Fourth Edition

Plant Roots

The Hidden Half

Fourth Edition

Plant Roots

The Hidden Half

Edited by

Amram Eshel • Tom Beeckman

CRC Press
Taylor & Francis Group
Boca Raton London New York

CRC Press is an imprint of the
Taylor & Francis Group, an **informa** business

Front Cover Credit: photographer, Claire Bouleau and artist, Dayla Luttwak.

CRC Press
Taylor & Francis Group
6000 Broken Sound Parkway NW, Suite 300
Boca Raton, FL 33487-2742

Printed on acid-free paper
Version Date: 20130304

International Standard Book Number-13: 978-1-4398-4648-3 (Hardback)

Visit the Taylor & Francis Web site at
http://www.taylorandfrancis.com

and the CRC Press Web site at
http://www.crcpress.com

Contents

PART III Regulation of Root Growth

PART IV Soil Resource Acquisition

PART V Root Response to Stress

PART VI Root–Rhizosphere Interactions

PART VII Modern Research Techniques

Preface to the Fourth Edition

This edition of *Plant Roots: The Hidden Half* is dedicated to the memory of Yoav Waisel (1931–2010), the initiator of this tome, the senior editor of its previous three editions, and the originator of its catchy name. It was his vision and tenacity that made this work such a great success among root scientists the world over. His untimely death took place a few months after he signed the contract to edit this fourth edition. He passed away suddenly, at the age of 79, having many plans for a very active future of writing and research. Yoav will be remembered affectionately by all who knew him and benefitted from his work.

In this edition, we present yet another stage in the never-ending saga of root research. The decade that has passed since the third edition of this book was published has been an era of great progress in biology in general, including plant sciences. The fundamental advancement brought about by the sequencing of whole genomes of model organisms, and the development of "omics" techniques—genomics, proteomics, and metabolomics—has affected root research as much as all other fields of biology.

Roots are still "the hidden half" of plants, and their research lags behind that of the aboveground parts of the plant. Nevertheless, it is evident from the content of this volume that root scientists have taken advantage of the great discoveries made in all other aspects of plant molecular biology. The content of this book reflects this transformation of methodologies and manners of biological studies, as it is reflected in studies of root biology at all levels, from the ecological to the cellular, and beyond.

We have endeavored to maintain the overall structure that our readers may be familiar with, presenting a broad range of topics related to root biology, yet having the individual chapters present a full, updated picture of their topic. We have a mixture of "old" and "new" subjects, but the new contributors greatly outnumber the ones who continued from the previous edition, reflecting the large change in modes of research and research topics that took place in the previous decade.

Part I—The Evolution and Genomics of Roots—consists of Chapters 1 and 2, which define root research in a wider context and that are characteristic of and meaningful for the recent developments in root research. Chapter 1, which describes where roots have come from, is a thorough update of the chapter about the evolution of roots and discusses their role in the global context of soil development and atmosphere composition. Chapter 2 portrays root research in the frame of the "new" biology of the postgenomic era, illustrating the important possibilities genomic research is currently creating for root research. Through the increasingly accumulating data on genome sequences throughout the plant kingdom, including representatives from primitive rootless plants, it will be tempting to speculate on how the development of roots became possible and to identify the genetic basis that was required for this event. We believe that this chapter will become a point of reference for future research in this area.

Part II—Root Structure—is a combination of nine chapters: Some are updates of previous topics, some are long-sought former omissions, and some are new topics that have emerged in recent years, shading a whole new perspective on our understanding of the mechanisms that determine root structure.

Part III—Regulation of Root Growth—includes nine chapters devoted to plant hormone action and signaling pathways that control root development: Some are updates of traditional topics written by former or by new contributors, while the ones concerning strigolactones and brassinosteroids reflect the recent advances in this field by defining these groups of compounds as proper plant hormones.

Part IV—Soil Resource Acquisition—includes ten chapters, which treat the subject both from the agricultural perspective and the ecological one. It is evident that agriculture and commercial growing of plants should apply the insight gained from studies of the plant root systems in their natural environment. The long appreciated chapter about roots as a source of food is also included in this part for this reason.

Part V—Root Response to Stress—includes seven chapters and is again a combination of updates and new topics, each representing the current state of the art on the subject. The impact of the genomic revolution on these topics is clearly reflected in the content of these chapters.

Part VI—Root-Rhizosphere Interactions—covers the topic from the beneficial microorganisms to the detrimental nematodes.

Part VII—Modern Research Techniques—includes two measuring techniques that hold great future for root research. One pertains to field research while the other must be used in the protected laboratory setting.

Keeping with the tradition of this series, the contributors did not only present a clear summation of their topic but also their vision of what is expected to be the course of development in the years to come. These views reflect the position of root research within the field of biology and science in general. Future prospects of global climate change and its impact on root growth are therefore also considered.

The 43 chapters that make up this book present a very wide coverage of themes related to root study and its application. Nevertheless, there will always be some topics that should have been included but did not find their place here due to a variety of reasons. We still hope that this book will serve well all those fascinated by plant roots, from basic or practical motives.

The cover was designed by Dalya Luttwak, who is probably the only sculpturer in the world that creates artistic interpretations of plant root systems. Her monumental metal sculptures were displayed in museums in the United States and in the Venice Biennale in Italy. In 2010, she had an exhibition called "Roots: The Hidden Half in Black and White," the name of which was derived from the previous edition of this book. We are grateful to Luttwak for her creative design.

It is our pleasure to thank all the contributors and Taylor & Francis Group's production team who were instrumental in the publication of this book.

Amram Eshel
Tom Beeckman

Editors

Amram Eshel received his BSc in biology in 1968 and his MSc cum laude in botany in 1971 (thesis entitled "Differences in Sodium Uptake Along the Primary Root of Corn Seedlings," under the supervision of Prof. Yoav Waisel), both from Tel Aviv University. He then received his PhD in botany in 1978 (thesis entitled "Studies of Mineral Relations of Plants by the Electron Probe," under the supervision of Prof. Yoav Waisel) from the same university. Dr. Eshel was a member of the academic staff at Tel Aviv University from 1980 until his retirement in 2012 and served as an associate professor in the Department of Plant Sciences since 1997. During those years, he held short visiting appointments at Ohio Agricultural Research Center in Wooster, Ohio, Michigan State University, Penn State, CNRS at Gif-Sur-Yvette, and University of Western Australia. His main research interests are in the area of plant physiological ecology, especially as related to mineral nutrition and water relations. He currently serves as the manager of the Sarah Racine Root Research Laboratory at Tel Aviv University.

Tom Beeckman received his master's degree in biology from the University of Ghent (1981–1985). He then completed his post-graduation in marine biology at Belgian Interuniversitary (1987–1989) and received his PhD in biology from the University of Ghent (1991–1997). He also studied botany at Ghent University in Belgium and after performing postdoctoral research at the university's Molecular Genetics Department, he became group leader of the Root Development Group at the Flanders Institute of Biotechnology (VIB) in 2001. In 2007, he became professor at Ghent University, teaching plant developmental biology. He devoted a considerable part of his research to understand how cell division is integrated into plant developmental processes, especially during the branching of roots.

Contributors

Mohamad Abu-Abied
The Institute of Plant Sciences
Volcani Center
Agricultural Research Organization
Bet Dagan, Israel

Nadim Alkharouf
Department of Computer and Information Sciences
Towson University
Towson, Maryland

Daniel F. Austin
Center for Sonoran Desert Studies
Arizona-Sonora Desert Museum
Tucson, Arizona

Mark R. Bakker
Bordeaux Sciences Agro
UMR 1220, TCEM
Gradignan, France

and

Institut National de la Recherche Agronomique
UMR 1220, TCEM
Villenave d'Ornon, France

Tom Beeckman
Integrative Plant Biology Division
Department of Plant Systems Biology
Flanders Institute for Biotechnology
and
Department Plant Biotechnology and Bioinformatics
Ghent University
Gent, Belgium

Philip N. Benfey
Department of Biology
and
Duke Center for Systems Biology
Duke University
Durham, North Carolina

A. Glyn Bengough
The James Hutton Institute

and

Division of Civil Engineering
University of Dundee
Dundee, United Kingdom

Malcolm J. Bennett
School of Biosciences
Centre for Plant Integrative Biology and Plant and Crop Sciences Division
University of Nottingham
Loughborough, United Kingdom

Nirit Bernstein
Institute of Soil Water and Environmental Sciences
Agricultural Research Organization
Volcani Center
Bet Dagan, Israel

Anthony Bishopp
School of Biosciences
Centre for Plant Integrative Biology and Plant and Crop Sciences Division
University of Nottingham
Loughborough, United Kingdom

Olga B. Blokhina
Division of Plant Biology
Department of Biosciences
University of Helsinki
Helsinki, Finland

Stephan Blossfeld
IBG-2 Plant Sciences
Forschungszentrum Jülich GmbH
Jülich, Germany

Ana I. Caño-Delgado
Department of Molecular Genetics
Center for Research in Agricultural Genomics
Barcelona, Spain

Nigel Chaffey
Department of Science
Bath Spa University
Bath, United Kingdom

François Chaumont
Institute of Life Sciences
Catholic University of Louvain
Louvain-la-Neuve, Belgium

Vincent Chochois
Plant Industry
Commonwealth Scientific and Industrial Research Organisation
Canberra, Australian Capital Territory, Australia

Yoan Coudert
Université Montpellier 2
CIRAD UMR DAP
Montpellier, France

Frédéric Danjon
Institut National de la Recherche Agronomique
UMR1202 BIOGECO
Cestas, France

and

Université de Bordeaux
UMR1202 BIOGECO
Talence, France

Ive De Smet
Division of Plant and Crop Sciences
School of Biosciences
University of Nottingham
Loughborough, United Kingdom

Marlies Demeulenaere
Integrative Plant Biology Division
Department of Plant Systems Biology
Flanders Institute for Biotechnology
and
Department Plant Biotechnology and Bioinformatics
Ghent University
Gent, Belgium

Xavier Draye
Earth and Life Institute
Catholic University of Louvain
Louvain-la-Neuve, Belgium

Quanying Du
Intercollege Graduate Degree Program in Ecology
The Pennsylvania State University
University Park, Pennsylvania

Joseph G. Dubrovsky
Departamento de Biologia Molecular de Plantas
Instituto de Biotecnología
Universidad Nacional Autónoma de México
Cuernavaca, Mexico

David M. Eissenstat
Department of Ecosystem Science and Management
The Pennsylvania State University
University Park, Pennsylvania

Sedeer El-Showk
Institute of Biotechnology and Department of Biosciences
University of Helsinki
Helsinki, Finland

Jhonathan E. Ephrath
French Associate Institute for Agriculture and Biotechnology of Drylands
Jacob Blaustein Institutes for Desert Research
Ben-Gurion University of the Negev
Beersheba, Israel

Amram Eshel
The George S. Wise Faculty of Life Sciences
Department of Molecular Biology and Ecology of Plants
Tel Aviv University
Tel Aviv, Israel

Kurt V. Fagerstedt
Division of Plant Biology
Department of Biosciences
University of Helsinki
Helsinki, Finland

Marc Faget
IBG-2 Plant Sciences
Forschungszentrum Jülich GmbH
Jülich, Germany

Guenter Feix
Institute of Biology III
University of Freiburg
Freiburg, Germany

Kimberly L. Gallagher
Lynch Laboratories
University of Pennsylvania
Philadelphia, Pennsylvania

Pascal Gantet
Université Montpellier 2
Montpellier, France

and

IRD, UMR DIADE, LMI RICE
Agricultural Genetics Institute
and
Department of Biotechnology and Pharmacology
University of Science and Technology of Hanoi
Hanoi, Vietnam

Jóska Gerendás
K+S KALI GmbH
Kassel, Germany

Simon Gilroy
Department of Botany
University of Wisconsin
Madison, Wisconsin

Alain Gojon
Biochimie et Physiologie Moléculaire des Plantes
Institut National de la Recherche Agronomique
Montpellier, France

Mary-Paz González-García
Department Molecular Genetics
Center for Research in Agricultural Genomics
Barcelona, Spain

Christoph R. Grünig
Microsynth AG
Balgach, Switzerland

Mohammad Tanbir Habib
Department of Forest Botany and Tree Physiology
Büsgen-Institut
Georg-August University of Göttingen
Göttingen, Germany

Charles Hachez
Institute of Life Sciences
Catholic University of Louvain
Louvain-la-Neuve, Belgium

Ykä Helariutta
Institute of Biotechnology and Department of Biosciences
University of Helsinki
Helsinki, Finland

Till Heller
Department of Forest Botany and Tree Physiology
Büsgen-Institut
Georg-August University of Göttingen
Göttingen, Germany

Christian Hermans
Laboratory of Plant Physiology and Molecular Genetics
Université libre de Bruxelles
Brussels, Belgium

Ko Hirano
Bioscience and Biotechnology Center
Nagoya University
Nagoya, Japan

Frank Hochholdinger
Institute of Crop Science and Resource Conservation
University of Bonn
Bonn, Germany

Parsa Hosseini
Soybean Genomics and Improvement Laboratory
United States Department of Agriculture
Agricultural Research Service
Beltsville, Maryland

and

Department of Computer and Information Sciences
Towson University
Towson, Maryland

Gregor Huber
IBG-2 Plant Sciences
Forschungszentrum Jülich GmbH
Jülich, Germany

Heba M.M. Ibrahim
Soybean Genomics and Improvement Laboratory
United States Department of Agriculture
Agricultural Research Service
Beltsville, Maryland

and

Faculty of Agriculture
Department of Genetics
Cairo University
Giza, Egypt

Siegfried Jahnke
IBG-2 Plant Sciences
Forschungszentrum Jülich GmbH
Jülich, Germany

Leentje Jansen
Integrative Plant Biology Division
Department of Plant Systems Biology
Flanders Institute for Biotechnology
and
Department Plant Biotechnology and Bioinformatics
Ghent University
Gent, Belgium

Robert Jarret
United States Department of Agriculture
Agricultural Research Service
Plant Genetic Resources
Griffin, Georgia

Mathieu Javaux
Earth and Life Institute
Catholic University of Louvain
Louvain-la-Neuve, Belgium

and

IBG-2 Plant Sciences
Forschungszentrum Jülich GmbH
Jülich, Germany

Paul Kenrick
Department of Earth Sciences
Natural History Museum
London, United Kingdom

Hinanit Koltai
Institute of Plant Sciences
Agricultural Research Organization
Volcani Center
Bet Dagan, Israel

Ingrid Kottke
Institute of Evolution and Ecology, Plant Evolutionary Ecology
Eberhard-Karls University Tübingen
Tübingen, Germany

Gábor M. Kovács
Department of Plant Anatomy
Institute of Biology
Eötvös Loránd University
Budapest, Hungary

Van Anh Le Thi
Université Montpellier 2
Montpellier, France

and

IRD, UMR DIADE, LMI RICE
Agricultural Genetics Institute
and
Department of Biotechnology and Pharmacology
University of Science and Technology of Hanoi
Hanoi, Vietnam

Daniel R. Lewis
Department of Biology
Wake Forest University
Winston-Salem, North Carolina

Louisa M. Liberman
Department of Biology
and
Duke Center for Systems Biology
Duke University
Durham, North Carolina

Guillaume Lobet
Earth and Life Institute
Catholic University of Louvain
Louvain-la-Neuve, Belgium

Jonathan P. Lynch
Department of Horticulture
The Pennsylvania State University
University Park, Pennsylvania

Ahmad M. Manschadi
Department of Crop Sciences
University of Natural Resources and Life Sciences
Vienna, Austria

Günther G.B. Manske
Center for Development Research
University of Bonn
Bonn, Germany

Hideaki Matsumoto
Institute of Plant Science and Resources
Okayama University
Okayama, Japan

Benjamin F. Matthews
Soybean Genomics and Improvement Laboratory
United States Department of Agriculture
Agricultural Research Service
Beltsville, Maryland

M. Luke McCormack
Intercollege Graduate Degree Program in Ecology
The Pennsylvania State University
University Park, Pennsylvania

Eric S. McLamore
Department of Agricultural and Biological Engineering
University of Florida
Gainesville, Florida

Chris J. Meyer
Department of Molecular and Cellular Biology
University of Guelph
Guelph, Ontario, Canada

Laila Moubayidin
Laboratory of Functional Genomics and Proteomics of Model Systems
Department of Biology and Biotechnology
Sapienza University of Rome
Rome, Italy

Gloria K. Muday
Department of Biology
Wake Forest University
Winston-Salem, North Carolina

Kerstin A. Nagel
IBG-2 Plant Sciences
Forschungszentrum Jülich GmbH
Jülich, Germany

Savithiry S. Natarajan
Soybean Genomics and Improvement Laboratory
United States Department of Agriculture
Agricultural Research Service
Beltsville, Maryland

Eric S. Ober
Department of Plant Biology and Crop Sciences
Rothamsted Research
Harpenden, United Kingdom

Pierdomenico Perata
Plant Lab
Sant'Anna School of Advanced Studies
Pisa, Italy

Serena Perilli
Laboratory of Functional Genomics and Proteomics of Model Systems
Department of Biology and Biotechnology
Sapienza University of Rome
Rome, Italy

Catherine Perrot-Rechenmann
Institut des Sciences du Végétal
CNRS UPR2355
Saclay Plant Sciences
Paris, France

Carol A. Peterson
Department of Biology
University of Waterloo
Waterloo, Ontario, Canada

Andrea Polle
Department of Forest Botany and Tree Physiology
Büsgen-Institut
Georg-August University of Göttingen
Göttingen, Germany

D. Marshall Porterfield
Department of Agricultural and Biological Engineering
Weldon School of Biomedical Engineering
Purdue University
West Lafayette, Indiana

Chiara Pucciariello
Plant Lab
Sant'Anna School of Advanced Studies
Pisa, Italy

R. George Ratcliffe
Department of Plant Sciences
University of Oxford
Oxford, United Kingdom

Boris Rewald
Forest Ecology, Department of Forest and Soil Sciences
University of Natural Resources and Life Sciences
Vienna, Austria

Joseph Riov
The Robert H. Smith Institute of Plant Sciences and Genetics in Agriculture
The Robert H. Smith Faculty of Agriculture, Food and Environment
The Hebrew University of Jerusalem
Rehovot, Israel

Sabrina Sabatini
Laboratory of Functional Genomics and Proteomics of Model Systems
Department of Biology and Biotechnology
Sapienza University of Rome
Rome, Italy

Einat Sadot
Institute of Plant Sciences
Agricultural Research Organization
Volcani Center
Bet Dagan, Israel

John Schiefelbein
Department of Molecular, Cellular, and Developmental Biology
University of Michigan
Ann Arbor, Michigan

Ulrich Schurr
IBG-2 Plant Sciences
Forschungszentrum Jülich GmbH
Jülich, Germany

Robert E. Sharp
Division of Plant Sciences
University of Missouri
Columbia, Missouri

Svetlana Shishkova
Departamento de Biologia Molecular de Plantas
Instituto de Biotecnología
Universidad Nacional Autónoma de México
Cuernavaca, Mexico

Thomas N. Sieber
Institute of Integrative Biology
ETH (Swiss Federal Institute of Technology)
Zurich, Switzerland

Moshe Silberbush
J. Blaustein Institute for Desert Research
Ben-Gurion University of the Negev
Beersheba, Israel

Alexia Stokes
Institut National de la Recherche Agronomique
Botany and Computational Plant Architecture (AMAP)
Montpellier, France

Sarah Swanson
Department of Botany
University of Wisconsin
Madison, Wisconsin

Ranjan Swarup
School of Biosciences
Centre for Plant Integrative Biology and Plant and Crop Sciences Division
University of Nottingham
Loughborough, United Kingdom

David Szwerdszarf
Institute of Plant Sciences
Agricultural Research Organization
Volcani Center
Bet Dagan, Israel

Eiichi Tanimoto
Department of Information and Biological Sciences
Graduate School of Natural Sciences
Nagoya City University
Nagoya, Japan

Jaimie M. Van Norman
Department of Biology
and
Duke Center for Systems Biology
Duke University
Durham, North Carolina

Nathalie Verbruggen
Laboratory of Plant Physiology and Molecular Genetics
Université libre de Bruxelles
Brussels, Belgium

Josep Vilarrasa-Blasi
Department of Molecular Genetics
Center for Research in Agricultural Genomics
Barcelona, Spain

Paul L.G. Vlek
Department of Ecology and Natural Resources Management
Center for Development Research
University of Bonn
Bonn, Germany

Anton P. Wasson
Plant Industry
Commonwealth Scientific and Industrial Research Organisation
Canberra, Australian Capital Territory, Australia

Michelle Watt
Plant Industry
Commonwealth Scientific and Industrial Research Organisation
Canberra, Australian Capital Territory, Australia

Darren M. Wells
School of Biosciences
Centre for Plant Integrative Biology and Plant and Crop Sciences Division
University of Nottingham
Loughborough, United Kingdom

W. Richard Whalley
Rothamsted Research
Hertfordshire, United Kingdom

Yana M. Wieckowski
Department of Molecular, Cellular, and Developmental Biology
University of Michigan
Ann Arbor, Michigan

Yoko Yamamoto
Institute of Plant Science and Resources
Okayama University
Okayama, Japan

Hanma Zhang
College of Life Sciences
Chongqing Normal University
Chongqing, People's Republic of China

I

The Evolution and Genomics of Roots

1

The Origin of Roots

Paul Kenrick
Natural History Museum

I. Introduction

Roots were an early development in plant life, evolving on land during the Devonian Period, 416 million to 360 million years ago (Gensel et al. 2001; Raven and Edwards 2001; Kenrick 2002; Boyce 2005). This was a time of enormous change, which witnessed the evolution of forest ecosystems from an earlier diminutive herbaceous vegetation of small rootless and leafless plants (Gensel and Edwards 2001; Gensel 2008; Meyer-Berthaud et al. 2010). From the outset, symbiotic associations with fungi were important (Taylor et al. 2004; Strullu-Derrien and Strullu 2007; Bonfante and Genre 2008), and it is clear that mycorrhizae and plant roots have coevolved in many different ways (Brundrett 2002; Wang and Qiu 2006; Taylor et al. 2009a). Roots combined with a fully integrated vascular system were essential to the evolution of large plants, enabling them to meet the requirements of anchorage and the acquisition of water and nutrients (Boyce 2005). Beginning in the Middle Devonian (ca 392 Ma), the earliest forest ecosystems already displayed an astonishing diversity of rooting systems encompassing extinct forms and others that are comparable in many ways to those of modern tree ferns and gymnosperms (Driese et al. 1997; Algeo and Scheckler 1998; Soria and Meyer-Berthaud 2004; Stein et al. 2007). The combined weight of evidence from both fossils and living plants demonstrates that once plants made the transition to the land, rooting organs evolved in a piecemeal fashion independently in several different clades, rapidly acquiring and extending functionality and complexity.

The evolution of rooting systems in plants during the Devonian Period had consequences that reached far beyond the plants themselves. Roots influenced the development of soils (Retallack 2001) and thereby indirectly the faunal component, which is known to have comprised a diversity of arthropods (Labandeira 2005, 2007) as well as nematodes (Poinar et al. 2008). Early terrestrial soil ecosystems were predominantly detritivore based and in this respect resembled the soils of today (Shear and Selden 2001). The physical effect of roots, in particular their ability to enhance the weathering of calcium-magnesium silicates, is thought to have had an enormous impact on key Earth systems, in particular the long-term or geochemical carbon cycle (Berner 1998; Algeo et al. 2001; Berner and Kothavala 2001; Bergman et al. 2004). Over millions of years, the cumulative effect of the evolution of roots and their action measured on a global scale was to draw down carbon dioxide from the atmosphere at an unprecedented rate. In this respect, their presence influenced on a grand scale the geochemistry of atmosphere and ocean, which through feedback mechanisms operating over millions of years, greatly influenced the evolution of other plant organ and tissue systems (Beerling and Berner 2005).

This chapter reviews the origin and early evolution of roots in land plants, focusing on new data from the fossil record of vascular plants interpreted in a phylogenetic context. In addition to providing a narrative of what happened and when, the evolution of root systems and their homologies is considered in detail. One aim is to provide the experimental biologist with a rationale for choosing organisms for research in root developmental biology. A second aim is to provoke the reader into considering how the evolution of roots in plants impacted on key Earth systems (e.g., carbon cycle).

II. Roots in the Fossil Record

Paleobotanists are hindered in their attempts to reconstruct plants from the past because much of the fossil record comprises broken or disarticulated remains preserved in sediments deposited in rivers or lakes. Piecing together plants to form a conceptual whole is a difficult and painstaking task, and frequently one must be satisfied with an incomplete organism, simply because parts are missing (Forey et al. 2004; Bateman and Hilton 2009). These missing parts are often the roots. Despite the difficulties inherent in reassembling whole plants, roots are occasionally preserved in fossils that have literally been uprooted and transported some distance. Also, it is possible to find environments where plants have been fossilized at their sites of growth. Under circumstances such as these, preservation of roots can be exceptionally good. Frequently, roots are the only parts of the plant that is preserved, and it is often possible to observe them as distinctive features in paleosols (Retallack 2001).

The oldest evidence of rooting structures that are probably attributable to land plants comes from Upper Silurian (Ludlow Series, 423 Ma) paleosols bearing small rhizome-like traces (Driese and Mora 2001; Retallack 2001) and from stem-group lycopods in similar age sediments (Kotyk et al. 2002). Paleosols containing root systems of more modern aspect, including downward-directed bifurcating branches with progressive decreases in root diameter, first became abundant during the Lower Devonian (Driese and Mora 2001; Hillier et al. 2008). Because rooted vegetation stabilizes riverbanks, it has been suggested that prior to the evolution of roots, river systems would have been braided rather than meandering (Davies and Gibling 2010a), thereby providing an indirect sedimentological indicator for the widespread development of rooting system on land. Also, the presence of binding root networks had other consequences for sediment deposition (Retallack 2001), encouraging, for example, lateral accretion in meandering channels. A recent survey of Cambrian to Devonian fluvial successions has documented the absence of meandering systems and lateral accretion sets prior to the Upper Silurian (Ludlow Series) (Davies and Gibling 2010b). Observations based on paleosols and on fluvial sedimentology therefore provide a broadly consistent picture of the development of shallow but extensive specialized rooting systems capable of stabilizing riverbanks by the early part of the Devonian Period.

A. Early Rooting Systems

1. Rhynie Chert: An Early Terrestrial Biota

The earliest direct evidence of rooting structures preserved in growth position and with preservation of detail at the cellular level comes from the 407.1 ± 2.2-million-year-old Rhynie Chert, Scotland (Mark et al. 2011). This sequence of fossiliferous cherts—rocks composed of finely crystalline silica—provides a window onto an early terrestrial environment, capturing a period when plant life on land was at an early stage of development (Rice et al. 1995; Trewin and Rice 2004). Studies of the depositional environment of the Rhynie Chert show that plants grew on sandy substrates in and around the margins of ephemeral ponds and lakes on an alluvial plain (Trewin et al. 1994; Fayers and Trewin 2004). The cherts formed as siliceous sinters, and they were deposited during multiple episodes of hot spring activity (Rice et al. 2002). This resulted in inundation and preservation of whole plants in their growth positions and the underlying soil. Petrographic thin sections are the method most widely employed to investigate and to reconstruct the plants, and they reveal amazing details of cell structure (e.g., Edwards 2004; Kerp et al. 2004). The Rhynie Chert thus provides unparalleled insights into rooting structures at an early stage of plant evolution.

2. Stems Functioning as Roots

The Rhynie Chert plants were all small with prostrate rhizomes. The anatomy and growth form of the rhizome have been documented in detail for *Nothia aphylla* (Kerp et al. 2001), a possible basal tracheophyte (zosterophyll) (Kenrick and Crane 1997), and they are also known for several other species of protracheophyte (Remy and Hass 1996; Edwards 2004). In *Nothia aphylla*, the rhizome is a system comprising several orders of lateral branching. The basic patterning of tissues resembles that of the aerial stems. In transverse section, one can recognize epidermis, hypodermis, cortex, and a vascular core. There is no endodermis. Along the underside of the rhizome, there is a conspicuous ventral ridge of rhizoids, which imparts a pronounced dorsiventral symmetry (Kerp et al. 2001). The rhizoidal ridge connects to the vascular core by distinctive connective tissue that differs from the cortical tissues in lack of intercellular spaces. The rhizome is thought to have been subterranean, but shallowly so, to depths on a millimeter scale. Uniquely among early plants, *Nothia aphylla* is interpreted as a geophyte, with short-lived probably seasonal aerial parts and a persistent rhizome. Another plant, *Horneophyton lignieri*, is known to have produced a lobed corm-like rooting structure rather than an extended rhizome (Kerp et al. 2001). Rhizomes however appear to have been the norm both in the Rhynie Chert and at other early sites (Gensel et al. 2001).

The phylogenetic placement of these early fossils demonstrates that rhizomatous growth was characteristic of the early vascular plants. It has been documented in protracheophytes such as *Aglaophyton major* (Edwards 1986; Remy and Hass 1996) and within the vascular plants in basal members of lycophyte (Gensel et al. 2001) and the euphyllophyte clades (Doran 1980; Li and Hsü 1987; Hao and Beck 1993). The earliest rooting structures in vascular plants were therefore aerial stems that were modified by the presence of rhizoids. They were superficial or shallowly subterranean. They undoubtedly served to anchor the plant, and in some they may also have functioned as perennating organs. It is likely that this form of growth would have led to the formation of extensive thickets of genetically identical individuals.

3. Rhizoids

Rhizoids are seldom preserved in fossils except under exceptional conditions such as those found in the Rhynie Chert (Edwards 2004) or when they are especially robust (e.g., Hueber and Banks 1979; Hernick et al. 2008; Figure 1.1). Their form is quite conservative in early fossils, but their developmental position on rhizomes and associated tissues exhibits variation and flexibility. All rhizoids documented to date developed from epidermal cells, and they were unicellular and smooth walled. In two Rhynie Chert tracheophytes, the rhizoids were quite short, and they were present on all surfaces of the rhizome (i.e., *Trichopherophyton*,

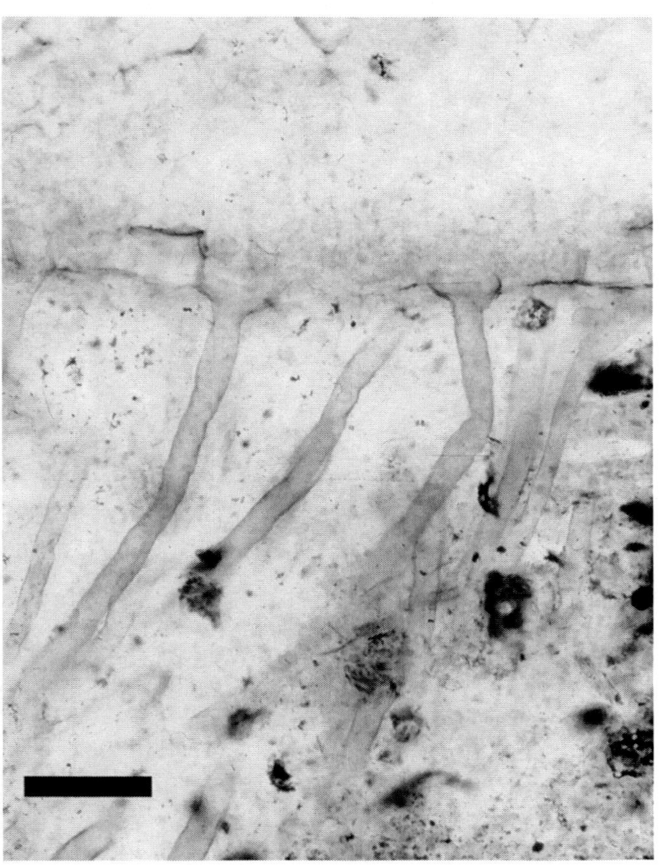

FIGURE 1.1 Unicellular rhizoids on the root system of a fossil plant from Rhynie Chert, Scotland (ca 407 Ma). The Rhynie plants are silicified and have been fossilized in their growth positions. Scale bar 100 μm.

Ventarura: Lyon and Edwards 1991). In other early plants, the rhizoids were much longer, and they appear to be associated with conspicuous preexisting structures and with specific tissue differentiation within the rhizome. Tufts of rhizoids developed on the distinctive hemispherical projections in the lower parts of the aerial axes and rhizome of *Rhynia gwynne-vaughanii*, but they were not exclusively associated with these (Edwards 2004). In another species, they were associated with the development of a conspicuous ridge of tissue on the lower surface of the rhizome (*Nothia*: Kerp et al. 2001), and in others their distribution was possibly also localized, but further detailed study is required (*Horneophyton*: Edwards 2004). In the fossil protracheophyte *Aglaophyton*, rhizoids developed from stomatal subsidiary cells. This was accompanied by cell divisions in underlying hypodermis creating bulges of rhizoid-bearing tissues on the lower surfaces of the rhizomes. Furthermore, underlying tissue differentiation produced apparent transfusion tissues linking vascular system to rhizoidal bulge (Remy and Hass 1996). Yet another variation was observed in the zosterophyll *Serrulacaulis*. In this plant, large deltoid spines were borne on stems of all sizes, but rhizoids developed on and among the spines on prostrate stems. These observations indicate that among early vascular plants, rhizoids were capable of being induced generally on aerial stems and associated appendages when in proximity to the soil. Furthermore, their induction was associated with some additional differentiation in underlying tissue systems, indicating that rhizoids played a role in absorption.

When conditions favored the preservation of minute cellular structures, rhizoids are observed in the great majority of species. Furthermore, they were expressed in both the haploid (gametophyte) and diploid (sporophyte) phases of the life cycle (e.g., *Aglaophyton*: Remy and Hass 1996), whereas in living vascular plants, where present, they are confined to the gametophyte. Rhizoids were apparently absent from one of the Rhynie Chert plants (Edwards 2004). Demonstrating absence conclusively in fossils is difficult because lack of evidence may simply reflect unsuitable conditions of preservation or incomplete fossils missing the key parts. In the Rhynie Chert, all the conditions necessary to preserve such delicate structures are met, and the root of the lycopod *Asteroxylon* has been studied in some detail, yet rhizoids have never been observed. This has led to the conclusion that in this plant they were truly absent (Edwards 2004), indicating that their presumed functions as anchoring or absorptive cells were not essential to all early vascular plants.

4. Branches with Uniquely Rooting Function

Rhizoid-bearing rhizomes that were very similar to aerial stems were common among the earliest vascular plants (Figure 1.2), but an increased level of specialization is notable

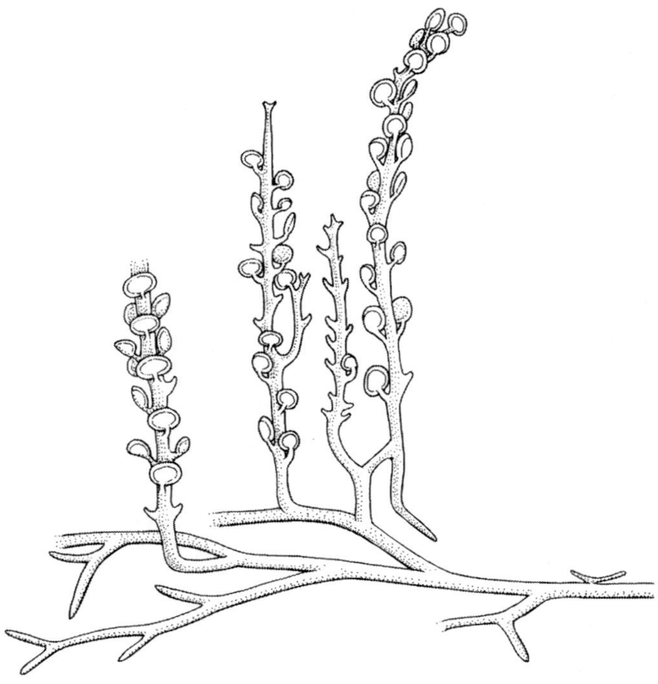

FIGURE 1.2 Growth form of clubmoss relative from early part of the Devonian Period (ca 410 Ma). Prostrate rhizome with upright branches bearing reniform sporangia. (Redrawn from Gerrienne, P., *Rev. Pal. Pal.*, 55, 317, 1988.)

among several early groups. Branches with a uniquely rooting function are first observed among relatives of the lycopods during the Early Devonian (416–398 Ma) (Gensel et al. 2001). These plants bore distinctive branches that were derived from aerial stems but were typically narrower and shorter, irregularly bifurcating, often with a sinuous appearance, and without emergences or leaves. A rooting function is further suggested by branch orientation, which is opposite to or perpendicular to the aerial stems. Evidence from several compression fossils shows roots emerging from points along prostrate stems (Gensel et al. 2001). Whereas in some species of *Zosterophyllum*, the aerial stems are all upright and roots emerged from the base in a dense tuft (Gensel et al. 2001; Hao et al. 2010). Because these fossils preserve overall form but not anatomical detail, it is not known whether rooting branches were endogenous. Similarly, the presence or absence of root caps and root hairs cannot be determined. Where there is anatomical preservation, as in *Asteroxylon mackiei* from the Rhynie Chert, the roots appear to be exogenous, arising probably as dichotomies with the main root deriving ultimately from the apical meristem of the aerial axis (Raven and Edwards 2001). Anatomically, they are clearly vascular, but neither root nor stem stele appears to be delimited by an endodermis. Branches such as these were nevertheless clearly functioning as roots, and they provide the earliest evidence of a positive geotropic response in vascular plants.

Dormant meristematic regions that had the potential to develop into new stems and perhaps roots were common to many early vascular plants (e.g., zosterophylls). These were most frequent on rhizomes, but they were also present on erect stems, where they were usually confined to the axillary position of lateral branches (Kenrick and Crane 1997; Gensel et al. 2001). In some species, they extended over a greater portion of the erect branching system (Edwards and Kenrick 1986; Li 1992). These structures were vascular and exogenous, and they are known to have developed into new aerial branches in some species, whereas in others it is possible that they developed into roots (Edwards and Kenrick 1986; Remy et al. 1986; Li 1992). The pattern of distribution and the developmental plasticity of these enigmatic meristematic regions are notable similarities to the widely discussed rhizophore of living Selaginellaceae. Phylogenetic evidence indicates, however, that homology is highly unlikely (Kenrick and Crane 1997). The reason for this is that Selaginellaceae are phylogenetically remote from those zosterophylls with dormant meristems (Figure 1.3). Assuming homology of these structures would therefore be highly unparsimonious because it would imply multiple losses of dormant meristems/rhizophores in intervening groups. It has been suggested that the function of dormant meristems in the life cycle of the plant was to facilitate vegetative growth or reproduction. Sedimentological evidence indicates that many such plants inhabited disturbed marginal aquatic settings where they were prone to catastrophic burial by flood (Edwards and Kenrick 1986). These dormant meristematic regions may therefore have facilitated regeneration after burial.

5. Bipolar Growth

The rooting systems considered thus far are essentially shoot-borne roots. Plants with single upright stems and downwardly directed bifurcating rooting systems were present in the fossil record by the mid-Devonian (398–385 Ma). One example is the Cladoxylopsida. These are an extinct group of fernlike plants of uncertain systematic position. They are clearly euphyllophytes (i.e., members of the group that contains modern ferns, horsetails, and seed plants), but their precise relationships are poorly understood (Figure 1.3). The group has been implicated in the origins of both ferns and horsetails (Hilton et al. 2003). One uprooted specimen, *Lorophyton goense*, provides an insight into the morphology of the whole plant. *L. goense* had a 2 cm diameter trunk that bore tufts of leaflike branches from the apex and numerous bifurcating roots from a slightly flared base. The whole plant is estimated to have been about 40 cm in height (Fairon-Demaret and Li 1993). Schweitzer and Li (1996) documented a similar growth form in the 50 cm tall lycopsid *Chamaedendron* by the early part of the Late Devonian (Frasnian Stage 385–375 Ma). The roots of *Chamaedendron* bifurcated at least four times and were at least 15 cm long (Figure 1.4). These fossils mark an early departure from shoot-borne rooting systems to a form of bipolar growth. Roots of these sorts were clearly differentiated as such, and they did not function as stems at any time during their ontogeny.

B. Roots of Trees

Trees first appeared in the fossil record during the mid-Devonian (Givetian Stage 392–385 Ma) (Stein et al. 2007), and it is clear from phylogenetic studies that arborescence evolved independently in several major clades of plants including seed plants, clubmosses, ferns, horsetails, and other groups of uncertain affinity (e.g., Cladoxylopsida) (Kenrick and Crane 1997; Kenrick 2000). This dramatic increase in size was in most groups a consequence of the evolution of the cambium, which is an innovation that also led to the development of much more extensive rooting systems. In tree ferns, rooting systems themselves contributed to the construction of the trunk. The rooting systems of early members of these clades are considered in the following text.

1. Horsetails

Paleozoic tree horsetails bore roots that were very similar to their modern relatives, which are all herbaceous (Taylor et al. 2009b). The rhizome of living *Equisetum* is a modified stem that produces aerial branches and adventitious roots. During the Carboniferous Period (360–299 Ma), the tree horsetail *Calamites* grew to a height of 20 m. The rooting system was comparatively shallow and comprised an enormous 40 cm diameter horizontal rhizome that bore adventitious roots in whorls at the nodes, much like branches on the erect stems. Roots also developed from the bases of upright branches. Anatomically, the calamite rhizome was very similar to the

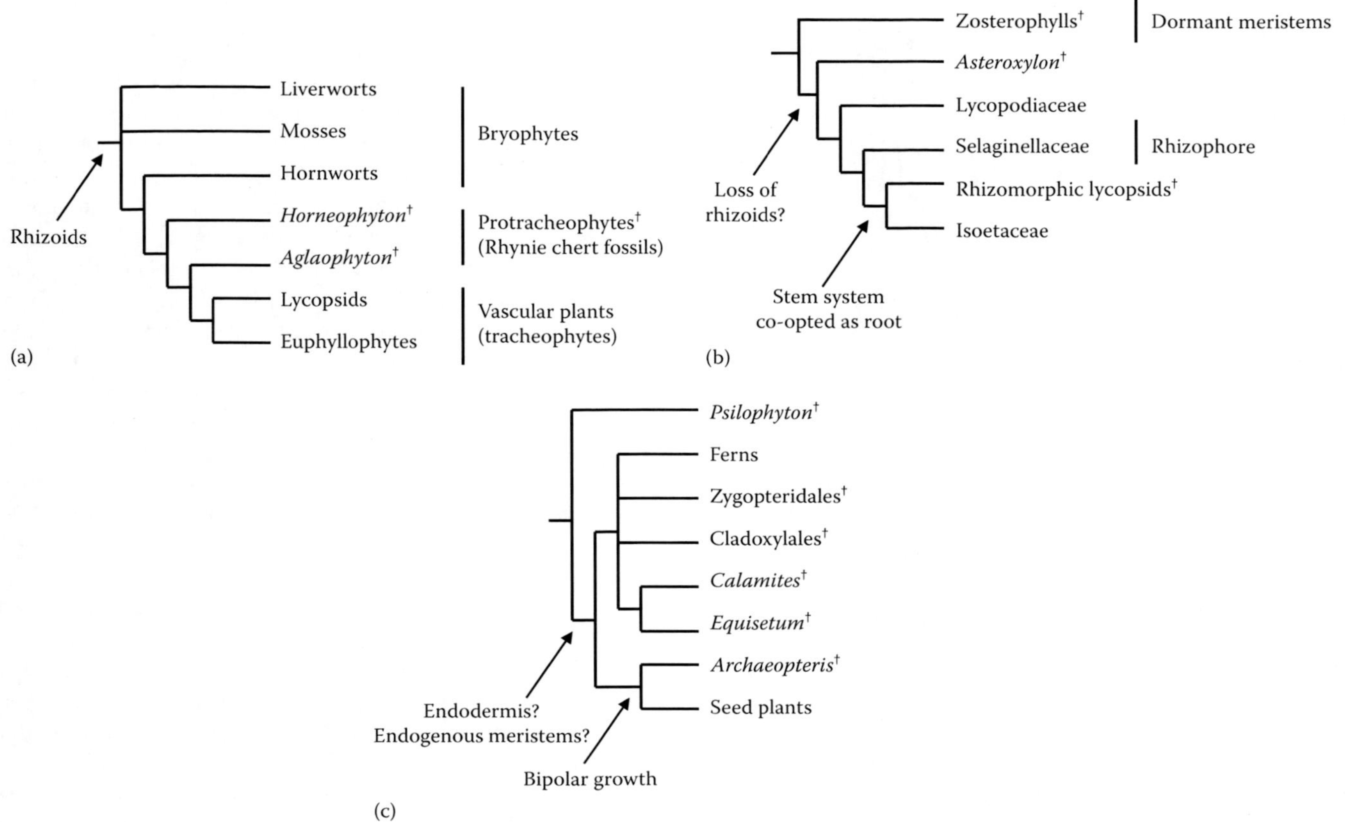

FIGURE 1.3 Phylogenetic relations among early land plant summarized and simplified, with aspects of root evolution depicted.† denotes extinct. Relations of some key early fossils and living plants in (a) basal land plants, (b) lycopods and (c) euphyllophytes. (Modified after Kenrick, P. and Crane, P.R., The origin and early diversification of land plants: A cladistic study, *Smithsonian Series in Comparative Evolutionary Biology*, Smithsonian Institution Press, Washington, DC, 1997; Kenrick, P., *Phil. Trans. R. Soc. Lond.*, B355, 847, 2000; Pryer, K.M. et al., *Am. J. Bot.*, 91, 1582, 2004; Rothwell, G.W. and Nixon, K.C., *Int. J. Plant Sci.*, 167, 737, 2006; Qiu, Y.L., *J. Syst. Evol.*, 46, 287, 2008; Karol, K.G. et al., *BMC Evol. Biol.*, 10, 321, 2010.)

stem, and it contained a large amount of secondary wood. The older and somewhat smaller horsetail *Archaeocalamites* (Carboniferous, Tournaisian, 345–359 Ma) was of similar construction but had a greatly expanded wooden rootstock at the base of a stem that bore numerous adventitious roots (Bateman 1991). Scrambling or climbing forms such as *Sphenophyllum* produced adventitious roots at nodes, often with leaves (Taylor et al. 2009b). The rhizome-based root system of horsetails probably originated from herbaceous Middle to Late Devonian (398–359 Ma) plants such as *Metacladophyton*, in which roots (8 cm long, 2.5 mm wide) were borne along one side of a horizontal rhizome (1 cm diameter) that gave rise to upright stems (Wang and Geng 1997; Wang and Lin 2007). Despite the great difference in size separating herbaceous from arborescent forms, the rooting system of Paleozoic Era horsetails was remarkably conservative.

2. Clubmosses

Paleozoic Era tree clubmosses had highly distinctive roots that are termed rhizomorphs. The rhizomorph was a determinate structure that branched dichotomously to form a shallow but laterally extensive rooting system that would have exceeded 10 m in diameter for the larger trees (Taylor et al. 2009b). Root apices terminated abruptly, and they bore a rim-like apical groove (Rothwell 1984). Rootlets were cylindrical and helically arranged, developed near the apical meristem, and are thought to have been exogenous (Rothwell and Erwin 1985). Each rootlet contained a large, air-filled cavity. No evidence of root caps or root hairs has been forthcoming, but this may reflect poor preservation and lack of data from rootlet apices. On older portions of the rooting system, secondary growth of cortex and stele caused rootlets to be abscised. Other Late Paleozoic Era arborescent clubmosses had rather different rooting structures. Instead of the extensive branched rhizomorph, plants such as *Chaloneria* and *Paurodendon* had unbranched rootstocks that were comparatively small and rounded and that had a lobed cormose base (Pigg and Rothwell 1983; Rothwell and Erwin 1985). Rootlets of both the rhizomorphic and cormose clubmosses share unique anatomical similarities, notably related to the location of the stele within a large cavity. This is one of several striking similarities to the roots of their living relatives, *Isoetes* (Stewart 1947).

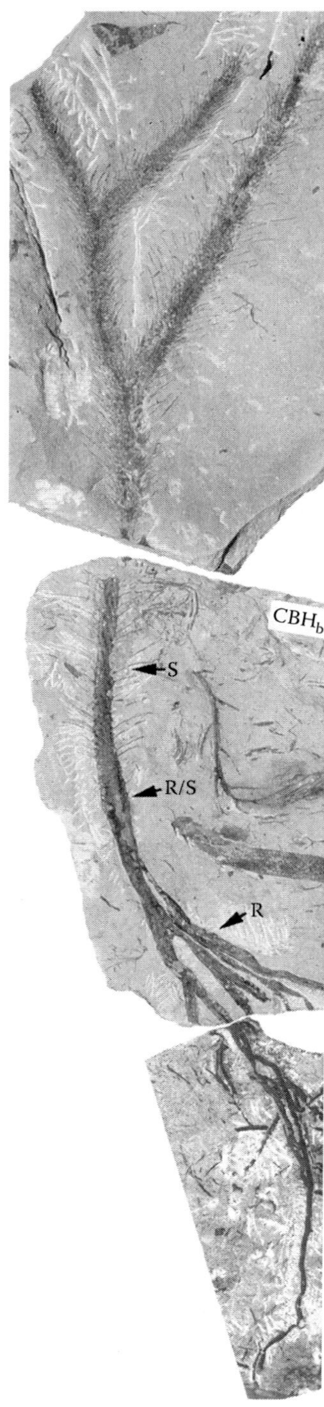

FIGURE 1.4 Uprooted fossil. A clubmoss (lycopsid) from China (Schweitzer and Li 1996) (ca 377 Ma). The specimen illustrated in 35 cm long, and it exhibits a bifurcating rooting system (R) and stem (S) bearing numerous microphylls and a slightly expanded root/stem junction (R/S).

3. Ferns

Ferns originated during the Devonian Period, and several distinctive extinct groups have been implicated as potential sister-group or stem-group members. These include the Zygopteridales (Phillips and Galtier 2005), Cladoxylales (Stein et al. 2007) and Stauropteridales (Rothwell and Nixon 2006; Taylor et al. 2009b; Figure 1.3). Zygopterids bore fronds with a unique 3D morphology in part and with pinnules of modest size. The pinnules had little or no laminae in the early members. Secondary growth has also been documented in several, and growth forms in different species encompassed both rhizome and erect trunks. In early zygopterids, the roots generally possessed a diarch xylem, and they developed endogenously from a rhizome. This contrasts with those species of living and extinct ferns that bear erect stems. In these, the site of root development is typically closely associated with leaf trace vasculature (Phillips and Galtier 2005). Roots also played a major structural role as a mantel that enclothed the trunk. In *Symplocopteris*, one of the oldest zygopterids, small roots covered with root hairs developed endogenously from large roots that also had a role in structural support. The large roots developed exogenously along the stems and are interpreted as modified branches with a strong positive geotropism (Phillips and Galtier 2005). The tree fernlike Cladoxylales were important understory and canopy elements in the earliest known forests, and some exceeded 8 m in height. These plants were anchored by numerous narrow roots that emerged either from the flared base of the trunk (Fairon-Demaret and Berry 2000; Stein et al. 2007) or from along the stem forming an extensive root mantle (Soria and Meyer-Berthaud 2004).

4. Progymnosperms

Progymnosperms were a grade of extinct, free-sporing plants that were closely related to seed plants (Figure 1.3; Hilton and Bateman 2006). In many aspects of their morphology, progymnosperms were intermediate between pteridophytes and gymnosperms. Plants bore fernlike foliage, and they ranged from small- or medium-size homosporous shrubs (Aneurophytales) to large heterosporous trees (Archaeopteridales) (Taylor et al. 2009b). Unlike ferns, progymnosperms produced gymnospermous wood from a bifacial cambium—one that produces both secondary xylem and phloem (Rothwell and Lev-Yadun 2005). Similarities with gymnosperms extended also to the roots (Scheckler 1995). Aneurophytales—the first progymnosperms to appear in the Middle Devonian (Eifelian 398–392 Ma)—had bifurcating roots that were probably determinate. Most roots were endogenous and some were adventitious. Cambial activity is evident in roots and rootlets of all sizes. Roots possessed cortical lacunae, periderm, persistent root hairs, and possibly an exodermis. There is no evidence of mycorrhizae (Scheckler 1995). In the progymnosperm *Eddya sullivanensis*, which is possibly a juvenile *Archaeopteris*, there was a complex taproot system, comprising a strong main root that gave rise to a profusion of small, much branched, probably endogenous laterals (Beck 1967). Evidence from petrified trunks indicates that in addition to a well-developed root system with secondary wood, *Archaeopteris* had adventitious latent primordial similar to those produced by some living trees, which eventually develop into roots on stem cuttings (Meyer-Berthaud et al. 1999; 2000).

5. Seed Plants

The earliest seed plants—also known as seed ferns or pteridosperms—were small narrow-stemmed, gracile shrubs, many of which had a scrambling or semi-self-supporting habit (DiMichele et al. 2006; Hilton and Bateman 2006). In this respect, they were very similar to the ancestral aneurophytalean progymnosperms. On these plants, roots had the capacity to develop throughout the aerial systems, frequently being borne along the stems (Taylor et al. 2009b) and sometimes organized into vertical rows or associated with other organs such as leaves (Meyer-Berthaud and Stein 1995). From these early forms, pteridosperms soon diversified to become prominent elements of tropical ecosystems evolving into large, dominant canopy trees, prominent understory trees, scrambling ground cover, thicket formers, and liana-like plants and vines (DiMichele et al. 2006). The climbers employed a remarkable diversity of adaptations including hooks, leaflet tendrils, tendrils terminating in adhesive pads, and aerial adventitious roots (Krings et al. 2003). It seems unlikely, however, that the pteridosperms produced any true root climbers, like modern angiosperms such as *Hedera* or *Monstera* (Krings et al. 2003).

In addition to the slender scrambling and climbing forms, Early Carboniferous Period (Mississippian Series 359–318 Ma) pteridosperms also included larger self-supporting or semi-self-supporting arborescent plants with upright woody trunks bearing large, bifurcating fernlike fronds (DiMichele et al. 2006). One common early form is *Medullosa*, which was a small tree. Roots developed adventitiously over an extended region at the base of the trunk, where they probably played a role in buttressing. The roots of *Medullosa* are a common component of coal balls (common spheroidal carbonate petrifactions in coal seams). Preservation of cellular-level details is exceptionally good, enabling the documentation of the anatomy in detail and observation of different developmental stages (Rothwell and Whiteside 1974; Taylor et al. 2009b). Older roots (~2.5 cm diameter) contained abundant secondary xylem and phloem and a wide band of periderm. The xylem was an actinostele with exarch development. Lateral roots originated from a thin-walled pericycle. Surrounding the pericycle was a narrow endodermis and parenchymatous cortex (Taylor et al. 2009b). Anatomically, roots were broadly similar to stems in the range of cell types expressed, but their stellar anatomy was simpler due to the absence of leafy appendages.

Early pteridosperms occupied a broad range of environments including peat-forming swamps, floodplains, extrabasin lowlands, and probably also higher ground (DiMichele et al. 2006). In some swamp-dwelling forms, there is evidence of adaptation to growth under very wet conditions. *Cordaites* was a group of scrambling shrubs and small trees with upright trunks (Taylor et al. 2009b). Some Carboniferous Period species had highly branched stilt roots. Anatomically, these ranged from diarch to pentarch, and secondary tissues were well developed in the larger roots, including secondary xylem and phloem, aerenchymatous phelloderm, and a compact layer of phellem with lenticels. The combination of aerenchyma, medullated protosteles, periderm, and lenticels has been likened to the stilt roots of plants growing in modern mangrove environments (Cridland 1964). In some species, lateral roots were borne in large clusters from phellem-covered protuberances. They exhibit no definite arrangement but appear to be borne on only one side of the primary root. They possessed the same complement of tissues as the main roots and had in addition a well-developed endodermis. Endophytic fungi believed to be endomycorrhizal have been reported, but it is difficult to be sure about whether these were true symbionts and not perhaps parasites or saprophytes (Taylor et al. 2009b).

III. Discussion

The early fossil record of land plants documents a continuum of variation in the evolution of rooting systems that began with simple rhizoid-bearing stems and developed, over a period of 40 million years, into a broad range of complex multicellular organs specialized in anchorage and nutrient acquisition. To define the term "root," one must draw an arbitrary line in this morphological continuum, and in applying the term to fossils, further complications arise due to missing details (Gensel et al. 2001; Boyce 2005). Frequently, data on the presence/absence of root cap and root hairs and on root development (endogenous vs. exogenous) are not available. Here, roots are defined simply as multicellular appendages that function in anchorage and in the acquisition of solutes and nutrients.

A. Root Origins

Fossil evidence demonstrates that roots evolved on land and that they were an early innovation of plant life. Roots, as defined here, are unique to vascular plants, but simpler functionally equivalent systems are more widespread. Rhizoid-bearing stems occur in living bryophytes and even in some green algae (e.g., Charales). Evidence from complete plants fossilized in growth position in the Rhynie Chert confirms the presence of similar systems in the earliest vascular plants. In these plants, rhizoids probably developed in response to physical contact of the stem with a substrate. The widespread occurrence of simple rhizoid-based systems in land plants (bryophytes, early fossil vascular plants) indicates that this form of rooting system preceded the evolution of true roots and may well have been a legacy inherited from green algal ancestors (Gensel et al. 2001; Kenrick 2002). Recently, striking confirmation of the shared evolutionary history of rhizoids has come from a comparison of rhizoid development in the moss *Physcomitrella patens* with root hair formation in the flowering plant *Arabidopsis thaliana*, demonstrating a highly conserved molecular pathway controlling development in both plants (Menand et al. 2007).

Shallow, rhizoid-bearing stems are adequate for small plants (30–50 cm tall) with modest transpirational needs and a simple vascular system, but also present in the Rhynie Chert were plants with organs that were clearly and uniquely specialized for rooting. *Asteroxylon* and its cosmopolitan relative *Drepanophycus* possessed small roots that were positively geotropic (Gensel et al. 2001). These plants are related to clubmosses, and based on their phylogenetic relations with other known species at the time, Kenrick (2002) argued that this implied that the roots of lycopods are uniquely derived and are therefore not homologous with those of other vascular plants. Recent discoveries of apparently similar roots in the earliest zosterophylls (Hao et al. 2010) now call this conclusion into question in so far as it applies to the rooting systems of basal herbaceous euphyllophytes and lycophytes.

B. Root Diversity in Early Terrestrial Ecosystems

By the Late Devonian and Early Carboniferous, an enormous variety of rooting structures had evolved, and roots had been co-opted to an equally broad range of additional functions. The evolution of large erect plants and in particular trees placed increasing demands on rooting systems, and these were solved in a variety of ways. From the outset, differentiation of tissues in the roots of arborescent plants followed closely patterns of tissue differentiation in erect trunks. One of the major sources of divergence in root and stem anatomy resulted from the influence of lateral appendages (e.g., the position and number of leaves in the shoots) on auxin-mediated stele morphology (Boyce 2005). In addition to their roles in water and nutrient uptake, roots played a key role in structural support. Large trees required an extensive rooting system to anchor the plant. The tree lycopods possessed very large bifurcating rooting systems that extended at shallow depths horizontally forming a stable platform for the pole-like trunk. It is likely that in early growth stages, development of the rooting system preceded that of the shoot system (Phillips and DiMichele 1992). Roots were co-opted as buttresses or mantles in the formation of the false trunks of tree ferns. These subsequently provided a substrate for vascular plant epiphytes and climbing or scrambling ferns, which are first observed in the fossil record on tree ferns during the Carboniferous Period (Rößler 2000).

In some groups, the evolution of arborescence did not involve major changes in growth form from the herbaceous antecedents. Arborescent horsetails maintained a rhizomatous rooting system that simply increased in girth through the addition of secondary tissues. Rooting systems became more highly modified in other groups. Progymnosperms developed bipolar growth, where shoot and specialized root were determined early in ontogeny as in their living relatives, the seed plants. Unlike the horsetails with their large horizontal rhizomes, the arborescent species of progymnosperm and pteridosperm had the ability to produce deeply penetrating roots. Bipolar growth was also a feature of the extinct tree lycopods, but phylogenetic and ontogenetic studies indicate that this evolved independently to seed plants (Rothwell and Erwin 1985; Bateman et al. 1992).

C. Origins of Roots in Clubmosses (Lycopods): The Modified Shoot Hypothesis

The origin of roots provides some remarkable examples of the co-option of organ systems to completely different roles. Perhaps the most striking of these is the hypothesized origin of the root system in extinct tree lycopods. The roots of these mainly arborescent plants are commonly associated with coal deposits of the Carboniferous Period, where they formed a massive but shallow bifurcating root system that was laterally extensive and could exceed 10 m in diameter (Taylor et al. 2009b). It has long been suspected that this rooting system—the so-called rhizomorph—is in fact a shoot modified for rooting (Rothwell and Erwin 1985; Rothwell 1995). The modified shoot hypothesis is based on the similarities between root and shoot anatomy, organization of appendages (rootlets, leaves), and evidence from a preserved early developmental stage (Stubblefield and Rothwell 1981). The main trunk of the rhizomorph is interpreted as a transformed stem and the rootlets as modified leaves. Under this interpretation, the whole shoot has become inverted and modified for anchorage and absorption of water and nutrients.

According to the modified shoot hypothesis, the rooting system of the tree lycopods is not homologous to that in other plants. Yet, despite this different evolutionary origin, recent research suggests that it may share similarities with other plants in the way that its development was regulated. Polar auxin transport is one such mechanism. Polar auxin flow can be recognized in certain types of fossil wood where lateral organs such as buds or branches disrupt flow, inducing the differentiation of distinctive patterns of circular tracheary elements in secondary xylem (Rothwell and Lev-Yadun 2005). Similar patterns have been observed in the root systems of tree lycopods demonstrating both the action of auxin transport and its direction of flow within the plant. As with modern seed plants, auxin flow was basipetal from shoot apex to root apex. These observations reveal that similar physiological mechanisms underlie the convergent evolution of bipolar growth in both plant groups (Sanders et al. 2011). The modified shoot hypothesis is derived from observations of long extinct fossils, but this once diverse clade of plants possesses living relatives in the family Isoetaceae, providing further potential to explore the developmental controls of this novel rooting system.

D. Mycorrhizae: An Early Symbiosis

Fungal symbioses involving mycorrhizae characterize over 90% of living plant species (Wang and Qiu 2006). The arbuscular mycorrhizae (Glomeromycota, Schüßler et al. 2001) form associations widely in liverworts, in hornworts, and in the roots of vascular plants (Smith and Read 1997; Read et al. 2000). Recent inferences from phylogenetic studies and direct observations of early fossils demonstrate that mycorrhizal associations are ancient and that arbuscular mycorrhizae are the ancestral type (Brundrett 2002; Taylor et al. 2004; Taylor and Krings 2005; Berbee and Taylor 2007; Krings et al. 2007; Strullu-Derrien and Strullu 2007). Much of the early fossil evidence comes from the Rhynie Chert. Arbuscules have been observed in both life cycle phases of one stem-group vascular plant (Remy et al. 1994). An extensive intercellular hyphal network and accompanying arbuscules formed within the cortex, and they were strikingly similar to those of modern *Glomus*. Furthermore, infection occurred very early in the plant's ontogeny, while the gametophyte was still attached to its spore. In this early fossil, the mycorrhizal infection was also clearly more extensive than that in living vascular plants. Arbuscules developed in the cortical tissues of both aerial and subaerial parts, with infection extending to the tips in both regions (Taylor et al. 2004). This extension of the infection to encompass the entire photosynthetic system is a key point of difference between the fossil *Aglaophyton* and modern vascular plants (Taylor et al. 2004; Strullu-Derrien and Strullu 2007).

Three fungal endophytes have been documented in the rhizome of the Rhynie Chert plant *Nothia aphylla*. These include narrow hyphae-producing clusters of small spores, large spherical spores or zoosporangia, and wide aseptate hyphae that formed intercellular vesicles within the cortex (Berbee and Taylor 2007; Krings et al. 2007). A range of host responses have also been demonstrated. These include bulges on infected rhizoids, encasement of intracellular hyphae, and partitioning of infected from uninfected tissues by secondarily thickened cell walls. The infection pathway in one of the *Nothia* endophytes is different to that in *Aglaophyton*. In *Nothia*, one fungus entered as an intracellular endophyte and then became intercellular within the cortex-forming vesicles, whereas in *Aglaophyton*, the fungus was an intercellular endophyte that became intracellular only in the mycorrhizal arbuscule zone.

These remarkable early fossils provide direct evidence of fungal associations and host plant responses that are in many respects similar to the mutualistic association of fungi in the roots of modern vascular plants. They indicate that arbuscular mycorrhizae and many features of the plant host response to fungal symbiosis are very ancient, implying a shared common origin. This is consistent with recent discoveries of similar signaling transduction pathways (*sym* genes) and cellular mechanisms of host response across modern land plants (Bonfante and Genre 2008; Bonfante and Selosse 2010; Wang et al. 2010).

E. Roots, Soils, and Early Phanerozoic Ecosystems

The Devonian Period witnessed a progressive increase in root biomass (Figure 1.5) that had an enormous impact on the environment, notably on the development of soils and on the evolution of the atmosphere. Prior to the Devonian Period, soils, if developed at all, are thought to have been predominantly thin and of microbial origin (Retallack 2001). The rhizomatous root systems of the earliest land plants were shallow and essentially superficial. They were unlikely to have contributed much to soil formation or to the weathering of the land surface. By the latter part of the Early Devonian, the development of positively geotropic roots coupled with a progressive increase in plant size led to more extensive and penetrating rooting systems (Elick et al. 1998). These nurtured an environment suitable for the stabilization of shallow soils. A further significant step in the development of soils was the evolution of the cambium, which from the outset is a tissue

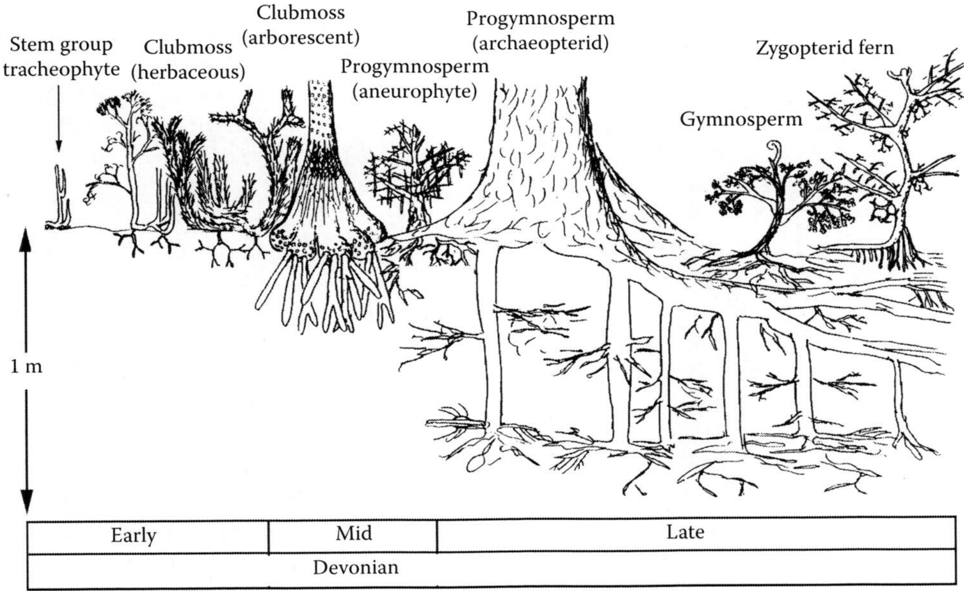

FIGURE 1.5 Relative size, morphology, and penetration depth of root systems from selected Devonian plants. (Modified after Algeo, T.J. and Scheckler, S.E., *Phil. Trans. R. Soc. Lond.*, B353, 113, 1998. With permission.)

that was expressed in both shoots and roots. Cambial activity led to a significant increase in plant size and in the extent and development of roots. Soil penetration depths were shallow (<20 cm) during the Eifelian–Givetian (386–377 Ma) but increased (80–100 cm) as archaeopterid trees spread during the Frasnian–Famennian (377–363 Ma) (Algeo et al. 2001).

The diversity of soils also increased during the Devonian Period (Retallack 2001). This phenomenon was the result of the interaction of several different factors including the increasing plant size, diversification of plant communities, exploitation of new habitats, and differences at a local scale in geology and topography. Roots increased rates of weathering by facilitating the movement of solutes through soils and through the release of humic acids. The effectiveness of soil mixing was boosted through anatomical innovations (Algeo and Scheckler 1998; Algeo et al. 2001). The evolution of the cambium initiated continuous perennial root growth and long-term survival of roots. The development of endogenous and adventitious roots, particularly in progymnosperms, enabled repeated penetration of a given soil volume. Plant communities had diversified in terms of both species and ecology. Forests contained trees comparable in size to modern forms. Landscape-level habitat diversity encompassed waterlogged peat-forming wetlands through well-drained higher-elevation hinterland forests. By the end of the Devonian Period, the biomass of continental ecosystems had increased enormously as had the structure and diversity of soils. Forest soils with modern profiles are recognizable at this time (Retallack 2001).

Deeper soils and larger plants go hand in hand, but the evolution of both may have been more intimately connected than previously supposed. One of the most remarkable phenomena of the Devonian Period was the independent and more or less concurrent evolution of trees in several plant groups (Kenrick and Davis 2004). Mosbrugger (1990) argued that this phenomenon may in fact have been caused by positive feedback between soil and plant, mediated by the roots. In essence, the idea is based on the observation that larger plants generally require more stable and comparatively deeper soils and that plants themselves contribute to soil formation. Under circumstances such as these, increments in plant size are expected to lead to increments in the quality and depth of the soil profile, which in turn would provide soils suitable for even larger plants. Plant size and soil depth therefore increased in parallel until other limiting factors came into play. Feedbacks such as this between plants and aspects of their environment are increasingly seen as important and pervasive in early terrestrial ecosystems. For example, the initial evolution of leaves was influenced by falling concentrations of atmospheric CO_2 during the Devonian Period, which itself was driven by plant-induced surficial weathering of rocks and the sequestration of carbon in sediments and soils (Osborne et al. 2004). Plant, soil, and atmosphere are therefore linked in an intricate network of geophysiological feedbacks (Beerling and Berner 2005).

The broader relevance of the root/soil relationship to the evolution of terrestrial and marine ecosystems during the Paleozoic Era is now widely appreciated (Berner and Kothavala 2001; Bergman et al. 2004; Beerling and Berner 2005; Berner et al. 2007; Beerling 2012; Kenrick et al. 2012). One area of particular importance relates to the effects of plants on the composition of the atmosphere through geological time. Living organisms are part of the biogeochemical carbon cycle, in which CO_2 is drawn down from the atmosphere through two main processes. One involves photosynthesis followed by burial of organic matter in sediments. The second depends on root-mediated weathering of calcium and magnesium silicates in surface rocks and soils (Berner and Kothavala 2001). This weathering process incorporates many steps, including the synthesis of plant- and atmosphere-derived carbonic or organic acids, the conversion of CO_2 to HCO_3^- in soil and groundwater, transport of HCO_3^- through rivers to the sea, and its eventual precipitation as a component of limestone or dolomite on the ocean floor. A vegetation of large plants with extensive root systems and associated mycorrhizal fungi greatly enhances the weathering effect (Berner and Kothavala 2001; Taylor et al. 2009a). Carbon sequestered in these ways is released back into the atmosphere slowly over tens or hundreds of millions of years through oxidative weathering of buried carbon and through the thermal breakdown of marine carbonates, resulting in degassing to the Earth's surface. Considered over geological timescales, the net effect of the presence of plants on land is to draw CO_2 out of the atmosphere, and this is thought to have had a major effect on climate, contributing to the glaciations of the latter part of the Paleozoic Era. The root-mediated weathering of silicates is therefore an important variable in biogeochemical models (Berner and Kothavala 2001; Bergman et al. 2004; Berner et al. 2007), but the effects of different types of vegetation are still poorly understood. The rooting systems of large vascular plants (i.e., trees) are thought to exert the most significant effect, but a recent laboratory-based study has shown that mosses also considerably enhanced silicate weathering over background abiotic levels (Lenton et al. 2012). The evolution of even modest rhizoid-based rooting systems in land plants may therefore have had a major effect on the chemistry of the atmosphere and on climate.

IV. Leading Topics, Gaps, and the Future

The origins and early evolution of roots went hand in hand with the development of shoots. Specialized rooting systems were necessary for the evolution of increasingly larger and more sophisticated plants capable of exploiting a broad range of habitats. Fossils provide an important source of evidence indicating that certain aspects of root morphology and development (e.g., bipolar growth, polar auxin transport, vascular cambium) evolved in a piecemeal fashion in various groups of vascular plants. Other features, such as rhizoids, appear to be homologous across land plants, and fossils from the Rhynie Chert provide new insights such as the presence of rhizoids in vascular plant sporophytes.

Other aspects of root evolution are still very poorly understood, including the early development of endogenous meristems, the root cap, and the endodermis (Gensel et al. 2001; Boyce 2005). Exceptional sites of fossil preservation such as the Rhynie Chert could provide insights into the evolution of these features. Also of key importance is the use of molecular methods to investigate the developmental controls of key cell types, tissues, and organs (Menand et al. 2007; Nishiyama 2007).

Fungi in general and mycorrhizae in particular were a ubiquitous component of early terrestrial ecosystems, but their study as fossils has received much less attention than the associated plants. We are beginning to understand the complexity of the relationships between mycorrhizal fungi and early plants through detailed studies of the Rhynie Chert (Taylor et al. 2004). One unexpected outcome has been the recognition that mycorrhizal infection in at least one plant encompassed the entire photosynthetic shoot system. Also, host plant responses are being recognized and documented (Krings et al. 2007). These discoveries are significant and relevant to understanding the early evolution of signaling transduction pathways and cellular mechanisms of host response across modern land plants (Bonfante and Genre 2008; Bonfante and Selosse 2010). Further detailed study of plant-fungal associations in the Rhynie Chert plants will undoubtedly yield additional significant insights.

Plants are a key component of the biogeochemical carbon cycle, and the advent of a land flora is thought to have had a huge impact on the evolution of atmosphere and climate through geological time. Roots influence the development of soils and they effect the deposition and chemistry of sediments (Retallack 2001). One of the most significant effects of plants is root-mediated weathering of calcium and magnesium silicates, which over geological time altered in major ways the composition of the atmosphere (Berner and Kothavala 2001). Rates of weathering are still rather poorly understood for different vegetation types and plant groups, especially for bryophytes and herbaceous vascular flora. Further experimental research is needed to evaluate the rates of various root- and rhizoid-mediated weathering (Lenton et al. 2012). Also needed is a more accurate understanding of the time of land colonization by plants and the time of acquisition of those key features of rooting systems (Kenrick et al. 2012).

References

Algeo TJ, Scheckler SE. 1998. Terrestrial-marine teleconnections in the Devonian: Links between the evolution of land plants, weathering processes, and marine anoxic events. *Phil Trans R Soc Lond* B353:113–130.

Algeo TJ, Scheckler SE, Maynard JB. 2001. Effects of the middle to late Devonian spread of vascular land plants on weathering regimes, marine biotas, and global climate. In: *Plants Invade the Land: Evolutionary and Environmental Consequences*, eds. PG Gensel, D Edwards, pp. 213–236. New York: Columbia University Press.

Bateman RM. 1991. Palaeobiological and phylogenetic implications of anatomically-preserved *Archaeocalamites* from the Dinantian of Oxroad Bay and Loch Humphrey Burn, southern Scotland. *Palaeontographica* B223:1–59.

Bateman RM, DiMichele WA, Willard DA. 1992. Experimental cladistic analysis of anatomically preserved lycopsids from the Carboniferous of Euramerica: An essay on paleobotanical phylogenetics. *Ann Mo Bot Gdn* 79:500–559.

Bateman RM, Hilton J. 2009. Palaeobotanical systematics for the phylogenetic age:applying organ- species, form-species and phylogenetic species concepts in a framework of reconstructed fossil and extant whole-plants. *Taxon* 58:1254–1280.

Beck CB. 1967. *Eddya sullivanensis*, gen. et sp. nov., a plant of gymnospermic morphology from the Upper Devonian of New York. *Palaeontographica* B121:1–21.

Beerling DJ. 2012. Atmospheric carbon dioxide: A driver of photosynthetic eukaryote evolution for over a billion years. *Phil Trans R Soc Lond* B367:477–482.

Beerling DJ, Berner RA. 2005. Feedbacks and the coevolution of plants and atmospheric CO_2. *Proc Natl Acad Sci USA* 102:1302–1305.

Berbee ML, Taylor JW. 2007. Rhynie chert: A window into a lost world of complex plant-fungus interactions. *New Phytol* 174:475–479.

Bergman NM, Lenton TM, Watson AJ. 2004. COPSE: A new model of biogeochemical cycling over Phanerozoic time. *Am J Sci* 304:397–437.

Berner RA. 1998. The carbon cycle and CO_2 over Phanerozoic time: The role of land plants. *Phil Trans R Soc Lond* B353:75–82.

Berner RA, Kothavala Z. 2001. GEOCARB III; a revised model of atmospheric CO_2 over Phanerozoic time. *Am J Sci*, 301:182–204.

Berner RA, VandenBrooks JM, Ward PD. 2007. Oxygen and evolution. *Science* 316:557–558.

Bonfante P, Genre A. 2008. Plants and arbuscular mycorrhizal fungi: An evolutionary-developmental perspective. *Trends Plant Sci* 13:492–498.

Bonfante P, Selosse M-A. 2010. A glimpse into the past of land plants and of their mycorrhizal affairs: From fossils to evo-devo. *New Phytol* 186:267–270.

Boyce KC. 2005. The evolutionary history of roots and leaves. In: *Vascular Transport in Plants*, eds. NM Holbrook, MA Zwieniecki, pp. 479–500. San Diego, CA: Elsevier.

Brundrett MC. 2002. Coevolution of roots and mycorrhizas of land plants. *New Phytol* 154:275–304.

Cridland AA. 1964. *Amyelon* in American coal balls. *Palaeontology* 7:186–209.

Davies NS, Gibling MR. 2010a. Cambrian to Devonian evolution of alluvial systems: The sedimentological impact of the earliest land plants. *Earth Sci Rev* 98:171–200.

Davies NS, Gibling MR. 2010b. Paleozoic vegetation and the Siluro-Devonian rise of fluvial lateral accretion sets. *Geology* 38:51–54.

DiMichele WA, Phillips TL, Pfefferkorn HW. 2006. Paleoecology of late Paleozoic pteridosperms from tropical Euramerica. *J Tor Bot Soc* 133:83–118.

Doran JB. 1980. A new species of *Psilophyton* from the lower Devonian of northern New Brunswick, Canada. *Can J Bot* 58:2241–2262.

Driese SG, Mora, CI. 2001. Diversification of Siluro-Devonian plant traces in paleosols and influence on estimates of paleoatmospheric CO_2 levels. In: *Plants Invade the Land: Evolutionary and Environmental Perspectives*, eds. PG Gensel, D Edwards, pp. 237–253. New York: Columbia University Press.

Driese SG, Mora CI, Elick JM. 1997. Morphology and taphonomy of root and stump casts of the earliest trees (middle to late Devonian), Pennsylvania and New York, U.S.A. *Palaios* 12:524–537.

Edwards DS. 1986. *Aglaophyton major*, a non-vascular land plant from the Devonian Rhynie Chert. *Bot J Lin Soc* 93:173–204.

Edwards D. 2004. Embryophytic sporophytes in the Rhynie and Windyfield cherts. *Trans R Soc Edin Earth Sci* 94:397–410.

Edwards D, Kenrick P. 1986. A new zosterophyll from the lower Devonian of Wales. *Bot J Lin Soc* 92:269–283.

Elick JM, Driese SG, Mora CI. 1998. Very large plant and root traces from the early to middle Devonian; implications for early terrestrial ecosystems and atmospheric pCO_2. *Geology* 26:143–146.

Fairon-Demaret M, Berry CM. 2000. A reconsideration of *Hyenia elegans* Kräusel et Weyland and *Hyenia 'complexa'* leclercq: Two middle Devonian cladoxylopsids from western Europe. *Int J Plant Sci* 161:473–494.

Fairon-Demaret M, Li C-S. 1993. *Lorophyton goense* gen. et sp. nov. from the lower Givetian of Belgium and a discussion of the middle Devonian Cladoxylopsida. *Rev Pal Pal* 77:1–22.

Fayers SR, Trewin NH. 2004. A review of the palaeoenvironments and biota of the Windyfield Chert. *Trans R Soc Edin Earth Sci* 94:325–339.

Forey P, Fortey R, Kenrick P, Smith AB. 2004. Taxonomy and fossils: A critical appraisal. *Phil Trans R Soc Lond* 359:639–653.

Gensel PG. 2008. The earliest land plants. *Annu Rev Ecol Evol Syst* 39:459–477.

Gensel PG, Edwards D (eds.) 2001. Plants invade the land: Evolutionary and environmental perspectives. In: *Critical Moments and Perspectives in Earth History and Paleobiology Series*. New York: Columbia University Press.

Gensel PG, Kotyk ME, Basinger JF. 2001. Morphology of above- and below-ground structures in early Devonian (Pragian-Emsian) plants. In: *Plants Invade the Land: Evolutionary and Environmental Perspectives*, eds. PG Gensel, D Edwards, pp. 83–102. New York: Columbia University Press.

Gerrienne P. 1988. Early Devonian plant remains from Marchin (north of Dinant Synclinorium, Belgium), I. *Zosterophyllum deciduum* sp. nov. *Rev Pal Pal* 55:317–335.

Hao S-G, Beck CB. 1993. Further observations on *Eophyllophyton bellum* from the Lower Devonian (Siegenian) of Yunnan, China. *Palaeontographica* B230:27–41.

Hao SG, Xue JZ, Guo DL, Wang DM. 2010. Earliest rooting system and root: Shoot ratio from a new *Zosterophyllum* plant. *New Phytol* 185:217–225.

Hernick LV, Landing E, Bartowski KE. 2008. Earth's oldest liverworts—*Metzgeriothallus sharonae* sp. nov. from the Middle Devonian (Givetian) of eastern New York, USA. *Rev Pal Pal* 148:154–162.

Hillier RD, Edwards D, Morrissey LB. 2008. Sedimentological evidence for rooting structures in the early Devonian Anglo-Welsh Basin (UK), with speculation on their producers. *Palaeogeog Palaeoclim Palaeoecol* 270:366–380.

Hilton J, Bateman RM. 2006. Pteridosperms are the backbone of seed-plant phylogeny. *J Tor Bot Soc* 133:119–168.

Hilton J, Geng B, Kenrick P. 2003. A novel late Devonian (Frasnian) woody Cladoxylopsid from China. *Int J Plant Sci* 164:793–805.

Hueber FM, Banks HP. 1979. *Serrulacaulis furcatus* gen. et sp. nov., a new zosterophyll from the lower Upper Devonian of New York State. *Rev Pal Pal* 28:169–189.

Karol KG, Arumuganathan K, Boore JL et al. 2010. Complete plastome sequences of *Equisetum arvense* and *Isoetes flaccida*: Implications for phylogeny and plastid genome evolution of early land plant lineages. *BMC Evol Biol* 10:321.

Kenrick P. 2000. The relationships of vascular plants. *Phil Trans R Soc Lond* B355:847–855.

Kenrick P. 2002. The origin of roots. In: *Plant Roots: The Hidden Half*, eds. Y Waisel, A Eshel, U Kafkafi, pp. 1–13. New York: Marcel Dekker, Inc.

Kenrick P, Crane PR. 1997. The origin and early diversification of land plants: A cladistic study. *Smithsonian Series in Comparative Evolutionary Biology*. Washington, DC: Smithsonian Institution Press.

Kenrick P, Davis PG. 2004. *Fossil Plants*. London, U.K.: The Natural History Museum.

Kenrick P, Wellman CH, Schneider H, Edgecombe GD. 2012. A timeline for terrestrialization: Consequences for the carbon cycle in the Palaeozoic. *Phil Trans R Soc Lond* B367:519–536.

Kerp H, Hass H, Mosbrugger V. 2001. New data on *Nothia aphylla* Lyon 1964 ex El-Saadawy et Lacey 1979, a poorly known plant from the lower Devonian Rhynie Chert. In: *Plants Invade the Land: Evolutionary and Environmental Perspectives*, eds. PG Gensel, D Edwards, pp. 52–82. New York: Columbia University Press.

Kerp H, Trewin NH, Hass H. 2004. New gametophytes from the early Devonian Rhynie Chert. *Trans R Soc Edin Earth Sci* 94:411–428.

Kotyk ME, Basinger JF, Gensel PG, de Freitas TA. 2002. Morphologically complex plant macrofossils from the late Silurian of Arctic Canada. *Am J Bot* 89:1004–1013.

Krings M, Kerp H, Taylor TN, Taylor EL. 2003. How Paleozoic vines and lianas got off the ground:on scrambling and climbing Carboniferous-Early Permian pteridosperms. *Bot Rev* 69:204–224.

Krings M, Taylor TN, Hass H, Kerp H, Dotzler N, Hermsen EJ. 2007. Fungal endophytes in a 400-million-yr-old land plant: Infection pathways, spatial distribution, and host responses. *New Phytol* 174:648–657.

Labandeira CC. 2005. Invasion of the continents: Cyanobacterial crusts to tree-inhabiting arthropods. *Trends Ecol Evol* 20:253–262.

Labandeira C. 2007. The origin of herbivory on land: Initial patterns of plant tissue consumption by arthropods. *Insect Sci* 14:259–275.

Lenton TM, Crouch M, Johnson M, Pires N, Dolan L. 2012. First plants cooled the Ordovician. *Nat Geosci* 5:86–89.

Li C-S. 1992. *Hsua robusta*, an early Devonian plant from Yunnan Province, China and its bearing on some structures of early land plants. *Rev Pal Pal* 71:121–147.

Li C-S, Hsü J. 1987. Studies on a new Devonian plant *Protopteridophyton devonicum* assigned to primitive fern from south China. *Palaeontographica* B207:111–131.

Lyon AG, Edwards D. 1991. The first zosterophyll from the lower Devonian Rhynie Chert, Aberdeenshire. *Trans R Soc Edin Earth Sci* 82:323–332.

Mark DF, Rice CM, Fallick AE, Trewin NH, Lee MR, Boyce A, Lee JKW. 2011. 40Ar/39Ar dating of hydrothermal activity, biota and gold mineralization in the Rhynie hot-spring system, Aberdeenshire, Scotland. *Geochim. Cosmochim. Acta* 75:555–569.

Menand B, Yi K, Jouannic S et al. 2007. An ancient mechanism controls the development of cells with a rooting function in land plants. *Science* 316:1477–1480.

Meyer-Berthaud B, Scheckler SE, Bousquet J-L. 2000. The development of *Archaeopteris*: New evolutionary characters from the structural analysis of an early Famennian trunk from southeast Morocco. *Am J Bot* 87:456–468.

Meyer-Berthaud B, Scheckler SE, Wendt J. 1999. *Archaeopteris* is the earliest known modern tree. *Nature* 398:700–701.

Meyer-Berthaud B, Soria A, Decombeix AL. 2010. The land plant cover in the Devonian: A reassessment of the evolution of the tree habit. In: *The Terrestrialization Process: Modelling Complex Interactions at the Biosphere-Geosphere Interface*, eds. M Vecoli, G Clément, B Meyer-Berthaud, pp. 59–70. London, U.K.: The Geological Society.

Meyer-Berthaud B, Stein WE. 1995. A reinvestigation of *Stenomyelon* from the late Tournaisian of Scotland. *Int J Plant Sci* 156:863–895.

Mosbrugger V. 1990. The tree habit in land plants. *Lect Notes Earth Sci* 28:1–161.

Nishiyama T. 2007. Evolutionary developmental biology of non-flowering land plants. *Int J Plant Sci* 168:37–47.

Osborne CP, Beerling DJ, Lomax BH, Chaloner WG. 2004. Biophysical constraints on the origin of leaves inferred from the fossil record. *Proc Nanl Acad Sci USA* 101:10360–10362.

Phillips TL, DiMichele, WA. 1992. Comparative ecology and life-history biology of arborescent lycopsids in late Carboniferous swamps of Euramerica. *Ann Mo Bot Gdn* 79:560–588.

Phillips TL, Galtier J. 2005. Evolutionary and ecological perspectives of late Paleozoic ferns. Part I. Zygopteridales. *Rev Pal Pal* 135:165–203.

Pigg KB, Rothwell GW. 1983. *Chaloneria* gen. nov.; heterosporous lycophytes from the Pennsylvanian of North America. *Bot Gaz* 144:132–147.

Poinar G, Kerp H, Hass H. 2008. *Palaeonema phyticum* gen. n., sp n. (Nematoda: Palaeonematidae fam. n.), a Devonian nematode associated with early land plants. *Nematology* 10:9–14.

Pryer KM, Schuettpelz E, Wolf PG, Schneider H, Smith AR, Cranfill R. 2004. Phylogeny and evolution of ferns (monilophytes) with a focus on the early leptosporangiate divergences. *Am J Bot* 91:1582–1598.

Qiu YL. 2008. Phylogeny and evolution of charophytic algae and land plants. *J Syst Evol* 46:287–306.

Raven JA, Edwards D. 2001. Roots: Evolutionary origins and biogeochemical significance. *J Exp Bot* 52:381–401.

Read DJ, Duckett JG, Francis R, Ligrone R, Russell A. 2000. Symbiotic fungal associations in 'lower' land plants. *Phil Trans R Soc Lond* B355:815–831.

Remy W, Hass H. 1996. New information on gametophytes and sporophytes of *Aglaophyton major* and inferences about possible environmental adaptations. *Rev Pal Pal* 90:175–193.

Remy W, Hass H, Schultka S. 1986. *Anisophyton potoniei* nov. spec. aus den Kühlbacher Schichten (Emsian) vom Steinbruch Ufersmühle, Wiehltalsperre. *Argumenta Palaeobotanica* 7:123–138.

Remy W, Taylor TN, Hass H, Kerp H. 1994. Four hundred-million-year-old vesicular arbuscular mycorrhizae. *Proc Natl Acad Sci USA* 91:11841–11843.

Retallack GJ. 2001. *Soils of the Past - An Introduction to Paleopedology*. Hoboken, NJ: John Wiley & Sons.

Rice CM, Ashcroft WA, Batten DJ et al. 1995. A Devonian auriferous hot spring system, Rhynie, Scotland. *J Geol Soc Lond* 152:229–250.

Rice CM, Trewin NH, Anderson LI. 2002. Geological setting of the early Devonian Rhynie cherts, Aberdeenshire, Scotland; an early terrestrial hot spring system. *J Geol Soc Lond* 159:203–214.

Rößler R. 2000. The late Palaeozoic tree fern *Psaronius*; an ecosystem unto itself. *Rev Pal Pal* 108:55–74.

Rothwell GW. 1984. The apex of *Stigmaria* (Lycopsida), rooting organ of Lepidodendrales. *Am J Bot* 71:1031–1034.

Rothwell GW. 1995. The fossil history of branching: Implications for the phylogeny of land plants. In: *Experimental and Molecular Approaches to Plant Biosystematics*, eds. PC Hoch, AG Stephenson, pp. 71–86. St. Louis, MO: Missouri Botanical Garden.

Rothwell GW, Erwin DM. 1985. The rhizomorph apex of *Paurodendron*: Implications for homologies among the rooting organs of lycopsida. *Am J Bot* 72:86–98.

Rothwell GW, Lev-Yadun S. 2005. Evidence of polar auxin flow in 375 million-year-old fossil wood. *Am J Bot* 92:903–906.

Rothwell GW, Nixon KC. 2006. How does the inclusion of fossil data change our conclusions about the phylogenetic history of euphyllophytes? *Int J Plant Sci* 167:737–749.

Rothwell GW, Whiteside KL. 1974. Rooting structures of the Carboniferous medullosan pteridosperms. *Can J Bot* 52:97–102.

Sanders H, Rothwell GW, Wyatt SE. 2011. Parallel evolution of auxin regulation in rooting systems. *Plant Syst Evol* 291:221–225.

Scheckler SE. 1995. Progymnosperms have gymnospermous roots. In: *The Evolution of Plant Architecture*, Abstract volume, eds. AR Hemsley, MH Kurmann. London, U.K.: Linnean Society London/Royal Botanic Gardens, Kew.

Schüßler A, Schwarzott D, Walker C. 2001. A new fungal phylum, the Glomeromycota: Phylogeny and evolution. *Mycol Res* 105:1413–1421.

Schweitzer H-J, Li C-S. 1996. *Chamaedendron* nov. gen., eine multisporangiate lycophyte aus dem Frasnium südchinas. *Palaeontographica* B238:45–69.

Shear WA, Selden PA. 2001. Rustling in the undergrowth: Animals in early terrestrial ecosystems. In: *Plants Invade the Land: Evolutionary and Environmental Perspectives*, eds. PG Gensel, D Edwards, pp. 29–51. New York: Columbia University Press.

Smith SE, Read DJ. 1997. *Mycorrhizal Symbiosis*. London, U.K.: Academic Press.

Soria A, Meyer-Berthaud B. 2004. Tree fern growth strategy in the late Devonian cladoxylopsid species *Pietzschia levis* from the study of its stem and root system. *Am J Bot* 91:10–23.

Stein WE, Mannolini F, Hernick LV, Landing E, Berry CM. 2007. Giant cladoxylopsid trees resolve the enigma of the Earth's earliest forest stumps at Gilboa. *Nature* 446:904–907.

Stewart WN. 1947. A comparative study of stigmarian appendages and *Isoetes* roots. *Am J Bot* 34:315–324.

Strullu-Derrien C, Strullu DG. 2007. Mycorrhization of fossil and living plants. *Comptes Rendus Palevol* 6:483–494.

Stubblefield SP, Rothwell GW. 1981. Embryogeny and reproductive biology of *Bothrodendrostrobus mundus* (Lycopsida). *Am J Bot* 68:625–634.

Taylor TN, Klavins SD, Krings M, Taylor EL, Kerp H, Hass H. 2003. Fungi from the Rhynie chert: A view from the dark side. *Trans R Soc Edin Earth Sci* 94:457–473.

Taylor TN, Krings M. 2005. Fossil microorganisms and land plants: Associations and interactions. *Symbiosis* 40:119–135.

Taylor LL, Leake JR, Quirk J, Hardy K, Banwart SA, Beerling DJ. 2009a. Biological weathering and the long-term carbon cycle: Integrating mycorrhizal evolution and function into the current paradigm. *Geobiology* 7:171–191.

Taylor TN, Taylor EL, Krings M. 2009b. *Paleobotany: The Biology and Evolution of Fossil Plants*. Boston, MA: Academic Press.

Trewin NH, Rice CM (eds.) 2004. The Rhynie Chert host-spring system: Geology, biota and mineralization. *Trans R Soc Edin Earth Sci* 94:283–521.

Trewin NH, Rolfe, WDI, Clarkson ENK, Panchen AL. 1994. Depositional environment and preservation of biota in the lower Devonian hot-springs of Rhynie, Aberdeenshire, Scotland. *Trans R Soc Edin Earth Sci* 84:433–442.

Wang Z, Geng B-Y. 1997. A new middle Devonian plant: *Metacladophyton tetraxylum* gen. et sp. nov. *Palaeontographica* B243:85–102.

Wang DM, Lin YJ. 2007. A new species of *Metacladophyton* from the late Devonian of China. *Int J Plant Sci* 168:1067–1084.

Wang B, Qiu YL. 2006. Phylogenetic distribution and evolution of mycorrhizas in land plants. *Mycorrhiza* 16:299–363.

Wang B, Yeun LH, Xue J-Y, Liu Y, Ané J-M, Qiu Y-L. 2010. Presence of three mycorrhizal genes in the common ancestor of land plants suggests a key role of mycorrhizas in the colonization of land by plants. *New Phytol* 186:514–525.

2

Arabidopsis Root: A Model Organ in Plant Genomics

Jaimie M. Van Norman
Duke University

Louisa M. Liberman
Duke University

Philip N. Benfey
Duke University

I. Introduction: The Genomic Sequencing Revolution in Plants

Given their subterranean existence, roots could be considered among the most inaccessible organs for biological study. Fortunately, roots are extremely amenable to growth in the absence of soil; currently, *Arabidopsis thaliana* seedlings are extensively studied in non-soil media. This dicotyledonous seedling has several attributes that facilitate studies of root biology including rapid germination, small size, and simple root architecture. By the late 1980s, *Arabidopsis* was being widely used as a research model for plant biology. The *Arabidopsis* root was emerging as a particularly powerful and elegant model in cellular and organ developmental biology and environmental response. In the decade prior to publication of the *Arabidopsis* genome, functionally important molecules in root biology were identified through genetic screens. However, the difficulty of cloning the identified loci was a significant roadblock in understanding the molecular mechanisms driving root growth and development. The attributes that initially made the *Arabidopsis* root a good physiological and developmental model organ have made it invaluable in the genomics era.

In the 1980s, it became clear that selection of a model plant would be useful for several reasons including the possibility of acquiring the entire sequence of a plant genome. The speed of sequencing was relatively slow and computational tools were in their infancy; thus, *Arabidopsis thaliana* was a practical candidate for sequencing as it possesses a relatively small genome compared to the average flowering plant and was amenable to molecular genetics methods (Leutwiler et al. 1984; Pruitt and Meyerowitz 1986). Many features of *Arabidopsis* made it advantageous as a genomic model plant, despite its lack of overt economic value. Key among these were the ease of transformation, its largely self-fertilizing nature, and the growing *Arabidopsis* research community who sought tools to identify the genes in their favorite quantitative trait loci (QTL) and developmental mutants (Meinke et al. 1998; Somerville and Koornneef 2002). The completion of the *Arabidopsis* genome paved the way for the genomic era in plants. Rapid technological improvements since 2000 have facilitated genome sequencing of several agriculturally relevant plant species, including rice and corn.

In this chapter, we describe some of the advances enabled by the genomic revolution as they pertain to plant and, more specifically, root biology. The sequenced genome greatly facilitated forward genetic analyses and led to new experimental approaches, collectively termed reverse genetics. Moreover, quantitative and association genetics were simplified and accelerated by a fully sequenced genome. Comparison between plant genomes can provide information regarding the genomic composition and evolutionary divergence of plants and can also provide insights into the origins of specialized metabolic pathways or organs, such as roots. Genomic approaches can be used to understand the unique qualities of plant biology and harness those qualities to maximize production of food, fiber, and fuel for global consumption. Importantly, the genomic sequence also led to the generation of invaluable tools for plant and root molecular biology, genetics, and systems biology approaches. Genome-wide transcriptional profiling, for example, enabled researchers to address the function of organs and tissues by determining which genes are being expressed. In roots, these studies have provided unprecedented resolution of transcriptional information and revealed unique cell type-specific characteristics as well as dynamic gene expression throughout developmental time. Additionally, transcriptional profiling of roots under stress demonstrated that cell identity mediates the root's response to abiotic stress. Finally, the future of genome-scale investigation, which is likely to require integration of transcriptional profiling with the proteomic and metabolomic landscapes, is discussed.

II. First Plant Genome: *Arabidopsis*

The start of the new century brought the completion of the first plant genome sequence and only the third completely sequenced multicellular eukaryotic genome. The ~125 megabase *Arabidopsis* genome had the largest gene set compared to *Drosophila melanogaster* and *Caenorhabditis elegans* but with just over 25,000 predicted genes had a gene density similar to *C. elegans* (*Arabidopsis* Genome Initiative 2000). The larger genome size of *Arabidopsis* is likely due to the number of duplicated genes and segmental repeats. Once the human genome was completed, comparison to the *Arabidopsis* genome revealed a similar number of genes, despite the human genome having more than 20 times the number of nucleotides (Venter et al. 2001).

Because a single-celled eukaryote was the last common ancestor between plants and animals, these genomes ultimately represent two independent solutions to the challenges associated with multicellularity. Comparisons of worm, fly, and plant genomes provided insight into the similarities and differences in genomic composition between these divergent groups (*Arabidopsis* Genome Initiative 2000). Conservation of predicted proteins in functional categories such as protein synthesis, energy, and metabolism indicated parallels in basic cellular functions. Additionally, characterization of genes conserved between *Arabidopsis* and humans can provide insights into the molecular mechanisms of human disease (Jones et al. 2008). The genomic differences between members of these two eukaryotic kingdoms can be used to examine the molecular nature of "plantness." For example, only about half of the putative plant transcription factors have counterparts in the other organisms, suggesting plant-specific evolution of many transcriptional regulators (Riechmann et al. 2000). Additionally, proteins predicted to function in pathways involved in perception and signal transduction of exogenous cues were vastly more abundant in plants (*Arabidopsis* Genome Initiative 2000). These differences likely reflect molecular solutions to the constraints of a sessile lifestyle and lack of cellular movement in plants and also may reveal plant-specific solutions to common problems of multicellularity, such as cell-to-cell communication. The *Arabidopsis* genome sequence provided a new tool to test hypotheses about plant genome evolution and gene function and also broke ground for the development of new experimental approaches to address questions in plant and root biology.

A. Forward and Reverse Genetic Approaches

Publication of the *Arabidopsis* genome sequence led to a major windfall in the number of identified genes with roles in the root and also in the types of molecular genetics analyses that could be applied to the study of root biology (reviewed in Ishida et al. 2008; Petricka and Benfey 2008). The genome sequence gave researchers access to an array of molecular markers and DNA clones that rapidly accelerated map-based cloning of genes based on mutant phenotypes. Greater efficiency in the progress from mutant phenotype to gene identity led to the rapid identification of key regulators of root development and the beginning of genetic pathway construction for these processes (example reviewed by Guimil and Dunand 2007). Innovative types of forward genetic screens could be conducted including the use of enhancer trap constructs to identify genes expressed in specific cell types and mutagenesis of reporter lines to isolate genes functioning upstream of a specific gene or developmental process (Aida et al. 2004; Willemsen et al. 2008). Despite the important advances brought about by the genome sequence itself, perhaps the most important information was the list of predicted proteins encoded by the genome. With these in hand, root biologists were able to use reverse genetic approaches to target proteins with specific functional domains or from specific gene families for genetic analyses by methods such as RNA interference. These types of experiments enabled the comparison of proteins with evolutionarily conserved functional domains and the functional dissection of gene families in *Arabidopsis* (reviewed by Alonso and Ecker 2006). Understanding the shared or specialized functions of individual gene family members is significant not only for studies of growth and development, but also in understanding genome evolution.

B. Quantitative and Association Genetic Approaches

Although forward and reverse genetic approaches are invaluable in characterizing gene function, natural variation offers an alternative to laboratory-induced mutations to understand root biology (Alonso-Blanco and Koornneef 2000). A key asset of *Arabidopsis* as a plant model system is the phenotypic variation between different, reproductively and often geographically isolated naturally occurring accessions. Phenotypes or traits that vary within and between populations are observed across a phenotypic continuum and thus described as quantitative traits. Quantitative traits often result from interactions between the genome and the environment. These types of traits can be of agricultural importance as they often directly lead to changes in plant growth and productivity. Therefore, characterization of quantitative traits especially in roots is increasingly valuable as the global demand for food, fiber, and fuel increases, while the availability of arable land declines.

Quantitative trait analyses aim to identify the genomic (molecular) basis of phenotypic variation, termed quantitative trait loci (QTL) (Alonso-Blanco and Koornneef 2000). Methods of mapping QTLs are similar to those utilized in forward genetic approaches to identify genes and were greatly expedited with genome sequence data. Although the *Arabidopsis* genome sequence was derived from a single accession (Col-0), molecular markers were identified in other accessions by sequencing smaller genomic fragments (Schmid et al. 2003; Nordborg et al. 2005). A mapped QTL, however, may contain hundreds or thousands of candidate genes, depending on the size of the mapping population and the extent of genotyping information. Additionally, quantitative traits are often multigenic, meaning that several genes may contribute to phenotypic variability. These difficulties make identification of the specific genomic change(s) responsible for the phenotypic variation at a QTL tedious and time-consuming.

The generation of recombinant inbred lines (RILs) has helped mitigate some of the difficulties associated with QTL mapping. RILs are the product of a cross between two accessions, in which individual progeny are self-fertilized over many generations until nearly all genomic loci are homozygous (Maloof 2003). Identification of several root QTLs has been aided by availability of the genome sequence and use of RILs in *Arabidopsis*. For example, the Umkirch-1 accession has a very short primary root. Mapping of a QTL from RILs resulted in identification of a genomic locus containing several genes. Genome sequence comparison between the parental accession and Columbia revealed a mutation in a single gene, *BREVIS RADIX*, which accounts for the majority of the variance observed in Umkirch-1 root length (Mouchel et al. 2004). However, there is an important caveat to these studies; an RIL population may reveal a genomic locus that strongly contributes to a trait; however, that locus may not be relevant to natural phenotypic variability.

To identify genomic loci that contribute to phenotypic variability from an evolutionary perspective, genome-wide association (GWA) mapping is a good alternative. GWA mapping looks for a relationship between a trait (or phenotype) and genomic sequence variation between individuals within a population. This requires that multiple individuals are genotyped at many loci, potentially revealing associated genomic loci at a higher resolution (Weigel and Mott 2009). The genotypes of natural populations reflect recombination events over evolutionary time-scales and thus represent an evolutionary perspective on the alleles contributing to the variation observed in a given trait. Resources for GWA analyses in *Arabidopsis* began with the direct sequencing of ~1000 loci distributed within the genomes of 96 unique *Arabidopsis* individuals from stock center collections and wild (natural) populations (Nordborg et al. 2005). Twenty wild accessions from this group were resequenced with high-density oligonucleotide arrays revealing substantial differences between the wild genomes and reference genome (Col-0) (Clark et al. 2007). Surprisingly, approximately 25% of the reference genome is missing or highly divergent in at least one wild accession, suggesting that functionally important genes are missing or nonfunctional in the reference genome. These observations suggest that it would be informative to sequence more genomes from naturally occurring populations.

Toward this end, a group of researchers has initiated a project to sequence 1001 *Arabidopsis* genomes with the aim of capturing the sequence variation in *Arabidopsis thaliana* at the level of the species (www.1001genomes.org). This sequencing effort is hierarchically structured with a small number of individual genomes being sequenced with a similar quality as the Col-0 reference genome. The majority of the genomes will be sequenced with less coverage allowing the sequencing of many more genomes. Ten individuals from 10 distinct populations in each of 10 geographical regions in Eurasia will be sequenced, plus 1 accession from North Africa (Weigel and Mott 2009). Sequencing data for >210 accessions have already been released with some of these genomes instantly applicable to QTL identification (Cao et al. 2011; Schneeberger and Weigel 2011). However, the primary application of these data will be in GWA studies. A recent GWA study examining 107 aboveground phenotypes in 199 plant lines showed that even with a relatively small sample size, common alleles with large effects could be identified (Atwell et al. 2011). Extension of GWA studies to include root system phenotypes will provide a more complete picture of the alleles that contribute both to root traits and overall adaptive traits in plants.

III. Comparative Genomics and Genetics

The increasing speed of sequencing coupled with decreasing cost and improvement of analytical tools led to genomic sequencing of plants with genomes that were once considered prohibitively large, particularly crop species such as rice, corn, and soybean. With more than a decade since the first plant genome was sequenced, the genomes of dozens of plant species are now sequenced and additional species are currently in progress (see Table 2.1). Genomic sequences are available for plants spanning the evolutionary tree from single-celled algae to nonvascular and vascular mosses to angiosperms (Figure 2.1). Comparative genomics among plants with distinct biochemical, developmental, or morphological features can provide insight into the molecular evolution of key plant traits and facilitate understanding of their specialized features (Paterson et al. 2000; Paterson 2006). It is now possible to assess how gene function might have evolved for specialized or novel processes and how gene regulation has

TABLE 2.1 Plant Genomes Completed and in Progress

Scientific Name	Common Name	Size (Mbp)	Status	Website
Arabidopsis lyrata	Rock cress	230	Published: Hu et al. (2011)	www.phytozome.net/alyrata
Arabidopsis thaliana	Mustard weed	125	Published: AGI (2000)	*arabidopsis.org/*
Aquilegia formosa	Columbine		Unpublished	www.phytozome.net/aquilegia.php
Brachypodium distachyon	Purple false brome	300	Published: IBI (2010)	www.brachypodium.org/
Brassica rapa	Field mustard	550	Unpublished	www.plantgdb.org/BrGDB/
Cannabis sativa	Marijuana	400	Unpublished	www.medicinalgenomics.com/ the-c-sativa-genome/
Carica papaya	Transgenic papaya, "Scv.unUp"	372	Published: Ming et al. (2008)	asgpb.mhpcc.hawaii.edu/papaya/
Citrus clementina	Clementine	295	In progress	www.citrusgenome.ucr.edu/
Citrus sinensis	Orange	319	In progress	www.phytozome.net/citrus.php
Cucumis sativus	Cucumber, IL 9930	244	Published: Huang et al. (2009)	www.phytozome.net/cucumber.php
Eucalyptus grandis	Eucalyptus		Unpublished	www.phytozome.net/eucalyptus.php
Fragaria vesca	Woodland strawberry	240	Published: Shulaev (2010)	www.strawberrygenome.org/
Glycine max	Soybean, cv. Williams 82	1100	Published: Schmutz (2010)	soybeangenome.org/
Gossypium spp.	Cotton	880	Unpublished	cottondb.org/
Hordeum vulgare	Barley		In progress	wheat.pw.usda.gov/ggpages/bgn/31/ druka
Lotus japonicus	Trefoil, cv. Miyakojima MG-20	472	Published: Sato (2009)	www.kazusa.or.jp/lotus/
Malus x domestica Borkh	Apple, cv. Golden delicious	742	Published: Velasco et al. (2010)	genomics.research.iasma.it/gb2/ gbrowse/apple/
Manihot esculenta	Cassava, cv. AM560-2	770	In progress	www.phytozome.net/cassava
Medicago truncatula	Barrel medic, cv. Jemalong A17	500	Unpublished	www.medicago.org/
Mimulus guttatus	Common monkeyflower	430	In progress	www.phytozome.net/mimulus
Oryza sativa ssp. *indica*	Rice, cv. 93–11	466	Published: Ouyang et al. (2007)	rice.plantbiology.msu.edu/
Physcomitrella patens	Moss	500	Published: Rensing et al. (2007)	genome.jgi-psf.org/physcomitrella/ physcomitrella.home
Populus trichocarpa	Black cottonwood	485	Published: Tuskan (2006)	www.phytozome.net/poplar.php
Prunus persica	Peach, cv. Lovell	220	Unpublished	www.rosaceae.org/peach/genome
Ricinus communis	Castor bean, cv. Hale	400	Published: Chan (2010)	castorbean.jcvi.org/index.php
Saccharum officinarum	Sugarcane	930	Unpublished	sugarcanegenome.org/
Selaginella moellendorffii	Spikemoss		Unpublished	genome.jgi-psf.org/Selmo1/Selmo1. download
Setaria italica	Foxtail millet	406	Unpublished	foxtailmillet.genomics.org.cn/
Solanum lycopersicum	Common tomato	950	In progress	sgn.cornell.edu/about/ tomato_project
Solanum tuberosum	Potato, DM1-3-516 R44	840	In progress	www.potatogenome.net
Sorghum bicolor	Sorghum	770	Published: Paterson (2009)	www.phytozome.net/sorghum
Theobroma cacao	Chocolate	327	Published: Argout et al. (2010)	genomevolution.org/CoGe/ OrganismView.pl?dsgid=10997
Vitis vinifera	Grapevine, Pinot noir cv. PN40024	487	Published: Jaillon (2007)	www.phytozome.net/grape.php
Zea mays ssp. *mays*	Maize, cv. B73	2600	Published: Schnable (2009)	maizesequence.org/

changed over evolutionary time. For example, plant comparative genomics provides the opportunity to address questions regarding the nature of root morphology and development.

A. Ancient Plants

Modern nonvascular and vascular mosses are considered representatives of early land plants. When plants colonized land approximately 450 million years ago, they encountered many challenges associated with a nonaquatic environment. Some multicellular algae developed rootlike structures, called rhizoids, which anchor these plants to rocky substrates in tidal zones. However, the migration of plants out of the water required development of specialized structures to acquire water and dissolved nutrients from terrestrial environments (Raven and Edwards 2001). The nonvascular moss *Physcomitrella patens* (Hedw.) Bruch & Schimp. is well positioned on the phylogenetic tree to inform our understanding of early land plants before the evolution of vasculature over 400 million years ago. *P. patens* has no true leaves or roots and requires a very moist environment. The rhizoid in moss participates in water and nutrient uptake but is not specialized for this function. Instead, the *P. patens* rhizoid is primarily involved

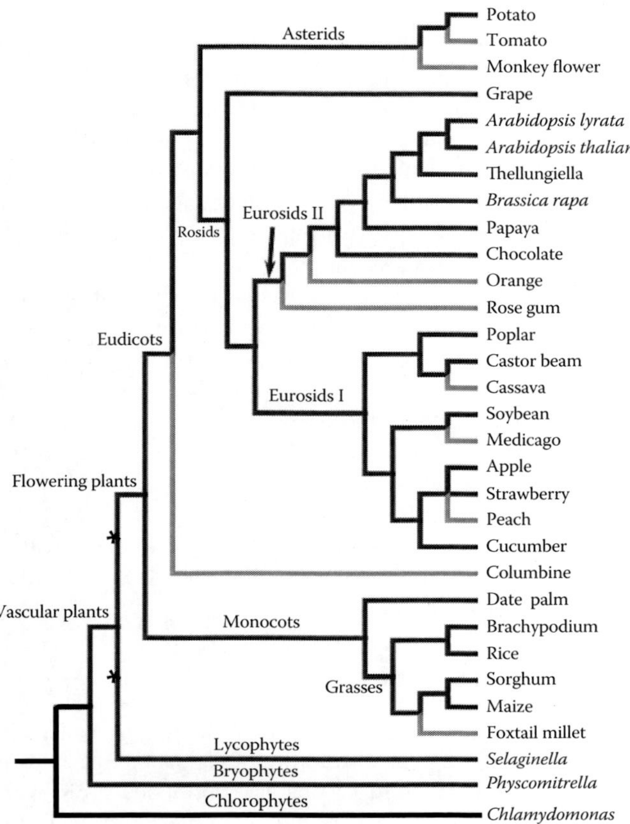

FIGURE 2.1 Phylogeny of plants with sequenced genomes. Branches in black are published and branches in gray are incomplete or not publically available. Asterisks mark the evolution of roots. Branch lengths are not proportional to time or evolutionary distance. (Figure adapted from James Schnable, CoGePedia, http:/genomevolution.org/wiki/index.php/Sequenced_plant_genomes)

in anchorage (Raven and Edwards 2001). Nonetheless, comparative genomic analyses, in the context of root biology, have revealed some compelling functional similarities among related moss and angiosperm genes. For example, *RHD SIX-LIKE1* (*PpRSL1*) and *PpRSL2* are required for rhizoid development in *P. patens*, and the related genes in *Arabidopsis*, *AtRHD6* and *AtRSL1*, function in hair development on the root epidermis. Despite this similarity, the transcriptional response of these genes to the plant hormone auxin is reversed in their respective organisms (Jang et al. 2011). Thus, *Physcomitrella* homologues of key *Arabidopsis* root developmental genes exist, and their comparative functional analysis provides insight into the molecular evolution of roots.

Completion of the first lycophyte genome provides an important evolutionary link between nonvascular mosses, like *Physcomitrella,* and modern vascular plants, like angiosperms. The spikemoss, *Selaginella moellendorffii* Hieron., is a non-seed vascular plant, which lacks true leaves but, by some criteria, has true roots. The roots of *Selaginella* originate from a rhizophore, a root-bearing organ that later differentiates into a root with a root cap and root hairs. *Selaginella* roots do not produce lateral roots like other vascular plants but instead branch from a bifurcation at the root tip (Banks 2009). It is hypothesized that roots evolved independently in the lycophyte and seed plant/fern lineages (see Figure 2.1). Thus, genomic comparisons between these two lineages will provide insight into two different solutions for the development of rooting structures. The *Selaginella* genome contains a similar number of genes (~22,000) as other plants, but it is the only known plant genome with no evidence of a genome duplication event (Banks et al. 2011). The massive proliferation and diversity of vascular land plants, the majority of which have roots, suggests that plants with vascular tissues and roots had an evolutionary advantage on land.

B. Sequencing Plants of Agronomic Importance

Angiosperms (flowering plants) are the most diverse group of land plants; today they dominate the landscape and are a key to human life. The two groups of angiosperms, monocots and dicots, are the primary caloric resource for many animal species, including humans, and are the foundation of modern agriculture. Many key crop species are monocots and chief among them are the cereal grains (e.g., rice, maize, wheat, sorghum), which are among the most economically important crops globally. Genome sequences for several monocots have been completed, including rice (International Rice Genome 2005), sorghum (Paterson et al. 2009), and maize (Schnable et al. 2009). Rice is one of the most important global food crops and has the smallest genome of the major cereals. Despite this, rice is predicted to have over 37,000 protein-coding genes with approximately 90% of *Arabidopsis* genes having a putative rice homologue. The remaining rice genes are predicted to have monocot- or species-specific functions (International Rice Genome 2005). The availability of sequence information for two subgroups (subspecies) of rice offers the additional possibility of intraspecies genomic comparison in a crop plant (International Rice Genome 2005; Yu et al. 2005). Recent GWA analyses combining phenotyping data and sequence information in many rice varieties have revealed genetic changes linked to complex aboveground traits (Zhao et al. 2011). High-throughput phenotyping of root architecture in 12 rice varieties (genotypes) has already demonstrated that root architecture traits are under strong genetic control (Iyer-Pascuzzi et al. 2010). Together, root phenotyping, genomics, and genetic variation provide powerful methods to identify genes controlling root system growth and development in crop plants.

Sorghum also serves as an attractive genomic model for cereal crops with over 90% of the predicted sorghum coding regions having orthologs in *Arabidopsis*, rice, or poplar (the model organism for trees, see below). As an inbreeding crop, sorghum is useful for studying agronomically important traits by genetic association studies; key among these traits in sorghum are drought and metal toxicity tolerance (Paterson et al. 2009). Although drought tolerance certainly has a root component, very little is known about how sorghum root systems participate in drought adaption. Critical progress has been made toward the understanding of sorghum's tolerance to toxic levels of aluminum in soils (Magalhaes 2006). Identification and cloning of an aluminum-activated citrate transporter has wide ramifications for future global crop improvement efforts, many of which will likely occur through modulation of root system traits.

The maize (B73) genome is the largest plant genome sequenced to date at 2.3 gigabase pairs. Maize is predicted to have only ~32,500 protein-coding genes, with the vast majority of its genome comprised of transposable elements. Maize is among the best-studied and genetically tractable cereals making it an important model in crop biology. Recently, a population of 5000 maize RILs was generated from diverse parental lines crossed to B73. This nested association mapping (NAM) population was established to map the genes that underlie complex, quantitative traits in maize. This population has already been successfully analyzed for two aboveground traits (Buckler et al. 2009; Tian et al. 2011) and could be used to identify genes with roles in maize root system architecture in the future.

With an increasing world population and demand for fuel, the necessity for alternative energy sources is attracting increasing attention. The model grass, *Brachypodium distachyon* (L.) P. Beauv., is at the forefront of biofuel research. As a member of the Pooideae subfamily of grasses, it was chosen as a model grass, in addition to rice, because of its rapid generation time, small size, and relatively small genome compared to its closely related polyploid relatives, wheat and barley. *Brachypodium* is also an excellent model for examining complex root systems as its root architecture is relatively compact, but its root development is similar to that of wheat (Watt et al. 2009). Using *Brachypodium* as a biofuel model enables researchers to more easily study methods for increasing plant biomass.

Black cottonwoods, or poplars, are the model organism for forest trees, which are currently a major source of lumber, fiber, and fuel. Poplar has 19 chromosomes and over 45,000 putative protein-coding genes indicating at least three whole-genome duplication events since the evolutionary divergence from its last common ancestor with *Arabidopsis* (Tuskan et al. 2006). Poplar has been demonstrated to be a good candidate for phytoremediation of contaminated soils. Genes for tolerance of toxic substances have been identified including cytochrome P-450s and glutathione S-transferases (GSTs) based on studies in *Arabidopsis*. Furthermore, expression of a GST has been demonstrated to be upregulated in poplar roots when the plants are exposed to munitions waste (Tanaka et al. 2007). Studying genes with upregulated expression in response to toxins should provide key information as to how the roots of poplar function to permit plant survival and rehabilitate waste areas.

Comparing the genomes of different species or accessions can provide insight into the ancestral relationships between diverse plant species and how evolution acts on genome structure and function. Additionally, comparative genomics can be used to address more specific developmental or physiological questions. In root biology, rice and maize have vastly more complex root systems comprised of more types of roots than *Arabidopsis*. Nevertheless, genes functioning in *Arabidopsis* root development and root system architecture can provide a straightforward starting point for genetic analysis of other root systems. Already, similar roles for related genes have been demonstrated in *Arabidopsis*, rice, and maize roots (reviewed by Coudert et al. 2011). With genomic sequence available for nearly 30 plant species and the imminent completion of 1001 *Arabidopsis* genomes, there is great potential for major advances in plant and root biology both in terms of understanding basic biological questions and in more applied agronomic improvement endeavors. Additionally, higher-resolution and higher-throughput approaches in phenotyping more complex root systems are being developed (Perret et al. 2007; Armengaud et al. 2009; French et al. 2009; Iyer-Pascuzzi et al. 2010). These approaches, together with genomic resources, will allow better characterization of root traits and detailed analyses of mutants or natural variants, which would lead to identification of the molecular mechanisms that regulate formation of root types and overall root architecture in these important crop species. These tools and the information they generate will contribute not only to our understanding of root biology but can be utilized to support crop improvement. These advances are already beginning to come to fruition, particularly in root biology, with research being conducted on plant biology's first genomic model, *Arabidopsis*.

IV. *Arabidopsis* Root as a Model Organ

Roots are essential for sensing and uptake of water and nutrients; the root system also plays a fundamental role in plant–environment interactions. Owing to continuous growth and development, plant root architecture, and morphology are particularly responsive to external stimuli, such as touch, gravity, and various biotic and abiotic conditions (Okada and Shimura 1990; Malamy 2005). However, characterization of a root system's response to rhizosphere conditions is difficult because of its subterranean location. Root observations are complicated by the loss of information upon extraction of roots from soil and by the large physical size and complexity of many crop root systems. The *Arabidopsis* root system provides a responsive yet tractable root system for biological study. Growth of *Arabidopsis* roots along the surface of agar plates allows detailed characterization of their previously buried growth and development (Schiefelbein and Benfey 1991; Armengaud et al. 2009; French et al. 2009). Microscopic observation of *Arabidopsis* roots is also straightforward as they are relatively thin (100–150 μm) and nearly translucent. These characteristics make *Arabidopsis* particularly attractive to plant biologists and facilitate not only developmental studies but also characterization of the root's response to a variety of specific conditions, such as osmotic stress and nutrient deficiencies and toxicities. *Arabidopsis* has also been useful for examination of biotic interactions between roots and other organisms, such as with nematodes and microbes (Wang et al. 2005; Lugtenberg and Kamilova 2009). Recent studies indicate that roots, including those of *Arabidopsis*, maize, and soybean, function in plant–plant interactions (Biedrzycki et al. 2009; Fang et al. 2011). Thus, *Arabidopsis* has served as a valuable model for questions pertaining to both basic and more applied plant biology.

The *Arabidopsis* root is a particularly tractable system for developmental biologists due to its simple morphological and cellular organization (reviewed in Benfey and Scheres 2000). One of the fundamental problems faced by developmental biologists in any multicellular system is that organ development occurs in four dimensions, that is, organs form in the three spatial dimensions and over time. Additionally, analysis of organogenesis can be complicated by the location in which it occurs, for example, development occurs in utero in many vertebrate animals. By contrast, in plants, the vast majority of organ development is postembryonic and continuous over the plant's life. During *Arabidopsis* embryogenesis, two populations of stem cells are formed, which will postembryonically give rise to the organs of the shoot and root systems (Jurgens 2001). In plant shoots, older aerial organs largely conceal both the region of stem cell activity and the newly forming organs. However, activity of the root stem cell population and its daughters remains accessible at the tip of the root.

The remarkable feature of continuous development in plants is easily observed in the *Arabidopsis* root. The initial (stem) cells form a single cell layer surrounding the quiescent center; these cells are collectively called the stem cell niche (Figure 2.2A, box). Asymmetric cell division of the initial cells at the root tip maintains the stem cell population while producing daughter cells that will undergo further replicative divisions, prior to expansion and differentiation (Dolan et al. 1993; Benfey and Scheres 2000). As cell division proceeds at the root apex, older cells are displaced shootward along the root's longitudinal axis. Consequently, cells become progressively more mature the further they are from the root tip, making the root's longitudinal axis a living developmental timeline. This timeline has historically been divided into three main zones (meristematic, elongation, and differentiation) based on the overall developmental activity of the region (Figure 2.2A). The progeny of each initial

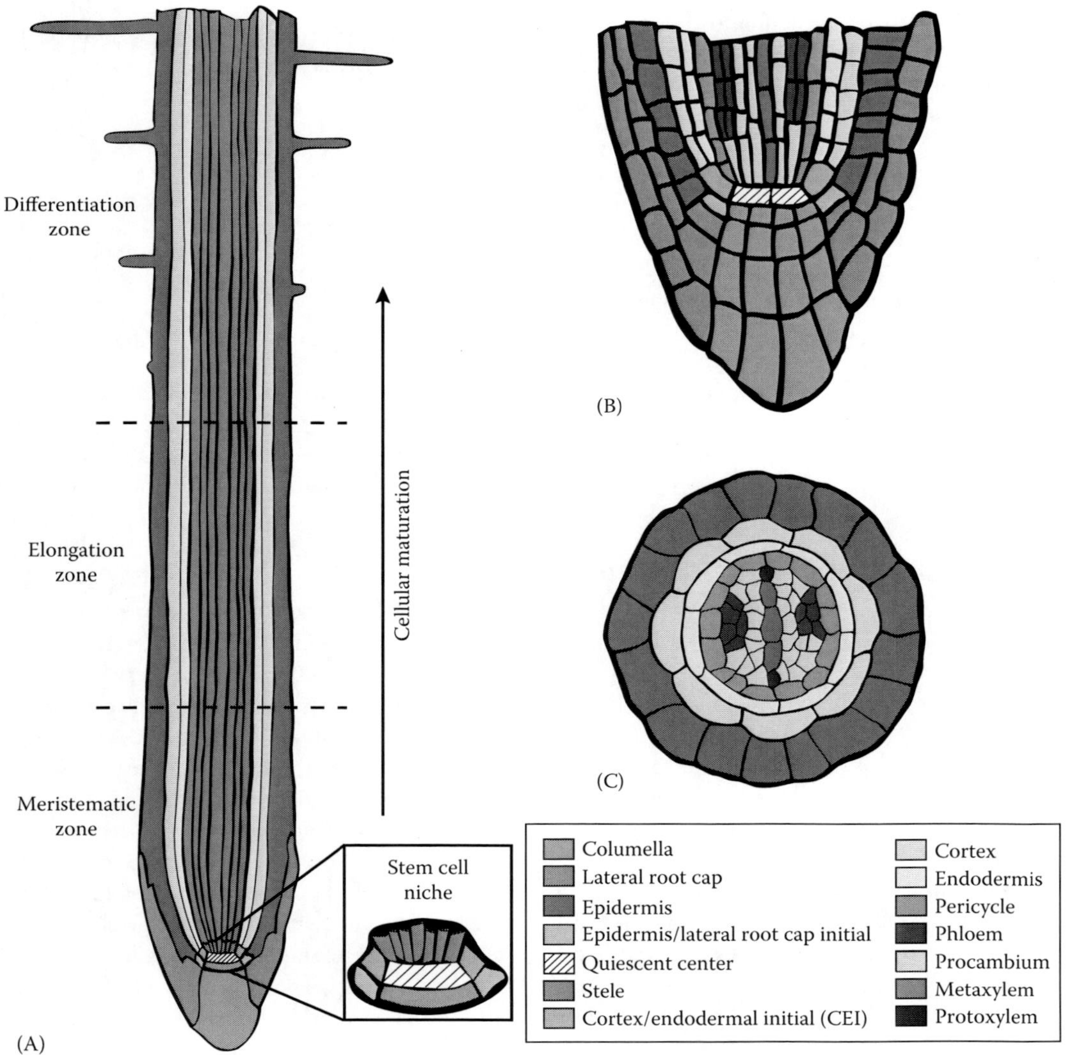

FIGURE 2.2 **(See color insert.)** Schematic of the cellular organization of the *Arabidopsis* root. (A) Median longitudinal section of a length of the root. The different developmental stages and the direction of cellular maturation are noted. Inset: the stem cell niche includes the quiescent center and the adjacent initial cells. (B) Median longitudinal section of the root tip. (C) Transverse cross section within the differentiation zone. The different cell types comprising the stele (A) are detailed in (B) and (C). This drawing is not to scale.

cell is typically restricted to a specific cellular lineage and cell file, meaning that each cell in the root can be traced back to a specific initial cell (Figure 2.2B). Root cell types are organized with the outer epidermal, cortical, and endodermal tissues forming concentric layers around a central stele (Dolan et al. 1993; Scheres 1994). Cell types within the stele, the pericycle, and the vascular tissues show a bilaterally symmetric organization (Figure 2.2C; Parizot et al. 2008). Thus, the largely radially symmetric organization of the root allows for the conceptual reduction of its development from four dimensions to two, with cell types in the radial axis and developmental time along the longitudinal axis.

The utility of genomic sequence information is not limited to studies of DNA sequence or structure, as the genome sequence predicts RNA transcripts and protein products, which are particularly interesting to many biologists. Knowledge of the mRNAs (transcripts) or proteins present in a tissue or cell holds the possibility of providing the unprecedented ability to characterize their function. Moreover, monitoring transcript levels over time or in response to treatments allows investigation into both developmental and response processes. However, examination of transcripts of individual genes provides limited information as to the function or state of the tissue or organ examined. Genome-wide examination of RNA transcripts or protein products can be used to uncover the function or state of a cell or tissue and to identify genes that likely function together in a given process. These types of studies represent the functional output of the genome and provide insight into how genome sequence is translated into biological function. The simple morphological and cellular organization of the *Arabidopsis* root has made it an ideal organ for functional genomics studies both at cellular and temporal resolution and under various conditions.

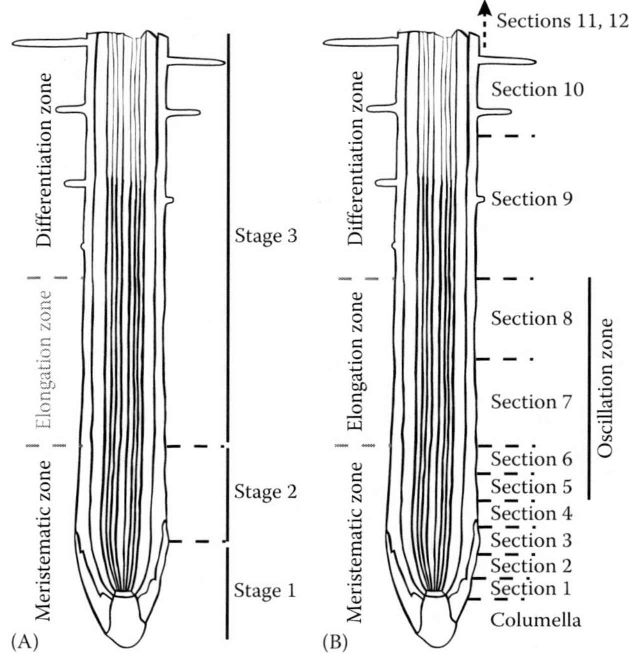

FIGURE 2.3 Schematic depicting the positions of the transverse sections sampled along the longitudinal axis for transcriptional profiling of the root's developmental stages. (A) The three transverse sections sampled by Birnbaum et al., 2003 and (B) the thirteen transverse sections sampled from individual roots by Brady et al., 2007 as related to the developmental zone boundaries. The region of the root encompassing the OZ is also indicated in (B). This drawing is not to scale.

V. Transcriptional Profiling of the *Arabidopsis* Root in Time and Space

A. Transcriptional Profiling in Developmental Time

To better understand the molecular programs underlying the developmental progression of cells in the *Arabidopsis* root, genome-wide transcriptional profiling using microarrays was performed on three serial transverse sections in the root's longitudinal axis (Birnbaum et al. 2003). The meristematic zone was divided into two sections (stages 1 and 2) and the elongation and the early differentiation zones comprised the third section (stage 3) (Figure 2.3A). As predicted, based on the developmental activity in the meristematic zone, genes involved in cell division-related processes showed peaks of expression in stages 1 and 2. Additionally, expression of a significant number of genes predicted to encode transcription factors and kinases peaked in stage 3, implying an important role for signaling in cellular maturation (Birnbaum et al. 2003). These results revealed the presence of complex transcription patterns in the developing root and suggested that specific transcriptional patterns occurred in smaller domains than those delineated by the three main developmental zones.

The analysis of gene expression at a higher spatial resolution was essential to address the intricacies of the transcriptional mechanisms underlying root development. Brady et al. (2007) microdissected a single *Arabidopsis* primary root into 13 transverse sections composed of 3–5 cell lengths each and profiled the mRNA expression by microarray (Figure 2.3B). A set of 40 dominant longitudinal expression patterns was identified (Brady et al. 2007). Most of these patterns showed a peak of expression across at least two adjacent sections. This is an important observation as it suggests that the dissected sections were smaller than the developmental regions under common transcriptional regulation and indicated a sufficient resolution for detecting specific expression patterns over developmental time. Confirmation that these expression patterns accurately reflect longitudinal region-specific biological processes can be visualized in the gene expression profiles associated with the columella. The columella functions in gravity sensing through activity of starch-containing amyloplasts, and expression in this longitudinal section was enriched for genes involved in starch catabolism (Brady et al. 2007). Because gene expression data from the transverse sections in the longitudinal axis accurately reflected a known function of specific root cells, these data can be used to infer novel biological functions occurring at particular developmental times.

A surprising outcome of these analyses is that nearly half of the longitudinal expression patterns peaked in nonadjacent developmental regions. These apparent fluctuations in expression were unexpected as developmental time is generally considered to be a linear, progressive process. The only gene expression pattern with multiple peaks that could be ascribed to known developmental events was that of genes functionally linked to the cell cycle. The first peak in the meristem sections (e.g., sections 1–4, Figure 2.3B) is consistent with proliferative cell divisions, and the second peak, in the shootward-most maturation sections (e.g., Sections 10–12, Figure 2.3B), is likely due to the cell divisions involved in formation of lateral root primordia. The majority of the expression patterns that appear to fluctuate in developmental time do not have clear developmental roles as yet. However, most of these patterns contain genes associated with metabolic or transport functions (Brady et al. 2007). Insight into the specific functions of these genes or pathways in the root will be necessary to comprehend the rationale behind their apparent repeated expression observed in developmental time.

Another key question in gene expression analyses over developmental time is the variability of expression among individual roots. To determine the reproducibility of the dominant expression patterns, a second root was examined. For nearly half of the dominant expression patterns, coexpression of greater than 90% of the constituent genes was observed. In patterns with reduced coexpression values, a subset of the genes often showed striking levels of coexpression (Brady et al. 2007). Patterns with limited coexpression often had apparent phasing or shifting of the peak of expression in developmental time. These shifts are hypothesized to represent gene expression patterns with a temporal component that is distinct from the developmental timescale assayed along the root's longitudinal axis. A computational method developed to identify genes with similar but temporally shifted expression patterns classified nearly 6000 genes as potentially acting at distinct timescales in the root (Orlando et al. 2010). These patterns might reflect dynamic gene expression patterns in the root similar to the oscillating gene expression patterns observed in other developmental processes such as vertebrate somitogenesis (reviewed by Dequeant and Pourquie 2008) or lateral root formation (Moreno-Risueno et al. 2010). Thus, comparison of gene expression in two individual roots revealed both highly reproducible patterns and temporally dynamic expression across the longitudinal axis.

Recently, periodic gene expression in the *Arabidopsis* root was linked to lateral root formation. Characterization of the transcriptional reporter, pDR5:GUS, suggested that there was a temporal component to this marker's expression near the root tip. Additionally, the expression of this marker at temporal intervals was correlated to lateral root development (De Smet et al. 2007). To examine the expression of this marker nondestructively and in real time, a transcriptional reporter with the luciferase enzyme was constructed (pDR5:Luciferase) (Moreno-Risueno et al. 2010). In vivo analyses revealed a rhythmic pulsing of DR5 expression near the root tip that behaved as an oscillating wave propagating along the root's axis. Following each wave of expression, a static point of DR5 expression that reflected the peak of the most recent wave was observed. These static points, termed prebranch sites, mark the positions at which lateral roots initiate later in development. These observations led to the hypothesis that an oscillating transcriptional mechanism operates to specify competence to form the next lateral root. Microdissection of this region, termed the oscillation zone (OZ) (Figure 2.3B), followed by genome-wide transcriptional profiling revealed nearly 3500 genes with oscillatory expression behavior (Moreno-Risueno et al. 2010). These results provide an entry point into a novel way of thinking about how gene expression controls temporally regulated developmental events in plants.

B. Transcriptional Profiling at Cell Type Resolution

A major puzzle in developmental biology is how cells become different from each other and what molecular mechanisms drive diverse cellular functions. Understanding these differences is made more challenging by the fact that cells are difficult to separate from each other in the context of a whole organism or organ. The largely radial organization of the *Arabidopsis* root allows for almost all the cell types to be observed in a median longitudinal section (Figure 2.2A and B). Coupling expression of GFP reporter constructs within specific root cell populations with fluorescence-activated cell sorting (FACS) and microarrays, gene expression profiles of different cell types could be examined and questions could be asked about their transcriptional similarities and differences (Birnbaum et al. 2003; Brady et al. 2007). These experiments provided new insights into cell-specific functions based on expression of genes associated with specific pathways. These experiments were the most comprehensive efforts to investigate gene expression in an organ at the level of individual cell types.

Cell type-specific gene expression profiling has been conducted in nearly all the cell populations that comprise the root (Birnbaum et al. 2003; Brady et al. 2007; Carlsbecker et al. 2010; Sozzani et al. 2010). These data include gene expression from the cortex–endodermal initials (CEI), the first plant initial cell population to be specifically isolated and transcriptionally profiled. Together these data revealed there are substantial differences in gene expression among cell types, correlating with the fact that different cell types carry out distinct functions. When differentially expressed genes were grouped by their presumptive biological function, or gene ontology terms, enrichment for many of these terms was found in single cell types. This result further demonstrated the unique expression profiles and the possibility for uncovering novel functions within each cell type. For example, root hairs are polarized epidermal cells that require cellular remodeling to develop. Expression of genes known to be involved in root hair differentiation, such as those encoding kinases and cell wall-loosening enzymes, was specifically enriched in this cell type. Additionally, examination of genes specifically expressed in root hairs also showed enrichment for gene ontology categories related to calcium ion transport and vesicle docking during exocytosis, processes which had been previously linked to root hair cell morphogenesis (Brady et al. 2007; Jones et al. 2008). This analysis confirmed previous studies of root hair function while additionally lending insight into novel functions of underexplored cell types, such as those that comprise the xylem.

In addition to finding gene expression differences between cell types, this analysis revealed 34 gene expression patterns associated with multiple cell types (Brady et al. 2007). These patterns of expression were associated with diverse gene ontology terms describing their putative functions. Many of the terms associated with patterns spanning multiple cell types describe general cellular processes that would be considered common among cell types, such as membrane localization. In other cases where expression patterns occur in neighboring tissues, the gene ontology terms suggest that these neighboring tissues may share a common function. However, some patterns of expression exist in unrelated tissues, which may be the result of similar biological processes occurring in disparate cell types. For example, both root hairs and xylem cells experience cell wall deposition upon terminal differentiation, and in both cell types, genes involved in cell wall deposition are enriched. Additionally, expression of genes with unknown function in the same pattern as those with known biological function allows for hypotheses to be made about the function of the novel genes. These data show the importance of profiling at the resolution of individual cell types both for confirming known and proposing novel cell type and gene functions.

C. Identification of Spatiotemporal Transcriptional Patterns

Gene expression in the longitudinal and radial root axes represents the transcriptional state of root cells in developmental time and space, respectively. However, the analyses of specific cellular populations often occurred over several developmental time points, while the longitudinal axis data are specific to a developmental time point but included multiple cell types. A combined analysis of gene expression in both time and space is potentially a powerful way to reveal biological processes likely to occur with both cell type and developmental stage specificity. Compilation of data from the dominant radial and longitudinal expression patterns uncovered peaks of relative expression that were restricted to a given tissue at a specific developmental time. However, some transcriptional information was excluded from these combined analyses based on criteria used to establish statistically significant relative gene expression values. Conversion of the expression data to a simple ON/OFF designation and requiring them to be ON in both datasets resulted in 158 groups of genes with expression restricted in both time and space (Brady et al. 2007). This binary analysis revealed distinct expression patterns that are not detectable by examining the relative expression data, for example, genes with expression restricted to the main developmental zones, or genes whose expression appears excluded from a particular cell type at all times (Brady et al. 2007).

Collectively these data comprise the most complete gene expression map of any organ to date with transcriptional information in both developmental time and at cell type-specific resolution. Nevertheless, not all of the cell types could be individually assayed (Brady et al. 2007). Ideally, a root gene expression map would measure gene expression with specificity in both cell type and developmental time simultaneously. To this end a computational method was developed to infer the gene expression information for specific cell types at each developmental time point (Cartwright et al. 2009). This method was used to identify a gene with a predicted peak of expression in the procambium cells, which had not been previously profiled, within the meristematic zone. In vivo validation with a transcriptional reporter showed that the predicted spatial and temporal expression pattern of this gene was accurate. Together these transcriptional analyses expose an abundance of unique gene expression patterns in both space and time and a surprising level of transcriptional complexity in the root. By integrating expression data both in specific cell types and in developmental time, novel cellular functions can be predicted. For example, these data can provide insight into the progression of initial cells to mature cells that are functionally specialized.

VI. Transcriptional Responses Link Development and Stress Responses in Root Cell Types

Plants are capable of dramatic phenotypic plasticity in the presence of diverse environmental conditions. Although this plasticity has been well documented at the whole plant and organ level, questions remain about how individual cells sense and respond to different stressors. To assess the effect of abiotic stress on different cell types and different developmental stages, salt toxicity and iron-deficient conditions were examined and compared (Dinneny et al. 2008). These studies revealed that cell identity was a key determinant in the root's transcriptional response to stress.

High salinity is a detrimental soil contaminant that adversely affects plant productivity. A phenotypic analysis using different salt concentrations was done to identify a concentration that elicited a strong phenotypic response but did not completely inhibit root growth. Additionally, to assess the timing of the root's transcriptional response to high salt, a time course after exposure was performed. These preliminary studies revealed that significant changes were observed after 1 h of exposure to 140 mM NaCl. To understand the transcriptional response at higher spatial and temporal resolution, four regions along the longitudinal axis and five cell populations including the epidermis, endodermis, cortex, columella root cap, and stele were profiled after 1 h of exposure to 140 mM NaCl (Figure 2.4). Increasing the resolution of the assay to longitudinal and cell type datasets revealed increased numbers of differentially expressed genes. The most profound

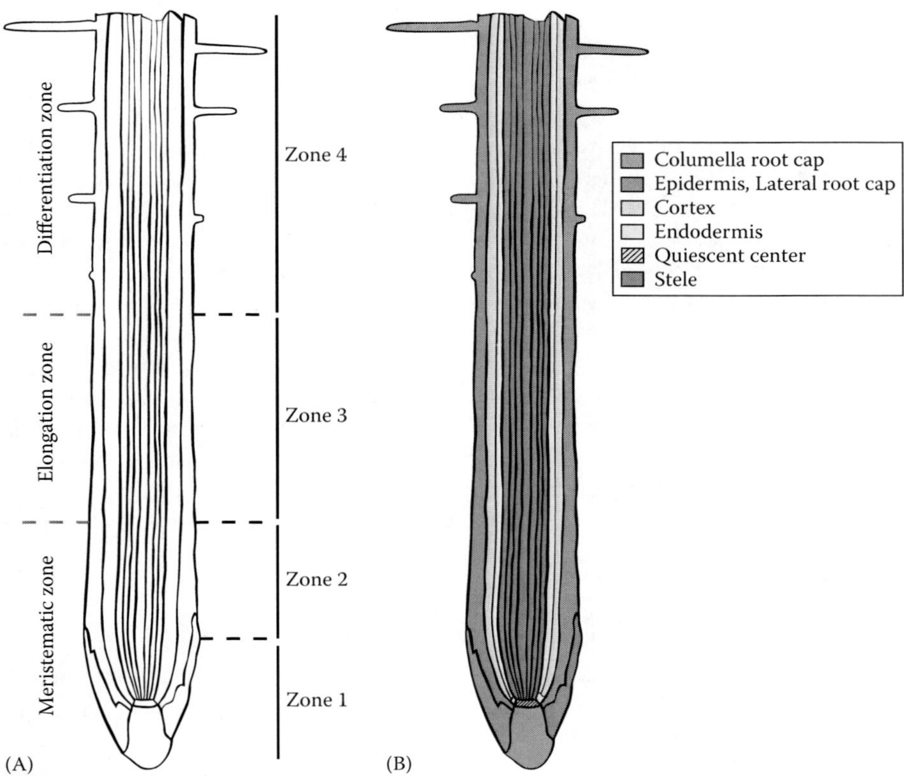

FIGURE 2.4 (See color insert.) Schematic depicting the transverse sections in the longitudinal axis and cell types sampled for transcriptional profiling of the *Arabidopsis* root's response to stress. (A) Four transverse sections and their relationship to the developmental zones and (B) five cell populations each profiled after exposure to abiotic stress. This drawing is not to scale.

result was that differential gene expression in response to salt stress occurred at the level of specific cell types, with the cortex being the most transcriptionally responsive cell type (Dinneny et al. 2008). This was unexpected as it indicates that transcriptional response to stress takes place in a cell type-dependent fashion.

This high-resolution dataset provided unprecedented spatial and temporal resolution for deciphering the root's response to salt toxicity. Radial swelling of the outer root tissues occurs in response to salt stress, a phenotype similar to cell wall biogenesis and tubulin mutants. Genes involved in cell wall biogenesis and tubulin were downregulated in the cortex and epidermis upon exposure to high salt, implicating these genes in the phenotypic response to salt stress. To investigate this hypothesis, plants with a loss-of-function mutation in a gene involved in cellulose biosynthesis, *COBRA*, were phenotypically examined under salt stress conditions. *cobra* mutant roots displayed increased radial swelling in the cortex under lower salt concentrations than wild type. This indicated that reduced expression of cell wall biogenesis genes contributed to salt-induced radial swelling phenotypes (Dinneny et al. 2008).

To determine whether the transcriptional response to salt stress was comparable to other stresses, differential gene expression in response to iron deficiency was examined. Similar to salt stress, increasing numbers of genes were differentially expressed with increasing spatial and temporal resolution, and cell type-specific gene expression changes were observed under iron-deficient conditions (Dinneny et al. 2008). As might be expected, genes involved in iron transport, storage, chelation, and metabolism were differentially regulated under low iron conditions. Additionally, root hairs were malformed and smaller than under standard conditions. Consistent with this phenotype, genes associated with cell wall biosynthesis and root hair morphogenesis were downregulated in epidermal cells under iron stress.

Comparing the gene expression responses to two abiotic stresses in different cell types and over the developmental zones revealed that gene expression within a given cell type was highly dependent on environmental conditions. For example, although there was a 20% overlap in genes that were differentially expressed under both stresses, only half had similar transcriptional changes in all the cell types profiled. Additionally, only a small percentage of biological functions that are enriched under standard conditions are similarly enriched under stressed conditions. These results indicate a high level of transcriptional plasticity in root cell types and are likely to reflect the necessity for plants to execute a fine-tuned response to variable environmental conditions. The next logical step was to ask what genes remain constant. A set of genes was found that showed consistent cell type-specific expression, despite environmental perturbations. The stable expression of certain genes under changing environmental conditions led to the hypothesis that these genes function to maintain cell identity and/or function (Dinneny et al. 2008).

The concept of cell type- and stress-specific transcriptional responses to abiotic stressors (Dinneny et al. 2008) was surprising given the evidence for a proposed universal stress response (Swindell 2006; Ma and Bohnert 2007; Walther et al. 2007). Whole-genome transcriptional data of entire organs or seedlings exposed to various stresses had identified a number of genes that appeared universally responsive. However, the complex patterns of gene expression revealed by higher spatial and temporal resolution studies suggested that many aspects of transcriptional complexity were largely obscured by analyses at the whole organ level (Birnbaum et al. 2005; Brady et al. 2007; Dinneny et al. 2008).

To further investigate cellular responses to various stresses, a meta-analysis was conducted on transcriptional data from whole roots under 14 stress conditions and from 4 stresses at cell type- and developmental stage-specific resolution (Iyer-Pascuzzi et al. 2011). Analyses designed to reveal genes that were universally responsive to stress showed that each stress elicited differential expression of genes associated with distinct gene ontology categories and only nine genes could be classified as universally responsive in the whole root. If requirements were relaxed, with genes responding in only half of the stress conditions, a set of so-called "common stress response" genes could be identified. Unexpectedly, when expression was examined at a higher spatial resolution, most of these common stress response genes displayed cell type-specific expression. Additionally, this transcriptional response was not simply observed in a single cell type under stress but showed specificity to both the cell type and the stress. The most transcriptionally responsive cell type could often be associated with a stress-induced phenotypic change (Dinneny et al. 2008; Iyer-Pascuzzi et al. 2011). This indicates that in the root, each stress elicits a distinct transcriptional response and this can be linked to morphological changes.

Common stress response genes were predicted to be broadly expressed; therefore, cell type-specific expression of these genes was unexpected. This specific expression might be linked to stress-induced changes in the expression of known cell identity regulators, as these genes were differentially expressed under different stresses (Iyer-Pascuzzi et al. 2011). Among the common stress response genes, those involved in responses to the plant hormone abscisic acid, a stress response hormone, were enriched (Iyer-Pascuzzi et al. 2011). The authors hypothesized that if cell identity regulators also functioned to mediate common stress responses, then mutations in these regulators would show defects in common stress responses, such as hypersensitivity to exogenous abscisic acid. Examination of plants with mutations in these cell identity genes revealed phenotypes consistent with altered responses to abscisic acid. These results suggest cell type-specific transcriptional response of common stress response genes is mediated through regulatory genes functioning in cell identity and identifies key points of interaction between developmental and stress response pathways.

The unique transcriptional response of various cell types and developmental regions to stress indicates that the identity and developmental stage of a cell dictates its response to a stress. These data can also provide functional insight into the genes whose expression remains enriched in a specific cell type regardless of the stress. This group of genes was designated as core marker genes, as several known cell identity regulators were found among them (Dinneny et al. 2008; Iyer-Pascuzzi et al. 2011). The remaining core marker genes are predicted to function in cell identity. Mutant analyses of several of these functionally uncharacterized genes supported this hypothesis. Surprisingly, in response to stress, these core marker genes showed differential expression in cell types other than the core cell type; however, the core marker gene's absolute expression level was always the highest in the core cell type. This suggests a mechanism to ensure that expression of cell identity genes is maintained at its highest level in the appropriate cell type, regardless of whether these genes respond to stress in another cell type. Together these data reveal that complex transcriptional networks function in the root to coordinate stress response with ongoing developmental processes.

VII. Identification of Novel Regulatory Modules from High-Resolution Transcriptional Data

Examination of the root's transcriptional responses to environmental conditions over time can be used to identify novel transcriptional modules that couple root development and environmental response. To address the transcriptional regulation of the root's physiological and developmental responses to iron deficiency (-Fe), genes encoding putative transcriptional regulators were selected for reverse genetic analysis (Dinneny et al. 2008; Long et al. 2010). Two novel proteins, POPEYE (PYE) and BRUTUS (BTS), were identified, and plants harboring mutations in these genes showed abnormal phenotypes under -Fe conditions. *pye* mutants have no detectable abnormal phenotype under iron-sufficient conditions; however under -Fe conditions, they show strong root growth inhibition, reduced root hair emergence, and swelling of epidermal and cortical cells in lateral roots (Long et al. 2010). Despite their phenotypic severity under -Fe conditions, *pye* plants have weaker iron starvation responses than wild type, yet *pye* mutants have increased iron accumulation under both iron-sufficient and iron-deficient conditions suggesting defects in iron homeostasis. Transcriptional analyses further supported this notion as genes involved in iron storage, uptake, and intercellular movement were misregulated in *pye* mutants under -Fe conditions (Long et al. 2010). These results implicate PYE as a regulator in a transcriptional module that regulates root growth and development in response to external conditions by monitoring and maintaining *in planta* iron levels.

Nuclear accumulation of PYE under -Fe conditions implies that it could act as a transcriptional regulator. However, its role as a direct transcriptional regulator of iron homeostasis was unclear. Genome-wide analysis of PYE–promoter interactions under -Fe conditions revealed that PYE interacts with the promoters of many genes, including several involved in iron homeostasis. PYE can therefore be placed in a gene regulatory network that functions directly in iron homeostasis. One of the genes tightly co-regulated with *PYE* was *BTS*. Like *PYE*, *BTS* expression is induced under -Fe conditions in the pericycle; however, in contrast to *pye*, *bts* mutants are less sensitive to -Fe growth conditions. Protein interaction assays revealed that while PYE and BTS do not physically interact, PYE and BTS do interact with other PYE-related proteins (Long et al. 2010). Together these data demonstrate one path from high-resolution spatiotemporal transcriptional information to identification of a novel regulatory module in organ development and physiology.

The use of cell type-specific transcriptional profiling was recently combined with a temporal analysis of gene expression following induction of transcription factors. Two transcriptional regulators, SHORTROOT (SHR) and SCARECROW (SCR), function in cortex/endodermal initials to regulate the asymmetric cell division that forms two distinct cell types, the cortex and endodermis (Figure 2.2B; Benfey et al. 1993; Pysh et al. 1999). To learn more about the transcriptional dynamics leading to this key cell division in root patterning, SHR and SCR were induced in their respective mutant backgrounds; then the ground tissue layer was transcriptionally profiled at specific time points. Key among the findings were that known direct targets of SHR and SCR and cell cycle-related genes were differentially expressed at the time coincident with asymmetric division of the cells. These data indicated that SHR and/or SCR may directly control cell division. To test this, chromatin immunoprecipitation of SHR followed by microarray revealed that direct targets of SHR include a D-type cyclin as well as other cell division-related genes (Sozzani et al. 2010). These experiments revealed an unexpected regulatory module that directly linked transcriptional regulation of root patterning and cell cycle control.

The availability of plant genomic sequences, particularly in *Arabidopsis*, has enabled rapid progress using diverse approaches to understand root biology. Genome data greatly facilitate reductionist approaches and have opened the door to large-scale system-wide approaches. With these tools in hand, root biologists are beginning to synthesize, integrate, and ultimately comprehend the interactions of the multiple components of the root collectively as a biological system. Understanding the properties of the root from a systems biology perspective will necessitate the future inclusion of additional system components.

VIII. Future Directions in Genomic Approaches

A. RNA Sequencing

Concurrent with the reduction in sequencing costs has come an increased interest in developing new and powerful sequencing methods and techniques. RNA sequencing (RNA-seq) is one such advance, which relies on generating cDNA libraries from mRNA or small RNAs that can be sequenced to determine the transcriptional profile of a given organism, tissue, or cell type. RNA-seq technology is rapidly changing and constantly increasing the number and length of sequencing reads that are possible to obtain with high fidelity. The advantages of this technology over hybridization techniques, such as the microarray, are numerous. RNA-seq allows for gene discovery because the nucleotides being sequenced do not have to be known a priori. Additionally, novel splice junctions and transcriptional isoforms can be identified using RNA-seq. These advances make it possible to not only determine which genes are expressed but also the relative abundance or the population distribution of isoforms associated with each open reading frame (Trapnell et al. 2010). Additionally noncoding RNAs such as microRNAs, other small RNAs, and long noncoding RNAs can all be assayed using RNA-seq (reviewed in Wang et al. 2009). The sensitivity of RNA detection using RNA-seq is also substantially higher than with microarray platforms, enabling greater dynamic range of quantification and a higher confidence when detecting low-abundance transcripts; these two factors greatly improve data quality. Ultimately, these data will be useful for understanding the transcriptional state of an organism, tissue, or cell type with unprecedented clarity.

Direct sequencing of RNA molecules is the next advance toward profiling the complete transcriptional landscape of roots and other organs. These technologies will soon revolutionize the way researchers profile transcriptomes and the way gene expression is quantified. The objective is to detect and quantify every transcript and RNA molecule in a tissue or even a single cell.

B. Proteomic and Metabolomic Analysis

Detecting and analyzing the protein populations within cells has proved challenging. Proteins can be considered the key functional units that carry out the specific structural and enzymatic activities required for cell function. Thus, proteomic information is necessary for a complete understanding of the functional readout of the genome. There are two primary types of genome-wide proteomic profiling: shotgun proteomics (e.g., liquid chromatography–tandem mass spectrometry) and quantitative proteomics (e.g., isobaric tags for relative and absolute quantitation [iTRAQ], label-free quantification, stable isotope labeling with amino acids in cell culture [SILAC]). Each assay has specific advantages, for example, shotgun proteomics can detect the largest number of different molecules, while quantitative proteomics detects the amounts of proteins at the expense of breadth of coverage. One of the barriers to achieving high resolution in proteomic approaches is that they generally require a large amount of biological material, a reality that is prohibitive for analyzing low-abundance cell types, like stem cells. Additionally, the poor sensitivity for detecting both low-abundance proteins and low molecular weight proteins is widely recognized as an area requiring improvement in the field (Domon and Aebersold 2010; Beck et al. 2011).

It has been shown that mRNA and protein levels do not strictly correlate by numerous different analyses (de Groot et al. 2007; de Godoy et al. 2008; de Sousa Abreu et al. 2009). There are multiple possible explanations for the low correlations observed when comparing mRNA and protein levels. Since mRNA and proteins have inherently different rates of synthesis, processing, and degradation, these processes should be considered when comparing these different molecules. For example, an mRNA and its corresponding protein may have different stability or half-life (Beck et al. 2011). A recent study designed to address mRNA and protein correlation in mammalian cells found that proteins are as much as five times more stable than mRNAs (Schwanhäusser et al. 2011). This information should be taken into consideration whenever possible in the design of future studies that aim to compare mRNA and protein abundances. As the technology for protein detection improves, the use of proteomic data to examine protein movement and degradation rates will be an invaluable component of studies on dynamic cell behaviors.

Proteomic analysis in plants also has great potential for unveiling protein modifications, members of protein complexes, and binding partners. Proteomics is the most reliable method for revealing the many posttranslational modifications that occur during the activation and regulation of proteins (reviewed in Mann and Jensen 2003). These modifications are subject to temporal and/or cell type-specific changes after protein synthesis and are not present in the mRNA molecules themselves. To date, protein phosphorylation has been extensively studied in *Arabidopsis* and a few other plants (reviewed in Jorrín-Novo et al. 2009; Kersten et al. 2009). These data have been compiled into a database and were used to create a phosphorylation site predictor for plants called PhosPhAt (http://phosphat.mpimp-golm.mpg.de) (Heazlewood et al. 2008). Additionally, work has begun to characterize redox regulation at the proteome level (reviewed by Wormuth et al. 2007). The use of these approaches on proteins from the root will expand our understanding of the role of posttranslational modifications in root biology.

To date approximately 50% of all known *Arabidopsis* proteins can be detected using mass spectrometry protocols (Baerenfaller et al. 2008). The gene ontology terms related to proteins found in the root included intracellular protein transport, response to oxidative stress, and toxin catabolic process. These terms were distinct from those enriched in other organs and tissues profiled in the plant, suggesting the diverse protein populations are functionally relevant to their locations. These data show that even with just 50% of the proteins accounted for, there is a wealth of information that can be gleaned from comprehensive proteomic analyses. A recent study examined the proteins in the root in a cell type-specific manner (Petricka et al. 2012). These data revealed many proteins are cell type-specific suggesting the spatial specificity of protein expression. This indicates that proteomic studies at cellular resolution in multicellular organisms are necessary in order to understand genetic networks.

In addition to RNA and protein, metabolites are a third class of molecules for which global information would be useful. Metabolites are not directly encoded by genes and thus are not as easily predicted from the genome sequence; nonetheless they play a vital role in carrying out cellular processes. There are estimated to be over 5000 metabolites in *Arabidopsis*, over half of which are currently detectable (Saito and Matsuda 2010). Determining the metabolic profile of root cell types will be useful for understanding the type and quantity of small molecules that are involved in root homeostasis. Kusano et al. (2011) looked at metabolite profiles to examine changes upon exposure to ammonium as a nitrogen source in rice. They observed ammonium accumulation in the roots and shoots of mutants defective in the conversion of ammonium to glutamine (Kusano et al. 2011). This study supports the use of metabolite profiling when trying to elucidate a global understanding of how roots respond to environmental factors. One issue with metabolomic profiling is that it is very difficult to identify novel metabolites or to characterize metabolites in an unbiased way. Additionally, the current level of detection for metabolites falls far short of a comprehensive analysis of all the small molecules in the plant. As detection capabilities improve, cell type-specific metabolite profiling will enhance our understanding of the unique functions of different cell types and the differential use of small molecules to execute particular cellular functions. It will be exciting to see the cell type-specific metabolomic profiles in roots and learn whether the metabolomic profiles are as distinct as the transcriptomic and proteomic profiles.

Although many methods exist for examining and profiling the transcriptome, proteome, or metabolome of plants, the next challenge will be to integrate the information provided by these analyses. This integration should ultimately provide a complete molecular description of the root. These descriptive analyses will lead to hypotheses for further study and more directed experiments to understand root biology.

IX. Conclusions and Perspectives

In the last decade, the pace and depth of our understanding of root biology has greatly increased. The rapidly increasing availability of plant genome sequences and the various tools that have arisen from these resources have propelled progress in both basic and applied fields. In particular, the diverse molecular genetics tools and research advantages of the *Arabidopsis* root have advanced our knowledge of the molecular mechanisms regulating root physiology and development. The imminent sequencing of hundreds of diverse *Arabidopsis* plant genomes, coupled with high-throughput phenotyping, will reveal novel loci controlling key root traits. By querying the vast natural variation of *Arabidopsis* ecotypes with modern technologies, quantitative genetic studies are poised to further revolutionize studies of root biology.

The availability of genomic resources for crop species such as maize and rice along with the development of novel phenotyping approaches will facilitate the identification of molecular mechanisms regulating root system architecture in crops. Additionally, genomic comparisons among diverse plants can be used as an entry point to the molecular underpinnings of unique features of form and function, such as floral morphology and root system architecture. Genomic comparisons across kingdoms can provide important insights into yet broader evolutionary questions. Because the plant and animal kingdoms have a single-celled common ancestor, the evolutionary success of these two independent groups of multicellular life forms was likely built from a similar molecular toolbox. Understanding similarities and differences in the molecular mechanisms underlying the evolution of these two groups may reveal underlying constraints on biological systems. If plants and animals arrived at similar solutions for a common problem, what can that tell us about the landscape of possible solutions? Limitations on the number of successful solutions that have persisted through evolutionary time might reveal fundamental principles underlying all biological systems.

Access to plant genome sequences is revolutionizing the way plant biologists approach research. Possibly the most dramatic change is the shift from investigating the function of single genes to examining multiple components at a genome-wide or system-wide level. Although the highly reductionist studies of previous decades were invaluable to plant and root biology, completely sequenced genomes now allow for gene-by-gene analyses to be placed into a broader context and linked together in the form of genetic networks. For example, in the *Arabidopsis* root, gene expression studies in specific cell types and in developmental time revealed that the level of gene regulation is more complex and dynamic than was previously predicted. Furthermore, similar analyses of gene expression under stress conditions did not support the prevailing hypothesis that plants have a universal stress response and instead indicate that the transcriptional responses elicited by abiotic stressors are specific to both the stress and the cell type. These large-scale datasets can be used to elaborate transcriptional modules as a first step toward building quantitative models of transcriptional networks that regulate root biology. The future incorporation of other functional components, such as proteins and metabolites, into these networks will facilitate a comprehensive understanding of cellular function. The ways in which these types of data are used to formulate new hypotheses and drive further innovation will shape the future of root biology.

References

Aida M, Beis D, Heidstra R et al. 2004. The PLETHORA genes mediate patterning of the *Arabidopsis* root stem cell niche. *Cell* 119:109–120.

Alonso JM, Ecker JR. 2006. Moving forward in reverse: Genetic technologies to enable genome-wide phenomic screens in *Arabidopsis*. *Nat Rev Genet* 7:524–536.

Alonso-Blanco C, Koornneef M. 2000. Naturally occurring variation in *Arabidopsis*: An underexploited resource for plant genetics. *Trends Plant Sci* 5:22–29.

Arabidopsis Genome Initiative. 2000. Analysis of the genome sequence of the flowering plant *Arabidopsis thaliana*. *Nature* 408:796–815.

Armengaud P, Zambaux K, Hills A et al. 2009. EZ-Rhizo: Integrated software for the fast and accurate measurement of root system architecture. *Plant J* 57:945–956.

Atwell S, Huang YS, Vilhjalmsson BJ et al. 2011. Genome-wide association study of 107 phenotypes in *Arabidopsis thaliana* inbred lines. *Nature* 465:627–631.

Baerenfaller K, Grossmann J, Grobei MA et al. 2008. Genome-scale proteomics reveals *Arabidopsis thaliana* gene models and proteome dynamics. *Science* 320:938–941.

Banks JA. 2009. *Selaginella* and 400 million years of separation. *Annu Rev Plant Biol* 60:223–238.

Banks JA, Nishiyama T, Hasebe M et al. 2011. The *Selaginella* genome identifies genetic changes associated with the evolution of vascular plants. *Science* 332:960–963.

Beck M, Claassen M, Aebersold R. 2011. Comprehensive proteomics. *Curr Opin Biotechnol* 22:3–8.

Benfey PN, Linstead PJ, Roberts K et al. 1993. Root development in *Arabidopsis*: Four mutants with dramatically altered root morphogenesis. *Development* 119:57–70.

Benfey PN, Scheres B. 2000. Root development. *Curr Biol* 10: R813–R815.

Biedrzycki ML, Jilany TA, Dudley SA, Bais HP. 2009. Root exudates mediate kin recognition in plants. *Commun Integr Biol* 3:28–35.

Birnbaum K, Jung JW, Wang JY et al. 2005. Cell type-specific expression profiling in plants via cell sorting of protoplasts from fluorescent reporter lines. *Nat Methods* 2:615–619.

Birnbaum K, Shasha DE, Wang JY et al. 2003. A gene expression map of the *Arabidopsis* root. *Science* 302:1956–1960.

Brady SM, Orlando DA, Lee J-Y et al. 2007. A high-resolution root spatiotemporal map reveals dominant expression patterns. *Science* 318:801–806.

Buckler ES, Holland JB, Bradbury PJ et al. 2009. The genetic architecture of maize flowering time. *Science* 325:714–718.

Cao J, Schneeberger K, Ossowski S et al. 2011. Whole-genome sequencing of multiple *Arabidopsis thaliana* populations. *Nat Genet* 43:956–963.

Carlsbecker A, Lee J-Y, Roberts C J et al. 2010. Cell signalling by microRNA165/6 directs gene dose-dependent root cell fate. *Nature* 465:316–321.

Cartwright DA, Brady SM, Orlando DA, Sturmfels B, Benfey PN. 2009. Reconstructing spatiotemporal gene expression data from partial observations. *Bioinformatics* 25:2581–2587.

Clark RM, Schweikert G, Toomajian C et al. 2007. Common sequence polymorphisms shaping genetic diversity in *Arabidopsis thaliana*. *Science* 317:338–342.

Coudert Y, Bès M, Van Anh Le T et al. 2011. Transcript profiling of crown rootless1 mutant stem base reveals new elements associated with crown root development in rice. *BMC Genomics* 12:387.

De Smet I, Tetsumura T, De Rybel B et al. 2007. Auxin-dependent regulation of lateral root positioning in the basal meristem of *Arabidopsis*. *Development* 134:681–690.

Dequeant ML, Pourquie O. 2008. Segmental patterning of the vertebrate embryonic axis. *Nat Rev Genet* 9:370–382.

Dinneny JR, Long TA, Wang JY et al. 2008. Cell identity mediates the response of *Arabidopsis* roots to abiotic stress. *Science* 320:942–945.

Dolan L, Janmaat K, Willemsen V et al. 1993. Cellular organisation of the *Arabidopsis thaliana* root. *Development* 119:71–84.

Domon, B, Aebersold R. 2010. Options and considerations when selecting a quantitative proteomics strategy. *Nat Biotechnol* 28:710–721.

Fang S, Gao X, Deng Y, Chen X, Liao H. 2011. Crop root behavior coordinates phosphorus status and neighbors: From field studies to three-dimensional in situ reconstruction of root system architecture. *Plant Physiol* 155:1277–1285.

French A, Ubeda-Tomas S, Holman TJ, Bennett MJ, Pridmore T. 2009. High-throughput quantification of root growth using a novel image-analysis tool. *Plant Physiol* 150:1784–1795.

de Godoy LMF, Olsen JV, Cox J et al. 2008. Comprehensive mass-spectrometry-based proteome quantification of haploid versus diploid yeast. *Nature* 455:1251–1254.

de Groot MJ. L, Daran-Lapujade P, van Breukelen B et al. 2007. Quantitative proteomics and transcriptomics of anaerobic and aerobic yeast cultures reveals post-transcriptional regulation of key cellular processes. *Microbiology* 15:3864–3878.

Guimil, S, Dunand C. 2007. Cell growth and differentiation in *Arabidopsis* epidermal cells. *J Exp Bot* 58:3829–3840.

Heazlewood JL, Durek P, Hummel J et al. 2008. PhosPhAt: A database of phosphorylation sites in *Arabidopsis thaliana* and a plant-specific phosphorylation site predictor. *Nucleic Acids Res* 36 (Database issue):D1015–D1021.

International Rice Genome. 2005. The map-based sequence of the rice genome. *Nature* 436:793–800.

Ishida T, Kurata T, Okada K, Wada T. 2008. A genetic regulatory network in the development of trichomes and root hairs. *Annu Rev Plant Biol* 59:365–386.

Iyer-Pascuzzi AS, Jackson T, Cui H et al. 2011. Cell identity regulators link development and stress responses in the *Arabidopsis* root. *Dev Cell* 21:770–782.

Iyer-Pascuzzi AS, Symonova O, Mileyko Y et al. 2010. Imaging and analysis platform for automatic phenotyping and trait ranking of plant root systems. *Plant Physiol* 152:1148–1157.

Jang G, Yi K, Pires ND, Menand B, Dolan L. 2011. RSL genes are sufficient for rhizoid system development in early diverging land plants. *Development* 138:2273–2281.

Jones AM, Chory J, Dangl JL et al. 2008. The impact of *Arabidopsis* on human health: Diversifying our portfolio. *Cell* 133:939–943.

Jorrín-Novo JV, Maldonado AM, Echevarría-Zomeño S et al. 2009. Plant proteomics update (2007–2008): Second-generation proteomic techniques, an appropriate experimental design, and data analysis to fulfill MIAPE standards, increase plant proteome coverage and expand biological knowledge. *J Proteomics* 72:285–314.

Jurgens G. 2001. Apical-basal pattern formation in *Arabidopsis* embryogenesis. *Embo J* 20:3609–3616.

Kersten B, Agrawal GK, Durek P et al. 2009. Plant phosphoproteomics: An update. *Proteomics* 9:964–988.

Kusano M, Tabuchi M, Fukushima A et al. 2011. Metabolomics data reveal a crucial role of cytosolic glutamine synthetase 1;1 in coordinating metabolic balance in rice. *Plant J* 66:456–466.

Leutwiler LS, Hough-Evans BR, Meyerowitz EM. 1984. The DNA of *Arabidopsis thaliana. Mol Gen Genet* 194:15–23.

Long TA, Tsukagoshi H, Busch W et al. 2010. The bHLH transcription factor POPEYE regulates response to iron deficiency in *Arabidopsis* roots. *Plant Cell* 22:2219–2236.

Lugtenberg B, Kamilova F. 2009. Plant-growth-promoting rhizobacteria. *Annu Rev Microbiol* 63:541–556.

Ma, S, Bohnert HJ. 2007. Integration of *Arabidopsis thaliana* stress-related transcript profiles, promoter structures, and cell-specific expression. *Genome Biol* 8:R49.

Magalhaes JV. 2006. Aluminum tolerance genes are conserved between monocots and dicots. *Proc Natl Acad Sci USA* 103:9749–9750.

Malamy JE. 2005. Intrinsic and environmental response pathways that regulate root system architecture. *Plant Cell Environ* 28:67–77.

Maloof JN. 2003. Genomic approaches to analyzing natural variation in *Arabidopsis thaliana. Curr Opin Genet Dev* 13:576–582.

Mann M, Jensen ON. 2003. Proteomic analysis of post-translational modifications. *Nat Biotechnol* 21:255–261.

Meinke DW, Cherry JM, Dean C et al. 1998. *Arabidopsis thaliana*: A model plant for genome analysis. *Science* 282:662–682.

Moreno-Risueno MA, Van Norman JM, Moreno A et al. 2010. Oscillating gene expression determines competence for periodic *Arabidopsis* root branching. *Science* 329:1306–1311.

Mouchel CF, Briggs GC, Hardtke CS. 2004. Natural genetic variation in *Arabidopsis* identifies BREVIS RADIX, a novel regulator of cell proliferation and elongation in the root. *Genes Dev* 18:700–714.

Nordborg M, Hu TT, Ishino Y et al. 2005. The pattern of polymorphism in *Arabidopsis thaliana. PLoS Biol* 3:e196.

Okada K, Shimura Y. 1990. Reversible root tip rotation in *Arabidopsis* seedlings induced by obstacle-touching stimulus. *Science* 250:274–276.

Orlando DA, Brady SM, Fink TM, Benfey PN, Ahnert SE. 2010. Detecting separate time scales in genetic expression data. *BMC Genomics* 11:381.

Parizot B, Laplaze L, Ricaud L et al. 2008. Diarch symmetry of the vascular bundle in *Arabidopsis* root encompasses the pericycle and is reflected in distich lateral root initiation. *Plant Physiol* 146:140–148.

Paterson AH. 2006. Leafing through the genomes of our major crop plants: Strategies for capturing unique information. *Nat Rev Genet* 7:174–184.

Paterson AH, Bowers JE, Bruggmann R et al. 2009. The *Sorghum bicolor* genome and the diversification of grasses. *Nature* 457:551–556.

Paterson AH, Bowers JE, Burow MD et al. 2000. Comparative genomics of plant chromosomes. *Plant Cell* 12:1523–1540.

Perret JS, Al-Belushi ME, Deadman M. 2007. Non-destructive visualization and quantification of roots using computed tomography. *Soil Biol Biochem* 39:391–399.

Petricka JJ, Benfey PN. 2008. Root layers: Complex regulation of developmental patterning. *Curr Opin Genet Dev* 18:354–361.

Petricka JJ, Schauer MA, Megraw M. et al. 2012. The protein landscape of the *Arabidopsis* root. *Proc Natl Acad Sci* 109:6811–6818.

Pruitt RE, Meyerowitz EM. 1986. Characterization of the genome of *Arabidopsis thaliana. J Mol Biol* 187:169–183.

Pysh LD, Wysocka-Diller JW, Camilleri C, Bouchez D, Benfey PN. 1999. The GRAS gene family in *Arabidopsis*: Sequence characterization and basic expression analysis of the SCARECROW-LIKE genes. *Plant J* 18 (1):111–119.

Raven JA, Edwards D. 2001. Roots: Evolutionary origins and biogeochemical significance. *J Exp Bot* 52:381–401.

Riechmann JL, Heard J, Martin G et al. 2000. *Arabidopsis* transcription factors: Genome-wide comparative analysis among eukaryotes. *Science* 290:2105–2110.

Saito K, Matsuda F. 2010. Metabolomics for functional genomics, systems biology, and biotechnology. *Annu Rev Plant Biol* 61:463–489.

Scheres B, Wolkenfelt H, Willemsen V et al. 1994. Embryonic origin of the *Arabidopsis* primary root and root meristem initials. *Development* 120:2475–2487.

Schiefelbein JW, Benfey PN. 1991. The development of plant roots: New approaches to underground problems. *Plant Cell* 3:1147–1154.

Schmid KJ, Sorensen TR., Stracke R et al. 2003. Large-scale identification and analysis of genome-wide single-nucleotide polymorphisms for mapping in *Arabidopsis thaliana. Genome Res* 13:1250–1257.

Schnable PS, Ware D, Fulton RS et al. 2009. The B73 maize genome: Complexity, diversity, and dynamics. *Science* 326:1112–1115.

Schneeberger K, Weigel D. 2011. Fast-forward genetics enabled by new sequencing technologies. *Trends Plant Sci* 16 (5):282–288.

Schwanhäusser B, Busse D, Li N et al. 2011. Global quantification of mammalian gene expression control. *Nature* 473 (7347):337–342.

Somerville C, Koornneef M. 2002. A fortunate choice: The history of *Arabidopsis* as a model plant. *Nat Rev Genet* 3:883–889.

de Sousa Abreu R, Penalva LO, Marcotte EM, Vogel C. 2009. Global signatures of protein and mRNA expression levels. *Mol Biosyst* 5:1512–1526.

Sozzani R, Cui H, Moreno-Risueno MA et al. 2010. Spatiotemporal regulation of cell-cycle genes by SHORTROOT links patterning and growth. *Nature* 466:128–132.

Swindell WR. 2006. The association among gene expression responses to nine abiotic stress treatments in *Arabidopsis thaliana. Genetics* 174:1811–1824.

Tanaka S, Brentner LB, Merchie KM et al. 2007. Analysis of gene expression in poplar trees (*Populus deltoides x nigra*, DN34) exposed to the toxic explosive hexahydro-1,3,5-trinitro-1,3,5-triazine (RDX). *Int J Phytorem* 9:15–30.

Tian F, Bradbury PJ, Brown PJ et al. 2011. Genome-wide association study of leaf architecture in the maize nested association mapping population. *Nat Genet* 43:159–162.

Trapnell C, Williams BA, Pertea G et al. 2010. Transcript assembly and quantification by RNA-Seq reveals unannotated transcripts and isoform switching during cell differentiation. *Nat Biotechnol* 28:511–515.

Tuskan GA, Difazio S, Jansson S et al. 2006. The genome of black cottonwood, *Populus trichocarpa* (Torr. & Gray). *Science* 313:1596–1604.

Venter JC, Adams MD, Myers EW et al. 2001. The sequence of the human genome. *Science* 291:1304–1351.

Walther D, Brunnemann R, Selbig J. 2007. The regulatory code for transcriptional response diversity and its relation to genome structural properties in *A. thaliana. PLoS Genet* 3:e11.

Wang Z, Gerstein M, Snyder M. 2009. RNA-Seq: A revolutionary tool for transcriptomics. *Nat Rev Genet* 10:57–63.

Wang X, Mitchum MG, Gao B et al. 2005. A parasitism gene from a plant-parasitic nematode with function similar to CLAVATA3/ESR (CLE) of *Arabidopsis thaliana. Mol Plant Pathol* 6:187–191.

Watt M, Schneebeli K, Dong P, Wilson IW. 2009. The shoot and root growth of *Brachypodium* and its potential as a model for wheat and other cereal crops. *Funct Plant Biol* 36:960–969.

Weigel D, Mott R. 2009. The 1001 genomes project for *Arabidopsis thaliana. Genome Biol* 10:107.

Willemsen V, Bauch M, Bennett T et al. 2008. The NAC domain transcription factors FEZ and SOMBRERO control the orientation of cell division plane in *Arabidopsis* root stem cells. *Dev Cell* 15:913–922.

Wormuth D, Heiber I, Shaikali J et al. 2007. Redox regulation and antioxidative defence in *Arabidopsis* leaves viewed from a systems biology perspective. *J Biotechnol* 129:229–248.

Yu J, Wang J, Lin W et al. 2005. The genomes of *Oryza sativa*: A history of duplications. *PLoS Biol* 3:e38.

Zhao K, Tung C-W, Eizenga GC et al. 2011. Genome-wide association mapping reveals a rich genetic architecture of complex traits in *Oryza sativa. Nat Commun* 2:467.

II

Root Structure

3

Cellular Patterning of the Root Meristem: Genes and Signals

Kimberly L. Gallagher
University of Pennsylvania

I. Introduction

It is easy to see how an infant child relates to the parent. With respect to overall pattern, it is essentially the adult in miniature: 2 arms, 2 legs, 10 fingers, and 10 toes. The same is not true when one compares a newly emerged plant seedling to the adult. The adult plant body plan is often a highly elaborated structure with multiple lateral appendages and organs. These structures may be underdeveloped or completely lacking in the newly germinated seedling. This is because much of the adult plant body is patterned postembryonically through the action of coordinated groups of stem cells in the shoot and root apices: the shoot and root apical meristems (SAM and RAM, respectively). Postembryonic development of the adult body plan allows the plant the plasticity to adjust its overall structure, to optimize the acquisition of resources like light and water, and to respond appropriately to biotic and abiotic signals.

As the SAM and RAM collectively give rise to all of the tissues of the adult plant, much attention has been given to how they are structured, how they function, and how they form. In previous editions of this book, this chapter discussed the physical and mechanical properties of the RAM and provided an excellent comparative analysis of RAM form and function considering multiple different taxa. In this edition, I will focus on the regulation and maintenance of the RAMs of angiosperms. Particular attention is paid to *Arabidopsis* development as much of the molecular and genetic analysis of RAM formation and maintenance has been done in this system. When possible, comparisons are drawn to other systems. As there are separate chapters in this book that concentrate entirely on vascular patterning of the root and hormone signaling, these topics are given only modest coverage here, although they clearly play very important roles in the patterning of the root. Prior to discussing the RAM, I provide definitions of anatomical terms and a brief description of the SAM as a basis for comparison to the RAM.

II. Anatomical Coordinates

There is inconsistent use of anatomical terminology in the literature particularly with respect to defining relative directions of growth and cell division (Baluska et al. 2005). Therefore, these terms will be defined here. By convention the tip of the shoot and the tip of the root are referred to as the shoot and root "apex," respectively (Figure 3.1A). Consequently in both the root and shoot, the term "apical" describes a location that, relative to the structure or feature being described, is toward the tip. For example, the epidermal cell labeled "a" in Figure 3.1B is apical to the cells labeled "b" and "c." Conversely "basal" is

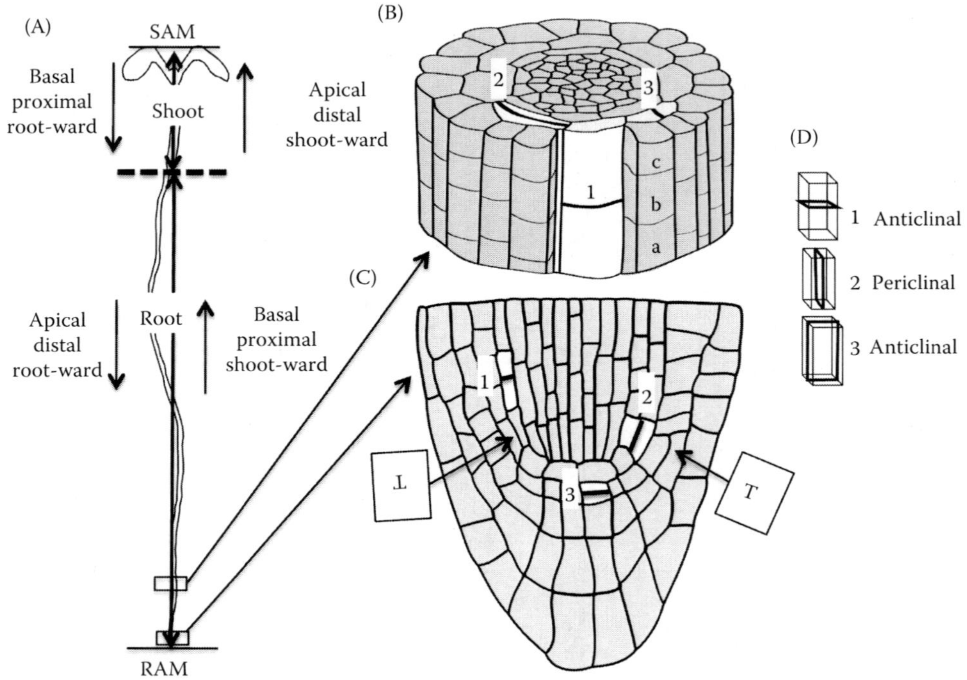

FIGURE 3.1 Anatomical coordinates. (A) Cartoon image of a hypothetical seedling. The junction between the root and shoot is indicated by the dashed line. Directional terms are labeled. (B) A transverse slice through an *Arabidopsis* root. (C) Medial longitudinal cross section through the *Arabidopsis* root meristem. The lowercase labels "a," "b," and "c" on the column of cells in (B) are for orientation and are referenced in the text. In both (B) and (C), oriented divisions are shown as bold black lines in the cells in white. The numbers associated with these divisions refer to the images in (D). "T" cell divisions are indicated in (C). The division on the left contributes to the plant body, whereas the one on the right adds to the root cap.

away from the apex and toward the point where the root joins the shoot (dotted line in Figure 3.1A; referred to by Dolan as the collet zone in *Arabidopsis*). The terms "proximal" and "distal" are also used to describe the position of a feature relative to the junction of the root and the shoot (usually the root–hypocotyl junction). "Proximal" indicates toward the root–shoot junction, whereas away is "distal." In Figure 3.1B, cell "c" is proximal to "b." Recently some have adopted the terms "root-ward" and "shoot-ward" in favor of proximal and distal (Baskin et al. 2010). In the root, shoot-ward is proximal, whereas in the shoot, proximal is root-ward.

With respect to cell division, the terms "anticlinal" and "periclinal" (Figure 3.1D) are used to describe the orientation of cell division relative to the surface of the organ, usually the epidermis. Anticlinal cell divisions create a new cell wall that is perpendicular (at a right angle) to the organ surface. In the root, anticlinal cell divisions can occur in two orientations: perpendicular or parallel to the axis of root elongation. The divisions labeled "1" in Figure 3.1B and C have occurred perpendicular to the epidermis and the primary axis of root growth. The type of anticlinal cell division increases the number of cells in the proximal–distal axis. In the *Arabidopsis* root, periclinal cell divisions largely occur within the plane of cell elongation and increase the number of cell layers in the organ. While the terms periclinal and anticlinal are very helpful in describing the orientation of cell divisions in the root meristem, they can be confusing when referring to divisions when cells are not cuboidal or when divisions are not clearly oriented parallel or perpendicular to the organ surface. For example, the apical cell in *Azolla filiculoides* Lam. (discussed in more detail later; see Figure 3.3E and F) is a tetrahedral and most of the formative cell divisions are oblique to the root surface. In this case, descriptors other than periclinal and anticlinal are useful in explaining the orientation of cell division, and the orientation may be described relative to the axis of the cell.

III. Definition of the Term "Meristem"

The term meristem was coined in 1858 by Carl Wilhelm von Nägeli to refer to populations of cells in the plant that are able to give rise to entire organs, for example, the SAM and RAMs (von Nägeli 1858). Part of the motivation for developing this new term was to distinguish populations of cells that produced an entire organ from those that added new tissues to an existing organ, like those of the vascular and cork cambia. Therefore, in its original conception, the term meristem would not have been applied to the vascular and cork cambia (which add to the vasculature and bark, respectively), although today both are generally classified as types of meristems. In defining a more modern definition of the term meristem, Esau suggested that the meristem is composed of a group of mitotically active cells that divide throughout the life of the plant to produce new cells, tissues, or organs. Esau points out that meristem cells are distinct from cells that divide occasionally within an organ to produce new cells or small clones of cells in that the meristem is maintained in a mitotically competent and relatively undifferentiated state throughout the life of the organ (Esau 1965). In this way, meristem cells are similar to stem cells in animals, which will be discussed later in this chapter.

IV. SAM

All of the aboveground tissues in the plant are produced through the action of the SAM. The SAM is located at the tip of the shoot axis and is generally a flat to dome-shaped structure that lies in between developing organ primordia (Esau 1965). Depending on the plant and the stage of development, the organs formed may be leaves, thorns, axillary, or floral meristems. The SAM of most angiosperms is organized into two to three distinct layers that are generally labeled the L1, L2, and L3, with the L1 being the outermost cell layer. In general the L1 and L2 layers are single sheets of cells, whereas the L3 layer is less organized and may contain multiple layers of cells. The L1 layer generally gives rise to the epidermis and the L3 layer, the vascular tissue (reviewed by Tooke and Battey 2003). However, this is not due to inherent developmental differences between the cells in these layers. Instead, experiments with cell ablations and L1–L2 chimeras have shown that cellular identity is neither predetermined nor invariant in the formation of lateral organs. Instead, cell fate is determined by its position. Indeed when examined, it appears as though most cell fate decisions in the plant are determined by position rather than lineage (reviewed by Szymkowiak and Sussex 1996).

Although cell fate is not invariant in the SAM, the formation of lateral organs and tissues generally proceeds in a predictable way. The regularity of SAM development has lead to the establishment of two models: the tunica-corpus model (originated by Schmidt 1924) and the meristem zonation model (from Foster 1943) that are widely used today to describe the organization and behavior of the SAM. In the tunica–corpus model, the meristem is divided, based upon the orientation of cell divisions, into two regions: the tunica and the corpus. In *Arabidopsis* the L1 and L2 layers of the SAM comprise the tunica (literally the tunic)—the outer covering of the meristem. The L3 comprises the corpus (literally the body). Cells in the tunica tend to divide anticlinally, and in doing so, they maintain single L1 and L2 cell layers. In contrast, the orientation of divisions in the corpus is irregular so a central mass of cells is created.

Overlaid upon the tunica-corpus organization are three different zones of cellular organization—the central zone (CZ), the peripheral zone (PZ), and the rib zone (RZ) (Medford 1992). These zones were originally based upon histological differences in gymnosperms (Foster 1943) but can be used to describe difference in cell division and gene expression in angiosperms. The central zone of the meristem extends over a portion of the cells in the tunica and corpus layers. The cells of the CZ were once thought to be largely quiescent, akin to the quiescent center cells found in the RAM of many species. However, it is now clear that the cells of the CZ do divide and in doing so displace their daughter cells toward the periphery of the CZ into the PZ, where organ initiation takes place. Cells in the peripheral zone divide more rapidly than their sister cells in the CZ. As these cells divide, their daughter cells are pushed further from the CZ and eventually differentiate as organ primordia. The RZ subtends the central and peripheral zones and divides at a rate that is intermediate between the central and peripheral zones. Divisions in the RZ increase the height of the SAM and contribute cells primarily to the stem tissues (Kerstetter and Hake 1997). In this way, the height of the plant is increased and lateral appendages are produced at the flanks of the meristem (reviewed by Traas and Vernoux 2002; Tooke and Battey 2003).

V. Root System

In most plants, the first structure to emerge from the germinating seed is the radicle (primary root) (Clowes 1961). In dicots, the primary root is often long lived, forming a prominent taproot that may continue to grow throughout the life of the plant. A primary root that continues to grow (or at least maintains the ability to grow) throughout the life of the plant is said to be indeterminate (Sinnot 1960). Truly indeterminate roots are probably rare (Shishkova et al. 2008). Lateral roots are generally smaller than the taproot and emerge from within the pericycle layer of the parent root (see Chapter 6). In monocots, the primary root is often short lived; it grows for a limited period of time and then growth ceases (determinate growth). Cessation of growth is often correlated with the emergence of multiple lateral and adventitious roots that form a fibrous root system composed of multiply branched lateral roots (Aloni et al. 2006). A prime example of this is rice. Upon germination the rice seminal (embryonic) root emerges. The seminal root is short lived, persisting only through the seedling stage of growth. Crown roots are prevalent in rice and emerge from the stem of the plant as opposed to root tissue and therefore are considered adventitious roots. Both the seminal and the crown roots generally give rise to lateral roots (both large and slender), some of which can themselves generate additional laterals (up to fifth-order branching is observed in rice root systems). As seminal, lateral, and crown roots have different developmental origins, there are specific mutations that inhibit formation of each type of root. For example, there are mutations that inhibit the formation of crown roots, but have little or no effect on the formation of the seminal root or the lateral roots (reviewed by Rebouillat et al. 2009). However independent of origin, once formed the growth of all of these roots relies on divisions in the root apical meristem (RAM).

Unlike the SAM, the RAM is a subapical structure that is covered at its apex by protective layers of cells that comprise the root cap (false colored in green in Figure 3.2). The RAM makes no lateral organs; instead, when cells in the RAM

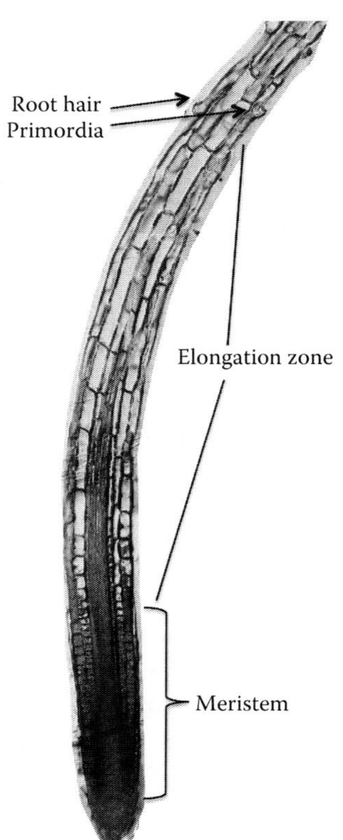

FIGURE 3.2 **(See color insert.)** Developmental zones in the root. Toluidine blue–stained single medial section through an *Arabidopsis* root (kindly provided by Shuang Wu). The different zones of the root are labeled. The root cap has been false colored in green, and the root initial cells are false colored in red.

divide, their daughter cells are displaced either apically to contribute to the root cap or basally to contribute to the body of the root (Clowes 1961). As cells are displaced basally into the body of the root, they generally divide one to several times (similar to the behavior of cells in the peripheral zone of the SAM) before eventually exiting the root meristem and entering into the root elongation zone, where rapid cell expansion occurs. Basal to the elongation zone is the cell differentiation zone, which is the region in which cells begin to adopt their respective developmental fates. For example, root hairs emerge from the epidermis (Figure 3.2) in the differentiation zone and the Casparian strips form in the endodermis. Cells in the root apex that are displaced apically during cell division enter the root cap. In *Arabidopsis*, these cells do not undergo a series of divisions before differentiation. Instead, they immediately differentiate as columella. As new columella cells are added through divisions in the RAM, the most apical cells of the root cap are sloughed off, resulting in the maintenance of a relatively constant number of root cap cells. The root then is comprised of a continuum of cells in which the least mature cells occupy a subapical position below the root cap with more mature cells situated both proximally in the root body and distally in the root cap/columella (Dolan et al. 1993; Baum and Rost 1996; Groot et al. 2003; Rost 2010).

VI. Structure of the RAM

As the RAM gives rise to all of the cells of the primary root and once formed the lateral roots, there has been much interest in describing its development and behavior. One of the early descriptions of root growth has been likened to the tunica–corpus model of SAM organization because it separates the root into two domains on the basis of the prevailing orientation of cell divisions. This is the Körper–Kappe model (Clowes 1961). To describe how the root grows, Schuepp (1917) examined fixed longitudinal sections through various different roots and noted that when two cell files were continuous with one cell file, the convergence of these cell files resulted in the formation of a T-junction (Figure 3.1C). These T-junctions were apparently formed by the reorientation of cell division from transverse to longitudinal in a series of clonally related cells. Based upon the orientation of this junction, Schuepp divided the root into the Körper (root body) and the Kappe (root cap). Divisions in the root body produced a T in which the crosshatch pointed toward the root apex, whereas in the cap, the T pointed toward the base. Schuepp found that in some plants, the boundary between the Korper and the cap was clear and consistent, whereas in others the boundary differed from root to root or changed with development. Schuepp also noted that Körper-type divisions increased the diameter of the root, whereas Kappe-type divisions did not, as these cells were regularly sloughed off as the root grew (Schuepp 1917; Clowes 1961).

The separation of the root apex into the root body and the root cap was also important in early descriptions by Guttenberg (1960). Based upon whether the root cap and cortex were clonally related, Guttenberg classified roots as either "closed" or "open." Roots with a closed meristem had clonally distinct ground tissue and root cap cell layers. Roots with an open meristem did not; instead, single progenitor cells in the root tip would divide to produce daughters that added both apically to the cap and basally to the root body. Recent work by Heimsch and Seago (2008) expanded upon the work of Guttenberg examining "over 45 orders and 132 families of basal angiosperms, monocots, and eudicots." From these comparisons, Heimsch and Seago found that angiosperms were more likely to have an exclusively closed organization than other groups. However, there were examples of both open and closed meristem organization in many families. Likewise Heimsch and Seago extended Guttenberg's classification, breaking out the closed and open organization into fifteen different subclassifications based upon how distinct the root body is from the root cap and the degree of organization in the root apex. As many basal angiosperms tend to have open meristems, Heimsch and Seago suggest that open may be the ancestral state.

In longitudinal sections through the RAM of many species that have closed meristems such as *Arabidopsis*, rice, and maize, it is possible to trace back individual cell files to a point of convergence (i.e., the head or crosshatch of Scheupp's T) that is distal to the tip of the stele (stellar pole). This observation led to the proposal that a subapical population of dedicated progenitor cells exists at the tip of these roots, similar to the CZ population of cells in the SAM (Clowes 1961; Barlow 1976). Hanstein labeled these progenitor cells histogens and proposed that three distinct histogens exist: the dermatogen, the periblem, and the plerome. These histogens provided a continuous source of new cells that produced the epidermis, cortex, and stele tissues, respectively (Hanstein 1870). In conception, the histogens were similar to the three primary germ layers in animals. Later Janczewski added a fourth histogen, the calyptrogens (Janczewski 1874; Clowes 1961). In cases where the root cap and the epidermis were clonally related, a calyptro-dermatogen was proposed. As illustrated in the following example (Figure 3.3A and B), a calyptro-dermatogen is common in dicots like *Arabidopsis*.

Implicit in Hanstein's theory of histogenesis was the concept that cell lineage determines cell fate. While a strict relationship between cell lineage and cell identity has not been experimentally borne out, in many roots a localized region in the apex has been identified that maintains the cell layers of the growing root. In the absence of perturbation, these cells divide in a predictable way to give rise to a constant number of cell files with predictable positions within the organ and predictable cell identities (Figure 3.3; Clowes 1961). These cells are now generally referred to as initials (Esau 1965). The initial cells have been variously defined based upon both position in the root and behavior. The definition that will be used here is that of Scheres et al. (1996) as it combines essential ideas from Esau (1965), Seago (1969), and Dolan et al. (1993). Scheres et al. define the root meristem initials as the cells at the end points of the linear cell files that with each asymmetric cell division add one cell to the plant body or the root cap, while the initial cell retains its position between the stele and the root cap. The daughter that is displaced from the center of the meristem may itself also divide asymmetrically (in Körper T divisions) to generate additional cell layers or divide symmetrically in transit-amplifying divisions that increase the length of the root. The behavior of the initial cells in *Arabidopsis thaliana* (L.) Heynh., *Oryza sativa* L. (rice), and *Azolla filiculoides* (water fern) provide good examples of how the cortex, epidermis, and root cap cells are generated through the asymmetric divisions of the initial cells and their daughters (Figures 3.3 and 3.4).

Upon germination the primary root of *Arabidopsis thaliana* is composed of single concentric layers of epidermis (p), cortex (c), endodermis (e), and pericycle. These four cell layers surround a central cylinder of vascular tissues, which along with the pericycle comprise the stele (S). At the root apex is the root cap, which is composed of the collumella and the lateral root cap (LRC)

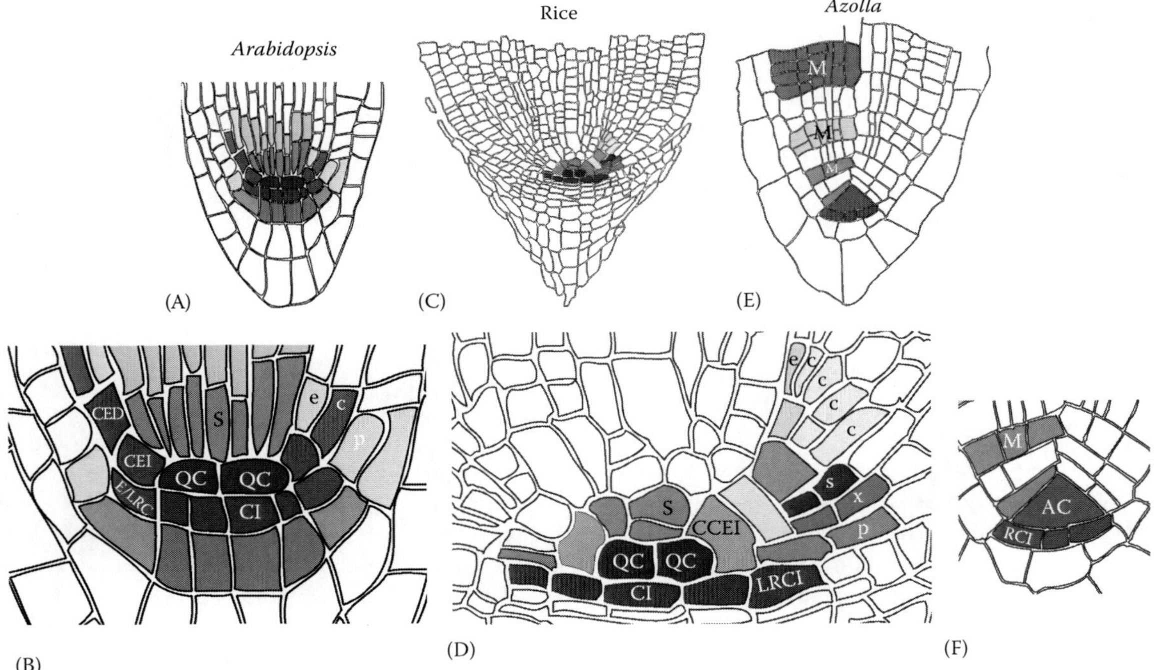

FIGURE 3.3 (See color insert.) Cellular organization of three root meristems. *Arabidopsis* is shown in (A) and cropped and enlarged to show the initial cells in (B). The rice meristem is shown in (C) and enlarged in (D). The structures of the *Arabidopsis* and rice roots are similar. However, the rice root has multiple different specialized cells in the ground tissue including the endodermis (e), schlerenchyma (s), and the exodermis (x) and multiple cortex (c) layers. In rice the ground tissue and the epidermis (p) are derived from the CCEI. In contrast, in *Arabidopsis*, these tissue are distinct, and the epidermis and lateral root cap share an initial (E/LRC) (Images based upon Courdert et al. 2010). The tip of an *Azolla* root is shown in (E) and close-up in (F). The apical cell (AC) is shown in green, and three merophytes are shown in different colors and labeled with an (M). Root cap initials (RCI) below the apical cell give rise to the root cap. (Based upon Gunning, B. et al., *Planta*, 143, 121, 1978.). Initial cells are labeled with capital letters. Cell types are indicated by lowercase. All common cell types are labeled with the same color. QC, quiescent center; CI, columella initial; S, stele initial; CEI, cortical endodermal initial; LRCI, lateral root cap initial.

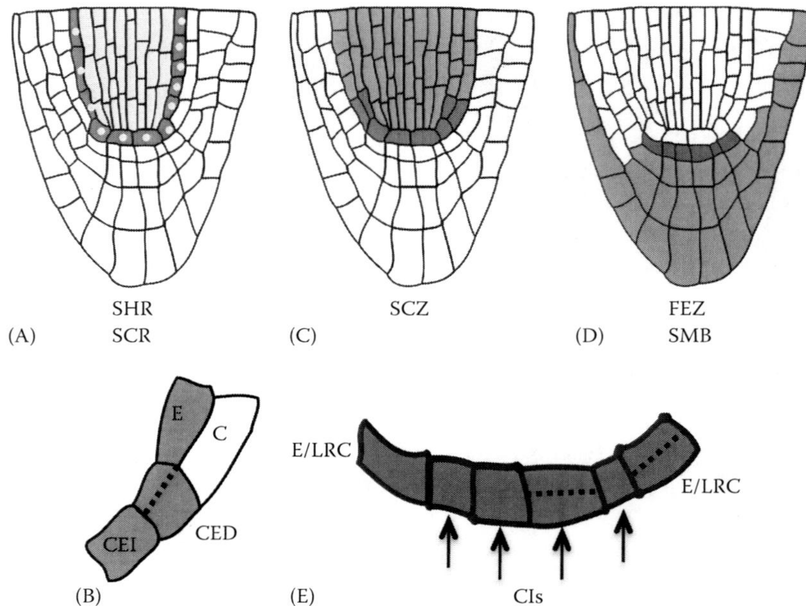

FIGURE 3.4 **(See color insert.)** Proteins involved in *Arabidopsis* root patterning. (A) SHR (yellow) and SCR (green) as SHR is transcribed in the stele and moves into the endodermis; the yellow dots in the endodermis indicate SHR movement. (B) The asymmetric division (dotted line) shown in the cortical endodermal daughter (CED) requires both SHR and SCR activity. (C) *SCZ* expression is highest in the QC and initials (dark orange) and present throughout the stele and ground tissue (lighter orange). (D) Expression of FEZ (magenta) and SMB (blue). (E) FEZ expression in the initials promotes periclinal cell division, whereas SMB inhibits periclinal division. E/LRC, epidermis lateral root cap initial; CI, columella initial; CEI, cortical endodermal initial.

(Dolan et al. 1993). The cellular pattern of the root is generated during embryogenesis (Scheres et al. 1994) and then maintained postembryonically through a series of stereotypical cell divisions that occur in the root apex (Dolan et al. 1993; Scheres et al. 1994, 1996; Baum and Rost 1996; Benfey and Scheres 2000; Baum et al. 2002). The cells primarily responsible for maintaining the pattern of the root are the initial cells and their immediate daughters (Figure 3.3A and B). The initial cells are arranged into three tiers that are located between the stele and the root cap and encircle the quiescent center (QC; as defined by Dolan et al. 1993). Periclinal and anticlinal divisions of dedicated initial cells in the upper and lower tiers of initials give rise to the columella and stele cell files, respectively. In contrast, common initials in the second and third tier give rise to the cortex and endodermis and the epidermis and the lateral root cap. Following these formative cell divisions, divisions (anticlinal) in the transit-amplifying cells transiently increase the number of cells in the meristematic zone.

In longitudinal cross sections through the root tip, the cortical endodermal initials (CEIs) are visible on the flanks of the QC (Figures 3.3A and B and 3.4A and B). The CEIs are the ultimate source of the cortex and endodermis. Generally each CEI divides once anticlinally to produce a daughter cell, the CED (cortical endodermal daughter), which is displaced proximally from the QC and divides periclinally (a Körper T division) to produce separate endodermal and cortical cell layers. Once formed, cells in the endodermis and cortex will continue to divide transversely and symmetrically in the meristem until they are displaced into the elongation zone. In some ecotypes, there can be a delay before the CED cell divides periclinally so that there are two or more undivided daughters in the cell file. However, in other ecotypes, the CEI itself may divide periclinally to produce separate initials for endodermis and cortex (Figures 3.3B and 3.4B; Dolan et al. 1993).

Genetic analysis of root patterning has shown that two related GRAS family transcription factors, *SHORT-ROOT* (*SHR*) and *SCARECROW* (*SCR*), promote the asymmetric divisions of the CEDs (Figure 3.4A and B; Di Laurenzio et al. 1996; Helariutta et al. 2000). The SHR protein is also required independently for specification of the endodermis (discussed in more detail Section XIV; Helariutta et al. 2000; Nakajima et al. 2001). *SCHIZORIZA* (*SCZ*) (Figure 3.4C), a protein with similarity to heat shock transcription factors, is required for specifying the cortex. Cortex specification prevents cells in the outer ground tissue layer from differentiating as epidermis (Doerner 2010; Pernas et al. 2010; ten Hove et al. 2010). In the absence of *SCZ*, root hairs emerge from inside of the root, dramatically tearing through the epidermis (Mylona et al. 2002). A role for small RNAs in radial patterning has also been suggested as mutations in *ARGONAUTE 1* (*AGO1*) or *HYL1* cause extra ground tissue layers in the root (Miyashima et al. 2009).

In *Arabidopsis* the root cap is composed of both the columella and the lateral root cap. The columella cells are produced by the transverse divisions of initial cells (CIs) that are immediately distal to the QC (Figures 3.3B and 3.4D and E) (Dolan et al. 1993; Baum and Rost 1996). Following the division of the

CI, the daughter cells that are in contact with the QC remain as initial cells, whereas the apically displaced daughter cells differentiate as columella. This is evidenced by the accumulation of starch granules in these cells. There are no transit-amplifying divisions in the production of the columella in *Arabidopsis* (Dolan et al. 1993; Willemsen et al. 2008); however, in other species, there are proliferative divisions in the root cap. For example, in both *Helianthus annuus* L. and *Zea mays* L., there is a gradation of cell divisions in the columella region of the root cap with the lowest rate of division generally near the QC, which suggests that the columella initials are not the only source of new columella cells in the root caps of these species (Clowes 1981).

In contrast to the dedicated initials that produce the columella, the cells that give rise to the lateral root cap (the epidermal lateral root cap initials; E/LRCIs) produce both the lateral root cap and the epidermis (Figures 3.3 and 3.4). The E/LRCI cells are apical to the QC and encircle the CIs. For the most part, periclinal cell divisions of the E/LRCI cells produce the LRC, whereas anticlinal divisions produce the epidermis (Dolan et al. 1993; Baum and Rost 1996). However, the timing of these divisions is not invariant. During time-lapse imaging periclinal cell divisions in cells one would have predicted to develop as epidermis resulted in the formation of new LRC cells (Campilho et al. 2006). There is therefore some flexibility in the sequence of divisions that leads to the formation of these two cell types. Once formed, transit-amplifying divisions occur in the LRC and the epidermis to increase the number of cells in these files. With continued division, cells in the epidermis will eventually be displaced into the elongation zone, whereas cells of the root cap will be lost as the root grows through the soil. Interestingly although the LRC and the COL are not clonally related, they share regulatory pathways. Willemsen et al. (2008) showed that two NAC domain-containing transcription factors, *FEZ* and *SOMBRERO* (*SMB*) (Figure 3.4D and E), regulate divisions in both the columella and E/LRC initials. The FEZ protein is expressed in both the columella and E/LRC initials and promotes periclinal cell division. SMB is expressed specifically in the CI and E/LRC daughter cells. The primary function of SMB appears to be inhibition of *FEZ* in cells outside of the initials, as ectopic expression of *FEZ* or loss of *SMB* activity results in ectopic periclinal cell divisions and extra cell layers in the columella and LRC. SMB along with two other NAC-domain proteins, BEARSKIN1 and BEARSKIN2, also controls the maturation of the root cap (Bennett et al. 2010). In contrast, FEZ appears to not play a direct role in cell identity (Willemsen et al. 2008).

Similar to the *Arabidopsis* root, rice roots are also organized into rings of cells that surround the vascular cylinder (Figure 3.3C and D). In rice however, there are multiple layers of cortex (the number of cortex layers varies between the different root types), and the ontogeny of the cortical, epidermal, and root cap cells differs between *Arabidopsis* and rice as it does between most dicots and monocots (Clowes 2000). In rice, the cortex and epidermal cell layers are clonally related, and the root cap is entirely separate from all other cell layers of the root (Clowes 1994). The initial cells in rice are arranged in three tiers that surround the QC. The cortical epidermal cell initial (common endodermis epidermis initial, CEEI) occupies the same tier as the QC cells. To form the epidermis and cortex, the CEEI divides anticlinally to produce a daughter cell that divides periclinally to generate the epidermal and the endodermal precursor cells. The resulting cell in the epidermal cell file divides anticlinally in a series of transit-amplifying cell divisions to produce the epidermal cell lineage. In contrast, the cell occupying the endodermal position divides periclinally several times to produce all of the specialized cortical cell layers of the root: exodermis (x), schlerenchyma (s), mesodermis (cortex, which will generally differentiate into aerenchyma in all roots with the exception of some laterals), and endodermis. These cell layers help the root to survive in water-saturated soils (Coudert et al. 2010). It is unclear how many initial cells are required to sustain the stele cell population or how these divisions are oriented. However, some work has been done to describe how various cell types arise in the vascular cylinder (reviewed by Rebouillat et al. 2009). In the mature rice roots, there appear to be separate initial cells for the LRC and columella cells.

In contrast to *Arabidopsis* and rice, all of the cell types (epidermis, cortex, and stele) of the *A. filiculoides* (water fern) root are derived from a single initial cell (Figure 3.3E and F; Gifford and Polito 1981a,b). At its apex, the *A. filiculoides* root is covered by a root cap. Proximal to the root cap are the root cap initial cells and a single pyramidal-shaped "apical cell" that gives rise to all cells of the plant body, except the root cap. It is thought that early in the development of the root, the apical cell divides along its distal face to produce the root cap initials so that for most of its growth, the *A. filiculoides* root has a closed meristem. However ferns of the genus *Marsilea* have an open meristem; there is no separate group of initials for the root cap. Instead, the apical cell divides continuously throughout the growth of the root along its apical face to maintain a steady population of root cap cells (Clowes 1961). Therefore, all cells in these roots are clonally related—produced by the regular divisions of a single initial cell.

The cells of the *A. filiculoides* root body are produced by divisions of the apical cell in which the new cell walls are oriented parallel to the three proximal faces of the mother cell. These formative cell divisions produce what are referred to as merophytes (essentially packets of cells) that divide periclinally several times (T divisions) to increase the number of cell files and then ultimately divide anticlinally in a series of transit-amplifying divisions to increase the number of cells in the file. The first division of the apical daughter cell (dark green in Figure 3.3E and F) produces an inner and an outer cell. The inner cell will give rise to the vascular tissue including the pericycle, the endodermis, and the inner cortex. The outer cell will give rise to the root hairs, the epidermis, and the outer cortex (Gunning et al. 1978; Gifford and Polito 1981b). Based upon experiments in other root systems that have a single apical cell at the root tip, some have contended that the apical cell of *Azolla* is not the source of all of the cells of the root body, but is instead more akin to the QC cells of rice and *Arabidopsis* (reviewed by Gifford 1983). The thought was that the large apical cell of *A. filiculoides* was largely quiescent and the first merophytes were the actual initials. Experiments by Clowes (1956a, 1958b) and Nitayangkura et al. (1980) however have shown this to be incorrect. The apical cell is mitotically active and provides all of the cells of the adult root. This has also been shown for the apical cells in *Equisetum scirpoides* Michx (Gifford and Kurth 1982) and *Marsilea vestita* Hook and Grev. (Kurth 1981). *Azolla* roots therefore lack a QC.

VII. Significance of an Open versus a Closed Meristem

The classification of roots based upon the organization of their meristem as either open or closed coveys information on how the cells in the meristem are arranged and potentially on how they divide to contribute cells to the root body and the root cap. The initial cells in closed meristems are often arranged in discrete tiers of cells; whereas the organization of open meristems is less well defined, and there is not a clear separation between the root body and cap. Likewise in roots with an open meristem, the quiescent center is often not as well defined as it is in roots with a closed meristem, and there is often a gradation of quiescence (Clowes 1961, 1981). However, these differences do not appear to correlate with any differences in growth capacity (Clowes 1961). So what then is the advantage of one configuration versus the other? Hamamoto et al. (2006) suggest that it may be the ability to produce border cells, individual free-living cells that are derived from the root cap through the activity of cell-wall-degrading enzymes. Roots with open meristems produce and shed considerably (7.5–200 times) more live border cells into the environment than do roots with a closed meristem. Bingham et al. (1993, 1995) has shown in pea that upon release from the root, border cells change their pattern of gene expression to produce antibiotics and even aluminum binding mucilage. It has been suggested that one of the functions of border cells is to act as guards in the soil—protecting the intact root from damage as it grows (Hawes et al. 1998). Roots with open meristems therefore may have an advantage to those with closed meristems in terms of resistance to biotic and abiotic attack. Interestingly one of the additional observations that Clowes (1981) made in examining *Helianthus* and maize roots was that due to differences in the distribution of mitotically active cells in their respective RAMs, *Helianthus* is expected to replace its root cap approximately every 1.5 days, whereas in maize this process would take 3 days. This may contribute to a larger pool of potential border cells in *Helianthus* as compared to maize.

VIII. Control of Meristem Size

The root apex is divided into the meristematic, elongation, and differentiation zones, with a region of transition between the meristem and elongation zones (Figure 3.5B; Baluska et al. 1990, 2010). For the size of each of these individual zones to remain constant, the rate of cell division in the meristem must equal the rate at which cells "enter" each of the other two zones. Often this is not the

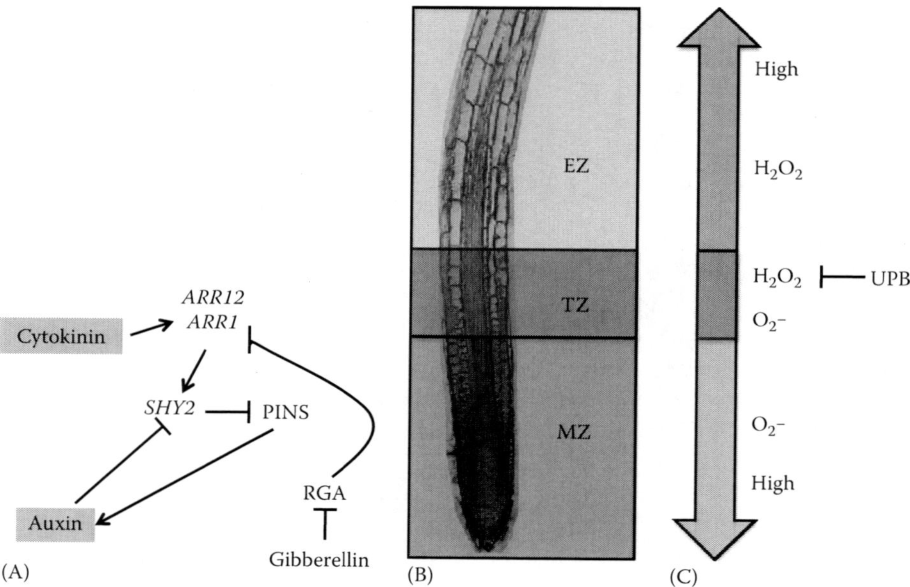

FIGURE 3.5 Regulation of meristem size. (A) The relative actions of cytokinin and auxin control the size of the meristem (MZ) in (B). In the transition zone (TZ) where cytokinin predominates cells are predisposed toward differentiation and auxin signaling is suppressed. In the MZ, auxin promotes cell division and inhibits cytokinin. Early in the growth of the meristem, high levels of gibberellins inhibit cytokinin signals, and the size of the meristem increases. Later in root growth, gibberellin levels fall off, and the size of the meristem is kept relatively constant. (C) In the elongation zone, a gradient of H_2O_2 (with levels increasing basally) promotes differentiation. A mobile protein, UPB1, facilitates this gradient by limiting H_2O_2 levels in the transition zone. An O_2^- gradient in the meristem promotes cell division. The overlap between these two ROS gradients defines the transition zone and the switch from cell division to cell elongation.

case, and the relative sizes of each of the regions change throughout the development of the root. For example, in examining the meristematic and differentiation zones of the pea root, which has an open meristem, Rost and Baum (1988) found that early in the development of the root when divisions were frequent, the height of the meristem (the distance between the root cap and the elongation zone) increased from 1.9 to 2.7 mm. Later in development as the mitotic index decreased, the meristem height returned to 2.0 mm. Interestingly, the height of the meristem was inversely correlated with the distance to the differentiation zone so that the differentiation zone was closer to the tip in slowly dividing roots. That is, as the rate of cell division decreased in the meristem, the differentiation zone moved apically toward the root tip. The apical movement of the differentiation zone was not directly equivalent to the decrease in the height of the meristem; instead, for every 0.19 mm change in the height of the meristem, they found a 1 mm change in the position of the differentiation zone. This effect could be exaggerated by treating the roots with an inhibitor of both cell division and cell elongation, indicating that the size of the meristem is dynamic and that mitotic activity in the tip may delay differentiation.

In *Arabidopsis* the size of the meristem is generally measured by counting the number of unexpanded cortex or epidermal cells on one side of a medial longitudinal cross section through the root tip. Therefore, factors that either promote divisions in the meristem or slow the progression of cells into the expansion zone increase the size of the meristem. The primary regulators of meristem size in *Arabidopsis* are the phytohormones, in particular auxin, cytokinin, and gibberellin (Figure 3.5A; reviewed by Galinha et al. 2009; Durbak et al. 2012). In the meristem, high levels of auxin and low levels of cytokinin promote cell division. In contrast, outside of the meristem, in the transition zone high levels of cytokinin promote cell differentiation and inhibit cell division. Both auxin and cytokinin signal through *SHORT HYPOCOTYL2* (*SHY2*), a member of the Aux/IAA family of auxin-induced genes, which inhibits the expression of PIN proteins and consequently polar auxin transport. Cytokinin promotes expression of *SHY2* (through the *ARR1* and *ARR12* transcription factors), whereas auxin inhibits *SHY2* expression. Gibberellin feeds into this pathway by promoting the degradation of the repressors of gibberellic acid (RGAs), which in turn results in decreased *SHY2* expression and decreased expression of cyclin-dependent kinase inhibitors (Dello Ioio et al. 2007; Achard et al. 2009). During the first few days of growth, gibberellin levels are high in the root, and cell division predominates over cell differentiation. However, by day 5 after germination, gibberellin levels begin to fall, and *SHY2* expression reaches its maximum (Dello Ioio et al. 2007). At this stage in root development, cell division matches cell differentiation, and the size of the meristem is largely stable. However, later in the growth of the root, cell divisions decrease, and the meristem shrinks in size. This often precedes a complete cessation of growth (Zhu et al. 1998a,b; Shishkova et al. 2008).

In *Arabidopsis* there is a natural variation between different ecotypes in the size of the root meristem. Mouchel et al. (2004) were able to exploit this natural variation, identifying a single locus that accounts for approximately 80% of the differences in root lengths between the individuals examined. They named this locus *BREVIS RADIX* (*BRX*). Homozygosity for the *brx* allele reduces the size of the elongation zone and the meristem by approximately 50% and 70%, respectively, when compared to plants with the *BRX* allele. The *BRX* gene was shown to function in the biosynthesis of brassinosteroids through upregulation of the *CONSTITUTIVE PHOTOMORPHOGENESIS AND DWARF* (*CPD*) gene. Consistent with this finding, application of brassinolide to *brx* roots was able to partially rescue root length; auxin was not. However, auxin-responsive gene expression (upon application of exogenous auxin) was almost entirely lost in the *brx* roots, an effect that was mimicked by treating *BRX* roots with brassinosteroid inhibitors. These results suggest that *BRX* activity is required for auxin signaling. Indeed, the production of brassinosteroids through *BRX* appears to be the "rate-limiting step" in auxin-induced gene expression. Auxin also increases the expression of *BRX* (Mouchel et al. 2006) and promotes its translocation into the nucleus, which is required for its function (Scacchi et al. 2009). In contrast, *BRX* expression is negatively regulated (albeit moderately) by the application of brassinolide and by auxin-induced degradation of BRX in the nucleus (Scacchi et al. 2009), indicating that *BRX* is at the center of a feedback loop between auxins and brassinosteroids, which have both positive and negative effects on *BRX* activity (Mouchel et al. 2004; Briggs et al. 2006; Scacchi et al. 2009).

Early work by Sachs (reviewed by Peters and Tomos 1996) suggested that some tissues exert more control over the development of the plant or organ than others. Sach's findings may result from differential mechanical signals or constraints on growth between different tissues that are imposed by the rigid network of cell walls that encase all cells within the plant. For example, the localized application or expression in the outer layers of the SAM of EXPANSINS, which reduces tension between cells by promoting cell-wall "loosening" is sufficient to induce leaf formation in tomato and tobacco plants (Fleming et al. 1997, 1999; Pien et al. 2001), suggesting that the L1 and L2 layers may constrain growth. Alternatively the primacy of a particular tissue in the development of an organ may be the result of the cell-to-cell transport or perception of tissue-specific growth factors. In *Arabidopsis* the dwarf phenotype of the *cpd* and *brassinosteroid-insensitive 1* (*bri1*) (brassinosteroid receptor deficient) mutants is rescued by the cell autonomous restoration of brassinosteroid perception in the epidermis (but not in the ground tissue), suggesting that the epidermis is the primary tissue regulating brassinosteroid-dependent above ground organ growth (Fleming et al. 1997, 1999; Pien et al. 2001). Similar experiments have been conducted to determine the primary tissues that regulate meristem growth in the root. In analyzing tissue-specific requirements for BRX, Mouchel et al. (2006) concluded that expression throughout the root vasculature was sufficient for rescue of the *brx* roots. In contrast, signal perception appears to be regulated by the epidermis as epidermal-specific restoration of the BRI1 receptor is sufficient to rescue *bri1* roots

(Hacham et al. 2011). Therefore, brassinosteroid production in the vasculature via BRX1 may signal to the epidermis via BRI1. In contrast, the critical site of action for gibberellins appears to be neither the stele nor the epidermis, but instead the endodermis. Expression of a dominant (degradation insensitive) allele of GAI in the endodermis results in a decrease in the size of the root meristem (Ubeda-Tomas et al. 2008) indicating that reduced growth of the endodermis may limit the rate of growth in other layers of the root and therefore the entire meristem.

IX. Changes in the Patterning of the Meristem

Not only does the size of the meristem change with development but also the patterning of the meristem. Many have argued that all roots in their mature embryonic form have a closed meristem and that the open arrangement is generated after germination. Heimsch and Seago convincingly dispute this contention citing multiple examples of roots that indeed start out with an open meristem configuration (Heimsch and Seago 2008). What this argument highlights however is that the organization of the RAM is not static; instead, it changes in response to both the environment and developmental signals. In examining twelve different species of roots, Armstrong and Heimsch (1976) found frequent changes in meristem organization during root growth, with examples of both apparently open meristems changing to closed and more frequently closed meristems becoming open. The conversion of a closed meristem into an open arrangement was often correlated with periclinal divisions in the cortex cell layers giving rise to columella cells so that the cortex and cap cells had a common cell lineage. It was unclear what changes led to the conversion of an open meristem into a closed one (Clowes 1981).

Experiments by Chapman et al. (2003) suggest that changes in the organization of the meristem, particularly from a closed to an open conformation, may indicate a change in the growth potential of the root and signal the eventual differentiation of the meristem. Using a labeling scheme that distinguishes between two different types of open meristem—basic open and intermediate open (Groot et al. 2004)—Chapman et al. (2003) showed that in five different species (*Clarkia unguiculata* Lindl., *Oxalis corniculata* L., *Dianthus caryophyllus* L., *Blumenbachia hieronymi* Urb., and *Salvia farinacea* Epling) with initially a closed meristem, the organization changed to an intermediate open prior to cessation of root growth. What they observed in all five species was an initial burst of growth upon germination that was then followed by a gradual deceleration of growth that preceded root determinacy. In all species, there was a wide-ranging shift in the organization of the meristem in which the size of the root cap generally decreased, cell shapes became irregular, cells became less densely cytoplasmic, and the clear distinction between the root cap and the root body was lost. This transformation took between 14 and 41 days depending upon the root species. These results indicate a connection between an open meristem organization and a limited growth potential but only in roots that started out with a closed meristem. Therefore an acquired open meristem may signal a switch to determinate root growth.

Examination of multiple different species indicates that the organization of the meristem and the potential for root growth is under both endogenous and environmental control. Shishkova et al. (2008) point out that for most roots, there is an initial phase of development where root growth is rapid and "indeterminate." During this phase, cell divisions within the meristem are able to replenish the cells that are lost to the elongation and the differentiation zones, and overall root growth is maintained. This indeterminate phase may then be followed by a period in which cell divisions in the meristem and growth cease either due to genetic programming or due to unfavorable growth conditions. In some cases, the cessation of growth correlates with root "dormancy" (as used by Clowes et al. 1967) in which the cells in the apex are still meristematic but simply arrested until conditions or programming signals reactivation. In contrast, cessation of growth may indicate determinate root growth and the eventual differentiation of the root apex. In this context, Shishkova et al. (2008) defined two types of determinate root growth: constitutive and nonconstitutive. For constitutively determinate roots, the loss of meristematic potential "is a natural part of root development" that occurs independently of the environment. Nonconstitutive "determinacy is induced usually by an environmental factor." However, even for constitutively determinate roots like those of pea, growth conditions can accelerate the loss of meristematic activity.

The root of the desert cactus, *Stenocereus gummosus* (Engelm.), provides a particularly interesting example of constitutively determinate root growth. The primary root of *S. gummosus* has an open meristem; cells in the apex divide to produce the root cap and root body. Upon germination, the body of the root is composed of an epidermal cell layer, 4–5 ground tissue layers, and a central stele. The initial cells are present at the base of the stele, and the meristem lacks any obvious QC. As the root grows, the size of the meristem decreases; the initial cells become enlarged and vacuolated. During this process, the size of the root cap also decreases until it is eventually lost. As a consequence of the loss of the meristem, the root tip differentiates, and root hairs form at the apex (Rodríguez-Rodríguez et al. 2003). This process is recapitulated in roots of *S. gummosus* that are generated from callus tissue, indicating a strong genetic component to determinate growth in this species (Shishkova et al. 2007).

When grown on rich medium in tissue culture, the primary root of *Arabidopsis* is maintained for more than 1 month with no obvious changes in the organization of the meristem. When grown in sand, the RAM of *Arabidopsis* seedlings switches from closed to open at about 28 days, which coincides with a cessation of root growth (Zhu et al. 1998a,b). Growing *Arabidopsis* roots on low-phosphorus medium can further accelerate this process (Sanchez-Calderon et al. 2005). By day 12, the meristems of *Arabidopsis* roots, grown on low-phosphate medium, were disorganized, the cells were highly vacuolated, there was no visible

QC, and root hairs had formed close to the tip, indicating a loss of meristematic activity and an apical migration of the differentiation zone similar to what was observed in pea (Rost and Baum 1988). Examination of mitotic activity in the low-phosphate plants using the *CycB1;1:uidA* marker revealed a significant decrease in the rate of cell division by day 4 after germination. By day 14, no expression of *CycB1;1:uidA* was observed. If plants were shifted back to high-phosphate medium at days 6 or 7, when roots still had between 11 and 13 *CycB1;1:uidA*-expressing cells in the meristem, root growth could be recovered. However, if roots were shifted to high-phosphate medium at day 8, when approximately 10 cells still expressed *CycB1;1:uidA*, no recovery was observed (Sanchez-Calderon et al. 2005), perhaps indicating a that a critical number of cells are required in the meristem to maintain root growth.

One way to understand the factors that control root growth is to look for roots that have precocious determinate growth. Null alleles of either *SHR* or *SCR* result in roots whose meristems become disorganized (Di Laurenzio et al. 1996; Helariutta et al. 2000). The *shr* and *scr* mutants initially have a closed meristem. However, as they grow, the division of the initial cells becomes irregular, the QC is lost, and differentiation proceeds to the root tip. This process takes 3–5 days in *shr-2* mutants and 5–7 days in *scr-4* mutants when grown on standard plant growth medium. The other most obvious defect in *shr-2* and *scr-4* mutants, which is apparent at the time of germination, is the presence of only one ground tissue layer instead of two in between the pericycle and the epidermis (Di Laurenzio et al. 1996; Helariutta et al. 2000). In the experiments of Rost and Baum (1988), the diameter of the pea root also decreased as the meristematic activity and meristem height decreased, perhaps indicating a decrease in the number of ground tissue layers. Rodríguez-Rodríguez et al. (2003) noted a decrease in the number of cortex cell layers from four to one, in *S. gummosus* so that a single endodermal and cortical cell layer comprised the ground tissue at the time that root growth ceased. As *SHR* and *SCR* are found in many plant species (Bolle 2004), perhaps changes in the levels of SHR and/or SCR play a role in regulating the switch to determinate growth in desert cactus and/or pea.

One consequence of determinate root growth in the desert cactus, the low-phosphate-grown *Arabidopsis*, and the *shr* and *scr* mutants is the emergence of lateral and adventitious roots (Rodríguez-Rodríguez et al. 2003; Sanchez-Calderon et al. 2005; Lucas et al. 2011). These results suggest that determinate growth causes a release of apical dominance, akin to cutting off the root tip (Zhang and Hasenstein 1999; Aloni et al. 2006). Therefore, constitutive determinacy (as in the case of *S. gummosus*) may be a timing mechanism that allows for the establishment of the primary root, which terminates early so that lateral roots are produced close to the soil surface for better access to water. In the case of *Arabidopsis* roots grown on low phosphate, determinate root growth may divert resources away from the primary root to allow the lateral roots to explore and potentially access more favorable soils. In plants like rice, which have short-lived determinate primary roots, the production of lateral and adventitious roots may create a multiply branched root that mechanically supports the plant better in wet soils (Rebouillat et al. 2009; Coudert et al. 2010). For *shr* and *scr* mutants, the emergence of laterals may help support plant growth in the presence of a smaller primary root (Lucas et al. 2011).

Extreme examples of mutations that cause precocious determinate root growth are *root meristemless* (*rml1* and *rml2*) of *Arabidopsis* (Cheng et al. 1995). Upon germination *Arabidopsis* roots contain an average of 17 epidermal and 17 cortical cells per cell file. At 4 days after germination, there was no increase in the number of cells in the *rml1* mutants and a very limited increase in *rml2* (approximately onefold compared to 10-fold in the wild type) indicating limited functionality of the meristems in these mutants. Both mutants also showed highly vacuolated cells and differentiation of the root tip as evidenced by the apical formation of vascular tissue and root hair emergence. Only *rml1* was able to form lateral roots, and the laterals that it formed terminated after the same number of divisions as the primary root, indicating a strong genetic component to the number of root cells. When the *RML1* gene was cloned, it was shown to encode γ-glutamylcysteine synthetase (Vernoux et al. 2000), the first enzyme in glutathione biosynthesis. Glutathione is a reducing agent that acts as a redox buffer. Either genetic or pharmacological inhibition of glutathione biosynthesis causes cell cycle arrest and defects in auxin signaling that affect QC maintenance, suggesting that the QC may be important for maintaining cell divisions in the root.

X. Importance of the QC in Maintaining Root Growth

The QC, as defined by Clowes (1956a,b, 1958a), is a region of low mitotic activity that lies at the apical pole of the stele at the point of convergence between the endodermal and cortical cell layers. The QC can be distinguished from other cells in the root on the basis of their weak (or slow) incorporation of radiolabeled nucleotides or nucleic acid precursors, indicating low mitotic activity (Clowes 1956). In some roots with large QCs like those of *Zea mays*, the number of QC cells can range from 200 to nearly 1000. In contrast, the number of QC cells in *Trifolium repens* L. is usually small, 2 cells on average (Clowes 1984), and in *Arabidopsis* 4–7 cells comprise the QC (Dolan et al. 1993). The relative degree of quiescence varies between different species of roots, between different roots on the same plant, and even between cells in the same QC, with cells at the center of the QC showing the highest degree of quiescence. In general the QC cells of roots with open meristems divide more frequently (normalized to the root cap initials) than those of closed meristems. Likewise as a trend, there is higher mitotic activity in small QCs than in large. The size of the QC may change throughout the life of the root (Clowes 1984). In some cases, changes in the size of the QC correlate with changes in the patterning of the root, particularly the vasculature (Feldman and Torrey 1976).

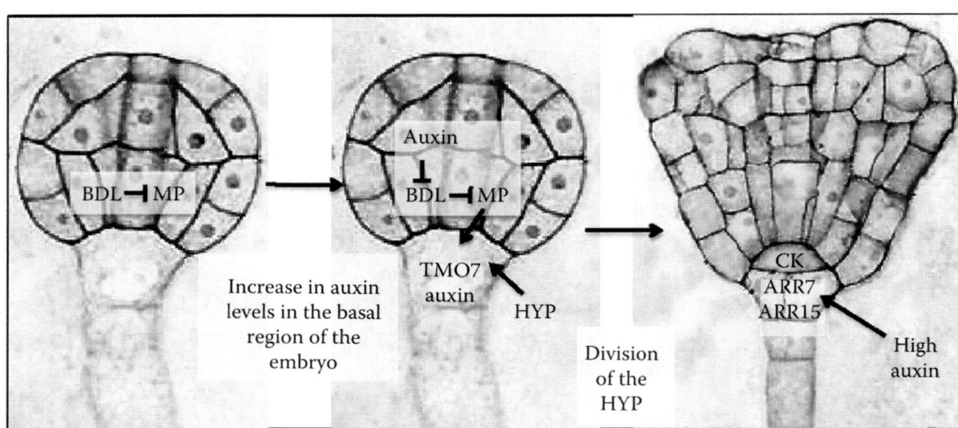

FIGURE 3.6 Specification of the hypophysis in the *Arabidopsis* embryo. An increase in auxin in the basal region of the embryo relieves repression of MP by BDL. This allows expression of TMO7, which moves into the top cell of the suspensor. This along with high auxin specifies this cell as the hypophysis (HYP). Later in growth of the embryo, when the hypophysis divides, high auxin is segregated to the lower daughter. High auxin promotes expression of ARR7 and ARR15 inhibiting cytokinin (CK) signaling. Cytokinin is required in the upper lens-shaped cell to specify the QC. (Images modified from Scheres, B. et al., *Development*, 120, 2475, 1994.)

It has been suggested that all roots have, at least at some point in their development, a QC (Jiang and Feldman 2005). In *Arabidopsis* the QC forms during the late globular stage of embryo development (Figure 3.6; Scheres et al. 1994). After fertilization, the *Arabidopsis* embryo divides asymmetrically to produce a basal cell, which forms the extra-embryonic suspensor and an apical cell, which will give rise to all cells of the embryo proper with the exception of the QC and the CIs. The QC and CIs form from the hypophysis, which is recruited from the suspensor. The recruitment of the hypophysis correlates with the establishment of an auxin maximum in the basal domain of the embryo. Factors that disrupt auxin signaling interfere with specification of the hypophysis (Christensen et al. 2000; Benjamins et al. 2001; Friml et al. 2004; Michniewicz et al. 2007). The auxin maximum at the base of the embryo results in the degradation of BODENLOS (BDL). Degradation of BDL frees the MONOPTEROS (MP) transcription factor from repression allowing the expression of the TARGET OF MONOPTEROUS 7 (TMO7) protein in the basal region of the embryo proper. Upon expression, TMO7 moves from the basal region of the embryo into the apical cell of the suspensor to specify this cell as the hypophysis. The hypophysis then divides to produce an apical lens-shaped cell and a basal cell (Schlereth et al. 2010). During this division, high auxin levels seem to segregate to the basal cell, whereas cytokinin predominates in the apical cells. Auxin in the basal cell promotes expression of ARR7 and AAR15, which inhibit cytokinin signaling (Muller and Sheen 2008). This partitioning of auxin and cytokinin signaling is required for the proper specification of the basal cell as columella and the apical cell as QC. *Arabidopsis* mutants (e.g., dominant alleles of BDL or loss-of-function alleles of MP) that fail to form a hypophysis fail to make a root.

Not all species specify the QC through recruitment of a hypophyseal cell. Instead, in some plants, the meristem and QC form from a small group of (apparently) disorganized cells at the root pole (von Guttenberg 1968; Jiang and Feldman 2005). In other species, all cells in the root meristem are mitotically active at the time of radicle emergence, and the QC forms post-embryonically (Clowes 1958a,b; Alfieri and Evert 1968). Roots that show constitutively or nonconstitutively determinate root growth either lack a QC entirely or they progressively lose their QC as the root grows (Shishkova et al. 2008). For example, in *S. gummosus* (Rodríguez-Rodríguez et al. 2003) and *A. filiculoides*, a QC is never formed (Clowes 1956, 1958b; Nitayangkura et al. 1980), and root growth is constitutively determinate. When Sanchez-Calderon et al. (2005) grew *Arabidopsis* seedlings on low-phosphate medium, the roots initially had recognizable QC cells and maintained some degree of root growth. The loss of QC markers by day 8 correlated with the inability to rescue root growth when returning the root to normal-phosphate medium. Likewise markers of the QC are either absent or misexpressed in both the *shr* and *scr* mutants (Di Laurenzio et al. 1996; Helariutta et al. 2000; Nakajima et al. 2001) and in other *Arabidopsis* single and double mutants that show constitutively determinate root growth. These results suggest that the presence of a QC is a prerequisite for indeterminate growth.

Two mechanisms have been proposed for the role of the QC in supporting indeterminate growth. The first suggests that the QC serves as a reservoir for cell replacement. During the growth of the root, initial cells are lost either due to wounding or due to the normal aging process. In order for the root to continue to grow, these cells must be replaced. The QC is a potential source of replacement cells. Experiments by Clowes (1967) in which root growth was perturbed through drought, wounding, or cold treatment found that after a state of imposed quiescence (referred to by Clowes as dormancy), growth could resume when the plants were transferred back to normal growing conditions. The basis of this recovery was traced back to cell divisions within the QC. Likewise similar results were found when roots were exposed to ionizing radiation (Clowes 1963, 1965) or in the roots of incense cedar during overwintering (Wilcox 1962). The thought is that stressors like cold treatment, drought, or ionizing radiation preferentially damage cells that are actively dividing. This means that the initial cells and their daughters would suffer the most damage. Since the QC has low mitotic activity, these cells are in a sense protected from loss. When the stressor is released, the QC can then divide to replace any injured cells and reform a functional meristem.

The second mechanism by which the QC may promote or maintain indeterminate growth is based largely upon the observation that in all *Arabidopsis* mutants, which lack a functional QC, the surrounding initial cells cease divisions and terminally differentiate. These results suggest that the QC acts as an organizing center that maintains the initial cell population in an undifferentiated state. The assumption is that the QC produces a short-range signal that affects cell behavior in a concentration-dependent manner such that the cells that are in direct contact with the QC receive the highest concentration of this signal and are therefore specified as initials. When the initial cells divide, their daughter cells are displaced from the QC and hence undergo some degree of differentiation. Cells in the larger meristem (those which undergo transit-amplifying cell divisions) receive even less of the QC-derived signal and finally cease division once they are significantly removed from the influence of the QC and begin the process of cell expansion and differentiation. Therefore, distance from the QC should affect both cell fate and competence. If the QC was lost, one would predict that the meristem is also lost.

To test this hypothesis, the four cells that comprise the QC were ablated in the *Arabidopsis* root. Following ablation, the QC was rapidly replaced by cells in the stele, in a region that correlated with the new auxin maximum (van den Berg and Willemsen 1997; Xu et al. 2006). Similar results were found when the QC cells were surgically excised in maize roots (Feldman 1976). However, if only one of the four QC cells was specifically ablated in the *Arabidopsis* root, replacement was delayed. In the time frame before replacement, van den Berg and Willemsen (1997) found that the columella initials that were in contact with the ablated QC failed to divide and instead differentiated as columella, whereas those in contact with the intact QC cells behaved normally. CEI cells in contact with an ablated QC cell continued to divide but did so periclinally in a manner similar to the CEDs, suggesting that the presence of the contacting QC cells prevents the CEI cells from adopting the CED cell fate. These roots now had essentially dedicated initials (as opposed to a common cortical endodermal initial) that gave rise separately to cortex and endodermis. As this is something that is observed in older roots, perhaps as the root ages either the QC signal wanes in intensity or the ability of the surrounding cells to respond to the signal weakens. From these ablation studies, it was concluded that a QC-derived signal maintains the initial cell population.

The concept of the QC as an "organizer" has also been tested genetically. In *scr* mutants both the primary and lateral roots show constitutively determinate root growth. These roots lack an endodermal cell layer and fail to maintain a QC. Upon the switch to determinate root growth, QC markers are lost, and both the former QC cells and the surrounding initials divide abnormally and differentiation eventually proceeds to the root tip (Di Laurenzio et al. 1996). In wild-type roots, the SCR protein is expressed in the endodermis, the QC, the CEIs, and their daughter cells (Malamy and Benfey 1997; Gallagher et al. 2004). To test which of these cell types is responsible for maintaining root growth, Sabatini et al. (2003) replaced wild-type *SCR* activity in either the QC or the CEI and their descendants. Restoration of *SCR* activity in the QC, but not in the CEI and the ground tissue, enabled reestablishment of indeterminate growth, suggesting that a wild-type QC is required to maintain root growth. In the roots with a rescued QC, the rate of root growth and the overall size of the meristem were decreased relative to a completely wild-type root; however, differentiation of the initial cells and QC was inhibited. These results suggest that the loss of the QC is the primary defect that results in precocious determinate growth in *scr* plants.

More recent results suggest that the requirement for *SCR* in root maintenance can be partially bypassed by the downregulation of retinoblastoma-related (RBR) activity (Wildwater et al. 2005). RBR both promotes cellular differentiation and inhibits E2F transcription factors, which promote mitosis. Although *RBR* is uniformly expressed in the root meristem, RNAi against *RBR* can partially rescue the stem cell maintenance phenotype of the *scr-4* mutant. These results indicate that one of the roles of SCR in the QC is to reduce RBR activity (through a yet unidentified mechanism) and therefore inhibit differentiation. In the maize root, low levels of *ZmRBR2;1* are found in the QC as compared to the root cap or the proximal stem cells, indicating that RBR activity in maize may also correlate with QC identity and that in maize RBR activity may be regulated at the level of transcription (Rymen et al. 2007; Jiang et al. 2010).

A prediction of the QC as "organizer" model is that initial cells achieve their progenitor cell status in the meristem based upon position relative to the QC and not based upon some inherent quality of the initial cell. This hypothesis was also tested using cell ablations. If CEI cells were ablated, cells from the adjacent pericycle layer could divide periclinally and invade the position formerly occupied by that ablated CEI. In this position, the pericycle daughter cells adopted the behavior of a CEI. They divided anticlinally to produce a CED; the CED in turn divided periclinally to produce functional endodermis and cortex cell layers. Likewise cortex cells were able to divide and functionally replace E/LRCI cells that were ablated, indicating that position relative to the QC maintains the initial cell status (van den Berg et al. 1995; van den Berg and Willemsen 1997). It is noteworthy that in these experiments, there was no indication of division by the QC to replace the lost initial cells, as has been suggested in other systems. This may be due to the limited damage caused by the ablation experiments. It may be that the QC only divides if a significant population of initial cells is lost. For example, the initial cells may provide a mobile signal that inhibits division of the QC, and loss of one to two cells does not significantly reduce this signal.

XI. Reformation and Maintenance of a QC

The finding that the QC is replaced when all four cells are ablated (van den Berg and Willemsen 1997; Xu et al. 2006) suggests that position rather than lineage also determines which cells become part of the QC. One of the key signals for QC formation and maintenance in the mature root is auxin, as it is during embryonic formation of the QC. An auxin maximum, which correlates with the position of the QC is maintained at the root tip by the cell-to-cell movement of auxin (via auxin efflux—PIN and influx—AUX carriers) from the shoot (Blilou et al. 2005; Grieneisen et al. 2007) and localized production of auxin in the root tip (Petersson et al. 2009). Genetic or pharmacological disruption of this auxin signal results in either loss of the QC or respecification of the QC at the position of the new auxin maximum, generally proximal to the original QC (Xu et al. 2006). This indicates that a continuous auxin signal is required to maintain the root meristem. The primacy of this auxin signal then sets up the potential for a self-organizing system in which auxin specifies the QC and the QC in turn recruits the initial cells (Grieneisen et al. 2007). Indeed, self-organization of the root meristem has been demonstrated in experiments in which the tip (QC and initials) is excised from the maize root (Feldman 1976). In these experiments, cell divisions are initiated in the stump (predominately in the pericycle cells), and a new meristem reforms. However, this meristem is only active following the establishment of a new QC. Similar results are seen in pea (Rost and Jones 1988). Interestingly the ability of auxin to respecify the meristem has its limits. If in addition to the root cap, the QC, and the initials, the transit-amplifying cells are also removed, a new meristem is not specified. Instead, lateral roots form at right angles to the root stump (Feldman 1976; Rost and Jones 1988). These results indicate that not all cells are able to respond to the auxin maximum with reformation of a QC.

Downstream of auxin in the formation and maintenance of the QC are the SHR and SCR genes (discussed previously) (Di Laurenzio et al. 1996; Helariutta et al. 2000; Sabatini et al. 2003), the *WUSCHEL RELATED HOMEOBOX 5* (*WOX5*) (Haecker et al. 2004), and *PLETHORA* (*PLT*) (Aida et al. 2004; Galinha et al. 2007) pathways (Xu et al. 2006). WOX5 is transcription factor in the *WUSCHEL* (*WUS*) gene family. The founding member of this family, *WUS*, is required for maintenance of both the shoot and floral meristems. *WUS* is expressed in a defined region in the center of the SAM referred to as the organizing center (OC). The function of the OC is similar to the QC in that signals from the OC maintain the surrounding stem cells in a mitotically active and undifferentiated state. *WOX5* is the functional equivalent of *WUS*. Correct expression of *WUS* in the root can rescue *wox5* mutants and expression of *WOX5* in the OC can rescue *wus* mutants (Sarkar et al. 2007). *WOX5* expression is initiated during embryogenesis in the upper lens-shaped cell that will form the QC immediately after the asymmetric division of the hypophysis. In the mature root, *WOX5* is expressed in the QC (Haecker et al. 2004). Loss of *WOX5* function results in an abnormal QC and differentiation of the distal stem cells as columella, suggesting a non-cell autonomous role for WOX5 in maintaining the fate of the columella initial cells in an undifferentiated state (Sarkar et al. 2007). There are mild effects of *wox5* on the structure of the proximal stem cells; however, root growth is not precociously determinate, suggesting that other gene may redundantly function in maintenance of the proximal initials. As WOX7 is the most closely related homolog of WOX5 in *Arabidopsis*, this is a likely candidate (Nardmann and Werr 2007; Nardmann et al. 2007).

There are homologs of *WOX5/WOX7* in rice, maize, brachypodium, sorghum, and poplar (Nardmann and Werr 2007; Nardmann et al. 2007). Analysis of the expression of the rice *WOX5/WOX7* homolog, *QHB* in rice, as well as ectopic expression studies suggests that the role of *QHB* in rice root development is similar to that of *WOX5* in *Arabidopsis*. In *Arabidopsis WOX5* expression is reduced or lost in *shr-1* and *scr-4* mutants, indicating that *WOX5* is downstream of both SHR and SCR (Haecker et al. 2004). This regulation is likely to be intact in rice, as *QHB* and *SCR* expression largely overlap, particularly in four cells at the tip of the root (Kamiya et al. 2003a,b; Haecker et al. 2004). However, the overlapping region of *SCR* and *QHB* may not define the entire QC region in rice. Analysis by Clowes (1956a, 1971, 1984) of QC structure in grasses suggests a large QC. Likewise examination of markers of cell divisions (CDKs) by Umeda et al. (1999a,b) in the rice roots showed broad expression in the meristem with a region (containing more than four cells) at the tip that is largely devoid of expression. This region has been interpreted as the QC. The region in which *SCR* and *QHB* overlap (Kamiya et al. 2003a) corresponds to approximately the center of the region defined by Umeda et al. (1999a,b). This may indicate a region of increased quiescence in this central region of the QC with reduced quiescence at the periphery.

The *PLETHORA* (*PLT*) genes are also downstream of auxin in maintenance of the QC but are largely independent of the *WOX5* and *SHR/SCR* pathways (Aida et al. 2004). *PLT1, PLT2, PLT3,* and *PLT4* are AP2 class transcription factors that are expressed in partially overlapping domains that converge upon the QC and the surrounding initial cells in the root meristem (Aida et al. 2004; Galinha et al. 2007). *PLT1* and *PLT2* are the most similar of the four genes. Loss of both *PLT1* and *PLT2* expression results in the loss of QC markers, differentiation of the initial cells, and formation of root hairs at the root tip, indicating that *PLT* functions in both the maintenance and the activity of the QC and initials (Aida et al. 2004). Triple mutants between *plt1, plt2,* and *plt3* are similar to the *rml1/2* mutants with no post-embryonic root growth. Quadruple mutants between *plt1, plt2, plt3,* and *plt4* (*babyboom*) resemble *mp* (nulls) or *BDL* (gain of function) mutants, entirely lacking a root. Analysis of *PLT1* and *PLT2* expression in the presence of exogenously applied auxin or in the *mp, non-phototropic hypocotyl 4* (NPH4) double mutants shows that *PLT1* and *PLT2* act downstream of auxin. In turn the *PLTs* also affect auxin activity through upregulation of the

PIN1, PIN3, and PIN4 proteins suggesting feedback between *PLT* and auxin levels. However, neither *PLT1* nor *PLT2* functions directly downstream of auxin as there is a significant delay in *PLT* expression after auxin application. One of the effects of ectopic expression of *PLT1* is the induction of ectopic meristems that show both expression of the QC-25 marker and the formation of columella cells in the absence of auxin accumulation. Collectively these results suggest that the *PLT* genes mediate QC specification downstream of auxin and that ectopic expression of *PLT* can bypass the need for an auxin maximum in formation of the meristem (Galinha et al. 2007). Examination of *PLT* genes in rice (Li and Xue 2011) and maize (Jiang and Feldman 2010) suggest similar role for the *PLT*s in maintaining the QC in these species.

The function of the PLTs is not restricted to the QC; they also function in the initial and transit-amplifying cells. Interestingly the function of the PLTs depends upon concentration. *PLT1*, *PLT2*, *PLT3*, and *PLT4* are all expressed in the RAM with the highest levels of expression in the QC and the initials. In the QC and the surrounding initials, high concentrations of *PLT2* promote stem cell fate. In the transit-amplifying cells, moderate levels of *PLT* promote mitotic activity, and even lower levels of *PLT* are required for differentiation. Based on the effect of *PLT* levels on cell behavior in *Arabidopsis*, an obvious candidate for regulation by *PLT* is RBR; however, downregulation of *RBR* is not sufficient to rescue *plt1* mutants (Galinha et al. 2007). *PLT* has been shown to upregulate expression of *HIGH PLOIDY2* (*HPY2*), a SUMO E3 ligase whose expression inhibits endoreduplication and cell differentiation. *HPY2* is expressed downstream of both *PLT1* and *PLT2*. Loss of HPY2 expression leads to a reduction in both B-type cyclins and cyclin-dependent kinases suggesting a mechanism by which *PLT*s function in the root meristem (Ishida et al. 2009). It is not known precisely what creates the gradient of *PLT* expression, but expression of *PLT1* and *PLT2* is regulated in part by the expression of both *HISTONE ACETYLTRANSFERASE* (*HAG1 a.k.a. GCN5*) and *ALTERATION/DEFICIENCY IN ACTIVATION 2B* (*ADA2b*) as loss-of-function alleles reduce the expression levels of *PLT* (*1* and *2*) and abolish the PLT gradient producing root meristem defects (Kornet and Scheres 2009). In maize the ratio of *zmPLT*, expression in the QC relative to the initials is approximately 4:1. Based upon the high levels of *zmPLT* in the QC relative to the proximal initials in both *Arabidopsis* and maize, it is tempting to speculate that the highest PLT levels promote quiescence. However, upon decapping of the maize root, the levels of *zmPLT* expression actually increase in the QC concomitant with activation of mitosis in these cells, indicating that high levels of PLT, at least in the maize root, do not promote quiescence (Jiang et al. 2010).

XII. Stem Cell in the Root Meristem

Based upon the different activities of the QC, some classify the QC cells as stem cells, whereas others associate it with the niche (reviewed by Jiang and Feldman 2005). The difference between these two groups lies in their definition of stem and niche cells (see Section XIII for more details). In a review article on stem cells, Morrison et al. (1997) came to the conclusion that "since different people define stem cells in different ways (for examples see Hall andWatt 1989; Potten and Loeffler 1997), formulating a generally acceptable definition can lead to a conclusion similar to that of U.S. Supreme Court Justice Stewart in regard to pornography: "It's hard to define, but I know it when I see it." However, there are two features that universally appear in definitions of stem cells. They are the ability to divide and to self-maintain so that with each division the stem cell is able to "pass on" stem cell identity to at least one of its daughter cells. Other characteristics that are often ascribed to stem cells are the ability to regenerate an organ or tissue after wounding, a low rate of cell division compared to its daughter cells and a relatively undifferentiated state with respect to other cells within the tissue or organ (Potten and Loeffler 1997; Morrison and Spradling 2008; Li and Clevers 2010). By this definition, both the QC cells and the initial cells in the root meristem could be classified as stem cells. However, it is clear that they are not identical populations of cells and there are probably differences in their "stemness."

In the context of the QC versus initial cells, it is useful to look at how Potten and Loeffler (1997) regard stem cells. In assigning the label stem cell, they point out that as stem cells are defined based upon their potential (i.e., how they will act in some uncertain future state), it is useful to distinguish between two different types of stem cells: actual stem cells and potential stem cells. Actual stem cells are ones that at the time of examination fulfill the key characteristic of a stem cell—they are undifferentiated, proliferative, and actively maintaining the stem cell population. In contrast, potential stem cells have the ability to behave as stem cells, but at the time examined, they are largely quiescent. There are multiple examples of these in animals, and often these cells exist within or near a population of actual stem cells. For example, within an intestinal crypt, actively cycling crypt-based columnar (CBC) cells exist at the base of the crypt among Paneth cells. CBC cells are multipotent stem cells that can reform a mini-gut in culture. Above the CBC cells are the label-retaining cells (LRCs) so named due to their retention of a pulse of bromodeoxyuridine (BrDU). As cell divisions would dilute the BrDU staining, these cells are relatively quiescent. It is thought that the LRC cells are potential stem cells that may be activated due to loss of or damage to the CBCs (reviewed by Potten and Loeffler 1997; Crosnier et al. 2006; Li and Clevers 2010).

Experiments in which the cells of the intestinal crypt are irradiated have shown that there are indeed functionally different populations of stem cells within the crypt (Ijiri and Potten 1986). With moderate to low doses of radiation, the actively cycling stem cells are killed, but at least two potential stem cell populations still exist, which become activated to repopulate the crypt. Higher doses of radiation kill the second population of cells, and still higher doses are required to deplete all mitotically competent cells. It is thought that the first cells to be lost in the crypt are those undergoing transit amplification, followed by the CBC cells, and then the LRCs (Ijiri and Potten 1986; Potten and Loeffler 1997; Barker et al. 2008). In the framework of the root, the crypt irradiation experiments (at least on their surface) are very similar to the root irradiation experiments performed by Clowes (1963, 1965). That is, after irradiation, the initial cells are lost, and the QC activates to repopulate the root meristem. Based upon this behavior, it is tempting to adopt the animal nomenclature and classify initial cells as actual stem cells and the QC cells as potential stem cells. Indeed, in nomenclature proposed by Barlow, the actively dividing initial cells that surround the QC are classified as functional initials and the cells in the QC, structural initials as indicating that these cells are variations of the same general type (reviewed by Jiang and Feldman 2005).

One of the problems in defining stem cells in plants has been the idea that all plant cells are essentially potential stem cells—under appropriate conditions able to regenerate either the entire plant or specific organs. This is in contrast to animal cells, where regenerative potential is limited to small pockets of stem cells found within defined regions of the animal (Sugimoto et al. 2011). The notion of the totipotent plant cell is largely based upon two observations: (1) the ability to regenerate entire seedlings from callus tissue in culture and (2) the capacity of wounded organs to regenerate following injury. During the formation of callus in tissue culture (as opposed to after wounding), mature plant tissues are treated with the appropriate cocktail of hormones to produce callus. The callus tissue is then coaxed (again through the application of the appropriate concentrations of phytohormones) into producing an entirely new plant. This progression is often thought of as a process of cellular dedifferentiation (from mature leaf or root) back to an embryonic or naïve state (callus) followed by redifferentiation. However, when the actual cells participating in the generation of callus were examined in *Arabidopsis*, they tended to be either xylem pole pericycle cells in the root or cells associated with the xylem poles in organs that lack pericycle. Interestingly, the cells that give rise to crown roots in rice are also associated with the xylem poles, perhaps suggesting a special role for these cells in plant growth (Atta et al. 2009; Rebouillat et al. 2009). When the expression profiles of the callus tissues were examined, they were not undifferentiated cells; instead, they resembled, independent of the source of the callus tissue (root, cotyledon, or petal), cells in the root meristem, with subepidermal expression of *WOX5, SCR, SHR,* and *PLT1.* Interestingly callus formation was blocked in plants harboring a mutation in the *ABERRANT LATERAL ROOT FORMATION* (*ALF4*) gene, which is required for the initiation of lateral roots. ALF4 is a nuclear-localized protein that maintains the pericycle cells in a "mitotically competent" state, which again suggests that the pericycle plays a key role in the formation of callus (Sugimoto et al. 2010). The pericycle therefore may represent a large population of potential stem cells that are able to generate de novo meristems not only during the process of lateral root formation but also in response to stress or damage. (For more details about the pericycle role in lateral roots development, see Chapter 6.) These results suggest that plant cells may not be all that different than animal cells in their developmental potential and that only specific cells within the plant have regenerative potential.

Stem cells are defined not only based upon their future behavior but also relative to the behavior of their daughter or neighboring cells. Potten and Loeffler (1997) suggest that within an organ, there may exist a continuum of different cellular identities. At one end of the spectrum is the naïve stem cell that has a sustained capacity for proliferation and at the other is the mature differentiated cell that has lost the ability to divide. In between these two extremes are transit cells, which have some aspects of stem cells but lack the ability to self-maintain. These cells are thought to be on a pathway that will lead to differentiation. Along this pathway, these cells divide one to several times to amplify the number of cells in the tissue. Some of these divisions may be asymmetric resulting in daughter cells that themselves may also divide asymmetrically to adopt different cell fates as is seen in the development of the ground tissue layers and the epidermis in the rice root. As transit cells are proliferative even in the absence of stem cells, they have some tissue regenerative potential. By assigning a proliferative function to the transit cells, the number of times that the actual stem cells need to divide is reduced; this may reduce the genetic load of the stem cells. A similar benefit is seen to having a population of quiescent, potential stem cells. Because these cells are not mitotically active, they are less sensitive to cellular damage and mutation. Recent papers looking at DNA damage (Curtis and Hays 2007; Yoshiyama et al. 2009), chiefly double-strand breaks (DSBs), showed that in the RAM the initial cells and their daughters are particularly sensitive to DSB and that they respond by activating a cell death pathway. During this process, the QC cells are spared (Fulcher and Sablowski 2009). Therefore, QC cells may represent a protected population of stem cells with less genetic load than the initial cells.

XIII. Stem Cell Niche Concept in the Root

The concept of a stem cell "niche" was first suggested by Schofield (1978) with respect to hematopoietic stem cells (HSCs). He proposed that within the organism, stem cells are found in association with other cells that "determine [their] behavior." He saw these associated cells as preventing the maturation of the stem cells. By preventing maturation, the niche cells played a role in maintaining the proliferative potential of the stem cell. As cells are displaced from the niche, they lose their stem cell identity and become first-generation colony-forming cells (what Potten and Loeffler (1997) would refer to as a transit cell). However, susceptible cells, which have not undergone terminal differentiation may also enter the niche and may be induced into a stem cell fate. This appears to be the case for the transit-amplifying cells in the intestinal crypt; if they are transplanted back into the base of the crypt, they can adopt the properties of intestinal stem cell (Barker et al. 2008). Likewise during both oogenesis and spermatogenesis in Drosophila, the progeny of the germ line stem cells (GSCs) retain the ability to serve as replacement cells for at least three generations after their production (Xie and Spradling 2000; Morrison and Spradling 2008). Therefore, the role of the niche is both to maintain the stem cell population and to recruit new cells upon loss of the native stem cell population. Repopulation of the niche in animals may occur via migration of transit cells back into the niche or through division of active or quiescent stem cells (Morrison and Spradling 2008). Interestingly Wilson and Trumpp (2006) suggest that the niche may at the same time promote both a quiescent (potential) and active stem cell population. At the center of the niche where inhibitory signals are highest, the quiescent stem cell population would be maintained. However, the actual stem cells would reside at the edges of the niche where the signals inhibiting mitosis are weaker. In the root, there is both an auxin and the *PLT* gradient with maxima at the QC. These two signals maintain both the QC and initial cell populations in the root apex.

Repopulation of a stem cell niche is classically illustrated by the ability of germ line stem cells (GSC) in contact with cap cells in the *Drosophila* germarium to change their pattern of cell division and replace ablated GSCs during oogenesis. It is thought that the cap cells comprise the niche and through direct signaling maintain the GSC population. However, signaling between the GSC is also important for stem cell maintenance, as the presence of a neighboring GSC affects the orientation of cell division and hence the fate of the daughter cells. This is similar to the results of van Den Berg et al. where the ablation of a root initial cell affected the orientation of division in the contacting initial so that new stem cells were generated (van den Berg et al. 1995; van den Berg and Willemsen 1997).

As the process of oogenesis in *Drosophila* is strikingly similar in its geometry to the process of root development in *Arabidopsis*, analogies have been drawn between the different cell types. The cap cells in the ovariole have been likened to the QC cells in *Arabidopsis* and the GSCs to the initials (Xie and Spradling 2000). Like the cap cells, the QC appears to have both stem cell maintenance and inductive capacity, as cells that come into contact with the QC through asymmetric division of the initials adopt a stem cell fate. This signaling function has led some to consider the QC as a niche as opposed to a stem cell population (reviewed by Jiang and Feldman 2005; Ivanov 2007). However, the presence of a signaling function does not necessitate the classification of the QC as part of the niche. Li and Clevers (2010) suggest that mutual signaling between active and quiescent stem cells may in fact reinforce differences between these two populations. In addition, there are examples of signaling between stem cells in several systems where neither cell type is considered a niche cell. Jiang and Feldman (2005) classify the QC as both stem cells (founder cells) and niche cells suggesting that the two populations are largely overlapping in the root meristem. However, it should be noted that the definition by Jiang and Feldman deviates from the stem cell niche concept as defined by Xie and Spradling (2000) who stated that "a requirement for intercellular signals does not by itself indicate the presence of a niche. A true niche should function independently of resident stem cells." While the QC may not significantly contribute to the initial cell population in *Arabidopsis*, it does divide, and in other systems, the contribution to root growth may be significant. Therefore, if the terms stem cell and niche cell are applied to the root meristem in the same way that they are in animal systems, the QC should be classified as a stem cell population.

What is the niche then in the root? In animals a distinction is made between stromal and epithelial niches (Scadden 2006; Morrison and Spradling 2008). In "stromal" niches, distinct and specific cells in the niche (e.g., the cap cells in the anterior end of the ovariole) control the fate of the resident stem cells. These cells are separate from the stem cells, and they exist in the absence of an active stem cell population. In contrast, "epithelial" niches are "devoid of specialized cell types" that control stemness. Instead, a localized domain defines the niche. Scadden (2006) suggests that some niches may be "composed of extracellular matrix and other non-cellular constituents" that control stemness. For example, ovarian follicle cells (FSCs) exist in a dynamic environment in which they maintain no continuous contacts with any specific cell type (Nystul and Spradling 2007). As an auxin maximum is the primary signal for formation and maintenance of the root meristem (Grieneisen et al. 2007), it may be useful to think of the root stem cell niche as an epithelial niche that induces the activity of the QC and surrounding stem cells. Cells with the highest levels of auxin express *WOX5*, *SCR*, and high levels of *PLT* and therefore become quiescent stem cells (QC), whereas the surrounding initials experience lower levels of auxin and become active stem cells. Homogeneous interactions between the quiescent and active stem cells then help to orchestrate the behavior of the entire stem cell population and maintain growth. It has been suggested in animals that the lack of a requirement for heterologous specialized cell types in an epithelial niche may lead to more flexibility in the positioning of the niche and the ability of the niche to move (Scadden 2006; Morrison and Spradling 2008). This may be particularly true in plants where relative movement of cells within an intact organ is severely limited.

Considering the root meristem as an epithelial niche may also help to explain recent results by Sena et al. (2009) who surgically ablated the tips of wild-type *Arabidopsis* roots at different positions in the root meristem. As was seen in experiments with maize and pea (Feldman 1976; Rost and Jones 1988), regeneration of the primary root was unsuccessful if the ablation removed the entire meristem. However, if the root cap, QC, and proximal meristem were removed, the entire root tip could regenerate. During the regeneration process, cells from all layers within the stump divided and reformed the tip. In maize, regeneration of the root resulted in the establishment of the QC prior to reformation of the meristem (Feldman 1976). This was not the case in *Arabidopsis*. In fact, root tip regeneration was possible (although slightly reduced) in genotypes that fail to maintain a QC. The authors interpret this as evidence that root regeneration is possible in the absence of a stem cells niche (Sena et al. 2009; Sena and Birnbaum 2010). However, if one considers the QC as a potential stem cell population (as discussed earlier) as opposed to a signaling component of a "stromal" niche, then the absence of a niche is not demonstrated by the lack of a QC. In the context of an epithelial niche concept, following ablation auxin levels rapidly increased at the tip. Once the auxin maximum is achieved, it induces stem cell identity in responsive cells, and the root tip reforms. In the time frame prior to the establishment of a new auxin maximum, the transit-amplifying cell is able to undergo limited divisions. In wild-type plants, a functional meristem and niche are regenerated. In the *scr* or *plt* mutants, the niche reforms, but growth is determinate due to defects in stem cell responses mediated by *PLT* and *SCR*. Considering the root meristem as an epithelial niche also eliminates the need for distinct new definitions of the niche in plants as compared to animals as was suggested by Tullio et al. (2010) who stated that "… QC cells derivatives can serve as a source of new initials. This feature of the QC is a remarkable difference in comparison to animal 'stem cell niches,' which, as far as we know, cannot replace stem cells proper." In the case of the epithelial niche concept, the QC cells are part of the stem cell population not the niche cells. In animals there are clear examples of stem cells generating new stem cells and the de novo generation of new niches (reviewed by Jones and Wagers 2008).

XIV. Importance of Cell Signaling in Root Growth

In both the shoot and the root, cell-to-cell signaling is required to maintain the normal function of the stem cells and their daughters. Perhaps the best understood example of this is the *CLAVATA* (*CLV*)–*WUS* signaling pathway. The primary signaling molecules in these pathways are the CLEs, secreted peptides that in their processed and functional form are approximately 13 amino acids long. In the CZ of the SAM, the *CLV3* peptide is expressed in the L1, L2, and upper region of the L3 layers. The *CLV3* peptide signals to cells in the L3 that express the RPK2, CLV2/CRN, and CLV1 receptors and repress the expression of *WUS* and therefore both the number of stem cells and the size of the OC in the SAM (for recent review, see Katsir et al. 2011).

In the *Arabidopsis* root, it has long been suspected that CLE peptides also play a role in maintaining the RAM, as ectopic expression of *CLV3*, *CLE19*, or *CLE40* (or application of the purified peptides encoded by these genes) leads to termination of root growth. Mutations in *clv2* ameliorate this phenotype (Fiers et al. 2004, 2005). Recently CLE40 has been identified as the endogenous CLV signal in the root (Stahl et al. 2009). *CLE40* is expressed in the differentiated cells in the columella (Figure 3.7A). The CLE40 peptide acts through the ARABIDOPSIS CRINKLY 4 (ACR4) receptor to restrict *WOX5* expression to the quiescent center cells (QC). Loss of *cle40* or *ACR4* leads to an expansion of the columella initials and an expansion of *WOX5* expression beyond the QC, but has no effect on the proximal initials. An important difference between the CLV–WUS pathways in the shoot and the root is that in the SAM, the *CLV3* and *WUS* domains partially overlap and *WUS* (which itself has recently been shown to move between cells [Yadav et al. 2011]) is required to turn on *CLV3*. In the root, *WOX5* and *CLV40* are in distinct domains. In addition in the shoot, stem cells produce *CLV3*, while in the root, differentiated cells produce *CLV40*. These results suggest that shoot and roots may have differently adapted the use of the CLV–WUS pathway. Remarkably this pathway may have been co-opted by cyst-forming parasitic nematodes as well. While CLE peptides are largely plant specific, there are non-plant species that express and release CLE peptides during the process of colonization. Expression of the CLE encoding *Hg-SYV46* gene from *Heterodera glycines* Ichinohe is able to rescue the *clv3* mutant, suggesting that the CLE peptides secreted by these plant pathogens are functional in the CLV–WUS pathway (Wang et al. 2005).

Another recently discovered class of signaling peptides in root development is the root meristem growth factors (RGFs): three functionally redundant homologous peptides that undergo posttranslational modification by tyrosylprotein sulfotransferase (TPST) (Figure 3.7B; Matsuzaki et al. 2010; Zhou et al. 2010). In situ hybridization revealed that *RGF1* is confined to the QC and columella stem cells, while RGF2 and RGF3 are mainly expressed in the innermost layer of central columella

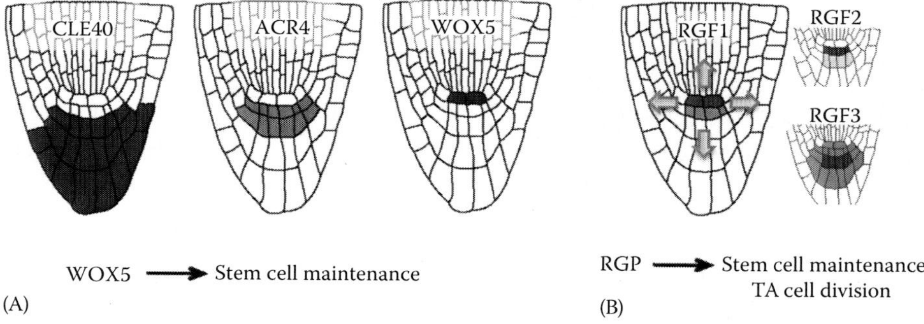

FIGURE 3.7 Peptide signals in the root meristem. (A) The CLE40 peptide is expressed in the columella and moves basally into the columella initials where it binds to the ACR4 receptor restricting WOX5 expression to the QC cells. (B) RGFs are recently identified peptides that maintain root growth. RGF1 is expressed in the QC and columella initials and moves out from these cells forming a gradient (indicated by the arrows). The expression patterns of *RGF2* and *RGF3* are also shown. Protein localization is not known for these two RGFs. Darker colors indicate higher expression.

cells. However, immunostaining showed that RGFs diffuse into the meristematic region (Matsuzaki et al. 2010). Genetic analysis revealed that the RGFs and the associated TPST1 are required for maintenance of the root stem cell niche in part through regulation of the auxin (PIN3 and PIN7) and PLT pathways. Loss of either the *RGFs* or the *TPST* results in a stunted root with a reduced meristem. This phenotype can be partially rescued by either external application of the purified sulfonated peptides or ectopic expression of *PLT2*.

The secretion of mobile peptides is only one of the signaling pathways that functions in root development. The other prevalent pathway is signaling through plasmodesmata (PD). In both *Arabidopsis* and in *Azolla*, a switch to determinate growth is correlated with a decrease in plasmodesmatal connectivity, suggesting that symplastic signaling may be a requirement for meristem maintenance (Gunning 1978; Zhu et al. 1998a; Baum et al. 2002). Plasmodesmata (PD) are membrane-lined intercellular channels that connect the cytoplasm of neighboring cells. Primary PD are formed during cytokinesis by the deposition of new cell wall and membrane material around endoplasmic reticulum derived tubules. All clonally related cells therefore are (at least potentially) connected by primary PD. Secondary PD form after cell division and are inserted into already existing cell walls. Thus, symplastic continuity is possible between both related and unrelated cells. The network and degree of connectivity between cells creates symplastic domains. The formation and maintenance of these domains are under developmental control. Likewise the structure and transport capacities of PD change over developmental time and are responsive to environmental conditions. For example, exposure of plants to aluminum induces the closure of PD (Sivaguru et al. 2000) and decreases both root elongation and mitotic activity in the root meristem.

In *Arabidopsis* Zhu et al. (1998a,b) examined PD densities (PDD) between the walls of all cells in the RAM as the root aged. They found that clonally related cells within the stele and within the ground tissue had the highest density of PD. In all tissues, the density of PD increased significantly from the time of germination up to 1 week. This increase in PDD preceded a rapid increase in the rate of root growth that was seen between week 1 and week 2 post germination. Likewise a significant decrease in PDD preceded the decline in root growth that was observed between weeks 2 and 4, with the lowest PDD detected at week 4. These results suggest that a decrease in symplastic connectivity may be one of the factors contributing to a switch to determinate growth in *Arabidopsis* roots. This is likely the case for roots of *A. filiculoides* as well. Gunning (1978) showed that the apical cell of *Azolla* divided 55 times (on average) before growth ceased. After the 35th division, he noted that the apical cells failed to maintain the same density of PD as earlier divisions. The same observations were made of the merophytes, indicating that as in *Arabidopsis*, decreased PDD precedes determinate root growth.

Symplastic signaling may also affect meristem organization. The presence of a single apical cell in the root meristem of ferns and lower vascular plants is often mirrored in the organization of the SAM. For example, in *Equisetum* both the SAM and the RAM contain prominent mitotically active apical cells that give rise to merophytes. Likewise the SAM of *A. filiculoides* contains a single tetrahedral-shaped apical cell that divides along its faces to produce all cells of the shoot. This similarity extends to angiosperms where plants have multiple initials in the RAMs and the SAMs (Imaichi and Hiratsuka 2007). The structural similarities between the SAM and the RAM may suggest a common origin for both of these stem cell populations (Bennett and Scheres 2010). However, it may also reflect an inherent limitation in these groups that restricts the size and number of initial cells in the meristem. For example, Imaichi and Hiratsuka (2007) provide a compelling explanation that is based upon the need for cells to communicate during development. In examining SAM structure in angiosperms, gymnosperms, pteridophytes, and lycopods, a strong correlation was found between the organization of the SAM and the plasmodesmatal network. Angiosperms can make both primary PD (between clonally related cells) and secondary PD (between both clonally related and nonrelated cells) and they have multi-initial SAMs. In contrast, all of the pteridophytes examined by Imaichi and Hiratsuka (2007) made only primary PD and had a single apical cell in their SAM. The lycopods were divided in their ability to make secondary PD. Interestingly, lycopods that made both primary and secondary PD had angiosperm-like SAMs with multiple initial cells, while those that were only able to make primary PD had "fern-like" meristems with a single apical cell. The authors suggest that the inability to make secondary PD and therefore to directly communicate between non-sister cells limits the meristem to a single apical cell that is highly connected to all of its sisters by primary PD.

There are a range of different molecules that are able to pass through PD, from small metabolites to mRNAs, microRNA, and proteins. In an effort to determine what fraction of root-expressed transcription factors move between cells, Lee et al. (2006) examined the domain of transcription and domain of protein localization for 61 tissue-specific transcription factors. They found evidence for cell-to-cell movement for approximately 16% of these proteins. Movement occurred between multiple different tissue types. The capacity for movement was also widespread with many different types of transcription factors including Dof, NAC, bZIP, MADS box, and MYB domain (Lee et al. 2006) proteins moving between cells. These proteins therefore represent a large number of potential signaling molecules in root development and homeostasis.

Perhaps one of the best-characterized symplastic signals in *Arabidopsis* root growth is the SHR transcription factor (Nakajima et al. 2001; Gallagher et al. 2004; Cui et al. 2007; Welch et al. 2007; Gallagher and Benfey 2009; Koizumi et al. 2011). As mentioned earlier, the roots of *shr-2* mutants show precocious determinate growth and defects in cellular patterning. The *shr-2* mutants lack an endodermal cell layer, their QC is disrupted, and metaxylem cells replace protoxylem cells (Helariutta et al. 2000; Carlsbecker et al. 2010). The *SHR* transcript is expressed exclusively in the stele; however, the protein is present not only in the stele but also in the endodermis and the QC, indicating that SHR moves from the stele into these neighboring cell layers (Figures 3.4A and 3.8). In the neighboring cell layer, SHR upregulates expression of the SCR transcription factor, which in turn maintains its own expression in the QC (Nakajima et al. 2001). In the CEI and CED cells, both SHR and SCR activate the expression of a D-type cyclin, CYCD6;1. Expression of CYCD6;1 in the CED triggers the asymmetric periclinal cell division that results in the formation of a separate cortex and endodermis. Later in root development, *CYCD6;1* expression is upregulated during middle cortex formation (Sozzani et al. 2010).

SHR homologs are found in multiple species of plants. The expression patterns of *OsSHR1* and *OsSCR* in rice along with interaction data suggest that this pathway is intact in the radial patterning of rice where the cortex and the epidermis are clonally related (Kamiya et al. 2003; Bolle 2004). In *Pinus radiata* D. Don, *PrSHR* is expressed in a pattern similar to *Arabidopsis* (Sole et al. 2008). In *Pinus sylvestris* L. and *Pisum sativum* L., the homologs of SCR (*PsySCR* and *PsSCR*, respectively) are both expressed in an arc between the stele tissue and the lateral root cap, which includes the endodermis. These results suggest that the SHR/SCR signaling pathway functions in monocots and in roots with an open meristem configuration (Sassa et al. 2001; Laajanen et al. 2007). The SHR signaling pathway is also not specific to the root, as experiments in leaf tissue also show movement of SHR and SHR-dependent regulation of the cell cycle (Gardiner et al. 2011).

Movement of SHR is required for its function. Nonmobile forms of SHR rescue neither the radial nor the vascular patterning defects seen in the *shr-2* mutants (Gallagher et al. 2004; Gallagher and Benfey 2009; Carlsbecker et al. 2010). This is because an endodermal dependent signal is required for patterning of the xylem. In the endodermis, SHR initiates expression of miRNA165/6, which in turn moves into the stele tissue where it inhibits its target mRNA, *PHABULOSA*. Dominant alleles of *PHB* with mutations in the miRNA binding site phenocopy the metaxylem defects in *shr-2*. SHR therefore is at the center of a reciprocal signaling pathway in which movement of SHR from the stele activates a microRNA which itself is a mobile signal that functions in the patterning of the vascular tissue (Carlsbecker et al. 2010; Vaten et al. 2011). miRNA 165/6 isoforms are found widely in plants (PMRD database, http://bioinformatics.cau.edu.cn/PMRD/) perhaps suggesting conserved patterning pathways. It is unclear what the adaptive value of this reciprocal signal is as opposed to SHR directly activating miRNA 165/6 in the stele. However, results from Carlsbecker et al. (2010) suggest that miRNA165/6 acts in a dosage-dependent manner. Graded movement is one way to create a gradient of active miRNA 165/6. One could speculate that such a system also provides the root with a feedback loop that speeds the formation of mature, thickened vascular tissue in the absence of an endodermal layer, which itself is both a suberized and lignified tissue that may provide some structural support to the root.

Research into the mechanisms by which SHR moves has revealed interesting sets of SHR-dependent positive and negative feedback loops that control SHR movement (Figure 3.8; Sena et al. 2004; Cui et al. 2007; Welch et al. 2007; Gallagher and Benfey 2009; Koizumi et al. 2011). In stele cells, the SHR protein is present in both the nucleus and the cytoplasm and moves into the endodermis. In the endodermis, SHR is exclusively nuclear localized and moves neither back into the stele nor into the endodermis (Gallagher et al. 2004). In the endodermis two SHR-dependent transcription factors, SCR and MAGPIE (MGP), and one SHR-independent transcription factor JACKDAW (JKD) inhibit movement of SHR either back into the stele or into the cortex, presumably through trapping SHR in the nuclei of endodermal cells. Interestingly, the stele expression of a SHR-interacting protein called SIEL (for SHR INTERACTING EMBRYONIC LETHAL) is downstream of both SHR and SCR. SIEL is an endosome-associated protein that interacts with SHR in the cytoplasm of stele cells and promotes movement of SHR into the endodermis (Koizumi et al. 2011). These results indicate that in the stele, SHR promotes its own movement, whereas in the endodermis, it inhibits it. This then may provide a mechanism for concentrating SHR in the nuclei of endodermal cells. Mutations that reduce movement of SHR or SHR activity (e.g., *siel-4* homozygotes or *shr-2* heterozygotes) result in ectopic divisions in the ground tissue (Koizumi et al. 2011). Interestingly conditions that increase SHR movement (e.g., homozygosity for *jkd-4* or SCR RNAi) (Cui et al. 2007; Welch et al. 2007) also increase the number of cell layers in the root. Regulation of SHR movement therefore may be a mechanism by which roots control the numbers of ground tissue layers that are formed.

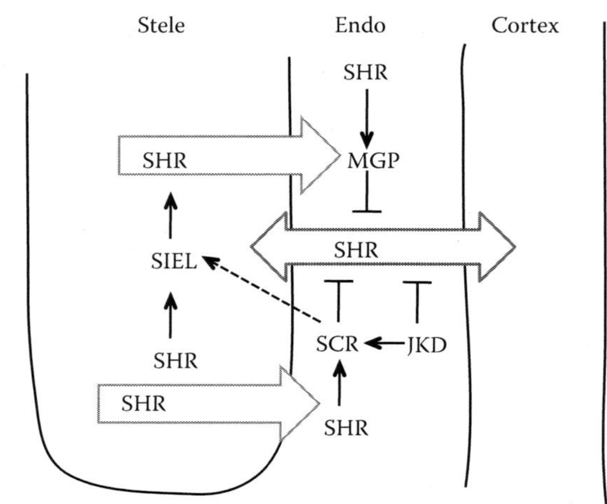

FIGURE 3.8 SHR signaling. The SHR protein moves out of the stele into the endodermis (endo) where it turns on both SCR and MGP. SCR, MGP, and JKD inhibit movement of SHR out of the endodermis and promote nuclear localization of SHR potentially increasing SHR concentrations in the endodermis. In the stele, SHR and SCR (though an unknown mechanism, indicated by the dashed arrow) promote the expression of the SIEL protein, which promotes SHR movement.

One of the latest mobile transcription factors shown to play a role in root patterning is a novel bHLH (subfamily 14) transcription factor named *UPBEAT1* (*UPB1*) (Tsukagoshi et al. 2010; Wells et al. 2010). UPB1 regulates root growth and meristem size through the regulation of the levels of reactive oxygen species (ROS). In the root, there are two major ROS, hydrogen peroxides (H_2O_2) and radical oxygens (O_2^-). O_2^- levels are highest in the meristem, whereas H_2O_2 is highest in the root elongation zone (Figure 3.5C). The overlap of O_2^- and H_2O_2 constitutes the transition zone (Dunand and Penel 2007; Dunand et al. 2007). *UPB1* mRNA is present in the vascular tissue as well as in the lateral root cap (LRC) of *Arabidopsis* roots, while the UPB1 protein is detected primarily in the nuclei of cells in the transition and elongation zones, suggesting that UPB moves between cells. Loss-of-function mutations in UPB1 affect ROS gradients in the root resulting in an expansion of the root meristem and a concomitant decrease in the elongation zone (Tsukagoshi et al. 2010). Examination of *UPB1* targets revealed that UPB1 repressed the expression of several peroxidases. Tsukagoshi et al. (2010) propose a model in which high levels of O_2^- promote cellular proliferation, whereas lower levels of O_2^- and high levels of H_2O_2 promote differentiation. As O_2^- and H_2O_2 form inverse but partially overlapping gradients, cell behavior can be controlled by subtle changes in the relative levels of the two ROS. Once the ratio of H_2O_2 to O_2^- reaches its maximal level, cell proliferation stops entirely, and cells become differentiated (Tsukagoshi et al. 2010).

UPB1 is not alone in regulating the levels of ROS in the root. One of the ways that auxin is thought to signal in the root is also through the generation of ROS. In the elongation zone, auxin induces cell-wall loosening through the formation of hydroxyl radicals that can readily react to form H_2O_2. Auxin can also interact with O_2 to form O_2^-. Therefore, depending upon the circumstances auxin could promote either cell division or cell elongation (Mori et al. 2009). It is interesting that mobile signals, like UPB1 and auxin, would function through the regulation of ROS as the generation of ROS is known to induce callose formation and block PD (Benitez-Alfonso and Jackson 2009; Benitez-Alfonso et al. 2011). Auxins move between cells via plasma membrane localized influx and efflux carriers. It is not known if they can also move via PD. Nonetheless, it is possible that one of the effects of auxin and UPB1 on ROS may be changes in PD permeability. This is probably not the only role for ROS in development however, as roles for ROS in the regulation of growth in animals have also been shown. For example, hematopoietic stem cells (HSC) are kept in a relatively quiescent state in the bone marrow by limiting the concentration of O_2-. Quiescence can be reversed by altering the levels of ROS (reviewed by Eliasson and Jonsson 2010) indicating a role for ROS in regulating animal as well as plant stem cells.

XV. Conclusions

In plants, much of the patterning of the organism occurs postembryonically through the action of meristems. The organization of the root meristem differs between different species of plants. However, in all meristems, there are populations of mitotically active cells termed initials. The initial cells in roots with a closed meristem divide in stereotypical ways so that the organization of the root is maintained relatively constant and cellular identities are stable and predictable. Initial cells have key properties that are associated with stem cells in animals including self-maintenance and a prolonged mitotic capacity. Like stem cells in animals, initial cells in the root create clones of cells that themselves may divide asymmetrically to create new cell layers or divide symmetrically (often in a series of transit-amplifying cell divisions) to increase the number of cells in the organ. Stemness is maintained through the action of both endogenous and exogenous cues, as is the size of the root meristem, elongation, and differentiation zone. The ability of the root to respond to extrinsic signals allows the root to conform its growth to suit the environment and in some cases entirely terminate growth of the primary meristem to allow the activity of lateral meristems. Genetic analysis, largely in *Arabidopsis*, has begun to uncover the genes and regulatory mechanisms involved in the cellular patterning of the meristems and root growth. Several of these pathways function not only in the root but also in the shoot meristem, suggesting similar points of control and regulation.

References

Aloni R, Aloni E, Langhans M et al. 2006. Role of cytokinin and auxin in shaping root architecture: Regulating vascular differentiation, lateral root initiation, root apical dominance and root gravitropism. *Ann Bot* 97: 883–893.

Armstrong JE, Heimsch C. 1976. Ontogenetic reorganization of root meristem in compositae. *Amer J Bot* 63: 212–219.

Atta R, Laurens L, Boucheron-Dubuisson E et al. 2009. Pluripotency of *Arabidopsis* xylem pericycle underlies shoot regeneration from root and hypocotyl explants grown in vitro. *Plant J* 57: 626–644.

Baluska F, Barlow PW, Baskin TI et al. 2005. What is apical and what is basal in plant root development? *Trends Plant Sci* 10: 409–411.

Baluska F, Kubica S, Hauskrecht M. 1990. Postmitotic isodiametric cell-growth in the maize root apex. *Planta* 181: 269–274.

Baluska F, Mancuso S, Volkmann D et al. 2010. Root apex transition zone: A signalling-response nexus in the root. *Trends Plant Sci* 15: 402–408.

Barker N, van de Wetering M, Clevers H. 2008. The intestinal stem cell. *Gene Dev* 22: 1856–1864.

Barlow. 1976. Towards an understanding of the behaviour of root meristems. *J Thoer Biol* 57: 433–451.

Baskin TI, Peret B, Baluska F et al. 2010. Shootward and rootward: Peak terminology for plant polarity. *Trends Plant Sci* 15: 593–594.

Baum S, Dubrovsky J, Rost T. 2002. Apical organization and maturation of the cortex and vascular cylinder in *Arabidopsis thaliana* (Brassicaceae) roots. *Am J Bot* 89: 908.

Baum S, Rost T. 1996. Root apical organization in *Arabidopsis thaliana* 1. Root cap and protoderm. *Protoplasma* 192: 178–188.

Benfey P, Scheres B. 2000. Root development. *Curr Biol* 10: 813–815.

Benitez-Alfonso Y, Jackson D. 2009. Redox homeostasis regulates plasmodesmal communication in *Arabidopsis* meristems. *Plant Signal Behav* 4: 655–659.

Benitez-Alfonso Y, Jackson D, Maule A. 2011. Redox regulation of intercellular transport. *Protoplasma* 248: 131–140.

Benjamins R, Quint A, Weijers D et al. 2001. The PINOID protein kinase regulates organ development in *Arabidopsis* by enhancing polar auxin transport. *Development* 128: 4057–4067.

Bennett T, Scheres B. 2010. Root development-two meristems for the price of one? *Curr Top Dev Biol* 91: 67–102.

Bennett T, van den Toorn A, Sanchez-Perez GF et al. 2010. SOMBRERO, BEARSKIN1, and BEARSKIN2 regulate root cap maturation in *Arabidopsis*. *Plant Cell* 22: 640–654.

van den Berg C, Willemsen V. 1997. Short-range control of cell differentiation in the *Arabidopsis* root meristem. *Nature* 390: 287–289.

van den Berg C, Willemsen V, Hage W et al. 1995. Cell fate in the *Arabidopsis* root meristem determined by directional signalling. *Nature* 378: 62–65.

Blilou I, Xu J, Wildwater M et al. 2005. The PIN auxin efflux facilitator network controls growth and patterning in *Arabidopsis* roots. *Nature* 433: 39–44.

Bolle C. 2004. The role of GRAS proteins in plant signal transduction and development. *Planta* 218: 683–692.

Briggs GC, Mouchel CF, Hardtke CS. 2006. Characterization of the plant-specific BREVIS RADIX gene family reveals limited genetic redundancy despite high sequence conservation. *Plant Physiol* 140: 1306–1316.

Brigham LA, Nicoll SM, Hawes MC. 1993. Isolation of cell-specific cDNAs from a pea root border cell library. *Plant Physiol* 102: 151–151.

Brigham LA, Woo HH, Nicoll SM et al. 1995. Differential expression of proteins and messenger-RNAs from border cells and root-tips of pea. *Plant Physiol* 109: 457–463.

Campilho A, Garcia B, Toorn HV et al. 2006. Time-lapse analysis of stem-cell divisions in the *Arabidopsis thaliana* root meristem. *Plant J* 48: 619–627.

Carlsbecker A, Lee J, Roberts C et al. 2010. Cell signalling by microRNA165/6 directs gene dose-dependent root cell fate. *Nature* 465: 316–321.

Chapman K, Groot E, Nichol S et al. 2003. Primary root growth and the pattern of root apical meristem organization are coupled. *J Plant Gr Reg* 21: 287–295.

Cheng JC, Seeley KA, Sung ZR. 1995. Rml1 and Rml2, *Arabidopsis* genes required for cell-proliferation at the root-tip. *Plant Physiol* 107: 365–376.

Christensen SK, Dagenais N, Chory J et al. 2000. Regulation of auxin response by the protein kinase PINOID. *Cell* 100: 469–478.

Clowes FAL. 1956a. Localization of nucleic acid synthesis in root meristems. *J Exp Bot* 7: 307–312.

Clowes FAL. 1956b. Nucleic acids in root apical meristems of *Zea*. *New Phytol* 55: 29–34.

Clowes FAL. 1958a. Development of quiescent centers in root meristems. *New Phytol* 57: 85–88.

Clowes FAL. 1958b. Protein synthesis in root meristems. *J Exp Bot* 9: 229–238.

Clowes FAL. 1961. *Apical Meristems*. Oxford, U.K.: Blackwell Science.

Clowes FAL. 1963. X-irradiation of root meristems. *Ann Bot* 27: 344–352.

Clowes FAL. 1965. Meristems and the effect of radiation on cells. *Endeavour* 24: 8–12.

Clowes FAL. 1971. The proportion of cells that divide in root meristems of *Zea mays* L. *Ann Bot* 35: 249–261.

Clowes FAL. 1981. The difference between open and closed meristems. *Ann Bot* 48: 761–767.

Clowes FAL. 1984. Size and activity of quiescent centers of roots. *New Phytol* 96: 13–21.

Clowes FAL. 1994. Origin of the epidermis in root-meristems. *New Phytol* 127: 335–347.

Clowes FAL. 2000. Pattern in root meristem development in angiosperms. *New Phytol* 146: 83–94.

Clowes FAL, Stewart HE. 1967. Recovery from dormancy in roots. *New Phytol* 66: 115–118.

Coudert Y, Périn C, Courtois B et al. 2010. Genetic control of root development in rice, the model cereal. *Trends Plant Sci* 15: 219–226.

Crosnier C, Stamataki D, Lewis J. 2006. Organizing cell renewal in the intestine: Stem cells, signals and combinatorial control. *Nat Rev Genet* 7: 349–359.

Cui H, Levesque M, Vernoux T et al. 2007. An evolutionarily conserved mechanism delimiting SHR movement defines a single layer of endodermis in plants. *Science* 316: 421–425.

Curtis MJ, Hays JB. 2007. Tolerance of dividing cells to replication stress in UVB-irradiated *Arabidopsis* roots: Requirements for DNA translesion polymerases eta and zeta. *DNA Repair (Amst)* 6: 1341–1358.

Dello Ioio R, Linhares FS, Scacchi E et al. 2007. Cytokinins determine *Arabidopsis* root-meristem size by controlling cell differentiation. *Curr Biol* 17: 678–682.

Di Laurenzio L, Wysocka-Diller J, Malamy JE et al. 1996. The SCARECROW gene regulates an asymmetric cell division that is essential for generating the radial organization of the *Arabidopsis* root. *Cell* 86: 423–433.

Doerner P. 2010. Plant development: Confused root cells have mixed identity in SCHIZORIZA. *Curr Biol* 20: R245–248.

Dolan L, Janmaat K, Willemsen V et al. 1993. Cellular organisation of the *Arabidopsis thaliana* root. *Development* 119: 71–84.

Dunand C, Crevecoeur M, Penel C. 2007. Distribution of superoxide and hydrogen peroxide in *Arabidopsis* root and their influence on root development: Possible interaction with peroxidases. *New Phytol* 174: 332–341.

Dunand C, Penel C. 2007. Localization of superoxide in the root apex of *Arabidopsis*. *Plant Signal Behav* 2: 131–132.

Durbak A, Yao H, McSteen P. 2012. Hormone signaling in plant development. *Curr Opin Plant Biol* 15: 92–96.

Eliasson P, Jonsson JI. 2010. The hematopoietic stem cell niche: Low in oxygen but a nice place to be. *J Cell Physiol* 222: 17–22.

Esau K. 1965. *Plant Anatomy*. New York: John Wiley & Sons.

Feldman L. 1976. The de novo origin of the quiescent center regenerating root apices of *Zea mays*. *Planta* 128: 207–212.

Feldman LJ, Torrey JG. 1976. The quiescent center and primary vascular tissue pattern formation in cultured roots of *Zea*. *Can J Bot* 53: 2796–2803.

Fiers M, Hause G, Boutilier K et al. 2004. Mis-expression of the CLV3/ESR-like gene CLE19 in *Arabidopsis* leads to a consumption of root meristem. *Gene* 327: 37–49.

Fiers M, Golemiec E, Xu J et al. 2005. The 14-amino acid CLV3, CLE19, and CLE40 peptides trigger consumption of the root meristem in *Arabidopsis* through a CLAVATA2-dependent pathway. *Plant Cell* 17: 2542–2553.

Fleming AJ, Caderas D, Wehrli E et al. 1999. Analysis of expansin-induced morphogenesis on the apical meristem of tomato. *Planta* 208: 166–174.

Fleming AJ, McQueenMason S, Mandel T et al. 1997. Induction of leaf primordia by the cell wall protein expansion. *Science* 276: 1415–1418.

Foster AS. 1943. Zonal structure and growth of the shoot apex in *Microcycas calocoma*. *Am J Bot* 30: 56–73.

Friml J, Yang X, Michniewicz M et al. 2004. A PINOID-dependent binary switch in apical-basal PIN polar targeting directs auxin efflux. *Science* 306: 862–865.

Fulcher N, Sablowski R. 2009. Hypersensitivity to DNA damage in plant stem cell niches. *Proc Natl Acad Sci USA* 106: 20984–20988.

Galinha C, Bilsborough G, Tsiantis M. 2009. Hormonal input in plant meristems: A balancing act. *Stem Cell Dev Biol* 20: 1149–1156.

Galinha C, Hofhuis H, Luijten M et al. 2007. PLETHORA proteins as dose-dependent master regulators of *Arabidopsis* root development. *Nature* 449: 1053–1057.

Gallagher KL, Benfey PN. 2009. Both the conserved GRAS domain and nuclear localization are required for SHORT-ROOT movement. *Plant J* 57: 785–797.

Gallagher KL, Paquette AJ, Nakajima K et al. 2004. Mechanisms regulating SHORT-ROOT intercellular movement. *Curr Biol* 14: 1847–1851.

Gardiner J, Donner TJ, Scarpella E. 2011. Simultaneous activation of SHR and ATHB8 expression defines switch to preprocambial cell state in *Arabidopsis* leaf development. *Dev Dyn* 240: 261–270.

Gifford EM. 1983. Concept of apical cells in bryophytes and pteridophytes. *Annu Rev Plant Physiol Plant Mol Biol* 34: 419–440.

Gifford EM, Kurth E. 1982. Quantitative studies of the root apical meristem of *Equisetum scirpoides*. *Am J Bot* 69: 464–473.

Gifford EM, Polito VS. 1981a. Growth of *Azolla filiculoides*. *BioScience* 31: 526–527.

Gifford EM, Polito VS. 1981b. Mitotic-activity at the shoot apex of *Azolla filiculoides*. *Am J Bot* 68: 1050–1055.

Grieneisen VA, Xu J, Marée AFM et al. 2007. Auxin transport is sufficient to generate a maximum and gradient guiding root growth. *Nature* 449: 1008–1013.

Groot EP, Doyle JA, Nichol SA et al. 2003. Phylogenetic distribution and evolution of root apical meristem organization in dicotyledonous angiosperms. *Int J Plant Sci* 165: 97–105.

Gunning B. 1978. Age-related and origin-related control of the numbers of plasmodesmata in cell walls of developing *Azolla* roots. *Planta* 143: 181–190.

Gunning B, Hughes J, Hardham A. 1978. Formative and proliferative cell divisions, cell differentiation, and developmental changes in the meristem of *Azolla* roots. *Planta* 143: 121–144.

von Guttenberg H. 1960. *Rundzuge der Histogenese hoherer Pflanzen. I. Die Angiospermen*. Berlin, Germany: Gebruder Borntraeger.

von Guttenberg H. 1968. *Der Primare Bau der Angiospermenwurzel*. Berlin, Germany: Gebruder Borntraeger.

Hacham Y, Holland N, Butterfield C et al. 2011. Brassinosteroid perception in the epidermis controls root meristem size. *Development* 138: 839–848.

Haecker A, Gross-Hardt R, Geiges B et al. 2004. Expression dynamics of WOX genes mark cell fate decisions during early embryonic patterning in *Arabidopsis thaliana*. *Development* 131: 657–668.

Hall PA, Watt FM. 1989. Stem cells: The generation and maintenance of cellular diversity. *Development* 106: 619–633.

Hamamoto L, Hawes M, Rost T. 2006. The production and release of living root cap border cells is a function of root apical meristem type in dicotyledonous angiosperm plants. *Ann Bot* 97: 917–923.

Hanstein J. 1870. *Die Entwicklung des Keimes der Monocotylen und Dikotylen*. Bonn, Germany: Marcus A.

Hawes MC, Brigham LA, Wen F et al. 1998. Function of root border cells in plant health: Pioneers in the rhizosphere. *Annu Rev Phytopath* 36: 311–327.

Heimsch C, Seago Jr J. 2008. Organization of the root apical meristem in angiosperms. *Am J Bot* 95: 1–21.

Helariutta Y, Fukaki H, Wysocka-Diller J et al. 2000. The SHORT-ROOT gene controls radial patterning of the *Arabidopsis* root through radial signaling. *Cell* 101: 555–567.

ten Hove C, Willemsen V, de Vries W et al. 2010. SCHIZORIZA encodes a nuclear factor regulating asymmetry of stem cell divisions in the *Arabidopsis* root. *Curr Biol* 20: 452–457.

Ijiri K, Potten CS. 1986. Radiation-hypersensitive cells in small intestinal crypts; their relationships to clonogenic cells. *Br J Cancer Suppl* 7: 20–22.

Imaichi R, Hiratsuka R. 2007. Evolution of shoot apical meristem structures in vascular plants with respect to plasmodesmatal network. *Am J Bot* 94: 1911–1921.

Ishida T, Fujiwara S, Miura K et al. 2009. SUMO E3 ligase HIGH PLOIDY2 regulates endocycle onset and meristem maintenance in *Arabidopsis*. *Plant Cell* 21: 2284–2297.

Ivanov V. 2007. Stem cells in the root and the problem of stem cells in plants. *Rus J Dev Biol* 38: 338–349.

Janczewski E. 1874. Recherches sur l'accroissement terminal des racines dans les phanerogames. *Ann Sci Nat 5eme Ser Bot* 20: 162–201.

Jiang K, Feldman LJ. 2005. Regulation of root apical meristem development. *Annu rev Cell Dev Biol* 21: 485–509.

Jiang K, Feldman L. 2010. Positioning of the auxin maximum affects the character of cells occupying the root stem cell niche. *Plant Signal Behav* 5: 202–204.

Jiang K, Zhu T, Diao Z et al. 2010. The maize root stem cell niche: A partnership between two sister cell populations. *Planta* 231: 411–424.

Jones DL, Wagers AJ. 2008. No place like home: Anatomy and function of the stem cell niche. *Nat Rev Mol Cell Biol* 9: 11–21.

Kamiya N, Itoh J, Morikami A et al. 2003b. The SCARECROW gene's role in asymmetric cell divisions in rice plants. *Plant J* 36: 45–54.

Kamiya N, Nagasaki H, Morikami A et al. 2003a. Isolation and characterization of a rice WUSCHEL-type homeobox gene that is specifically expressed in the central cells of a quiescent center in the root apical meristem. *Plant J* 35: 429–441.

Katsir L, Davies KA, Bergmann DC et al. 2011. Peptide signaling in plant development. *Curr Biol* 21: R356–364.

Kerstetter RA, Hake S. 1997. Shoot meristem formation in vegetative development. *Plant Cell* 9: 1001–1010.

Koizumi K, Wu S, MacRae-Crerar A et al. 2011. An essential protein that interacts with endosomes and promotes movement of the SHORT-ROOT transcription factor. *Curr Biol* 21: 1559–1564.

Kornet N, Scheres B. 2009. Members of the GCN5 histone acetyltransferase complex regulate PLETHORA-mediated root stem cell niche maintenance and transit amplifying cell proliferation in *Arabidopsis*. *Plant Cell* 21: 1070–1079.

Kurth E. 1981. Mitotic activity in theroot apex of the water fern *Marsilea vestita*. *Am J Bot* 68: 881–896.

Laajanen K, Vuorinen I, Salo V et al. 2007. Cloning of *Pinus sylvestris* SCARECROW gene and its expression pattern in the pine root system, mycorrhiza and NPA-treated short roots. *New Phytol* 175: 230–243.

Lee J, Colinas J, Wang J et al. 2006. Transcriptional and posttranscriptional regulation of transcription factor expression in *Arabidopsis* roots. *Proc Nat Acad Sci USA* 103: 6055–6060.

Li L, Clevers H. 2010. Coexistence of quiescent and active adult stem cells in mammals. *Science* 327: 542–545.

Li P, Xue H. 2011. Structural characterization and expression pattern analysis of the rice PLT gene family. *Acta Biochim Biophys Sin* 43: 688–697.

Lucas M, Swarup R, Paponov IA et al. 2011. Short-root regulates primary, lateral, and adventitious root development in *Arabidopsis*. *Plant Physiol* 155: 384–398.

Malamy JE, Benfey PN. 1997. Analysis of SCARECROW expression using a rapid system for assessing transgene expression in *Arabidopsis* roots. *Plant J* 12: 957–963.

Matsuzaki Y, Ogawa-Ohnishi M, Mori A. 2010. Secreted peptide signals required for maintenance of root stem cell niche in *Arabidopsis*. *Science* 329: 1065–1067.

Medford JI. 1992. Vegetative apical meristems. *Plant Cell* 4: 1029–1039.

Michniewicz M, Zago MK, Abas L et al. 2007. Antagonistic regulation of PIN phosphorylation by PP2A and PINOID directs auxin flux. *Cell* 130: 1044–1056.

Miyashima S, Hashimoto T, Nakajima K. 2009. ARGONAUTE1 acts in *Arabidopsis* root radial pattern formation independently of the SHR/SCR pathway. *Plant Cell* Physiol 50: 626–634.

Mori IC, Murata Y, Uraji M. 2009. Integration of ROS and hormone signaling. In: *Reactive Oxygen Species in Plant Signaling*, eds. LA del Rio, A Puppo, pp. 25–43. New York: Springer-Verlag.

Morrison SJ, Shah NM, Anderson DJ. 1997. Regulatory mechanisms in stem cell biology. *Cell* 88: 287–298.

Morrison SJ, Spradling AC. 2008. Stem cells and niches: Mechanisms that promote stem cell maintenance throughout life. *Cell* 132: 598–611.

Mouchel CF, Briggs GC, Hardtke CS. 2004. Natural genetic variation in *Arabidopsis* identifies BREVIS RADIX, a novel regulator of cell proliferation and elongation in the root. *Gene Dev* 18: 700–714.

Mouchel CF, Osmont KS, Hardtke CS. 2006. BRX mediates feedback between brassinosteroid levels and auxin signalling in root growth. *Nature* 443: 458–461.

Muller B, Sheen J. 2008. Cytokinin and auxin interaction in root stem-cell specification during early embryogenesis. *Nature* 453: 1094–1097.

Mylona P, Linstead P, Martienssen R et al. 2002. SCHIZORIZA controls an asymmetric cell division and restricts epidermal identity in the *Arabidopsis* root. *Development* 129: 4327–4334.

von Nägeli C. 1858. *Beiträge zur Wissenschaftlichen Botanik*. Leipzig, Germany: von Wilhelm Engelmann.

Nakajima K, Sena G, Nawy T et al. 2001. Intercellular movement of the putative transcription factor SHR in root patterning. *Nature* 413: 307–311.

Nardmann J, Werr W. 2007. The evolution of plant regulatory networks: What *Arabidopsis* cannot say for itself. *Curr Opin Plant Biol* 10: 653–659.

Nardmann J, Zimmermann R, Durantini D et al. 2007. WOX gene phylogeny in Poaceae: A comparative approach addressing leaf and embryo development. *Mol Biol Evol* 24: 2474–2484.

Nitayangkura S, Gifford Jr EM, Rost TL. 1980. Mitotic activity in the root apical meristem of *Azolla filiculoides* Lam., with special reference to the apical cell. *Am J Bot* 1484–1492.

Nystul T, Spradling A. 2007. An epithelial niche in the *Drosophila* ovary undergoes long-range stem cell replacement. *Cell Stem Cell* 1: 277–285.

Pernas M, Ryan E, Dolan L. 2010. SCHIZORIZA controls tissue system complexity in plants. *Curr Biol* 20: 818–823.

Peters WS, Tomos AD. 1996. The history of tissue tension. *Ann Bot* 77: 657–665.

Petersson SV, Johansson AI, Kowalczyk M et al. 2009. An auxin gradient and maximum in the *Arabidopsis* root apex shown by high-resolution cell-specific analysis of IAA distribution and synthesis. *Plant Cell* 21: 1659–1668.

Pien S, Wyrzykowska J, McQueen-Mason S et al. 2001. Local expression of expansin induces the entire process of leaf development and modifies leaf shape. *Proc Natl Acad Sci USA* 98: 11812–11817.

Potten CS, Loeffler M. 1997. *Stem Cells and Cellular Pedigrees-Conceptual Introduction*. London, U.K.: Academic Press.

Rebouillat J, Dievart A, Verdeil J et al. 2009. Molecular genetics of rice root development. *Rice* 2: 15–34.

Rodríguez-Rodríguez J, Shishkova S, Napsucialy-Mendivil S et al. 2003. Apical meristem organization and lack of establishment of the quiescent center in Cactaceae roots with determinate growth. *Planta* 217: 849–857.

Rost TL. 2010. The organization of roots of dicotyledonous plants and the positions of control points. *Ann Bot* 107: 1213–1222.

Rost TL, Baum S. 1988. On the correlation of primary root length, meristem size and protoxylem tracheary element position in pea-seedlings. *Am J Bot* 75: 414–424.

Rost T, Jones T. 1988. Pea root regeneration after tip excisions at different levels: Polarity of new growth. *Ann Bot* 61: 513.

Rymen B, Fiorani F, Kartal F et al. 2007. Cold nights impair leaf growth and cell cycle progression in maize through transcriptional changes of cell cycle genes. *Plant Physiol* 143: 1429–1438.

Sabatini S, Heidstra R, Wildwater M. 2003. SCARECROW is involved in positioning the stem cell niche in the *Arabidopsis* root meristem. *Gen Dev* 17:354–358.

Sanchez-Calderon L, Lopez-Bucio J, Chacon-Lopez A et al. 2005. Phosphate starvation induces a determinate developmental program in the roots of *Arabidopsis thaliana. Plant Cell Physiol* 46: 174–184.

Sarkar AK, Luijten M, Miyashima S et al. 2007. Conserved factors regulate signalling in *Arabidopsis thaliana* shoot and root stem cell organizers. *Nature* 446: 811–814.

Sassa N, Matsushita Y, Nakamura T et al. 2001. The molecular characterization and in situ expression pattern of pea SCARECROW gene. *Plant Cell Physiol* 42: 385–394.

Scacchi E, Osmont KS, Beuchat J et al. 2009. Dynamic, auxin-responsive plasma membrane-to-nucleus movement of *Arabidopsis* BRX. *Development* 136: 2059–2067.

Scadden DT. 2006. The stem-cell niche as an entity of action. *Nature* 441: 1075–1079.

Scheres B, McKhann HI, Van Den Berg C. 1996. Roots redefined: Anatomical and genetic analysis of root development. *Plant Physiol* 111: 959–964.

Scheres B, Wolkenfelt H, Willemsen V et al. 1994. Embryonic origin of the *Arabidopsis* primary root and meristem initials. *Development* 120: 2475–2487.

Schlereth A, Moller B, Liu W et al. 2010. MONOPTEROS controls embryonic root initiation by regulating a mobile transcription factor. *Nature* 464: 913–916.

Schmidt A. 1924. Histologisehe Studien ari phanerogamen Vegetation punkt. *Bot Arch* 8: 345–404.

Schofield R. 1978. The relationship between the spleen colony-forming cell and the haemopoietic stem cell. *Blood Cells* 4: 7–25.

Schuepp O. 1917. Untersuchungen uber Wachstum und Formwechsel von Vegetationspunkten. *Jahrb Wiss Bot* 57: 17–79.

Seago JL, Heimsch C. 1969. Apical organization in roots of the Convolvulaceae. *Am J Bot* 56: 131–138.

Sena G, Birnbaum K. 2010. Built to rebuild: In search of organizing principles in plant regeneration. *Curr Opin Genet Dev* 20: 460–465.

Sena G, Jung JW, Benfey PN. 2004. A broad competence to respond to SHORT ROOT revealed by tissue-specific ectopic expression. *Development* 131: 2817–2826.

Sena G, Wang X, Liu HY et al. 2009. Organ regeneration does not require a functional stem cell niche in plants. *Nature* 457: 1150–1153.

Shishkova S, Garcia-Mendoza E, Castillo-Diaz V et al. 2007. Regeneration of roots from callus reveals stability of the developmental program for determinate root growth in Sonoran Desert Cactaceae. *Plant Cell Rep* 26(5): 547–557.

Shishkova S, Rost T, Dubrovsky J. 2008. Determinate root growth and meristem maintenance in angiosperms. *Ann Bot* 101: 319.

Sinnot EW. 1960. *Plant Morphogenesis*. New York: McGraw-Hill Book Company.

Sivaguru M, Fujiwara T, Samaj J et al. 2000. Aluminum-induced 1 → 3-beta-D-glucan inhibits cell-to-cell trafficking of molecules through plasmodesmata. A new mechanism of aluminum toxicity in plants. *Plant Physiol* 124: 991–1006.

Sole A, Sanchez C, Vielba JM et al. 2008. Characterization and expression of a *Pinus radiata* putative ortholog to the *Arabidopsis* SHORT-ROOT gene. *Tree Physiol* 28: 1629–1639.

Sozzani R, Cui H, Moreno-Risueno M et al. 2010. Spatiotemporal regulation of cell-cycle genes by SHORTROOT links patterning and growth. *Nature* 466: 128–132.

Stahl Y, Wink RH, Ingram GC et al. 2009. A signaling module controlling the stem cell niche in *Arabidopsis* root meristems. *Curr Biol* 19: 909–914.

Sugimoto K, Gordon SP, Meyerowitz EM. 2011. Regeneration in plants and animals: Dedifferentiation, transdifferentiation, or just differentiation? *Trends Cell Biol* 21: 212–218.

Sugimoto K, Jiao Y, Meyerowitz EM. 2010. *Arabidopsis* regeneration from multiple tissues occurs via a root development pathway. *Dev Cell* 18: 463–471.

Szymkowiak EJ, Sussex IM. 1996. What chimeras can tell us about plant development. *Annu Rev Plant Physiol Plant Mol Biol* 47: 351–376.

Tooke F, Battey N. 2003. Models of shoot apical meristem function. *New Phytol* 159: 37–52.

Torrey JG, Feldman LJ. 1976. The organization and function of the root apex. *Am Sci* 65: 334–344.

Traas J, Vernoux T. 2002. The shoot apical meristem: The dynamics of a stable structure. *Phil Trans Royal Soc London. Series B Biol Sci* 357: 737–747.

Tsukagoshi H, Busch W, Benfey PN. 2010. Transcriptional regulation of ROS controls transition from proliferation to differentiation in the root. *Cell* 143: 606–616.

Tullio MCD, Jiang K, Feldman LJ. 2010. Redox regulation of root apical meristem organization: Connecting root development to its environment. *Plant Physiol Biochem* 48: 328–336.

Ubeda-Tomas S, Swarup R, Coates J et al. 2008. Root growth in *Arabidopsis* requires gibberellin/DELLA signalling in the endodermis. *Nat Cell Biol* 10: 625–628.

Umeda M, Iwamoto N, Umeda-Hara C et al. 1999b. Molecular characterization of mitotic cyclins in rice plants. *Mol Gen Genet* 262: 230–238.

Umeda M, Umeda-Hara C, Yamaguchi M et al. 1999a. Differential expression of genes for cyclin-dependent protein kinases in rice plants. *Plant Physiol* 119: 31–40.

Vaten A, Dettmer J, Wu S et al. 2011. Callose biosynthesis regulates symplastic trafficking during root development. *Dev Cell* 21: 1144–1155.

Vernoux T, Wilson RC, Seeley KA et al. 2000. The ROOT MERISTEMLESS1/CADMIUM SENSITIVE2 gene defines a glutathione-dependent pathway involved in initiation and maintenance of cell division during postembryonic root development. *Plant Cell* 12: 97–110.

Wang X, Mitchum MG, Gao B et al. 2005. A parasitism gene from a plant-parasitic nematode with function similar to CLAVATA3/ESR (CLE) of *Arabidopsis thaliana. Mol Plant Pathol* 6: 187–191.

Welch D, Hassan H, Blilou I et al. 2007. *Arabidopsis* JACKDAW and MAGPIE zinc finger proteins delimit asymmetric cell division and stabilize tissue boundaries by restricting SHORT-ROOT action. *Genes Dev* 21: 2196–2204.

Wells DM, Wilson MH, Bennett MJ. 2010. Feeling UPBEAT about growth: Linking ROS gradients and cell proliferation. *Dev Cell* 19: 644–646.

Wilcox H. 1962. Growth studies of the root of incense cedar, *Libocedrus decurrens*. I. The origin and development of primary tissues. *Am J Bot* 49: 221–236.

Wildwater M, Campilho A, Perez-Perez JM et al. 2005. The RETINOBLASTOMA-RELATED gene regulates stem cell maintenance in *Arabidopsis* roots. *Cell* 123: 1337–1349.

Willemsen V, Bauch M, Bennett T et al. 2008. The NAC domain transcription factors FEZ and SOMBRERO control the orientation of cell division plane in *Arabidopsis* root stem cells. *Dev Cell* 15: 913–922.

Wilson A, Trumpp A. 2006. Bone-marrow haematopoietic-stem-cell niches. *Nat Rev Immunol* 6: 93–106.

Xie T, Spradling AC. 2000. A niche maintaining germ line stem cells in the *Drosophila* ovary. *Science* 290: 328–330.

Xu J, Hofhuis H, Heidstra R et al. 2006. A molecular framework for plant regeneration. *Science* 311: 385–388.

Yadav RK, Perales M, Gruel J et al. 2011. WUSCHEL protein movement mediates stem cell homeostasis in the *Arabidopsis* shoot apex. *Genes Dev* 25: 2025–2030.

Yoshiyama K, Conklin PA, Huefner ND et al. 2009. Suppressor of gamma response 1 (SOG1) encodes a putative transcription factor governing multiple responses to DNA damage. *Proc Natl Acad Sci USA* 106: 12843–12848.

Zhang N, Hasenstein KH. 1999. Initiation and elongation of lateral roots in *Lactuca sativa. Int J Plant Sci* 160: 511–519.

Zhou W, Wei L, Xu J et al. 2010. *Arabidopsis* tyrosylprotein sulfotransferase acts in the Auxin/PLETHORA pathway in regulating postembryonic maintenance of the root stem cell niche. *Plant Cell* 22: 3692–3709.

Zhu T, Lucas W, Rost T. 1998a. Directional cell-to-cell communication in the *Arabidopsis* root apical meristem I. An ultrastructural and functional analysis. *Protoplasma* 203: 35–47.

Zhu T, O'Quinn R, Lucas W et al. 1998b. Directional cell-to-cell communication in the *Arabidopsis* root apical meristem II. Dynamics of plasmodesmatal formation. *Protoplasma* 204: 84–93.

4

Cellular Patterning in the Root Epidermis

Yana M. Wieckowski
University of Michigan

John Schiefelbein
University of Michigan

I. Patterning the Root Epidermis: *Arabidopsis* as a Model

Patterning of distinct cell types is a fundamental process in the formation of multicellular organisms. Accordingly, a large number of researchers have focused on uncovering the intricacies of cell patterning using a variety of model organisms and experimental approaches. In plants, the patterning of the cell types in the root epidermis has been used for more than a century to study this topic because these cells are easy to observe and to manipulate. In particular, the *Arabidopsis* root epidermis has emerged over the last 20 years as one of the leading models for studying cell pattern formation (Grierson and Schiefelbein 2002; Ueda et al. 2005; Guimil and Dunand 2006; Schellmann et al. 2007; Schiefelbein et al. 2009).

The *Arabidopsis* root epidermis contains two types of cells, root-hair cells and nonhair cells, which arise in a stereotyped pattern influenced by cell position. Because only two cell types are specified, root epidermal pattern formation is reduced to an either-or decision, making it an ideal system for studying cell-fate specification during development. Furthermore, root hairs are visible shortly after germination, and complete loss of hairs does not affect viability or fertility, making it a rapid and robust model system. An extensive body of research has characterized the development of *Arabidopsis* root epidermal cells from origin to maturity (Dolan et al. 1993, 1994; Scheres et al. 1994), providing a solid foundation for the study of cell-fate specification. New root epidermal cells are generated continuously in cell files. This means that a complete developmental timeline may be observed along the axis of growth at any point in time. An extensive array of genetic, molecular, and genomic resources, including transcriptome data from specific cell types and mutants, are available to aid the identification and analysis of genes and proteins. The study of root epidermal cell patterning in *Arabidopsis* has enriched our understanding of molecular-genetic regulation of cell-fate specification, and ongoing research continues to glean new information.

II. *Arabidopsis* Root Postembryonic Growth and Development

The *Arabidopsis* primary root is elegantly organized in concentric rings of cells surrounding a central vascular region. A transverse root section reveals four layers (from outside inward): the epidermis, cortex, endodermis, and pericycle surrounding the vascular tissue (phloem, xylem, and procambium) (Figure 4.1; Dolan et al. 1993). Stereotyped divisions of a small population of stem cells at the tip of the root combined with lack of cell movement give rise to clonally related cells organized in files along the axis of root growth (Dolan et al., 1993).

Epidermal cells arise from a group of 16 epidermal/lateral root cap (LRC) stem cells (Dolan et al. 1993,1994) maintained by four adjacent cells, called the quiescent center (QC), which serve as the stem cell niche (Figure 4.2; van den Berg et al. 1997). Each epidermal/LRC stem cell divides periclinally, generating an outer cell (the LRC daughter cell) as well as an inner cell. The inner cell divides transversely generating the epidermal daughter cell as well as regenerating the stem cell (Figure 4.2; Dolan et al. 1993, 1994). After epidermal cells are generated, they undergo rapid cell division in the meristem division zone (Figure 4.3).

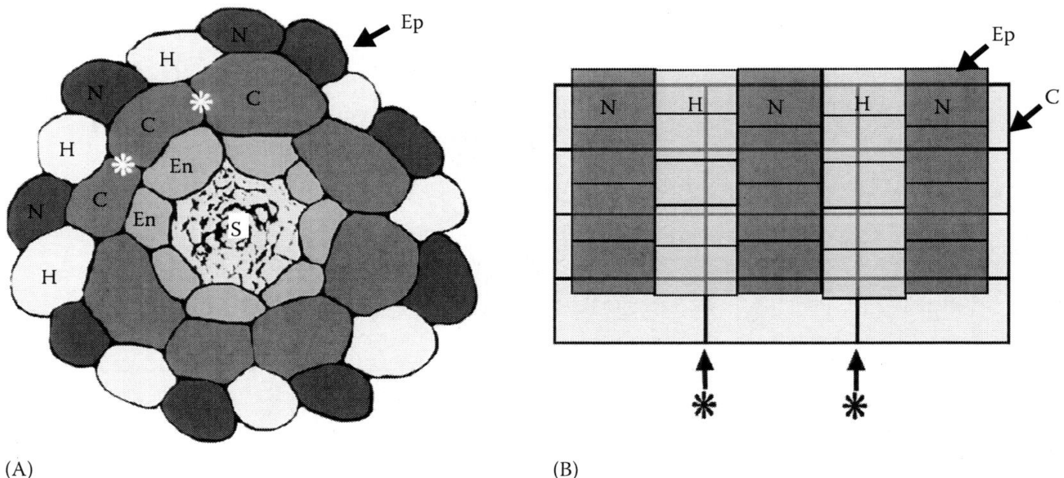

FIGURE 4.1 Position-dependent cell-fate specification. (A) Diagram of a transverse section of an *Arabidopsis* root meristem. The primary root is organized in concentric rings of cells including (from outward in) the epidermis [Ep], cortex [C], endodermis [En], and stele [S]. Cells that will develop a hair [H] are located over an ACCW (indicated as *), while cells that will develop into nonhair cells [N] are located over a single cortical cell. (B) Diagram of a surface view of a section of the root epidermis and underlying cortical cell. Note that the ACCW is microscopically visible through the epidermal layer allowing for determination of cell position relative to cortical cell boundary.

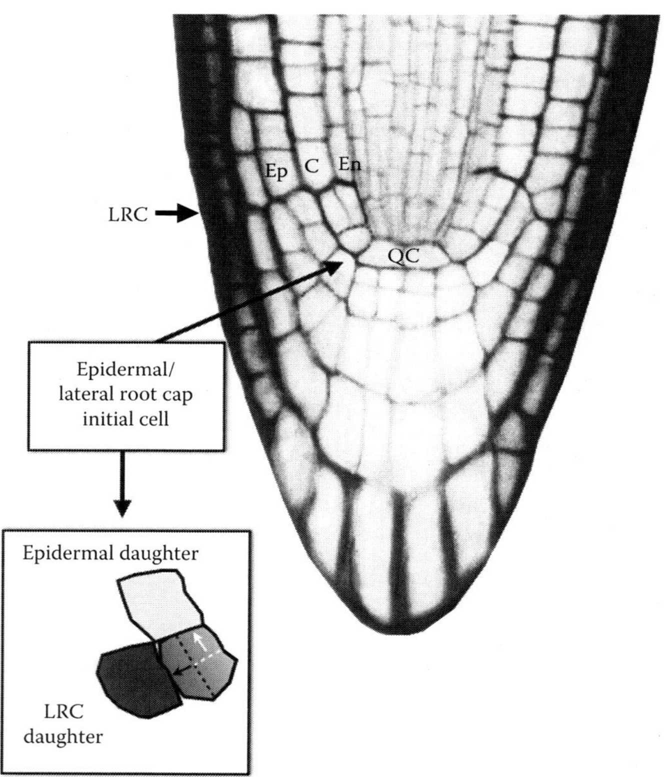

FIGURE 4.2 Cell types and organization of the *Arabidopsis* root meristem. Longitudinal section of the tip of an *Arabidopsis* root. Notice the structure organization of cell layers including the epidermis [Ep], cortex [C], endodermis [En], and LRC. Epidermal cells are derived from the epidermal/LRC initial cells (inset). The QC maintains the stem cells required to generate each cell layer.

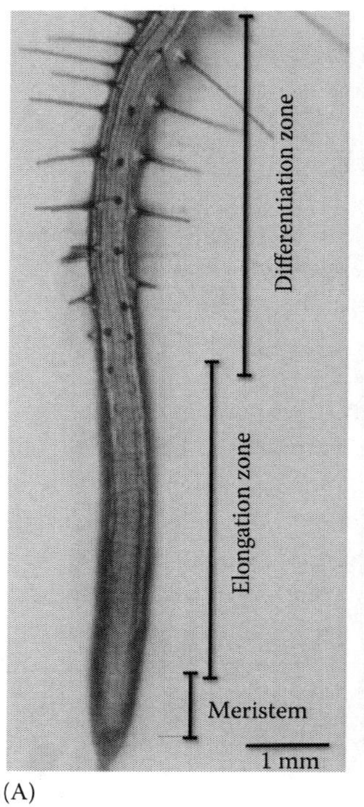

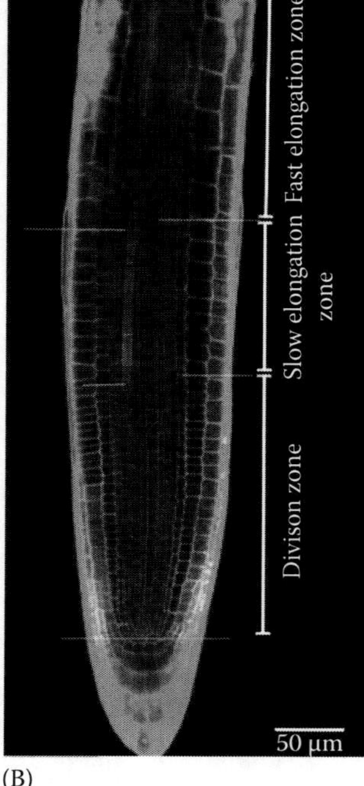

FIGURE 4.3 Root growth zones. Developmental zones of the *Arabidopsis* root and root meristem. (A) Low-magnification image of a root tip showing the progressive development of root hairs. (B) Confocal microscopy cross section of the root meristem depicting specific regions of growth within the root meristem.

Each meristematic epidermal cell divides approximately five to six times in a transverse orientation, which serves to generate additional cells in each file. Typically, a seedling root will have 16–24 epidermal cell files—one file from each of the 16 original stem cells and additional files generated by occasional longitudinal divisions (Dolan et al. 1994; Berger et al. 1998b).

Upon exiting the meristematic region, root cells cease division and rapidly elongate in the direction of root growth until their final cell dimensions are reached (Figure 4.3). When elongation ceases, root hairs emerge from the epidermis (Dolan et al. 1994). Mature hair cells possess long, tubular outgrowths approximately 22 microns in diameter and 1 mm or more in length (Figure 4.3; Grierson and Schiefelbein 2002). Root hairs are not multicellular structures but rather are tip-growing protrusions that emerge at the end of the cell nearest the root tip (Grierson and Schiefelbein 2002). Root hairs vastly increase the surface area of the root and are thought to aid plants in nutrient acquisition, water uptake, anchorage, and microbe interaction (Hofer 1991).

III. Root Epidermal Patterning: To Be or Not To Be a Hair Cell

Most angiosperms possess both hair cells and nonhair cells in their root epidermis. Among these species, three patterns of epidermal cell types have been identified: alternating pattern, random pattern, and position-dependent striped pattern (reviewed in Dolan and Costa 2001). The alternating pattern, which is widespread among monocots but present in only a few dicot species, results from asymmetric epidermal cell divisions that give rise to a large nonhair cell and a small hair cell (Clowes 2000). The random pattern, widespread among dicots and also found in some monocots, is characterized by a variable number of randomly distributed hair cells and nonhair cells—the final distribution is influenced primarily by environmental conditions (Cormack 1937, 1947; Clowes 2000; Pemberton et al. 2001). Lastly, the position-dependent striped pattern, in which hair cell files alternate with nonhair cell files, is found in a few families of dicots, including Brassicaceae (Cormack 1935; Clowes 2000; Dolan and Costa 2001; Pemberton et al. 2001).

A. Hair and Nonhair Cells: How Are They Different?

Epidermal cells in *Arabidopsis,* a model plant from the family Brassicaceae, are organized in position-dependent stripes relative to the anticlinal cortical cell wall (ACCW), the boundary between adjacent cortical cell files. Cells located over the ACCW (H-position cells) form root hairs, and those located over a single cortical cell file (N-position cells) do not (Figure 4.1; Dolan et al. 1993; Galway et al. 1994). This striped pattern has been shown by both clonal analysis (Berger et al. 1998b) and laser-ablation studies (Berger et al. 1998a) to be regulated by positional information as apposed to lineage.

Morphological differences between the two types of epidermal cells are not limited to the simple presence or absence of a root hair and are in fact discernible long before any evidence of root-hair emergence can be observed. In *Arabidopsis*, H-position cells have greater cytoplasmic density (Dolan et al. 1994; Galway et al. 1994), variant cell surface features (Dolan et al. 1994; Freshour et al. 1996), delayed vacuolation (Galway et al. 1994), and distinct chromatin organization (Xu et al. 2005; Costa and Shaw 2006), all characteristics attributed to greater metabolic activity involved in production of the root hair (Cutter 1978). Additionally, H files contain more cells with reduced cell length compared to cells in N files, a consequence of a greater rate of cell division (Dolan et al. 1994; Massucci et al. 1996; Berger et al. 1998b). Taken together, it is evident that epidermal cells are programmed early during postembryonic development to follow their distinct developmental paths. Nevertheless, until they exit from the meristem and begin elongation, epidermal cells are capable of switching fates if their position relative to underlying cortical cells changes (Berger et al. 1998a).

B. *Arabidopsis* Genes Controlling Root Epidermal Cell-Fate Specification

Molecular genetics research has unveiled an intricate regulatory network that guides *Arabidopsis* root meristem epidermal cell-fate specification. With the influence of positional information, this molecular machinery utilizes lateral inhibition and feedback mechanisms to regulate cell-type-specific expression of root-hair morphogenesis genes. Current knowledge of root epidermal cell-fate specification in *Arabidopsis* indicates that *GLABRA2* (*GL2*) is the most downstream player and it is often used as a readout of the upstream regulatory system. GL2 is a homeodomain transcription factor (Rerie et al. 1994) that is required for nonhair cell differentiation (Massucci et al. 1996). Loss of *GL2* expression results in specification of root-hair cells in all positions, producing a hairy root phenotype (Figure 4.4; Di Cristina et al. 1996; Massucci et al. 1996). Interestingly, loss of *GL2* does not affect all aspects of hair cell differentiation—cells develop early characteristics (vacuolation, cell size, cytoplasmic density) in accordance with their position (Massucci et al. 1996). The *GL2* gene is expressed in a position-dependent manner in developing nonhair cells already from the first or second cell above the epidermal/LRC initial cell (Figure 4.5; Massucci et al. 1996). Preferential *GL2* expression in N-position cells appears to be critical for nonhair specification, as *GL2* expression in the cells in the H position causes them to develop as nonhair cells. The GL2 protein negatively regulates transcription of root-hair morphogenesis genes and positively regulates nonhair cell-specific genes to promote the nonhair cell fate (Massucci et al. 1996; Lee and Schiefelbein 1999, 2002). Putative GL2 targets include *ROOT HAIR DEFECTIVE 6* (*RHD6*), which encodes a basic helix-loop-helix (bHLH) transcription factor involved in root-hair initiation (Menand et al. 2007), and *PHOSPHOLIPASE Dζ1* (*PLDζ1*), which plays a role in hair outgrowth and elongation (Ohashi et al. 2003). Binding of GL2 to the *PLDζ1* promoter (in gel shift assays) and inhibition of *PLDζ1* expression (*in planta*) (Ohashi et al. 2003) support GL2's role in downregulation of hair gene expression.

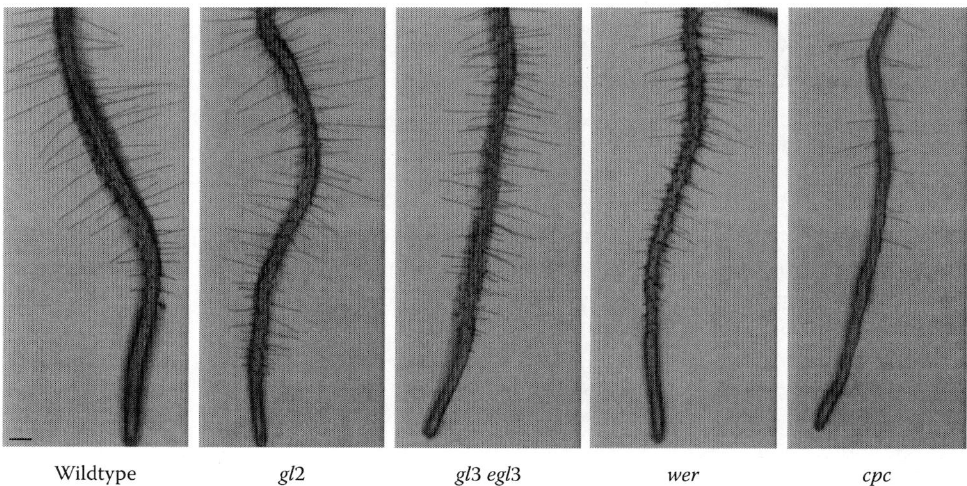

FIGURE 4.4 Root-hair phenotype of patterning gene mutants. Root-hair phenotypes of 4-day-old seedlings. Compared to wild type, the *gl2*, *gl3*, *egl3*, and *wer* mutants have increased root-hair cell frequency, while the *cpc* mutant has reduced hair cell frequency.

FIGURE 4.5 Position-dependent promoter activity of patterning genes *GL2*, *CPC*, *EGL3*, and *WER* during root development. The *GL2*, *CPC*, and *WER* promoter-reporter fusion constructs are preferentially expressed in the N-position cells in the meristematic region of the *Arabidopsis* root. *EGL3* promoter activity is localized mainly in developing H-position cells. Asterisks (*) indicate the position of H-cell files. Scale bars = 50 μm.

Four genes, *WEREWOLF* (*WER*), *GLABRA3* (*GL3*), *ENHANCER OF GLABRA3* (*EGL3*), and *TRANSPARENT TESTA GLABRA 1* (*TTG1*), are positive regulators of the nonhair cell fate and are essential for all aspects of nonhair cell development as loss-of-function mutations result in every epidermal cell developing complete characteristics of hair cells (Figure 4.4; Galway et al. 1994; Massucci et al. 1996; Lee and Schiefelbein 1999; Bernhardt et al. 2003). The four encoded proteins preferentially accumulate in N-position cells and are thought to act together as a core transcriptional complex to generate a spatially distinct expression pattern of *GL2* in cells in the N position.

The WER protein belongs to the R2R3 family of MYB-domain transcription factors and contains two MYB repeats and an acidic transcriptional activation domain (Lee and Schiefelbein 1999).

WER is expressed highly in developing root-hair and nonhair cells, later becoming restricted mainly to nonhair cells (Figure 4.5; Lee and Schiefelbein 1999). WER is nuclear localized and has been shown to influence *GL2* expression by binding to MYB-binding sites in the *GL2* promoter (Koshino-Kimura et al. 2005). Accordingly, the *WER* loss-of-function mutation effectively abolishes *GL2* promoter activity and transcript accumulation (Lee and Schiefelbein 1999). *WER* also plays a role in positively reinforcing the core transcriptional complex in developing nonhair cells through direct transcriptional activation of a highly related R2R3 MYB gene *MYB23* (Kang et al. 2009). The MYB23 protein is functionally equivalent to WER and is thought to contribute to increased production of the WER/MYB23-GL3/EGL3-TTG transcriptional complex in developing nonhair cells (Kang et al. 2009). This positive feedback loop is likely to ensure robust formation of the epidermal cell-type pattern.

The core transcriptional complex genes *GL3* and *EGL3*, which encode highly related bHLH proteins, are also necessary for *GL2* expression and nonhair cell-fate specification. Interestingly, the TTG-GL3/EGL3-WER core transcriptional complex suppresses expression of *GL3* and *EGL3* in cells in the N position, restricting promoter activity (Figure 4.5) and RNA accumulation to developing H-position cells (Bernhardt et al. 2005). However GL3 protein preferentially accumulates in developing nonhair cells, indicating that GL3 (and possibly EGL3 as well) moves from the H-position cells to the N-position cells, potentially as an additional feedback/lateral inhibition mechanism between adjacent epidermal cells (Bernhardt et al. 2005).

It is interesting to note that while GL3 and EGL3 are functionally similar in many ways, they also exhibit significant differences. For instance, the *gl3* single mutant has a moderate increase in hair cell frequency, whereas the *egl3* single mutant has no significant change in hair cell frequency, relative to the wild type (Bernhardt et al. 2003). Furthermore, the effect of overexpression of *GL3* is not as strong as *EGL3* (Bernhardt et al. 2003), which is an unexpected result considering the significance of their single mutant patterning defects. In addition, E*GL3* plays a role in regulation of seed coat mucilage and anthocyanin production, whereas *GL3* is not required for these biological functions (Zhang et al. 2003).

The third core transcriptional complex member is TTG1, a small protein containing WD40 repeats (Walker et al. 1999), that is, essential for specification of the nonhair fate. Overexpression of *GL3* and *EGL3* can overcome the effect of the *ttg1* mutation, but not the *wer* mutation (Bernhardt et al. 2003), suggesting that *GL3* and *EGL3* act in parallel with *TTG1* and potentially downstream of *WER* in the nonhair specification pathway. *TTG1* has also been shown to be important for trichome formation, anthocyanin production, and seed coat mucilage (Koornneef 1981, 1990). Interestingly, *TTG1*'s role in these varied biological processes all involves both an MYB and bHLH protein, suggesting a common mechanism for TTG1 action (Zhang et al. 2003).

While the importance of the TTG-GL3/EGL3-WER core transcriptional complex for nonhair cell-fate specification is most apparent, it also influences root-hair cell fate by promoting transcription of a three single-repeat R3 MYB-domain proteins CAPRICE (CPC), TRIPTYCHON (TRY), and ENHANCER OF TRY AND CPC (ETC1), which are necessary for root-hair cell-fate specification (Wada et al. 1997; Lee and Schiefelbein 2002; Schellmann et al. 2002; Wada et al. 2002; Bernhardt et al. 2003; Kirik et al. 2004; Simon et al. 2007). WER, TTG, and GL3/EGL3 are necessary for *CPC*, *TRY*, and *ETC1* gene expression (Lee and Schiefelbein 2002; Wada et al. 2002; Bernhardt et al. 2003; Simon et al. 2007), which is likely mediated through binding of WER to MYB-binding sites within the promoters (as has been shown for *CPC*) (Koshino-Kimura et al. 2005; Ryu et al. 2005). The *cpc* single mutant has a moderate reduction in root-hair cell frequency (Figure 4.4; Wada et al. 1997), and the *try* and *etc1* single mutants have slight to no effect on root-hair frequency (Schellmann et al. 2002; Kirik et al. 2004). However, the *cpc try* double (Schellmann et al. 2002) and *cpc try etc1* triple mutants completely lack root hairs, with the *etc1* enhancing the *cpc try* root-hair defective phenotype at the hypocotyl-root junction (Kirik et al. 2004). Furthermore, functional *CPC*, *TRY*, and *ETC1* genes are necessary to restrict *GL2* expression to cells in the N position. *GL2* expression is expanded moderately in *cpc* (Lee and Schiefelbein 2002; Wada et al. 2002), slightly in *try* and *etc1*, and to almost every epidermal cell in the *cpc try* double and *cpc try etc1* triple mutants (Simon et al. 2007). Conversion of all epidermal cells to nonhair cells is *WER*-dependent, as the *wer cpc try* triple mutant resembles the *wer* single mutant phenotypically, possessing hairs on every cell and lacking detectable *GL2::GUS* expression (Simon et al. 2007). This suggests that these CPC, TRY, and ETC1 act downstream of the TTG-GL3/EGL3-WER core transcriptional complex to influence hair cell fate by downregulating expression of *GL2* in H-position cells.

Current research supports the role of the R3 MYB-domain proteins as part of a lateral inhibition pathway between epidermal cell types. While *CPC*, *TRY*, and *ETC1* are necessary for specifying the hair cell fate, mRNA expression is detected only in developing cells in the N position (Schellmann et al. 2002; Wada et al. 2002; Kirik et al. 2004; Simon et al. 2007). CPC protein is found in nuclei of both N-position and H-position cells, indicating that CPC (and possibly TRY and ETC1 as well) moves from one cell type to the other likely through the plasmodesmata (Wada et al. 2002; Kurata et al. 2005). Altogether, it appears that the TTG-GL3/EGL3-WER core transcriptional complex promotes transcription of *CPC*, *TRY*, and *ETC1* in N-position cells and that these R3 MYB proteins then move to H-position cells acting as lateral inhibitors to prevent specification of the nonhair fate in cells in the H position.

While MYB proteins are known to bind both DNA and proteins in plants, loss of conserved amino acids required for MYB-protein binding to DNA indicates that CPC, TRY, and ETC1 likely function through protein-protein interactions (Wada et al. 2002). Furthermore, the *CPC*, *TRY*, and *ETC1* gene products are highly related to the R2R3 MYB-domain protein WER, except for the absence of the acidic transcriptional activation domain (Wada et al. 1997; Schellmann et al. 2002; Kirik et al. 2004). By constructing CPC-WER chimeric proteins, Tominaga et al. (2007)

showed that the WER R3 domain could functionally replace the CPC R3 domain as well as confer *GL2* promoter binding activity to the CPC protein. Together, molecular genetics and biochemical research suggests that through evolution the R3-single-repeat MYB-domain proteins CPC, TRY, and ETC1 lost functionality as activation-competent transcription factors.

C. Positional Information: How Epidermal Cells Know Where They Are

SCRAMBLED (SCM), a leucine-rich repeat receptor-like kinase (LRR-RLK), is necessary for the position dependence of root epidermal cell-fate specification. Loss-of-function mutations in *SCM* make a similar frequency of hair cells as wild type; however, there is a remarkable disconnect between cell fate and position (Kwak et al. 2005). Interestingly, the epidermal cell molecular network establishes cell identity in the *scm* mutant as it does in wild type; however, transcription factor expression (and subsequently cell fate) is no longer directly connected to the cell position relative to underlying cortical cell junctions (Kwak et al. 2005; Kwak and Schiefelbein 2007). Still, the pattern of epidermal cells is not completely random, even in the presumed *scm* null mutant. A tendency for the cells to adopt fate based on their position hints at an auxiliary or parallel mechanism to the SCM-signaling pathway influencing positional patterning.

SCM gene expression, mRNA accumulation, and protein localization occur throughout the developing root tissue, with the exception of the LRC (Kwak et al. 2005; Kwak and Schiefelbein 2008). In epidermal cells, the SCM protein is localized on the plasma membrane of both cell types but is enriched in H-position cells in the late division and early elongation zones (Kwak and Schiefelbein 2008). Using tissue-specific promoters to drive RNA interference of SCM in wild-type plants and to drive SCM gene expression in *scm* mutant plants, Kwak and Schiefelbein (2008) found that SCM expression and accumulation in epidermal cells is both sufficient and necessary for epidermal patterning. Furthermore, to fully complement the *scm* cell-fate patterning defect, preferential *SCM* expression in H-position cells is necessary, as driving SCM expression using nonhair cell promoters caused aberrant patterning (Kwak and Schiefelbein 2008).

Detailed genetic and expression studies indicate that the SCM-signaling pathway influences epidermal cell fate through preferential inhibition of *WER* transcription in the H cell position. *WER* expression is moderately increased in the *scm* mutant background and in the *cpc scm* mutant background, in which the damping effect of the CPC feedback loop is removed, there is an even greater increase in *WER* expression (Kwak and Schiefelbein 2007). Additionally, over-expressing *SCM* results in a significant reduction in *WER* mRNA accumulation (Kwak and Schiefelbein 2007). These and other findings imply that SCM influences the downstream transcription factor network by repressing *WER* transcription in H-position cells (Kwak and Schiefelbein 2007).

As *SCM* encodes a receptor-like kinase localized to the plasma membrane of epidermal cells, it is likely to receive a positional signal and translate it to underlying epidermal cells (Kwak et al. 2005). Surprisingly, biochemical and genetic evidence indicate that while the carboxyl-terminal kinase domain is necessary for function, it most likely lacks catalytical activity (Chevalier et al. 2005). Thus, SCM may act as an atypical receptor-like kinase, functioning through protein-protein interactions with an additional active kinase (Chevalier et al. 2005).

Although the identity of the putative SCM ligand is unknown, it appears to be related to the action of JACKDAW (JKD), a zinc-finger protein that affects patterning of several root tissues (Hassan et al. 2010). The appropriate expression of the *GL2*, *CPC*, and *WER* genes was shown to be JKD-dependent, and JKD acts in a non-cell-autonomous fashion from the underlying cortical cell layer to influence root epidermal patterning (Hassan et al. 2010). Given that JKD is a putative transcription factor, it may regulate the expression of a signaling gene (e.g., the SCM ligand) that controls the position-dependent patterning.

D. Chromatin Modifications Influence Root Epidermal Cell-Fate Specification

Chromatin modification may act as a sensor switch to modulate patterning signals. Initial evidence from Shu-Nong Bai and colleagues came in the way of altered position-dependent patterning after application of histone deacetylase compound TSA (Xu et al. 2005). In addition, they specifically implicated a histone deacetylase protein *HDA18*, in mediating epidermal patterning during development (Xu et al. 2005).

Additional support for the importance of chromatin modifications in epidermal patterning came from Costa and Shaw (2006) through their use of three-dimensional fluorescence in situ hybridization (3D FISH) on intact root tissue. They found that wild-type plants display an "open" chromatin organization around the *GL2* locus in N-position cells, where *GL2* is normally expressed, and a "closed" conformation in H-position cells, where *GL2* expression is repressed. Furthermore, they observed that specific chromatin states around the *GL2* locus are not inherited but rather are reset during mitosis and respecified in the following G1 phase of the cell cycle in response to local positional information. This suggests that root meristem cells are constantly assessing their position relative to underlying cortical cells and rapidly respond to changes in position by remodeling chromatin around patterning gene loci (Costa and Shaw 2006).

Position-dependent chromatin modifications may be mediated by the GEM1 protein, which was found to interact directly with TTG1 and affect histone methylation within the *GL2* and *CPC* promoters (Caro et al. 2007). Analysis of the phenotypic effect of overexpression and knockout of *GEM1* suggests a possible direct role of repressing *GL2* and *CPC* expression, potentially in H-position cells, to negatively influence the nonhair cell fate in H-position cells. It is tempting to speculate that the "open" and "closed" chromatin state around the *GL2* locus identified by Costa and Shaw may be regulated by GEM1.

E. Molecular Model of Position-Dependent Epidermal Patterning

Molecular genetics and biochemical evidence to date suggests a model for position-dependent fate specification in the *Arabidopsis* root epidermis (Figure 4.6). Essential to becoming a nonhair cell is accumulation of the TTG-GL3/EGL3-WER core transcriptional activation complex. On the other hand, an epidermal cell will become a hair cell if it accumulates enough CPC/TRY/ETC1 to prevent core transcriptional complex formation by preventing WER from binding to TTG-GL3/EGL3. Therefore, a cell with greater relative abundance of WER than CPC/TRY/ETC1 will develop as a nonhair cell, and one with greater CPC will develop as a hair cell. In developing nonhair cells, the core transcriptional complex (WER-GL3/EGL3-TTG) promotes nonhair cell fate (through expression of *GL2*), lateral inhibition (through expression of CPC/TRY/ETC1), downregulation of SCM accumulation, and intracellular feedback (through expression of *MYB23*). In developing hair cells, the lateral inhibitors (CPC/TRY/ETC1) promote hair cell fate by competing with WER for binding to TTG-GL3/EGL3 (Lee and Schiefelbein 1999, 2002; Tominaga et al. 2007), positively regulating *GL3/EGL3* expression, as well as strengthening preferential H-position accumulation of SCM (Kwak and Schiefelbein 2008).

At the core of the cell-fate specification model is the interaction between the different patterning gene products. Genetic evidence supports the role of WER, TTG1, and GL3/EGL3 as a core transcriptional activation complex and suggests that competition between WER and CPC/TRY/ETC1 for binding to TTG-GL3/EGL3 complex may be essential for epidermal patterning (Larkin et al. 2003; Pesch and Hulskamp 2004; Ueda et al. 2005). To support this model, protein interaction and competition have been shown both in yeast and plant cells. The GL3 and EGL3 proteins have been shown to interact with CPC, TRY, WER, and TTG1 and are thought to bridge the interaction between the MYB-domain and WD40-repeat proteins (Lee and Schiefelbein 1999; Payne et al. 2000; Wada et al. 2002; Bernhardt et al. 2003; Zhang et al. 2003). In support of this notion, the MYB-domain proteins WER, CPC, and TRY interact with a different region of the GL3 protein than TTG1 (Payne et al. 2000; Zhang et al. 2003). Additionally, TTG1 and the WER-equivalent MYB-domain protein GL1 cannot interact with each other (Payne et al. 2000; Zhang et al. 2003); however, they each interact with GL3 and EGL3.

Furthermore, protein interaction analysis in yeast supports the model of the single-repeat MYB-domain proteins disrupting WER binding to the core transcriptional activation complex in cells in the H position. Tominaga et al. (2007) used a yeast-three-hybrid experiment to demonstrate that the presence of CPC inhibits WER interaction with GL3 or EGL3. Thus, it is likely that CPC, TRY, and ETC1 promote the cell fate by preventing WER from binding to TTG1-EGL3/GL3.

Feedback mechanisms also play an essential role in epidermal patterning by ensuring positional accuracy of patterning gene expression and localization. One important feedback loop exists between GL3/EGL3 and CPC/TRY. GL3/EGL3

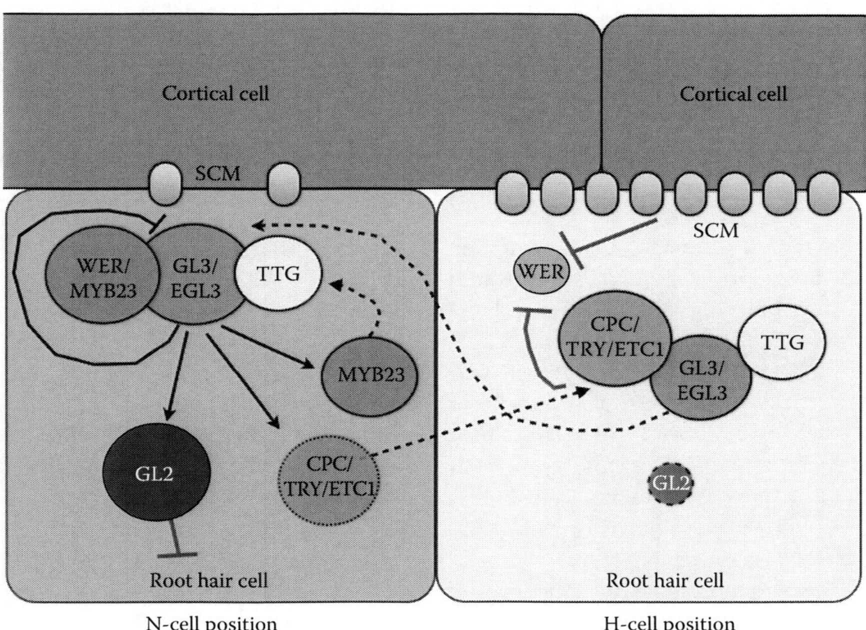

FIGURE 4.6 Model of the molecular network involved in position-dependent root epidermal cell-fate specification. The SCM receptor-like kinase is proposed to influence the downstream patterning network by repressing *WER* in H-position cells. Preferential accumulation of the activation complex (TTG1-GL3/EGL3-WER) in N-position cells leads to *GL2* expression (and subsequent inhibition of hair morphogenesis gene expression) in addition to expression of lateral inhibitors CPC/TRY/ETC1. CPC/TRY/ETC1 protein preferentially accumulates in H-position cells, where they compete with WER for binding to TTG1-GL3/EGL3, therefore preventing *GL2* expression in H-position cells. Note: Solid lines represent gene regulation, and dashed lines indicate protein movement.

localization in N-position cells is essential for *CPC* (and probably *TRY* and *ETC1*) expression (Bernhardt et al. 2003), and CPC/TRY are essential in H-position cells for *GL3/EGL3* expression (Bernhardt et al. 2005), generating a positive intercellular feedback loop. Additionally, feedback is necessary for establishing preferential hair cell localization of the SCM receptor (Kwak and Schiefelbein 2008). The core transcriptional activation complex negatively regulates *SCM* expression in N-position cells, and CPC/TRY positively regulate *SCM* expression in H-position cells (Kwak and Schiefelbein 2008). Together, this intricate molecular network within epidermal cells establishes cell fate based on position and then strengthens cellular differences between cells in each position through lateral inhibition and multiple feedback loops.

IV. Developmental Timeline of Epidermal Patterning

Patterning genes are expressed in the proper position-dependent location beginning as early as the heart stage of the embryo, suggesting that embryonic root epidermal cells have the same striped pattern before and after germination. *GL2* is first expressed throughout the epidermis of developing roots at late heart stage and progressively becomes restricted to N-position cells by the end of embryogenesis (Costa and Dolan 2003). This suggests that pattern formation occurs between late heart stage and the mature stage of embryogenesis. Interestingly, in situ analysis does not detect *WER* or *CPC* expression at the heart stage (Lin and Schiefelbein 2001; Costa and Dolan 2003) but rather late torpedo stage (Costa and Dolan, 2003). Functional WER TTG1 (Lin and Schiefelbein 2001), GL3/EGL3 (Bernhardt et al. 2005), and CPC/TRY (Simon et al. 2007) are required for N-cell-specific *GL2* expression in the embryo, suggesting similar molecular-genetic regulation in both the embryonic and postembryonic root.

Unexpectedly, the *scm* mutant does not exhibit an alteration in expression of *GL2* in mature embryos or emerging roots from germinating seeds (which are from embryonic origin), implying that *SCM* is not required for embryonic epidermal patterning (Kwak and Schiefelbein 2007). In fact, *GL2* expression patterning in the *scm* mutant background does not deviate from wild type until 3 days postgermination (Kwak and Schiefelbein 2007) or about the stage when root tip cells are derived from postembryonic meristematic activity (Scheres et al. 1994). This implicates an additional unknown mechanism in embryonic position-dependent epidermal patterning.

Postembryonic patterning and positional information are continuously generated and processed, as a change in position late in the root meristem can result in a change in fate (Berger et al. 1998a). SCM is necessary for establishing the pattern of meristem epidermal cells, not just maintaining a preset epidermal pattern established during embryogenesis, as may be implied from its lack of a role in embryonic root patterning (Kwak and Schiefelbein 2007). Pattern analysis of epidermal clones, clonally related files of cells generated by longitudinal division of a single epidermal cell, found the *scm* mutant to be significantly less effective at establishing two distinct cell fates in clones derived from developing hair cells (in which the clones would contain one new hair file and one nonhair file) (Kwak and Schiefelbein 2007). Thus, epidermal cells in the meristem continuously assess their position via the SCM-mediated signaling pathway during postembryonic root development.

V. Conclusion and Future Directions

The study of root epidermal patterning in *Arabidopsis* has provided a great deal of insight into the molecular-genetic regulation of cell-fate specification. Nevertheless, additional research is needed to address many unanswered questions.

To fully understand the regulation of position-dependent pattern, it is essential to identify the putative SCM ligand. Based on the specific organization of hair and nonhair cells relative to the underlying cortical cells, the SCM ligand is hypothesized to be asymmetrically distributed around the epidermis. However, the nature of the signal, where it is generated, and how it influences epidermal cells are currently unknown. In this regard, the further characterization of JKD and its role in root epidermal patterning may be informative. Additionally, it will be interesting to define the mechanism of signaling through the SCM pathway, as SCM has been shown to lack kinase activity (Chevalier et al. 2005).

Also, in view of the fact that SCM is not required for position-dependent cell patterning in the embryonic epidermis and the *scm* mutant does not entirely abolish the cell pattern in the postembryonic epidermis, additional patterning mechanisms beyond the SCM-signaling pathway are thought to exist (Kwak and Schiefelbein 2007). The recent discovery of *scm*-like mutants *qky*, *doq*, and *zet* may help answer some of these questions (Fulton et al. 2009); however, further research is needed to determine their precise role in root epidermal patterning.

Also of interest are the specific roles of redundant genes *GL3/EGL3* and *CPC/TRY/ETC1*. Not only do their loss-of-function root epidermal fate specification phenotypes differ (Schellmann et al. 2002; Bernhardt et al. 2003; Kirik et al. 2004; Bernhardt et al. 2005; Simon et al. 2007), but new developments suggest that their genetic regulation (Simon et al. 2007) and functions within the epidermal fate-specification pathway (Kwak and Schiefelbein 2008) may also have diverged. Analysis at the transcriptome level may provide interesting insight to their slightly divergent, yet significantly overlapping functions.

In a broader sense, it will be interesting to learn whether the same molecular components and regulatory interactions employed for root epidermal patterning in *Arabidopsis* are also used by other plant species. In particular, it is unknown whether plants that exhibit the alternating pattern or the random pattern utilize homologs of the *Arabidopsis* patterning genes, such as *WER*, *GL2*, and *CPC*. Comparative molecular studies of these patterning processes may shed light on the evolutionary origin of new patterning mechanisms in plants.

Acknowledgments

We thank Su-Hwan Kwak, Angela Bruex, and other members of the Schiefelbein Lab for helpful comments. Research on *Arabidopsis* root epidermal development in the Schiefelbein Lab is supported by grants from the National Science Foundation (IOS-0744599 and IOS-0723493).

References

Berger F, Haseloff J, Schiefelbein J, Dolan L. 1998a. Positional information in root epidermis is defined during embryogenesis and acts in domains with strict boundaries. *Curr Biol* 8:421–430.

Berger F, Hung CY, Dolan L, Schiefelbein J. 1998b. Control of cell division in the root epidermis of *Arabidopsis thaliana. Dev Biol* 194:235–245.

Bernhardt C, Lee MM, Gonzalez A, Zhang F, Lloyd A, Schiefelbein J. 2003. The bHLH genes *GLABRA3* (*GL3*) and *ENHANCER OF GLABRA3* (*EGL3*) specify epidermal cell fate in the *Arabidopsis* root. *Development* 130:6431–6439.

Bernhardt C, Zhao M, Gonzalez A, Lloyd A, Schiefelbein J. 2005. The bHLH, genes *GL3* and *EGL3* participate in an intercellular regulatory circuit that controls cell patterning in the *Arabidopsis* root epidermis. *Development* 132:291–298.

Caro E, Castellano MM, Gutierrez C. 2007. A chromatin link that couples cell division to root epidermis patterning in *Arabidopsis. Nature* 4:47.

Chevalier D, Batoux M, Fulton L, Pfister K, Yadav RK, Schellenberg M, Schneitz K. 2005. STRUBBELIG, defines a receptor kinase-mediated signaling pathway regulating organ development in *Arabidopsis. Proc Natl Acad Sci USA* 102:9074–9079.

Clowes F. 2000. Pattern in root meristem development in angiosperms. *New Phytol* 146:83–94.

Cormack R. 1935. Investigations on the development of root hairs. *New Phytol* 34:30–54.

Cormack R. 1937. The development of root hairs by *Elodea canadensis. New Phytol* 36:19–25.

Cormack R. 1947. A comparative study of developing epidermal cells in white mustard and tomato roots. *Am J Bot* 34:310–314.

Costa S, Dolan L. 2003. Epidermal patterning genes are active during embryogenesis in *Arabidopsis. Development* 130:2893–2901.

Costa S, Shaw PJ. 2006. Chromatin organization and cell fate switch respond to positional information in *Arabidopsis. Nature* 439:493–496.

Cutter E. 1978. The epidermis. In *Plant Anatomy.* London, U.K.: Clowes & Sons, pp. 94–106.

Di Cristina M, Sessa G, Dolan L, Linstead P, Baima S, Ruberti I, Morelli G. 1996. The *Arabidopsis* Athb-10 (GLABRA2) is an HD-Zip protein required for regulation of root hair development. *Plant J* 10:393–402.

Dolan L, Costa S. 2001. Evolution and genetics of root hair stripes in the root epidermis. *J Exp Bot* 52:413–417.

Dolan L, Duckett CM, Grierson C et al. 1994. Clonal relationships and cell patterning in the root epidermis of *Arabidopsis. Development* 120:2465–2474.

Dolan L, Janmaat K, Willemsen V, Linstead P, Poethig S, Roberts K, Scheres B. 1993. Cellular organisation of the *Arabidopsis thaliana* root. *Development* 119:71–84.

Freshour G, Clay RP, Fuller MS, Albersheim P, Darvill AG, Hahn MG. 1996. Developmental and tissue-specific structural alterations of the cell-wall polysaccharides of *Arabidopsis thaliana* roots. *Plant Physiol* 110:1413–1429.

Fulton L, Batoux M, Vaddepalli P et al. 2009. DETORQUEO, QUIRKY, and ZERZAUST, represent novel components involved in organ development mediated by the receptor-like kinase STRUBBELIG, in *Arabidopsis thaliana. PLoS Genet* 5(1):e1000355.

Galway ME, Massucci JD, Lloyd A, Walbot V, Davis RW, Schiefelbein J. 1994. The *TTG,* gene is required to specify epidermal cell fate and cell patterning in the *Arabidopsis* root. *Dev Biol* 166:740–754.

Grierson C, Schiefelbein J. 2002. Root hairs. *Arabidopsis Book* 1:e0060.

Guimil S, Dunand C. 2006. Patterning of *Arabidopsis* epidermal cells: epigenetic factors regulate the complex epidermal cell fate pathway. *Trends Plant Sci* 11:601–609.

Hassan H, Scheres B, Blilou I. 2010. JACKDAW, controls epidermal patterning in the *Arabidopsis* root meristem through a non-cell-autonomous mechanism. *Development* 137:1523–1529.

Hofer R-M. 1996. Root hairs. In: *Plant Roots: The Hidden Half,* eds. Y Waisel, A Eshel, U Kafkafi, 2nd edn., pp. 111–126. New York: Marcel Dekker, Inc.

Kang YH, Kirik V, Hulskamp M, Nam K, Hagely K, Lee MM, Schiefelbein J. 2009. The *MYB23* gene provides a positive feedback loop for cell fate specification in the *Arabidopsis* root epidermis. *Plant Cell* 21:1080–1094.

Kirik V, Simon M, Huelskamp M, Schiefelbein J. 2004. The *ENHANCER, OF, TRY, AND, CPC1* gene acts redundantly with *TRIPTYCHON,* and *CAPRICE,* in trichome and root hair cell patterning in *Arabidopsis. Dev Biol* 268:506–513.

Koornneef M. 1981. The complex syndrome of *ttg* mutants. *Arabidopsis Inf Serv* 45–51.

Koornneef M. 1990. Mutations affecting the testa color in *Arabidopsis. Arabidopsis Inf Serv* 27:94–97.

Koshino-Kimura Y, Wada T, Tachibana T, Tsugeki R, Ishiguro S, Okada K. 2005. Regulation of *CAPRICE,* transcription by MYB, proteins for root epidermis differentiation in *Arabidopsis. Plant Cell Physiol* 46:817–826.

Kurata T, Ishida T, Kawabata-Awai C et al. 2005. Cell-to-cell movement of the CAPRICE, protein in *Arabidopsis* root epidermal cell differentiation. *Development* 132:5387–5398.

Kwak S-H, Schiefelbein J. 2007. The role of the SCRAMBLED, receptor-like kinase in patterning the *Arabidopsis* root epidermis. *Dev Biol* 302:118–131.

Kwak S-H, Schiefelbein J. 2008. A feedback mechanism controlling SCRAMBLED, receptor accumulation and cell-type pattern in *Arabidopsis. Curr Biol* 18:1949–1954.

Kwak S-H, Shen R, Schiefelbein J. 2005. Positional signaling mediated by a receptor-like kinase in *Arabidopsis. Science* 307:1111–1113.

Larkin JC, Brown ML, Schiefelbein J. 2003. How do cells know what they want to be when they grow up? *Annu Rev Plant Biol* 54:403–430.

Lee MM, Schiefelbein J. 1999. WEREWOLF, a MYB-Related protein in *Arabidopsis* is a position-dependent regulator of epidermal cell patterning. *Cell* 99:473–483.

Lee MM, Schiefelbein J. 2002. Cell pattern in the *Arabidopsis* root epidermis determined by lateral inhibition with feedback. *Plant Cell* 14:611–618.

Lin Y, Schiefelbein J. 2001. Embryonic control of epidermal cell patterning in the root and hypocotyl of *Arabidopsis. Development* 128:3697–3705.

Massucci JD, Rerie WG, Foreman DR, Zhang M, Galway ME, David Marks M, Schiefelbein J. 1996. The homeobox gene *GLABRA2* is required for position-dependent cell differentiation in the root epidermis of *Arabidopsis thaliana. Development* 122:1253–1260.

Menand B, Yi K, Jouannic S, Hoffmann L, Ryan E, Linstead P, Schaefer DG, Dolan L. 2007. An ancient mechanism controls the development of cells with a rooting function in land plants. *Science* 316:1477–1480.

Ohashi Y, Oka A, Rodrigues-Pousada R, Possenti M, Ruberti I, Morelli G, Aoyama T. 2003. Modulation of phospholipid signaling by *GLABRA2* in root-hair pattern formation. *Science* 300:1427–1430.

Payne C, Zhang F, Lloyd AM. 2000. *GL3* encodes a bHLH, protein that regulates trichome development in *Arabidopsis* through interaction with GL1 and TTG1. *Genetics* 156:1349–1362.

Pemberton LMS, Tsai S, Lovell PH, Harris PJ. 2001. Epidermal patterning in seedling roots of eudicotyledons. *Ann Bot* 87:649–654.

Pesch M, Hulskamp M. 2004. Creating a two-dimensional pattern de novo during *Arabidopsis* trichome and root hair initiation. *Curr Opin Genet Dev* 14:422–427.

Rerie W, Feldmann K, David Marks M. 1994. The *GLABRA2* gene encodes a homeodomain protein required for normal trichome development in *Arabidopsis. Genes Dev* 8:1388–1399.

Ryu KH, Kang YH, Park Y-H, Hwang I, Schiefelbein J, Lee MM. 2005. The WEREWOLF, MYB, protein directly regulates CAPRICE, transcription during cell fate specification in the *Arabidopsis* root epidermis. *Development* 132:4765–4775.

Schellmann S, Hulskamp M, Uhrig J. 2007. Epidermal pattern formation in the root and shoot of *Arabidopsis. Biochem Soc Trans* 35:146–148.

Schellmann S, Schnittger A, Kirik V, Wada T, Okada K, Beermann A, Thumfahrt J, Jürgens G, Hülskamp M. 2002. TRIPTYCHON, and CAPRICE, mediate lateral inhibition during trichome and root hair patterning in *Arabidopsis. EMBO J* 21:5036–5046.

Scheres B, Wolkenfelt H, Willemsen V, Terlouw M, Lawson E, Dean C, Weisbeek P. 1994. Embryonic origin of the *Arabidopsis* primary root and root meristem initials. *Development* 120:2475–2475.

Schiefelbein J, Kwak S-H, Wieckowski Y, Barron C, Bruex A. 2009. The gene regulatory network for root epidermal cell-type pattern formation in *Arabidopsis. J Exp Bot* 60:1515–1521.

Simon M, Lee MM, Lin Y, Gish L, Schiefelbein J. 2007. Distinct and overlapping roles of single-repeat MYB, genes in root epidermal patterning. *Dev Biol* 311:566–578.

Tominaga R, Iwata M, Okada K, Wada T. 2007. Functional analysis of the epidermal-specific MYB, genes *CAPRICE,* and *WEREWOLF,* in *Arabidopsis. Plant Cell* 19:2264–2277.

Ueda M, Koshino-Kimura Y, Okada K. 2005. Stepwise understanding of root development. *Curr Opin Plant Biol* 8:71–76.

Van Den Berg C, Willemsen V, Hendriks G, Weisbeek P, Scheres B. 1997. Short-range control of cell differentiation in the *Arabidopsis* root meristem. *Nature* 390:287–289.

Wada T, Kurata T, Tominaga R et al. 2002. Role of a positive regulator of root hair development C*APRICE,* in *Arabidopsis* root epidermal cell differentiation. *Development* 129:5409–5419.

Wada T, Tachibana T, Shimura Y, Okada K. 1997. Epidermal cell differentiation in *Arabidopsis* determined by a Myb homolog C*PC. Science* 277:1113–1116.

Walker AR, Davison PA, Bolognesis-Winfield AC et al. 1999. The *TRANSPARENT, TESTA, GLABRA1* locus which regulates trichome differentiation and anthocyanin biosynthesis in *Arabidopsis* encodes a WD40 repeat protein. *Plant Cell*:11:1337–1349.

Xu C, Liu C, Wang Y, Li L-C, Chen W-Q, Xu Z-H, Bai S-N. 2005. Histone acetylation affects expression of cellular patterning genes in the *Arabidopsis* root epidermis. *Proc Natl Acad Sci USA* 102:14469–14474.

Zhang F, Gonzalez A, Zhao M, Payne C, Lloyd A. 2003. A network of redundant bHLH, proteins functions in all TTG1-dependent pathways of *Arabidopsis. Development* 130:4859–4869.

5

Structure and Function of Three Suberized Cell Layers: Epidermis, Exodermis, and Endodermis*

Chris J. Meyer
University of Guelph

Carol A. Peterson
University of Waterloo

I. Introduction

For a complete understanding of a biological system, it is advisable to integrate structure and function, as the two are inextricably linked. In studies of roots, however, these are often separated. Notable exceptions are the wide-ranging studies of McCully and Canny (see McCully 1995) and the numerous investigations of Clarkson, Sanderson, and Robards who considered the structure and function of roots with respect to ion and water uptake (see Clarkson 1996).

In the majority of textbooks, the reader is presented with cross sections of roots with mature primary development. This is a useful beginning, but it is, by no means, the whole story of a root system (see Waisel and Eshel 2002). Because roots have tip growth, there is an age continuum along an individual root—from the youngest region at the tip to the oldest at the base (adjacent to the stem). Along these roots, lateral and in some cases tertiary roots may develop, and in older regions of eudicot roots, secondary (thickening) growth can occur. Roots of various species have constitutive anatomical differences; as well, changes to their structure may occur in response to environmental conditions (see Enstone et al. 2003). As pointed out by McCully (1995), the anatomy of field-grown roots can differ significantly from that of laboratory-grown roots. All these considerations make it a daunting task to describe the microscopic structure of an entire root system, yet this is what is ultimately required to connect structure with function in the field.

As an initial step, various roots have been examined in laboratory settings to probe the physiological roles of certain cellular structures. From such studies, it has become clear that suberin in the walls of cells exerts a major influence on ion and water uptake, and protection against environmental stresses and pathogens. In this chapter, we have elected to describe three specialized layers of the root—the epidermis, exodermis, and endodermis—all of which are characterized by suberin deposits in the walls of their component cells.

* Since the writing of this chapter, Niko Geldner and associates (University of Lausanne, Switzerland) have published information concerning the plasma membrane proteins associated with Casparian band formation and the chemical nature of the band in *Arabidopsis thaliana* endodermis. In Naseer et al. (2012), the authors conclude that the Casparian band consists of lignin and not suberin based on a chemical analysis of the phenolic contents of the band. However, it is premature to claim definitively that the band lacks lipid and/or wax components in the absence of similar analyses for these chemicals. Two reviews of their work are currently available (i.e., Alassimone et al. 2012; Roppolo and Geldner 2012).

II. Suberin Forms: Chemistry, Anatomy, and Detection In Situ

Suberin is a complex biopolymer of fatty acids and phenolics with a known monomeric profile (reviewed in Kolattukudy 1980, 1984; Bernards 2002). The polymer can take several forms in plant roots—Casparian bands, suberin lamellae, diffuse, and wound suberin—and these vary either in composition or abundance. Each form has its unique anatomical and physiological features. There is no single histochemical test that can be used to detect suberin—three are required. Since suberin is a polymer composed of aliphatic and aromatic components, it can be stained for the former with dyes that partition into lipids such as the Sudan dyes or fluorol yellow 088 (Jensen 1962; Brundrett et al. 1991) and for the latter by the Hoepfner–Vorsatz test (Reeve 1951; Ling-Lee et al. 1977) or by observing autofluorescence (Figure 5.5A; Lulai and Morgan 1992). Since both lipid and phenolic components could be present but in a nonpolymerized condition, to prove the presence of the suberin polymer, it is necessary to show resistance to digestion by concentrated sulfuric acid (Johansen 1940).

Certainly the most trustworthy indicator of the presence of suberin is extraction, followed by analyses using gas chromatography coupled with mass spectrometry (GC-MS). However, visual observation of suberin in situ is useful as it clarifies the position and type of suberin being studied.

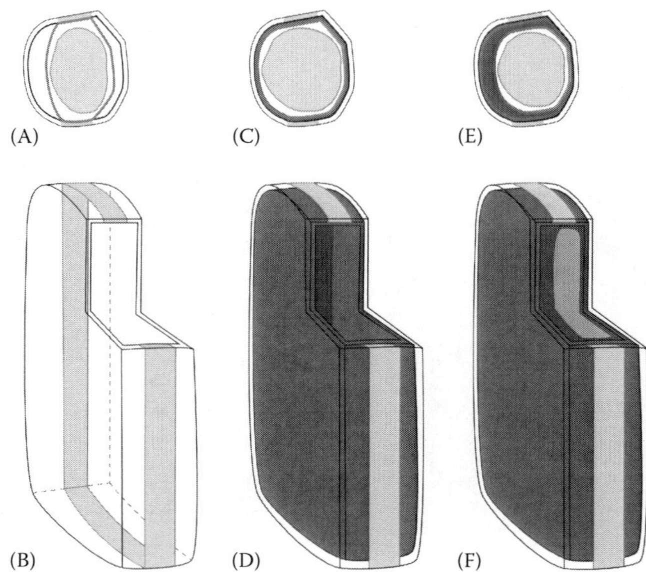

FIGURE 5.1 (See color insert.) Illustrations of endodermal cells (not to scale) displaying their states of maturity (A,C,E) in cross section and (B,D,F) in three dimensions. (A,B) State I, Casparian bands in anticlinal walls. (A) Band plasmolysis is depicted. (C,D) State II, suberin lamella between the cell wall and plasma membrane. (E,F) State III, tertiary wall thickening between the suberin lamella and plasma membrane. Black lines, cell wall borders; dark gray lines, plasma membranes; light gray lines, tonoplasts; yellow shading, Casparian bands; red shading, suberin lamellae; blue shading, tertiary wall thickenings. (Images (B,D,F) modified from Raven, P.H. et al., *Biology of Plants*, 6th edn., WH Freeman and Company, New York, 1999. With permission.)

A. Casparian Bands

1. Chemistry

The suberin in endodermal Casparian bands is mainly composed of poly(phenolics) (5%–6% of the total isolated cell wall fraction) with lesser amounts of aliphatics (0.1%–1%) (Schreiber et al. 1994; Schreiber 1996; Zeier and Schreiber 1997, 1998). The abundant poly(phenolics) are lignin-like in composition, except that a significant amount of hydroxycinnamic acids forms part of the polymer, along with the usual monolignol building blocks of lignin (Bernards, personal communication). Casparian band composition was determined in roots of three species (*Clivia miniata* [Lindl.] Regel, *Monstera deliciosa* Liebm., and *Glycine max* (L.) Merr.) where the endodermis contained Casparian bands but lacked suberin lamellae (Zeier and Schreiber 1997, 1998; Thomas et al. 2007). Such measurements require weeks of dedication to obtain enough isolated Casparian band material to conduct the analytical analyses. It is not possible to measure the chemical composition of exodermal Casparian bands because they develop concurrently with exodermal suberin lamellae (Enstone et al. 2003). The substructure of the Casparian band polymer, including its interactions with the cell wall and suberin lamellae, is still unknown.

2. Anatomy and Histochemical Detection

Casparian bands have long been known to occur in the endodermis of vascular plants with only a few exceptions (Damus et al. 1997) and more recently in the exodermis of many angiosperm species (Perumalla et al. 1990; Peterson and Perumalla 1990; Meyer et al. 2009). Casparian bands are embedded within the primary walls of cells, i.e., within their intermicrofibrillar spaces (see Steudle and Peterson 1998). The bands are typically located in the radial and transverse (end) walls (i.e., anticlinal walls) of the cells (Figure 5.1A and B) and form a continuous layer across the middle lamella so that they influence apoplastic radial movement of solutes between the cortex and the stele in the case of the endodermis (Figure 5.2A) and between the epidermis and the central cortex in the case of the exodermis (Figure 5.2B). Endodermal Casparian bands are often small and are, therefore, best seen with fluorescence microscopy due to the greatly increased contrast that a dark field affords. Specifically, the fluorochrome berberine hemisulfate followed by a counterstain of aniline blue or toluidine blue O is useful for this (Figure 5.3A; Brundrett et al. 1988; Lux et al. 2005), as is staining with periodic acid–Schiff's (PAS) reagent prior to observation with ultraviolet light (Watt et al. 1996), or PAS followed by fluorol yellow (Wang et al. 1995). Although the Casparian bands of the exodermis are larger than those of the endodermis, their presence can be masked by adjacent suberin lamellae; the procedure of Brundrett et al. (1988) that dampens the fluorescence of the lamellae is also recommended for observing these Casparian bands (Figure 5.5B). Using transmission electron microscopy (TEM), Casparian bands are often not conspicuous, although in some cases they can be recognized as a more or less dense

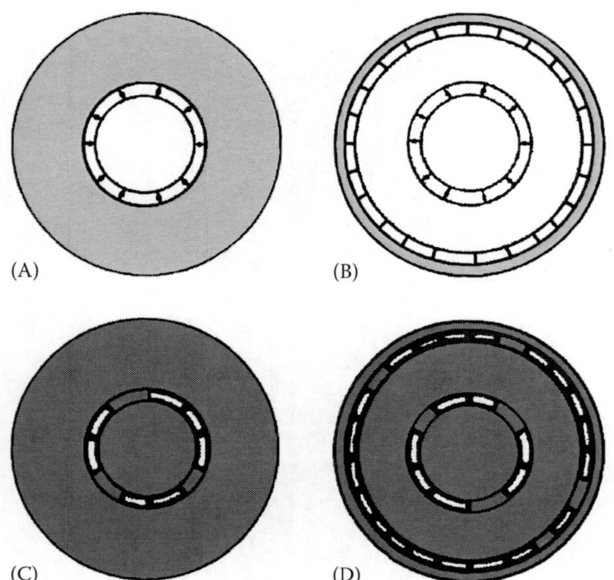

FIGURE 5.2 Diagrams of roots in cross section illustrating the differing contributions of (A,B) Casparian bands and (C,D) suberin lamellae to apoplastic isolation. (A) In a non-exodermal root, substances may flow apoplastically through the gray-shaded area as far as the endodermal Casparian bands. The stele (non-shaded) is apoplastically isolated from the central cortex and epidermis. (B) In an exodermal root, the exodermal Casparian band confines substances that flow apoplastically to the epidermis (gray-shaded area). (C) In a non-exodermal root, the protoplasts of endodermal cells with suberin lamellae (non-shaded areas) are isolated from the surrounding apoplast. (D) In an exodermal root, the protoplasts of both exodermal and endodermal cells with suberin lamellae (non-shaded areas) are isolated from the apoplast. (Modified from Peterson, C.A. and Cholewa, E., *Zeitsch. Pflanzen. Boden.*, 161, 521, 1998. With permission.)

collection of small, electron-opaque dots (Bonnett 1968). When the cells are without suberin lamellae, Casparian bands can be indicated indirectly by band plasmolysis (Bryant 1934; Bonnett 1968; Haas and Carothers 1975; Enstone and Peterson 1997).

In addition to the earlier histochemical tests, Casparian bands can be located by functional means, as they are virtually impermeable to ions and apoplastic fluorescent tracers. It is possible to test the permeability of the bands to ions by compartmental elution, a technique that can be used to measure the volume of the free space, i.e., the space into and out of which substances can move by diffusion. If Casparian bands are impermeable to an externally applied ion, its free space will consist of the wall spaces of the cells located external to the band. Conversely, if the bands are permeable to the ion, its free space will include all wall spaces (and lumens of dead cells) both internal and external to the bands. This technique has indicated that the Casparian bands of the exodermis are impermeable to calcium and phosphate ions (Peterson 1987; Cholewa and Peterson 2004; Waduwara 2007) and those of the endodermis to sodium ions (Baker 1971). A more rapid test is the blockage of apoplastic dye movement (Figure 5.5E; Peterson et al. 1978; Enstone and Peterson 1992; Alassimone et al. 2010), although it must always be remembered that these substances have wider molecular diameters than those of the ions that might be of actual interest. Despite this caveat, dye movement has, to date, been accurate in predicting the locations of Casparian bands.

To sum up, a positive test for a Casparian band consists of all three of the following: (1) positive histochemical reactions for lipids and phenolics, (2) resistance to strong acid digestion, and (3) prevention of apoplastic movement of substances.

B. Suberin Lamellae

1. Chemistry

The chemistry of suberin lamellae has been determined primarily from analyses of *Solanum tuberosum* L. tuber periderm. A suberin lamella is considered to be composed of two spatially distinct but covalently linked domains, i.e., the poly(phenolic) domain (SPPD, after Bernards 2002) embedded in the primary cell wall and the poly(aliphatic) domain (SPAD, after Bernards 2002) located between the cellulosic cell wall and plasma membrane. This two-domain model had previously been inferred from histochemical (Lulai and Morgan 1992) and nuclear magnetic resonance studies (Stark and Garbow 1992; Lopes et al. 2000a,b) and more recently from differential scanning calorimetry (Mattinen et al. 2009). When observed as osmium-stained suberin lamellae with a TEM, they appear as alternating bands of translucent (electron-light) and opaque (electron-dense) compounds. According to Bernards (2002), the translucent bands are composed of aliphatic monomers, whereas the opaque bands are presumably composed of a mixture of aromatic monomers, glycerol, and ester-linked portions of the aliphatic monomers. The typical monomeric composition of the two domains has been determined by analysis of depolymerized compounds isolated from mature tissues (Kolattukudy 1980, 1984; Graça and Pereira 2000a,b; Bernards 2002). Monomers are traditionally quantified and identified using GC-MS. It has always been assumed that these analyses refer to suberin lamellae, but it is possible that the periderm may also contain Casparian bands as these were found in *Pelargonium* × *hortorum* L.H. Bailey (Meyer and Peterson 2011).

In suberin lamellae, the monomer compositions of the SPAD and SPPD vary among species and cell types (Holloway 1983; Matzke and Riederer 1991; Zeier and Schreiber 1998, 1999; Zeier et al. 1999a,b). In general, the SPAD is lipophilic, containing variable amounts of fatty acids, ω-OH fatty acids, α,ω-dioic acids, and intercalated waxes. The SPPD consists mainly of hydroxycinnamic acid derivatives with lesser amounts of monolignols and tyramine (Borg-Olivier and Monties 1993; Negrel et al. 1996; Bernards and Lewis 1998).

2. Anatomy and Histochemical Detection

Suberin lamellae are easily seen with TEM as alternating electron-dense and electron-lucent layers (Sitte 1962). They are deposited against the face of the wall by the cell protoplast (Figure 5.1C and D). Since they are larger than Casparian bands, they can be detected by lipid stains, using bright-field (Figures 5.3B and 5.5C;

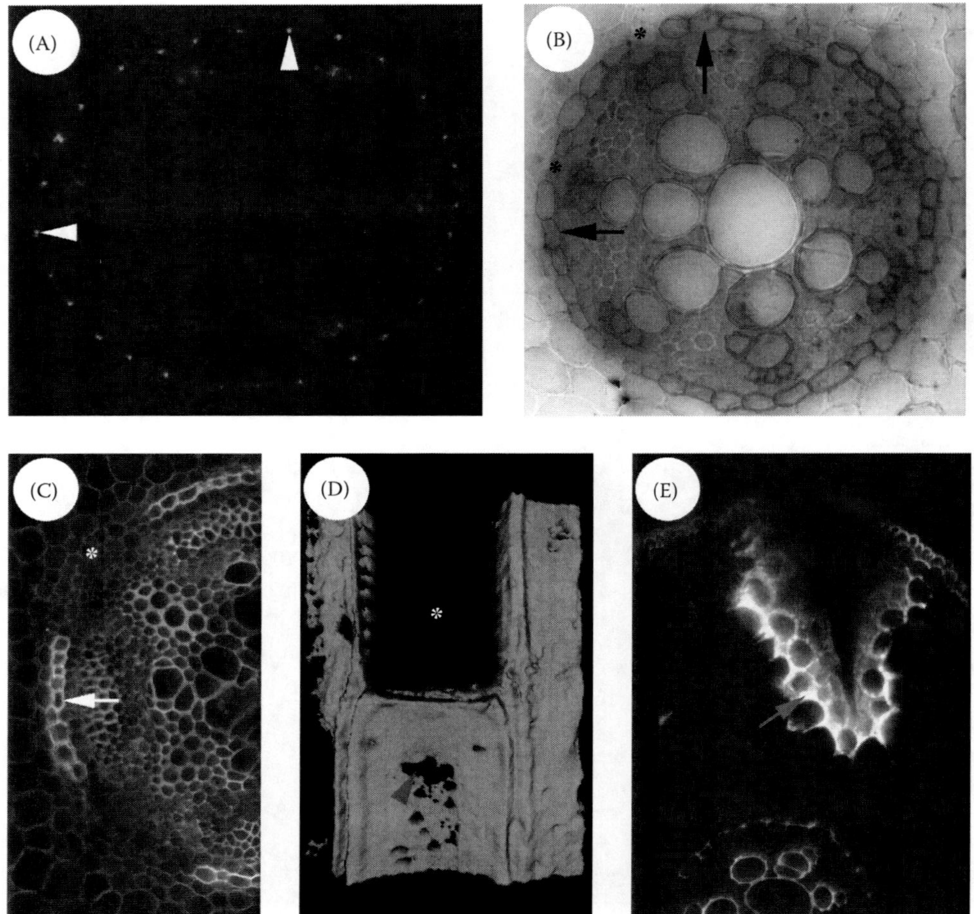

FIGURE 5.3 **(See color insert.)** Images of the endodermis and stele in (A–C) cross and (D) longitudinal views. (A) Stained with berberine hemisulfate–aniline blue. The endodermis has Casparian bands (white arrowheads) that appear dot-like in the anticlinal walls of *Allium cepa*. (B) Stained with Sudan red 7B. Mature endodermal cells contain suberin lamellae (black arrows) in *A. cepa*. Passage cells (asterisks) are located near the xylem poles. (C) Stained with fluorol yellow 088. Mature endodermal cells have suberin lamellae (white arrow), alternating with patches of passage cells (asterisk) in *Glycine max*. (D) Portions of *A. cepa* endodermal cells from a three-dimensional reconstruction using confocal laser scanning microscopy. Angle of rotation on the vertical axis was 5.6°. Asterisk indicates a passage cell (without suberin lamellae). The red arrowhead indicates a large opening in the suberin lamellae. Green color was assigned artificially for grayscale images. (E) Cross section of *A. cepa* root that had developed autofluorescent wound suberin following injury. Magenta arrow indicates suberin in an air space. (Images (A,B): courtesy of Ishari Waduwara; image (C): courtesy of Raymond Thomas; image (D): from Waduwara, C.I. et al., *Can. J. Bot.*, 86, 623, 2008. Reproduced with permission from the NRC Research Press.)

Sudan red 7B) as well as fluorescence microscopy (Figures 5.3C and 5.5D; fluorol yellow) as described by Brundrett et al. (1991).

C. Diffuse Suberin

1. Chemistry

This type of suberin is not as well known as those of Casparian bands and suberin lamellae. Diffuse suberin in the epidermis of soybean (*Glycine max* (L.) Merr.) roots consists of poly(aliphatics) (fatty acids, ω-OH fatty acids, α,ω-dioic acids), poly(phenolics) (vanillin and syringin), and associated waxes (Thomas et al. 2007). This type of suberin is similar in composition to that of endodermal suberin lamellae, but the amounts are substantially lower than those of the lamellae and Casparian bands. For instance, the amount of diffuse suberin in the epidermis of a young soybean root is about 4-fold lower than in the Casparian band (Thomas et al. 2007). Further, since the wall volume of the epidermis is much greater than the volume of the Casparian band, the density of diffuse suberin in the epidermis will be substantially lower than that of the band. Certainly this low-density suberin is an important feature for epidermal function (see Section III.B).

2. Anatomy and Histochemical Detection

Diffuse suberin was first identified by Peterson et al. (1978) as being non-lamellar and has since been detected in the epidermal cells of several species such as *Allium cepa* L. and *G. max* (Wilson and Peterson 1983; Brundrett et al. 1988; Thomas et al. 2007). Diffuse suberin can be

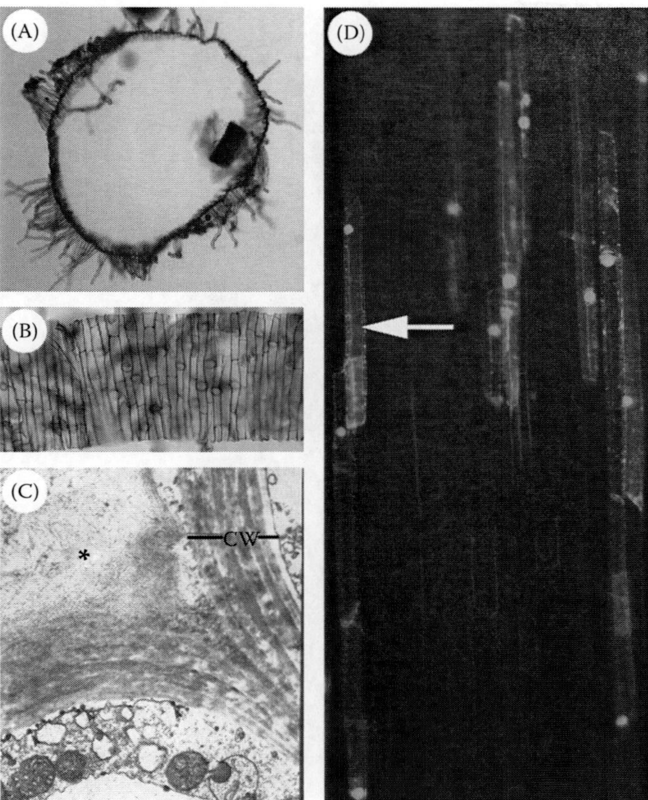

FIGURE 5.4 (See color insert.) Photomicrographs of the epidermis. (A) Cross section of *Glycine max* root treated with concentrated sulfuric acid. The epidermal walls resisted digestion. (B) Acid-digested epidermal cell walls in face view in *G. max*. (C) TEM image of the outer tangential and radial walls of epidermal cells in *Allium cepa*. Diffuse suberin in the cell walls (CW) is evident. Extracellular material (*) covering the outer tangential walls of epidermal cells is present. (D) Stained with uranin. Living epidermal cells accumulate the dye in the nucleus and cytoplasm (white arrow) in *A. cepa*. (Images (A,B): courtesy of Alice Fang; image (C): from Peterson, C.A. et al., *Protoplasma*, 96, 1, 1978. Reproduced with permission from Springer Science; image (D): courtesy of Ishari Waduwara.)

seen with TEM as inconspicuous bands within the wall (Figure 5.4C; Peterson et al. 1978; Fang 2006). The histochemical tests described for suberin lamellae, and acid digestion described for Casparian bands can be applied. However, unlike the Casparian band, this form of suberin is permeable to apoplastic dyes and smaller molecules (Figure 5.5E). It develops close to the root tip where the phenolic component was detected first, followed by the aliphatic component at 20 mm from the tip (Thomas et al. 2007).

D. Wound-Induced Suberin

Following wounding, plants react by synthesizing suberin. Initially, it is deposited in living cells adjacent to the wound (i.e., boundary zone). In eudicots, a new periderm can be initiated near the boundary zone. This periderm produces cells with suberin lamellae and possibly Casparian bands as described earlier.

1. Chemistry

Schreiber et al. (2005b) measured the suberin composition and amounts in native and wound-induced periderm (that would include the boundary zone) in *S. tuberosum* tubers. The composition of suberin was similar between native and wound periderm. On the other hand, the amounts of suberin and associated waxes were 40%–50% greater in native periderm compared with wound periderm (Schreiber et al. 2005b). This difference in suberin and wax amounts had a significant effect on water permeability. However, to the best of our knowledge, the chemistry of suberin in a boundary zone alone has never been purposefully investigated. Such an investigation could be carried out easily in monocot species since they do not form periderms. For example, Moon et al. (1984) observed the development of a boundary zone in wounded *A. cepa* roots.

2. Anatomy and Histochemical Detection

The synthesis of wound-induced suberin in the boundary zone of *A. cepa* roots occurs over a period of 4 days and begins with the deposition of phenolics followed by the lipid component (Moon et al. 1984). The suberin polymer is located within the walls of living cells adjacent to the wound, and it also infills adjacent air spaces (Figure 5.3E). Like Casparian bands, the wounded area is a barrier to the movement of apoplastic dyes, a feature that only occurs when the suberin resists acid digestion. Wound suberin can be detected with the same histochemical methods used for suberin lamellae (see Section II.B.2).

III. Epidermis

As aptly expressed by Clarke et al. (1979), "The epidermal cells of young roots occupy the key position at the root-soil interface. Not only are they the initial site of water and ion uptake but they are also (together with the root cap) the site of secretion of materials from root to rhizosphere." The root epidermis has a multitude of functions but, for the purpose of this chapter, we will focus on the occurrence and significance of suberin in this layer and on the viability of its cells.

A. Early Development and Permeability

The epidermis develops from the protoderm, a uniseriate layer of cells near the root apical meristem (Cutter 1971). In general, walls of very young root cells have a low permeability to solutes such as ions and tracer dyes (Lüttge and Weigl 1962; Rasmussen 1968; Enstone and Peterson 1992). It has been proposed that this low permeability is due to a lack of intermicrofibrillar spaces (through which dissolved substances move) in the wall and that these spaces form as the walls elongate, increasing their permeability (Enstone and Peterson 1992). However, even near the root tip, permeability studies with a fluorescent fabric-brightener dye showed that the outer tangential walls of the epidermal cells in *A. cepa* and *Zea mays* L. are permeable, although the radial and interior walls are not (Peterson and Perumalla 1984). This means that the soil solution can make contact with the plasma membranes

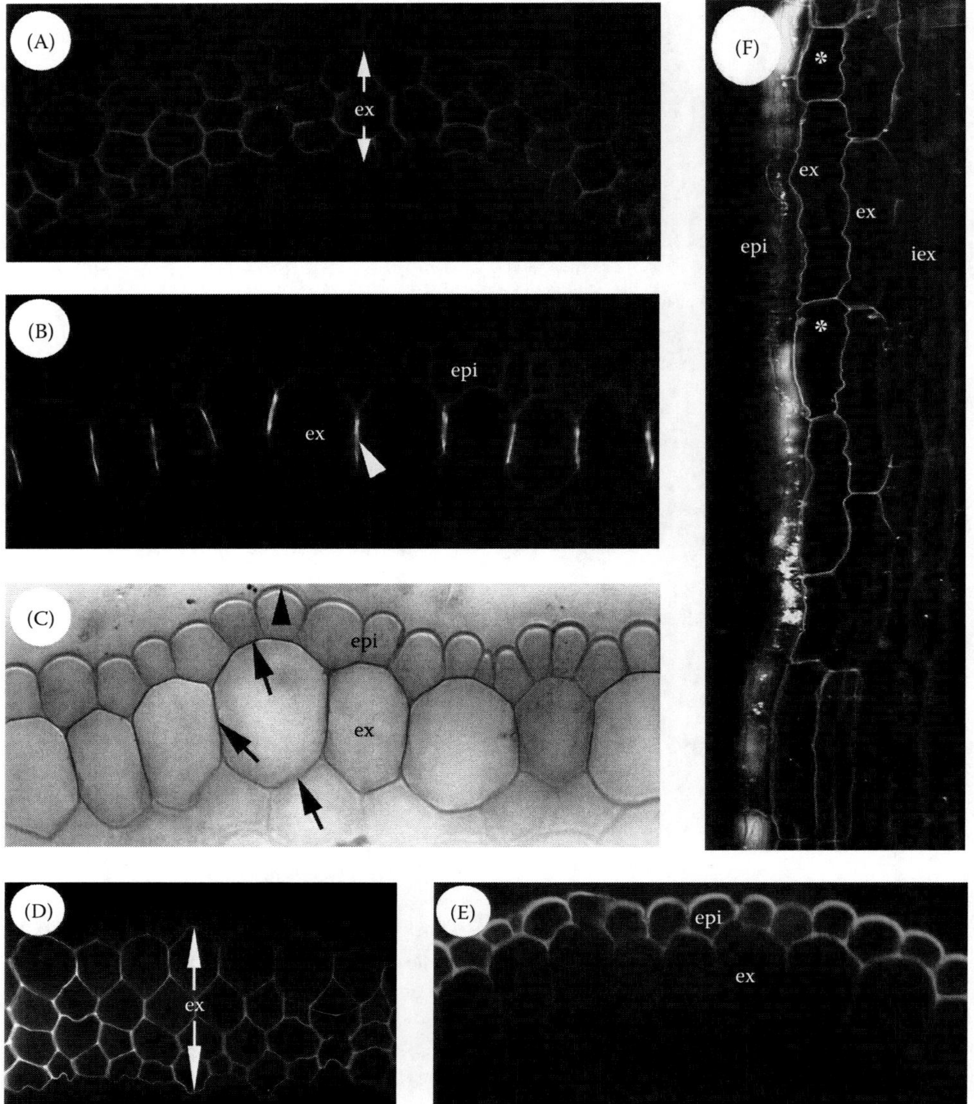

FIGURE 5.5 **(See color insert.)** Photomicrographs of the outer part of roots in (A–E) cross and (F) longitudinal section. (A) Autofluorescence with ultraviolet light. Walls of the MEX of *Iris germanica* are faint blue. (B) Stained with berberine hemisulfate–aniline blue. Casparian bands (white arrowhead) fluoresce yellow and occupy the anticlinal walls of the outermost exodermal layer in *I. germanica*. (C) Stained with Sudan red 7B. Suberin lamellae (black arrows) appear as red outlines in exodermal walls of *Allium cepa*. Also, diffuse suberin in the epidermis stains red (black arrowhead). (D) Stained with fluorol yellow 088. Suberin lamellae fluoresce yellow in exodermal cell walls of *I. germanica*. (E) Exodermal Casparian band prevents externally applied berberine (green fluorescence) from permeating the exodermis and central cortex in *A. cepa*. The walls of the epidermis are permeable to berberine. (F) Epidermal cells containing berberine thiocyanate crystals (yellow) and a dimorphic, biseriate exodermis with blue, autofluorescent walls in *I. germanica*. Asterisks indicate short cells. epi, epidermis; ex, mature exodermis; iex, immature exodermis. (Images (C, E): courtesy of Ishari Waduwara; image (F): Meyer, C.J. et al., *Ann. Bot.*, 103, 687, 2009. Reproduced with permission from Oxford University Press.)

of the epidermal cells in the young root zones, allowing direct uptake of ions at this location. In older zones further from the root tip, the radial and inner tangential walls of the epidermis became permeable to the dye (Peterson and Perumalla 1984).

Epidermal walls may be modified early in development by the presence of phenolic (Clarke et al. 1979) and aliphatic precursors (Wilson and Peterson 1983; Ranathunge et al. 2008). In the case of *G. max* (which lacks an exodermis), the histochemical tests for suberin have been verified by chemical analysis (Thomas et al. 2007). In *A. cepa* and *G. max*, using TEM, the suberin appears as somewhat electron-dense bands within the wall (Figure 5.4C; Peterson et al. 1987; Fang 2006), a formation that is termed "diffuse suberin" (see Section II.C.2). According to a histochemical study of 27 species, 22 tested positive for suberin in their epidermal walls, suggesting that this feature might be commonplace in roots (Wilson and Peterson 1983). A more extensive survey would be necessary to test this idea.

B. Significance of Epidermal Suberin for Root Function

The presence of suberin within all the walls of the epidermis raises the question "How do water and ions permeate these structures?" Tests of roots with an exodermis showed that the epidermis was permeable to ions (ferrous/ferric, De Rufz de Lavison 1910; sulfate, Peterson 1987; phosphate, Waduwara 2007) and apoplastic tracers like fabric-brightener dyes (Peterson et al. 1978, 1982) and berberine (Figure 5.5E; Enstone and Peterson 1992). These same substances were blocked from moving through the anticlinal walls of the exodermis. Such results demonstrate that the walls of the epidermis are permeable to molecules 961 Da in size; thus, water and ions of physiological significance would be able to pass through them unless they were bound by electrical charges.

We are used to thinking of suberin in the form of a lamella, in which there is a continuous sheet of the substance (Sitte 1962), or in the form of Casparian bands where the polymer apparently clogs up the intermicrofibrillar spaces of a specific part of the wall to form an effective barrier to apoplastic dyes and ions (see Steudle and Peterson 1998; Enstone et al. 2003). In the case of the more permeable epidermis, one could imagine that suberin is present but in a lesser density so that some of the intermicrofibrillar spaces remain open or are at least not totally obstructed.

Two initial studies indicate that epidermal suberin plays a key role in the resistance to root rot caused by *Phytophthora* (Thomas et al. 2007; Ranathunge et al. 2008). These studies featured *G. max*, a species with non-exodermal roots and a partial resistance to *Phytophthora sojae* (Perumalla et al. 1990; Schmitthenner 1985). Two contrasting genotypes were compared—Conrad with strong resistance and OX760-6 with weak resistance. According to a quantitative chemical analysis of their epidermal suberin, Conrad had twice as much suberin as did OX760-6 (Thomas et al. 2007). Total root suberin was positively correlated with plant survival in the field for nine varieties and also for 32 recombinant inbred lines between Conrad and OX 760-6. Because it was necessary to depolymerize suberin into its constituent monomers prior to GC-MS analysis, it was possible to refine the correlative analyses to domains (i.e., phenolic and aliphatic), substance classes (i.e., fatty acids, ω-hydroxy fatty acids, dioic fatty acids, and 1-alkanols), and even individual monomers. The domain of suberin that best correlated with strength of the resistance was the lipid. Within this, the greatest correlation was with the ω-hydroxy fatty acid substance class (Thomas et al. 2007). A comparison of the infection process in Conrad and OX760-6 provided insight into the role of the epidermis in *Phytophthora* resistance (Ranathunge et al. 2008). It is known that *G. max* roots are attacked by zoospores of *P. sojae* that are attracted to the youngest area of the root. These zoospores adhere to the root surface and encyst, and the pathogen enters the root by means of a germ tube that grows between the epidermal cells as illustrated by Enkerli et al. (1997). Do the middle lamellae that bind epidermal cells together contain suberin? Apparently the answer is "yes" because after digestion in concentrated sulfuric acid, the epidermal cells remain attached to each other (Figure 5.4A and B; Ranathunge et al. 2008). The crucial difference in infection between Conrad and OX760-6 was that the germ tubes experienced a 2–3 h delay in penetrating the epidermis in Conrad compared with OX760-6. In addition to the amounts of suberin in the epidermal middle lamellae, part of this delay may have been due to differences in suberin-filled intercellular spaces between the epidermis and the cortex. In these young roots, the endodermis had Casparian bands but not suberin lamellae. Nevertheless, compared with OX760-6, the oomycete was delayed in getting though the endodermis of Conrad, in which hyphal coils were seen in the central cortex near the endodermis. This sequence of events indicates that in Conrad, the delay at the epidermis had allowed the root to mount its chemical defenses, ultimately resulting in a less severe case of the disease.

C. Viability

The viability of epidermal cells in roots that develop an exodermis is an important issue. Ion uptake from the soil solution into the symplast has the potential to occur in living cells when the solution infiltrates their walls and comes in contact with their plasma membranes. In non-exodermal roots and young regions of exodermal roots where the exodermis is still immature, this potential for ion uptake extends to the cells of the central cortex and endodermis (see Kamula et al. 1994). However, when the exodermis matures, its cells have Casparian bands, and depending on the exodermal type (see Section IV.B.1), the majority or all of its cells develop suberin lamellae. The exodermis of *Iris germanica* L. represents an extreme case. It is a mixed, multiseriate type, in which all cells develop Casparian bands and suberin lamellae (Figure 5.6D). According to the early microautoradiographic work of Ziegler et al. (1963) and later work with the root pressure probe (Meyer et al. 2011), the exodermis of this species is highly impermeable to both apoplastic and transcellular flow of sulfate and sodium ions. In root regions with such a mature exodermis, the capacity for ion uptake depends on the viability of the epidermal cells and symplastic ion movement.

Whether an epidermal cell is dead or alive is often not readily apparent. Because the walls are thickened and the outer walls, especially in monocots, are arched and mechanically strong, the cells typically do not collapse when they die. Therefore, it is necessary to test for their viability, and there are several potentially suitable methods for this purpose (Stadelmann and Kinzel 1972). As one example, whole or bisected roots can be given a short exposure to uranin (disodium fluorescein) in an acidic buffer. After rinsing off the excess and mounting the root on a slide in a chamber (high enough to avoid damaging the uppermost root surface) filled with the buffer, one can focus on the outermost layer with an epifluorescence microscope. The epidermis can be identified by its position in the focal plane and also by the narrow dimensions of its cells compared with those of the underlying cortical cells. Uranin accumulates in the nuclei and cytoplasms of living but not dead cells (Figure 5.4D). Alternatively, nuclei can be

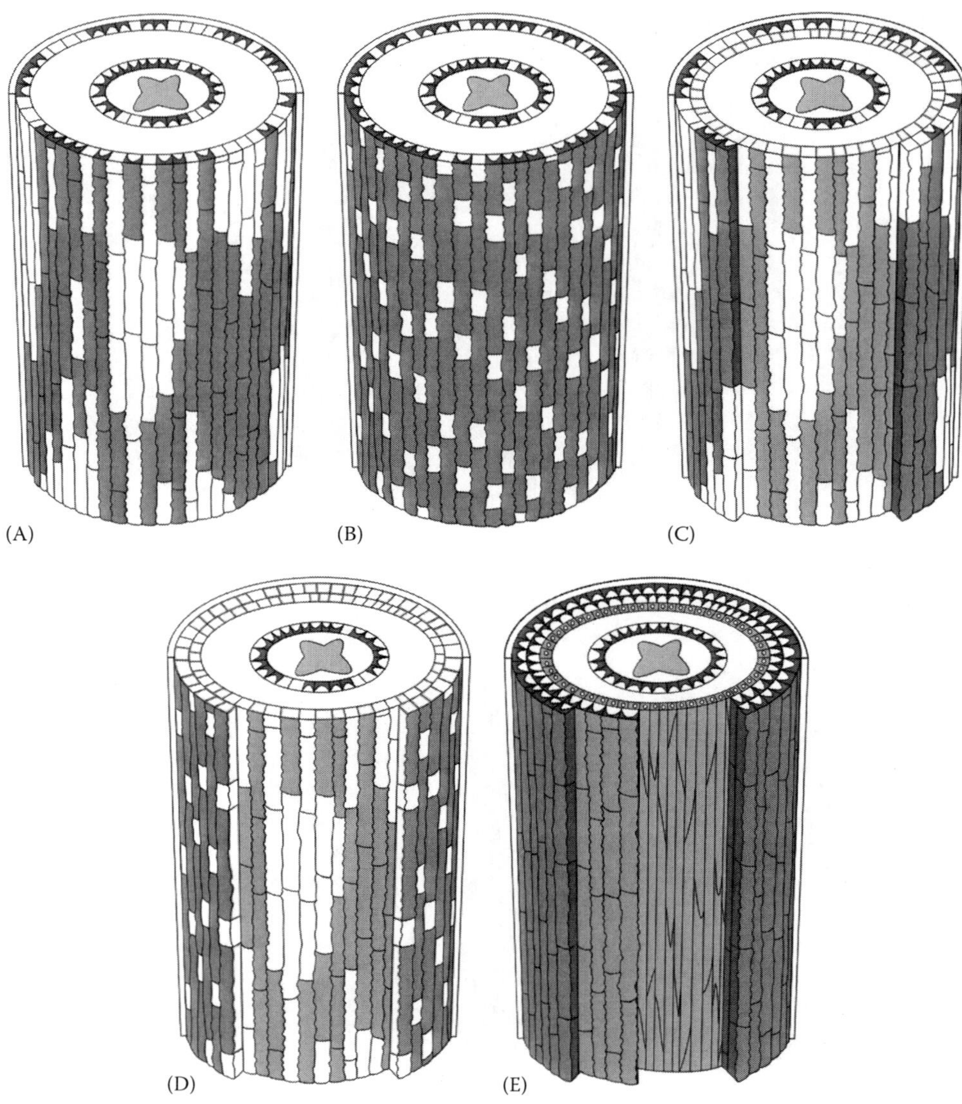

FIGURE 5.6 **(See color insert.)** Illustrations of four exodermal subtypes and one exodermal variation, shown in cross and longitudinal section. (A) Uniseriate, uniform. (B) Uniseriate, dimorphic. (C) Multiseriate, uniform. (D) Multiseriate, dimorphic (or mixed). (E) Reinforced exodermal variation, with sclerenchyma (blue cells). Red shading, suberin lamellae in cells of the first exodermal layer; brownish-red shading, suberin lamellae in cells of the second exodermal layer; deep red shading, tertiary walls. (Reproduced from Peterson, C.A., The exodermis and its interactions with the environment, in *Radical Biology: Advances and Perspectives on the Function of Plant Roots, Proceedings of the 11th Annual Penn State Symposium in Plant Physiology*, May 22–24, 1997, eds. Flores, H.E., Lynch, J.P., Eissenstat, D., American Society of Plant Biologists, Rockville, MD, 1997. With permission.)

stained with propidium iodide and visualized with fluorescence microscopy (Wang et al. 1995).

Does the maturation of the exodermis have a negative effect on the viability of epidermal cells? This question was investigated by Barrowclough and Peterson (1994) in *A. cepa*, a species with a dimorphic exodermis, in which the long cells make suberin lamellae and die a few days later (Ma and Peterson 2000). The short cells, on the other hand, make their lamellae later or not at all (Kroemer 1903; von Guttenberg 1968; Kamula et al. 1994), and the cells without lamellae remain alive. According to Barrowclough and Peterson (1994), the epidermal cells tend to die as the root ages. However, no increase in cell death rate occurs when the underlying exodermis matures. These somewhat counterintuitive results could be explained by looking at the orientations of the epidermal and exodermal cells (Figure 5.7; Barrowclough and Peterson 1994). As many as 98% of the epidermal cells overlaid an exodermal short cell at some point along their length. The few that did not (i.e., that overlaid only a long cell) died early.

It is logical to expect that epidermal cell viability will depend on a symplastic connection with cells of the cortex and stele, since the source of photosynthates for growth and maintenance of the root is delivered by the phloem to the stele. Data on the occurrence of plasmodesmata in the exodermis are limited, but

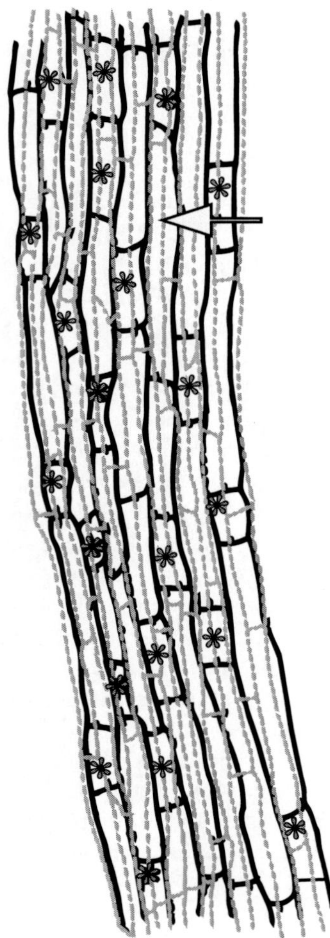

FIGURE 5.7 Camera lucida drawing of a portion of the epidermis (light, dashed lines) and exodermis (dark, solid lines) in *Allium cepa* root in face view. Most epidermal cells are subtended by short exodermal cells (asterisks). It is rare for an epidermal cell (arrow) to lie entirely over only long exodermal cells. (Modified from Barrowclough, D.E. and Peterson, C.A., *Physiol. Plant*, 92, 343, 1994. With permission.)

the few examples that are known indicate a variety of situations. In *A. cepa*, plasmodesmata connect the short cells without suberin lamellae (but not the long cells with suberin lamellae) to the cortical and epidermal cells (Ma and Peterson 2001b, 2003). What would be the case for a uniform exodermis without short cells? In field-grown *Z. mays,* many epidermal cells of the lateral roots remain alive and are connected to the underlying exodermal cells by plasmodesmata (Wang et al. 1995). Earlier, Clarkson et al. (1987) reported plasmodesmata connecting the cytoplasm of the mature exodermis to that of the epidermis and to that of the central cortex in the same species. In *I. germanica*, strong indirect evidence indicates that the plasmodesmata remain intact from the central cortex to the exodermis and epidermis (Meyer et al. 2009). Half of the epidermal cells remained alive after 14 days in a humid air gap (Meyer et al. 2011). In contrast, exposing *Z. mays* roots to a 2 day humid air treatment proved lethal to its epidermis (Enstone and Peterson 1998).

Roots, especially those growing in the field or in stressed conditions, often lose their epidermis and perhaps some adjoining tissues as well. The normal root development of trees (e.g., *Pinus banksiana* Lamb., *Eucalyptus pilularis* Sm., and *Pinus taeda* L.) and some herbaceous plants (e.g., *Hordeum vulgare* L., *Triticum aestivum* L., and *Avena sativa* L.) involves dieback to the endodermis (Wenzel and McCully 1991; McKenzie and Peterson 1995; Enstone et al. 2001). Under water stress, laboratory-grown roots of *H. vulgare* and *Lolium perenne* L. die back to the endodermis (Shone and Flood 1983; Jupp and Newman 1987). Laboratory-grown roots of the exodermal species *Z. mays* die back to the endodermis during drought stress, but in field-grown roots, only the epidermis is sloughed (Stasovski and Peterson 1991; Wenzel and McCully 1991). Laboratory-grown roots of *A. cepa* die back to the exodermis during drought stress (Stasovski and Peterson 1993). In *Oryza sativa* L. roots, both the epidermis and exodermis are sloughed off, leaving a sclerified layer at the root edge (Clark and Harris 1981). The effects of these dramatic anatomical changes on water loss from the root and on water and ion uptake would be fruitful areas of research that have scarcely been touched (see Section IV.D).

IV. Exodermis and Endodermis

A. Three States of Development

The cortex of the root originates from the ground meristem, and the exodermis and endodermis are the outermost and innermost parts of this tissue, respectively. In this chapter, the term "central cortex" will be used to refer to that part of the cortex located between the exodermis and endodermis.

Both the exodermis (i.e., a hypodermis with a Casparian band) and endodermis may pass through three states of maturity as reviewed by Ma and Peterson (2003). The first (primary) developmental state—State I—is reached when Casparian bands are deposited in the intermicrofibrillar spaces of radial and transverse cell walls (Figure 5.1A and B). In exodermal cells, the Casparian band typically fills the majority of the anticlinal wall space (Enstone et al. 2003; Meyer et al. 2009). In contrast, in the endodermis, the band fills only a small fraction of the anticlinal walls (Enstone et al. 2003). At this stage, a tight connection exists between the modified wall and the adjacent plasma membrane (Bonnett 1968; Karahara and Shibaoka 1992).

The second developmental state—State II—is marked by the deposition of a suberin lamella between the cell wall and plasma membrane (Figure 5.1C and D). This lamella severs the tight connection between the Casparian band and plasma membrane (Robards et al. 1973; Haas and Carothers 1975; Ma and Peterson 2001a). Often, mature suberin lamellae are perforated where plasmodesmata are located; hence they stay intact and the cells remain alive. Such perforations have been observed in the endodermal cells of all species so far investigated and are important because the intact plasmodesmata provide a symplastic connection between the pericycle and the innermost cells of the central cortex. In exodermal cells, suberin lamella perforations have been observed but to date only in *O. sativa* and *Z. mays* (Clark and Harris 1981; Clarkson et al. 1987; Wang et al. 1995). Conversely, in the long exodermal cells of *A. cepa*, suberin lamellae sever the plasmodesmata thereby interrupting symplastic transport, and these cells soon die (Ma and Peterson 2000). In addition to the small perforations associated with plasmodesmata, suberin lamellae in the endodermis and exodermis of *Z. mays* have larger gaps (see Section IV.D.1).

Lastly, in the third (tertiary) developmental state—State III—cellulosic wall thickenings that are often later embedded with lignin and sometimes also suberin are deposited along the radial, tangential, and transverse walls (Figure 5.1E and F). These tertiary walls are often U-shaped (Van Fleet 1961; Esau 1965; Clarkson et al. 1987; Zeier and Schreiber 1998). In the endodermis, the thicker wall is typically the innermost tangential, whereas in the exodermis it is the outermost tangential; in this respect, the endodermis and exodermis are mirror images of each other. In both layers, these tertiary walls can be pitted, thus preserving a population of short plasmodesmata connecting the protoplasts of the cells with those of their radially adjacent cells (Wang et al. 1995; Clarkson 1996; Ma and Peterson 2001b). There are species in which tertiary walls do not form in the exodermis. One example is *A. cepa*; its exodermis lacks tertiary walls because the suberin lamellae sever the plasmodesmata leading to death of the cells (Ma and Peterson 2001b). On the other hand, the mature exodermises of *Z. mays*, *Typha* spp., and *I. germanica* do possess tertiary walls (Wang et al. 1995; Seago et al. 1999; Meyer et al. 2009), indicating that the symplastic pathway remained open and that the cells were alive after their suberin lamellae were laid down.

Information regarding the evolution of the exodermis and endodermis is still relatively scarce. Key references on this topic are Damus et al. (1997), Raven and Edwards (2001), Brundrett (2002), and Tomescu (2007).

B. Exodermis

The general term "hypodermis" refers to one or more cell layers situated adjacent to the epidermis of any organ and is composed of cells that differ from those of the internally adjacent layer. In roots, a hypodermis is the outermost part of the cortex. The exodermis is defined as a type of hypodermis in which the cells have Casparian bands (Peterson and Perumalla 1990). This supersedes an earlier definition by von Guttenberg (1968) who referred to the exodermis as a hypodermis with suberin lamellae. Although the exodermis does not originate as the outermost root layer, it often assumes this position following sloughing off of the epidermis (Shishkoff 1986; McCully 1999), making it the root's first contact with soil particles, water, and dissolved minerals.

The exodermis is commonly but not invariably present in angiosperm roots, being present in over 90% of the 200 species surveyed (Perumalla and Peterson 1990; Perumalla et al. 1990). The same survey indicated that its distribution is related to taxonomy. The exodermis is structurally variable, as described and classified by Kroemer (1903).

1. Types of Exodermis

a. Uniseriate

The majority of species with an exodermis have a single-layered (uniseriate) type. Well-studied species include *A. cepa*, *O. sativa*, and *Z. mays* (Perumalla et al. 1990; see Enstone et al. 2003, and references therein). Within this type are two subtypes: (1) A "uniform exodermis" (Einheitliche Interkutis; after Kroemer 1903) consists of cells that are all of similar length, such as those in *Z. mays* (illustrated in Figure 5.6A). In this case, exodermal cell maturation can be irregular resulting in regions containing a combination of immature, unmodified cells along with cells that have reached State II. The former cells lack Casparian bands and, therefore, do not fit the definition of passage cells, i.e., cells with Casparian bands but no suberin lamellae (Peterson and Enstone 1996; Enstone and Peterson 1997); and (2) A "dimorphic exodermis" (Kurzzellen-Interkutis; after Kroemer 1903) has cells of two distinct lengths, i.e., short and long cells, such as found in *A. cepa* (Figure 5.6B). In this case, the short cells are passage cells because in mature root regions they possess Casparian bands but have delayed suberin lamella deposition in comparison with the long cells (von Guttenberg 1968; Kamula et al. 1994; Ma and Peterson 2001a). Typical Casparian bands are deposited only in the anticlinal walls, while the suberin lamellae are encrusted on the inner surfaces of all the walls.

b. Multiseriate

An additional type of exodermis is the multilayered (multiseriate) exodermis (MEX), which refers to an exodermis consisting of two or more cell layers. Development of a MEX occurs in approximately 18% of tested angiosperms with an exodermis (Meyer et al. 2009). The MEX develops centripetally from periclinal cell divisions of the outermost ground meristem layer (Seago and Marsh 1989; Peterson and Perumalla 1990; Seago et al. 1999). Species with a MEX include *I. germanica* (Kroemer 1903; Shishkoff 1986; Peterson and Perumalla 1990; Zeier and Schreiber 1998; Meyer et al. 2009), *Typha* spp. (Seago and Marsh 1989; Seago et al. 1999), and *Phragmites australis* (Cav.) Trin. ex Steud. (Armstrong et al. 2000; Soukup et al. 2002). Within the multiseriate type of exodermis, there are two subtypes: (1) a "uniform MEX" (Einheitliche mehrschichtigen Interkutis; after Kroemer 1903) in which the cells in all layers have similar lengths (Figure 5.6C) (examples of species that develop a uniform MEX are *Typha* spp. and *P. australis*), and (2) a "mixed MEX" (Gemischte mehrschichtigen Interkutis; after Kroemer 1903) in which the outermost exodermal layer is dimorphic, but all underlying layers are uniform (Figure 5.6D). To date, the mixed MEX has been observed in 14 species of various genera but all within the order Asparagales. One such species is *I. germanica* (Figure 5.5F; Kroemer 1903; Shishkoff 1986; Meyer et al. 2009).

An unusual Casparian band that occupies both the anticlinal and tangential cell walls develops in the MEX. This atypical structure forms a continuous layer around the root circumference and was termed a "continuous circumferential Casparian band" by Meyer et al. (2009) in a study of *I. germanica*. This type of Casparian band has also been observed in the MEX of *Typha* spp. (Seago and Marsh 1989; Seago et al. 1999) and *P. australis* (Soukup et al. 2002, 2007).

c. *Reinforced Exodermis*

One notable variation on the exodermal types listed earlier is the "reinforced exodermis" (verstärkte Interkutis, as described by Kroemer [1903]). A reinforced exodermis refers to a uniseriate or MEX with an underlying layer of sclerenchyma (Figure 5.6E). Such a structure occurs in the uniseriate exodermis of *O. sativa* roots (Ranathunge et al. 2003). It is more appropriate to define the reinforced exodermis as a variation rather than a type or subtype because the sclerenchyma layer lacks Casparian bands and thus is not a true part of the exodermis. The sclerenchyma layer is included as part of the hypodermis, which includes the exodermis.

2. Effects of Abiotic Stresses on Exodermal Development

Although exodermal development is constitutive in many angiosperms, the timing and rate of development can be altered when plants are exposed to differing environmental conditions. In particular, alterations occur in the regulation of when and how quickly the exodermis and its wall-modifying structures are synthesized or modified. In one set of examples, Clarkson et al. (1987) and Enstone and Peterson (1998) exposed the basal parts of *Z. mays* roots to humid air inside hydroponic chambers. Within the region of humid air, perhaps a combination of the lower water potential and/or increased capacity for gas exchange compared with completely submerged roots led to acceleration in exodermal suberin lamellae deposition. A similar reaction was observed for MEX maturation in *I. germanica* roots that were exposed to humid air (Meyer et al. 2009). In addition, acceleration in exodermal suberization occurred when *Z. mays* was grown in aeroponics, in vermiculite, or in a stagnant (oxygen-deficient) solution, compared with aerated nutrient solution (Zimmermann and Steudle 1998; Enstone and Peterson 2005).

Alternatively, the rate of exodermal maturation may not change, but root growth rate may decline. This is a common occurrence in roots exposed to drought or salt stress and results in a greater proportion of the root surface area being suberized (see Enstone et al. 2003, and references therein). Interestingly, in *A. cepa* adventitious roots that were drought stressed for 200 days, the short cells of the dimorphic exodermis remained alive, whereas the long exodermal cells and epidermis had died (Stasovski and Peterson 1993). These viable short exodermal cells maintained a symplastic continuity between the exodermis and central cortex, that was in turn symplastically connected to the endodermis and stele.

When roots are exposed to salt stress, often the abundance of suberin-associated fatty acids in the exodermis increases and in special cases an exodermis develops. For example, Krishnamurthy et al. (2009) detected significantly greater amounts of ω-hydroxy fatty acids in the exodermis of a salt-tolerant cultivar of *O. sativa* when exposed to salt stress (50–100 mM NaCl) for 7 days, compared with non-stressed plants. Also, Schreiber et al. (2005a), using *Ricinus communis* L. roots, measured increased SPAD deposition in the exodermis of NaCl-stressed roots (100 mM, 30 days) compared with their non-stressed counterparts. One notable but rare reaction is the initiation of exodermal development in response to salt stress. This was observed by Reinhardt and Rost (1995) who exposed non-exodermal roots of *Gossypium hirsutum* L. to 200 mM NaCl for 5 days. The development of an exodermis coupled with an increased density of monomers in the Casparian bands and suberin lamellae could enhance the restriction of Na^+ flow by clogging more of the intermicrofibrillar spaces, by blocking access to plasma membranes, or possibly by severing plasmodesmata. This enhanced restriction to Na^+ flow would be beneficial for roots of *Z. mays* or *O. sativa* because the uniseriate exodermises of these species are not nearly as resistant to Na^+ flow as the MEX of *I. germanica* (Steudle et al. 1993; Ranathunge et al. 2003; Meyer et al. 2011).

3. Molecular Mechanisms That May Regulate Exodermal Development

Recent studies of root radial patterning have begun to reveal the mechanisms behind the development of an endodermis in angiosperms. Interestingly, Cui et al. (2007) have demonstrated the importance of a direct interaction between scarecrow (SCR) and short-root (SHR) proteins for the development of a uniseriate endodermis in *Arabidopsis thaliana* (L.) Heynh. roots. SHR had previously been shown to induce an asymmetric cell division of cortical initials and was necessary for endodermal specification (Helariutta et al. 2000). In subsequent work, Cui et al. (2007) reported that SCR sequestered SHR in endodermal nuclei, preventing SHR from flowing into the central cortex. Furthermore, the SHR-SCR protein complex drives the transcription of *SCR* genes to ensure a steady and large supply of SCR for the interaction with and sequestration of SHR (i.e., a positive feedback loop). When SCR production was reduced in RNA interference mutants, SHR was not sequestered and supernumerary endodermal layers were formed. Therefore, the interaction of these proteins prevented the differentiation of additional endodermal layers (Cui et al. 2007). It is possible that similar mechanisms are involved in the development of uniseriate and multiseriate exodermal layers. Perhaps an SHR-like protein is generated in the protoderm cells (that later form the epidermis), which then interacts with an SCR-like protein in immature exodermal cells. Species with roots that have a uniseriate exodermis may well have an SCR-SHR developmental mechanism similar to that observed for the endodermis. For roots with a MEX, either a lack of SCR-like protein or an excess of SHR-like protein could induce additional periclinal cell divisions in underlying ground meristem cells. In roots that lack an exodermis, such as *A. thaliana* and *G. max*, the genes expressing SCR-like and SHR-like proteins may be absent or repressed. In the future, it would be interesting to elucidate the molecular mechanisms that regulate exodermal development.

C. Endodermis

The structure and significance of the endodermis has long been recognized (see reviews by Enstone et al. 2003; Ma and Peterson 2003). Casparian bands develop near the root tip, usually in a small area of the anticlinal walls of the cells. This may be in the center of the wall or displaced to either the inside or outside (see Wu et al. 2005; Meyer et al. 2009). In *A. cepa* and *Z. mays*, the Casparian bands enlarge to fill almost the entirety of the anticlinal walls when suberin lamellae form (Barnabas and Peterson 1992; Enstone and Peterson 1998).

A variety of cellular events are associated with Casparian band development. It is now known that five endodermis-specific, transmembrane proteins are involved in this development (Roppolo et al. 2011). Using young *A. thaliana* roots, Alassimone et al. (2010) and Roppolo et al. (2011) describe the following sequence. (1) Synthesis of "Casparian band membrane domain proteins" (CASPs) and their insertion throughout the entire endodermal plasma membrane; (2) The following events occur over a short time: repositioning of CASP1 to the membranes adjacent to the site of the future Casparian band; formation of a "Casparian band membrane domain" that prevents lateral diffusion of other transmembrane proteins; membrane adhesion to the wall at the site of the Casparian band; and insertion of phenolic compounds into the cell wall, identified by Casparian band autofluorescence; and (3) Lastly, the appearance of a mature Casparian band that blocks the apoplastic movement of propidium iodide. The mechanism whereby the cells produce the Casparian band (and its associated membrane features) in such a specific location remains a mystery, as neither microtubules nor microfilaments are involved (Haas and Carothers 1975; Karahara and Shibaoka 1992; Alassimone et al. 2010). Also still to be discovered are the enzymes involved in the biosynthesis and polymerization of the Casparian band. It has been suggested that the CASPs may form a scaffold to support these enzymes (Roppolo et al. 2011).

In addition to Casparian bands, endodermal cells often develop suberin lamellae. Unlike the exodermis where these two structures usually develop at about the same time, in the endodermis, the lamellae are typically deposited some time after the Casparian bands. There are reports that the suberin lamellae in *A. cepa* possess holes (see Section IV.D.1.a). Frequently, endodermal cells near the xylem poles are delayed in forming suberin lamellae although this pattern is not evident at the outset (Figure 5.8). Unlike lignin, suberin lamellae can stretch. This is evident when roots begin secondary growth and cell divisions of the vascular cambium (originating in the pericycle) cause an initial horizontal stretching of the tangential walls (Weerdenburg and Peterson 1984).

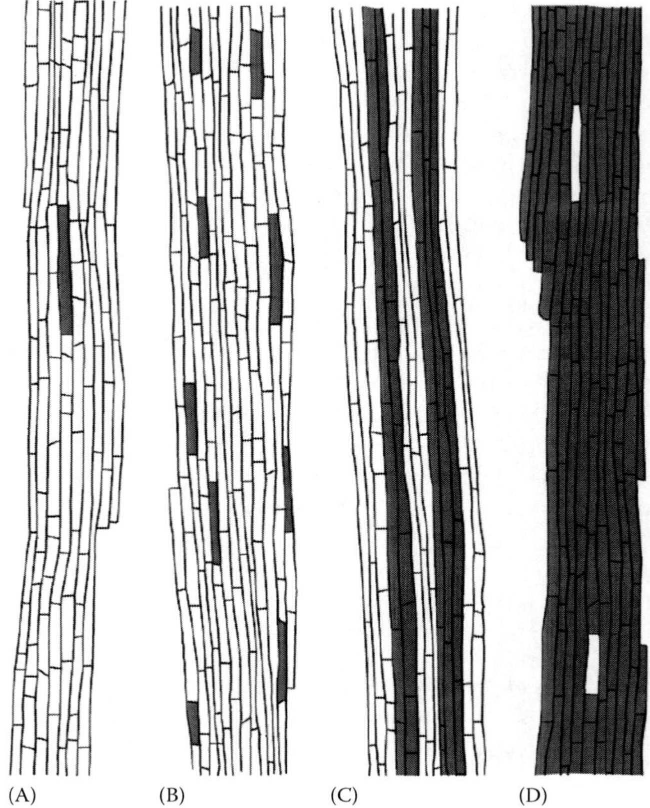

FIGURE 5.8 Drawings from montages of endodermal cells illustrating suberin lamellae deposition in progressively older regions of *Allium cepa* roots. Cells shaded with gray contain suberin lamellae. Distances from the root tip: (A) 5–35 mm, (B) 80–110 mm, (C) 125–155 mm, and (D) 255–285 mm. (From Waduwara, C.I. et al., *Can. J. Bot.*, 86, 623, 2008. Reproduced with permission from the NRC Research Press.)

D. Functions of the Exodermis and Endodermis: A Brief Overview

The following subsections cover the importance of the exodermis and endodermis for regulating ion and water transport, tolerating drought and salt stress, preventing radial oxygen loss, and resisting microorganism entry. For a recent review of these and additional exodermal and endodermal functions, see Ranathunge et al. (2010).

1. Radial Movement of Ions and Water

Water and ions move radially across roots from the soil solution to the tracheary elements of the xylem, and this movement can occur through three pathways (Figure 5.9). These are (1) the apoplastic path (through the cell walls, specifically through water-filled, intermicrofibrillar channels with diameters ranging from 5 to 30 nm (Nobel 2005), and lumens of dead cells), (2) the symplastic path (through cytoplasms of cells and their connecting plasmodesmata [functional diameter for transport about 2.5 nm (Evert 2006)]), and (3) the transcellular path (through cell walls, cytoplasms, and vacuoles). The symplastic and transcellular pathways are often collectively referred to as the cell-to-cell

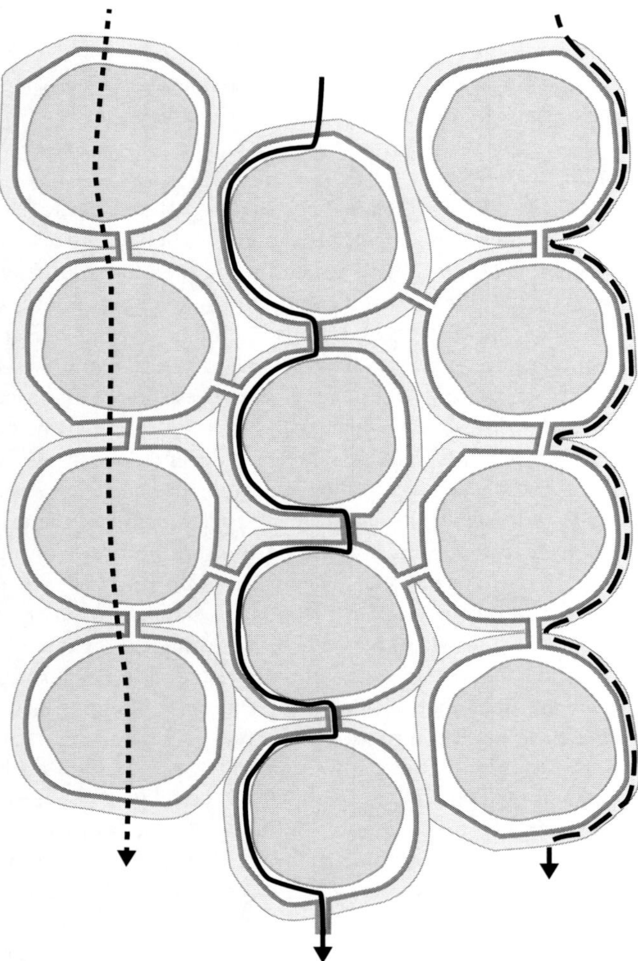

FIGURE 5.9 Illustration of the three parallel, radial transport pathways across unmodified parenchyma cells. The dotted, downward-directed line on the left side of the illustration indicates transcellular transport across cell walls, cytoplasms, and vacuoles. The solid, downward-directed line in the middle of the illustration depicts symplastic transport in cytoplasms of neighboring cells connected by plasmodesmata. The dashed line on the right side displays apoplastic transport in cell walls. Cell shading: outer light gray regions, cell walls; white regions internal to the walls, cytoplasms; gray regions internal to the cytoplasms, vacuoles. (Modified from Steudle, E. and Peterson, C.A., *J. Exp. Bot.*, 49, 775, 1998. With permission.)

pathway because, for water flow, it is not possible to differentiate the extent to which each is being used.

a. Ion Movement

In the absence of wall-modifying structures and uptake into the cytoplasm, the radial transport of ions occurs through the apoplast (see Steudle and Peterson 1998). This pathway is strongly affected by the development of Casparian bands in both the endodermis and exodermis. There is abundant evidence that these bands are virtually impermeable to ions. An initial impression of this can be obtained by observing blockage of apoplastic dye tracer movement, but definitive proof has come from experiments with the ions themselves. For the endodermis, evidence is available from compartmental elution of sodium (Baker 1971) and several studies with iron or heavy metals that can be seen with TEM (see Enstone et al. 2003 and references therein). Exceptions are rare (Ranathunge et al. 2005) and unlikely to be of physiological significance. For the exodermis, the evidence of apoplastic blockage of ion movement comes from compartmental elution studies with sulfate, calcium, and phosphate ions (Peterson 1987; Cholewa and Peterson 2004; Waduwara 2007) and direct observations of iron precipitates (de Rufz de Lavison 1910).

The impermeability of Casparian bands to ions prevents their apoplastic movement into major areas of the roots, thus limiting the locations where ion uptake into the symplast can occur. The endodermis prevents apoplastic ion movement into the stele (Figure 5.2A) so that the living cells of the root external to the Casparian bands can regulate the types of ions and their amounts that could be delivered to the stele. Endodermal Casparian bands play a second important role in delivering ions to the shoot. After passing symplastically past the Casparian bands, ions to be transported in the xylem are moved from the symplast to the apoplast (see Cholewa and Peterson [2004] for calcium and Alassimone et al. [2010] and references therein for boron). Interestingly, with respect to boron transporters, Alassimone et al. (2010) documented a very early polarity in the endodermal plasma membrane. These ions are trapped in the stele by the Casparian band, increasing the efficiency of their movement into the xylem. In contrast to the endodermis, the exodermis prevents apoplastic ion movement into a much larger area, i.e., the central cortex as well as the stele (Figures 5.2B and 5.10). Then ion uptake would be confined to the epidermis and living cells of the exodermis without suberin lamellae. It is astonishing that such a small deposition of suberin can have such a large impact on the area of the root where ion uptake can occur. In adventitious roots grown from *A. cepa* bulbs, maturation of the exodermis reduces the potential absorbing plasma membrane surface area from an initial 91 mm^2/mm root length to 16 mm^2, and death of the epidermis reduces this further to an area between 0.21 and 0.18 mm^2 depending on how many short cells remain without suberin lamellae (Kamula et al. 1994). The consequence of this drastic reduction in absorbing plasma membrane surface area for the uptake of specific ions by the root has, to date, not been tested. Because of the large number of species with a uniseriate, dimorphic exodermis (Peterson and Perumalla 1990; Perumalla et al. 1990; Meyer et al. 2009), such information would be of major importance in understanding root function.

It is interesting to note that the impact of suberin lamellae on ion movement, despite the larger quantity of poly(aliphatic) suberin they contain (Zeier et al. 1999b), is more limited than that of the Casparian band. Since a suberin lamella is located between the primary wall and plasma membrane of an individual cell and not in the wall itself, the lamella is not in a position to block the apoplastic path. Rather, it would impede the movement of substances into the individual cell it surrounds (Figure 5.2C and D). The assumption of suberin lamella impermeability

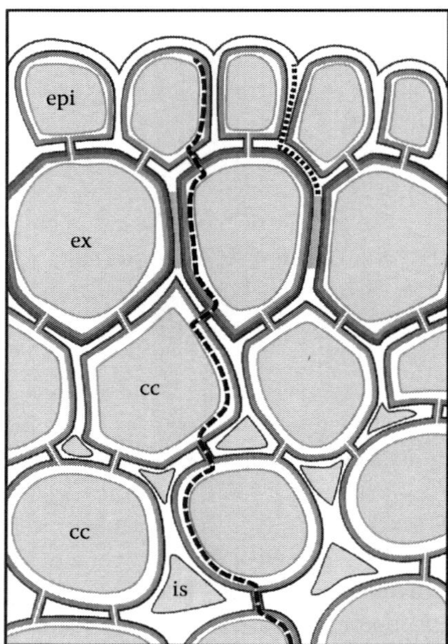

FIGURE 5.10 Drawing of the outer part of a root with a uniseriate exodermis, in cross section. Thicknesses of cell walls, cytoplasms, and plasmodesmata are exaggerated. Cell shading: outermost white regions bordered by thin black lines, cell walls; thick gray lines internal to cell walls, plasma membranes; white regions internal to the plasma membranes, cytoplasms; thin light gray lines internal to cytoplasms, tonoplast membranes; large gray regions internal to the tonoplasts, vacuoles. Wall-modifying structures: dark gray lines between the cell wall and plasma membrane in exodermal cells, suberin lamellae; striated bands in the radial walls of exodermal cells, Casparian bands. Transport pathways: long dashed line, symplastic transport through plasmodesmata; short dotted line, apoplastic transport blocked by a Casparian band. Abbreviations: epi, epidermis; ex, exodermis; cc, central cortex; is, intercellular space.

to ions rests on scant evidence (Evert et al. 1985). According to a TEM study by Wang et al. (1995), the suberin lamellae of *Z. mays* endodermis and exodermis have discontinuities ranging from 0.1 to 0.7 μm. Later in a study using confocal laser scanning microscopy to observe isolated *A. cepa* endodermis, Waduwara et al. (2008) noted various-sized pores (Figure 5.3D; average diameter 1.5 μm, average density 5 pores per 100 μm^2 suberin lamella surface area). In *Z. mays*, these pores sometimes occur in primary pit fields (Wang et al. 1995). However, in *H. vulgare*, the suberin lamellae were continuous across pit fields (Robards et al. 1973). Currently it is not clear if the pores are normally present, or if they are consistently associated with any endodermal features. The extent of their contribution to functions such as water and ion transport is also not known. Clearly, we have much to learn about the structure and function of suberin lamellae.

The earlier discussion presupposes that the roots are uninjured and perfectly healthy. This, however, may not be the case in nature—or in laboratory hydroponic experiments where flexing of lateral roots during solution changes, rinsing, etc. can open up apoplastic bypasses at the junctions of main and lateral roots (Moon et al. 1986; Peterson and Moon 1993). It is also possible that the permeability of apparent Casparian bands can vary. For example, "Casparian bands" in the endodermis of *Pinus bungeana* Zucc. ex Endl. needles are permeable to berberine, whereas those of the root are not (Wu et al. 2005).

b. Water Movement

Radial water flow in plants is usually powered by tension (suction) brought about by transpiration. Although radial water movement through roots is generally assumed to be diffusive, there is some evidence that it can occur by bulk flow in the apoplast under conditions of high transpiration (Aloni and Peterson 1998). Since water flow follows Ohm's law, the pathway(s) it takes across the root will depend on the resistances and areas of the paths. Thus, some water will move in all three paths but in varying amounts (Figure 5.9; see Steudle and Peterson 1998): (1) Apoplastic path. Unmodified cell walls present a low resistance to water flow. However, the Casparian bands of the exodermis and endodermis are in a position to limit radial water transport since the polymer is deposited in the intermicrofibrillar channels through which the water is moving. It has been impossible to quantify the water flow through the Casparian bands for technical reasons; (2) Symplastic path. Because the combined functional diameters of the openings in plasmodesmata are small compared to the areas of the walls or of the plasma membranes, this pathway should be minimally used by water; and (3) Transcellular path. Of all the cellular structures, membranes present the greatest resistance to water movement. However, aquaporins in the membranes reduce their resistance and allow water to flow more easily through this path. The resistance of this pathway is variable, depending on the presence and gating of the aquaporins (see Maurel et al. 2008). For this reason, it has proved difficult to determine the fractions of water moving in the various pathways. When suberin lamellae are present, they are in a position to hinder flow through the transcellular pathway (Figure 5.10). (It is sometimes assumed that the suberin lamellae interrupt the radial apoplastic pathway [e.g., Lux et al. 2011], but this is incorrect.) Perhaps the porous nature of the suberin lamella (Wang et al. 1995; Waduwara et al. 2008) coupled with the presence of aquaporins in the membranes of the mature exodermis and endodermis (Hachez et al. 2006) could partially compensate for this apparent sealing effect.

2. Drought and Salt Resistance

Root structure plays a critical role in resisting abiotic stresses such as drought and salt. Specifically, increased suberization of the exodermis and endodermis is important (see Sections III.C and IV.B.2); however, the involvement of these structures has been largely ignored in recent reviews of these subjects. An example of extreme drought tolerance has been thoroughly described for roots of *Agave deserti* Engelm. by North and Nobel (1998, 2000). Prolonged drought (45 days) accelerated endodermal maturation in *A. deserti* root regions more than 40 mm from the shoot base. In fact, most main roots of *A. deserti* remain alive after 180 days of drought (perhaps due to the stored water from the succulent shoot). Then when these roots were rewetted, new laterals were initiated and even the tip region of the main roots started elongating again (observed 7–11 days after rewetting).

Under favorable conditions when roots are growing rapidly, the exodermis matures much later than the endodermis, so there is a distal region where the cortical cells have access to ions from the soil. However, under unfavorable conditions, root growth slows down or stops and the exodermis matures closer to the tip (in the most extreme cases under the root cap; Perumalla and Peterson 1986). In this way, environmental conditions have a major impact on reducing the ion-absorbing areas while increasing protection against water loss.

3. Resistance to Oxygen Loss in Submerged Roots

It is well known that a mature exodermis is very restrictive to radial oxygen loss (ROL) from the aerenchyma to the rhizosphere (see Colmer 2003). This is particularly important for plants growing in submerged, oxygen-limiting substrates. Oxygen that moves into the root system through aerenchyma is used for respiratory processes. However, oxygen can diffuse out of the root into a medium with low oxygen. The exodermis functions as the primary barrier to reduce this oxygen loss. In fact, when roots are grown in stagnant conditions, the mature exodermis reduces ROL by more than 90% as measured in *P. australis* (Armstrong et al. 2000; Soukup et al. 2007), *Hordeum marinum* Huds. (Garthwaite et al. 2008), and *O. sativa* (Kotula et al. 2009). This reduction is positively correlated with a significant increase in the amount of aliphatic suberin monomers in the exodermis developed under stagnant conditions compared to the amount developed under aerated conditions. Considering the whole root, growth under stagnant conditions reduces the root growth rate so that more of it possesses a mature exodermis (Kotula et al. 2009). It has also been suggested that stagnant conditions lead to a more effective sealing around lateral roots (Soukup et al. 2007).

A fascinating trait in the primary roots of some species is the presence of so-called "windows" in the mature exodermis. Exodermal windows are essentially groups of cells in mature zones of the root that do not contain suberin lamellae and are located where lateral roots will emerge; it is unknown whether or not these cells contain Casparian bands. It is not known how ubiquitous these windows are in the angiosperms, but they have been observed in the roots of species that live at least part of their life in submerged conditions, including *P. australis* and *O. sativa* (Soukup et al. 2002; Armstrong and Armstrong 2005). The exodermal windows in aquatic roots provide low resistance pathways for radial water influx but are also susceptible to ROL (Armstrong et al. 2000).

4. Control of Microorganism Entry

The composition and location of the poly(phenolic) component of suberin lamellae make them important for limiting the penetration of pathogenic bacteria, fungi, and oomycetes into roots (Kolattukudy 1980, 1984; Lulai and Corsini 1998). This domain is thought to be linked covalently to primary cell wall carbohydrates (Bernards 2002). On the other hand, the aliphatic component of diffuse epidermal suberin is the most effective in preventing *Phytophthora sojae* (oomycete) invasion (Thomas et al. 2007; see Section III.B).

The efficacy of a mature exodermis in resisting the penetration of pathogenic and mycorrhizal fungal hyphae into the central cortex is known. For example, in *H. vulgare* roots where the exodermis is mature, the pathogen *Chaetomium globosum* Kunze is confined to the epidermis (Reissinger et al. 2003). Similarly, ectomycorrhizal fungi cannot penetrate the mature exodermis, but these specialized fungi instead develop an association with the epidermis (Massicotte et al. 1987). However, in the case of a dimorphic exodermis, pathogenic and vesicular–arbuscular mycorrhizal fungi enter the root through the passage cells (Hussey 1982; Peterson 1992; Kamula et al. 1995; Matsubara et al. 1999; Sharda and Koide 2008; Yu et al. 2001).

V. Concluding Remarks

The authors hope that readers of this chapter will be encouraged to include correlative anatomical studies with their investigations of physiology and/or molecular biology. The diversity of plants and their responses to environmental conditions makes it necessary to analyze the specific material of interest. The work of Lukas Schreiber and his associates (University of Bonn) that combines whole-plant physiology, anatomy, biochemistry and molecular biology is an example of a remarkable combination.

Since rather few plant biologists are being specifically trained in plant anatomy and related techniques, the first impediment for many researchers is a need for directions on how to proceed. The good news is that much anatomical information can be gained quickly using simple techniques. For example, the polychromatic stain toluidine blue O can be applied directly to fresh sections, making it possible to obtain specimens within minutes (O'Brien et al. 1964). Directions for this and many other simple and useful techniques can be found in Peterson et al. (2008). Most of the procedures involve making freehand sections of the material. Although this sectioning is a rapid method, it can be a challenge in the case of many roots because of their small diameters. However, it is possible to section these using supporting material or, if the roots are too small, a vibrating, freezing microtome.

Contrary to a somewhat general belief, not everything is known about plant anatomy. The following are some outstanding questions related to the information in this chapter.

1. Is it generally true that roots have suberin in their epidermis, and if so, is it effective against diseases other than *Phytophthora* root rot?
2. Is it generally true that suberin lamellae have pores, and if so, what are the consequences for ion and water uptake?
3. The tertiary walls of the endodermis and exodermis are lignified and usually have an arched shape, both of which suggest that they may have a mechanical role. Do these layers resist root shrinkage during drought?
4. How are the suberin constituents in the Casparian band organized? It is not obvious why these bands are so effective in preventing the passage of ions when they contain a minor amount of aliphatic suberin.

5. How is the Casparian band deposited in such a specific region of the walls?
6. What is the chemical makeup of wound suberin? Does it resemble that of Casparian bands?
7. How do the various types of exodermis and/or epidermal death affect ion and water uptake?

Indeed there is much still to be learned about root structure and function.

Acknowledgments

The authors thank Alice Fang, Dr. Ishari Waduwara, and Dr. Raymond Thomas for contributing some photomicrographs; Jesseline Gough-Meyer for graphic illustrations; and Dr. Mark Bernards for helpful discussions concerning suberin chemistry.

References

Alassimone J, Naseer S, Geldner N. 2010. A developmental framework for endodermal differentiation and polarity. *Proc Natl Acad Sci USA* 107:5214–5219.

Alassimone J, Roppolo D, Geldner N, Vermeer JEM. 2012. The endodermis – development and differentiation of the plant's inner skin. *Protoplasma* 249:433–443.

Aloni R, Peterson CA. 1998. Indirect evidence for bulk water flow in root cortical cell walls of three dicotyledonous species. *Planta* 207:1–7.

Armstrong J, Armstrong W. 2005. Rice: Sulfide-induced barriers to root radial oxygen loss, Fe^{2+} and water uptake, and lateral root emergence. *Ann Bot* 96:625–638.

Armstrong W, Cousins D, Armstrong J, Turner DW, Beckett PM. 2000. Oxygen distribution in wetland plant roots and permeability barriers to gas-exchange with the rhizosphere: A microelectrode and modeling study with *Phragmites australis*. *Ann Bot* 86:687–703.

Baker DA. 1971. Barriers to the radial diffusion of ions in maize roots. *Planta* 98:285–293.

Barnabas AD, Peterson CA. 1992. Development of Casparian bands and suberin lamellae in the endodermis of onion roots. *Can J Bot* 70:2233–2237.

Barrowclough DE, Peterson CA. 1994. Effects of growing conditions and development of the underlying exodermis on the vitality of the onion root epidermis. *Physiol Plant* 92:343–349.

Bernards MA. 2002. Demystifying suberin. *Can J Bot* 80:227–240.

Bernards MA, Lewis NG. 1998. The macromolecular aromatic domain in suberized tissue: A changing paradigm. *Phytochemistry* 47:915–933.

Bonnett Jr HT. 1968. The root endodermis: Fine structure and function. *J Cell Biol* 37:199–205.

Borg-Olivier O, Monties B. 1993. Lignin, suberin, phenolic acids and tyramine in the suberized, wound-induced potato periderm. *Phytochemistry* 32:601–606.

Brundrett MC. 2002. Coevolution of roots and mycorrhizas of land plants. *New Phytol* 154:275–304.

Brundrett MC, Enstone DE, Peterson CA. 1988. A berberine–aniline blue fluorescent staining procedure for suberin, lignin and callose in plant tissue. *Protoplasma* 146:133–142.

Brundrett MC, Kendrick B, Peterson CA. 1991. Efficient lipid staining in plant material with Sudan red 7B or Fluorol yellow 088 in polyethylene glycol–glycerol. *Biotech Histochem* 66:111–116.

Bryant AE. 1934. A demonstration of the connection of the protoplasts of the endodermal cells with the Casparian strips in the roots of barley. *New Phytol* 33:230–231.

Cholewa E, Peterson CA. 2004. Evidence for symplastic involvement in the radial movement of calcium in onion roots. *Plant Physiology* 134:1793–1802.

Clark LH, Harris WH. 1981. Observations on the root anatomy of rice (*Oryza sativa* L.). *Am J Bot* 68:154–161.

Clarke KJ, McCully ME, Miki NK. 1979. A developmental study of the epidermis of young roots of *Zea mays* L. *Protoplasma* 98:283–309.

Clarkson DT. 1996. Root structure and sites of ion uptake. In: *Plant Roots: The Hidden Half*, eds. Y Waisel, A Eshel, U Kafkafi, 2nd edn., pp. 483–510. New York: Marcel Dekker, Inc.

Clarkson DT, Robards AW, Stephens JE, Stark M. 1987. Suberin lamellae in the hypodermis of maize (*Zea mays*) roots; development and factors affecting the permeability of hypodermal layers. *Plant Cell Environ* 10:83–93.

Colmer TD. 2003. Long-distance transport of gases in plants: A perspective on internal aeration and radial oxygen loss from roots. *Plant Cell Environ* 26:17–36.

Cui H, Levesque MP, Vernoux T, Jung JW, Paquette AJ, Gallagher KL, Wang JY, Blilou I, Scheres B, Benfey PN. 2007. An evolutionarily conserved mechanism delimiting SHR movement defines a single layer of endodermis in plants. *Science* 316:421–425.

Cutter EG. 1971. *Plant Anatomy: Experiment and Interpretation. Part 2 Organs*. London, U.K.: Edward Arnold.

Damus M, Peterson RL, Enstone DE, Peterson CA. 1997. Modifications of cortical cell walls in roots of seedless vascular plants. *Bot Acta* 110:190–195.

De Rufz de Lavison J. 1910. Du mode de pénétration de quelques sels dans la plante vivante. Role de l'endoderme. *Rev gén Bot* 22:225–241.

Enkerli K, Hahn MG, Mims CW. 1997. Ultrastructure of compatible and incompatible interactions of soybean roots infected with the plant pathogenic oomycete *Phytophthora sojae*. *Can J Bot* 75:1493–1508.

Enstone DE, Peterson CA. 1992. The apoplastic permeability of root apices. *Can J Bot* 70:1502–1512.

Enstone DE, Peterson CA. 1997. Suberin deposition and band plasmolysis in the corn (*Zea mays* L.) root exodermis. *Can J Bot* 75:1188–1199.

Enstone DE, Peterson CA. 1998. Effects of exposure to humid air on epidermal viability and suberin deposition in maize (*Zea mays* L.) roots. *Plant Cell Environ* 21:837–844.

Enstone DE, Peterson CA. 2005. Suberin lamella development in maize seedling roots grown in aerated and stagnant conditions. *Plant Cell Environ* 28:444–455.

Enstone DE, Peterson CA, Hallgren SW. 2001. Anatomy of seedling tap roots of loblolly pine (*Pinus taeda* L.). *Trees Struct Funct* 15:98–111.

Enstone DE, Peterson CA, Ma F. 2003. Root endodermis and exodermis: Structure, function, and responses to the environment. *J Plant Gr Regul* 21:335–351.

Esau K. 1965. *Plant Anatomy*, 2nd edn. New York: John Wiley & Sons.

Evert RF. 2006. *Esau's Plant Anatomy*, 3rd edn. Hoboken, NJ: John Wiley & Sons.

Evert RF, Botha CEJ, Mierzwa RJ. 1985. Free-space marker studies on the leaf of *Zea mays* L. *Protoplasma* 126:62–73.

Fang X. 2006. Chemical composition of soybean root epidermal cell walls. MSc thesis. Department of Biology, University of Waterloo, Waterloo, Ontario, Canada.

Garthwaite AJ, Armstrong W, Colmer TD. 2008. Assessment of O_2 diffusivity across the barrier to radial O_2 loss in adventitious roots of *Hordeum marinum*. *New Phytol* 179:405–416.

Graça J, Pereira H. 2000a. Methanolysis of bark suberins: Analysis of glycerol and acid monomers. *Phytochem Anal* 11:45–51.

Graça J, Pereira H. 2000b. Suberin structure in potato periderm: Glycerol, long-chain monomers, and glyceryl and feruloyl dimers. *J Agr Food Cem* 48:5476–5483.

von Guttenberg, H. 1968. Der primäre Bau der Angiospermenwurzel. In: *Handbuch der Pflanzenanatomie*, ed. K Linsbauer, Vol. 8, pp. 141–159. Berlin, Germany: Gebrüder Borntraeger.

Haas DL, Carothers ZB. 1975. Some ultrastructural observations on endodermal cell development in *Zea mays* roots. *Am J Bot* 62:336–348.

Hachez C, Moshelion M, Zelazny E, Cavez D, Chaumont F. 2006. Localization and quantification of plasma membrane aquaporin expression in maize primary root: A clue to understanding their role as cellular plumbers. *Plant Mol Biol* 62:305–323.

Helariutta Y, Fukaki H, Wysocka-Diller JW et al. 2000. The SHORT-ROOT gene controls radial patterning of the *Arabidopsis* root through radial signaling. *Cell* 101:555–567.

Holloway P. 1983. Some variations in the composition of suberin from the cork layers of higher plants. Phytochemistry 22:495–502.

Hose E, Clarkson DT, Steudle E, Schreiber L, Hartung W. 2001. The exodermis—A variable apoplastic barrier. *J Exp Bot* 52:2245–2264.

Hussey RB. 1982. Interactions between VA mycorrhizal fungi and *Asparagus officinalis* L. roots. MSc thesis. Department of Botany, University of Guelph, Guelph, Ontario, Canada.

Jensen WA. 1962. *Botanical Histochemistry—Principles and Practice*. San Francisco, CA: W.H. Freeman.

Johansen DA. 1940. *Plant Microtechnique*. New York: McGraw-Hill.

Jupp AP, Newman EI. 1987. Morphological and anatomical effects of severe drought on the roots of *Lolium perenne* L. *New Phytol* 105:393–402.

Kamula SA, Peterson CA, Mayfield CI. 1994. The plasmalemma surface area exposed to the soil solution is markedly reduced by maturation of the exodermis and death of the epidermis in onion roots. *Plant Cell Environ* 17:1183–1193.

Karahara I, Shibaoka H. 1992. Isolation of Casparian strips from pea roots. *Plant Cell Physiol* 33:555–561.

Kolattukudy PE. 1980. Biopolyester membranes of plants: Cutin and suberin. *Science* 208:990–1000.

Kolattukudy PE. 1984. Biochemistry and function of cutin and suberin. *Can J Bot* 62:2918–2933.

Kotula L, Ranathunge K, Schreiber L, Steudle E. 2009. Functional and chemical comparison of apoplastic barriers to radial oxygen loss in roots of rice (*Oryza sativa* L.) grown in aerated or deoxygenated solution. *J Exp Bot* 60:2155–2167.

Krishnamurthy P, Ranathunge K, Franke R, Prakash HS, Schreiber L, Mathew MK. 2009. The role of root apoplastic transport barriers in salt tolerance of rice (*Oryza sativa* L.). *Planta* 230:119–134.

Kroemer K. 1903. Wurzelhaut, Hypodermis und Endodermis der Angiospermenwurzel. *Bibliotheca Botanica* 59:1–148.

Lopes MH, Gil AM, Silvestre ADJ, Neto CP. 2000a. Composition of suberin extracted upon gradual methanolysis of *Quercus suber* L. cork. *J Agr Food Cem* 48:383–391.

Lopes MH, Sarychev A, Neto CP, Gil AM. 2000b. Spectra editing of 13C CP/MAS NMR spectra of complex systems: Application to the structural characterization of cork cell walls. *Solid State NMR* 16:109–121.

Lulai E, Corsini D. 1998. Differential deposition of suberin phenolic and aliphatic domains and their roles in resistance to infection during potato tuber (*Solanum tuberosum* L.) wound-healing. *Physiol Mol Plant Pathol* 53:209–222.

Lulai EC, Morgan WC. 1992. Histochemical probing of potato periderm with neutral red: A sensitive cytofluorochrome for the hydrophobic domain of suberin. *Biotech Histochem* 67:185–195.

Lüttge U, Weigl J. 1962. Microautoradiographische Untersuchungen der Aufnahme und des Transportes von $^{35}SO_4^{--}$ und $^{45}Ca^{++}$ in Keimwurzeln von *Zea mays* und *Pisum sativum* L. *Planta* 58:113–126.

Lux A, Martinka M, Vaculik M, White PJ. 2011. Root responses to cadmium in the rhizosphere: A review. *J Exp Bot* 62:21–37.

Lux A, Morita S, Abe J, Ito K. 2005. An improved method for clearing and staining free-hand sections and whole-mount samples. *Ann Bot* 96:989–996.

Ma F, Peterson CA. 2000. Plasmodesmata in onion (*Allium cepa* L.) roots: A study enabled by improved fixation and embedding techniques. *Protoplasma* 211:103–115.

Ma F, Peterson CA. 2001a. Development of cell wall modifications in the endodermis and exodermis of *Allium cepa* roots. *Can J Bot* 79:621–634.

Ma F, Peterson CA. 2001b. Frequencies of plasmodesmata in *Allium cepa* L. roots: Implications for solute transport pathways. *J Exp Bot* 52:1051–1061.

Ma F, Peterson CA. 2003. Current insights into the development, structure, and chemistry of the endodermis and exodermis of roots. *Can J Bot* 81:405–421.

Massicotte HB, Peterson RL, Ackerley CA, Ashford AE. 1987. Ontogeny of *Eucalyptus pilularis–Pisolithus tinctorius* ectomycorrhizae. II. Transmission electron microscopy. *Can J Bot* 65:1940–1947.

Matsubara Y, Uetake Y, Peterson RL. 1999. Entry and colonization of *Asparagus officinalis* roots by arbuscular mycorrhizal fungi with emphasis on changes in host microtubules. *Can J Bot* 77:1159–1167.

Mattinen M, Filpponen I, Järvinen R, Li B, Kallio H, Lehtinen P, Argyropoulos D. 2009. Structure of the polyphenolic component of suberin isolated from potato (*Solanum tuberosum* var. Nikola). *J Agr Food Cem* 57:9747–9753.

Matzke K, Riederer M. 1991. A comparative study into the chemical constitution of cutins and suberins from *Picea abies* (L.) Karst., *Quercus robur* L., and *Fagus sylvatica* L. *Planta* 185:233–245.

McCully ME. 1995. How do real roots work? Some new views of root structure. *Plant Physiol* 109:1–6.

McCully ME. 1999. Roots in soil: Unearthing the complexities of roots and their rhizospheres. *Annu Rev Plant Physiol Plant Mol Biol* 50:695–718.

McKenzie BEM, Peterson CA. 1995. Root browning in *Pinus banksiana* Lamb. and *Eucalyptus pilularis* Sm. 1. Anatomy and permeability of the white and tannin zones. *Bot Acta* 108:127–137.

Meyer CJ, Peterson CA. 2011. Casparian bands occur in the periderm of *Pelargonium hortorum* stem and root. *Ann Bot* 107:591–598.

Meyer CJ, Peterson CA, Steudle E. 2011. Permeability of *Iris germanica*'s multiseriate exodermis to water, NaCl and ethanol. *J Exp Bot* 62:1911–1926.

Meyer CJ, Seago JL, Peterson CA. 2009. Environmental effects on the maturation of the endodermis and multiseriate exodermis of *Iris germanica* roots. *Ann Bot* 103:687–702.

Moon GJ, Clough BF, Peterson CA, Allaway WG. 1986. Apoplastic and symplastic pathways in *Avicennia marina* (Forsk.) Vierh. roots revealed by fluorescent tracer dyes. *Aust J Plant Physiol* 13:637–648.

Moon GJ, Peterson CA, Peterson RL. 1984. Structural, chemical, and permeability changes following wounding in onion roots. *Can J Bot* 62:2253–2259.

Naseer S, Lee Y, Lapierre C, Franke R, Nawrath C, and Geldner N. 2012. Casparian strip diffusion barrier in Arabidopsis is made of a lignin polymer without suberin. *Proc Nat Acad Sci* USA 109:10101–10106.

Negrel J, Pollet B, Lapierre C. 1996. Ether-linked ferulic acid amides in natural and wound periderms of potato tuber. *Phytochemistry* 43:1195–1199.

Nobel PS. 2005. *Physicochemical and Environmental Plant Physiology*, 3rd edn. San Diego, CA: Elsevier Academic Press.

North GB, Nobel PS. 1998. Water uptake and structural plasticity along roots of a desert succulent during prolonged drought. *Plant Cell Environ* 21:705–713.

North GB, Nobel PS. 2000. Heterogeneity in water availability alters cellular development and hydraulic conductivity along roots of a desert succulent. *Ann Bot* 85:247–255.

O'Brien TP, Feder N, McCully ME. 1964. Polychromatic staining of plant cell walls by Toluidine Blue O. *Protoplasma* 59:368–373.

Perumalla CJ, Peterson CA. 1986. Deposition of Casparian bands and suberin lamellae in the exodermis and endodermis of young corn and onion roots. *Can J Bot* 64:1873–1878.

Perumalla CJ, Peterson CA, Enstone DE. 1990. A survey of angiosperm species to detect hypodermal Casparian bands. I. Roots with a uniseriate hypodermis and epidermis. *Bot J Linn Soc* 103:93–112.

Peterson CA. 1987. The exodermal Casparian band of onion blocks apoplastic movement of sulphate ions. *J Exp Bot* 32:2068–2081.

Peterson RL. 1992. Adaptations of root structure in relation to biotic and abiotic factors. *Can J Bot* 70:661–675.

Peterson CA. 1997. The exodermis and its interactions with the environment. In: *Radical Biology: Advances and Perspectives on the Function of Plant Roots. Proceedings of the 11th Annual Penn State Symposium in Plant Physiology*, May 22–24, 1997, eds. HE Flores, JP Lynch, D Eissenstat, pp. 131–138. Rockville, MD: American Society of Plant Biologists.

Peterson CA, Cholewa E. 1998. Structural modifications in the apoplast and their potential impact on ion uptake. *Zeitsch Pflanzen Boden* 161:521–531.

Peterson CA, Emanuel ME, Wilson C. 1982. Identification of a Casparian band in the hypodermis of onion and corn roots. *Can J Bot* 60:1529–1535.

Peterson CA, Enstone DE. 1996. Functions of passage cells in the endodermis and exodermis of roots. *Physiol Plant* 97:592–598.

Peterson CA, Moon GJ. 1993. The effect of lateral root outgrowth on the structure and permeability of the onion root exodermis. *Bot Acta* 106:411–418.

Peterson CA, Murrmann M, Steudle E. 1993. Location of the major barriers to water and ion movement in young roots of *Zea mays* L. *Planta* 190:127–136.

Peterson CA, Perumalla CJ. 1984. Development of the hypodermal Casparian band in corn and onion roots. *J Exp Bot* 35:51–57.

Peterson CA, Perumalla CJ. 1990. A survey of angiosperm species to detect hypodermal Casparian bands. II. Roots with a multiseriate hypodermis or epidermis. *Bot J Linn Soc* 103:113–125.

Peterson RL, Peterson CA, Melville LH. 2008. *Teaching Plant Anatomy through Creative Laboratory Exercises*. Ottawa, Ontario, Canada: NRC Press.

Peterson CA, Peterson RL, Robards AW. 1978. A correlated histochemical and ultrastructural study of the epidermis and hypodermis of onion roots. *Protoplasma* 96:1–21.

Ranathunge K, Schreiber L, Franke R. 2010. Suberin research in the genomics era – new interest for an old polymer. *Plant Sci* 180:399–413.

Ranathunge K, Steudle E, Lafitte R. 2003. Control of water uptake by rice (*Oryza sativa* L.): Role of the outer part of the root. *Planta* 217:193–205.

Ranathunge K, Steudle E, Lafitte R. 2005. A new precipitation technique provides evidence for the permeability of Casparian bands to ions in young roots of corn (*Zea mays* L.) and rice (*Oryza sativa* L.). *Plant Cell Environ* 28:1450–1462.

Ranathunge K, Thomas RH, Fang X, Peterson CA, Gijzen M, Bernards MA. 2008. Soybean root suberin and partial resistance to root rot caused by *Phytophthora sojae*. *Phytopathology* 98:1179–1189.

Rasmussen HP. 1968. Entry and distribution of aluminum in *Zea mays*. *Planta* 81:28–37.

Raven JA, Edwards D. 2001. Roots: Evolutionary origins and biogeochemical significance. *J Exp Bot* 52:381–401.

Raven PH, Evert RF, Eichhorn SE. 1999. *Biology of Plants*, 6th edn. New York: WH Freeman and Company.

Reinhardt DH, Rost TL. 1995. Salinity accelerates endodermal development and induces an exodermis in cotton seedling roots. *Environ Exp Bot* 35:563–574.

Reissinger A, Winter S, Steckelbroeck S, Hartung W, Sikora RA. 2003. Infection of barley roots by *Chaetomium globosum*: Evidence for a protective role of the exodermis. *Mycological Res* 107:1094–1102.

Robards AW, Jackson SM, Clarkson DT, Sanderson J. 1973. The structure of barley roots in relation to the transport of ions into the stele. *Protoplasma* 77:291–311.

Roppolo D, Geldner N. 2012. Membrane and walls: who is master, who is servant? *Curr Opin Plant Biol* doi:10.1016/j.pbi.2012.09.009.

Roppolo D, De Rybel B, Tendon VD, Pfister A, Alassimone J, Vermeer JE, Yamazaki M, Stierhof YD, Beeckman T, Geldner N. 2011. A novel protein family mediates Casparian strip formation in the endodermis. *Nature* 473:380–383.

Schmitthenner AF. 1985. Problems and progress in control of *Phytophthora* root rot of soybean. *Plant Dis* 69:362–368.

Schreiber L. 1996. Chemical composition of Casparian strips isolated from *Clivia miniata* Reg. roots: Evidence for lignin. *Planta* 199:596–601.

Schreiber L, Breiner H, Riederer M, Düggelin M, Guggenheim R. 1994. The Casparian strip of *Clivia miniata* Reg. roots: Isolation, fine structure and chemical nature. *Bot Acta* 107:353–361.

Schreiber L, Franke R, Hartmann K. 2005a. Effects of NO_3 deficiency and NaCl stress on suberin deposition in rhizo- and hypodermal (RHCW) and endodermal cell walls (ECW) of castor bean (*Ricinus communis* L.) roots. *Plant Soil* 269:333–339.

Schreiber L, Franke R, Hartmann K. 2005b. Wax and suberin development of native and wound periderm of potato (*Solanum tuberosum* L.) and its relation to peridermal transpiration. *Planta* 220:520–530.

Seago JL Jr, Marsh LC. 1989. Adventitious root development in *Typha glauca*, with emphasis on the cortex. *Am J Bot* 76:909–923.

Seago JL Jr, Peterson CA, Enstone DE, Scholey CA. 1999. Development of the endodermis and hypodermis of *Typha glauca* Godr. and *Typha angustifolia* L. roots. *Can J Bot* 77:122–134.

Sharda JN, Koide RT. 2008. Can hypodermal passage cell distribution limit root penetration by mycorrhizal fungi? *New Phytol* 180:696–701.

Shishkoff N. 1986. The dimorphic hypodermis of plant roots: Its distribution in the angiosperms, staining properties, and interaction with root-invading fungi. PhD thesis, Cornell University, Ithaca, NY.

Shone MGT, Flood AV. 1983. Effects of periods of localized water stress on subsequent nutrient uptake by barley roots and their adaptation by osmotic adjustment. *New Phytol* 94:561–572.

Sitte P. 1962. Zum Feinbau der Suberinschichten im Flaschenkork. *Protoplasma* 54:555–559.

Soliday CL, Kolattukudy PE, Davis RW. 1979. Chemical and ultrastructural evidence that waxes associated with the suberin polymer constitute the major diffusion barrier to water vapor in potato tuber (*Solanum tuberosum* L.). *Planta* 146:607–614.

Soukup A, Armstrong W, Schreiber L, Franke R, Votrubová O. 2007. Apoplastic barriers to radial oxygen loss and solute penetration: A chemical and functional comparison of the exodermis of two wetland species, *Phragmites australis* and *Glyceria maxima*. *New Phytol* 173:264–278.

Soukup A, Votrubová O, Čížková H. 2002. Development of anatomical structure of roots of *Phragmites australis*. *New Phytol* 153:277–287.

Stadelmann EJ, Kinzel H. 1972. Vital staining of plant cells. In: *Methods in Cell Physiology*, ed. DM Prescott, Vol. 5, pp. 325–372. New York: Academic Press.

Stasovski E, Peterson CA. 1991. The effects of drought and subsequent rehydration on the structure and vitality of *Zea mays* seedling roots. *Can J Bot* 69:1170–1178.

Stasovski E, Peterson CA. 1993. Effects of drought and subsequent rehydration on the structure, viability, and permeability of *Allium cepa* adventitious roots. *Can J Bot* 71:700–707.

Stark RE, Garbow JR. 1992. Nuclear magnetic resonance relaxation studies of plant polyester dynamics. 2. Suberized potato cell wall. *Macromolecules* 25:149–154.

Steudle E, Murrmann M, Peterson CA. 1993. Transport of water and solutes across maize roots modified by puncturing the endodermis. Further evidence for the composite transport model of the root. *Plant Physiol* 103:335–349.

Steudle E, Peterson CA. 1998. How does water get through roots? *J Exp Bot* 49:775–788.

Thomas R, Fang X, Ranathunge K, Anderson TR, Peterson CA, Bernards MA. 2007. Soybean root suberin: Anatomical distribution, chemical composition, and relationship to partial resistance to *Phytophthora sojae*. *Plant Physiol* 144:299–311.

Tomescu AMF. 2007. The endodermis: A horsetail's tale. *New Phytol* 177:291–295.

Van Fleet DS. 1961. Histochemistry and function of the endodermis. *Bot Rev* 27:165–221.

Vogt E, Schönherr J, Schmidt HW. 1983. Water permeability of periderm membranes isolated enzymatically from potato tubers (*Solanum tuberosum* L.). *Planta* 158:294–301.

Waduwara CI. 2007. Onion root anatomy and the uptake of sulphate and phosphate ions. MSc thesis. Department of Biology, University of Waterloo, Waterloo, Ontario, Canada.

Waduwara CI, Walcott SE, Peterson CA. 2008. Suberin lamellae of the onion root endodermis: Their pattern of development and continuity. *Can J Bot* 86:623–632.

Waisel Y, Eshel A. 2002. Functional diversity of various constituents of a single root system. In: *Plant Roots: The Hidden Half*, eds. Y Waisel, A Eshel, U Kafkafi, 3rd edn., pp. 157–174. New York: Marcel Dekker, Inc.

Wang XL, McCully ME, Canny MJ. 1995. Branch roots of *Zea* V: Structural features that may influence water and nutrient transport. *Bot Acta* 108:209–219.

Watt M, vanderWelle CM, McCully ME, Canny MJ. 1996. Effects of local variations in soil moisture on hydrophobic deposits and dye diffusion in corn roots. *Bot Acta* 109:492–501.

Weerdenburg CA, Peterson CA. 1984. Effect of secondary growth on the conformation and permeability of the endodermis of broad bean (*Vicia faba*), sunflower (*Helianthus annuus*) and garden balsam (*Impatiens balsamina*). *Can J Bot* 62:907–910.

Wenzel CL, McCully ME. 1991. Early senescence of cortical cells in the roots of cereals. How good is the evidence? *Am J Bot* 78:1528–1541.

Wilson CA, Peterson CA. 1983. Chemical composition of the epidermal, hypodermal, endodermal and intervening cortical cells walls of various plant roots. *Ann Bot* 51:759–769.

Wu X, Lin J, Lin Q, Wang J, Schreiber L. 2005. Casparian strips in needles are more solute permeable than endodermal transport barriers in roots of *Pinus bungeana*. *Plant Cell Physiol* 46:1799–1808.

Yu T, Nassuth A, Peterson RL. 2001. Characterization of the interaction between the dark septate fungus *Phialocephala fortinii* and *Asparagus officinalis* roots. *Can J Microbiol* 47:741–753.

Zeier J, Goll A, Yokoyama M, Karahara I, Schreiber L. 1999a. Structure and chemical composition of endodermal and rhizodermal/hypodermal walls of several species. *Plant Cell Environ* 22:271–279.

Zeier J, Ruel K, Ryser U, Schreiber L. 1999b. Chemical analysis and immunolocalisation of lignin and suberin in endodermal and hypodermal/rhizodermal cell walls of developing maize (*Zea mays* L.) primary roots. *Planta* 209:1–12.

Zeier J, Schreiber L. 1997. Chemical composition of hypodermal and endodermal cell walls and xylem vessels isolated from *Clivia miniata*: Identification of the biopolymers lignin and suberin. *Plant Physiol* 113:1223–1231.

Zeier J, Schreiber L. 1998. Comparative investigation of primary and tertiary endodermal cell walls isolated from the roots of five monocotyledoneous species: Chemical composition in relation to fine structure. *Planta* 206:349–361.

Zeier J, Schreiber L. 1999. Fourier transform infrared-spectroscopic characterisation of isolated endodermal cell walls from plant roots: Chemical nature in relation to anatomical development. *Planta* 209:537–542.

Ziegler H, Weigl J, Lüttge U. 1963. Mikroautoradiographischer Nachweis der Wanderung von $^{35}SO_4^-$ durch die Tertiärendodermis der Iris-Wurzel. *Protoplasma* 56:362–370.

Zimmermann HM, Steudle E. 1998. Apoplastic transport across young maize roots: Effect of the exodermis. *Planta* 206:7–19.

6

Lateral Root Development

Leentje Jansen
Flanders Institute for Biotechnology
Ghent University

Marlies Demeulenaere
Flanders Institute for Biotechnology
Ghent University

Tom Beeckman
Flanders Institute for Biotechnology
Ghent University

I. Introduction

Distribution and availability of water and nutrients in the soil vary with time and space (Hodge 2006). The adaptation of a plant's root architecture by the initiation and elongation of lateral roots is an important strategy used by plants, being sessile organisms, to cope with the changing nutritional conditions (Malamy 2005; Hodge et al. 2009). The de novo formation of lateral roots allows the plant to explore the soil and to proliferate into nutrient-rich patches. Alternatively, plants can save energy by inhibiting elongation of lateral roots in regions where water and nutrients are scarce, while maintaining primary root growth to reach underground resources (Deak and Malamy 2005; Malamy 2005). When competing with other plants for the acquisition of immobile ions as phosphate, extensive branching offers great advantage (Fitter et al. 2002).

Lateral root development starts with the division of the founder cells, a developmental step that is known as "lateral root initiation." A clear definition and description of lateral root initiation can be found in the concerned chapters of the previous editions of this book (Charlton 1996; Lloret and Casero 2002). We refer the reader to these excellent and still pertinent reviews that include many references to lateral root initiation in a variety of plant species. In the third edition, Lloret and Casero (2002) already highlighted an ongoing evolution in the methodology used by the root developmental research community characterized by a gradual increase in the number of studies making use of molecular, genetic, and novel microscopy techniques that are mainly conducted in the model species *Arabidopsis thaliana*. During the last decade, this trend has only intensified, and genomic approaches such as large-scaled transcriptome analyses have provided the root community with a wealth of new information for future research. Today, we are faced with a significant progress in our understanding of the molecular mechanisms that are involved in root branching. In this context, we will focus particularly on these new insights dealing with intrinsic signaling pathways involved in lateral root development, mainly in *Arabidopsis*, but where possible a link with other species will be made.

A. Short Guide to Lateral Root Development

As part of the post-embryonic root system, lateral roots are initiated endogenously from cells within the parent root. These so-called founder cells undergo a series of highly organized divisions, leading to the formation of a new primordium (De Smet et al. 2006a). Founder cells are located at specific positions relative to the vascular tissues of the parent root. In *Arabidopsis*, lateral roots are formed from pericycle cells adjacent to the protoxylem poles (Dolan et al. 1993; Malamy and Benfey 1997b). This is also the case for onion, radish, and sunflower (Casero et al. 1995; Lloret and Casero 2002), but in carrot, where the pericycle consists of two layers, lateral roots are formed at the phloem poles (Lloret and Casero 2002). In maize and rice, both the pericycle and endodermis participate in the formation of the new primordium, which is initiated between protoxylem vessels, adjacent to phloem poles (Casero et al. 1995; Rebouillat et al. 2009; Jansen et al. 2012). In ferns, lateral root initiation differs from that of angiosperms in that it takes place in the endodermis (Hou et al. 2004). In the diarch vascular system of *Arabidopsis*, lateral roots are formed in two longitudinal rows, alternating between the two xylem poles (Dubrovsky et al. 2006). When grown on vertical agar plates in laboratory conditions, roots display a wavy growth pattern with the lateral roots forming at the convex side of the bends (Figure 6.1A; De Smet et al. 2007). Most plants have however a poly-arch vascular symmetry, and

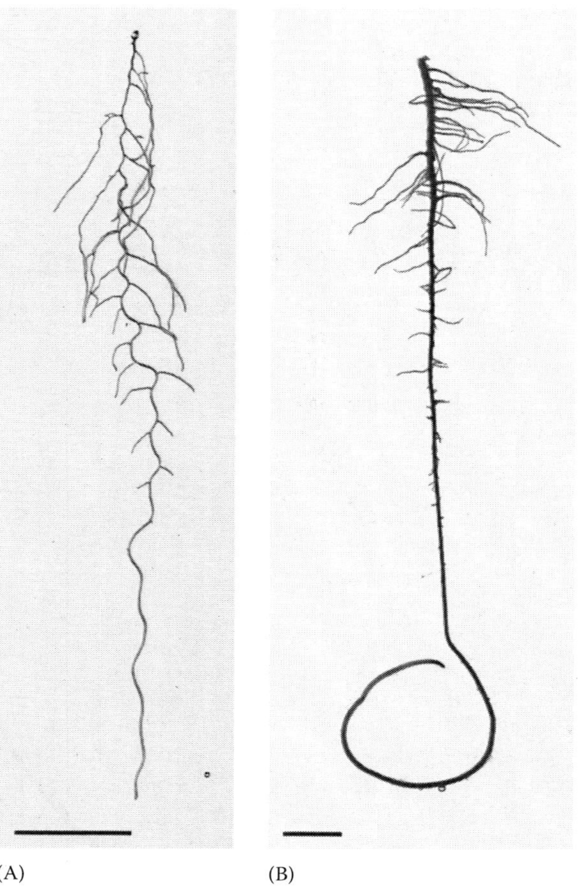

FIGURE 6.1 Lateral root growth in seedlings of *Arabidopsis thaliana* (A) and *Zea mays* (B). Bar = 1 cm.

although some regularity in the spacing of lateral roots within and between ranks has been demonstrated in certain plant species, the distribution of lateral roots in poly-arch root systems is clearly more complex compared to *Arabidopsis* (Figure 6.1B; Charlton 1987; Draye et al. 1999; Draye 2002).

Long before the first divisions become visible in the differentiation zone, pre-initiation events take place in the basal meristem, a transition zone at the root tip where cell division diminishes and cell size increases (Beemster et al. 2003; De Smet et al. 2007, 2008; De Rybel et al. 2010; Moreno-Risueno et al. 2010). In *Arabidopsis*, discrete xylem-pole pericycle cells become specified as pre-branch sites, a process called "priming" (De Smet et al. 2007; Péret et al. 2009; De Rybel et al. 2010; Moreno-Risueno et al. 2010). The first morphological sign of the initiation of a lateral root, which takes place after the cells reach the differentiation zone, are asymmetric, anticlinal divisions of founder cells (Dubrovsky et al. 2001). Further successive periclinal and anticlinal divisions will eventually give rise to the organized primordium (Laskowski et al. 1995; Sussex et al. 1995; Malamy and Benfey 1997b). While growing, the primordium protrudes through the different cell layers of the parent root, a process described as lateral root emergence (Malamy and Benfey 1997b). Once emerged, the meristem of the new lateral root is activated and further growth occurs through divisions of initial cells in the apical meristem (Malamy and Benfey 1997b). Lateral roots are formed in a clear chronosequence, meaning that the early stages are located closer to the root tip. When the initiation occurs strictly rootward, as is the case in *Arabidopsis* under controlled growth conditions, lateral roots are initiated close to the root tip, and no new lateral roots will be formed between already existing primordia (Dubrovsky et al. 2006). In more complex systems such as the maize root system, the spatial frame in which lateral roots can be initiated seems to be much more extensive than in *Arabidopsis*, which allows initiation in these species to occur in a region where primordia are already developing (MacLeod 1990).

In virtually all aspects of lateral root formation, from priming to the growth of the emerged lateral root, the plant hormone auxin plays an important role. Although other hormones are involved, auxin is by far the best studied when it comes to lateral root development. Auxin produced in the leaves is transported to the root tip through the vascular tissues (reviewed by Overvoorde et al. 2010), where it is laterally redistributed and transported shootward through cortex and epidermis (Swarup and Bennett 2003; Blilou et al. 2005). Besides the shoot, also the root tip and lateral roots are sites of auxin production (Ljung et al. 2005). A complex equilibrium of auxin biosynthesis and transport creates local auxin gradients needed for correct lateral root formation (Benková et al. 2003; see also Chapter 19). Signaling cascades link auxin distribution to specific developmental processes. Interfering with auxin biosynthesis, transport or response often leads to alterations in lateral root formation: plants overproducing auxin have higher numbers of lateral roots, while mutants impaired in auxin transport or one of the components of the signaling cascade form fewer lateral roots (reviewed by De Smet et al. 2006a). Researchers have taken advantage of these properties of auxin to develop a lateral root inducible system for *Arabidopsis.* Germinating the seeds in the presence of the auxin transport inhibitor N-1-naphthylphthalamic acid (NPA) prevents lateral root initiation, and subsequent transfer to naphthalene acetic acid (NAA) leads to a synchronized induction of a large number of lateral roots (Himanen et al. 2002).

Although regulated by intrinsic mechanisms, external signals have an important impact on the distribution, emergence, and elongation of lateral roots. Not only localization and concentration of water and nutrients, but also factors such as light, gravity, physical barriers, and interaction with other organisms influence root system architecture and thus help adjusting its function to the environmental conditions (Lucas et al. 2007; Ditengou et al. 2008; Nibau et al. 2008; Richter et al. 2009).

II. Birth of a New Organ

A. Pericycle: A Complex Tissue

Although lateral roots originate from the pericycle, not all pericycle cells can be primed for lateral root initiation, suggesting there are differences between individual cells within this tissue layer. Although formerly considered as one homogeneous tissue, there are increasing indications that the pericycle consists of different cell populations, which are closely associated with the vasculature and its differentiation (Parizot et al. 2008, 2010; Atta et al. 2009; Cui et al. 2011).

First, heterogeneity in the pericycle can be distinguished at the anatomical level. Cell width for instance is larger for cells prone to lateral root initiation in *Arabidopsis* (Laskowski et al. 1995) but smaller in onion when compared to other pericycle cells (Casero et al. 1989). There can also be a difference in distribution of plasmodesmata (Wright and Oparka 1997), and the arabinogalactan proteins at the cell surface vary between the different pericycle cell populations (Casero et al. 1998). Pericycle cells competent for lateral root initiation are often also shorter than other cells within the same layer (Casero et al. 1989; Lloret et al. 1989; Laskowski et al. 1995; Dubrovsky et al. 2000). This might indicate they continue dividing after leaving the meristem (Dubrovsky et al. 2000) or at least undergo extra divisions within the meristem (Beeckman et al. 2001).

The meristematic properties are a second way to distinguish pericycle cell populations. As lateral root initiation occurs in the differentiation zone, the founder cells have to resume cell cycle. Therefore, *Arabidopsis* xylem-pole pericycle cells leave the apical meristem in the G2 phase and retain their ability to divide, forming an extended meristem (Dubrovsky et al. 2000; Beeckman et al. 2001; Casimiro et al. 2003). Dense cytoplasm with small vacuoles and a high number of ribosomes, combined with a 4C cell content, point to the meristematic character of xylem-pole pericycle cells, while pericycle cells at the phloem poles have only a 2C cell content, do not show any meristematic activity, and have often only one large, central vacuole (Beeckman et al. 2001; Parizot et al. 2008). Treatments affecting the hormonal homeostasis within the root influence the meristematic activity of the different pericycle cell populations in opposite ways. NPA causes xylem-pole pericycle cells to lose their meristematic features, while cells at the phloem poles form a dense cytoplasm, show loss of the central vacuole, and occasionally even divide, although these divisions remain proliferative and will never lead to lateral root formation (Parizot et al. 2008). Transfer to NAA shifts the situation back to normal in a way that xylem-pole pericycle cells will divide, and this time, divisions are formative, meaning they will eventually form a lateral root primordium (Parizot et al. 2008). This difference between proliferative and formative divisions also appears when regenerating shoots from *Arabidopsis* root explants. Shoot formation takes place at the xylem poles starting from pericycle cells. Although the phloem pole pericycle cells have the competence to divide, they will only produce a file of short cells (Atta et al. 2009).

The pluripotent character of the xylem-associated pericycle cells suggests they are also distinct from other pericycle cells at the genetic level. This hypothesis was confirmed by the identification of two GAL4 enhancer trap lines with robust *GFP* expression in the three pericycle cell files in contact with a xylem pole (Laplaze et al. 2005; Parizot et al. 2008). In the line J0121, expression of *GFP* is turned on in the root elongation zone and disappears at places where a lateral root is forming (Laplaze et al. 2005). In a second enhancer trap line, Rm1007, GFP can be detected as early as the pericycle initials, right above the quiescent center, while it fades away gradually in older parts of the root (Parizot et al. 2008). NPA or NAA treatments have no effect on *GFP* expression in either of these lines, but on the other hand, aberrations in the expression pattern in a mutagenized J0121 line are associated with irregular lateral root phenotypes (Parizot et al. 2008). *GFP* expression in these lines is thus closely linked to their competence to initiate lateral roots. Remarkably, Rm1007 expression is visible in young embryonic stages, suggesting an early distinction between xylem-pole and phloem-pole pericycle cells early during development.

B. Regulation of Lateral Root Spacing

In recent years, different concepts have been proposed concerning the mechanisms which determine the positioning of lateral roots along the primary root. Several authors observed an oscillating auxin signaling maximum in protoxylem strands at the level of the basal meristem, which was linked to lateral root initiation (De Smet et al. 2007; De Rybel et al. 2010; Moreno-Risueno et al. 2010). However, different hypotheses have been published regarding the origin of this recurrent auxin maximum and its exact role in lateral root initiation.

Based on the localization of auxin transport proteins and computer simulations, Laskowski et al. (2008) suggested a self-organizing system, in which small alterations in cell shape, for instance as a consequence of root bending, result in changes in auxin distribution in these cells. These in turn trigger changes in the expression and localization of auxin transport proteins, which result in auxin accumulation in the pericycle cells at the outside of a bend in the differentiation zone (Laskowski et al. 2008). In this context, the authors suggested that the oscillating auxin response maxima observed in the basal meristem are the consequence of previously formed lateral roots, rather than the cause of the induction of a new branch, and that lateral root initiation is not preceded by priming in the basal meristem. The longitudinally spaced left–right positioning of lateral roots could as such be induced solely by the curvature of the main root (Laskowski et al. 2008).

Other authors however favor the concept of pre-branch site formation in the basal meristem (De Smet et al. 2007; Nieuwland et al. 2009; De Rybel et al. 2010; Moreno-Risueno et al. 2010). The auxin response maxima in protoxylem cells result in the priming of neighboring pericycle cells that will initiate a lateral root in more mature zones of the root. Thus, according to this model, priming of pericycle cells and lateral root initiation are separated both in time and space in the growing root (Overvoorde et al. 2010). However, the mechanism by which the auxin signal in the protoxylem cells is translated into pre-branch site formation in the neighboring pericycle cells is still unknown. Also different hypotheses have been proposed regarding the origin of the recurrent auxin response maxima. First, De Smet and colleagues (2007) suggested a link between the oscillating auxin response in the basal meristem and the gravitropic response of the root. Shoot-derived auxin is redistributed in the columella root cap cells according to the gravity vector and transported shootward through the lateral root cap and epidermis cells. At the level of the basal meristem, auxin coming from the root tip is redirected to the stele by differential localization of PIN-FORMED1 and 2 (PIN1 and PIN2) auxin efflux proteins (Blilou et al. 2005; Leyser 2006; Ditengou et al. 2008), which might cause the observed auxin response maxima (De Smet et al. 2007; Lucas et al. 2007, 2008). Lucas et al. (2007) confirmed that root gravitropism and lateral root initiation are co-regulated in *Arabidopsis*. This hypothesis is also in line with reports of other authors describing lateral root initiation to be dependent on auxin derived from the root tip (Casimiro et al. 2001; Bhalerao et al. 2002; Lucas et al. 2007) and could explain the fact that mutants defective in gravitropic response also have disturbed lateral root initiation (De Smet et al. 2007 and references therein). Moreno-Risueno et al. (2010) however propose a different mechanism controlling the periodicity of root branching and bending, based on the activity of an endogenous clock. According to their data, auxin pulses are not sufficient to cause priming. Instead, specification of pre-branch sites would be obtained by the oscillating expression of transcriptional regulators, among which there are several MADS-box transcription factors such as SHATTERPROOF1 and 2 (SHP1 and SHP2), SEEDSTICK (STK), and AGAMOUS-LIKE20 (AGL20) (Moreno-Risueno et al. 2010).

Recent identification of the GATA23 transcription factor and its role in founder cell specification argues in favor of the notion that priming takes place in the basal meristem (De Rybel et al. 2010). Patches of *GATA23* expression, which appear in the xylem-pole pericycle cells just above the basal meristem, are correlated with the oscillating auxin response in the basal meristem as reported by De Smet et al. (2007). GATA23 induces founder cell identity through an IAA28-controlled auxin response pathway (Figure 6.2). Auxin/INDOLE-3-ACETIC ACID (IAA) proteins are inhibitors of AUXIN RESPONSE FACTOR (ARF) transcription factors. Increasing auxin levels trigger the degradation of IAAs and allow the transcription of genes involved in auxin response (reviewed by Leyser 2006). Expression of a stabilized version of IAA28 indeed reduces the number of lateral roots formed (Rogg et al. 2001). Although IAA28 interacts with different ARFs, ARF7 and ARF19 are probably most relevant for the control of founder cell specification (De Rybel et al. 2010). Also CRANE/IAA18 might be involved in this auxin module, as it is expressed in the basal meristem (Ploense et al. 2009), and a gain-of-function mutation results in a reduced number of lateral roots (Uehara et al. 2008).

In addition to the endogenous mechanism that is controlling the regular spacing of lateral roots under controlled growth conditions, lateral roots can also be induced by external factors such as gravistimulation and mechanical bending. Already at the start of the twentieth century, Noll (1900) observed that lateral roots form preferentially at the convex side of a bend. This observation was confirmed over the years

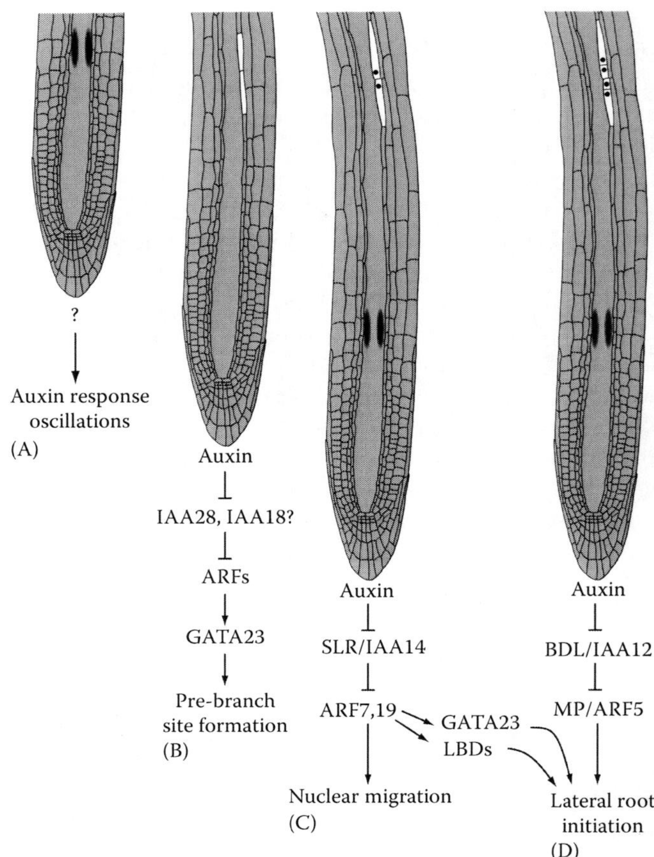

FIGURE 6.2 Specification of lateral root founder cells leading to lateral root initiation. (A) Auxin response oscillations in the protoxylem strands are an inducing factor for founder cell specification (priming). (B) An IAA28-dependent auxin signaling pathway leads to the induction of the transcription factor GATA23 in the adjacent xylem-pole associated pericycle cells, which results in the specification of these cells into lateral root founder cells. (C) The nuclei of the two founder cells from each cell file move to their common cell wall. This process is dependent on the SLR/IAA14-ARF7-ARF19 module. (D) A second module consisting of BDL/IAA12 and MP/ARF5 is activated. Together, these two modules are responsible for the initiation and regulation of the first asymmetric division of the founder cells.

by different studies using various setups (Fortin et al. 1989; Lucas et al. 2007; Ditengou et al. 2008; Richter et al. 2009). Reorientation of primary root growth upon gravistimulation depends on differential redistribution of auxin fluxes in the root tip (Ottenschlager et al. 2003; Swarup et al. 2005). Although an auxin maximum at the lower side of the root tip inhibits cell elongation at the concave side of the bend (Ottenschlager et al. 2003; Swarup et al. 2005), PIN1 relocation in a group of protoxylem cells at the convex side results in an auxin maximum in the protoxylem cells at the outside of the bend, which probably determines the site of lateral root initiation (Ditengou et al. 2008). In case of mechanical bending, founder cell specification is induced by temporary increases in cytoplasmic Ca^{2+} levels in the cell layers at the convex side of the bend, which are probably induced by the stretching of cells (Monshausen et al. 2009; Richter et al. 2009).

Redistribution of auxin fluxes and concentrations are mechanisms recurrently proposed in these different hypotheses to explain the determination of the lateral root initiation sites. It is therefore not surprising that interfering with auxin transport, for instance through addition of NPA, prevents the formation of auxin maxima and as such inhibits lateral root initiation (Casimiro et al. 2001; De Smet et al. 2007).

C. Lateral Root Initiation

As lateral root initiation takes place in the differentiation zone of the root, pericycle cells have to dedifferentiate or maintain their mitotic status once they leave the root apical meristem (Malamy and Benfey 1997a,b). Several reports, based on the expression of cell cycle regulators, point to the last option. First, pericycle cells seem to leave the meristem later than other cell files, and the expression of *CYCLIN-DEPENDENT KINASE AI* (CDKAI) suggests these cells maintain their cell division competence also in older root segments (Beeckman et al. 2001). Further, it was shown that xylem-pole pericycle cells are tetraploid (Beeckman et al. 2001). This was confirmed by the expression pattern of *CYCLIN A2;1* (CYCA2;1) whose transcript level rises during S phase and peaks at the end of G2 (Shaul et al. 1996). In the root, *CYCA2;1* expression disappears above the root apical meristem but reappears later on specifically in pericycle cells at the xylem poles (Burssens et al. 2000; Beeckman et al. 2001). The zone with reduced *CYCA2;1* expression corresponds to a region where all pericycle cells are in G1, while in the zone where *CYCA2;1* expression reappears in the xylem-pole pericycle, cell cycle proceeds toward G2 phase (Shaul et al. 1996; Beeckman et al. 2001). *CYCB1;1* is expressed during the G2 and M phase of the cell cycle (Hemerly et al. 1992; Doerner et al. 1996; Segers et al. 1996; Shaul et al. 1996), and its transcription precedes the completion of the first cell division cycle at the site of lateral root initiation, prior to any visual sign of primordium development (Ferreira et al. 1994; Malamy and Benfey 1997a; Gray et al. 1999; Beeckman et al. 2001). Therefore, *CYCB1;1* expression is commonly used as a marker for lateral root initiation (Himanen et al. 2002, 2004; Vanneste et al. 2005). While cell cycle progression is stimulated at xylem-pole pericycle cells and cell division takes place in founder cells, *KIP-RELATED PROTEIN2* (KRP2) expression blocks the G1-S transition in pericycle cells at the phloem poles in older parts of the roots and in cells opposite lateral root initiation sites, thereby preventing the formation of lateral roots at these sites (Himanen et al. 2002; Verkest et al. 2005).

Auxin plays a role in cell cycle regulation and thus lateral root initiation (Himanen et al. 2002). First, when plants are treated with NPA, the cell cycle is arrested at G1, and *KRP2* is strongly expressed in the pericycle (Himanen et al. 2004). Further, in the *solitary root1* (*slr1*) mutant, which expresses a stabilized form of SLR/IAA14, cell division in the pericycle is blocked (Fukaki et al. 2002; Vanneste et al. 2005), arguing for a direct link between SLR/IAA14-dependent auxin signaling and cell cycle activation (Vanneste et al. 2005). Nevertheless, pathways independent of auxin also seem to play a role in the regulation of cell division competence. *ABERRANT LATERAL ROOT FORMATION4* (*ALF4*) is not induced by auxin and seems to be involved in maintaining the meristematic state of the xylem-pole pericycle cells (Celenza et al. 1995; DiDonato et al. 2004). Although cells can still acquire founder cell identity in the *alf4* mutant, they will not divide, and therefore no lateral roots are formed in the absence of ALF4 (DiDonato et al. 2004; Dubrovsky et al. 2008).

In most plant species, lateral root initiation starts with the asymmetric division of founder cells (Casero et al. 1993, 1995; Malamy and Benfey 1997a). In *Arabidopsis*, the asymmetric divisions occur synchronously in one or two abutting pericycle cells and in three adjacent pericycle cell files (Dubrovsky et al. 2001). Two successive rounds of asymmetric cell divisions result in a file of small cells flanked by larger pericycle cells (De Smet et al. 2008). Asymmetric cell division during lateral root initiation is preceded by the migration of the nuclei of two abutting founder cells toward their common cell wall (De Smet et al. 2007; De Rybel et al. 2010). Prior to migration, auxin accumulates in the founder cells (Benková et al. 2003; Dubrovsky et al. 2008; De Rybel et al. 2010), and auxin signaling based on degradation of IAA28 and SLR/IAA14, and activation of ARF7 and ARF19, appeared to be necessary for coordinated nuclear migration (Figure 6.2; De Rybel et al. 2010). A general aspect of asymmetric cell divisions is that the daughter cells can have different cell fates (Scheres and Benfey 1999). In case of lateral root formation, these formative divisions are associated with the expression of the receptor-like kinase *ARABIDOPSIS CRINKLY4* (*ACR4*) specifically in the small daughter cells (De Smet et al. 2008). ACR4 is probably involved in the correct specification of the daughter cells and controls the number of pericycle cells involved in lateral root initiation by installing cell-to-cell communication with the neighboring pericycle cells thereby repressing formative divisions at these sites (De Smet et al. 2008). A similar dual role, combining a cell autonomous with a cell non-autonomous function, has been described for the maize homolog CRINKLY CRINKLY4 (CR4) during leaf development (Becraft et al. 2001).

The fact that asymmetric cell division is a prerequisite for the formation of lateral roots was demonstrated by the identification of *DIAGEOTROPICA* (*DGT*) in tomato (*Solanum lycopersicum* L.) (Ivanchenko et al. 2006). Although stretches of short pericycle cells are formed in the *dgt* mutant, indicating that cell cycle progression is not disturbed, these symmetric divisions are merely proliferative, and no lateral root primordia are formed (Ivanchenko et al. 2006). A similar phenomenon was observed in *slr1* plants in which cell cycle arrest was overcome by overexpression of core cell cycle regulators (Vanneste et al. 2005; De Smet et al. 2010). A disturbed PIN-dependent auxin distribution appeared to be responsible for the lateral-rootless phenotype. In the absence of the natural auxin distribution mechanism, the formation of auxin maxima is compromised (Okushima et al. 2005; Vanneste et al. 2005; De Smet et al. 2010). The need for a correct PIN localization is further demonstrated in plants with a partial loss-of-function allele of *GNOM*, a regulator of intracellular vesicle trafficking involved in the recycling of PIN1 to the plasma membrane (Steinmann et al. 1999; Geldner et al. 2003). Heterozygous *gnom*^{R5} mutants fail to establish the proper auxin maxima needed for formative divisions, resulting in random pericycle cell proliferation upon auxin application (Geldner et al. 2004).

Downstream of SLR/IAA14, ARF7 and ARF19, a second auxin signaling module is activated during lateral root initiation, consisting of BODENLOS (BDL)/IAA12 and MONOPTEROS (MP)/ARF5 (Figure 6.2; De Smet et al. 2010). Both BDL/IAA12 and MP/ARF5, which interact physically (Hamann et al. 2002), are expressed at the lateral root initiation sites (De Smet et al. 2010), and expression of *MP* in *slr1* mutants restores lateral root initiation and development (De Smet et al. 2010).

Auxin-induced release of ARF7 and ARF19 further leads to a plethora of responses, such as the transcription of members of the *LATERAL ORGAN BOUNDARIES DOMAIN/ASYMMETRIC LEAVES2-LIKE* (*LBD/ASL*) family including *LBD16*, *LBD18* and *LBD29* (Figure 6.2; Okushima et al. 2007). Expression of these genes rescued the lateral-rootless phenotype of *arf7arf19* double mutants (Okushima et al. 2007; Lee et al. 2009). LBD18 interacts with LBD33 and activates the expression of E2Fa, a positive regulator of the cell cycle. LBD18 thereby forms a link between auxin response and cell cycle regulation that leads to lateral root initiation (De Veylder et al. 2002; Berckmans et al. 2011). In rice, an orthologue of *LBD16* and *LBD29*, *CROWN ROOTLESS1/ADVENTITIOUS ROOTLESS1* (*CRL1/ARL1*), regulates the formation of adventitious and lateral roots (Inukai et al. 2005; Liu et al. 2005, 2010). Its expression is induced by auxin, and mutants show auxin-related defects in root development, including reduced numbers of lateral roots (Inukai et al. 2005; Liu et al. 2005). Also in maize two homologous LOB-domain proteins, ROOTLESS CONCERNING CROWN AND SEMINAL ROOTS (RTCS) and RTCS-LIKE (RTCL), have been shown to be auxin inducible and to be involved in the initiation and formation of seminal and shoot-borne roots (Taramino et al. 2007), pointing to conserved mechanisms between monocot and dicot species.

Apart from the promotive role of auxin in lateral root initiation, interplay with other hormones is also part of the regulation. Lateral root initiation is inhibited in mutants with increased ethylene levels (Nodzon et al. 2004; Ivanchenko et al. 2008; Prasad et al. 2010). Abscisic acid (ABA) also antagonizes the effect of auxin by inducing the expression of KRPs, thereby blocking cell cycle G1-S transition (Wang et al. 1998; De Smet et al. 2006b). Cytokinin on the other hand inhibits lateral root initiation by blocking the G2-M transition (Li et al. 2006). In normal conditions, cytokinin signaling is specifically inhibited in the protoxylem and the neighboring pericycle cells (Mähönen et al. 2006). Localized expression of a cytokinin biosynthesis gene in the xylem-pole pericycle cells reduces lateral root formation and disrupts the cell division pattern, while cytokinin degradation in these cells results in increased lateral root initiation (Laplaze et al. 2007). The addition of cytokinin negatively influences the expression of *PIN* genes and induces the degradation of PIN1 proteins (Laplaze et al. 2007; Marhavy et al. 2011), thereby preventing the formation of an auxin maximum.

D. Organogenesis: Shaping of the New Primordium

Only if the conditions described in the previous part are met (i.e., the combination of asymmetric cell divisions, cell specification, and a correct hormone homeostasis), a functional lateral root primordium can be formed. Three founder cells could be sufficient for the formation of a primordium in *Arabidopsis* (Dubrovsky et al. 2001). Other reports however estimate the number of founder cells involved to be 11 in *Arabidopsis* (Laskowski et al. 1995), and up to 40 in radish (Sussex et al. 1995). Although multiple cell files take part in the first steps of organogenesis, they are the founder cells from the central file that will contribute most to the formation of the primordium (Kurup et al. 2005).

After a set of asymmetric, anticlinal divisions, the smaller daughter cells expand radially followed by a periclinal division, forming a second cell layer. These early stages of lateral root formation seem quite robust, as they are conserved between different plant species (Lloret et al. 1989; Casero et al. 1995; Malamy and Benfey 1997b; Dubrovsky et al. 2000). In *Arabidopsis*, also subsequent divisions occur in a rather organized way, described by Malamy and Benfey (1997b) as a succession of stages going from a set of short pericycle cells after the asymmetric divisions (stage I), to the formation of a structure resembling that of a root meristem (Figure 6.3). At this point, mitotic activity decreases, and expansion of the basal cells drives the lateral root growth and emergence from the parent root. It is only after emergence that the meristem of the new lateral root is activated, and from that moment on the meristem initials are responsible for maintaining the growth (Malamy and Benfey 1997b; Ohtani et al. 2010).

Although histologically lateral root organogenesis is well described, little is known about the genetic pathways which take part in the process. Most mutants with abnormal lateral root phenotypes have defects at the level of founder cell specification or the first asymmetric division, making downstream events hard to elucidate. The few examples of lines with known aberrations in cell patterning all point to an important role of auxin and more particularly its distribution within the growing primordium (Benková et al. 2003; Ohtani et al. 2010). Already at stage II, an auxin gradient is established which ultimately forms a maximum at the primordium tip (Figure 6.3; Benková et al. 2003). This auxin distribution is realized by the combination of a localized expression of auxin transport proteins and their polar localization. At least six members of the *PIN* family are expressed in a specific pattern during primordium formation (Benková et al. 2003). Ectopic expression or mutations in these genes have an impact on the establishment of the auxin maximum and result in malformed primordia (Benková et al. 2003). Also the localization of PIN proteins within the cells is important, as shown by the *gnom* mutants. Although these mutants do not initiate lateral roots, treatment with NAA induces the formation of multilayered structures in large parts of the root, but they do not succeed in forming organized primordia (Geldner et al. 2004).

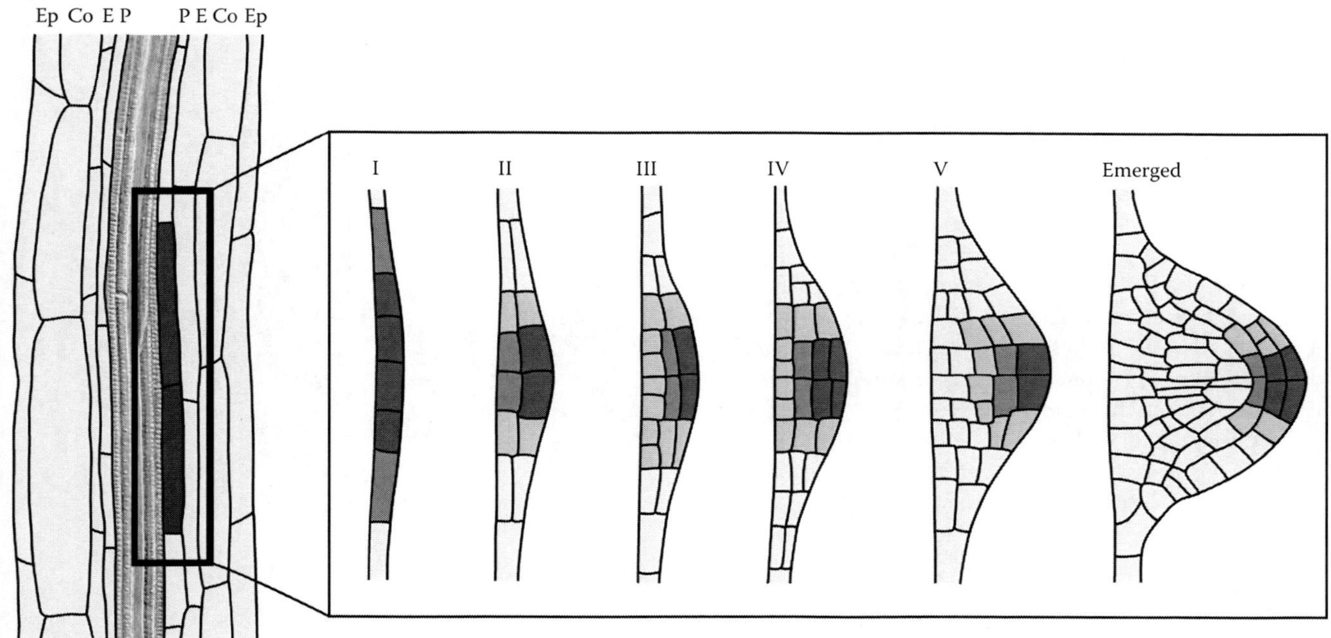

FIGURE 6.3 Lateral root organogenesis. Successive stages of lateral root primordium formation are illustrated, as described by Malamy and Benfey (1997b). Gray intensity corresponds to increasing auxin levels. Ep, epidermis; Co, cortex; E, endodermis; P, pericycle.

The initiation of lateral roots and the first steps of their formation depend on auxin coming from the root apical meristem (Bhalerao et al. 2002), and the role for auxin uptake into the primordium is ascribed to the auxin import facilitator AUXIN RESISTANT1 (AUX1). Its expression is induced just before the first periclinal divisions, the stage in which primordia of *aux1* mutants are blocked (Marchant et al. 2002). Once the new primordium has about three to five cell layers, it becomes in itself a source of auxin and it can grow independently from the parent root (Laskowski et al. 1995; Sussex et al. 1995; Ljung et al. 2005). This indicates that AUX1 together with the different PIN proteins is responsible for the correct auxin homeostasis.

The auxin response necessary for lateral root initiation occurs through SLR/IAA14-ARF7-ARF19 (Fukaki et al. 2002; Okushima et al. 2005). To study a possible role of the same module in later stages of lateral root formation, the lack of lateral root initiation sites in the *slr1* mutant was overcome by expressing a dominant negative form of SLR/IAA14 in young primordia. Only few aberrant lateral roots emerged, while most of the primordia remained arrested, indicating that SLR/IAA14-dependent auxin signaling is also needed for correct development (Fukaki et al. 2005). A gene which possibly acts downstream of this pathway is *PUCHI*, a transcription factor regulating the patterning of the lateral root primordium through restriction of the cell proliferation zone (Hirota et al. 2007). In the absence of functional PUCHI proteins, additional divisions start to occur in stage II primordia, which results in the formation of lateral roots which are swollen at the base due to excessive divisions (Hirota et al. 2007). Two other genes, to which a function similar to that of *PUCHI* was attributed, are expressed at later stages at the margins of the lateral root primordium. The expression of *CUP-SHAPED COTYLEDON3* (*CUC3*) and *LATERAL ORGAN BOUNDARIES* (*LOB*) leads to the formation of a boundary between the parent root and the developing lateral root and may also function in delimiting cell division (Shuai et al. 2002; Vroemen et al. 2003). ABA also seems to play a role at this stage. Two genes involved in ABA biosynthesis are expressed in pericycle and cortex cells surrounding lateral roots and at the base of lateral roots (Tan et al. 2003). As ABA slows down cell cycle activity (Wang et al. 1998), it might take part in defining the boundaries of the developing lateral root (De Smet et al. 2006b).

Two other hormones that influence primordia formation, through crosstalk with auxin, are cytokinin and ethylene. Cytokinin-overproducing plants or treatment with exogenous cytokinin not only reduces the lateral root density but also impairs patterning of the primordia (Laplaze et al. 2007; Kuderová et al. 2008). It seems however that only the founder cells are sensitive to cytokinin, as localized cytokinin production in young primordia did not have any effect on lateral root development (Laplaze et al. 2007). Moreover, *ISOPENTENYL TRANSFERASE5* (*IPT5*), which codes for a protein from the cytokinin biosynthesis pathway, is expressed in young lateral root primordia, suggesting that cytokinin is produced in developing primordia (Miyawaki et al. 2004). Therefore, aberrations in patterning must be a consequence of cytokinin response during lateral root initiation (Kuderová et al. 2008). Application of ethylene results in the formation of wider primordia due to cell enlargement and extra divisions at the apex of the lateral root and thus promotes its emergence (Ivanchenko et al. 2008).

E. On the Way to the Outer World

As cells undergo successive divisions, overlaying cell layers have to give way to the growing primordium. Although in *Arabidopsis* the number of these layers is limited to three, in plants such as maize or rice, the primordium encounters up to 16 layers on its way to the outer world (Feldman 1994; Hochholdinger et al. 2004). It implies that these plants might use different strategies to guide the young primordium through the mother root tissues. In maize, cortex cells collapse and their cytoplasm is broken down, while the lignified epidermis stays intact till it is pushed apart by the primordium (Bell and McCully 1970; Park et al. 2004). In *Arabidopsis*, however, cells of the endodermis, cortex, and epidermis remain intact during the whole emergence process and are separated through degradation of their middle lamellae (Laskowski et al. 2006). During the last decades, several hypotheses were developed trying to explain the process of lateral root emergence, going from solely mechanical pressure by the lateral root primordium to the secretion of enzymes, which has been demonstrated in different plant species (overview in Charlton 1996; Lloret and Casero 2002). In recent years, using *Arabidopsis* as a model, these theories were further adapted and elaborated, leading to the identification of molecular mechanisms involved in the process.

The development of a lateral root primordium in *Arabidopsis* coincides with the expression of cell wall–modifying enzymes which assist the emergence by loosening cellular interactions in overlaying tissues. These include *PGAZAT/ADPG2*, which codes for a polygalacturonase, *PLA1* and *PLA2*, two pectin lyases, and *PME1*, a pectin methylesterase (Laskowski et al. 2006; González-Carranza et al. 2007; Swarup et al. 2008). Pectin methyl esterase is believed to demethylate the pectin present in the middle lamellae, which is then degraded by polygalacturonase and pectin lyase, facilitating cell separation. The newly formed cell walls of the primary root meristem are highly methylated, which protects them from degradation (Torki et al. 2000; Laskowski et al. 2006). Also other genes may help the emergence process. Expression of *AUXIN-INDUCED IN ROOT CULTURES3* (AIR3), a subtilisin-like protease, and *XTR6*, a xyloglucan/xyloglucosyl transferase, was demonstrated in the outer layers of the mother root at the level of a developing primordium (Neuteboom et al. 1999; Swarup et al. 2008). Further, transcription of some expansins is highly induced during lateral root development (Laskowski et al. 2006; Swarup et al. 2008), but although expansins have been shown to be involved in cell separation, the increased expression may also, in part, be due to cell expansion in the growing primordium (Cho and Cosgrove 2000; Laskowski et al. 2006).

These cell wall–modifying genes were identified as being auxin inducible, suggesting also that the emergence of lateral roots is coordinated by auxin. This was confirmed by the observation that shoot-derived auxin and rootward auxin transport are needed for the emergence of lateral roots in young seedlings (Bhalerao et al. 2002; Wu et al. 2007). Auxin transported to the developing lateral roots also reaches the surrounding tissues of the parent root, where it induces the changes necessary for outgrowth (Bhalerao et al. 2002; Wu et al. 2007; Swarup et al. 2008).

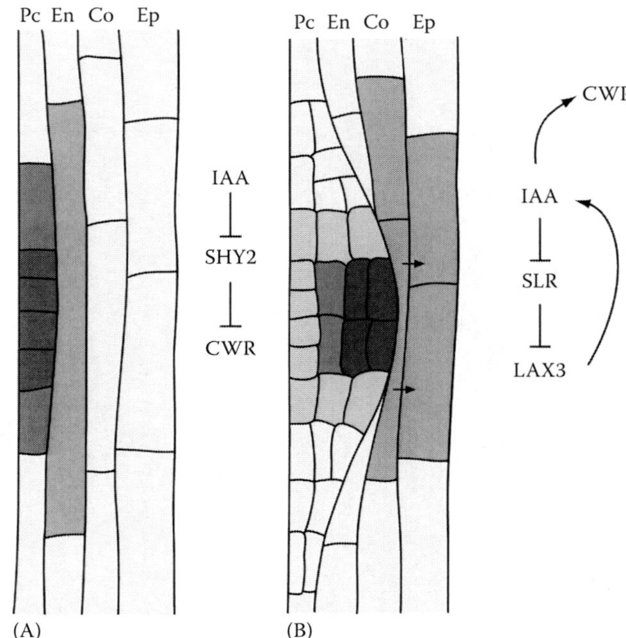

FIGURE 6.4 Lateral root emergence in *Arabidopsis*. Cell separation is accomplished through the activity of CWR enzymes. (A) Expression of CWRs in the endodermis is activated through auxin-induced degradation of SHY2/IAA3. (B) In cortex and endodermis, auxin signaling occurs through the SLR/IAA14-ARF7-ARF19 module to induce the LAX3 auxin import protein. Gray intensity corresponds to increasing auxin concentrations and arrows indicate auxin transport. Pc pericycle; En, endodermis; Co, cortex; Ep, epidermis.

The first layer encountered by the developing primordium is the endodermis. In this layer, auxin response is regulated by *SHORT HYPOCOTYL2* (SHY2)/IAA3, which is expressed in cells adjacent to the lateral root primordia (Figure 6.4A; Tian and Reed 1999; Swarup et al. 2008). *shy2* loss-of-function mutants have an increased number of emerged lateral roots, while in plants with a stabilized form of SHY2, lateral root emergence is delayed (Tian and Reed 1999; Swarup et al. 2008). When auxin from the growing primordium reaches the cortex and epidermis, *LIKE AUX1 3* (*LAX3*) transcription is induced through *degradation* of SLR/IAA14 and release of ARF7 and ARF19 (Figure 6.4B; Swarup et al. 2008). In turn, LAX3 acts as an auxin import carrier protein to increase the cellular auxin content (Bainbridge et al. 2008; Swarup et al. 2008), thus creating a very localized pool of auxin (Swarup et al. 2008). There are indications that both SHY2/IAA3 and LAX3 regulate the expression of a partially overlapping set of cell wall–remodeling (CWR) genes (Swarup et al. 2008).

F. Meristem Activation and Maintenance

Once emerged, divisions in the meristem of the new latera root have to be activated and maintained, so lateral root growth can be sustained (Malamy and Benfey 1997b). Knowledge about these processes is only fragmentary, and although there are indications that some mechanisms are lateral root specific (Wu et al. 2007), significant overlap with the maintenance of the primary root meristem was demonstrated (Celenza et al. 1995; Cheng et al. 1995). ROOT MERISTEMLESS1 (RML1) for instance is necessary for the activation of cell division in the embryonic primary as well as the lateral root meristems (Cheng et al. 1995). Although patterning occurs normally in the absence of RML1, the roots fail to elongate due to a block of the cell cycle in their meristem, a phenotype that could not be overcome by the addition of auxin (Cheng et al. 1995; Vernoux et al. 2000). In contrast to most other aspects of lateral root formation, meristem activation after emergence does not seem to be controlled by auxin but is influenced by ABA (De Smet et al. 2003). Application of ABA inhibits activation of the meristem in a reversible way, and the arrest cannot be overcome by increased auxin biosynthesis in the plant or external addition of NAA (De Smet et al. 2003).

However, different mutants have shown that a correct auxin homeostasis remains important also post-emergence (Celenza et al. 1995; Wu et al. 2007). Without application of exogenous auxin, lateral root primordia in *aberrant lateral root formation3* (*alf3*) mutant plants arrest growth after a couple of millimeters, and as cell death occurs, they lose their viability (Celenza et al. 1995). Although its exact function remains unknown, authors suggest ALF3 is part of the IAA biosynthetic pathway or plays a role in auxin transport in lateral root primordia (Celenza et al. 1995). Impaired auxin transport can indeed reduce lateral root elongation. MDR1 is a member of the multidrug resistance (MDR)-like ABC transporters. It participates in rootward auxin transport, and although expressed throughout the primary and lateral roots, its function is dispensable for primary root growth or early lateral root formation (Wu et al. 2007). In *MDR1*-deficient plants, the auxin response maximum at the root tip is reduced, and laterals are shorter compared to those of control plants (Wu et al. 2007).

III. Concluding Remarks

Increasing information on genomes of crop species as rice and maize will soon lead to new tools and implementation of knowledge from *Arabidopsis* in economically more relevant plants is becoming within reach. Although there might be differences in the individual pathways among species, one can expect that key mechanisms of lateral root formation are largely conserved between related species (Dubrovsky et al. 2006; Grunewald et al. 2007).

One common interspecific theme related to lateral root development is the role of plant growth regulators predominated by auxin. Unraveling the molecular pathways behind this hormone and the complex interplay with other hormones is currently a major theme and will gradually lead to the identification of molecular players in all stages of lateral root development. Computational modeling helps predicting the behavior of auxin and its effect on cellular processes (Laskowski et al. 2008; Lucas et al. 2008). It will represent an indispensable discipline to obtain a holistic view on this process. One other major challenge for future research is to integrate environmental conditions and their known impact on root system architecture into the current models for lateral root formation.

Acknowledgments

The authors thank Gert Van Isterdael for assistance with the artwork. M.D. is indebted to the Agency for Innovation by Science and Technology (IWT) for a PhD grant. Work in the Beeckman lab is supported by the Interuniversity Attraction Poles Programme (IUAPVI/33) initiated by the Belgian State Science Policy Office.

References

Atta R, Laurens L, Boucheron-Dubuisson E et al. 2009 Pluripotency of *Arabidopsis* xylem pericycle underlies shoot regeneration from root and hypocotyl explants grown in vitro. *Plant J* 57:626–644.

Bainbridge K, Guyomarc'h S, Bayer E et al. 2008 Auxin influx carriers stabilize phyllotactic patterning. *Genes Dev* 22:810–823.

Becraft PW, Kang SH, Suh SG 2001 The maize CRINKLY4 receptor kinase controls a cell-autonomous differentiation response. *Plant Physiol* 127:486–496.

Beeckman T, Burssens S, Inzé D 2001 The peri-cell-cycle in *Arabidopsis*. *J Exp Bot* 52:403–411.

Beemster GT, Fiorani F, Inzé D 2003 Cell cycle: the key to plant growth control? *Trends Plant Sci* 8:154–158.

Bell JK, Mccully ME 1970 A histological study of lateral root initiation and development in *Zea mays*. *Protoplasma* 70:179–205.

Benková E, Michniewicz M, Sauer M et al. 2003 Local, efflux-dependent auxin gradients as a common module for plant organ formation. *Cell* 115:591–602.

Berckmans B, Vassileva V, Schmid SP et al. 2011 Auxin-dependent cell cycle reactivation through transcriptional regulation of *Arabidopsis* E2Fa by lateral organ boundary proteins. *Plant Cell* 23:3671–3683.

Bhalerao RP, Eklöf J, Ljung K et al. 2002 Shoot-derived auxin is essential for early lateral root emergence in *Arabidopsis* seedlings. *Plant J* 29:325–332.

Blilou I, Xu J, Wildwater M et al. 2005 The PIN auxin efflux facilitator network controls growth and patterning in *Arabidopsis* roots. *Nature* 433:39–44.

Burssens S, Himanen K, Van De Cotte B et al. 2000 Expression of cell cycle regulatory genes and morphological alterations in response to salt stress in *Arabidopsis thaliana*. *Planta* 211:632–640.

Casero P, Casimiro I, Knoxm J 1998 Occurrence of cell surface arabinogalactan-protein and extensin epitopes in relation to pericycle and vascular tissue development in the root apex of four species. *Planta* 204:252–259.

Casero P, Casimiro I, Lloret P 1995 Lateral root initiation by asymmetrical transverse divisions by pericycle cells in four plant species: *Raphanus sativus*, *Helianthus annuus*, *Zea mays*, and *Daucus carota*. *Protoplasma* 188:49–58.

Casero PJ, Casimiro I, Rodríguez-Gallardo L, Martín-Partido G, Lloret PG 1993 Lateral root initiation by asymmetrical transverse divisions of pericycle cells in adventitious roots of *Allium cepa*. *Protoplasma* 176:138–144.

Casero P, García-Sánchez C, Lloret P, Navascués J 1989 Morphological features of pericycle cells in relation to their topographical location in onion adventitious roots. *New Phytol* 112:527–532.

Casimiro I, Beeckman T, Graham N et al. 2003 Dissecting *Arabidopsis* lateral root development. *Trends Plant Sci* 8:165–171.

Casimiro I, Marchant A, Bhalerao RP et al. 2001 Auxin transport promotes *Arabidopsis* lateral root initiation. *Plant Cell* 13:843–852.

Celenza JL, Jr., Grisafi PL, Fink GR 1995 A pathway for lateral root formation in *Arabidopsis thaliana*. *Genes Dev* 9:2131–2142.

Charlton WA 1987 Relationship between lateral root primordia in different ranks. *Ann Bot* 60:455–458.

Charlton WA 1996 Lateral root initiation. In *Plant Roots: The Hidden Half*, 2nd edn., eds. Y Waisel, A Eshel, U Kafkafi, pp. 149–173. New York: Marcel Dekker.

Cheng JC, Seeley KA, Sung ZR 1995 RML1 and RML2, *Arabidopsis* genes required for cell proliferation at the root tip. *Plant Physiol* 107:365–376.

Cho HT, Cosgrove DJ 2000 Altered expression of expansin modulates leaf growth and pedicel abscission in *Arabidopsis thaliana*. *Proc Natl Acad Sci USA* 97:9783–9788.

Cui H, Hao Y, Kovtun M et al. 2011 Genome-wide direct target analysis reveals a role for short-root in root vascular patterning through cytokinin homeostasis. *Plant Physiol* 157:1221–1231.

De Rybel B, Vassileva V, Parizot B et al. 2010 A novel aux/IAA28 signaling cascade activates GATA23-dependent specification of lateral root founder cell identity. *Curr Biol* 20:1697–1706.

De Smet I, Lau S, Voß U et al. 2010 Bimodular auxin response controls organogenesis in *Arabidopsis*. *Proc Natl Acad Sci USA* 107:2705–2710.

De Smet I, Signora L, Beeckman T et al. 2003 An abscisic acid-sensitive checkpoint in lateral root development of *Arabidopsis*. *Plant J* 33:543–555.

De Smet I, Tetsumura T, De Rybel B et al. 2007 Auxin-dependent regulation of lateral root positioning in the basal meristem of *Arabidopsis*. *Development* 134:681–690.

De Smet I, Vanneste S, Inze D, Beeckman T 2006a Lateral root initiation or the birth of a new meristem. *Plant Mol Biol* 60:871–887.

De Smet I, Vassileva V, De Rybel B et al. 2008 Receptor-like kinase ACR4 restricts formative cell divisions in the *Arabidopsis* root. *Science* 322:594–597.

De Smet I, Zhang H, Inzé D, Beeckman T 2006b A novel role for abscisic acid emerges from underground. *Trends Plant Sci* 11:434–439.

De Veylder L, Beeckman T, Beemster GT et al. 2002 Control of proliferation, endoreduplication and differentiation by the *Arabidopsis* E2Fa-DPa transcription factor. *Embo J* 21:1360–1368.

Deak KI, Malamy J 2005 Osmotic regulation of root system architecture. *Plant J* 43:17–28.

Didonato RJ, Arbuckle E, Buker S et al. 2004 *Arabidopsis ALF4* encodes a nuclear-localized protein required for lateral root formation. *Plant J* 37:340–353.

Ditengou FA, Teale WD, Kochersperger P et al. 2008 Mechanical induction of lateral root initiation in *Arabidopsis thaliana*. *Proc Natl Acad Sci USA* 105:18818–18823.

Doerner P, Jørgensen JE, You R, Steppuhn J, Lamb C 1996 Control of root growth and development by cyclin expression. *Nature* 380:520–523.

Dolan L, Janmaat K, Willemsen V et al. 1993 Cellular organisation of the *Arabidopsis thaliana* root. *Development* 119:71–84.

Draye X 2002. Consequences of root growth kinetics and vascular structure on the distribution of lateral roots. *Plant Cell Environ* 25:1463–1474.

Draye X, Delvaux B, Swennen R 1999 Distribution of lateral root primordia in root tips of *Musa*. *Ann Bot* 84:393–400.

Dubrovsky JG, Doerner PW, Colon-Carmona A, Rost TL 2000 Pericycle cell proliferation and lateral root initiation in *Arabidopsis*. *Plant Physiol* 124:1648–1657.

Dubrovsky JG, Gambetta GA, Hernández-Barrera A, Shishkova S, González I 2006 Lateral root initiation in *Arabidopsis*: developmental window, spatial patterning, density and predictability. *Ann Bot* 97:903–915.

Dubrovsky JG, Rost TL, Colon-Carmona A, Doerner P 2001 Early primordium morphogenesis during lateral root initiation in *Arabidopsis thaliana*. *Planta* 214:30–36.

Dubrovsky JG, Sauer M, Napsucialy-Mendivil S, et al. 2008 Auxin acts as a local morphogenetic trigger to specify lateral root founder cells. *Proc Natl Acad Sci USA* 105:8790–8794.

Feldman L 1994 The maize root. In *The Maize Handbook*, eds. M Freeling, V Walbot, pp. 29–37. New York: Springer-Verlag.

Ferreira PC, Hemerly AS, Engler JD et al. 1994 Developmental expression of the arabidopsis cyclin gene cyc1At. *Plant Cell* 6:1763–1774.

Fitter A, Williamson L, Linkohr B, Leyser O 2002 Root system architecture determines fitness in an *Arabidopsis* mutant in competition for immobile phosphate ions but not for nitrate ions. *Proc Biol Sci* 269:2017–2022.

Fortin MC, Pierce FJ, Poff KL 1989 The pattern of secondary root formation in curving roots of *Arabidopsis thaliana* (L.) Heynh. *Plant Cell Environ* 12:337–339.

Fukaki H, Nakao Y, Okushima Y, Theologis A, Tasaka M 2005 Tissue-specific expression of stabilized SOLITARY-ROOT/IAA14 alters lateral root development in *Arabidopsis*. *Plant J* 44:382–395.

Fukaki H, Tameda S, Masuda H, Tasaka M 2002 Lateral root formation is blocked by a gain-of-function mutation in the SOLITARY-ROOT/IAA14 gene of *Arabidopsis*. *Plant J* 29:153–168.

Geldner N, Anders N, Wolters H et al. 2003 The *Arabidopsis* GNOM ARF-GEF mediates endosomal recycling, auxin transport, and auxin-dependent plant growth. *Cell* 112:219–230.

Geldner N, Richter S, Vieten A et al. 2004 Partial loss-of-function alleles reveal a role for GNOM in auxin transport-related, post-embryonic development of *Arabidopsis*. *Development* 131:389–400.

González-Carranza ZH, Elliott KA, Roberts JA 2007 Expression of polygalacturonases and evidence to support their role during cell separation processes in *Arabidopsis thaliana*. *J Exp Bot* 58:3719–3730.

Gray WM, Del Pozo JC, Walker L et al. 1999 Identification of an SCF ubiquitin-ligase complex required for auxin response in *Arabidopsis thaliana*. *Genes Dev* 13:1678–1691.

Grunewald W, Parizot B, Inzé D, Gheysen G, Beeckman T 2007 Developmental biology of roots: one common pathway for all angiosperms? *Int J Plant Dev Biol* 1:212–225.

Hamann T, Benkova E, Bäurle I, Kientz M, Jürgens G 2002 The *Arabidopsis BODENLOS* gene encodes an auxin response protein inhibiting MONOPTEROS-mediated embryo patterning. *Genes Dev*16:1610–1615.

Hemerly A, Bergounioux C, Van Montagu M, Inze D, Ferreira P 1992 Genes regulating the plant cell cycle: isolation of a mitotic-like cyclin from *Arabidopsis thaliana*. *Proc Natl Acad Sci USA* 89:3295–3299.

Himanen K, Boucheron E, Vanneste S et al. 2002 Auxin-mediated cell cycle activation during early lateral root initiation. *Plant Cell* 14:2339–2351.

Himanen K, Vuylsteke M, Vanneste S et al. 2004 Transcript profiling of early lateral root initiation. *Proc Natl Acad Sci USA* 101:5146–5151.

Hirota A, Kato T, Fukaki H, Aida M, Tasaka M 2007 The auxin-regulated AP2/EREBP gene PUCHI is required for morphogenesis in the early lateral root primordium of *Arabidopsis*. *Plant Cell* 19:2156–2168.

Hochholdinger F, Woll K, Sauer M, Dembinsky D 2004 Genetic dissection of root formation in maize (*Zea mays*) reveals root-type specific developmental programmes. *Ann Bot* 93:359–368.

Hodge A 2006. Plastic plants and patchy soils. *J Exp Bot* 57:401–411.

Hodge A, Berta G, Doussan C, Merchan F, Crespi M 2009 Plant root growth, architecture and function. *Plant Soil* 321:153–187.

Hou G, Hill JP, Blancaflor EB 2004 Developmental anatomy and auxin response of lateral root formation in *Ceratopteris richardii*. *J Exp Bot* 55:685–693.

Inukai Y, Sakamoto T, Ueguchi-Tanaka M et al. 2005 Crown rootless1, which is essential for crown root formation in rice, is a target of an AUXIN RESPONSE FACTOR in auxin signaling. *Plant Cell* 17:1387–1396.

Ivanchenko MG, Coffeen WC, Lomax TL, Dubrovsky JG 2006 Mutations in the Diageotropica (Dgt) gene uncouple patterned cell division during lateral root initiation from proliferative cell division in the pericycle. *Plant J* 46:436–447.

Ivanchenko MG, Muday GK, Dubrovsky JG 2008 Ethylene-auxin interactions regulate lateral root initiation and emergence in *Arabidopsis thaliana*. *Plant J* 55:335–347.

Jansen L, Roberts I, De Rycke R, Beeckman T 2012 Phloem-associated auxin response maxima determine radial positioning of lateral roots in maize. *Philos Trans R Soc Lond B Biol Sci* 367(1595):1525–1533.

Kuderová A, Urbánková I, Válková M et al. 2008. Effects of conditional IPT-dependent cytokinin overproduction on root architecture of *Arabidopsis* seedlings. *Plant Cell Physiol* 49:570–582.

Kurup S, Runions J, Kohler U et al. 2005 Marking cell lineages in living tissues. *Plant J* 42:444–453.

Laplaze L, Benkova E, Casimiro I et al. 2007 Cytokinins act directly on lateral root founder cells to inhibit root initiation. *Plant Cell* 19:3889–3900.

Laplaze L, Parizot B, Baker A et al. 2005 GAL4-GFP enhancer trap lines for genetic manipulation of lateral root development in *Arabidopsis thaliana*. *J Exp Bot* 56:2433–2442.

Laskowski M, Biller S, Stanley K, Kajstura T, Prusty R 2006 Expression profiling of auxin-treated *Arabidopsis* roots: toward a molecular analysis of lateral root emergence. *Plant Cell Physiol* 47:788–792.

Laskowski M, Grieneisen VA, Hofhuis H et al. 2008 Root system architecture from coupling cell shape to auxin transport. *PLoS Biol* 6:e307.

Laskowski MJ, Williams ME, Nusbaum HC, Sussex IM 1995 Formation of lateral root meristems is a two-stage process. *Development* 121:3303–3310.

Lee HW, Kim NY, Lee DJ, Kim J 2009 LBD18/ASL20 regulates lateral root formation in combination with LBD16/ASL18 downstream of ARF7 and ARF19 in *Arabidopsis*. *Plant Physiol* 151:1377–1389.

Leyser O 2006 Dynamic integration of auxin transport and signaling. *Curr Biol* 16:R424–R433.

Li X, Mo X, Shou H, Wu P 2006 Cytokinin-mediated cell cycling arrest of pericycle founder cells in lateral root initiation of *Arabidopsis*. *Plant Cell Physiol* 47:1112–1123.

Liu HL, Wang GC, Feng Z, Zhu J 2010 Screening of genes associated with dedifferentiation and effect of LBD29 on pericycle cells in *Arabidopsis thaliana*. *Plant Growth Reg* 62:127–136.

Liu H, Wang S, Yu X et al. 2005 ARL1, a LOB-domain protein required for adventitious root formation in rice. *Plant J* 43:47–56.

Ljung K, Hull AK, Celenza J et al. 2005 Sites and regulation of auxin biosynthesis in *Arabidopsis* roots. *Plant Cell* 17:1090–1104.

Lloret PG, Casero PJ 2002 Lateral root initiation. In *Plant Roots: The Hidden Half*, eds. Y Waisel, A Eshel, U Kafkafi, 3rd edn., pp. 127–155. New York: Marcel Dekker, Inc.

Lloret P, Casero P, Pulgarín A, Navascués J 1989 The behaviour of two cell populations in the pericycle of *Allium cepa*, *Pisum sativum*, and *Daucus carota* during early lateral root development. *Ann Bot* 63:465–475.

Lucas M, Godin C, Jay-Allemand C, Laplaze L 2007 Auxin fluxes in the root apex co-regulate gravitropism and lateral root initiation. *J Exp Bot* 59:55–66.

Lucas M, Guedon Y, Jay-Allemand C, Godin C, Laplaze L 2008 An auxin transport-based model of root branching in *Arabidopsis thaliana*. *PLoS One* 3:e3673.

Macleod RD 1990 Lateral root primordium inception in *Zea mays*. *Environ Exp Bot* 30:225–234.

Mähönen AP, Bishopp A, Higuchi M et al. 2006 Cytokinin signaling and its inhibitor AHP6 regulate cell fate during vascular development. *Science* 311:94–98.

Malamy JE 2005 Intrinsic and environmental response pathways that regulate root system architecture. *Plant Cell Environ* 28:67–77.

Malamy JE, Benfey PN 1997a Down and out in *Arabidopsis*: the formation of lateral roots. *Trends Plant Sci* 2:390–396.

Malamy JE, Benfey PN 1997b Organization and cell differentiation in lateral roots of *Arabidopsis thaliana*. *Development* 124:33–44.

Marchant A, Bhalerao R, Casimiro I et al. 2002 AUX1 promotes lateral root formation by facilitating indole-3-acetic acid distribution between sink and source tissues in the *Arabidopsis* seedling. *Plant Cell* 14:589–597.

Marhavy P, Bielach A, Abas L et al. 2011 Cytokinin modulates endocytic trafficking of PIN1 auxin efflux carrier to control plant organogenesis. *Dev Cell* 21:796–804.

Miyawaki K, Matsumoto-Kitano M, Kakimoto T 2004 Expression of cytokinin biosynthetic isopentenyltransferase genes in *Arabidopsis*: tissue specificity and regulation by auxin, cytokinin, and nitrate. *Plant J* 37:128–138.

Monshausen GB, Bibikova TN, Weisenseel MH, Gilroy S 2009 Ca^{2+} regulates reactive oxygen species production and pH during mechanosensing in *Arabidopsis* roots. *Plant Cell* 21:2341–2356.

Moreno-Risueno MA, Van Norman JM, Moreno A et al. 2010 Oscillating gene expression determines competence for periodic *Arabidopsis* root branching. *Science* 329:1306–1311.

Neuteboom LW, Veth-Tello LM, Clijdesdale OR, Hooykaas PJ, Van Der Zaal BJ 1999 A novel subtilisin-like protease gene from *Arabidopsis thaliana* is expressed at sites of lateral root emergence. *DNA Res* 6:13–19.

Nibau C, Gibbs DJ, Coates JC 2008 Branching out in new directions: the control of root architecture by lateral root formation. *New Phytol* 179:595–614.

Nieuwland J, Maughan S, Dewitte W et al. 2009 The D-type cyclin CYCD4;1 modulates lateral root density in *Arabidopsis* by affecting the basal meristem region. *Proc Natl Acad Sci USA* 106:22528–22533.

Nodzon LA, Xu WH, Wang Y et al. 2004 The ubiquitin ligase XBAT32 regulates lateral root development in *Arabidopsis*. *Plant J* 40:996–1006.

Noll F 1900. *Über den bestimmenden Einfluss von Wurzelkrümmungen auf Entotenhung und Anordnung der Seitenwurzeln*. Berlin: P. Parey.

Ohtani M, Demura T, Sugiyama M 2010 Particular significance of SRD2-dependent snRNA accumulation in polarized pattern generation during lateral root development of *Arabidopsis*. *Plant Cell Physiol* 51:2002–2012.

Okushima Y, Fukaki H, Onoda M, Theologis A, Tasaka M 2007. ARF7 and ARF19 regulate lateral root formation via direct activation of *LBD/ASL* genes in *Arabidopsis*. *Plant Cell* 19:118–130.

Okushima Y, Overvoorde PJ, Arima K et al. 2005 Functional genomic analysis of the AUXIN RESPONSE FACTOR gene family members in *Arabidopsis thaliana*: unique and overlapping functions of ARF7 and ARF19. *Plant Cell* 17:444–463.

Ottenschlager I, Wolff P, Wolverton C et al. 2003 Gravity-regulated differential auxin transport from columella to lateral root cap cells. *Proc Natl Acad Sci USA* 100:2987–2991.

Overvoorde P, Fukaki H, Beeckman T 2010. Auxin control of root development. *Cold Spring Harb Perspect Biol* 2:a001537.

Parizot B, De Rybel B, Beeckman T 2010 VisuaLRTC: a new view on lateral root initiation by combining specific transcriptome data sets. *Plant Physiol* 153:34–40.

Parizot B, Laplaze L, Ricaud L et al. 2008 Diarch symmetry of the vascular bundle in *Arabidopsis* root encompasses the pericycle and is reflected in distich lateral root initiation. *Plant Physiol* 146:140–148.

Park WJ, Hochholdinger F, Gierl A 2004 Release of the benzoxazinoids defense molecules during lateral- and crown root emergence in *Zea mays*. *Plant Physiol* 161:981–985.

Péret B, De Rybel B, Casimiro I et al. 2009 *Arabidopsis* lateral root development: an emerging story. *Trends Plant Sci* 14:399–408.

Ploense SE, Wu MF, Nagpal P, Reed JW 2009 A gain-of-function mutation in *IAA18* alters *Arabidopsis* embryonic apical patterning. *Development* 136:1509–1517.

Prasad ME, Schofield A, Lyzenga W, Liu H, Stone SL 2010 *Arabidopsis* RING E3 ligase XBAT32 regulates lateral root production through its role in ethylene biosynthesis. *Plant Physiol* 153:1587–1596.

Rebouillat J, Dievart A, Verdeil JL et al. 2009 Molecular genetics of rice root development. *Rice* 2:15–34.

Richter GL, Monshausen GB, Krol A, Gilroy S 2009 Mechanical stimuli modulate lateral root organogenesis. *Plant Physiol* 151:1855–1866.

Rogg LE, Lasswell J, Bartel B 2001 A gain-of-function mutation in IAA28 suppresses lateral root development. *Plant Cell* 13:465–480.

Scheres B, Benfey PN 1999 Asymmetric cell division in plants. *Annu Rev Plant Physiol Plant Mol Biol* 50:505–537.

Segers G, Gadisseur I, Bergounioux C et al. 1996 The *Arabidopsis* cyclin-dependent kinase gene *cdc2bAt* is preferentially expressed during S and G_2 phases of the cell cycle. *Plant J* 10:601–612.

Shaul O, Mironov V, Burssens S, Van Montagu M, Inzé D 1996 Two *Arabidopsis* cyclin promoters mediate distinctive transcriptional oscillation in synchronized tobacco BY-2 cells. *Proc Natl Acad Sci USA* 93:4868–4872.

Shuai B, Reynaga-Peña CG, Springer PS 2002 The lateral organ boundaries gene defines a novel, plant-specific gene family. *Plant Physiol* 129:747–761.

Steinmann T, Geldner N, Grebe M et al. 1999 Coordinated polar localization of auxin efflux carrier PIN1 by GNOM ARF GEF. *Science* 286:316–318.

Sussex IM, Godoy JA, Kerk NM et al. 1995 Cellular and molecular events in a newly organizing lateral root meristem. *Philos Trans R Soc Lond B Biol Sci* 350:39–43.

Swarup K, Benková E, Swarup R et al. 2008 The auxin influx carrier LAX3 promotes lateral root emergence. *Nat Cell Biol* 10:946–954.

Swarup R, Bennett M 2003 Auxin transport: the fountain of life in plants? *Dev Cell* 5:824–826.

Swarup R, Kramer EM, Perry P et al. 2005 Root gravitropism requires lateral root cap and epidermal cells for transport and response to a mobile auxin signal. *Nat Cell Biol* 7:1057–1065.

Tan BC, Joseph LM, Deng WT et al. 2003 Molecular characterization of the *Arabidopsis* 9-cis epoxycarotenoid dioxygenase gene family. *Plant J* 35:44–56.

Taramino G, Sauer M, Stauffer JL Jr. et al. 2007 The maize (*Zea mays* L.) *RTCS* gene encodes a LOB domain protein that is a key regulator of embryonic seminal and post-embryonic shoot-borne root initiation. *Plant J* 50:649–659.

Tian Q, Reed JW 1999 Control of auxin-regulated root development by the *Arabidopsis thaliana* SHY2/IAA3 gene. *Development* 126:711–721.

Torki M, Mandaron P, Mache R, Falconet D 2000 Characterization of a ubiquitous expressed gene family encoding polygalacturonase in *Arabidopsis thaliana. Gene* 242:427–436.

Uehara T, Okushima Y, Mimura T, Tasaka M, Fukaki H 2008 Domain II mutations in CRANE/IAA18 suppress lateral root formation and affect shoot development in *Arabidopsis thaliana. Plant Cell Physiol* 49:1025–1038.

Vanneste S, De Rybel B, Beemster GT et al. 2005 Cell cycle progression in the pericycle is not sufficient for SOLITARY ROOT/IAA14-mediated lateral root initiation in *Arabidopsis thaliana. Plant Cell* 17:3035–3050.

Verkest A, Weinl C, Inzé D, De Veylder L, Schnittger A 2005 Switching the cell cycle. Kip-related proteins in plant cell cycle control. *Plant Physiol* 139:1099–10106.

Vernoux T, Wilson RC, Seeley KA et al. 2000 The ROOT MERISTEMLESS1/CADMIUM SENSITIVE2 gene defines a glutathione-dependent pathway involved in initiation and maintenance of cell division during postembryonic root development. *Plant Cell* 12:97–110.

Vroemen CW, Mordhorst AP, Albrecht C, Kwaaitaal MA, De Vries SC 2003 The CUP-SHAPED COTYLEDON3 gene is required for boundary and shoot meristem formation in *Arabidopsis. Plant Cell* 15:1563–1577.

Wang H, Qi Q, Schorr P et al. 1998 ICK1, a cyclin-dependent protein kinase inhibitor from *Arabidopsis thaliana* interacts with both Cdc2a and CycD3, and its expression is induced by abscisic acid. *Plant J* 15:501–510.

Wright KM, Oparka K 1997 Metabolic inhibitors induce symplastic movement of solutes from the transport phloem of *Arabidopsis* roots. *J Exp Bot* 48:1807–1814.

Wu G, Lewis DR, Spalding EP 2007 Mutations in *Arabidopsis* multidrug resistance-like ABC transporters separate the roles of acropetal and basipetal auxin transport in lateral root development. *Plant Cell* 19:1826–1837.

7

Vascular Development in *Arabidopsis* Roots

Anthony Bishopp
University of Nottingham

Sedeer El-Showk
University of Helsinki

Ykä Helariutta
University of Helsinki

I. Introduction: Why Study Vascular Development?

The development of a vascular system has been instrumental to the evolution of vascular plants and their subsequent domination of land. Nonvascular land plants (such as mosses and liverworts) are restricted to moist environments and are unable to attain large sizes due to an inability to effectively transport water and nutrients between organs. The evolution of a vascular system has not only provided plants with a transport mechanism by which water and nutrients can be conducted between organs but also provided them with a rigid structure. This has enabled vascular plants to grow tall enough to capture light and develop roots to search out water below the soil surface. Consequently, vascular land plants have been able to colonize habitats throughout most of the Earth.

Recent demand for biofuels has also sparked interest in vascular development. Although bioethanol can be made by fermentation from a variety of grains, this process requires a large amount of energy, and environmental life cycle analysis models have shown modest reductions in CO_2 emissions for grain-based ethanol production (California Energy Commission 2007). Recent research has focused on the development of cellulosic ethanol production where sugars can be extracted from plant waste products by either enzymatic or chemical hydrolysis. The majority of plant sugars are present as cellulose, hemicellulose, and lignin, and the greater part of these essential materials is locked within the vascular tissues. It is hoped that by engineering plants with alterations in the properties and numbers of vascular cells, we could produce either plants that have an increased amount of sugars stored within the vascular tissues or plants where the sugars are easier to extract. Therefore, a greater knowledge of vascular development would be an important asset in designing crops for the future.

One factor which has hampered efforts to study vascular development is the fact that the vascular tissues lie deep within the plant and imaging them can be extremely challenging. For this reason, the thin roots of the simple weed *Arabidopsis* have provided an excellent model with which to study vascular development. This chapter will therefore focus on the advances in our understanding of vascular development in *Arabidopsis*. We hope that by understanding the processes regulating vascular development in a model system, we will be able to transfer this knowledge to crop plants, such as cereals and trees.

II. Formation of the Vascular Cylinder during Embryogenesis

The *Arabidopsis* embryo forms through a regular pattern of stereotypical asymmetric cell divisions. Initially, an egg cell and one of the nuclei from a pollen grain fuse to create a zygote. The zygote then elongates and divides asymmetrically to create an upper cell that will give rise to the majority of the embryo and a lower cell that will give rise to the suspensor and also contribute toward the formation of the root. The upper cell undergoes a series of orchestrated longitudinal and transverse divisions to form the eight-cell proembryo. The apical-basal polarity is already defined by this stage (Figure 7.1). The upper tier of cells will form the shoot meristem and most of the cotyledons, while the lower tier will go on to form the abaxial cotyledons, the hypocotyl, and the root meristem. As the eight-cell proembryo divides periclinally, the radial axis forms, and this defines protoderm and ground tissue cell lineages (Jürgens et al. 1995). Several important events occur at the globular stage of embryogenesis (Figure 7.1). The root meristem precursor cells form through the division of the hypophysis (the uppermost suspensor cell). As the embryo transits between the globular and the late globular stage, the two cells comprising the vascular primordium divide periclinally to form the pericycle cell lineages and the vascular cell initials (Figure 7.1). Radial symmetry is broken in the apical portion of the embryo as cotyledons arise from the flanks of the apical domain.

The hormone auxin is instrumental in coordinating cell divisions and pattern formation during embryogenesis, and mutations in genes involved in auxin biosynthesis, transport, and signaling cause major patterning defects during embryogenesis. These mutations affect four distinct processes: the orientation of the first division of the apical daughter cell that gives rise to the two-cell embryo, resulting in failure to establish apical-basal polarity; the division of the hypophysis, causing a failure to establish a root meristem and resulting in rootless seedlings; the division of the

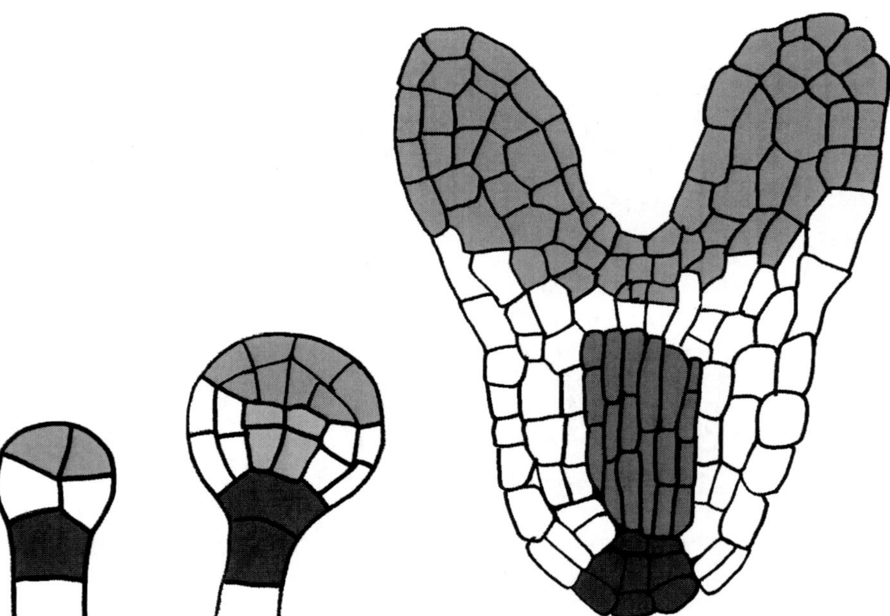

FIGURE 7.1 **(See color insert.)** The development of vascular cells during embryogenesis. Left is the eight-cell embryo. For simplicity, not all of the suspensor cells are shown. Already at this stage, the expression of marker genes indicates that the apical-basal identity is set. The uppermost suspensor cell, which will go on to form the root meristem, is shown in brown, and the cells which will form the shoot are shown in green. In the center is an early globular stage embryo. As cell identity is specified, the vascular primordium, marked in pink, will undergo a programmed set of divisions to form both the vascular initials and the pericycle cells. By the late torpedo stage (right), the vascular initial cells are clearly defined (red). Although these remain undifferentiated, tissue-specific marker genes are expressed. For example, at this stage, *AHP6* is already specifically expressed in two files that will later form the protoxylem.

vascular initial cells; and the initiation, growth, and separation of the cotyledons (Berleth and Jürgens 1993; Friml et al. 2003). In this chapter, we will concentrate on the events leading to the establishment and maintenance of correct cell division in the vascular cells.

Auxin is actively transported within the plant, and the formation of local auxin response maxima at discrete locations initiates many developmental processes. The direction of auxin transport is controlled by the asymmetric subcellular localization of the PIN class of auxin efflux proteins (Friml et al. 2003). Four PIN proteins (PIN1, PIN3, PIN4, and PIN7) are present during embryogenesis (Friml et al. 2003). During the formation of the apical-basal axis, PIN7 is localized to the apical membrane of the basal cell and forces auxin into the apical cell. This apical subcellular localization is maintained until the globular stage when it switches dynamically to the basal membrane to force auxin into the suspensor cells. PIN1 is expressed in an apolar fashion during early embryogenesis but becomes basally localized in the provascular cells next to the hypophysis to force auxin into the hypophysis. Quadruple *pin1 pin3 pin4 pin7* mutants display highly abnormal embryo formation. In these instances, it is not possible to separate the effects on vascular cell division from the effects of earlier defective cell divisions, and the mutants either are embryo lethal or develop into seedlings with severe apical defects and a nonfunctional root (Friml et al. 2003).

Regions of elevated auxin levels are perceived by the binding of auxin to the TIR1 family of auxin receptors (Dharmasiri et al. 2005a,b; Kepinski and Leyser 2005). When the TIR1 family receptors bind auxin, they stabilize the interaction between TIR1 and Aux/IAA proteins leading to ubiquitination and subsequent degradation of the Aux/IAAs (Tan et al. 2007). The Aux/IAAs act as inhibitors of auxin response by forming heterodimers with the ARF transcription factors, and this heterodimerization prevents the activation of auxin response genes (Abel et al. 1995). Degradation of Aux/IAAs releases the ARF proteins from this inhibitory effect and thus initiates the transcription of downstream response genes (Figure 7.2) (Mockaitis and Estelle 2008). These genes are part of large gene families (e.g., there are 23 ARF proteins and 29 Aux/IAA proteins in *Arabidopsis*), and their effects are often masked by genetic redundancy. However, mutations in some components result in strong phenotypes. There are several examples of dominant mutations in *Aux/IAA* genes, such as *axr3-1* (discussed later), that stabilize these proteins and render them resistant to TIR1-mediated degradation, causing them to confer a constitutive repression of auxin signaling. Additionally, mutations in certain *ARF* genes such as *MONOPEROS/ARF5* (*MP*) affect specific aspects of auxin response.

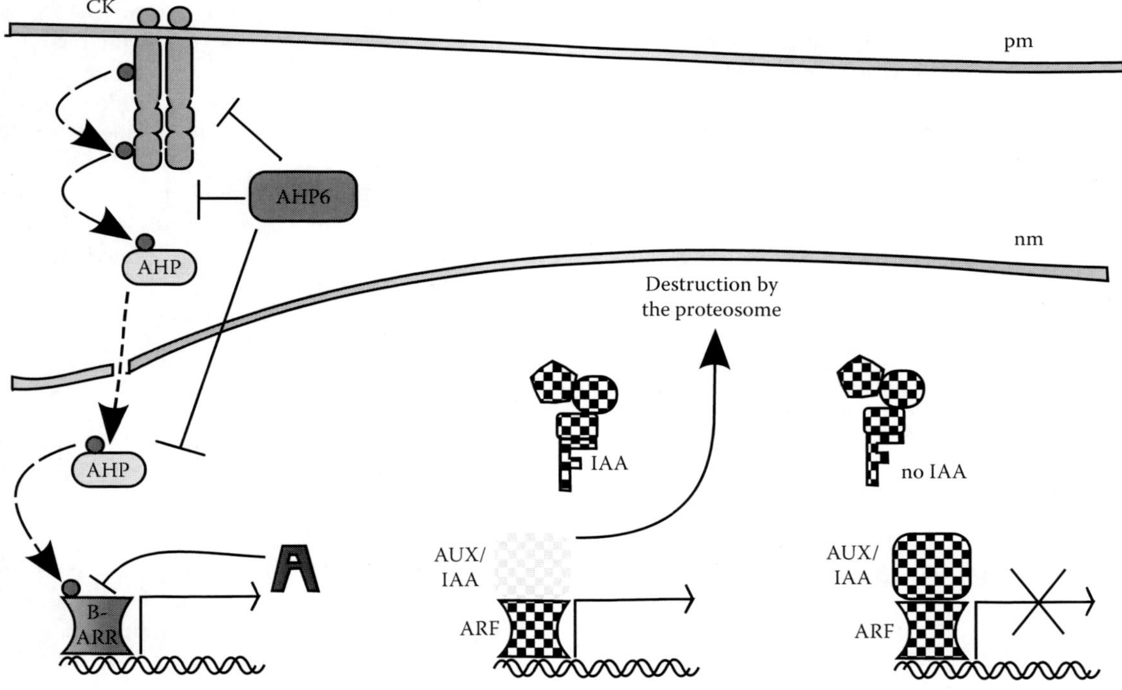

FIGURE 7.2 Hormonal perception: cytokinin and auxin signaling pathways. The cytokinin signaling pathway is shown with solid shading. Cytokinin is perceived by the CRE1 family of cytokinin receptors located on the plasma membrane (pm). This initiates autophosphorylation of the receptors, followed by the transfer of a phosphoryl group (small circle) to the phosphotransfer proteins (AHPs). When phosphorylated, these can move through the nuclear membrane (nm) and phosphorylate the type-B response regulators. This leads to transcription of the cytokinin response genes, including the type-A response regulators which act as inhibitors of cytokinin signaling. The pseudophosphotransfer protein AHP6 inhibits multiple stages of this phosphorelay. The auxin signaling components are shown with hatched marking. Under low-auxin conditions, the transcription of auxin response genes by the ARF proteins is blocked by the activity of the Aux/IAA proteins. When auxin is bound by a complex involving the TIR1-family of auxin receptors, this complex interacts with the Aux/IAAs and promotes the destruction of the Aux/IAAs by the 26S proteasome. This TIR1-mediated Aux/IAA degradation releases the ARFs from their transcriptional inhibition and allows the transcription of auxin response genes.

A wide range of alleles causing reduction of MP function to various degrees have provided some insight into the role of auxin during the earliest stages of vascular morphogenesis. Strong *mp* alleles display abnormal development as early as the eight-cell stage and fail to form a root at the early heart stage (Berleth and Jürgens 1993). In plants with *mp* alleles that do not prevent root development, the embryonic cell divisions occurring in the basal region of *mp* embryos proceed abnormally. Between the late heart and the torpedo stage of embryogenesis, the vascular initial cells and pericycle cells perform additional divisions to complete the radial pattern seen in the postembryonic root (Figure 7.1) (Scheres et al. 1994). At this stage, the number of vascular initials is already approaching the cell number of vascular initials observed in the postembryonic root; for example, at the late torpedo stage of embryogenesis, the stele typically comprises of about 9 pericycle cells compared with 12–14 in the postembryonic root meristem. By comparison, the basal portion of *mp* embryos contain relatively few indistinct cells indicating that little growth has occurred between the late heart and midtorpedo stages (Berleth and Jürgens 1993).

The ontogeny of root and hypocotyl cell lineages has been studied during embryogenesis by clonal activation of the β-glucuronidase marker gene (Scheres et al. 1994). This technique marks an individual cell in which the β-glucuronidase marker gene is activated by transposon excision along with all its daughter cells. This process gives rise to populations of marked cells (known as sectors) which can be used to follow the origin of these cells. This analysis confirmed that the cells in the vascular initials give rise to the vascular tissues of mature plants. However, in the mature root, these sectors always contained more than one cell in the radial direction and sometimes extended to ¼ of the circumference of the root, indicating that there is later division of the vascular cells in subsequent stages in which the radial pattern is refined. During embryogenesis, the individual vascular cells remain undifferentiated, and characteristic cellular properties (such as modification of the cell wall) are not visible until postembryonic growth occurs (Dolan et al. 1993).

III. Vascular Anatomy in the Growing Root

A zone of undifferentiated stem cells surrounds a central group of organizing cells in the quiescent center (QC) which produces and specifies cell identity in two directions resulting in respectively the columella root cap below and the vascular tissues above the QC. The QC and surrounding stem cells form the meristematic zone where cells are actively dividing (with the exception of the QC). Shootward of the meristem is a zone in which cell division ceases and the cells elongate, termed the elongation zone; shootward of the elongation zone is the differentiation zone, in which cells differentiate according to their specified fate.

The postembryonic root vascular tissue is comprised of several tissue types: xylem, phloem, and procambium. Together, these support the root system and function as a conductive network allowing the transport of water, nutrients, and chemical messengers between organs. In all angiosperms, the vascular tissue is contained within an outer boundary of pericycle cells, and together these form a region known as the stele. The endodermis and cortex ground tissues lie immediately outside of the pericycle cells. There is considerable variation in the vascular pattern both among species and even between individuals. Although the root vascular anatomy has been studied in a wide variety of species, the developmental mechanisms that specify the various cell types and which pattern the stele are only well understood in the model plant *Arabidopsis*.

Root growth is characterized by two distinct phases. During primary growth, the root elongates and the xylem and phloem cell lineages differentiate. The primary root of *Arabidopsis* is arranged according to a bisymmetric pattern. This means that there are two planes of symmetry located at 90° to each other. The first plane of symmetry runs through a central one-cell-wide xylem axis extending from the pericycle through the vascular cylinder (Figure 7.3). Two different types of xylem cell lineages characterize this axis. Protoxylem cell lineages differentiate at the marginal positions (i.e., adjacent to the pericycle cells), and metaxylem cells differentiate in the central positions. This axis is flanked on both sides by the intervening procambial cells; these provide a population of undifferentiated cells that later form the vascular cambium. Two phloem poles are located at 90° to the xylem axis, and the second plane of symmetry bisects these cells.

Dicotyledonous plants also undergo a secondary growth process where the vascular cambium divides to form secondary xylem and phloem. This secondary growth is responsible for the thickening of plant roots and stems. This thickening is most obvious in woody plants but also occurs in the *Arabidopsis* root.

A. Cambium

In contrast to animals, plants show indeterminate growth. As differentiated cells are rarely potent to divide or to redifferentiate with new cell identities, this requires the plants to maintain pools of self-renewing undifferentiated cells at the sites of active growth. These occur not only at the root and shoot apices, where they promote growth of the plant and allow new organs to form, but also in the cambium, where they allow the proliferation of vascular tissues and determine vascular cell fate.

During embryonic development, vascular initial cells are the only cell types present in the vascular cylinder. Soon after germination, the QC coordinates the differentiation of a subset of the vascular initials into either xylem or phloem. The remaining cells remain undifferentiated and are known as intervening procambial cells. Later these will form the lateral meristem, known as the cambium. At the moment, we have relatively little knowledge concerning how the transition from procambium to cambium occurs. During secondary development, both the

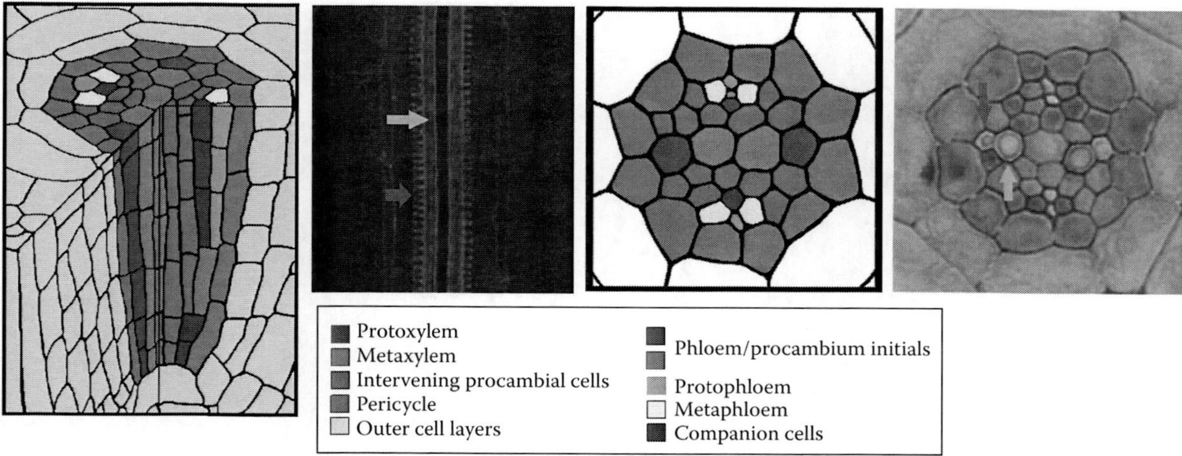

FIGURE 7.3 **(See color insert.)** Tools of the trade: vascular cell types in the mature root and their visualization. The schematic images show two different views of the vascular cells in the *Arabidopsis* root with the cell lineages marked. The center right panel shows a cross section taken approximately 50 µm above the QC. The cell layer immediately outside of the pericycle is the endodermis. Visualizing the vascular cells can be challenging. The center left panel shows a confocal image of a root stained with fuchsin-red dye. This stains lignified tissue, making the protoxylem and metaxylem cells easy to visualize. Protoxylem cells are characterized by a helical pattern on the cell wall (dark-blue arrow), while metaxylem cells stain more solidly and have a pitted pattern (light-blue arrow). A cross section through the root stained with toluidine-blue dye shows other cell types clearly. Here the xylem cells can be recognized by their thick cell walls (blue arrows). The proto- and metaphloem can be identified as they are small white cells at 90° to the xylem and the intervening procambial cells are dark blue.

cambial and pericycle cells undergo periclinal divisions. The cambial cells proliferate and give rise to secondary phloem centrifugally (outward) and xylem centripetally (inward). See also Chapter 8.

B. Xylem

The word "xylem" originates from the Greek *xylon* meaning "wood," and indeed xylem is the tissue that forms woody structures in the stems and roots and gives support to the plant. In addition to providing strength and structure to plants, xylem forms the conduit through which water, minerals and signaling molecules are transported from the roots to the rest of the plant. During primary root growth, protoxylem is the first xylem cell type to differentiate, but both proto- and metaxylem differentiate before the onset of secondary development. Protoxylem cells are characterized by being narrower and through their lignification, which occurs in an annular or a helical pattern (Figure 7.3). These cells typically mature before the adjacent tissues have elongated, and it is thought that the helical structure of their secondary cell wall allows them to stretch to accommodate the elongation of other tissues. Metaxylem cells are typically wider, and lignin is deposited across the cell wall in a pitted or reticulate manner. As the plant grows, the metaxylem cells are maintained, and they form tracheary elements and vessels, which are the water-conducting organs of the mature plant (Esau 1977).

There has been considerable effort to understand how lignin and cellulose are incorporated into the structure of xylem cell walls (Lacayo et al. 2010) and how these sugars can be most efficiently extracted. These studies have been mostly based on cell cultures, and there has been comparatively little research on the mechanisms underlying xylem development and how this can be manipulated to produce greater biomass for fuel production.

C. Phloem

The word "phloem" is derived from the Greek *phloos*, meaning "bark," since phloem forms the innermost layer of bark in trees. Phloem, which transports a wide range of substances, including sugars, micronutrients, amino acids, and hormones, plays a major role in interorgan communication and signaling within plants. It consists of several different cell types: enucleate sieve elements, which form the conductive network; companion cells, which maintain the sieve elements; and phloem fibers and phloem parenchyma (Esau 1977). In the *Arabidopsis* primary root, the phloem poles, which consist only of sieve elements and their adjoining companion cells, are located perpendicular to the xylem axis. The phloem and a subset of intervening procambial cells arise from the same precursor initial cells by several asymmetric cell divisions which take place within a distance of 50 µM from the QC (Mähönen et al. 2000). By about 70 µM above the QC, differentiation of the protophloem sieve elements is apparent (Bonke et al. 2003).

IV. Hormonal Signaling Establishes the Radial Vascular Pattern

The first indications that cytokinin signaling regulates vascular cell identity arose from the cloning and characterization of the *woodenleg* (*wol*) mutant. In addition to the embryonic defects, the *wol* mutant also has a short primary root that shows determinate growth with root growth aborting about 7 days after

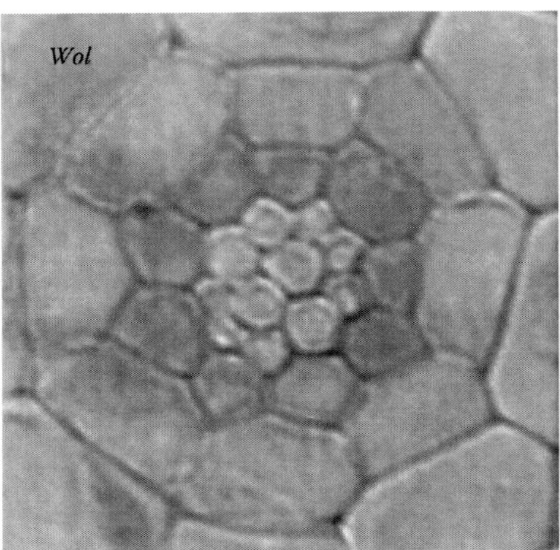

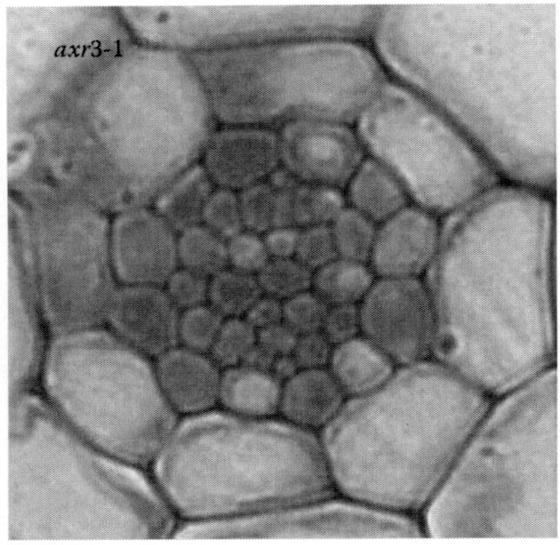

FIGURE 7.4 **(See color insert.)** The level of hormonal signaling regulates the extent of protoxylem differentiation. In the absence of cytokinin signaling (such as in the *wol* mutant), all vascular cells acquire protoxylem cell fates. Note the thick secondary cell walls in the left image. When auxin signaling is severely compromised (such as in the *axr*3-1 mutant), no xylem differentiation occurs. Note that no secondary walls are visible in the right image and all vascular cells are darkly stained. (Images courtesy of Hanna Help, University of Helsinki, Helsinki, Finland.)

germination. In the primary root, there are fewer cells within the vascular cylinder compared with wild type, and all these cells differentiate as xylem in a radially symmetric pattern (Figure 7.4) (Scheres et al. 1995). The number of vascular cells increases in the upper parts of the hypocotyl; both phloem and procambial cells are present in this zone. Despite the defects in the primary root, this mutant is not lethal as growth is rescued by the formation of adventitious roots from the hypocotyl. Cloning of this gene revealed that it encoded a two-component signaling molecule.

A. Two-Component Cytokinin Signaling Specifies Protoxylem versus Procambial Cell Fate

Independently from this effort to characterize *wol*, several research groups were investigating components of the cytokinin signaling pathway and identified the *CYTOKININ RESPONSE 1* (*CRE1*) and *ARABIDOPSIS HIS-KINASE 4* (*AHK4*) genes as receptors for cytokinin (Inoue et al. 2001; Suzuki et al. 2001; Ueguchi et al. 2001; Yamada et al. 2001). These efforts came together when it was discovered that *CRE1*, *AHK4*, and *WOL* loci are allelic. The *wol* mutation is a single-nucleotide substitution within the cytokinin-binding domain of CRE1/AHK4/WOL. While in vitro assays have been used to demonstrate the cytokinin-binding activity of CRE1, this binding activity is lost when the *wol* mutation is introduced (Yamada et al. 2001).

Analysis of plants with altered cytokinin levels strengthened the link between cytokinin signaling and the *wol* phenotype. Transgenic plants were generated in which cytokinin levels were depleted in the vascular cells by overexpressing the cytokinin depleting enzyme CYTOKININ OXIDASE (CKX). These lines phenocopy *wol* mutants as all vascular cells differentiate as xylem (Mähönen et al. 2006a). Conversely, when roots are treated with exogenous cytokinin, protoxylem identity is lost (Mähönen et al. 2006a). Together, these experiments demonstrate a role for cytokinin signaling in specifying the identity of vascular cells and show that in the absence of cytokinin signaling, protoxylem is the "default" identity.

In the following years, additional components of the cytokinin signaling pathway were discovered, and the mechanism through which it operates was elucidated. Cytokinin is initially perceived by a group of CRE1-like receptors located on the plasma membrane. These receptors then propagate this message to nuclear-localized response regulators via a phosphorylation cascade similar to that observed in bacterial two-component signaling systems. In this process, an environmental stimulus is perceived by the bacterial receptor, and this initiates the autophosphorylation of a conserved histidine residue (His) on the receptor. This phosphate group is then transferred from the His residue on the receptor to an aspartic acid residue (Asp) on a downstream response gene. This phosphotransfer modulates the activity of the response regulator and allows it to activate the transcription of downstream targets. A more complex variation of this process exists in plants and fungi where two additional phosphorelay steps are included and the phosphorelay occurs in the sequence His>Asp>His>Asp. It is this system that has been recruited for cytokinin signaling in plants (Figure 7.2).

The binding of cytokinin triggers the autophosphorylation of a His residue in the kinase domain of the cytokinin receptor. This phosphate group is then transferred to an Asp residue on the receiver domain of the cytokinin receptor and on to a His residue of a histidine phosphotransfer protein (HPt). When phosphorylated, the HPts can translocate to the nucleus and relay the phosphate group to a conserved Asp residue on a group of type-B response regulators. Phosphorylated type-B response regulators activate transcription of downstream targets of cytokinin response. Among these downstream targets is a group of cytokinin-responsive type A response regulators which act to inhibit cytokinin signaling by an unknown mechanism.

CRE1/AHK4/WOL belongs to a small gene family of cytokinin receptors that comprises two other members, *AHK2* and *AHK3* (Hwang et al. 2002). Surprisingly, when new supposedly null alleles of *CRE1* were identified, these did not show defects in vascular growth or development. At the same time, a genetic screen for modifiers of *wol* revealed some surprising results. Several mutants were identified which suppressed the *wol* phenotype. These included an extragenic suppressor mutant (*ahp6*—discussed later) and several intragenic suppressor mutants including some introducing premature stop codons (Mähönen et al. 2006b). This demonstrates that the *wol* mutant does not harbor a loss-of-function mutation but displays a dose-dependent negative effect on vascular cell proliferation and specification. However, *cre1 ahk2 ahk3* plants, which lack all three cytokinin receptors, show determinate root growth coupled with reduced vascular cell number of which all cells differentiate as xylem. This phenotype is stronger than *wol* because root growth is never rescued by the formation of adventitious roots (Mähönen et al. 2006b). This supports the observation that cytokinin signaling is required to restrict the specification of xylem genes but raises the new question of how the *wol* mutant exerts such a strong gain-of-function effect on cell proliferation and specification.

In vitro analysis of the phosphotransfer revealed that CRE1 exhibits not only a kinase activity to phosphorylate HPts in the presence of cytokinin but also a phosphatase activity which dephosphorylates HPts in the absence of cytokinin. As the *wol* mutation lies in the cytokinin-binding domain, it renders CRE1 to be constantly in a cytokinin-unbound state and exhibit constitutive phosphatase activity. This constitutive phosphatase activity has the effect of removing phosphate from the HPts and counteracting the kinase activities of the other receptors (Mähönen et al. 2006b).

Mutations have been identified and characterized in other genes involved in the phosphorelay. In *Arabidopsis*, the HPts are represented by a family of five members which all contain the conserved histidine necessary to act as phosphotransfer protein and one pseudomember which lacks this residue (*AHP6*). The individual HPt mutants do not display discernable phenotypes, but the quintuple *ahp1-5* mutant which lacks all five HPts has a strong vascular phenotype which copies *wol* or the triple cytokinin receptor mutant (Hutchison et al. 2006). In *Arabidopsis*, the cytokinin response regulators comprise large gene families. There are 11 type-B ARRs that contain both a phospho-receiving Asp residue and GARP DNA binding domain capable of initiating transcription of cytokinin target genes. Compared with other components of the cytokinin signaling machinery, there is greater specificity in the expression patterns of type-B ARRs; only *ARR1*, *ARR2*, *ARR10*, and *ARR12* are expressed strongly in the primary root tip (Mason et al. 2004). Although the single mutants do not display discernable vascular phenotypes, the *arr1 arr10 arr12* triple mutant phenocopies both the *cre1 ahk2 ahk3* triple mutant and the *ahp1-5* quintuple mutant (Argyros et al. 2008; Ishida et al. 2008). Together, these findings demonstrate a crucial role for cytokinin signal transduction via two-component signaling in the specification of vascular cell fate.

B. Spatially Specific AHP6 Expression Restricts Protoxylem to the Correct Position

The domain of high cytokinin signaling can be visualized by following the expression of primary response genes such as *ARR15* (a type-A ARR). *ARR15* is expressed in the intervening procambial cells adjacent to the xylem axis and is excluded from the protoxylem position (Mähönen et al. 2006a). When plants are treated with exogenous cytokinin, this domain of high cytokinin signaling expands into the protoxylem position at the same frequency with which protoxylem identity is lost. This raises the question: what factor normally restricts cytokinin signaling at this position to allow protoxylem differentiation?

An answer to this question arose from the genetic screen for modifiers of the *wol* phenotype. As well as the intragenic suppressors mentioned earlier, an additional extragenic locus (*AHP6*) was discovered which partially suppressed the *wol* phenotype. *AHP6* shares homology with the other five HPts but lacks the conserved His residue necessary for phosphotransfer. In vitro biochemical studies not only confirmed that AHP6 is incapable of acting as a phosphotransfer protein but also showed that AHP6 acts as an inhibitor of cytokinin signaling by interfering with the phosphorelay machinery (Mähönen et al. 2006a).

AHP6 is expressed specifically at the protoxylem position and in the two accompanying pericycle cells. When *AHP6* is knocked out, the domain of cytokinin signaling expands into the protoxylem position, and procambial cells are often present in the position normally occupied by protoxylem cells; *ahp6* mutants typically display a phenotype where protoxylem identity is missing from long stretches of one or both protoxylem cell lineages.

C. Auxin Signaling Promotes AHP6 Expression and Protoxylem Identity in the Correct Position

If AHP6 acts specifically at the protoxylem position to restrict the domain of cytokinin signaling, this raises the question of how *AHP6* expression is driven specifically in this domain.

Both synthetic auxin reporter genes (*DR5*) and the expression of an endogenous *Aux/IAA* gene (*IAA2*) have been used to follow auxin response in the vascular cylinder of the primary root meristem (Bishopp et al. 2011a). Close to the QC, the maximal auxin response was seen throughout the xylem axis, and later during meristem growth (approximately 40 μm from the QC), this became restricted to two maxima at the protoxylem positions. This pattern is completely nonoverlapping with the cytokinin response pattern. Experimental evidence suggests that *AHP6* is a direct target of the auxin signaling machinery. *AHP6* transcription is rapidly induced in auxin-treated roots, and a synthetic version of the *AHP6* promoter which lacks putative auxin binding sites fails to drive expression in the protoxylem. Additionally, transgenic lines in which auxin signaling is impaired (the stabilized *Aux/IAA axr3-1* was driven ectopically) and chemical treatments with auxin transport inhibitor NPA also lead to loss of *AHP6* expression at this position.

D. Mutually Inhibitory Interaction between Auxin and Cytokinin Specifies the Radial Vascular Pattern

There is a well-established model of auxin flow through the root. In this the rootward flow of auxin through the vascular cylinder to the root meristem is coordinated with the shootward flow of auxin through the epidermis and cortex. This provides a "reverse fountain" of auxin that maintains an auxin response maximum at the QC (Swarup and Bennett 2003; Blilou et al. 2007; Grieneisen et al. 2007). The auxin maximum in the xylem axis is maintained through the bisymmetric pattern of three PIN proteins (Bishopp et al. 2011a). *PIN7* is expressed in the intervening procambial cells that flank the xylem axis. The PIN7 protein is localized to lateral as well as rootward-facing plasma membranes. *PIN1* is expressed throughout the vasculature, but while it is only expressed in rootward-facing membranes in the xylem initials, it is additionally expressed on lateral membranes in the intervening procambial cells. Close to the QC, *PIN3* is expressed in the protoxylem-associated pericycle cells; higher up, this expression pattern expands to include the central cells of the xylem axis. Collectively these three PIN proteins direct the rootward flow of auxin through the xylem axis.

Cytokinin signaling is required for the radial transport of auxin from the intervening procambial cells into the xylem axis (Bishopp et al. 2011a). This can be seen in both *wol* mutants where an even pattern of auxin response is present in all vascular cells and in roots treated with cytokinin in which the domain of auxin response shrinks and can no longer be observed in cells at the protoxylem position. This effect is mediated through modulating the expression and possibly the activity of PIN proteins. PIN7 is completely absent from the vascular cylinder of *wol*, PIN3 levels are severely reduced, and PIN1 becomes exclusively localized on rootward-facing membranes in all vascular cells.

Together, these two domains of high hormonal signaling exert a mutually inhibitory effect on each other. Auxin signaling promotes *AHP6* expression, and this inhibits cytokinin signaling to enforce a domain of high auxin response (Figure 7.5). High cytokinin signaling in the intervening procambial cells promotes the radial transport of auxin out of those cells to enforce a domain of high cytokinin response. This mutually inhibitory effect is able to specify the boundary between the two hormonal domains and specify the differentiation of protoxylem in a bisymmetric pattern. Dramatic alterations in hormonal signaling such as the *wol* or *axr3-1* (which almost completely abolish cytokinin/auxin signaling within the vascular cylinder) cause either all-protoxylem or no-protoxylem phenotypes, respectively (Figure 7.4). Smaller alterations, such as the *cre1 ahk3* double mutant which has reduced but still active cytokinin signaling, result in a smaller domain of cytokinin signaling, which is typically reduced by one cell. The auxin domain expands accordingly, and this is accompanied by a comparable increase in protoxylem, such that these mutants display one or two ectopic files of protoxylem adjacent to the regular position.

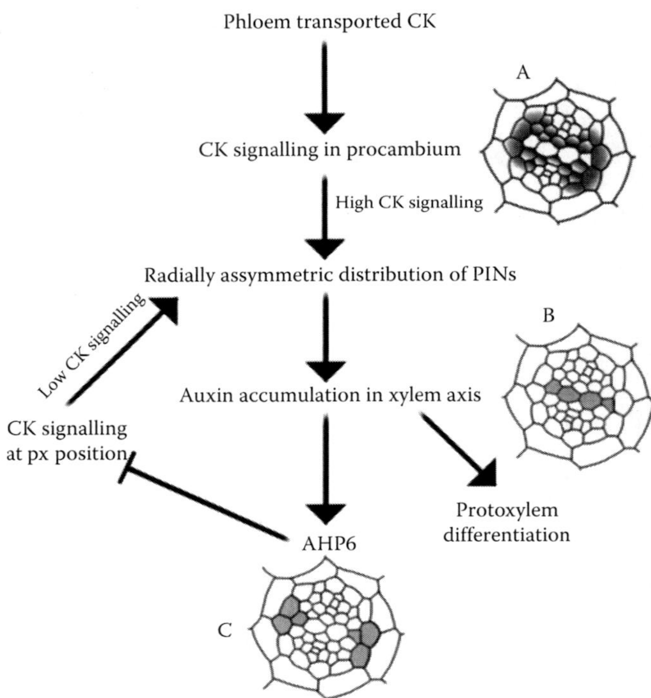

FIGURE 7.5 Mutually inhibitory domains of auxin and cytokinin pattern the vasculature. There is a cytokinin response maximum in the intervening procambial cells (A). In the mature root, the phloem is a source for at least some of this cytokinin. High cytokinin response at this position promotes the bisymmetric distribution of the PIN class of auxin efflux proteins. This forces the auxin into the xylem axis, and an auxin response maximum can be seen at this position (B). Auxin signaling at this position promotes the differentiation of protoxylem and the expression of *AHP6* (C) at the marginal positions of the xylem axis. AHP6 inhibits cytokinin signaling in the protoxylem position which reinforces this high auxin response.

Several lines of evidence suggest that it is high auxin at the protoxylem position (as opposed to low cytokinin) that specifies protoxylem identity. While manipulating cytokinin levels or signaling causes increased protoxylem specification, this has never been observed independently from alterations in auxin signaling. For example, when cytokinin levels are depleted by the ectopic expression of CKX, any increase in protoxylem differentiation is always accompanied by a comparable increase in auxin signaling. However, when auxin signaling is reduced in the *wol* background by ectopic expression of the stabilized *axr3-1* gene, the all-protoxylem phenotype of *wol* is somewhat suppressed, indicating that low cytokinin signaling is not sufficient for the specification of protoxylem identity. In addition, high auxin signaling at the protoxylem position initiates the transcription of the xylem identity gene *AtHB8* (see Section V; Donner et al. 2009; Carlsbecker et al. 2010; Bishopp et al. 2011a). While these observations suggest that it is high auxin which specifies protoxylem fate, further research will be needed to uncover whether cytokinin signaling promotes the transcription of a unique, as yet unidentified, set of genes required for maintaining intervening procambial identity and to establish phloem identity.

E. Top-Down Signaling Establishes the Initial Vascular Pattern

The mutually inhibitory interaction described earlier is capable of resetting vascular pattern; when plants are grown for many days on the auxin transport inhibitor NPA, the vascular cell number increases and multiple ectopic xylem poles form (Mattson et al. 1999; Bishopp et al. 2011a). However, it is very likely that top-down migration of auxin response from the cotyledons establishes the initial bisymmetric pattern during embryogenesis (Bishopp et al. 2011a), and it is certain that top-down hormonal transport maintains this pattern (Figure 7.5). It has been well documented that the top-down transport of auxin through polar auxin transport provides a substantial source of auxin to the meristem (Blilou et al. 2005; Grieneisen et al. 2007). There is also a top-down transport of cytokinin occurring through the phloem (Bishopp et al. 2011b). This cytokinin transport mechanism is required to maintain the cytokinin signaling domain in the intervening procambial cells of the root meristem, and mutants with impaired flow of cytokinin through the phloem show unstable hormonal signaling domains in the root and erratic differentiation of protoxylem (Bishopp et al. 2011b).

V. SHR, MicroRNA Movement, and HD-ZIP Gene Dosage Control the Centripetal Pattern of Protoxylem versus Metaxylem Specification

While the degree of hormonal signaling restricts the radial pattern of protoxylem specification, relatively minor changes in cytokinin levels or signaling do not alter the centripetal pattern of protoxylem (i.e., ectopic protoxylem is always restricted to the marginal positions of the vascular cylinder). This raises the question: how is the centripetal pattern set?

A. Level of Class III HD-ZIP Activity Determine the Centripetal Arrangement of Xylem Cells

The HD-ZIPs are a group of transcription factors unique to plants. Although all members contain a DNA binding homeodomain and a leucine zipper domain, this group can be further subdivided based on protein structure. In *Arabidopsis*, there are five members of the Class III group, *PHABULOSA* (*PHB*), *PHAVOLUTA* (*PHV*), *CORONA* (*CNA*), *REVOLUTA* (*REV*), and *AtHB8*. The function of these genes is regulated by the microRNAs miR165/6 which can cleave the mRNA of all five HD-ZIP genes (Zhong and Ye 2007). These genes have previously been shown to play important roles in the morphogenesis of the apical portion of the embryo, the shoot apical meristem, and the abaxial-adaxial patterning of lateral organs (Otsuga et al. 2001; Emery et al. 2003; Prigge et al. 2005).

Researchers identified the *phb-1d* mutant in a genetic screen for plants with altered vascular development as the mutant was short rooted and metaxylem differentiated in the place of protoxylem (Carlsbecker et al. 2010). This turned out to be a gain-of-function mutant with a point mutation in the miR165/6 targeting site. Normally, the expression of *PHB* mRNA is restricted by the action of miR165/6 to the metaxylem initial cells and adjacent procambial cells with a low level of expression in the protoxylem initials. However, the *phb-1d* mutation renders the *PHB* mRNA unresponsive to microRNA regulation, and the domain of *PHB* mRNA is expanded throughout the vascular cylinder and adjacent cells (Carlsbecker et al. 2010).

In addition to PHB, other members of the Class III HD-ZIP genes are expressed in the root vascular cylinder (Carlsbecker et al. 2010). *CNA* is expressed in the metaxylem initial cells and adjacent procambial cells. *AtHB8* is expressed specifically in the xylem axis. *REV* has a broad vascular expression pattern close to the QC. Genetic data support the idea that *PHV* is expressed in the root vasculature, but its expression is likely to be low and has so far been undetected. These genes act redundantly to specify xylem patterning, and loss-of-function mutants in any of these genes show normal xylem anatomy. However, when four of these genes are knocked out, protoxylem cells differentiate in the place of metaxylem (Figure 7.6). When all five Class III HD-ZIP genes are knocked out, there is no differentiation of either proto- or metaxylem (Figure 7.6). Together, this demonstrates that a gradient of Class III HD-ZIP activity controls the differentiation of the different xylem cell types in the correct centripetal pattern. HD-ZIP activity is required for xylem differentiation; at the lowest levels in the periphery of the vascular cylinder, it promotes the specification of protoxylem, and in the central regions where HD-ZIP activity is highest, it promotes the specification of metaxylem. How then is this gradient of HD-ZIP activity maintained?

B. SHR and MicroRNA Movement Act Non-Cell-Autonomously to Restrict HD-ZIP Expression

SHORT ROOT (SHR) and SCARECROW (SCR) are both members of the GRAS family of transcription factors and are required to specify the endodermis and cortex cell identity. Both these cell lineages are formed from a common initial cell that undergoes an asymmetric cell division and are collectively known as ground tissue. Mutants in either *shr* or *scr* have a single-cell lineage of ground tissue. In *shr* this has cortex features, and in *scr* this has features common to both cortex and endodermis. *SHR* mRNA is expressed within the vascular cylinder, but the protein moves into the adjacent cell layers (Helariutta et al. 2000; Nakajima et al. 2001). In the adjacent cell layer, SHR interacts with SCR and is sequestered into the nucleus. SHR and SCR activate downstream responses which specify the asymmetric cell division of the common initial cell to form endodermal and cortical cell lineages (Cui et al. 2007).

More recently, it was realized that these genes also regulate pattern formation in the vascular cylinder as in both the *shr* and *scr* mutants metaxylem differentiates ectopically in the

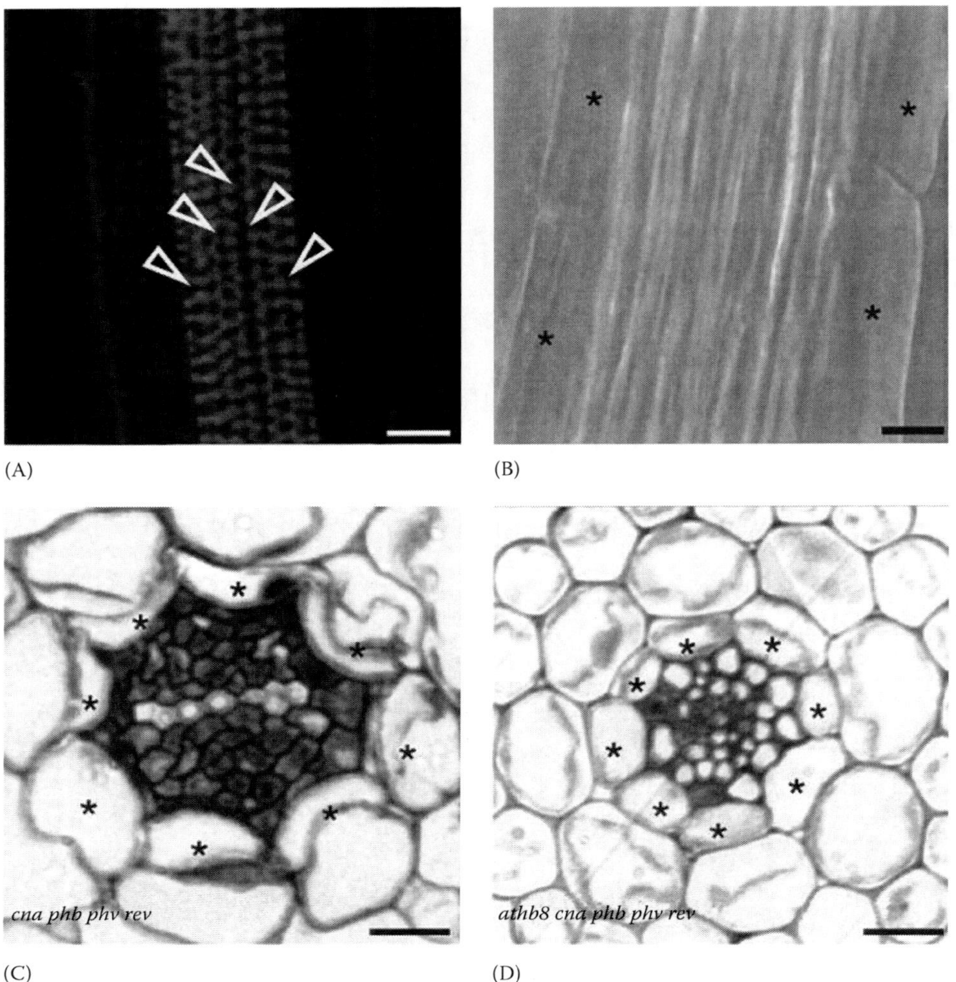

FIGURE 7.6 (See color insert.) HD-ZIP gene dosage controls the centripetal patterning of the xylem axis. When the gene dosage of class III HD-ZIP genes is low all cells differentiate as protoxylem. This can be seen here through Fuchsin staining (A) and cross sections (C) in the quadruple *can phb phv rev* mutant. The white arrowheads show protoxylem cell lines. When there is no class III HD-ZIP gene activity in the *athb8 can phb phv rev* quintuple mutant, no protoxylem develops. This can be seen in Nomarski images (B) and cross section below (D). It is not appropriate to perform Fuchsin stainings when there is no xylem as there will be no incorporation of the stain. The pericycle cells have been marked with asterisks in order to orientate the reader to the location where xylem would normally develop. (This image has been reproduced by permission from Macmillan Publishers Ltd., *Nature*, Carlsbecker, A., Lee, J.Y., Roberts, C.J. et al., Cell signalling by microRNA165/6 directs gene dose-dependent root cell fate, 465, 316–321, 2010. Copyright 2010.)

place of protoxylem. Genome-wide analysis of SHR and SCR targets revealed that all five class III HD-ZIP genes were upregulated in the *shr* and *scr* backgrounds. This increased HD-ZIP expression results in the xylem patterning defects in *shr*, and this phenotype can be rescued in *shr phb* double mutants. SHR protein is normally present in both the vascular cylinder and the endodermis, but the effects on xylem pattern require both SHR and SCR activity in the endodermis. The xylem patterning defects in *shr* can be complemented when an immobile version of SHR is driven in the endodermis but not when it is driven only in the vascular cylinder. It has been shown that SHR activates miR165/6 in the endodermis; miR165/6 levels are dramatically reduced in both *shr* and *scr* mutants, and SHR has been shown to bind to the promoters of several miR165/6 species. In order for the microRNA produced in the endodermis to encounter the HD-ZIPs in the vascular cylinder, there must be a mobile signal, and there is a considerable body of evidence to suggest that it is these microRNA species themselves that move.

The centripetal pattern of xylem specification is controlled by a gradient of gene dosage of the class III HD-ZIP genes. The expression of these genes is controlled by cell-to-cell communication between the vascular cylinder and the endodermis in which the transcription factor SHR is expressed in the vascular cylinder and then moves into the endodermis where, together with the transcription factor SCR, it activates the expression of mobile microRNAs. These microRNAs degrade the HD-ZIPs in the endodermis and vascular cylinder to create a gradient of

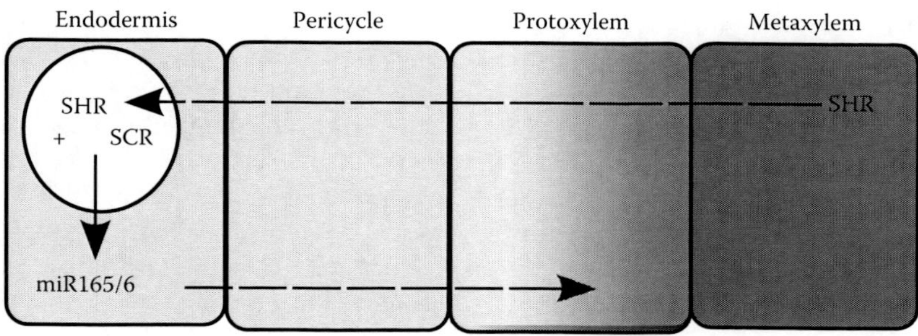

FIGURE 7.7 Cell-to-cell signaling from the endodermis specifies proto- versus metaxylem cell fate. SHR moves from the vascular cylinder to the endodermis. In the endodermis, it is sequestered into the nucleus where it interacts with the SCR transcription factor to activate the microRNA species 165 and 166. These species then move back into the vascular cylinder where they degrade the class III HD-ZIP transcription factors in the endodermis and the cells in the periphery of the stele. This creates a gradient of microRNA (shown in light grey) and a gradient of HD-ZIP expression (shown in dark grey). The HD-ZIP gene dosage specifies xylem cell fate so that metaxylem forms in the central cells where gene dosage is highest and protoxylem forms in the marginal position where gene dosage is lower.

HD-ZIP activity as described earlier (Figure 7.7). Auxin directly promotes the expression of the class III HD-ZIP gene *AtHB8* (Donner et al. 2009; Bishopp et al. 2011a). Future studies should reveal whether expression of other HD-ZIP genes is regulated by either cytokinin or auxin signaling.

VI. NAC Genes Oversee the Differentiation of Xylem Cells

A. Xylem Differentiation Is Regulated by VND6 and VND7

In order to identify genes involved in xylem morphogenesis, researchers used an in vitro system which allowed them to induce the formation of xylem in cell cultures (Kubo et al. 2005). By treating an *Arabidopsis* suspension culture with appropriate hormones, they were able to induce ~50% of the cells to differentiate into xylem elements within 7 days. Using a microarray to track changes in gene expression levels on a genomic level, they were able to identify 23 clusters of coregulated genes; of these, three clusters (comprising 224 genes) showed upregulation 6 days after induction, coincident with active xylem formation. While these included many genes encoding proteins that would be expected to play a role in developmental events, such as proteins involved in cell wall synthesis and programmed cell death, several putative transcription factors were also identified, including four NAC-domain transcription factors. These four transcription factors showed significant similarity to another NAC-domain transcription factor, *Z567*, previously seen to be upregulated during xylem formation in cell cultures of *Zinnia elegans* (Demura et al. 2002). Based on sequence similarity to *Z567*, three additional NAC-domain transcription factors were identified in the *Arabidopsis* genome; although they had not been recovered in the clusters, these three genes were also found to be upregulated in the microarray data. Collectively, these transcription factors were designated *VASCULAR-RELATED NAC-DOMAIN* (*VND*) *1–7*.

All of the *VND*s show a vascular-specific expression pattern in the shoot and root, with the exception of *VND1*, *4*, and *6*, which do not seem to be expressed in the shoot. Within the root vascular tissue, the VND transcription factors show variable expression patterns; *VND1*, *2*, and *3* are preferentially expressed in the intervening procambial cells adjacent to the root meristem, while *VND4–7* are observed primarily in mature xylem which does not yet show obvious secondary cell wall thickening. Within the xylem, *VND6* and *VND7* are nuclear localized in the central metaxylem and immature protoxylem, respectively.

While overexpression of *VND1-5* did not result in any morphological changes, overexpression of *VND6* or *VND7* resulted in transdifferentiation of various nonvascular cells into xylem in both poplar and *Arabidopsis*. In plants overexpressing *VND6*, the transdifferentiated cells had reticulate, pitted cell walls similar to those found in metaxylem; in those overexpressing *VND7*, the transdifferentiated cells had annular cell walls with a spiral thickening pattern, similar to protoxylem cells. In addition, several xylem-specific genes were found to be ectopically expressed in the *VND7* overexpression plants. By contrast, while knockout lines of *VND6* and *VND7* showed no morphological defects, the dominant repression of either gene (by overexpression of the gene fused to a strong repression domain) resulted in shorter roots with defects in the formation of metaxylem and protoxylem, respectively.

Both *VND6* and *VND7* appear to respond strongly to several hormones. Hypocotyls cultured with both auxin and cytokinin showed higher levels of *VND6* and *7* expressions than hypocotyls grown on hormone-free medium. The highest level of expression was found when plants were grown in the presence of auxin, cytokinin, and brassinosteroids.

B. VND7 Forms a Heterodimer with VNI2, Which Inhibits Upregulation of VND7 Targets

Yamaguchi et al. (2008) showed that VND7 can form homo- and heterodimers with the other NAC-domain proteins, including the VNDs; in addition, the stability of VND7 appears to be regulated by proteasome-mediated degradation. Taken together, these data suggest that the transcriptional activity of VND7 may be regulated by its interactions with other proteins. In later work, Yamaguchi et al. (2010, 2011) began to elucidate these interactions and identify other components regulated by VND7.

Using a yeast two-hybrid screen, they identified two NAC-domain proteins which interacted with VND7, dubbed *VND-INTERACTING* (*VNI*) *1* and *2*. Binding assays demonstrated that VNI2 effectively interacts with VND7, as well as with the other VND proteins with a lower affinity. Furthermore, *VNI2* is expressed in the xylem and phloem precursor cells in the root and shoot; this expression pattern overlaps with that of *VND7*.

Contrary to *VND7*, overexpression of *VNI2* results in discontinuities in protoxylem formation. Based on a transient *Luc* reporter assay, it was determined that *VNI2* is a transcriptional repressor. Given the overlapping expression and interaction with VND7, it seems likely that VNI2 affects xylem development by inhibiting the upregulation of *VND7* targets; indeed, a transient *Luc* reporter assay demonstrated just such repression. Although the *vni2* mutant shows no visible phenotype, the expression level of several genes involved in xylem formation is elevated in the mutant, consistent with *VNI2* acting as a transcriptional repressor during xylem development.

Finally, a transcriptome analysis by Yamaguchi et al. (2011) identified 63 putative targets of *VND7*. These targets include a broad range of proteins, such as transcription factors and proteolytic enzymes; this suggests the existence of a transcriptional network downstream of *VND7* regulating xylem differentiation. Several recently published studies have identified genes involved in processes such as secondary wall biosynthesis which may play a role in this regulatory network (see Yamaguchi et al. 2010 for a list).

VII. Long-Range Cell-to-Cell Communication through CLE Peptides Regulates Xylem Vessel Formation

A. CLE Peptides Inhibit Xylem Differentiation and Promote Procambial Proliferation

The CLAVATA3/EMBRYO SURROUNDING REGION-related (CLE) peptides are a family of intracellular signaling comprising 32 members in *Arabidopsis*. Tracheary element differentiation inhibiting factor (TDIF) is a dodecapeptide isolated from *Zinnia elegans* mesophyll cell cultures, where it inhibits tracheary element differentiation; this peptide sequence was found to be the same as the C-terminal amino acids in *Arabidopsis* CLE41 and CLE44 (Ito et al. 2006). Although overexpression of CLE19 has an effect on xylem development (Fiers et al. 2004), the pleiotropic nature of the phenotype in CLE19-overexpressing plants makes it difficult to distinguish between direct effects and secondary effects because of changes in cell proliferation.

In addition to suppressing differentiation into xylem, application of TDIF increased the proliferation of cambial cells. By screening for mutants insensitive to TDIF, researchers were able to identify the receptor, designated as *TDIF receptor* (*TDR*). Plants with a mutation in *tdr* had significantly fewer procambial cells than wild-type plants, though no significant difference was observed in the number of phloem and xylem cells (Hirakawa et al. 2008). It has also been shown that cotreatment with any of several CLE peptides (CLV3, CLE6, and CLE19) can enhance the vascular proliferation induced by CLE41 (Whitford et al. 2008).

Both *CLE41* and *CLE44* are expressed in the phloem and neighboring pericycle cells throughout the plant; while *TDR* is also expressed in vascular tissues throughout the plant, its expression is limited to intervening procambial cells. Based on these expression data, a model has been proposed wherein phloem and its neighboring cells produce and secrete TDIF, which is perceived by procambial cells through *TDR*, where it is thought to promote the proliferation of procambial cells and to prevent their differentiation into xylem (Hirakawa et al. 2008).

B. CLE Peptides Regulate Xylem Differentiation by Modulating Cytokinin Signaling

By exogenous application of different CLE peptides in the growth medium, researchers were able to determine that most (17 of 26) of the CLEs were capable of inhibiting the formation of protoxylem but not of metaxylem in a dose-dependent manner (Figure 7.8). Although all 17 CLEs also inhibit root growth, the researchers concluded that the protoxylem defects were not an indirect result of the altered root growth because two additional CLE peptides (CLE25 and CLE26) cause reduced root growth without affecting protoxylem development (Kondo et al. 2011).

Both *CLE9* and *CLE10* are expressed in a stele-specific manner in the differentiation and elongation zones of the root, as well as near the root meristem; both cause protoxylem defects when applied exogenously. In order to confirm that the effect of the exogenous application reflected in vivo activity, Kondo et al. constructed lines overexpressing *CLE10* in an inducible manner and observed similar protoxylem defects. *VND7* expression was also found to be reduced by the application of CLE10; given its roles as a master regulator of protoxylem differentiation, this indicates that *CLE10* inhibits protoxylem development at an early stage.

A microarray analysis was used to identify genes with expression levels altered by CLE10 treatment. By using CLE25 as a control, the researchers were able to distinguish the effect of CLE10 on root growth and isolate a set of genes involved in protoxylem regulation which responded to CLE10 treatment. Four genes were found to be significantly downregulated: *ARR5, ARR6, IPT7,* and *CKX3.* Although all of the downregulated genes play a role in cytokinin signaling or metabolism, no cytokinin-related genes were found to be upregulated by CLE10 treatment. Both *ARR5* and *ARR6* are type A ARRs, which negatively regulate cytokinin signaling; *IPT7* and *CKX3* play a role in cytokinin synthesis and degradation, respectively. Taken together, these results suggested that the effect of CLE10 on protoxylem may be mediated by a modulation of cytokinin signaling and homeostasis.

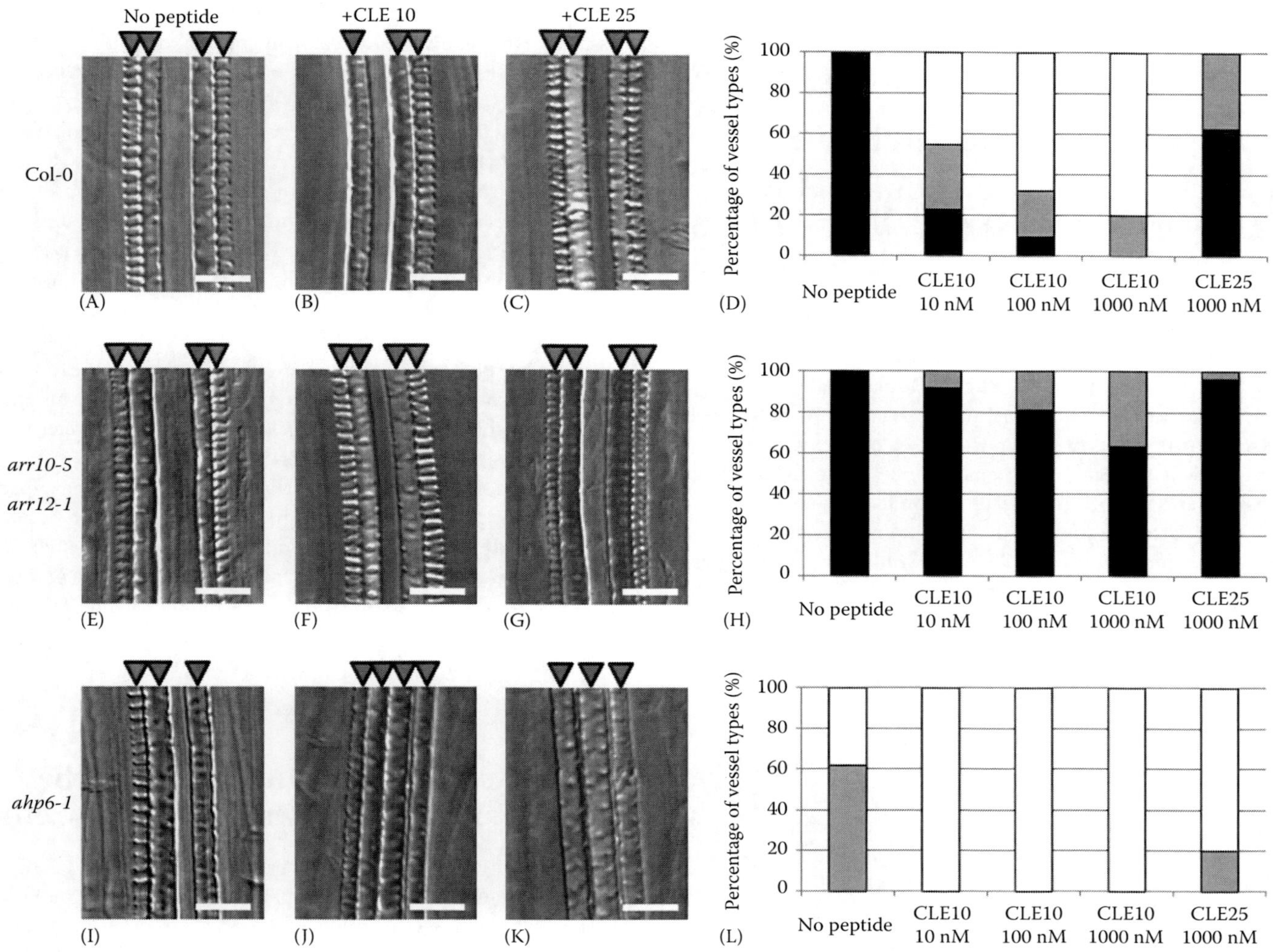

FIGURE 7.8 CLE10 suppresses protoxylem formation via cytokinin signaling. The application of CLE10, but not CLE25, causes protoxylem defects in wild type (A–D) but not in the *arr10 arr12* mutant (E–H) and enhances the loss of protoxylem in the *ahp6* mutant (I–L). As can be seen in the graphs, the effect of CLE10 is dose dependent. In the graph, black represents normal protoxylem development with two files, gray represents an abnormal phenotype with one of the two files interrupted or missing, and white represents a severely abnormal phenotype with both files missing or interrupted. (Reproduced from Kondo, Y. et al., *Plant Cell Physiol*, 52, 37, 2011. With permission from Oxford University Press.)

Because the *ahk2 ahk4* mutant, which is defective in two of the three cytokinin receptors, retains sensitivity to CLE10 treatment, it seems likely that CLE10 affects cytokinin status primarily by modulating signaling rather than metabolism. While no phenotype was observed in the protoxylem of *arr5* and *arr6* roots, the *arr5 arr6* double mutant exhibited an occasional loss of protoxylem and was found to be hypersensitive to CLE10, indicating that CLE10 may act to effectively increase cytokinin signaling by reducing the expression level of these type-A ARRs.

In order to confirm the activity of *CLE10* through cytokinin signaling, Kondo et al. examined the effect of CLE10 on plants defective in the type-B ARRs ARR10 and ARR12, which have been shown to be key mediators of the effect of cytokinin on root vascular development (see earlier). CLE10 treatment did not cause protoxylem defects in *arr10 arr12* plants, although root growth was still inhibited. Given that type-A and type-B ARRs are likely to compete for phosphotransfer, this suggests that the downregulation of *ARR5* and *ARR6* by *CLE10* may lead to increased *ARR10* and *ARR12* levels resulting in defects in protoxylem development. Because CLE10 treatment enhances the protoxylem phenotype in *ahp6* plants, it also seems likely that its action is at least partly independent of AHP6.

CLV2 encodes a CLE receptor; plants defective in *CLV2* are insensitive to the effect of *CLE10* on protoxylem formation and on *ARR5* expression. Because it lacks an intracellular kinase domain, *CLV2* cannot act as alone as a receptor. *CORYNE/suppressor of LLP2* (*CRN/SOL2*) is expressed in the root stele and has been proposed to form a heterodimer with *CLV2* in the shoot. Consistent with this, the *sol2-1* mutant showed insensitivity to CLE10 treatment (Kondo et al. 2011).

Taken together, these results suggest that *CLE9/10* signaling, mediated by a heterodimer of *CLV2* and *CRN*, regulates cytokinin signaling by modulating the expression levels of the type-A

ARRs *ARR5* and *ARR6* which compete with the type-B ARRs *ARR10* and *ARR12* for phosphotransfer, thus affecting protoxylem formation at a relatively early stage.

VIII. Differentiation of Phloem

A. APL Promotes Phloem Differentiation and Inhibits Protoxylem Differentiation

The *Altered Phloem Development* (*APL*) gene encodes an MYB coiled-coil transcription factor which is required for phloem development; it was identified from a mutant screen because the mutant produced short, determinate primary roots and had arrested shoot development (Bonke et al. 2003). *APL* is expressed in the cotyledons and hypocotyls of nearly mature *Arabidopsis* embryos. After the onset of *APL* expression, the anatomical pattern of the embryonic stele only occasionally resembles that of the postembryonic root, suggesting that phloem patterning is initiated late in embryogenesis and continues postembryonically. Postembryonically, *APL* is expressed throughout the phloem in the shoots and roots of seedlings. Close to the root meristem, its expression is limited to the protophloem sieve elements; higher up, expression expands into the developing companion cells and metaphloem sieve elements (Bonke et al. 2003).

Although *APL* is only expressed in developing phloem and does not appear to be expressed during the earlier asymmetric divisions, these divisions are delayed in the seedling-lethal *apl* mutant. *apl* plants also fail to form sieve elements and companion cells; instead, ectopic tracheary element-like cells are often found at the phloem position (Bonke et al. 2003).

Ectopic expression of *APL* throughout the vascular cylinder delays metaxylem differentiation and prevents cells at the protoxylem position from differentiating into tracheary elements. The undifferentiated protoxylem cells fail to undergo lignification and autolysis; although these cells do undergo a tangential division similar to phloem precursors, they nevertheless retain a nucleus, indicating that APL alone is not sufficient to result in ectopic phloem development (Bonke et al. 2003). The xylem defects seen in these lines seem to indicate that, in addition to specifying phloem, *APL* plays a role in inhibiting xylem differentiation at the phloem poles.

B. Brassinosteroids Induce Xylem Differentiation and Repress Phloem Differentiation

Cano Delgado et al. (2004) have identified a role for brassinosteroids in phloem development in the shoot. *BRL1–3* were identified based on sequence similarity with *BRI1*, a leucine-rich-repeat receptor kinase which is essential for the perception of brassinolide. All four proteins consist of a kinase domain connected to a single transmembrane pass followed by an LRR motif. When ectopically expressed under the *BRI1* promoter, *BRL1* and *BRL3* rescued *bri1* plants, indicating that they, too, are functional in brassinosteroid perception; immunoprecipitation assays further confirmed the affinity for brassinolide.

Unlike *BRI1*, the expression of both *BRL1* and *BRL3* is enriched in vascular tissues. *BRL1* is expressed in the columella and vascular initials of the embryonic root tip; postembryonically, expression continues in the columella, while it is also seen throughout the stele in the differentiation zone, continuing up through the hypocotyl into the midvein of the cotyledons. By contrast, *BRL3* is only expressed postembryonically and is found in the two protophloem files in the elongation zone of the root, as well as in phloem cells in the shoot.

Both *brl1-1* and *bri1-5* mutant plants exhibit an increased number of phloem cells and fewer xylem cells; the double mutant exhibits a stronger phenotype than either single mutant. The triple mutant *bri1-10 brl1-2 brl3* showed even stronger vascular defects, as well as being infertile.

A notable similarity between *APL* and the *BRL*s is the apparent connection between phloem determination and xylem repression, despite the different domains and opposite effects of these genes. These results suggest that the differentiation of these cell types may be somehow linked in a manner which remains unclear. Further research will most likely identify other components involved in these regulatory networks and clarify the interaction between them.

IX. Cytokinin Signaling Controls the Secondary Growth of Cambial Cells

The information presented so far was mostly related to primary root growth in which the cellular pattern is set up by the molecular mechanism described and root growth occurs through elongation. However, as roots grow larger, they present 3D growth during which, in addition to elongating, they undergo a process of thickening known as secondary growth. This process occurs mainly through thickening of the vascular cambium caused by cell proliferation. Cells within the cambium remain undifferentiated until environmental and internal cues trigger them to differentiate as either xylem or phloem cells. For detailed account of these processes, see Chapter 8. Although *Arabidopsis* plants only grow for 2 months and do not grow very large, the thickening of the root tissue is evident at the age of about 10 days and can be seen by an increase in the number of vascular cells.

Plant hormones such as auxin, gibberellin, and cytokinin have been implicated in the process of promoting division and differentiation of the cambium cells. In tree trunks, there is an auxin maximum in the cambial region of the trunk, and inhibition of this pattern leads to reduced cambial cell divisions and a reduction in xylem cell proliferation (Uggla et al. 1998, 2001; Nilsson et al. 2008). Overexpression of a rate-limiting enzyme in gibberellin biosynthesis, GA20 oxidase, in poplar increases cambial activity leading to an increase in stem height and thickness (Eriksson et al. 2000). However, there is little data that indicates whether these hormones play a role in promoting cambial development in roots.

As cytokinins are known to regulate cell division, it was proposed that they could also regulate cell divisions in the cambium. To test this, researchers examined *Arabidopsis* plants harboring mutations in key genes in the cytokinin biosynthesis pathway. Although there are many forms of cytokinin species in plants, the most biologically important fall into two classes: the isopentyladenine type (iP) and the *trans*-zeatin type (*t*-z). The *ATP/ADP isopentenyltransferase* genes (*IPTs*) catalyze the first step in the synthesis of cytokinin species from both of these classes (Kakimoto 2001; Miyawaki et al. 2004). In *Arabidopsis*, this gene family is represented by seven members. When the four most abundant members are knocked out in the *ipt1 ipt3 ipt5 ipt7* quadruple mutant, levels of both iP- and *t*-z-type cytokinins are severely reduced (Miyawaki et al. 2006).

The *ipt1 ipt3 ipt4 ipt7* quadruple mutants have slightly longer roots but show striking defects in root thickening compared with the wild type (Matsumoto-Kitano et al. 2008). In these mutants, the vascular cambium is almost completely missing, and there is no secondary growth until 21 days after germination. The decreased secondary growth of the *ipt* mutants can be overcome in a dose-dependent manner by the application of exogenous cytokinin, indicating that reduced cytokinin levels are responsible for this phenotype. Conversely, addition of cytokinin to wild-type plants or overexpression of IPT genes results in an enhancement of secondary growth. In both cases, only cell proliferation and not patterning is altered. Grafting of a wild-type shoot onto the root of the *ipt1 ipt3 ipt5 ipt7* mutant is sufficient to restore the levels of the iP class of cytokinins and the secondary growth of the root. Cytokinin species have been identified in the xylem and phloem sap (Kudo et al. 2009), and it has been shown that at least iP-type cytokinins travel through the phloem (Bishopp et al. 2011b). It is likely that, collectively, the transport of cytokinins through the vascular tissues acts as a mobile communication signal between the shoot and the root. It will be interesting to find out how much of an impact shoot-to-root signaling plays in controlling the secondary thickening of other more long-lived plant species.

In addition to this work performed in roots, there have been some interesting discoveries concerning secondary development during hypocotyl formation, and this has interesting implications for root development. Initially, secondary growth in the *Arabidopsis* hypocotyl involves proliferation of all cell types, but this is followed by a switch to xylem expansion and fiber differentiation. Sibout and colleagues (2008) investigated this phenomenon in a spectrum of *Arabidopsis* accessions from around the world by investigating the ratio of xylem to phloem. They identified a surprising degree of variation and investigated the genetic basis underlying this using quantitative trait loci (QTL) analysis. They discovered that the major loci regulating xylem-to-phloem ratio coincided with those that regulate flowering time. By studying gene expression data, they were able to show that xylem expansion occurs after floral induction but before the inflorescences emerge. They confirmed this by inducing the expression of the flowering regulator *CONSTANS* (*CO*). By doing so, they discovered that secondary development was enhanced in these lines, and a greater xylem proportion was observed. These data demonstrate a shoot-to-hypocotyl communication that regulates the onset of secondary development. It is likely that such a communication would extend also to the root as in many accessions studied hypocotyl and root thickness were correlated.

In the *Arabidopsis* hypocotyl, the proliferation of procambial cells and the suppression of xylem formation are controlled by a cell-to-cell communication mechanism involving the TDIF peptide and the TDR/PXY receptor (Hirakawa et al. 2008; Etchells and Turner 2010). This signaling pathway directly targets the *WUSCHEL*-related homeobox gene *WOX4* (Hirakawa et al. 2010), and *WOX4* expression is upregulated in a TDR-dependent manner following treatment with the TDIF peptide. Both the *tdr* and *wox4* mutants exhibited reduced secondary thickening in hypocotyls, although this was more pronounced in *tdr wox4* compared with *tdr* alone (Hirakawa et al. 2010). However, while the *tdr* mutants showed defects in maintaining a continuous ring of cambium, the *wox4* mutants did not. These data suggest that two signaling pathways may diverge after the recognition of TDIF by TDR and these may regulate cell proliferation and xylem differentiation separately. In addition, TDIF and TDR regulate the orientation of cell divisions (Etchells and Turner 2010). As all these components are also found in the root, it is very possible that they may regulate secondary development in the root as well.

X. Putting the Pattern Together

Most of the phenotypes described earlier have been affected in either acquiring the identity of specific vascular cell types (e.g., *ahp6*) or in the differentiation of those cell types (e.g., *VND* overexpression lines). However, there is also an additional class of vascular mutations that specifically affect the patterning. The basic helix-loop-helix (bHLH) transcription factor *LONESOME HIGHWAY* (*LHW*) does not promote the specification of any particular cell type but regulates vascular cell number and patterning (Ito and Bergmann 2007). During embryogenesis, there is little difference in development between *lhw* and wild type. The only difference observed during early embryogenesis was a slight delay in cell divisions in the basal part of the embryo. At the torpedo stage of embryogenesis, the vascular cylinder forms normally in *lhw* but is slightly narrower than that of wild type. In post-embryogenesis, there is a clear difference in cell number. While the number of cells in the endodermis and cortex is normal, *lhw* mutants have approximately half the number of pericycle and vascular cells than wild type. Striking alterations in symmetry are coupled with this reduced cell number; *lhw* mutants display only one pole of protoxylem and one pole of phloem which are separated by fewer metaxylem cells than normal.

In order to determine how LHW is related to known processes that regulate vascular patterning, the relationship between LHW and hormonal signaling was investigated. *lhw* mutants were able to respond to exogenous auxin indicating that *lhw* is unlikely to form a core component of the auxin signaling machinery. Additionally, when the auxin response was followed using the DR5 reporter, no differences were observed between the auxin response in *lhw* and wild type. However, *lhw* roots were resistant to long-term NPA treatment. In wild-type roots, there was a dramatic increase in cell number coupled with the resetting of vascular pattern to incorporate a greater number of vascular cells; however, this cell proliferation was greatly reduced in *lhw*, and ectopic xylem poles were never observed. The *wol lhw* double mutant had a further reduced cell number than either of the single mutants, but *lhw* was found to be epistatic to *wol* for vascular patterning. All cells in the *wol lhw* differentiated as protoxylem indicating that there is no requirement for LHW in the determination of protoxylem cell fate, and the double mutant phenotype suggests that LHW regulates cell proliferation independently from cytokinin signaling. There is some evidence that LHW might regulate stem cell maintenance at the root tip. The expression of some QC markers, such as *SCR*, is compromised in *lhw*. Future research will elucidate whether LHW exerts its effect on vascular patterning through maintaining the stem cell population in the root tip. If this effect is achieved through regulating the number of vascular cells, it may provide a platform for studying vascular patterning in species with much larger vascular cylinders than *Arabidopsis*. If so, this would be of great benefit in transferring information gleaned from *Arabidopsis* to crop plants.

XI. Prospects for the Future

In the last few years, there have been considerable advances in our understanding of vascular development in *Arabidopsis*. What originally started as isolated studies into the development of the individual tissue types is now moving toward a clear holistic picture. There are several examples of interconnections between the processes described earlier and of the identity of one vascular cell type regulating the identity of others. For example, there is cross talk between phloem and other tissues as TDIF produced in the phloem promotes the proliferation of procambial cells; additionally, phloem transport controls the supply of cytokinin into the system which affects the boundary between domains of high auxin and high cytokinin signaling in the root tip. A strong auxin signal directly promotes the expression of at least one HD-ZIP gene, *AtHB8*. The VNDs probably act downstream of these processes and have been shown to be mis-regulated following the application of CLE peptides. Further research should reveal whether there are additional layers of cross talk between these components as well as the full transcriptional network between components. For example, while auxin regulates protoxylem identity, the auxin response maximum does not overlap with the *VND7* expression pattern well, and it is likely that there is an as yet unidentified transcriptional network connecting these components.

This chapter has focused on the vascular development of the primary *Arabidopsis* root. However, there is still considerable research to be done. It would be fruitful to observe these processes during embryogenesis and in lateral roots where the vascular patterns are created de novo rather than based on prepatterning from earlier embryonic templates. There is a great deal of variation in vascular pattern within flowering plants, with different species typically having between 1 and 6 xylem poles. It will be exciting to learn how these patterns are created; it is possible that through computational modeling of the processes occurring in *Arabidopsis*, as well as the investigation of mutants with altered cell numbers, we can make informed extrapolations of how the vascular pattern formation may occur in other species. In addition, understanding the evolutionary history and dynamics of these networks will enable us to generalize the knowledge gained from studying *Arabidopsis* to other species more effectively. Through such studies, we may gain insight into how vascular plants came to dominate our planet and how we can better manipulate them to serve our increasing need for food, fuel, and biomaterials.

References

Abel S, Nguyen MD, Theologis A 1995 The PS-IAA4/5-like family of early auxin-inducible mRNAs in *Arabidopsis thaliana*. *J Mol Biol* 251:533–549.

Argyros RD, Mathews DE, Chiang YH et al. 2008 Type B response regulators of *Arabidopsis* play key roles in cytokinin signaling and plant development. *Plant Cell* 20:2102–2116.

Berleth T, Jurgens G 1993 The role of the MONOPTERIS gene in organising the basal body region of the *Arabidopsis* embryo. *Development* 118:575–587.

Bishopp A, Help H, El-Showk S et al. 2011a A mutually inhibitory interaction between auxin and cytokinin specifies root vascular pattern. *Curr Biol* 21:917–926.

Bishopp A, Lehesranta S, Vatén A et al. 2011b Phloem-transported cytokinin regulates polar auxin transport and maintains vascular pattern in the root meristem. *Curr Biol* 21:927–932.

Bonke M, Thitamadee S, Mähönen AP, Hauser MT, Helariutta Y 2003 APL regulates vascular tissue identity in *Arabidopsis*. *Nature* 426:181–186.

Caño-Delgado A, Yin Y, Yu C, Vafeados D, Mora-García S, Cheng J-C 2004 BRL1 and BRL3 are novel brassinosteroid receptors that function in vascular differentiation in *Arabidopsis*. *Development* 131:5341–5351.

Carlsbecker A, Lee JY, Roberts CJ et al. 2010 Cell signalling by microRNA165/6 directs gene dose-dependent root cell fate. *Nature* 465:316–321.

Cui H, Levesque MP, Vernoux T et al. 2007 An evolutionarily conserved mechanism delimiting SHR movement defines a single layer of endodermis in plants. *Science* 316:421–425.

Demura T, Tashiro G, Horiguchi G et al. 2002 Visualization by comprehensive microarray analysis of gene expression programs during transdifferentiation of mesophyll cells into xylem cells. *Proc Natl Acad Sci USA* 99:15794–15799.

Dharmasiri N, Dharmasiri S, Estelle M 2005a The F-box protein TIR1 is an auxin receptor. *Nature* 435:441–445.

Dharmasiri N, Dharmasiri S, Weijers D et al. 2005b Plant development is regulated by a family of auxin receptor F box proteins. *Dev Cell* 9:109–119.

Dolan L, Janmaat K, Willemsen V et al. 1993 Cellular organisation of the *Arabidopsis thaliana* root. *Development* 119:71–84.

Donner TJ, Sherr I, Scarpella E 2009 Regulation of preprocambial cell state acquisition by auxin signaling in *Arabidopsis* leaves. *Development* 136:3235–3246.

Emery JF, Floyd SK, Alvarez J et al. 2003 Radial patterning of *Arabidopsis* shoots by class III HD-ZIP and KANADI genes. *Curr Biol* 20:1768–1774.

Eriksson ME, Israelsson M, Olsson O, Moritz T 2000 Increased gibberellin biosynthesis in transgenic trees promotes growth biomass production and xylem fiber length. *Nat Biotechnol* 18:784–788.

Esau K 1977 *Anatomy of Seed Plants*, 2nd edn. New York: Wiley.

Fiers M, Hause G, Boutilier K, Casamitjana-Martinez E, Weijers D, Offringa R 2004 Mis-expression of the CLV3/ESR-like gene CLE19 in *Arabidopsis* leads to a consumption of root meristem. *Gene* 327:37–49.

Friml J, Vieten A, Sauer M et al. 2003 Efflux-dependent auxin gradients establish the apical-basal axis of *Arabidopsis. Nature* 426:147–153.

Grieneisen VA, Xu J, Maree AF, Hogeweg P, Scheres B 2007 Auxin transport is sufficient to generate a maximum and gradient guiding root growth. *Nature* 449:1008–1013.

Helariutta Y, Fukaki H, Wysocka-Diller J et al. 2000 The SHORT-ROOT gene controls radial patterning of the *Arabidopsis* root through radial signaling. *Cell* 101:555–567.

Hirakawa Y, Kondo Y, Fukuda H 2010 TDIF peptide signaling regulates vascular stem cell proliferation via the WOX4 homeobox gene in *Arabidopsis. Plant Cell* 22:2618–2629.

Hirakawa Y, Shinohara H, Kondo Y, Inoue A, Nakanomyo I, Ogawa M 2008 Non-cell-autonomous control of vascular stem cell fate by a CLE peptide/receptor system. *Proc Natl Acad Sci USA* 105:15208–15213.

Hutchison CE, Li J, Argueso C et al. 2006 The *Arabidopsis* histidine phosphotransfer proteins are redundant positive regulators of cytokinin signaling. *Plant Cell* 18:3073–3087.

Inoue T, Higuchi M, Hashimoto Y et al. 2001 Identification of CRE1 as a cytokinin receptor from *Arabidopsis. Nature* 409:1060–1063.

Ishida K, Yamashino T, Yokoyama A, Mizuno T 2008 Three type-B response regulators ARR1 ARR10 and ARR12 play essential but redundant roles in cytokinin signal transduction throughout the life cycle of *Arabidopsis thaliana. Plant Cell Physiol* 49:47–57.

Ito Y, Nakanomyo I, Motose H et al. 2006 Dodeca-CLE peptides as suppressors of plant stem cell differentiation. *Science* 313:842–845.

Jurgens G 1995 Axis formation in plant embryogenesis: cues and clues. *Cell* 81:467–470.

Kakimoto T 2001 Identification of plant cytokinin biosynthetic enzymes as dimethylallyl diphosphate:ATP/ADP isopentenyltransferases. *Plant Cell Physiol* 42:677–685.

Kepinski S, Leyser O 2005 The *Arabidopsis* F-box protein TIR1 is an auxin receptor. *Nature* 435:446–451.

Kondo Y, Hirakawa Y, Kieber JJ, Fukuda H 2011 CLE peptides can negatively regulate protoxylem vessel formation via cytokinin signaling. *Plant Cell Physiol* 52:37–48.

Kubo M, Udagawa M, Nishikubo N et al. 2005 Transcription switches for protoxylem and metaxylem vessel formation. *Genes Dev* 19:1855–1860.

Kudo T, Kiba T, Sakakibara H 2010 Metabolism and long-distance translocation of cytokinins. *J Integ Plant Biol* 52:53–60.

Lacayo CI, Malkin AJ, Holman HY et al. 2010 Imaging cell wall architecture in single *Zinnia elegans* tracheary elements. *Plant Physiol* 154:121–133.

Mähönen AP, Bishopp A, Higuchi M et al. 2006a Cytokinin signaling and its inhibitor AHP6 regulate cell fate during vascular development. *Science* 311:94–98.

Mähönen AP, Bonke M, Kauppinen L, Riikonen M, Benfey PN, Helariutta Y 2000 A novel two-component hybrid molecule regulates vascular morphogenesis of the *Arabidopsis* root. *Genes Dev* 14:2938–2943.

Mähönen AP, Higuchi M, Tormakangas K et al. 2006b Cytokinins regulate a bidirectional phosphorelay network in *Arabidopsis. Curr Biol* 16:1116–1122.

Mason MG, Li J, Mathews DE, Kieber JJ, Schaller GE 2004 Type-B response regulators display overlapping expression patterns in *Arabidopsis. Plant Physiol* 135:927–937.

Matsumoto-Kitano M, Kusumoto T, Tarkowski P et al. 2008 Cytokinins are central regulators of cambial activity. *Proc Natl Acad Sci USA* 105:20027–20031.

Mattsson J, Sung ZR, Berleth T 1999 Responses of plant vascular systems to auxin transport inhibition. *Development* 126:2979–2991.

Miyawaki K, Matsumoto-Kitano M, Kakimoto T 2004 Expression of cytokinin biosynthetic isopentenyltransferase genes in *Arabidopsis*: tissue specificity and regulation by auxin cytokinin and nitrate. *Plant J* 37:128–138.

Miyawaki K, Tarkowski P, Matsumoto-Kitano M et al. 2006 Roles of *Arabidopsis* ATP/ADP isopentenyltransferases and tRNA isopentenyltransferases in cytokinin biosynthesis. *Proc Natl Acad Sci USA* 103:16598–16603.

Mockaitis K, Estelle M 2008 Auxin receptors and plant development: a new signaling paradigm. *Annu Rev Cell Dev Biol* 24:55–80.

Nakajima K, Sena G, Nawy T, Benfey PN 2001 Intercellular movement of the putative transcription factor SHR in root patterning. *Nature* 413:307–311.

Nilsson J, Karlberg A, Antti H et al. 2008 Dissecting the molecular basis of the regulation of wood formation by auxin in hybrid aspen. *Plant Cell* 20:843–855.

Ohashi-Ito K, Bergmann DC 2007 Regulation of the *Arabidopsis* root vascular initial population by LONESOME HIGHWAY. *Development* 134:2959–2968.

Otsuga D, DeGuzman B, Prigge MJ, Drews GN, Clark SE 2001 REVOLUTA regulates meristem initiation at lateral positions. *Plant J* 25:223–236.

Prigge MJ, Otsuga D, Alonso JM, Ecker JR, Drews GN, Clark SE 2005 Class III homeodomain-leucine zipper gene family members have overlapping antagonistic and distinct roles in *Arabidopsis* development. *Plant Cell* 17:61–76.

Scheres B, Dilaurenzio L, Willemsen V et al. 1995 Mutations affecting the radial organization of the *Arabidopsis* root display specific defects throughout the embryonic axis. *Development* 121:53–62.

Scheres B, Wolkenfelt H, Willemsen V et al. 1994 Embryonic origin of the *Arabidopsis* primary root and root meristem initials. *Development* 120:2475–2487.

Suzuki T, Miwa K, Ishikawa K, Yamada H, Aiba H, Mizuno T 2001 The *Arabidopsis* sensor His-kinase AHk4 can respond to cytokinins. *Plant Cell Physiol* 42:107–113.

Swarup R, Bennett M 2003 Auxin transport: the fountain of life in plants? *Dev Cell* 5:824–826.

Tan X, Calderon-Villalobos LI, Sharon M et al. 2007 Mechanism of auxin perception by the TIR1 ubiquitin ligase. *Nature* 446:640–645.

Ueguchi C, Sato S, Kato T, Tabata S 2001 The AHK4 gene involved in the cytokinin-signaling pathway as a direct receptor molecule in *Arabidopsis thaliana*. *Plant Cell Physiol* 42:751–755.

Uggla C, Magel E, Moritz T, Sundberg B 2001 Function and dynamics of auxin and carbohydrates during earlywood/latewood transition in scots pine. *Plant Physiol* 125:2029–2039.

Uggla C, Mellerowicz EJ, Sundberg B 1998 Indole-3-acetic acid controls cambial growth in scots pine by positional signaling. *Plant Physiol* 117:113–121.

Whitford R, Fernandez A, De Groodt R, Ortega E, Hilson P 2008 Plant CLE peptides from two distinct functional classes synergistically induce division of vascular cells. *Proc Natl Acad Sci USA* 105:18625–18630.

Yamada H, Suzuki T, Terada K et al. 2001 The *Arabidopsis* AHK4 histidine kinase is a cytokinin-binding receptor that transduces cytokinin signals across the membrane. *Plant Cell Physiol* 42:1017–1023.

Yamaguchi M, Goue N, Igarashi H et al. 2010 VASCULAR-RELATED NAC-DOMAIN6 and VASCULAR-RELATED NAC-DOMAIN7 effectively induce transdifferentiation into xylem vessel elements under control of an induction system. *Plant Physiol* 153:906–914.

Yamaguchi M, Kubo M, Fukuda H, Demura T 2008 Vascular-related NAC-DOMAIN7 is involved in the differentiation of all types of xylem vessels in *Arabidopsis* roots and shoots. *Plant J* 55:652–664.

Yamaguchi M, Ohtani M, Mitsuda N et al. 2010 VND-INTERACTING2 a NAC domain transcription factor negatively regulates xylem vessel formation in *Arabidopsis*. *Plant Cell* 22:1249–1263.

Zhong R, Ye ZH 2007 Regulation of HD-ZIP III Genes by MicroRNA 165. *Plant Sig Behav* 2:351–353.

8

Secondary Growth of Tree Roots: Cytoskeletal Perspectives

Nigel Chaffey
Bath Spa University

I. Introduction

The primary apical meristems of the root and shoot produce the cells and tissue systems of the primary plant body, such as the epidermis, cortex, and primary vascular tissues, and give rise to elongation growth or extension of the relevant plant part (Evert and Eichhorn 2006). Secondary growth (synonymous with secondary thickening) is the province of the two laterally situated secondary meristems, the vascular cambium (herein simply referred to as the cambium) and the cork cambium, and results in increase in girth of the organ (Evert and Eichhorn 2006). Products of the vascular and cork cambia together constitute the secondary plant body—secondary xylem and secondary phloem—and periderm, respectively (Evert and Eichhorn 2006).

The terms primary and secondary merely indicate the order in which these two forms of growth occur during plant growth; they should not be seen as an indication of the relative importance of each. Indeed, both are necessary for the successful development of the majority of vascular plants, and secondary growth is essential to the development of the tree habit. Furthermore, both types of growth can occur at the same time at different places in the plant.

At the heart of secondary growth is the cambium and its capacity to divide and generate new cells. Notwithstanding the numerous reviews of this meristem (although almost solely of the shoot cambium) over the last 80 years (reviewed in Catesson 1974, 1994; Larson 1994; Lachaud et al. 1999), there has been comparatively little new cambial cell biology (i.e., employing molecular approaches) to review since the heyday of cambial ultrastructural studies in the 1960s and early 1970s. That contrasts markedly with the situation for the primary plant body, which in the past 30–40 years has benefited greatly from the development of new and still-emerging techniques in molecular biology (e.g., *The Arabidopsis Book* 2011). The application of those techniques, largely concentrated upon the "model plant" *Arabidopsis thaliana* (e.g., Koornneef and Meinke 2010), has generated a tremendous amount of information concerning plant development. Unfortunately, because of problems—either real or perceived—inherent in studying the cambium, particularly that of trees (see Chaffey 2002c,f), there appears to have been a reluctance to pursue such investigations, and it is only relatively recently that the techniques of modern cell biology have begun to be applied to this secondary meristem (e.g., Chaffey 2002a).

II. Scope of This Review

Aspects of primary growth and the biology of the primary body of roots are considered elsewhere in this volume. Background information on the anatomy and secondary growth of roots can be found in most standard plant anatomy texts (e.g., Evert and Eichhorn 2006). General aspects of the secondary growth of roots have previously been reviewed, most notably by Fayle (1968), Wilson (1975), and Woods (1991). Understandably, those reviews reflected the prevailing knowledge and interests of the time and generally stood outside of the root "looking in." However, since the last of those reviews was published, new insights into the cell biology of secondary growth in roots have been obtained, and we can now consider the "inner workings" of the system. Since currently most is known about the cytoskeleton during secondary vascular differentiation of the roots—and shoots—of the angiosperm trees, *Aesculus hippocastanum* L. (horse chestnut) (e.g., Chaffey et al. 1999, 2000a) and *Populus tremula* L. x *P. tremuloides* Michx. (hybrid aspen) (e.g., Chaffey et al. 2002a), this review will concentrate upon those topics in those taxa.

While root research is beginning to benefit from application of the techniques of "modern" cell biology, this is happening comparatively slowly. Consequently, although "molecular" studies will be referred to where appropriate, in large part, this review is firmly rooted in what some might call "old-fashioned" cell biology which emphasizes structures and their functions—known or inferred. This is inevitable because we know most about secondary growth at that level. Until the essential work of integrating knowledge at all the different levels of enquiry is further advanced—or, in some cases, even begun!—we will not have anything like a full picture and will not know where the gaps in knowledge and understanding are. Inevitably, there is a strong descriptive component to this contribution. This chapter will attempt to review the present state of knowledge of aspects of the cell biology of the secondary vascular system (SVS, the assemblage of secondary xylem, vascular cambium, and secondary phloem) of tree roots. In so doing, it is hoped to highlight some of the interesting features that make its further study highly desirable and rewarding! But first, some basic biology by way of setting the scene for those not familiar with the SVS.

III. Secondary Growth: An Introduction

The most cursory glance at a transverse section of a tree root (Figure 8.1a, b; see also Figure 12.1B in Evert and Eichhorn 2006) shows the essential features of its SVS, concentric tissue zones—the phloem toward the outside, the cambial zone, and the xylem toward the interior—and radial files of ray cells that traverse those zones. These radial and concentric features are both products of periclinal (synonyms in the literature: tangential, additive) cell division of the cambial initials (Figure 8.2) and of their daughter cells. Subsequently, these cambial derivatives grow and differentiate as phloem cells or xylem cells in a position-dependent manner.

Several subdivisions of the tissue zones are recognized in arboreal plants. For example, Bailey (1952) identified six zones related to wood formation: the cambial zone and the zones of cell enlargement, cell maturation, mature sapwood, sapwood-heartwood transition, and the inner core of heartwood. Broadly, those same zones are still recognized today (e.g., Déjardin et al. 2010), which is testament to the insight of those earlier workers, on the one hand. On the other, it is an acknowledgement of how little our understanding of the SVS has advanced in the intervening 6 decades. Undoubtedly, as our methods of analysis and enquiry become more refined, we will have cause, and need, to reevaluate those zones and define them more specifically and precisely by use of "molecular markers."

The powerhouse that drives all this growth activity within the SVS is the cambium. Unfortunately, although the cambium undeniably exists, there are currently no known markers that allow us unambiguously to identify true cambial initial cells (e.g., Catesson

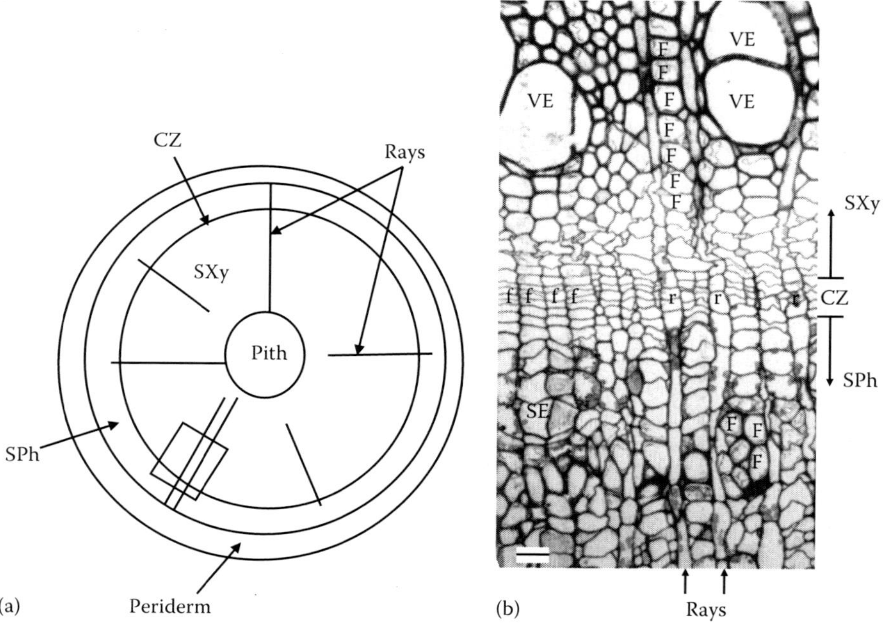

FIGURE 8.1 Root anatomy showing the concentric zones of tissues of the SVS as seen in transverse view in (a) whole root diagram and (b) in a light micrograph of hybrid aspen root (corresponding to rectangular area outlined in a). Key: CZ, cambial zone; f, fusiform cambial cell; F, fiber; r, ray cambial cell; SE, sieve elements; SPh, secondary phloem; SXy, secondary xylem; VE, vessel element; scale bar (for b) = 20 μm.

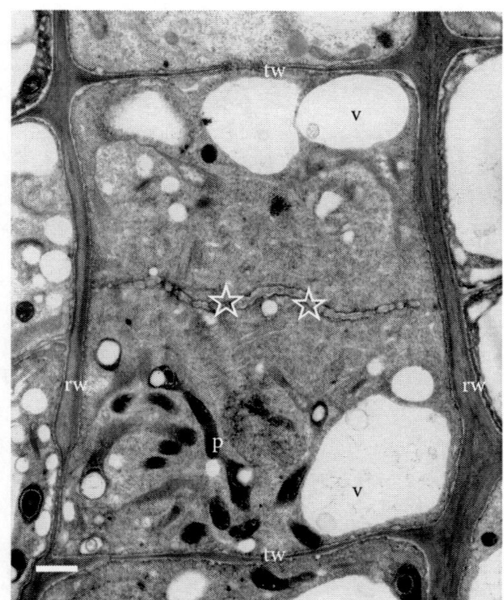

FIGURE 8.2 Transmission electron micrograph of a transverse section of the cambial zone of the taproot of horse chestnut showing a recently periclinally divided ray cambial cell. Key: p, plastid; rw, radial cell wall; tw, tangential cell wall; v, vacuole; star, nascent division wall; scale bar = 2.5 μm.

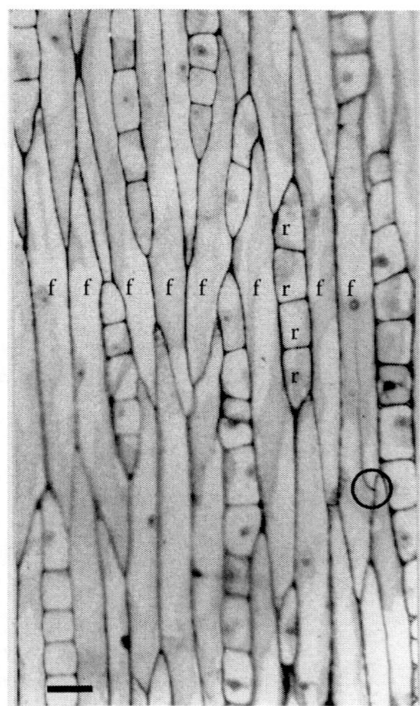

FIGURE 8.3 Light micrograph of a tangential longitudinal section through the cambial zone of the taproot of horse chestnut showing ray (r) and fusiform (f) cambial cells; the encircled region indicates a relatively recent pseudotransverse division wall; scale bar = 20 μm.

1994). Indeed, it is still a matter of debate as to whether the cambium proper is comprised of one or more tiers of cells (cf. Catesson 1974; Larson 1994; Iqbal 1995; Barlow 2005). For that reason, most workers are content to identify a region of similar-looking cells—the "cambial zone"—wherein the true cambium lies. But work by Fuchs et al. (2010a), which shows that plasmodesmata are retained, but at low density, throughout the cambial seasonal cycle of activity *only within the radial walls within the putative initial cells,* is probably the closest we have yet come to a cytologically distinct criterion for those elusive progenitors.

The established cambium consists of two morphologically distinct cell types: fusiform initials (which are greatly elongated in the axis of the root) and ray initials (which are cuboid) (Figure 8.3). Both are thin-walled, highly vacuolated cells (Figure 8.4) with the usual complement of organelles one would expect for a cell in the root (see Table 8.1). Although the cambium is a meristem, its cells are unlike those which typify primary meristems (e.g., Cutter 1978); for example, there are two distinctly different cambial cell types, both of which are more highly vacuolated and, certainly in the case of the fusiform cells, much more elongated than those of primary meristems.

Fusiform initials give rise to the cells of the vertical, or axial, component of the SVS, which, in horse-chestnut and hybrid aspen roots, are axial parenchyma, fibers, sieve elements, companion cells, and vessel elements (Table 8.2). Note, however, that transverse cell divisions of axial cambial derivatives within the cambial zone are necessary to produce axial parenchyma cells in both the xylem and phloem.

The ray initials produce the ray cells (Figure 8.1), which, in the xylem, can be further distinguished as two subtypes: contact and isolation ray cells (Braun 1984). Contact ray cells, where they lie next to vessel elements, have large-diameter pits, which are mirrored in the cell wall of the adjacent vessel element. Isolation ray cells, which may also lie next to vessel elements, have only small, elliptical simple pits (as for fibers and xylem axial parenchyma). All phloem ray cells appear to be of the same type. Although the

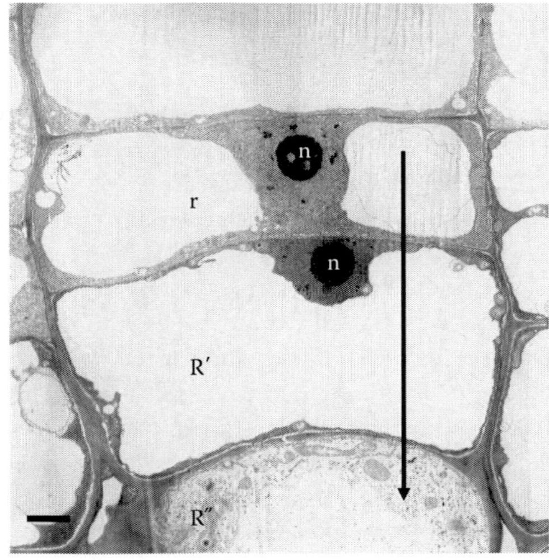

FIGURE 8.4 Transmission electron micrograph of a transverse section of the taproot of horse chestnut showing a file of phloem ray cells. Key: n, nucleus; r, ray cambial cell; R′ and R″, successive stages in differentiation of phloem ray cells; arrow indicates direction of differentiation; scale bar = 5 μm.

TABLE 8.1 Seasonal Changes in Ultrastructure of Fusiform and Ray Cambial Cells within Taproots of *Aesculus hippocastanum* L.

Feature	Active Cambium	Dormant Cambium
Cell walls	Thin, amorphous	Thick, lamellate
Nucleus[a]	Single	Single
Mitochondria	++	++
Plastids	++ (little starch)	++ (more starch)
Microtubules[b]	Random[c]	Oblique and helical[c]
Microfilaments[b]	Axial[d]	Axial[d]
Vacuome	Single, large vacuole	Dispersed, small vacuoles
Endoplasmic reticulum	++ (rough only)	++ (rough mainly, some smooth[e])
Dictyosomes	++ (active)	++ (less active)
Secretory vesicles	++	++
Coated vesicles	++	+(+)
PCR/TGN[f]	+	+
Plasmalemma	Undulating	Undulating
Oleosomes	+	++
Microbodies	+	+(+)
Ribosomes[g]	++	++

[a] Occasionally containing MT-like structures.
[b] Cortical elements.
[c] Endoplasmic MTs particularly prominent in ray cells.
[d] Random, but rare in ray cells.
[e] Smooth endoplasmic reticulum only seen in fusiform cells.
[f] Partially coated reticulum/*trans* Golgi network-like structure (Chaffey et al. 1997c).
[g] Frequently in polysomal configurations; +, +(+), ++, increasing abundance.

TABLE 8.2 Comparison of Some of the Characteristics of Cells of the SVS in the Taproot of *Aesculus hippocastanum*

Tissue/Cell Type	Width[a]	Length[a]	Wall[a]; ± Lignified; Type of Elaboration
Phloem			
Companion cell	Wider	Shorter	Thicker, nonlignified, pitted
Sieve element	Wider	Equal	Thicker, nonlignified, sieve areas[b]
Fiber	Wider	Longer	Thicker, lignified, simple pits, obliquely angled[c]
Axial parenchyma	Wider	Shorter	Thicker, nonlignified, pitted
Ray cell	Equal[d]	Longer[e]	Thicker, nonlignified, pitted
Cambium			
Ray cell	N/A	N/A	Thin, nonlignified, pit fields only
Fusiform cell	N/A	N/A	Thin, nonlignified, pit fields only
Xylem			
Fiber	Wider	Longer	Thicker, lignified, simple pits, obliquely angled[c]
Axial parenchyma	Wider	Shorter	Thicker, lignified, simple pits, transversely angled[c]
Vessel element	Wider	Equal	Thicker, lignified, bordered and contact pits, tertiary thickenings, simple perforations
Contact ray cell	Equal[d]	Longer[e]	Thicker, lignified, contact pits, interray cell pits
Isolation ray cell	Equal[d]	Longer[e]	Thicker, lignified, simple pits, interray cell pits

[a] Relative to precursor cambial cell.
[b] At sieve plate and within lateral cell walls.
[c] Orientation of long axis of elliptical pit relative to cell's long axis.
[d] Measured axially.
[e] Measured radially.

rays are predominantly uniseriate, biseriate rays are occasionally found in the xylem and phloem of hybrid aspen (particularly in the root). Additionally, phloem rays are subject to dilatation growth, which effectively makes the rays multiseriate within this tissue (Fig. 4D in Chaffey and Barlow 2001).

At maturity, and relative to the precursor cambial cell, all secondary vascular cells have different shapes, and their cell walls are thicker, variously elaborated, and, in the case of all xylem cells and phloem fibers, also lignified (Table 8.2; Chaffey 2002b). Many ultrastructural changes accompany secondary vascular differentiation (Iqbal 1995). However, it is really only now, with the availability of recently developed correlative cell biological techniques, such as immunolocalization, that the significance of some of those changes is being appreciated and understood.

IV. Root and Shoot Cambia Are Alike

Notwithstanding important differences between roots and shoots—for example, in terms of their origin and development (e.g., Evert and Eichhorn 2006), structure and function (e.g., Raven et al. 1999; Smith et al. 2010), and exposure to environmental and biotic factors (e.g., Hodge et al. 2009)—superficially at least, secondary growth in roots appears the same as it does in shoots, and for many years it has been almost an article of faith that root and shoot cambia were the same. To my knowledge, the only work on the cell biology of tree root cambia is the series of articles by Chaffey and coworkers (e.g., Chaffey et al. 1997a,b,c, 1998, 1999, 2000a,b; Chaffey and Barlow 2001, 2002), which examined aspects of the ultrastructure and dynamics of the cytoskeleton within taproots (*and* shoots) of *Aesculus hippocastanum*, and Chaffey and Barlow (2001, 2002) and Chaffey et al. (2002a) which extended those studies to hybrid aspen (*Populus tremula* x *P. tremuloides*). Very little difference between the cambia of roots and shoots *within* horse chestnut and *within* hybrid aspen was found in the ultrastructure of ray and cambial cells and in their seasonal cycle (Table 8.1) and the behavior of their microtubule (MT) and microfilament (MF) cytoskeleton, both during the cambial seasonal cycle and during differentiation of secondary xylem and phloem. Furthermore, very little difference appears to exist *between* these two taxa in respect of the cell features studied.

The major benefit of this revelation is that important cell biological aspects of shoot and root cambia do indeed appear to be similar, and information derived from one can be integrated with that derived from the other. This is comforting since more work is carried out on the easier-to-access aboveground SVS of shoots. However, lest we be tempted to ignore root cambia completely, it should be stressed that so few taxa have been studied that to conclude that *all* shoot and root cambia are the same may be premature. Certainly, much more work is needed to cover the range of tree and wood types, which covers angiosperms and gymnosperms, temperate and tropical, ring-porous and diffuse-porous, and storied and nonstoried cambia.

V. Cambial Cell Division and Growth

A. Arithmetic Cambium: Division and Multiplication

As a result of the combination of meristematic activity of the cambium and the growth of the daughter cells, the cambial cylinder moves outward. If radial growth were to continue unchecked, it is likely that the pressures so generated within the cambium and the phloem would result in these tissues being torn apart. Clearly, this does not happen; the radial expansion is somehow accommodated. The integrity of the SVS is maintained by at least three sites of cell division/growth in the radial plane, which together lead to an increase in girth. The main "pressure-relief valve" is presumed to be anticlinal (synonyms in the literature: radial, multiplicative) division of cambial initials to produce new cambial initials circumferentially, which subsequently grow and develop their own radial files of secondary vascular cells (e.g., Barlow 2005).

In nonstoried cambia, such as those of horse chestnut and hybrid aspen, where the ends of adjacent cambial cells are at different levels, these anticlinal divisions vary from nearly transverse to nearly longitudinal and are termed "pseudotransverse" (Evert and Eichhorn 2006). Because the division wall is offset from the vertical, both daughter cells are shorter than the mother cell. Thus, to increase the circumference of the cambial cylinder, not only do the daughters have to increase in diameter, but they also have to elongate to regain the original parental length. This contrasts with periclinal cell division, which is essentially vertical, longitudinal division that gives rise to two daughter cells of approximately equal length (Evert and Eichhorn 2006). In storied cambia, such as that of *Robinia*, where all fusiform cambial cells are the same length, anticlinal cell division is truly longitudinal, and two equal-length daughters are produced (Evert and Eichhorn 2006), which only have to increase in diameter to achieve increase in cambial circumference.

A second line of defense against unchecked radial expansion, but which is more pronounced in the stem, which increases in girth much more than the root, is so-called dilatation growth (e.g., Larson 1994; Evert and Eichhorn 2006) of phloem rays. In hybrid aspen, the typically uniseriate rays undergo anticlinal radial divisions and cell growth to broaden into multiseriate rays. Thirdly, the combination of anticlinal cell division in the cork cambium (Y. Waisel, personal communication, 2000) and cell growth also serves to relieve any buildup of pressure.

The conditions that promote and sustain these anticlinal cell divisions are presently unknown, but it is conceivable that some sort of pressure/stretch-detection mechanism exists at the cellular level, which subsequently sets off an appropriate number of rounds of cell division/enlargement sufficient to dissipate the immediate danger. However, a further consideration is that in dilatation growth, parenchymatous phloem ray cells, which are differentiated cambial derivatives, can subsequently be encouraged to undergo cell division. How similar is this cell division to that which takes place within the cambium proper? Do the differentiated ray cells have to dedifferentiate to a meristematic state before they can undergo division? Is this cell division also influenced by auxin flow as considered to be the case for the cambium (e.g., Aloni et al. 2000)? If so, how is the appropriate level/flow of auxin generated in this specialized region of the phloem? The presence of a radial gradient of auxin across the tissues of the SVS in stems of angiosperm and gymnosperm trees has been established (Sundberg et al. 2000), and its implications for regulation of wood formation examined (Nilsson et al. 2008). Extension of this study to a consideration of root cambia, dilatation growth, and the much understudied cork cambium should prove quite rewarding.

Mere production of more initial cells by anticlinal cell division in the cambium is not enough. The right sort of cambial initial has to be produced. A certain ratio of fusiform:ray cells needs to be maintained for proper functioning of the SVS (Larson 1994). To maintain that ratio, it will be necessary to produce more ray cells. Ray cells can be produced in several ways. Transformative cell divisions can occur in which either a part or the whole of a fusiform cambial cell becomes subdivided, by anticlinal, transverse cell divisions, to produce ray initial(s) (e.g., Phillips 1976). Additionally—or alternatively—a most unequal lateral cell division of a fusiform cambial cell can occur whereby a ray initial is cut off from the side of the fusiform cell (e.g., Phillips 1976). Understanding the conditions that promote the production of ray initials represents a considerable challenge.

B. Breaking the Rules

Of the two types of cambial initials, cell division of the fusiform cells is the most problematic. They are long cells (165–329 μm in horse-chestnut root, Chaffey et al. [1999], but can be up to several mm long in conifers, Larson [1994]), whose periclinal or pseudotransverse divisions "do not obey the usual laws of cell division" (Cutter 1975). For example, they violate Errera's Law (Cutter 1978), which states that cells divide by a wall of minimal surface area. As if that were not intriguing enough, they also appear to lack a preprophase band (PPB). The PPB is a circumferential aggregation of cortical MTs that appears to define both the plane of subsequent cell division and the site of attachment on the parental cell wall of the new division wall (e.g., van Damme et al. 2007). However, ultrastructural study of cell division in stem cambia of *Robinia* (Farrar and Evert 1997) failed to identify a PPB in fusiform initials, as has extensive search by immunofluorescent techniques for MTs in roots of horse chestnut and hybrid aspen (Chaffey and Barlow, unpublished).

If the PPB performs the role ascribed to it, how do fusiform initials manage without it? What guides the advancing cell plate to the correct place on the parental cell wall? Indeed, is the cell plate "guided" in the usual sense? Ultrastructural study of cell division in fusiform cambial cells of *Fraxinus* (Goosen-de Roo et al. 1984) identified a previously unsuspected class of MTs, within the phragmosome, which advances before the phragmoplast. They suggested that these longitudinally oriented (i.e., parallel to the developing cell wall) MTs guide the growing cell plate to the "parental cell wall site previously marked by the preprophase band of microtubules" (Goosen-de Roo et al. 1984). However, immunofluorescent studies on *Aesculus* and *Populus*—admittedly different species and using a different visualization technique—have not demonstrated any MTs in this location in dividing fusiform cells. Perhaps the best we can say at present is that it is uncertain whether a PPB is present in such cells or whether such phragmosomal MTs can perform the role suggested by Goosen-de Roo et al. (1984).

Of course, it might be that a PPB is unnecessary and a wide variety of daughter cell lengths can be tolerated—ranging from those that result from near-transverse to those from near-longitudinal variants of pseudotransverse cell division. The system then would be reliant upon elongation growth of both daughter cells to reconstitute the original cell length, rather than the accurate placement of the division wall at a specific site on the parental wall. On this basis, it may be proposed that a PPB is only essential to cells in situations where it is important to preserve a particular cell shape or to maintain specific cell-cell geometry. An example of this in the SVS could be the retention of the regular, radial symplasmic transport route, which is provided by the xylem and phloem rays. Consistent with this suggestion, a PPB has been identified in ray cambial cells by immunolocalization techniques (Chaffey and Barlow, unpublished). Nevertheless, it is necessary to establish in a more rigorous way whether a PPB really is absent from fusiform cells.

What do we currently know about the cell biology of cambial cell division in the root? So far, cytoskeletal dynamics have only been examined during periclinal cell division and the transverse cell divisions that generate axial phloem parenchyma cells in roots. Those studies, which have so far considered immunofluorescent localization of tubulin (for MTs), actin (for MFs), unconventional myosin VIII ("myosin") (characterized by Reichelt et al. 1999), and callose, have revealed that karyokinesis is accompanied by development of a mitotic spindle, which contains tubulin, but in which neither myosin nor actin has yet been detected. During subsequent cytokinesis, both tubulin and actin are immunolocalized at either side of the new division wall (Figure 8.5a,b), within

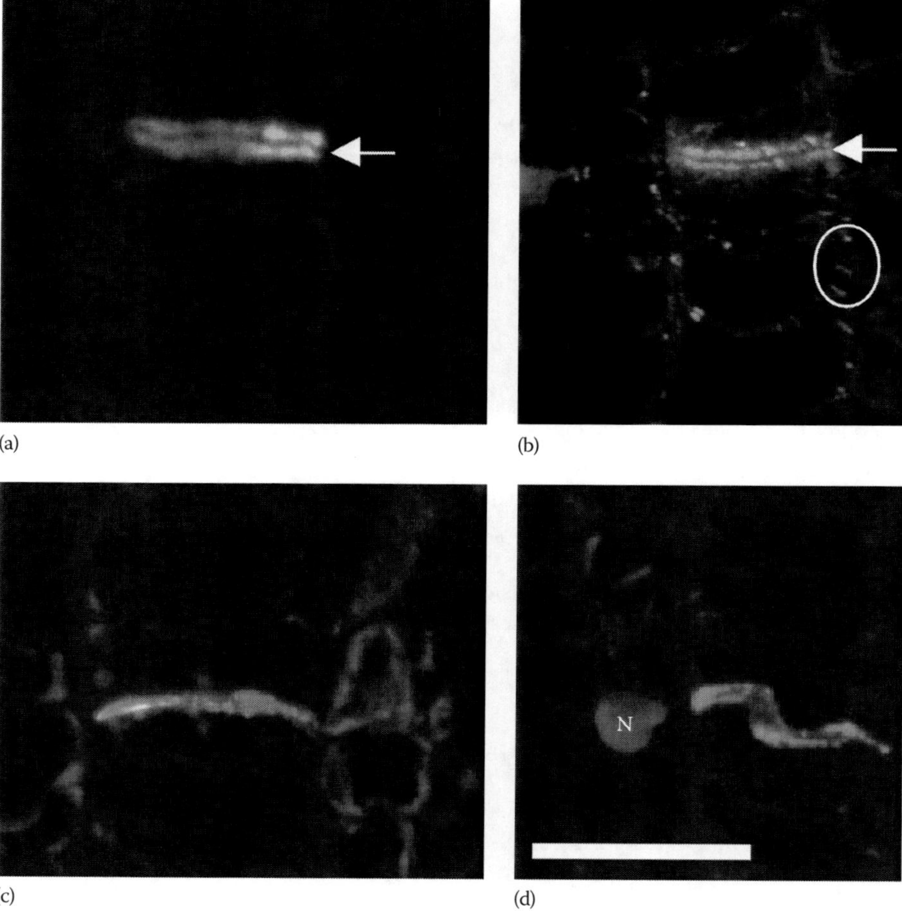

FIGURE 8.5 **(See color insert.)** Indirect immunofluorescent localization of various macromolecules involved in cell-wall formation within fusiform cambial cells in a transverse section of hybrid aspen root. (a) MTs and (b) MFs, within the phragmoplast; note the nonfluorescing central dark line, which corresponds to the site of the nascent division wall; encircled region shows axially oriented MF bundles in these obliquely cut cells; (c) myosin and (d) callose, at the site of the nascent division wall; N, nucleus; scale bar (in d) = 20 μm.

the phragmoplast, whereas myosin and callose are found within the wall (Figure 8.5c,d). Interestingly, cell walls at this very early stage of development are apparently unstained by Calcofluor (Chaffey 2002d), indicating that they are almost devoid of cellulose and, in agreement with earlier reports, of a callosic phase in cell-wall development (Waterkeyn 1967).

As for cell division in other cell systems, in longitudinal section, it is evident that the tubulin and actin are concentrated within the phragmosomes at the two ends of the centrifugally advancing cell plate. The MTs here, which lie at right angles to the plane of the developing cell wall (Figure 8.6), can be viewed as providing guides for transport of secretory vesicles to the developing cell wall, as can the associated MFs. Once vesicles arrive at the phragmoplast, the dense MT-MF array could help to trap such vesicles there, ensuring that their contents are released at the correct place. Immunolocalization of actin, and possibly tubulin, is also evident along the nascent cell wall between these two phragmoplasts, as is myosin and callose within the wall. The actin and myosin can be envisaged as acting in concert here as a motile cytoskeletal system (e.g., Reichelt and Kendrick-Jones 2000), which could translocate those dictyosome-derived secretory vesicles that arrive at the developed cell wall to the phragmoplasts, where they can contribute to the development of the nascent division wall.

Soon after this stage, anticallose antibodies no longer localize to the wall, but the cell walls stain strongly with Calcofluor, indicating the presence of cellulose (Chaffey 2002d). Whether callose stainability disappears because the callose is replaced or altered in some way so that its antigenicity is masked or lost is not known. In line with current interest in the chemistry of cambial cell walls and the enzymes that modify it (e.g., Follet-Gueye et al. 2000; Micheli et al. 2002), further study of this phenomenon is clearly warranted. However, neither callose nor myosin is

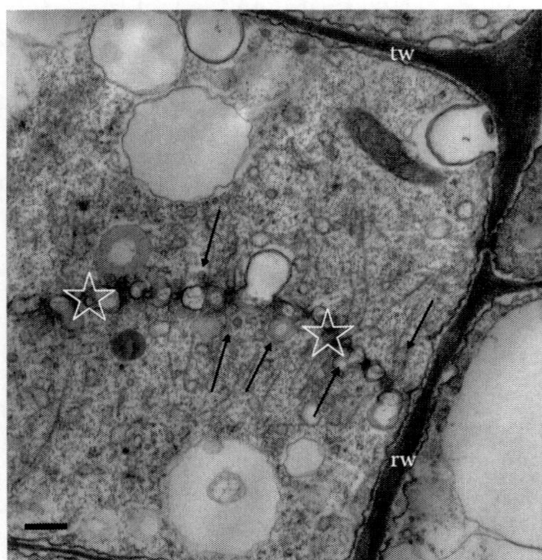

FIGURE 8.6 Transmission electron micrograph of a transverse section of the cambial zone of the taproot of horse chestnut showing a recently periclinally divided fusiform cambial cell undergoing cell-wall formation. Key: rw, radial cell wall; tw, tangential cell wall; star, nascent division wall; arrows, MTs; scale bar = 2.5 μm.

completely absent from the mature cell wall; both are retained at the pit fields (Chaffey and Barlow 2001), presumably in connection with the plasmodesmata at those sites (e.g., Radford et al. 1998; Radford and White 1998, 2011).

VI. Cytoskeletal Considerations of the Secondary Vascular System

Although many changes take place as cambial derivatives differentiate to become secondary vascular cells (Table 8.2; Iqbal 1995; Chaffey 2002b), alterations of their size, shape, and walls are among the most obvious. In view of the roles ascribed to cellulose microfibrils in influencing the direction of cell expansion (cf. Green 1969) and to the cytoplasmic MTs in influencing the arrangement of those microfibrils within the cell wall (reviewed by Baskin 2001; demonstrated by Paredez et al. 2006), to date, the cell walls and the cytoskeleton are the most studied aspects of root cambial cell biology. Using a combination of conventional transmission electron microscopy (TEM) and immunofluorescence (IFF) localization of the cytoskeletal proteins, tubulin (for MTs), actin (for MFs), and unconventional myosin VIII, the changes taking place to these cell components during secondary vascular differentiation have been examined in roots of *A. hippocastanum* and *P. tremula* x *P. tremuloides* (Chaffey et al. 1999, 2000a, 2002a; Chaffey and Barlow 2001, 2002; summarized for MTs and MFs in wood formation in Table 8.3).

TABLE 8.3 Behavior of the Cortical MT and MF Components of the Cytoskeleton during the Cambial Seasonal Cycle and Xylogenesis in the Taproot of *Aesculus hippocastanum*

Cell Type	Microtubules	Microfilaments
Ray cambial (active)	Random	Random
Ray cambial (dormant)	Parallel-helical	Random
Fusiform cambial (active)	Random	Axial
Fusiform cambial (dormant)	Parallel-helical	Axial
Fiber (early stages—first phase)	Parallel-helical, wavy	Axial[a]
Fiber (later stages—second phase)	Parallel-helical, dense	Axial[b]
Fiber (maturity)	Absent	Absent
Axial parenchyma (early stages)	Parallel-helical	Axial[a]
Axial parenchyma (maturity)	Axial	Axial[b]
Vessel element (early stages)	Random	Axial
Vessel element (middle stages)	Various[c]	Axial,[b] various[c,d]
Vessel element (maturity)	Absent	Absent
Contact ray (early stages)	Random, rings[e]	Axial, rings[e]
Contact ray (maturity)	Axial	Axial[b]
Isolation ray (early stages)	Random	Axial[a]
Isolation ray (maturity)	Axial	Axial[b]

[a] Ellipse associated with developing simple pits.
[b] Branched, but overall net-axial orientation maintained.
[c] Rings at peripheries of bordered and contact pits, and perforation plate, and at aperture of developing bordered pit, parallel-aligned arrays associated with developing tertiary thickenings.
[d] Meshwork over perforation plate.
[e] Rings at periphery of developing contact pit.

A. Cambium

1. Nondividing Cells

In nondividing cells of active cambia, the cortical MTs of both ray and fusiform cambial cells are randomly organized and overlap. Although cambial cell-wall structure has not specifically been examined in horse chestnut or in hybrid aspen, the literature shows that the primary walls of secondary xylem cells—which are derived from the cambial cell wall—contain disorganized cellulose microfibrils (e.g., Awano et al. 2000). Thus, the similarly randomly arranged MTs can be seen as circumstantial evidence for the idea that MTs influence the orientation of microfibrils.

Perhaps of more immediate significance is that the random arrangement of MTs-microfibrils in both cambial cell types potentially permits growth in any direction, that is, isotropically. However, cambial derivatives grow anisotropically, but this takes place principally along one or both of two axes—longitudinal and radial—and the combination of both varies quite dramatically depending upon the cell type. Thus, fibers grow considerably along both axes, sieve and vessel elements grow more radially than longitudinally, and ray cells grow almost solely radially. Further complications arise from the facts that ray cambial cells can "transform" into fusiform cambial cells (e.g., Larson 1994). Fusiform cambial cells can give rise to ray cells; daughters of pseudotransverse division of a fusiform initial must grow in both axes (although considerably more longitudinally) to regain the parental cell volume. Alternatively, fusiform cambial cells can undergo transverse divisions to produce axial parenchyma cells. In other words, the randomness of the MTs-microfibrils within the cambial cells can perhaps be interpreted as an indication that these cells are as yet uncommitted to a particular cell fate. Potentially, each cambial derivative could follow one of several pathways of differentiation; the observed random arrangement of MTs merely reflects this "keeping one's options open." Nevertheless, it is unusual to find random MTs in a cell whose shape is other than that of a sphere (cf. cultured tobacco protoplasts—Vissenberg et al. 2000b).

By contrast, the prominent MFs within fusiform cambial cells are found in axially oriented bundles throughout the cambial seasonal cycle. In ray cambial cells, however, the MF bundles are much less common and apparently randomly oriented, although they too persist in these cells throughout cambial activity and dormancy. Nowhere in cambial cells have MFs been observed cooriented with MTs. In both cambial cell types, the MF bundles are implicated in cytoplasmic streaming during the active part of the cambial seasonal cycle (Chaffey et al. 2000a).

The third cytoskeletal component so far identified in interphase cambial cells is myosin, whose localization seems to be restricted to sites of pit fields in thinned regions of the primary cell wall (Chaffey and Barlow 2001). This immunolocalization of myosin is mirrored by that of callose (which has been used as a marker for sites of plasmodesmata—e.g., Radford and White 1998) and has been interpreted to indicate that myosin is associated with the plasmodesmata that are concentrated at those wall sites (Chaffey and Barlow 2001, 2002). What property or process determines where and how many plasmodesmata are laid down is as yet unknown, but its study—and elucidation!—is of considerable relevance to attempts to understand cell-cell communication possibilities within and between the established secondary vascular tissues (e.g., Fuchs et al. 2011).

2. Dividing Cells

The cambium is intimately concerned with the dual processes of cell division and cell-wall formation. Much work has been devoted to an understanding of the involvement of the cytoskeleton in wall formation, particularly the potential role of MTs in orienting cellulose microfibrils in the developing cell wall (for a review, see, e.g., Baskin 2001). Although studies of the cambium provide strong evidence for the involvement of MTs in the general phenomenon of wall formation, as yet they do not give any insight into whether MT orientation influences microfibril orientation (or vice versa). However, what is known about the localization of tubulin and F-actin during wall formation in the cambium suggests that it is similar to that found in connection with primary meristems (e.g., Jürgens 2005; Cyr 2007).

Both MTs and F-actin (possibly as MFs) are found either side of the developing cell plate within the phragmoplast. Once the cell plate has developed to become a callose-rich structure, MTs are no longer associated with it; instead, they (and F-actin) continue to be localized at the edge of the expanding phragmosome disc. It is, however, noteworthy that MFs have been identified extending alongside the maturing, callose-rich cell plate apparently connecting the diametrically opposite edges of the expanding phragmosome. Furthermore, the callose-rich cell plate also contains myosin. This latter association has led to the suggestion that both MFs and myosin may act in concert to facilitate the movement of secretory vesicles—containing cell plate/wall precursors—away from the more mature callose-rich cell plate toward the edge of the phragmosome disc where new cell plate is being formed (see Chaffey and Barlow 2002).

There is clearly much more work to do in this area (which is fundamental to an appreciation of both the generation and maintenance of the SVS), but already some inconsistencies are apparent. In a TEM study of *Fraxinus excelsior* L., Goosen-de Roo et al. (1983) demonstrated the presence of MTs oriented parallel to the developing cell wall/plate. They suggested that these MTs were involved with guiding the developing cell wall to its appropriate site of union with the parent cell wall. This is attractive as a method of overcoming the problems anticipated in the absence of a PPB in the fusiform cambial cells. Unfortunately, indirect IIF study—albeit not in *Fraxinus*—has not so far demonstrated the presence of this MT array (Chaffey and Barlow, unpublished), nor has the work by Rensing et al. (2002) using cryofixation procedures and TEM to investigate cell division of fusiform cambial cells in *Pinus* spp.

Although MTs are associated with cell plate formation at the phragmoplast, their absence from the region between the advancing edges of the phragmoplast implies that they do not have a role in maturation of the early cell-wall proper—at least before it is attached to the parent cell wall. And, since such young walls are callose-rich—indeed, cellulose appears to be absent—this supports the notion that MTs are only needed where cellulose microfibrils are being incorporated into cell walls. These observations are in agreement with those of Scherp et al. (2002) on callose formation in cell plates and juvenile cell walls of the liverwort *Riella helicophylla* Bory et Mont.

Mirroring the distribution of callose in the cell plate is that of myosin. Initially, present throughout the whole length of the cell plate, myosin becomes localized to discrete regions in the wall proper—apparently at sites of plasmodesmata as inferred from the similar callose immunolocalization pattern (Chaffey and Barlow 2001). In a similar way, F-actin becomes localized to pit fields in the established cell wall (as well as being present as MF bundles within the cytoplasm), possibly in association with callose and myosin at the plasmodesmata. Because of their perceived role in cell-cell communication, the detailed structure of plasmodesmata has been a subject of intense research interest for many years (e.g., chapters in Baluška et al. 2006; Maule 2008; Bell and Oparka 2011), particularly in relation to rates of transport of organic compounds within the ray system of trees (see Section VI.D).

B. Early Stages of Secondary Vascular Differentiation

As cambial derivatives differentiate, their MTs generally adopt new arrangements, and it is tempting to suggest that this may indicate commitment of a cambial derivative to a particular cell fate (e.g., Chaffey et al. 1997b). However, since the earliest stages of differentiation are more easily identified by the increase in cell diameter and relocation of the nucleus to a tangential wall (e.g., Figure 8.4; Chaffey et al. 1999), it is not yet possible to decide whether changes to the MT cytoskeleton are cause or effect of the differentiation process, or without involvement therein.

1. Fibers

a. Biphasic Fiber Growth

Although both normal wood and gelatinous fibers (fibers with so-called gelatinous layers—an innermost secondary wall layer that can be distinguished from the outer secondary wall layer(s) by its high cellulose content and lack of lignin—Evert and Eichhorn 2006) are found within roots and shoots of hybrid aspen, only the cytoskeletal dynamics of normal wood fibers have so far been studied in such roots. Accordingly, only normal wood fibers are considered further.

Two phases of development of wood fibers are apparent. The first phase commences with the establishment of a parallel alignment of MTs, but although the individual MTs themselves appear wavy and may overlap, cross-links between MTs and the plasmalemma have not been observed. It is suggested that this phase is concerned with cell growth, hence with production of primary cell wall, and takes place while the cells are within the enlargement zone of the secondary xylem.

The succeeding phase shows an increase in the number of MTs relative to the first phase, but they are still present as a layer that is almost exclusively one MT deep. The MTs are all at the same angle within a cell (Figure 8.7a) and in places observed to be cross-linked to the plasmalemma, but not to each other. It is suggested that this phase represents fiber maturation, when fibers have ceased enlargement and are undergoing the main stage of cell-wall thickening. Presumably, these cross-links act to hold the MTs in a specific orientation to each other, which thereby promotes a more ordered orientation to the cellulose microfibrils (or to other components of the cell wall). It is widely acknowledged that fibers, and other wood cells, have highly ordered microfibrils in the various secondary wall layers. Presence of plasmalemmal cross-linked MTs in fibers during the second phase of growth—but not during the first phase or

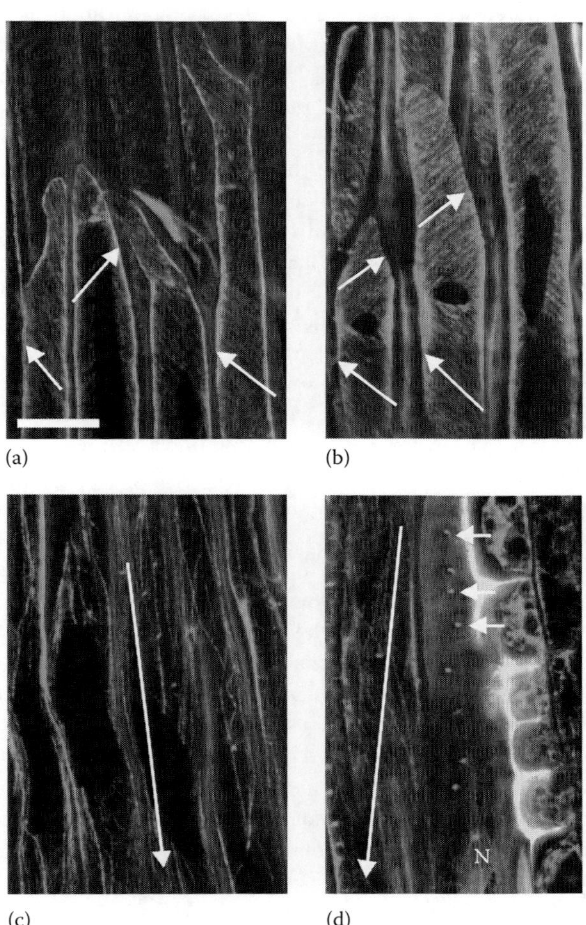

FIGURE 8.7 **(See color insert.)** Indirect immunofluorescent localization of cytoskeletal proteins in developing fibers (at second phase of development) seen in radial longitudinal sections of hybrid aspen root. (a) Xylem and (b) phloem, fibers showing parallel-aligned, obliquely oriented MTs (arrows); note that both the front and back walls of some of the phloem fibers are in view; (c) xylem and (d) phloem, fibers showing net-axially oriented MF bundles (arrows); note also the sites of simple pit formation (short arrows) in the phloem fibers; scale bar (in a) = 20 μm.

in fusiform cambial cells—can be related to this apparent prerequisite for a more highly ordered secondary cell wall, the better to perform its strengthening role. By contrast, for a primary cell wall—which is laid down in a cell that is still growing—the emphasis is more on flexibility and the ability of the wall to accommodate further increase in cell volume. In such a case, a less ordered microfibril arrangement will perhaps be more accommodating and may indeed be facilitated by the presence of nonplasmalemmal-cross-linked MTs.

If we accept that MT orientation has a role in determining the arrangement of cell-wall microfibrils, then observed variation in MT orientation from fiber to fiber, which is inferred to relate to successive stages of fiber differentiation, could be related to the deposition of different layers of the secondary cell wall (S_1, S_2, or S_3) or to successive development of different lamellae within a single wall layer (Barnett et al. 1998).

It is noteworthy that development of phloem fibers appears to be similar to that of their xylem counterparts. Certainly, they undergo a growth phase that is similar to the second phase of xylem fibers' development (Figure 8.7b). However, what we don't yet know is the full sequence of phloem fiber differentiation. This largely relates to the fact that more effort has been concentrated upon understanding xylogenesis in the SVS and partly to the apparent inability to identify phloem fibers at earlier stages of their differentiation (e.g., Chaffey et al. 2000b).

Parallels are also evident between development of xylem fibers and cotton "fibers" (which, strictly speaking, are seed-coat trichomes—e.g., Evert and Eichhorn 2006). Detailed ultrastructural work by Seagull (1992) with in vitro-grown cotton fibers has established that MTs are generally randomly oriented during fiber initiation and early elongation, reorienting to shallow-pitched (near-transverse) helices as elongation and primary wall deposition continue. Finally, they give way to steeply pitched (near-longitudinal) MT helices during secondary wall deposition (which orientation is similar to that of MTs in gelatinous wood fibers deemed to be laying down their characteristic cellulose-rich G-layer (Chaffey and Barlow, unpublished)). Accompanying these changes were increases in MT length, number, proximity to the plasmalemma, and a decrease in variability of MT orientation. With the exception of a shallow-helical phase and in the absence of data regarding MT length, the similarity of cotton-fiber and wood-fiber MT dynamics and the associated phases of growth is remarkable. Although the economic importance of cotton means that research on its trichomes is more fundable than that on wood fibers, in view of the similarity of the cell biology of growth between the two cell types, a bonus of the research on cotton may be a greater understanding of wood-fiber biology.

b. Bipolar Tip Growth

Over many years, it has been established that elongation growth of wood fibers occurs primarily, if not solely, at their tips (Larson 1994). Although the mechanism of this tip growth is not yet known, it has been suggested that it may be facilitated by secretion of wall-degrading enzymes at the tips (Wenham and Cusick 1975). Throughout all phases of fiber development, the MFs retain their net-axial orientation (Figure 8.7c, d). This is consistent with a role for them in supplying precursors or secretory vesicles, either indirectly via a role in cyclosis or by directed transport (Boevink et al. 1998), containing enzymes to the growing tips of these cells. Support for this view has been provided by the polar distribution of expansin mRNAs found in putative xylem fiber cells in stems of *Zinnia elegans* (Im et al. 2000). Along with xyloglucan endotransglycosylases (Vissenberg et al. 2000a), expansins are a class of enzymes that are implicated in "wall-loosening" events (Cosgrove 1998), such as may be necessary to facilitate tip growth of xylem fibers. Citing work of Roberts and Uhnak (1998) using the in vitro *Zinnia* system, in which leaf mesophyll cells are induced to differentiate as tracheary elements, Im et al. (2000) proposed that longitudinally oriented MTs might play a role in the translocation of the mRNAs to the fiber tips. *In planta,* in cells of the SVS of angiosperm trees, longitudinally oriented MTs have only so far been seen in gelatinous fibers of hybrid aspen (see also ultrastructural studies in *Populus* [Mia 1968; Fujita et al. 1974] and *Salix* [Robards and Kidwai 1972] and the immunofluorescent studies of *Fraxinus* [Prodhan et al. 1995]). The axially oriented MFs would appear to be a better candidate for any polar distribution of mRNAs that may be found in angiosperm trees, which would also be consistent with the role of MFs in vesicle delivery in growing pollen tubes (Kroeger et al. 2009). This interpretation also emphasizes the need for caution in making inferences about the natural system from an in vitro one.

Although fibers can increase in length quite markedly compared to their cambial precursors (e.g., to 359 μm from 247 μm in *A. hippocastanum* taproots—Chaffey et al. 1999), is the growth of fibers really "tip growth"? How does it compare with that in the more usual tip-growing systems of fungal hyphae, pollen tubes, root hairs, and fern protonemata (Geitman and Emons 2000)? The straightforward answer is that wood-fiber growth has not yet been studied in sufficient detail to permit a proper comparison with other tip-growing systems. However, some observations are pertinent. Wood fibers are single cells, both ends of which grow (i.e., they undergo bipolar growth—Chaffey et al. 1999), and whose MTs (Figure 8.7a) and MFs appear to extend to the tips of the cell. Other such systems are unipolar, and their apical regions usually have different MT and MF distributions to the rest of the cell. The growing tip of a wood fiber is long tapered and more varied in shape than the "hemisphere" of, for example, a pollen tube, and may bifurcate (Chaffey et al. 1999). Wood fibers bear wall elaborations—pits—which are absent from fungal hyphae or root hairs. During early development, wood fibers are in contact with other wood cells via plasmodesmata; all the other tip-growing systems are essentially isolated from other cells. The wall of fibers is probably much more structurally elaborate than that of most of the other systems and is lignified.

Taken together, wood fibers have a number of features that serve to distinguish them from the unipolar tip-growing systems. However, fibers are such an important economic cell type—for example, for wood pulp—that further study of their mode of growth is warranted and may even contribute to a greater understanding and appreciation of more conventional tip growth. It would be interesting—and no doubt instructive—to study wood-fiber development in a more detailed way, akin to that carried out by Cai et al. (2011) who examined the roles of MFs and MTs in the distribution of callose, sucrose, and cellulose synthases in tobacco pollen tubes.

c. Fiber Pits

The only deviation from the axial MF array in wood and phloem fibers is seen in the association of MFs with sites of simple pit formation in these cells (Figure 8.7d), where ellipses of antiactin reactive material are found. This is similar to the situation in axial and isolation ray parenchyma in both *Aesculus* and hybrid aspen. It may be significant that these ellipses are not seen in isolation, but are usually attached to MF bundles. Such an association might indicate that the MF bundles act as guides for the transport of enzymes of cell-wall modification, or precursors for cell-wall synthesis, to the site of pit development. The fact that no specific array of MTs has been seen at these sites implies that this cytoskeletal component is not directly involved with either the positioning or the development of the pits. Furthermore, the absence of MTs from the pit membrane of simple pits—and all other pit types in the secondary xylem, which remain as regions of unthickened primary wall-middle lamella—is consistent with the view that presence of MTs is required for secondary wall thickening to take place (Chaffey et al. 1999, 2002a).

d. Plasmatubules

Notwithstanding approximately 70 years of ultrastructural study of the SVS, new discoveries continue to be made. A good example of this is the presence of plasmatubules (PTs) in wood cells. PTs (Harris et al. 1982) are small-diameter, tubular evaginations of the plasmalemma found at sites where there is an inferred, but relatively short-term, high flux between the apoplasm and the symplasm (Chaffey and Harris 1985; Harris and Chaffey 1985, 1986). During ultrastructural studies of the SVS of *Aesculus* root, PTs were only seen in differentiating xylem cells (Chaffey et al. 1999). The plasmalemma-bound symplasmic compartment of an individual differentiating xylem cell is surrounded by a large apoplasmic space, which consists of its own cell wall, the walls of other cells, and the lumina of fully differentiated fibers and vessel elements. The observed distribution of PTs can thus be related to their inferred role in symplasm-apoplasm exchange, facilitating the resorption of the products of xylem cell lysis from the extracellular medium that might otherwise be lost. In this way, PTs may act as an adjunct to the symplasmic intercellular fluxes within and between the long-lived ray and axial parenchyma cells.

This notion of PT-facilitated recycling receives experimental support from a study of the uptake of radioactively labeled lysine by dwarf mistletoe (*Korthalsella lindsayi* (Oliver ex Hooker f.) Engler) from its host (Coetzee and Fineran 1989), in which PTs were implicated. Furthermore, absence of PTs from differentiating phloem cells is consistent with this suggestion for xylem, since here symplasmic routes between adjacent cells are maintained via plasmodesmata. However, even in situations within the phloem where such symplasmic continuity is lost, for example, during the differentiation of bast fibers, PTs have been recorded (Sal'nikov et al. 1993). Elsewhere, PTs have been illustrated and discussed in the study of tracheid differentiation in *Cryptomeria* sp. (Takabe and Harada 1986) and differentiating tracheary elements of *Salix dasyclados* (Sennerby-Forsse and von Fircks 1987). PTs are also evident in xylem fibers of *Acer* (Cronshaw 1965) and *Aesculus* (Barnett 1981).

2. Vessels and Vessel Elements

Vessel elements possess several distinct arrays of MTs and MFs in association with the numerous wall elaborations found in this cell type (Table 8.3); in view of space constraints, only development of the perforation plate and bordered pits will be considered further here.

a. Perforation Plate

The perforation plate is a highly specialized region of the vessel element wall. It is an MT-free zone (Chaffey et al. 1999) that is devoid of callose (Chaffey, Barlow and Barnett, unpublished observations), almost devoid of cellulose (Benayoun et al. 1981), and which neither secondarily thickens nor becomes lignified (Chaffey et al. 1999). Since removal of the perforation plate facilitates long-distance water transport throughout the tree, as individual vessel elements become vessels proper, it has been much studied (Meylan and Butterfield 1981; Butterfield 1995), although the details of its formation and subsequent loss are not fully known.

From a cytoskeletal point of view, we know that in roots (and shoots) of *Aesculus*, the periphery of the perforation plate is marked by a ring of both MTs and MFs, although in hybrid aspen, the MF ring appears to be absent, and a dense meshwork of MFs lies over the perforation plate itself. To date, and in the context of formation of the perforation plate, most significance is attached to the MF meshwork. In accordance with the view that enzymatic dissolution may play a role in the removal of the perforation plate (Benayoun 1983), it is suggested that transport of the secretory vesicles containing the necessary enzymes to the perforation plates is facilitated by the net-axial MF bundles of these cells (as occurs in pollen tube growth—e.g., Kroeger et al. 2009). However, once there, the MF meshwork effectively traps them so that their lytic cargoes are released only at the perforation plate. Clearly, work needs to be undertaken to test this hypothesis, but it has the merit of attempting to interpret structure and function in this cell.

b. Bordered Pits

Developing bordered pits were first detected as MT-free regions within the otherwise random array of MTs derived from the precursor fusiform cambial cell (Chaffey et al. 1997a). Subsequently, a polygonal array of MTs formed whose appearance was associated with deposition of cell wall, in a similar pattern to that of the MTs. Thereafter, a ring of MTs appeared, whose diameter decreased as the overarching pit border developed to leave a small elliptical aperture. A corresponding ring of MFs—strictly speaking actin since that is what the antibody used detects—also developed at the periphery of the bordered pit, whose diameter similarly decreased as the overarching pit border formed. The subsequent immunolocalization of myosin in association with the actin and tubulin at the developing bordered pits (Chaffey and Barlow 2002) led to the intriguing possibility that an actomyosin contractile system—a "plant muscle"—might be involved with the contraction of the MT-bounded pit aperture. This notion—as with much else concerning the cell biology of the tree SVS—remains to be fully explored, but it is not without commercial potential or relevance. For example, if the aperture of the bordered pit could be reduced further, it might hinder cell-cell passage of pathogenic organisms that might harm the tree. Alternatively, if the bordered pit aperture could be made wider (i.e., not reduced as much), this might facilitate infiltration of preservatives into any timber produced from the tree.

3. Multipurpose Microfilaments

A net-axial orientation of MFs is retained in the fusiform cambial cells and the axial components of the secondary xylem. However, there is a pronounced increase in the degree of branching of the MF network in axial xylem cells as cells differentiate from fusiform cambial cells into fibers and vessel elements. Thus, there appears to be a correlation between the degree of radial growth of the cell and the extent of branching of the MF network. In keeping with the notion that axial MFs are related to tip-directed cell-wall growth/elaboration/modification, it is suggested that the branched MF elements are involved with delivery of enzymes (such as expansins or xyloglucan endotransglycosylases) and/or wall precursors to specific sites on the side walls. There it may be necessary to break intermolecular bonds, such as xyloglucan cross-links (Fujino et al. 2000), so facilitating wall growth, or to permit wall elaboration such as development of simple pits in fiber cells. The degree of branching can be viewed as being at a maximum for the axial cell components in the case of transversely oriented MFs that are seen in vessel elements. Although MFs appear not to have a role in the change of orientation of MTs (Chaffey et al. 1997b), their putative involvement in formation of all pit types, perforations, and in directional cell expansion suggests that they have important roles in cell differentiation, generally, and secondary vascular differentiation specifically.

C. Seasonal Cycle of Cambial Activity-Dormancy

It is well known that cambial cells of the stems of temperate tree species undergo a seasonal cycle of activity-dormancy, which is paralleled by changes in cell-wall thickness (Catesson 1994; Chen et al. 2010). This was investigated within root cambia of horse chestnut (Chaffey et al. 1998), and a similar seasonal cycle was found. However, associated with the changes in cambial wall thickness was a previously unexpected cycle of MT reorientation. During cambial dormancy, when the cell walls are thickened, the MTs are arranged in a parallel-aligned helical orientation. When growth resumes and the walls become thin again, the random array of MTs reappears. These observations raise a number of questions.

First, how do the MTs, which appear to persist throughout the life of the cambial cells, survive dormancy? Although it is widely reported in the literature that MTs are depolymerized by low temperatures, their presence in cambia of roots *from frozen soil* (Chaffey et al. 1998) contradicts the view that MTs are absent from dormant cambia (Savidge 1993). Are the MTs in both active and dormant cambial cells naturally resistant to cold depolymerization? Or are the MTs of dormant cambial cells comprised of a cold-resistant type of tubulin, either an isoform of α- or β-tubulin, heterodimers of which constitute the MT, or a posttranslationally modified form of tubulin? Although preliminary immunofluorescent examination showed presence of acetylated and tyrosinated tubulins in the stem SVS of horse chestnut and hybrid aspen (Chaffey and Barlow, unpublished observations), no similar studies have yet been undertaken with roots (to my knowledge).

What is the significance of the helical orientation of the MTs during dormancy? Certainly, there appears to be an association throughout the SVS of parallel, helical MTs with wall thickening. It is seen during differentiation of fibers, vessel elements, and phloem cells. Therefore, by comparison with secondary wall formation in those other cells of the SVS and in keeping with the definition of a secondary wall as that which is laid down over an existing primary wall in a nongrowing cell, this "dormancy wall" of the cambium is considered a secondary wall (Chaffey et al. 1998).

Although it seems likely that the change of orientation of MTs from random to helical is related to the wall thickening during dormancy, whether this orientation of cytoplasmic MTs is mirrored by the orientation of cellulose microfibrils within the cell wall remains to be determined. It is tempting to suggest (Chaffey et al. 1998) that the dormancy-associated wall differs somehow from the primary wall which surrounds active cambial cells and that it is this difference that allows this "dormancy wall" to be lysed during spring cambial reactivation (Funada and Catesson 1991) but prevents the "active cambial primary wall" from suffering a similar fate.

However, once wall thickening has taken place, the MTs persist: Why? Do they serve other functions unrelated to wall elaboration? For example, do they act as a protein store that can be subsequently degraded and reutilized? This suggestion is superficially attractive for horse chestnut, which, unlike hybrid aspen, appears not to store protein in the SVS.

What seems less questionable is that the partial wall lysis in springtime will have several effects: It will reduce a major physical constraint to cell division/expansion, and it will also produce oligosaccharides. These sugars could serve as substrates for energy production and synthesis of cell components and/or act as solutes to raise cell turgor facilitating cell growth (Boldingh et al. 2000). It is also possible that oligosaccharins may be produced which can act as signaling molecules, which can have a number of roles in plant growth and development (Dumville and Fry 2000; Field 2009). Indeed, the simple sugars produced generally as part of wall digestion could have profound roles in aspects of growth and differentiation far beyond their use as substrates (Sheen et al. 1999; Smeekens 2000) and, potentially, far beyond the immediate vicinity of the cambium. In this regard, it is worth recalling the role attributed to a "secreted factor," possibly an oligosaccharide, in coordinating cell expansion and differentiation in the *Zinnia* mesophyll system (Roberts et al. 1997). Superficially, the enzymatic digestion that appears to accompany cambial reactivation has similarities with some aspects of the breaking of seed dormancy and fruit ripening. A more detailed comparison of all three phenomena is likely to provide further insights into cambial dormancy and spring activation.

Another important question arises in the context of cambial dormancy. Should dormant cambial cells be considered cambial cells? By definition, the cambium is a meristematic tissue. Clearly, dormant cambium is not meristematic, nor does it look like active cambium. This question is more than a semantic consideration, since it raises the further question of whether dormant cambial cells undergo a more profound "transdifferentiation" to become meristematic again, rather than simply wall thinning and vacuolation. If this is the case, arguably there is closer analogy between the "tree SVS system" with the transdifferentiation of mesophyll cells to generate tracheary elements in the *Zinnia* model system (Demura et al. 2002) than those who work with wood might feel comfortable with.

D. Continuity, Coordination, and Communication

To maintain the meristematic activity of the cambium and the growth and differentiation of its derivatives, input of materials is needed, as respirable substrates to support the high energy demand of cell division and growth, as structural raw materials, and as molecules that coordinate or orchestrate the process(es). Although hydrolysis of reserve materials (e.g., Fuchs et al. 2010b) and cambial wall lysis in situ may contribute to the cambium's demand for substrates, it is assumed that the majority of the required material is ultimately derived from photosynthesis, within the leaves for roots. Additionally, for stem cambia, it is also likely that local bark—or even pith—photosynthesis can supplement this leaf supply of photosynthate (Pfanz 2008). However, and regardless of the source of photosynthate, it must get into the cambial cells and their derivatives in order to fuel growth. Although it is well established that the *axial* pathway of photosynthate transport is within the sieve tubes of the phloem and that the rays are implicated in phloem-xylem exchange (van Bel 1990), the *radial* pathway from sieve tube to cambium is less clear (Sauter 2000; van Bel and Ehlers 2000).

Examination of the cytoskeleton within the long-lived cells of the SVS, particularly the ray cells, has suggested a role for this cell component in relation to photosynthate transfer to the cambial sink (Chaffey and Barlow 2001). At the final stages of xylem ray cell differentiation, when wall thickening and elaboration is complete, MFs and MTs are both bundled and axially oriented. At maturity, the MFs and MTs of axial xylem parenchyma cells are similarly oriented, parallel to the long cell axis. However, since the long axes of axial parenchyma and ray cells are at right angles to each other, so are the axes of orientation of their MT and MF cytoskeletons. Similarly, in mature phloem ray and axial parenchyma cells, although the MTs are more helically oriented (Figure 8.8a, b), a net-axial orientation of both MTs and MFs (Figure 8.8c) is attained. Since cell differentiation has finished and given that such ray cells live for many months (up to several years in the xylem), what is the role of MTs and MFs here? In view of the established importance of the long-lived ray cells as the symplasmic transport pathway between xylem and phloem (van Bel 1990) and their role in storage of reserve products (e.g., Sauter and van Cleve 1990; Höll 2000), Chaffey and Barlow (2001) proposed involvement of both cytoskeletal components in intracellular transport.

Additionally, unconventional myosin VIII and callose (a marker for plasmodesmata) have been immunolocalized to the pit fields between adjacent cells within the rays of the phloem and xylem and between ray cells and adjacent axial parenchyma (Chaffey and Barlow 2001). This evident plasmodesmatal coupling of living parenchyma cells thus provides an extensive 3-dimensional symplasmic network, which extends radially from the central pith to outer bark, circumferentially within the axial parenchyma of the xylem and phloem, and axially via the sieve tubes of the phloem, which ramifies throughout the tree.

The apparent colocalization of actin and myosin to the plasmodesmata of the membranes of the pits has led to further speculation that intercellular transport, between ray and axial parenchyma cells, may be mediated by plasmodesmata whose diameter can be increased/reduced via relaxation/contraction of an actomyosin complex (Chaffey and Barlow 2001). It is thus envisaged that this network might be involved in not only transport of photosynthate and other materials around the tree but also—by appropriate degrees of opening/closing of plasmodesmata—the establishment of symplasmically isolated domains (Lucas et al. 1993), which may have importance for coordination of developmental events throughout the tree (Chaffey and Barlow 2001).

This view of the radial-axial symplasmic pathway also suggests an additional role for the dilatation growth that can occur within phloem rays (Chaffey and Barlow 2001). Viewed in a transverse section of the root (or shoot), dilatation growth appears like a triangle (see, e.g., Fig. 14.8 in Barlow 2005), with its base toward the outside and its apex pointing toward the center of the root or shoot. In that way, it can be considered to act like a funnel, receiving photosynthate and other materials (such as hormones or other signaling molecules), from the sieve elements and associated cells and channeling it toward the bottom

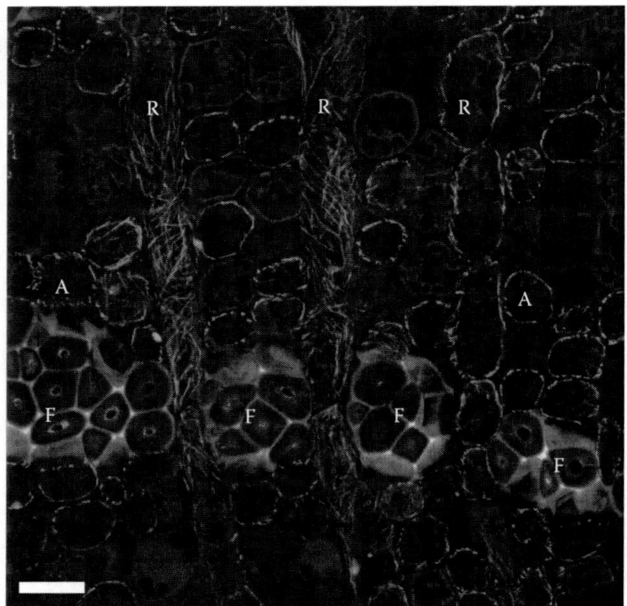

(a)

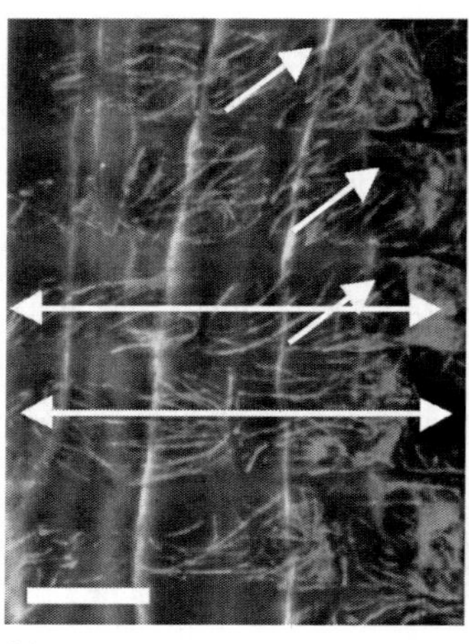

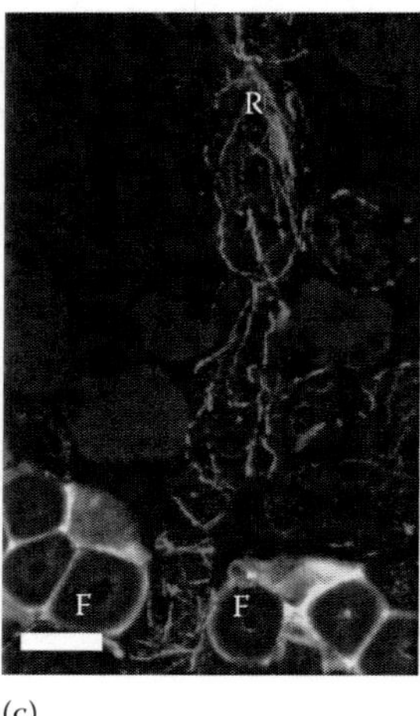

(b) (c)

FIGURE 8.8 (See color insert.) Indirect immunofluorescent localization of cytoskeletal proteins in phloem rays in hybrid aspen root. (a) Transverse section showing net-axial-helically oriented MTs in three rays (R) passing between axial parenchyma cells (A) and bundles of fibers (F); scale bar = 20 μm; (b) radial longitudinal section showing net-axially oriented MTs in ray cells (arrows) traversing a fiber bundle; scale bar = 10 μm; (c) transverse section showing net-axially oriented MF bundles in ray cells (R) passing between axial parenchyma cells and bundles of fibers (F); scale bar = 20 μm.

of the funnel to enter the uniseriate ray that then extends from the phloem into the cambial zone.

The cytoskeleton-facilitated solute transport so envisaged is expected to diminish as the helical/axial arrays of MTs and MFs are lost within the cells of the cambial zone, that is, in the cells where the transported material is required to fuel respiration and cell growth. One of the attractions of this idea is that transport can be bidirectional, from either the phloem side or the xylem side. Potentially, cambial reactivation in spring can be "kick started" by release of reserves from the axial and ray parenchyma cells of the xylem; thereafter, leaf photosynthate can be channeled to the cambium via the phloem route. Furthermore, while the gating of plasmodesmata could influence short-term flux of material between cells, mediated intracellularly via the cytoskeleton, the seasonal changes in plasmodesmatal frequency identified by Fuchs et al. (2010a) add another important dimension to coordination and communication within and between different tissues of the SVS.

VII. Where Next? (Ascendancy of *Arabidopsis*? Or the Tyranny of Thale Cress?)

When this question was posed previously (Chaffey 2002g), it was concluded that—given our then comparative ignorance about all aspects of "cell" biology of tree root cambia—further study in almost any direction will generate useful information. That is still the case; we know too little about what makes roots work, especially tree roots. It was also then acknowledged that one way to boost work with tree roots was to exploit the wood-forming capability of *Arabidopsis*, the model angiosperm (Koornneef and Meinke 2010). For many years, it has been known that *Arabidopsis* undergoes some degree of secondary thickening in its roots (Dolan and Roberts 1995); roots of *Arabidopsis* have been even been proposed as a model to study *plant* development (Scheres and Wolkenfeldt 1998). With demonstration that the hypocotyl of *Arabidopsis* could be encouraged to undergo quite extensive secondary growth under appropriate environmental conditions (Chaffey et al. 2002b), attention was directed away from subterranean organs toward the potential of aerial ones. In terms of research effort, there are many reviews and reports that extol the similarities between herbs and woody plants (e.g., Turner and Sieberth 2003; Nieminen et al. 2004; Groover 2005; Déjardin et al. 2010) and promote the use of *Arabidopsis* as a "proxy" for understanding the tree wood-forming system. Although the relevance of *Arabidopsis* to "proper" wood growth in trees was undermined somewhat by the failure to observe ray cells (Chaffey et al. 2002b), Mazur and Kurczynska (2011) have recently demonstrated not only ray formation but also intrusive growth of cambial cells in the stem inflorescence of *Arabidopsis*. Furthermore, ultrastructurally, the cambial cells of *Arabidopsis* appear like those of angiosperm tree roots and shoots, and the fibers and vessel elements of its wood are morphologically similar to those of hybrid aspen and horse chestnut (Chaffey et al. 2002b). The case for use of *Arabidopsis* to understand the SVS is certainly strong.

But, on the other hand, there are differences between the SVS of *Arabidopsis* and trees, such as in stress-responsive microRNAs (Lu et al. 2005) and the growth pattern and network of plasmodesmata (Fuchs et al. 2011). Interestingly, while Fuchs et al. (2011) rule out the usefulness of *Arabidopsis* for understanding *organizational and regulatory aspects* of wood formation, they do still envisage a role for the herb in regard to molecular approaches to answer question of xylogenesis and phloem formation.

Acknowledging that additional model systems are needed to cope with organismal diversity (Abzhanov et al. 2008), poplars—*Populus* spp.—can rightly lay claim to be the model angiosperm tree (Chaffey 2002c,f; Jansson and Douglas 2007; Jung et al. 2008). Indeed, poplar has been referred to as an "Arabidopsis for forestry" (Taylor 2002). Hearteningly, progress has been made in molecular studies of roots of this genus, for example, Kohler et al. (2003) advanced the notion of the root transcriptome by producing ESTs (expressed sequence tags) from hybrid cottonwood roots. Unfortunately, those ESTs are not from wood-forming tissues (see the images in http://mycor.nancy.inra.fr/IMGC/PoplarGenome/PoplarDB/FEBS/Rooting_stages.pdf). Hopes that this work might have inspired a new generation to undertake root SVS research are still to be realized.

So, a decade on, the research community still seems reluctant to "go underground." The temptation not to dig deeper (in both senses of the word!) is strong, and excuses can always be made for rhizo-averse behavior. But, do we need to study the tree root SVS? Are shoots and roots of trees sufficiently alike that the former can be used as a guide to the latter? Unless more workers take up the challenge of root work, we'll never know!

VIII. Concluding Comments

To study everything one should just about the cell biology of the tree SVS would be much too big an undertaking. Of necessity, one concentrates on those aspects that are sufficiently appealing to one's own interests, temperament, etc. Accordingly, my main focus has been on the role of the cytoskeleton in cambial biology and wood formation. Of course, the cytoskeleton is not the be-all and end-all of cell biology; it is only one aspect among many. Originally, the justification for examining the cytoskeleton in xylogenesis was the association of MT arrangement and microfibril orientation in the cell wall, which might influence cell morphogenesis. Associated with differentiation of all wood-cell types is a wide variety of changes to the cell wall. Thus, a complete wood-cell biology must include examination of wall polysaccharides (e.g., Grünwald et al. 2002), their enzymes of formation and modification (e.g., Bourquin et al. 2002; Micheli et al. 2002; Mellerowicz and Sundberg 2008), and enzymes of lignification (e.g., Šamaj and Boudet 2002). It is important therefore to explore and exploit all of the latest techniques to increase knowledge—and hopefully understanding of cells of the SVS. This even extends as far as to challenge our established notions of cell ultrastructure, which are usually derived from "conventional" chemical fixation protocols (e.g., Chaffey 2002e), by developing new techniques, such as high-pressure freezing/freeze substitution (Rensing 2002), which can often give a quite different view of cell structure (e.g., Kaneda et al. 2010). Continued advances in tissue preparation techniques are likely to shake up established views of cell biology of the SVS in a way similar to that which our ability to use a wide range of fluorescent probes has caused us to reevaluate many aspects of cell "behavior" (e.g., Mathur 2007).

And study of "cell biology" is not enough. To that must be added the study of the overall coordination of cambial and SVS activity from such factors as hormones (e.g., Elo et al. 2009) and transcription factors and genetic regulators (e.g., Du and Groover 2010; Risopatron et al. 2010), which intimately influence the final form and pattern of cells within the SVS. While it seems increasingly likely that work on *Arabidopsis* will continue to dominate in the short term, I am ever hopeful that more workers will eventually realize that trees are trees—and *Arabidopsis* is not—and start working with trees.

Furthermore, wood (secondary xylem) is not the only product of the tree SVS; secondary phloem ("inner bark") is also produced. Although, from a commercial stance, wood is probably the more important product, from a biological perspective, one must acknowledge that the wood could not be made if it were not for the transport of photosynthate to the differentiating cambial derivatives via the phloem to fuel the formation of the wood cells. But simply, no phloem, no wood. Thus, there is a great impetus to understand not only the cell—and molecular—biology of secondary phloem formation but also the interactions between secondary phloem and secondary xylem, which may be mediated via the living ray parenchyma that connects the two tissues through the vascular cambium (e.g., Chaffey and Barlow 2001). The tree SVS remains a challenging aspect of plant biology. It will not go away. Its study is important and will richly reward all those who rise to the challenge.

References

Abzhanov A, Extavour CG, Groover A et al. 2008 Are we there yet? Tracking the development of new model systems. *Trends Genet* 24:353–360.

Aloni R, Feigenbaum P, Kalev N, Rozovsky S 2000 Hormonal control of vascular differentiation in plants: the physiological basis of cambium ontogeny and xylem evolution. In *Cell and Molecular Biology of Wood Formation*, eds. R Savidge, JR Barnett, R Napier, pp. 223–236. Oxford, U.K.: Bios Scientific Publishers.

The Arabidopsis Book (2011) Accessible at http://www.bioone.org/page/arbo.j/aims

Awano T, Takabe K, Fujita M, Daniel G 2000 Deposition of glucuronoxylans on the secondary cell wall of Japanese beech as observed by immuno-scanning electron microscopy. *Protoplasma* 212:72–79.

Bailey IW 1952 Biological processes in the formation of wood. *Science* 115:255–259.

Baluška F, Volkmann D, Barlow PW, eds 2006 *Cell-Cell Channels*. New York: Landescience and Springer Science + Science Media.

Barlow P 2005 From cambium to early cell differentiation within the secondary vascular system. In *Vascular Transport in Plants*, eds. NM Holbrook, MA Zwienicki, pp. 279–306. San Diego, CA: Elsevier Academic Press.

Barnett JR 1981 Secondary xylem cell development. In *Xylem Cell Development*, ed. JR Barnett, pp. 47–95. Tunbridge Wells, U.K.: Castle House Publications.

Barnett JR., Chaffey NJ, Barlow PW 1998 Cortical microtubules and microfibril angle. In *Microfibril Angle in Wood*, ed. BG Butterfield, pp. 253–271. Christchurch, New Zealand: IAWA/IUFRO.

Baskin TI 2001 On the alignment of cellulose microfibrils by cortical microtubules: a review and a model. *Protoplasma* 215:150–171.

van Bel AJE 1990. Xylem-phloem exchange via the rays: the undervalued route of transport. *J Exp Bot* 41:631–644.

van Bel AJE, Ehlers K 2000 Symplasmic organization of the transport phloem and the implications for photosynthate transfer to the cambium. In *Cell and Molecular Biology of Wood Formation*, eds. R Savidge, JR Barnett, R Napier, pp. 85–99. Oxford, U.K.: Bios Scientific Publishers.

Bell K, Oparka K 2011 Imaging plasmodesmata. *Protoplasma* 248:9–25.

Benayoun J 1983 A cytochemical study of cell wall hydrolysis in the secondary xylem of poplar (*Populus italica* Moench). *Ann Bot* 52:189–200.

Benayoun J, Catesson AM, Czaninski Y 1981 A cytochemical study of differentiation and breakdown of vessel end walls. *Ann Bot* 47:687–698.

Boevink P, Oparka K, Cruz SS, Martin B, Betteridge A, Hawes C 1998 Stacks on tracks: the plant Golgi apparatus traffics on an actin/ER network. *Plant J* 15:441–447.

Boldingh H, Smith GS, Klages K 2000. Seasonal concentrations of non-structural carbohydrates of five *Actinidia* species in fruit, leaf and fine root tissue. *Ann Bot* 85:469–476.

Bourquin V, Nishikubo N, Abe H et al. 2002 Xyloglucan endotransglycosylases have a function during the formation of secondary cell walls of vascular tissues. *Plant Cell* 14:3073–3088.

Braun HJ 1984 The significance of the accessory tissues of the hydrosystem for osmotic water shifting as the second principle of water ascent, with some thoughts concerning the evolution of trees. *IAWA Bull ns* 5:274–294.

Butterfield BG 1995 Vessel element differentiation. In *The Cambial Derivatives*, ed. M Iqbal, pp. 93–106. Berlin, Stuttgart, Germany: Gebrüder Borntraeger.

Cai G, Faleri C, Del Casino C, Emons AMC, Cresti M 2011 Distribution of callose synthase, cellulose synthase, and sucrose synthase in tobacco pollen tube is controlled in dissimilar ways by actin filaments and microtubules. *Plant Physiol* 155:1169–1190.

Catesson A-M 1974 Cambial cells. In *Dynamic Aspects of Plant Ultrastructure*, ed. AW Robards, pp. 358–390. London, U.K.: McGraw-Hill Book Co.

Catesson A-M 1994 Cambial ultrastructure and biochemistry: changes in relation to vascular tissue differentiation and the seasonal cycle. *Int J Plant Sci* 155:251–261.

Chaffey NJ, ed. 2002a *Wood Formation in Trees: Cell and Molecular Biology Techniques*. Amsterdam, the Netherlands: Taylor & Francis.

Chaffey NJ 2002b Introduction. In *Wood Formation in Trees: Cell and Molecular Biology Techniques*, ed. NJ Chaffey, pp. 1–8. Amsterdam, the Netherlands: Taylor & Francis.

Chaffey NJ 2002c An introduction to the problems of working with the tree secondary vascular system. In *Wood Formation in Trees: Cell and Molecular Biology Techniques*, ed. NJ Chaffey, pp. 9–16. Amsterdam, the Netherlands: Taylor & Francis.

Chaffey NJ 2002d Wood microscopical techniques. In *Wood Formation in Trees: Cell and Molecular Biology Techniques*, ed. NJ Chaffey, pp. 17–40. Amsterdam, the Netherlands: Taylor & Francis.

Chaffey NJ 2002e Conventional (chemical-fixation) transmission electron microscopy and cytochemistry of angiosperm trees. In *Wood Formation in Trees: Cell and Molecular Biology Techniques*, ed. NJ Chaffey, pp. 41–64. Amsterdam, the Netherlands: Taylor & Francis.

Chaffey NJ 2002f Why is there so little research into the cell biology of the secondary vascular system of trees? *New Phytolt* 153:213–223.

Chaffey NJ 2002g. Secondary growth of roots: a cell biological perspective. In *Plant Roots: The Hidden Half*, eds. Y Waisel, A Eshel, U Kafkafi, 3rd edn., pp. 93–111. New York: Marcel Dekker, Inc.

Chaffey NJ, Barlow PW 2001 The cytoskeleton facilitates a three-dimensional symplasmic continuum in the long-lived ray and axial parenchyma cells of angiosperm trees. *Planta* 213:811–823. [Erratum, *Planta* 214:330–331.]

Chaffey NJ, Barlow PW 2002 Myosin, microtubules, and microfilaments: co-operation between cytoskeletal components during cambial cell division and secondary vascular differentiation in trees. *Planta* 214:526–536.

Chaffey NJ, Barlow PW, Barnett JR 1997a Formation of bordered pits in secondary xylem vessel elements of *Aesculus hippocastanum* L.: an electron and immunofluorescent microscope study. *Protoplasma* 197:64–75.

Chaffey NJ, Barlow PW, Barnett JR 1997b Microtubules rearrange during differentiation of vascular cambial derivatives, microfilaments do not. *Trees* 11:333–341.

Chaffey NJ, Barlow PW, Barnett JR 1998 A seasonal cycle of cell wall structure is accompanied by a cyclical rearrangement of cortical microtubules in fusiform cambial cells within taproots of *Aesculus hippocastanum* L. (Hippocastanaceae). *New Phytol* 139:623–635.

Chaffey NJ, Barlow PW, Barnett JR 2000a A cytoskeletal basis for wood formation in angiosperm trees: the involvement of microfilaments. *Planta* 210:890–896.

Chaffey NJ, Barlow PW, Barnett JR 2000b. Structure-function relationships during secondary phloem development in *Aesculus hippocastanum*: microtubules and cell walls. *Tree Physiol* 20:777–786.

Chaffey NJ, Barlow PW, Sundberg B 2002a Understanding the role of the cytoskeleton in wood formation in angiosperm trees: hybrid aspen (*Populus tremula* x *P. tremuloides*) as the model species. *Tree Physiol* 22:239–249.

Chaffey NJ, Barnett JR, Barlow PW 1997c Endomembranes, cell walls and cytoskeleton: aspects of the biology of the vascular cambium of *Aesculus hippocastanum* L. *Int J Plant Sci* 158:97–109.

Chaffey NJ, Barnett JR, Barlow PW 1999 A cytoskeletal basis for wood formation in angiosperm trees: the involvement of cortical microtubules. *Planta* 208:19–30.

Chaffey NJ, Cholewa E, Regan S, Sundberg B 2002b Secondary xylem development in *Arabidopsis*: a model for wood formation. *Physiol Plant* 114:594–600.

Chaffey NJ, Harris N 1985. Plasmatubules: fact or artefact? *Planta* 165:185–190.

Chen H-M, Han J-J, Cui K-M, He X-Q 2010 Modification of cambial cell wall architecture during cambium periodicity in *Populus tomentosa* Carr. *Trees* 24:533–540.

Coetzee J, Fineran BA 1989 Translocation of lysine from the host *Melicope simplex* to the parasitic dwarf mistletoe *Korthalsella lindsayi* (Viscaceae). *New Phytol* 112:377–381.

Cosgrove DJ 1998 Cell wall loosening by enzymes. *Plant Physiol* 118:333–339.

Cronshaw J 1965 The organization of cytoplasmic components during the phase of cell wall thickening in differentiating cambial derivatives of *Acer rubrum*. *Can J Bot* 43:1401–1407.

Cutter EG 1975 *Plant Anatomy: Experiment and Interpretation, Part 2, Organs*. London, U.K.: Edward Arnold.

Cutter EG 1978 *Plant Anatomy, Part 1, Cells and Tissues*, 2nd edn. London, U.K.: Edward Arnold.

Cyr R 2007. Plant mitosis, cytokinesis and cell plate formation. *Handbook of Plant Science* Vol. 1, ed. K Roberts, pp. 294–300. Chichester, U.K.: John Wiley.

Déjardin A, Laurans F, Arnaud D, Breton C, Pilate G, Leplé 2010 Wood formation in angiosperms. *C R Biologies* 333:325–334.

Demura T, Tashiro G, Horiguchi G et al. 2002 Visualization by comprehensive microarray analysis of gene expression programs during transdifferentiation of mesophyll cells into xylem cells. *Proc Natl Acad Sci USA* 99:15794–15799.

Dolan L, Roberts K 1995 Secondary thickening in roots of *Arabidopsis thaliana*. *New Phytol* 131:121–128.

Du J, Groover A 2010 Transcriptional regulation of secondary growth and wood formation. *J Integ Plant Biol* 52:17–27.

Dumville JC, Fry SC 2000 Uronic acid-containing oligosaccharins: their biosynthesis, degradation and signalling roles in non-diseased plant tissues. *Plant Physiol Biochem* 38:125–140.

Elo A, Immanen J, Nieminen K, Helariutta Y 2009 Stem cell function during plant vascular development. *Semin Cell Dev Biol* 20:1097–1106.

Evert RF, Eichhorn SE 2006. *Esau's Plant Anatomy Meristems, Cells, and Tissues of the Plant Body: Their Structure, Function, and Development*, 3rd edn. Hoboken, NJ: Wiley-Interscience.

Farrar JJ, Evert RF 1997 Seasonal changes in the ultrastructure of the vascular cambium of *Robinia pseudoacacia*. *Trees* 11:191–202.

Fayle DCF 1968 Radial growth of tree roots. *Fac For Univ Toronto Tech Rep* 9:1–183.

Field RA 2009 Oligosaccharide signalling molecules. In *Plant-Derived Natural Products*, eds. AE Osbourn, V Lanzotti, pp. 349–359. New York: Springer.

Follet-Gueye ML, Ermel FF, Vian B, Catesson AM, Goldberg R 2000 Pectin remodelling during cambial derivative differentiation. In *Cell and Molecular Biology of Wood Formation*, eds. R Savidge, JR Barnett, R Napier, pp. 289–294. Oxford, U.K.: Bios Scientific Publishers.

Fuchs M, van Bel AJE, Ehlers K 2010a Season-associated modifications in symplasmic organization of the cambium in *Populus nigra*. *Ann Bot* 105:375–387.

Fuchs M, van Bel AJE, Ehlers K 2011 Do symplasmic networks in cambial zones correspond with secondary growth patterns? *Protoplasma* 238:141–151.

Fuchs M, Ehlers K, Will T, van Bell AJE 2010b Immunolocalization indicates plasmodesmal trafficking of storage protein during cambial reactivation in *Populus nigra*. *Ann Bot* 106:385–394.

Fujino T, Sone Y, Mitsuishi Y, Itoh T 2000 Characterization of cross-links between cellulose microfibrils, and their occurrence during elongation growth in pea epicotyl. *Plant Cell Physiol* 41:486–494.

Fujita M, Saiki H, Harada H 1974 Electron microscopy of microtubules and cellulose microfibrils in secondary wall formation of poplar tension wood fibers. *Mokuzai Gakkaishi* 20:147–156.

Funada R, Catesson A-M 1991 Partial cell wall lysis and the resumption of meristematic activity in *Fraxinus excelsior* cambium. *IAWA Bull ns* 12:439–444.

Geitman A, Emons AMC 2000 The cytoskeleton in plant and fungal cell tip-growth. *J Microsc* 198:218–245.

Goosen-de Roo L, Bakhuizen R, van Spronsen PC, Libbenga KR 1984 The presence of extended phragmosomes containing cytoskeletal elements in fusiform cambial cells of *Fraxinus excelsior* L. *Protoplasma* 12:145–152.

Goosen-de Roo L, Burggraaf PD and Libbenga KR 1983 Microfilament bundles associated with tubular endoplasmic reticulum in fusiform cells in the active cambial zone of *Fraxinus excelsior* L. *Protoplasma* 116:204–208.

Green PB 1969 Cell morphogenesis. *Annu Rev Plant Physiol* 20:365–394.

Groover AT 2005 What genes make a tree a tree? *Trends Plant Sci* 10:210–214.

Grünwald C, Ruel K, Kim YS, Schmitt U 2002 On the cytochemistry of cell wall formation in poplar trees. *Plant Biol* 4:13–21.

Harris N, Chaffey NJ 1985 Plasmatubules in transfer cells of pea (*Pisum sativum* L.). *Planta* 165:191–196.

Harris N, Chaffey NJ 1986. Plasmatubules—real modifications of the plasmalemma. *Nordic J Bot* 6:599–607.

Harris N, Oparka KJ, Walker-Smith DW 1982 Plasmatubules: an alternative to transfer cells? *Planta* 156:461–465.

Hodge A, Berta G, Doussan C, Merchan F, Crespi M 2009 Plant root growth, architecture and function. *Plant Soil* 321:153–187.

Höll W 2000 Distribution, fluctuation and metabolism of food reserves in the wood of trees. In *Cell and Molecular Biology of Wood Formation*, eds. R Savidge, JR Barnett, R Napier, pp. 347–362. Oxford, U.K.: Bios Scientific Publishers.

Im K-H, Cosgrove DJ, Jones AM 2000 Subcellular localization of expansin mRNA in xylem cells. *Plant Physiol* 123:463–470.

Iqbal M, ed 1995 *The Cambial Derivatives*. Berlin, Stuttgart, Germany: Gebrüder Borntraeger.

Jansson S, Douglas CJ 2007 *Populus*: a model system for plant biology. *Annu Rev Plant Biol* 58:435–58.

Jung J-H, Kim S-G, Seo PJ, Park C-M 2008 Molecular mechanisms underlying vascular development. *Adv Bot Res* 48:1–68.

Jürgens G 2005 Cytokinesis in higher plants. *Annu Rev Plant Biol* 56:281–299.

Kaneda M, Rensing K, Samuels L 2010 Secondary cell wall deposition in developing secondary xylem of poplar. *J Integ Plant Biol* 52:234–243.

Kohler A, Delaruelle C, Martin D, Encelot N, Martin F 2003 The poplar root transcriptome: analysis of 7000 expressed sequence tags. *FEBS Lett* 542:37–41.

Koornneef M, Meinke D 2010 The development of *Arabidopsis* as a model plant. *Plant J* 61:909–921.

Kroeger JH, Daher FB, Grant M, Geitmann A 2009 Microfilament orientation constrains vesicle flow and spatial distribution in growing pollen tubes. *Biophys J* 97:1822–1831.

Lachaud S, Catesson A-M, Bonnemain J-L 1999 Structure and function of the vascular cambium. *CR Acad Sci, Paris, Sér III, Sci de la vie* 322:633–650.

Larson PR 1994 *The Vascular Cambium: Development and Structure*. Berlin: Springer.

Lu S, Sun Y-H, Shi R, Clark C, Li L, Chiang VL 2005 Novel and mechanical stress-responsive microRNAs in *Populus trichocarpa* that are absent from *Arabidopsis*. *Plant Cell* 17:2186–2203.

Lucas WJ, Ding B, van der Schoot C 1993 Plasmodesmata and the supracellular nature of plants. *New Phytol* 125:435–476.

Mathur J 2007 The illuminated plant cell. *Trends Plant Sci* 12:506–513.

Maule A 2008 Plasmodesmata: structure, function and biogenesis. *Curr Opin Plant Biol* 11:680–686.

Mazur E, Kurczynska EU 2012 Rays, intrusive growth, and storied cambium in the inflorescence stems of *Arabidopsis* (L.) Theynh. *Protoplasma* 249:217–220.

Mellerowicz E, Sundberg B 2008 Wood cell walls: biosynthesis, developmental dynamics and their implications for wood properties. *Curr Opin Plant Biol* 11:293–300.

Meylan BA, Butterfield BG 1981 Perforation plate differentiation in the vessels of hardwoods. In *Xylem Cell Development*, ed. JR Barnett, pp. 96–114. Tunbridge Wells, U.K.: Castle House Publications.

Mia AJ 1968. Organization of tension wood-fibers with special reference to the gelatinous layer in *Populus tremuloides* Michx. *Wood Sci* 1:105–115.

Micheli F, Ermel FE, Bordenave M, Richard L, Goldberg R 2002. Cell walls of woody tissues: cytochemical, biochemical and molecular analysis of pectins and pectin methylesterases. In *Wood Formation in Trees: Cell and Molecular Biology Techniques*, ed. NJ Chaffey, pp. 179–200. Amsterdam, the Netherlands: Taylor & Francis.

Nieminen KM, Kauppinen L, Helariutta Y 2004. A weed for wood? *Arabidopsis* as a genetic model for xylem development. *Plant Physiol* 135:653–659.

Nilsson J, Karlberg Å, Antti H et al. 2008 Dissecting the molecular basis of the regulation of wood formation by auxin in hybrid aspen. *Plant Cell* 20:843–855.

Paredez AR, Somerville CR, Ehrhardt DW 2006 Visualization of cellulose synthase demonstrates functional association with microtubules. *Science* 312:1491–1495.

Pfanz H 2008 Bark photosynthesis. *Trees* 22:137–138.

Phillips IDJ 1976 The cambium. In *Cell Division in Higher Plants*, ed. MM Yeoman, pp. 347–390. London, U.K.: Academic Press.

Prodhan AKMA, Funada R, Ohtani J, Abe H, Fukazawa K 1995 Orientation of microfibrils and microtubules in developing tension-wood-fibers of Japanese ash (*Fraxinus mandshurica* var. *japonica*). *Planta* 196:577–585.

Radford JE, Vesk M, Overall RL 1998 Callose deposition at plasmodesmata. *Protoplasma* 201:30–37.

Radford JE, White RG 1998 Localization of a myosin-like protein to plasmodesmata. *Plant J* 14:743–750.

Radford JE, White RG 2011 Inhibitors of myosin, but not actin, alter transport through *Tradescantia* plasmodesmata. *Protoplasma* 248:205–216.

Raven PH, Evert RF, Eichhorn SE 1999 *Biology of Plants*, 6th edn. New York: WH Freeman and Company/Worth Publishers.

Reichelt S, Kendrick-Jones J 2000 Myosins. In *Actin: A Dynamic Framework for Multiple Plant Cell Functions*, eds. C Staiger, F Baluška, D Volkmann, PW Barlow, pp. 29–44. Dordrecht, the Netherlands: Kluwer.

Reichelt S, Knight AE, Hodge TP, Baluška F, Šamaj J, Volkmann D, Kendrick-Jones J 1999 Characterization of the unconventional myosin VIII in plant cells and its localization at the post-cytokinetic cell wall. *Plant J* 19:555–567.

Rensing K 2002 Chemical and cryo-fixation for transmission electron microscopy of gymnosperm cambial cells. In *Wood Formation in Trees: Cell and Molecular Biology Techniques*, ed. NJ Chaffey, pp. 65–81. Amsterdam, the Netherlands: Taylor & Francis.

Rensing KH, Samuels AL, Savidge RA 2002 Ultrastructure of vascular cambial cell cytokinesis in pine seedlings preserved by cryofixation and substitution. *Protoplasma* 220:39–49.

Risopatron JPM, Sun Y, Jones BJ 2010 The vascular cambium: molecular control of cellular Structure. *Protoplasma* 247:145–161.

Robards AW, Kidwai P 1972 Microtubules and microfibrils in xylem fibers during secondary cell wall formation. *Cytobiologie* 6:1–21.

Roberts AW, Donovan SG, Haigler CH 1997 A secreted factor induces cell expansion and formation of metaxylem-like tracheary elements in xylogenetic suspension cultures of *Zinnia*. *Plant Physiol* 115:683–692.

Roberts A, Uhnak KS 1998 Tip growth in xylogenic suspension cultures of *Zinnia elegans*: implications for the relationship between cell shape and secondary cell wall pattern in tracheary elements. *Protoplasma* 204:103–113.

Sal'nikov VV, Ageeva MV, Yumashev VN, Lozovaya VV 1993 Ultrastructural analysis of bast fibers. *Russian Plant Physiol* 40:416–421.

Šamaj J, Boudet AM 2002 Immunolocalisation of enzymes of lignification. In *Wood Formation in Trees: Cell and Molecular Biology Techniques*, ed. NJ Chaffey, pp. 201–214. Amsterdam, the Netherlands: Taylor & Francis.

Sauter JJ 2000 Photosynthate allocation to the vascular cambium: facts and problems. In *Cell and Molecular Biology of Wood Formation*, eds. R Savidge, JR Barnett, R Napier, pp. 71–83. Oxford, U.K.: Bios Scientific Publishers.

Sauter JJ, van Cleve B 1990 Biochemical, immunochemical, and ultrastructural studies of protein storage in poplar (*Populus* x *canadensis* 'robusta') wood. *Planta* 183:92–100.

Savidge AR 1993 Formation of annual rings in trees. In *Oscillations and Morphogenesis*, ed. L Rensing, pp. 434–463. New York: Marcel Dekker, Inc.

Scheres B, Wolkenfelt H 1998 The *Arabidopsis* root as a model to study plant development. *Plant Physiol Biochem* 36:21–32.

Scherp P, Grotha R, Kutschera U 2002 Interactions between cytokinesis-related callose and cortical microtubules in dividing cells of the liverwort *Riella helicophylla. Plant Biol.* 4:619–624.

Seagull RW 1992 A quantitative electron microscopic study of changes in microtubule arrays and wall microfibril orientation during in vitro cotton fiber development. *J Cell Sci* 101:561–577.

Sennerby-Forsse L, von Fircks HA 1987 Ultrastructure of cells in the cambial region during winter hardening and spring dehardening in *Salix dasyclados* Wim. grown at two nutrient levels. *Trees* 1:151–163.

Sheen J, Zhou L, Jang J-C 1999 Sugars as signalling molecules. *Curr Opin Plant Biol* 2:410–418.

Smeekens S 2000 Sugar-induced signal transduction in plants. *Annu Rev Plant Physiol Plant Mol Biol* 51:49–81.

Smith AM, Coupland G, Dolan L et al. 2010 *Plant Biology*. New York: Garland Science.

Sundberg B, Uggla C, Tuominen H 2000 Cambial growth and auxin gradients. In *Cell and Molecular Biology of Wood Formation*, eds. R Savidge, JR Barnett, R Napier, pp. 169–188. Oxford, U.K.: Bios Scientific Publishers.

Takabe K, Harada H 1986 Polysaccharide deposition during tracheid wall formation in *Cryptomeria. Mokuzai Gakkaishi* 32:763–769.

Taylor G 2002 *Populus*: *Arabidopsis* for forestry. Do we need a model tree? *Ann Bot* 90:681–689.

Turner S, Sieburth LE 2003 Vascular patterning. *The Arabidopsis Book* 2: e0073 10.1199/tab.0073. http://my.aspb.org/members/group_content_view.asp?group=68456&id=110592.

Van Damme D, Vanstraelen M, Geelen D 2007 Cortical division zone establishment in plant cells. *Trends Plant Sci* 12:458–464.

Vissenberg K, Martinez-Vilchez IM, Verbelen J-P, Miller JG, Fry SC 2000a In vivo colocalization of xyloglucan endotransglycosylase activity and its donor substrate in the elongation zone of roots. *Plant Cell* 12:1229–1238.

Vissenberg K, Quwlo A-H, van Gestel K, Olyslaegers G, Verbelen J-P 2000b From hormone signal, via the cytoskeleton, to cell growth in single cells of tobacco. *Cell Biol Int* 24:343–349.

Waterkeyn L 1967 Sur l'existence d'un "stade callosique", présenté par la paroi cellulaire, au cours de la cytocinèse. *CR Acad Sci, Paris* 265:1792–1794.

Wenham MW, Cusick F 1975. The growth of secondary wood-fibres. *New Phytol* 74:247–261.

Wilson BF 1975. Distribution of secondary thickening in tree root systems. In *The Development and Function of Roots*, eds. JG Torrey, DT Clarkson, pp. 197–219. New York: Academic Press.

Woods FW 1991. Cambial activity of roots. In *Plant Roots: The Hidden Half*, eds. Y Waisel, A Eshel, U Kafkafi, pp. 149–160. New York: Marcel Dekker, Inc.

9

Rice: A Model Plant to Decipher the Hidden Origin of Adventitious Roots

Yoan Coudert
Université Montpellier 2

Van Anh Le Thi
Agricultural Genetics Institute
University of Science and Technology of Hanoi
Université Montpellier 2

Pascal Gantet
Agricultural Genetics Institute
University of Science and Technology of Hanoi
Université Montpellier 2

I. Introduction

In flowering plants, the primary root system is generated by the root apical meristem (RAM), a meristem that is formed during embryogenesis. The RAM will generate the seminal root (SR), which can branch to generate postembryonic lateral roots (LRs). These LRs can branch to generate other LRs, which result in additional branching levels. Some plant species can also generate other postembryonic roots, called adventitious roots (ARs), from the stem. Because ARs generally develop from aerial axes, they are easily observable until they reach the ground. Paradoxically, their cellular origin and the mechanisms controlling their formation remain hidden and are much less known than those governing the genesis of underground LR. In most monocot species, ARs develop constitutively and constitute the main part of the fibrous root system. In dicots, ARs can also develop constitutively or in response to abiotic or biotic stresses conferring the plant some adaptive advantage (Barlow 1986). The induction of AR formation in response to biotic stress is illustrated by the induction of AR formation from any part of the plant by the pathogenic bacterium *Agrobacterium rhizogenes* through the transfer of genes from the bacterial genome that modify hormonal balance of the plant (Figure 9.1A). These ARs are agravitropic, highly branched, and fast growing and synthesize compounds that sustain the bacteria. These roots can be isolated and easily cultured in vitro because they do not require any hormonal supply to maintain their growth. For this reason they are used as an artificial source of valuable secondary metabolites that are naturally synthesized by roots of medicinal plants (Guillon et al. 2006a,b). Some phytopathogenic fungi or viruses are also known to promote the formation of AR after infection of their host; however, the biological significance of this morphological response is not always clear (Dimond and Waggoner 1953; Rasa and Esau 1961). The AR of *Sesbania* sp. can develop nitrogen-fixing nodules when infected by symbiotic *Rhizobium* sp. bacteria (Duboux 1969). AR formation is also stimulated by wounding or mechanical stresses and often allows the plant to proliferate by vegetative propagation. This is the case for plants that are propagated by layering. The formation of AR after stem cutting or in vitro culture is the basis of clonal propagation of valuable genotypes. This can be a rate-limiting step for the multiplication and propagation of interesting varieties and thus stimulates research on the genetic and physiological determinants of AR formation (Geiss et al. 2009; see also Chapter 11).

In addition to these inducible processes of AR formation, many dicot and monocot species produce constitutively a profusion of AR (Figure 9.1B). These ARs can have particular functions such as the exploration and exploitation of the humus niches found in the aerial part of the canopy that are not accessible to the underground primary root system. In some cases, AR can serve to reinforce the trunk by fusing with it (Figure 9.1C) or by developing buttresses (Figure 9.1D). ARs can also form pillars that support the horizontal development of master branches

FIGURE 9.1 **(See color insert.)** Some examples of the diversity of ARs. (A) Hairy root developing from a *Catharanthus roseus* leaf after wounding and inoculating the vein with an *Agrobacterium rhizogenes* suspension (scale bar: 1 cm), (B) ARs formed from a *Ficus* sp. tree (scale bar: 1 m), (C) ARs fusing with the trunk of a *Ficus* sp. tree (scale bar: 30 cm), (D) buttress ARs in *Tetrameles nudiflora* (scale bar: 20 cm), (E) pillar ARs supporting a branch in *Ficus* sp. (scale bar: 50 cm), (F) ARs forming a protective cage around the trunk in *Pandanus* sp. (scale bar: 1 m), (G): clasping ARs and feeding ARs in a *Philodendron* sp. (scale bar: 5 cm), (H): aerial ARs in an epiphyte orchid (scale bar: 10 cm). (Photographs D, F, G, and H are from Claude Edelin, CNRS, Montpellier, France.)

(Figure 9.1E) or produce a protective cage around the trunk (Figure 9.1F). Semi-epiphytes such as *Philodendron* sp. can develop two kinds of AR: clasping roots that anchor the plant on the trunk and feeding roots that grow vertically to the ground (Figure 9.1G). Epiphytic orchids can develop aerial ARs that trap rain and dew humidity (Figure 9.1H).

In several plants, ARs grow from basal stem nodes or from rhizomes underground to form a major part of the plant root system. This is the case for monocot species such as rice or maize, for which ARs, also named crown roots (CRs), develop continuously during the plant growth to form a typical fasciculate root system (Rebouillat et al. 2009) (illustrated in Figure 9.2 for bamboo and rice). The rice root system first develops by forming the SR from the RAM and five embryonic CR that emerge from the coleoptile. More CR develop postembryonically from the nodes of the stem. SR and CR can branch to generate long LRs that are indeterminate and geotropic and short LRs that are determinate and ageotropic. These roots are characterized by a special radial differentiation. There are no layers of cortical cells in short LRs, while 3, 5, and 10 cortical cell layers are found in respectively long LRs, SRs, and CR (Rebouillat et al. 2009).

FIGURE 9.2 **(See color insert.)** CRs in monocots. (A) Bamboo stem with CRs developing at each node (scale bar: 15 cm), (B) detail of a crown of CR apexes emerging from a node of a bamboo stem (scale bar: 4 cm), (C) general view of the CR-made fibrous root system of rice (scale bar: 6 cm), (D) detail of the stem base of a wild-type rice showing the emergence of numerous CRs (scale bar: 5 mm), (E) detail of the stem base of the *crown rootless 1* rice mutant devoid of CRs, only the SR and some of its LRs can be observed (scale bar: 5 mm).

Rice is one of the most important cereals for human feeding (Khush 2005). For this reason the scientific community has chosen it as the model plant for monocots since 1985. Its genome was sequenced (Yu et al. 2002; Itoh et al. 2007) and several mutant collections were generated (Krishnan et al. 2009). In addition, cultivated rice has been adapted to many culture conditions, and a wide range of genetically and phenotypically diverse varieties (around 100,000) are available (McNally et al. 2009). The development of rice functional genomics allowed to start to decipher the molecular mechanisms controlling the differentiation of CR (Coudert et al. 2010). In this chapter, using rice as a reference, we will describe the histological origin of the CR and the hormonal and molecular genetics control of their formation. Based on the comparison of what is known for the formation of LR in *Arabidopsis* (Peret et al. 2009; see also Chapter 6), the model plant for dicots, we will discuss the differences and similarities among the ontogenetic and genetic mechanisms that control CR and LR differentiation, from stem and roots in monocots and dicots, respectively.

II. Initiation of CRs Occurs from a Ground Meristem inside the Stem

In rice, most of the postembryonic roots are derived from the stem and are named CR. Several articles describe the histogenesis of the rice stem and the formation of CR (Kaufman 1959) and in particular differentiation and organization of the CR primordium (Kawata and Harada 1975; Itoh et al. 2005). CR initial cells result from periclinal and anticlinal divisions of a portion of a meristematic layer called the ground meristem (GM). The GM is derived from the shoot apical meristem and generates the internal stem tissues (Kaufman 1959; Kawata and Harada 1975; Itoh et al. 2005). First, a shoot endodermis (starch sheath) and a meristematic cell layer are formed from the GM. This GM cell layer may be referred to as a shoot pericycle by analogy with the root pericycle in terms of position (inner to the endodermis and outer to vascular tissues of the stem) and function (rhizogenic capacity). The analogy between the GM in stem and the pericycle in roots will be further discussed in the light of recent findings concerning the genetic control of CR and LR development. The CR initiation process is divided into seven stages. The first stage (named Cr1) corresponds to the establishment of CR initial cells (Kawata and Harada 1975; Itoh et al. 2005). A dome-shape cell mass differentiating from the shoot pericycle can be distinguished from adjacent cells; this is the earliest detectable event of CR initiation (Figure 9.3A). At the second stage (Cr2), epidermis-endodermis (EE) and root cap (RC) initial cells of the future RM become

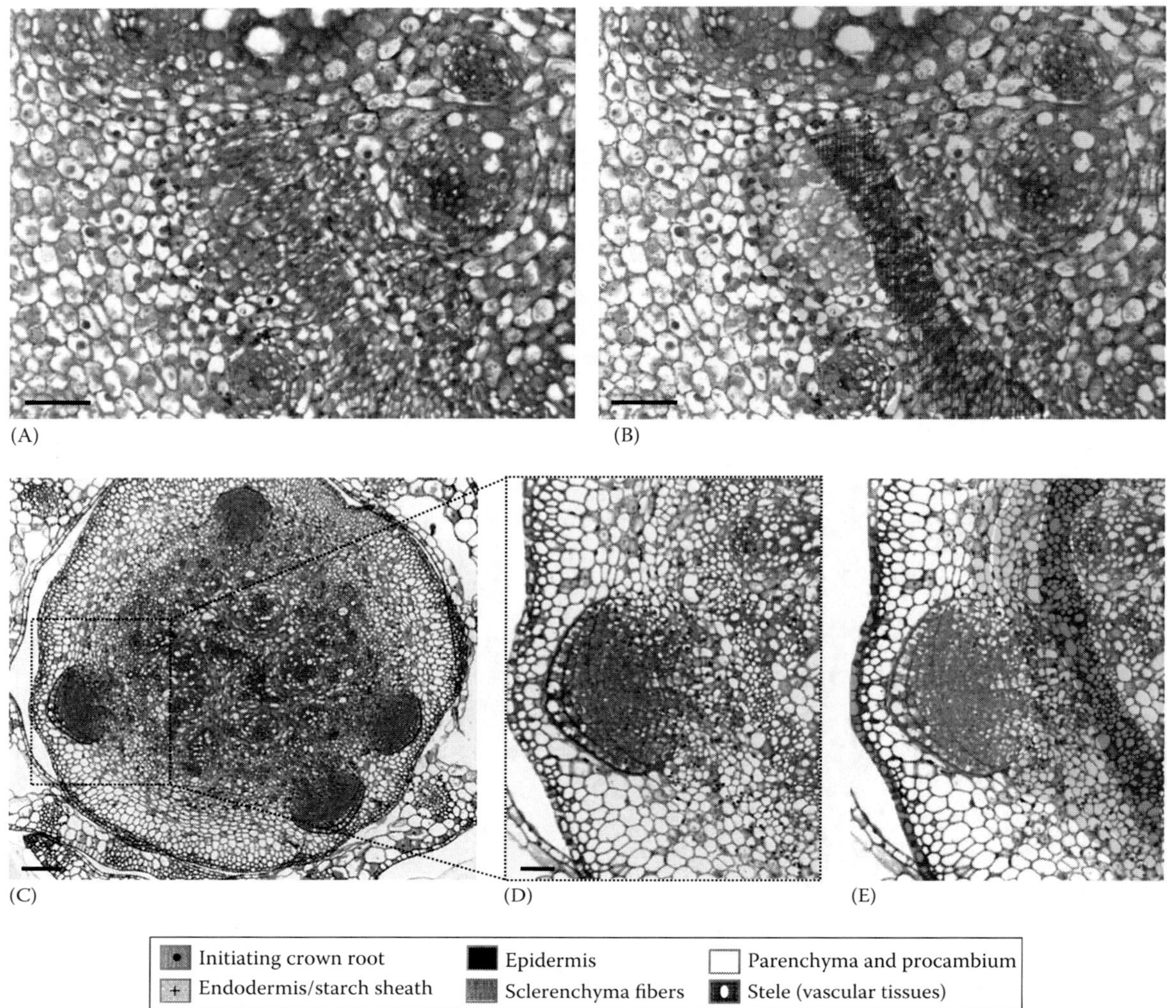

FIGURE 9.3 **(See color insert.)** Histological sections of rice stem base showing different stages of CR initiation. (A–B) early stages of CR initiation. (C–E) CR primordia after differentiation of a typical root meristem organization. (D and E) enlargements of the zone marked in picture C with dotted lines. (A, C, and D) periodic acid-Schiff and naphthol blue black-stained sections. (B and E) captioned sections. Scale bar: A, B, D, and E, 50 μm; C, 200 μm.

visible (Figure 9.3B) (Kawata and Harada 1975; Itoh et al. 2005). RC initials may become evident at the edge of the primordium when only two or three cell layers are formed (Kaufman 1959). Following stages of CR meristem formation correspond to the differentiation of epidermis and endodermis (Cr3), the differentiation of cortex (Cr4), and the establishment of fundamental root organization (RC reaches the stem epidermis) (Cr5). At this stage, the CR meristem is completely organized and reaches the stem epidermis. After that, in response to different developmental and external signals, we can observe an onset of elongation and vacuolation of the meristem cells (Cr6) followed by the emergence and the development of CR outside the stem (Cr7) (Kamiya et al. 2003; Itoh et al. 2005). When the CR has acquired a typical root organization (Cr5), the surrounding tissues in the stem are mostly differentiated and the stem has a typical radial histological structure constituted, from the periphery to the center, by the epidermis, the cortical parenchyma, the endodermis, a parenchyma layer containing procambium strands, a sclerenchyma fiber ring, and the stele (Figure 9.3C through E).

III. Initiation of CR Is Regulated by Auxin and Cytokinins and by Specific Transcription Factors

The recent discoveries and analyses of CR-less mutants in rice provided evidence that CR initiation is regulated by auxin and cytokinins (CKs) and involves different specific transcription factors (TFs). In two of these mutants, *crown rootless 4* (*crl4*) (Kitomi et al. 2008) and *Osgnom1* (Liu et al. 2009), the gene orthologous to the *Arabidopsis GNOM1* is inactivated, and initiation of CR is absent. GNOM1 is a membrane-associated guanine-nucleotide exchange factor of the ADP-ribosylation factor

G protein (ARF-GEF) (Steinmann et al. 1999). It regulates the intracellular trafficking of PIN1 (PINFORMED 1) auxin efflux carrier. In *Arabidopsis*, this process allows local regulation of auxin flux that is required for the proper asymmetric division of the two pericycle cells that may generate an LR primordium (Geldner et al. 2004; Peret et al. 2009). In rice, the expression pattern of *CRL4* coincides with the expression pattern of *OsPIN1* (Kitomi et al. 2008). Nevertheless, in plants where *OsPIN1* is downregulated by RNA interference, CR primordia can be initiated normally, and only their further development is inhibited (Xu et al. 2005). This suggests that in rice, other auxin efflux carriers than OsPIN1 are likely to be involved in the CR initiation process. These could be *OsPIN2*, *OsPIN5b*, and *OsPIN9* that are downregulated in the stem base of the *Osgnom1* mutant (Liu et al. 2009). These data show that an appropriate auxin transport mediated by *CRL4/OsGNOM1* is required to induce the divisions of the CR initial cells in the stem GM.

Auxin is also necessary for the induction of specific TFs that are indispensable for CR initiation. Two CR-less mutants, *adventitious rootless 1* (*arl1*) (Liu et al. 2005) and *crown rootless 1* (*crl1*) (Inukai et al. 2005), whose CR initiation is impaired, are affected in a gene encoding a Lateral Organ Boundary Domain (LBD) TF (Husbands et al. 2007; Matsumura et al. 2009). The expression of *ARL1/CRL1* is induced in response to exogenous auxin treatment, but auxin treatment cannot restore CR initiation in *crl1* mutant. The expression of *ARL1/CRL1* coincides with the auxin distribution pattern in the stem bases where CRs are initiated. This expression pattern requires an auxin-responsive element (ARE) in the *ARL1/CRL1* promoter. This ARE interacts in vitro with the auxin response factor OsARF16 TF. In addition, the induction of *ARL1/CRL1* expression by auxin is lost in plants overexpressing a ubiquitin-insensitive mutated AUXIN/INDOLE-3-ACETIC ACID protein. Taken together, these data suggest that *ARL1/CRL1* expression is induced by auxin via a transduction pathway involving the TRANSPORT INHIBITOR RESPONSE1/AUXIN RECEPTOR F-BOX PROTEIN1-3 (TIR1/AFB1-3). This auxin receptor (Dharmasiri et al. 2005; Kepinski and Leyser 2005) targets the degradation of an AUX/IAA regulatory protein, which leads to the activation of OsARF16, that on its turn binds the ARE element of *ARL1/CRL1* promoter and finally activates the transcription of the *ARL1/CRL1* gene. Interestingly such a regulatory cascade is reminiscent to the one involved in the control of LR in *Arabidopsis* that also starts with TIR1/AFB1-mediated auxin perception; AUX/IAA degradation; activation of ARF7 and ARF19, the two *Arabidopsis* orthologs of OsARF16; and finally the expression induction of *LBD16* and *LBD19*, which are phylogenetically closely related to *ARL1/CRL1* (Laplaze et al. 2005; Wilmoth et al. 2005; Okushima et al. 2007; Wang et al. 2007). In maize, the *ROOTLESS CONCERNING CROWN and SEMINAL ROOT* (*RTCS*) and its duplicated homologous gene *RTCS-LIKE* are also both involved in the control of CR initiation (Taramino et al. 2007). These genes are orthologous to *ARL1/CRL1*, they are both auxin regulated, and they both contain an ARE element in their promoter. This emphasizes that the developmental pathway involving auxin regulation of specific LBD TF for CR and LR initiation is well conserved in plants. The downregulation in the *arl1* mutant of the rice *WUSCHEL-TYPE HOMEOBOX* gene *QUIESCENT-CENTER-SPECIFIC HOMEOBOX* (*QHB*) and *OsSCARECROW* (*OsSCR*) suggests that these genes could be direct or indirect targets of ARL1/CRL1 TF (Kamiya et al. 2003).

Another TF was shown to regulate CR initiation in rice: WUSCHEL-RELATED HOMEOBOX 11, WOX11 (Zhao et al. 2009). The *wox11* mutant has a reduced number of CR, whereas *WOX11* overexpressing lines present an increased number of CR. *WOX11* expression is stimulated by both auxin and CK and is normally detected in the early CR primordia, suggesting that this TF may integrate auxin and CK signaling for the regulation of CR initiation. This TF downregulates the *type-A RESPONSE REGULATOR 2* (*ARR2*) gene expression and binds directly to its promoter. The expression of other *ARRs* genes is downregulated in *WOX11* overexpressing lines. ARRs are CK responsive and are known to inhibit the CK response (To et al. 2004). In *Arabidopsis*, CK disrupts LR initiation and patterning (Laplaze et al. 2007; Dello Ioio et al. 2008; Benkova and Hejatko 2009; Peret et al. 2009). This suggests that a fine control of the auxin/CK balance is required for a proper CR or LR initiation in rice and *Arabidopsis*, respectively.

In WT, *WOX11* is also expressed in the cell division zone of the CR meristem, suggesting that this gene is also involved in the development phase of CR in response to auxin and CK signals.

IV. Ethylene, Nitric Oxide, Gibberellic Acid, and Abscisic Acid Contribute to the Regulation of the Emergence and the Development of CR

After initiation, CRs develop inside the stem and stop to grow. Later CR growth is reactivated, and they emerge from the stem epidermis. CR emergence occurs regularly during plant growth and is synchronized with shoot development. In the rice stem, two crowns of CR primordia are formed, at two adjacent internodes. When the leaf of the node N emerges, the CRs of the node N-3 start to grow out (Fujii 1961; Kawata et al. 1963). In addition to this constitutive development, the emergence and the development of CR can also be stimulated by external signals. For instance, submergence induces CR emergence and is, at least in part, mediated by the accumulation of ethylene (Lorbiecke and Sauter 1999). CR emergence is promoted by exogenous ethylene treatment but not by gibberellic acid, auxin, or CK (Lorbiecke and Sauter 1999). Ethylene is necessary for the induction of cell cycle regulatory genes in CR primordia (Lorbiecke and Sauter 1999) and also promotes CR emergence by inducing epidermal cell death at the site of CR emergence (Mergemann and Sauter 2000). Gibberellic acid acts in synergy with ethylene to promote epidermal cell death, CR emergence, and development, whereas abscisic acid inhibits them (Steffens et al. 2006). Another stress hormone, nitric oxide, is also essential for CR emergence (Xiong et al. 2009). Also for auxin, evidence was found for its potential role during CR

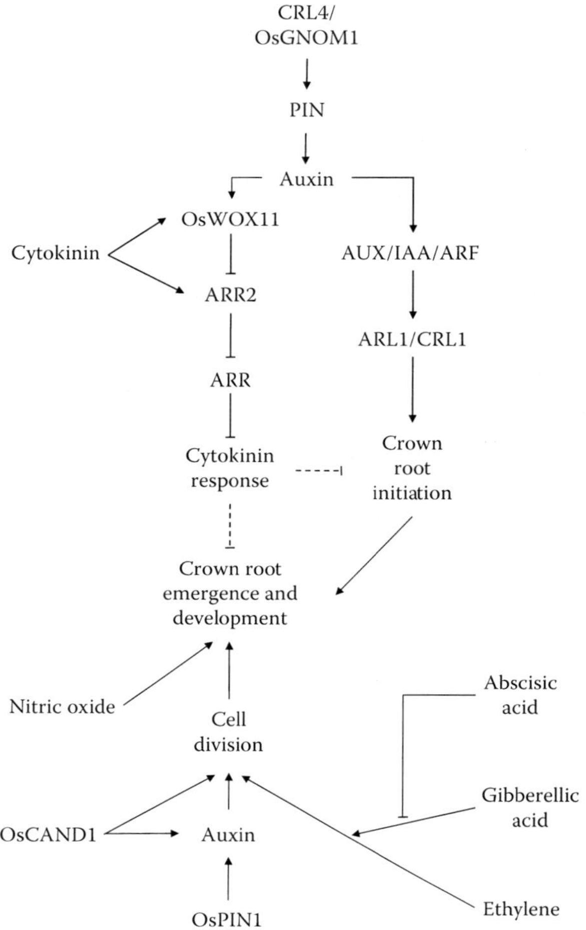

FIGURE 9.4 Gene regulatory network controlling CR initiation and development. Arrows represent the positive regulatory action of one element of the network on another one. A line ending with a trait represents the negative regulatory action of one element of the network on another one. Dotted line represent hypothetical link between two elements. Abbreviations: ARF, auxin response factor; ARL, adventitious rootless; ARR, type-A RESPONSE REGULATOR2; AUX/IAA, AUXIN/INDOLE-3-ACETIC ACID; CAND1, cullin-associated and neddylation-dissociated 1 (*CAND1*) SCF^{TIR1} ubiquitin ligase involved in auxin signaling via the AUX/IAA/ARF pathway; CR, crown root; CRL, crown rootless; GNOM1, membrane-associated guanine-nucleotide exchange factor of the ADP-ribosylation factor G protein (ARF-GEF) that regulates the traffic of PINFORMED1 (PIN1) auxin efflux carrier proteins; WOX11, WUSHEL-Related Homeobox 1.

emergence and development. OsPIN1 is a putative auxin efflux carrier in rice. In plants where *OsPIN1* expression was inhibited by RNA interference, CR emergence and development were strongly inhibited (Xu et al. 2005). This phenotype was rescued when *OsPIN1*-downregulated plants were treated with the auxin analog α-naphthyl-acetic acid and was mimicked when wild-type plants were treated with N-1-naphthylphalamic acid, an auxin transport inhibitor. A regulatory role of auxin in the emergence of CR is also attested by the characterization of the *Oscand1* mutant (Wang et al. 2010). *OsCAND1* is orthologous to *CAND1* from *Arabidopsis*. CAND1 encodes a cullin-associated and neddylation-dissociated 1 (*CAND1*) SCF^{TIR1} ubiquitin ligase involved in auxin signaling via the AUX/IAA/ARF pathway (Gray et al. 2001; Cheng et al. 2004; Chuang et al. 2004). The *Oscand1* mutant presents normal CR primordia, but CRs do not emerge. This phenotype can be reversed at least partially by exogenous auxin treatment. Transformation with the auxin-responsive DR5::GUS construction revealed that CR primordia of *Oscand1* mutants do not accumulate auxin in the central cylinder and that the distribution of auxin is diffuse in the RC, showing that auxin flux is affected in the mutant. Interestingly the expression of genes associated with the G2/M transition phase of the cell cycle is inhibited in *Oscand1* CR primordia. This indicates that *OsCAND1* is involved in the organization of auxin fluxes and in the regulation of cell division that allow proper CR emergence.

A summary of the current knowledge of the hormone and genetic network regulating CR initiation and emergence in rice is presented in Figure 9.4.

V. Similarities between Initiation Mechanisms and Tissue Origin of Root-Derived Roots and Stem-Derived Roots May Give Information about the Common Telomic Origin of Plant Axes

Knowledge acquired on *Arabidopsis* root-derived LR and rice stem-derived CR development suggests similarities between the genetic control and the cellular origin of these two postembryonic root types (Peret et al. 2009; Coudert et al. 2010). The *WUSCHEL*-related *WOX5*, a quiescent center (QC)-specific gene in *Arabidopsis*, is necessary for root stem cell maintenance (Breuninger et al. 2008; Ditengou et al. 2008). *WOX5* is expressed in the LR primordium a few hours after LR induction. In rice, *QHB*, a gene orthologous to *WOX5*, is also expressed in the root QC, in particular in CR (Kamiya et al. 2003). Genes encoding AS2/LOB-domain TFs, *LBD16* and *LBD29* in *Arabidopsis* and *CRL1* in rice, are involved in the control of postembryonic root initiation in both species (Inukai et al. 2005; Laplaze et al. 2005; Wilmoth et al. 2005; Okushima et al. 2007; Wang et al. 2007). These three genes are regulated by auxin via the AUX/IAA/ARF transduction pathway. Thus, key elements involved in the initiation of CR in rice and of LR in *Arabidopsis*, such as auxin signaling, AS2/LOB-domain, and WUSCHEL-like TFs, are well conserved or highly similar. This suggests that gene regulatory networks controlling postembryonic root initiation from stem or from roots may be well conserved in flowering plants. It can be hypothesized that these gene regulatory networks differentiated from a primitive one, which already controlled the initiation of roots from the telome of early land plants. The identification and comparison of LBD16, LBD29, and CRL1 target genes should allow a better understanding of the differences and the similarities in the gene regulatory networks controlling the initiation of root-derived and stem-derived roots.

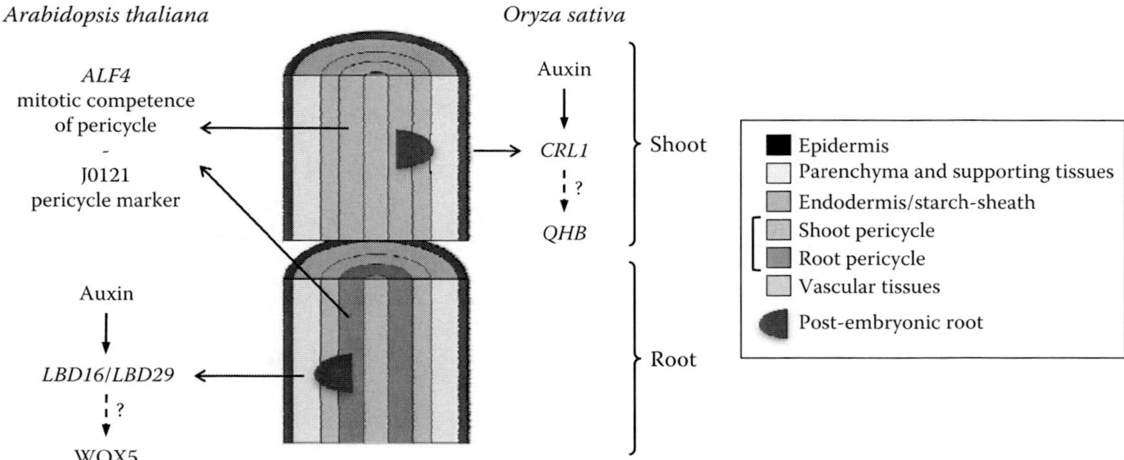

FIGURE 9.5 (See color insert.) Diagram representing genetic mechanisms related to postembryonic root development and conserved between shoot and root in *Arabidopsis* and rice. Abbreviations: ALF4, ABERRANT LATERAL ROOT FORMATION 4; CRL1, crown rootless1; GNOM1, membrane-associated guanine-nucleotide exchange factor of the ADP-ribosylation factor G protein (ARF-GEF) that regulates the traffic of PINFORMED1 (PIN1) auxin efflux carrier proteins; J0121, GFP enhancer trap line; QHB, WUSCHEL-TYPE HOMEOBOX GENE QUIESCENT-CENTER-SPECIFIC HOMEOBOX; LBD, Lateral Organ Boundary Domain; WOX4, WUSHEL-Related Homeobox 4.

Root and shoot are conceptually opposed to one another because of their obvious morphological differences resulting from their functional specifications. But interestingly, this overall accepted disparity is counterbalanced by genetic findings showing that same genes control the development of similar tissues in both root and shoot. This supports a common fundamental nature of these organs that are proposed to derive from the same telomeric ancestral structure. For example, *Arabidopsis* LRs are derived from root pericycle founder cells (Peret et al. 2009; see also Chapter 6). The pericycle, the outer limit of the stele that includes the vascular tissues, is inner to the endodermis and is generated from the RAM. In rice, CRs are derived from the GM, a meristematic tissue derived from the shoot apical meristem (Itoh et al. 2005). When CR primordia are formed, the GM is located between the central zone that contains the vascular tissues and the starch sheath that is constituted of cells rich in starch granules. The starch sheath corresponds to the shoot endodermis (Fukaki et al. 1998; Morita et al. 2007). Considering the relative histological location of the GM and its ability to initiate cell divisions to form CR meristems, it might be referred to as a shoot pericycle. Recent molecular evidence supports this hypothesis. In *Arabidopsis*, the GFP enhancer trap line J0121 specifically marks the pericycle at the xylem poles in the root (Sugimoto et al. 2010). This marker is also expressed around the vasculature in *Arabidopsis* aerial organs, cotyledons, and petals, suggesting the existence of cells with a pericycle-like identity in the shoot. In *Arabidopsis*, the *ABERRANT LATERAL ROOT FORMATION 4* (*ALF4*) gene is necessary to keep the mitotic capacity of root pericycle cells. In particular, *Arabidopsis alf4* mutants are devoid of LR, and AR formation is impaired in *alf4* hypocotyl (DiDonato et al. 2004). In response to an auxin treatment in vitro, the shoot pericycle-like cells marked in the J0121 line and the root pericycle cells can divide to generate calli (Sugimoto et al. 2010). These calli express *WOX5* suggesting that they constitute rootlike cell masses. In *alf4* mutants, this capacity to generate calli is lost in both root and aerial organs (Sugimoto et al. 2010). These results suggest that roots or rootlike structures can develop through an ALF4-dependent genetic pathway and from a pluripotent cell layer located in a similar position in the root and in the shoot. These data support the idea that the cell layer marked in the shoot of J0121 line has a molecular identity and organogenesis potential similar to the root pericycle and thus the existence of a shoot pericycle in *Arabidopsis*. *ALF4* has a unique ortholog in rice whose function has not been characterized yet. This gene represents a good candidate to study the conservation of the pericycle identity of the GM in rice. These data concerning the ontogenetic and genetic origin of postembryonic roots from stem or root are summarized in Figure 9.5.

VI. Conclusions: Adventitious Roots and Plant Improvement

With the model plant rice, significant progress has been made since 2005 in the understanding of the genetic and physiological determination of CR development. Key genes involved in this process have been identified that allow drawing a first outline of the regulatory network involved in the formation of roots from stem. As discussed earlier, these mechanisms can now be compared to those involved in LR initiation and development. This opens a new perspective to study how the different parts of the plant axis have differentiated during evolution.

In addition to the data acquired from functional studies of crown rootless mutants, we can expect that other approaches will contribute further to the deciphering of the mechanisms that control CR initiation and development in rice. Several studies have been conducted in rice to identify QTL involved in root development. A recent study used these data to conduct a meta-QTL analysis that allowed reducing the confidence interval of their position (Courtois et al. 2009). Some of them now include only a few genes. This is likely to allow in the near future the identification of new genes important for CR development. This will also benefit from the development of a root-specific transcriptome atlas. In rice, a transcriptome atlas produced by laser capture microdissection (LCM) covers 40 cell types including most cell types of seedling roots (Jiao et al. 2009), and other studies report on the expression of all auxin-responsive genes in various organs (Opanowicz et al. 2008; Jain and Khurana 2009). In addition, the development of sequencing projects and genomic resources for other major cereals such as wheat and maize and for the wild grass *Brachypodium distachyon* should contribute to the discovery of new genes involved in CR development (Hochholdinger and Tuberosa 2009; Reynolds et al. 2009). Because CR constitute the major part of the cereal root system, we can hope that the identification of the mechanisms involved in their initiation and development will help breeders to select root systems and to create new varieties better adapted to lower input systems (water, fertilizers), which is the major challenge for agriculture in a near future (Herder et al. 2010).

The progress made on the molecular mechanisms of constitutive CR initiation and development in rice is also likely to help decipher the mechanisms of AR in other plant species. We have seen that some major elements of the CR developmental network are well conserved not only between cereal species (e.g., *ARL1/CRL1* in rice versus *RTCS* in maize) but also between monocots and dicots when CR and LR gene regulatory networks are compared (e.g., *ARL1/CRL1* in rice versus *LBD16* and *LBD29* in *Arabidopsis*). In dicots, the induced AR rooting process has been studied in a large variety of species (reviewed in [Geiss et al. 2009]). Most of the mutants modulating inducible AR development in *Arabidopsis* affect genes such as *superroot1*, *superroot2*, or *yucca* that are different from the genes involved in the CR gene regulatory network in rice (Boerjan et al. 1995; Delarue et al. 1998; Zhao et al. 2001). The development of AR in *Arabidopsis* does not occur constitutively and remains anecdotic in normal conditions. We can hypothesize that the *Arabidopsis* AR mutants that are characterized by a capacity to constitutively produce AR from hypocotyls mostly reveal mechanisms involved in the induction of AR initiation rather than in the initiation process itself. From this point of view, the two model plants, *Arabidopsis* and rice, are complementary. Nevertheless, some regulatory signals are constant in CR development in rice and in AR formation in dicots such as auxin and ethylene (Geiss et al. 2009). For this reason, we assume that the study of CR development in rice, which occurs systematically during development, will help to discover in dicots species new key genes involved in inducible AR development via a candidate gene approach. This perspective will be precious to help the selection of elite cultivars with improved rooting capacity for plants that are vegetatively propagated.

References

Barlow PW 1986 Adventitious roots of whole plants: their forms, functions and evolution. *Dev Plant Soil Sci* 20:67–110.

Benkova E, Hejatko J 2009 Hormone interactions at the root apical meristem. *Plant Mol Biol* 69:383–396.

Boerjan W, Cervera MT, Delarue M et al. 1995 Superroot, a recessive mutation in *Arabidopsis*, confers auxin overproduction. *Plant Cell* 7:1405–1419.

Breuninger H, Rikirsch E, Hermann M, Ueda M, Laux T 2008 Differential expression of WOX genes mediates apical-basal axis formation in the *Arabidopsis* embryo. *Dev Cell* 14:867–876.

Cheng Y, Dai X, Zhao Y 2004 AtCAND1, a HEAT-repeat protein that participates in auxin signaling in *Arabidopsis*. *Plant Physiol* 135:1020–1026.

Chuang HW, Zhang W, Gray WM 2004 *Arabidopsis* ETA2, an apparent ortholog of the human cullin-interacting protein CAND1, is required for auxin responses mediated by the SCF(TIR1) ubiquitin ligase. *Plant Cell* 16:1883–1897.

Coudert Y, Perin C, Courtois B, Khong NG, Gantet P 2010 Genetic control of root development in rice, the model cereal. *Trends Plant Sci* 15:219–226.

Courtois B, Ahmadi N, Khowaja F et al. 2009 Rice root genetic architecture: meta-analysis from a drought QTL database. *Rice* 2:115–128.

Delarue M, Prinsen E, Onckelen HV, Caboche M, Bellini C 1998 Sur2 mutations of *Arabidopsis thaliana* define a new locus involved in the control of auxin homeostasis. *Plant J* 14:603–611.

Dello Ioio R, Nakamura K, Moubayidin L et al. 2008 A genetic framework for the control of cell division and differentiation in the root meristem. *Science* 322:1380–1384.

Dharmasiri N, Dharmasiri S, Estelle M 2005 The F-box protein TIR1 is an auxin receptor. *Nature* 435:441–445.

DiDonato RJ, Arbuckle E, Buker et al. 2004 *Arabidopsis* ALF4 encodes a nuclear-localized protein required for lateral root formation. *Plant J* 37:340–353.

Dimond AE, Waggoner PE 1953 The cause of epinastic symmetry in Fusarium wilt of tomato. *Phytopathology* 43:663–669.

Ditengou FA, Teale WD, Kochersperger P et al. 2008 Mechanical induction of lateral root initiation in *Arabidopsis thaliana*. *Proc Natl Acad Sci USA* 105:18818–18823.

Duboux E 1969 Ontogenèse des nodules caulinaires du *Sesbania rostrata* (Légumineuses). *Can J Bot* 62:982–994.

Fujii Y 1961 Studies on the regularity of root growth in rice and wheat plants. *Agric Bull Saga Univ* 12:1–117.

Fukaki H, Wysocka-Diller J, Kato T, Fujisawa H, Benfey PN, Tasaka M 1998 Genetic evidence that the endodermis is essential for shoot gravitropism in *Arabidopsis thaliana*. *Plant J* 14:425–430.

Geiss G, Gutierrez L, Bellini C 2009 Adventicious root formation: new insights and perspectives. *Annu Plant Rev* 37:127–156.

Geldner N, Richter S, Vieten A et al. 2004 Partial loss-of-function alleles reveal a role for GNOM in auxin transport-related, post-embryonic development of *Arabidopsis*. *Development* 131:389–400.

Gray WM, Kepinski S, Rouse D, Leyser O, Estelle M 2001 Auxin regulates SCF(TIR1)-dependent degradation of AUX/IAA proteins. *Nature* 414:271–276.

Guillon S, Tremouillaux-Guiller J, Pati PK, Rideau M, Gantet P 2006a Harnessing the potential of hairy roots: dawn of a new era. *Trends Biotechnol* 24:403–409.

Guillon S, Tremouillaux-Guiller J, Pati PK, Rideau M, Gantet P 2006b Hairy root research: recent scenario and exciting prospects. *Curr Opin Plant Biol* 9:341–346.

Herder GD, Van Isterdael G, Beeckman T, De Smet I 2010 The roots of a new green revolution. *Trends Plant Sci* 15:600–607.

Hochholdinger F, Tuberosa R 2009 Genetic and genomic dissection of maize root development and architecture. *Curr Opin Plant Biol* 12:172–177.

Husbands A, Bell EM, Shuai B, Smith HM, Springer PS 2007 LATERAL ORGAN BOUNDARIES defines a new family of DNA-binding transcription factors and can interact with specific bHLH proteins. *Nucleic Acids Res* 35:6663–6671.

Inukai Y, Sakamoto T, Ueguchi-Tanaka M et al. 2005 Crown rootless1, which is essential for crown root formation in rice, is a target of an AUXIN RESPONSE FACTOR in auxin signaling. *Plant Cell* 17:1387–1396.

Itoh J, Nonomura K, Ikeda K et al. 2005 Rice plant development: from zygote to spikelet. *Plant Cell Physiol* 46:23–47.

Itoh T, Tanaka T, Barrero RA et al. 2007 Curated genome annotation of *Oryza sativa* ssp. japonica and comparative genome analysis with *Arabidopsis thaliana*. *Genome Res* 17:175–183.

Jain M, Khurana JP 2009 Transcript profiling reveals diverse roles of auxin-responsive genes during reproductive development and abiotic stress in rice. *Febs J* 276:3148–3162.

Jiao Y, Tausta SL, Gandotra N et al. 2009 A transcriptome atlas of rice cell types uncovers cellular, functional and developmental hierarchies. *Nat Genet* 41:258–263.

Kamiya N, Nagasaki H, Morikami A, Sato Y, Matsuoka M 2003 Isolation and characterization of a rice WUSCHEL-type homeobox gene that is specifically expressed in the central cells of a quiescent center in the root apical meristem. *Plant J* 35:429–441.

Kaufman, PB 1959 Development of the shoot of *Oryza sativa* L. III. early stages in histoenesis of the stem and ontogeny of the adventitious root. *Phytomorphology* 9:382–404.

Kawata S, Harada J 1975 On the development of the crown root primordia in rice plants. *Proc Crop Sci Soc Jpn* 44:438–457.

Kawata S, Yamazaki K, Ishihara H et al. 1963 Studies on root system formation in rice plants in a paddy field: an example of its relationship with the growth stage. *Proc Crop Sci Soc Jpn* 33:168–180.

Kepinski S, Leyser O 2005 The *Arabidopsis* F-box protein TIR1 is an auxin receptor. *Nature* 435:446–451.

Khush GS 2005 What it will take to feed 5.0 billion rice consumers in 2030. *Plant Mol Biol* 59:1–6.

Kitomi Y, Ogawa A, Kitano H, Inukai Y 2008 *CRL4* regulates crown root formation through auxin transport in rice. *Plant Root* 2:19–28.

Krishnan A, Guiderdoni E, An G et al. 2009 Mutant resources in rice for functional genomics of the grasses. *Plant Physiol* 149:165–170.

Laplaze L, Benkova E, Casimiro I et al. 2007 Cytokinins act directly on lateral root founder cells to inhibit root initiation. *Plant Cell* 19:3889–3900.

Laplaze L, Parizot B, Baker A et al. 2005 GAL4-GFP enhancer trap lines for genetic manipulation of lateral root development in *Arabidopsis thaliana*. *J Exp Bot* 56:2433–2442.

Liu S, Wang J, Wang L et al. 2009 Adventitious root formation in rice requires OsGNOM1 and is mediated by the OsPINs family. *Cell Res* 19:1110–1119.

Liu H, Wang S, Yu X et al. 2005 ARL1, a LOB-domain protein required for adventitious root formation in rice. *Plant J* 43:47–56.

Lorbiecke R, Sauter M 1999 Adventitious root growth and cell-cycle induction in deepwater rice. *Plant Physiol* 119:21–30.

Matsumura Y, Iwakawa H, Machida Y, Machida C 2009 Characterization of genes in the ASYMMETRIC LEAVES2/LATERAL ORGAN BOUNDARIES (AS2/LOB) family in *Arabidopsis thaliana*, and functional and molecular comparisons between AS2 and other family members. *Plant J* 58:525–537.

McNally KL, Childs KL, Bohnert R et al. 2009 Genomewide SNP variation reveals relationships among landraces and modern varieties of rice. *Proc Natl Acad Sci USA* 106:12273–12278.

Mergemann H, Sauter M 2000 Ethylene induces epidermal cell death at the site of adventitious root emergence in rice. *Plant Physiol* 124:609–614.

Morita MT, Nakano A, Tasaka M 2007 *endodermal-amyloplast less 1* is a novel allele of *SHORT-ROOT*. *Adv Space Res* 39:1127–1133.

Okushima Y, Fukaki H, Onoda M, Theologis A, Tasaka M 2007 ARF7 and ARF19 regulate lateral root formation via direct activation of LBD/ASL genes in *Arabidopsis*. *Plant Cell* 19:118–130.

Opanowicz M, Vain P, Draper J, Parker D, Doonan JH 2008 *Brachypodium distachyon*: making hay with a wild grass. *Trends Plant Sci* 13:172–177.

Peret B, De Rybel B, Casimiro I et al. 2009 *Arabidopsis* lateral root development: an emerging story. *Trends Plant Sci* 14:399–408.

Rasa EA, Esau K 1961 Anatomic effects of curly top and aster yellows viruses on tomato. *Hilgardia* 30:469–515.

Rebouillat A, Dievart A, Verdeil JL et al. 2009 Molecular genetic of rice root development. *Rice* 2:15–34.

Reynolds M, Foulkes MJ, Slafer GA et al. 2009 Raising yield potential in wheat. *J Exp Bot* 60:1899–1918.

Steffens B, Wang J, Sauter M 2006 Interactions between ethylene, gibberellin and abscisic acid regulate emergence and growth rate of adventitious roots in deepwater rice. *Planta* 223:604–612.

Steinmann T, Geldner N, Grebe M et al. 1999 Coordinated polar localization of auxin efflux carrier PIN1 by GNOM ARF GEF. *Science* 286:316–318.

Sugimoto K, Jiao Y, Meyerowitz EM 2010 *Arabidopsis* regeneration from multiple tissues occurs via a root development pathway. *Dev Cell* 18:463–471.

Taramino G, Sauer M, Stauffer JL et al. 2007. The maize (*Zea mays* L.) RTCS gene encodes a LOB domain protein that is a key regulator of embryonic seminal and post-embryonic shoot-borne root initiation. *Plant J* 50:649–659.

To JP, Haberer G, Ferreira FJ et al. 2004 Type-A *Arabidopsis* response regulators are partially redundant negative regulators of cytokinin signaling. *Plant Cell* 16:658–671.

Wang D, Pei K, Fu Y et al. 2007 Genome-wide analysis of the auxin response factors (ARF) gene family in rice (*Oryza sativa*). *Gene* 394:13–24.

Wang XF, He FF, Ma XX et al. 2011 OsCAND1 is required for crown root emergence in rice. *Mol Plant* 4:289–299.

Wilmoth JC, Wang S, Tiwari SB et al. 2005 NPH4/ARF7 and ARF19 promote leaf expansion and auxin-induced lateral root formation. *Plant J* 43:118–130.

Xiong J, Lu H, Lu K, Duan Y, An L, Zhu C 2009 Cadmium decreases crown root number by decreasing endogenous nitric oxide, which is indispensable for crown root primordia initiation in rice seedlings. *Planta* 230:599–610.

Xu M, Zhu L, Shou H, Wu 2005 A PIN1 family gene, OsPIN1, involved in auxin-dependent adventitious root emergence and tillering in rice. *Plant Cell Physiol* 46:1674–1681.

Yu J, Hu S, Wang J et al. 2002 A draft sequence of the rice genome (*Oryza sativa* L. ssp. *indica*). *Science* 296:79–92.

Zhao Y, Christensen SK, Fankhauser C et al. 2001 A role for flavin monooxygenase-like enzymes in auxin biosynthesis. *Science* 291:306–309.

Zhao Y, Hu Y, Dai M, Huang L, Zhou DX 2009 The WUSCHEL-related homeobox gene WOX11 is required to activate shoot-borne crown root development in rice. *Plant Cell* 21:736–748.

10

Genetic Analysis of Maize Root Development

Frank Hochholdinger
University of Bonn

Guenter Feix
University of Freiburg

I. Introduction

Although maize has been a favored object for genetic analyses since the 1920s (Coe 2001), a systematic genetic analysis of the underground root system has only been initiated in the last decade (reviewed in Hochholdinger et al. 2004a,b; Hochholdinger and Zimmermann 2008; Hochholdinger and Tuberosa 2009; Hochholdinger 2009). The genetic dissection of the complex root structure of monocotyledonous plants and its formation is largely hampered by the difficulty of observing versatile underground root systems. Furthermore, the tremendous plasticity of root formation and the significant influence of environmental changes on root system architecture complicate the identification of mutants in their natural habitat. Therefore, until the mid-1990s, only one maize root mutant, *rt1* (Jenkins 1930), had been indirectly identified by its reduced lodging resistance associated with defects in shoot-borne root formation. In recent years, a set of monogenic mutants affecting different aspects of maize root development has been isolated (Wen and Schnable 1994; Hetz et al. 1996; Hochholdinger and Feix 1998a; Hochholdinger et al. 2001; Woll et al. 2005). Meanwhile, some of the affected genes have been cloned paving the way for elucidating the molecular basis of maize root formation (Wen et al. 2005; Taramino et al. 2007; Hochholdinger et al. 2008; von Behrens et al. 2011). For the identification and correct phenotypic assessment of such mutants, detailed knowledge of the structure and development of the root system was essential (cf. Feldman 1994; Kiesselbach 1999). As this aspect remains important for further mutant work, a short structural account of the root system of maize is given in the following text.

II. Maize Root System

A. Root Types and Their Formation during Development

The root system of maize consists of roots that are laid down during embryogenesis and roots that are formed during the postembryonic development after germination (Esau 1977; Fahn 1990; Kiesselbach 1999). The embryonic root system of maize consists of a single primary and a variable number of seminal roots, while postembryonic root system is composed of shoot-borne crown and brace roots and lateral roots that emerge from pericycle cells of all other roots (Figure 10.1A).

The primary root is endogenously initiated in the developing embryo (Yamashita 1991). It emerges from the basal pole of the seed and is essential during the early stages of development. Depending on the genetic background, primary root growth ceases (Larson and Hanway 1977; Feldman 1994) or continues until maturity (Kausch 1967; McCully and Canny 1988; Hetz et al. 1996; Kiesselbach 1999).

Seminal roots are also of embryonic origin, and their primordia are laid down at the scutellar node during the later stages of embryogenesis. The number of the seminal roots is controlled by several genes and is highly variable between different genotypes (Sass 1977; Feldman 1994; Kiesselbach 1999). Seminal roots together with the primary root are major determinants for the early vigor of the seedlings.

Shoot-borne crown roots are formed in variable numbers at consecutive underground stem nodes and are initiated at the

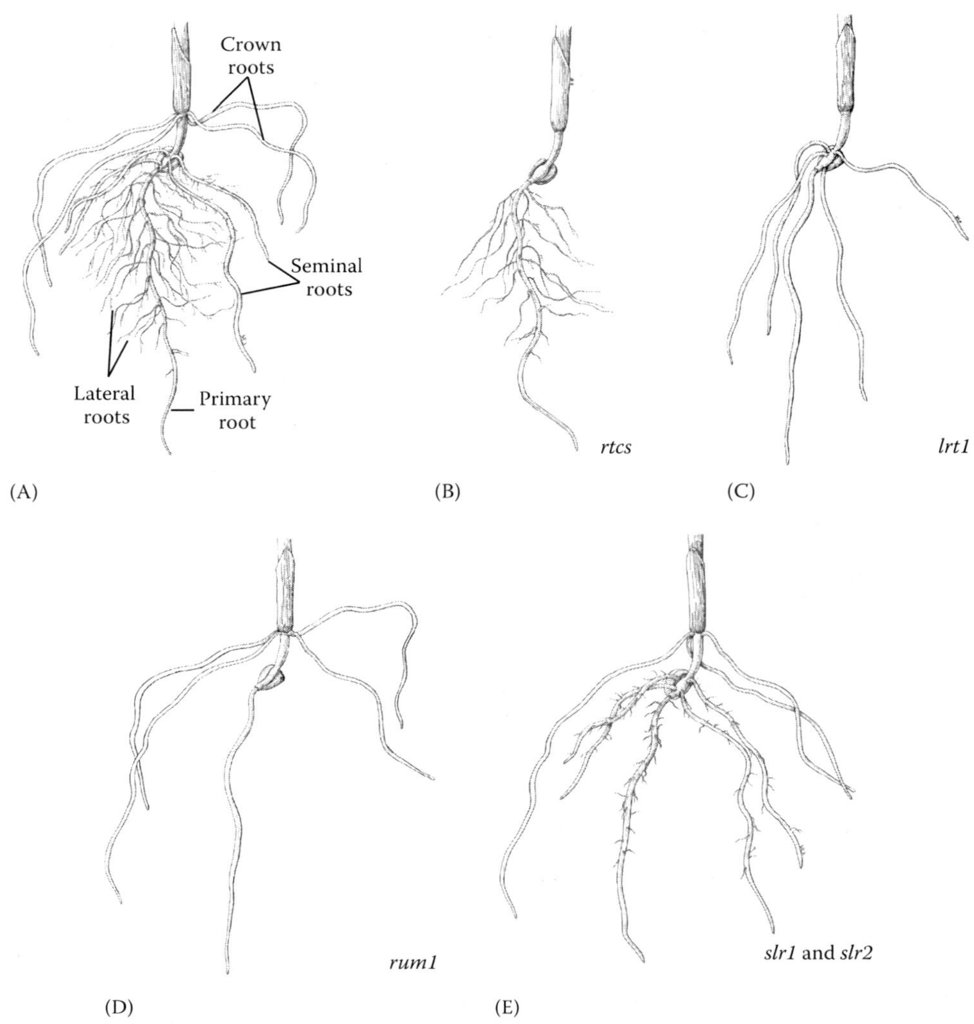

FIGURE 10.1 (A) Wild-type seedling root system. (B) The shoot-borne root initiation mutant *rtcs*. (C) The lateral root initiation mutant *lrt1*. (D) The lateral and seminal root initiation mutant *rum1*. (E) The lateral root elongation mutants *slr1* and *slr2*.

outermost cortical cell layers (Martin and Harris 1976; Varney and McCully 1991; Varney et al. 1991; Wang et al. 1991, 1994). The number and diameter of crown roots increases on higher shoot nodes. Crown roots represent the backbone of the overall root architecture of maize are therefore essential for the stability of the growing plant (McCully and Canny 1988). Shoot-borne roots formed from aboveground stem nodes are designated brace roots (Feldman 1994).

Lateral roots are initiated from pericycle and endodermis cells in the differentiation zone of roots by a multistep mechanism (Bell and McCully 1970; Esau 1977). Lateral roots are much thinner than crown roots (Peterson 1991) and can lead to the formation of secondary and higher order branches which have a great influence on the architecture of the root system (Lynch 1995). Lateral roots increase the absorbing surface of the root system and are essential for water and nutrient uptake.

Root hairs are unicellular epidermal extensions that are formed in the maturation zone of all roots and function in nutrient and water uptake by further increasing the absorbing surface of the root (Gilroy and Jones 2000). Despite these important functions, some maize mutants displaying reduced root hair elongation showed normal growth and development (Wen and Schnable 1994). A summary of the unique properties of the various root types is given in Table 10.1.

TABLE 10.1 Properties of the Maize Root System

Root Type	Site of Root Emergence	Time of Primordia Initiation	Number of Roots
Primary	Basal end of the seed	Early in embryogenesis	1
Seminal	Scutellar node of the seed	Late in embryogenesis	0–12
Crown	Underground shoot nodes	Whole postembryonic development	4–6 whorls
Brace	Aboveground shoot nodes	Late in plant development	2–3 whorls
Lateral	Pericycle cells of all roots	Whole postembryonic development	Many

B. Structure of Individual Roots

Although the overall structure of maize (Avery 1930; Sass 1977; Feldman 1994; Ishikawa and Evans 1995) and *Arabidopsis* roots (Dolan et al. 1994; Benfey and Schiefelbein 1994; Scheres and Wolkenfelt 1998) is similar, maize roots display some unique anatomical features (Hochholdinger and Zimmermann 2008). Among other differences, maize contains several hundred cells in the quiescent center and the root apical meristem (Feldman 1994) compared to only a few such cells in *Arabidopsis*. Moreover, radially maize contains 10–15 cortical cell files (Feldman 1994), while *Arabidopsis* displays a single-layer cortex consisting of eight cell files surrounding the stele. Furthermore, while in *Arabidopsis* root hair-forming trichoblasts can be predicted by their position relative to the underlying cortex cells, maize root hairs do not follow such a stereotypic pattern (Hochholdinger and Zimmermann 2008). Finally, while lateral roots of maize are formed adjacent to the primary phloem poles, these roots are initiated in *Arabidopsis* adjacent to the primary xylem poles (Hochholdinger and Zimmermann 2008).

C. Transition from Early to Late Root Architecture

Maize root system architecture is determined by the transition from early embryonic primary and seminal roots to postembryonic shoot-borne root structures. The early primary and seminal roots together with their lateral roots are essential for the development and vigor of the young seedling (Hochholdinger 2009). The shape of the early root system prior to seminal root emergence is comparable to the root system of the dicot model plant *Arabidopsis*. About 2 weeks after germination, postembryonic shoot-borne roots start to develop and gradually determine root system architecture in maize. The final shape of the root system is governed by the number and growth direction of the postembryonic crown roots together with their lateral roots (Lynch 1995). The development of the root system is strongly influenced by environmental factors such as water, oxygen, and nutrient availability, and the interaction with the biotic soil environment. Those can determine the rate of root initiation and growth and are the major causes for the plasticity of root system architecture (Aiken and Smucker 1996; McCully 1999).

III. Genetic Analysis of Maize Root Formation

A. Isolation of Maize Root Mutants

To identify genes involved in root formation and morphogenesis, a search for monogenic mutants was undertaken with the subsequent goal of isolating the mutated genes (reviewed in Hochholdinger et al. 2004a,b). Many traits contributing to the natural variation of maize root system architecture are quantitatively controlled by several genes (Hochholdinger et al. 2004a). Therefore, it is often very challenging to use these genotypes as starting material for the identification of the individual genes involved in these processes. Nevertheless, recently, several quantitative trait loci (QTL) linked to maize root traits have been mapped (Landi et al. 2010; Ruta et al. 2010). Since spontaneous mutations occur only very rarely, novel mutations related to maize root development were induced by chemical mutagens such as ethyl methanesulfonate (EMS) treatment of pollen (Hochholdinger and Feix 1998) or transposon tagging with the *Mutator* system (Wen and Schnable 1994; Hochholdinger et al. 2001; Woll et al. 2005). Two test systems were applied for the screening of large numbers of segregating F_2 families for monogenic mutants. The search for mutants with aberrations in the formation of the early root system was performed in a paper roll test (Hetz et al. 1996). This system allowed for the screening of a large number of seedlings under controlled conditions in a limited space. In brief, surface-sterilized kernels were placed on properly arranged filter paper sheets, which were rolled up and put with their lower part into beakers containing water only. The kernels were then left to germinate for up to 2 weeks. The developing seedling root system was visually inspected for aberrations after opening the rolled up sheets. Root mutants with defects at later stages of root formation can be indirectly identified in field tests of segregating F_2 populations based on characteristics of their aboveground parts as, for instance, reduced lodging resistance. This test is more demanding because only in rare instances a direct correlation of the observed changes with root system characteristics can be shown. This approach was successful in the case of a root lodging-resistant mutant (see the isolation of the *rtcs* mutant in the following text). After the identification of tentative mutant candidates, careful analyses of subsequent generations with respect to the expected Mendelian segregation ratios need to be performed also including outcrosses in different genetic backgrounds (cf. Woll et al. 2005).

The major classes of maize root mutants isolated thus far are characterized by the absence or reduction of specific root types thus affecting the overall root architecture (Table 10.2). A detailed account of the developmental stages affected by the different mutations is provided in Figure 10.2.

B. Root Initiation Mutants

The monogenic recessive mutant *rtcs* (*rootless for crown and seminal roots*) does not initiate seminal and crown roots leading to a phenotype that resembles the *Arabidopsis* root system with its simple primary root (Hetz et al. 1996; Figure 10.1B). The primary root of the mutant *rtcs* shows a reduced gravitropic response, while its elongation, lateral root density, and reaction to exogenously applied auxin are not affected (Taramino et al. 2007). The *rtcs* mutant displays reduced lodging resistance but can be propagated when supported by a stick and adequately provided with water and nutrients. The absence of cyclin expression (Hochholdinger and Feix 1998b) indicated that this mutation does not allow any cell divisions in the regions of primordia

TABLE 10.2 Characteristics of Monogenic Maize Root Mutants

	Affected Root Types						
Mutant	P	S	C	B	L	Histological Defect	References
rt1		x	x	x		Primordia initiation	Jenkins (1930)
rtcs		x	x	x		Primordia initiation	Hetz et al. (1996); Taramino et al. (2007)
lrt1	x	x	x		x	Primordia initiation	Hochholdinger and Feix (1998a)
rum1	x	x			x	Primordia initiation	Woll et al. (2005); von Behrens et al. (2011)
slr1	x	x			x	Cell elongation	Hochholdinger et al. (2001)
slr2	x	x			x	Cell elongation	Hochholdinger et al. (2001)
rth1	x	x	x	x	x	Root hair elongation	Wen and Schnable (1994); Wen et al. (2005)
rth2	x	x	x	x	x	Root hair elongation	Wen and Schnable (1994)
rth3	x	x	x	x	x	Root hair elongation	Wen and Schnable (1994); Hochholdinger et al. (2008)

P, primary root; S, seminal roots; C, crown roots; B, brace roots; L, lateral roots.
x indicates root types that are affected in the mutant.

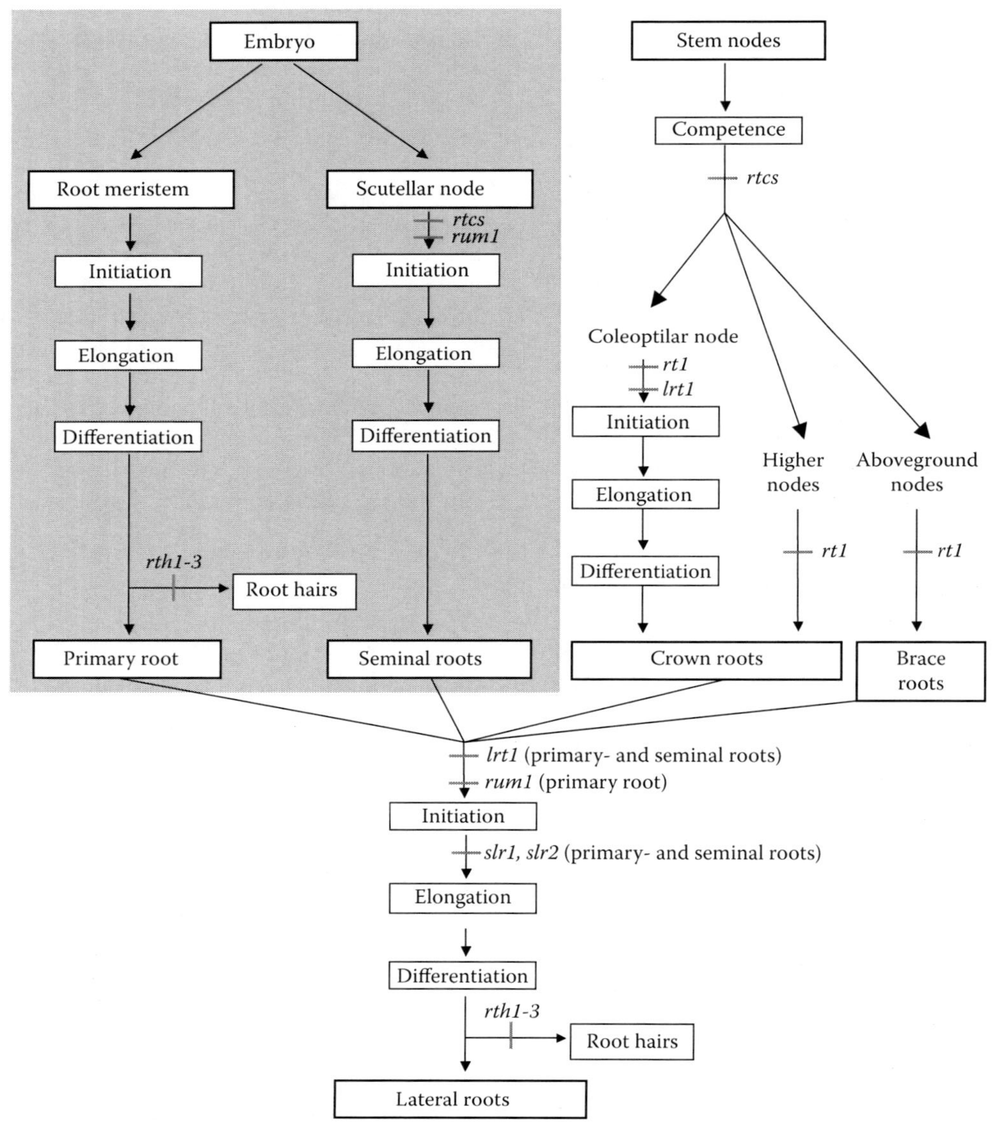

FIGURE 10.2 Known genetic checkpoints of maize root development.

formation. Map-based cloning revealed that *Rtcs*, which encodes a 26 kDa LOB (Lateral Organ Boundaries) domain protein, is located on chromosome 1S (Taramino et al. 2007). *Rtcs* gene expression in coleoptilar nodes is confined to emerging shoot-borne root primordia as illustrated by in situ hybridization experiments. The *Rtcs* promoter contains auxin response elements and is auxin inducible. Hence, *Rtcs* is an auxin responsive gene involved in the early events that lead to the initiation and maintenance of seminal and shoot-borne root primordia formation.

The monogenic recessive mutant *lrt1* (*lateral rootless 1*; Hochholdinger and Feix 1998a) lacks postembryonic root formation at the early seedling stage and does not form any lateral roots in the primary root and the seminal roots as demonstrated by microscopic analyses (Figure 10.1C). The defect of *lrt1* operates very early in root initiation before primordia are formed. Furthermore, the wild-type root phenotype cannot be rescued in mutants by the application of auxin. Moreover, the mutant does not form an elongated mesocotyl in the dark, although its photomorphogenic response appears to be normal (Hochholdinger and Feix 1998a). Double mutants of *lrt1* and *rtcs* show a strictly additive behavior suggesting that these genes are acting independently from each other. The *Lrt1* gene has not been cloned yet.

The semidominant mutant *rum1* (rootless with undetectable meristems 1) is deficient in the initiation of the embryonic seminal roots and the postembryonic lateral roots at the primary root, while lateral root initiation at the shoot-borne roots is not affected (Woll et al. 2005; Figure 10.1D). In the primary root, the mutant *rum1* displays reduced polar auxin transport and a delayed gravitropic response. Map-based cloning revealed that *Rum1*, which encodes a 28 kDa auxin-inducible monocot-specific Aux/IAA protein (von Behrens et al. 2011), is located on chromosome 3. RUM1 is an unstable protein localized in the nucleus which interacts with ZmARF25 and ZmARF34 (*Zea mays* AUXIN RESPONSE FACTOR 25 and 34; von Behrens et al. 2011).

C. Root Elongation Mutants

In the mutants *slr1* and *slr2* (*short lateral roots 1* and *2*), the mechanism of cell elongation of lateral roots is disturbed leading to lateral roots of about one-fourth the length of wild-type roots (Hochholdinger et al. 2001; Figure 10.1E) and very limited root hair formation because of their incomplete cell elongation. The two mutants display a very similar phenotype but proved to be different by allelism tests. A particular feature of these mutants is the strict root type specificity, since only lateral roots of the embryonic primary and seminal roots are affected but not lateral roots at the postembryonic shoot-borne roots. The shorter lateral roots are caused by smaller root cells compared to wild-type lateral roots as demonstrated by confocal laser scanning microscopy (Hochholdinger et al. 2001). In the double mutant *slr1/slr2*, the defect does not only influence lateral root-specific cell elongation but also leads to disarranged longitudinal cellular patterns in the primary and seminal roots indicating that the two loci affected in *slr1* and *slr2* are cooperating in the establishment of the lateral root specificity during early root development (Hochholdinger et al. 2001). The affected genes have not been cloned yet.

D. Root Maturation Mutants

Root hair formation takes place in the root maturation zone. Unicellular root hairs elongate via localized tip growth, a process mediated by polar exocytosis of secretory vesicles. Thus far, the three monogenic recessive mutants *rth1* to *rth3* (*roothairless 1 to 3*), which are defective in the morphology of root hairs of the primary root, have been isolated by transposon tagging (Wen and Schnable 1994). These mutants are of particular interest, since root hairs play an important role in plant nutrition by facilitating the uptake of water and nutrients (Gilroy and Jones 2000). The *rth1* and *rth2* mutants form normal root hair initials. However, the root hairs of the mutant *rth1* hardly elongate and reach only one-twentieth to one-tenth of the length of the root hairs of wild-type plants. The root hairs of the mutant *rth2* show elongation to about one-fifth to one-fourth of those of the wild type. In contrast to *rth1* and *rth2*, the mutant *rth3* displays abnormal root hair initial already at a very early stage of development as observed by scanning electron microscopy eventually leading to very short root hairs similar to *rth1*. The mutant *rth1* shows pleiotropic nutritional defects leading to stunted plants with purplish leaves that never form ears and only rarely produce tassels. In contrast to *rth1*, the plants of the mutants *rth2* and *rth3* show a normal growth performance during their entire development despite their drastically impaired root hair morphology.

Rth1 encodes a 100 kDa *SEC3* homolog (Wen et al. 2005) that is located on chromosome 1L. In yeast and mammals, *SEC3* is a subunit of the exocyst complex, which tethers exocytotic vesicles prior to their fusion. Cloning of the *Rth1* gene supported the hypothesis that exocyst-like proteins are involved in plant exocytosis which has been subsequently supported by additional experiments (Hala et al. 2008). Proteomic analyses of primary roots identified four proteins that accumulated to different levels in wild type and *rth1*. The preferential accumulation of a negative regulator in the *rth1* mutant proteome of the cell cycle (a prohibitin) may at least partially explain the delayed development and flowering of the *rth1* mutant. The *Rth2* gene has not been cloned yet.

Rth3 encodes a 71 kDa monocot-specific COBRA-like protein that displays structural features of a glycosylphosphatidylinositol protein (GPI anchor) (Hochholdinger et al. 2008) and maps to chromosome 1S. Members of the COBRA family are involved in various types of cell expansion and cell wall biosynthesis. In situ hybridization experiments showed that *Rth3* expression is confined in primary roots to root hair-forming trichoblasts and to lateral root primordia. Remarkably, in replicated field trials involving near-isogenic lines, the *rth3* mutant conferred significant losses in grain yield (Hochholdinger et al. 2008).

IV. Functional Genomic Analysis of Maize Root Development

To functionally characterize the molecular pathways that are affected in the diverse root mutants, transcriptomic and proteomic approaches were applied.

A. Transcriptomic Dissection of Maize Pericycle Cells

Transcriptome profiling via microarray analyses in combination with tissue-specific isolation techniques such as laser capture microdissection (LCM) allows for the analysis of a large number of genes in parallel and provides a powerful technique for the identification of specific gene expression profiles (Schnable et al. 2004). Each root cell type expresses a unique transcriptome. While in wild-type roots a subset of pericycle cells has the competence to initiate lateral roots, none of the pericycle cells of the mutant *rum1* has the capacity to divide and form lateral roots. Hence, comparing pericycle populations of wild-type and *rum1* roots allowed identifying gene expression patterns related to lateral root initiation. Comparative cell-type-specific analyses of the maize root pericycle transcriptomes of wild-type and *rum1* primary roots 64 h after germination, hence prior to lateral root initiation, revealed genes involved in signal transduction, transcription, and the cell cycle that will be helpful for the identification of checkpoints involved in lateral root formation downstream of *Rum1*. Similarly, pericycle cells of the inbred line B73 isolated via laser capture microdissection identified genes preferentially expressed in pericycle versus all other root cells of the differentiation zone of maize primary roots (Dembinsky et al. 2007). Among those, transcripts related to protein synthesis and cell fate were significantly enriched in pericycle versus non-pericycle cells. In summary, these pericycle specific gene expression experiments defined the molecular events during the specification of cell-cycle-competent pericycle cells prior to their first division and demonstrate that pericycle specification and lateral root initiation might be controlled by a different set of genes.

B. Proteomic Dissection of Maize Root Development

Typically, proteome comparisons of maize root types during development are based on two-dimensional protein separations and subsequent mass spectrometric analyses of differentially accumulated proteins (Hochholdinger et al. 2006). To begin with, this technique revealed significant changes in the composition of the maize primary root proteome during the early stages of development between 5 and 9 days after germination (Hochholdinger et al. 2005). Moreover, by analyzing 5-day-old primary roots of *lrt1* versus wild-type seedlings, it was demonstrated that lateral roots affect the protein composition of the primary root (Hochholdinger et al. 2004c). Furthermore, the soluble proteins related to shoot-borne root initiation in the coleoptilar node (Sauer et al. 2006) and seminal root initiation in embryos (Muthreich et al. 2010) were compared between wild type and the mutant *rtcs* and revealed that distinct molecular pathways are involved in the formation of these root types. Similarly, proteins related to lateral root initiation in the primary root have been identified by comparing the most abundant proteins of primary roots of wild type and *rum1* (Liu et al. 2006). In addition, tissue-specific analyses of cortex versus stele tissues of the maize primary root including the mutant *rum1* have been performed and revealed unique and conserved functions between these tissues (Saleem et al. 2009, 2010). Recently, proteome analyses of captured pericycle cells revealed the most abundant pericycle proteins (Dembinsky et al. 2007) and also proteome differences in wild-type versus *rum1* pericycle cells prior to lateral root formation (Liu et al. 2010). In summary, these comparative analyses identified differentially accumulated candidate proteins that might be involved in different aspects of root development. More recently, shotgun sequencing of protein samples became available which significantly enhanced the resolution of proteome studies and allowed in the case of constitutively secreted extracellular proteins from the maize primary root tip for the identification of 2848 distinct proteins (Ma et al. 2010).

In summary, transcriptome and proteome analyses revealed unique expression patterns for the different root and tissue types at the various developmental stages, suggesting complex regulatory mechanisms involved in maize root system development.

V. Conclusions and Outlook

The complex maize root system is composed of different embryonic and postembryonic root types formed at different stages of development. Recent progress in the genetic analysis of maize root development revealed, for example, that genes controlling maize root system architecture often affect the formation of multiple root types (Hochholdinger et al. 2004a) and that some of these genes such as *Lrt1* act only transiently at a specific developmental phase (Hochholdinger and Feix 1998a). Mapping of the root-related loci did not reveal any genomic hot spots involved in root development but rather a random distribution throughout the maize genome. The relatively small number of monogenic maize root mutants currently available underpins the difficulty of dissecting maize root development by classical genetic analyses considering the multitude of genes and their interactions anticipated to be involved in the development of the root system. The recent cloning of some of the genes identified by mutations provides therefore support for the ongoing genetic analysis. In this context, the cloning of *Rtcs*, *Rth1*, *Rth3*, and *Rum1* provided first insights into the molecular framework underlying the formation of the complex maize root system (Wen et al. 2005; Taramino et al. 2007; Hochholdinger et al. 2008; von Behrens et al. 2011). Furthermore, with the cloning of these genes and the availability of the maize genome sequence (Schnable et al. 2009), a systematic analysis for homologous sequences related to these gene families can be envisaged. In addition, transcriptome and proteome analyses might allow the reconstruction of pathways involved in the formation of the different maize root types by revealing putative downstream targets of the affected genes.

In the future, novel mutants, for instance, affected in lateral root formation of shoot-borne roots and cloning of additional genes involved in the determination of root system architecture will help to understand the molecular framework underlying root stock development in maize. Moreover, identification of direct molecular interactions of genes and proteins involved in root system formation by techniques such as bimolecular fluorescence complementation, yeast-two-hybrid experiments, and GST and other pull-down assays will allow for deeper insights in the molecular networks involved in root formation (von Behrens et al. 2011). Furthermore, GFP-marker lines will not only allow for tissue-specific and subcellular localizations of proteins but can also be used for analyzing the stability of proteins (von Behrens et al. 2011). In addition, the focus of maize root research will not be limited to analyses of genes regulating the endogenously determined basic structure of the maize root system. It will be extended to genes involved in the adaptive responses of roots to adverse environmental conditions such as drought or biotic interactions. To this end, high-throughput phenotyping platforms that allow for the precise analysis of adult root systems in their soil environment will be instrumental. Finally, analyses of natural variation by association mapping, QTL analyses, and marker-assisted breeding will support the search for favorable alleles and contribute to progress in maize breeding by combining genomic and quantitative genetic approaches.

References

Aiken RM, Smucker AJM 1996 Root system regulation of whole plant growth. *Annu Rev Phytopathol* 34:325–346.

Avery GS 1930 Comparative anatomy and morphology of embryos and seedlings of maize, oats and wheat. *Bot Gaz* 89:1–39.

von Behrens I, Komatsu M, Zhang Y et al. 2011 *Rootless with undetectable meristem1* encodes a monocot-specific AUX/IAA protein that controls embryonic seminal and postembryonic lateral root initiation in maize. *Plant J* 66:341–353.

Bell JK, McCully ME 1970 A histological study of lateral root initiation and development in *Zea mays. Protoplasma* 70:179–205.

Benfey PN, Schiefelbein JW 1994 Getting to the root of plant development: the genetics of *Arabidopsis* root formation. *Trends Genet* 10:84–88.

Coe EH 2001 The origins of maize genetics. *Nat Rev Genet* 2:898–905.

Dembinsky D, Woll K, Saleem M et al. 2007 Transcriptomic and proteomic analyses of pericycle cells of the maize (*Zea mays* L.) primary root. *Plant Physiol* 145:575–588.

Dolan L, Duckett CM, Grierson C et al. 1994 Clonal relationships and cell patterning in the root epidermis of *Arabidopsis. Development* 120:2465–2474.

Esau K 1977 *Plant Anatomy*, 2nd edn. New York: John Wiley & Sons, Inc.

Fahn A 1990 *Plant Anatomy*, 4th edn. Oxford, U.K.: Pergamon press.

Feldman L 1994 The maize root. In *The Maize Handbook*, eds. M Freeling, V Walbot, pp. 29–37. New York: Springer.

Gilroy S, Jones DL 2000 Through form to function: root hair development and nutrient uptake. *Trends Plant Sci* 5:56–60.

Hala M, Cole R, Synek L et al. 2008 An exocyst complex functions in plant cell growth in *Arabidopsis* and tobacco. *Plant Cell* 20:1330–1345.

Hetz W, Hochholdinger F, Schwall M, Feix G 1996 Isolation and characterisation of *rtcs*, a mutant deficient in the formation of nodal roots. *Plant J* 10:845–857.

Hochholdinger F 2009 The maize root system: morphology, anatomy and genetics. In *The Handbook of Maize*, eds. J Bennetzen, S Hake, pp. 145–160. New York: Springer.

Hochholdinger F, Feix G 1998a Early post-embryonic root formation is specifically affected in the maize mutant *lrt1. Plant J* 16:247–255.

Hochholdinger F, Feix G 1998b Cyclin expression is completely suppressed at the site of crown root formation in the nodal region of the maize root mutant *rtcs. J Plant Physiol* 153:425–429.

Hochholdinger F, Guo L, Schnable PS 2004c Lateral roots affect the proteome of the primary root of maize (*Zea mays* L.). *Plant Mol Biol* 56:397–412.

Hochholdinger F, Park WJ, Feix G 2001 Cooperative action of SLR1 and SLR2 is required for lateral root specific cell-elongation in maize. *Plant Physiol* 125:1529–1539.

Hochholdinger F, Park WJ, Sauer M, Woll K 2004a From weeds to crops: genetic analysis of root development in cereals. *Trends Plant Sci* 9:42–48.

Hochholdinger F, Sauer M, Dembinsky D, Hoecker N, Muthreich N, Saleem M, Liu Y 2006. Proteomic dissection of plant development. *Proteomics* 6:4076–4083.

Hochholdinger F, Tuberosa R 2009 Genetic and genomic dissection of maize root development and architecture. *Curr Opin Plant Biol* 12:172–177.

Hochholdinger F, Wen TJ, Zimmermann R et al. 2008 The maize (*Zea mays* L.) *roothairless3* gene encodes a putative GPI-anchored, monocot-specific, COBRA-like protein that significantly affects grain yield. *Plant J* 54:888–898.

Hochholdinger F, Woll K, Guo L, Schnable PS 2005 Analysis of the soluble proteome of maize (*Zea mays* L.) primary roots reveals drastic changes in protein composition during early development. *Proteomics* 18:4885–4893.

Hochholdinger F, Woll K, Sauer M, Dembinsky D 2004b Genetic dissection of root formation in maize (*Zea mays*) reveals root-type specific developmental programs. *Ann Bot* 93:359–368.

Hochholdinger F, Zimmermann R 2008 Conserved and diverse mechanisms in root development. *Curr Opin Plant Biol* 11:70–74.

Ishikawa H, Evans ML 1995 Specialized zones of development in roots. *Plant Physiol* 109:725–727.

Jenkins MT 1930 Heritable characters of maize XXXIV-rootless. *J Hered* 21:79–80.

Kausch W 1967 Lebensdauer der Primaerwurzel von Monokotyledonen. *Naturwissenschaften* 54:475.

Kiesselbach TA 1999 *The Structure and Reproduction of Corn*, 50th Anniversary edition. New York: Cold Spring Harbor Laboratory Press.

Landi P, Giuliani S, Salvi S, Ferri M, Tuberosa R, Sanguineti MC 2010 Characterization of root-yield-1.06, a major constitutive QTL for root and agronomic traits in maize across water regimes. *J Exp Bot* 61:3553–3562.

Larson WE, Hanway JJ 1977 Corn production. In *Corn and Corn Improvement*, ed. GF Spargue, pp. 625–669. Madison, WI: American Society of Agronomy.

Liu Y, Lamkemeyer T, Jakob A, Mi G, Zhang F, Nordheim A, Hochholdinger F 2006 Comparative proteome analyses of maize (*Zea mays* L.) primary roots prior to lateral root initiation reveal differential protein expression in the lateral root initiation mutant *rum1*. *Proteomics* 6:4300–4308.

Liu Y, von Behrens I, Muthreich N, Schütz W, Nordheim A, Hochholdinger F 2010 Regulation of the pericycle proteome in maize (*Zea mays* L.) primary roots by *RUM1* which is required for lateral root initiation. *Eur J Cell Biol* 89:236–241.

Lynch J 1995 Root architecture and plant productivity. *Plant Physiol* 109:7–13.

Ma W, Muthreich N, Liao C et al. 2010 The mucilage proteome of maize (*Zea mays* L.) primary roots. *J Proteome Res* 9:2968–2976.

Martin EM, Harris WM 1976 Adventitious root development from the coleoptilar node in *Zea mays* L. *Am J Bot* 63:890–897.

McCully ME 1999 Roots in soil: unearthing the complexities of roots and their rhizospheres. *Annu Rev Plant Physiol Plant Mol Biol* 50:695–718.

McCully ME, Canny MJ 1988 Pathways and processes of water and nutrient movements in roots. *Plant Soil* 111:159–170.

Muthreich N, Schützenmeister A, Schütz W et al. 2010 Regulation of the maize (*Zea mays* L.) embryo proteome by *RTCS* which controls seminal root initiation. *Eur J Cell Biol* 89:242–249.

Peterson RL 1991 Adaptations of root structure in relation to biotic and abiotic factors. *Can J Bot* 70:661–675.

Ruta N, Liedgens M, Fracheboud Y, Stamp P, Hund A 2010 QTLs for the elongation of axile and lateral roots of maize in response to low water potential. *Theor Appl Genet* 120:621–631.

Saleem M, Lamkemeyer T, Schützenmeister A, Fladerer C, Piepho H-P, Nordheim A, Hochholdinger F 2009. Tissue specific control of the maize (*Zea mays* L.) embryo, cortical parenchyma, and stele proteomes by RUM1 which regulates seminal and lateral root initiation. *J Proteome Res* 8:2285–2297.

Saleem M, Lamkemeyer T, Schützenmeister A et al. 2010 Specification of cortical parenchyma and stele of maize (*Zea mays* L.) primary roots by asymmetric levels of auxin, cytokinin and cytokinin-regulated proteins. *Plant Physiol* 152:4–18.

Sass JE 1977 Morphology. In: *Corn and Corn Improvement*, ed. GF Spargue, pp. 89–110. Madison, WI: American Society of Agronomy.

Sauer M, Jakob A, Nordheim A, Hochholdinger F 2006 Proteomic analysis of shoot-borne root initiation in maize (*Zea mays* L.). *Proteomics* 6:2530–2541.

Scheres B, Wolkenfelt H 1998 The *Arabidopsis* root as a model to study plant development. *Plant Physiol Biochem* 36:21–32.

Schnable PS, Hochholdinger F, Nakazono M 2004 Global expression profiling applied to plant development. *Curr Opin Plant Biol* 7:50–56.

Schnable PS, Ware D, Fulton RS et al. 2009 The B73 maize genome: complexity, diversity, and dynamics. *Science* 326:1112–1115.

Taramino G, Sauer M, Stauffer J, Multani D, Niu X, Sakai H, Hochholdinger F 2007 The *rtcs* gene in maize (*Zea mays* L.) encodes a lob domain protein that is required for post-embryonic shoot-borne and embryonic seminal root initiation. *Plant J* 50:649–659.

Varney GT, Canny MJ, Wang XL, McCully ME 1991 The branch roots of *Zea*. First order branches, their number, sizes and division into classes. *Ann Bot* 67:357–364.

Varney GT, McCully ME 1991 The branch roots of *Zea*. II. Developmental loss of the apical meristem in field-grown roots. *New Phytol* 118:535–546.

Wang XL, Canny MJ, McCully ME 1991 The water status of the roots of soil-grown maize in relation to the maturity of their xylem. *Physiol Plant* 82:157–162.

Wang XL, McCully ME, Canny MJ 1994 The branch roots of *Zea*. IV. The maturation and openness of xylem conduits in first-order branches of soil-grown roots. *New Phytol* 126:21–29.

Wen T-J, Hochholdinger F, Sauer M, Bruce W, Schnable PS 2005 The *roothairless1* gene of maize (*Zea mays*) encodes a homolog of *sec3*, which is involved in polar exocytosis. *Plant Physiol* 138:1637–1643.

Wen T-J, Schnable PS 1994 Analyses of mutants of three genes that influence root hair development in *Zea mays* (Gramineae) suggest that root hairs are dispensable. *Am J Bot* 81:833–843.

Woll K, Borsuk L, Stransky H, Nettleton D, Schnable PS, Hochholdinger F 2005 Isolation, characterization and pericycle specific transcriptome analyses of the novel maize (*Zea mays* L.) lateral and seminal root initiation mutant *rum1*. *Plant Physiol* 139:1255–1267.

Yamashita T 1991 Ist die Primaerwurzel bei Samenpflanzen exogen oder endogen? *Beitr Bio Pflanzen* 66:371–379.

11

Molecular Mechanisms Involved in Adventitious Root Formation

Joseph Riov
The Hebrew University of Jerusalem

David Szwerdszarf
Agricultural Research Organization

Mohamad Abu-Abied
Agricultural Research Organization

Einat Sadot
Agricultural Research Organization

I. Introduction

Rooting capacity is one of the economically important factors lost during the juvenile to mature phase change. The difficulties in propagation of promising clones of fruit trees, rootstocks, ornamental woody plants, and forest trees are a serious obstacle in breeding programs that are generally based on the production of rooted cuttings. Understanding the molecular basis of adventitious roots (AR) formation will contribute to our ability to design biotechnological solutions, not only for the benefit of agriculture but also for biomass, renewable energy, and biofuel production. For more detailed reviews on AR formation, the reader may refer to the following references: Dech (2009), Geiss et al. (2009), Lanteri et al. (2009), Ludwig-Müller (2009), and Pijut et al. (2011). In this chapter, we mainly summarize the latest research results, with a glance on some previous works performed on woody plants as the assay system. While woody plants are those that suffer the most from difficulties in rooting, they are not an easy system for biochemical and molecular studies. On the other hand, the research on AR formation using herbaceous model plants has elucidated several potential molecular mechanisms although these plants are practically easy to root. Therefore, this review combines and compares data obtained from both plant systems.

II. Process of AR Formation

AR formation is a complex process, in which roots differentiate and regenerate from stem tissues (Casson and Lindsey 2003). This process depends on multiple factors, such as the genetic background, physiology, and developmental stage of the mother plant, hormones, secondary metabolites, and environmental conditions, such as temperature, humidity, and oxygen pressure. (Geiss et al. 2009; Lanteri et al. 2009; Ludwig-Müller 2009). The differentiation of AR is often compared to that of lateral roots, since both are initiated from postembryonic tissues (Casson and Lindsey 2003). Despite the similarities between the two processes, many differences between the formation of lateral and AR have been observed. AR can either be induced de novo following wounding or other stimuli, but they can also be inherently preformed in intact plants during their normal development. Preformed root primordia can be found in woody plants, such as willow (*Salix fragilis* L.), *Thuja* species, and plants from the Malaceae (reviewed in Blakesley et al. 1991) as well as in herbaceous plants such as sweet potato (*Ipomoea batatas* (L.) Lam.) (Belehu et al. 2004; Firon et al. 2009) and oregano (*Origanum vulgare* L.) (Kuris et al. 1981). Preformed root primordia are found in nodal or in internodal regions, where they are associated with the phloem rays or the cambium (Blakesley et al. 1991;

Belehu et al. 2004; Firon et al. 2009). As mentioned previously, de novo AR are usually formed following wounding, but in some cases, they can be formed without wounding. Willow (*Salix cordata* Michx.), and poplar (*Populus deltoides* W. Bartram ex Marshall, *Populus balsamifera* L.) that are tolerant to flooding or burial (e.g., under sand dunes), can grow AR above the original root system in order to compensate for the lack of oxygen (Dech 2009). Spontaneous formation of AR occurs also in hypocotyls of *Arabidopsis* seedlings transferred from dark to light (Sorin et al. 2005) and in the collet, which is the transition zone between the root and hypocotyl (Casimiro et al. 2001; Smith and Fedoroff 1995).

AR formation in cuttings is often described to occur in four steps: (1) cell dedifferentiation, (2) cell division, (3) development of root primordia, and (4) root emergence (Blakesley et al. 1991; De Klerk et al. 1999). While molecular mechanistic data regarding these steps during AR formation start to accumulate, considerable data have been obtained regarding parallel steps during the formation of lateral roots, which might serve as guidelines for studies of AR formation. Nevertheless, it is already clear that major differences exist in the previously mentioned step 1 between lateral and AR. It was shown that in *Arabidopsis*, pre-initiation of lateral roots takes place in the basal meristem, very close to the root tip, as an inherent stage of the developmental program of the root (De Smet et al. 2007). Thereafter, initiation of root primordia occurs from the pericycle cell layer, 3–8 mm from the root tip (Casson and Lindsey 2003; Benkova and Bielach 2010), and, more precisely, from the pericycle cells adjacent to the xylem pole (Benkova and Bielach 2010). In contrast, stem tissues from which AR differentiate are variable (Blakesley et al. 1991) and can be either various established cell types or callus tissue that is formed as a result of wounding (Blakesley et al. 1991). The types of established cells which undergo dedifferentiation as AR induction starts are also

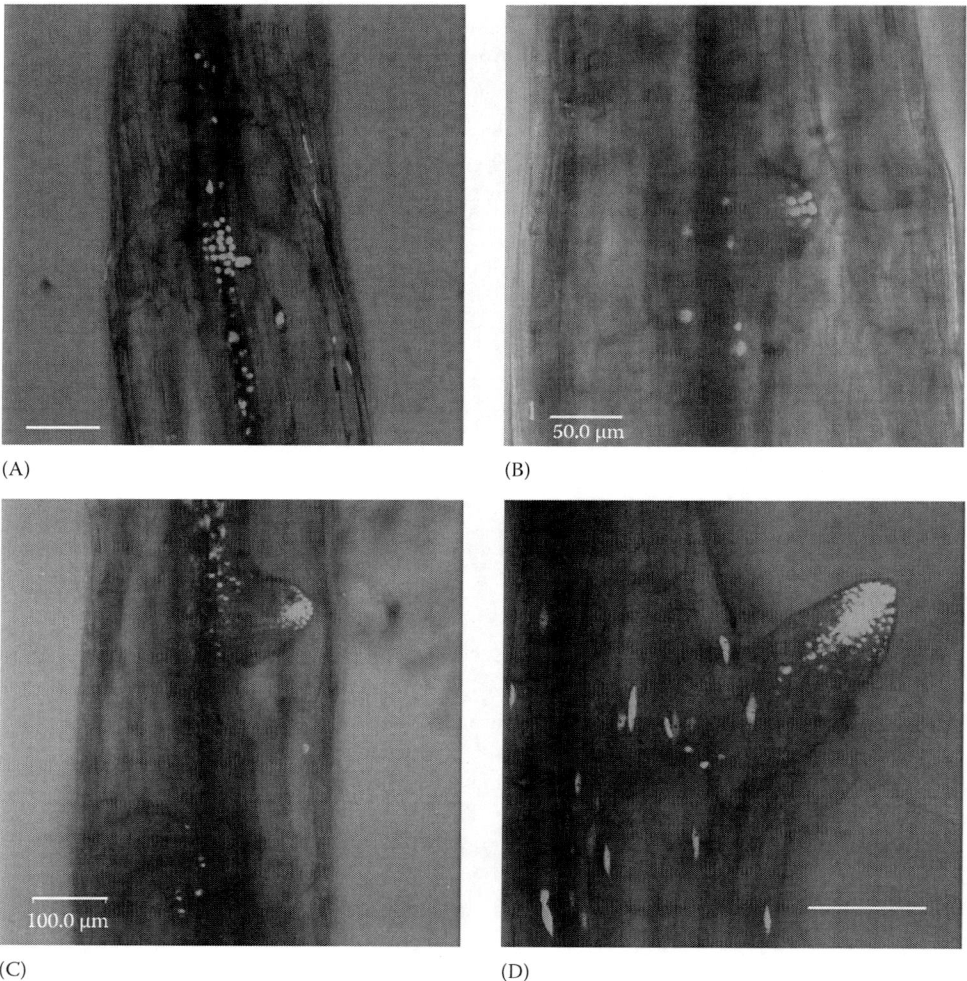

FIGURE 11.1 **(See color insert.)** Auxin accumulation during AR formation. *Arabidopsis* seedlings expressing the auxin responsive promoter DR5, linked to YFP with a nuclear localization signal (NLS), DR5::Venus marker (Laskowski et al. 2008), were grown in the dark. Hypocotyls were excised and incubated with 10 μM IBA. Auxin accumulation is shown during four stages of AR formation: (A) near the vascular tissue, (B) at the tip of an initial root primordia, (C) at the tip of a more developed primordium, and (D) at the tip of the root after emergence. Scale bars in A, C, D = 100 μm, in B = 50 μm. (From O. Rogovoy-Stelmakh, M. Abu-Abied, and E. Sadot, unpublished data.)

variable in different species and can be the cambium, phloem parenchyma, inner cortical parenchyma, non-differentiated secondary phloem and cambium between the vascular bundles, or other cell types (for additional information see Fahn 1990 and Ludwig-Müller 2009). After cell dedifferentiation occurs, the cells start to divide. In lateral roots, a few pericycle cells that acquire the feature of founder cells divide asymmetrically (Malamy and Benfey 1997; Dubrovsky et al. 2008). These cells, with additional flanking cells, undergo coordinated cell division and differentiation to form the lateral root primordium (Malamy and Benfey 1997; Benkova and Bielach 2010). The primordia continue to grow and eventually emerge through the endodermis, cortex, and epidermal layers of the primary root. The apical meristem of lateral roots is established only after root emergence (Malamy and Benfey 1997; Benkova and Bielach 2010). For lateral root emergence in *Arabidopsis*, it has been shown that auxin produced by the lateral root primordia induces the expression of LAX3, an auxin influx carrier in adjacent cells. In turn, LAX3 promotes the expression of auxin-dependent cell wall remodeling enzymes that lead to the separation of cells, which could otherwise block the emergence of the new roots (Swarup et al. 2008; Peret et al. 2009a,b). In contrast, AR emergence in rice (*Oryza sativa* L.), for example, depends on cell death of epidermal cells situated above the root primordium. Cell death is induced by ethylene, mediated by hydrogen peroxide (H_2O_2), and dependent on signaling through G proteins (Steffens and Sauter 2009a,b).

Many details are still missing regarding AR formation in cuttings: the particular pattern of auxin accumulation (initial data shown in Figure 11.1), its specific spatial and temporal regulators, transducers and cofactors, initial pattern of cell division and differentiation, the machinery of root emergence through the woody or half-woody layers, and the timing of establishment of the new root apical meristem.

III. Juvenile to Mature Phase Change in Relation to the Potential of AR Formation

A. Phase Change

Maturation in plant species is a process in which progressive changes in various developmental and morphological traits take place. Four phases of maturation have been characterized: (1) the embryonic phase, (2) the postembryonic juvenile vegetative phase, (3) the mature vegetative phase, and (4) the mature reproductive phase (Greenwood 1987; Poethig 1990). In both woody and herbaceous plants, flowering capacity indicates that the maturation stage has been achieved, but other traits, such as leaf shape and phyllotaxis, photosynthesis efficiency, shoot orientation, stem pigmentation, thorniness, chemical composition, disease and insect resistance, and the capability to form AR, are modified through the phase change (Hackett 1985; Poethig 1990, 2010; McGowran et al. 1998; Day et al. 2002). In woody plants, the juvenile traits, including rooting capability, are expressed at the bottom of the stem throughout the life of the plant and are sharply or gradually being replaced by maturity traits, including rooting incapability, toward the upper part of the main stem and the branches (Poethig 1990). Intermediate traits regulated by overlapping programs can be found along the stem and lateral shoots (Poethig 1990).

What are the mechanisms that regulate the juvenile to mature phase change? Rejuvenation of mature herbaceous shoots can be achieved by re-rooting, while mature woody shoots can sometimes return to juvenile phase by repeated graftings on juvenile rootstocks (Poethig 1990; Huang et al. 1992), suggesting that the controlling signal is encoded in the root system. Another model argues that a diffusible factor, which is produced outside the shoot apical meristem, underlies this process (Poethig 1990). Indeed, it was found that maturation of *Eucalyptus grandis* W. Hill seedlings was accompanied by the accumulation of a substance, named factor G, in the tissues comprising the base of the cuttings and the leaves, concomitantly with the decreased rooting ability (Paton et al. 1970). The G factor was later discovered to be composed of three rooting inhibitors, G1, G2, and G3, having a molecular skeleton which consists of a 2,3-dioxabicyclo[4.4.0] decene system (Sterns 1971). The G factor inhibited rooting of mung bean cuttings at high concentrations but promoted rooting at lower concentrations (Dhawan et al. 1979). The previous observations suggest that more than one mechanism is involved in the loss of rooting ability during the phase change.

One of the earliest signs of the differential response of juvenile and mature cuttings to root induction was revealed by comparative anatomical studies. Characterization of the process of AR formation in juvenile and mature cuttings of chestnut (*Castanea sativa* Mill.) treated with indole-3-butyric acid revealed that in both types of cuttings, induction of cell division occurred following auxin treatment; however, root primordia were formed only in juvenile cuttings (Ballester et al. 1999). Similar findings were observed in oak (*Quercus robur* L.) cuttings (Vidal et al. 2003). In pine (*Pinus taeda* L.), hypocotyls from 50-day-old seedlings readily formed AR, whereas epicotyls from these seedlings hardly formed AR (Greenwood et al. 2001). Yet, when 1 mm explants of both types of cuttings were cultured in the presence of different auxins, callus was equally formed initially from the cambium and later from the outer cortex. This indicates that cell division occurred in both types of cuttings, but differentiation of root primordia took place only in cuttings possessing juvenile traits. These observations suggest that one of the limiting factors in mature cuttings is a signal enabling a cross talk between dividing cells, which is necessary for the organization of root primordia.

B. Molecular Mechanism Underlying the Juvenile to Mature Phase Change in Woody Plants

What might be the juvenile specific signals for the formation of root primordia? Attempts were made during the years to identify molecules with differential expression between juvenile and mature tissues of various woody plants. Specific proteins that were differentially expressed in juvenile and mature tissues were identified in sequoia (*Sequoiadendron giganteum* (Lindl.) J. Buchholz) (Bon 1988), plum (*Prunus avium* L.) (Besford et al. 1996), and chestnut (*Castanea sativa*) (Amo-Marco et al. 1993). Polyamines are known to be involved in cell division and synthesis of DNA, RNA, and proteins (Takahashi and Kakehi 2010). It has been found that in both pine (*Pinus radiata*) and plum (*Prunus persica*), the ratio between the concentration of free polyamines and low molecular weight polyamine conjugates was more than one in juvenile plants but less than one in adult plants (Fraga et al. 2004). But when the concentration of total polyamines was measured, no difference was found between juvenile and mature plants of pine (*Pinus radiate* D. Don), while an increase during maturation was observed in plum (*Prunus persica* (L.) Batsch) (Fraga et al. 2004). In hazelnut (*Corylus avellana* L.), the concentration of free putrescine was relatively high in juvenile plants, but spermidine and spermine concentrations were higher in mature tissues (Rey et al. 1994). It may be concluded that polyamines might have specific roles in the juvenile to mature phases change in different woody plants.

Other studies have examined the involvement of epigenetic factors in the determination of the juvenile to mature phase change. Differences in DNA methylation between mature and juvenile or rejuvenated tissues were found in hazelnut (*Corylus avellana*) (Diaz-Sala et al. 1995), pine (*Pinus radiata*) (Fraga et al. 2002a,b), and acacia (*Acacia mangium*) (Baurens et al. 2004). It was shown that the degree of DNA methylation was higher in juvenile explants than in mature ones (Fraga et al. 2002a,b; Baurens et al. 2004). Phenotypic and biochemical differences that have been observed between juvenile and mature trees are most likely the cause of differential gene expression. During the years, attempts were made to characterize these expression differences in various woody plants. It was found that in ivy (*Hedera helix* L.), dihydroflavonol reductase (Murray and Hackett 1991) and a chlorophyll a/b binding-like protein (Woo et al. 1994) were upregulated in juvenile tissues, while a cell wall proline-rich protein was upregulated in mature tissues (Woo et al. 1994). Expression of a chlorophyll a/b-binding protein in larch (*Larix laricina* (Du Roi) K. Koch) was also higher in juvenile plants than that in mature ones (Hutchison et al. 1990). Transmembrane domains containing protein was found to be exclusively induced by auxin in juvenile cuttings of pine (*Pinus taeda* L.) (Busov et al. 2004), and a secreted cell wall protein rich in glycine and histidine was found to be preferentially expressed in micro-shoots derived from mature tissue of oak (*Quercus robur* L.) (Gil et al. 2003). Suppression of members of the poplar CENTRORADIALIS (CEN)/TERMINAL FLOWER 1 (TFL1) encoding genes that regulate shoot meristem identity accelerated the onset of mature characteristics of transgenic poplar (*Populus tremula* L. X *Populus alba* L.) trees (Mohamed et al. 2010). The previously mentioned findings might be unrelated or directly or indirectly related to the loss of AR formation in mature woody plants.

C. Molecular Mechanism of the Juvenile to Mature Phase Change in Herbaceous Plants

The molecular mechanisms underlying juvenile to mature phase change were more thoroughly studied in herbaceous model plants. Although the life cycle of woody plants is different from that of herbaceous plants, similar mechanisms underlying the phase change might exist in both types of plants. Initial indications of the molecular control of vegetative phase change (stage 2–3 mentioned previously) came from analysis of maize (*Zea mays* L.) mutants. The *Teopod1* and *Teopod2* mutations led to extended expression of juvenile traits, among which was an increase in the number of nodes bearing AR (Poethig 1988). In contrast, *Glossy15* mutation in maize caused an acceleration of adult cell characteristics in juvenile leaves (Evans et al. 1994; Moose and Sisco 1994). Later, the *Glossy15* gene was found to be an APETALA2-like encoding gene that controls leaf epidermal cell morphogenesis (Moose and Sisco 1996), which is regulated by miR172 (Lauter et al. 2005). *Corngrass1* (*Cg1*) is a maize mutant, in which juvenile traits are being retained during the adult reproductive phase (Chuck et al. 2007). The molecular basis of *Cg1* is overexpression of miR156, that is, concomitant with lower levels of miR172 (Chuck et al. 2007). It is interesting to note that the *Cg1* mutant exhibits AR formation in the upper nodes (Chuck et al. 2007), similar to the *Teopod1* and *Teopod2* mutants (Poethig 1988). Similarly, tomato (*Solanum lycopersicum* L. cv. 'Ailsa Craig') plants overexpressing miR156 exhibited extensive development of stem AR (Zhang et al. 2010). Recently, it has been discovered that miR156 and miR172 coordinate the juvenile to mature phase change in *Arabidopsis*, by a negative feedback loop involving the transcription factors SQUAMOSA PROMOTER BINDING PROTEIN LIKE SPL9 and SPL10 (Wu et al. 2009).

A study of miR156 and miR172 revealed similar changes in the expression of these genes in various woody plants, including *Acacia confuse* Merr., *Acacia colei* Maslin and L.A.J.Thomson, *Eucalyptus globules* Labill., *Hedera helix*, *Quercus acutissima* Carruth., and *Populus x Canadensis* Moench, during the juvenile to mature phase change (Wang et al. 2011). Overexpression of miR156 in transgenic poplar (*Populus x canadensis*) reduced the expression of miR156-targeted *SPL* genes and miR172 and prolonged the juvenile phase (Wang et al. 2011). Taken together, these data indicate that miR156 seems to be an evolutionarily conserved regulator of the vegetative phase change in both herbaceous and woody plants, which is involved in AR formation (Chuck et al. 2007; Zhang et al. 2010; Wang et al. 2011). MicroRNAs (miRNAs) form an abundant class of 21–24-nucleotide small RNAs that are common to diverse species of multicellular life. In plants, they are major regulators of development and other processes (reviewed by Zhang et al. 2006).

Another group of small regulatory RNAs is the group of trans-acting small (TAS) interfering RNAs (tasiRNAs). This is a special subgroup of 21–22-nucleotide endogenous siRNAs that are encoded by three loci, TAS1, TAS2, and TAS3 (Peragine et al. 2004; Vazquez et al. 2004; Adenot et al. 2006;

Fahlgren et al. 2006). It was shown that TAS3-derived tasiRNAs normally suppress the juvenile to mature phase transition via the downregulation of auxin response factor 3 (Fahlgren et al. 2006). In addition, *Arabidopsis* mutants with defects in RNA-dependent RNA polymerase 6 (RDR6) and suppressor of gene silencing 3 (SGS3) lack TAS3-derived tasiRNAs and exhibit accelerated transition from the juvenile to the mature phase during vegetative development (Peragine et al. 2004). Trans-acting small interfering RNAs were also found to regulate the juvenile to mature phase change in the moss *Physcomitrella patens* (Hedw.) Bruch and Schimp. (Talmor-Neiman et al. 2006). There are other effectors involved in the vegetative phase change in *Arabidopsis*; the *SERRATE* locus was shown to be required for the development of juvenile leaves while suppressing the development of mature leaves and inflorescences (Clarke et al. 1999). Loss of function mutations in *SQUINT* reduced the number of juvenile leaves but similar to the *SERRATE* mutation, had no effect on the flowering timing (Clarke et al. 1999; Berardini et al. 2001). Loss of function mutations of *HASTY* affected several aspects of plant development, among which was acceleration of the vegetative phase change (Telfer and Poethig 1998; Bollman et al. 2003). A double mutant of *SQUINT* and *HASTY* is shorter than each of the single mutants and exhibits stronger indications for mature traits on juvenile leaves, such as trichomes on the abaxial side of leaves 1 and 2, suggesting that the two genes act in parallel to regulate juvenile to mature phase change (Bollman et al. 2003). While *SQUINT* encodes for the *Arabidopsis* homolog of cyclophilin40, which is a protein found in association with the chaperon Hsp90 complex (Berardini et al. 2001), HASTY is the plant exportin-5 homolog (Bollman et al. 2003), which is involved in the translocation of the processed miRNA molecules from the nucleus (Park et al. 2005). Interestingly, the *HASTY* mutant has a shorter main root and more lateral roots than wt plants (Bollman et al. 2003).

Taken together, the juvenile to mature phase change in herbaceous plants, like in woody plants, is regulated by multiple factors, some unrelated and others directly or indirectly related to AR formation.

IV. Gene Context Favorable for AR Formation

A. Woody Plants

AR formation is a complex process, which is orchestrated by multiple genes, the products of which are organized in various regulatory layers and hierarchies. Knowledge from woody plants is starting to accumulate, while at present, the more detailed information comes from model plants. Several recent experiments applying DNA chip analysis were performed in order to define the molecular events underling AR formation in woody plants. A DNA chip analysis was performed in poplar (*Populus tremula X Populus alba*) to determine the early events after the excision of cuttings (Ramirez-Carvajal et al. 2009). Analysis of gene expression was carried out 0, 6, 24, and 48 h after excision. Since no auxin was necessary for AR formation in this species, changes in gene expression reflected the initial events leading to AR formation (after 10–14 days). It was found that genes of the ethylene biosynthesis pathway were upregulated, the expression of family members of the auxin response factors (ARF) or IAA families was either up- or downregulated, and cytokinin-regulated transcripts were decreased in accordance with their role in the process of AR formation. In addition, gene profiles in transgenic lines expressing a constitutively active form of the *Populus* type B cytokinin response regulator PtRR13, in which AR formation is decreased, were compared to those in a non-transformed line. The expression of only 11 genes was significantly different between the two plant populations during all time points tested. The PLEIOTROPIC DRUG RESISTANCE TRANSPORTER 9 (PDR9) encoding gene, which is involved in auxin efflux, and the gene encoding FIMBRIN-LIKE2 (FIM2), an actin-binding protein, were upregulated in the mutant. Other genes misregulated in the mutants were those encoding for CONTINUOUS VASCULAR RING1 and two APETALA2/ETHYLENE RESPONSE FACTOR genes (Ramirez-Carvajal et al. 2009).

DNA chip analysis was also performed in pine (*Pinus contorta* Douglas ex Loudon) hypocotyls induced to develop AR by auxin application. Profiles of gene expression were determined at several time points from day 0 till 33 days after root induction. Genes involved in cell division and cell wall weakening and genes related to water stress were upregulated during the initial stages. During later stages, genes involved in cell replication and stress were downregulated, suggesting maturation and functioning of the new roots (Brinker et al. 2004). During the phase of root meristem formation, the expression levels of genes involved in auxin transport and auxin responsive transcription increased, which is in agreement with the known role of auxin in this process.

In another study, gene expression analysis in rooting-competent pine (*Pinus radiata*) and chestnut (*Castanea sativa*) cuttings revealed the induction of *SCARECROW-LIKE* (*SCL*) genes 24 h after auxin application (Sanchez et al. 2007). These genes are members of the GRAS family of putative transcription factors responsible for the establishment of the root meristem (Helariutta et al. 2000; Sabatini et al. 2003; Long and Benfey 2006; Petricka and Benfey 2008). SCARECROW (*SCR*) is involved in the regulation of periclinal asymmetric cell division in the root meristem (Di Laurenzio et al. 1996) and is active in quiescent center cells (Sabatini et al. 2003). *SHORT ROOT* (*SHR*) is another member of the GRAS family that has been demonstrated to specify a single layer of endodermis via its movement from the central vascular tissue into neighboring cells, in which its interaction with SCR leads to its sequestration into the nucleus (Long and Benfey 2006; Petricka and Benfey 2008). In *Pinus radiata*, a *SHR*-like gene was found to be expressed in the cambial region of rooting competent cells of hypocotyl cuttings within the first 24 h after the initiation of rooting, before the activation of cell division and the organization of the AR meristem occurred (Sole et al. 2008). This suggests that *Pinus radiata SHR*-like transcription factor plays a role in AR formation. In *Arabidopsis*, *SHR* has recently been shown to regulate the indeterminate growth of primary, lateral, and AR by regulating cell division in the root apical meristem (Lucas et al. 2011).

B. Herbaceous Plants

Recently, Solexa deep sequencing analysis was performed to compare profiles of gene expression between stem node tissues without and with just-emerged brace roots, which are AR in maize (Li et al. 2011). The data demonstrated numerous differences in gene expression with prominent changes in auxin signaling pathways and cell wall metabolism.

Using mutants of rice and maize, it was found that the genes for *ARL1* (*adventitious rootless1*)/*CRL1* (*crown rootless1*) and *RTCS* (*rootless concerning crown and seminal root*), the ortholog of *ARL1/CRL1* from maize, which all encode conserved (lateral organ boundary) LOB-domain transcription factors, are involved in AR initiation and development in response to auxin (Inukai et al. 2005; Liu et al. 2005; Taramino et al. 2007; Coudert et al. 2010). In rice, two additional *crown rootless* mutants were found, *crl4* (*crown rootless4*) and *OsGnom1*, which are mutated in the gene ortholog of *GNOM1* in *Arabidopsis* that regulates polar auxin transport (Kitomi et al. 2008; Liu et al. 2009). AR formation studies in *Arabidopsis* confirmed the involvement of multiple pathways, with that of auxin signaling being the major one. Several *Arabidopsis* ecotypes were analyzed for differences in AR formation in hypocotyl sections following auxin treatment. It was found that rooting was controlled by several genes acting independently in additive-dominance fashion (King and Stimart 1998). Indeed, when another attempt was made to study the molecular basis for AR formation in *Arabidopsis*, several temperature-sensitive mutants were isolated, in which root formation was inhibited in various stages (Konishi and Sugiyama 2003; Sugiyama 2003). Three mutants (impaired in root redifferentiation), *rrd1,2,4,* had defects in cell proliferation (Sugiyama 2003). Five mutants (*root initiation defective*), *rid1–5,* failed to initiate root primordia. In another mutant (*root primordium defective*), *rpd1,* the development of root primordia was inhibited, and in three other mutants (*root growth defective*), *rgd1–3,* root growth after the establishment of the root apical meristem was arrested (Konishi and Sugiyama 2003). Interestingly, the *rid5* mutant required higher auxin concentrations, than wt plants, for rooting at the restrictive temperature. It was found that *rid5* resembles *mor1,* a temperature-sensitive mutant in a microtubule-associated protein (Whittington et al. 2001). This suggests an involvement of microtubules in auxin signaling. The gene *rpd1,* which is novel and specific to the plant kingdom, was found to function in maintaining cell proliferation (Konishi and Sugiyama 2006). Although the preceding data indicate that AR formation is regulated by multiple genes, there are studies demonstrating that single genes are capable of inducing AR formation. One of such genes is *superroot,* which is able to induce excessive lateral and AR formation (Boerjan et al. 1995). *Superroot 1* and *2* were found to encode for enzymes involved in auxin homeostasis (Delarue et al. 1998; Seo et al. 1998), the mutation of which led to IAA overproduction. When these mutants were compared to *ago1,* a mutant in *ARGONAUTE1* that barely forms AR, it was found that the loss of rooting capability was correlated with induction of *auxin response factor 17* (*arf17*) expression and reduction in the expression of *GH3* genes that are predicted to be involved in auxin conjugation with amino acids (Sorin et al. 2005, 2006). Of note, ARGONAUTE1 is a central component of the silencing machinery (Zhang et al. 2006). Later, when the involvement of the silencing machinery in AR formation was studied, it was confirmed that *arf17,* a target of miR160, is a negative regulator, and *arf6* and *arf8,* targets of miR167, are positive regulators of AR formation (Gutierrez et al. 2009).

In conclusion, the genetic studies in woody and herbaceous plants revealed parts of the genetic networks underlying the evolutionary conserved role of auxin, its metabolism, signaling pathways, and cross talk with other hormones in the induction of AR formation.

V. Cross Talk of Phytohormones during AR Formation

A. Auxin/Cytokinins

The inhibitory effect of cytokinins on AR formation is known for many years (Humphries 1960; Geiss et al. 2009). Roots are the major source of cytokinin synthesis, and according to one hypothesis, removal of roots induces AR formation due to the elimination of root apical dominance. Zeatin riboside was found to be a major component suppressing AR formation in a crude xylem sap collected from roots of squash (*Cucurbita maxima* Duchesne x *C. moschata* Duchesne) (Kuroha et al. 2002). This suggests that cytokinins might be involved in the root apical dominance phenomenon. Transgenic *Arabidopsis* plants overexpressing family members of the cytokinin oxidase/dehydrogenase (AtCKX), and as a result contained lower levels of cytokinins, had an increased number of AR (Werner et al. 2003). Interestingly, an *Arabidopsis* mutant in one of the cytokinin receptors AHK4 (named *wol-3*) and a triple mutant, *ahk2/ahk3/ahk4*, developed aberrant vascular systems in the primary and lateral roots but a normal vascular system in AR (Kuroha et al. 2006). Therefore, it has been suggested that cytokinins are necessary for the development of auxin transporting vascular system in the root, but not in AR. The balance between levels of auxin and cytokinins has been found to play an important role in root development. For example, in the root meristem of *Arabidopsis*, cytokinins promote cell differentiation by inhibiting both auxin signaling and transport, while auxin contributes to the maintenance of the root meristem by promoting cell division (Moubayidin et al. 2009; Werner and Schmulling 2009). In founder cells of lateral roots, cytokinins perturb the expression of PIN encoding genes, thus preventing the accumulation of auxin that is required for the formation of lateral root primordia (Laplaze et al. 2007).

B. Auxin/Ethylene

Since the first report on the stimulation of AR formation by ethylene (Zimmerman and Hitchcock 1933), numerous studies have been conducted to study the effect of this hormone on rooting. It was reported that a large number of plant species exhibited increased AR formation in response to ethylene (Mudge 1988). However, reports that not all species responded positively to ethylene and the sometimes contradictory results observed in the same species led Mudge (1988) to suggest that ethylene might not be as important as auxin in inducing AR formation. Additional evidence for the stimulative role of ethylene in AR formation came from studies demonstrating that inhibitors of ethylene biosynthesis and action inhibited rooting of various species (Wang and Pan 2006).

The observations that auxin stimulates ethylene production have led to a large number of studies, which initially examined whether there is a correlation between the level of auxin-induced ethylene production and the rate of AR formation. While such correlation was observed in some studies (Riov and Yang 1989), others failed to detect it (Mullins 1972; Geneve and Hauser 1982). Support for the role of auxin-induced ethylene production was provided by studies demonstrating that inhibitors of ethylene biosynthesis and action reduced the stimulative effect of auxin on rooting (Liu and Reid 1992).

The mechanism by which ethylene stimulates AR formation and the cross talk between auxin and ethylene in this process have been studied by several research groups. Since ethylene had no effect on auxin transport and metabolism nor on auxin levels in the rooting zone, Liu and Reid (1992) raised the possibility that ethylene affects AR formation by increasing the sensitivity of the tissues to auxin. Later studies adopted a genetic approach. It was shown that applied auxin was much less effective in inducing AR formation in ethylene-insensitive mutants of tomato (*Lycopersicon esculentum*) and petunia (*Petunia* x *hybrid* hort. ex Vilm.) than in wild type plants (Clark et al. 1999). The *Never ripe* tomato mutant had also less aboveground AR roots, a phenomenon which was also observed by others (Kim et al. 2008). Based on these observations, Clark et al. (1999) suggested that ethylene has a significant role in AR formation and that the stimulative effect of auxin is affected by ethylene responsiveness. Decreased AR formation in the tomato *Never ripe* mutant with or without auxin application was observed by others (Negi et al. 2010). Conversely, the *epinastic* mutant, which exhibits elevated ethylene levels and constitutive ethylene signaling, produced a higher number of AR. The observations that ethylene inhibited auxin transport in the hypocotyl cuttings and that it had no effect on IAA content in the cutting base suggest that the stimulative effect of ethylene might have resulted from increased sensitivity to auxin.

The role of ethylene in AR formation and the cross talk between auxin and ethylene have also been studied in plants subjected to flooding, which is known to enhance AR formation. It was shown that both auxin and ethylene acted to stimulate AR formation in flooded *Rumex palustris* plants (Visser et al. 1996). Since endogenous auxin concentration did not change during flooding-induced rooting, the authors suggested that the elevated ethylene production in flooded plants, which precedes the rooting response, increases the sensitivity of the tissues to auxin. Vidoz and colleagues observed that inhibition of ethylene biosynthesis or auxin transport resulted in a reduction in AR formation in flooded tomato plants (Vidoz et al. 2010). They also observed that ethylene stimulated auxin transport. Based on these observations, the authors proposed the following scenario to describe the cross talk between auxin and ethylene during flooding-induced increased AR formation in tomato: ethylene entrapment by water acts as the first warning signal of waterlogging, stimulating auxin transport. In this way, auxin accumulates in the stem and triggers additional ethylene production, which further stimulates the transport of auxin to the flooded parts of the plant. The resulted auxin accumulation in the base of plant then induces AR formation.

In conclusion, although there are sometimes variable observations regarding the role of ethylene in AR formation, most data indicate that ethylene, together with auxin, plays an important role in this processes. The nature of the cross talk between auxin and ethylene in AR formation has been examined only in a very few studies, which yielded different results. Clearly, there is a need for additional research to study this issue.

VI. Chemical Enhancers for AR Formation

A large number of chemicals, some of which are presented in Table 11.1, have been shown to stimulate AR formation.

In the following, we focus on nitric oxide (NO), which participates in the auxin signaling pathway and has gained an increasing interest in the last decade.

A. NO Metabolism and Role in Plants

In animal systems, NO is synthesized predominantly by the enzyme NO synthase (NOS) (Bredt and Snyder 1990). NOS converts L-arginine into L-citrulline in an NADPH-dependent reaction that releases one molecule of NO for each molecule of L-arginine. Assays of arginine to citrulline conversion and compounds that inhibit mammalian NOS have been used to demonstrate that NO synthesis by a NOS-like enzyme occurs in plants (Durner et al. 1998; Foissner et al. 2000; Moreau et al. 2010). Nevertheless, a canonical plant NOS has not yet been found (Moreau et al. 2010), except for the green alga *Ostreococcus tauri* C. Courties and M.-J. Chrétiennot-Dinet (Foresi et al. 2010). A putative NOS was initially characterized in plants (Guo et al. 2003), but several groups have later shown that this protein is not a genuine NOS but might be involved in NO accumulation, and thus, its name was changed to NO-associated protein (NOA1) (Crawford et al. 2006; Zemojtel et al. 2006; Moreau et al. 2010). NO is produced also by nitrate reductase (NR), which reduces nitrate to nitrite,

TABLE 11.1 Chemicals that Induce AR Formation

Chemical	Activity	Plant[a]	References
SNP	NO donor	Cucumber	Pagnussat et al. (2002)
SNAP	NO donor	Cucumber	Pagnussat et al. (2002)
8-Br-cGMP	Permeable cGMP	Cucumber	Pagnussat et al. (2003)
Sildenafil citrate	Phosphodiesterase inhibitor	Cucumber	Pagnussat et al. (2003)
Hematin/hemin	CO donor	Cucumber	Xuan et al. (2008)
Hydrogen peroxide	Active oxygen species	Mung bean, olive	Li et al. (2009); Sebastiani and Tognetti (2004)
Sodium hydrosulfide	Hydrogen sulfide donor	Soybean, sweet potato	Zhang et al. (2009)
Urea derivatives	Cytokinin-like	Pine	Ricci et al. (2008)
Sirtinol	Activation of auxin responsive genes	*Arabidopsis*	Zhao et al. (2003)
Auxin conjugates	Slow release of auxin		Riov (1993)
Paclobutrazol	GA biosynthesis inhibitor, enhancing IBA effect and sink capacity	Mung bean	Weisman and Riov (1994)
ACC/ethephon	Ethylene	Mung bean	Riov and Yang (1989)

[a] Some compounds were shown to affect various other plants not mentioned here.

but can further reduce nitrite to NO (Yamasaki et al. 1999; Yamasaki and Sakihama 2000; Rockel et al. 2002). Indeed, a triple mutant of the two *Arabidopsis* NR encoding genes, *Nia1* and *Nia2,* and of *NOA1, nia1/nia2/noa1,* exhibits no detectable NO production (Lozano-Juste and Leon 2010). Nitrite reduction to NO by nitrate reductase is not efficient, being only about 1% of its activity and requires low oxygen environment and higher nitrite than nitrate levels (Rockel et al. 2002). Reduction of nitrite to NO was exclusively demonstrated in root cells by other plant enzymes (Gupta et al. 2005; Stohr and Stremlau 2006). In addition, nonenzymatic reduction of nitrite to NO was detected (Bethke et al. 2004). This reaction requires acidic pH, and thus, it is restricted to the apoplasts of seeds and roots (Bethke et al. 2004; Stohr and Stremlau 2006). NO is unstable and reacts with O_2 to form both nitrite and nitrate (Wilson et al. 2008). It also reacts with reactive oxygen species (ROS) that modulate its levels and signaling (Wilson et al. 2008).

NO plays a role in multiple plant processes, such as germination, leaf expansion, lateral root development, flowering, stomatal closure, cell death, and defense against biotic and abiotic stresses (Besson-Bard et al. 2008; Wilson et al. 2008; Leitner et al. 2009; Moreau et al. 2010), and AR formation (Pagnussat et al. 2003; 2004; Lanteri et al. 2006, 2008, 2009).

B. NO Donors as AR Enhancers

NO donors that are mostly used in vitro are SNP (sodium nitroprusside, $Na_2[Fe(CN)_5NO]$), SNAP (*S*-nitroso *N*-acetyl penicillamine), and GSNO (*S*-nitrosoglutathione) (Floryszak-Wieczorek et al. 2006). In biological systems, NO release from its donors may occur both enzymatically and nonenzymatically by reducing agents or light (Wang et al. 2002). Importantly, while SNP releases mainly the cation NO^+, SNAP and GSNO release predominantly NO in the redox form of a free radical (Wang et al. 2002). Since the reduction and subsequent decomposition of SNP is accompanied by the release of cyanide, pronounced cellular toxicity is often observed (Wang et al. 2002; Floryszak-Wieczorek et al. 2006). Differences in the rate of NO release by light from the three donors and their half-life have been demonstrated. While the peak of NO release in SNP solution occurred 2 h after light induction, NO release in SNAP solution reached a maximum immediately after light induction. In GSNO solution, maximum NO release was recorded after 5 h. The half-life of the three donors was 12, 3, and 7 h for SNP, SNAP, and GSNO, respectively (Floryszak-Wieczorek et al. 2006). In 1997, it was reported that NO-releasing agents (SNP and others) promoted maize root elongation (Gouvêa et al. 1997). Five years later, it was shown that SNP and SNAP promoted AR formation in cucumber explants (Pagnussat et al. 2002). The effect of these NO donors on AR formation was comparable and synergistic to that of IAA. The specific NO scavenger cPTIO delayed the differentiation of AR promoted by IAA, suggesting that NO acts as a messenger in the auxin signaling pathway controlling AR formation. It was also found that IAA induced a transient increase in NO in the plants. It was further found that while permeable cGMP derivative induced AR formation, inhibitors of guanylate cyclase and phosphodiesterase inhibited the induction of AR formation by IAA and SNP, suggesting that cGMP is involved in this pathway (Pagnussat et al. 2003). Other components that have been suggested to be involved in the auxin/NO pathway promoting AR formation are mitogen-activated protein kinase (MAPK) (Pagnussat et al. 2004), calcium and calcium-dependent protein kinase (Lanteri et al. 2006), and phosphatidic acid and phospholipase D (Lanteri et al. 2008). Other plants in which NO donors were reported to promote AR development are rice (*Oryza sativa*) (Xiong et al. 2009), sunflower (*Helianthus annuus* L. cv. Morden) (Yadav et al. 2010), lavender (*Lavandula dentata* L.), cypress (*Chamaecyparis* spp.) (Lanteri et al. 2009), *Eucalyptus grandis*, and *Eucalyptus brachyphylla* C.A. Gardner (Abu-Abied et al. 2012). Of note, hydrogen peroxide (H_2O_2) was shown to

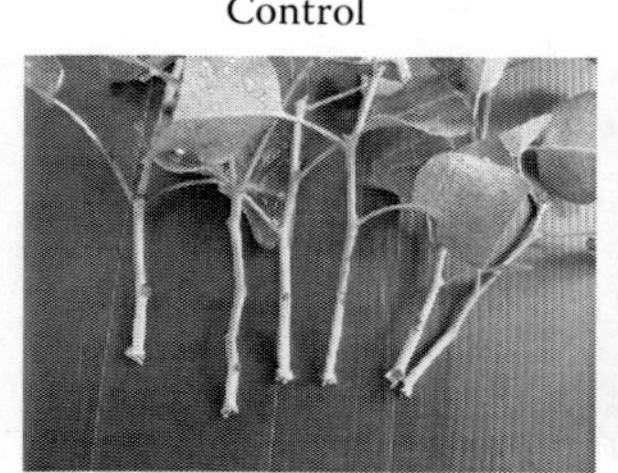

FIGURE 11.2 Effect of hydrogen peroxide on AR formation in *Eucalyptus gomphocephala* (cultivar clone). Cuttings were treated with 5000 ppm K-IBA with or without 1% H_2O_2 for 30 s and then rooted in vermiculite:crushed polystyrene (1:1) for 40 days. (From I. Mordehaev, M. Abu-Abied, J. Riov, and E. Sadot, unpublished data.)

promote NO synthesis that induced lateral root development (Wang et al. 2010) and AR formation in *Eucalyptus gomphocephala* (cultivar clone) (Figure 11.2).

VII. Conclusions

Taken together, the data suggest that while auxin is the major factor inducing AR formation, there are still open questions regarding its specific transducers, as well as the involvement of other cofactors which are relevant to AR formation. In addition, while it is clear that juvenile cuttings root more readily in many plant species, it is still to be verified which of the signaling pathways that dictate juvenility are relevant for AR formation and what are the limiting or inhibitory factors that prevent rooting in mature cuttings.

References

Abu-Abied M, Szwerdszarf D, Mordehaev I, Levy A, Rogovoy (Stelmakh) O, Belausov E, Yaniv Y, Uliel S, Katzenellenbogen M, Riov J, Ophir R, and Sadot E. 2012. Microarray analysis revealed upregulation of nitrate reductase in juvenile cuttings of *Eucalyptus grandis* which correlated with increased nitric oxide production and adventitious root formation. *Plant J* 71(5):787–799.

Adenot X, Elmayan T, Lauressergues D et al. 2006. DRB4-dependent TAS3 trans-acting siRNAs control leaf morphology through AGO7. *Curr Biol* 16:927–932.

Amo-Marco JB, Vidal N, Vieitez AM, Ballester A. 1993. Polypeptide markers differentiating juvenile and adult tissues in chestnut. *J Plant Physiol* 142:117–119.

Ballester A, San-Jose MC, Vidal N, Fernandez-Lorenzo JL, Vieitez AM. 1999. Anatomical and biochemical events during in vitro rooting of microcuttings from juvenile and mature phases of chestnut. *Ann Bot* 83:619–629.

Baurens FC, Nicolleau J, Legavre T, Verdeil JL, Monteuuis O. 2004. Genomic DNA methylation of juvenile and mature *Acacia mangium* micropropagated in vitro with reference to leaf morphology as a phase change marker. *Tree Physiol* 24:401–407.

Belehu T, Hammes PS, Robbertse PJ. 2004. The origin and structure of adventitious roots in sweet potato (*Ipomoea batatas*). *Aus J Bot* 52:551–558.

Benkova E, Bielach A. 2010. Lateral root organogenesis - from cell to organ. *Curr Opin Plant Biol* 13:677–683.

Berardini TZ, Bollman K, Sun H, Poethig RS. 2001. Regulation of vegetative phase change in *Arabidopsis thaliana* by cyclophilin 40. *Science* 291:2405–2407.

Besford RT, Hand P, Peppitt SD, Richardson CM, Thomas B. 1996. Phase change in *Prunus avium*: Differences between juvenile and mature shoots identified by 2-dimensional protein separation and in vitro translation of mRNA. *J Plant Physiol* 147:534–538.

Besson-Bard A, Pugin A, Wendehenne D. 2008. New insights into nitric oxide signaling in plants. *Annu Rev Plant Biol* 59:21–39.

Bethke PC, Badger MR, Jones RL. 2004. Apoplastic synthesis of nitric oxide by plant tissues. *Plant Cell* 16:332–341.

Blakesley D, Weston GD, Hall JF. 1991. The role of endogenous auxin in root initiation Part I. Evidence from studies on auxin application, and analysis of endogenous levels. *Plant Growth Regul* 10:341–353.

Boerjan W, Cervera MT, Delarue M et al. 1995. *Superroot*, a recessive mutation in *Arabidopsis*, confers auxin overproduction. *Plant Cell* 7:1405–1419.

Bollman KM, Aukerman MJ, Park MY et al. 2003. HASTY, the *Arabidopsis* ortholog of exportin 5/MSN5, regulates phase change and morphogenesis. *Development* 130:1493–1504.

Bon MC. 1988. J 16: An apex protein associated with juvenility of *Sequoiadendron giganteum*. *Tree Physiol* 4:381–487.

Bredt DS, Snyder SH. 1990. Isolation of nitric oxide synthetase, a calmodulin-requiring enzyme. *Proc Natl Acad Sci USA* 87:682–685.

Brinker M, van Zyl L, Liu W et al. 2004. Microarray analyses of gene expression during adventitious root development in *Pinus contorta*. *Plant Physiol* 135:1526–1539.

Busov VB, Johannes E, Whetten RW et al. 2004. An auxin-inducible gene from loblolly pine (*Pinus taeda* L.) is differentially expressed in mature and juvenile-phase shoots and encodes a putative transmembrane protein. *Planta* 218:916–927.

Casimiro I, Marchant A, Bhalerao RP et al. 2001. Auxin transport promotes *Arabidopsis* lateral root initiation. *Plant Cell* 13:843–852.

Casson SA, Lindsey K. 2003. Genes and signalling in root development. *New Phytol* 158:11–38.

Chuck G, Cigan AM, Saeteurn K, Hake S. 2007. The heterochronic maize mutant *Corngrass1* results from overexpression of a tandem microRNA. *Nat Genet* 39:544–549.

Clark DG, Gubrium EK, Barrett JE, Nell TA, Klee HJ. 1999. Root formation in ethylene-insensitive plants. *Plant Physiol* 121:53–60.

Clarke JH, Tack D, Findlay K, Van Montagu M, Van Lijsebettens M. 1999. The SERRATE locus controls the formation of the early juvenile leaves and phase length in *Arabidopsis. Plant J* 20:493–501.

Coudert Y, Perin C, Courtois B, Khong NG, Gantet P. 2010. Genetic control of root development in rice, the model cereal. *Trends Plant Sci* 15:219–226.

Crawford NM, Galli M, Tischner R et al. 2006. Response to Zemojtel et al: Plant nitric oxide synthase: Back to square one. *Trends Plant Sci* 11:526–527.

Day ME, Greenwood MS, Diaz-Sala C. 2002. Age- and size-related trends in woody plant shoot development: Regulatory pathways and evidence for genetic control. *Tree Physiol* 22:507–513.

De Klerk GJ, Van der Krieken W, de Jong JC. 1999. The formation of adventitious roots: New concepts, new possibilities. *In Vitro Cell Dev Biol Plant* 35:189–199.

De Smet I, Tetsumura T, De Rybel B et al. 2007. Auxin-dependent regulation of lateral root positioning in the basal meristem of *Arabidopsis. Development* 134:681–690.

Dech JP. 2009. Adventitious root production in woody plants of coastal dune ecosystems: Selective pressures, adaptive value and potential applications. In: *Adventitious Root Formation of Forest Trees and Horticultural Plants—From Genes to Applications*, ed. K Niemi, C Scagel, pp. 385–408. Trivandrum, India: Research Signpost.

Delarue M, Prinsen E, Onckelen HV, Caboche M, Bellini C. 1998. *Sur2* mutations of *Arabidopsis thaliana* define a new locus involved in the control of auxin homeostasis. *Plant J* 14:603–611.

Dhawan AK, Paton DM, Willing RR. 1979. Occurrence and bioassay responses of G. *Planta* 146:419–422.

Di Laurenzio L, Wysocka-Diller J, Malamy JE et al. 1996. The SCARECROW gene regulates an asymmetric cell division that is essential for generating the radial organization of the *Arabidopsis* root. *Cell* 86:423–433.

Diaz-Sala C, Rey M, Boronat A, Besford R, Rodriguez R. 1995. Variations in the DNA methylation and polypeptide patterns of adult hazel (*Corylus avellana* L.) associated with sequential in vitro subcultures. *Plant Cell Rep* 21:9–12.

Dubrovsky JG, Sauer M, Napsucialy-Mendivil S et al. 2008. Auxin acts as a local morphogenetic trigger to specify lateral root founder cells. *Proc Natl Acad Sci USA* 105:8790–8794.

Durner J, Wendehenne D, Klessig DF. 1998. Defense gene induction in tobacco by nitric oxide, cyclic GMP, and cyclic ADP-ribose. *Proc Natl Acad Sci USA* 95:10328–10333.

Evans MM, Passas HJ, Poethig RS. 1994. Heterochronic effects of *glossy15* mutations on epidermal cell identity in maize. *Development* 120:1971–1981.

Fahlgren N, Montgomery TA, Howell MD et al. 2006. Regulation of AUXIN RESPONSE FACTOR3 by TAS3 ta-siRNA affects developmental timing and patterning in *Arabidopsis. Curr Biol* 16:939–944.

Fahn A. 1990. *Plant Anatomy*, 4th edn. New York: Pergamon Press.

Firon N, Villordon A, LaBonte D et al. 2009. The sweet potato-botany and physiology: Storage root formation and development. In: *The Sweetpotato*, eds. G Loebenstein, G Thottappilly, pp. 13–26. Dordrecht, The Netherlands: Springer.

Floryszak-Wieczorek J, Milczarek G, Arasimowicz M, Ciszewski A. 2006. Do nitric oxide donors mimic endogenous NO-related response in plants? *Planta* 224:1363–1372.

Foissner I, Wendehenne D, Langebartels C, Durner J. 2000. In vivo imaging of an elicitor-induced nitric oxide burst in tobacco. *Plant J* 23:817–824.

Foresi N, Correa-Aragunde N, Parisi G et al. 2010. Characterization of a nitric oxide synthase from the plant kingdom: NO generation from the green alga *Ostreococcus tauri* is light irradiance and growth phase dependent. *Plant Cell* 22:3816–3830.

Fraga MF, Berdasco M, Diego LB, Rodriguez R, Canal MJ. 2004. Changes in polyamine concentration associated with aging in *Pinus radiata* and *Prunus persica. Tree Physiol* 24:1221–1226.

Fraga MF, Canal MJ, Rodriguez R. 2002a. Phase-change related epigenetic and physiological changes in *Pinus radiata* D. Don. *Planta* 215:672–678.

Fraga MF, Rodriguez R, Canal MJ. 2002b. Genomic DNA methylation-demethylation during aging and reinvigoration of *Pinus radiata. Tree Physiol* 22:813–816.

Geiss G, Gutierrez L, Bellini C. 2009. Adventitious root formation: New insights and perspectives. *Annu Plant Rev* 37:127–156.

Geneve RL, Hauser CW. 1982. The effect of IAA, IBA, NAA and 2,4-D on root formation and ethylene evolution in *Vigna radiata* cuttings. *J Am Soc Hort Sci* 202–205.

Gil B, Pastoriza E, Ballester A, Sanchez C. 2003. Isolation and characterization of a cDNA from *Quercus robur* differentially expressed in juvenile-like and mature shoots. *Tree Physiol* 23:633–640.

Gouvêa CMCP, Souza JF, Magalhães ACN, Martins IS. 1997. NO-releasing substances that induce growth elongation in maize root segments. *Plant Growth Reg* 21:183–187.

Greenwood MS. 1987. Rejuvenation of forest trees. *Plant Growth Reg* 6:1–12.

Greenwood MS, Cui X, Xu F. 2001. Response to auxin changes during maturation-related loss of adventitious rooting competence in loblolly pine (*Pinus taeda*) stem cuttings. *Physiol Plant* 111:373–380.

Guo FQ, Okamoto M, Crawford NM. 2003. Identification of a plant nitric oxide synthase gene involved in hormonal signaling. *Science* 302:100–103.

Gupta KJ, Stoimenova M, Kaiser WM. 2005. In higher plants, only root mitochondria, but not leaf mitochondria reduce nitrite to NO, in vitro and in situ. *J Exp Bot* 56:2601–2609.

Gutierrez L, Bussell JD, Pacurar DI et al. 2009. Phenotypic plasticity of adventitious rooting in *Arabidopsis* is controlled by complex regulation of AUXIN RESPONSE FACTOR transcripts and microRNA abundance. *Plant Cell* 21:3119–3132.

Hackett WP. 1985. Juvenility, maturation and rejuvenation in woody plants. *Hort Rev* 7:109–155.

Helariutta Y, Fukaki H, Wysocka-Diller J et al. 2000. The SHORT-ROOT gene controls radial patterning of the *Arabidopsis* root through radial signaling. *Cell* 101:555–567.

Huang LC, Lius S, Huang BL et al. 1992. Rejuvenation of *Sequoia sempervirens* by repeated grafting of shoot tips onto juvenile rootstocks in vitro: Model for phase reversal of trees. *Plant Physiol* 98:166–173.

Humphries EC. 1960. Inhibition of root development on petioles and hypocotyls of dwarf bean (*Phaseolus vulgaris*) by kinetin. *Physiol Plant* 13:659–663.

Hutchison KW, Sherman CD, Weber J et al. 1990. Maturation in larch: II. Effects of age on photosynthesis and gene expression in developing foliage. *Plant Physiol* 94:1308–1315.

Inukai Y, Sakamoto T, Ueguchi-Tanaka M et al. 2005. *Crown rootless1*, which is essential for crown root formation in rice, is a target of an AUXIN RESPONSE FACTOR in auxin signaling. *Plant Cell* 17:1387–1396.

Kim HJ, Lynch JP, Brown KM. 2008. Ethylene insensitivity impedes a subset of responses to phosphorus deficiency in tomato and petunia. *Plant Cell Environ* 31:1744–1755.

King JJ, Stimart DP. 1998. Genetic analysis of variation for auxin-induced adventitious root formation among eighteen ecotypes of *Arabidopsis thaliana* L. Heynh. *J Hered* 89:481–487.

Kitomi Y, Ogawa A, Kitano H, Inukai I. 2008. *CRL4* regulates crown root formation through auxin transport in rice. *Plant Root* 2:19–28.

Konishi M, Sugiyama M. 2003. Genetic analysis of adventitious root formation with a novel series of temperature-sensitive mutants of *Arabidopsis thaliana*. *Development* 130:5637–5647.

Konishi M, Sugiyama M. 2006. A novel plant-specific family gene, ROOT PRIMORDIUM DEFECTIVE 1, is required for the maintenance of active cell proliferation. *Plant Physiol* 140:591–602.

Kuris A, Altman A, Putievsky E. 1981. Vegetative propagation of spice-plants: Root formation in oregano stem cuttings. *Sci Hort* 14:151–156.

Kuroha T, Kato H, Asami T et al. 2002. A trans-zeatin riboside in root xylem sap negatively regulates adventitious root formation on cucumber hypocotyls. *J Exp Bot* 53:2193–2200.

Kuroha T, Ueguchi C, Sakakibara H, Satoh S. 2006. Cytokinin receptors are required for normal development of auxin-transporting vascular tissues in the hypocotyl but not in adventitious roots. *Plant Cell Physiol* 47:234–243.

Lanteri ML, Laxalt AM, Lamattina L. 2008. Nitric oxide triggers phosphatidic acid accumulation via phospholipase D during auxin-induced adventitious root formation in cucumber. *Plant Physiol* 147:188–198.

Lanteri ML, Pagnussat GC, Lamattina L. 2006. Calcium and calcium-dependent protein kinases are involved in nitric oxide- and auxin-induced adventitious root formation in cucumber. *J Exp Bot* 57:1341–1351.

Lanteri ML, Pagnussat GC, Laxalt AM, Lamattina L. 2009. Nitric oxide is down stream of auxin and is required for inducing adventitious root formation in herbaceous and woody plants. In: *Adventitious Root Formation o Forest Trees and Horticultural Plants—From Genes to Applications*, eds. K Niemi, C Scagel, pp. 31–50. Trivandrum, India: Research Signpost.

Laplaze L, Benkova E, Casimiro I et al. 2007. Cytokinins act directly on lateral root founder cells to inhibit root initiation. *Plant Cell* 19:3889–3900.

Laskowski M, Grieneisen VA, Hofhuis H et al. 2008. Root system architecture from coupling cell shape to auxin transport. *PLoS Biol* 6:e307.

Lauter N, Kampani A, Carlson S, Goebel M, Moose SP. 2005. microRNA172 down-regulates *glossy15* to promote vegetative phase change in maize. *Proc Natl Acad Sci USA* 102:9412–9417.

Leitner M, Vandelle E, Gaupels F, Bellin D, Delledonne M. 2009. NO signals in the haze: Nitric oxide signalling in plant defence. *Curr Opin Plant Biol* 12:451–458.

Li YJ, Fu YR, Huang JG, Wu CA, Zheng CC. 2011. Transcript profiling during the early development of the maize brace root via Solexa sequencing. *FEBS J* 278:156–166.

Li SW, Xue LG, Xu SJ, Feng HY, An LZ. 2009. Hydrogen peroxide acts as a signal molecule in the adventitious root formation of mung bean seedlings. *Envron Exp Bot* 65:63–71.

Liu JH, Reid DM. 1992. Auxin and ethylene stimulated adventitious rooting in relation to tissue sensitivity to auxin and ethylene production in sunflower hypocotyls. *J Exp Bot* 43:1191–1198.

Liu S, Wang J, Wang L et al. 2009. Adventitious root formation in rice requires OsGNOM1 and is mediated by the OsPINs family. *Cell Res* 19:1110–1119.

Liu H, Wang S, Yu X et al. 2005. ARL1, a LOB-domain protein required for adventitious root formation in rice. *Plant J* 43:47–56.

Long TA, Benfey PN. 2006. Transcription factors and hormones: New insights into plant cell differentiation. *Curr Opin Cell Biol* 18:710–714.

Lozano-Juste J, Leon J. 2010. Enhanced abscisic acid-mediated responses in *nia1nia2noa1-2* triple mutant impaired in NIA/NR- and AtNOA1-dependent nitric oxide biosynthesis in *Arabidopsis*. *Plant Physiol* 152:891–903.

Lucas M, Swarup R, Paponov IA et al. 2011. Short-Root regulates primary, lateral, and adventitious root development in *Arabidopsis*. *Plant Physiol* 155:384–398.

Ludwig-Müller J. 2009. Molecular basis for the role of auxin in adventitious rooting. In: *Adventitious Root Formation of Forest Trees and Horticultural Plants - From Genes to Applications*, eds. K Niemi, C Scagel, pp. 1–29. Trivandrum, India: Research Signpost.

Malamy JE, Benfey PN. 1997. Organization and cell differentiation in lateral roots of *Arabidopsis thaliana*. *Development* 124:33–44.

McGowran E, Douglas GC, Parkinson M. 1998. Morphological and physiological markers of juvenility and maturity in shoot cultures of oak (*Quercus robur* and *Q. petraea*). *Tree Physiol* 18:251–257.

Mohamed R, Wang CT, Ma C et al. 2010. *Populus CEN/TFL1* regulates first onset of flowering, axillary meristem identity and dormancy release in *Populus*. *Plant J* 62:674–688.

Moose SP, Sisco PH. 1994. Glossy15 controls the epidermal juvenile-to-adult phase transition in maize. *Plant Cell* 6:1343–1355.

Moose SP, Sisco PH. 1996. Glossy15, an *APETALA2*-like gene from maize that regulates leaf epidermal cell identity. *Genes Dev* 10:3018–3027.

Moreau M, Lindermayer C, Durner J, Klessig DF. 2010. NO synthesis and signaling in plants—Where do we stand? *Physiol Plant* 138:372–383.

Moubayidin L, Di Mambro R, Sabatini S. 2009. Cytokinin-auxin crosstalk. *Trends Plant Sci* 14:557–562.

Mudge KW. 1988. Effect of ethylene on rooting. In: *Adventitious Root Formation in Cuttings*, eds. TD Davis, BE Haissig, N Sankhla, pp. 150–161. Portland, OR: Dioscorides Press.

Mullins JW. 1972. Auxin and ethylene in adventitious root formation in *Phaseolus aureus* (Roxb.). In: *Plant Growth Substances 1970*, ed. DJ Carr, pp. 526–533. Berlin, Germany: Springer.

Murray JR, Hackett WP. 1991. Dihydroflavonol reductase activity in relation to differential anthocyanin accumulation in juvenile and mature phase *Hedera helix* L. *Plant Physiol* 97:343–351.

Negi S, Sukumar P, Liu X, Cohen JD, Muday GK. 2010. Genetic dissection of the role of ethylene in regulating auxin-dependent lateral and adventitious root formation in tomato. *Plant J* 61:3–15.

Pagnussat GC, Lanteri ML, Lamattina L. 2003. Nitric oxide and cyclic GMP are messengers in the indole acetic acid-induced adventitious rooting process. *Plant Physiol* 132:1241–1248.

Pagnussat GC, Lanteri ML, Lombardo MC, Lamattina L. 2004. Nitric oxide mediates the indole acetic acid induction activation of a mitogen-activated protein kinase cascade involved in adventitious root development. *Plant Physiol* 135:279–286.

Pagnussat GC, Simontacchi M, Puntarulo S, Lamattina L. 2002. Nitric oxide is required for root organogenesis. *Plant Physiol* 129:954–956.

Park MY, Wu G, Gonzalez-Sulser A, Vaucheret H, Poethig RS. 2005. Nuclear processing and export of microRNAs in *Arabidopsis*. *Proc Natl Acad Sci USA* 102:3691–3696.

Paton DM, Willing RR, Nichols W, Pryor LD. 1970. Rooting of stem cuttings of *Eucalyptus*: A rooting inhibitor in adult tissue. *Aust J Bot* 18:175–183.

Peragine A, Yoshikawa M, Wu G, Albrecht HL, Poethig RS. 2004. SGS3 and SGS2/SDE1/RDR6 are required for juvenile development and the production of trans-acting siRNAs in *Arabidopsis*. *Genes Dev* 18:2368–2379.

Peret B, De Rybel B, Casimiro I et al. 2009a. *Arabidopsis* lateral root development: An emerging story. *Trends Plant Sci* 14:399–408.

Peret B, Larrieu A, Bennett MJ. 2009b. Lateral root emergence: A difficult birth. *J Exp Bot* 60:3637–3643.

Petricka JJ, Benfey PN. 2008. Root layers: Complex regulation of developmental patterning. *Curr Opin Genet Dev* 18:354–361.

Pijut PM, Woeste KE, Michler CH. 2011. Promotion of adventitious root formation of difficult-to-root hardwood tree species. *Hort Rev* 38:213–251.

Poethig RS. 1988. Heterochronic mutations affecting shoot development in maize. *Genetics* 119:959–973.

Poethig RS. 1990. Phase change and the regulation of shoot morphogenesis in plants. *Science* 250:923–930.

Poethig RS. 2010. The past, present, and future of vegetative phase change. *Plant Physiol* 154:541–544.

Ramirez-Carvajal GA, Morse AM, Dervinis C, Davis JM. 2009. The cytokinin type-B response regulator PtRR13 is a negative regulator of adventitious root development in *Populus*. *Plant Physiol* 150:759–771.

Rey M, Tiburcio AF, Diaz-Sala C, Rodriguez R. 1994. Endogenous polyamine concentrations in juvenile, adult and in vitro reinvigorated hazel. *Tree Physiol* 14:191–200.

Ricci A, Rolli E, Dramis L, Diaz-Sala C. 2008. *N,N′*-bis-(2,3-Methylenedioxyphenyl) urea and *N,N′*-bis-(3,4-methylenedioxyphenyl) urea enhance adventitious rooting in *Pinus radiata* and affect expression of genes induced during adventitious rooting in the presence of exogenous auxin. *Plant Sci* 356–363.

Riov J. 1993. Endogenous and exogenous auxin conjugates in rooting of cuttings. *Acta Hort* 329:284–288.

Riov J, Yang SF. 1989. Ethylene and auxin-ethylene interaction in adventitious root formation in mung bean (*Vigna radiata*) cuttings. *J Plant Growth Reg* 8:131–141.

Rockel P, Strube F, Rockel A, Wildt J, Kaiser WM. 2002. Regulation of nitric oxide (NO) production by plant nitrate reductase in vivo and in vitro. *J Exp Bot* 53:103–110.

Sabatini S, Heidstra R, Wildwater M, Scheres B. 2003. *SCARECROW* is involved in positioning the stem cell niche in the *Arabidopsis* root meristem. *Genes Dev* 17:354–358.

Sanchez C, Vielba JM, Ferro E et al. 2007. Two *SCARECROW-LIKE* genes are induced in response to exogenous auxin in rooting-competent cuttings of distantly related forest species. *Tree Physiol* 27:1459–1470.

Sebastiani L, Tognetti R. 2004. Growing season and hydrogen peroxide effects on root induction and development in *Olea europaea* L. (cvs 'Frantoio' and 'Gentile di Larino') cuttings. *Sci Hort* 100:75–82.

Seo M, Akaba S, Oritani T et al. 1998. Higher activity of an aldehyde oxidase in the auxin-overproducing *superroot1* mutant of *Arabidopsis thaliana*. *Plant Physiol* 116:687–693.

Smith DL, Fedoroff NV. 1995. LRP1, a gene expressed in lateral and adventitious root primordia of *Arabidopsis*. *Plant Cell* 7:735–745.

Sole A, Sanchez C, Vielba JM et al. 2008. Characterization and expression of a *Pinus radiata* putative ortholog to the *Arabidopsis SHORT-ROOT* gene. *Tree Physiol* 28:1629–1639.

Sorin C, Bussell JD, Camus I et al. 2005. Auxin and light control of adventitious rooting in *Arabidopsis* require ARGONAUTE1. *Plant Cell* 17:1343–1359.

Sorin C, Negroni L, Balliau T et al. 2006. Proteomic analysis of different mutant genotypes of *Arabidopsis* led to the identification of 11 proteins correlating with adventitious root development. *Plant Physiol* 140:349–364.

Steffens B, Sauter M. 2009a. Epidermal cell death in rice is confined to cells with a distinct molecular identity and is mediated by ethylene and H_2O_2 through an autoamplified signal pathway. *Plant Cell* 21:184–196.

Steffens B, Sauter M. 2009b. Heterotrimeric G protein signaling is required for epidermal cell death in rice. *Plant Physiol* 151:732–740.

Sterns M. 1971. Crystal and molecular structure of a root inhibitor from *Eucalyptus grandis*, 4-ethyl-l-hydroxy-4,8,8,10,lO-pentamethyi-7,9-dioxo-2,3-dioxabicyclo [4.4.0]decene-5. *J Cryst Mol Struct* 1:373–381.

Stohr C, Stremlau S. 2006. Formation and possible roles of nitric oxide in plant roots. *J Exp Bot* 57:463–470.

Sugiyama M. 2003. Isolation and initial characterization of temperature-sensitive mutants of *Arabidopsis thaliana* that are impaired in root redifferentiation. *Plant Cell Physiol* 44:588–596.

Swarup K, Benkova E, Swarup R et al. 2008. The auxin influx carrier LAX3 promotes lateral root emergence. *Nat Cell Biol* 10:946–954.

Takahashi T, Kakehi J-I. 2010. Polyamines: Ubiquitous polycations with unique roles in growth and stress responses. *Ann Bot* 105:1–6.

Talmor-Neiman M, Stav R, Klipcan L et al. 2006. Identification of trans-acting siRNAs in moss and an RNA-dependent RNA polymerase required for their biogenesis. *Plant J* 48:511–521.

Taramino G, Sauer M, Stauffer JL Jr et al. 2007. The maize (*Zea mays* L.) *RTCS* gene encodes a LOB domain protein that is a key regulator of embryonic seminal and post-embryonic shoot-borne root initiation. *Plant J* 50:649–659.

Telfer A, Poethig RS. 1998. *HASTY*: A gene that regulates the timing of shoot maturation in *Arabidopsis thaliana*. *Development* 125:1889–1898.

Vazquez F, Vaucheret H, Rajagopalan R et al. 2004. Endogenous trans-acting siRNAs regulate the accumulation of *Arabidopsis* mRNAs. *Mol Cell* 16:69–79.

Vidal N, Arellano G, San-Jose MC, Vieitez AM, Ballester A. 2003. Developmental stages during the rooting of in-vitro-cultured *Quercus robur* shoots from material of juvenile and mature origin. *Tree Physiol* 23:1247–1254.

Vidoz ML, Loreti E, Mensuali A, Alpi A, Perata P. 2010. Hormonal interplay during adventitious root formation in flooded tomato plants. *Plant J* 63:551–562.

Visser E, Cohen JD, Barendse G, Blom C, Voesenek L. 1996. An ethylene-mediated increase in sensitivity to auxin induces adventitious root formation in flooded *Rumex palustris* Sm. *Plant Physiol* 112:1687–1692.

Wang P, Du Y, Li Y, Ren D, Song CP. 2010. Hydrogen peroxide-mediated activation of MAP kinase 6 modulates nitric oxide biosynthesis and signal transduction in *Arabidopsis*. *Plant Cell* 22:2981–2998.

Wang J, Pan R. 2006. Effect of ethylene on adventitious root formation. In: *Ethylene Action in Plants*, ed. NA Khan, pp. 69–79. Berlin, Germany: Springer.

Wang JW, Park MY, Wang LJ et al. 2011. MiRNA control of vegetative phase change in trees. *PLoS Genet* 7:e1002012.

Wang PG, Xian M, Tang X et al. 2002. Nitric oxide donors: Chemical activities and biological applications. *Chem Rev* 102:1091–1134.

Weisman Z, Riov J. 1994. Interaction of paclobutrazol and indole-3-butyric acid in relation to rooting of mung bean (*Vigna radiata)* cuttings. *Physiol Plant* 92:608–612.

Werner T, Motyka V, Laucou V et al. 2003. Cytokinin-deficient transgenic *Arabidopsis* plants show multiple developmental alterations indicating opposite functions of cytokinins in the regulation of shoot and root meristem activity. *Plant Cell* 15:2532–2550.

Werner T, Schmulling T. 2009. Cytokinin action in plant development. *Curr Opin Plant Biol* 12:527–538.

Whittington AT, Vugrek O, Wei KJ et al. 2001. MOR1 is essential for organizing cortical microtubules in plants. *Nature* 411:610–613.

Wilson ID, Neill SJ, Hancock JT. 2008. Nitric oxide synthesis and signalling in plants. *Plant Cell Environ* 31:622–631.

Woo H-H, Hackett WR, Das A. 1994. Differential expression of a chlorophyll *a/b* binding protein gene and a proline rich protein gene in juvenile and mature phase English ivy (*Hedera helix*). *Physiol Plant* 92:69–78.

Wu G, Park MY, Conway SR et al. 2009. The sequential action of miR156 and miR172 regulates developmental timing in *Arabidopsis*. *Cell* 138:750–759.

Xiong J, Lu H, Lu K et al. 2009. Cadmium decreases crown root number by decreasing endogenous nitric oxide, which is indispensable for crown root primordia initiation in rice seedlings. *Planta* 230:599–610.

Xuan W, Zhu FY, Xu S et al. 2008. The heme oxygenase/carbon monoxide system is involved in the auxin-induced cucumber adventitious rooting process. *Plant Physiol* 148:881–893.

Yadav S, David A, Bhatla SC. 2010. Nitric oxide modulates specific steps of auxin-induced adventitious rooting in sunflower. *Plant Sig Behav* 5:1–4.

Yamasaki H, Sakihama Y. 2000. Simultaneous production of nitric oxide and peroxynitrite by plant nitrate reductase: In vitro evidence for the NR-dependent formation of active nitrogen species. *FEBS Lett* 468:89–92.

Yamasaki H, Sakihama Y, Takahashi S. 1999. An alternative pathway for nitric oxide production in plants: New features of an old enzyme. *Trends Plant Sci* 4:128–129.

Zemojtel T, Frohlich A, Palmieri MC et al. 2006. Plant nitric oxide synthase: A never-ending story? *Trends Plant Sci* 11:524–525.

Zhang B, Pan X, Cobb GP, Anderson TA. 2006. Plant microRNA: A small regulatory molecule with big impact. *Dev Biol* 289:3–16.

Zhang H, Tang J, Liu XP et al. 2009. Hydrogen sulfide promotes root organogenesis in *Ipomoea batatas, Salix matsudana* and *Glycine max*. *J Int Plant Biol* 51:1086–1094.

Zhang X, Zou Z, Zhang J et al. 2011. Over-expression of sly-miR156a in tomato results in multiple vegetative and reproductive trait alterations and partial phenocopy of the sft mutant. *FEBS Lett* 585:435–439.

Zhao Y, Dai X, Blackwell HE, Schreiber SL, Chory J. 2003. SIR1, an upstream component in auxin signaling identified by chemical genetics. *Science* 301:1107–1110.

Zimmerman PW, Hitchcock AE. 1933. Initiation and stimulation of adventitious roots caused by unsaturated hydrocarbon gases. *Contrib Boyce Thompson Inst* 5:3513–3569.

III

Regulation of Root Growth

12

Auxin Signaling in Primary Roots

Catherine Perrot-Rechenmann
Institut des Sciences du Végétal

I. Introduction

Primary root growth results from a continuous production of new cells, sustained by cell division within the root apical meristem. After several rounds of cell division, root meristematic cells progressively exit the cell cycle, elongate, and enter into differentiation. These successive stages confer a longitudinal organization to the root (Figure 12.1). Close to the root tip, the stem cell niche is formed by the quiescent center (QC) and a small group of stem cells also named initials as they give rise to each root cell file. At the tip of the root, stem cell daughters rapidly differentiate into columella cells containing statoliths that are involved in gravity sensing. More laterally, stem cells divide and give rise to the lateral root cap. Other stem cells give rise to epidermis, cortex, endodermis, pericycle, and stele cells. The number of cells of each type varies by plant species, but the radial organization into concentric circles is conserved in all land plants. In the model plant *Arabidopsis*, only one layer of endodermal and cortical cells is present, and the stele includes two opposite poles of protophloem and two poles of protoxylem. The radial structure of the primary root is inherited from embryogenesis; cell specification and patterning are maintained postembryonically, thanks to the stem cell niche and the involvement of hormonal and peptidic regulators (Willemsen and Scheres 2004; De Smet et al. 2008; Matsuzaki et al. 2010; see also Chapter 3).

The phytohormone auxin is known to be critical for root patterning, primary root growth, and root architecture (in particular, lateral root development) (Tanaka et al. 2006; Vanneste and Friml 2009; Overvoorde et al. 2010; see also Chapter 6). Because root embryogenesis and lateral root development are covered by other contributions, the focus here is on auxin signaling and primary root growth.

II. Auxin Distribution in the Primary Root

Within the last 10 years, experimental and computer-based data highlighted the importance of auxin distribution and intercellular auxin transport for primary root growth. Auxin differentially accumulates within individual cells or group of cells to control auxin-dependent responses. Through a complex network of auxin biosynthetic and transport mechanisms, cells can transiently accumulate auxin or be depleted of auxin, thus generating conditions for modulations of auxin-mediated responses. A dynamic proximal-distal auxin gradient, exhibiting an auxin maximum in the QC, is superimposed on the longitudinal organization of the root apex (Figure 12.2A). A causal relationship between the putative auxin gradient and the physiological status of a given cell or its ability to respond is however not clearly established.

Auxin is mainly synthetized in young leaves and cotyledons and is transported to the root sink via the phloem and within the root tip via the combined action of polarly localized AUXIN RESISTANT 1/LIKE-AUXIN RESISTANT (AUX1/LAX) auxin influx and PIN-FORMED (PIN) efflux carriers (Swarup et al. 2001; Tanaka et al. 2006; Robert and Friml 2009). Members of the P-GLYCOPROTEIN/MULTIDRUG RESISTANT ABC transporter family (PGP/MDR) are also involved in the transport of auxin (Geisler and Murphy 2006). In addition, maintenance of the auxin gradient across the root meristem is supported by root-synthesized auxin (Stepanova et al. 2008; Petersson et al. 2009). Auxin flux converges to the QC and is redistributed through the lateral root cap and the epidermis to the basal part of the meristem. The basal meristem overlaps the transition zone that is defined as the area where cells progressively exit the cell cycle and start to elongate and enter

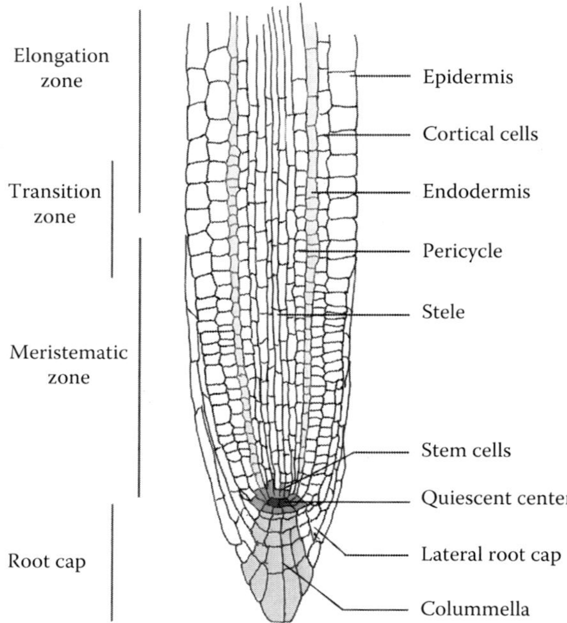

FIGURE 12.1 Primary root organization in the model plant *Arabidopsis thaliana*. The scheme represents a longitudinal and median section of the root, across the QC. Cell files derived from each stem cell are identified as well as the main zone of the apical root meristem. At a more distal position, not represented on the scheme, is the zone of differentiation.

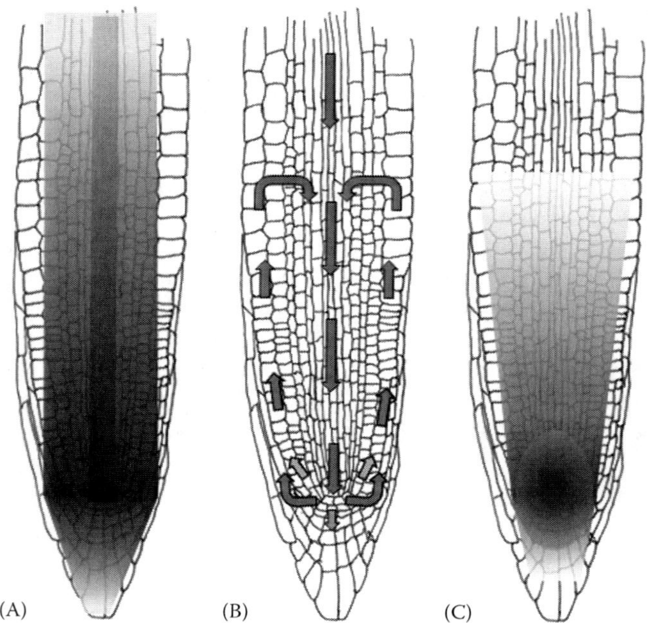

FIGURE 12.2 Informative gradients and auxin flow at the primary root. (A) Simplified representation of the auxin gradient at the root apex with auxin maximum in the QC. (B) Illustration of the inverted fountain model for auxin transport. (C) Schematic representation of the informative PLETHORA gradient at the root meristem. High levels contribute to the maintenance of the root stem cell niche, whereas distal decrease is required for transition to elongation.

into differentiation. At the transition zone, the auxin flux is reoriented back to the stele and the root tip, across the cortex, the endodermis, and the pericycle, thus contributing to the decrease of auxin concentration in more distal and peripheral tissues. Imprinting of lateral root initiation seems to occur at the basal meristem even if the first division of pericycle cells giving rise to lateral root primordia is observed at a more distal position (De Smet et al. 2007). This 3D loop of auxin redistribution at the root tip is often reported as the inverted fountain model (Swarup and Bennett 2003) (Figure 12.2B).

Our present understanding of auxin redistribution and auxin content within the primary root is mainly deduced from analysis of the asymmetric cellular localization of auxin carriers, visualization of auxin response maxima using transcriptional reporters, and more recently from modeling of auxin fluxes (Kramer and Bennett 2006; Grieneisen et al. 2007; Laskowski et al. 2008; Krupinski and Jonsson 2010; Kleine-Vehn et al. 2011). In the model plant *Arabidopsis*, high-sensitive measurements of endogenous indole-3-acetic acid (IAA) in various root cell types were used to map IAA distribution in the root tip (Petersson et al. 2009). Cell-specific GFP-expressing lines were used to prepare protoplasts and isolate specific cell types by fluorescence-activated cell sorting prior to quantification of IAA by mass spectrometry. All data confirm an IAA maximum in the QC. In the root meristem, IAA measurements suggest high levels of auxin in the stele, pericycle, endodermis, and cortical cells and relatively less auxin in the epidermis (Petersson et al. 2009), a distribution that is more or less in accordance with reported auxin response maxima using the recently developed DII-VENUS reporter (Brunoud et al. 2012). At the transition zone, the longitudinal auxin gradient is assumed to decrease as a result of the flow back to the stele, thus favoring cell elongation and differentiation. In addition, cytokinin signaling contributes to the regulation of the transition between the meristem and the elongation through regulation of SHY2/IAA3 repressor (Dello Ioio et al. 2008).

III. Auxin Perception

Whatever the auxin content, auxin-mediated responses require auxin perception and signaling. There are at least two classes of auxin receptors. One is the AUXIN BINDING PROTEIN 1 (ABP1), a low abundant protein, unique to the plant kingdom and identified in the 1980s on the basis of its capacity to bind auxin (Löbler and Klämbt 1985). The others are coreceptors resulting from the auxin-induced interaction of F-box proteins, TRANSPORT INHIBITOR RESISTANT 1 and related AUXIN F-BOX (AFBs), and their substrates Aux/IAA transcriptional repressors (Chapman and Estelle 2009).

A. ABP1 Auxin Receptor

A whole range of functional evidence supporting a role of ABP1 in auxin sensing and early responses indicates that the protein acts at the plasma membrane (PM) despite the absence of hydrophobic domains or posttranslational modification promoting a direct anchor of the protein within the membrane (Tromas et al. 2010; Sauer and Kleine-Vehn 2011). Part of the protein is located in the lumen of the endoplasmic reticulum (ER), thanks to the presence of a KDEL retention motif at the C-terminus. To date, nothing is known on the mechanisms controlling the targeting of the protein to the PM, and the relative distribution of ABP1 between the ER and the PM is still a matter of debate. The molecular interaction of ABP1 with membrane components also remains to be identified.

In the ER, the amount of the protein might be regulated by a RING ubiquitin ligase E3, RING MEMBRANE-ANCHOR 2 (Son et al. 2010). No experimental evidence supports a function of ABP1 in the ER; both regulation of protein half-life and KDEL-mediated retention of ABP1 in the ER compartment might be very efficient processes for controlling the relative amount of ABP1 that reaches the PM. Transcriptomic analysis revealed that ABP1 gene is ubiquitously and constantly expressed in plant tissues including root cells (Birnbaum et al. 2005; Schmid et al. 2005; Brady et al. 2007). ABP1 is essential from embryogenesis to root and shoot growth and development, thus throughout the whole plant life (Chen et al. 2001; Braun et al. 2008; Tromas et al. 2009).

ABP1 binds auxin at physiologically relevant auxin concentrations, but its binding affinity is tightly dependent on the pH. In vitro binding assays showed a strict pH-dependent binding affinity with an optimal pH around 5–5.5, that is, compatible with apoplastic pH, but a drastic decrease of affinity within one pH unit and virtually no binding at pH 7 (estimated pH in the ER) (Löbler and Klämbt 1985; Shimomura et al. 1986; Tian et al. 1995). Resolution of ABP1 structure confirmed the identification of a hydrophobic auxin binding pocket and revealed the presence of a zinc atom facilitating the positioning of the carboxyl group of auxin within the binding site (Woo et al. 2000, 2002).

Auxin binding to ABP1 induces PM hyperpolarization via modulation of ion fluxes across the PM, especially activation of proton pump ATPase and modulation of K^+ channels in a dose-dependent manner (Venis et al. 1992; Thiel et al. 1993; Bauly et al. 2000).

Recent work has established that ABP1 is required for clathrin-dependent endocytosis in root cells, acting as a positive factor in clathrin recruitment to the PM (Robert et al. 2010). Data revealed a constitutive function for ABP1 on endocytosis independent of any auxin stimulus (Figure 12.3). Interestingly, this effect is inhibited by the binding of auxin to ABP1 (Robert et al. 2010). Many PM proteins are subject to regular recycling via the clathrin-dependent mechanism, and disruption of this process might impair the recycling, the relative amount, and potentially the activity of a vast number of critical PM proteins. With the exception of PIN auxin efflux carriers, little is known about the consequences of ABP1-mediated inhibition of endocytosis in response to auxin. The binding of auxin to ABP1 inhibits PIN recycling without slowing down exocytosis; thus, the number of PIN proteins at the PM is increased and the cell capacity of auxin efflux is enhanced (Dhonukshe et al. 2007; Robert et al. 2010). This effect might contribute to the decrease of the intracellular content of auxin and together with auxin compartmentation, conjugation, and catabolism to restore auxin homeostasis.

ABP1 was shown also to mediate auxin-dependent activation of two antagonist Rho-GTPases influencing local organization of either F-actin or microtubules for ROP2 or ROP6 activation, respectively (Xu et al. 2010). This effect was reported in cotyledon pavement cells, and, to date, no experimental evidence supports a link between ABP1 with Rho-GTPase activity in roots, but it would be interesting to determine whether this effect is restricted to pavement cell development or is a more general process. In yeast, actin cytoskeleton is essential for clathrin-dependent endocytosis, and increasing evidence supports the importance of actin assembly and network in this process in mammals (Galletta et al. 2010). To date, it is not clear whether auxin/ABP1-mediated inhibition of endocytosis involves alteration of actin network via the regulation of Rho-GTPases, and this question will merit further investigations. It is important to notice that confusion should be avoided between ABP1 receptor and involvement of actin binding protein in the organization of actin network.

Alteration of ABP1 activity in plants is also related to modulations of transcriptional regulations by auxin (Braun et al. 2008; Tromas et al. 2009; Effendi et al. 2011). ABP1 acting primarily at the PM, it is assumed that the reported effects on gene expression require the involvement of downstream signaling components coordinating auxin sensing outside of the cell, early responses at the PM and auxin perception, and transcriptional responses in the nucleus. The signaling pathway downstream of ABP1 with or without auxin stimulus still has to be elucidated. Various secondary messengers have been identified as components of auxin signaling pathway, but their relative role remains to be determined (Kovtun et al. 1998; Mockaitis and Howell 2000; Strader et al. 2008; Lee et al. 2009; Xu et al. 2010).

B. TIR1/AFBs and Aux/IAA Nuclear Coreceptors

The regulation of protein degradation is pivotal for the control of multiple developmental and adaptive processes, including the subtle regulation of various hormonal responses among the auxin responses.

TIR1 and AFB1 to AFB5 proteins belong to the leucine-rich repeat (LRR) class of F-box proteins. TIR1 was initially identified from the *tir1-1 Arabidopsis* mutant isolated on a genetic screen based on root resistance to naphthylphthalamic acid (NPA), an inhibitor of polar auxin transport (Ruegger et al. 1998). *tir1* allelic mutants exhibit a weak resistance to NPA and are resistant to auxin, in particular to the synthetic auxin 2,4-D (Ruegger et al. 1998). TIR1 and AFBs are involved in SCF-type E3 ubiquitin ligase complex (SCF for SKP1-CUL1-Fbox) (Gray et al. 1999; Greenham et al. 2011). The SCF complex is composed of four major subunits: Cullin 1 (CUL1), SUPPRESSOR OF KINETOCHORE PROTEIN 1 (SKP1), RING-BOX 1 (RBX1)/REGULATOR OF CULLINS 1 (ROC1), and an F-box protein. Within such complex,

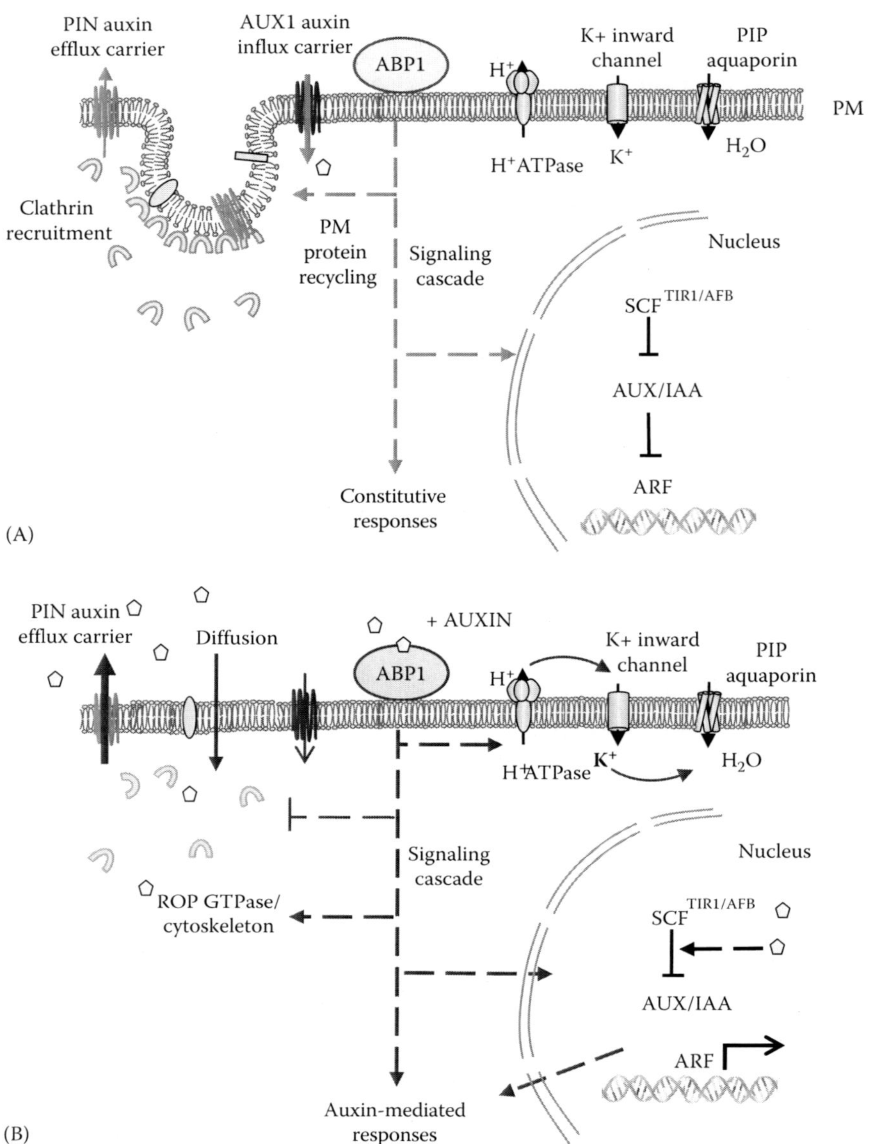

FIGURE 12.3 Schematic model integrating ABP1 signaling events. (A) Simulation of ABP1 downstream signaling in the presence of low auxin content. Constitutive ABP1-mediated responses, potentially independent of auxin binding, are a positive effect on clathrin recruitment and endocytosis and modulation of transcriptional responses. (B) Simulation of ABP1 downstream signaling after perception of auxin in increased auxin levels. Identified downstream targets of ABP1 action are activation of transporters and channels at the plasma membrane, activation of Rho-GTPases (at least in cotyledon epidermal cells), modulation of transcriptional responses, and inhibition of endocytosis (small pentagons represent auxin molecules).

the F-box protein provides the substrate specificity to the SCF complex for ubiquitination. Polyubiquitination of the substrate promotes its degradation by the 26S proteasome (Lechner et al. 2006; Vierstra 2009; Santner and Estelle 2010). For SCF^{TIR1}, the identified substrates are members of the Aux/IAA protein family (Figure 12.4). Aux/IAA proteins are transcriptional repressors that interact with activators of the AUXIN RESPONSE FACTOR (ARF) transcription factor family. ARF activators bind to auxin regulatory elements (AuxRE) found as single or multiple motifs within promoters of primary auxin responsive genes (Guilfoyle and Hagen 2007). Both Aux/IAAs and ARFs contain conserved protein-protein interaction domains, named domains III and IV, within their C-terminal half. Binding of Aux/IAA to ARF through these domains inhibits transcription either by direct inhibition of ARF activity, sequestration of ARF that avoids their binding to regulatory sequences, or recruitment of additional repressors as TOPLESS that was shown to interact with domain I of the Aux/IAA BODENLOS/IAA12 protein during embryogenesis (Szemenyei et al. 2008). Several members of the Aux/IAA family are acknowledged to be short-lived nuclear proteins. Auxin promotes their degradation, which enables transcriptional activation. Auxin directly promotes the interaction

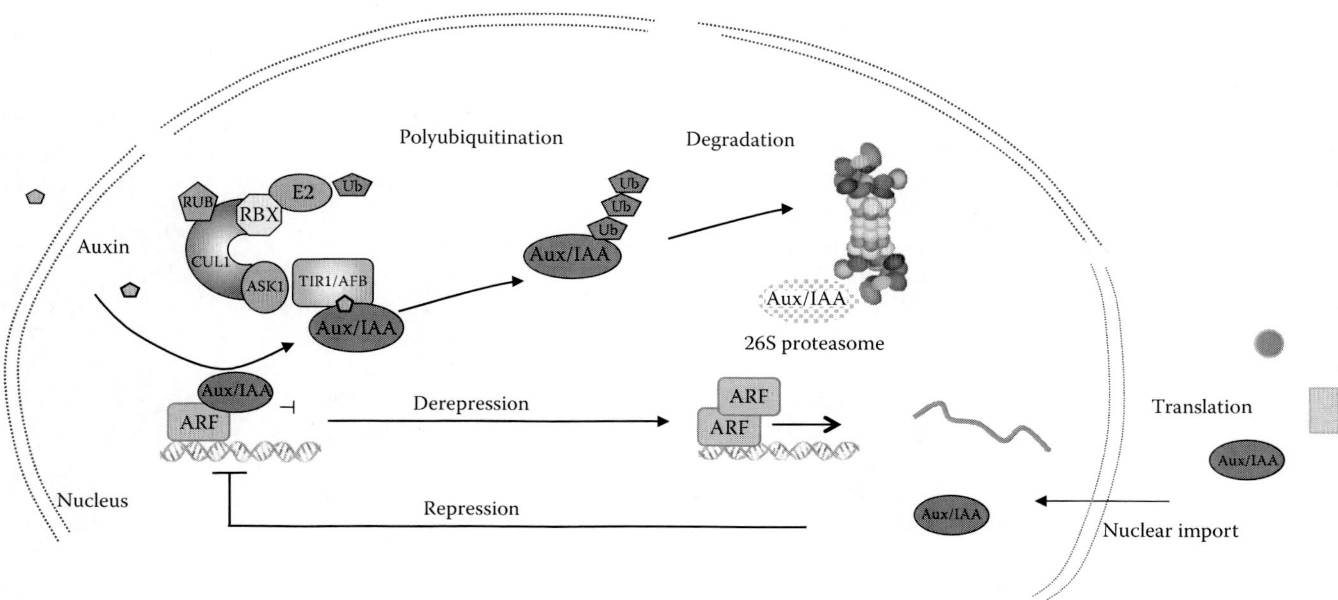

FIGURE 12.4 Current model for auxin-mediated transcriptional regulation by TIR1/AFB and Aux/IAA coreceptors. Under low-auxin conditions, auxin response genes are repressed via Aux/IAA repressors. In the presence of auxin into the nucleus, auxin promotes interaction between TIR1/AFB F-box proteins and Aux/IAA repressors. Aux/IAA proteins are polyubiquitinated by $SCF^{TIR1/AFB}$ E3 ubiquitin ligase and are degraded by the 26S proteasome. ARF transcriptional factors activate the transcription of auxin response genes. Among these, some encode Aux/IAA repressors that can thus exert repression.

between TIR1 and Aux/IAA, leading to their ubiquitination and degradation (Dharmasiri et al. 2005a; Kepinski and Leyser 2005; Maraschin Fdos et al. 2009). TIR1 and Aux/IAA thus behave as auxin coreceptors (Dharmasiri et al. 2005a; Kepinski and Leyser 2005). The binding of auxin to TIR1 or Aux/IAA alone is not documented. TIR1 and Aux/IAA can interact in the absence of auxin but with a low affinity. Auxin acts as a molecular glue increasing the interaction between TIR1 and Aux/IAA (Tan et al. 2007). For a coreceptor formed by TIR1 and AXR2/IAA7, the Kd is around 80 nM (Dharmasiri et al. 2005a; Kepinski and Leyser 2005). Domain II of Aux/IAA proteins is the domain of interaction with auxin and TIR1. A large number of auxin resistant mutants were found to exhibit mutations within this domain II that impaired interaction with the F-box and led to enhanced stability of the repressor. Resolution of TIR1 structure, in the presence of auxin and a synthetic domain II peptide, revealed the presence of a molecule of inositol hexakisphosphate (InsP6 or phytate) that is critical for stabilization of the LRR structure involved in the binding of auxin (Tan et al. 2007). Considering that at least TIR1, AFB1, AFB2, and AFB3 F-box proteins can potentially interact with up to 29 members of the Aux/IAA family, more than one hundred auxin coreceptors can theoretically be involved in the transcriptional control of auxin responsive genes. With such diversity, a complex picture of transcriptional responses to auxin has to be considered as the relative affinity of each F-box and Aux/IAA pair for each other and for auxin is a priori not similar, giving rise to a whole range of high- and low-affinity coreceptors (Parry et al. 2009). *In planta*, the pattern and timescale of expression of each member reduce the number of combinations within a given cell, but at least partial redundancy can be deduced from mutant analysis (Dharmasiri et al. 2005b; Parry et al. 2009).

TIR1 and *AFB* genes are rather broadly transcribed during plant development, but mRNA accumulation of *TIR1*, *AFB2*, and *AFB3*, but not *AFB1*, is restricted through the action of miR393 microRNA (Navarro et al. 2006; Parry et al. 2009). Little is known about the transcriptional regulation of TIR1 and AFB genes, except that TIR1 was shown to be induced by Pi deficiency (or partially repressed by increased Pi availability) (Perez-Torres et al. 2008). At the protein level, TIR1 and AFB F-box proteins are themselves susceptible to polyubiquitination and degradation by the 26S proteasome probably together with their Aux/IAA substrate (Stuttmann et al. 2009).

C. Other Putative Receptor

Reminiscent to the mechanisms reported earlier for TIR1/AFBs and Aux/IAA coreceptors, another F-box protein was recently reported to be an auxin receptor. This protein is S-Phase Kinase-Associated Protein 2A (SKP2A), an F-box protein involved in the polyubiquitination and 26S proteasome-mediated protein degradation of cell cycle transcription factors. The estimated Kd is around 200 mM for an *E. coli*-produced recombinant protein fused to maltose binding protein (Jurado et al. 2010).

SKP2A F-box is expressed from late S-phase to the mitotic phase of the cell cycle and mediates the degradation of identified substrates, ADENOVIRUS E2 PROMOTER BINDING FACTOR C (E2FC), and DIMERIZING PROTEIN (DPB) (del Pozo et al. 2002). E2FC and DPB act as heterodimers and function as transcriptional repressors for a subset of cell cycle-regulated genes (del Pozo et al. 2002, 2006). SKP2A is considered a positive regulator of cell division, whereas E2FC and DPB are involved in the regulation of the balance between proliferation and endoreduplication together with other E2F and DP dimers (De Veylder et al. 2002; del Pozo et al. 2006). Loss of function of SKP2A results in accumulation of increased levels of E2FC and DPB. Similarly, expression of an SKP2A protein mutated in the auxin binding domain cannot promote degradation of E2FC and DPB (Jurado et al. 2010). Binding of auxin to SKP2A triggers its own degradation and enhances E2FC and DPB degradation (Jurado et al. 2010). It is not clear whether auxin can promote degradation of SKP2A without the presence of the E2FC or DPB substrates or if their degradation is concomitant. With the effect of auxin on SKP2A, E2FC, and DPB half-life, a direct link is thus established between the regulation of cell division, more specifically the G2/M transition, and auxin-mediated proteolysis. Further investigation on the precise timing of these events during the progression of the cell cycle would help to better understand the importance of auxin in the regulation of SKP2A, E2FC, and DPB degradation.

Among the huge family of F-box proteins in plants, about one-third (around 200) contain LRR motifs (Gagne et al. 2002). This characteristic is of course not sufficient to enable auxin-promoted interaction with their substrates and auxin binding, but the possibility that other pairs of F-box and substrate have also the capacity to bind auxin cannot be excluded. Interestingly, the CORONATIVE INSENSITIVE 1 F-box protein (COI1) that belongs to the same clade as TIR1 and AFBs and is involved in jasmonic acid signaling is unable to bind auxin (Tan et al. 2007), whereas SKP2A, that is, much more distant from TIR1/AFB, seems to adopt a TIR1-like structure favoring auxin binding. The number of F-box/substrate auxin coreceptors might thus not end with those reported earlier.

IV. Auxin-Mediated Responses in Roots

Auxin is a major regulator of cell division and cell elongation, two critical and sequential cellular processes sustaining continuous root growth. Despite significant progress made within the last 10 years on molecular mechanisms linking auxin action and transcriptional regulation (Chapman and Estelle 2009) and recent data highlighting the importance of ABP1 for a large range of auxin responses (Tromas et al. 2010; Sauer and Kleine-Vehn 2011; Shi and Yang 2011), our present understanding of auxin control of root growth is paradoxically rather fragmentary with the notable exception of lateral root formation (Swarup et al. 2008; Peret et al. 2009; De Rybel et al. 2010; Dubrovsky et al. 2011; see also Chapter 6).

A. Primary Root Meristem and Auxin Signaling

The meristematic zone is characterized by active division of cells. Auxin is a permissive signal for cell division, and various core cell cycle regulators were shown to be regulated directly or indirectly by auxin either at the transcriptional (potentially involving TIR1/AFB and AUX/IAA coreceptors) or at the protein levels (Perrot-Rechenmann 2010). Molecular mechanisms of auxin action on cell cycle are still poorly understood, but ABP1 and SKP2A are likely to play major roles.

1. ABP1 Receptor and Meristem Maintenance

Characterization of conditional loss of function for ABP1 demonstrated that the protein acts at the G1 to S transition on regulators of the CYCLIN DEPENDENT KINASE/CYCLIN D/RETINOBLASTOMA pathway and is required for the entry and maintenance of root meristematic cells into the cell cycle (Tromas et al. 2009) (Figure 12.5B). In addition to its effect on cell division, ABP1 is essential for root meristem maintenance by acting on the gradient of PLETHORA genes (PLT). PLTs belong to AP2 transcription factor family and contribute to define root stem cell niche. PLTs act in a dose-dependent manner on the competence of the cells to be maintained within the meristem or to leave the meristem and start to elongate; thus, they are important regulators that mediate the site of the transition zone (Galinha et al. 2007) (Figure 12.2C). A loss of function in ABP1 results primarily in cell cycle arrest followed by a severe

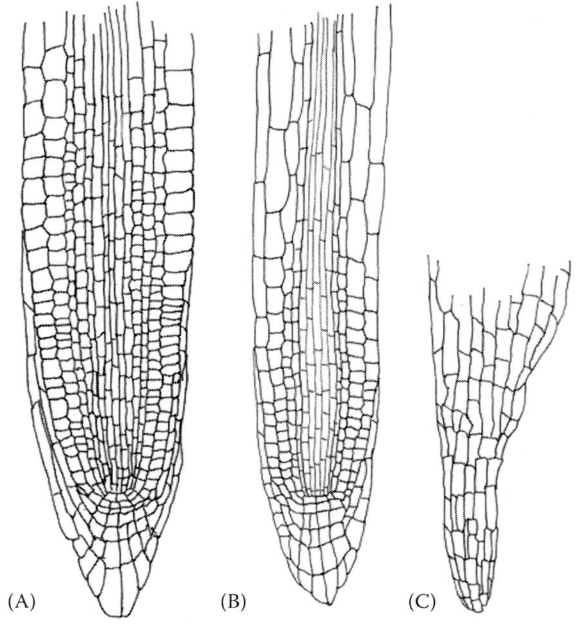

FIGURE 12.5 Comparison of longitudinal sections of primary root apex of various genotypes. (A) Primary root of 4-day-old wild-type seedling. (B) Primary root of a 4-day-old seedling inactivated for ABP1 after germination. After 3 days of inactivation of ABP1, the root meristem is severely reduced compared to wild-type or noninduced plants. Elongated and differentiated cells can be observed closer to the organizing QC. (C) Aborted primary root of class I *tir1afb1afb2afb3* quadruple mutant. The upper part corresponds to the basis of the hypocotyl; no quiescent center can be detected in these mutants unlike class III quadruple mutants that exhibit only minor alterations of the primary root apex (not shown).

decrease of PLETHORA gradients, thereby inducing a severe decrease of the root meristem size and anticipated elongation of meristematic cells (Tromas et al. 2009). Several recent reviews cover various aspects of ABP1 function, including the effect on cell cycle that are thus not further developed here (Perrot-Rechenmann 2010; Tromas et al. 2010; Sauer and Kleine-Vehn 2011; Shi and Yang 2011). ABP1 acts on the CDK/CYCLIN D/RETINOBLASTOMA pathway, but the precise target of this action is unknown, and components of the signaling pathway connecting ABP1 activity to cell cycle regulation remain to be identified.

Interestingly, differential endocytosis in the root meristem and at the transition zone modulates auxin response by providing specific positional information (Santuari et al. 2011). However, the contribution of ABP1 in this process has not been investigated yet.

2. SKP2A and Cell Division

SKP2A is expressed in the whole root at various stages of root growth. After 3–5 days of plant growth, the expression is limited to the root cap and to vascular tissues of differentiated roots (Jurado et al. 2008). Ectopic overexpression of SKP2A in roots results in a slight increase of the number of cell divisions in the meristem inducing a slight increase of root length (Jurado et al. 2008). A more obvious effect of SKP2A overexpression was reported on the number of lateral root primordia that is about twice that of wild-type roots. Enhanced lateral root primordia initiation is however dependent from auxin-mediated degradation of SOLITARY ROOT(SLR)/IAA14 by $SCF^{TIR1/AFBs}$ (Jurado et al. 2008). Overexpression of a mutated SKP2A protein, unable to bind auxin, has very little effect on the number of meristematic cells and root length. Loss of function of SKP2A, in the *skp2a* mutant, results in accumulation of increased levels of E2FC and DPB, a slight reduction of root meristem size, and a weak auxin resistance to the synthetic auxin 2,4-D that is additive to the resistance of *tir1-1* (Jurado et al. 2010). Phenotypes of *skp2a* and SKP2A overexpressor are much more subtle than phenotypes resulting from overexpression of E2FC and DPB or reduced activity of E2FC, respectively (del Pozo et al. 2002, 2006; Perrot-Rechenmann 2010), both affecting more drastically the size of the root meristem. This suggests that SKP2A might not be the sole protein involved in the degradation of E2FC and DPB transcriptional repressors especially in meristematic cells; it is likely that other critical regulators are required to control cell division within the meristem of the primary root.

B. Auxin Control of Cell Elongation

Exogenous application of auxin inhibits root growth by inhibiting cell elongation of all root cell layers. The inhibitory concentration of auxin varies according to the plant species and to the auxin molecule that is applied, but in *Arabidopsis* it is around 10^{-8} M for the natural auxin IAA. Lower concentrations potentially promote enhanced elongation of root cells, but this effect can be experimentally difficult to reveal. Auxin content in the elongation zone is close to the optimal concentration for promoting longitudinal cell elongation. Whereas ABP1 loss-of-function plants were shown to exhibit severe alteration of cell expansion in shoot tissues, no inhibition of longitudinal cell elongation in roots is observed in these plants even in the presence of exogenous auxin. Similarly, *tir1* mutants also exhibit normal longitudinal cell elongation. Conversely, both TIR1 and ABP1 were demonstrated to be required for radial cell expansion in roots.

Interestingly, trichoblastic cells (epidermal cells forming root hairs) respond to the auxin-mediated inhibition of cell elongation as other root cells but the same amount of auxin promotes enhanced elongation of root hairs. Endogenous auxin in these cells is clearly infraoptimal for root hair growth. This example illustrates that most cellular responses to auxin are dose dependent and that the optimal auxin concentration required for an optimal response varies according to the response even within the same cell and in the same time frame. This suggests the involvement of high- and low-affinity auxin signaling pathways triggering specific or combined responses. Increasing experimental evidence reveals that differential auxin responses are observed at the resolution of one cell or a small number of cells, with differential involvement of various ARF, Aux/IAA, and TIR1/AFB combinations (Weijers et al. 2006; De Smet et al. 2008; Mockaitis and Estelle 2008; Swarup et al. 2008).

Sensitivity of root cell elongation to exogenous auxin was broadly used to identify mutants resistant to auxin. Many of these mutants are gain of function mutants accumulating mutated Aux/IAA transcriptional repressors unable to be degraded by TIR1/AFB F-box proteins. The inhibitory effect of exogenous auxin on cell elongation is strongly reduced in these mutants (e.g., *axr2/iaa7, axr3/iaa17, msg2/iaa19*), and for those that accumulate the mutated protein in root epidermis, an agravitropic phenotype was also reported, as well as a reduction in the number or length of root hairs (Liscum and Reed 2002; Tatematsu et al. 2004). Many of these mutants also exhibit a reduced number of lateral roots, in the first place *solitary root, slr/iaa14*, that does not develop any lateral roots (Fukaki et al. 2002; Vanneste et al. 2005).

Major root developmental defects of *aux/iaa* gain-of-function mutants relate to cell elongation, gravitropism, root hair formation, and lateral root formation, whereas possible effects on cell division and meristem maintenance are poorly documented. Accumulation of a stable Aux/IAA repressor in gain-of-function mutants can displace the endogenous protein-protein equilibrium, and it is often difficult to conclude on a specific role for a given Aux/IAA. A better understanding of the respective role of Aux/IAA-mediated repression in auxin responses would come from loss-of-function mutant characterization, but it unfortunately has been compromised by a relative redundancy within members of the family mutants. Recently detailed characterization of a loss-of-function mutant for SHY2/IAA3 brought evidence supporting a role of this protein in the control of meristem size by regulating the transition zone under the control of cytokinin (Dello Ioio et al. 2008).

The *tir1* mutants exhibit a slightly longer primary root than control plants and less lateral roots (Ruegger et al. 1998). Single *afb* mutants are not significantly affected, but all exhibit resistance to 2,4-D (Dharmasiri et al. 2005b). TIR1 and AFB2 proteins seem to be the most important coreceptors in roots even if no severe alteration of primary root growth was shown for the double *tir1afb2* mutant (Dharmasiri et al. 2005b; Parry et al. 2009). Triple or quadruple *tir1* and *afb* mutants show more severe phenotypes suggesting that they act redundantly. A rootless phenotype is reported for a constant proportion of class I triple or quadruple *tir/afb* mutants (Figure 12.5C), thus mimicking the phenotype of *bdl/iaa12* and *monopteros (mp)/arf5* mutants (Hardtke and Berleth 1998; Hamann et al. 1999, 2002). BDL/IAA12 and MP/ARF5 define a functional auxin response module that is critical for root embryogenesis (Hamann et al. 2002; Weijers et al. 2005; Schlereth et al. 2010). The limitation of functional TIR1/AFB F-box proteins in the triple and quadruple mutants impairs the auxin-induced activation of ARF5-regulated genes and establishment of root stem cells during embryogenesis. A small proportion of the triple or quadruple mutants develop a short and agravitropic root (class II), and 50% to 40% of the plants develop a long primary root with alteration of gravitropism and inhibition of hair elongation (class III) (Dharmasiri et al. 2005b). This range of phenotypes suggests that a minimal threshold of Aux/IAA degradation is stochastically maintained within the triple and quadruple mutants to bypass embryogenesis defects; this residual auxin F-box activity might result from AFB4 or AFB5 proteins even if their capacity to target TIR1/AFB1–3 substrates has not been demonstrated (Walsh et al. 2006; Greenham et al. 2011).

V. Conclusion

Molecular mechanisms sustaining the ability of a cell to sense, interpret, and respond to a local change in auxin result from the involvement of an incredibly complex combination of multiple actors that are likely not all identified. Receptor, multiple auxin coreceptors, and downstream signaling components act in concert to promote or inhibit growth via modulation of auxin responses in the context of development or adaptation to environmental stimuli. Integrating the coordinated activity of all auxin receptors generates highest levels of complexity for the array of combinatorial regulations that underlie the pleiotropic effects of auxin. Whereas the mechanisms sustaining some specific and local auxin responses start to be elucidated, we are far from understanding how cells and complex tissues integrate, buffer, and coordinate activation or inhibition of distinct auxin signaling pathways. Other hormones interfering with auxin signaling in root growth as cytokinins, brassinosteroids, or gibberellic acid bring an additional level of complexity. Modeling of auxin transport has started to consider high level of complexity especially in roots, but perception, signaling, and cross talks have not been taken into account yet. It will certainly be one of the challenges of future research developments in auxin and root biology, ideally with a reciprocal benefit of modeling to biology, and vice versa.

Acknowledgments

The author is supported by the Centre National de la Recherche Scientifique and the French Research Agency (ANR, programme blanc). Choices had to be made to prepare this chapter, and I apologize to all authors whose exciting work is not referred in this chapter.

References

Bauly JM, Sealy IM, Macdonald H et al. 2000 Overexpression of auxin-binding protein enhances the sensitivity of guard cells to auxin. *Plant Physiol* 124:1229–1238.

Birnbaum K, Jung JW, Wang JY et al. 2005 Cell type-specific expression profiling in plants via cell sorting of protoplasts from fluorescent reporter lines. *Nat Methods* 2:615–619.

Brady SM, Orlando DA, Lee JY et al. 2007 A high-resolution root spatiotemporal map reveals dominant expression patterns. *Science* 318:801–806.

Braun N, Wyrzykowska J, Muller P et al. 2008 Conditional repression of AUXIN BINDING PROTEIN1 reveals that it coordinates cell division and cell expansion during post-embryonic shoot development in *Arabidopsis* and tobacco. *Plant Cell* 10:2746–2762.

Brunoud G, Wells DM, Oliva M et al. 2012 A novel sensor to map auxin response and distribution at high spatio-temporal resolution. *Nature* 482:103–106.

Chapman EJ, Estelle M 2009 Mechanism of auxin-regulated gene expression in plants. *Annu Rev Genet* 43:265–285.

Chen JG, Ullah H, Young JC, Sussman MR, Jones AM 2001 ABP1 is required for organized cell elongation and division in *Arabidopsis* embryogenesis. *Genes Dev* 15:902–911.

De Rybel B, Vassileva V, Parizot B et al. 2010 A novel aux/IAA28 signaling cascade activates GATA23-dependent specification of lateral root founder cell identity. *Curr Biol* 20:1697–1706.

De Smet I, Tetsumura T, De Rybel B et al. 2007 Auxin-dependent regulation of lateral root positioning in the basal meristem of *Arabidopsis*. *Development* 134:681–690.

De Smet I, Vassileva V, De Rybel B et al. 2008 Receptor-like kinase ACR4 restricts formative cell divisions in the *Arabidopsis* root. *Science* 322:594–597.

De Veylder L, Beeckman T, Beemster GT et al. 2002 Control of proliferation, endoreduplication and differentiation by the *Arabidopsis* E2Fa-DPa transcription factor. *Embo J* 21:1360–1368.

del Pozo JC, Boniotti MB, Gutierrez C 2002 *Arabidopsis* E2Fc functions in cell division and is degraded by the ubiquitin-SCF(AtSKP2) pathway in response to light. *Plant Cell* 14:3057–3071.

del Pozo JC, Diaz-Trivino S, Cisneros N, Gutierrez C 2006 The balance between cell division and endoreplication depends on E2FC-DPB, transcription factors regulated by the ubiquitin-SCF^{SKP2A} pathway in *Arabidopsis*. *Plant Cell* 18:2224–2235.

Dello Ioio R, Nakamura K, Moubayidin L et al. 2008 A genetic framework for the control of cell division and differentiation in the root meristem. *Science* 322:1380–1384.

Dharmasiri N, Dharmasiri S, Estelle M 2005a The F-box protein TIR1 is an auxin receptor. *Nature* 435:441–445.

Dharmasiri N, Dharmasiri S, Weijers D et al. 2005b Plant development is regulated by a family of auxin receptor F box proteins. *Dev Cell* 9:109–119.

Dhonukshe P, Aniento F, Hwang I et al. 2007 Clathrin-mediated constitutive endocytosis of PIN auxin efflux carriers in *Arabidopsis*. *Curr Biol* 17:520–527.

Dubrovsky JG, Napsucialy-Mendivil S, Duclercq J et al. 2011 Auxin minimum defines a developmental window for lateral root initiation. *New Phytol* 191:970–983.

Effendi Y, Rietz S, Fischer U, Scherer GF 2011 The heterozygous abp1/ABP1 insertional mutant has defects in functions requiring polar auxin transport and in regulation of early auxin-regulated genes. *Plant J* 65:282–294.

Fukaki H, Tameda S, Masuda H, Tasaka M 2002 Lateral root formation is blocked by a gain-of-function mutation in the SOLITARY-ROOT/IAA14 gene of *Arabidopsis*. *Plant J* 29:153–168.

Gagne JM, Downes BP, Shiu SH, Durski AM, Vierstra RD 2002 The F-box subunit of the SCF E3 complex is encoded by a diverse superfamily of genes in *Arabidopsis*. *Proc Natl Acad Sci USA* 99:11519–11524.

Galinha C, Hofhuis H, Luijten M et al. 2007 PLETHORA proteins as dose-dependent master regulators of *Arabidopsis* root development. *Nature* 449:1053–1057.

Galletta BJ, Mooren OL, Cooper JA 2010 Actin dynamics and endocytosis in yeast and mammals. *Curr Opin Biotechnol* 21:604–610.

Geisler M, Murphy AS 2006 The ABC of auxin transport: the role of p-glycoproteins in plant development. *FEBS Lett* 580:1094–1102.

Gray WM, del Pozo JC, Walker L et al. 1999 Identification of an SCF ubiquitin-ligase complex required for auxin response in *Arabidopsis thaliana*. *Genes Dev* 13:1678–1691.

Greenham K, Santner A, Castillejo C et al. 2011 The AFB4 auxin receptor is a negative regulator of auxin signaling in seedlings. *Curr Biol* 21:520–525.

Grieneisen VA, Xu J, Maree AF, Hogeweg P, Scheres B 2007 Auxin transport is sufficient to generate a maximum and gradient guiding root growth. *Nature* 449:1008–1013.

Guilfoyle TJ, Hagen G 2007 Auxin response factors. *Curr Opin Plant Biol* 10:453–460.

Hamann T, Benkova E, Baurle I, Kientz M, Jurgens G 2002 The *Arabidopsis BODENLOS* gene encodes an auxin response protein inhibiting MONOPTEROS-mediated embryo patterning. *Genes Dev* 16:1610–1615.

Hamann T, Mayer U, Jurgens G 1999 The auxin-insensitive *bodenlos* mutation affects primary root formation and apical-basal patterning in the *Arabidopsis* embryo. *Development* 126:1387–1395.

Hardtke CS, Berleth T 1998 The *Arabidopsis* gene *MONOPTEROS* encodes a transcription factor mediating embryo axis formation and vascular development. *EMBO J* 17:1405–1411.

Jurado S, Abraham Z, Manzano C, Lopez-Torrejon G, Pacios LF, Del Pozo JC 2010 The *Arabidopsis* cell cycle F-box protein SKP2A binds to auxin. *Plant Cell* 22:3891–3904.

Jurado S, Trivino SD, Abraham Z, Manzano C, Gutierrez C, Del Pozo C 2008 SKP2A protein, an F-box that regulates cell division, is degraded via the ubiquitin pathway. *Plant Sig Behav* 3:810–812.

Kepinski S, Leyser O 2005 The *Arabidopsis* F-box protein TIR1 is an auxin receptor. *Nature* 435:446–451.

Kleine-Vehn J, Wabnik K, Martiniere A et al. 2011 Recycling, clustering, and endocytosis jointly maintain PIN auxin carrier polarity at the plasma membrane. *Mol Syst Biol* 7:540.

Kovtun Y, Chiu WL, Zeng W, Sheen J 1998 Suppression of auxin signal transduction by a MAPK cascade in higher plants. *Nature* 395:716–720.

Kramer EM, Bennett MJ 2006 Auxin transport: a field in flux. *Trends Plant Sci* 11:382–386.

Krupinski P, Jonsson H 2010 Modeling auxin-regulated development. *Cold Spring Harb Perspect Biol* 2:a001560.

Laskowski M, Grieneisen VA, Hofhuis H et al. 2008 Root system architecture from coupling cell shape to auxin transport. *PLoS Biol* 6:e307.

Lechner E, Achard P, Vansiri A, Potuschak T, Genschik P 2006 F-box proteins everywhere. *Curr Opin Plant Biol* 9:631–638.

Lee JS, Wang S, Sritubtim S, Chen JG, Ellis BE 2009 *Arabidopsis* mitogen-activated protein kinase MPK12 interacts with the MAPK phosphatase IBR5 and regulates auxin signaling. *Plant J* 57:975–985.

Liscum E, Reed JW 2002 Genetics of Aux/IAA and ARF action in plant growth and development. *Plant Mol Biol* 49:387–400.

Löbler M, Klämbt D 1985 Auxin-binding protein from coleoptile membranes of corn (*Zea mays* L.). I. Purification by immunological methods and characterization. *J Biol Chem* 260:9848–9853.

Maraschin Fdos S, Memelink J, Offringa R 2009 Auxin-induced, SCF(TIR1)-mediated poly-ubiquitination marks AUX/IAA proteins for degradation. *Plant J* 59:100–109.

Matsuzaki Y, Ogawa-Ohnishi M, Mori A, Matsubayashi Y 2010 Secreted peptide signals required for maintenance of root stem cell niche in *Arabidopsis*. *Science* 329:1065–1067.

Mockaitis K, Estelle M 2008 Auxin receptors and plant development: a new signaling paradigm. *Annu Rev Cell Dev Biol* 24:55–80.

Mockaitis K, Howell SH 2000 Auxin induces mitogenic activated protein kinase (MAPK) activation in roots of *Arabidopsis* seedlings. *Plant J* 24:785–796.

Navarro L, Dunoyer P, Jay F et al. 2006 A plant miRNA contributes to antibacterial resistance by repressing auxin signaling. *Science* 312:436–439.

Overvoorde P, Fukaki H, Beeckman T 2010 Auxin control of root development. *Cold Spring Harb Perspect Biol* 2:a001537.

Parry G, Calderon-Villalobos LI, Prigge M et al. 2009 Complex regulation of the TIR1/AFB family of auxin receptors. *Proc Natl Acad Sci USA* 106:22540–22545.

Peret B, De Rybel B, Casimiro I et al. 2009 *Arabidopsis* lateral root development: an emerging story. *Trends Plant Sci* 14:399–408.

Perez-Torres CA, Lopez-Bucio J, Cruz-Ramirez A et al. 2008 Phosphate availability alters lateral root development in *Arabidopsis* by modulating auxin sensitivity via a mechanism involving the TIR1 auxin receptor. *Plant Cell* 20:3258–3272.

Perrot-Rechenmann C 2010 Cellular responses to auxin: division versus expansion. *Cold Spring Harb Perspect Biol* 2:a001446.

Petersson SV, Johansson AI, Kowalczyk M et al. 2009 An auxin gradient and maximum in the *Arabidopsis* root apex shown by high-resolution cell-specific analysis of IAA distribution and synthesis. *Plant Cell* 21:1659–1668.

Robert HS, Friml J 2009 Auxin and other signals on the move in plants. *Nat Chem Biol* 5:325–332.

Robert S, Kleine-Vehn J, Barbez E et al. 2010 ABP1 mediates auxin inhibition of clathrin-dependent endocytosis in *Arabidopsis*. *Cell* 143:111–121.

Ruegger M, Dewey E, Gray WM, Hobbie L, Turner J, Estelle M 1998 The TIR1 protein of *Arabidopsis* functions in auxin response and is related to human SKP2 and yeast Grr1p. *Gen Dev*12:198–207.

Santner A, Estelle M 2010 The ubiquitin-proteasome system regulates plant hormone signaling. *Plant J* 61:1029–1040.

Santuari L, Scacchi E, Rodriguez-Villalon A et al. 2011 Positional information by differential endocytosis splits auxin response to drive *Arabidopsis* root meristem growth. *Curr Biol* 21:1918–1923.

Sauer M, Kleine-Vehn J 2011 AUXIN BINDING PROTEIN1: The outsider. *Plant Cell* 23:2033–2043.

Schlereth A, Moller B, Liu W et al. 2010 MONOPTEROS controls embryonic root initiation by regulating a mobile transcription factor. *Nature* 464:913–916.

Schmid M, Davison TS, Henz SR et al. 2005 A gene expression map of *Arabidopsis thaliana* development. *Nat Genet* 37:501–506.

Shi JH, Yang ZB 2011 Is ABP1 an auxin receptor yet? *Mol Plant* 4:635–640.

Shimomura S, Sotobayashi T, Futai M, Fukui T 1986 Purification and properties of an auxin-binding protein from maize shoot membranes. *J Biochem* 99:1513–1524.

Son O, Cho SK, Kim SJ, Kim WT 2010 in vitro and in vivo interaction of AtRma2 E3 ubiquitin ligase and auxin binding protein 1. *Biochem Biophys Res Comm* 393:492–497.

Stepanova AN, Robertson-Hoyt J, Yun J et al. 2008 TAA1-mediated auxin biosynthesis is essential for hormone crosstalk and plant development. *Cell* 133:177–191.

Strader LC, Monroe-Augustus M, Bartel B 2008 The IBR5 phosphatase promotes *Arabidopsis* auxin responses through a novel mechanism distinct from TIR1-mediated repressor degradation. *BMC Plant Biol* 8:41.

Stuttmann J, Lechner E, Guerois R et al. 2009 COP9 signalosome- and 26S proteasome-dependent regulation of SCF^{TIR1} accumulation in *Arabidopsis*. *J Biol Chem* 284:7920–7930.

Swarup K, Benkova E, Swarup R et al. 2008 The auxin influx carrier LAX3 promotes lateral root emergence. *Nat Cell Biol* 10:946–954.

Swarup R, Bennett M 2003 Auxin transport: the fountain of life in plants? *Dev Cell* 5:824–826.

Swarup R, Friml J, Marchant A et al. 2001 Localization of the auxin permease AUX1 suggests two functionally distinct hormone transport pathways operate in the *Arabidopsis* root apex. *Genes Dev* 15:2648–2653.

Szemenyei H, Hannon M, Long JA 2008 TOPLESS mediates auxin-dependent transcriptional repression during *Arabidopsis* embryogenesis. *Science* 319:1384–1386.

Tan X, Calderon-Villalobos LI, Sharon M et al. 2007 Mechanism of auxin perception by the TIR1 ubiquitin ligase. *Nature* 446:640–645.

Tanaka H, Dhonukshe P, Brewer PB, Friml J 2006 Spatiotemporal asymmetric auxin distribution: a means to coordinate plant development. *Cell Mol Life Sci* 63:2738–2754.

Tatematsu K, Kumagai S, Muto H et al. 2004 *MASSUGU2* encodes Aux/IAA19, an auxin-regulated protein that functions together with the transcriptional activator NPH4/ARF7 to regulate differential growth responses of hypocotyl and formation of lateral roots in *Arabidopsis thaliana*. *Plant Cell* 16:379–393.

Thiel G, Blatt MR, Fricker MD, White IR, Millner P 1993 Modulation of K^+ channels in Vicia stomatal guard cells by peptide homologs to the auxin-binding protein C terminus. *Proc Natl Acad Sci USA* 90:11493–11497.

Tian H, Klambt D, Jones AM 1995 Auxin-binding protein 1 does not bind auxin within the endoplasmic reticulum despite this being the predominant subcellular location for this hormone receptor. *J Biol Chem* 270:26962–26969.

Tromas A, Braun N, Muller P et al. 2009 The AUXIN BINDING PROTEIN 1 is required for differential auxin responses mediating root growth. *PLoS One* 4:e6648.

Tromas A, Paponov I, Perrot-Rechenmann C 2010 AUXIN BINDING PROTEIN 1:Functional and evolutionary aspects. *Trends Plant Sci* 15:436–446.

Vanneste S, De Rybel B, Beemster GT et al. 2005 Cell cycle progression in the pericycle is not sufficient for SOLITARY ROOT/IAA14-mediated lateral root initiation in *Arabidopsis thaliana*. *Plant Cell* 17:3035–3050.

Vanneste S, Friml J 2009 Auxin: a trigger for change in plant development. *Cell* 136:1005–1016.

Venis MA, Napier RM, Barbier-Brygoo H, Maurel C, Perrot-Rechenmann C, Guern J 1992 Antibodies to a peptide from the maize auxin-binding protein have auxin agonist activity. *Proc Natl Acad Sci USA* 89:7208–7212.

Vierstra RD 2009 The ubiquitin-26S proteasome system at the nexus of plant biology. *Nat Rev Mol Cell Biol* 10:385–397.

Walsh TA, Neal R, Merlo AO et al. 2006 Mutations in an auxin receptor homolog AFB5 and in SGT1b confer resistance to synthetic picolinate auxins and not to 2,4-dichlorophenoxy-acetic acid or indole-3-acetic acid in *Arabidopsis. Plant Physiol* 142:542–552.

Weijers D, Benkova E, Jager KE et al. 2005 Developmental specificity of auxin response by pairs of ARF and Aux/IAA transcriptional regulators. *Embo J* 24:1874–1885.

Weijers D, Schlereth A, Ehrismann JS, Schwank G, Kientz M, Jurgens G 2006 Auxin triggers transient local signaling for cell specification in *Arabidopsis* embryogenesis. *Dev Cell* 10:265–270.

Willemsen V, Scheres B. 2004 Mechanisms of pattern formation in plant embryogenesis. *Annu Rev Genet* 38:587–614.

Woo EJ, Bauly J, Chen JG et al. 2000 Crystallization and preliminary X-ray analysis of the auxin receptor ABP1. *Acta Crystallogr D Biol Crystallogr* 56:1476–1478.

Woo EJ, Marshall J, Bauly J et al. 2002 Crystal structure of auxin-binding protein 1 in complex with auxin. *Embo J* 21:2877–2885.

Xu T, Wen M, Nagawa S et al. 2010 Cell surface- and rho GTPase-based auxin signaling controls cellular interdigitation in *Arabidopsis. Cell* 143:99–110.

13

Role of Gibberellins in Root Growth

Eiichi Tanimoto
Nagoya City University

Ko Hirano
Nagoya University

I. Introduction

Gibberellin (GA) is one group of plant hormones that has a chemical structure of gibberellane skeleton (Figure 13.1). GA is involved in almost all phases of plant growth and development. It promotes germination, elongation growth, flowering, and fruit development. Although several books and reviews have been published on the physiological role of GA, relatively limited references are available for roots, since GA does not strongly promote root elongation of many plants. GA was identified as a fungal toxin to cause unusual shoot elongation of rice plants, and it was further recognized as a plant hormone to regulate specifically shoot growth. However, by using GA-deficient mutants and inhibitors of GA biosynthesis, it is now evident that GA controls root growth at a low concentration range. Furthermore, molecular mechanism of GA signal transduction has recently been unveiled, and the sequential events occurring in shoots were found to take place also in roots. A short history and the research literatures of GA discovery were described in the previous edition (Tanimoto 2002).

Gibberellin was named after the Latin name of the fungus *Gibberella fujikuroi* (Saw.) Wr. that causes abnormal elongation of rice plants (Hori 1898). In 1954–1955, three research groups reported the isolation of active principle from the fungus. Soon after the discovery of GA as a fungal toxin, it was also isolated from higher plants. In consequence of the diverse structural flexibility of the gibberellane skeleton, 136 kinds of GAs have been identified to date from higher plants and from microorganisms; GA_{133}–GA_{136} were identified from loquat fruit (*Eriobotrya japonica* Lindl.) (Phuoc et al. 2008). The most updated information is available at http://www.plant-hormones.info/

II. Biosynthesis and Metabolism

A. Biosynthesis of Active GAs and Their Inactivation

The biosynthetic pathway of GA has been extensively studied in *Gibberella fujikuroi*, and the interconversion pathways of these GAs in plants have been investigated (Davies 1995; Frankenberger and Arshad 1995). The biosynthetic pathway and the mutational effect of enzymes involved in this pathway on GA production have been reviewed (Hedden and Kamiya 1997; Hedden 1999; Hedden and Proebsting 1999; Yamaguchi 2008). Of the 136 GAs identified to date, relatively few are thought to be physiologically active. The most known bioactive GAs are GA_1, GA_3, and GA_4, which are found in most of the higher plants (Figure 13.1).

The whole GA biosynthetic pathway is separated into three steps as shown in Figure 13.2. The first step is the formation of *ent*-kaurene. The *ent*-kaurene is made from isopentenyldiphosphate (IPP). It has long been believed that IPP is synthesized from mevalonic acid. In 1997, Lichtenthaler et al. found that IPP is formed from pyruvate and glyceraldehyde 3-phosphate via 1-deoxyxylulose in plastids. Hedden (1999) has reviewed the nonmevalonate pathway to IPP. Copalyl diphosphate synthase

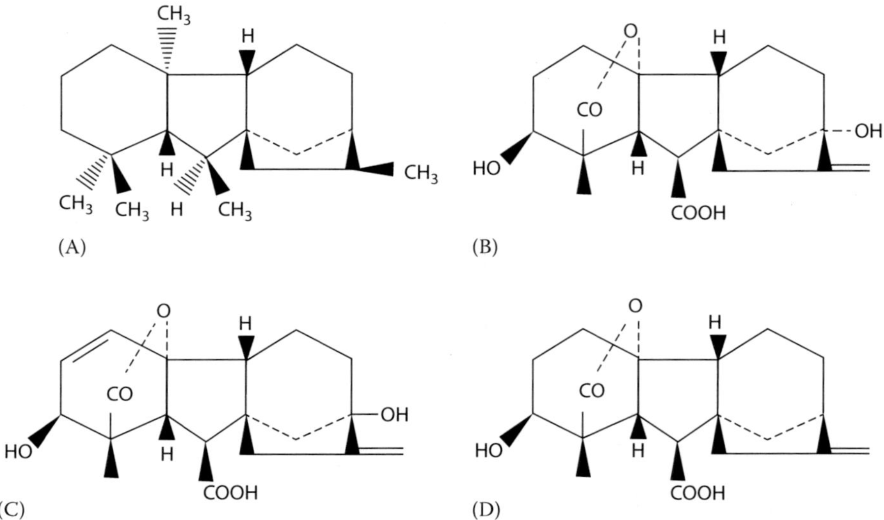

FIGURE 13.1 Structure of (A) *ent*-gibberellane, (B) GA_1, (C) GA_3, and (D) GA_4.

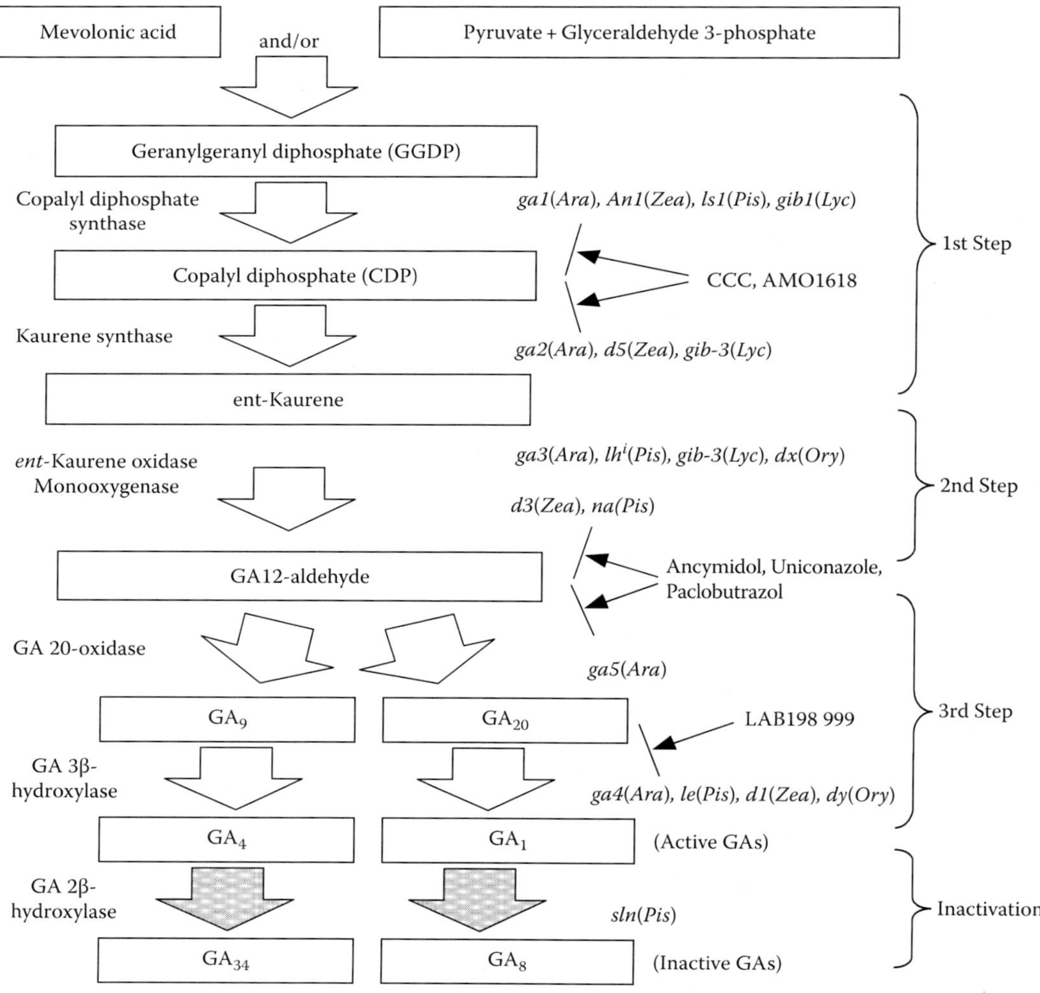

FIGURE 13.2 GA biosynthetic pathway, enzymes, corresponding mutant genes, and the site of inhibitor action.

(CPS) is thought to be the rate-limiting step of *ent*-kaurene synthesis (West et al. 1982), and the expression of CPS (*ga1*) gene was highest in shoot apices, root tips, and developing flowers in *Arabidopsis* (Silverstone et al. 1997). These data indicate that not only the shoot apices but also root tips are the source of the GA precursor.

The second step in the GA pathway is oxidation of *ent*-kaurene to GA_{12}-aldehyde. The *ent*-kaurene is oxidized by membrane-bound monooxygenases. Several GA-deficient dwarf mutants of *Arabidopsis*, pea, tomato, and rice are defective in *ent*-kaurene oxidase activity (Figure 13.2). The third and final step is the formation of the bioactive hormones GA_1 and GA_4 from GA_{12}-aldehyde. These reactions involve oxidative removal of C-20 to give a lactone between C-19 and C-10 and hydroxylation at the C-3β position. The former step is catalyzed by soluble oxidases that use 2-oxoglutarate as a cosubstrate. The latter reaction is the last and critical reaction to produce active GA_1 and GA_4. Further hydroxylation at C-2β position results in inactive GA_8 and GA_{34}. Not only GA biosynthesis but also the conversion to inactive form determines the active GA level (Figure 13.2). The genetic deletion of this inactivation step results in the accumulation of active GA that makes the mutant slender and tall like *sln* mutant of pea. Enzymes participating in hydroxylation at these C-2β and C-3β positions are under the control of auxin coming down from the shoot apical meristem (Figure 13.3) (Ross et al. 2000, 2002). The gene expression of enzymes participating in active GA biosynthesis, such as GA_{20}ox and GA_3ox, is also feedback regulated by endogenous level of active GA (Bidadi et al. 2010). The majority of genes encoding enzymes participating in GA biosynthesis and their cellular localization have been elucidated in *Arabidopsis thaliana* and *Oryza sativa* (Yamaguchi 2008).

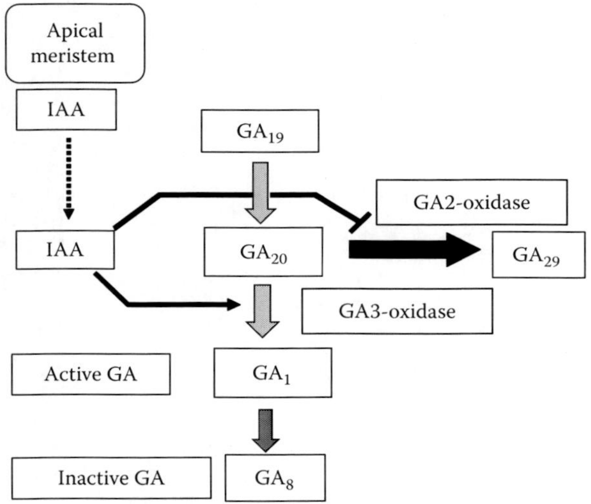

FIGURE 13.3 Regulation of GA_1 level by auxin (IAA) transported from apical meristem. IAA transported down from apical meristem promotes GA3-oxidase gene and represses GA2-oxidase gene, resulting in the increase in GA_1 level.

B. Site of GA Synthesis and Transport

GAs are known to move relatively freely from shoots to roots (Prochazka 1981; Matthysse and Scott 1984). There are no specific transporters reported for GAs (Reviews Leyser 2002; Tanimoto 2005; Kerr and Bennett 2007). Although at least some GAs seem to be transported from the roots to the shoots, a gradient pattern of endogenous GA concentration suggests that majority of GAs are transported from young developing tissue to older tissues. Since the enzymes of early GA production such as CPS are expressed in root tips, roots are, at least in part, a source of GA precursors. The gene for 3β-hydroxylase is also expressed in meristematic cells including root tips (Itoh et al. 1999). However, little information is available regarding the importance of these enzymes that are active in the roots in providing GAs to the shoots and the roots. The 3β-hydroxylase genes are suggested to be separately regulated in roots and stems in pea, since *le*-mutant with stunted stems caused by a mutation in 3β-hydroxylase genes has long roots (Figure 13.4) and contains sufficient amounts of active GA_1 in roots (Yaxley et al. 2001).

Growth-promoting GA_3 is mobile, since cotyledon- or shoot-applied GA_3 exerted growth promotion of roots, whereas root-applied GA_3 exerted growth promotion of shoots, in pea (Tanimoto 1994). Shoot-applied GA_3 also promoted root elongation in *Arabidopsis* (Bidadi et al. 2010). Additionally, [H^3] GA_1 is readily taken up and distributed throughout the whole plant (Davies and Rappaport 1975). By grafting experiments of GA-deficient mutants, the role of root-originated GA in shoot growth was evaluated (Dodd 2005). The ability to restore stem

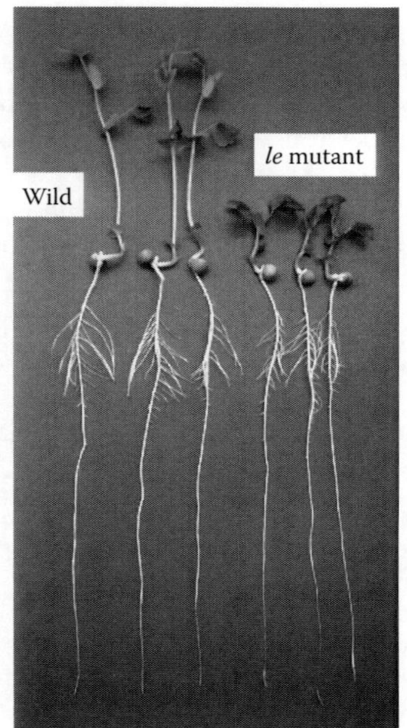

FIGURE 13.4 Ten-day-old seedlings of dwarf *le* mutant of pea and its wild counterparts. Stems are stunted but not the roots. (Pea seeds were kindly supplied by Dr. J. B. Reid, University Tasmania, Australia).

elongation of a GA-deficient scion is dependent on the nature of the malfunction in the GA biosynthetic pathway. WT rootstocks restored stem elongation of *nana* pea scions (Proebsting et al. 1992), but did not restore the growth of *le* scions (Reid et al. 1983). More recent gene-expression studies suggest that the sites of GA biosynthesis and response are mutually correlated, that is, young growing tissues show high expression of genes participating in GA biosynthesis (Kaneko et al. 2003; Yamaguchi 2008). Further quantitative measurement of organ-to-organ and tissue-to-tissue transport of GA_1 and its precursors is necessary for understanding the precise function of GA in a whole plant body.

C. GA in Soil

Soil-applied GA promotes shoot and root growth of dwarf maize (Frankenberger and Arshad 1995) revealing that GAs in the soil can affect plant growth. There is considerable evidence for the microbial production of GAs in soil (Frankenberger and Arshad 1995). GA-producing pathogenic fungi such as *Gibberella fujikuroi* strongly affect plant growth. Other soil microorganisms can also synthesize GAs and/or GA-like substances that may influence plants via the rhizosphere. But data regarding the stability of soil GA and its availability to plants are little known.

D. Chemical Inhibitors of GA Biosynthesis

Three types of inhibitors are known: onium type (CCC, AMO-1618), nitrogen-containing cycle type (ancymidol, paclobutrazol, uniconazole), and cyclohexantrion type (LAB198999) (Figure 13.5) (see Rademacher 1991). The action sites of these inhibitors are shown in Figure 13.2. These chemicals are being used for growth retardation of plants, for cultivation of tulips and chrysanthemums with short flower stalks, and for production of small plants for decoration.

In addition to these chemical inhibitors, GA-deficient dwarf cultivars were also employed to investigate the growth-enhancing activity of GA, since these plants produce only a small amount of endogenous GA and show high response to externally applied GA. Many of these cultivars were found to be mutants of GA biosynthesis. These mutants and cDNA clones for GA biosynthetic enzymes were reviewed by Hedden and Kamiya (1997). Some of these mutant genes are also indicated in Figure 13.2. Combination of these GA-less mutants with the chemical inhibitors of GA biosynthesis provided experimental plant material extremely sensitive to GA, for bioassay studies of GA quantification (Nishijima and Katsura 1989) and for root growth experiments, as shown in Figure 13.8.

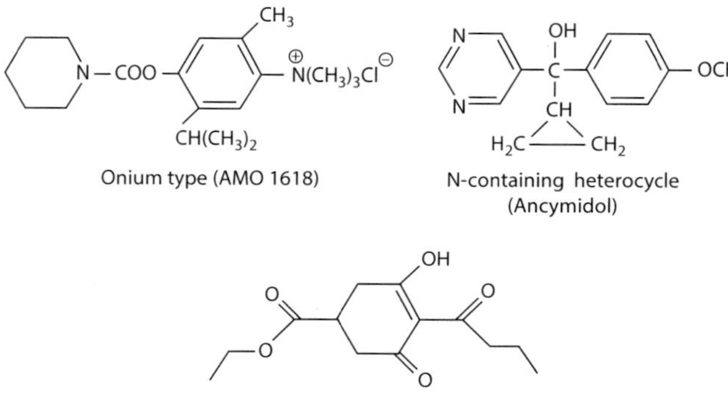

Onium type (AMO 1618)

N-containing heterocycle (Ancymidol)

Cyclohexanetrione (LAB 198999)

FIGURE 13.5 Synthetic plant growth retardants inhibiting GA biosynthesis. Site of action of these compounds is shown in Figure 13.2.

III. GA Functions in Roots

A. GA-Mediated Growth Regulation of Roots

There are conflicting reports of GA effect on root growth (Table 13.1) (Torrey 1976; Feldman 1984; Phinney 1984). However, basic information regarding GA functions on root growth has been established at least for some plant species such as peas and lettuce (reviewed by Tanimoto 2005).

Although externally applied, GA showed little effect on root elongation, the requirement of GA for root growth was suggested by the use of inhibitors of GA biosynthesis, such as CCC, AMO-1618, and ancymidol (Figure 13.5), which suppressed root growth. However, the requirement of endogenous GA for root growth has been debated since the suppression of root growth by some of these inhibitors was not reversed by the addition of external GA (Sachs and Kofranek 1963; Dunberg and Eliasson 1972; Crozier et al. 1973) in contrast to the case of shoot growth where growth inhibition was overcome by externally applied GA. But higher concentrations of the inhibitor were required for inhibition of root growth than for such inhibition of the shoots (Figure 13.6).

TABLE 13.1 Examples of Conflicting Effect of Externally Applied GA on Root Growth

Plant	Promotion	Inhibition	No Effect
Lactuca sativa L.	Aspinall et al. (1967)	Krekule and Ullman (1959)	Sawhney and Srivastava (1974)
Lycopersicon esculentum Mill.	Butcher and Street (1960)	Tognoni et al. (1967)	Finnie and Van Staden (1985)
Oryza sativa L.	Suge (1985)		Ogawa et al. (1976)
	Radi and Maeda (1988)		Katayama and Akita (1989)
Pisum sativum L.	Pecket (1960)		Manos (1961)
Triticum aestivum L.	Burström (1960)	Burström (1960)	
Zea mays L.	Whaley and Kephart (1957); Mertz (1966)	Konings and Wolf (1984); Svensson (1972)	Svensson (1972) Hirota (1980)

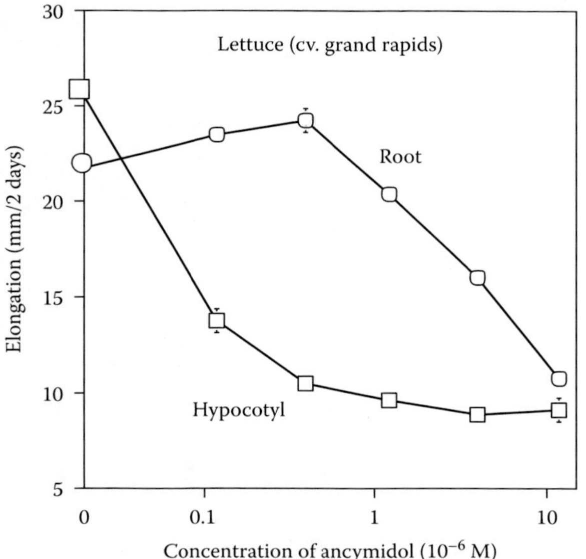

FIGURE 13.6 Inhibition of root and hypocotyl elongation by ancymidol in lettuce (*Lactuca sativa* L.). Hypocotyl was stunted at lower concentration than root. (Adapted from Tanimoto, E., *Plant Cell Physiol.*, 28, 963, 1987. With permission.)

In addition, recovery of root growth by GA addition was observed at a lower concentration range than those required for shoot growth (Figure 13.7). The elongation process of roots was followed during the inhibition and recovery phases. The complete recovery of the elongation rate was recorded by a rhizometer with minimum mechanical contact to roots (Figure 13.8) (Tanimoto and Watanabe 1986; Tanimoto 1988). When ancymidol-pretreated seedlings of lettuce were treated

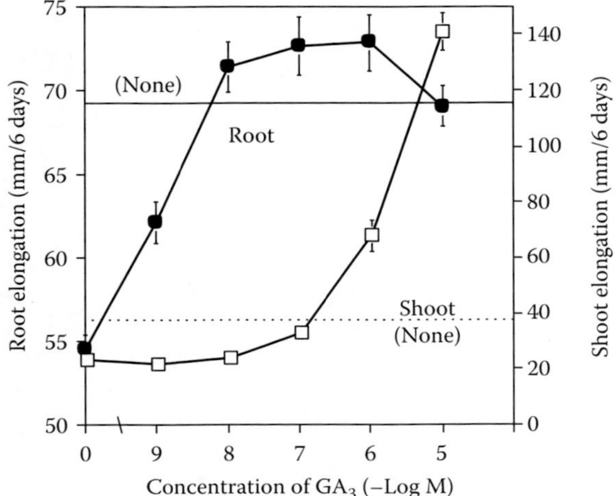

FIGURE 13.7 Promotion of root and shoot elongation of dwarf pea (cv. Little Marvel) by GA_3 in the presence of 3 * 10^{-6} M ancymidol. Root elongation was promoted at 1 nM GA_3 and saturated at 0.1 µM GA_3, whereas shoot elongation was enhanced at 1 µM and higher GA_3 concentrations. (Adapted from Tanimoto, E., *Plant Cell Physiol.*, 35, 1019, 1994. With permission.)

with two different concentrations of GA_3, 1 nmolar GA_3 promoted only root elongation, whereas 1 µmolar GA_3 enhanced both root and shoot growth (Figure 13.8).

The conflicting results of externally applied GA on root growth were suggested to have originated by the interaction with other hormones, auxin, ethylene, and cytokinins (Tanimoto 2005). As for the interaction of GA and auxin, Willige et al. (2011) recently reported that auxin transport is reduced in the inflorescences of *Arabidopsis thaliana* mutants deficient in GA biosynthesis and signaling. Their molecular investigations demonstrated that this reduced auxin transport correlates with a reduction in the abundance of PIN-FORMED (PIN) auxin efflux facilitators in GA-deficient plants and that PIN protein levels recover to wild-type levels following GA treatment. GA plays a role in keeping a normal status of PIN proteins and in supporting normal auxin transport from the shoot apex down to the root tips.

The mode of root elongation in GA-deficient dwarf mutants has also been an important reason why the role of GA in root elongation has been underestimated. It is well known that roots of dwarf peas (*le, na*) and dwarf maize (*d1, d5*) were not stunted, and the root length was comparable to that of the wild-type counterparts (Figure 13.4). As indicated in Figures 13.7 and 13.8, GA level required for root elongation is quite low so that roots can elongate while shoot elongation is suppressed at low GA level. Endogenous GA_1 content in the roots was compared between *le*-mutant and wild counterpart and was found not to be affected by *le*-mutation (Yaxley et al. 2001). Thus, the final step of GA_1 biosynthesis by 3β-hydroxylase is thought to be separately regulated in pea root and stem, probably by gene redundancy.

As shown in Figure 13.6, inhibitors of GA biosynthesis such as ancymidol and uniconazole promote root elongation at low concentration range in lettuce (Tanimoto 1987) and in *Arabidopsis* (Bidadi et al. 2010). The mechanism of this promotion is not clear but may involve a feedback regulation of GA biosynthesis responding to the shortage of GA (Bidadi et al. 2010).

B. GA-Mediated Control of Root Thickness

When treated with the inhibitors of GA biosynthesis, the elongation zone of the roots becomes thick mainly due to the expansion of cortex cells, that is, hypertrophy in cortex (Figure 13.10) (Tanimoto 1987; 1988). The inhibitor-induced thickening was reversed by GA (Figures 13.9 and 13.10) (Tanimoto 1988). The GA-mediated control of thickness was ascribed to the control of cellulose microfibril orientation in the cell walls. The orientation of cortical microtubules (CMT) and cellulose microfibrils is known to be transverse to the axis of roots (Hogetsu 1986; Hogetsu and Oshima 1986). That orientation causes root cells to elongate, resulting in the slender roots. However, once this GA function was disturbed, CMT orient themselves obliquely or longitudinally to the cell axis as observed in onion leaf sheath cells (Mita and Shibaoka 1984a,b).

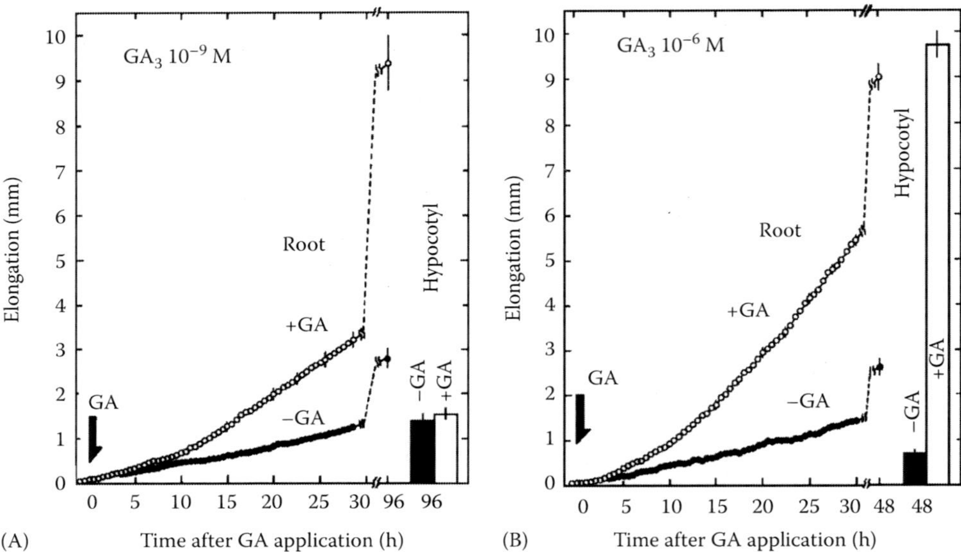

FIGURE 13.8 Differential promotion of root and hypocotyl growth by two concentrations of GA_3 in the presence of $4 * 10^{-6}$ M ancymidol in lettuce seedlings. Time course of root elongation was recorded every 30 min for 2 days in a rhizometer (From Tanimoto, E., *Plant Cell Physiol.*, 27, 1475, 1986), and the final length of roots and hypocotyls were measured after 4 days (GA_3 10^{-9} M) and after 2 days (GA_3 10^{-6} M). Whole seedlings were repeatedly dipped for 2 min in hydroponic solution containing growth regulators every 30 min. Ancymidol-suppressed growth of roots was promoted by 1 nM GA_3, whereas hypocotyl growth was not enhanced (A). GA_3 at 1 μM promoted root and hypocotyl growth in the presence of ancymidol (B).

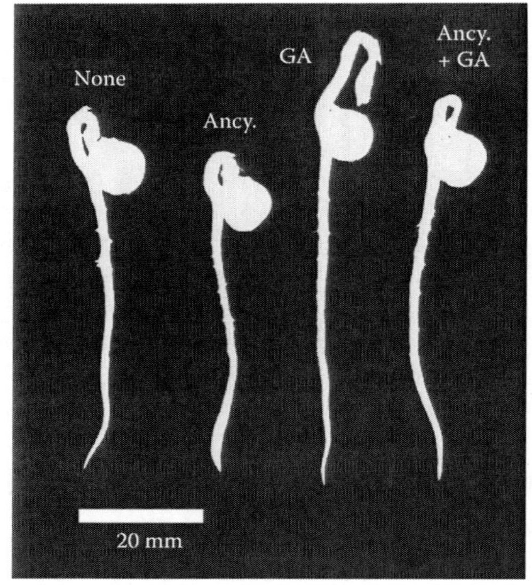

FIGURE 13.9 Shadowgraphs of pea seedlings treated with 19.5 μM ancymidol and/or 10 μM GA_3. Apical growing part of root is thickened by ancymidol, and GA_3 reversed it. (Adapted from Tanimoto, E., *Plant Cell Physiol.*, 29, 269, 1988. With permission.)

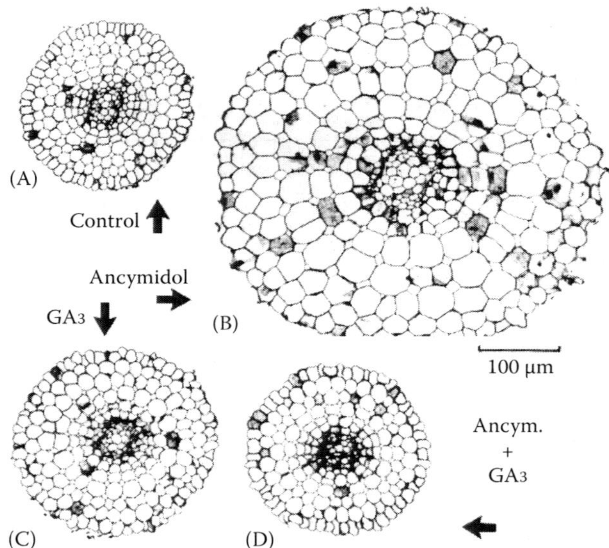

FIGURE 13.10 Ancymidol-induced thickening of lettuce root and its prevention by GA_3. (A) No hormone treatment, (B) Ancymidol-treatment, (C) GA_3-treatment, (D) Ancymidol + GA_3-treatment. (Adapted from Tanimoto, E., *Plant Cell Physiol.*, 28, 963, 1987. With permission.)

The same function of GA was discovered in roots by Baluška et al. (1993). They observed a GA-mediated transverse orientation of CMT in the distal elongation zone of the dwarf maize *d5*. GA-mediated control of root thickness was also observed in tea tree, *Camellia sinensis*, which has woody cell walls with mechanically harder properties than that of herbs and grasses (Tanimoto et al. 2004a,b). Thus, GA control of root thickening seems to be a common phenomenon in many plant families.

C. GA-Mediated Growth and Cell Wall Extension

Extension growth of plant cell is controlled by the rigidity of cell walls. The mechanical extensibility of cell walls is thought to participate in GA-mediated root elongation since GA increased the cell wall extensibility of pea root cells (reviewed by Tanimoto 2002).

D. GA-Mediated Gene Expression in Roots

The transcription of several genes was found to be upregulated by GAs, but the function of many of the genes is still unknown. *GASA* (*Gibberellic Acid Stimulated Arabidopsis*) gene family in *Arabidopsis* (Aubert et al. 1998) and *GAST* (*Gibberellic Acid Stimulated Transcript*) *1, 3, 4* in tomato (Taylor and Scheuring 1994) are expressed upon GA application in seedlings and roots. *pGASA4*-GUS expression in transgenic *Arabidopsis* was found primarily in all meristematic regions, including vegetative, inflorescence, and floral meristems, as well as in primary and lateral root tips. The expression pattern suggests that the *AtGASA4* plays a role in dividing cells rather than in elongating cells. GAST-like gene family was found to participate in the early lateral root development in rice and maize regulated by GA (Zimmermann et al. 2010).

Further studies on the products of these GA-regulated genes will unravel the GA-regulated growth and development of roots. For details of these molecular events, see Section IV.

E. Effect of GA on Rooting

Auxin strongly promotes rooting of cuttings and is thought to be essential for the differentiation of root primordia. In contrast, GA is rather inhibitory for rooting (Goldfarb et al. 1997). In fact, GA (Goldfarb et al. 1997) does not enhance the expression of Loblolly Pine (*Pinus taeda* L.) *Early Auxin-induced* (LPEA3) gene, which is thought to be involved in auxin-induced rooting, above the basal level. The findings that growth retardants such as paclobutrazol promote rooting of cuttings but simultaneously applied GA prevented the rooting (Porlingis and Koukourikou-Petridou 1996) suggest that GA is less required for rooting and rather inhibitory for auxin-promoted rooting.

IV. Signal Transduction of GA Actions

A. Early GA-Signaling Events

In contrast to the long research history of GA biosynthesis pathway, most of our knowledge on GA perception has been obtained during this past decade. Important signaling factors identified are GA receptor GID1 (GA INSENSITIVE DWARF1) (Ueguchi-Tanaka et al. 2005, Nakajima et al. 2006), DELLA proteins (Peng et al. 1997; Silverstone et al. 1998; Ikeda et al. 2001), and the F-box proteins GID2 (rice GA INSENSITIVE DWARF2)/SLY1 (*Arabidopsis* SLEEPY1) (McGinnis et al. 2003; Sasaki et al. 2003). We will briefly summarize how these factors function in GA signaling in the following.

The key event in GA signaling is the degradation of DELLA proteins, which are repressors of GA action. When GA receptor GID1 binds to GA (Ueguchi-Tanaka et al. 2005), GID1 forms a complex with DELLA proteins. The DELLA protein within the complex is then recognized and ubiquitinated by the $SCF^{GID2/SLY1}$ (Skp-CULLIN-GID2/SLY1 F-box protein) to be targeted for the 26S proteasome degradation (Figure 13.11) (Fu et al. 2002; McGinnis et al. 2003; Sasaki et al. 2003; Wang et al. 2009; Hirano et al. 2010). As a consequence, repressive activity of DELLA protein is relieved, and GA action such as cell elongation and induction of flowering occurs (Figure 13.11). Apart from the above mentioned model, it is now assumed that the repressive activity of DELLA protein can at least in part be inactivated merely by

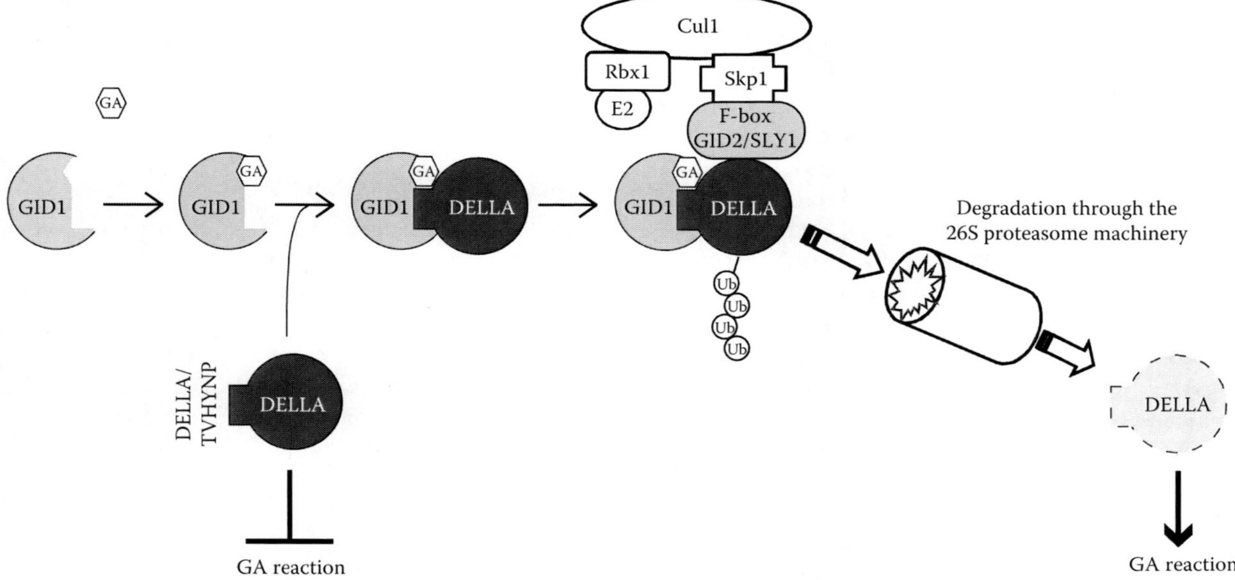

FIGURE 13.11 Model of GA signaling pathway in vascular plants. In the presence of GA, GID1 binds to DELLA, a negative regulator of GA action, to form the GA-GID1-DELLA complex. DELLA protein within the complex is recognized and ubiquitinated by the $SCF^{GID2/SLY1}$ complex and subsequently degraded through the 26S proteasome pathway. As a consequence, GA actions occur. (Adapted and modified from Hirano, K. et al., *Plant Cell*, 19, 3058, 2007.)

interacting with GID1 without requiring DELLA-protein degradation (Ariizumi et al. 2008; Ueguchi-Tanaka et al. 2008).

All of the current genetic, biochemical, and molecular experimental data support the earlier model of GA signaling. Of note, recent x-ray crystallography of GID1 has provided structure-based evidence that GA, GID1, and DELLA protein do interact (Murase et al. 2008; Shimada et al. 2008). It is assumed that upon GA-GID1 binding, the N-terminal lid region of GID1 covers the GA binding site. This conformational change of GID1 possibly allows DELLA protein to interact with the lid region of GID1, which could explain why GA is required for GID1 and DELLA protein to interact. Apart from the earlier GA-signaling factors, SPINDLY (SPY), a nucleocytoplasmic localized O-linked GlcNAc transferase, is known to function negatively in GA signaling by increasing the suppressive activity of DELLA protein (Olszewski et al. 2002).

Events that occur downstream of DELLA proteins are also beginning to be unraveled. For more details, we refer to some of the recent reviews on these topics (Hartweck 2008; Sun 2010).

B. GA Signaling in the Root System

1. Role of GA in the *Arabidopsis* Middle Cortex Formation

In *Arabidopsis* root, the endodermis and cortex each exist as a single layer at the early stage of development. At later stages (more than 2 weeks after germination), an additional longitudinal asymmetric cell division is often observed. It results in the formation of a second layer of cortex defined as the middle cortex. SCARECROW and LHP1 (LIKE HETEROCHROMATIN PROTEIN 1, an *Arabidopsis* homolog of chromosomal protein HP1) are assumed to repress this second cell division, since middle cortex in both the *scr* and the *lhp1* mutants appear earlier than the wild-type plants (Cui and Benfey 2009). Treatment of GA suppresses this mutant phenotype, whereas many of the GA-signaling-deficient mutants form a precocious middle cortex suggesting that DELLA protein functions to induce middle cortex formation. The only exception is that *spy* mutant which is expected to enhance GA signaling also forms precocious middle cortex, which cannot be explained at the moment. Very recently, mutation in the *Arabidopsis SCARECROW LIKE 3* (*SCL3*) was shown to result in precocious middle cortex formation, while constitutive expression of *SCL3* delayed the formation of middle cortex (Heo et al. 2011). Further studies suggested that SCL3 functions positively in GA signaling by attenuating DELLA protein (Heo et al. 2011; Zhang et al. 2011), which could explain why constitutive expression of SCL3 shows delayed middle cortex formation similar to that observed for GA treatment.

2. GA and Root Elongation

Arabidopsis root elongation growth depends on GA. Root length of GA-deficient *ga1-3* mutant (having mutation in the *ent*-CPS gene GA1, Figures 13.2 and 13.12) is shorter than that of the wild type, but this phenotype is rescued by GA application (Fu and Harberd 2003). Similarly, *ga1-3/gai-t6/rga24* triple mutant

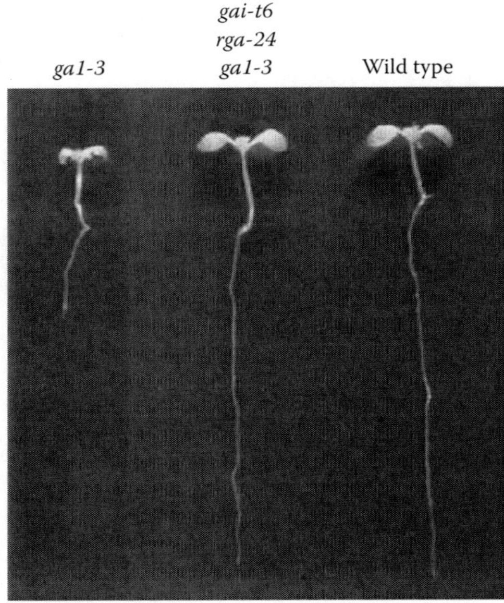

FIGURE 13.12 Representative 5-day-old primary seedling roots of *gai-t6/rga-24/ga1-3*, and the wild type. (Adapted from Fu, X. and Harberd, N.P., *Nature*, 421, 740, 2003. With permission.).

lacking GA1 and the DELLA proteins, GAI and RGA, have longer roots than *ga1-3* plants indicating that GA-enhanced root elongation is DELLA dependent (Figure 13.12) (Fu and Harberd 2003). To gain insight into the site of GA action in roots, Ubeda-Tomás et al. (2008) selectively expressed the nondegradable form of GAI (ndGAI) in each of the root tissues and observed the effect on root elongation zone (EZ) length. Results showed that root elongation was inhibited only when ndGAI was expressed in the endodermis, implying that regulation of root elongation by GA is governed primarily by the endodermis. Addition to the function of SCL3 in middle cortex formation mentioned earlier, SCL3 also positively controls root cell elongation in the EZ possibly through the regulation of GA signaling (Heo et al. 2011; Zhang et al. 2011).

3. GA and Root Meristem Size

Apart from cell elongation in the EZ, several groups found that root meristem size in *Arabidopsis* is also regulated by GA (Achard et al. 2009; Ubeda-Tomás et al. 2009; Moubayidin et al. 2010). Compared to the wild type, GA-deficient mutants such as *ga1-3* and *ga3ox1/ga3ox2* (double mutant of the *GA3ox* genes which encode enzymes catalyzing the last step of active GA biosynthesis, Figure 13.2) or GA biosynthesis inhibitor paclobutrazol-treated plants exhibit shorter proximal meristems (PM) due to reduced cell number and reduced cell length (Figure 13.13) (Ubeda-Tomás et al. 2009). As was described for the EZ in the previous section, the endodermis seems to be the primary GA-responsive tissue in PM cells, since only when the ndGAI was expressed in the endodermis, it showed reduced meristem size. Furthermore, GA-deficient and signaling mutants had reduced cyclin *B1;1* expression in the PM, and *ga1-3* mutant had increased expression of putative cyclin-dependent kinase

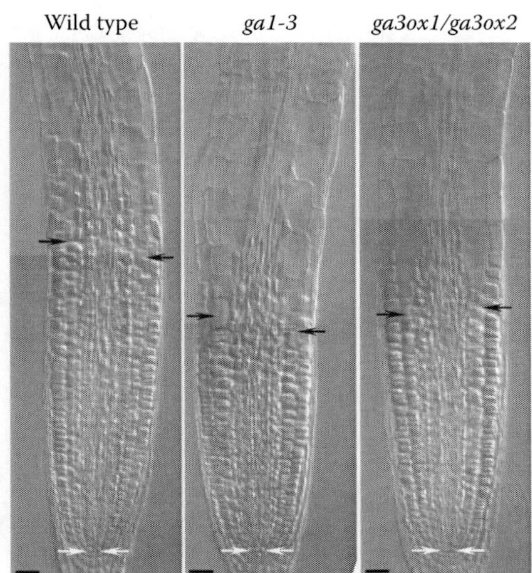

FIGURE 13.13 Bright-field microscopy images of *ga1-3* and *ga3ox1/ga3ox2* root meristem (both in Col-0 background) showing a reduction in meristem size compared to wild-type Col-0 of 4 days-after-germination seedlings. White and black arrowheads indicate quiescent center (QC) and transition zone (TZ) position, respectively. Scale bar represents 25 µm. (Adapted from Ubeda-Tomás, S. et al., *Curr. Biol.*, 19, 1194, 2009. With permission.)

inhibitors which inhibit cell-cycle progression suggestive for a role of GA in controlling the PM size through regulating cell division (Achard et al. 2009). Unlike animal cells, cell division and cell elongation in plants are restrained by adjacent tissues since plant cells are tightly connected with each other by rigid cell walls. Ubeda-Tomás et al. (2009) hypothesized that for roots to grow, cell expansion in the endodermis must first take place subsequently allowing the surrounding tissues to divide or elongate. A similar role in tissue restraining was suggested for the epidermal cells in the aerial part of the plant (Savaldi-Goldstein et al. 2007).

Moubayidin et al. (2010) also demonstrated the involvement of GA in regulating root meristem size possibly through a different mechanism. The size of the PM is determined by the balance between cell division at the PM and cell differentiation into the rapidly elongating cells in the meristem transition zone (TZ) (Figures 13.13 and 13.14A). In *Arabidopsis*, cell division rate prevails over cell differentiation rate until up to 5 days after germination, and therefore PM size increases at this stage. From 5 days onward, the meristem size is stabilized due to equivalent rate between division and differentiation. It has been well known that auxin and cytokinin antagonistically function in this event, auxin enhancing cell division whereas cytokinin, cell differentiation. An elegant work by Dello Ioio et al. (2008) revealed how these two phytohormones affect PM size. Cytokinin-response factors ARR1 and ARR12 directly enhance the expression of the Aux/IAA family gene, *SHY2* (Figure 13.14). *SHY2* is expressed in the vascular tissues of TZ and acts to reduce the expression of *PIN* auxin-transport facilitator genes. This causes reduced auxin redistribution and is suggested to lower the auxin level in the TZ and thus suppresses cell division rate. A recent follow-up paper by the same group (Moubayidin et al. 2010) elucidated that GA is at least in part involved in this auxin-cytokinin regulatory circuit (Figure 13.14). It was shown that the DELLA protein RGA can enhance the expression of the *ARR1* gene. The expression of *ARR1* is detectable only from 5 days after germination, a stage at which the final meristem size has already been determined. This is in contrast to the expression of GA nonregulated *ARR12* gene in which its expression is detectable immediately after germination. They also suggested that soon after germination, high levels of GA reduce the expression of *ARR1* by targeting DELLA protein for degradation, resulting in prevalence of cell division over differentiation (Moubayidin et al. 2010). At 5 days after germination, reduced GA level enhances *ARR1* expression, resulting in an equivalent rate of division and differentiation. As described in Section III, Willige et al. (2011) reported that auxin transport is reduced in mutants deficient in GA biosynthesis and signaling. They further revealed that GA deficiency (and thus possibly DELLA protein) promotes PIN2 protein for vacuolar degradation, implying that auxin transport in roots is controlled by various factors.

4. GA and Lateral Root Development

As stated in Section III.D, GAST-like gene family in rice and maize was found to participate in the early root development. This gene family encodes small polypeptides with 12 perfectly conserved Cys residues in the C-terminal domain, whereas their N-terminal domains are variable and contain a signal sequence characteristic of extracellular matrix secreted proteins. Recently, *Arabidopsis* GASA4 was shown to promote GA responses and suggested that these responses are dependent on the redox activity of GASA4, which requires the Cys-rich domain (Rubinovich and Weiss 2010).

5. GA Sensitivity in the Root

As mentioned in Section III, at least in lettuce and dwarf pea plants, the root possessed higher GA sensitivity than the hypocotyl and the shoot. This phenomenon might be explained by the different types of GID1 receptor functioning preferentially in the roots. *Arabidopsis* possesses three GID1 genes (*AtGID1a*, *1b*, and *1c*) with *AtGID1b* preferentially being expressed in the root and *AtGID1a* and *1c* in other organs (Griffiths et al. 2006). AtGID1b is different from the two others because it can interact with the DELLA protein in the absence of GA in yeast cells (Yamamoto et al. 2010). This suggests that AtGID1b may possess higher GA sensitivity *in planta* which could explain why roots have higher GA sensitivity compared to other organs. Actually, overexpression of *AtGID1b* in GA-deficient *ga1–3* showed longer roots compared to *AtGID1a* and *1c* overexpressed plants at low GA_4 concentration of 10^{-10} M (Ariizumi et al. 2008).

In summary, recent studies have started to unravel the molecular mechanism of GA signaling in the roots. However, spatial and temporal distribution of endogenous GA in the root has not yet been investigated in detail. Combining these data with molecular and biochemical results should facilitate our understanding of GA signaling in the roots.

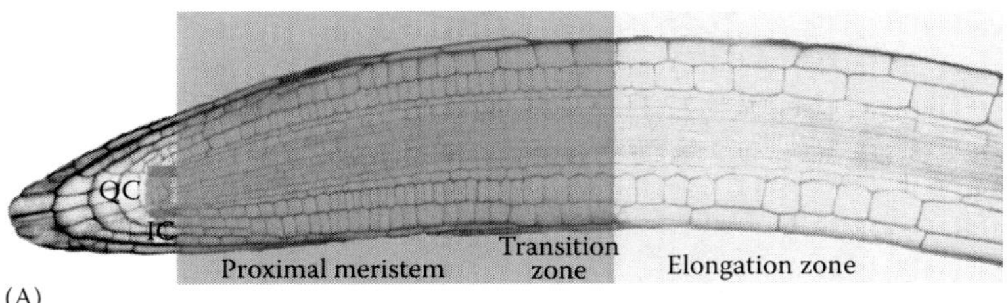

(A)

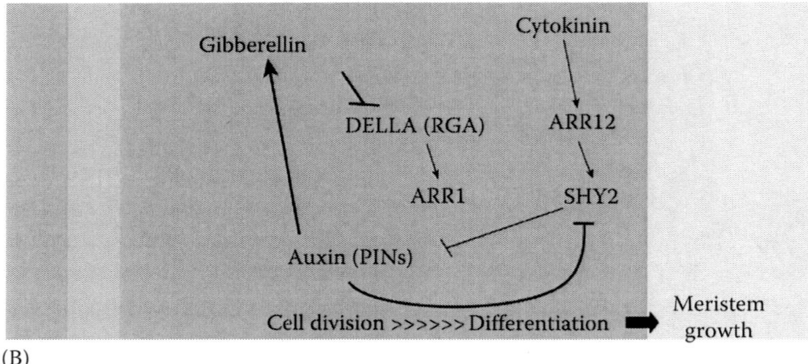

(B)

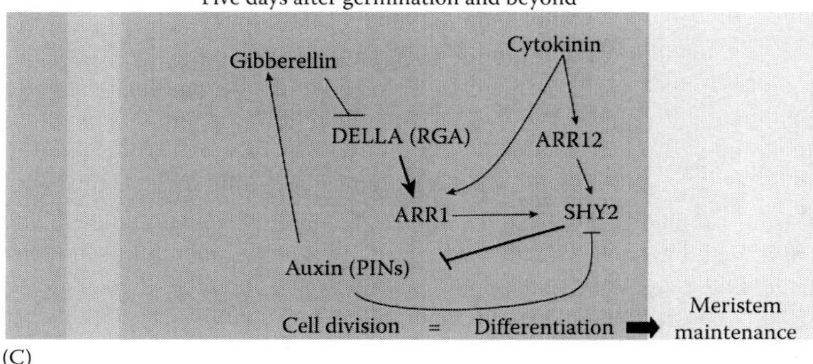

(C)

FIGURE 13.14 GA, cytokinin, and auxin signals control *Arabidopsis* root apical meristem size following germination. (A) Cross section of *Arabidopsis* root apical tissues proximal meristem is shown in dark grey, whereas the elongation zone is in light grey. Transition Zone is shown at the border of these two zones. QC (quiescent center), IC (Initial cells). (B) Schematic diagram illustrating the integration of the hormone signals and their pathways in root apical zones up to 5 days after germination. At this stage, cell division exceeds differentiation resulting in meristem growth. (C) Schematic diagram illustrating the integration of the hormone signals and their pathways in root apical zones from 5 days after germination. At this stage, balance between cell division and differentiation results in maintenance of meristem size. (Adapted from Ubeda-Tomás, S. and Bennett, M.J., *Curr. Biol.*, 20, R511, 2010. With permission.)

V. Concluding Remarks

The physiological function of GA has, since its discovery, been recognized to regulate the plant height which is determined by the elongation growth of stems or shoots. Breeding and genetic manipulation of GA biosynthesis have focused on the production of new cultivars of cereals with higher grain production, since plant height strongly affects the agricultural production. A typical success story is the Green Revolution, that is, the breeding of semidwarf rice (*Oryza sativa*) and wheat (*Triticum aestivum*) cultivars (Sasaki et al. 2002; Hedden 2003). Control of stem elongation by GA biosynthesis has also been conducted in horticulture to obtain short flowers and/or small plants by the application of GA biosynthesis inhibitors. Throughout this research, quantitative evaluation of GA functions in root growth has not well been documented. However, by using effective inhibitors of GA biosynthesis (Figure 13.2) and GA-less mutants, the essential role of GA for root growth was elucidated as shown in Figures 13.5 through 13.7. These dose-response experiments indicated that root elongation requires lower GA concentration than that of stems or hypocotyls (Figures 13.7 and 13.8).

Physiological importance of such difference between roots and stems at the level of GA sensitivity could be extended to naturally occurring dwarf plants. Dwarf plants with short stem may have agricultural and horticultural advantages only if root growth is not stunted, that is, shortage in GA biosynthesis in shoots may not strongly affect root growth which requires smaller amount of GA. Additionally, there are many plant species in nature with short stems but long roots. These are rosette plants, such as *Arabidopsis* lettuce (*Lactuca sativa*), spinach (*Spinacia oleracea* L.), and dandelion (*Taraxacum ruderalia* Kirschner, H. Øllg. & Stepanek). Their growth regulation is well adapted to survive the winter season. It seems detrimental for herbs and grasses to have elongated stems in winter, whereas root growth is a prerequisite to support aboveground organs. In contrast to auxin, overproduction of GA in the aboveground organ may not affect root growth, since GA is a saturating type of plant hormone with no significant suppressive effect at a higher concentration range (Tanimoto 2005). This GA property may allow the aboveground stem elongation, by producing large amount of GA, without suppressing root growth.

Molecular analyses of GA-signaling pathway elucidated that GA function is to inactivate the DELLA repressor proteins resulting in the expression of several genes (Figure 13.11). Although the functions of all DELLA-protein-suppressed genes are not fully understood, these genes play a role in cell division and cell differentiation in concert with auxin and cytokinin as shown in Figure 13.14.

Extreme shortage in endogenous GA causes the thickening of roots as shown in Figures 13.9 and 13.10. This abnormal thickening is mainly brought about by the expansion of cortical cells and is suggested to be conducted by cell-wall architecture. Control of cellulose-microfiber orientation is also conducted by GA. This GA function is responsible for the slenderness of roots and stems. However, further molecular mechanisms of GA-regulated cell elongation and cell thickening are currently under investigation.

As described in the last part of Section IV, molecular mechaism for high sensitivity of roots to GA is suggested to be shared by the GA receptor GID1b. Physiological function of this molecule is also under investigation in the author's laboratory.

In addition to auxin and cytokinin, indicated in Figures 13.3 and 13.14, another growth-promoting hormone, brassinosteroid, is also known to regulate root and shoot growth (Nomura et al. 1997; Yokota 1997; Nomura et al. 1999; see also Chapter 17).

The ratio of the aboveground part of a plant to the belowground counterpart may hopefully be manipulated by changing the endogenous level of GA and brassinosteroid together with sensitivity regulation through the quantitative control of gene expression in the future.

References

Achard P, Gusti A, Cheminant S et al. 2009 Gibberellin signaling controls cell proliferation rate in *Arabidopsis*. *Curr Biol* 19:1188–1193.

Ariizumi T, Murase K, Sun TP, Steber CM 2008 Proteolysis-independent down regulation of DELLA repression in *Arabidopsis* by the gibberellin receptor GIBBERELLIN INSENSITIVE DWARF1. *Plant Cell* 20:2447–2459.

Aspinall D, Palag LG, Addicot FT 1967 Abscisin II and some hormone-regulated plant responses. *Aust J Biol Sci* 20:869–882.

Aubert D, Chevillard M, Dorne A-M, Arlaud G, Herzog M 1998 Expression pattern of GASA genes in *Arabidopsis thaliana*: The GASA4 gene is up-regulated by gibberellins in meristematic regions. *Plant Mol Biol* 36:871–883.

Baluška F, Parker JS, Barlow PW 1993 A role for gibberellic acid in orienting microtubules and regulating cell growth polarity in maize root cortex. *Planta* 191:149–157.

Bidadi H, Yamaguchi S, Asahina M, Satoh S 2010 Effects of shoot-applied gibberellin/gibberellin-biosynthesis inhibitors on root growth and expression of gibberellin biosynthesis genes in *Arabidopsis thaliana*. *Plant Root* 4:4–11.

Burström H 1960 Influence of iron and gibberellic acid on the light sensitivity of roots. *Physiol Plant* 13:597–614.

Butcher DN, Street HE 1960 The effects of gibberellins on the growth of excised tomato roots. *J Exp Bot* 11:206–216.

Crozier A, Reid DM, Reeve DR 1973 Effects of AMO 1618 on growth morphology, and gibberellin content of *Phaseolus coccineus* seedlings. *J Exp Bot* 24:923–934.

Cui H, Benfey PN 2009 Interplay between SCARECROW, GA and LIKE HETEROCHROMATIN PROTEIN 1 in ground tissue patterning in the *Arabidopsis* root. *Plant J* 58:1016–1027.

Davies PJ 1995 *Plant Hormones: Physiology, Biochemistry and Molecular Biology*, 2nd edn., Dordrecht, the Netherlands: Kluwer.

Davies LJ, Rappaport L 1975 Metabolism of tritiated gibberellins in d-5 dwarf maize. I. In excised tissues and intact dwarf and normal plants. *Plant Physiol* 55:620–625.

Dello Ioio R, Nakamura K, Moubayidin L et al. 2008 A genetic framework for the control of cell division and differentiation in the root meristem. *Science* 322:1380–1384.

Dodd IC 2005 Root-to-shoot signaling: Assessing the roles of 'up' in the up and down world of long-distance signalling *in planta*. *Plant Soil* 274:251–270.

Dunberg A, Eliasson L 1972 Effects of growth retardants on Norway spruce (*Picea abies*). *Physiol Plant* 26:302–305.

Feldman LJ 1984 Regulation of root development. *Annu Rev Plant Physiol* 35:223–242.

Finnie JF, Van Staden J 1985 Effect of seaweed concentrate and applied hormones on in vitro cultured tomato roots. *J Plant Physiol* 120:215–222.

Frankenberger Jr. WT, Arshad M 1995 *Phytohormones in Soils, Microbial Production and Function*. New York: Marcel Dekker, Inc.

Fu X, Harberd NP 2003 Auxin promotes *Arabidopsis* root growth by modulating gibberellin response. *Nature* 421:740–743.

Fu X, Richards DE, Ait-Ali T et al. 2002 Gibberellin-mediated proteasome-dependent degradation of the barley DELLA protein SLN1 repressor. *Plant Cell* 14:3191–3200.

Goldfarb B, Lian Z, Lanz-Garcia C, Whetten R 1997 Auxin-induced gene expression during rooting of Loblolly Pine stem cuttings. In *Biology of Root Formation and Development*, eds. A Altman, Y Waisel, pp. 163–167. New York: Plenum.

Griffiths J, Murase K, Rieu I et al. 2006 Genetic characterization and functional analysis of the GID1 gibberellin receptors in *Arabidopsis. Plant Cell* 18:3399–3414.

Hartweck LM 2008 Gibberellin signaling. *Planta* 229:1–13.

Hedden P 1999 Recent advances in gibberellin biosynthesis. *J Exp Bot* 50:553–563.

Hedden P 2003 The genes of the green revolution. *Trends Genet* 19:5–9.

Hedden P, Kamiya Y 1997 Gibberellin biosynthesis: Enzymes, genes and their regulation. *Annu Rev Plant Physiol Plant Mol Biol* 48:431–460.

Hedden P, Proebsting WM 1999 Genetic analysis of gibberellin biosynthesis. *Plant Physiol* 119:365–370.

Heo JO, Chang KS, Kim IA, Lee MH, Lee SA, Song SK, Lee MM, Lim J 2011 Funneling of gibberellin signaling by the GRAS transcription regulator scarecrow-like 3 in the *Arabidopsis* root. *Proc Natl Acad Sci USA* 108:2166–2171.

Hirano K, Asano K, Tsuji H et al. 2010 Characterization of the molecular mechanism underlying gibberellin perception complex formation in rice. *Plant Cell* 22:2680–2696.

Hirano K, Nakajima M, Asano K et al. 2007 The GID1-mediated gibberellin perception mechanism is conserved in the Lycophyte *Selaginella moellendorffii* but not in the Bryophyte *Physcomitrella patens. Plant Cell* 19:3058–3079.

Hirota H 1980 Endogenous factors affecting the cultured growth of seminal roots of *Zea mays* L. seedlings grown in liquid culture. *Plant Cell Physiol* 21:961–968.

Hogetsu T 1986 Orientation of wall microfibril deposition in root cells of *Pisum sativum* L. var Alaska. *Plant Cell Physiol* 27:947–951.

Hogetsu T, Oshima Y 1986 Immunofluorescence microscopy of microtubule arrangement in root cells of *Pisum sativum* L. var. Alaska. *Plant Cell Physiol* 27:939–945.

Hori S 1898 Some observations on "Bakanae" disease of the rice plant. *Memo Agric Res Sta* 12:110–119.

Ikeda A, Ueguchi-Tanaka M, Sonoda Y et al. 2001 Slender rice, a constitutive gibberellin response mutant, is caused by a null mutation of the SLR1 gene, an ortholog of the height-regulating gene GAI/RGA/RHT/D8. *Plant Cell* 13:999–1010.

Itoh H, Tanaka-Ueguchi M, Kawaide H, Chen X, Kamiya Y, Matsuoka M 1999 The gene encoding tobacco gibberellin 3β-hydroxylase is expressed at the site of GA action during stem elongation and flower organ development. *Plant J* 20:15–24.

Kaneko M, Itoh H, Inukai Y et al. 2003 Where do gibberellin biosynthesis and gibberellin signaling occur in rice plants? *Plant J* 35:104–115.

Katayama K, Akita S 1989 Effect of exogenously applied gibberellic acid (GA_3) on initial growth of rice cultivars. *Jpn J Crop Sci* 58:217–224.

Kerr ID, Bennett MJ 2007 New insight into the biochemical mechanisms regulating auxin transport in plants. *Biochem J* 401:613–622.

Konings H, de Wolf A 1984 Promotion and inhibition by plant growth regulation of parenchyma formation in seedling roots of *Zea mays. Physiol Plant* 60:309–314.

Krekule J, Ullman J 1959 The influence of gibberellic acid on the growth of overground parts and roots of wheat, lettuce and oats. *Biol Plant* 1:22–30.

Leyser O 2002 Molecular genetics of auxin signaling. *Annu Rev Plant Biol* 53:377–398.

Lichtenthaler HK, Rohmer M, Schwender J 1997 Two independent biochemical pathways for isopentenyl diphosphate biosynthesis in higher plants. *Physiol Plant* 101:643–652.

Manos GE 1961 The effects of growth substances on attached and detached root tips of *Pisum sativum* L. *Physiol Plant* 14:697–711.

Matthysse AG, Scott TK 1984 Function of hormones at the whole plant level of organization. In *Hormonal Regulation of Development II Encyclopedia of Plant Physiol, New Series*, Vol. 10. ed. TK Scott, pp. 219–243. Berlin, Germany: Springer.

McGinnis KM, Thomas SG, Soule JD, Strader LC, Zale JM, Sun T, Steber CM 2003 The *Arabidopsis* SLEEPY1 gene encodes a putative F-box subunit of an SCF E3 ubiquitin ligase. *Plant Cell* 15:1120–1130.

Mertz D 1966 Hormonal control of root growth I. *Plant Cell Physiol* 7:125–135.

Mita T, Shibaoka H 1984a Effects of root excision on swelling of leaf sheath cells and on the arrangement of cortical microtubules in onion seedlings. *Plant Cell Physiol* 25:1521–1529.

Mita T, Shibaoka H 1984b Effects of S-3307, an inhibitor of gibberellin biosynthesis, on swelling of leaf sheath cells and on the arrangement of cortical microtubules in onion seedlings. *Plant Cell Physiol* 25:1531–1539.

Moubayidin L, Perilli S, Dello Ioio R, Di Mambro R, Costantino P, Sabatini S 2010 The rate of cell differentiation controls the *Arabidopsis* root meristem growth phase. *Curr Biol* 20:1138–1143.

Murase K, Hirano Y, Sun TP, Hakoshima T 2008 Gibberellin-induced DELLA recognition by the gibberellin receptor GID1. *Nature* 456:459–463.

Nakajima M, Shimada A, Takashi Y et al. 2006 Identification and characterization of *Arabidopsis* gibberellin receptors. *Plant J* 46:880–889.

Nishijima T, Katsura N 1989 A modified micro-drop bioassay using dwarf rice for detection of femtomol quantities of gibberellins. *Plant Cell Physiol* 30:623–627.

Nomura T, Kitasaka Y, Takatsuto S, Reid JB, Fukami M, Yokota T 1999 Brassinosteroid/sterol synthesis and plant growth as affected by lka and lkb mutations of pea. *Plant Physiol* 119:1517–1526.

Nomura T, Nakayama M, Reid JB, Takeuchi Y, Yokota T 1997 Blockage of brassinosteroid biosynthesis and sensitivity causes dwarfism in garden pea. *Plant Physiol* 113:31–37.

Ogawa M, Matsunaga E, Yamazaki Y, Pyamada K, Matsui T, Tobitsuka J 1976 A synergistic effect of 2-ethyl-1-isopropoxycarbonyl-3-(4-tolycarbamoyl)isourea with gibberellic acid on growth of rice seedlings. *Plant Cell Physiol* 17:743–749.

Olszewski N, Sun TP, Gubler F 2002 Gibberellin signaling: Biosynthesis, catabolism, and response pathways. *Plant Cell* 14: S61–S80.

Pecket RC 1960 Effects of gibberellic acid on excised pea roots. *Nature* 185:114–115.

Peng J, Carol P, Richards DE, King KE, Cowling RJ, Murphy GP, Harberd NP 1997 The *Arabidopsis* GAI gene defines a signaling pathway that negatively regulates gibberellin responses. *Genes Dev* 11:3194–3205.

Phinney BO 1984 Gibberellin A_1, dwarfism and the control of shoot elongation in higher plants. In *The Biosynthesis and Metabolism of Plant Hormones*, eds. A Crozier, JR Hillman, pp.17–41. London, U.K.: Cambridge University Press.

Phuoc LT, Mander LN, Koshioka M, Oyama-Okubo N, Nakayama M, Ito A 2008 Confirmation of structure and synthesis of three new 11b-OH C20 gibberellins from loquat fruit. *Tetrahedron* 64:4835–4851.

Porlingis IC, Koukourikou-Petridou MA 1996 Promotion of adventitious root formation in mung bean cuttings by four triazole growth retardants. *J Hort Sci* 71:573–579.

Prochazka S 1981 Translocation of growth regulators from roots in relation to the stem apical dominance in pea (*Pisum sativum* L.) seedlings. In *Structure and Function of Plant Roots*. eds. R Brouwer et al. pp. 407–409. The Hague, the Netherlands: Martinus Nijhoff/Dr W. Junk.

Proebsting WM, Hedden P, Lewis MJ, Croker SJ, Proebsting LN 1992 Gibberellin concentration and transport in genetic lines of pea: Effects of grafting. *Plant Physiol* 100:1354–1360.

Rademacher W 1991 Inhibitors of gibberellin biosynthesis: Applications in agriculture and horticulture. In *Gibberellins* eds. N Takahashi, BO Phinney, J MacMillan, T Daigaku, pp. 296–310. New York: Springer.

Radi SH, Maeda E 1988 Effect of brassinolide on the cultured rice root growth as modified by Figaron and gibberellic acid. *Jpn J Crop Sci* 57:191–198.

Reid JB, Murfet IC, Potts WC 1983 Internode length in *Pisum*. 2. Additional information on the relationship and action of loci *le, la, cry, na* and *lm*. *J Exp Bot* 34:349–364.

Ross JJ, O'Neill DP, Smith JJ, Kerckhoffs LHJ, Elliot RC 2000 Evidence that auxin promotes gibberellin A1 biosynthesis in pea. *Plant J* 21:547–552.

Ross JJ, O'Neill DP, Wolbang CM, Symons GM, Reid JB 2002 Auxin-gibberellin interactions and their role in plant growth. *J Plant Growth Regul* 20:346–353.

Rubinovich L, Weiss D 2010 The *Arabidopsis* cysteine-rich protein GASA4 promotes GA responses and exhibits redox activity in bacteria and in planta. *Plant J* 64:1018–1027.

Sachs RM, Kofranek AM 1963 Comparative cytohistological studies on inhibition and promotion of stem growth in *Chrysanthemum morifolium*. *Am J Bot* 50:772–779.

Sasaki A, Ashikari M, Ueguchi-Tanaka M et al. 2002 Green revolution: A mutant gibberellin-synthesis gene in rice: New insight into the rice variant that helped to avert famine over thirty years ago. *Nature* 416:701–702.

Sasaki A, Itoh H, Gomi K et al. 2003 Accumulation of phosphorylated repressor for gibberellin signaling in an F-box mutant. *Science* 299:1896–1898.

Savaldi-Goldstein S, Peto C, Chory J 2007 The epidermis both drives and restricts plant shoot growth. *Nature* 446:199–202.

Sawhney VK, Srivastava LM 1974 Cytochalasin-B-induced inhibition of root-hair growth in lettuce seedlings and its reversal by benzyladenine. *Planta* 119:165–168.

Scott TK 1984 Hormonal regulation of development. In *Encyclopedia of Plant Physiology, New Series*, Vol. 10. Berlin, Germany: Springer.

Shimada A, Ueguchi-Tanaka M, Nakatsu T et al. 2008 Structural basis for gibberellin recognition by its receptor GID1. *Nature* 456:520–523.

Silverstone A, Chang CW, Krol E, Sum TP 1997 Developmental regulation of the gibberellin biosynthetic gene GA1 in *Arabidopsis thaliana*. *Plant J* 12:9–19.

Silverstone AL, Ciampaglio CN, Sun T 1998 The *Arabidopsis* RGA gene encodes a transcriptional regulator repressing the gibberellin signal transduction pathway. *Plant Cell* 10:155–169.

Suge H 1985 Ethylene and gibberellin: Regulation of internode elongation and nodal root development in floating rice. *Plant Cell Physiol* 26:607–614.

Sun TP 2010 Gibberellin-GID1-DELLA: A pivotal regulatory module for plant growth and development. *Plant Physiol* 154:567–570.

Svensson SB 1972 A comparative study of the changes in root growth induced by coumarin, auxin, ethylene, kinetin and gibberellic acid. *Physiol Plant* 26:115–135.

Tanimoto E 1987 Gibberellin-dependent root elongation in *Lactuca sativa*: Recovery from growth retardant-suppressed elongation with thickening by low concentration of GA_3. *Plant Cell Physiol* 28:963–973.

Tanimoto E 1988 Gibberellin regulation of root growth with changes in galactose contents of the cell walls in *Pisum sativum*. *Plant Cell Physiol* 29:269–280.

Tanimoto E 1994 Interaction of gibberellin A_3 and ancymidol in the growth and cell-wall extensibility of dwarf pea roots. *Plant Cell Physiol* 35:1019–1028.

Tanimoto E 2002 Gibberellins. In *Plant Roots: The Hidden Half*, eds. Y Waisel A Eshel, U Kafkafi, 3rd edn., pp. 405–416. New York: Marcel Dekker, Inc.

Tanimoto E 2005 Regulation of root growth by plant hormones: Roles for auxin and gibberellin. *Crit Rev Plant Sci* 24:249–265.

Tanimoto E, Homma T, Lux A, Luxova M, Matsuo K, Abe J, Morita S, Inanaga S 2004b Phytohormone gibberellin regulates vegetative growth and flowering of hydroponics-grown tea trees. In *The Second International Conference on O-Cha(tea) Culture and Science*. Shizuoka, Japan, November 4–6, 2004.

Tanimoto E, Homma T, Matsuo K, Hoshino T, Lux A, Luxova M 2004a Root structure and cell-wall extensibility of adventitious roots of tea (*Camellia sinensis* cv. Yabukita). *Biologia, Bratislava, 59/s* 13:57–66.

Tanimoto E, Watanabe J 1986 Automated recording of lettuce root elongation as affected by indole-3-acetic acid and acid pH by a new rhizometer with minimum mechanical contact to root. *Plant Cell Physiol* 27:1475–1487.

Taylor BH and Scheuring CF 1994 A molecular marker for lateral root initiation: The RSI-1 gene of tomato (*Lycopersicon esculentum* Mill) is activated in early lateral root primordia. *Mol Gen Genet* 243:148–157.

Tognoni F, Halevy AH, Wittwer SH 1967 Growth of bean and tomato plants as affected by root absorbed growth substances and atmospheric carbon dioxide. *Planta* 72:43–52.

Torrey, JG 1976 Root hormones and plant growth. *Annu Rev Plant Physiol* 27:435–459.

Ubeda-Tomás S, Bennett MJ 2010 Plant development: Size matters, and it's all down to hormones. *Curr Biol* 20:R511–R513.

Ubeda-Tomás S, Federici F, Casimiro I et al. 2009 Gibberellin signaling in the endodermis controls *Arabidopsis* root meristem size. *Curr Biol* 19:1194–1199.

Ubeda-Tomás S, Swarup R, Coates J et al. 2008 Root growth in *Arabidopsis* requires gibberellin/DELLA signaling in the endodermis. *Nat Cell Biol* 10:625–628.

Ueguchi-Tanaka M, Ashikari M, Nakajima M et al. 2005 GIBBERELLIN INSENSITIVE DWARF1 encodes a soluble receptor for gibberellin. *Nature* 437:693–698.

Ueguchi-Tanaka M, Hirano K, Hasegawa Y, Kitano H, Matsuoka M 2008 Release of the repressive activity of rice DELLA protein SLR1 by gibberellin does not require SLR1 degradation in the gid2 mutant. *Plant Cell* 20:2437–2446.

Wang F, Zhu D, Huang X et al. 2009 Biochemical insights on degradation of *Arabidopsis* DELLA proteins gained from a cell-free assay system. *Plant Cell* 21:2378–2390.

West CA, Shen-Miller J, Railton ID 1982 Regulation of kaurene synthetase. In *Plant Growth Substances*, ed. PF Wareing, pp. 81–90. London, U.K.: Academic Press.

Whaley WG, Kephart J 1957 Effect of gibberellic acid on growth of maize roots. *Science* 125:234–235.

Willige BC, Isono E, Richter R, Zourelidou M, Schwechheimer C 2011 Gibberellin regulates PIN-FORMED abundance and is required for auxin transport-dependent growth and development in *Arabidopsis thaliana*. *Plant Cell* 23:2184–2195.

Yamaguchi S 2008 Gibberellin metabolism and its regulation. *Annu Rev Plant Biol* 59:225–251.

Yamamoto Y, Hirai T, Yamamoto E et al. 2010 A rice gid1 suppressor mutant reveals that gibberellin is not always required for interaction between its receptor, GID1, and DELLA proteins. *Plant Cell* 22:3589–3602.

Yaxley JR, Ross JJ, Sherriff LJ, Reid JB 2001 Gibberellin biosynthesis mutations and root development in pea. *Plant Physiol* 125:627–633.

Yokota T 1997 The structure, biosynthesis and function of brassinosteroids. *Trends Plant Sci* 2:137–143.

Zhang ZL, Ogawa M, Fleet CM, Zentella R, Hu J, Heo JO, Lim J, Kamiya Y, Yamaguchi S, Sun TP 2011 Scarecrow-like 3 promotes gibberellin signaling by antagonizing master growth repressor DELLA in *Arabidopsis*. *Proc Natl Acad Sci USA* 108:2160–2165.

Zimmermann R, Sakai H, Hochholdinger F 2010 The gibberellic acid stimulated-like gene family in maize and its role in lateral root development. *Plant Physiol* 152:356–365.

14

Molecular Basis of Cytokinin Action during Root Development

Serena Perilli*
Sapienza University of Rome

Laila Moubayidin*
Sapienza University of Rome

Sabrina Sabatini
Sapienza University of Rome

I. Introduction

In higher plants, development of all the adult organs and structures occurs mainly postembryonically and depends on the activity of the meristems. Meristems are small, specialized regions holding a stem-cell niche (SCN), a microenvironment where extracellular signals control stem-cell division and prevent their differentiation, thus allowing continuous generation of new cells sustaining indeterminate growth throughout the plant life span.

In *Arabidopsis*, the best characterized SCNs are within the shoot apical meristem and the root apical meristem (Figure 14.1A). Both shoot and root apical meristems are specified during embryogenesis and share a common functional organization: an organizing center that functions non-cell-autonomously to maintain the adjacent stem cells which, on their turn, give rise to a transit-amplifying population of cells in the meristem from which different tissues and organs are initiated (Sablowski 2010).

The *Arabidopsis* root apical meristem is located at the tip of the growing root and originates from the distal SCN (Figure 14.1A and B), where a group of four mitotically inactive cells, the quiescent center (QC), acts as an organizer to maintain the surrounding stem cells in an undifferentiated status (van Den Berg et al. 1997). Each stem cell gives rise to a specific tissue, generating a cylinder with distinct cell types organized in concentric layers: from outside to inside, epidermis, cortex, endodermis, and pericycle surround the central vascular bundle (Figure 14.1B) (Dolan et al. 1993).

Along the longitudinal axis, the root apex can be divided into three different developmental zones (Figure 14.1B): the SCN, the proximal meristem (PM), and the elongation/differentiation zone (EDZ).

In the root meristem, stem cells divide producing daughter cells, which undergo a finite number of cell divisions in the PM (the division zone) until they leave the meristem, rapidly expand, and differentiate to reach maturity (in the EDZ). Cell differentiation is initiated at the transition zone (TZ) encompassing the boundaries between dividing and expanding cells in the different files (Figure 14.1B). The TZ is different for each cell type, giving a jagged shape to the boundary between dividing and expanding cells (Figure 14.1B) (Dello Ioio et al. 2007; Scheres 2007).

The overall rate of root growth and the maintenance of root meristem size and activity depend on the relative rates of cell proliferation in the meristem and the extent of cell elongation/differentiation at the TZ (Beemster and Baskin, 1998). Upon seed germination, cell division in the meristem prevails over cell differentiation, allowing root apical meristem to grow and to increase in size. Few days after germination, the rate of cell division is balanced with the rate of cell differentiation, and final root meristem size is set, thus ensuring meristem maintenance and a constant rate of root growth (Dello Ioio et al. 2007).

* The authors Serena Perilli and Laila Moubayidin contributed equally to this chapter.

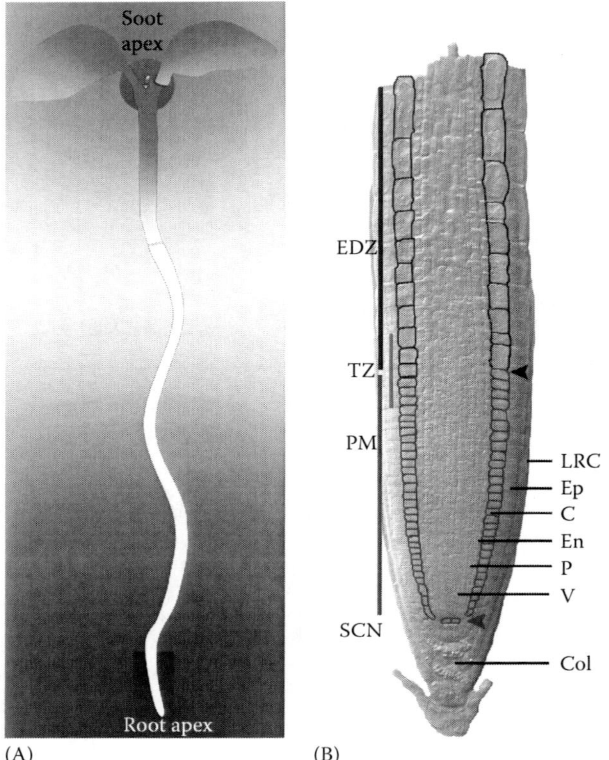

FIGURE 14.1 (See color insert.) Anatomy of the *Arabidopsis thaliana* root apical meristem. (A) Schematic view of *Arabidopsis thaliana* seedling, with highlighted shoot apex (blue circle) and root apex (brown rectangle). (B) Longitudinal view of the *Arabidopsis* root meristem, where the SCN is marked in light green, the QC in blue, a cortex cell file of the PM is marked in green, and a cortex cell file of the EDZ is marked in dark green. Black arrowhead indicates the cortex TZ (red line), where cells leave the meristem and enter the EDZ. Blue arrowhead indicates the QC. Col: columella; V: vascular tissue; P: pericycle; En: endodermis; C: cortex; Ep: epidermis; LRC: lateral root cap.

How this balance is achieved is a central question in plant developmental biology, and it has been demonstrated that it depends on the antagonistic interaction between two plant hormones: auxin and cytokinin (Dello Ioio et al. 2008; Moubayidin et al. 2010).

In this chapter, we will describe the role of the plant hormone cytokinin during root development, focusing on how cytokinin interacts with other hormones to control meristem growth and maintenance.

II. Molecular Basis of Cytokinins Function

Cytokinins are N^6-prenylated adenine derivatives involved in the control of several aspects of plant growth and development such as meristem activity, vascular differentiation, lateral root initiation, nodulation, as well as response to various biotic and abiotic stresses (Frugier et al. 2008; Argueso et al. 2010; Perilli et al. 2010). The effects of cytokinins on plant development and physiology have long been known, but in the past decade genetic and molecular analysis of mutants has provided new insights into the molecular mechanisms of cytokinin action *in planta* (see later).

A. Cytokinin Homeostasis

In higher plants, predominant cytokinins, known as the isoprenoid cytokinins, are isopentenyladenine (iP), trans-zeatin (tZ), cis-zeatin, and dihydrozeatin (Sakakibara 2006), whose homeostasis depends on the balance between synthesis and catabolism, a process that is strictly regulated spatially and temporally (Sakakibara 2006; Werner et al. 2003; Miyawaki et al. 2004). Among isoprenoid cytokinins, iP and tZ are major forms in *Arabidopsis thaliana* (Sakakibara 2006).

The initial, rate-limiting step in cytokinin biosynthesis is catalyzed by the *ATP/ADP-isopentenyltransferase* (*IPT*) gene family. In *Arabidopsis thaliana*, there are seven *IPT* genes (*AtIPT1*, *AtIPT3-8*), whose expression pattern reveals that cytokinins are synthesized throughout the plant, including roots, shoots, and immature seeds but with tissue- and organ-specific patterns of expression (Miyawaki et al. 2004).

IPTs catalyze the transfer of an isopentenyl group from dimethylallyl diphosphate to an adenine nucleotide (ATP, ADP, or AMP) (Kakimoto 2001; Takei et al. 2001). Multiple combinations of *ipt* mutants display drastic decreased levels of different cytokinins and severely inhibited shoot growth, while root growth is enhanced (Miyawaki et al. 2006), suggesting that cytokinins act as shoot-growth-promoting factors and negative regulators of root development.

In order to finely regulate steady-state levels of active cytokinins *in planta*, biosynthesis has to be tuned up with cytokinin degradation. In *Arabidopsis*, cytokinin breakdown is controlled by the activity of seven *CYTOKININ oxidase/dehydrogenase* (*CKX*) genes (*CKX1-7*), which catalyze the irreversible degradation of cytokinins by cleavage of the side chain (Mok and Mok 2001; Schmulling et al. 2003). Like *IPTs*, the *CKX* genes display different patterns of expression during plant development, showing highest activity in regions of active growth such as the shoot meristem, the root meristem, and emerging leaves (Werner et al. 2003). Overexpression of these genes phenocopies the effect of *ipt* mutations, confirming that cytokinins play a central role in different organs during various stages of plant development (Werner et al. 2003; Miyawaki et al. 2006).

The expression of key genes for cytokinin biosynthesis and homeostasis such as *IPTs* and *CKXs* is regulated by phytohormones including cytokinins, auxin, and abscisic acid. In *Arabidopsis*, auxin promotes the accumulation of the transcripts of *AtIPT5* and *AtIPT7* in roots, whereas cytokinins inhibit transcription of *AtIPT1*, *AtIPT3*, *AtIPT5*, and *AtIPT7* (Miyawaki et al. 2004). In maize, *CKX* genes are upregulated by cytokinins and abscisic acid (Brugiere et al. 2003).

A third fundamental process that must be properly regulated to control cytokinin action is the interconversion between inactive and active forms of cytokinins. In rice, a novel component of the cytokinin biosynthesis pathway has been isolated, the

LONELY GUY (*LOG*) gene (Kurakawa et al. 2007). *LOG* encodes an enzyme with phosphoribohydrolase activity that converts inactive cytokinin nucleotides into a free-base, biologically active form. This gene is specifically expressed in rice shoot meristem tips, and *log* mutants display smaller shoot meristem as a result of diminished active cytokinins at the top of the shoot meristem, where stem cells reside (Kurakawa et al. 2007). Recently, nine homologs of the rice *LOG* gene have been identified in *Arabidopsis* (*AtLOG1-9*), seven of which had enzymatic activities both in vitro and in vivo equivalent to that of rice *LOG* (Kuroha et al. 2009). Analysis of spatial distribution patterns of *AtLOGs* revealed that these genes are transcribed throughout the plant during development, with different and overlapping domains of expression for different *AtLOGs* (Kuroha et al. 2009). Moreover, some of the expression sites of *AtLOGs* overlap with those of *IPTs*, like in lateral root primordia and root vascular tissues, suggesting that cytokinins are synthesized and subsequently activated locally by *AtLOGs* to act as autocrine or paracrine signals. On the other hand, expression of some *AtLOGs* was observed in certain tissues (root hairs, the epidermis of the root elongation zone) in which no expression of *IPTs* was observed (Kuroha et al. 2009; Miyawaki et al. 2004). This suggests that a portion of cytokinin precursors is translocated from cell to cell and/or over a long range before becoming active (Kuroha et al. 2009). Multiple mutants of *AtLOGs* show a decreased sensitivity to cytokinins in terms of lateral root formation and inhibition of primary root growth (Kuroha et al. 2009), resembling the phenotype of plants overexpressing *CKXs* (Werner et al. 2003) and multiple loss-of-function *ipt* mutants (Miyawaki et al. 2006). This strongly suggests that *AtLOGs* are required in order to drive cytokinin activation for different developmental processes.

B. Cytokinin Perception and Signaling

Cytokinin signaling involves a multistep phosphorelay cascade similar to the bacterial two-component system (Kakimoto 2003; Argueso et al. 2010). In *Arabidopsis thaliana*, there are three transmembrane histidine kinase receptors, the ARABIDOPSIS HIS KINASE 2 (AHK2), AHK3, and AHK4/WOL1 (WOODENLEG 1)/CRE1 (CYTOKININ RESPONSE 1) (Inoue et al. 2001; Hwang and Sheen 2001), with largely overlapping expression patterns and partially redundant functions in cytokinin perception (Higuchi et al. 2004; Nishimura et al. 2004; Riefler et al. 2006). Cytokinin binding to the receptor induces autophosphorylation of the AHKs on a conserved His residue within their kinase domain (Inoue et al. 2001). A phosphate group is then transferred to a conserved Asp residue within the receiver domain of these AHK proteins and subsequently transferred to members of the ARABIDOPSIS HIS PHOSPHOTRANSFER PROTEINS (AHPs) family. Using *Arabidopsis* mesophyll protoplasts and T87 *Arabidopsis* tissue culture cells, it has been shown that some AHP proteins are translocated from the cytosol into the nucleus in response of exogenous cytokinin (Hwang and Sheen, 2001). However, it has been recently observed that *in planta*, AHPs are continuously transported in and out of the nucleus independently from their phosphorylation status and not in response to cytokinins (Punwani et al. 2010).

In the nucleus, phosphorylated AHPs transfer the phosphate group to members of the ARABIDOPSIS RESPONSE REGULATORS (ARRs) protein family that are classified into different types, type A and type B, depending on their C-terminal domain. Type-A ARRs (ARR3–ARR9, ARR15–ARR17) possess short C-termini, while type-B ARRs (ARR1, ARR2, ARR10–ARR14, ARR18–ARR21) have longer C-termini containing a conserved GARP DNA binding domain and function as transcriptional activators. All ARRs share a similar receiver domain with conserved residues targeted for phosphorylation (To and Kieber 2008).

AHPs mediate accumulation in the nucleus of another family of cytokinin-responsive genes, the CYTOKININ RESPONSE FACTORS (CRFs), members of the *Arabidopsis* AP2 family of transcription factors (Rashotte et al. 2006). Activated type-B ARRs and CRFs mediate cytokinin-regulated gene expression of common and unique targets (Rashotte et al. 2006), including type-A *ARRs* (Hwang and Sheen 2001; Rashotte et al. 2006). Activated type-A ARR proteins become more stable in response to cytokinin (To et al. 2007) and negatively regulate cytokinin signaling (To et al. 2004) via an as yet unknown mechanism. Since most type-A ARRs are predominantly localized in the nucleus, it has been suggested that they inhibit cytokinin signaling competing with type-B ARRs for phosphorylation.

Although the individual role of the different elements of the two-component cytokinin-signaling pathway is well understood, little is known regarding the way these molecules mediate a specific cytokinin-dependent developmental output. Nonetheless, there has been some progress in explaining the molecular mechanisms underlying cytokinin function in shaping root architecture (Perilli et al. 2010), as it will be described in the following sections.

III. Cytokinins and Root Development

Cytokinins have long been known as negative regulators of root development: exogenous cytokinin treatment has been shown to inhibit root elongation (Beemster and Baskin 1998), while reduction of endogenous cytokinin levels results in increased primary root elongation (Werner et al. 2003). Furthermore, several signal transduction and biosynthesis genes are preferentially expressed in the root (Higuchi et al. 2004; Mason et al. 2004; To et al. 2004), suggesting that cytokinins play a fundamental role during root development.

A. Cytokinins Control Cell Differentiation Rate of Meristematic Cells

It has been shown that during root development, cytokinins control cell differentiation rate of meristematic cells, thus determining root meristem size and root growth (Dello Ioio et al. 2007). Exogenous application of cytokinins inhibits root

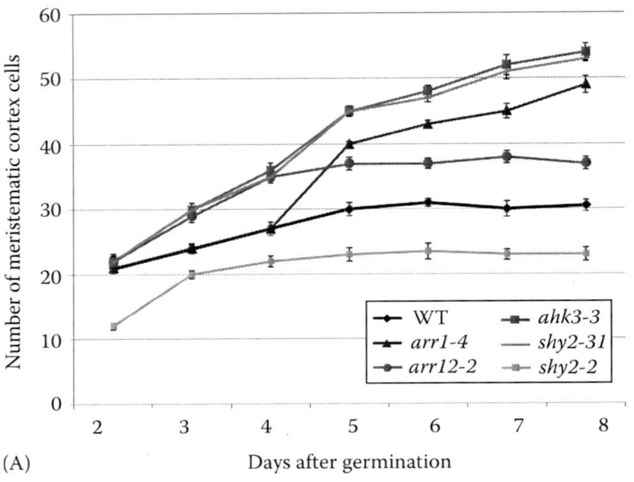

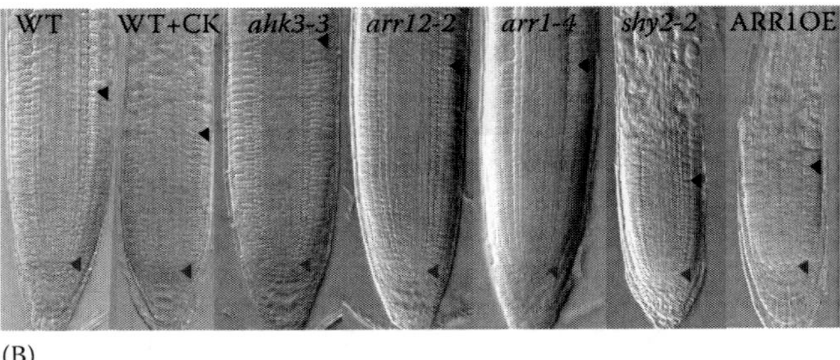

FIGURE 14.2 Cytokinins determine *Arabidopsis* root meristem size by controlling cell differentiation. (A) Time course of root meristem growth of wild-type (WT) and cytokinin-signaling mutants. Note that from 5 days after germination onward, WT meristems have a fixed number of cells while cytokinin-signaling mutants, with the exception of *arr12*, keep on growing accumulating meristematic cells. On the other hand, the *shy2-2* gain-of-function mutant displays a short meristem with a fixed number of cells already 3 days after germination. (B) Root meristems of WT and cytokinin-signaling mutants. Note that ARR1 overexpressor (ARR1OE) displays a short root meristem as the *shy2-2* gain-of-function mutant, mimicking the phenotype of root meristems treated with exogenous cytokinins (CK). Black and grey arrowheads indicate, respectively, the cortex TZ and the QC. Root meristem cell number is expressed as the number of cortex cells in a file extending from the QC to the first elongated cortex cell, excluded (from grey to black arrowhead).

growth and reduces root meristem size (Figure 14.2B) without altering stem-cell activity or cell division in the PM (Dello Ioio et al. 2007). Conversely, cytokinin biosynthesis mutants, mutations in the *AHK3* cytokinin receptor gene and in the type-B *ARR1* and *ARR12* genes, display strongly enhanced root growth and enlarged meristems (Figure 14.2A and B) (Dello Ioio et al. 2007, 2008). Interestingly, depletion of cytokinins, via *CKX* overexpression, specifically in the vascular tissue at the TZ, is sufficient to affect meristem size, while reduction of cytokinin levels in the whole meristem or in other root tissues at the TZ has no such effect (Dello Ioio et al. 2007).

Based on these data, it has been concluded that an AHK3/ARR1/ARR12 two-component cytokinin-signaling pathway specifically acts at the vascular tissue TZ to regulate the differentiation rate of meristematic cells, thus controlling root meristem size and root growth (Dello Ioio et al. 2007).

B. Cytokinins Mediate Cell Differentiation and Determine Final Root Meristem Size by Antagonizing the Auxin-Dependent Cell Division Input

In the meristem, rootward polar auxin transport and lateral distribution of auxin, mediated by the PINFORMED (PIN) auxin efflux facilitators, are crucial for the control of cell division and root meristem size (Blilou et al. 2005). Accordingly, exogenous application of auxin during root growth increases root meristem size (Dello Ioio et al. 2007) while multiple combinations of *pin* mutants display shorter root meristems and delayed root growth (Blilou et al. 2005). It has been shown that final root meristem size and root growth are determined by the antagonistic effects of cytokinins, which promote cell differentiation, and auxin, which promotes cell division (Dello Ioio et al. 2008). In particular, these two hormones control in opposite ways the abundance of the auxin-sensitive SHORT HYPOCOTYL 2 (SHY2) protein, a member of the auxin/indole-3-acetic acid (Aux/IAA) auxin-inducible family of genes (Tian and Reed 1999; Tian et al. 2003) that act as auxin-response inhibitors by preventing activation of auxin-responsive genes (Tian et al. 2002; Benjamins and Scheres 2008). Cytokinins, through the AHK3/ARR1 pathway, directly activate transcription of the *SHY2* gene in the vascular tissue at the TZ (Figure 14.3A) (Taniguchi et al. 2007; Dello Ioio et al. 2008), whereas auxin directs degradation of the SHY2 protein via the SCF^{TIR1} ubiquitin–ligase complex (Tian et al. 2003). SHY2 is necessary and sufficient to control root meristem size in response to cytokinin (Dello Ioio et al. 2008): *shy2-2* gain-of-function mutants display a short root meristem and impaired root growth, mimicking the effect of cytokinin application, whereas *shy2-31* loss-of-function mutants display an enlarged root meristem closely resembling *ahk3* and *arr1* mutant phenotype (Figure 14.2A) (Dello Ioio et al. 2008). Cytokinin-mediated activation of SHY2, in turn, leads to negative regulation of *PINs* expression at the vascular tissue TZ, limiting auxin transport and distribution and allowing cell differentiation of all tissues (Figure 14.3B) (Dello Ioio et al. 2008). Indeed, decreased expression of PINs was observed in *shy2-2* gain-of-function background (Dello Ioio et al. 2008), while a broader expression domain of these proteins was observed in the *shy2-31* loss-of-function mutant (Dello Ioio et al. 2008). The SHY2 protein represses auxin signaling and transport, on the one hand, and, on the other hand, it controls cytokinin biosynthesis by downregulating the *IPT5* gene, thus conferring robustness to this control circuit (Miyawaki et al. 2004; Dello Ioio et al. 2008).

C. Rate of Cell Differentiation Controls Root Meristem Growth

The molecular circuit described earlier explains how the balance between cell division and cell differentiation is achieved and maintained over time, but it does not clarify how a final root

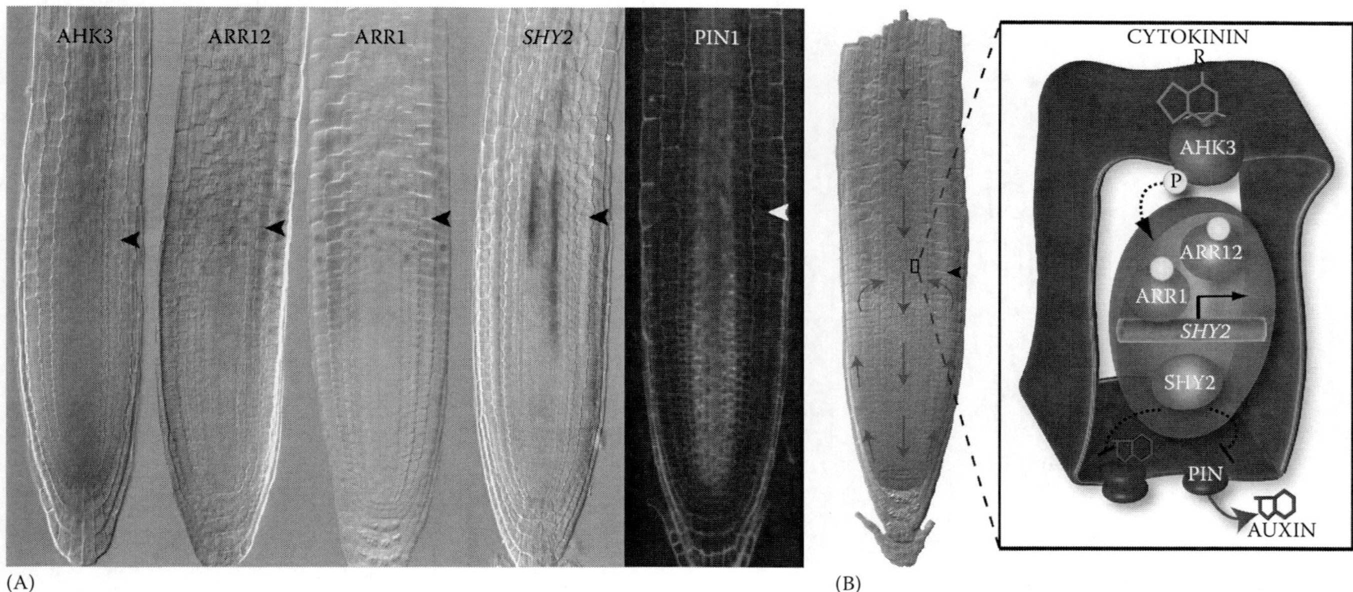

FIGURE 14.3 **(See color insert.)** Expression pattern of cytokinin-sensitive genes involved in root meristem size determination. (A) Expression of AHK3 in roots of *AHK3:GUS* plants, of ARR12 in roots of *ARR12:GUS* plants, of ARR1 in roots of *ARR1:GUS* plants, of *SHY2* in roots of *pSHY2:GUS* plants, and of PIN1 in roots of *PIN1:GFP* plants. Black and white arrowheads indicate the cortex TZ. (B) Longitudinal view of the *Arabidopsis* root meristem, where the domain of auxin activity is marked in blue, while the domain of cytokinin activity is marked in red. Arrows indicate direction of polar auxin transport. In the single-cell blowup schematic representation of cytokinin-signaling pathway acting during root meristem size determination. Cytokinins are perceived by the plasma membrane AHK3 receptor, which initiates a phosphorylation (P in the yellow circle) cascade ending with the activation, in the nucleus, of type-B ARR12 and ARR1 cytokinin-response transcription factors. Upon activation, ARR12 and ARR1 proteins directly bind the *SHY2* promoter inducing its transcription. SHY2 protein, in turn, downregulates *PIN* genes expression, limiting polar auxin transport.

meristem size is reached, that is, how a change in the relative rates of cell division and differentiation causes meristem growth to stop.

It has been shown recently that, after seed germination, root meristem growth is controlled by an increase in the rate of cell differentiation relative to cell division. The completion of growth and the achievement of a balance between cell division and cell differentiation, which are necessary in order to maintain the final meristem size, are brought about by the upregulation of *SHY2* caused by both ARR12 and ARR1 transcription factors (Moubayidin et al. 2010). The mRNA levels of the *SHY2* gene increase over time, reaching a maximum of expression at 5 days after germination (dpg), when ARR1 is activated. Moreover, transient expression of a nondegradable version of the SHY2 protein, during root meristem growth phase, is sufficient to reduce and prematurely set root meristem size, suggesting that a peak of SHY2 is necessary and sufficient to determine the final size of the root meristem (Moubayidin et al. 2010). Molecular as well as genetic analyses of double mutant combinations have demonstrated that during meristem growth (up to 5 dpg), expression of *SHY2* is controlled only by ARR12 that maintains SHY2 at a relatively low level, thus allowing prevalence of cell division over cell differentiation and the building up of the meristem (Figure 14.4A) (Moubayidin et al. 2010). At five dpg, upon activation of ARR1, the joint action of both ARR12 and ARR1 ensures optimal SHY2 level and sets the final root meristem size (Figure 14.4B) (Moubayidin et al. 2010).

As mentioned earlier, ARR1 and ARR12 exhibit different patterns of temporal expression. ARR12 is expressed at the root TZ (Figure 14.3A) following germination, whereas ARR1 is activated at the root TZ only 5 days after germination (Figure 14.3A) (Dello Ioio et al. 2007; Moubayidin et al. 2010), when root meristem size is set. It has been shown that the plant hormones gibberellins are responsible for selective repression of *ARR1* expression at early stages of meristem development (during meristem growth phase) and that activation of *ARR1* transcription 5 days after germination is due to a decrease in gibberellin activity, mediated by the DELLA protein *REPRESSOR OF GA 1-3* (RGA) (Figure 14.4A and B) (Moubayidin et al. 2010).

Similarly to auxin, gibberellins have been shown to act as positive regulators of root growth and meristem size by sustaining cell division (Ubeda-Tomás et al. 2008; Achard et al. 2009; Ubeda-Tomás et al. 2009). Exogenous application of gibberellins increases root meristem size and downregulates ARR1 expression 5 dpg, without affecting ARR12 levels (Moubayidin et al. 2010). In addition, inhibition of gibberellin biosynthesis allows early ARR1 expression at 3 dpg, during the root meristem growth phase, and results in a decrease of root meristem size (Moubayidin et al. 2010).

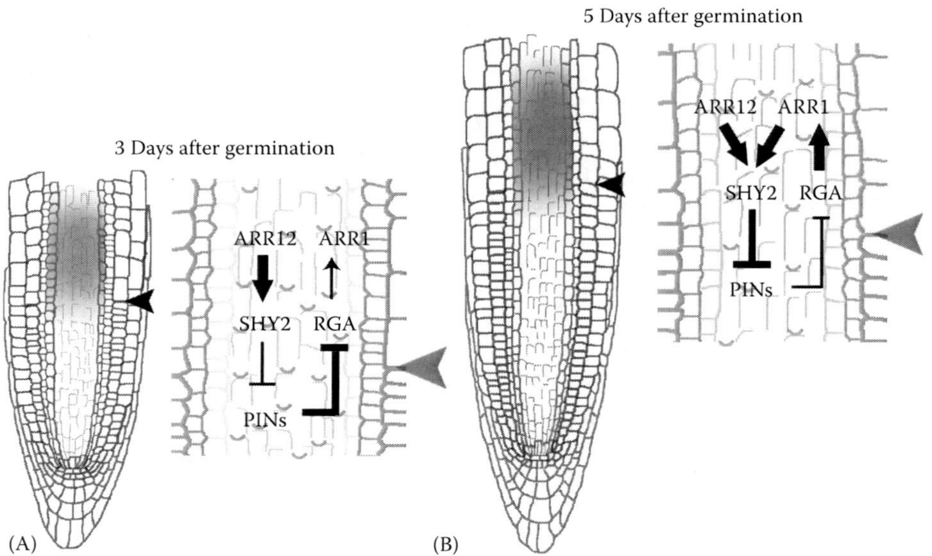

FIGURE 14.4 **(See color insert.)** The rate of cell differentiation controls the *Arabidopsis* root meristem growth phase. (A) The regulatory circuit controlling meristem growth 3 days after germination, when cell division overcomes cell differentiation. ARR12 drives low level of *SHY2* expression, sustaining PIN-mediated polar transport of auxin, which in turn supports gibberellin biosynthesis. High levels of gibberellin repress the activity of the DELLA protein RGA, thus suppressing *ARR1* transcription. (B) At five dpg, a decrease of gibberellin biosynthesis stabilizes the RGA protein, allowing transcriptional activation of *ARR1*, which eventually joins ARR12 in increasing *SHY2* expression, slowing PIN-mediated polar auxin transport and gibberellin biosynthesis. This results in an increase in cell differentiation that balances it with cell division and stops meristem growth. Black and gray arrowheads indicate the cortex TZ. Green area indicates the domain of *SHY2* expression, and different color gradations highlight changes in SHY2 levels.

Gibberellins promote degradation of DELLA proteins (Dill et al. 2001, 2004; Silverstone et al. 2001; Fu and Harberd 2003; Davière et al. 2008), which function as growth repressors during *Arabidopsis* seedling development (Ueguchi-Tanaka et al. 2007; Davière et al. 2008; Itoh et al. 2008). The DELLA protein *RGA* has been shown to positively control ARR1 expression, as loss-of-function *rga* mutants display lower levels of both *ARR1* and *SHY2*, and result in an enlarged root meristem resembling *arr1* and *shy2* phenotypes (Moubayidin et al. 2010). Interestingly, RGA protein levels increase during root meristem development specifically at the TZ following a decrease in gibberellins levels, 5 days after germination, as suggested by the reduced expression of several genes encoding rate-limiting enzymes in gibberellin biosynthesis (Moubayidin et al. 2010).

Therefore, during the root meristem growth phase, high level of gibberellins in the root apical meristem initially represses ARR1 expression by targeting the DELLA protein RGA for degradation (Moubayidin et al. 2010) (Figure 14.4A). In contrast, ARR12 is not regulated by gibberellin, enabling a low level of *SHY2* expression to be maintained (Figure 14.4A). Thus, PIN expression is elevated, enhancing auxin flow and gibberellin biosynthesis (Figure 14.4A) (Frigerio et al. 2006). As a result, the rate of cell division exceeds the rate of cell differentiation, allowing root meristem to build up (Figure 14.4A). At 5 days after germination, a reduction in gibberellin levels causes RGA to be stabilized, leading to ARR1 activation and *SHY2* upregulation (Figure 14.4B). SHY2, in turn, represses PIN expression, causing auxin redistribution; as a consequence, cell differentiation is enhanced and balanced with cell division, thereby setting root meristem size (Figure 14.4B) (Moubayidin et al. 2010).

D. Cytokinins and Root Embryo Development

The cytokinin and auxin antagonistic interaction has also been shown to play a crucial role in specifying root stem-cell niche during early stages of *Arabidopsis* embryo development (Müller and Sheen 2008) (Figure 14.5A). The *Arabidopsis* root SCN is laid down during early embryogenesis and starts with the specification of a single cell, the hypophysis, that asymmetrically divides to generate an apical cell, precursor of the QC cell, and a basal cell, precursor of the suspensor (Capron et al. 2009). Using a synthetic reporter to visualize cytokinin output, the two-component output sensor, and inducible genetic manipulations, it has been shown that an auxin maximum at the basal cell activates transcription of the type-A *ARR7* and *ARR15* genes, repressors of cytokinin signaling (Müller and Sheen 2008). Lack of both *ARR7* and *ARR15* genes leads to ectopic phosphorelay output in the basal daughter cell that prevents the establishment of a normal embryonic root SCN (Müller and Sheen 2008). Thus, auxin-mediated suppression of cytokinin signaling—via *ARR7* and *ARR15*—in the embryonic basal cell lineage is required in order to ensure the establishment of the root stem-cell niche (Figure 14.5A) (Müller and Sheen 2008). This mechanism is spatially and temporally confined to a defined developmental window, as expression of *ARR7* and *ARR15* is not detectable during late stages of embryo development any more (Müller and Sheen 2008).

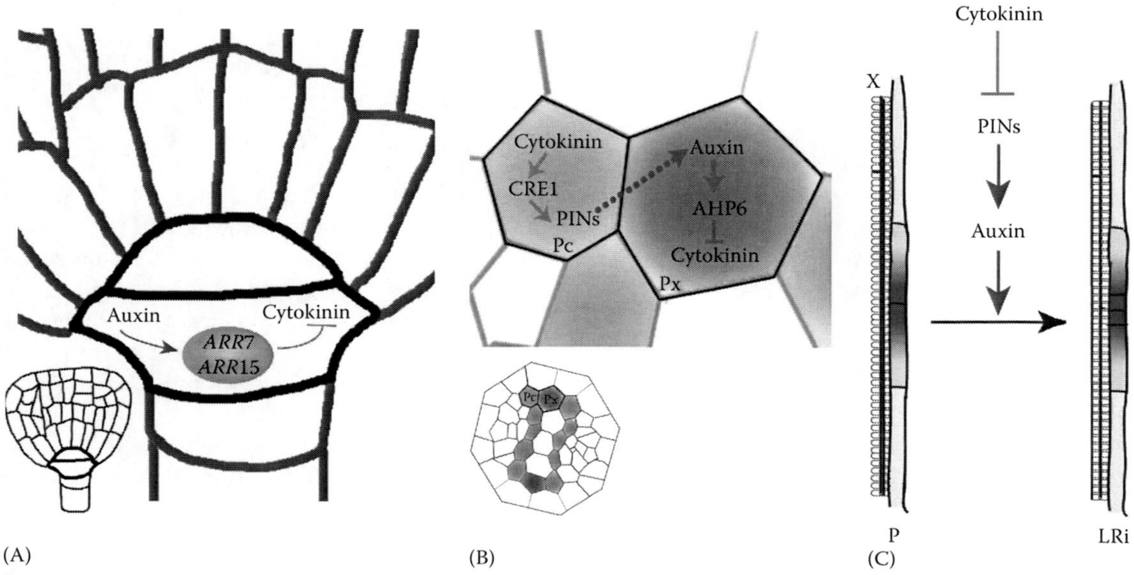

FIGURE 14.5 (See color insert.) Cytokinins are involved in different aspects of root development. (A) During early stages of embryo development, auxin activates type-A ARR7 and ARR15, negative regulators of cytokinin signaling, driving root SCN specification. (B) Schematic representation of a root apex cross section highlighting the procambium (Pc, in red) and the protoxylem (Px, in blue). Cytokinins, through the CRE1 receptor, maintain procambial cell identity promoting the localization of the PIN proteins on the lateral membranes of procambial cells, thus forcing auxin flow toward protoxylem (blue dotted arrow). Accumulation of auxin in the protoxylem induces expression of the pseudophosphotransfer AHP6 protein that promotes protoxylem differentiation by suppressing cytokinin signaling via an unknown mechanism. (C) Schematic representation of the initial stages of lateral root formation. Cytokinins negatively control lateral root initiation by downregulating *PIN* expression, thus preventing the establishment of the auxin maximum (in blue) in the xylem pole pericycle cells required for normal lateral root initiation. X: xylem; P: pericycle; LRi: lateral root initiation.

E. Cytokinins and Root Vascular Development

Cytokinins have been also implicated in the specification and maintenance of root vascular development, positively controlling procambial cell identity (Figure 14.5B). The root vascular cylinder has a bisymmetric pattern, with a central axis of xylem cell files consisting of protoxylem at marginal positions and metaxylem at central positions. This axis is flanked by a second plane, located 90° to this, running through the two phloem poles and intervening procambial cell files. The morphogenesis of vascular tissue in the root region is established through a set of asymmetric cell divisions requiring *CRE1* activity (Scheres et al. 1995; Mähönen et al. 2000; Nishimura et al. 2004). Mutations of the *CRE1* gene do not affect plant growth or development, with the exception of the *woodenleg* (*wol*) allele, that produces a CRE1 receptor lacking cytokinin binding activity and acting as a constitutively active phosphatase (Mähönen et al. 2006b).

wol plants have a determinate root growth phenotype resulting from a reduced number of cell divisions in the procambium and an abnormal radially symmetric pattern of differentiation of all vascular cells into protoxylem (Scheres et al. 1995; Mähönen et al. 2000, 2006a). Depletion of cytokinins in the *CRE1* domain of expression phenocopies the *wol* mutant and results in differentiation of all cell files within the root vascular cylinder into protoxylem. This indicates that cytokinin signaling is required for procambium maintenance during vascular development (Mähönen et al. 2006a). Analysis of an extragenic second-site suppressor mutation of the *wol* allele identified the *AHP6* locus, encoding for a "pseudo-HPT" that has been demonstrated to act as a negative regulator of cytokinin signaling (Mähönen et al. 2006a). A regulatory circuit between CRE1 and AHP6 has been proposed to regulate the balance between procambial cells and the differentiation of protoxylem elements in a bisymmetric pattern (Figure 14.5B) (Mähönen et al. 2006a). AHP6 promotes the differentiation of procambial cells into protoxylem by default. On the other hand, cytokinins that are responsible for procambial cell identity maintenance restrict *AHP6* expression to the protoxylem and the adjacent pericycle cells (Figure 14.5B). In these tissues, AHP6 inhibits cytokinin signaling, thus allowing differentiation of protoxylem elements in a bisymmetric pattern (Figure 14.5B) (Mähönen et al. 2006a).

In the procambial cells, high level of cytokinin transported through the phloem has been recently demonstrated to be required for the positioning of the PIN1, PIN3, and PIN7 auxin efflux proteins on the lateral membranes of these cells (Bishopp et al. 2011a,b). Reduction of cytokinin levels in the phloem alters PIN7 distribution and *AHP6* expression, resulting in the unstable formation of protoxylem files (Bishopp et al. 2011b).

This cytokinin-dependent PIN positioning forces auxin out of the procambial cells and its accumulation within protoxylem cells (Figure 14.5B). Here, the subsequent increase of auxin signaling promotes transcription of *AHP6*, which in turn inhibits cytokinin signaling, thus allowing the specification of protoxylem fate (Figure 14.5B) (Bishopp et al. 2011a). This model explains the all-protoxylem phenotype of the *wol* mutant: the auxin-response maximum expands in a radially symmetric fashion, and it can be suppressed by inhibiting auxin signaling within the CRE1 domain of expression (Bishopp et al. 2011a).

Type-B ARR1, ARR10, and ARR12 have been proposed to control transcription of genes regulating root vascular differentiation, since strong *arr1,10,12* loss-of-function mutants produce a short root and a *wol*-like phenotype (Yokoyama et al. 2007; Argyros et al. 2008; Ishida et al. 2008).

F. Cytokinins and Lateral Root Development

In *Arabidopsis*, lateral roots are initiated from a few pericycle cells at the periphery of the xylem poles that acquire the fate of lateral root founder cells (De Smet et al. 2007; Dubrovsky et al. 2008). Accumulation of auxin, mediated by both the auxin influx AUXIN RESISTANT 1 (AUX1) and auxin efflux PIN proteins (Benkova et al. 2003; Vieten et al. 2005; Ditengou et al. 2008), is necessary for the specification of the founder cells that subsequently divide and give rise to the lateral root primordium (Dubrovsky et al. 2008).

As in the case of primary root growth, several facts suggest that cytokinins are endogenous negative regulators of lateral root formation. Indeed, exogenous cytokinins inhibit lateral root initiation in *Arabidopsis* (Li et al. 2006; Laplaze et al. 2007; Kuderova et al. 2008) and in rice (Rani Debi et al. 2005). Moreover, several type-B *ARR* and *AHK* mutants as well as *CKX* overexpressing plants exhibit enhanced lateral root formation (Werner et al. 2001, 2003; Lohar et al. 2004; Mason et al. 2005; Riefler et al. 2006), whereas in several type-A *ARR* mutants, the number of lateral roots decreases (To et al. 2004).

The xylem-pole pericycle is the site where cytokinins act to inhibit lateral root initiation. Expression of *IPTs* in these cells has been demonstrated to inhibit initiation of lateral roots, a process that conversely can be enhanced by expressing the *CKX1* gene in xylem-pole pericycle cells (Laplaze et al. 2007).

Recently, the mechanism through which cytokinins interfere with lateral root initiation has been partially unveiled (Figure 14.5C). It has been demonstrated that cytokinin—through the CRE1 receptor (Li et al. 2006) and AHPs (Hutchison et al. 2006)—inhibits lateral root initiation by blocking division of pericycle cells, which will acquire founder cells identity (Li et al. 2006). Cytokinins have been shown to interfere with pericycle cell division by downregulating *PIN* gene expression, thus preventing the establishment of the auxin maximum necessary to induce cell fate respecification of divided pericycle cells into lateral root founder cells (Figure 14.5C) (Laplaze et al. 2007). Accordingly, overexpression of *IPTs* affects early lateral primordium patterning, as well as auxin distribution during lateral root primordium development (Kuderova et al. 2008).

Cytokinins also inhibit lateral root formation in the legume *Medicago truncatula* Gaertn. Silencing of the cytokinin receptor homolog CYTOKININ RESPONSE1 (MtCRE1) leads to cytokinin-resistant roots, with an increased number of lateral roots (Gonzalez-Rizzo et al. 2006), indicating a common role for cytokinin in regulating formation of lateral roots in higher plants.

IV. Cytokinins and Root Nodule Organogenesis

Root architecture (i.e., root system and its spatial configuration in terms of number and length of lateral organs) is a feature greatly depending on plant species, soil composition, and particularly on the availability of water and mineral nutrients (Lòpez-Bucio et al. 2003; Malamy 2005). In legumes, there are two major determinants of root architecture: soil conditions and interaction with symbiotic microorganisms. Besides lateral roots, legumes are able to develop a second type of lateral organ: the nitrogen-fixing nodule (Crespi and Frugier 2008). Morphogenesis of this root-derived organ is the consequence of a symbiotic interaction between soil *Rhizobium* bacteria and their legume hosts (Jones et al. 2007; Oldroyd and Downie 2008). Organogenesis of nitrogen-fixing root nodules (nodulation) starts at regions where root hairs develop and results from an interplay between bacterial and plant signals. The plant host mainly produces flavonoids, whereas bacteria synthesize and secrete lipochitooligosaccharidic signaling molecules, called Nod factors, that are perceived by the host root and act as signal molecules to initiate the nodule formation programmed in the host plant as well as to trigger the infection process (Crespi and Frugier 2008).

At the cellular level, infection of model legumes *Lotus japonicus* (Regel) K. Larsen and *Medicago truncatula* is initiated when bacteria are entrapped by a curled root hair. Concomitantly, Nod factors induce cytoskeletal rearrangement and transient proliferation of cells of the root pericycle layer. Protoxylem-pole cortical cells close to the site of bacterial infection dedifferentiate and reinitiate cell division, giving rise to the nodule primordium. At the same time, bacteria progress across root tissues toward the primordium inside channels called infection threads, whose formation is primarily host driven. Formation of infection threads is allowed by local cell wall hydrolysis and invagination of the plasma membrane. Infection threads then progress into the root cortex, branch, and ramify in the nodule primordium. Finally, Rhizobia are internalized into the nodule primordium, where they differentiate into a nitrogen-fixing form, the bacteroids, that can convert atmospheric nitrogen into ammonia (Stacey et al. 2006).

The first indication that cytokinins could be involved in nodule organogenesis came from the observation that rhizobial bacteria, such as *Rhizobium leguminosarum* Frank and *Bradyrhizobium japonicum* Kirchner, were able to secrete cytokinin-like compounds into the medium at concentrations that could affect root development (Phillips and Torrey 1972; Sturtevant and Taller 1989). Then, it was discovered that *Rhizobium* strains unable to synthesize Nod factors but capable of secreting tZ could induce uncolonized nodule-like structures in *Medicago sativa* L. roots, underlining the importance of this hormone during symbiosis (Cooper and Long 1994). The role of cytokinin as a key differentiation signal for nodule organogenesis was further confirmed by the observation that overexpression of heterologous *CKX* genes in *Lotus japonicus* (Regel) K. Larsen is sufficient to reduce nodulation (Lohar et al. 2004). In addition, expression analysis of the cytokinin-responsive gene *ARR5* in *L. japonicus* revealed that this gene is rapidly activated upon rhizobial inoculation in responding root hairs and dividing cortical cells (Lohar et al. 2004).

Only in recent years, the cytokinin-signaling components necessary to induce symbiotic nodulation have been identified (Gonzalez-Rizzo et al. 2006; Murray et al. 2007; Tirichine et al. 2007). Loss-of-function mutation of the *LOTUS HISTIDINE KINASE 1* (*LHK1*) gene, homolog of the *Arabidopsis* cytokinin receptor CRE1 (Tirichine et al. 2007), dramatically inhibits nodulation (Murray et al. 2007), whereas gain-of-function mutation in the same gene is sufficient to trigger cortical cell activation and spontaneous nodulation, suggesting that cytokinins are necessary and sufficient to initiate nodule organogenesis (Tirichine et al. 2007). Genetic analysis showed that the LHK1 pathway acts downstream of Nod factor perception and upstream of the activation of early nodulation-related transcription factors (Tirichine et al. 2007), and this LHK1-dependent pathway has been associated to nodule organogenesis acting in the cortex (Madsen et al. 2010).

Similarly, in *M. truncatula* lowering levels of the *LHK1*-orthologous *MtCRE1* blocks early stages of symbiotic interaction, such as growth of infection threads in the epidermis and induction of cortical cell division (Gonzalez-Rizzo et al. 2006). Expression analyses showed that MtCRE1 is initially expressed in cortical cells of the Nod factor responsive zone, then in the developing nodule primordium, and finally shifts to the nodule meristematic region (Lohar et al. 2006). Moreover, *MtCRE1/LHK1* cytokinin receptors mediate the activation of the early nodulation marker *NODULE INCEPTION* (*NIN*), a putative transcription factor essential for the formation of infection threads and bacterial entry (Schauser et al. 1999; Gonzalez-Rizzo et al. 2006; Marsh et al. 2007; Murray et al. 2007).

Nodule formation has been shown to depend also on auxin, because silencing of several PINs results in reduced nodulation (Huo et al. 2006). As in the *Arabidopsis* cytokinin-signaling mutants, in the *M. truncatula cre1* mutants, transcripts of several *MtPIN* genes are elevated (Plet et al. 2011), suggesting that the MtCRE1-dependent cytokinin signaling acts upstream of auxin-related pathways that are crucial for nodule organogenesis.

Future investigation will clarify the molecular basis of such a complex scenarios.

V. Concluding Remarks

Since their original isolation in autoclaved products of herring sperm DNA as cell division promoting factors (Miller et al. 1955), cytokinins have been implicated in the regulation of many plant developmental processes. In recent years, the identification of the molecular component of the cytokinin-signaling pathways and their target genes highlighted the importance of the interaction between cytokinin and other hormones, above all auxins. The importance of the cytokinin–auxin interaction during plant development is known since 1957, when the pioneer in vitro experiments performed by Skoog and Miller demonstrated that exposing callus cultures to a high auxin:cytokinin ratio resulted in root formation, whereas a low ratio of these hormones promotes shoot development (Skoog and Miller 1957). Since then, the cytokinin–auxin interaction has proved to be central, in vivo, for several developmental processes such as embryo development, root and shoot meristem maintenance, vascular development, as well as lateral root development and nodule organogenesis. Cytokinins can antagonize auxin activity or act in concert with it depending on the tissues or organs. For example, during root meristem development, cytokinins induce cell differentiation antagonizing the auxin-mediated cell division input, thus setting meristem size. On the other hand, during shoot apical meristem development, cytokinins act in synergy with auxin to maintain stem-cell activity (Zhao et al. 2010).

Future challenge will be to dissect the molecular basis of such differences to further understand how cytokinin/auxin interaction shapes plant development.

References

Achard P, Gusti A, Cheminant S et al. 2009 Gibberellin signaling controls cell proliferation rate in *Arabidopsis*. *Curr Biol* 19:1188–1193.

Argueso CT, Raines T, Kieber JJ 2010 Cytokinin signaling and transcriptional networks. *Curr Opin Plant Biol* 13:533–539.

Argyros RD, Mathews DE, Chiang YH et al. 2008 Type B response regulators of *Arabidopsis* play key roles in cytokinin signaling and plant development. *Plant Cell* 20:2102–2116.

Beemster GTS, Baskin TI 1998 Analysis of cell division and elongation underlying the developmental acceleration of root growth in *Arabidopsis thaliana*. *Plant Physiol* 116:1515–1526.

Benjamins R, Scheres B 2008 Auxin: The looping star in plant development. *Annu Rev Plant Biol* 59:443–465.

Benkova E, Michniewicz M, Sauer M et al. 2003 Local, efflux-dependent auxin gradients as a common module for plant organ formation. *Cell* 115:591–602.

Bishopp A, Help H, El-Showk S et al. 2011a A mutually inhibitory interaction between auxin and cytokinin specifies vascular pattern in roots. *Curr Biol* 21:917–926.

Bishopp A, Lehesranta S, Vatén A et al. 2011b Phloem-transported cytokinin regulates polar auxin transport and maintains vascular pattern in the root meristem. *Curr Biol* 21:927–932.

Blilou I, Xu J, Wildwater M et al. 2005 The PIN auxin efflux facilitator network controls growth and patterning in *Arabidopsis* root. *Nature* 433:39–44.

Brugiere N, Jiao S, Hantke S et al. 2003 Cytokinin oxidase gene expression in maize is localized to the vasculature, and is induced by cytokinins, abscisic acid, and abiotic stress. *Plant Physiol* 132:1228–1240.

Capron A, Chatfield S, Provart N et al. 2009 Embryogenesis: Pattern formation from a single cell. *The Arabidopsis Book*, 7:e0126. doi:10.1199/tab.0126.

Cooper JB, Long SR 1994 Morphogenetic rescue of *Rhizobium meliloti* nodulation mutants by *trans*-zeatin secretion. *Plant Cell* 6:215–225.

Crespi M, Frugier F 2008 De novo organ formation from differentiated cells: Root nodule organogenesis. *Sci Signal* 1:re11.

Davière JM, de Lucas M, Prat S 2008 Transcriptional factor interaction: A central step in DELLA function. *Curr Opin Genet Dev* 18:295–303.

De Smet I, Tetsumura T, De Rybel B et al. 2007 Auxin-dependent regulation of lateral root positioning in the basal meristem of *Arabidopsis*. *Development* 134:681–690.

Dello Ioio R, Linhares FS, Scacchi E et al. 2007 Cytokinins determine *Arabidopsis* root-meristem size by controlling cell differentiation. *Curr Biol* 17:678–682.

Dello Ioio R, Nakamura K, Moubayidin L et al. 2008 A genetic framework for the auxin/cytokinin control of cell division and differentiation in the root meristem. *Science* 322:1380–1384.

Dill A, Jung HS, Sun TP 2001 The DELLA motif is essential for gibberellin-induced degradation of RGA. *Proc Natl Acad Sci USA* 98:14162–14167.

Dill A, Thomas SG, Hu J et al. 2004 The *Arabidopsis* F-box protein SLEEPY1 targets gibberellin signaling repressors for gibberellin-induced degradation. *Plant Cell* 16:1392–1405.

Ditengou FA, Teale WD, Kochersperger P et al. 2008 Mechanical induction of lateral root initiation in *Arabidopsis thaliana*. *Proc Natl Acad Sci USA* 105:18818–18823.

Dolan L, Janmaat K, Willemsen V et al. 1993 Cellular organisation of the *Arabidopsis thaliana* root. *Development* 119:71–84.

Dubrovsky JG, Sauer M, Napsucialy-Mendivil S et al. 2008 Auxin acts as a local morphogenetic trigger to specify lateral root founder cells. *Proc Natl Acad Sci USA* 105:8790–8794.

Frigerio M, Alabadì D, Pérez-Gòmez J et al. 2006 Transcriptional regulation of gibberellin metabolism genes by auxin signaling in *Arabidopsis*. *Plant Physiol* 142:553–563.

Frugier F, Kosuta S, Murray JD et al. 2008 Cytokinin: Secret agent of symbiosis. *Trends Plant Sci* 13:115–120.

Fu X, Harberd NP 2003 Auxin promotes *Arabidopsis* root growth by modulating gibberellin response. *Nature* 421:740–743.

Gonzalez-Rizzo S, Crespi M, Frugier F 2006 The *Medicago truncatula* CRE1 cytokinin receptor regulates lateral root development and early symbiotic interaction with *Sinorhizobium meliloti*. *Plant Cell* 18:2680–2693.

Higuchi M, Pischke MS, Mähönen AP et al. 2004 *In planta* functions of the *Arabidopsis* cytokinin receptor family. *Proc Natl Acad Sci USA* 101:8821–8826.

Huo X, Schnabel E, Hughes K et al. 2006 RNAi phenotypes and the localization of a protein: GUS fusion imply a role for *Medicago truncatula* PIN genes in nodulation. *J Plant Growth Regul* 25:156–165.

Hutchison CE, Li J, Argueso C et al. 2006 The *Arabidopsis* histidine phosphotransfer proteins are redundant positive regulators of cytokinin signaling. *Plant Cell* 18:3073–3087.

Hwang I, Sheen J 2001 Two-component circuitry in *Arabidopsis* cytokinin signal transduction. *Nature* 413:383–389.

Inoue T, Higuchi M, Hashimoto Y et al. 2001 Identification of CRE1 as a cytokinin receptor from *Arabidopsis*. *Nature* 409:1060–1063.

Ishida K, Yamashino T, Yokoyama A et al. 2008 Three type-B response regulators, ARR1, ARR10 and ARR12, play essential but redundant roles in cytokinin signal transduction throughout the life cycle of *Arabidopsis thaliana*. *Plant Cell Physiol* 49:47–57.

Itoh H, Ueguchi-Tanaka M, Matsuoka M 2008 Molecular biology of gibberellins signaling in higher plants. *Int Rev Cell Mol Biol* 268:191–221.

Jones KM, Kobayashi H, Davies BW et al. 2007 How rhizobial symbionts invade plants: The *Sinorhizobium-Medicago* model. *Nat Rev Microbiol* 5:619–633.

Kakimoto T 2001 Identification of plant cytokinin biosynthetic enzymes as dimethylallyl diphosphate: ATP/ADP isopentenyltransferases. *Plant Cell Physiol* 42:677–685.

Kakimoto T 2003 Perception and signal transduction of cytokinins. *Annu Rev Plant Biol* 54:605–627.

Kuderova A, Urbankova I, Valkova M et al. 2008 Effects of conditional IPT-dependent cytokinin overproduction on root architecture of *Arabidopsis* seedlings. *Plant Cell Physiol* 49:570–582.

Kurakawa T, Ueda N, Maekawa M et al. 2007 Direct control of shoot meristem activity by a cytokinin-activating enzyme. *Nature* 445:652–655.

Kuroha T, Tokunaga H, Kojima M et al. 2009 Functional analyses of LONELY GUY cytokinin-activating enzymes reveal the importance of the direct activation pathway in *Arabidopsis*. *Plant Cell* 21:3152–3169.

Laplaze L, Benkova E, Casimiro I et al. 2007 Cytokinins act directly on lateral root founder cells to inhibit root initiation. *Plant Cell* 19:3889–3900.

Li X, Mo X, Shou H et al. 2006 Cytokinin-mediated cell cycling arrest of pericycle founder cells in lateral root initiation of *Arabidopsis*. *Plant Cell Physiol* 47:1112–1123.

Lohar DP, Schaff JE, Laskey JG et al. 2004 Cytokinins play opposite roles in lateral root formation, and nematode and Rhizobial symbioses. *Plant J* 38:203–214.

Lòpez-Bucio J, Cruz-Ramìrez A, Herrera-Estrella L 2003 The role of nutrient availability in regulating root architecture. *Curr Opin Plant Biol* 6:280–287.

Madsen LH, Tirichine L, Jurkiewicz A et al. 2010 The molecular network governing nodule organogenesis and infection in the model legume *Lotus japonicus*. *Nat Commun* 1:1–12.

Mähönen AP, Bishopp A, Higuchi M et al. 2006a Cytokinin signaling and its inhibitor AHP6 regulate cell fate during vascular development. *Science* 311:94–98.

Mähönen AP, Bonke M, Kauppinen L et al. 2000 A novel two-component hybrid molecule regulates vascular morphogenesis of the *Arabidopsis* root. *Genes Dev* 14:2938–2943.

Mähönen AP, Higuchi M, Tormakangas K et al. 2006b Cytokinins regulate a bidirectional phosphorelay network in *Arabidopsis. Curr Biol* 16:1116–1122.

Malamy JE 2005 Intrinsic and environmental response pathways that regulate root system architecture. *Plant Cell Environ* 28:67–77.

Marsh JF, Rakocevic A, Mitra RM et al. 2007 *Medicago truncatula* NIN is essential for rhizobial-independent nodule organogenesis induced by autoactive calcium/calmodulin-dependent protein kinase. *Plant Physiol* 144:324–335.

Mason MG, Li J, Mathews DE et al. 2004 Type B response regulators display overlapping expression patterns in *Arabidopsis. Plant Physiol* 135:1–11.

Mason MG, Mathews DE, Argyros DA et al. 2005 Multiple type-B response regulators mediate cytokinin signal transduction in *Arabidopsis. Plant Cell* 17:3007–3018.

Miller CO, Skoog F, Von Saltza MH et al. 1955 Kinetin, a cell division factor from deoxyribonucleic acid. *J Am Chem Soc* 77:1392.

Miyawaki K, Matsumoto-Kitano M, Kakimoto T 2004 Expression of cytokinin biosynthetic isopentenyltransferase genes in *Arabidopsis*: Tissue specificity and regulation by auxin, cytokinin, and nitrate. *Plant J* 37:128–138.

Miyawaki K, Tarkowski P, Matsumoto-Kitano M et al. 2006 Roles of *Arabidopsis* ATP/ADP isopentenyltransferases and tRNA isopentenyltransferases in cytokinin biosynthesis. *Proc Natl Acad Sci USA* 103:16598–16603.

Mok DWS, Mok MC 2001 Cytokinin metabolism and action. *Annu Rev Plant Physiol Plant Mol Biol* 52:89–118.

Moubayidin L, Perilli S, Dello Ioio R et al. 2010 The rate of cell differentiation controls the *Arabidopsis* root meristem growth phase. *Curr Biol* 20:1138–1143.

Müller B, Sheen J 2008 Cytokinin and auxin interaction in root stem cell specification during early embryogenesis. *Nature* 453:1094–1097.

Murray JD, Karas BJ, Sato S et al. 2007 A cytokinin perception mutant colonized by *Rhizobium* in the absence of nodule organogenesis. *Science* 315:101–104.

Nishimura C, Ohashi Y, Sato S et al. 2004 Histidine kinase homologs that act as cytokinin receptors possess overlapping functions in the regulation of shoot and root growth in *Arabidopsis. Plant Cell* 16:1365–1377.

Oldroyd GE, Downie JA 2008 Coordinating nodule morphogenesis with rhizobial infection in legumes. *Annu Rev Plant Biol* 59:519–546.

Perilli S, Moubayidin L, Sabatini S 2010 The molecular basis of cytokinin function. *Curr Opin Plant Biol* 13:21–26.

Phillips DA, Torrey JG 1972 Studies on cytokinin production by *Rhizobium. Plant Physiol* 49:11–15.

Plet J, Wasson A, Ariel F et al. 2011 MtCRE1-dependent cytokinin signaling integrates bacterial and plant cues to coordinate symbiotic nodule organogenesis in *Medicago truncatula. Plant J* 65:622–633.

Punwani JA, Hutchison CE, Schaller GE, Kieber JJ. 2010 The subcellular distribution of the Arabidopsis histidine phosphotransfer proteins is independent of cytokinin signaling. *Plant J* 62:473–482.

Rani Debi B, Taketa S, Ichii M 2005 Cytokinin inhibits lateral root initiation but stimulates lateral root elongation in rice (*Oryza sativa*). *J Plant Physiol* 162:507–515.

Rashotte AM, Mason MG, Hutchison CE et al. 2006 A subset of *Arabidopsis* AP2 transcription factors mediates cytokinin responses in concert with a two-component pathway. *Proc Natl Acad Sci USA* 103:11081–11085.

Riefler M, Novak O, Strnad M et al. 2006 *Arabidopsis* cytokinin receptor mutants reveal functions in shoot growth, leaf senescence, seed size, germination, root development, and cytokinin metabolism. *Plant Cell* 18:40–54.

Sablowski R 2010 Plant stem cell niches: From signaling to execution. *Curr Opin Plant Biol* 14:4–9.

Sakakibara H 2006 Cytokinins: Activity, biosynthesis, and translocation. *Annu Rev Plant Biol* 57:431–449.

Schauser L, Roussis A, Stiller J et al. 1999 A plant regulator controlling development of symbiotic root nodules. *Nature* 402:191–195.

Scheres B 2007 Stem-cell niches: Nursery rhymes across kingdoms. *Nat Rev Mol Cell Biol* 8:345–354.

Scheres B, Di Laurenzio L, Willemsen V et al. 1995 Mutations affecting the radial organization of the *Arabidopsis* root display specific defects throughout the embryonic axis. *Development* 121:53–62.

Schmulling T, Werner T, Riefler M et al. 2003 Structure and function of cytokinin oxidase/dehydrogenase genes of maize, rice, *Arabidopsis* and other species. *J Plant Res* 116:241–252.

Silverstone AL, Jung HS, Dill A et al. 2001 Repressing a repressor: Gibberellin-induced rapid reduction of the RGA protein in *Arabidopsis. Plant Cell* 13:1555–1566.

Skoog F, Miller C 1957 Chemical regulation of growth and organ formation in plant tissue cultured in vitro. *Symp Soc Exp Biol* 11:118–131.

Stacey G, Libault M, Brechenmacher L et al. 2006 Genetics and functional genomics of legume nodulation. *Curr Opin Plant Biol* 9:110–121.

Sturtevant DB, Taller BJ 1989 Cytokinin production by *Bradyrhizobium japonicum. Plant Physiol* 89:1247–1252.

Takei K, Sakakibara H, Sugiyama T 2001 Identification of genes encoding adenylate isopentenyltransferase, a cytokinin biosynthesis enzyme, in *Arabidopsis thaliana. J Biol Chem* 276:26405–26410.

Taniguchi M, Sasaki N, Tsuge T et al. 2007 ARR1 directly activates cytokinin response genes that encode proteins with diverse regulatory functions. *Plant Cell Physiol* 48:263–277.

Tian Q, Nagpal P, Reed JW 2003 Regulation of *Arabidopsis* SHY2/IAA3 protein turnover. *Plant J* 36:643–651.

Tian Q, Reed JW 1999 Control of auxin-regulated root development by the *Arabidopsis thaliana SHY2/IAA3* gene. *Development* 126:711–721.

Tian Q, Uhlir NJ, Reed JW 2002 *Arabidopsis* SHY2/IAA3 inhibits auxin-regulated gene expression. *Plant Cell* 14:301–319.

Tirichine L, Sandal N, Madsen LH et al. 2007 A gain-of-function mutation in a cytokinin receptor triggers spontaneous root nodule organogenesis. *Science* 315:104–107.

To JPC, Deruère J, Maxwell BB et al. 2007 Cytokinin regulates type-A *Arabidopsis* Response Regulator activity and protein stability via two-component phosphorelay. *Plant Cell* 19:3901–3914.

To JPC, Haberer G, Ferreira FJ et al. 2004 Type-A *Arabidopsis* response regulators are partially redundant negative regulators of cytokinin signaling. *Plant Cell* 16:658–671.

To JPC, Kieber JJ 2008 Cytokinin signaling: two-components and more. *Trends Plant Sci* 13:85–92.

Ubeda-Tomás S, Federici F, Casimiro I et al. 2009 Gibberellin signaling in the endodermis controls *Arabidopsis* root meristem size. *Curr Biol* 19:1194–1199.

Ubeda-Tomás S, Swarup R, Coates J et al. 2008 Root growth in *Arabidopsis* requires gibberellin/DELLA signalling in the endodermis. *Nat Cell Biol* 10:625–628.

Ueguchi-Tanaka M, Nakajima M, Motoyuki A et al. 2007 Gibberellin receptor and its role in gibberellin signaling in plants. *Annu Rev Plant Biol* 58:183–198.

Van den Berg C, Willemsen V, Hendriks C et al. 1997 Short-range control of cell differentiation in the *Arabidopsis* root meristem. *Nature* 390:287–289.

Vieten A, Vanneste S, Wisniewska J et al. 2005 Functional redundancy of PIN proteins is accompanied by auxin-dependent cross-regulation of PIN expression. *Development* 132:4521–4531.

Werner T, Motyka V, Laucou V et al. 2003 Cytokinin-deficient transgenic *Arabidopsis* plants show multiple developmental alterations indicating opposite functions of cytokinins in the regulation of shoot and root meristem activity. *Plant Cell* 15:2532–25550.

Werner T, Motyka V, Strnad M et al. 2001 Regulation of plant growth by cytokinin. *Proc Natl Acad Sci USA* 98:10487–10492.

Yokoyama A, Yamashino T, Amano Y et al. 2007 Type-B ARR transcription factors, ARR10 and ARR12, are implicated in cytokinin-mediated regulation of protoxylem differentiation in roots of *Arabidopsis thaliana*. *Plant Cell Physiol* 48:84–96.

Zhao Z, Andersen SU, Ljung K et al. 2010 Hormonal control of the shoot stem-cell niche. *Nature* 465:1089–1092.

15

Ethylene Regulates Root Growth and Development

Daniel R. Lewis
Wake Forest University

Gloria K. Muday
Wake Forest University

I. Introduction

Ethylene is a gaseous plant hormone that has profound effects on many aspects of plant growth and development. The role of changing levels of ethylene in modulating fruit ripening, seed germination, hypocotyl and root elongation, abscission, and responses to biotic and abiotic stresses has been well described (Abeles et al. 1992; Kendrick and Chang 2008). The best studied ethylene-regulated processes are hypocotyl elongation, particularly in dark-grown seedlings, and fruit ripening (as reviewed in Klee 2004; Giovannoni 2007; Kendrick and Chang 2008). Altered ethylene response in these two growth processes are the phenotypes used for isolation of mutants in ethylene signaling and synthesis, predominantly in *Arabidopsis thaliana* and *Lycopersicum esculentum* (tomato) (Klee 2004; Giovannoni 2007; Kendrick and Chang 2008). In recent years, the effect of ethylene on root growth and development has received substantially more attention, with identification of both inhibitory effects of ethylene on root elongation, gravitropism, and lateral root development and stimulatory effects of ethylene on root hair initiation. These recent studies using the plethora of ethylene signaling and synthesis mutants have identified mechanisms by which ethylene regulates root growth and development and have provided strong evidence for ethylene exerting its effect on root growth through interactions with auxin. This chapter will introduce the basics of ethylene signaling and how ethylene affects root growth and development, followed by sections that detail the mechanisms by which ethylene interacts with auxin signaling, transport, and synthesis to drive ethylene-dependent root growth and development.

II. Ethylene Signaling Pathway

The ethylene signal-transduction pathway has been identified through genetic analyses, and the details of this pathway have been reviewed elsewhere (Giovannoni 2007; Kendrick and Chang 2008; Stepanova and Alonso 2009). Important proteins in the ethylene signaling and synthesis pathways are briefly summarized here, as plants with mutations in these genes have been used to understand how ethylene affects root growth. Figure 15.1 contains a simplified schematic of the *Arabidopsis* ethylene signaling pathway with these proteins highlighted. *Arabidopsis ein2* and *etr1* mutants were isolated for loss of the ethylene-induced triple response, a process which is characterized by inhibition of hypocotyl elongation, enhancement in hook curvature, and hypocotyl swelling (Bleecker et al. 1988; Guzman and Ecker 1990; Chang et al. 1993). Additional screens identified mutants with enhanced ethylene response, including constitutive triple response (*ctr1*) (Kieber et al. 1993; Huang et al. 2003) and ethylene overproduction (*eto1*) (Guzman and Ecker 1990; Wang et al. 2004). The molecular defects in these mutants are now known, and the signaling proteins have been placed in the ethylene signal-transduction pathway (Kendrick and Chang 2008). Identification of the molecular basis of a number of tomato mutants with defects in ethylene-dependent fruit ripening illustrates substantial similarity in the signaling pathways between these two species (Klee 2004; Giovannoni 2007).

Physiological and genetic characterization of ethylene mutants has revealed a linear signaling pathway that begins with ethylene binding to and turning off its receptor proteins, including ETR1 and its tomato ortholog, NEVER-RIPE (NR) (Chang et al. 1993;

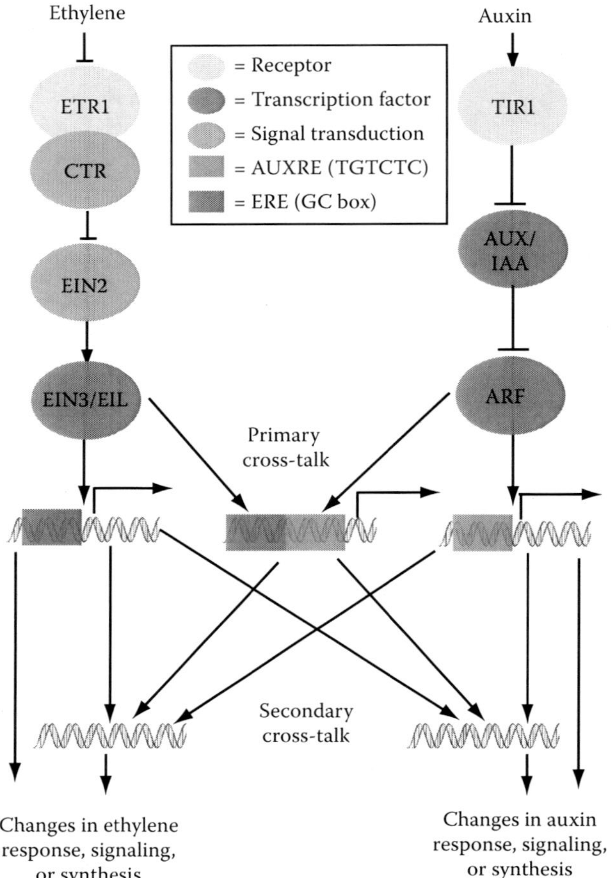

FIGURE 15.1 Model of auxin/ethylene cross talk. In *Arabidopsis*, ethylene and auxin responses are initiated through the ETR1 and TIR1 receptors, respectively. Ethylene binds to and inhibits ETR1 activity, which in turn leads to inhibition of the CTR kinase, a negative regulator of EIN2 activity. EIN2 activates the EIN3 and EIN3-like (EIL) family of transcription factors, which in turn promote transcription of genes containing an ethylene-responsive element (ERE) in their promoter region. Auxin signaling is mediated by proteasome-dependent degradation of AUX/IAA transcriptional repressors, which release ARF transcription factors to activate transcription of genes with auxin-responsive elements (AUXRE) in their regulatory region. Primary cross talk occurs by activation of genes that contain both AUXRE and ERE in their promoter region allowing both signaling pathways to directly regulate transcription. Secondary cross talk occurs through expression of genes that are either auxin or ethylene responsive, but whose activities control expression of genes that regulate the synthesis, signaling, or response to the other hormone. (Reprinted from Muday, GK. et al., *Trends Plant Sci.*, 17, 181, 2012.)

Wilkinson et al. 1995), as shown in Figure 15.1. CTR1, a protein kinase with sequence similarity to the catalytic domain of RAF protein kinase (a mitogen-activated protein kinase kinase kinase), is downstream from ETR1 and functions as a negative regulator of signaling (Kieber et al. 1993; Huang et al. 2003). EIN2 is an essential, positive modulator of ethylene signaling (Alonso et al. 1999) that, either directly or indirectly, controls the activity of transcription factors, including EIN3 and EIN3-like (EIL) proteins, whose targets include *ERF1* (*ETHYLENE RESPONSE FACTOR1*) and *EDF1* though 4 (*ETHYLENE RESPONSE DNA BINDING FACTOR1* through *4*) (Solano 1998; Alonso 2003). As a result of this hierarchical transcriptional cascade, ethylene either positively or negatively regulates diverse genes encoding proteins that mediate the growth response to ethylene (as reviewed by Stepanova et al. 2007; Kendrick and Chang 2008). This detailed understanding of ethylene signaling, and the associated genetic resources, provide an excellent framework in which the role of ethylene in root growth and development can be examined.

III. Ethylene Alters Root Growth and Development

The ability of ethylene to alter root growth and developmental processes is illustrated in Figure 15.2. These ethylene effects include modulation of primary root elongation, lateral root formation, root hair formation, root waving, and the gravitropic response. The inhibition of root elongation in the presence of high levels of ethylene or its precursor, 1-aminocyclopropane-1-carboxylic acid (ACC), has been documented in numerous species (Abeles et al. 1992; Le et al. 2001; Rahman et al. 2001; Swarup et al. 2002; Negi et al. 2010). Mutants with reduced sensitivity to ethylene, including *etr1*, *ein2*, *ein3*, and *eil1* in *Arabidopsis* and *never ripe* (*nr*) and *green ripe* (*gr*) in tomato, have elevated primary root growth rates demonstrating that ethylene inhibition of root growth is mediated by similar signaling pathways as ethylene's effects on hypocotyl elongation and fruit ripening (Barry and Giovannoni 2006; Ruzicka et al. 2007; Stepanova et al. 2007; Swarup et al. 2007; Negi et al. 2010). Roots of seedlings with enhanced ethylene signaling or synthesis, *ctr1* and *eto1*, respectively, exhibit reductions in the rate of root elongation (Kieber et al. 1993).

The inhibition of root elongation by ethylene has been examined with high temporal and spatial resolution and occurs within 5 min of treatment (Le et al. 2001). Exogenous ethylene treatment and the *eto1* mutation inhibit elongation predominantly in the central root elongation zone (EZ), found 500–1700 μm from the *Arabidopsis* root tip (Swarup et al. 2007; Ruzicka et al. 2007; Strader et al. 2010). There have been conflicting reports on whether the effects of ethylene on root growth also include an inhibition of cell division (Ortega-Martinez et al. 2007; Ruzicka et al. 2007). A particularly important feature of the growth inhibition by ethylene is that it closely mirrors the spatial and temporal effects of elevated auxin, inhibiting growth with similar changes in cell elongation in the primary root (Rahman et al. 2007).

The differential growth required for root gravitropism and root waving is also altered by ethylene. Root gravitropism is the reestablishment of a new direction of root growth that occurs when roots have been reoriented relative to the gravity vector, while root waving is sigmoidal growth that occurs when a root is grown on hard agar tilted at 45° (Okada and Shimura 1990). Ethylene treatment inhibits the gravitropic response in maize and *Arabidopsis* and increases the amplitude and frequency of

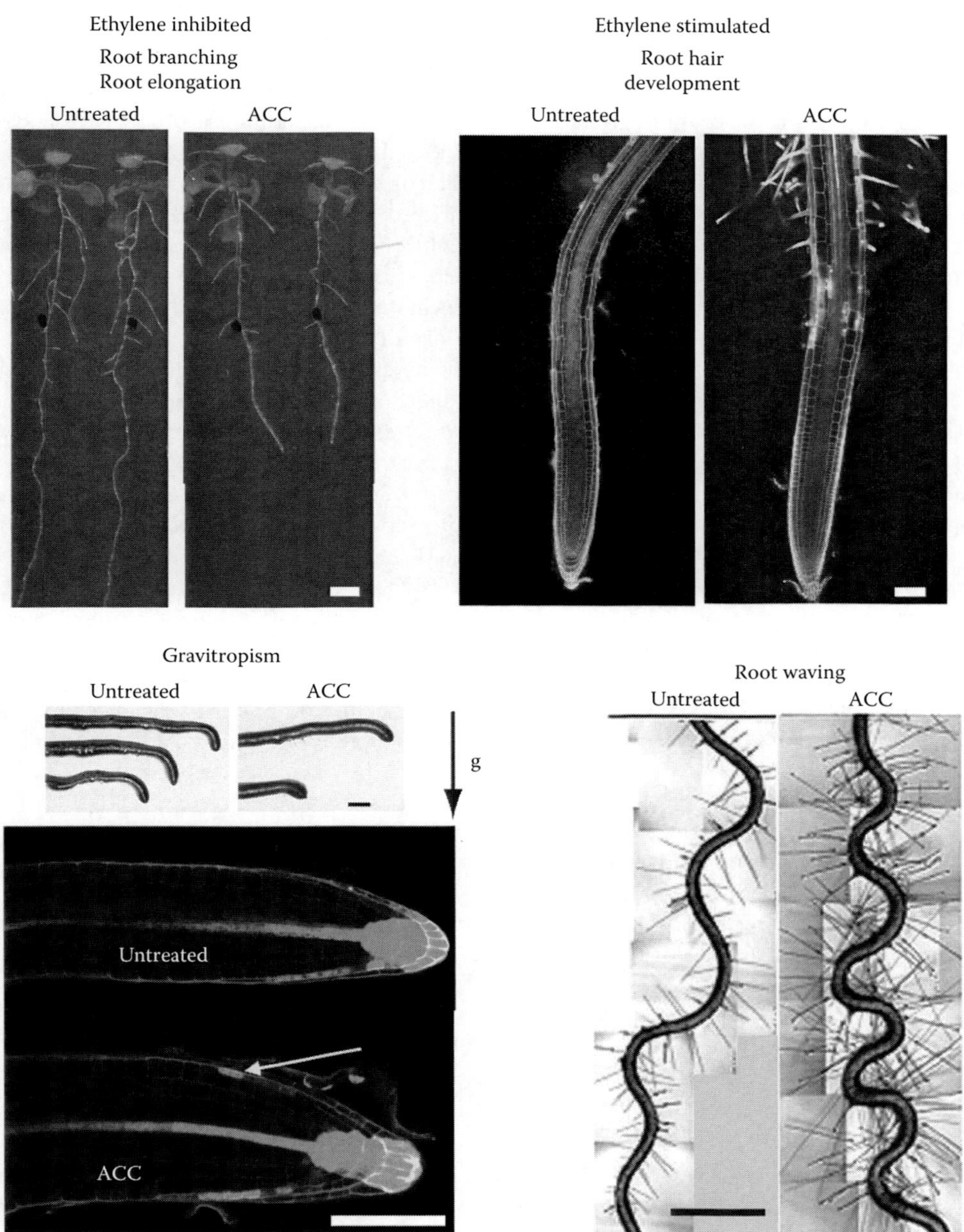

FIGURE 15.2 **(See color insert.)** Effects of ethylene on root growth and development. Treatment of roots with either ethylene or its biosynthetic precursor, ACC, alters root growth and development. The inhibitory effect of ACC on root elongation, lateral root development, and gravitropism is shown on the left. The images of roots undergoing gravitropism include both graviresponding roots at low magnification and confocal images of roots at 8 h after reorientation 90° relative to the gravity vector. The angle of gravity is indicated by an arrow. The confocal image shows individual cells after propidium iodide staining and the asymmetric expression of the auxin-responsive DR5-revGFP reporter across the untreated root. In contrast, ACC treatment leads to DR5-revGFP fluorescence on the upper side, which then minimizes the auxin gradient across the root reducing gravitropic curvature. The effect of ACC on stimulation of root hair formation and alteration of the periodicity of root waving is also shown. Scale bars = 100 μm for all micrographs except those depicting lateral root development and root waving, which are 1 mm. (Images of root waving are reprinted with permission from Buer et al. 2003, *Plant Physiol.* 132: 1085–1096.)

the root waving response (Lee et al. 1990; Buer et al. 2003, 2006; Chang et al. 2004; Lewis et al. 2011b). The inhibition of gravitropism by ethylene is lost in *ein2* and *etr1* mutants (Buer et al. 2006), suggesting an intact ethylene signaling pathway is necessary for the effect of ethylene on differential growth.

Recent genetic studies in *Arabidopsis* and tomato have shown that ethylene negatively regulates lateral root formation (Ivanchenko et al. 2008; Negi et al. 2008, 2010). Treatments or mutations to elevate ethylene levels inhibit lateral root formation, and *Arabidopsis* and tomato ethylene-insensitive mutants exhibit elevated numbers of lateral roots (Negi et al. 2008, 2010; Strader et al. 2010). Although most of these studies examined root growth on agar medium, the increase in lateral root formation seen in *nr* is even more striking in seedlings grown in soil for several weeks (Negi et al. 2010), suggesting that ethylene may have even more profound effects on roots during standard cultivation. Ethylene reduces lateral root development at the earliest stage of lateral root initiation, as determined by microscopic examination of lateral root primordia formation (Ivanchenko et al. 2008).

The effect of ethylene on adventitious root formation is less clear, as ethylene increases formation in some species (reviewed by De-Klerk et al. 1999) while decreasing adventitious root formation in others (Coleman et al. 1980; Nordstrom and Eliasson 1984). The effect of ethylene on adventitious root formation has only been examined in genetic models in a small number of cases (Clark et al. 1999; Kim et al. 2008; Negi et al. 2010; Sukumar 2010). In tomato, elevated endogenous or exogenous ethylene levels increase adventitious root formation, while ethylene-insensitive *never ripe* produces fewer adventitious roots (Clark et al. 1999; Kim et al. 2008; Negi et al. 2010). In contrast in *Arabidopsis*, ACC treatment, as well as the *eto1* and *ctr1* mutations, results in reduced adventitious root formation (Sukumar 2010).

In contrast to the negative effects of ethylene on root elongation and lateral root formation, root hair development is synergistically induced by ethylene and auxin (Kieber et al. 1993; Pitts et al. 1998; Schiefelbein 2000; Rahman et al. 2002). Auxin is able to rescue root hair elongation defects in ethylene-insensitive mutants, and inhibition of auxin influx exacerbates the *ein2* root hair phenotype (Rahman et al. 2002). Interestingly, ethylene also affects the positioning of root hair outgrowth, with ethylene application reinforcing the basal positioning of root hairs and ethylene-insensitive mutants having apically shifted root hairs (Fischer et al. 2007).

Elucidating the mechanisms by which ethylene regulates root growth and development is now an active area of study. Root elongation, gravitropism, waving, lateral root development, and root hair development are all auxin-dependent processes, suggesting that mechanisms for ethylene control of these processes may involve cross talk between these two hormones. The extent of auxin response in roots is defined by the amount of auxin synthesis, transport, and signaling activity. The following sections examine the experimental evidence for ethylene altering these three features of auxin action and how these mechanisms in turn may control the root elongation, gravitropism, and lateral root responses to ethylene.

IV. Examination of Cross Talk between Auxin and Ethylene Signaling Pathways

Auxin-dependent gene expression is strongly enhanced by elevated levels of ethylene. The auxin signaling machinery that is required for changes in auxin-modulated gene expression has been well characterized (as reviewed by Chapman and Estelle 2009). These auxin signaling components are briefly described in Figure 15.1. ARF (auxin response factors) are transcription factors that regulate expression of auxin-responsive genes (Ulmasov et al. 1997a), including the *SAURs*, *GH3*, and *AUX/IAA* gene families (Chapman and Estelle 2009). The AUX/IAA proteins function as transcriptional repressors (Ulmasov et al. 1997b), which are proteolytically destroyed when auxin levels rise (Tiwari et al. 2001). Forward genetic screens have identified mutants that are less sensitive to the negative effect of auxin on root elongation including *aux1, axr1, axr2, axr3, axr4,* and *slr1,* whose functions in auxin signaling are now being elucidated (as reviewed in Mockaitis and Estelle 2008). A related screen, focused on altered response to inhibitors of auxin transport, identified the *tir1* (*transport inhibitor resistant1*) mutant, which has a primary defect in auxin perception. *TIR1* encodes an F-box protein, which has been shown to function as an auxin receptor and to mediate proteolytic cleavage of transcriptional repressors, including the AUX/IAA proteins (Dharmasiri et al. 2005; Kepinski and Leyser 2005; Tan et al. 2007), to induce transcription of auxin-responsive genes. Figure 15.1 contains a summary of the auxin signaling pathway and the position of these proteins in the regulatory cascade driving expression of auxin-responsive genes.

The ability of ethylene to induce auxin signaling is readily observed after ethylene or ACC treatment or in the *eto1* background by increased expression of auxin-inducible reporters in the root EZ, including DR5-GUS (Ruzicka et al. 2007; Stepanova et al. 2007; Negi et al. 2008; Strader et al. 2010), DR5-GFP (Ruzicka et al. 2007; Lewis et al. 2011a), DR5-vYFP (Lewis et al. 2011b), and IAA2-GUS (Swarup et al. 2007). An example of DR5-vYFP induction after ACC treatment is shown in Figure 15.3. A functional auxin signaling network is required for these ethylene-dependent growth and gene expression changes (Ruzicka et al. 2007; Stepanova et al. 2007; Swarup et al. 2007). The auxin transcriptional machinery that controls this response is not yet clear. Two studies have examined the role of *arf7/arf19* in ethylene-mediated primary root growth inhibition with contradictory findings (Li et al. 2006; Ruzicka et al. 2007).

The evidence is much less clear on whether a fully functional ethylene signaling pathway is required for maximal auxin-dependent root growth inhibition. Two groups report different growth effects of exogenous auxin on the ethylene-insensitive mutants, *ein2* and *etr1*, with one report of insensitivity to NAA (Ruzicka et al. 2007) and another report of normal response to IAA (Stepanova et al. 2007). Finally, *etr1* and *ein2* mutants exhibit a wild-type response to IAA when transcript abundance of genes encoding auxin transport proteins and flavonol biosynthetic genes are examined, even though these mutations block the transcriptional responses to ACC (Lewis et al. 2011a,b). These results suggest that ethylene signaling is not necessary for either the effect of IAA on transcription of target genes or on growth inhibition.

The interactions between auxin and ethylene signaling pathways have also been examined in control of lateral root formation, a process in which auxin and ethylene have antagonistic effects. Although the expression of the DR5rev:GFP reporter increases in the root tip in response to ACC treatment, consistent

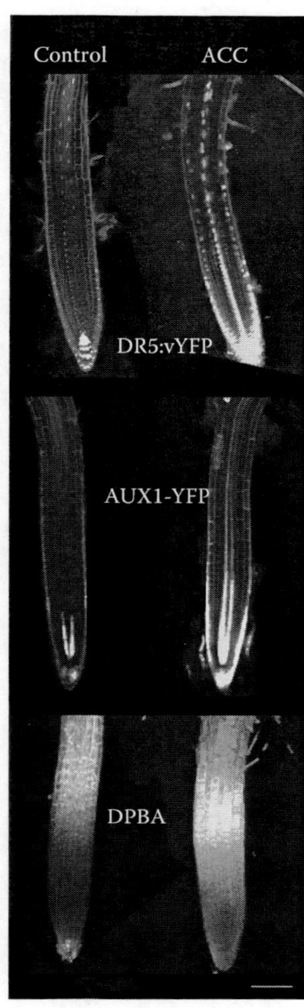

FIGURE 15.3 **(See color insert.)** Flavonoid accumulation, AUX1-YFP fluorescence, and auxin signaling are all elevated in the EZ after ACC treatment. In untreated roots, shown in the left images, DR5-vYFP expression (yellow) is highly expressed in the root columella cells at the root tip and in the central cylinder. After ACC treatment, there is increased expression in both of these tissues and in the epidermal tissues. In untreated roots, AUX1-YFP is expressed in pericycle and columella cells, and in both tissues there is strong enhancement after ACC treatment. The accumulation of flavonols was visualized by staining with diphenylboric acid 2-aminoethyl ester (DPBA), a dye that binds kaempferol and quercetin with different fluorescent properties, which were spectrally separated into kaempferol-DPBA (green) and quercetin-DPBA (yellow). Both compounds accumulate in the EZ after ACC treatment. All ACC treatments were with 1 μM and were for either 8 h (flavonol accumulation) or 12 h for DR5-vYFP and AUX1-YFP. Scale bars = 100 μm.

with the synergistic inhibition of growth in this region, in the mature regions of the root where lateral roots form, DR5rev:GFP fluorescence was reduced after ACC treatment with opposite responses in both regions to the ethylene synthesis inhibitor, aminoethoxyvinylglycine (AVG) (Lewis et al. 2011a). The ACC-dependent auxin transport protein transcript changes are lost in ethylene signaling mutants, *ein2* and *etr1*, while IAA-induced changes are not (Lewis et al. 2011a). Similarly, in *tir1*, auxin-induced gene expression changes are lost, but ACC-induced transcription and repression of root branching are maintained (Lewis et al. 2011a). These results are consistent with independent auxin and ethylene signaling pathways controlling transcription of the same target genes.

A model depicting the potential layers of cross talk that may control transcriptional responses to auxin and ethylene is shown in Figure 15.1. Although auxin and ethylene may both enhance expression of certain genes, these transcriptional effects may operate through independent signaling pathways and lead to primary transcriptional cross talk. Consistent with this possibility, examination of the upstream regulatory region of genes encoding enzymes in flavonol biosynthesis that are induced by both auxin and ethylene identified regions with sequence similarity to both AuxRE and ERE, sites of ARF or EIN3/EIL binding, respectively, in all the genes that were induced by both hormones. In contrast, the upstream regulatory region of one flavonol biosynthesis gene (*TT7*), which was induced by auxin, but not ethylene, had a potential AuxRE, but no identifiable ERE-like sequence (Lewis et al. 2011b).

To dissect the interactions between auxin and ethylene at the level of transcription more globally, a microarray analysis of dark-grown *Arabidopsis* roots was performed using wild type, *ein2*, and the auxin-insensitive *aux1* mutant with and without auxin and ethylene treatment (Stepanova et al. 2007). This experiment uncovered numerous genes that were only regulated by auxin or ethylene but also found that 33% of ethylene-regulated genes and 23% of auxin-regulated genes were in both data sets (Stepanova et al. 2007). Of the transcripts whose expression changed after auxin treatment, 38% required EIN2 function for this induction, while 28% of ethylene altered transcripts required AUX1 (Stepanova et al. 2007). Interpretation of the later result is complex, since AUX1 is an auxin transport protein and mutations in *AUX1* cause auxin insensitivity as a secondary phenotype. When the 5′-UTR of the genes in this large data set was examined, ARF binding sites were found to be enriched in the genes whose transcripts increased after auxin treatment, especially those that were not induced by ethylene. In contrast, EIN3/EIL and AP2/EREBP binding sites were not enriched in any of the populations identified in this study. The authors concluded that other types of transcription factors may be necessary for the ethylene response. They also concluded that ethylene and auxin may mostly act independently to control transcription (Stepanova et al. 2007). This possibility is also indicated in Figure 15.1 as a secondary cross talk, in which targets of ethylene and auxin cross talk are further downstream from primary auxin- or ethylene-responsive genes.

To fully understand the cross talk between auxin and ethylene in regulation of gene expression, additional experimentation is required. The experiment described earlier (Stepanova et al. 2007) and many others focused on ethylene effects on gene expression and used dark-grown plants in which root development is minimal. In contrast, experiments focused on auxin-induced transcript changes use light-grown seedlings and have demonstrated strong interconnected networks of light and auxin signaling (Cluis et al. 2004; Sibout et al. 2006). Furthermore, many of the microarray studies of auxin response were performed with RNA extracted from whole seedlings, rather than root tissue, further complicating interpretation of transcriptional events linked to root growth and development. It is clear that understanding this cross talk will require additional genome-wide transcript studies to understand these transcriptional interactions. Together, these studies show that auxin signaling networks are required for ethylene effects on some growth and gene expression responses, but additional experiments provided evidence that the cross talk between auxin and ethylene also occurred at other levels including regulation of the synthesis and transport of auxin.

V. Ethylene Modulates Auxin Transport

Ethylene regulates elongation growth, gravitropism, and lateral root development in part by modulating auxin transport. Early investigations into the mechanisms of ethylene action focused predominately on plant shoots, with several implicating auxin transport regulation as part of ethylene's mode of action in a range of plant species (Burg and Burg 1967; Beyer and Morgan 1971; Suttle 1988). More recent studies have focused on roots of genetically tractable species, including *Arabidopsis* and tomato. These studies have shown that elevated ethylene levels caused by treatment with the ethylene precursor, ACC, or by elevated synthesis in the *eto1* and *epi* mutants positively regulate both shootward and rootward auxin transport (Negi et al. 2008, 2010). The elevated IAA transport after ACC treatment is lost in the *Arabidopsis* and tomato ethylene signaling mutants *etr1* and *ein2* and *Never ripe*, respectively (Negi et al. 2008, 2010). Plants with mutations in genes encoding auxin transport proteins were used to identify the site of ethylene-regulated auxin transport (Negi et al. 2008; Lewis et al. 2011a).

The polar transport of auxin depends on three major classes of transmembrane proteins, PIN, ABCB, and AUX/LAX proteins, which have either influx or efflux activity, distinct membrane localization, and cell-type expression (as reviewed by Zazimalova et al. 2010). PIN proteins catalyze auxin efflux in heterologous systems (Petrasek et al. 2006), and *pin* mutants have reductions in auxin transport (Okada and Shimura 1992; Chen et al. 1998; Gälweiler et al. 1998). PIN1 is localized to the rootward face of cell membranes in pericycle and provascular cells of the root apex, while PIN2/AGR1/EIR1/WAV6 accumulates on the shootward face of epidermal cells (Gälweiler et al. 1998; Müller et al. 1998). PIN3 and PIN7 accumulate in the root apex and in pericycle and vascular cells along the whole root, but with less striking membrane asymmetry (Blilou et al. 2005; Wisniewska et al. 2006) and mutations in these two genes reduced rootward auxin transport and lateral root formation (Benkova et al. 2003; Blilou et al. 2005; Lewis et al. 2011a). In contrast, *pin2* mutant roots have reduced shootward auxin transport (Rashotte et al. 2000) and are completely agravitropic (Chen et al. 1998), consistent with the essential role of this polar auxin transport stream in mediating asymmetric auxin distribution in response to root reorientation relative to the gravity vector (as reviewed in Muday and Rahman, 2008).

ABCB/PGP/MDR proteins are ATP binding cassette transporters that have also been shown to mediate auxin transport and dependent processes, through mutant analyses, heterologous transport experiments, and fluorescent protein-reporter localization (Noh et al. 2001; Lin and Wang 2005; Terasaka et al. 2005; Wu et al. 2007). ABCB19 is detectable on the membrane in the pericycle, provascular, and cortical cells of primary and lateral roots, where it is required for straight primary root growth and lateral root elongation, respectively (Lewis et al. 2007; Wu et al. 2007). ABCB4 is localized in the epidermal cells of roots, in the elongation and differentiation zones, and regulates root gravitropism (Terasaka et al. 2005; Lewis et al. 2007). Some experimental results have suggested that ABCB proteins function in complex with PIN proteins (Blakeslee et al. 2007; Titapiwatanakun et al. 2008), while others find no evidence for such interaction (Mravec et al. 2008). Understanding the complete structure of auxin efflux complexes awaits additional experimentation (as reviewed in Peer et al. 2011).

AUX/LAX proteins mediate auxin uptake by plant cells (Marchant et al. 1999) and when expressed in heterologous systems (Yang et al. 2006). AUX1 is required for shootward auxin transport in root epidermal cells and is necessary for gravitropism (Bennett et al. 1996; Swarup et al. 2001). *aux1* mutants also have reduced rootward auxin transport under some environmental conditions (Negi et al. 2008; Lewis et al. 2011a), consistent with its expression in the central and epidermal tissues of the root (Marchant et al. 1999; Marchant et al. 2002; Negi et al. 2008). Single mutations in *lax1,2*, and *3* have very weak phenotypes, while *aux/lax* double mutants have more striking transport and root growth and lateral root developmental defects than *aux1* alone (Swarup et al. 2008; Lewis et al. 2011a).

The rapid growth inhibition caused by ethylene or ACC treatment is dependent in part upon regulation of auxin transport within the root tip. Forward genetic screens have identified ethylene-insensitive mutant alleles of *PIN2* and *AUX1*, suggesting normal auxin transport capacity was necessary for the full inhibition of growth by ethylene (Pickett et al. 1990; Luschnig et al. 1998). This response is specific to this set of transport proteins, as *pin1*, *pin4*, and *pin7* are normally sensitive to ethylene's effect on root elongation (Ruzicka et al. 2007). Ethylene increases the accumulation of *AUX1* and *PIN2* transcripts and promoter or protein fusion reporters (Ruzicka et al. 2007; Lewis et al. 2011a), which is linked to enhanced auxin accumulation at the root tip, also shown with AUX1-YFP and DR5-vYFP reporters in Figure 15.3. Treatment of *aux1* with low levels of NAA, which bypasses AUX/LAX proteins (Yamamoto and Yamamoto 1998), partially reverses ethylene insensitivity (Rahman et al. 2001), while IAA, which requires influx proteins for entry into cells, does not. This result suggests that auxin influx proteins are essential for ethylene inhibition of elongation. Reduced cell expansion may be caused by the enhanced accumulation of auxin in the EZ mediated by elevated expression of PIN2 and AUX1 in epidermal cells as shown in Figure 15.3.

Ethylene reduces root elongation and gravitropism at least partially through modulation of auxin transport by inducing the synthesis of flavonols, which are negative regulators of auxin transport (Buer et al. 2006; Lewis et al. 2011b). Flavonols have been shown to regulate auxin transport in vivo, by demonstration of elevated auxin transport in the *tt4* mutant which synthesizes no flavonols (Brown et al. 2001; Buer and Muday 2004; Peer et al. 2004) and in vitro by inhibition of auxin efflux either in tissue segments (Jacobs and Rubery 1988) or through ABCB proteins expressed in a heterologous system (Geisler et al. 2005; Bailly et al. 2008). The full inhibition of gravitropism by ethylene requires the presence of flavonols, with genetic evidence indicating that quercetin is the active molecule (Buer et al. 2006; Lewis et al. 2011b). Treatment of *Arabidopsis* seedlings with ethylene induces flavonol accumulation (Buer et al. 2006; Lewis et al. 2011b), as shown in Figure 15.3, in an *ETR1*- and *EIN2*-dependent fashion (Lewis et al. 2011b). Taken together, these data support the model shown in Figure 15.4 in which the coordinated increases in auxin transport from the root tip to the EZ (via PIN2 and AUX1) and inhibition of auxin transport within the EZ (by flavonols) together elevate the auxin concentration in expanding root cells to growth-inhibiting ranges.

The mechanisms by which ethylene regulates auxin transport and inhibits root branching have recently been identified. Ethylene increases rootward auxin transport and decreases lateral root number in *Arabidopsis* and tomato (Negi et al. 2008, 2010; Lewis et al. 2011a), a surprising result given the generally positive correlation between shoot to root transport of auxin and root branching. *pin3, pin7, aux1,* and *lax3* are less sensitive to the effect of ethylene on lateral root development and rootward auxin transport. In contrast, neither *abcb19* nor *pin2* shows any difference in ethylene responsiveness in either process. Additionally, PIN3, PIN7, and AUX1 transcript abundance and fluorescent protein reporters decreased in the presence of the ethylene synthesis inhibitor, AVG, and increased after ACC treatment, as illustrated in the model in Figure 15.4. Specifically, local depletions of PIN3 and PIN7 protein below (or on the rootward side) of developing lateral root primordia have been suggested to lead to local auxin maxima that drive lateral root formation (Laskowski et al. 2008). The local reduction in PIN3 and PIN7-GFP fluorescence is abolished after treatment with ACC, replaced by global elevation of PIN3 and PIN7 abundance through the whole root that correlates with enhanced long-distance auxin movement (Lewis et al. 2011a). While this enhanced rootward transport depletes auxin in the regions from which lateral roots initiate, it likely contributes to the elevated auxin levels in the root tip, which inhibit root elongation. This effect is also shown schematically in Figure 15.4.

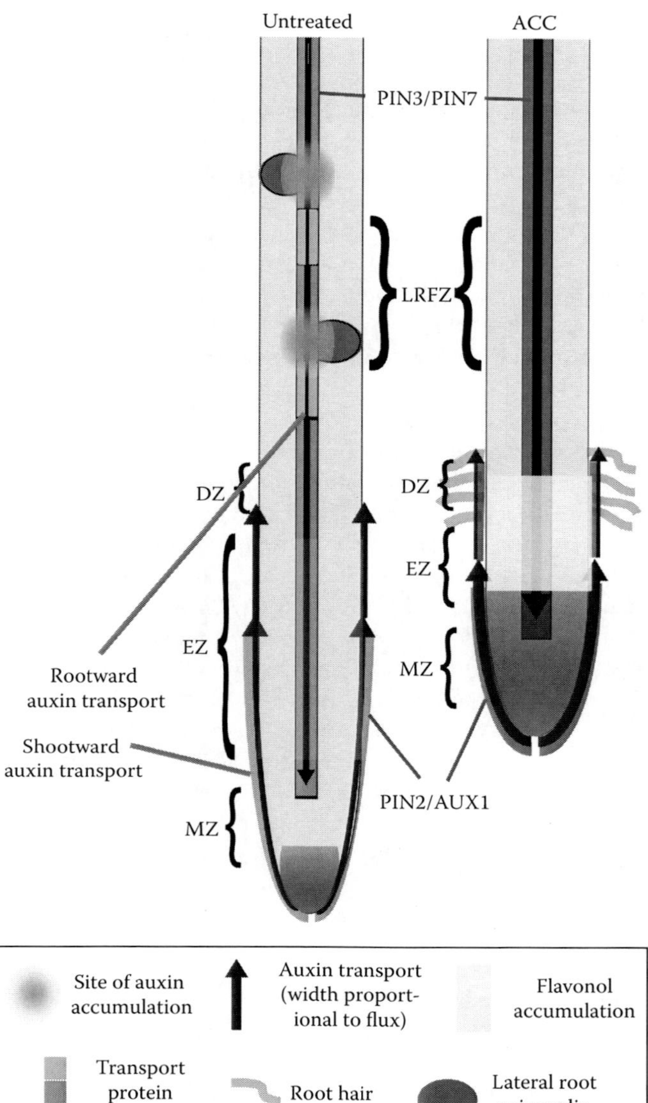

FIGURE 15.4 (See color insert.) Model of ethylene effects on root growth and development. In untreated roots, lateral root formation occurs in the lateral root-forming zone (LRFZ) as a result of localized auxin maxima. This accumulation may occur as a result of depletion of PIN3 and PIN7 proteins that prevents auxin movement away from these points, enhancing auxin accumulation. Ethylene increases rootward auxin transport by increasing transcription and translation of PIN3 and PIN7 in the central cylinder, which may then deplete the lateral root-forming region of auxin, while increasing auxin accumulation in the root apex or meristematic zone (MZ). This effect may be responsible for ethylene's negative regulation of lateral root formation. At the root tip, increases in transcription and translation of AUX1 and PIN2 enhance shootward transport of the auxin into the EZ. Ethylene also increases the accumulation of flavonols in the EZ, as shown by the increased yellow shading, which then traps auxin in the EZ through flavonol inhibition of auxin transport. This excess auxin accumulation in the EZ reduces cell expansion and enhances root hair development in the DZ. (Reprinted from Muday, GK. et al., *Trends Plant Sci.*, 17, 181, 2012.)

VI. Ethylene and Auxin Reciprocally Regulate Each Other's Synthesis

Several lines of evidence suggest that ethylene increases IAA accumulation at the root tip by enhancing the abundance of free IAA and that auxin induces ethylene synthesis in roots. The rate-limiting step in ethylene biosynthesis is catalyzed by ACC synthase (ACS), with multiple ACS genes identified in *Arabidopsis* (Wang et al. 2005). Ethylene synthesis has long been known to be induced by auxin treatment (Yang and Hoffman 1984; Woeste et al. 1999), and recent results have demonstrated that four *ACS* genes are induced by auxin and that at least three of these have conserved ARF binding sites (Stepanova et al. 2007).

The best evidence for increased IAA synthesis driving ethylene-mediated growth inhibition comes from a screen for ethylene-insensitive mutants (*wei* mutants), which identified several genes encoding proteins that function in ethylene-induced auxin synthesis, including WEI2 and WEI7 (Stepanova et al. 2005, 2008). Additionally, free IAA content at the root tip changes proportionally with ethylene accumulation (Ruzicka et al. 2007; Swarup et al. 2007). However, studies following de novo auxin biosynthesis in the presence of elevated ethylene levels have provided conflicting results (Ruzicka et al. 2007; Swarup et al. 2007). These results suggest regulation of WEI2- and WEI7-dependent IAA biosynthesis plays a role in the inhibition of root growth by ACC (Stepanova et al. 2008). In contrast, ethylene-induced increases in IAA synthesis do not seem to play a role in the inhibition of lateral root formation, as reduced levels of free IAA are detected in ACC-treated whole roots (Negi et al. 2008).

VII. Conclusions

It is now evident that a number of features of root growth and development are modulated by ethylene. Ethylene mediates these effects on root growth by acting through the canonical ethylene signaling pathways that were identified to have a role in other growth processes. The action of ethylene in mediating root elongation, gravitropism, and branching all appear to occur through cross talk with the hormone auxin, long known for its control of root growth and development. Ethylene modulates auxin synthesis, transport, and signaling with distinctly different effects when it synergistically combines with auxin to inhibit root elongation and when it antagonizes the positive effect of auxin on lateral root initiation. In particular, ethylene and auxin act through complex and often independent transcriptional cascades to enhance the synthesis of proteins needed for hormonal response, hormone synthesis, and auxin transport. Further understanding of these transcriptional networks and how auxin and ethylene signaling pathways control them is needed to completely understand the mechanisms by which these signaling pathways coordinate the growth and development of roots and other plant tissues.

References

Abeles F, Morgan P, Saltveit Jr. M 1992 *Ethylene in Plant Biology*. San Diego, CA: Academic Press, Inc.

Alonso JM, Hirayama T, Roman G, Nourizadeh S, Ecker JR 1999 EIN2, a bifunctional transducer of ethylene and stress responses in *Arabidopsis*. *Science* 284:2148–2152.

Bailly A, Sovero V, Vincenzetti V et al. 2008 Modulation of P-glycoproteins by auxin transport inhibitors Is mediated by interaction with immunophilins. *J Biol Chem* 283:21817–21826.

Barry CS, Giovannoni JJ 2006 Ripening in the tomato *Green-ripe* mutant is inhibited by ectopic expression of a protein that disrupts ethylene signaling. *Proc Natl Acad Sci USA* 103:7923–7928.

Benkova E, Michniewicz M, Sauer M et al. 2003 Local, efflux-dependent auxin gradients as a common module for plant organ formation. *Cell* 115:591–602.

Bennett MJ, Marchant A, Green HG et al. 1996 *Arabidopsis AUX1* gene: A permease-like regulator of root gravitropism. *Science* 273:948–950.

Beyer EM, Morgan PW 1971 Abscission: The role of ethylene modification of auxin transport. *Plant Physiol* 48:208–212.

Blakeslee JJ, Bandyopadhyay A, Lee OR et al. 2007 Interactions among PIN-FORMED and P-glycoprotein auxin transporters in *Arabidopsis*. *Plant Cell* 19:131–147.

Bleecker A, Estelle M, Magidin M, Somerville C, Kende H 1988 Insensitivity to ethylene conferred by a dominant mutation in *Arabidopsis thaliana*. *Science* 241:1086–1089.

Blilou I, Xu J, Wildwater M, Willemsen V et al. 2005 The PIN auxin efflux facilitator network controls growth and patterning in *Arabidopsis* roots. *Nature* 433:39–44.

Brown DE, Rashotte AM, Murphy AS et al. 2001 Flavonoids act as negative regulators of auxin transport in vivo in *Arabidopsis*. *Plant Physiol* 126:524–535.

Buer CS, Muday GK 2004 The *transparent testa4* mutation prevents flavonoid synthesis and alters auxin transport and the response of *Arabidopsis* roots to gravity and light. *Plant Cell* 16:1191–1205.

Buer CS, Sukumar P, Muday GK 2006 Ethylene modulates flavonoid accumulation and gravitropic responses in roots of *Arabidopsis*. *Plant Physiol* 140:1384–1396.

Buer CS, Wasteneys GO, Masle J 2003 Ethylene modulates root-wave responses in Arabidopsis. *Plant Physiol* 132:1085–1096.

Burg SP, Burg EA 1967 Inhibition of polar auxin transport by ethylene. *Plant Physiol* 42:1224–1228.

Chang SC, Kim YS, Lee JY et al. 2004 Brassinolide interacts with auxin and ethylene in the root gravitropic response of maize (*Zea mays*). *Physiol Plant* 121:666–673.

Chang C, Kwok SF, Bleecker AB, Meyerowitz EM 1993 *Arabidopsis* ethylene-response gene *ETR1*: Similarity of product to two-component regulators. *Science* 262:539–544.

Chapman EJ, Estelle M 2009 Mechanism of auxin-regulated gene expression in plants. *Annu Rev Genet* 43:265–285.

Chen RJ, Hilson P, Sedbrook J, Rosen E, Caspar T, Masson PH 1998 The *Arabidopsis thaliana AGRAVITROPIC 1* gene encodes a component of the polar-auxin-transport efflux carrier. *Proc Natl Acad Sci USA* 95:15112–15117.

Clark DG, Gubrium EK, Barrett JE, Nell TA, Klee HJ 1999 Root formation in ethylene-insensitive plants. *Plant Physiol* 121:53–60.

Cluis CP, Mouchel CF, Hardtke CS 2004 The *Arabidopsis* transcription factor HY5 integrates light and hormone signaling pathways. *Plant J* 38:332–347.

Coleman W, Huxter T, Reid D, Thrope T 1980 Ethylene as an endogenous inhibitor of root regeneration in tomato leaf disc cultures in vitro. *Physiol Plant* 48:519–525.

De-Klerk G, Krieken W, DeJong J 1999 The formation of adventitious roots: New concepts, new possibilities. *In Vitro Cell Dev Biol-Plant* 35:189–199.

Dharmasiri N, Dharmasiri S, Estelle M 2005 The F-box protein TIR1 is an auxin receptor. *Nature* 435:441–445.

Fischer U, Ikeda Y, Grebe M 2007 Planar polarity of root hair positioning in *Arabidopsis*. *Biochem Soc Trans* 35:149–151.

Gälweiler L, Guan C, Muller A et al. 1998 Regulation of polar auxin transport by *AtPIN1* in *Arabidopsis* vascular tissue. *Science* 282:2226–2230.

Geisler M, Blakeslee JJ, Bouchard R et al. 2005 Cellular efflux of auxin catalyzed by the *Arabidopsis* MDR/PGP transporter AtPGP1. *Plant J* 44:179–194.

Giovannoni JJ 2007 Fruit ripening mutants yield insights into ripening control. *Curr Opin Plant Biol* 10:283–289.

Guzman P, Ecker JR 1990 Exploiting the triple response of *Arabidopsis* to identify ethylene-related mutants. *Plant Cell* 2:513–523.

Huang Y, Li H, Hutchison CE, Laskey J, Kieber JJ 2003 Biochemical and functional analysis of CTR1, a protein kinase that negatively regulates ethylene signaling in *Arabidopsis*. *Plant J* 33:221–233.

Ivanchenko MG, Muday GK, Dubrovsky JG 2008 Ethylene-auxin interactions regulate lateral root initiation and emergence in *Arabidopsis thaliana*. *Plant J* 55:335–347.

Jacobs M, Rubery PH 1988 Naturally-occurring auxin transport regulators. *Science* 241:346–349.

Kendrick MD, Chang C 2008 Ethylene signaling: new levels of complexity and regulation. *Curr Opin Plant Biol* 11:479–485.

Kepinski S, Leyser O 2005 The *Arabidopsis* F-box protein TIR1 is an auxin receptor. *Nature* 435:446–451.

Kieber JJ, Rothenberg M, Roman G, Feldmann KA, Ecker JR 1993 CTR1, a negative regulator of the ethylene response pathway in *Arabidopsis*, encodes a member of the Raf family of protein kinases. *Cell* 72:427–441.

Kim HJ, Lynch JP, Brown KM 2008 Ethylene insensitivity impedes a subset of responses to phosphorus deficiency in tomato and petunia. *Plant Cell Environ* 31:1744–1755.

Klee HJ 2004 Ethylene signal transduction. Moving beyond *Arabidopsis*. *Plant Physiol* 135:660–667.

Laskowski M, Grieneisen VA, Hofhuis H et al. 2008 Root system architecture from coupling cell shape to auxin transport. *PLoS Biol* 6:(12):e307.

Le J, Vandenbussche F, Van Der Straeten D, Verbelen JP 2001 In the early response of *Arabidopsis* roots to ethylene, cell elongation is up- and down-regulated and uncoupled from differentiation. *Plant Physiol* 125:519–522.

Lee JS, Chang WK, Evans ML 1990 Effects of ethylene on the kinetics of curvature and auxin redistribution in gravistimulated toots of *Zea mays*. *Plant Physiol* 94:1770–1775.

Lewis DR, Miller ND, Splitt BL, Wu G, Spalding EP 2007 Separating the roles of acropetal and basipetal auxin transport on gravitropism with mutations in two *Arabidopsis* multidrug resistance-like ABC transporter genes. *Plant Cell* 19:1838–1850.

Lewis DR, Negi S, Sukumar P, Muday GK 2011a Ethylene inhibits lateral root development, increases IAA transport and expression of PIN3 and PIN7 auxin efflux carriers. *Development* 138:3485–3495.

Lewis DR, Ramirez MV, Miller ND et al. 2011b Auxin and ethylene induce flavonol accumulation through distinct transcriptional networks. *Plant Physiol* 156:144–164.

Li J, Dai X, Zhao Y 2006 A role for auxin response factor 19 in auxin and ethylene signaling in *Arabidopsis*. *Plant Physiol* 140:899–908.

Lin R, Wang H 2005 Two homologous ATP-binding cassette transporter proteins, AtMDR1 and AtPGP1, regulate *Arabidopsis* photomorphogenesis and root development by mediating polar auxin transport. *Plant Physiol* 138:949–964.

Luschnig C, Gaxiola RA, Grisafi P, Fink GR 1998 EIR1, a root-specific protein involved in auxin transport, is required for gravitropism in *Arabidopsis thaliana*. *Genes Dev* 12:2175–2187.

Marchant A, Bhalerao R, Casimiro I et al. 2002 AUX1 promotes lateral root formation by facilitating indole-3-acetic acid distribution between sink and source tissues in the *Arabidopsis* seedling. *Plant Cell* 14:589–597.

Marchant A, Kargul J, May ST et al. 1999 *AUX1* regulates root gravitropism in *Arabidopsis* by facilitating auxin uptake within root apical tissues. *EMBO J* 18:2066–2073.

Mravec J, Kubes M, Bielach A et al. 2008 Interaction of PIN and PGP transport mechanisms in auxin distribution-dependent development. *Development* 135:3345–3354.

Muday GK, Rahman A. 2008. Auxin transport and the integration of gravitropic growth. In plant Tropisms, S Gilroy, P Masson eds. pp. 47–48. Oxford, UK: Blackwell Publishing.

Muday GK, Rahman A, Binder BM 2012 Auxin and ethylene: collaborators or competitors? *Trends Plant Sci* 17:181–195.

Müller A, Guan C, Galweiler L et al. 1998 *AtPIN2* defines a locus of *Arabidopsis* for root gravitropism control. *EMBO J* 17:6903–6911.

Negi S, Ivanchenko MG, Muday GK 2008 Ethylene regulates lateral root formation and auxin transport in *Arabidopsis thaliana*. *Plant J* 55:175–187.

Negi S, Sukumar P, Liu X, Cohen JD, Muday GK 2010 Genetic dissection of the role of ethylene in regulating auxin-dependent lateral and adventitious root formation in tomato. *Plant J* 61:3–15.

Noh B, Murphy AS, Spalding EP 2001 *Multidrug resistance-like* genes of *Arabidopsis* required for auxin transport and auxin-mediated development. *Plant Cell* 13:2441–2454.

Nordstrom A-C, Eliasson L 1984 Regulation of root formation by auxin-ethylene interaction in pea stem cuttings. *Physiol Plant* 61:298–302.

Okada K, Shimura Y 1990 Reversible root tip rotation in *Arabidopsis* seedlings induced by obstacle-touching stimulus. *Science* 250:274–276.

Okada K, Shimura Y 1992 Mutational analysis of root gravitropism and phototropism of *Arabidopsis thaliana* seedlings. *Aust J Plant Physiol* 19:439–448.

Ortega-Martinez O, Pernas M, Carol RJ, Dolan L 2007 Ethylene modulates stem cell division in the *Arabidopsis thaliana* root. *Science* 317:507–510.

Peer WA, Bandyopadhyay A, Blakeslee JJ et al. 2004 Variation in expression and protein localization of the PIN family of auxin efflux facilitator proteins in flavonoid mutants with altered auxin transport in *Arabidopsis thaliana*. *Plant Cell* 16:1898–1911.

Peer WA, Blakeslee JJ, Yang H, Murphy AS 2011 Seven things we think we know about auxin transport. *Mol Plant* 4:487–504.

Petrasek J, Mravec J, Bouchard R, Blakeslee JJ, Abas M, Seifertova D, Wisniewska J, Tadele Z, Kubes M, Covanova M, Dhonukshe P et al. 2006 PIN proteins perform a rate-limiting function in cellular auxin efflux. *Science* 312:914–918.

Pickett FB, Wilson AK, Estelle M 1990 The *aux1* mutation of *Arabidopsis* confers both auxin and ethylene resistance. *Plant Physiol* 94:1462–1466.

Pitts RJ, Cernac A, Estelle M 1998 Auxin and ethylene promote root hair elongation in *Arabidopsis*. *Plant J* 16:553–560.

Rahman A, Amakawa T, Goto N, Tsurumi S 2001 Auxin is a positive regulator for ethylene-mediated response in the growth of *Arabidopsis* roots. *Plant Cell Physiol* 42:301–307.

Rahman A, Bannigan A, Sulaman W, Pechter P, Blancaflor EB, Baskin TI 2007 Auxin, actin and growth of the *Arabidopsis* thaliana primary root. *Plant J* 50:514–528.

Rahman A, Hosokawa S, Oono Y, Amakawa T, Goto N, Tsurumi S 2002 Auxin and ethylene response interactions during *Arabidopsis* root hair development dissected by auxin influx modulators. *Plant Physiol* 130:1908–1917.

Rashotte AM, Brady SR, Reed RC, Ante SJ, Muday GK 2000 Basipetal auxin transport is required for gravitropism in roots of *Arabidopsis*. *Plant Physiol* 122:481–490.

Ruzicka K, Ljung K, Vanneste S et al. 2007 Ethylene regulates root growth through effects on auxin biosynthesis and transport-dependent auxin distribution. *Plant Cell* 19:2197–2212.

Schiefelbein JW 2000 Constructing a plant cell. The genetic control of root hair development. *Plant Physiol* 124:1525–1531.

Sibout R, Sukumar P, Hettiarachchi C, Holm M, Muday GK, Hardtke CS 2006 Opposite root growth phenotypes of *hy5* versus *hy5 hyh* mutants correlate with increased constitutive auxin signaling. *Plos Genet* 2:e202.

Stepanova AN, Alonso JM 2009 Ethylene signalling and response: where different regulatory modules meet. *Curr Opin Plant Biol* 12:548–555.

Stepanova AN, Hoyt JM, Hamilton AA, Alonso JM 2005 A Link between ethylene and auxin uncovered by the characterization of two root-specific ethylene-insensitive mutants in *Arabidopsis*. *Plant Cell* 17:2230–2242.

Stepanova AN, Robertson-Hoyt J, Yun J et al. 2008 TAA1-mediated auxin biosynthesis is essential for hormone crosstalk and plant development. *Cell* 133:177–191.

Stepanova AN, Yun J, Likhacheva AV, Alonso JM 2007 Multilevel interactions between ethylene and auxin in *Arabidopsis* roots. *Plant Cell* 19:2169–2185.

Strader LC, Chen GL, Bartel B 2010 Ethylene directs auxin to control root cell expansion. *Plant J* 64:874–884.

Sukumar P 2010 Role of auxin and ethylene in adventitious root formation in *Arabidopsis* and tomato. PhD thesis, Wake Forest University, Winston Salem, NC.

Suttle JC 1988 Effect of ethylene treatment on polar IAA transport, net IAA uptake and specific binding of N-1-Naphthylphthalamic acid in tissues and microsomes isolated from etiolated pea Epicotyls. *Plant Physiol* 88:795–799.

Swarup K, Benkova E, Swarup R et al. 2008 The auxin influx carrier LAX3 promotes lateral root emergence. *Nat Cell Biol* 10:946–954.

Swarup R, Friml J, Marchant A et al. 2001 Localization of the auxin permease AUX1 suggests two functionally distinct hormone transport pathways operate in the *Arabidopsis* root apex. *Genes Dev* 15:2648–2653.

Swarup R, Parry G, Graham N, Allen T, Bennett M 2002 Auxin cross-talk: integration of signalling pathways to control plant development. *Plant Mol Biol* 49:411–426.

Swarup R, Perry P, Hagenbeek D et al. 2007 Ethylene upregulates auxin biosynthesis in *Arabidopsis* seedlings to enhance inhibition of root cell elongation. *Plant Cell* 19:2186–2196.

Tan X, Calderon-Villalobos LI, Sharon M et al. 2007 Mechanism of auxin perception by the TIR1 ubiquitin ligase. *Nature* 446:640–645.

Terasaka K, Blakeslee JJ, Titapiwatanakun B et al. 2005 PGP4, an ATP Binding Cassette P-Glycoprotein, catalyzes auxin transport in *Arabidopsis thaliana* roots. *Plant Cell* 17:2922–2939.

Titapiwatanakun B, Blakeslee JJ, Bandyopadhyay A et al. 2008 ABCB19/PGP19 stabilises PIN1 in membrane microdomains in *Arabidopsis*. *Plant J* 57:27–44.

Tiwari SB, Wang XJ, Hagen G, Guilfoyle TJ 2001 AUX/IAA proteins are active repressors, and their stability and activity are modulated by auxin. *Plant Cell* 13:2809–2822.

Ulmasov T, Hagen G, Guilfoyle TJ 1997a ARF1, a transcription factor that binds to auxin response elements. *Science* 276:1865–1868.

Ulmasov T, Murfett J, Hagen G, Guilfoyle TJ 1997b Aux/IAA proteins repress expression of reporter genes containing natural and highly active synthetic auxin response elements. *Plant Cell* 9:1963–1971.

Wang NN, Shih MC, Li N 2005 The GUS reporter-aided analysis of the promoter activities of *Arabidopsis* ACC synthase genes *AtACS4, AtACS5*, and *AtACS7* induced by hormones and stresses. *J Exp Bot* 56:909–920.

Wang KL, Yoshida H, Lurin C, Ecker JR 2004 Regulation of ethylene gas biosynthesis by the *Arabidopsis* ETO1 protein. *Nature* 428:945–950.

Wilkinson JQ, Lanahan MB, Yen HC, Giovannoni JJ, Klee HJ 1995 An ethylene-inducible component of signal transduction encoded by *Never-ripe*. *Science* 270:1807–1809.

Wisniewska J, Xu J, Seifertova D et al. 2006 Polar PIN localization directs auxin flow in plants. *Science* 312:883.

Woeste K, Vogel J, Kieber J 1999 Factors regulating ethylene biosynthesis in etiolated *Arabidopsis thaliana* seedlings. *Physiol Plant* 105:478–484.

Wu G, Lewis DR, Spalding EP 2007 Mutations in *Arabidopsis* multidrug resistance-like ABC transporters separate the roles of acropetal and basipetal auxin transport in lateral root development. *Plant Cell* 19:1826–1837.

Yamamoto M, Yamamoto KT 1998 Differential effects of 1-naphthaleneacetic acid, indole-3-acetic acid and 2,4-dichlorophenoxyacetic acid on the gravitropic response of roots in an auxin-resistant mutant of arabidopsis, aux1. *Plant Cell Physiol* 39:660–664.

Yang Y, Hammes UZ, Taylor CG, Schachtman DP, Nielsen E 2006 High-affinity auxin transport by the AUX1 influx carrier protein. *Curr Biol* 16:1123–1127.

Yang S, Hoffman N 1984 Ethylene biosynthesis and its regulation in higher plants. *Annu Rev Plant Physiol* 35:155–189.

Zazimalova E, Murphy AS, Yang HB, Hoyerova K, Hosek P 2010 Auxin Transporters - Why So Many? *Cold Spring Harb Perspect Biol* 2:a00152.

16

Abscisic Acid in Root Growth and Development

Ive De Smet
University of Nottingham

Hanma Zhang
Chongqing Normal University

I. Introduction

Abscisic acid (ABA) is a carotenoid-derived molecule that was first isolated from cotton plants and named abscisin (Liu and Carnsdagger 1961; Ohkuma et al. 1963) and from birch leaves and named dormin (Eagles and Wareing 1963). The name of this endogenous growth inhibitor in plants later became ABA (Addicott et al. 1968), derived from its initially assumed role in abscission of fruit and leaves. But this phytohormone also plays a major role in the response to stress and in fine-tuning growth and development (Lumba et al. 2010). ABA controls stomatal aperture, and based on its involvement in germination, bud growth, and lateral root outgrowth, it is also regarded as a dormancy hormone (Rohde et al. 2002; De Smet et al. 2006; Holdsworth et al. 2008). ABA does not appear to play a major role in organ development but rather in regulating the transitions between developmental stages, interacting with various other hormones to regulate plant growth and development (De Smet et al. 2003; Achard et al. 2006; Zhang et al. 2009, 2010). Here, we will describe the physiology and functions of ABA in roots, with a particular emphasis on its role in root growth, root branching, and root–shoot communication.

II. ABA Biosynthesis and Signaling

ABA is synthesized in both roots and leaves, and key enzymes have been identified in several species (see reviews by Taylor et al. 2000; Schwartz et al. 2003; Xiong and Zhu 2003; Nambara and Marion-Poll 2005). Its biosynthesis is triggered in response to heat, cold, drought, and other abiotic stress situations, and early synthesis steps take place in plastids and later in the cytosol. In higher plants, ABA is formed via the indirect pathway from a C40 carotenoid precursor (see reviews by Taylor et al. 2000; Seo and Koshiba 2002; Schwartz et al. 2003). The precursor is first converted into xanthoxin in the plastids. The first step of this conversion is the epoxidation of zeaxanthin and antheraxanthin to violaxanthin and is catalyzed by a zeaxanthin epoxidase (ZEP) (Marin et al. 1996). This is followed by the cleavage of violaxanthin to yield a C15 intermediate, xanthoxin, and this reaction is mediated by a 9-CIS-EPOXYCAROTENOID DIOXYGENASE (NCED) (Schwartz et al. 1997) and is considered to be a rate-limiting step in ABA biosynthesis (Thompson et al. 2000; Iuchi et al. 2001; Qin and Zeevaart 2002; Aswath et al. 2005). Xanthoxin is then exported to the cytosol and converted to ABA aldehyde through the activity of a SHORT-CHAIN ALCOHOL DEHYDROGENASE/REDUCTASE (SDR) and finally to ABA through the action of an ABA ALDEHYDE OXIDASE (AAO) (Rook et al. 2001; Cheng et al. 2002; Gonzalez-Guzman et al. 2004). AAO requires a sulfurylated molybdenum cofactor (MoCo) for its activity and the sulfurylation of the cofactor is mediated by a MoCo sulfurase (Xiong et al. 2001).

Over the years, several ABA receptors have been identified, but for most, this was met with controversy (McCourt and Creelman 2008; Klingler et al. 2010; Joshi-Saha et al. 2011). However, the most recent additions to this list—a family of soluble proteins called PYRABACTIN/PYRABACTIN-LIKE/REGULATOR COMPONENT OF ABA RECEPTOR (PYR/PYL/RCAR)—appear to convincingly function as ABA receptors (Ma et al. 2009; Melcher et al. 2009; Miyazono et al. 2009; Nishimura et al. 2009; Park et al. 2009; Santiago et al. 2009). In brief, the core pathway is initiated upon binding of ABA within the ligand-binding pockets of PYR/PYL/RCAR, which triggers a conformational change in the gate and latch loops of the receptors. The ABA-bound receptors subsequently interact with clade A PP2C protein phosphatases and inhibit their activity by docking within the PP2C active site. This allows accumulation of active, phosphorylated members of the SNF1-RELATED PROTEIN KINASE (SnRK) family, which phosphorylate and thereby modulate the activity of downstream (transcription) factors (Umezawa et al. 2009; Hubbard et al. 2010; Weiner et al. 2010; Fujii et al. 2011; Hauser et al. 2011; Figure 16.1). In contrast to some other hormones, which rely on protein turnover for their action, ABA signal transduction relies on phosphorylation as the primary mechanism (Lumba et al. 2010).

Subsequently, ABA signaling relies on transcriptional changes, and this is illustrated by the dramatic changes in gene expression that occur upon ABA treatment (Hoth et al. 2002; Seki et al. 2002). ABA-dependent gene expression is regulated by ABA-RESPONSIVE ELEMENT BINDING FACTORS/PROTEINS (ABFs/AREBs) (Figure 16.1), such as the bZIP-type transcription factor ABA INSENSITIVE 5 (ABI5) (Choi et al. 2005; Finkelstein et al. 2005). For ABA-dependent gene expression, the ABA-responsive element (or ABRE with the ACGTGT consensus sequence) is a major cis-regulatory element (Yamaguchi-Shinozaki and Shinozaki 2005), but ABA-mediated gene expression also involves other binding elements (Niu et al. 2002; Narusaka et al. 2003).

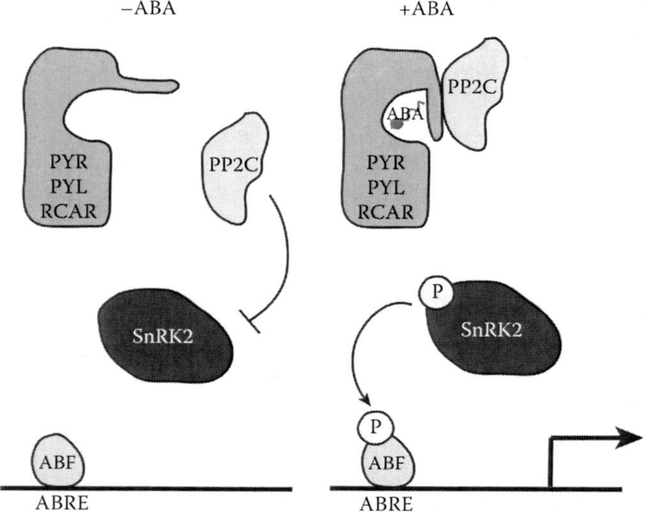

FIGURE 16.1 Schematic view of the core ABA signaling complex (see text for details).

III. ABA and Root Physiology

A. ABA Content and Distribution in Roots

ABA was detected in plant roots soon after its discovery in the 1960s (Tietz 1971; Milborrow and Robinson 1973). The ability of roots to synthesize ABA was first demonstrated by Hartung and Abou-Mandour (1980) in isolated root cultures of *Phaseolus coccineus* L. It is now well established that plant roots are an important site of ABA biosynthesis and accumulation. ABA in roots can be synthesized locally or is imported from shoots.

ABA content in roots varies considerably in different plant species, ranging from ~50 to 1890 pmol (g DW)$^{-1}$ (Hose et al. 2002; Table 16.1) and usually increases under stress conditions, such as water shortage (Zhang and Davies 1989), high salinity, and osmotic stress (Wolf et al. 1990). ABA content in roots is also affected by other environmental factors, such as nutrients (Peuke et al. 1994; Jeschke et al. 1997), rhizobacteria (Grossmann and Hansen 2001; Arkhipova et al. 2005), and hemiparasite plant infection (Jiang et al. 2004).

The spatial distribution of ABA within a root has been studied in different plant species. Rivier et al. (1977) reported a longitudinal ABA concentration gradient in maize roots, with the highest concentration around the root apex. Behl et al. (1981) observed a similar ABA concentration gradient in barley roots.

Table 16.1 ABA Content of Roots Under Different Stress Factors in Different Plant Species

	ABA Content		
Plant Species	Nonstressed Roots	Stressed Roots	Stresses
Zea mays	7.4 pmol (g FW)$^{-1}$	39.2 pmol (g FW)$^{-1}$	PEG
Zea mays	9.7 pmol (g FW)$^{-1}$	34.9 pmol (g FW)$^{-1}$	Salt
Anastatica hierochuntica	21 pmol (g FW)$^{-1}$	301 pmol (g FW)$^{-1}$	Salt
Chamaegigas intrepidus	56 pmol (g DW)$^{-1}$	1266 pmol (g DW)$^{-1}$	Drought
Xerophyta dasylirioides	90 pmol (g DW)$^{-1}$	297 pmol (g DW)$^{-1}$	Drought
Lycopersicon esculentum	110 pmol (g DW)$^{-1}$	210 pmol (g DW)$^{-1}$	Salt
Chamaegigas lanceolatum	200 pmol (g DW)$^{-1}$	6261 pmol (g DW)$^{-1}$	Drought
Myrothamnus moschata	483 pmol (g DW)$^{-1}$	2808 pmol (g DW)$^{-1}$	Drought
Simmondsia chinensis	780 pmol (g DW)$^{-1}$	120 pmol (g DW)$^{-1}$	Salt
Helianthus annuus	950 pmol (g DW)$^{-1}$	5680–7570 pmol (g DW)$^{-1}$	PEG

Source: Hose, E. et al., Abscisic acid in roots—Biochemistry and physiology, in *Plant Roots: The Hidden Half*, eds. Waisel, Y., Eshel, A., Kafkafi, U., pp. 435–448, Marcel Dekker, Inc., New York, 2002.

They incubated barley roots in solutions containing radioactively labeled ABA and found a radioactivity gradient along the longitudinal axis of the roots with the highest radioactivity at the root tips. Christmann et al. (2005) examined the tissue context of ABA distribution in *Arabidopsis* plants using noninvasive and cell-autonomous reporters and found that the highest level of physiologically active ABA in roots under nonstressed conditions was detected in the columella root cap and QC cells.

B. Factors That Regulate ABA Accumulation in Roots

In addition to the species-related and environmental factors, ABA accumulation in roots is also controlled by a number of internal factors, such as *de novo* synthesis, catabolism, transport, and exudation to the soil.

1. ABA Biosynthesis in Roots

ABA is likely to be synthesized in root tips, and under drought conditions, this source of ABA has been proposed to signal the water status to the shoot to control leaf stomatal conductance (Seo and Koshiba 2002; Schachtman and Goodger 2008; see Section VI). There is indeed strong biochemical and molecular evidence that the ABA biosynthesis pathway described in Section II operates in roots. Feldman et al. (1985) detected high concentrations of two intermediates of this pathway, xanthoxin and violaxanthin, in maize roots and found that the spatial distribution of violaxanthin matched with that of ABA in roots when grown in the dark. Parry and Horgan (1992) found all potential ABA precursors linked to this pathway in roots of several plant species including *Arabidopsis thaliana* (L.) Heynh., *Nicotiana plumbaginifolia* Viv., *Phaseolus vulgaris* L., *Pisum sativum* L., and *Lycopersicon esculentum* Mill. Expression studies in *Arabidopsis* have also shown that genes coding for the enzymes required for this pathway are all expressed in roots (Xiong et al. 2001; Xiong et al. 2002; Xiong and Zhu 2003). ZEP, SDR, and MoCo sulfurase are each encoded by a single gene in *Arabidopsis*, that is, *AtZEP*, *AtABA2*, and *AtABA3*, respectively. *AtZEP* and *AtABA3* are ubiquitously expressed throughout the whole plant including roots (Xiong et al. 2001, 2002). *AtABA2* has a restricted spatial expression pattern in roots, with strong expression at the branching points between lateral and primary roots and weak expression at root tips in nonstressed plants (Cheng et al. 2002). There are nine *NCED*-related genes in the *Arabidopsis* genome and at least five of them (*AtNCED2, 3, 5, 6,* and 9) are involved in ABA biosynthesis based on sequence and expression analyses (Iuchi et al. 2001; Schwartz et al. 2003). Two such genes, *AtNCED2* and *AtNCED3*, are strongly expressed in roots (Tan et al. 2003). The expression pattern of *AtNCED3* is of particular interest as it matches more or less with the pattern of ABA distribution in roots, with strong expression in root tips and vasculature in both primary and lateral roots (Tan et al. 2003). *AtNCED2* is expressed in lateral roots, the pericycle and the cortex cells of primary roots surrounding new lateral roots (Tan et al. 2003). Of the four *Arabidopsis* AAO genes, three (*AtAAO1, AtAAO2,* and *AtAAO3*) are expressed in roots (Seo et al. 2000; Koiwai et al. 2004).

2. ABA Catabolism in Roots

ABA can be catabolized through two different pathways: hydroxylation and conjugation. Hydroxylation can occur at three different positions on the ABA ring, but predominately at the C-8′ position (Zhou et al. 2004; Nambara and Marion-Poll 2005). This reaction is mediated by a subgroup of cytochrome P450 monooxygenases, CYP707As. The *Arabidopsis* genome contains four CYP707A-encoding genes (Kushiro et al. 2004; Saito et al. 2004). Genetic and functional analyses with one such gene, *CYP707A2*, have demonstrated a negative correlation between the activities of CYP707As and the level of ABA in plants (Kushiro et al. 2004). CYP707A-related sequences have been found in many different plants (Kushiro et al. 2004), and at least two of the *Arabidopsis CYP707A* genes, *CYP707A1* and *CYP707A3*, are known to be expressed in roots (Saito et al. 2004).

ABA can also form conjugates with sugars (Boyer and Zeevaart 1982; Hansen and Dörffling 1999). ABA-glucosyl ester (ABA-GE) is the most widespread conjugate and has been detected in roots and xylem sap in many plant species (Hansen and Dörffling 1999; Hartung et al. 2002; Sauter et al. 2002; Wilkinson and Davies 2002). ABA-GE is physiologically inactive (Bray and Zeevaart 1985). ABA glucosylation is mediated by a subset of glucosyltransferases, ABA glucosyltransferases (AOG) (Lehmann and Schütte 1980; Schwarzkopf and Miersch 1992), and *AOG* genes have been identified in adzuki bean (*Vigna angularis* (Willd.) Ohwi & Ohashi) (Xu et al. 2002) and *Arabidopsis* (Lim et al. 2005; Priest et al. 2006).

Despite progresses in the identification of genes involved in the hydroxylation and conjugation pathways, very little is known about the exact contribution of the pathways, or any of the identified genes, in controlling ABA content in roots.

3. ABA Transport between Roots and Shoots

ABA transport between roots and shoots also has an important impact on ABA accumulation in either organ. ABA, either as itself or as ABA-GE, can be exported from roots to shoots, imported from shoots to roots, and recirculated between roots and shoots (Jeschke et al. 1997; Hartung et al. 2002). Each of the transport routes consists of both long-distance and short-range intercellular movements. The long-distance movements take place in vascular vessels, that is, the xylem for the export route and phloem for the import route, and their rates are primarily determined by the mass flow in the respective vascular vessels. ABA intercellular movement is mediated by either diffusion or carrier-mediated transport. Because ABA is a weak acid, diffusion is considered to be the primary driving force for ABA intercellular movement, and membrane permeability is a key determining factor for this mode of transport. Physiological studies show significant variations in ABA permeability among different cell types or organelles (Hose et al. 2002). In maize and runner beans, root cells have lower ABA permeability than leaf mesophyll cells, and within an individual cell, the tonoplast membrane usually has lower ABA permeability than the plasma membrane (Daeter and Hartung 1990; Jovanovic et al. 1992). The low ABA permeability of the root plasma membrane makes ABA distribution within roots slow in comparison with the situation in leaves and therefore a likely rate-limiting step in ABA transport between roots and shoots, and, in addition, low ABA permeability of the tonoplast membrane can also reduce ABA accumulation in vacuoles. ABA can also be lost via exudation to the soil. This loss could be of significant amounts in alkaline soils (Slovik et al. 1995; Hose et al. 2002), but the low permeability in roots likely reduces ABA loss to the surrounding soil (Hose et al., 2002).

The existence of ABA carriers has been implicated in early physiological studies (Windsor et al. 1992; Wilkinson and Davies 1997), but molecular identification of such carriers was only achieved within the last year when two ABA carriers, PLEIOTROPIC DRUG RESISTANCE12 (PDR12)/ABCG40, and ABCG25, have been identified in *Arabidopsis* (Kang et al. 2010; Kuromori et al. 2010). The carriers belong to the same family of proteins known as the ATP-binding cassette (ABC) transporters, and there is evidence that they are located on the plasma membrane. PDR12/ABCG40 is believed to be an ABA importer. When PDR12/ABCG40 is expressed in yeast and bacteria suspension cells, it enhances ABA uptake into the cells. *Arabidopsis* plants with no PDR12/ABCG40 activity display slower responses to exogenously applied ABA (Kang et al. 2010). ABCG25 is proposed as an ABA exporter (Kuromori et al. 2010). Membrane vesicles derived from ABCG25-expressing insect cells showed ATP-dependent ABA transport activities. ABCG25-overexpressing plants showed higher leaf temperatures, implying a function in stomatal regulation. Early physiological evidence suggests that carrier-mediated ABA transport only occurred in specific tissues, such as root tips, and under acid conditions (Astle and Rubery 1983). However, both *PDR12/ABCG40* and *ABCG25* are expressed in a broad range of tissues (Kang et al. 2010; Kuromori et al. 2010), indicating that their encoded proteins act in many different tissues. Nevertheless, there is no direct functional evidence for the involvement of the identified ABA carriers in ABA transport in the roots or between the roots and the shoots.

IV. ABA and Root Growth

A. ABA and Inhibition or Promotion of Root Growth

ABA has been generally regarded as a growth inhibitor of roots. This conclusion is based on experimental evidence obtained with exogenous ABA applications. In many plant species, exogenously applied ABA above a certain threshold causes inhibition of root growth (Newton 1974; Pilet 1975; Spollen et al. 2000). The precise cellular and molecular mechanisms of this ABA action have not yet been fully established. Early studies suggest that the inhibition is due to reduced cell elongation (Pilet 1975). More recent data show that ABA inhibits cell division (Wang et al. 1998; Zhang et al. 2010). Wang et al. (1998) proposed a possible mechanism for the latter mode of action and suggested that ABA might act by upregulating the expression of *INTERACTOR OF CDC2 KINASE1 (ICK1)/KIP-RELATED PROTEIN1 (KRP1)*, which encodes a protein that shares sequence similarity with mammalian cyclin-dependent kinase (CDK) inhibitors and has been shown to physically interact with core cell cycle regulators such as CELL DIVISION CONTROL 2A (CDC2A)/CYCLIN DEPENDENT KINASE A1 (CDKA1), and CYCLIN D3 (CYCD3) in in vitro binding assays (Wang et al. 1997). This proposal is based on the observation that ABA induces the expression of *ICK1/KRP1* in leaves (Wang et al. 1998). However, it is not clear whether *ICK1/KRP1* is also induced by ABA in roots and if the level of ICK1/KRP1 activity correlates with root growth in ABA-treated plants. There is also evidence that ABA inhibits root growth by modulating ethylene synthesis or signaling (Cheng et al. 2002). For example, blocking ethylene signaling was found to partially restore root growth in *aba2* mutants (Cheng et al. 2002). A recent study by Wang et al. (2011) reveals a cross talk between ABA and auxin in regulating root growth. The authors identified two ABA overly sensitive mutant alleles in the *AUXIN RESPONSE FACTOR2* (*ARF2*) gene and established that ABA treatment induced the expression of *ARF2*, repressed the ARF2 target *HOMEOBOX PROTEIN33* (*HB33*) expression, and altered auxin distribution and signaling in the primary root tips. The *arf2* mutants and plants overexpressing *HB33* display an increased sensitivity to ABA in both root growth and seed germination, whereas plants with reduced *HB33* expression show the opposite phenotype.

Despite the overwhelming physiological evidence in support of an inhibitory effect of ABA on root growth, none of the currently known ABA-deficient or ABA-insensitive mutants actually show enhanced root growth. This suggests that the inhibitory effect observed in experiments using exogenous ABA application may not necessarily reflect the function of endogenously produced ABA under nonstressed conditions. In fact, some ABA-deficient or ABA-insensitive mutants actually show retarded growth (Cheng et al. 2002; Lin et al. 2007; Fujii and Zhu 2009; Nakashima et al. 2009; Fujii et al. 2011). For example, mutants that lack ABA2 activity display retarded root and shoot growth, and this phenotype can be rescued by exogenous application of ABA (Cheng et al. 2002; Lin et al. 2007). Loss-of-function mutants of several SnRK2 kinases, which are a key component of the recently identified core ABA signaling cascade, also display retarded root/shoot growth even under high humidity (Fujii and Zhu 2009; Nakashima et al. 2009). The SUMO E3 ligase SIZ1 negatively regulates ABA signaling by sumoylation of ABI5, causing primary root growth arrest (Miura et al. 2009). In addition, the RING E3 Ligase KEEP ON GOING is involved in ABA signaling through maintaining low levels of ABI5 in the absence of stress, also affecting main root growth (Stone et al. 2006). Furthermore, an ABA-insensitive mutant, *abi8*, displays severe defect in root growth and root meristem maintenance (Brocard-Gifford et al. 2004).

The ABI8/ELONGATION DEFECTIVE 1 (ELD1)/KOBITO 1 (KOB1) protein is involved in root growth, as the mutant displays terminal differentiation of the RAM and cessation of cell division in the root tip (Cheng et al. 2000; Pagant et al. 2002; Brocard-Gifford et al. 2004). However, the protein is placed in a network of signaling factors, and its amino acid sequence does not reveal any domains of known biochemical function (Brocard-Gifford et al. 2004).

The retarded growth phenotypes of the deficient and insensitive mutants suggest that basal levels of ABA are required for maintaining normal root and plant growth. This interpretation is consistent with numerous, but often overlooked, observations that low levels of ABA, either endogenously produced or exogenously applied, promote root and plant growth (Koornneef 1986; Zeevaart and Creelman 1988; Rock and Zeevaart 1991; Finkelstein and Zeevaart 1994; Vartanian et al. 1994; Nambara et al. 1998; Zhou et al. 1998; Ephritikhine et al. 1999; Sharp et al. 2000; LeNoble et al. 2004; Barrero et al. 2005). ABA therefore appears to be playing two opposite roles in plant and root growth depending on its concentrations, inhibiting growth at high concentrations and promoting growth at low concentrations. There is some evidence that ethylene may play a role in ABA-induced promoting of root growth under certain conditions (Sharp 2002). For example, Spollen et al. (2000) found that ABA-deficient plants displayed slower root growth in comparison with control plants under water stress and this phenotype was linked to excessive production of ethylene. Both the root growth phenotype and the enhanced ethylene production disappeared when ABA content in roots was restored with exogenous ABA application, indicating that ABA limits ethylene production in water-stressed plants and thereby promotes root growth (Spollen et al. 2000).

B. ABA and the Primary Root Meristem

Arabidopsis root meristems contain a well-defined stem cell niche, which includes the quiescent center and the surrounding initials (van den Berg et al. 1997; Scheres 2007), and maintenance of this niche is essential for the function and development of the root. While the role of auxin and various other regulators in maintenance of QC quiescence and suppression of stem cell differentiation has been studied extensively (Ortega-Martinez et al. 2007; Sarkar et al. 2007; Scheres 2007), the involvement of ABA has been largely overlooked. Using the widely used ABA biosynthesis inhibitor fluridone, Zhang et al. (2010) concluded that ABA promotes QC quiescence and suppresses cell differentiation in the stem cell niche (Han et al. 2010). The ABA-mediated regulation of QC division appears to be independent of ethylene, but ABI1, ABI2, ABI3, ABI5, and ENHANCED RESPONSE TO ABA 1 (ERA1) appear to play some role (Zhang et al. 2010). Importantly, the regulation of stem cell differentiation by ABA in the root meristem requires the function of WUSCHEL-RELATED HOMEOBOX 5 (WOX5) (Zhang et al. 2010), a major regulator of stem cell maintenance (Sarkar et al. 2007). In addition, MP seems to be involved in the regulation of stem cell differentiation by ABA in the root meristem (Zhang et al. 2010). Together with earlier observations on ABA-dependent effects on root meristems in *Arabidopsis* and *Medicago truncatula* Gaertn. (Cheng et al. 2002; Barrero et al. 2005; Liang et al. 2007), the recent results by Zhang et al. (2010) provide a cellular mechanism for a positive role for ABA in promoting root meristem maintenance and root growth.

V. ABA and Lateral Roots

Before lateral roots become visible upon emergence from the main root, they have progressed through several developmental stages such as initiation, primordium development, and meristem activation (Laskowski et al. 1995; Malamy and Benfey 1997; Figure 16.2). At present, there is no convincing molecular evidence that ABA plays a role in the early stages of lateral root initiation and development. Nevertheless, some observations point in the direction that ABA could negatively regulate cell cycle progression and cell division events associated with lateral root initiation (Figure 16.2). For example, the progression through the G1–S checkpoint is partially regulated by the level of the ICKs/KRPs, a family of cell cycle inhibitors (Wang et al. 2000; De Veylder et al. 2001; Verkest et al. 2005), and this is linked to lateral root initiation as plants overproducing *ICK2/KRP2* display a reduction in lateral root number (Himanen et al. 2002). While it remains to be shown whether auxin and ABA act antagonistically at the control of the G1 to S transition in pericycle cells during lateral root initiation by regulating KRP levels, it is intriguing that generally auxin downregulates *KRP* levels (Himanen et al. 2002), whereas ABA upregulates them (Wang et al. 1998). In addition, some *NCED* genes, which are involved in ABA biosynthesis (Taylor et al. 2000), were reported to be expressed in pericycle cells surrounding lateral root initiation sites (Tan et al. 2003). During later stages of lateral root formation, the expression domains of these *NCED* genes are restricted to the base of the primordia (Tan et al. 2003).

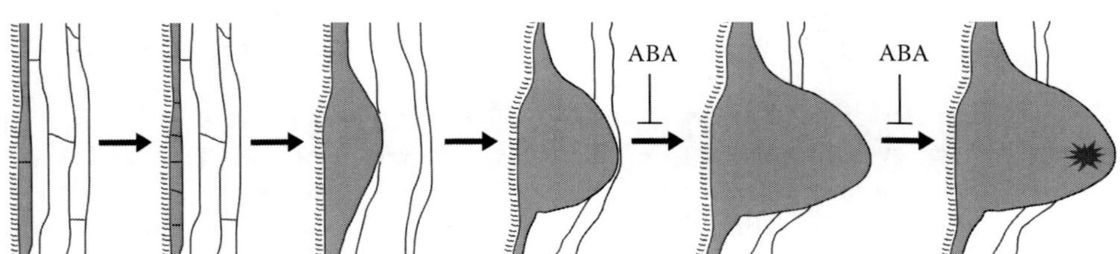

FIGURE 16.2 Key steps in lateral root development, with ABA acting at lateral root emergence and meristem activation (dark gray star).

Given that ABA slows down cell cycle activity, restraining cell proliferation in neighboring cells could be essential to define the organ boundaries for future lateral roots.

In agreement with its general function in regulating seed and bud dormancy (Rohde et al. 2002; Ruttink et al. 2007; Holdsworth et al. 2008), ABA has been put forward as being involved in "lateral root dormancy" (De Smet et al. 2006). This reversible stage can be seen as ABA-induced suppression of the activation of newly formed lateral root meristems and is intimately connected to nitrate (see further); implying that lateral root meristems are activated when soil conditions are favorable (De Smet et al. 2003; Figure 16.2). Several components of ABA signaling, such as ABI1, ABI2, and ABI3, were shown to play some role in the ABA-induced lateral root dormancy (De Smet et al. 2003). In further support of this, ABI3 is auxin inducible in lateral root primordia and is already expressed early on in lateral root development (Brady et al. 2003). Furthermore, the maize VIVIPAROUS1 (VP1) gene, which is believed to be the ABI3 ortholog in *Arabidopsis*, seems to mediate a specific interaction between auxin and ABA signaling (Suzuki et al. 2001). However, the number of lateral roots in *abi3* loss-of-function mutants is similar to that in the wild type, indicating that VP1/ABI3 has only a subtle role in lateral root development (Brady et al. 2003). The VP1/ABI3 transcription factor is regulated upstream by the ERA1 protein, which might have a direct role in downregulating the transcription of ABI3 in lateral roots. In agreement with the repressive effect of ERA1 on ABI3 and the positive role of ABI3 in lateral root development, the lateral root number of the *era1-2* mutant is increased compared with that of the wild type (Brady et al. 2003). The previously described control mechanism on meristem activation is of crucial importance during the development of the root system upon changing soil environment or stress conditions, as will become clear in the next section.

VI. ABA, Roots, and Environment

A. ABA-Mediated Control on Root System Architecture in Changing Environment

Root system architecture, the spatial configuration of the root system in the soil, is strongly affected by water and nutrient availability, and it responds to drought stress, salt stress, nutrient scarceness, etc. In such challenging environments, plants still need to produce an optimal root system to survive and grow. For example, in soybean, the protein kinase GmWNK1 seems to play a role in fine-tuning ABA-dependent ABA homeostasis in response to stress and as such regulating lateral root development (Wang et al. 2010).

One way nitrogen availability affects lateral root development is through a systemic inhibitory effect that is most apparent at high nitrate concentrations and that occurs immediately after the emergence of the lateral root primordia from the parent root (Zhang et al. 1999; Zhang and Forde 2000). While the regulatory role of nitrate is likely mediated through several pathways, the inhibitory effect is significantly reduced in several ABA synthesis mutants (*aba1-1*, *aba2-3*, *aba2-4*, and *aba3-2*) and two ABA-insensitive mutants (*abi4* and *abi5*) (Signora et al. 2001), and high nitrate concentrations mimic the effect of exogenous ABA on lateral root meristem activation (De Smet et al. 2003). These data strongly support that ABA plays a major role in mediating the effect of nitrate on lateral root growth.

Plant roots have the ability to grow toward the direction of high water availability and away from that of high osmolarity (hydrotropism). Under drought conditions that induce ABA synthesis, *Arabidopsis* seedlings produce short, hairless lateral roots. This response is dramatically reduced in the ABA-deficient *aba-1* and ABA-insensitive *abi1-1* mutants (Vartanian et al. 1994). In addition, inhibition of lateral root development is a typical adaptive response of roots to drought stress, which appears to mainly take place at the level of meristem activation and root elongation, where ABA seems to play an important role, and not at the level of initiation (Xiong et al. 2006).

Osmotica downregulate a post-initiation step in lateral root formation, at the level of meristem activation, and ABA is a critical component of the mechanism that regulates this response (Deak and Malamy 2005). Increasing concentrations of osmotica strongly repress or delay the formation of emerged lateral roots with activated meristems. As *aba2-1* and *aba3-1* mutant seedlings grown under osmotic stress have a dramatically larger root system than that of wild-type seedlings, ABA is clearly involved in this process (Deak and Malamy 2005). *LATERAL ROOT DEVELOPMENT2* (*LRD2*), which encodes LONG CHAIN ACYL-COA SYNTHETASE 2 (LACS2), was shown to be part of the mechanism that represses lateral root formation in response to osmotic stress, and LRD2 affects lateral root development in an ABA-dependent way (Macgregor et al. 2008). The *lrd2* mutant develops more lateral roots under repressive osmotic conditions than the wild type (Deak and Malamy 2005).

In addition, the R2R3-type MYB transcription factor MYB96, which is expressed in developing lateral root primordia and is induced by auxin and drought, plays a role in meristem activation, for instance during ABA-mediated drought stress response, and MYB96 possibly interacts with auxin homeostasis and auxin response during adaptive stress responses (Seo et al. 2009). In further support of the cross talk between ABA and auxin (De Smet et al. 2003; Seo et al. 2009), the transcription factor VP1 (also known as ABI3), which is auxin-inducible in lateral root primordia, might define another node of interaction between ABA and auxin signaling during lateral root development (Brady et al. 2003). In addition, the *aberrant lateral root formation3* (*alf3*) mutant displays a lateral root arrest after the emergence of lateral root primordia (Celenza et al. 1995), which resembles lateral root arrest upon ABA treatment and drought stress (De Smet et al. 2003; Deak and Malamy 2005). However, while ABA-induced lateral root arrest cannot be rescued by an increased supply of auxin (De Smet et al. 2003), exogenously supplied auxin rescues the ALF3-dependent developmental arrest (Celenza et al. 1995) and the osmotic repression of lateral root development (Deak and Malamy 2005), suggesting additional auxin-independent pathways. In addition to a potential cross talk between ABA and auxin at the signaling level, ABA might also influence lateral root development by interfering with polar auxin transport (Belin et al. 2009; Shkolnik-Inbar and Bar-Zvi 2010). This is further supported by some ABA-responsive mutants, such as *abi3*, *lrd2*, *aba2-1*, and *aba3-1*, which have a reduced response to polar auxin transport inhibitors (Brady et al. 2003; Deak and Malamy 2005).

B. ABA and Adaptation to Water Stress

In addition to its local role in regulating root growth and development, root-derived ABA can also act as a long-distance signal to regulate processes in leaves. An example is its involvement in the regulation of stomata closure in leaves in response to water stress in roots. When the root system of a plant encounters drying soil, it produces some specific signals which move to the leaves where it induces stomata closure to reduce water loss through transpiration. ABA plays a critical role in this adaptive response and is a key component of the signal for the root-to-shoot signaling process. But, ABA is likely not the only component, and further evidence is needed to elucidate these signals. Water stress increases ABA biosynthesis in roots (Griffiths et al. 1996) and ABA export to shoots (Tardieu and Davies 1992; Griffiths et al. 1996; Tardieu et al. 1996; Wilkinson et al. 1998; Borel et al. 2001). Zhang and Davies (1990, 1991) demonstrated that ABA from roots was required for the water stress–induced stomata closure in leaves, as removing ABA from the xylem sap prevented stomatal closure in water-stressed plants and increasing ABA content in xylem sap, by either exogenous application of ABA in soil or direct injection into xylem, induced stomata closure in plants in the absence of water stress. Once in the leaves, ABA can act in two different capacities to promote stomatal closure. First, it can act as a signal to induce the expression of ABA biosynthesis genes and thereby increase ABA content in leaves, and there is a general agreement in the literature about this role (Dodd 2005). Secondly, it can also act together with leaf-produced ABA to promote stomatal closure. There are contradictory reports about the exact contributions of root-synthesized ABA in this latter role. There are many reports that ABA from roots alone is insufficient to induce stomata closure and other water-stress responses in leaves (Sauter et al. 2001; Holbrook et al. 2002; Thompson et al. 2007). On the other hand, there are other reports that root synthesized ABA is capable of inducing water-stress responses in leaves including stomata closure (Tal et al. 1970; Jones et al. 1987; Fambrini et al. 1995; Chen et al. 2003). For example, when scions of two tomato ABA-deficient mutants, *flacca* and *sitiens*, which are deficient at the last step in ABA biosynthesis (Sindhu and Walton 1988; Taylor et al. 1988), were grafted onto wild-type (WT) root stocks, water-stress responses in the mutant scions were similar to that in WT control (Jones et al. 1987), indicating that ABA from wild-type roots is sufficient to induce water-stress responses in leaves. However, a recent study argues against the involvement of ABA in the root-to-shoot communication of water-stress signals (Christmann et al. 2007). The authors found that the stomatal response in the shoot to limited soil water supply was not affected by the capacity of the root to synthesize ABA and proposed that the water-stress signal was communicated from the root to the shoot via a hydraulic signal (Christmann et al. 2007).

VII. Model for ABA Function in Plant Growth and Development: A Tale Emerging from Studies in Roots

Studies in roots have revealed different regulatory properties of ABA in growth and development. At the morphological level, it stimulates growth at low concentrations, inhibits growth at high concentrations, and promotes dormancy in meristems. At the cellular level, it inhibits cell division and promotes stem cells in root meristems. Figure 16.3 presents a hypothetic model which connects the regulatory properties of ABA at both the cellular and the morphological levels. It proposes inhibition of cell division and suppression of stem cell differentiation as two major regulatory properties of this hormone. Both properties are dose dependent and can be translated to growth promotion, growth inhibition, or dormancy depending on ABA concentrations and developmentally associated sensitivity. At very low ABA concentrations, meristems become defective due to the loss of stem cells. Under this condition, increasing ABA can have a positive impact on root growth by helping the maintenance of the stem cell population in root meristems. At the intermediate and high concentration ranges, ABA inhibits growth by suppressing cell division in meristems. In seeds or newly formed meristems, ABA keeps meristems in dormant state by preserving stem cells

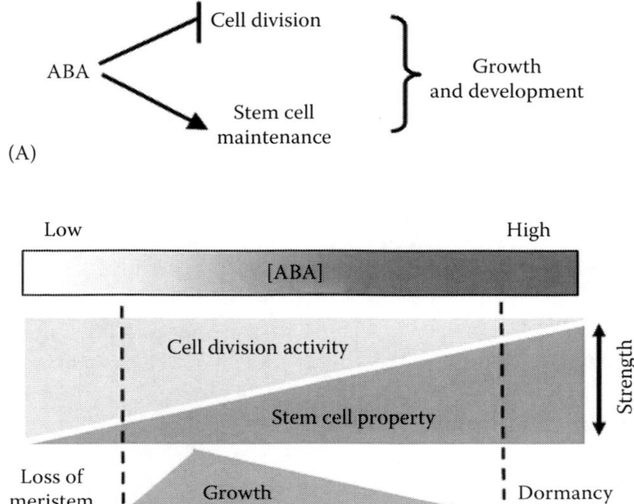

FIGURE 16.3 A model for ABA functions in plant growth and development. (A) Two regulatory properties of ABA at the cellular level. (B) A model showing how the regulatory properties are translated into growth promotion, growth inhibition, or meristem dormancy.

and inhibiting cell division. Although the model is based on observations in roots, it can be applied to shoots as ABA induces similar morphological regulations in both roots and shoots (Sharp et al. 2000; Finkelstein et al. 2002). It is obvious that ABA does not act alone in controlling plant growth and development, almost certainly acting together with other regulators, and the model is therefore oversimplified. Nevertheless, it can serve as a working hypothesis for future studies.

VIII. Concluding Remarks

ABA plays a major role in controlling root system architecture, enabling plant growth under various conditions. While in the past decade enormous progress was made in our understanding of the core signaling mechanism, how ABA specifically controls root branching and how it interacts with other hormone pathways in the root still need to be addressed.

Acknowledgments

We thank Daniela Dietrich for critical reading and useful suggestions. This work was supported by a BBSRC David Phillips Fellowship to IDS and BBSRC (24/P19323), Royal Society (25/R4-JP), and EU grants (RAM-235121) to HZ.

References

Achard P, Cheng H, De Grauwe L et al. 2006. Integration of plant responses to environmentally activated phytohormonal signals. *Science* 311:91–94.

Addicott FT, Lyon JL, Ohkuma K et al. 1968. Abscisic acid: a new name for abscisin II (dormin). *Science* 159:1493.

Arkhipova TN, Veselov SU, Melantiev AI, Marty NEV, Kudoyerova GR. 2005. Ability of bacterium *Bacillus* to produce cytokinins and to influence the growth and endogenous hormone content of lettuce plants. *Plant Soil* 272:201–209.

Astle MC, Rubery PH. 1983. Carriers for abscisic acid and indole-3-acetic acid in primary roots: their regional localisation and thermodynamic driving forces. *Planta* 157:53–63.

Aswath CR, Kim SH, Mo SY, Kim DH. 2005. Transgenic plants of creeping bent grass harboring the stress inducible gene, 9-cis-epoxycarotenoid dioxygenase, are highly tolerant to drought and NaCl stress. *Plant Growth Regul* 47:129–139.

Barrero JM, Piqueras P, Gonzalez-Guzman M et al. 2005. A mutational analysis of the ABA1 gene of *Arabidopsis thaliana* highlights the involvement of ABA in vegetative development. *J Exp Bot* 56:2071–2083.

Behl R, Jeschke WD, Hartung W. 1981. A compartmental analysis of abscisic acid in roots of *Hordeum distichon*. *J Exp Bot* 32:889–897.

Belin C, Megies C, Hauserova E, Lopez-Molina L. 2009. Abscisic acid represses growth of the *Arabidopsis* embryonic axis after germination by enhancing auxin signalling. *Plant Cell* 21:2253–2268.

van den Berg C, Willemsen V, Hendriks G, Weisbeek P, Scheres B. 1997. Short-range control of cell differentiation in the *Arabidopsis* root meristem. *Nature* 390:287–289.

Borel C, Frey A, Marion-Poll A, Simmoneau T, Tardieu F. 2001. Does engineering abscisic acid biosynthesis in *Nicotiana plumbaginifolia* modify stomatal response to drought? *Plant Cell Environ* 24:477–489.

Boyer GL, Zeevaart JA. 1982. Isolation and quantitation of beta-d-glucopyranosyl abscisate from leaves of *Xanthium* and spinach. *Plant Physiol* 70:227–231.

Brady SM, Sarkar SF, Bonetta D, McCourt P. 2003. The ABSCISIC ACID INSENSITIVE 3 (ABI3) gene is modulated by farnesylation and is involved in auxin signalling and lateral root development in *Arabidopsis*. *Plant J* 34:67–75.

Bray EA, Zeevaart JA. 1985. The Compartmentation of abscisic acid and beta-d-glucopyranosyl abscisate in mesophyll cells. *Plant Physiol* 79:719–722.

Brocard-Gifford I, Lynch TJ, Garcia ME, Malhotra B, Finkelstein RR. 2004. The *Arabidopsis thaliana* ABSCISIC ACID-INSENSITIVE8 encodes a novel protein mediating abscisic acid and sugar responses essential for growth. *Plant Cell* 16:406–421.

Celenza JL, Jr., Grisafi PL, Fink GR. 1995. A pathway for lateral root formation in *Arabidopsis thaliana*. *Genes Dev* 9:2131–2142.

Chen GX, Fu XP, Lips SH M S. 2003. Control of plant growth resides in the shoot, and not in the root, in reciprocal grafts of *flacca* and wild-type tomato (*Lycopersicon esculentum*), in the presence and absence of salinity stress. *Plant Soil* 256:205–215.

Cheng WH, Endo A, Zhou L et al. 2002. A unique short-chain dehydrogenase/reductase in *Arabidopsis* glucose signalling and abscisic acid biosynthesis and functions. *Plant Cell* 14:2723–2743.

Cheng JC, Lertpiriyapong K, Wang S, Sung ZR. 2000. The role of the *Arabidopsis* ELD1 gene in cell development and photomorphogenesis in darkness. *Plant Physiol* 123:509–520.

Choi HI, Park HJ, Park JH et al. 2005. *Arabidopsis* calcium-dependent protein kinase AtCPK32 interacts with ABF4, a transcriptional regulator of abscisic acid-responsive gene expression, and modulates its activity. *Plant Physiol* 139:1750–1761.

Christmann A, Hoffmann T, Teplova I, Grill E, Muller A. 2005. Generation of active pools of abscisic acid revealed by in vivo imaging of water-stressed *Arabidopsis*. *Plant Physiol* 137:209–219.

Christmann A, Weiler EW, Steudle E, Grill E. 2007. A hydraulic signal in root-to-shoot signalling of water shortage. *Plant J* 52:167–174.

Daeter W, Hartung W. 1990. Compartmentation and transport of abscisic acid in mesophyll cells of intact leaves of *Valerianella locusta*. *J Plant Physiol* 136:306–312.

De Smet I, Signora L, Beeckman T, Inze D, Foyer CH, Zhang H. 2003. An abscisic acid-sensitive checkpoint in lateral root development of *Arabidopsis*. *Plant J* 33:543–555.

De Smet I, Zhang H, Inze D, Beeckman T. 2006. A novel role for abscisic acid emerges from underground. *Trends Plant Sci* 11:434–439.

De Veylder L, Beeckman T, Beemster GT et al. 2001. Functional analysis of cyclin-dependent kinase inhibitors of *Arabidopsis. Plant Cell* 13:1653–1668.

Deak KI, Malamy J. 2005. Osmotic regulation of root system architecture. *Plant J* 43:17–28.

Dodd IC. 2005. Root-to-shoot signalling: Assessing the roles of 'up' in the up and down world of long-distance signalling in planta. *Plant Soil* 274:251–270.

Eagles CF, Wareing PF. 1963. Dormancy regulators in woody plants: Experimental induction of dormancy in *Betula pubescens. Nature* 199:874–875.

Ephritikhine G, Fellner M, Vannini C, Lapous D, Barbier-Brygoo H. 1999. The sax1 dwarf mutant of *Arabidopsis thaliana* shows altered sensitivity of growth responses to abscisic acid, auxin, gibberellins and ethylene and is partially rescued by exogenous brassinosteroid. *Plant J* 18:303–314.

Fambrini M, Vernieri P, Toncelli ML, Rossi VD, Pugliesi C. 1995. Characterization of a wilty sunflower (*Helianthus annuus* L) mutant. 3. Phenotypic interaction in reciprocal grafts from wilty mutant and wild-type plants. *J Exp Bot* 46:525–530.

Feldman LJ, Arroyave NJ, Sun PS. 1985. Abscisic acid, xanthoxin and violaxanthin in the caps of gravistimulated maize roots. *Planta* 166:483–489.

Finkelstein R, Gampala SS, Lynch TJ, Thomas TL, Rock CD. 2005. Redundant and distinct functions of the ABA response loci ABA-INSENSITIVE(ABI)5 and ABRE-BINDING FACTOR (ABF)3. *Plant Mol Biol* 59:253–267.

Finkelstein RR, Gampala SS, Rock CD. 2002. Abscisic acid signalling in seeds and seedlings. *Plant Cell* 14:S15-S45.

Finkelstein RR, Zeevaart JA. 1994. Gibberellin and abscisic acid biosynthesis and response. In *The Arabidopsis Book*, eds. CR Somerville, EM Meyerowitz, Cold Spring Harbor, NY: Cold Spring Harbor Laboratory Press.

Fujii H, Verslues PE, Zhu JK. 2011. *Arabidopsis* decuple mutant reveals the importance of SnRK2 kinases in osmotic stress responses in vivo. *Proc Natl Acad Sci USA* 108:1717–1722.

Fujii H, Zhu JK. 2009. *Arabidopsis* mutant deficient in 3 abscisic acid-activated protein kinases reveals critical roles in growth, reproduction, and stress. *Proc Natl Acad Sci USA* 106:8380–8385.

Gonzalez-Guzman M, Abia D, Salinas J, Serrano R, Rodriguez PL. 2004. Two new alleles of the abscisic aldehyde oxidase 3 gene reveal its role in abscisic acid biosynthesis in seeds. *Plant Physiol* 135:325–333.

Griffiths A, Parry AD, Jones HG, Tomos AD. 1996. Abscisic acid and turgor pressure regulation in tomato roots. *J Plant Physiol* 149:372–376.

Grossmann K, Hansen H. 2001. Ethylene-triggered abscisic acid: A principle in plant growth regulation? *Physiol Plant* 113:9–14.

Han W, Zhang H, Wang MH. 2010. Fluridone affects quiescent centre division in the *Arabidopsis thaliana* root stem cell niche. *BMB Rep* 43:813–817.

Hansen H, Dörffling K. 1999. Changes of free and conjugated abscisic acid and phaseic acid in xylem sap of drought-stressed sunflower plants. *J Exp Bot* 50:1599–1605.

Hartung W, Abou-Mandour AA. 1980. Abscisic acid in root cultures of *Phaseolus occineus* L. *Z Pflanzenphysiol* 97:265–270.

Hartung W, Sauter A, Hose E. 2002. Abscisic acid in the xylem: where does it come from, where does it go to? *J Exp Bot* 53:27–32.

Hauser F, Waadt R Schroeder JI. 2011. Evolution of abscisic acid synthesis and signalling mechanisms. *Curr Biol* 21:R346–R355.

Himanen K, Boucheron E, Vanneste S, de Almeida Engler J, Inze D, Beeckman T. 2002. Auxin-mediated cell cycle activation during early lateral root initiation. *Plant Cell* 14:2339–2351.

Holbrook NM, Shashidhar VR, James RA, Munns R. 2002. Stomatal control in tomato with ABA-deficient roots: Response of grafted plants to soil drying. *J Exp Bot* 53:1503–1514.

Holdsworth MJ, Bentsink L, Soppe WJ. 2008. Molecular networks regulating *Arabidopsis* seed maturation, after-ripening, dormancy and germination. *New Phytol* 179:33–54.

Hose E, Sauter A, Hartung W. 2002. Abscisic acid in roots—Biochemistry and physiology. In *Plant Roots: The Hidden Half*, eds. Y Waisel, A Eshel, U Kafkafi, pp. 435–448. New York: Marcel Dekker, Inc.

Hoth S, Morgante M, Sanchez JP, Hanafey MK, Tingey SV, Chua NH. 2002. Genome-wide gene expression profiling in *Arabidopsis thaliana* reveals new targets of abscisic acid and largely impaired gene regulation in the abi1-1 mutant. *J Cell Sci* 115:4891–4900.

Hubbard KE, Nishimura N, Hitomi K, Getzoff ED, Schroeder JI. 2010. Early abscisic acid signal transduction mechanisms: Newly discovered components and newly emerging questions. *Genes Dev* 24:1695–1708.

Iuchi S, Kobayashi M, Taji T et al. 2001. Regulation of drought tolerance by gene manipulation of 9-cis-epoxycarotenoid dioxygenase, a key enzyme in abscisic acid biosynthesis in *Arabidopsis. Plant J* 27:325–333.

Jeschke WD, Peuke AD, Pate JS, Hartung W. 1997. Transport, synthesis and catabolism of abscisic acid (ABA) in intact plants of castor bean (*Ricinus communis* L.) under phosphate deficiency and moderate salinity. *J Exp Bot* 48:1737–1747.

Jiang F, Jeschke WD, Hartung W. 2004. Abscisic acid (ABA) flows from *Hordeum vulgare* to the hemiparasite *Rhinanthus minor* and the influence of infection on host and parasite abscisic acid relations. *J Exp Bot* 55:2323–2329.

Jones HG, Sharp CS, Higgs KH. 1987. Growth and water elations of wilty mutants of tomato (*Lycopersicon esculentum* Mill.). *J Exp Bot* 38:1848–1856.

Joshi-Saha A, Valon C, Leung J. 2011. Abscisic acid signal off the STARting block. *Mol Plant* 4:562–580.

Jovanovic L, Daeter W, Hartung W. 1992. Compartmental analysis of abscisic acid in root segments of two maize lines differing in drought susceptibility. *J Exp Bot* 43:S37.

Kang J, Hwang JU, Lee M et al. 2010. PDR-type ABC transporter mediates cellular uptake of the phytohormone abscisic acid. *Proc Natl Acad Sci USA* 107:2355–2360.

Klingler JP, Batelli G, Zhu JK. 2010. ABA receptors: The START of a new paradigm in phytohormone signalling. *J Exp Bot* 61:3199–3210.

Koiwai H, Nakaminami K, Seo M, Mitsuhashi W, Toyomasu T, Koshiba T. 2004. Tissue-specific localization of an abscisic acid biosynthetic enzyme, AAO3, in *Arabidopsis*. *Plant Physiol* 134:1697–1707.

Koornneef M. 1986. Genetic aspects of abscisic acid. In *A Genetic Approach to Plant Biochemistry*, eds. AD Blonstein, PJ King, pp. 35–54. Vienna, Austria: Springer.

Kuromori T, Miyaji T, Yabuuchi H et al. 2010. ABC transporter AtABCG25 is involved in abscisic acid transport and responses. *Proc Natl Acad Sci USA* 107:2361–2366.

Kushiro T, Okamoto M, Nakabayashi K et al. 2004. The *Arabidopsis* cytochrome P450 CYP707A encodes ABA 8′′-hydroxylases: Key enzymes in ABA catabolism. *EMBO J* 23:1647–1656.

Laskowski MJ, Williams ME, Nusbaum HC, Sussex IM. 1995. Formation of lateral root meristems is a two-stage process. *Development* 121:3303–3310.

Lehmann H, Schütte HR. 1980. Purification and characterization of an abscisic acid glucosylating enzyme from cell suspension cultures of *Macleaya microcarpa*. *Z Pflanzenphysiol* 96:277–280.

LeNoble ME, Spollen WG, Sharp RE. 2004. Maintenance of shoot growth by endogenous ABA: Genetic assessment of the involvement of ethylene suppression. *J Exp Bot* 55:237–245.

Liang Y, Mitchell DM, Harris JM. 2007. Abscisic acid rescues the root meristem defects of the *Medicago truncatula latd* mutant. *Dev Biol* 304:297–307.

Lim EK, Doucet CJ, Hou B, Jackson RG, Abramsn SR, Bowles DJ. 2005. Resolution of (+)-abscisic acid using an *Arabidopsis* glycosyltransferase. *Tetrahedron-Asymmetr* 16:143–147.

Lin PC, Hwang SG, Endo A, Okamoto M, Koshiba T, Cheng WH. 2007. Ectopic expression of ABSCISIC ACID 2/GLUCOSE INSENSITIVE 1 in *Arabidopsis* promotes seed dormancy and stress tolerance. *Plant Physiol* 143:745–758.

Liu WC, Carnsdagger HR. 1961. Isolation of abscisin, an abscission accelerating substance. *Science* 134:384–385.

Lumba S, Cutler S, McCourt P. 2010. Plant nuclear hormone receptors: A role for small molecules in protein-protein interactions. *Annu Rev Cell Dev Biol* 26:445–469.

Ma Y, Szostkiewicz I, Korte A et al. 2009. Regulators of PP2C phosphatase activity function as abscisic acid sensors. *Science* 324:1064–1068.

Macgregor DR, Deak KI, Ingram PA, Malamy JE. 2008. Root system architecture in *Arabidopsis* grown in culture is regulated by sucrose uptake in the aerial tissues. *Plant Cell* 20:2643–2660.

Malamy JE, Benfey PN. 1997. Organization and cell differentiation in lateral roots of *Arabidopsis thaliana*. *Development* 124:33–44.

Marin E, Nussaume L, Quesada A et al. 1996. Molecular identification of zeaxanthin epoxidase of *Nicotiana plumbaginifolia*, a gene involved in abscisic acid biosynthesis and corresponding to the ABA locus of *Arabidopsis thaliana*. *EMBO J* 15:2331–2342.

McCourt P, Creelman R. 2008. The ABA receptors: We report you decide. *Curr Opin Plant Biol* 11:474–478.

Melcher K, Ng LM, Zhou XE et al. 2009. A gate-latch-lock mechanism for hormone signalling by abscisic acid receptors. *Nature* 462:602–608.

Milborrow BV, Robinson DR. 1973. Factors affecting the biosynthesis of abscisic acid. *J Exp Bot* 24:537–548.

Miura K, Lee J, Jin JB, Yoo CY, Miura T, Hasegawa PM. 2009. Sumoylation of ABI5 by the *Arabidopsis* SUMO E3 ligase SIZ1 negatively regulates abscisic acid signalling. *Proc Natl Acad Sci USA* 106:5418–5423.

Miyazono K, Miyakawa T, Sawano Y et al. 2009. Structural basis of abscisic acid signalling. *Nature* 462:609–614.

Nakashima K, Fujita Y, Kanamori N et al. 2009. Three *Arabidopsis* SnRK2 protein kinases, SRK2D/SnRK2.2, SRK2E/SnRK2.6/OST1 and SRK2I/SnRK2.3, involved in ABA signalling are essential for the control of seed development and dormancy. *Plant Cell Physiol* 50:1345–1363.

Nambara E, Kawaide H, Kamiya Y, Naito S. 1998. Characterization of an *Arabidopsis thaliana* mutant that has a defect in ABA accumulation: ABA-dependent and ABA-independent accumulation of free amino acids during dehydration. *Plant Cell Physiol* 39:853–858.

Nambara E, Marion-Poll A. 2005. Abscisic acid biosynthesis and catabolism. *Annu Rev Plant Biol* 56:165–185.

Narusaka Y, Nakashima K, Shinwari ZK et al. 2003. Interaction between two cis-acting elements, ABRE and DRE, in ABA-dependent expression of *Arabidopsis* rd29A gene in response to dehydration and high-salinity stresses. *Plant J* 34:137–148.

Newton RJ. 1974. Abscisic acid effects on growth and metabolism in the roots of *Lemna minor*. *Physiol Plant* 80:108–112.

Nishimura N, Hitomi K, Arvai AS et al. 2009. Structural mechanism of abscisic acid binding and signalling by dimeric PYR1. *Science* 326:1373–1379.

Niu X, Helentjaris T, Bate NJ. 2002. Maize ABI4 binds coupling element1 in abscisic acid and sugar response genes. *Plant Cell* 14:2565–2575.

Ohkuma K, Lyon JL, Addicott FT, Smith OE. 1963. Abscisin II, an abscission-accelerating substance from young cotton fruit. *Science* 142:1592–1593.

Ortega-Martinez O, Pernas M, Carol RJ, Dolan L. 2007. Ethylene modulates stem cell division in the *Arabidopsis thaliana* root. *Science* 317:507–510.

Pagant S, Bichet A, Sugimoto K et al. 2002. KOBITO1 encodes a novel plasma membrane protein necessary for normal synthesis of cellulose during cell expansion in *Arabidopsis*. *Plant Cell* 14:2001–2013.

Park SY, Fung P, Nishimura N et al. 2009. Abscisic acid inhibits type 2C protein phosphatases via the PYR/PYL family of START proteins. *Science* 324:1068–1071.

Parry AD, Horgan R. 1992. Abscisic acid biosynthesis in roots. I The identification of potential abscisic acid precursors, and other carotenoids. *Planta* 187:185–191.

Peuke AD, Jeschke WD, Hartung W. 1994. The uptake and flow of C, N and ions between roots and shoots in *Ricinus communis* L III. Long-distance transport of abscisic acid depending on nitrogen nutrition and salt stress. *J Exp Bot* 45:741–747.

Pilet PE. 1975. Abscisic acid as a root growth inhibitor: physiological analyses. *Planta* 122:299–302.

Priest DM, Ambrose SJ, Vaistij FE et al. 2006. Use of the glucosyltransferase UGT71B6 to disturb abscisic acid homeostasis in *Arabidopsis thaliana*. *Plant J* 46:492–502.

Qin X, Zeevaart JA. 2002. Overexpression of a 9-cis-epoxycarotenoid dioxygenase gene in *Nicotiana plumbaginifolia* increases abscisic acid and phaseic acid levels and enhances drought tolerance. *Plant Physiol* 128:544–551.

Rivier L, Milon H, Pilet P-E. 1977. Gas chromatography mass spectrometric determinations of abscisic acid levels in the cap and the apex of maize roots. *Planta* 134:23–27.

Rock CD, Zeevaart JA. 1991. The aba mutant of *Arabidopsis thaliana* is impaired in epoxy-carotenoid biosynthesis. *Proc Natl Acad Sci USA* 88:7496–7499.

Rohde A, Prinsen E, De Rycke R, Engler G, Van Montagu M, Boerjan W. 2002. PtABI3 impinges on the growth and differentiation of embryonic leaves during bud set in poplar. *Plant Cell* 14:1885–1901.

Rook F, Corke F, Card R, Munz G, Smith C, Bevan MW. 2001. Impaired sucrose-induction mutants reveal the modulation of sugar-induced starch biosynthetic gene expression by abscisic acid signalling. *Plant J* 26:421–433.

Ruttink T, Arend M, Morreel K et al. 2007. A molecular timetable for apical bud formation and dormancy induction in poplar. *Plant Cell* 19:2370–2390.

Saito S, Hirai N, Matsumoto C et al. 2004. *Arabidopsis* CYP707As encode (+)-abscisic acid 8''-hydroxylase, a key enzyme in the oxidative catabolism of abscisic acid. *Plant Physiol* 134:1439–1449.

Santiago J, Dupeux F, Round A et al. 2009. The abscisic acid receptor PYR1 in complex with abscisic acid. *Nature* 462:665–668.

Sarkar AK, Luijten M, Miyashima S et al. 2007. Conserved factors regulate signalling in *Arabidopsis thaliana* shoot and root stem cell organizers. *Nature* 446:811–814.

Sauter A, Davies WJ, Hartung W. 2001. The long-distance abscisic acid signal in the droughted plant: The fate of the hormone on its way from root to shoot. *J Exp Bot* 52:1991–1997.

Sauter A, Dietz KJ, Hartung W. 2002. A possible stress physiological role of abscisic acid conjugates in root-to-shoot signalling. *Plant Cell Environ* 25:223–228.

Schachtman DP, Goodger JQ. 2008. Chemical root to shoot signalling under drought. *Trends Plant Sci* 13:281–287.

Scheres B. 2007. Stem-cell niches: Nursery rhymes across kingdoms. *Nat Rev Mol Cell Biol* 8:345–354.

Schwartz SH, Qin X, Zeevaart JA. 2003. Elucidation of the indirect pathway of abscisic acid biosynthesis by mutants, genes, and enzymes. *Plant Physiol* 131:1591–1601.

Schwartz SH, Tan BC, Gage DA, Zeevaart JA, McCarty DR. 1997. Specific oxidative cleavage of carotenoids by VP14 of maize. *Science* 276:1872–1874.

Schwarzkopf E, Miersch O. 1992. in vitro glucosylation of dihydrojasmonic acid and abscisic acid. *Biochem Physiol Pflanz* 188:57–65.

Seki M, Ishida J, Narusaka M et al. 2002. Monitoring the expression pattern of around 7,000 *Arabidopsis* genes under ABA treatments using a full-length cDNA microarray. *Funct Integr Genomics* 2:282–291.

Seo M, Koiwai H, Akaba S et al. 2000. Abscisic aldehyde oxidase in leaves of *Arabidopsis thaliana*. *Plant J* 23:481–488.

Seo M, Koshiba T. 2002. Complex regulation of ABA biosynthesis in plants. *Trends Plant Sci* 7:41–48.

Seo PJ, Xiang F, Qiao M et al. 2009. The MYB96 transcription factor mediates abscisic acid signalling during drought stress response in *Arabidopsis*. *Plant Physiol* 151:275–289.

Sharp RE. 2002. Interaction with ethylene: changing views on the role of abscisic acid in root and shoot growth responses to water stress. *Plant Cell Environ* 25:211–222.

Sharp RE, LeNoble ME, Else MA, Thorne ET, Gherardi F. 2000. Endogenous ABA maintains shoot growth in tomato independently of effects on plant water balance: Evidence for an interaction with ethylene. *J Exp Bot* 51:1575–1584.

Shkolnik-Inbar D, Bar-Zvi D. 2010. ABI4 mediates abscisic acid and cytokinin inhibition of lateral root formation by reducing polar auxin transport in *Arabidopsis*. *Plant Cell* 22:3560–3573.

Signora L, De Smet I, Foyer CH, Zhang H. 2001. ABA plays a central role in mediating the regulatory effects of nitrate on root branching in *Arabidopsis*. *Plant J* 28:655–662.

Sindhu RK, Walton DC. 1988. Xanthoxin metabolism in cell-free preparations from wild type and wilty mutants of tomato. *Plant Physiol* 88:178–182.

Slovik S, Daeter W, Hartung W. 1995. Compartmental redistribution and long-distance transport of abscisic acid (ABA) in plants as influenced by environmental changes in the rhizosphere—A biomathematical model. *J Exp Bot* 46:881–894.

Spollen WG, LeNoble ME, Samuels TD, Bernstein N, Sharp RE. 2000. Abscisic acid accumulation maintains maize primary root elongation at low water potentials by restricting ethylene production. *Plant Physiol* 122:967–976.

Stone SL, Williams LA, Farmer LM, Vierstra RD, Callis J. 2006. KEEP ON GOING, a RING E3 ligase essential for *Arabidopsis* growth and development, is involved in abscisic acid signalling. *Plant Cell* 18:3415–3428.

Suzuki M, Kao CY, Cocciolone S, McCarty DR. 2001. Maize VP1 complements *Arabidopsis* abi3 and confers a novel ABA/auxin interaction in roots. *Plant J* 28:409–418.

Tal M, Imber D, Itai C. 1970. Abnormal stomatal behavior and hormonal imbalance in flacca, a wilty mutant of tomato: I. root effect and kinetin-like activity. *Plant Physiol* 46:367–372.

Tan BC, Joseph LM, Deng WT et al. 2003. Molecular characterization of the *Arabidopsis* 9-cis epoxycarotenoid dioxygenase gene family. *Plant J* 35:44–56.

Tardieu F, Davies WJ. 1992. Stomatal response to abscisic acid is a function of current plant water status. *Plant Physiol* 98:540–545.

Tardieu F, Lafarge T, Simonneau T. 1996. Stomatal control by fed or endogenous xylem ABA in sunflower: Interpretation of correlations between leaf water potential and stomatal conductance in anisohydric species. *Plant Cell Environ* 19:75–84.

Taylor IB, Burbidge A, Thompson AJ. 2000. Control of abscisic acid synthesis. *J Exp Bot* 51:1563–1574.

Taylor IB, Linforth RST, Al-Naieb RJ, Bowman WR, Marples BA. 1988. The wilty tomato mutants flacca and sitiens are impaired in the oxidation of ABA-aldehyde to ABA. *Plant Cell Environ* 11:739–745.

Thompson AJ, Andrews J, Mulholland BJ et al. 2007. Overproduction of abscisic acid in tomato increases transpiration efficiency and root hydraulic conductivity and influences leaf expansion. *Plant Physiol* 143:1905–1917.

Thompson AJ, Jackson AC, Symonds RC et al. 2000. Ectopic expression of a tomato 9-cis-epoxycarotenoid dioxygenase gene causes over-production of abscisic acid. *Plant J* 23:363–374.

Tietz A. 1971. Nachweis yon Abscisinssäure in Wurzeln. Planta (Berl.) 93:93–96. Umezawa T, Sugiyama N, Mizoguchi M et al. 2009. Type 2C protein phosphatases directly regulate abscisic acid-activated protein kinases in *Arabidopsis*. *Proc Natl Acad Sci USA* 106:17588–17593.

Umezawa T, Sugiyama N, Mizoguchi M et al. 2009. Type 2C protein phosphatases directly regulate abscisic acid-activated protein kinases in *Arabidopsis*. *Proc Natl Acad Sci USA* 106:17588–17593.

Vartanian N, Marcotte L, Giraudat J. 1994. Drought rhizogenesis in *Arabidopsis thaliana* (differential responses of hormonal mutants). *Plant Physiol* 104:761–767.

Verkest A, Weinl C, Inze D, De Veylder L, Schnittger A. 2005. Switching the cell cycle. Kip-related proteins in plant cell cycle control. *Plant Physiol* 139:1099–1106.

Wang H, Fowke LC, Crosby WL. 1997. A plant cyclin-dependent kinase inhibitor gene. *Nature* 386:451–452.

Wang L, Hua D, He J, Duan Y, Chen Z, Hong X, Gong Z. 2011. *Auxin Response Factor2 ARF2* and its regulated homeodomain gene *HB33* mediate abscisic acid response in *Arabidopsis*. *PLoS Genet* 7, e1002172.

Wang H, Qi Q, Schorr P, Cutler AJ, Crosby WL, Fowke LC. 1998. ICK1, a cyclin-dependent protein kinase inhibitor from *Arabidopsis thaliana* interacts with both Cdc2a and CycD3, and its expression is induced by abscisic acid. *Plant J* 15:501–510.

Wang Y, Suo H, Zheng Y et al. 2010. The soybean root-specific protein kinase GmWNK1 regulates stress-responsive ABA signalling on the root system architecture. *Plant J* 64:230–242.

Wang H, Zhou Y, Gilmer S, Whitwill S, Fowke LC. 2000. Expression of the plant cyclin-dependent kinase inhibitor ICK1 affects cell division, plant growth and morphology. *Plant J* 24:613–623.

Weiner JJ, Peterson FC, Volkman BF, Cutler SR. 2010. Structural and functional insights into core ABA signalling. *Curr Opin Plant Biol* 13:495–502.

Wilkinson S, Corlett JE, Oger L, Davies WJ. 1998. Effects of xylem pH on transpiration from wild-type and flacca tomato leaves. A vital role for abscisic acid in preventing excessive water loss even from well-watered plants. *Plant Physiol* 117:703–709.

Wilkinson S, Davies WJ. 1997. Xylem sap pH increase: A drought signal received at the apoplastic face of the guard cell that involves the suppression of saturable abscisic acid uptake by the epidermal symplast. *Plant Physiol* 113:559–573.

Wilkinson S, Davies WJ. 2002. ABA-based chemical signalling: the co-ordination of responses to stress in plants. *Plant Cell Environ* 25:195–210.

Windsor ML, Milborrow BV, McFarlane IJ. 1992. The Uptake of (+)-S- and (−)-R-abscisic acid by suspension culture cells of hopbush (*Dodonaea viscosa*). *Plant Physiol* 100:54–62.

Wolf O, Jeschke WD, Hartung W. 1990. Long-distance transport of abscisic acid in salt-stressed *Lupinus albus* plants. *J Exp Bot* 41:593–600.

Xiong L, Ishitani M, Lee H, Zhu JK. 2001. The *Arabidopsis* LOS5/ABA3 locus encodes a molybdenum cofactor sulfurase and modulates cold stress- and osmotic stress-responsive gene expression. *Plant Cell* 13:2063–2083.

Xiong L, Lee H, Ishitani M, Zhu JK. 2002. Regulation of osmotic stress-responsive gene expression by the LOS6/ABA1 locus in *Arabidopsis*. *J Biol Chem* 277:8588–8596.

Xiong L, Wang RG, Mao G, Koczan JM. 2006. Identification of drought tolerance determinants by genetic analysis of root response to drought stress and abscisic acid. *Plant Physiol* 142:1065–1074.

Xiong L, Zhu JK. 2003. Regulation of abscisic acid biosynthesis. *Plant Physiol* 133:29–36.

Xu ZJ, Nakajima M, Suzuki Y, Yamaguchi I. 2002. Cloning and characterization of the abscisic acid-specific glucosyltransferase gene from adzuki bean seedlings. *Plant Physiol* 129:1285–1295.

Yamaguchi-Shinozaki K, Shinozaki K. 2005. Organization of cis-acting regulatory elements in osmotic- and cold-stress-responsive promoters. *Trends Plant Sci* 10:88–94.

Zeevaart JA, Creelman RA. 1988. Metabolism and physiology of abscisic acid. *Annu Rev Plant Physiol Plant Mol Biol* 39:439–473.

Zhang S, Cai Z, Wang X. 2009. The primary signalling outputs of brassinosteroids are regulated by abscisic acid signalling. *Proc Natl Acad Sci USA* 106:4543–4548.

Zhang J, Davies WJ. 1989. Abscisic acid produced in dehydrating roots may enable the plant to measure the water status of the soil. *Plant Cell Environ* 12:73–81.

Zhang J, Davies WJ. 1990. Changes in the concentration of ABA in xylem sap as a function of changing soil water status can account for changes in leaf conductance and growth. *Plant Cell Environ* 13:277–285.

Zhang J, Davies WJ. 1991. Antitranspirant activity in xylem sap of maize plants. *J Exp Bot* 42:317–321.

Zhang H, Forde BG. 2000. Regulation of *Arabidopsis* root development by nitrate availability. *J Exp Bot* 51:51–59.

Zhang H, Han W, De Smet I et al. 2010. ABA promotes quiescence of the quiescent centre and suppresses stem cell differentiation in the *Arabidopsis* primary root meristem. *Plant J* 64:764–774.

Zhang H, Jennings A, Barlow PW, Forde BG. 1999. Dual pathways for regulation of root branching by nitrate. *Proc Natl Acad Sci USA* 96:6529–6534.

Zhou R, Cutler AJ, Ambrose SJ et al. 2004. A new abscisic acid catabolic pathway. *Plant Physiol* 134:361–369.

Zhou L, Jang JC, Jones TL, Sheen J. 1998. Glucose and ethylene signal transduction crosstalk revealed by an *Arabidopsis* glucose-insensitive mutant. *Proc Natl Acad Sci USA* 95:10294–10299.

17

Brassinosteroid Signaling in Root Development

Josep Vilarrasa-Blasi
Center for Research in Agricultural Genomics

Mary-Paz González-García
Center for Research in Agricultural Genomics

Ana I. Caño-Delgado
Center for Research in Agricultural Genomics

I. Introduction

Brassinosteriods (BRs) are polyhydroxylated triterpenoids essential for plant growth. BRs participate in a myriad of developmental processes, such as seed germination, pollen tube growth, male fertility, vascular development, flowering time, and senescence. During the plant life cycle, BRs modulate the plant response to environmental factors such as light, temperature, salt, and pathogens among others (Bajguz and Hayat 2009). BRs were originally discovered in *Brassica napus* L. pollen (Grove 1979). The past two decades, genetic and biochemical analyses have identified the main BR signaling and synthesis components in the plant model species *Arabidopsis* and rice (Vert et al. 2005; Kim and Wang 2010; Li 2010). The BR pathway is currently among the most highly studied signal transduction pathways in plants (Vert et al. 2005; Kim and Wang 2010; Clouse 2011; Figure 17.1). In aerial plant organs, the growth-promoting properties of BRs are exemplified by their positive effect on cell elongation. Recently, the comprehensive characterization of BR contribution to root growth has shown that in addition to cell elongation, BR-mediated cell cycle progression is central for growth and meristem maintenance in the primary root. This chapter summarizes the state of knowledge on BR signaling in *Arabidopsis* and the current understanding of BR action in root development.

II. Brassinosteroid Signaling Pathway

Plant and animal steroids share high structural similarity although the cellular mechanisms for BR signaling in plants differs from the signaling based on nuclear-localized receptors found in animals (Thummel and Chory 2002). BRs are perceived by a plasma membrane–localized leucine-rich-repeat receptor-like-kinase (LRR-RLK) protein, BRASSINOSTEROID INSENSITIVE 1 (BRI1), that belongs to a family of protein kinases that contain more than 200 members (Shiu and Bleecker 2001). The *BRI1* gene was identified as a putative BR receptor through a genetic screen for BR-insensitive mutants in *Arabidopsis* (Li and Chory 1997) and was reported to be ubiquitously expressed in the plant (Friedrichsen et al. 2000). An intensive study of its structure has indicated that BRI1 has an extracellular domain consisting of 24 LRR domains interrupted by a 70-amino acid island domain (ID) placed between 20th and 21st LRR, a surface pocket for brassinolide (BL) binding, a transmembrane domain, a cytoplasmic serine/threonine kinase domain (KD), a juxtamembrane domain, and a short C-terminal domain (Li and Chory 1997; Friedrichsen et al. 2000; Wang et al. 2005; Hothorn et al. 2011). In the presence of BL, homodimerized BRI1 receptors interact with coreceptors of the SOMATIC EMBRYOGENESIS RECEPTOR KINASE (SERK) family that are required for signaling, of

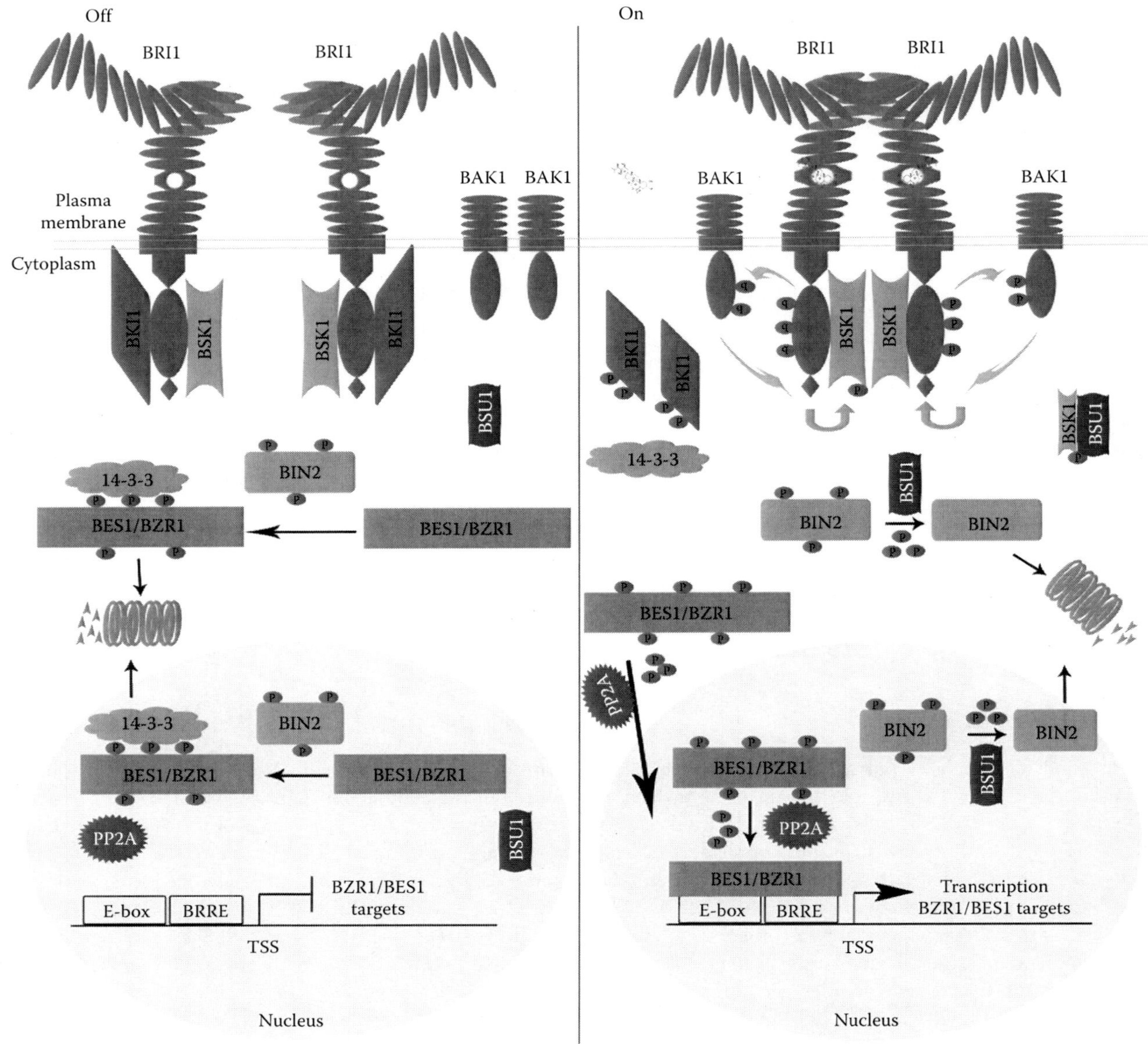

FIGURE 17.1 **(See color insert.)** BR signaling pathway in *Arabidopsis*. Schematic representation of BR signaling pathway when is inactive (off) and active (on) in the left and right panels, respectively. In the absence of BL, BKI1 binds to the inactive form of BRI1 receptor together with BSKs at the cell surface; thus, BSU remains inactive. Inside the cell, BIN2 kinase phosphorylates BES1 and BZR1 transcription factors, promoting their cytoplasm retention thought the binding of 14-3-3 phosphoproteins and their subsequent proteasome-mediated degradation. In consequence, no changes in gene expression are observed (left panel). Upon BR-ligand binding to the extracellular domain of BRI1 receptor, BRI1 kinase phosphorylates BKI1 promoting its release to the cytoplasm and letting BRI1/BAK1 heterodimerization to occur. Transphosphorylation between BRI1 and BAK1 coreceptor activates the signaling complex, mediating phosphorylation of BSKs. Then BSK will bind to BSU, which in turn mediates BIN2 proteasome degradation by dephosphorylation of a conserved Tyr residue. The negative regulation of BIN2 upon BR perception together with the PP2A phosphatase activity mediates dephosphorylation of BES1 and BZR1 transcription factors. Dephosphorylated BES1 and BZR1 transcription factors in the nucleus control BR-mediated genome transcription and cellular responses in the plant (right panel).

which SERK3/BAK1 is the most prominent (Li et al. 2002; Nam and Li 2002; Figure 17.1). In this scenario, BR binds to the extracellular domain of BRI1 (Kinoshita et al. 2005), and the kinase of BRI1 phosphorylates tyrosine residues of the negative regulator BRI1 kinase inhibitor1 (BKI1) (Jaillais et al. 2011b). This promotes BRI1 heterodimerization with its coreceptor BRI1 ASSOCIATED KINASE-1 (BAK1), activating the signaling complex that inactivates BRASSINOSTEROID INSENSITIVE 2 (BIN2) (a homolog to mammalian GSK3 kinases) through BR-SIGNALING KINASES (BSKs) and *bri1* SUPPRESSORS 1 (BSU1) (Mora-García et al. 2004; Peng et al. 2008; Tang et al. 2008; Kim et al. 2010). A dynamic and transient interplay between the main ligand-binding receptors and the coreceptors has recently been proposed (Jaillais et al. 2011a).

BR-ligand perception at the plasma membrane triggers distinct transcriptomic responses in the plant nucleus that enables the plant to adapt to internal cues and major environmental conditions (Goda et al. 2002; Nemhauser et al. 2006). This differential gene expression involves a stabilization of at least two transcription factors specific to plants BZR1 (BRASSINAZOLE RESISTANT 1) and BES1 (BRI1 EMS SUPRESSOR 1, also designated BZR2, BRASSINAZOLE RESISTANT 2) (Wang et al. 2002; Yin et al. 2002). Both can be phosphorylated by BIN2 kinase (He et al. 2002). The phosphorylated forms of BES1/BZR1 interact with 14-3-3 phosphoproteins promoting proteasome-mediated degradation, thus attenuating the signal (Gampala et al. 2007; Ryu et al. 2007). BIN2 kinase activity is negatively regulated by BSU1-mediated dephosphorylation and subsequent degradation in a proteasome-dependent manner (Mora-García et al. 2004). BIN2 inactivation leads to dephosphorylation of BZR1/BES1 by protein phosphatase 2A (PP2A) and its translocation into the nucleus. There it can regulate BR-responsive gene expression (Tang et al. 2011).

BES1 and BZR1 transcription factors of *Arabidopsis* share 88% amino acid sequence identity. Genome-wide identification of BZR1 and BES1 target genes led to the identification of complex regulatory networks that integrate with hormonal and light signaling pathways during plant growth. BZR1 and BES1 proteins can bind to both E-boxes (CANNTG) and BRRE (BR response element, CGTGT/CG) and function as transcriptional activators or repressors depending on their specific target (Sun et al. 2010; Yu et al. 2011). The large number of nonoverlapping target genes for BZR1 and BES1 further evidence the tissue-specific nature of these responses (Yin et al. 2005; Li et al. 2009; Gudesblat and Russinova 2011). Future studies should uncover whether BRs can modulate specific cellular responses without the involvement of BES1/BZR1.

The modulation of BRs signaling events in the cell is generally accomplished by BRI1 vesicle trafficking (Geldner et al. 2007; Viotti et al. 2010). Originally, a hypothesis of BRI1 complex endocytosis upon BRI1 receptor activation was proposed. These studies were based on the use of green fluorescent protein (GFP)-tagged BRI1 and different pharmacological inhibitors to describe the subcellular dynamics of the receptor in *Arabidopsis* roots. Upon blocking the transition between early- and late-endosome mediated by Brefeldin A, BRI1 was accumulating, resulting in enhanced signaling responses (Russinova et al. 2004; Geldner et al. 2007; Viotti et al. 2010). A recent study that implemented the use of a fluorescently labeled BR, Alexa Fluor 647 castasterone (AFCS), has facilitated the visualization of BRI1-AFCS endocytosis in living cells. This innovative study demonstrates that the BRI1 receptor is active at the plasma membrane and addresses a role for BRI1-mediated endocytosis in BR signal attenuation (Irani et al. 2012).

Based on the BRI1 ligand-binding domain characteristics, three additional BRI1-like receptor homologs have been identified in *Arabidopsis* (BRI1-like genes) (Caño-Delgado et al. 2004; Ceserani et al. 2009) and rice (Yamamuro et al. 2000). Of those, the BRL1 and BRL3 have true receptor functions and play specific roles in vascular development, where their transcriptional expression is enriched (Caño-Delgado et al. 2004). Studies aimed at the identification of BRL1 and BRL3 receptor complexes, and their downstream signaling targets should be very important in the coming years for understanding how BRs adapt environmental responses into cell-specific cellular growth and differentiation programs.

III. Brassinosteroids Contribute to Important Aspects of Plant Postembryonic Development

Plants use BRs to translate environmental stimuli into developmental responses. BR-deficient mutants and plants treated exogenously with BL- or BR-synthesis inhibitors are dramatically affected in seed germination (Steber and McCourt 2001), photomorphogenesis (Li et al. 1996), organ size and architecture (Clouse et al. 1996), apical dominance (Clouse et al. 1996), senescence (Li and Chory 1997), vascular development (Szekeres et al. 1996, Caño-Delgado et al. 2004), male fertility (Ye et al. 2010), pollen tube growth (Grove 1979), and flowering time (Li and Chory 1997; Domagalska et al. 2007; Figure 17.2). At the cellular level, BRs mediate cell growth by controlling cell division, elongation, and differentiation activities. The ability of BR hormones to promote cell growth in a variety of plant species has been described (Fujioka and Sakurai 1997). The most classical example is the elongation of the pollen tube, a fundamental step toward reproductive success in flowering plants (Grove 1979). In *Arabidopsis*, genetic analysis has shown that strong BR mutants are male sterile, and the mutant defects in pollen tube elongation were found to correlate with a reduction in the expression of a set of genes required for microspore mother cell development (*SPL/NZZ*), for microspore development (*TDF1*, *AMS*, and *MYB103*), and for tapetal development and pollen wall formation (*MS1/MS2*). Those genes are direct targets of BES1 (Ye et al. 2010).

Decades of physiological studies in several plant species have shown that BL application promotes cell elongation regulating the expression of primary cell wall–loosening enzymes such as xyloglucan endotransglycosylases (XETs) (Zurek et al. 1994). Upon germination, BR-deficient mutants display severe growth defects in the size of hypocotyls in seedlings that can be reversed by exogenous application of BL (Szekeres et al. 1996; Azpiroz et al. 1998). Recently, a mechanism for BR-mediated promotion of cell elongation has been proposed based on the targeted expression of cell wall biosynthesis enzymes by BES1 and BZR1 detected in two parallel chromatin immunoprecipitation (ChIP)-chip experiments (Sun et al. 2010; Yu et al. 2011). One further example is the demonstration by ChIP experiments that the BR-activated BES1 can bind to the promoters of a group

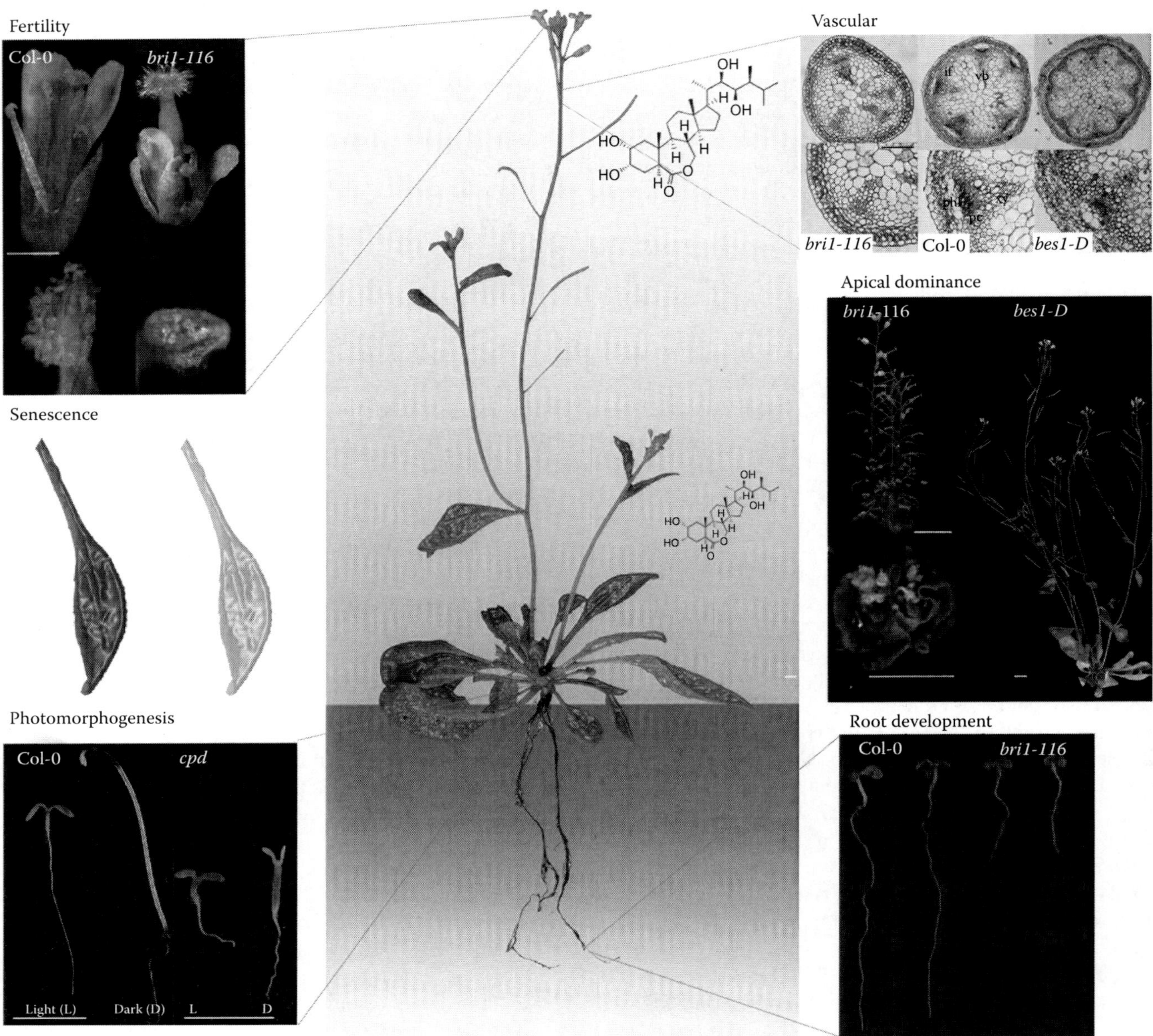

FIGURE 17.2 **(See color insert.)** BR contribution to plant development. The identification of BR mutants by genetic analysis has permitted to assign important roles for BRs in plant development. *Fertility*: loss-of-function BR mutants exhibit short stamens and defective pollen grain production that result in male-sterile plants (Ye et al. 2010). *Senescence*: exogenous application of active BRs was found to promote senescence, and BRs mutants show impaired senescence programs (Li and Chory 1997). *Photomorphogenesis*: *Arabidopsis* seedlings grow in darkness show elongated hypocotyls and closed cotyledons, in contrast mutants defective in BR mutants (e.g., *constitutive photomorphogenesis and dwarfism*, *cpd*) that show photomorphogenic responses with open cotyledons and short hypocotyls (Szekeres et al. 1996). *Vascular*: BR contribution to vascular bundle number in the shoot inflorescence stem by promoting early provascular divisions (Caño-Delgado et al. 2004; Ibañes et al. 2009). *Apical dominance*: loss-of-function BR mutants are dwarf and exhibit a loss of apical dominance. Gain-of-function *bes1-D* mutant (Li and Chory 1997, Yin et al. 2002). *Root development*: BR mutants show reduced root length caused by both alterations in cell-cycle progression and cell elongation (González-García et al. 2011; Hacham et al. 2011) (Vb; vascular bundle, If; interfascicular fibers, phl; phloem, pc; procambium, xy; xylem). (From González-García, M.P. et al., *Development*, 138, 849, 2011. With permission.)

of *Arabidopsis* cellulose synthase genes (*AtCESA*) that control primary cell wall extension (Xie et al. 2011).

BRs mutants exhibit an array of cellular defects in leaf growth and development. BRs loss-of-function mutants have small and round-shaped leaves with short petioles (Szekeres et al. 1996; Li and Chory 1997; Choe et al. 1998; Friedrichsen et al. 2000; Li et al. 2001), whereas BR gain-of-function mutants exhibit enlarged leaves with elongated petioles (Choe et al. 2001; Wang et al. 2002; Yin et al. 2002; Oh et al. 2011; Figure 17.2). The dwarfed leaf phenotype of the BR loss-of-function mutants has been attributed to impaired cell growth (Chory et al. 1991; Clouse et al. 1996; Kauschmann et al. 1996; Szekeres et al. 1996; Fujioka and Sakurai 1997; Azpiroz et al. 1998). In addition, several *ucu1/bin2* mutant alleles were shown to display leaves rolling spirally downward concomitant with a reduced cell expansion of abaxial epidermal cells (Perez-Perez et al. 2002). A recent analysis has shown that increased BRI1 levels or modification of Tyr-phosphorylation properties of BRI1 receptor produces plants with increased leaf size (Gonzalez et al. 2010; Oh et al. 2011). The analysis of these transgenic plants indicated that BRs contribute to leaf growth by promoting cell division. Despite the results described earlier, a reduction in cell division could be compensated for by an increase in cell area not necessarily resulting in larger leaves (Aguirrezabal et al. 2006), which has hampered a clear dissection between primary and secondary BR-mediated growth defects in this organ. Currently, the precise contribution of BRs to cell division or elongation activities during leaf growth and the molecular mechanisms involved in this developmental process remain unknown.

In general, because BR mutants generally show alterations in organ size, it has been difficult to establish whether the observed phenotypes are directly caused by overall cellular defects (cell division and cell elongation), by deregulation of cell-type and/or stage-specific signaling programs (cell differentiation), or a combination of both effects. A merge of quantitative phenotypic analyses with the identification of a cell-based specific analysis of BR responses will advance our understanding of tissue-specific BR signaling.

The study of the role of BRs in vascular development in *Arabidopsis* comes from the original characterization of BR-deficient mutants *cpd* and *dwf7* (Szekeres et al. 1996; Choe et al. 2001) and BR perception mutants (Caño-Delgado et al. 2004). The specificity of the BRI1 homolog receptors BRL1 and BRL3, expressed in the vascular tissues and the mutant phenotypes in the inflorescence stems, has allowed the proposal of a role for BRs in modulating the xylem/phloem differentiation ratio (Caño-Delgado et al. 2004). A more recent analysis of BR mutants has identified a role for BRs in promoting vascular bundle formation (Ibañes et al. 2009; Caño-Delgado et al. 2010). Mutants with reduced BRI1 receptor activity, signaling, or BR levels (Szekeres et al. 1996; Li and Chory 1997) were found to have a reduced number of vascular bundles, whereas mutants with enhanced BR signaling (Wang et al. 2002; Yin et al. 2002) or BR levels (Choe et al. 2001) exhibited an increased number of vascular bundles. While the mechanisms for BR-mediated provascular divisions and cell differentiation are unknown, the available transcriptomic studies in *Zinnia* and *Arabidopsis* cell cultures point to members of class III homeodomain transcription factors as candidate regulators of this process (Ohashi-Ito et al. 2002; Ohashi-Ito and Fukuda 2003). The combination of computational approaches with molecular analyses will help to clarify the role of BR signaling in provascular divisions during vascular development (Fabregas et al. 2010).

IV. Brassinosteroid Contribution to Root Growth and Development

The simple and stereotyped organization of the *Arabidopsis* primary root has motivated the research of fundamental biological questions in plant organogenesis and development. The study of the primary root as a model is widely extended in plant biology, as it is elegantly gathered in the present book. The *bri1* mutants were originally described by showing a short-root phenotype in the presence of BL (Clouse et al. 1996). Since then, the analyses of root phenotypes in several BR-deficient and insensitive mutants have assigned a role of BRs in distinct aspects of root growth and development, such as gravitropism (Kim et al. 2000), lateral root formation (Bao et al. 2004), root-hair cell fate (Kuppusamy et al. 2009), meristem size, and distal stem cell differentiation (González-García et al. 2011; Hacham et al. 2011). In root growth inhibition assays, BR-deficient roots are more resistant to heavy metal cadmium (Cd) than wild-type plants, indicating BRs play a role in the plant response to Cd (Villiers et al. 2012). Physiological studies have proven valuable to our understanding of how BRs modulate root developmental processes in concerted action with other plant hormones, such as auxin and cytokinin (Scacchi et al. 2010). Experiments using exogenous hormone application have shown that BRs can stimulate root gravitropism independently of auxin (Kim et al. 2000; Li et al. 2005). In contrast, the characterization and study of *bri1* and BR-synthesis mutants *dwarf4* resulted in the notion that at low levels, both hormones synergistically promote lateral root emergence (Bao et al. 2004), and root elongation where BR action has been proposed to be subordinated to that of auxin (Yoshimitsu et al. 2011). Also in root growth, the functional analysis of BREVIS RADIX (BRX) has connected BRs synthesis and auxin signaling (Mouchel et al. 2006). At the protophloem cells of the root apical meristem, the concerted action of BRX, MONOPTEROS (MP), and SHORT HYPOCOTYL 2 (SHY2) establishes a feedback loop that regulates PIN3 in the control of root meristem size (Scacchi et al. 2010). Contrary to that, BR signaling in the root was reported not to have a significant effect on auxin efflux carriers PIN1, 3, and 7 transcription (Hacham et al. 2011), whereas it regulates PIN2 and PIN4 accumulation in the epidermis, cortex, and columella cells at both transcriptional and posttranscriptional levels (Hacham et al. 2012). Certainly, the role of auxin gradients in BR-mediated cell-cycle progression and differentiation deserves further analysis.

A. BRs at the Root Epidermis

The characterization of *bri1* epidermal patterning defects together with the expression of hair-fate reporters assigned a role for BR signaling in providing positional information to root epidermal cells (Kuppusamy et al. 2009). BRs control the expression of key regulators of epidermal patterning, such as WEREWOLF (WER) and its downstream target GLABLA2 (GL2). Exogenous BL application increases the transcript levels of these two genes (>2 FC), (Nemhauser et al. 2006). In addition, the root-hair-specific GL2:GUS transcriptional reporter appeared to be misexpressed in the *bri1* mutant roots, which exhibited radial patterning defects in the epidermis. Although neither *GL2* nor *WER* genes were among the list of BZR1/BES1 high-fidelity targets recently released (Sun et al. 2010; Yu et al. 2011), the identification of downstream signals transmitted from the BRI1 receptor in this root-hair-fate response yet deserves further study. The presence of the BRI1 receptor in the epidermis was shown to be related to the promotion of growth of shoots and roots (Savaldi-Goldstein et al. 2007; Hacham et al. 2011). In the root, driving the expression of BRI1 receptor under GL2 promoter recovers the root meristem defects of *bri1* null mutants, whereas specific expression of BRI1 in inner cell layers failed to do that. Based on these results, a signaling mechanism from the epidermis to the inner meristematic cells has been suggested, although the molecular nature of the moving signal remains to be identified.

B. Root Analysis for the Study of BRs in Cell-Cycle Progression and Differentiation

The role of BRs in promoting cell elongation in the primary root has been demonstrated at both the physiological and the genetic level (Szekeres et al. 1996; Müssig et al. 2003; Mouchel et al. 2006). However, the role of BRs in cell-cycle progression has remained controversial (Clouse et al. 1996; Miyazawa et al. 2003). The cellular analysis of *bri1* roots has shown that

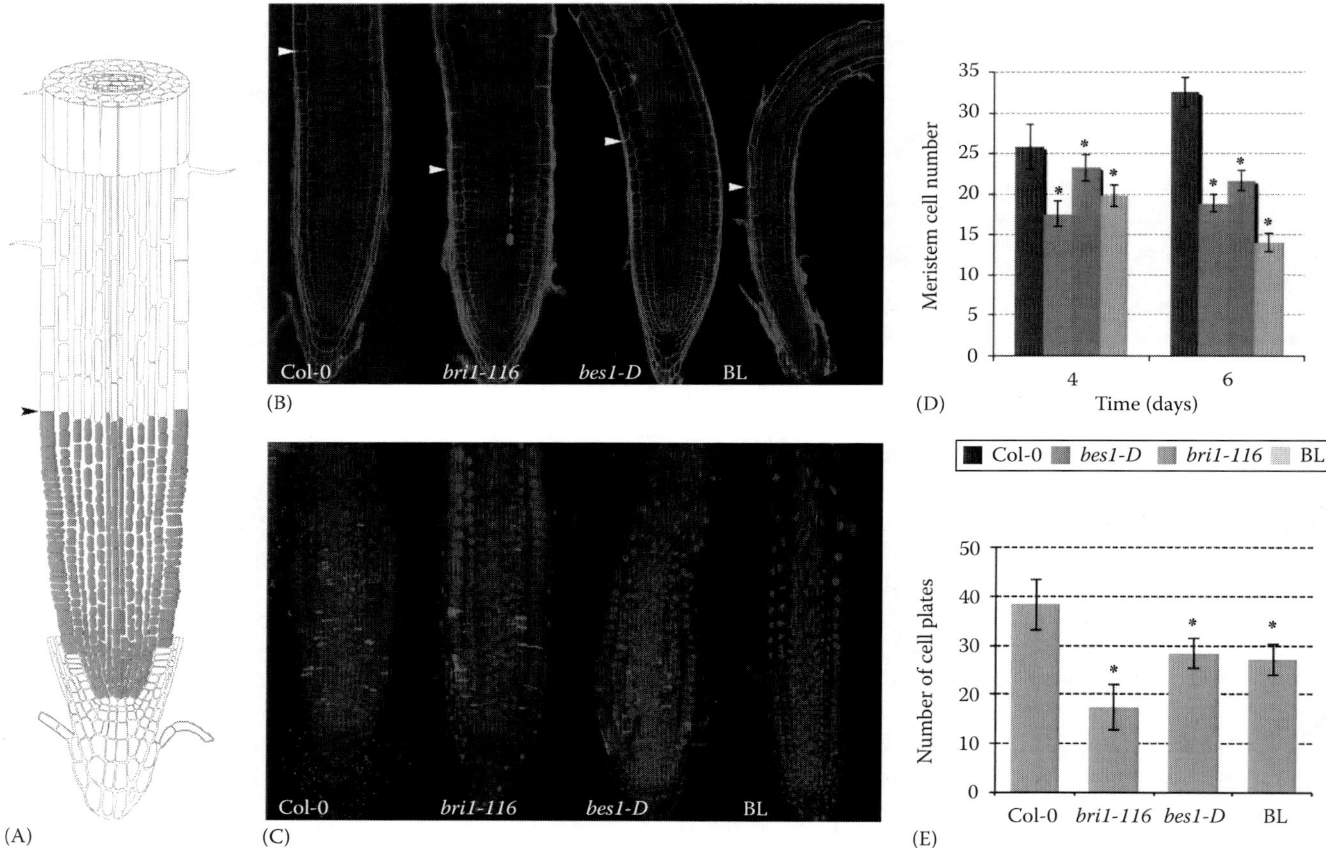

FIGURE 17.3 **(See color insert.)** BRs control root growth by maintaining the normal cell-cycle progression in the meristem. (A) Schematic representation of a 6-day-old *Arabidopsis* primary root. Green cells represent the meristematic region (dividing cells). (B) Root meristems of Col-0, *bri1-116* (loss-of-function mutant), *bes1-D* (gain-of-function mutant), and 6-day-old plants continuously treated with 4 nM BL. Counterstained roots with propidium iodide (PI) staining (red), a white arrow points the end of the meristematic zone. (D) Representation of the number of cells within the meristematic region. (C) Immunolocalization of KNOLLE protein (pink) and cell nucleus stained with DAPI (blue) in different BR mutant background and BL treatment. (E) Representation of the number of cell plates for each genotype/physiological treatment (pink). Note that the differences in the number of meristematic cells and the number of KNOLLE cell plates in BR mutants and plant exogenously treated with BL in comparison to Col-0 wild type. (From González-García, M.P. et al., *Development*, 138, 849, 2011. With permission.)

short *bri1* roots were caused by both defects in cell expansion and in the normal progression of the cell cycle resulting in a short meristem size (González-García et al. 2011; Hacham et al. 2011). Intriguingly, plants treated with BL or mutants with enhanced BR signaling such as *bes1-D* displayed a reduced root meristem, suggesting that balanced levels of the BRI1 receptor and downstream signaling are needed in order to maintain normal cell division activities in the root meristem (González-García et al. 2011). These findings prompted the study of BR-mediated root growth from a different perspective (Figure 17.3).

The effects of BRs on meristem division have been analyzed using different cell-cycle reporter lines. The analysis was done using a destruction box containing *pCYCB1;1:GUS* and a *pCYCB1;1:GFP* marker that visualizes cells at the G2-M phase of the cell cycle (Colon-Carmona et al. 1999), a reporter of the plant specific cell-cycle inhibitor ICK2/KRP2 (De Veylder et al. 2001), and immunofluorescence of KNOLLE to label the cell plates of dividing cells (Volker et al. 2001). Different BR mutant backgrounds and BL-treated plants were analyzed in the genetic background of these reporters. In the *bri1* mutants, the misexpression of CYB1;1, ICK2/KRP2, and immunofluorescence analysis of KNOLLE (Volker et al. 2001) revealed a reduced number of cell plates in the root meristem pointing to an arrested or slower cell cycle. Remarkably, the defects observed in *bri1* root apical meristems were recovered by overexpression of the CYCD3;1 gene (González-García et al. 2011), previously reported to increase cell proliferation (Riou-Khamlichi et al. 1999). These results together with the premature appearance of epidermal root hairs in *bri1* mutants indicated that BRI1 signaling is required to control not only cell elongation but also the balance between cell division and differentiation in the primary root.

Conversely, several lines of experimental evidence support the idea that increased BR signaling levels can lead to an acceleration of cell cycle during primary root growth: (1) a reduction of CYCB1;1 expression; (2) a reduction in the number of KNOLLE-labeled cell plates; (3) the reduced ICK2/KRP2 levels; (4) an increased number of cells that expressed QC identity markers, such as *pSCR:GFP* at the root stem cell niche; (5) an accelerated cellular differentiation revealed by the differentiation of root hairs in the proximal side of the meristem; and (6) an increased distal stem cell differentiation, characterized by deposition of starch granules in the former distal columella stem cells (CSC) (Figure 17.4).

These observations suggest that BRI1 signaling would be required for root growth by controlling the normal cell-cycle progression of meristematic cells including the almost nondividing QC cells (González-García et al. 2011). In agreement, a recent quantitative confocal microscopy study has estimated that BRI1 receptor density is significantly higher in cells that retain division activity compared to those cells of the QC (van Esse et al. 2011). Together, these root analyses open a new window to study how cells behave in response to BRs. Root growth dynamic studies using a combination of mathematical modeling

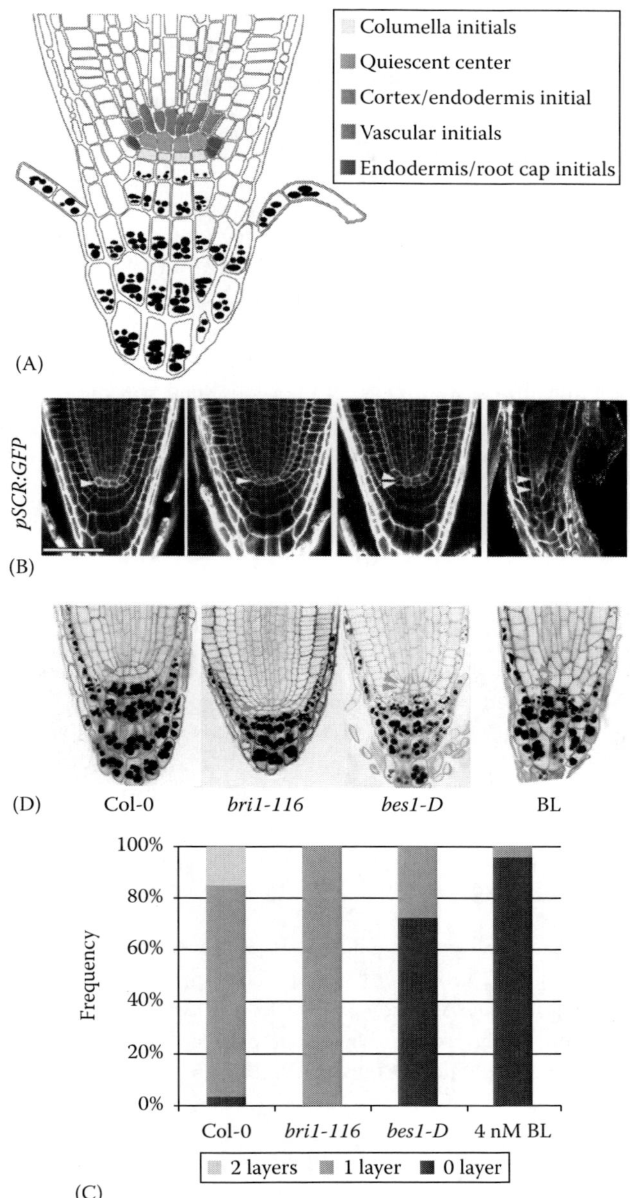

FIGURE 17.4 (See color insert.) BRs promote distal stem cell differentiation in *Arabidopsis* primary root. (A) Cartoon representing the root stem cell niche in the root apex (colored area). The quiescent center (QC) cells maintain the surrounding stem cells that give rise to all cell types present in the *Arabidopsis* root (van den Berg et al. 1997). Below the QC, the differentiated columella cells accumulate starch granules (black dots). (B) Six-day-old roots show the expression of the QC identity marker *pSCR:GFP* in the genetic background of several BR mutants. (C) Modified pseudo-Schiff PI staining (mPS-PI) of mutant roots reveals a duplication of the QC cells in both *bes1-D* (gain-of-function) and in the exogenous BL treatment (white and green arrows) and defects on the number of columella stem cell layers. (D) Quantification of the number of columella stem cell (CSC) layers in 6-day-old roots. *Bes1-D* mutants and BL-treated plants show an increased CSC differentiation; CSC layers disappear, and the frequency of CSC divisions in *bri1* mutants is uncoupled compared to that of the Col-O wild type. (From González-García, M.P. et al., *Development*, 138, 849, 2011. With permission.)

with a spatial/temporal analysis of BR action in root growth will be very helpful in understanding how BRs balance cell division, elongation, and differentiation in the plant.

V. Concluding Remarks

The last two decades of genetic and biochemical studies in the plant model system *Arabidopsis* have served to learn how plants perceive BRs and transduce the signals from the plasma membrane to the nuclei into transcriptional responses. It is currently clear that BRs can modulate distinct developmental processes in the plant by simultaneously acting on cell division, elongation, and differentiation, very likely in a cell-specific manner. The recent root analysis of BR mutants provides a starting point to investigate how BRs control plant growth and development. The unraveling of the action of BRI1-like members and the signal specificity of BR-mediated responses are challenges for the next decade.

Acknowledgments

We acknowledge M. Philips for comments on the manuscript. JV-B is funded by a PhD fellowship from the Generalitat de Catalunya. AIC-D is supported by the Marie Curie Initial Training Network "BRAVISSIMO" (grant no. PITN-GA-2008-215118). M.G.-G is funded by a grant from the Spanish Ministry of Education and Science (BIO2008/00505) to AIC-D.

References

Aguirrezabal L, Bouchier-Combaud S, Radziejwoski A, Dauzat M, Cookson SJ, Granier C. 2006. Plasticity to soil water deficit in *Arabidopsis thaliana*: Dissection of leaf development into underlying growth dynamic and cellular variables reveals invisible phenotypes. *Plant Cell Environ* 29:2216–2227.

Azpiroz R, Wu YW, Locascio JC, Feldmann KA. 1998. An *Arabidopsis* brassinosteroid-dependent mutant is blocked in cell elongation. *Plant Cell* 10:219–230.

Bajguz A, Hayat S. 2009. Effects of brassinosteroids on the plant responses to environmental stresses. *Plant Physiol Biochem* 47:1–8.

Bao F, Shen J, Brady SR, Muday GK, Asami T, Yang Z. 2004. Brassinosteroids interact with auxin to promote lateral root development in *Arabidopsis*. *Plant Physiol* 134:1624–1631.

Caño-Delgado A, Lee J-Y, Demura T. 2010. Regulatory mechanisms for specification and patterning of plant vascular tissues. *Annu Rev Cell Dev Biol* 26:605–637.

Caño-Delgado A, Yin Y, Yu C et al. 2004. BRL1 and BRL3 are novel brassinosteroid receptors that function in vascular differentiation in *Arabidopsis*. *Development* 131:5341–5351.

Ceserani T, Trofka A, Gandotra N, Nelson T. 2009. VH1/BRL2 receptor-like kinase interacts with vascular-specific adaptor proteins VIT and VIK to influence leaf venation. *Plant J* 57:1000–1014.

Choe S, Dilkes BP, Fujioka S, Takatsuto S, Sakurai A, Feldmann KA. 1998. The DWF4 gene of *Arabidopsis* encodes a cytochrome P450 that mediates multiple 22alpha-hydroxylation steps in brassinosteroid biosynthesis. *Plant Cell* 10:231–243.

Choe S, Fujioka S, Noguchi T, Takatsuto S, Yoshida S, Feldmann KA. 2001. Overexpression of DWARF4 in the brassinosteroid biosynthetic pathway results in increased vegetative growth and seed yield in *Arabidopsis*. *Plant J* 26:573–582.

Chory J, Nagpal P, Peto CA. 1991. Phenotypic and genetic analysis of det2, a new mutant that affects light-regulated seedling development in *Arabidopsis*. *Plant Cell* 3:445–459.

Clouse SD. 2011. Brassinosteroid signal transduction: From receptor kinase activation to transcriptional networks regulating plant development. *Plant Cell* 23:1219–1230.

Clouse SD, Langford M, Mcmorris TC. 1996. A brassinosteroid-insensitive mutant in *Arabidopsis thaliana* exhibits multiple defects in growth and development. *Plant Physiol* 111:671–678.

Colon-Carmona A, You R, Haimovitch-Gal T, Doerner P. 1999. Technical advance: Spatio-temporal analysis of mitotic activity with a labile cyclin-GUS fusion protein. *Plant J* 20:503–508.

De Veylder L, Beeckman T, Beemster GT et al. 2001. Functional analysis of cyclin-dependent kinase inhibitors of *Arabidopsis*. *Plant Cell* 13:1653–1668.

Domagalska MA, Schomburg FM, Amasino RM, Vierstra RD, Nagy F, Davis SJ. 2007. Attenuation of brassinosteroid signaling enhances FLC expression and delays flowering. *Development* 134:2841–2850.

Fabregas N, Ibanes M, Cano-Delgado AI. 2010. A systems biology approach to dissect the contribution of brassinosteroid and auxin hormones to vascular patterning in the shoot of *Arabidopsis thaliana*. *Plant Signal Behav* 5:903–906.

Friedrichsen DM, Joazeiro CA, Li J, Hunter T, Chory J. 2000. Brassinosteroid-insensitive-1 is a ubiquitously expressed leucine-rich repeat receptor serine/threonine kinase. *Plant Physiol* 123:1247–1256.

Fujioka S, Sakurai A. 1997. Brassinosteroids. *Nat Prod Rep* 14:1–10.

Gampala SS, Kim T-W, He J-X et al. 2007. An essential role for 14–3–3 proteins in brassinosteroid signal transduction in *Arabidopsis*. *Dev Cell* 13:177–189.

Geldner N, Hyman DL, Wang X, Schumacher K, Chory J. 2007. Endosomal signaling of plant steroid receptor kinase BRI1. *Genes Dev* 21:1598–1602.

Goda H, Shimada Y, Asami T, Fujioka S, Yoshida S. 2002. Microarray analysis of brassinosteroid-regulated genes in *Arabidopsis*. *Plant Physiol* 130:1319–1334.

Gonzalez N, De Bodt S, Sulpice R et al. 2010. Increased leaf size: Different means to an end. *Plant Physiol* 153:1261–1279.

González-García MP, Vilarrasa-Blasi J, Zhiponova M et al. 2011. Brassinosteroids control meristem size by promoting cell cycle progression in *Arabidopsis* roots. *Development* 138:849–859.

Grove MD, Spencer GF, Rohwedder WK, Mandava N et al. 1979. Brassinolide, a plant growth-promoting steroid isolated from *Brassica napus* pollen. *Nature* 281:216–217.

Gudesblat GE, Russinova E. 2011. Plants grow on brassinosteroids. *Curr Opin Plant Biol* 14:530–537.

Hacham Y, Holland N, Butterfield C et al. 2011. Brassinosteroid perception in the epidermis controls root meristem size. *Development* 138:839–848.

Hacham Y, Sela A, Friedlander L, Savaldi-Goldstein S. 2012. BRI1 activity in the root meristem involves post-transcriptional regulation of PIN auxin efflux carriers. *Plant Signal Behav* 7:68–70.

He JX, Gendron JM, Yang Y, Li J, Wang ZY. 2002. The GSK3-like kinase BIN2 phosphorylates and destabilizes BZR1, a positive regulator of the brassinosteroid signaling pathway in *Arabidopsis*. *Proc Natl Acad Sci USA* 99:10185–10190.

Hothorn M, Belkhadir Y, Dreux M et al. 2011. Structural basis of steroid hormone perception by the receptor kinase BRI1. *Nature* 474:467–471.

Ibañes M, Fàbregas N, Chory J, Caño-Delgado AI. 2009. Brassinosteroid signaling and auxin transport are required to establish the periodic pattern of *Arabidopsis* shoot vascular bundles. *Proc Natl Acad Sci USA* 106:13630–13635.

Irani N, Di Rubbo S, Mylle E et al. 2012. Fluorescently labeled castasterone reveals control of brassinosteroid signaling from the plasma membrane by clathrin- and ARF GEF-dependent endocytosis. *Nat Chem Biol* 8:583–589.

Jaillais Y, Belkhadir Y, Balsemao-Pires E, Dangl JL, Chory J. 2011a. Extracellular leucine-rich repeats as a platform for receptor/coreceptor complex formation. *Proc Natl Acad Sci USA* 108:8503–85037.

Jaillais Y, Hothorn M, Belkhadir Y, Dabi T, Nimchuk ZL, Meyerowitz EM, Chory J. 2011b. Tyrosine phosphorylation controls brassinosteroid receptor activation by triggering membrane release of its kinase inhibitor. *Genes Dev* 25:232–237.

Kauschmann A, Jessop A, Koncz C, Szekeres M, Willmitzer L, Altmann T. 1996. Genetic evidence for an essential role of brassinosteroids in plant development. *Plant J* 9:701–713.

Kim SK, Chang SC, Lee EJ, Chung WS, Kim YS, Hwang S, Lee JS. 2000. Involvement of brassinosteroids in the gravitropic response of primary root of maize. *Plant Physiol* 123:997–1004.

Kim SY, Kim BH, Lim CJ, Lim CO, Nam KH. 2010. Constitutive activation of stress-inducible genes in a brassinosteroid-insensitive 1 (bri1) mutant results in higher tolerance to cold. *Physiol Plant* 138:191–204.

Kim T-W, Wang Z-Y. 2010. Brassinosteroid signal transduction from receptor kinases to transcription factors. *Annu Rev Plant Biol* 61:681–704.

Kinoshita T, Cano-Delgado A, Seto H, Hiranuma S, Fujioka S, Yoshida S, Chory J. 2005. Binding of brassinosteroids to the extracellular domain of plant receptor kinase BRI1. *Nature* 433:167–171.

Kuppusamy KT, Chen AY, Nemhauser JL. 2009. Steroids are required for epidermal cell fate establishment in *Arabidopsis* roots. *Proc Natl Acad Sci USA* 106:8073–8076.

Li J. 2010. Regulation of the nuclear activities of brassinosteroid signaling. *Curr Opin Plant Biol* 13:540–547.

Li J, Chory J. 1997. A putative leucine-rich repeat receptor kinase involved in brassinosteroid signal transduction. *Cell* 90:929–938.

Li J, Nagpal P, Vitart V, Mcmorris TC, Chory J. 1996. A role for brassinosteroids in light-dependent development of *Arabidopsis*. *Science* 272:398–401.

Li J, Nam, KH, Vafeados, D Chory J. 2001. BIN2, a new brassinosteroid-insensitive locus in *Arabidopsis*. *Plant Physiol* 127:14–22.

Li J, Wen J, Lease KA, Doke JT, Tax FE, Walker JC. 2002. BAK1, an *Arabidopsis* LRR receptor-like protein kinase, interacts with BRI1 and modulates brassinosteroid signaling. *Cell* 110:213–222.

Li L, Xu J, Xu Z-H, Xue H-W. 2005. Brassinosteroids stimulate plant tropisms through modulation of polar auxin transport in *Brassica* and *Arabidopsis*. *Plant Cell* 17:2738–2753.

Li L, Yu X, Thompson A et al. 2009. *Arabidopsis* MYB30 is a direct target of BES1 and cooperates with BES1 to regulate brassinosteroid-induced gene expression. *Plant J* 58:275–286.

Miyazawa Y, Nakajima N, Abe T et al. 2003. Activation of cell proliferation by brassinolide application in tobacco BY-2 cells: Effects of brassinolide on cell multiplication, cell-cycle-related gene expression, and organellar DNA contents. *J Exp Bot* 54:2669–2678.

Mora-García S, Vert G, Yin Y, Caño-Delgado A, Cheong H, Chory J. 2004. Nuclear protein phosphatases with Kelch-repeat domains modulate the response to brassinosteroids in *Arabidopsis*. *Genes Dev* 18:448–460.

Mouchel CF, Osmont KS, Hardtke CS. 2006. BRX mediates feedback between brassinosteroid levels and auxin signalling in root growth. *Nature* 443:458–461.

Müssig C, Shin G-H, Altmann T. 2003. Brassinosteroids promote root growth in *Arabidopsis*. *Plant Physiol* 133:1261–1271.

Nam KH, Li J. 2002. BRI1/BAK1, a receptor kinase pair mediating brassinosteroid signaling. *Cell* 110:203–212.

Nemhauser JL, Hong F, Chory J. 2006. Different plant hormones regulate similar processes through largely nonoverlapping transcriptional responses. *Cell* 126:467–475.

Oh MH, Sun J, Oh DH, Zielinski RE, Clouse SD, Huber SC. 2011. Enhancing *Arabidopsis* leaf growth by engineering the BRASSINOSTEROID INSENSITIVE1 receptor kinase. *Plant Physiol* 157:120–131.

Ohashi-Ito K, Demura T, Fukuda H. 2002. Promotion of transcript accumulation of novel Zinnia immature xylem-specific HD-Zip III homeobox genes by brassinosteroids. *Plant Cell Physiol* 43:1146–1153.

Ohashi-Ito K, Fukuda H. 2003. HD-zip III homeobox genes that include a novel member, ZeHB-13 (*Zinnia*)/AtHB-15 (*Arabidopsis*), are involved in procambium and xylem cell differentiation. *Plant Cell Physiol* 44:1350–1358.

Peng P, Yan Z, Zhu,Y, Li J. 2008. Regulation of the *Arabidopsis* GSK3-like kinase BRASSINOSTEROID-INSENSITIVE 2 through proteasome-mediated protein degradation. *Mol Plant* 1:338–346.

Perez-Perez JM, Ponce MR, Micol JL. 2002. The UCU1 *Arabidopsis* gene encodes a SHAGGY/GSK3-like kinase required for cell expansion along the proximodistal axis. *Dev Biol* 242:161–173.

Riou-Khamlichi C, Huntley R, Jacqmard A, Murray JA. 1999. Cytokinin activation of *Arabidopsis* cell division through a D-type cyclin. *Science* 283:1541–1544.

Russinova E, Borst J-W, Kwaaitaal M, Caño-Delgado A, Yin Y, Chory J, De Vries SC. 2004. Heterodimerization and endocytosis of *Arabidopsis* brassinosteroid receptors BRI1 and AtSERK3 (BAK1). *Plant Cell* 16:3216–3229.

Ryu H, Kim K, Cho H, Park J, Choe S, Hwang I. 2007. Nucleocytoplasmic shuttling of BZR1 mediated by phosphorylation is essential in *Arabidopsis* brassinosteroid signaling. *Plant Cell* 19:2749–2762.

Savaldi-Goldstein S, Peto C, Chory J. 2007. The epidermis both drives and restricts plant shoot growth. *Nature* 446:199–202.

Scacchi E, Salinas P, Gujas B et al. 2010. Spatio-temporal sequence of cross-regulatory events in root meristem growth. *Proc Natl Acad Sci USA* 107:22734–2279.

Shiu S, Bleecker A. 2001. Receptor-like kinases from *Arabidopsis* form a monophyletic gene family related to animal receptor kinases. *Proc Natl Acad Sci USA* 98:10763–10768.

Steber CM, Mccourt P. 2001. A role for brassinosteroids in germination in *Arabidopsis*. *Plant Physiol* 125:763–769.

Sun Y, Fan X-Y, Cao D-M et al. 2010. Integration of brassinosteroid signal transduction with the transcription network for plant growth regulation in *Arabidopsis*. *Dev Cell* 19:765–777.

Szekere M, Németh K, Koncz-Kálmán Z et al. 1996. Brassinosteroids rescue the deficiency of CYP90, a cytochrome P450, controlling cell elongation and de-etiolation in *Arabidopsis*. *Cell* 85:171–182.

Tang W, Kim T-W, Oses-Prieto JA et al. 2008. BSKs mediate signal transduction from the receptor kinase BRI1 in *Arabidopsis*. *Science* 321:557–560.

Tang W, Yuan M, Wang R et al. 2011. PP2A activates brassinosteroid-responsive gene expression and plant growth by dephosphorylating BZR1. *Nat Cell Biol* 13:124–131.

Thummel CS, Chory J. 2002. Steroid signaling in plants and insects—Common themes, different pathways. *Genes Dev* 16:3113–3129.

Van Den Berg C, Willemsen V, Hendriks G, Weisbeek P, Scheres B. 1997. Short-range control of cell differentiation in the *Arabidopsis* root meristem. *Nature* 390:287–289.

Van Esse GW, Westphal AH, Surendran RP et al. 2011. Quantification of the brassinosteroid insensitive1 receptor in planta. *Plant Physiol* 156:1691–1700.

Vert G, Nemhauser JL, Geldner N, Hong F, Chory J. 2005. Molecular mechanisms of steroid hormone signaling in plants. *Annu Rev Cell Dev Biol* 21:177–201.

Villiers F, Jourdain A, Bastien O et al. 2012. Evidence for functional interaction between brassinosteroids and cadmium response in *Arabidopsis thaliana*. *J Exp Bot* 63:1185–1200.

Viotti C, Bubeck J, Stierhof Y-D et al. 2010. Endocytic and secretory traffic in *Arabidopsis* merge in the trans-Golgi network/early endosome, an independent and highly dynamic organelle. *Plant Cell* 22:1344–1357.

Volker A, Stierhof YD, Jurgens G. 2001. Cell cycle-independent expression of the *Arabidopsis* cytokinesis-specific syntaxin KNOLLE results in mistargeting to the plasma membrane and is not sufficient for cytokinesis. *J Cell Sci* 114:3001–3012.

Wang X, Li X, Meisenhelder J et al. 2005. Autoregulation and homodimerization are involved in the activation of the plant steroid receptor BRI1. *Dev Cell* 8:855–865.

Wang Z-Y, Nakano T, Gendron J et al. 2002. Nuclear-localized BZR1 mediates brassinosteroid-induced growth and feedback suppression of brassinosteroid biosynthesis. *Dev Cell* 2:505–513.

Xie L, Yang C, Wang X. 2011. Brassinosteroids can regulate cellulose biosynthesis by controlling the expression of CESA genes in *Arabidopsis*. *J Exp Bot* 62:4495–4506.

Yamamuro C, Ihara Y, Wu X et al. 2000. Loss of function of a rice brassinosteroid insensitive1 homolog prevents internode elongation and bending of the lamina joint. *Plant Cell* 12:1591–1606.

Ye Q, Zhu W, Li L, Zhang S, Yin Y, Ma H, Wang X. 2010. Brassinosteroids control male fertility by regulating the expression of key genes involved in *Arabidopsis* anther and pollen development. *Proc Natl Acad Sci USA* 107:6100–6105.

Yin Y, Vafeados D, Tao Y, Yoshida S, Asami T, Chory J. 2005. A new class of transcription factors mediates brassinosteroid-regulated gene expression in *Arabidopsis*. *Cell* 120:249–259.

Yin Y, Wang Z-Y, Mora-García S et al. 2002. BES1 accumulates in the nucleus in response to brassinosteroids to regulate gene expression and promote stem elongation. *Cell* 109:181–191.

Yoshimitsu Y, Tanaka K, Fukuda W et al. 2011. Transcription of DWARF4 plays a crucial role in auxin-regulated root elongation in addition to brassinosteroid homeostasis in *Arabidopsis thaliana*. *PLoS One* 6:e23851.

Yu X, Li L, Zola J et al. 2011. A brassinosteroid transcriptional network revealed by genome-wide identification of BESI target genes in *Arabidopsis thaliana*. *Plant J* 65:634–646.

Zurek DM, Rayle DL, Mcmorris TC, Clouse SD. 1994. Investigation of gene expression, growth kinetics, and wall extensibility during brassinosteroid-regulated stem elongation. *Plant Physiol* 104:505–513.

18

Role of Strigolactones in Root Development and Communication

Hinanit Koltai
Agricultural Research Organization

I. Introduction

Strigolactones (SLs) are carotenoid-derived terpenoid lactones (Matusova et al. 2005). They were identified 40 years ago as germination stimulants for the parasitic plants *Striga* and *Orobanche* (e.g., Cook et al. 1966, 1972; Yokota et al. 1998; Matusova et al. 2005; Akiyama and Hayashi 2006; Xie et al. 2007, 2008a,b, 2009; Goldwasser et al. 2008; Gomez-Roldan et al. 2008; recently reviewed by Xie et al. 2010) and later, as stimulants of hyphal branching of the symbiotic arbuscular mycorrhizal fungi (AMF; reviewed by Xie et al. 2010). More recently, SLs have been found to act as long-distance branching factors that suppress growth of preformed axillary buds (Gomez-Roldan et al. 2008; Umehara et al. 2008). Thus, they fit the characteristics of a novel branch-inhibiting signal acting as a long-distance signal transmitter (e.g., Gomez-Roldan et al. 2008; Umehara et al. 2008; Brewer et al. 2009; Ferguson and Beveridge 2009; reviewed by Dun et al. 2009). SLs or their derivatives have therefore been defined as a new group of plant hormones.

Several studies have suggested an additional role for SLs in plant development as regulators of root development, and this is the focus of this chapter. I will present evidence of the effect of SLs on root development and growth and of the suggested network of plant hormones that are interconnected with these effects. SL regulation of root development suggests that they are mediators of plant response to nutritional conditions and coordinators of root and shoot development. On the other hand, SLs function also as signal molecules that facilitate an ancient plant-symbiotic interaction by enhancing hyphal branching of the symbiotic AMF. SLs may have evolved as regulators of plant development and co-opted to be a signal for beneficial communication between plant roots and symbiotic organisms.

II. Strigolactones: Biosynthesis and Composition

SLs are suggested to be cleavage products of carotenoids or carotenoid-derived molecules via carotenoid-cleavage dioxygenase (CCD) enzyme activity (Umehara et al. 2008). Indeed, in several plant species, including *Arabidopsis thaliana*, rice (*Oryza sativa* L.), petunia (*Petunia hybrida* D. Don ex Louden), pea (*Pisum sativum* L.), tomato (*Solanum lycopersicum* L.), and chrysanthemum (*Dendranthema grandiflorum* (Ramat. Kitamura)) (e.g., Drummond et al. 2009; Liang et al. 2010; Vogel et al. 2010; reviewed by Dun et al. 2009; Leyser 2009), mutations in CCDs have been associated with a hyperbranching phenotype that exhibits reduced SL levels. These CCDs include CCD7, also identified as MAX3,

RMS5, and HTD1/D17 (Booker et al. 2004; Johnson et al. 2006; Zou et al. 2006), and CCD8, also identified as MAX4, RMS1, D10, and DAD1 (Sorefan et al. 2003; Bainbridge et al. 2005; Snowden et al. 2005; Arite et al. 2007). CCD7 and CCD8 are suggested to catalyze successive carotenoid-cleavage reactions (Schwartz et al. 2004; Auldridge et al. 2006). Former and later step in SL biosynthesis are suggested to be executed by DWARF27 (D27), a carotenoid isomerase and MAX1, a cytochrome P450 monooxygenase, respectively (Booker et al. 2004; Gomez-Roldan et al. 2008; Alder et al. 2012). Additional steps in SL synthesis are anticipated but still unknown (Matusova et al. 2005).

The presence of SLs has been demonstrated in several plant species. In each of these, a mixture of several SL compounds was found, whereas similar or even identical SL compounds were shown to exist in more than one plant species. All natural SLs contain a tricyclic lactone (ABC part) that connects via an enol ether bridge to a butenolide group (D ring) (Figure 18.1); they include one or more hydroxyl or acetyloxyl groups in the A/B ring moiety and one or two methyl groups on the A ring. 5-Deoxystrigol is thought to be the common precursor of these SLs (reviewed by Xie et al. 2010).

Plant species found to produce SLs include cotton (*Gossypium hirsutum* L.), sorghum (*Sorghum bicolor* (L.) Moench), maize (*Zea mays* L.), common millet (*Panicum miliaceum* L.), cowpea (*Vigna unguiculata* (L.) Walp.), red clover (*Trifolium pratense* L.), garden pea (*P. sativum* L.), rice (*O. sativa* cv. Nipponbare), tobacco (*Nicotiana tabacum* L.), tomato (*S. lycopersicum* L.) and *A. thaliana* (reviewed by Xie et al. 2010). In addition, the ratios of the different SLs vary with plant growth conditions, age, and cultivar (reviewed by Yoneyama et al. 2008, 2009).

SL biosynthesis has been shown to occur mainly in the plant roots. This was deduced from experiments performed with several plant species, which demonstrated that grafting of wild-type (WT) roots to an SL-synthesis-mutant shoot was sufficient to reverse the mutant phenotype to that of the WT. SL synthesis has also been suggested to occur in the lower part of the shoot. Interstock grafting of the WT lower-shoot part between mutant roots and shoot was sufficient to revert the SL-synthesis-mutant shoot phenotype to that of the WT (Napoli 1996; Foo et al. 2001; Booker et al. 2004; Gomez-Roldan et al. 2008; Umehara et al. 2008; Koltai et al. 2010b; reviewed by Dun et al. 2009). However,

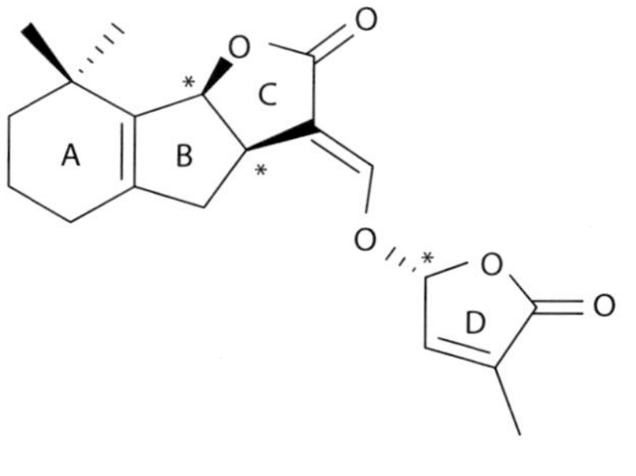

FIGURE 18.1 A schematic representation of the basic chemical structure of natural strigolactones (SLs). All natural SLs contain a tricyclic lactone (ABC part) that connects via an enol ether bridge to a butenolide group (D ring). For activity, stereochemistry at C* is important. (Illustrated by Dr. Cristina Prandi.)

to confer a significant reduction in shoot branching (Foo et al. 2001; Brewer et al. 2009; Ferguson and Beveridge 2009), SLs, their metabolites, or other unknown secondary messengers are suggested to move in the root-to-shoot direction (e.g., Napoli 1996; Beveridge et al. 2000; Koltai et al. 2010b; reviewed by Dun et al. 2009). The apparent presence of the SL orobanchol was detected in *Arabidopsis* xylem sap, indicating root-derived SL transport to the shoot (Kohlen et al. 2011).

The signal-transduction pathway of SLs is still mostly unknown. However, MAX2, an F-box protein, and DWARF14 (D14) have been suggested to be components of SL response and to function in ubiquitin-mediated degradation of as yet unknown target proteins (Stirnberg et al. 2007; Umehara et al. 2008; Waters et al. 2012).

III. Effects of Strigolactones on Shoot Development

The first indication of SLs' hormonal activity came from a study of shoot branching. Some years ago, a particular class of mutants was identified, which displayed an increase in bud outgrowth that could not be entirely explained by the presence or activity of cytokinin or auxin. Rather, it was shown that this class of mutants is defective in synthesizing or responding to a novel branch-inhibiting signal, SMS (named after mobile/cell phone text messages). SMS was suggested to act as a long-distance cue and to be involved in branching control. It was also suggested to be regulated by auxin and another, unknown, feedback signal (reviewed by Dun et al. 2009).

The cross talk between SLs and auxin was extensively examined in the shoot, giving rise to several models describing this association. SLs were suggested to be an auxin-promoted secondary messenger that moves up into the buds to repress their outgrowth (Brewer et al. 2009, Ferguson and Beveridge 2009, reviewed by Dun et al. 2009). Alternatively, restrained bud outgrowth was suggested to be a result of a SL-mediated reduction in the capacity of the main shoot for polar auxin transport from the apical meristem; this reduction was suggested to lead to inhibition of polar auxin transport from the buds and thereby to restrain bud outgrowth (e.g., Bennett et al. 2006; Mouchel and Leyser 2007; Ongaro and Leyser 2008; Crawford et al. 2010; Leyser 2010). Finally, for the coordinated control of axillary branching, both auxin and SLs were suggested to have a dynamic feedback loop (Hayward et al. 2009). Apically derived auxin was shown to induce SL synthesis in the root via the AXR1/TIR1 signal-transduction pathway, involving stability of the Aux/IAA protein, IAA12, and induction of MAX3 and MAX4 expression (reviewed by Beveridge and Kyozuka 2010). A direct feedback effect of SL response on its own biosynthesis was demonstrated in rice roots: D10 (MAX4 homolog) expression was upregulated in SL pathway mutants, whereas SL application led to restoration of its expression to wild-type levels in the biosynthesis mutant but not the response mutant (Umehara et al. 2008).

Other aboveground roles suggested for SLs included positive regulation of light harvesting (Mayzlish-Gati et al. 2010) and mimicking of light-adapted seedling growth (Tsuchiya et al. 2010). Additional information on SL regulation of shoot development can be found in, for example, Beveridge and Kyozuka (2010) and Leyser (2009). This chapter, however, will focus on the recently identified role of SLs in root development and communication.

IV. Strigolactones Regulate Root Development

SLs have been recently shown to play a role in the development of plant roots, as well as shoots. Their regulating function was shown to affect both root architecture and root-hair elongation (Figure 18.2) and to be dependent on growth conditions.

A. Primary Root Formation

Under conditions of carbohydrate limitation that usually lead to a reduction in primary root (PR) length (Jain et al. 2007), the PR of *Arabidopsis* mutants flawed in either SL response (*max2*) or synthesis (*max1* and *max4*) was shorter than that of the WT, suggesting that SLs are positive regulators of PR length. Supplementation of GR24 (a bioactive synthetic SL; Johnson et al. 1981) under these conditions extended the PR length of the WT and SL-deficient mutants, but not of the SL-response mutant, suggesting that the positive effect of SLs on PR length is mediated via MAX2 signaling (Ruyter-Spira et al. 2011).

However, under conditions of sufficient sugar, only relatively lower concentrations of GR24 led to a MAX2-dependent increase in PR length and then only in 8-day-old seedlings; the opposite effect, inhibition of PR length, was recorded for treatments with relatively higher doses of GR24, in a MAX2-independent fashion (Ruyter-Spira et al. 2011). The reduction of PR length in the *max* mutants was accompanied by a reduction in the number of PR meristem cells, which could be rescued by application of GR24 in SL-deficient, but not SL-insensitive mutants. Together, the results suggested that SLs are positive regulators of PR elongation, dependent on growth conditions (Ruyter-Spira et al. 2011). Similarly, in rice, mutants that are SL-deficient or SL-insensitive exhibited a shorter crown root phenotype, which could be restored by GR24 treatments to that of the WT in the SL-deficient but not in the SL-insensitive mutants (Arite et al. 2012).

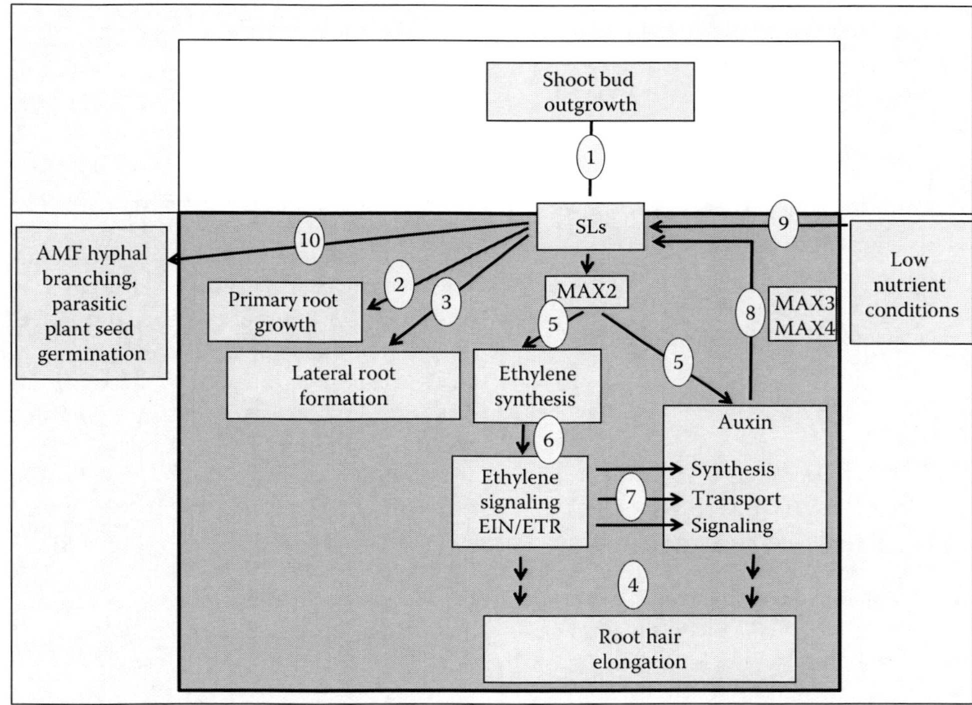

FIGURE 18.2 A schematic representation of strigolactones' (SLs) regulation of root development and communication and the hormonal network in which SLs are involved. In the shoot (represented by white section), SLs repress bud outgrowth (1), whereas in the root (gray section), they are regulators of primary root growth (2), lateral root formation (3), and root-hair elongation (4). The latter is effected via suggested cross talk of SLs with ethylene and auxin: SL signaling, conveyed via MAX2 activity, leads to induction of auxin signaling and/or ethylene synthesis (5), which in turn leads to ethylene signaling, conveyed via EIN and ETR proteins (6). Ethylene signaling positively affects auxin synthesis, transport, and signaling (7). Auxin and ethylene signaling, both individually and together, lead to root-hair elongation (4), while auxin positively regulates SL synthesis by induction of MAX3 and MAX4 expression (8). Low-nutrient growth conditions induce SL production and secretion (9), and SL secretion, in turn, enhances hyphal branching of arbuscular mycorrhizal fungi (AMF) and germination of parasitic plant seeds (10).

B. Lateral Root Formation

Under conditions of sufficient phosphate (Pi), SLs were shown to negatively regulate lateral root (LR) formation in *Arabidopsis*. First, mutants deficient in SL biosynthesis (i.e., *max3* and *max4*) and response (i.e., *max2*) showed an increased number of LRs in comparison with the WT (Kapulnik et al. 2011a; Ruyter-Spira et al. 2011). Second, treatments of seedlings with GR24 under sufficient Pi conditions led to a reduction in LR number in the WT and in the SL-synthesis mutants, but not in the SL-response mutant (Kapulnik et al. 2011a; Ruyter-Spira et al. 2011). However, under Pi-limiting conditions, GR24 application increased LR formation in the WT and in the SL-synthesis mutants, but not in the SL-response mutant (Ruyter-Spira et al. 2011). These results suggested that SLs are mediators of LR formation, with a negative effect under high-Pi conditions and a positive effect under low-Pi conditions, in both cases via the MAX2-dependent signaling (Kapulnik et al. 2011a; Ruyter-Spira et al. 2011).

C. Root-Hair Elongation

An increase in root-hair (RH) length of WT and SL-deficient mutants, but not of the SL-insensitive mutant *max2*, was recorded with GR24 treatments as early as 24 h following GR24 application. The *max2* mutants remained resistant to GR24 in terms of RH elongation even when relatively high concentrations of GR24 were applied. Hence, SLs were suggested to have a positive effect on RH length, which is mediated via MAX2 (Kapulnik et al. 2011a).

The dependence of the ability of SLs to regulate PR and LR development on sucrose and Pi conditions, respectively, as well as RH elongation, raises the possibility that SLs are regulators of plant responses to adverse growth conditions. Moreover, phytohormones other than SLs regulate PR, LR, and RH development, suggesting the existence of a network consisting of SLs and other plant hormones that act to regulate root development. These issues are discussed later.

V. Strigolactones Interact with Other Plant Hormones in the Regulation of Root Development

As mentioned, SLs and auxin are thought to interact in the shoot. SLs were suggested to be either effectors of auxin canalization or secondary messengers of auxin, thereby repressing shoot-bud outgrowth. In the root, however, it has been suggested that SLs, auxin, and ethylene are involved in intimate cross talks (Figure 18.2).

A. Strigolactone–Auxin Cross Talk

Similar to the situation suggested in the shoot, it may be that SLs are involved in RH elongation via modulation of auxin flux in the roots. Evidence for this notion includes the SLs' suggested interference with the activity of auxin-efflux carriers in the presence of exogenously applied auxin, thereby affecting PR growth, cell elongation, and RH elongation in tomato roots (Koltai et al. 2010a). In addition, upon GR24 treatment, PIN1-, PIN3-, and PIN7-GFP intensities in the primary vascular tissue of the PR tip decreased (Ruyter-Spira et al. 2011). Moreover, RH elongation was shown to be a result of increased auxin levels in the epidermal cells (e.g., Pitts et al. 1998), which was dependent on auxin transport. AUX1-dependent auxin transport through non-hair epidermal cells is necessary for RH elongation (Jones et al. 2009). Hence, the positive effect of SLs on RH elongation (Kapulnik et al. 2011a) might be mediated through modulation of auxin flux.

The effect of SLs on LR formation may also be explained, to a certain extent, by SLs' postulated ability to modulate auxin flux. LR formation is dependent, at least in part, on PIN1 auxin-efflux activity, and is suggested to be regulated by, among other things, local auxin levels, sensitivity, or both (reviewed by Péret et al. 2009). Indeed, in correlation with its negative regulation of LR formation, GR24 treatment resulted in decreased PIN1-GFP intensities, suggesting PIN1's involvement in the GR24-mediated reduction of LR formation. However, when auxin levels were increased by exogenous application, no reduction was seen in PIN1-GFP intensities upon GR24 application. Under these conditions, GR24 application increased, rather than reduced, LR formation (Ruyter-Spira et al. 2011).

It could be that by being modulators of auxin flux, SLs change the auxin concentrations present locally for LR formation since under relatively low auxin levels, SLs reduce auxin flux to the root, leading to inhibition of LR formation. Whereas under high auxin levels, this SL-mediated reduction of auxin flux allows generation of auxin optima and induction of LR formation (Ruyter-Spira et al. 2011). Another possible explanation for the effect of SLs on LR formation may be associated with modulation of auxin sensitivity. This is because under low-Pi conditions, increased auxin sensitivity, mediated via the TIR1 auxin receptor, is suggested to lead to changes in root-system architecture (Pérez-Torres et al. 2008). Accordingly, auxin signaling was shown to be involved in the RH elongation induced by SL: Under low GR24 concentrations, the auxin receptor mutant *tir1-1* (Dharmasiri et al. 2005) was less responsive to SLs than the WT (Kapulnik et al. 2011b). However, SL signaling was shown not to be involved in RH elongation induced by auxin: the SL-insensitive mutant *max2* was as responsive to auxin as the WT (Kapulnik et al. 2011b). Also, it was suggested that GR24-dependent reduction of apical auxin production might contribute to the SLs' ability to suppress LR production (Ruyter-Spira et al. 2011).

B. Strigolactone–Ethylene Cross Talk

A cross talk has been demonstrated between SLs and ethylene, with respect to the RH-elongation response. Although the SL-response mutant was responsive to an ethylene precursor, suggesting that SL signaling is not necessary for the ethylene response, ethylene was suggested to be involved in the SL response based on three lines of evidence. One, the ethylene-signaling mutants *etr* and *ein* (Stepanova and Alonso 2009) showed markedly reduced RH elongation upon GR24 treatment, suggesting the involvement of ethylene signaling in the SL response (Kapulnik et al. 2011b). Two, blockage of ethylene biosynthesis by the ethylene-synthesis inhibitor 2-aminoethoxyvinylglycine (AVG) eliminated SLs' effect on RH elongation, suggesting that the SL response required ethylene biosynthesis. Three, GR24 treatments were shown to elevate *At-acs2* transcription (Kapulnik et al. 2011b), *acs* genes encode key rate-determining enzymes in ethylene synthesis, and their transcription is correlated to the level of ethylene synthesis (e.g., Yamamoto et al. 1995; Barry et al. 2000; Yamagami et al. 2003). Taken together, SLs are suggested to induce ethylene biosynthesis, and this induction leads to activation of ethylene signaling which is necessary (although perhaps not exclusively) for the RH response to SLs. This implies that ethylene and SLs are in the same hormonal pathway regulating RH elongation, where ethylene is epistatic to SLs (Kapulnik et al. 2011b; Koltai 2011).

Moreover, ethylene was shown to regulate auxin synthesis and transport, including auxin efflux and PIN protein expression. It has also been shown to modulate auxin response and distribution in roots (e.g., Ruzicka et al. 2007; Swarup et al. 2007; Ivanchenko et al. 2008; Negi et al. 2008, 2010; reviewed by Yoo et al. 2009). Since ethylene was shown to be epistatic to SLs and its biosynthesis to be essential for the RH response to SLs (Kapulnik et al. 2011b), the SL and auxin pathways may also converge through that of ethylene for SL-mediated regulation of RH elongation (Kapulnik et al. 2011b; Koltai 2011).

VI. Strigolactones May Be Mediators of Root Responses to Nutritional Status

RHs are directly associated with the plant's ability to absorb nutrients from the soil (reviewed by Gilroy and Jones 2000). Moreover, formation and elongation of RHs increase under conditions of nutrient deficiency, including Pi, nitrogen (N), and iron (Fe) (reviewed by López-Bucio et al. 2003; Muller and Schmidt 2004). In addition, the development of primary and lateral roots has been shown to be affected by Pi levels (e.g., López-Bucio et al. 2002) and to be important for plant adaptive growth (reviewed by Hell and Hillebrand 2001; Osmont et al. 2007). LR initiation and PR elongation are also largely affected by levels of N, Fe, aluminum (Al), calcium (Ca), and sulfur (S) (reviewed by Osmont et al. 2007; López-Bucio et al. 2003).

Phytohormones have been shown to be involved in some of the alterations in RH, PR, and LR development induced by nutritional status. For example, RH elongation in response to low Fe required both ethylene and auxin signaling (reviewed by López-Bucio et al. 2003; Rubio et al. 2009). Under low-Pi conditions, the changes in LR formation in *Arabidopsis* were also suggested to be a result of increased auxin sensitivity, mediated via the TIR1 auxin receptor (Pérez-Torres et al. 2008).

On the other hand, hormone synthesis has been shown to be affected by nutrient status. For example, higher levels of auxin accumulated in the roots of plants grown under low nitrate conditions vs. high nitrate conditions (Caba et al. 2000; Walch-Liu et al. 2006).

Along the same lines, the ability to affect RH elongation, as well as LR and PR development, may position SLs as mediators of the root response to low-nutrient conditions. SLs positively regulate RH elongation (Kapulnik et al. 2011a), and RHs may enhance nutrient absorption by the plant. Under low-Pi conditions, SLs enhance LR formation (Ruyter-Spira et al. 2011), and increased LR formation may be important for plant growth under those conditions (Al-Ghazi et al. 2003; reviewed by Osmont et al. 2007). Furthermore, low-Pi and low-N conditions induced SL production (López-Ráez and Bouwmeester 2008; López-Ráez et al. 2008; Kohlen et al. 2011) and secretion (Yoneyama et al. 2007a,b) in plants. Morever, SLs were shown to be necessary for seedling response to low-Pi conditions (Mayzlish-Gati et al. 2012).

Taken together, these data indicate that SLs, like other plant hormones, are regulators of the plant root response to low-nutrient conditions, while their synthesis is regulated by the plant's nutritional status. This nutritional regulation may lead to the changes in root development and thus to a response to growth conditions.

VII. Strigolactones May Be Coordinators of Root-to-Shoot Development in Response to Pi Starvation

Plant growth is highly regulated throughout development: a global growth coordination of the different organs is needed for optimal plant function. For example, the production of photosynthetic assimilation products in leaves has to be fine-tuned to the availability of water and nutrients supplied by the root, necessitating linking between shoot and root growth (Paul and Foyer 2001).

Pi starvation leads to enhanced root-to-shoot ratio (Drew 1975; Ericsson 1995): the number of shoot branches is decreased (Troughton 1977), whereas in the root, the outgrowth of LRs is increased and that of the PR is decreased under these conditions (Williamson et al. 2001).

In rice and *Arabidopsis*, the reduction in bud outgrowth under low Pi concentrations in the media was associated with an increase in SL levels in the roots (Umehara et al. 2010; Kohlen et al. 2011). These results suggest that SLs are regulators of shoot architecture in response to Pi status. Since SLs are also suggested to regulate plant root response to low Pi conditions (Mayzlish-Gati et al. 2012; Ruyter-Spira et al. 2011), they could potentially be modulators of the coordinated development of roots and shoots in response to Pi starvation.

VIII. Strigolactones Are Signaling Molecules of Beneficial Plant Communication

The first indication of SLs' biological role, long before their identification as plant hormones, was their stimulation of seed germination of the parasitic plants *Striga* and *Orobanche* (e.g., Cook et al. 1972; Yokota et al. 1998; Matusova et al. 2005; Akiyama and Hayashi 2006; Xie et al. 2007, 2008a,b, 2009; Goldwasser et al. 2008; Gomez-Roldan et al. 2008). SLs were also identified as stimulants of hyphal branching of the symbiotic AMF (e.g., Akiyama et al. 2005; Besserer et al. 2006, 2008; Gomez-Roldan et al. 2008; Yoneyama et al. 2008). Thus, at least two plant–fungus and plant–plant associations evolved to include SLs as signals for host recognition. The role of SLs as signal molecules of parasitic weed–plant interactions is presented in several recent reviews (e.g., Xie et al. 2010) and will not be discussed here. The role of SLs in the beneficial plant–fungus association is presented in the following.

The arbuscular mycorrhizal (AM) symbiosis is an association between soil AMF and roots of higher plants. Under suitable conditions, symbiotic associations are formed with most terrestrial vascular flowering plants (Smith and Read 2008) and are therefore thought to be the most prevalent and ancient symbiosis on earth (Remy et al. 1994). This symbiosis was suggested to have played a crucial role in the plants' success in colonizing the land (Simon et al. 1993).

The AMF–host association begins with a "presymbiotic stage," in which the fungal spore germinates in the soil and the hyphae develop but exhibit limited growth. In the presence of a host root, the hypha starts to branch (e.g., Gianinazzi-Pearson et al. 1989; Giovannetti et al. 1996; Buée et al. 2000; Nagahashi and Douds 2000). Once a contact is formed between the hypha and a host root epidermis, the fungus forms an appressorium, and a prepenetration apparatus is formed by the host root (Siciliano et al. 2007).

In the symbiotic stage, the fungus penetrates the root epidermis and grows within the root cortex. However, despite penetration through the cell wall, the arbuscular hyphae remain in the apoplastic compartment of the root cortex, surrounded by an interfacial matrix and the periarbuscular membrane. The fungi form morphologically distinct, specialized structures within the root, consisting of arbuscules in the *Arum*-type association, or extensive intracellular coiled hyphae in the *Paris*-type association (Smith and Read 2008). Completing the fungal life cycle, the fungus produces an extraradical mycelium from which spores are eventually formed (Smith and Read 2008).

Both partners of the AM symbiosis benefit from the association, bidirectional exchange between the plant and fungus is suggested to take place mainly through the arbuscules (reviewed by Bucher 2007). AMF enhance the plant's ability to bridge the "depletion zone" established in the soil once mineral concentrations decrease in close proximity to root surfaces. Thus, AMF permeate a greater soil volume around the root and consequently increase the amount of nutrients available to the plant. The availability and uptake of Pi was shown to be promoted by the AM symbiosis, via both the greater root absorbing surface provided by the hyphae and active import, accumulation and storage of Pi in the fungus vacuoles (e.g., Harrison and van Buuren 1995; Maldonado-Mendoza et al. 2001; Smith et al. 2003; Benedetto et al. 2005; Bucher 2007).

In return, the fungus receives fixed carbon from the host. Between 4% and 20% of the plant's total photosynthetic products have been suggested to be provided to the fungi during the AM association, and fungal carbon demand has been shown to affect carbon partitioning in the plant (reviewed by Bago et al. 2000).

Despite spontaneous spore germination and limited development of the fungal hyphae in the absence of a host (reviewed by Giovannetti et al. 2010), AM hyphae undergo extensive branching only in the presence of host roots, and hyphal branching was shown to be induced by root exudates (e.g., Gianinazzi-Pearson et al. 1989; Giovannetti et al. 1996; Buée et al. 2000; Nagahashi and Douds 2000). These branching factors were first identified as a group of lipophilic compounds (Nagahashi and Douds 2000) and later specified to be SLs. Branching of the AMF *Gigaspora margarita* W. N. Becker and I. R. Hall, was shown to be induced by SLs purified from root exudates of *Lotus japonicus* (Regel) K. Larsen (Akiyama et al. 2005; Akiyama and Hayashi 2006), and the synthetic SL GR24 was shown to effectively induce AMF branching at 10^{-8} M (Gomez-Roldan et al. 2008). In addition, root exudates of SL-deficient mutants of pea and tomato showed a reduced level of AMF hyphal branching compared to the WT (e.g., Gomez-Roldan et al. 2008; Koltai et al. 2010b). Finally, SLs were shown to induce mitochondrial activation, energy metabolism, and mitosis in AMF (Besserer et al. 2006, 2008).

The exchange of signals between the host and fungi is suggested to promote successful mycorrhization, including the promotion of AMF hyphal branching in response to root exudates (Giovannetti et al. 1996; Buée et al. 2000). AMF hyphal branching in response to SLs thus positions these compounds as important signaling molecules for this association. However, SLs are probably not essential for the AM symbiosis, because both spore germination and some degree of hyphal development and branching occur even without the host presence or its root exudates (reviewed by Giovannetti et al. 2010). These spontaneous events may provide the opportunity for encounters between the developing hyphae and the host root, even in the absence of SLs.

In addition, it is still unclear whether SLs determine host specificity. On the one hand, root exudates of non-AMF-host plants were shown to be unable to induce AMF hyphal branching (Buée et al. 2000; Nagahashi and Douds 2000), suggesting a role for root exudates in the determination of interaction specificity between the plant host and the AMF. On the other hand, similar SLs were identified in root exudates of AMF–host plants and in *A. thaliana* (Brassicaceae) and lupin (Fabaceae), two non-AMF–host plants (Goldwasser et al. 2008; Yoneyama et al. 2008). With regard to the low-specificity nature of the AM association, species- and even cultivar-specific combinations of SLs were identified (Yoneyama et al. 2008; reviewed by Xie et al. 2010), arguing against a role for SLs in determining host specificity.

Interestingly, mycorrhized plants showed reduced susceptibility to *Striga* or *Orobanche* due to a lower level of induction of parasitic seed germination, suggesting they produce and/or exude less SLs (e.g., Lendzemo et al. 2009; Fernández-Aparicio et al. 2010). Accordingly, SL production was shown to be significantly reduced upon AM symbiosis in tomato (López-Ráez et al. 2011). It could be that AMF and SLs are associated by a negative feedback loop, that is, SLs enhance AMF colonization, and AMF colonization then leads to a reduction in SL synthesis or exudation.

IX. Are Strigolactones Actively Secreted by Roots?

Passive SL dispersal, that is, via diffusion or leakage from the plant root, might support a main role for SLs in plant morphogenesis, while the AMF and parasitic plants, interacting with the host plant, simply exploited the leaking substance from the roots for selection of their host. Active secretion from the roots by a controlled transport mechanism might support a main role for SLs in the establishment of symbioses.

Several studies have suggested higher secretion/production of SLs when both mycotrophic and non-mycotrophic plants were exposed to low-Pi conditions (Yoneyama et al. 2007a,b; López-Ráez and Bouwmeester 2008; López-Ráez et al. 2008; Kohlen et al. 2011). Under these conditions, both SL production and secretion were suggested to increase (Yoneyama et al. 2007a). However, root exudation of amino acids and reducing sugars also increases in plants under conditions of Pi deficiency. Changes in the membrane permeability of Pi-deficient roots were suggested to be the cause for the increase in root exudation, rather than changes in the content of these compounds in the root (Graham et al. 1981).

Hence, it is possible that instead of there being a well-controlled and specific mechanism for SL secretion, increased secretion of SLs under low-Pi conditions is a result of both a higher level of SL biosynthesis by the plant and higher membrane permeability of the root cells under these conditions. However, recently a SLs - specific ABC transporter (PDRI) was identified in petunia, and it was suggested that by functioning as a cellular transporter it has a role in both shoot branching and mycorrhiza association (Kretzschmar et al. 2012).

X. Strigolactones' Primary Biological Role

SLs play a role in both beneficial AMF–plant communication and plant development. However, it is not yet clear whether SLs were first produced by the plant to control organogenesis or to promote symbiosis.

SLs appear to have a pivotal role in determining the architecture of both root and shoot. This role may be crucial for plant survival and productivity. First, SL regulation of shoot branching may enable the plant to respond to, for example, physical damage or changing light conditions (reviewed by Leyser 2009 and by Waldie et al. 2010). Secondly, the effect of SLs on roots is suggested to be associated with the plant ability to respond to different nutritional levels, and, thirdly, SLs may be modulators of the coordinated development of root and shoot in response to Pi starvation. Such pivotal roles for SLs may have been the major force in their evolution, suggesting that the primary role of SLs is in plant morphogenesis, rather than in enhancement of beneficial plant interactions.

Moreover, SL-biosynthesis enzymes were identified in ferns (*Selaginella moellendorfii* Hieron.), bryophytes (*Physcomitrella patens* (Hedw.) Bruch & Schimp.) the Charales and the liverwort (*Marchantia polymorpha* L., Arite et al. 2007; Rensing et al. 2008; Delaux et al. 2012); in *P. patens* and the Charales, SLs were shown to be produced and released into the medium (Proust et al. 2011; Delaux et al. 2012). These results suggest that SLs were produced by primitive plants very early in evolution. These primitive plants possess characters significantly different from higher plants, for example, they lack seeds, roots, and axillary meristems (Bowman et al. 2007). Therefore, SLs probably predate the development of these plant organs and might have primarily emerged in plants to control other ancient biological processes (e.g., cell division or elongation).

A clue as to the primitive role of SLs comes from Proust et al. (2011) study in *P. patens*. SLs were shown in *P. patens* to control branching of protonemal filaments and colony extension. Moreover, the presence of neighboring colonies, which in wild-type plants causes the arrest of colony extension, was not detected by Ppccd8 mutant colonies, leading to their continuous expansion. These results suggest a role for SLs in bryophytes as signaling factors controlling development; it is suggested that they may be reminiscent of bacterial-quorum-sensing molecules, evolved in bryophytes to be used for communication between colonies to control coordinated colony growth (Proust et al. 2011).

On the other hand, the involvement of SLs in AMF symbiosis may suggest an ancient role as facilitators of this plant–fungus interaction. The AM symbiosis is thought to date as far back as 400 million years, as evident by the discovery of an arbuscular fossil in *Aglaophyton major* D. S. Edwards (Remy et al. 1994), a Devonian-period plant. Moreover, Ordovician rocks in Wisconsin dating back 460 million years were found to contain fossilized fungal hyphae and spores that strongly resemble those of the modern genus *Glomus*. These fossilized fungi might have been free living, but they may also have formed mycorrhizal-like relationships with the bryophytes (Redecker et al. 2000). It has been suggested that AM symbiosis was crucial for the colonization and life habit of plant life on land (Pirozynski and Malloch 1975). This might designate the AM symbiotic association as a crucial step in evolution and SLs as its promoters.

However, comparative analysis of the CCD7, CCD8, and MAX2 protein sequences between multiple plant species (Plaza 2.0; http://bioinformatics.psb.ugent.be/plaza/) identified the presence of CCD7 and CCD8 homologues not only in land plants (embryophytes, including bryophytes and vascular plants) but also in green algae of the chlorophytes (Figure 18.3; Chlorophytes descend from green algae-like ancestors; Lewis and McCourt 2004). Also, the Charales, the freshwater green algal closely related to land plants, produce and exude strigolactones (Delaux et al. 2012). Therefore, the ability to biosynthesize SL is present also in aquatic plants, and thus, in the green algae-like ancestor of both embryophytes and chlorophytes, predating AM symbiosis.

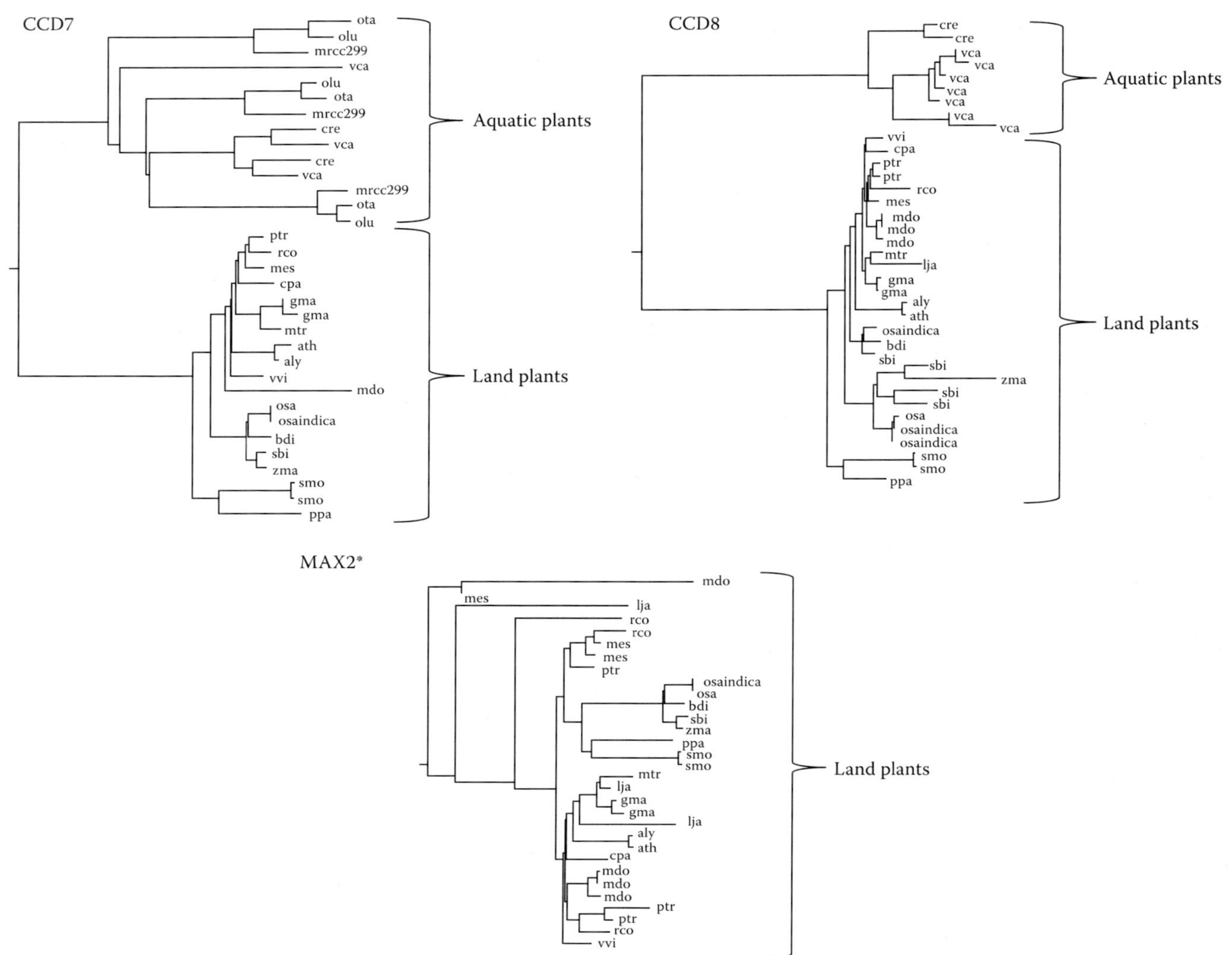

FIGURE 18.3 Phylogenetic trees (phylograms) of homologous gene family of CCD7 (AT2G44990 in *Arabidopsis thaliana*), CCD8 (AT4G32810 in *A. thaliana*), and MAX2 (AT2G42620 in *A. thaliana*) in land and aquatic plant species, from genome sequences available at http://bioinformatics.psb.ugent.be/plaza/ (Proost et al. 2009). Abbreviations: *Chlamydomonas reinhardtii* (cre), *Volvox carteri* (vca), *Micromonas* sp. RCC299 (mrcc299), *Ostreococcus tauri* (ota), *Ostreococcus lucimarinus* (olu), *Physcomitrella patens* (ppa), *Selaginella moellendorffii* (smo), *Zea mays* (zma), *Sorghum bicolor* (sbi), *Brachypodium distachyon* (bdi), *Oryza sativa* ssp. *Indica* (osaindica), *Oryza sativa* ssp. *Japonica* (osa), *Vitis vinifera* (vvi), *Carica papaya* (cpa), *Arabidopsis lyrata* (aly), *Arabidopsis thaliana* (ath), *Populus trichocarpa* (ptr), *Ricinus communis* (rco), *Manihot esculenta* (mes), *Malus domestica* (mdo), *Glycine max* (gma), *Medicago truncatula* (mtr), *Lotus japonicus* (lja). The phylogenetic trees of the gene families were created using PhyML (a maximum-likelihood method for phylogenetic inference). * For MAX2, the stripped multiple sequence alignment used to infer the phylogenetic tree is only 94 amino acids long.

Interestingly, MAX2, putatively involved in SLs response seems not to be present in chlorophytes (Figure 18.3). The presence of additional yet unknown components of SLs response and additional, multicellular chlorophyte genomes (when available) might need to be examined. However, these results may suggest that although the ability to synthesize SLs was present in land plants' ancestors, a pathway of SL response was developed only with the establishment of terrestrial plant life habit (and is therefore not present in chlorophytes).

One of the key questions in understanding the adaptation of aquatic green plants to life on land is how programs for development and growth were changed to allow development from apical meristem, and the patterning and production of tissues (Bowman et al. 2007). As mentioned earlier, the SLs' response is associated with the development and architecture of both shoot and root in plant lands. It might be that the evolution of the SL response, which might be absent in aquatic ancestors but present in land plants, has played a role in the process of coordinated development of tissues, to allow terrestrial life habit.

The involvement of SLs in plant parasitism, however, probably provides an example of plant compounds dedicated to a particular function (i.e., plant development or symbiosis establishment) that might be exploited by another organism to serve an alternative function (i.e., parasitic plant seed germination). Another explanation for the exploitation of SLs for plant parasitism is that SLs' ability to induce parasitic plant seed germination was derived from a preexisting mechanism of self-regulation of plant germination by SLs (reviewed by Westwood et al. 2010). This suggestion is supported by the findings of a positive role for SLs in *Arabidopsis* seed germination (Tsuchiya et al. 2010; Nelson et al. 2011).

XI. Conclusions and Perspective

The recently identified roles for SLs as regulators of plant development position them as newly identified players in a long-known game, which is governed by the activity of several different phytohormones. It is not surprising, therefore, to find SLs in an intimate relationship with the other plant hormones that regulate development. Moreover, SLs' putative roles as modulators of plant responses to nutritional conditions, and as coordinators of shoot and root development, suggest a pivotal role for SLs in plant growth and development, which might have played a part in the adaptation of plants to terrestrial life habit.

However, curiously, both SL-deficient and SL-insensitive mutants are able to grow, develop, and reproduce, at least under controlled conditions as usually applied in laboratories or experimental fields. Moreover, these mutants may have some evolutionary advantage in not being infected by parasitic plants. Nevertheless, SLs have been detected in a diverse array of land plant species, suggesting they have an important role. It might be that SLs are involved in the fine-tuning of plant development, such as coordination of root and shoot development under conditions of Pi deficiency. This fine-tuning of plant development may be common to all or many plant species and may be one of the mechanisms that allow plants to respond to their environment.

References

Akiyama K, Hayashi H. 2006. Strigolactones: Chemical signals for fungal symbionts and parasitic weeds in plant roots. *Ann Bot* 97:925–931.

Akiyama K, Matsuzaki K-I, Hayashi H. 2005. Plant sesquiterpenes induce hyphal branching in arbuscular mycorrhizal fungi. *Nature* 435:824–827.

Al-Ghazi Y, Muller B, Pinloche S et al. 2003. Temporal responses of *Arabidopsis* root architecture to phosphate starvation: Evidence for the involvement of auxin signalling. *Plant Cell Environ* 26:1053–1066.

Alder A, Jamil M, Marzorati M et al. 2012. The path from {beta}-carotene to carlactone, a strigolactone-like plant hormone. *Science* 335:1348–1351.

Arite T, Iwata H, Ohshima K et al. 2007. DWARF10, an RMS1/MAX4/DAD1 ortholog, controls lateral bud outgrowth in rice. *Plant J* 51:1019–1029.

Arite T, Kameoka H, Kyozuka J. 2012. Strigolactone positively controls crown root elongation in rice. *J Plant Growth Regul* 31:165–172.

Auldridge ME, Block A, Vogel JT et al. 2006. Characterization of three members of the *Arabidopsis* carotenoid cleavage dioxygenase family demonstrates the divergent roles of this multifunctional enzyme family. *Plant J* 45:982–993.

Bago B, Pfeffer PE, Shachar-Hill Y. 2000. Carbon metabolism and transport in arbuscular mycorrhizas. *Plant Physiol* 124:949–958.

Bainbridge K, Sorefan K, Ward S et al. 2005. Hormonally controlled expression of the *Arabidopsis* MAX4 shoot branching regulatory gene. *Plant J* 44:569–580.

Barry CS, Llop-Tous MI, Grierson D. 2000. The regulation of 1-aminocyclopropane-1-carboxylic acid synthase gene expression during the transition from system-1 to system-2 ethylene synthesis in tomato. *Plant Physiol* 123:979–986.

Benedetto A, Magurno F, Bonfante P et al. 2005. Expression profiles of a phosphate transporter gene (*GmosPT*) from the endomycorrhizal fungus *Glomus mosseae*. *Mycorrhiza* 15:620–627.

Bennett T, Sieberer T, Willett B et al. 2006. The *Arabidopsis* MAX pathway controls shoot branching by regulating auxin transport. *Curr Biol* 16:553–563.

Besserer A, Becard G, Jauneau A et al. 2008. GR24, a synthetic analog of strigolactones, stimulates the mitosis and growth of the arbuscular mycorrhizal fungus *Gigaspora rosea* by boosting its energy metabolism. *Plant Physiol* 148:402–413.

Besserer A, Puech-Pagès V, Kiefer P et al. 2006. Strigolactones stimulate arbuscular mycorrhizal fungi by activating mitochondria. *PLoS Biol* 4:e226.

Beveridge CA, Kyozuka J. 2010. New genes in the strigolactone-related shoot branching pathway. *Curr Opin Plant Biol* 13:34–39.

Beveridge CA, Symons GM, Turnbull CGN. 2000. Auxin inhibition of decapitation-induced branching is dependent on graft-transmissible signals regulated by genes Rms1 and Rms2. *Plant Physiol* 123:689–698.

Booker J, Auldridge M, Wills S et al. 2004. MAX3/CCD7 is a carotenoid cleavage dioxygenase required for the synthesis of a novel plant signalling molecule? *Curr Biol* 14:1232–1238.

Bowman JL, Floyd SK, Sakakibara K. 2007. Green genes-comparative genomics of the green branch of life. *Cell* 129:229–234.

Brewer PB, Dun EA, Ferguson BJ et al. 2009. Strigolactone acts downstream of auxin to regulate bud outgrowth in pea and *Arabidopsis*. *Plant Physiol* 150:482–493.

Bucher M. 2007. Functional biology of plant phosphate uptake at root and mycorrhiza interfaces. *New Phytol* 173:11–26.

Buée M, Rossignol M, Jauneau A et al. 2000. The pre-symbiotic growth of arbuscular mycorrhizal fungi is induced by a branching factor partially purified from plant root exudates. *Mol Plant Microbe Interact* 13:693–698.

Caba JM, Centeno ML, Fernández B et al. 2000. Inoculation and nitrate alter phytohormone levels in soybean roots: Differences between a supernodulating mutant and the wild type. *Planta* 211:98–104.

Cook CE, Whichard LP, Turner B et al. 1966. Germination of witchweed (*Striga lutea* Lour.): Isolation and properties of a potent stimulant. *Science* 154:1189–1190.

Cook CE, Whichard LP, Wall M et al. 1972. Germination stimulants. II. Structure of strigol, a potent seed germination stimulant for witchweed (*Striga lutea*). *J Am Chem Soc* 94:6198–6199.

Crawford S, Shinohara N, Sieberer T et al. 2010. Strigolactones enhance competition between shoot branches by dampening auxin transport. *Development* 137:2905–2913.

Delaux P-M, Xie X, Timme RE, Puech-Pages V, Dunand C, Lecompte E, Delwiche CF, Yoneyama K, Bécard G, Séjalon-Delmas N. 2012. Origin of strigolactones in the green lineage. *New Phytol* 195:849–863.

Dharmasiri N, Dharmasiri S, Estelle M. 2005. The F-box protein TIR1 is an auxin receptor. *Nature* 435:441–445.

Drew MC. 1975. Comparison of the effects of a localized supply of phosphate, nitrate, ammonium and potassium on the growth of the seminal root system, and the shoot, in barley. *New Phytol* 75:479–490.

Drummond RSM, Martinez-Sanchez NM, Janssen BJ et al. 2009. *Petunia hybrida* CAROTENOID CLEAVAGE DIOXYGENASE7 is involved in the production of negative and positive branching signals in petunia. *Plant Physiol* 151:1867–1877.

Dun EA, Brewer PB, Beveridge CA. 2009. Strigolactones: Discovery of the elusive shoot branching hormone. *Trend Plant Sci* 14:364–372.

Ericsson T. 1995. Growth and shoot: Root ratio of seedlings in relation to nutrient availability. *Plant Soil* 168–169:205–214.

Ferguson BJ, Beveridge CA. 2009. Roles for auxin, cytokinin, and strigolactone in regulating shoot branching. *Plant Physiol* 149:1929–1944.

Fernández-Aparicio M, García-Garrido JM, Ocampo JA et al. 2010. Colonisation of field pea roots by arbuscular mycorrhizal fungi reduces *Orobanche* and *Phelipanche* species seed germination. *Weed Res* 50:262–268.

Foo E, Turnbull CGN, Beveridge CA. 2001. Long-distance signalling and the control of branching in the rms1 mutant of pea. *Plant Physiol* 126:203–209.

Gianinazzi-Pearson V, Branzanti B, Gianinazzi S. 1989. in vitro enhancement of spore germination and early hyphal growth of a vesicular-arbuscular mycorrhizal fungus by host root exudates and plant flavonoids. *Symbiosis* 7:243–255.

Gilroy S, Jones DL. 2000. Through form to function: Root hair development and nutrient uptake. *Trend Plant Sci* 5:56–60.

Giovannetti M, Avio L, Sbrana C. 2010. Fungal spore germination and pre-symbiotic mycelial growth—Physiological and genetic aspects. In *Arbuscular Mycorrhizas: Physiology and Function*, eds. H Koltai, Y Kapulnik, 2nd edn., pp. 3–32. Dordrecht, the Netherlands: Springer.

Giovannetti M, Sbrana C, Citernesi AS et al. 1996. Analysis of factors involved in fungal recognition responses to host-derived signals by arbuscular mycorrhizal fungi. *New Phytol* 133:65–71.

Goldwasser Y, Yoneyama K, Xie X et al. 2008. Production of strigolactones by *Arabidopsis thaliana* responsible for *Orobanche aegyptiaca* seed germination. *Plant Growth Regul* 55:21–28.

Gomez-Roldan V, Fermas S, Brewer PB et al. 2008. Strigolactone inhibition of shoot branching. *Nature* 455:189–194.

Graham JH, Leonard RT, Menge JA. 1981. Membrane-mediated decrease in root exudation responsible for phorphorus inhibition of vesicular-arbuscular mycorrhiza formation. *Plant Physiol* 68:548–552.

Harrison MJ, van Buuren ML. 1995. A phosphate transporter from the mycorrhizal fungus *Glomus versiforme*. *Nature* 378:626–629.

Hayward A, Stirnberg P, Beveridge C et al. 2009. Interactions between auxin and strigolactone in shoot branching control. *Plant Physiol* 151:400–412.

Hell R, Hillebrand H. 2001. Plant concepts for mineral acquisition and allocation. *Curr Opin Biotechnol* 12:161–168.

Ivanchenko MG, Muday GK, Dubrovsky JG. 2008. Ethylene-auxin interactions regulate lateral root initiation and emergence in *Arabidopsis thaliana*. *Plant J* 55:335–547.

Jain A, Poling MD, Karthikeyan AS et al. 2007. Differential effects of sucrose and auxin on localized phosphate deficiency-induced modulation of different traits of root system architecture in *Arabidopsis*. *Plant Physiol* 144:232–247.

Johnson X, Brcich T, Dun EA et al. 2006. Branching genes are conserved across species. Genes controlling a novel signal in pea are coregulated by other long-distance signals. *Plant Physiol* 142:1014–1026.

Johnson AW, Gowada G, Hassanali A et al. 1981. The preparation of synthetic analogues of strigol. *J Chem Soc Perkin Trans* 1:1734–1743.

Jones AR, Kramer EM, Knox K et al. 2009. Auxin transport through non-hair cells sustains root-hair development. *Nat Cell Biol* 11:78–84.

Kapulnik Y, Delaux P-M, Resnick N et al. 2011a. Strigolactones affect lateral root formation and root-hair elongation in *Arabidopsis*. *Planta* 233:209–216.

Kapulnik Y, Resnick N, Mayzlish-Gati E et al. 2011b. Strigolactones interact with ethylene and auxin in regulating root-hair elongation in *Arabidopsis*. *J Exp Bot* 62(8):2915–2924.

Kohlen W, Charnikhova T, Liu Q et al. 2011. Strigolactones are transported through the xylem and play a key role in shoot architectural response to phosphate deficiency in non-AM host *Arabidopsis thaliana*. *Plant Physiol* 155:974–987.

Koltai H. 2011. Strigolactones are regulators of root development. *New Phytol* 190:545–549.

Koltai H, Dor E, Hershenhorn J et al. 2010a. Strigolactones' effect on root growth and root-hair elongation may be mediated by auxin-efflux carriers. *J Plant Growth Regul* 29:129–136.

Koltai H, Lekkala SP, Bhattacharya C et al. 2010b. A tomato strigolactone-impaired mutant displays aberrant shoot morphology and plant interactions. *J Exp Bot* 61:1739–1749.

Kretzschmar T, Kohlen W, Sasse J et al. 2012. A petunia ABC protein controls strigolactone-dependent symbiotic signalling and branching. *Nature* 483:341–344.

Lendzemo V, Kuyper T, Vierheilig H. 2009. *Striga* seed-germination activity of root exudates and compounds present in stems of *Striga* host and nonhost (trap crop) plants is reduced due to root colonization by arbuscular mycorrhizal fungi. *Mycorrhiza* 19:287–294.

Lewis LA, McCourt RM. 2004. Green algae and the origin of land plants. *Am J Bot* 91:1535–1556.

Leyser O. 2009. The control of shoot branching: an example of plant information processing. *Plant Cell Environ* 32:694–703.

Leyser O. 2010. The power of auxin in plants. *Plant Physiol* 154:501–505.

Liang J, Zhao L, Challis R et al. 2010. Strigolactone regulation of shoot branching in chrysanthemum (*Dendranthema grandiflorum*). *J Exp Bot* 61:3069–3078.

López-Bucio J, Cruz-Ramírez A, Herrera-Estrella L. 2003. The role of nutrient availability in regulating root architecture. *Curr Opin Plant Biol* 6:280–287.

López-Bucio J, Hernandez-Abreu E, Sanchez-Calderon L et al. 2002. Phosphate availability alters architecture and causes changes in hormone sensitivity in the *Arabidopsis* root system. *Plant Physiol* 129:244–256.

López-Ráez JA, Bouwmeester H. 2008. Fine-tuning regulation of strigolactone biosynthesis under phosphate starvation. *Plant Signaling Behav* 3:963–965.

López-Ráez JA, Charnikhova T, Fernández I et al. 2011. Arbuscular mycorrhizal symbiosis decreases strigolactone production in tomato. *J Plant Physiol* 168:294–297.

López-Ráez JA, Charnikhova T, Gómez-Roldán V et al. 2008. Tomato strigolactones are derived from carotenoids and their biosynthesis is promoted by phosphate starvation. *New Phytol* 178:863–874.

Maldonado-Mendoza IE, Dewbre GR, Harrison MJ. 2001. A phosphate transporter gene from the extra-radical mycelium of an arbuscular mycorrhizal fungus *Glomus intraradices* is regulated in response to phosphate in the environment. *Mol Plant Microbe Interact* 14:1140–1148.

Matusova R, Rani K, Verstappen FWA et al. 2005. The strigolactone germination stimulants of the plant-parasitic *Striga* and *Orobanche* spp. are derived from the carotenoid pathway. *Plant Physiol* 139:920–934.

Mayzlish-Gati E, Lekkala SP, Resnick N et al. 2010. Strigolactones are positive regulators of light-harvesting genes in tomato. *J Exp Bot* 61:3129–3136.

Mayzlish-Gati E, De-Cuyper C, Goormachtig S et al. 2012. Strigolactones are involved in root response to low phosphate conditions in *Arabidopsis*. *Plant Physiol*. doi:10.1104/pp.112.202358.

Mouchel CF, Leyser O. 2007. Novel phytohormones involved in long-range signalling. *Curr Opin Plant Biol* 10:473–476.

Muller M, Schmidt W. 2004. Environmentally induced plasticity of root hair development in *Arabidopsis*. *Plant Physiol* 134:409–419.

Nagahashi G, Douds Jr DD. 2000. Partial separation of root exudate components and their effects upon the growth of germinated spores of AM fungi. *Mycol Res* 104:1453–1464.

Napoli C. 1996. Highly branched phenotype of the Petunia dad1-1 mutant is reversed by grafting. *Plant Physiol* 111:27–37.

Negi S, Ivanchenko M, Muday G. 2008. Ethylene regulates lateral root formation and auxin transport in *Arabidopsis thaliana*. *Plant J* 55:175–187.

Negi S, Sukumar P, Liu X et al. 2010. Genetic dissection of the role of ethylene in regulating auxin-dependent lateral and adventitious root formation in tomato. *Plant J* 61:3–15.

Nelson DC, Scaffidi A, Dun EA, Waters MT, Flematti GR, Dixon KW, Beveridge CA, Ghisalberti EL, Smith SM. 2011. F-box protein MAX2 has dual roles in karrikin and strigolactone signalling in *Arabidopsis thaliana*. *Proc Natl Acad Sci USA* 108:8897–8902.

Ongaro V, Leyser O. 2008. Hormonal control of shoot branching. *J Exp Bot* 59:67–74.

Osmont KS, Sibout R, Hardtke CS. 2007. Hidden branches: Developments in root system architecture. *Annu Rev Plant Biol* 58:93–113.

Paul MJ, Foyer CH. 2001. Sink regulation of photosynthesis. *J Exp Bot* 52:1383–1400.

Péret B, De Rybel B, Casimiro I et al. 2009. *Arabidopsis* lateral root development: An emerging story. *Trends Plant Sci* 14:399–408.

Pérez-Torres C-A, Lopez-Bucio J, Cruz-Ramirez A et al. 2008. Phosphate availability alters lateral root development in *Arabidopsis* by modulating auxin sensitivity via a mechanism involving the TIR1 auxin receptor. *Plant Cell* 20:3258–3272.

Pirozynski KA, Malloch DW. 1975. The origin of land plants: A matter of mycotrophism. *Biosystems* 6:153–164.

Pitts RJ, Cernac A, Estelle M. 1998. Auxin and ethylene promote root hair elongation in *Arabidopsis*. *Plant J* 16:553–560.

Proost S, Van Bel M, Sterck L, Billiau K, Van Parys T, Van de Peer Y, Vandepoele K. 2009. PLAZA: A comparative genomics resource to study gene and genome evolution in plants. *Plant Cell* 21:3718–3731.

Proust H, Hoffmann B, Xie X, Yoneyama K, Schaefer DG, Yoneyama K, Nogué F, Rameau C. 2011. Strigolactones regulate protonema branching and act as a quorum sensing-like signal in the moss *Physcomitrella patens*. *Development* 138:1531–1539.

Redecker D, Kodner R, Graham LE. 2000. Glomalean fungi from the Ordovician. *Science* 289:1920–1921.

Remy W, Taylor TN, Hass H et al. 1994. Four hundred-million-year-old vesicular arbuscular mycorrhizae. *Proc Natl Acad Sci U S A* 91:11841–11843.

Rensing SA, Lang D, Zimmer AD et al. 2008. The *Physcomitrella* genome reveals evolutionary insights into the conquest of land by plants. *Science* 319:64–69.

Rubio V, Bustos R, Irigoyen M et al. 2009. Plant hormones and nutrient signalling. *Plant Mol Biol* 69:361–373.

Ruyter-Spira C, Kohlen W, Charnikhova T et al. 2011. Physiological effects of the synthetic strigolactone analog GR24 on root system architecture in *Arabidopsis*: Another below-ground role for strigolactones? *Plant Physiol* 155:7217–7234.

Ruzicka K, Ljung K, Vanneste S et al. 2007. Ethylene regulates root growth through effects on auxin biosynthesis and transport-dependent auxin distribution. *Plant Cell* 19:2197–2212.

Schwartz SH, Qin X, Loewen MC. 2004. The biochemical characterization of two carotenoid cleavage enzymes from *Arabidopsis* indicates that a carotenoid-derived compound inhibits lateral branching. *J Biol Chem* 279:46940–46945.

Siciliano V, Genre A, Balestrini R et al. 2007. Transcriptome analysis of arbuscular mycorrhizal roots during development of the prepenetration apparatus. *Plant Physiol* 144:1455–1466.

Simon L, Bousquet J, Levesque RC et al. 1993. Origin and diversification of endomycorrhizal fungi and coincidence with vascular land plants. *Nature* 363:67–69.

Smith SE, Read DJ. 2008. *Mycorrhizal Symbiosis*, 3rd edn. Burlington, MA: Elsevier.

Smith SE, Smith FA, Jakobsen I. 2003. Mycorrhizal fungi can dominate phosphate supply to plants irrespective of growth responses. *Plant Physiol* 133:16–20.

Snowden KC, Simkin AJ, Janssen BJ et al. 2005. The decreased apical dominance1/*Petunia hybrida* CAROTENOID CLEAVAGE DIOXYGENASE8 gene affects branch production and plays a role in leaf senescence, root growth, and flower development. *Plant Cell* 17:746–759.

Sorefan K, Booker J, Haurogné K et al. 2003. MAX4 and RMS1 are orthologous dioxygenase-like genes that regulate shoot branching in *Arabidopsis* and pea. *Genes Dev* 17:1469–1474.

Stepanova AN, Alonso JM. 2009. Ethylene signalling and response: Where different regulatory modules meet. *Curr Opin Plant Biol* 12:548–555.

Stirnberg P, Furner IJ, Ottoline Leyser HM. 2007. MAX2 participates in an SCF complex which acts locally at the node to suppress shoot branching. *Plant J* 50:80–94.

Swarup R, Perry P, Hagenbeek D et al. 2007. Ethylene upregulates auxin biosynthesis in *Arabidopsis* seedlings to enhance inhibition of root cell elongation. *Plant Cell* 19:2186–2196.

Troughton A. 1977. The effect of phosphorus nutrition upon the growth and morphology of young plants of *Lolium perenne* L. *Ann Bot* 41:85–92.

Tsuchiya Y, Vidaurre D, Toh S et al. 2010. A small-molecule screen identifies new functions for the plant hormone strigolactone. *Nat Chem Biol* 6:741–749.

Umehara M, Hanada A, Magome H et al. 2010. Contribution of strigolactones to the inhibition of tiller bud outgrowth under phosphate deficiency in rice. *Plant Cell Physiol* 51:1118–1126.

Umehara M, Hanada A, Yoshida S et al. 2008. Inhibition of shoot branching by new terpenoid plant hormones. *Nature* 455:195–200.

Vogel JT, Walter MH, Giavalisco P et al. 2010. SlCCD7 controls strigolactone biosynthesis, shoot branching and mycorrhiza-induced apocarotenoid formation in tomato. *Plant J* 61:300–311.

Walch-Liu P, Ivanov II, Filleur S et al. 2006. Nitrogen regulation of root branching. *Ann Bot* 97:875–881.

Waldie T, Hayward A, Beveridge C. 2010. Axillary bud outgrowth in herbaceous shoots: How do strigolactones fit into the picture? *Plant Mol Biol* 73:27–36.

Waters MT, Nelson DC, Scaffidi A et al. 2012. Specialisation within the DWARF14 protein family confers distinct responses to karrikins and strigolactones in *Arabidopsis*. *Development* 139:1285–1295.

Westwood JH, Yoder JI, Timko MP et al. 2010. The evolution of parasitism in plants. *Trends Plant Sci* 15:227–235.

Williamson L, Ribrioux S, Fitter A, Leyser O. 2001. Phosphate availability regulates root system architecture in *Arabidopsis*. *Plant Physiol* 126:1–8.

Xie X, Kusumoto D, Takeuchi Y et al. 2007. 2′-Epi-orobanchol and solanacol, two unique strigolactones, germination stimulants for root parasitic weeds, produced by tobacco. *J Agric Food Chem* 55:8067–8072.

Xie X, Yoneyama K, Harada Y et al. 2009. Fabacyl acetate, a germination stimulant for root parasitic plants from *Pisum sativum*. *Phytochemistry* 70:211–215.

Xie X, Yoneyama K, Kusumoto D et al. 2008a. Isolation and identification of alectrol as (+)-orobanchyl acetate, a germination stimulant for root parasitic plants. *Phytochemistry* 69:427–431.

Xie X, Yoneyama K, Kusumoto D et al. 2008b. Sorgomol, germination stimulant for root parasitic plants, produced by *Sorghum bicolor*. *Tetrahedron Lett* 49:2066–2068.

Xie X, Yoneyama K, Yoneyama K. 2010. The strigolactone story. *Annu Rev Phytopathol* 48:93–117.

Yamagami T, Tsuchisaka A, Yamada K et al. 2003. Biochemical diversity among the 1-amino-cyclopropane-1-carboxylate synthase isozymes encoded by the *Arabidopsis* gene family. *J Biol Chem* 278:49102–49112.

Yamamoto M, Miki T, Ishiki Y et al. 1995. The synthesis of ethylene in melon fruit during the early stage of ripening. *Plant Cell Physiol* 36:591–596.

Yokota T, Sakai H, Okuno K et al. 1998. Alectrol and orobanchol, germination stimulants for *Orobanche minor*, from its host red clover. *Phytochemistry* 49:1967–1973.

Yoneyama K, Xie X, Kusumoto D et al. 2007a. Nitrogen deficiency as well as phosphorus deficiency in sorghum promotes the production and exudation of 5-deoxystrigol, the host recognition signal for arbuscular mycorrhizal fungi and root parasites. *Planta* 227:125–132.

Yoneyama K, Xie X, Sekimoto H et al. 2008. Strigolactones, host recognition signals for root parasitic plants and arbuscular mycorrhizal fungi, from Fabaceae plants. *New Phytol* 179:484–494.

Yoneyama K, Xie X, Yoneyama K et al. 2009. Strigolactones: Structures and biological activities. *Pest Manage Sci* 65:467–470.

Yoneyama K, Yoneyama K, Takeuchi Y et al. 2007b. Phosphorus deficiency in red clover promotes exudation of orobanchol, the signal for mycorrhizal symbionts and germination stimulant for root parasites. *Planta* 225:1031–1038.

Yoo S-D, Cho Y, Sheen J. 2009. Emerging connections in the ethylene signalling network. *Trends Plant Sci* 14:270–279.

Zou J, Zhang S, Zhang W et al. 2006. The rice HIGH-TILLERING DWARF1 encoding an ortholog of *Arabidopsis* MAX3 is required for negative regulation of the outgrowth of axillary buds. *Plant J* 48:687–698.

19

Root Gravitropism

Ranjan Swarup
University of Nottingham

Darren M. Wells
University of Nottingham

Malcolm J. Bennett
University of Nottingham

I. Introduction

Plants adapt their development in response to environmental signals in order to optimize their chances of survival. These environmental cues include directional signals such as gravity and light. Plant organs adjust their position relative to such signals by employing tropic responses to modify their direction of growth. For example, roots employ directional signals like gravity to explore the soil environment and acquire anchorage and resources.

Tropic responses represent key adaptive traits throughout root development. During germination of some species, primary roots develop the ability to respond to gravity prior to emergence from the seed coat (Ma and Hasenstein 2006), thereby aiding seedling establishment. When secondary roots emerge from the primary root, they initially grow out horizontally (Mullen and Hangarter 2003), while higher orders of lateral roots often exhibit randomized directions. These diagravitropic and plagiotropic patterns of root growth greatly facilitate the acquisition of water and nutrients (Basu et al. 2007), particularly in the topsoil where phosphorus availability is the greatest (Lynch and Brown 2001). While root angle is primarily regulated by the gravitropic response, other directional signals like touch and water/oxygen gradients also modify root angle by inducing thigmo-, hydro-, and oxytropic responses when roots encounter compacted, drying, and waterlogged soil, respectively (Porterfield 2002).

Tropic responses have fascinated scientists for over 300 years. Hydrotropism was initially proposed by Dodart (1700) as a mechanism for roots to grow downward in response to soil water content. Root gravitropism was first demonstrated by Knight (1806) after observing that young roots reorient downward irrespective of the angle the germinating seed was positioned. Darwin (1881) first described root thigmotropism in his book "The power of movement in plants," while oxytropism was initially described by Molisch (1884) after observing that submerged roots grew toward the water–air interface. Despite the large intervening period of time, our understanding about the mechanisms regulating many of these root tropisms remains patchy. In contrast, the last 15 years have seen several important breakthroughs in our understanding about the molecular components, signals, and tissues that function during root gravitropism.

Root gravitropism can be divided into three spatially and temporally distinct phases: gravity perception, gravitropic signal transmission, and gravitropic response (reviewed by Swarup and Bennett 2009). In brief, a compelling body of genetic and physiological evidence has demonstrated the importance of starch-filled plastids (termed statoliths) in columella cells at the root apex for gravity perception (see Figure 19.1). The gravitropic signal transmission phase is triggered following the repositioning of statoliths within gravity-sensing columella cell causing an asymmetric release of the growth regulator auxin (Rashotte et al. 2001; Boonsirichai et al. 2003; Ottenschlager et al. 2003; see Figure 19.1) that is transported from columella cells via the lateral root cap to the gravitropic responsive tissues in the elongation zone (Swarup et al. 2005). The resulting lateral auxin gradient drives the gravitropic response phase

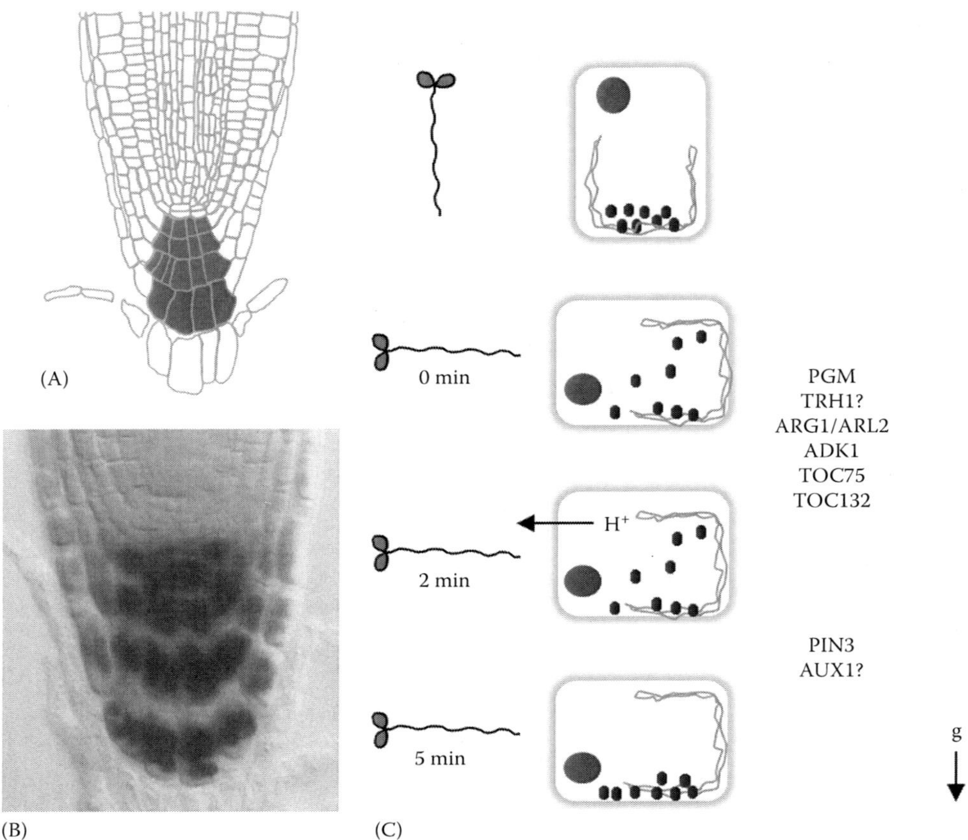

FIGURE 19.1 (See color insert.) Gravity perception. (A) A cartoon of an *Arabidopsis* root with columella cells marked in gray. (B) Image of an *Arabidopsis* root stained for starch. (C) A schematic representation of gravity perception in *Arabidopsis*. Change in gravity vector results in sedimentation of amyloplasts in the columella cells that initiates a cascade of events including alkalinization of cytoplasm and repositioning of auxin efflux carrier PIN3 to the new bottom of the cell by a yet unknown mechanism.

(Chavarría-Krauser et al. 2008), by triggering differential growth where cell expansion on the lower side of the elongation zone is reduced relative to the upper side, causing the root to bend downward.

This chapter will primarily focus on recent advances in our knowledge about the molecular and cellular mechanisms that regulate each of these three phases of the root gravitropic response in the primary root of the model plant *Arabidopsis thaliana*. For information about gravitropism in other plant species or other root tropic responses, readers are directed to a number of excellent reviews (Konings 1995; Muday 2001; Tasaka et al. 2001; Boonsirichai et al. 2002; Blancaflor and Masson 2003; Perbal and Driss-Ecole 2003; Morita and Tasaka 2004; Esmon et al. 2005; Porterfield 2002; Molas and Kiss 2009; Perbal 2009; Morita 2010).

II. Gravity Perception in the *Arabidopsis* Root

A. Columella Represents the Site of Gravity Perception in the *Arabidopsis* Root

Experiments involving surgical removal of the root cap that resulted in loss of gravitropic response show that the root cap is the primary site of graviperception (Barlow 1974; Blancaflor and Masson 2003; Morita and Tasaka 2004). The *Arabidopsis* root cap is comprised of lateral root cap cells and several stories of columella cells termed S1, S2, S3, etc. (Figure 19.1A and B). Experiments using laser ablation (Blancaflor et al. 1998) or heavy-ion microbeam irradiation (Tanaka et al. 2002) revealed that within the *Arabidopsis* root cap, columella cells are the primary site of graviperception. By ablating different stories of columella cells, Blancaflor et al. (1998) showed that S1 and S2 columella cells had maximum impact on the gravitropic behavior of roots. Roots in which S1 and S2 layers had been ablated showed greatest reduction in root curvature and deviation from vertical growth after gravistimulation. These roots also exhibited longest presentation times (i.e., the minimum exposure time to a 1 g field to induce a response; Sack 1991). Interestingly, ablation of S3 cells resulted in reduction in root gravitropism without affecting the presentation time, suggesting an impaired transmission of the gravity signal, rather than a major reduction in sensitivity to the gravity stimulus. These results imply that columella cells within S1 and S2 layers represent the primary site of gravity sensing in the *Arabidopsis* root. There is some evidence in maize that suggests that there may be additional gravity-sensing sites outside of the root cap (Wolverton et al. 2002b). The authors used a special rotating stage to maintain the root tip at a set angle, while selected regions in the elongation zone were kept at a gravistimulated angle. Under these conditions, maize roots continue to show some differential growth suggesting the presence of minor graviperception sites outside the root cap (Wolverton et al. 2002b). However, the tissue location of such secondary sites for graviperception remains to be identified.

B. Statoliths Function as Gravity Sensors

Gravity has been omnipresent ever since life evolved on earth. Plants and other living organisms have evolved various systems for the perception of gravity. Cell organelles such as mitochondria (Myxomycetes), nuclei (fungi and sphenopsids), and sedimentable barium sulfate crystals (characean algae) have all been reported to act as gravisensors (Barlow 1995). In higher plant roots, starch-filled amyloplasts function as gravity sensors in columella cells. Columella cells are ideally suited to this task as ultrastructural studies show that these cells lack a central vacuole, their nucleus and ER are localized at the cell periphery, and they have freely sedimentable amyloplasts (Sack 1991; Zheng and Staehelin 2001).

Sedimentation of statoliths activates an, as yet, unknown receptor triggering a signal transduction pathway to promote the development of a gravitropic curvature. Support for the starch–statolith hypothesis (Sack 1997; Figure 19.1C) comes from several genetic studies that reported on starch-deficient mutants exhibiting defects in gravity perception (Kiss et al. 1996; MacCleery and Kiss 1999; Fitzelle and Kiss 2001). Kiss et al. (1996) discovered that there is a direct correlation between starch content and graviperception and 51%–60% of the wild-type level of starch is near the threshold amount needed for full gravitropic sensitivity. Experiments involving hypergravity or high-gradient magnetic fields to mimic a gravitational field provide further support for the starch–statolith hypothesis (Kuznetsov and Hasenstein 1996; Fitzelle and Kiss 2001). Kuznetsov and Hasenstein (1996) showed that *Arabidopsis* wild-type roots but not starchless *TC7* mutants could respond to high magnetic fields. Similarly, Fitzelle and Kiss (2001) showed that hypergravity conditions can cause statolith sedimentation in starch-deficient mutants thereby rescuing the root gravitropic defects. Fitzelle and Kiss (2001) experiments also suggest that starch is not essential for gravity sensing as long as amyloplasts can sediment, a view also shared by Morita (2010). Though there is an overwhelming support and experimental evidence for the starch–statolith hypothesis, the possibility of other mechanisms operating in parallel cannot be ruled out (Wolverton et al. 2002a,b; LaMotte and Pickard 2004).

C. Translating Statolith Sedimentation into a Physiological Signal(s)

Modeling studies suggest that statolith sedimentation is required to generate a lateral auxin gradient (Band et al. 2012). However, early events of gravity perception are not well understood. It has been proposed that sedimenting amyloplasts may activate mechanosensitive ion channels through distortion of the actin cytoskeleton (Sievers et al. 1991; Perbal and Driss-Ecole 2003). Despite the fact that amyloplasts in columella cells have been shown to be surrounded by a fine network of actin filaments (Collings et al. 2001), the involvement of actin in graviperception remains ambiguous. Treatment of *Arabidopsis* roots with the actin cytoskeleton inhibitor latrunculin B (Lat B) has been observed to result in enhanced gravitropic curvature (Hou et al. 2004). Similarly mutations in *DIS1* and *DIS2* genes, which encode proteins of the ARP (actin-related protein) 2/3 complex, show enhanced gravitropic response (Reboulet et al. 2010). These results may suggest that actin is not involved in the early stages of gravity perception. Recent high-resolution electron tomography studies indicate that the weight of statoliths is sufficient to locally deform the ER (Leitz et al. 2009). This may provide a more direct mechanism to trigger mechanosensitive ion channels. However, until more direct evidence is available, this question remains unanswered.

Despite our lack of understanding of early events of graviperception, work in the past 10 years has shed light on the events following perception. Soon after a gravity stimulus, there is a rapid change in the pH of root cap cells (Scott and Allen 1999; Fasano et al. 2001; Hou et al. 2004). These pH changes appear to be linked with statolith sedimentation since they were not observed in the starchless mutant *pgm*. Cytoplasmic pH in columella cells has been shown to increase from 7.2 to 7.6 within the first minute and was followed by the sustained acidification of the apoplast (Fasano et al. 2001). Significantly, disrupting the pH shift in columella cells by acidifying or alkalinizing agents altered the gravitropic responses, suggesting that these pH changes are of functional importance and form components of the gravitropic signaling machinery. Mutations in the *ARG1* (*ALTERED RESPONSE TO GRAVITY*) gene confer an agravitropic root phenotype and block cytosolic alkalinization of columella cells (Boonsirichai et al. 2003). Expressing the *ARG1* gene in columella cells only is sufficient to rescue the *arg1* gravitropic defect, suggesting that ARG1 plays a key role in early stages of gravity sensing.

ARG1 and its closely related gene *ARL2* (*ARG1-like 2*) encode DnaJ-like peripheral membrane proteins (Sedbrook et al. 1999; Boonsirichai et al. 2003; Guan et al. 2003). ARG1 is associated with microsomal membranes and the cytoskeleton and is proposed to be involved in regulating the trafficking or function of membrane proteins needed for the early phases of gravity signal transduction such as PIN3 (Boonsirichai et al. 2003).

Recently, a role for plastid TOC (translocon of outer membrane of chloroplasts) complexes is proposed for ARG1-mediated gravity signal transduction (Stanga et al. 2009). In a screen for enhancers of *arg1*, *modifier of arg1* (*mar1*) and *mar2* mutations were identified. *MAR1* and *MAR2* encode different components of the TOC complex. What the plastid import machinery has to do with gravitropism is not clear. Stanga et al. (2009) speculate that a direct or indirect interaction of the TOC complex on the amyloplast envelope with a transducer on the PM or ER with ARG1 may modulate the activity/localization of this transducer. Jouhet and Gray (2009) suggest that the pea homolog of MAR2 can bind to actin. This could provide a direct link between statolith sedimentation and the actin cytoskeleton.

III. Root Gravitropic Signal Transduction

A. Auxin Represents the Primary Gravitropic Signal

While gravity is perceived in the columella cells, the gravi-response (bend) takes place in the elongation-zone tissues (Figure 19.2A). To explain this, Cholodny (1927) and Went (1926) proposed the movement of a gravitropic signal(s) from root cap to the elongation zone. Auxin represents a strong candidate for the gravitropic signal, but several other signals have also been linked with gravitropism (reviewed by Philosoph-Hadas et al. 2005) including brassinosteroids (Li et al. 2005), ethylene (Buer et al. 2006), cytokinin (Aloni et al. 2004), and abscisic acid (Han et al. 2009). But in the absence of compelling genetic evidence, the role for these other signals in regulating root gravitropism is still unresolved. In contrast, there is overwhelming genetic, pharmacological, and molecular evidence linking auxin to root gravitropism. For example, many auxin transport and response mutants show agravitropic roots (Leyser et al. 1993; Timpte et al. 1994; Bennett et al. 1996; Leyser et al. 1996; Nagpal et al. 2000; Swarup et al. 2001, 2004, 2005, Dharmasiri et al. 2006; Table 19.1) and auxin transport inhibitors disrupt root gravitropism (Marchant et al. 1999). Mutation in the *AUX1* gene results in severely agravitropic roots (Bennett et al. 1996). Later AUX1 was shown to function as a high-affinity auxin influx carrier (Yang et al. 2006). Swarup et al. (2001) showed that movement of auxin from columella cells to the elongation zone (shootward auxin transport) was disrupted in *aux1* mutants underlining the importance of auxin transport for root gravitropism.

Swarup et al. (2005) mapped the root tissues required to transport auxin during a gravitropic response. This work showed that root gravitropism is dependent on auxin being transported via the lateral root cap and elongating epidermal cells and provided conclusive evidence that auxin functions as the primary gravitropic signal linking gravisensing and graviresponsive tissues. Auxin-responsive reporters (Rashotte et al. 2001; Boonsirichai et al. 2003; Ottenschlager et al. 2003; Swarup and Bennett 2009; Figure 19.3) exhibit asymmetric changes in their patterns of expression on the lower side of the gravistimulated roots consistent with the

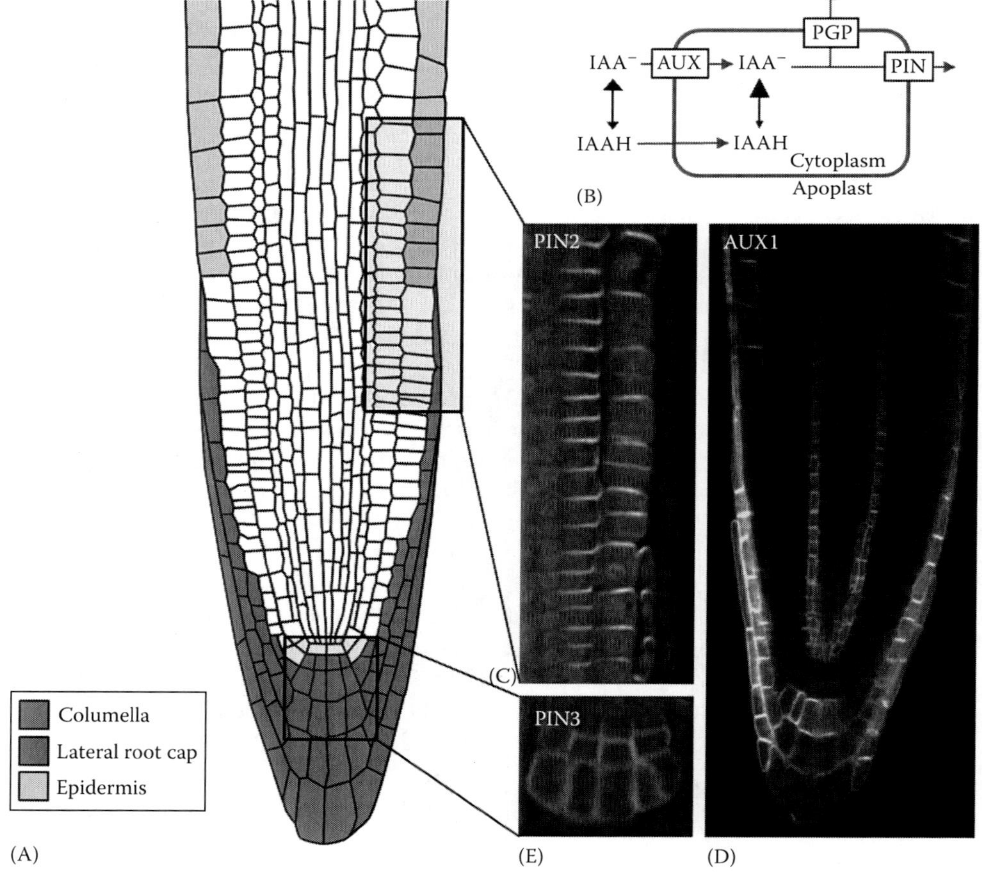

FIGURE 19.2 **(See color insert.)** Gravity signal transmission. (A) A cartoon of an *Arabidopsis* root marking tissues involved in gravipercep-tion (blue), signal transmission (gray), and response (sky blue). (B) Auxin influx and efflux carriers facilitate auxin movement following a gravity stimulus. (C) Confocal image showing PIN2 (green) localization in root apex. (D) Confocal image showing AUX1 (yellow) localization in root apex. (E) Confocal image showing PIN3 (green) localization in columella cells.

TABLE 19.1 Summary of the Key Genes That Regulate Root Gravitropism in *Arabidopsis*

Gene Name	Gene/Protein	Mutant Phenotype	References
Graviperception			
PGM	Phosphoglucomutase	Reduced root and shoot gravitropism	Caspar and Pickard (1989)
ARG1	Dna J protein	Reduced root and shoot gravitropism	Sedbrook et al. (1999); Boonsirichai et al. (2003)
ARL2	Dna J protein	Reduced root and shoot gravitropism	Guan et al. (2003)
ADK1	Adenosine kinase	Root cap morphogenesis and altered gravisensitivity	Young et al. (2006)
TOC75-III & TOC132	Components of the translocon of outer membrane of chloroplasts (TOC) complex	Modifiers of *arg1*	Stanga et al. (2009)
Gravisignal transmission			
PIN3	Auxin efflux carrier	Reduced root gravitropism	Friml et al. (2002)
PIN2/EIR1/AGR1	Auxin efflux carrier	Agravitropic roots	Müller et al. (1998); Paciorek et al. (2005); Abas et al. (2006)
AUX1	Auxin influx carrier	Agravitropic roots	Bennett et al. (1996); Swarup et al. (2001, 2004, 2005); Yang et al. (2006)
PGP1	ABC transporter family	Reduced root gravitropism	Lin and Wang (2005)
GNOM	ARF-GEF	Reduced root gravitropism in weak alleles	Geldner et al. (2003, 2004)
PGP4	ABC transporter family	Reduced root gravitropism	Terasaka et al. (2005)
AXR4	Novel protein	Agravitropic roots	Hobbie and Estelle (1995); Dhramasiri et al. (2006)
RCN1	PP2A	Reduced root gravitropism	Garbers et al. (1996); Deruere et al. (1999)
TT4	Chalcone synthase	Altered root gravitropism	Buer et al. (2006)
TRH1	Potassium transporter	Altered root gravitropism	Vicente-Agullo et al. (2004)
Graviresponse			
AXR1	Ubiquitin-activating enzyme	Agravitropic roots	Leyser et al. (1993)
AXR2	AUX/IAA protein	Agravitropic roots (GoF)	Timpte et al. (1994); Nagpal et al. (2000)
AXR3	AUX/IAA protein	Agravitropic roots (GoF)	Leyser et al. (1996); Rouse et al. (1998)
IAA14/SLR1	AUX/IAA protein	Agravitropic roots (GoF)	Fukaki et al. (2002), (2005)
IAA3/SHY2	AUX/IAA protein	Agravitropic roots (GoF)	Weijers et al. (2005)
ARF7,19	Auxin-response factors	Reduced root gravitropism	Okushima et al. (2005)
TIR1 (AFB2-4)	F box protein (auxin receptor)	Reduced root gravitropism	Ruegger et al. (1998); Dharmasiri et al. (2005a,b); Kepinski and Leyser (2005)
PLD Zeta2	Phospholipase D	Reduced root gravitropism	Li and Xue (2007)

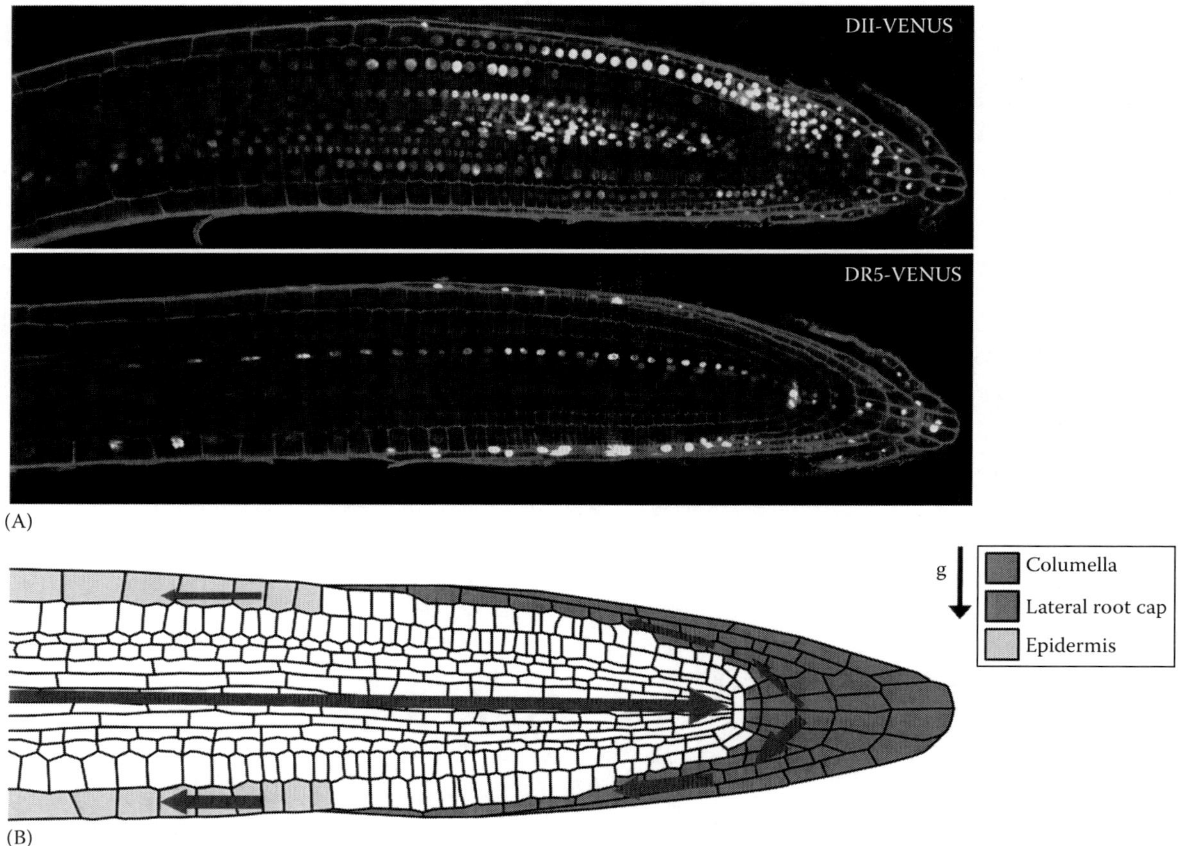

FIGURE 19.3 **(See color insert.)** Formation of an auxin-response gradient. (A) Confocal image of a gravistimulated *Arabidopsis* root showing differential expression of auxin-response reporter DII-VENUS (upper) and DR5:VENUS (lower). Note the complimentary expression patterns of the two reporters. Fluorescence asymmetry becomes apparent in DII-VENUS roots much earlier (1 h) following a gravistimulus than in DR5:VENUS roots (3 h). (B) An interpretation of (A) showing differential auxin movement following gravity perception. Upon gravity perception, asymmetric localization of PIN3 in the columella cells results in differential auxin movement between upper and lower side of a gravistimulated root as seen by asymmetric localization of auxin-response reporter DR5:VENUS.

transmission of a lateral auxin gradient to the lower side of the root. Starchless *pgm* mutants fail to exhibit this asymmetric auxin gradient up to 6 h following reorientation thus directly linking statolith sedimentation with the generation of asymmetric auxin gradient (Wolverton et al. 2011; Band et al. 2012).

B. PIN3: Creating a Lateral Auxin Gradient

Friml et al. (2002) showed that the auxin efflux carrier PIN3 is relocalized to the lower face of columella cells within 5 min of a gravistimulus (Figure 19.2E) providing an elegant molecular mechanism for the establishment of a lateral auxin-response gradient (Rashotte et al. 2001; Ottenschlager et al. 2003; Swarup and Bennett 2009; Figure 19.3). Surprisingly, the *pin3* agravitropic phenotype is very mild (Friml et al. 2002). Kleine-Vehn et al. (2010) have recently shown that this is simply due to the redundancy with PIN7. Trafficking of PIN proteins is regulated by the ARF-GEF protein, GNOM (Geldner et al. 2003). The *GNOM* gene is strongly expressed in columella cells (Geldner et al. 2004) and is required for PIN3 polarization (Kleine-Vehn et al. 2010). Partial loss-of-function alleles of *GNOM* disrupt root gravitropism as a result of impaired relocalization of PIN proteins.

C. AUX1/PIN2: Channeling the Auxin Gradient via the Lateral Root Cap to Elongation Zone

Once a PIN3-dependent lateral auxin gradient has been generated in the columella cells due to asymmetric localization of PIN3, the auxin influx carrier AUX1 and the efflux carrier PIN2 move this auxin via the lateral root cap to expanding cells in the elongation zone (LRC; Figure 19.2A). This AUX1-/PIN2-driven shootward movement of auxin is critical for root gravitropism as shown by genetic, pharmacological, and cell biology-based studies (Rashotte et al. 2000; Swarup et al. 2001, 2005, 2007; Ottenschlager et al. 2003). In addition to AUX1, another putative auxin influx carrier AtPGP4 is also expressed in LRC cells (Santelia et al. 2005; Terasaka et al. 2005). However, genetic evidence suggests that this auxin carrier plays only a minor role in the redistribution of the gravity-induced lateral auxin gradient.

D. AUX1/PIN2: Delivering Auxin to Elongation-Zone Tissues

Auxin influx and efflux carrier activities are also required to establish the lateral auxin gradient in the elongation-zone tissues. *AUX1* (Swarup et al. 2001, 2005) and *PIN2* (Luschnig et al. 1998; Blilou et al. 2005) are both expressed in the elongation-zone tissues and show root gravitropic defects. In contrast to *aux1*, *pin2* mutants only exhibit a weak agravitropic phenotype (Chen et al. 1998; Luschnig et al. 1998; Blakeslee et al. 2007) and a 30% reduction in shootward transport (Rashotte et al. 2000; Shin et al. 2005). The auxin efflux carriers *AtPGP1* and *AtPGP19* are also expressed in the root distal elongation zone in epidermal and cortical cells, respectively (Blakeslee et al. 2007). Though single or double *pgp1* and *pgp19* mutants do not have strong agravitropic phenotype, a triple *pin2 pgp1pgp19* mutant exhibited an agravitropic root phenotype as severe as that of *aux1* (Blakeslee et al. 2007). This additive phenotype suggests that both PIN and PGP classes of auxin efflux carriers are required to transport the lateral auxin gradient during a root gravitropic response.

E. PIN2: Providing Directionality to Auxin Movement

PIN2 is asymmetrically localized at the shootward PM face of LRC and epidermal cells (Figure 19.2C) facilitating shootward transport of auxin (Wiśniewska et al. 2006). Interestingly, PIN2 is localized on the rootward face of cortical cells of the meristem and is involved in the recycling of auxin back toward the root apex (Blilou et al. 2005). This auxin reflux loop appears to be important for root gravitropism as altering the polarity of PIN2 in the cortical cells results in diminished gravitropism (Rahman et al. 2010).

Posttranslational modifications and regulation of PIN trafficking appear to play a key role in PIN localization and polarity. For example, like other PIN proteins, phosphorylation appears to be crucial in determining PIN2 polarity (Sukumar et al. 2009; Zourelidou et al. 2009; Dhonukshe et al. 2010; Huang et al. 2010; Zhang et al. 2010). Auxin has also been demonstrated to effect endocytosis of PIN2 (Paciorek et al. 2005). Abas et al. (2006) showed that differential regulation of PIN2 trafficking between upper and lower side of a gravistimulated root may be crucial for the maintenance of the asymmetric auxin gradient established by PIN3 in columella cells. These posttranslational mechanisms have important functional consequences during root gravitropism. On the lower side, endocytosis of PIN2 is blocked resulting in more PIN2 on the PM of expanding epidermal cells, while on the upper side, PIN2 is rapidly internalized and degraded. The importance of PIN2 in maintaining this differential auxin gradient is evident in the *wav5–62* allele of PIN2 where a single glycine to glutamic acid substitution makes this pin2*wav6–52* protein resistant to degradation and results in an agravitropic phenotype. Disruption of PIN2 cycling by mutations in the *phospholipase Dζ2* gene or the PLD-inhibitor 1-butanol also reduces root gravitropism (Li and Xue 2007).

F. AUX1: Facilitating Efficient Auxin Uptake

While PIN2 provides directionality of auxin movement, AUX1 (Figure 19.2D) appears essential for the efficient uptake of auxin by expanding epidermal cells. Computer simulations of auxin fluxes through elongation-zone tissues suggest that expression of AUX1 and PIN2 in the epidermis minimizes the effect of radial diffusion while facilitating shootward auxin transport (Swarup et al. 2005).

A recent report by Monshausen et al. (2011) provides further insight into the importance of AUX1 in root gravitropism. Using confocal microscopy and automated computer vision-based image analysis, the authors showed that there is an increase in surface pH on the lower side of a gravistimulated root. This pH shift was not seen in *aux1* mutants underlining the importance of AUX1 in this process. They propose that an increase in apoplastic pH on the lower side of a gravistimulated root shifts the IAA^-/IAAH equilibrium toward IAA^- thus reducing the level of membrane permeable IAAH. IAA^- cannot enter the cells by diffusion and requires carrier (AUX1)-mediated uptake. Thus, during a graviresponse, both AUX1 and PIN2 play crucial roles in moving the lateral auxin gradient through the LRC cells to the cells in the elongation zone, and auxin promotes its own carrier-mediated transport via AUX1 (Monshausen et al. 2011) and PIN2 (Robert and Offringa 2008). The work of Monshausen et al. (2011) also clarifies the common misconception that because protonated IAA is membrane permeable, influx carriers play only a supplemental role in auxin distribution. Instead, root epidermal cells expressing *AUX1* are estimated to have a carrier-mediated IAA influx 15 times greater than the diffusive contribution (Kramer and Bennett 2006; Swarup et al. 2005).

The *Arabidopsis* primary root elongation zone contains approximately 16 rapidly expanding epidermal cells in each longitudinal file of cells (Beemster and Baskins 2001). Swarup et al. (2005) expressed *AUX1* in variable number of elongation-zone epidermal cells and showed that expressing *AUX1* in up to the first six of these rapidly expanding epidermal cells was not sufficient to rescue the *aux1* agravitropic phenotype. This led the authors to conclude that the root gravitropic response requires *AUX1* to be expressed in every expanding epidermal cell, presumably to ensure that the lateral auxin gradient persists through the elongation zone and drives asymmetric root growth.

IV. Root Gravitropic Response

A. Root Gravitropism Represents an Auxin-Dependent Differential Root Growth Response

Detailed growth measurements of gravitropic root growth in wild-type *Arabidopsis* reveal that initial root curvature results from differential inhibition of expansion in the distal elongation zone (DEZ), followed by an acceleration of elongation along the top side of the DEZ (Ishikawa and Evans 1997; Mullen et al. 1998a,b). This pattern of differential growth is consistent with the Cholodny–Went hypothesis that predicts that differential growth patterns are induced by a gravity-induced auxin gradient (Band et al. 2012; Brunoud et al. 2012).

B. Epidermis Represents the Primary Target for Auxin Action

The lateral auxin gradient is presumed to directly act on expanding cells in all elongation-zone tissues. However, simulations using a 3D root elongation-zone model suggest that the lateral auxin gradient accumulates 10- to 20-fold more in the epidermis than in the underlying cortical and endodermal tissues (Swarup et al. 2005) consistent with recent observations using the new auxin reporter, DII-VENUS (Band et al. 2012; Brunoud et al. 2012; Figure 19.3). Partitioning the lateral auxin gradient between root tissues in such a manner is predicted to result in root gravitropic bending being driven primarily by differential cell elongation in the epidermis. We directly tested the predicted functional importance of the epidermis to root gravitropism by selectively disrupting the auxin response in this tissue using the stabilized auxin repressor protein axr3–1 (Rouse et al. 1998). Results from our axr3–1-targeted expression studies demonstrate that the auxin response in expanding epidermal cells (but not the underlying tissues) is essential for root gravitropism (Swarup et al. 2005). In contrast, the basal rate of root growth is only slightly reduced when the auxin response in the epidermis is disrupted. These contrasting growth effects could be explained if the epidermis primarily regulated auxin-dependent differential root growth, whereas basal root growth was driven by the expansion of the inner tissues. In such a model, the epidermis would restrain basal root growth in response to a lateral auxin gradient. As the outermost root tissue, the epidermis is ideally positioned to regulate organ bending (Kutschera 1989).

C. Cellular Targets for Auxin

1. Early Events in Graviresponse Are Likely to Be Nongenomic

Detailed physiological and imaging experiments have shown that the root gravitropic response is detectable within 10 min after a gravity stimulus (Ishikawa and Evans 1997; Monshausen et al. 2011). The rapidity of this response suggests that early phase of gravitropic response is likely to be caused by nongenomic effects on auxin distribution. Monshausen et al. (2011) recently showed that there are changes in the surface pH of root elongation-zone cells within minutes of a gravistimulus. This AUX1-dependent alkalinization of apoplast was also observed in single (*tir1*) or multiple auxin receptor (*tir1afb2afb3*) mutant backgrounds clearly indicating that early events in a gravitropic response are not dependent on the TIR1-mediated auxin-response machinery that modifies gene expression. It is not clear how this TIR1-independent auxin-response module promotes pH-dependent differential growth. ABP1 was one of the first proteins shown to bind to auxin, but until recently, its role in auxin-regulated plant development has been unclear (Tromas et al. 2009, 2010; Robert et al. 2010; Xu et al. 2010; Effendi et al. 2011). Although *abp1* knockouts are embryo lethal, ABP1/abp1 heterozygotes are reported to be agravitropic (Effendi et al. 2011). ABP1 has been shown to promote differential growth by inhibiting auxin-sensitive clathrin-mediated endocytosis (Robert et al. 2010), hence is well positioned to control nongenomic gravitropic response at the plasma membrane.

It is also not very clear how changes in apoplastic pH result in differential growth, but calcium (Ca^{2+}) is thought to be involved in translating this local auxin response into a physiological signal. Several studies have implicated changes in Ca^{2+} fluxes in root gravitropism (Lee et al. 1983, 1984; Evans 1986). For example, asymmetric Ca^{2+} gradient has been reported in gravistimulated pea and maize roots (Lee et al. 1984), and application of Ca^{2+} chelators results in loss of gravitropic sensitivity of maize roots (Lee et al. 1983). However, until recently, it was not clear if Ca^{2+} changes were associated with gravity perception, signaling, or response. Using the FRET-based Ca^{2+} sensor YC3.6 and high-resolution imaging, Monshausen et al. (2011) observed that local auxin application at the root tip results in a shootward wave of Ca^{2+}. This Ca^{2+} wave was disrupted by Ca^{2+} channel blocker and led them to propose that auxin perception results in activation of a calcium channel at the PM resulting in an increase in cytosolic Ca^{2+}. This increase in Ca^{2+} is likely to activate a plasma membrane H/OH conductance that results in the alkalinization of the apoplast. Ca^{2+}-induced alkalinization has also been associated with tip growth in pollen, root hairs, and touch stimulation (Feijo et al. 1999; Monshausen et al. 2007, 2008). Changes in cell wall pH have been proposed to play a key role in cell expansion (Cosgrove 1998, 2000). It is likely that alkalinization of apoplast impacts cell wall rigidity and therefore growth by promoting intermolecular cross-links (Brady and Fry 1997). Besides, pH changes and Ca^{2+} fluxes, reactive oxygen species (ROS) are also likely to play a role in root gravitropism (Joo et al. 2001, 2005). ROS has been shown to accumulate on the lower side of gravity-stimulated roots and appear to act downstream of auxin as application of hydrogen peroxide (H_2O_2) can induce curvature even in NPA-treated roots (Joo et al. 2005). Auxin induces ROS production during root gravitropism by activating phosphatidylinositol 3-kinase (PtdIns 3 kinase), and pretreatment with an inhibitor of PtdIns 3-kinase activity blocked auxin-mediated ROS generation and reduced root gravitropic curvature (Joo et al. 2005).

Auxin has also been shown to regulate root gravitropism via nitric oxide (NO) and cGMP (Hu et al. 2005; Perera et al. 2006). Like ROS, NO has been reported to accumulate in the tissues on the lower side of gravity-stimulated roots (Hu et al. 2005). This pattern of NO accumulation can be blocked by addition of the auxin transport inhibitor NPA and restored by asymmetric application of auxin. Manipulating NO levels using either NO scavengers or inhibitors of NO synthase resulted in reduction in both NO accumulation and gravitropic bending underlining the physiological importance of NO in root gravitropism. NO can either signal by directly modifying proteins via nitrosylation (Durner and Klessig 1999) or elevating cGMP levels (Durner et al. 1998). During root gravitropism, auxin-induced NO appears to signal via cGMP because unilateral application of a membrane permeable analogue 8-Br-cGMP can restore curvature of NPA-treated roots (Hu et al. 2005). Nevertheless, treatment with NO or cGMP inhibitors did not completely block root gravitropism, suggesting that additional signals are required.

2. Later Events in Graviresponse Are Genomic

Though early events in gravitropism appear to be regulated by nongenomic mechanisms, the later phase of gravity response is dependent on auxin-regulated transcription (Abel and Theologis 1996) as several auxin-signaling mutants are agravitropic (Leyser 2006). The output of the TIR1 auxin-signaling pathway is dependent on the dynamic equilibrium of two key families of proteins. The first one includes auxin-response factors (ARFs) that are the transcription factors that bind to the auxin-response elements (AuxRE) in the promoters of auxin-regulated genes. The second one includes Aux/IAA proteins that are repressors of auxin responses (Tiwari et al. 2001, 2003; Leyser 2006, 2011; Kieffer et al. 2010). Aux/IAA proteins bind to the ARFs and keep them in an inactive state. In the presence of auxin, Aux/IAA proteins are targeted for degradation by the ubiquitin proteosome pathway mediated by the $SCF^{TIR1/AFB}$ ubiquitin-ligase complex (Dharmasiri et al. 2005a,b; Kepinski and Leyser 2005). Once Aux/IAA proteins are degraded, ARFs can initiate transcription from auxin-regulated promoters. Both ARFs and Aux/IAA proteins belong to large multigene families, and genetic, genomic, and molecular studies have identified the key molecular players regulating gravitropism. ARF7 and ARF19 appear to be the key ARFs regulating root gravitropism (Okushima et al. 2005). Single mutants in either are gravitropic, but double *arf7arf19* mutants are agravitropic, implying functional redundancy. As expected, auxin-regulated gene expression is severely impaired in the *arf7arf19* double mutant (Okushima et al. 2005). Similarly dominant gain of function mutants in aux/IAA proteins axr2, axr3, shy2, and slr1 are agravitropic (Leyser et al. 1996; Tian and Reed 1999; Nagpal et al. 2000; Fukaki et al. 2002; Weijers et al. 2005). Molecular studies have shown that ARF7 and ARF19 interact with IAA14/SLR1 and IAA3/SHY2 (Fukaki et al. 2005; Weijers et al. 2005). At this stage, it is not clear which genes are target for regulation. It appears that these target genes are required for the persistence of the asymmetric auxin gradient. For example, Yamada et al. (2009) showed that *TIR2*, which encodes a tryptophan aminotransferase (a key enzyme in auxin biosynthesis), is asymmetrically expressed on the lower side of the elongation zone of gravistimulated roots. Mutations in TIR2 reduce root gravitropism. These results suggest that local auxin biosynthesis may help to maintain the lateral auxin gradient between lower and upper side of a gravistimulated root. On the other hand, transcriptome analysis suggests that several cell wall-related genes including *AtXTH18* and *AtXTH19* are under ARF7 and ARF19 regulation in expanding root epidermal cells (Wilson and Bennett unpublished results). *AtXTH18* and *AtXTH19* encode cell wall enzymes that cross-link cell wall components causing inhibition of cell expansion (Vissenberg et al. 2005). It is tempting to speculate that initially through its nongenomic effect, auxin brings about rapid differential cell expansion, while later on, through its genomic effects, auxin regulates cell wall biosynthesis and remodeling.

V. Conclusion and Perspectives

There has been a significant increase in our understanding of root gravitropism, but many important questions remain to be answered. The gravity receptor detecting statolith sedimentation has yet to be identified. How is PIN3 asymmetrically localized in response to statolith sedimentation? What is the precise role of signals such as pH, NO, ROS, and IP3 during root gravitropism? What is the nature of the receptor that brings about the rapid nongenomic effects of auxin during gravitropism? Which genes are crucial targets for genomic regulation?

Though genetics has been and will remain at the forefront of our arsenal of experimental approaches to study root gravitropism, newer approaches and technologies need to be adopted to get deeper mechanistic insights. The development of novel imaging devices such as ROTATO which combine real-time image analysis with a computer-controlled clinostat has produced valuable information as to the threshold stimulus angle and presentation time and provided the first direct link between auxin flux and statolith sedimentation (Mullen et al. 2000; Wolverton et al. 2002b, 2011). While transcriptomic approaches will continue to underpin our knowledge of the later events of gravitropism (Zhu et al. 2002; Kimbrough et al. 2004), considering the speed at which gravitropic bending takes place, proteomics-based approaches are likely to unravel novel regulators during early stages of gravity perception and response (Young et al. 2006). In addition, chemical biology approaches may uncover novel inhibitors of root gravitropism (Yamazoe et al. 2005; Na et al. 2011).

In addition to defining the key components and intercellular pathways, efforts have also been made in recent years to develop a quantitative, rather than just a qualitative, understanding of root gravitropic signaling processes. Improvement in detection methods using machine vision and computation and development of novel auxin sensors (Figure 19.3) and image analysis tools have started to provide a more detailed quantitative insight into root gravitropism (French et al. 2009; Brooks et al. 2010; Miller et al. 2010; Band et al. 2012, Brunoud et al. 2012) and have contributed to the identification of novel gravitropic mutants (Miller et al. 2010).

In addition, advances in confocal microscopy that offer improved cellular and subcellular resolution with development of software to segment and track individual cells will add a new dimension to gravitropic research (Sethuraman et al. 2009) that has focused too much on the gene and organ scales in the past. Interpretation of such data will require the development of novel mathematical approaches (Chavarría-Krauser 2006; Chavarría-Krauser et al. 2008; Band et al. 2012) and modeling frameworks that can reflect the complex morphological dynamics of the graviresponse (e.g., OpenAlea; http://openalea.gforge.inria.fr). Such approaches will lead to development of experimentally testable, mathematically sound, and realistic multiscale models (Swarup et al. 2005; Laskowski et al. 2008; Kondrachuk and Starkov 2011) that will provide deeper insights into the mechanisms regulating root gravitropism.

Acknowledgments

The authors would like to acknowledge research support from the BBSRC, EPSRC, and European Space Agency. The authors would also like to acknowledge the BBSRC professorial fellowship to MJB and CISB award to CPIB.

References

Abas L, Benjamins R, Malenica N et al. 2006. Intracellular trafficking and proteolysis of the *Arabidopsis* auxin-efflux facilitator PIN2 are involved in root gravitropism. *Nat Cell Biol* 8: 249–256.

Abel S, Theologis A. 1996. Early genes and auxin action. *Plant Physiol* 111: 9–17.

Aloni R, Langhans M, Aloni E, Ullrich CI. 2004. Role of cytokinin in the regulation of root gravitropism. *Planta* 220: 177–182.

Band LR, Wells DM, Larrieu A et al. 2012. Root gravitropism is regulated by a transient lateral auxin gradient controlled by a novel tipping point mechanism. *Proc Natl Acad Sci USA* 109: 4668–4673.

Barlow PW. 1974. Regeneration of cap of primary roots of Zea-mays. *New Phytol* 73: 937.

Barlow PW. 1995. Gravity perception in plants: A multiplicity of systems derived by evolution? *Plant Cell Environ* 18: 951–962.

Basu P, Zhang YJ, Lynch JP, Brown KM. 2007. Ethylene modulates genetic, positional, and nutritional regulation of root plagiogravitropism. *Funct Plant Biol* 34: 41–51.

Beemster G, Baskins T. 2001. Stunted Plant1 mediates effects of cytokinin, not auxin, on cell division and expansion in the root of *Arabidopsis*. *Plant Physiol* 124: 1718–1727.

Bennett MJ, Marchant A, Green HG et al. 1996. *Arabidopsis AUX1* gene: A permease-like regulator of root gravitropism. *Science* 273: 948–950.

Blakeslee JJ, Bandyopadhyay A, Lee OR et al. 2007. Interactions among PIN-FORMED and P-glycoprotein auxin transporters in *Arabidopsis*. *Plant Cell* 19: 131–147.

Blancaflor EB, Fasano JM, Gilroy S. 1998. Mapping the functional roles of cap cells in the response of *Arabidopsis* primary roots to gravity. *Plant Physiol* 116: 213–222.

Blancaflor EB, Masson PH. 2003. Plant gravitropism. Unraveling the ups and downs of a complex process. *Plant Physiol* 133: 1677–1690.

Blilou I, Xu J, Wildwater M et al. 2005. The PIN auxin efflux facilitator network controls growth and patterning in *Arabidopsis* roots. *Nature* 433: 39–44.

Boonsirichai K, Guan C, Chen R, Masson PH. 2002. Root gravitropism: An experimental tool to investigate basic cellular and molecular processes underlying mechanosensing and signal transmission in plants. *Annu Rev Plant Biol* 53: 421–447.

Boonsirichai K, Sedbrook JC, Chen RJ, Gilroy S, Masson PH. 2003. Altered response to gravity is a peripheral membrane protein that modulates gravity-induced cytoplasmic alkalinization and lateral auxin transport in plant statocytes. *Plant Cell* 15: 2612–2625.

Brady J, Fry SC. 1997. Formation of Di-isodityrosine and loss of isodityrosine in the cell walls of tomato cell-suspension cultures treated with fungal elicitors or H_2O_2. *Plant Physiol* 115: 87–92.

Brooks T, Miller N, Spalding EP. 2010. Plasticity of *Arabidopsis* root gravitropism throughout a multidimensional condition space quantified by automated image analysis. *Plant Physiol* 152: 206–216.

Brunoud G, Wells DM, Oliva M et al. 2012 A novel sensor to map auxin response and distribution at high spatio-temporal resolution. *Nature* 482: 103–108.

Buer CS, Sukumar P, Muday, GK. 2006. Ethylene modulates flavonoid accumulation and gravitropic responses in roots of *Arabidopsis*. *Plant Physiol* 140: 1384–1396.

Chavarría-Krauser A. 2006. Quantification of curvature production in cylindrical organs such as roots and hypocotyls. *New Phytol* 171: 633–641.

Chavarría-Krauser A, Nagel KA, Palme K, Schurr U, Walter A, Scharr H. 2008. Spatio-temporal quantification of differential growth processes in root growth zones based on a novel combination of image sequence processing and refined concepts describing curvature production. *New Phytol* 177: 811–821.

Chen RJ, Hilson P, Sedbrook J, Rosen E, Caspar T, Masson PH. 1998. The *Arabidopsis thaliana AGRAVITROPIC1* gene encodes a component of the polar-auxin-transport efflux carrier. *Proc Natl Acad Sci USA* 95: 15112–15117.

Cholodny N. 1927. Wuchshormone und Tropismen bei den Panzen. *Biol Zent* 47: 604–626.

Collings DA, Zsuppan G, Allen NS, Blancaflor EB. 2001. Demonstration of prominent actin filaments in the root columella. *Planta* 212: 392–403.

Cosgrove DJ. 1998. Cell wall loosening by expansins. *Plant Physiol* 118: 333–339.

Cosgrove DJ. 2000. Loosening of plant cell walls by expansions. *Nature* 407: 321–326.

Darwin C. 1881. *The Power of Movement in Plants*. London, U.K.: John Murray.

Deruere J, Jackson K, Garbers C, Soll D, DeLong A. 1999. The RCN1-encoded a subunit of protein phosphatase 2A increases phosphatase activity *in vivo*. *Plant J* 20: 389–399.

Dharmasiri N, Dharmasiri S, Estelle M. 2005a. The F-box protein TIR1 is an auxin receptor. *Nature* 435: 441–445.

Dharmasiri N, Dharmasiri S, Weijers D et al. 2005b. Plant development is regulated by a family of auxin receptor f box proteins. *Dev Cell* 9: 109–119.

Dharmasiri S, Swarup R, Mockaitis K et al. 2006. AXR4 is required for localization of the auxin influx facilitator AUX1. *Science* 312: 1218–1220.

Dhonukshe P, Huang F, Galvan-Ampudia C et al. 2010. Plasma membrane-bound AGC3 kinases phosphorylate PIN auxin carriers at TPRXS(N/S) motifs to direct apical PIN recycling. *Development* 137: 3245–3255.

Dodart L. 1700. Sur l' affection de la perpendiculaire remarquable dans toutes les tiges, dans plusieurs raciness, et autant qu'il est possible dans toutes les branches des plantes. *Hist Acad Roy Sci* 61:47–63.

Durner J, Klessig DF. 1999. Nitric oxide as a signal in plants. *Curr Opin Plant Biol* 2: 396–374.

Durner J, Wendehenne D, Klessig DF. 1998. Defense gene induction in tobacco by nitric oxide, cyclic GMP, and cyclic ADPribose. *Proc Natl Acad Sci USA* 95: 10328–10333.

Effendi Y, Rietz S, Fischer U, Scherer G. 2011. The heterozygous abp1/ABP1 insertional mutant has defects in functions requiring polar auxin transport and in regulation of early auxin-regulated genes. *Plant J* 65: 282–294.

Esmon CA, Pedmale UV, Liscum E. 2005. Plant tropisms: Providing the power of movement to a sessile organism. *Int J Dev Biol* 49: 665–674.

Evans ML. 1986. Role of calcium in gravity perception of plant roots. *Adv Space Res* 6: 61–65.

Fasano JM, Swanson SJ, Blancaflor EB, Dowd PE, Kao TH, Gilroy S. 2001. Changes in root cap pH are required for the gravity response of the *Arabidopsis* root. *Plant Cell* 13: 907–921.

Feijo JA et al. 1999. Growing pollen tubes possess a constitutive alkaline band in the clear zone and a growth-dependent acidic tip. *J Cell Biol* 144: 483–496.

Fitzelle KJ, Kiss JZ. 2001. Restoration of gravitropic sensitivity in starch-deficient mutants of *Arabidopsis* by hypergravity. *J Exp Bot* 52: 265–275.

French A, Ubeda-Tomás S, Holman TJ, Bennett MJ, Pridmore T. 2009. High-throughput quantification of root growth using a novel image-analysis tool. *Plant Physiol* 150: 1784–1795.

Friml J, Wisniewska J, Benkova E, Mendgen K, Palme K. 2002. Lateral relocation of auxin efflux regulator PIN3 mediates tropism in *Arabidopsis*. *Nature* 415: 806–809.

Fukaki H, Nakao Y, Okushima Y, Theologis A, Tasaka M. 2005. Tissue-specific expression of stabilized SOLITARY-ROOT/IAA14 alters lateral root development in *Arabidopsis*. *Plant J* 44: 382–395.

Fukaki H, Tameda S, Masuda H, Tasaka M. 2002. Lateral root formation is blocked by a gain-of-function mutation in the SOLITARY-ROOT/IAA14 gene of *Arabidopsis*. *Plant J* 29: 153–168.

Garbers C, DeLong A, Deruere J, Bernasconi P, Soll D. 1996. A mutation in protein phosphatase 2A regulatory subunit A affects auxin transport in *Arabidopsis*. *EMBO J* 15: 2115–2124.

Geldner N, Anders N, Wolters H et al. 2003. The *Arabidopsis* GNOM ARF-GEF mediates endosomal recycling, auxin transport, and auxin-dependent plant growth. *Cell* 112: 219–230.

Geldner N, Richter S, Vieten A et al. 2004. Partial loss-of-function alleles reveal a role for *GNOM* in auxin transport-related, post-embryonic development of *Arabidopsis*. *Development* 131: 389–400.

Guan CH, Rosen ES, Boonsirichai K, Poff KL, Masson PH. 2003. The ARG1-LIKE2 gene of *Arabidopsis* functions in a gravity signal transduction pathway that is genetically distinct from the PGM pathway. *Plant Physiol* 133: 100–112.

Han W, Rong H, Zhang H, Wanga M-H. 2009. Abscisic acid is a negative regulator of root gravitropism in *Arabidopsis thaliana*. *Biochem Biophys Res Commun* 378: 695–700.

Hobbie L, Estelle M. 1995. The *axr4* auxin-resistant mutants of *Arabidopsis thaliana* define a gene important for root gravitropism and lateral root initiation. *Plant J* 7: 211–220.

Hou GC, Kramer VL, Wang YS et al. 2004. The promotion of gravitropism in *Arabidopsis* roots upon actin disruption is coupled with the extended alkalinization of the columella cytoplasm and a persistent lateral auxin gradient. *Plant J* 39: 113–125.

Hu X, Neill SJ, Tang Z, Cai W. 2005. Nitric oxide mediates gravitropic bending in Soybean roots. *Plant Physiol* 137: 663–670.

Huang F, Zago M, Abas L, Marion A, Galván-Ampudia C, Offringa R. 2010. Phosphorylation of conserved pin motifs directs *Arabidopsis* PIN1 polarity and auxin transport. *Plant Cell* 22: 1129–1142.

Ishikawa H, Evans ML. 1997. Novel software for analysis of root gravitropism: Comparative response patterns of *Arabidopsis* wild-type and *axr1* seedlings. *Plant Cell Environ* 20: 919–928.

Joo JH, Bae YS, Lee JS. 2001. Role of auxin-induced reactive oxygen species in root gravitropism. *Plant Physiol* 126: 1055–1060.

Joo JH, Yoo HJ, Hwang I, Lee JS, Nam KH, Bae YS. 2005. Auxin-induced reactive oxygen species production requires the activation of phosphatidylinositol 3-kinase. *FEBS Lett* 579: 1243–1248.

Jouhet J, Gray JC. 2009. Interaction of actin and the chloroplast protein import apparatus. *J Biol Chem* 284: 19132–19141.

Kepinski S, Leyser O. 2005. The *Arabidopsis* F-box protein TIR1 is an auxin receptor. *Nature* 435: 446–451.

Kieffer M, Neve J, Kepinski S. 2010. Defining auxin response contexts in plant development. *Curr Opin Plant Biol* 13: 12–20.

Kimbrough JM, Salinas-Mondragon R, Boss WE, Brown CS, Sederoff HW. 2004. The fast and transient transcriptional network of gravity and mechanical stimulation in the *Arabidopsis* root Apex. *Plant Physiol* 136: 2790–2805.

Kiss JZ, Wright, JB, Caspar T. 1996. Gravitropism in roots of intermediate-starch mutants of *Arabidopsis. Physiol Plant* 97: 237–244.

Kleine-Vehn J, Dinga Z, Jones R, Tasaka M, Morita M, Friml J. 2010. Gravity-induced PIN transcytosis for polarization of auxin fluxes in gravity-sensing root cells. *Proc Natl Acad Sci USA* 107: 22344–22349.

Knight TA. 1806. On the direction of the radicle and germen during the vegetation of seeds. *Phil Trans R Soc Lond B* 99: 108–120.

Kondrachuk AV, Starkov VN. 2011. Modeling the kinetics of root gravireaction. *Microgr Sci Technol* 23: 221–225.

Konings H. 1995. Gravitropism of roots—An evaluation of progress during the last 3 decades. *Acta Bot Neerl* 44: 195–223.

Kramer EM, Bennett MJ. 2006. Auxin transport: A field in flux. *Trends Plant Sci* 11: 382–386.

Kutschera U. 1989. Tissue stresses in growing plant organs. *Physiol Plant* 77: 157–163.

Kuznetsov OA, Hasenstein KH. 1996. Intracellular magnetophoresis of amyloplasts and induction of root curvature. *Planta* 198: 87–94.

LaMotte CE, Pickard BG. 2004. Control of gravitropic orientation. II. Dual receptor model for Gravitropism. *Funct Plant Biol* 31: 109–120.

Laskowski M, Grieneisen VA, Hofhuis H et al. 2008. Root system architecture from coupling cell shape to auxin transport. *PLoS Biol* 6: e307.

Lee J, Mulkey T, Evans ML. 1983. Gravity-induced polar transport of calcium across root tips of maize. *Plant Physiol* 73: 874–876.

Lee J, Mulkey T, Evans ML. 1984. Inhibition of polar calcium movement and gravitropism in roots treated with auxin-transport inhibitors. *Planta* 160: 536–543.

Leitz G, Kang B-H, Schoenwaelder M, Staehelin A. 2009. Statolith sedimentation kinetics and force transduction to the cortical endoplasmic reticulum in gravity-sensing *Arabidopsis* columella cells. *Plant Cell* 21: 843–860.

Leyser O. 2006. Dynamic integration of auxin transport and signalling. *Curr Biol* 16: R424–R433.

Leyser O. 2011. Auxin, self-organization, and the colonial nature of plants. *Curr Biol* 21: R331–R337.

Leyser HMO, Lincoln CA, Timpte C, Lammer D, Turner J, Estelle M. 1993. *Arabidopsis* auxin-resistance gene *AXR1* encodes a protein related to ubiquitin-activating enzyme-E1. *Nature* 364: 161–164.

Leyser HMO, Pickett FB, Dharmasiri S, Estelle M. 1996. Mutations in the AXR3 gene of *Arabidopsis* result in altered auxin response including ectopic expression from the SAUR-AC1 promoter. *Plant J* 10: 403–413.

Li L, Xu J, Xu ZH, Xue HW. 2005. Brassinosteroids stimulate plant tropisms through modulation of polar auxin transport in *Brassica* and *Arabidopsis*. *Plant Cell* 17: 2738–2753.

Li G, Xue HW. 2007. *Arabidopsis* PLDζ2 regulates vesicle trafficking and is required for auxin response. *Plant Cell* 19: 281–295.

Lin RC, Wang HY. 2005. Two homologous ATP-binding cassette transporter proteins, AtMDR1 and AtPGP1, regulate *Arabidopsis* photomorphogenesis and root development by mediating polar auxin transport. *Plant Physiol* 138: 949–964.

Luschnig C, Gaxiola R, Grisafi P, Fink G. 1998. EIR1, a root-specific protein involved in auxin transport, is required for gravitropism in *Arabidopsis thaliana. Genes Dev* 12: 2175–2187.

Lynch JP, Brown KM. 2001. Topsoil foraging—An architectural adaptation of plants to low phosphorus availability. *Plant Soil* 237: 225–237.

Ma Z, Hasenstein KH. 2006. The onset of gravisensitivity in the embryonic root of flax. *Plant Physiol* 140: 159–166.

MacCleery SA, Kiss JZ. 1999. Plastid sedimentation kinetics in roots of wild-type and starch-deficient mutants of *Arabidopsis. Plant Physiol* 120: 183–192.

Marchant A, Kargul J, May ST et al. 1999. AUX1 regulates root gravitropism in *Arabidopsis* by facilitating auxin uptake within root apical tissues. *EMBO J* 18: 2066–2073.

Miller ND, Brooks T, Assadi A, Spalding EP. 2010. Detection of a gravitropism phenotype in glutamate receptor-like 3.3 mutants of *Arabidopsis thaliana* using machine vision and computation. *Genetics* 186: 585–593.

Molas ML, Kiss JZ. 2009. Phototropism and gravitropism in plants. *Adv Bot Res* 49: 1–34.

Molisch H. 1884. Ueber die Ablenkung der Wurzeln von ihrer normalen Wachsthumsrichtung durch Gase (Aerotropismus). *Ber Dtsch Bot Ges* 2: 160–169.

Monshausen GB, Bibikova TN, Messerli MA, Shi C, Gilroy S. 2007. Oscillations in extracellular pH and reactive oxygen species modulate tip growth of *Arabidopsis* root hairs. *Proc Natl Acad Sci USA* 104: 20996–21001.

Monshausen GB, Messerli MA, Gilroy S. 2008. Imaging of the Yellow Cameleon 3.6 indicator reveals that elevations in cytosolic Ca^{++} follow oscillating increases in growth in root hairs of *Arabidopsis. Plant Physiol* 147: 1690–1698.

Monshausen GB, Miller ND, AS, Gilroy S. 2011. Dynamics of auxin-dependent Ca^{2+} and pH signaling in root growth revealed by integrating high-resolution imaging with automated computer vision-based analysis. *Plant J* 65: 309–318.

Morita MT. 2010 Directional gravity sensing in gravitropism. *Annu Rev Plant Biol* 61: 705–720.

Morita MT, Tasaka M. 2004. Gravity sensing and signalling. *Curr Opin Plant Biol* 7: 712–718.

Muday GK. 2001. Auxins and tropisms. *J Plant Growth Regul* 20: 226–243.

Mullen JL, Hangarter RP. 2003. Genetic analysis of the gravitropic setpoint angle in lateral roots of *Arabidopsis*. *Adv Space Res* 31: 2229–2236.

Mullen JL, Ishikawa H, Evans ML. 1998a. Analysis of changes in relative elemental growth rate patterns in the elongation zone of *Arabidopsis* roots upon gravistimulation. *Planta* 206: 598–603.

Mullen JL, Turk E, Johnson K et al. 1998b. Root-growth behavior of the *Arabidopsis* mutant rgr1—Roles of gravitropism and circumnutation in the waving/coiling phenomenon. *Plant Physiol* 118: 1139–1145.

Mullen JL, Wolverton C, Ishikawa H, Evans ML. 2000. Kinetics of constant gravitropic stimulus responses in *Arabidopsis* roots using a feedback system. *Plant Physiol* 123: 665–670.

Müller A, Guan C, Gälweiler L et al. 1998. *AtPIN2* defines a locus of *Arabidopsis* for root gravitropism control. *EMBO J* 17: 6903–6911.

Na X, Hu Y, Yue K et al. 2011. Narciclasine modulates polar auxin transport in *Arabidopsis* roots. *J Plant Physiol* 168: 1149–1156.

Nagpal P, Walker LM, Young JC et al. 2000. *AXR2* encodes a member of the Aux/IAA protein family. *Plant Physiol* 123: 563–573.

Okushima Y, Overvoorde PJ, Arima K et al. 2005. Functional genomic analysis of the auxin response factor gene family members in *Arabidopsis thaliana*: Unique and overlapping functions of ARF7 and ARF19. *Plant Cell* 17: 444–463.

Ottenschlager I, Wolff P, Wolverton C et al. 2003. Gravity-regulated differential auxin transport from columella to lateral root cap cells. *Proc Natl Acad Sci USA* 100: 2987–2991.

Paciorek T, Zazimalova E, Ruthardt N et al. 2005. Auxin inhibits endocytosis and promotes its own efflux from cells. *Nature* 435: 1251–1256.

Perbal G. 2009. From roots to GRAVI-1: Twenty five years for understanding how plants sense gravity. *Microgravity Sci Technol* 21: 3–10.

Perbal G, Driss-Ecole D. 2003. Mechanotransduction in gravisensing cells. *Trends Plant Sci* 8: 498–504.

Perera IY, Hung CY, Brady S, Muday GK, Boss WF. 2006. A universal role for inositol 1,4,5-trisphosphate-mediated signaling in plant gravitropism. *Plant Physiol* 140: 746–760.

Philosoph-Hadas S, Friedman H, Meir S. 2005. Gravitropic bending and plant hormones. *Vitam Horm* 72: 31–78.

Porterfield DM. 2002. The biophysical limitations in physiological transport and exchange in plants grown in microgravity. *J Plant Growth Regul* 21: 177–190.

Rahman A, Takahashi M, Shibasaki A et al. 2010. Gravitropism of *Arabidopsis thaliana* roots requires the polarization of PIN2 toward the root tip in meristematic cortical cells. *Plant Cell* 22: 1762–1776.

Rashotte AM, Brady SR, Reed RC, Ante SJ, Muday GK. 2000. Basipetal auxin transport is required for gravitropism in roots of *Arabidopsis*. *Plant Physiol* 122: 481–490.

Rashotte AM, DeLong A, Muday GK. 2001. Genetic and chemical reductions in protein phosphatase activity alter auxin transport, gravity response and lateral root growth. *Plant Cell* 13: 1683–1697.

Reboulet J, Kumar P, Kiss J. 2010. DIS1 and DIS2 play a role in tropisms in *Arabidopsis thaliana*. *Env Exp Bot* 67: 474–478.

Robert S, Kleine-Vehn J, Barbez Elke et al. 2010. ABP1 mediates auxin inhibition of clathrin dependent endocytosis in *Arabidopsis*. *Cell* 143: 111–121.

Robert HS, Offringa R. 2008. Regulation of auxin transport polarity by AGC kinases. *Curr Opin Plant Biol* 11: 495–502.

Rouse D, Mackay P, Stirnberg P, Estelle M, Leyser, HMO. 1998. Changes in auxin response from mutations in an *AUX/IAA* gene. *Science* 279: 1371–1373.

Ruegger M, Dewey E, Gray WM, Hobbie L, Turner J, Estelle M. 1998. The TIR1 protein of *Arabidopsis* functions in auxin response and is related to human SKP2 and yeast Grr1p. *Genes Dev* 12: 198–207.

Sack FD. 1991. Plant gravity sensing. *Int Rev Cytol* 127: 193–252.

Sack FD. 1997. Plastids and gravitropic sensing. *Planta* 203: S63–S68.

Santelia D, Vincenzetti V, Azzarello E et al. 2005. MDR-like ABC transporter AtPGP4 is involved in auxin-mediated lateral root and root hair development. *FEBS Lett* 579: 5399–5406.

Scott AC, Allen NS. 1999. Changes in cytosolic pH within *Arabidopsis* root columella cells play a key role in the early signaling pathway for root gravitropism. *Plant Physiol* 121: 1291–1298.

Sedbrook JC, Chen RJ, Masson PH. 1999. ARG1 (Altered Response to Gravity) encodes a DnaJ-like protein that potentially interacts with the cytoskeleton. *Proc Natl Acad Sci USA* 96: 1140–1145.

Sethuraman V, Taylor S, Pridmore T, French A, Wells D. 2009. Segmentation and tracking of confocal images of *Arabidopsis thaliana* root cells using automatically-initialized network snakes. In *Proceedings of 3rd International Conference on Bioinformatics and Biomedical Engineering*, June 11–13, 2009, pp. 2243–2246. Piscataway, NJ: Institute of Electrical and Electronics Engineers.

Shin HS, Guo ZB, Blancaflor EB, Masson PH, Chen RJ. 2005. Complex regulation of *Arabidopsis* AGR1/PIN2-mediated root gravitropic response and basipetal auxin transport by cantharidin-sensitive protein phosphatases. *Plant J* 42: 188–200.

Sievers A, Buchen B, Volkman D, Hejnowicz Z. 1991. Role of cytoskeleton in gravity perception. In *The Cytoskeletal Basis of Plant Growth and Form* (ed.), CW Lloyd, pp. 169–182. London, U.K.: Academic Press.

Stanga J, Boonsirichai K, Sedbrook J, Otegui M, Masson PH. 2009. A role for the TOC complex in *Arabidopsis* root gravitropism. *Plant Physiol* 149: 1896–1905.

Sukumar P, Edwards K, Rahman A, DeLong A, Muday GK. 2009. Pinoid kinase regulates root gravitropism through modulation of PIN2-dependent basipetal auxin transport in *Arabidopsis. Plant Physiol* 150: 722–735.

Swarup R, Bennett MJ. 2009. Root gravitropism. *Ann Plant Rev* 37: 157–174.

Swarup R, Friml J, Marchant A et al. 2001. Localization of the auxin permease AUX1 suggests two functionally distinct hormone transport pathways operate in the *Arabidopsis* root apex. *Genes Dev* 15: 2648–2653.

Swarup R, Kargul J, Marchant A et al. 2004. Structure-function analysis of the presumptive *Arabidopsis* auxin permease AUX1. *Plant Cell* 16: 3069–3083.

Swarup R, Kramer EM, Perry P et al. 2005. Root gravitropism requires lateral root cap and epidermal cells for transport and response to a mobile auxin signal. *Nat Cell Biol* 7: 1057–1065.

Swarup R, Perry P, Hagenbeek D et al. 2007. Ethylene up regulates auxin biosynthesis in *Arabidopsis* seedlings to enhance inhibition of root cell elongation. *Plant Cell* 19: 2186–2196.

Tanaka A, Kobayashi Y, Hase Y, Watanabe H. 2002. Positional effect of cell inactivation on root gravitropism using heavy-ion microbeams. *J Exp Bot* 53: 683–687.

Tasaka M, Kato T, Fukaki H. 2001. Genetic regulation of gravitropism in higher plants. *Int Rev Cytol* 206: 135–154.

Terasaka K, Blakeslee JJ, Titapiwatanakun B et al. 2005. PGP4, an ATP binding casette p-glycoprotein, catalyzes auxin transport in *Arabidopsis thaliana* roots. *Plant Cell* 17: 2922–2939.

Tian Q, Reed JW. 1999. Control of auxin-regulated root development by the *Arabidopsis thaliana* SHY2/IAA3 gene. *Development* 126: 711–721.

Timpte C, Wilson AK, Estelle M. 1994. The *AXR2-1* mutation of *Arabidopsis thaliana* is a gain-of-function mutation that disrupts an early step in auxin response. *Genetics* 138: 1239–1249.

Tiwari SB, Hagen G, Guilfoyle T. 2003. The roles of auxin response factor domains in auxin-responsive transcription. *Plant Cell* 15: 533–543.

Tiwari SB, Wang X-J, Hagen G, Guilfoyle TJ. 2001. AUX/IAA proteins are active repressors, and their stability and activity are modulated by auxin. *Plant Cell* 13: 2809–2822.

Tromas A, Braun N, Muller P et al. 2009. The auxin binding protein 1 is required for differential auxin responses mediating root growth. *PLoS One* 4: e6648.

Tromas A, Paponov I, Perrot-Rechenmann C. 2010. Auxin binding protein 1: Functional and evolutionary aspects. *Trends Plant Sci* 15: 436–446.

Vicente-Agullo F, Rigas S, Desbrosses G, Dolan L, Hatzopoulos P, Grabov A. 2004. Potassium carrier TRH1 is required for auxin transport in *Arabidopsis* roots. *Plant J* 40: 523–535.

Vissenberg K, Oyama M, Osato Y, Yokoyama R, Verbelen J-P, Nishitani K. 2005. Differential expression of AtXTH17, AtXTH18, AtXTH19 and AtXTH20 genes in *Arabidopsis* roots. Physiological roles in specification in cell wall construction. *Plant Cell Physiol* 46: 192–200.

Weijers D, Benkova E, Jager KE et al. 2005. Developmental specificity of auxin response by pairs of ARF and Aux/IAA transcriptional regulators. *EMBO J* 24: 1874–1885.

Went F. 1926. On growth accelerating substances in the coleoptile of *Avena sativa. Proc Kon Ned Akad Weten* 30: 1019.

Wiśniewska J, Xu J, Seifertová D et al. 2006. Polar PIN localization directs auxin flow in plants. *Science* 312: 883.

Wolverton C, Ishikawa H, Evans ML. 2002a. The kinetics of root gravitropism: Dual motors and sensors. *J Plant Growth Regul* 21: 102–112.

Wolverton C, Mullen JL, Ishikawa H, Evans ML. 2002b. Root gravitropism in response to a signal originating outside of the cap. *Planta* 215: 153–157.

Wolverton C, Paya AM, Toska J. 2011. Root cap angle and gravitropic response rate are uncoupled in the *Arabidopsis pgm-1* mutant. *Physiol Plant* 141: 373–382.

Xu T, Wen M, Nagawa S, Fu Y et al. 2010. Cell surface- and rho GTPase-based auxin signaling controls cellular interdigitation in *Arabidopsis.* Cell 143: 99–110.

Yamada Y, Greenham K, Prigge MJ, Jensen PJ, Estelle M. 2009. The transport inhibitor response 2 gene is required for auxin synthesis and diverse aspects of plant development. *Plant Physiol* 151: 168–179.

Yamazoe A, Hayashi K, Kepinski S, Leyser O, Nozaki H. 2005. Characterization of terfestatin A, a new specific inhibitor for auxin signalling. *Plant Physiol* 139: 779–789.

Yang YD, Hammes UZ, Taylor CG, Schachtman DP, Nielsen E. 2006. High-affinity auxin transport by the AUX1 influx carrier protein. *Curr Biol* 16: 1123–1127.

Young LS, Harrison BR, Narayana MUM, Moffatt BA, Gilroy S, Masson PH. 2006. Adenosine kinase modulates root gravitropism and cap morphogenesis in *Arabidopsis. Plant Physiol* 142: 564–573.

Zhang J, Nodzynski T, Pencík A, Rolcík J, Friml J. 2010. PIN phosphorylation is sufficient to mediate PIN polarity and direct auxin transport. *Proc Natl Acad Sci USA* 107: 918–922.

Zheng HQ, Staehelin LA. 2001. Nodal endoplasmic reticulum, a specialized form of endoplasmic reticulum found in gravity-sensing root tip columella cells. *Plant Physiol* 125: 252–265.

Zhu T, Chang HS, Wang X, Feldman LJ. 2002. Transcription profiling of the early gravitropic response in *Arabidopsis* using high-density oligonucleotide probe microarrays. *Plant Physiol* 130: 720–728.

Zourelidou M, Müller I, Willige BO, Nill C, Jikumaru Y, Li H, Schwechheimer C. 2009. The polarly localized D6 protein kinase is required for efficient auxin transport in *Arabidopsis thaliana. Development* 136: 627–636.

20

Calcium: From Root Macronutrient to Mechanical Signal

Sarah Swanson
University of Wisconsin

Simon Gilroy
University of Wisconsin

I. Calcium as a Macronutrient

Calcium is an essential plant macronutrient (Maathuis 2009). Intracellularly, Ca^{2+} regulates the activity of a host of enzymes and regulatory proteins (Dodd et al. 2010), and in the apoplast, this ion has long been recognized as a key structural component of the pectin-rich regions of the cell wall. Here it plays a role in coordinating between the acidic groups on pectin to increase wall rigidity (Jones and Lunt 1967). Pectins make up to 35% of the primary cell wall polymers in dicots (2%–10% in the grasses) and approximately 5% in woody tissues (Mohnen 2008). Thus, pectin structure has an important impact on the mechanical properties of the wall. Roots do show a great deal of variability in pectin content related to both genotypic and environmental factors. For example, closely related buckwheat and maize cultivars can exhibit significant differences in amounts of pectin in the root that change in response to stresses such as the presence of the toxic metal Al^{3+} (Dejene et al. 2005; Eticha et al. 2005; Yang et al. 2011). Such regulation of the amount and placement of pectin provides the plant with one mechanism to control wall properties. However, variations in the availability of apoplastic Ca^{2+} and in the masking of the carboxylic groups on the pectin polymer by esterification and de-esterification (pectin methylesterases) provide another avenue for the plant to modulate the rigidity of the wall. The dynamic nature of such changes in Ca^{2+} levels and esterification makes these Ca^{2+}-pectin interactions an important feature in the remodeling of wall properties essential for the regulation of turgor-driven growth (Wolf et al. 2009). Not surprisingly therefore, mutants in this regulatory system, such as those in the pectin methylesterases, have defects on growth and developmental patterning (e.g., Jiang et al. 2005; Bosch and Hepler 2006; Tian et al. 2006; Peaucelle et al. 2008).

Calcium also plays other key large-scale structural roles in the plant. For example, Ca^{2+} is essential in stabilizing membranes through coordinating with the phosphates of the phospholipid bilayer (Papahadjopoulos and Ohki 1969), and reducing apoplastic Ca^{2+} availability rapidly leads to solute leakage from the cells of the plant. These widespread structural roles explain (1) the high Ca^{2+} requirement to support plant growth and development and (2) the need for a high-capacity Ca^{2+} delivery mechanism to supply the Ca^{2+} extracted from the soil by the roots to both their own cells and those throughout the rest of the plant body.

Although Ca^{2+} is relatively immobile in the plant, being rapidly sequestered, it is readily redistributed from root to shoot through the xylem. The extremely low Ca^{2+} levels found in the phloem mean this tissue is not likely to be a major highway for Ca^{2+} movement. Xylem Ca^{2+} arrives from the soil via both apoplastic and symplastic routes, although the precise relative contribution of each remains unknown (White 2001; White and Broadley 2003; Cholewa and Peterson 2004; Hayter and Peterson 2004). The symplastic route requires initial channel-mediated uptake across root cell plasma membranes that is thought to occur through Ca^{2+}-selective and unselective channels (Demidchik and Maathuis 2007). Such symplastic uptake/transport supports both the Ca^{2+} requirement of the cells of the root and possible transport to the xylem for export to the shoot. Although symplastic cell-to-cell transport should allow Ca^{2+} to permeate to the stele, the highly immobile nature of cytoplasmic Ca^{2+} and the extremely low cytosolic Ca^{2+} concentrations suggest this could be a slow process. Sustaining a significant flux by this mechanism would require complex coordination and integration of Ca^{2+} transporters at the plasma and internal membranes across the root. The apoplastic component of root Ca^{2+} transport, with its relatively high ion mobility, therefore likely supports much of the flow to the stele. Ca^{2+} movement through the apoplastic space also allows the root to balance uptake and use by root cells (the symplastic route) with the delivery of the large amounts of Ca^{2+} needed to be exported through the xylem to sustain shoot function. Here Ca^{2+} would permeate the apoplast at relatively high concentrations in equilibrium with the soil up to the Casperian strip of the endodermis. At this point, Ca^{2+} may continue in a wholly apoplastic route to the xylem or be forced to cross the endodermal cell plasma membrane and transit through the cytoplasm of these cells before entering the vasculature. Specialized endodermal cells called passage cells are thought to be important for this process. This is because, although they do form a casperian strip, passage cells do not undergo the further extensive deposition of suberin and lignin seen elsewhere in the endodermis (White 2001). These cells therefore retain a plasma membrane that is relatively exposed to the apoplastic space that is in contact with the soil. This symplastic excursion would provide a point at which Ca^{2+} fluxes into the plant can be monitored and controlled but the capacity of such a transport pathway to deliver enough Ca^{2+} to the vasculature has been debated (White 2001; Hayter and Peterson 2004). Although the Ca^{2+} transporters and regulators responsible for the flux to the xylem remain unknown, a strong candidate for part of this system has recently emerged in *Arabidopsis* in the Ca^{2+}-binding protein MCA2 (Yamanaka et al. 2010). Although MCA2 may not directly be an influx channel itself, it does appear to be a component that controls Ca^{2+} uptake capacity. MCA2 is discussed in more detail in Section III.A of this chapter.

Calcium levels fluctuate in the plant dependent on soil and environmental conditions and the Ca^{2+} usage related to organ growth and development. Considering the essential role that Ca^{2+} plays, not surprisingly the plant checks and controls these changes in Ca^{2+} supply and redistribution. For example, the diurnal fluctuations in the rate of Ca^{2+} delivery from the roots to shoots appear to be closely monitored by stomata in the aerial parts of the plant in order to coordinate Ca^{2+} uptake with cellular requirements (Tang et al. 2007). This Ca^{2+} sensing operates through a plastid-localized extracellular Ca^{2+} sensor (CAS; Han et al. 2003; Nomura et al. 2008) that likely modulates H_2O_2- and NO-dependent signaling cascades in the guard cell (Wang et al. 2012b). These signaling events open and close the stomata and so control rates of transpiration and their associated rates of Ca^{2+} delivery. Although this extracellular Ca^{2+} monitoring function of CAS is strongly supported by the literature, precisely how a protein found in the inner thylakoid membrane can sense extracellular Ca^{2+} levels to trigger this guard cell signaling network remains obscure.

Further clues as to the molecular components of plant Ca^{2+} sensing/adaptation networks come from the Ca^{2+}-hypersensitive phenotypes seen in some Ca^{2+} channel mutants such as GLR1.1 (Kang and Turano 2003) and CNGC2 (Chan et al. 2003). Ca^{2+} channel knockouts might be expected to have impaired Ca^{2+} uptake capacities and to show resistance to elevated extracellular Ca^{2+}. Their actual hypersensitive phenotypes hint at a possible role in Ca^{2+} sensing. For example, CNGC2 is a cyclic nucleotide-gated Ca^{2+} channel (Leng et al. 1999) that plays roles in NO-mediated defense signaling (Ali et al. 2007), but a knockout line also shows hypersensitivity to the Ca^{2+} levels (10 mM) that support vigorous growth in wild-type plants (Chan et al. 2003). Plants exhibit complex transcriptional changes to increased or decreased Ca^{2+} supply (Maathuis et al. 2003; Chan et al. 2008). As expected, ion transporters represent a major class of genes that change in response to such alterations in Ca^{2+} availability, but increased Ca^{2+} was also accompanied by the induction of genes thought to be markers of a wide array of abiotic stresses (Chan et al. 2008). This pattern was disrupted in the *cngc2* mutant. These widespread transcriptional changes imply a complex signaling network is acting to tailor myriad responses to ambient Ca^{2+} levels in the soil and/or in the plant body.

Such a sensory system that coordinates transport and demand for Ca^{2+} is extremely important as the relatively immobile nature of Ca^{2+} in the plant means that rapidly growing tissues can readily run out of this key macronutrient. Such Ca^{2+} deficiency is associated with a range of plant diseases such as bitter pit in apples and blossom-end rot in tomatoes where the lack of adequate Ca^{2+} supply to these fast-growing organs is thought to lead to a loss of cellular integrity and subsequent tissue necrosis (Ho and White 2005).

II. Calcium and Signaling Networks

Paradoxically, despite the essential role of Ca^{2+} in plant biology described earlier, Ca^{2+} is also a cytotoxin. The phosphate-based energy metabolism that lies at the heart of cellular metabolism means that when Ca^{2+} levels become elevated in the cytoplasm, Ca^{2+} and the relatively high levels of free phosphate in the cell react and precipitate as insoluble calcium phosphate. This causes irreversible damage to cellular biochemistry. Thus, although total Ca^{2+} levels are high in plant tissues, often reaching mM levels in the apoplast and in organelles such as the vacuole, the free levels in the cytoplasm are kept 10,000-fold lower, at approximately 100 nM. This extremely low resting Ca^{2+} level is maintained by the activity of a series of ATP-powered pumps and cotransporters on the plasma and organellar membranes that actively pump Ca^{2+} from the cytosol (Dodd et al. 2010; Kudla et al. 2010; Pittman 2011).

The large electrochemical gradient between cytosolic and apoplast/organelle Ca^{2+} levels, with very low, highly regulated, and stable resting intracellular levels, means that opening a Ca^{2+} channel leads to rapid Ca^{2+} influx. Increases in cytosolic concentrations to micromolar levels are rapidly attained, but only low absolute numbers of calcium ions need enter the cell to cause this change. The resulting elevation in Ca^{2+} tends to be highly localized as the relatively few ions that have entered are quickly pumped out or sequestered, protecting the cytoplasm from the toxic effects of sustained high Ca^{2+} levels. This feature of rapid, reversible, localized increases against an actively maintained, low resting background has been capitalized upon by the cell to yield a Ca^{2+}-dependent signaling system that regulates a vast array of plant processes (Dodd et al. 2010; Kudla et al. 2010). Thus, many stimuli lead to Ca^{2+} influx to the cytosol which itself contains a host of proteins responsive to the micromolar Ca^{2+} levels that can be safely maintained for the duration of these influx/export cycles. Further, the amplitude, duration, and localization of Ca^{2+} changes, the so-called Ca^{2+} signature, is thought to carry regulatory information about the signals that triggered the change, making Ca^{2+} a ubiquitous cellular regulator capable of simultaneously carrying information about a very diverse range of stimuli and plant processes (Dodd et al. 2010; Kudla et al. 2010).

III. Calcium Signaling/Regulatory Networks in Roots

In roots, Ca^{2+} signals have been characterized in response to a host of environmental and endogenous signals ranging from stresses such as salt, cold, and osmotic shock (Figure 20.1; Kiegle et al. 2000) to the hormones such as auxin that integrate root growth and development (Monshausen et al. 2011), interactions with endosymbiotic microbes (Capoen et al. 2011), and even the biophysical forces related to growth through the soil (Monshausen et al. 2009; Richter et al. 2009).

A. Root Calcium Signaling and the Mechanical Environment of the Soil

For the root, growth through the soil presents a wealth of mechanical stimuli including rocks, hardpan layers, and even the mechanical friction of moving past soil particles (Fasano et al. 2002; Massa et al. 2003; Chehab et al. 2009). The ability of the root to navigate and penetrate this mechanically stimulus-rich environment varies between plants, and the mapping of QTLs for soil penetration in rice suggests a complex underlying genetic network (Ali et al. 2000; Zheng et al. 2000). However, a common theme across species seems to be the conserved nature of the signaling that is linked to initial touch perception. Thus, touch has been well characterized as involving Ca^{2+}-dependent signaling across kingdoms, and there is a plethora of data on Ca^{2+}-dependent mechanical signaling in plants. This link

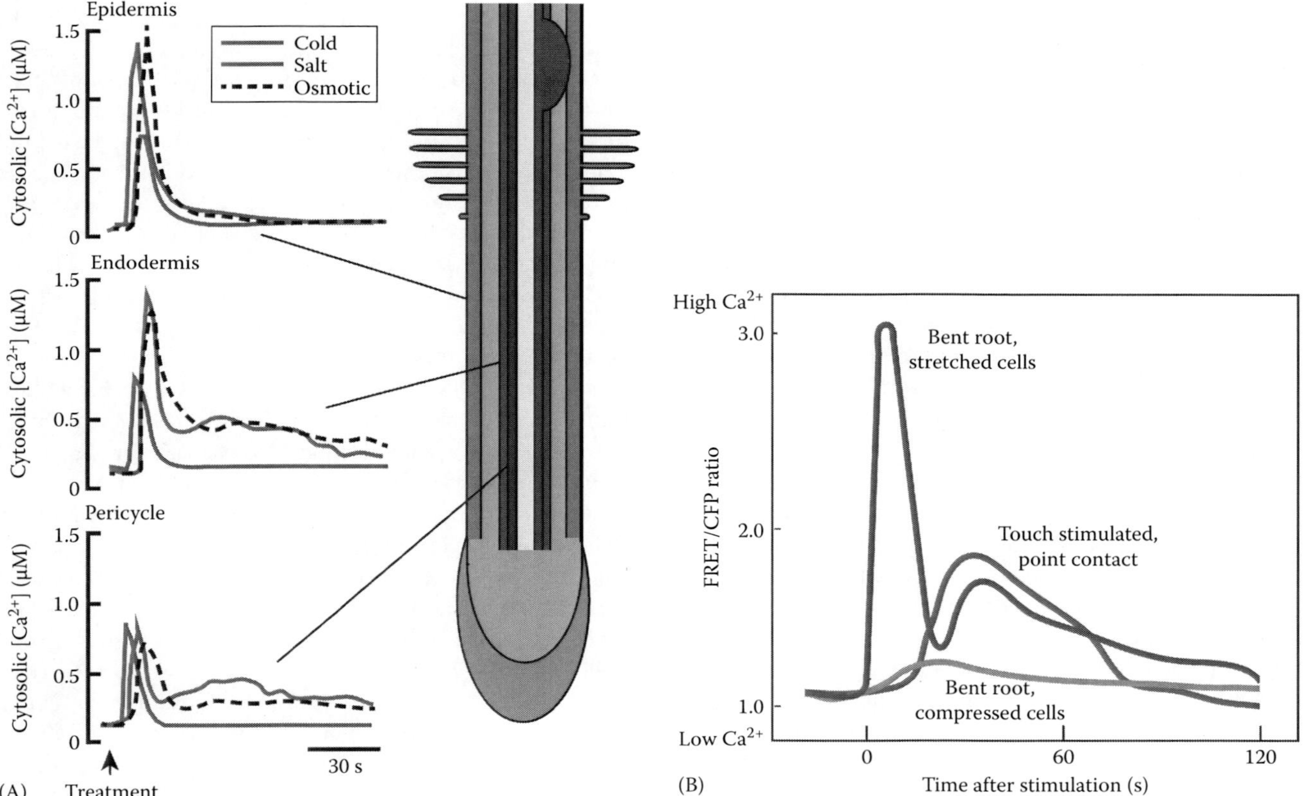

FIGURE 20.1 Cell-type-specific Ca^{2+} changes in roots responding to cold, salt, and osmotic stress and to mechanostimulation. (A) The luminescent Ca^{2+} sensing protein aequorin was targeted to the cell types outlined using tissue-specific promoters. Ca^{2+} was then monitored at a whole plant level in seedlings subjected to cold shock, 220 mM NaCl, or 440 nM mannitol. (Redrawn from Kiegle, E. et al., *Plant J.*, 23, 267, 2000.) (B) Mechanically induced Ca^{2+} changes in the root tip in response to point contact with a micropipette, or in cells under tension and compression as a root bends. Ca^{2+} levels were monitored using the FRET-based Ca^{2+} sensor YC3.6, where an increase in the ratio of FRET signal to that of the CFP in the sensor reflects an increase in Ca^{2+}. (Redrawn from Monshausen, G.B. et al., *Plant Cell*, 21, 2341, 2009.)

between Ca^{2+} and touch was initially made at the whole plant level from observations using the Ca^{2+}-sensitive luminescent photoprotein aequorin. These kinds of measurements showed that in response to local touch and mechanical perturbation such as by wind, cytosolic Ca^{2+} levels transiently rose in the plant (e.g., Knight et al. 1991b, 1992). Measurements made with fluorescent Ca^{2+}-sensitive dyes followed and allowed responses to be localized down to the cellular level. These kinds of experiments revealed touch-induced Ca^{2+} changes throughout the root, with the root cap showing particularly high sensitivity (Legue et al. 1997). Increased mechanical sensitivity of cap cells would fit well with this organ being the first to encounter new environments and obstacles as the root forces its way through the soil.

More recently, the touch-induced Ca^{2+} responses of the root have been monitored using genetically encoded, green-fluorescent protein-based sensors of the Cameleon family (Figure 20.1; Miyawaki et al. 1997, 1999; Nagai et al. 2004). These probes have allowed researchers for the first time to make noninvasive measurements of the dynamics of Ca^{2+} in the roots of plants stably transformed with sensors that permit the subcellular resolution afforded by confocal microscopy. These kinds of measurements are beginning to reveal a tremendously rich Ca^{2+} signaling landscape within the plant as described in the following.

Mechanical stimulation can be applied to the root in many forms that try to mimic different natural sources of mechanical stimulation, and these varied stimuli are now known to elicit Ca^{2+} changes that show qualitatively and quantitatively different Ca^{2+} signatures. Thus, point contacts have been shown to elicit local, rapid, monophasic (i.e., single peak) Ca^{2+} increases in root cells (Figure 20.1B; Legue et al. 1997; Monshausen et al. 2009) that may reflect responses normally seen to herbivore or fungal attack (Hardham et al. 2008). Such local touch stimulation can be accompanied by accumulation of a network of actin microfilaments, depolymerization of microtubules, and recruitment of ER and peroxisomes (Hardham et al. 2008), potentially all related to structural aspects of mounting a localized defense response, that may itself be coordinated by the initial Ca^{2+} signal. In roots, local touch stimulation also leads to rapid, Ca^{2+}-dependent production of apoplastic ROS and a simultaneous alkalinization of the wall and acidification of the cytoplasm (Monshausen et al. 2009). The coordinated cytoplasmic and wall pH fluxes strongly suggest that Ca^{2+} has triggered proton influx across the plasma membrane, although the molecular nature of the transporters involved remains unknown. The increased pH and ROS levels in the wall may both be related to localized modification of their rigidity of the wall, with alkaline pH inhibiting the slippage of wall polymers normally related to acid growth (Cosgrove 2005) and ROS potentially cross-linking and/or severing components of the wall matrix (Swanson and Gilroy 2010).

A similar suite of responses is seen upon applying organ-wide mechanical stress imposed as the root bends. Here a biphasic Ca^{2+} increase is elicited but only in cells under tension (Figure 20.1B; Monshausen et al. 2009; Richter et al. 2009). This observation implies the triggering of a stretch sensor in the cells that is either directly or indirectly linked to the gating of a Ca^{2+} influx channel. The Ca^{2+} channel blockers La^{3+}, Gd^{3+}, and ruthenium red can inhibit such mechanically induced Ca^{2+} signals (e.g., Knight et al. 1991a, 1992; Legue et al. 1997; Monshausen et al. 2009), tentatively suggesting that both influx across the plasma membrane (blocked by La^{3+} and Gd^{3+}) and influx from internal sources (sensitive to ruthenium red) are involved. La^{3+} and Gd^{3+} block both peaks of the biphasic Ca^{2+} increase, and ruthenium red also attenuates both changes. In addition, at a cellular level, the Ca^{2+} elevation spreads throughout the stimulated cell cytoplasm and persists for many seconds. Taken together, these observations of extended cellular dynamics and pharmacological sensitivities imply that Ca^{2+} influx across the plasma membrane may prime the cell for response but that an amplification system such as Ca^{2+}-induced Ca^{2+} release (CICR) from internal stores may also be involved in sustaining these initial changes. CICR is well known as a signal amplifier in animal cells as part of sustained Ca^{2+} spiking and proliferation of Ca^{2+} waves within the cytoplasm (Berridge 2002; Falik et al. 2005). However, such conclusions from pharmacological interventions must always be treated with caution as the site(s) of action of many of these agents remain poorly defined in plants. For example, at high levels, La^{3+} has been reported to interfere with anion channels (Lewis and Spalding 1998) that are unlikely to reflect the stretch-activated Ca^{2+} channels that are the targets of the root touch response experiments. Similarly, in addition to its effects on intracellular Ca^{2+} stores, ruthenium is also known to bind to a host of proteins other than Ca^{2+}-selective ion channels, such as tubulin (Deinum et al. 1985) and calmodulin (Sasaki et al. 1992), and the effects of these possible alternate sites of action on plant cell function and signaling are poorly characterized.

Clearly, many questions as to how these mechanically induced signals are generated and propagated will only be answered from direct structure-function studies on the plant mechanoreceptor(s). However, the molecular identity of these (presumably) proteins remains elusive. Although mechanoresponsive ion channels have been characterized in plant membranes by electrophysiology (discussed in Monshausen et al. 2008a,b), they have yet to be identified at the molecular level. For example, there are no plant homologs of the animal transient receptor potential channels (TRP channels) that represent one well-defined class of mechanosensors in other organisms (Lin and Corey 2005), and homologs of the well-studied MEC mechanosensor complex from *C. elegans* (Arnadottir and Chalfie 2010) are not found in the sequenced plant genomes. However, plants do possess proteins that either structurally or functionally mimic mechanosensors found in *E. coli* and yeast.

Perhaps the best studied mechanosensitive channel is the Mechanosensitive channel of Small conductance (MscS) of bacteria (Kung et al. 2010). These channels act as safety valves to rapidly expel solutes should the bacterium experience hypoosmotic shock. The channel is made of a series of transmembrane domains that form an iris-like structure that opens as the tension

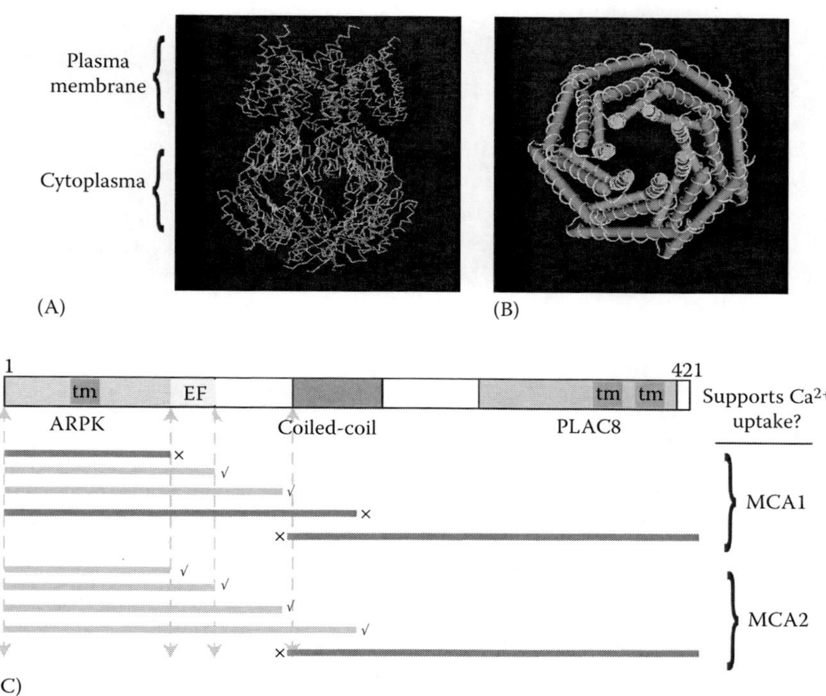

FIGURE 20.2 The structure of the bacterial MscS mechanosensory channel and the *Arabidopsis* proteins MCA1. (A) MscS exists as a homoheptamer, with each subunit having three transmembrane domains. The C-terminal domains combine to form a large hollow cage or gondola with seven openings. (B) Top view from outside the cell of the iris structure showing transmembrane pore. Tension in the membrane pulls on the intramembrane domains to open the pore. The cytoplasmic domains in the following have been removed from this view for clarity. MscS structures were rendered using Cn3D from the National Center for Biotechnology Information (Wang et al. 2000). (C) Domain structure of MCA1. Truncations in MCA1 and MCA2 are shown that do (light grey) and do not (dark grey) support enhanced Ca^{2+} uptake when expressed in yeast (Nakano et al. 2011).

in the membrane that develops during mechanical/osmotic stress pulls on the sides of the channel (Figure 20.2A and B). On the cytoplasmic side of the membrane, the protein also forms a cage-like structure around the mouth of the pore. This may act as a mechanical filter to prevent large cell contents from exiting through the channel when it opens. *Arabidopsis* has 10 MscS-like genes (the *AtMSLs*), and rice has 6 (Haswell and Meyerowitz 2006). *AtMsl2* and 3 are most closely related to the bacterial MscS and play roles in plastid function such as influencing the placement of the chloroplast division ring (Haswell and Meyerowitz 2006; Wilson et al. 2011). However, there are 5 MSLs expressed in *Arabidopsis* roots (AtMSL4, 5, 6, 9, and 10), and these are slightly more divergent from the prototypical bacterial channels. From the electrophysiological phenotypes of their knockout mutants, MSL 9 and 10 are thought to play a largely redundant role as stretch-activated Cl^- channels, with MSL 4, 5, and 6 having a similar but more minor activity (Haswell et al. 2008). Unexpectedly, the quintuple knockout in all five genes shows no overt root developmental phenotype and no alteration in response to osmotic, salt, and dehydration/rehydration stresses nor to mechanical stimuli (Haswell et al. 2008; Monshausen and Gilroy, unpublished). Thus, although the MSL genes remain clear candidates for encoding root mechanosensors, their precise physiological role has yet to be defined. In addition, although permeable to Ca^{2+}, these channels are much more likely to be transporting Cl^-, and so they do not seem to represent the mechanosensitive Ca^{2+} channels likely responsible for the touch-related Ca^{2+} changes outlined earlier.

Another strong candidate for direct links to mechanical Ca^{2+} response came from a screen of *Arabidopsis* cDNAs that could rescue the phenotype of a knockout in the yeast Mid1 gene (Nakagawa et al. 2007). Mid1 one forms part of a yeast mechanosensitive Ca^{2+} channel along with the product of the Cch1 gene. Therefore, *Arabidopsis* cDNAs that rescued the yeast mutant were hypothesized to encode for an equivalent function in the plant. The rescuing clone, named *MCA1*, caused elevated Ca^{2+} uptake when expressed in yeast and led to a novel stretch-activated Ca^{2+} conductance being measured when expressed in mammalian cells. Overexpression of *MCA1* also led to increased basal levels of Ca^{2+} uptake and increased magnitude of a Ca^{2+} signal in response to osmotic shock in *Arabidopsis* (Nakagawa et al. 2007) as did the tobacco homologs when expressed in tobacco BY-2 cells (Yamanaka et al. 2010; Kurusu et al. 2011). Interestingly, *mca1* knockout lines have a reduced ability to penetrate hard agar (Nakagawa et al. 2007), suggesting a possible mechanical sensing-related lesion. Lastly, MCA1 is found in the plasma membrane. Thus, the characterization of MCA1 appears highly consistent with it playing a Mid1-like role in plants and gating Ca^{2+} entry upon mechanostimulation.

However, MCA1 shows only 10% amino acid homology to Mid1, making interpretation of its role more complex. Unfortunately, the structure of the protein gives few clues to function (Figure 20.2C). The N-terminal region of MCA1 has a domain similar to putative regulatory domains found in rice protein kinases (including a possible transmembrane domain) and a Ca^{2+}-binding EF-hand domain. These are separated from the C-terminus of the protein by a coiled-coil domain that may play a role in interactions with partner proteins. The C-terminal portion has two putative transmembrane domains in a PLAC8 motif, a cysteine-rich domain of unknown function (also known as DUF614) that is found in many placental-enriched genes (Galaviz-Hernandez et al. 2003). MCA1 (and MCA2; see the following) forms homodimers and tetramers (Yamanaka et al. 2010; Nakano et al. 2011).

MCA1 has a single paralog, MCA2 (72.7% amino acid identity, 89.4% similarity), which can also rescue the yeast Mid1 knockout. Both MCA1 and MCA2 are ubiquitously expressed, especially in the vasculature and endodermis but not in cortical and epidermal cells. MCA2 is excluded from the root cap and the root elongation zone (Yamanaka et al. 2010). MCA2-null lines showed reduced Ca^{2+} uptake by roots, suggesting the protein may play a role in regulating normal root Ca^{2+} influx. Its restricted expression to the endodermis and vasculature would then point to a role in loading Ca^{2+} from the soil to the transpiration stream, consistent with ideas outlined earlier that the endodermis may play a major role in regulating Ca^{2+} entry to the xylem (White 2001; White and Broadley 2003; Cholewa and Peterson 2004; Hayter and Peterson 2004). *mca2* null mutants do not show the altered penetration of hard agar seen in *mca1* knockouts, implying there may be different roles for each protein, with perhaps MCA1 playing a larger role in mechanical responses. Consistent with these ideas about nonoverlapping function, the double mutant behaves as the single *mca1* null in these assays (Yamanaka et al. 2010). Expressing the N-terminal of MCA1 up to its coiled-coil domain showed stimulated Ca^{2+} uptake activity when expressed in yeast, as did a similar construct for MCA2 (Nakano et al. 2011). However, including the coiled-coil domain abolished Ca^{2+} uptake activity associated with MCA1 but not MCA2 (Figure 20.2C), suggesting different structure-function constraints on the two proteins. Neither C-terminal portion from each gene had Ca^{2+} uptake activity in yeast. In the full-length protein, stimulating Ca^{2+} uptake is dependent on the presence of the first N-terminal transmembrane domain (Nakano et al. 2011), again consistent with its role as a plasma membrane protein.

Thus, although MCA1 and MCA2 represent strong candidates for a component of a mechanosensitive Ca^{2+} channel complex, this activity resides in the N-terminal part of the protein and is associated with a region containing a single EF-hand domain and a single transmembrane motif. Thus, even as a homotetramer, there is a question as to whether this structure could form a pore across the membrane that would be needed for channel activity. MCA1 and 2 may represent regulatory components of a channel complex, but this will remain speculation until either they or their associated channel passes the "gold standard" of channel identification in showing the requisite electrophysiological signature when heterologously expressed. What seems to have emerged from these studies is that MCA2 is very likely a component of the system regulating Ca^{2+} loading from the apoplastic route to the xylem for redistribution to the aerial parts of the plant.

In addition to direct effects on membrane tension and possible gating of ion channel activity, mechanical signals will also likely affect wall properties, and so candidate mechanosensors have also been sought in this region of the plant cell. Wall-associated protein kinases have recently come to the fore as candidates for wall integrity sensors (Cheung and Wu 2011) that could play a role in mechanosensing. These kinases possess an extracellular, likely sensory, domain, a single transmembrane span, and then an intracellular serine-threonine protein kinase domain. From this family, the THESEUS1 and HERCULES1 receptor-like kinases have been shown to play important roles in the vegetative development in *Arabidopsis*, including root cell expansion (Hematy et al. 2007; Guo et al. 2009). The close relative of THESEUS1, a wall kinase called FERONIA, not only acts in this THESEUS1/HERCULES1 system to support vegetative cell elongation such as root hair tip growth but also plays critical roles in other growth-related processes such as pollen tube function and pathogen resistance (Huck et al. 2003; Rotman et al. 2003; Escobar-Restrepo et al. 2007; Sandaklie-Nikolova et al. 2007; Kessler et al. 2010). In these cases, the kinase appears to monitor the wall status in the "host" tissues as the pollen tube or hypha advances through them. In the pollen tube itself, there are close relatives of FERONIA in the ANXUR kinases. The proteins seem to play an analogous role in the pollen tube to that FREONIA plays in the female reproductive tissues, in monitoring cell wall status to regulate growth versus cell bursting (Boisson-Dernier et al. 2009; Miyazaki et al. 2009). The extracellular domains of these kinases likely interact with wall polymers, translating changes in structure to cytosolic signaling cascades triggered through their cytoplasmic kinase domains. In the case of FERONIA, further elements of the downstream network have also been identified in the RAC/ROP GTPases and changes in ROS production (Cheung and Wu 2011). Interactions of their extracellular domain with wall pectins have been directly shown for the wall-associated kinases WAK1 and WAK2, and the affinity of these proteins for the wall polymers was increased by the presence of Ca^{2+} and de-esterification of the pectin (Decreux and Messiaen 2005; Kohorn et al. 2009). These WAKs are also expressed in the root, and their suppression leads to reduced root growth (Kohorn et al. 2006), but whether these kinases truly transfer information as to the mechanical status of the wall related to growth remains to be defined.

A further candidate for a sensor of mechanical stress is AtHK1, a histidine kinase from *Arabidopsis* that operates as the sensor element of a two-component signaling relay. This protein is found most abundantly in the plasma membrane of root cells, and its expression is highly upregulated in response to osmotic stress (Urao et al. 1999). AtHK1 is able to rescue knockouts in the yeast osmosensing kinase SLN1, and in *Arabidopsis*, it appears not only to act as an osmosensor in vegetative growth, acting through a downstream MAP kinase cascade (Droillard et al. 2002), but also to play a key role in desiccation during seed development (Wohlbach et al. 2008). Whether sensitivity to osmotic stress could also translate into a protein sensitive to mechanical forces in the membrane remains to be determined.

B. Molecular Responses to Mechanical Stimulation

At the molecular level, mechanical stimulation has been shown to trigger a host of events ranging from ethylene and jasmonate production (Tretner et al. 2008) to a range of rapid and widespread transcriptional changes (e.g., Braam and Davis 1990; Kimbrough et al. 2004; Lee et al, 2005; Walley et al. 2007; Leblanc-Fournier et al. 2008). Prominent among the genes that are upregulated are those encoding Ca^{2+}-related proteins. Indeed, of the first four *Arabidopsis* genes identified as being upregulated in response to touch (*TCH1–4*), *TCH1* encodes a calmodulin (CaM 2) and *TCH2* and *3*, calmodulin-like proteins (CML24 and 12, respectively; Braam and Davis 1990). These observations suggest that a Ca^{2+}-dependent mechanical response system may be rewiring itself after an initial Ca^{2+} signal to allow for adaptation to/memory of the initial mechanical stimulus. Consistent with such ideas, the transcription of *TCH* genes has been shown to respond to alterations in Ca^{2+} level (Braam 1992a).

The precise roles of many of these *TCH* genes in the mechanical response have proven difficult to define. A notable exception is *TCH4*, which encodes a xyloglucan endotransglycosylase that has been demonstrated to play an important role in cell wall remodeling (Antosiewicz et al. 1997; Thompson and Fry 2001). This activity would fit well with changes in the cell wall needed to adapt to mechanical stress. However, even though *TCH1-3* are known Ca^{2+}-related proteins and there has been much discussion of their potential to transduce mechanical signals (e.g., Braam and Davis 1990; Knight et al. 1991b; Braam 1992b; Telewski 2006; Monshausen and Gilroy 2009a,b), direct evidence for their role in mechanical perception or response has been lacking. *TCH2/CML24* has been shown to control flowering time through an NO-dependent pathway (Tsai et al. 2007), but until recently, a functional relationship to touch sensing had not been revealed. However, *TCH2/CML24* has now been shown to play a role in modulating root growth that may relate to mechanosensitivity. Thus, *cml24-2* and *cml24-4* mutants show reduced root growth, altered cell file rotation, and enhanced skewing (Wang et al. 2012a,b). These phenotypes only emerge when the roots are growing on the surface of hard agar, suggesting some relationship to the mechanical environment of growth. Consistent with this idea, *cml24-2* and *cml24-4* show an aberrant thigmotropic response to growing along the surface of a glass barrier (Wang et al. 2012a). Cortical microtubules were also misaligned in the mutants, which themselves showed enhanced sensitivity to microtubule depolymerizing drugs. Microtubule orientation has long been known to be sensitive to mechanical force in a range of plant systems from protoplasts to intact meristems (Wymer et al. 1996; Hamant et al. 2008). Cortical microtubules are also known to reflect the ordering of processes intimately linked to the control of growth, such as the complex feedback mechanisms behind the alignment of cellulose synthesis to the wall (Paredez et al. 2008; Fujita et al. 2011, 2012) or auxin transporter positioning (Heisler et al. 2010). Taken together, these observations suggest that TCH2/CML24 may play a role in controlling the orientation of cortical microtubules required for the normal regulation of root growth. Whether TCH2/CML24 acts directly on microtubules or through some other microtubule-associated proteins and whether such regulation is controlled by Ca^{2+} levels is currently unknown. However, Ca^{2+} and calmodulin have previously been characterized as regulators of plant microtubule function, likely through other microtubule-associated proteins (Cyr 1991; Fisher et al. 1996), suggesting Ca^{2+}-dependent regulation of microtubule alignment through TCH2/CML24 is a distinct possibility. TCH2/CML24 has also been shown to interact with the IQ domain of myosin VIII in mammalian cells (Abu-Abied et al. 2006) raising the possibility of a role in actin function as well. Microfilaments and microtubules are known to show complex interactions (Schwab et al. 2003), and so CML24 may be playing a role in coordinating these two components of the cytoskeleton.

IV. Obstacle Avoidance and Root System Architecture

As roots navigate the soil, a further mechanical cue they constantly need to sense and respond to is obstacles to the direction of growth such as rocks and hardpan layers. Roots may be able to predict when they are coming close to an obstacle through a local increase in the accumulation of their own secreted compounds and elicit an avoidance response prior to contact (Falik et al. 2005). However, it is clear that roots can also directly sense and respond to the mechanical environment of the soil to elicit highly adaptive changes in growth patterns and root system architecture to avoid such obstacles. Experimental evidence from *Arabidopsis* primary roots suggests that initially upon contact with a barrier such as a stone, the root may try to force its way past the obstacle. Should this fail, the tensions in the tissue generated to try and force past the object cause the root to change its growth pattern toward growing round the barrier (Massa and Gilroy 2003a; Monshausen et al. 2009). The lateral progress over the surface of the object appears to represent a compromise between touch and gravity-directed growth, with touch stimulation downregulating gravitropic signaling at the level of inhibiting sensory events in the columella cells of the root cap (Massa and Gilroy, 2003a,b).

The bending of the root upon contact with the obstacle also seems to trigger a range of other responses linked to remodeling the branching pattern/architecture of the root system to take into account the presence of the obstacle. The branching pattern of the root system reflects the positioning of lateral root primordia that will eventually form the new lateral root, and such positioning is governed at multiple levels. There appears to be an endogenous pattern of auxin signaling laid down in the basal meristem of the root that may prepattern primordia positioning (De Smet et al. 2007; see also Chapter 6), and this is coupled to oscillations in transcriptional regulation that predefine branching sites (De Smet et al. 2007; Moreno-Risueno et al. 2010). Together, these mechanisms provide an inherent, genetically preprogrammed root system architecture. However, superimposed on this predefined patterning is a set of environmental response systems, such as to the nutrient levels around the root, that can radically modify this prepatterning (Malamy 2005). One of these modifiers is mechanical signaling.

It has long been known that lateral roots tend to form in the convex sides of curves in the root system whether such curves form from physical bending, root waving, gravitropic reorientation, or navigating barriers in the soil (Noll 1900; Goss and Russell 1979; Fortin et al. 1989; De Smet et al. 2007; O'Brien et al. 2007; Ditengou et al. 2008; Laskowski et al. 2008; Lucas et al. 2008a,b). Lateral root primordia form from divisions of specific pericycle cells in a process intimately related to auxin-dependent signaling (Casimiro et al. 2003). One model for how bends can recruit lateral root placement therefore relates to the altered geometry in the bend causing auxin accumulation over several hours to levels above the threshold for lateral root specification (Laskowski et al. 2008). However, an additional mechanism is suggested from the observation that the bending-induced placement of lateral roots is able to suppress the reduction in lateral root formation seen in auxin-related mutants such as *aux1* and *tir1*. Thus, the induction pathway may (1) lie in parallel or upstream of these auxin-related events or (2) alter the sensitivity of the existing auxin-dependent pathway to possibly operate at ambient auxin levels (Ditengou et al. 2008; Richter et al. 2009). The trigger also does not appear to be linked to possible export of gravitropic signals from the root cap that might be expected to occur as the orientation of the root changes during the production of a bend (Ditengou et al. 2008; Richter et al. 2009).

The mechanically induced Ca^{2+} increase described earlier, however, does seem intrinsic to this process of reprogramming the pericycle to lateral root founder cell fate. Thus, transiently bending roots for different periods of time established that the inductive period for this lateral root recruitment to the curve was a bend that was maintained for only 20 s (Ditengou et al. 2008; Richter et al. 2009). After this period, the signal was established, and lateral root primordium production occurred even in roots that had been returned to straight growth. Formation of lateral root primordia is visible about 12 h after curve formation, but the 20 s time frame needed for induction matches the duration of the bending Ca^{2+} signal (Figure 20.1B). In addition, this curve-related Ca^{2+} change is seen throughout the stretched cells of the root, including the pericycle where the initial divisions to produce the lateral root primordium will occur. Blocking these changes with La^{3+} also selectively blocked curve-induced lateral root production (Richter et al. 2009). Pericycle cells are also known to be able modulate their Ca^{2+} responses in a tissue-specific manner independently of the other cell types in the root (Figure 20.1; Kiegle et al. 2000), suggesting they may possess a closely regulated Ca^{2+} signaling system. However, the lateral root-induction response must be more complex than simply raising pericycle Ca^{2+} as there is a limited region of the root where such lateral root induction can occur (Richter et al. 2009), and simply raising pericycle Ca^{2+} levels, e.g., by cold shock, does not lead to this same kind of lateral root induction. Taken together, these observations suggest that a program of endogenous developmental patterning is still overlaid on this environmental response pathway to lateral root induction.

Defining whether the mechanically induced pericycle Ca^{2+} signal truly reflects the trigger for developmental reprogramming will clearly require identifying further molecular components of this system. Mechanical stimulation is also known to trigger local fluxes of protons and production of ROS (Monshausen et al. 2009) that could represent important further elements of this system. Thus, as noted earlier, touch or bending yields both cytosolic and wall ROS production in many plant systems, and in roots, this ROS production appears to be triggered by the mechanical Ca^{2+} signal. Although originally thought of as toxic by-products of metabolism, ROS are now recognized as key cellular regulators acting both through the modulation of a range of proteins and through modifications of the structure of the wall itself (Swanson and Gilroy 2010). In the root, ROS have recently been highlighted as a potential switch from division to differentiation through the action of a transcription factor called UPBEAT that coordinates peroxidase activities (Tsukagoshi et al. 2010), reinforcing the possible functional role of the mechanically elicited ROS in developmental programming. The changes in pH seen upon mechanostimulation also have regulatory potential. Swings as large as 0.4 and 3 pH units were recorded in the cytosol and wall respectively upon touch stimulation (Monshausen et al. 2009). Such large-scale changes in pH would be expected to have major effects on enzyme activities in the cell and in the wall, and so these pH changes could well act in the global reprogramming of the cell.

In addition to arising from the environmental, mechanical stimulation is inherent in the process of turgor-driven growth, where monitoring and responding to the stress and strain experienced as cells expand is thought to be a fundamental aspect of plant cell growth control (Dutta and Robinson 2004; Monshausen and Gilroy 2009a). The bend-induced lateral root recruitment and its associated Ca^{2+} signal, pH changes, and ROS production are also seen in roots naturally making curves as they circumvent objects (Richter et al. 2008; Monshausen et al. 2009), suggesting this response is likely a normal aspect of root system development and capable of being triggered by the endogenous forces of growth. It is very likely that curve induction of lateral root formation to the convex side of the bend allows mechanical signaling to promote the production of branches oriented away from where the main root tip is growing, setting the root system architecture to optimally explore and exploit the entire soil volume.

V. Parallels between Mechanical Signaling and Root Hair Growth

One theme that emerges from the earlier discussion is that there appears to be a strong correlation between Ca^{2+} signals and subsequent change in pH and ROS in roots. Indeed, in root hairs, the link between these three signaling elements strongly correlates to growth. Thus, root hairs grow in an oscillatory pattern with cyclical periods of slower and faster growth. In *Arabidopsis*, this cycle appears to be about 20 s (Monshausen et al., 2007, 2008a,b). The tip growth in these cells is associated with an oscillating Ca^{2+} gradient (Figure 20.3A; Monshausen et al. 2008a,b) thought to drive Ca^{2+}-dependent elements of the tip growth machinery and both pH and ROS oscillate along with this process (Monshausen et al. 2007). When the precise timing of the oscillations of each of these molecules was mapped onto growth, it showed that growth preceded Ca^{2+} influx by 5 s and ROS and pH changes occurred simultaneously but 2 s after the Ca^{2+} (Figure 20.3B; Monshausen et al. 2007, 2008a,b). This has led to a model of regulation where stretch-activated channels were opened by growth (i.e., the endogenous mechanical forces of turgor-driven expansion), followed by Ca^{2+} influx triggering proton fluxes and ROS production (in this case, mutant analysis showed the ROS production is mediated by the NAPDH oxidase AtRbohC) (Foreman et al. 2003; Monshausen et al. 2007). The role of the ROS and pH changes may well be to modify wall properties to help prevent runaway expansion and cell bursting. A complex feedback loop appears to exist between Ca^{2+} influx via hyperpolarization- and ROS-gated channels, mechanically gated Ca^{2+} influx, and Ca^{2+}-dependent ROS production (Foreman et al. 2003; Takeda et al. 2008) that may help integrate the growth-driven Ca^{2+} influx via mechanically sensitive channels to the activity of intracellular signaling cascades themselves linked to ROS and Ca^{2+}. Similar links have been forged between growth, Ca^{2+} fluxes, pH changes, and interactions in the wall in pollen tubes (Robinson and Messerli 2002; Hepler et al. 2012), suggesting the potentially widespread nature of a conserved response cassette of Ca^{2+}, ROS, and pH.

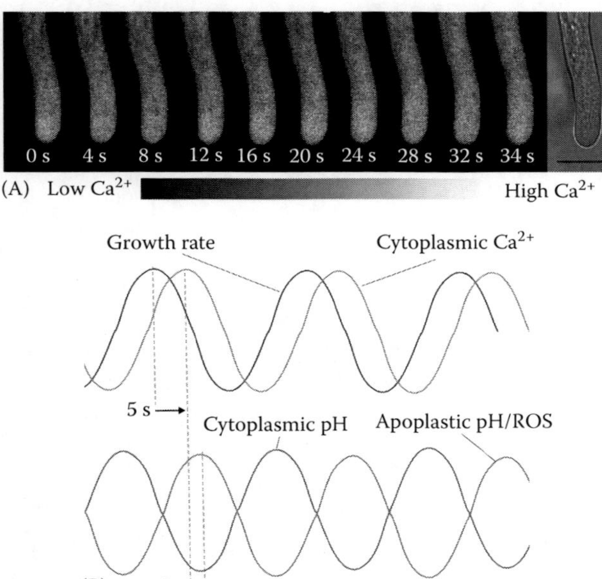

FIGURE 20.3 Ca^{2+} changes associated with root hair growth in *Arabidopsis*. (A) Ca^{2+} changes at the tip of a growing root hair of *Arabidopsis* visualized by confocal ratio imaging of the Ca^{2+} sensor YC3.6. Scale bar represents 10 μm. Imaging was performed as described in (Monshausen et al. 2008) and Ca^{2+} levels coded according to the inset scale. (B) Relationship between oscillations in growth, cytosolic Ca^{2+}, apoplastic ROS/pH changes, and cytosolic pH. Cross-correlation analysis was used to infer the most likely relationship between the timing of the peaks of each of these parameters to each other and to growth. (From Monshausen, G.B. et al., *Proc. Natl. Acad. Sci. USA*, 104, 20996, 2007; Monshausen, G.B. et al., *Plant Physiol.*, 147, 1690, 2008a; Monshausen, G. et al., Touch sensing and thigmotropism, in *Plant Tropisms*, eds., Gilroy, S. and Masson, P.H, Blackwell Publishing, Ames, IA, pp. 91–122, 2008b.)

VI. Conclusions and Perspectives

There are clearly many gaps in our knowledge of these Ca^{2+}-mechanical signaling systems in the root. A prime example of such limitations is highlighted by our understanding of how another component known to be linked to mechanical signaling, ATP in the root apoplast (Weerasinghe et al. 2009), acts to mediate cellular responses. Extracellular ATP is well characterized as a signal in animals and is emerging as an important regulator in plants as well (Tanaka et al. 2010a,b; Clark and Roux 2011). Although extracellular ATP is known to elicit Ca^{2+} signaling in the root that appears to involve heterotrimeric G proteins in its action (Tanaka et al. 2010a,b), key elements of this system that would enable us to understand how it contributes to root function, such as its receptors and downstream signaling components, have yet to be well defined.

Ca^{2+} signals are seen in roots in response to many different stimuli, and the discussion earlier indicates that even within the mechanical response, there appear to be many processes linked to Ca^{2+}-dependent regulation. Teasing apart these networks will require precise measurements of each Ca^{2+} signaling signature to define how unique each response really is, coupled to defining shared and unique downstream molecular players. The ubiquitous nature of the Ca^{2+} signal in the root may also be hinting at a role for Ca^{2+} in integrating multiple responses. Thus, shared components between these varied Ca^{2+} signaling systems may well be pointing us toward important conserved regulatory cassettes in root growth and development.

Acknowledgments

This work was supported by grants from the National Science Foundation and the National Aeronautics and Space Agency.

References

Abu-Abied M, Golomb L, Belausov E et al. 2006. Identification of plant cytoskeleton-interacting proteins by screening for actin stress fiber association in mammalian fibroblasts. *Plant J* 48: 367–379.

Ali R, Ma W, Lemtiri-Chlieh F, Tsaltas D, Leng Q, von Bodman S, Berkowitz GA. 2007. Death don't have no mercy and neither does calcium: *Arabidopsis* cyclic nucleotide gated channel2 and innate immunity. *Plant Cell* 19: 1081–1095.

Ali ML, Pathan MS, Zhang J, Bai G, Sarkarung S, Nguyen HT. 2000. Mapping QTLs for root traits in a recombinant inbred population from two indica ecotypes in rice. *Theor Appl Genet* 101: 756–766.

Antosiewicz DM, Purugganan MM, Polisensky DH, Braam J. 1997. Cellular localization of *Arabidopsis xyloglucan* endotransglycosylase-related proteins during development and after wind stimulation. *Plant Physiol* 115: 1319–1328.

Arnadottir J, Chalfie M. 2010. Eukaryotic mechanosensitive channels. *Annu Rev Biophys* 39: 111–137.

Berridge MJ. 2002. The endoplasmic reticulum: A multifunctional signaling organelle. *Cell Calcium* 32: 235–249.

Boisson-Dernier A, Roy S, Kritsas K, Grobei MA, Jaciubek M, Schroeder JI, Grossniklaus U. 2009. Disruption of the pollen-expressed feronia homologs anxur1 and anxur2 triggers pollen tube discharge. *Development* 136: 3279–3288.

Bosch M, Hepler PK. 2006. Silencing of the tobacco pollen pectin methylesterase NtPPME1 results in retarded in vivo pollen tube growth. *Planta* 223: 736–745.

Braam J. 1992a. Regulated expression of the calmodulin-related TCH genes in cultured *Arabidopsis* cells: Induction by calcium and heat shock. *Proc Natl Acad Sci USA* 89: 3213–3216.

Braam J. 1992b. Regulation of expression of calmodulin and calmodulin-related genes by environmental stimuli in plants. *Cell Calcium* 13: 457–463.

Braam J, Davis RW. 1990. Rain-, wind-, and touch-induced expression of calmodulin and calmodulin-related genes in *Arabidopsis*. *Cell* 60: 357–364.

Capoen W, Sun J, Wysham D et al. 2011. Nuclear membranes control symbiotic calcium signaling of legumes. *Proc Natl Acad Sci USA* 108: 14348–1453.

Casimiro I, Beeckman T, Graham N et al. 2003. Dissecting *Arabidopsis* lateral root development. *Trends Plant Sci* 8: 165–171.

Chan CW, Schorrak LM, Smith RK, Jr., Bent AF, Sussman MR. 2003. A cyclic nucleotide-gated ion channel, CNGC2, is crucial for plant development and adaptation to calcium stress. *Plant Physiol* 132: 728–731.

Chan CW, Wohlbach DJ, Rodesch MJ, Sussman MR. 2008. Transcriptional changes in response to growth of *Arabidopsis* in high external calcium. *FEBS Lett* 582: 967–976.

Chehab EW, Eich E, Braam J. 2009. Thigmomorphogenesis: A complex plant response to mechano-stimulation. *J Exp Bot* 60: 43–56.

Cheung AY, Wu HM. 2011. Theseus 1, Feronia and relatives: A family of cell wall-sensing receptor kinases? *Curr Opin Plant Biol* 14: 632–641.

Cholewa E, Peterson CA. 2004. Evidence for symplastic involvement in the radial movement of calcium in onion roots. *Plant Physiol* 134: 1793–1802.

Clark G, Roux SJ. 2011. Apyrases, extracellular ATP and the regulation of growth. *Curr Opin Plant Biol* 14: 700–706.

Cosgrove DJ. 2005. Growth of the plant cell wall. *Nat Rev Mol Cell Biol* 6: 850–861.

Cyr RJ. 1991. Calcium calmodulin affects microtubule stability in lysed protoplasts. *J Cell Sci* 100: 311–317.

Decreux A, Messiaen J. 2005. Wall-associated kinase WAK1 interacts with cell wall pectins in a calcium-induced conformation. *Plant Cell Physiol* 46: 268–278.

De Smet I, Tetsumura T, De Rybel B et al. 2007. Auxin-dependent regulation of lateral root positioning in the basal meristem of *Arabidopsis*. *Development* 134: 681–690.

Deinum J, Wallin M, Jensen PW. 1985. The binding of Ruthenium red to tubulin. *Biochim Biophys Acta* 838: 197–205.

Demidchik V, Maathuis FJ. 2007. Physiological roles of nonselective cation channels in plants: From salt stress to signalling and development. *New Phytol* 175: 387–404.

Ditengou FA, Teale WD, Kochersperger P et al. 2008. Mechanical induction of lateral root initiation in *Arabidopsis thaliana*. *Proc Natl Acad Sci USA* 105: 18818–18823.

Dodd AN, Kudla J, Sanders D. 2010. The language of calcium signaling. *Annu Rev Plant Biol* 61: 593–620.

Droillard M, Boudsocq M, Barbier-Brygoo H, Lauriere C. 2002. Different protein kinase families are activated by osmotic stresses in *Arabidopsis thaliana* cell suspensions. Involvement of the MAP kinases AtMPK3 and AtMPK6. *FEBS Lett* 527: 43–50.

Dutta R, Robinson KR. 2004. Identification and characterization of stretch-activated ion channels in pollen protoplasts. *Plant Physiol* 135: 1398–1406.

Escobar-Restrepo JM, Huck N, Kessler S et al. 2007. The Feronia receptor-like kinase mediates male-female interactions during pollen tube reception. *Science* 317: 656–660.

Eticha D, Stass A, Horst WJ. 2005. Cell-wall pectin and its degree of methylation in the maize root-apex: Significance for genotypic differences in aluminium resistance. *Plant Cell Environ* 11: 1410–1420.

Falik O, Reides P, Gersani M, Novoplansky A. 2005. Root navigation by self inhibition. *Plant Cell Environ* 28: 562–569.

Fasano JM, Massa GD, Gilroy S. 2002. Ionic signaling in plant responses to gravity and touch. *J Plant Growth Regul* 21: 71–88.

Fisher DD, Gilroy S, Cyr RJ. 1996. Evidence for opposing effects of calmodulin on cortical microtubules. *Plant Physiol* 112: 1079–1087.

Foreman J, Demidchik V, Bothwell JH et al. 2003. Reactive oxygen species produced by NADPH oxidase regulate plant cell growth. *Nature* 422: 442–446.

Fortin MC, Pierce FJ, Poff KL. 1989. The pattern of secondary root formation in curving roots of *Arabidopsis thaliana* L. Heynh. *Plant Cell Environ* 12: 337–339.

Fujita M, Himmelspach R, Hocart CH, Williamson RE, Mansfield SD, Wasteneys GO. 2011. Cortical microtubules optimize cell-wall crystallinity to drive unidirectional growth in *Arabidopsis*. *Plant J* 66: 915–928.

Fujita M, Lechner B, Barton DA, Overall RL, Wasteneys GO. 2012. The missing link: Do cortical microtubules define plasma membrane nanodomains that modulate cellulose biosynthesis? *Protoplasma* 249 (Suppl 1): 59–67.

Galaviz-Hernandez C, Stagg C, de Ridder G, Tanaka TS, Ko MS, Schlessinger D, Nagaraja R. 2003. Plac8 and Plac9, novel placental-enriched genes identified through microarray analysis. *Gene* 309: 81–89.

Goss MJ, Russell RS. 1979. Effects of mechanical impedance on root growth in barley. *J Exp Bot* 31: 577–588.

Guo H, Li L, Ye H, Yu X, Algreen A, Yin Y. 2009. Three related receptor-like kinases are required for optimal cell elongation in *Arabidopsis thaliana*. *Proc Natl Acad Sci USA* 106: 7648–7653.

Hamant O, Heisler MG, Jonsson H et al. 2008. Developmental patterning by mechanical signals in *Arabidopsis*. *Science* 322: 1650–1655.

Han S, Tang R, Anderson LK, Woerner TE, Pei ZM. 2003. A cell surface receptor mediates extracellular Ca^{2+} sensing in guard cells. *Nature* 425: 196–200.

Hardham AR, Takemoto D, White RG. 2008. Rapid and dynamic subcellular reorganization following mechanical stimulation of *Arabidopsis* epidermal cells mimics responses to fungal and oomycete attack. *BMC Plant Biol* 8: 63.

Haswell ES, Meyerowitz EM. 2006. MscS-like proteins control plastid size and shape in *Arabidopsis thaliana*. *Curr Biol* 16: 1–11.

Haswell ES, Peyronnet R, Barbier-Brygoo H, Meyerowitz EM, Frachisse JM. 2008. Two MscS homologs provide mechanosensitive channel activities in the *Arabidopsis* root. *Curr Biol* 18: 730–734.

Hayter ML, Peterson CA. 2004. Can Ca^{2+} fluxes to the root xylem be sustained by Ca^{2+}-ATPases in exodermal and endodermal plasma membranes? *Plant Physiol* 136: 4318–4325.

Heisler MG, Hamant O, Krupinski P et al. 2010. Alignment between PIN1 polarity and microtubule orientation in the shoot apical meristem reveals a tight coupling between morphogenesis and auxin transport. *PLoS Biol* 8: e1000516.

Hematy K, Sado PE, Van Tuinen A et al. 2007. A receptor-like kinase mediates the response of *Arabidopsis* cells to the inhibition of cellulose synthesis. *Curr Biol* 17: 922–931.

Hepler PK, Kunkel JG, Rounds CM, Winship LJ. 2012. Calcium entry into pollen tubes. *Trends Plant Sci* 17: 32–38.

Ho LC, White PJ. 2005. A cellular hypothesis for the induction of blossom-end rot in tomato fruit. *Ann Bot* 95: 571–581.

Huck N, Moore JM, Federer M, Grossniklaus U. 2003. The *Arabidopsis* mutant feronia disrupts the female gametophytic control of pollen tube reception. *Development* 130: 2149–2159.

Jiang L, Yang SL, Xie LF et al. 2005. Vanguard1 encodes a pectin methylesterase that enhances pollen tube growth in the *Arabidopsis* style and transmitting tract. *Plant Cell* 17: 584–596.

Jones RGW, Lunt OR. 1967. Function of calcium in plants. *Bot Rev* 33: 407–426.

Kang J, Turano FJ. 2003. The putative glutamate receptor 1.1 AtGLR1.1 functions as a regulator of carbon and nitrogen metabolism in *Arabidopsis thaliana*. *Proc Natl Acad Sci USA* 100: 6872–6877.

Kessler SA, Shimosato-Asano H, Keinath NF et al. 2010. Conserved molecular components for pollen tube reception and fungal invasion. *Science* 330: 968–971.

Kiegle E, Moore CA, Haseloff J, Tester MA, Knight MR. 2000. Cell-type-specific calcium responses to drought, salt and cold in the *Arabidopsis* root. *Plant J* 23: 267–278.

Kimbrough JM, Salinas-Mondragon R, Boss WF, Brown CS, Sederoff HW. 2004. The fast and transient transcriptional network of gravity and mechanical stimulation in the *Arabidopsis* root apex. *Plant Physiol* 136: 2790–2805.

Knight MR, Campbell AK, Smith SM, Trewavas AJ. 1991a. Recombinant aequorin as a probe for cytosolic free Ca^{2+} in *Escherichia coli*. *FEBS Lett* 282: 405–408.

Knight MR, Campbell AK, Smith SM, Trewavas AJ. 1991b. Transgenic plant aequorin reports the effects of touch and cold-shock and elicitors on cytoplasmic calcium. *Nature* 352: 524–526.

Knight MR, Smith SM, Trewavas AJ. 1992. Wind-induced plant motion immediately increases cytosolic calcium. *Proc Natl Acad Sci USA* 89: 4967–4971.

Kohorn BD, Johansen S, Shishido A et al. 2009. Pectin activation of MAP kinase and gene expression is WAK2 dependent. *Plant J* 60: 974–982.

Kohorn BD, Kobayashi M, Johansen S et al. 2006. An *Arabidopsis* cell wall-associated kinase required for invertase activity and cell growth. *Plant J* 46: 307–316.

Kudla J, Batistic O, Hashimoto K. 2010. Calcium signals: The lead currency of plant information processing. *Plant Cell* 22: 541–563.

Kung C, Martinac B, Sukharev S. 2010. Mechanosensitive channels in microbes. *Annu Rev Microbiol* 64: 313–329.

Kurusu T, Yamanaka T, Nakano M et al. 2011. Involvement of the putative Ca^{2+}-permeable mechanosensitive channels, NtMCA1 and NtMCA2, in Ca^{2+} uptake, Ca^{2+}-dependent cell proliferation and mechanical stress-induced gene expression in tobacco *Nicotiana tabacum* BY-2 cells. *J Plant Res* 125: 555–568.

Laskowski M, Grieneisen VA, Hofhuis H et al. 2008. Root system architecture from coupling cell shape to auxin transport. *PLoS Biol* 6: e307.

Leblanc-Fournier N, Coutand C, Crouzet J et al. 2008. Jr-ZFP2, encoding a Cys2/His2-type transcription factor, is involved in the early stages of the mechano-perception pathway and specifically expressed in mechanically stimulated tissues in woody plants. *Plant Cell Environ* 31: 715–726.

Lee D, Polisensky DH, Braam J. 2005. Genome-wide identification of touch- and darkness-regulated *Arabidopsis* genes: A focus on calmodulin-like and XTH genes. *New Phytol* 165: 429–444.

Legue V, Blancaflor E, Wymer C, Perbal G, Fantin D, Gilroy S. 1997. Cytoplasmic free Ca^{2+} in *Arabidopsis* roots changes in response to touch but not gravity. *Plant Physiol* 114: 789–800.

Leng Q, Mercier RW, Yao W, Berkowitz GA. 1999. Cloning and first functional characterization of a plant cyclic nucleotide-gated cation channel. *Plant Physiol* 121: 753–761.

Lewis BD, Spalding EP. 1998. Nonselective block by La^{3+} of *Arabidopsis* ion channels involved in signal transduction. *J Membr Biol* 162: 81–90.

Lin SY, Corey DP. 2005. TRP channels in mechanosensation. *Curr Opin Neurobiol* 15: 350–357.

Lucas M, Godin C, Jay-Allemand C, Laplaze L. 2008a. Auxin fluxes in the root apex co-regulate gravitropism and lateral root initiation. *J Exp Bot* 59: 55–66.

Lucas M, Guedon Y, Jay-Allemand C, Godin C, Laplaze L. 2008b. An auxin transport-based model of root branching in *Arabidopsis thaliana*. *PLoS One* 3: e3673.

Maathuis FJ. 2009. Physiological functions of mineral macronutrients. *Curr Opin Plant Biol* 12: 250–258.

Maathuis FJ, Filatov V, Herzyk P, Krijger GC, Axelsen KB, Chen S, Green BJ et al. 2003. Transcriptome analysis of root transporters reveals participation of multiple gene families in the response to cation stress. *Plant J* 35: 675–692.

Malamy JE. 2005. Intrinsic and environmental response pathways that regulate root system architecture. *Plant Cell Environ* 28: 67–77.

Massa GD, Fasano JM, Gilroy S. 2003. Ionic signaling in plant gravity and touch responses. *Gravit Space Biol Bull* 16: 71–82.

Massa GD, Gilroy S. 2003a. Touch and gravitropic set-point angle interact to modulate gravitropic growth in roots. *Adv Space Res* 31: 2195–2202.

Massa GD, Gilroy S. 2003b. Touch modulates gravity sensing to regulate the growth of primary roots of *Arabidopsis thaliana*. *Plant J* 33: 435–445.

Miyawaki A, Griesbeck O, Heim R, Tsien RY. 1999. Dynamic and quantitative Ca^{2+} measurements using improved cameleons. *Proc Natl Acad Sci USA* 96: 2135–2140.

Miyawaki A, Llopis J, Heim R, McCaffery JM, Adams JA, Ikura M, Tsien RY. 1997. Fluorescent indicators for Ca^{2+} based on green fluorescent proteins and calmodulin. *Nature* 388: 882–887.

Miyazaki S, Murata T, Sakurai-Ozato N, Kubo M, Demura T, Fukuda H, Hasebe M. 2009. ANXUR1 and 2, sister genes to FERONIA/SIRENE, are male factors for coordinated fertilization. *Curr Biol* 19: 1327–1331.

Mohnen D. 2008. Pectin structure and biosynthesis. *Curr Opin Plant Biol* 11: 266–277.

Monshausen GB, Bibikova TN, Messerli MA, Shi C, Gilroy S. 2007. Oscillations in extracellular pH and reactive oxygen species modulate tip growth of *Arabidopsis* root hairs. *Proc Natl Acad Sci USA* 104: 20996–21001.

Monshausen GB, Bibikova TN, Weisenseel MH, Gilroy S. 2009. Ca^{2+} regulates reactive oxygen species production and pH during mechanosensing in *Arabidopsis* roots. *Plant Cell* 21: 2341–2356.

Monshausen GB, Gilroy S. 2009a. Feeling green: Mechanosensing in plants. *Trends Cell Biol* 19: 228–235.

Monshausen GB, Gilroy S. 2009b. The exploring root–root growth responses to local environmental conditions. *Curr Opin Plant Biol* 12: 766–772.

Monshausen GB, Messerli MA, Gilroy S. 2008a. Imaging of the Yellow Cameleon 3.6 indicator reveals that elevations in cytosolic Ca^{2+} follow oscillating increases in growth in root hairs of *Arabidopsis*. *Plant Physiol* 147: 1690–1698.

Monshausen GB, Miller ND, Murphy AS, Gilroy S. 2011. Dynamics of auxin-dependent Ca^{2+} and pH signaling in root growth revealed by integrating high-resolution imaging with automated computer vision-based analysis. *Plant J* 65: 309–318.

Monshausen G, Swanson S, Gilroy S. 2008b. Touch sensing and thigmotropism. In *Plant Tropisms*. eds. S Gilroy, PH Masson, pp. 91–122. Ames, IA: Blackwell Publishing.

Moreno-Risueno MA, Van Norman JM, Moreno A, Zhang J, Ahnert SE, Benfey PN. 2010. Oscillating gene expression determines competence for periodic *Arabidopsis* root branching. *Science* 329: 1306–1311.

Nagai T, Yamada S, Tominaga T, Ichikawa M, Miyawaki A. 2004. Expanded dynamic range of fluorescent indicators for Ca^{2+} by circularly permuted yellow fluorescent proteins. *Proc Natl Acad Sci USA* 101: 10554–10559.

Nakagawa Y, Katagiri T, Shinozaki K et al. 2007. *Arabidopsis* plasma membrane protein crucial for Ca^{2+} influx and touch sensing in roots. *Proc Natl Acad Sci USA* 104: 3639–3644.

Nakano M, Iida K, Nyunoya H, Iida H. 2011. Determination of structural regions important for Ca^{2+} uptake activity in *Arabidopsis* MCA1 and MCA2 expressed in yeast. *Plant Cell Physiol* 52: 1915–1930.

Noll F. 1900. Uber den bestimmenden Einfluss von Wurzelkruemmengen auf Entstehung und Anordnung der Seitcnwurzeln. *Landtw Jahrb* 29: 361–426.

Nomura H, Komori T, Kobori M, Nakahira Y, Shiina T. 2008. Evidence for chloroplast control of external Ca^{2+}-induced cytosolic Ca^{2+} transients and stomatal closure. *Plant J* 53: 988–998.

O'Brien EE, Brown JS, Moll JD. 2007. Roots in space: A spatially explicit model for below-ground competition in plants. *Proc Biol Sci* 274: 929–934.

Papahadjopoulos D, Ohki S. 1969. Stability of asymmetric phospholipid membranes. *Science* 164: 1075–1077.

Paredez AR, Persson S, Ehrhardt DW, Somerville CR. 2008. Genetic evidence that cellulose synthase activity influences microtubule cortical array organization. *Plant Physiol* 147: 1723–1734.

Peaucelle A, Louvet R, Johansen JN, Hofte H, Laufs P, Pelloux J, Mouille G. 2008. *Arabidopsis* phyllotaxis is controlled by the methyl-esterification status of cell-wall pectins. *Curr Biol* 18: 1943–1948.

Pittman JK. 2011. Vacuolar Ca^{2+} uptake. *Cell Calcium* 50: 139–146.

Richter GL, Monshausen GB, Krol A, Gilroy S. 2009. Mechanical stimuli modulate lateral root organogenesis. *Plant Physiol* 151: 1855–1866.

Robinson KR, Messerli MA. 2002. Pulsating ion fluxes and growth at the pollen tube tip. *Sci STKE* 2002: pe51.

Rotman N, Rozier F, Boavida L, Dumas C, Berger F, Faure JE. 2003. Female control of male gamete delivery during fertilization in *Arabidopsis thaliana*. *Curr Biol* 13: 432–436.

Sandaklie-Nikolova L, Palanivelu R, King EJ, Copenhaver GP, Drews GN. 2007. Synergid cell death in *Arabidopsis* is triggered following direct interaction with the pollen tube. *Plant Physiol* 144: 1753–1762.

Sasaki T, Naka M, Nakamura F, Tanaka T. 1992. Ruthenium red inhibits the binding of calcium to calmodulin required for enzyme activation. *J Biol Chem* 267: 21518–21523.

Schwab B, Mathur J, Saedler R, Schwarz H, Frey B, Scheidegger C, Hulskamp M. 2003. Regulation of cell expansion by the distorted genes in *Arabidopsis thaliana*: Actin controls the spatial organization of microtubules. *Mol Genet Gen* 269: 350–360.

Swanson S, Gilroy S. 2010. ROS in plant development. *Physiol Plant* 138: 384–392.

Takeda S, Gapper C, Kaya H, Bell E, Kuchitsu K, Dolan L. 2008. Local positive feedback regulation determines cell shape in root hair cells. *Science* 319: 1241–1244.

Tanaka K, Gilroy S, Jones AM, Stacey G. 2010a. Extracellular ATP signaling in plants. *Trends Cell Biol* 20: 601–608.

Tanaka K, Swanson SJ, Gilroy S, Stacey G. 2010b. Extracellular nucleotides elicit cytosolic free calcium oscillations in *Arabidopsis*. *Plant Physiol* 154: 705–719.

Tang RH, Han S, Zheng H et al. 2007. Coupling diurnal cytosolic Ca^{2+} oscillations to the CAS-IP_3 pathway in *Arabidopsis*. *Science* 315: 1423–1426.

Telewski FW. 2006. A unified hypothesis of mechanoperception in plants. *Am J Bot* 93: 1466–1476.

Thompson JE, Fry SC. 2001. Restructuring of wall-bound xyloglucan by transglycosylation in living plant cells. *Plant J* 26: 23–34.

Tian GW, Chen MH, Zaltsman A, Citovsky V. 2006. Pollen-specific pectin methylesterase involved in pollen tube growth. *Dev Biol* 294: 83–91.

Tretner C, Huth U, Hause B. 2008. Mechanostimulation of *Medicago truncatula* leads to enhanced levels of jasmonic acid. *J Exp Bot* 59: 2847–2856.

Tsai YC, Delk NA, Chowdhury NI, Braam J. 2007. *Arabidopsis* potential calcium sensors regulate nitric oxide levels and the transition to flowering. *Plant Signal Behav* 2: 446–454.

Tsukagoshi H, Busch W, Benfey PN. 2010. Transcriptional regulation of ROS controls transition from proliferation to differentiation in the root. *Cell* 143: 606–616.

Urao T, Yakubov B, Satoh R, Yamaguchi-Shinozaki K, Seki M, Hirayama T, Shinozaki K. 1999. A transmembrane hybrid-type histidine kinase in *Arabidopsis* functions as an osmosensor. *Plant Cell* 11: 1743–1754.

Walley JW, Coughlan S, Hudson ME et al. 2007. Mechanical stress induces biotic and abiotic stress responses via a novel cis-element. *PLoS Genet* 3: 1800–1812.

Wang Y, Geer LY, Chappey C, Kans JA, Bryant SH. 2000. Cn3D: Sequence and structure views for Entrez. *Trends Biochem Sci* 25: 300–302.

Wang Y, Wang B, Gilroy S, Chehab EW, Braam J. 2012a. CML24 is involved in root mechanoresponses and cortical microtubule orientation in *Arabidopsis*. *J Plant Growth Regul* 30: 467–479.

Wang WH, Yi XQ, Han AD et al. 2012b. Calcium-sensing receptor regulates stomatal closure through hydrogen peroxide and nitric oxide in response to extracellular calcium in *Arabidopsis*. *J Exp Bot* 63: 177–190.

Weerasinghe RR, Swanson SJ, Okada SF et al. 2009. Touch induces ATP release in *Arabidopsis* roots that is modulated by the heterotrimeric G-protein complex. *FEBS Lett* 583: 2521–2526.

White PJ. 2001. The pathways of calcium movement to the xylem. *J Exp Bot* 52: 891–899.

White PJ, Broadley MR. 2003. Calcium in plants. *Ann Bot* 92: 487–511.

Wilson ME, Jensen GS, Haswell ES. 2011. Two mechanosensitive channel homologs influence division ring placement in *Arabidopsis* chloroplasts. *Plant Cell* 23: 2939–2949.

Wohlbach DJ, Quirino BF, Sussman MR. 2008. Analysis of the *Arabidopsis* histidine kinase ATHK1 reveals a connection between vegetative osmotic stress sensing and seed maturation. *Plant Cell* 20: 1101–1117.

Wolf S, Mouille G, Pelloux J. 2009. Homogalacturonan methyl-esterification and plant development. *Mol Plant* 2: 851–860.

Wymer CL, Wymer SA, Cosgrove DJ, Cyr RJ. 1996. Plant cell growth responds to external forces and the response requires intact microtubules. *Plant Physiol* 110: 425–430.

Yamanaka T, Nakagawa Y, Mori K et al. 2010. MCA1 and MCA2 that mediate Ca^{2+} uptake have distinct and overlapping roles in *Arabidopsis*. *Plant Physiol* 152: 1284–1296.

Yang JL, Zhu XF, Zheng C, Zhang YJ, Zheng SJ. 2011. Genotypic differences in Al resistance and the role of cell-wall pectin in Al exclusion from the root apex in *Fagopyrum tataricum*. *Ann Bot* 107: 371–378.

Zheng HG, Babu RC, Pathan MS, Ali L, Huang N, Courtois B, Nguyen HT. 2000. Quantitative trait loci for root-penetration ability and root thickness in rice: Comparison of genetic backgrounds. *Genome* 43: 53–61.

IV

Soil Resource Acquisition

21

Root-Based Solutions to Increasing Crop Productivity

Michelle Watt
Commonwealth Scientific and Industrial Research Organisation

Anton P. Wasson
Commonwealth Scientific and Industrial Research Organisation

Vincent Chochois
Commonwealth Scientific and Industrial Research Organisation

I. Introduction

Advances in agriculture last century enabled global food production to keep up with a near fourfold increase in human population. From the 1950s, arable land for agriculture barely increased, making this scientific feat even more impressive (Evans 1993). Many improvements involved soil management, and thus indirectly the roots of new varieties empirically selected by breeders on yield alone. Examples from Australian wheat cropping systems include the addition of phosphorus fertilizer early in the century, addition of lime and manganese to support nitrogen fixation by legumes in acidic soils mid-century, rotations with canola and legumes to break root disease cycles and allow response to nitrogen application in the 1980s, and the widespread adoption of zero tillage late in the century which probably enabled survival of farms through the millennium drought at the turn of this century (Angus 2001; Kirkegaard and Hunt 2010). Global food production must double again over the next 100 years to meet population growth and growing affluence and diet shifts toward meat in the most populous countries, China and India. In the case of cereals, which provide over half the calories to humans, this can be achieved without an increase in food price if the yield of crops increases annually by about 1.5% (50% faster than the current rate), and this yield is reached with less water, less fertilizer, and less fertile land for agriculture (Fischer and Edmeades 2010). Therefore, productivity (production per land area per resource input) must increase. Here we address this challenge in terms of opportunities from the genetic improvement and agronomic management of roots using ideotypes or traits suited to environments or farming systems (Figure 21.1). We bias our examples from the literature on evaluations in soil in controlled conditions or in the field in farming systems. The scope of the review is crop root systems that have a role in water and nutrient uptake, anchorage and soil amendment, but are not harvested for food. Root crops are covered in Chapter 30.

II. Water

Agricultural productivity can be increased by developing new varieties and management practices that encourage root systems to capture more water and allow this water to be utilized more efficiently for biomass and yield (Passioura 1977; Passioura and Angus 2010). A number of examples exist where increased water capture by roots was achieved indirectly by breeding for tolerance to root diseases (e.g., Ogbonnaya et al. 2001), or soil toxicities (discussed in the following), thus greatly increasing root length and performance (e.g., see Figure 21.2A for the impact of disease on root development). The importance of overcoming soil borne constraints with root genetics or management to increase water capture and use efficiency cannot be overemphasized. In the short term, this approach will continue to provide the greatest gains in crop productivity.

Many scientists have selected for high yielding varieties in water-scarce environments to incorporate beneficial root traits into breeding programs (Bruce et al. 2002). It is from these studies that varieties with potentially valuable root traits have been identified (e.g., Champoux et al. 1995; Ali et al. 2000; Shen et al. 2001). Direct selection of the root traits by ideotype breeding (targeting a developmental or physiological trait of potential

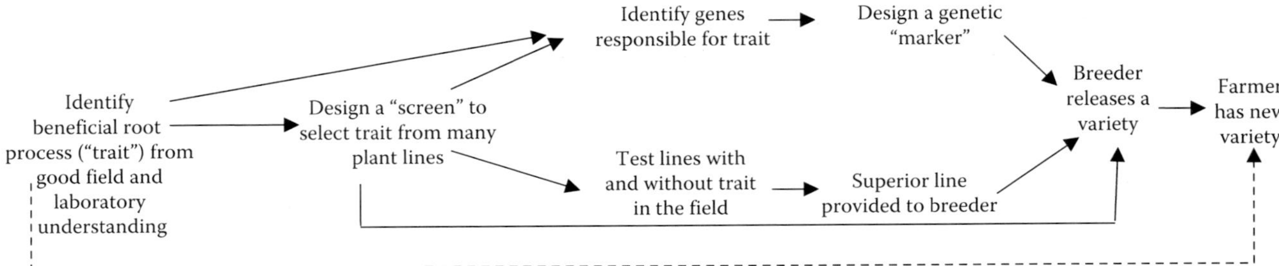

FIGURE 21.1 Ideotype or trait breeding to deliver a new variety to farmers with improved roots (solid line arrows). It may also be possible to use a management practice to generate the desirable root trait in the field (dashed line arrow). Fastest productivity gains will arise from combining new varieties and practices.

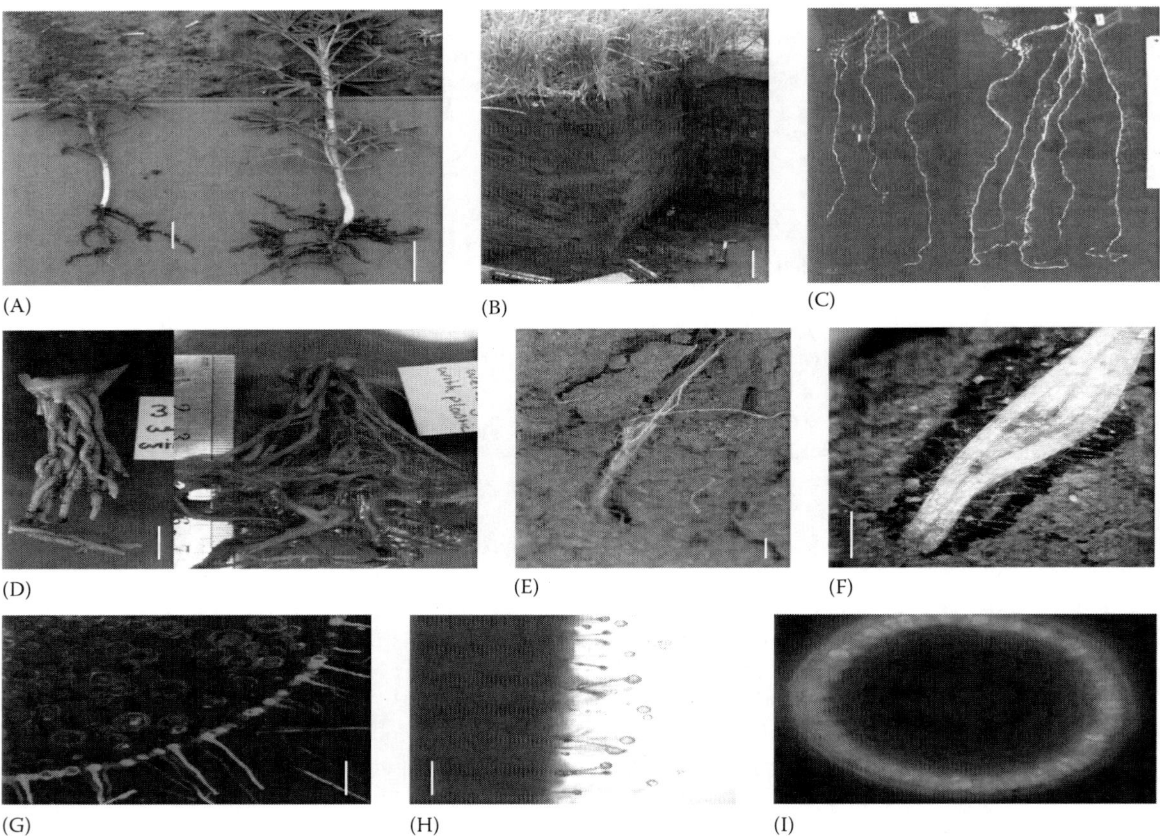

FIGURE 21.2 **(See color insert.)** (A) Lupin roots from the field. Left: infected with rhizoctonia fungal root disease and root length severely stunted and nodule number reduced. Right: healthy without rhizoctonia root disease. Bar = 2 cm. (B) Pit dug within a wheat crop at flowering and grain development in southeastern Australia. Maximum root depth ~150 cm. Bar = 20 cm. (C) Wheat genotypes of same age with genetic variation in root system vigor. (Reproduced from Gewin, Nature, 466, 552, 2010.) Ruler at right = 30 cm. (D) Sorghum nodal roots growing at two angles from the crown; wide angle on left is in moist soil; narrow angle on right is in dry soil. Bar = 1 cm. (Photo courtesy of M. Rostamza.) (E) Roots clumped in a pore in the pit shown in (B). Bar = 2 mm. (F) Wheat roots clumped in a soil pore with root hairs contacting sides of pore, enabling hydraulic conductivity with soil. Bar = 2 mm. (Photo courtesy of M.E. McCully.) (G) Phosphate transporter expressed specifically in rice root hairs (*Pht1;1 promoter::gfp* gene). Bar = 50 µm. (Reproduced from Schunmann, P.H.D. et al, *J. Exp. Bot.*, 55, 855, 2004.) (H) Sorgoleone exudate (yellow drops) released from sorghum root hairs. Exudate is a biocide and electron transport inhibitor in plants that take it up. Bar = 50 µm. (Photo courtesy of K. Schneebeli.) (I) Avenacin (blue) localized to the epidermal cells of oat. Compound toxic to wheat Take-All fungus. Bar = 50 µm.

value) or marker-assisted breeding (using their underlying genomic regions with molecular markers) can speed up genetic gain in water-limiting environments, compared to selecting on yield alone (Richards et al. 2010; Figure 21.1). However, the traits must be highly heritable and relatively simple (Richards 2006). There are few examples of root traits directly selected by ideotype breeding to increase water capture.

The work of Hurd and colleagues in the 1960 in Canada is one of the earliest examples, coupling root studies in controlled conditions and field trials. They linked the performance of wheat cultivars under drought conditions in the field with the growth of those varieties' root systems in glass-faced root boxes under varying levels of soil moisture. They identified genotypic variation in rooting patterns that were expressed in a variety of soil-water patterns (Hurd 1964). The key traits identified as being of value were an early root vigor, surface root density, and, when stressed, root density at depth. Continuing studies into the mid-1970s identified a drought-resistant variety that exhibited substantial root mass irrespective of water treatment and a high rate of vertical root growth (Hurd 1974). Importantly, these studies were extended into the field with the use of soil coring, which was found to agree well with the glass root boxes.

Work conducted by Taylor and coworkers on soybean in Iowa advanced the same approach further (Taylor 1980). They started with an extensive analysis of the edaphic and climatological factors involved in soybean production and concluded that soil water in the farming system was recharged annually, but not fully exploited. Root length density to a depth of 150 cm was adequate; however, deeper root length density, increased by a higher vertical descent rate for the roots, was modeled to increase both the total water uptake and the water-use efficiency. The arrival of the roots at the underexploited 150 cm layer would coincide with reproductive development, so the additional water would be partitioned into grain production. Greenhouse studies revealed variation in taproot growth rates of between 3.1 and 4.3 cm/day over a 27 day period (Taylor 1980). In the field, it was demonstrated that this variation was dependent on soil temperature and decreased with time (Kaspar et al. 1981; Stone and Taylor 1983).

These classic examples of selection for roots with greater water uptake pointed to the importance of greater root length at depth (see Figure 21.2B for illustration of the depth of wheat roots in the field). The value of root depth traits, however, should not be overstated. A root depth trait will not contribute to increased water uptake where water is not available and recharged into the deep layers annually, and there are subsoil constraints such as harsh pH or nutrient deficiencies (Passioura 1983, 2006a; Yambao et al. 1992). Tennant and Hall (2001) reviewed crop water use in Western Australia and concluded that root depth was a function of soil type, overriding the capacity of crops for deeper rooting (Dracup et al. 1992; Gregory et al. 1992). Kirkegaard and Lilley (2007) found that root depth was a function of soil wetting, soil type, and vegetative growth duration. Deep-reaching roots that extract more water are a trait that would be valuable in deeper soils with a high water holding capacity and in a climate with terminal drought stress (Wasson et al. 2012). Kirkegaard et al. (2007) demonstrated the value of accessing deep water by using rainout shelters and irrigation in the field to determine the efficiency with which different sources of water were converted to wheat grain yield. They showed that an additional 10.5 mm of subsoil water absorbed after anthesis from 1.35 to 1.85 m depth would increase grain yield by 0.62 ton/ha, equating to 59 kg/ha for every additional mm. This study supported that of Manschadi et al. (2006) in large root boxes, which found that greater root length at depth of wheat contributed to increased yield by allowing better water extraction during grain filling: each additional mm of water extracted after anthesis generated an extra 55 kg/ha of grain yield (see Chapter 22). In rice, deep rooting is considered a drought resistance trait and, where constitutively expressed, a drought avoidance mechanism (Mambani and Lal 1983; Kondo et al. 2003; Bernier et al. 2008). Among upland rice varieties, a deeper root system allows higher yields under drought, and a high deep-root-to-shoot-weight ratio is linked to increased drought resistance (Fukai and Cooper 1995).

Root processes linked to greater deep root length include vigor and angle (see Figure 21.2C and D for illustration of the variation vigor and angle). Manschadi et al. found a wide variation in root angle among diverse Australian wheats, with lines with narrow seedling angles originating from northern soils with stored, deep water, and those with wider spreading angles from Mediterranean environments that receive surface showers during the wheat growing season (Manschadi et al. 2008). Japanese winter wheats adapted to a drier environment had a narrower angle, linking that to better depth penetration (Oyanagi et al. 1993). Variation in root angle will need to be combined with root vigor at depth (Palta and Watt 2009). Wide variation in seeding vigor exists in wheat, and this may be related to differences in rooting depth at the end of the season (Richards et al. 2007).

Greater water uptake by roots in different soil layers could be improved with increased root and root–soil hydraulic conductivity. It was proposed that an increase in xylem size in rice would lead to a decrease in axial resistance and better water uptake from deeper soil layers (Nguyen et al. 1997). Some studies linked root thickness and drought resistance (Ekanayake et al. 1985). A study in rice identified QTL for basal root thickness and showed a significant 0.33 correlation with yield in upland conditions but no significant correlation in lowland conditions (Li et al. 2005). Yambao et al. (1992) showed a correlation between root thickness and xylem size for thin fibrous roots, but they couldn't conclude that xylem size was an important factor in drought resistance. Selection pressure for rice with deep root systems and large xylem capacities could be exerted with infrequent irrigations that keep the deep soil layers at field capacity but which allowed the shallower layer to dry out (Bernier et al. 2008).

Resistance at the interface between soil and root has been implicated as the cause of incomplete utilization of subsoil water (Stirzaker and Passioura 1996). Rowse and Goodman (1981) concluded that in common bean radial resistance was a greater determinant of water uptake than axial resistance. This could be overcome in crops with selection for anatomical variation or expression of aquaporins (Bramley et al. 2009; see also Chapter 24). Alternatively, breeding could target root hairs, as there is evidence that they increase water uptake (see also Section IV.B). Water uptake in barley mutants lacking root hairs was analyzed over a 2 day period, showing less water uptake and less uptake per unit root length, despite greater root branching (Segal et al. 2008). There is significant evidence that roots do not grow through soil uniformly, but rather exploit preexisting pores and channels (Passioura 1991; see Figure 21.2E). In pores, the roots branch less but have more root hairs making contact with the pore walls (White and Kirkegaard 2010; see Figure 21.2F). Improved root hair contact may lead to an increase in root–soil hydraulic conductivity.

Research and physiological breeding conducted in the 1970s is still the best known example of incorporating a root trait into a crop to improve water-use efficiency. Passioura demonstrated that by forcing wheat plants to rely on a single seminal root (by excision), they used less water prior to flowering, the higher resistance to water flow effectively metering out the soil-water supply (Passioura 1972). This physiological effect was then combined with two key observations. Firstly, wheat receiving small irrigations over the course of an entire season yielded better than varieties that received four times as much water in a single basal irrigation at the start of the season, a key indication that physiologically relevant timing of water availability is a key component of improved water-use efficiency. Secondly, there was a correlation between harvest index and the amount of water used by a variety after flowering as a percentage of its total water use, with the varieties with harvest indices of 0.4–0.5 using 30% of their water after flowering, whereas those with harvest indexes of <0.1 using only 10% after flowering. The implication of these finding was that where stored soil moisture was the primary source of a wheat plant's water uptake, increased axial resistance might increase the efficiency with which that water was used.

Richards and Passioura (1981a,b) investigated increasing axial resistance either by reducing the number of seminal roots or reducing the xylem vessel diameter of those roots, opting to focus on xylem diameter by virtue of there being a source of natural variation in wheat and because the trait had a higher heritability. Incorporation of the trait did produce modest improvement in grain yield in years of water stress, but no yield penalties in years of abundant rainfall, when nodal roots developed to exploit the additional rainfall and reliance on the seminal roots was reduced.

Thus, good candidates for the basic developmental and physiological determinants of root water uptake and water-use efficiency are well established. However, root characteristics have not been incorporated into commercial breeding programs based on root phenotyping because of the basic difficulties in selecting for root traits (Nguyen et al. 1997). Emphasis has been placed on using marker-assisted selection (MAS) to incorporate root traits using QTLs. Shen et al. (2001) developed near-isogenic lines (NILs) of IR64 with QTLs from the rice line Azucena. Four QTLs respectively from chromosomes 1, 2, 7, and 9 were identified. One of three NILs containing the chromosome 1 allele had improved root traits; three of seven NILs for the chromosome 7 allele had improved root mass at depth (but not other traits); four of six NILs for the chromosome 9 allele had improved maximum root length (but not other traits); of the five NILs carrying the chromosome 2 target, none had significantly improved root phenotypes. It is surprising to find out how few of the lines actually had improved roots. Whether this is an issue with MAS (e.g., distance between the marker and the desired gene) or indicative of the complexity underlying the genetic control of root traits is unclear. Courtois et al. (2003) identified 28 QTLs for 10 root traits in rice, or 1–4 main effect QTLs per trait, including root length, root thickness, and root dry weight in various layers (of a 1 m tube). Another QTL study sought to introduce the 4 QTLs from Azucena line into the Indian upland rice variety Kalinga III, which had not previously been used for QTL mapping. The four QTLs were pyramided, requiring 6 years, 8 generations, 3000 marker assays, and 323 lines. In field experiments, 22 NILs were evaluated, showing that the chromosome 9 QTL was associated with increased root length under both irrigated and drought stress treatments. No significant effects were observed for the other targets (Steele et al. 2006). A rice allele derived from Way Rarem (an upland-adapted, drought susceptible rice variety from Indonesia), *qtl12.1*, was identified in a screen for grain yield in drought conditions and was shown to confer a 7% advantage in water uptake in water-limited conditions, enough to confer a yield advantage (Bernier et al. 2009). A regression analysis showed that both the root length beneath 30 cm and the maximum root depth traits were correlated to increased water uptake—lines with the Way Rarem allele have 18% greater root length at depth. A similar chromosomal region was identified by Yue et al. (2006) in a tube screen for drought tolerance traits that sought to apply the drought stress during the same developmental stage for all lines, so as to separate drought escape from the drought avoidance and drought tolerance traits, including root traits. The 12 root traits gave 74 QTL which could be assigned to 36 genomic regions. However, these examples have revealed multiple QTL and illustrate the challenges when dealing with QTL in breeding for water-limited environments (Passioura et al. 2007).

III. Root Traits to Increase Yield Potential

Analyses of potential yields of the major crops rice, wheat, and maize last century suggest increases were achieved mainly by increasing the biomass of the shoot and the ratio of harvested grain to shoot biomass, called the harvest index (Fischer and Edmeades 2010). What was happening to the root systems during this time? Given that these increases were often associated with a trend in earlier flowering, decline in days between germination and yield, it seems feasible that breeders in some environments have been inadvertently selecting for more functional and possibly more efficient root systems (e.g., increased uptake and transport per unit carbon). Carbon allocated to the roots is used for building cell components, generating root length, surface area, vascular capacity and strength, and for respiration for synthesis of compounds and powering transport. On average, 5%–10% of net carbon fixed by the plant is released to the soil in exudates (reviewed in Farrar et al. 2003); similarly, associations with soil microorganisms such as mycorrhizae can "cost" the plant 20% of carbon fixed in the leaves (Graham 2000). Is there scope to reduce carbon allocation to roots for further genetic gains?

Modeling and some experimental evidence suggest efficiencies can be made with allocation of belowground carbon, but these depend on limiting water and nutrient resources, and their spatial and temporal distribution in the soil through the season (Lynch 2007a). For example, some bean and wheat genotypes have lower root respiration rates per unit uptake of nutrients per length of root than others when either phosphorus or nitrogen is limiting (Nielsen et al. 2001; Palta and Watt 2009). However, quantification of root carbon in soil and the field is lacking as measurement is technically very difficult, if not impossible.

In a recent view point on the importance of root system size in dry environments, Palta et al. (2011) found no evidence that extra root length occurred at a cost to vegetative biomass production and yield. Similarly, modeling of wheat root growth suggested that faster overall root system growth and colonization of soil layers were associated with yield increases in most environments, including wetter conditions (Lilley and Kirkegaard 2011). Modeling in bean and maize, however, suggests the architecture of the root system is important. In bean, root structural carbon was best allocated into an architecture that favors top soil root length to capture limiting phosphorus, and deep root length to capture deep water at the end of the season (Ho et al. 2005; see also Chapter 26). Modeling of maize crops in the midwest of the United States also suggested that narrow, deep penetrating, root system architecture is partly responsible for gains in yield (Hammer et al. 2009). Thus, it would seem that root growth rate and direction of growth relative to limiting resources are important. Both these traits are being investigated by research groups worldwide, require field evaluation, and very likely will need to be combined in new varieties (Palta and Watt 2009).

One characteristic of cereal root systems relies on the presence of a second, morphologically and anatomically distinct type of dense, fibrous roots, called nodal, adventitious, or crown roots. These postembryonic roots emerge later than the seminal root systems and can capture, in some conditions, an important part of the resources allocated to the roots, as their dry weight can be greater than that of the seminal roots. Although wheat and barley nodal roots absorbed respectively up to three to four times more water, and two to three more salts than seminal roots in hydroponic conditions, their excision did not affect significantly grain yield of both cereals (Krassovsky 1926), as opposed to seminal root excision. Moreover, as concluded in this study, the removal of nodal roots was compensated by an increased growth of the seminal roots, which are able to absorb twice as much water as the nodal roots per unit of dry weight. One way suggested by Richards (1991) to improve crops, and especially water-use efficiency of crops, would be to reduce the root biomass (Richards 1991). It has been discussed that cereals root systems are too big for the water they extract (Passioura 1983; Richards 1991) and that their reduction may free some resources to make more seminal root branching at depth in order to capture more water during dry seasons. Interestingly, a 40% increase in maize yield under severe stress was achieved by a 35% reduction in root mass in the shallowest 50 cm of soil (Bolanos et al. 1993). It is thought that this reduced the stress signals caused by intermittent soil drying, reducing stomatal closing, and increasing transpiration.

In that context, genes affecting nodal root emergence and growth have been identified in maize and rice, such as RTCS1 (Hetz et al. 1996; Taramino et al. 2007) also called ARL1 (Liu et al. 2005)/CRL1 (Inukai 2005) in rice, or OsGNOM1/CRL4 in rice (Liu et al. 2009). It could be interesting to investigate these genes, as well as any QTL affecting the abundance of nodal roots relative to seminal roots, in field experiments in order to evaluate their impact on grain yield, especially in drought conditions, and in environments where crops grow mostly on stored soil water rather than on rainwater. As mentioned later, a knockout mutant would probably not be the best candidate to investigate for this trait: RTCS1, the maize knockout mutant devoid of crown roots, was identified in the field through its lack of lodging resistance. One can imagine that this kind of phenotype is not suited in the field as it obviously weakens the mechanical resistance to winds and can even alter both grain yield and quality of the crop (Easson et al. 1993; Berry et al. 2004; Reynolds et al. 2011). Thus, a line with reduced, rather than totally abolished, crown root system may be a better candidate in terms of coping with field conditions and farmers' demands, especially if better yields, hence heavier heads, are expected.

This could be achieved by reducing the number of nodal root axes emerging or hampering the nodal root growth specifically, without affecting seminal root growth. The first approach implies focusing on root emergence processes, which involve mostly auxin-responsive genes mentioned earlier (RTCS1-ARL1/CRL1, OsGNOM1/CRL4). Few genes affecting specifically nodal root growth (and not emergence) have been identified so far. Genes involved in root meristem maintenance and function such as genes affecting or responsible for cytokinin transport and signaling may be a lead. In rice, the gene WOX11 has been shown to respond both to auxin and cytokinin to initiate and pursue crown root growth (Zhao et al. 2009). This gene is suggested to be an integrator of auxin and cytokinin signaling which regulates cell proliferation during crown root development.

An innovative approach to decreasing root "cost" was taken by Zhu et al. (2010a). They have been selecting for maize genotypes with more cortical aerenchyma under drought conditions and finding them to be more tolerant. The idea is that breaking down cortical cells, albeit after they are built, saves on their maintenance respiration, freeing up carbon to build more length and surface area (see also Chapter 26).

It is attractive to think that savings in carbon could be made by selecting for decreased net carbon exuded to the rhizosphere or decreased carbon transferred to fungi such as mycorrhizae. However, some exudates play important roles increasing the availability of phosphorus or nitrogen to roots; others prevent toxicity by elements such as Al, as discussed in the following. The function of others such as the hexose sugar glucose is less clear. Studies with labeled carbon report that exudation is increased by the presence of microorganisms, suggesting that exudates could be a cost to yield unless they provide a benefit in return (Martin 1977). Processes in the rhizosphere mediated by exudates and microbial associations are not well understood in field conditions; however, there are encouraging signs that they can be manipulated by management or genetics to have a large impact on productivity (Watt et al. 2006a,b; Richardson and Simpson 2011).

Any trade-off in carbon allocation between roots and shoots becomes critical when carbon from photosynthesis and storage is limiting, around flowering and development of the fruit or grain when shoots are senescing. Generally in annual crops, although rarely measured, root growth and function ceases around that time. A simple way to manage water and nutrient uptake by roots is to change flowering time, either agronomically, by altering the sowing time, or genetically, by breeding for altered flowering times. For example, delaying flowering extends time for root growth and uptake of phosphorus (Nord and Lynch 2008). When water is limiting at grain development, extra water can be accessed with extra root length, to either maintain photosynthesis or mobilize plant carbohydrates to the grain (Kirkegaard et al. 2007). Lopes and Reynolds (2010) found in the field that wheat genotypes with more root growth at depth had cooler canopies, form more green leaf area, or have higher transpiration and greater yield. Carbon allocation to roots either during the season or post flowering appeared to be at no cost to yield in the dry environment of the trial. The authors suggest a trade-off between carbon allocated to roots and stem storage carbohydrates.

IV. Nutrients

A. Nitrogen

The vast bulk of nitrogen fixation occurs biologically, with atmospheric nitrogen converted to ammonia by nitrogen-fixing bacteria (Postgate 1998). This is the best example of a root-based process that has been exploited successfully in agriculture. Symbiotic biological nitrogen fixation in legumes, between the plant and N-fixing rhizobia, is a root trait (organ and symbiosis) that contributes directly to the N requirements of the plant: for every ton of shoot dry matter produced by crop legumes, the symbiosis contributed 30–40 kg of nitrogen (Peoples et al. 2009). However, the contribution to the agricultural system of rotational cropping, at least in terms of N, may not be that significant, as the majority of that N is removed with the harvested seed—the true benefits of rotational cropping may come from the effect of legumes breaking disease cycles and on soil biology (Angus 2001; Peoples et al. 2009). There are opportunities to improve the fixation of nitrogen by legumes under water deficit. For example, Sinclair et al. (2000, 2007) identified a drought-tolerant variety of soybean with nitrogen fixation, which was then crossed to a high yielding variety. The progeny was assessed in the greenhouse for nitrogen fixation under mild drought stress, and two of the lines were shown to outperform commercial cultivars in the field.

Nitrogen fertilizer use is widespread in agriculture but relatively inefficient with only 33% of the N applied being harvested in grain (Raun and Johnson 1999). The remaining applied nitrogen is lost, either through leaching and runoff, volatilization, or denitrification by soil microbes (Vitousek et al. 1997). Improved nitrogen use efficiency (NUE) equates to improved nitrogen uptake of applied fertilizer (uptake efficiency) and how well the nitrogen that is absorbed is utilized by the plant to produce the yield (utilization efficiency).

Altered root traits might contribute to improved uptake efficiency. Although nitrate is mobile in the soil, root architecture may not be the most relevant trait in ensuring the most efficient uptake (Burns 1980; Robinson and Rorison 1983). For example, Gallais and Coque (2005) did not identify a clear benefit to large root systems in terms of yield in a low-N environment. Similarly, in a high-N environment, root-to-shoot ratios are comparatively low. In maize, 79% of the N supply to the roots is through mass flow, 20% is through diffusion, and only 1% is through direct interception by the roots. Indeed, evolution appears to have delivered a bias against exploiting high levels of N with extensive root growth at a cost in carbohydrate: in *Arabidopsis*, lateral root growth and root length are reduced by high-nitrate treatments (Zhang and Forde 1998, 2000). Lateral roots also proliferate where they encounter nitrogen-rich patches, but this is unlikely to be relevant in the agricultural context where large amounts of fertilizer are being used and plants are usually grown in monoculture and root systems compete with each other (Walch-Liu et al. 2006).

Early root vigor and a high root length density may be a useful trait in conditions where nitrogen is likely to be leached from the soil. Liao et al. (2004a, 2006) showed that early root growth improved nitrogen uptake, even though the root-to-shoot ratio was ultimately unaltered. Furthermore, vigorous growth was associated with a higher root length density (measured as the length of root in a volume of soil), where few thick roots were replaced by many thinner roots without a necessary increase in carbon allocation. This leads to higher NO_3^- uptake in conditions where it is likely to be leached (Wiesler and Horst 1993, 1994).

Where ammonium, rather than nitrate, is the plant's primary source of nitrogen exploration may be a valuable trait, given that it is significantly less mobile. There may also be a benefit associated with greater root density at depth, allowing more nitrogen to be intercepted as it leaches from the surface layer (Gastal and Lemaire 2002). Increasing the surface area of roots through increased root hair growth is a known response to nutrient stress (Robinson and Rorison 1987; Gilroy and Jones 2000). However, where there is an abundance of N, increased root hair length does not seem to enhance N acquisition. Modeling of nutrient acquisition has shown that increased root diameter does not significantly increase the uptake capacity of NO_3^- or NH_4^+ (Robinson and Rorison 1983, 1987; Clarkson 1985). The incorporation of a biological nitrification inhibiting exudate into crop roots could also improve nitrogen use efficiency from the rhizosphere, especially if combined with longer root hairs (Subbarao et al. 2009).

A longer term prospect for improving nitrogen acquisition would be the genetic modification of nonlegumes to engage in nodule development and symbiosis with rhizobia, a prospect that is becoming more feasible as more is learned about the molecular basis of the rhizobium-legume symbiosis (Charpentier and Oldroyd 2010). However, the value of these traits is yet to be demonstrated empirically.

B. Phosphorus

Phosphorus limitation reduces growth of the shoot and reproductive organs and is a constraint to yield in many agronomic systems (Marschner 1995). Phosphorus and crop productivity has two main constraints. First, many crops require phosphorus fertilizer as most soil P is bound to mineral and organic compounds and is not readily available to roots. Second most phosphorus fertilizer is mined from phosphate rock, and predictions suggest world reserves will be depleted by the end of this century (Cordell et al. 2009). Phosphorus efficiencies ultimately need to come in two ways; plants grow more biomass and yield at a lower level of critical tissue phosphorus, and phosphorus in residues and waste is saved and recycled (Cornish 2010). To date, there has been little practical success in developing crop plants with a lower critical phosphorus requirement, although native plants from very-low-phosphorus regions have mechanisms to grow with extremely low phosphorus in leaves and efficient recycling strategies that may be exploited in the future (Lambers et al. 2011). The second option is feasible, but current technologies are too expensive. However, crop improvement research has had success exploiting plant strategies to increase access to bound soil phosphorus, more effectively taking up applied phosphorus fertilizer or mining P reserves in soil (Marschner 1995; Richardson 2009; Richardson et al. 2009). Strategies include modification of root architecture to increase growth of roots where phosphorus is highest, increased root hair length and density to increase surface area to grab phosphorus in solution, and increased release of exudates to access bound phosphorus. The following examples deal with experiments that were conducted both in soil and in the field and are offering promise to farmers.

Perhaps one of the most successful examples of a root-based solution to crop productivity involves breeding for more surface roots to access phosphorus accumulated at the top of the soil profile (Lynch and Brown 2001; Lynch 2007b). Lynch and his team began in the old, acidic, and highly P-"fixing" soils of Colombia, South America (Yan et al. 1995). There they identified a bean variety with the ability to yield as high as other varieties with less phosphorus fertilizer and studied it for mechanisms that could explain its superior ability to access phosphorus. The root architecture of the P-efficient variety was more horizontal than the other varieties. Branch roots emerged and grew from the base of the tap root at a more horizontal angle close to the soil surface, placing greater surface area of the axes and subtending branches within the surface soil where phosphorus from fertilizer and dead plant matter accumulated. A breeding "screen" for branch root angle was used in pouches with large sheets of germination paper where root systems grew in a horizontal plane and angles were measured (Bonser et al. 1996; Liao et al. 2001). QTLs were identified that are associated with the wide root angle; however, because the phenotypic screen is quick, easy, and directly associated with the trait, it may be a more economic approach than marker-assisted selection (Liao et al. 2004b; Beebe et al. 2006). Field studies in Central America showed that varieties with horizontal roots produce more shoot biomass and prevent phosphorus runoff in farming systems, and thus have higher phosphorus use efficiency (Henry et al. 2010). Lynch and colleagues now train breeders in China to screen for root angle and incorporate that trait into new varieties for farmers (Lynch, personal comm.). New varieties with wider root angle should lead to increases in P use efficiency in China, and this will be valuable to global P supplies and prices in the future. Recent impressive gains in grain yield in China are correlated tightly and positively with increased phosphorus and nitrogen fertilizer application and irrigation (Davies, personal comm.).

Root hairs greatly increase surface area and reduce the diffusion distance between phosphate and the root, at little or no metabolic cost to whole plant growth (Bates and Lynch 2001; Zhu et al. 2010b; see Figure 21.2G for illustration of the importance of root hairs to phosphate uptake). Many plant species produce more and longer root hairs when phosphorus becomes limiting. Gahoonia et al. (1997) found a wide variation in root hair length and density among wheat and barley genotypes and associated longer, denser hairs with greater P uptake in low-P soils. The importance of root hairs for P uptake and yield was then demonstrated using a mutant of barley. Trials in soil showed that the wild type with hairs had greater yield in lower phosphorus conditions compared to the mutant without hairs (Gahoonia and Nielsen 2004). These studies suggest that having longer and denser root hairs is a valuable trait for selection to increase phosphorus use.

Direct measurement of root hairs is time consuming and tricky because they are highly variable along the root, and therefore, the use of molecular markers is likely to be the best way breeders can incorporate the trait. Zhu et al. (2005) identified a QTL associated with root hair growth in response to phosphorus in maize that could be used in marker-based selection. Candidate genes, such as RSL4 identified in an *Arabidopsis* genotype that, when constitutively expressed, produced longer root hairs, could form the basis of molecular markers in crops in the future (Yi et al. 2010).

Roots of many plant species release a range of compounds to the rhizosphere, exudates, that solubilize P from bound mineral and organic compounds (Randall et al. 2001). Roots exude hydrogen ions and organic acids, which lower the soil pH and increase P availability. Organic anions can be exuded, competitively binding to elements such as aluminum and releasing phosphate ions. Likewise, phosphatase enzymes are released, which cleave phosphate from organic molecules. Other exudates likely operate indirectly, such as those that suppress microorganisms competing for phosphate made available in soil solution by other directly solubilizing exudates (Martinoia et al. 2006) or those that stimulate microorganisms that release phosphate from substrates unavailable to plants (Richardson et al. 2001b). The synthesis, membrane transport, timing of release, and localization of P-related exudates in roots and rhizosphere have received substantial physiological research, and efforts have been under way to identify, through phenotypic screens, regulatory genes for use in breeding for about 20 years.

A novel approach was taken by Richardson and his team using genetic modification. *Arabidopsis*, clover, and tobacco plants were transformed with a fungal gene that encodes a phytase enzyme, rendering them the ability to access phosphate from phytate (Richardson et al. 2001a; George et al. 2004, 2005b). Most plants have no or very little phytase naturally, and this limits their ability to access phosphate from the organic molecule phytate, which can be a substantial component of total soil phosphorus. The group found, however, that the growing medium determined how well the plants performed. Transformed plants performed well both on agar and on soils to which additional phytate had been added, but in most soils, they were not effective. Importantly, the group went on to understand the limitations of enzyme function in soil and discovered that pH is critical (George et al. 2005a). At low pH, the phytase was quickly adsorbed to soil surfaces and rendered ineffective within minutes of application; at high pH (7.5), the phytate remained in solution and functional for days. These observations highlight the need to evaluate genetic variation of root traits in soil because even simple substrate mechanisms present in soils are not always occurring under in vitro or controlled environments (e.g., George et al. 2008).

V. Land

Much of the world's best land for farming surrounds cities and towns, and it continues to be taken over by buildings and roads. To maintain production with declining good land for agriculture, options are to (1) move to more marginal lands with new varieties or management practices, (2) maintain or improve the properties of current land by developing new varieties with roots adapted to soil conservation practices, and (3) intensify agriculture on current land with more yield per land per time with new crops and more frequent rotations. Each option involves improvement of the roots.

A. Roots Adapted to Marginal, Hostile Soils

New varieties have been developed with root traits adapted to marginal, hostile soils. Examples are provided for salty and acidic soils, which cover a large portion of the earth's surface.

Soil salinity from irrigation or the subsoil is a major constraint to crop production worldwide (Rengasamy 2010). Durum wheat, used for pasta, is particularly sensitive to salt in the root zone, compared to bread wheat. Physiology, breeding, and molecular biology were used by Munns and her team to increase the salinity tolerance of durum wheat (reviewed by Passioura 2010). First, a collection of durum wheat plants and their wild relatives was made from around the world. This collection was screened in a gravel culture system whereby the roots were regularly flushed with salty solution. With physiological knowledge of plant response to salt stress, it was decided that the salt concentration of the expanding young leaves would be used to identify lines that excreted salt and thus avoided toxicity (Munns and James 2003). A wild relative was identified with very low levels of salt accumulation in the leaves. Crosses were made to a sensitive durum cultivar that accumulated salt, providing the population to conduct the genetics to identify two genomic regions associated with the leaf salt, and the mapping to identify two genes, Nax1 and Nax2 (Munns et al. 2003). Both encode sodium transporters that export salt out of the cells (Huang et al. 2008). Nax1 is located in the leaf sheath, preventing accumulation in the leaf blade, and Nax2 is located in the roots in the parenchyma cells, where it pumps Na^+ out of the xylem (James et al. 2006). Nax2 has been most effective in the field to date, and trials on salty soils in northern Australia using lines with and without the gene have demonstrated a 20% yield advantage over the current durum variety (Munns et al. 2012). This fine example of physiology, classical breeding, genetics, and molecular biology resulted in a new wheat genotype with higher yield on marginal lands within 10 years, from 2001 to 2011, building on stress physiology and membrane research from the 1970s (Munns and Tester 2008). There are opportunities to incorporate this gene into durum and bread wheat in other salt-prone regions and countries through gene selection using breeding markers or phenotypic selection (James et al. 2011). In the future, it may be possible to incorporate root traits to maintain root elongation in salty subsoils (Rahnana et al. 2011).

A similar combination of physiology, breeding, and molecular biology took place in the identification of aluminum tolerance to acid soils (Ryan et al. 2011). Wheat plants derived of parents from Brazil, a region of the world with highly acidic soils, were collected and shown to have higher yields in acidic soils in southern Australia (Fisher and Scott 1987). Delhaize, applying physiological knowledge of plant responses to acid soils, showed that the tolerant wheats exuded malate from their root tips in response to the element Al (Delhaize et al. 1993). The main constraint to crop growth in acidic soils is the toxic element Al, which severely inhibits plant growth by entering the root tip and disrupting cell division and preventing root elongation (see Chapter 33). Malate binds to the Al around or within the apoplast of the root tips, allowing the tip to continue to elongate. By carefully subdividing the root zones, Ryan and colleagues showed that the efflux was localized to the root tip and that it was mediated by an Al-activated channel (Ryan et al. 1995, 1997). With near-isogenic material and knowledge of the trigger and location of efflux, Sasaki et al. (2004) went on to use a subtractive approach to identify a cDNA that corresponded to a gene encoding an Al-triggered transporter, ALMT1. Molecular markers, a phenotypic screen (root elongation or malate efflux in response to Al), and germplasm sources of the gene are available to breeders to develop wheat varieties tolerant to acid soils (Sasaki et al. 2006). This gene was transformed into barley using genetic engineering. Barley, normally very sensitive to Al, was rendered tolerant to acid soils (Delhaize et al. 2009). This and other Al tolerance genes can be used in conventionally selected and genetically modified crops in the future (Ryan et al. 2011).

B. Roots Adapted to Farming Practices That Conserve Soil Properties

In the 1970s, conservation farming practices were developed to prevent soil erosion. Since, they are used in various forms depending on region and farming system. An underlying feature of these practices is minimum soil cultivation and disturbance. Seeds are sown with tines or disks without plowing to maintain soil structure and reduce erosion. These practices save farmers fuel for plowing, increase the porosity of the soil (by preventing the destruction of natural pores and channels in the soil), increase carbon, and increase the diversity and numbers of microorganisms and macroflora (Liebig et al. 2004; Zhang et al. 2007; reviewed by Watt et al. 2006; Table 3). Despite these benefits, downsides in some regions include poor early shoot growth leading to lower yield, deep drainage of nitrogen and water, and increased herbicide resistance (herbicide spraying increases as an alternative to weed control through cultivation). Farmers will be attracted to these practices if yields increase and their production costs go down. A valuable approach is to develop new varieties suited to the soil conditions of these management practices (Richards et al. 2007; Kirkegaard and Hunt 2010). Here we provide examples of root traits in wheat to increase yields in conservation systems.

Slow shoot growth, sometimes referred to as "poor early vigor," can occur in conservation systems around the world (Kirkegaard 1995; Lekberg and Koide 2005; Thomas et al. 2007). Studies with wheat showed that early vigor is increased if unplowed field soil is either cultivated *or* fumigated (Chan et al. 1987). In other words, an interaction between soil structure and microorganisms can restrict early shoot growth in unplowed soils. Simpfendorfer et al. (2002) found that inhibitory *Pseudomonas* bacteria were the microorganisms responsible for slowing wheat shoot growth in unplowed soils in southeastern Australia. Watt et al. (2003, 2005) undertook studies in controlled and field conditions to understand how structure and microorganisms were causing the poor early shoot vigor. Roots in unplowed soil tended to grow in cracks and pores between harder peds and were 30% shorter and three times more distorted than roots in plowed soil. The slow-growing root tips accumulated higher numbers of bacteria than the faster-growing root tips in cultivated soil, and *Pseudomonas* were in higher numbers around the tips than other bacteria. These studies suggested that wheat genotypes with inherently faster root growth may be better adapted to unplowed conditions than those with inherently slower root growth. To test this, a wheat genotype selected for high shoot vigor was compared with a wheat cultivar available to farmers that had shown poor early vigor in conservation systems. The vigorous wheat had 64% faster shoot growth and 39% faster root growth than the conventional cultivar in unplowed soil. Early vigorous growth in conservation systems can be achieved by using varieties selected with faster root or shoot growth.

A second advantage of varieties with higher root and shoot vigor in conservation systems may be that they capture water and nitrogen that would drain below the root zone of less vigorous genotypes. The infiltration rate of water in soils with long-term conservation practices can be three times greater than that in soil in plowed systems due to the development of channels and pores in the soil by roots and other organisms (Liebig et al. 2004; Zhang et al. 2007). If the current crop roots have not colonized the soil sufficiently or leaf area and transpiration are too low, greater movement of water through the profile results in greater drainage of water and nitrogen below the root depth. For example, Kirkegaard et al. (2001) reported that 76 mm of water drained below the root zone in the unplowed system compared to 26 mm in the plowed system in Australia in a season when the total rainfall was 256 mm. The 50 mm that was lost for the crop would have contributed to greater shoot biomass during the season and to greater mobilization of carbohydrates into the grain at the end of the season if it could have been captured. An extra 10 mm of water captured by the roots during grain development can result in an extra half ton of wheat yield (Kirkegaard et al. 2007). Thus, conservation practices which cause more water to reach the deep soil layers combined with varieties with more root length at depth that captures that water would be of great benefit to productivity. Substantial variation in mature root length and depth has been observed in wheat (Richards et al. 2007). This variation is available for selection in breeding programs targeted at conservation farming systems, provided that phenotypic screens or genomic regions for molecular markers become available to breeders (Richards et al. 2010).

Weeds are one of the largest constraints to agricultural productivity globally (Fischer et al. 2009). They use water and nutrients during and between the crop growing season and occupy land area. Plowing the soil is one of the most effective means to reduce weeds, and therefore, conservation systems rely heavily on herbicides. Herbicide resistance can develop causing farmers to resort to plowing. A way to reduce weeds in conservation systems is to develop crops that compete effectively against weeds for light, water, and nutrients and suppress weed seed production. Root exudates inhibitory to other plants, called "allelochemicals," have received substantial attention as a useful crop characteristic to suppress weeds (Bertin et al. 2003; Weston and Duke 2003). Sorghum and fine fescue (*Festuca rubra* L.) provide striking examples. Some sorghum genotypes are grown as cover crops in subtropical areas to reduce weed infestations. These genotypes produce and release "sorgoleone" specifically from their root hairs (Figure 21.2H) (easy to see in water because it is colored and hydrophobic; Watt and Weston 2009). This highly toxic compound is a hydrophobic hydroquinone, lasts a long time in soil, and has similar properties to synthetic herbicides, inhibiting electron transport in photosynthesis and respiration in plants that take it up (Czarnota et al. 2001). Certain fine fescues are used in temperate roadsides, pastures, and reclamation sites to minimize annual weed infestations. These fescues produce a nonprotein amino acid called meta-tyrosine that is highly inhibitory to the growth of other plants including weeds (Bertin et al. 2007). The compound appears to be specifically synthesized and released from epidermal layers close to root tips (Watt and Weston 2009). Attempts are underway to select the sorghum and fescue allelochemicals intentionally in breeding programs (Weston, personal comm.). Their potency, localization to specific root tissues, and visibility make them ideal for phenotypic selection. A gene associated with sorgoleone production, SOR1, is identified and available to develop a molecular marker to select for sorgoleone in new sorghum varieties (Yang et al. 2004).

Encouraging research conducted under controlled conditions suggests that crops such as wheat produce and exude phenolic allelochemicals that suppress weeds (e.g., Wu et al. 2001). However, these genotypes have yet to be tested and applied in the field.

C. Roots Adapted to Agricultural Intensification on Current Land

The third option for increasing production from existing fertile land is to increase cropping intensity, production per land area per time. This can be achieved by (1) sowing earlier or harvesting later and extending the growing duration of a single crop; (2) introducing more crops in a given season, shortening the seasons of current crops, or removing any fallow periods; and (3) harvesting more from a single crop. A striking example of the final point is "dual-purpose crops," sown, grazed by animals or mechanically removed to provide feed during the vegetative phase, and then allowed to go to yield (Kirkegaard et al. 2008; Kelman and Dove 2009). These options for crop intensification on a given area of land require both new management practices and new crop genetics and can be very profitable, provided that inputs such as fertilizer and water increase minimally.

Breeders and agronomists are aware that crop phenologies, such as flowering time, must be optimized in new farming systems. Relatively less attention has been paid to opportunities for selecting the best root systems, although the "break crop" benefits of the roots of legumes, canola, and non-wheat cereals for reducing disease and increasing the nitrogen availability to these crops are well recognized (Passioura and Angus 2010). There are bound to be new opportunities from root-based traits, notably to increase water and fertilizer use efficiencies in these more intense systems, to avoid root disease problems, and to increase soil carbon. For example, oats produces avenacin specifically in the epidermal cells around the root tips (Figure 21.2I). It is toxic to the Take-All fungus and can be exploited to inhibit root diseases (Watt and Weston 2009). Another example of management to intensify agriculture is earlier sowing—an effective way to extend the vegetative period and use more water that would be lost to evaporation early in the season. In India in the drier, limited irrigation areas, Indian farmers are keen to sow early. However, sowing soils are very hot, and seed must be planted deep. Traits that enable good root establishment while resources are allocated to the emerging shoot, and water uptake at high temperatures are desirable. Innovative new farming systems combined with new crop root genetics will be the fastest route to more profitable and productive agriculture in the future. These will require an integrated, laboratory and field research team, with modelers, geneticists, physiologists, and agronomists (Kirkegaard and Hunt 2010).

VI. Challenge for the Future: Linking Laboratory to the Field

Laboratory-based discoveries tend to remain in the lab, and field-related work tends to remain in the field. Several dozen genes and QTLs affecting root emergence, growth, and vigor have been identified by molecular or at least laboratory-based approaches (Hochholdinger and Zimmermann 2008; Hochholdinger and Tuberosa 2009; Coudert et al. 2010). However, examples where these discoveries have been evaluated in the field, in terms of the impact of the selected trait on roots in field conditions, as well as its impact on grain yield or quality, are quite rare.

Field-based experiments have the advantage of enabling large-scale experiments, such as screening for a trait in several thousands of individuals. Hence, the maize mutant RTCS1, which fails in initiating crown and seminal roots, was identified in a field experiment (Hetz et al. 1996). But this example is probably not a general rule, and most mutants affected on root traits have been isolated using small-scale, seedling stage screenings (Hochholdinger et al. 2001) involving water cultures (Inukai et al. 2001), tissue culture-derived screens (Liu et al. 2005), or paper rolls (Woll et al. 2005).

Similarly, potentially valuable root traits identified in laboratory studies are rarely evaluated in field trials. One can cite the work of Li et al. (2010), who overexpressed the gene ZmPTF1 in maize and assessed the yield of the transgenic maize lines in a low-Pi soil. The transgenic lines displayed an improved root growth and branching as well as a higher grain yield per plant in hydroponics, sand pots, and low-Pi pots. Although the switch from hydroponics and sand to field-originated soil is obviously laudable, the experiments have not been conducted directly in a field but under controlled growth conditions in pots. Conditions such as soil temperature and compaction as well as the soil volume available for root exploration are very different from an actual field, and thus, field behavior cannot be extrapolated directly from these results.

The work of Steele and collaborators (Steele et al. 2006, 2007) is a nice example of field evaluation of identified QTLs. Four rice QTLs related to root growth (and a fifth related to aroma) were introgressed from Azucena into Kalinga III cultivar through glasshouses selections in the UK. After the selection process, seeds were exported to India where five field experiments were conducted in irrigated and drought conditions to assess the impact of the introgressed QTLs on drought resistance. Even though the first field trial was performed in pots in order to enable a comprehensive root analysis (Steele et al. 2006), subsequent on-farm experiments were conducted on these lines to assess the impact of the QTLs on straw and grain yield (Steele et al. 2007). The study led to the conclusion that the presence of several introgressed QTLs improved both grain and straw yield, and the two most promising lines are tested in advanced yield trials for future release in medium upland of eastern India. It is interesting to note that the lines were originally designed with the assumption that bigger root systems would trigger a better resistance to drought and that the final results suggest that the new lines performed better than Kalinga III only in favorable conditions (where the mean yield is greater than 1 ton/ha). This clearly shows how important it is to test root traits in real field conditions, as the benefit might not be the one expected. The wheat-rye translocations 1RS.1BL are another example of field evaluation of a genomic region associated with root traits (Ehdaie et al. 2003; Sharma et al. 2009). This translocation is associated with root traits and greater wheat yields, which explains why it has been abundantly used and studied, including in field conditions (Lukaszewski 1990, 1993; Villareal et al. 1997, 1998; Kumar et al. 2003; Monneveux et al. 2003). But examples where the consequences of a modified gene expression (silenced, constitutively expressed, or overexpressed) have been evaluated in the field are scarce.

The constitutive expression of the rice gene OsNAC10 is a good example (Jeong et al. 2010). This gene is a member of a largely uncharacterized family of transcription factors gathering 140 genes in rice. The authors cloned the gene under either a whole plant constitutive promoter or under a root-specific constitutive promoter. These transgenic lines were then tested in fields under drought or well-watered conditions. Interestingly, only the root-specific expression of this gene resulted in a better grain filling, increasing the grain yield by 25%–42% in drought conditions and by 5%–14% in normal conditions. These lines, expressing OsNAC10 constitutively in roots, showed enlarged stele, cortex, and epidermis resulting in 25% increase of root diameter compared to the nontransgenic control line and the whole plant promoter construct.

The scarcity of examples where gene discoveries have been applied or assessed in the field can be explained by several reasons. Firstly, a limitation to validating discoveries in the field is the fact that field conditions may be very different from pot experiments in growth cabinets, glasshouses, or even outside in birdcages or even field sites. It is difficult to obtain similar conditions on several experiments due to the extreme variability of environmental conditions. Obviously, pot experiments are clearly favored by root scientists as it is much easier to obtain the whole root system from a plant, avoiding any loss of material and more convenient to control watering and growth conditions. It is also possible to presterilize the soil if the project involves a pathogen or symbiotic bacterial or fungal communities. The experiments are thereby repeatable and comparable with others when performed in a similar set of conditions. These normalized growth conditions enable comparison of several experiments, but one must keep in mind that field conditions are very different from pot conditions in terms of soil temperature (especially when these pots are black), compaction, water content, and aeration (Passioura 2006b). These contrasting soil conditions can be responsible for unexpected, even surprising results in the field, which may differ largely from the results obtained in pots, possibly making the trait less interesting. For example, the dwarfing genes *Rht* displayed increased root length in gel cultures but decreased root length in soil-filled columns and field experiment (Wojciechowski et al. 2009).

Secondly, it is extremely complicated to get and analyze an entire root system from a plant which has grown directly in the soil. This approach forces the scientist to extrapolate the whole root system from a sample. Obtaining this sample of root system is labor intensive. Even if the results reflect the potential of the trait in real conditions, they are much more difficult to obtain and analyze than pot experiments.

Thirdly, molecular laboratories don't have necessarily access to fields. This reflects the gap between applied and basic research. Indeed, field trials require specific knowledge, competence, and background that a purely laboratory-based team may not have. If the research institution does not possess a field trial area and the associated infrastructures, discoveries may not leave the lab unless collaborations are initiated with other institutions or research organizations which have access to fields.

Finally, traits which may be interesting for breeding purposes are usually the result not of a single gene mutation or overexpression but of complex polygenic combinations which can be difficult to generate and understand using reverse genetics techniques. Moreover, valuable traits are often the result of subtle variations of gene(s) expression, whereas the classical method to identify and characterize genes consists in strongly silencing or overexpressing them one by one, in order to quantify the phenotypes associated with it. Although very powerful to identify potential candidates for a trait, this sort of approach is not suitable for commercial crops for which the quality or ease of growth in the field may be altered by the affected gene(s).

References

Ali M, Pathan MS, Zhang J, Bai G, Sarkarung S, Nguyen HT. 2000. Mapping QTLs for root traits in a recombinant inbred population from two *indica* ecotypes in rice. *Theor Appl Gen* 101:756–766.

Angus JF. 2001. Nitrogen supply and demand in Australian agriculture. *Aust J Exp Agric* 41:277–288.

Bates TR, Lynch JP. 2001. Root hairs confer a competitive advantage under low phosphorus availability. *Plant Soil* 236:243–250.

Beebe SE, Yan M, Blair X et al. 2006. Quantitative trait loci for root architecture traits correlated with phosphorus acquisition in common bean. *Crop Sci* 46:413–423.

Bernier J, Atlin GN, Serraj R, Kumar A, Spaner D. 2008. Breeding upland rice for drought resistance. *J Sci Food Agric* 88:927–939.

Bernier J, Serraj R, Kumar A et al. 2009. The large-effect drought-resistance QTL qtl12. 1 increases water uptake in upland rice. *Field Crop Res* 110:139–146.

Berry PM, Sterling M, Spink JH et al. 2004. Understanding and reducing lodging in cereals. *Adv Agric* 84:217–271.

Bertin C, Weston LA, Huang T, Jander G, Owens T, Meinwald J, Schroeder FC. 2007. Grass roots chemistry: Meta-Tyrosine, an herbicidal nonprotein amino acid. *Proc Natl Acad Sci USA* 104:16964–16969.

Bertin C, Yang XH, Weston LA. 2003. The role of root exudates and allelochemicals in the rhizosphere. *Plant Soil* 256:67–83.

Bolanos J, Edmeades GO, Martinez L. 1993. Eight cycles of selection for drought tolerance in lowland tropical maize. III. Responses in drought-adaptive physiological and morphological traits. *Field Crop Res* 31:269–286.

Bonser AM, Lynch JP, Snapp S. 1996. Effect of phosphorus deficiency on growth angle of basal roots of *Phaseolus vulgaris* L. *New Phytol* 132:281–288.

Bramley H, Turner NC, Turner DW, Tyerman SD. 2009. Roles of morphology, anatomy, and aquaporins in determining contrasting hydraulic behavior of roots. *Plant Physiol* 150:348–364.

Bruce WB, Edmeades GO, Barker TC. 2002. Molecular and physiological approaches to maize improvement for drought tolerance. *J Exp Bot* 53:13–25.

Burns IG. 1980. Influence of the spatial distribution of nitrate and the uptake of N by plants: A review and a model for rooting depth. *J Soil Sci* 31:155–173.

Champoux M, Wang G, Sarkarung S et al. 1995. Locating genes associated with root morphology and drought avoidance in rice via linkage to molecular markers. *Theor Appl Gen* 90:969–981.

Chan KY, Mead JA, Roberts WP. 1987. Poor early growth of wheat under direct drilling. *Aust J Agric Res* 38:791–800.

Charpentier M, Oldroyd G. 2010. How close are we to nitrogen-fixing cereals? *Curr Opin Plant Biol* 13:556–564.

Clarkson DT. 1985. Factors affecting mineral nutrient acquisition by plants. *Ann Rev Plant Physiol* 36:77–116.

Cordell D, Drangert JO, White S. 2009. The story of phosphorus: Global food security and food for thought. *Glob Environ Change Human Policy Dimens* 19:292–305.

Cornish P. 2010. A postscript to "Peak P"—An agronomist's response to diminishing P reserves. In *Proceedings of the 15th Agronomy Conference*, Lincoln, New Zealand. Food Security from Sustainable Agriculture. http://www.regional.org.au/au/asa/2010/plenary/sustainability/7452_cornish.htm

Coudert Y, Perin C, Courtois B, Khong NG, Gantet P. 2010. Genetic control of root development in rice, the model cereal. *Trends Plant Sci* 15:219–226.

Courtois B, Shen L, Petalcorin W, Carandang S, Mauleon RP, Li ZK. 2003. Locating QTLs controlling constitutive root traits in the rice population IAC 165 × Co39. *Euphytica* 134:335–345.

Czarnota MA, Paul RN, Dayan FE, Nimbal CI, Weston LA. 2001. Mode of action, localisation, chemical composition and activity of sorgoleone: A potent PSII inhibitor in Sorghum spp. root exudates. *Weed Technol* 15:813–825.

Delhaize E, Ryan PR, Randall PJ. 1993. Aluminum tolerance in wheat (*Triticum aestivum* L.) 2. Aluminum-stimulated excretion of malic acid from root apices. *Plant Physiol* 103:695–702.

Delhaize E, Taylor P, Hocking PJ, Simpson RJ, Ryan PR, Richardson AE. 2009. Transgenic barley (*Hordeum vulgare* L.) expressing the wheat aluminium resistance gene (TaALMT1) shows enhanced phosphorus nutrition and grain production when grown on an acid soil. *Plant Biotechnol J* 7:391–400.

Dracup M, Belford RK, Gregory PJ. 1992. Constraints to root growth of wheat and lupin crops in duplex soils. *Aust J Exp Agric* 32:947–961.

Easson DL, White EM, Pickles SJ. 1993. The effects of weather, seed rate and cultivar on lodging and yield in winter-wheat. *J Agric Sci* 121:145–156.

Ehdaie B, Whitkus RW, Waines JG. 2003. Root biomass, water-use efficiency, and performance of wheat-rye translocations of chromosomes 1 and 2 in spring bread wheat 'Pavon'. *Crop Sci* 43:710–717.

Ekanayake I, O'Toole, JC, Garrity DP, Masajo TM. 1985. Inheritance of root characters and their relations to drought resistance in rice. *Crop Sci* 25:927–933.

Evans LT. 1993. *Crop Evolution, Adaptation and Yield*. London, U.K.: Cambridge University Press.

Farrar J, Hawes M, Jones D, Lindow S. 2003. How roots control the flux of carbon to the rhizosphere. *Ecology* 84:827–837.

Fischer RA, Byerlee D, Edmeades GO. 2009. Can technology deliver on the yield challenge to 2050? *Expert Meeting on How to Feed the World in 2050*. Food and Agriculture Organization of the United Nations Economic and Social Development Department, Rome, Italy. http://www.fao.org/wsfs/forum2050/wsfs-background-documents/wsfs-expert-papers/en

Fischer RA, Edmeades GO. 2010. Breeding and cereal yield progress. *Crop Sci* 50:S85–S98.

Fisher JA, Scott BJ. 1987. Response to selection for aluminium tolerance. In *Priorities in Soil/Plant Relations: Research for Plant Production*, eds. PGE Searle, BG Davey, pp. 135–137. Sydney, Australia: The University of Sydney.

Fukai S, Cooper M. 1995. Development of drought-resistant cultivars using physiomorphological traits in rice. *Field Crop Res* 40:67–86.

Gahoonia TS, Care D, Nielsen NE. 1997. Root hairs and phosphorus acquisition of wheat and barley cultivars. *Plant Soil* 191:181–188.

Gahoonia TS, Nielsen NE. 2004. Barley genotypes with long root hairs sustain high grain yields in low-P field. *Plant Soil* 262:55–62.

Gallais A, Coque M. 2005. Genetic Variation and selection for nitrogen use efficiency in maize: A synthesis. *Maydica* 50:531–547.

Gastal F, Lemaire G. 2002. N uptake and distribution in crops: An agronomical and ecophysiological perspective. *J Exp Bot* 53:789–799.

George TS, Gregory PJ, Hocking P, Richardson AE. 2008. Variation in root-associated phosphatase activities in wheat contributes to the utilization of organic P substrates in vitro, but does not explain differences in the P-nutrition of plants when grown in soils. *Environ Exp Bot* 64:239–249.

George TS, Richardson AE, Hadobas PA, Simpson RJ. 2004. Characterization of transgenic *Trifolium subterraneum* L. which expresses phyA and releases extracellular phytase: Growth and P nutrition in laboratory media and soil. *Plant Cell Environ* 27:1351–1361.

George TS, Richardson AE, Simpson RJ. 2005a. Behaviour of plant-derived extracellular phytase upon addition to soil. *Soil Biol Biochem* 37:977–988.

George TS, Simpson RJ, Hadobas PA, Richardson AE. 2005b. Expression of a fungal phytase gene in *Nicotiana tabacum* improves phosphorus nutrition of plants grown in amended soils. *Plant Biotechnol J* 3:129–140.

Gewin V. 2010. Food: An underground revolution. *Nature* 466:552–553

Gilroy S, Jones DL. 2000. Through form to function: Root hair development and nutrient uptake. *Trends Plant Sci* 5:56–60.

Graham JH. 2000. Assessing costs of arbuscular mycorrhizal symbiosis in agroecosystems. In *Current Advances in Mycorrhizae Research*, eds. GK Podila, DD Douds, pp. 111–126. St. Paul, MN: APS Press.

Gregory PJ, Tennant D, Hamblin AP, Eastham J. 1992. Components of the water balance on duplex soils in Western Australia. *Aust J Exp Agric* 32:845–855.

Hammer GL, Dong ZS, McLean G et al. 2009. Can changes in canopy and/or root system architecture explain historical maize yield trends in the US corn belt? *Crop Sci* 49:299–312.

Henry A, Chaves NF, Kleinman PJA, Lynch JP. 2010. Will nutrient-efficient genotypes mine the soil? Effects of genetic differences in root architecture in common bean (*Phaseolus vulgaris* L.) on soil phosphorus depletion in a low-input agro-ecosystem in Central America. *Field Crop Res* 115:67–78.

Hetz W, Hochholdinger F, Schwall M, Feix G. 1996. Isolation and characterization of rtcs, a maize mutant deficient in the formation of nodal roots. *Plant J* 10:845–857.

Ho MD, Rosas JC, Brown KM, Lynch JP. 2005. Root architectural tradeoffs for water and phosphorus acquisition. *Funct Plant Biol* 32:737–748.

Hochholdinger F, Park WJ, Feix GH. 2001. Cooperative action of SLR1 and SLR2 is required for lateral root-specific cell elongation in maize. *Plant Physiol* 125:1529–1539.

Hochholdinger F, Tuberosa R. 2009. Genetic and genomic dissection of maize root development and architecture. *Curr Opin Plant Biol* 12:172–177.

Hochholdinger F, Zimmermann R. 2008. Conserved and diverse mechanisms in root development. *Curr Opin Plant Biol* 11:70–74.

Huang S, Spielmeyer W, Lagudah ES, Munns R. 2008. Comparative mapping of HKT genes in wheat, barley, and rice, key determinants of Na^+ transport, and salt tolerance. *J Exp Bot* 59:927–937.

Hurd EA. 1964. Root study of three wheat varieties and their resistance to drought and damage by cracking. *Can J Plant Sci* 44:240–248.

Hurd EA. 1974. Phenotype and drought tolerance in wheat. *Agric Meteorol* 14:39–55.

Inukai Y. 2005. Crown rootless1, which is essential for crown root formation in rice, is a target of an AUXIN RESPONSE FACTOR in auxin signaling. *Plant Cell Online* 17:1387–1396.

Inukai Y, Miwa M, Nagato Y, Kitano H, Yamauchi A. 2001. Characterization of rice mutants deficient in the formation of crown roots. *Breed Sci* 51:123–129.

James RA, Blake C, Byrt CS, Munns R. 2011. Major genes for Na^+ exclusion, Nax1 and Nax2 wheat HKT1;4 and HKT1;5, decrease Na^+ accumulation in bread wheat leaves under saline and waterlogged conditions. *J Exp Bot* 62:2939–2947.

James RA, Davenport RJ, Munns R. 2006. Physiological characterization of two genes for Na^+ exclusion in durum wheat, Nax1 and Nax2. *Plant Physiol* 142:1537–1547.

Jeong JS, Kim YS, Baek KH, Jung H, Ha SH, Do Choi Y, Kim M, Reuzeau C, Kim JK. 2010. Root-specific expression of OsNAC10 improves drought tolerance and grain yield in rice under field drought conditions. *Plant Physiol* 153:185–197.

Kaspar TC, Woolley DG, Taylor HM. 1981. Temperature effect on the inclination of lateral roots of soybeans. *Agric J* 73:383–385.

Kelman WM, Dove H. 2009. Growth and phenology of winter wheat and oats in a dual-purpose management system. *Crop Past Sci* 60:921–932.

Kirkegaard JA. 1995. A review of trends in wheat yield responses to conservation cropping in Australia. *Aust J Exp Agric* 35:835–848.

Kirkegaard JA, Howe GN, Simpfendorfer S, Angus JF, Gardner PA, Hutchinson P. 2001. Poor wheat yield response to conservation cropping—Causes and consequences during 10 years of the Harden tillage trial. In *Proceedings of the 10th Australian Agronomy Conference*, Hobart, Australia.

Kirkegaard JA, Hunt JR. 2010. Increasing productivity by matching farming system management and genotype in water-limited environments. *J Exp Bot* 61:4129–4143.

Kirkegaard JA, Lilley JM. 2007. Root penetration rate—A benchmark to identify soil and plant limitations to rooting depth in wheat. *Aust J Exp Agric* 47:590–602.

Kirkegaard JA, Lilley JM, Howe GN, Graham JM. 2007. Impact of subsoil water use on wheat yield. *Aust J Agric Res* 58:303–315.

Kirkegaard JA, Sprague SJ, Dove H et al. 2008. Dual-purpose canola—A new opportunity in mixed farming systems. *Aust J Agric Res* 59:291–302.

Kondo M, Pablico PP, Aragones DV et al. 2003. Genotypic and environmental variations in root morphology in rice genotypes under upland field conditions. *Plant Soil* 255:189–200.

Krassovsky I. 1926. Physiological activity of the seminal and nodal roots of crop plants. *Soil Sci* 21:307–325.

Kumar S, Kumar N, Balyan HS, Gupta PK. 2003. 1BL.1RS translocation in some Indian bread wheat genotypes and strategies for its use in future wheat breeding. *Caryologia* 56:23–30.

Lambers H, Finnegan PM, Laliberte E et al. 2011. Phosphorus nutrition of proteaceae in severely phosphorus-impoverished soils: Are there lessons to be learned for future crops? *Plant Physiol* 156:1058–1066.

Lekberg Y, Koide RT. 2005. Is plant performance limited by abundance of arbuscular mycorrhizal fungi? A meta-analysis of studies published between 1988 and 2003. *New Phytol* 168:189–204.

Li ZX, Gao QA, Liu YZ, He CM, Zhang XR, Zhang JR. 2010. Overexpression of transcription factor ZmPTF1 improves low phosphate tolerance of maize by the regulation of carbon metabolism and root growth. *In Vitro Cell Dev Biol Animal* 46:180–180.

Li Z, Mu P, Li C, Zhang H, Li Z, Gao Y, Wang X. 2005. QTL mapping of root traits in a doubled haploid population from a cross between upland and lowland japonica rice in three environments. *Theor Appl Gen* 110:1244–1252.

Liao M, Fillery I, Palta J. 2004a. Early vigorous growth is a major factor influencing nitrogen uptake in wheat. *Funct Plant Biol* 31:121–129.

Liao M, Palta JA, Fillery IRP. 2006. Root characteristics of vigorous wheat improve early nitrogen uptake. *Aust J Agric Res* 57:1097–1107.

Liao H, Rubio G, Yan X, Cao A, Brown K, Lynch J. 2001. Effect of phosphorus availability in basal root shallowness in common bean. *Plant Soil* 232:69–79.

Liao H, Yan XL, Rubio G, Beebe SE, Blair MW, Lynch JP. 2004b. Genetic mapping of basal root gravitropism and phosphorus acquisition efficiency in common bean. *Funct Plant Biol* 31:959–970.

Liebig MA, Tanaka DL, Wienhold BJ. 2004. Tillage and cropping effects on soil quality indicators in the northern Great Plains. *Soil Till Res* 78:131–141.

Lilley JM, Kirkegaard JA. 2011. Benefits of increased soil exploration by wheat roots. *Field Crop Res* 122:118–130.

Liu S, Wang J, Wang L et al. 2009. Adventitious root formation in rice requires OsGNOM1 and is mediated by the OsPINs family. *Cell Res* 19:1110–1119.

Liu H, Wang S, Yu X et al. 2005. ARL1, a LOB-domain protein required for adventitious root formation in rice. *Plant J* 43:47–56.

Lopes MS, Reynolds MP. 2010. Partitioning of assimilates to deeper roots is associated with cooler canopies and increased yield under drought in wheat. *Funct Plant Biol* 37:147–156.

Lukaszewski AJ. 1990. Frequency of 1rs 1al and 1rs 1bl translocations in United States wheats. *Crop Sci* 30:1151–1153.

Lukaszewski AJ. 1993. Reconstruction in wheat of complete chromosome-1b and chromosome-1r from the 1rs.1bl translocation of Kavkaz origin. *Genome* 36:821–824.

Lynch JP. 2007a. Rhizoeconomics: The roots of shoot growth limitations. *Hortscience* 42:1107–1109.

Lynch JP. 2007b. Roots of the second green revolution. *Aust J Bot* 55:493–512.

Lynch JP, Brown KM. 2001. Topsoil foraging—An architectural adaptation of plants to low phosphorus availability. *Plant Soil* 237:225–237.

Mambani B, Lal R. 1983. Response of upland rice varieties to drought stress: II Screening rice varieties by means of variable moisture regimes along a toposequence. *Plant Soil* 73:73–94.

Manschadi AM, Christopher J, Devoil P, Hammer GL. 2006. The role of root architectural traits in adaptation of wheat to water-limited environments. *Funct Plant Biol* 33:823–837.

Manschadi AM, Hammer GL, Christopher JT, DeVoil P. 2008. Genotypic variation in seedling root architectural traits and implications for drought adaptation in wheat (*Triticum aestivum* L.). *Plant Soil* 303:115–129.

Marschner H. 1995. *Mineral Nutrition of Higher Plants*, 2nd edn. London: Academic Press.

Martin JK. 1977. Factors influencing the loss of organic carbon from wheat roots. *Soil Biol Biochem* 9:1–7.

Martinoia E, Weisskopf L, Abou-Mansour E et al. 2006. White lupin has developed a complex strategy to limit microbial degradation of secreted citrate required for phosphate acquisition. *Plant Cell Environ* 29:919–927.

Monneveux P, Villareal R, Reynolds MP, Mujeeb-Kazi A. 2003. Effect of T1BL.1RS chromosome translocation on agronomic traits and transpiration efficiency in bread and durum wheat under drought conditions. *Cereal Res Commun* 31:363–370.

Munns R, James RA. 2003. Screening methods for salinity tolerance: A case study with tetraploid wheat. *Plant Soil* 253:201–218.

Munns R, Rebetzke GJ, Husain S, James RA, Hare RA. 2003. Genetic control of sodium exclusion in durum wheat. *Aust J Agric Res* 54:627–635.

Munns R, Tester M. 2008. Mechanisms of salinity tolerance. *Ann Rev Plant Biol* 59:651–681.

Munns R, James RA, Xu B, Athman A, Conn SJ, Jordans C et al. 2012. Wheat grain yield on saline soils is improved by an ancestral Na+ transporter gene. *Nature Biotech* 30: 360-U173.

Nguyen HT, Babu RC, Blum A. 1997. Breeding for drought resistance in rice: Physiology and molecular genetics considerations. *Crop Sci* 37:1426–1434.

Nielsen KL, Eshel A, Lynch JP. 2001. The effect of phosphorus availability on the carbon economy of contrasting common bean (*Phaseolus vulgaris* L.) genotypes. *J Exp Bot* 52:329–339.

Nord EA, Lynch JP. 2008. Delayed reproduction in *Arabidopsis thaliana* improves fitness in soil with suboptimal phosphorus availability. *Plant Cell Environ* 31:1432–1441.

Ogbonnaya FC, Subrahmanyam NC, Moullet O et al. 2001. Diagnostic DNA markers for cereal cyst nematode resistance in bread wheat. *Aust J Agric Res* 52:1367–1374.

Oyanagi A, Nakamoto T, Wada M. 1993. Relationship between root growth angle of seedlings and vertical distribution of roots in the field in wheat [*Triticum aestivum*] cultivars. *Jpn J Crop Sci* 64:565–570.

Palta JA, Chen X, Milroy SP, Rebetzke GJ, Dreccer MF, Watt M. 2011. Large root systems: Are they useful in adapting wheat to dry environments? *Funct Plant Biol* 38:347–354.

Palta J, Watt M. 2009. Crop roots systems form and function: Improving the capture of water and nutrients with vigorous root systems. In *Crop Physiology: Applications for Genetic Improvement and Agronomy*, eds. VO Sadras, D Calderini, pp. 309–325. Burlington, Ontario, Canada: Elsevier.

Passioura J. 1972. The effect of root geometry on the yield of wheat growing on stored water. *Aust J Agric Res* 23:745–752.

Passioura JB. 1977. Grain yield, harvest index, and water use of wheat. *J Aust Inst Agric Sci* 43:117–120.

Passioura JB. 1983. Roots and drought resistance. *Agric Water Manage* 7:265–280.

Passioura JB. 1991. Soil structure and plant growth. *Aust J Soil Res* 29:717–728.

Passioura JB. 2006a. Increasing crop productivity when water is scarce-from breeding to field management. *Agric Water Manage* 80:176–196.

Passioura JB. 2006b. The perils of pot experiments. *Funct Plant Biol* 33:1075–1079.

Passioura JB. 2010. Scaling up: The essence of effective agricultural research. *Funct Plant Biol* 37:585–591.

Passioura JB, Angus JF. 2010. Improving productivity of crops in water-limited environments. *Adv Agric* 106:37–75.

Passioura JB, Spielmeyer W, Bonnett DG. 2007. Requirements for success in marker-assisted breeding for drought-prone environments. In *Advances in Molecular Breeding Toward Drought and Salt Tolerant Crops*, ed. MA Jenks et al., pp. 479–500. Dordrecht, the Netherland: Springer.

Peoples MB, Brockwell J, Herridge DF et al. 2009. The contributions of nitrogen-fixing crop legumes to the productivity of agricultural systems. *Symbiosis* 481–3:1–17.

Postgate J. 1998. *Nitrogen Fixation*, 3rd edn. Cambridge, U.K.: Cambridge University Press.

Rahnama A, Munns R, Poustini K, Watt M. 2011. A screening method to identify genetic variation in root growth response to a salinity gradient. *J Exp Bot* 62:69–77.

Randall PJ, Hayes JE, Hocking PJ, Richardson AE. 2001. Root exudates in phosphorus acquisition by plants. *Plant Nutr Acquis* 71:100–520.

Raun WR, Johnson GV. 1999. Improving nitrogen use efficiency for cereal production. *Agron J* 91:357–363.

Rengasamy P. 2010. Soil processes affecting crop production in salt-affected soils. *Funct Plant Biol* 37:613–620.

Reynolds M, Bonnett D, Chapman SC et al. 2011. Raising yield potential of wheat. I. Overview of a consortium approach and breeding strategies. *J Exp Bot* 62:439–452.

Richards RA. 1991. Crop improvement for temperate Australia—Future opportunities. *Field Crop Res* 26:141–169.

Richards RA. 2006. Physiological traits used in the breeding of new cultivars for water-scarce environments. *Agric Water Manage* 80:197–211.

Richards RA, Passioura JB. 1981a. Seminal root morphology and water use of wheat. II. Environmental effects. *Crop Sci* 21:249–252.

Richards RA, Passioura JB. 1981b. Seminar root morphology and water use of wheat. II. Genetic variation. *Crop Sci* 21:253–255.

Richards RA, Rebetzke GJ, Watt M, Condon AG, Spielmeyer W, Dolferus R. 2010. Breeding for improved water productivity in temperate cereals: Phenotyping, quantitative trait loci, markers and the selection environment. *Funct Plant Biol* 37:85–97.

Richards RA, Watt M, Rebetzke GJ. 2007. Physiological traits and cereal germplasm for sustainable agricultural systems. *Euphytica* 154:409–425.

Richardson AE. 2009. Regulating the phosphorus nutrition of plants: Molecular biology meeting agronomic needs. *Plant Soil* 322:17–24.

Richardson AE, Hadobas PA, Hayes JE. 2001a. Extracellular secretion of *Aspergillus* phytase from *Arabidopsis* roots enables plants to obtain phosphorus from phytate. *Plant J* 25:641–649.

Richardson AE, Hadobas PA, Hayes JE, O'Hara CP, Simpson RJ. 2001b. Utilization of phosphorus by pasture plants supplied with myo-inositol hexaphosphate is enhanced by the presence of soil micro-organisms. *Plant Soil* 229:47–56.

Richardson AE, Hocking PJ, Simpson RJ, George TS. 2009. Plant mechanisms to optimise access to soil phosphorus. *Crop Past Sci* 60:124–143.

Richardson AE, Simpson RJ. 2011. Soil microorganisms mediating phosphorus availability. *Plant Physiol* 156:989–996.

Robinson D, Rorison IH. 1983. Relationships between root morphology and nitrogen availability in a recent theoretical model describing nitrogen uptake from soil. *Plant Cell Environ* 6:641–647.

Robinson D, Rorison IH. 1987. Root hairs and plant-growth at low nitrogen availabilities. *New Phytol* 107:681–693.

Rowse HR, Goodman D. 1981. Axial resistance to water movement in broad bean (*Vicia faba*) roots. *J Exp Bot* 32:591–8.

Ryan PR, Delhaize E, Randall PJ. 1995. Malate efflux from root apices and tolerance to aluminum are highly correlated in wheat. *Aust J Plant Physiol* 22:531–536.

Ryan PR, Skerrett M, Findlay GP, Delhaize E, Tyerman SD. 1997. Aluminum activates an anion channel in the apical cells of wheat roots. *Proc Natl Acad Sci USA* 94:6547–6552.

Ryan PR, Tyerman SD, Sasaki T et al. 2011. The identification of aluminium-resistance genes provides opportunities for enhancing crop production on acid soils. *J Exp Bot* 62:9–20.

Sasaki T, Ryan PR, Delhaize E et al. 2006. Sequence upstream of the wheat (*Triticum aestivum* L.) ALMT1 gene and its relationship to aluminium resistance. *Plant Cell Physiol* 47:1343–1354.

Sasaki T, Yamamoto Y, Ezaki B et al. 2004. A wheat gene encoding an aluminum-activated malate transporter. *Plant J* 37:645–653.

Schunmann PHD, Richardson AE, Smith FW et al. 2004. Characterization of promoter expression patterns derived from the Pht1 phosphate transporter genes of barley (*Hordeum vulgare* L.). *J Exp Bot* 55:855–865.

Segal E, Kushnir T, Mualem Y, Shani U. 2008. Water uptake and hydraulics of the root hair rhizosphere. *Vadose Zone J* 7:1027–1034.

Sharma S, Bhat PR, Ehdaie B, Close TJ, Lukaszewski AJ, Waines JG. 2009. Integrated genetic map and genetic analysis of a region associated with root traits on the short arm of rye chromosome 1 in bread wheat. *Theor Appl Gen* 119:783–793.

Shen L, Courtois B, McNally K L, Robin S, Li Z. 2001. Evaluation of near-isogenic lines of rice introgressed with QTLs for root depth through marker-aided selection. *Theor Appl Gen* 103:75–83.

Simpfendorfer S, Kirkegaard JA, Heenan DP, Wong PTW. 2002. Reduced early growth of direct drilled wheat in southern New South Wales—Role of root inhibitory pseudomonads. *Aust J Agric Res* 53:323–331.

Sinclair TR, Purcell LC, King CA, Sneller CH, Chen P, Vadez V. 2007. Drought tolerance and yield increase of soybean resulting from improved symbiotic N2 fixation. *Field Crop Res* 101:68–71.

Sinclair TR, Purcell LC, Vadez V, Serraj R, King CA, Nelson R. 2000. Identification of soybean genotypes with N2 fixation tolerance to water deficits. *Crop Sci* 60:1803–1809.

Steele KA, Price AH, Shashidhar HE, Witcombe JR. 2006. Marker-assisted selection to introgress rice QTLs controlling root traits into an Indian upland rice variety. *Theor Appl Gen* 112:208–221.

Steele KA, Virk DS, Kumar R, Prasad SC, Witcombe JR. 2007. Field evaluation of upland rice lines selected for QTLs controlling root traits. *Field Crop Res* 101:180–186.

Stirzaker RJ, Passioura JB. 1996. The water relations of the root–soil interface. *Plant Cell Environ* 19:201–208.

Stone JA, Taylor HM. 1983. Temperature and the development of the taproot and lateral roots of four indeterminate soybean cultivars. *Agron J* 75:613–618.

Subbarao GV, Kishii M, Nakahara K et al. 2009. Biological nitrification inhibition (BNI)—Is there potential for genetic interventions in the Triticeae? *Breed Sci* 59:529–545.

Taramino G, Sauer M, Stauffer JL et al. 2007. The maize (*Zea mays* L.) RTCS gene encodes a LOB domain protein that is a key regulator of embryonic seminal and post-embryonic shoot-borne root initiation. *Plant J* 50:649–659.

Taylor HM. 1980. Postponement of severe water stress in soybeans by rooting modifications: A progress report. In *Proceeding of the World Soybean Research Conference II*, ed. FT Corbin, pp. 160–178. Boulder, CO: Westview Press.

Tennant D, Hall D. 2001. Improving water use of annual crops and pastures—Limitations and opportunities in Western Australia. *Aust J Agric Res* 52:171–182.

Thomas GA, Titmarsh GW, Freebairn DM, Radford BJ. 2007. No-tillage and conservation farming practices in grain growing areas of Queensland—A review of 40 years of development. *Aust J Exp Agric* 47:887–898.

Villareal RL, Banuelos O, Mujeeb-Kazi A. 1997. Agronomic performance of related durum wheat (*Triticum turgidum* L.) stocks possessing the chromosome substitution T1BL.1RS. *Crop Sci* 37:1735–1740.

Villareal RL, Banuelos O, Mujeeb-Kazi A, Rajaram S. 1998. Agronomic performance of chromosomes 1B and T1BL.1RS near-isolines in the spring bread wheat Seri M82. *Euphytica* 103:195–202.

Vitousek PM, Mooney HA, Lubchenco J, Melillo JM. 1997. Human domination of Earth's ecosystems. *Science* 277:494–499.

Walch-Liu P, Ivanov II, Filleur S, Gan Y, Remans T, Forde BG. 2006. Nitrogen regulation of root branching. *Ann Bot* 97:875–881.

Wasson AP, Richards RA, Chatrath R, Misra SC, Sai Prasad SV, Rebetzke GJ, Kirkegaard JA, Christopher J, Watt M. 2012. Traits and selection strategies to improve root systems and water uptake in water-limited wheat crops. *J Exp Bot* 63:3485–3498.

Watt M, Kirkegaard JA, Passioura JB. 2006. Rhizosphere biology and crop productivity—A review. *Aust J Soil Res* 44:299–317.

Watt M, Kirkegaard JA, Rebetzke GJ. 2005. A wheat genotype developed for rapid leaf growth copes well with the physical and biological constraints of unploughed soil. *Funct Plant Biol* 32:695–706.

Watt M, McCully ME, Kirkegaard JA. 2003. Soil strength and rate of rot elongation alter the accumulation of Pseudomonas spp. and other bacteria in the rhizosphere of wheat. *Funct Plant Biol* 30:483–491.

Watt M, Silk WK, Passioura JB. 2006b. Rates of root and organism growth, soil conditions, and temporal and spatial development of the rhizosphere. *Ann Bot* 97:839–855.

Watt M, Weston LA. 2009. Specialised root adaptations display cell-specific developmental and physiological diversity. *Plant Soil* 322:39–47.

Weston LA, Duke SO. 2003. Weed and crop allelopathy. *Crop Rev Plant Sci* 22:367–389.

White RG, Kirkegaard JA. 2010. The distribution and abundance of wheat roots in a dense, structured subsoil—Implications for water uptake. *Plant Cell Environ* 33:133–148.

Wiesler F, Horst WJ. 1993. Differences among maize cultivars in the utilization of soil nitrate and the related losses of nitrate through leaching. *Plant Soil* 151:193–203.

Wiesler F, Horst WJ. 1994. Root growth and nitrate utilization of maize cultivars under field conditions. *Plant Soil* 163:267–277.

Wojciechowski T, Gooding MJ, Ramsay L, Gregory PJ. 2009. The effects of dwarfing genes on seedling root growth of wheat. *J Exp Bot* 60:2565–2573.

Woll K, Borsuk LA, Stransky H, Nettleton D, Schnable PS, Hochholdinger F. 2005. Isolation, characterization, and pericycle-specific transcriptome analyses of the novel maize lateral and seminal root initiation mutant rum1. *Plant Physiol* 139:1255–1267.

Wu H, Haig T, Pratley J, Lemerle D, An M. 2001. Allelochemicals in wheat (*Triticum aestivum* L.): Cultivar difference in exudation of phenolic acids. *J Agric Food Chem* 49:3742–3745.

Yambao E, Ingram K, Real J. 1992. Root xylem influence on the water relations and drought resistance of rice. *J Exp Bot* 43:925–932.

Yan XL, Beebe SE, Lynch JP. 1995. Genetic-variation for phosphorus efficiency of common bean in contrasting soil types. 2. Yield response. *Crop Sci* 35:1094–1099.

Yang XH, Scheffler BE, Weston LA. 2004. SOR1, a gene associated with bioherbicide production in sorghum root hairs. *J Exp Bot* 55:2251–2259.

Yi K, Menand B, Bell E, Dolan L. 2010. A basic helix-loop-helix transcription factor controls cell growth and size in root hairs. *Nat Genet* 42:264–267.

Yue B, Xue W, Xiong L et al. 2006. Genetic basis of drought resistance at reproductive stage in rice: Separation of drought tolerance from drought avoidance. *Genetics* 172:1213–1228.

Zhang GS, Chan KY, Oates A, Heenan DP, Huang GB. 2007. Relationship between soil structure and runoff/soil loss after 24 years of conservation tillage. *Soil Till Res* 92:122–128.

Zhang H, Forde BG. 1998. An *Arabidopsis* MADS box gene that controls nutrient-induced changes in root architecture. *Science* 279:407–409.

Zhang H, Forde BG. 2000. Regulation of *Arabidopsis* root development by nitrate availability. *J Exp Bot* 51:51–59.

Zhao Y, Hu Y, Dai M, Huang L, Zhou DX. 2009. The WUSCHEL-related homeobox gene WOX11 is required to activate shoot-borne crown root development in rice. *Plant Cell* 21:736–748.

Zhu JM, Brown KM, Lynch JP. 2010a. Root cortical aerenchyma improves the drought tolerance of maize (*Zea mays* L.). *Plant Cell Environ* 33:740–749.

Zhu JM, Kaeppler SM, Lynch JP. 2005. Mapping of QTL controlling root hair length in maize (*Zea mays* L.) under phosphorus deficiency. *Plant Soil* 270:299–310.

Zhu JM, Zhang CC, Lynch JP. 2010b. The utility of phenotypic plasticity of root hair length for phosphorus acquisition. *Funct Plant Biol* 37:313–322.

22

Root Architecture and Resource Acquisition: Wheat as a Model Plant

Ahmad M. Manschadi
University of Natural Resources and Life Sciences

Günther G.B. Manske
University of Bonn

Paul L.G. Vlek
University of Bonn

I. Introduction

Producing 70% more food for an additional 2.3 billion people by 2050 is one of the main challenges world agriculture will face in the coming decades (FAO 2009). Considering the escalating scarcity of natural resources, increasing water and nutrient use efficiency in crop plants is a key component of the strategy to sustainably enhance agricultural productivity and food production (Vlek et al. 1997; Tilman et al. 2002; Lynch 2007; Fageria et al. 2008; Falkenmark et al. 2009; Passioura and Angus 2010). The global average water use efficiency (grain yield per unit of seasonal evapotranspiration) of rain-fed wheat (*Triticum aestivum* L.), for instance, is currently only 32%–44% of attainable efficiency, as a result of deficiencies in soil mineral nutrients and inadequate crop/soil management practices (Sadras and Angus 2006). The nutrient use efficiencies are currently only 30%–50% of applied nitrogen (N) and around 45% of phosphorus (P) fertilizers (Raun and Johnson 1999; Tilman et al. 2002). This means that a significant amount of applied N and P is lost from agricultural fields due to surface runoff, leaching, microbial denitrification, and volatilization. Such nutrient losses harm neighboring ecosystems and diminish water quality through over-enrichment, eutrophication, hypoxia, and decline in biodiversity (Tilman et al. 2002). Phosphorus fertilizers are produced from nonrenewable resources, and the global P reserves are being rapidly depleted. The existing rock phosphate reserves could be exhausted in the next 50–100 years (Steen 1998; Cordell et al. 2009). Furthermore, inefficiencies in nutrient use are associated with substantial economic losses. Nitrogen fertilizers, for instance, represent about one-third of the costs of cereal production (Le Gouis et al. 2000; Fageria and Baligar 2005). Improving crop/soil management practices and developing more efficient crop cultivars can make a significant contribution to meet the challenges of increasing food production while reducing the adverse environmental impacts of agriculture and reducing farmers' costs.

Plant roots are of fundamental importance to soil resource acquisition. Exploiting genetic diversity in root traits and acquisition of scarce soil resources can significantly enhance resource use efficiency in crop plants. There is widespread evidence for genotypic variation in the root characteristics of many crop species (O'Toole and Bland 1987; Ludlow and Muchow 1990). In wheat, such traits include the depth of rooting, root elongation rate, root distribution at depth, xylem vessel diameter, and root to shoot dry matter ratio (Hurd 1968; O'Brien 1979; Richards and Passioura 1989; Siddique et al. 1990; Gregory 1994; Manschadi et al. 1998; Hoad et al. 2001; Manske et al. 2002; Manschadi et al. 2006).

Root architecture is defined as the in situ space-filling properties or the spatial distribution of the root system within the rooting volume. It is known to play an important role in the ability of crop plants to capture soil resources and hence largely determines their productivity and adaptation to suboptimal soil conditions (Lynch 1995; Fitter 2002; Doussan et al. 2003; Ho et al. 2005; Manschadi et al. 2006; Hodge et al. 2009).

This chapter provides a review of root architectural traits and their role in acquisition of soil water, P, and N resources. It highlights the existence of genotypic variation for the key traits that determine root architecture and discusses the implications of these root attributes for crop adaptation to nutrient-deficient and drought-prone environments. While this review draws on published research regarding the root system of many crop species, it focuses mainly on the root architecture of wheat as a model plant.

II. Morphology of Wheat Root System

The root system of cereals is composed of two root types, the seminal roots (also called seed-borne or primary roots), which develop at the scutellar and epiblast nodes of the embryonic hypocotyl of the germinating caryopsis, and adventitious roots (also called shoot-borne, nodal, secondary, or crown roots), which subsequently emerge from the coleoptile tiller bud and the basal nodes of the main shoot and tillers (Klepper 1984). The number of seminal root axes in spring wheat cultivars is 3–5 (O'Brien 1979; Manschadi et al. 2008). Winter wheat varieties grown in temperate environments develop on average six seminal axes (Gregory et al. 1978). Seed size and seminal root number are positively correlated, although in some genotypes this is not the case (O'Toole and Bland 1987). The seminal roots penetrate the soil earlier and deeper than the adventitious roots.

The crown root system of wheat develops on the lower shoot internodes, typically, 1–2 cm beneath the soil surface. The development of these nodal roots in wheat is linearly related to the number of leaves on the culm (Gregory et al. 1978; Klepper et al. 1984; Klepper 1991) so that genotypes with faster leaf appearance rate and higher tillering capacity are expected to produce more nodal roots and hence greater root-length density (RLD, unit root length per volume of soil) in the topsoil. Indeed, results of experiments with semidwarf bread wheat cultivars grown under P deficiency in acid soils showed that RLD in the topsoil was positively correlated with the number of tillers (Manske et al. 2000). The RLD of wheat varies broadly between 2 and 10 cm cm^{-3} soil depending on the genotype, stage of plant development, soil depth, and environmental factors.

The average root radius of wheat plants ranges between 0.07 and 0.15 mm. The horizontal spread of wheat roots is usually between 30 and 60 cm (Weaver 1926; Manschadi et al. 2006). However, under deficient moisture conditions, the roots may grow in lower soil horizons and have a greater vertical distribution (Mishra et al. 1999). Wheat roots can be abundant at soil depths of over 100 cm, with some roots even reaching below 200 cm (Gregory et al. 1978; Hoad et al. 2001).

III. Root Architectural Traits

The architecture of plant root systems is the result of interplay among a number of signals from various sources. Given the high spatiotemporal variability in soil water and mineral nutrients, root systems generally exhibit a high degree of developmental plasticity as the result of interactions between genetic endogenous characteristics (intrinsic pathways) and biotic and abiotic environmental stimuli (extrinsic pathways) (Malamy 2005). The following developmental mechanisms enable plants to optimize their root architecture—initiating lateral root primordia, modifying growth rate of primary and lateral roots, and regulating root gravitropism (Malamy 2005; Hodge et al. 2009). Substantial variation in root architecture has been reported both among plant species (Kutschera 1960; Fitter and Stickland 1992; Bouma et al. 2001) and within genotypes of crop species (Masi and Maranville 1998; Liao et al. 2001; Lynch and Brown 2001; McPhee 2005; Manschadi et al. 2006).

The root architectural traits can be divided into geometric properties determining the shape of the root system (e.g., the growth angle of root axes, lateral root expansion, root depth) and traits describing the root structure (e.g., topological traits describing the pattern of root branching). Together with the extension rate of individual root axes, these traits determine the in situ space-filling properties of a root system within the soil (i.e., root architecture), with the implication that the overall geometric configuration has some functional significance for the temporal and spatial patterns of water and nutrient acquisition (Fitter 1987; Lynch 1995; Fitter 2002; Hodge et al. 2009; see also Chapter 26).

The growth angle of root axes is a principal component of root system architecture that has been strongly associated with acquisition efficiency of soil resources in many crop species.

In common bean (*Phaseolus vulgaris* L.), for instance, the angle of basal roots (the main lateral branches from the hypocotyl in a dicot seedling, Lynch 1995) is the major determinant of root architecture, with genotypes exhibiting a wider basal root angle appearing to develop a shallower root system (Lynch and van Beem 1993; Nielsen et al. 1999; Liao et al. 2001; Lynch and Brown 2001). Likewise, Kato et al. (2006) demonstrated that the growth angle of nodal roots in rice (*Oryza sativa* L.) affects vertical root distribution and rooting depth. Bengough et al. (2004) observed a relatively narrower angular spread of seminal roots in wild barley when compared with modern cultivars. The authors hypothesized that narrower angular spread and, hence, a deeper root system was an advantage for plant species such as wild barley that had evolved in water-limiting conditions where obtaining water from depth was important for survival. Moreover, it can be assumed that such a trait would affect fecundity of the plants, thus driving their spread in the wild population.

A. Genotypic Variation in Wheat Seminal Root Characteristics

In spite of growing evidence that root angle plays an important role in shaping the architecture of root systems in many crop species, little is known about the extent of genotypic variability for the growth angle of seminal roots in wheat germplasm. Nakamoto and Oyanagi (1994) reported that deeply rooted Japanese wheat genotypes exhibit a narrower angle of seminal roots, while genotypes with a shallower root system tend to grow their seminal roots more horizontally. Manschadi et al. (2008) investigated the seminal root characteristics of a set of wheat genotypes in small gel-filled chambers. At 8 days after seed transfer, the seminal roots visible through the clear glass were scanned using a flat bed scanner. All wheat genotypes developed the primary seminal root (or radical) and the second and third root axis, counting from right to left (Figure 22.1). The fourth and fifth seminal roots were also observed in some genotypes. For each wheat seedling, the growth angle of individual root axes was measured at 3 cm distance from the seed relative to a vertical line passing through the stem base. The primary seminal root appeared to grow almost vertically, while the fourth and fifth seminal roots elongated almost horizontally (78.4° ± 6.79°) and did not differ significantly among the wheat genotypes. The average growth angle of the second and third seminal root axes, however, showed significant genotypic variation among the wheat cultivars with the drought-tolerant line SeriM82 exhibiting the smallest growth angle (Figure 22.2). Based on these

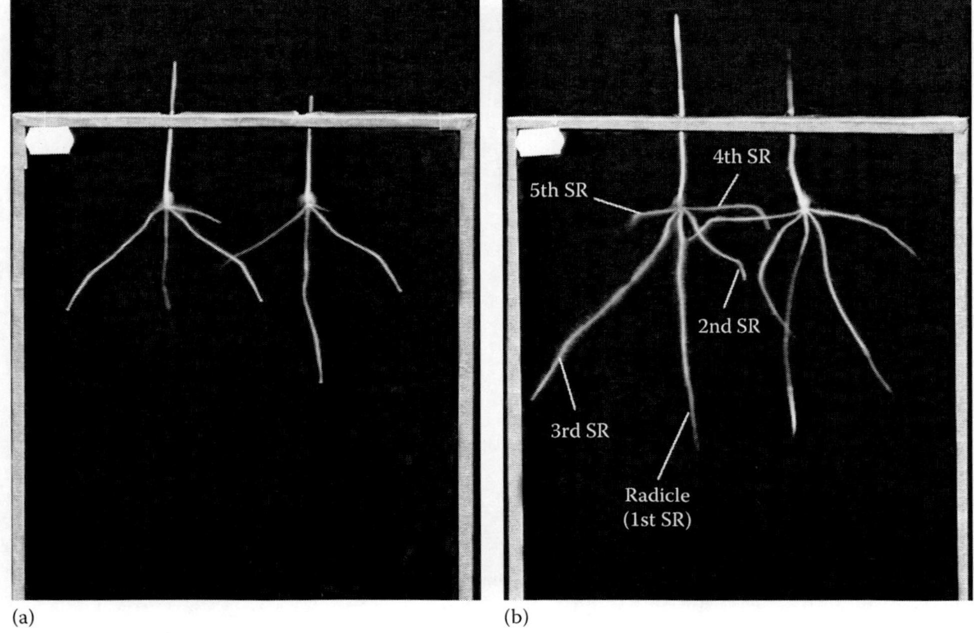

FIGURE 22.1 Scans of gel-filled chambers showing seminal roots of wheat genotypes Hartog (a) and SeriM82 (b) at 8 days after seed transfer. SR, seminal root.

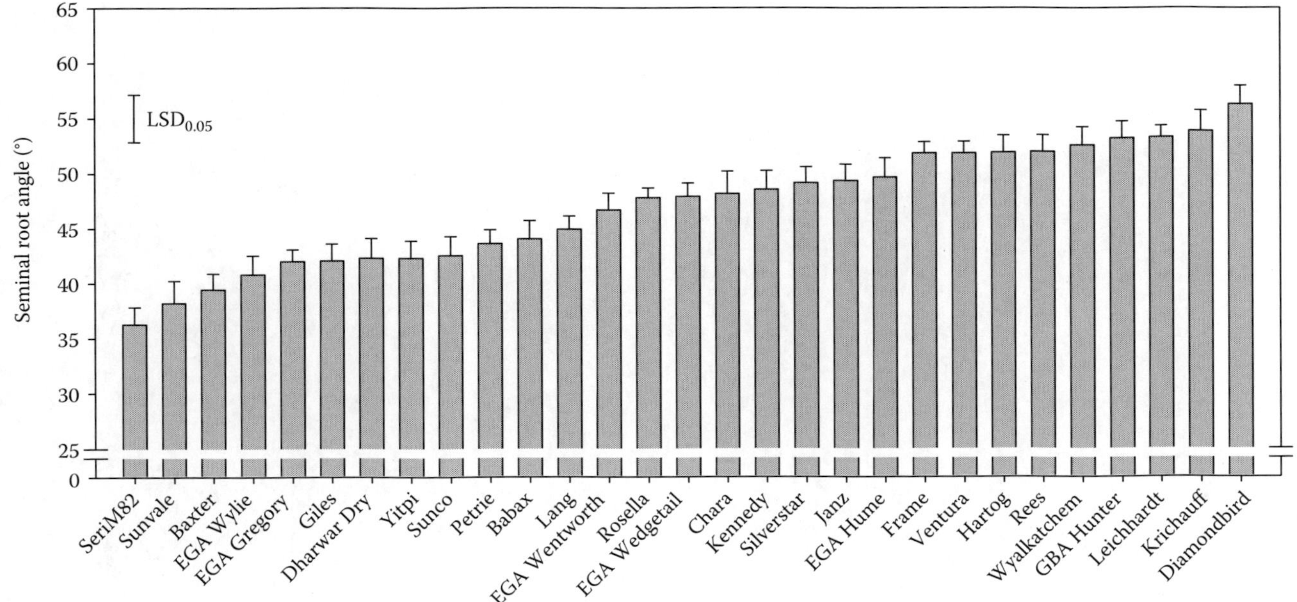

FIGURE 22.2 Growth angle (degrees) of wheat seminal roots in a set of CIMMYT and widely grown Australian genotypes. The bars indicate standard error of the mean; the vertical bar represents LSD ($p = 0.05$) for comparing the mean values. (From Manschadi, A.M. et al., *Plant Soil*, 303, 115, 2008.)

studies with a limited set of wheat genotypes, the values for the growth angle of the second and third seminal roots ranged from 36° to 56° (Nakamoto and Oyanagi 1994; Manschadi et al. 2008).

The wheat genotypes also exhibit substantial variation in the number of seminal roots. While the Australian spring wheat cultivars produce three to five seminal root axes (O'Brien 1979; Manschadi et al. 2008), winter wheat varieties grown in temperate environments develop on average six seminal axes (Gregory et al. 1978). Furthermore, experimental results suggest that the growth angle and number of seminal root axes do not correlate (O'Brien 1979; Manschadi et al. 2008).

B. Quantitative Description of Root Architecture

The root architecture of plants at early developmental stages can be relatively easily quantified based on the growth angle and number of seminal or basal roots. With the increase in rooting depth, initiation and growth of nodal roots, and appearance of lateral root branches, the quantification of root system architecture becomes increasingly difficult. So far, only a few attempts have been made to describe genotypic variation in root architecture of older plants at the quantitative level.

The topological approach based on the number and spatial arrangement of root links can yield valuable parameters for quantitative characterization of root branching, but the requirement for detailed data on root links, which are difficult to acquire, renders this method less suitable for large-scale analysis of root architecture (Fitter 1987; Fitter and Stickland 1991; Berntson 1997; Bouma et al. 2001; Fitter 2002). Fractal geometry is a quantitative method of describing complex natural objects with non-integer dimensions (Mandelbrot 1983). Root systems appear to approximate fractal objects over a finite range of scales because the repetitive branching of roots leads to a certain degree of self-similarity, which is a fundamental characteristic of fractal objects (Tatsumi et al. 1989; Fitter and Stickland 1992; Berntson 1994, 1996; Eshel 1998). In root architectural analysis, a fractal dimension (*D*) is calculated from the root system images using the box-counting method. This procedure is based on superimposing a grid on the root system image and counting the numbers of squares intercepted by roots for various square sizes (Tatsumi et al. 1989; Nielsen et al. 1997; Manschadi et al. 2008). By definition, if a 2D object is fractal, the value of *D* must be greater than 1 and less than or equal to 2. Variations in *D* of root architecture in response to genotype, water, and nutrient availability have been reported in a number of crop species (Eghball et al. 1993; Lynch and van Beem 1993; Masi and Maranville 1998; Nielsen et al. 1999).

In a recent study, Manschadi et al. (2008) investigated the fractal dimension of a set of wheat genotypes contrasting in drought tolerance. The wheat seedlings grown for 6 weeks (four to five leaf stage) in soil-filled root chambers showed significant variation in the lateral extension and branching intensity of roots (Figure 22.3). The drought-tolerant wheat SeriM82 exhibited the most compact root system with the main axes showing a near vertical growth orientation. The values of *D* calculated for the entire root system varied substantially among the genotypes. The lowest value of *D* (1.62 ± 0.043) was obtained for SeriM82 and the highest value for the standard wheat cultivar Hartog (1.77 ± 0.019). Compared to wheat genotypes, the spring barley cultivar Mackay had the most vigorous root system with a highly intensive branching structure and a *D* value of (1.84 ± 0.040). The greater *D* value of the barley cultivar reflects the relatively more vigorous and highly branched root system of this species. This is consistent with other reports demonstrating that the mature root system of barley tends to explore the topsoil more efficiently by producing greater root length and mass compared to wheat (Gregory et al. 1992; López-Castañeda and Richards 1994).

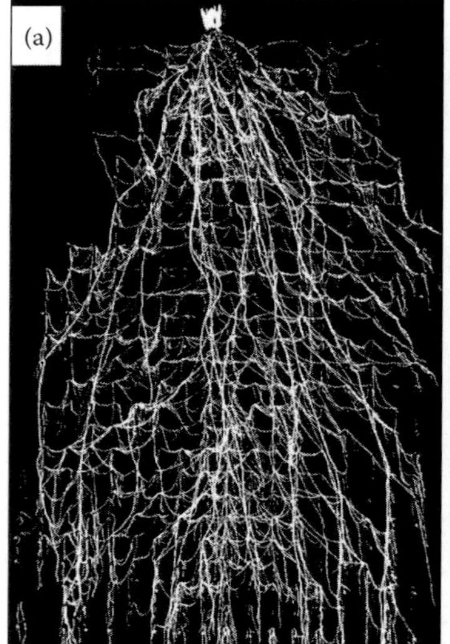

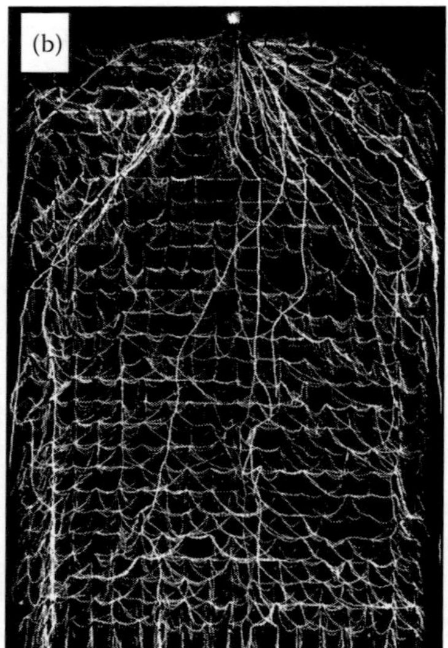

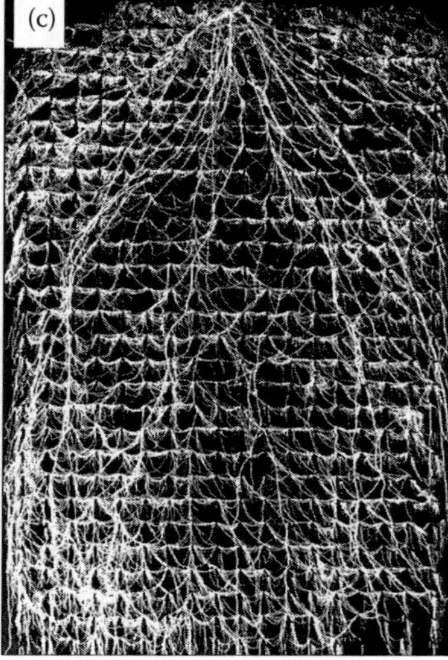

FIGURE 22.3 Root system images at 33 days after emergence for wheat genotypes SeriM82 (a) and Hartog (b) and barley cv. Mackay (c) grown in soil-filled root observation chambers. (From Manschadi, A.M. et al., *Plant Soil*, 303, 115, 2008.)

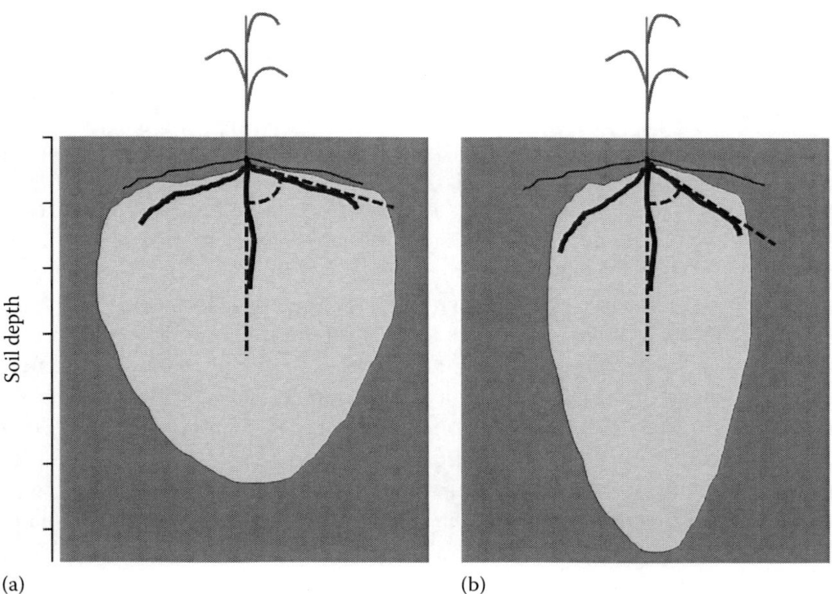

FIGURE 22.4 Schematic representation of the relationship between seminal root angle and root architecture in wheat genotypes contrasting in drought tolerance; (a) wider seminal root angle and a large and shallow root system (cv. Hartog) and (b) narrow seminal root angle and a compact and deep root system (SeriM82).

Although fractal dimension reflects genotypic differences in root typology and branching pattern, the values of *D* may vary among various studies due to differences in plant age, growth conditions, and techniques used to determine *D*, as well as species-specific differences in root architecture. For instance, the *D* values reported by Manschadi et al. (2008) for wheat are similar to those presented by Masi and Maranville (1998) for sorghum but generally greater than those measured for some other species (Tatsumi et al. 1989; Fitter and Stickland 1992; Eghball et al. 1993; Berntson 1994; Eshel 1998; Nielsen et al. 1999). But generally, it can be concluded that the value of *D* represents an index of the "space-filling" properties of the root system and encompasses both topological and geometric root characteristics. Variation in *D* among the wheat cultivars suggests significant genotypic differences in root development and branching structure which may explain their adaptation to water stress conditions.

C. Relationship between Wheat Seminal Root Angle and Root Architecture

The observed concentration of roots in the soil volume directly underneath the SeriM82 plant (see Figure 22.3) can be attributed to the growth angle of seminal root axes. In fact, SeriM82 exhibits a narrower seminal root growth angle compared to the standard wheat cultivar Hartog (see Figure 22.2). In a detailed study of root growth and water extraction pattern of single plants grown in large root observation chambers, Manschadi et al. (2006) found that, compared to Hartog, the mature root system of the drought-tolerant genotype SeriM82 was less laterally spread and more compact. The maximum lateral root extension measured from the stem base was approximately 45 cm for SeriM82 and 60 cm for Hartog. This resulted in significantly greater root-length production at depth and hence higher water extraction from deeper soil layers in SeriM82. While little is known about the mature root system architecture of field-grown wheat genotypes, the findings of Manschadi et al. (2006) and the observations of Oyanagi (1994) clearly suggest that wheat root system architecture is closely linked to the angle of seminal root axes at the seedling stage and that the early rooting characteristics determine the architecture and functioning of the mature root system later in the season. It means that wheat genotypes exhibiting a narrow seminal root angle are likely to develop a compact and deep root system, whereas those with wider angle of seminal axes tend to form a broad, shallow root system (Figure 22.4). Similar relationships between gravitropic growth of the root axes and spatial deployment of root systems have also been reported in other crops such as rice (Kato et al. 2006) and common bean (Lynch and van Beem 1993; Nielsen et al. 1999; Liao et al. 2001).

IV. Implications of Root Architectural Traits for Resource Acquisition

Architectural traits play a crucial role in determining the temporal and spatial distribution of the root systems in soil, thereby affecting acquisition efficiency of water and nutrient resources.

Nitrogen (N) and phosphorus (P) are the nutrients that most frequently limit crop growth and yield, especially in low-input agricultural systems in the tropics and subtropics. The nutrient use efficiency of cereal crops is generally defined as grain production per unit of nutrient available (from the soil and/or fertilizer). Variation in nutrient use efficiency among crop genotypes can be attributed to the efficiency of absorption (acquisition efficiency) and the efficiency with which the absorbed nutrient is utilized to produce the grains (utilization efficiency) (Moll et al. 1982; Manske et al. 2001; Wang et al. 2010b).

Thus, acquisition efficiency refers to the ability of crops to take up nutrient from soils (absorbed nutrient per total nutrient in the rooting volume), while utilization efficiency is largely determined by re-translocation of nutrient from vegetative plant tissues to generative organs (grain yield per unit of total plant nutrient). Research in many crops has shown that under conditions of nutrient deficiency, genotypic variation in nutrient use efficiency is largely due to variation in acquisition efficiency, whereas utilization efficiency of accumulated nutrients is more important when soil supply is adequate (Le Gouis et al. 2000; Manske et al. 2001; Wang et al. 2010a). Nutrient deficiency in agricultural soils may occur as a result of insufficient quantity and low mobility of nutrients and/or poor solubility of the abundant chemical forms in the soil. Thus, the nutrient-efficient genotypes may have an increased capacity to exploit the soil and/or convert non-available nutrient forms into plant-available ones (Rengel and Marschner 2005). The herringbone root architecture (branches from a single main root axis) seems to be more efficient for the uptake of mobile nutrients due to the reduced overlapping of depletion zones, whereas the dichotomous architecture type (highly branched root system) has a better ability to acquire immobile nutrients (Fitter and Stickland 1991; Fitter et al. 1991). Considering the central role of root system characteristics in the capture of soil resources, this chapter focuses on acquisition efficiency.

A. Nutrient Acquisition Efficiency

1. Phosphorus

Compared to other essential mineral nutrients, P is the least mobile and least available to plants in many soils. As the contribution of transpiration-driven mass flow to satisfy crop P demand is typically very small, almost all of the required P must reach the root surface via diffusion (Föhse et al. 1991). The diffusion coefficient for inorganic P (Pi), the only form of P that is taken up by roots, is low. Therefore, plants can extract such immobile nutrients only from the soil volume close to the root surface. The quantities that can be extracted are limited by the nutrient concentration at the root–soil interface. The soil solution concentrations of phosphate are small ($<0.05\ \mu g\ g^{-1}$) as compared to that of nitrate-N ($100\ \mu g\ g^{-1}$). Traits of crop plants that will increase the availability and uptake of immobile nutrients, such as P, are (1) increase of root surface area through specific root architectural, morphological, and anatomical characteristics, (2) exudation of organic compounds and protons, and (3) formation of symbioses with microorganisms such as arbuscular mycorrhizae (Rengel and Marschner 2005; Lambers et al. 2006; Lynch 2007; Ramaekers et al. 2010; Wang et al. 2010c). Changes in root morphology, such as root hair formation, and exudation of nutrient mobilizing compounds are localized to the microenvironments determined by root proliferation and distribution, and therefore root architecture. Thus, root architecture can be considered as a higher-order organismic trait within which traits at the organ and tissue level operate (Lynch and Brown 2001).

Phosphorus availability usually declines substantially with depth in agricultural soils because of fertilizer application and cultivation of topsoil and inherently poor mobility of P in the soil profile. Several studies have shown that root architectural traits in legume crops, such as common bean (*P. vulgaris* L.) and soybean (*Glycine max* L.), are critically important for adaptation to P-deficient soils. The root system of many annual dicots, including common bean and soybean, is composed of three main types of axes: a taproot, "basal roots" emerging from the basal end of the taproot, and adventitious roots emerging from the hypocotyl below the soil surface (Walk et al. 2006). Shallower growth of basal roots, increased adventitious rooting, and greater dispersion of lateral roots enable root foraging in the topsoil where P availability is high and contribute to enhanced P efficiency in these species (Ge et al. 2000; Lynch and Brown 2001; Miller et al. 2003; Liao et al. 2004b, Ho et al. 2005; Ao et al. 2010; Wang et al. 2010c). In maize (*Zea mays* L.), genotypes with enhanced or sustained lateral rooting appear to exhibit greater P acquisition, biomass accumulation, and relative growth rate at low P availability as compared to other genotypes (Zhu and Lynch 2004).

Genotypic variation in P acquisition efficiency has also been reported for wheat, with P-efficient genotypes producing higher shoot biomass and grain yield (Fageria and Baligar 1999; Manske et al. 2000; Ozturk et al. 2005). In contrast to extensive research on the role of root architecture in adaptation to P deficiency in legume species (Lynch 2007), relatively few studies have addressed the implications of genetic diversity in root architectural traits for P acquisition by wheat. The studies in wheat have so far focused on the effects of RLD and dry weight in topsoil on P uptake. The density of roots in the upper soil layers seems to be the most important trait associated with improved acquisition of relatively immobile nutrients such as phosphorus.

Manske et al. (2000) reported that wheat genotypes with higher RLD are able to take up more phosphorus. At low P supply, there was a strong correlation between RLD in topsoil and P uptake and grain yield, whereas under adequate P supply this correlation is lost. However, where P is available in subsoil (residual or native P), the roots of P-stressed wheat plants proliferated in the deeper soil horizons (Manske et al. 2000). Many wild forms and landraces of wheat possess large root systems but tend to lodge because of their tall culms (Manske 1989; Vlek et al. 1996). Compared with landraces, root size of modern cultivars is small and may be too small for optimum uptake of water and nutrients and maximum grain yield (Waines and Ehdai 2007). There is a concern that decades of breeding for improved yield potential under high-input conditions may have resulted in wheat germplasm with smaller root systems. This was investigated in 44 historically important lines released by CIMMYT (International Maize and Wheat Improvement Center) and grown under conditions of adequate nutrient and water supply. Results of this study indicated that wheat grain yield had increased on

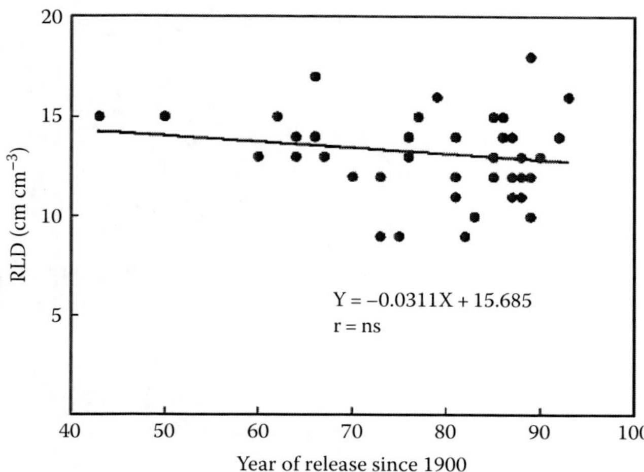

FIGURE 22.5 The effect by the year of release on root length density (RLD) in the upper soil layer (0–20 cm) of 44 important CIMMYT wheat lines released since 1942 grown on an alkaline soil with nutrients and water amply supplied, means of three replicated plots. (From Manske, G.G.B., Utilization of the genotypic variability of VAM-symbiosis and root length density in breeding phosphorus efficient wheat cultivars at CIMMYT, Final report of a special project, CIMMYT, de México, Mexico, 1997.)

average by 50.4 kg ha^{-1} year^{-1} over the years, but the absolute values of RLD had not significantly changed (Figure 22.5; Manske 1997).

Root diameter and root hair abundance are other important determinants of P acquisition efficiency in wheat (Jones et al. 1989) because they modify the P depletion profiles in the rhizosphere (Gahoonia et al. 1997). Genotypes with thinner roots showed improved P uptake (Manske et al. 1996). At high P and ample water supply, root hairs are rudimentary, whereas under deficient conditions long root hairs are abundant (Föhse et al. 1991). Root hair density, scored in plants grown in pot cultures with quartz sand, varied considerably among 19 bread wheat genotypes and was positively correlated with P uptake at anthesis when plants were grown on a P-deficient calcareous Aridisol (Manske 1997).

Vesicular-arbuscular and arbuscular mycorrhizal fungi (AMF) are also important for the uptake of nutrients that are relatively immobile, such as P (Hayman and Mosse 1971; Bucher 2007; Adesemoye and Kloepper 2009), Cu (Gildon and Tinker 1983), and Zn (Swaminathan and Verma 1979). The AMF symbiosis improves P influx (P uptake per unit root length), but increased fertilizer doses reduce the infection by native AMF. With reduced fertilizer application in an irrigated field trial, improved AMF infection at the stage of tillering resulted in higher wheat grain yields (Manske et al. 1995). In a P-fixing acid Andisol, AMF infection rate was positively correlated with P uptake (Manske et al. 2000). However, AMF explained only 25% of the variance in P uptake. Other traits such as RLD and phosphatase excretion by the roots were also important.

Screening experiments have shown considerable genotypic variability in AMF colonization and efficiency in wheat (Bertheau et al. 1980; Manske 1990; Kapulnik and Kushnir 1991; Hetrick et al. 1992; Osborne and Rengel 2002). Old wheat cultivars rely more on symbiosis than modern wheat (Manske 1990; Yücel et al. 2009). Both the extent of AMF infection and the degree of benefit from AMF appear to be heritable traits (Manske 1990; Vlek et al. 1996; Pandey et al. 2006) and could be used to breed new varieties for reduced-input agriculture (Sawers et al. 2010). High AMF efficiency of a wheat variety can be transferred by crossing to a non-efficient one (Manske 1990; Diederichs and Manske 1991). Depending on the crossing combination and environment, one could observe recessive, intermediate, or dominant inheritance modi, plasmatic effects, effects of heterosis, and even a change of dominance within the same genotype under different environments (Manske 1989, 1990; Diederichs and Manske 1991).

Roots can also induce changes in the pH of the rhizosphere. Carboxylic acids, especially citric and malic acid, are the major compounds in exudates of wheat roots that are responsible for this process. Compared with other monocots, wheat is highly resistant to lime-induced chlorosis caused by Fe deficiency. Under conditions of Fe deficiency, the roots release high amounts of chelating exudates (phytosiderophores), which form plant-available Fe–phytosiderophore complexes and transport Fe to the roots (Marschner et al. 1986; Neumann and Römheld 2002). Exudation promotes the dissolution of poorly available nutrients such as P (Neumann and Römheld 1999) and Mn (Godo and Reisenhauer 1980). Poorly available organic P in the soil, which commonly accounts for 40%–50% of the total P supply to plants, can be utilized with the help of root phosphatases (Helal 1990). Genotypic differences in root phosphatase, excreted or bound at the root surface, were shown by McLachlan (1980). The acid phosphatase activity at the surface of intact wheat roots was determined for 4-week-old plantlets of 42 bread wheat genotypes from the CIMMYT collection grown in P-free nutrient solution by the *p*-nitrophenyl assay (Tabatabai and Bremner 1969). Phosphatase activity was positively correlated with P uptake of those genotypes grown in the field on a P-deficient Andisol (Manske 1997; Portilla et al. 1998). This indicates that the wheat plants could utilize organic P from the soil.

2. Nitrogen

Although there is a growing understanding of the importance of root growth and architecture for P acquisition (reviewed by Lynch 2007), less is known about the effects of root morphology and architecture on N uptake. Extensive field studies with winter wheat grown under various N supply treatments suggest that there is a significant genotypic variation in N acquisition efficiency and hence the total N uptake in shoot tissues (Le Gouis et al. 2000; Barraclough et al. 2010). However, a multisite comparison of N use efficiency between old and modern winter wheat varieties in the United Kingdom indicated that on-farm N use efficiency has not increased over the years. Increases in wheat grain yield have rather been accompanied by increases in N fertilizer amounts than by improvements in N acquisition (Sylvester-Bradley and Kindred 2009).

Inorganic N in the soil occurs mainly as nitrate (NO_3-N) and ammonium (NH_4-N). NO_3-N is a highly mobile, non-adsorbing ion, which moves readily toward the roots via both mass flow and diffusion. Root vigor, rooting depth, RLD, and root longevity (especially for post-anthesis N uptake) appear to be important attributes for soil N acquisition (Hirel et al. 2007; Foulkes et al. 2009; Garnett et al. 2009). As N fertilizers are commonly applied to the topsoil, wheat genotypes with extensive root growth in the upper soil layers appear to have higher efficiency of fertilizer-N uptake. Liao et al. (2006) reported that the early and more extensive horizontal growth of the roots in the top 70 cm of the soil profile was responsible for the superior N acquisition by vigorous wheat breeding lines. Similarly, Palta et al. (2007) found that the earlier and more abundant root branching and growth in vigorous wheat lines, and the associated increase in RLD, resulted in higher N uptake.

Nitrate easily moves with water and hence can be leached down the soil profile when soil N supply exceeds plant uptake capacity. This can occur during early crop development stages when soil N supply from applied fertilizers is substantially higher than N uptake capacity by the plants. Under such conditions, root growth and distribution at depth and post-anthesis root longevity seem to be critical for sufficient uptake of N from subsoil at the later stages of plant development. An anatomical study of the number and arrangement of xylem tracheary elements of field-grown wheat roots indicated that all roots occurring in subsoil were seminal axes and their branches, and no nodal roots were present at depth greater than 60 cm (Watt et al. 2008). Thus, a greater number of seminal root axes may result in more intensive root branching and RLD at depth, which substantially enhances the capacity of root systems to acquire N and water from the subsoil. Wheat genotypes exhibiting a high number of seminal roots combined with a narrow growth angle may be ideally suited to soils with high potential for leaching. In wheat, genotypic variation in root distribution at depth and its implications for water uptake has been reported both under controlled (Manschadi et al. 2006) and field conditions (Reynolds et al. 2007; Christopher et al. 2008). Wheat genotypes with superior ability to extract water from the subsoil may have also higher nutrient acquisition efficiency in soils with high potential for leaching. Root growth often ceases post-anthesis due to internal plant competition for assimilates, which are preferentially allocated to reproductive organs. Thus, crop phenology, especially the onset of anthesis, is a major determinant of maximum root depth potential.

B. Water Accessibility and Drought Tolerance

Drought is generally regarded as the most widespread limitation to crop productivity and yield stability in rain-fed production systems. Consequently, developing cultivars with enhanced drought adaptation and higher yield has been the focus of many crop improvement programs. Root traits can moderate the effects of drought either by increasing the rate of water uptake through increased RLD or by allowing a greater amount of water extraction through increased root-length distribution at depth (Passioura 1983; Gregory 1989; Manschadi et al. 2006; Hammer et al. 2009). Increased rooting depth and production of greater length of roots in the deeper soil layers in response to soil drying in surface layers have been reported as important drought-adaptive root trait in many crop species (O'Toole and Bland 1987) including wheat (Hurd 1968; Asseng et al. 1998; Xue et al. 2003).

In a field experiment, wheat genotypes grown on residual moisture demonstrated significant variation in vertical root distribution in the soil profile (Table 22.1). The drought-tolerant semidwarf genotypes produced more roots in the deeper layers (Pastor, Sujata, Dharwar, Synthetic 2; SeriM82), whereas the non-tolerant samples (Sonalika, Tevee, Baviacora and Pavon) had fewer roots in the deep soil. Scores for stay green (maintaining green leaf area longer during the grain-filling period) were positively correlated with RLD in deep soil layer (80–100 cm) (Manske and Vlek 2002).

The ability of SeriM82 to maintain green leaf area longer after anthesis (stay green) was also observed in a multiyear field experiment conducted in a summer-rainfall-dominant location in Australia (Christopher et al. 2008).

The yield of SeriM82 was 6%–28% greater than that of the standard cultivar Hartog (a Pavon-type cultivar) due to its stay-green phenotype, which was expressed in those years when it yielded significantly more than Hartog. However, where the availability of deep soil moisture was limited, SeriM82 failed to exhibit significantly greater yield or to express the stay-green

TABLE 22.1 Grain Yield, Scores for Stay Green, and RLD in Different Soil Layers of 12 Wheat Genotypes Contrasting in Drought Tolerance Grown on Residual Moisture (One Single Irrigation at Sowing) in an Arid Environment in Northwest Mexico (Means of Six Replications)

			RLD in Soil Layers (cm cm^{-3})		
Genotypes	Grain Yield (t ha^{-1})	Score[a] Stay Green	0–100 cm	0–20 cm	80–100 cm
Sonalika	2.29	1	0.95	1.92	0.40
Baviacora	2.51	3	0.98	1.85	0.43
Pavon	2.15	1	1.11	2.45	0.45
Nesser	2.07	2	1.00	1.82	0.45
Tevee	2.29	3	0.88	1.92	0.49
SeriM82	1.76	2	1.08	2.09	0.56
Check 3	0.71	5	1.14	2.71	0.61
Synthetic 1	2.44	4	0.92	1.76	0.65
Synthetic 2	2.15	5	1.04	1.89	0.68
Dharwar	2.20	3	1.09	2.25	0.70
Sujata	2.65	5	0.88	1.42	0.71
Pastor	2.52	3	1.01	1.62	0.72
Mean	2.15	3	1.01	1.97	0.57
LSD 5%	0.38	0.93	0.18	0.72	0.24

Source: Manske, G.G.B. and Vlek, P.L.G., Root architecture—Wheat as a model plant, in *Plant Roots: The Hidden Half*, eds., Waisel, Y., Eshel, A., and Kafkafi, U., 3rd edn., Marcel Dekker, Inc., New York, pp. 249–259, 2002.

[a] Score stay green after anthesis, 1 < 5 for increasing green.

phenotype. Thus, the stay-green phenotype in SeriM82 appears to be closely related to its deep root architecture and hence the ability to extract greater amount of water from depth after anthesis (Christopher et al. 2008).

Studies on the root structural traits such as topology and branching pattern their implications for soil resource acquisition are limited, due to difficulties in observation and quantification of these architectural traits. Angadi and Entz (2002) attributed the relatively greater soil moisture depletion by a standard height sunflower cultivar to more uniform root distribution, as the RLDs did not differ between the cultivars compared. In wheat, Manschadi et al. (2006) observed a more even root distribution with depth in the drought-tolerant genotype SeriM82 compared to the standard variety Hartog. More even root-branching structure results in a more uniform root distribution in the soil and hence greater amount of water extraction, although the total root length was similar in both genotypes (see Fig. 5 in Manschadi et al. 2006).

1. Adaptation to Contrasting Drought Environments

The wheat production environments differ markedly in the temporal pattern of rainfall distribution and hence the timing and severity of drought stress during the crop life cycle.

According to a framework introduced by CIMMYT, the wheat-growing regions of the world are divided into 12 mega-environments (MEs) (Braun et al. 1996; Christopher et al. 2005; Ortiz et al. 2008). An ME is defined as a broad, often noncontiguous or transcontinental area with similar biotic and abiotic stresses, cropping systems, consumer preferences, and volume of production. Drought stress is the major yield-limiting factor in ME4A, ME4B, and ME4C, where rain-fed wheat is grown under low rainfall combined with temperate or hot temperature regimes. These MEs differ significantly in terms of the temporal pattern of water stress relative to wheat development. In the winter rainfall, Mediterranean environments of ME4A, wheat crops often suffer from post-anthesis moisture stress, while in ME4B, pre-anthesis water shortage is the dominant drought stress pattern. Residual drought stress occurs in ME4C, where wheat is grown on stored soil moisture. The CIMMYT representative locations/regions for ME4A are Aleppo, Syria, and Settat, Morocco, for ME4B Marco Juárez, Argentina, and ME4C Indore, India (Braun et al. 1996; Ortiz et al. 2008).

The overall advantage of the architectural root traits must, therefore, be interpreted in the context of the type of drought environment in which the wheat crops are grown. In low-rainfall environments with a summer-dominant rainfall pattern (ME4C), such as southeastern Australia and Indore in India, wheat crops are grown largely on water stored in the subsoil with a substantial risk that water may run out before the completion of grain filling. In contrast, genotypes adapted to the winter-rainfall-dominant Mediterranean environments, for example, in the Mediterranean region and southwestern Australia, rely heavily on in-season rainfall (Nix 1975).

Root architecture is likely to be associated with genotypic adaptation to these contrasting environments. While developing a compact and deep root system to increase access to water from the deeper soil layers during the critical grain-filling phase appears to be desirable in ME4C, possessing a large and shallow root system with greater potential for water extraction early in the season in order to reduce unproductive soil evaporation is more suitable for Mediterranean environments (ME4A).

The growth angle of seminal axes is one of the principal determinants of wheat root system architecture (see Figure 22.4). In a study of genotypic variation in growth angle and number of seminal roots among 27 current Australian and 3 CIMMYT wheat genotypes, Manschadi et al. (2008) found that wheat cultivars grown in the Mediterranean environments of southern and western Australia tended to have a wide root angle, whereas drought-tolerant genotypes with higher yield in the summer-dominant rainfall regions exhibited narrower angular spread of seminal axes. Similarly, Oyanagi (1994) reported that Japanese winter wheat cultivars adapted to drier environments possess a narrow seminal root angle and consequently a deeper root system, whereas genotypes developed for more favorable, wetter environments express a more horizontal seminal root growth and superficial root system.

In addition to growth angle, the number of seminal roots may also affect the degree of adaptation to drought stress. Seminal roots tend to grow in deeper soil horizons and can therefore make a significant contribution to water uptake from subsoil (Watt et al. 2008). Thus, a greater number of seminal root axes may result in more intensive root branching and RLD at depth, which substantially enhances the capacity of root systems to explore the deep soil layers more effectively. Genotypes expressing high number of seminal roots combined with a narrow growth angle may be ideally suited to environments where plants rely largely on subsoil water.

The architecture of a mature wheat root system is not only determined by the number and growth angle of seminal roots but is also affected by the development of nodal or adventitious roots. Little is known about genotypic variation in the number and gravitropic response of nodal axes in wheat. However, because nodal roots are concentrated in the top soil layers, they may play a major role in water extraction from the shallow soil layers particularly in Mediterranean-type environments, where due to substantial in-season rainfall, rewetting of the topsoil occurs more frequently.

It is important to note that drought tolerance is a complex character resulting from the interaction of a range of root and shoot traits (Ludlow and Muchow 1990; Richards et al. 2002, 2010; Passioura and Angus 2010). Root and shoot growth are inextricably linked because root system growth and maintenance depend on the assimilate supply by shoot. Concurrently, the size and activity of the root system determines the rate at which the shoot system can produce photosynthates. Therefore, in evaluating the implications of root architectural traits for drought adaptation, it is important to consider the attributes governing seedling establishment and vigor, leaf canopy growth and development, timing of anthesis, and biomass allocation to the root system. For instance, the roots of the drought-tolerant wheat line SeriM82 exhibit less lateral spread and a more uniform and deep root architecture in the post-anthesis phase (Manschadi et al. 2006). Compared to the standard wheat cultivar Hartog, SeriM82 is less vigorous early in the season with significantly lower shoot and root dry weights but substantially higher R:S (Root to Shoot) ratio (Manschadi et al. 2008). Allocating relatively more biomass into the root system early in the life cycle combined with lower early shoot vigor may be an adaptive advantage of SeriM82, as this will reduce pre-anthesis water use and, hence, improve post-anthesis water availability. Small amounts of water "shifted" from pre- to post-anthesis in this way can have major effects on grain yield in water-limited situations.

In a simulation study with the Agricultural Production Systems Simulator model (APSIM) (Keating et al. 2003; http://www.apsim.info/apsim/), Manschadi et al. (2006) examined the likely effects of modified root traits on wheat growth and yield formation across a wide range of summer-dominant rainfall environments in southern Queensland (QLD), Australia. The APSIM simulations were run with 100 years of daily weather data from three locations contrasting in the soil type and the maximum amount of plant-available water content (PAWC) in the profile. The three locations were Roma, representing a low-yielding, Oakey, a high-yielding, and Goondiwindi, an intermediate-yielding wheat environment, respectively. In each year of simulation, the standard wheat cultivar Hartog was "sown" on June 1 at a density of 100 plants m^{-2} under non-limiting nitrogen supply. To generate a wide range of standard starting soil water conditions, the soil profile at sowing was assumed to be recharged to one-third, two-thirds, and full level of PAWC, each year at each site. In the second set of simulations, instead of Hartog, a root-modified (RM) wheat genotype was sown. The RM genotype represented a hypothetical wheat line with the aboveground characteristics of Hartog but root attributes of the drought-tolerant genotype SeriM82, that is, narrower root system architecture with the ability to extract more soil moisture per soil volume, particularly deep in the profile late in the growing season.

Analysis of the simulation results indicated a mean relative yield benefit of 14.5% in water-deficit seasons for the RM genotype and that each additional millimeter of water extracted during grain filling generated on average an extra 55 kg ha^{-1} of grain yield (Figure 22.6). Considering that the seasonal water use efficiency for grain production in well-managed wheat crops is about 20–25 kg ha^{-1} mm^{-1} (French and Schultz 1984; Sadras and Angus 2006), the marginal water use efficiency of extra soil moisture that becomes available post-anthesis is about two times higher than that calculated over the whole growing season. This means that small amounts of additional moisture extracted post-anthesis can lead to relatively large differences in grain yield. Similar to these simulation results, Kirkegaard et al. (2007) measured an efficiency of 59 kg wheat grain ha^{-1} for each extra millimeter of water extracted by wheat roots from subsoil late in the season.

It seems that by linking this strategy, that is, shifting water use from pre- to post-anthesis, with drought-adaptive root traits such as narrower growth angle and higher number of seminal axes, wheat genotypes such as SeriM82 are likely to possess an optimal combination of attributes for adaptation to summer-dominant rainfall environments, where effective utilization of stored soil moisture is critical for improved yields under drought stress. In contrast, wheat genotypes with increased early shoot vigor and greater root growth and branching intensity have been shown to be better adapted to winter-dominant rainfall environments. The yield advantage of vigorous wheat genotypes has been attributed to greater water use and aboveground biomass production early in the season when vapor

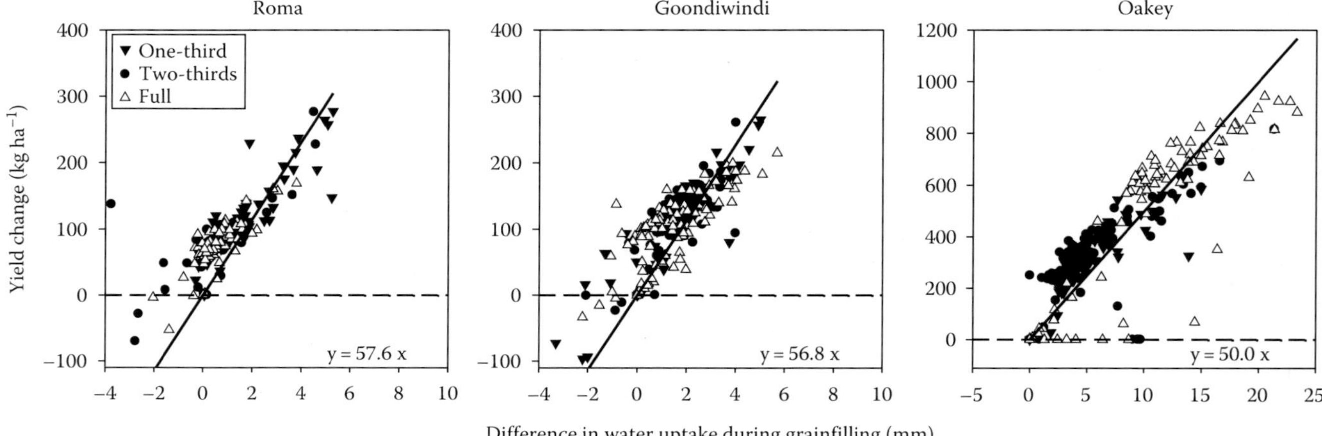

FIGURE 22.6 Simulated yield difference between the standard wheat cultivar Hartog and the RM genotype plotted against difference in water uptake during the grain-filling phase. One-Third, Two-Thirds, and Full indicate the initial soil water content at sowing as the fraction of total plant available soil water in the profile; the slope of regression line $y = ax$ represents water use efficiency for grain yield at each location. (From Manschadi, A.M. et al., *Funct. Plant Biol.*, 33, 823, 2006.)

pressure deficit is low, reduced soil evaporative loss of water, and greater water use efficiency (Rebetzke and Richards 1999; Liao et al. 2004a, 2006).

C. Root Architectural Trade-Offs for Soil Resources Acquisition

A common response of crop plants to low availability of soil resources is increasing biomass allocation to the root system. Increased carbon allocation to roots bears the cost of reduced allocation to photosynthetic shoot tissues and reproductive organs. If plants are considered to optimize the allocation of resources among various organs, their investment in new growth structures, organs, and processes should be maintained at a level where the marginal cost equals the marginal benefit (Lynch and Ho 2005). Estimating carbon budgets can help in assessing the importance of increased carbon allocation to roots for the adaptation of plants to low soil nutrient and water availability. In common bean, for instance, the root respiration rate per unit of root growth is substantially lower in P-efficient genotypes, enabling them to sustain a greater root biomass at equivalent carbon cost (Nielsen et al. 2001).

Another adaptation mechanism to low-P soil in common bean is shifting biomass allocation to more metabolically efficient root types, such as adventitious roots, which have greater specific root length (SRL, root length per unit biomass, m g^{-1}) and less construction cost than tap and basal roots (Miller et al. 2003). In a simulation analysis of trade-offs between adventitious and basal root growth and P acquisition in common bean, Walk et al. (2006) found that increased carbon allocation to adventitious roots in stratified soil reduced the growth of tap and basal lateral roots, yet P acquisition increased by up to 10%. Furthermore, greater formation of aerenchyma (large air spaces) in the root cortex has also been reported to reduce the metabolic costs of root growth by increasing SRL (Lynch and Ho 2005).

The previous sections discussed the role and importance of root morphology and architecture for the acquisition of individual soil resources P, N, and water. However, crop plants grown under field conditions typically encounter multiple stresses during their life cycle and must co-optimize their resource allocation for acquisition of several limiting resources. In addition, many soil resources are unevenly distributed in space and time. Thus, root architectural traits that increase the acquisition efficiency for one soil resource may incur trade-offs for the acquisition of other resources. For instance, root architectures that are more efficient in exploiting topsoil resources may be advantageous in low-P soils, but they may not be adapted to drought environments due to their reduced access to water stored in subsoil. Ho et al. (2005) investigated the root architectural trade-offs for P and water acquisition in bean genotypes grown under vertically stratified P stress, water stress, and combined P and water stress. The shallow-rooted genotypes had greater biomass and total P accumulation under P stress, and deep-rooted genotypes performed better under terminal drought stress with sufficient P supply. Interestingly, the bean genotype with best performance under combined P and water stress was the one with a large dimorphic root system that permitted vigorous rooting both in the surface and deep soil horizons. This genotype had also the most vigorous shoot because increased diversion of biomass to the root system was not at the expense of aboveground growth.

A vigorous root system may also be important for the acquisition of soluble nutrients, especially NO_3-N. Indeed, vigorous wheat breeding lines appear to have superior N acquisition due to their large root systems (Liao et al. 2006; Palta et al. 2007). Dunbabin et al. (2003) used a 3D root architectural model to assess the efficiency of a range of root architectural types, from the classic herringbone to the dichotomous type, in NO_3-N capture from a coarse-textured soil of high leaching potential under Mediterranean climatic conditions. The simulated root systems were considered to have an equivalent carbon cost due to maintaining an equal total root volume at each point in time. Results of this study showed that the root architecture likely to maximize NO_3-N capture under these conditions is one that rapidly develops a high density of roots in the topsoil, thereby reducing NO_3-N leached with the early season rains, along with a deep, vigorous taproot, enabling access to leached NO_3-N and water in the subsoil later in the season. Therefore, such root architecture would not only increase N acquisition efficiency and reduce NO_3-N leaching but would also enhance terminal drought adaptation by improving access to water stored in subsoil. These simulation and empirical studies suggest that root architectural attributes that enhance P acquisition may also be beneficial for the acquisition of other immobile soil nutrients, whereas traits optimizing water uptake would also increase the capture of soluble, mobile nutrients such as NO_3-N. However, the co-optimization of root architecture for the acquisition of multiple soil resources will remain a challenging problem in root research (Lynch 2007).

Acquisition of P and N by plant roots is strongly affected by soil moisture content. In agricultural fields, nutrients are commonly concentrated in the topsoil due to the application of fertilizers. Topsoil, however, dries out frequently, reducing the mobility of soil N and P. Because diffusion of inorganic P in dry soil is very low, P uptake declines with decreasing soil moisture content. However, the root system can redistribute soil water by taking up water from moist deep soil layers and releasing it into shallower dry soil at night, a process termed "hydraulic lift" (Caldwell and Richards 1989; Burgess et al. 2000). Therefore, nutrient uptake from dry topsoil is expected to increase when the soil is moistened due to hydraulic lift. In wheat, Valizadeh et al. (2003) found that P banded in dry topsoil was taken up by plants when roots had access to moist subsoil and hydraulically lifted water was released into dry soil. The importance of hydraulic lift for nutrient uptake from topsoil in field-grown crops has not been quantified.

V. Selecting for Desirable Root Traits

Crop improvement programs in the past few decades have largely focused on enhancing yield potential and resistance to biotic stresses under high-input conditions. Direct selection for adaptation to low soil fertility and water stress has received relatively less attention, although genetic variation in root characteristics and its implications for nutrient and water acquisition efficiency has been identified in many crop species.

The reasons for neglecting direct selection for nutrient- and drought-adaptive root traits lie, among others, in the difficulty of isolating specific edaphic stresses from other co-occurring stresses, confounding effects of biotic stresses affecting roots, difficulty of directly evaluating root phenotypes in the field, large phenotypic plasticity of root traits in response to soil conditions, and lack of rapid and cost-effective screening methods (Sinclair et al. 2004; Lynch 2007; Richards et al. 2010).

Desirable root traits that are usually expressed at later stages of crop development, such as greater distribution of roots at depth and uniform root-branching pattern, appear to be less suitable for large-scale cost-effective screening programs, unless they can be reliably linked with surrogate measures like canopy temperature, stomatal conductance, grain carbon isotope discrimination, shoot nutrient content, and stay green (Fischer et al. 1998; Reynolds et al. 1999; Christopher et al. 2008; Lopes and Reynolds 2010). In addition, root characteristics exhibit a high degree of phenotypic plasticity in response to temporal and spatial variation in rooting environment (Poorter and Nagel 2000; Fitter 2002; Hodge 2006), which may complicate the identification of genotypic variation in adaptive root traits (Price et al. 2002). Given the obstacles to selection for traditional root traits, such as rooting depth, architectural root properties that are expressed at early stages of crop development and determine the growth and functioning of the mature root system later in the season may be more suitable as selection criteria in crop improvement programs (Manschadi et al. 2006). The angle and number of seminal, basal, and nodal roots appear to be such a trait, as it seems to strongly influence the root architecture and thus the timing and amount of nutrient and water uptake. Genotypic variation in angular spread of root axes has been reported in many crop species including barley (Hargreaves et al. 2009), common bean (Bonser et al. 1996; Lynch and Brown 2001; Liao et al. 2004b), rice (*O. sativa* L.) (Kato et al. 2006; Norton and Price 2009), sorghum (*Sorghum bicolor* L.), maize (*Z. mays* L.) (Giuliani et al. 2005; Singh et al. 2010), and wheat (Oyanagi 1994; Manschadi et al. 2008). Rapid and cost-effective screening for this trait can be undertaken using growth pouches or small gel-filled root chambers (Bengough et al. 2004; Lynch 2007; Manschadi et al. 2008). Coupled with digital scanners/cameras and image analysis software, these screening methods allow rapid characterization of seedling root traits of a large number of lines and genotypes (Manschadi et al. 2008; Yazdanbakhsh and Fisahn 2009; Le Bot et al. 2010). Screening techniques for analyzing roots grown in soil or other solid media are also emerging. The x-ray microtomography, for instance, is suitable for measuring the 3D configuration of the root system and root–soil interactions in a smaller number of genotypes. Combination of simple 2D root screening methods (e.g., Petri dishes, pouches, agar plates) with x-ray microtomography seems to be an efficient strategy to enhance the selection of desirable root traits (Gregory 2009). In addition to efficient and accurate phenotype screening, QTL (quantitative trait loci) mapping of root architecture, marker-assisted selection, functional-structural modeling of plant (root) systems, and close collaboration between breeders, geneticists, crop physiologists, soil scientists, and agronomists will be needed for successful development of water- and nutrient-efficient crop cultivars adapted to target environments (de Dorlodot et al. 2007; Su et al. 2009).

VI. Summary and Conclusions

Enhancing water and nutrient use efficiency in crop plants is a key component of the strategy to sustainable increase of agricultural productivity, reduction of the adverse environmental impacts of agriculture, and meeting the increasing food demand in the coming decades. Root system characteristics play a major role in soil resource acquisition and crop productivity in drought-prone and nutrient-deficient environments. The architecture of plant root systems appears to be particularly important for crop adaptation to multiple abiotic stresses occurring in such environments.

The root system architecture is largely determined by the growth angle of seminal, basal, and nodal roots, which has been strongly associated with temporal and spatial acquisition efficiency of soil water and P. There is extensive evidence of genotypic variation in the number and growth angle of root axes in many crop species. In wheat, the root system architecture is closely linked to the angle of seminal root axes expressed at the seedling stage. Wheat genotypes exhibiting a narrow seminal root angle are likely to develop a compact and deep root system, whereas those with wider angle of seminal axes tend to form a broad, shallow root system. Similar relationships between gravitropic growth of the root axes and spatial deployment of root systems have been reported in barley, common bean, maize, rice, and sorghum.

Selection for growth angle and number of root axes may help to identify genotypes with a root system architecture that is better adapted to specific drought and nutrient-deficient conditions.

Given the existence of genetic variability, expression at a very early stage of crop development, and recent advances in developing relatively simple, rapid, and cost-effective techniques for screening large number of plants, the number and angular distribution of root axes appear to represent promising traits that can potentially be exploited in crop improvement programs. However, due to plastic root responses to multiple soil stresses and the complex interactions between root and shoot growth, the high-throughput screening systems for seedling root characteristics need to be combined with more in-depth examination of root system growth and architecture under field conditions.

The advantage of the architectural root traits must be interpreted in the context of the type of environment in which the crops are grown. Developing a compact and deep root system to increase access to water and N from the deeper soil layers during the post-anthesis phase appears to be desirable in low-rainfall environments with a summer-dominant rainfall pattern, where crops are grown largely on water stored in the subsoil with a substantial risk that water may run out before the completion of grain filling. In contrast, possessing a large and shallow root system with greater potential for water extraction early in the season in order to reduce unproductive soil evaporation is more suitable for winter-rainfall-dominant Mediterranean environments. Optimizing root architecture to increase the acquisition efficiency for one soil resource may incur trade-offs for the acquisition of other resources. For instance, root architectures that are more efficient in exploiting topsoil resources may be advantageous in soils with low P bioavailability, but they may not be adapted to drought environments due to their reduced access to water stored in subsoil. The co-optimization of root architecture for the acquisition of multiple soil resources requires further research.

Characterization of the crop environment is an important step in evaluating the implications of desirable root traits for improving water and nutrient acquisition and crop productivity. This is particularly important for drought adaptation, as crops grown in water-limited environments may experience significant inter-seasonal variation in timing and severity of drought stress. Crop simulation models linked to long-term weather data provide a valuable tool for environmental characterization and quantitative assessment of the impact of specific root traits on resource use efficiency and crop yield in a much larger sample of environments than is possible experimentally.

Considering the complex interactions between root and shoot growth, it is evident that improving water and nutrient use efficiency and crop yield will require a whole plant growth and functioning approach. It is also important to note that enhancing the acquisition and utilization of soil resources requires not only genotypes with improved root and shoots traits but also agronomic practices that aim to maximize resource availability. Thus, successful exploitation of genetic variability in root architectural traits to increase water and nutrient use efficiency in crop plants will greatly depend on the collaborations between, among others, crop physiologists, modelers, breeders, geneticists, soil scientists, and agronomists.

References

Adesemoye AO, Kloepper JW. 2009. Plant–microbes interactions in enhanced fertilizer-use efficiency. *Appl Microbiol Biotechnol* 85:1–12.

Angadi SV, Entz MH. 2002. Root system and water use patterns of different height sunflower cultivars. *Agron J* 94:136–145.

Ao J, Fu J, Tian J, Yan X, Liao H. 2010. Genetic variability for root morph-architecture traits and root growth dynamics as related to phosphorus efficiency in soybean. *Funct Plant Biol* 37:304–312.

Asseng S, Ritchie JT, Smucker AJM, Robertson MJ. 1998. Root growth and water uptake during water deficit and recovering in wheat. *Plant Soil* 201:265–273.

Barraclough PB, Howarth JR, Jones J, Lopez-Bellido R, Parmar S, Shepherd CE, Hawkesford MJ. 2010. Nitrogen efficiency of wheat: Genotypic and environmental variation and prospects for improvement. *Eur J Agron* 33:1–11.

Bengough AG, Gordon DC, Al-Menaie H et al. 2004. Gel observation chamber for rapid screening of root traits in cereal seedlings. *Plant Soil* 262:63–70.

Berntson GM. 1994. Root systems and fractals: How reliable are calculations of fractal dimensions? *Ann Bot* 73:281–284.

Berntson GM. 1996. Fractal geometry, scaling and the description of plant root architecture. In *Plant Roots: The Hidden Half*, eds. Y Waisel, A Eshel, U Kafkafi, 2nd edn., pp. 259–272. New York: Marcel Dekker, Inc.

Berntson GM. 1997. Topological scaling and plant root system architecture: Developmental and functional hierarchies. *New Phytol* 135:621–634.

Bertheau Y, Gianinazi-Pearson V, Gianinazi S. 1980. Development and expression of endomycorrhizal associations in wheat. I. Evidence of varietal effects. *Ann Amélioration Plantes* 30:67–78.

Bonser AM, Lynch J, Snapp S. 1996. Effect of phosphorus deficiency on growth angle of basal roots in *Phaseolus vulgaris*. *New Phytol* 132:281–288.

Bouma TJ, Nielsen KL, Hal JV, Koutstaal B. 2001. Root system topology and diameter distribution of species from habitats differing in inundation frequency. *Funct Ecol* 15:360–369.

Braun HJ, Rajaram S, van Ginkel M. 1996. CIMMYT's approach to breeding for wide adaptation. *Euphytica* 92:175–183.

Bucher M. 2007. Functional biology of plant phosphate uptake at root and mycorrhiza interfaces. *New Phytol* 173:11–26.

Burgess SSO, Pate JS, Adams MA, Dawson TE. 2000. Seasonal water acquisition and redistribution in the Australian woody phreatophyte, *Banksia prionotes*. *Ann Bot* 85:215–224.

Caldwell MM, Richards JH. 1989. Hydraulic lift: Water efflux from upper roots improves effectiveness of water uptake by deep roots. *Oecologia* 79:1–5.

Christopher JT, Manschadi AM, Borrell AK, Hammer GL. 2005. A stay-green wheat with high yield under rain fed conditions in sub tropical Australia. In *Proceedings of the Second International Conference on Integrated Approaches to Sustain and Improve Plant Production Under Drought Stress, InterDrought II*, pp. 209, Rome, Italy, September 24–28.

Christopher JT, Manschadi AM, Hammer GL, Borrell AK. 2008. Developmental and physiological traits associated with high yield and stay-green phenotype in wheat. *Aust J Agric Res* 59:354–364.

Cordell D, Drangert J-O, White S. 2009. The story of phosphorus: Global food security and food for thought. *Global Environ Change* 19:292–305.

de Dorlodot S, Forster B, Pagès L, Price A, Tuberosa R, Draye X. 2007. Root system architecture: Opportunities and constraints for genetic improvement of crops. *Trends Plant Sci* 12:474–481.

Diederichs C, Manske GGB. 1991. The role of VA mycorrhiza for crop nutrition in warmer regions. In *Wheat for the Nontraditional Warm Areas*, ed. DA Saunders, pp. 352–371. de México, Mexico: CIMMYT.

Doussan C, Pagès L, Pierret A. 2003. Soil exploration and resource acquisition by plant roots: An architectural and modelling point of view. *Agronomie* 23:419–431.

Dunbabin V, Diggle A, Rengel Z. 2003. Is there an optimal root architecture for nitrate capture in leaching environments? *Plant Cell Environ* 26:835–844.

Eghball B, Settimi JR, Maranville JW, Parkhurst AM. 1993. Fractal analysis for morphological description of corn roots under nitrogen stress. *Agron J* 85:287–289.

Eshel A. 1998. On the fractal dimensions of a root system. *Plant Cell Environ* 21:247–251.

Fageria NK, Baligar VC. 1999. Phosphorus-use efficiency in wheat genotypes. *J Plant Nutr* 22:331–340.

Fageria NK, Baligar VC. 2005. Enhancing nitrogen use efficiency in crop plants. *Adv Agron* 88:97–185.

Fageria NK, Baligar VC, Li YC. 2008. The role of nutrient efficient plants in improving crop yields in the twenty first century. *J Plant Nutr* 31:1121–1157.

Falkenmark M, Rockström J, Karlberg L. 2009. Present and future water requirements for feeding humanity. *Food Security* 1:59–69.

FAO. 2009. *How to Feed the World in 2050*. pp. 1–35, Rome, Italy: FAO.

Fischer RA, Rees D, Sayre KD, Lu Z-M, Condon AG, Saavedra AL. 1998. Wheat yield progress associated with higher stomatal conductance and photosynthetic rate, and cooler canopies. *Crop Sci* 38:1467–1475.

Fitter AH. 1987. An architectural approach to the comparative ecology of plant root systems. *New Phytol* 106:61–77.

Fitter AH. 2002. Characteristics and functions of root systems. In *Plant Roots: The Hidden Half*, eds. Y Waisel, A Eshel, U Kafkafi, 3rd edn., pp. 249–259. New York: Marcel Dekker, Inc.

Fitter AH, Stickland TR. 1991. Architectural analysis of plant root systems 2. Influence of nutrient supply on architecture in contrasting plant species. *New Phytol* 118:383–389.

Fitter AH, Stickland TR. 1992. Fractal characterization of root system architecture. *Funct Ecol* 6:632–635.

Fitter AH, Stickland TR, Harvey ML, Wilson GW. 1991. Architectural analysis of plant root systems 1. Architectural correlates of exploitation efficiency. *New Phytol* 118:375–382.

Föhse D, Claassen N, Jungk A. 1991. Phosphorus efficiency of plants. II. Significance of root radius, root hairs and cation-anion balance for phosphorus influx in seven plant species. *Plant Soil* 132:261–272.

Foulkes MJ, Hawkesford MJ, Barraclough PB, Holdsworth MJ, Kerr S, Kightley S, Shewry PR. 2009. Identifying traits to improve the nitrogen economy of wheat: Recent advances and future prospects. *Field Crops Res* 114:329–342.

French RJ, Schultz JE. 1984. Water use efficiency of wheat in a Mediterranean-type environment. I. The relation between yield, water use and climate. *Aust J Agric Res* 35:743–764.

Gahoonia TS, Care D, Nielsen NE. 1997. Root hairs and phosphorus acquisition of wheat and barley cultivars. *Plant Soil* 191:181–188.

Garnett T, Conn V, Kaiser BN. 2009. Root based approaches to improving nitrogen use efficiency in plants. *Plant Cell Environ* 32:1272–1283.

Ge Z, Rubio G, Lynch JP. 2000. The importance of root gravitropism for inter-root competition and phosphorus acquisition efficiency: Results from a geometric simulation model. *Plant Soil* 218:159–171.

Gildon A, Tinker PB. 1983. Interactions of vesicular-arbuscular mycorrhizal infection and heavy metals in plants. I. The effects of heavy metals on the development of vesicular-arbuscular mycorrhizas. *New Phytol* 95:247–261.

Giuliani S, Sanguineti MC, Tuberosa R, Bellotti M, Salvi S, Landi P. 2005. Root-ABA1: A major constitutive QTL, affects maize root architecture and leaf ABA concentration at different water regimes. *J Exp Bot* 56:3061–3070.

Godo GH, Reisenhauer HM. 1980. Plant effects on soil manganese availability. *Soil Sci Soc Am J* 44:993–995.

Gregory PJ. 1989. Concepts of water use efficiency. In *Soil and Crop Management for Improved WUE in Rainfed Areas. Proceedings of an International Workshop*, eds. HC Harris, PJM Cooper, P Pala, pp. 9–20. Ankara, Turkey, May 15–19.

Gregory PJ. 1994. Root growth and activity. In *Physiology and Determination of Crop Yield*, eds. KJ Boote, JM Bennett, TR Sinclair, GM Paulsen, pp. 65–93. Madison, WI: ASA, CSSA, and SSSA.

Gregory PJ. 2009. Root phenomics of crops: Opportunities and challenges. *Funct Plant Biol* 36:922–929.

Gregory PJ, McGowan M, Biscoe PV, Hunter B. 1978. Water relations of winter wheat. 1. Growth of the root system. *J Agric Sci* 91:91–102.

Gregory PJ, Tennant D, Belford RK. 1992. Root and shoot growth, and water and light use efficiency of barley and wheat crops grown on a shallow duplex soil in a Mediterranean-type environment. *Aust J Agric Res* 43:555–573.

Hammer GL, Dong Z, McLean G et al. 2009. Can changes in canopy and/or root system architecture explain historical maize yield trends in the U.S. corn belt? *Crop Sci* 49:299–312.

Hargreaves CE, Gregory PJ, Bengough AG. 2009. Measuring root traits in barley (*Hordeum vulgare* ssp. *vulgare* and ssp. *spontaneum*) seedlings using gel chambers, soil sacs and x-ray microtomography. *Plant Soil* 316:285–297.

Hayman DS, Mosse B. 1971. Plant growth response to vesicular-arbuscular mycorrhiza. 1. Growth of Endogone inoculated plants in phosphate deficient soils. *New Phytol* 70:19–27.

Helal HM. 1990. Varietal differences in root phosphatase activity as related to the utilization of organic phosphates. *Plant Soil* 123:161–163.

Hetrick BDA, Wilson GWT, Cox TS. 1992. Mycorrhizal dependence of modern wheat varieties, landraces and ancestors. *Can J Bot* 70:2032–2040.

Hirel B, Le Gouis J, Ney B, Gallais A. 2007. The challenge of improving nitrogen use efficiency in crop plants: Towards a more central role for genetic variability and quantitative genetics within integrated approaches. *J Exp Bot* 58:2369–2387.

Ho MD, Rosas JC, Brown KM, Lynch JP. 2005. Root architectural tradeoffs for water and phosphorus acquisition. *Funct Plant Biol* 32:737–748.

Hoad SP, Russell G, Lucas ME, Bingham IJ. 2001. The management of wheat, barley, and oat root systems. *Adv Agron* 74:193–246.

Hodge A. 2006. Plastic plants and patchy soils. *J Exp Bot* 57:401–411.

Hodge A, Berta G, Doussan C, Merchan F, Crespi M. 2009. Plant root growth, architecture and function. *Plant Soil* 321:153–187.

Hurd EA. 1968. Growth of roots of seven varieties of spring wheat at high and low moisture levels. *Agron J* 60:201–205.

Jones GPD, Blair GJ, Jessop RS. 1989. Phosphorus efficiency in wheat—A useful selection criteria. *Field Crops Res* 21:257–264.

Kapulnik Y, Kushnir U. 1991. Growth dependency of wild, primitive and modern cultivated wheat lines on vesicular-arbuscular mycorrhizal fungi. *Euphytica* 56:27–36.

Kato Y, Abe J, Kamoshita A, Yamagishi J. 2006. Genotypic variation in root growth angle in rice (*Oryza sativa* L.) and its association with deep root development in upland fields with different water regimes. *Plant Soil* 287:117–129.

Keating BA, Carberry PS, Hammer GL, Probert ME, Robertson MJ, Holzworth D, Huth NI, Hargreaves JNG, Meinke H, Hochman Z. 2003. An overview of APSIM, a model designed for farming systems simulation. *Eur J Agron* 18:267–288.

Kirkegaard JA, Lilley JM, Howe GN, Graham JM. 2007. Impact of subsoil water use on wheat yield. *Aust J Agric Res* 58:303–315.

Klepper B. 1984. Response of the wheat plant to non-animal stresses. In *Proceedings of National Wheat Pasture Symposium*, ed. GW Horn, pp. 235–243. Misc. Publ. 115 of the Oklahoma Agric. Exp. Sta., Stillwater, OK.

Klepper B. 1991. Root-shoot relationships. In *Plant Roots: The Hidden Half*, eds. Y Waisel, A Eshel, U Kafkafi, pp. 265–286. New York: Marcel Dekker, Inc.

Klepper B, Belford RK, Rickman RW 1984 Root and shoot development in winter wheat. *Agron J* 76:117–122.

Kutschera L. 1960. *Wurzelatlas Mitteleuropaeischer Ackerunkraeuter und Kulturpflanzen*, Frankfurt am Main, Germany: DLG Verlag.

Lambers H, Shane MW, Cramer MD, Pearse SJ, Veneclaas EJ. 2006. Root structure and functioning for efficient acquisition of phosphorus: Matching morphological and physiological traits. *Ann Bot* 98:693–713.

Le Bot J, Serra V, Fabre J, Draye X, Adamowicz S, Pagès L. 2010. DART: A software to analyse root system architecture and development from captured images. *Plant Soil* 326:261–273.

Le Gouis J, Béghin D, Heumez E, Pluchard P. 2000. Genetic differences for nitrogen uptake and nitrogen utilisation efficiencies in winter wheat. *Eur J Agron* 12:163–173.

Liao M, Fillery IRP, Palta JA. 2004a. Early vigorous growth is a major factor influencing nitrogen uptake in wheat. *Funct Plant Biol* 31:121–129.

Liao M, Palta JA, Fillery IRP. 2006. Root characteristics of vigorous wheat improve early nitrogen uptake. *Aust J Agric Res* 57:1097–1107.

Liao H, Rubio G, Yan X, Cao A, Brown KM, Lynch JP. 2001. Effect of phosphorus availability on basal root shallowness in common bean. *Plant Soil* 232:69–79.

Liao H, Yan X, Rubio G, Beebe SE, Blair MW, Lynch JP. 2004b. Genetic mapping of basal root gravitropism and phosphorus acquisition efficiency in common bean. *Funct Plant Biol* 31:959–970.

Lopes MS, Reynolds MP. 2010. Partitioning of assimilates to deeper roots is associated with cooler canopies and increased yield under drought in wheat. *Funct Plant Biol* 37:147–156.

López-Castañeda C, Richards RA. 1994. Variation in temperate cereals in rain-fed environments. I. Grain yield, biomass and agronomic characteristics. *Field Crops Res* 37:51–62.

Ludlow MM, Muchow RC. 1990. A critical evaluation of traits for improving crop yields in water-limited environments. *Adv Agron* 43:107–153.

Lynch JP. 1995. Root architecture and plant productivity. *Plant Physiol* 109:7–13.

Lynch JP. 2007. Roots of the second green revolution. *Aust J Bot* 55:493–512.

Lynch JP, Brown KM. 2001. Topsoil foraging—An architectural adaptation of plants to low phosphorus availability. *Plant Soil* 237:225–237.

Lynch JP, Ho MD. 2005. Rhizoeconomics: Carbon costs of phosphorus acquisition. *Plant Soil* 269:45–56.

Lynch JP, van Beem JJ. 1993. Growth and architecture of seedling roots of common bean genotypes. *Crop Sci* 33:1253–1257.

Malamy JE. 2005. Intrinsic and environmental response pathways that regulate root system architecture. *Plant Cell Environ* 28:67–77.

Mandelbrot BB. 1983. *The Fractal Geometry of Nature*, New York: W.H. Freeman.

Manschadi AM, Christopher J, deVoil P, Hammer GL. 2006. The role of root architectural traits in adaptation of wheat to water-limited environments. *Funct Plant Biol* 33:823–837.

Manschadi AM, Hammer GL, Christopher JT, deVoil P. 2008. Genotypic variation in seedling root architectural traits and implications for drought adaptation in wheat (*Triticum aestivum* L.). *Plant Soil* 303:115–129.

Manschadi AM, Sauerborn J, Stützel H, Göbel W, Saxena MC. 1998. Simulation of faba bean (*Vicia faba* L.) root system development under Mediterranean conditions. *Eur J Agron* 9:259–272.

Manske GGB. 1989. The efficiency of the inoculation by the VA mycorrhizal fungi *Glomus manihotis* in spring wheat genotypes and its inheritance in F1- and R1-generations at different phosphate forms applied and different weather conditions. PhD dissertation in German, with English abstract, University of Göttingen, Göttingen, Germany.

Manske GGB. 1990. Genetical analysis of the efficiency of VA Mycorrhiza with spring wheat. I. Genotypical differences and reciprocal cross between an efficient and non-efficient variety. In *Genetic Aspects of Plant Mineral Nutrition*, eds. N El Bassam, M Dambroth, BC Loughman, pp. 397–405. Dordrecht, the Netherlands: Kluwer.

Manske GGB. 1997. Utilization of the genotypic variability of VAM-symbiosis and root length density in breeding phosphorus efficient wheat cultivars at CIMMYT. Final report of a special project. pp. 304, de México, Mexico: CIMMYT.

Manske GGB, Lüttger AB, Behl RK, Vlek PLG. 1995. Nutrient efficiency based on VA mycorrhizae and total root length of wheat cultivars grown in India. *Angew Bot* 69:108–110.

Manske GGB, Ortiz-Monasterio JI, van Ginkel M, González RM, Fischer RA, Rajaram S, Vlek PLG. 2001. Importance of P uptake efficiency versus P utilization for wheat yield in acid and calcareous soils in Mexico. *Eur J Agron* 14:261–274.

Manske GGB, Ortiz-Monasterio JI, Van Ginkel M, González RM, Rajaram S, Molina E, Vlek PLG. 2000. Traits associated with improved P-uptake efficiency in CIMMYT's semidwarf spring bread wheat grown on an acid Andisol in Mexico. *Plant Soil* 221:189–204.

Manske GGB, Ortiz-Monasterio JI, van Ginkel M, Gonzalez R, Vlek PLG. 1996. Phosphorus uptake, utilization efficiency and grain yield of semidwarf wheat grown in acid or alkaline, P deficient soils. In *Fifth International Wheat Conference*, Ankara, Turkey, June 10–14.

Manske GGB, Ortiz-Monasterio JI, van Ginkel RM, Rajaram S, Vlek PLG. 2002. Phosphorus use efficiency in tall, semi-dwarf and dwarf near-isogenic lines of spring wheat. *Euphytica* 125:113–119.

Manske GGB, Vlek PLG. 2002. Root architecture—Wheat as a model plant. In *Plant Roots: The Hidden Half*, eds. Y Waisel, A Eshel, U Kafkafi, 3rd edn., pp. 249–259. New York: Marcel Dekker, Inc.

Marschner H, Kissel M, Römheld V. 1986. Different strategies in higher plants in mobilization and uptake of iron. *J Plant Nutr* 9:695–713.

Masi CEA, Maranville JW. 1998. Evaluation of sorghum root branching using fractals. *J Agric Sci* 131:259–265.

McLachlan KD. 1980. Acid phosphatase activity of intact roots and phosphorus nutrition of plants. I. Assay conditions and phosphatase activity. *Aust J Agric Res* 31:429–440.

McPhee K. 2005. Variation for seedling root architecture in the core collection of pea germplasm. *Crop Sci* 45:1758–1763.

Miller CR, Ochoa I, Nielsen KL, Beck D, Lynch JP. 2003. Genetic variation for adventitious rooting in response to low phosphorus availability: Potential utility for phosphorus acquisition from stratified soils. *Funct Plant Biol* 30:973–985.

Mishra HS, Rathore TR, Tomar VS. 1999. Root growth, water potential and yield of irrigated wheat. *Irrig Sci* 18:117–123.

Moll RH, Kamprath EJ, Jackson WA. 1982. Analysis and interpretation of factors which contribute to efficiency and nitrogen utilization. *Agron J* 74:562–564.

Nakamoto T, Oyanagi A. 1994. The direction of growth of seminal roots of *Triticum aestivum* L. and experimental modification thereof. *Ann Bot* 73:363–367.

Neumann G, Römheld V. 1999. Root excretion of carboxylic acids and protons in phosphorus-deficient plants. *Plant Soil* 211:121–130.

Neumann G, Römheld V. 2002. Root-induced changes in the availability of nutrients in the rhizosphere. In *Plant Roots: The Hidden Half*, eds. Y Waisel, A Eshel, U Kafkafi, 3rd edn., pp. 617–649. New York: Marcel Dekker, Inc.

Nielsen KL, Eshel A, Lynch JP. 2001. The effect of phosphorus availability on the carbon economy of contrasting common bean (*Phaseolus vulgaris* L.) genotypes. *J Exp Bot* 52:329–339.

Nielsen K, Lynch J, Weiss H. 1997. Fractal geometry of bean root systems: Correlations between spatial and fractal dimension. *Am J Bot* 84:26–33.

Nielsen KL, Miller CR, Beck D, Lynch JP. 1999. Fractal geometry of root systems: Field observations of contrasting genotypes of common bean (*Phaseolus vulgaris* L.) grown under different phosphorus regimes. *Plant Soil* 206:181–190.

Nix HA. 1975. The Australian climate and its effect on grain yield and quality. In *Australian Field Crops, Wheat and Other Temperate Cereals*, eds. A Lazenby, EM Matheson, pp. 183–226. Sydney, Australia: Angus and Robertson.

Norton GJ, Price AH. 2009. Mapping of quantitative trait loci for seminal root morphology and gravitropic response in rice. *Euphytica* 166:229–237.

O'Brien L. 1979. Genetic variability of root growth in wheat (*Triticum aestivum* L.). *Aust J Agric Res* 30:587–595.

Ortiz R, Sayre KD, Govaerts B et al. 2008. Climate change: Can wheat beat the heat? *Agric Ecosyst Environ* 126:46–58.

Osborne LD, Rengel Z. 2002. Genotypic differences in wheat for uptake and utilisation of P from iron phosphate. *Aust J Agric Res* 53:837–844.

O'Toole JC, Bland WL. 1987. Genotypic variation in crop plant root systems. *Adv Agron* 41:91–145.

Oyanagi A. 1994. Gravitropic response growth angle and vertical distribution of roots of wheat (*Triticum aestivum* L.). *Plant Soil* 165:323–326.

Ozturk L, Eker S, Torun B, Cakmak I. 2005. Variation in phosphorus efficiency among 73 bread and durum wheat genotypes grown in a phosphorus-deficient calcareous soil. *Plant Soil* 269:69–80.

Palta JA, Fillery IRP, Rebetzke GJ. 2007. Restricted-tillering wheat does not lead to greater investment in roots and early nitrogen uptake. *Field Crops Res* 104:52–59.

Pandey R, Singh B, Nair TVR. 2006. Effect of arbuscular-mycorrhizal inoculation in low phosphorus soil in relation to P-utilization efficiency of wheat (*Triticum aestivum*) genotypes. *Indian J Agric Sci* 76:349–353.

Passioura JB. 1983. Roots and drought resistance. *Agric Water Manage* 7:265–280.

Passioura J, Angus JF. 2010. Improving productivity of crops in water-limited environments. *Adv Agron* 106:37–75.

Poorter H, Nagel O. 2000. The role of biomass allocation in the growth response of plants to different levels of light, CO_2; nutrients and water: A quantitative review. *Aust J Plant Physiol* 27:595–607.

Portilla CI, Molina GE, Cruz-Flores G, Ortiz-Monasterio I, Manske GGB. 1998. Colonizacion micorrizica arbuscular, actividad fosfatasica y longitud radical como respuesta a estres de fosforo en trigo y triticale cultivados en un andisol. *Terra* 16:55–60.

Price AH, Steele KA, Gorham J et al. 2002. Upland rice grown in soil-filled chambers and exposed to contrasting water-deficit regimes: I. Root distribution, water use and plant water status. *Field Crops Res* 76:11–24.

Ramaekers L, Remans R, Rao IM, Blair MW, Vanderleyden J. 2010. Strategies for improving phosphorus acquisition efficiency of crop plants. *Field Crops Res* 117:169–176.

Raun WR, Johnson GV. 1999. Improving nitrogen use efficiency for cereal production. *Agron J* 91:357–363.

Rebetzke GJ, Richards RA. 1999. Genetic improvement of early vigour in wheat. *Aust J Agric Res* 50:291–302.

Rengel Z, Marschner P. 2005. Nutrient availability and management in the rhizosphere: Exploiting genotypic differences. *New Phytol* 168:305–312.

Reynolds M, Dreccer F, Trethowan R. 2007. Drought-adaptive traits derived from wheat wild relatives and landraces. *J Exp Bot* 58:177–186.

Reynolds MP, Rajaram S, Sayre KD. 1999. Physiological and genetic changes of irrigated wheat in the post-green revolution period and approaches for meeting projected global demand. *Crop Sci* 39:1611–1621.

Richards RA, Passioura JB. 1989. A breeding program to reduce the diameter of the major xylem vessel in the seminal roots of wheat and its effect on grain yield in rain-fed environments. *Aust J Agric Res* 40:943–950.

Richards RA, Rebetzke GJ, Condon AG, van Herwaarden AF. 2002. Breeding opportunities for increasing the efficiency of water use and crop yield in temperate Cereals. *Crop Sci* 42:111–121.

Richards RA, Rebetzke GJ, Watt M, Condon AGT, Spielmeyer W, Dolferus R. 2010. Breeding for improved water productivity in temperate cereals: Phenotyping, quantitative trait loci, markers and the selection environment. *Funct Plant Biol* 37:85–97.

Sadras VO, Angus JF. 2006. Benchmarking, water-use efficiency of rainfed wheat in dry environments. *Aust J Agric Res* 57:847–856.

Sawers RJH, Gebreselassie MN, Janos DP, Paszkowski U. 2010. Characterizing variation in mycorrhiza effect among diverse plant varieties. *Theor Appl Genet* 120:1029–1039.

Siddique KHM, Belford RK, Tennant D. 1990. Root: Shoot ratios of old and modern, tall and semi-dwarf wheats in a Mediterranean environment. *Plant Soil* 121:89–98.

Sinclair TR, Purcell LC, Sneller CH. 2004. Crop transformation and the challenge to increase yield potential. *Trends Plant Sci* 9:70–75.

Singh V, van Oosterom EJ, Jordan DR, Messina CD, Cooper M, Hammer GL. 2010. Morphological and architectural development of root systems in sorghum and maize. *Plant Soil* 333:287–299.

Steen I. 1998. Phosphorus availability in the 21st century: Management of non-renewable resource. *Phosphorus Potassium* 217:25–31.

Su JY, Zheng Q, Li HW, Li B, Jin RL, Tong YP, Li ZS. 2009. Detection of QTLs for phosphorus use efficiency in relation to agronomic performance of wheat grown under phosphorus sufficient and limited conditions. *Plant Sci* 176:824–836.

Swaminathan K, Verma BC. 1979. Reponses of three crop species to vesicular-arbuscular mycorrhizal infection on zinc-deficient Indian soils. *New Phytol* 82:481–487.

Sylvester-Bradley R, Kindred DR. 2009. Analysing nitrogen responses of cereals to prioritize routes to the improvement of nitrogen use efficiency. *J Exp Bot* 60:1939–1951.

Tabatabai MA, Bremner JM. 1969. Use of p-nitrophenylphosphate for assay of soil phosphatase activity. *Soil Biol Biochem* 1:301–307.

Tatsumi J, Yamauchi A, Kono Y. 1989. Fractal analysis of plant root systems. *Ann Bot* 64:499–503.

Tilman D, Cassman KG, Matson PA, Naylor R, Polasky S. 2002. Agricultural sustainability and intensive production practices. *Nature* 418:671–677.

Valizadeh GR, Rengel Z, Rate AW. 2003. Response of wheat genotypes efficient in P utilisation and genotypes responsive to P fertilisation to different P banding depths and watering regimes. *Aust J Agric Res* 54:59–65.

Vlek PLG, Kühne RF, Denich M. 1997. Nutrient resources for crop production in the tropics. *Philos Trans R Soc London, Ser B* 352:975–985.

Vlek PLG, Lüttger AB, Manske GGB. 1996. The potential contribution of arbuscular mycorrhiza to the development of nutrient and water efficient wheat. In *The Ninth Regional Wheat Workshop for Eastern, Central and Southern Africa*, eds. DG Tanner, TS Payne, OS Abdalla, pp. 28–46. Addis Ababa, Ethiopia: CIMMYT, October 2–6.

Waines JG, Ehdai B. 2007. Domestication and crop physiology: Roots of green-revolution wheat. *Ann Bot* 100:991–998.

Walk TC, Jaramillo R, Lynch JP. 2006. Architectural tradeoffs between adventitious and basal roots for phosphorus acquisition. *Plant Soil* 279:347–366.

Wang L, Chen F, Zhang F, Mi G. 2010a. Two strategies for achieving higher yield under phosphorus deficiency in winter wheat grown in field conditions. *Field Crops Res* 118:36–42.

Wang X, Shen J, Liao H. 2010b. Acquisition or utilization, which is more critical for enhancing phosphorus efficiency in modern crops? *Plant Sci* 179:302–306.

Wang X, Yan X, Liao H. 2010c. Genetic improvement for phosphorus efficiency in soybean: A radical approach. *Ann Bot* 106:215–222.

Watt M, Magee LJ, McCully ME. 2008. Types, structure and potential for axial water flow in the deepest roots of field-grown cereals. *New Phytol* 178:135–146.

Weaver JE. 1926. *Root Development of Field Crops*, New York: McGraw-Hill Book Company.

Xue Q, Zhu Z, Musick JT, Stewart BA, Dusek DA. 2003. Root growth and water uptake in winter wheat under deficit irrigation. *Plant Soil* 257:151–161.

Yazdanbakhsh N, Fisahn J. 2009. High throughout phenotyping of root growth dynamics, lateral root formation, root architecture and root hair development enabled by PlaRoM. *Funct Plant Biol* 36:938–946.

Yücel C, Özkan H, Ortaş I, Yağbasanlar T. 2009. Screening of wild emmer wheat accessions (*Triticum turgidum* subsp. *dicoccoides*) for mycorrhizal dependency. *Turk J Agric For* 33:513–523.

Zhu J, Lynch JP. 2004. The contribution of lateral rooting to phosphorus acquisition efficiency in maize (*Zea mays*) seedlings. *Funct Plant Biol* 31:949–958.

23

Root pH Regulation

Jóska Gerendás
K+S KALI GmbH

R. George Ratcliffe
University of Oxford

I. pH Values and Their Measurement

Proton concentrations are major determinants of cell function, in part because the properties of many biological molecules are pH dependent, but also because protons are directly involved in cellular processes. Thus, many proteins have pH-dependent conformations, leading to pH-dependent enzyme activities, and indeed the pH at which a protein is most stable correlates with the pH of its subcellular location (Chan and Warwicker 2009). Moreover, cellular processes such as metabolism and transport are necessarily pH dependent if they require protons or hydroxide ions as substrates. It follows that uncontrolled variations in pH might be disruptive, implying the need to regulate intracellular pH values for optimal cell function. The set point for regulation is likely to vary between subcellular compartments; indeed, differences in pH across membranes play a crucial role in both bioenergetics and nutrient uptake, and the preferred value for a specific compartment may also vary according to the physiological state of the tissue. Moreover, the finite capacity of the regulation system and/or its response time may permit transient changes in pH that can then be perceived as signals by the perturbed cell.

Given the central role of the proton in biology, it is not surprising that considerable efforts have been directed toward measuring intracellular pH values, to analyzing pH regulation, and to testing the significance of the changes that occur in particular circumstances. Measurements in and around roots established many years ago that soil pH usually falls in the range 3–9 (Brady and Weil 2008), that the root apoplast (Grignon and Sentenac 1991; Yu et al. 2000) and the root vacuoles (Guern et al. 1991) are typically mildly acidic, and that the pH of the root cytoplasm is usually in the range 7.2–7.6 (Guern et al. 1991). Microelectrodes, fluorescence probes, and nuclear magnetic resonance (NMR) spectroscopy are the principal measurement techniques that are used to measure intracellular and extracellular pH values. These methods have been crucial in exploring the fundamental role of pH in plant biology (Rengel 2002), and a brief survey of their present status in root research is given in the following paragraphs.

A. Microelectrodes

Microelectrodes allow fast, real-time monitoring of the pH in the cytoplasm (pH_{cyt}) and vacuole (pH_{vac}) of impaled cells (Felle 1993), as well as pH measurements of the apoplast (Felle 1998; Yu et al. 2000) and the rhizosphere (Zieschang et al. 1993). Microelectrode measurements of intracellular pH are unavoidably invasive, but the method has the advantage that it provides simultaneous measurements of the functionally related membrane potential, and by using a triple barreled electrode, it is possible to make simultaneous measurements of both pH and other ions, for example, pH and potassium, or pH and nitrate (Carden et al. 2003). In fact, most recent microelectrode studies on roots focus on these other ions, rather than on protons, with the pH measurements providing evidence for the cytoplasmic or vacuolar location of the electrode (Miller et al. 2001).

Microelectrodes are particularly suitable for measuring pH gradients in the rhizosphere, either with electrode arrays (Fischer et al. 1989) or by making multiple measurements with stationary electrodes (Newman et al. 1987) or by using ion-selective vibrating electrodes (Kochian et al. 1992). Vibrating electrodes have considerable advantages over their stationary counterparts because of better sensitivity and temporal resolution. As well as providing pH maps around the root surface, spatially resolved pH measurements in the rhizosphere can be used to quantify H^+ fluxes into and out of roots (Newman et al. 1987; Miller et al. 1991; Kochian et al. 1992; Shabala et al. 1997; Monshausen et al. 2007; Sun et al. 2009; see also Chapter 43). A further option is to combine microelectrode measurements with colorimetric or fluorescence methods in which the roots are grown in an agar medium containing a colorimetric pH indicator such as bromocresol purple (Marschner and Römheld 1983; Gollany and Schumacher 1993) or a pH-sensitive fluorescent dye such as fluorescein dextran (Monshausen et al. 2007). Colorimetric methods provide accurate measurements of the pH gradients that develop around roots (Jaillard et al. 1996), and there is good agreement between the H^+ fluxes calculated from colorimetric and potentiometric data (Plassard et al. 1999).

B. Fluorescent Probes

Probes with pH-dependent spectral properties that report directly from specific intracellular locations have many advantages for pH determination in vivo (Khramtsov 2005; Han and Burgess 2010). The use of pH-dependent fluorescent probes, whether dyes or proteins, for the measurement of extracellular and intracellular pH values in roots is particularly attractive because of the high sensitivity and high spatial resolution with which the probes can be detected (Swanson et al. 2011). The so-called ratiometric method depends on introducing a fluorescent probe with a suitable pH dependence into the region or compartment of interest and then deducing the pH from the intensity ratio of the fluorescence observed either at two excitation frequencies (e.g., 2′,7′-bis-(2-carboxyethyl)-5-(and 6-)carboxy fluorescein, BCECF) or at two emission frequencies (e.g., SNARF-1). It is also possible to use a pseudoratiometric method in which the pH is deduced from the fluorescence of two separate probes, one that is pH sensitive and one that is not (Swanson et al. 2011).

Fluorescent dyes can be introduced by microinjection into specific cells, for example, into root hairs, or by incubating the roots with the dye or a more lipophilic precursor. The latter approach is preferable, since it is noninvasive and easy to implement, but it can be difficult to control and this may lead to uncertainty about the location of the dye. For example, the commonly used pH probe BCECF is often loaded by incubating roots with the acetomethyl ester (BCECF-AM). This procedure can lead to cytoplasmic and vacuolar pools of BCECF (Brauer et al. 1996), and it may be difficult to distinguish between them. Short loading times (Plieth et al. 1999) or microinjection with dextran-linked BCECF (Scott and Allen 1999) minimize the sequestration of the dye into the vacuole. Calibration curves are usually based on in situ measurements, to avoid the uncertainty of inferring pH from an in vitro calibration that may not reflect the intracellular environment, although sometimes this procedure may be difficult to implement successfully (Bibikova et al. 1998).

As an alternative to fluorescent dyes, pH-sensitive variants of green fluorescent protein (GFP) can provide convenient pH sensors in transformable systems (Moseyko and Feldman 2001; Swanson et al. 2011). An important advantage of this approach is that the GFP probes can be targeted to specific subcellular locations, although currently these probes are only used to measure apoplastic and cytosolic pH values in roots. For example, pHluorins have been targeted to the apoplast by fusing a secretory sequence to the gene, allowing measurements of the apoplastic pH in *Arabidopsis* roots (Gao et al. 2004). Similarly, the H148D variant of GFP can be used to monitor the dynamics of cytosolic pH in *Arabidopsis* root hairs in response to mechanical stimulation and growth (Monshausen et al. 2007, 2009, 2011). Using high-resolution confocal laser scanning microscopy, images can be collected on a timescale of a few seconds allowing excellent temporal resolution for analyzing the dynamics of the observed pH changes. This can lead to prodigious quantities of data, which in turn are driving the development of automated techniques for extracting the pH information from such images (Monshausen et al. 2011).

Fluorescence measurements can also be used to measure pH in the rhizosphere using pH optodes (Blossfeld and Gansert 2007; Blossfeld et al. 2010). This noninvasive method depends on the measurement of the fluorescence decay time of pH-sensitive dyes incorporated into a sensor that is fixed to the internal surface of a transparent rhizotron. This allows the measurement of a 2D array of pH values from a plane in the rhizosphere defined by the position of the sensor. The method is particularly suitable for long-term measurements of the spatial and temporal dynamics of pH at the root–soil interface because of the stability of the system over timescales of weeks (Blossfeld and Gansert 2007).

C. NMR Spectroscopy

Like the fluorescence method, in vivo NMR measurements of pH depend on the detection of tissue NMR signals with pH-dependent properties (Ratcliffe 1994; Khramtsov 2005). In its most widely applied form, ^{31}P NMR is used to detect the cytoplasmic and vacuolar pools of inorganic phosphate (P_i), and the position of the corresponding signals in the NMR spectrum, that is, the chemical shift of each signal, is used to deduce pH_{cyt} and pH_{vac} on the basis of a suitable pH calibration curve. Although the effective pK_a of P_i at 6.8 is rather too high for the accurate determination of acidic pH values, NMR has been used extensively for measurements of both pH_{cyt} and pH_{vac} in roots and other plant tissues (Ratcliffe et al. 2001). Analysis of exceptionally well-resolved ^{31}P NMR spectra from *Acer pseudoplatanus* L. cell suspensions suggests that the cytoplasmic P_i signal can be split into a cytosolic signal, corresponding to a pH of 7.4, and an organellar signal from the mitochondria and plastids, corresponding to a pH of 7.55 (Pratt et al. 2009). Recording NMR spectra at the highest available field strength should make it easier to resolve the cytosolic P_i signal, but comparable observations have yet to be reported for roots, perhaps because of the cellular heterogeneity within the root tissues that are used for NMR measurements.

In vivo NMR measurements of pH can sometimes be made from pH-dependent signals of metabolites that do not contain a phosphate group. For example, the lower pK_a values of the organic acids make them more sensitive probes for acidic pH values than P_i, allowing pH_{vac} to be measured from the malate signals observed in the ^{13}C NMR spectra of ^{13}C-labeled roots (Chang and Roberts 1989). Similarly, it has been possible to exploit the pH dependence of the fine structure of the ^{1}H-coupled ^{14}N NMR signal of ammonium to measure a vacuolar pH of 3.9 in the roots of Norway spruce (*Picea abies* (L.) Karst.) (Aarnes et al. 2007). Alternatively, pH values can be determined from the NMR signals of exogenous pH probes that are readily taken up by root tissues, for example, methyl phosphonate (Couldwell et al. 2009; Pratt et al. 2009), detectable by ^{31}P NMR, and α-methylfluorinated alanines (Pfeffer et al. 2004), detectable by ^{19}F NMR. The use of trifluoromethyl alanine in maize roots revealed a lower pH_{vac} (4.6) than commonly reported from ^{31}P NMR measurements, as well as considerable heterogeneity, reflected in the width of the in vivo ^{19}F NMR signal.

Although versatile, the temporal resolution of NMR measurements of intracellular pH is generally poorer than that achieved with fluorescence and microelectrodes, and little, if any, spatial information can be deduced at the cellular level from the usual spectroscopic measurements. In principle, NMR imaging, which has found a number of specialized applications in plants (Köckenberger et al. 2004; Van As 2007), might be able to solve this problem, since spatially resolved pH measurements have been performed with NMR imaging in animal models (Khramtsov 2005), but the technical complexity of this approach to pH measurement probably explains why it has not been applied to roots or indeed other plant tissues. In fact, the literature suggests that NMR methods are being used less frequently for pH measurements in root tissues than previously, despite the metabolic information that NMR often provides simultaneously with the pH measurement, and it seems that the incentive to develop spatially resolved NMR measurements of pH has been diminished by the increasing power of the fluorescence approach.

II. Factors Influencing pH

It has been known for many years that the pH values in and around roots are not necessarily constant, and changes in pH are frequently observed. Many of these effects have been described in detail, and it is convenient to distinguish the changes in pH that occur during normal growth and development, from those that arise under adverse growth conditions.

A. pH Changes during Normal Growth and Development

The marked shift in metabolism that occurs following the onset of illumination causes several well-characterized pH changes in leaves, including acidification of the thylakoid lumen and alkalinization of the chloroplast stroma in the light (Heldt et al. 1973; Kramer et al. 1999) and acidification of the vacuole in the leaves of CAM plants in the dark (Stidham et al. 1983). In contrast, most roots find themselves in a more stable environment, and pH fluctuations due to diurnal changes in the immediate environment of a root are less likely.

In fact, a more important influence on the pH values in and around roots is growth itself, since it is commonly observed that hydroponic growth on nitrate tends to alkalinize the growth medium, whereas supplying ammonium leads to acidification (Kirkby and Mengel 1967; Marschner 2012). The outcome depends on the mineral nutrients available to the plant, and the impact of the nitrogen source reflects in part the proton balance for the assimilation of nitrate and ammonium into amino acids (Raven and Smith 1976; Raven 1986; Raven and Wollenweber 1992; Britto and Kronzucker 2005). A similar effect can be observed in the rhizosphere (Marschner and Römheld 1983), although nitrogen nutrition is only one of many factors that can influence rhizosphere pH (Hinsinger et al. 2003), and it is superimposed on a spatial variation along the root axis that tends to ensure that the pH at the root tip is relatively alkaline (Fischer et al. 1989; Miller et al. 1991; Taylor and Bloom 1998). However, although rather large pH differences, typically 1–2 pH units, can develop in the growth medium or the rhizosphere as a result of differences in the nitrogen supply, only small differences are observed between the intracellular pH values for the roots of nitrate- and ammonium-grown plants (Gerendás et al. 1990).

The growth-related pH changes that occur in the rhizosphere can have a profound effect on nutrient availability (Marschner 2012). Thus, the acidification of the rhizosphere and root apoplast that accompanies growth on ammonium, or to a lesser extent nitrogen fixation, favors the uptake of a range of nutrients, including both macronutrients, such as phosphate (Ding et al. 2011), and minor nutrients, such as boron, iron, and manganese. Moreover, a shortage of certain nutrients can trigger specific pH changes, as in the acidification of the rhizosphere that occurs when organic acids are released in response to a shortage of phosphorus (Neumann and Römheld 1999) and the acidification of the rhizosphere (Santi and Schmidt 2009) and the apoplast (Kosegarten et al. 2004) that occurs in response to iron deficiency. The magnitudes of the resulting pH changes are likely to be very dependent on the conditions experienced by the plant, and it is more difficult, for example, to reduce the apoplastic and rhizosphere pH values in high-carbonate soils (Hauter and Mengel 1988). In fact, the optimum rhizosphere pH for growth is a function of the soil type and the nutrient requirements of the plant, and the increased availability of certain nutrients at acidic pH values may also lead to toxic levels for others. Thus, iron and manganese toxicity can be a significant problem in acid soils (Gupta and Gupta 1998), whereas on other soil types it may be advantageous to use an ammonium fertilizer to increase the availability of the same nutrients (Cummings and Xie 1995; Malhi et al. 2000). The availability of aluminum also increases at acidic pH values, and aluminum toxicity is the most important constraint on plant growth in acidic soils (Liao et al. 2006; see also Chapter 33). Growth-related alkalinization of the rhizosphere also occurs, and this can be important in conferring resistance to toxic levels of aluminum (Degenhardt et al. 1998) and in determining the availability of copper (Bravin et al. 2009).

The normal growth and development of a plant also requires localized or cell-specific pH changes within the roots. The acid growth theory of cell elongation emphasizes the acidification of the apoplast (Rayle and Cleland 1992) and its role in increasing the activity of cell-wall expansins (Cosgrove 2000). Spatially resolved measurements of apoplastic pH in roots have provided numerous instances of correlations between localized pH changes in the cell wall and growth responses (Taylor et al. 1996; Bibikova et al. 1998; Fan and Neumann 2004; Monshausen et al. 2007, 2011; Staal et al. 2011), and a peptide that is essential for regulating apoplastic pH during root hair development has been identified (Wu et al. 2007). Localized acidification of the apoplast is sometimes correlated with localized alkalinization of the cytoplasm (Bibikova et al. 1998; Monshausen et al. 2007), and the dynamics of the pH changes can be followed with excellent time resolution using fluorescent probes and laser scanning confocal microscopy (Monshausen et al. 2007). Alkalinization of the root cell wall can also be observed, for example, in response to mechanical stimulation, where the increase in pH in a stretched surface is expected to work against the deformation by increasing the rigidity of the cell wall (Monshausen et al. 2009). Again alkalinization of the cell wall correlates with a localized acidification of the cytoplasm, implying a mechanistic coupling of the two events at the level of the plasma membrane ion transporters (Figure 23.1). Other examples of localized intracellular pH changes associated with developmental events in roots include the acidification of specific cortical cells that precedes cell death and aerenchyma formation in rice (*Oryza sativa* L.) roots (Kawai et al. 1998) and the alkalinization of the root hair cells of alfalfa (*Medicago sativa* L.) in response to *Rhizobium meliloti* Nod factors (Felle et al. 1996). It may also be noted that the pH difference between the acidic apoplast and the slightly alkaline cytoplasm favors the diffusion of undissociated auxin into cells and thus contributes to polar auxin transport, a process that is critical for growth and development (Kramer and Bennett 2006).

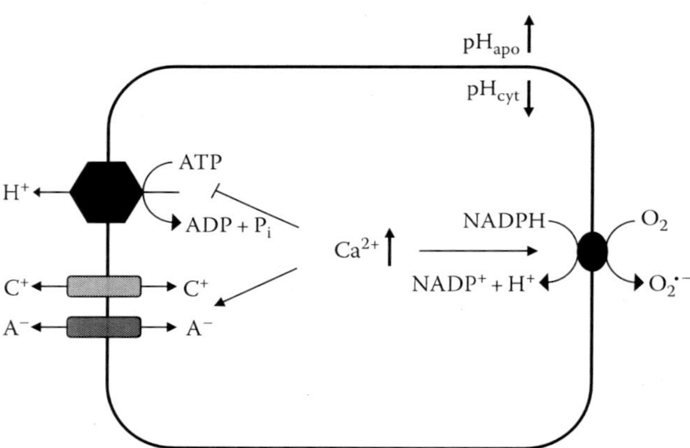

FIGURE 23.1 Processes leading to cytosolic acidification and cell-wall alkalinization following mechanical stimulation. An initial increase in cytosolic Ca^{2+} inhibits the plasma membrane H^+-ATPase while simultaneously activating the plasma membrane NADPH oxidase and increasing the transport activity of cation and/or anion channels (Figure based on Monshausen et al. 2009). The superoxide released in the cell wall can be converted into hydrogen peroxide, which can act as a signaling molecule after transport into the cytoplasm (Bienert et al. 2006). Thus, the Ca^{2+} signal triggers transient intracellular and extracellular changes in both pH and reactive oxygen species. (Modified from Monshausen, G.B. et al., *Plant Cell*, 21, 2341, 2009.)

B. pH Changes under Adverse Conditions

In the laboratory, apoplastic pH can be altered by incubating roots in buffered solutions (e.g., Winch and Pritchard 1999), while intracellular pH values can be manipulated by incubating roots with permeable weak acids and bases (e.g., Walker et al. 1998; Gerendás and Ratcliffe 2000). Weak acids tend to accumulate preferentially in the more alkaline compartments of roots, and weak bases in the acidic compartments. However, since the accumulation of either type of compound tends to reduce the pH differences between the compartments, the intracellular partitioning of these compounds usually depends on the loading conditions and the duration of the experiment. Manipulations of this kind emphasize the finite capacity of the pH regulatory mechanisms, and although primarily of interest in a laboratory context, they can also be relevant in the field. For example, the use of ammonia-based fertilizers can lead to both high soil pH values and high ammonium content (Liu et al. 1995), and this combination has the potential to perturb root pH values.

Root pH values also respond to several environmental stresses, including flooding, drought, salt stress, and unfavorable soil pH values. The effect of oxygen deprivation on flooding-intolerant plants is particularly severe, since it causes an acidification of the cytoplasm that can lead to cell death (Roberts et al. 1984a,b). This phenomenon has been extensively documented in the roots of maize (Saint-Ges et al. 1991; Roberts et al. 1992; Fox et al. 1995; Libourel et al. 2006) and other species (Felle 1996; Stoimenova et al. 2003; Tournaire-Roux et al. 2003; Kulichikhin et al. 2007). Typically, there is an immediate decrease of 0.5–0.6 pH units in pH_{cyt} following the imposition of anoxia, followed by a period of stabilization or partial recovery, before the onset of a further acidification that eventually proves lethal (Roberts et al. 1984a). There is some uncertainty as to whether the fall in pH_{cyt} under anoxia is the cause, rather than a consequence, of cell death with a critical review concluding that the evidence was inconclusive (Greenway and Gibbs 2003). It has also been argued that the stabilization of pH_{cyt} at a pH lower than the normal aerobic value—a phenomenon that has been observed over extended periods in stem tissues from anoxia-tolerant *Potamogeton* species (Summers et al. 2000; Dixon et al. 2006; Koizumi et al. 2011)—should be regarded as a set point that is optimal for anaerobic metabolism (Felle 2005). The response of the pH to anoxia is influenced by the external pH experienced by the roots (Fox et al. 1995; Xia and Roberts 1996), with lower external pH values promoting cytoplasmic acidosis, and the acidification is reduced if the roots are first acclimatized to a low oxygen level by a hypoxic pretreatment (Xia and Roberts 1994). This latter observation is particularly relevant to field conditions where the onset of anaerobiosis is likely to be gradual (Drew 1997; See also Chapter 32).

The effects of other abiotic stresses on root pH values are generally less marked than the effect of oxygen deprivation. Hyperosmotic shock caused only small increases in pH_{cyt} and pH_{vac} in excised maize root tips, with exposure to an osmotic potential of −1.35 MPa resulting in an alkalinization of the cytoplasm of just 0.1 pH units after 3 h (Spickett et al. 1992). The effect of sodium chloride on pH_{cyt} is similar to that of nonionic osmotica at comparable osmotic potentials, with sodium chloride having a negligible effect at concentrations up to 100 mM (Spickett et al. 1993; Halperin et al. 2003), but the influx of sodium into the vacuole can lead to a substantial increase in pH_{vac} (Fan et al. 1989; Spickett et al. 1993; Katsuhara et al. 1997). For example, a vacuolar alkalinization of 0.6 pH units was observed when maize root tips were exposed to sodium chloride concentrations greater than 200 mM, but the effect was considerably smaller (0.3 pH units) in the root tips of the halophyte *Spartina anglica* C. Hubb (Spickett et al. 1993). Similarly hyperosmotic shock had a smaller effect on the vacuolar pH in drought-resistant pearl millet (*Pennisetum americanum* (L.) Leeke) roots than in maize roots (Nagarajan et al. 2001). Drought stress also increases xylem pH in both roots and shoots, favoring the release of abscisic acid (ABA) into the xylem and increasing ABA delivery to the shoot (Wilkinson 1999).

Intracellular pH values are also sensitive to the external pH, but observations on plant cell suspensions (Fox and Ratcliffe 1990; Gout et al. 1992) suggest that this factor is only likely to be important at extreme pH values. Thus, varying the external pH between 8 and 9.25 had no effect on pH_{cyt} and pH_{vac} in maize root tips (Gerendás and Ratcliffe 2000), and increasing the external pH from 6 to 10 only caused an increase of 0.1 pH units in the pH_{cyt} of *Chamaegigas intrepidus* Dinter ex Heil roots (Schiller et al. 1998). Similarly, modest decreases in the external pH from around neutral to 4.5 or 4 have little or no effect on pH_{cyt} in maize root tissues (Gerendás et al. 1990), although there is good evidence from observations on cell suspensions that the passive influx of protons at or below pH 4.5 can overwhelm the capacity of the plasma membrane H^+-ATPase, leading to acidification of the cytoplasm (Gout et al. 1992). In fact, poor root growth at acidic pH values correlated with a net decrease in proton release in maize and broad bean (*Vicia faba* L.) (Yan et al. 1992), and upregulation of the plasma membrane H^+-ATPase occurs during adaptation of roots to low pH values (Yan et al. 1998; Zhu et al. 2009), but studies of the relationship between growth and external pH rarely include measurements of the intracellular pH values. In an exception to this generalization, an analysis of the effect of external pH on the root pH values in chickpea (*Cicer arietinum* L.) and lupin (*Lupinus angustifolius* L.) showed only small changes over the pH range 4.8–8, and it was concluded that this factor could not explain the growth differences between the two species on acid and alkaline soils (Hartung et al. 2002). See also Chapter 32 by Fagerstedt et al. in this volume.

III. pH Regulation

The inherent pH dependence of many cellular events, which arises from the prevalence of ionizable groups among the metabolites and macromolecules that make up the cell, indicates that pH regulation must be an essential feature of all living systems. The need for pH regulation is further emphasized by the fact that even normal cell growth can lead to the net production or consumption of protons (Raven 1986). So while there are many situations in which intracellular pH values might change, these changes actually occur against a general background of pH homeostasis. This is achieved by maintaining an interactive balance between the H^+-consuming and H^+-generating processes within the cell. Moreover, if these processes are to act as pH regulatory mechanisms, then their activity has to be modulated, either directly or indirectly, by the pH changes that are occurring. There are two general solutions to the problem of pH regulation—proton transport (biophysical pH regulation) or proton metabolism (biochemical pH regulation)—and these mechanisms have to be deployed in such a way as to allow functionally important pH changes to occur at specific locations at specific times.

A. Contribution of Membrane Transport to pH Regulation

Membrane transport contributes to pH regulation by providing pH-dependent mechanisms for the movement of ions and metabolites between compartments (Figure 23.2). While attention focuses on the plasma membrane and tonoplast H^+-ATPases, their contribution to pH regulation depends on charge-compensating movements of other ions. It follows that biophysical regulation of pH depends on the coordination of primary and secondary membrane transport mechanisms. However, the potential importance of the H^+-ATPases is reinforced, because many of the secondary fluxes are themselves strongly influenced by the proton gradients established by the electrogenic pumps.

The likely involvement of multiple membrane transport processes in pH regulation is further emphasized by a rigorous analysis of the factors that determine pH. This shows that the pH of a solution in equilibrium with a gas phase containing carbon dioxide (CO_2) depends on the partial pressure of the CO_2 (pCO_2), the concentration of the weak acids in the solution ($[A_{1,tot}]$, $[A_{2,tot}]$, …, $[A_{i,tot}]$, …), and the strong ion difference ([SID]) of the solution, that is, the net charge carried by the

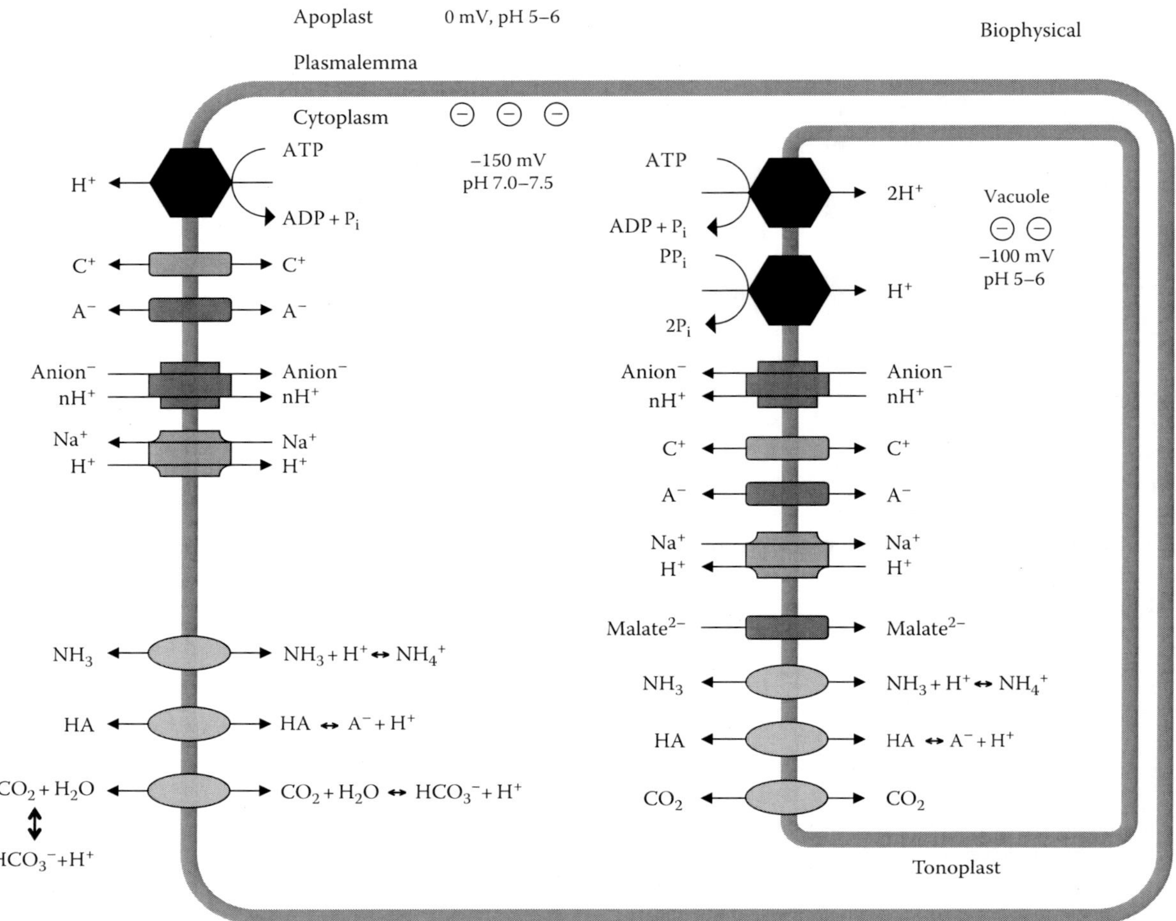

FIGURE 23.2 Transport processes across the plasma membrane and tonoplast that can contribute to the biophysical regulation of the apoplastic, cytoplasmic, and vacuolar pH values in a root. Proton pumping from the cytoplasm is accompanied by the movement of strong ions through a variety of channels and carriers. The net change in ion balance, together with any change in the weak acid content of the compartment of interest, determines the overall effect on the pH.

ions that are fully dissociated under the prevailing conditions (Stewart 1983). This analysis leads to an equation for [H$^+$] that takes the form of a fourth-order polynomial:

$$[H^+]^4 + a[H^+]^3 + b[H^+]^2 + c[H^+] + d = 0$$

where a, b, c, and d are determined by the dissociation constants for the ionic equilibria involving water, weak acids, and CO_2 and by the values of [SID], [$A_{i,tot}$], and pCO_2 (Stewart 1983; Gerendás and Schurr 1999). This equation, which can be solved numerically if sufficient analytical information is available, can be used to calculate the pH of the solutions in intracellular and extracellular compartments, for example, in the xylem and phloem (Gerendás and Schurr 1999). The equation also has important implications for an understanding of pH regulation since it indicates that changes in pH can only occur through changes in the independent variables [SID], [$A_{i,tot}$], and pCO_2. Thus, biophysical pH regulation is usually the result of a multiplicity of membrane transport processes leading to a net change in [SID] and/or [$A_{i,tot}$] (Figure 23.2). Moreover, proton transport alone is insufficient to cause a change in pH, because the transfer of charge by an electrogenic proton pump has to be balanced by other charge transfers to avoid changes in the membrane potential (Gerendás and Schurr 1999). So while a proton cotransport system, such as a Na$^+$/H$^+$ antiport (Bassil et al. 2011; Figure 23.2), can change pH, proton pumps can only be effective in tandem with other transport processes.

Subject to this constraint, all three plant proton pumps—the plasma membrane H$^+$-ATPase, the tonoplast H$^+$-ATPase, and the tonoplast H$^+$-pyrophosphatase—are expected to play a role in pH regulation in the cytoplasm (Figure 23.2), with the plasma membrane H$^+$-ATPase also contributing to pH regulation in the apoplast and xylem vessels. While huge advances have been made in the molecular characterization of these pumps (Gaxiola et al. 2007), this has not led so far to a deeper understanding of the contribution of the pumps to pH regulation. Plasma membrane H$^+$-ATPases are largely regulated by posttranslational modification, with activation occurring through phosphorylation-dependent 14–3–3 protein binding (Morsomme and Boutry 2000; Palmgren 2001; Gaxiola et al. 2007). This process does not appear to have been linked directly to cytoplasmic pH regulation, beyond speculation that 14–3–3 protein interactions might allow the coordinated regulation of all three pumps to ensure cytoplasmic pH homeostasis (Gaxiola et al. 2007), but the increase in ATPase activity that occurs as pH_{cyt} falls below its normal level (Luo et al. 1999; Morsomme and Boutry 2000) is certainly consistent with a role for the plasma membrane H^+-ATPase in cytoplasmic pH regulation. Any increase in proton pumping into the apoplast has to be matched by enhanced cation uptake and/or enhanced anion efflux to avoid hyperpolarization of the membrane (Gerendás and Schurr 1999), and a good illustration is provided by the activation of potassium influx that occurred in barley roots in response to acidification of the apoplast (Amtmann et al. 1999).

The role of the plasma membrane H^+-ATPase in opposing the passive influx of protons across the plasma membrane was elegantly demonstrated in a sycamore (*A. pseudoplatanus*) cell suspension (Gout et al. 1992). Cytoplasmic pH regulation failed when the external pH was reduced below 4.5, and at this critical pH, it was observed that the oxygen consumption rate had risen to its maximum possible value and that the ATP level had fallen to less than 50% of its normal steady-state concentration. It was concluded that the plasma membrane H^+-ATPase was no longer able to compensate for the influx of protons at the low external pH because of a shortage of ATP. This conclusion was confirmed by performing experiments on phosphate-starved cells, with low levels of ATP, and observing that pH_{cyt} only recovered to its normal value of 7.5 at external pH values of 4.5 and 6.0 after the resupply of P_i allowed the ATP level to recover to its normal value (Gout et al. 1992). The importance of the plasma membrane H^+-ATPase in establishing and maintaining the normal steady-state value of pH_{cyt} is further demonstrated by the changes that occur during the adaptation of maize roots to low external pH (Yan et al. 1998). Adaptation increased the proton pumping activity of the H^+-ATPase, and it was concluded that this effect was a significant factor in the survival of the roots at low pH.

Biophysical pH regulation is also important in controlling the pH in the cytoplasm and apoplast of roots in many other situations. For example, the assimilation of ammonium releases protons (Raven and Smith 1976; Gerendás and Ratcliffe 2000), and since there are only limited possibilities for restoring the proton balance metabolically (Raven 1986), it has been found that the regulation of pH_{cyt} during growth is dominated by proton transport (Raven and Smith 1976; Allen and Raven 1987). This increases the likelihood of adverse effects at acid pH values, and so one response to growth on ammonium is to increase the activity of the plasma membrane H^+-ATPase (Yamashita et al. 1995; Schubert and Yan 1997; Zhu et al. 2009). In contrast, in nitrate-grown plants, both the theoretical analysis (Raven and Smith 1976) and the experimental evidence (Allen and Raven 1987) suggests that the contribution of proton transport to pH regulation is supplemented by biochemical mechanisms, resulting in tighter pH regulation in nitrate-grown roots (Gerendás et al. 1990).

The role of membrane transport processes in opposing the acidification of the cytoplasm under anaerobiosis has been discussed at length (Ratcliffe 1999; Greenway and Gibbs 2003; Felle 2005). Substrate limitation of the H^+-ATPases was advanced as a possible cause of the acidification of the cytoplasm in anoxic maize root tips (Saint-Ges et al. 1991), but subsequent work on similar tissues suggested that pH regulation was unaffected by changes in the ATP level within the normal physiological range (Xia et al. 1995; Ratcliffe 1999). In fact, the significance of these experiments really depends on the extent to which the plasma membrane H^+-ATPase functions under anoxia (Ratcliffe 1999; Greenway and Gibbs 2003; Felle 2005). Microelectrode data indicated a depolarization of alfalfa root hairs under anoxia (Felle 1996), suggesting that the proton pump was deactivated, and it seems certain that the pump is inhibited to a significant extent (Felle 2005). On the other hand, the residual activity responds to the presence of fusicoccin, an activator of the plasma membrane H^+-ATPase, hyperpolarizing alfalfa root hairs in the presence of respiratory inhibitors (Felle 1996) and stimulating the release of protons from maize root tips under anoxia (Xia and Roberts 1996). For maize root tips, it was concluded that proton pumping (with the necessary charge compensation by the transport of other ions) made little contribution to cytoplasmic pH regulation in anoxic maize root tips at an external pH of around 6 (Xia and Roberts 1996) and a similar conclusion was reached for alfalfa root hairs (Felle 1996). Thus, as emphasized by Felle (2005), the significance of the residual plasma membrane H^+-ATPase activity under anoxia probably lies in maintaining a proton gradient to drive the uptake of metabolizable substrates.

The contribution of the proton pumps in the tonoplast to pH regulation under anoxia is also relevant (Greenway and Gibbs 2003), both because of the potential significance of the vacuole as a source or sink for transportable ions and metabolites (Figure 23.2) and because the expression and activity of the tonoplast H^+-pyrophosphatase increase under anoxia (Carystinos et al. 1995; Liu et al. 2010). Fluorescence and NMR measurements on maize root hairs led to the conclusion that pH_{vac} is maintained by the tonoplast H^+-ATPase under aerobic conditions and by the tonoplast H^+-pyrophosphatase under anoxia (Brauer et al. 1997), a result that is consistent with the view that plant cells shift to the increased utilization of pyrophosphate as an energy source under anoxia (Mustroph et al. 2005; Huang et al. 2008).

Lactate is commonly produced, at least initially, as a glycolytic end product during oxygen deficiency, and this is often accompanied by lactate efflux into the apoplast and external medium. The potential impact on pH_{cyt} of several possible mechanisms that could support the export of lactate and/or lactic acid has been discussed in detail (Greenway and Gibbs 2003; Felle 2005), and it was concluded that the contribution of these processes to pH regulation is uncertain. Hypoxic pretreatment of maize roots increased lactate efflux from anoxic root tips (Xia and Saglio 1992), and it also improved cytoplasmic pH regulation under anoxia (Xia and Roberts 1994), but subsequent work showed that the greater lactate efflux from the acclimated maize root tips was accompanied by a smaller net efflux of protons (Xia and Roberts

1996). Thus, lactate efflux did not make a major contribution to the observed changes in the external pH, and this is difficult to reconcile with a significant role for the process in intracellular pH regulation.

B. Contribution of Metabolism to pH Regulation

Metabolism contributes to pH regulation by providing pH-dependent pathways that allow a shift toward proton consumption as pH falls and toward proton production as pH increases (Figure 23.3; Davies 1986; Raven 1986). This biochemical mechanism of pH regulation depends on the existence of enzymes with pH-dependent activities that catalyze H^+-consuming or H^+-producing reactions. The relationship between the fluxes through these pathways in a metabolic network and the activities of the enzymes with nonzero elasticity coefficients for H^+ is generally unpredictable (Fell 1997), and so experimental verification is essential to demonstrate that a particular flux is indeed responding to a change in intracellular pH in a way that contributes to pH regulation. Moreover, in analyzing the contribution that such fluxes might make to pH regulation, it is necessary to consider the impact of the step of interest within the context of metabolism as a whole, and this entails a consideration of the pathways that supply the substrate, as well as the pathways that allow the regeneration of any coenzymes that might be involved in the reaction (Figures 23.4 and 23.5).

The most commonly encountered example of biochemical pH regulation is based on the pH-dependent balance between the synthesis and degradation of malate (Figure 23.4; Davies 1986; Sakano 1998). The synthesis of malate from carbon skeletons derived ultimately from glucose generates protons:

$$C_6H_{12}O_6 + 2CO_2 \rightarrow 2C_4H_4O_5^{2-} + 4H^+$$

and the step catalyzed by PEP carboxylase (PEPC) has an alkaline pH optimum; while the oxidative decarboxylation of malate to pyruvate, and the subsequent regeneration of NAD^+ using oxygen as a terminal electron acceptor, consumes protons:

$$C_4H_4O_5^{2-} + H^+ + \frac{1}{2}O_2 \rightarrow C_3H_3O_3^- + H_2O + CO_2$$

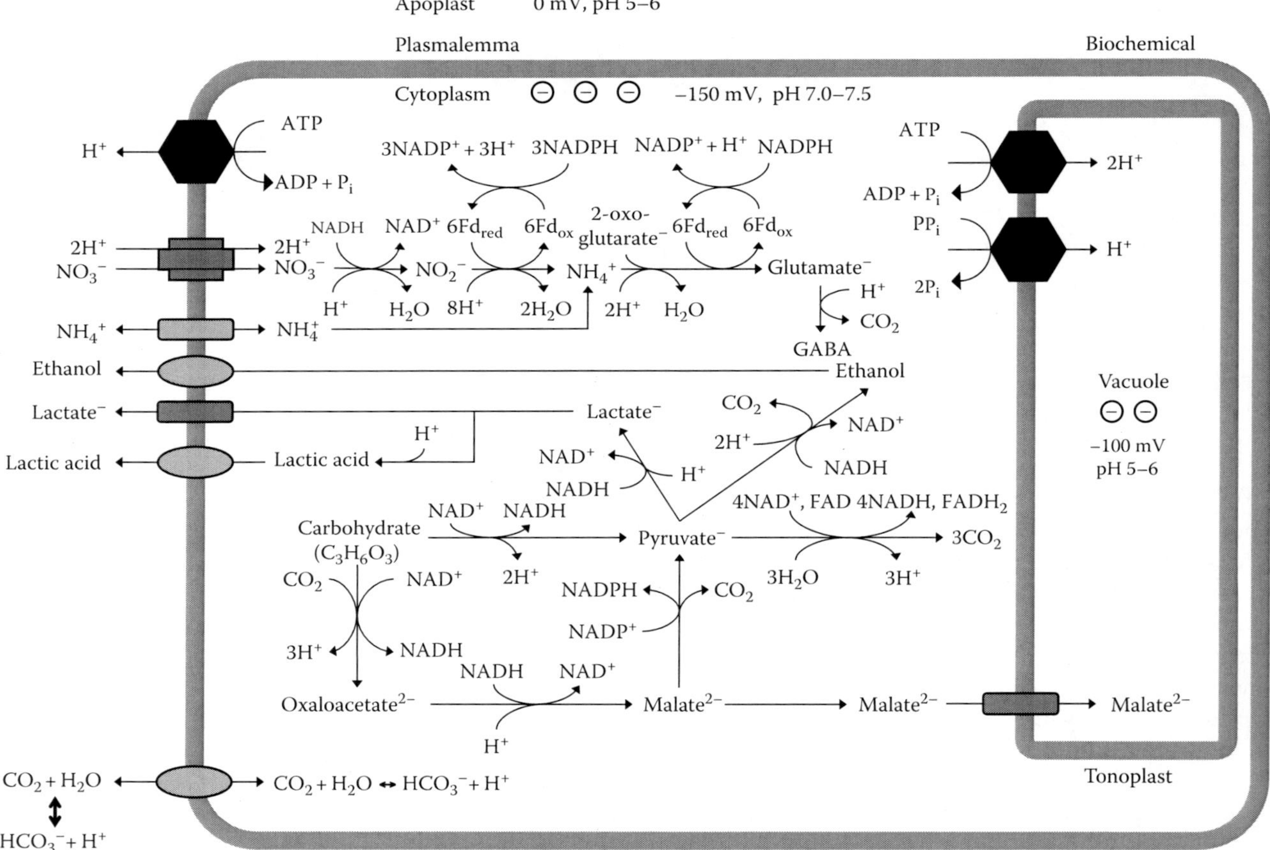

FIGURE 23.3 Metabolic processes that can contribute to the biochemical regulation of pH_{cyt} in a root. The diagram provides balanced equations for the aerobic and anaerobic oxidation of carbohydrate, the synthesis and degradation of malate, the reduction of nitrate, and the subsequent assimilation of ammonium into glutamate. These pathways contribute to pH regulation through the consumption and production of protons, but their net effect depends on the pathways available for the recycling of reducing power (Figures 23.4 and 23.5). The decarboxylation of oxaloacetate by PEPK and the metabolic involvement of other carboxylates in pH regulation, for example, citrate and oxalate, are omitted for reasons of clarity.

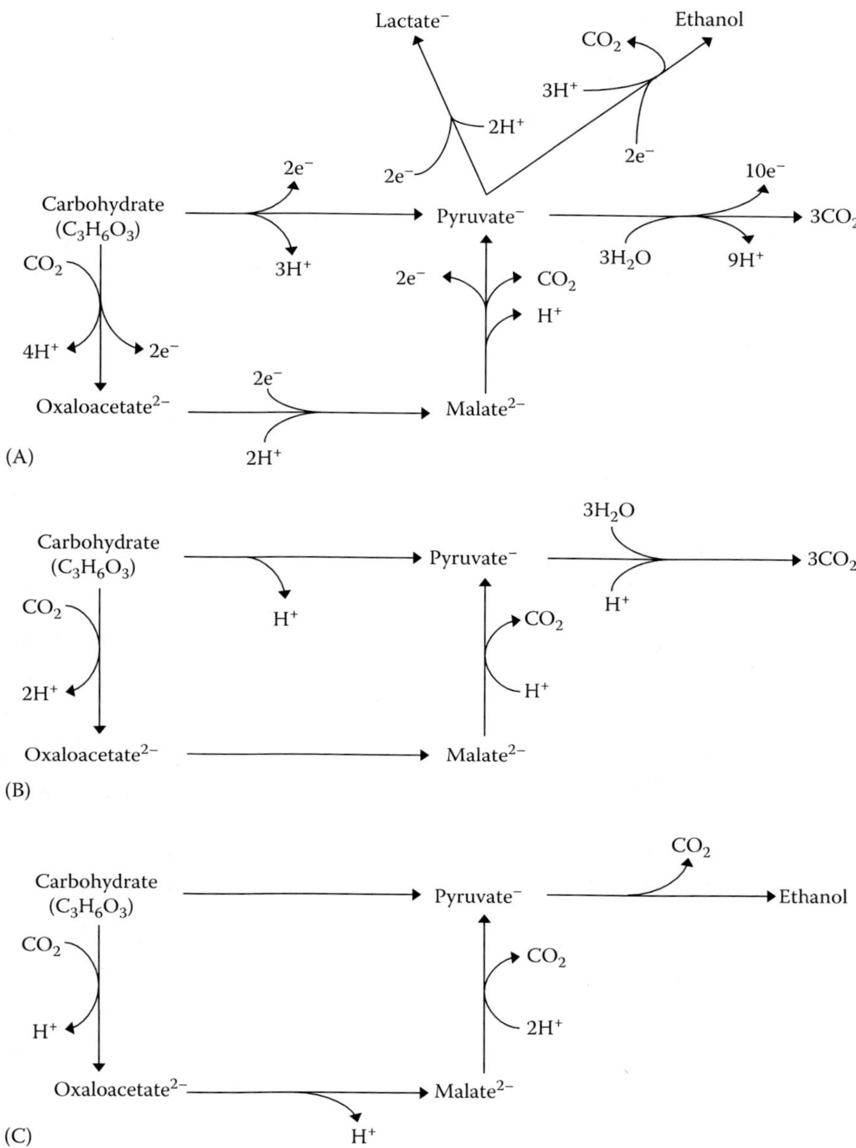

FIGURE 23.4 Proton consumption and production during the aerobic and anaerobic oxidation of carbohydrate and the synthesis and degradation of malate. (A) The proton balance before allowing for changes in oxidation state; (B) the proton balance under aerobic conditions, assuming that electrons are supplied by the complete oxidation of glucose according to the equation $C_6H_{12}O_6 + 3O_2 \rightarrow 6CO_2 + 12H^+ + 12e^-$ and consumed by the reduction of oxygen according to the equation $O_2 + 4H^+ + 4e^- \rightarrow 2H_2O$; and (C) the proton balance under anaerobic conditions, assuming that electrons are supplied by the conversion of carbohydrate to pyruvate according to the equation $C_6H_{12}O_6 \rightarrow 2C_3H_3O_3^- + 6H^+ + 4e^-$ and consumed by the conversion of pyruvate to ethanol according to the equation $C_3H_3O_3^- + 3H^+ + 2e^- \rightarrow C_2H_6O + CO_2$. The proton balance of the steps involved in the synthesis and degradation of malate is different under aerobic (B) and anaerobic (C) conditions, although the net effect is qualitatively the same with malate synthesis releasing H^+ and malate degradation consuming H^+. The route from malate to pyruvate via the decarboxylation of oxaloacetate has been omitted for clarity, but the proton balance for the conversion of malate to pyruvate via PEPCK is the same as the route shown here. (Walker and Chen 2002.)

and the step catalyzed by malic enzyme has an acidic pH optimum. Thus, malate metabolism can respond to pH changes in a way that compensates for them, and as an example, this effect has long been invoked to explain the higher organic acid content of the roots and other tissues of nitrate-grown plants (Kirkby and Mengel 1967; Raven 1986), since nitrate assimilation consumes protons (Figure 23.5). It may also be noted that malate synthesis is potentially unlimited in photosynthetically active plants with stores of carbohydrate, and this may make alkalinization of the cytoplasm easier to control than acidification (Allen and Raven 1987; Gerendás et al. 1990).

PEP carboxykinase (PEPCK), which catalyzes the decarboxylation of oxaloacetate and thus provides an alternative route for the degradation of malate to pyruvate, may also

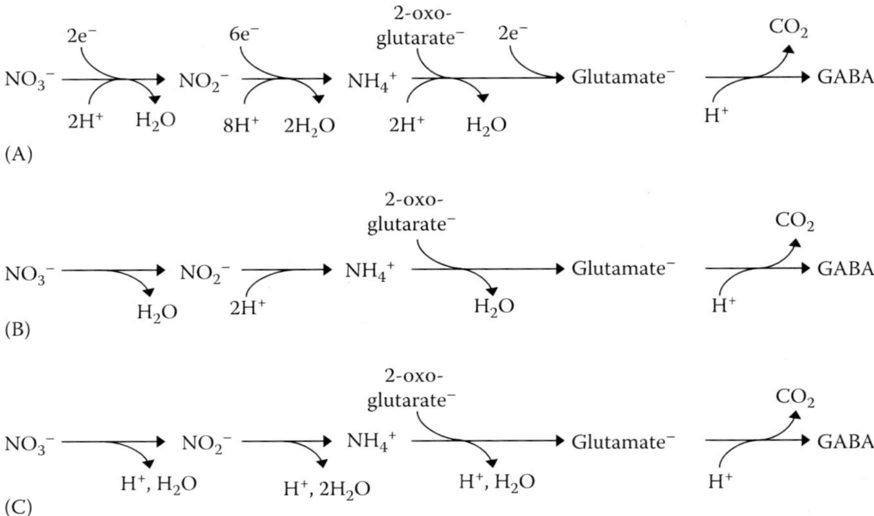

FIGURE 23.5 Proton consumption and production during the conversion of nitrate to GABA. (A) The proton balance before allowing for the provision of reducing power and a source of carbon skeletons; (B) the proton balance under aerobic conditions, assuming that electrons are supplied by the complete oxidation of glucose according to the equation $C_6H_{12}O_6 + 3O_2 \rightarrow 6CO_2 + 12H^+ + 12e^-$; and (C) the proton balance under anaerobic conditions, assuming that electrons are supplied by the conversion of carbohydrate to pyruvate according to the equation $C_6H_{12}O_6 \rightarrow 2C_3H_3O_3^- + 6H^+ + 4e^-$. A comparison of (B) and (C) shows that nitrate assimilation consumes protons under aerobic conditions and releases them under anaerobic conditions (Stoimenova et al. 2003). The overall proton balance for the incorporation of ammonium into glutamate is further influenced by the pathway used to replenish the 2-oxoglutarate pool. (Gerendás and Ratcliffe 2000.)

contribute to the operation of the malate biochemical pH-stat (Walker and Chen 2002). PEPCK activity increases in roots and other tissues in response to treatments that cause acidification (Chen et al. 2004), and when maize roots were incubated with ammonium at an acidic pH, the level of PEPCK increased while the level of malic enzyme declined (Walker et al. 2001). Since the effect on the proton balance of converting malate to pyruvate is the same irrespective of the pathway, these observations are consistent with a role for PEPCK in place of malic enzyme in the biochemical pH-stat (Walker et al. 2001; Walker and Chen 2002).

More recently, it has been shown in *Arabidopsis* that the expression levels of the PEPC kinases that activate PEPC by phosphorylation and the phosphorylation state of PEPC increase in response to alkalinization of the cytosol, while at the same time, the expression of PEPCK decreases (Chen et al. 2008). This led to the conclusion that the PEPC kinases and PEPCK should be considered as key enzymes in the pH-dependent synthesis and degradation of malate alongside PEPC and malic enzyme (Chen et al. 2008). These observations on *Arabidopsis* have also extended the concept of the biochemical pH-stat from its original focus on metabolic enzymes, to include the pH regulation of the genetic processes that determine the level of such enzymes.

While the malate biochemical pH-stat is often invoked in discussions of proton balance during nitrogen assimilation, its relevance during nitrogen acquisition has been challenged on the grounds that (1) it ignores the proton transport that accompanies the transfer of nitrate and ammonium to the cytosol from outside the cell, (2) much of the supporting evidence from tissue measurements of ions and organic acids is not directly interpretable in terms of events in the cytosol, and (3) nitrate nutrition is often not associated with a higher PEPC activity than ammonium nutrition (Britto and Kronzucker 2005). Ignoring several complications, including the subcellular compartmentation of nitrate reduction, the storage of nitrate in root vacuoles, the export of imported nitrate to the shoot, and the differential utilization of nitrate and ammonium in vascular land plants, it is straightforward to compare the H^+ balance associated with the uptake and assimilation of nitrate and ammonium on a molar basis as shown in the next paragraph.

Assuming that the uptake of nitrate by a $2H^+/NO_3^-$ transporter is charge compensated (rectified) by proton pumping by the plasma membrane H^+-ATPase, then

$$NO_3^-(\text{ext}) + H^+(\text{ext}) \rightarrow NO_3^-(\text{cyt}) + H^+(\text{cyt})$$

The conversion of nitrate to ammonium is a reduction process (Figure 23.5A):

$$NO_3^-(\text{cyt}) + 8e^- + 10H^+(\text{cyt}) \rightarrow NH_4^+(\text{cyt}) + 3H_2O$$

and assuming that the reducing power for nitrate reduction is provided by the complete oxidation of glucose to CO_2

$$C_6H_{12}O_6(\text{cyt}) + 3O_2 \rightarrow 6CO_2 + 12H^+(\text{cyt}) + 12e^-$$

then the overall equation for the uptake and reduction of nitrate to ammonium is

$$6NO_3^-(ext) + 6H^+(ext) + 4C_6H_{12}O_6 + 12O_2 + 6H^+(cyt)$$
$$\rightarrow 6NH_4^+(cyt) + 24CO_2 + 18H_2O$$

This equation shows that the overall process alkalinizes both the external medium, as often observed (Kirkby and Mengel 1967; Marschner 2012), and the cytoplasm, where a very minor alkalinization has been observed in short-term nitrate uptake and assimilation experiments on maize roots (Espen et al. 2004). However, if it is also assumed that ammonium uptake is rectified by proton pumping, that is, that the uptake of each ammonium ion is charge compensated by the release of one proton, then the provision of ammonium follows the equation:

$$6NH_4^+(ext) + 6H^+(cyt) \rightarrow 6NH_4^+(cyt) + 6H^+(ext)$$

This shows that the observed acidification of the external medium is associated with an alkalinization of the cytosol that is exactly the same as that expected for the uptake and assimilation of nitrate. The underlying assumptions in this analysis are discussed elsewhere (Britto and Kronzucker 2005), but it can be concluded that since there is no difference in proton balance between the two processes, it is not necessary to invoke a biochemical pH-stat to explain the observed difference in organic acid content of tissues grown on different nitrogen sources. It is then argued that the higher organic acid content in nitrate-grown plants reflects the greater tissue concentration of cations, particularly potassium and calcium in such plants.

Aside from these compelling theoretical considerations, there is also experimental evidence that changes in PEPC activity are not necessarily significant for cytoplasmic pH regulation. For example, an ammonium-induced increase in pH_{cyt} in maize root tips stimulated carboxylate synthesis, but the contribution of this process to pH regulation during and after the ammonium treatment was quantitatively insignificant (Gerendás and Ratcliffe 2000). Similarly, an investigation of pH regulation in sycamore cells showed that the onset of carboxylate synthesis in response to a change in external pH was too slow to account for the stabilization of pH_{cyt} (Gout et al. 1992). Moreover, on switching the external pH back to its original value, pH_{cyt} returned to its normal value before any detectable fall in the malate and citrate levels. Thus, while the stimulation of carboxylate synthesis by elevated pH_{cyt} would have influenced pH homeostasis during the continued exposure of the sycamore cells to alkaline pH values, it did not play a critical role in stabilizing the pH during the periods in which the cells were responding to an alteration in the pH of the external medium (Gout et al. 1992). These examples emphasize the point that the contribution of biochemical pH-stats to pH regulation can only be established through experiment.

Biochemical mechanisms of pH regulation are commonly invoked in studies of root pH regulation during anaerobiosis (Figures 23.3 through 23.5; Ratcliffe 1999; Greenway and Gibbs 2003; Felle 2005). As an example, there is a continuing interest in the link between the acidification of the cytoplasm that occurs shortly after the onset of anaerobiosis and the commonly observed switch from lactate to ethanol production (Kulichikhin et al. 2009). According to the biochemical pH-stat model of pH regulation under anoxia (Davies et al. 1974), lactate dehydrogenase, with an alkaline pH optimum, is inhibited by the acidification, while pyruvate decarboxylase, with an acidic pH optimum, is activated. The net result is that lactate production is expected to be transient, with the switch from a H^+-generating fermentation pathway that produces lactate

$$C_6H_{12}O_6 \rightarrow 2CH_3CH(OH)COO^- + 2H^+$$

to a H^+-neutral pathway that produces ethanol

$$C_6H_{12}O_6 \rightarrow 2CH_3CH_2OH + 2CO_2$$

stabilizing pH_{cyt} (Figure 23.4). NMR methods have been used extensively to probe the validity of this model in roots (Ratcliffe 1997, 1999), providing strong evidence for the role of pH as an in vivo regulator of pyruvate decarboxylase (Roberts et al. 1984b; Fox et al. 1995). While the quantitative correlation between lactate fermentation and cytoplasmic acidification is species dependent, it seems clear that the pH-dependent switch from lactate to ethanol production can be an important factor in pH regulation under anoxia (Ratcliffe 1999, Greenway and Gibbs 2003; Felle 2005).

There are several other H^+-consuming pathways that can be activated by a fall in pH_{cyt} and which could therefore be important in pH regulation under anoxia. First, it has been shown that the decarboxylation of malate (Figure 23.4) contributes to pH regulation in maize root tips during the anoxic response (Roberts et al. 1992; Edwards et al. 1998) and that it is the combined activation of pyruvate decarboxylase and malic enzyme that is responsible for the stabilization and partial recovery of pH_{cyt} that occurs following the onset of oxygen deprivation (Roberts et al. 1992). Second, pH is a regulator of glutamate decarboxylase in vivo, with a fall in pH leading to an increase in the H^+-consuming synthesis of γ-aminobutyrate (GABA) from glutamate (Carroll et al. 1994; Crawford et al. 1994). Although GABA synthesis is a commonly observed response to oxygen deprivation and is linked to the hypoxia-induced accumulation of alanine in *Arabidopsis* roots (Miyashita and Good 2008), it was insignificant during the initial stabilization of pH_{cyt} in maize root tips (Roberts et al. 1992) indicating that it could only contribute to pH homeostasis during the subsequent metabolism of the anaerobic root tissue. Both of these examples illustrate the point that while in vitro enzyme activities may be informative (e.g., Kulichikhin et al. 2009), conclusive evidence for a potential pH-stat usually requires measurements of metabolite levels and metabolic fluxes.

The role of nitrate reduction in pH regulation under anoxia has been examined by comparing the effect of oxygen deprivation on a tobacco transformant that lacked nitrate reductase in the roots (Stoimenova et al. 2003). Root segments from the transformant showed a greater acidification of the cytoplasm under anoxia than those from wild-type plants, but this was caused by the higher metabolic rate in the roots lacking nitrate reductase rather than by the simple cessation of nitrate reduction. In fact, in most plant tissues, nitrate is only reduced as far as nitrite under anoxia, and in contrast to aerobic conditions, this reaction produces protons when the only source of NADH is glycolysis (Figure 23.5):

$$C_6H_{12}O_6 + 2NO_3^- \rightarrow 2C_3H_3O_3^- + 2NO_2^- + 2H_2O + 2H^+$$

This analysis shows that nitrate reduction to nitrite cannot function as a biochemical pH-stat for regulating pH_{cyt} under anoxia, despite the activation of nitrate reductase at acidic pH values (Botrel et al. 1996), and thus cannot provide an explanation for the beneficial effect of nitrate on cytoplasmic pH regulation (Libourel et al. 2006). The reason for the greater acidification in tobacco roots lacking nitrate reductase was traced to the much higher fermentation rate in the transformants than the wild type (Stoimenova et al. 2003). Nitrate metabolism also features in an alternative fermentation pathway that exploits the high oxygen affinity of hypoxia-induced class 1 hemoglobin to regenerate NAD^+ under hypoxia in the presence of nitrate (Igamberdiev and Hill 2004). This process recycles nitrate via nitrite and nitric oxide and is as acidifying as the normal fermentation pathway to lactate, producing two protons per glucose molecule:

$$C_6H_{12}O_6 + O_2 \rightarrow 2C_3H_3O_3^- + 2H_2O + 2H^+$$

This equation assumes that the NADH for the reduction of the residual oxygen present under hypoxia is obtained from glycolysis, but while the cycle provides a mechanism for regenerating NAD^+ in the presence of nitrate, it does not provide an explanation for improved cytoplasmic pH regulation under those conditions (Libourel et al. 2006).

IV. Physiological Consequences of pH Changes

Although the pH regulatory mechanisms available to plants are powerful, they have a finite capacity, and there are many situations in which the pH may deviate from the notional normal value. Thus, the contribution of the plasma membrane H^+-ATPase to pH regulation is inevitably limited by both kinetic and thermodynamic considerations, and a biochemical mechanism such as the decarboxylation of malate can only consume H^+ until the available malate is exhausted. Moreover, it is possible to envisage cell- and tissue-specific changes in the regulatory mechanisms that lead to different pH values in the same compartment under different circumstances (Felle 2005). So changes in intracellular and apoplastic pH values do occur, and these changes often have biochemical and physiological consequences. The link between pH and cell function can be investigated by manipulating intracellular pH values, and a change in pH that triggers a change in function may be considered to be a signal and/or messenger (Felle 2001). While changes in root pH values certainly trigger local events in root cells, the distance over which pH signals can be transmitted is unclear. For example, although long-distance transmission of pH signals from root to shoot via the xylem sap has been proposed as a mechanism for the stress-induced alkalinization of the leaf apoplast and the subsequent redistribution of ABA in the response of leaves to drought (Wilkinson 1999; Schachtman and Goodger 2008), the direct transfer of information through the long-distance propagation of pH changes seems unlikely to be the mechanism that causes the apoplastic pH change (Felle 2001; Jackson et al. 2003; Felle et al. 2005). It seems more likely that root-to-shoot communication reflects the apoplastic transport of the stress factor itself or of a compound formed in the root in response to the stress (Felle et al. 2005; Geilfus and Mühling 2012).

The pH sensitivity of enzymes and transporters ensures that changes in pH can influence biochemical activity in vivo, and there is good evidence for the pH-dependent activation of several pathways in roots, for example, the fermentation pathway to ethanol (Roberts et al. 1984b; Fox et al. 1995) and the conversion of glutamate to GABA (Ford et al. 1996). Transport steps can also be affected by pH changes, and pH-dependent gating of aquaporins provides an explanation for the reduction in water uptake observed during anoxia (Tournaire-Roux et al. 2003). Reductions in cytoplasmic pH caused by oxygen deprivation, respiratory inhibitors, and permeable weak acids all caused a decrease in water permeability in the roots of *Arabidopsis thaliana*, and the underlying mechanism of the pH response was established by heterologous expression of aquaporins in *Xenopus* oocytes (Tournaire-Roux et al. 2003). This work provided an explanation for the reduction of root permeability under anoxia and an unequivocal example of the way in which a change in root pH can have an important physiological consequence. However, pH is unlikely to be solely responsible for regulating root water transport, since plant aquaporins are also sensitive to calcium (Gerbeau et al. 2002; Alleva et al. 2006) and changes in root pH_{cyt} are often associated with changes in cytosolic calcium (Subbaiah et al. 1994; Plieth et al. 1999). Interactions between calcium and pH signaling complicate the assessment of the importance of pH changes (Felle 2001), and in general the regulation of important physiological processes might be expected to be under the control of multiple effectors (Figure 23.1).

Changes in apoplastic and/or cytoplasmic pH are also implicated in a wide range of developmental processes in roots, including root growth (Taylor et al. 1996; Winch and Pritchard 1999), root hair development (Bibikova et al. 1998; Monshausen et al. 2007), gravitropism (Scott and Allen 1999; Fasano et al. 2001; Monshausen et al. 2009, 2011), root–microbe interactions (Ramos et al. 2009a,b), and cell death (Kawai et al. 1998). While the involvement of pH changes and/or proton fluxes across the plasma membrane in these processes is usually readily demonstrated, it can be more difficult to obtain a full mechanistic explanation of the observed effects. For example, understanding the interaction between changes in pH, calcium, reactive oxygen species, and auxin distribution during the gravitropic response is a considerable challenge (Peer et al. 2011), and the speed of the initial gravitropic response, which occurs on a timescale of seconds to minutes, requires detailed experimental analysis to capture the dynamics of the signaling processes (Monshausen et al. 2011). Such an analysis shows that the apoplastic alkalinization that occurs in gravitropism is both auxin and calcium dependent (Monshausen et al. 2011). Changes in root apoplastic pH also have the potential to cause extensive changes in gene expression, including numerous genes associated with calcium signaling and cell-wall modification (Lager et al. 2010), again indicating that developmental changes associated with changes in pH may only be partly attributable to the pH sensitivity of biochemical processes.

V. Concluding Remarks

The interplay between the biochemical and biophysical mechanisms of pH regulation, coupled with the ubiquity of the proton and the need to permit changes in particular circumstances, can make it difficult to quantify the contribution of specific mechanisms to the overall process of pH control in roots. While the molecular tools exist to dissect these mechanisms in some detail, current research tends to focus on the implications of pH changes for cell function rather than on the processes that permit the pH changes to occur. Thus, fundamental questions, such as what determines the set point for pH regulation, or which features of the regulatory system permit oscillations in pH, or what is the balance between biophysical and biochemical processes in a particular stress response, are largely ignored in favor of establishing the dynamics of localized pH changes and their temporal relationship to other events in signaling cascades. Arguably, modern plant science is better equipped to deal with these functional questions than with wrestling with the complexities that arise in the interpretation of such a readily measurable parameter as pH.

References

Aarnes H, Eriksen AB, Petersen D, Rise F. 2007. Accumulation of ammonium in Norway spruce (*Picea abies*) seedlings measured by in vivo ^{14}N-NMR. *J Exp Bot* 58:929–934.

Allen S, Raven JA. 1987. Intracellular pH regulation in *Ricinus communis* grown with ammonium or nitrate as N source: The role of long distance transport. *J Exp Bot* 38:580–596.

Alleva K, Niemietz CM, Sutka M et al. 2006. Plasma membrane of *Beta vulgaris* storage root shows high water channel activity regulated by cytoplasmic pH and a dual range of calcium concentrations. *J Exp Bot* 57:609–621.

Amtmann A, Jelitto TC, Sanders D. 1999. K^+-selective inward-rectifying channels and apoplastic pH in barley roots. *Plant Physiol* 120:331–338.

Bassil E, Tajima H, Liang YC et al. 2011. The *Arabidopsis* Na^+/H^+ antiporters NHX1 and NHX2 control vacuolar pH and K^+ homeostasis to regulate growth, flower development, and reproduction. *Plant Cell* 23:3482–3497.

Bibikova TN, Jacob T, Dahse I, Gilroy S. 1998. Localized changes in apoplastic and cytoplasmic pH are associated with root hair development in *Arabidopsis thaliana*. *Development* 125:2925–2934.

Bienert GP, Schjoerring JK, Jahn TP. 2006. Membrane transport of hydrogen peroxide. *Biochim Biophys Acta* 1758:994–1003.

Blossfeld S, Gansert D. 2007. A novel non-invasive optical method for quantitative visualization of pH dynamics in the rhizosphere of plants. *Plant Cell Environ* 30:176–186.

Blossfeld S, Perriguey J, Sterckeman T, Morel JL, Lösch R. 2010. Rhizosphere pH dynamics in trace-metal-contaminated soils, monitored with planar pH optodes. *Plant Soil* 330:173–184.

Botrel A, Magné C, Kaiser WM. 1996. Nitrate reduction, nitrite reduction and ammonium assimilation in barley roots in response to anoxia. *Plant Physiol Biochem* 34:645–652.

Brady NC, Weil RR. 2008. *The Nature and Properties of Soils*. 14th edn. Upper Saddle River, NJ: Prentice Hall.

Brauer D, Uknalis J, Triana R, Shachar-Hill Y, Tu SI. 1997. Effects of bafilomycin A_1 and metabolic inhibitors on the maintenance of vacuolar acidity in maize root hair cells. *Plant Physiol* 113:809–816.

Brauer D, Uknalis J, Triana R, Tu SI. 1996. Subcellular compartmentation of different lipophilic fluorescein derivatives in maize root epidermal cells. *Protoplasma* 192:70–79.

Bravin MN, Martí AL, Clairotte M, Hinsinger P. 2009. Rhizosphere alkalisation—A major driver of copper bioavailability over a broad pH range in an acidic, copper-contaminated soil. *Plant Soil* 318:257–268.

Britto DT, Kronzucker HJ. 2005. Nitrogen acquisition, PEP carboxylase, and cellular pH homeostasis: New views on old paradigms. *Plant Cell Environ* 28:1396–1409.

Carden DE, Walker DJ, Flowers TJ, Miller AJ. 2003. Single-cell measurements of the contributions of cytosolic Na^+ and K^+ to salt tolerance. *Plant Physiol* 131:676–683.

Carroll AD, Fox GG, Laurie S, Phillips R, Ratcliffe RG, Stewart GR. 1994. Ammonium assimilation and the role of γ-aminobutyric acid in pH homeostasis in carrot cell suspensions. *Plant Physiol* 106:513–520.

Carystinos GD, MacDonald HR, Monroy AF, Dhindsa RS, Poole RJ. 1995. Vacuolar H^+-translocating pyrophosphatase is induced by anoxia or chilling in seedlings of rice. *Plant Physiol* 108:641–649.

Chan P, Warwicker J. 2009. Evidence for the adaptation of protein pH-dependence to subcellular pH. *BMC Biol* 7:69.

Chang K, Roberts JKM. 1989. Observation of cytoplasmic and vacuolar malate in maize root tips by ^{13}C NMR spectroscopy. *Plant Physiol* 89:197–203.

Chen ZH, Jenkins GI, Nimmo HG. 2008. pH and carbon supply control the expression of phosphoenolpyruvate carboxylase kinase genes in *Arabidopsis thaliana*. *Plant Cell Environ* 31:1844–1850.

Chen ZH, Walker RP, Técsi LI, Lea PJ, Leegood RC. 2004. Phosphoenolpyruvate carboxykinase in cucumber plants is increased both by ammonium and acidification, and is present in the phloem. *Planta* 219:48–58.

Cosgrove DJ. 2000. Loosening of plant cell walls by expansins. *Nature* 407:321–326.

Couldwell DL, Dunford R, Kruger NJ, Lloyd DC, Ratcliffe RG, Smith AMO. 2009. Response of cytoplasmic pH to anoxia in plant tissues with altered activities of fermentation enzymes: Application of methyl phosphonate as an NMR pH probe. *Ann Bot* 103:249–258.

Crawford LA, Bown AW, Breitkreuz KE, Guinel FC. 1994. The synthesis of γ-aminobutyric acid in response to treatments reducing cytosolic pH. *Plant Physiol* 104:865–871.

Cummings GA, Xie HS. 1995. Effect of soil pH and nitrogen source on the nutrient status in peach: II. Micronutrients. *J Plant Nutr* 18:553–562.

Davies DD. 1986. The fine control of cytosolic pH. *Physiol Plant* 67:702–706.

Davies DD, Grego S, Kenworthy P. 1974. The control of the production of lactate and ethanol by higher plants. *Planta* 118:297–310.

Degenhardt J, Larsen PB, Howell SH, Kochian LV. 1998. Aluminum resistance in the *Arabidopsis mutant alr*-104 is caused by an aluminum-induced increase in rhizosphere pH. *Plant Physiol* 117:19–27.

Ding X, Fu L, Liu C et al. 2011. Positive feedback between acidification and organic phosphate mineralization in the rhizosphere of maize (*Zea mays* L.). *Plant Soil* 349:13–24.

Dixon MH, Hill SA, Jackson MB, Ratcliffe RG, Sweetlove LJ. 2006. Physiological and metabolic adaptations of *Potamogeton pectinatus* L. tubers support rapid elongation of stem tissue in the absence of oxygen. *Plant Cell Physiol* 47:128–140.

Drew MC. 1997. Oxygen deficiency and root metabolism: Injury and acclimation under hypoxia and anoxia. *Annu Rev Plant Physiol Plant Mol Biol* 48:223–250.

Edwards S, Nguyen B, Do B, Roberts JKM. 1998. Contribution of malic enzyme, pyruvate kinase, *phosphoenolpyruvate* carboxylase, and the Krebs cycle to respiration and biosynthesis and to intracellular pH regulation during hypoxia in maize root tips observed by nuclear magnetic resonance and gas chromatography-mass spectrometry. *Plant Physiol* 116:1073–1081.

Espen L, Nocito FF, Cocucci M. 2004. Effect of NO_3^- transport and reduction on intracellular pH: An in vivo NMR study in maize roots. *J Exp Bot* 55:2053–2061.

Fan TWM, Higashi RM, Norlyn J, Epstein E. 1989. In vivo ^{23}Na and ^{31}P NMR measurements of a tonoplast Na^+/H^+ exchange process and its characteristics in two barley cultivars. *Proc Natl Acad Sci U S A* 86:9856–9860.

Fan L, Neumann PM. 2004. The spatially variable inhibition by water deficit of maize root growth correlates with altered profiles of proton flux and cell wall pH. *Plant Physiol* 135:2291–2300.

Fasano JM, Swanson SJ, Blancafor EB, Dowd PE, Kao T, Gilroy S. 2001. Changes in root cap pH are required for the gravity response of the *Arabidopsis* root. *Plant Cell* 13:907–921.

Fell D. 1997. *Understanding the Control of Metabolism*. London, U.K.: Portland Press.

Felle HH. 1993. Ion-selective microelectrodes: Their use and importance in modern plant cell biology. *Bot Acta* 106:5–12.

Felle HH. 1996. Control of cytoplasmic pH under anoxic conditions and its implication for plasma membrane proton transport in *Medicago sativa* root hairs. *J Exp Bot* 47:967–973.

Felle HH. 1998. The apoplastic pH of the *Zea mays* root cortex as measured with pH-sensitive microelectrodes: Aspects of regulation. *J Exp Bot* 49:987–995.

Felle HH. 2001. pH: Signal and messenger in plant cells. *Plant Biol* 3:577–591.

Felle HH. 2005. pH regulation in anoxic plants. *Ann Bot* 96:519–532.

Felle HH, Herrmann A, Hückelhoven R, Kogel KH. 2005. Root-to-shoot signalling: Apoplastic alkalinization, a general stress factor and defence response in barley (*Hordeum vulgare*). *Protoplasma* 227:17–24.

Felle HH, Kondorosi E, Kondorosi A, Schultze M. 1996. Rapid alkalinization in alfalfa root hairs in response to rhizobial lipochitooligosaccharide signals. *Plant J* 10:295–301.

Fischer WR, Flessa H, Schaller G. 1989. pH values and redox potentials in microsites of the rhizosphere. *Z Pflanzenernähr Bodenk* 152:191–195.

Ford YY, Ratcliffe RG, Robins RJ. 1996. Phytohormone-induced GABA production in transformed root cultures of *Datura stramonium*: An in vivo ^{15}N NMR study. *J Exp Bot* 47:811–818.

Fox GG, McCallan NR, Ratcliffe RG. 1995. Manipulating cytoplasmic pH under anoxia: A critical test of the role of pH in the switch from aerobic to anaerobic metabolism. *Planta* 195:324–330.

Fox GG, Ratcliffe RG. 1990. ^{31}P NMR observations on the effect of the external pH on the intracellular pH values in plant cell suspension cultures. *Plant Physiol* 93:512–521.

Gao D, Knight MR, Trewavas AJ, Sattelmacher B, Plieth C. 2004. Self-reporting *Arabidopsis* expressing pH and $[Ca^{2+}]$ indicators unveil ion dynamics in the cytoplasm and in the apoplast under abiotic stress. *Plant Physiol* 134:898–908.

Gaxiola RA, Palmgren MG, Schumacher K. 2007. Plant proton pumps. *FEBS Lett* 581:2204–2214.

Geilfus CM, Mühling KH (2012) Transient alkalinization in the leaf apoplast of *Vicia faba* L. depends on NaCl stress intensity: An *in situ* ratio imaging study. *Plant Cell Environ.* 35:578–587.

Gerbeau P, Amodeo G, Henzler T, Santoni V, Ripoche P, Maurel C. 2002. The water permeability of *Arabidopsis* plasma membrane is regulated by divalent cations and pH. *Plant J* 30:71–81.

Gerendás J, Ratcliffe RG. 2000. Intracellular pH regulation in maize root tips exposed to ammonium at high external pH. *J Exp Bot* 51:207–219.

Gerendás J, Ratcliffe RG, Sattelmacher B. 1990. ^{31}P nuclear magnetic resonance evidence for differences in intracellular pH in the roots of maize seedlings grown with nitrate or ammonium. *J Plant Physiol* 137:125–128.

Gerendás J, Schurr U. 1999. Physicochemical aspects of ion relations and pH regulation in plants—A quantitative approach. *J Exp Bot* 50:1101–1114.

Gollany HT, Schumacher TE. 1993. Combined use of colorimetric and microelectrode methods for evaluating rhizosphere pH. *Plant Soil* 154:151–159.

Gout E, Bligny R, Douce R. 1992. Regulation of intracellular pH values in higher plant cells. Carbon-13 and phosphorus-31 nuclear magnetic resonance studies. *J Biol Chem* 267:13903–13909.

Greenway H, Gibbs J. 2003. Mechanisms of anoxia tolerance in plants. II. Energy requirements for maintenance and energy distribution to essential processes. *Funct Plant Biol* 30:999–1036.

Grignon C, Sentenac H. 1991. pH and ionic conditions in the apoplast. *Annu Rev Plant Physiol Plant Mol Biol* 42:103–128.

Guern J, Felle H, Mathieu Y, Kurkdjian A. 1991. Regulation of intracellular pH in plant cells. *Int Rev Cytol* 127:111–173.

Gupta UC, Gupta SC. 1998. Trace element toxicity relationships to crop production and livestock and human health: Implications for management. *Commun Soil Sci Plant Anal* 29:1491–1522.

Halperin SJ, Gilroy S, Lynch JP. 2003. Sodium chloride reduces growth and cytosolic calcium, but does not affect cytosolic pH, in root hairs of *Arabidopsis thaliana* L. *J Exp Bot* 54:1269–1280.

Han J, Burgess K. 2010. Fluorescent indicators for intracellular pH. *Chem Rev* 110:2709–2728.

Hartung W, Leport L, Ratcliffe RG, Sauter A, Duda R, Turner NC. 2002. Abscisic acid concentration, root pH and anatomy do not explain growth differences of chickpea (*Cicer arietinum* L.) and lupin (*Lupinus angustifolius* L.) on acid and alkaline soils. *Plant Soil* 240:191–199.

Hauter R, Mengel K. 1988. Measurement of pH at the root surface of red clover (*Trifolium pratense*) grown in soils differing in proton buffer capacity. *Biol Fert Soils* 5:295–298.

Heldt HW, Werdan K, Milovancev M, Geller G. 1973. Alkalization of the chloroplast stroma caused by light-dependent proton flux into the thylakoid space. *Biochim Biophys Acta* 314:224–241.

Hinsinger P, Plassard C, Tang C, Jaillard B. 2003. Origins of root-mediated pH changes in the rhizosphere and their responses to environmental constraints: A review. *Plant Soil* 248:43–59.

Huang SB, Colmer TD, Millar AH. 2008. Does anoxia tolerance involve altering the energy currency towards PPi? *Trends Plant Sci* 13:221–227.

Igamberdiev AU, Hill RD. 2004. Nitrate, NO and haemoglobin in plant adaptation to hypoxia: An alternative to classic fermentation pathways. *J Exp Bot* 55:2473–2482.

Jackson MB, Saker LR, Crisp CM, Else MA, Janowiak F. 2003. Ionic and pH signalling from roots to shoots of flooded tomato plants in relation to stomatal closure. *Plant Soil* 253:103–113.

Jaillard B, Ruiz L, Arvieu JC. 1996. pH mapping in transparent gel using color indicator videodensitometry. *Plant Soil* 183:85–95.

Katsuhara M, Yazaki Y, Sakano K, Kawasaki T. 1997. Intracellular pH and proton-transport in barley root cells under salt stress: In vivo P-31 NMR study. *Plant Cell Physiol* 38:155–160.

Kawai M, Samarajeewa PK, Barrero RA, Nishiguchi M, Uchimiya H. 1998. Cellular dissection of the degradation pattern of cortical cell death during aerenchyma formation of rice shoots. *Planta* 204:277–287.

Khramtsov VV. 2005. Biological imaging and spectroscopy of pH. *Curr Org Chem* 9:909–923.

Kirkby EA, Mengel K. 1967. Ionic balance in different tissues of the tomato plant in relation to nitrate, urea or ammonium nutrition. *Plant Physiol* 42:6–14.

Kochian LV, Shaff JE, Kühtreiber WM, Jaffe LF, Lucas WJ. 1992. Use of an extracellular, ion-selective, vibrating microelectrode system for the quantification of K^+, H^+, and Ca^{2+} fluxes in maize roots and maize suspension cells. *Planta* 188:601–610.

Köckenberger W, De Panfilis C, Santoro D, Dahiya P, Rawsthorne S. 2004. High resolution NMR microscopy of plants and fungi. *J Microsc* 214:182–189.

Koizumi Y, Hara Y, Yazaki Y, Sakano K, Ishizawa K. 2011. Involvement of plasma membrane H^+-ATPase in anoxic elongation of stems in pondweed (*Potamogeton distinctus*) turions. *New Phytol* 190:421–430.

Kosegarten H, Hoffmann B, Rroco E, Grolig F, Glüsenkamp KH, Mengel K. 2004. Apoplastic pH and Fe^{III} reduction in young sunflower (*Helianthus annuus*) roots. *Physiol Plant* 122:95–106.

Kramer EM, Bennett MJ. 2006. Auxin transport: A field in flux. *Trends Plant Sci* 11:382–386.

Kramer DM, Sacksteder CA, Cruz JA. 1999. How acidic is the lumen? *Photosynth Res* 60:151–163.

Kulichikhin KY, Aitio O, Chirkova TV, Fagerstedt KV. 2007. Effect of oxygen concentration on intracellular pH, glucose-6-phosphate and NTP content in rice (*Oryza sativa*) and wheat (*Triticum aestivum*) root tips: In vivo ^{31}P-NMR study. *Physiol Plant* 129:507–518.

Kulichikhin KY, Chirkova TV, Fagerstedt KV. 2009. Activity of biochemical pH-stat enzymes in cereal root tips under oxygen deficiency. *Russ J Plant Physiol* 56:377–388.

Lager I, Andréasson O, Dunbar TL, Andréasson E, Escobar MA, Rasmusson AG. 2010. Changes in external pH rapidly alter gene expression and modulate auxin and elicitor responses. *Plant Cell Environ* 33:1513–1528.

Liao H, Wan H, Shaff J, Wang X, Yan X, Kochian LV. 2006. Phosphorus and aluminum interactions in soybean in relation to aluminum tolerance. Exudation of specific organic acids from different regions of the intact root system. *Plant Physiol* 141:674–684.

Libourel IGL, van Bodegom PM, Fricker MD, Ratcliffe RG. 2006. Nitrite reduces cytoplasmic acidosis under anoxia. *Plant Physiol* 142:1710–1717.

Liu ZJ, Clay SA, Clay DE, Harper SS. 1995. Ammonia fertilizer influences atrazine adsorption-desorption characteristics. *J Agric Food Chem* 43:815–819.

Liu Q, Zhang Q, Burton RA, Shirley NJ, Atwell BJ. 2010. Expression of vacuolar H^+-pyrophosphatase (*OVP3*) is under control of an anoxia-inducible promoter in rice. *Plant Mol Biol* 72:47–60.

Luo H, Morsomme P, Boutry M. 1999. The two major types of plant plasma membrane H^+-ATPases show different enzymatic properties and confer differential pH sensitivity of yeast growth. *Plant Physiol* 119:627–634.

Malhi SS, Harapiak JT, Nyborg M, Gill KS. 2000. Effects of long-term applications of various nitrogen sources on chemical soil properties and composition of bromegrass hay. *J Plant Nutr* 23:903–912.

Marschner P. 2012. *Marschner's Mineral Nutrition of Higher Plants*. 3rd edn. London, UK.: Academic Press.

Marschner H, Römheld V. 1983. In vivo measurement of root induced pH changes at the soil-root interface: Effect of plant species and nitrogen source. *Z Pflanzenphysiol* 111:241–251.

Miller AJ, Cookson SJ, Smith SJ, Wells DM. 2001. The use of microelectrodes to investigate compartmentation and the transport of metabolized inorganic ions in plants. *J Exp Bot* 52:541–549.

Miller AL, Smith GN, Raven JA, Gow NAR. 1991. Ion currents and the nitrogen status of roots of *Hordeum vulgare* and non-nodulated *Trifolium repens*. *Plant Cell Environ* 14:559–567.

Miyashita Y, Good AG. 2008. Contribution of the GABA shunt to hypoxia-induced alanine accumulation in roots of *Arabidopsis thaliana*. *Plant Cell Physiol* 49:92–102.

Monshausen GB, Bibikova TN, Messerli MA, Shi C, Gilroy S. 2007. Oscillations in extracellular pH and reactive oxygen species modulate tip growth of *Arabidopsis* root hairs. *Proc Natl Acad Sci U S A* 104:20996–21001.

Monshausen GB, Bibikova TN, Weisenseel MH, Gilroy S. 2009. Ca^{2+} regulates reactive oxygen species production and pH during mechanosensing in *Arabidopsis* roots. *Plant Cell* 21:2341–2356.

Monshausen GB, Miller ND, Murphy AS, Gilroy S. 2011. Dynamics of auxin-dependent Ca^{2+} and pH signaling in root growth revealed by integrating high-resolution imaging with automated computer vision-based analysis. *Plant J* 65:309–318.

Morsomme P, Boutry M. 2000. The plant plasma membrane H^+-ATPase: Structure, function and regulation. *Biochim Biophys Acta* 1465:1–16.

Moseyko N, Feldman LJ. 2001. Expression of pH-sensitive green fluorescent protein in *Arabidopsis thaliana*. *Plant Cell Environ* 24:557–563.

Mustroph A, Albrecht G, Hajirezaei M, Grimm B, Biemelt S. 2005. Low levels of pyrophosphate in transgenic potato plants expressing *E. coli* pyrophosphatase lead to decreased vitality under oxygen deficiency. *Ann Bot* 96:717–726.

Nagarajan S, Dijkema C, Van As H. 2001. Metabolic response of roots to osmotic stress in sensitive and tolerant cereals—Qualitative in vivo ^{31}P nuclear magnetic resonance study. *Indian J Biochem Biophys* 38:149–152.

Neumann G, Römheld V. 1999. Root excretion of carboxylic acids and protons in phosphorus-deficient plants. *Plant Soil* 211:121–130.

Newman IA, Kochian LV, Grusak MA, Lucas WJ. 1987. Fluxes of H^+ and K^+ in corn roots: Characterization and stoichiometries using ion-selective microelectrodes. *Plant Physiol* 84:1177–1184.

Palmgren MG. 2001. Plant plasma membrane H^+-ATPases: Powerhouses for nutrient uptake. *Annu Rev Plant Physiol Plant Mol Biol* 52:817–845.

Peer WA, Blakeslee JJ, Yang H, Murphy AS. 2011. Seven things we think we know about auxin transport. *Mol Plant* 4:487–504.

Pfeffer PE, Shachar-Hill Y, Tu SI, Brauer D. 2004. Measurement of intracellular pH in maize root tissue with α-methyl fluorinated alanines and in-vivo ^{19}F NMR spectroscopy. *Physiol Plant* 122:373–379.

Plassard C, Meslem M, Souche G, Jaillard B. 1999. Localization and quantification of net fluxes of H^+ along maize roots by combined use of pH indicator dye videodensitometry and H^+-selective microelectrodes. *Plant Soil* 211:29–39.

Plieth C, Sattelmacher B, Hansen UP, Knight MR. 1999. Low pH mediated elevations in cytosolic calcium are inhibited by aluminium: A potential mechanism for aluminium toxicity. *Plant J* 18:643–650.

Pratt J, Boisson AM, Gout E, Bligny R, Douce R, Aubert S. 2009. Phosphate (Pi) starvation effect on the cytosolic Pi concentration and Pi exchanges across the tonoplast in plant cells: An in vivo ^{31}P nuclear magnetic resonance study using methylphosphonate as a Pi analog. *Plant Physiol* 151:1646–1657.

Ramos AC, Lima PT, Dias PN, Kasuya MCM, Feijó JA. 2009a. A pH signaling mechanism involved in the spatial distribution of calcium and anion fluxes in ectomycorrhizal roots. *New Phytol* 181:448–462.

Ramos AC, Martins MA, Okorokova-Façanha AL et al. 2009b. Arbuscular mycorrhizal fungi induce differential activation of the plasma membrane and vacuolar H^+ pumps in maize roots. *Mycorrhiza* 19:69–80.

Ratcliffe RG. 1994. *In vivo* NMR studies of higher plants and algae. *Adv Bot Res* 20:43–123.

Ratcliffe RG. 1997. *In vivo* NMR studies of the metabolic response of plant tissues to anoxia. *Ann Bot* 79 (Suppl. A):39–48.

Ratcliffe RG. 1999. Intracellular pH regulation in plants under anoxia. In *Regulation of Acid-Base Status in Animals and Plants*, eds. Egginton S, Taylor EW, Raven JA, pp. 193–213. Cambridge, UK.: Cambridge University Press.

Ratcliffe RG, Roscher A, Shachar-Hill Y. 2001. Plant NMR spectroscopy. *Progr Nucl Magn Reson Spectrosc* 39:267–300.

Raven JA. 1986. Biochemical disposal of excess H^+ in growing plants? *New Phytol* 104:175–206.

Raven JA, Smith FA. 1976. Nitrogen assimilation and transport in vascular land plants in relation to intracellular pH regulation. *New Phytol* 76:415–431.

Raven JA, Wollenweber B. 1992. Temporal and spatial aspects of acid-base regulation. *Curr Topics Plant Biochem Physiol* 11:270–294.

Rayle DL, Cleland RE. 1992. The acid growth theory of auxin-induced cell elongation is alive and well. *Plant Physiol* 99:1271–1274.

Rengel Z. (ed.) 2002. *Handbook of Plant Growth: pH as the Master Variable*. New York: Marcel Dekker, Inc.

Roberts JKM, Callis J, Jardetzky O, Walbot V, Freeling M. 1984a. Cytoplasmic acidosis as a determinant of flooding intolerance in plants. *Proc Natl Acad Sci U S A* 81:6029–6033.

Roberts JKM, Callis J, Wemmer D, Walbot V, Jardetzky O. 1984b. Mechanism of cytoplasmic pH regulation in hypoxic maize root tips and its role in survival under hypoxia. *Proc Natl Acad Sci U S A* 81:3379–3383.

Roberts JKM, Hooks MA, Miaullis AP, Edwards S, Webster C. 1992. Contribution of malate and amino acid metabolism to cytoplasmic pH regulation in hypoxic maize root tips studied using nuclear magnetic resonance spectroscopy. *Plant Physiol* 98:480–487.

Saint-Ges V, Roby C, Bligny R, Pradet A, Douce R. 1991. Kinetic studies of the variations of cytoplasmic pH, nucleotide triphosphates (^{31}P NMR) and lactate during normoxic and anoxic transitions in maize root tips. *Eur J Biochem* 200:477–482.

Sakano K. 1998. Revision of biochemical pH-stat: Involvement of alternative pathway mechanisms. *Plant Cell Physiol* 39:467–473.

Santi S, Schmidt W. 2009. Dissecting iron deficiency-induced proton extrusion in *Arabidopsis* roots. *New Phytol* 183:1072–1084.

Schachtman DP, Goodger JQD. 2008. Chemical root to shoot signalling under drought. *Trends Plant Sci* 13:281–287.

Schiller P, Hartung W, Ratcliffe RG. 1998. Intracellular pH stability in the aquatic resurrection plant *Chamaegigas intrepidus* in the extreme environmental conditions that characterize its natural habitat. *New Phytol* 140:1–7.

Schubert S, Yan F. 1997. Nitrate and ammonium nutrition of plants: Effects on acid/base balance and adaptation of root cell plasmalemma H^+ ATPase. *Z Pflanzenernähr Bodenk* 160:275–281.

Scott AC, Allen NS. 1999. Changes in cytosolic pH within *Arabidopsis* root columella cells play a key role in the early signaling pathway for root gravitropism. *Plant Physiol* 121:1291–1298.

Shabala SN, Newman IA, Morris J. 1997. Oscillations in H^+ and Ca^{2+} ion fluxes around the elongation region of corn roots and effects of external pH. *Plant Physiol* 113:111–118.

Spickett CM, Smirnoff N, Ratcliffe RG. 1992. Metabolic response of maize roots to hyperosmotic shock: An in vivo ^{31}P nuclear magnetic resonance study. *Plant Physiol* 99:856–863.

Spickett CM, Smirnoff N, Ratcliffe RG. 1993. An in vivo nuclear magnetic resonance investigation of ion transport in maize (*Zea mays*) and *Spartina anglica* roots during exposure to high salt concentrations. *Plant Physiol* 102:629–638.

Staal M, de Cnodder T, Simon D et al. 2011. Apoplastic alkalinization is instrumental for the inhibition of cell elongation in the *Arabidopsis* root by the ethylene precursor 1-aminocyclopropane-1-carboxylic acid. *Plant Physiol* 155:2049–2055.

Stewart PA. 1983. Modern quantitative acid-base chemistry. *Can J Physiol Pharmacol* 61:1444–1461.

Stidham MA, Moreland DE, Siedow JN. 1983. ^{13}C nuclear magnetic resonance studies of crassulacean acid metabolism in intact leaves of *Kalanchoë tubiflora*. *Plant Physiol* 73:517–520.

Stoimenova M, Libourel IGL, Ratcliffe RG, Kaiser WM. 2003. The role of nitrate reduction in the anoxic metabolism of roots. II. Anoxic metabolism of tobacco roots with or without nitrate reductase activity. *Plant Soil* 253:155–167.

Subbaiah CC, Zhang J, Sahs MM. 1994. Involvement of intracellular calcium in anaerobic gene expression and survival of maize seedlings. *Plant Physiol* 105:369–376.

Summers JE, Ratcliffe RG, Jackson MB. 2000. Anoxia tolerance in the aquatic monocot *Potamogeton pectinatus*: Absence of oxygen stimulates elongation in association with an unusually large Pasteur effect. *J Exp Bot* 51:1413–1422.

Sun J, Chen S, Dai S et al. 2009. NaCl-induced alternations of cellular and tissue ion fluxes in roots of salt-resistant and salt-sensitive poplar species. *Plant Physiol* 149:1141–1153.

Swanson S, Choi WG, Chanoca A, Gilroy S. 2011. In vivo imaging of Ca^{2+}, pH, and reactive oxygen species using fluorescent probes in plants. *Annu Rev Plant Biol* 62:273–297.

Taylor AR, Bloom AJ. 1998. Ammonium, nitrate and proton fluxes along the maize root. *Plant Cell Environ* 21:1255–1263.

Taylor DP, Slattery J, Leopold AC. 1996. Apoplastic pH in corn root gravitropism: A laser scanning confocal microscopy measurement. *Physiol Plant* 97:35–38.

Tournaire-Roux C, Sutka M, Javot H et al. 2003. Cytosolic pH regulates root water transport during anoxic stress through gating of aquaporins. *Nature* 425:393–397.

Van As H. 2007. Intact plant MRI for the study of plant-water relations, membrane permeability, cell-to-cell and long-distance water transport. *J Exp Bot* 58:743–756.

Walker DJ, Black CR, Miller AJ. 1998. The role of cytosolic potassium and pH in the growth of barley roots. *Plant Physiol* 118:957–964.

Walker RP, Chen ZH. 2002. Phosphoenolpyruvate carboxykinase: Structure, function and regulation. *Adv Bot Res* 22:93–189.

Walker RP, Chen ZH, Johnson KE, Famiani F, Tecsi L, Leegood RC. 2001. Using immunochemistry to study plant metabolism: The examples of its use in the localization of amino acids in plant tissues, and of phosphoenolpyruvate carboxykinase and its possible role in pH regulation. *J Exp Bot* 52:565–576.

Wilkinson S. 1999. pH as a stress signal. *Plant Growth Regul* 29:87–99.

Winch S, Pritchard J. 1999. Acid-induced wall loosening is confined to the accelerating region of the root growing zone. *J Exp Bot* 50:1481–1487.

Wu J, Kurten EL, Monshausen G, Hummel GM, Gilroy S, Baldwin IT. 2007. NaRALF, a peptide signal essential for the regulation of root hair tip apoplastic pH in *Nicotiana attenuata*, is required for root hair development and plant growth in native soils. *Plant J* 52:877–890.

Xia JH, Roberts JKM. 1994. Improved cytoplasmic pH regulation, increased lactate efflux, and reduced cytoplasmic lactate levels are biochemical traits expressed in root tips of whole maize seedlings acclimated to a low oxygen environment. *Plant Physiol* 105:651–657.

Xia JH, Roberts JKM. 1996. Regulation of H^+ extrusion and cytoplasmic pH in maize root tips acclimated to a low oxygen environment. *Plant Physiol* 111:227–233.

Xia JH, Saglio PH. 1992. Lactic acid efflux as a mechanism of hypoxic acclimation of maize root tips to anoxia. *Plant Physiol* 100:40–46.

Xia JH, Saglio P, Roberts JKM. 1995. Nucleotide levels do not critically determine survival of maize root tips acclimated to a low oxygen environment. *Plant Physiol* 108:589–595.

Yamashita K, Kasai M, Ezaki B et al. 1995. Stimulation of H^+ extrusion and plasma membrane H^+-ATPase activity of barley roots by ammonium treatment. *Soil Sci Plant Nutr* 41:133–140.

Yan F, Feuerle, R, Schäffer S, Fortmeier H, Schubert S. 1998. Adaptation of active proton pumping and plasmalemma ATPase activity of corn roots to low root medium pH. *Plant Physiol* 117:311–319.

Yan F, Schubert S, Mengel K. 1992. Effect of low root medium pH on net proton release, root respiration, and root growth of corn (*Zea mays* L.) and broad bean (*Vicia faba* L.). *Plant Physiol* 99:415–421.

Yu Q, Tang C, Kuo J. 2000. A critical review on methods to measure apoplastic pH in plants. *Plant Soil* 219:29–40.

Zhu Y, Di T, Xu G et al. 2009. Adaptation of plasma membrane H^+-ATPase of rice roots to low pH as related to ammonium nutrition. *Plant Cell Environ* 32:1428–1440.

Zieschang HE, Köhler K, Sievers A. 1993. Changing proton concentrations at the surfaces of gravistimulated *Phleum* roots. *Planta* 190:546–554.

24

Root Water Uptake and Water Flow in the Soil–Root Domain

Guillaume Lobet
Catholic University of Louvain

Charles Hachez
Catholic University of Louvain

François Chaumont
Catholic University of Louvain

Mathieu Javaux
Catholic University of Louvain
Forschungszentrum Jülich GmbH

Xavier Draye
Catholic University of Louvain

I. Introduction

Water is by far the resource that plants absorb in largest amounts from their environment. However, only a tiny fraction of the water taken up is used for the production of new assimilates. The largest part is lost to the atmosphere as stomata open to allow diffusion of carbon dioxide into leaves. For maize alone, this expensive exploit represents annually ca. 3×10^{14} L of water worldwide, enough to fill 250 million Olympic swimming pools (London 2012 Olympic Games had two of them) (FAO 2012). Although transpiration prevents excessive heating of sunlit leaves and ensures long-distance transport of nutrients from the bulk soil to the upper plant organs, the contrasting transpiration efficiencies of C3 and C4 species is an indication that actual crop water use often exceeds theoretical requirements. However, the fact that breeding for reduced crop water use has often led to negative results (Blum 2005) stresses that a more detailed consideration of the dynamics of water fluxes in the soil–plant–atmosphere system is needed (Gregory et al. 2005).

In a transpiring plant, the diffusion of vapor to the atmosphere through stomata leads to the evaporation of water from the mesophyll cell walls and the displacement of the water–air interface to narrower interstitial spaces between wall fibrils. This increases capillary forces at the wall surface and the tension at the leaf side of the soil–plant hydraulic continuum. In the absence of any resistance to water flow along the continuum, this tension would lead to an immediate displacement of water from the soil and maintain the system in a steady state. However, water movement is impeded by several features of both the soil matrix and plant structures, which complicate water entry into the plant and decrease the water potential in the whole continuum (Figure 24.1A). The master control, played by the stomata, which operates where the water potential gradient is the largest, is therefore subordinate to the ability of the upstream continuum to cope with the transpirational demand. If the impedance between the bulk soil and the stomata gets so large that the plant water potential decreases to damaging levels, the plant will wilt and die unless it closes its stomata.

In the absence of transpiration, the xylem tension disappears and the residual water potential gradient between the soil and the root is dominated by the osmotic component (Javot and Maurel 2002). In such conditions, the driving force to water flow relies on the difference in solute concentration between the xylem and the soil solution, resulting from the active pumping of solutes from the soil to the root. The resulting water flow can generate a small positive pressure in the xylem, which can assist the refilling of embolized vessels (see succeeding text) but does not

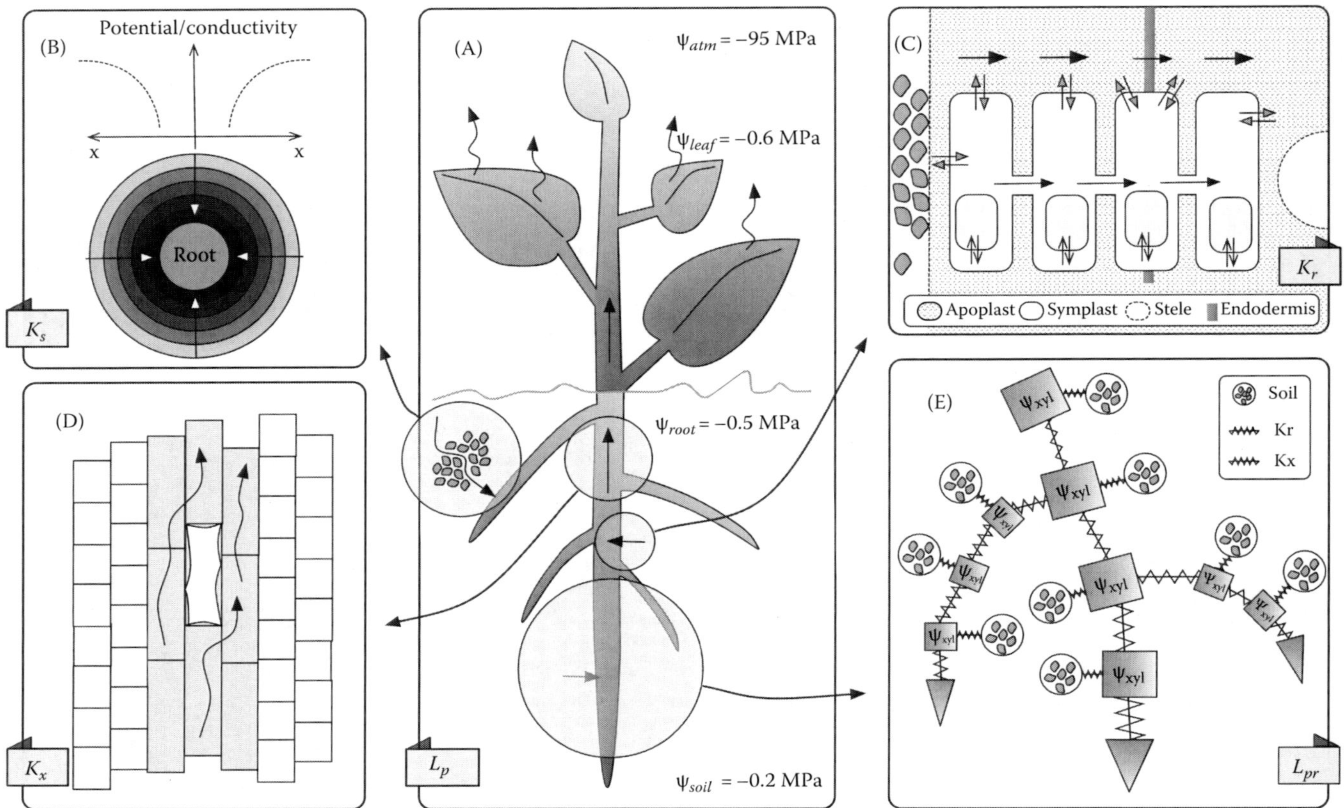

FIGURE 24.1 Water flow in the SPAC. (A) Cohesion tension mechanism. The water flows along the SPAC following the gradient of water potential between the different compartments. (B) Water movement in the soil. The water flows in the soil following local gradient of matrix forces. Close to the root surface, the radial nature of the water flow induces a strong water potential gradient. (C) Radial movement of water in the root. Water entering the roots has to traverse several cell layers using a combination of symplastic, cell-to-cell, and apoplastic pathway. Dark arrows represent pressure-driven convection, while grey arrows represent osmotically driven flow. (D) Axial movement of water in the root. Water moves in xylem vessels following water potential gradients between roots and leaves. The path can be interrupted locally at embolized vessels (cavitation). (E) Importance of the root system architecture as an integrative element. The analogy between the root system hydraulic architecture and an electric network makes it possible to predict the sites of water uptake within the root system.

contribute to the massive absorption of water mentioned earlier. For this reason, the scenario of root pressure–driven flow is only briefly mentioned in the section dealing with xylem refilling.

Biologists often use the analogy with the Ohm's law to express the dependency between water potential, water flux, and hydraulic resistance (van den Honert 1948). At the whole-plant scale, the electrical analogy is expressed as $E = (\psi_{atm} - \psi_{soil}) * L_p$, where E is the total water flow, ψ_{atm} is the water potential of the atmosphere, and L_P and ψ_{soil} are aggregated values of the plant hydraulic conductivity and soil water potential. As will be explained throughout this chapter, the complex hydraulic architecture of the soil–plant system and the values of all resistances impeding water movement determine the water potential, hydraulic conductivity, and water flow at every location in the system, at any given transpiration rate. It is their integration at a higher scale that sets the spatial distribution and amount of water E flowing through the plant, referred to as "uptake."

Excellent reviews of root water uptake have been published already (Jackson et al. 2000; Javot et al. 2002; Draye et al. 2010; Maurel et al. 2010; Aroca et al. 2011). This chapter focuses on the mechanisms or principles that impede (and, thereby, control) water flow on its way from the bulk soil to the shoot of transpiring plants and on novel ways of analyzing how water uptake patterns emerge from the integration of several hydraulic controls in the soil–plant system.

II. Water Moves in a Complex System

Despite a general agreement on the driving forces of water flow in the soil–plant–atmosphere continuum (SPAC), on the possible routes taken by water, and on the local mechanisms that control its flow, it remains difficult to predict quantitatively the contribution of specific mechanisms to the overall regulation of root water uptake. This is largely due to the composite pathway and the spatially distributed and temporally variable nature of the hydraulic controls.

Before entering the root, water flows in the bulk soil toward the soil–root interface. Its movement follows ψ_{soil} gradient and is impeded by friction against the solid phase. Because soils are comprised of an ensemble of pores of different sizes, their hydraulic conductivity decreases nonlinearly with water content (Figure 24.1B), viz., with water getting restricted to narrower pores. As mentioned earlier, transpiration decreases the amount of water and thus the water potential in the immediate vicinity around the roots. In a homogeneous domain, this is likely to cause a drop of the soil conductivity, leading to a redistribution of the sites of water uptake. In the corresponding electrical analogy, the soil domain is represented by spatially distributed capacitance and resistance elements that are physically linked to the soil water potential.

When entering the root, water flows into a structure comprised of several layers of differentiated cells. As stated in the composite transport model (Steudle et al. 1998), water crosses the successive layers using pressure-driven apoplastic (cell wall and intercellular spaces) and symplastic (plasmodesmata) pathways or osmotically driven transcellular pathways (water channels). As water flows preferably in routes of low resistance, the contribution of the three pathways is ultimately controlled by the geometry, cell wall properties, and membrane permeability of each layer (Steudle et al. 1998; Steudle 2000) (Figure 24.1C). The most obvious controlling features are the number of cell layers laid down by the meristem (organ level), the irreversible deposition of hydrophobic barriers in the apoplast (tissue level; see chapter by Meyer and Peterson in this volume; Enstone et al. 2003; Ranathunge et al. 2011), and the dynamically regulated activity of water channels (cell level, Maurel et al. 1993; Javot et al. 2002; Chaumont et al. 2005). The electrical analogy could be used to integrate cell-, tissue-, and organ-level effects on the control of the radial flow. It would comprise resistances to flow between the different compartments (cell walls, intercellular space, and cytoplasm), connected in agreement with the radial anatomy and subjected to a difference of water potential between the root surface and the xylem lumen.

Once in the xylem, water flows over long distances through the network of vessels (Figure 24.1D). Because xylem conductivity is by several order of magnitude higher than that of tissues along the radial path, it is often assumed that the xylem route imposes a negligible impedance to water flow in the SPAC (Steudle et al. 1998). However, this route is by large the longest one taken by water within the plant. In a maize plant, for example, water runs for several meters in the xylem but only a few millimeters from the root surface to the xylem and from the xylem to the mesophyll cells, viz., a difference of three orders of magnitude. There is also abundant evidence that the xylem route can be the limiting route in the system as the likelihood of cavitation increases with path length (Peirce 1936; Tyree et al. 1989; Sperry et al. 2002). Frictional drag and cavitation, the major limitations to water flow in the xylem, obey two very different laws, with friction being predicted by Poiseuille's law, while cavitation is, at best, only predicted by probability models. As previously mentioned, an electrical analogy could be used to work out the detailed flow of water through the network of xylem conduits. Here, however, in addition to using resistances where friction is operating, on/off switches with partially stochastic behaviors would be needed at the level of each xylem vessel.

The obvious need to integrate different types and scales of control at the suborgan level also applies at the whole-plant level. Indeed, water flows simultaneously from many different places in the bulk soil and enters the plant through many different roots, following trajectories that are not hydraulically equivalent. Even for plants growing in uniform hydroponic systems, water uptake is unlikely to be uniform over the root surface. The structure and geometry of the soil/root system place therefore additional constraints on water flow and uptake. It is certainly at this level that the electrical analogy has been most used (Doussan et al. 1998), probably due to the simplicity of the circuit, comprising a simple tree of nodes connected by axial resistances and linked to their local environment by radial resistances (Figure 24.1E).

Looking at the whole array of conductivities of the media or structures in which water is flowing, some appear to be variable on long time scales, while others are highly dynamic. Some appear to vary following simple physical laws, while others follow intrinsic or responsive regulatory pathways. In the next section, we review our understanding of the phenomena responsible for the variation of these conductivities, considering successively those that operate in the soil, in the radial path, in the xylem network, and at the system level.

III. Water Flow Control in the Soil Matrix

A. Soil Hydraulic Properties That Affect Water Flow to the Roots

In soils, water follows passively the water potential gradient to move from the bulk soil to the root surface. Accordingly, the soil water potential gradients and the soil hydraulic conductivity will, at any given time, control the supply of water to the plant. As discussed in Section III.B, these hydraulic properties of the soil compartment display highly dynamic behavior, which is most likely to determine the patterns of water availability and uptake. A crucial feature of soils is their ability to store water, which is due to their porous structure. However, the amount of energy needed to extract water from the soil increases as the soil water content decreases, because capillary and surface-binding forces reduce the free energy of water retained in pores (matric potential). At equilibrium, the matric potential is related to the volumetric soil water content by a water retention (or release) curve, which is principally determined by the pore size distribution and the wetting properties of the pore surface. Water retention curves also display hysteresis according to the wetting or drying cycle of the soil.

A second crucial feature of soils is their ability to conduct water when subjected to a gradient of water potential. The soil hydraulic conductivity is inversely related to friction and is therefore affected by fluid properties (viscosity and density) and soil structure, that is, the size, shape, tortuosity, and connectivity of the pore network filled with water. Even in fully saturated conditions, the conductivity value (K_{sat}) may vary by several orders of magnitude between fine- and coarse-textured soils.

At the field scale, values of the saturated conductivity in certain horizons exhibit a coefficient of variation up to 900%, mainly determined by structural features like macropores (Mallants et al. 1997). When the soil desaturates, the hydraulic conductivity decreases dramatically because flow paths become increasingly tortuous and drag forces between the fluid and the solid phases increase. This steep curve induces ratios between saturated and unsaturated conductivity values at 1.5 MPa ranging between 10^6 (for clay) and 10^{21} (for sand). The hydraulic conductivity curve characterizes this dependence of the hydraulic conductivity on soil moisture.

B. Soil Water Dynamics in the Presence of Roots

The Richards equation (Richards 1931) is usually used to model soil water flow, taking into account the nonlinear water release and hydraulic conductivity behaviors of soils. At the root scale, considering a radial geometry, the solution of this equation close to the root surface indicates that, under constant uptake, the soil matrix potential decreases nonlinearly toward the root surface, which results in a higher resistance to water flow compared to the bulk soil (Gardner 1965). In conditions of high transpiration rate and medium to dry soil moisture, the decreased soil conductivity near the root surface may limit root water uptake, despite the presence of sufficient water in the bulk soil (Schröder et al. 2009).

The redistribution of soil water can also affect root water uptake. At night, as long as the plant hydraulic conductivity is large enough and no capillary or hydrophobic barrier develop during the previous day, soil hydraulic potential tends to even out, restoring the water content around roots and in the water-depleted zones. The intensity of this lateral and vertical redistribution depends on soil hydraulic properties and on the soil and plant hydraulic status. At the field scale, it also depends on soil hydraulic properties that are shaped by natural (e.g., soil horizons due to pedogenesis) and anthropic processes (e.g., soil tillage) but also by soil topography (through runoff) and row/inter-row patterns.

C. Hydraulic Properties of the Rhizosphere

The rhizosphere is the soil zone that surrounds and is affected by the presence and activity of roots. Its boundaries are diffuse and its spatial extent ranges from sub-micron to supra-centimeter scales, depending on the process of interest (Darrah 1993; Hinsinger et al. 2005). In the rhizosphere, soil hydraulic properties can be profoundly modified by the presence and activity of roots and microorganisms. The combination of structural, chemical, and biological effects is relatively complex and not well understood. In addition, the hydraulic properties of the rhizosphere are difficult to measure with conventional techniques. It follows that different and sometimes contradicting results were obtained with respect to the impact of rhizosphere processes on soil hydraulic properties (Moradi et al. 2012).

Root growth causes compaction of the surrounding soil in front of the root apex (Dexter 1987). This increases the bulk density (Guidi et al. 1985) and reorganizes the soil pore distribution and continuity. While Whalley et al. (2004) suggest that this would generate a decrease of the infiltration, Aravena et al. (2011) show with modeling that this could increase the soil conductivity by raising soil water connectivity.

Exudates from roots and microbial and fungal communities also affect the hydraulic properties of the rhizosphere. Mucilage principally contains polysaccharides that retain high water content at low water potentials (Watt et al. 1994; McCully et al. 1997; Read et al. 1999). Exudates contain also surfactants that reduce the surface tension of soil solution (Read et al. 2003) and other hydrophobic compounds that increase the soil water repellency as compared to bulk soil (Hallett et al. 2003), potentially generating a slower rate of infiltration in drier soils (Czarnes et al. 2000). Carminati et al. (2011) combined the observations of higher water-holding capacity and unsaturated conductivity into a simple model and proposed that water availability to plants would increase under unsaturated conditions.

The geometry of the root–soil contact in itself can also influence the conductivity of the rhizosphere. In young root segments, this is commonly modified by the production of hairs that extend into the rhizosphere, arising from selected cells of the root epidermis (or rhizodermis [Esau 1965]). Root hairs increase the apparent conductivity of the soil–root contact by increasing the surface through which water flows into the root (Gilroy et al. 2000; Bibikova et al. 2003; Segal et al. 2008). Root hair formation and length are tightly controlled in response to local environmental conditions.

Another modification of the root–soil geometry consists in the shrinkage of roots (and soil), which generates air gaps at the soil–root interface (Veen et al. 1992; Carminati et al. 2009). This phenomenon, which occurs typically under very dry conditions, can be seen as a safety mechanism leading to a dramatic reduction of the transfer of water between the soil and the root.

D. Soil, Constitutive Part of the Soil–Plant–Atmosphere Continuum

Because the water supply from the bulk soil to the root surface depends on soil properties that are shaped by prior water uptake, the nonlinear water retention and hydraulic conductivity behaviors of the soil, combined with the array of rhizosphere processes, are likely to generate complex patterns of water uptake at the whole-plant scale.

As the soil has long been considered a continuous medium, the analogy with the electrical circuit has never been used in soil physics. Unlike plants, the whole soil domain can be approached with a unique equation, such as Richards equation, that can be solved numerically.

The complexity of the soil compartment has been largely neglected in the past and water uptake has been assumed to be mostly determined by root length density (RLD). This simplified view is supported by several reports showing, by experiments or calculations, that under wet conditions, the soil conductivity is much higher than the root radial conductivity (Gardner et al. 1962; Newman 1969a,b; Arya et al. 1975a,b; Passioura 1980). However, under low moisture conditions that are often encountered, the soil conductivity may become limiting for water uptake (Draye et al. 2010). Interestingly, this suggests that our current understanding of water capture largely ignores many aspects of its dynamics.

IV. Regulation and Control of Radial Water Flow

A. Radial Water Movement

In response to the hydraulic gradients, water molecules that enter the root have to flow through the complex anatomical structure of roots, as described by the composite transport model (Steudle et al. 1998). Depending on the tissue anatomy, developmental stages, and environmental conditions, water will flow through a combination of paths that offers the lowest hydraulic resistance to its passage. It was generally assumed that under transpiring conditions, when hydrostatic water gradients are dominant, the apoplastic path would be dominant (Steudle et al. 1998). This simplified view has been recently challenged by reports that water flowing via the cell-to-cell path could account for almost the whole radial root water transport, even while the plant is transpiring (Bramley et al. 2009; Knipfer et al. 2010, 2011). In addition, the relative contribution of cell-to-cell and apoplastic paths to the total hydraulic conductivity depends on the species and even on the ecotype (Bramley et al. 2009; Sutka et al. 2011). Nowadays, the common view is that there is neither a purely apoplastic nor a pure cell-to-cell root water movement but rather a combination of both, due to the local variations of water potentials, cell wall and cell membrane hydraulic resistances.

B. Modification of Anatomical Structures

The conductivity of the root apoplasm is primarily determined by the overall structure of the cortex, viz., the number of cell layers, wall thickness, and cell size. This basic organization can be further modified by highly selective cell death and dissolution in the cortex leading to the formation of large air cavities (aerenchyma) within the root cortical cylinder. This results in a reduction of root conductivity (L_{pr}), which is at least partly due to a drastic reduction of the cumulated membrane surface and of the aquaporin-mediated radial transport of water (Fan et al. 2007; Yang et al. 2012).

The hydraulic conductivity of the apoplasm is locally disrupted by the deposition of lignin and suberin compounds within the cell wall matrix, which occlude wall pores otherwise filled with water (reviewed by Meyer and Peterson in this book and by Schreiber and Franke 1999). The endodermis is a major apoplastic barrier that develops with a very regular maturation pattern in most angiosperm species (Perumalla et al. 1986). It contributes to the prevention of water loss (cutting off apoplastic bypasses), the maintenance of root pressure (Peterson et al. 1993), and the protection against pathogen and detrimental environmental conditions (Hose et al. 2001; Enstone et al. 2003; Thomas et al. 2007).

The exodermis (defined as a hypodermis with Casparian bands) fulfills similar functions (with the exception of root pressure maintenance). However, the extent, composition, and rate at which exodermal apoplastic barriers develop strongly depend on environmental cues such as drought, anoxia, salinity, heavy metals, or nutrient stresses (Hose et al. 2001; Enstone et al. 2003, 2005). The cultivation of maize seedlings in aeroponic conditions induces the formation of an exodermis in primary roots (Zimmermann et al. 2000; Hachez et al. 2006, 2012), leading to a reduction of L_{pr} by a factor 1.5–3.6 compared to hydroponic culture. Furthermore, water deficit is known to promote or intensify the development of root apoplastic barriers (both endodermis and exodermis) that tend to develop closer to the tip (Perumalla et al. 1986; Hose et al. 2001; Karahara et al. 2004; Enstone et al. 2005; Vandeleur et al. 2009).

C. Regulation by Water Channels

Plants have the ability to rapidly alter their L_{pr} in a span of a few minutes to a few hours (Moshelion et al. 2002; Javot et al. 2003; Lopez et al. 2003; Cochard et al. 2007b; Hachez et al. 2012). This fine regulation is under metabolic control and may be accounted for by changes in cell membrane permeability triggered by the expression and specific activation of water channel proteins known as aquaporins (Maurel et al. 2001; Cochard et al. 2007b; Maurel et al. 2008). Angiosperm species possess the highest multiplicity of aquaporin isoforms of all kingdoms of life with 33, 35, 36, 55, and 71 homologues in rice, *Arabidopsis*, maize, poplar, and cotton, respectively, falling into 4–5 phylogenetic subfamilies (Chaumont et al. 2001; Johanson et al. 2001; Sakurai et al. 2005; Gupta et al. 2009; Park et al. 2010). Plant aquaporins are found in all subcellular compartments forming or derived from the secretory pathway (Maurel et al. 2008) as well as in some organelles such as chloroplasts (Maurel et al. 2008; Uehlein et al. 2008). Among aquaporin subfamilies, plasma membrane intrinsic proteins (PIPs) mostly localize in the plasma membrane where they control the bulk of the cell membrane water movement as shown by numerous pharmacological and reverse genetics evidences (Martre et al. 2002; Siefritz et al. 2002; Tournaire-Roux et al. 2003; Hachez et al. 2006).

In roots, the contribution of aquaporins to the total root water flow ranges from 20% to 80% (Maurel et al. 2001; Javot et al. 2003) as assessed by the use of aquaporin inhibitors such as silver (as $AgNO_3$ or silver sulfadiazine), gold (as $HAuCl_4$), mercurial compounds (Tyerman et al. 1999; Niemietz et al. 2002; Bramley et al. 2009), acid load (Tournaire-Roux et al. 2003), and hydrogen peroxide (Ye et al. 2006) or by reverse genetics approaches (reviewed by Hachez et al. 2006). Reverse genetics has turned particularly useful to characterize the role of specific isoforms, although this approach presents clear limitations due to the multigenic nature of aquaporin subfamilies and compensation effects by close homologues of the gene whose expression has been altered (reviewed by Hachez et al. 2006).

In plants, silencing of PIP isoforms (by RNAi or knockout mutants) generally results in a lowered hydraulic conductivity of root cell plasma membrane (denoted as L_{pcell}) (Kaldenhoff et al. 1998; Martre et al. 2002; Siefritz et al. 2002; Javot et al. 2003). *Arabidopsis* transgenic lines in which expression of several PIP isoforms had been knocked out compensated for a three- to five-fold decrease in L_{pcell} by increasing the root/leaf dry mass ratio, so that the overall L_{pr} tended to remain the same (Kaldenhoff et al. 1998; Martre et al. 2002). However, an increase of the root/shoot mass ratio was not found to be a general trend as different plant species used different ways to cope with such reduction in L_{pcell} (reviewed by Hachez et al. 2006).

Root water flow is generally higher during the day than at night due to the high evaporative demand inherent to the evapotranspiration process. One of the plants short-term resources to increase their root system conductance is to decrease the cell-to-cell water resistance, which is achieved through a tight control of aquaporin expression and/or activity. Diurnal expression of aquaporin transcripts and/or proteins has been reported in a wide variety of plant species like *Zea mays* L. (Lopez et al. 2003; Hachez et al. 2012), *Samanea saman* (Jacq.) Merr. (Moshelion et al. 2002), *Lotus japonicus* (Regel) K. Larsen, walnut tree (Cochard et al. 2007b), *Helianthus annuus* L., *Vitis vinifera* L., or rice (Sakurai-Ishikawa et al. 2011). Aquaporin levels and activity are usually higher during the day than at night, although exceptions to this trend exist for some specific isoforms. The overall higher aquaporin expression and activity observed during the day facilitates the water movement via the cell-to-cell path by significantly increasing the L_{pcell} compared to nighttime values (Hachez et al. 2012).

L_{pcell} also tends to increase from the root surface toward the endodermis and this further supports the importance of the cell-to-cell path in mediating radial root water flow (Bramley et al. 2009). This increase in conductance of the cell-to-cell path could serve as a compensatory mechanism to the decreasing apoplastic space available as water radially progresses toward the stele.

Regulation of aquaporin expression and activity is of prime importance as it allows a fast adaptation of the cell membrane hydraulics to the cell's ever-changing surrounding environment. Multiple levels of regulation have been unraveled and novel mechanisms are still to be discovered. The reader can refer to thorough reviews by Chaumont et al. (2005) and Maurel et al. (2010).

Expression of aquaporins is under a tight transcriptional control by a variety of environmental or hormonal cues of biotic or abiotic origin. Such transcriptional control allows the plant to cope with adverse or changing environmental conditions that impact soil water availability, evapotranspiration rate, or plant water uptake ability (Maathuis et al. 2003; Alexandersson et al. 2005). Long-term abiotic stresses such as drought, cold, and salinity usually induce a marked drop in L_{pr}, mostly achieved through downregulation of aquaporin expression and activity although some upregulated isoforms have been identified (Jang et al. 2004; Alexandersson et al. 2005).

Whereas concomitant downregulation of L_{pr} and L_{pcell} is usually expected upon stress, this is actually not always the case and both parameters can vary independently (Vandeleur et al. 2009; Sutka et al. 2011; Hachez et al. 2012). In response to stress, L_{pcell} may even increase, as a result of an increase in aquaporin expression and activity, while L_{pr} may decrease or remain unchanged (Vandeleur et al. 2009; Hachez et al. 2012). One plausible explanation for such discrepancies between L_{pcell} and L_{pr} behavior would be that the increase in cortex cell conductivity is insufficient to compensate for the drop in L_{pr} due to the presence of other control points acting as bottlenecks to water flow, most likely at locations where membranes are crossed (as discussed by Hachez et al. 2012; see also succeeding text).

Once synthesized, PIP protein amount in the plasma membrane is modulated by regulation of their subcellular localization. Constitutive cycling of PIPs between the plasma membrane and endomembrane compartments has been demonstrated (Li et al. 2011; Luu et al. 2011). Modulation of this cycling presents a way to rapidly modulate PIP abundance at the plasma membrane in response to stress (Boursiac et al. 2008; Li et al. 2011). In addition to that, PIP subcellular localization can also be regulated by specific physical interaction between different isoforms (Zelazny et al. 2007).

The opening or closure of the aquaporin pore (gating) is also controlled by cytosolic calcium or pH, both of which trigger conformational changes occluding the pore (Tournaire-Roux et al. 2003; Törnroth-Horsefield et al. 2006; Boursiac et al. 2008). Other factors possibly affecting the gating behavior involve methylation, acetylation, phosphorylation, heteromerization, pressure pulses, solute gradients, and temperature (for review, see Chaumont et al. 2005; Maurel et al. 2010).

Finally, hormonal signals also control aquaporin expression and L_{pr}. Abscisic acid (ABA), a well-known stress hormone, has long-lasting effects on plant water status (Nagel et al. 1994; Thompson et al. 2007; Parent et al. 2009). Root hydraulic properties indeed respond to application of exogenous ABA, the most common effect being an increase in L_{pr} (Zhang et al. 1995; Aroca et al. 2006; Mahdieh et al. 2009; Ruiz-Lozano et al. 2009). As inferred from a study on transgenic maize lines impaired in the regulation of ABA biosynthesis, this increase of L_{pr} occurs via the upregulation of aquaporin activity. However, ABA can actually trigger either an up- or a downregulation of aquaporin expression/activity, depending on time, dose (exogenously applied or not), or species (Hose et al. 2000; Martinez-Ballesta et al. 2003; Parent et al. 2009; reviewed by Maurel et al. 2010).

D. Interplay between Root Anatomy and Aquaporin Activity

The changes in root anatomy described earlier may be accompanied by a concomitant adjustment of aquaporin expression/activity. Expression of root aquaporins has been detected in the vicinity of apoplastic barriers where they presumably facilitate a transmembrane flow of water (Hachez et al. 2006; Vandeleur et al. 2009; Hachez et al. 2012). Such a phenomenon has been recently reported in maize where a correlation between exodermis suberization and PIP protein localization could be detected (Hachez et al. 2012). Plants that were grown under aeroponic conditions and that developed a hypodermis with Casparian bands, effectively forcing more water to flow through the membranous uptake path, showed increased levels of some PIP isoforms in the vicinity of the apoplastic barrier when compared to roots lacking such suberized exodermis. However, a similar correlation between suberization pattern and root hydraulic properties of the endodermis could not be observed in *Arabidopsis* (Sutka et al. 2011).

Interestingly ABA has been shown to induce suberin biosynthesis in *Arabidopsis* roots, and the genes involved in this process were upregulated by this hormone in *Agrobacterium*-induced tumors developing a suberization of their cell walls (Duan et al. 2005; Efetova et al. 2007). These observations suggest that root suberization might be induced by ABA, whose role in regulating aquaporin expression level and activity has been demonstrated, as discussed in the previous section. The interplay between endogenous ABA levels, extent of root suberization, and aquaporin expression and activity clearly deserves further characterization.

Root radial water conductivity is therefore controlled by several processes acting either on the long term (developmentally driven or in response to lasting adverse environmental conditions) or on the short term (e.g., in response to diurnal variation to transpiration demand). The interplay between both types of processes and their relative contributions seems to be highly dependent on the plant species and the growing conditions. Altogether, this multilevel regulation of root water conductivity could confer a high plasticity and adaptation potential of the root water uptake mechanisms.

V. Control of the Axial Water Flow

A. Cavitation Onset

As stated earlier, most of the plant water movement takes place in the interconnected network of xylem vessels. The hydraulic continuity of this vascular network can be locally disrupted, as xylem vessels under tension are threatened by the nucleation of small gas bubbles whose rapid expansion (embolism) leads to cavitation. Because the conductivity of embolized vessels drops to zero, the movement of water is forced to flow through the remaining adjacent and non-cavitated vessels (Peirce 1936; Sperry et al. 1988b; Tyree et al. 1989). As a result, the flow rate and friction in the remaining vessels increase, and by virtue of the Ohm's analogy, the water potential downstream of the cavitation site decreases. This might increase the likelihood of cavitation in the system in a feedforward way, unless the remaining vessels are less prone to cavitation. As long as the tension in the xylem is maintained (or increases, due to a lesser availability of water in the soil), an increasing number of vessels are likely to become more prone to embolization, as shown on Figure 24.1D. The relation between xylem tension and the proportion of embolized vessels describes a plant's susceptibility to cavitation.

The susceptibility to cavitation has been shown to have various effects on the plant water status, ranging from changes in leaf conductivity (Johnson et al. 2012) and stomatal conductance (Cochard 2002; Zufferey et al. 2011) to lowering plant yield (Cochard et al. 2007a). It remains controversial whether the axial water movement is the weak link of the hydraulic flow in the SPAC. It is indeed often assumed that in crop plants, the limiting factor is the radial component (Steudle et al. 1998). More experiments are therefore required to clarify the functional importance of cavitation events.

B. Cavitation Avoidance

Although xylem cavitation is essentially a passive phenomenon, plants use two strategies to keep some level of control. The first strategy embraces all plant features that influence its ability to prevent cavitation. The existence of a control at this level is demonstrated by the variability of the susceptibility to cavitation between plant species (Sperry et al. 1997; Pockman et al. 2000; Cochard et al. 2008), cultivars (Cochard et al. 2007a; Li et al. 2009; Rewald et al. 2011a), and even organs, where roots have been shown to be more susceptible to cavitation than stem (Hacke et al. 1996; Sperry et al. 1997; Hacke et al. 2000).

In several instances, differences in cavitation susceptibility have been linked to differences in xylem wall (and pits) structure in both angiosperms (Herbette et al. 2010; Christman et al. 2012) and gymnosperms (Hacke et al. 2009; Delzon et al. 2010). Such structural features are thought to influence gas leakage from an embolized vessel and the formation of nucleation sites in functional adjacent vessels (Tyree et al. 1989; Jarbeau et al. 1995; Delzon et al. 2010). A recent hypothesis states that the total surface of the pit membrane is negatively correlated with cavitation avoidance (Christman et al. 2012). In addition, cavitation can weaken xylem vessels and render them more prone to further cavitation, a phenomenon that does not occur to the same extent in all plant species (Hacke et al. 2001; Stiller et al. 2002).

Other options to restrict cavitation susceptibility are to avoid excessive increase in xylem tension. The surface ratio between the roots and the leaves appears to be a key factor controlling the maintenance of a low tension. For example, an undersized root system (relative to the leaf evaporative surface) would lead to an increase in xylem tension and a greater susceptibility to cavitation (Sperry et al. 1998; Hacke et al. 2000; Sperry et al. 2002). Similarly, isohydric stomatal behaviors preventing the variation of the plant water potential (Tardieu et al. 1998) are also likely to reduce the risk of cavitation.

C. Cavitation Recovery

The second strategy relates to the restoration of the lost hydraulic conductivity through the refilling of embolized vessels. The osmotically driven pressure buildup that can occur in the root at very low transpiration flow (viz., during the night or winter) is the first mechanism supporting this strategy. Provided the xylem pressure rises above a threshold value that depends on the xylem diameter (Yang et al. 1992), the gas present in the embolized vessel may dissolve into the water that is pushed into the vessels until they are refilled (Sperry et al. 1987, 1988a,b, 1994). The extent and frequency of pressure-driven refilling vary widely between species.

Xylem refilling under tension (i.e., while the plant is transpiring) has been observed in several species and environmental conditions (McCully et al. 1998; McCully 1999; Zufferey et al. 2011). This discovery caused lively debates within the scientific community, but it seems now accepted that "novel refilling" (Hacke et al. 2003) does indeed occur. Nonetheless, the mechanisms remain unclear and several theories coexist (Holbrook and Zwieniecki 1999; Clearwater et al. 2005). Figure 24.2 illustrates the conceptual framework proposed by Zwieniecki and Holbrook (2009) to "provoke discussion, organize existing information and provide a useful guideline for future studies of xylem refilling under tension." In this framework, active transport of solutes from neighboring parenchyma cells to the embolized vessels generates an osmotic flow of water into the vessel forming high osmotic droplets at the wall surface. The expression of specific aquaporin isoforms appears to be involved in this osmotic flow (Secchi et al. 2010, 2011; Secchi and Zwieniecki 2011). The partially hydrophobic xylem walls prevent water from the embolized vessel to leak to the neighboring vessel under tension, while small gas-filled channels in the vessel wall enable the exit of gas trapped in the embolized vessel. Finally, additional supply of water in vapor phase from the neighboring vessel is proposed as a novel pathway for refilling.

The hydraulic conductivity of the network of xylem vessels appears therefore as a complex and dynamic variable in the soil–plant continuum. Its complexity lays in the very different nature of the many components that contribute to its regulation. The active solute transport and the multiple paths for water entry and gas exit during refilling deserve more attention as these processes are extremely important in controlling plant long distance water movement.

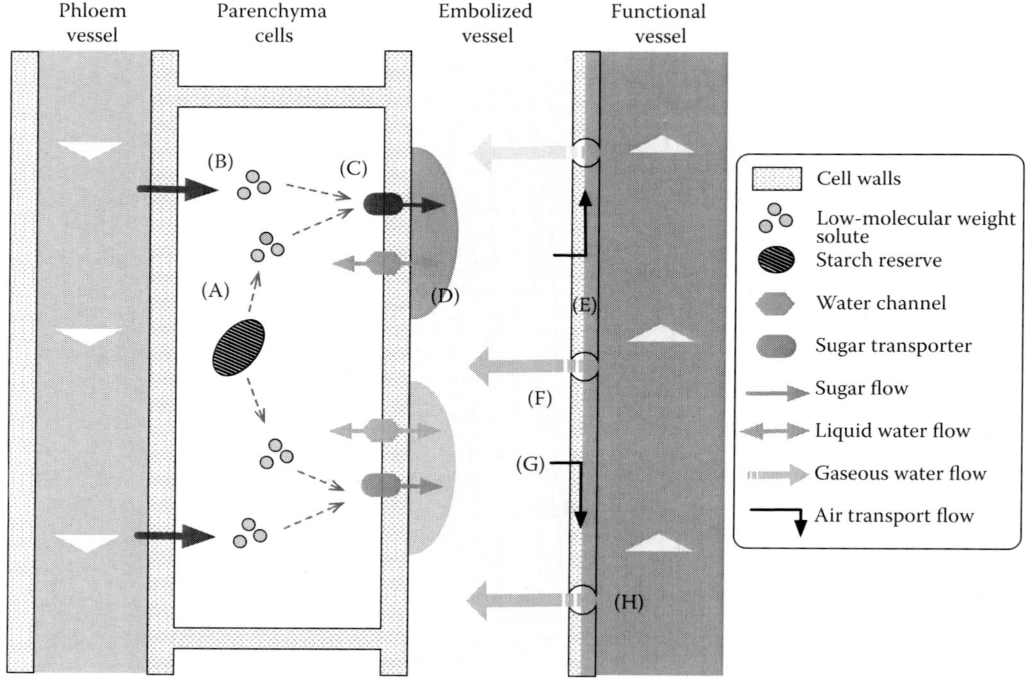

FIGURE 24.2 Scenario for xylem refilling under tension. Starch (A) and phloem (B) serve as sources of low-molecular-weight solutes that are actively transported into the embolized vessel (C). The accumulation of solutes results in water movement from xylem parenchyma cells by osmosis, forming droplets with high osmotic activity on internal vessel walls (D). The partially non-wettable walls of xylem conduits prevent these droplets from being removed by suction from still-functioning vessels (E). Condensation of water vapor provides a second pathway by which water refills cavitated conduits, allowing adjacent conduits to provide water for refilling (F). As the high osmotic droplets grow inside the vessel, the embolus is removed by forcing gas into solution and by pushing gas through small pores through the vessel walls to intercellular spaces (G). The flared opening of the bordered pit chamber acts like a check valve until the lumen is filled, thus preventing contact with the highly wettable bordered pit membranes (H). Reconnection occurs once the pressure in the lumen exceeds that of the entry threshold into the bordered pit chambers. A hydrophobic layer within pit membranes might provide the needed connectivity among multiple bordered pits. (Reproduced from *Trends Plant Sci.*, 14, Zwieniecki, M.A. and Holbrook, N.M., Confronting Maxwell's demon: Biophysics of xylem embolism repair, 530–534, Copyright 2009, with permission from Elsevier.)

VI. Root Architecture

A. Root System Architecture and the Potential for Water Uptake

Root systems are structured and ever changing populations of roots of different types (Eshel and Waisel 1996). Much of their structure arise from growth and branching processes, the latter being responsible for their treelike topological structure. The mechanisms that lead to the formation of a meristem or a new branch are under strong genetic control (Péret et al. 2009; see also the chapter by Jansen et al. in this volume). However, growth and branching respond to global and local environmental cues, including soil water potential (Pregitzer et al. 1993; Hodge 2004; Zhu et al. 2005), which provides an additional control of water flow. There are indeed several ways by which root architecture influences the ability of root systems to take up water.

Firstly, the treelike topological structure of root systems impacts how water tension generated in the shoot organs propagates simultaneously to thousands of individual roots, which influences their hydraulic conductivities (see preceding sections) and the local driving force for water uptake.

Secondly, roots belong to a hierarchy of root types whose features (e.g., diameter, axial conductance, or growth rate are often assumed to decrease with the root order) affect their impedance to radial and axial water flow (Rewald et al. 2011b and references cited therein). As root architecture is developing, the total root size increases, but the relative proportions of the different root types change as well, leading to a complex evolution of the root systems' hydraulic properties.

Thirdly, roots comprise a succession of segments of increasing age from the tip to the base. As root segments get older, the deposition of apoplastic barriers decreases their radial conductivity, while the maturation of xylem vessels increases their axial conductivity (Steudle et al. 1989; Doussan et al. 1998; Hachez et al. 2006). Young segments located behind the elongation zone appear to be sites of peak absorption, while older segments are thought to be mainly conductive pipes for the water absorbed by more distal segments and branches (Boyer 1995). One has to note, however, that the relation between segment age and segment position is not absolute and depends on root growth rate (Lecompte et al. 2001).

The coupling of root architecture with the hydraulic features of root segments has been referred to as the "hydraulic architecture" of a root system (Doussan et al. 1998). It has been shown that radial and axial water flows in individual segments can be estimated simultaneously for all segments of a root system using the Ohm's analogy, which turns into a system of multiple linear equations whose structure is set by the hydraulic architecture of the root system (Landsberg et al. 1978; Doussan et al. 1998). This finding led to a family of models that formalize the strong relationship between root architecture, hydraulic properties, and water uptake capacity of root systems (Doussan et al. 2006; Javaux et al. 2008).

B. Root Placement and the Availability of Water

In addition to size and topology, root architecture also embraces the 3D localization of root segments in the soil. Root placement is at the core of the water flow dynamics from the soil to the plant, because soil moisture is usually not uniform and varies in space and time under the influence of the environment and of the root system itself (Draye et al. 2010).

The most recognized contribution of root system architecture to water uptake is certainly the root system ability to explore deep soil layers that contain water reserves during critical phenological phases. In a crop stand, where strong horizontal competition with neighboring plants is occurring, it is generally assumed that a deeper root system gives access to more water (King et al. 2003), although the optimal exploration strategy should also depend on the scenario of climate, soil water, and plant water use over the whole season. Indeed, multiple studies have highlighted the link between rooting depth and drought resistance in a panel of species such as vegetables (Johnson et al. 2000), cereals (Shen et al. 2001; Reynolds et al. 2007; Steele et al. 2007; Bernier et al. 2009; Henry et al. 2011; Wasson et al. 2012), trees (Nagarajah et al. 1981; Pinheiro et al. 2005), grasses (Marcum et al. 1995; Hammer et al. 2009), or legumes (Kashiwagi et al. 2006).

Several traits contribute to rooting depth, including root insertion angles and gravitropism (Hammer et al. 2009), timing of axial root emission, and penetration ability of hard subsoil (Lynch et al. 2012). Recent work has also shown that an increased proportion of root cortical aerenchyma leads to a substantial decrease in root maintenance respiration (Zhu et al. 2010), which is thought to account for more than 50% of the global assimilate consumption (Nielsen et al. 1998). By decreasing the maintenance cost, an increased allocation of assimilates to primary root growth can be achieved, leading to deeper rooting and improved resistance to water deficit under both control and drought conditions. Traits influencing root depth appear to be under the control of multiples genes (Johnson et al. 2000; Shen et al. 2001; de Dorlodot et al. 2007; Reynolds et al. 2007; Steele et al. 2007; Bernier et al. 2009), indicating some potential for improvement.

A less recognized contribution of root system architecture to water uptake relates to the ability of root systems to shape soil moisture gradients in a way that optimizes water flow from the bulk soil to the root. During the growing season, evaporation and drainage create pronounced vertical water potential gradients that are modified by the spatial patterns of root water uptake in a way that orients water flow to the depleted areas close to roots. However, parts of the root system that are surrounded by zones with high RLDs can fail to attract water from the bulk soil if they are unable to decrease the soil water potential below that of the surrounding zone. In this context, the gravitropic behavior of crown roots in cereals may contribute to keep vertical "low-RLD channels" between crown roots, in which capillary rise occurs and is capable of supplying water to the upper layers. Finally, root system extension also contributes to the placement of young segments in new soil regions and ensures strong soil–root water potential gradients near the most conductive root segments.

VII. Integration of Water Flow in the Soil–Plant Domain

Our understanding of the hydraulic behavior at the whole-plant (or crop) scale is much less advanced. The complexity of the whole system raised numerous discussions about the main resistance to water flow in the soil–plant domain. Analyses of the plant hydraulic conductance in soils coupled with simplistic or inaccurate assumptions led to apparently contradictory conclusions on the main resistance (in the cortex, the xylem, or the rhizosphere). Several authors found that under wet soil conditions, soil conductivity is much higher than root radial conductivity, and vice versa (Gardner et al. 1962; Newman 1969b; Arya et al. 1975; Passioura 1980). A comparison of root and soil conductances suggested that there is an important range of soil moisture where both of them can locally limit water uptake (Figure 24.3; Draye et al. 2010). Therefore, understanding how these controls cooperate (e.g., additive, feedback of feedforward actions) has become pivotal for predicting the patterns of water uptake and the development of soil moisture gradients.

The major perspective in the area of root water uptake is therefore to bridge the gap between local controls or regulatory mechanisms and water flow within the soil–plant domain. For this to happen, we need to build a "systems view" integrating the regulation of the driving force, the hydraulics of the soil–plant domain at different scales, and the additional regulation layer provided by shoot–root signaling. Hopefully, novel tools are being developed that will move research in this direction. This section introduces some of these tools and illustrates with a few examples how such integration can help in elaborating and validating new hypotheses.

A. Novel Techniques Supporting Integration

The practical difficulties in capturing the water dynamics of the whole system, which are largely responsible for the slow pace of progress at the system scale, are now being addressed with novel experimental approaches that support the integration of root architecture with 2- or 3D soil water content. Neutron tomography (Carminati et al. 2010; Esser et al. 2010) and nuclear magnetic resonance (Jahnke et al. 2009; Borisjuk et al. 2012) but also x-ray computer tomography (Mairhofer et al. 2012) and light transmission imaging enable the visualization of root growth and the impact of water uptake on the soil properties and soil moisture distribution. Coupled with tracer experiments and detailed measurements of the plant water potential and xylem fluxes, these techniques are expected to help analyze how plant and soil resistances are distributed, develop, and affect water flow.

In parallel to experimental techniques, novel integrated modeling approaches have been recently proposed. These rely on the coupling of a 3D physical model of water flow in the soil with a biophysical model of plant water flow based on the root hydraulic architecture (Javaux et al. 2011). The R-SWMS model

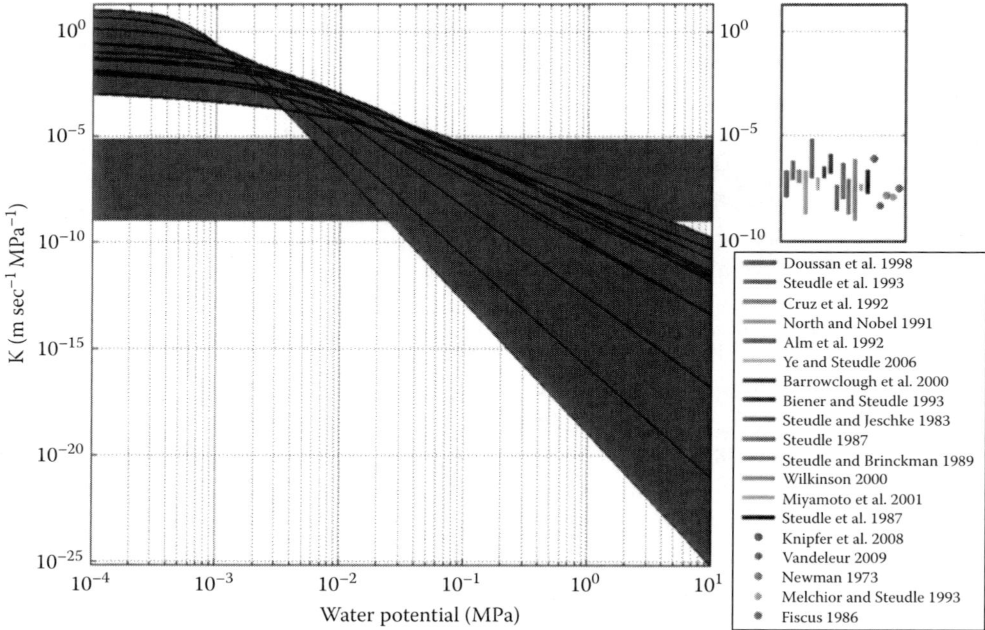

FIGURE 24.3 Envelopes of typical soil conductivity curves (dark area between curves) and apparent root conductivity values (dark rectangle area) redrawn from the literature. The upper right panel represents root conductivity values from 19 (listed in order in the reference list on the lower right panel). (Reproduced from Draye, X. et al., Model-assisted integration of physiological and environmental constraints affecting the dynamic and spatial patterns of root water uptake from soils, *J. Exp. Bot.*, 2010, 61, 2145, by permission of Oxford University Press.)

(Javaux et al. 2008) performs this coupling using Richards equation for the soil compartment (see preceding text) and defining the sink term for each soil voxel (typically having a volume lower than 1 cm^3) based on the solution of Doussan's equations for the root segments present in the voxel. The latter equations are solved using, for each segment, the local values of the soil water potential that satisfy Richards equations. An iterative numerical algorithm is used to find the 3D water flows in the whole soil–plant domain that satisfy both models. Using this type of model, it is possible to consider the different controls that have been discussed earlier. For example, the radial conductivity can be set to change with root order and segment age (Doussan et al. 1998), or the axial conductivity can decrease with xylem water potential using a susceptibility function (Li et al. 2009).

B. Need for Integration in Plant Physiology Studies

The appeal of deciphering cellular mechanisms underlying local conductivities should not draw the attention away from the fact that their effect on the local water flow is largely dependent on the hydraulic state of the whole system. For example, the effect of aquaporin activity on the radial conductivity of deep roots is likely to be marginal under well-watered conditions in the upper soil layers, while it would make a strong contribution to water uptake when water has been depleted from the upper soil layers. Understanding the regulation of local controls may thus require consideration of other parts of the system.

Recent experiments on ABA signaling during root zone drying indicated that the response of root xylem ABA concentration to decreasing soil water potential varied with the texture of the soil substrate (Figure 24.4A; Dodd et al. 2010). The mechanisms responsible for this substrate dependence were not clearly understood. Interestingly, independent simulation experiments with the R-SWMS model indicated that the concentration of a stress factor (called "ABA analogue") in the xylem sap would follow very similar trends and substrate dependency (Figure 24.4B; Draye et al. 2010). The explanation proposed in the modeling study relied on the hydraulic specificities of soil substrates (see preceding text). In particular, the conductivity of the sand substrate is likely to drop before that of the clay loam substrate in the vicinity of the roots, leading to a local water deficit (and ABA production) at higher soil water potential in the sand substrate. Although this hypothesis remains to be tested, this example indicates that an integration of the soil and plant hydraulic properties may provide new hypotheses for plant physiological experiments.

C. Integration Helps in Understanding Water Uptake Patterns

When soil moisture content is unevenly distributed and relatively low, a small part of the root system located in a wetter zone may acquire a large part of the soil water, essentially because the soil hydraulic conductivity and soil–root potential gradient are higher in the wetter zones. In this situation, water

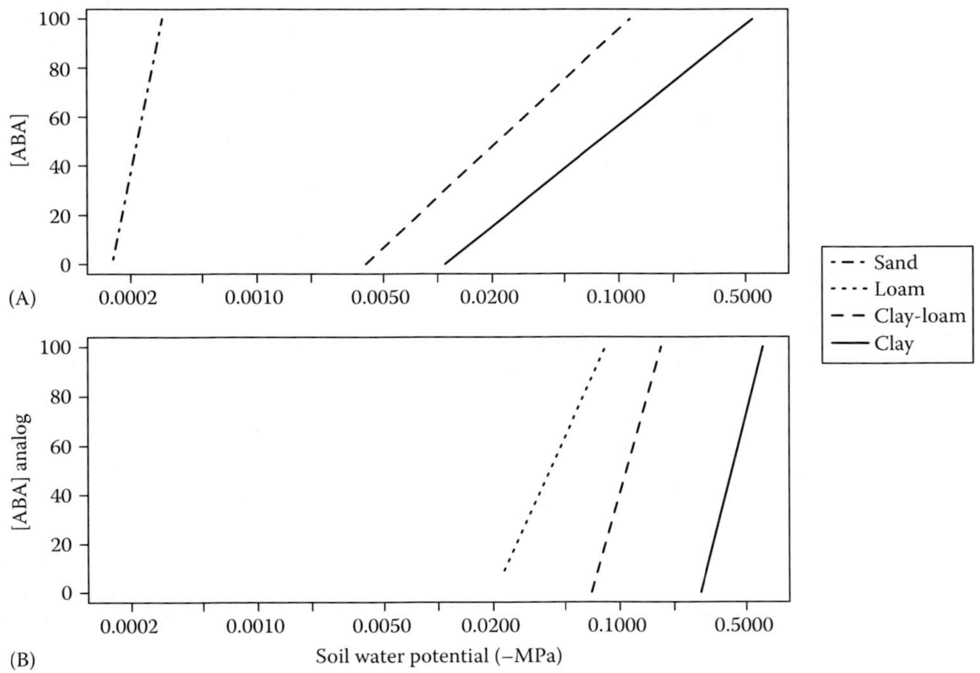

FIGURE 24.4 Evolution of xylem concentration of ABA (A, experimental) (Redrawn from Dodd, I.C. et al., *J. Exp. Bot.*, 61, 3543, 2010.) and a hypothetical stress factor (B, simulation) with decreasing soil water potential. (Redrawn from Draye, X. et al., *J. Exp. Bot.*, 61, 2145, 2010.)

is not extracted proportionally to RLD and it is said that a lack of uptake in certain root zones is compensated for by increased uptake in wetter zone (Jarvis 2010). Similar deviations from the proportionality of uptake to RLD are expected to occur when the xylem conductivity is reduced (Draye et al. 2010).

This compensation phenomenon is a passive process that results from the redistribution of soil and plant water and arises intrinsically from the physical laws underlying water flow in the SPAC (as discussed in the previous sections). However, such compensation is also determined indirectly by root placement and hydraulic architecture. It follows that differences between water extraction patterns of different genotypes that result from compensation and can be explained by pure physical laws should not require complex biological explanations (see examples in Draye et al. 2010). Using an integrated approach, it becomes possible to extract the part of the variation that can be accounted for by simple compensation and to isolate the part of variation for which a specific biological explanation is needed.

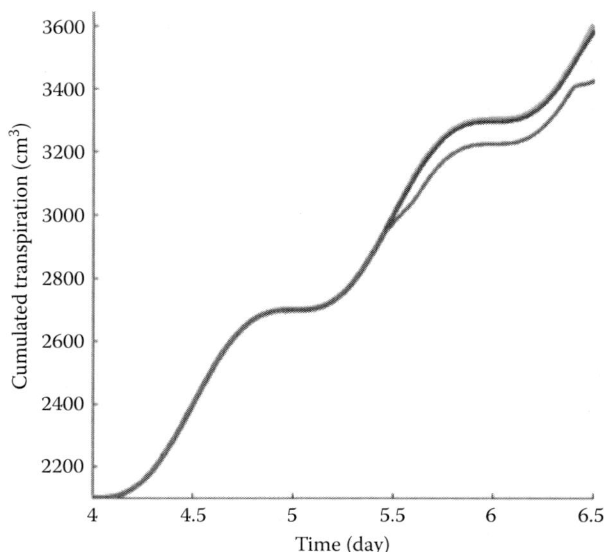

FIGURE 24.5 Cumulated transpiration rate for a plant with (grey line, lowest values) and without (dark line) cavitation of the xylem vessels, as compared to potential transpiration (grey line, highest values).

D. Integration Helps to Find Where and When Control Is Effective

As mentioned earlier, the radial flow of water is unlikely to be uniform over the entire root system surface, nor is the axial flow through all root segments. It follows that the many controls or regulatory processes that have been discussed in this chapter are unlikely to operate equally at all locations within the soil–plant domain. Depending on the situation, this heterogeneity may or may not be relevant. The case of xylem cavitation, which impedes the rate of water uptake under dry soil conditions and high evaporative demand, is one example where this heterogeneity matters. The hydraulic effect of xylem embolism has mostly been analyzed using the susceptibility to cavitation curves that quantify the loss of axial conductance, usually at the shoot or whole-plant level. However, because the soil–plant system is not hydraulically uniform, cavitation may be more important in some parts of the system than in other, and this is likely to be reflected in the patterns of water uptake and in the evolution of soil water potential.

To illustrate the benefits of integration, xylem cavitation was implemented in the R-SWMS model using a Weibull function that simulates the loss of xylem conductivity in response to the xylem water potential (Li et al. 2009). Simulations with maize and with a day–night sinusoidal transpiration demand in a globally wet soil reproduced the expected decrease of transpiration (Figure 24.5). Interestingly, the simulations indicated that cavitation occurred preferentially in the distal part of deep crown roots (not shown), which was not an intuitive result. Obviously, additional experiments are needed to determine whether this prediction matches the reality, or if our understanding of soil–plant hydraulics (as formalized in the model) needs further refinement. Nevertheless, if the prediction turned to be experimentally valid, failure to account for this asymmetric occurrence of cavitation would lead to inaccurate predictions on the spatial distribution of water uptake.

E. Future Outlooks

The last decade has seen an increasing awareness that soil and plant hydraulics (and not only RLD) play an important quantitative role in water capture, especially in drought-prone environments (de Dorlodot et al. 2007; Lynch 2007; Bernier et al. 2009). This shift of emphasis was made possible thanks to the accumulation of new data on the molecular and cellular processes controlling water flow in the plant and to the development of novel tools allowing the integration of water flow in all compartments and at different scales.

The major perspective in the area of root water uptake is therefore to rely on this development to bridge the gap between local controls or regulatory mechanisms and water flow within the soil–plant domain. For this to happen, we need to build a "systems view" integrating the regulation of the driving force, the hydraulics of the soil–plant domain at different scales, and the additional regulatory layer provided by shoot–root signaling. In particular, there is a need for plant biologists and soil hydrologists to develop together their understanding of water flow in the soil–plant system.

In this context, developing a holistic view of the mechanisms controlling the distribution of water uptake (spatial and temporal) will be crucial for our understanding of plant water use and soil water availability, mainly because water uptake at any given time modifies water availability in the next time step (feedforward). Being able to separate between the contributions of the many controls to water flow during the crop cycle will be especially important in predicting the water availability to plants with promising traits.

Functional–structural plant models are likely to play a pivotal role in future research, especially because they offer a unique mathematical framework to integrate, in a quantitative way, molecular, physiological, biophysical, and hydrological data relating to important aspects of plant water use as phenology, assimilation, growth, and long-distance signaling (Godin et al. 2005). The relevance of models considering the flow of water to individual rootlets of a complete root system has long been questioned in soil physics, primarily because the precise geometry of the root system is impossible to capture (Molz et al. 1970; King et al. 2003). As exploration tools, however, such models have a great potential since they enable visual, intuitive, and quantitative appreciation of the hydraulic behavior of root systems in their soil environment.

References

Alexandersson E, Fraysse L, Sjovall-Larsen S et al. 2005. Whole gene family expression and drought stress regulation of aquaporins. *Plant Mol Biol* 59:469–484.

Aravena JE, Berli M, Ghezzehei TA, Tyler SW. 2011. Effects of root-induced compaction on rhizosphere hydraulic properties—X-ray microtomography imaging and numerical simulations. *Environ Sci Technol* 45:425–431.

Aroca R, Ferrante A, Vernieri P, Chrispeels MJ. 2006. Drought, abscisic acid and transpiration rate effects on the regulation of PIP aquaporin gene expression and abundance in *Phaseolus vulgaris* plants. *Ann Bot* 98:1301–1310.

Aroca R, Porcel R, Ruiz-Lozano JM. 2011. Regulation of root water uptake under abiotic stress conditions. *J Exp Bot* 63:43–57.

Arya L, Blake G, Farrell D. 1975. Field study of soil-water depletion patterns in presence of growing soybean roots. 2. Effect of plant-growth on soil-water pressure and water-loss patterns. *Soil Sci Soc Am J* 39:430–436.

Arya L, Farrell D, Blake G. 1975. Field study of soil-water depletion patterns in presence of growing soybean roots. 1. Determination of hydraulic properties of soil. *Soil Sci Soc Am J* 39:424–430.

Bernier J, Serraj R, Kumar A et al. 2009. The large-effect drought-resistance QTL qtl12.1 increases water uptake in upland rice. *Field Crops Res* 110:139–146.

Bibikova T, Gilroy S. 2003. Root hair development. *J Plant Growth Regul* 21:383–415.

Blum A. 2005. Drought resistance, water-use efficiency, and yield potential—Are they compatible, dissonant, or mutually exclusive? *Aust J Agric Res* 56:1159–1168.

Borisjuk L, Rolletschek H, Neuberger T. 2012. Surveying the plant's world by magnetic resonance imaging. *Plant J* 70:129–146.

Boursiac Y, Boudet J, Postaire O et al. 2008. Stimulus-induced downregulation of root water transport involves reactive oxygen species-activated cell signalling and plasma membrane intrinsic protein internalization. *Plant J* 56:207–218.

Boyer JS. 1995. *Measuring the Water Status of Plants and Soils.* New York: Academic Press.

Bramley H, Turner N, Turner D, Tyerman SD. 2009. Roles of morphology, anatomy, and aquaporins in determining contrasting hydraulic behavior of roots. *Plant Physiol* 150:348–364.

Carminati A, Moradi AB, Vetteriein D et al. 2010. Dynamics of soil water content in the rhizosphere. *Plant Soil* 332:163–176.

Carminati A, Schneider CL, Moradi AB et al. 2011. How the rhizosphere may favor water availability to roots. *Vadose Zone J* 10:988–998.

Carminati A, Vetterlein D, Weller U, Vogel H-J, Oswald SE. 2009. When roots lose contact. *Vadose Zone J* 8:805–809.

Chaumont F, Barrieu F, Wojcik E, Chrispeels M, Jung R. 2001. Aquaporins constitute a large and highly divergent protein family in maize. *Plant Physiol* 125:1206–1215.

Chaumont F, Moshelion M, Daniels MJ. 2005. Regulation of plant aquaporin activity. *Biology Cell* 97:749–764.

Christman MA, Sperry JS, Smith DD. 2012. Rare pits, large vessels and extreme vulnerability to cavitation in a ring-porous tree species. *New Phytol* 193:713–720.

Clearwater MJ, Goldstein G. 2005. Embolism repair and long distance water transport. In *Vascular Transport in Plants*, eds. NM Holbrook, MA Zwieniecki, pp. 375–399. London, U.K.: Elsevier.

Cochard H. 2002. Xylem embolism and drought-induced stomatal closure in maize. *Planta* 215:466–471.

Cochard H, Barigah ST, Kleinhentz M, Eshel A. 2008. Is xylem cavitation resistance a relevant criterion for screening drought resistance among *Prunus* species? *J Plant Physiol* 165:976–982.

Cochard H, Casella E, Mencuccini M. 2007a. Xylem vulnerability to cavitation varies among poplar and willow clones and correlates with yield. *Tree Physiol* 27:1761–1767.

Cochard H, Venisse J-S, Barigah TS, Brunel N et al. 2007b. Putative role of aquaporins in variable hydraulic conductance of leaves in response to light. *Plant Physiol* 143:122–133.

Czarnes S, Hallett PD, Bengough AG, Young IM. 2000. Root- and microbial-derived mucilages affect soil structure and water transport. *Eur J Soil Sci* 51:435–443.

Darrah PR. 1993. The rhizosphere and plant nutrition: A quantitative approach. *Plant Soil* 155–156:1–20.

Delzon S, Douthe C, Sala A, Cochard H. 2010. Mechanism of water-stress induced cavitation in conifers: Bordered pit structure and function support the hypothesis of seal capillary-seeding. *Plant Cell Environ* 33:2101–2111.

Dexter A. 1987. Compression of soil around roots. *Plant Soil* 97:401–406.

Dodd IC, Egea G, Watts CW, Whalley WR. 2010. Root water potential integrates discrete soil physical properties to influence ABA signalling during partial rootzone drying. *J Exp Bot* 61:3543–3551.

de Dorlodot S, Forster B, Pagès L et al. 2007. Root system architecture: Opportunities and constraints for genetic improvement of crops. *Trends Plant Sci* 12:474–481.

Doussan C, Pagès L, Vercambre G. 1998a. Modelling of the hydraulic architecture of root systems: An integrated approach to water absorption—Model description. *Ann Bot* 81:213–223.

Doussan C, Pierret A, Garrigues E, Pagès L. 2006. Water uptake by plant roots: II—Modelling of water transfer in the soil root-system with explicit account of flow within the root system—Comparison with experiments. *Plant Soil* 283:99–117.

Doussan C, Vercambre G, Pagès L. 1998b. Modelling of the hydraulic architecture of root systems: And integrated approach to water absorption—Distribution of axial and radial conductances in maize. *Ann Bot* 81:225–232.

Draye X, Kim Y, Lobet G, Javaux M. 2010. Model-assisted integration of physiological and environmental constraints affecting the dynamic and spatial patterns of root water uptake from soils. *J Exp Bot* 61:2145–2155.

Duan H, Schuler MA. 2005. Differential expression and evolution of the *Arabidopsis* CYP86A subfamily. *Plant Physiol* 137:1067–1081.

Efetova M, Zeier J, Riederer M et al. 2007. A central role of abscisic acid in drought stress protection of *Agrobacterium*-induced tumors on *Arabidopsis*. *Plant Physiol* 145:853–862.

Enstone DE, Peterson CA. 2005. Suberin lamella development in maize seedling roots grown in aerated and stagnant conditions. *Plant Cell Environ* 28:444–455.

Enstone DE, Peterson CA, Ma F. 2003. Root endodermis and exodermis: Structure, function, and responses to the environment. *J Plant Growth Regul* 21:335–351.

Esau K. 1965. *Plant Anatomy*, 2nd edn., New York: John Wiley & Sons.

Eshel A, Waisel Y. 1996. Multiform and multifunction of various constituents of one root system. In *Plant Roots: The Hidden Half*, eds. Y Waisel, A Eshel, U Kafkafi, 2nd edn., pp. 175–192. New York: Marcel Dekker, Inc.

Esser HG, Carminati A, Vontobel P, Lehmann EH, Oswald SE. 2010. Neutron radiography and tomography of water distribution in the root zone. *J Plant Nutr Soil Sci* 173:757–764.

Fan M, Bai R, Zhao X, Zhang J. 2007. Aerenchyma formed under phosphorus deficiency contributes to the reduced root hydraulic conductivity in maize roots. *J Integr Plant Biol* 49:598–604.

FAO, ed. 2012. FAO STAT. Available from: http://faostat.fao.org/ [accessed on April 2012].

Gardner WR. 1965. Dynamic aspects of soil-water availability to plants. *Annu Rev Plant Physiol* 16:323–342.

Gardner W, Ehlig C. 1962. Some observations on movement of water to plant roots. *Agric J* 54:453–456.

Gilroy S, Jones DL. 2000. Through form to function: Root hair development and nutrient uptake. *Trends Plant Sci* 5:56–60.

Godin C, Sinoquet H. 2005. Functional-structural plant modelling. *New Phytol* 166:705–708.

Gregory PJ, Ingram JSI, Brklacich M. 2005. Climate change and food security. *Philos Trans R Soc B Biol Sci* 360:2139–2148.

Guidi G, Poggio G, Petruzzelli G. 1985. The porosity of soil aggregates from bulk soil and from soil adhering to roots. *Plant Soil* 87:311–314.

Gupta AB, Sankararamakrishnan R. 2009. Genome-wide analysis of major intrinsic proteins in the tree plant *Populus trichocarpa*: Characterization of XIP subfamily of aquaporins from evolutionary perspective. *BMC Plant Biol* 9:134.

Hachez C, Moshelion M, Zelazny E, Cavez D, Chaumont F. 2006. Localization and quantification of plasma membrane aquaporin expression in maize primary root: A clue to understanding their role as cellular plumbers. *Plant Mol Biol* 62:305–323.

Hachez C, Veselov D, Ye Q et al. 2012. Short-term control of maize cell and root water permeability through plasma membrane aquaporin isoforms. *Plant Cell Environ* 35:185–198.

Hacke UG, Jansen S. 2009. Embolism resistance of three boreal conifer species varies with pit structure. *New Phytol* 182:675–686.

Hacke UG, Sauter JJ. 1996. Drought-induced xylem dysfunction in petioles, branches, and roots of *Populus balsamifera* L. and *Alnus glutinosa* (L.) Gaertn. *Plant Physiol* 111:413–417.

Hacke UG, Sperry JS. 2003. Limits to xylem refilling under negative pressure in *Laurus nobilis* and *Acer negundo*. *Plant Cell Environ* 26:303–311.

Hacke UG, Sperry JS, Ewers BE et al. 2000. Influence of soil porosity on water use in *Pinus taeda*. *Oecologia* 124:495–505.

Hacke UG, Stiller V, Sperry JS, Pittermann JJ, McCulloh KA. 2001. Cavitation fatigue. Embolism and refilling cycles can weaken the cavitation resistance of xylem. *Plant Physiol* 125:779–786.

Hallett PD, Gordon DC, Bengough AG. 2003. Plant influence on rhizosphere hydraulic properties: Direct measurements using a miniaturized infiltrometer. *New Phytol* 157:597–603.

Hammer G, Dong Z, Mclean G et al. 2009. Can changes in canopy and/or root system architecture explain historical maize yield trends in the US corn belt? *Crop Sci* 49:299–312.

Henry A, Gowda VRP, Torres RO, McNally KL, Serraj R. 2011. Variation in root system architecture and drought response in rice (*Oryza sativa*): Phenotyping of the OryzaSNP panel in rainfed lowland fields. *Field Crops Res* 120:205–214.

Herbette S, Cochard H. 2010. Calcium is a major determinant of xylem vulnerability to cavitation. *Plant Physiol* 153:1932–1939.

Hinsinger P, Gobran GR, Gregory PJ, Wenzel WW. 2005. Rhizosphere geometry and heterogeneity arising from root-mediated physical and chemical processes. *New Phytol* 168:293–303.

Hodge A. 2004. The plastic plant: Root responses to heterogeneous supplies of nutrients. *New Phytol* 162:9–24.

Holbrook NM, Zwieniecki M. 1999. Embolism repair and xylem tension: Do we need a miracle? *Plant Physiol* 120:7–10.

van den Honert TH. 1948. Water transport in plants as a catenary process. *Discuss Faraday Soc* 3:146–153.

Hose E, Clarkson D, Steudle E, Schreiber L, Hartung W. 2001. The exodermis: A variable apoplastic barrier. *J Exp Bot* 52:2245–2264.

Hose E, Steudle E, Hartung W. 2000. Abscisic acid and hydraulic conductivity of maize roots: A study using cell-and root-pressure probes. *Planta* 211:874–882.

Jackson RB, Sperry JS, Dawson TE. 2000. Root water uptake and transport: using physiological processes in global predictions. *Trends Plant Sci* 5:482–488.

Jahnke S, Menzel MI, van Dusschoten D et al. 2009. Combined MRI-PET dissects dynamic changes in plant structures and functions. *Plant J* 59:634–644.

Jang JY, Kim DG, Kim YO, Kim JS, Kang H. 2004. An expression analysis of a gene family encoding plasma membrane aquaporins in response to abiotic stresses in *Arabidopsis thaliana*. *Plant Mol Biol* 54:713–725.

Jarbeau JA, Ewers FW, Davis SD. 1995. The mechanism of water-stress-induced embolism in two species of chaparral shrubs. *Plant Cell Environ* 18:189–196.

Jarvis N. 2010. Comment on "Macroscopic root water uptake distribution using a matric flux potential approach." *Vadose Zone J* 9:499.

Javaux M, Draye X, Doussan C, Vanderborght J, Vereecken H. 2011. Root water uptake: Toward 3-D functional approaches. In *Encyclopedia of Agrophysics*, eds. J Gliński, J Horabik, J Lipiec, pp. 717–722. Dordrecht, the Netherlands: Springer.

Javaux M, Schroeder T, Vanderborght J, Vereecken H. 2008. Use of a three-dimensional detailed modeling approach for predicting root water uptake. *Vadose Zone J* 7:1079–1088.

Javot H, Lauvergeat V, Santoni V et al. 2003. Role of a single aquaporin isoform in root water uptake. *Plant Cell* 15:509–522.

Javot H, Maurel C. 2002. The role of aquaporins in root water uptake. *Ann Bot* 90:301–313.

Johanson U, Karlsson M, Johansson I et al. 2001. The complete set of genes encoding major intrinsic proteins in *Arabidopsis* provides a framework for a new nomenclature for major intrinsic proteins in plants. *Plant Physiol* 126:1358–1369.

Johnson W, Jackson L, Ochoa O. 2000. Lettuce, a shallow-rooted crop, and *Lactuca serriola*, its wild progenitor, differ at QTL determining root architecture and deep soil water exploitation. *Theor Appl Genet* 101:1066–1073.

Johnson DM, McCulloh KA, Woodruff DR, Meinzer FC. 2012. Evidence for xylem embolism as a primary factor in dehydration-induced declines in leaf hydraulic conductance. *Plant Cell Environ* 35:760–769.

Kaldenhoff R, Grote K, Zhu JJ, Zimmermann U. 1998. Significance of plasmalemma aquaporins for water-transport in *Arabidopsis thaliana*. *Plant J* 14:121–128.

Karahara I, Ikeda A, Kondo T, Uetake Y. 2004. Development of the Casparian strip in primary roots of maize under salt stress. *Planta* 219:41–47.

Kashiwagi J, Krishnamurthy L, Crouch JH, Serraj R. 2006. Variability of root length density and its contributions to seed yield in chickpea (*Cicer arietinum* L.) under terminal drought stress. *Field Crops Res* 95:171–181.

King J, Gay A, Sylverster-Bradley R et al. 2003. Modelling cereal root systems for water and nitrogen capture: Towards an economic optimum. *Ann Bot* 91:383–390.

Knipfer T, Besse M, Verdeil J-L, Fricke W. 2011. Aquaporin-facilitated water uptake in barley (*Hordeum vulgare* L.) roots. *J Exp Bot* 62:4115–4126.

Knipfer T, Fricke W. 2010. Root pressure and a solute reflection coefficient close to unity exclude a purely apoplastic pathway of radial water transport in barley (*Hordeum vulgare*). *New Phytol* 187:159–170.

Landsberg J, Fowkes N. 1978. Water movement through plant roots. *Ann Bot* 42:493–508.

Lecompte F, Ozier-Lafontaine H, Pagès L. 2001. The relationships between static and dynamic variables in the description of root growth. Consequences for field interpretation of rooting variability. *Plant Soil* 236:19–31.

Li Y, Sperry JS, Shao M. 2009. Hydraulic conductance and vulnerability to cavitation in corn (*Zea mays* L.) hybrids of differing drought resistance. *Environ Exp Bot* 66:341–346.

Li X, Wang X, Yang Y et al. 2011. Single-molecule analysis of PIP2;1 dynamics and partitioning reveals multiple modes of *Arabidopsis* plasma membrane aquaporin regulation. *Plant Cell* 23:3780–3797.

Lopez F, Bousser A, Sissoëff I et al. 2003. Diurnal regulation of water transport and aquaporin gene expression in maize roots: Contribution of PIP2 proteins. *Plant Cell Physiol* 44:1384–1395.

Luu DT, Martiniere A, Sorieul M, Runions J, Maurel C. 2011. Fluorescence recovery after photobleaching reveals high cycling dynamics of plasma membrane aquaporins in *Arabidopsis* roots under salt stress. *Plant J* 69:894–905.

Lynch JP. 2007. Roots of the second green revolution. *Aust J Bot* 55:493–512.

Lynch JP, Brown KM. 2012. New roots for agriculture: Exploiting the root phenome. *Philos Trans R Soc B Biol Sci* 367:1598–1604.

Maathuis FJ, Filatov V, Herzyk P et al. 2003. Transcriptome analysis of root transporters reveals participation of multiple gene families in the response to cation stress. *Plant J* 35:675–692.

Mahdieh M, Mostajeran A. 2009. Abscisic acid regulates root hydraulic conductance via aquaporin expression modulation in *Nicotiana tabacum*. *J Plant Physiol* 166:1993–2003.

Mairhofer S, Zappala S, Tracy SR et al. 2012. RooTrak: Automated recovery of three-dimensional plant root architecture in soil from X-ray microcomputed tomography images using visual tracking. *Plant Physiol* 158:561–569.

Mallants D, Mohanty BP, Vervoort A, Feyen J. 1997. Spatial analysis of saturated hydraulic conductivity in a soil with macropores. *Soil Technol* 10:115–131.

Marcum KB, Morton S, Engelke M, White R. 1995. Rooting characteristics and associated drought resistance of zoysia grasses. *Agric J* 87 (3):534–538.

Martinez-Ballesta MC, Aparicio F, Pallas V, Martinez V, Carvajal M. 2003. Influence of saline stress on root hydraulic conductance and PIP expression in *Arabidopsis*. *J Plant Physiol* 160:689–697.

Martre P, Morillon R, Barrieu F et al. 2002. Plasma membrane aquaporins play a significant role during recovery from water deficit. *Plant Physiol* 130:2101–2110.

Maurel C, Chrispeels MJ. 2001. Aquaporins. A molecular entry into plant water relations. *Plant Physiol* 125:135–138.

Maurel C, Reizer J, Schroeder J, Chrispeels M. 1993. The vacuolar membrane protein gamma-TIP creates water specific channels in *Xenopus* oocytes. *EMBO J* 12:2241–2247.

Maurel C, Simonneau T, Sutka M. 2010. The significance of roots as hydraulic rheostats. *J Exp Bot* 61:3191–3198.

Maurel C, Verdoucq L, Luu D, Santoni V. 2008. Plant aquaporins: Membrane channels with multiple integrated functions. *Annu Rev Plant Biol* 59:595–624.

McCully M. 1999. Root xylem embolisms and refilling. Relation to water potentials of soil, roots, and leaves, and osmotic potentials of root xylem sap. *Plant Physiol* 199:1001–1008.

McCully ME, Boyer JS. 1997. The expansion of maize root-cap mucilage during hydration. 3 Changes in water potential and water content. *Physiol Plant* 99:169–177.

McCully ME, Huang CX, Ling LEC. 1998. Daily embolism and refilling of xylem vessels in the roots of field-grown maize. *New Phytol* 138:327–342.

Molz F, Remson I. 1970. Extraction term models of soil moisture use by transpiring plants. *Water Resour Res* 6:1346.

Moradi AB, Carminati A, Lamparter A et al. 2012. Is the rhizosphere temporarily water repellent? *Vadose Zone J* 11(3):Doi 10.2136/Vzj 2011.0120.

Moshelion M, Becker D, Biela A et al. 2002. Plasma membrane aquaporins in the motor cells of Samanea saman: Diurnal and circadian regulation. *Plant Cell* 14:727–739.

Nagarajah S, Ratnasuriya GB. 1981. Clonal variability in root growth and drought resistance in tea (*Camellia sinensis*). *Plant Soil* 60:153–155.

Nagel OW, Konings H, Lambers H. 1994. Growth rate, plant development and water relations of the ABA deficient tomato mutant *sitiens*. *Physiol Plant* 92:102–108.

Newman EI. 1969a. Resistance to water flow in soil and plant. II A review of experimental evidence on rhizosphere resistance. *J Appl Ecol* 6:261–272.

Newman EI. 1969b. Resistance to water flow in soil and plant. I soil resistance in relation to amounts of root–theoretical estimates. *J Appl Ecol* 6:1–12.

Nielsen K, Bouma T, Lynch J, Eissenstat DM. 1998. Effects of phosphorus availability and vesicular–arbuscular mycorrhizas on the carbon budget of common bean (*Phaseolus vulgaris*). *New Phytol* 139:647–656.

Niemietz CM, Tyerman SD. 2002. New potent inhibitors of aquaporins: Silver and gold compounds inhibit aquaporins of plant and human origin. *FEBS Lett* 531:443–447.

Parent B, Hachez C, Redondo E et al. 2009. Drought and abscisic acid effects on aquaporin content translate into changes in hydraulic conductivity and leaf growth rate: A trans-scale approach. *Plant Physiol* 149:2000–2012.

Park W, Scheffler B, Bauer P. 2010. Identification of the family of aquaporin genes and their expression in upland cotton (*Gossypium hirsutum* L.). *BMC Plant Biol* 10:142.

Passioura JB. 1980. The transport of water from soil to shoot in wheat seedlings. *J Exp Bot* 31:333–345.

Peirce GJ. 1936. The state of water in ducts and tracheids. *Plant Physiol* 11:623–628.

Perumalla CJ, Peterson CA. 1986. Deposition of Casparian bands and suberin lamellae in the exodermis and endodermis of young corn and onion roots. *Can J Bot* 38:1872–1878.

Péret B, de Rybel B, Casimiro I et al. 2009. *Arabidopsis* lateral root development: An emerging story. *Trends Plant Sci* 14:399–408.

Peterson CA, Murrmann M, Steudle E. 1993. Location of the major barriers to water and ion movement in young roots of *Zea mays* L. *Planta* 190:127–136.

Pinheiro HA, DaMatta F, Chaves ARM, Loureiro ME, Ducatti C. 2005. Drought tolerance is associated with rooting depth and stomatal control of water use in clones of *Coffea canephora*. *Ann Bot* 96:101–108.

Pockman WT, Sperry JS. 2000. Vulnerability to xylem cavitation and the distribution of Sonoran Desert vegetation. *Am J Bot* 87:1287–1299.

Pregitzer KS, Hendrick RL, Fogel R. 1993. The demography of fine roots in response to patches of water and nitrogen. *New Phytol* 125:575–580.

Ranathunge K, Schreiber L, Franke R. 2011. Suberin research in the genomics era—New interest for an old polymer. *Plant Sci* 180:399–413.

Read DB, Bengough AG, Gregory PJ et al. 2003. Plant roots release phospholipid surfactants that modify the physical and chemical properties of soil. *New Phytol* 157:315–326.

Read DB, Gregory PJ, Bell AE. 1999. Physical properties of axenic maize root mucilage. *Plant Soil* 211:87–91.

Rewald B, Ephrat JE, Rachmilevitch S. 2011b. A root is a root is a root? Water uptake rates of *Citrus* root orders. *Plant Cell Environ* 34:33–42.

Rewald B, Leuschner C, Weisman Z, Ephrat JE. 2011a. Influence of salinity on root hydraulic properties of three olive varieties. *Plant Biosys* 145:12–22.

Reynolds M, Dreccer F, Trethowan R. 2007. Drought-adaptive traits derived from wheat wild relatives and landraces. *J Exp Bot* 58:177–186.

Richards LA. 1931. Capillary conduction of liquids through porous mediums. *J Appl Physics* 1:318–333.

Ruiz-Lozano JM, Alguacii MDM, Barzana G, Vemieri P, Aroca R. 2009. Exogenous ABA accentuates the differences in root hydraulic properties between mycorrhizal and non mycorrhizal maize plants through regulation of PIP aquaporins. *Plant Mol Biol* 70:565–579.

Sakurai J, Ishikawa F, Murai-Hatano M, Hayashi H et al. 2011. Transpiration from shoots triggers diurnal changes in root aquaporin expression. *Plant Cell Environ* 34:1150–1163.

Sakurai J, Ishikawa F, Yamaguchi T, Uemura M, Maeshima M. 2005. Identification of 33 rice aquaporin genes and analysis of their expression and function. *Plant Cell Physiol* 46:1568–1577.

Schreiber L, Hartmann K, Skrabs M, Zeier J. 1999. Apoplastic barriers in roots: Chemical composition of endodermal and hypodermal cell walls. *J Exp Bot* 50:1267–1280.

Schröder T, Javaux M, Vanderborght J, Körfgen B, Vereecken H. 2009. Implementation of a microscopic soil–root hydraulic conductivity drop function in a three-dimensional soil–root architecture water transfer model. *Vadose Zone J* 8:783–792.

Secchi F, Gilbert ME, Zwieniecki MA. 2011. Transcriptome response to embolism formation in stems of *Populus trichocarpa* provides insight into signaling and the biology of refilling. *Plant Physiol* 157:1419–1429.

Secchi F, Zwieniecki MA. 2010. Patterns of PIP gene expression in *Populus trichocarpa* during recovery from xylem embolism suggest a major role for the PIP1 aquaporin subfamily as moderators of refilling process. *Plant Cell Environ* 33:1285–1297.

Secchi F, Zwieniecki MA. 2011. Sensing embolism in xylem vessels: The role of sucrose as a trigger for refilling. *Plant Cell Environ* 34:514–524.

Segal E, Kushnir T, Mualem Y, Shani U. 2008. Water uptake and hydraulics of the root hair rhizosphere. *Vadose Zone J* 7:1027.

Shen L, Courtois B, McNally KL, Robin S, Li Z. 2001. Evaluation of near-isogenic lines of rice introgressed with QTLs for root depth through marker-aided selection. *Theor Appl Genet* 103:75–83.

Siefritz F, Tyree MT, Lovisolo C, Schubert A. 2002. PIP1 plasma membrane aquaporins in tobacco. *Plant Cell Online* 14:869–876.

Sperry JS, Adler FR, Campbell GS, Comstock JP. 1998. Limitation of plant water use by rhizosphere and xylem conductance: Results from a model. *Plant Cell Environ* 21:347–359.

Sperry JS, Donnelly JR, Tyree MT. 1988a. A method for measuring hydraulic conductivity and embolism in xylem. *Plant Cell Environ* 11:35–40.

Sperry JS, Donnelly JR, Tyree MT. 1988b. Seasonal occurrence of xylem embolism in sugar maple (*Acer saccharum*). *Am J Bot* 75:1212–1218.

Sperry JS, Holbrook NM, Zimmermann MH, Tyree MT. 1987. Spring filling of xylem vessels in wild grapevine. *Plant Physiol* 83:414–417.

Sperry JS, Ikeda T. 1997. Xylem cavitation in roots and stems of Douglas fir and white fir. *Tree Physiol* 17:275–280.

Sperry JS, Saliendra NZ. 1994. Intra- and inter-plant variation in xylem cavitation in *Betula occidentalis. Plant Cell Environ* 17:1233–1241.

Sperry JS, Stiller V, Hacke UG. 2002. Soil water uptake and water transport through root systems. In *Plant Roots: The Hidden Half*, eds. Y Waisel, A Eshel, U Kafkafi, 3rd edn., pp. 663–681. New York: Marcel Dekker, Inc.

Steele K, Virk D, Kumar R, Prasad S, Witcombe J. 2007. Field evaluation of upland rice lines selected for QTLs controlling root traits. *Field Crops Res* 101:180–186.

Steudle E. 2000. Water uptake by plant roots: An integration of views. *Plant Soil* 226:45–56.

Steudle E, Frensch J. 1989. Osmotic responses of maize roots. *Planta* 177:281–295.

Steudle E, Peterson CA. 1998. How does water get through roots? *J Exp Bot* 49:775–788.

Stiller V, Sperry JS. 2002. Cavitation fatigue and its reversal in sunflower (*Helianthus annuus* L.). *J Exp Bot* 53:1155–1161.

Sutka M, Li G, Boudet J et al. 2011. Natural variation of root hydraulics in *Arabidopsis* grown in normal and salt-stressed conditions. *Plant Physiol* 155:1264–1276.

Tardieu F, Simonneau T. 1998. Variability among species of stomatal control under fluctuating soil water status and evaporative demand: Modelling isohydric and anisohydric behaviours. *J Exp Bot* 49:419–432.

Thomas R, Fang X, Ranathunge K et al. 2007. Soybean root suberin: Anatomical distribution, chemical composition, and relationship to partial resistance to *Phytophthora sojae. Plant Physiol* 144:299–311.

Thompson A, Andrews J, Mulholland B et al. 2007. Overproduction of abscisic acid in tomato increases transpiration efficiency and root hydraulic conductivity and influences leaf expansion. *Plant Physiol* 143:1905.

Törnroth-Horsefield S, Wang Y, Hedfalk K et al. 2006. Structural mechanism of plant aquaporin gating. *Nature* 439:689–694.

Tournaire-Roux C, Sutka M, Javot H et al. 2003. Cytosolic pH regulates root water transport during anoxic stress through gating of aquaporins. *Nature* 425:393–397.

Tyerman SD, Bohnert HJ, Maurel C, Steudle E, Smith JAC. 1999. Plant aquaporins: Their molecular biology, biophysics and significance for plant water relations. *J Exp Bot* 50:1055–1071.

Tyree MT, Sperry JS. 1989. Vulnerability of xylem to cavitation and embolism. *Annu Rev Plant Physiol Plant Mol Biol* 40:19–36.

Uehlein N, Otto B, Hanson DT et al. 2008. Function of Nicotiana tabacum aquaporins as chloroplast gas pores challenges the concept of membrane CO_2 permeability. *Plant Cell* 20:648–657.

Vandeleur RK, Mayo G, Shelden MC et al. 2009. The role of plasma membrane intrinsic protein aquaporins in water transport through roots: diurnal and drought stress responses reveal different strategies between isohydric and anisohydric cultivars of grapevine. *Plant Physiol* 149:445–460.

Veen BW, Van Noordwijk M, De Willigen P, Boone F, Kooistra M. 1992. Root-soil contact of maize, as measured by a thin-section technique. *Plant Soil* 139:131–138.

Wasson AP, Richards RA, Chatrath R et al. 2012. Traits and selection strategies to improve root systems and water uptake in water-limited wheat crops. *J Exp Bot* 63:3485–3498.

Watt M, McCully M, Canny M. 1994. Formation and stabilization of rhizosheaths of *Zea mays* L. *Plant Physiol* 106:179–186.

Whalley WR, Riseley B, Leeds-Harrison PB et al. 2004. Structural differences between bulk and rhizosphere soil. *Eur J Soil Sci* 56:353–360.

Yang X, Li Y, Ren B et al. 2012. Drought-induced root aerenchyma formation restricts water uptake in rice seedlings supplied with nitrate. *Plant Cell Physiol* 53:495–504.

Yang S, Tyree MT. 1992. A theoretical model of hydraulic conductivity recovery from embolism with comparison to experimental data on *Acer saccharum. Plant Cell Environ* 15:633–643.

Ye Q, Steudle E. 2006. Oxidative gating of water channels (aquaporins) in corn roots. *Plant Cell Environ* 29:459–470.

Zelazny E, Borst JW, Muylaert M et al. 2007. FRET imaging in living maize cells reveals that plasma membrane aquaporins interact to regulate their subcellular localization. *Proc Natl Acad Sci U S A* 104:12359–12364.

Zhang J, Zhang X. 1995. Exudation rate and hydraulic conductivity of maize roots are enhanced by soil drying and abscisic acid treatment. *New Phytol* 131:329–336.

Zhu J, Brown KM, Lynch JP. 2010. Root cortical aerenchyma improves the drought tolerance of maize (*Zea mays* L.). *Plant Cell Environ* 33:740–749.

Zhu J, Kaeppler SM, Lynch JP. 2005. Mapping of QTLs for lateral root branching and length in maize (*Zea mays* L.) under differential phosphorus supply. *Theor Appl Genet* 111:688–695.

Zimmermann HM, Hartmann K, Schreiber L, Steudle E. 2000. Chemical composition of apoplastic transport barriers in relation to radial hydraulic conductivity of corn roots (*Zea mays* L.). *Planta* 210:302–311.

Zufferey V, Cochard H, Améglio T, Spring JL, Viret O. 2011. Diurnal cycles of embolism formation and repair in petioles of grapevine (*Vitis vinifera* cv. Chasselas). *J Exp Bot* 62:3885–3894.

Zwieniecki MA, Holbrook NM. 2009. Confronting Maxwell's demon: Biophysics of xylem embolism repair. *Trends Plant Sci* 14:530–534.

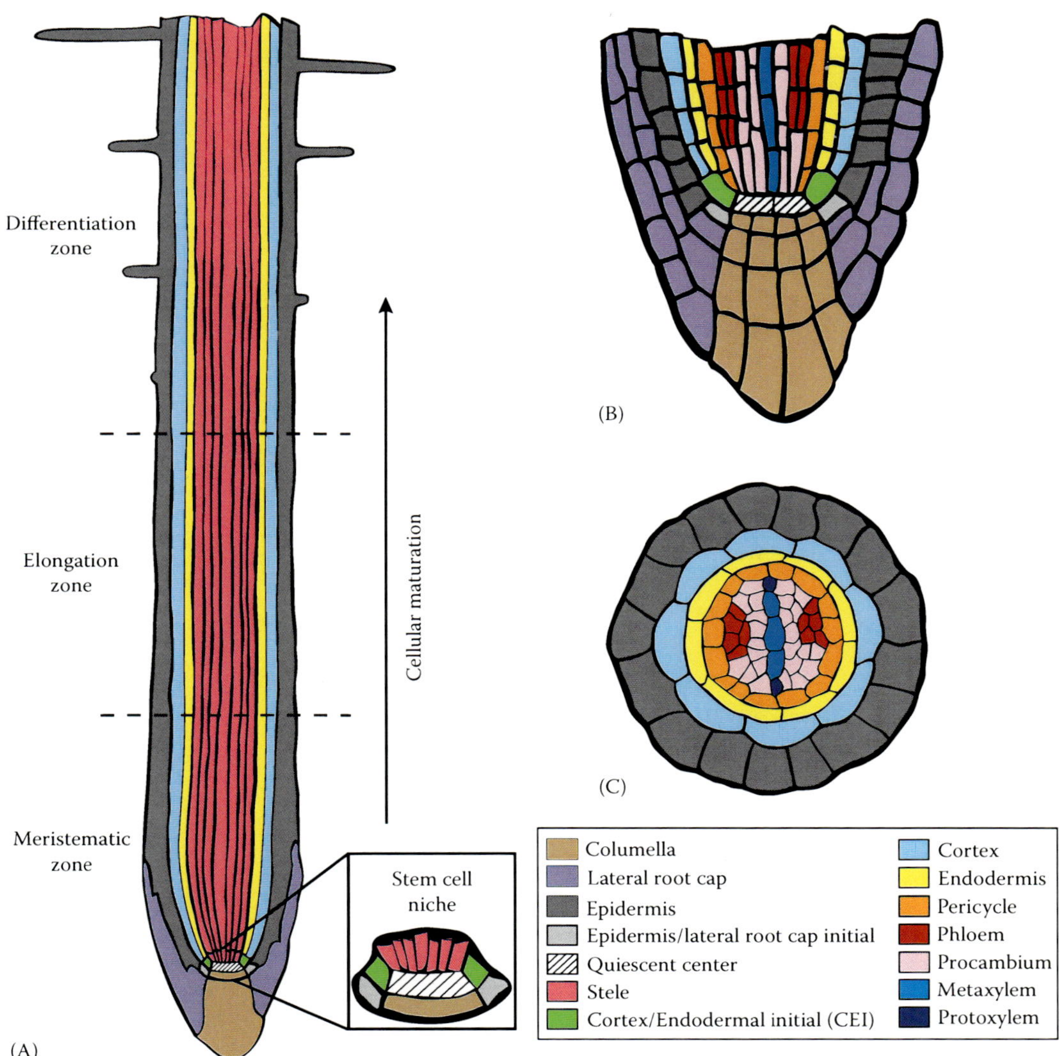

FIGURE 2.2 Schematic of the cellular organization of the *Arabidopsis* root. (A) Median longitudinal section of a length of the root. The different developmental stages and the direction of cellular maturation are noted. Inset: the stem cell niche includes the quiescent center and the adjacent initial cells. (B) Median longitudinal section of the root tip. (C) Transverse cross section within the differentiation zone. The different cell types comprising the stele (A) are detailed in (B) and (C). This drawing is not to scale.

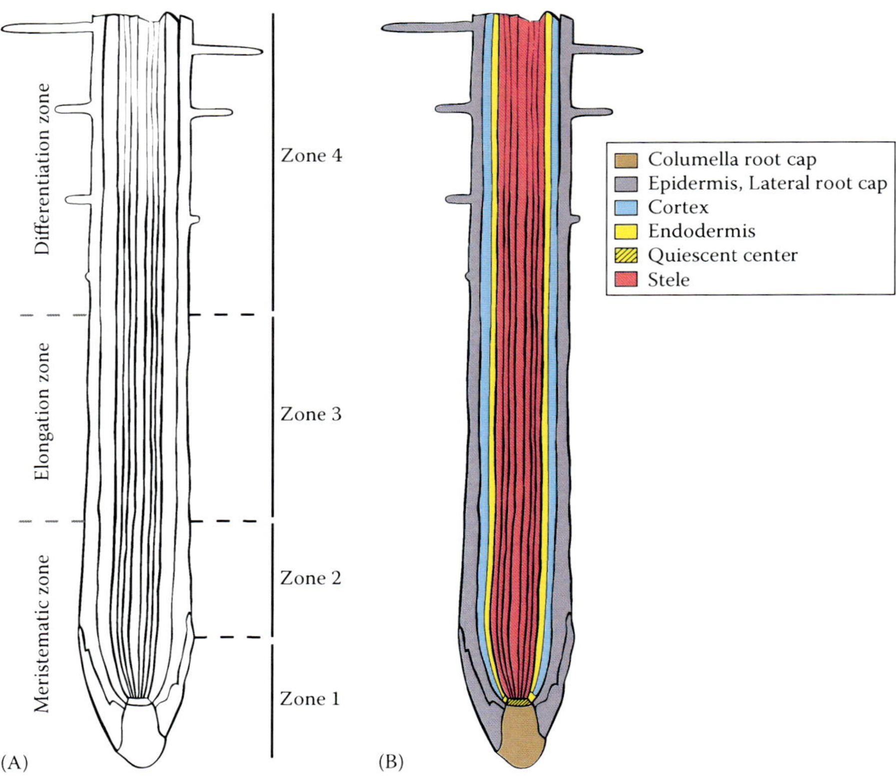

FIGURE 2.4 Schematic depicting the transverse sections in the longitudinal axis and cell types sampled for transcriptional profiling of the *Arabidopsis* root's response to stress. (A) Four transverse sections and their relationship to the developmental zones and (B) five cell populations each profiled after exposure to abiotic stress. This drawing is not to scale.

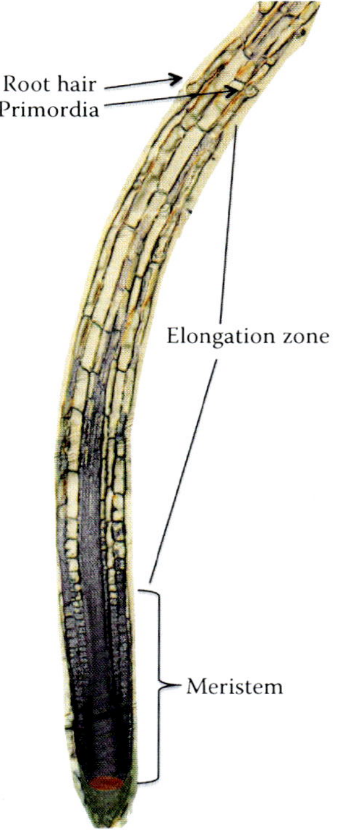

FIGURE 3.2 Developmental zones in the root. Toluidine blue–stained single medial section through an *Arabidopsis* root (kindly provided by Shuang Wu). The different zones of the root are labeled. The root cap has been false colored in green, and the root initial cells are false colored in red.

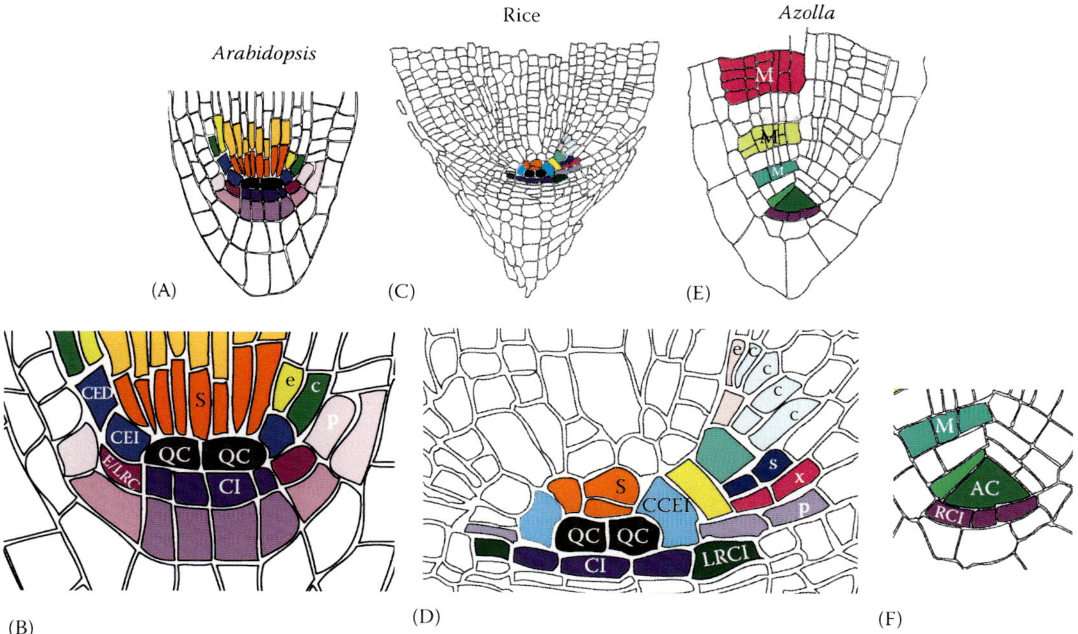

FIGURE 3.3 Cellular organization of three root meristems. *Arabidopsis* is shown in (A) and cropped and enlarged to show the initial cells in (B). The rice meristem is shown in (C) and enlarged in (D). The structures of the *Arabidopsis* and rice roots are similar. However, the rice root has multiple different specialized cells in the ground tissue including the endodermis (e), schlerenchyma (s), and the exodermis (x) and multiple cortex (c) layers. In rice the ground tissue and the epidermis (p) are derived from the CCEI. In contrast, in *Arabidopsis*, these tissue are distinct, and the epidermis and lateral root cap share an initial (E/LRC) (Images based upon Courdert et al. 2010). The tip of an *Azolla* root is shown in (E) and close-up in (F). The apical cell (AC) is shown in green, and three merophytes are shown in different colors and labeled with an (M). Root cap initials (RCI) below the apical cell give rise to the root cap. (Based upon Gunning, B. et al., *Planta*, 143, 121, 1978.). Initial cells are labeled with capital letters. Cell types are indicated by lowercase. All common cell types are labeled with the same color. QC, quiescent center; CI, columella initial; S, stele initial; CEI, cortical endodermal initial; LRCI, lateral root cap initial.

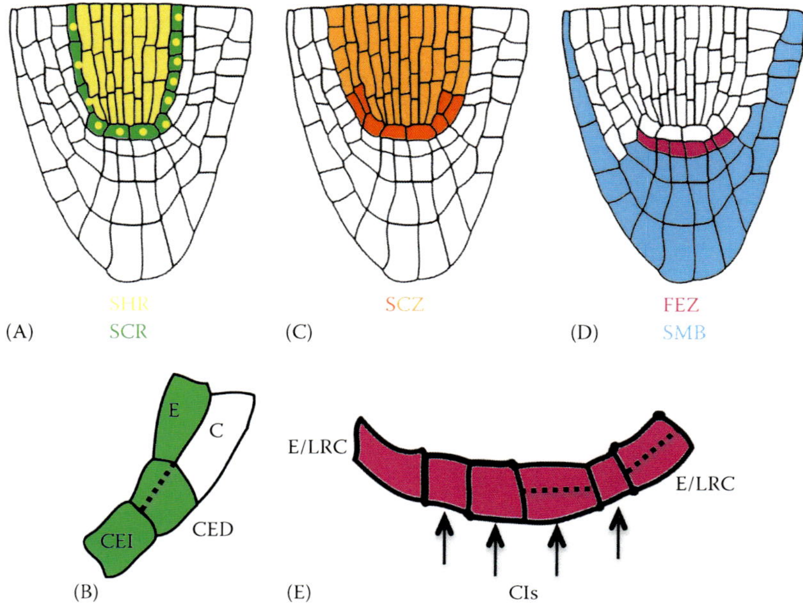

FIGURE 3.4 Proteins involved in *Arabidopsis* root patterning. (A) SHR (yellow) and SCR (green) as SHR is transcribed in the stele and moves into the endodermis; the yellow dots in the endodermis indicate SHR movement. (B) The asymmetric division (dotted line) shown in the cortical endodermal daughter (CED) requires both SHR and SCR activity. (C) *SCZ* expression is highest in the QC and initials (dark orange) and present throughout the stele and ground tissue (lighter orange). (D) Expression of FEZ (magenta) and SMB (blue). (E) FEZ expression in the initials promotes periclinal cell division, whereas SMB inhibits periclinal division. E/LRC, epidermis lateral root cap initial; CI, columella initial; CEI, cortical endodermal initial.

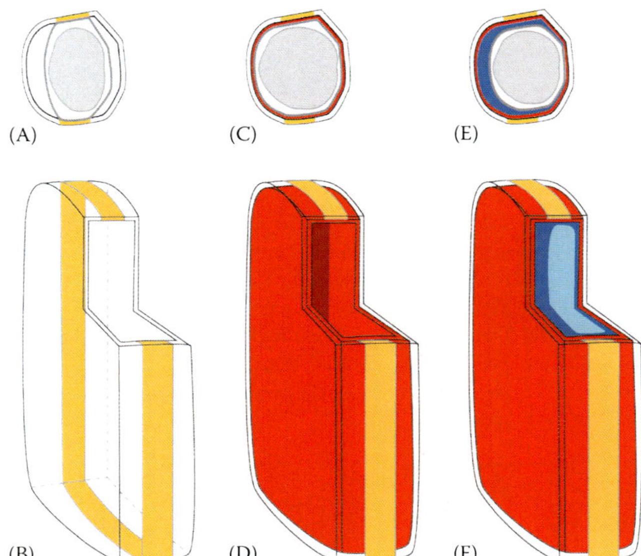

FIGURE 5.1 Illustrations of endodermal cells (not to scale) displaying their states of maturity (A,C,E) in cross section and (B,D,F) in three dimensions. (A,B) State I, Casparian bands in anticlinal walls. (A) Band plasmolysis is depicted. (C,D) State II, suberin lamella between the cell wall and plasma membrane. (E,F) State III, tertiary wall thickening between the suberin lamella and plasma membrane. Black lines, cell wall borders; dark gray lines, plasma membranes; light gray lines, tonoplasts; yellow shading, Casparian bands; red shading, suberin lamellae; blue shading, tertiary wall thickenings. (Images (B,D,F) modified from Raven, P.H. et al., *Biology of Plants*, 6th edn., WH Freeman and Company, New York, 1999. With permission.)

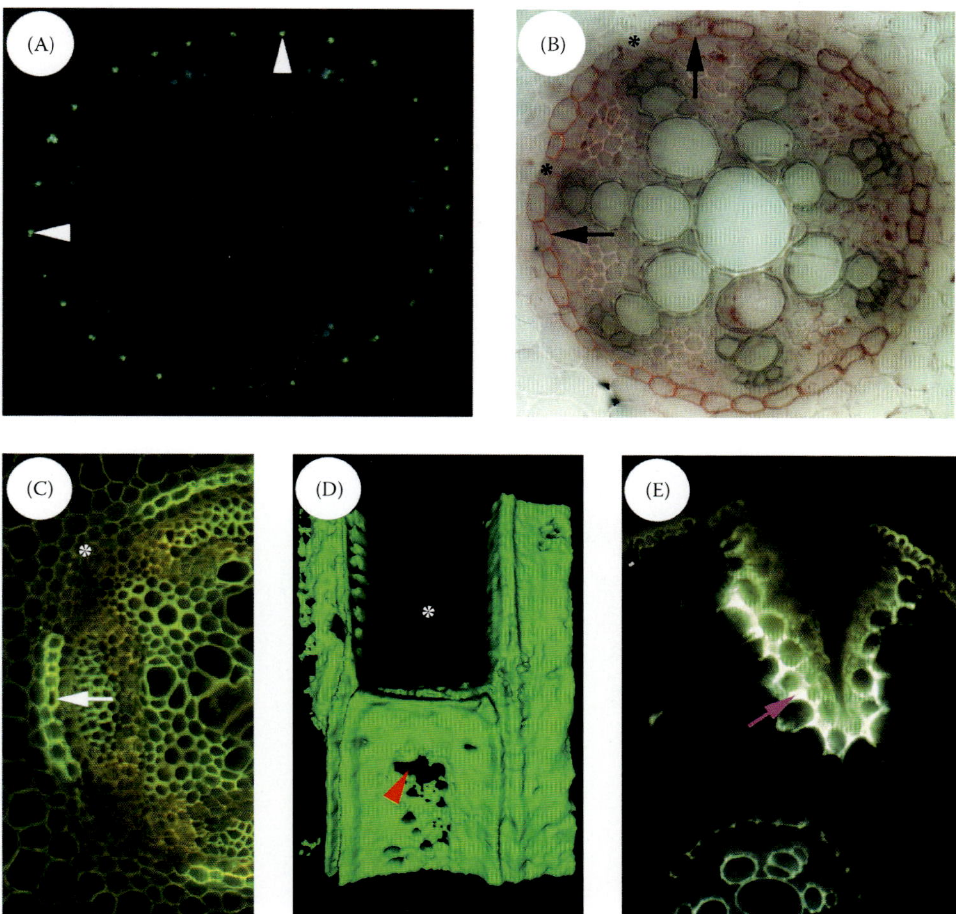

FIGURE 5.3 Images of the endodermis and stele in (A–C) cross and (D) longitudinal views. (A) Stained with berberine hemisulfate–aniline blue. The endodermis has Casparian bands (white arrowheads) that appear dot-like in the anticlinal walls of *Allium cepa*. (B) Stained with Sudan red 7B. Mature endodermal cells contain suberin lamellae (black arrows) in *A. cepa*. Passage cells (asterisks) are located near the xylem poles. (C) Stained with fluorol yellow 088. Mature endodermal cells have suberin lamellae (white arrow), alternating with patches of passage cells (asterisk) in *Glycine max*. (D) Portions of *A. cepa* endodermal cells from a three-dimensional reconstruction using confocal laser scanning microscopy. Angle of rotation on the vertical axis was 5.6°. Asterisk indicates a passage cell (without suberin lamellae). The red arrowhead indicates a large opening in the suberin lamellae. Green color was assigned artificially for grayscale images. (E) Cross section of *A. cepa* root that had developed autofluorescent wound suberin following injury. Magenta arrow indicates suberin in an air space. (Images (A,B): courtesy of Ishari Waduwara; image (C): courtesy of Raymond Thomas; image (D): from Waduwara, C.I. et al., *Can. J. Bot.*, 86, 623, 2008. Reproduced with permission from the NRC Research Press.)

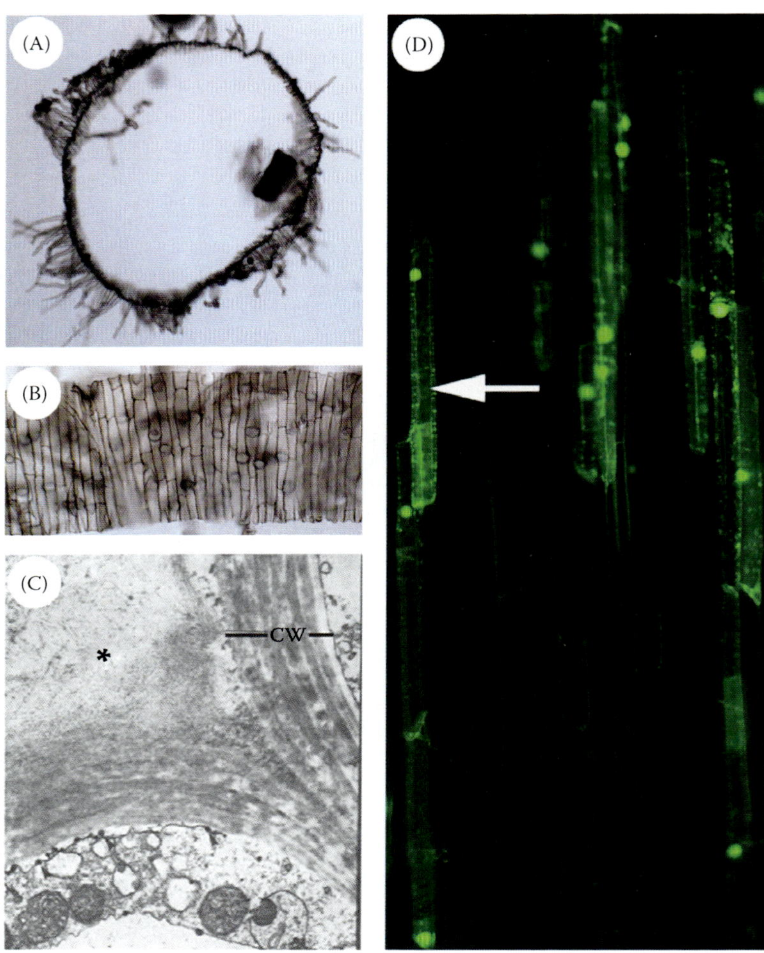

FIGURE 5.4 Photomicrographs of the epidermis. (A) Cross section of *Glycine max* root treated with concentrated sulfuric acid. The epidermal walls resisted digestion. (B) Acid-digested epidermal cell walls in face view in *G. max*. (C) TEM image of the outer tangential and radial walls of epidermal cells in *Allium cepa*. Diffuse suberin in the cell walls (CW) is evident. Extracellular material (*) covering the outer tangential walls of epidermal cells is present. (D) Stained with uranin. Living epidermal cells accumulate the dye in the nucleus and cytoplasm (white arrow) in *A. cepa*. (Images (A,B): courtesy of Alice Fang; image (C): from Peterson, C.A. et al., *Protoplasma*, 96, 1, 1978. Reproduced with permission from Springer Science; image (D): courtesy of Ishari Waduwara.)

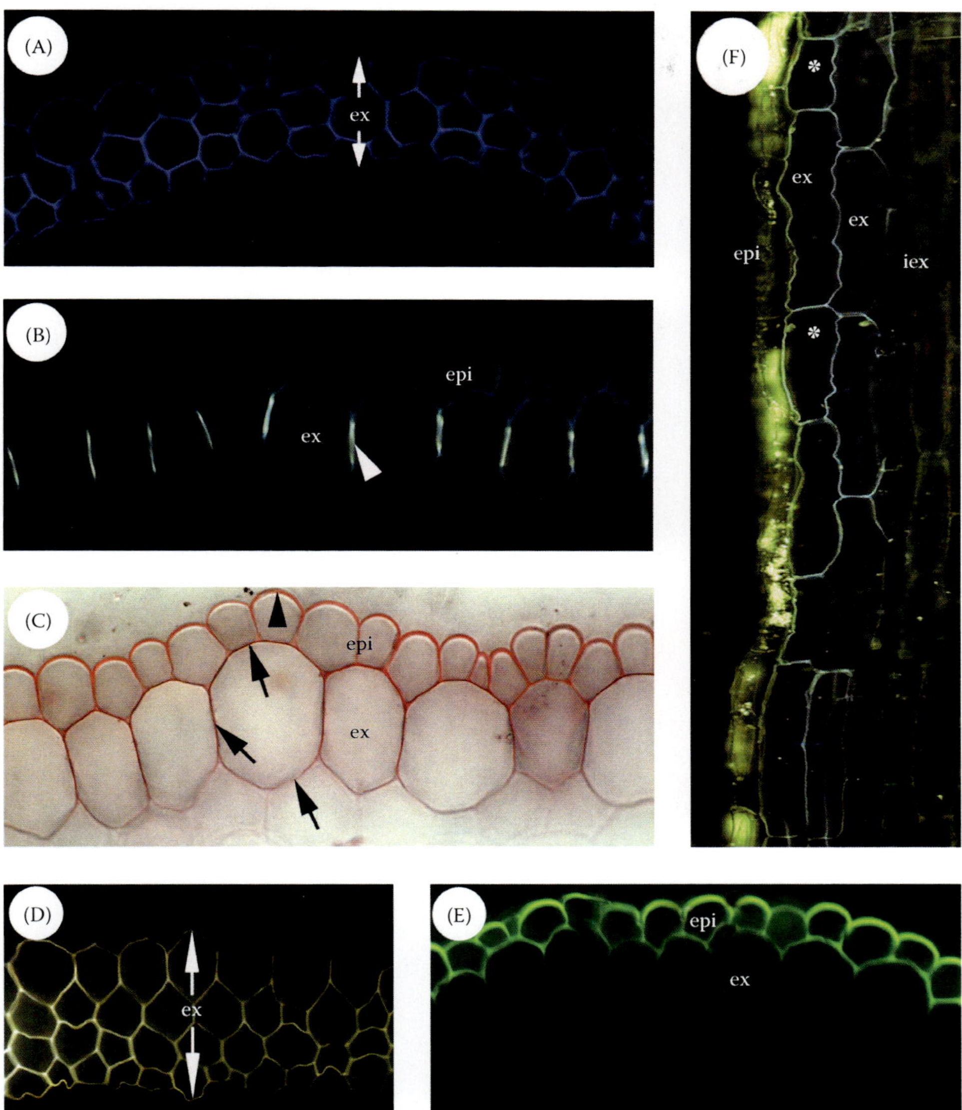

FIGURE 5.5 Photomicrographs of the outer part of roots in (A–E) cross and (F) longitudinal section. (A) Autofluorescence with ultraviolet light. Walls of the MEX of *Iris germanica* are faint blue. (B) Stained with berberine hemisulfate–aniline blue. Casparian bands (white arrowhead) fluoresce yellow and occupy the anticlinal walls of the outermost exodermal layer in *I. germanica*. (C) Stained with Sudan red 7B. Suberin lamellae (black arrows) appear as red outlines in exodermal walls of *Allium cepa*. Also, diffuse suberin in the epidermis stains red (black arrowhead). (D) Stained with fluorol yellow 088. Suberin lamellae fluoresce yellow in exodermal cell walls of *I. germanica*. (E) Exodermal Casparian band prevents externally applied berberine (green fluorescence) from permeating the exodermis and central cortex in *A. cepa*. The walls of the epidermis are permeable to berberine. (F) Epidermal cells containing berberine thiocyanate crystals (yellow) and a dimorphic, biseriate exodermis with blue, autofluorescent walls in *I. germanica*. Asterisks indicate short cells. epi, epidermis; ex, mature exodermis; iex, immature exodermis. (Images (C, E): courtesy of Ishari Waduwara; image (F): Meyer, C.J. et al., *Ann. Bot.*, 103, 687, 2009. Reproduced with permission from Oxford University Press.)

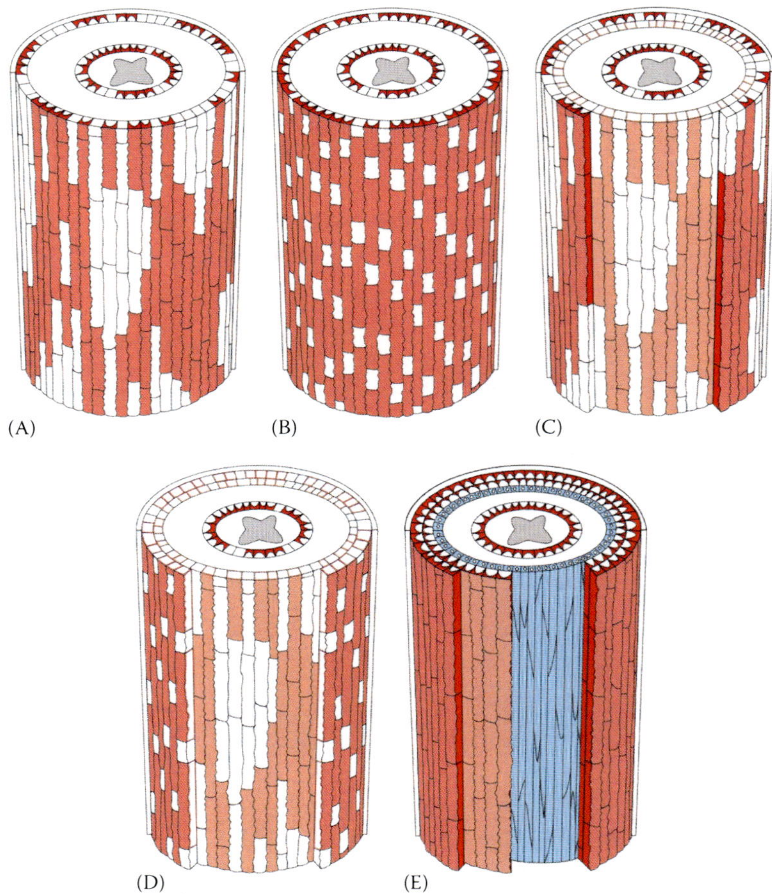

FIGURE 5.6 Illustrations of four exodermal subtypes and one exodermal variation, shown in cross and longitudinal section. (A) Uniseriate, uniform. (B) Uniseriate, dimorphic. (C) Multiseriate, uniform. (D) Multiseriate, dimorphic (or mixed). (E) Reinforced exodermal variation, with sclerenchyma (blue cells). Red shading, suberin lamellae in cells of the first exodermal layer; brownish-red shading, suberin lamellae in cells of the second exodermal layer; deep red shading, tertiary walls. (Reproduced from Peterson, C.A., The exodermis and its interactions with the environment, in *Radical Biology: Advances and Perspectives on the Function of Plant Roots, Proceedings of the 11th Annual Penn State Symposium in Plant Physiology*, May 22–24, 1997, eds. Flores, H.E., Lynch, J.P., Eissenstat, D., American Society of Plant Biologists, Rockville, MD, 1997. With permission.)

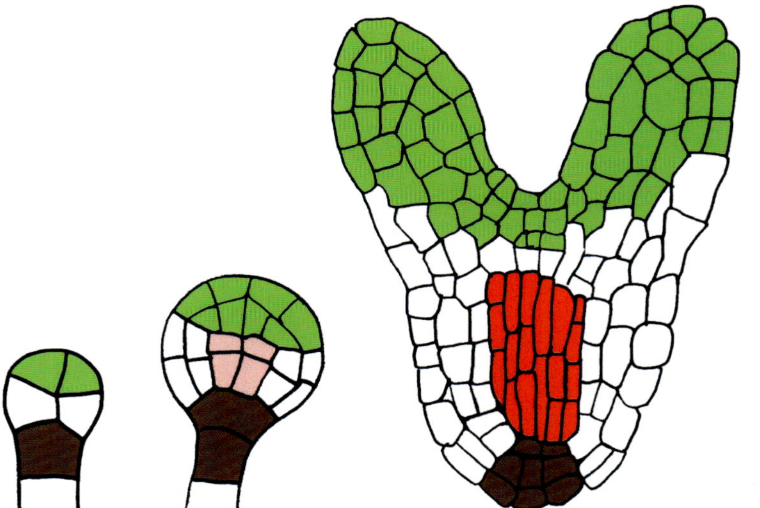

FIGURE 7.1 The development of vascular cells during embryogenesis. Left is the eight-cell embryo. For simplicity, not all of the suspensor cells are shown. Already at this stage, the expression of marker genes indicates that the apical-basal identity is set. The uppermost suspensor cell, which will go on to form the root meristem, is shown in brown, and the cells which will form the shoot are shown in green. In the center is an early globular stage embryo. As cell identity is specified, the vascular primordium, marked in pink, will undergo a programmed set of divisions to form both the vascular initials and the pericycle cells. By the late torpedo stage (right), the vascular initial cells are clearly defined (red). Although these remain undifferentiated, tissue-specific marker genes are expressed. For example, at this stage, *AHP6* is already specifically expressed in two files that will later form the protoxylem.

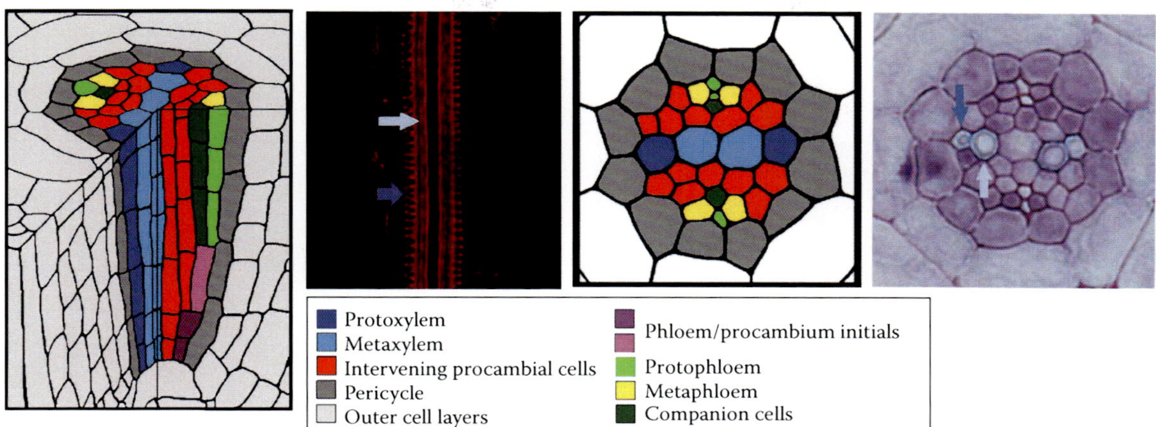

FIGURE 7.3 Tools of the trade: vascular cell types in the mature root and their visualization. The schematic images show two different views of the vascular cells in the *Arabidopsis* root with the cell lineages marked. The center right panel shows a cross section taken approximately 50 μm above the QC. The cell layer immediately outside of the pericycle is the endodermis. Visualizing the vascular cells can be challenging. The center left panel shows a confocal image of a root stained with fuchsin-red dye. This stains lignified tissue, making the protoxylem and metaxylem cells easy to visualize. Protoxylem cells are characterized by a helical pattern on the cell wall (dark-blue arrow), while metaxylem cells stain more solidly and have a pitted pattern (light-blue arrow). A cross section through the root stained with toluidine-blue dye shows other cell types clearly. Here the xylem cells can be recognized by their thick cell walls (blue arrows). The proto- and metaphloem can be identified as they are small white cells at 90° to the xylem and the intervening procambial cells are dark blue.

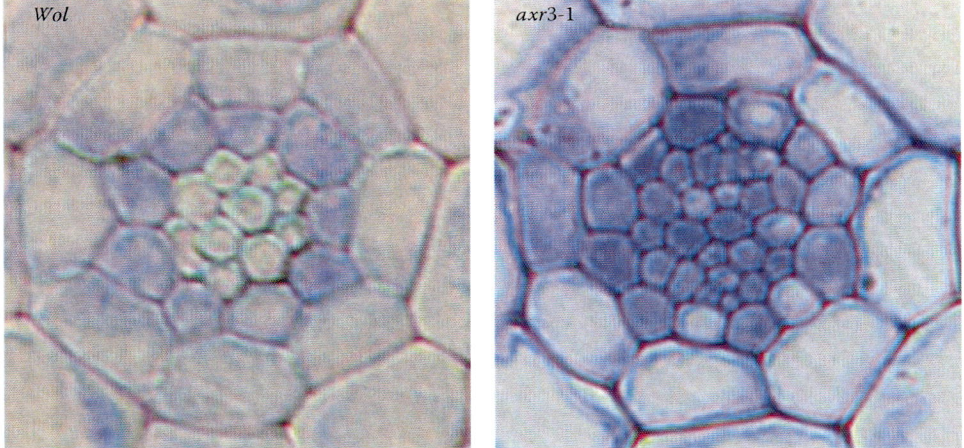

FIGURE 7.4 The level of hormonal signaling regulates the extent of protoxylem differentiation. In the absence of cytokinin signaling (such as in the *wol* mutant), all vascular cells acquire protoxylem cell fates. Note the thick secondary cell walls in the left image. When auxin signaling is severely compromised (such as in the *axr*3-1 mutant), no xylem differentiation occurs. Note that no secondary walls are visible in the right image and all vascular cells are darkly stained. (Images courtesy of Hanna Help, University of Helsinki, Helsinki, Finland.)

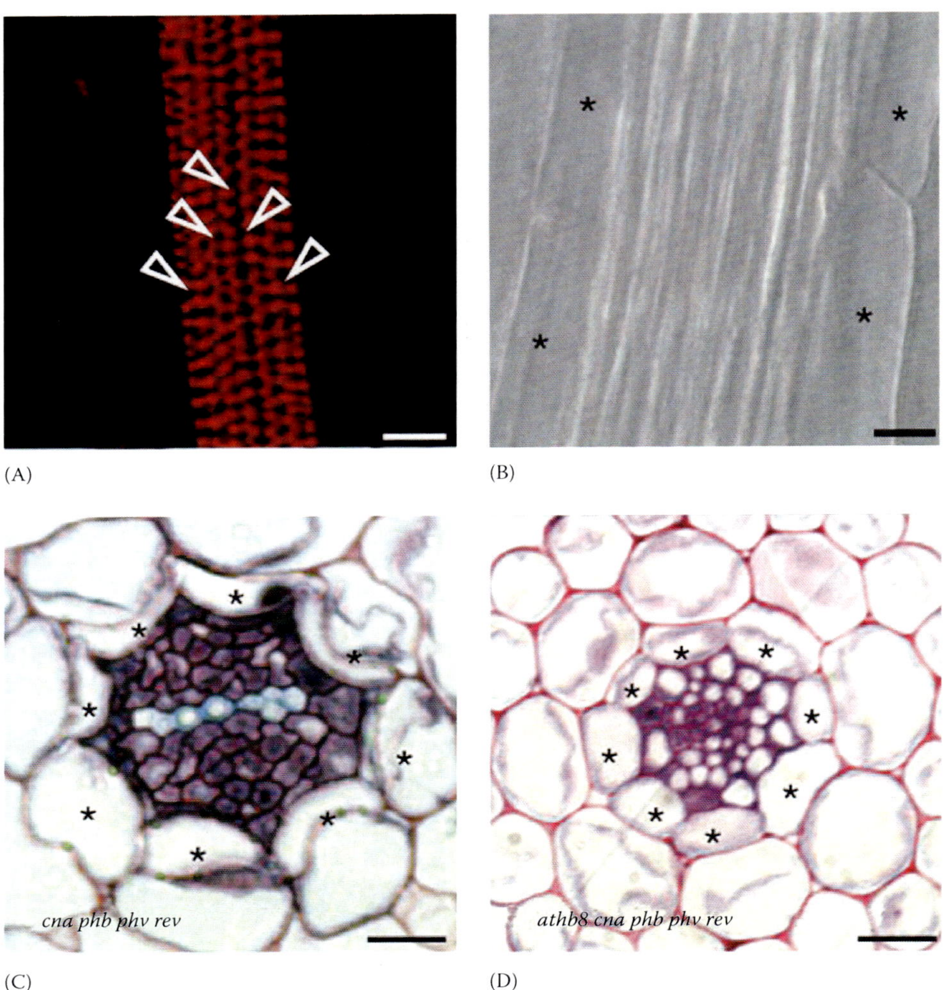

FIGURE 7.6 HD-ZIP gene dosage controls the centripetal patterning of the xylem axis. When the gene dosage of class III HD-ZIP genes is low all cells differentiate as protoxylem. This can be seen here through Fuchsin staining (A) and cross sections (C) in the quadruple *can phb phv rev* mutant. The white arrowheads show protoxylem cell lines. When there is no class III HD-ZIP gene activity in the *athb8 can phb phv rev* quintuple mutant, no protoxylem develops. This can be seen in Nomarski images (B) and cross section below (D). It is not appropriate to perform Fuchsin stainings when there is no xylem as there will be no incorporation of the stain. The pericycle cells have been marked with asterisks in order to orientate the reader to the location where xylem would normally develop. (This image has been reproduced by permission from Macmillan Publishers Ltd., *Nature*, Carlsbecker, A., Lee, J.Y., Roberts, C.J. et al., Cell signalling by microRNA165/6 directs gene dose-dependent root cell fate, 465, 316–321, 2010. Copyright 2010.)

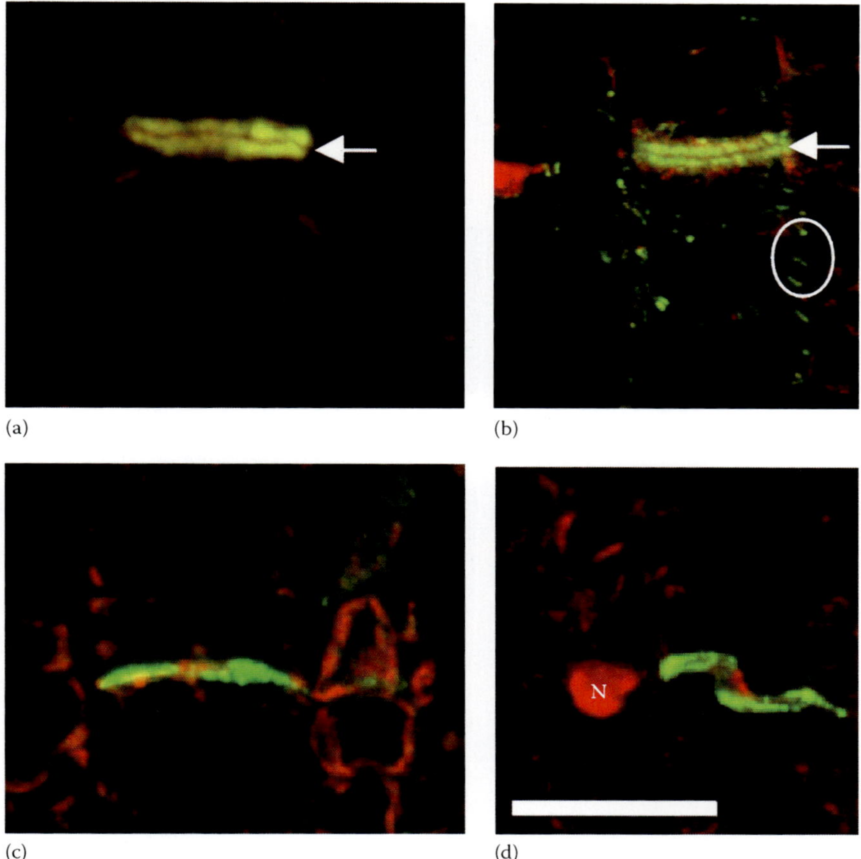

FIGURE 8.5 Indirect immunofluorescent localization of various macromolecules involved in cell-wall formation within fusiform cambial cells in a transverse section of hybrid aspen root. (a) MTs and (b) MFs, within the phragmoplast; note the nonfluorescing central dark line, which corresponds to the site of the nascent division wall; encircled region shows axially oriented MF bundles in these obliquely cut cells; (c) myosin and (d) callose, at the site of the nascent division wall; N, nucleus; scale bar (in d) = 20 μm.

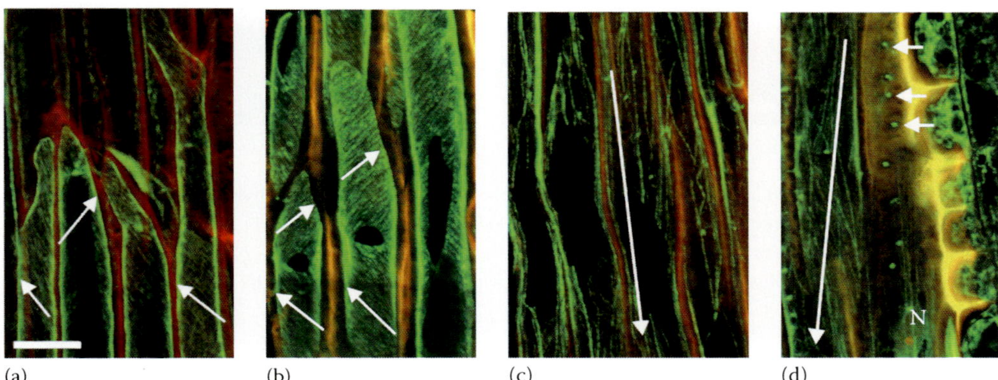

FIGURE 8.7 Indirect immunofluorescent localization of cytoskeletal proteins in developing fibers (at second phase of development) seen in radial longitudinal sections of hybrid aspen root. (a) Xylem and (b) phloem, fibers showing parallel-aligned, obliquely oriented MTs (arrows); note that both the front and back walls of some of the phloem fibers are in view; (c) xylem and (d) phloem, fibers showing net-axially oriented MF bundles (arrows); note also the sites of simple pit formation (short arrows) in the phloem fibers; scale bar (in a) = 20 μm.

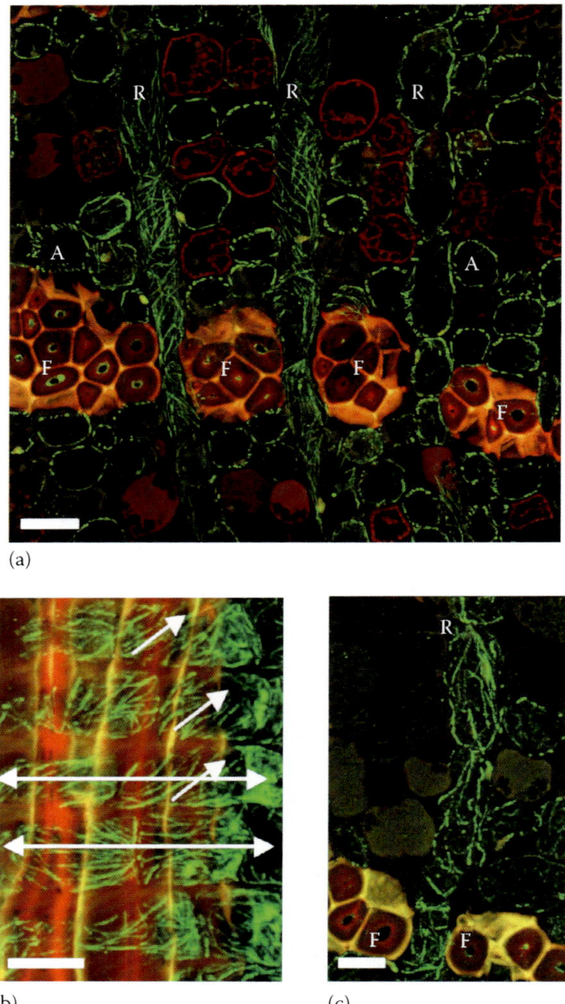

FIGURE 8.8 Indirect immunofluorescent localization of cytoskeletal proteins in phloem rays in hybrid aspen root. (a) Transverse section showing net-axial-helically oriented MTs in three rays (R) passing between axial parenchyma cells (A) and bundles of fibers (F); scale bar = 20 μm; (b) radial longitudinal section showing net-axially oriented MTs in ray cells (arrows) traversing a fiber bundle; scale bar = 10 μm; (c) transverse section showing net-axially oriented MF bundles in ray cells (R) passing between axial parenchyma cells and bundles of fibers (F); scale bar = 20 μm.

FIGURE 9.1 Some examples of the diversity of ARs. (A) Hairy root developing from a *Catharanthus roseus* leaf after wounding and inoculating the vein with an *Agrobacterium rhizogenes* suspension (scale bar: 1 cm), (B) ARs formed from a *Ficus* sp. tree (scale bar: 1 m), (C) ARs fusing with the trunk of a *Ficus* sp. tree (scale bar: 30 cm), (D) buttress ARs in *Tetrameles nudiflora* (scale bar: 20 cm), (E) pillar ARs supporting a branch in *Ficus* sp. (scale bar: 50 cm), (F) ARs forming a protective cage around the trunk in *Pandanus* sp. (scale bar: 1 m), (G): clasping ARs and feeding ARs in a *Philodendron* sp. (scale bar: 5 cm), (H): aerial ARs in an epiphyte orchid (scale bar: 10 cm). (Photographs D, F, G, and H are from Claude Edelin, CNRS, Montpellier, France.)

FIGURE 9.2 CRs in monocots. (A) Bamboo stem with CRs developing at each node (scale bar: 15 cm), (B) detail of a crown of CR apexes emerging from a node of a bamboo stem (scale bar: 4 cm), (C) general view of the CR-made fibrous root system of rice (scale bar: 6 cm), (D) detail of the stem base of a wild-type rice showing the emergence of numerous CRs (scale bar: 5 mm), (E) detail of the stem base of the *crown rootless 1* rice mutant devoid of CRs, only the SR and some of its LRs can be observed (scale bar: 5 mm).

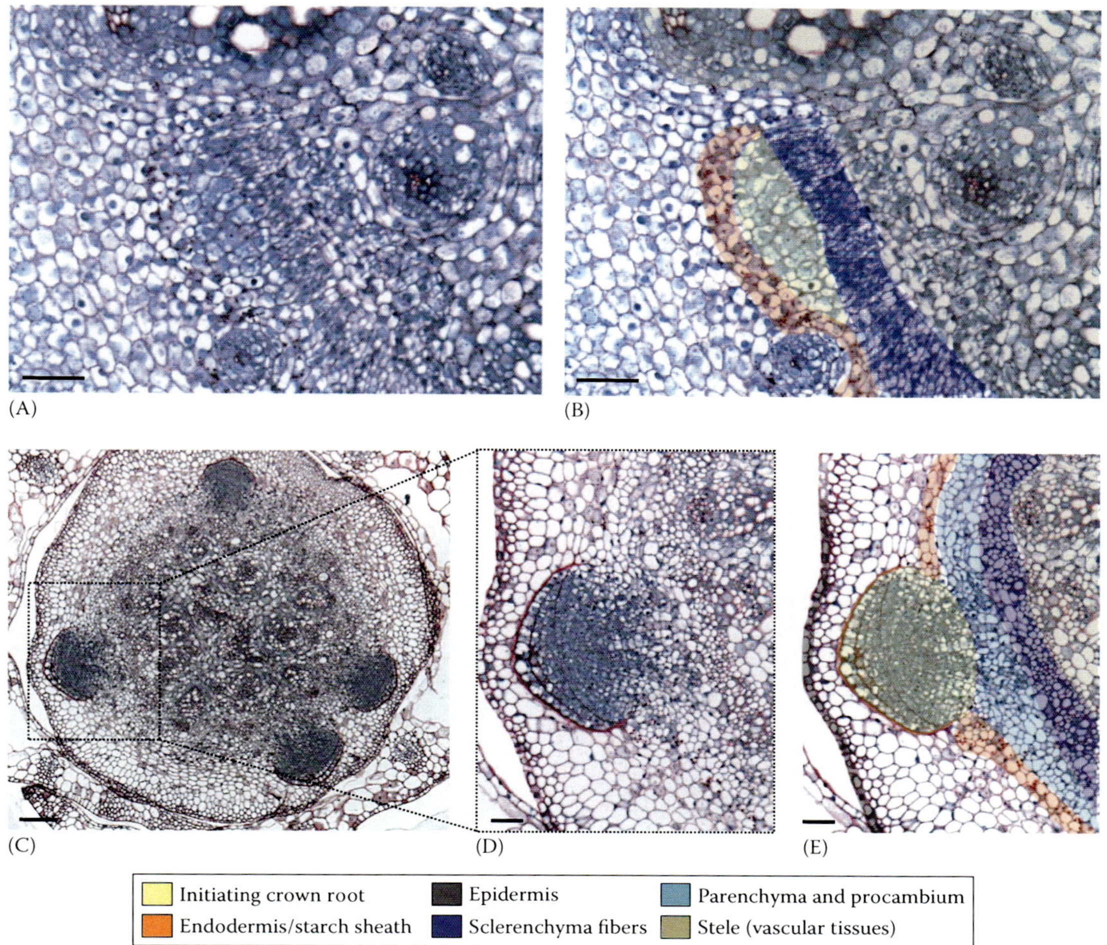

FIGURE 9.3 Histological sections of rice stem base showing different stages of CR initiation. (A–B) early stages of CR initiation. (C–E) CR primordia after differentiation of a typical root meristem organization. (D and E) enlargements of the zone marked in picture C with dotted lines. (A, C, and D) periodic acid-Schiff and naphthol blue black-stained sections. (B and E) captioned sections. Scale bar: A, B, D, and E, 50 μm; C, 200 μm.

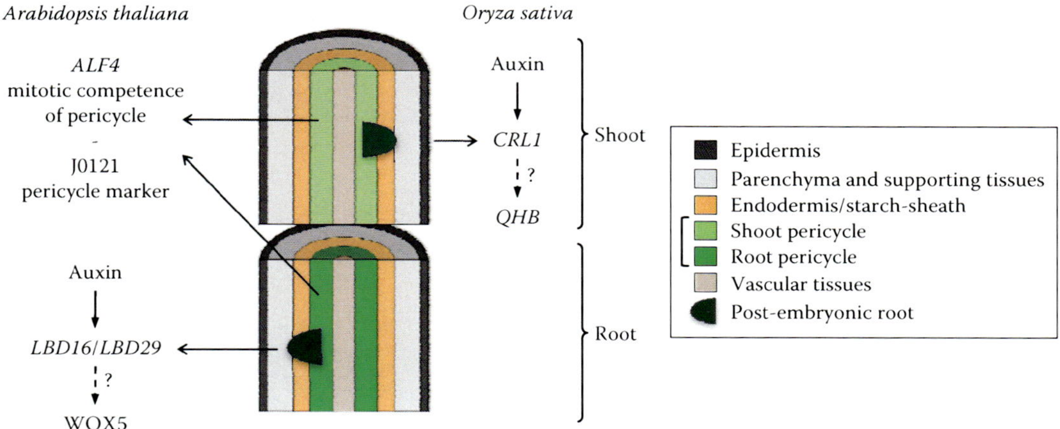

FIGURE 9.5 Diagram representing genetic mechanisms related to postembryonic root development and conserved between shoot and root in *Arabidopsis* and rice. Abbreviations: ALF4, ABERRANT LATERAL ROOT FORMATION 4; CRL1, crown rootless1; GNOM1, membrane-associated guanine-nucleotide exchange factor of the ADP-ribosylation factor G protein (ARF-GEF) that regulates the traffic of PINFORMED1 (PIN1) auxin efflux carrier proteins; J0121, GFP enhancer trap line; QHB, WUSCHEL-TYPE HOMEOBOX GENE QUIESCENT-CENTER-SPECIFIC HOMEOBOX; LBD, Lateral Organ Boundary Domain; WOX4, WUSHEL-Related Homeobox 4.

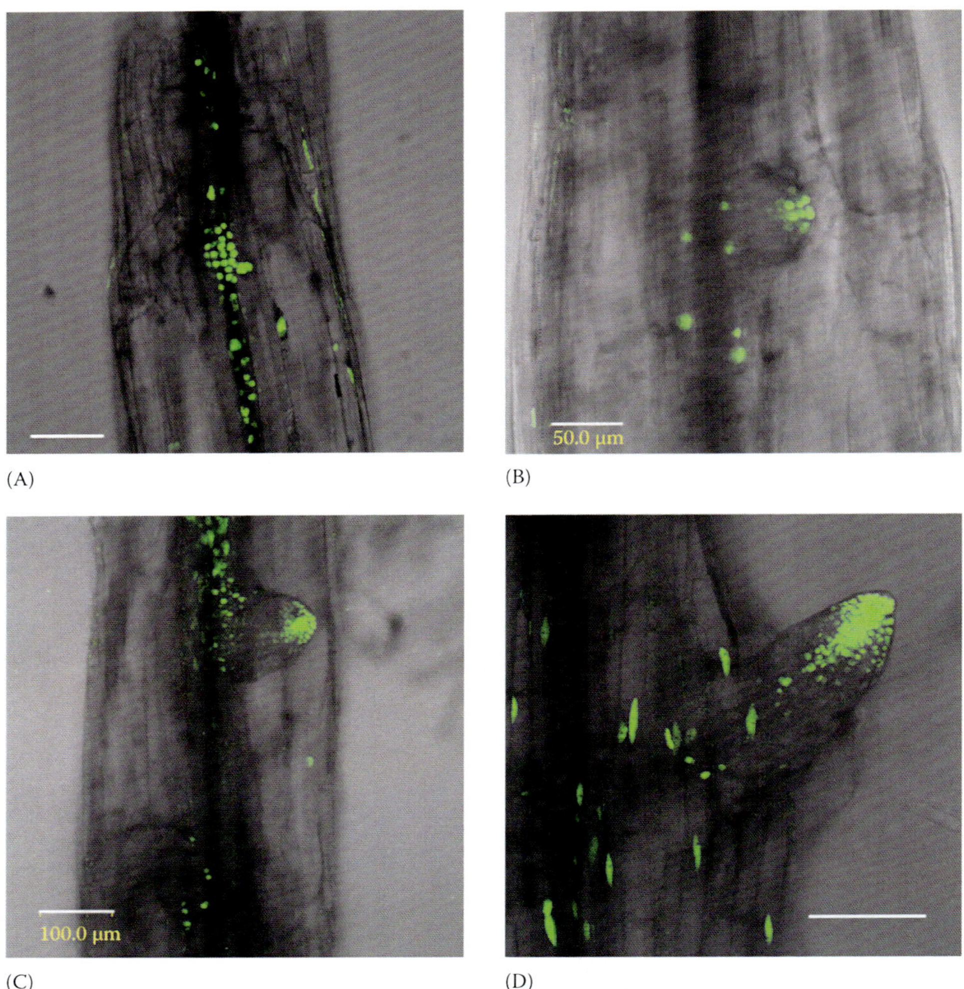

FIGURE 11.1 Auxin accumulation during AR formation. *Arabidopsis* seedlings expressing the auxin responsive promoter DR5, linked to YFP with a nuclear localization signal (NLS), DR5::Venus marker (Laskowski et al. 2008), were grown in the dark. Hypocotyls were excised and incubated with 10 μM IBA. Auxin accumulation is shown during four stages of AR formation: (A) near the vascular tissue, (B) at the tip of an initial root primordia, (C) at the tip of a more developed primordium, and (D) at the tip of the root after emergence. Scale bars in A, C, D = 100 μm, in B = 50 μm. (From O. Rogovoy-Stelmakh, M. Abu-Abied, and E. Sadot, unpublished data.)

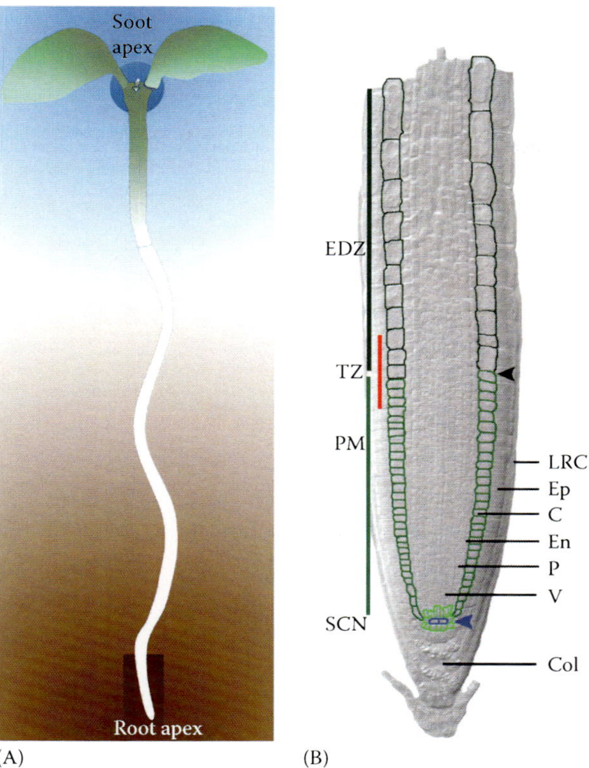

FIGURE 14.1 Anatomy of the *Arabidopsis thaliana* root apical meristem. (A) Schematic view of *Arabidopsis thaliana* seedling, with highlighted shoot apex (blue circle) and root apex (brown rectangle). (B) Longitudinal view of the *Arabidopsis* root meristem, where the SCN is marked in light green, the QC in blue, a cortex cell file of the PM is marked in green, and a cortex cell file of the EDZ is marked in dark green. Black arrowhead indicates the cortex TZ (red line), where cells leave the meristem and enter the EDZ. Blue arrowhead indicates the QC. Col: columella; V: vascular tissue; P: pericycle; En: endodermis; C: cortex; Ep: epidermis; LRC: lateral root cap.

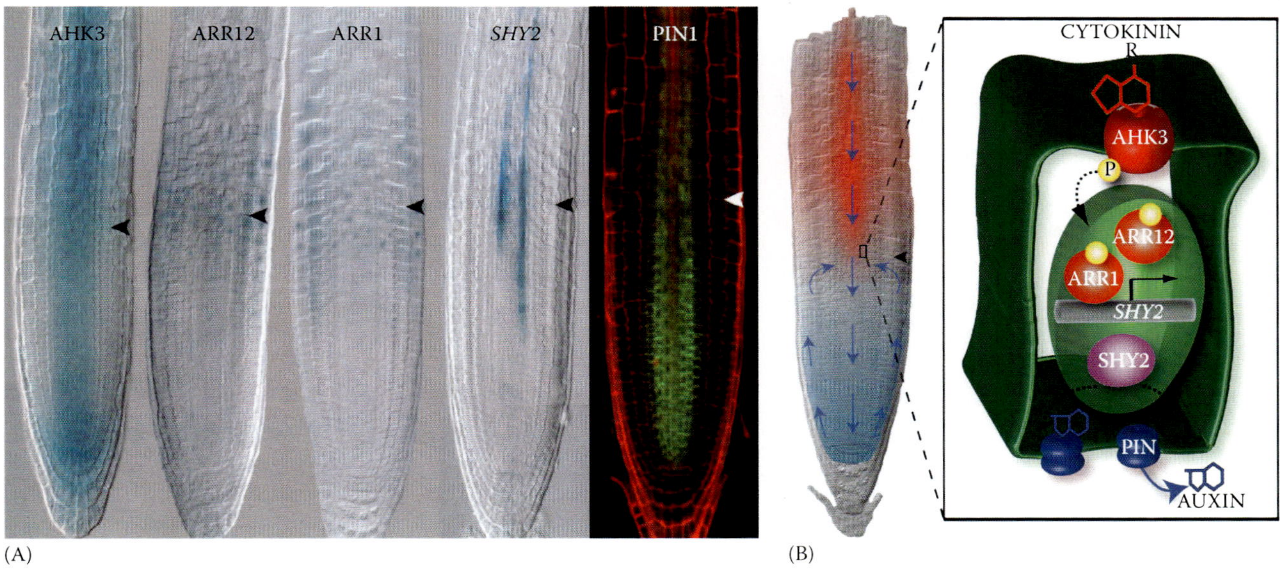

FIGURE 14.3 Expression pattern of cytokinin-sensitive genes involved in root meristem size determination. (A) Expression of AHK3 in roots of *AHK3:GUS* plants, of ARR12 in roots of *ARR12:GUS* plants, of ARR1 in roots of *ARR1:GUS* plants, of *SHY2* in roots of *pSHY2:GUS* plants, and of PIN1 in roots of *PIN1:GFP* plants. Black and white arrowheads indicate the cortex TZ. (B) Longitudinal view of the *Arabidopsis* root meristem, where the domain of auxin activity is marked in blue, while the domain of cytokinin activity is marked in red. Arrows indicate direction of polar auxin transport. In the single-cell blowup schematic representation of cytokinin-signaling pathway acting during root meristem size determination. Cytokinins are perceived by the plasma membrane AHK3 receptor, which initiates a phosphorylation (P in the yellow circle) cascade ending with the activation, in the nucleus, of type-B ARR12 and ARR1 cytokinin-response transcription factors. Upon activation, ARR12 and ARR1 proteins directly bind the *SHY2* promoter inducing its transcription. SHY2 protein, in turn, downregulates *PIN* genes expression, limiting polar auxin transport.

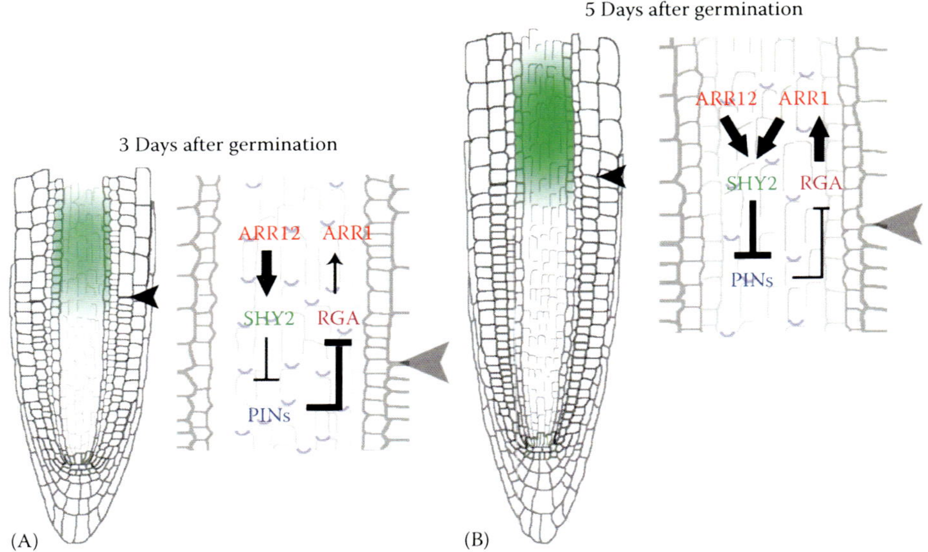

FIGURE 14.4 The rate of cell differentiation controls the *Arabidopsis* root meristem growth phase. (A) The regulatory circuit controlling meristem growth 3 days after germination, when cell division overcomes cell differentiation. ARR12 drives low level of *SHY2* expression, sustaining PIN-mediated polar transport of auxin, which in turn supports gibberellin biosynthesis. High levels of gibberellin repress the activity of the DELLA protein RGA, thus suppressing *ARR1* transcription. (B) At five dpg, a decrease of gibberellin biosynthesis stabilizes the RGA protein, allowing transcriptional activation of *ARR1*, which eventually joins ARR12 in increasing *SHY2* expression, slowing PIN-mediated polar auxin transport and gibberellin biosynthesis. This results in an increase in cell differentiation that balances it with cell division and stops meristem growth. Black and gray arrowheads indicate the cortex TZ. Green area indicates the domain of *SHY2* expression, and different color gradations highlight changes in SHY2 levels.

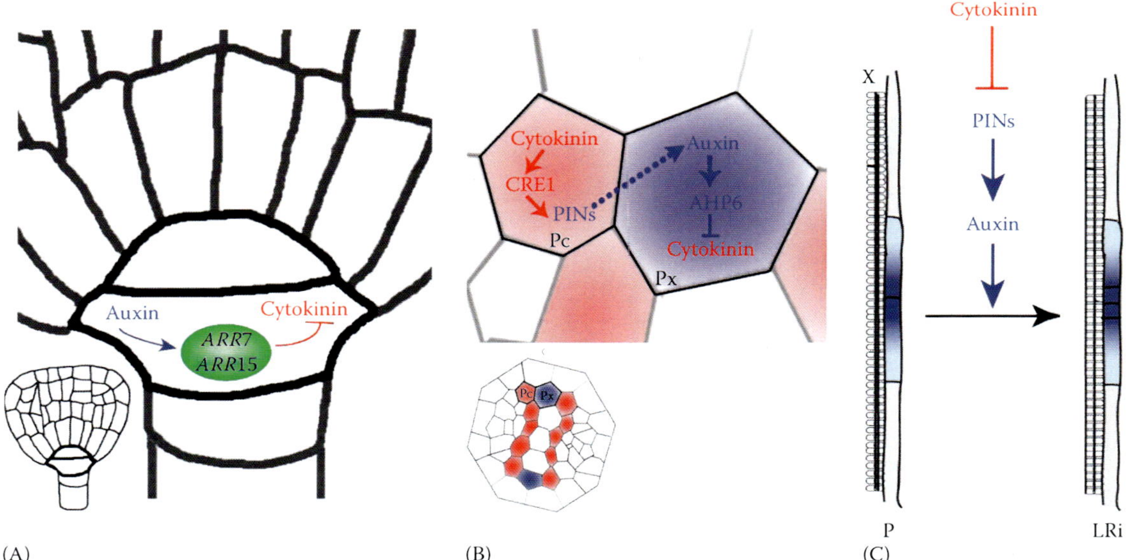

FIGURE 14.5 Cytokinins are involved in different aspects of root development. (A) During early stages of embryo development, auxin activates type-A ARR7 and ARR15, negative regulators of cytokinin signaling, driving root SCN specification. (B) Schematic representation of a root apex cross section highlighting the procambium (Pc, in red) and the protoxylem (Px, in blue). Cytokinins, through the CRE1 receptor, maintain procambial cell identity promoting the localization of the PIN proteins on the lateral membranes of procambial cells, thus forcing auxin flow toward protoxylem (blue dotted arrow). Accumulation of auxin in the protoxylem induces expression of the pseudophosphotransfer AHP6 protein that promotes protoxylem differentiation by suppressing cytokinin signaling via an unknown mechanism. (C) Schematic representation of the initial stages of lateral root formation. Cytokinins negatively control lateral root initiation by downregulating *PIN* expression, thus preventing the establishment of the auxin maximum (in blue) in the xylem pole pericycle cells required for normal lateral root initiation. X: xylem; P: pericycle; LRi: lateral root initiation.

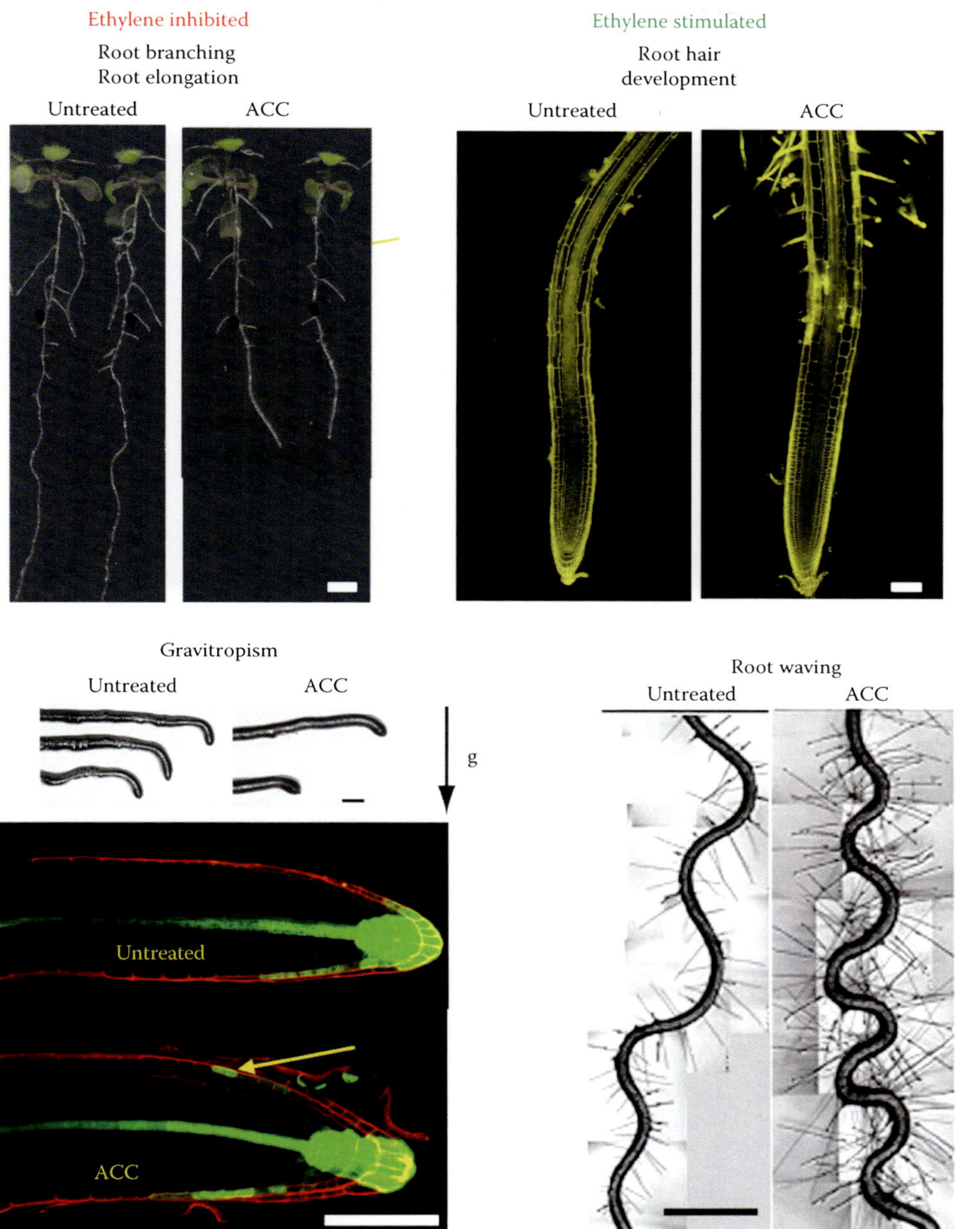

FIGURE 15.2 Effects of ethylene on root growth and development. Treatment of roots with either ethylene or its biosynthetic precursor, ACC, alters root growth and development. The inhibitory effect of ACC on root elongation, lateral root development, and gravitropism is shown on the left. The images of roots undergoing gravitropism include both graviresponding roots at low magnification and confocal images of roots at 8 h after reorientation 90° relative to the gravity vector. The angle of gravity is indicated by an arrow. The confocal image shows individual cells after propidium iodide staining and the asymmetric expression of the auxin-responsive DR5-revGFP reporter across the untreated root. In contrast, ACC treatment leads to DR5-revGFP fluorescence on the upper side, which then minimizes the auxin gradient across the root reducing gravitropic curvature. The effect of ACC on stimulation of root hair formation and alteration of the periodicity of root waving is also shown. Scale bars = 100 μm for all micrographs except those depicting lateral root development and root waving, which are 1 mm. (Images of root waving are reprinted with permission from Buer et al. 2003, *Plant Physiol.* 132: 1085–1096.)

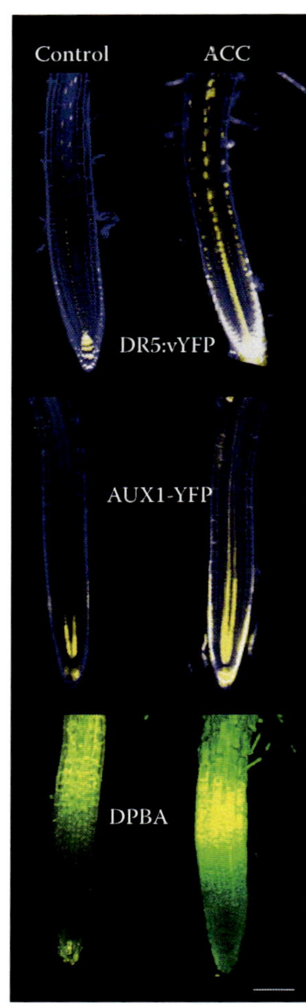

FIGURE 15.3 Flavonoid accumulation, AUX1-YFP fluorescence, and auxin signaling are all elevated in the EZ after ACC treatment. In untreated roots, shown in the left images, DR5-vYFP expression (yellow) is highly expressed in the root columella cells at the root tip and in the central cylinder. After ACC treatment, there is increased expression in both of these tissues and in the epidermal tissues. In untreated roots, AUX1-YFP is expressed in pericycle and columella cells, and in both tissues there is strong enhancement after ACC treatment. The accumulation of flavonols was visualized by staining with diphenylboric acid 2-aminoethyl ester (DPBA), a dye that binds kaempferol and quercetin with different fluorescent properties, which were spectrally separated into kaempferol-DPBA (green) and quercetin-DPBA (yellow). Both compounds accumulate in the EZ after ACC treatment. All ACC treatments were with 1 μM and were for either 8 h (flavonol accumulation) or 12 h for DR5-vYFP and AUX1-YFP. Scale bars = 100 μm.

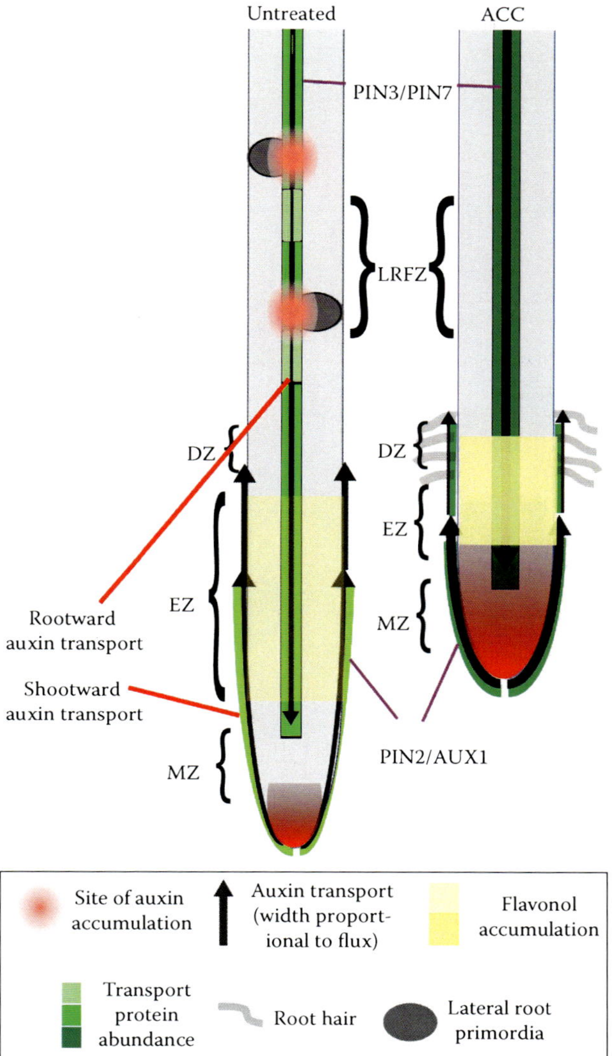

FIGURE 15.4 Model of ethylene effects on root growth and development. In untreated roots, lateral root formation occurs in the lateral root-forming zone (LRFZ) as a result of localized auxin maxima. This accumulation may occur as a result of depletion of PIN3 and PIN7 proteins that prevents auxin movement away from these points, enhancing auxin accumulation. Ethylene increases rootward auxin transport by increasing transcription and translation of PIN3 and PIN7 in the central cylinder, which may then deplete the lateral root-forming region of auxin, while increasing auxin accumulation in the root apex or meristematic zone (MZ). This effect may be responsible for ethylene's negative regulation of lateral root formation. At the root tip, increases in transcription and translation of AUX1 and PIN2 enhance shootward transport of the auxin into the EZ. Ethylene also increases the accumulation of flavonols in the EZ, as shown by the increased yellow shading, which then traps auxin in the EZ through flavonol inhibition of auxin transport. This excess auxin accumulation in the EZ reduces cell expansion and enhances root hair development in the DZ. (Reprinted from Muday, GK. et al., *Trends Plant Sci.*, 17, 181, 2012.)

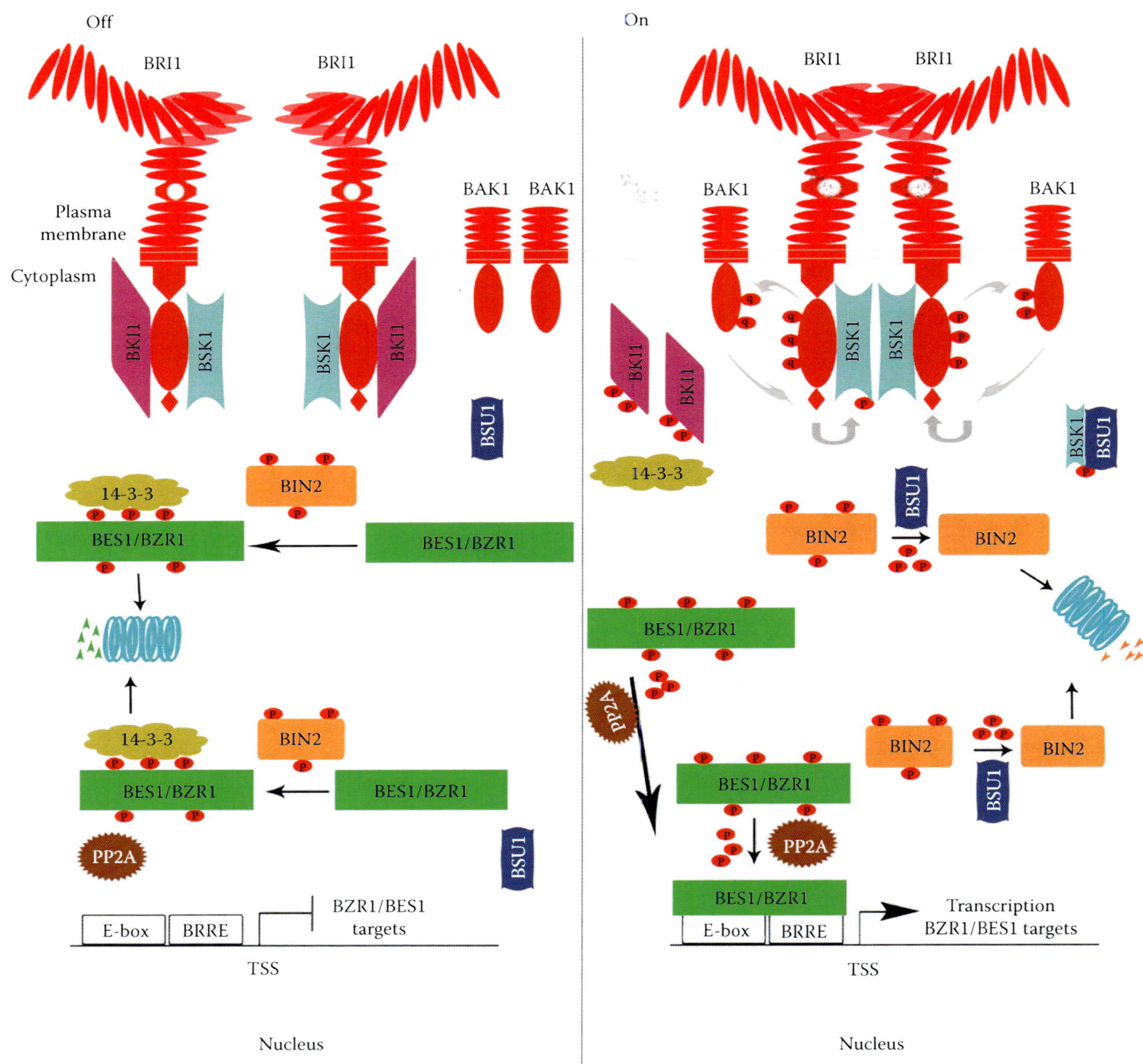

FIGURE 17.1 BR signaling pathway in *Arabidopsis*. Schematic representation of BR signaling pathway when is inactive (off) and active (on) in the left and right panels, respectively. In the absence of BL, BKI1 binds to the inactive form of BRI1 receptor together with BSKs at the cell surface; thus, BSU remains inactive. Inside the cell, BIN2 kinase phosphorylates BES1 and BZR1 transcription factors, promoting their cytoplasm retention thought the binding of 14-3-3 phosphoproteins and their subsequent proteasome-mediated degradation. In consequence, no changes in gene expression are observed (left panel). Upon BR-ligand binding to the extracellular domain of BRI1 receptor, BRI1 kinase phosphorylates BKI1 promoting its release to the cytoplasm and letting BRI1/BAK1 heterodimerization to occur. Transphosphorylation between BRI1 and BAK1 coreceptor activates the signaling complex, mediating phosphorylation of BSKs. Then BSK will bind to BSU, which in turn mediates BIN2 proteasome degradation by dephosphorylation of a conserved Tyr residue. The negative regulation of BIN2 upon BR perception together with the PP2A phosphatase activity mediates dephosphorylation of BES1 and BZR1 transcription factors. Dephosphorylated BES1 and BZR1 transcription factors in the nucleus control BR-mediated genome transcription and cellular responses in the plant (right panel).

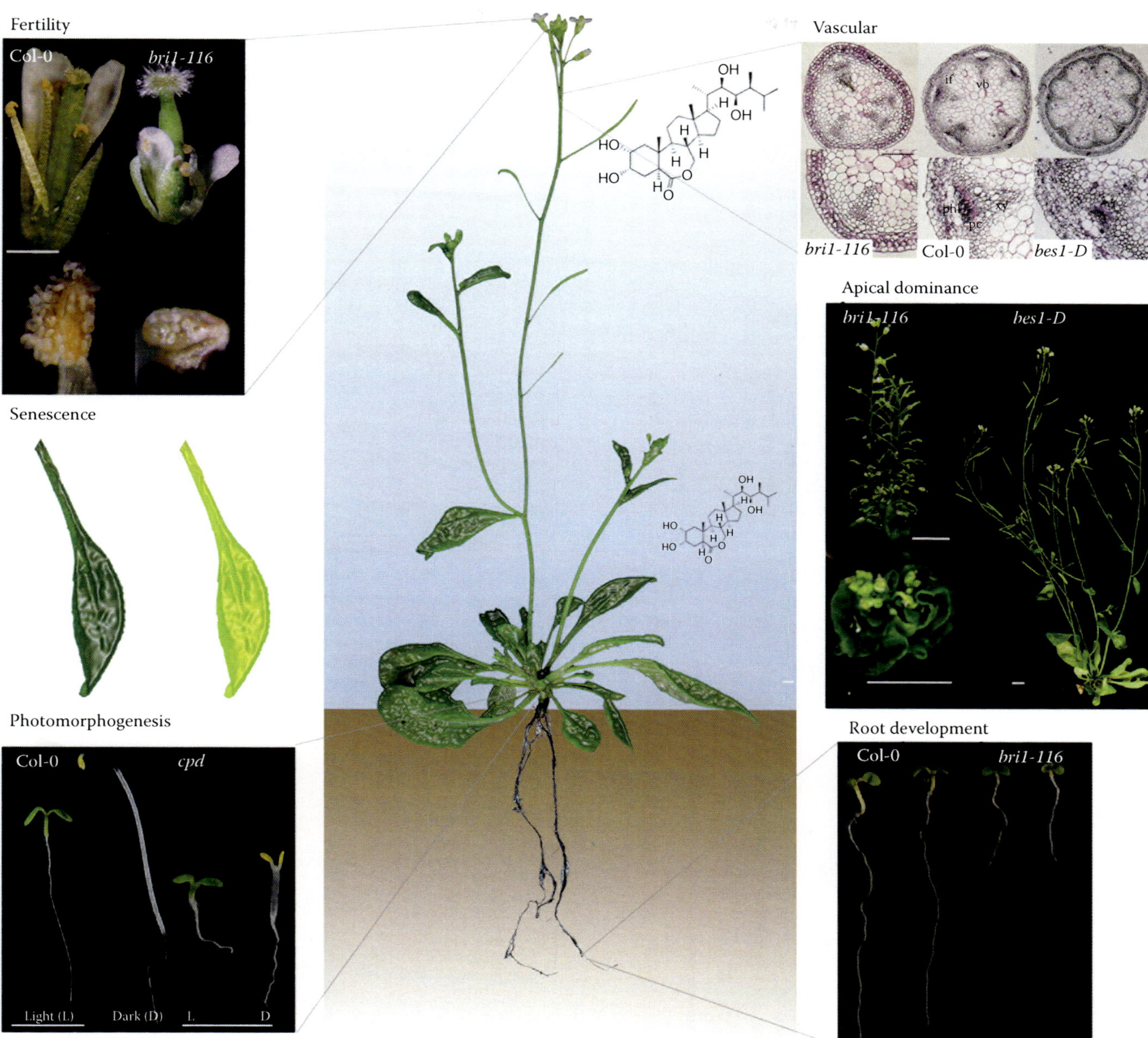

FIGURE 17.2 BR contribution to plant development. The identification of BR mutants by genetic analysis has permitted to assign important roles for BRs in plant development. *Fertility*: loss-of-function BR mutants exhibit short stamens and defective pollen grain production that result in male-sterile plants (Ye et al. 2010). *Senescence*: exogenous application of active BRs was found to promote senescence, and BRs mutants show impaired senescence programs (Li and Chory 1997). *Photomorphogenesis*: *Arabidopsis* seedlings grow in darkness show elongated hypocotyls and closed cotyledons, in contrast mutants defective in BR mutants (e.g., *constitutive photomorphogenesis and dwarfism*, *cpd*) that show photomorphogenic responses with open cotyledons and short hypocotyls (Szekeres et al. 1996). *Vascular*: BR contribution to vascular bundle number in the shoot inflorescence stem by promoting early provascular divisions (Caño-Delgado et al. 2004; Ibañes et al. 2009). *Apical dominance*: loss-of-function BR mutants are dwarf and exhibit a loss of apical dominance. Gain-of-function *bes1-D* mutant (Li and Chory 1997, Yin et al. 2002). *Root development*: BR mutants show reduced root length caused by both alterations in cell-cycle progression and cell elongation (González-García et al. 2011; Hacham et al. 2011) (Vb; vascular bundle, If; interfascicular fibers, phl; phloem, pc; procambium, xy; xylem). (From González-García, M.P. et al., *Development*, 138, 849, 2011. With permission.)

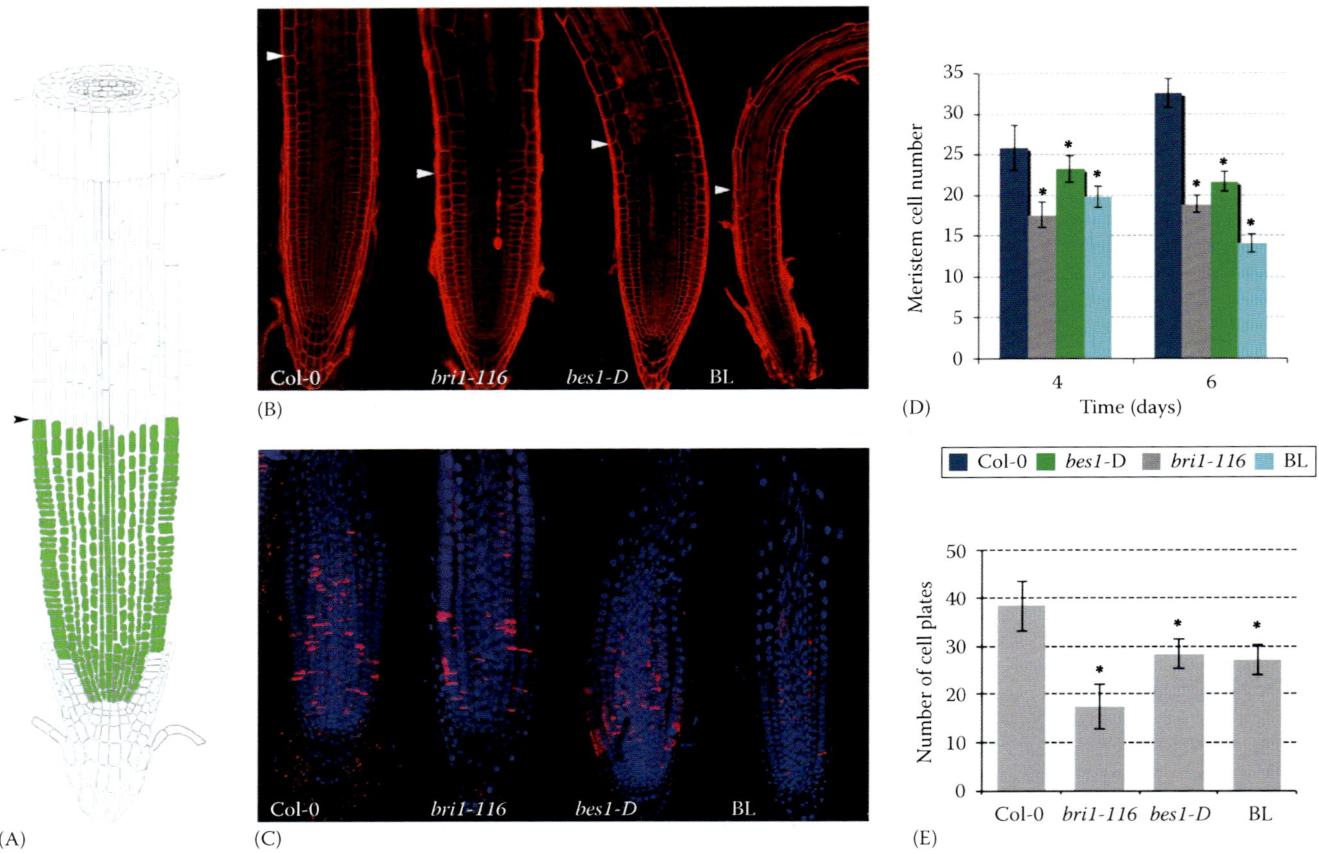

FIGURE 17.3 BRs control root growth by maintaining the normal cell-cycle progression in the meristem. (A) Schematic representation of a 6-day-old *Arabidopsis* primary root. Green cells represent the meristematic region (dividing cells). (B) Root meristems of Col-0, *bri1-116* (loss-of-function mutant), *bes1-D* (gain-of-function mutant), and 6-day-old plants continuously treated with 4 nM BL. Counterstained roots with propidium iodide (PI) staining (red), a white arrow points the end of the meristematic zone. (D) Representation of the number of cells within the meristematic region. (C) Immunolocalization of KNOLLE protein (pink) and cell nucleus stained with DAPI (blue) in different BR mutant background and BL treatment. (E) Representation of the number of cell plates for each genotype/physiological treatment (pink). Note that the differences in the number of meristematic cells and the number of KNOLLE cell plates in BR mutants and plant exogenously treated with BL in comparison to Col-0 wild type. (From González-García, M.P. et al., *Development*, 138, 849, 2011. With permission.)

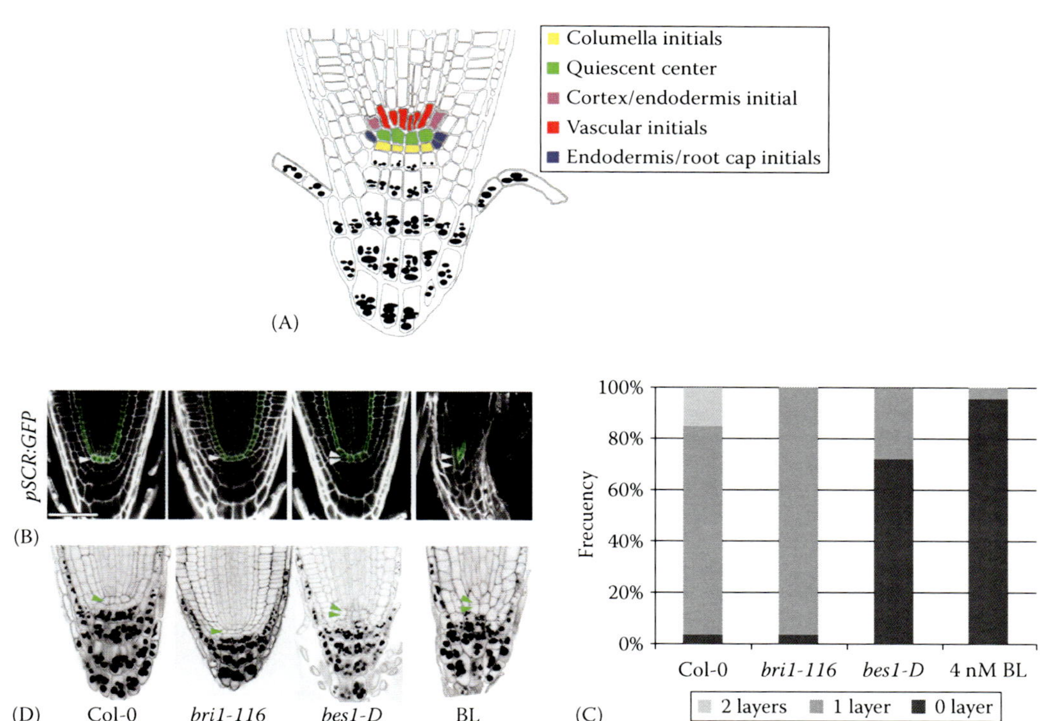

FIGURE 17.4 BRs promote distal stem cell differentiation in *Arabidopsis* primary root. (A) Cartoon representing the root stem cell niche in the root apex (colored area). The quiescent center (QC) cells maintain the surrounding stem cells that give rise to all cell types present in the *Arabidopsis* root (van den Berg et al. 1997). Below the QC, the differentiated columella cells accumulate starch granules (black dots). (B) Six-day-old roots show the expression of the QC identity marker *pSCR:GFP* in the genetic background of several BR mutants. (C) Modified pseudo-Schiff PI staining (mPS-PI) of mutant roots reveals a duplication of the QC cells in both *bes1-D* (gain-of-function) and in the exogenous BL treatment (white and green arrows) and defects on the number of columella stem cell layers. (D) Quantification of the number of columella stem cell (CSC) layers in 6-day-old roots. *Bes1-D* mutants and BL-treated plants show an increased CSC differentiation; CSC layers disappear, and the frequency of CSC divisions in *bri1* mutants is uncoupled compared to that of the Col-O wild type. (From González-García, M.P. et al., *Development*, 138, 849, 2011. With permission.)

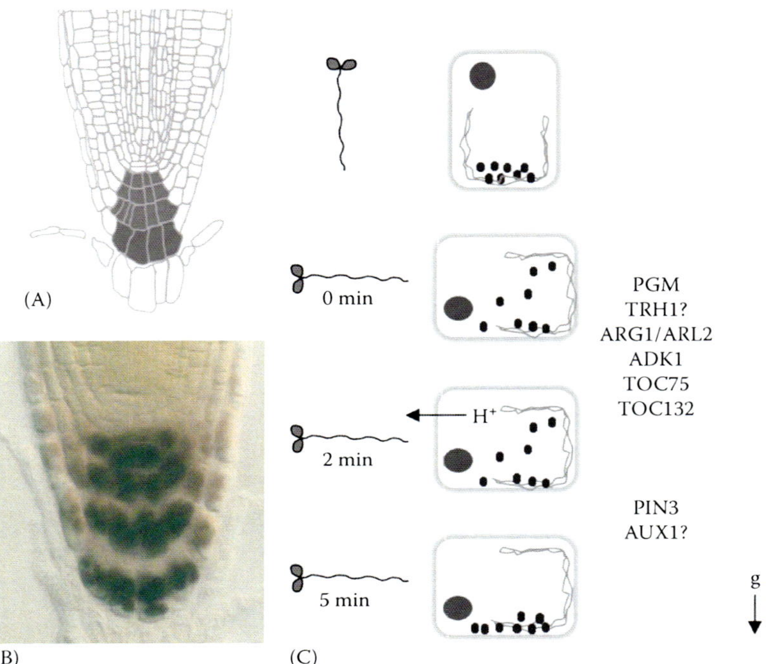

FIGURE 19.1 Gravity perception. (A) A cartoon of an *Arabidopsis* root with columella cells marked in gray. (B) Image of an *Arabidopsis* root stained for starch. (C) A schematic representation of gravity perception in *Arabidopsis*. Change in gravity vector results in sedimentation of amyloplasts in the columella cells that initiates a cascade of events including alkalinization of cytoplasm and repositioning of auxin efflux carrier PIN3 to the new bottom of the cell by a yet unknown mechanism.

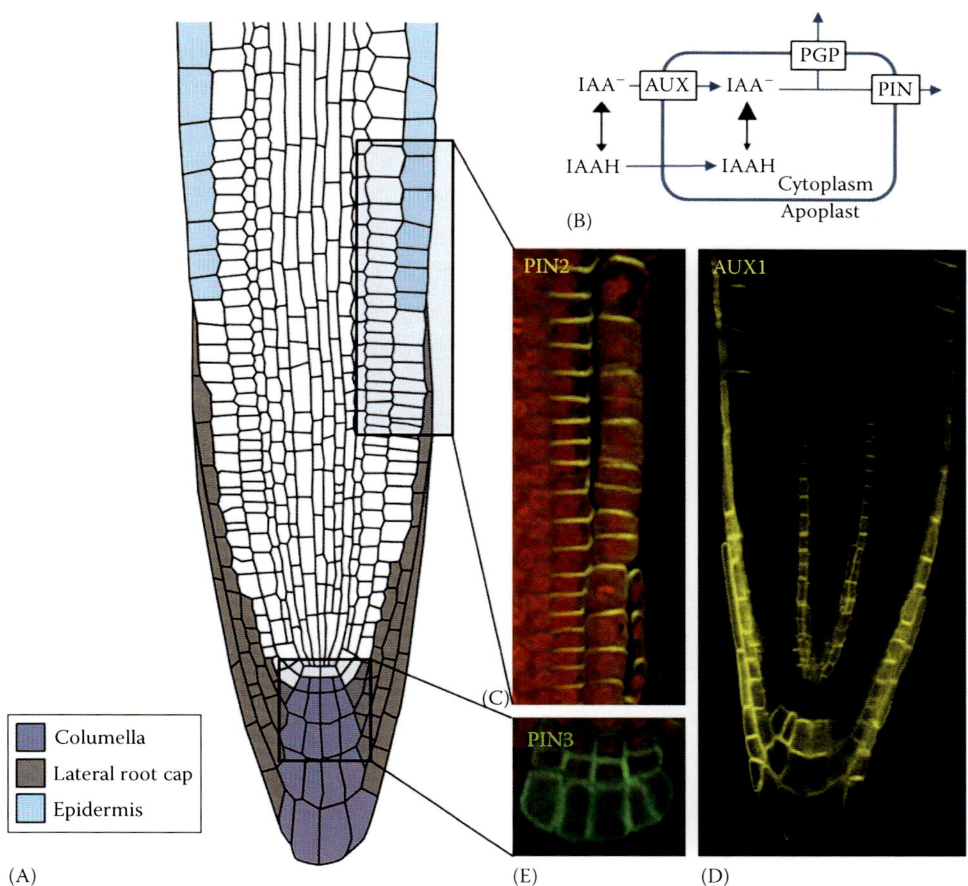

FIGURE 19.2 Gravity signal transmission. (A) A cartoon of an *Arabidopsis* root marking tissues involved in graviperception (blue), signal transmission (gray), and response (sky blue). (B) Auxin influx and efflux carriers facilitate auxin movement following a gravity stimulus. (C) Confocal image showing PIN2 (green) localization in root apex. (D) Confocal image showing AUX1 (yellow) localization in root apex. (E) Confocal image showing PIN3 (green) localization in columella cells.

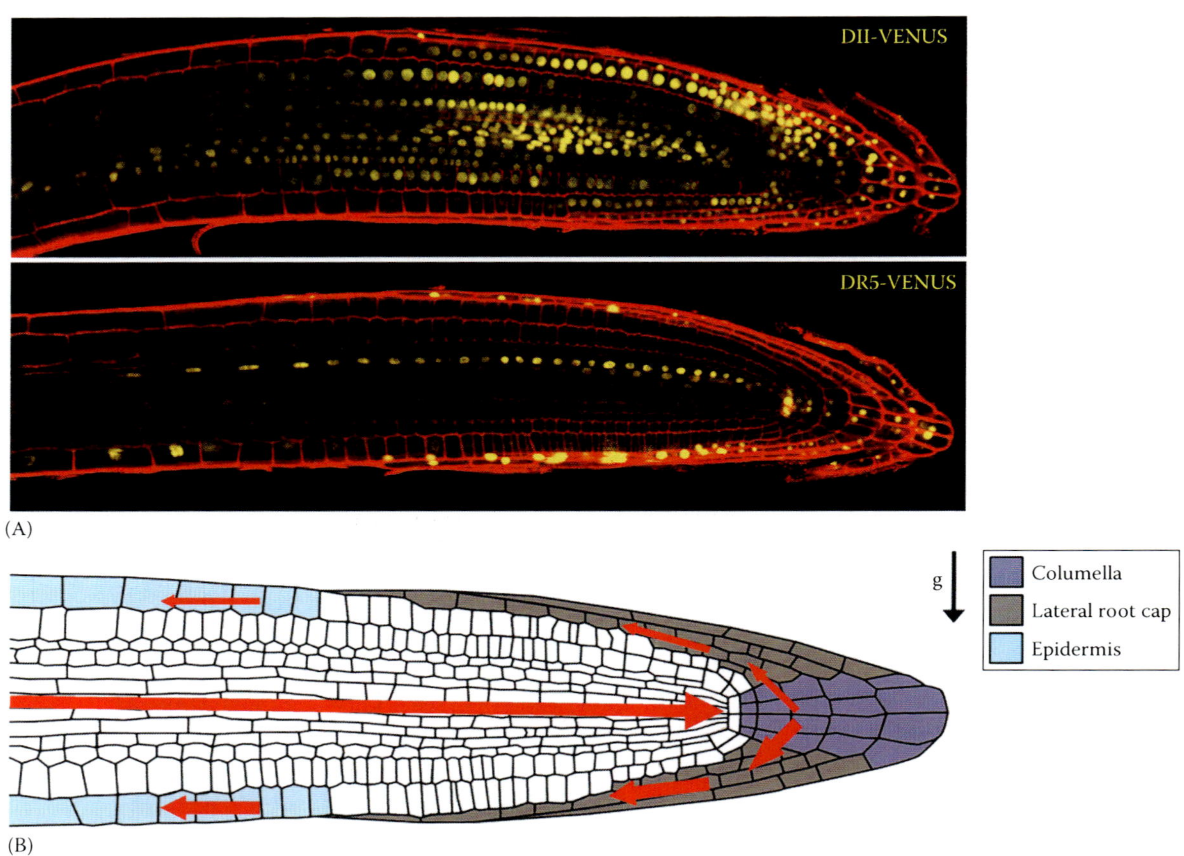

FIGURE 19.3 Formation of an auxin-response gradient. (A) Confocal image of a gravistimulated *Arabidopsis* root showing differential expression of auxin-response reporter DII-VENUS (upper) and DR5:VENUS (lower). Note the complimentary expression patterns of the two reporters. Fluorescence asymmetry becomes apparent in DII-VENUS roots much earlier (1 h) following a gravistimulus than in DR5:VENUS roots (3 h). (B) An interpretation of (A) showing differential auxin movement following gravity perception. Upon gravity perception, asymmetric localization of PIN3 in the columella cells results in differential auxin movement between upper and lower side of a gravistimulated root as seen by asymmetric localization of auxin-response reporter DR5:VENUS.

FIGURE 21.2 (A) Lupin roots from the field. Left: infected with rhizoctonia fungal root disease and root length severely stunted and nodule number reduced. Right: healthy without rhizoctonia root disease. Bar = 2 cm. (B) Pit dug within a wheat crop at flowering and grain development in southeastern Australia. Maximum root depth ~150 cm. Bar = 20 cm. (C) Wheat genotypes of same age with genetic variation in root system vigor. (Reproduced from Gewin, Nature, 466, 552, 2010.) Ruler at right = 30 cm. (D) Sorghum nodal roots growing at two angles from the crown; wide angle on left is in moist soil; narrow angle on right is in dry soil. Bar = 1 cm. (Photo courtesy of M. Rostamza.) (E) Roots clumped in a pore in the pit shown in (B). Bar = 2 mm. (F) Wheat roots clumped in a soil pore with root hairs contacting sides of pore, enabling hydraulic conductivity with soil. Bar = 2 mm. (Photo courtesy of M.E. McCully.) (G) Phosphate transporter expressed specifically in rice root hairs (*Pht1;1 promoter::gfp* gene). Bar = 50 μm. (Reproduced from Schunmann, P.H.D. et al, *J. Exp. Bot.*, 55, 855, 2004.) (H) Sorgoleone exudate (yellow drops) released from sorghum root hairs. Exudate is a biocide and electron transport inhibitor in plants that take it up. Bar = 50 μm. (Photo courtesy of K. Schneebeli.) (I) Avenacin (blue) localized to the epidermal cells of oat. Compound toxic to wheat Take-All fungus. Bar = 50 μm.

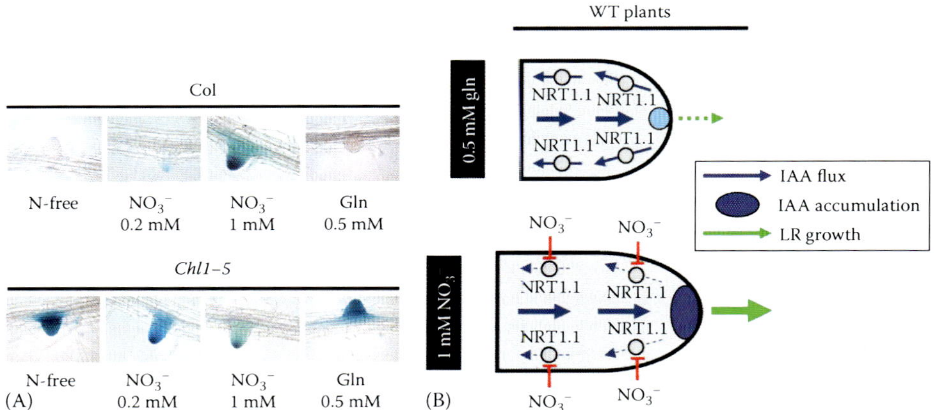

FIGURE 25.9 (A) Histochemical staining of GUS activity in newly emerged lateral roots of transgenic *Arabidopsis* plants expressing *DR5::GUS* (an auxin biomarker) in wild-type or *chl1-5* background. The plants were cultivated for 8 days on media containing nitrogen sources described in the figure. (B) Schematic model for NRT1.1 control of lateral root growth in response to nitrate. Two situations are shown to illustrate the specific effect of NO_3^- on lateral root growth, corresponding to plants supplied either with 0.5 mM glutamine or with 1 mM NO_3^- (1 mM external N in both cases). The model postulates that in the absence of NO_3^- (glutamine-fed plants), NRT1.1 favors basipetal transport of auxin in lateral roots, thus preventing auxin accumulation at the lateral root tip. This slows down outgrowth and elongation of lateral roots. At 1 mM NO_3^-, facilitation of basipetal auxin transport by NRT1.1 is inhibited, leading to auxin accumulation in the lateral root tip and accelerated growth of lateral root. (From Krouk, G. et al., *Dev. Cell*, 18, 927, 2010b.)

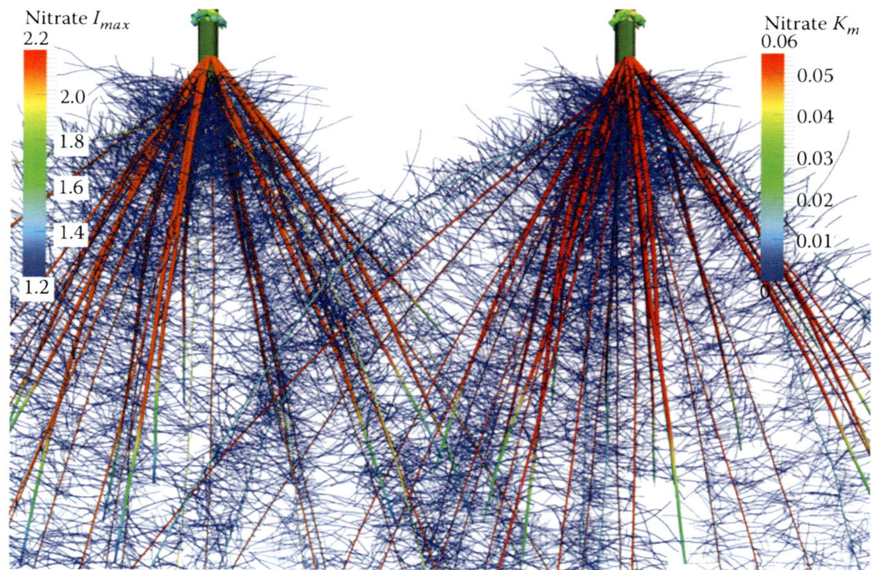

FIGURE 26.2 Spatial maps of Michaelis–Menton kinetic parameters (I_{max} [μmol·cm^{-2} d^{-1}], K_m [μM]) for nitrate in maize root systems 40 days after germination. (From M. Silberbush, J. Postma, and J.P. Lynch, unpublished data; Postma, J. and Lynch, J., *Ann. Bot.*, 107, 829, 2011. With permission.)

FIGURE 28.2 Root system development, seedlings, and adult plants of *Pachycereus pringlei* and *Stenocereus gummosus* from Sonoran Desert. (A) Adult plant of *P. pringlei*. Arrow indicates a massive root at the soil surface. (B) Root system of adult *P. pringlei* plant where a bayonet-like taproot (arrow) and shallow, apparently lateral, roots can be clearly recognized. (C) *P. pringlei* seedling with primary root that terminated its growth and four basal roots formed at the base of the primary root. (D) Adult *S. gummosus* plant. (E) *S. gummosus* seedlings whose primary root terminated growth. (F) *S. gummosus* seedlings that form compact root system as most lateral and basal roots also terminated growth. (G) Young *S. gummosus* plants with clearly defined taproots. Bars = 5 mm. (Kind donation of photo (B) by Dr. F. Molina-Freaner from the Universidad Nacional Autónoma de México, Mexico, is gratefully acknowledged. Other photos by the authors.)

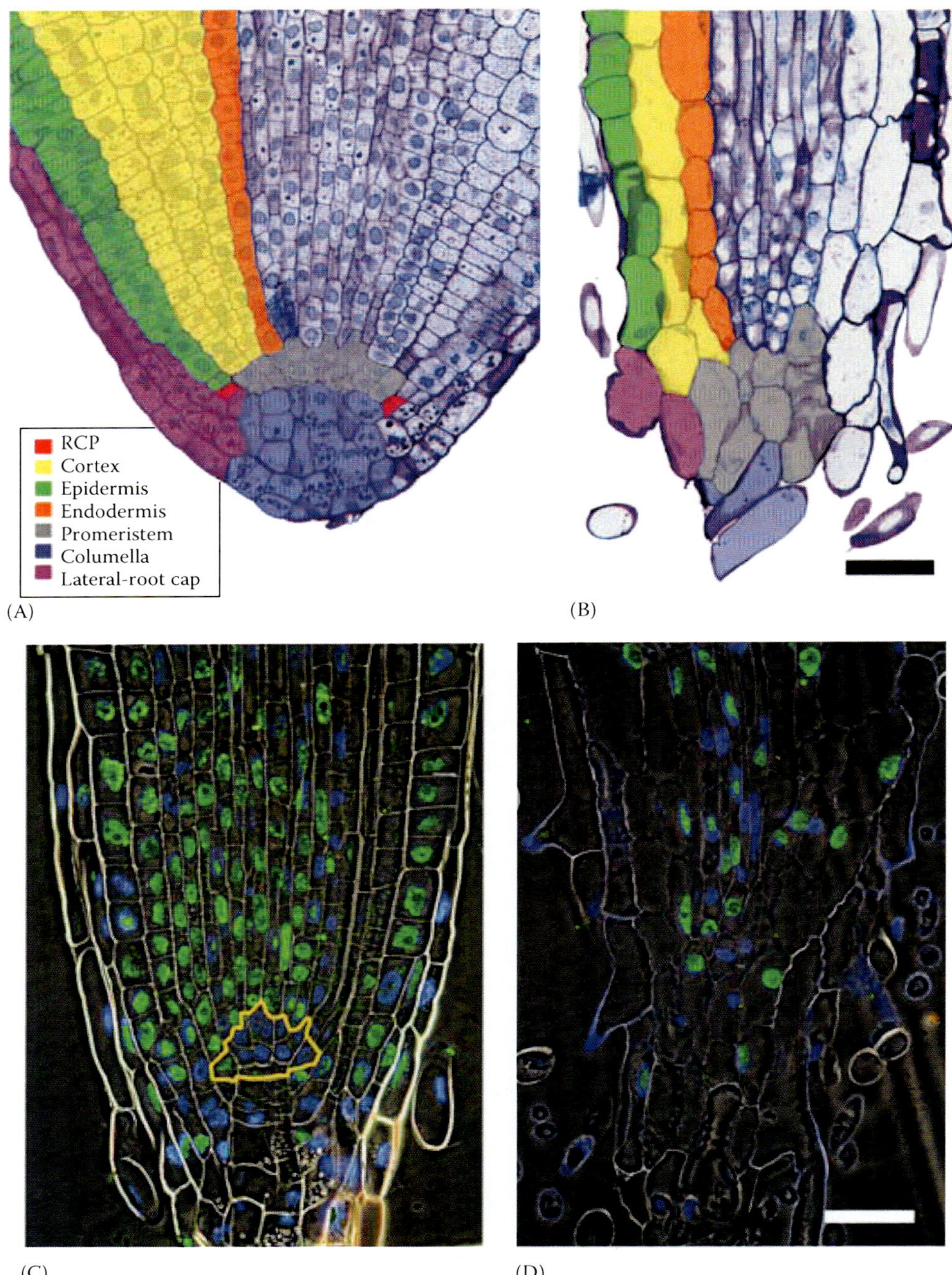

FIGURE 28.6 Primary root apical meristem organization and exhaustion in *S. gummosus* (A, B) and *P. pringlei* (C, D). (A) Open type of meristem organization in a seedling just after seed germination. Promeristem (gray-colored cells) is the zone where quiescent center cells could be established within the open-type meristem. (B) Primary root apex 3 days after germination. Note, almost all cells of the root cap are lost, cortex is formed by just one cell file, differentiation zone approached to the tip, and the root hairs are formed at the very tip. (C, D) Developmental changes in the primary roots of *P. pringlei* and immunofluorescence analysis of cell proliferation. DAPI (blue staining of nuclei) and FITC (green; immunological staining of nuclei that passed S-phase of the cell cycle during 24 h treatment with BrDU) images superimposed. (C) Primary root of recently germinated seedling. Quiescent center is outlined with yellow line. (D) Primary root at the end of meristem exhaustion. Note: No compact group of FITC-negative cells is detected indicating an absence of the quiescent center at this stage. All ex-meristematic cells are differentiated, and many of them passed S-phase (green FITC-positive cells). Bars = 50 μm. (Photo by the authors, Rodríguez-Rodríguez, J.F. et al., *Planta*, 217, 849, 2003 with modifications. Kind permission of Springer Science + Business Media.)

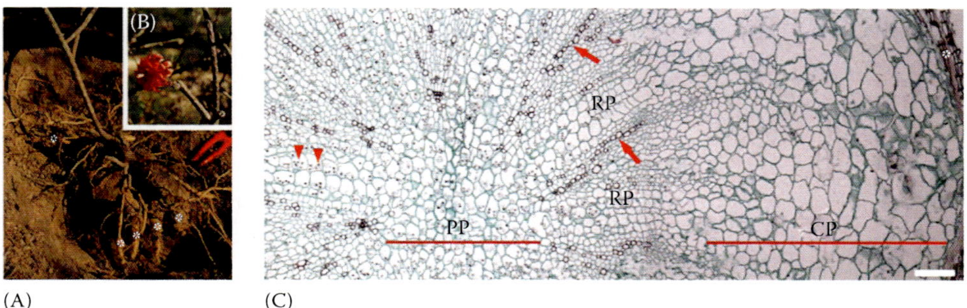

FIGURE 28.8 Tuberous root morphology, succulent stem, flower, and anatomical features of tuberous root in *Peniocereus viperinus* (Cactaceae). (A) In *P. viperinus* plant, several storage roots are formed during different years. Each tuberous root is marked with white asterisk. Note the perennial most massive tuberous root to the right of the image. (B) Aboveground vegetative and generative organs are produced when water becomes available. During this period, as in all geophytes, photosynthates are produced and transported to the tuberous root where they are accumulated in form of starch. During dry season, aboveground organs die. Note that stem is narrow and less succulent than in most other Cactaceae. (C) A section of young tuberous root. Note multilayered periderm (asterisk), abundant starch accumulation (arrowheads), xylem vessels (arrows), pith parenchyma (PP), ray parenchyma (RP), and cortical parenchyma (CP). Bar = 200 μm. (All images were kindly donated by Dr. T. Terrazas, Universidad Nacional Autónoma de México, Mexico.)

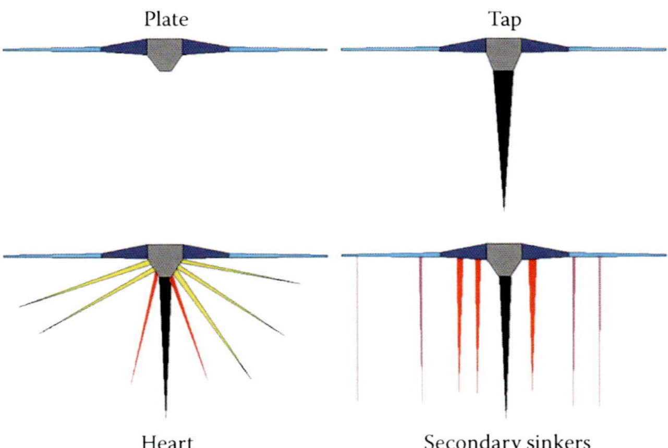

FIGURE 29.4 Schematic representation of the four classes of tree root architecture according to Köstler et al. (1968). Gray: stump, black: taproot, blue: ZRT, light blue: shallow roots beyond ZRT, red: proximal secondary sinkers, violet: distal secondary sinkers, yellow: oblique roots. Intermediate-depth roots and deep roots are not figured.

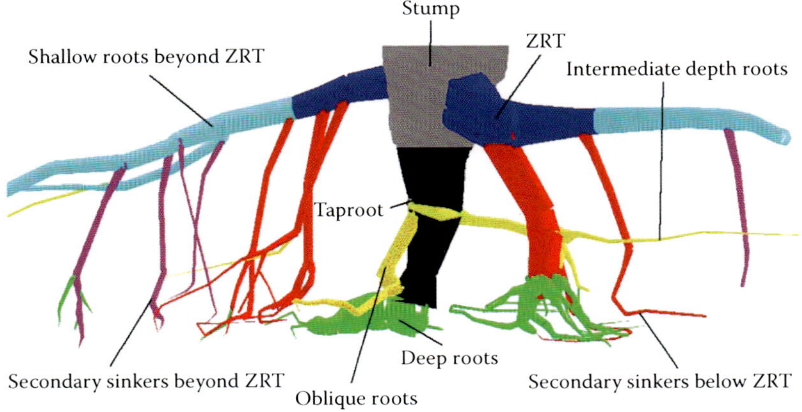

FIGURE 29.5 Architectural analysis in mature *P. pinaster*: Schematic representation of root types. Nine compartments were defined mainly as a function of their angle toward horizontal plane, depth of their initial branching point and tapering. (From Danjon, F. et al., *New Phytol.*, 168, 387, 2005. With permission.)

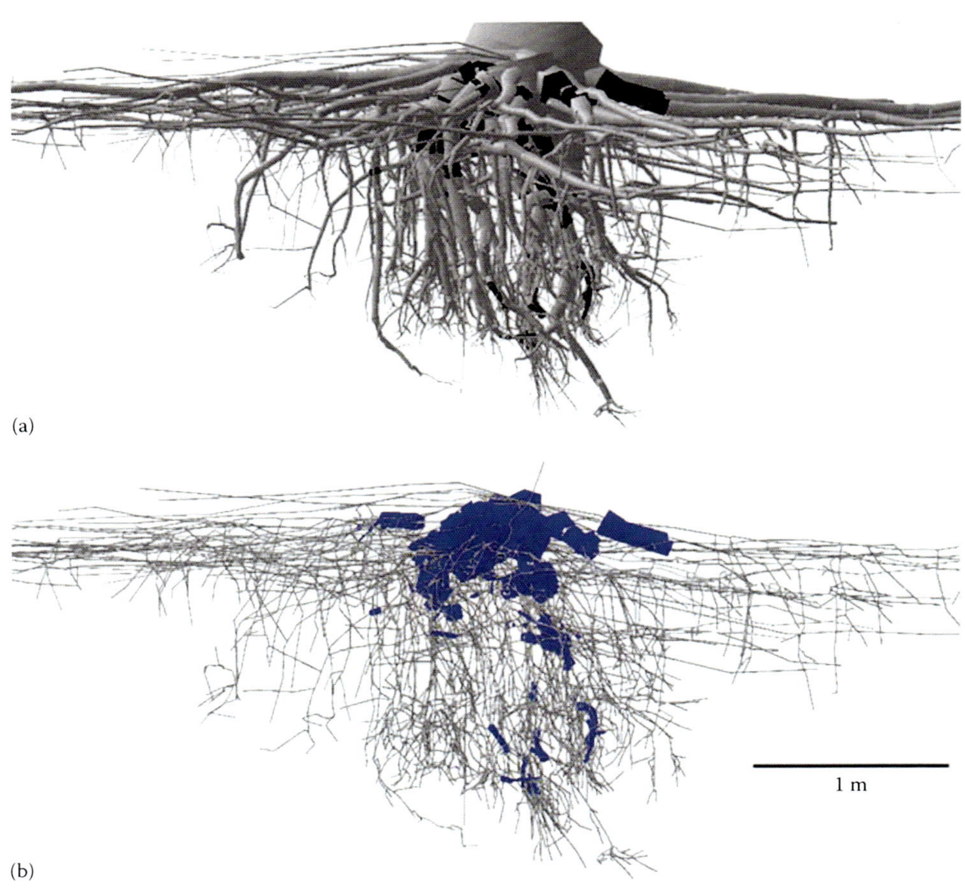

FIGURE 29.6 Root grafts in an 18-year-old *P. pinaster* root system measured by 3D digitizing. (a) Root grafts are in black. (b) Same tree, segments with a root graft drawn with their real diameter. The diameter of all other segments was set at 4 mm. (Data from Khlifa, R., Meredieu, C., Augusto, L., Trichet, P., Danjon, F. Unpublished data)

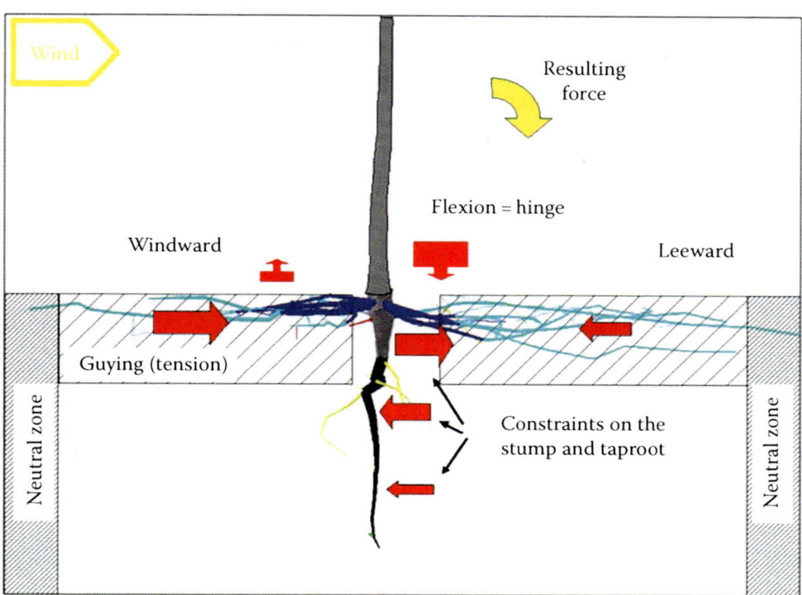

FIGURE 29.8 Schematic drawing of the forces experienced by a sapling with a taproot system in sandy soil, when the stem is pushed to the right by a strong wind. Wind damage could be toppling or stem breakage. (Reproduced from Danjon, F. and Fourcaud, T., L'arbre et son enracinement, *Actes du colloque CIAG Sylviculture Forêts et Tempêtes*, juin 30, 2009, Pessac, France, Revue Innovations Agronomiques INRA, Paris, France, Vol. 6, pp. 17–37, 2009. With permission.)

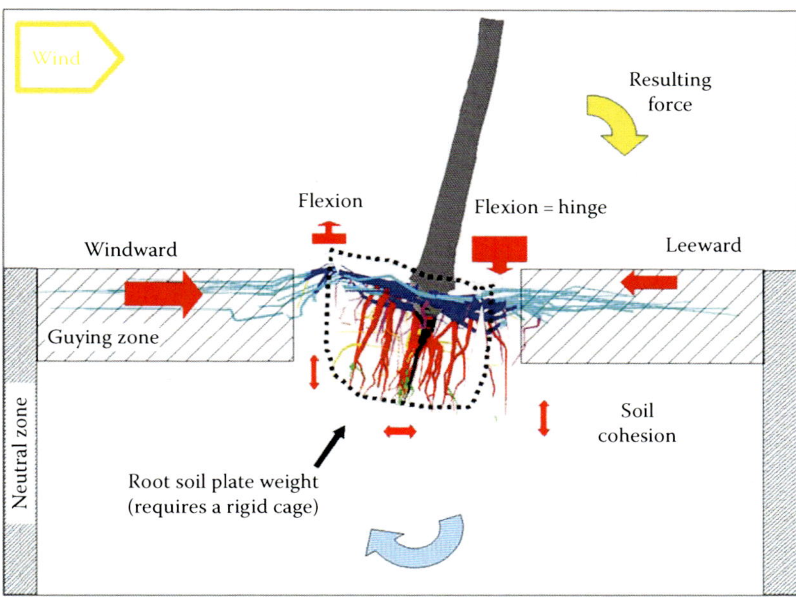

FIGURE 29.9 Schematic drawing of the forces experienced by a mature tree with secondary sinkers' root system in sandy soil, when the stem is pushed to the right by a strong wind. Wind damage could be uprooting with root soil plate or stem breakage. (Reproduced from Danjon, F. and Fourcaud, T., L'arbre et son enracinement, *Actes du colloque CIAG Sylviculture Forêts et Tempêtes*, juin 30, 2009, Pessac, France, Revue Innovations Agronomiques INRA, Paris, France, Vol. 6, pp. 17–37, 2009.)

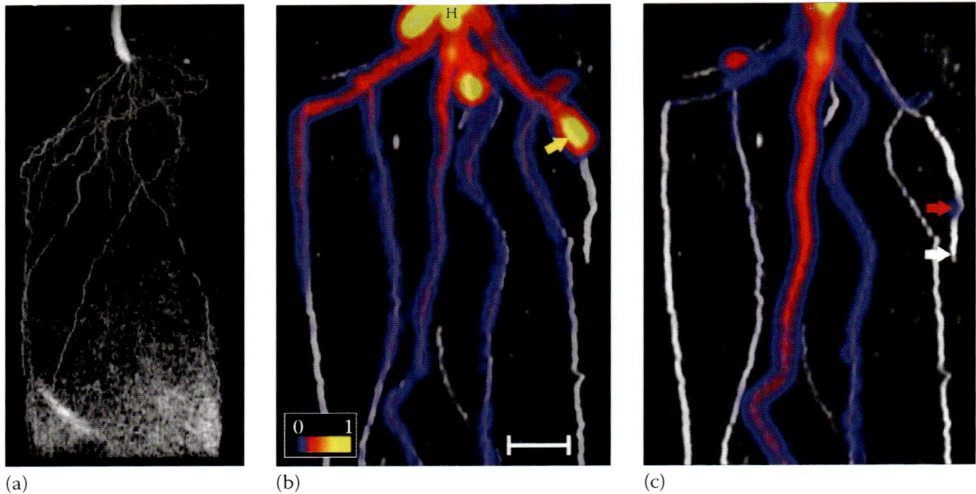

FIGURE 31.6 Noninvasive imaging of roots growing in substrates. (a) Wheat roots growing in loamy sand visualized with MRI (van Dusschoten, unpublished). (b, c) Maize roots in sand coregistered with MRI (grey) and PET (colored). Growth was exemplarily analyzed on a root as shown on the right-hand side of panels (b) and (c). The root tip position was visualized on day 15 (yellow arrow), day 16 (red arrow), and day 17 (white arrow) after sowing. Scale bar 1 cm. (Modified from Jahnke, S. et al., *Plant J.*, 59, 634, 2009.)

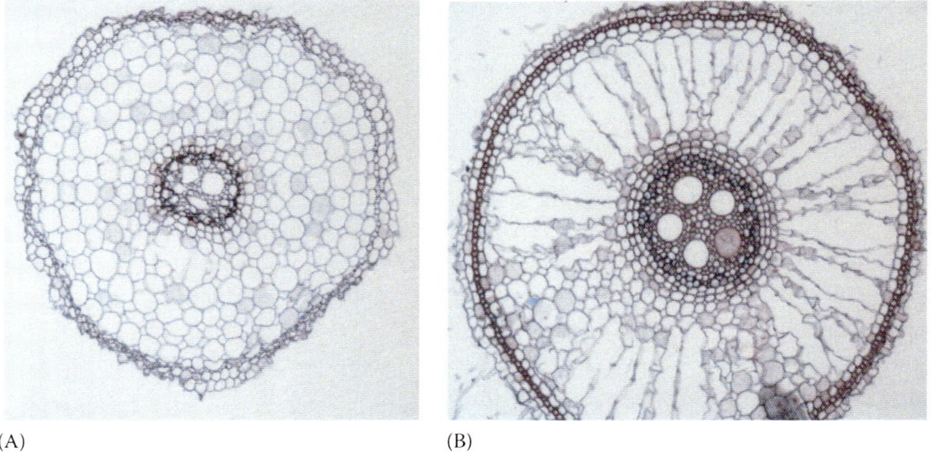

FIGURE 32.1 Development of aerenchyma in rice roots grown under stagnant solution culture conditions. (A) Section of a young rice root. (B) Section of a fully developed rice root. Note the thick-walled exodermis spanning two layers of cells, protecting the root tissue from oxygen loss to the stagnant environment. (Photo courtesy of Kurt V. Fagerstedt.)

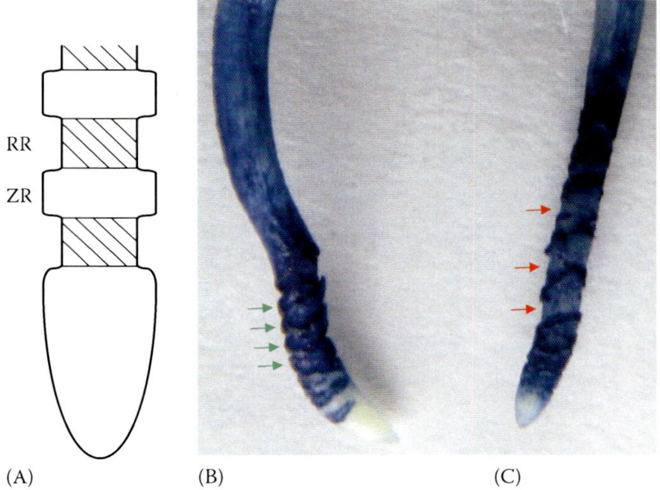

FIGURE 33.3 Changes in the morphology of the pea root apex during and after recovery from Al treatment. (A) Schematic diagram of the root apex structure induced by Al toxicity. RR: rupture region, ZR: zonary region stained with Evans blue. (B) Pea root treated with 40 μM Al for 24 h. (C) Pea root treated with 40 μM Al for 12 h and then in Al-free solution for 12 h. Both (B) and (C) show the image of Evans blue staining. Green arrows indicate the zones stained markedly with Evans blue corresponding to the ZR in (A). Red arrows indicate the rupture zone corresponding to the RR in (A). (From Motoda, H. et al., *Plant Signal. Behav.*, 6, 98, 2011. With the permission of Landes Bioscience.)

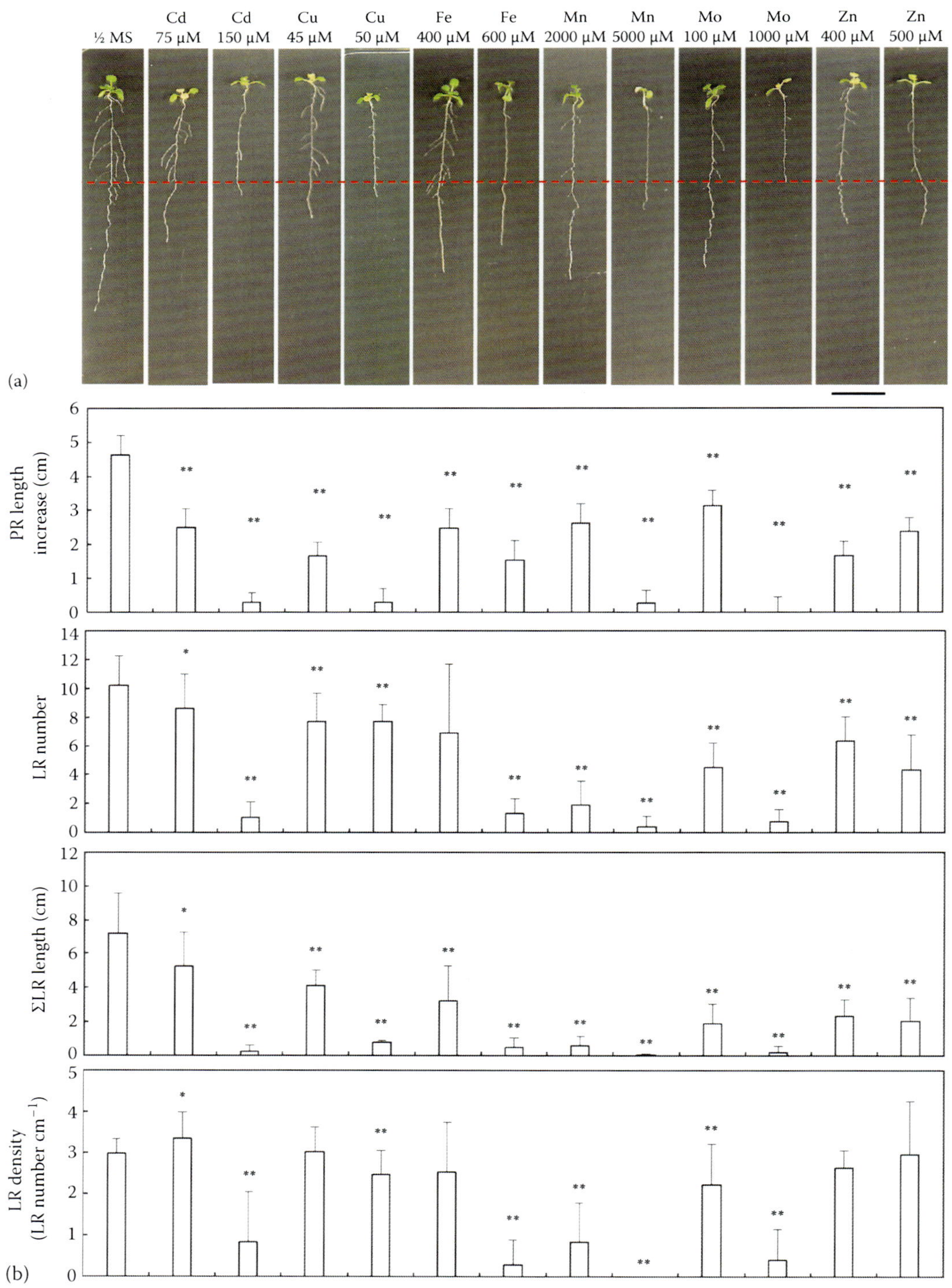

FIGURE 34.1 Seedlings were germinated for 6 days on 0.5× Murashige and Skoog medium and were transferred to media enriched with TMEs (75, 150 μM $CdSO_4$; 45, 50 μM $CuSO_4$; 400, 600 μM Fe-EDTA, 2000, 5000 μM $MnSO_4$; 100, 1000 μM MoO_3 and 400, 500 μM $ZnSO_4$). (a) Representative picture of seedlings one week after transfer. The dashed red lines indicate the average PR length upon transfer. Scale bar: 2 cm; (b) Root architecture parameters: primary root length (PR) after transfer, lateral root (visibly emerged > 1 mm) number (LR), sum of lateral root length (ΣLR) and lateral root density (LR density) calculated as the number of LR per root length portion comprised between the first and the last LR. N = 10–20 observations ±std. Stars indicate statistical difference between a given treatment and the control conditions (0.5× MS): $P < 0.10$ (*), $P < 0.01$ (**).

FIGURE 34.3 Photograph of a *N. caerulescens* plant growing in a rhizobox filled with heterogeneously contaminated soil. The rhizobox is divided (dashed line) into two equal left and right compartments, filled with uncontaminated soil (Ni 0) on the left, and with soil contaminated with nickel (Ni 500: 500 µg Ni g^{-1} dry soil) on the right. Foraging for Ni is highlighted by the higher proportion of roots developed in the right compartment enriched with Ni. (Courtesy of C. Dechamps.)

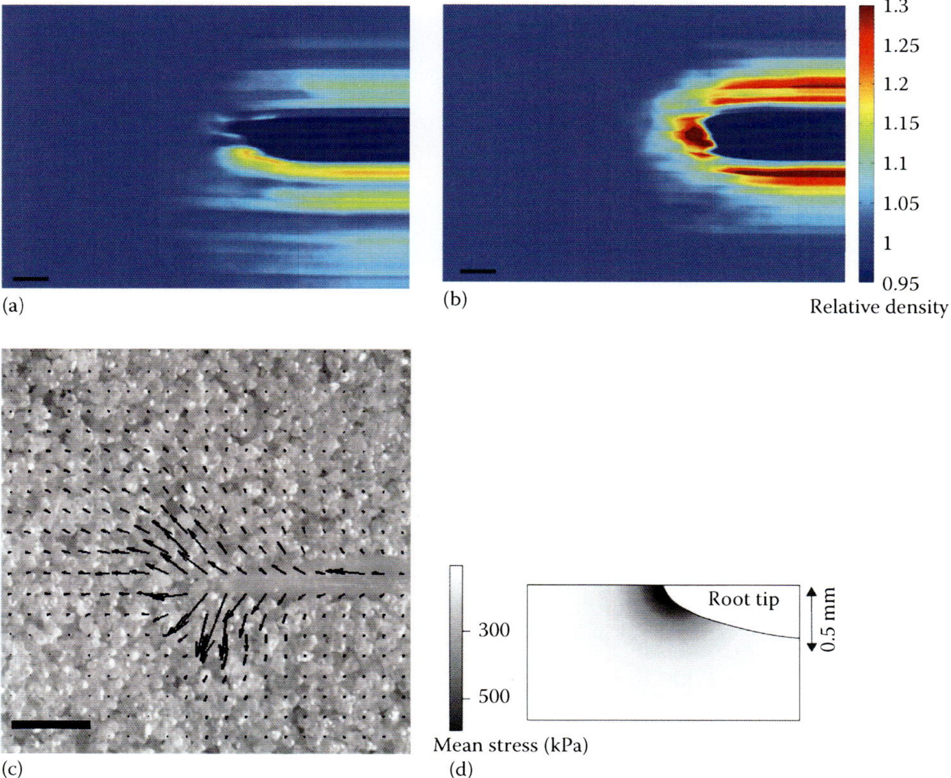

FIGURE 37.13 Soil displacement patterns around growing root tips. (a and b) Soil compression around wild type (KYS; a) or decapped mutant (*agt1*$_{dec}$; b) maize roots (Redrawn from Vollsnes, A.V. et al., *Eur. J. Soil Sci.*, 61, 926, 2010, Figures 9a and b.) estimated from (c) displacement fields in sand measured with particle image velocimetry. (Redrawn from Vollsnes, A.V. et al., *Eur. J. Soil Sci.*, 61, 926, 2010, Figures 6a.) (d) Predicted stress distribution around root tips of pea growing in sandy loam. Black scale bars are 2 mm long. (Redrawn Kirby, J.M. and Bengough, A.G., *Eur. J. Soil Sci.*, 53, 119, 2002, Figure 4. With permission.)

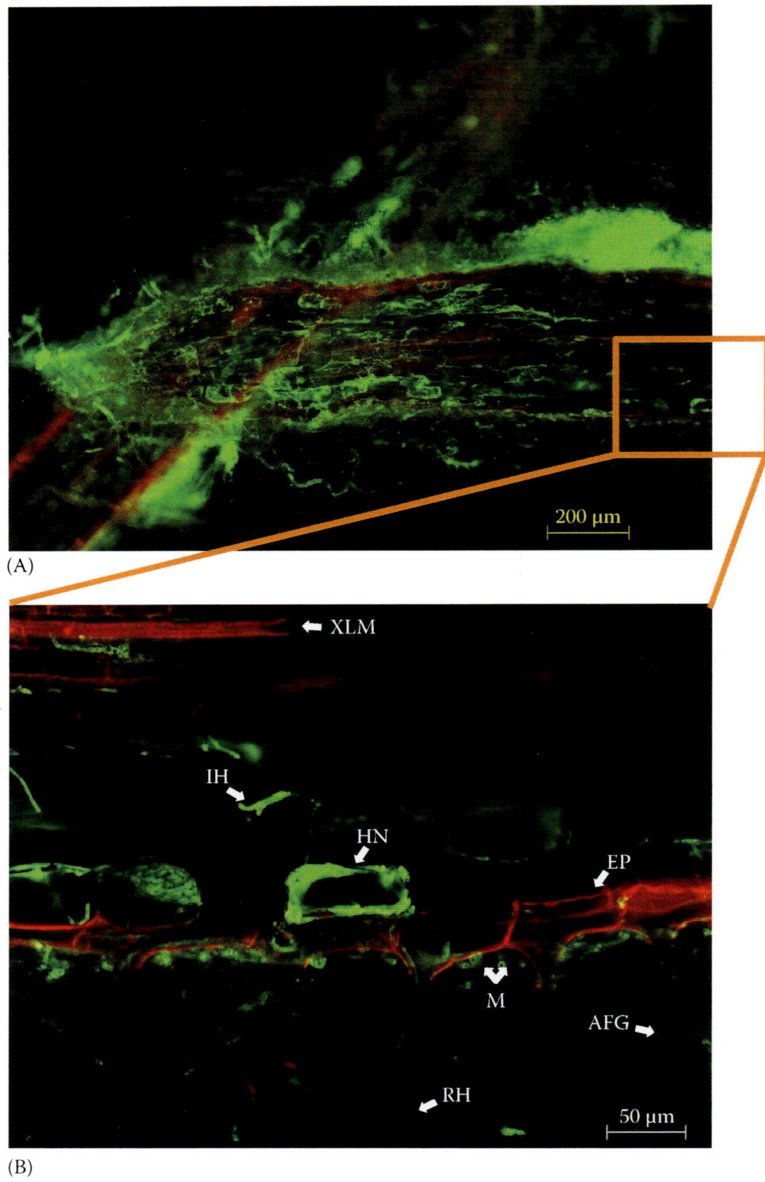

FIGURE 39.1 Morphological stages of *L. bicolor* on and in *Populus* x *canescens* roots. (A) Root tip ensheathed by a fungal mantle with emanating hyphae, (B) details of root cells in contact with hyphae forming a Hartig net. The organisms were cocultured for 5 weeks in a Petri dish on agar medium. Infected roots were fixed and 65–70 μm thick longitudinal sections were prepared with a sliding microtome. The fungal hyphae were stained with the chitin-specific florescent dye WGA-Alexa Fluor-488 (green), and the plant cell walls were stained with propidium iodide (red). Sample was visualized under a Zeiss fluorescent microscope. M = mantle, HN = Hartig net, RH = root hair, EP = epidermis, XLM = xylem, IH = invading hyphae, AFG = autofluorescence of glue.

FIGURE 40.3 Scheme indicating the extraordinary species-rich orchid communities as epiphytes (e.g., *Stelis* spp.) and terrestrials (e.g., *Epidendrum* spp., *Prostechea vespa*, *Sobralia rosea*) in the Andean tropical mountain rain forest supported by multiple, shared Tulasnellales (basidiomes on bark and rotten branches).

FIGURE 40.4 Scheme displaying orchids in shady ECM forests (Fagales, Caesalpiniaceae, Dipterocarpaceae, Pinaceae) as supported by ECMF mycobionts profiting from carbon supply by trees. Fungi indicate main groups as given in Table 40.2 (Thelephorales, Sebacinales group A, Russulales, Tuberaceae, Tulasnellales).

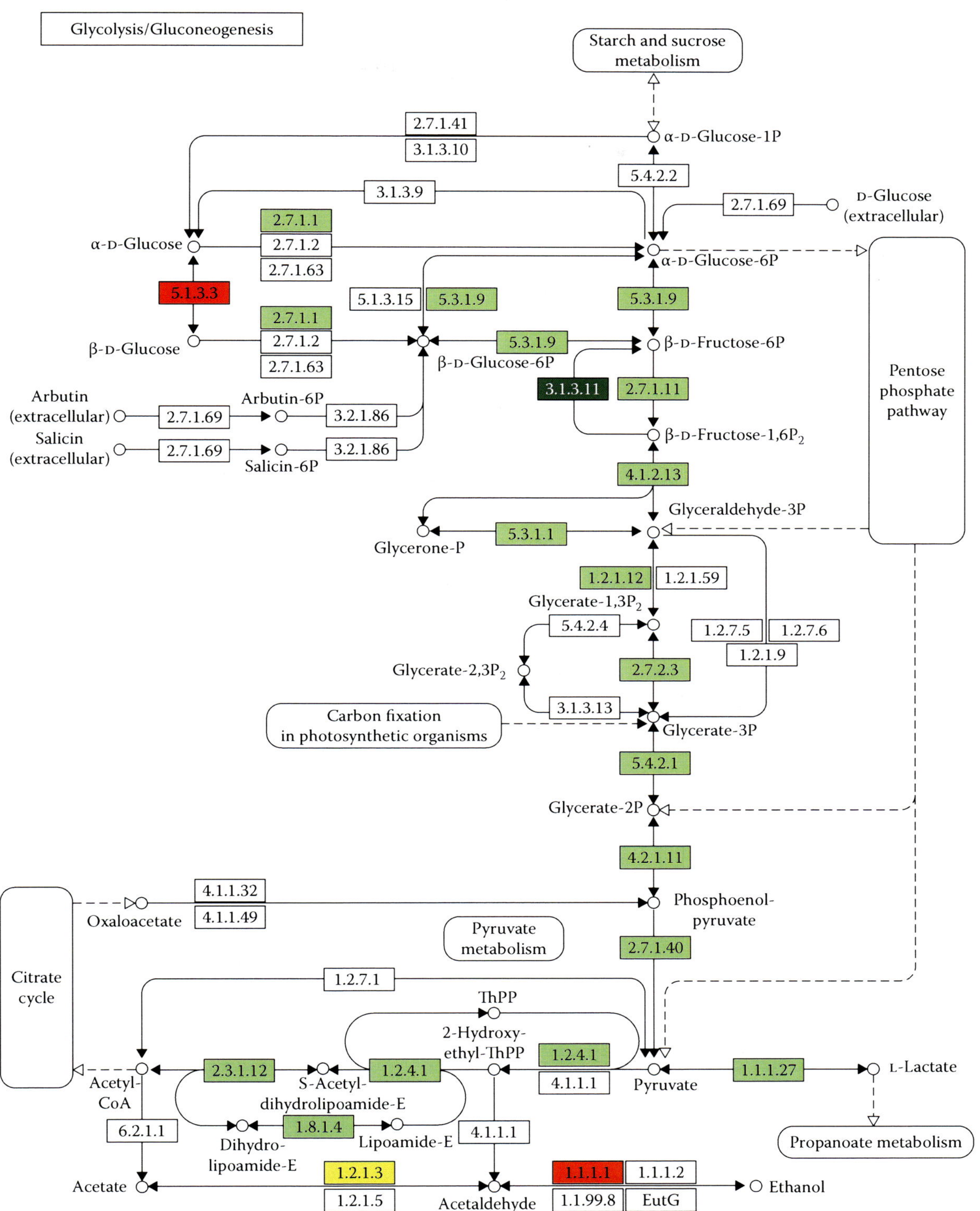

FIGURE 41.2 Expression profiles of the RNAs encoding enzymes in glycolysis/gluconeogenesis pathway in compatible syncytia 10 dai derived from data from Ithal et al. (2007a). Enzymes colored in red are encoded by downregulated genes. Enzymes colored in yellow are encoded by more than one gene and those different gene copies are up- and downregulated, respectively. Genes encoding enzymes colored in green are upregulated. Dark green indicates genes with transcripts accumulating more than 75% of the upregulated genes.

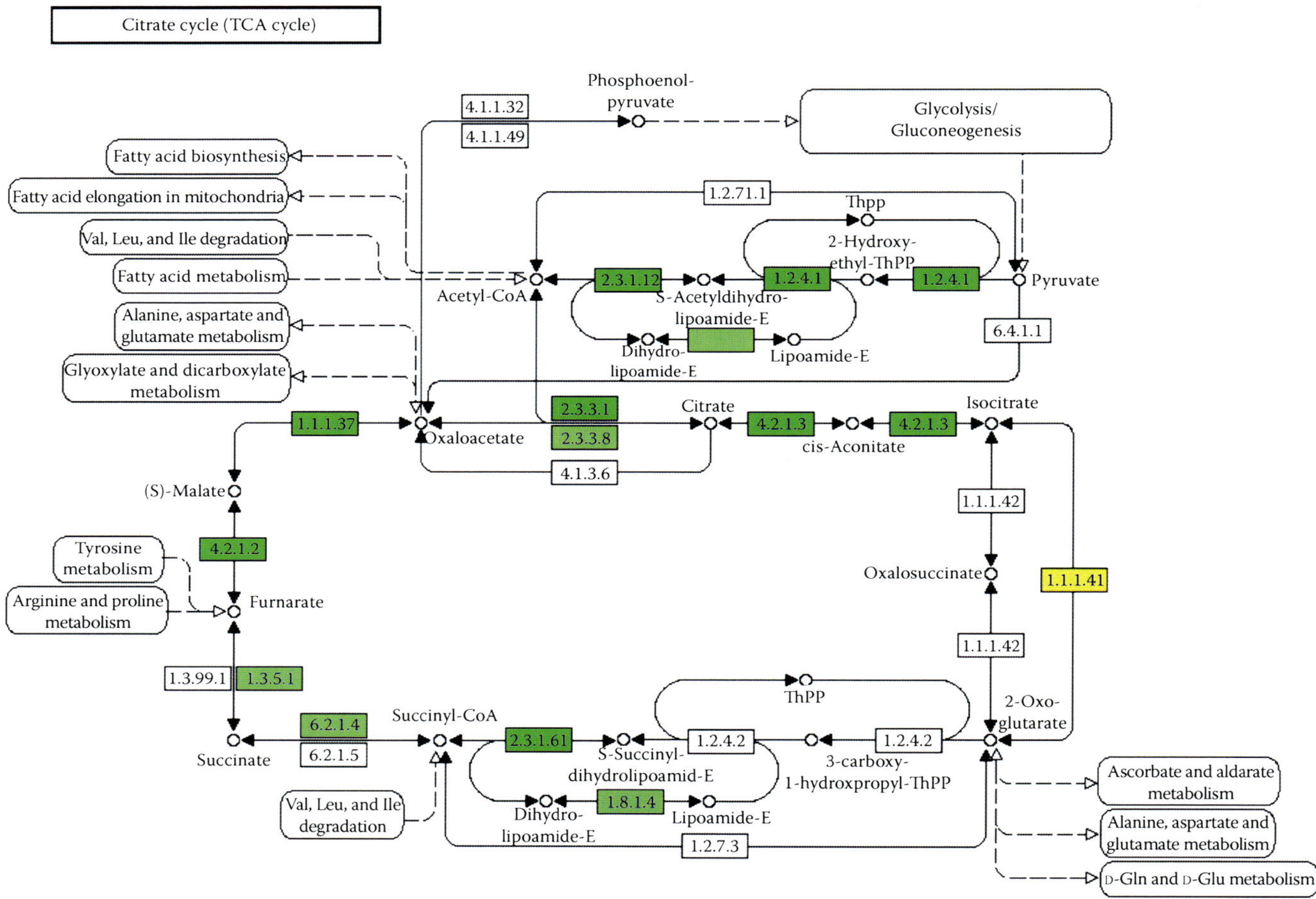

FIGURE 41.3 Expression profiles of the RNAs encoding enzymes in the TCA cycle as found in syncytia formed during a compatible reaction 10 dai. Enzymes colored in yellow are encoded by more than one gene and those different gene copies are up- and downregulated, respectively. Genes encoding enzymes colored in green are upregulated. (Data from Ithal N et al., *Mol. Plant-Microbe. Interact.*, 20, 293, 2007a.)

FIGURE 42.9 *Solanum lycopersicum* roots infested with *Orobanche aegyptiaca* tubercles. The picture was taken with a magnifying camera MR, Bartz, Carpinteria, CA.

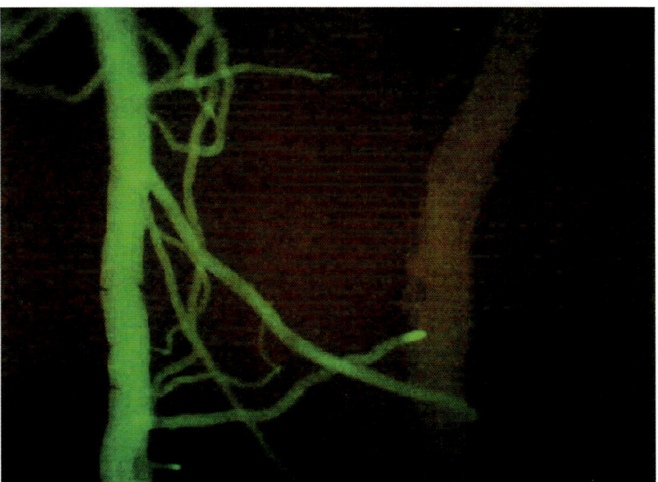

FIGURE 42.10 The fluorescent green roots belong to a genetically transformed *Zea mays* genotype expressing the GFP. The dark nonfluorescent roots belong to a non-GFP maize variety and can be clearly distinguished from the GFP roots. The picture was taken with a custom-made UV-MR camera system. (see Faget et al. [2009] for details; Image courtesy of M. Faget, M. Liedgens, P. Stamp, P. Flutsch and J.M. Herrera, Zurich, Switzerland.)

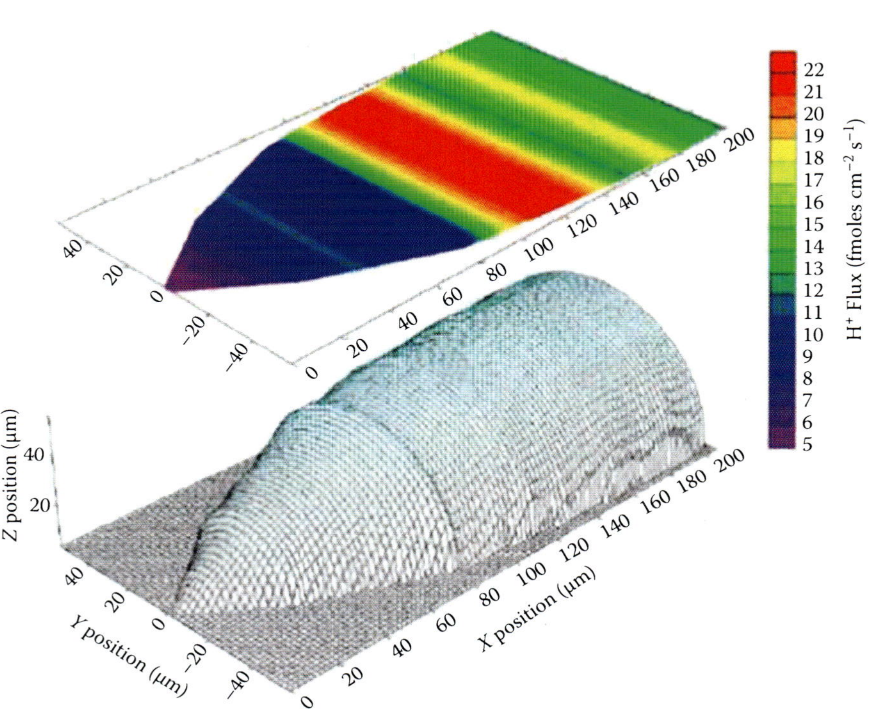

FIGURE 43.9 3D profile of proton efflux from a *T. latifolia* root. Increased proton flux occurs near the zone of elongation, where cells are rapidly metabolizing and driving polar root elongation ($\Delta X = 10\ \mu m$, $F = 0.30$ Hz). (Reproduced from Porterfield, D.M., Use of microsensors for studying the physiological activity of plant roots, in *Plant Roots: The Hidden Half*, eds. Y. Waisel, A. Eshel, U. Kafkafi, 3rd edn., Marcel Dekker, Inc., New York, pp. 333–347, 2002.)

25

Inorganic Nitrogen Acquisition and Signaling: Physiological and Molecular Aspects

Alain Gojon
Institut National de la Recherche Agronomique

I. Introduction

One of the most important roles of the roots is to acquire mineral resources from the soil. Among the various nutrients required by the plants, nitrogen (N) is the one needed in greatest amounts because it typically accounts for 1%–3% of the dry biomass (Marschner 1995). Most plants take up nitrogen in the form of nitrate (NO_3^-) and/or ammonium (NH_4^+) present in the soil solution. Indeed, only a small proportion of all plant species (notably the legumes) are able to set up symbioses that allow them to indirectly use the unlimited source of atmospheric N_2. For the vast majority of species root N acquisition is strongly constrained by the dramatic fluctuations that occur, in both time and space, for NO_3^- and NH_4^+ availability in the soils (Miller et al. 2007). Modern agriculture has solved this problem by a massive use of N fertilizers that in turn led to a major disturbance of the overall nitrogen cycle between the atmosphere, hydrosphere, and biosphere (Galloway et al. 2003). Both these harmful environmental consequences and the high economic and energetic costs of excessive inputs of N fertilizers are now concerning enough to raise the improvement of N use efficiency by crops as a major goal in agronomy and plant sciences (Good et al. 2004).

In this context, the root system is a key target of new breeding strategies as many root-related traits are of crucial importance for determining the efficiency of N acquisition from the soil (Garnett et al. 2009). However, these traits are not easy to identify, because they ultimately result from a complex interplay between structure and function of the roots and between the actions of internal and external signals that modulate N acquisition efficiency. For instance, the various transporter proteins involved in root NO_3^- or NH_4^+ uptake need to be identified and finely characterized not only to improve our basic knowledge of the N uptake processes by plants but also because enhancing their intrinsic activity and/or their expression level may appear as a promising way to increase N uptake. Nonetheless, there are many studies demonstrating that a refined characterization of the functional properties of transporter proteins is not sufficient to understand what actually determines N influx. The precise localization of these proteins within the root tissues is also crucial, because differences in functional properties between transporters only make sense when their specific localization pattern is taken into account (the NH_4^+ uptake system of *Arabidopsis thaliana* that is detailed later is a very illustrative example of this point). Furthermore, the size and architecture of the root system is at least as important as the activity of transport systems for determining nutrient capture from the soil. There are reports indicating that higher root density in soil (generally associated with more branched root systems) actually improves N uptake efficiency by crops (Garnett et al. 2009; see also Chapter 26). However, the role of root development in determining N acquisition is difficult to investigate, and our knowledge is much restricted to studies in laboratory setup that certainly poorly mimic actual field conditions. Finally, root N uptake is a very dynamic physiological function, strongly responsive to various endogenous or environmental factors, at both functional (regulation of transport systems) and morphological (plasticity of root system architecture) levels. This adaptability is due to the action of signaling mechanisms that target transporter expression/activity or key regulatory steps of root development. There is certainly a myriad of such signaling mechanisms that differ in nature (e.g., local vs. systemic) and act at different levels (e.g., transcriptional vs. posttranscriptional). However, they presumably cooperate in a very complex way to yield an integrated response of the root system for modulating N uptake efficiency as a function of the nutrient demand of the plant and of biotic and abiotic constraints (Forde 2002; De Kroon et al. 2009).

This chapter aims at addressing most of the previously mentioned aspects that need to be considered together for a better understanding of root N acquisition by plants. For simplicity, it will be focused on NO_3^- or NH_4^+ uptake systems and their regulation, as well as the effects of these ions (directly or through their assimilation into organic molecules) on root development and architecture. Recent insight into the molecular mechanisms underlying these functional or morphological properties will be presented, at least through selected examples out of many of them which can be found in the literature. This focus on physiological and molecular aspects is justified by the tremendous progress in their understanding that occurred during the past decade. However, a strong limitation related to this choice is that most of the information reviewed here is documented in model species only (and more particularly *A. thaliana*) and that we do not always know whether it is also valid for crop plants.

II. Identity and Function of Root NO_3^- and NH_4^+ Uptake Systems

A. Influx/Efflux

One surprising observation concerning NO_3^- or NH_4^+ uptake by plant roots is that it corresponds, as for many other mineral ions, to the balance between two opposite but concomitant unidirectional fluxes, namely, influx and efflux (Crawford and Glass 1998; Gojon et al. 2009). There are considerable discrepancies between estimates of the respective importance of influx and efflux, depending on species, environmental conditions, and methods used. Influx can only be measured through incorporation of isotopes (the short-lived radioactive ^{13}N or the stable ^{15}N) in very short-term (5–15 min) labeling experiments to minimize subsequent efflux of the tracer (Clarkson et al. 1996). Efflux may be assayed either by tracer leakage out of the roots or by difference between influx and net uptake rate, the latter being determined by ion depletion from the external medium or long-term (>1 h) isotope labeling. For NO_3^-, a classical outcome of many experiments with plants under steady-state nutrient supply is that efflux represents ~10%–30% of influx (Mackown et al. 1981; Delhon et al. 1995). However, since the efflux generally increases with increasing internal NO_3^- concentration in root tissues, it can rise to a much higher proportion of the influx in plants with ample stores of vacuolar NO_3^-, especially if these plants experience a low NO_3^- provision (Mackown et al. 1981). For NH_4^+, efflux to the external medium is often stronger than for NO_3^- (Morgan and Jackson 1988) and can even be quite considerable, amounting to up 80% of the influx (Britto et al. 2001). The physiological role of NO_3^- or NH_4^+ efflux has been a matter of debate for long, as it represents a loss of key nutrients by the plant. If NO_3^- efflux is sometimes considered as an unavoidable "leak," due to the electrochemical potential gradient favoring its transport out of the cells, this cannot hold true for NH_4^+ because in this case the electrochemical potential gradient is opposite and requires the involvement of an active (energy-consuming) transport system that secretes this ion into the medium. The high energy consumption associated with root NH_4^+ efflux has even been proposed to be one main reason for the toxicity of pure NH_4^+ nutrition (Britto et al. 2001). The recent evidence that NO_3^- transport out of the cell into the external medium is driven by specific transporter proteins that are different from those ensuring NO_3^- influx (see Section II.F) strongly suggests that NO_3^- efflux may indeed play a physiological role under particular conditions (Segonzac et al. 2007).

B. High- and Low-Affinity Transport Systems

The kinetics of root NO_3^- or NH_4^+ uptake (or influx) by the roots obeys to the general feature described for most inorganic ions and reveals two different phases of transport as a function of the external concentration (Marschner 1995; Crawford and Glass 1998). The high-affinity phase (low concentration range) is saturable and can be modeled using the Michaelis–Menten formalism, whereas the low-affinity phase (high concentration range) corresponds in most cases to a linear increase in root uptake (or influx) with increasing external concentration (Figure 25.1). This general pattern has been consistently observed in almost all species investigated to date and under most environmental conditions. However, the external concentration threshold for saturation of the high-affinity phase can vary considerably (generally from 0.1 to 1 mM). It has been postulated for long that these two phases of NO_3^- or NH_4^+ uptake are illustrative of the additive activities of two different influx transport systems, namely, the high-affinity and low-affinity transport systems (HATS and LATS, respectively). Given the much higher rate of uptake measured in the high as compared to the low concentration range, it has also been inferred that HATS and LATS are low- and high-capacity transport systems, respectively. Noteworthily, the recent molecular characterization of the NO_3^- and NH_4^+

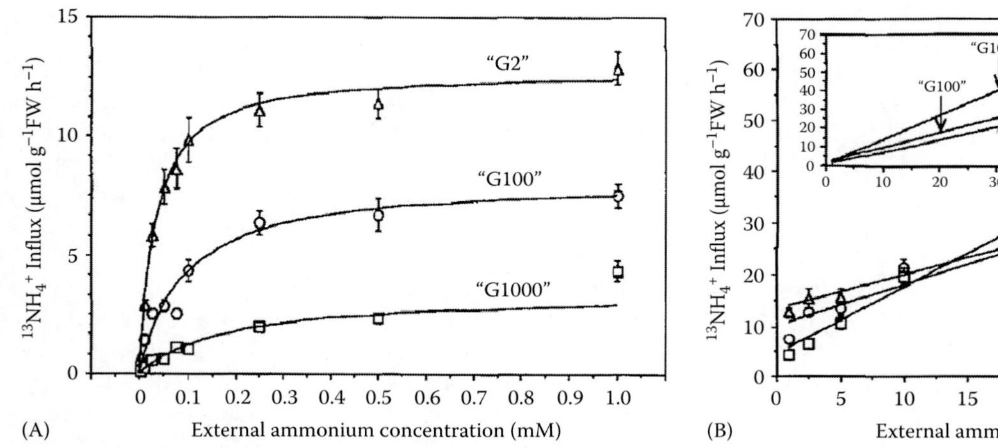

FIGURE 25.1 Influx of $^{13}NH_4^+$ into rice roots at low (A) or high (B) concentrations. Rice seedlings (3 weeks old) were grown at 2 (G2), 100 (G100), or 1000 (G1000) μM NH_4^+. Each data point is the mean of 16 (A) or more than 6 (B) replicates with SE. Inset in (B): estimated LATS fluxes after subtracting the V_{max} of the HATS from the measured influxes at high concentrations. (From Wang, M.Y. et al., *Plant Physiol.*, 103, 1259, 1993.)

transporter proteins has largely confirmed these physiological predictions (see Sections III and IV).

Given the low concentration of NH_4^+ found in most soils, it is quite obvious that the NH_4^+ HATS plays a predominant role over the LATS for nutrition of the plant (von Wirén et al. 2000). This is less clear in the case of NO_3^-, especially in agrosystems where fertilizer application may increase NO_3^- concentration in the soil solution up to 10 mM, i.e., in the range where the activity of the LATS far exceeds that of the HATS (Miller et al. 2007). However, soil NO_3^- concentration dramatically fluctuates in both time and space, and nitrogen acquisition by the plant does not remain constant at all developmental stages, making it difficult to anticipate the relative contributions of HATS and LATS to total root NO_3^- uptake by crops. Combining data on kinetic parameters of root NO_3^- uptake assayed in laboratory conditions and sequential measurements of soil NO_3^- concentrations during the growing season, Malagoli et al. (2004) calculated that the NO_3^- HATS has a major contribution to the overall root N acquisition by oilseed rape plants grown in the field under regular agricultural practice. Nevertheless, an additional complexity is that the kinetic parameters of root N uptake (especially those of the HATS) are also subjected to significant changes due to the regulation of transporter expression/activity (see Section III). For instance, both NH_4^+ and NO_3^- HATS are strongly repressed by high N status of the plant, suggesting that at unlimited availability of these ions in the soil, the LATS may play a predominant role not only because the external concentration of their substrates is high enough to sustain their transport activity but also because they are the only active uptake systems in roots of well-fed plants.

C. Transporter Proteins for NO_3^- and NH_4^+

The past decade has provided major insight into the individual NO_3^- and NH_4^+ carrier proteins responsible for N uptake by plant roots (Tsay et al. 2007; Gojon et al. 2009). This is particularly true concerning both NO_3^- and NH_4^+ HATS, for which the most important players have been identified and functionally characterized, thus providing a quite clear view of the general organization of these two transport systems. Furthermore, additional transporters have been isolated that participate in NO_3^- influx by the LATS and in NO_3^- efflux out of the roots or into the xylem sap. At the opposite, nothing is known at the molecular level concerning the LATS for NH_4^+ or the transport systems responsible for NH_4^+ efflux. Most advances were made in *A. thaliana*, but increasing evidence indicates that the model we now have for the molecular structure of the transport systems in this plant is also largely valid for many other species.

D. NO_3^- HATS

Plant high-affinity NO_3^- influx transporters were mostly found in the NRT2.1 family of NNP (nitrate–nitrite porters) transporters. Seven *NRT2* genes are present in the *Arabidopsis* genome (Orsel et al. 2002; Okamoto et al. 2003), but it is now clear that it is the *NRT2.1* gene product that plays a major role in the root NO_3^- HATS of this species. Indeed, knockout mutants for *NRT2.1* have lost of up to 80% of the HATS activity as compared to wild-type plants (Figure 25.2; Filleur et al. 2001; Li et al. 2007). Accordingly, NRT2.1 is predominantly localized in the plasma membrane of epidermis (including root hairs) and cortex cells of the mature part of the roots (Girin et al. 2007; Wirth et al. 2007), i.e., in the portion of the root system responsible for most of the ion uptake from the soil. Additional but minor components of the *Arabidopsis* NO_3^- HATS are probably the NRT2.2 and NRT1.1 (which has a dual affinity for NO_3^-) transporters that can make a small contribution to the overall high-affinity uptake of NO_3^- under specific conditions (Liu and Tsay 2003; Li et al. 2007).

NRT2 genes have been cloned from a large variety of species, including barley (Trueman et al. 1996; Vidmar et al. 2000), rice (Feng et al. 2011), soybean (Amarasinghe et al. 1998), *Medicago truncatula* (Ruffel et al. 2008), and *Nicotiana plumbaginifolia* (Krapp et al. 1998). Although the functional characterization of

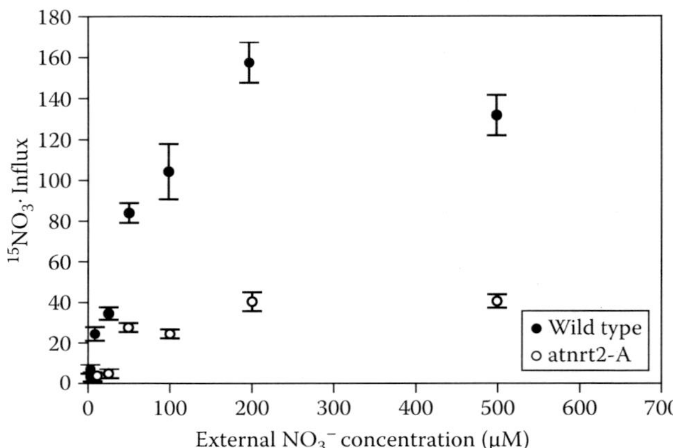

FIGURE 25.2 Kinetic characteristics of nitrate influx in wild-type and *atnrt2-A* mutant (deletion mutant of *NRT2.1*) plants. Plants were grown in hydroponic conditions in the presence of 1 mM NH_4NO_3 as sole nitrogen source and then transferred to 1 mM NO_3^+ for 7 days. Root influx is given in μmol $^{15}NO_3^-$ h^{-1} g^{-1} of root dry wt. Bars indicate standard error (n = 6). Root $^{15}NO_3^-$ influx was measured after 5 min labeling with a complete nutrient solution containing between 0.005 and 0.5 mM $^{15}NO_3^-$ to determine the nitrate influx mediated by the HATS. (From Filleur, S. et al. *FEBS Lett.* 489, 200, 2001.)

most of these genes remains to be done, one or several of them were found in each species to be predominantly expressed in the roots and to display a regulation pattern consistent with those known for NO_3^- HATS activity and for the *Arabidopsis NRT2.1* gene (e.g., induction by NO_3^-, repression by high N status of the plant, and stimulation by photosynthates; see Sections III and IV). This suggests that NRT2 transporters indeed play a key role in high-affinity NO_3^- uptake in most, if not all, plant species.

However, despite the experimentally proven or postulated key role of NRT2 transporters in NO_3^- HATS activity, most attempts to demonstrate the NO_3^- influx activity of these proteins in heterologous expression systems (such as *Xenopus* oocytes) have failed until it was realized that co-expression of NRT2s with another protein of the NAR2 family was required to yield NO_3^- transport (Tong et al. 2005; Orsel et al. 2006). This is quite unique among the plant nutrient transporters because NAR2 proteins have no known transport or catalytic activity, and their function is at present totally elusive. Furthermore, in *Arabidopsis,* the NAR2.1 protein is required not only for activity but also for expression of NRT2.1, because null *nar2.1* mutant plants have a strongly lowered level of NRT2.1 protein in the plasma membrane (Wirth et al. 2007). The reason why NAR2.1 regulates NRT2.1 expression is not known, but physical interaction at the plasma membrane has been demonstrated between the *Arabidopsis* NRT2.1 and NAR2.1 proteins, suggesting that the active transport system may in fact be an hetero-oligomer of NRT2 and NAR2 (Orsel et al. 2006; Yong et al. 2010). However, not all NRT2 proteins need a NAR2 partner to mediate NO_3^- transport, as shown by the fact that the *Arabidopsis* NRT2.7 yields NO_3^- influx in oocytes when expressed alone (Chopin et al. 2007). This makes it difficult to understand why a dual-component transport system has emerged as a main player in high-affinity root NO_3^- uptake.

E. NO_3^- LATS

Plant low-affinity NO_3^- transporters identified to date mostly belong to the large NRT1/PTR family (53 members in *A. thaliana*), which also comprises nitrite, peptide, and carboxylic acid transporters (Tsay et al. 2007). In *A. thaliana*, among the eight PTR proteins characterized as NO_3^- transporters, only two (NRT1.1 and NRT1.2) are thought to be involved in the root NO_3^- LATS.

NRT1.1 is an unusual dual-affinity transporter. It displays high-affinity transport activity when phosphorylated at the T101 residue, while the non-phosphorylated form is a low-affinity transporter (Liu and Tsay 2003). Thus, NRT1.1 has been proposed to be involved in both HATS and LATS, depending on its T101 phosphorylation status. NRT1.2 is a low-affinity NO_3^- transporter that is expressed constitutively in the rhizodermis and that actually participates in the LATS as evidenced by reduced NO_3^- uptake in the high concentration range found in *NRT1.2* under-expressors or knockout mutant as compared to wild-type plants (Huang et al. 1999; Krouk et al. 2006).

It is obvious that other contributors to the NO_3^- LATS in *A. thaliana* still await identification because neither NRT1.1 nor NRT1.2 appear to account for most of the activity of this transport system. Indeed, under-expression or knockout mutation of *NRT1.2* results only in limited inhibition of the LATS activity (Huang et al. 1999; Krouk et al. 2006). Likewise, a major role of NRT1.1 in the overall N acquisition by the roots is questioned for several reasons. First, the *NRT1.1* gene is predominantly expressed in root tips or in internal tissues of the root (e.g., in the stele), which make a relatively small contribution to the uptake of NO_3^- from the external medium. Second, defects in either HATS or LATS activity in mutants for *NRT1.1* (*chl1* mutants) are found only in very specific conditions, corresponding, for instance, to rare situations where *NRT2.1* is not expressed (for a role in the HATS) or to mixed NH_4^+/NO_3^- nutrition (for a role in the LATS) where NH_4^+ and not NO_3^- is actually the main N form taken up by the plant (Touraine and Glass 1997; Wang et al. 1998). Third, as will be detailed in the following, NRT1.1 is now considered to have a dual NO_3^- signaling/transport function, suggesting that its main role is rather to be a NO_3^- sensor.

F. NO_3^- Efflux Out of the Roots

The identification of NAXT1, an *A. thaliana* transporter of the NRT1/PTR family specifically involved in NO_3^- efflux to the external medium, has provided some understanding of what could be a physiological function for this transport process (Segonzac et al. 2007). NAXT1 is localized at the plasma membrane of cortex cells of the roots. Under normal physiological conditions, knockout mutants for *NAXT1* have no visible phenotype. However, in response to acid load treatments (exogenous supply of 10 mM propionate) that lead to an acidification of the cytoplasm, NO_3^- efflux out of the roots is strongly stimulated in wild-type plants, but not in *naxt1* mutants. The interpretation of these data is that NAXT1 is responsible for the acidification-induced root NO_3^- efflux, which may aim at neutralizing the electrical effects of the strong stimulation of H^+ excretion by the H^+-ATPase triggered by cytoplasm acidification. Accordingly, NAXT1 expression is markedly increased at the protein level in response to acid load (Segonzac et al. 2007). Thus, the assumption is that besides being a major N source for nutrition of the plant, NO_3^- is also a highly mobile anion that can easily be used as a counterion for balancing major changes in cation fluxes (H^+ in this case) and prevent excessive alteration of electrical potential gradients across membranes.

G. Radial NO_3^- Transport to the Xylem

Although not being *sensu stricto* involved in uptake from the external medium, there are several other steps of NO_3^- transport in the roots that play a key role in the N nutrition of the plant. Indeed, once taken up, NO_3^- may be either directly assimilated in the roots, stored in root vacuoles, or transported radially to the stele for secretion into the xylem sap. This last process is of major importance because the leaves are the predominant site for NO_3^- assimilation in many plant species (Andrews 1986). The radial transport of NO_3^- to the stele is thought to occur by diffusion along the symplasm (cytosolic continuum of the different cell layers connected by plasmodesmata), but its secretion into the extracellular space of the xylem vessels is a regulated process involving specific efflux transport systems. The NRT1.5 transporter of the NRT1/PTR family, located in the plasma membrane of pericycle and xylem parenchyma cells, has been shown to participate in this secretion in roots of *A. thaliana* (Lin et al. 2008). NRT1.5 can mediate bidirectional (influx or efflux) NO_3^- transport when expressed in *Xenopus* oocytes. However, knockout *nrt1.5* mutants display a reduced rate of NO_3^- translocation to the shoot and a lowered NO_3^- concentration in the xylem sap, indicating that under physiological conditions, the role of NRT1.5 is to contribute to NO_3^- efflux to the xylem. Root parenchyma cells of *A. thaliana* possess another NRT1 transporter, NRT1.8, which acts as a low-affinity influx transporter taking up NO_3^- from the xylem sap into the root stele (Li et al. 2010). This probably allows basal portions of the roots that are in contact with soil areas depleted from NO_3^- to retrieve this nutrient from the xylem, following its long-distance transport from more apical portions where uptake from the external medium takes place.

H. NH_4^+ HATS

High-affinity NH_4^+ transporters were isolated from the AMT/MEP/Rh superfamily, first by functional complementation of a yeast mutant strain deficient in NH_4^+ uptake in the low concentration range (Ninneman et al. 1994; Gazzarini et al. 1999). In *A. thaliana*, five *AMT1* (*AMT1.1* to *1.5*) genes and one more distantly related *AMT2.1* gene were found in the genome. Out of these six genes, at least four (*AMT1.1*, *1.2*, *1.3*, and *2.1*) are significantly expressed in the roots, and genetic evidence indicates that the NH_4^+ HATS relies on the additive activities of the AMT1.1, AMT1.2, and AMT1.3 transporters (Loqué et al. 2006; Yuan et al. 2007). Indeed, a quadruple knockout mutant (*qko*) for all four *AMT1.1-1.3* and *2.1* genes retains only a residual NH_4^+ HATS activity (5%–10% of the wild-type) that may be attributable to AMT1.5 (Yuan et al. 2007). The complementation of this *qko* with individual *AMT* genes showed that *AMT1.1*, *1.2*, or *1.3*, but not *AMT2.1*, could restore part of the wild-type HATS activity. Accordingly, single mutants for either *AMT1.1*, *1.2*, or *1.3* display a reduced HATS activity, whereas no defect in root NH_4^+ uptake could be found in an *AMT2.1* mutant. Altogether, these data suggest that the root high-affinity uptake of NH_4^+ is a function shared by the three AMT1.1, 1.2, and 1.3 transporters. Interestingly, these three AMT1 differ in both their apparent affinity for NH_4^+ and in their tissue localization in the roots. While AMT1.1 and 1.3 have a low Km for NH_4^+ (50 and 60 μM, respectively) and are predominantly found in the plasma membrane of epidermal cells and especially in root hairs, AMT1.2 has a much higher Km (>230 μM) and is localized in cortical and endodermal cells (Yuan et al. 2007). These differences certainly have a physiological meaning. Indeed, the concentration of NH_4^+ is generally very low in the external soil solution, but is thought to be higher in the apoplasm inside the roots, because of the electrostatic attraction by negatively charged cell walls. Thus, it makes sense to have the transporter with the lowest affinity localized in inner cell types of the root as compared to those with high affinity.

AMT1 transporters have been characterized in other plant species than *A. thaliana*, such as tomato (Lauter et al. 1996), rice (Kumar et al. 2003), and *Lotus japonicus* (D'Apuzzo et al. 2004). Interestingly, in tomato, *LeAMT1.1* and *LeAMT1.2* are preferentially expressed in root hairs, which confirms the key role of these sites for efficient uptake of cations like NH_4^+ that have a very limited mobility in the soil.

Finally, beside their obvious role in the primary acquisition of the soil NH_4^+ by the roots, the AMT1 transporters of the NH_4^+ HATS most probably also fulfill the critical function of retrieving the endogenous NH_4^+ that is secreted out of the roots. Indeed, even when NH_4^+ is not the major N source for nutrition of the plant, amino acid catabolism in the root cells generates a significant amount of free NH_4^+, part of which undergoes efflux to the external medium (Morgan and Jackson 1988). The NH_4^+ HATS is essential for ensuring reuptake of this NH_4^+ and thus preventing what could be a major nutrient loss. Although it is always difficult to extrapolate data from lower organisms for understanding higher plants physiology, the potential importance of this retrieval function is illustrated by the *Δmep1Δmep2Δmep3* triple mutant of yeast, lacking all three AMT/MEP transporters present in this species, and that is unable to grow properly on any N source because it constantly looses NH_4^+ in the external medium (Marini et al. 1997).

III. Regulation of Root NO_3^- and NH_4^+ Uptake Systems

The capacity of a plant to acquire NO_3^- or NH_4^+ from the soil solution varies during the developmental cycle and is strongly dependent on the environmental conditions. In particular, root N uptake capacity is finely tuned to match the N demand for growth or to compensate for changes in the external NO_3^- or NH_4^+ availability. Compelling evidence has suggested that most of the regulation of root NO_3^- and NH_4^+ uptake occurs at the level of influx rather than of efflux (Lee 1993; Wang et al. 1993; Delhon et al. 1995; Crawford and Glass 1998). As a consequence, the regulation pattern of the various influx transport systems has been extensively investigated and is now quite well characterized (especially for the HATS), at both physiological (transport system activity) and molecular (gene expression) levels. Several examples illustrating this general pattern will be described in the following. However, we still have very little information concerning the regulatory mechanisms themselves, from the whole-plant signaling cascades triggering the changes in expression/activity of the root transport systems to the most downstream events involved in the direct control of the expression of transporter genes or of the activity of transporter proteins.

There are many different mechanisms regulating root N uptake systems in plants. For simplicity, only three major regulations will be addressed in this chapter, namely, regulation by NO_3^-, feedback repression by the N status of the plant, and stimulation by photosynthates.

A. Regulation by NO_3^-

It is now well documented that NO_3^- is not only a nutrient but also a signal molecule *per se*, which is sensed by the plants and affects many aspects of their metabolism or development (Crawford 1995; Stitt 1999). One of the best-known examples of the signaling effect of NO_3^- is the induction of its own transport systems. This was first revealed by the observation that the uptake rate of NO_3^- in young wheat, corn, cotton, and tobacco seedlings markedly accelerated during a period of 10 h following the first supply of the ion (Jackson et al. 1972). This response seems to be a specific feature of the NO_3^- HATS, because the activity of the LATS appeared most often to be insensitive to prior NO_3^- provision to the plant (Crawford and Glass 1998). Accordingly, it is often considered that the NO_3^- HATS actually consists of two components, a constitutive HATS (cHATS), active regardless of whether NO_3^- was previously supplied or not, and an inducible HATS (iHATS) that requires a NO_3^- pretreatment to be expressed or activated. Induction of NO_3^- uptake by NO_3^- was found to require protein synthesis and to be independent of NO_3^- assimilation, as it is unaffected by knockout mutation of nitrate reductase (NR) which catalyzes the first step of NO_3^- reduction (Hole et al. 1990; King et al. 1993; Wang et al. 2004). Even very low concentrations of NO_3^- or short-term pulses of NO_3^- supply are able to trigger the induction process. Altogether, these observations have led to the hypothesis that NO_3^- itself acts as an inducer of its HATS in the roots.

The advent of molecular studies on NO_3^- transporter genes largely confirmed this hypothesis. It was indeed demonstrated in *A. thaliana* that root mRNA levels of NRT2.1, NRT2.2, NAR2.1, and NRT1.1, i.e., the only known components of the HATS in this species, were strongly increased upon first NO_3^- supply (Tsay et al. 1993; Filleur and Daniel-Vedele 1999; Okamoto et al. 2003). For both NRT1.1 and NRT2.1, promoter–reporter gene fusions indicated that at least part of this control acts at the transcriptional level (Guo et al. 2001; Girin et al. 2007). On the other hand, NRT1.2, which contributes specifically to the LATS, is not regulated by NO_3^- at the mRNA level (Huang et al. 1999). Numerous studies on other plant species also indicate NO_3^- regulation of a large number of NRT genes (Amarasinghe et al. 1998; Vidmar et al. 2000; Feng et al. 2011). Furthermore, NO_3^- transporter genes are only few of the many genes that are NO_3^- responsive. In *A. thaliana* for instance, at least several hundred genes, and possibly up to 10% of the transcriptome, are under direct control by NO_3^- signaling (Wang et al. 2004; Krouk et al. 2010a). Yet, despite its general occurrence, the induction of NO_3^- transporters by NO_3^- remains almost a unique feature, which departs from the general observation that most other ion transporters involved in root nutrient uptake are induced by the absence of their substrate rather than by its presence. Why many NO_3^- transporters have the unusual characteristic to be expressed only after NO_3^- supply? This question has no convincing answer yet.

The signaling pathways responsible for NO_3^- regulation of root transport systems are still mostly unknown at the molecular level. However, at least two key players of these pathways, namely, NLP7 and NRT1.1, have been recently identified in *Arabidopsis*. The NLP (NIN-like proteins) family of *Arabidopsis* comprises nine members that are believed to act as transcription factors. The interest in this particular family concerning NO_3^- signaling was motivated by the fact that (1) NIN proteins are involved in the regulation of symbiotic nitrogen fixation in legumes and (2) NIT2, a NLP of the green unicellular algae *Chlamydomonas reinhardtii*, is a major transcription factor involved in NO_3^- signaling in this species (Camargo et al. 2007). A functional genomic approach on the *NLP* gene family of *Arabidopsis* revealed that *NLP7* is required for correct NO_3^- signaling (Castaings et al. 2009). In particular, knockout mutants for *NLP7* show a marked attenuation of the NO_3^- induction of the NO_3^- transporter genes *NRT2.1* and *NRT2.2*.

The finding that the NRT1.1 dual-affinity NO_3^- transporter has a NO_3^- signaling role occurred almost by accident. Indeed, transcriptome analysis of a deletion mutant for *NRT1.1* (*chl1-5* mutant) unexpectedly revealed strong alterations in the response of several genes to N availability (Muños et al. 2004). In particular, the high-affinity NO_3^- transporter gene *NRT2.1* was found to become insensitive to repression by high N (NH_4NO_3) supply in response to *NRT1.1* mutation. Further investigations showed that this mis-regulation is due to the impairment of a specific NO_3^- signaling mechanism that down-regulates *NRT2.1* in response to high NO_3^- provision (Krouk et al. 2006). This provided molecular evidence to earlier predictions that NO_3^- can actually have two opposite actions, as an inducer or as a repressor of its own uptake systems, depending on the concentration and duration of NO_3^- treatments (King et al. 1993). This also suggested that NRT1.1 fulfills a dual transport/signaling role, possibly acting as a NO_3^- sensor in a

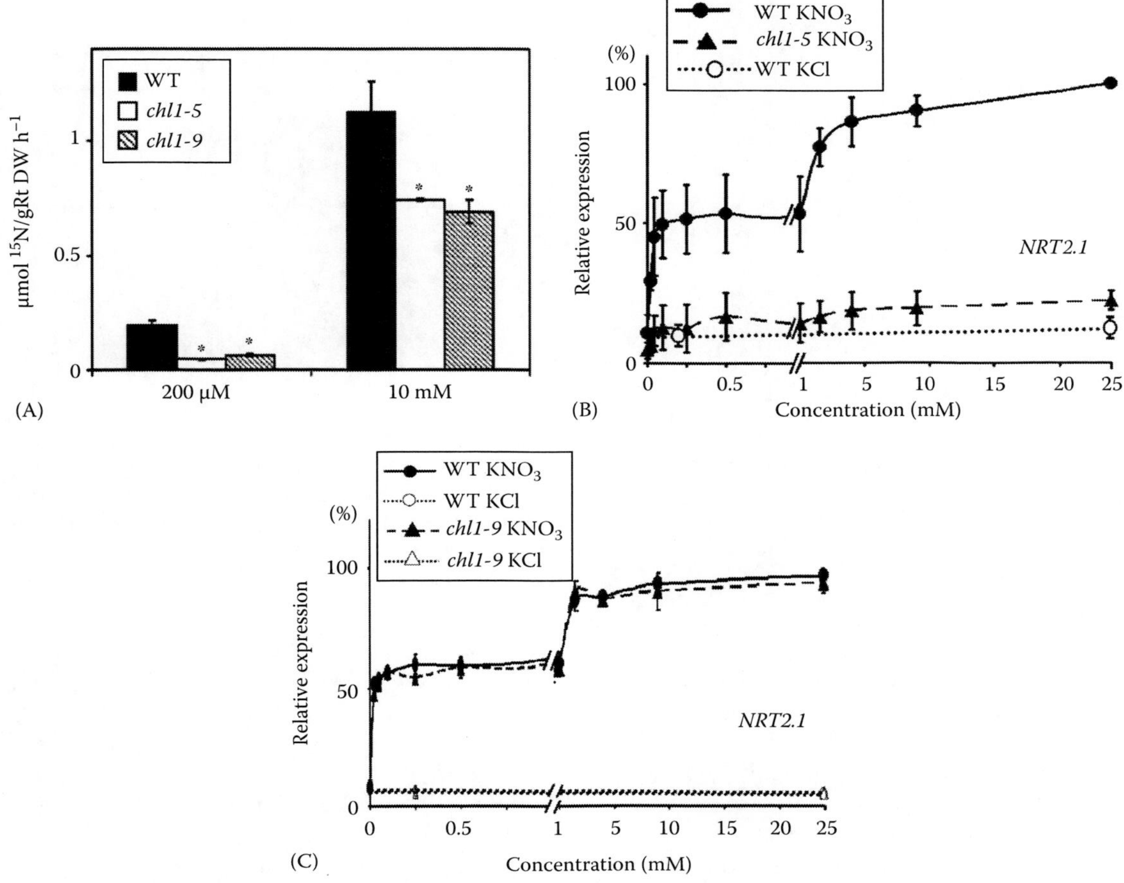

FIGURE 25.3 (A) Defective nitrate uptake in *chl1-5* (deletion of *NRT1.1*) and *chl1-9* (P492L substitution in *NRT1.1*) mutants of *NRT1.1*. Nitrate uptake measured using 10 mM or 200 μM $K^{15}NO_3$ for 30 min. The plants were grown for 10 days in NO_3^--free medium using NH_4^+ as the sole N source, treated with fresh medium overnight and again for an additional 3 h, and then exposed for 30 min to the indicated concentrations of KNO_3 or KCl. The data are the mean ±SD for three experiments; $p < 0.01$ compared to the wild-type (t test). (B,C) Concentration-dependent induction of *NRT2.1* expression (Q-PCR analysis) by nitrate in wild-type plants and the *chl1-5* mutant (B) or the *chl1-9* mutant (C). The relative expression is the expression normalized to that for *NRT2.1* in wild-type plants exposed to 25 mM KNO_3 for 30 min. The plants were grown as in (A). The data are the mean ±SD for three experiments. (From Ho, C.L. et al., *Cell* 138, 1184, 2009.)

similar way to the transceptor (transporter/receptor) proteins for inorganic P, NH_4^+, or amino acids in yeast (Holsbeeks et al. 2004; Gojon et al. 2011). Additional evidence of a dual transport/signaling function of NRT1.1 was provided by studies showing that (1) *NRT1.1* (as well as *NLP7*) were identified in a genetic screen for regulatory mutants altered in NO_3^- signaling (Wang et al. 2009), (2) NRT1.1 was required not only for NO_3^- repression of *NRT2.1* but also for its NO_3^- induction under particular experimental conditions (Ho et al. 2009), and (3) point mutations in the *NRT1.1* gene could uncouple NO_3^- transport and NO_3^- signaling by NRT1.1 (Ho et al. 2009). In particular, the P492L substitution affecting NRT1.1 in the *chl1-9* mutant results in the loss of the NO_3^- transport activity of the protein, without modifying the NRT1.1-dependent induction of NRT2.1 by NO_3^- (Figure 25.3). This direct evidence that the NO_3^- signaling function of NRT1.1 is independent from its transport activity is probably the strongest argument supporting the hypothesis that NRT1.1 is indeed a NO_3^- sensor. The mechanism involved in the NO_3^- regulation of *NRT2.1* by NRT1.1 is unknown, but it is postulated that NO_3^- binding at a specific recognition site on the NRT1.1 protein results in a conformational change that triggers transduction of the NO_3^- signal, possibly through interaction with other proteins (Ho et al. 2009).

B. Feedback Repression by the N Status of the Whole Plant

Considerable evidence indicates that the N uptake capacity of the roots is ultimately regulated by the N demand of the whole organism, in order to fit N acquisition with growth (Imsande and Touraine 1994; Gojon et al. 2009). This control is thought to be the result of a feedback repression exerted by signals of the N status of the plant. Under N satiety conditions (high external N provision for instance), both NO_3^- and NH_4^+ HATS display a very low capacity, whereas N deficiency (low N provision

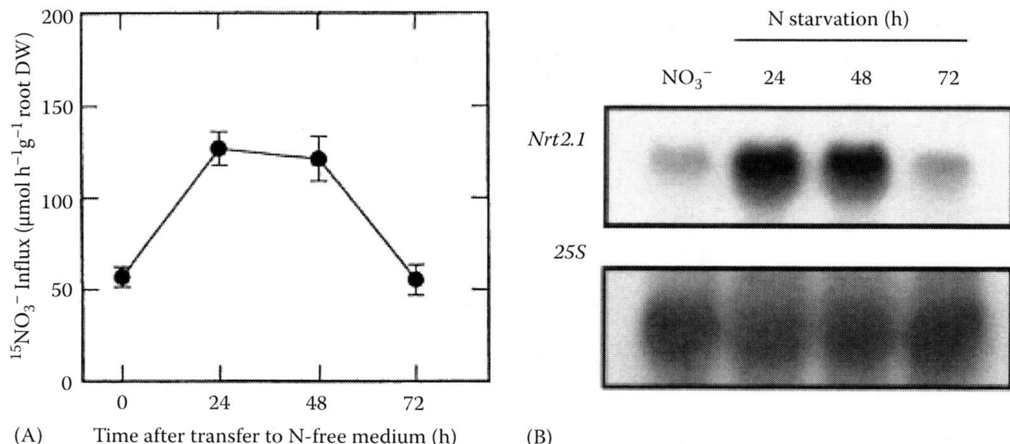

FIGURE 25.4 (A) Root $^{15}NO_3^-$ influx in 18-day-old hydroponically grown *A. thaliana* plants either supplied with a nutrient solution containing 1 mM KNO_3 or transferred from this solution to N-free medium for 24, 48, or 72 h. Root $^{15}NO_3^-$ influx was measured after 5 min labeling with a complete nutrient solution containing 0.2 mM $^{15}NO_3^-$. The values are the means of six replicates. Vertical bars indicate SE. (B) Northern blot of *NRT2.1* expression in the roots of plants grown and treated as in (A). (From Lejay, L. et al., *Plant J.*, 18, 509, 1999.)

or N-starvation treatment) leads after one or a few days to a marked increase in this capacity (Figure 25.4; see also Figure 25.1; Morgan and Jackson 1988; Hole et al. 1990; Gazzarini et al. 1999; Lejay et al. 1999). This suggests that both HATS can be reversibly repressed/derepressed, with almost all possible intermediate states between full repression and derepression, depending on the N status of the plant. Accordingly, most transporters involved in either NO_3^- or NH_4^+ HATS were shown to be subject to this repression/derepression mechanism. For instance, transfer of well-fed *Arabidopsis* plants to a N-deprived medium results after 1–4 days in a steep increase in root mRNA levels of *NRT2.1*, *NRT2.2*, *AMT1.1*, *AMT1.2*, and *AMT1.3* genes (Figure 25.4), suggesting that the increase in HATS capacity is due to a higher density of the corresponding proteins in the plasma membrane (Gazzarini et al. 1999; Lejay et al. 1999; Li et al. 2007; Yuan et al. 2007).

Split-root experiments have clearly shown, at least for NO_3^-, that feedback repression of the HATS by high N status results from the action of systemic signals. For instance, suppressing the NO_3^- source from one side of the split-root system of *Arabidopsis* plants results after 1–2 days in a stimulation of root NO_3^- uptake by the untreated side of the root system still fed with this ion (Figure 25.5; Gansel et al. 2001). Thus, although their own N supply remained unchanged, the untreated roots have reacted to the N deficiency experienced by the other organs of the plant (the N-starved roots and the shoots) to compensate for the sudden lack of NO_3^- acquisition by the treated roots. This indicates that a whole-plant signaling system acts to integrate the N status of the various organs and to convey this information to the roots for adequate tuning of the N uptake systems. This mechanism is of course necessary in the complex multicellular plant, to make sure that the roots fulfill their specific functions to the benefit of the whole organism. Surprisingly, only a few root N transporters have been shown to be regulated by the systemic signaling of the whole-plant N status. Among these, the NRT2 transporters involved in the NO_3^- HATS are certainly the most important ones, confirming the high flexibility of this transport system (Gansel et al. 2001). In contrast, the LATS for NO_3^- or NH_4^+ show little sensitivity, if any, to the feedback regulation exerted by N status of the plant (Wang et al. 1993; Lejay et al. 1999).

A very popular hypothesis postulates that the repressive signals related to the N status of the plant are downstream metabolites of the N assimilation pathway such as amino acids (Crawford and Glass 1998). When supplied exogenously, many amino acids but especially glutamine repress both the expression and activity of root NO_3^- and NH_4^+ HATS (Muller and Touraine 1992; Rawat et al. 1999; Zhuo et al. 1999; Nazoa et al. 2003). Furthermore, glutamine is a major N storage form in many plant species and accumulates in tissues under N satiety conditions. Thus, the rationale is that N satiety results in a repression of root NO_3^- and NH_4^+ HATS because it is associated with high levels of free glutamine in root cells that act as a suppressor of NO_3^- and NH_4^+ transporters. Conversely, N deprivation leads to a fast metabolization of glutamine to support protein synthesis relieving NO_3^- and NH_4^+ HATS from feedback repression by this metabolite. Finally, this mechanism can also account for the systemic nature of this regulation because extensive cycling of amino acids (in particular glutamine) between roots and shoot is expected to provide a means to informing the roots of the N status of the whole plant (Cooper and Clarkson 1989). However, this elegant hypothesis still awaits final confirmation because the underlying mechanisms remain totally obscure. Not a single gene involved in the feedback repression of root N acquisition by N metabolites has been identified to date. Hopefully, the recent isolation of several *Arabidopsis hni* (*high nitrogen insensitive*) mutants impaired in the systemic downregulation of *NRT2.1* by high N

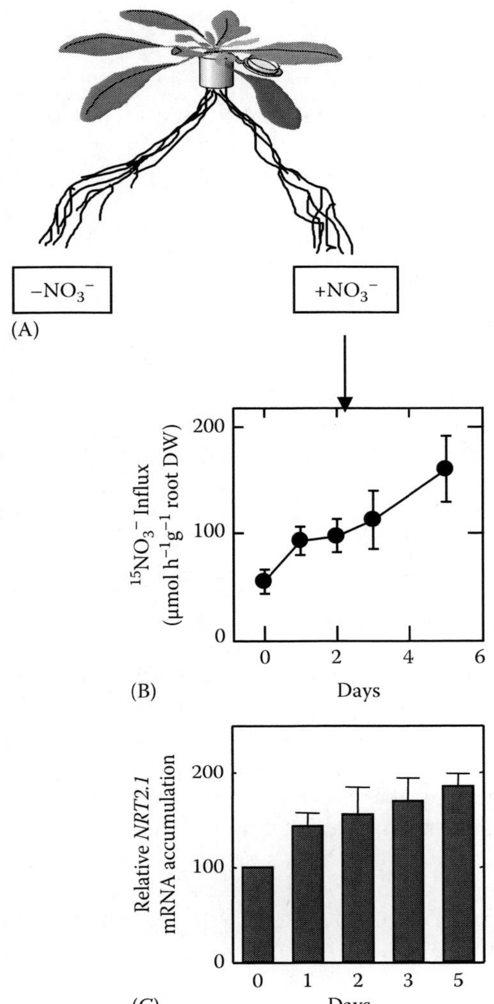

FIGURE 25.5 Response of root $^{15}NO_3^-$ influx and *NRT2.1* expression to localized supply of NO_3^-. (A) Principle of the treatment. After growth for 6 weeks on 1 mM NO_3^-, one side of the split-root system ($-NO_3^-$) was supplied with an N-free solution for 1, 2, 3, or 5 days, while the other side ($+NO_3^-$) remained on 1 mM NO_3^-. (B) Time course of $^{15}NO_3^-$ influx in the ($+NO_3^-$) side of the split-root system after the beginning of the localized starvation treatment. Data are means of six replicates ±SE. (C) Changes in *NRT2.1* transcript accumulation in the ($+NO_3^-$) side of the split-root system after the beginning of the localized starvation treatment. Relative quantification data of mRNA accumulation are means of values from three independent experiments ±SE. (From Gansel, X. et al., *Plant J.*, 26, 143, 2001.)

provision will allow further insight into the molecular aspects of this crucial regulation (Girin et al. 2010).

Surprisingly, despite the fact that the data concerning regulation of the NO_3^- and NH_4^+ HATS at the protein level are scarce, it is at the posttranslational level that a key mechanism has been deciphered for repression of high-affinity NH_4^+ uptake in response to high N supply (Loqué et al. 2007). This mechanism involves a conformational change at the cytoplasmic C terminus of AMT1 proteins providing the possibility to rapidly shut off the NH_4^+ transport activity. For AMT1.1, this conformational change is triggered by phosphorylation of the T460 residue in response to exogenous supply of NH_4^+ (Lanquar et al. 2009). Not only AMT1.1 but also AMT1.2 is subject to this allosteric inhibition of activity (Neuhäuser et al. 2007). This repressive mechanism is impressively efficient, for at least two reasons. First, AMT1 transport systems are actually trimers of AMT1 polypeptides, each having its own transport activity. Because upon phosphorylation the C terminus of any subunit may interact with its neighbors and to inactivate them, a single phosphorylation event of one subunit only is sufficient to suppress the transport activity of the whole transport system. Second, AMT1 phosphorylation in response to NH_4^+ is time- and concentration-dependent but occurs as soon as 5 min following exposure to NH_4^+, even if NH_4^+ is transiently supplied for only 3 min (Figure 25.6). Altogether, this certainly provides a means to quickly limit influx of large amounts of NH_4^+ in the cytoplasm, which can be toxic in many plant species (Britto et al. 2001). The signaling events involved in AMT1 posttranslational inactivation by NH_4^+ are unknown, but their connection with the systemic feedback control exerted by the N status of the plant is questioned. First, AMT1 phosphorylation probably occurs too fast to involve a whole-plant signaling cascade. Second, only exogenous NH_4^+ is apparently

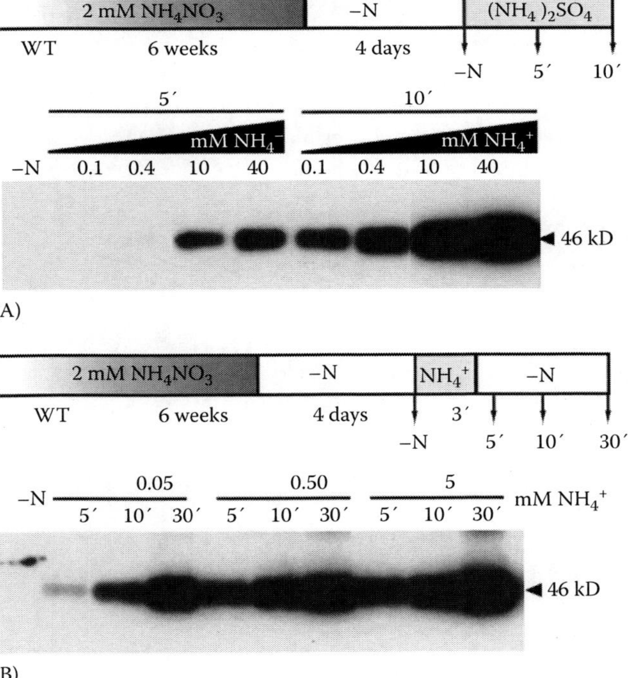

FIGURE 25.6 Time course of AMT1 phosphorylation of wild-type plants grown for 6 weeks in 2 mM NH_4NO_3 and nitrogen-starved (−N) for 4 days prior to ammonium addition. (A) Plants were exposed to NH_4^+ [as $(NH_4)_2SO_4$], and roots were collected after 5 and 10 min. (B) Plants were exposed to NH_4^+ [as $(NH_4)_2SO_4$] for 3 min and transferred back to nitrogen-free medium (−N). Samples were collected 5, 10, and 30 min after transfer. Protein gel blots were performed using the AMT1-P antiserum. (From Lanquar, V. et al., *Plant cell*, 21, 3610, 2009.)

able to trigger AMT1 inactivation, because high levels of cytoplasmic NH_4^+ resulting from inhibition of the NH_4^+ assimilation pathway by methionine sulfoximine failed to induce AMT1 phosphorylation (Lanquar et al. 2009). Therefore, it is postulated that this gating mechanism is activated by a membrane sensor of extracellular NH_4^+, either the NH_4^+ transporter itself (acting as a transceptor) or a separate protein, which is able to interact with AMT1 C terminus (e.g., a receptor kinase). Although no precise mechanism has been identified to date, there is also evidence that NRT2 transporters may be subject to strong posttranscriptional and/or posttranslational regulation for controlling NO_3^- HATS activity (Fraisier et al. 2000; Wirth et al. 2007).

IV. Stimulation by Photosynthates

Another major factor governing root acquisition of NO_3^- and NH_4^+ is photosynthesis. Uptake of both ions fluctuates diurnally, with an increase during the light period and a decrease at night (Figure 25.7; Clément et al. 1978; Delhon et al. 1995; Gazzarini et al. 1999; Lejay et al. 1999). Furthermore, shading or increased illumination of the shoot is followed within hours by a decrease or increase in root N uptake, respectively (Gastal and Saugier 1989; Delhon et al. 1996). Experiments under continuous light or with CO_2-free air have shown that it is not the endogenous clock nor the light *per se* which is responsible for these effects (Delhon et al. 1996; Lejay et al. 2003, 2008). Rather, it is postulated that the root uptake systems for NO_3^- and NH_4^+ are positively regulated by photosynthates transported from the shoot to the root, and thus react with some delay with respect to the modulation of photosynthesis (Delhon et al. 1996). Accordingly, decay of root N uptake at night can be prevented by exogenous sucrose supply to the roots (Delhon et al. 1996; Lejay et al. 1999, 2003). Interestingly, at least for NO_3^-, both HATS and LATS seem to be affected by this regulation (Lejay et al. 1999), which is interpreted as a means to coordinate N acquisition by the roots with C acquisition by the shoot. This coordination is made necessary by the fact that the vast majority of the N in a plant, and up to 40% of its C, are found in the same molecules, i.e., the proteins.

At the molecular level, many *Arabidopsis* NO_3^- or NH_4^+ transporter genes (e.g., *NRT1.1*, *NRT2.1*, *AMT1.2*, *AMT1.3*) are regulated by photosynthesis and sugars at the mRNA level in a very similar fashion as the HATS/LATS for these ions (Figure 25.7; Gazzarini et al. 1999; Lejay et al. 1999, 2003). The precise mechanisms of this regulation are unknown, but it has been shown that they involve sugar sensing and signaling pathways, in particular a yet uncharacterized signaling role of specific steps or metabolites of the oxidative pentose phosphate pathway (Lejay et al. 2008).

V. Nitrogen Regulation of the Root System Architecture

The overall N acquisition by the plant is not only controlled by the expression/activity of root NO_3^- or NH_4^+ transporters but also by the size and architecture of the root system that determine the total volume of soil explored by the plant and the size and location of the exchange surface between the soil and the plant. As for transporter expression/activity, both the size and the architecture of the root system are strongly dependent on external N availability and on internal N status of the plant (Zhang and Forde 2000). Root system architecture (RSA) has been reported to be particularly affected by NO_3^-, which has a

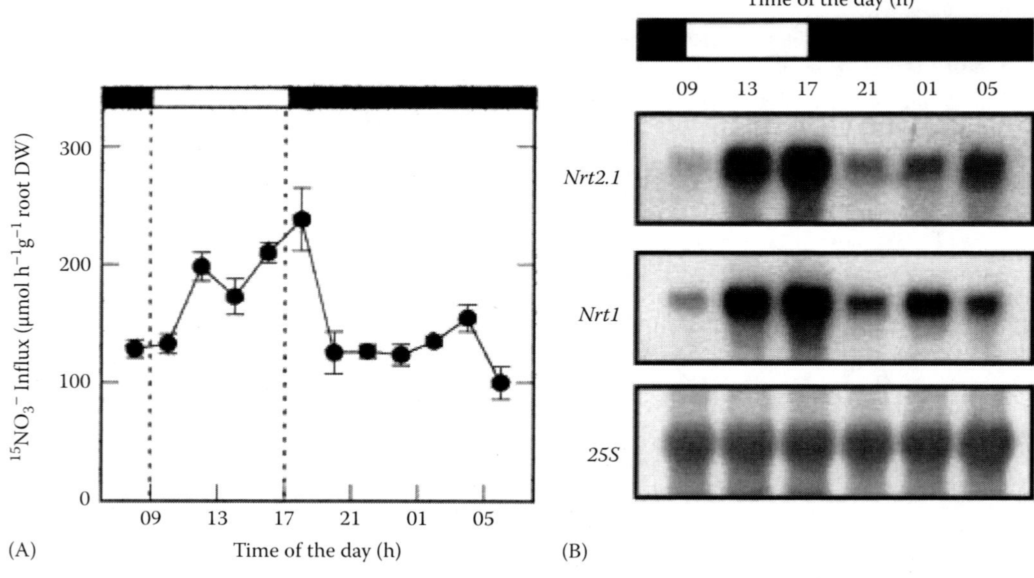

FIGURE 25.7 (A) Diurnal changes in root $^{15}NO_3^-$ influx of 6-week-old *A. thaliana* plants grown on complete nutrient solution containing 1 mM NO_3^-. Root $^{15}NO_3^-$ influx was measured after 5 min labeling with a complete nutrient solution containing 1 mM$^{15}NO_3^-$. The values are the means of 12 replicates. Vertical bars indicate SE. (B) Diurnal changes in expression of *NRT2.1* and *NRT1.1* in the roots of 6-week-old *A. thaliana* plants grown on complete nutrient solution containing 1 mM NO_3^-. (From Lejay, L. et al., *Plant J.*, 18, 509, 1999.)

limited impact on growth of the main root of species with a tap-root system, but strongly modulates root branching by directly regulating lateral root development and growth (Zhang and Forde 1998; Chapman et al. 2011). Surprisingly, at least until recently, RSA appeared relatively insensitive to NH_4^+ (Zhang et al. 1999). This is counterintuitive, since one can postulate that the plasticity of RSA is certainly more critical to optimize the acquisition of a slowly diffusible ion like NH_4^+, rather than that of a highly mobile ion like NO_3^-, which can move in the soil over large distances through convection and diffusion (Robinson 1996). One possible explanation to this unexpectedly high effect of NO_3^- on RSA may be that even for the uptake of diffusible nutrients, the ability of an individual plant to optimize soil colonization by the roots is crucial for competing with other plants or with microorganisms (Hodge et al. 1999).

An extensive analysis of the multiple responses of RSA to changes in N availability is out of the scope of this chapter, especially because unlike the transport systems that appear to be conserved among species, the plasticity of the root system in response to N can uncover many different responses, depending on the intrinsic morphology and anatomy of roots that are highly variable between species (Robinson 1994). However, three major aspects of the N regulation of root growth and development appear to be commonly observed and will be addressed in the following.

A. Shoot:Root Biomass Ratio

A decrease in the external N availability generally favors root over shoot growth, resulting in a decrease of the shoot:root biomass ratio, that can be quite dramatic in N-starved compared to well-fed plants (Robinson 1994). This does not necessarily mean that root biomass is increased in absolute terms by N deprivation. Root growth is often actually decreased by N-limitation, but to a much lesser extent than shoot growth. This very common adaptive response (seen in many plant species and also for other nutrients than N) aims at favoring the growth of the organ (the roots in this case) in charge of the acquisition of the element that is most limiting for growth (N in this case). However, little is known concerning the mechanisms responsible for it. Several hypotheses have been proposed, that may not exclude each other. First, N-limited plants accumulate carbohydrates at a high level in the leaves, which has been proposed to lead to an enhanced phloem transport downward to the roots, thus alleviating a C limitation of root growth occurring in non-starved plants (Rufty et al. 1988; Freixes et al. 2002). Nevertheless, it is unclear whether this is a cause or a consequence of favored root over shoot growth in N-limited plants. Second, both NO_3^- and NH_4^+ supply (but more strongly NO_3^-) were shown to stimulate cytokinin synthesis in the roots and subsequent xylem transport of these hormones to the shoot (Sakakibara et al. 2006). Because cytokinins have a strong positive effect on growth of the aerial organs, it is possible that the decrease in shoot:root ratio resulting from N deprivation may be partly due to the arrest of this cytokinin-induced stimulation of shoot growth. Third, there is convincing evidence that NO_3^- itself may be a signal regulating the allocation of resources between roots and shoot. Indeed, Scheible et al. (1997) reported that plants of tobacco lines under-expressing NR display, like wild-type plants, a high shoot:root ratio when fed with ample supply of NO_3^- as sole N source. This was unexpected because NR under-expressors take up and accumulate large amounts of NO_3^- but have a severely impaired organic N production that prevents them from growing properly. Thus, NR under-expressors grown on NO_3^- show at the same time the strong growth reduction associated with N limitation and a high shoot:root ratio associated with N sufficiency. This suggests that shoot:root ratio is actually determined by the internal stores of NO_3^- in the tissues, rather than by the organic N status of the plant. Finally, lateral root growth is specifically repressed by high N status of the plant (detailed in the following section), which contributes to increasing shoot:root ratio under N sufficiency.

B. Systemic Repression of Lateral Root Growth by High N Status of the Plant

Consistent with the increase in shoot:root biomass ratio observed in response to ample N supply, high N status of the whole plant has been reported to exert a systemic repression on lateral root growth. Best characterized in *Arabidopsis*, this mechanism acts when plants are supplied with a high NO_3^- concentration (>10 mM) and results in an inhibition of meristem activation in newly emerged lateral root primordia (Zhang et al. 1999). As a consequence, even if a high number of primordia may have been initiated, they are blocked in their development and fail to generate an elongating lateral root, thus markedly restricting the overall branching of the root system. Importantly, this inhibition is reversible. Transfer of the plants from high to low NO_3^- medium results in fast meristem activation in the quiescent primordia, leading to the generation of numerous lateral roots, generally interpreted as a foraging response. The repression of lateral root growth by high NO_3^- is systemic because it can be observed in split-root plants, where high NO_3^- supply on one portion of the root system is able to repress lateral root primordia in another portion of the root system fed with a low NO_3^- medium. In agreement with the hypothesis that NO_3^- may act as a signal favoring shoot over root growth (Scheible et al. 1997, see in prior text), systemic repression of lateral root primordia is maintained in NR-deficient mutants of *Arabidopsis* (Zhang et al. 1999). However, organic N sources such as glutamine also trigger inhibition of meristem activation in lateral root primordia. This shows that NO_3^- itself, together with downstream products of its assimilation, directly or indirectly participates in the long-distance signaling responsible for systemic repression of root growth by high N status of the plant.

Although the precise mechanisms responsible for the N-induced systemic repression of lateral root growth are unknown, several lines of evidence suggest a direct implication of hormone signaling pathways. Indeed, *Arabidopsis mutants* either insensitive to ABA (*abi4* and *abi5*) or deficient in ABA synthesis (*aba2* and *aba3*) were found to display a reduced inhibition of lateral root growth in response to high NO_3^- (Signora et al. 2001). Furthermore, exogenous ABA supply mimics this nitrate-induced inhibition, and mutants that still produce visible lateral roots in presence of 0.5 μM ABA (*labi* mutants) are also less sensitive to repression by high NO_3^- (Zhang et al. 2007).

C. Local Stimulation of Lateral Root Growth by NO_3^- and NH_4^+

In contrast with its systemic inhibitory effect on lateral root growth, NO_3^- locally promotes generation and/or elongation of lateral roots. This dual action of NO_3^- results in one of the most spectacular and probably ecologically relevant adaptive response of RSA, the so-called lateral root proliferation in NO_3^- rich patches (Zhang et al. 1999). Challenged by a strong spatial heterogeneity of NO_3^- concentration in the soil, many plant species react in specifically stimulating root colonization of the soil areas where NO_3^- is most abundant (Robinson 1994). This is generally associated with a dramatically increased generation and/or elongation of lateral roots in the NO_3^- rich patches (Figure 25.8) that improves the acquisition efficiency of this nutrient (Drew 1975; Zhang et al. 1999; Linkohr et al. 2002; Remans et al. 2006).

This adaptive response is allowed by the relief of the systemic inhibition described previously, because NO_3^- provision to only one part of the root system is generally not sufficient to activate the repression exerted by the N status of the plant. However, the most important mechanism driving preferential lateral root growth in NO_3^--rich patches is related to specific local NO_3^- signaling. Indeed, this response is seen in NR-deficient plants (Zhang and Forde 1998), which shows that lateral roots proliferate into NO_3^--rich patches not because they are well fed with N, but because they have detected the presence of NO_3^- in these patches. Local NO_3^- sensing promotes lateral root growth at two different levels: (1) stimulation of primordia emergence, resulting in an increased density of lateral roots (Linkohr et al. 2002; Krouk et al. 2010b), and (2) stimulation of the elongation of emerged lateral roots (Zhang et al. 1999; Remans et al. 2006). Several molecular components of this NO_3^- signaling pathway have now been identified in *Arabidopsis*. The ANR1 MADS-box putative transcription factor was shown to be required because ANR1 downregulated lines are almost unable to stimulate lateral root elongation in response to a localized NO_3^- supply (Zhang and Forde 1998). More recently, the NRT1.1 NO_3^- transporter was reported to also play a crucial role in the signaling process, possibly as an upstream membrane NO_3^- sensor, responsible for the initial perception of the external NO_3^- that triggers the whole pathway (Remans et al. 2006). Indeed, knockout mutants for *NRT1.1* show reduced preferential lateral root growth in a high NO_3^- patch, as the ANR1 downregulated lines. For instance, for 5 days after transfer of split-root plants to a heterogeneous medium, lateral roots have grown ca. fourfold faster in the high NO_3^- side as compared to the low NO_3^- side in wild-type plants, whereas this difference is only twofold in the *chl1-5* mutant of *NRT1.1* (Figure 25.8). Furthermore, *chl1* mutants have a markedly reduced ANR1 transcript accumulation, suggesting that NRT1.1 acts upstream of ANR1 in the signaling pathway (Remans et al. 2006).

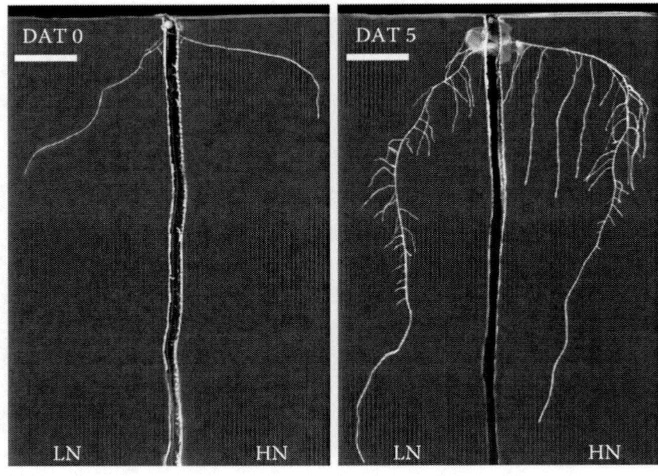

(A)

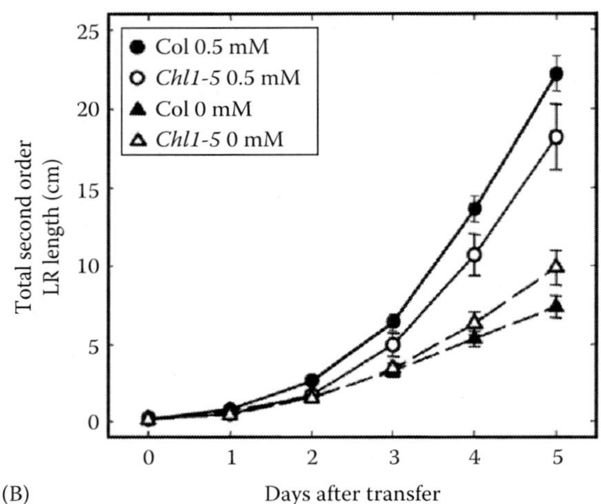

(B)

FIGURE 25.8 (A) Response of the root system architecture to a localized high NO_3^- supply (HN; 0.5 mM) in wild-type plants of *Arabidopsis*. LN, low NO_3^- medium (no added NO_3). DAT, day after transfer (Scale bars: 1 cm.). At DAT 0, the young seedlings with a root system pruned to two first-order lateral roots were transferred to a vertically segmented agar plate, with one first-order lateral root on the HN side, and the other one on the LN side. At DAT 5, many second-order laterals were generated but have markedly elongated in the HN side only. (B) Time course of second-order lateral root growth in HN (0.5 mM) or LN (0 mM) sides of the split-root system in Col wild-type and the *chl1-5* mutant, showing that mutation of *NRT1.1* alters preferential LR growth in the nitrate-rich patch. (From Remans, T. et al., *Proc. Natl. Acad. Sci. U S A*, 103, 19206, 2006.)

These data support the hypothesis that the NO_3^- sensing role of NRT1.1 is not restricted to the regulation of other NO_3^- transport systems (see in prior text) but also governs the root morphological changes triggered by changes in NO_3^- availability (Gojon et al. 2011). Unlike for the NRT1.1-dependent regulation of *NRT2.1* expression, a precise molecular mechanism has been proposed to account for the effect of NRT1.1 in the NO_3^- regulation of lateral root growth (Krouk et al. 2010b). This mechanism involves a direct connection between NO_3^- and auxin signaling, a hypothesis that has been put forward for long to explain the strong effects of NO_3^- on RSA (Zhang and Forde 2000). Indeed, the NO_3^- induced stimulation of lateral root emergence and growth is associated with an increased accumulation of auxin at the apex of primordia or newly emerged lateral roots (Krouk et al. 2010b). This local auxin maximum is crucial for efficient emergence and subsequent growth of these roots (Benkova et al. 2003). NRT1.1 is required for the locally increased accumulation of auxin in response to NO_3^-, not because it promotes this accumulation in presence of NO_3^- but because it prevents it in the absence of NO_3^- (Figure 25.9). Accordingly, primordia and newly emerged lateral roots of NRT1.1-deficient mutants display a constitutively high level of auxin that results in a constitutive high rate of primordia emergence and growth, regardless of the external NO_3^- concentration (Krouk et al. 2010b). Thus, it appears that NRT1.1 acts as a repressor of lateral root growth in the absence of or at low availability of NO_3^-. The reason why NRT1.1 controls the NO_3^--induced accumulation of auxin in lateral roots is that this protein is apparently able to facilitate the transport of auxin, in addition to that of NO_3^-, as shown in heterologous expression experiments in *Xenopus* oocytes or *Saccharomyces cerevisiae* (Krouk et al. 2010b). Furthermore, auxin transport facilitation by NRT1.1 is (1) inhibited by NO_3^- and (2) most likely contributing to the remobilization of the hormone out of the lateral root toward the parent root. Altogether, these data led to the proposal of a model for the mechanism of NRT1.1-dependent regulation of lateral root growth by NO_3^-. Briefly, this model postulates that in the absence of or at low external availability of NO_3^-, NRT1.1 behaves as an auxin transport facilitator that removes the hormone out of the lateral root, therefore repressing emergence and growth of this organ. At high NO_3^- concentration, auxin transport facilitation by NRT1.1 is inhibited, and auxin accumulates at the apex of the lateral root, which in turn promotes emergence and growth (Figure 25.9).

Two other recent studies reinforce a direct role of auxin signaling in the N regulation of RSA in *Arabidopsis*. First, Gifford et al. (2008) found that the organic forms of nitrogen, such as glutamine, are able to induce *AUXIN RESPONSE FACTOR 8* (*ARF8*) expression in the pericycle where lateral root primordia are initiated. Actually, high N supply represses miRNA167 in this tissue, leading to the accumulation of the transcript of its target *ARF8*. This results in a stimulation of lateral root primordia initiation, which is however more than counterbalanced by a concomitant inhibition of primordia emergence, ultimately restricting lateral root density in plants with high N provision. Second, Vidal et al. (2010) reported that NO_3^- transiently induces the expression of the auxin receptor gene *AFB3*, suggesting that NO_3^- not only modifies local auxin levels in the root tissues but also increases the sensitivity of these tissues to auxin. These authors found that *AFB3* was induced by NO_3^- in particular in the pericycle and that NO_3^- supply increased the density of total lateral roots (emerged plus non-emerged) in wild-type plants, but not in an *afb3* mutant, highlighting a crucial role of *AFB3* in the stimulation of lateral root initiation by NO_3^-.

The hypothesis that, in addition to NO_3^-, external NH_4^+ could also be sensed specifically by the roots and could trigger

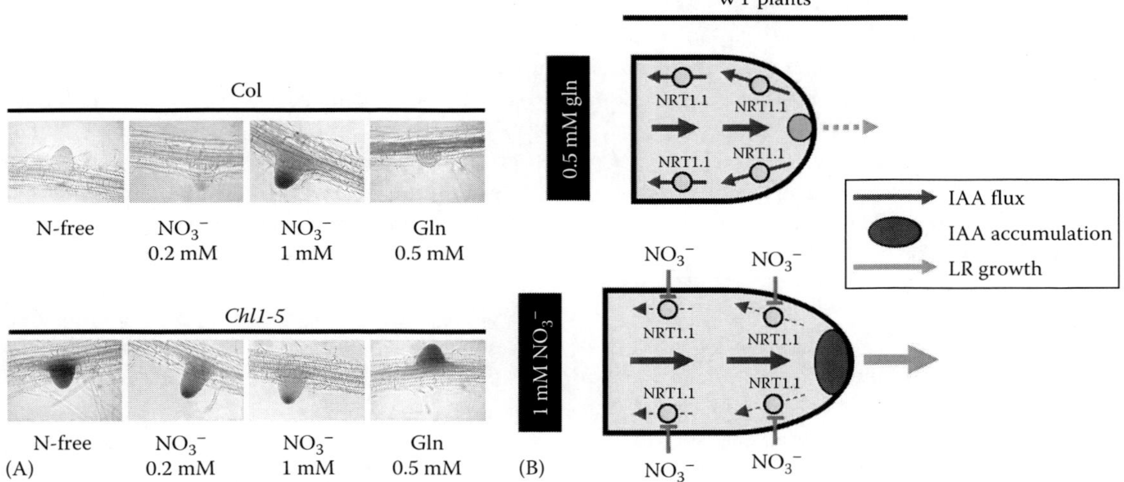

FIGURE 25.9 **(See color insert.)** (A) Histochemical staining of GUS activity in newly emerged lateral roots of transgenic *Arabidopsis* plants expressing *DR5::GUS* (an auxin biomarker) in wild-type or *chl1-5* background. The plants were cultivated for 8 days on media containing nitrogen sources described in the figure. (B) Schematic model for NRT1.1 control of lateral root growth in response to nitrate. Two situations are shown to illustrate the specific effect of NO_3^- on lateral root growth, corresponding to plants supplied either with 0.5 mM glutamine or with 1 mM NO_3^- (1 mM external N in both cases). The model postulates that in the absence of NO_3^- (glutamine-fed plants), NRT1.1 favors basipetal transport of auxin in lateral roots, thus preventing auxin accumulation at the lateral root tip. This slows down outgrowth and elongation of lateral roots. At 1 mM NO_3^-, facilitation of basipetal auxin transport by NRT1.1 is inhibited, leading to auxin accumulation in the lateral root tip and accelerated growth of lateral root. (From Krouk, G. et al., *Dev. Cell*, 18, 927, 2010b.)

adaptive RSA responses has been a matter of debate. In his pioneering work on this topic, Drew (1975) reported that heterogeneous NH_4^+ supply to sand-grown barley plants led to a similar localized proliferation of lateral roots as that observed with NO_3^-. However, because these experiments were not performed in sterile conditions, it was questioned whether this was actually a response due to NH_4^+ *per se* or to NO_3^- that could have been produced locally by nitrification. Moreover, experiments with split-root *Arabidopsis* plants indicated that the localized stimulation of lateral root elongation induced by NO_3^- could not be mimicked by NH_4^+ (Zhang et al. 1999; Remans et al. 2006). Nevertheless, more recent reports provide convincing evidence that NH_4^+ present in the external medium is indeed perceived by the roots as a signal molecule that can regulate lateral root development. First, the phosphorylation-dependent inactivation of AMT transporters by high NH_4^+ supply in *Arabidopsis* roots apparently relies on a sensing system specific for exogenous NH_4^+ (Lanquar et al. 2009, see in prior text). Second, detailed reconsideration of the action of NH_4^+ on RSA in *Arabidopsis* unraveled specific effects different from those of NO_3^- (Lima et al. 2010). Unlike NO_3^- that predominantly promotes lateral root emergence and elongation, localized NH_4^+ supply was found to mainly increase initiation of lateral root primordia of high-order laterals (second and mostly third order that are initiated on first- and second-order lateral roots, respectively). This effect was most probably missed by the other groups that investigated lateral root proliferation in response to localized N supply, simply because none of them considered third-order laterals. This NH_4^+-induced generation of higher-order lateral roots results in a much more "bushy" architecture than with NO_3^-, suggesting that these two ions have complementary effects on RSA, that were shown to be at least partly additive when NH_4^+ and NO_3^- are supplied together (Lima et al. 2010). Interestingly, the RSA responses to localized NH_4^+ supply required the presence of a functional AMT1.3 transporter (Lima et al. 2010), thus making an interesting parallel with the NRT1.1-dependent lateral root proliferation in response to localized NO_3^- supply. However, it is not known yet whether this is related to a putative NH_4^+ sensing role of AMT1.3 in the roots.

VI. Conclusions

Our knowledge concerning the physiological and molecular aspects of root N acquisition has considerably increased during the past decade. This is particularly exemplified by the molecular characterization of NH_4^+ and NO_3^- transporters, as we are now able to put protein names on specific transport processes that were only characterized at the functional level (i.e., flux measurements). The generalization of the use of reverse genetics has allowed drawing firm conclusions concerning the physiological importance of some of these proteins. For instance, knockout mutations of specific *AMT1* or *NRT2* genes yielded genotypes almost totally impaired in the NH_4^+ or NO_3^- HATS, respectively. This demonstrates that we have certainly identified the major players of these transport systems. Nonetheless, only a dozen (out of ~70 in total) of AMT and NRT transporters have been functionally characterized in *Arabidopsis*. In particular, both NH_4^+ and NO_3^- LATS still await a detailed molecular identification. Thus, a lot of work still remains to be done in order to have a general overview of the molecular structure of the various NH_4^+ or NO_3^- transport systems that are active in this model species and to answer in the case of N this general question: why are there multiple ion transporters in plants? This task is even more tremendous when considering crops, which may have larger transporter families than *Arabidopsis* (for instance, there are 53 NRT1/PTR genes in *Arabidopsis* but ~80 in rice).

Concurrently, the investigation of the N regulation of RSA has unraveled largely unexpected mechanisms (e.g., involvement of transporters as nutrient sensors, identification of specific regulatory miRNAs) that, however, seem to share a common feature, i.e., a direct crosstalk between N and hormone signaling pathways. No doubt that the near future will provide many more examples of such mechanisms that will help us understand how they interact to ensure the spectacular plasticity of RSA in response to changes in N availability in the soil. Transfer of the knowledge acquired in model species to crops will certainly be particularly difficult on this topic, especially because we desperately lack high-throughput phenotyping facilities adapted to large plants as compared to *Arabidopsis*.

A challenge will also certainly be to integrate studies on N transport systems and studies on RSA. These are two strongly interconnected aspects of root N acquisition that are still mostly investigated separately. There is little doubt that they now need to be considered together in future strategies aiming at improving N use efficiency.

Finally, the regulatory mechanisms that modulate root N acquisition are also beginning to be revealed. One of the most significant developments deals with the identification of putative sensing systems (e.g., NRT1.1) that trigger the signaling cascades controlling transport systems expression and/or activity or RSA. The example of NRT1.1 is illustrative of the idea that these sensing systems may drive both physiological and developmental responses to N. Putting more efforts in the elucidation of these sensing/signaling mechanisms is of high priority, because this is probably where we may expect to identify upstream "master" genes that coordinate a complex array of individual mechanisms to ultimately determine the efficiency of N acquisition by the roots.

References

Amarasinghe BHR, de Bruxelles GL, Braddon M et al. 1998. Regulation of *GmNRT2* expression and nitrate transport activity in roots of soybean (*Glycine max*). *Planta* 206:44–52.

Andrews M. 1986. The partitioning of nitrate assimilation between root and shoot of higher plants. *Plant Cell Environ* 9:511–519.

Benkova E, Michniewicz M, Sauer M et al. 2003. Local, efflux-dependent auxin gradients as a common module for plant organ formation. *Cell* 115:591–602.

Britto DT, Siddiqi MY, Glass ADM, Kronzucker HJ. 2001. Futile membrane NH_4^+ cycling: A cellular hypothesis to explain ammonium toxicity in plants. *Proc Natl Acad Sci U S A* 98:4255–4258.

Camargo A, Liamas A, Schnell RA et al. 2007. Nitrate signaling by the regulatory gene NIT2 in *Chlamydomonas*. *Plant Cell* 19:3491–3503.

Castaings L, Camargo A, Pocholle D et al. 2009. The nodule inception-like protein 7 modulates nitrate sensing and metabolism in *Arabidopsis*. *Plant J* 57:426–435.

Chapman N, Whalley WR, Lindsey K, Miller AJ (in press). Water supply and not nitrate concentration determines primary root growth in *Arabidopsis*. *Plant Cell Environ* 34:1630–1638 (doi: 10.1111/j.1365-3040.2011.02358.x).

Chopin F, Orsel M, Dorbe MF et al. 2007. The *Arabidopsis* AtNRT2.7 nitrate transporter controls nitrate content in seeds. *Plant Cell* 19:1590–1602.

Clarkson DT, Gojon A, Saker LR et al. 1996. Nitrate and ammonium influxes in soybean (*Glycine max*) roots: Direct comparison of ^{13}N and ^{15}N tracing. *Plant Cell Environ* 19:859–868.

Clément CR, Hopper MJ, Jones LHP, Leafe EL. 1978. The uptake of nitrate by *Lolium perenne* from flowing nutrient solution. II. Effect of light, defoliation, and relationship to CO_2 flux. *J Exp Bot* 29:1173–1183.

Cooper HD, Clarkson DT. 1989. Cycling of amino-nitrogen and other nutrients between shoots and roots in cereals—A possible mechanism integrating shoot and root in the regulation of nutrient uptake. *J Exp Bot* 40:753–762.

Crawford NM. 1995. Nitrate: Nutrient and signal for plant growth. *Plant Cell* 7:859–868.

Crawford NM, Glass ADM. 1998. Molecular and physiological aspects of nitrate uptake in plants. *Trends Plant Sci* 3:389–395.

D'apuzzo E, Rogato A, Simon-Rosin U et al. 2004. Characterization of three functional high-affinity ammonium transporters in *Lotus japonicus* with different transcriptional regulation and spatial expression. *Plant Physiol* 134:1763–1774.

De Kroon H, Visser EJW, Huber H, Mommer L, Hutchings MJ. 2009. A modular concept of plant foraging behaviour: The interplay between local responses and systemic control. *Plant Cell Environ* 32:704–712.

Delhon P, Gojon A, Tillard P, Passama L. 1995. Diurnal regulation of NO_3^- uptake in soybean plants. I. Changes in NO_3^- influx, efflux, and N utilization in the plant during the day/night cycle. *J Exp Bot* 46:1585–1594.

Delhon P, Gojon A, Tillard P, Passama L. 1996. Diurnal regulation of NO_3^- uptake in soybean plants. IV. Dependence on current photosynthesis and sugar availability to the roots. *J Exp Bot* 47:893–900.

Drew MC. 1975. Comparison of the effects of a localized supply of phosphate, nitrate, ammonium and potassium on the growth of the seminal root system, and the shoot, in barley. *New Phytol* 75:479–490.

Feng H, Yan M, Fan X et al. 2011. Spatial expression and regulation of rice high-affinity nitrate transporters by nitrogen and carbon status. *J Exp Bot* 62:2319–2332.

Filleur S, Daniel-Vedele F. 1999. Expression analysis of a high-affinity nitrate transporter isolated from *Arabidopsis thaliana* by differential display. *Planta* 207:461–469.

Filleur S, Dorbe MF, Cerezo M et al. 2001. An *Arabidopsis* T-DNA mutant affected in *Nrt2* genes is impaired in nitrate uptake. *FEBS Lett* 489:200–224.

Forde B. 2002. Local and long-range signaling pathways regulating plant responses to nitrate. *Annu Rev Plant Biol* 53:203–224.

Fraisier V, Gojon A, Tillard P, Daniel-Vedele F. 2000. Constitutive expression of a putative high-affinity nitrate transporter in *Nicotiana plumbaginifolia*: Evidence for post-transcriptional regulation by a reduced nitrogen source. *Plant J* 23:489–496.

Freixes S, Thibaud MC, Tardieu F, Muller B. 2002. Root elongation and branching is related to local hexose concentration in *Arabidopsis thaliana* seedlings. *Plant Cell Environ* 25:1357–1366.

Galloway JN, Aber JD, Erisman JW et al. 2003. The nitrogen cascade. *BioScience* 53:341–356.

Gansel X, Muños S, Tillard P, Gojon A. 2001. Differential regulation of the NO_3^- and NH_4^+ transporter genes *AtNrt2.1* and *AtAmt1.1* in *Arabidopsis*: Relation with long distance and local controls by N status of the plant. *Plant J* 26:143–156.

Garnett T, Conn V, Kaiser BN. 2009. Root based approaches to improving nitrogen use efficiency in plants. *Plant Cell Environ* 32:1272–1283.

Gastal F, Saugier B. 1989. Relationships between nitrogen uptake and carbon assimilation in whole plants of tall fescue. *Plant Cell Environ* 12:407–418.

Gazzarini S, Lejay L, Gojon A, Ninnemann O, Frommer WB, VonWirèn N. 1999. Three functional transporters for constitutive, diurnally regulated, and starvation-induced uptake of ammonium into *Arabidopsis* roots. *Plant Cell* 11:937–947.

Gifford ML, Dean A, Gutiérrez RA, Coruzzi G, Birnbaum K. 2008. Cell-specific nitrogen responses mediate developmental plasticity. *Proc Natl Acad Sci U S A* 105:803–808.

Girin T, El Kafafi ES, Widiez T et al. 2010. Identification of *Arabidopsis mutants* impaired in the systemic regulation of root nitrate uptake by the nitrogen status of the plant. *Plant Physiol* 153:1250–1260.

Girin T, Lejay L, Wirth J et al. 2007. Identification of a 150 bp cis-acting element of the *AtNRT2.1* promoter involved in the regulation of gene expression by the N and C status of the plant. *Plant Cell Environ* 30:1366–1380.

Gojon A, Krouk G, Perrine-Walker F, Laugier E. 2011. Nitrate transceptor(s) in plants. *J Exp Bot* 62:2299–2308.

Gojon A, Nacry P, Davidian JC. 2009. Root uptake regulation: A central process for NPS homeostasis in plants. *Curr Opin Plant Biol* 12:328–338.

Good AG, Shrawat AK, Muench DG. 2004. Can less yield more? Is reducing nutrient input into the environment compatible with maintaining crop production? *Trends Plant Sci* 9:597–605.

Guo FQ, Wang R, Chen M, Crawford NM. 2001. The *Arabidopsis* dual-affinity nitrate transporter gene AtNRT1.1 (CHL1) is activated and functions in nascent organ development during vegetative and reproductive growth. *Plant Cell* 13:1761–1777.

Ho CH, Lin SH, Hu HC, Tsay YF. 2009. CHL1 functions as a nitrate sensor in plants. *Cell* 138:1184–1194.

Hodge A, Robinson D, Griffiths DS, Fitter AH. 1999. Why plants bother: Root proliferation results in increased nitrogen capture from an organic patch when two grasses compete. *Plant Cell Environ* 22:811–820.

Hole DJ, Emran AM, Fares Y, Drew MC. 1990. Induction of nitrate transport in maize roots, and kinetics of influx, measured with nitrogen-13. *Plant Physiol* 93:642–647.

Holsbeeks I, Lagatie O, Van Nuland A, Van de Velde S, Thevelein JM. 2004. The eukaryotic plasma membrane as a nutrient-sensing device. *Trends Biochem Sci* 29:556–564.

Huang NC, Liu KH, Lo HJ, Tsay YF. 1999. Cloning and characterization of an *Arabidopsis* nitrate transporter gene that encodes a constitutive component of low-affinity uptake. *Plant Cell* 11:1381–1392.

Imsande J, Touraine B. 1994. N demand and the regulation of nitrate uptake. *Plant Physiol* 105:3–7.

Jackson WA, Volk RJ, Tucker TC. 1972. Apparent induction of nitrate uptake in nitrate-depleted plants. *Agron J* 64:518–521.

King BJ, Siddiqi MY, Ruth TJ, Warner RL, Glass ADM. 1993. Feedback regulation of nitrate influx in barley roots by nitrate, nitrite, and ammonium. *Plant Physiol* 102:1279–1286.

Krapp A, Fraisier V, Scheible WR et al. 1998. Expression studies of *Nrt2;1Np*, a putative high-affinity nitrate transporter: Evidence for its role in nitrate uptake. *Plant J* 14:723–731.

Krouk G, Crawford NM, Coruzzi GM, Tsay YF. 2010a. Nitrate signaling: Adaptation to fluctuating environments. *Curr Opin Plant Biol* 13:266–273.

Krouk G, Lacombe B, Bielach A et al. 2010b. Nitrate and auxin transport by NRT1.1 defines a mechanism for nutrient sensing in plants. *Dev Cell* 18:927–937.

Krouk G, Tillard P, Gojon A. 2006. Regulation of the high-affinity NO_3^- uptake system by NRT1.1-mediated NO_3^- demand signaling in *Arabidopsis*. *Plant Physiol* 142:1075–1086.

Kumar A, Silim SN, Okamoto M, Siddiqi MY, Glass ADM. 2003. Differential expression of three members of the *AMT1* gene family encoding putative high-affinity NH_4^+ transporters in roots of *Oryza sativa* subspecies *indica*. *Plant Cell Environ* 26:907–914.

Lanquar V, Loque D, Hormann F et al. 2009. Feedback inhibition of ammonium uptake by a phospho-dependent allosteric mechanism in *Arabidopsis*. *Plant Cell* 21:3610–3622.

Lauter FR, Ninnemann O, Bucher M, Riesmeier JW, Frommer WB. 1996. Preferential expression of an ammonium transporter and of two putative nitrate transporters in root hairs of tomato. *Proc Natl Acad Sci U S A* 93:8139–8144.

Lee RB. 1993. Control of net uptake of nutrients by regulation of influx in barley plants recovering from nutrient deficiency. *Ann Bot* 72:223–230.

Lejay L, Gansel X, Cerezo M et al. 2003. Regulation of root ion transporters by photosynthesis: Functional importance and relation with hexokinase. *Plant Cell* 15:2218–2232.

Lejay L, Tillard P, Lepetit M et al. 1999. Molecular and functional regulation of two NO_3^- uptake systems by N- and C-status of *Arabidopsis* plants. *Plant J* 18:509–519.

Lejay L, Wirth J, Pervent M, Cross JMF, Tillard P, Gojon A. 2008. Oxidative pentose phosphate pathway-dependent sugar sensing as a mechanism for regulation of root ion transporters by photosynthesis. *Plant Physiol* 146:2036–2053.

Li JY, Fu YL, Pike SM et al. 2010. The *Arabidopsis* nitrate transporter NRT1.8 functions in nitrate removal from the xylem sap and mediates cadmium tolerance. *Plant Cell* 22:1633–1646.

Li W, Wang Y, Okamoto M, Crawford NM, Siddiqi MY, Glass AD. 2007. Dissection of the *AtNRT2.1:AtNRT2.2* inducible high-affinity nitrate transporter gene cluster. *Plant Physiol* 143:425–433.

Lima JE, Kojima S, Takahashi H, von Wirén N. 2010. Ammonium triggers lateral root branching in *Arabidopsis* in an AMMONIUM TRANSPORTER1;3-dependent manner. *Plant Cell* 22:3621–3633.

Lin SH, Kuo HF, Canivenc G et al. 2008. Mutation of the *Arabidopsis* NRT1.5 nitrate transporter causes defective root-to-shoot nitrate transport. *Plant Cell* 20:2514–2528.

Linkohr BI, Williamson LC, Fitter AH, Leyser OHM. 2002. Nitrate and phosphate availability and distribution have different effects on root system architecture of *Arabidopsis*. *Plant J* 29:751–760.

Liu KH, Tsay YF. 2003. Switching between the two action modes of the dual-affinity nitrate transporter CHL1 by phosphorylation. *EMBO J* 22:1–9.

Loqué D, Lalonde S, Looger LL, Von Wiren N, Frommer WB. 2007. A cytosolic trans-activation domain essential for ammonium uptake. *Nature* 446:195–198.

Loqué D, Yuan L, Kojima S et al. 2006. Additive contribution of AMT1;1 and AMT1;3 to high-affinity ammonium uptake across the plasma membrane of nitrogen-deficient *Arabidopsis* roots. *Plant J* 48:522–534.

MacKown CT, Volk RJ, Jackson WA. 1981. Nitrate accumulation, assimilation and transport by decapited corn roots. Effects of prior nitrate nutrition. *Plant Physiol* 68:133–138.

Malagoli P, Laine P, Le Deunff E, Rossato L, Ney B, Ourry A. 2004. Modeling nitrogen uptake in oilseed rape cv Capitol during a growth cycle using influx kinetics of root nitrate transport systems and field experimental data. *Plant Physiol* 134:388–400.

Marini AM, Soussi-Boudekou S, Vissers S, Andre B. 1997. A family of ammonium transporters in *Saccharomyces cerevisiae*. *Mol Cell Biol* 17:4282–4293.

Marschner H. 1995. *Mineral Nutrition of Higher Plants*. London, U.K.: Academic Press.

Miller AJ, Fan X, Orsel M, Smith SJ, Wells DM. 2007. Nitrate transport and signalling. *J Exp Bot* 58:2297–2306.

Morgan MA, Jackson WA. 1988. Reciprocal ammonium transport into and out of plant roots: Modifications by plant nitrogen status and elevated root ammonium. *J Exp Bot* 40:207–214.

Muller B, Touraine B. 1992. Inhibition of NO_3^- uptake by various phloem-translocated amino acids in soybean seedlings. *J Exp Bot* 43:617–623.

Muños S, Cazettes C, Fizames C et al. 2004. Transcript profiling in the *chl1-5* mutant of *Arabidopsis* reveals a role of the nitrate transporter NRT1.1 in the regulation of another nitrate transporter, NRT2.1. *Plant Cell* 16:2433–2447.

Nazoa P, Vidmar JJ, Tranbarger TJ et al. 2003 Regulation of the nitrate transporter gene *AtNRT2.1* in *Arabidopsis thaliana*: Responses to nitrate, amino acids and developmental stage. *Plant Mol Biol* 52:689–703.

Neuhäuser B, Dynowski M, Mayer M, Ludewig U. 2007. Regulation of NH_4^+ transport by essential cross talk between AMT monomers through the carboxyl tails. *Plant Physiol* 143:1651–1659.

Ninnemann O, Jauniaux JC, Frommer WB. 1994. Identification of a high-affinity ammonium transporter from plants. *EMBO J* 13:3464–3471.

Okamoto M, Vidmar JJ, Glass ADM. 2003. Regulation of *NRT1* and *NRT2* gene families of *Arabidopsis thaliana*: Responses to nitrate provision. *Plant Cell Physiol* 44:304–317.

Orsel M, Chopin F, Leleu O et al. 2006. Characterization of a two-component high-affinity nitrate uptake system in *Arabidopsis*. Physiology and protein-protein interaction. *Plant Physiol* 142:1304–1317.

Orsel M, Krapp A, Daniel-Vedele F. 2002. Analysis of the NRT2 nitrate transporter family in *Arabidopsis*. Structure and gene expression. *Plant Physiol* 129:886–896.

Rawat SR, Silim SN, Kronzuker HJ, Siddiqi MY, Glass ADM. 1999. *AtAmt1.1* gene expression and NH_4^+ uptake in roots of *Arabidopsis thaliana*: Evidence for regulation by root glutamine levels. *Plant J* 19:143–152.

Remans T, Nacry P, Pervent M et al. 2006. The *Arabidopsis* NRT1.1 transporter participates in the signaling pathway triggering root colonization of nitrate-rich patches. *Proc Natl Acad Sci U S A* 103:19206–19211.

Robinson D. 1994. The responses of plants to non-uniform supplies of nutrients. *New Phytol* 127:635–674.

Robinson D. 1996. Resource capture by localized root proliferation: Why do plants bother? *Ann Bot* 77:179–185.

Ruffel S, Freixes S, Balzergue S et al. 2008. Systemic signaling of the plant nitrogen status triggers specific transcriptome responses depending on the nitrogen source in *Medicago truncatula*. *Plant Physiol* 146:2020–2035.

Rufty TW Jr, Huber SC, Volk RJ. 1988. Alteration in leaf carbohydrate metabolism in response to N stress. *Plant Physiol* 88:725–730.

Sakakibara H, Kentaro T, Naoya H. 2006. Interactions between nitrogen and cytokinin in the regulation of metabolism and development. *Trends Plant Sci* 11:440–448.

Scheible WR, Lauerer M, Schulze ED, Caboche M, Stitt M. 1997. Accumulation of nitrate in the shoots acts as a signal to regulate shoot-root allocation in tobacco. *Plant J* 11:671–691.

Segonzac C, Boyer JC, Ipotesi E et al. 2007. Nitrate efflux at the root plasma membrane: Identification of an *Arabidopsis* excretion transporter. *Plant Cell* 19:3760–3777.

Signora L, De Smet I, Foyer CH, Zhang, H. 2001. ABA plays a central role in mediating the regulatory effects of nitrate on root branching in *Arabidopsis*. *Plant J* 28:655–662.

Stitt M. 1999. Nitrate regulation of metabolism and growth. *Curr Opin Plant Biol* 2:178–186.

Tong Y, Zhou JJ, Li Z, Miller AJ. 2005. A two-component high-affinity nitrate uptake system in barley. *Plant J* 41:441–450.

Touraine B, Glass ADM. 1997. NO_3^- and ClO_3^- fluxes in the *chl1-5* mutant of *Arabidopsis thaliana*. Does the *CHL1-5* gene encode a low-affinity NO_3^- transporter? *Plant Physiol* 114:137–144.

Trueman LJ, Richardson A, Forde BG. 1996. Molecular cloning of higher plant homologues of the high-affinity nitrate transporters of *Chlamydomonas reinhardtii* and *Aspergillus nidulans*. *Gene* 175:223–231.

Tsay YF, Chiu CC, Tsai CB, Ho CH, Hsu PK. 2007. Nitrate transporters and peptide transporters. *FEBS Lett* 581:2290–2300.

Tsay YF, Schroeder JI, Feldmann KA, Crawford NM. 1993. The herbicide sensitivity gene *CHL1* of *Arabidopsis* encodes a nitrate-inducible nitrate transporter. *Cell* 72:705–713.

Vidal EA, Araus V, Lu C et al. 2010. Nitrate-responsive miR393/AFB3 regulatory module controls root system architecture in *Arabidopsis thaliana*. *Proc Natl Acad Sci U S A* 107:4477–4482.

Vidmar JJ, Zhuo D, Siddiqi MY, Glass ADM. 2000. Isolation and characterization of *HvNRT2.3* and *HvNRT2.4*, cDNAs encoding high-affinity nitrate transporters from roots of barley. *Plant Physiol* 122:783–792.

Wang R, Liu D, Crawford NM. 1998. The *Arabidopsis* CHL1 protein plays a major role in high-affinity nitrate uptake. *Proc Natl Acad Sci U S A* 95:15134–15139.

Wang MY, Siddiqi MY, Ruth TJ, Glass ADM. 1993. Ammonium uptake by rice roots. II. Kinetics of $^{13}NH_4^+$ influx across the plasmalemma. *Plant Physiol* 103:1259–1267.

Wang R, Tischner R, Gutierrez RA et al. 2004. Genomic analysis of the nitrate response using a nitrate reductase-null mutant of *Arabidopsis*. *Plant Physiol* 136:2512–2522.

Wang R, Xing X, Wang Y, Tran A, Crawford NM. 2009. A genetic screen for nitrate regulatory mutants captures the nitrate transporter gene NRT1.1. *Plant Physiol* 151:472–478.

von Wirèn N, Gazzarini S, Gojon A, Frommer WB. 2000. The molecular physiology of ammonium uptake and retrieval. *Curr Opin Plant Biol* 3:254–261.

Wirth J, Chopin F, Santoni V et al. 2007. Regulation of root nitrate uptake at the NRT2.1 protein level in *Arabidopsis thaliana. J Biol Chem* 282:23541–23552.

Yong Z, Kotur Z, Glass AD. 2010. Characterization of an intact two-component high-affinity nitrate transporter from *Arabidopsis* roots. *Plant J* 63:739–748.

Yuan L, Loqué D, Kojima S et al. 2007. The organization of high-affinity ammonium uptake in *Arabidopsis* roots depends on the spatial arrangement and biochemical properties of AMT1-type transporters. *Plant Cell* 19:2636–2652.

Zhang H, Forde BG. 1998. An *Arabidopsis* MADS-box gene controlling nutrient-induced changes in root architecture. *Science* 279:407–409.

Zhang H, Forde BG. 2000. Regulation of *Arabidopsis* root development by nitrate availability. *J Exp Bot* 51:51–59.

Zhang H, Jennings A, Barlow PW, Forde BG. 1999. Dual pathways for regulation of root branching by nitrate. *Proc Natl Acad Sci U S A* 96:6529–6534.

Zhang H, Rong H, Pilbeam D. 2007. Signalling mechanisms underlying the morphological responses of the root system to nitrogen in *Arabidopsis thaliana. J Exp Bot* 58:2328–2338.

Zhuo D, Okamoto M, Vidmar JJ, Glass ADM. 1999. Regulation of a putative high-affinity nitrate transporter (*Nrt2;1At*) in roots of *Arabidopsis thaliana. Plant J* 17:563–568.

26

Nutrient Uptake and Root System Architecture Modeling: Past and Prospects for the Future

Moshe Silberbush
Ben-Gurion University of the Negev

Amram Eshel
Tel-Aviv University

Jonathan P. Lynch
The Pennsylvania State University

I. Introduction

Modeling is an essential step in understanding plant root systems (Fitter 1991, 1996, 2002). The inherent difficulties in measuring root system structure and function make modeling uniquely important for this field of research. Traditionally, there has been a division between modeling root system shape (cf., Berntson 1996; Lynch and Nielsen 1996; Pagès 2002) and modeling of physiological processes carried out by plant roots (cf. Silberbush 1996, 2002; Sperry et al. 2002). In recent years, the tendency has shifted toward combined structural–functional models. These include an architectural element simulating the structural development of the root system with functional simulation of physiological and ecological processes at the root–soil interface. In this chapter, we review techniques used for modeling nutrient uptake and root–rhizosphere interactions. Later, we describe modeling of root system structure and function in 3D space using *SimRoot* as an example (Lynch et al. 1997). A brief account of future simulation needs concludes this chapter.

II. Modeling Nutrient Uptake

Modeling of nutrient uptake by roots was one of the earliest applications of computers in agriculture and plant sciences in general. The models were used for a number of purposes: analysis of the processes that govern plant nutrient uptake, predicting plant performance under varying soil conditions, establishing guidelines for fertilization regimes, and so forth. Therefore, model structure and level of detail differed widely among the various models published over the years. Some were empirical models, without any attempt to describe the mechanisms involved. A typical example to this type is QUEFTS (QUantitative Evaluation of the Fertility of Tropical Soils), a model that uses empirical equations and parameters to provide estimates of yield as a function of soil nitrogen (N)-phosphorus (P)-potassium (K) fertility indices (Smaling and Janssen 1993). Another example is APSIM (Agricultural Production systems SIMulator), which is a package of sub-models to simulate crop production, using terrestrial (Delve et al. 2009) or environmental (Ludwig et al. 2009) factors. The present review, however, is focused on mechanistic models of which a partial list is presented in Table 26.1. The models in this list were developed to solve specific cases, and they are arranged according to the modeled soil and root structure: 1D, 2D, 3D, radial flow, and different variations within each group.

A. One-Dimensional Models

Early crop root models were written in conjunction with crops such as small grains or alfalfa that are uniformly broadcast across the field at a certain average density (Hagin et al. 1976; Jemison et al. 1994a,b). Such models considered the rooting volume and the root distribution to be uniform along the horizontal axes with variation occurring along the depth (z) axis only (Silberbush 2002), hence the term "one-dimensional" (1D) models. The rooting volume was divided into a number of depth layers, each with its own thickness. The model followed the changes in each layer in terms of root length density (RLD, root length

TABLE 26.1 List of Water and Ion Uptake Models

Type of Model	Model/Authors	Characteristics	References
Single cell	SNAP	Passive uptake, pesticides	Behrendt et al. (1995)
1D[a]	Hagin–Amberger	N transformation and uptake	Hagin et al. (1976)
	Gardner	Water uptake	Gardner (1991)
	Rao–Mathur	Cd transformation and uptake	Rao and Mathur (1994)
	LEACHM	Chemical reactions and uptake	Hutson and Wagenet (1995), Jemison et al. (1994a,b)
	SNAPS	Pesticide uptake	Behrendt et al. (1995)
Single root	TROIKA	Radial water flow to a single root	Lambert and Penning de Vries (1973)
	Helyar–Munns	Single root with root hairs, P transformation and uptake	Helyar and Munns (1975)
	Nye–Marriott	Radial convection–diffusion flow and uptake by roots, extendible r_1[b]	Nye and Marriott (1969)
	Baldwin–Nye–Tinker	Radial convection–diffusion flow	Baldwin et al. (1973)
	Bhat–Nye–Baldwin	Baldwin's model + uptake by root hairs	Bhat et al. (1976)
	Claassen–Barber	Extendible r_1 (without competition)	Claassen and Barber (1976)
	Barber–Cushman ("Nutrient Uptake")	Constant r_1 (root competition), constant BP[c]	Cushman (1979), Barber and Cushman (1981), Oates and Barber (1987)
	Itoh–Barber	As Barber–Cushman, with root hairs	Itoh and Barber (1983)
	Bar-Yosef–Fishman–Talpaz	Zn uptake, different chemical forms, soil water depletion	Bar-Yosef et al. (1980)
	De Willigen–Van Noordwijk	Constant and zero sink for high and low concentrations, respectively	De Willigen and Van Noordwijk (1994a,b)
	Hoffland et al.	r_1 adjusted to root density	Hoffland et al. (1990)
	COMP8	As Baldwin–Nye–Tinker, variable BP, variable r_1, competition between plant species	Ibrikci et al. (1998), Smethurst and Comerford (1993)
	Cushman (1980)	Barber–Cushman + sink/source (root hairs or microorganisms)	Cushman (1980)
	Leadley–Reynolds–Chapin	As Barber–Cushman, N components' uptake	Leadley et al. (1997)
	Bouldin	Multi-ions without interaction, linear influx functions	Bouldin (1989)
	Reginato et al.	"Free boundary" (variable root radius)	Reginato et al. (1990)
	NST 3.0	Uptake by root hairs, variable BP	Claassen (1990)
	Silberbush–Sorek–Yakirevich	Adjusted r_1, multi-ion uptake, interactions among ions, salinity/toxicity	Silberbush et al. (1993)
	Yanai	Depletion and steady-state ion influx, root competition	Yanai (1994), Williams and Yanai (1996)
2D	2DSOIL	Modular set of models, includes transport, uptake, exchange between phases	Acock and Pachepsky (1996), Timlin et al. (1996)
	Geelhoed et al.	Exudation of protons and citrate	Geelhoed et al. (1999)
	Simunek–Hopmans	Passive and active uptake from high and low concentrations, respectively	Simunek and Hopmans (2009)
	GOSSYM	Cotton growth	Baker et al. (1983)
	SOYMOD/OARDC	Soybean growth	Meyer et al. (1979)
	Benjamin–Ahuja–Allmaras	Corn growth, water uptake, and nitrate leaching	Benjamin et al. (1996)
3D	Geelhoed et al.	Uptake by root hairs, zero sink	Geelhoed et al. (1997a,b)
	Somma–Hopmans–Clausnitzer	Solute transport and uptake, and root growth, using finite-element technique	Somma et al. (1998)
	Schnepf–Leitner	Finite elements, diffusion to roots	Schnepf and Leitner (2009)

[a] 1D(vertical) models.

[b] The external boundary of the soil rhizocylinder.

[c] BP, soil buffer power.

per unit soil volume), soil compaction, and water and mineral contents. Root growth includes both increases in RLD with time within each layer resulting from branching within the layer and extension of the roots down from the upper soil layers to deeper ones. Rooting depth and root distribution profiles represented the characteristics of the simulated root system. Soil physical (movement of water and solutes) and chemical (chemical interactions, absorption/desorption between clay surfaces and soil solution, dissolution) processes and rain or irrigation events as well as root absorption activity determined water and mineral content of each layer. These models usually also included a simulation of a certain temperature gradient based on known values for a certain depth, taken from meteorological measurements of a nearby weather station and a known temperature profile. Root growth and activity were a function of the RLD and soil conditions in each layer. The overall performance of the root system was represented by the sum of the activities in the layers from the top down to the current rooting depth. Such models were coupled with models of aboveground crop development. They were amenable to calibration and verification by standard soil coring methods.

B. Two-Dimensional Models

These models were associated with models of row crops such as cotton (Baker et al. 1983), soybean (Meyer et al. 1979), or maize (Benjamin et al. 1996). They typically assumed a uniform plant distribution along the row and a known row distance with bisymmetrical development at the two sides of the row. Therefore, calculations were made for root development and activity at one side of the row only, with roots from one side allowed to cross the midline between the rows and contribution from two sides taken into account. In such models, the rooting volume was represented by a plane of a certain thickness extending across the row and down in to the soil. Root distribution in both horizontal and vertical axes was simulated, hence the term "two-dimensional" (2D) models. The simulated rooting volume was divided into cubes arranged in rows and columns with a certain probability of movement of roots and water through the horizontal and vertical faces of the cube. The roots were again represented by a certain RLD at each cube that changed as a result of branching and extension from neighboring cubes, and with time. Higher probability of root penetration through the horizontal than the vertical faces of the cube accounted for the effect of gravitropism on root distribution. Here, also root growth and activity were a function of the RLD and soil conditions in each cube. The overall performance of the root system was represented by the sum of the activities in all parts of the rooting volume penetrated by the roots at the current time step, and the total sum was doubled to account for calculating one side of the row only.

C. Radial-Flow or Single-Root Models

This type of model represents the root system of the plant by a single cylinder acquiring water and nutrients from a uniform soil volume around it. The equations making up the model describe the radial movement of water and nutrients from the bulk soil to the root surface. The pioneering modeling work of P.H. Nye (e.g., Nye and Marriott 1969) reviewed by Tinker and Nye (2000) combined radial mass flow and diffusion toward a cylindrical root, a basic approach that has been widely used since then (e.g., Barber and Cushman 1981; Oates and Barber 1987; Claassen and Steingrobe 1999) and is still in use (Roose and Schnepf 2008). A brief description of this approach follows, based on Barber and Cushman (1981):

The radial flux J_r of a solute in the soil toward the root is driven by mass flow and diffusion, assuming constant soil moisture. The partial differential equation that describes this flow is

$$\frac{\partial C_l}{\partial t} = \frac{1}{r}\left(rD\frac{\partial C_l}{\partial r} + \frac{v_0 r_0 C_l}{b} \right) \quad (26.1)$$

where

C_l is the solute concentration in the soil solution
t is time
r the radial distance from the root axis
v_0 water velocity at r_0 root radius
D the effective diffusion coefficient in the soil
b the soil buffer power for the solute

The radial flow occurs between two boundaries:

The external boundary, at radial distance r_1:

$$J_r = 0 \quad r = r_1 \quad t > 0 \quad (26.2a)$$

The inner boundary, at the root surface:

$$Db\frac{\partial C_l}{\partial r} + v_0 C_l = \frac{J_{\max}(C_l - C_{\min})}{K_m + C_l - C_{\min}} \quad r = r_0 \quad t > 0 \quad (26.2b)$$

where $J_{\max}$, K_m, and $C_{\min}$ are the Michaelis–Menten influx kinetics parameters. In plain language, the flux occurs between the external boundary (where it equals zero) and the root surface, where the flow in the soil equals the uptake rate by the root. The soil buffer power b ($=\partial C_s/\partial C_l$, where C_s is the diffusible solute concentration) was either assumed to be constant (Barber and Cushman 1981; Oates and Barber 1987) or variable according to the solute adsorption isotherm kinetics (Van Rees et al. 1990):

$$b = \theta + \rho_b K_d \quad (26.3)$$

where

θ is soil volumetric moisture content
ρ_b soil bulk density
K_d the derivative (slope) of the adsorption isotherm as a function of C_l

This form was included as an option in the NST 3.0 model (Claassen and Steingrobe 1999; see Table 26.1).

Uptake by the whole root system, T, is obtained by dual integration with time of the flux to the root, and the integration with time along the growing root (Barber and Cushman 1981):

$$T = 2\pi r_0 L_0 \int_0^{t_m} J_r(r_0 s)ds + 2\pi r_0 \int_0^{t_m} \frac{dL}{dt} \int_0^{t_m - t} J_r(r_0 s)dsdt \qquad (26.4)$$

The first term in Equation 26.4 calculates uptake by already-existing roots, with length L_0; in the second term, the integration is of the derivative function of root growth with time, $L(t)$.

Single-root modeling was recently reviewed, and its parameters reevaluated (Roose and Schnepf 2008). Such models were applied to whole root systems assuming that they could be represented by uniform cylinders with average characteristics and a single/uniform RLD value, either constant (Barber and Cushman 1981) or changing with time as roots grow (Hoffland et al. 1990). One of the basic features of such a description is a smooth (i.e., non-branched) cylindrical root, with a uniform radius. Roose and Fowler (2004) tested the validity of these characteristics and showed, with several simplifying assumptions, that branching improves root foraging of the soil and thus should enhance the simulated uptake of barely mobile nutrients such as inorganic P (Figure 26.1). This correction should make the model more realistic, since the single-root models often underestimated P uptake by plants grown in low-P soils under field conditions (Silberbush and Barber 1984; Ernani et al. 1994; Lu and Miller 1994). These models predicted closely P and Mn uptake by plants grown in pots under controlled conditions (Barber 1995; Sadana and Claassen 2000) but systematically overestimated Mg uptake from an acid soil (Rengel 1991), which indicates that the use of real root architecture is just one part of the problem and the soil characteristics under extreme conditions should be more realistically simulated, in addition to the soil–root interface, which is far from being satisfactorily characterized, as will be discussed later.

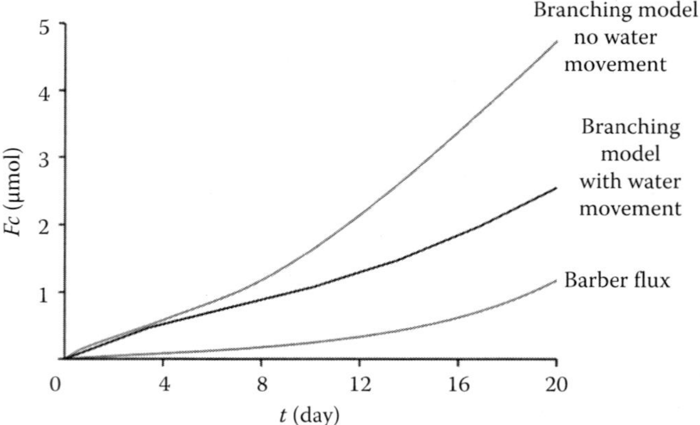

FIGURE 26.1 Phosphate uptake with time as simulated with a single-root (Barber) model and two cases of branched root systems: soil water content kept constant or water flow and depletion due to uptake by the roots. (Reprinted from *J. Theor. Biol.*, 228, Roose, T. and Fowler, A.C., A mathematical model for water and nutrient uptake by plant root systems, 173–184. Copyright 2004, with permission from Elsevier.)

Another issue is the accuracy of the single-root models in predicting uptake by woody plants and trees. Root systems of perennial plants are regarded as being in a steady state that can be represented by a constant RLD distribution in the rooting volume compared to those of annuals that develop during the growth season. A comparison between transient models (represented by NST 3.0) and two steady-state models indicated a systematic underprediction of water uptake by forest trees by the latter (Lin and Kelly 2010). The authors suspected that the steady-state models contain inherent artifacts, which leave the problem of modeling nutrient uptake by perennial plants still unresolved.

D. Demand by Plants

Scaife (1976) pointed out the problem in uptake modeling due to relating uptake merely to the concentration/availability in the growth medium: Differences in paces of root and shoot growth in different growth stages should lead to accumulation or deficiencies of various nutrients. There are two views with respect to predicting rates of ion and water uptake: One is based upon the "uptake capacity" as the controlling variable, namely, the predicted capability of each part of the root system to take up these soil resources based on its size, physiological characteristics, and ambient conditions. The other point of view focuses on the "nutrient demand" by the plant according to its size, where growth and transpiration are the leading factors that determine nutrient and water uptake. Dividing these two parameters one by the other yields a fraction that modulates the whole plant activity and attenuates it to the competence of the root system to support plant growth.

The uptake capacity of the root was quantified for K (Siddiqi and Glass 1986), nitrate-N (Hole et al. 1990), P (Xu et al. 2007), and sulfate-S (Maruyama-Nakashita 2008). This was correlated with the activation of the transporters responsible for the transport of these nutrients across the cell membrane, whose production and activity was controlled by their genes and may be quantified by the production of their mRNA by the appropriate genes (Glass 2002).

A change in the Michaelis–Menten kinetic parameters (increase in J_{max} and decrease in K_m) in plants that were previously deprived of these nutrients was shown for P by Drew et al. (1984) and for K by Siddiqi and Glass (1986). Steingrobe and Schenk (1994) reported an increase in J_{max} of nitrate with the relative growth rate (RGR) of lettuce. A list of high-affinity transporters of several nutrients and plant species is provided in Table 26.2. Deficiency of N would induce high-affinity nitrate transporter systems (HATS) in all parts of the plant, including roots; low-affinity transporters (LATS), on the other hand, will be induced only locally, as in the case of ammonium transporters (Epstein and Bloom 2005; Ruffel et al. 2008). The classical study of Drew and Saker (1975) showed an enhanced root growth in nitrate-rich patches. Recent studies (Little et al. 2005; Remans et al. 2006) found a link between enhanced lateral initiation and a specific nitrate transporter (*NRT2.1*) gene activity,

TABLE 26.2 List of High-Affinity Transporters of Nutrients Identified in Roots of Different Plant Genotypes

Nutrient/Ion	K_m	Transporter	Species	References[b]	Remarks
NO_3^-N	10–25 μM	AtNRT2	*Arabidopsis*	1	
	18.6 μM		Corn	6	
NH_4^+-N	~200 μM	AtAMT1	*Arabidopsis*	11	100–600 μM, depends on membrane voltage
Urea-N	<4 μM	AtDUR3	*Arabidopsis*	7	
K^+	15–24 μM	AtHAK5	*Arabidopsis*	5	
	6–10 μM	HvHAK1	Barley	4	
	196 μM	AtCHX13	*Arabidopsis*	16	
	6.2 μM	CaHAK1	Pepper	8	Inhibited by monovalent cations
o-phosphate-P	31 μM	LePT1	Tomato	3	
	3.1 μM	PHT1	*Arabidopsis*	10	
	0.225 μM	H+/Pi co-transporter	*Hakea sericea*	12	Proteoid roots, dual co-transporter
	40.8 μM				
	10 μM	na[a]	Corn	13	Up to 100 μM Pi
SO_4^{2-}S	na	SULTR1	*Arabidopsis*	9,15	
Fe^{2+}	54–93 nM	RIT1	Pea	2	
	4.7 μM	na	Barley	14	Interaction with Zn^{2+}

[a] na, not available.

[b] [1]Cerezo et al. (2001); [2]Cohen et al. (2004); [3]Daram et al. (1998); [4]Fulgenzi et al. (2008); [5]Gierth et al. (2005); [6]Hole et al. (1990); [7]Kojima et al. (2007); [8]Martinez-Cordero et al. (2005); [9]Rouached et al..(2008); [10]Mitsukawa et al. (1997); [11]Neuhäuser et al. (2007); [12]Sousa et al. (2007); [13]Xu et al. (2007); [14]Zaharieva and Römheld (2000); [15]Maruyama-Nakashita (2008); [16]Zhao et al. (2008).

which indicates a possible interrelationship between the signaling of nitrate availability and lateral root initiation, with possible involvement of auxin (Krouk et al. 2010).

There are several indications that different components of the root system possess different transport systems (Eshel and Waisel 1996; Waisel and Eshel 2002). Recently, different transporters were found in different tissues and root classes in rice, *Oryza sativa* (Ai et al. 2009). While the *OsPT2* seems to be a low-affinity transporter (K_m above 50 μM) for P translocation throughout the plant, the *OsPT6* is a high-affinity transporter (K_m 3–10 μM) located in roots but also in other tissues, although it seems to be the transporting system responsible for uptake from low external concentrations. Furthermore, while *OsPT2* is localized in the stele, *OsPT6* was expressed in both epidermal and cortical cells of young primary and lateral roots. Both transporters appeared to become activated by P deprivation.

Nitrate transporters are also active in different parts of the plant (Epstein and Bloom 2005). Up to 10 nitrate transporters were identified in different plants and organs. The high-affinity system *NRT2* is also activated by N deprivation. Recently, we measured different K_m values for nitrate influx to different root classes of young corn plants, which indicated that different transporters dominate uptake in different root classes (Silberbush and Lynch, unpublished; Postma et al. 2010). Nitrogen deprivation induces also interactions between nitrate and other nutrients and of water uptake (Cramer et al. 2009). The effect on water influx seems to be indirect and influenced by diverse factors such as aquaporin activity in roots and phytohormone regulation of leaf stomatal conductance, which affects transpiration.

III. Processes at the Root–Soil Interface

The term "rhizosphere" is used for that section of the soil that is affected by the root, to distinguish it from the bulk soil. Its characteristics are different from those of the bulk soil (Marschner 1998; Hinsinger et al. 2009). Root exudation of organic substances, pH changes due to anion–cation balance and H^+/OH^- secretion pumps (Marschner and Römheld 1996; Gerendás and Ratcliffe 2002), and depletion or accumulation of solutes due to uptake or efflux are the main causes of such differences. Also, the young root surfaces are accompanied by the content of cells, originated in the root cap and border cells, with the typical saprophytic and supportive microflora that feed on it (Sievers et al. 2002; Driouich et al. 2007). The root–soil interface is that part of the rhizosphere that is attached to the root surface, where the transition from fluxes of water and solutes in the soil to physiological mechanisms of uptake by the root takes place. The traditional single-root models use transport parameters measured in the bulk soil to quantify flows toward the root and uptake parameters separately measured in a nutrient solution. The mathematical treatment of the soil–root boundary is one of the crucial parts of the radial-flow models (Barber and Cushman 1981).

Root hairs, which are extensions of the root epidermis cells and mycorrhizal hyphae, increase the soil–root contact area but also change the geometry of the active surface (Geelhoed et al. 1997b) and consequently the conditions at the soil–root interface (Schilling et al. 1998). The uptake models failed so far to account for such changes at the soil–root interface governed by the living root and its associated microflora (Cushman 1980).

Exudation of organic compounds by roots, phytosiderophores, enables gramineous plants to enhance and control their supply of micronutrients metals like iron (Fe) and zinc (Zn), better than broadleaf plants (Gerendás and Ratcliffe 2002; Murata et al. 2008). Active roots of terrestrial plants invest matter and bioenergy in ameliorating their growth environment (McCully 1999; Pinton et al. 2007). The inadequate handling of this part of the system is one of the major sources of errors in existing uptake models. It has been difficult to mimic the living system accurately by mechanistic models. Whether this is an inherent trait or just a matter of a proper modification of these models is rather a philosophical question. Current plant nutrient uptake models do not provide sufficient solutions to this problem.

IV. Interactions with the Rhizosphere Biota

Another open issue is the microbiological activity at the soil–root interface (Pinton et al. 2007). Most studies on these topics were carried out in the lab and only a few in vivo, even there are some studies on trees grown in the field (Grayston et al. 1997). Bacterial abundance at the soil–root interface is enhanced by root exudation that, in turn, is affected the plant N nutrition (Liljeroth et al. 1990). Rhizosphere bacteria affect plant P nutrition by solubilization of barely soluble P forms (Wenzel et al. 1994). Bacterial siderophores improved iron uptake by cucumber and maize plants (Walter et al. 1994). The mutualism between roots and mycorrhizal fungi was studied extensively by several disciplines. Arbuscular mycorrhiza (AM), in particular, are most beneficial in improving uptake of hardly accessible soil nutrients as P (Bell et al. 1989; Jakobsen et al. 2005; Bucher 2007), macroelements such as calcium (Azcon and Barea 1992) and iron (Treeby 1992), and microelements like manganese (Kothari et al. 1991) and zinc (Bell et al. 1989). Early modeling attributed the contribution of AM to enhance P uptake by roots to their improved soil foraging, resulting from their large surface contact per unit of mass (Barber 1995; Schnepf and Roose 2006). Recent studies revealed that mycorrhiza may contribute to plant N nutrition by specific absorption of amino acids originated from decomposed organic forms (Nasholm et al. 2009). The ability of roots to release phosphatases and organic anions to enhance P uptake is also been affected by rhizosphere bacteria and mycorrhizae (Richardson et al. 2009). The inclusion of microorganism activities in modeling of the soil–root interface is still lagging behind the needs, although the importance of these interrelationships between plant roots and microorganisms is greater than previously evaluated (Grayston et al. 1997). Exudation by roots, which affects rhizosphere bacterial community and its activities, is directly linked to plant nutrient availability, both of trees and annual plants. The long time since the solitary attempt of Cushman (1980) to put these processes in exact terms indicates the gap between the needs and the practice of their inclusion in root modeling.

V. Modeling Root System Structure and Function

As computers and programming techniques developed, they have enabled more detailed modeling of root system structures, as a basis to functional models. These "root architecture" (also named "structural–functional") models are based on a detailed description of root system growth and development with time in a 3D space (Dupuy et al. 2010). Root systems are described as hierarchal arrays of different classes of roots, where the younger class would develop by branching from the former one. A comprehensive review on the structure and description of root systems was provided by Fitter (2002). Root systems vary in branching frequency, relative orientation of their different components (i.e., branching angle and direction), thickness, and primary and secondary vascular growth, just to mention a few. The basic root structure is governed by the plant genotype. Yet, the actual appearance of the root system is highly affected by the local growth conditions in the soil, like compaction and moisture, which determine the soil mechanical impedance and the consequent root penetrability, and the overall effect of gravitropism that makes growing root tips to incline downward. Also, changes in P (Camacho-Cristobal et al. 2008) and N form (Drew and Saker 1975; Liu et al. 2008) abundance can affect root branching by enhancing lateral root initiation. The mechanistic models that were developed to quantify root architectural characteristics rely on the genotypic expression of the plant (Lynch et al. 1997; Johnson et al. 2000), and changes in soil impedance may also been accounted for (Clausnitzer and Hopmans 1994). The actual description of a root system is much more complex and difficult to predict (Diggle 1988; Pagès et al. 2004). Still, species and plant life forms do follow inherited patterns.

The simulated root system in such models usually starts from the seedling tap root that elongates gravitropically. Laterals emerge at a certain frequency along the tap root at some spherical angles, and their growth direction is also affected by gravitropism. The actual location and direction of each root is stochastically determined from statistics of these parameters for the considered plant genotype. This way it is possible to simulate the effect of various gravitropic responses or other root characteristics (Pagès et al. 2004). The primary laterals give off secondary ones according to the same principles, each root category having its own characteristics in terms of radius, tapering angle, elongation rate, and specific density. Again, these model parameters allow for modeling of various genotypes and for studying projected effects of future genetic changes.

Another example is the model of Schnepf and Leitner (2009), who used a finite-element method (FEM) to describe a 3D distribution and growth of the root system with time. The same approach was used by Somma et al. (1998), who used a 3D FEM soil model, and added to it root distribution and a simplified sink term for uptake of water and nutrients (see the reference for details).

Physiological activity of the root system as a whole is the sum of the processes that take place in each of its members. Each root is assigned a certain ion and water uptake capability and respiration costs. Integrating the activity of all the roots in the model yields estimates of gains and losses associated with the whole system. Commonly, the soil is considered to be uniform in terms of its effect on root growth and ion and water content. However, in certain cases, the rooting volume can be divided into horizons, each with its own physical and chemical characteristics (e.g., the LEACHM model; see Table 26.1). Plant root systems, on the other hand, seem to be adapted to soil layering by altering the proportions between deep and lateral growth, like genotypic adaptation to phosphorus abundance in the upper soil (Lynch and Brown 2001; Liao et al. 2004; Beebe et al. 2006).

SimRoot is an example of a structural–functional plant model (SFPM) that has been useful in understanding the function of root traits for soil resource acquisition (Lynch et al. 1997; Postma and Lynch 2011). *SimRoot* uses empirical data for the number, orientation, elongation, and branching of various root classes to accurately simulate the growth and development of root systems (Figure 26.2). *SimRoot* has been parameterized for maize (*Zea mays* L.), common bean (*Phaseolus vulgaris* L.), squash (*Cucurbita* spp.), and white lupin (*Lupinus albus* L.) and is currently able to simulate acquisition of phosphorus, nitrogen, potassium, and water, incorporating growth and photosynthetic responses to resource limitation. Spatial variability of N and P uptake kinetics in the root system is explicitly modeled (Figure 26.2; Rubio et al. 2004). Both the Barber–Cushman model (Barber 1995) and Simunek's SWMS (Soil–Water Movement Simulating, Somma et al. 1998) can be used for simulating nutrient transport in the soil and acquisition by roots. Results from *SimRoot* agree well with results from empirical studies (Figure 26.3).

The purpose of *SimRoot* is to assist in the identification and evaluation of root traits for potential use in crop breeding programs to enhance soil resource acquisition. It has been useful in identifying several fractal properties of roots as a means to estimate root architecture in soil from limited sampling, including 1D and 2D fractal measures of 3D root architecture (Nielsen et al. 1997) and aggregate measures such as the combination of fractal abundance, fractal dimension, and lacunarity (Walk et al. 2004), and showed how these metrics might be related to the efficiency of soil exploration (Walk et al. 2004). *SimRoot* was useful in quantifying the effects of genotypic variation in basal root growth angle on phosphorus acquisition, considering spatial competition within and between root systems (Ge et al. 2000; Rubio et al. 2001). In these two applications, *SimRoot* was useful in estimating metrics that were difficult (fractal geometry in multiple dimensions) or infeasible (overlap of phosphorus depletion zones among neighboring roots) to measure directly.

Another utility of SFPMs is to employ their carbon and resource accounting functions to assess potential tradeoffs resulting from plant investment in specific root traits. An example of this application is the use of *SimRoot* to show that the utility of genotypic variation for adventitious rooting is dependent on the respiration rates of adventitious roots compared with those for laterals arising from basal roots (Walk et al. 2006). In some scenarios, increased adventitious rooting may be counterproductive for soil exploration, because of reduced lateral branching on basal root axes. Another example of the use of the resource accounting function of *SimRoot* is the recent demonstration that genotypic variation in the formation of root cortical aerenchyma (RCA) may substantially enhance phosphorus acquisition by reducing

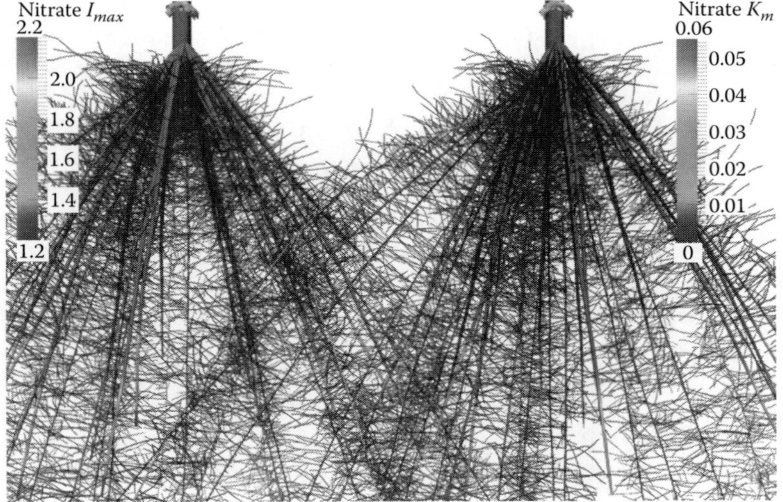

FIGURE 26.2 **(See color insert.)** Spatial maps of Michaelis–Menton kinetic parameters (I_{max} [μmol · cm^{-2} d^{-1}], K_m [μM]) for nitrate in maize root systems 40 days after germination. (From M. Silberbush, J. Postma, and J.P. Lynch, unpublished data; Postma, J. and Lynch, J., *Ann. Bot.*, 107, 829, 2011. With permission.)

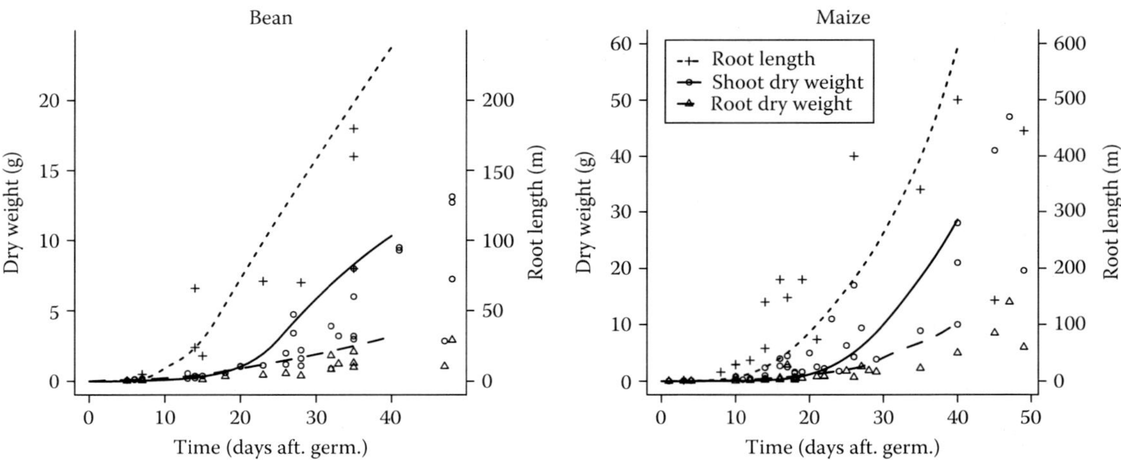

FIGURE 26.3 Verification of simulated data from *SimRoot* (lines). Data points were taken from 30 different publications, which were independent from the papers used to parametrize the model. (From Postma, J. and Lynch, J., *Ann. Bot.*, 107, 829, 2011. With permission.)

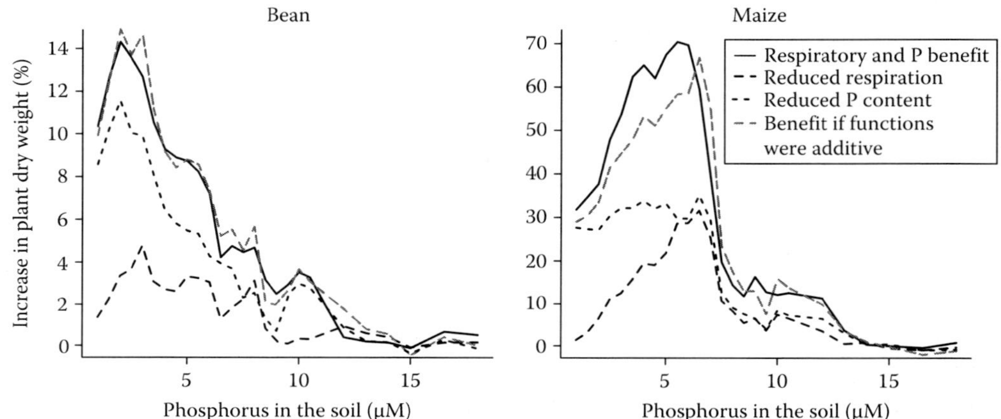

FIGURE 26.4 Relative benefit of root cortical aerenchyma for plant growth as a function of soil phosphorus bioavailability. (From Postma, J. and Lynch, J., *Ann. Bot.*, 107, 829, 2011.)

the metabolic costs of soil exploration (Postma and Lynch 2011). In this case, *SimRoot* was able to quantify the benefits of RCA for reduced maintenance respiration and reduced tissue phosphorus content independently (Figure 26.4).

A third type of application of SFPMs is in evaluating phenotypes that do not exist in nature. An example of this application is the use of *SimRoot* to identify morphological synergism among root hair length, density, and configuration for phosphorus acquisition (Ma et al. 2001). These traits are coordinated in natural phenotypes, but in *SimRoot* they could be independently varied to reveal their interactions.

A fourth type of application for SPFMs is the analysis of a large number of phenotypes *in silico* so that empirical studies can be focused on the most promising cases. At least 20 distinct root architectural and anatomical traits under genetic control are potential targets for crop breeding for enhanced nutrient acquisition. These traits may interact with each other to create functional synergism, as in the case of root hair traits for phosphorus acquisition, or may have negative interactions. In the simplest case, if each of these 20 traits existed in two possible states (e.g., shallow vs. deep growth angles, or long vs. short root hairs), we would have 2^{20} possible integrated phenotypes. This number is too large to evaluate in empirical studies, even if genotypes possessing all possible trait combinations could be obtained, but this number could readily be evaluated in a SFPM such as *SimRoot*. A related application is the evaluation of given phenotypes in multiple environments. For example, *SimRoot* predicts that RCA will increase the efficiency of nitrogen acquisition in coarse-textured soils, which experience greater leaching than finer soils (Figure 26.5). Similarly, RCA is more useful in regions with greater precipitation and leaching (data not shown).

Finally, a fifth type of application of SPFMs for understanding nutrient acquisition is in scaling processes in time and space and across levels of biological organization. For example, root proliferation in responses to nutrient patches may assist local nutrient acquisition but could actually decrease nutrient acquisition by the whole plant by diverting resources from soil exploration and therefore the acquisition of water or nutrients from

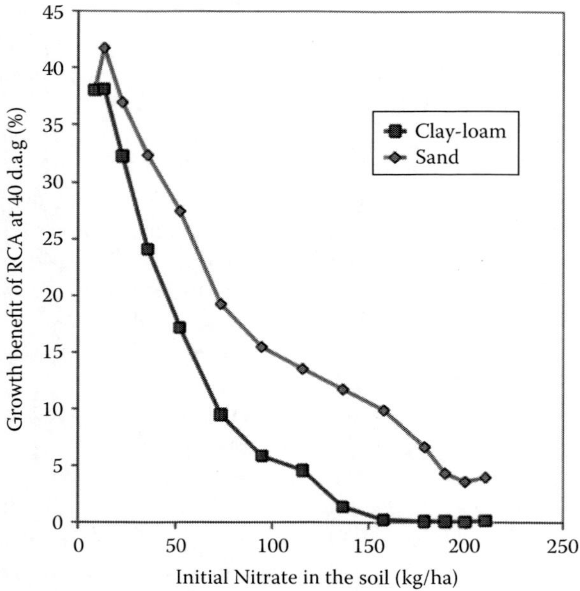

FIGURE 26.5 *SimRoot* predicts that root cortical aerenchyma could improve maize growth in low N soils by about 40%, and in coarse-textured soils could improve growth at high N. (From J. Postma and J.P. Lynch, unpublished data.)

other soil domains. At larger scales, a SPFM could be linked with regional climate models to predict nutrient uptake in future climates or across large spatial areas.

As these examples illustrate, modeling is not a replacement for empirical research but can substantially improve the speed, precision, and cost of empirical research, by focusing it in domains that are most likely to be important for the function of interest.

VI. Root System Modeling: The Next Generation

Evaluation of the predictive ability of 3D architecture root models has been hindered for a long time by the lack of appropriate data. In recent years, measurements of intact root system architecture and real growth patterns have achieved meaningful progress. A variety of methods have been developed for capturing the 3D structure of plant root systems. Excavated root systems of woody roots were measured and reconstructed quantitatively by a number of techniques reviewed in Tobin et al. (2007) and Danjon and Reubens (2008). A variety of noninvasive methods were developed in recent years for 3D analysis of root system structure and activity. Rice and soybean plants grown in a phytagel-based transparent medium scanned by a 3D laser scanner (Fang et al. 2009). X-ray microtomography was performed on two alder roots, grown for 4 months in a natural moraine soil, and, in order to obtain the data in the desired resolution, the roots had to be transferred into a Plexiglas receptacles filled with quartz sand for analysis (Kaestner et al. 2006). More recently, Hargreaves et al. (2009) used the same technique to study barley root systems grown in soil-filled sacs, and Tracy et al. (2010) used it to measure root systems of soil-grown wheat and maize and even *Arabidopsis*. Dhondt et al. (2010) have shown that new developments in high-resolution x-ray computed tomography make this technique amenable for studies of *Arabidopsis* root systems.

In attempts to combine measurement of root system architecture and function, Amato et al. (2009) used electrodes to measure soil resistivity as an indicator of soil moisture content and root biomass distribution in the rooting volume of alfalfa. A combination of magnetic resonance imaging (MRI) and positron emission (PET) of ^{11}C was used by Jahnke et al. (2009) to visualize the roots and to monitor photoassimilate transport through maize root system and sugar beet and radish storage roots.

As a result, the so-called hidden half of plants has become more and more "visible." Approaches developed for modeling single roots, which oversimplified the soil–root system, may be improved and updated. There are means to provide more realistic root architecture, including subunits like root hairs and mycorrhizal hyphae (Itoh and Barber 1983; Wulfsohn and Nyengaard 1999). Furthermore, the link between these traits and the activation of their controlling genes were also presented (Liao et al. 2004; Armengaud et al. 2009). Giving those traits, quantitative measures enable their inclusion in mechanistic models, as already shown by several models (Dunbabin et al. 2003; Liao et al. 2004; de Dorlodot et al. 2007). Also, these models may be coupled with advanced uptake models that deal with water and solute flows to gain the combined advantages (Somma et al. 1998).

Developments in root biology open new opportunities for uptake modeling. The more realistic, quantitative evaluation of the root system and the activity of its different components in uptake, together with the more accurate description of the soil, open new possibilities (Roose and Schnepf 2008). On the other hand, we know now more about the absorption systems in the plant: from speculations about the "true" model to quantify influx to roots, when the transport system was unknown (Jensén et al. 1987), the recent molecular studies on the structure of transporters open new options in studying uptake mechanisms by roots. Combining the two approaches is the natural development in uptake modeling. Yet, the gain from such an approach is not guaranteed. Sensitivity analyses of the existing uptake models indicated that, for the uptake of most nutrients, the limitation lays in the transport of ions in the rhizosphere toward the root, rather than the uptake by the root per se (Barber 1995). This is definitely true for less-mobile nutrients like phosphate and metallic microelements. The only major nutrient whose uptake modeling may be improved is nitrate: being a monovalent anion, nitrate does not interact with the soil matrix and moves via the soil solution mainly by mass flow (Barber 1995).

Root system modeling has also been used to simulate phenomena related to other root system functions such as tree anchorage, which depends on the physical contact between root surface and the adjacent soil volume (e.g., Dupuy et al. 2007). The effects of plant root systems on the stability of slopes are another area investigated through simulation (Van Beek et al. 2005; Danjon et al. 2008; Schwarz et al. 2010). We are still far from knowing how to quantify root effects on the surrounding soil. The interactions between the plant roots and the soil seem to be the major research effort in the years to come, both by plant biologists and soil scientists.

References

Acock B, Pachepsky YA 1996 Convective-diffusion model of two-dimensional root growth and proliferation. *Plant Soil* 180:231–240.

Ai PH, Sun SB, Zhao JN et al. 2009 Two rice phosphate transporters, OsPht1;2 and OsPht1;6, have different functions and kinetic properties in uptake and translocation. *Plant J* 57:798–809.

Amato M, Bitella G, Rossi R et al. 2009 Multi-electrode 3D resistivity imaging of alfalfa root zone. *Eur J Agron* 31:213–222.

Armengaud P, Zambaux K, Hills A et al. 2009 EZ-Rhizo: Integrated software for the fast and accurate measurement of root system architecture. *Plant J* 57:945–956.

Azcon R, Barea JM 1992 The effect of vesicular-arbuscular mycorrhizae in increasing Ca acquisition by alfalfa plants in calcareous soils. *Biol Fert Soils* 13:155–159.

Baker DN, Lambert JR, McKinion JM 1983 GOSSYM: A simulator of cotton crop growth and yield. *South Carolina Agric Expt Sta Tech Bull* #1089. Clemson, SC: Clemson University.

Baldwin JP, Nye PH, Tinker PB 1973. Uptake of solutes by multiple root systems from soil. III. A model for calculating the solute uptake by a randomly dispersed root system developing a finite volume of soil. *Plant Soil* 63:621–635.

Barber SA 1995 *Soil Nutrient Bioavailability: A Mechanistic Approach*, 2nd edn. New York: Wiley.

Barber SA, Cushman JH 1981 Nitrogen uptake model for agronomic crops. In *Modeling Wastewater Renovation Land Treatment*, ed. IK Iskandar, pp. 382–409. New York: Wiley.

Bar-Yosef B, Fishman S, Talpaz HA 1980 Model of zinc movement to single roots in soils. *Soil Sci Soc Am J* 44:1272–1279.

Beebe SE, Rojas-Pierce M, Yan XL et al. 2006 Quantitative trait loci for root architecture traits correlated with phosphorus acquisition in common bean. *Crop Sci* 46:413–423.

Behrendt H, Bruggemann R, Morgenstern M 1995 Numerical and analytical model of pesticide root uptake model comparison and sensitivity. *Chemosphere* 30:1905–1920.

Bell MJ, Middleton KJ, Thompson JP 1989 Effects of vesicular-arbuscular mycorrhizae on growth and phosphorus and zinc nutrition of peanut (*Arachis hypogaea* L.) in an Oxisol from subtropical Australia. *Plant Soil* 117:49–57.

Benjamin JG, Ahuja LR, Allmaras RR 1996 Modeling corn rooting patterns and their effects on water uptake and nitrate leaching. *Plant Soil* 179:223–232.

Berntson GM 1996 Fractal geometry, scaling, and the description of plant root architecture. In *Plant Roots: The Hidden Half*, eds. Y Waisel, A Eshel, U Kafkafi, 2nd edn., pp. 259–272. New York: Marcel Dekker, Inc.

Bhat KKS, Nye PH, Baldwin JP 1976 Diffusion of phosphate to plant roots in soil. IV. The concentration distance profile in the rhizosphere of roots with root hairs in a low-P soil. *Plant Soil* 40:309–319.

Bouldin DR 1989 A multiple ion uptake model. *J Soil Sci* 40:309–319.

Bucher M 2007 Functional biology of plant phosphate uptake at root and mycorrhiza interfaces. *New Phytol* 173:11–26.

Camacho-Cristobal JJ, Rexach J, Conejero G, Al-Ghazi Y, Nacry P, Doumas P 2008 PRD, an *Arabidopsis* AINTEGUMENTA-like gene, is involved in root architectural changes in response to phosphate starvation. *Planta* 228:511–522.

Cerezo M, Tillard P, Filleur S, Munos S, Daniel-Vedele F, Gojon A 2001 Major alterations of the regulation of root NO_3^- uptake are associated with the mutation of Nrt2.1 and Nrt2.2 genes in *Arabidopsis*. *Plant Physiol* 127:262–271.

Claassen N, Barber SA 1976. Simulation model for nutrient uptake from soil by a growing plant root system. *Agron J* 68:961–964.

Claassen N 1990 Die Aufnahme von Nährstoffen aus dem boden durch die höhere Pflanze als ergebins von Verfügbarkeit und Aneignungsvermögen. Göttingen, Germany: Severin.

Claassen N, Steingrobe B 1999 Mechanistic simulation models for a better understanding of nutrient uptake from soil. In *Mineral Nutrition of Crops: Fundamental Mechanisms and Implications*, ed. Z Rengel, pp. 327–367. New York: Haworth Press.

Clausnitzer V, Hopmans JW 1994 Simultaneous modeling of transient three-dimensional root growth and soil water flow. *Plant Soil* 164:299–314.

Cohen CK, Garvin DF, Kochian LV 2004 Kinetic properties of a micronutrient transporter from *Pisum sativum* indicate a primary function in Fe uptake from the soil. *Planta* 218:784–792.

Cramer MD, Hawkins HJ, Verboom GA 2009 The importance of nutritional regulation of plant water flux. *Oecologia* 161:15–24.

Cushman JH 1979 An analytical solution to solute transport near root surfaces for low initial concentration. I. Equation development. *Soil Sci Soc Am J* 43:1087–1090.

Cushman JH 1980 Analytical study of the effect of ion depletion (replenishment) caused by microbial activity near toots. *Soil Sci* 129:69–87.

Danjon F, Barker DH, Drexhage M, Stokes A 2008 Using three dimensional plant root architecture in models of shallow-slope stability. *Ann Bot* 101:1281–1293.

Danjon F, Reubens B 2008 Assessing and analyzing 3D architecture of woody root systems, a review of methods and applications in tree and soil stability, resource acquisition and allocation. *Plant Soil* 303:1–34.

Daram P, Brunner S, Persson BL, Amrhein N, Bucher M 1998 Functional analysis and cell-specific expression of a phosphate transporter from tomato. *Planta* 206:225–233.

de Dorlodot S, Forster B, Pagès L, Price A, Tuberosa R, Draye X 2007 Root system architecture: Opportunities and constraints for genetic improvement of crops. *Trends Plant Sci* 12:474–481.

De Willigen P, Van Noordwijk M 1994a Mass flow and diffusion of nutrients to a root with constant or zero-sink uptake. I. Constant uptake. *Soil Sci* 157:162–170.

De Willigen P, Van Noordwijk M 1994b Mass flow and diffusion of nutrients to a root with constant or zero-sink uptake. II. Zero-sink uptake. *Soil Sci* 157:171–175.

Delve RJ, Probert ME, Cobo JG et al. 2009 Simulating phosphorus responses in annual crops using APSIM: Model evaluation on contrasting soil types. *Nutr Cycling Agroecosyst* 84:293–306.

Dhondt S, Vanhaeren H, Van Loo D, Cnudde V, Inzé D 2010 Plant structure visualization by high-resolution X-ray computed tomography. *Trends Plant Sci* 15:419–422.

Diggle AJ 1988 ROOTMAP—A model in 3-dimensional coordinates of the growth and structure of fibrous root systems. *Plant Soil* 105:169–178.

Drew MC, Saker LR 1975 Nutrient supply and growth of seminal root system in barley. 2. Localized, compensatory increases in lateral root growth and rates of nitrate uptake when nitrate supply is restricted to only part of root system. *J Exp Bot* 26:79–90.

Drew MC, Saker LR, Barber SA, Jenkins W 1984 Changes in the kinetics of phosphate and potassium absorption in nutrient-deficient barley roots measured by a solution-depletion technique. *Planta* 160:490–499.

Driouich A, Durand C, Vicre-Gibouin M 2007 Formation and separation of root border cells. *Trends Plant Sci* 12:14–19.

Dunbabin V, Diggle A, Rengel Z 2003 Is there an optimal root architecture for nitrate capture in leaching environments? *Plant Cell Environ* 26:835–844.

Dupuy LX, Fourcaud T, Lac P, Stokes A 2007 A generic 3D finite element model of tree anchorage integrating soil mechanics and real root system architecture. *Am J Bot* 94:1506–1514.

Dupuy L, Gregory PJ, Bengough AG 2010 Root growth models: Towards a new generation of continuous approaches. *J Exp Bot* 61:2131–2143.

Epstein E, Bloom AJ 2005 *Mineral Nutrition of Plants: Principles and Perspectives*, 2nd edn. Sunderland, MA: Sinauer.

Ernani PR, Santos JCP, Kaminski J, Rheinheimer DS 1994 Prediction of phosphorus uptake by a mechanistic model in a low phosphorus highly weathered soil. *J Plant Nutr* 17:1067–1078.

Eshel A, Waisel Y 1996 Multiform and multifunction of various constituents of one root system. In *Plant Roots: The Hidden Half*, eds. Y Waisel, A Eshel, U Kafkafi, 2nd edn., pp. 175–192. New York: Marcel Dekker, Inc.

Fang S, Yan X, Liao H 2009 3D reconstruction and dynamic modeling of root architecture in situ and its application to crop phosphorus research. *Plant J* 60:1096–1108.

Fitter A 1991 Characteristics and functions of root systems. In *Plant Roots: The Hidden Half*, eds. Y Waisel, A Eshel, U Kafkafi, pp. 3–25. New York: Marcel Dekker, Inc.

Fitter A 1996 Characteristics and functions of root systems. In *Plant Roots: The Hidden Half*, eds. Y Waisel, A Eshel, U Kafkafi, 2nd edn., pp. 1–20. New York: Marcel Dekker, Inc.

Fitter A 2002 Characteristics and functions of root systems. In *Plant Roots: The Hidden Half*, eds. Y Waisel, A Eshel, U Kafkafi, 3rd edn., pp. 15–32. New York: Marcel Dekker, Inc.

Fulgenzi FR, Peralta ML, Mangano S et al. 2008 The ionic environment controls the contribution of the barley HvHAK1 transporter to potassium acquisition. *Plant Physiol* 147:252–262.

Gardner WR 1991 Modeling water uptake by roots. *Irrig Sci* 12:109–114.

Ge ZY, Rubio G, Lynch JP 2000 The importance of root gravitropism for inter-root competition and phosphorus acquisition efficiency: Results from a geometric simulation model. *Plant Soil* 218:159–171.

Geelhoed JS, Findenegg GR, Van Riemsdijk WH 1997a Availability to plants of phosphate adsorbed on goethite: Experiment and simulation. *Eur J Soil Sci* 48:473–481.

Geelhoed JS, Mous SLJ, Findenegg GR 1997b Modeling zero sink nutrient uptake by roots with root hairs from soil: Comparison of two models. *Soil Sci* 162:544–553.

Geelhoed JS, Van Riemsdijk WH, Findenegg GR 1999 Simulation of the effect of citrate exudation from roots on the plant availability of phosphate adsorbed on goethite. *Eur J Soil Sci* 50:379–390.

Gerendás J, Ratcliffe RG 2002 Root pH regulation. In *Plant Roots: The Hidden Half*, eds. Y Waisel, A Eshel, U Kafkafi, 3rd edn., pp. 553–570. New York: Marcel Dekker, Inc.

Gierth M, Maser P, Schroeder JI 2005 The potassium transporter AtHAK5 functions in K^+ deprivation-induced high-affinity K^+ uptake and AKT1 K^+ channel contribution to K^+ uptake kinetics in *Arabidopsis* roots. *Plant Physiol* 137:1105–1114.

Glass ADM 2002 Nutrient absorption by plant roots: Regulation of uptake to match plant demand. In *Plant Root: The Hidden Half*, eds. Y Waisel, A Eshel, U Kafkafi, 3rd edn., pp. 571–586. New York: Marcel Dekker, Inc.

Grayston SJ, Vaughan D, Jones D 1997 Rhizosphere carbon flow in trees, in comparison with annual plants: The importance of root exudation and its impact on microbial activity and nutrient availability. *Appl Soil Ecol* 5:29–56.

Hagin J, Amberger A, Kruh G, Segall E 1976. Outlines of a computer simulation model on residual and added nitrogen changes and transport in soils. *J Plant Nutr Soil Sci* 139:443–455.

Hargreaves CE, Gregory PJ, Bengough AG 2009 Measuring root traits in barley (*Hordeum vulgare* ssp. *vulgare* and ssp. *spontaneum*) seedlings using gel chambers, soil sacs and X-ray microtomography. *Plant Soil* 316:285–297.

Helyar KR, Munns DN 1975 Phosphate fluxes in the soil-plant system: A computer simulation. *Hilgardia* 43:103–130.

Hinsinger P, Bengough AG, Vetterlein D, Young IM 2009. Rhizosphere: Biophysics, biogeochemistry and ecological relevance. *Plant Soil* 321:117–152.

Hoffland E, Bloemhof HS, Leffelaar PA, Findenegg GR, Nelemans JA 1990 Simulation of nutrient uptake by a growing root system considering increasing root density and inter-root competition. In *Plant Nutrition—Physiology and Application*, ed. ML Van Beusichem, pp. 9–15. Amsterdam, the Netherlands: Kluwer.

Hole DJ, Emran AM, Fares Y, Drew MC 1990 Induction of nitrate transport in maize roots, and kinetics of influx, measured with nitrogen-13. *Plant Physiol* 93:642–647.

Hutson JL, Wagenet RJ 1995 An overview of LEACHM: A process based model of water and solute movement, transformations, plant uptake and chemical reactions in the unsaturated zone. In *Chemical Equilibrium and Reaction Models*, eds. RH Loepert, AP Schwab, S Goldberg. ASA Spec Publ No 42, pp. 97–112. Madison, WI: American Society of Agronomy.

Ibrikci H, Ulger AC, Cakir B, Buyuk G, Guzel N 1998. Modeling approach to nitrogen uptake by field-grown corn. *J Plant Nutr* 21:1943–1954.

Itoh S, Barber SA 1983 A numerical solution of whole plant nutrient uptake for soil-root systems with root hairs. *Plant Soil* 70:403–413.

Jahnke S, Menze MI, van Dusschoten D et al. 2009 Combined MRI–PET dissects dynamic changes in plant structures and functions. *Plant J* 59:634–644.

Jakobsen I, Chen BD, Munkvold L, Lundsgaard T, Zhu YG 2005 Contrasting phosphate acquisition of mycorrhizal fungi with that of root hairs using the root hairless barley mutant. *Plant Cell Environ* 28:928–938.

Jemison JM, Jabro JD, Fox RH 1994a Evaluation of LEACHM. I. Simulation of drainage, bromide leaching, and corn bromide uptake. *Agron J* 86:843–851.

Jemison JM, Jabro JD, Fox RH 1994b Evaluation of LEACHM. II. Simulation of nitrate leaching from nitrogen-fertilized and manured corn. *Agron J* 86:852–859.

Jensén P, Erdei L, Møller IM 1987 K^+ uptake in plant roots: Experimental approach and influx models. *Physiol Plant* 70:743–748.

Johnson WC, Jackson LE, Ochoa O et al. 2000 Lettuce, a shallow-rooted crop, and *Lactuca serriola*, its wild progenitor, differ at QTL determining root architecture and deep soil water exploitation. *Theor Appl Genet* 101:1066–1073.

Kaestner A, Schneebeli M, Graf F 2006 Visualizing three-dimensional root networks using computed tomography. *Geoderma* 136:459–469.

Kojima S, Bohner A, Gassert B, Yuan LX, von Wiren N 2007 AtDUR3 represents the major transporter for high-affinity urea transport across the plasma membrane of nitrogen-deficient *Arabidopsis* roots. *Plant J* 52:30–40.

Kothari SK, Marschner H, Römheld V 1991 Effect of vesicular-arbuscular mycorrhizal fungus and rhizosphere micro-organisms on manganese reduction in the rhizosphere and manganese concentration in maize (*Zea mays* L.). *New Phytol* 117:649–655.

Krouk G, Lacombe B, Bielach A et al. 2010. Nitrate-regulated auxin transport by NRT1.1 defines a mechanism for nutrient sensing in plants. *Dev Cell* 18:927–937.

Lambert JR, Penning de Vries FWT 1973 Dynamics of water in the soil-plant-atmosphere system: A model named Troika. In *Physical Aspects of Soil Water and Salts in Ecosystems*, eds. A Hadas, D Swartzendruber, PE Rijtema, M Fuchs, B Yaron, pp. 257–273. Ecological Studies. Analysis and Synthesis, Vol. 4. Berlin, Germany: Springer.

Leadley PW, Reynolds JF, Chapin FS 1997. A model of nitrogen uptake by *Eriophorum vaginatum* roots in the field: Ecological implications. *Ecol Monogr* 67:1–22.

Liao H, Yan XL, Rubio G, Beebe SE, Blair MW, Lynch JP 2004 Genetic mapping of basal root gravitropism and phosphorus acquisition efficiency in common bean. *Funct Plant Biol* 31:959–970 (Errata 33:207; 2006).

Liljeroth E, Baath E, Mathiasson I, Lundborg T 1990 Root exudation and rhizosphere bacterial abundance of barley (*Hordeum vulgare* L.) in relation to nitrogen fertilization and root growth. *Plant Soil* 127:81–89.

Lin W, Kelly M 2010 Nutrient uptake estimates for woody species as described by the NST 3.0, SSAND, and PACTS mechanistic nutrient uptake models. *Plant Soil* 335:199–212.

Little DY, Rao HY, Oliva S, Daniel-Vedele F, Krapp A, Malamy JE 2005 The putative high-affinity nitrate transporter NRT2.1 represses lateral root initiation in response to nutritional cues. *Proc Natl Acad Sci USA* 102:13693–13698.

Liu JX, Han LL, Chen FJ, Bao J, Zhang FS, Mi GH 2008 Microarray analysis reveals early responsive genes possibly involved in localized nitrate stimulation of lateral root development in maize (*Zea mays* L.). *Plant Sci* 175:272–282.

Lu S, Miller MH 1994 Prediction of phosphorus uptake by field-grown maize with the Barber-Cushman model. *Soil Sci Soc Am J* 58:852–857.

Ludwig F, Milroy SP, Asseng S 2009 Impacts of recent climate change on wheat production systems in Western Australia. *Clim Change* 92:495–517.

Lynch JP, Brown KM 2001 Topsoil foraging—An architectural adaptation of plants to low phosphorus availability. *Plant Soil* 237:225–237.

Lynch J, Nielsen KL 1996 Simulation of root system architecture. In *Plant Roots: The Hidden Half*, eds. Y Waisel, A Eshel, U Kafkafi, 2nd edn., pp. 247–257. New York: Marcel Dekker, Inc.

Lynch JP, Nielsen KL, Davis RD, Jablokow AG 1997 SimRoot: Modelling and visualization of root systems. *Plant Soil* 188:139–151.

Ma Z, Walk TC, Marcus A, Lynch JP 2001 Morphological synergism in root hair length, density, initiation and geometry for phosphorus acquisition in *Arabidopsis thaliana*: A modeling approach. *Plant Soil* 236:221–235.

Marschner H 1998 Soil-root interface: Biological and biochemical processes. In *Soil Chemistry and Ecosystem Health*, eds. PM Huang, DC Adriano, TJ Logan, RT Checkai, pp. 191–231. Madison, WI: Soil Science Society of America.

Marschner H, Römheld V 1996 Root induced changes in the availability of micronutrients in the rhizosphere. In *Plant Roots: The Hidden Half*, eds. Y Waisel, A Eshel, U Kafkafi, 2nd edn., pp. 557–579. New York: Marcel Dekker, Inc.

Martinez-Cordero MA, Martinez V, Rubio F 2005 High-affinity K+ uptake in pepper plants. *J Exp Bot* 56:1553–1562.

Maruyama-Nakashita A 2008 Transcriptional regulation of genes involved in sulfur assimilation in plants: Understanding from the analysis of high-affinity sulfate transporters. *Plant Biotechnol* 25:323–328.

McCully ME 1999 Roots in soil: Unearthing the complexities of roots and their rhizospheres. *Annu Rev Plant Physiol Plant Mol Biol* 50:695–718.

Meyer GE, Curry RB, Streeter JG, Mederski HJ 1979 *SOYMOD/OARDC—A Dynamic Simulator of Soybean Growth, Development, and Seed Yield. I. Theory, Structure, and Validation*. Wooster, OH: Ohio Agr Res Dev Cntr Res Bull # 1113.

Mitsukawa N, Okumura S, Shirano Y et al. 1997 Overexpression of an *Arabidopsis thaliana* high-affinity phosphate transporter gene in tobacco cultured cells enhances cell growth under phosphate-limited conditions. *Proc Natl Acad Sci USA* 94:7098–7102.

Murata Y, Harada E, Sugase K et al. 2008. Specific transporter for iron(III)-phytosiderophore complex involved in iron uptake by barley roots. *Pure Appl Chem* 80:2689–2697.

Nasholm T, Kielland K, Ganeteg U 2009 Uptake of organic nitrogen by plants. *New Phytol* 182:31–48.

Neuhäuser B, Dynowski M, Mayer M, Ludewig U 2007 Regulation of $NH4^+$ transport by essential cross talk between AMT monomers through the carboxyl tails. *Plant Physiol* 143:1651–1659.

Nielsen KL, Lynch JP, Weiss HN 1997 Fractal geometry of bean root systems: Correlations between spatial and fractal dimension. *Am J Bot* 84:26–33.

Nye PH, Marriott FHC 1969 A theoretical study of the distribution of substances around roots resulting from simultaneous diffusion and mass flow. *Plant Soil* 30:459–472.

Oates K, Barber SA 1987 Nutrient uptake: A microcomputer program to predict nutrient absorption from soil by roots. *J Agron Educ* 16:65–68.

Pagès L 2002 Modeling root system architecture. In *Plant Roots: The Hidden Half*, eds. Y Waisel, A Eshel, U Kafkafi, 3rd edn., pp. 359–382. New York: Marcel Dekker, Inc.

Pagès L, Vercambre G, Drouet JL et al. 2004 Root Typ: A generic model to depict and analyse the root system architecture. *Plant Soil* 258:103–119.

Pinton R, Varanini Z, Nannipieri P 2007. *The Rhizosphere: Biochemistry and Organic Substances at the Soil-Plant Interface*, 2nd edn. Boca Raton, FL: CRC.

Postma JA, Jaramillo RE, Lynch JP 2010 Towards modeling the function of root traits for enhancing water acquisition by crops. In *Response of Crops to Limited Water: Understanding and Modeling Water Stress Effects on Plant Growth Processes*. Adv. in Agricultural Systems Modeling Series 1, pp. 251–275. Madison, WI: ASA-CSSA-SSSA.

Postma J, Lynch J 2011 Theoretical evidence for the functional benefit of root cortical aerenchyma in soils with low phosphorus availability. *Ann Bot* 107:829–841.

Rao S, Mathur S 1994. Modeling heavy metal (cadmium) uptake by soil-plant root systems. *J Irrig Drain Eng ASAE* 120:89–96.

Reginato JC, Tarzia DA, Cantero A 1990 On the free boundary problem for the Michaelis-Menten absorption model for root growth. *Soil Sci* 150:722–728.

Remans T, Nacry P, Pervent M et al. 2006 A central role for the nitrate transporter NRT2.1 in the integrated morphological and physiological responses of the root system to nitrogen limitation in *Arabidopsis*. *Plant Physiol* 140:909–921.

Rengel Z 1991 Modeling magnesium uptake from an acid soil. (3) Determination of root magnesium concentration. *Soil Sci Soc Am J* 55:1612–1615.

Richardson AE, Hocking PJ, Simpson RJ, George TS 2009 Plant mechanisms to optimise access to soil phosphorus. *Crop Pasture Sci* 60:124–143.

Roose T, Fowler AC 2004 A mathematical model for water and nutrient uptake by plant root systems. *J Theor Biol* 228:173–184.

Roose T, Schnepf A 2008 Mathematical models of plant-soil interaction. *Phil Trans R Soc A* 366:4597–4611.

Rouached H, Wirtz M, Alary R et al. 2008 Differential regulation of the expression of two high-affinity sulfate transporters, SULTR1.1 and SULTR1.2, in *Arabidopsis*. *Plant Physiol* 147:897–911.

Rubio G, Sorgona A, Lynch J 2004 Spatial mapping of phosphorus influx in bean root systems using digital autoradiography. *J Exp Bot* 55:2269–2280.

Rubio G, Walk T, Ge ZY, Yan XL, Liao H, Lynch JP 2001 Root gravitropism and below-ground competition among neighbouring plants: A modelling approach. *Ann Bot* 88:929–940.

Ruffel S, Freixes S, Balzergue S et al. 2008. Systemic signaling of the plant nitrogen status triggers specific transcriptome responses depending on the nitrogen source in *Medicago truncatula*. *Plant Physiol* 146:2020–2035.

Sadana US, Claassen N 2000 Manganese dynamics in the photosphere and Mn uptake by different crops evaluated by a mechanistic model. *Plant Soil* 218:233–238.

Scaife MA 1976 The use of a simple dynamic model to interpret the phosphate response of plants grown in solution culture. *Ann Bot* 40:1217–1229.

Schilling G, Gransee A, Deubel A, Lezovic G, Ruppel S 1998 Phosphorus availability, root exudates, and microbial activity in the rhizosphere. *Z Pflanzenernähr Bodenkd* 161:465–478.

Schnepf A, Leitner D 2009 FEM simulation of below ground processes on a 3-dimensional root system geometry using Distmesh and Comsol multiphysics. In *Algoritmy 2009: 18th*

Conference on Scientific Computing, eds. A Handlovicova, P Frolkovic, K Mikula, D Sevcovic, pp. 321–330. Bratislava, Slovakia: Slovak University of Technology.

Schnepf A, Roose T 2006 Modelling the contribution of arbuscular mycorrhizal fungi to plant phosphate uptake. *New Phytol* 171:669–682.

Schwarz M, Lehmann P, Or D 2010 Quantifying lateral root reinforcement in steep slopes—From a bundle of roots to tree stands. *Earth Surf Process Landforms* 35:354–367.

Siddiqi MY, Glass ADM 1986 A model for the regulation of K^+ influx, and tissue potassium concentrations by negative feedback effects upon plasmalemma influx. *Plant Physiol* 81:1–7.

Sievers A, Braun M, Monshausen GB 2002 The root cap: Structure and function. In *Plant Roots: The Hidden Half*, eds. Y Waisel, A Eshel, U Kafkafi, 3rd edn., pp. 33–47. New York: Marcel Dekker, Inc.

Silberbush M 1996 Simulation of ion uptake from the soil. In *Plant Roots: The Hidden Half*, eds. Y Waisel, A Eshel, U Kafkafi, 2nd edn., pp. 643–658. New York: Marcel Dekker, Inc.

Silberbush M 2002 Simulation of ion uptake from the soil. In *Plant Roots: The Hidden Half*, eds. Y Waisel, A Eshel, U Kafkafi, 3rd edn., pp. 651–661. New York: Marcel Dekker, Inc.

Silberbush M, Barber SA 1984 Phosphorus and potassium uptake of field-grown soybean cultivars predicted by a simulation model. *Soil Sci Soc Am J* 48:592–596.

Silberbush M, Sorek S, Yakirevich A 1993 K+ uptake by root systems grown in soil under salinity. I. A mathematical model. *Transp Porous Media* 11:101–116.

Simunek J, Hopmans JW 2009 Modeling compensated root water and nutrient uptake. *Ecol Model* 220:505–521.

Smaling EMA, Janssen BH 1993 Calibration of QUEFTS, a model predicting nutrient uptake and yields from chemical soil fertility indices. *Geoderma* 59:21–44.

Smethurst PJ, Comerford NB 1993 Simulating nutrient uptake by single or competing and contrasting root systems. *Soil Sci Soc Am J* 57:1361–1367.

Somma F, Hopmans JW, Clausnitzer V 1998 Transient three-dimensional modeling of soil water and solute transport with simultaneous root growth, root water and nutrient uptake. *Plant Soil* 202:281–293.

Sousa MF, Facanha AR, Tavares RM, Lino-Neto T, Geros H 2007 Phosphate transport by proteoid roots of *Hakea sericea*. *Plant Sci* 173:550–558.

Sperry JS, Stiller V, Hacke UG 2002 Soil water uptake and water transport through root systems. In *Plant Roots: The Hidden Half*, eds. Y Waisel, A Eshel, U Kafkafi, 3rd edn., pp. 663–681. New York: Marcel Dekker, Inc.

Steingrobe B, Schenk MK 1994 A model relating the maximum nitrate inflow of lettuce (*Lactuca sativa* L.) to the growth of roots and shoots. *Plant Soil* 162:249–257.

Timlin D, Pachepsky YA, Acock B 1996 A design for a modular, generic soil simulator to interface with plant models. *Agron J* 88:162–169.

Tinker PB, Nye PH 2000. *Solute Movement in the Rhizosphere*, 2nd edn. Oxford, UK: Oxford Univ. Press.

Tobin B, Cermák J, Chiatante D et al. 2007 Towards developmental modelling of tree root systems. *Plant Biosys* 141:481–501.

Tracy SR, Roberts JA, Black CR, McNeill A, Davidson R, Mooney SJ 2010 The x-factor: Visualizing undisturbed root architecture in soils using x-ray computed tomography. *J Exp Bot* 61:311–313.

Treeby MT 1992 The role of mycorrhizal fungi and non-mycorrhizal micro-organisms in iron nutrition of citrus. *Soil Biol Biochem* 24:857–864.

Van Beek LPH, Wint J, Cammeraat LH, Edwards JP 2005. Observation and simulation of root reinforcement on abandoned Mediterranean slopes. *Plant Soil* 278:55–74.

Van Rees KCJ, Comerford NB, Rao PSC 1990 Defining soil buffer power: Implications for ion diffusion and nutrient uptake modeling. *Soil Sci Soc Am J* 54:1505–1507.

Waisel Y, Eshel A 2002 Functional diversity of various constituents of a single root system. In *Plant Roots: The Hidden Half*, eds. Y Waisel, A Eshel, U Kafkafi, 3rd edn., pp. 157–174. New York: Marcel Dekker, Inc.

Walk T, Jaramillo R, Lynch J 2006 Architectural tradeoffs between adventitious and basal roots for phosphorus acquisition. *Plant Soil* 279:347–366.

Walk TC, van Erp E, Lynch JP 2004 Modelling applicability of fractal analysis to efficiency of soil exploration by roots. *Ann Bot* 94:119–128.

Walter A, Römheld V, Marschner H, Crowley DE 1994 Iron nutrition of cucumber and maize: Effect of *Pseudomonas putida* YC 3 and its siderophore. *Soil Biol Biochem* 26:1023–1031.

Wenzel CL, Ashford AE, Summerell BA 1994 Phosphate-solubilizing bacteria associated with proteoid roots of seedlings of waratah [*Telopea speciosissima* (Sm.) R.Br.]. *New Phytol* 128:487–496.

Williams M, Yanai RD 1996. Multi-dimensional sensitivity analysis and ecological implications of a nutrient uptake model. *Plant soil* 180:311–324.

Wulfsohn D, Nyengaard JR 1999 Simple stereological procedure to estimate the number and dimensions of root hairs. *Plant Soil* 209:129–136.

Xu Z, Li K, Liu Z, Zhang, K, Zhang J 2007 Correlations between kinetic parameters of phosphate uptake and internal phosphorus concentrations in maize plants: Making it possible to estimate the status of phosphate uptake according to shoot phosphorus concentrations. *Commun Soil Sci Plant Anal* 38:1–15.

Yanai R 1994 A steady-state model of nutrient uptake accumulating for newly grown roots. *Soil Sci Soc Am J* 58:1562–1571.

Zaharieva T, Römheld V 2000 Specific Fe^{2+} uptake system in strategy I plants inducible under Fe deficiency. *J Plant Nutr* 23:1733–1744.

Zhao J, Cheng NH, Motes CM et al. 2008 AtCHX13 is a plasma membrane K+ transporter. *Plant Physiol* 148:796–807.

27

Global Change and Root Lifespan

David M. Eissenstat
The Pennsylvania State University

M. Luke McCormack
The Pennsylvania State University

Quanying Du
The Pennsylvania State University

I. Introduction

A. Importance of Root Lifespan

Like other plant organs, roots have a life history in which they pass from birth to death. The size and population structure of the root system is determined by the birth rate and death rate of the individual roots. The study of root population dynamics is of interest to many disciplines, including crop science, physiology, ecology, and soil science. For example, a better understanding of root lifespan could enable agronomists and horticulturalists to increase yields while reducing agrochemical inputs. Severe root losses, such as those caused by drought or pathogens, clearly are not conducive to crop production. Growing too many roots, however, may also be undesirable, since large amounts of carbohydrates and mineral nutrients are needed for root growth and maintenance that otherwise might be allocated to photosynthetic organs or harvested parts. An optimization approach suggests that, other things being equal, total plant growth should be greatest when a root system maximizes water and nutrient acquisition per unit resource supplied from the shoot (e.g., Thornley 1998). If roots are produced in the most favorable soil patches and shed when they are no longer efficient in water and nutrient absorption, then production, theoretically, should be maximized.

The birth and death of roots also influence plant competition. Root competition can be more intense than shoot competition (Wilson 1988). Just as perennial structures aboveground can give plants a competitive advantage for light capture, there may be advantages to long-lived roots in the capture of limited soil resources. Resource preemption can be an important component of competitive success. For example, in climates with winter precipitation, perennial grasses with an established root system are much more effective than seedlings of perennial grasses at competing with annual grass species during the spring and summer (Harris 1967). Clearly, root population dynamics can have important consequences on species distribution and abundance.

Root population dynamics also influences ecosystem processes associated with material and energy flows. Approximately 33% of global net primary production is used for fine-root production, based on fine-root biomass in 253 field studies in a wide range of ecosystems and assuming roots have a lifespan of 1 year, possibly a conservative estimate (Jackson et al. 1997). In other studies, belowground net primary productivity (BNPP) has been estimated to be at least as great as aboveground net primary productivity (Vogt et al. 1986; Caldwell 1987). While there is a wide agreement that the production and death of fine roots can have a substantial influence on ecosystem carbon and mineral nutrient cycling, these processes are not well understood. Many ecologists have been concerned with understanding how BNPP varies among ecosystems and predicting how it may change in response to global change, which is the focus of this chapter.

B. Ephemeral Portion of the Root System

The first roots of plants developing from seed are indeterminate, often extending greatly in length as the taproot or other seminal roots develop. The major laterals that first emerge from these primary roots and the adventitious or nodal roots that emerge from the stem base are also typically indeterminate, often extending decimeters or more in length. These indeterminate roots, often called pioneer roots in trees, form the basic framework of the root system and may live as long as the plant lives. These roots may have limited mycorrhizal colonization and several layers of hypodermis with limited passage cell development (Zadworny and Eissenstat 2011, see also Chapter 5). These roots are often more tolerant of soil moisture deficits and presumably other environmental stressors than similar-age fibrous roots (Polverigiani et al. 2011). Although pioneer roots may be common in ingrowth cores and other conditions where the root system is disturbed and may still be less than 1 mm in diameter, they should not be considered part of the ephemeral root system.

Ephemeral roots are often associated with the finest laterals of a root system, sometimes referred to as fibrous or feeder roots. Usually the annual production of fine laterals that typically emerge from the woody roots has only two or three orders of branching (Guo et al. 2008; Valenzuela-Estrada et al. 2008; Xia et al. 2010). In probably most woody species, these lower-order roots (using stream-based branching scheme) never undergo secondary development of the vascular tissue or develop a periderm (Brundrett and Kendrick 1988; Eissenstat and Achor 1999; Xia et al. 2010). This population of roots represents the ephemeral root system that typically turns over in a manner analogous to leaves and is the chief contributor of carbon and mineral nutrients by roots to the detritus pool. Note that many roots less than 2 mm in diameter, depending on the species, may not be ephemeral but woody semipermanent members of the root system (Guo et al. 2008; Valenzuela-Estrada et al. 2008). These roots are usually third, fourth, or higher order and have secondary development, when examined in cross section. In an extreme example, the ericoid species *Vaccinium corymbosum* L. has fourth-order woody roots that are only about 0.1 mm in diameter (Figure 27.1). Studies aimed at understanding the physiology, chemistry, or longevity of the ephemeral roots need to distinguish these small-diameter woody roots from the lower-order roots that represent the population of roots that turns over with high frequency and are typically the mycorrhizal absorptive roots of the plant. How the longevity of these ephemeral roots will be influenced by global change is only beginning to be understood.

II. Variation in Root Lifespan

A. Sources of Variation

Estimates of root lifespan vary widely. The median lifespan of the finest roots can range from about 20 days at some times of the year in fast-growing trees and deciduous fruit crops to longer than a year in slow-growing forest trees, according to studies using transparent windows in the soil (Eissenstat and Yanai 1997). In a data set containing 190 studies in nonagricultural ecosystems, based mainly on changes in biomass from sampling soil monoliths, soil cores, or ingrowth cores, average root lifespans ranged from about 290 days in tropical ecosystems to about 3 years in high-latitude ecosystems, with considerable variation within each ecosystem type (Gill and Jackson 2000). Studies based on tracer approaches have indicated that fine roots less than 2 mm in diameter, which may have included both

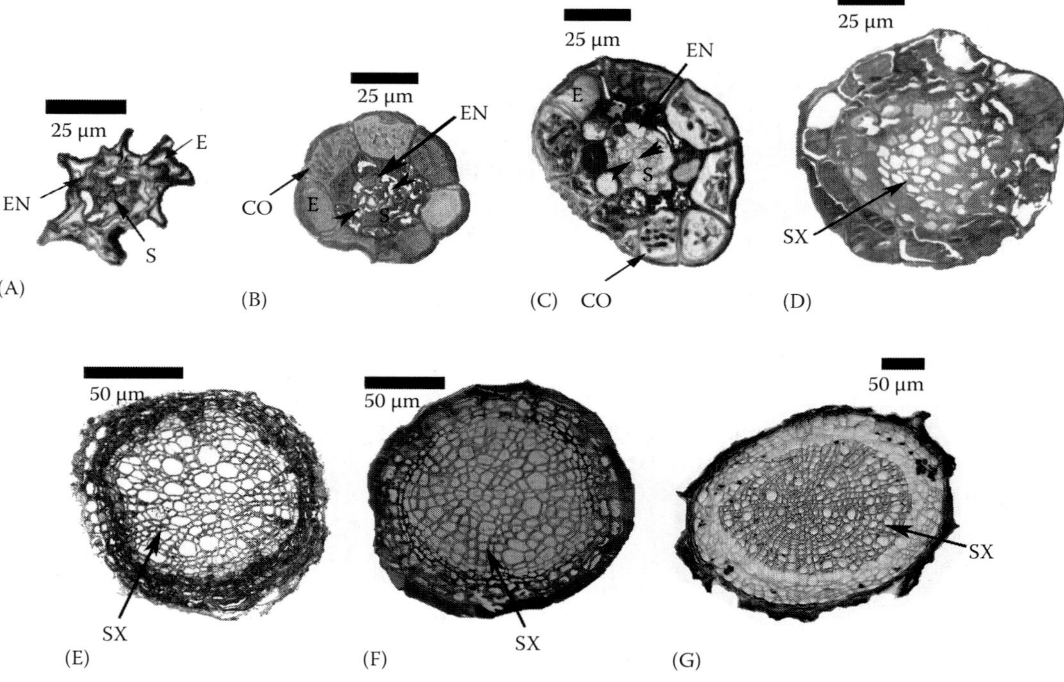

FIGURE 27.1 Cross sections of the first seven root orders of *V. corymbosum* from roots collected at 15–20 cm soil depth under field conditions. (A) Light micrograph of a first-order root about 500 μm from the root tip showing a single cell layer of epidermal cells (E), a single layer of endodermal cells (EN), and stele (S). Cross sections of (B) second- and (C) third-order roots with no secondary development showing some ericoid fungal coils (CO) in the epidermal cells and fourth- (D), fifth- (E), sixth- (F) and seventh-order roots (G) with secondary xylem development (SX). Note that scale bar changes in length depending on root order. (Reprinted from *Am. J. Bot.*, 95, Valenzuela-Estrada, L.R., Vera-Caraballo, V., Ruth, L.E., and Eissenstat, D.M., Root anatomy, morphology and longevity among root orders in *Vaccinium corymbosum* (Ericaceae), 1506–1514. Copyright 2008, with permission from Elsevier.)

ephemeral and woody fine roots, may live considerably longer—averaging 4–8 years in some temperate forests (Gaudinski et al. 2000; Matamala et al. 2003). Although differences in methods contribute substantially to differences in estimates of lifespan, as discussed previously (Eissenstat and Yanai 2002), undoubtedly much of the variation in reported root lifespan is caused by differences in environmental conditions and plant species.

B. Economic Approaches to Root Lifespan

The same theories that attempt to explain variation in leaf lifespan have been applied to roots (Grime 1977; Chapin 1980; Aerts 1995). These theories suggest, for example, that plants that have slow growth rates and are adapted to chronically low-nutrient sites should have long lifespan of the absorptive organs compared to more fertile sites. Tissue retention in nutrient-poor sites allows nutrients to be retained as well, which is important if root and shoot growth rates are restricted by nutrient limitations. There is considerable evidence that leaf longevity is consistent with this hypothesis (Reich et al. 1997; Wright et al. 2004), but roots have been less well studied. In a pot study, Black et al. (1998) showed that the species with the shortest root lifespan, *Prunus avium* L., had considerably faster-growing root and shoot systems than the species with longest root lifespan, *Picea sitchensis* (Bong.) Carrière. In a study comparing grasses from nutrient-poor and nutrient-rich habitats in pots in the field, the grasses from the nutrient-rich habitat had lost a greater percentage of their leaves and roots by the end of the second growing season (Ryser 1996). These results suggest that roots and leaves do have suites of traits associated with tissue longevity.

Other, more comprehensive studies have not found a correlation between leaf and root longevity. Although roots and leaves may have similar sets of traits associated with their respective lifespan, such as N concentration, thickness, and metabolic activity, both anecdotal and direct experimentation have often revealed that variation among species in leaf longevity may not be closely linked to variation in root longevity. For example, desert succulents have long-lived leaves but short-lived "rain" roots (Huang and Nobel 1992; North et al. 1993). In seasonal dry climates, cluster roots of evergreen woody plants, such as those in the Proteaceae (Lamont 1995) and ericoid mycorrhizal hair roots of plants in the Ericaceae (Smith and Read 1997), are often shed during extended dry periods. Withington et al. (2006) examined 11 temperate tree species in a common garden in Poland ranging in leaf lifespan from 0.5 to 6 years and in root lifespan from 0.5 to 2.5 years. While both root and leaf lifespan were negatively correlated with tissue N concentration, root lifespan and leaf lifespan were not correlated with each other. Trees with short-lived leaves like maples (*Acer* spp.) had long-lived roots and species with long-lived leaves like spruce (*Picea abies* (L.) H. Karst.) had relatively short-lived roots. Similar results have been observed in a comparison of 12 tree species in common garden in central Pennsylvania (McCormack et al. 2012), clearly indicating that the selection pressures on leaf longevity may act independently from those on root longevity. Because favorable soil environments for water and nutrient uptake may be temporally displaced from favorable conditions for photosynthesis by leaves, it seems reasonable that root and leaf traits are not necessarily correlated.

C. Difficulties and Definitions

The single greatest impediment to the study of root lifespan is the difficulty of studying roots in their natural environment. Many approaches have been taken with varying degrees of success. Often studies are not long enough to establish clear year-to-year variation or to have allowed the plants to fully adjust to installation of root measuring devices (e.g., minirhizotrons) or treatments. For example, fertilization studies are often conducted for only a few years, so they may not characterize steady-state responses to a new level of fertility.

The interpretation of estimates of root lifespan is hindered by inconsistencies in methods of reporting root dynamics. In many ecosystem studies, the main objective is to estimate BNPP (kg ha^{-1} year^{-1}). The term "root turnover" has often been used synonymously with annual root production or annual root mortality and thus has units such as kg ha^{-1} year^{-1}. Alternatively, "root turnover" may be used to describe the specific rate of root mortality, in units of year^{-1}. One way to report the specific rate of root mortality is the rate constant in exponential decay (Eissenstat and Yanai 2002). Root turnover rate is also commonly reported as annual root production or annual root mortality divided by root standing crop. Studies differ in whether minimum (Hendrick and Pregitzer 1993), average (Aber et al. 1985; Aerts et al. 1992), or maximum (Dahlman and Kucera 1965; Gill and Jackson 2000) root biomass are used to estimate standing crop. Gill and Jackson (2000) found about one-third of root turnover studies report only the mean standing crop. An important disadvantage of using minimum or maximum standing crop is that the minimum or maximum value in any distribution is dependent on the number of samples collected and the sampling error associated with sample measurement. Nonetheless, Gill and Jackson (2000) found that maximum standing crop could be accurately estimated by mean standing crop by a regression approach (r^2 = 0.90), based on 20 data sets that included both maximum and mean root biomass.

Root lifespan is inversely proportional to root turnover rate, with the constant of proportionality dependent on the definitions of turnover rate and lifespan. Many recent studies that follow the fate of individual roots with minirhizotrons report only median lifespan (or similarly half-life of the cohort), partly because many of the roots in the study have not died by the end of the study and partly because the median is a better estimator of the central location of a highly skewed distribution—a condition common to survivorship curves. Clearly, average lifespan may be considerably longer than median lifespan if an appreciable fraction of the population lives a very long time. Studies that follow individual roots typically report median lifespans of specific cohorts (roots born at the same point in time), because different cohorts may exhibit very different median lifespans (Kosola et al. 1995).

Lastly, one must exercise caution in extrapolating median root lifespan to estimates of annual turnover for two reasons. One is that the average may depart markedly from the median. Secondly, and more importantly, root production is not continuous throughout the year, especially in regions with a dormant season such as winter in the temperate and boreal region. A spring cohort of roots may have a median lifespan of only 90 days, but that does not mean turnover is 4 $year^{-1}$, if no roots are being produced for about 6 months during the dormant season. In this case, turnover would approximately be only 2 $year^{-1}$.

III. Effects of Global Change on Root Lifespan

Fine-root responses to global-change factors will be varied and complex depending on species, site, and magnitude of environmental change. Among important factors expected to change in the twenty-first century, most notable are atmospheric carbon dioxide concentration, temperature, nitrogen deposition, and precipitation. Changes will include both increases in magnitude such as with carbon dioxide and increases in variability such as precipitation. Discussions of global-change effects quickly become complex as one considers direct and indirect as well as short- and long-term effects. Moreover, often the effects of an environmental factor may have a nonlinear effect on lifespan. Clearly, very low and very high soil temperatures, for example, will likely negatively affect root lifespan, but the effects of moderate changes in mean soil temperature, say from 15°C to 18°C, are less clear. Our discussion will focus primarily on direct, shorter-term effects of global-change factors on fine-root lifespan as they are within the scope of feasible experimental manipulations at the plant to ecosystem scale.

A. Elevated Atmospheric CO_2

As a principal global-change factor, elevated CO_2 and its effects on ecosystem processes has received substantial interest and study over the past few decades. Because elevated atmospheric CO_2 effectively alters the C budget for whole ecosystems, it is not surprising that elevated CO_2 has frequently (Iversen et al. 2008; Jackson et al. 2009), but not always (Phillips et al. 2006), resulted in increased total root production and roughly parallel increases in mortality. Direct effects of elevated atmospheric CO_2 on soil concentrations of CO_2 are likely unimportant because of the considerably higher concentration in the soil atmosphere (soil concentrations can be on the order of 10,000 ppm) than in atmosphere surrounding the leaves. Rather, effects of elevated atmospheric CO_2 must act on fine-root lifespan through its effects on whole-plant physiology and indirectly through its effects on ecosystem properties.

The influence of elevated CO_2 on root behavior can be analyzed from a cost:benefit perspective (Eissenstat and Yanai 1997). Assuming elevated CO_2 reduces the expense of providing C to support the benefit roots provided in terms of water and nutrient acquisition, root C allocation should increase with increases in atmospheric CO_2. Thus plants under elevated CO_2 should exhibit greater ability to acquire water and nutrients, which is usually associated with greater root growth. For example, greater net primary productivity at the Duke, Oak Ridge, and Rhinelander FACE sites was linked to increased N uptake (Finzi et al. 2007), which may be partially explained by greater fine-root biomass (Drake et al. 2011). Meta-analysis of early studies examining effects of elevated CO_2 on whole-plant and root growth commonly showed an increase, but not always (Curtis and Wang 1998).

A simple economic analogy would lead to the prediction that elevated CO_2 should increase root lifespan because the cheaper C should reduce the cost of root maintenance, thus improving the marginal rate of return on any root investment. The real world, however, is not always that simple. Empirical evidence indicates no systematic increase in root lifespan under elevated atmospheric CO_2 (Table 27.1). Indeed, there are as many studies indicating a decrease in root lifespan as an increase (Table 27.2). Phillips et al. (2009) observed a 50% increase in fine-root lifespan of *Pinus ponderosa* P. Lawson & C. Lawson, while on the other hand, Thomas et al. (1999) observed 65% reduction in root lifespan of *Pinus radiata* D. Don grown with elevated CO_2. Moreover, many studies have found little to no effect of elevated atmospheric CO_2 on either whole-plant growth or fine-root lifespan, especially where soil fertility is low (Oren et al. 2001; Johnson et al. 2006). In these studies, CO_2 is not limiting so increases in its abundance do not influence either plant growth or the efficiency of nutrient acquisition. Thus, it is consistent with an economic analogy that there is no effect on lifespan. Other complicating factors may include roots of higher densities becoming less-efficient, higher soil moisture under elevated CO_2 (because of greater water use efficiency) leading to greater herbivory/pathogen pressure as well as shifting patterns of allocation between roots and mycorrhizal fungi. Clearly, more detailed analysis on direct causes of mortality in elevated CO_2

TABLE 27.1 Summary of the Number of Studies Reporting a Given Response of Fine-Root Lifespan and Relevant Root Dynamics to Global Change Effects

	Positive (+)	Negative (−)	No Effect (0)
Elevated CO_2	4	5	8
Rising temperature	0	5	2
Increased precipitation	1	5	3
Decreased precipitation	0	4	1
Increased N deposition	5	4	3

Whenever references reported multiple studies with different effects on different species, then both effects were counted. For example, Coleman unpublished reported a decrease in lifespan with elevated CO_2 for one species and no change for another, which resulted in counting one negative effect (−) and one no effect (0).

TABLE 27.2 Studies Reporting Effects of Elevated CO_2 (eCO_2) on Fine-Root Lifespan

Plant Type	Location	Length of Study	Method	Effect of eCO_2	Reference	Notes
Mixed grassland	Central UK	1.5 years	Chamber on soil bins, D	0, –	Fitter et al. (1998)	Study included two grassland types; only one showed a response to CO_2
Aspen seedlings	Michigan, United States	<1.0 year	OTC on soil bins, D	0	Kubiske et al. (1998)	May have observed differential effects of CO_2 with high and low available soil nitrogen
Pine saplings	New Zealand	2 years	OTC on natural soil, D	–	Thomas et al. (1999)	Median root lifespan decreased from 951 to 333 days. Elevated CO_2 increased both root production and mortality
Pine seedlings	California, United States	4 years	OTC on natural soil, D	+	Johnson et al. (2000)	Study also noted strong effect of soil temperature on root survival as well as a negative effect of nitrogen additions on root lifespan
Semiarid grassland	Colorado, United States	5 years	OTC on natural soil, D	+	Milchunas et al. (2005)	Elevated CO_2 also increased total root production and mortality in the semiarid shortgrass steppe grassland
Desert shrub	Nevada, United States	5 years	FACE on natural soil, I	0, –	Phillips et al. (2006)	Negative effect occurred in mixed community plots; plots with single species saw no change
Fir seedlings	Oregon, United States	3 years	Chamber on soil bins, I	0	Johnson et al. (2006)	No measured change in lifespan though rooting depth appeared to be altered
Sweet gum trees	Tennessee, United States	9 years	FACE on natural soil, I	+	Iversen et al. (2008)	CO_2 effect on root lifespan was inferred based on a decrease in fine-root turnover
Aspen trees	Wisconsin, United States	2 years	FACE on natural soil, D	0	Pregitzer et al. (2008)	Minirhizotron results only reported for years 2003 and 2004 and were mixed between the 2 years
Pine and decid. trees	North Carolina, United States	6 years	FACE on natural soil, D	–	Pritchard et al. (2008)	Elevated CO_2 decreased average root lifespan from 574 to 500 days
Pine seedlings	Oregon, United States	3 years	Chamber on soil bins, D	+	Phillips et al. (2009)	Elevated CO_2 increased average root lifespan and total root production
Mixed grassland	Texas, United States	2 years	Gradient tunnel, D	0	Anderson et al. (2010)	Overall effects of CO_2 on root lifespan were negligible. In cases of high root proliferation, elevated CO_2 decreased root lifespan
Pine/oak seedlings	Alabama, United States	3 years	OTC on soil bins, D	0	McCormack et al. (2010)	Effect of elevated CO_2 on roots appeared to be relatively small compared to measured effects on soil fungi
Scrub oak trees	Florida, United States	10 years	OTC on natural soil, D	0	Stover et al. (2010)	Evergreen scrub oak system with no effect of elevated CO_2 on fine-root lifespan

Note: Methods listed pertain to the experimental system used as well as whether direct observations (D) of fine-root lifespan were reported or if the effect on fine-root lifespan was inferred through indirect observations (I) or related evidence such as root turnover.

experiments is needed to improve our understanding of both direct and indirect effects of this global-change factor on root lifespan. We note that in studies where C exposure was for more than 5 years, generally elevated CO_2 effects on lifespan were modest or negligible. Thus, at least based on the limited research so far, it does not seem that compared to other global-change factors, elevated CO_2 will markedly shift root lifespan in most ecosystems unless there is substantial shifts in species composition of plant communities.

B. Altered Precipitation

Altered precipitation presents a particular challenge when considering how global-change factors will impact ecosystem dynamics like fine-root lifespan. Unlike atmospheric CO_2, change in the amount of precipitation an ecosystem receives will unlikely be a steady, monotonic increase through time. Instead, climate models consistently predict increased variability in patterns of precipitation with overall averages increasing in some areas and decreasing in others (IPCC 2007). As such, forecasting future fine-root lifespan requires understanding how root populations respond to both increases in water availability as well as soil moisture deficits. Furthermore, it will be important to know when ephemeral roots are most sensitive. For example, in areas that experience distinct changes in rainfall during seasonal dormancy, there may be little to no impact on root lifespan. In contrast, the impact of a severe drought every 5 or 10 years during the growing season may profoundly affect root lifespan.

Despite the potential complexity of changing patterns of precipitation, responses of fine-root lifespan to drought have been fairly consistent (Table 27.1). Though few studies have directly observed drought effects on fine-root lifespan, their results

TABLE 27.3 Studies Reporting Effects of Changing Soil Temperature on Fine-Root Lifespan

Plant Type	Location	Length of Study	Method	Rising Temp.	Reference	Notes
Ryegrass	UK	<0.5 year	Pots in greenhouse, D	–	Forbes et al. (1997)	Different temperature treatments included 15°C, 21°C, and 27°C. Differences only observed between 15°C and 27°C treatments
Mixed grassland	Central UK	2.5 years	Soil bins with heater cables, I	0	Fitter et al. (1999)	Heating caused increased mortality but no equivalent increase in root production
NA	NA	NA	I	–	Gill and Jackson (2000)	Based on literature review of global patterns of fine-root turnover
Ryegrass and clover	Italy and UK	<0.5 year	Comparison of two sites, D	–	Watson et al. (2000)	Study compared plant growth across one warmer site in Italy and a cooler site in the UK
Mixed decid. trees	New Hampshire, United States	2 years	Field observations, I	–	Tierney et al. (2003)	Based on variation in ambient temperature, mortality increased during warmer months. May be confounded with root phenology
Pine seedlings	Oregon, United States	3 years	Chamber on soil bins, I	0	Johnson et al. (2006)	No measured change in lifespan though rooting depth appeared to be altered
Mixed evrg. trees	California, United States	3 years	Field observations, I	–	Kitajima et al. (2010)	Based on variation in ambient temperature and precipitation
Increased variability						
Pine trees	Georgia, United States	1 year	Partial tree harvest, D	–	Jones et al. (2003)	Tree harvest increased variability in temperature. Also an interaction with increased water availability
Lower winter temperature						
Deciduous trees	New Hampshire, United States	2 years	Snow removal at field site, I	–	Tierney et al. (2001)	Snow removal in winter, which decreased winter soil temperatures
Spruce trees	SE Germany	1 year	Snow removal at field site, I	–	Gaul et al. (2008)	Snow removal in winter, which decreased winter soil temperatures

Note: Methods listed pertain to the experimental system used as well as whether direct observations (D) of fine-root lifespan were reported or if the effect on fine-root lifespan was inferred through indirect observations (I) or related evidence such root turnover.

taken with several other relevant studies support the expectation that drought stress tends to reduce root lifespan (Table 27.3). For example, Meier and Leuschner (2008) found that drought decreased fine-root lifespan by about 50% in beech seedlings (*Fagus sylvatica* L.) grown in outdoor soil bins. However, if soil moisture deficits are not severe, deep roots or roots in moist soil regions readily supply water to roots in dry soil layers at night, as found in grape (*Vitis* spp., Bauerle et al. 2008), leading to no effect of localized soil moisture deficits on root lifespan. Nocturnal redistribution of water is partly a function of the water available to the roots in the moist soil; the degree transpiration is limited at night and the plant's hydraulic architecture. Thus, the effects of soil moisture deficits on root lifespan may be negligible in citrus (Kosola and Eissenstat 1994) and grape (Anderson et al. 2003; Bauerle et al. 2008) but substantial in a species such as *Vaccinium* that have narrow xylem vessels (Valenzuela-Estrada et al. 2009). Root tolerance of desiccation may also be related to root anatomy, including development of prominent hypodermis (Eissenstat 1998). Other studies have observed increased mortality in fine roots during periods of severe drought (Mainiero and Kazda 2006; Mission et al. 2006), which may imply direct desiccation from low root water potentials or root shedding because of severe reductions in photosynthate causing selective mortality of roots providing the least benefit. Furthermore, many studies have reported longer root lifespan for roots produced deeper in the soil profile compared to those in shallow soil layers (Pritchard et al. 2008; Phillips et al. 2009; McCormack et al. 2010), which Peek et al. (2006) hypothesized could be due to more constant and typically higher soil moister levels deeper in the soil.

Unlike drought, increases in water availability appear to have both negative and positive effects on fine-root lifespan (Table 27.3). Of the four studies directly observing responses of fine-root lifespan to increased water availability, one found increased lifespan (Pregitzer et al. 1993), one decreased lifespan (Jones et al. 2003), and two remaining studies found that the fine-root lifespan of some species was decreased, while others were unaffected (Coleman unpublished results; Adams et al. unpublished results). Jones et al. (2003) observed reduced lifespan following a tree removal treatment that resulted in increased soil moisture available to the remaining trees as total system transpiration decreased. However, while the observed decrease in lifespan was significant, it is difficult to confidently say whether the decrease was driven by soil moisture or other factors that were influenced by the tree removal including variance in soil temperature. Similarly, Kitajima et al. (2010) observed increased turnover associated with reduced lifespan as water availability increased in a semiarid mixed-conifer forest in southern California, United States. Other studies (Coleman unpublished results; Adams et al. unpublished) found species variation in lifespan with water addition, with many species showing decreased lifespan, but some showing no effect. The only study to observe increased lifespan (Pregitzer et al. 1993) was conducted in a mixed temperate forest including *Populus grandidentata* Michx. along with *Quercus rubra* Michx., *Acer rubrum* L., and *Fagus grandifolia* Ehrh. However, it is possible that the relatively short length of observation (~100 days) in this study compared to other studies that ranged from 1 to more than 3 years in length only captured a transient boost in lifespan that would have changed given more time. It is also possible that there were competitive interactions with roots of longer-lived species dominating the wet patches leading to the perceived increase in root lifespan.

Understanding the underlying mechanisms driving responses of fine-root lifespan to altered precipitation is an important part of predicting responses of fine roots in the future. Such research would need to monitor whole-plant water relations, including nocturnal hydraulic redistribution in the plant, to better understand how different species cope with soil moisture deficits. Causes for shifts in root lifespan because of increased soil moisture availability are less clear, barring extreme increases such as flooding. Increased water availability may increase microbial activity, both bacterial and fungal, contributing to greater infection by pathogens. Root herbivores are also more likely to be active in moist soil. Increased soil moisture may also cause more rapid decomposition of dying roots. Because the assessment of root death using minirhizotron images often includes cues such as root shriveling and disappearance, rapid root decomposition may lead to shorter estimates of root lifespan. In addition, rapid decomposition of organic matter may increase nutrient availability, which can either increase or decrease root lifespan depending on several factors (see Section III.D). In summary, extremes in soil moisture availability, either as excess or as deficits, will likely diminish root lifespan in many species. More modest shifts in soil moisture on root lifespan from increased variability in precipitation will likely differ among species and be difficult to predict.

C. Temperature

While temperatures at the ecosystem level are likely to become more variable much like precipitation, models also predict that the overall averages will generally increase across the globe, with generally larger shifts occurring at higher latitudes. Only two studies have reported direct effects of warmer temperatures on fine-root lifespan and in both cases lifespan was decreased (Table 27.4; Forbes et al. 1997; Watson et al. 2000). However, both studies were relatively short (≤100 days) and have limitations. Forbes et al. (1997) compared fine-root lifespan of *Lolium perenne* L. grown in a greenhouse at 15°C, 21°C, and 27°C and only found a difference in lifespan at the highest temperature. While a 12°C daily and seasonal fluctuation is well within the range of temperature of many surface soils, it is substantially greater than the 1°C–3°C average increase expected for an annual average, making direct comparisons to global climate change difficult. In another study, Watson et al. (2000) reported decreased fine-root lifespan of *L. perenne* and *Trifolium repens* L. grown in the field. However, the plants were grown at two different locations; a cooler site in the United Kingdom and a warmer site in Italy. While differences in temperature were likely partly responsible for the effect on root lifespan, other possible factors cannot be excluded.

Further supporting an expected decrease in fine-root lifespan with increasing temperature are several reports highlighting increased mortality and turnover (inversely related to lifespan) with increasing temperature. Tierney et al. (2003) and Kitajima et al. (2010) both observed an increase in fine-root mortality and turnover with increasing soil temperature across multiple years. Both studies observed root dynamics in systems without experimental warming and depended on natural variation within and between seasons creating both an advantage and disadvantage over experimental soil-warming studies. Their main disadvantage rests with their potential to conflate responses to temperature with patterns of plant and root phenology across seasons. However, in studies with experimental warming, fine-root mortality tends to increase concomitantly with fine-root production (Fitter et al. 1999) making it difficult to draw conclusions regarding fine-root lifespan as increased mortality may simply mirror the greater number of roots produced and not a shortening of root longevity. Using another approach, Gill and Jackson (2000) also suggest that increasing temperature increases fine-root turnover. Reviewing 190 published studies, the authors found a positive relationship between temperature and root turnover in grasslands, in shrublands, and in the fine roots of forest systems. While the observed trends may be caused simply by changing suites of species along the temperature gradients, it is still consistent with the hypothesis that temperature plays a large role in mediating rates of fine-root turnover and potentially root lifespan.

Factors leading to decreased root lifespan from increasing temperature may be both endogenous and exogenous to the plant root. Increased temperature may directly increase root metabolism and respiration leading to faster root aging and decreased lifespan. The link between temperature and respiration has been shown in many studies and the link between respiration, root aging, and subsequent root mortality has been also been suggested. However, fine roots may also acclimate to increased temperature in a matter of days resulting in little or no significant increase in respiration. Bryla et al. (2001) showed that root respiration of citrus trees was significantly elevated when temperature was increased over relatively short time periods (1 h).

TABLE 27.4 Studies Reporting Effects of Increased and Decreased Precipitation on Fine-Root Lifespan

Plant Type	Location	Length of Study	Method	Effect of Precipitation		Reference	Notes
				Increase	*Decrease*		
Forest, deciduous		<0.5 year	Irrigation tubes in field, D	+	•	Pregitzer et al. (1993)	Study used direct observations through Biotron root windows and not minirhizotrons
NA	NA	NA	I	–	•	Gill and Jackson (2000)	Based on literature review of global patterns of fine rot turnover
Mixed decid. trees	Tennessee, United States	5 years	Throughfall displacement, I	0	0	Joslin et al. (2000)	Root turnover may have increased with increased water but was not significant
Pine trees	Georgia, United States	1 year	Partial tree harvest, D	–	•	Jones et al. (2003)	Tree harvest increased soil moisture and decreased root lifespan. Possible interaction with variability in soil temp
Beech trees	Southern Germany	1 year	Field observations, I	•	–	Mainiero and Kazda (2006)	Study observed increased mortality during natural drought
Pine trees	California, United States	1 year	Field observations, I	•	–	Misson et al. (2006)	Study observed increased mortality during natural drought
Desert shrub	Utah, United States	2 years	Field observations, I	•	–	Peek et al. (2006)	Observed increased mortality during drought. However, there was no explicit drought treatment
Beech seedlings	Germany	2 years	Controlled soil water in bins, D	•	–	Meier and Leuschner (2008)	Reduced root lifespan due to drought was consistent across seedlings of *F. sylvatica* from four different populations
Mixed evrg. trees	California, United States	3 years	Field observations, I	–	•	Kitajima et al. (2010)	Based on variation in ambient temperature and precipitation
Evrg/decid. seedlings	South Carolina, United States	6 years	Irrigated field plots, D	–, 0	•	Coleman, unpublished	Decreased lifespan observed in *Pinus taeda*, while *Po. deltoides* was unaffected
Deciduous tree planting	Pennsylvania, United States	3 years	Irrigation tubes in field, D	–, 0	•	Adams et al., unpublished	Lifespan decreased in three species, while one was unaffected

Note: Methods listed pertain to the experimental system used as well as whether direct observations (D) of fine-root lifespan were reported or if the effect on fine-root lifespan was inferred through indirect observations (I) or related evidence such root turnover.

However, when roots were allowed to acclimate over a period of 6 days, the effect of increased temperature was vastly reduced. Increased temperature may also indirectly increase root metabolism and aging through increased nutrient uptake. Similar to increased water availability, elevated temperatures may increase mineralization rates of many nutrients increasing their availability in the soil. Furthermore, increased temperature would likely increase whole-plant transpiration rate. Both factors would increase the specific root uptake rates of soil nutrients including nitrate (NO_3^-), which has a high metabolic cost for uptake. Still, in some systems, changes in temperature may have little to no effect on root lifespan. Johnson et al. (2006) found that much like elevated CO_2, elevated temperature had little to no effect on fine-root lifespan in their nutrient-poor system.

Most studies suggest that increasing temperature will have a negative effect on fine-root lifespan. However, some have found that temperature variability may also play a role in mediating root lifespan, while others have found unique interactions with winter snow removal, which reduces soil temperatures in winter, also potentially reducing lifespan. In addition to altering soil moisture by tree removal, Jones et al. (2003) increased the variability of soil temperature along the gap edges by allowing greater solar radiation to penetrate to the forest floor. Increased short-term variability in temperature was associated with a negative effect on fine-root lifespan despite maintaining a relatively constant longer-term average. While the direct mechanism is unknown, it seems plausible that the increased temperature variability reduced the ability of roots to acclimate to increases in temperature leading to consistently higher metabolism, respiration rates, and faster aging. Using a different experimental manipulation, two other studies have reported increased fine-root mortality following winter snow removal in forest systems (Tierney et al. 2001; Gaul et al. 2008). These experiments represent an interesting interaction between potential changes in winter precipitation and resulting decreases in soil temperature as the insulating snow layer was no longer present. Gaul et al. (2008) were also able to show increased variability in soil temperature following snow removal as well, which may have acted in a similar manner to the effects observed by Jones et al. (2003).

In summary, higher temperatures have occasionally, but not always, been associated with shorter root lifespan. It is difficult, however, to distinguish the direct effects of temperature on root lifespan from the numerous indirect effects that temperature can have on the abiotic and biotic factors that influence root longevity.

D. Soil Nitrogen

Atmospheric reactive nitrogen (N) emissions have increased as a result of N fertilization and fossil fuel combustion (Galloway et al. 2004, 2008). These emissions are deposited back onto terrestrial ecosystems, causing roughly 5- to 50-fold increases over preindustrial N deposition rates in parts of Asia, Europe, and North America (Dentener et al. 2006). In addition, soil warming can increase rates of organic matter mineralization, leading to increase in soil N (Melillo et al. 2002, 2011). The effects of increases in soil N on root lifespan are context dependent and may lead to increases, decreases, or no effect on root lifespan (Table 27.5).

Among studies that have tracked the fate of individual roots, increased nitrogen availability has been associated with decreased root lifespan in *Populus* (Pregitzer et al. 1995, 2000), *P. abies* (Majdi and Kangas 1997; Majdi and Andersson 2005), and *Pi. ponderosa* (Johnson et al. 2000). Nitrogen addition can lead to decreases in soil pH and reductions in base cation/aluminum ratios, which can also lead to reduced root longevity (Du 2011). Plot-level increases in soil N has also led to increases in root lifespan in *Populus deltoides* W. Bartram ex Marshall stands (Kern et al. 2004) and *Fraxinus mandshurica* Rupr. stands (Yu et al. 2009).

In an efficiency context, plants should optimize carbon expenditure for uptake of nutrients that limit growth. To predict how increased nutrient availability might affect optimal root lifespan requires information on root respiration and uptake capacity as a function of root age. In some plants, roots of high metabolic activity associated with high root N concentrations might be expected to exhibit rapid declines in uptake capacity with age and therefore earlier mortality than roots of lower metabolic activity (cf. Pregitzer et al. 1998). Strong reductions in root lifespan with N deposition might occur where plant growth is not limited by N (i.e., N saturated), because of increased costs without a concomitant increase in benefit.

Spatially localized nutrient enrichment can have different effects on root efficiency and root longevity than variation in site fertility, in which the whole root system is affected. Studies with trees in the field have demonstrated enhanced fine-root persistence in fertile patches (Pregitzer et al. 1993; Fahey and Hughes 1994; Wells 1999; Baldi et al. 2010; Adams et al. unpublished data). For example, greater root lifespan has been observed in localized fertile patches created with water and nitrogen in a forest stand dominated by *Po. grandidentata*

TABLE 27.5 Studies of at Least 2 Years Reporting Influence of Added N on Fine-Root Lifespan

Dominant Species	Location	Length of Addition; Fertilization Amount	Effect Size of N	References
Fr. mandshurica	Mt. Maoer, China	2 years; 100 kg $ha^{-1}year^{-1}$ N	+23.5%	Yu et al. (2009)
Prunus persica	Italy	5 years; compost (240 kg $ha^{-1}year^{-1}$ N)	+19%	Baldi et al. (2010)
P. abies	Southwest Sweden	4 years; 100 kg $ha^{-1}year^{-1}$ N	+12.5%	Majdi et al. (2001)
Po. deltoides	Wisconsin	2 years; 50 (N50) 100 (N100) 200 (N200) kg $ha^{-1}year^{-1}$ N	+10% (N50), +13% (N100), +16% (N200)	Kern et al. (2004)
Po. deltoides	South Carolina	6 years; 80 kg $ha^{-1}year^{-1}$ N	+5.7%	Coleman, unpublished
Larix gmelinii	Mt. Maoer, China	2 years; 100 kg $ha^{-1}year^{-1}$ N	No effect	Yu et al. (2009)
Pi. ponderosa	California	4 years; 100 and 200 kg $ha^{-1}year^{-1}$ N	No effect	Phillips et al. (2006)
Pinus palustris	Georgia	3 years; 50 kg $ha^{-1}year^{-1}$ N	No effect	Guo et al. (2008)
Pr. persica	Italy	5 years; mineral (70–130 kg $ha^{-1}year^{-1}$ N)	No effect	Baldi et al. (2010)
Pi. taeda	South Carolina	6 years; 80 kg $ha^{-1}year^{-1}$ N	−45%	Coleman, unpublished
Pi. ponderosa	California	3 years; 100 (N100) and 200 (N200) kg $ha^{-1}year^{-1}$ N	−34% (N100) −43% (N200)	Johnson et al. (2000)
P. abies	Northern Sweden	13 years; 92.3 kg $ha^{-1}year^{-1}$ N	−9%	Majdi and Andersson (2005)

Michx., *Prunus pensylvanica* L. f., and other second-growth hardwoods (Pregitzer et al. 1993); in a forest dominated by *Acer saccharum* Marshall, *F. grandifolia*, and *Betula alleghaniensis* Britton (Fahey and Hughes 1994); and in a nearly pure 60-year-old *A. saccharum* stand (Wells 1999).

In summary, there are conflicting results on the effects of soil nutrients on root lifespan. Inconsistencies may be partially related to indirect effects of fertility and soil acidity and to differences in methodology. More direct observations of the survival of individual roots along fertility gradients, in long-term fertilization trials, and in response to nutrient patches are needed before we can generalize about the effect of soil fertility on lifespan. As with our discussion on effects of other global-change factors, additional measurements of shoot growth, leaf N concentration, as well as root respiration will help in the determination whether the additional N improves root efficiency or not.

IV. Summary

The ability of a root system to forage efficiently for water and nutrients depends on the production and loss of individual roots in soil of spatially and temporally heterogeneous moisture and fertility and on the physiological activity of these roots, which changes with age. There can be a wide variation in root lifespan, and sources for the variation are not well understood. Predicting how global change may lead to changes in root lifespan is complicated by both species variation in how they may respond to a change in their environment and how the influence of a global-change factor such as precipitation or nitrogen depends on initial site conditions.

Both abiotic and biotic factors affect root lifespan and often these factors interact. Higher moisture, for example, may diminish root lifespan more by allowing for more root herbivores and pathogens then by directly affecting root maintenance costs or uptake. Moreover, any global-change factor that affects the plants ability to assimilate carbon may also affect the likelihood that root lifespan may increase or decrease. Despite these considerable challenges, a review of the literature suggests that higher soil temperatures are likely to diminish root lifespan. Strong excess or deficits in soil moisture are also likely to either not affect or decrease root lifespan, depending on the plant species. Available evidence suggests that extended exposure to elevated CO_2 may have only modest effects on root lifespan, but more study is needed. The influence of N deposition is particularly uncertain, as it can strongly decrease, increase, or not affect root lifespan. We expect some of the variation in N deposition experiments is due to the relatively short duration and variation in lithology affecting soil acidity and soil thickness. Additional studies examining factors influencing root lifespan combined with optimization modeling are needed to unravel the numerous controls and constraints on the influence of global change on the lifespan of plant roots.

References

Aber JD, Melillo JM, Nadelhoffer KJ, McClaugherty CA, Pastor J 1985 Fine root turnover in forest ecosystems I relation to quantity and form of nitrogen availability: A comparison of two methods. *Oecologia* 66:317–321.

Aerts R 1995 The advantages of being evergreen. *Trends Ecol Evol* 10:402–407.

Aerts R, Bakker C, De Caluwe H 1992 Root turnover as determinant of the cycling of C, N and P in a dry heathland ecosystem. *Biogeochemistry* 15:175–190.

Anderson LJ, Comas JH, Lakso AN, Eissenstat DM 2003 Multiple risk factors in root survivorship: A 4-year study in Concord grape. *New Phytol* 158:489–501.

Anderson LJ, Derners JD, Polley HW, Gordon WS, Eissenstat DM, Jackson RB 2010 Root responses along a subambient to elevated CO_2 gradient in a C_3-C_4 grassland. *Global Change Biol* 16:454–468.

Baldi E, Toselli M, Eissenstat DM, Marangoni B 2010. Organic fertilization leads to increased peach root production and lifespan. *Tree Physiol* 30:1373–1382.

Bauerle TL, Richards JH, Smart DR, Eissenstat DM 2008 Importance of internal hydraulic redistribution for prolonging the lifespan of roots in dry soil. *Plant Cell Environ* 31:177–186.

Black KE, Harbron CG, Franklin M, Atkinson D, Hooker JE 1998 Differences in root longevity of some tree species. *Tree Physiol* 18:259–293.

Brundrett MC, Kendrick B 1988 The mycorrhizal status, root anatomy, and phenology of plants in a sugar maple forest. *Can J Bot* 66:1153–1173.

Bryla DR, Bouma TJ, Hartmond U, Eissenstat DM 2001 Influence of temperature and soil drying on respiration of individual roots in citrus: Integrating greenhouse observations into a predictive model for the field. *Plant Cell Environ* 24:781–790.

Caldwell MM 1987 Competition between roots in natural communities. In *Root Development and Function*, eds. PJ Gregory, JV Lake, DA Rose, pp. 167–185. New York: Cambridge University Press.

Chapin III FS 1980 The mineral nutrition of wild plants. *Annu Rev Ecol Syst* 11:233–260.

Curtis PE, Wang X 1998 A meta-analysis of elevated CO_2 effects on woody plant mass, form, and physiology. *Oecologia* 113:299–313.

Dahlman RC, Kucera RL 1965 Root productivity and turnover in native prairie. *Ecology* 46:84–89.

Dentener F, Stevenson D, Ellingsen K et al. 2006 The global atmospheric environment or the next generation. *Environ Sci Technol* 40:3586–3594.

Drake JE, Gallet-Budynek A, Hofmockel KS et al. 2011 Increases in the flux of carbon belowground stimulate nitrogen uptake and sustain the long-term enhancement of forest productivity under elevated CO_2. *Ecol Lett* 14:349–357.

Du Q 2011 The influence of nitrogen deposition on root lifespan in Northeastern temperate forests. MS thesis. University Park, PA: The Pennsylvania State University.

Eissenstat DM 1998 Responses of fine roots to dry surface soil: A case study in citrus. In *Radical Biology: Advances and Perspectives in the Function of Plant Roots*, eds. HE Flores, JP Lynch, DM Eissenstat. Curr Topics Plant Physiol. Vol. 17, pp. 224–237. Rockville, MD: American Society of Plant Physiology.

Eissenstat DM, Achor DS 1999 Anatomical characteristics of roots of citrus rootstocks that vary in specific root length. *New Phytol* 141:309–321.

Eissenstat DM, Yanai RD 1997 The ecology of root lifespan. *Adv Ecol Res* 27:1–62.

Eissenstat DM, Yanai RD 2002 Root Lifespan, efficiency, and turnover. In *Plant Roots: The Hidden Half*, eds. Y Waisel, A Eshel, U Kafkafi, 3rd edn., pp. 221–238. New York: Marcel Dekker, Inc.

Fahey TJ, Hughes JW 1994 Fine root dynamics in a northern hardwood forest ecosystem, Hubbard Brook Experimental Forest, NH. *J Ecol* 82:533–548.

Finzi AC, Norby RJ, Calfapietra C et al. 2007 Increases in nitrogen uptake rather than nitrogen-use efficiency support higher rates of temperate forest productivity under elevated CO_2. *Proc Natl Acad Sci USA* 104:14014–14019.

Fitter AH, Graves JD, Self GK, Brown TK, Bogie DS, Taylor K 1998 Root production, turnover and respiration under two grassland types along an altitudinal gradient: Influence of temperature and solar radiation. *Oecologia* 114:20–30.

Fitter AH, Self GK, Brown TK, Bogie DS, Graves JD, Benham D, Ineson P 1999 Root production and turnover in an upland grassland subjected to artificial soil warming respond to radiation flux and nutrients, not temperature. *Oecologia* 120:575–581.

Forbes PJ, Black KE, Hooker JE 1997 Temperature-induced alteration to root longevity in *Lolium perenne*. *Plant Soil* 190:87–90.

Galloway JN, Dentener FJ, Capone DG, Boyer EW, Howrath RW, Seitzinger SP, Asner GP, Cleveland CC, Green PA, Holland EA 2004 Nitrogen cycles: Past, present and future. *Biogeochemistry* 70:153–226.

Galloway JN, Townsend AR, Erisman JW, Bekunda M, Cai ZC, Freney JR, Martinelli LA, Seitzinger SP, Sutton MA 2008 Transformation of the nitrogen cycle: Recent trends, questions, and potential solutions. *Science* 320:889–892.

Gaudinski JB, Trumbore SE, Davidson EA, Zheng S 2000 Soil carbon cycling in a temperate forest: Radiocarbon-based estimates of residence times, sequestration rates and partitioning of fluxes. *Biogeochemistry* 51:33–69.

Gaul D, Hertel D, Leuschner C 2008 Effects of experimental soil frost on the fine-root system of mature Norway spruce. *J Plant Nutr Soil Sci* 171:690–698.

Gill RA, Jackson RB 2000 Global patterns of root turnover for terrestrial ecosystems. *New Phytol* 147:13–31.

Grime JP 1977 Evidence for the existence of three primary strategies in plants and its relevance to ecological and evolutionary theory. *Am Nat* 111:1169–1194.

Guo D, Xia M, Wei X, Chang W, Liu Y, Wang Z 2008 Anatomical traits associated with absorption and mycorrhizal colonization are linked to root branch order in twenty-three Chinese temperate tree species. *New Phytol* 180:673–683.

Harris GA 1967 Some competitive relationships between *Agropyron spicatum* and *Bromus tectorum*. *Ecol Monogr* 37:89–111.

Hendrick RL, Pregitzer KS 1993 Patterns of fine root mortality in two sugar maple forests. *Nature* 361:59–61.

Huang B, Nobel PS 1992 Hydraulic conductivity and anatomy for lateral roots of *Agave deserti* during growth and drought-induced abscission. *J Exp Bot* 43:1441–1449.

Intergovernmental Panel on Climate Change 2007 Climate Change 2007: The Physical Science Basis—Contribution of Working Group I to the Fourth Assessment Report of the Intergovernmental Panel on Climate Change, eds. S. Solomon, et al. New York: Cambridge University Press.

Iversen CM, Ledford J, Norby RJ 2008 CO_2 enrichment increases carbon and nitrogen input from fine root in a deciduous forest. *New Phytol* 179:837–847.

Jackson RB, Cook CW, Pippen JS, Palmer SM 2009 Increased belowground biomass and soil CO_2 fluxes after a decade of carbon dioxide enrichment in a warm-temperate forest. *Ecology* 90:3352–3366.

Jackson RB, Mooney HA, Schulze ED 1997 A global budget of fine root biomass, surface area, and nutrient contents. *Proc Natl Acad Sci USA* 94:7362–7366.

Johnson MG, Phillips DL, Tingey DT, Storm MJ 2000 Effects of elevated CO_2, N-fertilization, and season on survival of ponderosa pine fine roots. *Can J For Res* 30:220–228.

Johnson MG, Rygiewicz PT, Tingey DT, Phillips DL 2006 Elevated CO_2 and elevated temperature have no effect on Douglas-fir fine-root dynamics in nitrogen-poor soil. *New Phytol* 170:345–356.

Jones RH, Mitchell RJ, Stevens GN, Pecot SD 2003 Controls of fine root dynamics across a gradient of gap sizes in a pine woodland. *Oecologia* 134:132–143.

Joslin JD, Wolfe MH, Hanson PJ 2000 Effects of altered water regimes on forest root systems. *New Phytol* 147:117–129.

Kern CC, Friend AL, Johnson J M-F, Coleman MD 2004 Fine root dynamics in a developing *Populus deltoides* plantation. *Tree Phys* 24:651–660.

Kitajima K, Anderson KE, Allen MF 2010 Effect of soil temperature and soil water content on fine root turnover rate in a California mixed conifer ecosystem. *J Geophys Res* 115:1–12.

Kosola KR, Eissenstat DM 1994 The fate of surface roots of citrus seedlings in dry soil. *J Exp Bot* 45:1639–1645.

Kosola KR, Eissenstat DM, Graham JH 1995 Root demography in mature citrus trees: The influence of *Phytophthora nicotianae*. *Plant Soil* 171:283–288.

Kubiske ME, Pregitzer KS, Zak DR, Mikan CJ 1998 Growth and C allocation of *Populus tremuloides* genotypes in response to atmospheric CO_2 and soil N availability. *New Phytol* 140:251–260.

Lamont BB 1995 Mineral nutrient relations in Mediterranean regions of California, Chile and Australia. In *Ecology and Biogeography of Mediterranean Ecosystems of Chile, California, and Australia*, eds. MTK Arroyo, PH Zedler, MD Fox, pp. 211–238. New York: Springer.

Mainiero R, Kazda M 2006 Depth-related fine root dynamics of *Fagus sylvatica* during exception drought. *For Ecol Manage* 237:135–142.

Majdi H, Andersson P 2005 Fine root production and turnover in a Norway spruce stand in Northern Sweden: Effects of nitrogen and water manipulation. *Ecosystems* 8:191–199.

Majdi H, Damm E, Nylund J-E 2001 Longevity of mycorrhizal roots depends on branching order and nutrient availability. *New Phytol* 150:195–202.

Majdi H, Kangas P 1997 Demography of fine roots in response to nutrient applications in a Norway spruce stand in southwestern Sweden. *Ecoscience* 4:199–205.

Matamala R, Gonzalez-Meler MA, Jastrow JD, Norby R, Schlesinger WH 2003 Impacts of fine root turnover on forest NPP and soil C sequestration potential. *Science* 302:1385–1387.

McCormack ML, Adams TS, Smithwick EAH, Eissenstat DM 2012 Predicting fine root lifespan from plant functional traits in temperate trees. *New Phytol* 195:823–831.

McCormack ML, Pritchard SG, Breland S et al. 2010 Soil fungi respond more strongly than fine roots to elevated CO_2 in a model regenerating longleaf pine-wiregrass ecosystem. *Ecosystems* 12:901–916.

Meier I, Leuschner C 2008 Genotypic variation and phenotypic plasticity in the drought response of fine roots of European beech. *Tree Physiol* 28:297–309.

Melillo JM, Butler S, Johnson J et al. 2011 Soil warming, carbon–nitrogen interactions, and forest carbon budgets. *Proc Natl Acad Sci USA* 108:9508–9512.

Melillo JM, Steudler PA, Aber JD et al. 2002 Soil warming and carbon-cycle feedbacks to the climate system. *Science* 298:2173–2176.

Milchunas DG, Morgan JA, Mosier AR, Lecain DR 2005 Root dynamics and demography in shortgrass steppe under elevated CO_2, and comments on minirhizotron methodology. *Global Change Biol* 11:1837–1855.

Misson L, Gershenson A, Tang J, McKay M, Cheng W, Goldstein A 2006 Influences of canopy photosynthesis and summer rain pulses on root dynamics and soil respiration in a young ponderosa pine forest. *Tree Physiol* 26:833–844.

North GB, Huang B, Nobel PS 1993 Changes in the structure and hydraulic conductivity of desert succulents as soil water status varies. *Bot Acta* 106:126–135.

Oren R, Ellsworth DS, Johnsen KH et al. 2001 Soil fertility limits carbon sequestration by forest ecosystems in a CO_2-enriched atmosphere. *Nature* 411:469–472.

Peek MS, Leffler AJ, Hipps L, Ivans S, Ryel RJ, Caldwell MM 2006 Root turnover and relocation in the soil profile in response to seasonal soil water variation in a natural stand of Utah juniper (*Juniperus osteosperma*). *Tree Physiol* 26:1469–1476.

Phillips DL, Johnson MG, Tingey DT, Catricala CE, Hoyman TL, Nowak, RS 2006 Effects of elevated CO_2 on fine root dynamics in a Mojave Desert community: A FACE study. *Global Change Biol* 12:61–73.

Phillips DL, Johnson MG, Tingey DT, Storm MJ 2009 Elevated CO_2 and O_3 effects on fine-root survivorship in ponderosa pine mesocosms. *Oecologia* 160:827–837.

Polverigiani S, McCormack ML, Mueller CW, Eissenstat DM 2011 Growth and physiology of olive pioneer and fibrous roots exposed to soil moisture deficits. *Tree Physiol* 31:1228–1237.

Pregitzer KS, Burton AJ, King JS, Zak DR 2008 Soil respiration, root biomass, and root turnover following long-term exposure of northern forests to elevated atmospheric CO_2 and tropospheric O_3. *New Phytol* 180:153–161.

Pregitzer KS, Hendrick RL, Fogel R 1993 The demography of fine roots in response to patches of water and nitrogen. *New Phytol* 125:575–580.

Pregitzer KS, Laskowski MJ, Burton AJ, Lessard VC, Zak DR 1998 Variation in sugar maple root respiration with root diameter and soil depth. *Tree Physiol* 18:665–670.

Pregitzer KS, Zak DR, Maziasz J, DeForest J, Curtis PS, Lussenhop J 2000 Interactive effects of atmospheric CO_2 and soil-N availability on fine roots of *Populus tremuloides*. *Ecol Appl* 10:18–33.

Pregitzer KS, Zak DR, Curtis PS, Kubiske ME, Teeri JA, Vogel CS 1995 Atmospheric CO_2, soil nitrogen and fine root turnover. *New Phytol* 129:579–585.

Pritchard SG, Strand AE, McCormack ML et al. 2008 Fine root dynamics in a loblolly pine forest are influenced by free-air-CO_2-enrichment: A six-year-minirhizotron study. *Global Change Biol* 14:588–602.

Reich PB, Walters MB, Ellsworth DS 1997 From tropics to tundra: Global convergence in plant functioning. *Proc Natl Acad Sci USA* 94:13730–13734.

Ryser P 1996 The importance of tissue density for growth and life span of leaves and roots: A comparison of five ecologically contrasting grasses. *Func Ecol* 10:717–723.

Smith SE, Read DJ 1997 *Mycorrhizal Symbiosis*, 2nd edn. San Diego, CA: Academic Press.

Stover DB, Day FP, Drake BG, Hinkle CR 2010 The long-term effects of CO_2 enrichment on fine root productivity, mortality, and survivorship in a scrub oak ecosystem at Kennedy Space Center, Florida, USA. *Environ Exp Bot* 69:214–222.

Tierney GL, Fahey TJ, Groffman PM, Hardy JP, Fitzhugh RD, Driscoll CT 2001 Soil freezing alters fine root dynamics in a northern hardwood forest. *Biogeochemistry* 56:175–190.

Tierney GL, Fahey TJ, Groffman PM, Hardy JP, Fitzhugh RD, Driscoll CT, Yavitt JB 2003 Environmental control of fine root dynamics in a northern hardwood forest. *Global Change Biol* 9:670–679.

Thomas SM, Whitehead D, Reid JB, Cook FJ, Adams JA, Leckie AC 1999 Growth, loss, and vertical distribution of *Pinus radiata* fine roots growing at ambient and elevated CO_2 concentration. *Global Change Biol* 5:107–121.

Thornley JHM 1998 Modelling shoot:root relations: The only way forward? *Ann Bot* 81:165–171.

Valenzuela-Estrada LR, Richards JH, Diaz A, Eissenstat DM 2009 Patterns of nocturnal rehydration in root tissues of *Vaccinium corymbosum* L. under severe drought conditions. *J Exp Bot* 60:1241–1247.

Valenzuela-Estrada LR, Vera-Caraballo V, Ruth LE, Eissenstat DM 2008 Root anatomy, morphology and longevity among root orders in *Vaccinium corymbosum* (Ericaceae). *Am J Bot* 95:1506–1514.

Vogt KA, Grier CC, Vogt DJ 1986 Production, turnover, and nutrient dynamics of above- and belowground detritus of world forests. *Adv Ecol Res* 15:303–377.

Watson CA, Ross JM, Bagnaresi U et al. 2000 Environment-induced modifications to root longevity in *Lolium perenne* and *Trifolium repens*. *Ann Bot* 85:397–401.

Wells CE 1999 Advances in the fine root demography of woody species. PhD thesis. University Park, PA: Pennsylvania State University.

Wilson JB 1988 Shoot competition and root competition. *J Appl Ecol* 25:279–296.

Withington JM, Reich PB, Oleksyn J, Eissenstat DM. 2006 Comparisons of structure and life span in roots and leaves among temperate trees. *Ecol Monogr* 76:381–397.

Wright IJ, Reich PB, Westoby M et al. 2004 The worldwide leaf economics spectrum. *Nature* 428:821–827.

Xia M, Guo D, Pregitzer KS 2010 Ephemeral root modules in *Fraxinus mandshurica*. *New Phytol* 188:1065–1074.

Yu SQ, Wang ZQ, Shi JW, Yu LZ, Quan XK 2009 Effects of nitrogen fertilization on fine root lifespan of *Fraxinus mandshurica* and *Larix gmelinii*. *J Appl Ecol* 20:2332–2338.

Zadworny M, Eissenstat DM 2011 Contrasting the morphology, anatomy, and fungal colonization of new pioneer and fibrous roots. *New Phytol* 190:213–221.

28

Developmental Adaptations in Roots of Desert Plants with Special Emphasis on Cacti

Joseph G. Dubrovsky
Universidad Nacional Autónoma de México

Svetlana Shishkova
Universidad Nacional Autónoma de México

I. Introduction

An important question in root developmental biology would be how adaptations acquired during millennia of plant evolution affect current root development, or, the subject of this chapter, which root developmental mechanisms become adaptive traits in desert plants. Therefore, we would elucidate a framework of developmental processes in roots of desert plants that can help us understand how they are related to plant adaptation to this specific environment. The concept of "developmental adaptations" implies those developmental programs that evolved in response to environmental conditions (Dubrovsky and North 2002). Here, we apply this concept to roots of desert plants. We will first give a short overview of desert environmental conditions; next, we will outline the main ecological groups of desert plants and then consider the main developmental adaptations of root systems in desert environments. These adaptations will be analyzed in various plant species but with special emphasis on Cactaceae as one of the best studied groups of desert plants in terms of root development. General reviews on roots of desert plants and their adaptive features can be found in the literature (Cannon 1911; Drew 1979; Rundel and Nobel 1991; Dubrovsky and North 2002; Nobel 2002).

II. Desert Environmental Conditions

Desert ecosystems are "water-controlled ecosystems with infrequent, discrete, and largely unpredictable water inputs" (Noy-Meir 1973). According to the widely accepted system of Meigs (1953), there are three types of arid, or desert, regions: (a) extremely arid lands that receive less than 60–100 mm of mean annual precipitation where no rainfall may occur the whole year; (b) arid zones that get less than 250 mm (60–100 to 150–250 mm) of annual rainfall; and (c) semiarid areas, sometimes called semideserts, that have a mean annual precipitation from 150–250 to 250–500 mm (Noy-Meir 1973; Walter 1985). Importantly, in deserts, much more water may be lost by evapotranspiration than is received in rainfall.

Hot deserts are usually located in the north or south subtropical areas. Desert regions near their equatorial margins receive most of their scarce rainfall in summer, while regions near the higher latitude margins receive precipitation, if ever, in winter. Intermediate regions have erratic infrequent rainfalls that can occur any time of the year (McGinnies 1979a). Temperate (cold) deserts are located at higher latitudes and are characterized by low winter temperatures and snowfalls, although summer can be rather hot. A world map of the major hot and temperate deserts can be found in the previous edition of this book (Nobel 2002), as well as in other specialized texts on deserts (McGinnies 1979b; Walton 2009). Freezing zones of the pole regions are sometimes called "polar deserts"; water is mainly unavailable there because it exists as ice and snow. As vegetation in these regions is rather limited and adaptations for survival are very different from that of hot or temperate desert plant species, root adaptations of polar desert plants are not covered in this chapter.

Whereas temperature and radiation fluctuate in deserts in predictable and similar patterns over the year and year by year, rainfall events can be very irregular and unpredictable and can vary to a great extent between years. Still, generally in deserts, there are from 10 to 50 rainy days each year. During the few hours when precipitation occurs, it may be rather heavy (Noy-Meir 1973). While hot deserts experience very high daytime temperatures, especially in summer, nighttime temperatures can fall drastically. Cloud-free sky leads to high daytime solar radiation and also permits rapid loss by long-wave radiation at night. Daytime temperature of the soil surface commonly surpasses that of the air. Many studies have shown that temperatures at the soil surface in deserts can exceed 70°C, whereas minimum temperature, for example, in the Sonoran Desert can be as low as 0°C (Jordan and Nobel 1984; Nobel 1988).

III. Main Groups of Desert Plants and General Features of Their Root Systems

The roots of different species are heavily intermingled in the upper horizon of the desert soils (Gibbens and Lenz 2001). The specific environmental conditions will affect competition between desert plant species. For example, optimal temperatures for root growth are distinct for different species (Drennan and Nobel 1996; Nobel et al. 1992). Heterogeneity of nutrients and water in soil patches (Rundel and Nobel 1991; Martre et al. 2002) is another important factor that affects root competition. Differences in nutrient uptake capacity among groups of desert plants, for example, between grasses and shrubs (Ivans et al. 2003) or between distinct types of shrubs (James and Richards 2007), are also important determinants of root competition. In spite of almost permanent water deficiency in deserts, most desert plants have roots colonized by mycorrhizal fungi (Bethlenfalvay et al. 1984; Jacobson 1997; Carrillo-Garcia et al. 1999; Collier et al. 2003; Tian et al. 2006), which is an important adaptation for desert soils that are commonly poor in mineral nutrients. In the following, we consider the general features of roots of main groups of desert plants.

A. Resurrection Plants

Desiccation-tolerant resurrection plants that form roots belong to Lycopodiophyta and Magnoliophyta divisions (Alpert 2000). A representative desert species of Pteridophytes, *Selaginella lepidophylla* (Hook. & Grev.) Spring, native to the Chihuahuan Desert, has a shallow root system (Nobel 2002). Among angiosperms, resurrection plants of South Africa can withstand close to 0% relative humidity in their tissues (Gaff 1971). Their habitat is not strictly desert, but can be very similar with rocky surfaces of granitic inselbergs and ferricretes at high elevation (Fischer 2004). A classical example of resurrection angiosperms is *Craterostigma plantagineum* Hochst. of the Scrophulariaceae family. This interesting species accumulates stachyose in the roots at the highest concentration recorded in a plant tissue (over 40% of the dry weight). This carbohydrate reserve is important for dehydration tolerance of the root tissues and is metabolized to sucrose during dehydration. Decrease in stachyose concentration correlates with accumulation of sucrose that has a protective role in dehydrated tissues (Norwood et al. 2003).

B. Ephemeral Plants

Ephemeral plants have a short life cycle of their aboveground organs and include two subgroups: annuals and ephemeroids. The only reserve of annuals is the seed whose germination depends exclusively on external water. Ephemeroids are perennials whose aboveground organs appear only for a short period of the year and their storage organs (bulbs and rhizomes) persist below the soil surface (Noy-Meir 1973). Root systems of annuals are generally shallow. For example, roots of Mojave Desert winter annuals are found mainly within 10–20 cm of the upper soil layer (Bell et al. 1979). However, under certain conditions, spring annuals that germinate in late fall overwinter, forming small rosettes, and have low evapotranspiration, while their roots continue growing into deeper layers and may reach depth of 1 m (Gutierrez and Whitford 1987). Therefore, they can have access to more nutrients. Overall root dry mass allocation ranges between 4% and 13% in eight annual species of Mojave Desert (Bell et al. 1979) indicating low carbon investment in root development. Roots of ephemeral plants, both annual and perennial, are colonized by arbuscular and other type of mycorrhizae. In a survey of 73 ephemerals of 20 families from Gurbantunggut Desert (Junggar Basin, northwest China), no species was recorded as nonmycorrhizal (Shi et al. 2006). In a detailed analysis of five ephemeral species, it was found that the proportion of root colonized ranged from 2% to 85% with an average of 19% (Shi et al. 2007).

C. Perennial Grasses

Root systems of perennial grasses reach on average 35 cm (Nobel 2002) and a maximum 1.6 m in soil depth (Gibbens and Lenz 2001). The roots of a single plant can often spread out to a significant distance around the aboveground organs. A Saharan desert grass, *Aristida pungens* Desf., extends its roots to 20 m (Price 1911). Perennial grasses are well adapted to harsh desert environments, and their optimal temperature for root growth is one of the highest among desert species. For example, for a Sonoran Desert species *Pleuraphis rigida* Thurb., optimal root growth temperature is 35°C–37°C (Drennan and Nobel 1996).

The soil depth at which optimal root growth rate is attained by *P. rigida* is 4–9 cm (Franco and Nobel 1989, 1990; Drennan and Nobel 1996).

D. Shrubs and Trees

Desert shrubs and trees, except phreatophytes (see Section III.E), have common characteristics in their root system distribution. The average root depth in different species of this group ranges from 20 to 80 cm (Nobel 2002). Roots of shrubs and trees penetrate relatively deep and are also widespread beneath the soil surface (Figure 28.1; Gibbens and Lenz 2001). Interestingly, in four different shrubs of the Mojave Desert, some degree of nonoverlapping spatial root segregation was found when fine root growth among different depths and microsites was analyzed (Wilcox et al. 2004). Richards and Caldwell (1987) showed that the roots of the Great Basin Desert shrub *Artemisia tridentata* Nutt. behave as conduits for hydraulic lift, which is a process of water movement by roots from relatively moist to relatively dry soil layers. Water absorbed during daytime by roots at all soil depths enters the transpiration stream. During nighttime, when transpiration is reduced, water efflux occurs passively by roots due to significant water potential differences between root tissues and neighboring dry soil. So, water is liberated to dryer soil layers and becomes available for nutrient uptake by shallow roots (Richards and Caldwell 1987; Caldwell and Richards 1989; Caldwell et al. 1998). Even though the phenomenon of hydraulic lift is known to also occur in nondesert plants, its role in a harsh desert environment is essential for water economy and root growth maintenance (Caldwell et al. 1998). The fact that fine root growth in different shrubs may predominantly occur at different soil depths (Wilcox et al. 2004) raises the question of how the phenomenon of hydraulic lift in one species can affect the water supply and root growth of the other species in the desert plant community.

E. Phreatophytes

Trees and shrubs whose roots reach deep underground water tables are considered phreatophytes. It is commonly accepted that these species have the deepest root systems. For example, *Boscia albitrunca* (Burch.) Gilg & Gilg-Ben. in the African Kalahari Desert reaches a depth of 68 m (Canadell et al. 1996), and *Prosopis juliflora* (Sw.) DC. from the Sonoran Desert penetrates to almost 53 m (Phillips 1963). Under some geological conditions, for example, in Central Asian Taklamakan Desert where annual precipitation is extremely low, about 35 mm per year, permanent groundwaters are not very deep and range between 2 and 24 m (Gries et al. 2003; Arndt et al. 2004). Under these hyperarid conditions, trees and shrubs survive only because of groundwater access. They apparently are not subjected to physiologically significant water stress and maintain vigorous growth (Gries et al. 2003). Mineral nutrients in deserts are another limiting factor (Bohrer et al. 2001), and the species in Taklamakan Desert obtain nitrates from groundwater though its concentration is very low (Arndt et al. 2004). Due to water availability, some phreatophytes from the Fabaceae family maintain their capability to atmospheric nitrogen fixation (Shearer et al. 1983; Arndt et al. 2004). Interestingly, when surface water becomes available, as a result of snowmelt in the adjacent mountains, many phreatophytic shrubs and trees do not use this additional source of water and no fine roots are produced near the soil surface (Arndt et al. 2004; Zeng et al. 2006). Similarly, low-level root branching is described for some phreatophytic shrubs of Kazakh Desert where underground waters are at about 6.5 m (Baytulin 1987). This illustrates that permanent water source is a dominant factor defining root architecture in this group of desert plants.

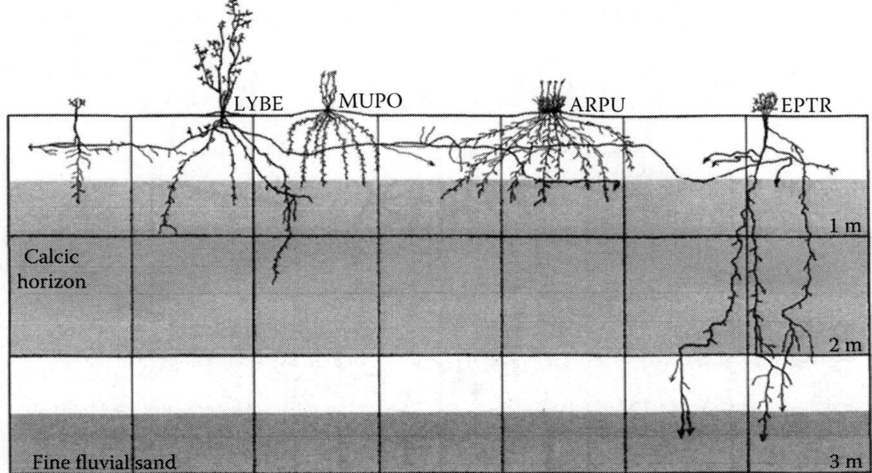

FIGURE 28.1 Roots systems of some shrubs and grasses from the Chihuahuan Desert. A wolfberry shrub (LYBE, *Lycium berlandieri*), two bunch grasses, bush muhly (MUPO, *Muhlenbergia porteri*), red threeawn (ARPU, *Aristida purpurea*), and an ephedra shrub (EPTR, *Ephedra trifurca*) growing in a fine-loamy soil. Roots ending in arrows were not followed further. Note that the roots of different species are heavily intermingled. (Illustration from Gibbens, R.P. and Lenz, J.M., *J. Arid Environ.*, 49, 221, 2001. Kind permissions to reproduce this figure granted by Mr. Lenz [USDA Agricultural Research Service, Jornada Experimental Range, Las Cruces, NM] and Elsevier are gratefully acknowledged.)

FIGURE 28.2 (See color insert.) Root system development, seedlings, and adult plants of *Pachycereus pringlei* and *Stenocereus gummosus* from Sonoran Desert. (A) Adult plant of *P. pringlei*. Arrow indicates a massive root at the soil surface. (B) Root system of adult *P. pringlei* plant where a bayonet-like taproot (arrow) and shallow, apparently lateral, roots can be clearly recognized. (C) *P. pringlei* seedling with primary root that terminated its growth and four basal roots formed at the base of the primary root. (D) Adult *S. gummosus* plant. (E) *S. gummosus* seedlings whose primary root terminated growth. (F) *S. gummosus* seedlings that form compact root system as most lateral and basal roots also terminated growth. (G) Young *S. gummosus* plants with clearly defined taproots. Bars = 5 mm. (Kind donation of photo (B) by Dr. F. Molina-Freaner from the Universidad Nacional Autónoma de México, Mexico, is gratefully acknowledged. Other photos by the authors.)

F. Succulents

Roots of desert succulents are usually nonsucculent and shallow (Figure 28.2A and B). In agaves and cacti, these are on average located only 9 cm in depth (Nobel 2002) with a recorded minimum of 2 cm for *Dudleya saxosa* (M. E. Jones) Britton & Rose (Nobel and Zutta 2007) and a recorded maximum of 115 cm for *Pachycereus pringlei* (S. Watson) Britton & Rose (Niklas et al. 2002), both Sonoran Desert species. Root-to-shoot dry weight ratio in desert succulents is conspicuously low, on average 0.12; this ratio generally is about 10 or more times greater in other groups of desert plants (Noy-Meir 1973; Nobel 2002). It is yet unknown whether these differences are related to relatively low investment in root construction or greater root turnover compared to other groups of desert plants. These possibilities are open because of shallow root system in succulents and known temperature extremes at soil surface (Jordan and Nobel 1984; Nobel 1988).

IV. Main Developmental Adaptations in Desert Plants

Commonly recognized adaptations of plants to desert environments are frequently considered physiological and are called "ecophysiological." As developmental changes and physiological adjustments are interrelated, the distinction between ecophysiological and developmental adaptations is somewhat artificial. Therefore, throughout this chapter, we use the term "developmental adaptations."

A. Root Tissue Differentiation and Resistance to Water Loss

As the main limiting factor for plants in the deserts is rainfall, the most important process for desert plants is the uptake of water and the minimization of water loss. Shallow roots of desert succulents allow rapid response even to a light rainfall. Nevertheless, during onset of drought, when the upper soil layers dry out more quickly than lower ones, a loss of water can occur through the shallow roots. Root systems of perennial desert succulents with shallow roots, such as *Ferocactus acanthodes* (Lem.) Britton & Rose, *Opuntia ficus-indica* (L.) Miller, and *Agave deserti* Engelm. from North America, have rectifier-like properties, that is, they permit the ready movement of water in one direction, but not in the other. Namely, roots rapidly take up water when soil is wet, but water loss by roots into the dry soil during drought is rather limited (Nobel and Sanderson 1984; North and Nobel 1992). Changes in root hydraulic conductivity in wet and dry soil are important for this phenomenon. Root hydraulic conductivity, which is a measure of the easiness of water entry and flow along the roots, can be separated into radial and axial conductivity (Huang and Nobel 1992, 1994; Nobel 1994; North and Nobel 1995). Root hydraulic conductivity of desert succulents and its changes in dehydrated and rehydrated roots are age dependent (Lopez and Nobel 1991). Thus, drying in soil for 30 days causes three- to fivefold decrease in the hydraulic conductivity of 3-month-old roots of *F. acanthodes* and *O. ficus-indica* and a two- to threefold decrease for 12-month-old roots (North and Nobel 1992). Nevertheless, lower root hydraulic conductivity during drought does not completely account for the rectifying properties of the root–soil continuum.

Differentiated tissues of *A. deserti* roots growing in drying soil develop more suberized and lignified hypodermis, endodermis, and sclerified cortex adjacent to the endodermis compared to roots growing under wet conditions. This adaptation decreases water loss from the root tissues (North and Nobel 1991, 1998, 2000). In *F. acanthodes*, as a part of secondary root growth, a periderm is formed that under wet conditions is composed of six or seven layers, two to three of which are suberized. The number of suberized layers is increasing two and three times in roots subjected to 18 and 35 days of drought, respectively. These developmental changes help to decrease water loss from roots (Nobel and Huang 1992). Also, in the species mentioned, air-filled lacunae are formed in the cortex caused by rupture of cortical cells in the dehydrated roots. Suberized and lignified hypodermis, endodermis, and sclerified cortex as well as cortical lacunae decrease significantly the radial water flow. In addition, embolism of xylem vessels in drought-subjected roots reduces the axial component of hydraulic conductivity (North and Nobel 1991, 1992).

Remarkably, the commonly observed gradient in the level of tissue differentiation (more mature root segment has more differentiated tissues) can be reversed if the younger root segment grows under drying condition and more mature one under wet conditions. This was shown experimentally on the same individual roots of *A. deserti*. In the younger root region growing in drying soil, the number of lignified cortical cell layers (sclerenchyma) around the endodermis can be as much as eight; however, in more mature root region grown in a wet substrate, these cell layers are absent. Similarly, the endodermis development is much more advanced in the younger root portion under drying condition than in the more mature root portion under wet conditions. Also, the endodermis inner cell walls are three times thicker in younger than in more mature root parts (North and Nobel 2000). These developmental adaptations occurring in response to available water are an important component of root plasticity in this desert species.

Abscission of lateral roots during drought decreases the root area and can further reduce the water loss for the entire root system, especially for agaves (Nobel and Huang 1992). Besides, roots can shrink radially due to the loss of water from root cells, and as a result, an air gap may develop between the root and the drying soil. In such a case, the contact between the roots and the soil is interrupted, and water must diffuse in the vapor phase to cross the air gap. Root-to-soil air gaps can thus help to limit water loss from roots to soil (Nobel and Cui 1992). Moreover, soil sheaths about 0.7 mm thick that surround young roots can contribute to the formation of the air gaps. These sheaths are composed of soil particles, root hairs, and mucilaginous substances exuded from the roots and found in some cactus species, such as *F. acanthodes* and *O. ficus-indica* (North and Nobel 1992; Huang et al. 1993), as well as in some monocots, such as southern rushes and Restionaceae (Shane et al. 2009). Not only young roots decrease in diameter more considerably than older roots but also the dehydration and contraction of mucilage cause the contraction of overall soil sheath. Therefore, a larger air gap is formed near sheathed roots than unsheathed roots. In addition, upon drying, the soil particles in the sheath aggregate more tightly, making the sheaths less permeable to water. That is why the water loss rate for roots with soil sheaths removed is greater than that for sheathed roots (Huang et al. 1993).

After soil rewetting, roots rapidly undergo structural changes that can increase the hydraulic conductivity and thus the water uptake from wet soil (Nobel 1988; Nobel and Huang 1992). The air bubbles in xylem can be redissolved, and the root–soil air gaps eliminated as a consequence of root swelling and increasing of the mucilage volume on the sheathed roots. In addition, the rehydration of cortex cells of the young roots reduces lacunae and facilitates water entry to the stele and xylem (Nobel and Sanderson 1984). In addition, upon soil rewetting, new roots are induced and existing roots elongate, thus increasing root surface area and water uptake capability of the whole root system (this feature is reviewed in the following). The contribution of these two processes, namely, the increase of hydraulic conductivity and new lateral root formation, can vary among desert species. Moreover, the first few days after rainfall, rapidly rehydrated existing roots are accountable for the majority of water uptake (Rundel and Nobel 1991).

As we mentioned earlier, both water uptake and resistance to water loss depend significantly on root tissue differentiation state, which is a dynamic feature related to both root age and water availability. Developmental processes in the whole root system such as new root formation and root shedding are important factors that define hydraulic conductivity of the whole root system. Therefore, root survival, growth, and tissue differentiation are tightly linked with the ability of the roots either to take up water or to prevent water loss.

B. Root Determinacy and Its Ecological Significance

Root growth is the basis for survival in desert plant species as the main root function is to provide and transport water and mineral nutrients. The only group of species that do not depend on this root function to this extent is desert epiphytes, for example, from genus *Tillandsia* of Bromeliaceae family. Desert *Tillandsia* species associate with free nitrogen-fixing bacteria (Brighigna et al. 1992; Puente and Bashan 1994) and develop rudimentary roots that serve mainly for attachment to the host trees' branches (Bernal et al. 2005). Actually, it is unknown how roots of these plant species stop growing, and it would be interesting to know whether the root apical meristem is maintained or lost soon after seed germination. The presence of an apical meristem in a developed root is a main criterion that defines whether the root growth is determinate or not (Shishkova et al. 2008). If the root has an active apical meristem, then indeterminate growth typical for most plant species, including desert ones, is observed. But if the meristem exhausts and root apical cells differentiate into mature cells, determinate root growth is recognized.

Norman Boke from the University of Oklahoma was one of the first to discover determinate growth of lateral roots in desert succulents and to explain its adaptive significance. He found that *Opuntia arenaria* Engelm. formed long feeder roots that in turn formed numerous laterals lacking secondary growth and periderm. These lateral roots successively produced clusters of short determinate roots that he called "spurs" (Boke 1979). Therefore, one type of determinate roots in this species is lateral root spurs. Root tips of longer feeder roots are active for a relatively short time; then, the apical meristems die and root tips become bract-like structures. A lateral root that develops behind the dead meristem permits continuation of growth of the root system, which forms a cluster of sympodially branched roots (Figure 28.3). However, the apical meristem of some laterals that were formed behind the dead meristem also dies soon afterward, and thus a terminal cluster of determinate roots, called terminal root spurs, forms (Boke 1979). All root-spur cells fully differentiate, and the epidermis produces long and densely spaced root hairs. Estimation based on original figures of root spurs (Boke 1979) shows high density of root-spur clusters, on average one cluster per mm of the parent root. Considering that each cluster consists of 4–5 spurs, we can expect abundant root hair formation. This feature is of much significance for water and mineral uptake. Lateral rootlets with

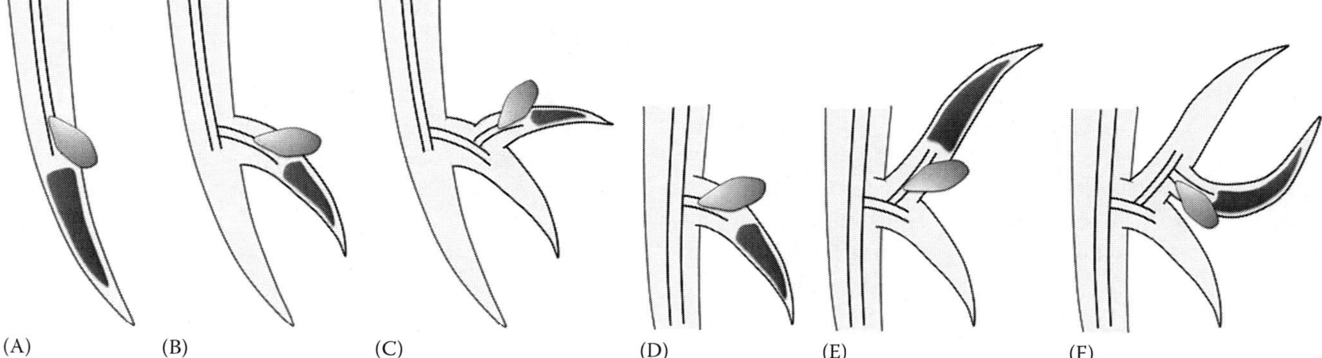

FIGURE 28.3 Schematic representation of developmental sequence of sympodially branched root spurs (root clusters) formation in some Cactaceae. (A–C) Terminal root clusters are formed as a result of full differentiation of the meristematic cells of the main roots and lateral root primordium formation in the subapical region, first in the main root (A) and then on the lateral roots (B, C) that also terminate growth. (D–F) Lateral root clusters are formed as a result of full differentiation of the meristematic cells of the first-order lateral root and formation of next order lateral root primordia; the lateral roots of each subsequent order also exhibit determinate root growth. Dark-shaded area at the tip indicates fully differentiated ex-meristematic cells. (Based on data presented in Boke, N.H., *Am. J. Bot.*, 66, 1085, 1979; Dubrovsky, J.G., *Planta*, 203, 85, 1997a; Dubrovsky, J.G., Determinate primary root growth in *Stenocereus gummosus* (Cactaceae), its organization and role in lateral root development. In *Biology of Root Formation and Development*, eds. A Altman, Y Waisel, pp. 13–20. New York: Plenum Publishing Corporation, 1997b; Dubrovsky, J.G. et al., *Am. J. Bot.*, 90, 823, 2003.)

determinate growth are also found in *O. tunicata* (Lehmann) Link & Otto *var. davisii* (Engelm. & Bigelow) L. Benson (Boke 1979) and in other Cactaceae (Dubrovsky 1997a; Dubrovsky and Gómez-Lomelí 2003).

Determinate growth of lateral roots is not uncommon among plants (Shishkova et al. 2008), but it is very rare in young *primary* roots, as this organ is the starting point of whole root system formation, at least in dicots. Some Sonoran Desert Cactaceae, such as *Stenocereus gummosus* (Engelm.) Gibson & Horak, *Ferocactus peninsulae* (F.A.C. Weber) Britton & Rose, and *P. pringlei*, are characterized by determinate growth of the primary root, which in different species stops growing 2–9 days after the beginning of germination. This growth arrest is accompanied by full differentiation of meristematic cells and formation of numerous root hairs at the very tip, the process called "meristem exhaustion" (Dubrovsky 1997a,b; Dubrovsky and Gómez-Lomelí 2003). This developmental pattern is extremely stable, and out of several thousand *S. gummosus* or *P. pringlei* seedlings analyzed and grown in a number of substrates, not a single plant was found that had indeterminate root growth (Dubrovsky and Shishkova, personal observation). Meristem exhaustion appears to act as a "physiological" root tip decapitation since the beginning of meristem exhaustion coincides with the timing of lateral root initiation, as documented for *S. gummosus* (Dubrovsky 1997a). Primary and lateral roots of these species also possess abundant root hairs. Interestingly, in *S. gummosus*, the root hairs start their formation during seed germination even before the mature embryo root starts its growth and without the usually required epidermal cell elongation; each root epidermal cell can form a root hair (Figure 28.4). These adaptive features increase root surface area and facilitate water and mineral uptake.

Let us consider now what ecological significance determinate primary root growth has at younger developmental stages. Optimal periods for seed germination are very short in the desert and depend on precipitation. Soil water potential becomes very low soon after a rain and seedling survival depends on plant biomass (Dubrovsky 1996). In the Cactaceae with determinate root growth, lateral roots develop very soon after primary root formation. For example, in *S. gummosus*, *Stenocereus thurberi* (Engelm.) Buxbaum, and *F. peninsulae*, lateral roots start to emerge 3–5 days after the start of radicle protrusion (Dubrovsky 1997a) indicating that root branching is a relatively rapid process in these cacti. As commonly lateral roots also have determinate growth, they terminate growth being relatively short (Figures 28.2F and 28.5A). As a result, a very compact root system is formed (Figure 28.2F) that permits the plant to economize the carbon cost of root system construction. It works efficiently as many root hairs promote water and mineral uptake resulting in rapid shoot biomass increase and successful seedling establishment (Dubrovsky 1998).

Next, we will consider what consequences determinate growth has for the root system formation during the transition from juvenile to adult stage. In the literature, the dominant root of many dicots is called a taproot and is usually believed to originate from the primary root (e.g., Mauseth 2009, p. 144). In *S. gummosus* and *P. pringlei* plants, the taproot is well defined (Figure 28.2B through G). The latter species is a large columnar cactus, and its bayonet-like taproot (Figure 28.2B) is important for mechanical support of the principal stem in vertical position and for wind resistance. Not only the taproot in these cacti develops from the primary root but also lateral roots contribute to its formation. When the root apical meristem becomes exhausted, the final primary root length can range from 1 to 9 mm in *S. gummosus* (Dubrovsky 1997a) and from 18 to 46 mm in *P. pringlei* (Dubrovsky and Gómez-Lomelí 2003). In the latter species, the taproot length of an adult plant is much more than one order of magnitude longer than that of the primary root (Niklas et al. 2002). This indicates that the primary root formed in the seedling is only a starting point of taproot formation (Figure 28.2C). After termination of the primary root growth, a lateral root is formed close to the primary (Figure 28.2F) or lateral root tip that grows in the same direction as the primary root (Dubrovsky, personal observation). This suggests that at a later stage, sympodially branched lateral roots fuse together with the primary root and this fusion becomes masked when secondary growth

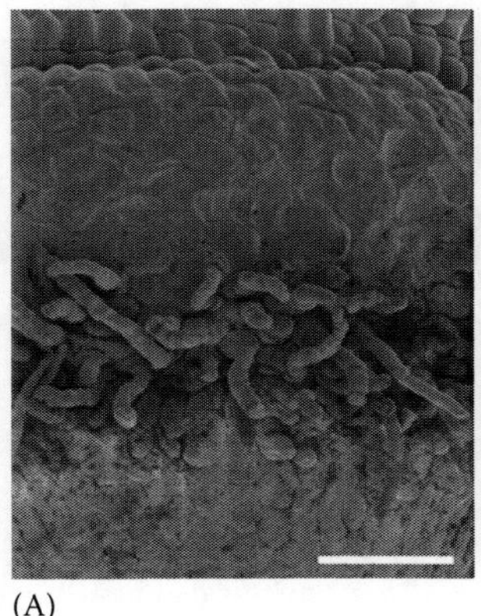
(A)

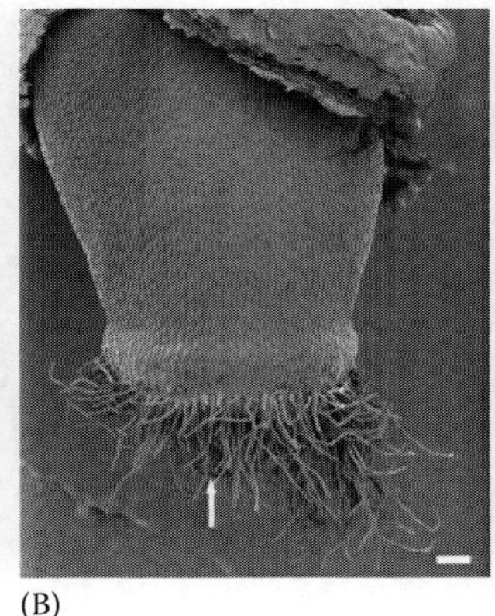
(B)

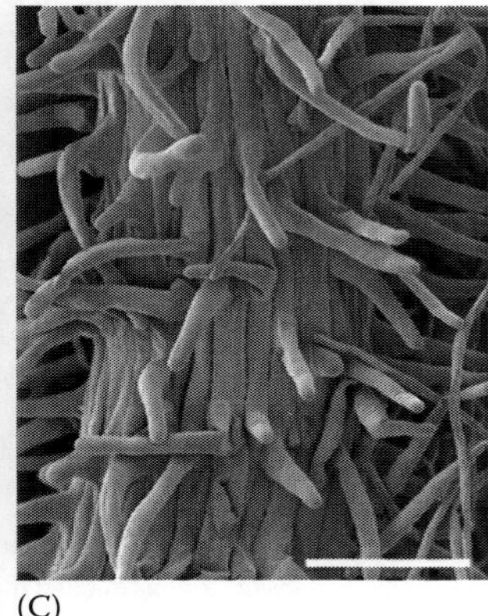
(C)

FIGURE 28.4 Root hair formation in *S. gummosus* seedling. (A) A collar of first root hairs at the base of the primary root from cells produced during embryogenesis and not subjected to elongation. (B) A young seedling with root hairs before the primary root (arrow) started its active growth. (C) Grown primary root with developed root hairs. Note that each epidermal cell forms a root hair. Bars = 100 μm. (Photo by the authors.)

takes place. Therefore, the taproot in these cacti is apparently a sequential mixture of primary and of different-order lateral roots. This developmental scenario is an important adaptation for formation of strong and woody taproot that provides the biomechanical properties necessary for supporting the heavy shoot (Niklas et al. 2002).

C. Developmental Bases and Mechanisms of Root Determinacy

The maintenance of the root apical meristem depends on the functioning and activity of the quiescent center and surrounding initial (stem) cells (see Chapter 3). In the Cactaceae with determinate primary root growth, activity of initial cells is barely detected. The quiescent center in *S. gummosus* is not established post-germination, and in *P. pringlei*, it is established for only a few days and later it cannot be detected (Figure 28.6; Rodríguez-Rodríguez et al. 2003). The absence or a limited functionality of the quiescent center post-germination in some Cactaceae can explain the primary root determinacy.

Developmental changes occurring in *S. gummosus* are strikingly rapid and dramatic. In the mature embryo, all tissues are well developed. During 72 h post-germination, the meristem gradually becomes exhausted, and all tissues at the root tip lose their meristematic characteristics and differentiate (Figures 28.5B and 28.6). The period of steady-state growth, in which the meristem maintains cell proliferation activity and does not change its length, is not longer than 36 h after the start of radicle protrusion; then, the meristem begins to sharply decrease (Dubrovsky 1997a). During meristem exhaustion, no events of programmed cell death (PCD) were detected in the root apical meristem of *S. gummosus* and *P. pringeli*. However, some root cap cells in these species undergo PCD, and later, when meristematic cells differentiate, the root cap is sloughed off (Figures 28.5B and 28.6). Also, PCD events were found in the root hairs of both species, and root hair death appears to be an important adaptation for root hair renovation that takes place in newly formed lateral roots (Shishkova and Dubrovsky 2005). Roots of these cacti regenerated from calli maintain determinate growth, suggesting that this developmental program is present in different root types (Shishkova et al. 2007).

D. Root Dormancy and Developmental Timing as Adaptation

Interestingly, while a strategy to exhaust the apical meristem and completely terminate root growth evolved in Cactaceae, another strategy was evolved in long-lived monocots of Western Australian Desert. Root apices of southern rushes, particularly *Lyginia barbata* R.Br (Restionaceae), cope with harsh environments by entering dormancy yet apparently maintaining their meristem. Summer dormant roots are covered by sand sheaths surrounding the tips, thereby protecting the live meristems (Shane et al. 2009). During this time, dormant root tips are characterized by diminished metabolism. For example, respiration rate, total protein, and water content decrease 10, 3.5, and 1.4 times compared to growing root tip, respectively (Shane et al. 2009). Differentiation processes continue in dormant roots and are found at 1–3 mm from the root apex, while in a growing root, this distance is at least ten times greater (Shane et al. 2010). Roots reestablish meristematic activity after about half a year dormancy period, when most of the hot and dry season is over. The root tip grows through the sand sheath, and the new roots extend to a maximal length of 0.5 m during the winter season, having grown at an average rate of 2.3 mm per day (Shane et al. 2009). Thus, individual roots have the cycles of growth, which are interrupted by dormancy periods. Dormancy of the root apical meristem enables root survival during the harsh season, thereby maintaining the whole root system activity. The intriguing question of how meristematic cells attain dormancy has yet

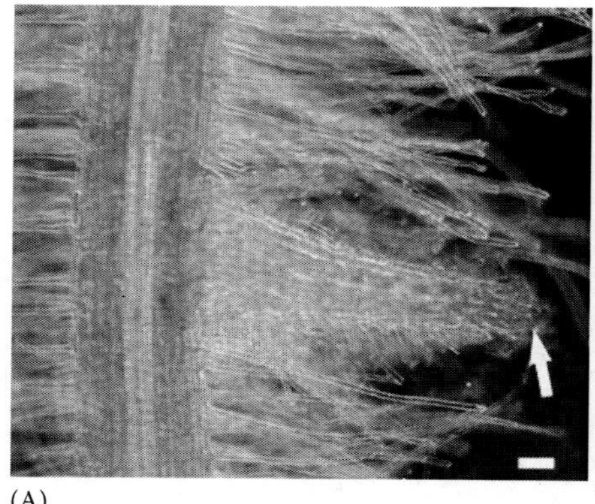

(A)

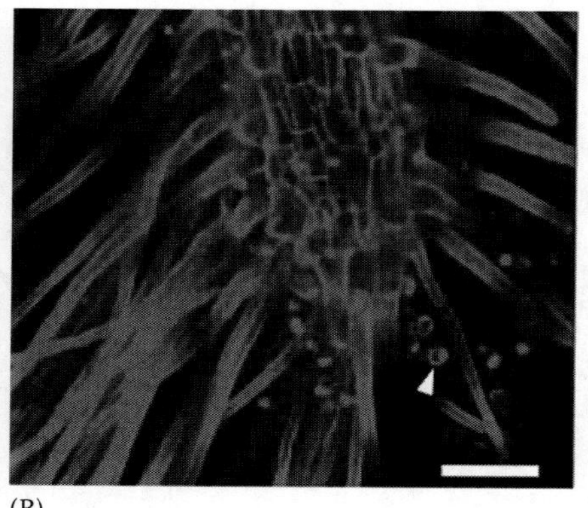

(B)

FIGURE 28.5 Determinate pattern of root growth in some Sonoran Desert Cactaceae. (A) Determinate root growth of a first-order lateral root of *P. pringlei* observed with Nomarski optics. Note that at the tip of the lateral root (arrow), root hairs are formed indicating complete meristem exhaustion. Also note that the lateral root is shorter than the root hairs formed on the primary root. (B) Live primary root of *S. gummosus* seedling 8 days after seed germination stained with propidium iodide and observed with a laser scanning confocal microscope. Note that the root tip cells are elongated, epidermal cells of the former meristem are completely differentiated, root hairs are formed, and the root cap is sloughed off. Arrowhead indicates nucleus of dead root cap cell. Bars = 100 μm. (Photo by the authors.)

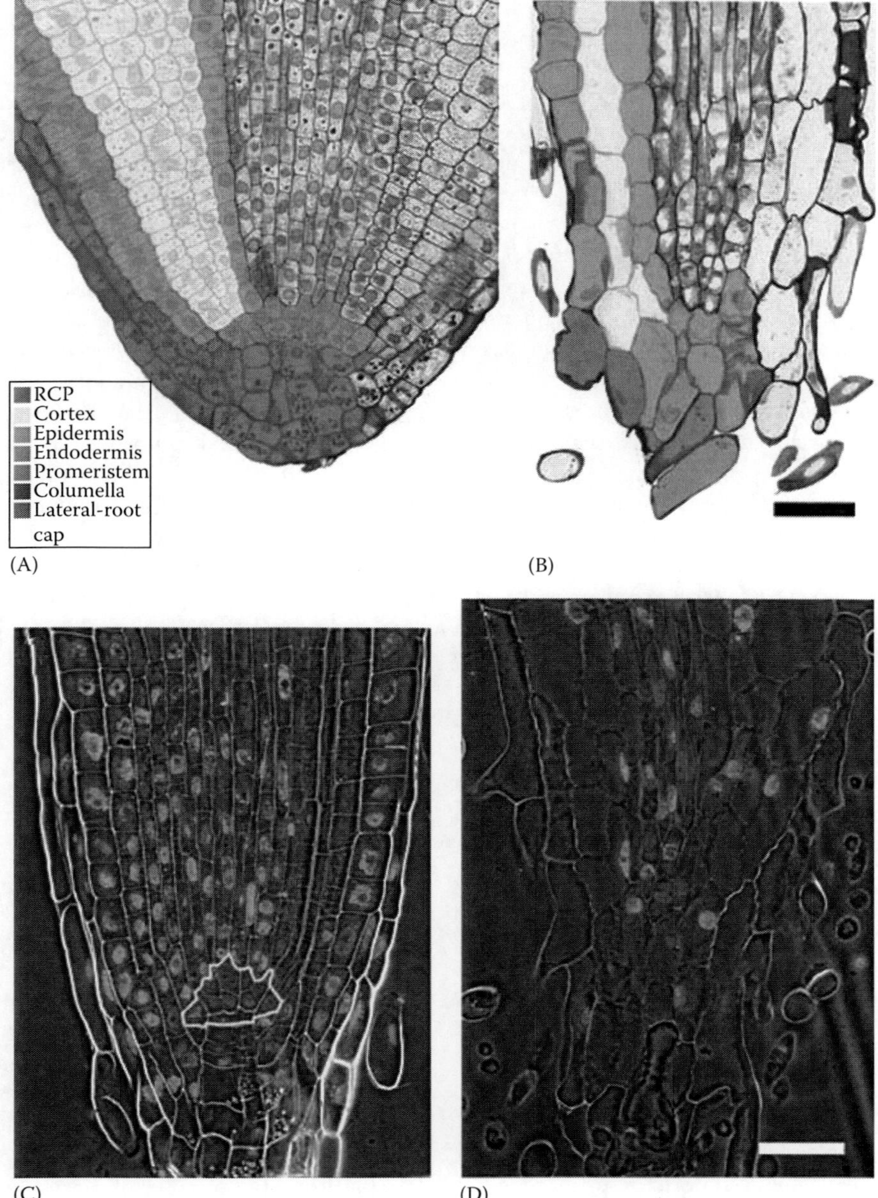

FIGURE 28.6 (See color insert.) Primary root apical meristem organization and exhaustion in *S. gummosus* (A, B) and *P. pringlei* (C, D). (A) Open type of meristem organization in a seedling just after seed germination. Promeristem (gray-colored cells) is the zone where quiescent center cells could be established within the open-type meristem. (B) Primary root apex 3 days after germination. Note, almost all cells of the root cap are lost, cortex is formed by just one cell file, differentiation zone approached to the tip, and the root hairs are formed at the very tip. (C, D) Developmental changes in the primary roots of *P. pringlei* and immunofluorescence analysis of cell proliferation. DAPI (blue staining of nuclei) and FITC (green; immunological staining of nuclei that passed S-phase of the cell cycle during 24 h treatment with BrDU) images superimposed. (C) Primary root of recently germinated seedling. Quiescent center is outlined with yellow line. (D) Primary root at the end of meristem exhaustion. Note: No compact group of FITC-negative cells is detected indicating an absence of the quiescent center at this stage. All ex-meristematic cells are differentiated, and many of them passed S-phase (green FITC-positive cells). Bars = 50 μm. (Photo by the authors, Rodríguez-Rodríguez, J.F. et al., *Planta*, 217, 849, 2003 with modifications. Kind permission of Springer Science + Business Media.)

to be answered. More studies of root apical meristem activity are also needed to find out how common is this phenomenon among desert plant species.

A different dormancy strategy evolved in desert cacti. In this plant group, substrate drying promotes lateral root primordia initiation and development (North et al. 1993; Dubrovsky et al. 1998a). During a dry period in a desert, which may last for a long time, the lateral root primordia do not emerge until a subsequent rain event. As soon as water becomes available, the primordia develop into roots. Therefore, during a relatively long time, the primordia are maintained in a dormant state within a parent root. Primordium dormancy is not well studied in desert plants, and we do not know whether all initiated primordia are capable of subsequently emerging as lateral roots. This question awaits further investigation. Primordium dormancy is a very interesting developmental adaptation as it may represent a large fraction of the root system in a latent form. There are indications that dormant primordia also develop in monocots, such as agaves (Nobel 1988).

Water available after a drought period induces amazingly rapid lateral root emergence. The roots formed in response to rain are called "rain roots" (Nobel 1988). These roots are usually short lived (Nobel 1988) and are also termed "ephemeral roots" (Dubrovsky and North 2002). In less than 8 h after a rainfall, a few millimeters long rain roots are formed in *Opuntia puberula* Pfeiff. and *F. acanthodes* (Kausch 1965; Nobel and Sanderson 1984; Nobel 1988). In *O. ficus-indica* after experimental drought and 20 h after rewetting, lateral roots emerged that were 2–4 mm long and on average 0.7 mm in diameter; the cell number in the meristem of these newly formed lateral roots was 40% less compared to that found in the parent root under wet conditions. During the second day after rewetting, they attain 80% of the maximum growth rate of the parent root (Dubrovsky et al. 1998a). Such rapid emergence and growth of rain roots is a vital developmental adaptation as new roots permit rapid reestablishment of water uptake and mineral nutrition. For example, rain roots of *A. deserti* and *F. acanthodes* have approximately twofold higher hydraulic conductivity than 4-month-old established roots (Nobel 2002).

The basis of such rapid lateral root emergence has yet to be studied. It is probable that rapid root emergence upon rewetting results from the drought-induced dormant primordia and rapid triggering of cell elongation and proliferation. Indeed, some published data support this hypothesis. For example, in *F. acanthodes*, primordia developed during drought have 10–20 cell layers indicating the advanced stage they reached before rewetting (North et al. 1993). In *O. ficus-indica*, the length of fully elongated cells in recently emerged rain roots reaches 87% of that in parent root growing under optimal conditions, and average cell cycle duration in the root apical meristem is about 20% shorter than in the meristem of the parent root (Dubrovsky et al. 1998a). In general, cell cycle duration in the Cactaceae root apical meristem is relatively short ranging from 14 to 28 h (Dubrovsky et al. 1998a,b). These features indicate that developmental timing of growth and morphogenetic processes are important developmental adaptations that increase plant fitness and are important for plant survival.

E. Stability of Developmental Processes

As mentioned earlier, during both gradual and rapid soil drying, new root primordia are formed even at greater extent than under wet conditions, as it is documented for *O. ficus-indica* (Dubrovsky et al. 1998a). This increase ranges from 1.5 to 2.2 times as compared to roots grown under wet conditions. Under gradual drying, the length of fully elongated cortical cells is about 70% of those under wet conditions (Dubrovsky et al. 1998a). It seems that the increase in lateral root primordium density is not a consequence of shorter cells as this increase is maintained if the analysis is performed on a per cell basis also (Dubrovsky et al. 2009). In another cactus species, *P. pringlei*, the density of roots and primordia increases under a lower substrate water potential, but because cell length decreases proportionally, there is no change in lateral root initiation per cellular basis along the primary root (Dubrovsky and Gómez-Lomelí 2003). This indicates that even though lateral root initiation is not increasing under water deficit, it is maintained stable in this desert species. This stability appears to be important for seedling establishment after seed germination and therefore for plant survival in the harsh desert environment.

F. Developmental Changes in the Root Apical Meristem during Drought

Periods of active root growth are infrequent in the desert and depend on the water availability. During periods of soil drying, root growth dynamics and meristem activity vary depending on soil depth. In a species characterized by indeterminate root growth, *O. ficus-indica*, rapid soil drying sharply inhibits root growth, and most of the root apices die within 5 days (Dubrovsky et al. 1998a). During gradual drying at deeper soil layers, root mortality is relatively low. Remarkably, the length of the meristem, the duration of the cell division cycle, and the length of the elongation zone are unchanged under rapid drying, but developmental changes do occur in gradually drying roots. For example, in the latter condition, the number of meristematic cells decreases by 30%, the duration of the cell cycle in the meristem increases by 36%, and fully elongated cells become 30% shorter than under conditions of active root growth (Dubrovsky et al. 1998a). This suggests that at deeper soil layers, roots acclimate and may have sufficient time for developmental changes to occur. These changes become critical for maintenance of root growth under drying conditions.

G. Changes in the Root Apical Meristem Depend on the Temperature

For Sonoran Desert species, the drought-deciduous C3 shrub, *Encelia farinosa* Torr & A. Gray, the C4 bunchgrass *P. rigida*, and the monocarpic CAM succulent, *A. deserti*, the length of meristematic cells is greater at optimal growth temperatures than at suboptimal temperatures. Also, the length of the cell division zone is greater at optimal than at suboptimal and supraoptimal temperatures. For example, in a perennial grass *P. rigida*, it was close to 1 mm in length at optimal temperature and was on average 790 and 840 μm at suboptimal and supraoptimal temperatures, respectively (Drennan and Nobel 1996). Similar trend can be found in the cactus *O. ficus-indica* (Drennan and Nobel 1998). Under suboptimal temperatures, cell division zone decreases 16%–54% in this species. This suggests that cell production should also be lower under suboptimal temperature. Notably, at supraoptimal temperatures, no decrease in the meristem length is observed (Drennan and Nobel 1996), indicating that the meristem may become dormant. Interestingly, the length of cell division zone at summer did not differ from that at winter. Altogether, considering that optimal temperatures and root distribution in soil depths are different for different species, temperature-dependent processes of root growth represent complex root adaptation at a level of a desert plant community that requires further investigation.

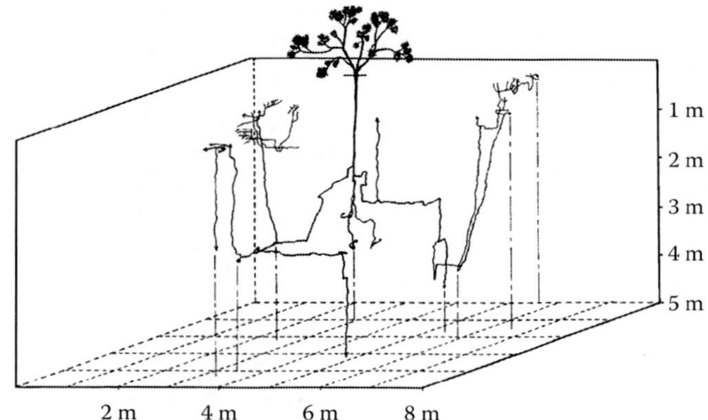

FIGURE 28.7 Root system of a large crucifixion thorn (*Koeberlinia spinosa*) shrub from the Chihuahuan Desert. Horizontal line at base of plant denotes soil surface, and plant top is drawn to same scale as roots. Roots ending in arrows were not followed further. Upward growing roots extended to within 10 cm of soil surface. (Illustration from Gibbens, R.P. and Lenz, J.M., *J. Arid Environ.*, 49, 221, 2001. Kind permissions to reproduce this figure granted by Mr. Lenz [USDA Agricultural Research Service, Jornada Experimental Range, Las Cruces, NM] and Elsevier.)

H. Upward Growing Roots

Since the beginning of the twentieth century, there were indications that a typical positive gravitropic root response could occasionally be reversed; for instance, roots can grow upslope. On several figures from the classical monograph by W. A. Cannon on roots of Sonoran Desert plants (1911), roots angled upward are depicted. Similar records can be found in several other studies. The most extensive of them describes root systems of various individuals for more than 40 shrub, grass, and forb species from the Chihuahuan Desert (Gibbens and Lenz 2001, including previous data from Gile et al. 1995, 1997, 1998). These authors showed that upward growing roots were common for most shrub species analyzed, such as creosote bush (*Larrea tridentata* (Sess. & Moc. ex DC.) Cov.), tarbush (*Flourensia cernua* DC.), wolfberry (*Lycium berlandieri* Dunal), winterfat (*Ceratoides lanata* (Pursh) J.T. Howell), mesquite (*Prosopis glandulosa* Torr.), longleaf ephedra (*Ephedra trifurca* Torr. ex S. Wats), soaptree yucca (*Yucca elata* Engelm.), and four-wing saltbush (*Atriplex canescens* (Pursh) Nutt.). Moreover, upward growing roots were also found in root systems of a rhizomatous perennial forb, leatherweed, (*Croton pottsii* (Klotzsch) Muell.-Arg.), and parasitic three fans (*Krameria lanceolata* Torr.). For different shrub species, either first-, second-, and greater-order lateral roots can grow upward. For example, a first-order lateral root of a tarbush grew upward from 60 to 19 cm depth and then grew horizontally (Gibbens and Lenz 2001). In many instances, the roots growing vertically or at an angle upward can nearly reach the surface. Thus, some creosote bush and tarbush roots angle upward within 5–8 cm of the soil surface. Some tarbush roots grew upward within about 2 cm of the soil surface under the canopy of another plant! A root system of a large crucifixion thorn was a remarkable example of many major roots growing vertically upward (Figure 28.7; Gibbens and Lenz 2001).

All these examples suggest that roots grow upward in a search of water, showing hydrotropic response. However, no reports on hydrotropism in roots of desert species are known. Although there are studies of root hydrotropic response since it was observed by Darwin and Darwin (1880) in the last nineteenth century, its genetic control is at the beginning of elucidation (Eapen et al. 2005; Miyazawa et al. 2009; Takahashi et al. 2009). It is most probable that in desert species, upward growing roots develop in response to more positive soil water potential in upper soil levels after small rainfalls, overriding positive gravitropic response. This phenomenon may have important adaptive significance and may be more widespread than is currently recognized.

I. Developmental Adaptations Dependent on Inter-Root Communication

Inter-root communication of desert species represents an important mechanism involved in control of root growth and soil exploration by individual roots. Roots of two individual plants of *Ambrosia dumosa* W. W. Payne shrub inhibit each other's root growth when in direct contact (growth rate 2 mm per day); however, other noncontacting roots continue to grow at normal rates (growth rate 13 mm per day, Mahall and Callaway 1991, 1992, 1996). The authors conclude that this constitutes a detection and avoidance mechanism by which soil volume commonly used by neighboring plants could be minimized. The fact that direct contact of roots results in a 6.5-fold decrease in root growth rate represents an interesting developmental adaptation that forces roots of two different plants to avoid direct contact. Another type of inter-root communication is found between shrubs of *L. tridentata* and *A. dumosa*. Even without direct contact, root of one *Larrea* plant inhibits growth of roots of other plants of the same species. Moreover, *Larrea* roots inhibit root growth of *Ambrosia* without direct contact, but the latter does not inhibit growth of *Larrea* (Mahall and Callaway 1991, 1992). These findings have important ecological and adaptive implications as due to inter-root communication, these shrubs can be distributed in a more regular way, and thereby competition for water and mineral resources would be decreased (Mahall and Callaway 1992). Future research is required to establish the nature of this adaptation (Callaway and Mahall 2007).

J. Specialized Root Types: Succulent, Tuberous Storage, and Contractile Roots

In some desert species, specialized types of roots have evolved that improve plant survival in harsh desert environment. Mostly, these specialized root types are known in the Cactaceae. Some species of the family maintain succulence of their root tissues that permits water storage similarly to succulent stems. Succulent roots were first reported a century ago for *Opuntia vivipara* Rose, which develops fleshy roots (Cannon 1911). Massive succulent roots and succulent shoot-to-root transition zones are found in a number of small cacti of different genera (reviewed by Dubrovsky and North 2002). In Cactaceae, primary cortex and secondary xylem are the main tissues, which develop root succulence (Gibson 1978). Succulent roots were also reported for *Yucca glauca* Nutt. of Agavaceae family (Markle 1917) and a relict gymnosperm of the Namib Desert, *Welwitschia mirabilis* Hook.f. (Gibson 1996; Kutschera et al. 1997).

Tuberous roots of desert plants mainly accumulate starch and frequently are also succulent, especially in genera *Echinocereus*, *Neoevansia*, *Peniocereus*, *Wilcoxia*, and *Pterocactus* of the Cactaceae family (Gibson 1978; Loza-Cornejo and Terrazas 1996; Herrera-Cárdenas et al. 2000). This developmental adaptation is of great significance in plant survival as tuberous roots accumulate a lot of reserves and may attain a significant volume. For example, root of *Peniocereus greggii* (Engelm.) Britton & Rose can reach up to 60 cm in diameter, 15–20 cm long, and a weight of 27–56 kg (Britton and Rose 1963). During optimal growth period, members of these genera develop succulent stems, accumulate starch in the root, and set fruits. During the driest part of the year, the aboveground stems die and live buds are maintained belowground. When water becomes available again, bud meristems are activated and form narrow succulent stems using most of the starch and other reserves accumulated in the tuberous roots. The latter are perennial and reach substantial size due to secondary growth. Important anatomical features of tuberous roots that have adaptive significance for maintaining storage root function are related to the prevention of water loss due to formation of a suberized periderm. These roots sustain secondary growth, during which a ray dilatation occurs that prevents internal tissue cracks and makes secondary tissues that are visible as radial rows without fibers. In tuberous roots, starch-containing tissues are mainly cortex and ray parenchyma and in some cases also pith-like parenchyma (Loza-Cornejo and Terrazas 1996; Herrera-Cárdenas et al. 2000). Some of these features can be observed in Figure 28.8.

Contractile roots constitute another interesting developmental adaptation evolved in some small desert Cactaceae, Agavaceae (North et al. 2008; Garrett et al. 2010 and references therein), and in some other taxa. The root of "living rock" cactus *Ariocarpus fissuratus* Schumann K. is capable of longitudinal shortening as a result of radial enlargement of basal root regions. Due to the root contraction, the stem moves down to the soil at a rate of 6–30 mm per year. Root contraction also occurs in ephemeral annual plants such as Asteraceae whose roots contract and permit shoot apex to deepen 10 mm during 50 days. This contraction permits formation of hypogeal inflorescence (Zamski et al. 1983). Also, in a rare Negev Desert lily, *Pancratium sickenbergeri* Asch. & Schweinf., root contraction helps to decrease the lily bulb herbivory (Ruiz et al. 2006). Root contraction in these species is a developmental adaptation that became essential for plant survival as it allows the plant to avoid lethal or unfavorable temperatures at the soil surface and significantly decreases herbivory.

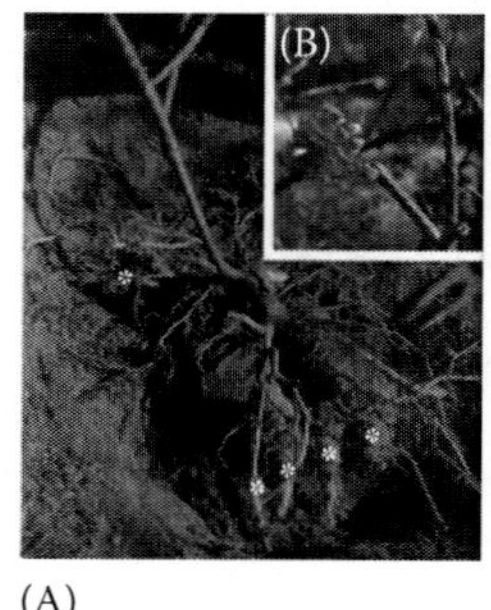

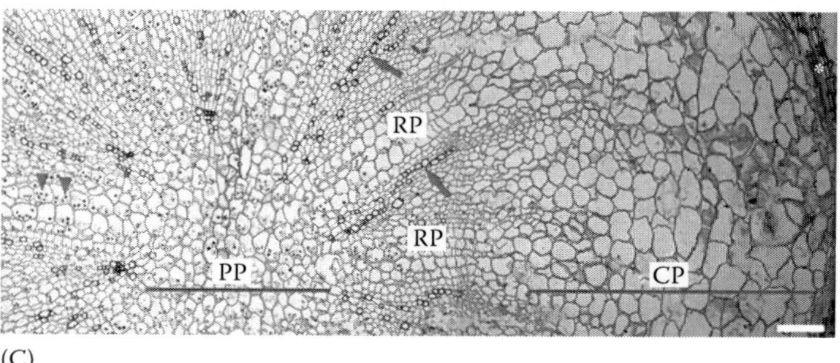

FIGURE 28.8 (See color insert.) Tuberous root morphology, succulent stem, flower, and anatomical features of tuberous root in *Peniocereus viperinus* (Cactaceae). (A) In *P. viperinus* plant, several storage roots are formed during different years. Each tuberous root is marked with white asterisk. Note the perennial most massive tuberous root to the right of the image. (B) Aboveground vegetative and generative organs are produced when water becomes available. During this period, as in all geophytes, photosynthates are produced and transported to the tuberous root where they are accumulated in form of starch. During dry season, aboveground organs die. Note that stem is narrow and less succulent than in most other Cactaceae. (C) A section of young tuberous root. Note multilayered periderm (asterisk), abundant starch accumulation (arrowheads), xylem vessels (arrows), pith parenchyma (PP), ray parenchyma (RP), and cortical parenchyma (CP). Bar = 200 μm. (All images were kindly donated by Dr. T. Terrazas, Universidad Nacional Autónoma de México, Mexico.)

V. Conclusions and Future Perspectives

In this review, we described a number of developmental adaptations that are essential for plant survival in desert. Taking into account all these adaptations together represents an integrative approach. A developmental adaptation at one level of organization underlies a similar adaptation at a higher level. For example, such developmental adaptations as a relatively short cell cycle in the root apical meristem and rapid cell elongation under optimal growth are the bases for rapid rain root emergence after drought. Quiescent center establishment and/or activity are crucial factors for determinate or indeterminate root growth pattern. Independent consideration of a particular developmental adaptation by itself is incomplete because each adaptation interacts with another adaptation. For example, formation of dormant lateral root primordia during drought serves as a basis for subsequent rain root formation and a significant increase of root hydraulic conductivity.

A number of developmental processes that have adaptive significance in roots were not considered here as the available information is scarce. This does not mean that other adaptive features are less important. For example, the fact that shoots are produced from roots in *Opuntia arbuscula* is known since the classical studies of Cannon (1911). Root buds were reported in *O. arenaria* (Boke 1979) and found in *Myrtillocactus geometrizans* (Mart. ex. Pfeiff.) Console (Dubrovsky, personal observation). These fragmented observations stimulate further studies of root bud formation as a developmental adaptation.

Different plant life forms, for example, a shrub, a perennial grass, and a succulent from Sonoran Desert, may have different thermal niches for optimal root growth that affect the timing and soil depth where roots predominantly grow (Drennan and Nobel 1996). Further studies could define how these niches are established and how they could be changed with global warming. Future research of developmental adaptations will help to better comprehend root plasticity. In general, studies of root adaptation in desert species will have increasing importance because of global warming (Etterson and Shaw 2001; Cushman and Borland 2002; Thomas et al. 2005; Nobel 2006).

The understanding of the species biology, particularly their roots, will be more complete when studies on ecological, physiological, developmental, hormonal, genetic, and molecular-biology levels will be combined. The interdisciplinary approach of ecological developmental biology ("eco-devo"), which focuses on understanding development in environmental conditions that vary in nature, has emerged in the last decade (Gilbert 2001). This discipline studies developmental responses to changing environmental conditions. Developmental plasticity, that is, the ability of a single genotype to produce a spectrum of phenotypes in different environmental conditions, is an important concept for "eco-devo" (Gilbert 2001; Sultan 2007). In order to respond to the environment, a plant should integrate multiple external and internal cues and adjust gene expression accordingly. As a result, multiple regulatory elements involving transcription factors, their target proteins, and hormones might be used by a plant to translate environmental signals to developmental modifications (Sultan 2010). Recently, a great advance in understanding of plant ecological development, from regulatory pathways to developmental response, was made for some models, for example, avoidance of neighbor shade (Sultan 2010). In this context, roots of desert plants would represent a convenient model, because a response to rainfall can be very rapid and notable. For instance, elucidation of the regulatory mechanisms involved in control of lateral root initiation in response to drought and emergency and elongation of lateral roots in response to rainfall, as well as of changes of root hydraulic conductivity in response to drought and soil rewetting, would represent radical advance in this field. Still, roots being a subterranean part of the plant represent a difficult model for "eco-devo" studies. Moreover, because of little advance of molecular genetics and genomics studies of desert plants, this kind of research of their roots is even more challenging. Nevertheless, recently emerging sequencing technologies, such as mRNA-Seq, and subsequent de novo assembly of transcriptome permit a global analysis of differential gene expression even when the genome is not sequenced (see rev. Ozsolak and Milos 2011). One early example of such an approach is the analysis of leaf transcriptome of angiosperm resurrection plant, *C. plantagineum,* in response to desiccation (Suarez-Rodriguez et al. 2010). The authors of this chapter are currently involved in a study of differential gene expression in root tip of *P. pringlei* (Cactaceae) during determinate primary root growth using mRNA-Seq and de novo transcript assembly methodology. This and rapidly emerging new approaches will permit to investigate and better understand the "eco-devo" aspects of root biology and specifically developmental adaptations of roots of desert species.

Acknowledgments

The authors are grateful to Dr. Teresa Terrazas from the Universidad Nacional Autónoma de México (UNAM), Mexico, and Dr. Gretchen B. North from the Occidental College, Los Angeles, for their valuable comments and suggestions that permitted to improve this chapter. Research in the authors' laboratory was supported by the Dirección General de Asuntos del Personal Académico—Programa de Apoyo a Proyectos de Investigación e Innovación Tecnológica, UNAM (grants IN204312 to J.G.D. and IN204912 to S.S.), and Consejo Nacional de Ciencia y Technología, Mexico (grants 127957 to J.G.D. and 79736 to S.S.).

References

Alpert P 2000 The discovery, scope, and puzzle of desiccation tolerance in plants. *Plant Ecol* 151:5–17.

Arndt SK, Kahmen A, Arampatsis C, Popp M, Adams M 2004 Nitrogen fixation and metabolism by groundwater-dependent perennial plants in a hyperarid desert. *Oecologia* 141:385–394.

Baytulin IO 1987 Stroenie i Rabota Kornevoy Sistemy Rasteniy (Structure and function of plant root system). Alma-Ata: Nauka Publishing of Kazakh Soviet Republic (In Russian).

Bell KL, Hiatt HD, Niles WE 1979 Seasonal changes in biomass allocation in eight winter annuals of the Mohave Desert. *J Ecol* 67:781–787.

Bernal R, Valverde T, Hernández-Rosas L 2005 Habitat preference of the epiphyte *Tillandsia recurvata* (Bromeliaceae) in a semi-desert environment in Central Mexico. *Can J Bot* 83:1238–1247.

Bethlenfalvay GJ, Dakessian S, Pacovsky RS 1984 Mycorrhizae in a southern California desert: Ecological implications. *Can J Bot* 62:519–524.

Bohrer G, Kagan-Zur V, Roth-Bejerano N, Ward D 2001 Effects of environmental variables on vesicular–arbuscular mycorrhizal abundance in wild populations of *Vangueria infausta*. *J Veg Sci* 12:279–288.

Boke NH 1979 Root glochids and root spurs of *Opuntia arenaria* (Cactaceae). *Am J Bot* 66:1085–1092.

Brighigna L, Montaini P, Favilli F, Trejo AC 1992 Role of the nitrogen-fixing bacterial microflora in the epiphytism of *Tillandsia* (Bromeliaceae). *Am J Bot* 79:723–727.

Britton NL, Rose JN 1963 *The Cactaceae*. New York: Dover.

Caldwell MM, Dawson TE, Richards JH 1998 Hydraulic lift: Consequences of water efflux from the roots of plants. *Oecologia* 113:151–116.

Caldwell MM, Richards JH 1989 Hydraulic lift: Water efflux from upper roots improves effectiveness of water uptake by deep roots. *Oecologia* 79:1–5.

Callaway RM, Mahall BE 2007 Plant ecology: Family roots. *Nature* 448:145–147.

Canadell J, Jackson RB, Ehleringer JR, Mooney HA, Sala OE, Schultze ED 1996 A global review of rooting patterns II. Maximum root depth. *Oecologia* 108:583–595.

Cannon WA 1911 *The Root Habits of Desert Plants*, Publication 131. Washington, DC: Carnegie Institution of Washington.

Carrillo-Garcia A, León de la Luz J-L, Bashan Y, Bethlenfalvay GJ 1999 Nurse plants, mycorrhizae, and plant establishment on a disturbed area of the Sonoran Desert. *Restor Ecol* 7:321–335.

Collier SC, Yarnes CT, Herman RP 2003 Mycorrhizal dependency of Chihuahuan Desert plants is influenced by life history strategy and root morphology. *J Arid Environ* 55:223–229.

Cushman JC, Borland AM 2002 Induction of Crassulacean acid metabolism by water limitation. *Plant Cell Environ* 25:295–310.

Darwin C, Darwin F 1880 *The Power of Movements in Plants*. London, UK: John Murray.

Drennan PM, Nobel PS 1996 Temperature influences on root growth for *Encelia farinosa* (Asteraceae), *Pleuraphis rigida* (Poaceae), and *Agave deserti* (Agavaceae) under current and doubled CO_2 concentrations. *Am J Bot* 83:133–139.

Drennan PM, Nobel PS 1998 Root growth dependence on soil temperature for *Opuntia ficus-indica*: Influences of air temperature and a doubled CO_2 concentration. *Funct Ecol* 12:959–964.

Drew MC 1979 Root development and activities. In *Arid-land Ecosystems: Structure, Functioning and Management*, eds. DW Goodall, RA Perry, KMW Howes, Vol. 1, pp. 573–607. Cambridge, U.K.: Cambridge University Press.

Dubrovsky JG 1996 Seed hydration memory in Sonoran Desert Cacti and its ecological implication. *Am J Bot* 83:624–632.

Dubrovsky JG 1997a Determinate primary-root growth in seedlings of Sonoran Desert Cactaceae; its organization, cellular basis, and ecological significance. *Planta* 203:85–92.

Dubrovsky JG 1997b Determinate primary root growth in *Stenocereus gummosus* (Cactaceae), its organization and role in lateral root development. In *Biology of Root Formation and Development*, eds. A Altman, Y Waisel, pp. 13–20. New York: Plenum Publishing Corporation.

Dubrovsky JG 1998 Determinate root growth as an adaptation to drought in Sonoran Desert Cactaceae. In *Radical Biology: Advances and Perspectives on the Function of Plant Roots*, eds. HE Flores, JP Lynch, D Eissenstat, pp. 471–474. Rockville, MD: American Society of Plant Physiologists.

Dubrovsky JG, Contreras-Burciaga L, Ivanov VB 1998b Cell cycle duration in the root meristem of Sonoran Desert Cactaceae as estimated by cell-flow and rate-of-cell production methods. *Ann Bot* 81:619–624.

Dubrovsky JG, Gómez-Lomelí LF 2003 Water deficit accelerates determinate developmental program of the primary root and does not affect lateral root initiation in a Sonoran Desert cactus (*Pachycereus pringlei*, Cactaceae). *Am J Bot* 90:823–831.

Dubrovsky JG, North GB 2002 Root structure and function in the Cactaceae. In *Cactus Biology and Uses*, ed. PS Nobel, pp. 41–56. Berkeley, CA: University of California Press.

Dubrovsky JG, North GB, Nobel PS 1998a Root growth, developmental changes in the apex, and hydraulic conductivity for *Opuntia ficus-indica* during drought. *New Phytol* 138:75–82.

Dubrovsky JG, Soukup A, Napsucialy-Mendivil S, Jeknic Z, Ivanchenko MG 2009 The lateral root initiation index: An integrative measure of primordium formation. *Ann Bot* 103:807–817.

Eapen D, Barroso ML, Ponce G, Campos ME, Cassab G 2005 Hydrotropism: Root growth responses to water. *Trends Plant Sci* 10:44–50.

Etterson JR, Shaw RG 2001 Constraint to adaptive evolution in response to global warming. *Science* 294:151–154.

Fischer E 2004 Scrophulariaceae. In *The Families and Genera of Vascular Plants, Flowering Plant Dicotyledons (except Acanthaceae including Avicenniaceae)*, ed. JW Kadereit, Vol. VII, pp. 333–432. Berlin, Germany: Springer.

Franco AC, Nobel PS 1989 Effect of nurse plants on the microhabitat and growth of cacti. *J Ecol* 77:870–886.

Franco AC, Nobel PS 1990 Influences of root distribution and growth on predicted water uptake and interspecific competition. *Oecologia* 82:151–157.

Gaff DF 1971 Desiccation-tolerant flowering plants in Southern Africa. *Science* 174:1033–1034.

Garrett TY, Huynh CV, North GB 2010 Root contraction helps protect the "living rock" cactus *Ariocarpus fissuratus* from lethal high temperatures when growing in rocky soil. *Am J Bot* 97:1951–1960.

Gibbens RP, Lenz JM 2001 Root systems of some Chihuahuan Desert plants. *J Arid Environ* 49:221–263.

Gibson AC 1978 Structure of *Pterocactus tuberosus*, a cactus geophyte. *Cactus Succ J* 50:41–43.

Gibson AC 1996 *Structure-Function Relations of Warm Desert Plants*. Berlin, Germany: Springer.

Gilbert SF 2001 Ecological developmental biology: Developmental biology meets the real world. *Dev Biol* 233:1–12.

Gile LH, Gibbens RP, Lenz JM 1995 Soils and sediments associated with remarkable, deeply-penetrating roots of crucifixion thorn (*Koeberlinia spinosa* Zucc.). *J Arid Environ* 31:137–151.

Gile LH, Gibbens RP, Lenz JM 1997 The near ubiquitous pedogenic world of mesquite roots in an arid basin floor. *J Arid Environ* 35:39–58.

Gile LH, Gibbens RP, Lenz JM 1998 Soil-induced variability in root systems of creosote bush (*Larrea tridentata*) and tarbush (*Flourensia cernua*). *J Arid Environ* 39:57–78.

Gries D, Zeng F, Foetzki A et al. 2003 Growth and water relations of *Tamarix ramosissima* and *Populus euphratica* on Taklamakan desert dunes in relation to depth to a permanent water table. *Plant Cell Environ* 26:725–736.

Gutierrez JR, Whitford WG 1987 Chihuahuan Desert annuals: Importance of water and nitrogen. *Ecology* 68:2032–2045.

Herrera-Cárdenas RT, Terrazas T, Loza-Cornejo S 2000 Anatomía comparada del tallo y de la raíz de las especies del género *Neoevansia Marshall* (Cactaceae). *Boletín de la Sociedad Botánica de México* 67:5–16.

Huang B, Nobel PS 1992 Hydraulic conductivity and anatomy for lateral roots of *Agave deserti* during root growth and drought-induced abscission. *J Exp Bot* 43:1441–1449.

Huang B, Nobel PS 1994 Root hydraulic conductivity and its components, with emphasis on desert succulents. *Agron J* 86:767–774.

Huang B, North GB, Nobel PS 1993 Soil sheaths, photosynthate distribution to roots, and rhizosphere water relations for *Opuntia ficus-indica*. *Int J Plant Sci* 154:425–431.

Ivans CI, Leffler AJ, Spaulding U, Stark JM, Ryel RJ, Caldwell MM 2003 Root responses and nitrogen acquisition by *Artemisia tridentate* and *Agropyron desertorum* following small summer rainfall events. *Oecologia* 134:317–324.

Jacobson KM 1997 Moisture and substrate stability determine VA-mycorrhizal fungal community distribution and structure in an arid grassland. *J Arid Environ* 35:59–75.

James JJ, Richards JH 2007 Influence of temporal heterogeneity in nitrogen supply on competitive interactions in a desert shrub community. *Oecologia* 152:721–727.

Jordan PW, Nobel PS 1984 Thermal and water relations of roots of desert succulents. *Ann Bot* 54:705–717.

Kausch W 1965 Beziehungen zwischen Wurzelwachstum, Transpiration und CO_2-Gaswechsel bei einigen Kakteen. *Planta* 66:229–238.

Kutschera L, Lichtenegger E, Sobotik M, Haas D 1997 *Die Wurzel, das neue Organ, ihre Bedeutung für das Leben von Welwitschia mirabilis und anderer Arten der Namib sowie von Arten angrenzender Gebiete, mit Erklärungen des geotropen Wachstums der Pflanzen*. Klagenfurt, Austria: Pflanzensoziologisches Institut.

Lopez FB, Nobel PS 1991 Root hydraulic conductivity of two cactus species in relation to root age, temperature, and soil water status. *J Exp Bot* 42:143–149.

Loza-Cornejo S, Terrasas T 1996 Anatomía del tallo y de la raíz de dos especies de *Wilcoxia Britton & Rose* (Cactaceae) del noroeste de México. *Boletin de la Sociedad Botánica de México* 59:13–23.

Mahall BE, Callaway RM 1991 Root communication among desert shrubs. *Proc Natl Acad Sci U S A* 88:874–876.

Mahall BE, Callaway RM 1992 Root communication mechanisms and intracommunity distributions of two Mojave Desert shrubs. *Ecology* 73:2145–2151.

Mahall BE, Callaway RM 1996 Effects of regional origin and genotype on intraspecific root communication in the desert shrub *Ambrosia dumosa* (Asteraceae). *Am J Bot* 83:93–98.

Markle MS 1917 Root systems of certain desert plants. *Bot Gaz* 64:177–205.

Martre P, North GB, Bobich EG, Nobel PS 2002 Root deployment and shoot growth for two desert species in response to soil rockiness. *Am J Bot* 89:1933–1939.

Mauseth JD 2009 *Botany: An Introduction to Plant Biology*, 4th edn. Sudbury, MA: Jones and Bartlett Publishers.

McGinnies WG 1979a General description of desert areas. In *Arid Land Ecosystems: Structure, Functioning, and Management*, eds. DW Goodall, RA Perry, pp. 5–20. Cambridge, U.K.: Cambridge University Press.

McGinnies WG 1979b Arid-land ecosystems—Common features throughout the world. In *Arid Land Ecosystems: Structure, Functioning, and Management*, eds. DW Goodall, RA Perry, Vol. 1, pp. 299–316. Cambridge, U.K.: Cambridge University Press.

Meigs P 1953 Design and use of homoclimatic maps: Dry climates of Israel as example. In *Desert Research. Proceedings of International Symposium*, Jerusalem, Israel, May 7–14, 1952, pp. 99–111. Jerusalem, Israel: Research Council of Israel, Special publication no. 2.

Miyazawa Y, Ito Y, Moriwaki T, Kobayashi A, Fujii N, Takahashi H 2009 A molecular mechanism unique to hydrotropism in roots. *Plant Science* 177:297–301.

Niklas KJ, Molina-Freaner F, Tinoco-Ojanguren C, Paolillo Jr DJ 2002 The biomechanics of *Pachycereus pringlei* root systems. *Am J Bot* 89:12–21.

Nobel PS 1988 *Environmental Biology of Agaves and Cacti*. New York: Cambridge University Press.

Nobel PS 1994 Root-soil responses to water pulses in dry environments. In *Exploitation of Environmental Heterogeneity by Plants*, eds. MM Caldwell, RW Pearcy, pp. 285–304. San Diego, CA: Academic Press.

Nobel PS 2002 Ecophysiology of roots of desert plants, with special emphasis on agaves and cacti. In *Plant Roots: The Hidden Half*, eds. Y Waisel, A Eshel, U Kafkafi, 2nd edn., pp. 961–971. New York: Marcel Dekker, Inc.

Nobel PS 2006 Environmental productivity indices and productivity for *Opuntia ficus-indica* under current and elevated atmospheric CO_2 levels. *Plant Cell Environ* 14:637–646.

Nobel PS, Cui MY 1992 Hydraulic conductances of the soil, the root soil air gap, and the root—Changes for desert succulents in drying soil. *J Exp Bot* 43:319–326.

Nobel PS, Huang B 1992 Hydraulic and structural changes for lateral roots of two desert succulents in response to soil drying and rewetting. *Int J Plant Sci* 153:163–170.

Nobel PS, Miller PM, Graham EA 1992 Influence of rocks on soil temperature, soil water potential, and rooting patterns for desert succulents. *Oecologia* 92:90–96.

Nobel PS, Sanderson J 1984 Rectifier-like activities of roots of two desert succulents. *J Exp Bot* 35:727–737.

Nobel PS, Zutta BR 2007 Rock associations, root depth, and temperature tolerances for the "rock live-forever," *Dudleya saxosa*, at three elevations in the north-western Sonoran Desert. *J Arid Environ* 69:15–28.

North GB, Brinton EK, Garrett TY 2008 Contractile roots in succulent monocots: Convergence, divergence and adaptation to limited rainfall. *Plant Cell Environ* 31:1179–1189.

North GB, Huang B, Nobel PS 1993 Changes in structure and hydraulic conductivity for root junctions of desert succulents as soil water status varies. *Bot Acta* 106:126–135.

North GB, Nobel PS 1991 Changes in hydraulic conductivity and anatomy caused by drying and rewetting roots of *Agave deserti* (Agavaceae). *Am J Bot* 78:906–915.

North GB, Nobel PS 1992 Drought-induced changes in hydraulic conductivity and structure in roots of *Ferocactus acanthodes* and *Opuntia ficus-indica*. *New Phytol* 120:9–19.

North GB, Nobel PS 1995 Hydraulic conductivity of concentric root tissues of *Agave deserti* Engelm. under wet and drying conditions. *New Phytol* 130:47–57.

North GB, Nobel PS 1998 Water uptake and structural plasticity along roots of a desert succulent during prolonged drought. *Plant Cell Environ* 21:705–713.

North GB, Nobel PS 2000 Heterogeneity in water availability alters cellular development and hydraulic conductivity along roots of a desert succulent. *Ann Bot* 85:247–255.

Norwood M, Toldi O, Richter A, Scott P 2003 Investigation into the ability of roots of the poikilohydric plant *Craterostigma plantagineum* to survive dehydration stress. *J Exp Bot* 54:2313–2321.

Noy-Meir I 1973 Desert ecosystems: Environment and producers. *Annu Rev Ecology Syst* 4:25–51.

Ozsolak F, Milos PM 2011 RNA sequencing: Advances, challenges and opportunities. *Nat Rev Genet* 12:87–98.

Phillips WS 1963 Depth of roots in soil. *Ecology* 44:424.

Price RS 1911 The roots of some North African desert-grasses. *New Phytol* 10:328–340.

Puente M-E, Bashan Y 1994 The desert epiphyte *Tillandsia recurvata* harbours the nitrogen-fixing bacterium *Pseudomonas stutzeri*. *Can J Bot* 72:406–408.

Richards JH, Caldwell MM 1987 Hydraulic lift: Substantial nocturnal water transport between soil layers by *Artemisia tridentata* roots. *Oecologia* 73:486–489.

Rodríguez-Rodríguez JF, Shishkova S, Napsucialy-Mendivil S, Dubrovsky G. 2003 Apical meristem organization and lack of establishment of the quiescent center in Cactaceae roots with determinate growth. *Planta* 217:849–857.

Ruiz RN, Ward D, Saltz D 2006 Population differentiation and the effects of herbivory and sand compaction on the subterranean growth of a desert lily. *J Hered* 97:409–416.

Rundel PW, Nobel PS 1991 Structure and function in desert root systems. In *Plant Root Growth: An Ecological Perspective*, ed. D Atkinson, pp. 349–378. London, U.K.: Blackwell Scientific Publications.

Shane MW, McCully ME, Canny MJ et al. 2009 Summer dormancy and winter growth: Root survival strategy in a perennial monocotyledon. *New Phytol* 183:1085–1096.

Shane MW, McCully ME, Canny MJ, Pate JS, Huang C, Ngo H, Lambers H 2010 Seasonal water relations of *Lyginia barbata* (Southern rush) in relation to root xylem development and summer dormancy of root apices. *New Phytol* 185:1025–1037.

Shearer G, Kohl DH, Virginia RA et al. 1983 Estimates of N_2-fixation from variation in the natural abundance of ^{15}N in Sonoran desert ecosystems. *Oecologia* 56:365–373.

Shi ZY, Feng G, Christie P, Li XL 2006 Arbuscular mycorrhizal status of spring ephemerals in the desert ecosystem of Junggar Basin, China. *Mycorrhiza* 16:269–275.

Shi ZY, Zhang LY, Li XL, Feng G, Tian CY, Christie P 2007 Diversity of arbuscular mycorrhizal fungi associated with desert ephemerals in plant communities of Junggar Basin, northwest China. *Appl Soil Ecol* 35:10–20.

Shishkova S, Dubrovsky JG 2005 Developmental programmed cell death in primary roots of Sonoran Desert Cactaceae. *Am J Bot* 92:1590–1594.

Shishkova S, García-Mendoza E, Castillo-Díaz V, Moreno NE, Arellano J, Dubrovsky JG 2007 Regeneration of roots from callus reveals stability of the developmental program for determinate root growth in Sonoran Desert Cactaceae. *Plant Cell Rep* 26:547–557.

Shishkova S, Rost TL, Dubrovsky JG 2008 Determinate root growth and meristem maintenance in angiosperms. *Ann Bot* 101:319–340.

Suarez-Rodriguez MCS, Edsgärd D, Hussain SS et al. 2010 Transcriptomes of the desiccation-tolerant resurrection plant *Craterostigma plantagineum*. *Plant J* 63:212–228.

Sultan SE 2007 Development in context: The timely emergence of eco-devo. *Trends Ecol Evol* 22:575–582.

Sultan SE 2010 Plant developmental responses to the environment: Eco-devo insights. *Curr Opin Plant Biol* 13:96–101.

Takahashi H, Miyazawa Y, Fujii N 2009 Hormonal interactions during root tropic growth: Hydrotropism versus gravitropism. *Plant Mol Biol* 69:489–502.

Tian C, Shi, Chen Z, Feng G 2006 Arbuscular mycorrhizal associations in the Gurbantunggut Desert. *Chin Sci Bull* 51(Suppl. I):140–146.

Thomas DSG, Knight M, Wiggs GFS 2005 Remobilization of southern African desert dune systems by twenty-first century global warming. *Nature* 435:1218–1221.

Walter H 1985 *Vegetation of the Earth and Ecological Systems of the Geo-Biosphere*, 3rd edn. Berlin, Germany: Springer.

Walton K 2009 *The Arid Zones*. Chicago, IL: Aldine Transaction.

Wilcox CS, Ferguson JW, Fernandez GCJ, Nowak RS. 2004 Fine root growth dynamics of four Mojave Desert shrubs as related to soil moisture and microsite. *J Arid Environ* 56:129–148.

Zamski E, Ucko O, Koller D 1983 The mechanism of root contraction in *Gymnarrhena micrantha*, a desert plant. *New Phytol* 95:29–35.

Zeng F, Bleby TM, Landman PA, Adams MA, Arndt SK 2006 Water and nutrient dynamics in surface roots and soils are not modified by short-term flooding of phreatophytic plants in a hyperarid desert. *Plant Soil* 279:129–139.

29

Root Systems of Woody Plants

Frédéric Danjon
Institut National de la Recherche Agronomique
Université de Bordeaux

Alexia Stokes
Institut National de la Recherche Agronomique

Mark R. Bakker
Bordeaux Sciences Agro
Institut National de la Recherche Agronomique

I. Introduction

Woody plants are dominant species covering a large proportion of the natural terrestrial biome; trees and shrubs are used intensively as crops in orchards, vineyards, and plantation forests. Globally, forest covers 31% of the earth's land surface and tree root systems account for nearly 20% of forest biomass (FAO 2010).

Root systems of all plants generally have the same major functions, that is, water and nutrient uptake, anchorage, carbohydrate storage and association with symbiotic organisms, deposition and excretion of biochemical compounds, as well as production of propagules (Pagès 2002). This chapter will focus on the specificity of woody plant root systems. Root systems of woody plants differ from those of herbaceous plants largely because of the framework of larger rigid woody roots supporting the finer roots. The limit between the two classes is generally set to 2 mm in diameter (Böhm 1979), but this limit is controversial because the response pattern of the finest diameter fractions (those lower than 0.5 mm in diameter) can greatly differ from the larger fractions (Zobel et al. 2007). Large and fine roots do not have the same functions and are typically investigated by different methods. Fine roots are often studied with regard to their resource uptake functions and within a context of seasonal variation, longevity, and mortality. Woody plants generally live longer than herbaceous species and thus exhibit a very high variability in root lifespan, ranging from several days for root hairs to sometimes several hundreds of years for the stump and main roots. The immediate environment around the tree root system is also altered over the years, in particular with regard to soil chemical and physical properties (Bowen 1985). The coarse roots of woody plants usually undergo radial growth, which results in rigid roots that are important for anchorage. The functions of any given root segment change significantly as the plant grows bigger. Even small trees can colonize a large volume of soil, for example, 50 m^3 for a 3-year-old *Prunus persica* (L.) Batsch (Vercambre et al. 2003). Woody plants can also grow to very large dimensions, requiring a large and mechanically efficient root system (Anten et al. 2005). Because of their size, roots of one large tree can often grow to several meters in depth and several dozens of meters laterally and therefore will encounter very variable soil conditions both spatially and temporally. Woody plants are usually perennials and are thus subject to variable and possibly harsh seasonal field conditions, for example, storms, freezing, flooding, and drought. Thus, root systems of woody plants have to demonstrate both stability (survival) and plasticity (resilience) (Bowen 1985). Woody plants can be studied out of the field (e.g., in pots, hydroponics, and gel-filled chambers) only at the seedling or young sapling stage. Mature root systems have to be studied in the field, where they often grow in complex ecosystems. Roots really form a "system" and their functioning depends on the entire 3D root system structure including both the topological arrangement of all the elements and their geometrical characteristics (Pagès 1999).

In-depth studies of coarse roots have largely been hindered over the years due to the lack of appropriate methods, which really take into account their structure. Such methods have been available since the late 1990s but are still rarely used (Danjon and Reubens 2008). Much more research has been devoted to the study of fine-root stocks and their turnover in forest systems.

II. Defining Fine and Coarse Roots

Fine roots (defined generally as roots having a diameter below a threshold root diameter, most commonly 1 or 2 mm but 3 or 5 mm in some older studies) are often not considered by those studying coarse-root systems (e.g., with regard to tree mechanical stability) and typically are those parts of root systems that are lost when whole-root system excavations are carried out using mechanical excavators. Alternatively, fine roots might thus be defined as those roots that break off when a mechanical excavation of a root system is carried out, and in turn, coarse roots being those roots that typically occur with a low incidence in auger samples collected for the study of fine roots. It is not so much the size of roots that determines their inclusion in a fine-/coarse-root class but the measuring method used and the functions investigated, for example, anchorage, absorption of nutrients and water, or symbiosis.

Brunner and Godbold (2007) estimated that fine roots represent 21.4% of total root carbon content in boreal forests, 19.0% in temperate forests, and 13.6% in tropical forests, that is, on average 18.0%, or between one-fifth and one-sixth of total root biomass. Nevertheless, fine roots can also account for 95% of total root length and surface area (Persson 2002). In contrast, coarse and fine roots contribute equally (20%) to the carbon flux in trees and soil in central European forests (Brunner and Godbold 2007).

III. Measurement Methods

Uniform standardized methods for investigating any given root parameter do not exist (see Böhm (1979); Smit et al. (2000); Danjon and Reubens (2008) for an extensive description of root methods), because roots exhibit a large variety in size and dimensions, that is, from millimeters to tens of centimeters in diameter and from millimeters to tens of meters in length (Reubens et al. 2007).

Plant roots are structured into systems. Coarse roots therefore have a very heterogeneous distribution at the stand scale and are generally concentrated around and beneath the stump (e.g., Danjon et al. 1999a). Coarse roots also have orientation and branching patterns that depend both upon plant genetics and soil conditions, as well as their interaction. Coarse roots have been studied in detail with regard to their anchorage functions, which is determined by the whole 3D architecture. This 3D architecture also determines partly the distribution of fine roots at the tree level. Therefore, coarse roots generally have to be studied by excavating/extracting whole-root systems and by measuring the biomass or root architecture (Cairns et al. 1997; Danjon and Reubens 2008) or by measuring them partially and reconstructing the whole-root systems through modeling techniques (e.g., Spek and van Noordwijk 1994; Pagès et al. 2004). Measurements of specific features like the number of first-order lateral roots in planting stock (Deans et al. 1992) have also been performed.

In homogeneous tree and shrub systems, for example, plantations, the distribution of fine roots and mycorrhizas at the stand level is fairly homogeneous in the horizontal plane. This distribution varies mainly as a function of depth, soil heterogeneity, and plant cover and also as a function of time, with major seasonal changes. Fine-root attributes, among which biomass, length, surface area, specific root length (SRL), specific root area, branching, root tip number, and diameter distribution are the most commonly recorded parameters, are generally assessed with regard to soil depth, soil volume, or stand surface area (Böhm 1979). Not all studies clearly indicate whether the reported values relate only to live fine roots or if they also include dead fine roots (but see studies by Persson and Stadenberg 2009, 2010; Persson 2012), as the evaluation of root vitality is not easy and is mainly based on morphological criteria (Vogt and Persson 1991; Persson and Stadenberg 2010). Various methods exist for estimating root vitality, such as staining root tissue using, for example, triphenyltetrazolium chloride (TTC) reduction (Steponkus and Lanphear 1967; Clemensson-Lindell 1994; Comas et al. 2000; Withington et al. 2006), or performing physiological measurements such as oxygen consumption (Richter 2007) and root respiration (Comas et al. 2000). However, these techniques are time-consuming and may not be applicable in routine root separations.

In most studies on fine roots harvested within a plant community where several species are present, roots are not sorted by species, due to the difficulties in identifying which roots belong to which species. In a meta-analysis on fine roots, fine-root biomass was assessed separately for trees in 70% of the cited references, in both boreal and temperate forest ecosystems (Finer et al. 2011a). In tropical forest, fine roots were measured unsorted in 88% of the references. Understory biomass averaged 31% of total fine-root biomass in boreal forests and 20% in temperate forest (Finer et al. 2011a). However, distinguishing roots of different species can now be performed through morphological characteristics, molecular tools (Jackson et al. 2000; Mommer et al. 2011), or infrared methods (Roumet et al. 2006; Lei and Bauhus 2010). Alternatively, Brunner et al. (2004) used microsatellite markers to investigate intraspecific overlap. The assessment of such attributes can be based on different collection methods (core sampling, monolith sampling), retrieval methods of previously installed sampling devices (e.g., ingrowth cores, ingrowth meshes, and ingrowth nets (Hirano et al. 2009; Lukac and Godbold 2010), or observation methods (minirhizotrons, rhizoboxes, and trench walls). These methods allow for the study of fine-root mass, turnover, or nutrient uptake. However, these methods, with the exception of trench walls, are unsuitable for studying coarse-root systems. The trench wall root impact count can be used in different ways to assess both fine- and coarse-root distribution (Böhm 1979; Smit et al. 2000; Levillain et al. 2011). Nevertheless, even if data provide indications concerning vertical coarse-root distribution, results may be biased by the position of the trench relative to neighboring trees, and very few data can be obtained with regard to root system architecture and biomass. Clearly, the methodology of studying coarse roots and fine roots is different, and as a result, there is fairly little overlap in the scientific communities studying either type of roots except for Oppelt et al. (2005a,b). Oppelt et al. (2005a,b) combined a root auger approach with manual excavation and 3D digitizing of coarse roots to assess the exploration, exploitation, and root distribution of both coarse and fine roots in fruit tree and shrub species in Botswana. The topological and geometrical relationships between coarse and fine roots were also studied recently by recording the position and length of fine-root branching from carefully excavated coarse roots (Khuder et al. 2006). Sap-flow measurements have also provided information concerning functional relationships between coarse and fine roots (Coners and Leuschner 2002).

IV. Fine Roots

A. Functions of Fine Roots

The main function that is traditionally attributed to fine roots is that of nutrient and water uptake (uptake function). Storage of starch and nutrients (storage function) and physical stabilization (anchorage function) by fine roots are considered of lesser importance relative to coarse roots. The uptake function has been studied in various ways, using controlled nutrient conditions, hydroponics, aeroponics, pot experiments, rhizotrons, root observation labs, and different in situ devices (root windows, mesocosms). These methods permit direct access to roots, for example, the application of tracers or the carrying out of nutrient mobilization studies in the plant. For example, bioassays on excised fine roots, which consist of immersing the fine roots into solutions containing isotopes, were used to assess the phosphorus (P) status of the soil environment as experienced by the plant (Dighton and Harrison 1983), with high uptake rates of ^{32}P indicative of low P status. Isotopes were also used in a Danish research forest to evaluate whether the nutrient uptake capacity of *Quercus robur* L. roots from different soil layers (organic topsoil at 5 cm depth, mineral soil at 50 cm depth) would differ for nitrogen (N) (supplied as $^{15}NH_4^+$), potassium (K) (supplied by the analogue $^{86}Rb^+$), and P (supplied as $H_2^{32}PO_4^-$). Results showed a physiological plasticity of uptake with depth (Göransson et al. 2007), where the uptake capacity from excised fine roots from the topsoil layers was higher for N and K and similar for P compared to the deeper soil layer, which is somewhat counterintuitive as *Q. robur* is considered a deep-rooted species (Göransson et al. 2007). In another experiment in the same research forest, using 40-year-old stands of *Q. robur*, *Fagus sylvatica* L., and *Picea abies* (L.) H. Karst., ^{15}N and cesium (Cs) (analogue for K) were applied in situ close to fine roots at 5 and 50 cm soil depths followed by the measurement of these tracers in the foliage and leaf buds. Results showed a higher recovery of both tracers from 5 cm soil depths than 50 cm soil depths for *Q. robur*, but no differences with soil depth for either *F. sylvatica* or *P. abies*. It was also shown that uptake capacity for both nutrients was poorly related to rooting density, suggesting that root density is not an ideal predictor for actual nutrient uptake (Göransson et al. 2008).

Following root differentiation over time, roots have often been divided into different classes, that is, non-suberized versus suberized (Kramer and Bullock 1966), white relative to brown (Sands et al. 1982), or nonwoody and woody (van Rees and Comerford 1990). Based on studies on *Pinus banksiana* Lamb., *Pinus taeda* L., and *Eucalyptus pilularis* Sm., Enstone et al. (2001) and Kumar et al. (2007) have proposed that a distinction in three anatomically different root zones—respectively, white, condensed tannin (CT), and cork zone—is more appropriate. The underlying reason for this is that the browning process, which occurs as the white region matures, would rather be caused by the deposition of condensed tannins in the walls of all cells external to the stele than by deposition of suberin in various tissues (Mckenzie and Peterson 1995). For the order of magnitude, in *P. taeda*, the white zones represented around 5% of root surface area, compared to around 70% for the CT zones, some 2.7%–5% for mycorrhizal short roots, and the cork zone, representing the oldest root parts, some 10%–20% (Kumar et al. 2007).

Regardless of the precise definition for different root zones, nutrient absorption by roots appears to be highest at the root apex or the as yet non-suberized parts close behind the apex (the white zones). In two experiments with *P. abies* seedlings, grown in hydroponics (Häussling et al. 1988) and in root boxes with soil (Dieffenbach et al. 1997), the uptake of K^+, Ca^{2+}, and Mg^{2+} was most effective at the root tips and root elongation zones. However, some uptake of Mg^{2+} still occurred at the distal parts of the root system when the roots became older (Dieffenbach et al. 1997), and significant absorption of nutrients and water was reported even through suberized and woody roots (Kramer and Bullock 1966; Chung and Kramer 1975; van Rees and Comerford 1990).

As a result of root functioning and growth, the biological, physical, and chemical properties in the zone around the roots can be altered. This zone, the so-called rhizosphere, can exhibit a wide range of spatial and temporal heterogeneities (Hinsinger et al. 2009). For example, for nutritional elements that are limiting growth (and that have a low mobility in the soil, in particular P), the rhizosphere can become depleted due to uptake, whereas those elements that can move freely with mass flow (generated by plant transpiration) may accumulate close to the root. In many cases, roots develop symbiosis with fungal partners (Smith and Read 1997), permitting the exploration and exploitation of larger soil volumes than by roots only (then referred to as mycorhizosphere). Up to 75% of the uptake potential of young *P. taeda* root systems was made up of ectomycorrhizal hyphae (Rousseau et al. 1994), and in *P. abies* forests in Sweden, the total amount of ectomycorrhizal mycelium (including mantles) was estimated at 700–900 kg ha^{-1}, that is, the same order of magnitude as fine-root biomass (Wallander et al. 2001). In mature *Pinus pinaster* Aiton stands, all root tips, regardless of soil depth, were colonized by mycorrhizal fungi. The overall number of live morphotypes for the top 60 cm was 379,000 m^{-2} for the humid site and 625,000 m^{-2} for the dry site (Bakker et al. 2006, 2009). The role of symbiosis has been shown to be significant for nutrient mobilization and plant uptake of elements such as P, N, K, Cu, and Zn (Marschner and Dell 1994). These root–rhizosphere or root–mycorhizosphere interactions are discussed in more detail in Part VI of this volume and will not be commented on any further in this chapter.

Fine roots can also fulfill the role of storage of starch and sugars and can contain a significant amount of nutrients (Jackson et al. 1997; Yuan and Chen 2010). Presumably this may be more important for the larger diameter fractions (1–2 mm diameter) than the smaller fractions (<1 mm) following the observations on functional differences between these fractions (e.g., Withington et al. 2006; Ostonen et al. 2007a; Zobel et al. 2007; Goebel et al. 2011). High levels of starch in fine roots may be present the whole year round (Ericsson and Persson 1980) and be used for root growth during periods of low supplies, when needs for growth cannot be fulfilled through photosynthesis, for example, during the winter months in montane and subalpine forest ecosystems (Mao et al. in press). Van Praag et al. (1988) attributed somes peaks in sequential coring studies to simple variations in starch content, whereas Marshall (1986) suggested that insufficient carbohydrate (starch and sugar) content was responsible for reductions in fine-root biomass that occurred during drought stress.

B. Variations in Fine Roots: Horizontally, Vertically, and Over Time

Variations in the size of root systems may be related to ontogeny or size-related differences that occur over time or among different growth conditions (Ovington 1957; Helmisaari et al. 2002; Ritson and Sochacki 2003; Zerihun and Montagu 2004; Coyle and Coleman 2005; Coyle et al. 2008). The root mass fraction (RMF) in *Pinus sylvestris* L. decreased from 46% at the age of 7 years to 22% at 55 years (Ovington 1957) in the United Kingdom and from 25% at the age of 15 years to 13% at 100 years in Finland (Helmisaari et al. 2002). As a result, in a majority of studies, an increase of fine-root biomass is reported in the earlier age stages until canopy closure (reported age for maximum values between 30 and 100 year of stand age), after which it levels off or even decreases in maturing stands or stands with gaps in the canopy (Chen and Popadiouk 2002). Such a pattern of change in fine-root biomass with stand age (Figure 29.1) was reported both for deciduous tree species (Idol et al. 2000; Le Goff and Ottorini 2001; Claus and George 2005; Bakker et al. 2008; Yuan and Chen 2010) and coniferous tree species (Vogt et al. 1987; Vanninen and Makela 1999; Helmisaari et al. 2002; Fujimaki et al. 2007; Yuan and Chen 2010), but in some cases comparisons between stands of different ages did not show differences in fine-root biomass with stand age. Not only modifications in root physiology occur with stand development over time, but the dimensions of individual trees and allocation proportions can also change, along with stand density. For example, 3 years after a thinning treatment, fine-root biomass of the trees in a thinned stand was 143 g m^{-2} compared

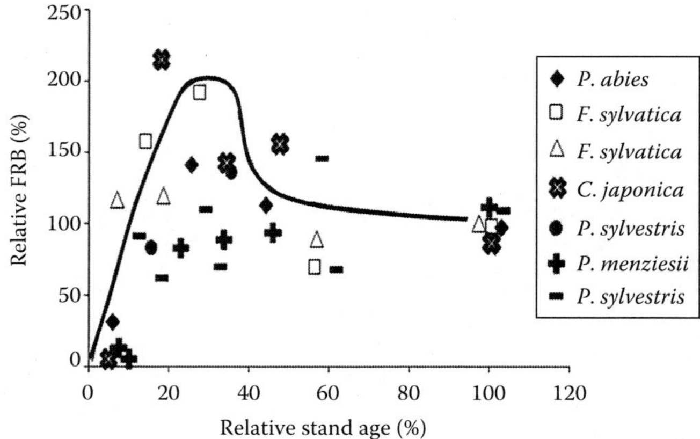

FIGURE 29.1 Hypothetical pattern of fine-root biomass with stand age (after a concept proposed by Claus and George 2005). The magnitude of fine-root biomass for each chronosequence stand is expressed relative to fine-root biomass in the oldest stand of the respective chronosequence. Relative stand age is expressed as a percentage of the oldest age in the chronosequence. Data from Claus and George (2005) for *P. abies* [♦] and *F. sylvatica* [□], Bakker et al. (2008) for *F. sylvatica* (Δ), Fujimaki et al. (2007) for *C. japonica* (✽), Helmisaari et al. (2002) for *P. sylvestris* (●), Vogt et al. (1987) for *P. menziesii* (✚), and Vanninen and Makela (1999) *P. sylvestris* (▬).

to 218 g m^{-2} for a control stand (Noguchi et al. 2011), suggesting that for a lower stem number (due to thinning or death of some individuals) that typically occurs with increasing stand age, fine-root biomass will decrease. This brings up a rather crucial issue in root studies: where coarse roots can often easily be assigned to individual trees, for fine roots this is not the case. Where tree circumference or diameter values of individual trees generally show good correlations with coarse-root or total root biomass (Tobin et al. 2007b), fine-root biomass is often badly predicted. Chen et al. (2004) proposed that a fine-root biomass inventory (at the stand level) be carried out and then aboveground characteristics used to calculate a fine-root biomass per tree for a "representative" tree size of mean basal area for each stand. These data would then enable significant correlations between fine-root biomass per tree and diameter at the ground surface to be derived for the forest types investigated. However, in a mature old-growth monospecific *F. sylvatica* forest, no indications were found for territoriality or asymmetric competition in the presence of neighbors, with no relationship between fine-root biomass and the distance from the stem (Lang et al. 2010), suggesting that it may not be so easy to assign fine roots to a particular tree. The use of microsatellite markers could be helpful tools as they were successfully used to assign fine-root fragments to individual trees in *Abies alba* Mill. in Switzerland, showing root overlap to occur both vertically and laterally (Brunner et al. 2004).

C. Plasticity and Resources

Fine-root biomass is generally lower at more productive forest sites (Keyes and Grier 1981; Vogt et al. 1987; Walters and Reich 1989; Olsthoorn 1991; Achat et al. 2008) or in stands benefiting from fertilization or irrigation (Axelsson and Axelsson 1986; Albaugh et al. 1998; Maier and Kress 2000; Bakker et al. 2009). In response to spatial and temporal variability in nutrient availability, trees can display morphological plasticity by selectively allocating biomass for growth within nutrient-rich patches (Jackson et al. 1990; Hutchings and De Kroon 1994; Robinson 1994; George et al. 1997; Hodge 2004), sometimes accompanied by a lower investment in root growth in the nutrient-poor patches (Hodge 2009). Root growth can be achieved by forming new roots ("birth of roots") to profit from nutrient-rich patches, but inversely, an increased root turnover ("root death") was also reported to be an efficient mechanism to increase P and K uptake (Steingrobe 2005). Root growth into nutrient patches can be increased by an increase in root elongation or by producing finer roots with a higher SRL (i.e., root length per unit dry mass) or a lower root tissue density (RTD). The extent to which this happens will depend on species traits, with fast-growing species showing a greater increase in SRL (George et al. 1997; Mou et al. 1997; Comas and Eissenstat 2004). This can also depend on the nutrient deficiency occurring (Zhang and George 2009). In a case of magnesium (Mg) deficiency in young *P. sylvestris* and *P. abies* seedlings, Mg uptake and accumulation in roots were increased but not root growth, in the vicinity of Mg-rich spots (Zhang and George 2009), suggesting a different response to Mg patches than to N and P patches. For sites with overall higher fertility or receiving fertilizers, a lower fine-root biomass can be coupled with a higher SRL compared to less productive sites or treatments (Bakker et al. 2009; Maurice et al. 2010). A study in New Zealand using the soil chronosequence approach showed that SRL was comparably high at the strongly P-limited late stages and at the youngest less P-limited site, but root diameters were smaller and tissue densities higher at the late stages (Holdaway et al. 2011). In a study comparing 11 tree species, root diameters were not significantly different between hardwood and conifer species, but hardwood roots had significantly greater SRL than the conifers (Withington et al. 2006), suggesting some overall difference in root dimensions or tissue density between hardwoods and conifers (Comas and Eissenstat 2004). Physiological adaptation for nutrient uptake can also take place (Fransen et al. 1999; Göransson et al. 2006), for example, by increasing uptake rate per unit of root length in the most favorable soil layers (Göransson et al. 2006). The optimum morphology and physiology of fine roots will ultimately depend on how a plant can best invest carbon to optimize benefits under given environmental conditions (Chapin et al. 1987; Eissenstat 1992; Ryser 2006). Also, root systems may perhaps perceive conflicting signals from different parts of the root system at the same time (Hodge 2009), yielding lower responses to heterogeneities in soils under natural conditions relative to more artificial experimental conditions. This may explain why there are ecological limits to plant plasticity (Valladares et al. 2007). Nutrient availability was shown to modify fine-root diameter distributions (Zobel et al. 2007) as well as RTD (Ostonen et al. 2007a). Any plastic responses of the root system in terms of changes in SRL or SRA will have consequences for root uptake, respiration, turnover, and decomposition, because these processes differ among root of different diameter size fractions or among root orders (Eissenstat and Yanai 1997; Pregitzer et al. 1998; King et al. 2002; Pierret et al. 2005; Goebel et al. 2011). Fine-root parameters, for example, root calcium (Ca):aluminum (Al) ratios in the case of acidity (Cronan and Grigal 1995; Vanguelova et al. 2007) or SRL changes due to fertilization levels, Al stress, light conditions, elevated temperature, and CO_2 (Ostonen et al. 2007b), were suggested as suitable indicators for environmental conditions or nutrient availability.

D. Soil Properties, Horizontal and Vertical Spread

Most fine and small roots are generally concentrated in the upper layers where most of the nutrients are located (Jackson et al. 1996; Parker and Van Lear 1996; Gwenzi et al. 2011), but depth distributions can be different through biomes (Gale and Grigal 1987; Jackson et al. 1996; Schenk and Jackson 2005). Maximum rooting depth is often assumed to be reached directly below the tree base via the taproot or sinker roots (Stone and Kalisz 1991), but for those species that can form obliquely descending roots or sinkers from lateral roots, roots may reach depths similar to or greater than those of the taproot and nearby sinkers (Stone and Kalisz 1991). The record values for maximum rooting depth ranges between 40 and 61 m (see Stone and Kalisz 1991), but maximum depth values generally do not exceed 20 m. Christina et al. (2011) found that in intensively managed *Eucalyptus* plantations on soils without any physical or chemical barrier, the rooting front depth was about 85% of mean tree height for stands with less than 20 m height. In a study by Schenk and Jackson (2005) on the global distribution of deep roots, about 10% of all the root systems analyzed were considered as deeply rooted, that is, either 5% of all roots were deeper than 2 m or at least a part of the roots were deeper than 4 m. The occurrence of deep roots was more likely to occur in seasonally dry, semiarid, or certain tropical regions with water infiltration depths and evaporative demand as the main climatic factors to determine vertical root distributions on a global scale (Schenk and Jackson 2005). Rooting depth may increase in response to higher atmospheric CO_2 concentration (Thomas et al. 1999; Iversen 2010). A small number of deep roots with small diameter may be sufficient to maintain transpiration during drought (Breda et al. 1993).

Hard soils are known to limit root growth or elongation (Unger and Kaspar 1994; Clark et al. 2003; Sudmeyer et al. 2004), and in such soils, root channels (macropore created by live roots or channel left after a root has died and decomposed) can often be found (Nambiar and Sands 1992). Roots can proliferate preferentially in channels (see Ghestem et al. 2011) and channels were found to occur on average 2.4 to 3.7 times per meter of trench wall (Figure 29.2) in studies on *F. sylvatica* (Bakker et al. 2008) and *P. pinaster* (Achat et al. 2008). Old decaying root systems (Van Lear et al. 2000) or scattered root channels containing several fine and small roots (van Noordwijk et al. 1991; Parker and Van Lear 1996; Pierret et al. 2005; Achat et al. 2008; Bakker et al. 2008)

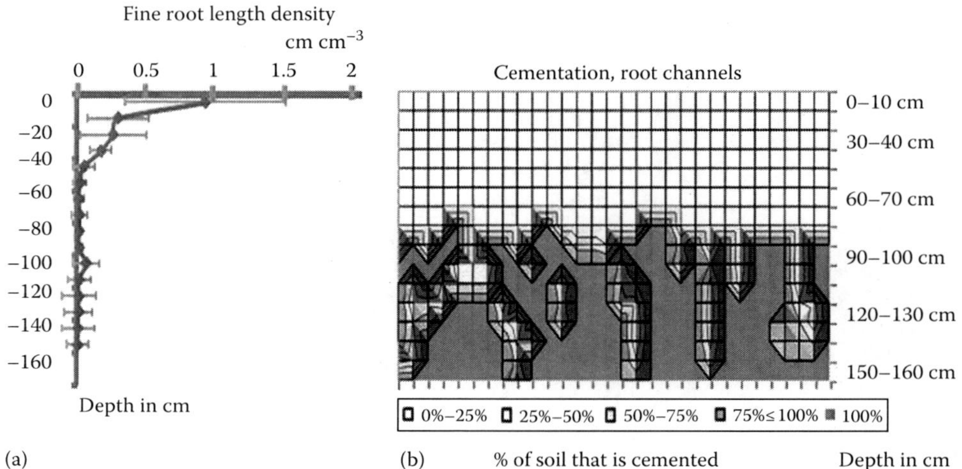

FIGURE 29.2 (a) Depth distribution of fine roots in a 82-year old *F. sylvatica* stand in Fongeres, France (fine-root length density) versus soil physical obstacles (b) where topsoil is without obstacles, subsoil with obstacles but featuring albeluvic tongues (data are means ± standard error, SE). Roots can develop freely above the dense subsoil (but see high SE in topsoils due to highly colonized nutrient-rich patches) and otherwise only in the root channels within the tongues of softer material (at greater depths higher SEs also occur; Figures made by M. Bakker from the original dataset).

enhanced root density by up to 20 times as compared to the soil matrix. These channels formed as a result of lower soil strength, better aeration, higher fertility, and more favorable moisture conditions. Subsoils with chemical constraints may also impede root growth into deep layers (Gwenzi et al. 2011). As pointed out by van Noordwijk et al. (1991), a coating of organic matter in the old root channels may help to complex aluminum (Al) and so reduce Al toxicity, which may be relevant under given conditions of soil acidity, at least at some soil depths.

The horizontal spread of root systems is suggested to be determined largely by the extension of the coarse-root system (Nygren et al. 2009), as fine roots or associated hyphae attached to the structural roots are generally short and do not extend far into the soil substrate. When the horizontal spread of fine roots is evaluated as a function of distance from the tree, a negative relationship between fine-root biomass per unit soil volume and the distance to the tree is generally assumed. However, this relationship is not always apparent, especially in studies where trench walls were used (Bouillet et al. 2002; Schmid and Kazda 2005; Achat et al. 2008; Bakker et al. 2008; Maurice et al. 2010), presumably because the dimensions of such trench wall pits are not adapted for such questions. Instead, at such a limited scale, the distance to the nearest coarse root may be a better predictor of local fine-root density. Using a ^{15}N tracer technique applied to boreal *P. abies* and *P. sylvestris* trees with about 20 cm DBH, Göttlicher et al. (2008) evaluated the lateral spread of the root system to be 4–5 m, but root systems also overlapped between trees. In their review on the maximum radial spread of root systems, Stone and Kalisz (1991) reported both examples where responses to applied fertilizers were detected at 3–6 m distance and examples where root damage occurred up to 40 m distance from the tree. The maximum radial spread they recorded was over 50 m. In a study evaluating both horizontal and vertical root proliferation in agroforestry systems with two *Eucalyptus* and two *Pinus* species in Australia, Sudmeyer et al. (2004) showed that the lateral extent of the roots was 10–44 m, which is equal to 1.5–2.5 (maximum of 4) times the height of the trees. Among the suggested factors affecting the variability of fine-root densities in the horizontal direction are genotypic differences, soil features such as slope direction and stoniness (Stone and Kalisz 1991), presence of understory species and irregularities at the forest floor surface (Achat et al. 2008), the presence of zones of preferential water flow (Bouillet et al. 2002), as well as the presence of nutrient patches or decaying stumps and coarse woody debris (Bengough et al. 2000; Van Lear et al. 2000; Achat et al. 2008). Variability in root intersects will be lower for increasing soil volumes or observation surface areas (Pierret et al. 2005; Achat et al. 2008) and will be larger for increasing soil depth as rooting gets sparser (Bengough et al. 2000). Overall, variations in water and nutrient variability often occur at scales of 20–50 m or much less, that is, within the range of lateral spread of individual trees or shrubs (Stone and Kalisz 1991).

Roots are often distributed in patches ("root clusters") rather than homogeneously in the soil. Trench wall fine-root counts showed that such clusters occurred 4–10 times per 1 m trench wall (120 cm deep), mostly in the top 50 cm, both in a chronosequence of *F. sylvatica* (Bakker et al. 2008) and in *P. pinaster* stands having contrasting site fertility (Achat et al. 2008). Bouillet et al. (2002) found root clusters in young *Eucalyptus* to be related to zones of preferential water flow rather than distance to the stem. Fleischer et al. (2006) reported that *F. sylvatica* formed weaker (i.e., less dense) small root clusters in larger cluster regions, avoiding inter-root overlapping compared to *P. abies* that possessed stronger clustering in small cluster areas. These small root clusters were inherently present in the top 20 cm and could not be clearly related to stand density (Schmid and Kazda 2005). Bengough et al. (2000) suggested that roots cluster in space because of their branched connections with daughter roots, as a result of induction by patches of nutrients or due to concentrations of roots in cracks or biopores in compacted soils. This suggests that clustering depends both on species (branching patterns) and site conditions (nutrient patches, soil compaction).

E. Fine-Root Turnover

Fine-root turnover is assumed to be a major pathway for carbon fluxes into the soil and belowground nutrient cycling (Jackson et al. 1997; Brunner and Godbold 2007) and can depend both on traits of the plant considered (e.g., root diameter, root function, root order, and carbohydrate reserves) and on environmental conditions such as soil type, soil depth, ambient temperature, and so on (Marshall 1986; Van Praag et al. 1988; Gill and Jackson 2000; Guo et al. 2008b; Yuan and Chen 2010; Goebel et al. 2011; Finer et al. 2011b). Various methods for the determination of fine-root turnover exist and include the use of sequential coring, ingrowth cores or meshes, image analysis based on minirhizotron images, root carbon age based on ^{14}C, and mass balance methods based on C or N budgets (Pritchard and Strand 2008). The resulting turnover values may vary substantially and studies comparing turnover methods are scanty and so far do not permit an integration of the different methods (King et al. 2002; Tierney and Fahey 2002; Guo et al. 2008a; Pritchard and Strand 2008; Strand et al. 2008; see also Chapter 27). Stocks of dead fine roots were reported somewhat higher than live fine roots (Persson and Stadenberg 2010); on average the live/dead ratio for <2 mm roots was 1.4.

F. Fine-Root Competition, Facilitation, Territoriality of Roots

When fine roots are separated from soil cores or recognized on images or trench wall profiles, it immediately becomes clear that roots of different species occur in the same samples, whether they be roots of different trees or roots of the understory mixed together with roots of overstory species. For example, roots of the target overstory *P. pinaster* represented only 10%–30% of all roots in the topsoil layers of a forest stand (Achat et al. 2008). These data lead to the question whether under such conditions, roots are likely to compete for resources (competition) or perhaps may benefit from each other's presence (facilitation)? Typically, for mixtures of trees and/or understory species, it is assumed that, for example, a deeper rooting pattern might occur for any one given species and shallower patterns for remaining species (Coomes and Grubb 2000; Schmid and Kazda 2005; Bolte and Villanueva 2006). Such vertical niche differentiation has been suggested to reduce competition for resources (Casper and Jackson 1997; Jose et al. 2006). However, no belowground overyielding in terms of fine-root biomass occurred for species-rich forest stands relative to species-poor stands (Meinen et al. 2009a,b), possibly related to an absence of vertical stratification of root systems of different species. Instead, a broad overlap of the root systems of neighboring trees of different species was prevailing both in soil cores and in situ root growth chambers (Rewald and Leuschner 2009), demonstrating a clearly asymmetric competition between species, with *Fagus* having the highest belowground competitive ability. In mixed stands of overstory *P. pinaster* and various understory species, root distributions were limited by a hard pan and high water tables and did not differ significantly between species (Achat et al. 2008), suggesting competition for resources from the shared soil horizons. Through the examination of root traits and competition in heterogeneous environments, Rajaniemi (2007) showed that root biomass and time to reach a nutrient-rich patch were less important traits to explain the competitive ability than the ability to quickly explore and fill a soil volume with roots. In some cases, new roots profit from old decaying root systems (van Noordwijk et al. 1991; Nambiar and Sands 1992; Parker and Van Lear 1996), facilitating their growth downward. Also, one species' root development may show a specific but positive response when roots of another species are encountered (de Kroon 2007). Messier et al. (2009) tested the effects of competition (for resources and for space) between grasses and tree species and showed that early successional tree species featured root avoidance or segregation and late successional tree species demonstrated root tolerance of competition as mechanisms for optimizing nutrient uptake.

V. Coarse Roots

A. Coarse-Root Measurements

Noninvasive techniques to investigate root system structure within the soil have been used successfully in small container plants using X-ray computed tomography (CT) or nuclear magnetic resonance (NMR) imaging. Such techniques can capture the dynamics of root growth because they are nondestructive (Lontoc-Roy et al. 2006). Discrimination between roots and soil can be improved using a root-tracing software (Tracy et al. 2010). In the field, available techniques, such as ground-penetrating radar (GPR) or electrical resistivity, can at best be used only in certain types of soil for biomass estimation after calibration, see for example, Butnor et al. (2003). Several methodological papers have been published concerning this topic, but as the signal varies as a function of soil water content and root wood density, it is very difficult to reconstitute root system architecture (Danjon and Reubens 2008).

When investigating coarse-root structure and function, the first step is therefore to access the root system. Manual excavation is laborious, whereas high-pressure air or water lances are efficient and do not harm ligneous roots. Once excavated, the root system can be measured in situ or brought back to the laboratory. Manual uprooting is suitable for smaller plants, but for larger plants, mechanical uprooting using a trestle, a lumbering crane, or a mechanical shovel is generally much more rapid. Due to the large variability in root system structure, it is often better to measure a large number of root systems obtained via uprooting than to spend a lot of time over a careful excavation and in situ measurement of a small number of trees, even if some roots are lost during uprooting (Danjon and Reubens 2008). A subsample of long shallow roots can be extracted over several meters before uprooting the root system. Before uprooting, it is recommended to remove the understory and litter and to free the large shallow roots with an air or water lance. Information about the position of the root system including north, soil level, and horizontal plane has to be firmly tagged on the root system before extraction in the case of *ex situ* measurements (Danjon and Reubens 2008).

One further problem when studying coarse-root systems is to determine the limit between roots and shoots and even the true soil level. As a root system grows, the soil level in the vicinity of the stem often rises due to the accretion of root volume. Conversely, the soil between trees can sink when heavy machinery is used in the forest, which results in soil compaction (Tobin et al. 2007b). Additionally, many tropical trees have large buttresses or aerial roots (Pavlis and Jenik 2000), and even temperate forest species can possess small buttresses in larger trees.

Root systems are often studied by taking into account only their RMF (also referred to as "root partitioning coefficient," RPC—Canham et al. 1996) (e.g., Coutand et al. 2008; Poorter et al. 2010). In the TRY plant trait database, roots are described by only two traits (mycorrhiza type and maximal rooting depth) compared to 50 traits for the aboveground parts (Kattge et al. 2011). Measurements can simply consist of cutting up the root system in segments and classifying them by diameter class or root type for biomass measurements (Le Goff and Ottorini 2001; Scanlan and Hinz 2010) and/or position of the root volume. An assessment of several aspects of root architecture can be performed without measurement of geometry by the following:

1. Fractal branching analysis, recording diameter before and after each branching point, and length between branching points (Spek and van Noordwijk 1994; Soethe et al. 2007). They can be used to reconstruct a whole-root system from the proximal diameter of lateral roots, considering that the fractal branching parameters (i.e., tapering at branching point, tapering between branching points, share of the largest root segment after branching in the sum of cross-sectional area (CSA) after the branching point, interlateral length) do not vary in the structure. Reconstructions have often not been accurate because the parameters vary as a function of root type, and geometry and/or the structuring in root types is not taken into account (Danjon and Reubens 2008).
2. Topological indices can also be computed to determine if the root system has a herringbone-like or a more dichotomous structure (Fitter 1987). A herringbone topology exploits more efficiently the soil near the root–stem base (stump), whereas dichotomous structures generally more efficiently explore the soil at a larger scale. However, root systems of larger woody plants can have a fairly complex structure with, for example, large herringbone shallow roots and dichotomous secondary sinkers (Danjon et al. 2005). Therefore, a whole-root system topology parameter is not always relevant in large woody plants.

Root system geometry is important for anchorage and resource capture. Partial measurements have been performed by measuring CSA of a root at given distances and depths from the collar in 2D (Coutts 1983) or 3D (Drexhage et al. 1999). This method can be used in a simple way for measuring just CSA and azimuth of all shallow second-order roots. However, more modern 3D digitizing can be used if further information is required (Danjon and Reubens 2008).

To perform 3D digitizing of root architecture, a coding system has to be used. In recent years, only the multiscale tree graph (mtg) coding (Godin et al. 1999) was used for root measurement (Danjon and Reubens 2008). This coding is simple and parsimonious; it can be easily corrected and adapted to objectives for which it was not initially designed (Godin 2000). Components of root segments are generally root axes divided into arbitrary segments chosen so as to most efficiently represent changes in direction and taper.

Three-dimensional root architecture, including both topology and geometry of a subpart or of a full root system, can be assessed by either (Danjon and Reubens 2008) of the following:

1. Manually using a tape, a protractor, and a compass (Dupuy et al. 2007) or a frame for small plants (Khuder et al. 2006)
2. Using a low magnetic field 3D digitizer driven by software devoted to the measurement of plant architecture (Danjon et al. 1999b)

Both types of measurement can be coded in the aforementioned mtg format.

Laser scanning is a rapid way to digitize a rigid uprooted root system (Wagner et al. 2010), but it cannot capture hidden parts in complex root systems and can only yield root volume location. Therefore this technique cannot be used directly for architectural analysis (Danjon and Reubens 2008) unless a program is written to determine automatically the topology from the laser data as suggested by Tracy et al. (2010).

Once the root architecture of a species has been characterized, rapid architecture measurements can be performed as a high-throughput phenotyping, measuring root diameters, lengths, and azimuths so as to answer a given specific scientific question. Such an analysis was carried out on 1200 root systems of *P. pinaster* saplings of different provenances (Danjon et al. 2009), in order to study anchorage and water capture potential. Less than 10 min per tree were spent on the measurement (including uprooting). For the dynamics of coarse-root growth, field rhizotrons can be used (Jourdan and Rey 1997a), but artifacts are expected (Lecompte et al. 2001). Coarse-root growth can also be partially inferred through a series of excavations, a chronosequence, and/or by annual ring measurements (Danjon and Reubens 2008). Age assessment in coarse roots is difficult because they form discontinuous rings, and in angiosperms, xylem structure is usually diffuse-porous, even if stems are ring-porous (see Stokes 2002).

VI. Architecture of Root Systems

Roots have a large number of functions but show little morphological variation (Fitter 1987). The growth of shoots of trees and shrubs is strongly constrained by genetics, that is, by the

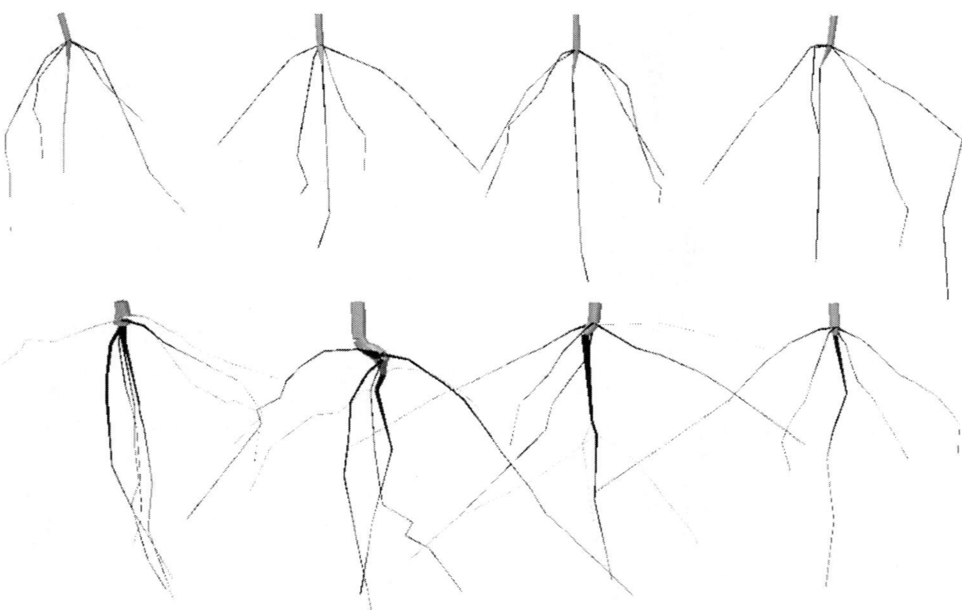

FIGURE 29.3 Graphical reconstruction of root system of *J. curcas* seedling grown from seed, measured by 3D digitizing. *J. curcas* seedlings grow a taproot and four oblique laterals from the stump. Stump and lateral roots in gray, taproot in black. Above: 2-week-old seedlings which differed on average by only 8% from the model. Below: 4-week-old seedlings, which differed on average by 18% from the model. (From Reubens, B. et al., *J. Arid Environ.* 75, 201, 2011.)

endogenous architectural model of the species. Nevertheless, growth and form are also influenced by the local environment, depending upon the plasticity of the genotype (Malamy 2005). The aim of architectural analysis is to identify the successive architectural phases of a species/genotype that constitutes its architectural model (Barthélémy and Caraglio 2007). Root systems also follow architectural models, which rely mainly on growth, branching, and morphological differentiation of axes (Atger and Edelin 1994b). However, shoots grow in a medium (air), which is much more homogeneous than soil and are therefore mainly shaped by neighboring plant competition and general resource availability. Most roots grow in soil, which is a complex medium with high spatial and temporal environmental variability (Harper et al. 1991). Soil is often heterogeneous with distinct soil layers, stones, hard pan, a water table, coarse woody debris, areas of toxicity, variability in compaction, and water and nutrient content. A further confounding parameter with regard to root architecture is that planting methods can perturb enormously the root system (Puhe 2003; Moore et al. 2008). Therefore, it is more difficult to see the underlying root architectural model of a given species compared to the shoot model. Root systems thus have to be more opportunistic than shoots, in order to capture efficiently resources or to achieve an optimal anchorage. With regard to architectural analyses over time, growth flushes and growth units generally cannot be retrieved in root systems due to the lack of morphological markers (Danjon and Reubens 2008). According to Köstler et al. (1968), the root architecture of a genotype is best expressed in a soil with no limiting factors. However, probably only a low number of species have a very characteristic root system structure, for example, *Jatropha curcas* L. When grown from seed, this species initially develops one taproot and four perpendicular oriented oblique laterals, branching from the stump (Reubens et al. 2011; Figure 29.3).

Zobel and Waisel (2010) proposed a standardized root nomenclature for individual plant roots with regard to four types: (1) the taproot or primary root that emerges from the seed and is generally negatively gravitropic, (2) lateral roots branching from the taproot and their branches, (3) shoot-borne roots, and (4) basal roots originating from the hypocotyl that is the organ between the base of the shoot and the base of the taproot (adventitious roots).

Cannon (1949) published a classification of root systems of herbaceous plants that could be partly used for woody plant seedlings. In this classification, root systems without adventitious roots mainly differ in the length or forking of the taproot and in the distribution and length of second-order roots along the taproot. In a major overview of root system architecture of German forest trees, Köstler et al. (1968) proposed a qualitative classification based on drawings and subjective classification. These authors defined four types of architecture (Figure 29.4): (1) plate root systems with only shallow roots ("Flachwurzelsystem"); (2) taproot systems with a taproot and shallow roots ("Pfahlwurzelsystem"); (3) heart root systems with shallow, oblique, and sinker roots branching from the stump ("Herzwurzelsystem"); and (4) root systems with shallow roots and secondary sinkers ("Senkerwurzelsystem"). Secondary sinkers are vertical downward roots branching from shallow laterals. Much more variability in rooting can be found in tropical trees, which in particular can show well-developed aerial roots, such as stilt roots, spine roots, buttresses, and pneumorhizae

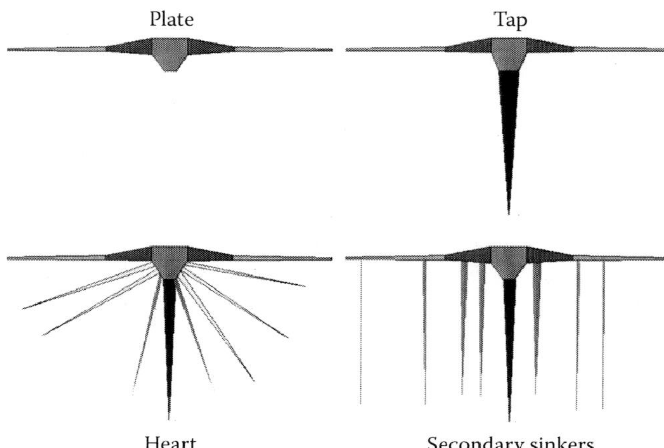

FIGURE 29.4 (See color insert.) Schematic representation of the four classes of tree root architecture according to Köstler et al. (1968). Gray: stump, black: taproot, blue: ZRT, light blue: shallow roots beyond ZRT, red: proximal secondary sinkers, violet: distal secondary sinkers, yellow: oblique roots. Intermediate-depth roots and deep roots are not figured.

(Pavlis and Jenik 2000). Some tropical species also only develop secondary root systems from the collar after death of the taproot apex (Kahn 1977, 1983).

Most plants repeat their architectural unit through "reiteration," which can be complete or partial, that is, the plant repeatedly expresses the juvenile growth pattern of the seedling (Barthélémy and Caraglio 2007). A branching root that is the same type as its mother root and branches near its apex can be defined as a reiteration (Collet et al. 2006). This type of reiteration may have occurred through a traumatic event, for example, after the death of the apex, in winter (Collet et al. 2006), or through wounding of an axis (Atger and Edelin 1994c), due to herbivory, machinery, wind damage, or soil movement. In the taproot and lateral surface roots of *Platanus* sp., Atger and Edelin (1994b) described sympodial branching originating from the death of root apices followed by the formation of several subterminal new axes, corresponding to a systematic forking ("immediate reiteration"). Edelin and Atger (1994) also reported that the perennial horizontal roots of *Cecropia obtusa* systematically and successively fork into three branches, and at a distance of 4 m from the trunk, one of these roots develops as a secondary sinker. This strategy allows the root system to spread out radially even with a low number of second-order roots branching from the stump. A delayed branching by formation of reiteration at the base of the stem (in *Cecropia*) or at the collar (in *Platanus*) is referred to as a "delayed reiteration," or ability to grow new roots on the framework of older roots or shoots. This reiteration permits the tree to once again explore and exploit the soil around the stump, where the originally established root system has little or no actively absorbing sections (Kahn 1983; Atger and Edelin 1994b). Systematic immediate reiteration was also observed in *Quercus rubra* (Lyford 1980). However, *Anaxagorea dolichocarpa*, a smaller tropical tree species, does not have reiterations in its root system, whereas its shoots are built up by systematic total reiteration (Atger and Edelin 1994b). Generally, branches originating from a fork are separated from each other by smaller angles than those between parent roots and their laterals (Lyford 1980).

Therefore, using the classifications of Köstler et al. (1968), Atger and Edelin (1994a), and Zobel and Waisel (2010), the endogenous root system architecture of a genotype can be characterized by (1) the ability to develop a substantial amount of oblique roots and/or secondary sinkers, (2) the level of immediate reiteration, (3) the ability to reiterate by retarded branching, and (4) the proportion of roots in the four classes defined by Zobel and Waisel (2010). These properties have a high incidence on root system functioning. Analogous to the absorbing capacity of herringbone versus dichotomous patterns (Fitter 1987), root systems without reiteration would be the most efficient for long-range exploration, root systems with immediate reiteration would be the most efficient for shorter-range exploration, and root systems with retarded reiterations would be the most efficient for exploitation over time (Atger and Edelin 1994a).

The branching order is rarely sufficient to describe the fairly complex root architecture in tree root systems (Jourdan and Rey 1997b). Hallé and Oldemann (1970) developed qualitative "architectural analysis" for shoots, which was adapted to root systems by Kahn (1977, 1983), Atger and Edelin (1994b), Leroux and Pagès (1994), and Atger and Edelin (1995). Quantitative 3D architectural analysis was initiated by de Reffye (1979) on shoots and adapted to roots through morphological modeling by Pagès and Aries (1988) on simple structures and Jourdan et al. (1995) on mature *Elaeis guineensis* Jacq. (oil palms).

Architectural analysis on root systems consists of defining types of axes composing the root system and assessing the properties of each type in terms of longitudinal and radial growth rate and tropism, apex length and diameter, branching rate, angle and type of branches carried, monopodial or sympodial, immediate or delayed branching, forking, and mortality (Atger and Edelin 1994b). The basic output of architectural analysis is a table of properties by root type and a schematic representation of stages of development of the species. See as an example Tables 1 and 2 in Pagès et al. (2004) for a list of parameters potentially describing root types and soil properties and Atger and Edelin (1994b) for schematic drawings. A sapling tree root system could be described using—five to six root types (Collet et al. 2006; Reubens et al. 2009) and a mature tree by about—seven to eight categories of roots (Atger and Edelin 1994b; Jourdan and Rey 1997a; Danjon et al. 2005).

Vercambre et al. (2003) and Collet et al. (2006) defined root types that correspond roughly to the branching order; two second-order types were defined by Collet et al. (2006), one with reiteration due to winter mortality and one without. A classification by Jourdan and Rey (1997a) was also based on branching orders but with subclasses for orientation (horizontal or vertical) or depth (shallow or deep).

Danjon et al. (2005) established another type of classification, to determine which part of the root system in mature

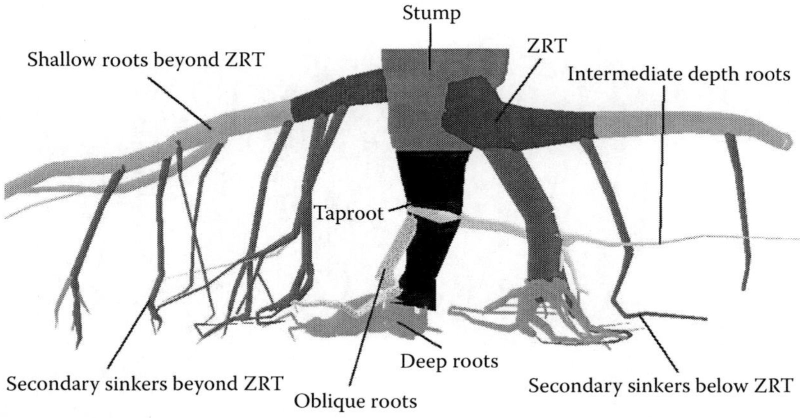

FIGURE 29.5 **(See color insert.)** Architectural analysis in mature *P. pinaster*: Schematic representation of root types. Nine compartments were defined mainly as a function of their angle toward horizontal plane, depth of their initial branching point and tapering. (From Danjon, F. et al., *New Phytol.*, 168, 387, 2005. With permission.)

P. pinaster is important for mechanical stability and to track structural adaptation to the dominant wind direction. Nine root compartments were defined (Figure 29.5). The first-order root was broken into two compartments, stump (or "root bole") and taproot (Nicoll et al. 1995). The stump is the upper portion of the taproot with a large diameter where most of the shallow lateral roots originate. The taproot is the largest vertical root originating directly from the stump. Shallow roots were divided in the zone of rapid taper (ZRT, Eis 1974) and shallow roots beyond ZRT. The ZRT is the proximal section of shallow roots near the stump that shows a mechanical reinforcement in secondary growth in response to wind movement (Coutts 1987). The above part of the ZRT can partially emerge from the soil as a buttress. The three other compartments were intermediate-depth roots, deep roots, and oblique roots. Relative or absolute limits were set arbitrarily to define these compartments. This classification was used successfully in other species and in seedlings and saplings (e.g., Khuder 2007; Reubens et al. 2009). Certain root compartments, like secondary sinkers in *P. pinaster*, possess very distinct characteristics, but Vercambre et al. (2003) observed a large variability within each root type in *Pr. persica* trees, resulting in a continuum between the two extreme root system types.

Once established, these classifications were used for developmental modeling of root systems (Tobin et al. 2007b). Collet et al. (2006) used this type of modeling to study the effect of grass competition on *Quercus* seedling root systems. In the same way, Lamanda et al. (2008) assessed the dynamics of above- and belowground competition in *Cocos nucifera* L. agroforestry stands using both 3D aerial and belowground architectural modeling. Architectural analysis on roots was first used by Danjon et al. (2005) to compare quantitatively various treatments: these authors compared root volume in each compartment in three circular sectors in windthrown versus undamaged root systems of mature *P. pinaster* trees. A more detailed analysis was made by examining several properties, for example, root length, diameter, angle to the soil surface, and interlateral length, in each root type in *Robinia* seedlings subject to four treatments (Khuder et al. 2006, 2007). The plasticity of root architecture was assessed through architectural analysis in *P. pinaster* grown in sandy spodosols in well-irrigated conditions (930 mm annual precipitation). This species has a very stable root architecture starting with a taproot system as a seedling and developing progressively to a secondary sinker system with deep branches in larger trees. Few oblique roots or intermediate-depth horizontal roots occur (Danjon et al. 2005). Even planted cuttings essentially only differed from planted seedlings by the number of second-order roots branching from the stump, of which there were only half as many in cuttings (Khuder et al. 2007). In shallower soils, *P. pinaster* grows large secondary sinkers more rapidly, but again, the same architectural model exists (Danjon et al. 2005). Different provenances also possessed the same architectural scheme but differed largely in the size or the number of taproots on one side and in the relative size of shallow roots on the other side, probably as a result of natural selection for mechanical stability versus growth in drier areas (Danjon et al. 2009).

It should be noted that most common European conifers grow a root system similar to *P. pinaster*, whereas European hardwood species more likely develop heart-type root systems (Köstler et al. 1968; Stokes 2002). In *Picea sitchensis* (Bong.) Carr., the ranking of main second-order shallow roots is generally stable; only roots with large diameter tips produce major structural roots with strong secondary growth (Coutts and Lewis 1983).

Intra- or inter-root system grafts are common in woody plants, as roots, once elongated, do not move in the soil. Grafts result from a morphological union of cambium xylem and phloem of previously distinct neighbor roots (Graham and Bormann 1966). As described by Lyford (1980) in *Q. rubra* and Danjon et al. (2005) in *P. pinaster* (Figure 29.6), root grafts are common in the central root system, where the density of roots is large, and therefore, the probability that two

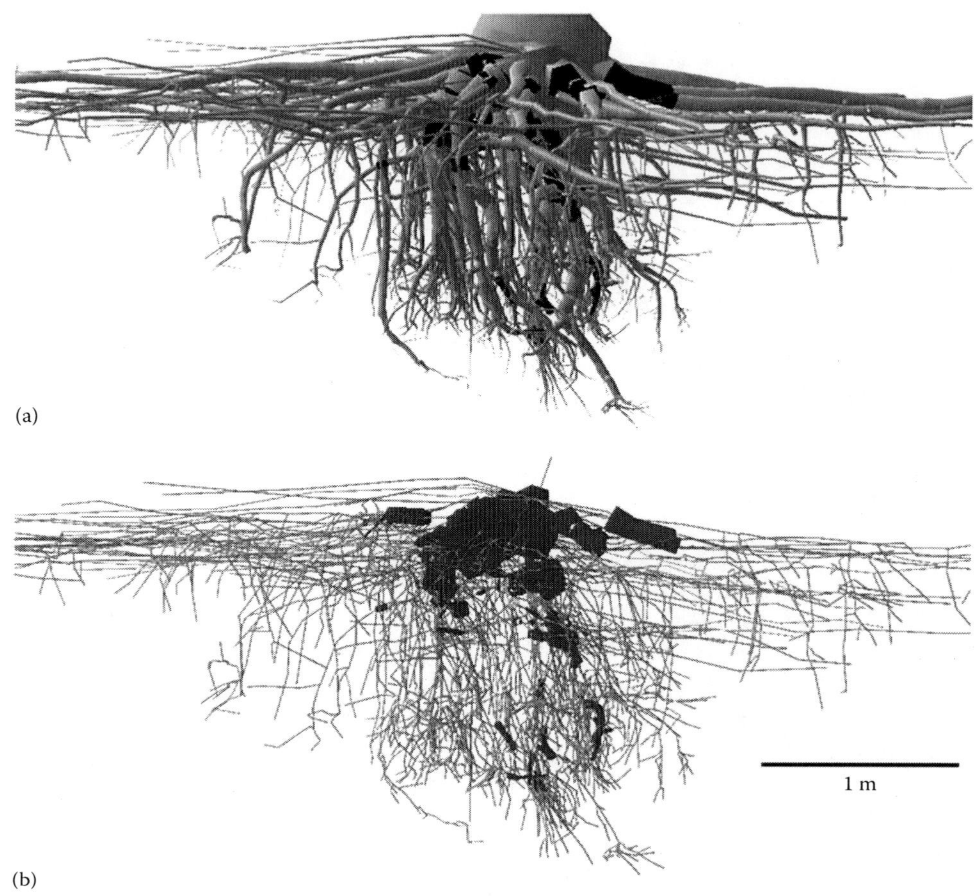

FIGURE 29.6 (See color insert.) Root grafts in an 18-year-old *P. pinaster* root system measured by 3D digitizing. (a) Root grafts are in black. (b) Same tree, segments with a root graft drawn with their real diameter. The diameter of all other segments was set at 4 mm. (Data from Khlifa, R., Meredieu, C., Augusto, L., Trichet, P., Danjon, F. Unpublished data)

roots cross is high. A large amount of grafts was also reported by Edelin and Atger (1994) in shallow roots on *Platanus hybrida* Brot., which form a sort of plinth in older trees with old and newborn roots. In the same way, intra-tree grafting is uncommon in the outer part of the root system, but *Q. rubra* and *P. pinaster* have displayed grafting to neighboring trees of the same species. Kulla and Lohmus (1999) reported that in *P. abies*, the number of inter-tree grafts increases when the distance between trees decreases or when the humus layer is narrower. Intra-tree grafts improve rigidity of the central root system and therefore improve resistance to windthrow. Inter-tree grafts can have a fairly large impact on ecosystems because when a tree is thinned or in case of stem breakage, its root system can partly survive, thanks to the still-standing tree to which it is grafted (Graham and Bormann 1966). Carbohydrates are transferred from the living trees to these stumps (Fraser et al. 2006, 2007; Tobin et al. 2007a; Tarroux and DesRochers 2010). In Ireland, 2%–11% of stumps in stands of *P. sitchensis* were found to be callusing, probably as a result of carbohydrate flux via root grafting (Tobin et al. 2007a). Root grafts are also a way through which pathogens can be transmitted (Epstein 1978). The benefits of root grafts are listed by Lev-Yadun (2011).

Forest stands are often found on shallow soils because deeper soils are used for agricultural purposes. The plasticity of root architecture with regard to soil depth can vary largely between species. According to Lebourgeois and Jabiol (2002), *Quercus petraea* (Matt.) Liebl. can develop deep root systems even in poor soils like heavy clays, horizons with coarse elements, and hydromorphic or compacted horizons. However, in *F. sylvatica* roots, downward growth is arrested by such poor conditions. *Q. petraea* does not demonstrate a particular acclimation of its root architecture in shallow soils, whereas *F. sylvatica* acclimates to shallow soils through greater extension of a plate root system. In *P. pinaster* (Figure 29.7), a similar acclimative response occurs, affecting in turn mechanical stability (see in succeeding text). Rooting depth and root system architecture also influence hydraulic redistribution (Brooksbank et al. 2011). Woody plants can have a major effect on ecosystems through hydraulic redistribution (Prieto et al. 2012), that is, the passive movement of water between diverse layers of soil via different roots within a root system.

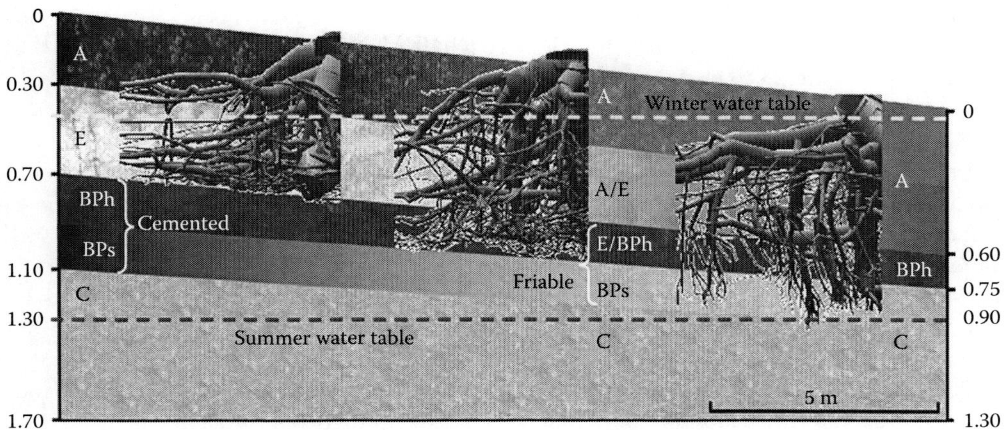

FIGURE 29.7 Diagram of a typical hydromorphic sandy spodosol toposequence with the corresponding rooting patterns of 38 cm mean DBH *P. pinaster* root systems. The horizontal scale is only for the toposequence. Same horizontal and vertical scale for root systems and the soil depth (m). Horizon names according to the World Reference Base for Soil Resources (WRB, Food and Agriculture Organization of the United Nations, Rome 1998). (Reproduced from Danjon, F. et al., *New Phytol.*, 168, 387, 2005. With permission.)

VII. Biomass Allocation and Carbon Content

From a major meta-analysis on forest ecosystem biomass, Cairns et al. (1997) established that root biomass generally ranges from 13% to 28% of the plant biomass, independent of latitude, soil texture, tree age, temperature, temperature:precipitation ratio, and plant type (angiosperm vs. gymnosperm). The RMF is slightly lower (19.3% vs. 20.6%) in tropical sites compared to temperate and boreal sites. Variability is higher in younger stands, in angiosperms, and in tropical sites. In a similar study, Kurz et al. (1996) found that in temperate and boreal forests, the RMF ranges between 7% and 23%, averaging 18.7%. A lack of trends in RMF with age was also found by Yang and Luo (2011) who carried out a meta-analysis on 16 chronosequences. Examining the relationships with the aforementioned variables, Cairns et al. (1997) showed that root biomass per hectare in forest ecosystems is only correlated to stand age. In areas with low-level background wind loading, the RMF tends to be larger at wider spacings (Ritson and Sochacki 2003; Barton and Montagu 2006). When arboreal plants with branches and secondary thickening age, they store biomass in the stem but shed periodically fine roots underground and branches by self-pruning aboveground, especially in dense stands. Conversely, the stump thickens at the same rate as the stem and shedding of coarse roots generally results from poor soil conditions and attacks by pathogens such as root rot (Puhe 2003). It should also be noted that coppicing removes the shoots but leaves the root system. According to Kurz et al. (1996) and Cairns et al. (1997), fine-root (<5 mm) mass fraction in roots decrease from 70% in young stands to less than 10% when the root biomass per hectare is larger than 75 Mg. Single-tree coarse-root biomass can be estimated from stem diameter at breast height (DBH) using species- and site-specific allometric equations (Tobin et al. 2007b).

Few data have been collected on coarse-root respiration. In *Eucalyptus* hybrids' plantations, coarse-root respiration per unit of external root surface area showed no relationship with root diameter (Marsden et al. 2008a). According to Chen et al. (2009), in *Acacia crassicarpa* A. Cunn. ex Benth. and *Eucalyptus urophylla* S.T. Blake plantations, at the stand level, both coarse and fine roots each contribute equally toward root respiration.

Taking into account the architecture of the tree, Bert and Danjon (2006) showed that in mature *P. pinaster*, carbon content in root wood varies mainly as a function of root diameter, ranging from about 53.7% in 1 cm diameter root segments to 51.5% in >5 cm diameter root segments. As a result, the weighted mean carbon concentration reached 51.7% in both stump and coarse-root wood. The quantity of carbon was greater in the root bark (54.5%), stem sapwood (53.3%), and stem heartwood (54.3%).

VIII. Anchorage Mechanics

A primary function of roots is to support the aboveground organs of plants, that is, to counteract the effect of gravity, and self-loading of the shoots. In most trees and woody plants, anchorage is a secondary consequence of competition for light above ground (Fitter 1987). In a forest, taller trees can shade shorter neighbors as well as being more efficient at capturing light (Anten et al. 2005). As root systems are usually constrained within the soil matrix, self-supporting mechanisms are not necessary. Therefore, mechanical loading on root systems is largely defined by a plant's aerial organs. Nevertheless, soil physical properties also influence root mechanical behavior. The mechanical state of woody plants is constantly being modified through both primary and secondary growth (Almeras and Fournier 2009). In an aerial environment that is relatively free from physical obstacles, tree trunks and branches may grow preferentially toward light patches, inducing stem lean. However, root systems remain fixed within the substrate and reactions to the aboveground physical environment will be slower. Therefore, the quality of anchorage can compromise the ability of stems to move away from the vertical axis, even when lean is advantage for maximizing light capture (Almeras and Fournier 2009).

The anchorage function largely drives the biomass allocation to roots as well as allocation within the root system (Stokes et al. 1997; Coutand et al. 2000; Danjon et al. 2005). The quality of anchorage will determine the ability of trees and shrubs to resist vertical forces (e.g., grazing; Ennos and Pellerin 2000), compressive forces (e.g., trampling; Striker et al. 2006), and lateral forces (e.g., substrate mass movement [Stokes et al. 2009] or flooding [Karrenberg et al. 2003]). Woody plants also have to resist overturning caused by strong winds. Storms are responsible for more than 50% of all primary abiotic and biotic damage (by volume) to European forests from catastrophic events (Gardiner et al. 2010). Even mild wind loading can affect tree growth through physical (abrasive) actions, alteration of the local atmospheric conditions around foliage (causing desiccation and affecting transpiration and photosynthesis), as well as inducing acclimative physiological responses. These responses can reduce stem height and volume, resulting in tapered or stunted trees (see Telewski 1995). Root anchorage may be improved through changes in biomass allocation within root systems, particularly when there is one dominant wind direction (Stokes and Guitard 1997).

The quality of anchorage in trees is usually measured through one of four methods:

1. Performing an inventory of damage after a storm event and measuring damage to trees, soil properties, and shoot and root–soil plate characteristics (Mason 1985; Harrington and De Bell 1996; Cucchi and Bert 2003).
2. Static bending tests whereby trees are winched sideways until stem, root system, or root–soil plate failure (Coutts 1983; Ray and Nicoll 1998; Cucchi et al. 2004). Static tests generally overestimate wind firmness because during storm events, wind gusts sway trees, weakening the root system (Peltola 2006; James and Kane 2008). Many older studies do not take into account dominant wind directions, which can bias results if tree acclimation to a given prevailing wind direction has occurred (Danjon et al. 2005).
3. Measurements of root system architecture in windthrown trees compared to trees that did not fail during a storm event (Harrington and De Bell 1996; Danjon et al. 2005). Indicators of stability, for example, stem lean, can be used (Danjon et al. 1999a) as well as static winching tests (Lindström and Rune 1999; Khuder et al. 2007; Gilman and Grabosky 2011).
4. A 2D or 3D numerical analysis of root anchorage including root architecture and root and soil properties. Numerical analysis is useful because it is difficult to measure both root and soil mechanical properties during uprooting (Dupuy et al. 2007).

Wind forces are extremely variable in space and time (Ennos 1999) and data on wind-induced movement in trees during wind events are rare (but see Watson 2000; James and Kane 2008). When a woody stem sways due to a strong wind, wind energy is partially absorbed by the shoots, and part is transmitted to the soil via the stump and roots. When the wind force along the stem (uprooting moment; a moment is a measure of the tendency of a force to rotate an object about an axis) exceeds the tree's resistive bending moment at a particular angle of deflection, the tree will deflect further. The tree will then start to fail if the wind force along the stem exceeds its maximum resistive bending moment. However, once anchorage has started to fail, the leaning stem becomes the second uprooting force, through the mass of deflecting stem and crown, regardless of wind loading (Ray and Nicoll 1998). The relative strengths of the stem and root system will then determine the exact mode of failure. Uprooting is thus a complex process that depends on wind speed, velocity, and turbulence; stem and root wood properties; root system architecture; and soil properties. During storms, soil water content is often high due to increased precipitation, reducing soil cohesion and weakening further root anchorage.

In most taprooted trees, when the stem is displaced sideways by wind loading, the stump and taproot push into the soil on the leeward side of the tree, with the top half rotating and the bottom half remaining reasonably well anchored (Crook and Ennos 1997). In some cases, the lower half of the taproot may make a semicircular movement and push into the soil on the windward side (Hintikka 1972). Shallow lateral roots are held in tension on the windward side of the tree and push into or buckle on the leeward side (Figure 29.8). If the shallow roots are well anchored, the entire taproot will rotate windward (Crook and Ennos 1997). In larger trees possessing a more platelike root system, during wind loading the root–soil plate rotates and is lifted upward on the

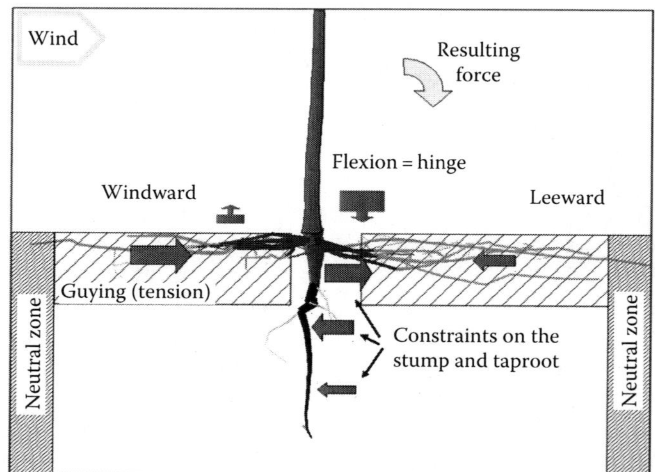

FIGURE 29.8 **(See color insert.)** Schematic drawing of the forces experienced by a sapling with a taproot system in sandy soil, when the stem is pushed to the right by a strong wind. Wind damage could be toppling or stem breakage. (Reproduced from Danjon, F. and Fourcaud, T., L'arbre et son enracinement, *Actes du colloque CIAG Sylviculture Forêts et Tempêtes*, juin 30, 2009, Pessac, France, Revue Innovations Agronomiques INRA, Paris, France, Vol. 6, pp. 17–37, 2009. With permission.)

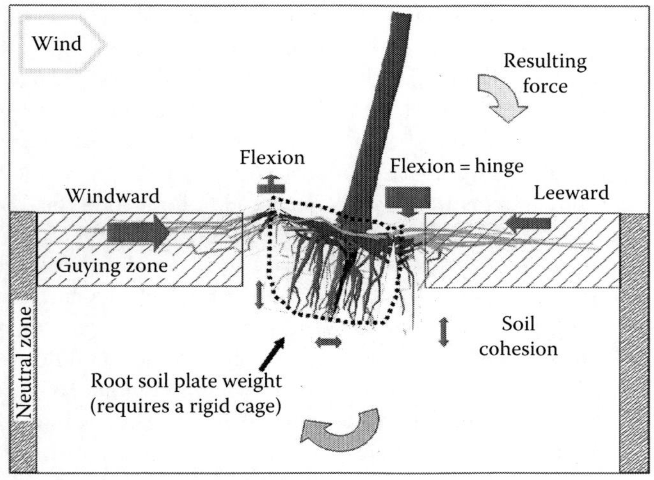

FIGURE 29.9 (See color insert.) Schematic drawing of the forces experienced by a mature tree with secondary sinkers' root system in sandy soil, when the stem is pushed to the right by a strong wind. Wind damage could be uprooting with root soil plate or stem breakage. (Reproduced from Danjon, F. and Fourcaud, T., L'arbre et son enracinement, *Actes du colloque CIAG Sylviculture Forêts et Tempêtes*, juin 30, 2009, Pessac, France, Revue Innovations Agronomiques INRA, Paris, France, Vol. 6, pp. 17–37, 2009.)

windward side (Figure 29.9). Lateral windward roots are subjected to tensile and possibly shearing forces (Coutts 1983). Leeward lateral roots are pushed onto the hard-bearing surface of the soil near the stump and are subjected to bending and compressive forces. Each root within a root system can act as a lever arm and resist bending at the leeward hinge, that is, the point about which a root plate will rotate during failure (Coutts 1983). In shallow root systems, doubling the distance between the hinge and the stump will double the turning moment required for uprooting (Coutts et al. 1999). The strength of the hinge can determine if the root–soil plate will slide into the soil or be lifted out of the ground (Figure 29.10). Roots perpendicular to the dominant wind direction will be subject to torsion, or twisting forces, and contribute little to the resistive turning moment of the tree. Additional resistance to uprooting is provided initially by root and soil strength around the edges of the root–soil plate and by the root–soil plate weight. The contribution of each component to anchorage has been determined in experimental and numerical studies (Coutts 1983, 1986; Crook and Ennos 1996; Fourcaud et al. 2008; Ghani et al. 2009) but varies as a function of root system architecture and soil structure (Danjon et al. 2005).

A. Mechanisms That Increase Anchorage Strength

Wind damage in smaller trees can cause toppling, that is, stem lean due to stem base/stump breakage or root failure near the stump; trees often survive with a permanently bowed stem (Coutts 1983; Moore et al. 2008). Moore et al. (2008) listed several causes of toppling of saplings, including a poor RMF, use of provenances with weak root development, root deformations in the container, or resulting from poor quality planting and weak structural roots at planting. In smaller trees prone to toppling, the root system design that provides maximum anchorage for minimum construction costs is generally a large and rigid taproot acting like a stake, guyed by lateral roots (Ennos 1993). Growing deeper lateral roots improves resistance to vertical uprooting and moves the center of rotation further in the soil, constraining taproot rotation (Ennos et al. 1993). In larger trees, it is more efficient to develop a wide central root system resulting in a large potential root–soil plate, because the weight of the root–soil plate rises with the fourth power of its linear dimension (Ennos 1993). Moore (2000) and Cucchi et al. (2004) showed that in *P. radiata* and *P. pinaster*, respectively, wider root–soil plates provided better resistance to uprooting. Failed root–soil plates are larger toward or perpendicular to the dominant wind direction and smaller on the leeward side, with large roots broken near the stump, often at major points of branching (Coutts 1983). Their windward radius ranges from approximately five times the DBH in smaller trees to two times the DBH in large temperate trees (Weber and Mattheck 2005). In *P. pinaster*, Danjon et al. (2005) found that the radius of the windward side of the root–soil plate corresponded to the extension of the ZRT. In taprooted systems, or root systems with a

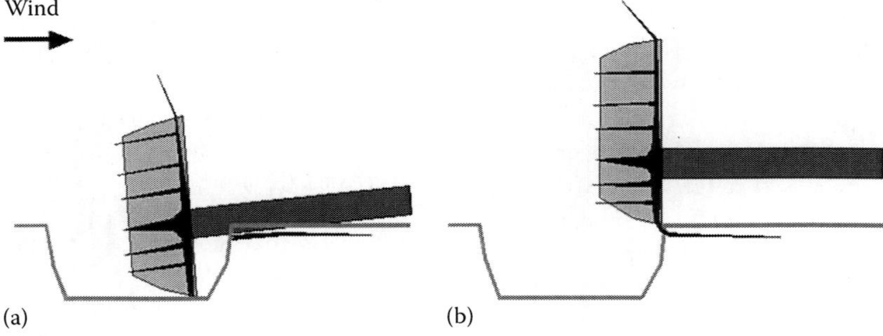

FIGURE 29.10 Uprooted tree with a root soil plate. (a) When the roots at the leeward hinge break, the root soil plate slides into the soil. (b) When the roots of the leeward hinge do not break, the root system is lifted out of the ground. (From Danjon, F. and Fourcaud, T., L'arbre et son enracinement, *Actes du colloque CIAG Sylviculture Forêts et Tempêtes*, juin 30, 2009, Pessac, France, Revue Innovations Agronomiques INRA, Paris, France, Vol. 6, pp. 17–37, 2009. With permission.)

large central mass of deep roots, it was also shown that lateral root ZRT potentially defined root plate size, but that anchorage could be influenced by the length of the deep central roots (Fourcaud et al. 2008; Ghani et al. 2009). Fourcaud et al. (2008) showed through numerical modeling that in low cohesion frictional sand-like soils, during uprooting, leeward shallow laterals penetrated little into the soil, whereas windward roots were lifted upward, resulting in a shallow and eccentric rotation axis to the leeward side. If central roots were longer than the windward radius of the plate (as defined by the ZRT), they were a major component of anchorage. If central roots were shorter than the length of the plate radius, it was the plate radius itself that governed anchorage. In high-cohesion saturated clay-like soils, the root–soil plate size was determined by the longest stiff root, leeward laterals were pushed into the soil, and the rotation axis was located below the stem. In both cases, root sections within the root–soil plate contributed little to anchorage; roots must cross the edge of the soil–root plate and be anchored to the distal substrate outside the plate in order to provide resistance to uprooting. In tropical trees, large buttresses acting in tension and compression can play a dominant role in anchoring the stem, along with the presence of aerial roots, especially in shallow soils (Crook et al. 1997; Pavlis and Jenik 2000).

In *P. sitchensis,* trees with a more restricted rooting depth developed both a larger root–soil plate horizontal area and allocated more biomass to roots (Ray and Nicoll 1998). A more rigid root–soil plate ensures a smaller early displacement during uprooting and may prevent initial failure in the soil (Coutts 1986). In both *P. sitchensis* (Coutts 1983) and *P. pinaster* (Cucchi et al. 2004) static winching tests, the maximal force to cause uprooting was recorded at a low displacement of the stem, when soil failed under tension because of its low elasticity. Therefore, soil resistance plays a key role in uprooting and could be manipulated to increase root anchorage in sensitive areas, for example, the urban environment.

The contribution of individual roots to anchorage depends in part on the physical properties of the root material and the geometry of that root (Coutts 1983). The bending stiffness (product of the elastic modulus of the beam material and the area moment of inertia of the beam cross section) of a circular beam is related to the fourth power of its diameter, which means that a root with a large CSA will be much stiffer than two roots with the same cumulative CSA. A wide angle between laterals shortens the lever arm provided by shallow laterals. Conversely, the strength of roots held in tension depends on the sum of their CSA but can be reduced by unequal loading, because failure often occurs through sequential breakage (Coutts 1983; Pollen and Simon 2005). An increase in the number of branches on a root improves its resistance to uprooting in tension (Dupuy et al. 2005a). Therefore, root topology and geometry can influence a tree's mechanical stability. In a numerical study, Dupuy et al. (2005b) compared the anchorage properties of several typical root system architectures (classified according to Köstler et al. 1968). These authors found that heart- and taproot systems in both clay- and sand-like soils possessed the most efficient anchorage due to a stiff central root system. In heart root systems, large, oblique roots also branched rapidly into smaller roots, thus holding a large mass of soil, the weight of which will help counteract the uprooting moment. The taproot system was more resistant to overturning in sand-like soil because of increased rooting depth. Shallow plate root systems were poorly anchored in both soil types. In an experimental study on mature *P. pinaster*, it was found that well-anchored root systems resemble a rigid cage-like structure (Danjon et al. 2005). This cage is formed through the development of a long and wide ZRT and numerous secondary sinkers holding together a large mass of soil. Shallow lateral roots beyond the ZRT guy the structure in place. Few oblique roots or sinkers form on the stump and there are few immediate reiterations and no retarded reiteration (Danjon et al. 2005). As found in many species (Persson 2002), *P. pinaster* root systems form a framework comprising about 10–20 main shallow second-order roots (Khuder et al. 2007). As a consequence, if a young tree cannot establish a homogeneous array of shallow roots branching from the stump, the root system will not regenerate roots in sectors without horizontal roots and therefore cannot develop secondary sinkers in these sectors. The root system will therefore possess an incomplete cage (Figure 29.11) resulting in poor mechanical stability (Danjon et al. 2005).

When a woody plant is subjected to frequent wind loads, even in mildly windy conditions, growth responses occur in

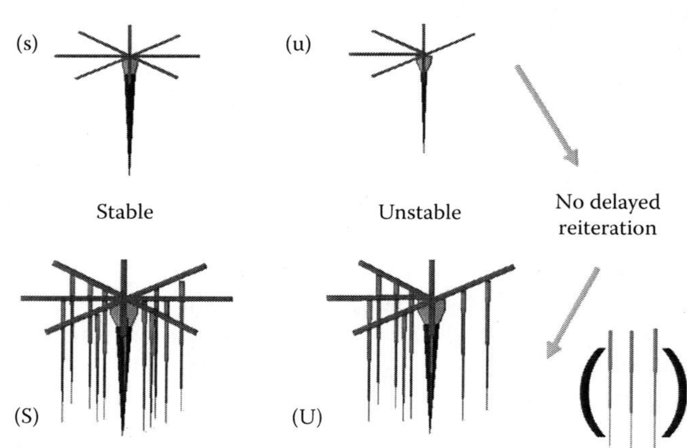

FIGURE 29.11 An example of the consequences of root system architectural type on tree mechanical stability. In *P. pinaster.* (s) mechanically stable sapling with a well-developed deep vertical taproot and a homogeneous array of shallow roots; (u) mechanically unstable sapling with a small taproot and a heterogeneous circular distribution of shallow laterals. (S) larger stable tree grown from (s) with a complete cage imprisoning a large mass of soil. (U) unstable large tree grown from (u); As *P. pinaster* does not demonstrate retarded reiteration, a tree like (u) cannot redevelop a homogeneous array of shallow root and therefore cannot grow secondary sinkers in a sector without laterals. (U) will be unstable because of an incomplete root system cage. (From Danjon, F. and Fourcaud, T., L'arbre et son enracinement, *Actes du colloque CIAG Sylviculture Forêts et Tempêtes*, juin 30, 2009, Pessac, France, Revue Innovations Agronomiques INRA, Paris, France, Vol. 6, pp. 17–37, 2009. With permission.)

the tree, permitting it to acclimate to the physical movement induced by wind loading. Tree stems may be more tapered and shorter, with differences in internal xylem structure (Telewski 1995). On the windward sides of trees, roots frequently held in tension may be thinner and more branched per unit area of soil (Stokes et al. 1995). As the tensile strength of roots increases with decreasing diameter (in roots smaller than 10 mm in diameter) (Genet et al. 2005), more numerous and thinner roots will increase the tensile strength of the root–soil matrix. The mean tensile strength of roots in forest trees usually varies from 6 MPa to around 60 MPa (see Table 2 in Stokes 2002), but surprisingly high values can also be found, for example, 730 MPa in *F. sylvatica* roots (Bischetti et al. 2005). The tensile strength of non-rooted soil is 3–5 orders of magnitude weaker than root wood (Coutts 1983).

In young trees and woody plants, as a root emerges from the central rooting zone, it will grow initially as a mechanically weak absorbing tip. The root will have a negligible role in anchorage mechanics with regard to wind loading. As the tree grows, the same root section may successively be part of the tension zone in the young tree, then located in the potential shear zone in an older tree, to eventually become included in the ZRT and absorbed in the stump in the mature tree (Weber and Mattheck 2005). This shifting in functions is mirrored in the shape of the internal xylem structure (Stokes and Mattheck 1996) and corresponding growth rings. Nicoll and Dunn (1999) recorded modifications in radial growth in shallow structural roots of *P. sitchensis* and attributed these changes to a response to frequent and strong wind loading. Secondary thickening in woody roots may occur asymmetrically, resulting in eccentrically shaped roots, sometimes forming "I"- or "T"-shaped cross sections. Extra wood is usually laid down in the parts of roots most subject to mechanical stress, for example, the upper and lower sides of lateral roots, or root bases near the stump (forming the ZRT). The extra woody tissue will increase rigidity along the bending axis and resulting root shapes are considered analogous to I-beams or T-beams in civil engineering (Nicoll and Ray 1996).

At the whole-root system scale, a distinctive acclimative growth response to prevailing winds was observed by Danjon et al. (2005) in *P. pinaster* trees. After an analysis of 50-year-old trees that had been damaged (or not) during a wind storm in 1999, it was found that undamaged trees possessed more root volume and more branched shallow windward roots beyond the ZRT. The ZRT itself possessed a large root volume and there was more secondary sinker root volume on the leeward side of undamaged trees. Similarly, mature *P. pinaster* trees growing at the windward edge of a stand possessed a 20% stronger anchorage than more sheltered trees inside the stand (Cucchi et al. 2004). From an analysis on winching tests of 2000 *P. sitchensis* in Great Britain, Nicoll et al. (2008) concluded that tree anchorage strength increases linearly with wind exposure. Wind exposure accounted for 5% of maximal force, much more than the soil depth (0.6%) and soil type (1.7%) but much less than stem weight (63.2%). The high contribution of stem weight can be explained in that anchorage strength is primarily a function of tree size; it simply scales with the drag force of wind on the crown and stem.

Acclimative growth in response to wind loading varies also as a function of landscape topology and substrate geometry: in mature *P. sitchensis* trees growing in a windy area, trees on flat ground possessed more root biomass in the leeward sector, whereas nearby trees on a 30° slope had more root biomass in the windward sector, which was itself perpendicular to the slope direction (Nicoll et al. 2006). In *Robinia pseudoacacia* L. seedlings grown in a greenhouse and subjected (or not) to mechanical perturbation, trees without mechanical perturbation had more lateral root biomass in sectors perpendicular to the slope direction, whereas seedlings subjected to nondirectional mechanical perturbation possessed more root biomass in upslope lateral roots (Khuder et al. 2006).

Knowledge about root biomechanics has slowly been put to use as an ecosystem service throughout recent years. Not only do root systems anchor a plant to substrate, but they can also "fix" that substrate in place. On hillslopes, riverbanks, and man-made slopes, vegetation is used increasingly to stabilize soil against shallow landslides or prevent against water erosion (see reviews by Reubens et al. 2007; Stokes et al. 2009). With regard to shallow landslides, plant roots need to traverse the potential slip surface (fragile zone beneath the soil surface where slope failure initiates) in order to contribute to slope stability. Plant root architectural traits and root strength are major parameters required in slope stability models incorporating vegetation into calculations of slope safety factors (Pollen and Simon 2005; Danjon et al. 2008). Currently, most of these models can only consider thin and fine roots acting in tension during a shallow landslide, even though thicker roots, which act like soil nails, are also a major component fixing the soil in place. Fewer studies have been carried out on root traits and erosion prevention (see Reubens et al. 2007), although the contribution of fine roots to aggregate stability (a measure of erodibility) through the binding network of roots and root–soil carbon exchange is becoming fairly well documented (De Baets et al. 2007; Fattet et al. 2011).

IX. Conclusions

Fine roots represent the lowest carbon pool but the highest carbon sink of root systems, yet the paucity of our knowledge on this rather small fraction of terrestrial carbon is appealing and its linkage to small and structural roots is insufficiently understood. Among the many issues poorly investigated so far, the following research challenges for the coming years can be mentioned with emphasis: (1) the relationship between carbon storage pools in roots of different sizes, the age of this carbon, and fine-root dynamics—Are carbon allocation, root production, and longevity very different between different root diameter classes? (2) With regard to the spatial and temporal functioning of fine roots, what drives physiological plasticity in response to nutrient or water uptake and growth, and are there differences in uptake between seasons or between different hours of the day? (3) Will root proliferate in deeper layers following increased climate change-induced nutrient and water demands? (4) Is there significant nutrient resorption from dead fine roots and from small or coarse roots, and do fine roots constitute a relevant nutrient pool? (5) What can be the effect of root litter on soil quality, on temporal variation of nutrient uptake, or on priming of old soil carbon? (6) What is the intensity of intra- and interspecific fine-root overlap and how meaningful is this in the light of facilitation or competition in complex environments? (7) What can be the belowground diversity in terms of fine-root species diversity and are there advantages in sharing nutrients, carbon, or water through mycorrhizal networks or root grafting, or are the disadvantages greater (e.g., pathogen pathway)? (8) How rigid or flexible are aboveground:belowground ratios including fine roots (LAI:RAI or leaf mass:fine-root mass) and how meaningful are these?

Research papers whereby the coarse woody root architecture between different treatments is compared have only recently appeared. Coarse roots are still often ignored or simply considered in terms of biomass and without a dimensional structure, even though roots are structured into systems composed of several root types. Root systems are usually very plastic and acclimative growth occurs in response to, for example, the local physical and chemical soil environment or external mechanical loading. Taking into account the structure through architectural analysis will allow for more powerful analyses in many fields (e.g., Danjon et al. 2005). In root biomechanics, the challenge will be to bring together information on anchorage provided through different methods of investigation. In particular, knowledge can be used to design more efficient numerical simulations of tree overturning, including, for example, heterogeneous soil and root properties, soil moisture content, and more realistic root systems. Fourcaud et al. (2008) even suggested designing process-based plant growth models to simulate the dynamics of acclimative growth at the whole-tree level. The contribution of each component to anchorage may then be better assessed as well as the consequences of, for example, perturbation to the architectural scheme through specific planting techniques. Soil reinforcement on slopes will also be better quantified taking into account the role of coarse-root architecture (Danjon et al. 2008).

Acknowledgment

Funding from the French government (ANR-2010-STRA-003-01 "Ecosfix" project) is gratefully acknowledged.

References

Achat DL, Bakker MR, Trichet P. 2008. Rooting patterns and fine root biomass of *pinus pinaster* assessed by trench wall and core methods. *J For Res* 13:165–175.

Albaugh T, Allen H, Dougherty P, Kress L, King J. 1998. Leaf area and above- and belowground growth responses of loblolly pine to nutrient and water additions. *For Sci* 44:317–328.

Almeras T, Fournier M. 2009. Biomechanical design and long-term stability of trees: Morphological and wood traits involved in the balance between weight increase and the gravitropic reaction. *J Theor Biol* 256:370–381.

Anten N, Casado-Garcia R, Nagashima H. 2005. Effects of mechanical stress and plant density on mechanical characteristics, growth, and lifetime reproduction of tobacco plants. *Am Nat* 166:650–660.

Atger C, Edelin C. 1994a. Stratégies d'occupation du milieu souterrain par les systèmes racinaires des arbres. *Revues d'écologie (Terre Vie)* 49:343–356.

Atger C, Edelin C. 1994b. Preliminary data on the comparative architecture of roots and crowns. *Can J Bot* 72:963–975.

Atger C, Edelin C. 1995. A case of sympodial branching based on-ground endogenous determinism in root-system—*Platanus hybrida Brot Acta Bot Gallica* 142:23–30.

Axelsson E, Axelsson B. 1986. Changes in carbon allocation patterns in spruce and pine trees following irrigation and fertilization. *Tree Physiol* 2:189–204.

Bakker MR, Augusto L, Achat DL. 2006. Fine root distribution of trees and understory in mature stands of maritime pine (*Pinus pinaster*) on dry and humid sites. *Plant Soil* 286:37–51.

Bakker MR, Jolicoeur E, Trichet P, Augusto L, Plassard C, Guinberteau J, Loustau D. 2009. Adaptation of fine roots to annual fertilization and irrigation in a 13-year-old *Pinus pinaster* stand. *Tree Physiol* 29:229–238.

Bakker MR, Turpault MP, Huet S, Nys C. 2008. Root distribution of *Fagus sylvatica* in a chronosequence in western France. *J For Res* 13:176–184.

Barthélémy D, Caraglio Y. 2007. Plant architecture: A dynamic, multilevel and comprehensive approach to plant form, structure and ontogeny. *Ann Bot* 99:375–407.

Barton C, Montagu K. 2006. Effect of spacing and water availability on root: shoot ratio in *Eucalyptus camaldulensis. For Ecol Manage* 221:52–62.

Bengough AG, Castrignano A, Pagès L, van Noordwijk M. 2000. Sampling strategies, scaling and statistics. In *Root Methods: A Handbook*, eds. A Smit, AG Bengough, C Engels, M van Noordwijk, S Pellerin, S Van de Geijn, pp. 147–174. Berlin, Germany: Springer.

Bert D, Danjon F. 2006. Carbon concentration variations in the roots, stem and crown of mature *Pinus pinaster* (Ait.). *For Ecol Manage* 222:279–295.

Bischetti GB, Chiaradia EA, Simonato T, Speziali B, Vitali B, Vullo P, Zocco A. 2005. Root strength and root area of forest species in Lombardy. *Plant Soil* 278:11–22.

Böhm W. 1979. Methods of Studying Root Systems. Ecological Studies no. 33. Berlin, Germany: Springer, 188 pp.

Bolte A, Villanueva I. 2006. Interspecific competition impacts on the morphology and distribution of fine roots in European beech (*Fagus sylvatica* L.) and Norway spruce (*Picea abies* (L.) Karst). *Eur J For Res* 125:15–26.

Bouillet J, Laclau J, Arnaud M, M'Bou A, Saint-André L, Jourdan C. 2002. Changes with age in the spatial distribution of roots of Eucalyptus clone in Congo—Impact on water and nutrient uptake. *For Ecol Manage* 171:43–57.

Bowen G. 1985. Roots as a component of tree productivity. In *Attributes of Trees as Crop Plants*, eds. M Cannell, J Jackson. Abbots Ripton, U.K.: Institute of Terrestrial Ecology Natural Environment Research Council.

Breda N, Cochard H, Dreyer E, Granier A. 1993. Water transfer in a mature oak stand (*Quercus petraea*)—Seasonal evolution and effects of a severe drought. *Can J For Res* 23:1136–1143.

Brooksbank K, White DA, Veneklaas EJ, Carter JL. 2011. Hydraulic redistribution in *Eucalyptus kochii* subsp *borealis* with variable access to fresh groundwater. *Tree Struct Funct* 25:735–744.

Brunner I, Godbold DL. 2007. Tree roots in a changing world. *J For Res* 12:78–82.

Brunner I, Ruf M, Luscher P, Sperisen C. 2004. Molecular markers reveal extensive intraspecific below-ground overlap of silver fir fine roots. *Mol Ecol* 13:3595–3600.

Butnor J, Doolittle J, Johnsen K, Samuelson L, Stokes T, Kress L. 2003. Utility of ground-penetrating radar as a root biomass survey tool in forest systems. *Soil Sci Soc Am J* 67:1607–1615.

Cairns M, Brown S, Helmer E, Baumgardner G. 1997. Root biomass allocation in the world's upland forests. *Oecologia* 111:1–11.

Canham C, Berkowitz A, Kelly V, Lovett G, Ollinger S, Schnurr J. 1996. Biomass allocation and multiple resource limitation in tree seedlings. *Can J For Res* 26:1521–1530.

Cannon W. 1949. A tentative classification of root systems. *Ecology* 30:452–458.

Casper B, Jackson R. 1997. Plant competition underground. *Annu Rev Ecol Syst* 28:545–570.

Chapin F, Bloom A, Field C, Waring R. 1987. Plant-responses to multiple environmental-factors. *Bioscience* 37:49–57.

Chen H, Popadiouk R. 2002. Dynamics of North American boreal mixedwoods. *Environ Rev* 10:137–166.

Chen W, Zhang Q, Cihlar J, Bauhus J, Price D. 2004. Estimating fine-root biomass and production of boreal and cool temperate forests using aboveground measurements: A new approach. *Plant Soil* 265:31–46.

Chen D, Zhang Y, Lin Y, Chen H, Fu S. 2009. Stand level estimation of root respiration for two subtropical plantations based on in situ measurement of specific root respiration. *For Ecol Manage* 257:2088–2097.

Christina M, Laclau J-P, Goncalves JLM, Jourdan C, Nouvellon Y, Bouillet J-P. 2011. Almost symmetrical vertical growth rates above and below ground in one of the world's most productive forests. *Ecosphere* 2:art27.

Chung H, Kramer P. 1975. Absorption of water and 32P through suberized and unsuberized roots of loblolly pine. *Can J For Res* 5:229–235.

Clark L, Whalley W, Barraclough P. 2003. How do roots penetrate strong soil? *Plant Soil* 255:93–104.

Claus A, George E. 2005. Effect of stand age on fine-root biomass and biomass distribution in three European forest chronosequences. *Can J For Res* 35:1617–1625.

Clemensson-Lindell A. 1994. Triphenyltetrazolium chloride as an indicator of fine-root vitality and environmental-stress in coniferous forest stands—Applications and limitations. *Plant Soil* 159:297–300.

Collet C, Löf M, Pagès L. 2006. Root system development of oak seedlings analysed using an architectural model. Effects of competition with grass. *Plant Soil* 279:367–383.

Comas L, Eissenstat D. 2004. Linking fine root traits to maximum potential growth rate among 11 mature temperate tree species. *Funct Ecol* 18:388–397.

Comas L, Eissenstat D, Lakso A. 2000. Assessing root death and root system dynamics in a study of grape canopy pruning. *New Phytol* 147:171–178.

Coners H, Leuschner C. 2002. In situ water absorption by tree fine roots measured in real time using miniature sap-flow gauges. *Funct Ecol* 16:696–703.

Coomes D, Grubb P. 2000. Impacts of root competition in forests and woodlands: A theoretical framework and review of experiments. *Ecol Monogr* 70:171–207.

Coutand C, Dupraz C, Jaouen G, Ploquin S, Adam B. 2008. Mechanical stimuli regulate the allocation of biomass in trees: Demonstration with young *Prunus avium* trees. *Ann Bot* 101:1421–1432.

Coutand C, Julien J, Moulia B, Mauget J, Guitard D. 2000. Biomechanical study of the effect of a controlled bending on tomato stem elongation: Global mechanical analysis. *J Exp Bot* 51:1813–1824.

Coutts M. 1983. Root architecture and tree stability. *Plant Soil* 71:171–188.

Coutts M. 1986. Components of tree stability in sitka spruce on peaty gley soil. *Forestry* 59:173–197.

Coutts M. 1987. Developmental processes in tree root systems. *Can J For Res* 17:761–767.

Coutts M, Lewis G. 1983. When is the structural root-system determined in Sitka spruce. *Plant Soil* 71:155–160.

Coutts M, Nielsen C, Nicoll B. 1999. The development of symmetry, rigidity and anchorage in the structural root system of conifers. *Plant Soil* 217:1–15.

Coyle D, Coleman M. 2005. Forest production responses to irrigation and fertilization are not explained by shifts in allocation. *For Ecol Manage* 208:137–152.

Coyle DR, Coleman MD, Aubrey DP. 2008. Above- and belowground biomass accumulation, production, and distribution of sweetgum and loblolly pine grown with irrigation and fertilization. *Can J For Res* 38:1335–1348.

Cronan C, Grigal D. 1995. Use of calcium aluminium ratios as indicators of stress in forest ecosystems. *J Environ Qual* 24:209–226.

Crook M, Ennos A. 1996. The anchorage mechanics of deep rooted larch *Larix europea x L. japonica*. *J Exp Bot* 47:1509–1517.

Crook M, Ennos A. 1997. The increase in anchorage with tree size in the tropical taprooted tree *Mallotus wrayi* King (Euphorbiaceae). In *Plant Biomechanics*, eds. G Jeronimidis, J Vincent. Reading, U.K.: Centre for Biomimetics.

Crook M, Ennos A, Banks J. 1997. The function of buttress roots: A comparative study of the anchorage systems of buttressed (*Aglaia* and *Nephelium ramboutan* species) and non-buttressed (*Mallotus wrayi*) tropical trees. *J Exp Bot* 48:1703–1716.

Cucchi V, Bert D. 2003. Wind-firmness in *Pinus pinaster* Ait. stands in Southwest France: Influence of stand density, fertilisation and breeding in two experimental stands damaged during the 1999 storm. *Ann For Sci* 60:209–226.

Cucchi V, Meredieu C, Stokes et al. 2004. Root anchorage of inner and edge trees in stands of Maritime pine (*Pinus pinaster* Ait.) growing in different podzolic soil conditions. *Tree Struct Funct* 18:460–466.

Danjon F, Barker DH, Drexhage M, Stokes A. 2008. Using three-dimensional plant root architecture in models of shallow-slope stability. *Ann Bot* 101:1281–1293.

Danjon F, Bert D, Godin C, Trichet P. 1999a. Structural root architecture of 5-year-old *Pinus pinaster* measured by 3D digitising and analysed with AMAPmod. *Plant Soil* 217:49–63.

Danjon F, Eveno E, Bernier F, Chambon J, Lozano P, Plomion C, Garnier-Géré P. 2009. Genetic variability in 3D coarse root architecture in *Pinus pinaster*. In *Proceedings of Second International Conference on Wind Effects on Trees*, eds. H Mayer, D Schindle, October 13–16, 2009, Vol. 19, pp. 155–161. Freiburg, Germany: Meteorologischen Instituts der Albert Ludwigs Universität.

Danjon F, Fourcaud T. 2009. L'arbre et son enracinement. *Actes du colloque CIAG Sylviculture Forêts et Tempêtes*, juin 30, 2009, Pessac, France. Revue Innovations Agronomiques INRA, Paris, France, Vol. 6, pp. 17–37.

Danjon F, Fourcaud T, Bert D. 2005. Root architecture and wind-firmness of mature *Pinus pinaster*. *New Phytol* 168:387–400.

Danjon F, Reubens B. 2008. Assessing and analyzing 3D architecture of woody root systems, a review of methods and applications in tree and soil stability, resource acquisition and allocation. *Plant Soil* 303:1–34.

Danjon F, Sinoquet H, Godin C, Colin F, Drexhage M. 1999b. Characterisation of structural tree root architecture using 3D digitising and AMAPmod software. *Plant Soil* 211:241–258.

De Baets S, Poesen J, Knapen A, Barbera GG, Navarro JA. 2007. Root characteristics of representative Mediterranean plant species and their erosion-reducing potential during concentrated runoff. *Plant Soil* 294:169–183.

Deans J, Mason W, Harvey F. 1992. Clonal differences in planting stock quality of Sitka spruce. *For Ecol Manage* 49:101–107.

Dieffenbach A, Göttlein A, Matzner E. 1997. In-situ soil solution chemistry in an acid forest soil as influenced by growing roots of Norway spruce (*Picea abies* [L.] Karst). *Plant Soil* 192:57–61.

Dighton J, Harrison A. 1983. Phosphorus-nutrition of lodgepole pine and sitka spruce stands as indicated by a root bioassay. *Forestry* 56:33–43.

Drexhage M, Chauvière M, Colin F, Nielsen C. 1999. Development of structural root architecture and allometry of *Quercus petraea*. *Can J For Res* 29:600–608.

Dupuy LX, Fourcaud T, Lac P, Stokes A. 2007. A generic 3D finite element model of tree anchorage integrating soil mechanics and real root system architecture. *Am J Bot* 94:1506–1514.

Dupuy L, Fourcaud T, Stokes A. 2005a. A numerical investigation into factors affecting the anchorage of roots in tension. *Eur J Soil Sci* 56:319–327.

Dupuy L, Fourcaud T, Stokes A. 2005b. A numerical investigation into the influence of soil type and root architecture on tree anchorage. *Plant Soil* 278:119–134.

Edelin C, Atger C. 1994. Stem and root tree architecture: Questions for plant biomechanics. *Biomimetics* 2:253–266.

Eis S. 1974. Root system morphology of western hemlock, western red cedar, and douglas fir. *Can J For Res* 4:28–38.

Eissenstat D. 1992. Costs and benefits of constructing roots of small diameter. *J Plant Nutr* 15:763–782.

Eissenstat D, Yanai R. 1997. The ecology of root lifespan. *Adv Ecol Res* 27:1–63.

Ennos A. 1993. The scaling of root anchorage. *J Theor Biol* 161:61–75.

Ennos A. 1999. The aerodynamics and hydrodynamics of plants. *J Exp Biol* 202:3281–3284.

Ennos A, Crook M, Grimshaw C. 1993. A comparative-study of the anchorage systems of Himalayan balsam *Impatiens glandulifera* and mature sunflower *Helianthus annuus*. *J Exp Bot* 44:133–146.

Ennos A, Pellerin S. 2000. Plant anchorage. In *Root Methods. A Handbook*, eds. A Smit, AG Bengough, C Engels, M van Noordwijk, S Pellerin, S van de Geijn, pp. 545–565. Berlin, Germany: Springer.

Enstone D, Peterson C, Hallgren S. 2001. Anatomy of seedling tap roots of loblolly pine (*Pinus taeda* L.). *Tree Struct Funct* 15:98–111.

Epstein A. 1978. Root graft transmission of tree pathogens. *Annu Rev Phytopathol* 16:181–192.

Ericsson A, Persson H. 1980. Seasonal changes in starch reserves and growth of fine roots of 20-year-old Scots pines. *Ecol Bull* 32:239–250.

FAO. 2010. Food and Agriculture Organization of the United Nations Global forest resources assessment. FAO Forestry Paper 147. Rome, Italy: FAO.

Fattet M, Fu Y, Ghestem M et al. 2011. Effects of vegetation type on soil resistance to erosion: Relationship between aggregate stability and shear strength. *Catena* 87:60–69.

Finer L, Ohashi M, Noguchi K, Hirano Y. 2011a. Factors causing variation in fine root biomass in forest ecosystems. *For Ecol Manage* 261:265–277.

Finer L, Ohashi M, Noguchi K, Hirano Y. 2011b. Fine root production and turnover in forest ecosystems in relation to stand and environmental characteristics. *For Ecol Manage* 262:2008–2023.

Fitter A. 1987. An architectural approach to the comparative ecology of plant-root systems. *New Phytol* 106:61–77.

Fleischer F, Eckel S, Schmid I, Kazda M. 2006. Point process modelling of root distribution in pure stands of *Fagus sylvatica* and *Picea abies*. *Can J For Res* 36:227–237.

Fourcaud T, Ji J-N, Zhang Z-Q, Stokes A. 2008. Understanding the impact of root morphology on overturning mechanisms: A modelling approach. *Ann Bot* 101:1267–1280.

Fransen B, Blijjenberg J, de Kroon H. 1999. Root morphological and physiological plasticity of perennial grass species and the exploitation of spatial and temporal heterogeneous nutrient patches. *Plant Soil* 211:179–189.

Fraser EC, Lieffers VJ, Landhausser SM. 2006. Carbohydrate transfer through root grafts to support shaded trees. *Tree Physiol* 26:1019–1023.

Fraser EC, Lieffers VJ, Landhausser SM. 2007. The persistence and function of living roots on lodgepole pine snags and stumps grafted to living trees. *Ann For Sci* 64:31–36.

Fujimaki R, Tateno R, Tokuchi N. 2007. Root development across a chronosequence in a Japanese cedar (*Cryptomeria japonica* D. Don) plantation. *J For Res* 12:96–102.

Gale M, Grigal D. 1987. Vertical root distributions of northern tree species in relation to successional status. *Can J For Res* 17:829–834.

Gardiner B, Blennow K, Carnus J-M et al. 2010. Destructive storms in European forests: Past and forthcoming impacts. EFI Atlantic, Final Report to European Commission—DG Environment.

Genet M, Stokes A, Salin F, Mickovski S, Fourcaud T, Dumail J, van Beek R. 2005. The influence of cellulose content on tensile strength in tree roots. *Plant Soil* 278:1–9.

George E, Seith B, Schaeffer C, Marschner H. 1997. Responses of *Picea, Pinus,* and *Pseudotsuga* roots to heterogeneous nutrient distribution in soil. *Tree Physiol* 17:39–45.

Ghani MA, Stokes A, Fourcaud T. 2009. The effect of root architecture and root loss through trenching on the anchorage of tropical urban trees (*Eugenia grandis* Wight). *Tree Struct Funct* 23:197–209.

Ghestem M, Sidle RC, Stokes A. 2011. The influence of plant root systems on subsurface flow: Implications for slope stability. *Bioscience* 61:869–879.

Gill R, Jackson R. 2000. Global patterns of root turnover for terrestrial ecosystems. *New Phytol* 147:13–31.

Gilman EF, Grabosky J. 2011. *Quercus virginiana* root attributes and lateral stability after planting at different depths. *Urban For Urban Green* 10:3–9.

Godin C. 2000. Representing and encoding plant architecture: A review. *Ann For Sci* 57:413–438.

Godin C, Costes E, Sinoquet H. 1999. A method for describing plant architecture which integrates topology and geometry. *Ann Bot* 84:343–357.

Goebel M, Hobbie SE, Bulaj B et al. 2011. Decomposition of the finest root branching orders: Linking belowground dynamics to fine-root function and structure. *Ecol Monogr* 81:89–102.

Göransson H, Fransson A-M, Jonsson-Belyazid U. 2007. Do oaks have different strategies for uptake of N K and P depending on soil depth? *Plant Soil* 297:119–125.

Göransson H, Ingerslev M, Wallander H. 2008. The vertical distribution of N and K uptake in relation to root distribution and root uptake capacity in mature *Quercus robur, Fagus sylvatica* and *Picea abies* stands. *Plant Soil* 306:129–137.

Göransson H, Wallander H, Ingerslev M, Rosengren U. 2006. Estimating the relative nutrient uptake from different soil depths in *Quercus robur, Fagus sylvatica* and *Picea abies*. *Plant Soil* 286:87–97.

Göttlicher SG, Taylor AFS, Grip H, Betson NR, Valinger E, Högberg MN, Högberg P. 2008. The lateral spread of tree root systems in boreal forests: Estimates based on (15)N uptake and distribution of sporocarps of ectomycorrhizal fungi. *For Ecol Manage* 255:75–81.

Graham BJ, Bormann F. 1966. Natural root grafts. *Bot Rev* 32:255–292.

Guo D, Li H, Mitchell RJ, Han W, Hendricks JJ, Fahey TJ, Hendrick RL. 2008a. Fine root heterogeneity by branch order: Exploring the discrepancy in root turnover estimates between minirhizotron and carbon isotopic methods. *New Phytol* 177:443–456.

Guo D, Mitchell RJ, Withington JM, Fan P-P, Hendricks JJ. 2008b. Endogenous and exogenous controls of root life span, mortality and nitrogen flux in a longleaf pine forest: Root branch order predominates. *J Ecol* 96:737–745.

Gwenzi W, Veneklaas EJ, Holmes KW, Bleby TM, Phillips IR, Hinz C. 2011. Spatial analysis of fine root distribution on a recently constructed ecosystem in a water-limited environment. *Plant Soil* 344:255–272.

Hallé F, Oldemann R. 1970. Essai sur l'architecture et la dynamique de croissance des arbres tropicaux. Paris, France: Masson.

Harper J, Jones M, Hamilton N. 1991. The evolution of roots and the problem of analysing their behaviour. In *Plant Root Growth: An Ecological Perspective*, ed. D Atkinson. Oxford, U.K.: Blackwell Scientific Publications.

Harrington C, De Bell D. 1996. Above- and below-ground characteristics associated with wind toppling in a young *Populus* plantation. *Tree Struct Funct* 11:109–118.

Häussling M, Jorns C, Lehmbecker G, Hechtbuchholz C, Marschner H. 1988. Ion and water-uptake in relation to root development in Norway spruce (*Picea abies* (L.) Karst). *J Plant Physiol* 133:486–491.

Helmisaari H, Makkonen K, Kellomaki S, Valtonen E, Malkonen E. 2002. Below- and above-ground biomass, production and nitrogen use in Scots pine stands in eastern Finland. *For Ecol Manage* 165:317–326.

Hinsinger P, Bengough AG, Vetterlein D, Young IM. 2009. Rhizosphere: Biophysics, biogeochemistry and ecological relevance. *Plant Soil* 321:117–152.

Hintikka V. 1972. Wind-induced movements in forest trees. *Commun Inst For Fenn* 76:1–56.

Hirano Y, Noguchi K, Ohashi M, Hishi T, Makita N, Fujii S, Finér L. 2009. A new method for placing and lifting root meshes for estimating fine root production in forest ecosystems. *Plant Root* 3:26–31.

Hodge A. 2004. The plastic plant: Root responses to heterogeneous supplies of nutrients. *New Phytol* 162:9–24.

Hodge A. 2009. Root decisions. *Plant Cell Environ* 32:628–640.

Holdaway RJ, Richardson SJ, Dickie IA, Peltzer DA, Coomes DA. 2011. Species- and community-level patterns in fine root traits along a 120 000-year soil chronosequence in temperate rain forest. *J Ecol* 99:954–963.

Hutchings M, De Kroon H. 1994. Foraging in plants—The role of morphological plasticity in resource acquisition. *Adv Ecol Res* 25:159–238.

Idol T, Pope P, Ponder F. 2000. Fine root dynamics across a chronosequence of upland temperate deciduous forests. *For Ecol Manage* 127:153–167.

Iversen CM. 2010. Digging deeper: Fine-root responses to rising atmospheric CO_2 concentration in forested ecosystems. *New Phytol* 186:346–357.

Jackson R, Canadell J, Ehleringer J, Mooney H, Sala O, Schulze E. 1996. A global analysis of root distributions for terrestrial biomes. *Oecologia* 108:389–411.

Jackson R, Manwaring J, Caldwell M. 1990. Rapid physiological adjustment of roots to localized soil enrichment. *Nature* 344:58–60.

Jackson R, Mooney H, Schulze E. 1997. A global budget for fine root biomass, surface area, and nutrient contents. *Proc Natl Acad Sci U S A* 94:7362–7366.

Jackson R, Sperry J, Dawson T. 2000. Root water uptake and transport: Using physiological processes in global predictions. *Trends Plant Sci* 5:482–488.

James KR, Kane B. 2008. Precision digital instruments to measure dynamic wind loads on trees during storms. *Agric For Meteorol* 148:1055–1061.

Jose S, Williams R, Zamora D. 2006. Belowground ecological interactions in mixed-species forest plantations. *For Ecol Manage* 233:231–239.

Jourdan C, Rey H. 1997a. Architecture and development of the oil-palm (*Elaeis guineensis* Jacq) root system. *Plant Soil* 189:33–48.

Jourdan C, Rey H. 1997b. Modelling and simulation of the architecture and development of the oil-palm (*Elaeis guineensis* Jacq) root system.1. The model. *Plant Soil* 190:217–233.

Jourdan C, Rey H, Guedon Y. 1995. Architectural analysis and modelling of the branching process of the young oil-palm root system. *Plant Soil* 177:63–72.

Kahn F. 1977. Analyse structurale des systèmes racinaires des plantes ligneuses de la forêt tropicale dense humide. *Candollea* 32:321–358.

Kahn F. 1983. Architecture comparée de forêts tropicales humides et dynamique de la rhizosphère. Thèse de doctorat d'état, Université de Montpellier II, Montpellier, France.

Karrenberg S, Blaser S, Kollmann J, Speck T, Edwards P. 2003. Root anchorage of saplings and cuttings of woody pioneer species in a riparian environment. *Funct Ecol* 17:170–177.

Kattge J, Diaz S, Lavorel S et al. 2011. TRY—A global database of plant traits. *Global Change Biol* 17:2905–2935.

Keyes M, Grier C. 1981. Above-ground and below-ground net production in 40-year-old douglas-fir stands on low and high productivity sites. *Can J For Res* 11:599–605.

Khuder H. 2007. Etude de l'effet d'une pente sur l'architecture et les propriétés mécaniques des systèmes racinaires de semis d'arbres. Thése de Doctorat Talence, France: Université de Bordeaux I.

Khuder H, Danjon F, Stokes A, Fourcaud T. 2006. Growth response and root architecture of black locust seedlings growing on slopes and subjected to mechanical perturbation. In *Proceedings of the Fifth Plant Biomechanics Conference*, August 28 to September 1, 2006, Stockholm, Sweden, pp. 299–303.

Khuder H, Stokes A, Danjon F, Gouskou K, Lagane F. 2007. Is it possible to manipulate root anchorage in young trees? *Plant Soil* 294:87–102.

King J, Albaugh T, Allen H, Buford M, Strain B, Dougherty P. 2002. Below-ground carbon input to soil is controlled by nutrient availability and fine root dynamics in loblolly pine. *New Phytol* 154:389–398.

Köstler J, Brückner E, Bibelriether H. 1968. *Die Wurzeln der Waldbäume*. Hamburg, Germany: Paul Parey.

Kramer P, Bullock H. 1966. Seasonal variations in the proportions of suberized and unsuberized roots of trees in relation to the absorption of water. *Am J Bot* 53:200–204.

de Kroon H. 2007. Ecology—How do roots interact? *Science* 318:1562–1563.

Külla T, Lõhmus K. 1999. Influence of cultivation method on root grafting in Norway spruce (*Picea abies* [L.] Karst). *Plant Soil* 217:91–100.

Kumar P, Hallgren SW, Enstone DE, Peterson CA. 2007. Root anatomy of *Pinus taeda* L.: Seasonal and environmental effects on development in seedlings. *Tree Struct Funct* 21:693–706.

Kurz W, Beukema S, Apps M. 1996. Estimation of root biomass and dynamics for the carbon budget model of the Canadian forest sector. *Can J For Res* 26:1973–1979.

Lamanda N, Dauzat J, Jourdan C, Martin P, Malezieux E. 2008. Using 3D architectural models to assess light availability and root bulkiness in coconut agroforestry systems. *Agrofor Syst* 72:63–74.

Lang C, Dolynska A, Finkeldey R, Polle A. 2010. Are beech (*Fagus sylvatica*) roots territorial? *For Ecol Manage* 260:1212–1217.

Le Goff N, Ottorini J. 2001. Root biomass and biomass increment in a beech (*Fagus sylvatica* L.) stand in North-East France. *Ann For Sci* 58:1–13.

Lebourgeois F, Jabiol B. 2002. Enracinements comparés du chêne sessile, du chêne pédonculé et du hêtre. Réflexions sur l'autécologie des essences. *Rev Forest Franç* 54:17–42.

Lecompte F, Ozier-Lafontaine H, Pagès L. 2001. The relationships between static and dynamic variables in the description of root growth. Consequences for field interpretation of rooting variability. *Plant Soil* 236:19–31.

Lei P, Bauhus J. 2010. Use of near-infrared reflectance spectroscopy to predict species composition in tree fine-root mixtures. *Plant Soil* 333:93–103.

Leroux Y, Pagès L. 1994. Development and polymorphism of roots in rubber tree seedlings (*Hevea brasiliensis*). *Can J Bot* 72:924–932.

Lev-Yadun S. 2011. Why should trees have natural root grafts? *Tree Physiol* 31:575–578.

Levillain J, M'Bou AT, Deleporte P, Saint-André L, Jourdan C. 2011. Is the simple auger coring method reliable for below-ground standing biomass estimation in *Eucalyptus* forest plantations? *Ann Bot* 108:221–230.

Lindström A, Rune G. 1999. Root deformation in plantations of container-grown Scots pine trees: Effects on root growth, tree stability and stem straightness. *Plant Soil* 217:29–37.

Lontoc-Roy M, Dutilleul P, Prasher SO, Han L, Brouillet T, Smith DL. 2006. Advances in the acquisition and analysis of CT scan data to isolate a crop root system from the soil medium and quantify root system complexity in 3-D space. *Geoderma* 137:231–241.

Lukac M, Godbold DL. 2010. Fine root biomass and turnover in southern taiga estimated by root inclusion nets. *Plant Soil* 331:505–513.

Lyford WH. 1980. *Development of the root system of northern red oak (Quercus rubra L)*. USDA Forest Service, *Harvard Forest Paper* 21:1–31.

Maier C, Kress L. 2000. Soil CO_2 evolution and root respiration in 11 year-old loblolly pine (*Pinus taeda*) plantations as affected by moisture and nutrient availability. *Can J For Res* 30:347–359.

Malamy J. 2005. Intrinsic and environmental response pathways that regulate root system architecture. *Plant Cell Environ* 28:67–77.

Mao Z, Bonis M, Rey H, Saint-André L, Stokes A, Jourdan C (in press) Which processes drive fine root elongation in a natural forest ecosystem? *Plant Ecol Divers.*

Marschner H, Dell B. 1994. Nutrient-uptake in mycorrhizal symbiosis. *Plant Soil* 159:89–102.

Marsden C, Nouvellon Y, Epron D. 2008a. Relating coarse root respiration to root diameter in clonal *Eucalyptus* stands in the Republic of the Congo. *Tree Physiol* 28:1245–1254.

Marshall J. 1986. Drought and shade interact to cause fine-root mortality in douglas fir seedlings. *Plant Soil* 91:51–60.

Mason E. 1985. Causes of juvenile instability of *Pinus radiata* in New Zealand. *NZ J For Sci* 15:263–280.

Maurice J, Laclau J-P, Scorzoni Re D et al. 2010. Fine root isotropy in *Eucalyptus grandis* plantations. Towards the prediction of root length densities from root counts on trench walls. *Plant Soil* 334:261–275.

Mckenzie B, Peterson C. 1995. Root browning in *Pinus banksiana* Lamb and *Eucalyptus pilularis* Sm.1. Anatomy and permeability of the white and tannin zones. *Bot Acta* 108:127–137.

Meinen C, Hertel D, Leuschner C. 2009a. Biomass and morphology of fine roots in temperate broad-leaved forests differing in tree species diversity: Is there evidence of below-ground overyielding? *Oecologia* 161:99–111.

Meinen C, Leuschner C, Ryan NT, Hertel D. 2009b. No evidence of spatial root system segregation and elevated fine root biomass in multi-species temperate broad-leaved forests. *Tree Struct Funct* 23:941–950.

Messier C, Coll L, Poitras-Lariviere A, Belanger N, Brisson J. 2009. Resource and non-resource root competition effects of grasses on early- versus late-successional trees. *J Ecol* 97:548–554.

Mommer L, Dumbrell AJ, Wagemaker CNAM, Ouborg NJ. 2011. Belowground DNA-based techniques: Untangling the network of plant root interactions. *Plant Soil* 348:115–121.

Moore J. 2000. Differences in maximum resistive bending moments of *Pinus radiata* trees grown on a range of soil types. *For Ecol Manage* 135:63–71.

Moore JR, Tombleson JD, Turner JA, Van der Colff M. 2008. Wind effects on juvenile trees: A review with special reference to toppling of radiata pine growing in New Zealand. *Forestry* 81:377–387.

Mou P, Mitchell R, Jones R. 1997. Root distribution of two tree species under a heterogeneous nutrient environment. *J Appl Ecol* 34:645–656.

Nambiar E, Sands R. 1992. Effects of compaction and simulated root channels in the subsoil on root development, water-uptake and growth of radiata pine. *Tree Physiol* 10:297–306.

Nicoll BC, Berthier S, Achim A, Gouskou K, Danjon F, van Beek LPH. 2006. The architecture of *Picea sitchensis* structural root systems on horizontal and sloping terrain. *Tree Struct Funct* 20:701–712.

Nicoll B, Dunn A. 1999. The effects of wind speed and direction on radial growth of structural roots. In *Proceedings of the Conference on the Supporting Roots of Trees and Woody Plants: Form, Function and Physiology*, Bordeaux, France, July 20–24, 1998, Series: Developments in Plant and Soil Sciences, ed. A Stokes, Vol. 87, pp. 61–76. Dordrecht, the Netherlands: Kluwer.

Nicoll B, Easton E, Milner A, Walker C, Coutts M. 1995. Distribution of biomass within root systems of sitka spruce clones. In *Wind Stability Factors in Tree Selection: Distribution of Biomass within Root Systems of Sitka Spruce Clones*, eds. M Coutts, J Grace. Cambridge, U.K.: Cambridge University Press.

Nicoll BC, Gardiner BA, Peace AJ. 2008. Improvements in anchorage provided by the acclimation of forest trees to wind stress. *Forestry* 81:389–398.

Nicoll B, Ray D. 1996. Adaptive growth of tree root systems in response to wind action and site conditions. *Tree Physiol* 16:891–898.

Noguchi K, Han Q, Araki MG, Kawasaki T, Kaneko S, Takahashi M, Chiba Y. 2011. Fine-root dynamics in a young hinoki cypress (*Chamaecyparis obtusa*) stand for 3 years following thinning. *J For Res* 16:284–291.

van Noordwijk M, Widianto, Heinen M, Hairiah K. 1991. Old tree root channels in acid soils in the humid tropics—Important for crop root penetration, water infiltration and nitrogen management. *Plant Soil* 134:37–44.

Nygren P, Lu M, Ozier-Lafontaine H. 2009. Effects of turnover and internal variability of tree root systems on modelling coarse root architecture: Comparing simulations for young *Populus deltoides* with field data. *Can J For Res* 39:97–108.

Olsthoorn A. 1991. Fine root density and root biomass of 2 douglas fir stands on sandy soils in the netherlands.1. Root biomass in early summer. *Neth J Agric Sci* 39:49–60.

Oppelt A, Kurth W, Godbold D. 2005a. Contrasting rooting patterns of some arid-zone fruit tree species from Botswana—II. Coarse root distribution. *Agrofor Syst* 64:13–24.

Oppelt A, Kurth W, Jentschke G, Godbold D. 2005b. Contrasting rooting patterns of some arid-zone fruit tree species from Botswana—I. Fine root distribution. *Agrofor Syst* 64:1–11.

Ostonen I, Lõhmus K, Helmisaari H-S, Truu J, Meel S. 2007a. Fine root morphological adaptations in Scots pine Norway spruce and silver birch along a latitudinal gradient in boreal forests. *Tree Physiol* 27:1627–1634.

Ostonen I, Püttsepp U, Biel C et al. 2007b. Specific root length as an indicator of environmental change. *Plant Biosyst* 141:426–442.

Ovington J. 1957. Dry-matter partitioning by *Pinus sylvestris* L. *Ann Bot* 21:287–314.

Pagès L. 1999. Why model root system architecture? In *Proceedings of the Conference on the Supporting Roots of Trees and Woody Plants: Form, Function and Physiology*, Bordeaux, France, July 20–24, 1998, Series: Developments in Plant and Soil Sciences, ed. A Stokes, Vol. 87, pp. 187–194. Dordrecht, the Netherlands: Kluwer.

Pagès L. 2002. Modeling root system architecture. In *Plant Roots: The Hidden Half*, eds. Y Waisel, A Eshel, U Kafkafi, 3rd edn., pp. 543–574. New York: Marcel Dekker, Inc.

Pagès L, Aries F. 1988. Sarah—A simulation model for growth, development and architecture of root systems. *Agronomie* 8:889–896.

Pagès L, Vercambre G, Drouet J, Lecompte F, Collet C, Le Bot J. 2004. Root Typ: A generic model to depict and analyse the root system architecture. *Plant Soil* 258:103–119.

Parker M, Van Lear D. 1996. Soil heterogeneity and root distribution of mature loblolly pine stands in piedmont soils. *Soil Sci Soc Am J* 60:1920–1925.

Pavlis J, Jenik J. 2000. Roots of pioneer trees in the Amazonian rain forest. *Tree Struct Funct* 14:442–455.

Peltola HM. 2006. Mechanical stability of trees under static loads. *Am J Bot* 93:1501–1511.

Persson A. 2012. The high input of soil organic matter from dead tree fine roots into the forest soil. *Int J For Res*. Article ID 217402.

Persson HA. 2002. Root system of arboreal plants. In *Plant Roots: The Hidden Half*, eds. Y Waisel, A Eshel, U Kafkafi, 3rd edn., pp. 287–313. New York: Marcel Dekker, Inc.

Persson HA, Stadenberg I. 2009. Spatial distribution of fine-roots in boreal forests in eastern Sweden. *Plant Soil* 318:1–14.

Persson HA, Stadenberg I. 2010. Fine root dynamics in a Norway spruce forest (*Picea abies* (L.) Karst) in eastern Sweden. *Plant Soil* 330:329–344.

Pierret A, Moran C, Doussan C. 2005. Conventional detection methodology is limiting our ability to understand the roles and functions of fine roots. *New Phytol* 166:967–980.

Pollen N, Simon A. 2005. Estimating the mechanical effects of riparian vegetation on stream bank stability using a fiber bundle model. *Water Resour Res* 41, W07025:11 pp.

Poorter H, Niinemets U, Walter A, Fiorani F, Schurr U. 2010. A method to construct dose-response curves for a wide range of environmental factors and plant traits by means of a meta-analysis of phenotypic data. *J Exp Bot* 61:2043–2055.

Pregitzer K, Laskowski M, Burton A, Lessard V, Zak D. 1998. Variation in sugar maple root respiration with root diameter and soil depth. *Tree Physiol* 18:665–670.

Prieto I, Armas C, Pugnaire FI. 2012. Water release through plant roots: new insights into its consequences at the plant and ecosystem level. *New Phytol* 193:830–841.

Pritchard SG, Strand AE. 2008. Can you believe what you see? Reconciling minirhizotron and isotopically derived estimates of fine root longevity. *New Phytol* 177:287–291.

Puhe J. 2003. Growth and development of the root system of Norway spruce (*Picea abies*) in forest stands—A review. *For Ecol Manage* 175:253–273.

Rajaniemi TK. 2007. Root foraging traits and competitive ability in heterogeneous soils. *Oecologia* 153:145–152.

Ray D, Nicoll B. 1998. The effect of soil water-table depth on root-plate development and stability of Sitka spruce. *Forestry* 71:169–182.

van Rees K, Comerford N. 1990. The role of woody roots of slash pine-seedlings in water and potassium absorption. *Can J For Res* 20:1183–1191.

de Reffye P. 1979. Modélisation de l'architecture des arbres par des processus stochastiques. Simulation spatiale des modèles tropicaux sous l'effet de la pesanteur. Application au *Coffea robusta*, Thèse de doctorat d'Etat. Université de Paris-Sud Orsay, France.

Reubens B, Achten WMJ, Maes WH, Danjon F, Aerts R, Poesen J, Muys B. 2011. More than biofuel? *Jatropha curcas* root system symmetry and potential for soil erosion control. *J Arid Environ* 75:201–205.

Reubens B, Pannemans B, Danjon F et al. 2009. The effect of mechanical stimulation on root and shoot development of young containerised *Quercus robur* and *Robinia pseudoacacia* trees. *Tree Struct Funct* 23:1213–1228.

Reubens B, Poesen J, Danjon F, Geudens G, Muys B. 2007. The role of fine and coarse roots in shallow slope stability and soil erosion control with a focus on root system architecture: A review. *Tree Struct Funct* 21:385–402.

Rewald B, Leuschner C. 2009. Belowground competition in a broad-leaved temperate mixed forest: Pattern analysis and experiments in a four-species stand. *Eur J For Res* 128:387–398.

Richter AK. 2007. *Fine root growth and vitality of european beech in acid forest soils with a low base saturation*. PhD Nr. 17496, Zurich, Switzerland: ETH.

Ritson P, Sochacki S. 2003. Measurement and prediction of biomass and carbon content of *Pinus pinaster* trees in farm forestry plantations, south-western Australia. *For Ecol Manage* 175:103–117.

Robinson D. 1994. The responses of plants to nonuniform supplies of nutrients. *New Phytol* 127:635–674.

Roumet C, Picon-Cochard C, Dawson L, Joffre R, Mayes R, Blanchard A, Brewer M. 2006. Quantifying species composition in root mixtures using two methods: Near-infrared reflectance spectroscopy and plant wax markers. *New Phytol* 170:631–638.

Rousseau J, Sylvia D, Fox A. 1994. Contribution of ectomycorrhiza to the potential nutrient-absorbing surface of pine. *New Phytol* 128:639–644.

Ryser P. 2006. The mysterious root length. *Plant Soil* 286:1–6.

Sands R, Fiscus E, Reid C. 1982. Hydraulic-properties of pine and bean roots with varying degrees of suberization, vascular differentiation and mycorrhizal infection. *Aust J Plant Physiol* 9:559–569.

Scanlan CA, Hinz C. 2010. Using radius frequency distribution functions as a metric for quantifying root systems. *Plant Soil* 332:475–493.

Schenk H, Jackson R. 2005. Mapping the global distribution of deep roots in relation to climate and soil characteristics. *Geoderma* 126:129–140.

Schmid I, Kazda M. 2005. Clustered root distribution in mature stands of *Fagus sylvatica* and *Picea abies*. *Oecologia* 144:25–31.

Smit A, Bengough AG, Engels C, van Noordwijk M, Pellerin S, van de Geijn S. 2000. *Root Methods: A Handbook*. Berlin, Germany: Springer.

Smith S, Read D. 1997. *Mycorrhizal Symbiosis*, 2nd edn. London, U.K.: Academic Press.

Soethe N, Lehmann J, Engels C. 2007. Root tapering between branching points should be included in fractal root system analysis. *Ecol Model* 207:363–366.

Spek L, van Noordwijk M. 1994. Proximal root diameter as predictor of total root size for fractal branching models 2. Numerical model. *Plant Soil* 164:119–127.

Steingrobe B. 2005. A sensitivity analysis for assessing the relevance of fine-root turnover for P and K uptake. *J Plant Nutr Soil Sci* 168:496–502.

Steponkus P, Lanphear F. 1967. Refinement of the triphenyl tetrazolium chloride method of determining cold injury. *Plant Physiol* 42:1423–1426.

Stokes A. 2002. Biomechanics of tree root anchorage. In *Plant Roots: The Hidden Half*, eds. Y Waisel, A Eshel, U Kafkafi, 3rd edn., pp. 269–286. New York: Marcel Dekker, Inc.

Stokes A, Atger C, Bengough AG, Fourcaud T, Sidle RC. 2009. Desirable plant root traits for protecting natural and engineered slopes against landslides. *Plant Soil* 324:1–30.

Stokes A, Fitter A, Coutts M. 1995. Responses of young trees to wind and shading—Effects on root architecture. *J Exp Bot* 46:1139–1146.

Stokes A, Guitard D. 1997. Tree root response to mechanical stress. In *The Biology of Root Formation and Development*, eds. A Altman, Y Waisel. New York: Plenum.

Stokes A, Mattheck C. 1996. Variation of wood strength in tree roots. *J Exp Bot* 47:693–699.

Stokes A, Nicoll B, Coutts M, Fitter A. 1997. Responses of young Sitka spruce clones to mechanical perturbation and nutrition: Effects on biomass allocation, root development, and resistance to bending. *Can J For Res* 27:1049–1057.

Stone E, Kalisz P. 1991. On the maximum extent of tree roots. *For Ecol Manage* 46:59–102.

Strand AE, Pritchard SG, McCormack ML, Davis MA, Oren R. 2008. Irreconcilable differences: Fine-root life spans and soil carbon persistence. *Science* 319:456–458.

Striker G, Insausti P, Grimoldi A, Leon R. 2006. Root strength and trampling tolerance in the grass *Paspalum dilatatum* and the dicot *Lotus glaber* in flooded soil. *Funct Ecol* 20:4–10.

Sudmeyer R, Speijers J, Nicholas B. 2004. Root distribution of *Pinus pinaster*, *P. radiata Eucalyptus globulus* and *E. kochii* and associated soil chemistry in agricultural land adjacent to tree lines. *Tree Physiol* 24:1333–1346.

Tarroux E, DesRochers A. 2010. Frequency of root grafting in naturally and artificially regenerated stands of *Pinus banksiana*: Influence of site characteristics. *Can J For Res* 40:861–871.

Telewski FW. 1995. Wind-induced physiological and developmental responses in trees. In *Wind and Trees*, eds. M Coutts, J Grace. Cambridge, U.K.: Cambridge University Press.

Thomas S, Whitehead D, Reid J, Cook F, Adams J, Leckie A. 1999. Growth, loss, and vertical distribution of *Pinus radiata* fine roots growing at ambient and elevated CO_2 concentration. *Global Change Biol* 5:107–121.

Tierney G, Fahey T. 2002. Fine root turnover in a northern hardwood forest: A direct comparison of the radiocarbon and minirhizotron methods. *Can J For Res* 32:1692–1697.

Tobin B, Black K, McGurdy L, Nieuwenhuis M. 2007a. Estimates of decay rates of components of coarse woody debris in thinned Sitka spruce forests. *Forestry* 80:455–469.

Tobin B, Cermak J, Chiatante D et al. 2007b. Towards developmental modelling of tree root systems. *Plant Biosyst* 141:481–501.

Tracy SR, Roberts JA, Black CR, McNeill A, Davidson R, Mooney SJ. 2010. The X-factor: Visualizing undisturbed root architecture in soils using X-ray computed tomography. *J Exp Bot* 61:311–313.

Unger P, Kaspar T. 1994. Soil compaction and root-growth—A review. *Agron J* 86:759–766.

Valladares F, Gianoli E, Gomez JM. 2007. Ecological limits to plant phenotypic plasticity. *New Phytol* 176:749–763.

Van Lear D, Kapeluck P, Carroll W. 2000. Productivity of loblolly pine as affected by decomposing root systems. *For Ecol Manage* 138:435–443.

Van Praag H, Sougnezremy S, Weissen F, Carletti G. 1988. Root turnover in a beech and a spruce stand of the Belgian Ardennes. *Plant Soil* 105:87–103.

Vanguelova EI, Hirano Y, Eldhuset TD et al. 2007. Tree fine root Ca/Al molar ratio—Indicator of Al and acidity stress. *Plant Biosyst* 141:460–480.

Vanninen P, Makela A. 1999. Fine root biomass of Scots pine stands differing in age and soil fertility in southern Finland. *Tree Physiol* 19:823–830.

Vercambre G, Pagès L, Doussan C, Habib R. 2003. Architectural analysis and synthesis of the plum tree root system in an orchard using a quantitative modelling approach. *Plant Soil* 251:1–11.

Vogt K, Persson H. 1991. Measuring growth and development of roots. In *Techniques and Approaches in Forest Tree Ecophysiology*, eds. J Lassoie, T Hinckley, pp. 477–501. Boca Raton, FL: CRC.

Vogt K, Vogt D, Moore E, Fatuga B, Redlin M, Edmonds R. 1987. Conifer and angiosperm fine-root biomass in relation to stand age and site productivity in douglas fir forests. *J Ecol* 75:857–870.

Wagner B, Gaertner H, Ingensand H, Santini S. 2010. Incorporating 2D tree-ring data in 3D laser scans of coarse-root systems. *Plant Soil* 334:175–187.

Wallander H, Nilsson L, Hagerberg D, Baath E. 2001. Estimation of the biomass and seasonal growth of external mycelium of ectomycorrhizal fungi in the field. *New Phytol* 151:753–760.

Walters M, Reich P. 1989. Response of ulmus-americana seedlings to varying nitrogen and water status.1. photosynthesis and growth. *Tree Physiol* 5:159–172.

Watson A. 2000. Wind-induced forces in the near-surface lateral roots of radiata pine. *For Ecol Manage* 135:133–142.

Weber K, Mattheck C. 2005. The double nature of the root plate. *Allg. Forst U. Jagdz* 176:77–85.

Withington JM, Reich PB, Oleksyn J, Eissenstat DM. 2006. Comparisons of structure and life span in roots and leaves among temperate trees. *Ecol Monogr* 76:381–397.

Yang Y, Luo Y. 2011. Isometric biomass partitioning pattern in forest ecosystems: Evidence from temporal observations during stand development. *J Ecol* 99:431–437.

Yuan ZY, Chen HYH. 2010. Fine root biomass production turnover rates, and nutrient contents in boreal forest ecosystems in relation to species climate fertility, and stand age: Literature review and meta-analyses. *Crit Rev Plant Sci* 29:204–221.

Zerihun A, Montagu K. 2004. Belowground to aboveground biomass ratio and vertical root distribution responses of mature *Pinus radiata* stands to phosphorus fertilization at planting. *Can J For Res* 34:1883–1894.

Zhang J, George E. 2009. Root proliferation of Norway spruce and Scots pine in response to local magnesium supply in soil. *Tree Physiol* 29:199–206.

Zobel RW, Kinraide TB, Baligar VC. 2007. Fine root diameters can change in response to changes in nutrient concentrations. *Plant Soil* 297:243–254.

Zobel RW, Waisel Y. 2010. A plant root system architectural taxonomy: A framework for root nomenclature. *Plant Biosyst* 144:507–512.

30

Roots as a Source of Food

Daniel F. Austin
Arizona-Sonora Desert Museum

Robert Jarret
United States Department of Agriculture

I. Why Do We Eat Roots?

Many people consider grasses such as rice, wheat, maize, barley, and oats to be the "staff of life." Grass seeds are rich in carbohydrates, which are major elements of the human diet; yet they are not the only sources of these nutrients. Many root crops (Table 30.1) are similarly rich in carbohydrates and contain other important nutrients as well.

Different ecological and climatic areas provide distinct resources and opportunities to grow food. For example, the unique ecology of the northern subtropical mountain belt of the Old World resulted in the dominance of seed crops, while the root and tuber crops were best developed in the New World tropical lowlands (cf. Sauer 1952; Hawkes 2007). There are also distinct environmental issues in these areas. In tropical forests with leaf-cutter ant invasions or in savannas prone to locusts, root crops are truly the "staff of life." Root crops provide calories and, more importantly to many people, a full stomach (Figure 30.1). For example, people in Amazonian Brazil with whom the first author has worked could not understand how a person might get a full meal without eating a sizeable portion of starchy *farinha* (manioc).

According to the OED (2009), the term "vegeculture" (or vegiculture; vegetative propagation) was coined by Day (1917). In fact, use of the word is older because it was used by Perrine (1840), although both authors used vegeculture in a sense different from Ames (1939) and Sauer (1952). Carl Sauer was among the first to ask which came first, the root or the seed. Sauer (1969) developed the theory that vegeculture preceded seed cultures. Later studies also indicated that many root crops and other vegetatively propagated plants were in cultivation before seed plants, especially in the tropics (e.g., Norman et al. 1984; Rindos 1984; Hather 1996; Shuji and Matthews 2002). Beaumont (1993) found that vegeculture was 11,000 years old in Southeastern Asia; seed culture was of a more recent origin. Dickau et al. (2007) showed that root crops were present in Panama between 9700 and 7800 cal BP and that maize did not appear until 7800–7000 cal BP. Heiser (1990), in his book *Seeds to Civilization*, asserted that seed crops allowed civilizations to develop. Some researchers question that assertion and suggest that cultures dependent on the consumption of roots arose first and that later cultures dependent on seed crops partially replaced them. Perhaps root crops provided sustenance to humans while they domesticated seed crops.

Hawkes (1983, 2007) and others suggested that neither seed culture nor vegeculture predominated uniformly in all areas simultaneously but that one or the other predominated in different geographical areas. Still, there are many people in the tropics who largely depend on "root crops" (e.g., Alcorn 1984; Denslow and Padoch 1988; Horton 1988; Lebot 2009). China and Japan make extensive use of the sweet potato even though these countries lie mostly within temperate zones. Apparently, cultures of Mediterranean and European origin have a different view as to the relative importance of root crops in relation to seed crops.

II. Are "Root Crops" Really Roots?

"Root" is often (mis)used in a generic sense when referring to crops. Technical papers sometimes imply that roots are "That part of a plant ... which is normally below the earth's surface" (OED 2009). However, even nonbotanical dictionaries like the OED recognize tubers ("An underground structure consisting

TABLE 30.1 Plants with Roots Used for Food in Various Parts of the World

Species	Common Names	Origin	Edible Part	Uses	Compounds	Cytology
Apios americana	American potato-bean, ground-bean	North America	Roots	Raw or cooked vegetable	Starch, protein	Diploid, 2n = 22; triploids, 2n = 33
Armoracia rusticana	Horseradish	Europe	Roots	Condiment	Glucoside sinigrin, isothiocyanates, indoles, vitamin C	2n = 32
Beta vulgaris	Beet, beet root	Eurasia	Roots, leaves	Cooked vegetable, pickles	Saccharose, betanins, geosmin	Diploid, 2n = 20
Brassica rapa ssp. *campestris*	Turnip	Europe	Roots, leaves	Fresh or cooked vegetable	Sugar, starch, isothiocyanates, indoles	Diploid, 2n = 20
Chaerophylum bulbosum	Turnip-rooted chervil, parsnip chervil	Europe	Roots	Vegetable	Starch	Unknown
Cichorium intybus	Chicory	Mediterranean to China	Roots, leaves	Root coffee substitute; leaves as greens	Lactucin, lactucopicrin, sugar, inulin andresins	Diploid, 2n = 16,18
Daucus carota	Carrot	Europe and Asia	Roots	Raw or cooked vegetable	Sugar, proteins, vitamins, anthocyanin, carotenes, terpenoids	Diploid, 2n = 18
Ipomoea batatas	Sweet potato, batata, camote	South America	Roots, leaves	Cooked vegetable; leaves as greens	Starch, sugar, vitamins, β-carotene, β-amylase	Tetraploid and hexaploid, 2n =60, 90
Manihot esculenta	Cassava, manioc, tapioca	South America	Roots, leaves in some strains	Cooked vegetable, cassava, farinha, manioc, tapioca, gari, fu-fu	Carbohydrate, protein, minerals, cyanogenic glucosides	Tetraploid, 2n = 36
Pastinaca sativa	Parsnip	Europe	Roots	Cooked vegetable	Sugar, starch	2n = 22
Petroselinium cripsum cv. *Tuberosum*	Hamburg parsley, turnip-rooted parsley	Europe	Roots, leaves in other cultivars	Raw or cooked vegetable for salads, soups, stews, etc.	Starch, apiol, myristicin, carotene, niacin, ascorbic acid, minerals	Diploid, 2n = 22
Raphanus sativus	Radish	Eurasia	Roots	Relish, appetizer, cooked vegetable	Carbohydrate, protein, vitamin C, niacin, isothiocyanates, indoles	Diploid, 2n = 18
Scolymus hispanicus	Spanish salsify	Europe	Roots	Vegetable	Starch, phenolic acids, flavonoids	2n = 20
Scorzonera hispanica	Scorzonera, black oyster plant	Europe	Roots, leaves	Root boiled, baked or in soup; rubber substitute; leaves as greens	Starch, inulin, sesquiterpene glucoside	Diploid and tetraploid, 2n = 14, 28
Sium sisarum	Skirret	China	Roots	Vegetable	Starch	Unknown; related species 2n = 12
Smyrnium olusatrum	Alexanders, wild celery, horse parsley	Mediterranean	Leaves, stems, roots	Blanched or boiled	Starch, essential oils, kaempferol, quercetin, flavonols, furocoumarins, acetylenic compounds, lactones (istanbulin-A, istanbulin-B), furanogermacrane glechomafuran (=alexandrofuran), sesquiterpenes, acetooxyfurano-endesmane	2n = 22
Trapogon porrifolius	Salsify, vegetable oyster	Europe	Roots	Root boiled, baked or in soup	Starch	Diploids, 2n = 12

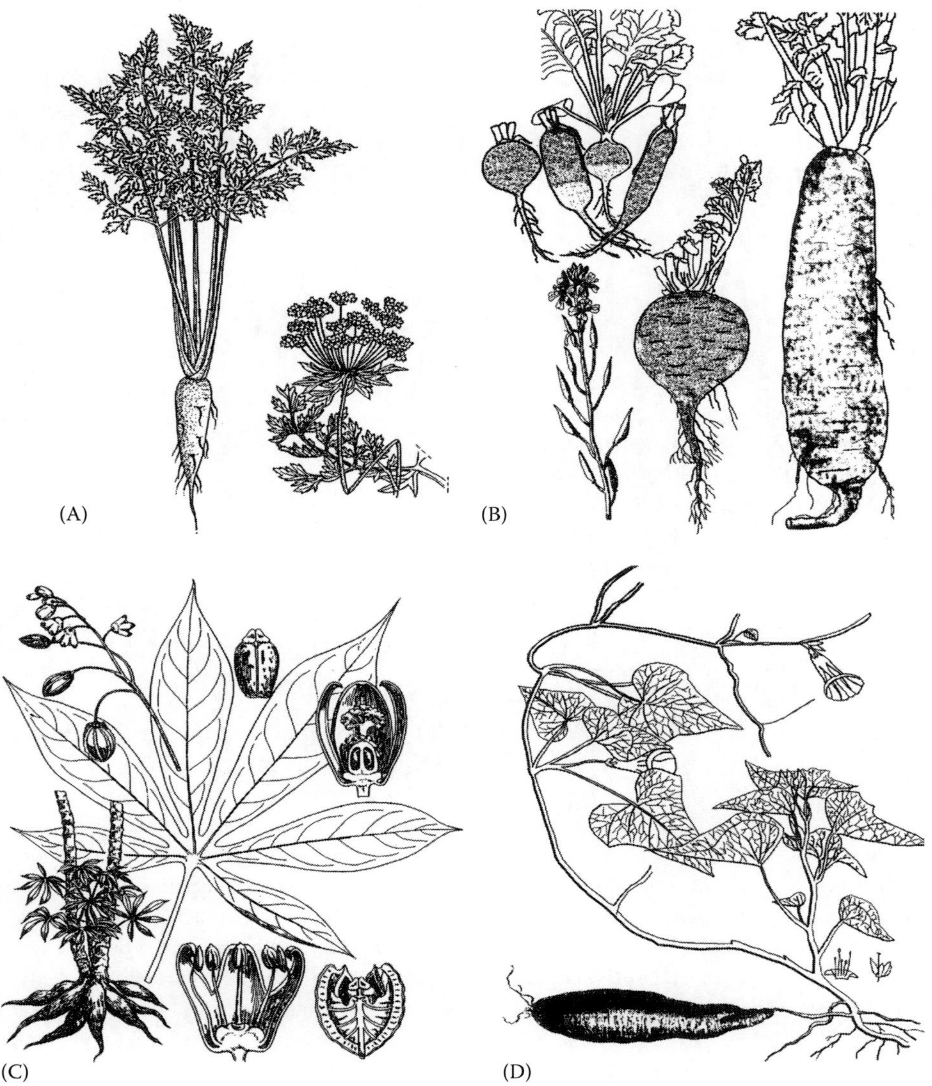

FIGURE 30.1 Some of the major root crops. (A) Carrot (*Daucus carota*). (From Siemonsma and Piluek 1993.) (B) Radish (*Raphanus sativus*). Six varieties are illustrated. (Adapted and redrawn from Harrison et al. 1981.) (C) Manioc (*Manihot esculenta*). (From Proctor et al. 1984.) (D) Sweet potato (*Ipomoea batatas*). (Adapted and redrawn from Velloso 1829; Nicholson 1885.)

of a solid thickened portion or outgrowth of a stem or rhizome, of a more or less rounded form, and bearing 'eyes' or buds from which new plants may arise"), "bulbs" ("The underground spheroidal portion of the stem ... 'a subterranean bud'"), and corms ("A short fleshy rhizome, or bulb-like subterraneous stem of a monocotyledonous plant"). Some elect to use the word "roots" because it is simpler (e.g., Onwueme 1978; Vietmeyer 1992). However, only a comparatively small number of the species cultivated can be considered root crops in the strictest sense. Most produce tubers or corms (see Armstrong 2012 for a partial list). An additional complicating factor is the dual use of the word "tuber." Some restrict the use of the term "tuber" to a modified stem serving as a storage organ. Others use the word to include certain storage roots (Bell 1991; Kays et al. 1992; Wilson and Wickham 1992a,b; Fitter 2002).

This discussion will exclude the non-root "root crops" such as ahipa, jicama, or yam beans (*Pachyrhizus* spp.); arracacha (*Arracacia xanthorrhiza*); chufa or rush nut (*Cyperus esculentus*); maca (*Lepidium meyenii*); mashua (*Tropaeolum tuberosum*); oca (*Oxalis tuberosa*); onion (*Allium cepa*); potato (*Solanum tuberosum*); taro or cocoyams (*Colocasia esculenta* and *Xanthosoma* spp.); ulluco (*Ullucus tuberosus*); yams (*Dioscorea* spp.); and water chestnut (*Trapa natans*). Many sources of information are available for those interested in plants with edible underground parts, including Purseglove (1968, 1972), Simmonds (1976), Onwueme (1978), Zohary and Hopf (2000), and Lebot (2009).

III. Major Root Crops

Common names in this discussion are restricted to those most often used in English. For additional common names, see Siemonsma and Piluek (1993), Rehm (1994), Brako et al. (1995), Quattrocchi (1999), and von Erhardt et al. (2010).

A. Tropical Root Crops

1. Manioc

a. Classification and Relatives

Manioc (*Manihot esculenta* Crantz, Euphorbiaceae) contains ~98 species native to the tropical and warm regions of the Americas (Figure 30.1C). The genus has been divided into 19 sections (Rogers and Appan 1973). Of those, these authors suggested that *Manihot aesculifolia* Pohl, *Manihot auriculata* McVaugh, *Manihot rubricaulis* I.M. Johnston, and *Manihot pringlei* S. Watson were closely allied with *M. esculenta* (Rogers and Appan 1973). Allem (1994, 1999) divided manioc into three subspecies, one cultivated (*M. esculenta* subsp. *esculenta*) and two wild (*M. esculenta* subsp. *flavelliformis* (Pohl) Ciferri and *M. esculenta* subsp. *peruviana* (Muell.-Arg.) Allem). Although these subspecies are apparently reproductively isolated in the wild, successful artificial hybridizations between them have been made (Rogers 1963; Jennings 1995).

b. Origin and Domestication

Manioc originated in South America, perhaps over 9000 years BP (Piperino and Pearsall 1998), arriving in Panama by ca 8000 cal BP (Dickau et al. 2007). De Candolle (1886), Vavilov (1951), and Onwueme (1978) favored a Brazilian origin; that region is supported by recent molecular genetics studies (Fregene et al. 1994; Roa et al. 1997; Chavarriaga-Aguirre et al. 1998; Olsen and Schaal 1999, 2006). Secondary centers of diversity have been recognized in Central America and NE, central, and SW Brazil (Rogers and Appan 1973; Nassar 2002). Allem (1987, 1994, 1999) found *M. esculenta* growing wild in Brazil and molecular genetics data support his interpretation (Olsen and Schaal 1999, 2001; Olsen 2002, 2004), although earlier RFLP data did not (Haysom et al. 1994).

There is evidence that the flour made from the roots of manioc was an important trade item as early as 3000 BCE (Simmonds 1976). Manioc was widely distributed in the New World at the time of European arrival, its range extending from Mexico to southern South America and arguably Easter Island (Langdon 1988). Manioc was discovered by Columbus in Hispaniola and recorded with the name *ñamé* (among other spellings) because he confused it with the true yam (*Dioscorea* spp.) of Africa (Sauer 1969). After manioc's arrival in Europe, the species quickly spread into other areas of the Old World (Prendergast et al. 1998). In some of those places, such as Africa, manioc has become more important in local diets than in its homeland.

Manioc at one time was thought to be composed of two species; now these are called "bitter" and "sweet" varieties (cf. Wilson and Dufour 2002). The varieties that are "bitter" contain substantial amounts of cyanogenic glucosides that must be removed to make the roots edible. A complex technique was developed among pre–Columbian American cultures to accomplish this extraction (Rogers 1963; Sauer 1969; Nye 1991). Technically, only one resultant product of this extraction is "cassava," but the name is sometimes applied to the whole plant (see Gade 2002 for complexities of names). "Sweet" varieties do not contain the cyanogenic glucosides in abundance (Nye 1991) and may be cooked directly and eaten without the elaborate extraction procedure.

Cultivated *Manihot* is a diploid or diploidized tetraploid with 2n = 36, as are all studied members of the genus (Awoleye et al. 1994; Jennings 1995; Okogbenin et al. 2006; Raji et al. 2009). Magoon et al. (1969) suggested that manioc is a segmental allotetraploid, although that has been called into question (Chavarriaga-Aguirre et al. 1998). There is a suggestion that the cultivated plants may be more variable because of genetic exchanges occurring between cultivated and wild plants (Fregene et al. 1994; Olsen and Schaal 2006). Global genetic resources of *Manihot* are maintained by the Consultative Group for International Agricultural Research.

c. Production

Manioc grows best within areas where temperatures range from 25°C to 29°C and rainfall is 100–150 cm/year. Plants are monoecious and pollinated by insects. Seeds, however, are not used in cultivation; propagation is by stem cuttings. Although manioc is a perennial, it is usually harvested during the first or the second year. When grown on fertile soils, typically few roots are produced and there is excessive vegetative growth. Thus, the crop has the advantage of producing best on soils of low fertility.

Onwueme (1978) devoted eight pages to the various diseases and pests of manioc or cassava. Cassava mosaic is the most important viral disease and whitefly (*Bemisia tabaci*) is its primary vector. Bacterial blight (*Xanthomonas manihotis*) and stem rot (*Erwinia* spp.) are also problems.

Fungal diseases include brown leaf spot (*Cercospora henningsii*), leaf spot (*Phyllosticta* spp.), white thread (*Fomes lignosus*), anthracnose (*Colletotrichum gloeosporioides*), and super elongation disease (*Sphaceloma* spp.). Insect pests include variegated grasshopper (*Zonocerus variegatus*), green spider mite (*Mononychellus tanajoa*), red spider mite (*Tetranychus telarius*), web mite (*Oligonychus* spp.), hornworm (*Erinnyis ello*), scale (*Aoindomytilus albus*), mealybug (*Phenacoccus manihoti*) (Schulthess et al. 1991), whiteflies, mites, and termites. Root knot is caused by the nematode *Meloidogyne incognita*.

Manioc contains ~35% starch, 1%–2% protein, and 1%–2% fiber. The important minerals are mostly phosphorus and iron. Vitamin C is usually ~0.35% of fresh weight and there are traces of niacin and vitamins A, B_1, and B_2. The content of thiamine and riboflavin is negligible. The root is rich in arginine but low in methionine, lysine, tryptophan, phenylalanine, and tyrosine. The cyanogenic glucosides linamarin and lotaustralin are present in varying quantities. Both glucosides hydrolyze to prussic acid.

Africa is the largest consumer of manioc and has been since shortly after World War II (Horton 1988). Africa alone has grown about one-third of the crop production for the world since 1984 (Nakamoto et al. 1994; Simpson and Conner-Ogorzaly 1995). Asia is next and South America third for the remainder (FAO 2010).

2. Sweet Potato

a. Classification and Relatives

The sweet potato (*Ipomoea batatas* (L.) Lam.) belongs to the Convolvulaceae. *I. batatas* is mostly called *batata* or *camote* in its homeland, and Austin (1988) has discussed the linguistic relationships between these American names (Figure 30.1D). *Ipomoea* contains about 600 species that are concentrated in the tropics (Austin and Huamán 1996). Currently there are 16 taxa (13 species, two varieties, and a named hybrid) in *Ipomoea* series *Batatas* (Jarret and Austin 1994; Austin and Huáman 1996). Indeed, there is evidence that the species of this series share unique gene sequences (Huang and Sun 2000). It has been suggested that the progenitor of the cultivated sweet potato is *Ipomoea trifida* (Kunth) G. Don, but various data indicate that this is not the case (cf. Austin 1988; Bohac et al. 1993; Shindo et al. 1999). *I. trifida* was probably involved in the origin of the sweet potato (Buteler et al. 1999), but it is far from clear that it is the sole ancestor. Regardless of its phyletic relationship, *I. trifida* has been useful in studies of self-incompatibility (Kowyama et al. 2008).

I. batatas is common in the wild in the American tropics. Moreover, wild tetraploid sweet potatoes are now known to extend from Mexico to Ecuador (Austin et al. 1993; Bohac et al. 1993).

b. Origin and Domestication

The sweet potato was brought into cultivation in the Americas, where all but one wild relative was endemic (Austin 1991). Sauer (1969), using one set of data, and Austin (1988), using another, suggested that the area of origin was in northwestern South America. Two regions of diversity exist—northwestern South America and southern Mexico through Mesoamerica (Austin 1983, 1988).

There are many cultivars of *I. Batatas*. The American diversity is now being studied by the International Potato Center (CIP) in Lima, Peru, which currently maintains more than 5000 lines (Huamán et al. 1999). A smaller collection is maintained in the United States (Jarret 1989). Pacific diversity appears to be a subset of the Americas (Yen 1974).

Columbus took the sweet potato with him back to Europe along with the Caribbean name *age* or *aje* (Sauer 1969). The species was spread into the Old World tropics in the early 1500s; it reached China by or before 1594 and Japan by 1674 (Austin 1988; Bohac et al. 1995).

The cultivated sweet potato is a hexaploid (2n = 90), whereas the wild sweet potato is tetraploid (2n = 60). Allied species are diploids (2n = 30) and tetraploids (Jones 1990; Bohac et al. 1993). Ozias-Akins and Jarret (1994) reported nuclear DNA contents and ploidy-level variation within the genus *Ipomoea*. Cytological data disagree on whether or not sweet potato is an autoploid (Shiotani 1988) or alloploid (Magoon et al. 1970). Morphological and molecular genetics data suggest that both *I. trifida* and *Ipomoea triloba* L. may have been involved in the alloploid origin of the sweet potato (Austin 1988; Shindo et al. 1999).

c. Production

I. batatas is a warm-weather crop growing best at temperatures above 24°C; below 10°C, growth is negligible. Sweet potatoes are perennials that are often handled in agricultural practices as annuals. The species is self-incompatible and allogamous. Pollination is entomophilous. Both honeybees (*Apis mellifera*) and bumblebees (*Bombus* spp.) are common pollinators. Propagation is either from roots or from stems, though stems are most commonly used. Although the plants will produce in poor soils, they respond well to fertilizer. Types and quantities of additives required depend on the local conditions, but excessive nitrogen results in vegetative and not root growth (Hill et al. 1992).

Perhaps the most damaging pest of sweet potato is the weevil (*Cylas formicarius*), which has now spread around the world (Austin et al. 1991). Snook et al. (1994) correlated levels of various esters of p-coumaric acid in sweet potato vine and root latex with the leaf feeding index values for *Cy. formicarius* and suggested a relationship between latex chemistry and insect resistance. Shahidul Islam et al. (2002) reported variation in foliar polyphenolic composition among 1389 sweet potato genotypes. However, a correlation between polyphenolic content and insect resistance was not established. To date, no conventional source of resistance to sweet potato weevil has been identified. Other damage-causing insect pests are the beetle *Eusepes* spp., a vine borer (*Omphisa anastomosalis*), a hawk moth (*Herse convolvuli*), wireworms (*Conoderus* spp.), white grubs (*Plectris aliena*), and flea beetles (*Chaetocnema confinis*). There are several fungal diseases that affect the sweet potato, including black rot (*Ceratocystis fimbriata*), scurf (*Moniliochaetes infuscans*), fusarium wilt (*Fusarium oxysporum* f. *batatas*), and soft rot (*Rhizopus stolonifer*). Nematodes are uncommon problems in sweet potato, but exceptions are sting nematode (*Belonolaimus gracilis*), root lesion nematode (*Pratylenchus* spp.), and root-knot nematode (*Meloidogyne* spp.). Jansson and Raman (1991) provide a modern discussion of pests in this species.

Numerous viruses infect the crop. Viral infection reduces yields and hinders the international exchange of plant material. Domola et al. (2008) identified nine sweet potato viruses in South Africa. Reduction in marketable yields of virus-infected plants was 21%–38% but can be much higher (Mwanga et al. 2002). Virus resistance has been identified in the *I. batatas* gene pool (Mwanga et al. 2002), and genetically modified (GM) plants are also being evaluated (Okada et al. 2001)—though not all reports are promising (Eicher et al. 2006). In addition to technical difficulties associated with the production of the GM plants, socioeconomic difficulties and those related to intellectual property rights, biosafety regulatory frameworks, expenditure to generate GE crops, and opposition by nongovernmental activists await a clear solution (Reddy et al. 2009). Nevertheless, research in this area continues (Sonada et al. 2009).

Sweet potatoes contain 44%–78% starch, 8%–27% sugar, 1%–11% protein, fiber, β-amylase, and up to 11% (11 g/100 g fresh weight) of β-carotene. There are also riboflavin, fair amounts of thiamine, calcium, iron, and vitamins A and C (Hill et al. 1992). Carotene and vitamin A are antioxidants that have been claimed to be associated with cancer prevention (Liu 2004; Nishino et al. 2009). Orange-red-fleshed roots contain higher levels of total carotenoids than white- and cream-fleshed lines (Wu et al. 2008). The predominant carotenoid is trans β-carotene. The beneficial effects of sweet potato in the diets of children are well documented (Low et al. 1997, 2007; van Jaarsveld et al. 2005). Garcia and Walter (1998) and Zhang and Oates (1999) noted large differences in starch structural characteristics among varieties in the size of amylase, gelatinization temperature, granule morphology, ash, N content, and amylose content. Variation among Caribbean sweet potato varieties was observed for pasting properties, amylase content, and reducing and nonreducing sugars (Aina et al. 2009). Root starch characteristics of cultivated *I. batatas* and the closely related *I. trifida* are quite distinct (Asante et al. 2006).

Studies suggest that compounds with antimutagenic activity are produced by sweet potato roots (Yoshimoto et al. 1998, 1999; Lee et al. 2007). Other reported medicinal properties associated with the consumption of purple-fleshed varieties (roots or leaves) include anti-atherosclerotic activity (Park et al. 2010), reduction in serum hepatic biomarkers (Suda et al. 2008), decreased lipid peroxidation (Chen et al. 2008), and amelioration of cognition deficits (Shan et al. 2009). Most, but not all (Miyazaki et al. 2005), of these appear to be associated with elevated levels of anthocyanins on antioxidant activity (Kano et al. 2005; Dini et al. 2009). Root periderm is reported to contain allelopathic substances (Peterson and Harrison 1995; Chon and Boo 2005).

Sweet potato is a convenient model for the study of environmental and chemical factors affecting storage root initiation and development (McSpence 1971; David and Alamu 1980; Eguchi et al. 1994; Toshihiko and Yoshida 2008; Firon et al. 2009). In vitro culture systems have been used (Eguchi and Yoshida 2008). In recognition of the potential of sweet potato as a "space food," hydroponic culture systems have also been developed (Eguchi et al. 1996; Uewada et al. 1992; Eguchi and Yoshida 2004; Kitaya et al. 2007). Anatomical aspects of sweet potato storage root formation were recently reviewed and examined by Villordon et al. (2009).

Asia and Africa grow more sweet potatoes than any other region. In the period between 1984 and 1990, these areas produced almost 93% of the crop (Simpson and Conner-Ogorzaly 1995). Most of that production came from China, Uganda, and Nigeria. China alone produced 75.8 million tons in 2007 (FAO 2010). The United States ranks 12th in terms of global sweet potato production/country (FAO 2010), and tropical production has been declining for some years. U.S. production has been increasing slowly in recent years possibly in response to the development of new uses for the crop and an expanding market in Europe.

B. Temperate Root Crops

1. Beet

a. Classification and Relatives

Beet (*Beta vulgaris* L., Chenopodiaceae)—the British call it "beetroot"—has 11–13 species in Europe and Asia (Letschert 1993; Mabberley 2009). There are four lineages within *B. vulgaris* whose common names have remained reasonably stable over time, while the scientific names associated with those cultigens have varied greatly. Ford-Lloyd (1995) and Lange et al. (1999) produced the most recent classifications. These four groups of *B. vulgaris*, using the more familiar system, are sugar beet (subsp. *vulgaris*); the fodder beet, forage beet, or mangelwurzel (subsp. *vulgaris*); the garden beet or red beet (subsp. *vulgaris*); and the spinach beet or Swiss chard (subsp. *cicla* (L.) Koch). Although as many as eight subspecies have been recognized within *B. vulgaris*, the most recent systems include all cultivated root beets in *B. vulgaris* L. subsp. *vulgaris* (Letschert 1993). All taxa are biennials (weedy races may be annuals, cf. Ford-Lloyd 1995). The wild sea beet is *B. vulgaris* subsp. *maritima* (L.) Arcang. plus other subspecies. There are also wild hybrid forms associated with cultivated plants (Boudry et al. 1993; Ford-Lloyd 1995). Kadereit et al. (2006) found that *Beta macrocarpa* Guss. (=*B. vulgaris* subsp. *macrocarpa* (Guss.) Thell.) is sister to *B. vulgaris*. Kadereit et al. (2006) suggested that, on the basis of ITS sequences, only *B. vulgaris* and *B. macrocarpa* should be included within section *Beta*.

Additional information on relationships is available in Kadereit et al. (2003), Müller and Borsch (2005), and Hohmann et al. (2006).

b. Origin and Domestication

Although no archaeological records exist for preclassical times, linguistic records place leafy forms of the cultivated beet to the eighth century BCE in Babylonia (Zohary and Hopf 2000). Synopses of the classical and modern history of the species in Europe and some other areas were given by de Blok (1985) and Ford-Lloyd (1995). Wild forms, often called the "wild sea beet," from which the cultigens may have been derived, are distributed throughout Europe, the Near East, and the Indian subcontinent (Doney et al. 1990; Letschert 1993; Ford-Lloyd 1995).

Many cultivated forms of the beet exist, including plants that produce storage roots and edible leaves. Desplanque et al. (1999, 2002) found little genetic diversity in the cultivated beet but great diversity in wild beets. In Europe, weedy forms are known to result from locally produced seed. This weed-seed origin accounts for the expansion of the distribution of these weedy annuals since the 1970s (Bartsch et al. 1999). Boudry et al. (1993) and Arnaud et al. (2003) examined the consequences of this finding in regard to the release of transgenic sugar beets and concluded that the use of transgenic strains would actually make the weed problem worse. There is such concern over these and other production-related issues in Europe that a meeting was held to address them (Frese et al. 2009). Gene pools of plants cultivated in the Americas are restricted in size and decline with time in cultivation (McGrath et al. 1999).

Chromosome numbers of beets are 2n = 18. There is a polyploid series in *Beta* including tetraploids (*Beta corolliflora* Zosimovic ex Buttler and *Beta patellaris* Moq., 2n = 36) and (a hexaploid *Beta trigyna* Waldst. and Kitt., 2n = 54).

c. Production

Beets are biennial and flowering requires vernalization. A period of 2 weeks at 4°C–10°C is required for temperate varieties, but shorter periods are sufficient for tropical ones. Flowers are self-incompatible and pollination is anemophilous. Propagation is always from seed. Tropical plantings usually require importation of seeds from temperate regions. Beets require ample fertilization for profitable yields, especially with nitrogen (Siemonsma and Piluek 1993).

Although not as susceptible to insect pests as some other crops, beets are subject to downy mildew (*Peronospora parasitica* and *Cercospora beticola*) of the leaves. *Phoma betae* and other fungi cause damping off. Beet mosaic virus also causes problems (Dusi and Peters 1999). Larvae of beet webworms (*Loxostege sticticalis*) feed on the leaves, and aphids and beet leaf miners also cause damage. Root-knot nematode (*Meloidogyne* sp.) affects the roots.

Beet roots typically contain 7%–10% carbohydrates (sucrose), 1.5%–2% protein, and small quantities of fat, ash, and fiber. However, roots of some cultivars of sugar beet may contain >18% sucrose (Ford-Lloyd et al. 1995). Roots contain a lower mineral and vitamin content than most other vegetables. The red color is produced by betanins (red betacyanins). Geosmin causes the earthy smell.

Europe is the largest producer of beets, with North and Central America in second place (Simpson and Conner-Ogorzaly 1995). Many other temperate areas cultivate the plants, mostly for the home or local markets. Europe, Asia, and North America produce almost 70% of the sugar beets. Similarly, Europe and North America grow most of the forage beets (Simpson and Conner-Ogorzaly 1995). Spinach beets are also grown and used in northern India and parts of Central and South America (Siemonsma and Piluek 1993).

2. Carrot

a. Classification and Relatives

Carrot (*Daucus carota* L., Apiaceae) is a complex of 13 wild and cultivated subspecies (Heywood 1983), with 12 of these occurring in Europe (Pankhurst 2010). The species includes the cultivated edible carrot (subsp. *sativus* (Hoffm.) Archangeli) and weedy forms (subsp. *carota*) locally referred to as "Queen Anne's lace" in the United States and "wild carrot" in Great Britain (Figure 30.1C). *Daucus* is a genus of ca. 22 species: 10 in Europe and the others spread through the Mediterranean, southwest and central Asia, tropical Africa, Australia, New Zealand, and the Americas (Sáenz Laín 1981; Heywood 1983; Mabberley 2009).

b. Origin and Domestication

Prior to Linnaeus fixing the Greek loanword *daucus* to the carrot, both that and Latin *pastinaca,* which Linnaeus applied to the parsnip (see succeeding text), were generic terms that included carrots, parsnips, and some other plants (Andrews 1949a). Efforts to determine the region of origin and document the spread of the cultivated carrot have been fraught with disagreement, largely because many botanists have ignored the linguistic literature. Zohary and Hopf (1993) did not discuss the species. Later, Zohary and Hopf (2000) commented that the archaeological data were "deplorably fragmentary" and that its history relies heavily on written records.

Banga (1957a,b) erroneously excluded the carrot from the early Greek literature and thought that it was introduced in the twelfth-century Spain by the Arabs. Zohary and Hopf (2000), apparently unaware of Andrews (1949a), accepted Körber-Grohne (1987) that the *staphylinos* of Dioscorides (40–90 CE) is carrot—especially since Dioscorides mentioned the characteristic trait of the dark central flower. Thus, the cultivated carrot has been in Greece since at least the time of Dioscorides (ca. 40–90 CE) and arrived later in some other parts of Europe. Recent archaeological study shows that cultivated carrots have been distributed throughout most of Europe since at least the Early Middle Ages (407–1000 CE), earlier in some areas (Bakels 2009). Spalik and Downie (2007) postulated a Mediterranean origin for *D. carota*, although they did not address the cultivated subspecies. Purple and yellow variants are thought to have spread into the Mediterranean and Western Europe in the twelfth to fourteenth centuries and to China, India, and Japan in the eighteenth century (Heywood 1983; Siemonsma and Piluek 1993; Riggs 1995). Laufer (1919) thought the carrot arrived in China during the Yuan dynasty (1260–1367 CE).

Carrot consists of the "eastern type" with mostly anthocyanin and branched roots and the "western type," having predominantly carotene and unbranched roots. There seems to be consensus that eastern carrots were first domesticated near Afghanistan and adjacent regions, perhaps as early as the tenth century in Persia (Matzkevitzh 1929; Banga 1957a; Heywood 1983; Siemonsma and Piluek 1993; Mabberley 2009). Western cultivars were perhaps derived from the yellow-rooted and white-rooted plants.

Carrots have 2n = 18 chromosomes. Whitaker (1949) thought that neither polyploidy nor structural changes were involved in speciation within *Daucus*. Molecular genetics studies of carrot cultivars distinguish between subsp. *sativus* and the other subspecies (Riggs 1995; Nakajima et al. 1998; Vivek and Simon 1998 1999; Lee et al. 2001). Moreover, these studies show closer relationships between cultivated and wild carrots than between other subspecies or other species of *Daucus*. The relationships between these plants are complex (Small 1978). *Daucus* species do not vary much in the number of rDNA loci despite variation in published chromosome numbers (Iovene et al. 2008).

The entire plastid genome of *D. carota* has been sequenced (Ruhlman et al. 2006) and this may stimulate additional research on the phylogeny of the crop. Genetic resources of *Daucus* and other umbelliferous crops are available in various gene banks (Astley 1997).

c. Production

Carrots are biennials that require vernalization before flowering. Plants at high latitudes bolt when exposed to temperatures of 2°C–6°C over 5–12 weeks. The species is protandrous and largely allogamous. Bees and flies are pollinators. Propagation is from seeds, and added nutrients are required for good production. Among the commonly applied fertilizers are potassium, nitrogen, phosphorus, calcium, and magnesium. The plants are more susceptible to diseases at high soil pH, especially in those with high chloride concentrations (Siemonsma and Piluek 1993).

A wide array of problems affects carrot production (Davis and Raid 2002). Among these are leaf blight (*Alternaria dauci* and *Cercospora carotae*) and root-knot nematodes (*Meloidogyne hapla*). Plants are also attacked by powdery mildews, white rust, and bacterial blights. Root rot is caused by several organisms, including *Botrytis cinerea, Fusarium* spp., *Sclerotinia sclerotiorum, Pythium violae*, and *Erwinia carotovora*. Insect pests include carrot root fly (*Psila rosae*), lygus bug (*Lygus hesperus* and *Lygus elisus*), leafhoppers (*Macrosteles fascifrons*), carrot weevil (*Listronatus oregonensis*), and armyworms (*Spodoptera* spp.). Aphids are vectors of at least 14 viral diseases in carrots.

Carrots are known for their high carotene content (~8285 μg/100 g), which some claim to prevent cancer. Between 2000 and May 2010, some 17,600 studies were published claiming some kind of impact of β-carotene on cancer: some negative, others positive. The consensus appears to be that a mixture of phytochemicals including β-carotene gives positive results against some cancers (e.g., Liu 2004; Nishino et al. 2009). The roots also contain 6%–9% carbohydrates (sugar), 1% protein, ash, 5–10 mg vitamin C, 40 mg calcium, and 1 mg iron per 100 g fresh weight. Terpenoids and other volatile compounds contribute to the taste of raw carrots (Siemonsma and Piluek 1993).

Commercial world production of carrots in the mid-1990s was more than 14 million metric tons yearly. This production was up from about 11 million in 1984. China (34% 2003–2005), Russia (7% 2003–2005), and the United States (7% 2003–2005) have led the remainder of the world in carrot production for several years. Except for China, carrot production in much of Asia is for local markets. In 2005, China produced about one-third of the global production of carrots, followed by Russia and the United States (FAO 2010; World Carrot Museum 2010).

3. Radish

a. Classification and Relatives

Radish (*Raphanus sativus* L.) belongs to the Brassicaceae (Figure 30.1B). The genus contains three species that are distributed from Western Europe to central Asia (Al-Shehbaz 1985; Mabberley 2009). Three species are generally recognized: *Raphanus sativus*; *Raphanus raphanistrum* L., including several subspecies (five in Flora Europaea online, Pankhurst 2010); and *Raphanus boissieri* Al-Shehbaz (Al-Shehbaz 1985).

b. Origin and Domestication

Radishes have been in cultivation at least since 2780 BCE (Al-Shehbaz 1985) and were known to the Assyrians, Chinese, Egyptians, Greeks, and Romans (Davidson 1999; Zohary and Hopf 2000; Mabberley 2009). The oldest crop form was drawn on the inner walls of the pyramids almost 4000 years ago (Al-Shehbaz 1985; Crisp 1995). From the Mediterranean, the radish spread to China by ~500 BCE and to Japan ~700 CE, where some radishes still retain the older sizes and shapes (Siemonsma and Piluek 1993).

R. sativus is usually considered to be of hybrid origin, although the ancestor(s) have yet to be identified. Several genetic studies, summarized by Lü et al. (2008), indicate that there are multiple haplotypes present in the various cultivars. *R. raphanistrum* is not the ancestor of the Eastern lineages of radish because it has a unique haplotype not in the cultivated plants (Yamane et al. 2005). The Eastern wild radish has contributed genes to the cultivated Asian plants, but it is not clear if it was through origin or subsequent gene flow. Lü et al. (2008) suggest that the Eastern black lineages (their var. *niger*) have an origin distinct from the European cultivars.

Several classification systems of *R. sativus* have been proposed (e.g., Kitamura 1958; Pistrick 1987; Siemonsma and Piluek 1993; Porcher 2010), none of them very satisfactory. Lü et al. (2008) used the system by Kitamura (1958), recognizing three groups—*R. sativus* L. var. *sativus* (European small radish), *R. sativus* var. *hortensis* Backer (E Asian big radish), *R. sativus* var. *niger* (Mill.) J. Kern (black radish). These three groups were supported by their molecular genetics analysis as distinct clades. However, Rabbani et al. (1998) and Liu et al. (2008), using other molecular markers, did not find the same groupings as Lü et al. (2008).

Chromosomes are 2n = 18 in radish and the related species.

c. Production

The radish is an annual and is grown from seeds. The edible part may be only the hypocotyl or both hypocotyl and upper part of the taproot. Flowers are allogamous and entomophilous. The crop requires more organic material in the soil than some others; best results require an application of incorporated nitrogen, phosphorus, and potassium in addition to surface-applied nitrogenous fertilizers (Siemonsma and Piluek 1993).

Cercospora brassicola (leaf spot) and *P. parasitica* (downy mildew) are common foliar diseases affecting the radish. Root diseases of the radish in temperate areas are black rot (*Aphanomyces raphani*) and yellows (*F. oxysporum*). Clubroot (*Plasmodiophora* spp.) also causes problems. Insect pests of the radish include flea beetles (*Phyllotreta* spp.), aphids, and mustard sawfly (*Athalia proxima*). Problems with root-knot nematodes (*Meloidogyne* spp.) are common.

Radish root contains ~5% carbohydrates, 0.6% protein, 32 mg calcium, 21 mg phosphorus, 0.6 mg iron per 100 mg fresh weight, and a trace of fat (Siemonsma and Piluek. 1993). Vitamins A, B1, and B2 are present in trace quantities. Vitamin C (25 mg) and niacin (0.3 mg) are also present. Vitamin A and the cruciferous isothiocyanates and indoles have been linked with possible cancer prevention (cf. Thornalley 2002; Amitabha 2005; Fimognari and Hrelia 2006).

The species occupies ~2% of the world production of vegetables with about 7 million tons/year (Siemonsma and Piluek 1993). Most of the world's radish production comes from the United States, China, Japan, and India (Meensalu 2010).

4. Turnip

a. Classification and Relatives

Turnip is the common name for *Brassica rapa* L. of the Brassicaceae. This plant continues to be called *Brassica campestris* in spite of its merger with *Brassica rapa* in the 1800s (Oost et al. 1987). According to the intraspecific classification developed by Toxopeus et al. (1984, 1985, 1988; Toxopeus in Siemonsma and Piluek 1993), the turnip belongs to the cultivar group Vegetable Turnip (synonyms: *B. rapa* L. subsp. *campestris* (L.) Clapham; *B. campestris* L. subsp. *rapa* (L.) Hook. f. and Anders.; *B. campestris* L. subsp. *rapifera* (Metzger) Sinsk.; *B. rapa* L. subsp. *rapa* sensu authors; *B. rapa* L. var. *rapa* sensu authors).

b. Origin and Domestication

The word turnip is derived from the English dialectic words *turn* = round and *nàep* or *nip* = root (OED online 2009). An archaeobotanical record of turnip is known from the Byzantine period of the fourth or fifth century CE (Hather et al. 1992), but the earliest linguistic record is from 1800 BCE for the word *laptu* in ancient Assyrian (Oppenheim et al. 1973 cited in Vogl-Lukasser et al. 2007).

Studies by Song et al. (1990, 1996) showed that the turnip is genetically close to other cultivars of *B. rapa* in India, China, and Japan. However, the root turnip is genetically and historically a species brought into cultivation in the Mediterranean (Song et al. 1990). *B. rapa* was either simultaneously brought into cultivation in the Mediterranean and China or has long and separate domestication histories in both regions (Zohary and Hopf 2000; Toxopeus and Baas 2004; Zhao et al. 2005).

Chromosomes are 2n = 300, of polyploid origin (Koch et al. 2003). *B. rapa* subsp. *campestris* was involved in the origin of the allotetraploid rutabaga (*Brassica napus* L., 2n = 38) (Palmer et al. 1983).

c. Production

Turnip is an annual or biennial herb with a stout taproot. Biennial types bolt after comparatively low temperatures. *B. rapa* consists of diverse crops that are completely cross compatible (Chrunga et al. 1999; Ochs et al. 1999). Propagation is from seeds. High nitrogen fertilization is required for good yields (Siemonsma and Piluek 1993).

Turnips are susceptible to the same pests as other members of *B. rapa*, such as soft rot (*E. carotovora*), downy mildew (*P. parasitica*), and clubroot (*Plasmodiophora brassicae*). Between 2002 and 2010, at least 4670 papers were devoted to the turnip mosaic virus (Google Scholar search, 3 January 2011); other viruses also cause production problems. Insect pests of turnips include caterpillars of white butterflies (*Pieris rapae* and *Pieris napi*), moths, webworms, leaf webber, aphids, turnip root fly (*Delia floralis*), red turnip beetle (*Entomoscelis americana*), and striped flea beetle (*Phyllotreta striolata*).

Both isothiocyanates and indoles in cruciferous vegetables have been associated with potential cancer prevention (cf. Thornalley 2002; Amitabha 2005; Fimognari and Hrelia 2006). China, Europe, and Japan are the leaders in production of this crop, although exact figures are combined with carrot production (FAO 2010).

IV. Minor Root Crops

A. Chicory

1. Classification and Relatives

Chicory (*Cichorium intybus* L., Asteraceae) and five other *Cichorium* species are native to Europe, the Mediterranean, and Ethiopia (Kiers 2000; Gemeinholzer and Bachmann 2005). Endive (*Cichorium endivia* L.) is a related species cultivated for its leaves.

2. Origin and Domestication

Chicory was probably brought into cultivation in the Mediterranean. Horace (65–8 BCE) wrote "*Me pascunt olivae, me cichorea, me malvae*" ("As for me, olives, chicory, and mallows provide sustenance"). Europeans have developed chicory roots of uniform size, shape, and time of maturation that are used as a coffee additive or substitute (Davidson 1999). Moreover, they have developed four cultivar lineages: Root Chicory group, Witloof group, Pain de Sucré group, and Radicchio group. Leaves of the Witloof, Pain de Sucré, and Radicchio groups are used in salads.

Roots contain the bitter sesquiterpene lactones (lactucin and lactucopicrin) and also sugar, inulin, resins, esculetin, esculin, cichoriin, umbelliferone, scopoletin, and 6.7-dihydrocoumarin, plus further sesquiterpene lactones and their glycosides (Van Beck et al. 1990; DeBruyn et al. 1992; Bais and Ravishankar 2001). Roots of chicory also contain the guaianolide phytoalexin cichoralexin (Monde et al. 1990). In medical experiments, root extract was shown to lower triglyceride and cholesterol levels in rats (Kaur et al. 1989; Uberoid 1991). Hepatoprotective activity has also been associated with chicory (Gilani et al. 1993; Ahmed et al. 2003). Inulin, besides being a sugar substitute, probiotic, and dietary fiber, has been associated with reducing the occurrence of tumors (Femia et al. 2002; Gibson et al. 2004; Schmidt et al. 2007; Topping 2007).

Chromosomes in this species are reported as 2n = 16, 18, 18 + 0-2B.

B. American Potato Bean Groundnut

1. Classification and Relatives

Groundnut (*Apios americana* Medik.) is a member of the Fabaceae (Figure 30.2C). *Apios* is a genus of North America and Eastern Asia with 10 species (Mabberley 2009). The groundnut, formerly called *Apios tuberosa* Moench, is one of two species of *Apios* native to North America (Woods 2005).

2. Origin and Domestication

Archaeological records from Missouri dating between 4500 and 3600 BCE indicate that the groundnut was used by the indigenous Americans (Harl 2009). *Apios* roots were later used by many indigenous peoples and European Americans (Austin 2004). Groundnut is now cultivated in the United States, Europe, and Japan. The species is notable for producing roots capable of containing >17% protein and 44.6% carbohydrates (Duke 1989).

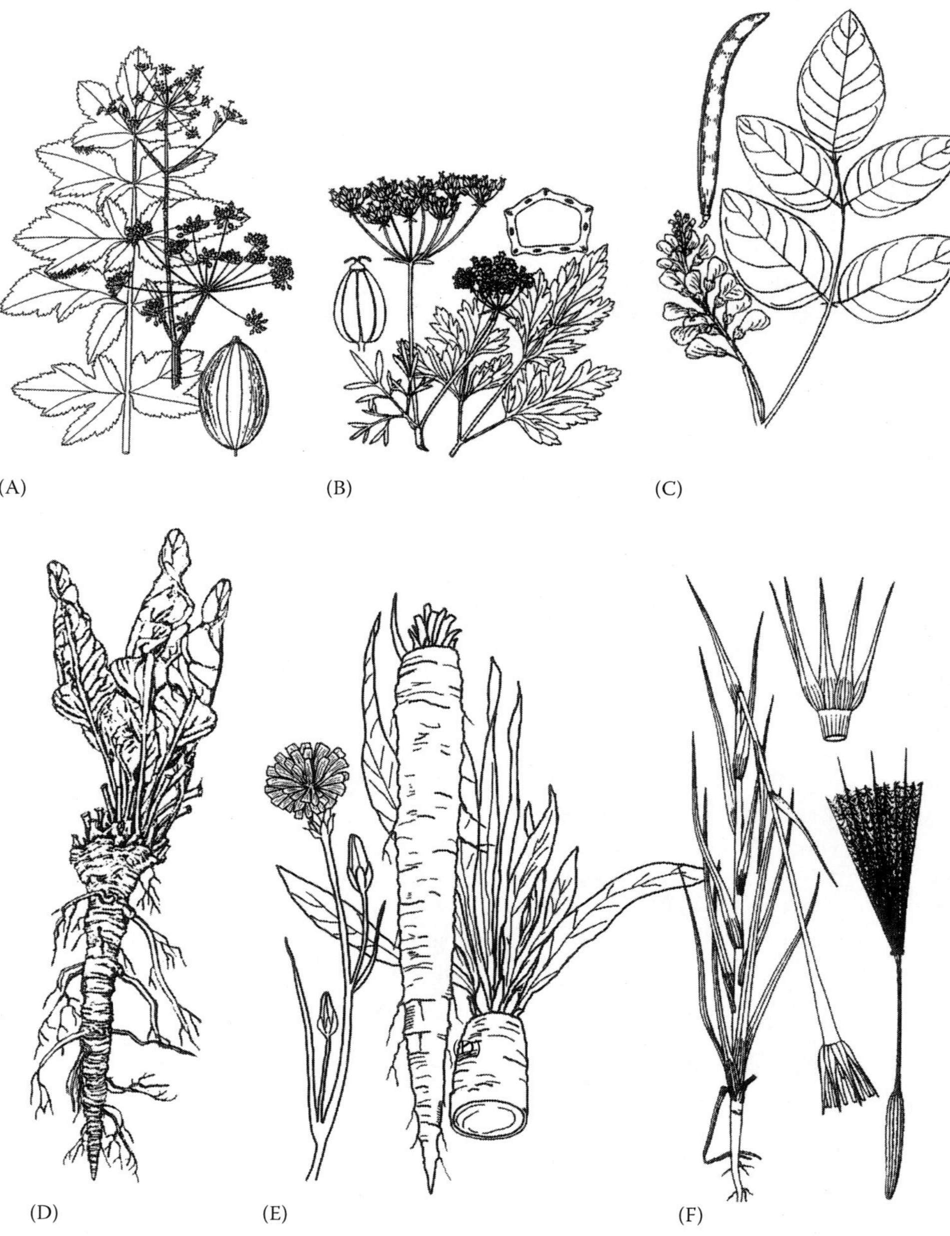

FIGURE 30.2 Some of the minor root crops. (A) Parsnip (*Pastinaca sativa*). (B) Parsley (*Petroselinum crispum*). (C) American groundnut (*Apios americana*). (D) Horseradish (*Armoracia rusticana*). (E) Scorzonera (*Scorzonera hispanica*). (F) Salsify (*Tragopogon porrifolius*). (A: From Steyermark 1963; B, C, F: From Diggs et al. 1999; D, E: Adapted and redrawn from Harrison et al. 1981.)

Biologically active compounds in the roots combat hypertension and hyperlipidemia, at least in laboratory rats (Iwai and Matsue 2007). While formerly believed to be pollinated by flies (Westerkamp and Paul 1993), the flowers are actually pollinated by megachilid bees (Bruneau and Anderson 1994).

Chromosome numbers of the diploids are 2n = 22. Some northern plants are triploids with 2n = 33 (Joly and Bruneau 2004).

C. Hamburg Parsley or Turnip-Rooted Parsley

1. Classification and Relatives

Parsley (*Petroselinum crispum* (Mill.) Fuss, Apiaceae) grows in the Mediterranean parts of Europe (Mabberley 2009; Figure 30.2B), along with the second species in the genus, *Petroselinum segetum* (L.) W.D.J.Koch. A related genus is *Apium* (Apieae) (Valiejo-Roman et al. 2006; Zhou et al. 2009). Three basic types of parsley are often recognized, curly leaved (var. *crispum*), flat-leaved or Italian parsley (var. *neapolitanum* Danert), and Hamburg parsley (var. *tuberosum* (Bernh.) Crov.).

2. Origin and Domestication

Parsley is a biennial plant that was used to flavor food by the Romans ~2000 years ago (Andrews 1949b; Simmonds 1976). The Greeks apparently did not eat parsley; some say it was medicinal (Aegineta 1844). Several sources state that parsley was used to crown winners of the Isthmian games (e.g., Tillyard 1913; Azeez and Parthasarathy 2008), but that is a faulty translation of σέλινον (*selinon*), which is wild celery (Andrews 1949b; Broneer 1962). Parsley was supposedly considered a symbol of death and put on tombs, but that too is dubious since the word used was *selinon* (Iverson 1976). Parsley spread into Western Europe in the fifteenth and sixteenth centuries (e.g., Aketer and Kühn 2008); from there it was introduced into the temperate parts of the world where it often became naturalized.

Hamburg parsley originated in Germany in the 1500s (Davidson 1999). Roots of this variety are used like those of the related celeriac (*Apium graveolens* L. var. *rapaceum* (Mill.) Gaudin)—boiled as a vegetable and put in soups, stews, or salads. Although fresh plants are seasonally available in some places in the United States, Hamburg parsley is seen more frequent in Europe. Leaves, seeds, and presumably roots of parsley contain the essential oils apiol and myristicin. Zhang et al. (2006), Edris (2007), and Mazzio and Soliman (2009) summarized the existing evidence that these compounds may reduce the chances for cancer. Furocoumarins and graveolone are found in the leaves (Beier et al. 1994); these substances cause photodermatitis (Egan and Sterling 1993). Polyacetylenic alcohols in the roots are effective in treating fungal infections (Nitz et al. 1990). Leaves and perhaps roots of parsley also contain carotene, niacin, ascorbic acid, iodine, iron, and calcium. Ascorbic acid and carotene are antioxidants that are said to be associated with cancer prevention (Liu 2004; Nishino et al. 2009). *Phytophthora* root rot was first reported in the species by Davis et al. (1994).

Chromosome number of parsley is 2n = 22.

D. Horseradish

1. Classification and Relatives

Horseradish (*Armoracia rusticana* P. Gaertn., B. Mey. and Scherb.) is a member of the Brassicaceae (Figure 30.2D). The genus has three species whose distributions range from southeastern Europe to Siberia (Al-Shehbaz 1988; Zhno et al. 2001).

2. Origin and Domestication

De Candolle (1886) concluded that the horseradish was native to Eastern Europe, based largely on linguistic relationships of the Slavic name *chren*. Horseradish has been cultivated for perhaps 2000 years (De Candolle 1886). While there are claims that *Armoracia* was known to the Egyptians in 1500 BC (Foster 2004; Goodwin and Collins 2008), the earliest documented mention was by Dioscorides, later by Pliny (Courter and Rhodes 1969). This is one of the "five bitter herbs" of the Jews (Grieve 1931). The pungent smell is due to the presence of the glucoside sinigrin (2-propenyl glucosinolate). The plant contains isothiocyanates and indoles. High vitamin C content also makes horseradish antiscorbutic. Horseradish peroxidase is known to degrade aflatoxin B_1 (Das and Mishra 2000).

Chromosome number is 2n = 32. Meiotic traits seem to indicate that cultigens are of hybrid origin (Stokes 1955).

E. Parsnip

1. Classification and Relatives

Parsnip (*Pastinaca sativa* L., Apiaceae) (Figure 30.2A) is one of the eight species in *Pastinaca* that are spread through temperate Eurasia (Menemen and Jury 2001). Three species occur in Europe, with four subspecies of *P. sativa* (spp. *latifolia* (Duby) DC., spp. *sativa*, spp. *sylvestris* (Mill.) Rouy and E.G.Camus, spp. *urens* (Req. ex Godr.) Čelak.) (see Loos 1993 and Pankhurst, 2010 for alternate taxonomy).

2. Origin and Domestication

Daucus and *Pastinaca* were originally generic terms applied to a number of plants but often included parsnips (see carrot for more). According to Andrews (1958), the Greek ἐλαφόβοσκον of Dioscorides and the Latin equivalent *elapoboskon* used by Pliny refer to the parsnip. He postulated that parsnip was brought into domestication in central Italy during the Augustan era (63 BCE–14 CE). Andrews' conclusion generally agrees with Zohary and Hopf (2000), although they suggest a "European and Western Asia element" and lament the meager archaeological data.

Parsnip was an important vegetable and fodder plant in Western and Central Europe from the Middle Ages into the eighteenth century (Hanelt et al. 2001). Simpson and Conner-Ogorzaly (1995) called parsnip "old-fashioned." The carrot is certainly now more popular, and Andrews (1958) suggests that has long been the case. Parsnips were used before the birth of Christ in the Mediterranean region, but good, fleshy forms were not developed before the Middle Ages (Hedrick 1919; McGee 1984). The main producers now are Great Britain, France, Canada, the United States, the Scandinavian countries, the Netherlands, and Hungary (Hanelt et al. 2001).

Parsnip was brought to the New World by the English pilgrims who settled at Plymouth, MA, in 1620. This was one of the few crops introduced by the Europeans that quickly became popular with the indigenous Americans (Simpson and Conner-Ogorzaly 1995), who, like pre-Columbian Europeans, had few sweet foods. Parsnip was introduced into the Caribbean in 1554 (Simmonds 1976). Soon after the species was brought to the New World, it escaped from cultivation and two varieties became weeds: subsp. *sativa* and subsp. *sylvestris* (Mill.) Rouy and E.G.Camus (var. *pratensis* Pers. of Kartesz 1994).

Chromosome are 2n = 22 (Van Loon and Kieft 1980).

F. Salsify or Vegetable Oyster

1. Classification and Relatives

Salsify (*Tragopogon porrifolius* L.) belongs to the Asteraceae (Figure 30.2F). This is one of ~110 *Tragopogon* species native to the temperate Mediterranean Eurasian region (Mabberley 2009).

2. Origin and Domestication

Theophrastus (372–288 BCE) and Pliny (23–79 CE) called these *tragopogon*, saying they were commonly eaten in Egypt (Bostock and Riley 1856; Flora of Australia online 2010). Later salsify was listed by Albertus Magnus in the thirteenth century under the name *Oculus porce* or *flos campi* (Sturtevant 1890) and mentioned through the 1500s and 1600s as being in gardens. Salsify was first listed in American gardens in 1806. This species and two of its congeners (*Tragopogon dubius* Scop. and *Tragopogon pratensis* L.) were introduced into North America, where all three escaped and became weeds (Kartesz 1994).

Salsify is diploid with 2n = 12 chromosomes. Ownbey (1950) reported that the North American species can hybridize with each other; hybrids have recently been synthesized (Tate et al. 2009).

G. Scorzonera or Black Oyster Plant

1. Classification and Relatives

Scorzonera (*Scorzonera hispanica* L.) is a member of the Asteraceae (Figure 30.2E). The genus contains ~175 species native to the Mediterranean and to central Asia (Mabberley 2009), although as now circumscribed, *Scorzonera* is polyphyletic (Mavrodiev et al. 2004).

2. Origin and Domestication

Scorzonera is based on the common names in both Romance and Germanic languages, *chersos* (serpent, toad, Latin), *escuerzo* (toad, Spanish), and *escurçó* (viper, Catalan), that allude to a use to treat snakebites (Mattioli 1561; Real Academia Española 2001). *Scorzonera* was possibly in cultivation by the 1540s when they were discussed by Fuchs' *De historia stirpium* (González Bueno 2008), although Hernández Bermejo and León (1994) did not think they were. Still, the ancient and current widespread use in southern Europe suggests a long human involvement (Rivera et al. 2006). *S. hispanica* was first cultivated in southern Europe and spread from there; it was taken from Spain to France early in the seventeenth century. The species was in the Americas by 1806 (Sturtevant 1890).

There are now about 24 named cultivars of the species (Hernández Bermejo and León 1994; Dolota and Dabrowska 2004). Some five species have been used either for edible roots, edible leaves, or rubber substitutes. The roots contain inulin (Dolota and Dabrowska 2004). Bryanskii et al. (1992) found a sesquiterpene glucoside in the plants. Bosch (2004) summarized the chemical composition.

Chromosome numbers in *S. hispanica* are 2n = 14, 28 (Hernández Bermejo and León 1994).

H. Skirret

1. Classification and Relatives

Skirret (*Sium sisarum* L., Apiaceae) has 8–14 related species in the Northern Hemisphere and Africa (Pu and Watson 2005; Spalik and Downie 2006; Downie et al. 2008; Mabberley 2009). The genus *Sium* is polyphyletic (Hardway et al. 2004; Spalik et al. 2009).

2. Origin and Domestication

Skirret is an Asian species grown for its roots that are eaten like salsify, used as a coffee substitute, or distilled into liquor (Pistrick 2002). The plant is native to Western Asia (Spalik and Downie 2006; Pankhurst 2010) and was introduced into Britain before 1548 (Paillieux and Bois 1885). *S. sisarum* was mentioned in Gerard's *Herbal* in 1597. The plant was unknown to the Greeks and Romans (Andrews 1958); it was in the Americas by 1775 (Sturtevant 1890).

As with many members of the family Apiaceae, *Sium* contains polyacetylenes (Crowden et al. 1969). However, the chemicals are present in such low concentrations that they cause no harm to humans eating the roots.

S. sisarum has 2n = 11 chromosomes (Bell and Constance 1966). The chromosome numbers of the related species *Sium latifolium* L. and *Sium suave* Walter are 2n = 10, 12, 22 (Maude 1939; Bell and Constance 1960; Crawford and Hartman 1972).

I. Spanish Salsify

1. Classification and Relatives

Spanish salsify (*Scolymus hispanicus* L.) is a member of the Asteraceae. The genus has three species in the Mediterranean, including Europe (Vázquez 2000; Strother 2005). *Scolymus* is closely related to *Sonchus* and *Lactuca* (Karis et al. 1992).

2. Origin and Domestication

Spanish salsify is from southern Europe, where it was mentioned by Theophrastus and Pliny (Hernández Bermejo and León 1994), and its roots are eaten like *Tragopogon*. The taste of the root is similar to that of salsify and is appreciated in Spain (Davidson 1999), although not everyone there has enjoyed it (Pardo-de-Santayana et al. 2006). *Sc. hispanicus* is sometimes cultivated in southern Italy (Gaetano 2009). Studies have shown a variety of chemicals and medicinal applications for *Sc. hispanicus* (Astudillo-Vázquez et al. 2009), and these plants have been used to treat cancer and other maladies (Pieroni et al. 2002; Alpinar et al. 2009).

Chromosomes are 2n = 20 (Kuzmanov and Georgieva 1980).

J. Turnip-Rooted Chervil or Parsnip Chervil

1. Classification and Relatives

Turnip-rooted chervil (*Chaerophyllum bulbosum* L.) is a member of the Apiaceae. The genus has ~30 species in the northern temperate regions (Spalik et al. 2001) and 12 species in Europe (Mabberley 2009). *Chaerophyllum tuberosum* Royle is cultivated in the Himalayas where it is called *sham*, and the roots are eaten (Hedrick 1919).

2. Origin and Domestication

Turnip-rooted chervil is native to Europe and naturalized in the United States (Mabberley 2009). The cultigen was introduced into Britain in 1726 and in 1864 was sent to the United States by F. Webster, the consul at Munich. Péron (1990) says that it was introduced to France in 1846, although some say it is native there (Pankhurst 2010). Apparently, the species was considered wild by Camerarius in 1588, Clusius in 1601, and Bauhin in 1623 (Hedrick 1919).

Cultivar development began in France in 1985 and there are now cultivars grown in that country (Péron 1990). Because there are polyacetylenes in the plants (Christensen et al. 2002; Christensen and Brandt 2006), some consider the leaves and stems poisonous (SHBL 2010), but the roots have such low concentrations of these chemicals that they pose no problem for human consumption (Crowden et al. 1969).

The species has 2n = 22 chromosomes (Kamari et al. 2004).

K. Alexanders, Wild Celery, or Horse Parsley

1. Classification and Relatives

Alexanders (*Smyrnium olusatrum* L., Apiaceae) is one of the seven Mediterranean species in the genus (Mabberley 2009). Perfoliate Alexanders (*Smyrnium perfoliatum* L.) is distributed through Central and southern Europe and southwest Asia. Like Alexanders, its blanched stems and leaves are eaten.

2. Origin and Domestication

Alexanders is native to Eurasia and naturalized in Great Britain, the Netherlands, and the United States (Randall 2003; Mabberley 2009). The plants were well known to the classical Greeks and Romans and their neighbors who ate the leaves, stems, and roots (Hedrick 1919; Leach 1982; Hernández Bermejo and León 1994).

Sm. olusatrum is thought to have been introduced into Britain by the Roman conquest that began AD 43 and declined in the fourth century (Hernández Bermejo and León 1994), particularly since macrofossils were recorded from Roman levels in the Welsh town of Caerwent (Dickson 1994). Others recording the species in early records include Apicius' *De re coquinaria* of the late fourth or early fifth century (Terrise 1990), Benedictus Crispus (died 725 or 735; Stannard 1966), and the *Saxon Leechdom* of ca. 940 (Cockayne 1864; OED 2009). Moreover, Alexanders was identified among Saxon and Norman garden remains dated from the ninth to the tenth centuries (Jones et al. 1991). Alexanders was in British gardens by or before Henricus Angelicus who mentioned cultivating it ca. 1235–1313 (Harvey 1987). Plants are still associated with archaeological sites inland in the British Isles (Feehan 2000).

Alexanders was commonly used for medicine and food until it was largely replaced when improved celery (*Ap. graveolens* L.) was introduced from the eastern Mediterranean. One of the reasons the herb was eaten is because of the aromatic herbage (*Smyrnium* is from Greek σμυρνα, myrrh), due to the essential oils (Randall 2003). Indeed, Alexanders contains numerous chemicals, including kaempferol, quercetin, flavonols, furocoumarins, acetylenic compounds, lactones (istanbulin-A, istanbulin-B), furanogermacrane glechomafuran (=alexandrofuran), sesquiterpenes, and acetooxyfuranoeudesmane (Mölleken et al. 1998; Randall 2003). Moreover, Alexanders was used to flavor metheglin (a Welsh word for spiced mead derived from *meddyglyn*, a compound of *meddyg*, "healer, doctor" + *llyn*, "liquor") (OED 2009).

Smyrnium is not mentioned by Zohary and Hopf (2000), but Hedrick (1919) gave a synoptic history from its dominance through decline. *Smyrnium* is still eaten and used as medicine from Turkey to Spain but as a minor resource (Ertuğ 2004; Tardío et al. 2006; Bullitta et al. 2007; Lentini and Venza 2007; Martignano et al. 2008).

The species has 2n = 22 chromosomes (Maude 1939; de Clavijo-J 1988).

V. Future of Root Crops

In the period between the 1970s and early 2000s, there were numerous predictions made about the importance of root crops in the future, and programs were organized to teach and learn more about them (e.g., Omawale 1979; Trèche 1996). In spite of these predictions and promises of the "Green Revolution" and the current "Biotechnology Revolution," little gain in crop production has been made in most of the world, particularly in developing countries. One study in Asia shows that, while the human population continues to grow at an alarming rate, increases in both grain crop and root and tuber crop yields have increased only slightly (Kumar 2006). Efforts to increase food production, particularly for the poor, languish, while the rich countries promote GM foods. The poor cannot afford the seeds, much less the chemicals to make GM crops grow and flourish (e.g., Hails 2002, Andersson and de Vincente 2010;

Taylor 2007; Cummings 2008). Instead of using species that historically have proven acceptable for consumption, there are calls for "new" crops (Glover et al. 2007). Janick (2007) pointedly illustrated the disconnection between plant biotechnologists and the people who actually grow the crops and eat them.

Surely, there will also be future shifts in major and minor crop dominance as in the past. There are even new uses for old crops:

1. Growing plant roots for phytoremediation (McBride 2003; Okoronkwo et al. 2005; Peuke and Rennenberg 2005; Russell. 2005; Swartjes et al. 2007; Turner 2009).
2. Molecular farming (Commandeur et al. 2003; Fischer and Schillberg 2004; Horn et al. 2004; Basaran and Rodríguez-Cerezo 2008; Spök and Karner 2008; Fischer et al. 2009).
3. Plant roots in culture provide an amazing array of compounds, including phenolics, steroidal saponins, coumarins, starch, essential oils, and many others. The literature on this subject is extensive. A few examples include Abbasi et al. (2007), Chang et al. (2005), Fu et al. (2006), Kim et al. (2004). Santos et al. (2005), Triplett et al. (2008), Sidwa-Gorycka et al. (2009), and Zhai et al. (2010). At least 2700 papers were published on the topic between 2009 and June 2010 (scholar-google.com), giving some indication of the emphasis being put on this subject. Root culture is not without problems or critics, but progress is being made (Fischer et al. 2004; Georgiev et al. 2007; Srivastava and Srivastava 2007).

In the 2002 edition of this book, the first author pointed out the inadequacy of the germplasm collections of root crops, their poorly understood systematic relationships, and the need to broaden the genetic base of the existing cultivars. While progress has occurred in some of these areas, others remain essentially unchanged.

Germplasm collections are still being amassed for some crops, while others struggle to maintain what has been collected due to inadequate resources. Ironically, the scientific community acknowledges that crop genetic diversity is disappearing. However, attracting resources to adequately maintain what has already been collected and to acquire new material continues to be a challenge (Tao et al. 1989).

The CIP assembled the largest collection of sweet potato germplasm in the world and then began eliminating clones based on their relationship with other clones, as determined by genetic analysis at a limited number of loci (Roca and Tay 2007). In contrast, other crops are still being actively collected (e.g., Rubenstein et al. 2006; Maxted et al. 2007; Malice and Baudoin 2009). Studies from various regions show that the loss of genetic diversity in many crops and their relatives is verging on irretrievable. For some, it may be too late. Nabhan (2008) revisited many of the areas proposed by Vavilov as important centers of crop diversity and found alarming genetic erosion. His is not a unique commentary on that problem (e.g., Bezançon et al. 2009; van Heerwaarden et al. 2009; CIP 2010; Lipper et al. 2010; Lockie and Carpenter 2010). Much of the crop genetic erosion that is occurring appears to be the result of international megabusinesses' promotion of GM crops that suppresses the ability or the desire of local farmers to grow traditional varieties (Cooney and Dickson 2006; Taylor 2007; Cummings 2008). Gene pools of crop-related species likewise are threatened and indeed are disappearing, as even remote areas are developed.

Knowledge of the phyletic relationships of root crops is necessary in order to identify and conserve materials that adequately represent/expand the existing crop gene pools. However, the phylogenies of many of the crop plants that sustain us are poorly understood. Although many tools are now available for testing and refining phylogenetic hypotheses (e.g., microsatellites, SNPs, DNA sequence databases), the funds for this type of research are often not available and the necessary plant materials may be inaccessible. Even when the funds or plant materials are available, some studies are performed using inadequate sample sizes and/or on materials that are misidentified.

There are serious questions about how effective ex situ germplasm collections are for preserving genetic diversity far from the centers of diversity. This topic has generated considerable discussion in the past few decades (e.g., Lacy 1987; Hammer 2003; Perales-R et al. 2003; Guerrant et al. 2004; Koo 2004; Meilleur and Hodgkin 2004; Gepts 2006; Ulloa et al. 2006). Many argue that there is higher diversity in crops cultivated in the field by native agriculturists (in situ) and lower in those of stored germplasm collections (ex situ). Due to genetic drift, gene banks can become markedly lower in diversity over time when compared to in situ collections. However, in terms of providing ready international access to germplasm in support of crop improvement and related research activities, no acceptable alternative currently exists—unless both are used.

The future of root crops as sustenance for the hungry must recognize the problems that inevitably arise with an ever-growing human population. As the need for food increases as a consequence of population growth, crop genetic diversity typically declines. After all, there is limited space to live and obtain resources. Increased mobility of crop diseases is also linked with larger human populations, and the existing narrow gene-based cultivars often grown in monoculture are inadequately prepared to cope with current or future plagues. As new or more virulent and/or resistant strains of current diseases and insect pests appear, crops will be even less capable of producing adequate yields. Rational preparation for such scenarios might include conserving and broadening the genetic base of root crops and supporting efforts to better understand their phyletic relationships. In the absence of sufficient genetic resources properly maintained and readily available, this goal is unlikely to be met.

Acknowledgments

Sandra Austin provided information or gave helpful suggestions on the original manuscript. Research on sweet potato and its relatives has been supported by grants from the Centro Internacional de la Papa, Lima, Peru; U.S. Department of Agriculture (grant 58-6659-1-102); USAID/USDA (grant 59-319R-4-001; D. F. Austin and Z. Huamán coinvestigators); and the National Geographic Society (grant 4478-91). Mardie Banks (Visual Resources, Florida Atlantic University) redrew and composed the figures. Ihsan Al-Shehbaz offered advice on the Brassicaceae.

References

Abbasi BH, Tian C-L, Murch SJ, Saxena PK, Liu C-Z. 2007. Light-enhanced caffeic acid derivatives biosynthesis in hairy root cultures of *Echinacea purpurea. Plant Cell Rep* 26:1367–1372.

Aegineta P. 1844. *The Seven Books of Paulus Aegineta: Translated from the Greek by Francis Adams*. London, U.K.: Sydenham Society, http://books.google.com/

Ahmed A, Al-Howiriny TA, Siddiquia AB. 2003. Antihepatotoxic activity of seeds of *Cichorium intybus. J Ethnopharm* 87:237–240.

Aina AJ, Falade KO, Akingbala JO, Titus P. 2009. Physicochemical properties of twenty-one Caribbean sweet potato cultivars. *Int J Food Sci Technol* 44:1696–1704.

Aketer O, Kühn M. 2008. Desiccated plant macrofossils from the medieval castle of Marmorera, Switzerland, with a note on the identification of leaves of Cyperaceae. *Environ Archaeol* 13:37–50.

Alcorn JB. 1984. *Huastec Mayan Ethnobotany*. Austin, TX: University of Texas Press.

Allem AC. 1994. The origin of *Manihot esculenta* Crantz (Euphorbiaceae). *Genet Resour Crop Evol* 41:133–150.

Allem AC. 1999. The closest wild relatives of cassava (*Manihot esculenta* Crantz). *Euphytica* 107:123–133.

Alpinar K, Özyürek M, Kolak U et al. 2009. Antioxidant capacities of some food plants wildly grown in Ayvalik of Turkey. *Food Sci Technol Res* 15:59–64.

Al-Shehbaz IA. 1985. The genera of Brassiceae (Cruciferae; Brassicaceae) in the Southeastern United States. *J Arnold Arbor* 66:279–351 (*Raphanus* pp. 327–334).

Al-Shehbaz IA. 1988. The genera of Arabideae (Cruciferae; Brassicaceae) in the Southeastern United States. *J Arnold Arbor* 66:85–166 (*Armoriaca* pp. 160–166).

Ames O. 1939. *Economic Annuals and Human Cultures*. Cambridge, MA: Botanical Museum of Harvard University.

Amitabha R. 2005. Cancer preventive role of selected dietary factors. *Ind J Cancer* 42:15–24.

Andersson MS, de Vincente MC. 2010. *Gene Flow between Crops and Their Wild Relatives*. Baltimore, MD: Johns Hopkins University Press.

Andrews AC. 1949a. The carrot as a food in the classical era. *Class Philol* 44:182–196.

Andrews AC. 1949b. Celery and parsley as foods in the Greco-Roman period. *Class Philol* 44:91–99.

Andrews AC. 1958. The parsnip as a food in the classical era. *Class Philol* 53:145–152.

Armstrong, W.P. 2012. Vegetables From Underground. http://waynesword.palomar.edu/vege1.htm (accessed December 20, 2012).

Arnaud J-F, Viard F, Delescluse M, Cuguen J. 2003. Evidence for gene flow via seed dispersal from crop to wild relatives in *Beta vulgaris* (Chenopodiaceae): Consequences for the release of genetically modified crop species with weedy lineages. *Proc R Soc Ser B* 270:1565–1571.

Asante SA, Yamada T, Hisamatsu M, Shiotani I. 2006. Studies on the properties of starch of diploid *Ipomoea trifida* (H.B.K.) Don. strains. *Starch* 45:299–306.

Astley D. 1997. *Report of a Workshop on Umbellifer Crops Genetic Resources*. Rome, Italy: International Plant Genetic Resources Institute, 55pp.

Astudillo-Vázquez A, Mata R, Navarrete A. 2009. El reino vegetal, fuente de agentes antiespasmódicos gastrointestinales y antidiarreicos. *Rev Latinoam Quím* 37:7–44.

Austin DF. 1983. Variability in sweet potato in America. *Proc Am Soc Hortic Sci* 27(Pt B):15–26.

Austin DF. 1988. The taxonomy, evolution and genetic diversity of sweet potatoes and related wild species. In: *Exploration, Maintenance, and Utilization of Sweet Potato Genetic Resources*, pp. 27–60. Report of the *First Sweet Potato Planning Conference, 1987*. Lima, Peru: Centro Internacional de la Papa.

Austin DF. 1991. *Ipomoea littoralis* (Convolvulaceae)—Taxonomy, distribution and ethnobotany. *Econ Bot* 45:251–256.

Austin DF. 2004. *Florida Ethnobotany*. Boca Raton, FL: CRC Press.

Austin DF, Huamán Z. 1996. A synopsis of *Ipomoea* (Convolvulaceae) in the Americas. *Taxon* 43:3–38.

Austin DF, Jansson RK, Wolfe GW. 1991. Convolvulaceae and *Cylas*: A proposed hypothesis on the origins of this plant: Insect relationship. *Trop Agric (Trinidad)* 68:162–170.

Austin DF, Jarret R, Tapia E, De La Puente F. 1993. Collecting tetraploid *I. batatas* (Linnaeus) Lamarck in Ecuador. *FAO/IBPGR Plant Genet Resour Newsl* 91/92:33–35.

Awoleye F, Van Duren M, Dolezel J, Novak FJ. 1994. Nuclear DNA content and in vitro induced somatic polyploidization cassava (*Manihot esculenta* Crantz) breeding. *Euphytica* 76:195–202.

Azeez S, Parthasarathy VA. 2008. Parsley. In: *Chemistry of Spices*, eds. VA Parthasarathy, B Chempakam, JT Zachariah, pp. 376–400. Wallingford, U.K.: CABI International.

Bais HP, Ravishankar GA. 2001. *Cichorium intybus* L—Cultivation, processing, utility, value addition and biotechnology, with an emphasis on current status and future prospects. *J Sci Food Agric* 81:467–484.

Bakels CC. 2009. *The Western European Loess Belt: Agrarian History, 5300 BC–AD 1000*. Dordrecht, the Netherlands: Springer.

Banga O. 1957a. The origin of the European cultivated carrot. *Euphytica* 6:54–63.

Banga O. 1957b. The development of the original European carrot material. *Euphytica* 6:64–76.

Bartsch D, Lehnen M, Clegg J, Pohl-Orf M, Schuphan I, Ellstrand NC. 1999. Impact of gene flow from cultivated beet on genetic diversity of wild sea beet populations. *Mol Ecol* 8:1733–1741.

Basaran P, Rodríguez-Cerezo E. 2008. Plant molecular farming. Opportunities and challenges. *Crit Rev Biotechnol* 28:153–172.

Beaumont P. 1993. Drylands: Environmental management and development. *Routledge Natural Environment Series. Japanese Studies Series*. New York: Nissan Institute/Routledge.

Beier RC, Ivie GW, Oertli EH. 1994. Linear furanocoumarins and graveolens from the common herb parsley. *Phytochemistry* 36:869–872.

Bell AD. 1991. Plant Form. *An Illustrated Guide to Flowering Plant Morphology*. Oxford, U.K.: Oxford University Press.

Bell CR, Constance L. 1960. Chromosome numbers in Umbelliferae. II. *Am J Bot* 47:24–32.

Bell CR, Constance L. 1966. Chromosome numbers in Umbelliferae. III. *Am J Bot* 53:512–520.

Bezançon G, Pham J-L, Deu M, Vigouroux Y, Sagnard F, Mariac C, Kapran I, Mamadou A, Gérard B, Ndjeunga J, Chantereau J. 2009. Changes in the diversity and geographic distribution of cultivated millet (*Pennisetum glaucum* (L.) R. Br.) and sorghum (*Sorghum bicolor* (L.) Moench) varieties in Niger between 1976 and 2003. *Genet Resour Crop Evol* 56:223–236.

Bohac JR, Austin DF, Jones A. 1993. Discovery of wild tetraploid sweetpotatoes. *Econ Bot* 47:193–201.

Bohac JR, Dukes PD, Austin DF. 1995. Sweetpotato. In: *Evolution of Crop Plants*, 2nd edn., eds. J Smartt, NW Simmonds, pp. 57–62. London, U.K.: Longman Group, Ltd.

Bosch CH. 2004. *Scorzonera hispanica* L. In: *Plant Resources of Tropical Africa 2. Vegetables*, eds. GJH Grubben, OA Denton, pp. 454–455. Wageningen: PROTA Foundation/ Backhuys Publishers.

Bostock J, Riley HT. 1856. *The Natural History of Pliny*, Vol. 4. Bohn's classical library. London, U.K.: G. Bell and Sons.

Boudry P, Morchen M, Saumitou-Laprade P. 1993. The origin and evolution of weed beets: Consequences for the breeding and release of herbicide-resistant transgenic sugar beets. *Theor Appl Genet* 87:471–478.

Brako L, Rossman AY, Farr DF. 1995. *Scientific and Common Names of 7,000 Vascular Plants in the United States*. St. Paul, MN: American Phytopathological Society.

Broneer O. 1962. The Isthmian victory crown. *Am J Archaeol* 66:259–263.

Bruneau A, Anderson GJ. 1994. To bee or not to bee? The pollination biology of *Apios americana* (Leguminosae). *Plant Syst Evol* 192:147–149.

Bryanskii OV, Tolstikhina VV, Zinchenko SV. 1992. A sesquiterpene glucoside from cultivated cells of *Scorzonera hispanica*. *Chem Nat Compd* 28:556–567.

Bullitta S, Piluzza G, Viegi L. 2007. Plant resources used for traditional ethnoveterinary phytotherapy in Sardinia (Italy). *Genet Resour Crop Evol* 54:1447–1464.

Buteler MI, Jarret RL, LaBonte DR. 1999. Sequence characterization of microsatellites in diploid and polyploid *Ipomoea*. *Theor Appl Genet* 99:123–132.

Chang C-K, Chang KS, Lin Y-C, Liu S-Y, Chen C-Y. 2005. Hairy root cultures of *Gynostemma pentaphyllum* (Thunb.) Makino: A promising approach for the production of gypenosides as an alternative of ginseng saponins. *Biotechnol Lett* 27:1165–1169.

Chavarriaga-Aguirre P, Maya MM, Bonierbale MW, Kresovich S, Fregene MA, Tohme J, Kochert G. 1998. Microsatellites in cassava (*Manihot esculenta* Crantz): Discovery, inheritance and variability. *Theor Appl Genet* 97:493–501.

Chen C-M, Lin Y-L, Chen C-Y, Hsu C-Y, Shieh M-J, Liu J-F. 2008. Consumption of purple sweetpotato leaves decreases lipid peroxidation and DNA damage in human. *Asia Pac J Clin Nut* 17:408–414.

Chon S-U, Boo H-O. 2005. Differences in allelopathic potential as influenced by root periderm colour of sweet potato (*Ipomoea batatas*). *J Agric Crop Sci* 191:75–80.

Christensen LP, Brandt K. 2006. Acetylenes and psoralens. In: *Plant Secondary Metabolites: Occurrence, Structure and Role in the Human Diet*, eds. A Crozier, MN Clifford, H Ashihara, pp. 137–173. New York: Wiley-Blackwell.

Christensen LP, Hansen SL, Purup S, Brandt K. 2002. Naturally occurring acetylenes in common food plants: Chemistry, occurrence and bioactivity. In: *Health Promoting Compounds in Vegetables and Fruits*. Proceedings of Workshop in Karrebaeksminde, Denmark, November 6–8, 2002, eds. K Brandt, B Akesson, pp. 54–68. Tjele, Denmark: DIAS Report. Horticulture no. 29.

Chrunga B, Verma N, Monanty A. Shivanna KR. 1999. Production and characterization of interspecific hybrids between *Brassica maurorum* and crop brassicas. *Theor Appl Genet* 98:608–613.

CIP. 2010. Sweetpotato/Genetic resources conservation. Accessed May 20, 2010. http://www.cipotato.org/sweetpotato/germplasm.asp

Cockayne TO (ed.). 1864. *Leechdoms, Wortcunning, and Starcraft of Early England: Being a Collection of Documents, for the Most Part Never before Printed, Illustrating the History of Science in This Country before the Norman Conquest*, Vol. 2. London, U.K.: Longman, Green, Longman, Roberts, and Green.

Commandeur U, Twyman RM, Fischer R. 2003. The biosafety of molecular farming in plants. *AgBiotechNet* 5:1–9.

Cooney R, Dickson B (eds.). 2006. *Biodiversity and the Precautionary Principle. Risk, Uncertainty and Practice in Conservation and Sustainable Use*. Sterling, VA: Stylus Publishing.

Courter JW, Rhodes AM. 1969. Historical notes on horseradish. *Econ Bot* 23:156–164.

Crawford DJ, Hartman RL. 1972. Chromosome numbers and taxonomic notes for Rocky Mountain Umbelliferae. *Am J Bot* 59:386–392.

Crisp P. 1995. Radish. *Raphanus sativus* (Cruciferae). In: *Evolution of Crop Plants*, 2nd edn., eds. J Smartt, NW Simmonds, pp. 86–88. London, U.K.: Longman Group, Ltd.

Crowden JB, Harborne JB, Heywood VH. 1969. Chemosystematics of the Umbelliferae—A general survey. *Phytochemistry* 8:1963–1894.

Cummings CH. 2008. *Uncertain Peril. Genetic Engineering and the Future of Seeds*. Boston, MA: Beacon Press.

Das C, Mishra HN. 2000. In vitro degradation of aflatoxin B_1 by horseradish peroxidase. *Food Chem* 68:309.

David CR, Alamu S. 1980. The effect of growth regulators on tuber initiation and growth in rooted leaves of two sweet potato cultivars. *Ann Bot* 45:363–364.

Davidson A. 1999. *The Oxford Companion to Food*. Oxford, U.K.: Oxford University Press.

Davis M, Raid R. 2002. *Compendium of Umbelliferous Diseases*. St. Paul, MN: APS Press.

Davis RM, Winterbottom CQ, Valencia J. 1994. First report of *Phytophthora* root rot of parsley. *Plant Dis* 78:1122.

Day HA. 1917. *Vegeculture: How to Grow Vegetables, Salads, and Herbs in Town and Country*. London, U.K.: Methuen & Co., Ltd.

de Blok TSM. 1985. The genus *Beta*: Domestication, taxonomy and interspecific hybridization for plant breeding. *Acta Hortic* 182:335–344.

de Candolle A. 1886. *Origin of Cultivated Plants*. Reprinted. New York: Hafner Publishing Co., 1967.

de Clavijo-J. ER. 1988. Números cromosomáticos de plantes occidentales, 452–465. *An Jard Bot Madr* 45:259–266.

DeBruyn A, Alvarez AP, Sandra P. 1992. Isolation and identification of O-beta-D-fructofuranosyl-(21)O-beta-D-fructofuranosyl-(21)-D-fructose, a product of the enzymic hydrolysis of the inulin from *Cichorium intybus*. *Carbohydr Res* 235:303–308.

Denslow JS, Padoch C (eds.). 1988. *People of the Tropical Rain Forest*. Berkeley, CA: University of California Press.

Desplanque B, Boudry P, Broomberg K, Saumitou-Laprade P, Cuguen J, Van Dijk H. 1999. Genetic diversity and gene flow between wild, cultivated and weedy forms of *Beta vulgaris* L. (Chenopodiaceae), assessed by RFLP and microsatellite markers. *Theor Appl Genet* 98:1194–1201.

Desplanque B, Hautekèete N, Van Dijk H. 2002. Transgenic weed beets: Possible, probable, avoidable? *J Appl Ecol* 39:561–571.

Dickau R, Ranere AJ, Cooke RG. 2007. Starch grain evidence for the preceramic dispersals of maize and root crops into tropical dry and humid forests of Panama. *Proc Natl Acad Sci U S A* 104:3651–3656.

Dickson C. 1994. Macroscopic fossils of garden plants from British Roman and medieval deposits. In: *Garden Plants, Species, Forms and Varieties from Pompeii to 1800*, eds. E Moe, JH Dickson, PM Jorgensen, pp. 47–72. Rixensaret, Belgium: PACT 42.

Diggs GMJ, Lipscomb BL, O'Kennon RJ. 1999. *Shinners & Mahler's Illustrated Flora of North Central Texas*. Fort Worth, TX: Issue 16. DOI.wiley.com/10.1002/asi.20463.

Dini I, Tenore GC, Dini A. 2009. Saponins in *Ipomoea batatas* tubers: Isolation, characterization, quantification and antioxidant properties. *Food Chem* 113:411–419.

Dolota A, Dabrowska B. 2004. Raw fibre and inulin content in roots of different scorzonera cultivars (*Scorzonera hispanica* L.) depending on cultivation method. *Folia Hortic* 16:31–37.

Domola M, Thompson GJ, Aveling TAS, Laurie SM, Strydom H, van den Berg AA. 2008. Sweet potato viruses in South Africa and the effect of viral infection on storage root yield. *Afr Plant Protect* 14:15–23.

Doney DK, Whitney ED, Terry J. 1990. The distribution and dispersal of *Beta vulgaris* L. subsp. *maritima* germplasm in England, Wales and Ireland. *J Sugar Beet Res* 27:29–38.

Downie SR, Katz-Downie DS, Sun F-J, Lee C-S. 2008. Phylogeny and biogeography of Apiaceae tribe Oenantheae inferred from nuclear rDNA ITS and cpDNA *psbI*-5′τρνK$^{(uuu)}$ sequences, with emphasis on the North American endemics clade. *Botany* 86:1039–1064.

Duke JA. 1989. *Apios americana* Medik. (Fabaceae)—Groundnut. In: *CRC Handbook of Nuts*, pp. 22–25. Boca Raton, FL: CRC Press.

Dusi AN, Peters D. 1999. Beet mosaic virus: Its vector and host relationships. *Phytopathol Zeitsch* 147:293–298.

Edris AE. 2007. Pharmaceutical and therapeutic potentials of essential oils and their individual volatile constituents: A review. *Phytother Res* 21:308–323.

Egan CL, Sterling G. 1993. Phytophotodermatitis: A visit to Margaritaville. *Cutis* 51:41–42.

Eguchi T, Kitano M, Eguchi H. 1994. Effect of root temperature on sink strength of tuberous root in sweet potato plants (*Ipomoea batatas* Lam.). *Biotronics* 23:75–80.

Eguchi T, Kitano K, Eguchi H. 1996. New system of hydroponics for growth analysis of sweet potato tuber. *Biotronics* 25:85–88.

Eguchi T, Yoshida S. 2004. A cultivation method to ensure tuberous root formation in sweetpotatoes (*Ipomoea batatas* (L.) Lam.). *Environ Control Biol* 42:259–266.

Eguchi T, Yoshida S. 2008. Effects of application of sucrose and cytokinin to roots on the formation of tuberous roots in sweetpotato (*Ipomoea batatas* (L.) Lam.). *Plant Root* 2:7–13.

Eicher CK, Maredia K, Sithole-Niang I. 2006. Crop biotechnology and the African farmer. *Food Policy* 31:504–527.

Ertuğ F. 2004. Wild edible plants of the Bodrum Area (Muğla, Turkey). *Turk J Bot* 28:161–174.

FAO. 2010. Food and Agricultural Organization of the United Nations FAOSTAT report. Accessed June 9, 2010. http://faostat.fao.org/site/339/default.aspx

Feehan J. 2000. A prior's herb garden. *Archaeol Ireland* 14:44.

Femia AP, Luceri C, Dolara P, Giannini A, Biggeri A, Salvadori M, Clune Y, Collins KJ, Paglierani M, Caderni G. 2002. Antitumorigenic activity of the prebiotic inulin enriched with oligofructose in combination with the probiotics *Lactobacillus rhamnosus* and *Bifidobacterium lactis* on azoxymethane-induced colon carcinogenesis in rats. *Carcinogenesis* 23:1953–1960.

Fimognari C, Hrelia P. 2006. Sulforaphane as a promising molecule for fighting cancer. *Rev Mutat Res* 635:90–104.

Firon N, LaBonte D, Villordon A, McGregor C, Kfir Y, Pressman E. 2009. Botany and physiology: Storage root formation and development. In: *The Sweetpotato*, eds. G Loebenstein, G Thottappilly, pp. 13–26. New York: Springer.

Fischer R, Schillberg S. 2004. *Molecular Farming: Plant-made Pharmaceuticals and Technical Proteins*. Weinheim, Germany: Wiley-VCH.

Fischer R, Schillberg S, Twyman RM. 2009. Molecular farming of antibodies in plants. In: *Recent Advances in Plant Biotechnology*, eds. A Irakosyan, PB Kaufman, pp. 35–63. New York: Springer.

Fischer R, Stoger E, Schillberg S, Christou P, Twyman RM. 2004. Plant-based production of biopharmaceuticals. *Curr Opin Plant Biol* 7:152–158.

Fitter A. 2002. Characteristics and functions of root systems. In: *Plant Roots: The Hidden Half*, eds. Y Waisel, A Eshel, U Kafkaki, 3rd edn., pp. 15–32. New York: Marcel Dekker, Inc.

Flora of Australia Online. 2010. Flora of Australia Online: Norfolk and Lord Howe Islands. Australian Biological Resources Study, Canberra. Accessed 20 May 2010. http://www.anbg.gov.au/abrs/online-resources/flora/stddisplay.xsql?pnid = 5927

Ford-Lloyd BV. 1995. Sugarbeet, and other cultivated beets. *Beta vulgaris* L. (Chenopodiaceae). In: *Evolution of Crop Plants*, 2nd edn., eds. J Smartt, NW Simmonds, pp. 35–40. London, U.K.: Longman Group, Ltd.

Foster RJ. 2004. *A Horse of Another Flavor*. Accessed May 19, 2010. http://www.foodproductdesign.com/articles/2004/09/food-product-design-ingredient-insight-septembe.aspx

Fregene MA, Vargas J, Ikea J, Angel F, Tohne J, Asiedu RA, Akoroda MO, Roca WN. 1994. Variability of chloroplast DNA and nuclear ribosomal DNA in cassava (*Manihot esculenta* Crantz) and its wild relatives. *Theor Appl Genet* 89:719–727.

Frese L, Maggioni L, Lipman E (eds.). 2009. Report of a Working Group on *Beta* and the World *Beta* Network. *Third Joint Meeting*, March 8–11, 2006, Puerto de la Cruz, Tenerife, Spain. Rome, Italy: Bioversity International.

Fu C-X, Xu Y-J, Zhao D-X, Ma FS. 2006. A comparison between hairy root cultures and wild plants of *Saussurea involucrata* in phenylpropanoids production. *Plant Cell Rep* 24:750–754.

Gade DW. 2002. Names for *Manihot esculenta*: Geographical variations and lexical clarification. *J Latin Am Geogr* 1:55–74.

Gaetano L. 2009. Microevolution of *Scolymus hispanicus* L. (Compositae) in southern Italy: From gathering of wild plants to some attempts of cultivation. *J Agric Rural Dev Trop Subtrop*, Suppl. 92:119–126.

Garcia AM, Walter Jr WM. 1998. Physicochemical characterization of starch from Peruvian sweetpotato selections. *Starch* 50:331–337.

Gemeinholzer B, Bachmann K. 2005. Examining morphological and molecular diagnostic character states of *Cichorium intybus* L. (Asteraceae) and *C. spinosum* L. *Plant Syst Evol* 253:105–123.

Georgiev MI, Pavlov AI, Bley T. 2007. Hairy root type plant in vitro systems as sources of bioactive substances. *J Appl Microbiol Biotechnol* 74:1175–1185.

Gepts P. 2006. Plant genetic resources conservation and utilization. The accomplishments and future of a societal insurance policy. *Crop Sci* 46:2278–2292.

Gibson GR, Probert HM, Loo JV, Rastall RA, Roberfroid MB. 2004. Dietary modulation of the human colonic microbiota: Updating the concept of prebiotics. *Nutr Res Rev* 17:259–275.

Gilani AN, Janbaz KN, Javed MG. 1993. Hepatoprotective activity of *Cichorium intybus*, and indigenous medicinal plant. *Med Sci Res* 21:151–152.

Glover JD, Cox CM, Reganold JP. 2007. Future farming: A return to roots. *Sci Am* 297:82–89.

González Bueno A. 2008. Los "Fuchs" castellanos impresos por los Birckmann: En torno a un tratado de Botánica renacentista conservado en la Bibliotech "Marqués de Valdecilla." *Pecia Complut* 5:46–72.

Goodwin K, Collins T. 2008. Horseradish. Accessed May 19, 2010. http://academics.hamilton.edu/foodforthought/Our_Research_files/parsnips.pdf

Grieve M. 1931. *A Modern Herbal: The Medicinal, Culinary, Cosmetic and Economic Properties, Cultivation and Folklore of Herbs, Grasses, Fungi, Shrubs, & Trees with All Their Modern Scientific Uses. 1971*. Reprinted. New York: Dover Publications.

Guerrant EO, Havens K, Maunder M. 2004. *Ex Situ Plant Conservation: Supporting Species Survival in the Wild*. Washington, DC: Island Press.

Hails RS. 2002. Assessing the risks associated with new agricultural practices. *Nature* 418:685–688.

Hammer K. 2003. A paradigm shift in the discipline of plant genetic resources. *Genet Resour Crop Evol* 50:3–10.

Hanelt P, Büttner R, Mansfeld R. 2001. *Mansfeld's Encyclopedia of Agricultural and Horticultural Crops*, Vol. 2. Berlin, Germany: Springer.

Hardway TM, Spalik K, Watson MF, Katz-Downie DS, Downie SR. 2004. Circumscription of Apiaceae tribe Oenantheae. *S Afr J Bot* 70:393–406.

Harl JL. 2009. Archaic period of East-Central Missouri. In: *Archaic Societies: Diversity and Complexity across the Midcontinent*, eds. TE Emerson, DL McElrath, AC Fortier, pp. 377–400. New York: SUNY Press.

Harrison BE, Maesfield GB, Wallis M, Nicholson BE. 1981. *The Oxford Book of Food Plants*. Oxford, UK: Oxford University Press.

Harvey JH. 1987. The square garden of Henry the Poet. *Garden Hist* 15:1–11.

Hather JG. 1996. The origins of tropical vegeculture: Zingiberaceae, Araceae, and Dioscoreaceae in Southeast Asia. In: *The Origins and Spread of Agriculture and Pastoralism in Eurasia*, ed. DR Harris, pp. 538–551. New York: UCL Press.

Hather JG, Peña-Chocarro L, Sidell EJ. 1992. Turnip remains from Byzantine Sparta. *Econ Bot* 46:395–400.

Hawkes JG. 1983. *The Diversity of Crop Plants*. Cambridge, MA: Harvard University Press.

Hawkes JG. 2007. The ecological background of plant domestication. In: *The Domestication and Exploitation of Plants and Animals*, eds. PJ Ucko, GW Dimbleby, pp. 17–30. Piscataway, NJ: Transaction Publishers.

Haysom HR, Chan TL, Hughs MA. 1994. Phylogenetic relationships of *Manihot* species revealed by restriction fragment length polymorphism. *Euphytica* 76:227–234.

Hedrick UP. 1919. *Sturtevant's Notes on Edible Plants*. Albany, NY: J.B. Lyon.

Heiser CB Jr. 1990. *Seeds to Civilization*. Cambridge, MA: Harvard University Press.

Hernández Bermejo JE, León J (eds.). 1994. *Neglected Crops: 1492 from a Different Perspective. Issue 26 of FAO Plant Production and Protection Series*. Rome, Italy: Food & Agriculture Organization.

Heywood VH. 1983. Relationships and evolution of the *Daucus carota* complex. *Isr J Bot* 35:51–65.

Hill WA, Bonsi CK, Loretan PA (eds.). 1992. *Sweetpotato Technology for the 21st Century*. Tuskegee, AL: Tuskegee University.

Hohmann S, Kadereit JW, Kadereit G. 2006. Understanding Mediterranean-Californian disjunctions: Molecular evidence from Chenopodiaceae-Betoideae. *Taxon* 55:67–78.

Horn ME, Woodard SL, Howard JA. 2004. Plant molecular farming: Systems and products. *Plant Cell Rep* 22:711–720.

Horton D. 1988. *Underground Crops*. Morrilton, AZ: Winrock International.

Huáman Z, Aguilar C, Ortiz R. 1999. Selecting a Peruvian sweetpotato core collection on the basis of morphological, ecogeographical, and disease pest reaction data. *Theor Appl Genet* 98:840–844.

Huang JC, Sun M. 2000. Genetic diversity and relationships of sweetpotato and its wild relatives in *Ipomoea* series *Batatas* (Convolvulaceae) as revealed by inter-simple sequence repeat (ISSR) and restriction analysis of chloroplast DNA. *Theor Appl Genet* 100:1050–1060.

Iovene M, Grzebelus E, Carputo D, Jiang J, Simon PW. 2008. Major cytogenetic landmarks and karytype analysis in *Daucus carota* and other Apiaceae. *Am J Bot* 95:793–804.

Iverson AM. 1976. The ancient Greek "death" aspect of spring in Mandel'štam's poetry. *Slav East Eur J* 20:34–39.

Iwai K, Matsue H. 2007. Ingestion of *Apios americana* Medikus tuber suppresses blood pressure and improves plasma lipids in spontaneously hypertensive rats. *Nutr Res* 27:218–224.

Janick J. 2007. The origins of horticultural technology and science. *Acta Hortic* 759:41–60.

Jansson RK, Raman KV (eds.). 1991. *Sweet Potato Pest Management. A Global Perspective*. Boulder, CO: Westview Press.

Jarret RL. 1989. A repository for sweet potato germplasm. *HortScience* 25:885–886.

Jarret RJ, Austin DF. 1994. Genetic diversity and systematic relationships in sweetpotato (*Ipomoea batatas* (L.) Lam.) and related species as revealed by RAPD analysis. *Genet Resour Crop Evol* 41:165–173.

Jennings DL. 1995. Cassava. *Manihot esculenta* (Euphorbiaceae). In: *Evolution of Crop Plants*, 2nd edn., eds. J Smartt, NW Simmonds, pp. 128–132. London, U.K.: Longman Group, Ltd.

Joly S, Bruneau A. 2004. Evolution of triploidy in *Apios americana* (Leguminosae) revealed by genealogical analysis of the histone *h3-d* gene. *Evolution* 58:284–295.

Jones A. 1990. Unreduced pollen in a wild tetraploid relative of sweetpotato. *J Am Soc Hort Sci* 115:512–516.

Jones G, Straker V, Davis A. 1991. Early Medieval plant use and ecology. In: *Aspects of Saxon and Norman London II: Finds and Environmental Evidence*, ed. AG Vince, pp. 347–385. London and Middlesex, U.K.: Archaeological Society Special Paper 12.

Kadereit G, Borsch T, Weising K, Freitag H. 2003. Phylogeny of Amaranthaceae and Chenopodiaceae and the evolution of C4 photosynthesis. *Int J Plant Sci* 164:959–986.

Kadereit G, Hohmann S, Kadereit JW. 2006. A synopsis of Chenopodiaceae subfam. Betoideae and notes on the taxonomy of *Beta*. *Willdenowia* 36:9–19.

Kamari G, Blanché C, Garbari F. 2004. Mediterranean chromosome number reports-14. *Flora Mediterr* 14:423–453.

Kano M, Takayanagi T, Harada K, Makino K, Ishikawa F. 2005. Antioxidative activity of anthocyanins from purple sweet potato, *Ipomoea batatas* cultivar Ayamurasaki. *Biosci Biotechnol Biochem* 69:979–988.

Karis PO, Kallersjo M, Bremer K. 1992. Phylogenetic analysis of the Cichorioideae (Asteraceae), with emphasis on the Mutisieae. *Ann Mo Bot Gard* 79:416–427.

Kartesz JT. 1994. *A Synonymized Checklist of the Vascular Flora of the United States, Canada and Greenland*, Vol. 1, Checklist, 2nd edn. Portland, OR: Timber Press.

Kaur N, Gupta AK, Saijpaul S. 1989. Triglyceride and cholesterol lowering effect of chicory roots in the liver of dexamethasone-injected rats. *Med Sci Res* 17:1009–1010.

Kays SJ, Collins WW, Bouwkamp JC. 1992. A response: The sweetpotato storage organ is a root, not a tuber. In: *Sweetpotato Technology for the 21st Century*, eds. WA Hill, CK Bonsi, PA Loretan, pp. 307–313. Tuskegee, AL: Tuskegee University.

Kiers AM. 2000. Endive, chicory, and their wild relatives: A systematic and phylogenetic study of *Cichorium* (Asteraceae). Doctoral dissertation published as Gorteria supplement 5:1–78, Universiteit Leiden.

Kim OT, Kim MY, Hong MH, Ahn JC, Hwang B. 2004. Stimulation of asiaticoside accumulation in the whole plant cultures of *Centella asiatica* (L.) Urban by elicitors. *Plant Cell Rep* 23:339–344.

Kitamura S. 1958. Varieties of radish and their transition. In: *Japanese Radish*, ed. I Nishiyama, pp. 1–19. Tokyo, Japan: Japanese Science Society Press. (in Japanese).

Kitaya Y, Hirai H, Wei X, Islam AFMS, Yamamoto M. 2007. Growth of sweetpotato cultured in the newly designed hydroponic system for space farming. *Adv Space Res* 41:730–735.

Koch M, Al-Shehbaz IA, Mummenhoff K. 2003. Molecular systematics, evolution, and population biology in the mustard family (Brassicaceae). *Ann Mo Bot Gard* 90:151–171.

Koo B. 2004. *Saving Seeds: The Economics of Conserving Crop Genetic Resources Ex Situ in the Future Harvest Centres of the CGIAR*. Wallingford, Oxfordshire, U.K.: CABI Publishing.

Körber-Grohne U. 1987. *Nutzpflanzen in Deutschland: Kulturgeschichte und Biologie*. Stuttgart, Germany: K. Theiss Verlag.

Kowyama Y, Tsuchiya T, Kakeda K. 2008. Molecular genetics of sporophytic self-incompatibility in *Ipomoea*, a member of the Convolvulaceae. In: *Self-Incompatibility in Flowering Plants Evolution, Diversity and Mechanisms*, ed. VE Franklin-Tong, pp. 259–274. Berlin, Germany: Springer.

Kumar BM. 2006. Agroforestry: The new old paradigm for Asian food security. *J Trop Agric* 44:1–14.

Kuzmanov B, Georgieva S. 1980. In Chromosome number reports LXIX, ed. Löve A. *Taxon* 29:703–730 (cf.715).

Lacy RC. 1987. Loss of genetic diversity from managed populations: Interacting effects of drift, mutation, immigration, selection and population subdivision. *Conserv Biol* 1:143–158.

Langdon R. 1988. Manioc, a long concealed key to the enigma of Easter Island. *Geogr J* 154:324–336.

Lange W, Brandenburg WA, de Bock TSM. 1999. Taxonomy and cultonomy of beet (*Beta vulgaris* L.). *Bot J Linn Soc* 130:81–96.

Laufer B. 1919. *Sino-Iranica*. Chicago, IL: Field Museum of Natural History, *Anthropology Series* 15(3), publ. 201.

Leach HM. 1982. On the origins of kitchen gardening in the Ancient Near East. *Gard Hist* 10:1–16.

Lebot, V. 2009. *Tropical Root and Tuber Crops. Cassava, Sweet Potato, Yams and Aroids*. Cambridge, MA: CABI North American Office.

Lee B-Y, Levin GA, Downie SA. 2001. Relationships within the spiny-fruited Umbellifers (Scandiceae subtribes Daucinae and Torilidinae) as assessed by phylogenetic analysis of morphological characters. *Syst Bot* 26:622–642.

Lee J-S, Shin M-J, Park Y-K, Ahn Y-S, Chung M-N, Kim H-S, Kim J-M. 2007. Antibacterial and antimutagenic effects of sweet-potato tips extract. *Korean J Crop Sci* 52:303–310.

Lentini F, Venza F. 2007. Wild food plants of popular use in Sicily. *J Ethnobiol Ethnomed* 3:15–27.

Letschert JPW. 1993. *Beta* section *Beta*: Biogeographical patterns of variation, and taxonomy. Thesis, Wageningen Agricultural University Papers 91:1–155.

Lipper L, Anderson CL, Dalton TJ (eds.). 2010. *Seed Trade in Rural Markets. Implications for Crop Diversity and Agricultural Development*. London, U.K.: The Food and Agricultural Organization of the United Nations and Earthscan.

Liu L-W, Zhao L-P, Gong Y-Q et al. 2008. DNA fingerprinting with genetic diversity analysis of late-bolting radish cultivars with RAPD, ISSR and SRAP markers. *Sci Hort* 116:240–247.

Liu RH. 2004. Potential synergy of phytochemicals in cancer prevention: Mechanism of action. *J Nutr* 134:3479S–3485S.

Lockie S, Carpenter D. 2010. *Agriculture, Biodiversity and Markets: Livelihoods and Agroecology in Comparative Perspective*. London, U.K.: Earthscan.

Loos GH. 1993. Zur taxonomie von *Pastinaca sativa* L. s. lat. *Floristische Rundbriefe* 27:16–19.

Low JW, Arimond M, Osman N, Cunguara B, Zano F, Tschirley D. 2007. A food-based approach introducing orange-fleshed sweet potatoes increased Vitamin A intake and serum retinol concentrations in young children in rural Mozambique. *Am J Nutr* 137:1320–1327.

Low JW, Kinyae P, Gichoki S, Oyunga MA, Hagenimana V, Kabria J. 1997. Combating vitamin A deficiency through the use of sweet potato: Results from phase I of an action research project in South Nyanza, Kenya. Lima, Peru: Centro Internacional de la Papa (CIP).

Lü N, Yamane K, Ohnishi O. 2008. Genetic diversity of cultivated and wild radish and phylogenetic relationships among *Raphanus* and *Brassica* species revealed by the analysis of trnK/matK sequence. *Breed Sci* 58:15–22.

Mabberley DJ. 2009. *The Plant-Book*, 3rd edn., with corrections. Cambridge, U.K.: Cambridge University Press.

Magoon ML, Krishnan R, Bai KV. 1969. Morphology of the pachytene chromosomes and meiosis in *Manihot esculenta*. *Cytologia* 34:612–624.

Magoon ML, Krishnan R, Bai KV. 1970. Cytological evidence on the origin of the sweet potato. *Theor Appl Genet* 40:360–366.

Malice M, Baudoin JP. 2009. Genetic diversity and germplasm conservation of three minor Andean tuber crop species. *Biotechnol Agric Soc Environ* 13:441–448.

Martignano F, Falco V, Traclò BRG, Hammer K. 2008. Agricultural biodiversity in Grecìa and Bovesìa, the two Griko-speaking areas in Italy. *PGR Newsl* 156:43–49.

Mattioli PA. 1561. *Epistolarum Medicinalium Libri Quinque*. Prague: Melantrichius.

Matzkevitzh VI. 1929. The carrot of Afghanistan. *Bull Appl Bot Genet Plant Breed* 20:517–562 (in Russian).

Maude PF. 1939. The Merton Catalogue. A list of the chromosome numerals of species of British flowering plants. *New Phytol* 38:1–31.

Mavrodiev EV, Edwards CE, Albach DC, Gitzendanner MA, Soltis PS, Soltis DE. 2004. Phylogenetic relationships in subtribe Scorzonerinae (Asteraceae: Cichorioideae: Cichorieae) based on ITS sequence data. *Taxon* 53:699–712.

Maxted N, Ford-Lloyd BV, Kell SP, Irinodo J, Dulloo E (eds.). 2007. *Crop Wild Relative Conservation and Use*. Wallingford, Oxfordshire, U.K.: CABI Publishing.

Mazzio EA, Soliman KFA. 2009. In vitro screening for the tumoricidal properties of international medicinal herbs. *Phytother Res* 23:385–398.

McBride MB. 2003. Toxic metals in sewage sludge-amended soils: Has promotion of beneficial use discounted the risks? *Adv Environ Res* 8:5–19.

McGee H. 1984. *On Food and Cooking. The Science and Lore of the Kitchen*. New York: Charles Scribner's.

McGrath JM, Derrico CA, Yu Y. 1999. Genetic diversity in selected, historical US sugarbeet germplasm and *Beta vulgaris* subsp. *maritima*. *Theor Appl Genet* 98:968–976.

McSpence JA. 1971. Cultivation of detached sweet potato (*Ipomoea batatas* (L.) Lam.) leaves with tuberous roots for photosynthetic studies. *Photosynthetica* 5:424–425.

Meensalu L. 2010. Radish. Accessed June 1, 2010. http://www.eestitoit.ee/?page_id = 412&language = en

Meilleur BA, Hodgkin T. 2004. In situ conservation of crop wild relatives: Status and trends. *Biodiv Conserv* 13:663–684.

Menemen Y, Jury SL. 2001. A taxonomic revision of the genus *Pastinaca* L. (Umbelliferae). *Isr J Plant Sci* 49:67–77.

Miyazaki Y, Kusano S, Doi H, Aki O. 2005. Effects on immune response of antidiabetic ingredients from white-skinned sweet potato (*Ipomoea batatas* L.). *Nutrition* 21:358–362.

Mölleken U, Sinnwell V, Kukbezka K-H. 1998. Essential oil composition of *Smyrnium olusatrum*. *Phytochemistry* 49:1709–1714.

Monde K, Oya T, Shirat A. 1990. A guaianolide phytoalexin, cichoralexin, from *Cichorium intybus*. *Phytochemistry* 29:3449–3452.

Müller K, Borsch T. 2005. Phylogenetics of Amaranthaceae based on matK/trnK sequence data. Evidence from parsimony, likelihood, and Bayesian analyses. *Ann Mo Bot Gard* 92:66–102.

Mwanga ROM, Yencho GC, Moyer JW. 2002. Diallel analysis of sweetpotato for resistance to sweetpotato virus disease. *Euphytica* 128:237–248.

Nabhan GP. 2008. *Where Our Food Comes From. Retracing Nikolay Vavilov's Quest to End Famine*. Washington, DC: Island Press/Shearwater Books.

Nakajima Y, Oeda K, Yamamoto T. 1998. Characterization of genetic diversity of nuclear and mitochondrial genomes in *Daucus* varieties by RAPD and AFLP. *Plant Cell Rep* 17:848–853.

Nakamoto ST, Wanitprapha K, Iwaoka W, Huang A. 1994. *Cassava, Ginger, Sweet Potato, and Taro Trade Statistics Research Extension Series 150*. Honolulu, HI: Institute of Tropical Agriculture and Human Resources.

Nassar NMA. 2002. Cassava, *Manihot esculenta* Crantz, genetic resources: Origin of the crop, its evolution and relationships with wild relatives. *Genet Mol Res* 1:298–305.

Nicholson G. (ed.). 1885–1889. *The Illustrated Dictionary of Gardening: A Practical and Scientific Encyclopedia of Horticulture for Gardeners and Botanists*. London, UK: L. Upcott Gill.

Nishino H, Murakoshi M, Tokuda H, Satomi Y. 2009. Cancer prevention by carotenoids. *Arch Biochem Biophys* 483:165–168.

Nitz S, Spraul MH, Drawert F. 1990. C_{17} polyacetylenic alcohols as the major constituents in roots of *Petroselinum crispum* Mill. ssp. *tuberosum*. *J Agric Food Chem* 38:1440–1447.

Norman MJT, Pearson CJ, Searle PGE. 1984. *The Ecology of Tropical Food Crops*. Cambridge, U.K.: Cambridge University Press.

Nye MM. 1991. The mis-measure of manioc (*Manihot esculenta* Euphorbiaceae). *Econ Bot* 45:47–57.

Ochs G, Schock G, Trischler M, Kosemund K, Wild A. 1999. Complexity and expression of the glutamine synthetase multigene family in the amphidiploid crop *Brassica napus*. *Plant Mol Biol* 39:395–405.

OED Online. 2009. *Oxford English Dictionary*. Oxford, U.K.: Oxford University Press.

Okada Y, Saito A, Nishiguchi M, Kimura T, Mori M, Hanada K, Sakai J, Miyazaki Y, Murata T. 2001. Virus resistance in transgenic sweetpotato [*Ipomoea batatas* L. (Lam.)] expressing the coat protein gene of sweetpotato feathery mottle virus. *Theor Appl Gen* 103:743–751.

Okogbenin E, Marin J, Fregene M. 2006. An SSR-based molecular genetic map of cassava. *Euphytica* 147:433–440.

Okoronkwo NE, Igwe JC, Onwuchekwa EC. 2005. Risk and health implications of polluted soils for crop production. *Afr J Biotechnol* 4:1521–1524.

Olsen KM. 2002. Population history of *Manihot esculenta* (Euphorbiaceae) inferred from nuclear DNA sequences. *Mol Ecol* 11:901–911.

Olsen KM. 2004. SNPs, SSRs and inferences on cassava's origin. *Plant Mol Biol* 56:517–526.

Olsen KM, Schaal BA. 1999. Evidence on the origin of cassava: Phytogeography of *Manihot esculenta*. *Proc Natl Acad Sci U S A* 96:5586–5591.

Olsen KM, Schaal BA. 2001. Microsatellite variation in cassava (*Manihot esculenta*, Euphorbiaceae) and its wild relatives: Further evidence for a southern Amazonian origin of domestication. *Am J Bot* 88:131–142.

Olsen KM, Schaal BA. 2006. DNA sequence data and inheritances on cassava's origin of domestication. In: *Documenting Domestication: New Genetic and Archaeological Paradigms*, ed. MA Zeder, pp. 123–133. Berkeley, CA: University of California Press.

Omawale. 1979. The nutritional significance of root and tuber crop development as staples in the Caribbean community. *Arch Latinoam Nutr* 29:311–325.

Onwueme IC. 1978. *The Tropical Tuber Crops*. Chicester, U.K.: John Wiley & Sons.

Oost EH, Brandenburg WA, Reuling GHM, Jarvis CE. 1987. Lectotypification of *Brassica rapa* L., *B. campestris* L. and neotypification of *B. chinensis* L. (Cruciferae). *Taxon* 36:625–634.

Oppenheim AL, Reiner E, Biggs RD, Renger JM, Slot M (eds.). 1973. *The Assyrian Dictionary of the Oriental Institute of the University of Chicago*, Vol. 9. Glückstadt, Germany: Augustin Verlagbuchhandlung.

Ownbey M. 1950. Natural hybridization and amphiploidy in the genus *Tragopogon*. *Am J Bot* 37:487–499.

Ozias-Akins P, Jarret RL. 1994. Nuclear DNA content and ploidy levels in the genus *Ipomoea*. *J Am Soc Hortic Sci* 119:110–115.

Paillieux A, Bois D. 1885. *Le Potager d'un curieux: histoire, culture et usages de 100 plantes comestibles peu connues ou inconnues*. Paris, France: Librarie agricole de la Maison rustique.

Palmer JD, Shields CR, Cohen DB, Otron TJ. 1983. An unusual mitochondrial DNA plasmid in the genus *Brassica*. *Nature* 301:725–728.

Pankhurst R. 2010. Flora Europaea online. http://rbg-web2.rbge.org.uk/FE/fe.html

Pardo-de-Santayana M, Tardío J, Heinrich M, Touwaide A, Morales R. 2006. Plants in the works of Cervantes. *Econ Bot* 60:159–181.

Park K-H, Kim J-R, Lee J-S, Lee H, Cho K-H. 2010. Ethanol and water extract of purple sweet potato exhibits anti-atherosclerotic activity and inhibits protein glycation. *J Med Food* 13:91–98.

Perales-R H, Brush SB, Qualset CO. 2003. Dynamic management of maize landraces in Central Mexico. *Econ Bot* 57:21–34.

Péron J-Y. 1990. Tuberous-rooted chervil: A new root vegetable for temperate climates. In: *Advances in New Crops*, eds. J Janick, JE Simon, pp. 422–423. Portland, OR: Timber Press.

Perrine H. 1840. Random records of tropical Florida. *Mag Hortic*. Reprinted in *Tequesta* 11:58–62. 1951. http://digitalcollections.fiu.edu/tequesta/files/1951/51_1_03.pdf

Peterson JK, Harrison HF Jr. 1995. Sweet potato allelopathic substance inhibits growth of purple nutsedge (*Cyperus rotundus*). *Weed Sci* 9:277–280.

Peuke AD, Rennenberg H. 2005. Phytoremediation. *EMBO Rep* 6:497–501.

Pieroni A, Janiak V, Dürr CM, Lüdeke S, Trachsel E, Heinrich M. 2002. In vitro antioxidant activity of non-cultivated vegetables of ethnic Albanians in Southern Italy. *Phytother Res* 16:467–473.

Piperino DR, Pearsall DM. 1998. *The Origins of Agriculture in the Lowland Neotropics*. San Diego, CA: Academic Press.

Pistrick K. 1987. Untersuchungen zur Systematik der Gattung *Raphanus* L. *Kulturpflanzen* 35:225–321.

Pistrick K. 2002. Notes on neglected and underutilized crops. Current taxonomical overview of cultivated plants in the families Umbelliferae and Labiatae. *Genet Resour Crop Evol* 49:211–221.

Porcher MH. 2010. Sorting *Raphanus* names. Multilingual Multiscript Plant Name Database. http://www.plantnames.unimelb.edu.au/new/Raphanus.html#sativus-sativus

Prendergast H, Etkin NL, Harris DR, Houghton PJ. 1998. *Plants for Food and Medicine*. London, U.K.: The Royal Botanic Gardens.

Proctor GR, Brunt MA, Lewis CB. 1984. *Flora of the Cayman Islands*. London: HMSO, Kew Bulletin Additional Series XI.

Pu F, Watson MF. 2005. *Sium* In: *Flora of China. Apiaceae through Ericaceae*, eds. Z-Y Wu, PH Raven, Vol. 14, pp. 115–116. St. Louis, MO: Missouri Botanical Garden Press.

Purseglove JW. 1968. *Tropical Crops. Dicotyledons*. 2 Vols. London, U.K.: Longmans Group.

Purseglove JW. 1972. *Tropical Crops. Monocotyledons*. London, U.K.: Longmans Group.

Quattrocchi U. 1999. *CRC World Dictionary of Plant Names. Common Names, Scientific Names, Eponyms, Synonyms, and Etymology*. Boca Raton, FL: CRC Press LLC.

Rabbani MA, Murakami Y, Kuginuki Y, Takayanagi K. 1998. Genetic variation in radish (*Raphanus sativus* L.) germplasm from Pakistan. *Genet Resour Crop Evol* 45:307–316.

Raji AAJ, Anderson JV, Kolade OA, Ugwu1 CD, Dixon AGO, Ingelbrecht IL. 2009. Gene-based microsatellites for cassava (*Manihot esculenta* Crantz): Prevalence, polymorphisms, and cross-taxa utility. *BMC Plant Biol* 9:1–11.

Randall RE. 2003. *Smyrnium olusatrum* L. *J Ecol* 91:325–340.

Real Academia Española. 2001. Diccionario de la lengua Española, Vigésima segunda edición. http://buscon.rae.es/draeI/ (accessed July 05, 2010).

Reddy DVR, Sudarshana MR, Fuchs M, Rao NC, Thottapilly G. 2009. Genetically engineered virus-resistant plants in developing countries: Current status and future prospects. *Adv Virus Res* 75:185–220.

Rehm S (ed.). 1994. *Multilingual Dictionary of Agronomic Plants*. Dordrecht, the Netherlands: Kluwer.

Riggs TJ. 1995. Carrot. *Daucus carota* (Umbelliferae). In: *Evolution of Crop Plants*, eds. J Smartt, NW Simmonds, 2nd edn., pp. 477–480. London, U.K.: Longman Group, Ltd.

Rindos D. 1984. *The Origins of Agriculture: An Evolutionary Perspective*. Orlando, FL: Academic Press.

Rivera D, Obón C, Heinrich M, Inocencio C, Verde A, Fajardo J. 2006. Gathered Mediterranean food plants—Ethnobotanical investigations and historical development. In: *Local Mediterranean Food Plants and Nutraceuticals. Forum of Nutrition*, Vol. 59, eds. M Heinrich, WE Müller, C Galli, pp. 18–74. Basel, Switzerland: Karger.

Roa AC, Maya MM, Duque MC, Tohme J, Allem AC, Bonierbale MW. 1997. AFLP analysis of relationships among cassava and other Manihot species. *Theor Appl Genet* 95:741–750.

Roca W, Tay D. 2007. Background on the development of the Global Strategy for Ex Situ conservation of sweetpotato genetic resources. Lima, Peru: Centro Internacional de la Papa (CIP).

Rogers DJ. 1963. Studies of *Manihot esculenta* Crantz (cassava) and related species. *Bull Torrey Bot Club* 90:43–54.

Rogers DJ, Appan SG. 1973. *Manihot, Manihotoides* (Euphorbiaceae), *Monograph 13*. New York: Flora Neotropica.

Rubenstein KD, Smale M, Widrlechner MP. 2006. Demand for genetic resources and the U.S. National Plant Germplasm System. *Crop Sci* 46:1021–1031.

Ruhlman T, Lee S-B, Jansen RK, Hostetler JB, Tallon LJ, Town CD, Daniell H. 2006. Complete plastid genome sequence of *Daucus carota*: Implications for biotechnology and phylogeny of angiosperms. *BMC Genomics* 7:222–235.

Russell K. 2005. The use and effectiveness of phytoremediation to treat persistent organic pollutants. Washington, DC: U.S. Environmental Protection Agency.

Sáenz Laín C. 1981. Research on *Daucus* L. (Umbelliferae). *An Jard Bot Madr* 37:481–534.

Santos PAG, Figueiredo AC, Oliveira MM, Barroso JG, Pedro LG, Deans SG, Scheffe JJC. 2005. Growth and essential oil composition of hairy root cultures of *Levisticum officinale* W.D.J. Koch (lovage). *Plant Sci* 168:1089–1096.

Sauer CO. 1952. *Agricultural Origins and Dispersals*. New York: American Geographic Society.

Sauer CO. 1969. *The Early Spanish Main*. Berkeley, CA: University of California Press.

Schmidt BM, Ilicb N, Pouleva A, Raskina I. 2007. Toxicological evaluation of a chicory root extract. *Food Chem Toxicol* 45:1131–1139.

Schulthess F, Baumgärtner JU, Delucchi V, Gutierrez AP. 1991. The influence of the cassava mealybug, *Phenacoccus manihoti* Mat.-Ferr. (Hom., Pseudococcidae), on yield formation of cassava, *Manihot esculenta* Crantz. *J Appl Entomol* 111:155–165.

Shan Q, Lu J, Zheng Y et al. 2009. Purple sweet potato color ameliorates cognition deficits and attenuates oxidative damage and inflammation in aging mouse brain induced by D-galactose. *J Biomed Biotechnol* Article ID 564737 (doi:10.1155/2009/564737).

SHBL. 2010. Cerfeuil tubéreux, *Chaerophyllum bulbosum*. Société d'Horticulture du Bas-léon (SHBL). Accessed June 3, 2010. http://hortimail.over-blog.com/article-18363187.html

Shahidul Islam MD, Yoshimoto M, Yahara S, Okuno S, Ishiguro K, Yamakawa O. 2002. Identification and characterization of foliar polyphenolic composition in sweetpotato (*Ipomoea batatas*) genotypes. *J Agric Food Chem* 50:3718–3722.

Shindo A, Tahara M, Tashiro R, Ochiai T. 1999. Characterization of sweetpotato chromosomes by genomic in situ hybridization. *XVI International Botanical Congress*, August 1–7, 1999, St. Louis, MO, Poster No. 1851.

Shiotani I. 1988. Genomic structure and the gene flow in sweet potato and related species. In: *Exploration, Maintenance, and Utilization of Sweet Potato Genetic Resources*, pp. 61–74. Report of the *First Sweet Potato Planning Conference, Lima-Perú 1987*. Lima, Peru: Centro Internacional de la Papa.

Shuji Y, Matthews PJ (eds.). 2002. *Vegeculture in Eastern Asia and Oceania*. Osaka, Japan: Japan Center for Area Studies.

Sidwa-Gorycka M, Krolicka A, Orlita A, Malinski E, Golebiowski M, Kumirska J, Chromik A, Biskup E, Stepnowski P, Lojkowska E. 2009. Genetic transformation of *Ruta graveolens* L. by *Agrobacterium rhizogenes*: Hairy root cultures a promising approach for production of coumarins and furanocoumarins. *Plant Cell Tissue Organ Cult* 97:59–69.

Siemonsma JJ, Piluek K (eds.). 1993. *Plant Resources of South-East Asia No. 8. Vegetables*. Wageningen, the Netherlands: Pudoc Scientific.

Simmonds NW (ed.). 1976. *Evolution of Crop Plants*. London, UK: Longmans.

Simpson BB, Conner-Ogorzaly M. 1995. *Econ Bot. Plants in Our World*, 2nd edn. New York: McGraw-Hill.

Small E. 1978. A numerical taxonomic analysis of the *Daucus carota* complex. *Can J Bot* 56:248–276.

Snook ME, Data ES, Kays SJ. 1994. Characterization and quantification of hexadecyl, octadecyl, and eicosyl esters of p-coumaric acid in the vine and root latex of sweetpotato [*Ipomoea batatas* (L.) Lam.]. *J Agric Food Chem* 42:2589–2595.

Song K, Osborn TC, Williams PH. 1990. *Brassica* taxonomy based on nuclear restriction fragment length polymorphisms (RFLPs). 3. Genome relationships in *Brassica* and related genera and the origin of *B. oleracea* and *B. rapa* (syn. *campestris*). *Theor Appl Genet* 79:497–506.

Song K, Tang K, Osborn TC, Ping L. 1996. Genome variation and evolution of *Brassica* amphidiploids. *Acta Hortic* 407:35–44.

Sonoda S, Mori M, Nishiguchi M. 2009. Homology-dependent virus resistance in transgenic plants with the coat protein gene of sweet potato feathery mottle potyvirus: Target specificity and transgene methylation. *Phytopathology* 89:385–391.

Spalik A, Wojewodzka A, Downie SR. 2001. Delimitation of genera in Apiaceae with examples from Scandiceae Subtribe Scandicinae. *Edinburgh J Bot* 58:331–346.

Spalik K, Downie SR. 2006. The evolutionary history of *Sium* sensu lato (Apiaceae): Dispersal, vicariance, and domestication as inferred from ITS rDNA phylogeny. *Am J Bot* 93:747–761.

Spalik K, Downie SR. 2007. Intercontinental disjunctions in *Cryptotaenia* (Apiaceae, Oenantheae): An appraisal using molecular data. *J Biogeogr* 34:2039–2054.

Spalik K, Downie SR, Watson WF. 2009. Generic delimitations within the *Sium* alliance (Apiaceae tribe Oenantherae) inferred from cpDNA rps16-5′τρνK(uuu) and nrDNA ITS sequences. *Taxon* 58:735–748.

Spök A, Karner S. 2008. *Plant Molecular Farming. Opportunities and Challenges*. JRC Scientific and Technical Reports. Luxembourg: Office for Official Publications of the European Communities.

Srivastava S, Srivastava AK. 2007. Hairy root culture for mass-production of high-value secondary metabolites. *Crit Rev Biotechnol* 27:29–43.

Stannard J. 1966. Benedictus Crispus, an eighth century medical poet. *J Hist Med Allied Sci* 21:24–46.

Stokes GW. 1955. Seed development and failure in horseradish. *J Heredity* 46:15–21.

Steyermark JA. 1963. *Flora of Missouri*. Ames: Iowa State University Press.

Strother JL. 2005. Scolymus. In: *Flora of North American North of Mexico*, Vol. 19, Asteridae, part 6: Asteraceae, part 1. Editorial Committee, 220. New York: Oxford University Press.

Sturtevant EL. 1890. History of garden vegetables (continued from page 332). *Am Nat* 24: 629–646; 719–744. *Scorzonera* pp. 643–644; *Sium* pp. 719–720.

Suda I, Ishikawa F, Hatakeyama M, Miyawaki M, Kudo T, Hirano K, Ito A, Yamakawa O, Horiuchi S. 2008. Intake of purple sweetpotato beverage affects on serum hepatic biomarker levels of healthy adult men with borderline hepatitis. *Eur J Clin Nutr* 62:60–67.

Swartjes FA, Dirven-Van Breemen EM, Otte PF, Van Beelen P, Rikken MGJ, Tuinstra J, Spijker J, Lijzen JPA. 2007. Human health risks due to consumption of vegetables from contaminated sites. RIVM report 71170140. Bilthoven, the Netherlands: RIVM [National Institute for Public Health and the Environment].

Tao K-L, Williams JT, Van Sloten DH. 1989. Base-collections of crop genetic resources: Their future importance in a man-dominated world. *Environ Conserv* 16:311–316.

Tardío J, Pardo-de-Santayana M, Morales R. 2006. Ethnobotanical review of wild edible plants in Spain. *Bot J Linn Soc* 152:27–71.

Tate JE, Symonds VV, Doust AN, Buggs RJA, Mavrodiev E, Majure LC, Soltis PS, Soltis DE. 2009. Synthetic polyploids of *Tragopogon miscellus* and *T. mirus* (Asteraceae): 60 Years after Ownbey's discovery. *Am J Bot* 96:979–988.

Taylor IEP (ed.). 2007. *Genetically Engineered Crops. Interim Policies, Uncertain Legislation*. Binghamton, NY: Haworth Food & Agricultural Products Press.

Terrise A. 1990. Note historico-nomenclaturale sur le maceron (*Smyrnium olusatrum* L.). *Bulletin de Societe Botanique de Centre-Ouest, n.s.* 21:71–74.

Thornalley PJ. 2002. Isothiocyanates: Mechanism of cancer chemopreventive action. *Anti-Cancer Drugs* 13:331–338.

Tillyard EMW. 1913. Theseus, Sinis, and the Isthmian games. *J Hell Stud* 33:296–312.

Topping D. 2007. Cereal complex carbohydrates and their contribution to human health. *J Cereal Sci* 46:220–229.

Toshihiko T, Yoshida S. 2008. Effects of application of sucrose and cytokinin to roots on the formation of tuberous roots in sweetpotato (*Ipomoea batatas* (L.) Lam.). *Plant Root* 2:7–13.

Toxopeus H, Baas J. 2004. *Brassica rapa* L. [Internet] Record from Protabase. GJH Grubben, OA Denton, eds. PROTA (Plant Resources of Tropical Africa/Ressources végétales de l'Afrique tropicale). Wageningen, the Netherlands. Accessed May 16, 2010. http://database.prota.org/search.htm

Toxopeus H, Oost EH. 1985. A cultivar group classification of *Brassica rapa* L. *Crucifer Newsl* 10:6–7.

Toxopeus H, Oost EH, Reuling G. 1984. Current aspects of the taxonomy of cultivated *Brassica* species. *Crucifer Newsl* 9:55–58.

Toxopeus H, Oost EH, Yamagishi H, Prescott-Allen R. 1988. Cultivar group classification of *Brassica rapa* L.: Update 1988. *Crucifer Newsl* 13:9–11.

Trèche S. 1996. Tropical root and tuber crops as human staple food. In: *Congresso latino americano de raizes tropicais. Montpellier: ORSTOM 1996*, 24 p. Congresso Latino Americano de Raizes Tropicais, 1. São Pedro (BRA) 1996/10/07–10. http://www.documentation.ird.fr/fdi/notice.php?ninv = fdi:010009053

Triplett BA, Moss SC, Bland JM, Bowd MK. 2008. Induction of hairy root cultures from *Gossypium hirsutum* and related *Gossypium barbadense* to produce gossypol and related compounds. *In vitro Cell Dev Biol Plant* 44:508–517.

Turner AH. 2009. *Urban Agriculture and Soil Contamination: An Introduction to Urban Gardening. Practice Guide #25*. Louisville, KY: Center for Environmental Policy and Management, University of Louisville. http://cepm.louisville.edu/Pubs_WPapers/practiceguides/PG25.pdf

Uberoid SK. 1991. Cholesterol lowering effect of chicory (*Cichorium intybus*) root in facceine-fed rats. *J Med Sci Res* 19:643–644.

Uewada T, Kiyota M, Kitaya Y, Aiga I. 1992. Hydroponic cultivation of sweetpotato. In: *Sweetpotato Technology for the 21st Century*, eds. WA Hill, CK Bonsi, PA Loretan, pp. 120–125. Tuskegee, AL: Tuskegee University.

Ulloa M, Stewart JMD, Garcia-CEA, Godoy-AS. 2006. Cotton genetic resources in the western states of Mexico: In situ conservation status and germplasm collection for ex situ preservation. *Genet Resour Crop Evol* 53:653–668.

Valiejo-Roman CM, Shneyer VS, Samiguillin TH, Terentieva EI, Pimenov MG. 2006. An attempt to clarify taxonomic relationships in "Verwandtschaftskreis der Gattung *Ligusticum*" (Umbelliferae-Apioideae) by molecular analysis. *Plant Syst Evltn* 257:25–43.

Van Beck TA, Maas P, King BM. 1990. Bitter sesquiterpene lactones from chicory roots. *J Agric Food Chem* 38:1035–1037.

van Heerwaarden J, Hellin J, Visser RF, van Eeuwijk FA. 2009. Estimating maize genetic erosion in modernized smallholder agriculture. *Theor Appl Genet* 119:875–888.

Van Jaarsveld PJ, Faber B, Tanumihardjo SA, Nestel P, Lombard CJ, Benade AJS. 2005. B-Carotene rich orange-fleshed sweet potato improves the vitamin A status of primary school children assessed with the modified-relative-dose-response test. *Am J Clin Nutr* 81:1080–1087.

Van Loon JC, Kieft B. 1980. In Löve A, ed. Chromosome number reports LXVIII. *Taxon* 29: 538–542 (cf. p. 540).

Vavilov NI. 1951. The origin, variation, immunity and breeding of cultivated plants (Translated by KS Chester). *Chron Bot* 13:1–366.

Vázquez FM. 2000. The genus *Scolymus* Tourn. ex L. (Asteraceae): Taxonomy and distribution. *An Jard Bot Madr* 58:83–100.

Velloso, JM da Conceição. 1829. *Flora Fluminensis*. Parisi: A. Senefelder.

Vietmeyer N. 1992. Forgotten roots of the Incas. In: *Chilies to Chocolate. Food the Americas Gave the World*, eds. N Foster, LS Cordell, pp. 95–104. Tucson, AZ: University of Arizona Press.

Villordon AQ, La Bonte DR, Firon N, Kfir Y, Pressman E, Schwartz A. 2009. Characterization of adventitious root development in sweetpotato. *HortScience* 44:651–655.

Vivek BS, Simon PW. 1998. Genetic relationships and diversity in carrot and other *Daucus* taxa based on nuclear restriction fragment length polymorphisms. *J Am Soc Hortic Sci* 123:1053–1057.

Vivek BS, Simon PW. 1999. Phylogeny and relationships in *Daucus* based on restriction fragment length polymorphisms (RFLPs) of the chloroplast and mitochondrial genomes. *Euphytica* 105:183–189.

Vogl-Lukasser B, Vogl CR, Reiner H. 2007. The turnip (*Brassica rapa* L. subsp. *rapa*) in Eastern Tyrol (Lienz district; Austria). *Ethnobot Res Appl* 5:305–317.

Von Erhardt W, Götz E, Bödeker N, Seybold S. 2010. *The Timber Press Dictionary of Plant Names*. Portland, OR: Timber Press, Inc.

Westerkamp C, Paul H. 1993. *Apios americana*, a fly-pollinated papilionaceous flower. *Plant Syst Evol* 187:135–144.

Whitaker TW. 1949. A note on the cytology and systematic relationship of the carrot. *Proc Am Soc Hortic Sci* 53:305–308.

Wilson LA, Wickham LD. 1992a. New perspectives on tuberization in sweetpotato. In: *Sweetpotato Technology for the 21st Century*, eds. WA Hill, CK Bonsi, PA Loretan, pp. 296–306. Tuskegee, AL: Tuskegee University.

Wilson LA, Wickham LD. 1992b. A rejoinder: Further perspective on tuberization. In: *Sweetpotato Technology for the 21st Century*, eds. WA Hill, CK Bonsi, PA Loretan, pp. 314–317. Tuskegee, AL: Tuskegee University.

Wilson WM, Dufour DL. 2002. Why "bitter" cassava? Productivity of "bitter" and "sweet" cassava in a Tukanoan Indian settlement in the northwest Amazon. *Econ Bot* 56:49–57.

Woods M. 2005. A revision of the North American species of *Apios* (Fabaceae). *Castanea* 70:85–100.

World Carrot Museum. 2010. http://www.carrotmuseum.co.uk/ (accessed July 05, 2010)

Yamane K, Lü N, Ohnishi O. 2005. Chloroplast DNA variations of cultivated radish and its wild relatives. *Plant Sci* 168:627–634.

Yen DE. 1974. *The Sweet Potato and Oceania*. Honolulu, HI: Bishop Museum Press.

Yoshimoto M, Okuno S, Yoshinana M, Yamakawa O. 1998. Antimutagenic activity of water extracts from sweetpotato. *Trop Agric* 75:308–313.

Yoshimoto M, Okuno S, Yoshinana M, Yamakawa O, Yamaguchi M, Yamada J. 1999. Antimutagenicity of sweetpotato (*Ipomoea batatas*) roots. *Biosci Biotechnol Biochem* 63:537–541.

Zhai DD, Jin HZ, Zhong JJ. 2010. A new sesquiterpene from hairy root cultures of *Artemisia annua*. *Chin Chem Lett* 21:590–592.

Zhang H, Chen F, Wang X, Yao H-Y. 2006. Evaluation of antioxidant activity of parsley (*Petroselinum crispum*) essential oil and identification of its antioxidant constituents. *Food Res Int* 39:833–839.

Zhang T, Oates C. 1999. Relationship between alpha-amylase degradation and physico-chemical properties of sweet potato starches. *Food Chem* 65:157–163.

Zhao J, Wang X, Deng B, Lou P, Wu J, Sun R, Xu Z, Vromans J, Koornneef M, Bonnema G. 2005. Genetic relationships within *Brassica rapa* as inferred from AFLP fingerprints. *Theor Appl Genet* 110:1301–1314.

Zhou J, Gong X, Downie SR, Peng H. 2009. Towards a more robust phylogeny of Chinese Apiaceae subfamily Apioideae: Additional evidence from nrDNA ITS and cpDNA intron (*rpl16* and *rps16*) sequences. *Molecular Phylogen Evol* 53:56–68.

Zhou T, Lou L, Yang G, Al-Shehbaz IA. 2001. *Armoriaca*. In Wu Zhengi-yi and RH Raven eds. *Flora of China*, Vol. 8, p. 86. St. Louis, MO: Missouri Botanical Garden Press.

Zohary D, Hopf M. 1993. *Domestication of Plants in the Old World*, 2nd edn. Oxford, U.K.: Clarendon Press.

Zohary D, Hopf M. 2000. *Domestication of Plants in the Old World*, 3rd edn. New York: Oxford University Press.

V

Root Response to Stress

31

Temperature Effects on Root Growth

Marc Faget
Forschungszentrum Jülich GmbH

Stephan Blossfeld
Forschungszentrum Jülich GmbH

Siegfried Jahnke
Forschungszentrum Jülich GmbH

Gregor Huber
Forschungszentrum Jülich GmbH

Ulrich Schurr
Forschungszentrum Jülich GmbH

Kerstin A. Nagel
Forschungszentrum Jülich GmbH

I. Introduction

Temperature is one of the major external inputs plants are exposed to from germination to senescence. Temperature variation has a strong influence on the establishment and function of plant root systems; furthermore, temperature can affect overall plant productivity. In the first edition of this book, Bowen (1991) covered the relationships of low and/or high temperatures on specific plant functions ranging from nutrient uptake and utilization to photosynthesis and carbon partitioning. In the second and third versions of this book, McMichael and Burke (1996, 2002) focused on optimal temperature for root growth, genetic diversity in the responses of plant roots to temperature, temperature stresses encountered in soil environments, changes in root metabolic activity, ultimate crop yields, and mycorrhizal associations. In this chapter, we update those topics as well as include root response to temperature gradients within the soil and particularly add a summary of the available noninvasive methods for studying temperature effects on root system development and function.

II. Dynamics of Soil Temperature

Most variations in soil temperature result from temperature variations at the surface. The main components of the balance between incoming and outgoing energy at the soil surface are daytime heating (mostly short-wave radiation from the sun but also long-wave radiation from the sky and from plant canopies), night-time cooling (long-wave radiation emitted from the surface), heat flow into the ground and air, and latent heat loss by evaporation (Marshall and Holmes 1996). Thus, heat is usually stored in the soil during daytime and released at night. The same mechanisms cause heat storage during spring and summer and release during autumn and winter. Soil temperature follows surface temperature with a time lag and progressive damping with depth. Any detailed modeling of soil temperature dynamics must involve liquid and gaseous water movement as well as heat transfer. In many cases, annual (Figure 31.1) as well as diurnal (Figure 31.2) fluctuations can be sufficiently approximated by a simple harmonic (sinusoidal) waveform, where the phase angle depends on soil characteristics and depth in the soil (West 1952; Fluker 1958; Penrod et al. 1960; de Vries 1975; Marshall and Holmes 1996; Hillel 2004; Nobel 2009). Decrease of temperature with depth (z) is typically modeled as an exponential function $e^{-z/d}$ governed by a damping depth, d. At a depth of $5d$, the temperature variations are reduced to less than 1% of the fluctuation at the surface, thus being almost damped out. The damping depth depends on the period of the temperature oscillation and the thermal properties of the soil (Hillel 2004; Nobel 2009). It increases with the thermal conductivity of the soil which in turn increases with bulk density and water content (Marshall and Holmes 1996). Diurnal fluctuations reach far less deeply into the soil than annual variations; due to the dependence on

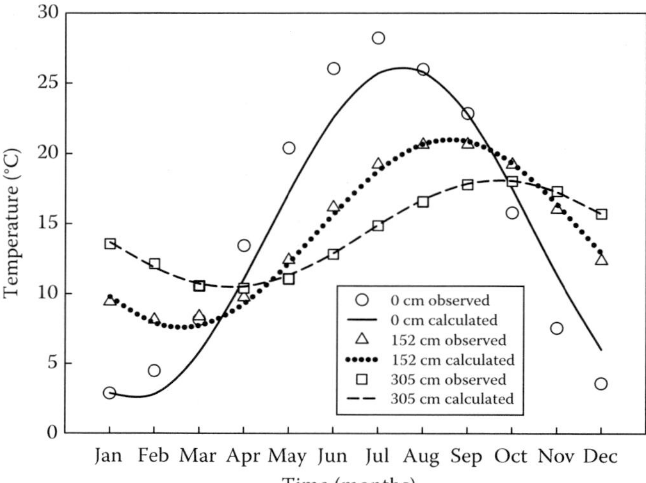

FIGURE 31.1 Monthly variation of surface and soil temperature at different depths for a 5-year norm. Calculation was done using a simple harmonic function. (From Penrod, E.B. et al., *Soil Sci.*, 90, 275, 1960.)

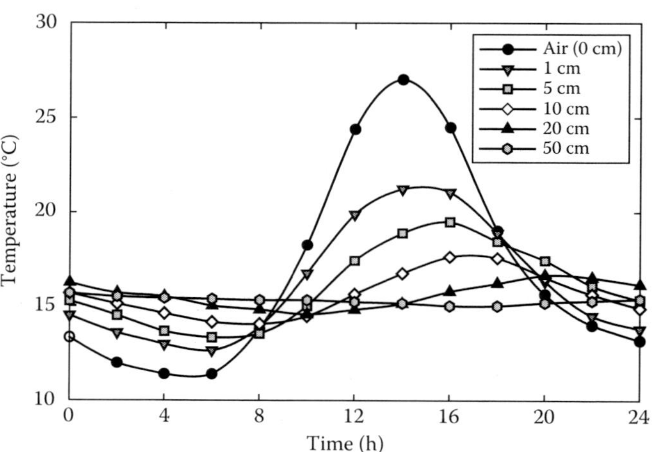

FIGURE 31.2 Diel cycle of air and soil temperature in different depths of a sandy cambisol. (From Walter, A. et al., *Annu. Rev. Plant Biol.*, 60, 279, 2009.)

oscillation period, the damping depth for diurnal fluctuations is typically more than one order of magnitude smaller than for annual variations (Nobel 2009).

There are many more factors affecting soil temperatures (Renaud et al. 2001; Baker and Baker 2002; Balland and Arp 2005; Saito et al. 2006; Bittelli et al. 2008; Kollet et al. 2009). Any factor directly influencing surface temperature, such as shading or soil cover, tends to have a strong impact on soil temperature (Paul et al. 2004).

On a global level, increases in soil temperature will follow climate changes. The Intergovernmental Panel on Climate Change (2007) reports that 29,000 observational data series from 75 studies show significant changes in many physical and biological systems, more than 89% as a response to global warming.

III. Strategies for Alteration of Soil Temperature

A. Alteration of Soil Temperature in the Field

Alteration of soil temperature in the field is a difficult task. The purpose is to create environmental conditions that match more favorably the characteristics of the plants. The choice of the approach to be taken depends on whether the goal is to cool or warm the soil. Some of the early work conducted with mulches showed that straw mulches, for example, tended to lower the soil temperature (Cooper 1973). Hartwig and Ammon (2002) showed that living mulches such as cover crops reduce the surface soil temperature as well as have an effect on soil erosion, nitrogen budget, and weed and pest control. Plastic mulches tend to raise soil temperature by changing the balance of long-wave radiation emitted from the surface and by reducing heat loss through evaporation and convection. The net effect of plastic mulches on the energy balance depends on the absorption properties of the plastic. Black mulches absorb short-wave radiation, and that energy is then transferred into the soil across the soil–mulch interface. For clear mulches, soil heating occurs by direct absorption of radiation at the soil surface. It was suggested that although the heat transfer mechanisms involved are different, the resulting soil heat fluxes may be similar (Ham and Kluitenberg 1994). Nevertheless, later studies show that different plastic mulches such as transparent, white, black, or white and black coextruded polyethylene have a significantly different effect on mean root temperatures (Baghour et al. 2002a,b,c) and yield (Ibarra-Jiménez et al. 2011).

Many studies which demonstrate the impact of plastic mulches on plant development have been performed. Mbagwu (1991), working with cassava plants, showed that yields were increased when the soil was covered by plastic mulches and that the magnitude of the increases was cultivar dependent; Tardieu and Pellerin (1991) showed in mulch and non-mulch experimentation that nodal root trajectory of field-grown maize is determined by soil temperature; Wien et al. (1993) also showed that root development in tomatoes was enhanced when the soil was covered by plastic mulches.

Other approaches to change the soil temperature include various tillage practices (Griffith et al. 1973; Gupta et al. 1982, 1983; Malhi and O'Sullivan 1990) and multicrop techniques such as varying row space. A suitable tillage method would increase the root zone temperature in spring in northern latitudes and decrease it at the onset of monsoons in tropical climates (Lal 1991).

B. Alteration of Soil Temperature in the Greenhouse and under Laboratory Conditions

Various techniques have been developed for studying the impact of soil temperature on plants or controlling plant performance in greenhouse practice. One technique to modify the root zone temperature in a more controlled environment is to switch on the greenhouse heating system before sowing and to suspend sheets of transparent polythene above the soil to trap solar energy (Monteith et al. 1983). Polystyrene tiles smeared with soil can be used to reduce the warming of the soil surface by the air (Gregory 1983).

It is of major importance to be able to control precisely the root zone temperature independently from the aboveground temperature under laboratory or greenhouse conditions. Water baths are often used to control the soil temperature of containers in scientific experiments (Marshall and Waring 1985; Engels and Marschner 1990; Engels et al. 1992), and they can be used in combination or individually to establish different temperatures in upper and lower soil levels, particularly when the goal is to create a temperature gradient more consistent with field conditions (Bryla et al. 1997; Sowinski et al. 1998). Fyfield and Gregory (1989) used an osmotic technique consisting of a semipermeable membrane sac containing a polyethylene glycol solution to study germination rate depending on temperature and water potential.

McMichael (1998) employed specially constructed cabinets that used forced air of a given temperature to control the temperature of roots growing in soil or polyvinylchloride tubes filled with nutrient solution independent of the aboveground temperature.

When plants are grown in a nutrient solution, it is possible to change the temperature of the medium by using a thermocirculator (Cumbus and Nye 1982) or to control thermostatically the desired temperature (Hurewitz and Janes 1983).

For plants grown in agar, soil, or solution, the use of temperature-controlled water or glycol circulating through heat exchangers in or surrounding the soil is also a good way to create parameters similar to field conditions (Gregory 1983; Bland et al. 1990; Greer et al. 2006; Nagel et al. 2009; Poire et al. 2010).

IV. Root Response to Various Temperatures

A. Response to Low Temperature

If the temperature of the soil differs significantly from the species-specific optimum (Table 31.1) (Blacklow 1972; Sattelmacher et al. 1990; Gladish and Rost 1993; Lyr and Garbe 1995; Seiler 1998; Wang and Camp 2000), then the structure and function of the root system can be altered. This might even result in a total loss of function or at least severe damage of the root system. The negative effects of root zone chilling can appear in a wide range of temperatures, from just above 0°C to higher temperatures of up to 20°C, depending on species-specific adaptation (Huang et al. 2005a). The response of plant roots to root zone chilling varies from one species to another and includes morphological and physiological responses (Figure 31.3). Root zone chilling is, of course, particularly important if tropical and subtropical plant

TABLE 31.1 Species-Specific Optimum Temperature for Root Growth

Plant Species	Optimum Temperature (°C)	Reference
Fragaria sp.	12–18	Wang and Camp (2000)
Pinus sylvestris	15	Lyr and Garbe (1995)
P. sativum	15	Gladish and Rost (1993)
Solanum tuberosum	15	Sattelmacher et al. (1990)
Tilia cordata	20	Lyr and Garbe (1995)
Fagus sylvatica	20	Lyr and Garbe (1995)
Quercus robur	25	Lyr and Garbe (1995)
Ipomoea batatas	25	Sattelmacher et al. (1990)
Helianthus sp.	25–30	Seiler (1998)
Zea mays	30	Blacklow (1972)
Manihot esculenta	35	Sattelmacher et al. (1990)

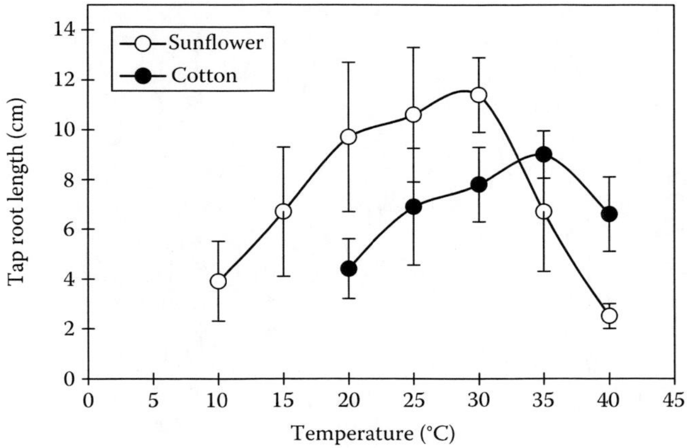

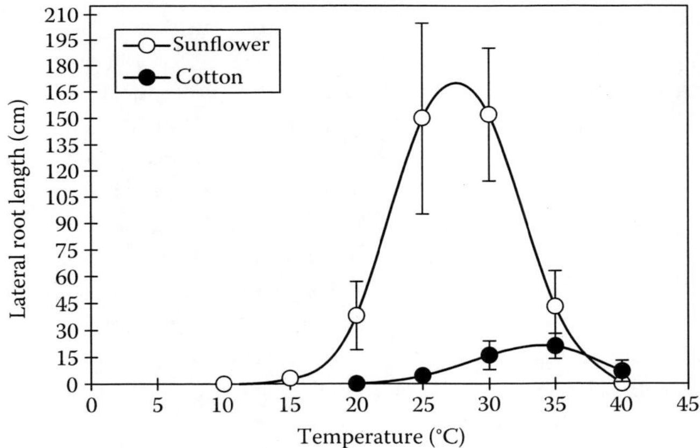

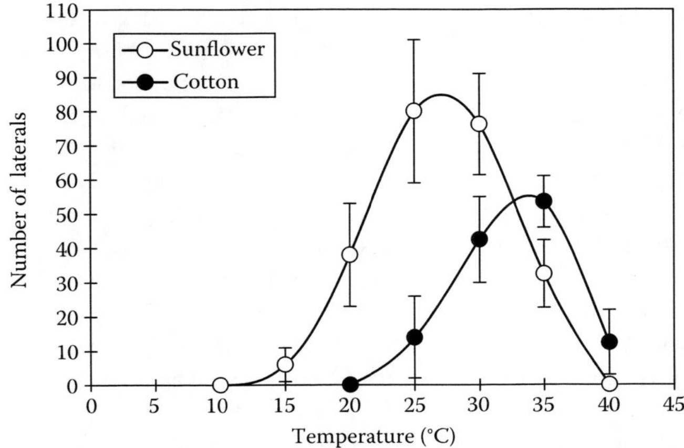

FIGURE 31.3 Development of root systems of 10-day-old cotton and sunflower seedlings as a function of temperature. (From McMichael, B.L. and Quisenberry, J.E., *Environ. Exp. Bot.*, 33, 53, 1993.)

species are cultivated in the temperate zone because they are generally not adapted to cooler temperatures (Allen and Ort 2001).

The most obvious morphological response to temperatures lower than the species-specific optimum is that the plants produce smaller root systems and roots of thinner diameters (Brouwer and Hoagland 1964; Pahlavanian and Silk 1988; Nagel et al. 2009). This phenotype is caused by a reduction of the relative elemental growth rates and cell elongation rates in the root growth zone. For example, maize roots showed a 40% reduction of growth rate at root zone temperatures of 16°C compared to 25°C (Nagel et al. 2009). Investigations of root systems via Magnetic Resonance Imaging (MRI) revealed that the biomass was reduced especially in the basal parts of the root systems as well as in the lateral roots (Nagel et al. 2009). Furthermore, root systems grown at low temperatures are less branched, leading to more compacted root systems and therefore to a reduction in the soil volume explored by the root system (Brouwer and Hoagland 1964; Cumbus and Nye 1982; Nagel et al. 2009). Of course, other factors such as the spatial distribution and availability of water and nutrients will also affect the branching of lateral roots and thus the compactness of a root system.

The most significant physiological response of plants subjected to chilling events is that the uptake of water and nutrients is reduced (Nielsen et al. 1960, 1974). This results from a decreased root hydraulic conductivity and a malfunction of the stomata in the leaves and appears to be the most important difference between chilling-sensitive and chilling-tolerant plant species (Aroca et al. 2001; Melkonian et al. 2004; Aroca et al. 2005). Aside from a decrease in water diffusion rates across the plasma membrane, the loss of aquaporin function causes a breakdown of the hydraulic conductivity in chilling-sensitive but not in chilling-tolerant plants (Bloom et al. 2004; Lee et al. 2005). Furthermore, the activity of the membrane-bound H^+-ATPase is crucial for an improved chilling tolerance. For example, Ahn et al. (1999) showed that H^+-ATPase degraded quickly in chilling-sensitive *Cucumis sativus* roots if a critical temperature threshold of 12°C was crossed; however, it increased constantly with decreasing temperature in chilling-tolerant *Cucurbita ficifolia* roots. Hence, as a major consequence of root zone chilling, the canopies of chilling-sensitive plant species dry out and eventually die.

The plasma membranes of root cells can also be directly affected by root zone chilling, that is, by reactive oxygen species (ROS) generated by the chilling events. For example, the formation of H_2O_2 reaches critical concentrations in chilling-sensitive plants, whereas chilling tolerant plants do not show high levels of H_2O_2 (Aroca et al. 2005; Zhang et al. 2007). A high content of ROS results in peroxidation of the membrane lipids and consequently in a decreased membrane fluidity and an increased membrane permeability. This can be the reason for increased electrolyte leakage from chilling-sensitive roots, which is often found in chilling-sensitive plant species (Ahn et al. 1999; Aroca et al. 2005; Popov et al. 2010).

Finally, chilling events also affect plant nutrition. For example, root zone chilling decreased the NH_4^+ and NO_3^- absorption by plant roots, affecting NO_3^- more strongly than NH_4^+ uptake (Clarkson et al. 1986; Smart and Bloom 1991; Clarkson et al. 1992; Bloom et al. 1998). Hence, low soil temperatures cause a preference for NH_4^+ over NO_3^- nutrition. However, the root zone temperature is not the only factor governing NO_3^- uptake during chilling events; internal nitrogen status is also a crucial parameter (Smart and Bloom 1991). Furthermore, the uptake of other nutrients, such as boron, can also be suppressed by root chilling and thus in a second step result in severe deficiency effects (Huang et al. 2005a).

B. Response to High Temperature

Global warming predicted by the IPCC (2007) for the next decades will affect soil temperature and consequently also plant productivity (Wang et al. 2010). Climate models indicate that the effect may be stronger in high latitudes, causing a shortening of the snow season, a speeding up of soil warming in spring, and consequently an increase in the length of the growing season (Dankers and Christensen 2005; Mellander et al. 2007). Generally, an increase in soil temperature is accompanied by an increase in root growth (Pahlavanian and Silk 1988; Misra 1999; Liu et al. 2011), but when root systems are exposed to temperatures that are higher than the optimum, plant roots generally show decreased elongation rates as a first response (Arndt 1937; Fortin and Poff 1991). However, due to species-specific adaptations, this critical temperature threshold varies as strongly as in the case of low temperature. For example, elongation rates of tomato roots were reduced at soil temperatures over 30°C, while elongation rates of white pine roots still increased at the same temperature range (Cooper 1973).

Similar to the responses of roots to low temperatures, heat-stressed roots can also show an increase in electrolyte leakage and lipid peroxidation, with stronger effects in heat-sensitive plants and when the plants are additionally subjected to drying soils (Liu and Huang 2000; Liu et al. 2002; Huang et al. 2005b). These responses, like those of cold-stressed roots, are linked to the formation of ROS.

Generally, the size of the root system is reduced by temperatures above the optimum. This size reduction is often due to a decreased number and length of lateral roots (Sattelmacher et al. 1990; Pardales et al. 1999). The duration of exposure to high temperatures also appears to have an impact on the growth and development of root systems. If the high temperature of the root zone is only temporary (i.e., for 1 day), then the lateral root growth of *Pisum sativum* is inhibited during the period of high temperature and restarts when the root zone temperature is favorable again (Gladish and Rost 1993). Furthermore, Eidsten and Gislerod (1986) showed that a relatively short (30 min) exposure to temperatures over 30°C retarded the growth of curly parsley roots and that a low daily average root zone temperature did not compensate for the damage caused by the short exposure to high temperatures (Figure 31.4).

Interestingly, in lettuce (*Lactuca sativa*), the negative effects of high temperatures (38°C) on root elongation can be reduced by inhibiting the biosynthesis of the phytohormone ethylene, whereas the application of ethylene precursors can induce high temperature reactions in lettuce roots even when they are growing at 20°C (Qin et al. 2007). This is due to the general functioning of ethylene in roots as a regulator of cell elongation

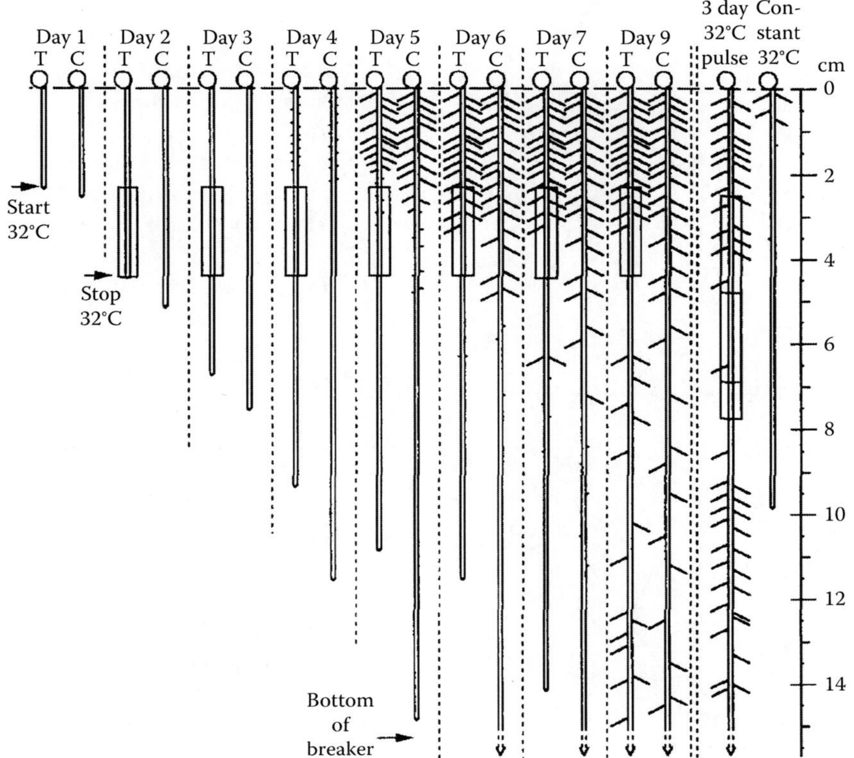

FIGURE 31.4 Schematic representation of 9 days of development of two pea roots, one grown at constant 25°C and one exposed to a 1-day 32°C pulse. Also depicted at the 9-day stage for comparison is a root exposed to a 3-day 32°C pulse and one grown at constant 32°C. Arrows indicate when 32°C treatment started and stopped and the location of the bottom of the container. T treatment plant (1 day 32°C pulse), C control plant (constant 25°C), box encloses length of primary root produced during 32°C exposure in 1 day's time. Three-day pulse example is shown with three boxes. (From Gladish, D.K. and Rost, T.L., *Environ. Exp. Bot.*, 33, 243, 1993.)

(Ruzicka et al. 2007). Other strong effects of phytohormones on high temperatures and heat shock responses by plant roots have been shown by applying synthetic cytokinin or abscisic acid (Bonham-Smith et al. 1988; Liu et al. 2002). Additional studies showed that high temperature can also promote an auxin-mediated hypocotyl elongation (Gray et al. 1998; Cheng et al. 2011); however, it seems to play no role in the developmental inhibition of *P. sativum* roots at high root zone temperatures (Gladish et al. 2000). Hence, the interplay between different phytohormones is crucial for the responses of root growth to high temperatures.

A strategy of plant roots to avoid temperatures above the species-specific optimum (Table 31.1) can be negative thermotropism. For example, when maize roots are subjected to temperatures that come close to the heat shock range, the gravitropic curvature can be modified or even reversed (Fortin and Poff 1991).

C. Response to Temperature Gradient versus Homogenous Soil Temperatures

Under natural conditions, plant roots are exposed to seasonal climate fluctuations which lead to temporal as well as spatial (depth) changes in soil temperature. Accordingly, roots normally grow in a vertical temperature gradient. It is well known that root growth as well as physiological and biochemical processes in the plant will be directly and/or indirectly affected by the root zone temperature. However, this knowledge has been gained from plant roots exposed to artificial uniform temperature regimes, and research on the influence of spatially heterogeneous soil temperatures on plant growth and performance is scarce. To simulate temperature gradients similar to field conditions, root zone temperature in labs or greenhouses must be controlled precisely; for example, one method uses tube systems circulating temperature-controlled water/glycol for cooling the bottom part of isolated pots, boxes, or columns (Bland et al. 1990; Sowinski et al. 1998; Nagel et al. 2009).

A vertical temperature gradient (20°C–10°C top to bottom) has a profound effect on the root system architecture of oilseed rape plants (Figure 31.5). The rooting depth is only slightly enlarged by the vertical gradient compared to roots exposed to a uniform root temperature of 15°C corresponding to the averaged root zone temperature of the gradient (Figure 31.5a). However, lateral root growth and especially the angle at which lateral roots branch from the primary roots increase in the gradient relative to the homogenous root zone temperature (Figure 31.5b,c). Consequently, these modifications of root system architecture

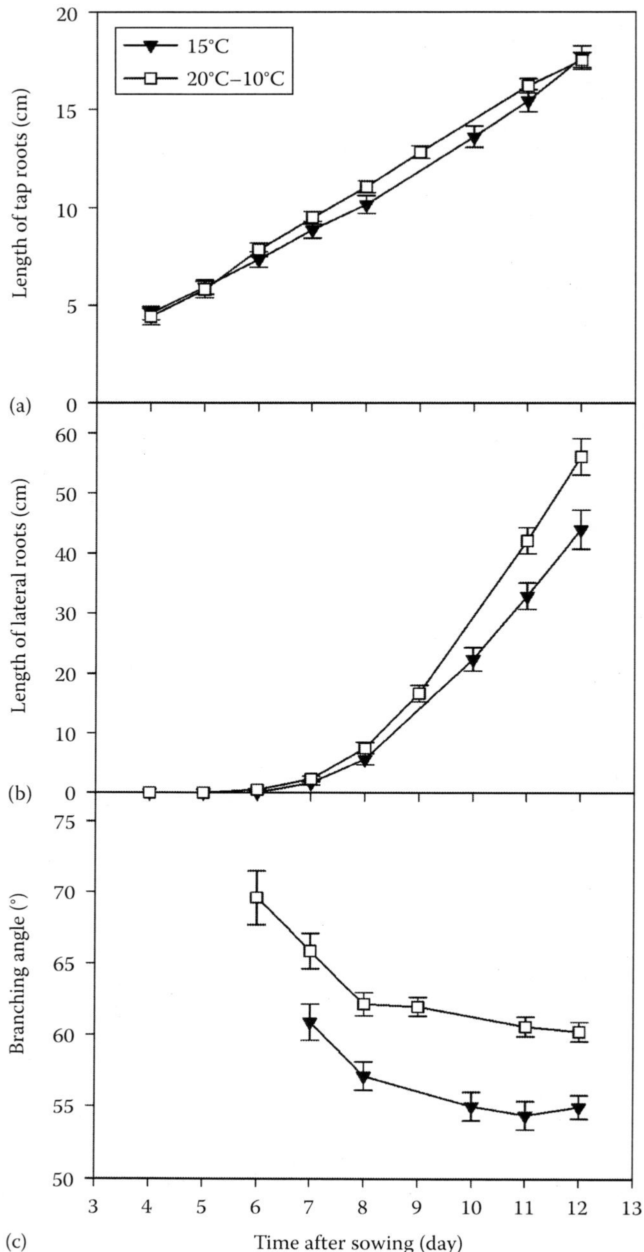

FIGURE 31.5 Effect of a vertical temperature gradient (20°C–10°C) versus a uniform root zone temperature (15°C) on root system architecture of oilseed rape plants: length of tap root (a) and lateral roots (b) and branching angle representing the averaged angle between the tap root and branched lateral roots (c). (Modified from Nagel, K.A. et al., *Funct. Plant Biol.*, 36, 947, 2009.)

lead to higher rooting densities and a larger exploited volume in regions with more favorable temperatures for root growth and nutrient uptake, which is in the top soil layers of the earlier mentioned vertical temperature gradient (Nagel et al. 2009). The ability of plants to adapt to a given vertical temperature gradient could increase the efficiency of the entire root system because the uptake and translocation of essential resources from root to shoot is strongly temperature dependent (Ching and Barber 1979; Engels and Marschner 1990; Ye et al. 2003). In the same way, the fraction of roots within a gradient exposed to suboptimal temperatures has an impact on assimilate transport from shoot to root (Sowinski et al. 1998). Thus, it is crucial for plant fitness and survival to place an adequate number of roots in upper soil layers especially during early growth in spring, when the air and topsoil are becoming warmer, while the subsoil temperature is still lower.

Because plants are naturally adapted to heterogeneous soil temperature conditions, a major source of discrepancies between findings from lab and field experiments might be the artificial homogeneous temperature regimes to which roots are mostly exposed during lab experiments. Therefore, one of the major current challenges of plant science is the simulation of field conditions under controlled lab or greenhouse conditions to improve the understanding of plant growth and performance in a changing environment.

V. Noninvasive Methods for Studying Temperature Effects on Root Growth

Roots growing in the soil have been traditionally studied by extracting them from soil samples of a defined volume (Box and Ramseur 1993; Samson and Sinclair 1994). Such destructive approaches do not yet allow researchers to address the geometry of a root system present in natural soil or to analyze dynamic processes, such as acclimation of root system architecture or growth to varying temperature conditions. To monitor the dynamics of root growth, noninvasive measurement techniques are required. In the field, observation of roots growing through structured soil has been achieved, for example, with a periscopic approach (Gregory 1979) or by using minirhizotron cameras (Williams and Weil 2004). The latter devices consist of transparent tubes inserted in the soil in which a small camera can be positioned manually to analyze root systems of multiple plants that randomly meet the walls of the minirhizotrons (see also Chapter 42). Drawbacks of this method are that only a limited part of the total root system can be observed; it is often difficult to distinguish roots from the soil background, and data evaluation requires extreme care. An alternative approach is plant cultivation in transparent soil columns or rhizoboxes (Thaler and Pagès 1996; Watt et al. 2006) or the cultivation of plant roots in a transparent medium, such as agar (Nagel et al. 2006). Software tools like GROWSCREEN-Root (Mühlich et al. 2008) or EZ-Rhizo (Armengaud et al. 2009) have been recently developed to quantify automatically the entire root system and branching patterns of roots.

The described optical methods are powerful tools but have some drawbacks because they are limited to 2D analysis, by which only part of a root system can be observed. Also, rhizotrons are generally tilted to direct root growth to the transparent plates at the lower side of the boxes and thus possibly affect root (and shoot) architecture. To identify 3D root system architecture under more natural conditions, noninvasive root imaging in soil under well-defined conditions is needed.

Methods previously used only for medical issues are now becoming applicable to plant studies, but studies using such advanced methods to analyze the effect of temperature on root growth are almost completely nonexistent. Neutron irradiation produced excellent images of a soybean root system when very flat (2 mm), sand-filled rhizotrons were used (Nakanishi et al. 2003). The authors also used x-ray computed tomography (CT) to image soybean roots growing in pots 3 cm in diameter. Although the quality of the images was affected by the water content of the sand, the authors found that water content of the substrate showed minimum values about 1 mm apart from the root surface. One can assume that this could be different at different soil temperatures, but it has not been studied yet. X-ray CT has been used also to analyze the growth velocity of potato tubers (Ferreira et al. 2010). Though the 3D geometry of a complete root system can be, in principle, analyzed with x-ray CT (Heeraman et al. 1997), due to attenuation problems, there is still a need to improve this technique, and the detection of fine roots particularly requires good methodology (Pierret et al. 2005).

High soil water content or soil inhomogeneities may affect also Nuclear Magnetic Resonance (NMR) imaging (MRI) of roots, but dedicated NMR analysis methods can help to improve signal-to-noise ratios allowing, for example, the imaging of roots growing in soil (Brown et al. 1990; Southon and Jones 1992). The relevance of substrate structure and density on gravitropic curvature of oat coleoptiles (and roots) was measured with MRI in soil or glass beads used as substrates (Antonsen et al. 1999). To visualize sugar beet or radish roots, or measure growth rates of maize roots in soil, MRI is a very versatile and appropriate method (Jahnke et al. 2009), and there is progress in detecting even thin wheat roots growing in loamy sand with MRI (Figure 31.6a).

Furthermore, noninvasive imaging of roots can be achieved by labeling plants with radioactive tracer molecules, in particular, positron-emitters such as ^{11}C, ^{13}N, or ^{15}O. This can be done by administering tracers directly to the roots or by measuring the transport of radioactive compounds delivered from radiolabeled source leaves to the roots. Emitted positrons may be detected by autoradiographic techniques allowing high spatial resolution as shown, for example, for uptake of ^{18}FDG by *Sorghum* roots (Hattori et al. 2008), but radioactivity is then measured only at the surface of root containing vessels because positrons have a range of no more than a few millimeters in plant or soil materials. However, the γ-radiation emitted during electron–positron annihilation can penetrate plant soil materials and can be quantitatively detected or imaged by using PET (positron emission tomography) technology. Thus far, in most studies using PET to visualize radiolabeled compound distribution and transport in roots, plants were grown hydroponically, which does not reflect root behavior under natural soil conditions; see, for example, uptake of $^{13}NH_4^+$- or ^{15}O-labeled water by rice roots (Kiyomiya et al. 2001) or iron (^{52}Fe) uptake of roots comparing Fe-sufficient and Fe-deficient barley plants (Tsukamoto et al. 2009). By combining MRI and PET, both structure and growth of maize roots growing in natural sand as well as transport of ^{11}C-labeled photoassimilates in the roots have been imaged and analyzed (Figure 31.6b,c).

To our knowledge, there is at this time only one report where MRI–PET imaging has been applied for a proof-of-principle study of temperature effects on root structure and transport of radiolabeled photoassimilates in the roots (Nagel et al. 2009). One can expect that combining the different techniques which are now available will lead to a better understanding of the mechanism of root structure and function in response to temperature conditions.

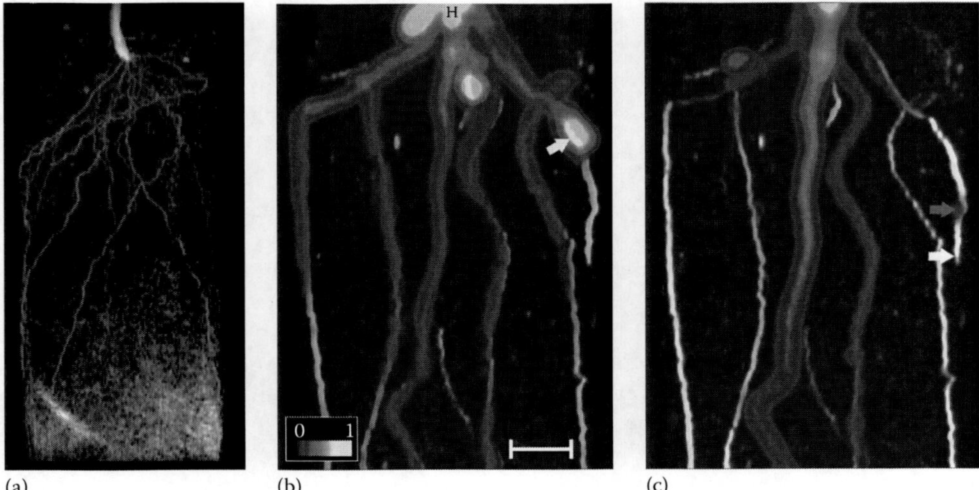

FIGURE 31.6 (See color insert.) Noninvasive imaging of roots growing in substrates. (a) Wheat roots growing in loamy sand visualized with MRI (van Dusschoten, unpublished). (b, c) Maize roots in sand coregistered with MRI (grey) and PET (colored). Growth was exemplarily analyzed on a root as shown on the right-hand side of panels (b) and (c). The root tip position was visualized on day 15 (yellow arrow), day 16 (red arrow), and day 17 (white arrow) after sowing. Scale bar 1 cm. (Modified from Jahnke, S. et al., *Plant J.*, 59, 634, 2009.)

VI. Conclusion/Outlook

Plant environments are subjected to diurnal and seasonal variations in temperature. The effects of temperature on root systems of different plant species differ and depend on the specific optimal conditions of each plant species. The major challenge for the future will be to establish or mimic natural field conditions under controlled lab or greenhouse conditions to extend our knowledge of plant growth and performance in dynamic environments. Combining different noninvasive phenotyping methods will enable us to study the responses of root system structure and function to dynamic changes in temperature regimes. This knowledge will facilitate plant breeding for better adapted crops with more efficient yield under extreme low or high temperatures.

References

Ahn SJ, Im YJ, Chung GC, Cho BH, Suh SR 1999 Physiological responses of grafted-cucumber leaves and rootstock roots affected by low root temperature. *Sci Hort* 81:397–408.

Allen DJ, Ort DR 2001 Impacts of chilling temperatures on photosynthesis in warm-climate plants. *Trends Plant Sci* 6:36–42.

Antonsen F, Johnsson A, Futsaether C, Krane J 1999 Nuclear magnetic resonance imaging in studies of gravitropism in soil mixtures. *New Phytol* 142:59–66.

Armengaud P, Zambaux K, Hills A et al. 2009 Ez-rhizo: Integrated software for the fast and accurate measurement of root system architecture. *Plant J* 57:945–956.

Arndt CH 1937 Water absorption in the cotton plant as affected by soil and water temperatures. *Plant Physiol* 12:703–720.

Aroca R, Amodeo G, Fernandez-Illescas S, Herman EM, Chaumont F, Chrispeels MJ 2005 The role of aquaporins and membrane damage in chilling and hydrogen peroxide induced changes in the hydraulic conductance of maize roots. *Plant Physiol* 137:341–353.

Aroca R, Tognoni F, Irigoyen JJ, Sanchez-Diaz M, Pardossi A 2001 Different root low temperature response of two maize genotypes differing in chilling sensitivity. *Plant Physiol Biochem* 39:1067–1073.

Baghour M, Moreno DA, Hernandez J, Castilla N, Romero L 2002a Influence of thermal regime of soil on the sulfur (S) and selenium (Se) concentration in potato plants. *J Environ Sci Health a-Toxic/Hazard Subst Environ Eng* 37:1075–1085.

Baghour M, Moreno DA, Villora G, Hernandez J, Castilla N, Romero L 2002b The influence of the root zone temperatures on the phytoextraction of boron and aluminium with potato plants growing in the field. *J Environ Sci Health a-Toxic/Hazard Subst Environ Eng* 37:939–953.

Baghour M, Moreno DA, Villora G et al. 2002c Root-zone temperature influences the distribution of Cu and Zn in potato-plant organs. *J Agric Food Chem* 50:140–146.

Baker JM, Baker DG 2002 Long-term ground heat flux and heat storage at a mid-latitude site. *Clim Change* 54:295–303.

Balland V, Arp PA 2005 Modeling soil thermal conductivities over a wide range of conditions. *J Environ Eng Sci* 4:549–558.

Bittelli M, Ventura F, Campbell GS, Snyder RL, Gallegati F, Pisa PR 2008 Coupling of heat, water vapor, and liquid water fluxes to compute evaporation in bare soils. *J Hydrol* 362:191–205.

Blacklow WM 1972 Influence of temperature on germination and elongation of the radicle and shoot of corn (*Zea mays* L.)1. *Crop Sci* 12:647–650.

Bland WL, Mesarch MA, Wolfe JE 1990 A controlled-temperature rhizotron. *Crop Sci* 30:1142–1145.

Bloom AJ, Randall LB, Meyerhof PA, St. Clair DA 1998 The chilling sensitivity of root ammonium influx in a cultivated and wild tomato. *Plant Cell Environ* 21:191–199.

Bloom AJ, Zwieniecki MA, Passioura JB, Randall LB, Holbrook NM, St. Clair DA 2004 Water relations under root chilling in a sensitive and tolerant tomato species. *Plant Cell Environ* 27:971–979.

Bonham-Smith PC, Kapoor M, Bewley JD 1988 Exogenous application of abscisic acid or triadimefon affects the recovery of *Zea mays* seedlings from heat shock. *Physiol Plant* 73:27–30.

Bowen GD 1991 Soil temperature, root growth and plant function. In: *Plant Roots: The Hidden Half*, eds. Y Waisel, A Eshel, U Kafkafi, pp. 309–330. New York: Marcel Dekker, Inc.

Box JE, Ramseur EL 1993 Minirhizotron wheat root data—Comparisons to soil core root data. *Agric J* 85:1058–1060.

Brouwer R, Hoagland A 1964 Responses of bean plants to root temperatures. Ii. Anatomical aspects. *Meded Inst biol Scheik Onderz Landb-Gewass* 236:23–31.

Brown JM, Kramer PJ, Cofer GP, Johnson GA 1990 Use of nuclear-magnetic resonance microscopy for noninvasive observations of root-soil water relations. *Theor Appl Climatol* 42:229–236.

Bryla DR, Bouma TJ, Eissenstat DM 1997 Root respiration in citrus acclimates to temperature and slows during drought. *Plant Cell Environ* 20:1411–1420.

Cheng N-H, Liu J-Z, Liu X et al. 2011 *Arabidopsis* monothiol glutaredoxin, AtGRXS17, is critical for temperature-dependent postembryonic growth and development via modulating auxin response. *J Biol Chem* 286:20398–20406.

Ching PC, Barber SA 1979 Evaluation of temperature effects on k uptake by corn. *Agric J* 71:1040–1044.

Clarkson DT, Hopper MJ, Jones LHP 1986 The effect of root temperature on the uptake of nitrogen and the relative size of the root system in lolium perenne. I. Solutions containing both NH_4^+ and NO_3^-. *Plant Cell Environ* 9:535–545.

Clarkson DT, Jones LHP, Purves JV 1992 Absorption of nitrate and ammonium ions by *Lolium perenne* from flowing solution cultures at low root temperatures. *Plant Cell Environ* 15:99–106.

Cooper AJ 1973 *Root Temperature and Plant Growth*. Slough, U.K.: Commonwealth Agricultural Bureaux.

Cumbus IP, Nye PH 1982 Root zone temperature effects on growth and nitrate absorption in rape (*Brassica napus* cv. Emerald). *J Exp Bot* 33:1138–1146.

Dankers R, Christensen OB 2005 Climate change impact on snow coverage, evaporation and river discharge in the sub-arctic tana basin, northern fennoscandia. *Clim Change* 69:367–392.

de Vries DA 1975 Heat transfer in soils. In: *Heat and Mass Transfer in the Biosphere*, eds. DA de Vries, NH Afgan, pp. 5–28. New York: Wiley.

Eidsten IM, Gislerod HR 1986 The effect of root temperature on growth of curled parsley. *Plant Soil* 92:23–28.

Engels C, Marschner H 1990 Effect of suboptimal root zone temperatures at varied nutrient supply and shoot meristem temperature on growth and nutrient concentrations in maize seedlings (*Zea mays* L.). *Plant Soil* 126:215–225.

Engels C, Munkle L, Marschner H 1992 Effect of root zone temperature and shoot demand on uptake and xylem transport of macronutrients in maize (*Zea mays* L.). *J Exp Bot* 43:537–547.

Ferreira SJ, Senning M, Sonnewald S, Kessling PM, Goldstein R, Sonnewald U 2010 Comparative transcriptome analysis coupled to X-ray CT reveals sucrose supply and growth velocity as major determinants of potato tuber starch biosynthesis. *BMC Genomics* 11.

Fluker BJ 1958 Soil temperatures. *Soil Sci* 86:35–46.

Fortin M-C, Poff KL 1991 Characterization of thermotropism in primary roots of maize: dependence on temperature and temperature gradient, and interaction with gravitropism. *Planta* 184:410–414.

Fyfield TP, Gregory PJ 1989 Effects of temperature and water potential on germination, radicle elongation and emergence of mung bean. *J Exp Bot* 40:667–674.

Gladish DK, Rost TL 1993 The effects of temperature on primary root growth dynamics and lateral root distribution in garden pea (*Pisum sativum* L., cv. "Alaska"). *Environ Exp Bot* 33:243–258.

Gladish DK, Sutter EG, Rost TL. 2000 The role of free indole-3-acetic acid (IAA) levels, iaa transport, and sucrose transport in the high temperature inhibition of primary root development in pea (P*isum sativum* L., cv. Alaska). *J Plant Gr Regul* 19:347–358.

Gray WM, Östin A, Sandberg G, Romano CP, Estelle M 1998 High temperature promotes auxin-mediated hypocotyl elongation in *Arabidopsis*. *Proc Natnl Acad Sci USA* 95:7197–7202.

Greer DH, Wunsche JN, Norling CL, Wiggins HN 2006 Root-zone temperatures affect phenology of bud break, flower cluster development, shoot extension growth and gas exchange of 'Braeburn' (*Malus domestica*) apple trees. *Tree Physiol* 26:105–111.

Gregory PJ 1979 Periscope method for observing root-growth and distribution in field soil. *J Exp Bot* 30:205–214.

Gregory PJ 1983 Response to temperature in a stand of pearl-millet (*Pennisetum typhoides*).3. Root development. *J Exp Bot* 34:744–756.

Griffith DR, Mannering JV, Galloway HM, Parsons SD, Richey CB 1973 Effect of eight tillage-planting systems on soil temperature, percent stand, plant growth, and yield of corn on five Indiana soils. *Agric J* 65:321–326.

Gupta SC, Larson WE, Linden DR 1983 Tillage and surface residue effects on soil upper boundary temperatures. *Soil Sci Soc Am J* 47:1212–1218.

Gupta SC, Radke JK, Larson WE, Shaffer MJ 1982 Predicting temperatures of bare- and residue-covered soils from daily maximum and minimum air temperatures. *Soil Sci Soc Am J* 46:372–376.

Ham JM, Kluitenberg GJ 1994 Modeling the effect of mulch optical-properties and mulch soil contact resistance on soil heating under plastic mulch culture. *Agric For Meteorol* 71:403–424.

Hartwig NL, Ammon HU 2002 Cover crops and living mulches. *Weed Sci* 50:688–699.

Hattori E, Uchida H, Harada N et al. 2008 Incorporation and translocation of 2-deoxy-2-[^{18}F]fluoro-d-glucose in *Sorghum bicolor* (L.) moench monitored using a planar positron imaging system. *Planta* 227:1181–1186.

Heeraman DA, Hopmans JW, Clausnitzer V 1997 Three dimensional imaging of plant roots in situ with X-ray computed tomography. *Plant Soil* 189:167–179.

Hillel D 2004 *Introduction to Environmental Soil Physics*. San Diego, CA: Academic Press.

Huang L, Ye Z, Bell RW, Dell B 2005a Boron nutrition and chilling tolerance of warm climate crop species. *Ann Bot* 96:755–767.

Huang X, Lakso AN, Eissenstat DM 2005b Interactive effects of soil temperature and moisture on concord grape root respiration. *J Exp Bot* 56:2651–2660.

Hurewitz J, Janes HW 1983 Effect of altering the root-zone temperature on growth, translocation, carbon exchange-rate, and leaf starch accumulation in the tomato. *Plant Physiol* 73:46–50.

Ibarra-Jiménez L, Lira-Saldivar RH, Valdez-Aguilar LA, Lozano-Del Río J 2011 Colored plastic mulches affect soil temperature and tuber production of potato. *Acta Agric Scan B Soil Plant Sci* 61:365–371.

IPCC 2007 *Climate change 2007: Contribution of Working Group I, II and III to the Fourth Assessment Report of the Intergovernmental Panel on Climate Change*. Cambridge, NY: Cambridge University Press.

Jahnke S, Menzel MI, van Dusschoten D et al. 2009 Combined MRI-PET dissects dynamic changes in plant structures and functions. *Plant J* 59:634–644.

Kiyomiya S, Nakanishi H, Uchida H et al. 2001 Real time visualization of ^{13}N-translocation inrrice under different environmental conditions using positron emitting tracer imaging system. *Plant Physiol* 125:1743–1753.

Kollet SJ, Cvijanovic I, Schuttemeyer D, Maxwell RM, Moene AF, Bayer P 2009 The influence of rain sensible heat and subsurface energy transport on the energy balance at the land surface. *Vadose Zone J* 8:846–857.

Lal R 1991 Tillage and agricultural sustainability. *Soil Till Res* 20:133–146.

Lee SH, Chung GC, Steudle E 2005 Gating of aquaporins by low temperature in roots of chilling-sensitive cucumber and chilling-tolerant figleaf gourd. *J Exp Bot* 56:985–995.

Liu Q, Yin H, Chen J et al. 2011 Belowground responses of *Picea asperata* seedlings to warming and nitrogen fertilization in the eastern Tibetan Plateau. *Ecol Res* 26:637–648.

Liu X, Huang B 2000 Heat stress injury in relation to membrane lipid peroxidation in creeping bentgrass. *Crop Sci* 40:503–510.

Liu X, Huang B, Banowetz G 2002 Cytokinin effects on creeping bentgrass responses to heat stress: I. Shoot and root growth. *Crop Sci* 42:457–465.

Lyr H, Garbe V 1995 Influence of root temperature on growth of *Pinus sylvestris, Fagus sylvatica, Tilia cordata* and *Quercus robur. Tree Struct Funct* 9:220–223.

Malhi SS, O'Sullivan PA 1990 Soil temperature, moisture and penetrometer resistance under zero and conventional tillage in central Alberta. *Soil Till Res* 17:167–172.

Marshall JD, Waring RH 1985 Predicting fine root production and turnover by monitoring root starch and soil-temperature. *Can J For Res* 15:791–800.

Marshall TJ, Holmes JW 1996 *Soil Physics*. New York: Cambridge University Press.

Mbagwu JSC 1991 Influence of different mulch materials on soil-temperature, soil-water content and yield of 3 cassava cultivars. *J Sci Food Agric* 54:569–577.

McMichael BL 1998 The influence of seed treatments on early growth in cotton under different environmental conditions. *Proc Beltwide Cotton Prod Res Conf*, National Cotton Council, Memphis, TN, Vol. 2, pp. 1410–1410.

McMichael BL, Burke JJ 1996 Temperature effects on root growth. In: *Plant Roots: The Hidden Half*, eds. Y Waisel, A Eshel, U Kafkafi, 2nd edn., pp. 383–396. New York: Marcel Dekker, Inc.

McMichael BL, Burke JJ 2002 Temperature effects on root growth. In: *Plant Roots: The Hidden Half*, eds. Y Waisel, A Eshel, U Kafkafi, 3rd edn., pp. 717–728. New York: Marcel Dekker, Inc.

McMichael BL, Quisenberry JE 1993 The impact of the soil environment on the growth of root systems. *Environ Exp Bot* 33:53–61.

Melkonian J, Yu L-X, Setter TL 2004 Chilling responses of maize (*Zea mays* L.) seedlings: Root hydraulic conductance, abscisic acid, and stomatal conductance. *J Exp Bot* 55:1751–1760.

Mellander P-E, Löfvenius M, Laudon H 2007 Climate change impact on snow and soil temperature in boreal scots pine stands. *Clim Change* 85:179–193.

Misra RK 1999 Root and shoot elongation of rhizotron-grown seedlings of *Eucalyptus nitens* and *Eucalyptus globulus* in relation to temperature. *Plant Soil* 206:37–46.

Monteith JL, Marshall B, Saffell RA et al. 1983 Environmental-control of a glasshouse suite for crop physiology. *J Exp Bot* 34:309–321.

Mühlich M, Truhn D, Nagel KA, Walter A, Scharr H, Aach T 2008 Measuring plant root growth. In: *Lecture Notes in Computer Science 5096*, ed. G Rigoll, pp. 497–506. Heidelberg, Germany: Springer.

Nagel KA, Kastenholz B, Jahnke S et al. 2009 Temperature responses of roots: Impact on growth, root system architecture and implications for phenotyping. *Funct Plant Biol* 36:947–959.

Nagel KA, Schurr U, Walter A 2006 Dynamics of root growth stimulation in *Nicotiana tabacuma* in increasing light intensity. *Plant Cell Environ* 29:1936–1945.

Nakanishi TM, Okuni Y, Furukawa J et al. 2003 Water movement in a plant sample by neutron beam analysis as well as positron emission tracer imaging system. *J Radioanalyt Nucl Chem* 255:149–153.

Nielsen KF 1974 Roots and root temperature. In: *The Plant Root and its Environment*, ed. EW Carson, pp. 293–335. Charlottesville,VA: University of Virginia Press.

Nielsen KF, Halstead RL, Maclean AJ, Holmes RM, Bourgest SJ 1960 The influence of soil temperature on the growth and mineral composition of oats. *Can J Soil Sci* 40:255–263.

Nobel PS 2009 *Physiochemical and Environmental Plant Physiol.* Oxford, U.K.: Academic Press.

Pahlavanian AM, Silk WK 1988 Effect of temperature on spatial and temporal aspects of growth in the primary maize root. *Plant Physiol* 87:529–532.

Pardales JRJ, Banoc DM, Yamauchi A, Iijima M, Kono Y 1999 Root system development of cassava and sweetpotato during early growth stage as affected by high root zone temperature. *Plant Prod Sci* 2:247–251.

Paul KI, Polglase PJ, Smethurst PJ, O'Connell AM, Carlyle CJ, Khanna PK 2004 Soil temperature under forests: A simple model for predicting soil temperature under a range of forest types. *Agric For Meteorol* 121:167–182.

Penrod EB, Elliott JM, Brown WK 1960 Soil temperature variation (1952–1956) at Lexington, Kentucky. *Soil Sci* 90:275–283.

Pierret A, Moran CJ, Doussan C 2005 Conventional detection methodology is limiting our ability to understand the roles and functions of fine roots. *New Phytol* 166:967–980.

Poire R, Schneider H, Thorpe MR, Kuhn AJ, Schurr U, Walter A 2010 Root cooling strongly affects diel leaf growth dynamics, water and carbohydrate relations in Ricinus communis. *Plant Cell Environ* 33:408–417.

Popov VN, Antipina OV, Trunova TI 2010 Lipid peroxidation during low-temperature adaptation of cold-sensitive tobacco leaves and roots. *Russ J Plant Physiol* 57:144–147.

Qin L, He J, Lee SK, Dodd IC 2007 An assessment of the role of ethylene in mediating lettuce (*Lactuca sativa*) root growth at high temperatures. *J Exp Bot* 58:3017–3024.

Renaud F, Scott HD, Brewer DW 2001 Soil temperature dynamics and heat transfer in a soil cropped to rice. *Soil Sci* 166:910–920.

Ruzicka K, Ljung K, Vanneste S et al. 2007 Ethylene regulates root growth through effects on auxin biosynthesis and transport-dependent auxin distribution. *Plant Cell* 19:2197–2212.

Saito H, Simunek J, Mohanty BP 2006 Numerical analysis of coupled water, vapor, and heat transport in the vadose zone. *Vadose Zone J* 5:784–800.

Samson BK, Sinclair TR 1994 Soil core and minirhizotron comparison for the determination of root length density. *Plant Soil* 161:225–232.

Sattelmacher B, Marschner H, Kühne R 1990 Effects of the temperature of the rooting zone on the growth and development of roots of potato (*Solanum tuberosum*). *Ann Bot* 65:27–36.

Seiler GJ 1998 Influence of temperature on primary and lateral root growth of sunflower seedlings. *Environ Exp Bot* 40:135–146.

Smart DR, Bloom AJ 1991 Influence of root NH_4^+ and NO_3^- content on the temperature response of net NH_4^+ and NO_3^- uptake in chilling sensitive and chilling resistant *Lycopersicon* taxa. *J Exp Bot* 42:331–338.

Southon TE, Jones RA 1992 NMR imaging of roots—Methods for reducing the soil signal and for obtaining a 3-dimensional description of the roots. *Physiol Plant* 86:322–328.

Sowinski P, Richner W, Soldati A, Stamp P 1998 Assimilate transport in maize (*Zea mays* L.) seedlings at vertical low temperature gradients in the root zone. *J Exp Bot* 49:747–752.

Tardieu F, Pellerin S 1991 Influence of soil-temperature during root appearance on the trajectory of nodal roots of field-grown maize. *Plant Soil* 131:207–214.

Thaler P, Pagès L 1996 Root apical diameter and root elongation rate of rubber seedlings (*Hevea brasiliensis*) show parallel responses to photoassimilate availability. *Physiol Plant* 97:365–371.

Tsukamoto T, Nakanishi H, Uchida H et al. 2009 ^{52}Fe translocation in barley as monitored by a positron-emitting tracer imaging system (PETIS): Evidence for the direct translocation of Fe from roots to young leaves via phloem. *Plant Cell Physiol* 50:48–57.

Walter A, Silk WK, Schurr U 2009 Environmental effects on spatial and temporal patterns of leaf and root growth. *Annu Rev Plant Biol* 60:279–304.

Wang J, Gao S, Lin J, Mu Y, Mu C 2010 Summer warming effects on biomass production and clonal growth of *Leymus chinensis. Crop Past Sci* 61:670–676.

Wang SY, Camp MJ 2000 Temperatures after bloom affect plant growth and fruit quality of strawberry. *Sci Hort* 85:183–199.

Watt M, Silk WK, Passioura JB 2006 Rates of root and organism growth, soil conditions, and temporal and spatial development of the rhizosphere. *Ann Bot* 97:839–855.

West ES 1952 A study of the annual soil temperature waves. *Aust J Sci Res* 5:303–314.

Wien HC, Minotti PL, Grubinger VP 1993 Polyethylene mulch stimulates early root-growth and nutrient-uptake of transplanted tomatoes. *J Am Soc Hort Sci* 118:207–211.

Williams SM, Weil RR 2004 Crop cover root channels may alleviate soil compaction effects on soybean crop. *Soil Sci Soc Am J* 68:1403–1409.

Ye ZQ, Huang LB, Bell RW, Dell B 2003 Low root zone temperature favours shoot B partitioning into young leaves of oilseed rape (*Brassica napus*). *Physiol Plant* 118:213–220.

Zhang Y, Zhang Y, Zhou Y, Yu J 2007 Adaptation of cucurbit species to changes in substrate temperature: Root growth, antioxidants, and peroxidation. *J Plant Biol* 50:527–532.

32

Flooding Tolerance Mechanisms in Roots

Kurt V. Fagerstedt
University of Helsinki

Olga B. Blokhina
University of Helsinki

Chiara Pucciariello
Sant'Anna School of Advanced Studies

Pierdomenico Perata
Sant'Anna School of Advanced Studies

I. Introduction

Flooding of native and agricultural lands is a major problem on Earth. On the whole world scale, the land area exposed to flooding is more than 17 million km^2 annually. Dramatic floods occur in all continents of our planet and result in an estimated damage of more than U.S. $80 billion annually (Global Register of Major Flood Events; www.dartmouth.edu/~floods/Archives/2005sum.htm). Furthermore, it has been predicted that the global climate change will lead into an increase in the frequency and in the severity of flooding events (Arnell and Liu 2001). Floods are changing the natural patterns of plant distribution and biodiversity (Silvertown et al. 1999), and as most of our food crop species (including some rice cultivars) are intolerant of flooding, waterlogging has a devastating impact on global food production (Normile 2008).

In this chapter, we are concentrating on the mechanisms that plants have, whether native plant species or agricultural cultivars, to overcome the problems inflicted on them by flooding of the soil. The adaptation mechanisms to the waterlogged environment include changes in the root morphology and anatomy as well as in the primary metabolism of root cells. Naturally, for all this to happen, plants need ingenious sensing, signaling, and execution systems to act appropriately in a changing environment. As these matters are not only of purely scientific interest but have a huge commercial background, the amount of publications in this field is enormous. In the following pages we have tried to express the latest views on the topics and collected the newest information with the expense of older articles no less important for the development of scientific knowledge.

II. Starting Point: Basic Anatomy of the Root

Plant roots have to serve two important functions: They anchor the plant to the soil and allow the uptake of important nutrients and water from the soil. To be able to fulfill these tasks, plant roots have to be able to grow into the soil to reach the nutrients and water, and to be able to do this, they need the energy provided by cellular respiration. For this they need a sufficient supply of oxygen that they must get from the air present in variable amounts in between the soil particles. Another problem may be the high CO_2 (>5 kPa) in waterlogged-flooded soils (Greenway et al. 2006).

Internal gas supply is another possibility, which plays an especially important role in the case of soil compaction or soil flooding by excess water, both of which hinder the diffusion of gases in the soil. During their evolution, some plant species have adapted well to soil flooding and this adaptation has been achieved by two different means: either by increasing the air spaces inside the roots thus allowing diffusion of gases or by changing their respiratory activity to reduce the consumption of oxygen while maintaining the supply of ATP for vital cellular maintenance.

In this chapter we will concentrate on the development of air spaces, i.e., aerenchyma in roots. There are several points that have to be taken into account: Is the aerenchyma formation induced or constitutive? How is the development regulated? What are the roles of plant hormones and reactive oxygen and nitrogen species (ROS and RNS) and free calcium ions? Which genes are acting in the regulation of the development? What kind of network is formed with all these physiological processes together? These are the points in aerenchyma development that we are concentrating on in the following pages. When talking about directions in the roots, we have adopted the latest fashion using the terms "rootward" and "shootward" (Baskin et al. 2010).

A. Basic Features of Root Anatomy: Where Is Aerenchyma Formed?

Plant root structure is a lot simpler than that of the shoot. This is mainly due to the fact that the shoot has a complicated branching pattern and bears leaves, flowers, and fruits, and often some additional structures such as thorns or hairs. Disregarding whether the plant is a dicotyledonous or monocot species, its root has a radially symmetric structure with the vascular tissues inside the root central cylinder. The outermost part of this cylinder is formed by the pericycle, a layer of cells from which the lateral roots will eventually emerge. Inside this cylinder we will find the actual vascular tissues, xylem, and phloem, also organized in a radial symmetric fashion: In monocotyledonous plants, there are normally several bundles of xylem and phloem; hence, the root is called polyarch, while in dicots and gymnosperms, the number of bundles is less than 8 leading to oligarch organization of the central cylinder.

The tissue outside the central cylinder is called the root cortex. The innermost layer of this is the endodermis, with its more or less pronounced cell wall thickenings and the Casparian band, which has a profound effect on the water and nutrient uptake by the root. The rest of the cortex is formed mainly of parenchymal tissue with small or larger schizogenous air-filled intercellular spaces where pectin in the middle lamella has been degraded. It is in this cortical tissue that aerenchyma is formed in plant roots. In wetland plants, cortical aerenchyma is formed constitutively, but it is even more pronounced in roots growing under oxygen-deprived conditions. Dryland plants do also develop root cortical aerenchyma in the case of a sudden flood or other circumstances where oxygen diffusion to the roots is hindered. Figure 32.1 shows cortical aerenchyma development in rice roots grown in stagnant nutrient solution culture. Rice roots develop aerenchyma also under aerated conditions, but it is only under oxygen deprivation that a barrier layer develops under the rhizodermis, which then prevents radial oxygen loss (ROL) from the roots into the medium. This barrier is discussed in Section II.C of this chapter.

B. Aerenchyma Formation and Programmed Cell Death Signaling

As root survival is of vital significance in all economically important cultivated plant species under flooding stress conditions, the anatomical structures in wetland plants that thrive in flooded habitats have interested man for long. Hence, the anatomical features of wetland plants with well-developed aerenchyma in the root cortex were described already at the end of the 1800s and at the beginning of the 1900s (Klinge 1879;

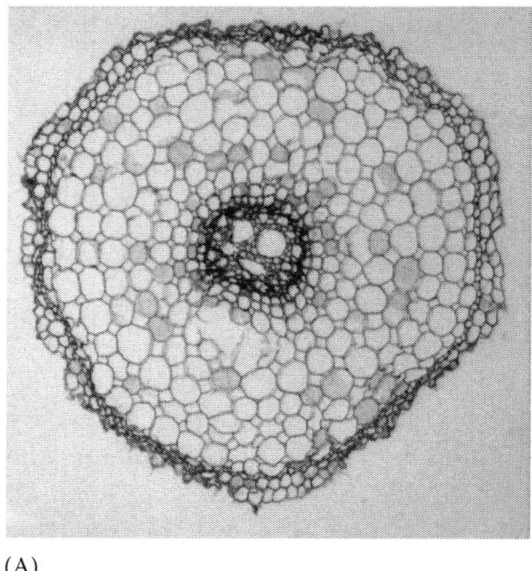

(A)

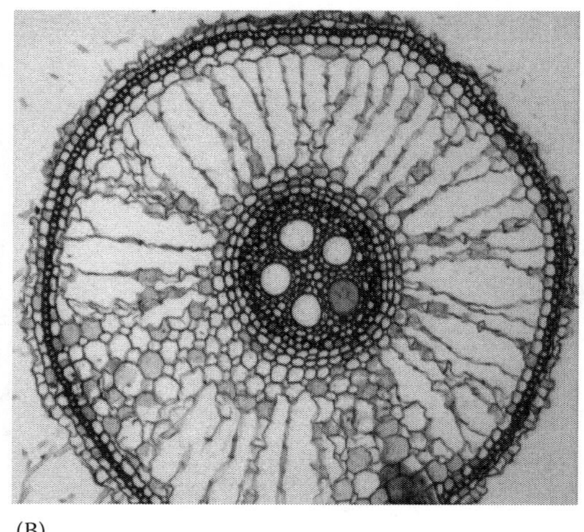

(B)

FIGURE 32.1 **(See color insert.)** Development of aerenchyma in rice roots grown under stagnant solution culture conditions. (A) Section of a young rice root. (B) Section of a fully developed rice root. Note the thick-walled exodermis spanning two layers of cells, protecting the root tissue from oxygen loss to the stagnant environment. (Photo courtesy of Kurt V. Fagerstedt.)

Plowman 1906; Kükenthal 1909; Wille 1926). During the twentieth century, the root anatomical structures were described in many wetland plants including the rather extreme developments in, i.e., *Scirpus* and *Carex* species (Crawford 1910; Kaul 1971; Fagerstedt 1992). In *Carex rostrata* Stokes, large portions of the root cortex die and the remaining tangential parts of the cell walls form sheetlike structures (Figure 32.2). In these roots, some radial rows of cortical cells still remain intact/viable connecting the rhizodermis and exodermis with the central cylinder allowing the passage of water and nutrients.

Aerenchyma can be formed by three very different processes: (1) It can develop when the intercellular spaces between cells enlarge through the breakdown of pectic substances in the middle lamellae. Such development is called schizogenous and can be seen, e.g., in *Rumex* species (Laan et al. 1989). This process may also involve PCD; (2) Another possibility is by programmed cell death, PCD, of cells in more or less large areas in the root cortex (Kawai et al. 1998). The air spaces formed by this kind of development are called lysigenous air spaces, the term meaning that cells have disintegrated during PCD; and (3) The third possibility is the development of expansigenous honeycomb aerenchyma, which is characterized by intercellular spaces that develop into lacunae through cell division and expansion. This type is typical for basal aquatic angiosperms (Nymphaeales) (Seago et al. 2005). All these events are developmentally controlled. In Figure 32.1 we can see an example of the type-two aerenchyma formation, where parenchymal cells in the rice root cortex gradually enlarge and then die during the formation of the radial air cavities in between the live cells, like the spokes of a bicycle wheel.

Even though the air spaces in roots, rhizomes, and stems take up a large part of the volume of the organs, the risk of water filling these spaces in the case of a wound does not seem to be great. There are two main factors preventing the water flowing in: The air cavities do not form continuous space through the different plant organs, but there are thin septa separating the root cavities from the shoot, and in the shoot, there are septa in each internode (Fagerstedt 1992). Another fact affecting the influx of water is the presence of hydrophobic compounds lining the walls of the air cavities (Woolley 1983).

What is the evidence we have that the death of cells in the cortical regions is due to the well-organized PCD? In plant cells, PCD is known to happen in a number of developmental events such as the classical development of tracheal elements in vascular tissues but more importantly during stresses, which are related to the availability of oxygen. These are the development of lysigenous air spaces in roots under mechanical pressure (He et al. 1996a), high-temperature stress, and nitrogen, phosphorus, and sulfur deficiencies (Konings and Verschuren 1980; Bouranis et al. 2003; Fan et al. 2003; Postma and Lynch 2011). How these are connected with oxygen is explained by the effects

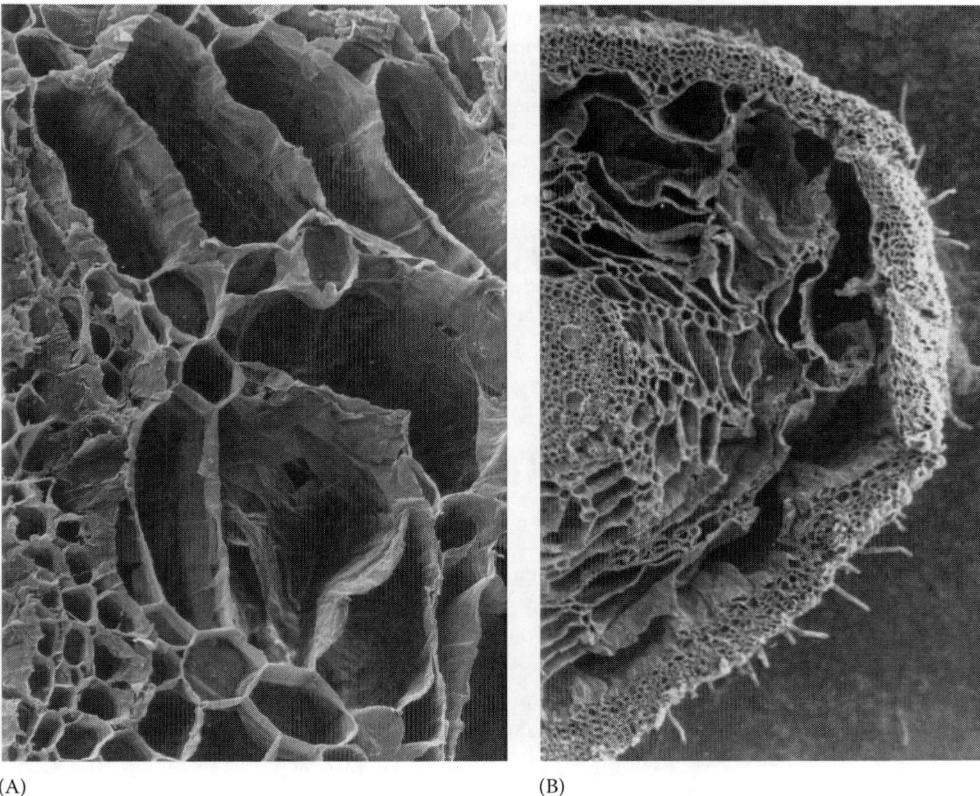

FIGURE 32.2 Scanning electron micrographs of a *C. rostrata* Stokes root showing the well-developed aerenchyma with a larger magnification (A) and the cortical region of the root shown in cross section (B). Note that the majority of the cortical cells have died and only their tangential cell walls remain thus creating large air-filled cavities. (Photo courtesy of Kurt V. Fagerstedt.)

on respiration/root surface area (or root volume) ratio: During nutrient starvation, the development of aerenchyma means less respiration and input of organic matter to a unit length of root, thus allowing for more root surface to scavenge nutrients in the soil (Fan et al. 2003, see also Chapter 26). However, there must be even more regulatory events involved as it has been measured that during phosphorus starvation, root respiration decreased by 70% while at the same time the amount of living cells in the roots decreased by 30% only (Fan et al. 2003).

Considering the cytological and molecular changes that take place during root PCD, we know that plant mitochondria have an important role at the beginning of the sequence of events. High mitochondrial matrix Ca^{2+} ion concentration, decreased inner membrane potential, and increased ROS level favor the opening of a mitochondrial permeability transition pore (PTP), which leads to the release of agents that initiate PCD (Virolainen et al. 2002). One of these agents known to act in plant cells is cytochrome *c*. As yet, we do not know nearly as many of the details in plant cells as are known in mammalian cell PCD, but we have some indications that many of the mammalian pro- and antiapoptotic proteins induce PCD in plant cells. For example, in tobacco leaves expressing the mammalian proapoptotic Bax protein, cell death was induced (Lacomme and Santa Cruz 1999; Mitsuhara et al. 1999). Several other key features of PCD have been shown in maize (*Zea mays* L.) root lysigenous aerenchyma formation including TUNEL (terminal deoxynucleotidyl transferase-mediated dUTP nick-end labeling) of nuclei and DNA laddering both indicating DNA degradation and cytoplasmic condensation (Gunawardena et al. 2001; Gunawardena 2008). In addition, characterization of caspase-specific protein fragmentation in apoptotic plant cells has been done, and caspase-3-like proteases have been detected in vivo, which imply their importance in the implementation of plant PCD (Chichkova et al. 2004; Zhang et al. 2009). Since then, several other PCD-related proteins such as vacuolar processing enzyme (VPE) have been found in plants (Zhang et al. 2010).

Lysigenous aerenchyma formation has been studied in detail and divided into the following steps (Bouranis et al. 2007): (1) the activation process during which the plant cell mitochondria play an important role, (2) the execution process when lytic enzymes are released to break down cellular structures, (3) the dissemination process during which the lytic cell cavity forms, and (4) the termination process during which the boundaries for the lytic cell cavity are formed.

The external stimuli initiating lysigenous aerenchyma formation includes a variety of factors such as low external oxygen concentrations (especially during flooding and soil compaction), increased ethylene concentration, and phosphate, nitrogen, or sulfur starvation. Naturally, increased ethylene can be formed in the root cells, but as ethylene is an easily permeable gas, it can also be generated in the soil by other plant or microbial material. Next we will have a look at all these matters individually.

Oxygen deprivation and its role in the initiation of aerenchyma formation have intrigued man for decades. It was only very recently that the oxygen sensing system was found in plant cells (Gibbs et al. 2011; Licausi et al. 2011). The plant system involves a ubiquitin-dependent N-end rule pathway for protein degradation. A similar system has been found earlier to function in mammals. Mammals also have another system, which is not present in plant cells and which involves a hypoxia-inducible factor-1 (Hif1-α, Bergeron et al. 1999). Also, quite recently a protein named the GRIM REAPER (GRI) that is sensitive to ROS has been found in plants, but it remains to be seen whether this plays a role in oxygen sensing in plants (Wrzaczek et al. 2010).

Once the signal has been perceived, whether it is low oxygen or mechanical pressure on the roots, the intracellular sequence of events are most probably rather similar. We know that the lowest oxygen concentrations are found inside the root just slightly shootward from the root tip (Gibbs et al. 1998). Low oxygen concentrations affect root metabolic events very quickly leading to the classical symptoms: drop in cytosolic pH, low adenylate energy charge (Fox et al. 1995), and also a general decline in protein synthesis. Only a small group of proteins are synthesized and they are related to sugar metabolism and glycolysis and to cell wall degradation (Subbaiah and Sachs 2003). We also know that within a few minutes, the low-oxygen signal leads to the opening of calcium channels, and hence, the cytosolic free Ca^{2+} ion concentration increases leading to the release of cytochrome *c* from plant mitochondria (Virolainen et al. 2002) and initiating PCD. Free calcium seems to be very important for aerenchyma development and we know that if calcium is bound (e.g., with EGTA) or if its release is hindered by ruthenium red, aerenchyma development is prevented in maize roots (He et al. 1996b; Drew et al. 2000). Does the rise in cytoplasmic free Ca^{2+} concentration follow that of H_2O_2? There are some indications that this may be the case (Rentel et al. 2004), but it is known that the Ca^{2+} rise induces the plasma membrane (PM) NADPH oxidase resulting in H_2O_2 production (Sagi and Fluhr 2001).

The role of ethylene in aerenchyma formation has been studied for a long time and we know already quite a lot of the details (e.g., in rice, see Fukao and Bailey-Serres 2008). Ethylene is not formed under totally anaerobic conditions because oxygen is needed in the last step catalyzed by 1-aminocyclopropane-1-carboxylate (ACC) oxidase. The sequence of events involving ethylene and leading into aerenchyma formation is as follows: First, under low oxygen concentrations, ACC synthase is induced (He et al. 1996a,b). Next, ACC oxidase catalyzes the formation of ethylene. As ethylene is formed in roots, it can immediately act in the root cortex and initiate PCD. As ethylene is an easily permeable gas, it may be important that a permeability barrier is formed under the rhizodermis preventing diffusion of ethylene out of the roots (Armstrong et al. 2000). This ROL barrier is discussed in Section II.C of this chapter.

Ethylene is then sensed by PM ethylene receptors, which act in a similar fashion as two-component histidine kinases (Bleecher and Kende 2000). These work together with a kinase (CTR1 in *Arabidopsis*, Kieber et al. 1993). Once ethylene is bound to the ETR2, the activation of CTR1 stops, and hence, EIN2 inactivation is released. This leads to activation of EIN3 and ERF1 transcription factors (TFs) (Solano et al. 1998). More recently another gene, LSD1, has been found (Mühlenbock et al. 2007), which operates together with EDS1 and PAD4 and is thought to prepare the cells to receive the ethylene signal. In addition, it has also been shown that MAP kinases (Hahn and Harter 2009) and G proteins (Steffens and Sauter 2010) play a role in ethylene signaling cascades to increase the signal strength as well as to activate ethylene biosynthesis.

Without going very much into the details of glycolysis and fermentation during hypoxia or anoxia (which are dealt in Section IV of this chapter), we can point out the metabolic facts that are needed for the initiation of aerenchyma development. First, we have seen that the mitochondrial PTP opening is an important event and it is not only caused by the increased concentration of free cytoplasmic Ca^{2+} ions. This is logical as the release of a proapoptotic signal from plant mitochondria is fatal to the cell and must not happen accidentally. Other conditions in the cytoplasm that favor the opening of the PTP are low adenylate energy charge, high Ca^{2+} in the mitochondrial matrix, and the presence of nitric oxide (NO) (Virolainen et al. 2002; Fagerstedt 2010).

ROS and RNS seem to play important parts on many levels in the initiation and proceeding of the formation of lysigenous air spaces. The PM NADPH oxidase produces superoxide radicals, which are quickly dismutated to H_2O_2 by the many superoxide dismutases (SODs) present in plant cells (Sang et al. 2001). Hence, they can take part in the signaling of low-oxygen conditions especially since it is known that NOX (NADPH oxidase) is indirectly activated by increased cytosolic concentration of Ca^{2+} ions (Keller et al. 1998). We also know that H_2O_2 levels are regulated (among other matters) by Rho-related small GTPases called ROPs (Agrawal et al. 2003; Bailey-Serres and Chang 2005).

Nitric oxide, on the other hand, is known to be produced in the same cellular sites as ROS (although the cellular processes for NO production are not all yet clear; for a review see Igamberdiev and Hill 2004) and can cause damage as well as the formation of nitrosothiols in proteins. It is tempting to hypothesize that these proteins are regulated by the formation of nitrosothiols as these same proteins are known to be modified by H_2O_2 (Lindermayr et al. 2005). While plant cells are protected against ROS by a wide array of antioxidants, they are protected against NO by hemoglobins (Hbs) (Igamberdiev et al. 2005). The role for NO in the development of aerenchyma is by no means clear as yet. It has been shown that alfalfa plants overexpressing class I Hb do not develop aerenchyma in their roots, while in the wild-type control plants, this did take place (Dordas et al. 2003). The picture is made more complicated by the fact that maize cell cultures with suppressed Hb expression produced more ethylene that the control plants (Manac'h-Little et al. 2005). In addition, it is known that nitric oxide (NO) can affect ethylene levels in plants through the inhibition of its synthesis (Lindermayr et al. 2006). *S*-nitrosylation of methionine adenosyltransferase, an enzyme responsible for the synthesis of ethylene precursor *S*-adenosyl methionine, represents an example of regulatory protein *S*-nitrosylation and supports the idea of antagonistic relationship between NO and ethylene (Arasimowicz and Floryszak-Wieczorek 2007).

The sequence of events starting the development of aerenchyma in roots is visualized in Figure 32.3, and the downstream

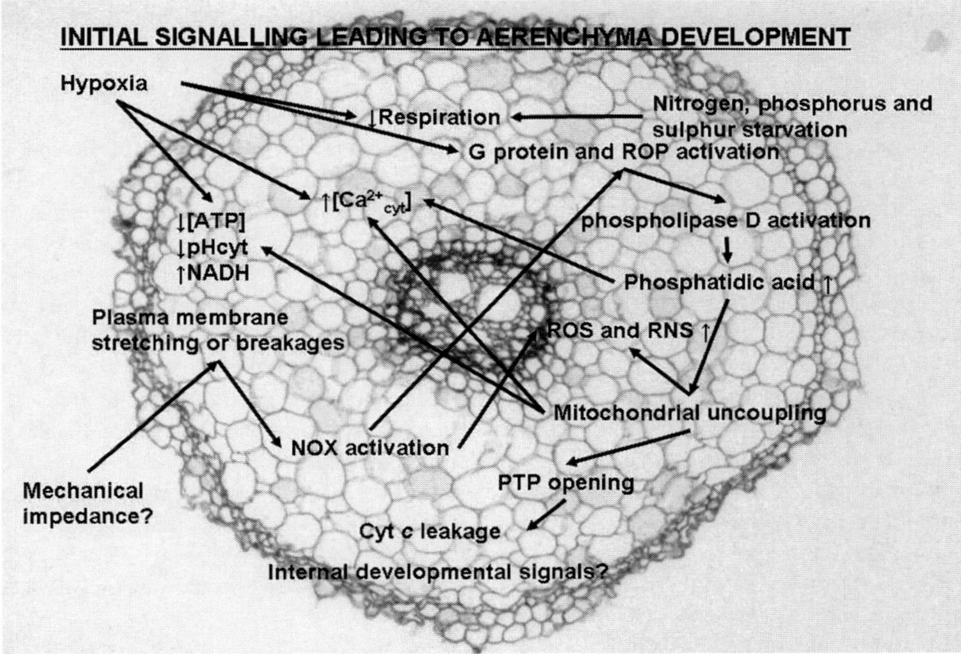

FIGURE 32.3 Initial signaling events of aerenchyma development and PCD immediately after induction by hypoxic environment. Background photomicrograph is of a cross section of a young rice root at the stage where cell death has not yet commenced. NOX, NADPH oxidase; RNS, reactive nitrogen species; ROS, reactive oxygen species; ROP, RHO-like GTPases of plants. (Reprinted with kind permission from Springer Science+Business Media: *Waterlogging Signaling and Tolerance in Plants*, Mancuso, S., Shabala, S., eds., Programmed cell death and aerenchyma formation under hypoxia, 2010, pp. 99–118, Fagerstedt, K.V.)

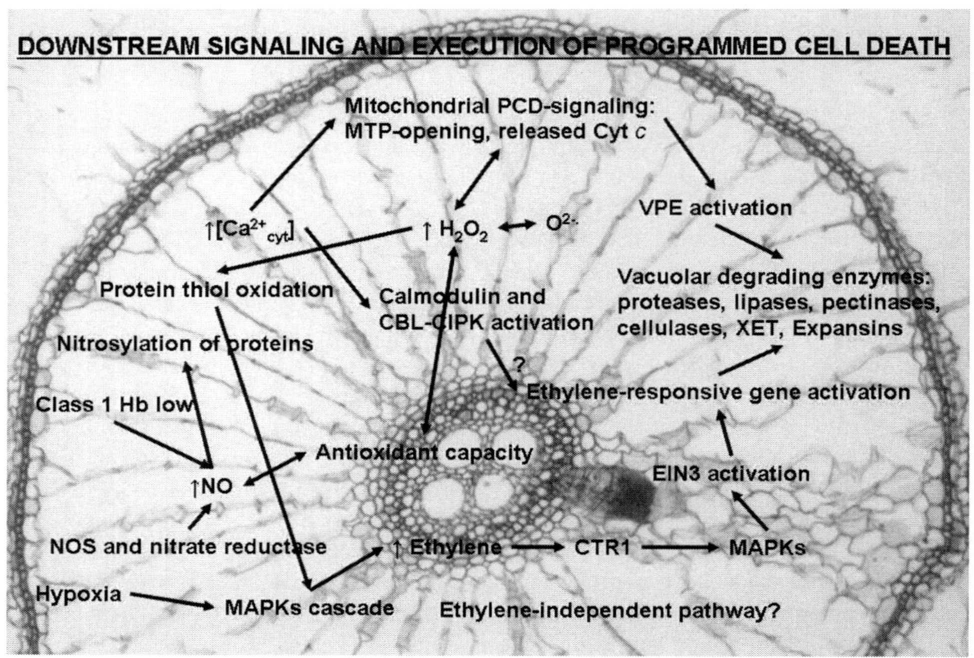

FIGURE 32.4 Downstream signaling leading to the execution of PCD during lysigenous aerenchyma formation. Several signals are in action simultaneously, and if certain thresholds are exceeded, the signal is passed on and eventually the vacuolar degrading enzymes are activated and released once the tonoplast ruptures. The ethylene-independent pathway, which seems to be in action in some wetland plants with constitutive aerenchyma formation, is not known yet. How certain cells are chosen to die and certain to survive is not known either. It may be that some cells are predestined to die and hence are more sensitive to the signals for PCD. CBL, calcineurin B-like proteins; CIPK, CBL-interacting protein kinases; CTR1, a negative regulator of the ethylene response pathway; EIN3, ethylene insensitive 3; MAPKs, mitogen-activated protein kinase; NO, nitric oxide; NOS, nitric oxide synthase; PCD, programmed cell death; XET, xyloglucan endotransglycosidase. (Reprinted with kind permission from Springer Science+Business Media: *Waterlogging Signaling and Tolerance in Plants*, Mancuso, S., Shabala, S., eds., Programmed cell death and aerenchyma formation under hypoxia, 2010, pp. 99–118, Fagerstedt, K.V.)

events have been collected in Figure 32.4. Once PCD has progressed, quite a few enzymes take part in the clearing of the remains of cells in the lysigenous air spaces. Once the vacuole has ruptured with concomitant drop in pH, hydrolytic enzymes are released and attack the remaining cellular organelles and DNA (Hara-Nishimura et al. 2005; Van Doorn and Woltering 2005). Proteins are degraded by the VPE, which is a caspase-like plant protease (Hatsugai et al. 2001). It is evident that there is a whole array of proteases at work during lysigenous aerenchyma formation. But so far we only have information indicating that in *Trifolium subterraneum* L. during hypoxia, the array of proteases changes and that there are specific lysis-related proteases at work (Aschi-Smiti et al. 2003). It has been shown that many cell wall-degrading enzymes are activated too. These include expansins, cellulases, xyloglucan endotransglycosidase (XET), and pectinases, many of which are induced by hypoxia and by ethylene (for more details see review by Jackson and Armstrong 1999).

During the recent decade, it has become clear that the schizogenous and lysigenous air space formation does not occur in the whole plant kingdom. In the basal angiosperms (the Nymphaeales and the Acorales), roots can develop expansigenous honeycomb aerenchyma. This type of aerenchyma develops by expansion of intercellular spaces into lacunae by cell division and cell expansion (Seago et al. 2005).

What are the determinants of the specific 3D structure of root aerenchyma in a given species? As in many instances concerning plant development, with perhaps the exceptions of flower and root tip development (e.g., Swarup et al. 2005; Péret et al. 2009), we do not have yet much information on the regulation of 3D development. In roots we can say that aerenchyma develops normally in midcortical regions, at least a few cell layers below the rhizodermis, and never in the vascular cylinder in the center of the root (Justin and Armstrong 1987). One of the determining facts could be the oxygen concentration inside the tissues, but detailed studies have shown that PCD is not initiated in the cells in the area with lowest oxygen concentrations (Evans 2003). Hence, it is difficult to say what determines the anatomical features of aerenchyma, but there are indications, such as the fact that aerenchyma is not initiated in places immediately outside the endodermis and pericycle where lateral roots emerge (Campbell and Drew 1983; Bouranis et al. 2006), that would imply cross talk with other developmental signals in addition to ethylene.

C. Development of the Radial Oxygen Loss Barrier

During the recent years, transport barriers in plant tissues formed out of cutinized, suberized, or otherwise waxy layers in plant cell walls have gathered new interest especially since we have realized how little we know of the placement of these compounds into specific sites in the cell walls (Schreiber 2010). In waterlogging-tolerant plant species, it has been noticed that a barrier layer develops under the rhizodermis, and this restricts the loss of oxygen from the roots into the surrounding medium. This barrier is called the ROL barrier (Armstrong et al. 2000; Visser et al. 2000; Colmer 2003; Shiono et al. 2011). In this area, the cell walls are impregnated with some substance that inhibits oxygen diffusion. Earlier it was shown that this barrier is formed out of suberized and/or lignified cell walls (Insalud et al. 2006; Kotula and Steudele 2009), but a recent and more detailed study has now indicated that this layer is probably developing earlier than the time the ROL hindrance can be measured (Shiono et al. 2011). In this investigation, deposition of some electron-dense substance occurs before suberin formation, coinciding with the loss of ROL (Shiono et al. 2011). However, it may be that in older roots, both layers are needed in order to keep the oxygen from diffusing out from the root tissues. In Figure 32.5, the formation of suberized cell walls inside of the rhizodermis can be seen stained with a fluorescent suberin dye (Fagerstedt, unpublished).

D. Root and Rhizome Ventilation during Waterlogging Enables Fast Exchange of Gases

From the text mentioned earlier, it has become clear that aerenchyma in the roots, rhizomes, and also aerial parts of the plant greatly affects diffusion of gases in and out of the roots. Nonetheless, it has been calculated and shown that diffusion alone cannot supply adequate oxygen for the needs of root cell respiration at any great distance. In pea, the distance was only 7–8 cm of root length (Armstrong et al. 1982, 1983), but we

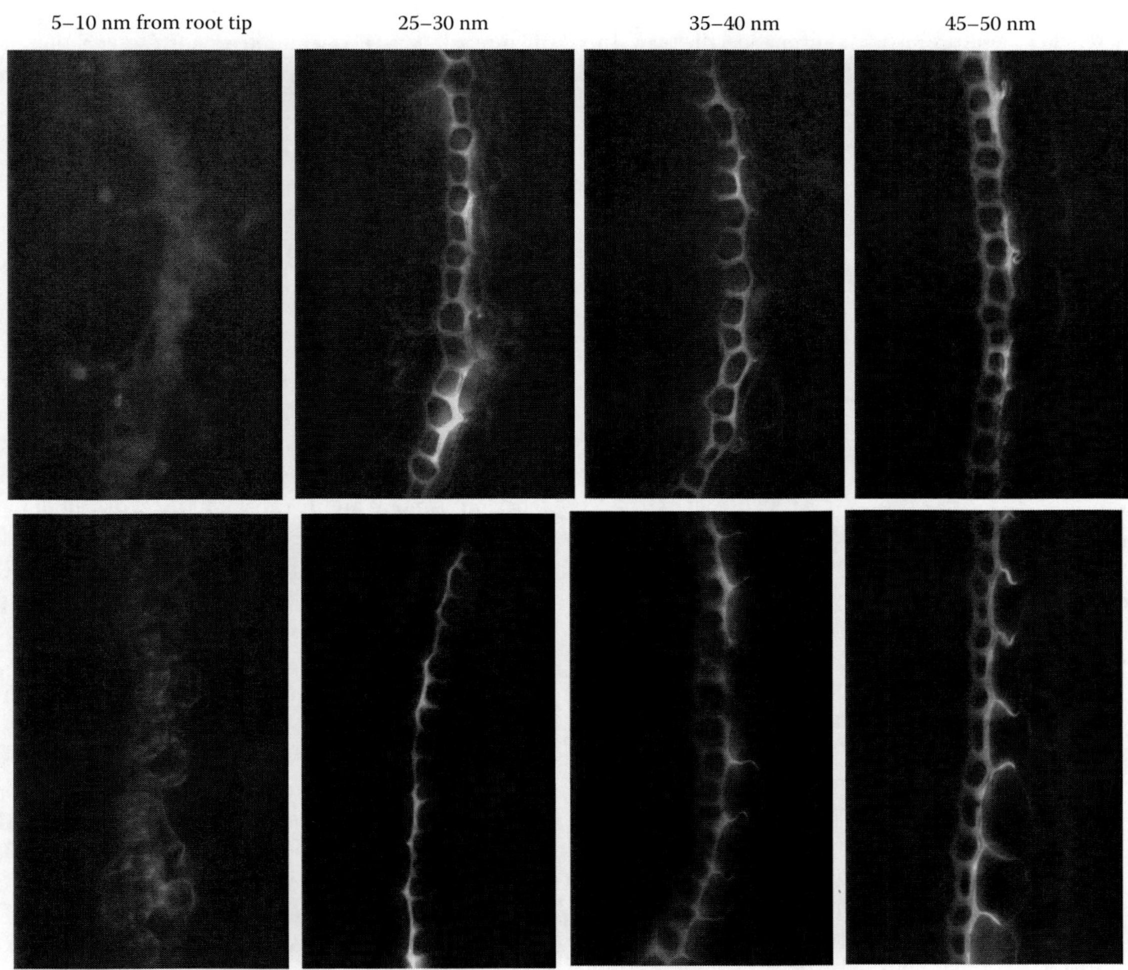

FIGURE 32.5 **(See color insert.)** Fluorescent microscope images of the exodermis of rice root segments (14 μm) taken at several distances from the tip of rice roots grown under aerobic (upper row) and stagnant (lower row) solution culture conditions and stained with berberin–aniline blue fluorescent dye (0.1% in water). The bright green fluorescence seen in the tangential cell walls and some of the radial cell walls indicates presence of suberin especially in roots grown under stagnant conditions. (Photo courtesy of Kurt V. Fagerstedt.)

know that in any larger wetland plants, such as *Phragmites australis* (Cav.) Trin. ex Steud., *Schoenoplectus lacustris* L., and *Iris pseudacorus* L., gas passages are measured in meters. This has directed the research efforts into various aspects of mass flow that enables much faster exchange of gases. It has been noticed that certain—but not all—wetland plants have convective flow of gases through the tissues (for a review, see Wegner 2010). There are several ways the convective flow can be achieved:

By humidity-induced pressurization, where the pressure needed is created by the water vapor pressure difference between the outside environment of the leaf and the inside compartment, where the air is saturated with water vapor. The pressure builds up in green tissues with stomata and the pressure drives mass flow of gases through rhizomes and out of, e.g., dead culms with open air spaces. A mathematical model created by Armstrong and coworkers (1996a) shows that this type of flow is greatly affected by relative air humidity and air temperature.

By thermal osmosis, where the energy for gas flow is created by the difference in temperature inside and outside the leaf. This idea was originally formulated theoretically and then tested experimentally (Grosse 1996).

By venturi-induced convection, where the pressure difference is created by wind passing over broken culms of, e.g., *P. australis*. This mechanism has been plausibly demonstrated by Armstrong and coworkers (1992, 1996b).

The mass flow of gases has been studied in great detail in many investigations mainly concerning aboveground stem and underground rhizome ventilation. Naturally, roots connected to the rhizome will benefit from the airflow that goes past their connection to the rhizome. The convective flow has been studied in a large array of plant species with the most exotic being various *Equisetum* species (Armstrong and Armstrong 2011).

E. Adventitious Root Formation

Adventitious roots are usually formed on stems or leaves in response to external stimuli, e.g., mechanical wounding or abiotic stress. It has been shown that they are formed in cereals under flooding stress and can improve the flooding tolerance of the plants probably since they grow near the soil surface where some oxygen is present (Fagerstedt and Crawford 1987). Unlike roots of embryonic origin, adventitious roots develop from non-meristematic tissue, from the phloem, or from the root pericycle (Wang and Pan 2006) and rely on complex signaling network for the induction.

Several investigations report the involvement of auxin and the two well-established hypoxic markers, NO and elevated Ca^{2+}, in adventitious root induction by phosphatidic acid. The pathway is active under normoxic (normal oxygen) conditions and relies on auxin-stimulated NO production and phospholipase D-derived phosphatidic acid accumulation (Lanteri et al. 2008) and is dependent on Ca^{2+} concentration (Lanteri et al. 2006). The requirement of NO accumulation for adventitious root formation has been further confirmed in arginase-deficient *Arabidopsis* mutants: When treated with an auxin analog, these plants produced more NO and exhibited enhanced growth of adventitious roots (Flores et al. 2008). Lipid-derived signals mediated by phospholipases and phosphatidic acid converge hormonal, ROS/NO signaling pathways and IP_3-dependent Ca^{2+} fluxes to regulate the polarized cell growth including adventitious root formation and root hair growth (Zhang et al. 2005).

Recent studies on posttranscriptional modification in root development have revealed implication of microRNA (miRNA) in the regulation of adventitious root formation and in stress response (Meng et al. 2010). Regulation based on miRNA represents a highly conserved pathway supporting cross talk between auxin signaling, nutritional status, and stress stimuli in plants. In *Arabidopsis*, miR167 induces adventitious root emergence and acts via targeting positive regulators, the auxin response factors ARF6 and ARF7, whereas miR160 targets ARF17—a negative regulator of adventitious root growth (Gutierrez et al. 2009). In turn ARF6, ARF7, and ARF17 have been shown to reciprocally regulate miRNA abundance and ARF17 can negatively affect its own regulator miR160 and has a positive effect on miR167, thus representing a complex cascade of feedback loop regulation in the process of adventitious root initiation (Gutierrez et al. 2009).

III. Uptake of H_2O and Minerals under Flooding

A. Aquaporins

Water is essential for virtually every metabolic reaction in living organisms. In higher plants, the root system effectively performs uptake of water and nutrients from the environment, redistributes it within the root (radial transport), and transports it to the shoot via xylem vessels (axial transport). The hydraulic conductivity of a root is a measure of membrane water permeability and relies on osmotic and hydrostatic forces (Kjellbom et al. 1999). A number of environmental factors such as oxygen deprivation, drought, nutrient deficiency, chilling, and high concentration of toxic ions can reduce the hydraulic conductivity of roots (Maurel et al. 2002; Bramley et al. 2007, and references therein). There are three distinct pathways for radial transport of water across the root to the stele: apoplastic (via cell walls and intercellular spaces), symplastic (via plasmodesmata), and transmembrane pathway (via aquaporins, APs) (Bramley et al. 2007). The symplastic and transmembrane water flow are mediated by a diverse group of APs, intrinsic membrane proteins (MIPs), which are able to transport water and other nonpolar molecules (Maurel et al. 2009). APs are ubiquitously expressed in plants and can constitute up to 20% of membrane proteins (Johansson et al. 1996).

Different groups of APs are localized at the PM (plasma membrane intrinsic proteins, PIPs), the tonoplast (tonoplast intrinsic proteins, TIPs), the endoplasmic reticulum (ER; small basic intrinsic proteins, SIPs), and the nodulin 26-like intrinsic proteins (NIPs, localized to internal membranes and the PM) (Maurel et al. 2002, 2009). Besides water, MIPs are able to transport a number of nonelectrolytes: CO_2, NH_3, H_2O_2, glycerol (aquaglyceroporins), boric acid, and urea (Weig and Jakob 2000; Flexas et al. 2006; Leshem et al. 2006; Tanaka et al. 2008; Ludewig and Dynowski 2009). Some of these specialized channels cannot transport water, as is the case with boron-transporting NIP in *Arabidopsis* (Tanaka et al. 2008). Interestingly, external H_2O_2 may exert a regulatory function on PIP channel activity, causing protein internalization (Maurel et al. 2009). The possibility exists, as has been shown for a mammalian cell culture system, that alongside with other signaling molecules, APs can also facilitate NO transport across biological membranes (Herrera et al. 2006). It is not yet clear how relevant the AP-dependent pathway is for signaling under stress conditions when NO accumulates. It has also been shown that NO can freely penetrate biological membranes (Lamattina et al. 2003; Pryor et al. 2006; Blokhina and Fagerstedt 2010b).

Another intrinsic property of plant APs relevant to both normal metabolism and hypoxia is the exclusion of protons. Due to small channel pore size, hydrophobicity, and a conserved critical arginine residue, plant APs are impermeable to protons and hence add to the preservation of pH gradients across PM and tonoplast (Ludewig and Dynowski 2009). Under hypoxia-induced cytoplasmic acidification, this would mean that no relief can come from functional APs to alleviate the pH decrease. However, AtNIP2;1 has been shown to contribute to lactic acid transport, preferably in the protonated form, in the roots of flooded *Arabidopsis* plants (Choi and Roberts 2007). Among nine *Arabidopsis* NIP genes, NIP 2;1 is extremely responsive to hypoxia: Its transcript level increases 70-fold in the roots after 1 h of waterlogging, and it is characterized by low permeability to water (Choi and Roberts 2007). A much stronger induction of NIP 2;1 is caused by total oxygen deprivation: Up to 300-fold accumulation of the corresponding transcript has been measured 2 h after the onset of anoxia. The authors consider this transport activity to be adaptive (Choi and Roberts 2007). In addition to lactic acid, there is a possibility that other small-size neutral molecules and weak acids, which accumulate under oxygen deprivation, e.g., alanine, acetaldehyde, acetic acid, γ-aminobutyric acid (GABA), and malic acid, may be transported via the same route (Bramley and Tyerman 2010).

The efficiency of water absorption by roots depends also on root anatomy and morphology of root system and on the abundance/activity of APs. Indeed, apoplastic radial water transport may be blocked by suberin layers in the exodermis and endodermis, e.g., as in anoxia-tolerant species with the O_2 impermeable barrier to prevent ROL (Colmer 2003; see Section II.C). The remaining possibilities, i.e., symplastic and transmembrane water transport, require that water molecules cross the PM and, in case of the transmembrane path, both the PM and the tonoplast (Bramley and Tyerman 2010). The efficiency of this route is affected by AP expression level and by AP gating. A number of physiological changes that occur under oxygen deprivation can regulate AP activity, cytoplasmic acidification being the most prominent factor. In *Arabidopsis* PIP, it has been shown that charged amino acids in the cytoplasmic loop (D loop, conserved histidine 197) are responsible for interaction with protons. The subsequent conformational change in the PIP protein results in pore closure and, hence, in reduced hydraulic conductivity of the roots under oxygen deprivation (Tournaire-Roux et al. 2003). Interestingly, changes in pH imposed from the apoplastic side of the PM did not affect PIP permeability, and APs of the tonoplast were insensitive to cytosolic pH changes (Tournaire-Roux et al. 2003). A hypothesis for the physiological role of anoxia-induced closure of APs and the resulting inhibition of water transport has been suggested: The reduced water flow through the root sustains accumulation of volatile ethylene, and in turn, the increased ethylene initiates a signaling cascade leading to aerenchyma formation and alleviation of oxygen deprivation (Holbrook and Zwieniecki 2003).

Free Ca^{2+} ions and phosphorylation/dephosphorylation of the pore protein can also affect PIP activity, and dephosphorylation favors pore opening (Tornroth-Horsefield et al. 2006; Bramley et al. 2007; Nyblom et al. 2009). The phosphorylation state of the protein does not change pH sensitivity of AP, as has been shown for tobacco PIP2;1 in yeast expression system (Fischer and Kaldenhoff 2008). The increase in cytosolic Ca^{2+}, characteristic for hypoxic conditions (Subbaiah et al. 1994a,b), may lead to AP activity inhibition through Ca^{2+} binding to a divalent cation binding site, resulting in conformational movement of loop D of the protein and subsequent channel blocking (Nyblom et al. 2009). Ca^{2+} may also affect the permeability of AP indirectly via activation of Ca^{2+}-dependent protein kinases and therefore sustain protein phosphorylation (Azad et al. 2004).

Alongside with posttranslational modification and regulation, AP activity is under transcriptional control. In a number of microarray studies on hypoxic response, APs have exhibited a complicated transcriptional pattern. A short-term hypoxic exposure (0.5–2 h) has resulted in rapid upregulation of NIP-type APs and tonoplast-localized TIPs (Klok et al. 2002; Branco-Price et al. 2005; Liu et al. 2005; Loreti et al. 2005; van Dongen et al. 2009). The latter may be needed in order to adjust to the cytoplasmic volume change that takes place due to enhanced water uptake from water-saturated apoplast during flooding (Tyerman et al. 2002; Bramley et al. 2007). Some of the repressed PIPs are predominantly expressed in roots (Bramley et al. 2007). On the whole, prolonged oxygen deprivation has been found to cause substantial downregulation of diverse AP types resulting in overall inhibition of water transport under oxygen deprivation.

B. Mineral Uptake and Ion Transport under Oxygen Shortage

Roots and the rhizosphere in the soil have a complex interaction: The soil provides the plant roots with vital nutrients and water, while the roots affect the rhizosphere with the diffusion of oxygen and carbon dioxide and with the many organic compounds leaking from the roots. Flooding of the soil disturbs this interaction in many ways and often leads to root death especially in flooding-intolerant plants. In the next few paragraphs, we will go through this interaction.

Waterlogging leads to multiple metabolic changes that affect nutrient uptake in a negative way through restricted energy supply for root growth and for transmembrane transport activities and through decreased transpiration and inhibited water transport via APs (Drew 1997; Bramley et al. 2007). Nutrient uptake rate in roots is regulated by many factors: by physical and chemical processes in the root and in the surrounding soil, by soil microorganisms, by the root system architecture, and by root metabolism (Marschner 1995; Elzenga and Veen 2010).

Changes in the rhizosphere imposed by waterlogging also have an impact on nutrient uptake: In the highly reduced waterlogged soils, the availability of ferrous (Fe^{2+}), phosphorus, manganese, and zinc increases because reduced forms of iron and manganese have higher solubility and, consequently, higher mobility in the soil (Marschner 1995; Plekhanova 2007). Hence, hypoxic plants may accumulate near-toxic levels of these ions (Plekhanova 2007; Elzenga and Veen 2010). The changes in the soil result from actions of anaerobic microorganisms, which in the absence of oxygen, use nitrate as terminal electron acceptor for respiration and reduce nitrate to nitrous oxides NO, N_2O, and N_2 in the process of denitrification (Marschner 1995). The depletion of nitrate in flooded soils is accompanied by conversion of organic nitrogen to ammonium (Ashraf and Habib-ur-Rehman 1999). During long-term flooding soil, doubling in NH_4^+ soil content has been observed (Ashraf and Habib-ur-Rehman 1999). The disproportion between nitrogen forms in waterlogged soils may be beneficial for hypoxic plants in terms of uptake energy costs. Nitrate uptake occurs via NO_3^-/H^+ symporter with subsequent H^+ export by proton-pumping ATPase consuming one ATP, while ammonia in the NH_3 form may passively cross the membrane, although it can add to metabolic acidification of the cytoplasm (Elzenga and Veen 2010). The increased ammonia content in hypoxic soil has been shown to exert a positive effect on root growth, as compared with nitrate. The roots formed when ammonia was the source of nitrogen appeared to be thinner and longer. This strategy has been considered as adaptive for nutrient acquisition (Elzenga and Veen 2010).

The reductive activity of microorganisms also results in the alkalinization of the soil under waterlogging. Indeed, the reduction of transition metals, such as iron, requires a proton to be absorbed from the environment. Secondly, many of the reduced compounds formed have alkaline properties: $NO_3^- \rightarrow N_2$, $Mn^{4+} \rightarrow Mn^{2+}$, $Fe^{3+} \rightarrow Fe^{2+}$, and $SO_4^{2-} \rightarrow S^{2-}$ (H_2S) (Marschner 1995; Plekhanova 2007). Besides, a complex relationship exists between different reduced species in the waterlogged soil: Reduction of iron in Fe(III)PO_4 (if present in the soil in substantial amounts) causes an increase in phosphate solubility and availability for plant roots, while reduction of sulfate to H_2S causes the formation of semisoluble sulfides of iron, zinc, and copper, therefore diminishing the pool available for uptake.

Cessation of root growth and decline in nutrient uptake are immediate responses to root hypoxia reflecting the energy-saving strategy of the flooded plant. In waterlogged soils, inhibition of root growth may be also caused by secondary microbial metabolites—monocarboxylic and phenolic acids (Pang et al. 2007). Several reasons can lead to hypoxia-induced inhibition of nutrient acquisition. First, the arrest of root growth decreases the interception of available soil nutrients (Marschner 1995). Secondly, the decrease in adenylate energy charge both affects root growth and the PM potential. The PM depolarization under hypoxia is well established and results from the inhibition of proton-pumping H^+-ATPases (Greenway and Gibbs 2003; Colmer and Greenway 2011). The outcome of impaired ATPase functioning is a decreased proton gradient across the PM and less negative PM potential. These effects, in turn, have an inhibitory effect on secondary membrane transport activities, including nutrient transport via symporters (Armstrong and Drew 2002; Elzenga and Veen 2010; Colmer and Greenway 2011). It has been suggested also that secondary metabolites produced by soil microorganisms under waterlogging such as formic, acetic, or propionic acid as well as phenolic acids can affect the net fluxes of K^+ and H^+ in a way that aggravates the effect of hypoxia (Pang et al. 2007). Treatment of barley roots with these compounds under normoxic conditions results in the depolarization of PM, increased K^+ efflux, and influx of H^+ (Pang et al. 2007).

Radial gradient of diminishing O_2 concentration that causes near-anoxic conditions in the stele is found in the roots exposed to low external O_2. Epidermal and cortex cells may experience moderate hypoxia, allowing for ion uptake, while the stele may experience severe hypoxia or anoxia. This results in differences in energy supply, in the activity of energy-dependent ion transporters, and in the degree of membrane depolarization between the cortex cells and the stele (Colmer and Greenway 2011). Indeed, the depletion of ATP in the xylem parenchymal cells will inhibit the H^+-ATPase function and reduce ion release into the xylem. The overall reduction in xylem ion loading may be exemplified by potassium fluxes. Under hypoxic conditions, K^+ accumulated by root epidermal and cortex cells may enter xylem parenchymal cells by diffusion via plasmodesmata (Colmer and Greenway 2011). Further K^+ loading through highly selective outward-rectifying K^+ SKOR channels in the xylem parenchymal cells will be repressed due to the channel closure under anoxic conditions. However, membrane depolarization will favor passive K^+ entrance into xylem (Pang and Shabala 2010). Additionally, K^+ and other ions may be released into the xylem through nonspecific outward-rectifying channels (NORC) (Pang and Shabala 2010). Thus, ion uptake under hypoxia should be considered as a two-step process: (1) uptake by epidermal and cortex cells and (2) xylem loading, with different regulatory factors affecting the efficiency (Colmer and Greenway 2011).

IV. Physiological Adaptations to Low O_2

Even though anatomical adaptations in roots and rhizomes play an important role in the survival of plants and tissues during low-oxygen conditions in both wetland and dryland plant species, there are instances such as total submergence, where anatomical adaptations are not enough. Still, we may find differences in tolerance between plant species. Such differences in sensitivity to flooding or anoxia are due to biochemical, not anatomical, features (Crawford 1978). In the following pages, we will concentrate on three main metabolic processes that are related to tolerance or intolerance of low-oxygen conditions: the consumption of carbohydrates, the action of plant mitochondria and cellular energetics, and the differences in accumulation of anoxia-specific metabolites. As cytoplasmic pH regulation is dealt with in Chapter 23, we do not deal with this topic here.

A. Starch Breakdown and Sugar Metabolism under Oxygen Deprivation

1. Carbohydrate Metabolism Is Strongly Affected by Low O_2

When O_2 levels fall below the requirements for aerobic respiration, the balance between carbon supply and demand depends on the complex control of the partitioning of assimilates into sucrose and starch. Under aerobic conditions, the fate of carbon assimilated through the Calvin cycle is split into two distinct pathways. A fraction is retained by the chloroplast for starch synthesis, and another fraction is exported to the cytosol and utilized for sucrose production. While sucrose provision is usually sufficient for immediate demand during the day, starch accumulation during the day sustains the metabolic requirements at the following night (Smith and Stitt 2007; Zeeman et al. 2007).

As soon as O_2 availability decreases, soluble sugars are rapidly channeled to the fermentative pathway for ATP generation (Figure 32.6), in an attempt to compensate for the lack of oxidative phosphorylation in the mitochondrial respiration (Perata and Alpi 1993). Early investigations of tolerant and sensitive plants under O_2 deprivation suggested a key role for carbohydrate metabolism in the adaptation to low O_2 (Guglielminetti et al. 1995a, 1997; Perata et al. 1996, 1998). An efficient use of the carbohydrate pool is a central trait in species that are well adapted

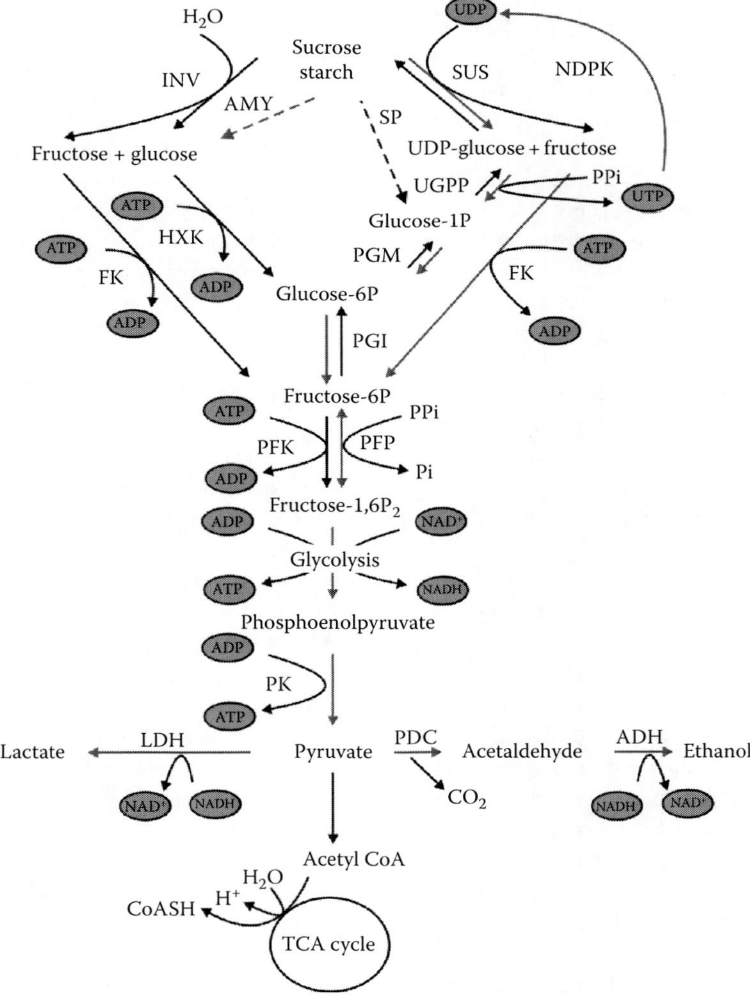

FIGURE 32.6 Multiple routes of sugar catabolism under O_2 deprivation. Grey arrows indicate reactions promoted during O_2 deprivation. ADH, alcohol dehydrogenase; AMY, amylases; FK, fructokinase; HXK, hexokinase; INV, invertase; NDPK, nucleoside diphosphate kinase; PFK, ATP-dependent phosphofructokinase; PDC, pyruvate decarboxylase; PGI, phosphoglucoisomerase; PGM, phosphoglucomutase; PK, pyruvate kinase; PFP, PPi-dependent phosphofructophosphatase; SP, starch Pase; SUS, sucrose synthase; UGPP, UDP-glucose pyrophosphatase. (Adapted from Bailey-Serres, J. and Voesenek, L.A.C.J., *Annu. Rev. Plant. Biol.*, 59, 313, 2008.)

to water submergence (Sachs and Vartapetian 2007). Tobacco roots overexpressing pyruvate decarboxylase (PDC) and therefore displaying higher ethanol production under anoxia showed the rapid use of carbohydrate reserves and premature cell death, while replenishment of carbohydrates through glucose feeding improved their survival (Tadege et al. 1998). However, root consumption of starch in the flooding-tolerant *Luffa cylindrica* Roem. and flooding-intolerant bitter melon (*Momordica charantia* L.) decreased rapidly in both species under flooding (Su et al. 1998). This happened despite increased sucrose, glucose, and fructose levels, suggesting that root sugar levels are not critical for the flooding tolerance of these species (Su et al. 1998).

With low O_2 concentrations, some plants accumulate sugars (Setter et al. 1987). Many species of *Poales* and *Asterales* accumulate intermediate-term storage carbohydrate pools such as fructans (Albrecht et al. 1993; Albrecht and Biemelt 1998). Flooding-tolerant *Iris* spp. display a high ability to store fructan as the main storage carbohydrate (Kubin 1992). Fructans have been found to accumulate in response to O_2 deficiency in both flooding-tolerant and flooding-intolerant species growing in flood-prone areas. However, there are higher absolute values and ratios between fructan and starch in the flooding-tolerant species (Albrecht et al. 1997a,b). In wheat roots, total carbohydrate content has been found to increase during hypoxia (Huang and Johnson 1995), and the increase is prominent in fructans (Albrecht et al. 2004). In contrast, anoxically treated wheat roots deplete all soluble carbohydrates and die within hours (Albrecht et al. 2004). Under low O_2 conditions, sugar and fructan accumulation occurs during the metabolic adjustment between respiration and fermentation toward a new balance (Albrecht et al. 2004). An investigation of prolonged hypoxia on the sugar uptake of tomato roots has revealed that sugar concentrations increased as well as the hexose uptake (Gharbi et al. 2009). Increased hexose transport has been found to be concomitant with the induction of the hexose transporter gene *LeHT2*, which is mainly expressed in sink organs (roots and green fruits) (Gear et al. 2000).

2. Starch Metabolism under Low O_2

Starchy seeds are often more tolerant to O_2 deprivation than fatty seeds (Raymond et al. 1985). This is a consequence of their ability to maintain a high energy metabolism even under anaerobiosis through starch catabolism, which does not require O_2 (Alpi and Beevers 1983). In cereal grains, starch stored in the endosperm represents a major reserve, which is hydrolyzed to soluble sugars during germination (Perata et al. 1992; Perata and Alpi 1993; Guglielminetti et al. 1995a,b; Loreti et al. 2003a). Soluble sugars are translocated to the embryonic axis to be used as carbon and energy sources for shoot and root growth. The ability to mobilize starch into soluble sugars under O_2 deprivation appears to be more important for tolerance than the availability of the carbohydrate reserve itself. Both potato tubers and *Acorus calamus* L. rhizomes have high carbohydrate reserves, of which only the latter are able to mobilize and thus showing a submergence-tolerant phenotype (Arpagaus and Braendle 2000).

Rice has the advantage over other cereals in harboring a complete set of starch-degrading enzymes such as α- and β-amylases, debranching enzyme and maltases that eventually function even under anoxia. Indeed, in germinating rice, α-amylases play a major role in degrading native granules of starch that provide sugars toward shoot and root (Murata et al. 1968; Dunn 1974; Sun and Henson 1991). However, root emergence is more sensitive than coleoptile to O_2 deficiency, and primary and lateral root growth is delayed under water (Kutschera et al. 1990; Lee et al. 2009). This suggests a preferential flux of sugars from endosperm toward the true green leaves under oxygen limitation. α-Amylases are endo-amylolytic enzymes that catalyze the hydrolysis of α-l,4-linked glucose polymers of starch in plants. Under O_2 deprivation, most of these enzymes are not produced in anoxia-intolerant cereals such as wheat, barley, oat, and rye. Consequently, they suffer from sugar starvation and eventually die when the oxygen availability is limited (Guglielminetti et al. 1995b; Perata et al. 1996). Indeed, while wheat fails to germinate under anoxia, addition of sugar induces germination and root protruding through the seed coat (Perata et al. 1992).

The rice genome encodes at least ten different isoforms of α-amylases that can be grouped into three subfamilies: Amyl (A–C), Amy2 (A), and Amy3 (A–F) (Rodriguez et al. 1992). *RAmy1A* is hormonally modulated by gibberellins (GAs) under aerobic conditions. During germination, GAs are synthesized in the embryo and diffuse through the starchy endosperm to reach the aleurone layer, where α-amylase gene induction takes place (Figure 32.7A; Fincher 1989). GAs also activate *RAmy1A* in the epithelium (Gubler et al. 1995). α-Amylases are then released into the endosperm, where they catalyze the hydrolysis of stored starch. *Ramy1A* has been detected in anaerobic samples (Perata et al. 1997); however, its induction is drastically slowed down by anoxia (Perata et al. 1993; Hwang et al. 1999). This amylase isoform is also present in rice roots (Thomas and Rodriguez 1994).

The major α-amylase isoform is encoded by *RAmy1A*, which may suggest that in the absence of GAs, rice germination is severely impaired. This assumption has been, however, dismissed by the evidence that the GA-deficient mutant *Tan-ginbozu* can readily germinate, in both air and anoxia (Loreti et al. 2003b). These results are explained by the presence of GA-independent α-amylase genes in rice, such as *RAmy3D* (Figure 32.7B), that have been shown to be present also in root tissues (Thomas and Rodriguez 1994). Indeed, *Ramy3D* expression has been observed in anoxic rice seedlings and, remarkably, is anoxia induced (Perata et al. 1997; Hwang et al. 1999). *Ramy3D* is not induced by GAs, since it does not possess the GA-responsive *cis*-acting element in its promoter region (Morita et al. 1998; Loreti et al. 2003b). Instead, *Ramy3D* is upregulated by sugar starvation, suggesting a direct link with diminished soluble sugar availability under anoxia, rather than a direct induction by anaerobiosis (Guglielminetti et al. 1995a; Perata et al. 1996; Loreti et al. 2003a).

In rice, sugar repression of *Ramy3D* and *Ramy3E* is controlled at a transcriptional level by a 100 bp sugar-responsive element (SRE) on the promoter region, containing three essential motifs, GC box, G box, and TA box, for the activation of the promoter

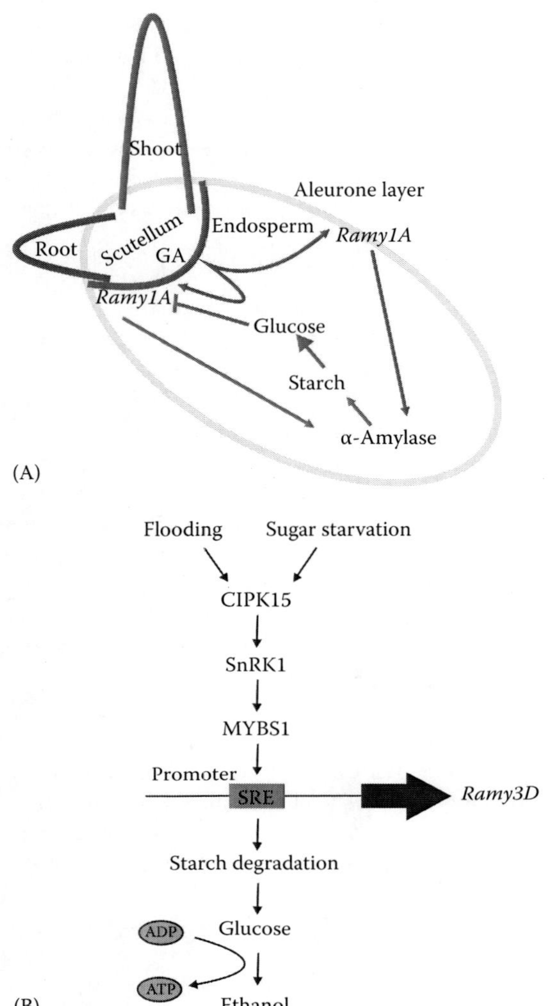

FIGURE 32.7 (A) Interaction between sugar and GA signaling versus *Ramy1A* expression sites during germination. (Adapted from Perata, P. et al., *Plant Cell*, 9, 2197, 1997; Kaneko, M. et al., *Plant Physiol.*, 128, 1264, 2002.) During rice seed germination, the embryo synthesizes GA, which diffuses to the endosperm, reaching the aleurone layer. *Ramy1A* GA-triggered induction takes place in both the epithelium and aleurone layers and induces α-amylase activity in the endosperm where starch is hydrolyzed to soluble sugars. The resulting soluble sugars repress *Ramy1A* expression in the epithelium but not in the aleurone. (B) The *Ramy3D* activation pathway that coordinates the rice seedling response to O_2 deprivation and sugar deficiency for tolerance to flooding. (Adapted from Lee, K.W. et al., *Sci. Signal.*, 2(91), ra61, 2009.) CIPK15 acts as the main upstream positive regulator of SnRK1 at a posttranscriptional level in response to sugar starvation and flooding. SnRK1 is necessary for the activation of MYBS1 that drives the *Ramy3D* expression binding to the SRE located in the promoter region of the *Ramy3D* gene.

under sugar starvation (Lu et al. 1998; Toyofuku et al. 1998; Chen et al. 2002). Sugar signaling also regulates α-amylase by controlling its mRNA stability, since in suspension-cultured rice cells, the half-life of α-amylase mRNA is longer after the depletion of sucrose (Sheu et al. 1996).

Lu et al. (2002) have demonstrated that a TF (MYBS1) binds specifically to the TA box in vivo and in vitro, thus functioning as a transcriptional activator of *Ramy3D* SRE under sugar depletion. Both *MYBS1* and *Ramy3D* expressions were shown to be regulated by the general regulator yeast sucrose nonfermenting 1-related protein kinase SnRK1A, which is expressed in various organ tissues, including young roots, and plays a central role in the sugar signaling pathway (Lu et al. 2007). In the rice T-DNA insertional mutant *snrk1a*, the root growth has been shown to be more severely retarded than shoot growth, suggesting a significant role of this signaling in this organ (Lu et al. 2007). Recently, cross talk between sugar and O_2-deficiency signaling in flooding tolerance was demonstrated in rice plants (Lee et al. 2009). The calcineurin B-like (CBL) interacting protein kinase 15 (CIPK15) has been indicated as the key regulator of carbohydrate catabolism and fermentation during rice germination, through the *SnRK1A*- and *MYBS1*-mediated sugar signaling pathway (Figure 32.7B; Lee et al. 2009). The allelic *cipk15* knockout rice mutant roots were on average 15% shorter than wild-type roots, regardless of sugar feeding. However, exogenous sucrose led to increased root growth in both wild-type and *cipk15* seedlings by 36% and 44%, respectively (Lee et al. 2009). CIPK proteins interact with CBL calcium sensor proteins in order to propagate plant Ca^{2+} signaling (Batistic and Kudla 2004). Calcium has been indicated as a regulator of multiple stress signaling, both biotic and abiotic (for a review see Lecourieux et al. 2006), and has also been suggested as triggering low O_2 signaling in plants. Increased cytosolic Ca^{2+} concentrations have been observed in maize roots and *Arabidopsis* seedlings under flooding (Subbaiah et al. 1994a,b; Sedbrook et al. 1996). The specific CBL that interacts with CIPK15 under O_2 deprivation has not yet been identified.

At the stages of germination and seedling growth, CIPK15 appears to play a major role in low O_2 tolerance in rice. The ethylene-responsive factor (ERF) *Submergence 1A* (*Sub1A*) gene is instead thought to play a key role in the submergence tolerance of mature rice plants (Fukao et al. 2006; Xu et al. 2006). *Sub1A* is part of the major Sub1 QTL located on chromosome 9, which encodes a subgroup of ERF genes involved in the quiescent adaptive mechanisms of lowland rice. The tolerance-related allele *Sub1A-1*, which is only present in a few rice accessions, is induced by ethylene in aerial tissues under submergence and restricts carbohydrate depletion through the negative regulation of *Ramy3D* expression probably via the *Sub1C* gene (Figure 32.8; Bailey-Serres and Voesenek 2008). In submerged plants, *Sub1C* is positively regulated by GA, thus implying hormonal control of the carbohydrate level (Fukao et al. 2006). The management and economy of carbohydrate under O_2 deprivation represents an energy reservoir for the regrowth of plants when the submergence water recedes. While limited information is available about the role of *Sub1A* in roots under low oxygen, it has been shown to have a role in acclimation to dehydration that paradoxically occurs in the natural progression of a flooding event (Fukao et al. 2011). A rapid accumulation of *Sub1A* has been observed in both shoots and roots under dehydration. However,

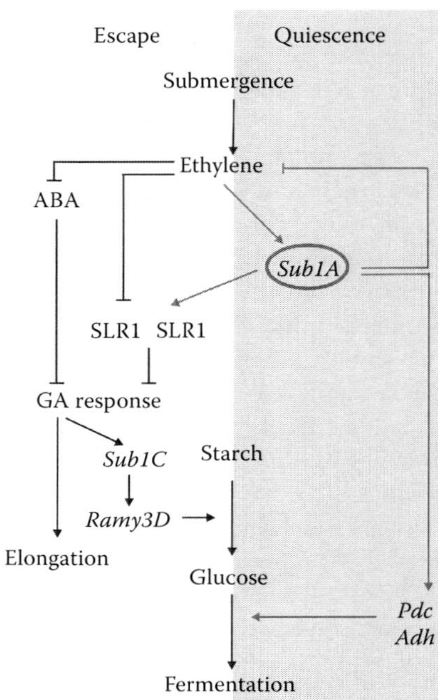

FIGURE 32.8 The carbohydrate metabolism pathways in lowland rice plants showing the "quiescence strategy" and the "escape strategy" under submergence. The escape strategy promotes GA-induced elongation with concomitant starch and soluble sugar catabolism. Submergence stimulates the production/entrapment of ethylene, which then activates *Sub1A* in rice varieties adopting the quiescence strategy. *Sub1A* positively regulates the two GA signaling suppressors SLR1-SLRL1. This inactivates GA signaling with reduced *Sub1C* expression and a concomitant reduction in carbohydrate consumption. *Sub1A* also increases mRNA accumulation of pyruvate decarboxylase (*Pdc*) and alcohol dehydroxygenase (*Adh*), enhancing fermentation. (Adapted from Bailey-Serres, J. et al., *Rice*, 3, 138, 2010.)

the inhibition of root growth was found to be less severe in M202(*Sub1*) rice plants, a near isogenic *Sub1* introgression line, indicating that SUB1A enhances tolerance to osmotic stress also in roots (Fukao et al. 2011).

Recently, the existence of an interrelation between Sub1A and CIPK15 signaling has been suggested in submerged rice seedlings, since a low *CIPK15* transcript level was observed in the *Sub1A* containing variety FR13A (Kudahettige et al. 2010).

Deepwater rice varieties do not respond to submergence by adopting a quiescence strategy. Instead, they grow rapidly, showing elongation of the internodes under water in an attempt to escape low O_2 (Bailey-Serres and Voesenek 2008). Moreover, the decline in ABA observed under submergence has been shown to augment the ethylene-dependent adventitious root formation (Steffens et al. 2006). Raskin and Kende (1984) have reported high carbohydrate consumption in the rapidly elongating internodes under submergence. Deepwater rice varieties survive submergence through an "escape strategy" as long as plant growth is fast enough to allow aerial tissue to get away from submergence and avoid sugar starvation (Figure 32.8; Bailey-Serres and Voesenek 2008). Shoot elongation is stimulated by the ethylene-regulated *Snorkel* genes *SK*1 and *SK2* (Hattori et al. 2009). How the *SK* genes regulate internode elongation is still unclear. However, the arrest of elongation, subsequent to deepwater plant treatment with uniconazole (inhibitor of GA biosynthesis), suggests the involvement of GA biosynthesis or signal transduction (Nagai et al. 2010). The organ-specific expression of *SK* genes has revealed their absence in root tissues (Hattori et al. 2009).

3. Sucrose Metabolism under Low O_2

Sucrose is a product of carbon fixation and is also the main form in which carbohydrates are transported to plant sink organs. Sucrose can be catabolized via two independent routes: through UDP-dependent reversible sucrose synthase (SUS) reaction, leading to UDP-glucose and fructose, and through invertase (INV) irreversible hydrolysis to glucose and fructose (Figure 32.6).

Under O_2 deprivation, the SUS route is upregulated, while INV gene expression and enzyme activities are repressed (Geigenberger 2003; Bologa et al. 2003). The energetic cost for sucrose entry into glycolysis as hexose phosphates via the INV way is in fact disadvantageous when compared to the SUS route. This is because the former reaction needs two ATP molecules per mol of sucrose catabolized and the latter reaction only needs one molecule of pyrophosphate (PPi) (van Dongen et al. 2003). This was elegantly demonstrated in transgenic potato tubers with high INV activities, which were unable to maintain ATP production under low O_2 (Bologa et al. 2003). Experimental evidence suggests the preferential activation of the SUS pathway under anaerobiosis in several plant species. In maize, *SUS* genes show upregulation under low O_2, while *INV* gene expression is repressed (Zeng et al. 1998, 1999). In rice germinated under anoxia, SUS activity increases under O_2 deprivation, while INV activity declines (Ricard et al. 1991; Guglielminetti et al. 1995a). *Arabidopsis* SUS is encoded by a small multigene family where different isoforms differentiate in spatial and temporal distribution, suggesting a similar function in different cell types or developmental stages (Bieniawska et al. 2007). Under hypoxic conditions, the levels of *SUS1* and *SUS4* genes increase in both roots and leaves, together with the amount of proteins and total SUS activity (Martin et al. 1993; Baud et al. 2004; Bieniawska et al. 2007). These two isoforms show 89% identity, but they are less than 68% identical to the other isoforms (Baud et al. 2004). An analysis of single *sus1* and *sus4* mutants has not revealed any unpaired growth during flooding. However, the double mutant is severely damaged under submergence (Bieniawska et al. 2007). In maize, the loss of both SUS isoforms SUS1 and SH1 impairs the ability of roots to survive anoxia and hypoxia (Ricard et al. 1998).

However, experimental evidence suggests that the role of SUS under O_2 deprivation is not to provide substrate for glycolysis (Guglielminetti et al. 1996; Biemelt et al. 1999; Subbaiah and Sachs 2001; Albrecht and Mustroph 2003a). Deficiency of *Shrunken-1*-encoded SUS in maize has been shown to have no effect on the rate of ethanol production under anaerobic conditions (Guglielminetti et al. 1996). Root extracts from *SUS* antisense plants, showing a reduced tolerance to hypoxia, do not show any difference in the level of glycolysis intermediates, thus suggesting that the limitation of glycolysis is not responsible for the observed sensitivity (Biemelt et al. 1999).

SUS has been suggested to have a role in channeling carbohydrate toward the cell wall polymers for later use (Biemelt et al. 1999). Indeed, increased SUS activity has been found to correlate with increased cellulose content and cell wall structures in the roots of wheat (Albrecht and Mustroph 2003a,b). In addition, in maize roots under O_2 deprivation, callose induction has been observed (Subbaiah and Sachs 2001). The specific role for preferential channeling of carbon to structural polysaccharide synthesis under O_2 deprivation is, however, still unclear.

B. Energy Metabolism and Plant Mitochondria under Oxygen Deprivation

Plant root cell mitochondria do not only play a central role in the energy metabolism of root cell but also provide carbon skeletons for amino acid and other biosyntheses. Depending on the severity of flooding stress, mitochondrial events can be hampered and may lead to increased ROS production adding to the stress.

1. Metabolic and Structural Alterations

Mitochondria, the main energy-producing organelle, rely on molecular oxygen as terminal electron acceptor, on the availability of respiratory substrates provided by tricarboxylic acid cycle (TCA), and on reducing power of NADH. NADH is utilized as a substrate by complex I and by external and internal NAD(P)H dehydrogenases unique to plant mitochondria (Rasmusson et al. 2008), therefore making them sensitive to cytoplasmic pyridine nucleotide status and to other metabolic factors controlling this status. Under oxygen deprivation, these factors may include glycolysis (NADH production), GABA shunt (NADH consumption in glutamate dehydrogenase reaction), nitrate reductase (NR) (NADP(H) consumption), and some other reactions (Greenway and Gibbs 2003; Bailey-Serres and Voesenek 2008). Under hypoxic stress such as waterlogging, mitochondria can encounter considerably different conditions in terms of O_2 concentration, dependent on their tissue localization. Steep O_2 gradients may exist in seeds, roots, and bulky organs (van Dongen et al. 2003, 2004; Benamar et al. 2008; Borisjuk and Rolletschek 2009). Under oxygen deprivation, roots have to cope with declining O_2 and energy restrictions and still sustain, at least to some extent, water transport (see Section III.A) and ion and nutrient uptake (see Section III.B).

Hypoxic mitochondria function in a special cytoplasmic environment: highly reduced due to NADH accumulation, acidic due to hypoxia-induced lactic acid production, and under diminishing concentration of terminal electron acceptor O_2. Although the activity of cytochrome oxidase (COX) is not really limited under hypoxic conditions because of its extremely low Km for oxygen (10 μM) (Affourtit et al. 2001), the respiration rate declines when O_2 concentration is well above Km. The degree of respiratory inhibition in the roots is under metabolic control from pyruvate fluxes, which are enhanced due to glycolysis activation (Zabalza et al. 2009). Recently, hypoxic operation of TCA cycle in a "split" mode, when only reactions 2-oxoglutarate → succinyl-CoA → succinate and oxaloacetate → malate are operational, has been suggested for *Lotus japonicus* (Regal) nodulated roots and leaves (Rocha et al. 2010). This metabolic alteration links glycolysis to TCA cycle controlling glycolytic pyruvate accumulation via alanine aminotransferase (AlaAT) with the production of alanine. Preventing pyruvate buildup is essential for diminishing the respiration rate and hence for avoidance of complete anoxia. 2-Oxoglutarate, which is also formed in AlaAT reaction, can enter a TCA cycle at the level 2-oxoglutarate dehydrogenase (OGDG) resulting in the synthesis of one ATP molecule and in succinate accumulation (Rocha et al. 2010). NAD^+ needed for OGDG reaction is provided from NADH by malate dehydrogenase converting oxaloacetate to malate. In turn, oxaloacetate is supplied by aspartate aminotransferase with concurrent production of glutamate needed as a substrate for pyruvate conversion by AlaAT (Rocha et al. 2010). The "non-classic" mode of TCA cycle may be operational not only under oxygen deprivation but also in illuminated leaves and in developing seeds (reviewed in Sweetlove et al. 2010).

The hypothesis of tight metabolic interaction between glycolysis and TCA cycle is further supported by the recently described glycolytic substrate channeling to mitochondria (Graham et al. 2007). The existing cytoplasmic pool of glycolytic enzymes can undergo rapid partition to the surface of mitochondria in response to respiratory increase (as is the case with hypoxia-accumulating pyruvate) to sustain effective channeling of the substrates. The following glycolytic enzymes have been found to associate with outer mitochondrial membrane: hexokinase, fructose-bisphosphate aldolase, triosephosphate isomerase, glyceraldehyde-3-phosphate dehydrogenase, phosphoglycerate kinase, enolase, and pyruvate kinase (Graham et al. 2007). The nature of this association is highly dynamic, occurs in the time range of minutes, and is tightly regulated by metabolite-induced conformational changes (Graham et al. 2007).

Alongside with metabolic adjustment and spatial association with the glycolytic pathway, mitochondria exhibit structural reorganization of protein complexes involved in oxidative phosphorylation (complexes I–IV and ATP synthase) resulting in the formation of respiratory supercomplexes—respirasomes (Dudkina et al. 2006, 2008, 2010; Bultema et al. 2009). The supercomplex organization has been suggested to enhance the electron transfer between complexes, to aid the substrate channeling, and to reduce the diffusion path for mobile electron carriers ubiquinone and cytochrome *c*. In plant mitochondria, association between the complex I and the dimer of complex III is thought to regulate alternative oxidase (AOX) through limiting enzyme access to the substrate ubiquinone (Dudkina et al. 2006, 2010; Heinemeyer et al. 2007).

2. ROS and NO Formation in Plant Mitochondria

Hypoxic mitochondria produce ROS and NO via multiple mechanisms (Andreyev et al. 2005; Kaiser et al. 2007; Lenaz et al. 2007; Schwarzländer et al. 2009; Blokhina and Fagerstedt 2010a; Gupta et al. 2011). Protein complexes of mitochondrial electron transport chain (ETC) are capable of ROS formation

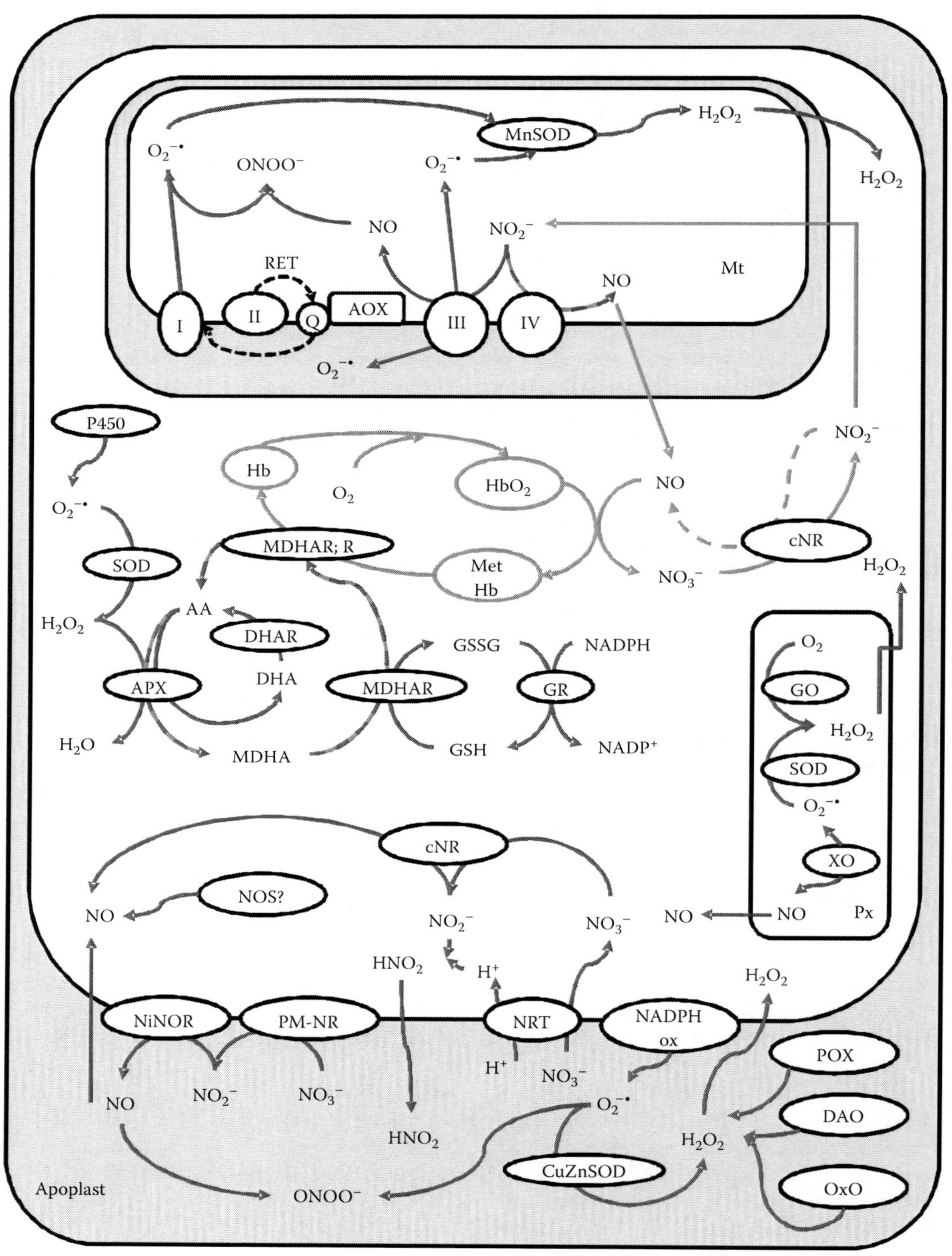

MnSOD
H_2O_2
$O_2^{-\bullet}$
$ONOO^-$
$O_2^{-\bullet}$
H_2O_2
NO
NO_2^-
RET
NO
Mt
I
II
Q
AOX
III
IV
$O_2^{-\bullet}$
P450
NO_2^-
Hb
HbO_2
NO
O_2
$O_2^{-\bullet}$
MDHAR; R
cNR
SOD
Met
Hb
NO_3^-
H_2O_2
AA
H_2O_2
DHAR
O_2
GSSG
NADPH
DHA
APX
MDHAR
GR
GO
H_2O_2
H_2O
MDHA
GSH
$NADP^+$
SOD
$O_2^{-\bullet}$
cNR
XO
NOS?
NO
NO_2^-
NO_3^-
NO
NO
Px
HNO_2
H^+
H_2O_2
NiNOR
PM-NR
NRT
NADPH
ox
POX
H^+
NO_3^-
NO
NO_2^-
NO_3^-
$O_2^{-\bullet}$
DAO
HNO_2
H_2O_2
CuZnSOD
Apoplast
$ONOO^-$
OxO

under normal and pathological conditions. This ROS formation is dependent on mitochondrial inner membrane potential, activity of ATP synthase, and the reduction state of the respiratory complex proteins. All these three parameters are inherently connected with each other and are affected under oxygen deprivation in a way that favors ROS formation (Adam-Vizi and Chinopoulos 2006; Igamberdiev and Hill 2009; Murphy 2009; Blokhina and Fagerstedt 2010a). The redox state of the ubiquinone radical is the key regulator of electron leakage from ETC and the formation of the superoxide radical (Lenaz et al. 2007; Murphy 2009). Under over-reduced conditions, univalent reduction of O_2 occurs at complex I, complex III, and in reverse electron flow from succinate dehydrogenase (complex II) to complex I (Figure 32.9; Turrens 2003; Andreyev et al. 2005).

In waterlogged soils, roots also encounter an environment with low redox potential where reduced substances such as Mn^{2+}, Fe^{2+}, H_2S, and nitrite can build up (Drew 1997). Roots have been shown to possess two redundant membrane-bound systems for metabolization of nitrate, nitrite, and NO under hypoxia (Stöhr et al. 2001; Gupta et al. 2005; Stöhr and Stremlau 2006; Igamberdiev et al. 2010): one being the PM-bound nitrite:NO reductase (NiNOR) (see later) (Stöhr and Stremlau 2006; Gupta et al. 2011) and the other the mitochondria-associated nitrite reductase activity capable of converting nitrite to NO and sensitive to respiratory inhibitors of complex III and AOX (Gupta et al. 2005; Kaiser et al. 2007; Igamberdiev and Hill 2009; Gupta and Kaiser 2010). It is not clear yet whether nitrite reduction is mediated directly by ETC complexes, or whether an additional nitrite reductase protein exists in mitochondria (Figure 32.9). Interestingly, only root mitochondria are able to produce NO via the nitrite-dependent reaction. The molecular basis and physiological significance of such diversification of function between the root and leaf mitochondria is not clear (Kaiser et al. 2007). The mitochondria in barley and rice roots have been shown to synthesize ATP anaerobically at a rate comparable with that of glycolysis, oxidizing NAD(P)H in the presence of nitrite with concurrent NO release (Stoimenova et al. 2007). The beneficial effect of NR under anoxia in roots has been confirmed in the experiments with transformed tobacco (*Nicotiana tabacum* L.) plants lacking NR: The transgenic plants were more sensitive to oxygen deprivation (Stoimenova et al. 2003).

In addition to the metabolic regulation of internal oxygen concentration via pyruvate described earlier (Rocha et al. 2010), hypoxically produced NO has been implicated in the regulation of respiration rate to sustain low O_2 levels (Borisjuk et al. 2007; Benamar et al. 2008). NO exerts a reversible inhibitory effect on COX, therefore lowering O_2 consumption by mitochondria. In turn, NO can be oxidized by COX to nitrite, and this would release COX inhibition and increase electron flow to O_2. It has been also shown that NO can increase Km of COX to oxygen making the enzyme more sensitive to fluctuations on O_2 concentration (Cooper et al. 2003). Fine-tuning of this cycle may also be achieved by additional factors such as NO scavenging by Hbs (Igamberdiev et al. 2005, 2010).

FIGURE 32.9 **(See color insert.)** Intracellular and extracellular sites of ROS and NO formation, scavenging by antioxidant systems and involvement of Hb–NO cycle in NO elimination under hypoxia. In the apoplast, NADPH oxidase is the main enzyme producing superoxide anion radical ($O_2^{-\bullet}$), which is dismutated by SOD (CuZn-SOD) to H_2O_2. Several other apoplastic enzymes such as POX, diaminooxidase (DAO), and germin-like OxO add to H_2O_2 formation in the apoplast. As a membrane-permeable compound, apoplastic H_2O_2 may affect cytosolic H_2O_2 level. Nitrate (NO_3^-) acquired as a nutrient from the soil can be converted to nitrite (NO_2^-) by apoplast-facing PM-NR or transported into the cytosol via nitrate–/nitrite–proton symporter (NRT). Apoplastic NO_2^- is further reduced to nitric oxide (NO) by root-specific PM-bound NiNOR. Additionally non-enzymatic reduction of NO_2^- to NO may occur, which is enhanced by acidic pH and phenolic compounds. NO formed freely penetrates the PM and enters the cytosol, or alternatively it may react in the apoplast with $O_2^{-\bullet}$ originated from NADPH oxidase reaction and form highly reactive peroxynitrite ($ONOO^-$). Multiple sites of $O_2^{-\bullet}$ formation inside the cell include cytochrome P450 enzymes in ER and cytoplasm, peroxisomal xanthine oxidoreductase (XO), and mitochondrial ETC complexes I and III. Additionally reverse electron transport from complex II via ubiquinone (Q) to complex I adds to $O_2^{-\bullet}$ formation at complex I site. The over-reduced state of Q favors electron leakage to O_2 and formation of $O_2^{-\bullet}$. AOX prevents ROS formation via competition for electrons and diminishing the over-reduction of ETC. $O_2^{-\bullet}$ formed is quickly dismutated to H_2O_2 by compartment-specific SODs (CuZnSOD in the cytosol and peroxisomes and MnSOD in mitochondria and peroxisomes). Peroxisomes produce H_2O_2 as a result of fatty acid β-oxidation and glycolate oxidation by GO. Accumulated H_2O_2 is detoxified by an array of antioxidant systems (omitted for clarity) with ascorbate–GSH cycle being the most versatile system present in cytoplasm (see figure), mitochondria, chloroplast, and peroxisomes (omitted). Under hypoxia, NO_2^- level builds up in the cytoplasm due to the induction of cytosolic nitrate reductase (cNR), different from PM-NR. Accumulating nitrite can escape to the apoplast in the protonated form HNO_2. The same enzyme, cNR, is capable of reducing NO_2^- to NO, using NADH (also accumulating under hypoxia) as an electron donor. Alternatively, L-arginine-dependent NO synthesis may occur via NOS-like enzyme. Under low oxygen, peroxisomal XO exhibits increased NO-producing activity. Mitochondrial respiratory complexes III and IV are able to use nitrite as an alternative electron acceptor under low-oxygen tension, releasing NO as a product. Mitochondrial nitrite reductase activity is stimulated by a decrease in pH. In turn, NO regulates the respiration rate and diminishes ROS formation via inhibition of cytochrome *c* oxidase. Hypoxically induced class 1 Hbs prevent NO accumulation in the cytoplasm via Hb–NO cycle: Oxygenated form of Hb (HbO_2) converts NO to nitrate (NO_3^-) with concurrent formation of MetHb. The latter is reduced to Hb by MDHAR or another putative MetHb reductase (R) at the expense of NADPH. MDHAR is an enzyme of H_2O_2-eliminating ascorbate–GSH cycle—a powerful antioxidant system. Nitrate formed after NO oxygenation by HbO_2 is reduced by cNR to nitrite (NO_2^-), which can once again enter the cycle after further reduction to NO by mitochondrial or cytoplasmic systems. Mt, mitochondria; Px, peroxisome; MDHA, monodehydroascorbate; DHA, dehydroascorbate; AA, ascorbate; GSH/GSSG, reduced and oxidized glutathione, respectively; APX, ascorbate peroxidase; DHAR, dehydroascorbate reductase; GR, glutathione reductase. Other abbreviations are explained in the text. Red lines denote routes of ROS formation; blue lines denote routes of NO formation, and green lines mark the reactions of Hb–NO cycle.

C. Differences in Accumulation of Anoxia-Specific Metabolites and Protein and Gene Expression Levels in Roots and Shoots

One can easily imagine that when complete anoxia is all of a sudden imposed on the roots, the effects are devastating. After a few minutes of chaos in the root cells, metabolic events are quickly reorganized in a way to maintain the adenylate energy charge as far as possible. Once the mitochondria are unable to produce ATP, the cells turn to fermentation, first to lactic acid and then to ethanolic fermentation. At the same time, protein synthesis (and most other syntheses as well) is slowing down to a halt, and only a few dozens of vital proteins, mostly related to glycolysis and fermentation, are maintained or even induced. The fundamental facts mentioned earlier are most probably valid in practically all plant species under anoxic conditions. Under hypoxia, things are different and depend largely on the amount of oxygen present in the tissues. All these events have been studied ever since the 1970s and reviewed at more or less regular intervals, the latest perhaps being the massive review by Greenway and Gibbs (2003) and the book edited by Mancuso and Shabala (2010), and hence, we will not dwell on them more in this chapter.

The recent developments in publicly available microarray data and metabolome- and translatome-level studies of the response to the lack of oxygen for several plant species have revealed a wealth of information on the complex signature and multilevel regulation of metabolic events (van Dongen et al. 2009; Mustroph et al. 2009, 2010). Whole-plant hypoxia stress response, however similar in its core, differs considerably between organs and tissues on the level of transcription, translation, and the metabolome (Mustroph et al. 2009). An assessment of multilevel response in *Arabidopsis* after 2 h hypoxia has been performed in root (proliferating cells, endodermis, vasculature, phloem companion cells, atrichoblast epidermis, cortex meristematic zone, cortex elongation, and maturation zone) and shoot (photosynthetic tissues, leaf epidermis, leaf guard cells, phloem companion cells, bundle sheath) tissues (Mustroph et al. 2009). According to this study, hypoxia-induced mRNAs were induced in all cell types examined, while hypoxia-reduced mRNAs were more cell-type specific.

A recent study on transcriptomic and metabolomic profiling with *Arabidopsis* root material treated with decreasing oxygen concentrations (21%, 8%, 4%, and 1% (v/v) external oxygen) for 0, 5, 2, and 48 h has shown that root growth was inhibited, and transcript and metabolite profiles were significantly altered in response to a moderate decrease in oxygen concentrations (van Dongen et al. 2009). The data indicate that low oxygen leads to a preferential upregulation of genes important to trigger adaptive responses. Even a moderate oxygen decrease led to a small but highly specific set of genes being induced very early in the response, and downregulated genes mainly encoded proteins involved in energy-consuming processes. On the whole, the results showed clearly that large-scale reprogramming of gene expression and metabolism takes place when oxygen concentration is decreased in a very narrow range.

A very interesting new investigation of the transcriptome of *Arabidopsis* and 20 other organisms from four kingdoms(!) was published in 2010 (Mustroph et al. 2010). It illuminated the broadly conserved low-oxygen stress responses (some of which were described earlier) while also indicating that plant-specific hypoxia-responsive unknown proteins (HUPs), of which 16 were studied further in this investigation, confirmed the earlier belief that HUPs with limited phylogenetic distribution influence low-oxygen stress endurance. This tolerance naturally means tightly co-regulated molecular events, where specific TFs play a role. The current knowledge of TFs concerning low-oxygen adaptation have been reviewed by Licausi (2010), including the oxygen-sensing mechanism described in Section II.B in this chapter (Gibbs et al. 2011; Licausi et al. 2011).

D. Changes in Cytosolic pH during Oxygen Deprivation Stress

During oxygen deprivation stress, a general phenomenon in the root cells is a decline in cytosolic pH, which then affects many metabolic events in cellular metabolism such as the decrease in lactic acid synthesis and increase in ethanolic fermentation. The actual cause of the change in pH can be due to several events concerning proton consumption and production during aerobic and anaerobic oxidation of carbohydrate and during the conversion of nitrate to GABA. These points are presented in Chapter 23. In addition, changes in pH also can have a signaling role, which has been reviewed recently by Felle (2010).

V. Oxidative and Nitrosative Stress and ROS and NO Signaling under Limited O_2

A. ROS Formation under Oxygen Deprivation

Generation of ROS is typical for all living tissues and their formation increases under practically all stress conditions—a phenomenon known as oxidative stress. It incorporates both excessive formation of ROS and accumulation and evolution of RNS and is characterized by insufficient antioxidant protection. Roots or the whole plant exposed to limited O_2 shows diverse structural and metabolic responses aimed to avoid further decline in O_2 or to adapt to the low oxygen partial pressure (Visser et al. 2000, 2003; Bailey-Serres and Voesenek 2008; Colmer and Voesenek 2009). Both these strategies also involve coping with oxidative stress, a general response under various stress situations, including O_2 deprivation. ROS may be formed even during hypoxia, as has been shown for the roots of cereals and rhizomes of *Iris* spp. (Biemelt et al. 2000; Blokhina et al. 2001). The possibility of hypoxia-induced ROS formation has been further supported in a study on peroxidative ethane emission from submerged rice seedlings and seedlings exposed to atmosphere with O_2 concentration below 1% (Santosa et al. 2007). However, the most profound accumulation of ROS occurs when O_2 supply is resumed.

Multiple enzymatic and nonenzymatic routes leading to ROS formation exist in plant cells (Figure 32.9). In the apoplast, H_2O_2 may be produced by pH-dependent peroxidases (POX), diamino oxidase, germin-like oxalate oxidase (OxO), Cu-Zn SOD, and NADPH oxidase (Bolwell and Wojtaszek 1997; Bolwell et al. 2002; Blokhina and Fagerstedt 2006). Intracellular systems responsible for the activation of O_2 are represented by ETCs of chloroplast and mitochondria and an array of enzymatic reactions: cytoplasmic and mitochondrial SODs, PRXs, xanthine oxidase, P450 enzymes, and peroxisomal glycolate oxidase (GO) (Mittler et al. 2004; Blokhina et al. 2003; Blokhina and Fagerstedt 2006; Van Breusegem et al. 2008). Gene expression profiling of hypoxic response supports earlier findings that ROS and NO are true hypoxic metabolites and that their levels are under the control of antioxidant systems. Several microarray studies of hypoxia-initiated transcriptional changes in *Arabidopsis* roots and whole seedlings have revealed that transcripts related to ROS and NO formation and detoxification are induced (Klok et al. 2002; Branco-Price et al. 2005; Liu et al. 2005; Loreti et al. 2005; van Dongen et al. 2009). Respiratory burst oxidase, some of the cytochrome P450 family proteins, several SODs, L-ascorbate PRX, monodehydroascorbate reductase (MDHAR), nonsymbiotic Hb1, GSH-metabolizing enzymes, enzymes related to lipid metabolism, and mitochondrial functions are all induced by oxygen deprivation (summarized in Blokhina and Fagerstedt 2010b).

B. NO: Origin and Sites of Action under Oxygen Deprivation Stress

Enzymatic production of NO is associated with both apoplastic and cytosolic compartments with multiple routes leading to NO synthesis (reviewed by Gupta et al. 2011). With nitrite as a substrate and NAD(P)H as an electron donor, NO can be synthesized by cytosolic nitrate reductase (cNR) and by root-specific PM-bound NiNOR (Yamasaki et al. 1999; Yamasaki and Sakihama 2000; Stöhr et al. 2001; Stöhr and Stremlau 2006). In the apoplast of root cells, nitrite for NiNOR reaction can be supplied by PM-bound apoplast-facing NR, thus creating a powerful system for NO synthesis and emission in roots (Figure 32.9; Stöhr et al. 2001; Stöhr and Stremlau 2006). In addition to this coupled apoplastic system, the activity of cNR is regulated by environmental factors. The activation of this enzyme under hypoxia and accumulation of nitrite has been shown in vitro and in vivo (Gupta et al. 2011).

The origin of NO in plant cells and tissues has been recently reviewed (Gupta et al. 2011). According to the authors, NO biosynthesis in plants can occur either oxidatively or reductively. The so far best-documented biosynthetic and reductive route goes through NR in the cytosol. This as well as the PM-bound nitrate reductase (PM-NR) and NiNOR and the mitochondrial nitrite reduction through cytochrome *c* oxidase and/or reductase all use nitrate and/or nitrite as the substrate (Stöhr et al. 2001; Gupta et al. 2011). The oxidative biosynthetic routes start either from L-arginine, polyamines, or hydroxylamine. Interestingly, the so far unknown plant nitric oxide synthase (NOS) protein and the corresponding gene have been identified in a green alga *Ostreococcus tauri* C. Courties and M.-J. Chrétiennot-Dinet (Foresi et al. 2010). This first plant NOS shares features with mammalian NOS enzymes, the protein sequence shares 45% similarity with mammalian isoforms, its catalytic domain structure resembles that of animal NOS, and the purified protein exhibits NO-producing activity in vitro with kinetic characteristics similar to its animal counterparts (Foresi et al. 2010).

NO-producing activity of NR has been carefully studied in several plant species by Rockel and coworkers (2002). The NO-synthesizing activity under normal aerobic conditions emitted from the leaves of nitrate-fertilized plants in light was in the region of 0.1%–0.01% of total NR activity. Interestingly, anoxic treatment of the plant resulted in leaves accumulating nitrite to concentrations exceeding those in normal illuminated leaves up to 100-fold, and NO production was drastically increased especially in the dark. It was concluded that NO can be produced in variable quantities by NR depending on the total NR activity, the NR activation state, and the cytosolic nitrite and nitrate concentration (Rockel et al. 2002). In addition, feedback regulation of NR by NO has been shown to take place (Rosales et al. 2011). In mammals it has been shown that nitration of proteins and lipids caused by NO can have a role in redox signaling and injury (Rubbo and Radi 2008). It remains to be seen whether this is the case also in plants.

The physiological effects of NO are based on its reduction of free metal ions or oxidation of metals in protein complexes such as Hb: Fe-nitrosyl formation resulting in activation of guanylate cyclase and hemoxygenase; inhibition of P450, cytochrome *c* oxidase, aconitase, and catalase (CAT); and downregulation of ferritin (Wink and Mitchell 1998; Navarre et al. 2000). NO intervention with TCA cycle enzyme would also have an impact on the energy status of the cell, already affected by COX inhibition. The NO inhibitory effect on COX in potato tuber mitochondria has been shown (de Oliveira et al. 2008). Therefore, not only the TCA cycle enzymes but also ETC are targets for NO regulation also under oxygen deprivation stress.

1. Interaction of Class I Hemoglobins and NO

In plants, there are three types of Hbs: symbiotic, nonsymbiotic and truncated Hbs. While the symbiotic Hbs facilitate oxygen diffusion in root nodules, the roles of nonsymbiotic and truncated Hbs have remained more obscure. Nonsymbiotic class I Hb overexpression has been associated with improved tolerance to hypoxia and to low temperature, NO exposure, and nutrient deprivation. The mechanism underlying the resulting improved stress performance is not completely understood, but it is not associated with O_2 delivery.

It is known, however, that reactions of NO with Hb allow the maintenance of NAD^+ levels for the needs of glycolysis under hypoxic conditions (Igamberdiev and Hill 2004; Igamberdiev et al. 2005). This alleviation of hypoxic stress has been observed in maize cell cultures where overexpression of the hypoxically inducible class I Hb resulted in the maintenance of the energy status, and these cells showed less induction of ADH and were able to maintain the redox status of the cells. Similarly, transformed alfalfa roots overexpressing Hb have lower levels of NO (Dordas et al. 2004; Dordas 2009). Hence, the regulation of NO level under oxygen deprivation can be achieved in plants via interaction with class 1 nonsymbiotic Hbs through several routes. This can happen in a reaction with oxyhemoglobin to form nitrate and methemoglobin (MetHb) (Fe^{3+}) with the latter being reduced to Hb (Fe^{2+}) in an NADPH-depending reaction (Figure 32.9; Igamberdiev et al. 2004), or through another route where NO interacts with deoxyhemoglobin to form nitrosylhemoglobin (Dordas et al. 2003). Under low-oxygen tension, nitrosylhemoglobin will represent a significant part of the Hb pool.

This process that also involves NADH recycling to NAD^+ and maintenance of redox state in mitochondria has been defined as Hb–NO cycle (Igamberdiev et al. 2005, 2007, 2010). NO reacts with oxygenated Hb to form nitrite at the expense of the reducing power of NADH (Figure 32.9). To complete the cycle, MetHb formed in the reaction has to be reduced to Hb (Igamberdiev et al. 2005). The presence or absence of appropriate reductant may hinder functionality of the Hb–NO cycle in vivo (Smagghe et al. 2008).

The Hb:NO story has helped solve the background for the age-old method of alleviating flooding damage of field crops: The beneficial role of nitrate and nitrite supplementation under anoxia has been attributed to cytoplasmic alkalinization and to NADH recycling (Stoimenova et al. 2003; Libourel et al. 2006); moreover, anoxic mitochondria have been shown to synthesize ATP in the presence of NAD(P)H and nitrite with concurrent production of NO (Stoimenova et al. 2007). However, the situation is not straightforward as has been indicated by a comparative analysis of primary metabolome profiling studies and expression profiles of low-oxygen-treated plant material where the Hb expression and NO signaling exhibited complex pattern across the compared species (Narsai et al. 2011).

Bioinformatic approach has revealed that there are NO-responsive genes and promoters in *Arabidopsis thaliana* (Grun et al. 2006; Palmieri et al. 2008). Earlier studies have indicated transcriptional changes induced by NO involved in such diverse areas as signal transduction, defense and cell death, transport, basic metabolism, and ROS production and degradation. Palmieri and coworkers (2008) continued on this basis and found eight families of common transcription factor-binding sites (TFBS) in promoter regions of NO-regulated genes, based on microarray analyses. Most of these TFBSs concerned genes involved in particular stress responses, such as jasmonic acid (JA) biosynthesis.

VI. Antioxidant Protection of Root Tissues with Special Emphasis on Low-Oxygen Conditions

Even though it may seem paradoxical, antioxidant protection of root tissues under oxygen deprivation stress is of paramount importance to the root tissues. This is due to a few main issues: During oxygen deprivation stress, the adenylate energy charge is often jeopardized and this leads to the slowing down of practically all biosyntheses of antioxidants and of their turnover. Of all the biosyntheses in plant cells, the synthesis of only a few proteins, mainly related to glycolysis and sugar metabolism, is maintained or induced during anoxia (Branco-Price et al. 2008). This means that the tissues are less protected against oxidative damage during the oxygen deprivation period, and moreover, once the oxygen concentration is restored to normoxic levels, e.g., after a flooding period, root cells may experience a sudden burst in ROS concentrations leading to oxidative damage. Such postanoxic damage has been shown in experiments where plants were placed under anoxic conditions for almost a lethal period and were then brought back to air (Monk et al. 1987; Van Toai and Bolles 1991). A positive effect of antioxidant supplementation and/or upregulation has been noticed in a number of similar studies (Monk et al. 1989; Noctor and Foyer 1998; Scebba et al. 1998; Blokhina et al. 1999, 2000, 2001, 2003; Blokhina and Fagerstedt 2006, 2010b; Boo and Jung 1999; Smirnoff 2000; Foyer and Noctor 2011). It has also been suggested that leaf and root cell antioxidant systems may react differentially to changes in oxygen concentration. In leaves, low-mass antioxidants would play the major role in ROS detoxification during postanoxia, while in roots, increased AOX capacity would support the antioxidant systems, possibly by preventing ROS formation in mitochondria (Skutnik and Rychter 2009). Naturally, both small molecular and enzymatic antioxidants show tissue-dependent concentration patterns as has been observed in onion root zones (del Carmen Cordoba-Pedregosa et al. 2003).

As the antioxidant arsenal of plant cells and the changes taking place during flooding or oxygen deprivation stress has been reviewed extensively (Blokhina et al. 2003; Blokhina and Fagerstedt 2006), we will only present the basics in this chapter. The small molecular antioxidants in plants consist of ascorbate, glutathione (GSH), thioredoxins (TRX), tocopherols, and carotenoids, and in addition, many phenolic compounds in plants have considerable antioxidant activity (Pietta 2000; Hernandez et al. 2009; Procházková et al. 2011). A whole array of enzymes is needed to recycle the antioxidants into their active form once they have been oxidized. These are dehydroascorbate reductase (DHAR), TRX reductase, GSH reductase, lipoamide dehydrogenase, and thiol transferase (glutaredoxin). Many plant-produced vitamins such as the water-soluble vitamins B and C as well as the lipid-soluble vitamins A, E, and K have antioxidant properties as reviewed recently by Asensi-Fabado and Munné-Bosch (2010).

To supplement the small molecular antioxidants, plants and animal have enzymes, which scavenge deleterious ROS. These include the many SODs, CATs, and PRX. Furthermore, there are also a number of enzymes detoxifying lipid peroxidation products: GSH *S*-transferases, phospholipid-hydroperoxide GSH PRX, and ascorbate PRX.

A. Small Molecular Antioxidants

GSH (glutamylcysteinylglycine) is a tripeptide, which is an abundant compound in plant tissues in the cytosol, ER, vacuole, and mitochondria (Jimenez et al. 1998). GSH has an important role together with its oxidized form (GSSG) in the maintenance of the cellular redox balance, and hence, it has a role in redox regulation of gene expression (Wingate et al. 1988; Alscher 1989) and also in the regulation of the cell cycle (Sanchez-Fernandez et al. 1997). During oxidative stress, GSH functions by scavenging H_2O_2 and reacts nonenzymatically with other ROS: singlet oxygen, superoxide radical, and hydroxyl radical (Larson 1988). GSH can also regenerate another powerful water-soluble antioxidant, ascorbic acid, via ascorbate–GSH cycle (Foyer and Halliwell 1976; Noctor and Foyer 1998).

Vitamin C or ascorbic acid is a powerful and omnipresent antioxidant in plant tissues (Noctor and Foyer 1998; Arrigoni and de Tullio 2000; Horemans et al. 2000). In addition to practically all plant cell types and cell organelles, ascorbate is also present in the apoplast. In non-stressed situations, a large proportion of ascorbic acid (90% of the ascorbate pool) exists in its reduced form in leaves and chloroplasts (Smirnoff 2000). The intracellular concentration of ascorbate is, e.g., 20 mM in the cytosol and 20–300 mM in the chloroplast stroma (Foyer and Lelandais 1996). Ascorbic acid can scavenge oxygen radicals such as superoxide and hydroxyl radicals and protect cells against other active forms of oxygen such as the singlet oxygen and H_2O_2 via ascorbate PRX reaction (Noctor and Foyer 1998), which is a part of the ascorbate–GSH cycle presented in Figure 32.9. Although ascorbic acid is water soluble, it helps in the protection of lipid membranes by regenerating tocopherol (vitamin E) from tocopheroxyl radicals (Thomas et al. 1992). In addition to redox reactions, dehydroascorbic acid has been shown to act as a signaling molecule in the regulation of stomatal closure (Fotopoulos et al. 2008).

The importance of the hydrophobic *Vitamin E or tocopherols and tocotrienols* lies in the antioxidant and non-antioxidant functions they have in biological membranes (Kagan 1989). During the postanoxic phase in plant tissues, vitamin E interferes with the chain reactions of lipid autoxidation (Serbinova and Packer 1994), which would otherwise destroy whole membranes. As described earlier, the regeneration of the tocopheroxyl radical to the reduced form can be achieved by vitamin C (ascorbate), reduced GSH (Fryer 1992), or coenzyme Q (Kagan 1989; Kagan et al. 2000).

Plants, and especially our native plant species apart from cultivars, synthesize a large number of different *phenolics* (aromatic nitrogen-containing compounds such as flavonoids and tannins) (Grace and Logan 2000). Many of the phenolics have free radical scavenging activity and are hence used as antioxidants even in our daily foodstuff. Phenolic compounds can also save cells from the oxidative action of hydrogen peroxide (Takahama and Oniki 1997). Biosynthesis of phenolic compounds has been shown to be induced under various stress conditions such as heavy metal stress (Sgherri et al. 2003; Mithöfer et al. 2004), and in root tissues, increased levels of phenolics have been determined under various biotic and abiotic stress factors (Saviranta et al. 2010).

B. Enzymatic Antioxidants

SODs protect plant tissues against superoxide, $O_2^{\bullet-}$, producing H_2O_2, which is then converted to water and oxygen by CATs. SODs are present in all cell compartments in the form of different isoenzymes, which are FeSOD (in prokaryotic organisms, in the chloroplast stroma), MnSOD (in prokaryotic organisms and in the mitochondrion of eukaryotes), and the structurally unrelated Cu/ZnSOD (in the cytosol and chloroplasts, in gram-negative bacteria) (Bowler et al. 1992). Increasing amounts of superoxide have been detected in plant roots under flooding stress (Yan et al. 1996), and this led into increased SOD activities in the roots (Kalashnikov et al. 1994; Biemelt et al. 2000). Also other stresses such as salt stress can lead into increased SOD levels (Attia et al. 2008). Recent research has concentrated on the induction mechanisms, while miRNA398 (Sunkar et al. 2006) and a mitochondrial protein AtMTM1 have been shown to affect SOD levels (Su et al. 2007). In addition, it seems that the mitochondrial redox balance may regulate the level of MnSOD in the roots of *A. thaliana* (Morgan et al. 2008).

Recently, an interesting new finding has indicated that oxidative modification of major antioxidant enzymes, e.g., by protein nitration due to NO accumulation under hypoxia, can lead to a decreased rate of ROS elimination. The nitration of tyrosine 34 in MnSOD (the mitochondrial isoform) has been shown to increase the energetic barrier for ligand (superoxide) access to the metal center of enzyme (Moreno et al. 2011).

CATs, PRXs, and ascorbate PRXs are important antioxidants not only for the protoplasts but also as they are present in the apoplast protecting cells from apoplastic H_2O_2. PRXs take part in the β-oxidation of fatty acids, and as a side product, H_2O_2 is produced. CATs then break it down to dioxygen and water. Under stressed conditions, CAT can also work in a peroxidatic reaction, where it produces hydroxyl radicals ($OH^{\bullet}$; Elstner and Osswald 1994), which are very strong oxidants and lead to radical chain reactions with organic molecules, particularly with polyunsaturated fatty acids (PUFA) in membrane lipids. This is why tocopherols and tocotrienols (vitamin E) are vitally important within the biological membranes repairing the damage and preventing lipid radical chain reactions (Serbinova and Packer 1994) in the roots, while ascorbate PRX and GSH have an important role in protecting the chloroplasts against ROS in the shoots (Asada 1999; De Tullio et al. 2004; De Tullio et al. 2007; Foyer and Noctor 2011).

The phospholipid hydroperoxide glutathione peroxidase (PHGPX) is a mitochondrial enzyme (Yang et al. 2006) known to be inducible under various stress conditions. It acts by catalyzing the regeneration of phospholipid hydroperoxides using the reducing power of GSH. As it has been implicated that the PHGPX gene is induced by increasing levels of ROS (Avsian-Kretchmer et al. 1999), it seems probable that it will also be induced under flooding stress in the roots, but so far there is no direct evidence on this.

VII. Summary and Future Perspectives

The kingdom of plants on the whole shows a variety of amazing tolerance limits toward flooded conditions and concomitantly an array of adaptation mechanisms, damage prevention, and repair systems that maintain plant life in harsh environments. If and when we can introduce these mechanisms to our food plant cultivars, mankind will be able to use large areas of flooded lands for food production. This is a great challenge to our researchers and we are already seeing the first results of this kind of work, e.g., in the production of the golden rice.

In nature, oxygen deprivation of the root system occurs during flooding season causing plant death or significant loss of biomass and yield in agriculturally important plants. The physiological response of the root, an organ that is most affected by flooding, has a major impact on plant survival under oxygen deprivation. During the last decades, extensive studies on plant adaptation to hypoxia have resulted in accumulation of knowledge on anatomical, physiological, metabolic, and gene expression alterations brought about by the lack of oxygen. Development of root aerenchyma, rapid stem elongation, adventitious root formation, and cell wall modification manifested by the ROL barrier are the main anatomical features sustaining the delivery of residual O_2 from the outside or prevention of its loss from the inside. On the cellular and subcellular levels, such metabolic pathways as starch breakdown, redundant routes for sucrose metabolization, activation of the glycolytic pathway, and the onset of cytoplasmic acidification as a result of fermentation processes underlie the accumulation of hypoxia-specific metabolites. This response is caused by profound changes on transcriptional and translational levels.

An important aspect that we may see advancing in the future is research of the regulation of events during flooding tolerance. We have already seen this concerning individual metabolic pathways or developmental events such as aerenchyma formation, but today our large genomic and transcriptomic datasets of a vast array of plants species make it possible to analyze developmental events also on the evolutionary scale. This may help us to understand how all the various adaptations we see in the flooding tolerance mechanisms of plants have developed and how they are related to each other. With the rapid development of the methods, we are entering an exciting period when also the fine molecular mechanisms behind flooding tolerance will be elucidated. One of the main challenges for the future research lie in the dissection of particular signaling pathways leading to the formation of distinct metabolic signature in anoxia-tolerant and anoxia-intolerant plants, adventitious root emergence, induction of ROL deposition, and aerenchyma formation via the induction of PCD. The latter is particularly important not only for the field of hypoxic research, but it is also of fundamental interest in plant physiology. So far, unlike in animal studies, very little is known about initiation and propagation of PCD in plants. These pathways are evolutionarily distinct and the homologues of animal PCD machinery are nonexistent in the plant kingdom with a few exceptions that make the picture even more complicated.

Recent developments in high-throughput techniques, e.g., standardization and wide availability of microarray technology, allow the researchers to monitor the expression pattern of virtually every gene. The number of plant genomes that has been sequenced so far includes 25 published genomes (November 2011), and sequencing of many more is on the way. The availability of new software for monitoring of root growth and morphology puts physiological studies on a higher level both qualitatively and quantitatively, allowing for high-throughput screening. At the same time, these approaches generate large amounts of data that need proper handling and analysis. One of the main challenges for researchers will be integration of available anatomical and biochemical data with the vast amount of information obtained during the post-genomic era. The most important tasks will be to dissect the complicated redundant pathways/networks of tolerance to O_2 deprivation.

One of the great missing facts in flooding tolerance, namely, the oxygen-sensing mechanism in plants, was published in *Nature* in 2011. However, we still do not know whether there are other sensing mechanisms for ROS and there are indications that such mechanisms may be present, as plant cells are sensitive to H_2O_2 and the superoxide radical at least. This discovery will undoubtedly lead into better understanding of the hypoxic response and allow its manipulation for improvement of tolerance.

The research efforts reviewed in this chapter show the great diversity in plant adaptation to flooded environments, aided by anatomical, physiological, and developmental features. This knowledge will be used in the breeding of agricultural crop cultivars for the use in the 17 million km^2 of fields under flood annually on the Earth.

References

Adam-Vizi V, Chinopoulos C. 2006. Bioenergetics and the formation of mitochondrial reactive oxygen species. *Trends Pharmacol Sci* 27:639–645.

Affourtit C, Krab K, Moore AL. 2001. Control of plant mitochondrial respiration [Review]. *Biochim Biophys Acta* 1504:58–69.

Agrawal GK, Iwahashi H, Rakwal R. 2003. Small GTPase 'Rop': Molecular switch for plant defense responses. *FEBS Lett* 546:173–180.

Albrecht G, Biemelt S. 1998. A comparative study on carbohydrate reserves and ethanolic fermentation in the roots of two wetland and non-wetland species after commencement of hypoxia. *Physiol Plant* 104:81–86.

Albrecht G, Biemelt S, Baumgartner S. 1997a. Accumulation of fructans following oxygen deficiency stress in related plant species with different flooding tolerances. *New Phytol* 136:137–144.

Albrecht G, Biemelt S, Wiedenroth EM, Praznik W. 1997b. Changes in carbohydrate content following oxygen deficiency stress in related *Senecio* species with different flooding tolerances. *Phyton Ann Rei Bot* 37:7–12.

Albrecht G, Kammerer S, Praznik W, Wiedenroth EM. 1993. Fructan content of wheat seedlings (*Triticum aestivum* L.) under hypoxia and following re-aeration. *New Phytol* 123:471–476.

Albrecht G, Mustroph A. 2003a. Sucrose utilization via invertase and sucrose synthase with respect to accumulation of cellulose and callose synthesis in wheat roots under oxygen deficiency. *Russ J Plant Physiol* 50:813–820.

Albrecht G, Mustroph A. 2003b. Localization of sucrose synthase in wheat roots: Increased in situ activity of sucrose synthase correlates with cell wall thickening by cellulose deposition under hypoxia. *Planta* 217:252–260.

Albrecht G, Mustroph A, Fox TC. 2004. Sugar and fructan accumulation during metabolic adjustment between respiration and fermentation under low oxygen conditions in wheat roots. *Physiol Plant* 120:93–105.

Alpi A, Beevers H. 1983. Effects of O_2 concentration on rice seedlings. *Plant Physiol* 71:30–34.

Alscher RG. 1989. Biosynthesis and antioxidant function of glutathione in plants. *Physiol Plant* 77:457–464.

Andreyev A, Kushnareva Y, Starkov A. 2005. Mitochondrial metabolism of reactive oxygen species. *Biochemistry (Moscow)* 70:200–214.

Arasimowicz M, Floryszak-Wieczorek J. 2007. Nitric oxide as a bioactive signalling molecule in plant stress responses. *Plant Sci* 172:876–887.

Armstrong J, Armstrong W. 2011. Reasons for the presence and absence of convective (pressurized) ventilation in the genus *Equisetum*. *New Phytol* 190:387–397.

Armstrong W, Armstrong J, Beckett PM. 1992. *Phragmites australis*: Venturi- and humidity-induced pressure flows enhance rhizome aeration and rhizosphere oxidation. *New Phytol* 120:197–207.

Armstrong W, Armstrong J, Beckett PM. 1996a. Pressurized ventilation in emergent macrophytes: The mechanism and mathematical modeling of humidity-induced convection. *Aquat Bot* 54:121–135.

Armstrong W, Armstrong J, Beckett PM, Halder JE, Lythe S, Holt R, Sinclair A. 1996b. Pathways for aeration and the mechanisms and beneficial effects of humidity- and venturi-induced convections in *Phragmites australis* (Cav.) Trin ex. Steud. *Aquat Bot* 54:177–197.

Armstrong W, Cousins D, Armstrong J, Turner DW, Beckett PM. 2000. Oxygen distribution in wetland plant roots and permeability barriers to gas exchange with the rhizosphere: A microelectrode and modelling study with *Phragmites australis*. *Ann Bot* 86:687–703.

Armstrong W, Drew MC. 2002. Root growth and metabolism under oxygen deficiency. In *Plant Roots: The Hidden Half*, 3rd edn., eds. Y Waisel, A Eshel, U Kafkafi, pp. 729–761. New York: Marcel Dekker, Inc.

Armstrong W, Healy MT, Lythe S. 1983. Oxygen diffusion in pea II. Oxygen concentration in the primary root apex as affected by growth, the production of laterals and radial oxygen loss. *New Phytol* 94:549–559.

Armstrong W, Healy MT, Webb T. 1982. Oxygen diffusion in pea I. Pore space resistance in the primary root. *New Phytol* 91:647–659.

Arnell N, Liu C. 2001. Hydrology and water resources. In *Climate Change 2001: Impacts, Adaption, and Vulnerability*, eds. JJ McCarthy, OF Canziani, NA Leary, DJ Dokken, KS White, pp. 191–233. Cambridge, U.K.: Cambridge University Press.

Arpagaus S, Braendle R. 2000. The significance of α-amylase under anoxia stress in tolerant rhizomes (*Acorus calamus* L.) and non-tolerant tubers (*Solanum tuberosum* L., var. Désirée). *J Exp Bot* 51:1475–1477.

Arrigoni O, de Tullio MC. 2000. The role of ascorbic acid in cell metabolism: Between gene-directed functions and unpredictable chemical reactions. *J Plant Physiol* 157:481–488.

Asada K. 1999. The water-water cycle in the chloroplasts: Scavenging of active oxygens and dissipation of excess photons. *Annu Rev Plant Physiol Plant Mol Biol* 50:601–639.

Aschi-Smiti S, Chaibi W, Brouquisse R, Ricard B, Saglio P. 2003. Assessment of enzyme induction and aerenchyma formation as mechanisms for flooding tolerance in *Trifolium subterraneum* 'Park'. *Ann Bot* 91:195–204.

Asensi-Fabado MA, Munné-Bosch S. 2010. Vitamins in plants: Occurrence, biosynthesis and antioxidant function. *Trends Plant Sci* 15:582–592.

Ashraf M, Habib-ur-Rehman. 1999. Interactive effects of nitrate and long-term waterlogging on growth, water relations, and gaseous exchange properties of maize (*Zea mays* L.). *Plant Sci* 144:35–43.

Attia H, Arnaud N, Karray N, Lachaal M. 2008. Long-term effects of mild salt stress on growth, ion accumulation and superoxide dismutase expression of *Arabidopsis* rosette leaves. *Physiol Plant* 132:293–305.

Avsian-Kretchmer O, Eshdat Y, Gueta-Dahan Y, Ben-Hayyim G. 1999. Regulation of stress-induced phospholipid hydroperoxide glutathione peroxidase expression in citrus. *Planta* 209:469–477.

Azad AK, Sawa Y, Ishikawa T, Shibata H. 2004. Phosphorylation of plasma membrane aquaporin regulates temperature-dependent opening of tulip petals. *Plant Cell Physiol* 45:608–617.

Bailey-Serres J, Chang R. 2005. Sensing and signalling in response to oxygen deprivation in plants and other organisms. *Ann Bot* 96:507–518.

Bailey-Serres J, Fukao T, Ronald P, Ismail A, Heuer S, Mackill D. 2010. Submergence tolerant rice: SUB1's journey from landrace to modern cultivar. *Rice* 3:138–147.

Bailey-Serres J, Voesenek LACJ. 2008. Flooding stress: Acclimations and genetic diversity. *Annu Rev Plant Biol* 59:313–339.

Baskin TI, Peret B, Baluška F. 2010. Shootward and rootward: Peak terminology for plant polarity. *Trends Plant Sci* 15:593–594.

Batistic O, Kudla J. 2004. Integration and channeling of calcium signaling through the CBL calcium sensor/CIPK protein kinase network. *Planta* 219:915–924.

Baud S, Vaultier MN, Rochat C. 2004. Structure and expression profile of the sucrose synthase multigene family in *Arabidopsis*. *J Exp Bot* 55:397–409.

Benamar A, Rolletschek H, Borisjuk L. 2008. Nitrite–nitric oxide control of mitochondrial respiration at the frontier of anoxia. *Biochim Biophys Acta Bioenerg* 1777:1268–1275.

Bergeron M, Yu AY, Solway KE, Semenza GL, Sharp FR. 1999. Induction of hypoxia-inducible factor-1 (HIF-1) and its target genes following focal ischaemia in rat brain. *Eur J Neurosci* 11:4159–4170.

Biemelt S, Hajirezaei MR, Melzer M, Albrecht G, Sonnewald U. 1999. Sucrose synthase activity does not restrict glycolysis in roots of transgenic potato plants under hypoxic conditions. *Planta* 210:41–49.

Biemelt S, Keetman U, Mock H, Grimm B. 2000. Expression and activity of isoenzymes of superoxide dismutase in wheat roots in response to hypoxia and anoxia. *Plant Cell Environ* 23:135–144.

Bieniawska Z, Barratt PDH, Garlick AP, Thole V, Kruger NJ, Martin C, Zrenner R, Smith AM. 2007. Analysis of the sucrose synthase gene family in *Arabidopsis*. *Plant J* 49:810–828.

Bleecher AB, Kende H. 2000. Ethylene: A gaseous signalling molecule in plants. *Annu Rev Cell Dev Biol* 16:1–18.

Blokhina O, Chirkova TV, Fagerstedt KV. 2001. Anoxic stress leads to hydrogen peroxide formation in plant cells. *J Exp Bot* 52:1179–1190.

Blokhina O, Fagerstedt KV. 2006. Oxidative stress and antioxidant defences in plants. In *Oxidative stress, Disease and Cancer*, ed. J Quek, pp. 151–199. Singapore: World Scientific, eBooks http://ebooks.worldscinet.com/ISBN/9781860948046/9781860948046_0004html.

Blokhina O, Fagerstedt KV. 2010a. Reactive oxygen species and nitric oxide in plant mitochondria: Origin and redundant regulatory systems. *Physiol Plant* 138:447–462.

Blokhina O, Fagerstedt KV. 2010b. Oxidative metabolism, ROS and NO under oxygen deprivation. *Plant Physiol Biochem* 48:359–373.

Blokhina OB, Fagerstedt KV, Chirkova TV. 1999. Relationships between lipid peroxidation and anoxia tolerance in a range of species during post-anoxic reaeration. *Physiol Plant* 105:625–632.

Blokhina O, Virolainen E, Fagerstedt KV. 2003. Antioxidants, oxidative damage and oxygen deprivation stress: A review. *Ann Bot* 91:179–194.

Blokhina O, Virolainen E, Fagerstedt KV, Hoikkala A, Wähälä K, Chirkova TV. 2000. Antioxidant status of anoxia-tolerant and -intolerant plant species under anoxia and reaeration. *Physiol Plant* 109:396–403.

Bologa KL, Fernie AR, Leisse A, Loureiro ME, Geigenberger P. 2003. A bypass of sucrose synthase leads to low internal oxygen and impaired metabolic performance in growing potato tubers. *Plant Physiol* 132:2058–2072.

Bolwell GP, Bindschedler LV, Blee KA. 2002. The apoplastic oxidative burst in response to biotic stress in plants: A three-component system. *J Exp Bot* 53:1367–1376.

Bolwell GP, Wojtaszek P. 1997. Mechanisms for the generation of reactive oxygen species in plant defence—A broad perspective. *Physiol Mol Plant Pathol* 51:347–366.

Boo YC, Jung J. 1999. Water deficit-induced oxidative stress and antioxidative defenses in rice plants. *J Plant Physiol* 155:255–261.

Borisjuk L, Macherel D, Benamar A, Wobus U, Rolletschek H. 2007. Low oxygen sensing and balancing in plant seeds: A role for nitric oxide. *New Phytol* 176:813–823.

Borisjuk L, Rolletschek H. 2009. The oxygen status of the developing seed. *New Phytol* 182:17–30.

Bouranis DL, Chorianopoulou SN, Kollias C, Maniou P, Protonotarios VE, Siyannis VF, Hawkesford MJ. 2006. Dynamics of aerenchyma distribution in the cortex of sulphate-deprived adventitious roots of maize. *Ann Bot* 97:695–704.

Bouranis DL, Chorianopoulou SN, Siyiannis VF, Protonotarios VE, Hawkesford MJ. 2003. Aerenchyma formation in roots of maize during sulphate starvation. *Planta* 217:382–391.

Bouranis DL, Chorianopoulou SN, Siyiannis VF, Protonotarios VE, Hawkesford MJ. 2007. Lysigenous aerenchyma development in roots—Triggers and cross-talks for a cell elimination program. *Int J Plant Dev Biol* 1:127–140.

Bowler C, van Montagu M, Inze D. 1992. Superoxide dismutase and stress tolerance. *Annu Rev Plant Physiol Plant Mol Biol* 43:83–116.

Bramley H, Turner DW, Tyerman SD, Turner NC. 2007. Water flow in the roots of crop species: The influence of root structure, aquaporin activity, and waterlogging. In *Advances in Agronomy*, ed. DL Sparks, pp. 133–196. New York: Academic Press.

Bramley H, Tyerman S. 2010. Root water transport under waterlogged conditions and the roles of aquaporins. In *Waterlogging Signalling and Tolerance in Plants*. eds. S Mancuso, S Shabala, pp. 151–180. Berlin, Germany: Springer.

Branco-Price C, Kaiser KA, Jang CJH, Larive CK, Bailey-Serres J. 2008. Selective mRNA translation coordinates energetic and metabolic adjustments to cellular oxygen deprivation and reoxygenation in *Arabidopsis thaliana*. *Plant J* 56:743–755.

Branco-Price C, Kawaguchi R, Ferreira RB, Bailey-Serres J. 2005. Genome-wide analysis of transcript abundance and translation in *Arabidopsis* seedlings subjected to oxygen deprivation. *Ann Bot* 96:647–660.

Bultema JB, Braun H, Boekema EJ, Kouřil R. 2009. Megacomplex organization of the oxidative phosphorylation system by structural analysis of respiratory supercomplexes from potato. *Biochim Biophys Acta Bioenerg* 1787:60–67.

Campbell R, Drew MC. 1983. Electron microscopy of gas space (aerenchyma formation) in adventitious roots of *Zea mays* L. subjected to oxygen shortage. *Planta* 157:350–357.

del Carmen Cordoba-Pedregosa M, Cordoba F, Villalba JM, Gonzalez-Reyes JA. 2003. Zonal changes in ascorbate and hydrogen peroxide contents, peroxidase, and ascorbate-related enzyme activities in onion roots. *Plant Physiol* 131:697–706.

Chen PW, Lu CA, Yu TS, Tseng TH, Wang CS, Yu SM. 2002. Rice α-amylase transcriptional enhancers direct multiple mode regulation of promoters in transgenic rice. *J Biol Chem* 277:13641–13649.

Chichkova NV, Kim SH, Titova ES, Kalkum M, Morozov VS, Rubtsov YP, Kalinina NO, Taliansky ME, Vartapetian AB. 2004. A plant caspase-like protease activated during the hypersensitive response. *Plant Cell* 16:157–171.

Choi W, Roberts DM. 2007. *Arabidopsis* NIP2;1, a major intrinsic protein transporter of lactic acid induced by anoxic stress. *J Biol Chem* 282:24209–24218.

Colmer TD. 2003. Aerenchyma and an inducible barrier to radial oxygen loss facilitate root aeration in upland, paddy and deep-water rice (*Oryza sativa* L.). *Ann Bot* 91:301–309.

Colmer TD, Greenway H. 2011. Ion transport in seminal and adventitious roots of cereals during O2 deficiency. *J Exp Bot* 62:39–57.

Colmer TD, Voesenek LACJ. 2009. Flooding tolerance: Suites of plant traits in variable environments. *Funct Plant Biol* 36:665–681.

Cooper CE, Davies NA, Psychoulis M. 2003. Nitric oxide and peroxynitrite cause irreversible increases in the K-m for oxygen of mitochondrial cytochrome oxidase: In vitro and in vivo studies. *Biochim Biophys Acta Bioenerg* 1607:27–34.

Crawford RMM. 1978. Metabolic adaptation to anoxia. In *Plant Life in Anaerobic Environments*, eds. DD Hook, RMM Crawford, p. 564. Ann Arbor, MI: Ann Arbor Science Publishers.

Crawford FC. 1910. *Anatomy of the British Carices*. Edinburgh, U.K.: Oliver and Boyd.

De Tullio MC, Ciraci S, Liso R, Arrigoni O. 2007. Ascorbic acid oxidase is dynamically regulated by light and oxygen. A tool for oxygen management in plants? *J Plant Physiol* 164:39–46.

De Tullio MC, Liso R, Arrigoni O. 2004. Ascorbic acid oxidase: An enzyme in search of a role. *Biol Plant* 48:161–166.

van Dongen JT, Frohlich A, Ramirez-Aguilar SJ et al. 2009. Transcript and metabolite profiling of the adaptive response to mild decreases in oxygen concentration in the roots of *Arabidopsis* plants. *Ann Bot* 103:269–280.

van Dongen JT, Roeb GW, Dautzenberg M. 2004. Phloem import and storage metabolism are highly coordinated by the low oxygen concentrations within developing wheat seeds. *Plant Physiol* 135:1809–1821.

van Dongen JT, Schurr U, Pfister M, Geigenberger P. 2003. Phloem metabolism and function have to cope with low internal oxygen. *Plant Physiol* 131:1529–1543.

Dordas C. 2009. Nonsymbiotic hemoglobins and stress tolerance in plants. *Plant Sci* 176:433–440.

Dordas C, Hasinoff B, Igamberdiev AU, Manac'h N, Rivoal J, Hill RD. 2003. Expression of a stress-induced haemoglobin affects NO levels produced by alfalfa under hypoxic stress. *Plant J* 35:763–770.

Dordas C, Hasinoff B, Rivoal J, Hill R. 2004. Class-1 hemoglobins, nitrate and NO levels in anoxic maize cell-suspension cultures. *Planta* 219:66–72.

Dordas C, Rivoal J, Hill RD. 2003. Plant haemoglobins, nitric oxide and hypoxic stress. *Ann Bot* 91:173–178.

Drew MC. 1997. Oxygen deficiency and root metabolism: Injury and acclimation under hypoxia and anoxia. *Annu Rev Plant Physiol Plant Mol Biol* 48:223–250.

Drew MC, He CJ, Morgan PW. 2000. Programmed cell death and aerenchyma formation in roots. *Trends Plant Sci* 5:123–127.

Dudkina NV, Heinemeyer J, Sunderhaus S, Boekema EJ, Braun H. 2006. Respiratory chain supercomplexes in the plant mitochondrial membrane. *Trends Plant Sci* 11:232–240.

Dudkina NV, Kouřil R, Peters K, Braun H, Boekema EJ. 2010. Structure and function of mitochondrial supercomplexes. *Biochim Biophys Acta Bioenerg* 1797:664–670.

Dudkina NV, Sunderhaus S, Boekema EJ, Braun HP. 2008. The higher level of organization of the oxidative phosphorylation system: Mitochondrial supercomplexes. *J Bioenerg Biomembr* 40:419–424.

Dunn G. 1974. A model for starch breakdown in higher plants. *Phytochemistry* 13:1341–1346.

Elstner EF, Osswald W. 1994. Mechanisms of oxygen activation during plant stress. *Proc R Soc Edinb* 102B:31–154.

Elzenga JTM, Veen H. 2010. Waterlogging and plant nutrient uptake. In *Waterlogging Signalling and Tolerance in Plants*, eds. S Mancuso, S Shabala, pp. 23–35. Berlin, Germany: Springer.

Evans DE. 2003. Aerenchyma development. *New Phytol* 161:35–49.

Fagerstedt KV. 1992. Development of aerenchyma in roots and rhizomes of *Carex rostrata* (*Cyperaceae*). *Nordic J Bot* 12:115–120.

Fagerstedt KV. 2010. Programmed cell death and aerenchyma formation under hypoxia. In *Waterlogging Signaling and Tolerance in Plants*, eds. S Mancuso, S Shabala, pp. 99–118. Berlin, Germany: Springer.

Fagerstedt KV, Crawford RMM. 1987. Is anoxia tolerance related to flooding tolerance? *Funct Ecol* 1:49–55.

Fan M, Zhu J, Richards C, Brown KM, Lynch JP. 2003. Physiological roles for aerenchyma in phosphorus-stressed roots. *Funct Plant Biol* 30:493–506.

Felle HH. 2010. pH signaling during anoxia. In *Waterlogging Signaling and Tolerance in Plants*, eds. S Mancuso, S Shabala, pp. 79–98. Berlin, Germany: Springer.

Fincher GB. 1989. Molecular and cellular biology associated with endosperm mobilization in germinating cereal grains. *Annu Rev Plant Physiol Plant Mol Biol* 40:305–346.

Fischer M, Kaldenhoff R. 2008. On the pH regulation of plant aquaporins. *J Biol Chem* 283:33889–33892.

Flexas J, Ribas-Carbó M, Hanson DT. 2006. Tobacco aquaporin NtAQP1 is involved in mesophyll conductance to CO_2 in vivo. *Plant J* 48:427–439.

Flores T, Todd CD, Tovar-Mendez A. 2008. Arginase-negative mutants of *Arabidopsis* exhibit increased nitric oxide signaling in root development. *Plant Physiol* 147:1936–1946.

Foresi N, Correa-Aragunde N, Parisi G, Calo G, Salerno G, Lamattina L. 2010. Characterization of a nitric oxide synthase from the plant kingdom: NO generation from the green alga *Ostreococcus tauri* is light irradiance and growth phase dependent. *Plant Cell* 22:3816–3830.

Fotopoulos V, De Tullio MC, Barnes J, Kanellis AK. 2008. Altered stomatal dynamics in ascorbate oxidase over-expressing tobacco plants suggests a role for dehydroascorbate signalling. *J Exp Bot* 59:729–737.

Fox GG, McCallan NR, Ratcliffe RG. 1995. Manipulating cytoplasmic pH under anoxia—A critical test of the role of pH in the switch from aerobic to anaerobic metabolism. *Planta* 195:324–330.

Foyer CH, Halliwell B. 1976. The presence of glutathione and glutathione reductase in chloroplasts: A proposed role in ascorbic acid metabolism. *Planta* 133:21–25.

Foyer CH, Lelandais MA. 1996. A comparison of the relative rates of transport of ascorbate and glucose across the thylakoid, chloroplast and plasmalemma membranes of pea leaves mesophyll cells. *J Plant Physiol* 148:391–398.

Foyer CH, Noctor G. 2011. Ascorbate and glutathione: The heart of the redox hub. *Plant Physiol* 155:2–18.

Fryer MJ. 1992. The antioxidant effects of thylakoid vitamin E (α-tocopherol). *Plant Cell Environ* 15:381–392.

Fukao T, Yeung E, Bailey-Serres J. 2011. The submergence tolerance regulator Sub1A mediates crosstalk between submergence and drought tolerance in rice. *Plant Cell* 23:412–427.

Fukao T, Bailey-Serres J. 2008. Ethylene—A key regulator of submergence responses in rice. *Plant Sci* 175:43–51.

Fukao T, Xu K, Ronald PC, Bailey-Serres J. 2006. A variable cluster of ethylene response factor-like genes regulates metabolic and developmental acclimation responses to submergence in rice. *Plant Cell* 18:2021–2034.

Gear ML, McPhillips ML, Patrick JW, McCurdy DW. 2000. Hexose transporters of tomato: Molecular cloning, expression analysis and functional characterization. *Plant Mol Biol* 44:687–697.

Geigenberger P. 2003. Response of plant metabolism to too little oxygen. *Curr Opin Plant Biol* 6:247–256.

Gibbs DJ, Lee SC, Isa NM, Gramuglia S, Fukao T, Bassel GW, Correia CS, Corbineau F, Theodoulou FL, Bailey-Serres J, Holdsworth MJ. 2011. Homeostatic response to hypoxia is regulated by the N-rule pathway in plants. *Nature* 479:415–418.

Gharbi I, Ricard B, Smiti S, Bizid E, Brouquisse R. 2009. Increased hexose transport in the roots of tomato plants submitted to prolonged hypoxia. *Planta* 230:441–448.

Gibbs J, Turner DW, Armstrong W, Darwent MJ, Greenway H. 1998. Response to oxygen deficiency in primary maize roots. I. Development of oxygen deficiency in the stele reduces radial solute transport to the xylem. *Aust J Plant Physiol* 25:745–758.

Grace S, Logan BA. 2000. Energy dissipation and radical scavenging by the plant phenylpropanoid pathway. *Trans R Soc Lond B* 355:1499–1510.

Graham JWA, Williams TCR, Morgan M, Fernie AR, Ratcliffe RG, Sweetlove LJ. 2007. Glycolytic enzymes associate dynamically with mitochondria in response to respiratory demand and support substrate channeling. *Plant Cell* 19:3723–3738.

Greenway H, Armstrong W, Colmer TD. 2006. Conditions leading to high CO_2 (>5 kPa) in waterlogged-flooded soils and possible effects on root growth and metabolism. *Ann Bot* 98:9–32.

Greenway H, Gibbs J. 2003. Mechanisms of anoxia tolerance in plants. II. Energy requirements for maintenance and energy distribution to essential processes. *Funct Plant Biol* 30:999–1036.

Grosse W. 1996. The mechanism of thermal transpiration (=thermal osmosis). *Aquat Bot* 54:101–110.

Grun S, Lindermayr C, Sell S, Durner J. 2006. Nitric oxide and gene regulation in plants. *J Exp Bot* 57:507–516.

Gubler F, Kalla R, Roberts JK, Jacobsen JV. 1995. Gibberellin-regulated expression of a *myb* gene in barley aleurone cells: Evidence for Myb transactivation of a high-pI α-amylase gene promoter. *Plant Cell* 7:1879–1891.

Guglielminetti L, Alpi A, Perata P. 1996. Shrunken−1-encoded sucrose synthase is not required for the sucrose-ethanol transition in maize under anaerobic conditions. *Plant Sci* 119:1–10.

Guglielminetti L, Yamaguchi J, Perata P, Alpi A. 1995a. Amylolytic activities in cereal seeds under aerobic and anaerobic conditions. *Plant Physiol* 109:1069–1076.

Guglielminetti L, Perata P, Alpi A. 1995b. Effect of anoxia on carbohydrate metabolism in rice seedlings. *Plant Physiol* 108:735–741.

Guglielminetti L, Wu Y, Boschi E, Yamaguchi J, Favati A, Vergara M, Perata P, Alpi A. 1997. Effects of anoxia on sucrose degrading enzymes in cereal seeds. *J Plant Physiol* 150:251–258.

Gunawardena AHLAN. 2008. Programmed cell death and tissue remodelling in plants. *J Exp Bot* 59:445–451.

Gunawardena AHLAN, Pearce DM, Jackson MB, Hawes CR, Evans DE. 2001. Characterisation of programmed cell death during aerenchyma formation induced by ethylene or hypoxia in roots of maize. *Planta* 212:205–214.

Gupta KJ, Fernie AR, Kaiser WM, van Dongen JT. 2011. On the origins of nitric oxide. *Trends Plant Sci* 16:160–168.

Gupta KJ, Kaiser WM. 2010. Production and scavenging of nitric oxide by barley root mitochondria. *Plant Cell Physiol* 51:576–584.

Gupta KJ, Stoimenova M, Kaiser WM. 2005. In higher plants, only root mitochondria, but not leaf mitochondria reduce nitrite to NO, in vitro and in situ. *J Exp Bot* 56:2601–2609.

Gutierrez L, Bussell JD, Pacurar DI, Schwambach J, Pacurar M, Bellini C. 2009. Phenotypic plasticity of adventitious rooting in *Arabidopsis* is controlled by complex regulation of AUXIN RESPONSE FACTOR transcripts and microRNA abundance. *Plant Cell* 21:3119–3132.

Hahn A, Harter K. 2009. Mitogen-activated protein kinase cascades and ethylene: Signaling, biosynthesis or both? *Plant Physiol* 149:1207–1210.

Hara-Nishimura I, Hatsugai N, Nakaune S, Kuroyanagi M, Nishimura M. 2005. Vacuolar processing enzyme, an executor of plant cell death. *Curr Opin Plant Biol* 8:404–408.

Hatsugai N, Kuroyanagi M, Yamada K, Meshi T, Tsuda S, Kondo M, Nishimura M, Hara-Nishimura I. 2001. A plant vacuolar protease, VPE, mediates virus-induced hypersensitive cell death. *Science* 305:855–858.

Hattori Y, Nagai K, Furukawa S et al. 2009. The ethylene response factors SNORKEL1 and SNORKEL2 allow rice to adapt to deep water. *Nature* 460:1026–1030.

He CJ, Finlayson SA, Drew MC, Jordan WR, Morgan PW. 1996a. Ethylene biosynthesis during aerenchyma formation in roots of maize subjected to mechanical impedance and hypoxia. *Plant Physiol* 112:1679–1685.

He CJ, Morgan PW, Drew MC. 1996b. Transduction of an ethylene signal is required for cell death and lysis in the root cortex of maize during aerenchyma formation induced by hypoxia. *Plant Physiol* 112:463–472.

Heinemeyer J, Braun H, Boekema EJ, Kouřil R. 2007. A structural model of the cytochrome c reductase/oxidase supercomplex from yeast mitochondria. *J Biol Chem* 282:12240–12248.

Hernandez I, Alegre L, Van Breusegem F, Sergi Munne-Bosch S. 2009. How relevant are flavonoids as antioxidants in plants? *Trends Plant Sci* 14:125–132.

Herrera M, Hong NJ, Garvin JL. 2006. Aquaporin-1 transports NO across cell membranes. *Hypertension* 48:157–164.

Holbrook NM, Zwieniecki MA. 2003. Plant biology: Water gate. *Nature* 425:361–361.

Horemans N, Foyer CH, Potters G, Asard H. 2000. Ascorbate function and associated transport systems in plants. *Plant Physiol Biochem* 38:531–540.

Huang B, Johnson JW. 1995. Root respiration and carbohydrate status of two wheat genotypes in response to hypoxia. *Ann Bot* 75:427–432.

Hwang YS, Thomas BR, Rodriguez RL. 1999. Differential expression of rice α-amylase genes during seedling development under anoxia. *Plant Mol Biol* 40:911–920.

Igamberdiev A, Baron K, Hill R. 2007. Nitric oxide as an alternative electron carrier during oxygen deprivation. In *Nitric Oxide in Plant Growth, Development and Stress Physiology*, eds. L Lamattina, J Polacco, pp. 255–268. Berlin, Germany: Springer.

Igamberdiev AU, Baron K, Manac'h-Little N, Stoimenova M, Hill RD. 2005. The haemoglobin/nitric oxide cycle: Involvement in flooding stress and effects on hormone signalling. *Ann Bot* 96:557–564.

Igamberdiev AU, Bykova NV, Shah JK, Hill RD. 2010. Anoxic nitric oxide cycling in plants: Participating reactions and possible mechanisms. *Physiol Plant* 138:393–404.

Igamberdiev AU, Hill RD. 2004. Nitrate, NO and haemoglobin in plant adaptation to hypoxia: An alternative to classic fermentation pathways. *J Exp Bot* 55:2473–2482.

Igamberdiev AU, Hill RD. 2009. Plant mitochondrial function during anaerobiosis. *Ann Bot* 103:259–268.

Igamberdiev A, Seregélyes C, Manac'h N, Hill R. 2004. NADH-dependent metabolism of nitric oxide in alfalfa root cultures expressing barley hemoglobin. *Planta* 219:95–102.

Insalud N, Bell RW, Colmer TD, Berkasem B. 2006. Morphological and physiological responses of rice (*Oryza sativa*) to limited phosphorus supply in aerated and stagnant solution culture. *Ann Bot* 98:995–1004.

Jackson MB, Armstrong W. 1999. Formation of aerenchyma and the processes of plant ventilation in relation to soil flooding and submergence. *Plant Biol* 1:274–287.

Jimenez A, Hernandez JA, Pastori G, del Río LA, Sevilla F. 1998. Role of the ascorbate-glutathione cycle of mitochondria and peroxisomes in the senescence of pea leaves. *Plant Physiol* 118:1327–1335.

Johansson I, Larsson C, Ek B, Kjellbom P. 1996. The major integral proteins of spinach leaf plasma membranes are putative aquaporins and are phosphorylated in response to Ca^{2+} and apoplastic water potential. *Plant Cell* 8:1181–1191.

Justin SHFW, Armstrong W. 1987. The anatomical characteristics of roots and plant response to soil flooding. *New Phytol* 106:465–495.

Kagan VE. 1989. Tocopherol stabilizes membrane against phospholipase A, free fatty acids, and lysophospholipids. *Ann New York Acad Sci* 570:121–135 (in *Vitamin E: Biochemistry and Health Implications*, eds. AT Diplock, J Machlin, L Packer, WA Pryor).

Kagan VE, Fabisiak JP, Quinn PJ. 2000. Coenzyme Q and vitamin E need each other as antioxidants. *Lipids* 214:11–18.

Kaiser W, Gupta K, Planchet E. 2007. Higher plant mitochondria as a source for NO. In *Nitric Oxide in Plant Growth, Development and Stress Physiolog*, eds. L Lamattina, J Polacco, pp. 1–14. Berlin, Germany: Springer.

Kalashnikov JE, Balakhnina TI, Zakrzhevsky DA. 1994. Effect of soil hypoxia on activation of oxygen and the system of protection from oxidative destruction in roots and leaves of *Hordeum vulgare. Russ J Plant Physiol* 41:583–588.

Kaneko M, Itoh H, Ueguchi-Tanaka M, Ashikari M, Matsuoka M. 2002. The α-amylase induction in endosperm during rice seed germination is caused by gibberellin synthesized in epithelium. *Plant Physiol* 128:1264–1270.

Kaul RB. 1971. Diaphragms and aerenchyma in *Scirpus validus. Am J Bot* 58:808–816.

Kawai M, Samarajeewa PK, Barrero RA, Nishiguchi M, Uchimiya H. 1998. Cellular dissection of the degradation pattern of cortical cell death during aerenchyma formation of rice roots. *Planta* 204:277–287.

Keller T, Damude HG, Werner D, Doerner P, Dixon RA Lamb C. 1998. A plant homolog of the neutrophil NADPH oxidase gp91ph_x subunit gene encodes a plasma membrane protein with Ca^{2+} binding motifs. *Plant Cell* 10:255–266.

Kieber JJ, Rothenberg M, Roman G, Feldman KA, Ecker JR. 1993. CTR1, a negative regulator of the ethylene-response pathway in *Arabidopsis*, encodes a member of the raf family of protein kinases. *Cell* 72:427–441.

Kjellbom P, Larsson C, Johansson I, Karlsson M, Johanson U. 1999. Aquaporins and water homeostasis in plants. *Trends Plant Sci* 4:308–314.

Klinge J. 1879. Vergleichend histologische Untersuchung der Gramineen- und Cyperaceen-wurzeln. Mémoires de L'Academie de Sciences de St. Pétersbourg, VIIE Serie, tome XXVI, No 12 St. Pétersbourg.

Klok EJ, Wilson IW, Wilson D. 2002. Expression profile analysis of the low-oxygen response in *Arabidopsis* root cultures. *Plant Cell* 14:2481–2494.

Konings H, Verschuren G. 1980. Formation of aerenchyma in roots of *Zea mays* in aerated solutions, and its relation to nutrient supply. *Physiol Plant* 49:265–270.

Kotula L, Steudele E. 2009. Measurements of oxygen permeability coefficients of rice (*Oryza sativa* L.) roots using a new perfusion technique. *J Exp Bot* 60:2155–2167.

Kubin P. 1992. Fructans: The role in adaptation to hypoxia. In *Second International Symposium on Fructans*, ed. CJ Pollock, pp. 2–13, Aberystwyth, U.K.

Kudahettige NP, Pucciariello C, Parlanti S, Alpi A, Perata P. 2010. Regulatory interplay of the Sub1A and CIPK15 pathways in the regulation of α-amylase production in flooded rice plants. *Plant Biol* 13:611–619.

Kükenthal G. 1909. Cyperaceae–Caricoideae. In *Das Planzenreich IV 20*, ed. A Engler, Vol. 38, pp. 1–814. Lepzig, Germany: Verlag von Wilhelm Engelmann.

Kutschera U, Siebert C, Masuda Y, Sievers A. 1990. Effects of submergence on development and gravitropism in the coleoptile of *Oryza sativa* L. *Planta* 183:112–119.

Laan P, Berrevoets MJ, Lythe S, Armstrong W, Blom CWPM. 1989. Root morphology and aerenchyma formation as indicators of the flood-tolerance of *Rumex* species. *J Ecol* 77:693–703.

Lacomme C, Santa Cruz S. 1999. Bax-induced cell death in tobacco is similar to the hypersensitive reaction. *Proc Natl Acad Sci U S A* 96:7956–7961.

Lamattina L, Garcia-Mata C, Graziano M, Pagnussat G. 2003. Nitric oxide: The versatility of an extensive signal molecule. *Annu Rev Plant Biol* 54:109–136.

Lanteri ML, Laxalt AM, Lamattina L. 2008. Nitric oxide triggers phosphatidic acid accumulation via phospholipase D during auxin-induced adventitious root formation in cucumber. *Plant Physiol* 147:188–198.

Lanteri ML, Pagnussat GC, Lamattina L. 2006. Calcium and calcium-dependent protein kinases are involved in nitric oxide- and auxin-induced adventitious root formation in cucumber. *J Exp Bot* 57:1341–1351.

Larson RA. 1988. The antioxidants of higher plants. *Phytochemistry* 27:969–978.

Lecourieux D, Ranjeva R, Pugin A. 2006. Calcium in plant defence signalling pathways. *New Phytol* 171:249–269.

Lee KW, Chen PW, Lu CA, Chen S, Ho THD, Yu SM. 2009. Coordinated responses to oxygen and sugar deficiency allow rice seedlings to tolerate flooding. *Sci Signal* 2(91):ra61.

Lenaz G, Fato R, Formiggini G, Genova ML. 2007. The role of coenzyme Q in mitochondrial electron transport. *Mitochondrion* 7:S8–S33.

Leshem Y, Melamed-Book N, Cagnac O. 2006. Suppression of *Arabidopsis* vesicle-SNARE expression inhibited fusion of H_2O_2-containing vesicles with tonoplast and increased salt tolerance. *Proc Natl Acad Sci U S A* 103:18008–18013.

Libourel IG, van Bodegom PM, Fricker MD, Ratcliffe RG. 2006. Nitrite reduces cytoplasmic acidosis under anoxia. *Plant Physiol* 142:1710–1717.

Licausi F. 2010. Regulation of the molecular response to oxygen limitations in plants. *New Phytol* 190:550–555.

Licausi F, Kosmacz M, Weits DA et al. 2011. Oxygen sensing in plants is mediated by an N-end rule pathway for protein destabilization. *Nature* 479:419–422.

Lindermayr C, Saalbach G, Bahnweg G, Durner J. 2006. Differential inhibition of *Arabidopsis* methionine adenosyltransferases by protein S-nitrosylation. *J Biol Chem* 281:4285–4291.

Lindermayr C, Saalbach G, Durner J. 2005. Proteomic identification of S-nitrosylated proteins in *Arabidopsis. Plant Physiol* 137:921–930.

Liu F, Van Toai T, Moy LP, Bock G, Linford LD, Quackenbush J. 2005. Global transcription profiling reveals comprehensive insights into hypoxic response in *Arabidopsis. Plant Physiol* 137:1115–1129.

Loreti E, Alpi A, Perata P. 2003a. α-Amylase expression under anoxia in rice seedlings: An update. *Russ J Plant Physiol* 50:737–742.

Loreti E, Poggi A, Novi G, Alpi A, Perata P. 2005. A genome-wide analysis of the effects of sucrose on gene expression in *Arabidopsis* seedlings under anoxia. *Plant Physiol* 137:1130–1138.

Loreti E, Yamaguchi J, Alpi A, Perata P. 2003b. Gibberellins are not required for rice germination under anoxia. *Plant Soil* 253:137–143.

Lu CA, Ho THD, Ho SL, Yu SM. 2002. Three novel MYB proteins with one DNA binding repeat mediate sugar and hormone regulation of α-amylase gene expression. *Plant Cell* 14:1963–1980.

Lu CA, Lim EK, Yu SM. 1998. Sugar response sequence in the promoter of a rice α-amylase gene serves as a transcriptional enhancer. *J Biol Chem* 273:10120–10131.

Lu CA, Lin CC, Lee KW et al. 2007. The SnRK1A protein kinase plays a key role in sugar signaling during germination and seedling growth of rice. *Plant Cell* 19:2484–2499.

Ludewig U, Dynowski M. 2009. Plant aquaporin selectivity: Where transport assays, computer simulations and physiology meet. *Cell Mol Life Sci* 66:3161–3175.

Manac'h-Little N, Igamberdiev AU, Hill RD. 2005. Hemoglobin expression affects ethylene production in maize cell cultures. *Plant Physiol Biochem* 43:485–489.

Mancuso S, Shabala S (eds.). 2010. *Waterlogging Signaling and Tolerance in Plants*. Berlin, Germany: Springer.

Marschner H. 1995. *Mineral Nutrition of Higher Plants*, 2nd edn. London, U.K.: Academic Press.

Martin T, Frommer WB, Salanoubat M, Willmitzer L. 1993. Expression of an *Arabidopsis* sucrose synthase gene indicates a role in metabolisation of sucrose both during phloem loading and in sink organs. *Plant J* 4:367–377.

Maurel C, Javot H, Lauvergeat V. 2002. Molecular physiology of aquaporins in plants. *Int Rev Cytol* 215:105–148.

Maurel C, Santoni V, Luu D, Wudick MM, Verdoucq L. 2009. The cellular dynamics of plant aquaporin expression and functions. *Curr Opin Plant Biol* 12:690–698.

Meng Y, Ma X, Chen D, Wu P, Chen M. 2010. MicroRNA-mediated signaling involved in plant root development. *Biochem Biophys Res Commun* 393:345–349.

Mithöfer A, Schulze B, Boland W. 2004. Biotic and heavy metal stress response in plants: Evidence for common signals. *FEBS Lett* 566:1–5.

Mitsuhara I, Malik KA, Miura M, Ohashi Y. 1999. Animal cell-death suppressors Bcl-xL and Ced–9 inhibit cell death in tobacco plants. *Curr Biol* 9:775–778.

Mittler R, Vanderauwera S, Gollery M, Van Breusegem F. 2004. Reactive oxygen gene network of plants. *Trends Plant Sci* 9:490–498.

Monk LS, Braendle R, Crawford RMM. 1987. Catalase activity and post-anoxic injury in monocotyledonous species. *J Exp Bot* 38:233–246, doi:10.1093/jxb/38.2.233.

Monk LS, Fagerstedt KV, Crawford RMM. 1989. Oxygen toxicity and superoxide dismutase as an antioxidant in physiological stress. *Physiol Plant* 76:456–459.

Moreno DM, Martí MA, De Biase PM. 2011. Exploring the molecular basis of human manganese superoxide dismutase inactivation mediated by tyrosine 34 nitration. *Arch Biochem Biophys* 507:304–309.

Morgan MJ, Lehmann M, Schwarzländer M et al. 2008. Decrease in manganese superoxide dismutase leads to reduced root growth and affects tricarboxylic acid cycle flux and mitochondrial redox homeostasis. *Plant Physiol* 147:101–114.

Morita A, Umemura T, Kuroyanagi M, Futsuhara Y, Perata P, Yamaguchi J. 1998. Functional dissection of a sugar-repressed α-amylase gene (*RAmy1A*) promoter in rice embryos. *FEBS Lett* 423:81–85.

Mühlenbock P, Plaszczyca M, Plaszczyca M, Mellerowicz E, Karpinski S. 2007. Lysigenous aerenchyma formation in *Arabidopsis* is controlled by LESION SIMULATING DISEASE1. *Plant Cell* 19:3819–3830.

Murata T, Akazawa T, Fukuchi S. 1968. Enzymic mechanism of starch breakdown in germinating rice seeds: 1 An analytical study. *Plant Physiol* 43:1899–1905.

Murphy MP. 2009. How mitochondria produce reactive oxygen species. *Biochem J* 417:1–17.

Mustroph A, Lee SC, Oosumi T. 2010. Cross-kingdom comparison of transcriptomic adjustments to low-oxygen stress highlights conserved and plant-specific responses. *Plant Physiol* 152:1484–1500.

Mustroph A, Zanetti ME, Jang CJH. 2009. Profiling translatomes of discrete cell populations resolves altered cellular priorities during hypoxia in *Arabidopsis. Proc Natl Acad Sci U S A* 106:18843–18848.

Nagai K, Hattori Y, Ashikari M. 2010. Stunt or elongate? Two opposite strategies by which rice adapts to floods. *J Plant Res* 123:303–309.

Narsai R, Rocha M, Geigenberger P, Whelan J, van Dongen JT. 2011. Comparative analysis between plant species of transcriptional and metabolic responses to hypoxia. *New Phytol* 190:472–487.

Navarre DA, Wendehenne D, Durner J, Noad R, Klessig DF. 2000. Nitric oxide modulates the activity of tobacco aconitase. *Plant Physiol* 122:573–582.

Noctor G, Foyer CH. 1998. Ascorbate and glutathione: Keeping active oxygen under control. *Annu Rev Plant Physiol Mol Biol* 49:249–279.

Normile D. 2008. Sowing the seeds of expertise. *Science* 18:332.

Nyblom M, Frick A, Wang Y. 2009. Structural and functional analysis of SoPIP2;1 mutants adds insight into plant aquaporin gating. *J Mol Biol* 387:653–668.

de Oliveira HC, Wulff A, Saviani EE, Salgado I. 2008. Nitric oxide degradation by potato tuber mitochondria: Evidence for the involvement of external NAD(P)H dehydrogenases. *Biochim Biophys Acta Bioenerg* 1777:470–476.

Palmieri MC, Sell S, Huang X. 2008. Nitric oxide-responsive genes and promoters in *Arabidopsis thaliana*: A bioinformatics approach. *J Exp Bot* 59:177–186.

Pang J, Cuin T, Shabala L, Zhou M, Mendham N, Shabala S. 2007. Effect of secondary metabolites associated with anaerobic soil conditions on ion fluxes and electrophysiology in barley roots. *Plant Physiol* 145:266–276.

Pang J, Shabala S. 2010. Membrane transporters and waterlogging tolerance. In *Waterlogging Signaling and Tolerance in Plants*, eds. S Mancuso, S Shabala, pp. 197–219. Berlin, Germany: Springer.

Perata P, Alpi A. 1993. Plant responses to anaerobiosis. *Plant Sci* 93:1–17.

Perata P, Geshi N, Akazawa T, Yamaguchi J. 1993. Effect of anoxia on the induction of α-amylase in cereal seeds. *Planta* 191:402–408.

Perata P, Guglielminetti L, Alpi A. 1996. Anaerobic carbohydrate metabolism in wheat and barley, two anoxia-intolerant cereal seeds. *J Exp Bot* 47:999–1006.

Perata P, Loreti E, Guglielminetti L, Alpi A. 1998. Carbohydrate metabolism and anoxia tolerance in cereal grains. *Acta Bot Neerl* 47:269–283.

Perata P, Matsukura C, Vernieri P, Yamaguchi J. 1997. Sugar repression of a gibberellin-dependent signaling pathway in barley embryos. *Plant Cell* 9:2197–2208.

Perata P, Pozueta-Romero J, Akazawa T, Yamaguchi J. 1992. Effect of anoxia on starch breakdown in rice and wheat seeds. *Planta* 188:611–618.

Péret B, De Rybel B, Casimiro I. 2009. *Arabidopsis* lateral root development: An emerging story. *Trends Plant Sci* 14:399–408.

Pietta PG. 2000. Flavonoids as antioxidants. *J Nat Prod* 63:1035–1042.

Plekhanova I. 2007. Transformation of Fe, Mn, Co, and Ni compounds in humic podzols at different moisture. *Biol Bull* 34:67–75.

Plowman AB. 1906. The comparative anatomy and phylogeny of the *Cyperaceae. Ann Bot* 20:1–30.

Postma J, Lynch J. 2011. Theoretical evidence for the functional benefit of root cortical aerenchyma in soils with low phosphorus availability. *Ann Bot* 107:829–841.

Procházková D, Boušová I, Wilhelmová N. 2011. Antioxidant and prooxidant properties of flavonoids. *Fitoterapia* 82:513–523.

Pryor WA, Houk KN, Foote CS. 2006. Free radical biology and medicine: It's a gas, man! *Am J Physiol Regul Integr Comp Physiol* 291:R491–R511.

Raskin I, Kende H. 1984. Effect of submergence on translocation, starch content and amylolytic activity in deep-water rice. *Planta* 162:556–559.

Rasmusson AG, Geisler DA, Møller IM. 2008. The multiplicity of dehydrogenases in the electron transport chain of plant mitochondria. *Mitochondrion* 8:47–60.

Raymond P, Al-Ani A, Pradet A. 1985. ATP production by respiration and fermentation, and energy-charge during aerobiosis and anaerobiosis in 12 fatty and starchy germinating-seeds. *Plant Physiol* 79:879–884.

Rentel MC, Lecourieux D, Quaked F et al. 2004. OXI1 kinase is necessary for oxidative burst-mediated signalling in *Arabidopsis. Nature* 427:858–861.

Ricard B, Rivoal J, Spiteri A, Pradet A. 1991. Anaerobic stress induces the transcription and translation of sucrose synthase in rice. *Plant Physiol* 95:669–674.

Ricard B, Van Toai T, Chourey P, Saglio P. 1998. Evidence for the critical role of sucrose synthase for anoxic tolerance of maize roots using a double mutant. *Plant Physiol* 116:1323–1331.

Rocha M, Licausi F, Araujo WL. 2010. Glycolysis and the tricarboxylic acid cycle are linked by alanine aminotransferase during hypoxia induced by waterlogging of *Lotus japonicus. Plant Physiol* 152:1501–1513.

Rockel P, Strube F, Rockel A, Wildt J, Kaiser WM. 2002. Regulation of nitric oxide (NO) production by plant nitrate reductase in vivo and in vitro. *J Exp Bot* 53:103–110.

Rodriguez RL, Huang N, Sutliff TD, Ranjhan S, Karrer E, Litts J. 1992. Organization, structure and expression of the rice α-amylase multigene family. In *Rice Genetics II: Proceedings of the Second International Rice Genetics Symposium*, pp. 417–429. Los Banos, Philippines: IRRI.

Rosales EP, Iannone MF, Groppa MD, Benavides MP. 2011. Nitric oxide inhibits nitrate reductase activity in wheat leaves. *Plant Physiol Biochem* 49:124–130.

Rubbo H, Radi R. 2008. Protein and lipid nitration: Role in redox signaling and injury. *Biochim Biophys Acta Gen Subj* 1780:1318–1324.

Sachs MM, Vartapetian BB. 2007. Plant anaerobic stress I. Metabolic adaptation to oxygen deficiency. *Plant Stress* 1:123–135.

Sagi M, Fluhr R. 2001. Superoxide production by plant homologues of the gp91phox NADPH oxidase. Modulation of activity by calcium and by tobacco mosaic virus infection. *Plant Physiol* 126:1281–1290.

Sanchez-Fernandez R, Fricker M, Corben LB et al. 1997. Cell proliferation and hair tip growth in the *Arabidopsis* root are under mechanistically different forms of redox control. *Proc Natl Acad Sci U S A* 94:2745–2750.

Sang Y, Cui D, Wang X. 2001. Phospholipase D and phosphatidic acid-mediated generation of superoxide in *Arabidopsis. Plant Physiol* 126:1449–1458.

Santosa I, Ram P, Boamfa E. 2007. Patterns of peroxidative ethane emission from submerged rice seedlings indicate that damage from reactive oxygen species takes place during submergence and is not necessarily a post-anoxic phenomenon. *Planta* 226:193–202.

Saviranta NM, Julkunen-Tiitto R, Oksanen E, Karjalainen RO. 2010. Red clover (*Trifolium pratense* L.) isoflavones: Root phenolic compounds affected by biotic and abiotic stress factors. *J Sci Food Agric* 90:418–423.

Scebba F, Sebastiani L, Vitagliano C. 1998. Changes in activity of antioxidative enzymes in wheat (*Triticum aestivum*) seedlings under cold acclimation. *Physiol Plant* 104:747–752.

Schreiber L. 2010. Transport barriers made of cutin, suberin and associated waxes. *Trends Plant Sci* 15:546–553.

Schwarzländer M, Fricker MD, Sweetlove LJ. 2009. Monitoring the in vivo redox state of plant mitochondria: Effect of respiratory inhibitors, abiotic stress and assessment of recovery from oxidative challenge. *Biochim Biophys Acta Bioenerg* 1787:468–475.

Seago JL, Marsh LC, Stevens KJ, Soukup A, Votrubová O, Enstone DE. 2005. A re-examination of the root cortex in wetland flowering plants with respect to aerenchyma. *Ann Bot* 96:565–579.

Sedbrook JC, Kronebusch PJ, Borisy GG, Trewavas AJ, Masson PH. 1996. Transgenic AEQUORIN reveals organ specific cytosolic Ca^{2+} responses to anoxia in *Arabidopsis thaliana* seedlings. *Plant Physiol* 111:243–257.

Serbinova EA, Packer L. 1994. Antioxidant properties of α-tocopherol and α-tocotrienol. *Methods Enzymol* 234:354–366.

Setter TL, Waters I, Atwell BJ, Kupkanchanakul T, Greenway H. 1987. Carbohydrate status of terrestrial plants during flooding. In *Plant Life in Aquatic and Amphibious Habitats*, ed. RMM Crawford, pp. 411–433. Oxford, U.K.: British Ecological Society, Blackwell Scientific Publishers.

Sgherri C, Cosi E, Navari-Izo F. 2003. Phenols and antioxidative status of *Raphanus sativus* grown in copper excess. *Physiol Plant* 118:21–28.

Sheu JJ, Yu TS, Tong WF, Yu SM. 1996. Carbohydrate starvation stimulates differential expression of rice α-amylase genes that is modulated through complicated transcriptional and posttranscriptional processes. *J Biol Chem* 271:26998–27004.

Shiono K, Ogawa S, Yamazaki S, Isoda H, Fujimura T, Nakazono M, Colmer TD. 2011. Contrasting dynamics of radial O_2-loss barrier induction and aerenchyma formation in rice roots of two lengths. *Ann Bot* 107:89–99.

Silvertown J, Dodd ME, Gowing DJG, Mountford JO. 1999. Hydrologically defined niches reveal a basis for species richness in plant communities. *Nature* 400:61–63.

Skutnik M, Rychter AM. 2009. Differential response of antioxidant systems in leaves and roots of barley subjected to anoxia and post-anoxia. *J Plant Physiol* 166:926–937.

Smagghe BJ, Trent III JT, Hargrove MS. 2008. NO dioxygenase activity in hemoglobins is ubiquitous in vitro, but limited by reduction *in vivo. PLoS One* 3:e2039.

Smirnoff N. 2000. Ascorbic acid: Metabolism and functions of a multi-faceted molecule. *Curr Opin Plant Biol* 3:229–235.

Smith AM, Stitt M. 2007. Coordination of carbon supply and plant growth. *Plant Cell Environ* 30:1126–1149.

Solano R, Stepanova A, Chao Q, Ecker JR. 1998. Nuclear events in ethylene signalling, a transcriptional cascade mediated by ETHYLENE-INSENSITIVE3 and ETHYLENE-RESPONSE-FACTOR1. *Genes Dev* 12:3703–3714.

Steffens B, Sauter M. 2010. G proteins as regulators in ethylene-mediated hypoxia signaling. *Plant Signal Behav* 5:375–378.

Steffens B, Wang J, Sauter M. 2006. Interactions between ethylene, gibberellin and abscisic acid regulate emergence and growth rate of adventitious roots in deepwater rice. *Planta* 223:604–612.

Stöhr C, Stremlau S. 2006. Formation and possible roles of nitric oxide in plant roots. *J Exp Bot* 57:463–470.

Stöhr C, Strube F, Marx G, Ullrich WR, Rockel P. 2001. A plasma membrane-bound enzyme of tobacco roots catalyses the formation of nitric oxide from nitrite. *Planta* 212:835–841.

Stoimenova M, Igamberdiev AU, Gupta K, Hill RD. 2007. Nitrite-driven anaerobic ATP synthesis in barley and rice root mitochondria. *Planta* 226:465–474.

Stoimenova M, Libourel IG, Ratcliffe RG, Kaiser WM. 2003. The role of nitrate reduction in the anoxic metabolism of roots II. Anoxic metabolism of tobacco roots with or without nitrate reductase activity. *Plant Soil* 253:155–167.

Su Z, Chai MF, Lu PL, An R, Chen J, Wang XC. 2007. AtMTM1, a novel mitochondrial protein, may be involved in activation of the manganese-containing superoxide dismutase in *Arabidopsis*. *Planta* 226:1031–1039.

Su PH, Wu TH, Lin CH. 1998. Root sugar level in luffa and bitter melon is not referential to their flooding tolerance. *Bot Bull Acad Sinica* 39:175–179.

Subbaiah CC, Bush DS, Sachs MM. 1994a. Elevation of cytosolic calcium precedes anoxic gene expression in maize suspension-cultured cells. *Plant Cell* 6:1747–1762.

Subbaiah CC, Sachs MM. 2001. Altered patterns of sucrose synthase phosphorylation and localization precede callose induction and root tip death in anoxic maize seedlings. *Plant Physiol* 125:585–594.

Subbaiah CC, Sachs MM. 2003. Molecular and cellular adaptations of maize to flooding stress. *Ann Bot* 91:119–127.

Subbaiah CC, Zhang J, Sachs MM. 1994b. Involvement of intracellular calcium in anaerobic gene expression and survival of maize seedlings. *Plant Physiol* 105:369–376.

Sun Z, Henson CA. 1991. A quantitative assessment of the importance of barley seed α-amylase, debranching enzyme, and α-glucosidase in starch degradation. *Arch Biochem Biophys* 284:298–305.

Sunkar R, Kapoor A, Zhu JK. 2006. Posttranscriptional induction of two Cu/Zn superoxide dismutase genes in *Arabidopsis* is mediated by downregulation of miR398 and important for oxidative stress tolerance. *Plant Cell* 18:2051–2065.

Swarup R, Kramer EM, Perry PJ, Knox K, Leyser HMO, Haseloff J, Beemster GTS, Bhalerao R, Bennett MJ. 2005. Root gravitropismrequireslateral root cap and epidermalcells fortransportandrespon seto amobileauxinsignal. *Nat Cell Biol* 7:1057–1065.

Sweetlove LJ, Beard KFM, Nunes-Nesi A, Fernie AR, Ratcliffe RG. 2010. Not just a circle: Flux modes in the plant TCA cycle. *Trends Plant Sci* 15:462–470.

Tadege M, Brändle R, Kuhlemeier C. 1998. Anoxia tolerance in tobacco roots: Effect of overexpression of pyruvate decarboxylase. *Plant J* 14:327–335.

Takahama U, Oniki T. 1997. A peroxide/phenolics/ascorbate system can scavenge hydrogen peroxide in plant cells. *Physiol Plant* 101:845–852.

Tanaka M, Wallace IS, Takano J, Roberts DM, Fujiwara T. 2008. NIP6;1 Is a boric acid channel for preferential transport of boron to growing shoot tissues in *Arabidopsis*. *Plant Cell* 20:2860–2875.

Thomas BR, Rodriguez RL. 1994. Metabolite signals regulate gene expression and source/sink relations in cereal seedlings. *Plant Physiol* 106:1235–1239.

Thomas CE, McLean LR, Parker RA, Ohlweiler DF. 1992. Ascorbate and phenolic antioxidant interactions in prevention of liposomal oxidation. *Lipids* 27:543–550.

Tornroth-Horsefield S, Wang Y, Hedfalk K. 2006. Structural mechanism of plant aquaporin gating. *Nature* 439:688–694.

Tournaire-Roux C, Sutka M, Javot H. 2003. Cytosolic pH regulates root water transport during anoxic stress through gating of aquaporins. *Nature* 425:393–397.

Toyofuku K, Umemura T, Yamaguchi J. 1998. Promoter elements required for sugar-repression of the *RAmy3D* gene for α-amylase in rice. *FEBS Lett* 428:275–280.

Turrens JF. 2003. Mitochondrial formation of reactive oxygen species. *J Physiol* 552:335–344.

Tyerman SD, Niemietz CM, Bramley H. 2002. Plant aquaporins: Multifunctional water and solute channels with expanding roles. *Plant Cell Environ* 25:173–194.

Van Breusegem F, Bailey-Serres J, Mittler R. 2008. Unraveling the tapestry of networks involving reactive oxygen species in plants. *Plant Physiol* 147:978–984.

Van Doorn WG, Woltering EJ. 2005. Many ways to exit? Cell death categories in plants. *Trends Plant Sci* 10:117–122.

Van Toai TT, Bolles CS. 1991. Postanoxic injury in soybean (*Glycine max*) seedlings. *Plant Physiol* 97:588–592.

Virolainen E, Blokhina O, Fagerstedt KV. 2002. Ca^{2+}-induced high amplitude swelling and cytochrome *c* release from wheat (*Triticum aestivum* L.) mitochondria under anoxic stress. *Ann Bot* 90:509–516.

Visser EJW, Colmer TD, Blom CWPM, Voesenek LACJ. 2000. Changes in growth, porosity, and radial oxygen loss from adventitious roots of selected mono- and dicotyledonous wetland species with contrasting types of aerenchyma. *Plant Cell Environ* 23:1237–1245.

Visser EJW, Voesenek LACJ, Vartapetian BB, Jackson MB. 2003. Flooding and plant growth. *Ann Bot* 91:107–109.

Wang J, Pan R. 2006. Effect of ethylene on adventitious root formation. In *Ethylene Action in Plants*, ed. NA Khan, pp. 69–79. Berlin, Germany: Springer.

Wegner LH. 2010. Oxygen transport in waterlogged plants. In *Waterlogging Signaling and Tolerance in Plants*, eds. S Mancuso, S Shabala, pp. 3–22. Berlin, Germany: Springer.

Weig AR, Jakob C. 2000. Functional identification of the glycerol permease activity of *Arabidopsis thaliana* NLM1 and NLM2 proteins by heterologous expression in *Saccharomyces cerevisiae*. *FEBS Lett* 481:293–298.

Wille F. 1926. Beiträge zur Anatomie des Cyperaceenrhizoms. *Beih Bot Zbl* 43:267–309.

Wingate VPM, Lawton MA, Lamb CJ. 1988. Glutathione causes a massive and selective induction of plant defense genes. *Plant Physiol* 87:206–210.

Wink DA, Mitchell JB. 1998. Chemical biology of nitric oxide: Insights into regulatory, cytotoxic, and cytoprotective mechanisms of nitric oxide. *Free Radic Biol Med* 25:434–456.

Woolley JT. 1983. Maintenance of air in intercellular spaces of plants. *Plant Physiol* 72:989–991.

Wrzaczek M, Brosche M, Kollist H, Kangasjärvi J. 2010. *Arabidopsis* GRI is involved in the regulation of cell death induced by extracellular ROS. *Proc Natl Acad Sci U S A* 106:5412–5417.

Xu K, Xu X, Fukao T et al. 2006. *Sub1A* is an ethylene-response-factor-like gene that confers submergence tolerance to rice. *Nature* 442:705–708.

Yamasaki H, Sakihama Y. 2000. Simultaneous production of nitric oxide and peroxynitrite by plant nitrate reductase: In vitro evidence for the NR-dependent formation of active nitrogen species. *FEBS Lett* 468:89–92.

Yamasaki H, Sakihama Y, Takahashi S. 1999. An alternative pathway for nitric oxide production in plants: New features of an old enzyme. *Trends Plant Sci* 4:128–129.

Yan B, Dai Q, Liu X, Huang S, Wang Z. 1996. Flooding-induced membrane damage, lipid oxidation and activated oxygen generation in corn leaves. *Plant Soil* 179:261–268.

Yang XD, Dong CJ, Liu JY. 2006. A plant mitochondrial phospholipid hydroperoxide glutathione peroxidase: Its precise localization and higher enzymatic activity. *Plant Mol Biol* 62:951–962.

Zabalza A, van Dongen JT, Froehlich A. 2009. Regulation of respiration and fermentation to control the plant internal oxygen concentration. *Plant Physiol* 149:1087–1098.

Zeeman SC, Smith SM, Smith AM. 2007. The diurnal metabolism of leaf starch. *Biochem J* 401:13–28.

Zeng Y, Wu Y, Avigne WT, Koch KE. 1999. Rapid repression of maize invertases by low oxygen. Invertase/sucrose synthase balance, sugar signaling potential, and seedling survival. *Plant Physiol* 121:599–608.

Zeng Y, Wu Y, Wayne TA, Koch KE. 1998. Differential regulation of sugar-sensitive sucrose synthase by hypoxia and anoxia indicate complementary transcriptional and posttranscriptional responses. *Plant Physiol* 116:1573–1583.

Zhang H, Dong S, Wang M, Wang W, Song W, Dou X, Zheng X, Zhang Z. 2010. The role of vacuolar processing enzyme (VPE) from *Nicotiana benthamiana* in the elicitor-triggered hypersensitive response and stomatal closure. *J Exp Bot* 61:3799–3812.

Zhang L, Xu Q, Xing D, Gao C, Xiong H. 2009. Real-time detection of caspase-3-like protease activation in vivo using fluorescence resonance energy transfer during plant programmed cell death induced by ultraviolet c overexposure. *Plant Physiol* 150:1773–1783.

Zhang W, Yu L, Zhang Y, Wang X. 2005. Phospholipase D in the signaling networks of plant response to abscisic acid and reactive oxygen species. *Biochim Biophys Acta Mol Cell Biol Lipids* 1736:1–9.

33

Plant Roots under Aluminum Stress: Toxicity and Tolerance

Hideaki Matsumoto
Okayama University

Yoko Yamamoto
Okayama University

I. Introduction: Aluminum Toxicity and Acid Soils

Aluminum toxicity is the primary problem limiting agricultural production in acid soils. The reason for this is because aluminum is one of the most prevalent minerals in the Earth's crust, and high acidity in soil can render aluminum soluble, thereby poisoning the water taken up by the roots. Acid soils occupy 3.95 billion ha (30%) of the world's ice-free land area (Baligar et al. 1998) comprising both the tropical and temperate belts, and as much as 50% of the world's potentially arable crops are negatively affected by acid soils (Matsumoto et al. 2008).

Acid soils inhibit crop production in many developing countries where food production is crucial. In addition to the natural factors that affect weathering, agricultural farming processes such as the excessive supply of inorganic fertilizers or removal of cations by harvest lower the soil pH. Furthermore, the acidity of environment is gradually increased due to environmental pollution and acid rain (Kopáček et al. 2009). Acid soils are infertile because they lack the basic nutrients, such as Ca^{2+}, Mg^{2+}, and K^{+} and are also characterized by high content of toxic elements such as Al, Mn, and Fe or deficiency of P. Most acid soils have low cation exchange capacity, leading to loss of essential minerals and to poor crop production.

II. Occurrence and Chemistry of Al

Exchangeable Al is, next to Mn, the major toxic elements in most acid soils. Al exists as insoluble aluminosilicates or oxides in the neutral or weakly acidic soil. At neutral pH, Al forms a poorly soluble Al–phosphatehydroxo complex with a complicated chemical form and biological function limiting solubility to ~20 μM (Yokel 2004). At pH below 5, Al^{3+} exists as the octahedral hexahydrate, $Al(H_2O)_6^{3+}$, often abbreviated as Al^{3+}. As the solution becomes less acidic, $Al(H_2O)_6^{3+}$ undergoes successive deprotonations to yield $Al(OH)^{2+}$ and $Al(OH)_2^+$. In neutral solution, $Al(OH)_3$ precipitates as gibbsite that redissolves in basic solutions owing to formation of tetrahedral $Al(OH)_4^-$ as aluminate anion. The nadir of aqueous Al solubility is at pH 6.2 (Harris et al. 1997), with free Al ion concentration at pH 4, 5, 6, and 7 being ~50 mM, 50 μM, 0.05 μM, and 0.05 nM, respectively (Yokel 2004).

Time-dependent formation of polynuclear species may also take place (Martin 1986). Since Al toxicity differs with the chemical form of Al, many studies have been done regarding this interaction, especially between Al^{3+} and mononuclear hydroxy–Al. Generally, Al^{3+} is more phytotoxic than $Al(OH)^{2+}$ or $Al(OH)_2^+$. Root elongation of wheat and clover was inhibited to less than 4% by 25 μM $Al(OH)_4^-$ at pH 8 but not at pH 8.9 where elongation was unaffected, suggesting the $Al(OH)_4^-$ is almost nontoxic. In the soil solution, Al^{3+} reacts not only with OH^- but also with phosphate, F^-, SO_4^{2-}, silicate, and a large number of organic ligands. Under specific conditions of OH/Al ratio, total Al, and stirring rate, $AlO_4Al_{12}(OH)_{24}(H_2O)_{12}^{7+}$ (Al_{13} polymer), which is highly toxic, can be formed (Parker et al. 1989; Kinraide 1990). Nevertheless, Al_{13} was not observed in soil solutions (Funakawa et al. 1993).

III. Inhibition of Root Elongation by Al

Inhibition of root elongation is the first visible symptom of Al stress. In most plant species, root elongation is markedly inhibited by Al^{3+} at the μmol level in a simple solution containing Ca^{2+} alone. Inhibition of root elongation of Al-sensitive maize occurred within 30 min of Al treatment (Llungany et al. 1995). Changes in the electric charge of the root surface by other ions, especially cations, affect the accessibility of Al^{3+}. Root elongation in Al-sensitive wheat (*Triticum aestivum*) cultivar, Scout 66, was apparently inhibited by a 3 h treatment with 5 μM Al, but that of Al-tolerant Atlas 66 was inhibited to the same degree only by a 10 times higher concentration of Al (Sasaki 1996). The root apex (root cap, meristem, distal transition zone [DTZ], and elongation zone) accumulated most of the Al and played a major role in the Al-perception mechanism (Zheng et al. 2005a). Indeed, only the apical 2–3 mm of maize and pea roots should be exposed to Al for the inhibition of root elongation to take place (Delhaize et al. 1995; Matsumoto et al. 1996). Sivaguru et al. (1998) found that the distal part of transition zone of the root apex of corn, where cells undergo a preparatory phase for rapid elongation, is the primary target of Al.

A. Morphological Changes of Intact Roots and Root Cells under Al Stress

A number of studies have shown that the rhizodermis of roots may rupture or crack when exposed to Al (Budikova and Ciamporova 1998; Yamamoto et al. 2001). Morphological changes in wheat roots were characterized by cracks on the root surface (Sasaki 1996). Shortening of the root elongation zone with the formation of cracks by Al is accompanied by an increase in the diameter and a decrease of the length of the cells in the outer layer of the cortex of the elongation zone of Atlas 66 wheat plants (Sasaki et al. 1996; Matsumoto 2000). The ratio of length to diameter of the cells in the control root was three to four times larger than that in the Al-treated roots, and cells in the second and third layers of the cortex were swollen laterally. Therefore, cracking might be caused by the outward pressure of the cells in outer cortex of wheat roots. Jones et al. (2006) reported that exposure to Al induced a rigidification of the cell wall in maize root. They suggested that Al-induced root inhibition in maize occurs by rigidification of the epidermal layers through the function of callose. The injuries to the root by the formation of ruptures or cracks are induced at root apex, especially epidermis. The order of appearance of cell injury with Al is in the order of accumulation with a gradual increase of Al: epidermis > outer cortex > middle cortex > inner cortex > stele (Ciamporova 2002).

Zhu et al. (2003) found that Al seriously inhibited the production and release of border cells, resulting in clumping of border cells in Al-sensitive wheat (Scout 66) but less clustering in Al-tolerant wheat (Atlas 66). The number of border cells released from roots treated with Al was significantly less than that from roots grown without Al treatment. The removal of border cells from root tips of both Atlas 66 and Scout 66 enhanced the Al-induced inhibition of root elongation concomitant with increased Al accumulation in the root. Later, Huang et al. (2009b) reported that morphological changes in root outer cell layers between Al-sensitive and wild-type rice protect the roots against the toxicity of Al and other metals by preventing metal penetration into the inner cells.

B. Hormone and Al Toxicity

Al has a localized effect on auxin transport, and unilateral application of Al caused root curvature. Application of Al to the root cap of Al-sensitive maize strongly promoted acropetal transport of auxin, reducing auxin transport polarity (basipetal transport divided by acropetal transport) from 6.3 to 2.1. Treatment of the root cap with Ca^{2+}-enhanced basipetal movement of auxin, increasing auxin transport polarity from 6.3 to 7.6, suggests that Al and Ca have opposite effects on this process (Hasenstein et al. 1988). Kollmeier et al. (2000) provided the evidence that the primary mechanism responsible for genotypic differences in Al resistance is located within DTZ and suggested that a signaling pathway in the root apex mediates the Al signal between the DTZ and the elongation zone through basipetal auxin transport.

Massot et al. (2002) reported that Al^{3+}-induced inhibition of root growth may be predicted by significant changes in cytokinin's content and composition and enhanced ethylene biosynthesis. It has been identified that increased ethylene production was closely associated with Al-induced root growth inhibition in lotus plants (Sun et al. 2007). Recently, Sun et al. (2010) investigated the role of ethylene and auxin in Al^{3+}-induced inhibition of root elongation in *Arabidopsis thaliana* using wild type and mutants defective in ethylene signaling (*etr1-3* and *ein2-1*) and auxin polar transporter (*aux-7* and *pin 2*). Their findings suggest that Al^{3+}-induced ethylene production is likely to act as a signal that alters auxin distribution in roots by disrupting AUX-1- and PIN2-mediated auxin polar transport, leading to the arrest of root elongation.

C. Impact of Al on Cytoskeletal Proteins

The Al-induced inhibition of longitudinal cell expansion and cell swelling in the elongation zone might be related to the disorder of the cytoskeletal network (Sivaguru et al. 1999a). The actin network plays an important role in the plant cell under Al stress (Frantzios et al. 2005; Ahad et al. 2007; Amenos et al. 2009). In soybean cells, the tension and organization of the actin network were modified by calcium-regulated phosphatase and kinase (Grabski et al. 1998). Al induced a significant increase in the tension within the transvacuolar actin network in soybean cells (Grabski et al. 1995). This may result from the formation of nonhydrolyzable (Al^{3+}–ATP) complexes whose binding to actin/myosin can modify filament contraction. The tension-inducing activity of Al in the presence of inhibitors of kinases and phosphatase was investigated (Grabski et al. 1995). Inhibitors of calmodulin (CaM) (W-7 and calmidazolium) and CaM-dependent kinases (KT 5926) such as CaM-like domain protein kinase (CDPK) and myosin light chain kinase (MLCK) induced a decrease in the tension within the actin network of soybean cells. Preincubation of cells with the CM inhibitors prevented the ability of Al to enhance the tension of the actin network. The loss of the tension-enhancement activity mediated by Al was also observed in cells preincubated with an inhibitor of MLCK, KT 5926.

Al treatment resulted in a reorganization of microtubules (MTs) in the inner cortex of the elongation zone of *Zea mays* (Blancaflor et al. 1998). The orientation of the MTs is closely related to cell expansion. Longitudinally elongating cells have transversely oriented MTs. MT-disrupting agents promote lateral expansion but inhibit longitudinal expansion. Cortical MTs are known to be involved in the orientation of cellulose microfibrils. Indeed, the disappearance of the cortical MTs in elongating cells of wheat roots that was observed under Al stress (Sasaki et al. 1997a) might be responsible for these changes in cell growth. Moreover, the time-dependent effect of Al on MT stability was correlated with the reduction of root growth. However, the orientation of MTs in the outer cortex and epidermis remained unchanged even after chronic symptoms of toxicity were evident. They also found that the auxin-induced reorientation and cold-induced depolymerization of MTs in the outer cortex were blocked by Al, suggesting that Al increased the stability of MTs in these cells. The changes of behavior of MTs under Al stress may depend on the growth phase of the cells (Sivaguru et al. 1999b).

D. Inhibition of Cell Division

Cell division in root meristems of several plants is inhibited by Al (Clarkson 1965; Morimura et al. 1978b). Cell division accounts for only 1%–2% of the overall root elongation, and cell cycle in plants takes about 1 day. However, the primal phenomena of Al toxicity are the inhibition of root elongation that occurs within hour(s) of Al treatment. Thus, attention has been largely paid to the inhibition of root cell elongation as the primary site of Al toxicity. However, the inhibition of cell elongation at the elongation zone is not fatal for plant growth as long as the cells can divide at the meristematic zone, suggesting that the lethal cause of Al toxicity might be inhibition of cell division (Matsumoto 2000).

Clarkson reported as early as 1965 a close correlation between the cessation of root elongation and disappearance of mitotic forms in the onion roots under Al stress. This phenomenon was further confirmed by a distinct localization of Al in the nuclei of the onion root meristematic zone (Morimura et al. 1978a), in the developing lateral roots of pea (*Pisum sativum* cv. Alaska) (Matsumoto et al. 1976), and in nuclei of root hair cells (Matsumoto et al. 1977b). Al entered into cells of the soybean (*Glycine max* L. Merr.) cv Young, an Al-sensitive genotype, and accumulated at nuclei in the meristematic region of the root tip (Silva et al. 2000). Unwinding of double strands of DNA is a prerequisite for expression of genetic information, but separation of double strands of DNA was interrupted by Al (Matsumoto 1991). Furthermore, the structural change of chromatin in pea roots treated with Al in vivo implied that Al induced the condensation and/or aggregation of chromatin (Matsumoto 1988). These results suggested that the inhibition of RNA synthesis was caused by Al (Morimura et al 1978a).

In suspension of tobacco cells, Sivaguru et al. (1999b) found that the actively dividing log-phase cells were characterized by faint and larger phragmoplasts and unusually enlarged daughter nuclei after 6 h of Al treatment. After a 24 h treatment, no phragmoplasts and spindle microtubules (SMTs) from cells having metaphase plate chromosomes were observed, suggesting that Al might block cell division directly at the metaphase. Doncheva et al. (2005) investigated that the short-term influence (5–180 min) of 50 μM Al on cell division in *Z. mays* L. varieties differs in Al resistance. Their observations suggest a fast change in cell patterning rather than a general karyotoxic effect after exposure to Al for a short time in Al-sensitive plants. No such changes were found in Al-resistant maize. The occurrence of different types of chromosomal aberrations, reduction in the amount of nuclear DNA, and persistence of the phytotoxic effects at the posttreatment stage suggest carcinogenic effects of Al on rice plants (Mohanty et al. 2004).

IV. Site of Al Toxicity

A. Cell Wall

Although there is disagreement with regard to the site of Al toxicity–namely, symplastic or apoplastic (Zhdesng et al. 2005a)–many investigators have stated that 30%–90% of the absorbed Al is localized in the apoplast (Tice et al. 1992; Rengel 1996). Al binding to the cell wall can be advantageous for plants because

Al is trapped and entry of toxic Al into cytosol could be inhibited (Wang et al. 2004; Hiradate et al. 2007). Regarding the genotypic differences in Al tolerance, evidence for a role of the cell wall negativity possessing higher capacity of Al binding was provided with maize and rice (Eticha et al. 2005; Yang et al. 2008).

As to the binding site of Al in the apoplast, pectin carboxyl was suggested as a plausible candidate, although almost no evidence has been found to show the binding of Al to pectin in vivo (Horst 1995; Matsumoto et al. 1977a; Eticha et al. 2005; Rangel et al. 2005). Although Al is bound by the negative charge of pectin, the binding capacity of pectin varies with the plant species, and the pectin content is extremely different between monocots and dicots. Even in the same species, the pectin content of the roots differs with the position on the root or with the chemical modification of pectin, such as methylation or demethylation changes with the physiological activity of the cell. Binding of Al to the cell wall pectin matrix would be important, and their methylation reaction could regulate the affinity for Al (Schmohl et al. 2000; Wehr et al. 2004; Mimmo et al. 2005). Le et al. (1994) found that Al increased pectin hemicellulose contents in the elongating zone of squash. These changes could protect the cells from Al toxicity because Al was trapped by pectin (Zheng et al. 2004).

In buckwheat (*Fagopyrum esculentum* Moench), formation of Al–P complexes in the cell wall, such as low-solubility $Al_4(PO_4)_3$, may be helpful by decreasing uptake of Al into cytosol (Zheng et al. 2005b). In maize, Al treatment greatly enhanced Si accumulation in the cell wall fraction, resulting in reducing mobility of apoplastic Al, thus detoxifying Al (Wang et al. 2004). The swollen cells of wheat roots were characterized by the drastic accumulation of lignin resulting in the rigidification of cell wall under Al stress (Sasaki et al. 1996).

The decrease of cell viability in the elongation zone of wheat roots coincided with the time required for the inhibition of the root elongation and lignin deposition (Sasaki et al. 1997b). Al affected the biochemical and biophysical nature of the cell wall. In Al-sensitive wheat (Scout 66), Al modified the metabolism of cell wall and makes the cell wall thick and rigid, resulting in growth inhibition (Tabuchi et al. 2001). In Al-tolerant wheat (Atlas 66), Al treatment decreased the osmotic potential of the root cells. This did not occur in Scout 66, suggesting that Al-tolerant Atlas 66 osmotically adapted to water uptake which allowed for the elongation of the cells under Al stress (Tabuchi et al. 2004).

Interactions of Al with other cell wall components, such as enzymes, extensin, xyloglucan, and plasma membrane, may also affect the functional integrity of the cell wall. Furthermore, the targeting of the cell wall and plasma membrane through the cytoskeleton proteins will become important because the signal of Al stress on the apoplast can be transferred into the symplast through the cytoskeleton even if the entry of Al into symplast is strictly limited (Horst et al. 1999). Al-induced cell wall–associated kinase receptors (WAKs) indicated that overexpression of this protein can induce Al tolerance (Sivaguru et al. 2003a). Al increased the level of covalently bound cell wall proteins in pea root (*P. sativum* cv. Alaska). In vitro and in vivo, Al-binding experiments suggested that extensin has the highest capacity to bind Al among cell wall proteins (Kenjebaeva et al. 2000).

B. Plasma Membrane

The plasma membrane is one of the first targets of Al (Haug 1984; Matsumoto 2000). Structural and functional changes of the plasma membranes are induced by Al binding (Chen et al. 1991; Nichol et al. 1993). The important change in the plasma membrane mediated by Al is the alteration of membrane potential (V_m) of the plasma membrane as well as the changes on surface potential (zeta potential). A change of membrane depolarization potential of root cap cells indicated Al tolerance in snap beans (*Phaseolus vulgaris* L.) (Olivetti et al. 1995; Lindberg et al. 1997; Papernik et al. 1997). Wherrett et al. (2005) reported that Al-activated ion fluxes might induce changes in the membrane potential, and that these responses would differ between wheat genotypes that differed with regard to Al tolerance.

The zeta potential of the cell membrane is known to regulate the accessibility of Al^{3+} inside of the cells (Kinraide et al. 1992). Al reduced the negative charge associated with phospholipids, that is, depolarization of zeta potential, and proteins by binding to these charged groups or shielding the surface charges. The role of zeta potential in terms of Al toxicity was also compared between Al-tolerant and Al-sensitive varieties (Wagatsuma et al. 1995). Ahn et al. (2001, 2002, 2004) and Ahn and Matsumoto (2006) conducted intensive research regarding changes in zeta potential and plasma membrane H^+-ATPase activity. Zeta potential might have a regulatory effect on the H^+-ATPase. The segmental analysis showed that the zeta potential was more negative at root tip than in other regions (Figure 33.1; Ahn et al. 2004). It was also suggested that the rate of depolarization of the plasma membrane zeta potential in the Al-sensitive wheat ES8 line was always higher than in the Al-tolerant ET8 line, at Al concentration in excess of 10 μM, which inhibited root elongation (Ahn et al. 2004). However, the varietal sensitivity to Al^{3+} was not always based on the difference in cell surface electrical potential (Kinraide et al. 1992). Further research will be needed to understand the relationship between ion flux and zeta potential under Al stress.

Changes in the plasma membrane lipid composition were found between Al-resistant (PT741) and Al-sensitive (Katepwa) cultivar of wheat (Zhang et al. 1997). One of the biochemical changes of the plasma membrane is the Al-dependent lipid peroxidation (for more details, see Section VII).

C. Aluminum Toxicity Recovery

Many studies of the mechanism of Al toxicity and tolerance have been performed. Root elongation completely stops in the presence of excess level of toxic Al, and plants begin dying without recovery. The root elongation in acid soil might not be uniform, and the inhibition and elongation and/or re-elongation including the recovery process might occur simultaneously or may

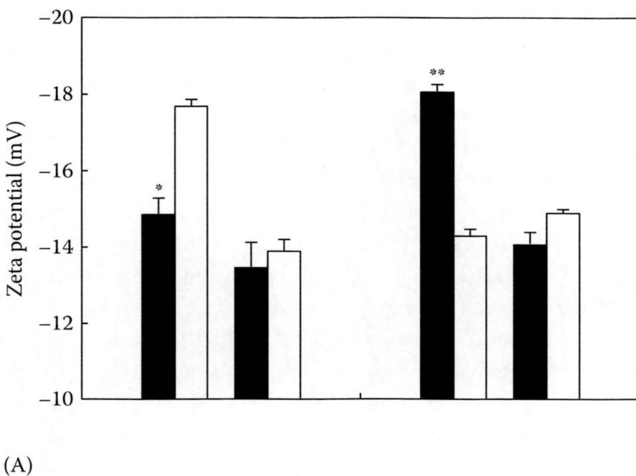

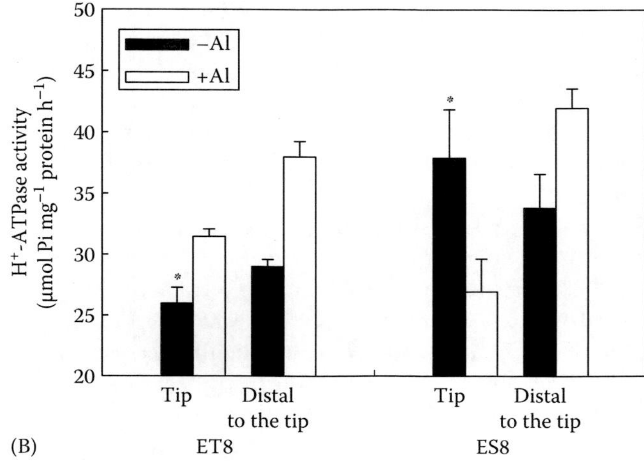

FIGURE 33.1 Effect of Al treatment (2.6 μM) on the surface potential (zeta potential) (A) and the H⁺-ATPase activity (B) of plasma membrane (PM) vesicles isolated from root tips (0–10 mm) and the region distal to the tip (10–20 mm) of ET8 and ES8 near-isogenic wheat lines. Plants were grown in 0.2 mM $CaCl_2$ solution for 5 days from germination and were then treated with 2.6 μM Al (pH 4.5) for 4 h before isolation of PM vesicles. The electrophoresis medium was free of Al. Values are means ±SE (n = 4) of three separate experiments. Asterisks indicate statistically different means between −Al (closed columns) and +Al (open columns) treatment: *, $P < 0.05$; **, $P < 0.025$ (Student's t-test). (From Ahn, S.J. et al., *New Phytol.*, 162, 71, 2004. With permission.)

be repeated under the limited concentration of Al (Matsumoto et al. 2012). However, research on the process of recovery from Al-induced inhibition of root elongation has received little attention (Tamas et al. 2003; Kikui et al. 2007). Morphological changes in the root apex together with the inhibition of root elongation are induced by Al (Eleftherios et al. 1993; Ciamporova 2000, 2002; Yamamoto et al. 2001; Budikova et al. 1998, 2004; Kopittke et al. 2008). Recovery process from Al-induced morphological change at root apex of pea was investigated (Motoda et al. 2010, 2011). The relative root growth (RRG), which is the ratio of root length, increased during a certain period in the Al-treated seedlings compared to the control seedlings and was 35% during Al treatment and 75% in the recovery stage (Figure 33.2). RRG was stable, and this treatment was used as a model system for the recovery process. As shown in Figure 33.3, root apex showed two different morphologies. The rupture region (RR) and belt-like

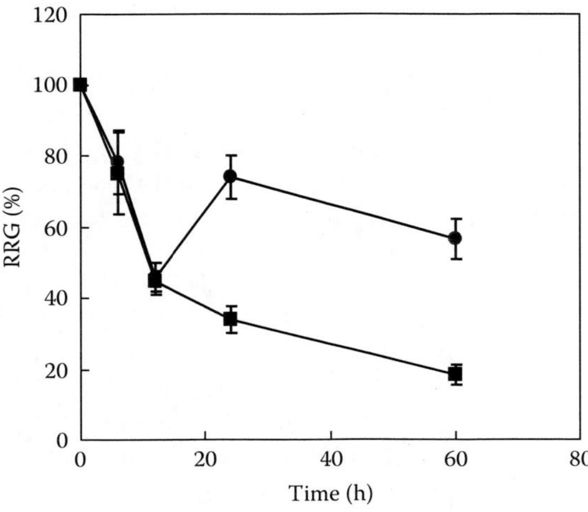

FIGURE 33.2 RRG of the pea root cultured in 40 μM Al solution for 12 h and then in Al-free solution (●) (50 μM $CaCl_2$ solution) or left in the same Al solution (■). RRG during 0–6, 6–12, 12–24, and 24–60 h was calculated. Experiments were repeated twice which showed similar tendency. Values represent means ±SEM (n = 8) is a typical set of experiment. (From Motoda, H. et al., *Plant Soil*, 333, 49, 2010. With permission.)

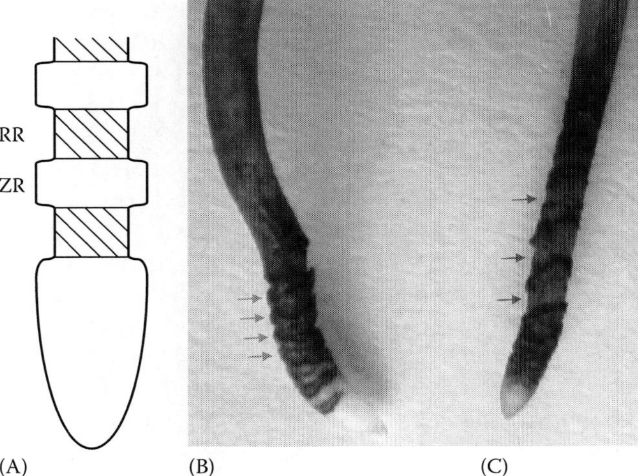

FIGURE 33.3 **(See color insert.)** Changes in the morphology of the pea root apex during and after recovery from Al treatment. (A) Schematic diagram of the root apex structure induced by Al toxicity. RR: rupture region, ZR: zonary region stained with Evans blue. (B) Pea root treated with 40 μM Al for 24 h. (C) Pea root treated with 40 μM Al for 12 h and then in Al-free solution for 12 h. Both (B) and (C) show the image of Evans blue staining. Green arrows indicate the zones stained markedly with Evans blue corresponding to the ZR in (A). Red arrows indicate the rupture zone corresponding to the RR in (A). (From Motoda, H. et al., *Plant Signal. Behav.*, 6, 98, 2011. With the permission of Landes Bioscience.)

zonary region (ZR) stained markedly with Evans blue were observed. ZR includes dead or injured cells because of staining with Evans blue and does not detach from the root even during the recovery process unlike rupture zone. The rupture zone extended during the recovery process (Figure 33.3C) suggesting the probable elongation of the central cylinder during the recovery process.

Several papers suggested that Al toxicity syndromes were caused by oxidative stress (Cakmak et al. 1991; Yamamoto et al. 2002; Matsumoto et al. 2012). The possible involvement of oxidative stress in the model system was investigated. The content of superoxide anion, H_2O_2, and lignin increased in the continued treatment but decreased during the recovery process (Matsumoto et al. 2012). The result suggested the rupture was induced by the increase of the rigidification in cell wall, and elongation of central cylinder was induced by reducing the penetration of Al during the recovery process through ZR. ZR may help re-elongate central cylinder. Thus, the programmed cell death (PCD)-like function of ZR during the recovery process was proposed (Matsumoto et al. 2012; for more details, see Section III).

V. Calcium

Ca^{2+} is an essential element for root growth, and much work was devoted to its role in Al toxicity (Blamey et al. 1993; Qifu et al. 2002). Al-induced changes in cell physiology, occurring in the cytoplasm and at the plasma membrane, might be caused by the disruption of Ca^{2+} homeostasis. An inhibitory effect of Al was observed not only on Ca^{2+} uptake but also on its translocation from the apical region. Results of a number of studies (Jacob and Northcote 1985; Sasaki et al. 1994; Huang et al. 1996) implied that Al blocks the Ca^{2+} channels on the plasma membrane.

The antagonistic effect between Ca^{2+} and Al^{3+} is well known. The question is how does Al inhibit the physiological functions of Ca^{2+}? Replacement of functional Ca^{2+} from the membranes and from the cell wall in the root might be one mechanism, although the evidence for such displacement in vivo is limited (Kinraide et al. 1992). Competition between Ca^{2+} and Al^{3+} is thought to weaken the cell walls either by reducing the number of Ca cross-links or by replacing Ca^{2+} cross-links with stronger Al^{3+}, making them too rigid for growth. Displacement of apoplastic Ca^{2+} by Al is, in part, due to the competition for ligands, such as pectin carboxyl residues in the cell wall (Reid et al. 1995). However, Ca^{2+} was not displaced by Al in wheat (Ryan et al. 1997a), but the signal initiated or disrupted by excess Al inhibited the growth in the meristem of *Allium cepa* (Schofield et al. 1998).

The disruption of Ca^{2+} metabolism by Al was debated. In wheat seedlings, root growth can be severely inhibited by Al^{3+} concentrations that do not affect Ca^{2+} uptake, while the addition of other ameliorating cations, for example, 30 mM Na^+, 3 mM Mg^{2+}, or 50 μM tris(ethylenediamine) cobalt (III), depressed Ca^{2+} uptake (Ryan et al. 1994). Phytotoxic effect of Al on the root hairs of *Arabidopsis* was not due to the blockage of Ca^{2+}-permeable channels required for Ca^{2+} influx into the cytoplasm (Jones et al. 1998). On the other hand, Zhang and Rengel (1999) found that the increase of cytoplasmic free Ca^{2+} ions ($[Ca^{2+}]c$) in root apical cells was higher in Al-sensitive wheat (ES8) than in Al-tolerant one (ET8). The Al-related increase in $[Ca^{2+}]c$ was correlated with inhibition of root growth and was reversible upon removing the ambient $AlCl_3$.

Recently, the dynamics of $[Ca^{2+}]c$ in different zones of *Arabidopsis* roots has been studied using the yellow cameleon (YC) Ca^{2+} sensor, a chimeric protein that relies on fluorescence resonance energy transfer (FRET) as an indicator of $[Ca^{2+}]c$ (Rincón-Zachary et al. 2010). In this system, although Al ion evoked increases in $[Ca^{2+}]c$ in the root transition zone, the $[Ca^{2+}]c$ transients were similarly observed in roots of Al-sensitive, Al-tolerant, and wild-type plants, indicating that early $[Ca^{2+}]c$ may not be tightly linked to Al toxicity. The effects of Al on Ca signaling were assessed in tobacco expressing a Ca^{2+}-monitoring luminescent protein, aequorin, as well as in a newly isolated putative plant Ca^{2+} channel protein from *A. thaliana*, AtTPC1 (two-pore channel 1). The TPC1 channels were demonstrated to be the only Al-sensitive channel which was involved in Ca signaling. Lin et al. (2005) reported the involvement of an Al-sensitive signaling pathway requiring TPC1-type channel-dependent Ca^{2+} influx in the presence of salicylic acid, a key plant defense-inducing agent. It was found that the stimulation of Ca^{2+} influx by both salicylic acid and cold shock was via an Al-sensitive pathway.

In conclusion, the important issue still remains to be solved: a role of the transient increase in $[Ca^{2+}]c$ in Al toxicity, namely, the possibility that an increase in $[Ca^{2+}]c$ could be the second messenger to evoke other Al responses (for more details, see Section VIII).

VI. Callose

Callose (β-1,3-glucan) synthesis is very sensitive to Al stress and has been considered a reliable marker for Al toxicity. The callose concentration in the 10–30 mm root tip of cowpea was inversely related to the root elongation rate, when the roots were subjected to Al concentration above 10 μM (Wissemeier et al. 1992). Furthermore, a negative correlation was found between the relative callose concentration and relative root elongation rates of three soybean genotypes differing in Al sensitivity. Callose is synthesized by β-1,3-glucan synthetase on the plasma membrane and activated by Ca^{2+}. It is still unknown where the Ca^{2+} required for callose synthesis comes from.

Ca^{2+} may not be the only signal for callose formation, and the alteration in the plasma membrane architecture might also be important for callose synthesis (Jacob et al. 1985). As callose is released into the apoplast, the cell wall of root cells of Al-treated plants most likely contain callose depositions. What is the inhibitory function of callose under Al stress? Callose can be considered as a sealing agent in plants. Callose is localized in the wall around the plasmodesmata, which appear to be structurally subdivided. Thus, the constricting force exerted by the accumulated callose would be transmitted to the plasmodesmal core

(Turner et al. 1994). This would inhibit the transport of cellular compounds through plasmodesmata under Al stress.

Pinpoint deposition of callose in Al-sensitive wheat root treated with Al was detected at the plasmodesmata using monoclonal antibodies against 1 → 3-β-D-glucose (callose). Such an increase of callose inhibited the cell-to-cell trafficking of molecules through the plasmodesmata, resulting in the inhibition of root elongation, since these events were markedly repressed in the presence of 2-deoxy-D-glucose which is a callose synthase inhibitor (Sivaguru et al. 2000). On the other hand, in an Al-tolerant *Arabidopsis* mutant, no direct relationship between Al uptake and callose formation was established (Larsen et al. 1996).

VII. Oxidative Stress and Al Toxicity

A. Lipid Peroxidation

Peroxidation of lipids is a typical symptom of oxidative stress, and the enhancement of peroxidation of lipids by Al has been reported in various systems (Rath et al. 2000). Iron is a transition metal having two common oxidation numbers (2, 3) and is well known to act as a catalyst for redox reactions both in vitro and in vivo. In conjunction with oxygen, Fe can act as a catalyst for the production of hydroxyl radicals, which initiate lipid peroxidation. On the other hand, Al is not a transition metal and has a fixed oxidation number of 3. Therefore, Al itself cannot catalyze redox reactions directly, and the Al-enhanced peroxidation of lipids should be caused indirectly.

In animals, if peroxidation of liposomes, erythrocytes, synaptosomes, myelin, low-density lipoproteins, or microsomes is stimulated by Fe^{2+}, the presence of Al ion increases the peroxidation rate. It was proposed that under such condition, Al ions bind to membranes and cause a subtle rearrangement of membrane lipids, which aids the propagation of lipid peroxidation via Fe-mediated production of hydroxyl radicals as well as alkoxyl or peroxyl radicals (Halliwell et al. 2007).

Similarly in plants, Al and Fe synergistically enhanced the peroxidation of lipids in cultured tobacco (*Nicotiana tabacum* L.) cells, which leads to the alterations in membrane permeability (Ono et al. 1995) and eventually caused apoptosis-like cell death (Yamaguchi et al. 1999). Under the same conditions, lipophilic antioxidants protected cells from both the peroxidation of lipids and the loss of viability (Yamamoto et al. 1997). Thus, the peroxidation of lipids synergistically enhanced by both Al and Fe is an actually toxic event leading to cell death. In whole-plant systems, the enhancement of peroxidation of lipids has been frequently reported (Cakmak et al. 1996). However, unlike the cultured cell systems described earlier, the enhancement of the peroxidation of lipids could be carried only by Al, without Fe supply. Furthermore, the peroxidation of lipids seems not to be the primary cause leading to inhibition of root elongation under Al stress (Horst et al. 1992; Yamamoto et al. 2001).

Yin et al. (2010a) reported a possible involvement of a part of the lipid peroxide–derived aldehydes, such as highly electrophilic alpha, beta-unsaturated aldehydes (2-alkeral), in the Al-induced root elongation inhibition in tobacco. Aluminum treatment markedly increased the contents of aldehydes. Compared to wild-type plants, the transgenic tobacco plants overexpressing *Arabidopsis* 2-alkenal reductase showed less accumulation of the aldehydes, less retardation of root elongation under Al stress, and higher regrowth after Al removal (Yin et al. 2010b). Thus, the authors suggested that the lipid peroxide–derived aldehydes such as 4-hydroxyl-(E)-2-nonenal and (E)-2-hexanal could injure root cells directly.

B. Reactive Oxygen Species Production

Superoxide anion (O_2^-) is the primary form of reactive oxygen species (ROS). In nonphotosynthetic tissue, O_2^- could be formed mainly via NADPH oxidase localized in the plasma membrane or via electron transfer in the inner membrane of mitochondria. Kawano et al. (2003) reported that the addition of high concentrations of Al (6 mM $AlCl_3$ as an optimal concentration) triggered the generation of O_2^- in cultured tobacco cells at pH 5.8, which seemed to stimulate the influx of Ca^{2+}. On the other hand, in cultured tobacco cells treated with relatively low concentrations of Al ($\leq$100 μM), the enhancement of ROS was related to mitochondrial dysfunction (Yamamoto et al. 2002). It seems that Al triggers mitochondrial dysfunctions followed by ROS production which is the toxic event that eventually leads to cell death. Interestingly, in seedlings of pea, Al similarly triggered ROS production, which was observed simultaneously with mitochondrial dysfunction and inhibition of root elongation (Yamamoto et al. 2002). In the root tips of *Cassia tora* L., activity of cell wall–bound peroxidases was increased by Al, which was clearly correlated with lignin accumulation and H_2O_2 production (Xue et al. 2008). The authors further suggested the possible involvement of jasmonate and nitric oxide (NO) in the regulation of the cell wall peroxidase activity and lignin synthesis. In barley roots exposed to Al, growth was inhibited, and DNA ladders occurred after 8 h treatments with 0.1–1.0 mM Al. The authors suggested that the Al ion induces cell death possibly via ROS-activated signal transduction pathway (Pan et al. 2001).

C. Genetic Evidence Supporting a Link between Al Stress and Oxidative Stress

Many lines of genetic evidence including transcriptomic analyses have supported the idea that Al stress causes oxidative stress in tobacco (Ezaki et al. 1995), rye (*Secale cereale* L.) (Milla et al. 2002), *Arabidopsis* (Ezaki et al. 2000; Kumari et al. 2008), wheat (Houde et al. 2008), *Medicago truncatula.* (Chandran et al. 2008a,b), and so on. In *Arabidopsis*, large-scale transcriptomic analysis of root responses to Al was performed (Kumari et al. 2008). Exposure to Al triggered changes in the transcript levels for several genes related to the oxidative stress pathway including ascorbate peroxidase, glutathione reductase, superoxide dismutase (SOD), and class III peroxidase. The biological role of Al stress-inducible genes was investigated in *Arabidopsis* by root elongation assay (Ezaki et al. 2000), suggesting that a part of the

Al-induced inhibition of root elongation is caused by oxidative stress. Ascorbic acid (ASA) is the most abundant antioxidant in plants and is regenerated by the action of dehydroascorbate reductase (DHAR). The overexpression of *Arabidopsis* cytosolic DHAR gene in tobacco leads to lower hydrogen peroxide level, higher ASA level, and better root growth under Al exposure than wild-type plants, indicating that the overexpression of DHAR confers Al tolerance. Taken together, these reports indicate that a part of the Al-induced inhibition of root elongation is caused by oxidative stress and that overexpression of the genes encoding antioxidant systems could confer Al tolerance in plants.

D. Nitric Oxide

NO is involved in ROS (Halliwell and Gutteridge 2007) and serves as a signaling molecule modulating numerous physiological processes in animals and plants. The relationship between NO production and the Al-induced inhibition of root elongation has been investigated in several plants (Wang et al. 2005). In the roots of *Arabidopsis*, it was reported that cells of the distal portion of the transition zone emitted NO, which was blocked by Al (Illes et al. 2006). Then, in the roots of *Hibiscus moscheutos* L., in addition to the ameliorative effect of exogenous NO on the Al-induced inhibition of root elongation, it was reported that Al inhibited the activity of NO synthase (NOS) and reduced NO concentrations (Tian et al. 2007). Since other treatments leading to the reduction of NO in root apical cells (NO scavenger, inhibitors of NOS, and nitrate reductase) caused the inhibition of root elongation, the authors suggested that the reduction of endogenous NO concentrations resulting from inhibition of NOS activity by Al could lead to inhibition of root elongation.

VIII. Signal Transduction and Al Signal

Several studies suggested that interactions of Al with the elements of signal transduction pathways in cells are apparently primary events (Haug et al. 1994; Panda et al. 2009). Special attention has been paid to phosphoinositide-associated signal transduction (Martinez-Esteveg et al. 2003). $AlCl_3$ and Al–citrate inhibited phospholipase C (PLC) of the microsomal membrane in a dose-dependent manner of wheat roots. I_{50} was observed at 15–20 μM Al in wheat roots (Jones et al. 1995).

Binding of Al to microsomes and liposomes was found to be lipid dependent with the signal transduction element PIP_2 having the highest affinity for Al with an Al/lipid stoichiometry of 1:1. Al disrupted production of second messengers such as inositol 1,4,5-triphosphate (IP3) and phosphatidic acid (PA) by blocking PLC in cell suspension (Chee-Gonzalez et al. 2009); however, phospholipase D (PLD) and diacylglycerol kinase (DGK) activities were stimulated by Al in *Coffea arabica* L. Recently, Pejchar et al. (2010) reported that Al inhibited the formation of diacylglycerol mediated by phosphatidylcholine-hydrolyzing PLC. These results suggest that the effect of Al on the signal transduction pathway is associated with the mechanism of Al toxicity (Illes et al. 2006; Ramos-Díaz et al. 2007; Pejchar et al. 2008).

How and by which receptors is the Al signal recognized and how is it transported into the cytoplasm at the root apices? Bennet and Breen (1991) proposed that the Al signal is received in the root cap of maize. Kasai et al. (1993) speculated that the transduction of Al signal in barley roots is related to an increase of abscisic acid (ABA), and Shen et al. (2004a) described the effects of ABA on the efflux of citrate from soybean root under Al stress. They suggested that ABA is involved in early response, after which a K-252a-sensitive protein kinase plays a key regulatory role in the activity of an anion channel within the plasma membrane. The transport of exogenously applied [^{3}H]indole-3-acetic acid to the meristematic zone was significantly inhibited by Al in maize roots. The signaling pathway in the root apex mediating the Al signal may be responsible for the genotypic difference in Al resistance (Kollmeier et al. 2000).

NO could be an important key signaling molecule mediating numerous physiological processes in plants including Al stress responses (see Section VII), but the biochemical mechanism in terms of the transduction of Al signal is poorly understood. In order to accurately delineate the signal transduction pathway, an identification of the events occurring at the early phase of Al treatment might be necessary.

In wheat, Al has been shown to induce a transient expression of a protein kinase which is prerequisite for the extrusion of malate within 30 s of exposure. K252a reduced an Al-dependent efflux of malate from Al-tolerant wheat apices, which implies that protein phosphorylation is involved in the response of Al of this species. Since this process occurs within several minutes, an Al-associated signal transduction process might be involved in the exudation of organic acids (Osawa et al. 2001; Shen et al. 2005). Ligaba et al. (2009) reported that S384 of TaALMT1, an Al-activated malate transporter (*ALMT1*) in wheat root apex, is an essential residue regulating basal transport as well as Al activation of transport activity in TaALMT1 via direct protein phosphorylation.

Furthermore, addition of a toxic concentration of Al to cell suspension cultures of *C. arabica* induced the rapid and transient activation of a protein kinase that phosphorylated myelin basic protein (Arroyo-Serralta et al. 2005). Sivaguru et al. (2003b) determined that Al depolymerized the MTs and depolarized the membranes in *Arabidopsis*, via the activity of Ca channel blockers. They proposed that Ca^{2+} influx might involve glutamate receptors, which in animals are ligand-gated cation channels and are also known to be present in the genome of *Arabidopsis*. They also demonstrated that glutamate depolymerized the MTs and depolarized the plasma membrane. These responses, as well as the inhibition of root elongation, occurred within the few minutes of Al treatment but were evoked more rapidly by glutamate than Al. MT depolymerization and membrane depolarization, whether induced by glutamate or by Al, could be blocked by a specific antagonist of ionotropic glutamate receptors, 2-amino-5-phosphonopenanate, whereas an

Al-gated anion channel antagonists blocked the two responses to Al but not to glutamate. They speculated that Al induces the secretion of glutamate, where it binds to its receptor and triggers the influx of Ca which results in the observed depolymerization of MTs and the depolarization of membrane.

IX. Metabolism Affected by Al

Treatment with 75 μM Al reduced O_2 uptake by excised wheat roots by 23% and 35% after 12 and 24 h, respectively. Mitochondria isolated from Al-treated roots had reduced oxidative capacity with supply of electrons to complexes I and II. It was found that initially, Al affected electron flow through complexes I and II and, after longer exposure, interacted with other sites in the mitochondria (de Lima and Copeland 1994). Al tolerance of snap beans (cv. Dade) was an inducible trait. In this cultivar, the resumption of root elongation during recovery from Al treatment was accompanied by increased rates of maintenance respiration, potentially reflecting diversion of energy to metabolic pathways that offset the adverse effects of Al toxicity (Cumming et al. 1992).

Carbon metabolism was also affected by Al. Al stress increased alcohol dehydrogenase, sucrose synthase, and lactate dehydrogenase activity in wheat (cv. Vulcan) roots (Copeland and de Lima 1992). The first two enzymes in the pentose phosphate pathway (glucose-6-phosphatedehydrogenase and 6-phosphogluconate dehydrogenase) decreased in Al-sensitive wheat cv. Grana. However, these two enzymes first increased but then decreased in Al-tolerant rye (Sláski et al. 1996). Kumari et al. (2008) reported a large-scale transcriptomic analyses of root responses to Al.

In general, a broader range of changes in transcript abundance was observed after 48 h (1114 genes) as compared to 6 h (401 genes) of Al treatment. Within the genes related to glycolytic pathway, transcript levels of one gene encoding fructose-bisphosphate aldolase and three genes encoding pyruvate kinase increased after 6 and 48 h exposure, respectively. Interestingly, among all TCA cycle genes, transcripts for only one gene, a mitochondrial malate dehydrogenase increased after 48 h. In the pentose phosphate pathway, increase in the abundance of transcripts of two genes encoding 6-phosphogluconate dehydrogenase was observed at 6 and 48 h exposure.

In tobacco suspension-cultured cells, Al reduced the uptake rate of sucrose within 3 h of treatment, which was related to decreases in fresh weight, soluble sugar content, and osmolality. Thus, it has been proposed that the inhibition of sucrose uptake by Al is a primary event responsible for lowered osmolality and hence lowered water uptake and the inhibition of root elongation (Abdel-Basset et al. 2010). On the other hand, the burst of ROS was observed in the relatively late phase after 6 h or more of Al treatment. The ROS production was strongly related to both the dysfunction of mitochondria as well as ATP depletion and the loss of growth capability. Thus, in order to understand the mechanism of Al-induced ROS production and cell death, the biochemical changes of sugar metabolism triggered by the prevention of sugar uptake at early phase of Al treatment should be elucidated.

X. Aluminum Tolerance

As mentioned at the beginning of this chapter, Al toxicity is a major factor inhibiting plant growth in acid soil. Al toxicity retards crop production 20% in East Asia, in sub-Saharan Africa, and North America, 31% in Latin America, and 38% in South East Asia (Wood et al. 2000). Attention was paid to improving the agricultural production in acid soils. Acid soil is a worldwide problem, and the use of Al-tolerant plants is a strategy to minimize the loss of crops in acid soil. So far, two mechanisms of Al tolerance have been proposed (Taylor 1991; Kochian 1995). One is the exclusion, or external tolerance mechanism, which was defined in case where Al was prevented from entering the plant cells, and the other is internal tolerance mechanism which was defined in case where Al enters the cells and tolerance is achieved by detoxification processes. After the previous edition of this book was published (Matsumoto 2002a), major developments were made regarding the mechanism of the major Al tolerance, that is, exclusion mechanism at the molecular level (Pellet et al. 2007). Especially the findings of new transporters encoded by known genes have contributed to our molecular understanding of the Al tolerance (Sasaki et al. 2004; Larsen et al. 2007; Magalhaes et al. 2007; Furuichi et al. 2010; Xia et al. 2010).

XI. Exclusion Mechanism

Since higher plants cannot move away from the acid soil, they have developed ways to reduce this edaphic stress. An effective strategy to reduce the stress is to chelate the toxic Al^{3+} with exuded organic anions in the rhizosphere rendering less toxicity of Al^{3+}.

A. Exudation of Organic Acids

The pathway involved in the release of organic acids has been examined in order to clarify the mechanisms of the release of organic acid which are native chelators into the rhizosphere of Al-resistant crop plants (Jones 1998; Ryan et al. 2001; Kochian et al. 2004, 2005). The release of organic acids occurs mainly at the root apex where the toxicity symptoms appear under Al stress (Ryan et al. 1993; Osawa et al. 2001). The exudation of organic acids may also contribute to the utilization of Al phosphate in the soil (Takita et al. 1999). It was generally accepted that the amount of organic acids at the root tip was similar in Al-tolerant and Al-sensitive cultivars. Miyasaka et al. (1991) first reported that Al-tolerant snap bean (*P. vulgaris*) exuded 70 times more citrate in the presence of Al than in its absence and 10 times more citrate than the sensitive cultivars.

The major organic acids exuded by roots of different plant species were malate, citrate, and oxalate (Ryan et al. 2001; Klug et al. 2010). The efficiency of Al-tolerance activity of each

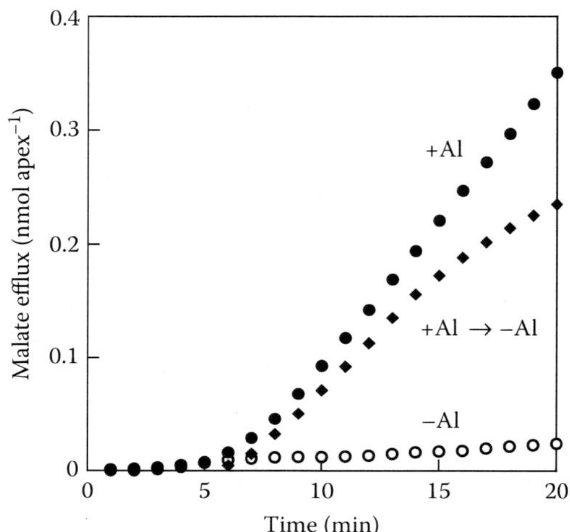

FIGURE 33.4 Malate efflux from root apexes of Al-resistant wheat (cv. Alaska) with lapse of time after exposure to 200 μM $AlCl_3$. Excised root apexes (2 mm in length from root apices) were exposed to 200 μM $CaCl_2$ (○; Ca solution) or the Ca solution containing 200 μM Al (pH 4.2) (●; Ca + Al solution) passing through at a flow rate of 0.5 ml min^{-1}. Root apices were exposed to the Ca + Al solution for 1 min (♦) followed by three rinses with the Ca solution and subsequent exposure to the Ca solution at a flow rate of 0.5 mL min^{-1} from 5 min after addition of the Ca + Al solution. Eluents passed through root apexes were collected every 1 min and assayed for the determination of malate concentration. (From Osawa, H. and Matsumoto, H., *Plant Physiol.*, 126, 411, 2001. With permission.)

organic acid depends on its chelating ability (Ryan et al. 2001), and citrate was the most powerful chelator. The time lag between the exposure to Al and the efflux of organic acids depends on crop species (Li et al. 2000b). Figure 33.4 shows the efflux of malate from the root tips of Al-tolerant wheat (Osawa et al. 2001). The efflux started after 5 min of exposure of the root tips to Al and continued for at least 15 min. However, the efflux of malate did not occur in the absence of Al. When the root tips were exposed to Al for only 1 min followed by the removal of Al, efflux started after 5 min with a gradual decrease rate of efflux. It is interesting that root tips can respond to the Al signal given for only 1 min by the later efflux of malate.

In more than 36 lines of wheat cultivars differing in Al resistance that were screened, Al-stimulated malate release was correlated with Al resistance (Ryan et al. 1995). Although it is generally accepted that the capacity of organic acid exudation correlated with Al tolerance, there are exceptions (Yang et al. 2008). After 20 h of Al exposure, root growth was resumed in the Al-tolerant soybean varieties concomitant with the marked exudation of citrate but remained severely inhibited in the Al-sensitive ones (Yang et al. 2000). The direct contact of Al with the root was essential for inducing the acid secretion (Yang et al. 2000). Dark treatment and the removal of shoots inhibited the citrate exudation markedly (Yang et al. 2001).

The effect of P deficiency on Al-induced citrate exudation was investigated in three soybean varieties differing in low-P tolerance. P starvation alone failed to induce secretion of organic acids from all three soybean varieties. However, P deficiency altered Al-induced citrate exudation over time, showing a complex interaction (Nian et al. 2003). In lupin (*Lupinus pilosus* L.), citrate and malate exudation was enhanced by P deficiency in the absence of Al than that in the presence of Al (Ligaba et al. 2004). The induction of 48 kD protein kinase within 30 s of exposure for the excretion of malate from wheat root (Osawa et al. 2001) and the involvement of ABA in the excretion of citrate from soybean root (Shen et al. 2004a) were prerequisite for secretion of the two acids.

Al-induced depolarization of root cap cell membrane potential is probably linked to, but is not sufficient to trigger, malate release in Al-tolerant wheat (Papernik et al. 1997). Both de novo synthesis and activation of an anion channel are needed for Al-induced secretion of citrate in *C. tora*, but in buckwheat, the plasma membrane protein responsible for oxalate secretion preexists (Yang et al. 2006). Shen et al. (2004b) reported that the anion channels of tap roots were greater in number and higher in activity than those of basal roots, thus allowing more citrate secretion and Al resistance in this root type of common bean (*P. vulgaris*). The charge balance caused by flux of either H^+ in case of citrate exudation (Ohno et al. 2003; Shen et al. 2004a) or K^+ in case of malate exudation was observed (Osawa et al. 2002). The regulatory mechanisms for the Al-induced efflux of K^+ and malate from the root apex of Al-resistant wheat (cv. Atlas) have been characterized. The results suggested that Al-induced K^+ efflux was not a prerequisite for the induction of the release of malate (Osawa and Matsumoto 2002).

B. Organic Anion Transporter

Ma et al. (2000) found that the short arm of chromosome 3R was essential for the exudation of organic acids in triticale. Many investigators reported that a channel blocker reduced the Al-induced exudation of organic acids, suggesting the exudation of organic acids through the organic anion transporter. Therefore, the most important mechanism of Al tolerance relies on the release of organic anions from roots which bind with toxic Al^{3+} and detoxify them. Ryan and Delhaize (2010) proposed the idea that mechanism of Al^{3+} resistance based on organic acids efflux from root cells evolved relatively recently from mutations that co-opted transport proteins from other locations or other functions to perform new roles.

The pathway involved in the release of organic acids has been examined in order to clarify the mechanism of organic acid release in Al-tolerant crop plants. Generally, the content of organic acids in the root was similar in Al-tolerant and Al-sensitive plants, indicating that transporters of organic acids through the plasma membrane control the exudation of organic acids (Dill et al. 1987; Barbier-Brygoo et al. 2000; Matsumoto 2000). Al-activated malate, citrate, and anion transporters have been reported from different species (Ryan et al. 1997b; Kollmeier et al. 2001; Piñeros et al. 2001; Zhang et al. 2001; Sasaki et al. 2004; Zhang et al. 2008).

XII. Genetic Basis for Al Tolerance

The bimodal distribution of phenotypes corresponding to a 3:1 segregation ratio for Al tolerance and sensitivity in populations derived from crosses between Al-tolerant and Al-sensitive wheat cultivars revealed the presence of single major genes for Al tolerance (Garvin 1998). In addition to major genes conferring major differences in Al tolerance, there is also some evidence that minor genes or modifier genes may play a role in modulating the effect of major Al tolerance genes. The D genome of wheat may determine the tolerance to acid soil and, consequently, contribute to the increased adaptation of hexaploid wheat during their evolution. Atlas 66 is a well-known Al-tolerant cultivar of wheat. However, not all the genes for tolerance to Al in Atlas 66 are located on the D genome chromosome (Berzonsky 1992). Furthermore, Al tolerance in wheat is a dominant trait, and the majority of observed variability could be explained by two or three gene pairs, each gene affecting the same character with complete dominance of each gene pair. Al tolerance in the ditelosomic line of the Chinese Spring wheat cultivar revealed that genes controlling this character were located on the short arm of chromosome 5A and the long arm of chromosome 2D and 4D (Delhaize et al. 1993a).

Conservation of the Al-tolerance gene by various species was investigated. RFLP markers for a major wheat Al-tolerance gene Alt_{BH} were found on the long arm of chromosome 4D, while in rye, the Al-tolerance gene was located on chromosome 4 which harbors chromosome segments homologous to regions of wheat chromosome 4D. In barley, the Al-tolerance gene, Al_p, is almost certainly orthologous to the wheat Alt_{BH} gene due to the fact that the relative positions of Al_p and Alt_{HB} with respect to a common set of molecular markers are virtually identical in both genomes (Berzonsky 1992). Ditelosomic substitution lines with the chromosomes of D genome of wheat cv. Chinese Spring substituted for their homeologues in *Triticum turgidum* L. cv. Langdon showed that substitution lines involving chromosome 4D were more Al tolerant than Langdon. The tolerance was found to be controlled by a single dominant gene, designated Alt2, located in the proximal region of the long arm of chromosome 4D (Luo and Dvořák 1996). A quantitative trait locus (QTL) analysis of the Al-tolerant wheat (Atlas 66) revealed one QTL on the distal region of chromosome arm 4D where a malate transporter gene was mapped (Ma et al. 2005). Similarly with Al-tolerant Chinese wheat landrace FSW, one major QTL was mapped on chromosome 4DL that cosegregated with Xups4, a marker for the promoter of the *ALMT1* gene. The other two QTLs were located on chromosomes 3BL and 2A, respectively (Cai et al. 2008).

Restriction fragment length polymorphism (RFLP) mapping of Alp suggested that it was localized to the long arm of chromosome 4H. The Al-tolerance gene Alt_{BH} was located on the long arm of chromosome 4D, suggesting the possibility that Al tolerance in barley and wheat may be due to the reaction of orthologous loci (Tang et al. 2000). With Al-tolerant (cv. Koshihikari) and Al-sensitive (cv. Kasalath) rice differing in the capability of secretion of citrate, three QTLs controlling Al tolerance were detected on chromosome 1, 2, and 6 (Ma et al. 2002).

Mao et al. (2004) investigated the cDNA-amplified fragment length polymorphism (cDNA-AFLP) using Al-tolerant (cv. Azucema) and Al-sensitive (cv. IR1552) rice. Nineteen function-known genes were found among 34 transcript-derived fragments (TDFs) regulated by Al stress. The results indicated that Al stress could induce the biosynthesis of lignin and other cell wall components in roots.

QTL for *A. thaliana* Al tolerance was analyzed using a recombinant inbred population (R2) derived from Landsberg and Columbia accessions. Two significant single-factor QTLs ($p < 0.05$) were detected by measuring relative root length (RRL) on chromosome 1 and 4, where the Columbia allele showed positive and negative effects on the Al tolerance. QTLs could explain about 43% of the total variation of Al tolerance among the recombinant inbred (RI) population (Kobayashi and Koyama 2002). With the same system, malate release explained nearly all (95%) of the variation in Al tolerance in the population, although only two of the QTLs have been identified (Hoekenga et al. 2003).

QTL analysis of Al tolerance was performed using *Ler*/Cvi RI lines of *A. thaliana*. Two QTLs were detected at the top of chromosome 1 and bottom of chromosome 3. These QTLs explained 40% and 16% of the phenotypic variation of Al tolerance and the positive effect of the Cvi allele. The QTL on chromosome 1 is possibly related to malate excretion (Kobayashi et al. 2005). The expression of genes on the short arm of triticale (*Triticosecale* Wittmark cv. Currency) chromosome 3R is induced by Al and turned out to be necessary for the release of organic acids (Ma et al. 2000).

In rye, Al tolerance is controlled by at least four independent loci, Alt1, Alt2, Alt3, and Alt4, located on chromosome arms 6RS, 3RS, 4RL, and 7RS, respectively (Miftahudin et al. 2002; Benito et al. 2010). Benito et al. (2010) proposed the presence of Alt3 and Alt4 (*ScALMT1*) orthologues in other cereals. As to the physiological functions of Al-tolerance genes, Delhaize et al. (1993b) found that Al tolerance, controlled by the Alt gene in wheat, appeared to be dominant across a range of Al concentrations based on identical Al tolerance in heterozygotes and Alt 1 homozygotes. This gene controls the excretion of malate upon Al stress.

XIII. Major Al Tolerance Genes

A. ALMT1 Family

Sasaki et al. (2004) isolated a novel gene, Al-activated malate transporter (*TaALMT1*), which encodes an *ALMT1* from wheat (*T. aestivum*) root apex under Al stress. Constitutive expression of *ALMT1* in Al-tolerant wheat (cv. ET8) was much higher than in sensitive wheat (cv. ES8). Base sequences of ALMT1 cDNA differed slightly between ET8 (*ALMT1-1*) and ES8 (*ALMT1-2*) at six nucleotides that encode two amino acids. Quantitative RT-PCR analysis suggested that ALMT1 co-segregates with the phenotypic traits such as their root elongation inhibition and malate exudation using F2 and F3 seeds of ES7 × ET7 under Al stress. *Xenopus laevis* oocyte system demonstrated that an inward current was specifically induced by Al and malate but not citrate. It is very important that Al but not La or other trivalent cations could activate the malate efflux, suggesting that the ALMT1 protein transports malate in the presence of Al only but not other trivalent cations. It is now clear that other members of the *ALMT* family have a similar function in other plant species including *Arabidopsis* (Hoekenga et al. 2006; Kobayashi et al. 2007; Kovermann et al. 2007; Liu et al. 2009), rape (*Brassica napus* L.) (Ligaba et al. 2006), rye (Fontecha et al. 2007; Collins et al. 2008), maize (Piñeros et al. 2008), and wheat (Raman et al. 2006).

TaALMT1 protein (Yamaguchi et al. 2005) and BnALMT1 protein (Ligaba et al. 2007) are located on the plasma membrane, and Raman et al. (2005) showed that it consists of six exons interrupted by five introns in bread wheat. They further identified molecular markers targeting insertion/deletion (indel) and SSR repeats within intron 3 region of the *ALMT1* gene. Both markers, *ALMT1-SSR3a* and *ALMT1-SSR3b*, based on repetitive indels, exhibited complete co-segregation with Al tolerance, malate efflux, and a cleavage amplified polymorphic sequence (CAPS) marker discriminating *ALMT1-1* and *ALMT1-2* alleles, in a doubled haploid population derived from Diamond bird (Al tolerant)/Janz (Al sensitive). The higher level of variation in intron 3 suggests that this genomic region has been constrained by indels, SSR, and single nucleotide polymorphisms (Raman et al. 2006). The possible molecular determinant for Al tolerance involving a homology of the wheat *TaALMT1* was found and named *AtALMT1* (At1go8430) in *A. thaliana*. Two *ALMT1* homologues, *BnALMT1* and *BnALMT2*, were also found to increase Al tolerance in rape (Ligaba et al. 2006).

Among cereal crops, rye is the most tolerant species, and an *ALMT* orthologue, *ScALMT1*, was identified. *ScALMT1* was primarily expressed in the root apex and upregulated when Al was present in the medium. Fivefold differences in the expression were found between the Al-tolerant and the Al-sensitive genotypes. Much higher absolute expression levels were detected in rye than in the moderately tolerant Chinese Spring wheat cultivar (Fontecha et al. 2007).

The finding of *ALMT1* is important milestone in genetic research of Al tolerance because (1) *ALMT1* encodes the malate transporter which exudes the malate from root apex in the presence of Al but not other trivalent ions and alleviates Al toxicity, (2) *ALMT1* is of plant origin and malate exuded through ALMT is completely nontoxic, and (3) transgenic barley expressing the wheat *ALMT1* gene can exude the malate and grow well in hydroponic solution containing Al and in acid soils (Delhaize et al. 2004; Figure 33.5). Therefore, the utilization of *ALMT1* gene is expected

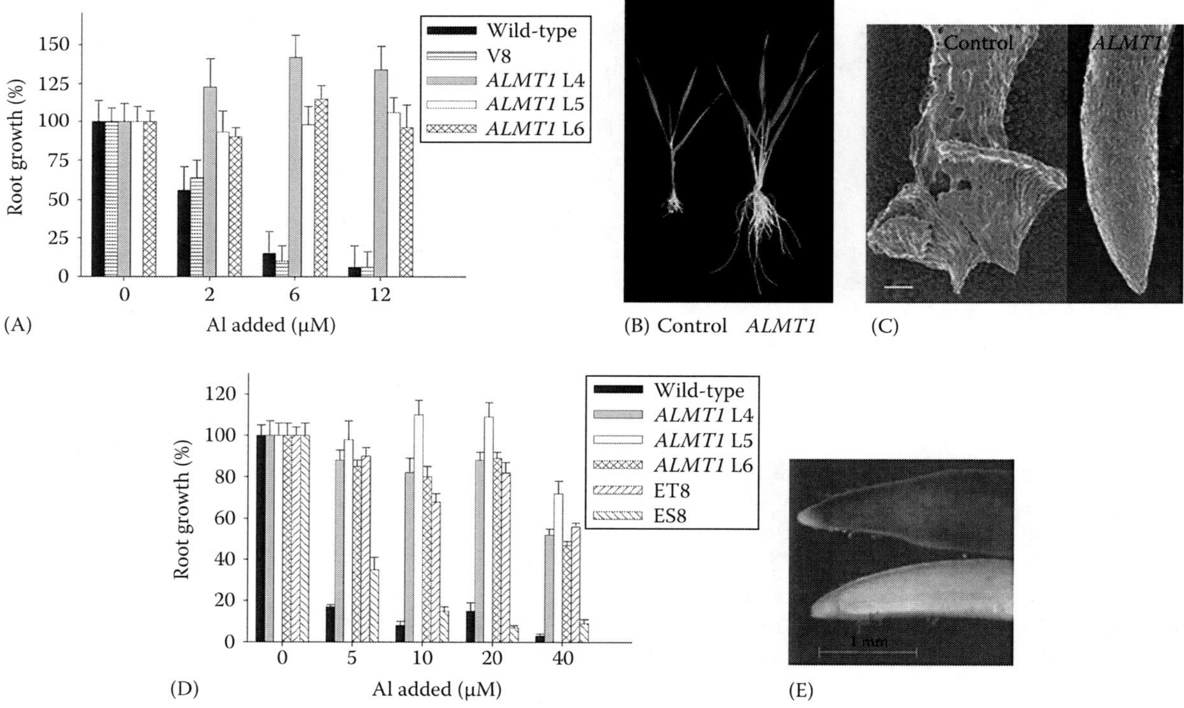

FIGURE 33.5 ALMT1 confers Al tolerance to barley grown in hydroponic culture. (A) Root elongation of T0 (primary transgenics; lines 4–6 denoted by L4–L6) barley lines grown in hydroponic culture. (B) Effect of 3 μM Al on growth over 10 days of the T0 generations of the control (empty vector) line and ALMT1 line 5. (C) Scanning electron micrograph showing the effect of Al (3 μM) on the morphology of the root apex from the control line (empty vector) and ALMT1 line grown for 10 days. (D) Root elongation of T2 homozygous barley lines grown in hydroponic culture. The wheat lines ET8 and ES8 are near-isogenic lines that differ in Al tolerance. (E) Roots stained with hematoxylin for detection of Al. The bottom root is derived from the T2 homozygous ALMT1 line 5, and the top root is from its azygous sister line. (From Delhaize, E. et al., *Proc. Natl. Acad. Sci. U S A*, 101, 15249, Copyright 2004, with the permission of National Academy of Sciences, U.S.A.)

to increase crop yield in acid soils in the future (Matsumoto and Sivaguru 2008). The malate transporting ability of *ALMT1-1* was similar to that of *ALMT1-2*, but the expression pattern between the two genes differed suggesting that the *ALMT*1 expression rate determines the Al tolerance (Sasaki et al. 2004).

With 69 Japanese and non-Japanese wheat lines, genomic regions upstream and downstream of *ALMT1* have been characterized. The first 1000 bp downstream of *ALMT1* was conserved, which did not correlate with Al tolerance, but the first 1000 bp upstream of the *ALMT1* coding region was more variable in six different patterns. Type II–VI had blocks of sequence which were duplicated and triplicated. In lines of non-Japanese origin, the number of repeats in the upstream region was positively correlated with the levels of *ALMT1* expression and Al resistance (Sasaki et al. 2006).

Vacuolar malate may reach very high concentrations and fluctuate rapidly, whereas cytosolic malate is kept at a constant level allowing optimal metabolism. *ALMT* gene family may encode a vacuolar malate channel, and *AtALMT9* is targeted to the vacuole (Kovermann et al. 2007). A transcription factor, zinc finger protein STOP1, was found to regulate the expression of multigenes that protect *Arabidopsis* from H^+ and Al toxicity (Iuchi et al. 2007; Sawaki et al. 2009). Similarly, Yamaji et al. (2009) found a C2H2-type zinc finger transcription factor Al-resistance transcription factor 1 (ART1), which specifically regulates the expression of genes related to Al tolerance in rice (*Oryza sativa*). Al-specific malate excretion is carried out by a combined regulation of *AtALMT*1 expression and activation of AtALMT1 protein. Reversible phosphorylation was important for the transcriptional and posttranslational regulation of *AtALMT1* (Kobayashi et al. 2007). In maize, not all ZmALMT1-type transporters mediate Al-activated organic acid responses but were implicated in the selective transport of anions involved in mineral nutrition and ion homeostasis processes, rather than mediating a specific Al-activated citrate exudation response at the rhizosphere of maize roots (Piñeros et al. 2008).

B. ABC Transporter Family

The ABC transporter genes encode a multidrug resistance (MDR) protein which functionally detoxifies organic and inorganic substances. MDR is known as a member of the ATP-binding cassette (ABC) protein superfamily. The hallmark of ABC transporters is their ability to derive energy from ATP hydrolysis to transport molecules through membranes. This requires the presence of an ATP hydrolyzing activity which is associated with the ABC domain of the ABC transporter protein. Full-length ABC transporters contain two of these ABC domains also called nucleotide-binding folds (NBF). These domains are also characterized by the presence of sequence motifs called Walker A box and Walker B box which border the ATP-binding consensus motif. The transporters found in prokaryotes are only half in size. Upon dimerization, they form the active transport complexes responsible for their functions (Schulz et al. 2006).

Larsen's group found two mutants of *Arabidopsis* (Al-sensitive 1 (*als1)* and Al-sensitive 3 (*als3*)), and those genes encode proteins of the ABC transporter family (Larsen et al. 2003, 2007; Gabrielson et al. 2006). ALS3 may function to redistribute accumulated Al away from sensitive tissues in order to protect the growing root from the toxic effects of Al. ALS1 accumulates at the tonoplast of root cells and may be important for movement of some substrate, possibly chelated Al, as part of a mechanism of Al sequestration. However, the transport proteins and their molecular function are still unknown.

Star 1 and *Star 2* genes (sensitive to Al rhizotoxicity 1 and 2) from rice encode ABC protein, which may protect plants from Al stress. STAR1 and STAR2 are responsible for Al tolerance in rice. STAR1 encodes a nucleotide-binding domain, while STAR2 encodes a transmembrane domain of a bacterial-type ABC transporter. Disruption of either gene resulted in hypersensitivity to Al. Both STAR1 and STAR2 are expressed mainly in the roots and are specifically induced by Al exposure. STAR1 interacts with STAR2 to form a complex that localizes to the vesicle membrane of all root cells. STAR1/2 shows efflux transport activity specific for UDP-glucose, which may be used to modify the cell wall under Al toxicity (Huang et al. 2009a).

C. MATE Family

MDR protein has the capacity to confer drug resistance to microbes by an efflux mechanism: thus proteins are named the multidrug and toxin extrusion (or efflux) as the acronym MATE (Delhaize et al. 2007). Recently, a novel Al-tolerance gene responsible for Alt_{SB} locus in sorghum (*Sorghum bicolor* (L.) Moench), which is a member of the MATE gene family (designated *SbMATE*) and encodes an Al-activated citrate efflux transporter, was identified. Al-inducible Alt_{SB} expression specifically in the root apices was responsible for sorghum Al tolerance (Magalhaes et al. 2007). *SbMATE* is not related to the ALMT family. A broader role for MATE family in providing Al tolerance not only in sorghum but also in the grass family as a whole will provide valuable molecular and genetic information for crop production in acid soils. Wang et al. (2007) identified *HvMATE* gene encoding a MATE protein, as a candidate controlling Al tolerance in barley. Relative expression of the *HvMATE* gene was 30-fold greater in Al-tolerant cv. Dayton than Al-sensitive cv. Gardiner. *HvMATE* expression was significantly correlated with Al-tolerance and Al-activated citrate efflux.

A gene belonging to the MATE family which correlates with the citrate efflux phenotype was found to be involved in Al resistance in wheat. Whole-genome linkage mapping using the F2 population derived from a cross between cv. Carazinho (citrate efflux) and the EGA-Burke (no citrate efflux) revealed a major locus on chromosome 4BL, $Xce_{c'}$, which accounts for more than 50% of the phenotypic variation in citrate efflux (Ryan et al. 2009). Furukawa et al. (2007) identified a gene (*HvAACT1*) responsible for the Al-activated citrate secretion from barley root under Al stress. This gene also belongs to the MATE family. A good correlation was found between the expression of *HvAACT1* and citrate secretion in 10 barley cultivars differing in Al resistance.

Al tolerance in barley is conditioned by the Alp locus on the long arm chromosome 4H which is associated with Al-activated release of citrate from roots (Wang et al. 2007). Yokosho et al. (2010) isolated two homologous genes of MATE family, ScFRDL1 and ScFRDL2, from rye. ScFRDL1 shared 94.2% identity with HvAACTi (Furukawa et al. 2007), and ScFRDL2 shared 80.6% identity with OsFRDL2, a putative Al-responsive protein in rice. ScFRD1 was involved in efflux of citrate into xylem for Fe translocation from the roots to the shoots, while ScFRDL2 was involved in Al-activated citrate secretion in rye. The expression of a citrate transporter MATE gene is crucial for citrate exudation in common bean. However, although the expression of the citrate transporter is a prerequisite for citrate exudation, genotypic Al resistance in common bean particularly depends on the capacity for maintaining the cytosolic pool that enables citrate exudation (Eticha et al. 2010).

XIV. Exudation of Nonorganic Anions

Many reports so far suggested that organic anion exudation correlates with Al tolerance (Ma et al. 2001; Delhaize et al. 2007). However, exceptions were observed in signal grass (*Brachiaria decumbens* Stapf) (Wenzl et al. 2001) and maize (Piñeros et al. 2005). These results support the suggestion that the exudation of nonorganic anions plays a role in Al resistance (Parker et al. 1998). In some plants, there is linkage between phosphorus deficiency and Al toxicity through the secretion of organic acids (Gaume et al. 2001; Dong et al. 2004). The exudation of phosphate from the root apex of Al-resistant maize and wheat cultivars alleviates Al toxicity (Pellet et al. 1995). The exudation of flavonoid-type phenols (catechin and quercetin) from maize root tips, which is enhanced by the silicon pretreatment, has a potential role in alleviating Al toxicity (Kidd et al. 2001). Nevertheless, the role of excreted organic acids as detoxifiers against Al toxicity is evident (Ryan et al. 2001; Matsumoto 2002b).

Basu et al. (1999) found a 23 kDa peptide in root exudate that segregated with the Al-resistance phenotype in F2 population of *T. aestivum* and had a significant Al-binding capacity. For the extremely high tolerance of the camphor tree (*Cinnamomum camphora* Meisn.) against Al toxicity, Osawa et al. (2011) found the important role of proanthocyanidin (PA)-accumulating cells on the epidermal apex. They proposed that outer PA-accumulating cells that constitutively proliferate until the end of early expansion may shield the inner epidermis cells against highly toxic Al. Then, the abrupt detachment of the PA cells could help replace the Al-damaged cell surface with the expanding epidermis cells.

XV. Internal Al-Tolerance Mechanism

The strategy of exclusion mechanism is to exclude the toxic Al into roots by the chelation of Al with exuded organic acids and other substances. However, some crops can retain a large amount of Al inside the cell without any growth reduction. This means that the cell has a mechanism for detoxifying Al after uptake. The leaf of hydrangea (*Hydrangea macrophylla* (Thunb.) Ser.) contains a larger amount of Al in the cytoplasm, which is chelated with citrate at a molar ratio 1:1 (Ma et al. 1997a). In the buckwheat, Al in the cytoplasm is combined with oxalate at a molar ratio of 1:3 and sequestered into vacuoles (Ma et al. 1997b, 1998). In these Al-accumulating plants, it seems that intracellular organic acids have a significant role in reducing Al toxicity.

Xylem transport of toxic Al from the root apex to the leaves and their sequestration into the vacuole associated with Al–citrate complex in buckwheat (Ma et al. 2001), as well as progressive vacuolation in Al-treated root tips of barley (Ikeda et al. 1993), have been reported. Xylem transport of the toxic Al from the root apex to leaves occurs in the form of Al–citrate complex (Ma et al. 2001). Phenolic substances which chelate Al play a role in the mechanism of Al tolerance (Ofei-Manu et al. 2001; Barceló et al. 2002; Tolra et al. 2005).

Amelioration of Al toxicity is induced by the formation of an aluminosilicate compound in the root apoplast (Hodson et al. 1993). The following genes and mechanisms may contribute to Al tolerance. Al stress induces the generation of ROS which are toxic for plant growth. Watt (2003) and Ezaki et al. (1996, 2000, 2001) reported that genes encoding enzyme proteins which ameliorate oxidative stress participate in the tolerance mechanism against Al stress. Al-tolerant tobacco cell lines contained antioxidant, like ascorbate and glutathione, at a much higher level than in Al-sensitive cell lines (Devi et al. 2003), suggesting that scavenging systems of ROS may play a role in Al tolerance.

Al-induced inhibition of root growth and accumulation of malondialdehyde (MDA; an indicator of lipid peroxidation) was lower in homozygous transgenic plants (T2) compared with WT plants. SOD activity was higher in homologous transgenic plants (T2) than in WT plants. These results suggest that sensitivity to Al toxicity can be reduced by overexpression of *WMnSOD1* (Basu et al. 2001). Alfalfa was engineered by introducing the *Pseudomonas aeruginosa* citrate synthase (CS) gene controlled by the *Arabidopsis* Act2 constitutive promoter. The transgenic events showed that bacterial CS overexpression can be a useful tool to help achieve Al tolerance (Barone et al. 2008).

Pyruvate phosphate dikinase catalyzes the phosphorylation of pyruvate yielding phosphoenolpyruvate which is a substrate of phosphoenolpyruvate carboxylase. Transgenic tobacco plant overexpressing pyruvate phosphate dikinase showed increased exudation of organic acids and decreased accumulation of Al in the roots (Trejo-Téllez et al. 2010). Mitochondrial enzymes including malate dehydrogenase and CS increased, but aconitase decreased under Al stress in soybean. Expression of *Gm-AICT*, a gene showing homology to Al-activated citrate transporter, was also induced under Al treatment. The Al-dependent changes in activity and expression of these enzymes are consistent with the assumption that they support the sustained release of citrate from soybean roots (Xu et al. 2010).

Similarly, overexpression of malate dehydrogenase gene in alfalfa enhanced organic acid production and confers tolerance to Al (Tesfaye et al. 2001). The plasma membrane is one of the major targets for toxic Al^{3+}. Functional damage of plasma membrane by Al is related to the denaturation of lipids together with reduced membrane fluidity. The transgenic *Arabidopsis* expressing the gene (*S851*) of the root of *Stylosanthes hamata* (L.) Taub. encoding Δ8 sphingolipid desaturase had different sphingolipid composition of cell membrane that could protect plants from Al stress (Ryan et al. 2007). Delhaize et al. (1999) cloned a wheat cDNA (*TaPPS1*) which encodes phosphatidylserine synthetase (PSS). The transgenic yeast overexpressing *TaPPS1* resulted in the increase in Al resistance, but a high level of *TaPSS1* expression in *Arabidopsis* and tobacco has been shown to result in the appearance of necrotic lesions on leaves.

A large amount of absorbed Al is located in the cell wall. Thus *WAK1* (cell wall–associated receptor kinase 1) may play an important role in the Al signal transduction via WAK. *WAK1* gene in the root showed a typical "on" and "off" pattern, suggesting that *WAK1* is a representative of the Al-induced early genes. WAK proteins are localized preferentially to the peripheries of cortical cell, within the elongation zone of *Arabidopsis*. Furthermore, *WAK1*-overexpressing transgenic *Arabidopsis* showed enhanced root growth under Al stress (Sivaguru et al. 2003a).

XVI. Mucilage

The meristem and cap region where Al toxicity is dominant are coated with a mucilage layer that ranges in thickness from 50 μM to 1 mm. Mucilage has various protective functions against toxic metals in the soil and has a high Al-binding capacity. Al bound to mucilage of wheat roots accounted for approximately 25%–35% of the Al remaining after desorption by citric acid (Archambault et al. 1996). In the rhizosphere, the Al is bound to mucilage that blocks its entry into the root. When the mucilage was periodically removed from the root tips of cowpea with a brush, inhibition of root elongation was more severe. Apparently, binding of Al to mucilage is a mechanism of Al tolerance. A good correlation exists between mucilage volume and Al tolerance. Organic acids released into the mucilage droplet would diffuse slowly; thus, the mucilage droplet would form a region of high concentration of organic acids where Al is captured before reaching the root surface.

Poschenrieder et al. (2008) found that supply of Mn in excess amounts can substantially ameliorate Al resistance. This amelioration was clearly related to higher mucilage production. Removal of mucilage significantly reduced Al accumulation in *Melastoma malabathricum* L. (Al accumulator). Al-adsorbed mucilage and bioassay with alfalfa seedlings indicated that the concentrated Al in the mucilage of *M. malabathricum* bound very weakly to cation exchange sites of the mucilage. The higher charge density in *M. malabathricum* mucilage, derived from unmethylated uronic acid, is assumed to be related to preferential adsorption of trivalent cation. Not only a higher degree of methylation in the uronic acid but also H^+ release from roots to the mucilage appears to be responsible for the loose binding of Al in *M. melabathricum* (Watanabe et al. 2008). However, a protective role of the mucilage against Al injury is difficult to reconcile with the lack of information about the effect of Al on mucilage excretion. On the contrary, disappearance of mucilage is one of the first visible symptoms of Al toxicity (Puthota et al. 1991).

The role of mucilage in the protection of roots against Al toxicity depends on the amount of mucilage excreted and how strongly Al bounds to mucilage. Mucilage from maize roots is strongly bound to Al but failed to prevent Al-induced inhibition of root elongation (Li et al. 2000a). Approximately 50% of the total Al of the root apices was located in the mucilage of cowpea, while only 9%–22% of Al in maize root was bound to mucilage. The binding is decreased by the lower content of uronic acids (3%) in maize mucilage as compared to 11.5% in cowpea mucilage (Li et al. 2000a). It will be necessary to determine the kinetic data of synthesis and excretion of mucilage in order to better understand the role of mucilage in Al resistance.

XVII. pH Regulation in the Rhizosphere

Solubility of Al depends strongly on pH suggesting that high-solution pH may reduce the solubility and toxicity of Al. An increase in the pH of dilute nutrient solution from 4.5 to 4.6 caused a 26% decline in soluble Al concentration (Blamey et al. 1983). This suggests that even a slight pH change can affect the toxicity of Al. Although Al stress is always a combination of two stresses, acidic and Al, the Al treatment has rarely been differentiated from the acidic stress (Koyama et al. 1995; Rangel et al. 2005; Babourina et al. 2006). A vibrating microelectrode was used to measure pH at a radial distance of 20 and 50 μm from the surface of the root tip of the wild-type and an Al-tolerant *Arabidopsis* (Degenhardt et al. 1998). The Al-tolerant *Arabidopsis* mutant (alt-104) showed a clear increase of pH in the rhizosphere at the presence of Al, but the wild type did not. The increased flux raised the root surface pH of alt-104 by 0.15 unit, suggesting that Al resistance in alt-104 is mediated by pH change in the rhizosphere. The difference in Al resistance between the wild type and alt-104 disappeared when roots were grown in pH-buffered medium. It is interesting that no difference in root H^+ fluxes between the wild type and alt-104 was detected in the absence of Al.

A correlation between impaired H^+-fluxes across the plasma membrane and Al-induced growth inhibition was observed in root apices of squash. Furthermore, the inhibition of H^+-pumping rate in the highly purified PM vesicles obtained from the Al-treated apical root portions coincided with the inhibition of root growth under Al stress. These results suggest that the changes in surface pH mediated by altered dynamics of H^+ efflux and influx across the root tip plasma membrane play an important role in root growth as affected by Al (Ahn et al. 2002). Root growth of *Arabidopsis* is inhibited by proton rhizotoxicity in low ionic strength media when the medium has a pH lower than 5.0. QTL analysis of *Arabidopsis* at pH 4.7 revealed that two major QTLs on chromosomes 2 and 5 and an additional six epistatic interacting loci pairs control proton resistance in the Ler/Col recombinant inbred population. These results suggest that these genetic factors are independently associated with proton resistance in comparison to the known Al-resistance QTL and epistases detected in the same RI population at 4 μM Al at pH 5.0. This indicates that different genetic factors regulate the mechanism of resistance to each stress in *Arabidopsis* (Ikka et al. 2007). Alt 1-1 (Al tolerant) obtained through EMS-induced mutagenesis of *Arabidopsis* mutant, als 3-1, suggests that this mutation positively impacts Al resistance in a manner dependent on pH adjustment rather than Al exclusion (Gabrielson et al. 2006).

XVIII. General Conclusion

Al is the major element in the soils, and solubilized Al^{3+} at pH lower than 4.5–5.0 inhibits root elongation markedly. Due to the increasing environmental problems and the anticipated population growth especially in the developing countries, attention has been paid to improve the agricultural production in acid soils. Under such situations, vast knowledge on the mechanism of Al toxicity and tolerance has been provided in past few decades.

Al^{3+} has a strong affinity to the cell constituents, and various cell functions including plasma membrane, cell wall, Ca, signal transduction, and cell division are adversely affected by Al toxicity. Exclusion or extracellular mechanism plays an important role in Al tolerance. The natures of various transporter for organic anions were characterized. Furthermore, genes encoding the transporter proteins have been isolated, and transgenic plant overexpressing those genes was evidenced to grow better in acid soil in the laboratory scale. Our final goal is to grant the increased demand for food by the improvement of plant growth in acid soils. We sincerely hope that the accumulation of new and valuable knowledge on the Al stress will help to accomplish the awaiting task leading to the final goal.

References

Abdel-Basset R, Ozuka S, Demiral T et al. 2010 Aluminum reduces sugar uptake in tobacco cell cultures: A potential cause of inhibited elongation but not of toxicity. *J Exp Bot* 61:1597–1610.

Ahad A, Nick P. 2007. Actin in bundled in activation-tagged tobacco mutants that tolerate aluminum. *Planta* 225: 451–468.

Ahn SJ, Matsumoto H. 2006. Review: The role of the plasma membrane in the response of plant roots to aluminum toxicity. *Plant Signal Behav* 1:37–45.

Ahn SJ, Rengel Z, Matsumoto H. 2004. Aluminum-induced plasma membrane surface potential and H+-ATPase activity in near-isogenic wheat lines differing in tolerance to aluminum. *New Phytol* 162:71–79.

Ahn SJ, Sivaguru M, Chung GC, Rengel Z, Matsumoto H. 2002. Aluminum-induced growth inhibition is associated with impaired efflux and influx of H^+ across the plasma membrane in root apices of squash (*Cucurbita pepo*). *J Exp Bot* 53:1959–1966.

Ahn SJ, Sivaguru M, Osawa H, Chung GC, Matsumoto H. 2001. Aluminum inhibits the H^+-ATPase activity by permanently altering the plasma membrane surface potential in squash roots. *Plant Physiol* 126:1381–1390.

Amenos M, Corrales I, Poschenrieder C et al. 2009. Different effects of aluminum on the actin cytoskeleton and brefeldin A-sensitive vesicle recycling in root apex cells of two maize varieties differing in root elongation rate and aluminum tolerance. *Plant Cell Physiol* 50:528–540.

Archambault DJ, Zhang G, Taylor GJ. 1996. Accumulation of Al in root mucilage of an Al-resistant and an Al-sensitive cultivar of wheat. *Plant Physiol* 112:1471–1478.

Arroyo-Serralta GA, Ku-Gonzalez A, Hernandez-Sotonayor SMT, Aguilar JJZ. 2005. Exposure to toxic concentration of aluminum activates a MAPK-like protein in cell suspension cultures of *Coffea arabica. Plant Physiol Biochem* 43:27–35.

Babourina O, Ozturk L, Cakmak I, Rengel Z. 2006. Reactive oxygen species production in wheat roots is not linked with changes in H^+ fluxes during acidic and aluminum stresses. *Plant Signal Behav* 1:70–75.

Baligar VC, Beaver WV, Ahlrichs JL. 1998. Nature and distribution of acid soils in the world. In *Proceedings of a Workshop to Develop a Strategy for Collaborative Research and Dissemination of Technology in Sustainable Crop Production in Acid Savannas and other Problem Soils of the World*, ed. RE Schaffert, pp. 1–12. May 4–6, 1998. Purdue, IN: Purdue University Press.

Barbier-Brygoo H, Vinauger M, Colombet J et al. 2000. Anion channels in higher plants: Functional characterization, molecular structure and physiological role. *Biochem Biophys Acta* 1465:199–218.

Barceló J, Poschenrieder C. 2002. Fast root growth responses, root exudates, and internal detoxification and cues to the mechanisms of aluminum toxicity and resistance: A review. *Environ Exp Bot* 48:75–92.

Barone P, Rosellini D, LaFayette P et al. 2008. Bacterial citrate synthase expression and soil aluminum tolerance in transgenic alfalfa. *Plant Cell Rep* 27:893–901.

Basu U, Goldbold D, Taylor GJ. 2001. Transgenic *Brassica napus* plants overexpressing aluminum-induced mitochondrial manganese superoxide dismutase cDNA are resistant to aluminum. *Plant Cell Environ* 24:1269–1278.

Basu U, Good AG, Aung T et al. 1999. A 23-kDa, root exudates polypeptide co-segregates with aluminum resistance in *Triticum aestivum. Physiol Plant* 106:53–61.

Benito C, Silva-Navas J, Fontecha G et al. 2010. Rye *Alt3* and *Alt4* aluminum tolerance loci to orthologous genes in other cereals. *Plant Soil* 327:107–120.

Bennet RJ, Breen CM. 1991. The aluminium signal: New dimensions to mechanisms of aluminium tolerance. In *Plant-Soil Interactions at Low pH*, eds. RJ Wright, VC Baligar, RP Murrmann, pp. 703–716. Dordrecht, the Netherlands: Kluwer.

Berzonsky WA. 1992. The genomic inheritance of aluminum tolerance in 'Atlas 66' wheat. *Genome* 35:689–693.

Blamey FPC, Asher CJ, Kerven GL, Edwards DG. 1993. Factors affecting aluminium sorption by calcium pectate. *Plant Soil* 149:87–94.

Blamey FPC, Edwards DG, Asher CJ. 1983. Effects of aluminium, OH:Al and P:Al molar ratios, and ionic strength on soybean root elongation in solution culture. *Soil Sci* 136:197–207.

Blancaflor EB, Jones DL, Gilroy S. 1998. Alterations in the cytoskeleton accompany aluminum-induced growth inhibition and morphological changes in primary roots of maize. *Plant Physiol* 118:159–172.

Budikova S, Ciamporova M. 1998. Growth and structural responses of maize roots to aluminium stress. In *Progress in Botanical Research*, eds. I Tsekos, M Moustakas, pp. 419–422. Dordrecht, the Netherlands: Kluwer.

Budikova S, Durcekova K. 2004. Aluminium accumulation in roots of Al-sensitive barley cultivar changes root cell structure and induces close synthesis. *Biologia* 59:215–220.

Cai S, Bai G-H, Zhang D. 2008. Quantitative trait loci for aluminum resistance in Chinese wheat landrace FSW. *Theor Appl Genet* 117:49–56.

Cakmak I, Horst WJ. 1991. Effect of aluminium on lipid peroxidation, superoxide dismutase, catalase, and peroxidase activities in root tips of soybean (*Glycine max*). *Physiol Plant* 83:463–468.

Chandran D, Sharopova N, Ivashuta S et al. 2008a. Transcriptome profiling identified novel genes associated with aluminum toxicity, resistance and tolerance in *Medicago truncatula*. *Planta* 228:151–166.

Chandran D, Sharopova N, VandenBosch KA, Garvin KA, Samac DA. 2008b. Physiological and molecular characterization of aluminum resistance in *Medicago truncatula*. *BMC Plant Biol* 8:89.

Chee-Gonzalez L, Munoz-Sanchez JA, Racagni-Di Palma G. 2009. Effect of phosphate n aluminium-inhibited growth and signal transduction pathways in *Coffea arabica* suspension cells. *J Inorg Biochem* 103:1497–1503.

Chen J, Sucoff E, Stadelmann EJ. 1991. Aluminium and temperature alteration of cell membrane permeability of *Quercus rubra*. *Plant Physiol* 96:644–649.

Ciamporova M. 2000. Diverse responses of root cell structure to aluminium stress. *Plant Soil* 226:113–116.

Ciamporova M. 2002. Morphological and structural response of plant roots to aluminum at organ, tissue and cellular levels. *Biol Plant* 45:161–171.

Clarkson DT. 1965. The effect of aluminium and some other trivalent metal cations on cell division in the root apices of *Allium cepa*. *Ann Bot* (N.S.) 29:309–315.

Collins NC, Shirley NJ, Saeed M. 2008. An *ALMT1* gene cluster controlling aluminum tolerance at the *Alt4* locus of rye (*Scale cereale* L.). *Genetics* 179:669–682.

Copeland L, de Lima ML. 1992. The effect of aluminum on enzyme activities in wheat roots. *J Plant Physiol* 140:641–645.

Cumming JR, Cumming AB, Taylor GJ. 1992. Patterns of root respiration associated with the induction of aluminium tolerance in *Phaseolus vulgaris* L. *J Exp Bot* 43:1075–1081.

Degenhardt J, Larsen PB, Howell SH, Kochian LV. 1998. Aluminum resistance in the *Arabidopsis* mutant alr-104 is caused by an aluminum-induced increase in rhizosphere pH. *Plant Physiol* 117:19–27.

Delhaize E, Benjamin D, Gruber D, Ryan PR. 2007. The roles of organic anion permease in aluminium resistance and mineral nutrition. *FEBS Lett* 581:2255–2262.

Delhaize E, Craig S, Beaton CD et al. 1993a. Aluminum tolerance in wheat (*Triticum aestivum* L.). *Plant Physiol* 103:685–693.

Delhaize E, Hebb DM, Richards KD et al. 1999 Cloning and expression of wheat (*Triticum aestivum* L.) phosphatidylserine synthase cDNA. *J Biol Chem* 274:7082–7088.

Delhaize E, Ryan PR. 1995. Aluminum toxicity and tolerance in plants. *Plant Physiol* 107:315–321.

Delhaize E, Ryan PR, Hebb DM. 2004. Engineering high-level aluminum tolerance in barley with the *ALMT1* gene. *Proc Natl Acad Sci U S A* 101:15249–15254.

Delhaize E, Ryan PR, Randall PJ. 1993b. Aluminum tolerance in wheat. II. Aluminum-stimulated excretion of malic acid from root apices. *Plant Physiol* 103:695–702.

Devi SR, Yamamoto Y, Matsumoto H. 2003. An intracellular mechanism of aluminum tolerance associated with high antioxidant status in cultured tobacco cells. *J Inorg Biochem* 97:59–68.

Dill ET, Holden MJ, Colombini M. 1987. Voltage gating in VDAC is markedly inhibited by micromolar quantities of aluminum. *J Membr Biol* 99:187–196.

Doncheva S, Amenos M, Poschenrieder C, Barceló C. 2005. Root cell patterning: A primary target for aluminium toxicity in maize. *J Exp Bot* 56:1213–1220.

Dong D, Peng X, Yan X. 2004. Organic acid exudation induced by phosphorus deficiency and/or aluminium toxicity in two contrasting soybean genotypes. *Physiol Plant* 122:190–199.

Eleftherios EP, Moustakas M, Fiagiskas N. 1993. Aluminate-induced changes in morphology and ultrastructure of *Thinopyrum* roots. *J Exp Bot* 44:427–436.

Eticha D, Stass A, Horst WJ. 2005. Cell-wall pectin and its degree of methylation in the maize root-apex: Significance for genotypic differences in aluminium resistance. *Plant Cell Environ* 28:1410–1420.

Eticha D, Zahn M, Bremer M et al. 2010. Transcriptomic analysis reveals differential gene expression in response to aluminium in common bean (*Phaseolus vulgaris*) genotypes. *Ann Bot* 105:1119–1128.

Ezaki B, Gardner RC, Ezaki Y, Matsumoto H. 2000. Expression of aluminum-induced genes in transgenic *Arabidopsis* plants can ameliorate aluminum stress and/or oxidative stress. *Plant Physiol* 122:657–665.

Ezaki B, Katsuhara M, Kawamura M, Matsumoto H. 2001. Different mechanisms of four aluminum (Al)-resistant transgenes for Al toxicity in *Arabidopsis*. *Plant Physiol* 127:918–927.

Ezaki B, Tsugita S, Matsumoto H. 1996. Expression of a moderately anionic peroxidase is induced by aluminium treatment in tobacco cells: Possible involvement of peroxidase isozymes in aluminium stress. *Physiol Plant* 96:21–28.

Ezaki B, Yamamoto Y, Matsumoto H. 1995. Cloning and sequencing of the cDNAs induced by aluminium treatment and Pi starvation in cultured tobacco cells. *Physiol Plant* 93:11–18.

Fontecha G, S-Navas J, Benito C et al. 2007. Candidate gene identification of aluminum-activated organic acid transporter gene at the *Alt4* locus for aluminum tolerance in rye (*Secale cereale* L.). *Theor Appl Genet* 114:249–260.

Frantzios G, Galatis B, Apostolakos P. 2005. Aluminum causes variable responses in actin filament cytoskeleton of the root tip cell of *Triticum turgidum*. *Protoplasma* 225:129–140.

Funakawa S, Hirai H, Kyuma K. 1993. Speciation of aluminum in soils solution from forest soils in northern Kyoto with special reference to their pedogenic process. *Soil Sci Plant Nutr* 39:281–290.

Furuichi J, Sasaki T, Tsuchiya Y et al. 2010. An extracellular hydrophilic carboxyl-terminal domain regulates the activity of TaALMT1, the aluminum-activated malate transport protein of wheat. *Plant J* 64:47–55.

Furukawa J, Yamaji N, Wang H et al. 2007. An aluminum-activated citrate transporter in barley. *Plant Cell Physiol* 48:1081–1091.

Gabrielson KM, Cancel JD, Morua LF, Larsen PB. 2006. Identification of dominant mutations that confer increased aluminium tolerance through mutagenesis of the Al-sensitive *Arabidopsis* mutant, *als3-1*. *J Exp Bot* 57:943–951.

Garvin DG. 1998. The genetic basis of aluminum tolerance in crops. In *Proceeding of a Workshop to Develop a Strategy for Collaborative Research and Dissemination of Technology in Sustainable Crop Production in Acid Savannas and other Problem Soils of the World*, ed. RE Schaffert, May 4–6, 1998, pp. 73–77. Purdue, IN: Purdue University Press.

Gaume A, Machler F, Fossard E. 2001. Aluminum resistance in two cultivars of *Zea mays* L.: Root exudation of organic acids and influence of phosphorus nutrition. *Plant Soil* 234:73–84.

Grabski S, Armoys E, Busch B. 1998. Regulation of actin tension in plant cells by kinase and phosphatase. *Plant Physiol* 116:279–290.

Grabski S, Schindler M. 1995. Aluminum induces rigor within the actin network of soybean cells. *Plant Physiol* 108:897–901.

Halliwell B, Gutteridge JMC. 2007. Free radicals in biology and medicine. Oxford, U.K.: Oxford University Press.

Harris WR, Berthon G, Day JP et al. 1997. Separation of aluminum in biological systems. In *Research Issues in Aluminum Toxicity*, eds. RA Yokel, MS Golub, pp. 91–116. Washington, DC: Taylor & Francis.

Hasenstein KH, Evans ML. 1988. Effects of cations on hormone transport in primary roots of *Zea mays*. *Plant Physiol* 86:890–894.

Haug A. 1984. Molecular aspects of aluminum toxicity. *Crit Rev Plant Sci* 1:345–373.

Haug A, Shi B, Vitorello V. 1994. Aluminum interaction with phosphoinositide-associated signal transduction. *Arch Toxicol* 68:1–7.

Hiradate S, Ma JF, Matsumoto H. 2007. Strategies of plants to adapt to mineral stresses in problem soils. *Adv Agron* 96:65–132.

Hodson MJ, Sangster AG. 1993. Interaction between silicon and aluminum in *Sorghum bicolor* (L) Moench: Growth analysis and X-ray microanalysis. *Ann Bot* 72:389–400.

Hoekenga OA, Maron L, Pineros MA et al. 2006. *AtALMT1*, which encodes a malate transporter, is identified as one of several genes critical for aluminum tolerance in *Arabidopsis*. *Proc Natl Acad Sci U S A* 103:9738–9743.

Hoekenga OA, Vision TJ, Shaff JE et al. 2003. Identification and characterization of aluminum tolerance loci in *Arabidopsis* (*Landsberg erecta* x Columbia) by quantitative trait locus mapping. A physiologically simple but genetically complex trait. *Plant Physiol* 132:936–948.

Horst WJ. 1995. The role of the apoplast in aluminium toxicity and distance of higher plants: A review. *Z Pflanzenernähr Bodenkd* 158:419–428.

Horst WJ, Asher CJ, Cakmak I. 1992. Short-term responses of soybean roots to aluminium. *J Plant Physiol* 140:174–178.

Horst WJ, Schmol N, Kollmeier H, Baluska F. 1999. Does aluminium affect root growth of maize through interaction with the cell wall-plasma membrane-cytoskeleton continuum? *Plant Soil* 215:163–174.

Houde M, Diallo AO. 2008. Identification of genes and pathways associated with aluminum stress and tolerance using transcriptome profiling of wheat near-isogenic lines. *BMC Genomics* 9:400.

Huang JW, Pellet DW, Papernick LA, Kochian LV. 1996. Aluminium interactions with voltage-dependent calcium transport in plasma membrane vesicles isolated from roots of aluminum-sensitive and tolerant wheat cultivars. *Plant Physiol* 110:561–569.

Huang CF, Yamaji N, Mitani N et al. 2009a. A bacterial-type transporter is involved in aluminum tolerance in rice. *Plant Cell* 21:655–667.

Huang CF, Yamaji N, Nishimura M, Tajima S, Ma JF. 2009b. A rice mutant sensitive to Al toxicity is defective in the specification of root outer cell layers. *Plant Cell Physiol* 50:976–985.

Ikeda H, Tadano T. 1993. Ultrastructural changes of the root tip cells in barley induced by a comparatively low concentration of aluminum. *Soil Sci Plant Nutr* 39:109–117.

Ikka T, Kobayashi Y, Iuchi S et al. 2007. Natural variation of *Arabidopsis thaliana* reveals that aluminum resistance and proton resistance are controlled by different genetic factors. *Theor Appl Genet* 115:709–719.

Illes P, Schlicht M, Pavlovkin J et al. 2006. Aluminum toxicity in plants: internalization of aluminium into cells of the transition zone in *Arabidopsis* root apices related to changes in plasma membrane potential endosomal behavior, and nitric oxide production. *J Exp Bot* 57:4205–4213.

Iuchi S, Koyama H, Iuchi A et al. 2007. Zinc finger protein STOP1 is critical for proton tolerance in *Arabidopsis* and coregulates a key gene in aluminum tolerance. *Proc Natl Acad Sci U S A* 104:9900–9905.

Jacob SR, Northcote DH. 1985. In vivo glucan synthesis by membranes of celery petioles: The role of the membrane in determining the type of linkage formed. *J Cell Sci* (Suppl. 2):1–11.

Jones DL. 1998. Organic acids in the rhizosphere—A critical review. *Plant Soil* 205:25–44.

Jones DL, Blancaflor EB, Kochian LV, Gilroy S. 2006. Spatial coordination of aluminum uptake, production of reactive oxygen species, callose production and wall rigidification in maize roots. *Plant Cell Environ* 29:1309–1318.

Jones DL, Gilroy S, Larsen PB, Howell SH, Kochian LV. 1998. Effect of aluminum on cytoplasmic Ca^{2+} homeostasis in root hairs of *Arabidopsis thaliana* (L.). *Planta* 206:378–387.

Jones DL, Kochian LV. 1995. Aluminum inhibition of the inositol 1,4,5-triphosphate signal transduction pathway in wheat roots: A role in aluminum toxicity? *Plant Cell* 7:1913–1922.

Kasai M, Sasaki M, Tanakamaru S, Yamamoto Y, Matsumoto H. 1993. Possible involvement of abscisic acid in increases in activities of two vacuolar H^+-pumps in barley roots under aluminum stress. *Plant Cell Physiol* 34:1335–1338.

Kawano T, Kadono T, Furuichi T et al. 2003. Aluminum-induced distortion in calcium signaling involving oxidative burst and channel regulation in tobacco BT-2 cells. *Biochem Biophys Res Commun* 308:35–42.

Kenjebaeva S, Yamamoto Y, Matsumoto H. 2000. The impact of aluminium on the distribution of cell wall glycoproteins of pea root tip and their Al-binding capacity. *Soil Sci Plant Nutr* 47:629–636.

Kidd PS, Llugany M, Poschenrieder C, Gunse B, Barcelo J. 2001. The role of root exudates in aluminium resistance and silicon-induced amelioration of aluminium toxicity in three varieties of maize (*Zea mays* L.) *J Exp Bot* 52:1339–1352.

Kikui S, Sasaki T, Osawa H, Matsumoto H, Yamamoto Y. 2007. Malate enhances recovery from aluminum-caused inhibition of root elongation in wheat. *Plant Soil* 290:1–15.

Kinraide TB. 1990. Assessing the rhizotoxicity of the aluminate ion, $Al(OH)_4^-$. *Plant Physiol* 93:1620–1625.

Kinraide TB, Ryan PR, Kochian LV. 1992. Interactive effects of Al^{3+}, H^+, and other cations on root elongation considered in terms of cell-surface electrical potential. *Plant Physiol* 99:1461–1468.

Klug B, Horst JW. 2010. Oxalate exudation into the root–tip water free space confers protection from aluminum toxicity and allows aluminum accumulation in the symplast in buckwheat (*Fagopyrum esculentum*). *New Phytol* 187:380–391.

Kobayashi Y, Furuta Y, Ohno T, Hara T, Koyama H. 2005. Quantitative trait loci controlling aluminium tolerance in two accessions of *Arabidopsis thaliana* (*Landsberg erecta* and Cape Verde Islands). *Plant Cell Environ* 28:1516–1524.

Kobayashi Y, Hoekenga OA, Itoh H et al. 2007. Characterization of AtALMT1 expression in aluminum-inducible malate release and its role for rhizotoxic stress tolerance in *Arabidopsis. Plant Physiol* 145:843–852.

Kobayashi Y, Koyama H. 2002. QTL analysis of Al tolerance in recombinant inbred lines of *Arabidopsis thaliana. Plant Cell Physiol* 43:1526–1533.

Kochian LV. 1995. Cellular mechanisms of aluminum toxicity and resistance in plants. *Annu Rev Plant Physiol Plant Mol Biol* 46:237–260.

Kochian LV, Hoekenga OA, Pineros MA. 2004. How do crop plants tolerate acid soil? Mechanism of aluminum tolerance and phosphorus deficiency. *Annu Rev Plant Biol* 55:459–493.

Kochian LV, Pineros MA, Hoekenga OA. 2005. The physiology, genetics and molecular biology of plant aluminum resistance and toxicity. *Plant Soil* 274:175–195.

Kollmeier M, Dietrich P, Bouer CS, Horst WJ, Herdrich R. 2001. Aluminum activates a citrate-permeable anion channel in the aluminum-sensitive zone of the maize root apex: A comparison between an aluminum-sensitive and an aluminum-resistant cultivar. *Plant Physiol* 126:397–412.

Kollmeier M, Felle HH, Horst WJ. 2000. Genotypical differences in aluminum resistance of maize are expressed in the distal part of the transition zone. Is reduced basipetal auxin flow involved in inhibition of root elongation by aluminum? *Plant Physiol* 122:945–956.

Kopáček J, Hejzlar J, Kaňa J et al. 2009. Trends in aluminium export from a mountainous area to surface waters, from deglaciation to the recent: Effects of vegetation and solid development, atmospheric acidification, and nitrogen-saturation. *J Inorg Biochem* 103:1439–1448.

Kopittke PM, Blamey FPC, Menzies NW. 2008. Toxicities of soluble Al, Cu. and La include rupture on rhizodermal and root critical cells of cowpea. *Plant Soil* 303:217–227.

Kovermann P, Meyer S, Hortensteiner S et al. 2007. The *Arabidopsis* vacuolar malate channel is a member of the ALMT family. *Plant J* 52:1169–1180.

Koyama H, Toda T, Yokota S, Dawair Z, Hara T. 1995. Effects of aluminum and pH on root growth and cell viability in *Arabidopsis thaliana* strain Landsberg in hydroponic culture. *Plant Cell Physiol* 36:201–205.

Kumari M, Taylor GJ, Deyholos MK. 2008. Transcriptomic responses to aluminum stress in roots of *Arabidopsis thaliana. Mol Genet Genomics* 279:339–357.

Larsen PB, Cancel, J, Rounds, M, Ochoa V. 2007. *Arabidopsis* ALS1 encodes a root tip and stele localized half type ABC transporter required for root growth in an aluminum toxic environment. *Planta* 225:1447–1458.

Larsen PB, Geisler MJ, Jones CA. 2003. *ALS3* encodes a phloem-localized ABC transporter-like protein that is required for aluminum tolerance in *Arabidopsis. Plant J* 42:353–363.

Larsen PB, Tai C-Y, Kochian LV, Howell SH. 1996. *Arabidopsis* mutants with increased sensitivity to aluminum. *Plant Physiol* 110:743–751.

Le VH, Kuraishi S, Sakurai N. 1994. Aluminum-induced rapid root inhibition and changes in cell-wall components of squash seedlings. *Plant Physiol* 106:971–976.

Li XF, Ma JF, Hiradate S, Matsumoto H. 2000a. Mucilage strongly binds aluminum but does not prevent root from aluminum injury in *Zea mays. Physiol Plant* 108:152–160.

Li XF, Ma JF, Matsumoto H. 2000b. Pattern of Al-induced secretion of organic acids differs between rye and wheat. *Plant Physiol* 123:1537–1543.

Ligaba A, Katsuhara M, Ryan PR, Sibasaka M, Matsumoto H. 2006. The *BnALMT1* and *BnALMT2* genes from *Brassica napus* L. encode aluminum-activated malate transporters that enhance the aluminum resistance of plant cell. *Plant Physiol* 142:1294–1303.

Ligaba A, Katsuhara M, Sakamoto W, Matsumoto H. 2007. The BnALMT1 protein that is an aluminum-activated malate transporter is localized in the plasma membrane. *Plant Signal Behav* 2:255–259.

Ligaba A, Kochian K, Piñeros M. 2009. Phosphorylation at S384 regulates the activity of the TaALMT1 malate transporter that underlies aluminum resistance in wheat. *Plant J* 60:411–423.

Ligaba A, Yamaguchi M, Shen H et al. 2004. Phosphorus deficiency enhances plasma membrane H^+-ATPase activity and citrate exudation in grater purple lupin (*Lupinus pilosus*). *Funct Plant Biol* 31:1075–1083.

de Lima ML, Copeland L. 1994. The effect of aluminium on respiration of wheat roots. *Physiol Plant* 90:51–58.

Lin C, Yu Y, Kadono T et al. 2005. Action of aluminum, novel TPC1-type channel inhibitor, against salicylate-induced and cold-shock-induce calcium influx in tobacco BY-2 cells. *Biochem Biophys Res Commun* 332:823–830.

Lindberg S, Strid H. 1997. Aluminium induces rapid changes in cytoplasmic pH and free calcium and potassium concentrations in root protoplast of wheat *(Triticum aestivum)*. *Physiol Plant* 99:405–414.

Liu J, Magalhaes JV, Shaff J, Kochian LV. 2009. Aluminum-activated citrate and malate transporters from the MATE and ALMT families function independently to confer *Arabidopsis* aluminum tolerance. *Plant J* 57:389–399.

Llungany M, Poschennieder C, Barceló J. 1995. Monitoring of aluminium-induced inhibition of root elongation in four maize cultivars differing in tolerance to aluminium and proton toxicity. *Physiol Plant* 93:265–271.

Luo M-C, Dvořák J. 1996. Molecular mapping of an aluminum tolerance locus on chromosome 4D of Chinese Spring wheat. *Euphytica* 91:31–35.

Ma H-X, Bai G-H, Carver BF, Zhou L-L. 2005. Molecular mapping of a quantitative trait locus for aluminum tolerance in wheat cultivar Atlas 66. *Theor Appl Genet* 112:51–57.

Ma JF, Hiradate S, Matsumoto H. 1998. High aluminum resistance in buckwheat. II. Oxalic acid detoxifies aluminum internally. *Plant Physiol* 117:753–759.

Ma JF, Hiradate S, Nomoto K, Iwashita T, Matsumoto H. 1997a. Internal detoxification mechanism of Al in *Hydrangea*. Identification of Al form in the leaves. *Plant Physiol* 113:1033–1039.

Ma JF, Ryan PR, Delhaize E. 2001. Aluminum tolerance in plants and the complexing role of organic acids. *Trends Plant Sci* 6:273–278.

Ma JF, Shen R, Zhao Z et al. 2002. Response of rice to Al stress and identification of quantitative trait loci for Al tolerance. *Plant Cell Physiol* 43:652–659.

Ma JF, Taketa S, Yang ZM. 2000. Aluminum tolerance genes on the short arm of chromosome 3R are linked to organic acid release in triticale. *Plant Physiol* 122:687–694.

Ma JF, Zheng SJ, Matsumoto H, Hiradate S. 1997b. Detoxifying aluminum with buckwheat. *Nature* 390:569–570.

Magalhaes J, Liu J, Guimarães CT et al. 2007. A gene in the multidrug and toxic compound extrusion (MATE) family confers aluminum tolerance in sorghum. *Nat Genet* 39:1156–1161.

Mao C, Yi K, Yang L et al. 2004. Identification of aluminium-regulated genes by cDNA-AFLP in rice (*Oryza sativa* L.): Aluminium-regulated genes for the metabolism of cell wall Components. *J Exp Bot* 55:137–143.

Martin RB. 1986. The chemistry of aluminum as related to biology and medicine. *Clin Chem* 32:1797–1806.

Martinez-Esteveg M, Racagni-Oi Palma G, Munoz-Sanchez JA et al. 2003. Aluminium differentially modifies lipid metabolism from the phosphoinositide pathway in *Coffea arabica* cells. *J Plant Physiol* 160:1297–1303.

Massot N, Nicander B, Barcelo J, Poschenrieder C, Tillberg E. 2002. A rapid increase in cytokinin level and enhanced ethylene evaluation precede Al^{3+}-induced inhibition of root growth in bean seedlings (*Phaseolus vulgaris* L.). *Plant Growth Regul* 37:105–112.

Matsumoto H. 1988. Changes of the structure of pea chromatin by aluminum. *Plant Cell Physiol* 29:281–287.

Matsumoto H. 1991. Biochemical mechanism of the toxicity of aluminium and the sequestration of aluminum in plant cells. In *Plant-Soil Interactions at Low pH*, eds. RJ Wright, VC Baligar, RP Murrmann, pp. 825–838. Dordrecht, the Netherlands: Kluwer.

Matsumoto H. 2000. Cell biology of aluminum toxicity and tolerance in higher plants. *Int Rev Cytol* 200:1–46.

Matsumoto H. 2002a. Plant roots under aluminum stress: toxicity and tolerance In *Plant Roots: The Hidden Half*, eds. Y Waisel, A Eshel, U Kafkafi, 3rd edn., pp. 821–838. New York: Marcel Dekker, Inc.

Matsumoto H. 2002b. Metabolism of organic acids and metal tolerance in plants exposed to aluminum. In *Physiology and Biochemistry of Metal Toxicity and Tolerance in Plants*, eds. MN Prasad, K Strzalka, pp. 95–109. New York: Marcel Dekker, Inc.

Matsumoto H, Hirasawa E, Torikai H, Takahashi E. 1976. Localization of absorbed Al in pea root and its binding to nucleic acids. *Plant Cell Physiol* 17:127–137.

Matsumoto H, Morimura S, Takahashi E. 1977a. Less involvement of pectin in the precipitation of aluminium in pea root. *Plant Cell Physiol* 18:325–335.

Matsumoto H, Morimura S, Takahashi E. 1977b. Binding of aluminium to DNA of DNP in pea root nuclei. *Plant Cell Physiol* 18:987–993.

Matsumoto H, Motoda H. 2012. Aluminum toxicity recovery processes in root apices. Possible association with oxidative stress. *Plant Sci* 185–186:1–8.

Matsumoto H, Senoo Y, Kasai M, Maeshima M. 1996. Response of the plant root to aluminum stress: Analysis of the inhibition of the root elongation and changes in membrane function. *J Plant Res* 109:99–105.

Matsumoto H, Sivaguru M. 2008. Advances in the aluminum toxicity and tolerance of plants for increased productivity in acid soils. In *Soil Contamination: New Research*, ed. AN Dubois, pp. 1–42. New York: Nova Science Publication.

Miftahudin, Scoles GJ, Gustafson JP. 2002. AFLP markers tightly linked to the aluminum-tolerance gene *Alt3* in rye (*Secale cereale* L.). *Theor Appl Genet* 104:626–631.

Milla MA, Butler E, Huete AR et al. 2002. Expressed sequence tag-based gene expression analysis under aluminum stress in rye. *Plant Physiol* 130:1706–1716.

Mimmo T, Marzadori C, Montecchio D, Gessa C. 2005. Characterization of Ca- and Al-pectate gels by thermal analysis and FT-IR spectroscopy. *Carbohydr Res* 340:2510–2519.

Miyasaka SC, Buta JG, Howell RK, Foy CD. 1991. Mechanism of aluminum tolerance in snap beans. Root exudation of citric acid. *Plant Physiol* 96:737–743.

Mohanty S, Das AB, Das P, Mohanty P. 2004. Effects of a low dose of aluminum on mitotic and meiotic activity, 4C DNA content, and pollen sterility in rice, *Oryza sativa* L. cv. Lalat. *Ecotoxicol Environ Saf* 59:70–75.

Morimura S, Matsumoto H. 1978a. Effect of aluminium on some properties and template activity of purified pea DNA. *Plant Cell Physiol* 19:429–436.

Morimura S, Takahashi E, Matsumoto H. 1978b. Association of aluminium with nuclei and inhibition of cell division in onion (*Allium cepa*) roots. *Z Pflanzenphysiol* 88:395–401.

Motoda H, Kano Y, Hiragami F, Kawamura K, Matsumoto H. 2010. Morphological changes in the apex of pea roots during and after recovery from aluminum treatment. *Plant Soil* 333:49–58.

Motoda H, Kano Y, Hiragami F, Kawamura K, Matsumoto H. 2011. Changes in rupture formation and zonary region stained with Evans blue during the recovery process from aluminum toxicity in the pea root apex. *Plant Signal Behav* 6:98–100.

Nian H, Ahn SJ, Yang ZM, Matsumoto H. 2003. Effect of phosphorus deficiency on aluminium-induced citrate exudation in soybean (*Glycine max*). *Physiol Plant* 117:229–236.

Nichol BE, Oliveria LA, Glass AD, Siddiqi MY. 1993. The effects of aluminum on the influx of calcium, potassium, ammonia, nitrate, and phosphate in an aluminum-sensitive cultivar of barley (*Hordeum vulgare* L.). *Plant Physiol* 101:1263–1266.

Ofei-Manu P, Wagatsuma T, Ishikawa S et al. 2001. The plasma membrane strength of the root-tip cells and root phenolic compounds are correlated with Al tolerance in several common woody plants. *Soil Sci Plant Nutr* 47:359–375.

Ohno T, Koyama H, Hara T. 2003. Characterization of citrate transport through the plasma membrane in carrot mutant cell line with enhanced citrate excretion. *Plant Cell Physiol* 44:146–162.

Qifu MA, Rengel Z, Kuo J. 2002. Aluminium toxicity in rye (*Secale cereale*): Root growth and dynamics of cytoplasmic Ca^{2+} in intact root tips. *Ann Bot* 89:241–244.

Olivetti GP, Cumming JR, Etherton B. 1995. Membrane potential depolarization of root cap cells precedes aluminum tolerance in snap bean. *Plant Physiol* 109:123–129.

Ono K, Yamamoto Y, Hachiya, Matsumoto H. 1995. Synergistic inhibition of growth by aluminum and iron of tobacco (*Nicotiana tabacum* L.) cells in suspension culture. *Plant Cell Physiol* 36:115.

Osawa H, Endo I, Hara Y, Matsushima Y, Tange T. 2011. Transient proliferation of proanthocyanidin-accumulating cells on the epidermal apex contributes to high aluminum-resistant root elongation in camphor tree. *Plant Physiol* 155:433–446.

Osawa H, Matsumoto H. 2001. Possible involvement of protein phosphorylation in aluminum-responsive malate efflux from wheat root apex. *Plant Physiol* 126:411–423.

Osawa H, Matsumoto H. 2002. Aluminium triggers malate-independent potassium release via ion channels from the root apex in wheat. *Planta* 215:405–412.

Pan J, Zhu M, Chen H. 2001. Aluminum-induced cell death in root-tip cells of barley. *Environ Exp Bot* 46:71–79.

Panda SK, Baluška F, Matsumoto H. 2009. Aluminum stress signaling in plants. *Plant Signal Behav* 4:592–597.

Papernik LA, Kochian LV. 1997. Possible involvement of Al-induced electrical signals in Al tolerance in wheat. *Plant Physiol* 115:657–667.

Parker DR, Kinraide TB, Zelazny LW. 1989. On the phytotoxicity of polynuclear hydroxyaluminum complexes. *Soil Sci Soc Am J* 53:789–796.

Parker DR, Pedler JF. 1998. Probing the malate hypothesis of differential aluminum tolerance in wheat by using other rhizotoxic ions as proxies for Al. *Planta* 205:389–396.

Pejchar P, Martin P, Zuzana N et al. 2010. Aluminium ions inhibit the formation of diacylglycerol generated by phosphatidylcholine-hydrolysing phospholipase C in tobacco cells. *New Phytol* 188:150–160.

Pejchar P, Pleskot R, Schwarzerová K et al. 2008. Aluminum ions inhibit phospholipase D in a microtubule-dependent manner. *Cell Biol Int* 35:554–556.

Pellet DM, Grunes DL, Kochian LV. 1995. Organic acid exudation as an aluminum-tolerance in maize (*Zea mays* L.). *Planta* 196:788–795.

Pellet DM, Papernick LA, Jones DL et al. 2007. Involvement of multiple aluminum exclusion mechanism in aluminum tolerance. *Plant Soil* 192:63–68.

Piñeros MA, Cançado GM, Maron LG et al. 2008. Not all ALMT1-type transporters mediate aluminum-activated organic acid responses: The case of *ZnALMT1*—An anion-selective transporter. *Plant J* 53:352–367.

Piñeros MA, Kochian LV. 2001. A patch-clamp study on the physiology of aluminum toxicity and aluminum tolerance in maize. Identification and characterization of Al^{3+}-induced anion channels. *Plant Physiol* 125:292–305.

Pinéros MA, Scaff JE, Manslank HS, Alves VMC, Kochin LV. 2005. Aluminum resistance in maize cannot be solely explained by root organic acids exudation. *Plant Physiol* 137:231–241.

Poschenrieder C, Gunse B, Corrales I, Barceló J. 2008. A glance into aluminum toxicity and resistance in plants. *Sci Total Environ* 400:356–368.

Puthota V, C-Ortega R, Johnson J, Ownby J. 1991. An ultrastructural study of the inhibition of mucilage secretion in the wheat root cap by aluminum. In *Plant-Soil Interactions at Low pH*, eds. Brito-Argáed. RJ Wright, VC Baligar, RP Murrmann, pp. 779–787. Dordrecht, the Netherlands: Kluwer.

Raman H, Raman R, Wood R, Martin P. 2006. Repetitive indel markers within the *ALMT1* gene conditioning aluminium tolerance in wheat (*Triticum aestivum* L.). *Mol Breeding* 18:171–183.

Raman H, Zhang KP, Cakir M et al. 2005. Molecular characterization and mapping of ALMT1, the aluminum-tolerant gene of bread wheat (*Triticum aestivum* L.). *Gene* 48:781–791.

Ramos-Díaz A, Brito-Argáez L, Munnik T, Hernández-Sotomayor S. 2007. Aluminum inhibits phosphatidic acid formation blocking the phospholipase C pathway. *Planta* 225:393–401.

Rangel AF, Mobin M, Rao IM. 2005. Proton toxicity interferes with the screening of common bean (*Phaseolus vulgaris* L.) genotypes for aluminium resistance in nutrient solution. *J Plant Nutr Soil Sci* 168:607–616.

Rath I, Barz W. 2000. The role of lipid peroxidation in aluminium toxicity in soybean cell suspension cultures. *Z Naturforsch C* 55:957–964.

Reid RJ, Tester MA, Smith A. 1995. Calcium/aluminium interactions in the cell wall and plasma membrane of *Chara*. *Planta* 195:362–368.

Rengel Z. 1996. Uptake of aluminium by plant cells. *New Phytol* 134:389–406.

Rincón-Zachary M, Teaster ND, Sparks JA et al. 2010. Fluorescence resonance energy transfer-sensitized emission of yellow cameleon 3.60 reveals root zone-specific calcium signatures in *Arabidopsis* in response to aluminum and other trivalent cations. *Plant Physiol* 152:1442–1458.

Ryan PR, Delhaize E. 2010. The convergent evolution of aluminium resistance in plants exploits a convenient currency. *Funct Plant Biol* 37:275–284.

Ryan PR, Delhaize E, Randall PJ. 1995. Malate efflux from root apices and tolerance to aluminium are highly correlated in wheat. *Aust J Plant Physiol* 22:531–536.

Ryan PR, Delahize E, Randall PJ. 2001. Function and mechanism of organic anion exudation from plant roots. *Ann Rev Plant Physiol Plant Mol Biol* 52:527–566.

Ryan PR, Ditomaso JM, Kochian LV. 1993. Aluminum toxicity in roots: An investigation of spatial sensitivity and the role of the root cap. *J Exp Bot* 44:437–446.

Ryan PR, Kinraide TB, Kochian LV. 1994. Al^{3+}-Ca^{2+}-interactions in aluminum rhizotoxicity I. Inhibition of root growth is not caused by reduction of calcium uptake. *Planta* 192:98–103.

Ryan PR, Liu Q, Sperling P. 2007. A higher plant Δ8 sphingolipid denaturase with a preference for (Z)-isomer formation confers aluminum tolerance to yeast and plants. *Plant Physiol* 144:1–11.

Ryan PR, Raman H, Gupta S et al. 2009. A second mechanism for aluminum resistance in wheat relies on the constitutive efflux of citrate from roots. *Plant Physiol* 149:340–351.

Ryan PR, Reid RJ, Smith FA. 1997a. Direct evaluation of the Ca^{2+}-displacement hypothesis for Al toxicity. *Plant Physiol* 113:1355–1357.

Ryan PR, Skerrett M, Findlay GP, Delhaize E, Tyernan SD. 1997b. Aluminum activates an anion channel in the apical cells of wheat roots. *Proc Natl Acad Sci U S A* 94:6547–6552.

Sasaki M. 1996. Study of aluminum toxicity on root growth in wheat (*Triticum aestivum* L.). PhD dissertation, Okayama University, Kurashiki, Okayama.

Sasaki M, Yamamoto Y, Matsumoto H. 1994. Putative Ca^{2+} channels of plasma membrane vesicles are not involved in the tolerance mechanism of aluminum in aluminum tolerant wheat (*Triticum aestivum* L.) cultivar. *Soil Sci Plant Nutr* 40:709–714.

Sasaki M, Yamamoto Y, Matsumoto H. 1996. Lignin deposition induced by aluminum in wheat (*Triticum aestivum*) roots. *Physiol Plant* 96:193–198.

Sasaki M, Yamamoto Y, Matsumoto H. 1997a. Aluminum inhibits growth and stability of cortical microtubules in wheat (*Triticum aestivum*) roots. *Soil Sci Plant Nutr* 43:469–472.

Sasaki M, Yamamoto Y, Ma JF, Matsumoto H. 1997b. Early events induced by aluminum stress in elongating cells of wheat root. *Soil Sci Plant Nutr* 43:1009–1014.

Sasaki T, Ryan PR, Delhaize E et al. 2006. Sequence upstream of the wheat (*Triticum aestivum* L.) ALMT1 gene and its relationship to aluminum resistance. *Plant Cell Physiol* 47:1343–1354.

Sasaki T, Yamamoto Y, Ezaki B et al. 2004. A wheat gene encoding an aluminum-activated malate transporter. *Plant J* 37:645–653.

Sawaki Y, Iuchi S, Kobayashi Y et al. 2009. STOP1 regulates multiple genes that protect *Arabidopsis* from proton and aluminum toxicities. *Plant Physiol* 150:281–294.

Schmohl N, Horst WJ. 2000. Pectin methylesterase modulates aluminum sensitivity in *Zea mays* and *Solanum tuberosum* L. *Physiol Plant* 109:119–127.

Schofield RMS, Pallon J, Fiskesjö G et al. 1998. Aluminum and calcium distribution patterns in aluminum-intoxicated roots of *Allium cepa* do not support the calcium-displacement hypothesis and indicate signal-mediated inhibition of root growth. *Planta* 205:175–180.

Schulz B, Üner Kolukisaoglu H. 2006. Genomics of plant ABC transporters: The alphabet of photosynthetic life forms or just holes in membrane? *FEBS Lett* 580:1010–1016.

Shen H, He LF, Sasaki T et al. 2005. Citrate secretion coupled with the modulation of soybean root tip under aluminum stress: Up-regulation of transcription, translation, and threonine-oriented phosphorylation of plasma membrane H^{+}-ATPase. *Plant Physiol* 138:287–296.

Shen H, Ligaba A, Yamaguchi M et al. 2004a. Effect of K-252a and abscisic acid on the efflux of citrate from soybean roots. *J Exp Bot* 55:663–671.

Shen H, Yan XL, Cai KZ et al. 2004b. Differential Al resistance and citrate secretion in the tap and basal roots of common bean seedlings. *Physiol Plant* 121:593–603.

Silva IR, Syth TJ, Moxley DF et al. 2000. Aluminum accumulation at nuclei of cells in the root tip. Fluorescent detection using lumogallion and confocal laser scanning microscopy. *Plant Physiol* 123:543–552.

Sivaguru M, Baluska F, Volkmann D, Felle HH, Horst WJ. 1999a. Impacts of aluminum on the cytoskeleton of the maize root apex. Short-term effects on the distal part of the transition zone. *Plant Physiol* 119:1073–1082.

Sivaguru M, Ezaki B, He ZH et al. 2003a. Aluminum-induced gene expression and protein localization of a cell wall-associated receptor kinase in *Arabidopsis. Plant Physiol* 132:1–11.

Sivaguru M, Fujiwara T, Šamaj J et al. 2000. Aluminum-induced 1,3-D-glucan inhibits cell-to-cell trafficking of molecules through plasmodesmata: A new mechanism of Al toxicity in plants. *Plant Physiol* 124:991–1005.

Sivaguru M, Horst WJ. 1998. The distal part of the transition zone is the most aluminum-sensitive apical zone of maize. *Plant Physiol* 116:155–163.

Sivaguru M, Pike S, Gassmann W, Baskin TI. 2003b. Aluminum rapidly depolymerizes cortical microtubules and depolarizes the plasma membrane: Evidence that these responses are mediated by a glutamate receptor. *Plant Cell Physiol* 44:667–675.

Sivaguru M, Yamamoto Y, Matsumoto H. 1999b. Differential impacts of aluminium on microtubule organization depends on growth phase in suspension-cultured tobacco cells. *Physiol Plant* 107:110–119.

Sláski JJ, Zhang G, Basu U, Stephens JL, Taylor GJ. 1996. Aluminum resistance in wheat (*Triticum aestivum*) is associated with rapid, Al-induced changes in activities of glucose-6- phosphate dehydrogenase and 6-phospho-gluconate dehydrogenase in root apices. *Physiol Plant* 98:477–484.

Sun P, Tian QY, Chen J, Zhang W-H. 2010. Aluminium-induced inhibition of root elongation in *Arabidopsis* is mediated by ethylene and auxin. *J Exp Bot* 61:347–356.

Sun P, Tian QY, Zhao MG et al. 2007. Aluminum-induced ethylene production is associated with inhibition of root elongation in *Lotus japonicus* L. *Plant Cell Physiol* 48:1229–1235.

Tabuchi A, Kikui S, Matsumoto H. 2004. Differential effects of aluminum on osmotic potential and sugar accumulation in the root of Al-resistant and Al-sensitive wheat. *Physiol Plant* 12:106–112.

Tabuchi A, Matsumoto H. 2001. Changes in cell-wall properties of wheat (*Triticum aestivum*) roots during aluminum-induced growth inhibition. *Physiol Plant* 112:353–358.

Takita E, Koyama H, Hara T. 1999. Organic acid metabolism in aluminum-phosphate utilizing cells of carrot (*Daucus carota* L.). *Plant Cell Physiol* 40:489–495.

Tamas L, Huttova J, Mistrik I. 2003. Inhibition of Al-induced rot elongation and enhancement of Al-induced peroxidase activity in Al-sensitive and Al-resistant barley cultivars are positively correlated. *Plant Soil* 250:193–200.

Tang Y, Sorreslls ME, Kochian LV, Gorvin DF. 2000. Identification of RFLP markers linked to the barley aluminum tolerance gene *Alp. Crop Sci* 40:778–782.

Taylor GJ. 1991. Current views of the aluminum stress response: The physiological basis of tolerance. *Curr Top Plant Biochem Physiol* 10:57–93.

Tesfaye M, Temple SJ, Allan DL et al. 2001. Overexpression of malate dehydrogenase in transgenic alfalfa enhances organic acid synthesis and confers tolerance to aluminum. *Plant Physiol* 127:1836–1844.

Tian Q-Y, Sun D-H, Zhan M-G, Zhang W-H. 2007. Inhibition of nitric oxide synthase (NOS) underlines aluminum-induced inhibition of root elongation in *Hibiscus moscheutos. New Phytol* 174:322–331.

Tice KR, Parker DR, DeMason DA. 1992. Operationally defined apoplastic and symplastic aluminum fractions in root tips of aluminum-intoxicated wheat. *Plant Physiol* 100:309–318.

Tolra RP, Poscenrieder C, Luppi B, Barceló J. 2005. Aluminium-induced changes in the profiles of both organic acids and phenolic substances underlie Al tolerance in *Rumex acetosa* L. *Environ Exp Bot* 54:231–238.

Trejo-Téllez LI, Stenzel FC, Gómez-Merino FC, Schmitt JM. 2010. Transgenic tobacco plants overexpressing pyruvate phosphate dikinase increase exudation of organic acids and decrease accumulation of aluminum in the roots. *Plant Soil* 326:187–198.

Turner A, Wells B, Roberts K. 1994. Plasmodesmata of maize root tips: Structure and composition. *J Cell Sci* 107:3351–3361.

Wagatsuma T, Jujo K, Ishikawa S, Nakashima T. 1995. Aluminum-tolerant protoplasts from roots can be collected with positively charged silica microbeads: A methods based on differences in surface negativity. *Plant Cell Physiol* 36:1493–1502.

Wang J, Raman H, Zhou M et al. 2007. High-resolution mapping of the *Alp* locus and identification of a candidate gene *HvMATE* controlling aluminium tolerance in barley (*Hordeum vulgare* L.). *Theor Appl Genet* 115:265–276.

Wang Y, Stass A, Horst WJ. 2004. Apoplastic binding of aluminum is involved in silicon-induced amelioration of aluminum toxicity in maize. *Plant Physiol* 136:3762–3770.

Wang Y-S, Yang ZM. 2005. Nitricoxide reduces aluminum toxicity by preventing oxidative stress in the root of *Cassia tora* L. *Plant Cell Physiol* 46:1915–1923.

Watanabe T, Misawa S, Hiradate S, Osaki M. 2008. Characterization of root mucilage from *Melastoma malabathricum,* with emphasis on its roles in aluminum accumulation. *New Phytol* 178:581–589.

Watt DA. 2003. Aluminium-responsive genes in sugarcane: Identification and analysis of expression under oxidative stress. *J Exp Bot* 54:1163–1174.

Wehr JB, Menzies NW, Blamey FPC. 2004. Inhibition of cell-wall autolysis and pectin degradation by cations. *Plant Physiol Biochem* 42:485–492.

Wenzl P, Patino GM, Chaves AL, Mayer JE, Rao IM. 2001. The high level of aluminum resistance in signal grass is not associated with known mechanisms of external aluminum detoxification in root apices. *Plant Physiol* 125:1473–1484.

Wherrett T, Ryan PR, Delhaize E, Shabala S. 2005. Effect of aluminum on membrane potential and ion fluxes at the apices of wheat roots. *Funct Plant Biol* 32:199–208.

Wissemeier AH, Dieming A, Hergenroder A et al. 1992. Callose formation as parameter for assessing genotypical plant tolerance of aluminium and manganese. *Plant Soil* 146:67–75.

Wood S, Sebastian K, Scherr SJ. 2000. *Pilot Analysis of Global Ecosystems: Agroecosystems.* Washington, DC: Rosen.

Xia JX, Kasai T, Yamaji N, Ma JF. 2010. Plasma membrane-associated transporter for aluminum in rice. *Proc Natl Acad Sci U S A* 107:18381–18385.

Xu M, You J, Hou N et al. 2010. Mitochondrial enzymes and citrate transporter contribute to the aluminum-induced citrate secretion from soybean (*Glycine max*) roots. *Funct Plant Biol* 37:285–295.

Xue YJ, Tao L, Yang ZM. 2008. Aluminum-induced cell wall peroxidase activity and lignin synthesis are differentially regulated by jasmonate and nitric oxide. *J Agric Food Chem* 56:9676–9684.

Yamaguchi M, Sasaki T, Sivaguru M et al. 2005. Evidence for the plasma membrane localization of Al-activated malate transporter (ALMT1). *Plant Cell Physiol* 46:812–816.

Yamaguchi Y, Yamamoto Y, Matsumoto H. 1999. Cell death process initiated by a combination of aluminum and iron in suspension-cultured tobacco cells (*Nicotiana tabacum*): Apoptosis-like cell death mediated by calcium and proteinase. *Soil Sci Plant Nutr* 45:647–657.

Yamaji N, Huang CF, Nagao S et al. 2009. A zinc finger transcription factor ART1 regulates multiple genes implicated in aluminum tolerance in rice. *Plant Cell* 21:3339–3349.

Yamamoto Y, Hachiya A, Matsumoto H. 1997. Oxidative damage to membrane by a combination of aluminum and iron in suspension-cultured tobacco cells. *Plant Cell Physiol* 38:1333–1339.

Yamamoto Y, Kobayashi Y, Matsumoto H. 2001. Lipid peroxidation is an early symptom triggered by aluminum, but not the primary cause of elongation inhibition in pea roots. *Plant Physiol* 125:199–208.

Yamamoto Y, Kobayashi Y, Rama Devi S, Rikiishi S, Matsumoto H. 2002. Aluminum toxicity is associated with mitochondrial dysfunction and the production of reactive oxygen species in plant cells. *Plant Physiol* 128:63–72.

Yang JJ, Li YY, Zhang YJ, Matsumoto H. 2008. Cell wall polysaccharides are specifically involved in the exclusion of aluminum from rice root apex. *Plant Physiol* 146:602–611.

Yang ZM, Nian H, Sivaguru M, Tanakamaru S, Matsumoto H. 2001. Characterization of aluminium-induced citrate secretion in aluminium-tolerant soybean (*Glycine max*) plants. *Physiol Plant* 113:64–71.

Yang ZM, Sivaguru M, Horst WJ, Matsumoto H. 2000. Aluminum tolerance is achieved by exudation of citric acid from roots of soybean (*Glycine max*). *Physiol Plant* 110:72–77.

Yang JL, Zheng SJ, He YF et al. 2006. Comparative studies on the effect of a protein-synthesis inhibitor on aluminium-induced secretion of organic acids from *Fagopyrum esculentum* Moench. and *Cassia tora* L. roots. *Plant Cell Environ* 29:240–246.

Yin L, Mano J, Wang S, Tsuji W, Tanaka K. 2010a. The involvement of lipid peroxide-derived aldehydes in aluminum toxicity of tobacco roots. *Plant Physiol* 152:1406–1417.

Yin L, Wang S, Eltayeb AE et al. 2010b. Overexpression of dehydroascorbate reductase, but not monodehydroascorbate reductase, confers tolerance to aluminum stress in transgenic tobacco. *Planta* 231:609–621.

Yokel RA. 2004. Aluminum. In *Elements and their Compounds in the Environment*, 2nd edn., eds. E Merian, M Anke, M Inhat, M Stoeppler, pp. 635–658. Weinheim, Germany: WILEY-VCH Verlag GmbH & Co. KGaA.

Yokosho K, Yamaji N, Ma JF. 2010. Isolation and characterization of two MATE genes in rye. *Funct Plant Biol* 37:296–303.

Zhang W-H, Rengel Z. 1999. Aluminium induces an increase in cytoplasmic calcium in intact wheat root apical cells. *Aust J Plant Physiol* 26:401–409.

Zhang W-H, Ryan PR, Sasaki T et al. 2008. Characterization of the TaALMT1 protein as an Al^{3+}-activated anion channel in transformed tobacco (*Nicotiana Tabacum* L.) cells. *Plant Cell Physiol* 49:1316–1330.

Zhang W-H, Ryan PR, Tyerman SD. 2001. Malate-permeable channels and cation channels activated by aluminum in the apical cells of wheat roots. *Plant Physiol* 125:1459–1472.

Zhang G, Slaski JJ, Archamtault DJ, Taylor GJ. 1997. Alteration of plasma membrane lipids in aluminum-resistant and aluminum-sensitive wheat genotypes in response to aluminum stress. *Physiol Plant* 99:302–306.

Zheng SJ, Xianyong L, Yang J, Lin Q, Tang C. 2004. The kinetics of aluminum adsorption and desorption by root cell walls of an aluminum resistant wheat (*Triticum aestivum* L.) cultivum. *Plant Soil* 261:85–90.

Zheng SJ, Yang JL. 2005a. Target sites of aluminum phytotoxicity. *Biol Plant* 49:321–331.

Zheng SJ, Yang JL, Ile Y et al. 2005b. Immobilization of aluminum with phosphorus in root is associated with high aluminum resistance in buckwheat. *Plant Physiol* 138:297–303.

Zhu MY, Ahn SJ, Matsumoto H. 2003. Inhibition of growth and development of root border cells in wheat by Al. *Physiol Plant* 117:359–367.

34

Root Responses to Trace Metallic Elements

Nathalie Verbruggen
Université libre de Bruxelles

Christian Hermans
Université libre de Bruxelles

I. General Introduction

The transfer of chemical elements from soil to plants is a fundamental process that controls life on Earth. Increased interest in plant nutrition and in trace elements in particular is due to growing understanding of their importance for human health and for a safe environment.

This chapter is devoted to root responses to trace metallic elements (TMEs), with two main aspects:

1. It reviews root morphological responses to TME excess or deficiencies.
2. It presents current knowledge on cellular responses to critical TME concentrations and tolerance mechanisms to TME excess, with a special emphasis on hyperaccumulators.

Before addressing these issues, we will define TMEs and present the factors affecting their bioavailability.

II. Classifications of Trace Metallic Elements

In the previous version of this chapter by Hagemeyer and Breckle (2002), the term "heavy metals" was used. That term often applies a group name for metals and metalloids associated with environmental pollution. Legal regulations still refer to "heavy metals" as a list of 11 elements with a density higher than five, soluble at physiological conditions, and relevant in the environmental context: arsenic (which is actually a metalloid), cadmium (Cd), chromium, cobalt (Co), copper (Cu), lead (Pb), mercury (Hg), nickel (Ni), tin, vanadium, and zinc (Zn). The term "trace elements" refers to chemical elements that occur in the Earth's crust in amounts less than 0.1% in mass. Since our review focuses on root responses, we will also consider those metals that are micronutrients, that is to say, which are present as trace amounts in plant organs (<0.01% plant dry weight [DW]) and fulfill biological functions. While iron (Fe) is found in abundance in the Earth's crust (about 5% in mass), it is present only in trace amounts in plant materials. Manganese (Mn) is not considered as a trace element according to the geochemical definition because its abundance reaches the 0.1% threshold, but in the same way as Fe, it is an essential micronutrient.

While the classification of metals according to their abundance is largely adopted, it does not allow to predict their behavior. Over 50 years ago, Ahrland et al. (1958) have proposed to classify metal ions according to their behavior as electron acceptors while forming a chemical bond, called Lewis acids, defined as elemental species with an available lowest unoccupied molecular orbital. That classification allows prediction of

TABLE 34.1 Characteristics of Considered TME

Metal Ion	Common Oxidation States	Hardness
Essential TMEs		
Copper	+2	Soft
	+3	Borderline
Iron	+2	Borderline
	+3	Hard
Manganese	+2	Borderline
	+3	Hard
Molybdenum	+4	Hard
	+6	Hard
Nickel	+2	Borderline
Zinc	+2	Borderline
Nonessential TMEs		
Cadmium	+2	Soft
Cobalt	+2	Borderline
	+3	Hard
Lead	+2	Soft
	+4	Borderline
Mercury	+2	Soft

Source: Adapted from Duffus, J.H., *Pure Appl. Chem.*, 74, 793, 2002; Andreini, C. et al., *J. Biol. Inorg. Chem.*, 13, 1205, 2008.

preferred ligands and metal complexes properties (reviewed by Duffus 2002). Metal ions can be classified as type A, type B, or borderline. Type A (hard) metals are nonpolarizable, preferentially form complexes with oxygen donors, and the bonding in these complexes is mainly ionic. Type B (soft) metals preferentially bind to soft ligands to give more covalent bonding. Borderline metals are intermediate, with variable coordination chemistry (Table 34.1; Duffus 2002; Andreini et al. 2008). That classification also allows interpretation of the biochemical basis for toxicity.

A classification of TMEs is also established according to the essential, beneficial, or nonessential nature of the element to plants.

1. *Essential TMEs*
   Of the 92 known elements, 17 elements are classified as *essential* because they are required to complete plant's life cycle (Marschner 1995). Among those, Cu, Fe, Mn, molybdenum (Mo), Ni, and Zn are metallic micronutrients. In a classification of essential elements according to their biochemical function, all TMEs belong to the same group of "nutrients that are involved in redox reactions", except Mn, which is in the group "nutrients that remain in ionic form" (as defined by Mengel and Kirkby 1987, cited by Taiz and Zeiger 2010). By analyzing the roles and the distribution of metal ions in enzymatic catalysis, Adreini et al. (2008) have concluded that redox-inert metal ions, such as Zn, are used in enzymes to stabilize negative charges and to activate substrates by virtue of their Lewis acid properties, whereas redox-active metal ions, such as Cu, Fe, Mn, Mo, and Ni, can be used as Lewis acids and as redox centers (Table 34.1).

Essential TMEs might play many different biological roles in plants (Hansch and Mendel 2009), and some important examples are given below.

Due to its ability to readily gain and lose an electron and consequently to cycle between the oxidized Cu(II) and reduced Cu(I) states, Cu serves as an important catalytic cofactor in redox reactions and is the cofactor of proteins involved in many biochemical processes including photosynthesis, respiration, removal of superoxide radicals, ethylene perception, cell wall modification, and Mo cofactor biosynthesis (Burkhead et al. 2009; Cohu and Pilon 2010).

As Cu, Fe is a cofactor of many ubiquitous proteins. The redox potential of Fe^{2+}/Fe^{3+} enables its inclusion in the form of heme or Fe–sulfur clusters, in a number of protein complexes, such as those involved in electron transfer of photosynthesis and respiration (Curie et al. 2003). Fe is also required for chlorophyll synthesis.

Mn has several oxidative states and can play the role of catalyst in electron transfer reactions. It is a component of the oxygen-evolving complex (OEC), also called water-splitting complex. OEC is associated with photosystem II and generates dioxygen from water using a catalytic Mn_4CaO_n cluster (n varies with the mechanism and nature of the intermediate) (Kanady et al. 2011). Another important Mn-containing enzyme is Mn superoxide dismutase (Mn-SOD), which is responsible for detoxifying superoxide radicals in mitochondria and peroxisomes. Mn also activates decarboxylases and dehydrogenases involved in the Krebs cycle. Furthermore, it is the cofactor of enzymes that are involved in the biosynthetic pathway, leading to aromatic amino acids, auxin, gibberellic acid, and chlorophyll precursors (Pittman 2005; Williams and Pittman 2010).

Mo is required by several enzymes participating in reduction and oxidation reactions, in which it is bound to pterin to form the Mo cofactor or Moco. Molybdoenzymes are involved in nitrate assimilation, sulfite detoxification, abscisic acid (ABA) biosynthesis, and purine degradation (Schwarz and Mendel 2006). Legumes need more Mo than other plants because it is required by nitrogenase in symbiotic bacteria for the fixation of atmospheric nitrogen in the nodules.

Ni was relatively recently established as an essential element for plants (Brown et al. 1987). It is only used as an integral component of the urease, which catalyzes the hydrolysis of urea to carbon dioxide and ammonia (Tejada-Jimenez et al. 2009). Ni deficiency results in decreased activity of urease and subsequently in toxic accumulation of urea (Eskew et al. 1983; Bai et al. 2006). Besides that, Ni also coordinates with S and O ligands of many enzymes or ligands of tetrapyrrole structures (reviewed by Yusuf et al. 2011). It is also reported that small quantities of Ni are essential for root nodule growth and hydrogenase activation in some legumes (Yussuf et al. 2011).

Zn is used as a catalytic or structural component in proteins and in protein–protein interactions (Frausto da Silva and Williams 2001; Adreini et al. 2008; Clemens 2010). The best-known example for a Zn-containing protein domain is the "Zn finger," which consists of antiparallel hairpin motifs. The majority of Zn-stabilized proteins are involved in gene regulation and interact with nucleic acids (reviewed by Clemens 2010). For example, more than one-fifth of *Arabidopsis* transcription factors are Zn-requiring proteins. Other main Zn-requiring enzymes are hydrolases, oxidoreductases, and transferases (reviewed by Clemens 2010).

2. *Beneficial elements* can promote growth or may be essential to particular taxa but not to all plants (Pilon-Smits et al. 2009). Up until now, Co is the only beneficial TME identified in plants (Pilon-Smits et al. 2009). It is an essential element in animals, being the central metal ion of cobalamin (vitamin B_{12}), which is only synthesized by bacteria. While Co has a role in legume nutrition as an essential element for symbiotic N_2-fixing microorganism, its importance to other plant species is still ambiguous.
3. *Nonessential elements* are those that are neither essential nor beneficial. The evolution of mineral elements as essential or nonessential depends on chemical properties but also on their abundance in the Earth's crust. For example, Cd is naturally found at low concentrations in the Earth's crust and may not have been recruited during evolution because of its lower abundance compared to Zn, which is a neighboring element in the periodic table (Clemens 2006). Therefore, some nonessential elements can be particularly toxic because they mimic essential ones. The list of nonessential TMEs is large, but for simplicity, we will consider those with physiological relevance: Cd, Hg, and Pb. Those metals are of major concern with respect to plant exposure as well as human food-chain accumulation (Clemens 2006). Because chromium is only slightly soluble in the soil solution and not easily taken up by plants (Kabata-Pendias 2010), it will not be considered in this chapter. Toxicity of TME can be due to oxidative damage (see Section IV.B.2). Regarding Cd and Hg, impaired metabolism is also due to the high affinity for sulfhydryl (–SH) groups of Cd(II) and Hg(II) (Patra and Sharma 2000; Patra et al. 2004; Verbruggen et al. 2009a). As a consequence of exposure to Cd and Hg, glutathione (γ-glutamate–cysteine–glycine tripeptide) depletion is an important source of oxidative stress. As mentioned earlier, Cd also mimics essential elements. It can compete with Zn, Fe, Ca, and other cations for transport as well as in binding to proteins (reviewed by Verbruggen et al. 2009a). The extent to which Hg is harmful depends on the Hg form. Organomercurial compounds (which can be found in fertilizers and pesticides) may be 200 times more toxic than inorganic Hg (Patra and Sharma 2000).

III. Availability of Trace Metallic Elements in Soils and Their Uptake by Plants

TME concentrations vary widely by geographic location, and their occurrence in the environment can be due to natural sources or to human activities. Those elements, even though normally found in trace quantities, can quickly become persistent environmental contaminants and toxic to life. The uptake of a given TME by plants depends not only on the soil elemental concentration *per se* but also on the chemical form(s), bioavailability and mobility of the element, and on the chemical and biological properties of the soil. TMEs move from the bulk soils to the root surface, are further transported into the root (which represents the first barrier to selective accumulation of metals from the soil solution), and are translocated from the root to the shoot (Alloway 1995). We will describe here soil and plant factors influencing TME uptake.

A. Occurrence of Trace Metallic Elements in Soils and Strategies to Address Human Health and Environmental Issues

Pedogenetic processes and anthropogenic activities can modify the occurrence of TMEs in natural and agricultural systems. Some lands naturally contain relatively high TME levels. For example, serpentine soils, which are derived from ultramafic rocks, are usually rich in Ni and Fe (Mesjasz-Przybyiowicz et al. 2007). Over the past 200 years, emissions of toxic TMEs by human activities have significantly exceeded those from natural sources for practically all metals (Clemens 2006). Extensive mining, agriculture, and industry have released enormous amounts of TMEs into the environment, which create worldwide environmental and health concerns (Cuypers et al. 2012). Cd contamination is widespread, particularly in soils containing waste materials from mining industry and in sludge-amended and Cd-rich phosphatic fertilized soils (Lugon-Moulin et al. 2006). Farmland may also contain elevated quantities of essential TMEs, due to the continuous spreading of Cu-rich manure (Bolan et al. 2003; Guan et al. 2011), Cu-based fungicides (Wightwick et al. 2008), or scattering of Zn-rich sewage sludge (Chaney 1993). To address environmental and human health issues related to contaminated ecosystems, phytoremediation (use of plants and their associated microbes) has attracted much attention during the last decades. The many possible ways in which plants can achieve this include phytoextraction, phytovolatilization, detoxification, and sequestration (reviewed by Chaney et al. 2007; Verbruggen and Leduc 2009; Zhao and McGrath 2009). The two first processes are particularly attractive in that their use has the potential of moving the contaminants from the local ecosystem altogether. In phytoextraction, plants take up the contaminant of interest from soil, sediment, or water and accumulate it. Phytoextraction is possible within a reasonable time frame in weakly to moderately degree of pollution, by using plants with efficient metal root-to-shoot translocation and high accumulation capacity in the shoot. In such a case, harvesting the aboveground biomass succeeds in removing the contaminant from the local ecosystem. Phytovolatilization is relevant to those trace element contaminants that can be metabolized to a volatile form (e.g., Hg, which is volatile in its elemental form). For more heavily contaminated soils, phytostabilization with tolerant plants aims at stabilization of the contaminated site and at lowering the risk of pollutants leaching. Hyperaccumulators (see Section V) can be directly used for phytoremediation strategies. However, most hyperaccumulators do not develop sufficient biomass, even in the absence of high concentrations of heavy metals to be useful in phytoextraction, and/or are not amenable to agronomic practices (mechanical harvest). To create suitable plants for phytoremediation, one possibility is the selection of high-biomass plants like poplar, Indian mustard, or tobacco with the highest tolerance and accumulation capacities, provided that the species displays sufficient genetic variation for those characteristics. An opportunity is also sought in the association of metal-tolerant microorganisms to enhance tolerance. Finally, genetic manipulation can improve accumulation, tolerance, and detoxification capacities of high-biomass and rapid-growing plants in order to optimize the phytoextraction process. Genetic determinants of those traits can be identified through the study of hyperaccumulators.

Besides TME excess, marginal and agricultural soils can also be depleted with essential TMEs. Fe, Mn, and Zn deficiencies are well-recognized limiting factors in crop production systems (White and Zasoski 1999; Cakmak 2002; Williams and Pittman 2010). Low TME content in plants can also have an impact on human health. Diets of populations subsisting on cereals, or inhabiting regions where soil mineral imbalances occur, often lack Cu, Fe, or Zn, among others (White and Broadley 2005). Fe deficiency is currently the most common and widespread nutritional disorder in the world (http://www.who.int/nutrition/topics/ida/en/index.html). In the frame of plant nutrition studies for global health, the increase of essential TME (mainly Fe and Zn) content in grains of staple food (= so called biofortification process) has become an important quality factor for human consumption and a strategy to alleviate mineral malnutrition, particularly in developing countries (Zhao and McGrath 2009; White and Broadley 2011). Current research aims at identifying physiological and genetic determinisms related to improved efficiency in the uptake by roots and use of TMEs. Undeniably hyperaccumulators (see Section V) are once more considered as valuable resources for this research area.

B. Soil Factors Influencing the Absorption of Trace Metallic Elements by Plants

TMEs present in the soil can be divided into different fractions: metals within the matrix of soil minerals, metal precipitates, metals sorbed to clays and organic matter, and soluble metals in the soil solution. Plants cannot access the total TME pool in soil but only the latter fraction. Soil factors as various as pH, redox potential (Eh), organic matter, clay, water regime, redox conditions, cation exchangeable capacity, and microbiological activity determine the proportion of metals in the soil solution (Micó et al. 2008; Kabata-Pendias 2010). Nonetheless, not all dissolved TMEs but only the bioavailable fraction, which mainly depends on metal speciation (chemical forms that an element takes in solution), can be directly absorbed by roots. Hence, soil solution pH strongly affects TME speciation and solubility. For example, most acid soils, which cover less than one-third of the Earth's surface, have an excess of Mn for plants (Kochian et al. 2004). The divalent form of Mn(II), which is the only assimilable Mn form to plants, predominates in the soil solution with low pH, while Mn(III) and Mn(IV) forms predominate at higher pH and cannot be accumulated in plants (Dučić and Polle, 2005). Calcareous soils, which represent 30% of world's cultivated surface, worsen Fe deficiency in plants (Chen and Barak 1982), despite the general abundance of Fe in soils. The concentrations of free Fe^{3+} and Fe^{2+} (the forms taken up by roots according to the plant species; see Section III.C) are nearly less than 10^{-15} M in well-aerated soils at neutral or basic pH because most of the Fe is present in insoluble forms (Kim and Guerinot 2007).

A last classification of TMEs can be done according to the mobility in soil (Kabata-Pendias 2010): (1) Hg and Pb are relatively strongly adsorbed elements by soil particles and are not readily transported to the plant aboveground parts; (2) Co, Cu, Mn, Mo, and Ni are mobile elements in soil and readily taken up by plants; and (3) Cd and Zn are very mobile elements in soil and easily bioaccumulated in plants.

C. Uptake Systems of Trace Metallic Elements in Plants

Plants possess highly effective uptake systems for TMEs in the soil solution, which are usually present in the submicromolar to low nanomolar concentration range. Tremendous progress has been achieved in the identification of TME transporters encoding genes. However, most of the studies of metal transporters lack direct functional analysis at the protein level. Rather most of the reports present gene expression data, mutant analysis, or heterologous expression in yeast, which do not necessarily identify the real in vivo substrates of those transporters. For most of the elements, the transport is not specific, enabling nonessential metals to compete for the absorption and enter the root cells as well. When essential TMEs become scarce in soil, or when the mineral balance is unfavorable, plants frequently modify their absorption through coordinated regulation of mineral uptake transporters both at the transcriptional and posttranscriptional levels. In addition to regulation according to element availability, expression of several genes coding for transport of macro- and micronutrients shows circadian rhythmicity (e.g., Fe, Duc et al. 2009; Cu, Andrés-Colás et al. 2010; Perea-García et al. 2010; for other elements, see Haydon et al. 2011). The clock is believed to regulate the fluxes and assimilation of nutrients, and in return, changes in nutrient concentration are believed to provide a feedback to modulate the behavior of the clock (Haydon et al. 2011).

Daily changes in redox conditions are believed to play a role in the rhythmically schedule of metal homeostasis (Puig and Peñarrubia 2009).

1. Copper

Cu can be taken up by the root by members of the COPPER TRANSPORTER PROTEINS (COPT) family: COPT1, which is expressed at the root tips and possibly COPT2 (Kampfenkel et al. 1995; Sancenón et al. 2003, 2004). Most Cu in soil is found in its oxidized form Cu(II), while COPT proteins transport the element in its reduced form Cu(I). Several lines of evidence suggest that FERRIC REDUCTASE OXIDASE (FRO) also reduces Cu(II) (Cohen et al. 1997; Mukherjee et al. 2006; Puig et al. 2007; Cohu and Pilon 2010). However, Cu(II) reduction by FRO has not been demonstrated. Interestingly, *FRO3* expression is increased in *Arabidopsis* roots upon both Fe and Cu shortage, supporting a role in Fe as well as in Cu acquisition from soil (Mukherjee et al. 2006). Some COPT family members are differentially regulated by Cu availability in *Arabidopsis*: *COPT1* is induced upon starvation, and *COPT1*, *COPT2*, and *COPT4* are induced upon excess (Wintz et al. 2003).

In addition to COPT transporters, several ZINC-REGULATED TRANSPORTER, IRON-REGULATED TRANSPORTER PROTEINS (ZIPs) may be involved in Cu uptake (Wintz et al. 2003). The Cu induction of the gene encoding the HEAVY METAL ATPase 5 (HMA5) in roots (maximum expression levels were already achieved at Cu concentrations as low as 10 μM) may result in an increase of Cu efflux, supporting a role in root detoxification (Andrés-Colás et al. 2006; del Pozo et al. 2010).

2. Iron

As already mentioned, Fe is chiefly present as insoluble forms (ferric hydroxide) in aerobic environments at neutral or basic pH. To mobilize that element, plants have developed two different strategies. The first strategy, utilized by dicots and non-graminaceous monocots, is based on the acidification of the rhizosphere by a proton pump to increase the solubility of Fe^{3+} complexes followed by the reduction of Fe^{3+} by ferric chelate reductases. Subsequently, Fe^{2+} ions are transported through the plasma membrane of root epidermis cells. It is referred to as the reduction strategy or strategy I. In *Arabidopsis*, acidification is mainly due to H^+-ATPASE 2 (AHA2) activity, and reduction occurs through FERRIC REDUCTION OXIDASES enzymes, such as FRO2, and transport of Fe^{2+} by the IRON-REGULATED TRANSPORTER 1 (IRT1) of the ZIP transporter family, which is the main transporter for high-affinity Fe uptake in roots (Eide et al. 1996; Pakrasi et al. 1999; Connolly et al. 2002, 2003; Vert et al. 2002; Guerinot 2010; Ivanov et al. 2012). FRO activity induced upon Fe shortage is mainly localized in young lateral roots, with highest activity at the root apex and subapical root regions (Römheld and Marschner 1981; Chen et al. 2010). In response to Fe deficiency, plants are known to accumulate other metals. In such cases, *Arabidopsis* IRT1 and its rice orthologue are likely to also transport Cd, Co, Mn, Ni, and Zn (Eide et al. 1996; Vert et al. 2002; Nakanishi et al. 2006; Nishida et al. 2011). In the second strategy developed by graminaceous monocots like maize, phytosiderophores (which are molecules of the mugineic acid family) are secreted outside the root and chelate Fe. The Fe(III)–phytosiderophore complex enters then the roots by a specific transporter. In maize, that transporter has been identified as YELLOW STRIPE 1 (YS1), a distant member of the oligopeptide transporter family (Curie et al. 2001). It is referred to as strategy II or the chelation strategy. Zn(II)-, Ni(II)-, and Cu(II)-phytosiderophores can also be transported through the Fe–phytosiderophore pathway in maize (Von Wiren et al. 1996; Schaaf et al. 2004); however, *YS1* expression data do not support a role in primary Cu or Zn transport (Roberts et al. 2004).

The dogma that plants evolve either strategy I or strategy II to acquire Fe collapsed with the identification of strategy I in addition to strategy II uptake system in rice (Ishimaru et al. 2006; Bashir et al. 2010).

3. Molybdenum

Plants take up Mo in the form of the molybdate anions MoO_4^{2-} and $HMoO_4^-$, which are the predominant species in soil solution (McGrath et al. 2010). The MOLYBDATE TRANSPORTER 1 (MOT1), which belongs to the sulfate transporter superfamily, has been proposed to be involved in Mo uptake from the soil (Tomatsu et al. 2007). However, while *AtMOT1* is expressed in every root cell layer, its expression is highest in endodermis and stele cells. Using a GFP fusion, Tomatsu et al. (2007) localized MOT1 in the plasma membrane and in endomembranes, whereas Baxter et al. (2008) localized the same transporter in the mitochondria, where the first committed step in molybdopterin biosynthesis takes place. The latter localization challenges the direct role of MOT1 in Mo uptake from the soil solution. Interestingly, *AtMOT1* was also identified as the causal gene driving the reduced shoot Mo content in *Arabidopsis* accessions, which suggests that the mitochondria could act as a control point to regulate whole-plant Mo content (Baxter et al. 2008). Equally interesting, the expression of *AtMOT1* does not seem to be affected upon Mo deficiency (Ide et al. 2011). This illustrates that all the permeases mediating the acquisition of TMEs are not always upregulated at the transcript level at least upon starvation conditions.

4. Manganese

Two pathways are known to allow the entry of Mn in roots: IRT1, as previously mentioned, and NATURAL RESISTANCE-ASSOCIATED MACROPHAGE PROTEIN 1 (NRAMP1). NRAMP1 was demonstrated to be the major high-affinity Mn^{2+} transporter in *Arabidopsis* (Cailliatte et al. 2010). It is localized to the plasma membrane, and the gene expression is restricted to the roots and is induced upon Mn depletion, in particular in the elongation zone. Tissue expression seems to vary with the age of the root, being first abundant in all cell layers, and decreasing as the root matures except in the vascular cells and in root hairs. The broad selectivity of *NRAMP1* was illustrated in vivo. When *NRAMP1* was expressed in the *irt1* mutant upon the activity of the Fe deficiency–inducible, root epidermis–specific

IRT1 promoter, it could restore the capacity to take up Fe and to a lesser extent Zn and Co, which is impaired in the mutant (Cailliatte et al. 2010). In rice, *NRAMP1* is mainly expressed in roots and is involved in Cd accumulation in the shoots (Takahashi et al. 2011).

5. Nickel

No specific uptake system for Ni in plants has been identified up to now, but several members of the ZIP, NRAMP, and YELLOW STRIPE-LIKE (YSL) families can transport Ni (Tejada-Jimenez et al. 2009). There is an interaction between Ni- and Fe-uptake systems (Akbas et al. 2009). For example, AtIRT1 can also transport Ni^{2+}, and Fe-deficiency symptoms are usually attributed to Ni excess that interrupts the Fe transport system and root-to-shoot translocation (Nishida et al. 2011).

6. Zinc

Zn is transported through several ZINC TRANSPORTERS (ZIPs) (Grotz et al. 1998; Colangelo and Guerinot 2006; Yang et al. 2010). However, the exact ZIP members involved in Zn uptake from the soil have still not been clearly identified. In response to Zn deficiency, approximately half of the ZIP family genes and in particular *AtZIP4* are induced (Grotz et al. 1998; Wintz et al. 2003; van de Mortel et al. 2006; Assunção et al. 2010). Supraoptimal concentrations of Zn interfere with the uptake and allocation of other elements, such as Cu, Fe, Mn, and P (Tewari et al. 2008; Sagardoy et al. 2009; Yang et al. 2011).

7. Beneficial and Nonessential Trace Metallic Elements

Nonessential TMEs enter roots through nonspecific transporters. Cd uptake can occur through Ca, Fe, and Zn transporters like ZIP, NRAMP, and other cation channels, such as depolarization-activated calcium channels (DACC) and hyperpolarization-activated calcium channels (HACC) and voltage-insensitive cation channels (VICC) (Verbruggen et al. 2009a; Lux et al. 2011; Mendoza-Cozatl et al. 2011). As already mentioned, the YS1 transporter of maize seems also to be able to transport Cd(II)–phytosiderophores but in a very inefficient way, making that contribution most probably insignificant (Schaaf et al. 2004). Actually phytosiderophores poorly mobilize Cd and therefore constitute an advantage relative to the reduction strategy to acquire Fe under Cd stress (Meda et al. 2007).

Part of Co toxicity to plants can be due to interference with Mn and Fe uptake (Liu et al. 2000). Mn concentration in soil seems to exert a major control on how plants take up and accumulate Co (Faucon et al. 2009; Collins and Kinsela 2011). Co competes with Fe to access the IRT1 transporter, whereas Co uptake is clearly dependent on the Fe status of the plant. In fact, Fe-deficient plants usually contain more Co in their tissues (Korshunova et al. 1999), and in such cases, *IRT1* expression is higher, which allows more Co to be taken up by the roots (Morrissey et al. 2009). FERROPORTIN 1 (FPN1) seems to be required for Co movement from root to shoot (Morrissey et al. 2009). For Hg and Pb, the identity of the transporters is not known. Although Hg is usually released in its metal or ionic form, it is readily methylated to methyl Hg in the environment by bacteria. Inorganic forms are thought to be more available to plants than are organic ones. Pb may enter roots through ionic channels. Inhibition of Pb absorption by calcium is well known, and Pb can be transported by Ca-permeable channels (Pourrut et al. 2011). Alternative nonselective pathways may be through cyclic nucleotide–gated ion channels and low-affinity cation transporters (reviewed by Pourrut et al. 2011).

D. Transport and Circulation of Trace Metallic Elements in Plants

After crossing the plasma membrane of epidermal cells through a metal uptake system, TME ions, as intermediate to soft Lewis acids, can bind carboxylic, amino, and thiol groups. As a result, free TME concentrations will be very low (they have been calculated at less than one Cu or Zn free ion per cell; Outten and O'Halloran 2001; Finney and O'Halloran 2003). Metal ions not occupying sites in proteins are expected to be bound to low-molecular-weight metal ligands (Haydon and Cobbett 2007). Important metal ligands for long-distance and intracellular transport include nicotianamine (NA), glutathione (GSH), phytochelatins (PCs), histidine (HIS), and citrate. Our knowledge about transport of ligands of metallic ions and ligand–metal complexes has been reviewed by Haydon and Cobbett (2007). There is good evidence that circulation of Fe, for example, occurs as Fe–citrate complexes in xylem and Fe–NA complexes in phloem (Haydon and Cobbett 2007; Guerinot 2010). Transport of the Fe ligand citrate is achieved by members of the MULTIDRUG AND TOXIC COMPOUND EXTRUSION (MATE) transporters: FERRIC REDUCTASE DEFICIENT 3 (AtFRD3) in *Arabidopsis* and FRD-LIKE 1 (OsFRDL1) in rice. YSL members seem to transport Fe–NA and other metal–NA complexes (Rogers and Guerinot 2002; Curie et al. 2009; Guerinot 2010; Ivanov et al. 2012). Fe concentration is highly regulated, and ferritins, which are Fe proteins localized in the plastids of plants, seem also to play a role in that process. Ferritins have been considered as storage and protective proteins to prevent Fe to react with oxygen (Briat et al. 2010). Ferritins may control the concentration and distribution of Fe. In roots, where Fe is less abundant than in shoots, ferritins mainly accumulate close to the root tip in the endodermis cell layer, and at the emergence of secondary roots, where they may buffer Fe flux between the cortex and the stele (reviewed by Briat et al. 2006). In regard to NA, several lines of evidence support that it chelates Cu in the xylem sap and, in addition, is involved in phloem loading and/or unloading of Cu, Fe, and Zn (Curie et al. 2009). Intracellular metal trafficking can involve metallochaperone proteins. So far, only Cu metallochaperones have been identified and shown to interact with the COPT1 transporter (reviewed by Puig and Peñarrubia 2009; Cohu and Pilon 2010). Existence of a Zn metallochaperone network probably does not exist because cellular Zn sites are abundant, about 20-fold more abundant than Cu sites (Clemens 2010).

The radial transport of TMEs through the root up to the xylem can take place either via a symplasmic pathway or an apoplasmic pathway in regions lacking a Casparian bands (usually at the extreme root tip and in regions in which lateral roots are being initiated) (reviewed by Haydon and Cobbett 2007). Efflux of several TMEs out of the cytoplasm of pericycle and xylem parenchyma cells into the apoplastic xylem vessels involves HEAVY METAL ATPASEs (HMAs). Xylem loading of Zn and Cd by HMA2 and HMA4 is the best understood (Hussain et al. 2004; Sinclair et al. 2007; Wong and Cobbet 2009). Besides HMA2 and HMA4, PLANT CADMIUM RESISTANCE 2 (PCR2) is also significantly involved in Zn translocation from the root to the shoot in *Arabidopsis* (Song et al. 2010a). In addition to its localization in xylem parenchyma, PCR2 is also present in the plasma membrane of root epidermis cells where it detoxifies Zn excess through efflux in the rhizosphere (Song et al. 2010a). HMA5, a Cu-transporting pump that is highly expressed in roots, is proposed to be involved in Cu loading into the xylem (Kobayashi et al. 2008; Cohu and Pilon 2010). However, that role is still a subject of discussion. Initial work on a *hma5* loss-of-function *Arabidopsis* mutant suggested a role in Cu detoxification in roots through the efflux across the root cell plasma membrane (Andrés-Colás et al. 2006). Compared to wild-type plants, *hma5* was hypersensitive to Cu excess, and it over accumulated Cu in roots without significant modification of the accumulation in shoots, which argued against a role for HMA5 in root-to-shoot translocation (Andrés-Colás et al. 2006). No other transporter for Cu loading in xylem has been proposed yet. Interestingly, *HMA5* was associated with a major QTL for Cu tolerance in *Arabidopsis* (Kobayashi et al. 2008). Because the major QTL for Cu tolerance was linked with differential Cu translocation from roots to shoots, the role of the candidate gene, *HMA5*, was proposed to be responsible for that process (Kobayashi et al. 2008). As the HMA5 cellular localization was not yet demonstrated, it has been speculated that HMA5 could transport Cu from the cytosol to the apoplasm, or to the endoplasmic reticulum.

Metal transport at the subcellular level is at the heart of homeostasis. Vacuolar transport is of particular importance in metal storage/delivery and detoxification. Vacuolar transport can occur in ionic forms or metal–chelate complexes. The best example is probably the detoxification as PC–metal chelates. Recently, the ATP-BINDING CASSETTE (ABC) ABCC1 and ABCC2 that transport PC–As in the vacuoles have been uncovered in *Arabidopsis* (Song et al. 2010b), which may also transport other PC complexes, in particular PC–Cd. HMA3 is involved in vacuolar influx of Cd/Zn/Co/Pb (Morel et al. 2009). In rice, high *HMA3* expression in roots is associated to reduced Cd flux to the shoots and seeds (Ueno et al. 2010). CATION DIFFUSION FACILITATORS/METAL TRANSPORTER PROTEINS (CDF/MTP) have been identified that show high substrate specificity upon heterologous expression in yeast, Zn (Blaudez et al. 2003), Ni (Persans et al. 2001), and Mn (Delhaize et al. 2003), but not always (Persans et al. 2001). CATION/H^+ EXCHANGER 2 and 4 (CAX2, CAX4) can transport Ca, Zn, Cd, Mn, and probably other cations against protons (reviewed by Manohar et al. 2011). NRAMP3 and NRAMP4 are responsible for Cd, Fe, Mn, and Zn vacuolar efflux and seem to have low substrate specificity (Thomine et al. 2000, 2003; Lanquar et al. 2010). Plastid and mitochondrial transports have been recently reviewed by Nouet et al. (2011) and will not be addressed here. Knowledge on transport to other cellular compartments is scarce.

E. Importance of Rhizosphere Activity

The rhizosphere encompasses the millimeters of soil surrounding a plant root (Bais et al. 2006). Very complex interactions occur between roots and the rhizosphere, in which root exudates (ions, free oxygen and water, enzymes, mucilage, and a diverse array of carbon-containing primary and secondary metabolites) play an important role (reviewed by Bais et al. 2006; see also Part VI). Some bacteria, known as "plant growth–promoting rhizobacteria" (PGPR), can increase root hair production and root surface area, thereby affecting TME absorption capacity (Dutta and Podile 2010). Different mechanisms have been identified that can explain that effect: production of plant growth regulators that cause root cell elongation like IAA (indole-3-acetic acid) or production of enzymes degrading ethylene (which inhibits root growth) (reviewed by Lugtenberg and Kamilova 2009). Furthermore, some bacteria can acidify the rhizosphere and produce high concentrations of chelating agents that can enhance the solubility of metals like Fe. A novel mechanism affecting Fe uptake by plant was identified in *Bacillus subtilis* GB03. That PGPR strain can activate the plant's own Fe acquisition machinery to increase the assimilation of metal ions (Zhang et al. 2009). Mechanistically, GB03 activates strategy I Fe uptake responses by inducing an increased transcript accumulation of the gene encoding the FE DEFICIENCY–INDUCED TRANSCRIPTION FACTOR (FIT) (see Section IV.B.3), which in turn is thought to upregulate the Fe uptake system I in *Arabidopsis* roots (Zhang et al. 2009). Endophytes, which live inside the plant, constitute another positive factor of plant growth and health (e.g., by reducing the negative impact of plant pathogen and by improving plant nutrition) and can have a certain impact on plant resistance to TME excess (Sheng et al. 2008; Mastretta et al. 2009; Weyens et al. 2009; Ma et al. 2011). Finally, mycorrhizae are usually of major importance in the capacity of plants to colonize soils contaminated by TME excess or deficient in nutrients. In harsh environments, positive roles of fungi on associated plants include nutrition improvement as well as defense against absorption of metal excess (Schützendübel and Polle 2002; Adriaensen et al. 2005; Gohre and Paszkowski 2006; Hildebrandt et al. 2006; Krznaric et al. 2009).

IV. Response of Roots to Trace Metallic Elements

Plants respond to fluctuations in mineral nutrient concentrations by altering the physiology and morphology of their roots. Essential TMEs are required by plants in balanced proportions.

TABLE 34.2 Critical Trace Metal Element Concentrations in Plants and Hyperaccumulation Thresholds

Element	Critical Level ($\mu g\ g^{-1}$)		Hyperaccumulation Concentration Criterion ($\mu g\ g^{-1}$)		Number of[c]	
	Deficiency	Toxicity	To Date	Newly Suggested	Taxa[c]	Families[c]
Cadmium	n.r.	6–10	>100		9	6
Cobalt	n.r.	0.4-several	>1,000	>300	(26)	(11)
Copper	1–5	20–30	>1,000	>300	(35)	(15)
Lead	n.r.	0.6–28	>1,000		(14)	(7)
Manganese	10–20	200–3,500	>10,000		10	6
Mercury	n.r.	below 10[a]	n.r.		0	
Nickel	0.002–0.004	10–50	>1,000		390	42
Zinc[b]	15–30	100–700	>10,000	>3,000	15	6

Source: Adapted from Verbruggen, N. et al., *Curr. Opin. Plant Biol.*, 12, 364, 2009a; Krämer, U., *Annu. Rev. Plant Biol.*, 61, 517, 2010.

n.r., not reported.

[a] Concentrations are not reported in the literature. As Hg is more toxic than Cd at similar bioaccumulation levels, its critical concentration level should be below 10 $\mu g\ g^{-1}$ (Schat, personal communication).

[b] White and Broadley (2011).

[c] Brackets indicate unconfirmed data.

Because they are needed in much smaller amounts than macronutrients, the concentration window between deficiency and toxicity is narrow (Table 34.2). Nonessential TMEs can rapidly become toxic at even lower concentrations. Deviation from equilibrium can result in drastic nutritional disorders and profound root morphological adaptation. Plants, like other organisms, possess homeostasis mechanisms to fine-tune adequate concentrations of essential TME in cellular compartments and to minimize the damage from exposure to nonessential ones. All plants possess basic mechanisms to cope with TME excess, which have been summarized by Hall (2002): restriction of metal movement to roots by mycorrhizae, binding to cell wall and root exudates, restricted influx across plasma membrane, active efflux into apoplast, chelation in cytosol by various ligands, repair and protection of plasma membrane, and detoxification by transport and accumulation of metals in vacuoles. There is a wide variation in those responses between and within plant species. This natural variation can be exploited to identify genetic determinants of root plasticity and metal homeostasis. This strategy has already been used in *Arabidopsis* and will be further illustrated in this chapter by the discussion of the extreme case of hyperaccumulators (Section V). First, we will summarize current knowledge about the perception and response of roots to fluctuations in TME concentrations.

A. Inventory of Root Morphological Responses to Critical Trace Metallic Element Concentrations

Plants usually compensate for the limitation of essential TMEs by engaging both morphological and metabolic changes in roots in order to improve metal acquisition. Accelerated root elongation and increased number and length of root hairs increase the surface area where roots contact the soil are such adaptive responses. However, enhanced growth is not observed under all micronutrient deficiencies. Other specific anatomical changes for enhancing metal acquisition can also be observed in certain species, such as the formation of transfer cells and of cluster roots, consisting of a large number of determinate branch roots over a very short distance of the main root axis. On the other hand, exposure to elevated concentrations of essential and nonessential TMEs can cause suberization of root tissues and development of apoplasmic barriers. The most frequently observed change in root morphology upon TME excess is undoubtedly an inhibition of the primary root (PR) growth, due partly to inhibition of cell divisions and enhancement of lateral branching to various degrees. The inhibition of root growth is a rapid visual response that is frequently used to evaluate plant tolerance. Indeed, the maintenance of root growth appears a very crucial trait in plant tolerance. The following order of metal toxicity has been observed in ryegrass Cu > Ni > Mn > Pb > Cd > Zn > Hg > Fe, which follows the order of metallic–organic complexes stability (Wong and Bradshow 1982). Observations on other plant species were similar except for Mn (Wong and Bradshow 1982). Measurement of root growth can be easily done in hydropony or in vitro on vertical plates. Figure 34.1 shows the root morphological adaptations of the model nontolerant species *Arabidopsis thaliana* exposed for 1 week to toxic concentrations of Cd, Cu, Fe, Mn, Mo, and Zn after transfer of 6-day-old plants. At TME concentrations that inhibit PR growth by about 50% lateral root (LR) number, LR density and Σ LR length are reduced. However, Cu (at the tested concentration which inhibits root growth by 70%) and Zn do not affect LR density but LR elongation. Those metal-induced changes in root morphology are considered to redirect root development away from a local elevated source of metal (Potters et al. 2007). However, in certain hypertolerant species, the root foraging capacity can be accelerated in TME-rich patches. The latter features will be further developed in Section V.

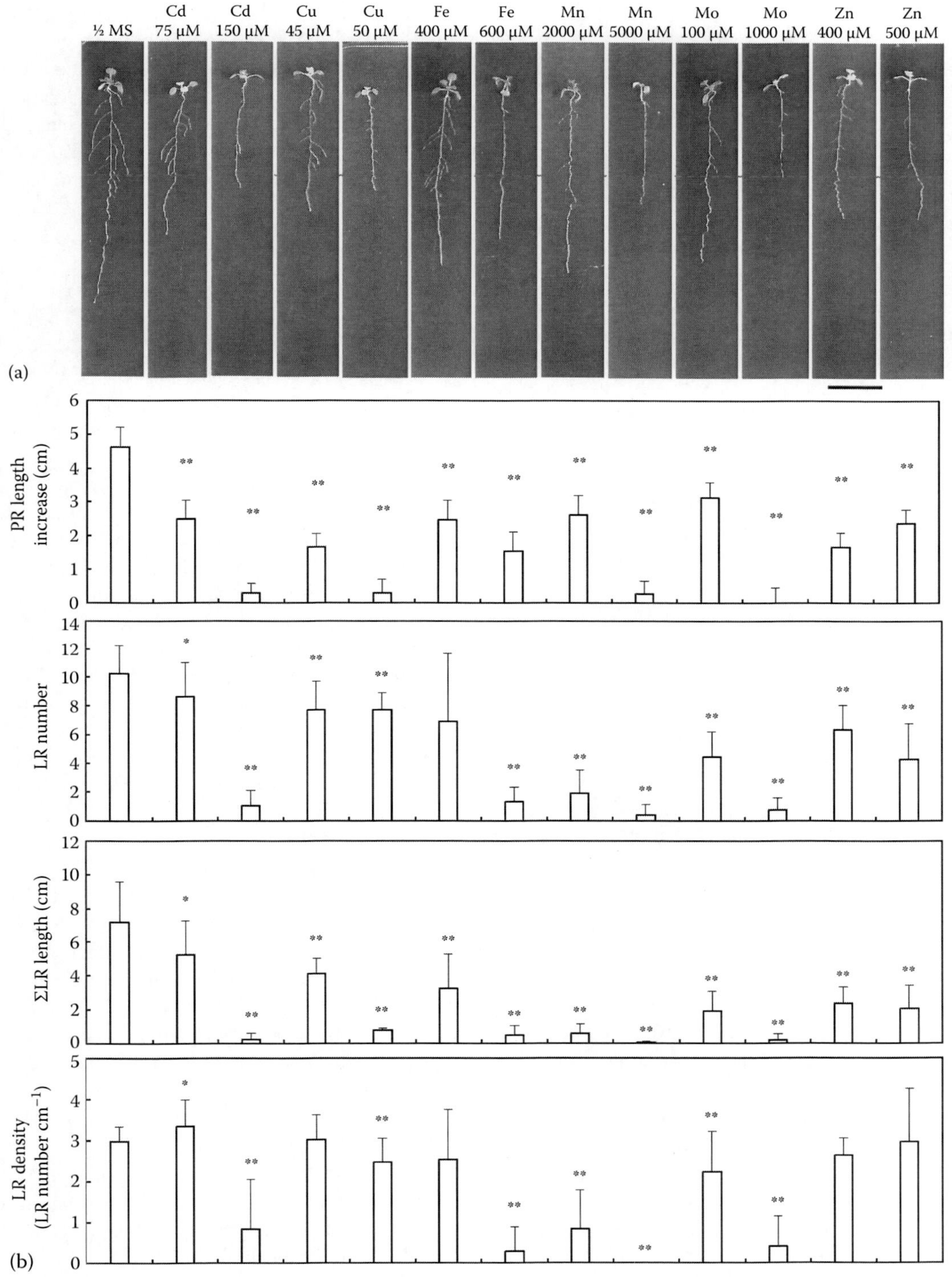

FIGURE 34.1 (See color insert.) Seedlings were germinated for 6 days on 0.5× Murashige and Skoog medium and were transferred to media enriched with TMEs (75, 150 μM $CdSO_4$; 45, 50 μM $CuSO_4$; 400, 600 μM Fe-EDTA, 2000, 5000 μM $MnSO_4$; 100, 1000 μM MoO_3 and 400, 500 μM $ZnSO_4$). (a) Representative picture of seedlings one week after transfer. The dashed red lines indicate the average PR length upon transfer. Scale bar: 2 cm; (b) Root architecture parameters: primary root length (PR) after transfer, lateral root (visibly emerged > 1 mm) number (LR), sum of lateral root length (ΣLR) and lateral root density (LR density) calculated as the number of LR per root length portion comprised between the first and the last LR. N = 10–20 observations ±std. Stars indicate statistical difference between a given treatment and the control conditions (0.5× MS): $P < 0.10$ (*), $P < 0.01$ (**).

1. Essential Trace Metallic Elements

a. Copper

Responses to Cu limitation observed upon in vitro culture (typically in the presence of bathocuproine disulfic acid, a Cu chelator) include accelerated root elongation of *Arabidopsis* (Sancenón et al. 2004; del Pozo et al. 2010; Garcia-Molina et al. 2011). Transgenic *Arabidopsis COPT1* antisense plants with lower expression of the Cu transporter gene also exhibited extended root growth compared to wild-type plants (Sancenón et al. 2004). Roots are the preferential accumulation site of that metal (Panou-Filotheou et al. 2004; Lequeux et al. 2010). Root morphology changes in response to excessive Cu concentrations range from disruption of the root cuticle and reduced root hair proliferation to severe deformation of root structure including thickening of the cortex layer (Arduini et al. 1995; Sheldon and Menzies 2005). Both root elongation and lateral branching are sensitive to Cu excess (Kim et al. 2007). Numerous stunted LRs are frequently observed in high-Cu-treated plants, suggesting that LR elongation is more sensitive to Cu toxicity than initiation (Lequeux et al. 2010). In addition, high Cu decreases root meristem cell proliferation (Liu et al. 2009), affects cell integrity, causing ion influx and thereby inhibiting cell elongation and damaging cell integrity in the root transition zone (Madejon et al. 2009), and ultimately results in root cell death (Yeh et al. 2003; Lequeux et al. 2010).

b. Iron

Typical morphological adaptations of roots to Fe shortage include the swelling of root tips, enhanced LR development, reduced LR growth, and the increase of the number and length of root hairs (Moog et al. 1995; Schmidt et al. 2000; Romera and Alcántra 2004; Muller and Schmidt 2004; Schmidt 2008). PR length is correlated with tissue concentration of Fe in *Arabidopsis* (Ward et al. 2008). The root hair zone and its preceding region seem to be the main Fe acquisition section along the root (Schmidt et al. 2000; Jakoby et al. 2004; Seguela et al. 2008; Ivanov et al. 2012). Only certain cells within the root epidermis, that is, those lying over anticlinal cortical cell walls (trichoblasts), are capable of producing root hairs, whereas other cells (atrichoblasts) remain hairless (Schmidt and Schikora 2001). Low Fe conditions induce the formation of ectopic hairs in positions that are normally occupied by non-hair cells (e.g., overlying periclinal cortical cell walls) (Schmidt and Schikora 2001). Other specific anatomical changes for enhancing Fe acquisition can also be observed in certain species, such as the formation of transfer cells (Krämer et al. 1980; Landsberg 1986; Schmidt and Bartels 1996) and of cluster roots (Hagström et al. 2001; Zaïd et al. 2003; Shane and Lambers 2005). Under Fe excess, the viability of root border cells issued from root cap zone can be reduced, which can cause their mechanically disengaged from the root tips. Those dead cells could protect the root from the damage caused by harmful Fe excess (Xing et al. 2008b). In anoxic conditions (favoring Fe^{2+} ions), a common feature of aquatic species is to develop on their root surface an Fe plaque (Moore et al. 2011). The diffusion of oxygen into the rhizosphere oxidizes Fe^{2+} to Fe^{3+}. Ferric Fe is then precipitated as Fe oxide/hydroxides (ferrihydrite, goethite, siderite) onto the root surface, forming the plaque (Hansel and Fendorf 2001). The Fe plaque may be amorphous or crystalline. For example, in rice, the Fe plaque is composed of fine needles about 10 nm thick and 300 nm long at the most, outside the plasma membranes of the epidermal cells (Moore et al. 2011). The presence of the plaque may also act as a barrier for the uptake and movement of other TMEs or metalloids (Ye et al. 1997; Moore et al. 2011).

c. Manganese

Mn deficiency reduces PR growth (Konno et al. 2003; Cailliatte et al. 2010). Interestingly, the formation of abundant root hairs is largely reported for Mn-deficient plants (Ma et al. 2001; Konno et al. 2003; Inoue et al. 2006; Schmidt et al. 2008; Yang et al. 2008) and differs from the Fe-deficiency root hair phenotype previously mentioned. Mn deficiency induces a unique root hair phenotype comprising changes in the length and in the position of root hairs relative to the underlying cortical cells (Yang et al. 2008). Upon Mn shortage, atrichoblasts redifferentiate, resulting in the formation of root hairs in ectopic positions. This indicates that Mn deficiency does not simply promote the elongation of hairs from trichoblasts, but alters the developmental program of rhizodermal cells (Yang et al. 2008). Ma et al. (2001) suggested that "the effect of Mn deficiency on root hair density is a secondary result of reduced root extension." High Mn suppressed root hair at low pH while inducing main root growth (Konno et al. 2003). Mn generally tends to accumulate predominantly in the shoots rather than in the roots (Millaleo et al. 2010). Interestingly, grape root growth was shown to be particularly tolerant to high Mn levels (Mou et al. 2011).

d. Molybdenum

Mo deficiencies are considered rare in most agricultural cropping areas (Kaiser et al. 2005). Mo in acid soils tends to be unavailable to plants. This is why most Mo deficiencies occur on acid soils. As Mn deficiency, Mo deficiency can increase root hair formation at pH 6 and suppress main root growth slightly (Konno et al. 2003). Information about Mo toxicity (more often observed in basic soils) on plants is scarce. Roots accumulate higher Mo concentration than the shoot, and root yield is more affected upon high Mo levels (reviewed by McGrath et al. 2010).

e. Nickel

Ni concentration in soil is usually beyond plant needs, and Ni-deficiency symptoms in roots have received little attention, but sporadic studies reported the reduced germination rate and vigor of Ni-deprived barley grains (Brown et al. 1987). Upon exposure to increasing doses of Ni, tissue concentration rises more rapidly in roots than in shoots (Schaaf et al. 2006; Nishida et al. 2011). It is reported that over 80% of the Ni in the root may be present in the vascular cylinder (reviewed by Chen et al. 2009b). In tobacco exposure to high Ni levels quickly represses root growth and induces dark brown color (Boominathan and Doran 2002). In maize roots, Ni excess induced an increased number of cortical cell layers (Maksimovic et al. 2007).

f. Zinc

Moderate Zn-deficient conditions increased root length in several plant species (Genc et al. 2007; Chen et al. 2009a), while impact on root hair density is very modest or not existent (Ma et al. 2001; Genc et al. 2007). Study of natural variation of the Arabidopsis root response to low and high Zn has recently highlighted an important role for Zn in the development of lateral roots (Richard et al. 2011). Zn toxicity is generally characterized by browning of the root system, a decreased number and shorter lateral roots, as well as deceleration of mitotic activity (Sagardoy et al. 2009; Jain et al. 2010). Interestingly, tolerance to Zn excess is partially correlated with Zn root-to-shoot translocation in *A. thaliana* (Richard et al. 2011).

2. Beneficial and Nonessential Trace Metallic Elements

a. Cadmium

Cd preferentially accumulates in the roots (Gong et al. 2003; Smeets et al. 2008; Li et al. 2010) and is translocated to the aerial parts where it causes chlorosis in young leaves and alters the photosynthetic processes (Baryla et al. 2001; Janik et al. 2010). Cd exposure induces oxidative stress and enhanced production of ROS, which are thought to induce part of the root changes. Cd poisoning causes a change of root tip color, decreased root tip diameter, alteration in cell shape, and an inhibition of root growth due to cell death, first in the meristematic zone and later in the elongation zone (Song et al. 2003; Suzuki 2005; Kim et al. 2007; Lux et al. 2011). Cd toxicity further induces the formation of lateral and adventitious roots (Xu et al. 2011). Interestingly, PR growth inhibition can be reversed by Ca application to Cd-stressed plants (Kim et al. 2002; Suzuki 2005). That antagonism is assumed to result from a competition for transporters between the two ions (Clemens 2006) and/or interaction in the signaling of stress (see Section IV.B). Radial movement of Cd is restricted by a thicker root cortex (Maksimovic et al. 2007). At the level of the roots, Cd movement to the xylem is impeded both in the symplast and in the apoplast, thereby reducing the translocation to the shoot where Cd can inhibit photosynthesis. Cd detoxification occurs mainly through vacuolar sequestration of PC–Cd complexes (reviewed by Verbruggen et al. 2009a; Mendoza-Cozatl et al. 2011). Cd-induced barrier in the apoplasmic pathway is manifested by the isolation of the stele through accelerated maturation of endodermis, exodermis, and endodermis cell wall impregnation with suberin (Lux et al. 2011).

b. Cobalt

The effect of Co addition on barley root elongation was shown to be mainly dependent on soil properties, such as the effective cation exchange capacity (Micó et al. 2008). Most of the studies on natural plant adaptation to Co excess have been done in the Katangan Cu belt (Democratic Republic of Congo) (Baker and Brooks 1989; Faucon et al. 2009). Mechanisms of Co tolerance and accumulation are still poorly understood.

c. Lead

Exposure to Pb provokes inhibition of root growth and blackening of the root tissues (Song et al. 2003; Patra et al. 2004; Sharma and Dubey 2005; Kim et al. 2007; John et al. 2009). In a recent study on *Acacia farnesiana* grown in vitro, roots were longer and thicker, with higher number of secondary roots when exposed to 250 and 500 mg Pb^{2+} L^{-1}, while they stopped growing at 1000 mg L^{-1} (90%–93% shorter than the ones without added Pb) (Maldonado-Magaña et al. 2011). Pb mainly accumulates in roots with a low translocation to the shoot (Maldonado-Magaña et al. 2011). At the subcellular level, Pb was localized mainly in the cell walls, intercellular spaces, and vacuoles of root cells (Phang et al. 2010). In the root meristem, Pb disturbs microtubule organization, leading to inhibition of cell division (Eun et al. 2000).

d. Mercury

The application of mercuric salts inhibited root growth in various species (Ganeshan and Manoharan 1983; Gupta 1991; Fargasova 1994) and also suppressed graviresponses in maize roots (Pilet and Versel 1981). Nonetheless, the absorption of organic and inorganic Hg from soil by plants is relatively low, and Hg accumulates preferentially in roots because there is a barrier to Hg translocation to the shoot (Patra and Sharma 2000). A large proportion of Hg is associated with cell walls. Up to 93% of the total Hg found in the roots was immobilized in the cell walls (Válega et al. 2009). A minor proportion of total Hg in roots, which is nevertheless important for tolerance to Hg, is associated with PCs (Carrasco-Gil et al. 2011; Válega et al. 2009). However, the main process damaged by excess of Hg is reported to be the photosynthesis (Ruiz et al. 2011 and references therein).

B. Stress Signaling Induced by Trace Metallic Elements

This part focuses on the signal pathways responsible for the sensing and transduction of stress induced by TME disequilibrium, ultimately driving the activation of transcription factors that modify expression of responsive genes, and allowing adaptive responses to take place.

1. Perception of Trace Metallic Element Concentration

The uptake and assimilation of nutrients must be coordinated with mineral supply and plant demand in a complex and regulated interacting network. Plants constantly monitor and respond to the availability of essential mineral nutrients in the soil (Williams and Pittman 2010). However, there is still extremely little knowledge about the mechanisms used by plants to gauge mineral concentrations. Roots are usually the first site for the perception of fluctuations in TME concentrations and for the transmission of a signal along transduction pathways through the plant vascular system.

Some knowledge exists for the response to Cu shortage: the SQUAMOSA PROMOTER BINDING PROTEIN–LIKE 7

(SPL7) transcription factor, which is mainly expressed in roots, is proposed to be involved in the detection of Cu availability and in the orchestration of whole-plant Cu delivery (see also Section IV.B.3; Yamasaki et al. 2009; Cohu and Pilon 2010).

In the case of Fe, the identity of a sensor in plants remains elusive. Many lines of evidence point to a systemic signaling of micronutrient shortage that more than likely adds to a root local sensing. This is also true for Fe. A spatial microarray analysis of the root response to Fe shortage was performed in *Arabidopsis* using cell sorting (Dinneny et al. 2008). Expression of genes related to metal-ion transport and NA biosynthesis was upregulated in the epidermis, while the stele response was enriched for stress-responsive and signaling genes, suggesting that Fe deficiency is sensed internally. Perception of micronutrient starvation or excess by other organs than the roots must be transmitted among the plant parts in order to regulate the uptake or reallocate mineral resources. NA, which is involved in long-distance metal (Co, Cu, Fe, Mn, Ni, and Zn) transport (Haydon and Cobbett 2007; see Section III.D), has been proposed as a kind of Fe sensor (Curie and Briat 2003). NA is thought to carry Fe in the phloem of monocots as well as of dicots (Haydon and Cobbett 2007; Guerinot 2010). In nongraminaceous plants where NA is known to accumulate upon Fe increase, NA has been proposed to provide a signal that will downregulate Fe uptake in roots (Curie and Briat 2003). The NA-deficient *chloronerva* tomato mutant is unable to sense its Fe status and shows a constitutive Fe-deficiency response (Curie and Briat 2003). Currently, the idea that NA is an Fe sensor is abandoned because NA is relatively abundant and only weakly regulated by Fe status (Curie, personal communication). ROS (see also Section IV.B.2) can also be considered as sensors of high Fe. Upon Fe increase, ferritin mRNAs are first induced and subsequently posttranscriptionally regulated through oxidative signals (Ravet et al. 2012).

As for Fe, several lines of evidence support a systemic response to Zn availability. For example, when Zn sequestration is enhanced in the shoot by increasing its transport to the vacuoles, Zn transporters in the root and shoot are upregulated (Gustin et al. 2009). The Zn-deficiency signal is currently unknown.

2. Role of Reactive Oxygen Species and Nitric Oxide in Signaling Trace Metallic Element Stress

a. Reactive Oxygen Species

Reactive oxygen species (ROS) usually increase after exposure to TME excess, as a result of the Haber–Weiss and Fenton reactions, which are catalyzed by redox metals, or as consequence of metabolism alterations (Verma and Dubey 2003; Ortega-Villasante et al. 2007; Zhang et al. 2008; Bi et al. 2009; Phang et al. 2010; Cuypers et al. 2012). In response to high Zn, accumulation of ROS in Arabidopsis roots was only observed in the mtp1 mutant devoid of a main vacuolar sequestration pathway and not in wild-type plants (Kawachi et al. 2009). Production of ROS seems to follow different waves, which may be of different origins. There is a plethora of contradictory data describing ROS production after TME (Cd mostly) excess, which may be reconciled by a careful analysis of production kinetics. Rapid production of superoxides can also be enzymatically catalyzed by NAPH oxidases. Evidence for upregulation of NAPH oxidases in roots exposed to TME excess was provided for Cu (Navari-Izzo et al. 2006), Cd (Remans et al. 2010), and Ni (Hao et al. 2006). In an elegant study on isolated plasma membranes, mitochondria, and intact roots, Heyno et al. (2008) provided evidence that Cd stimulates H_2O_2 production in mitochondria while inhibiting the one of O_2^- at plasma membranes. H_2O_2 can diffuse out of the cell to the apoplast and/or out of the roots. This was corroborated by the observation that H_2O_2 was detected inside and outside of Cd-treated roots (Ortega-Villasante et al. 2007). ROS production in response to micronutrient starvation is poorly documented, unlike deficiencies of macronutrients (Shin et al. 2005). The consequences of ROS-enhanced production in roots are not well understood. Apart from inducing an oxidative stress, ROS accumulation may be involved in inducing stress-responsive genes and has been proposed to be an upstream signaling mediator of such response (Cuypers et al. 2012 and references therein). It is believed that the subcellular localization, the quantity, and the type of ROS produced are involved in the specificity of the cellular response (Foyer and Noctor 2009). Oxidative signals are usually transmitted by MAP kinase (MAPK) cascades. In particular, two *Arabidopsis* MAPKs, AtMPK3 and AtMPK6, and their homologues in other plant species have been repeatedly reported to be induced by Cd or Cu excess (Jonak et al. 2004; Yeh et al. 2007; Liu et al. 2010a; Wang et al. 2010). Calcium may be part of the signal transduction cascade acting upstream and/or downstream of the ROS production. ROS have also been proposed to be involved in the Cu-induced increased lignification (Pasternak et al. 2005).

The consequence of increased ROS production may include changes in gene expression, direct alteration of the function of proteins through modification of thiols, as well as alterations in the redox state of the cell. Unlike many other signaling molecules, ROS are believed to participate at all stages of the stress response (signal perception, signal transduction, and amplification) (Foyer and Noctor 2009). Several lines of evidence support cross talk between ROS, NO, and plant hormones in response to TME excess, but the underlying mechanisms are poorly known.

b. Nitric Oxide

Nitric oxide (NO) is well documented as a signaling molecule during root organogenesis. NO is a free radical gas that promotes LR development, possibly in an auxin-dependent process (Correa-Aragunde et al. 2004). NO production is an essential step in response to Fe deficiency (Graziano and Lamattina 2007; Besson-Bard et al. 2009) by stabilizing the FIT transcription factor (see Section IV.B). This molecule is also involved in the root growth sensitivity to Cd in Arabidopsis (Besson-Bard et al. 2009) or in the tolerance to high-Zn exposure in potato (Xu et al. 2010). Short-term exposure to Cd was reported to induce NO production in *Arabidopsis* (Besson-Bard et al. 2009; De Michele et al. 2009), Indian mustard and pea (Bartha et al. 2005), while long-term exposure reduced NO production in roots and leaves of pea and rice (Rodriguez-Serrano et al. 2006, 2009; Xiong et al. 2009a,b). NO production was also reduced in the root elongation zone upon long-term Cu treatment (Peto et al. 2011). A model for the cellular response to long-term Cd exposure was proposed by Rodriguez-Serrano et al. (2009), consisting of a cross talk between Ca, ROS, and NO. In their model, the reduction in NO production is proposed to result from the inhibition of NO SYNTHASE (NOS) activity by the Cd-induced decrease in Ca levels. As NO can react with O_2^-, Cd-induced reduction of NO is thought to favor O_2^- accumulation, which may act as effectors for further signaling. Upon Ca addition to Cd-treated roots, NO production was increased, and O_2^- accumulation was reduced, supporting the link between ROS, NO, and Ca.

3. Transcription Factors Regulating Responses to Trace Metallic Element Shortage

Using homology to known transcription factors, mutant analysis, or nutrient-shortage-induced promoter, several transcription factors regulating responses to Cu, Fe, and Zn deficiency have been recently identified.

SPL7 has been proposed to be a master regulator of Cu homeostasis when in shortage (Yamasaki et al. 2009). *SPL7*, which is mainly expressed in the roots, may regulate several metal transporters and other components of metal homeostasis upon Cu limitation. In strong support, wild-type *Arabidopsis* plants unlike *spl7* mutants have higher transcript levels of *COPT1*, *COPT2*, *ZIP2*, *FRO3*, and *YSL2* upon Cu deficiency. Furthermore, it seems that SPL7 also orchestrates the downregulation through miRNAs of the expression of genes encoding Cu proteins, in order to save Cu in the cell and maintain plastocyanin levels in chloroplasts (see Section IV.B.4; Yamasaki et al. 2007).

In response to Fe limitation, *FE-INDUCED TRANSCRIPTION FACTOR* (FIT, previously named *AtFIT1, bHLH29*, or *FRU*) is largely expressed in roots and is considered as a major regulator of strategy I (Guerinot and Colangelo 2004; Jakoby et al. 2004; Guerinot 2010; Ivanov et al. 2012). FIT is a basic helix-loop-helix transcription factor orthologue of FER, previously identified in tomato (Ling et al. 2002). FIT is activated by a yet unknown mechanism (Meiser et al. 2011). FIT activation may occur by physical interaction with other transcription factors, such as bHLH38, bHLH39, bHLH100, and bHLH101. Corresponding genes are induced by Fe deficiency in a FIT-independent manner, and interaction between FIT and both bHLH38 and bHLH39 has been demonstrated (Yuan et al. 2008; Meiser et al. 2011). FIT is believed to regulate the Fe acquisition machinery at multiple levels. It promotes the transcription of *AHA2*, *FRO2*, and to a lesser extent that of *IRT1* (Colangelo and Guerinot 2004; Yuan et al. 2008; Ivanov et al. 2012). FIT seems instead to be necessary for the posttranscriptional regulation of IRT1. While *IRT1* mRNA accumulation was decreased by a factor of 2 but still present in *fit* mutants under Fe shortage, IRT1 protein abundance was extremely reduced (by a factor of 10) (Colangelo and Guerinot 2004; Seguela et al. 2008). FIT was proposed to regulate a factor required for IRT1 protein stability (Colangelo and Guerinot 2004). Interestingly, FIT itself is regulated at the posttranscriptional level, by NO and ethylene. In addition to FIT, another bHLH transcription factor, POPEYE (PYE), regulates Fe-deficiency responsive genes. PYE seems to be involved in the mobilization of Fe in the root and its transport to the shoot (Long et al. 2010). Interestingly, transcriptional response to Fe deficiency also includes changes in genes encoding putative Cu and Zn transporters (Buckhout et al. 2009; Yang et al. 2010). Three stele-expressed genes, which are believed to be nonspecific for Fe, *FRO3*, *NICOTINAMINE SYNTHASE 4* (*NAS4*), and *ZINC FACILITATOR 1* (*ZIF1*), are targets of PYE, which supports a role for this regulator in cross talk between metal homeostasis networks (Long et al. 2010; Yang et al. 2010; Ivanov et al. 2012). In particular, cross-homeostasis between Fe and Zn is frequently observed in plants (Buckhout et al. 2009; Shanmugam et al. 2011, 2012).

With regard to the adaptation to Zn deficiency, the first transcription factors regulating Zn acquisition, BASIC LEUCINE-ZIPPER 19 and 23 (AtbZIP19, 23), have been recently identified (Assunção et al. 2010). bZIP19 and bZIP23 were shown to be involved in the upregulation of different members of the ZIP transporter family (ZIP1, 2, 3, 4, 5, 9, 12 and IRT3) as well as other components of the Zn homeostasis network (Assunção et al. 2010).

4. Role of MicroRNAs in Signaling Trace Metallic Element Stress

Even though transcription factors are key players in stress responses, posttranscriptional processes such as those regulated by microRNAs (miRNAs) are crucial for the determination of the ultimate mRNA activity (Khan et al. 2011). Recently, specific miRNAs have emerged as a new and important type of mobile signals during nutrient deficiency responses (Chiou 2007; Sunkar et al. 2007; Khan et al. 2011). miRNAs are small (20–24 nucleotides) nonprotein-coding RNAs that are precisely excised as discrete species from the stem of an imperfect stem-loop pre-miRNA precursor (Voinnet 2009). miRNAs bind to complementary sequences on target messenger RNA transcripts (mRNAs), usually resulting in translational repression or target degradation and gene silencing (Voinnet 2009). Different miRNAs have been identified that seem to be transcribed under the activity of SPL7 under Cu deprivation (Sunkar et al. 2006; Yamasaki et al. 2007; Abdel-Ghany and Pilon 2008; Yamasaki et al. 2009). One of the SPL7 targets, miR398 mediates the degradation of *CDS1* and *CDS2* mRNAs encoding Cu/Zn SODs (Cu/Zn SODs), whose function is compensated by Fe SOD in low Cu conditions (Yamasaki et al. 2007; Abdel-Ghany and Pilon 2008). The *COPPER CHAPERONE FOR SOD1* (*CCS1*) gene encoding the chaperone delivering Cu to the Cu, Zn SODs is also a target of miR398. Interestingly, stresses inducing enhanced production of superoxide radicals were reported to repress the expression of miR398. A negative correlation was found between miR398 and *COPPER/ZINC SUPEROXIDE DISMUTASE 1, 2* (*CSD1, 2*) mRNAs under many abiotic and biotic stress conditions, such as Ni, Cu, or Fe excess, Cd stress, high light, paraquat, ozone, bacterial infections (repressing miR398), and low Cu levels (inducing miR398) (reviewed by

Zhu et al. 2011). While miR398 mainly accumulates in leaves, it was also abundant in phloem (Buhtz et al. 2010). Two other Cu deficiency–induced miRNAs, miR397 and miR408, accumulate in roots and leaves, respectively, as well as in phloem sap, and target laccases, which are Cu-containing enzymes involved in lignin synthesis (Buhtz et al. 2010).

There is little indication regarding the regulation of the Fe-deficiency response by miRNA. Interestingly, using hybridization to microarray containing all known plant miRNAs, six of them were shown to be downregulated in *Brassica juncea* upon Fe deficiency compared to full nutrient conditions (Buhtz et al. 2010). Among those, several were upregulated by Cu deficiency, such as miR398, suggesting that Fe deficiency led to a reduction of Cu-deficiency response. Only miR158, predicted to target a pentatricopeptide repeat–containing protein of unknown function, a lipase, and xyloglucan fucosyltransferases, was upregulated in phloem sap (Buhtz et al. 2010).

If the role of some miRNAs in regulating plant metabolism upon nutrient deficiency is now well established, however, their role in modulating the root system architecture remains elusive.

5. Roles of Hormonal and Plant Growth Regulators in Signaling Trace Metallic Element Stress

Hormone interactions during root growth and development are treated in other chapters (see Chapters 12 through 20). Here we emphasize the role of hormones in the morphological response to TME stress and the regulation of mineral acquisition. Morphological changes upon TME deficiency or excess frequently indicate alterations of the hormonal balance. To date, studies examining the hormone signaling in response to critical concentrations of TME (Cu, Fe, Cd, etc.) are scarce, and the mechanistic details underlying their role remain elusive.

a. Auxin

The availability of essential TMEs may interfere with hormonal signaling to reprogram root development for a more efficient acquisition of the elements. The best documented case is probably the influence of the Fe supply on auxin distribution in roots. Giehl et al. (2012) proposed an auxin-integrating model that accounts for the root foraging response toward Fe-rich sources under low-Fe background. Upon local supply, the increase of the Fe symplastic pool mediated by IRT1 triggers the expression of the auxin importer *AUXIN RESISTANT 1* (*AUX1*) to accumulate auxin in the LR apices and to further stimulate the elongation of emerged lateral roots. Besides, Fe deficiency can cause the formation of extra root hairs, via the well-known ethylene signaling cascade (see Section IV.B). However, a possible auxin response pathway including AUX1 and/or AXR1 was proposed by Schmidt and Schikora (2001) as another pathway in which ethylene acts. Later, Chen et al. (2010) demonstrated a major role of auxin in the induction of Fe acquisition system in *Arabidopsis*. The inhibition of polar auxin transport prevented the induction of *FIT* and *FRO2* genes. Using mutants in polar auxin transport and chemical treatments, the authors proposed a new model in which NO acts downstream of auxin to activate *FIT* and *FRO2* genes and subsequently root ferric chelate reductase activity under Fe deficiency (Chen et al. 2010). A direct implication of auxin in the induction of strategy I was also recently proposed in *Malus xiaojinensis*, in which the root response was blocked by preventing IAA transport from the shoot (Wu et al. 2012). Several lines of evidence support the view that in addition to a primary perception of Fe shortage in roots, systemic auxin signal accumulates in the shoot apex and is transported to the root to coordinate the Fe-deficiency responses (Wu et al. 2012). Those data corroborate earlier studies on the importance of the shoot in Fe-deficiency response (Landsberg 1984; Romera et al. 1992; Schmidt and Schikora 2001; Schmidt et al. 2003).

Zn may be directly involved in auxin signaling as a predominant metal ion binding to AUXIN BINDING PROTEIN 1 (ABP1) (Woo et al. 2002). In rice, root growth inhibition upon Zn shortage is mainly due to a decline in newly adventitious roots that led to reduced total root number rather than inhibition of longitudinal root growth (Widodo et al. 2010). Auxin is believed to control adventitious root formation, and several putative auxin-induced/responsive genes were differentially regulated upon Zn deficiency (Widodo et al. 2010), supporting a major role of that hormone in the response.

Root growth inhibition induced by TME excess has been associated with changes in auxin distribution, as supported by reporter genes studies. As shown in Figure 34.1, morphological changes of roots usually include an inhibition of PR growth as well as an inhibition of proper elongation of lateral roots. Auxin is believed to play a key role in the stress-induced morphology. Upon Cd exposure, auxin distribution is modified, with a higher concentration in root tissues and, in the most severe toxicity cases, a decrease in the root tip (Potters et al. 2007, 2009; Mei et al. 2009; Xu et al. 2011). Under high Cu, auxin accumulates in the root section just above the meristem, which is proposed to account for the formation of multiple LRs (Lequeux et al. 2010). Interestingly, auxin was also proposed to be involved in the reduced NO production observed in the elongation zone (Peto et al. 2011). Together with ethylene, auxin might also be involved in the formation of root hairs in the area adjacent to the Cu^{2+}-treated root tips (Pasternak et al. 2005). However, exposure to high levels of other essential TMEs, like Mn or Ni, does not seem to increase auxin levels (Mei et al. 2009).

The hormonal signaling mechanisms are unknown. Interestingly, auxin is thought to be involved in lignin formation, and enhanced lignification was observed after several TME excess, in particular Cu and Cd (Schützendübel et al. 2002; van de Mortel et al. 2008; Elobeid et al. 2012). A recent study in poplar using a *GH3::GUS* reporter gene supported high IAA conjugation rates in the vascular system of Cd-exposed plants and thus a depletion of active auxin pool. GUS activity was confined to the endodermis, pericycle, procambium, and primary phloem (Elobeid et al. 2012). High *GH3* expression was previously observed in a global transcriptomic study in *B. juncea* after exposure to Cd (Minglin et al. 2005).

b. Cytokinin

Cytokinins are negative regulators of root growth. Transgenic *Arabidopsis* and tobacco plants expressing *CYTOKININ OXIDASE/DEHYDROGENASE* (*CKX1* or *CKX3*) genes under the transcriptional control respectively of the *WRKY6* or *PYK10* promoter of *Arabidopsis* have been engineered. Those plants have enhanced root-specific degradation of cytokinin and form a larger root system, with a similar shoot than the wild-type plants (Werner et al. 2010). Increasing the root-to-shoot ratio improved the concentrations of several nutrients, including the TME Zn and Mn in the leaves (mineral profile was measured in rosettes only). On a contaminated soil, Cd shoot concentration was also increased in the *CKX1* or *CKX3* transgenics. The uptake systems of Mn and Zn were upregulated in the roots of the transgenic plants, contributing to the increase in nutrient accumulation. However, Fe concentration was decreased in the leaves of the transgenic lines, while *IRT1* and *FRO2* were upregulated. The latter observation was probably due to the relief of the negative regulation of Fe transport system I by cytokinin (Seguela et al. 2008). These results highlight the complexity of metal homeostasis and the difficult way to engineer crops biofortified in micronutrients.

In general, there is a poor knowledge about the regulation of CK metabolism and sensitivity upon TME stress. Seguela et al. (2008) have proposed that CKs control the root Fe uptake machinery through a root growth–dependent pathway in order to adapt nutrient uptake to the demand of the plant. They have shown that CK addition inhibited the Fe acquisition strategy I through a FIT-independent pathway and that inhibition depended on the HISTIDINE KINASE 3 (AHK3) and CYTOKININ RESPONSE 1 (CRE1) CK receptors. Reduced flow of cytokinin to the shoot would be expected to play an important role in the resource allocation shift in favor of more root growth. Decrease in CK content was however reported upon Cd treatment that inhibits root growth (Veselov et al. 2003), highlighting the importance of hormonal balance to account for the observed phenotype. Cd differentially affected the concentrations of trans-zeatin and N6-isopentenyladenine ribosides in the roots of two barley cultivars (Atanassova et al. 2000). The significance of those changes on the root responses is not known. Upon exposure to Cu concentration inhibiting the PR growth of *Arabidopsis* plants, a higher CK level, as deduced by the *ARR5::GUS* reporter gene expression, was observed along the root and could participate to the root growth inhibition (Lequeux et al. 2010). Those observations need to be confirmed by other experiments.

c. Abscisic Acid

Possible involvement of ABA in the root responses to critical TME concentrations is poorly documented. ABA application inhibited Fe deficiency–induced acidification of the rhizosphere (Landsberg et al. 1986). Several studies reported an increase in ABA content under exposure to Cd, Co, Cu, Ni, or Zn at concentrations inhibiting root growth (Monni et al. 2001; Hsu and Kao 2003, 2005; Zengin 2006). Furthermore, ABA is known to lead to activation of NADPH oxidase (Kwak et al. 2003), making a possible link with ROS signaling.

d. Ethylene

The relationship between TME availability and stress-induced ethylene production has been largely investigated. Gas emanation is rapidly triggered in roots upon exposure to Cd (Rodriguez-Serrano et al. 2006; Arteca and Arteca 2007), high Cu (Arteca and Arteca 2007), or low Fe concentrations (Waters and Blevins 2000; Romera and Alcántra 2004; Molassiotis et al. 2005). Ethylene has been depicted in morphological and physiological responses to cope with these nutrient constraints. Its involvement was demonstrated in the promotion of root hairs upon Fe (Moog et al. 1995; Schmidt et al. 2000; Romera and Alcántra 2004) and Mn (Yang et al. 2008) and in the expression of genes related to mineral acquisition and homeostasis such as upon Fe deficiency (Lucena et al. 2006; García et al. 2010; Lingam et al. 2011; Wu et al. 2011). In this well-documented case, Fe deficiency upregulates genes in ethylene biosynthesis and signaling, while Fe-related genes are upregulated by ethylene (García et al. 2010). In addition, FIT protein is stabilized by interacting with ETHYLENE INTENSIVE 3 (EIN3) and ETHYLENE INSENSITIVE 3-LIKE 1 (EIL1) transcription factors of the ethylene pathway, which further increases the Fe acquisition response (Lingam et al. 2011).

In the case of metallic excess, ethylene production is partly responsible for root growth inhibition. Plant growth–promoting bacteria with 1-aminocyclopropane-1-carboxylic acid (ACC) deaminase activity can increase plant tolerance to metal excess and promote root growth by decreasing ethylene accumulation (Burd et al. 1998; Gamalero et al. 2009).

e. Salicylic Acid

Salicylic acid (SA) is known as an endogenous signal mediating local and systemic plant defense responses against pathogens (Rivas-San Vicente and Plasencia 2011). In various plant species, that molecule is also implicated in response to various abiotic stresses including those induced by Cd, Co, Cu, Ni, and Pb (Landberg and Greger 2002; Metwally et al. 2003; Freeman et al. 2005; Hsu and Kao 2005; El-tayeb et al. 2006; Rodriguez-Serrano et al. 2006). TME exposure can increase free SA production in roots, and preexposure to SA was shown to alleviate metal toxicity symptoms (Metwally et al. 2003; Choudhury and Panda 2004; Freeman et al. 2005; Krantev et al. 2008; Popova et al. 2009; Belkhadi et al. 2010; He et al. 2010). In this sense, SA appears to be, just like in mammals, an "effective therapeutic agent" for plants (Rivas-San Vicente and Plasencia 2011). In plants, enhanced tolerance to abiotic stress by SA pretreatment seems to be due to increased antioxidant capacity (Horváth et al. 2007). Even if the mechanisms of SA-mediated increased resistance to TMEs have not been completely elucidated yet, they seem to share certain steps with pathogens response.

V. Tolerance and Hyperaccumulation of Trace Metallic Elements

A. Definition of Tolerance and Accumulation Strategies

When considering TME concentration in shoots as a function of the soil TME concentration, three general accumulation strategies can be distinguished in plants (Figure 34.2; Baker 1981): (1) exclusion, which is characterized by mechanisms acting to minimize metal accumulation in aboveground tissues; (2) indication with a positive correlation between shoot and soil element concentrations, in that case proportional relationships exist between metal levels in the soil, uptake and accumulation in plant parts (Baker 1981); and (3) accumulation, described by high metal concentration in the shoot, even when soil concentration is low.

Based on the concentration of metals in aboveground plant parts, Brooks et al. (1979) have further distinguished plants as either accumulators or as hyperaccumulators. Hyperaccumulators can accumulate exceptional concentrations of trace elements in their aerial parts without visible toxicity symptoms, usually reflected by a root-to-shoot ratio of metal content above 1. Foliar concentration thresholds for hyperaccumulation have been arbitrarily set to about 100 times the element mean concentrations in other plants grown in the same environment but are currently subject to revision (Table 34.2).

Over 450 plant species (about 0.2% of angiosperms) have been identified as hyperaccumulators of trace elements (reviewed by Verbruggen et al. 2009b; Krämer 2010).

Hyperaccumulators have recently gained considerable interest because of their potential use in phytoremediation (Raskin and Ensley 2000), phytomining, and food crop biofortification (Zhao and McGrath 2009; White and Broadley 2011). They also represent the extreme end of natural variation of the metal homeostasis network. Remarkable insights into the evolutionary potential of plants to respond to elevated soil TME have recently been made, in particular through the study of two species, *Noccaea* (previously named *Thlaspi*) *caerulescens* and *Arabidopsis halleri*. These two Brassicaceae are close relatives to *A. thaliana* that have become models to study metal homeostasis in plants. Both species seem to hyperaccumulate Zn constitutively at the species level (that is to say that all populations analyzed so far possess that trait). Some calamine populations of both species can also hyperaccumulate Cd, and for *Noccaea*, serpentine populations can hyperaccumulate Ni. Grafting experiments in different plant species support that accumulation of metal in the shoot is largely controlled by root activity (Wagner et al. 1988; Arao et al. 2008; Guimarães et al. 2009). Rootstocks and shoot scions of the Zn hyperaccumulator *N. caerulescens* and the nonaccumulator *Thlaspi perfoliatum* were reciprocally grafted (Guimarães et al. 2009). Those experiments established that the root is the primary driver of Zn hyperaccumulation and shoot Zn hypertolerance. On the contrary, split-root experiments in excluder species unambiguously show that plant tolerance is mainly a root property (Harmens et al. 1993). One of the mechanisms of Cu tolerance in excluders such as tolerant *Silene vulgaris* would be enhanced ATP-dependent Cu efflux across the root cell plasma membrane (Van Hoof et al. 2001).

The following sections focus on anatomical, physiological, and molecular mechanisms underlying the exceptional behavior of hyperaccumulators.

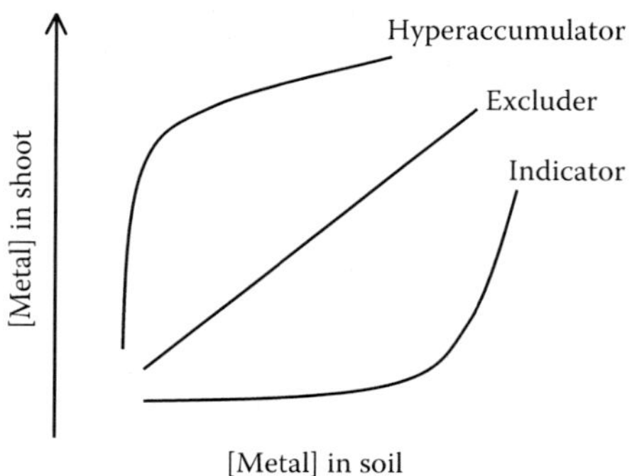

FIGURE 34.2 Plants have developed three main strategies for the management of potentially toxic TMEs. Excluders maintain constant elemental concentration in shoot over a wide range of soil metal content. Indicators increase shoot concentration along with soil mineral concentration. Accumulators accumulate metals in their aerial parts regardless of the concentration of the soil. (From Baker, A.J.M., *J. Plant Nutr.*, 3, 643, 1981.)

B. Root Morphological Adaptation to Cope with High Trace Metallic Element Concentrations in Hyperaccumulator Species

Hyperaccumulators are efficient to mobilize nonlabile trace element pools. The underlying mechanisms are still poorly understood, and there is a current lack of evidence for a role of root exudates. A possible role of bacteria and fungi of the rhizosphere has been recently examined by Alford et al. (2010). We will review here unique root morphological adaptations exhibited by some hyperaccumulators.

1. Foraging Capacity of Hyperaccumulators

In this chapter, the root morphological responses were already described of nontolerant species such as *A. thaliana* to conditions reflecting homogenous elevated TME concentrations (Figure 34.1), notwithstanding that the distribution of these elements is usually heterogeneous in soil. As pointed out for macronutrients elsewhere (see Chapter 25; Hermans et al. 2006), a foraging response to patchily distributed TME can be observed. Unlike to nonhyperaccumulator species, which generally restrict root growth to avoid undesired elevated metal conditions, some hyperaccumulators can actively proliferate in soil metal-rich hotspots. For example, *N. caerulescens* shows zincophilic root foraging capacity, increasing root length and root hair production in high-Zn-containing patches. Some populations of that species also allocate more root biomass into Cd- or Ni-enriched soil portions (Figure 34.3; Schwarz et al. 1999;

FIGURE 34.3 (See color insert.) Photograph of a *N. caerulescens* plant growing in a rhizobox filled with heterogeneously contaminated soil. The rhizobox is divided (dashed line) into two equal left and right compartments, filled with uncontaminated soil (Ni 0) on the left, and with soil contaminated with nickel (Ni 500: 500 μg Ni g^{-1} dry soil) on the right. Foraging for Ni is highlighted by the higher proportion of roots developed in the right compartment enriched with Ni. (Courtesy of C. Dechamps.)

Whiting et al. 2000; Haines 2002; Broadley et al. 2007; Dechamps et al. 2008). Root foraging for Zn and Cd requirement was also described in another Zn/Cd hyperaccumulator species, *Sedum alfredii* (Liu et al. 2010b). However, this positive chemotropism does not hold true for every hyperaccumulator species and every trace element. For instance, *Berkheya coddii* Rossler does not forage toward Ni-rich patches but still maintain the same root distribution pattern under uniformly high Ni concentrations (Moradi et al. 2009). Nevertheless, the principle of how heterogeneously distributed TME triggers these root development responses is poorly understood.

2. Difference in Root Structure between Hyperaccumulators and Nonhyperaccumulators

Difference in root structure between tolerant and sensitive plants can provide some answers to the question of how the root system has adapted under the selective pressure of high TMEs. Hyperaccumulation can also be associated with modification of the endodermis. In *N. caerulescens*, Casparian bands (the first stage of endodermal development) form <1 mm from the root tip, and suberin lamellae (the second stage of endodermal development) are formed in all endodermal cells ca 5–6 mm from the apex. In *Thlaspi arvense*, Casparian bands develop ca 2 mm from the root tip, and suberin lamellae are formed in all endodermal cells >10 mm from the root tip (Broadley et al. 2007; Zelko et al. 2008). A "peri-endodermal" layer of cells with irregularly thickened inner tangential walls extending to <1 mm from the root tip is another root particularity in *N. caerulescens* (Zelko et al. 2008). That layer is composed of secondary cell walls impregnated by suberin/lignin, forming a compact cylinder surrounding the endodermis from the outer side. This layer was not observed in the related nonaccumulator *T. arvense* (Zelko et al. 2008) or *A. thaliana* (van de Mortel et al. 2006) when compared with *N. caerulescens*. The function of that layer is not understood. As the Zn contents in the root apoplasm of *N. caerulescens* and *T. arvense* are similar (Lasat et al. 1998), the role of the layer is probably not associated with metal storage. The authors rather speculated an active role in efficient xylem loading, through the extended surface of the plasma membrane adhered to the irregular thickenings of the peri-endodermal layer in the root of *N. caerulescens*. A role in buffering the flux through the stele may not be excluded.

Another comparison of root structure between hyperaccumulating and nonhyperaccumulating plants was realized within the same species, *Senecio coronatus*, collected from their native environment (Mesjasz-Przybylowicz et al. 2007). The Ni-hyperaccumulator genotype differs from the nonhyperaccumulator one by distinct groups of specialized cells in the inner cortical region with an organelle-rich cytoplasm that produced many spherical bodies. Those cells had lower Ni concentrations than adjacent inner cortex and phloem. Casparian bands were observed in exodermal cell walls of both genotypes, with most probably higher suberin content in the nonhyperaccumulator (as revealed by fluorescence upon aniline blue staining). Together, these observations suggest anatomical adaptation underlying efficient Ni uptake by the root and translocation to the shoot in hyperaccumulator.

C. Metal Homeostasis Network in the Roots of Hyperaccumulator

The study of the two species *A. halleri* and *N. caerulescens* has tremendously advanced our molecular understanding of metal homeostasis in plants and, in particular, of metal transport. Comparisons between hyperaccumulators and nonhyperaccumulators highlight three main processes of metal homeostasis in the roots: (1) metal uptake, (2) metal mobility, and (3) metal loading into the apoplastic xylem. These three processes are more efficient in the roots of hyperaccumulators and account for the hyperaccumulation phenotype. In addition, adaptation to extreme metallic environments seems to also include enhanced protection against oxidative stress in hyperaccumulators (Boominathan and Doran 2002; Freeman et al. 2004; Freeman and Salt 2007).

1. Uptake of Trace Metallic Elements by Roots

Physiological experiments have clearly shown that the hyperaccumulation trait was associated with enhanced metal uptake by the roots (reviewed by Verbruggen et al. 2009b; Krämer 2010). Enhanced Zn uptake in *N. caerulescens* compared to the nonaccumulator *T. arvense* was proposed to be at least partly due

TABLE 34.3 Genes with a Putative Function in Metal Homeostasis and Overexpressed in the Roots of Hyperaccumulators *A. halleri* and/or *N. caerulescens* Compared to Nonaccumulator Relatives

Related Function	Name	Annotation	References
Metal uptake into cells	*ZIP4*	ZIP family of metal transporters	Becher et al. (2004); Weber et al. (2004); Hammond et al. (2006); van de Mortel et al. (2006)
	ZIP6	ZIP family of metal transporters	Becher et al. (2004); Filatov et al. (2006); Hammond et al. (2006); Talke et al. (2006)
	ZIP9	ZIP family of metal transporters	Weber et al. 2004; Hammond et al. (2006); Talke et al. (2006); van de Mortel et al. (2006)
	ZIP10	ZIP family of metal transporters	Hammond et al. (2006); Talke et al. (2006); van de Mortel et al. (2006)
	IRT1	ZIP family of metal transporters	Becher et al. (2004); van de Mortel et al. (2006)
	IRT3	ZIP family of metal transporters	Hammond et al. (2006); Talke et al. (2006); Lin et al. (2009)
Metal vacuolar sequestration	*MTP1*	Cation diffusion facilitator	Becher et al. (2004); Weber et al. (2004); Talke et al. (2006); van de Mortel et al. (2006)
	MTP8	Cation diffusion facilitator	Hammond et al. (2006); Talke et al. (2006); van de Mortel et al. (2006)
	CAX2	Ca^{++}: cation antiporter	Weber et al. (2004); Hammond et al. (2006)
	HMA3	P-type metal ATPase	Becher et al. (2004); Filatov et al. (2006); Hammond et al. (2006); van de Mortel et al. (2006); Ueno et al. (2011)
Metal remobilization from the vacuole	*NRAMP3*	Natural resistance-associated macrophage	Weber et al. (2004); Filatov et al. (2006); Talke et al. (2006)
Xylem loading/unloading of metal/ligands/metal–ligand complexes	*HMA4*	P-type metal ATPase	Becher et al. (2004); Weber et al. (2004); Filatov et al. (2006); Hammond et al. (2006); Talke et al. (2006); van de Mortel et al. (2006)
	FRD3	Multidrug and toxin efflux family transporter	Hammond et al. (2006); Talke et al. (2006); van de Mortel et al. (2006); van de Mortel et al. (2008)
	YSL3	Yellow stripe–like transporter	Gendre et al. (2007)
Synthesis of metal ligands	*NAS1*	Nicotianamine synthase	Hammond et al. (2006); van de Mortel et al. (2006)
	NAS2	Nicotianamine synthase	Becher et al. (2004); Weber et al. (2004); Talke et al. (2006); van de Mortel et al. (2006)
	NAS3	Nicotianamine synthase	Hammond et al. (2006); Talke et al. (2006); van de Mortel et al. (2006)
	NAS4	Nicotianamine synthase	Weber et al. (2004); Hammond et al. (2006); van de Mortel et al. (2006); van de Mortel et al. (2008)
	SAMS3	S-adenosylmethionine synthetase	Talke et al. (2006)
	AOSA2	Cysteine synthase	Weber et al. (2004); Talke et al. (2006)

Source: Adapted from Verbruggen, N. et al., *Curr. Opin. Plant Biol.*, 12, 364, 2009a.
Gene names refer to *A. thaliana.*

to a constitutive higher expression of the ZIP transporter ZNT1 (Lasat et al. 1996). Other overexpressed ZIP transporters have been identified from global analysis of Zn hyperaccumulator transcriptomes (Table 34.3). Yet there is no functional analysis determining the contribution of individual members to the Zn uptake from the soil solution to the root. In the case of Zn hyperaccumulators, the constitutive upregulation of Zn transporters seems to be related to the root sensing of Zn starvation due to efficient loading of Zn in the xylem. This has been demonstrated in *A. halleri* (Hanikenne et al. 2008; see Section V.C.3). In the case of Ni hyperaccumulation, the high uptake of Ni in the roots does not seem to be related to a Ni-deficiency response (Richau et al. 2009).

In the Zn hyperaccumulator *A. halleri*, exposure of roots to Zn excess does not induce the typical Fe-deficiency response as observed in *Arabidopsis*. Shanmugam et al. (2011) showed that Fe-uptake genes in roots of *A. halleri* and *A. thaliana* roots are differentially regulated. Similarly, the capacity to accumulate Cd in the shoot of *A. halleri* was positively correlated with the capacity to accumulate Fe, Zn, and K (Willems et al. 2010). These data point to the existence of an important homeostasis adaptation for elements other than the one being hyperaccumulated and, in particular, to the existence of a cross talk between Fe and Zn homeostasis as observed in global transcriptomics analysis (Hammond et al. 2006; Talke et al. 2006; van de Mortel et al. 2006).

2. Metal Mobility through the Roots

Hyperaccumulation of trace metals also depends on the limitation of root vacuolar sequestration, facilitating radial movement across the root to the xylem. On the other hand, enhanced metal vacuolar sequestration in the shoot is believed to be the main detoxification pathway and to allow the plant to become highly tolerant to internal metal concentrations, as supported by grafting experiments (Guimarães et al. 2009).

Lower accumulations of Zn in root vacuoles have been observed in *N. caerulescens* compared to *T. arvense* (about 2.5 times; Lasat et al. 1998) or in the Zn-hyperaccumulating ecotype of *S. alfredii* compared to the nonaccumulating one (about 2.7 times; Yang et al. 2006). Similar observations have been published for Cd (1.5-fold higher vacuolar Cd fraction in the roots in the *N. caerulescens* population) showing a lower capacity to accumulate high Cd concentrations in the shoot (Xing et al. 2008a). However, genes encoding transporters potentially involved in the transport of Zn or Cd into the vacuole were not less expressed compared to nonaccumulator-related species. On the contrary, *ZINC TRANSPORTER OF ARABIDOPSIS THALIANA 1 ZAT1/MPT1*, a member of the cation diffusion facilitator family, encoding a vacuolar transporter of Zn is highly expressed in the roots of *A. halleri* compared to Arabidopsis. However, the protein abundance was not verified, and posttranscriptional regulation may not be excluded. In the shoots, MTP1 is believed to be responsible for the detoxification of Zn and partially for the Zn hypertolerance trait in *A. halleri* (Talke et al. 2006; Willems et al. 2007; Shahzad et al. 2010). Similarly, *HMA3* encoding a vacuolar metal pump was constitutively highly expressed in the roots of *A. halleri* and *N. caerulescens*. In the latter species, natural variation for *HMA3* expression was found to be associated with Cd hyperaccumulation (Ueno et al. 2011). The authors show good evidence that higher *HMA3* expression allows the ecotype "Ganges" (South of France) to efficiently detoxify Cd through vacuolar sequestration in shoots (Ueno et al. 2011). In roots, *NcHMA3* was only expressed at pericycle cells (Ueno et al. 2011).

Other reported modifications that may influence metal mobility in hyperaccumulators include the higher production of metal ligands (Callahan et al. 2006) such as His (Krämer et al. 1996), NA (Weber et al. 2004), and GSH (Freeman et al. 2004; Meyer et al. 2011). In some cases, part of the hypertolerance could be mimicked in *A. thaliana* by overexpressing the corresponding biosynthetic enzyme genes (Ingle et al. 2005; Pianelli et al. 2005; Freeman and Salt 2007).

In Ni hyperaccumulators of the *Alyssum* genus, *Noccaea goesingense* and *N. caerulescens* species, free His is an important Ni binding ligand. Krämer et al. (1996) demonstrated that enhanced production of His was responsible for the Ni hyperaccumulation phenotype in *Alyssum*, that is to say, Ni tolerance and high Ni root-to-shoot translocation rates. However, His-overproducing transgenic *Arabidopsis* lines also displayed higher tolerance to Ni but did not exhibit increased Ni concentrations in either xylem sap or shoot tissue, suggesting that additional factors are necessary to recapitulate the complete hyperaccumulator phenotype (Ingle et al. 2005). Richau et al. (2009) showed that the high rate of root-to-shoot translocation of Ni in *N. caerulescens* compared with *T. arvense* depends on the combination of two distinct characters, that is, a greatly enhanced root His concentration and a strongly decreased ability to accumulate His-bound Ni in root cell vacuoles. The identity of the vacuolar transporter is not known.

In *N. caerulescens*, NA seems to be instrumental for the circulation of Ni. Exposure to Ni excess triggers the accumulation of NA in the roots. However, in that species, enhanced NA production occurs in the leaves and is transported to the roots through the phloem where it chelates part of the Ni. Ni–NA chelates are further translocated from the root to the shoot where Ni accumulates (Vacchina et al. 2003; Mari et al. 2006; Gendre et al. 2007; Curie et al. 2009). Interestingly, *N. caerulescens* roots highly express *YELLOW STRIPE-LIKE YSL3* and *YSL5* genes encoding putative NA and NA–metal complexes transporters (Gendre et al. 2007; Table 34.3). In particular, NcYSL3 was demonstrated to be an Fe/Ni–NA influx transporter (Gendre et al. 2007). So, at least in *N. caerulescens*, two mechanisms seem to coexist to efficiently transport metals to the aboveground tissues, one depending on HIS and another on NA. As some populations of that species are able to hyperaccumulate both Zn and Ni, these mechanisms may be involved in tolerance and/or accumulation traits of the two metals.

In *A. halleri* roots, enhanced NA content is locally produced by high NA synthase protein levels (Weber et al. 2004). This led the authors to propose an important role for NA in Zn hyperaccumulation in *A. halleri*. The identification of NA–Zn complexes in vivo further supported a direct role for NA in Zn homeostasis, in particular in the efficient root-to-shoot translocation (Trampczynska et al. 2010). This role was recently demonstrated by the generation of *AhNAS2-RNAi* plants that lost the capacity to hyperaccumulate Zn (Deinlein et al. 2012). In *A. halleri*, NA is proposed to form complexes with Zn(II) in root cells and to facilitate symplastic passage of Zn(II) toward the xylem (Deinlein et al. 2012).

GSH is a major cellular antioxidant and is also a precursor of the metal chelator PCs. Ni hypertolerance in various *Noccaea* hyperaccumulators seems to be associated to elevated levels of GSH and of SA, which is proposed to be involved in the regulation of high GSH levels in those species (Freeman et al. 2004, 2005). As there is no indication of speciation of Ni with S ligands, GSH is proposed to be involved in the protection against Ni-induced ROS (Freeman et al. 2004). Transgenic *Arabidopsis* overproducing GSH were more tolerant to Ni but did not contain more Ni in the shoot (Freeman et al. 2004). High GSH levels have also been measured in the roots of Zn, Cd hyperaccumulating populations of *N. caerulescens* and *A. halleri*, compared to *Arabidopsis* (Meyer et al. 2011). The putative role of GSH and PC in metal hypertolerance was investigated by inhibiting gamma-glutamylcysteine synthetase, a main enzyme of the biosynthetic pathway of GSH (Schat et al. 2002). That treatment did not decrease hypertolerance of different Cu, Cd, Ni, and Zn hypertolerant species including hyperaccumulators, which argues against a role for GSH or PCs in metal tolerance. Furthermore, lower PC levels were measured in the roots of Cd, Zn hyperaccumulators upon Cd stress than in *Arabidopsis* (Meyer et al. 2011). The higher levels of PC synthesis in *A. thaliana*, compared to *A. halleri* and *N. caerulescens*, most probably indicate a higher availability of metal ions at the site of PC synthesis. Possibly the lower activity of xylem loading and/or effluxing system for Cd in

A. thaliana compared to Cd hyperaccumulators favors the accumulation of high Cd concentrations in roots, which activate the synthesis of PC–Cd complexes and their subsequent sequestration in root vacuoles (Meyer et al. 2011).

3. Xylem Loading of Trace Metallic Elements and Root-to-Shoot Translocation

As already mentioned, the P1B-type ATPases, also known as the heavy metal transporting ATPases (HMAs), play an important role in transporting metal ions against their electrochemical gradient using the energy provided by ATP hydrolysis. HMA4 is localized in the plasma membrane (Courbot et al. 2007) of vascular tissues and is involved in the root-to-shoot translocation of Zn and Cd in plants (Verret et al. 2004; Wong and Cobbett 2008). *HMA4* is overexpressed both in *N. caerulescens* and *A. halleri* (Bernard et al. 2004; Hussain et al. 2004; Talke et al. 2006; Courbot et al. 2007). In *A. halleri*, *HMA4* was identified as a major determinant of Zn and Cd tolerance (Courbot et al. 2007; Willems et al. 2007) and of Zn and Cd hyperaccumulation (Frérot et al. 2010; Willems et al. 2010). These results refute the idea that the metal accumulation and tolerance are independent traits. In strong support, *HMA4* RNAi lines with a lower expression of *HMA4* translocated less Zn and Cd from the root to the shoot (Zn accumulation was shown in the pericycle cells) and were more sensitive to exposure to high Zn or Cd (Hanikenne et al. 2008). Furthermore, high *HMA4* expression was demonstrated to be directly responsible for the upregulation of Zn uptake transporters in *A. halleri*. Upregulation of the *IRT3* and *ZIP4* Zn transporters genes could be mimicked by the overexpression of *AhHMA4* (Hanikenne et al. 2008).

VI. Conclusion

Through roots, plants acquire water and nutrients from the complex chemical and biological environment of the soil. Research on plant nutrition is pivotal to crop productivity but also to the environment and to human health through enhancement of the food nutritional quality. Furthermore, improving nutrient use efficiency and, in particular, the uptake efficiency by roots may reduce the need for fertilizers and extend the cultivated land to marginal soils. Roots, which have been out of sight of breeders, are at the center of recent research efforts to produce crops with a better yield and with a lower input requirement (Gewin 2010). This type of research has recently benefitted from omics tools, and a picture is beginning to emerge for the regulation of some elements. Much still needs to be learnt, however, about sensing and signaling of changes in nutrient concentrations and responses to them. There is an important variability in plant responses and adaptation to nutrient stress and that variability may be exploited to understand regulatory mechanisms underlying plant nutrition. Such strategy was successful to elucidate the exceptional metal tolerance and accumulation traits of hyperaccumulators.

In short, "Nothing in biology makes sense except in the light of evolution" (Dobzhansky, 1973).

Acknowledgments

We apologize to those authors whose work has not been cited. We sincerely thank colleagues for critical reading of draft versions of the manuscript (S. Clemens; C. Curie; M. Hanikenne; P. Meerts; D. Mendoza-Cozatl; C.-L. Meyer and H. Schat).

References

Abdel-Ghany SE, Pilon M. 2008. MicroRNA-mediated systemic down-regulation of copper protein expression in response to low copper availability in *Arabidopsis*. *J Biol Chem* 283:15932–15945.

Adriaensen K, Vralstad T, Noben J-P et al. 2005. Copper-adapted *Suillus luteus*, a symbiotic solution for pines colonizing Cu mine spoils. *Appl Environ Microbiol* 71:7279–7284.

Ahrland S, Chatt J, Davies NR. 1958. The relative affinities of ligand atoms for acceptor molecules and ions. *Quart Rev Chem Soc* 12:265–276.

Akbas H, Dane F, Meric C. 2009. Effect of nickel on root growth and the kinetics of metal ions transport in onion (*Allium cepa*) root. *Ind J Biol Biophys* 46:332–336.

Alford ER, Pilon-Smits EAH, Paschke MW. 2010. Metallophytes-a view from the rhizosphere. *Plant Soil* 337:33–50.

Alloway BJ. 1995. Soil processes and the behaviour of heavy metals. In *Heavy Metals in Soils*, ed. BJ Alloway, pp. 11–37. London, U.K.: Blackie Academic and Professional.

Andreini C, Bertini I, Cavallaro G et al. 2008. Metal ions in biological catalysis: From enzyme databases to general principles. *J Biol Inorg Chem* 13:1205–1218.

Andrés-Colás N, Perea-Garcia A, Puig S et al. 2010. Deregulated copper transport affects *Arabidopsis* development especially in the absence of environmental cycles. *Plant Physiol* 153:170–184.

Andrés-Colás N, Sancenón V, Rodriguez-Navarro S et al. 2006. The *Arabidopsis* heavy metal P-type ATPase HMA5 interacts with metallochaperones and functions in copper detoxification of roots. *Plant J* 45:225–236.

Arao T, Takeda H, Nishihara E. 2008. Reduction of cadmium translocation from roots to shoots in eggplant (*Solanum melongena*) by grafting onto *Solanum torvum* rootstock. *Soil Sci Plant Nutr* 54:555–559.

Arduini I, Godbold DL, Onnis A. 1995. Influence of copper on root growth and morphology of *Pinus pinea* L. and *Pinus pinaster* Ait. seedlings. *Tree Physiol* 15:411–415.

Arteca RN, Arteca JM. 2007. Heavy-metal-induced ethylene production in *Arabidopsis thaliana*. *J Plant Physiol* 164:1480–1488.

Assunção AG, Herrero E, Lin YF et al. 2010. *Arabidopsis thaliana* transcription factors bZIP19 and bZIP23 regulate the adaptation to zinc deficiency. *Proc Natl Acad Sci U S A* 107:10296–10301.

Atanassova L, Vassilev A, Pissarska M et al. 2000. Cd-induced changes of root cytokinins from two barley cultivars. *Compt Rend Acad Bulg Sci* 53:91–94.

Bai C, Reilly CC, Wood BW. 2006. Nickel deficiency disrupts metabolism of ureides, amino acids, and organic acids of young pecan foliage. *Plant Physiol* 140:433–443.

Bais HP, Weir TL, Perry LG et al. 2006. The role of root exudates in rhizosphere interactions with plants and other organisms. *Annu Rev Plant Biol* 57:233–266.

Baker AJM. 1981. Accumulators and excluders—Strategies in the response of plants to heavy metals. *J Plant Nutr* 3:643–654.

Baker AJM, Brooks RR. 1989. Terrestrial higher plants which hyperaccumulate metallic elements—A review of their distribution, ecology and phytochemistry. *Biorecovery* 1:81–126.

Bartha B, Kolbert Z, Erdei L. 2005. Nitric oxide production induced by heavy metals in *Brassica juncea* L. Czern. and *Pisum sativum* L. *Acta Biol Szeged* 49:9–12.

Baryla A, Carrier P, Franck F et al. 2001. Leaf chlorosis in oilseed rape plants (*Brassica napus*) grown on cadmium-polluted soil: Causes and consequences for photosynthesis and growth. *Planta* 212:696–709.

Bashir K, Ishimaru Y, Nishizawa NK. 2010. Iron uptake and loading into rice grains. *Rice* 3:122–130.

Baxter I, Muthukumar B, Park HC et al. 2008. Variation in molybdenum content across broadly distributed populations of *Arabidopsis* thaliana is controlled by a mitochondrial molybdenum transporter (MOT1). *PLoS Genet* 4:e1000004.

Becher M, Talke IN, Krall L et al. 2004. Cross-species microarray transcript profiling reveals high constitutive expression of metal homeostasis genes in shoots of the zinc hyperaccumulator *Arabidopsis halleri*. *Plant J* 37:251–268.

Belkhadi A, Hediji H, Abbes Z et al. 2010. Effects of exogenous salicylic acid pre-treatment on cadmium toxicity and leaf lipid content in *Linum usitatissimum* L. *Ecotoxicol Environ Saf* 73:1004–1011.

Bernard C, Roosens N, Czernic P et al. 2004. A novel CPx-ATPase from the cadmium hyperaccumulator *Thlaspi caerulescens*. *FEBS Lett* 569:140–148.

Besson-Bard A, Gravot A, Richaud P et al. 2009. Nitric oxide contributes to cadmium toxicity in *Arabidopsis* by promoting cadmium accumulation in roots and by up-regulating genes related to iron uptake. *Plant Physiol* 149:1302–1315.

Bi YH, Chen WL, Zhang WN et al. 2009. Production of reactive oxygen species, impairment of photosynthetic function and dynamic changes in mitochondria are early events in cadmium-induced cell death in *Arabidopsis thaliana*. *Biol Cell* 101:629–643.

Blaudez D, Kohler A, Martin F et al. 2003. Poplar metal tolerance protein 1 (MTP1) confers zinc tolerance and is an oligomeric vacuolar zinc transporter with an essential leucine zipper motif. *Plant Cell* 15:2911–2928.

Bolan N, Adriano D, Mani S et al. 2003. Adsorption, complexation, and phytoavailability of copper as influenced by organic manure. *Environ Toxicol Chem* 22:450–456.

Boominathan R, Doran PM. 2002. Ni-induced oxidative stress in roots of the Ni hyperaccumulator, *Alyssum bertolonii*. *New Phytol* 156:205–215.

Briat J-F, Duc C, Ravet K et al. 2010. Ferritins and iron storage plants. *Biochim Biophys Acta* 1800:806–814.

Broadley MR, White PJ, Hammond JP et al. 2007. Zinc in plants. *New Phytol* 173:677–702.

Brooks RR, Morrison RS, Reeves RD et al. 1979. Hyperaccumulation of nickel by *Alyssum Linnaeus* (Cruciferae). *Proc R Soc Lond B* 203:387–403.

Brown PH, Welch RM, Cary EE. 1987. Nickel: A micronutrient essential for higher plants. *Plant Physiol* 85:801–803.

Buckhout TJ, Yang TJ, Schmidt W. 2009. Early iron-deficiency-induced transcriptional changes in *Arabidopsis* roots as revealed by microarray analyses. *BMC Genomics* 10:147.

Buhtz A, Pieritz J, Springer F et al. 2010. Phloem small RNAs, nutrient stress responses, and systemic mobility. *BMC Plant Biol* 10:64.

Burd GI, Dixon DG, Glick BR. 1998. A plant growth-promoting bacterium that decreases nickel toxicity in seedlings. *Appl Environ Microbiol* 64:3663–3668.

Burkhead JL, Reynolds KA, Abdel-Ghany SE et al. 2009. Copper homeostasis. *New Phytol* 182:799–816.

Cailliatte R, Schikora A, Briat JF et al. 2010. High-affinity manganese uptake by the metal transporter NRAMP1 is essential for *Arabidopsis* growth in low manganese conditions. *Plant Cell* 22:904–917.

Cakmak I. 2002. Plant nutrition research: Priorities to meet human needs for food in sustainable ways. *Plant Soil* 247:3–24.

Callahan DL, Baker AJM, Kolev SD et al. 2006. Metal ion ligands in hyperaccumulating plants. *J Biol Inorg Chem* 11:2–12.

Carrasco-Gil S, Alvarez-Fernandez A, Sobrino-Plata J et al. 2011. Complexation of Hg with phytochelatins is important for plant Hg tolerance. *Plant Cell Environ* 34:778–791.

Chaney RL. 1993. Zinc phytotoxicity. In *Zinc in Soils and Plants*, ed. AD Robson, pp. 135–150. Dordrecht, the Netherlands: Kluwer.

Chaney RL, Angle JS, Broadhurst CL, Peters CA, Tappero RV, Sparks DL. 2007. Improved understanding of hyperaccumulation yields commercial phytoextraction and phytomining technologies. *J Environ Qual* 36:1429–1443.

Chen Y, Barak P. 1982. Iron nutrition of plants in calcareous soils. *Adv Agron* 35:217–240.

Chen WR, He ZL, Yang XE et al. 2009a. Zinc efficiency is correlated with root morphology, ultrastructure, and antioxidative enzymes in rice. *J Plant Nutr* 32:287–305.

Chen C, Huang D, Liu J. 2009b. Functions and toxicity of nickel in plants: Recent advances and future prospects. *Clean* 37:304–313.

Chen WW, Yang JL, Qin C et al. 2010. Nitric oxide acts downstream of auxin to trigger root ferric-chelate reductase activity in response to iron deficiency in *Arabidopsis*. *Plant Physiol* 154:810–819.

Chiou T-J. 2007. The role of microRNAs in sensing nutrient stress. *Plant Cell Environ* 30:323–332.

Choudhury S, Panda SK. 2004. Role of salicylic acid in regulating cadmium induced oxidative stress in *Oryza sativa* L. roots. *Bulg J Plant Physiol* 30:95–110.

Clemens S. 2006. Toxic metal accumulation, responses to exposure and mechanisms of tolerance in plants. *Biochimie* 88:1707–1719.

Clemens S. 2010. Zn—A versatile player in plant cell biology. In *Cell Biology of Metals and Nutrients*, eds. R Hell, R-R Mendel, pp. 281–298. Berlin, Germany: Springer.

Cohen CK, Norvell WA, Kochian LV. 1997. Induction of the root cell plasma membrane ferric reductase (An exclusive role for Fe and Cu). *Plant Physiol* 114:1061–1069.

Cohu CM, Pilon M. 2010. Cell biology of copper. In *Cell Biology of Metals and Nutrients*, eds. R Hell, R-R Mendel, Plant Cell Monographs, Vol. 17, pp. 55–74. Berlin, Germany: Springer.

Colangelo EP, Guerinot ML. 2006. Put the metal to the petal: Metal uptake and transport throughout plants. *Curr Opin Plant Biol* 9:322–330.

Collins R, Kinsela A. 2011. Pedogenic factors and measurements of the plant uptake of cobalt. *Plant Soil* 339:499–512.

Connolly EL, Campbell NH, Grotz N et al. 2003. Overexpression of the FRO2 ferric chelate reductase confers tolerance to growth on low iron and uncovers posttranscriptional control. *Plant Physiol* 133:1102–1110.

Connolly EL, Fett JP, Guerinot ML. 2002. Expression of the IRT1 metal transporter is controlled by metals at the levels of transcript and protein accumulation. *Plant Cell* 14:1347–1357.

Correa-Aragunde N, Graziano M, Lamattina L. 2004. Nitric oxide plays a central role in determining lateral root development in tomato. *Planta* 218:900–905.

Courbot M, Willems G, Motte P et al. 2007. A major quantitative trait locus for cadmium tolerance in *Arabidopsis halleri* colocalizes with HMA4, a gene encoding a heavy metal ATPase. *Plant Physiol* 144:1052–1065.

Curie C, Briat J-F. 2003. Iron transport and signaling in plants. *Annu Rev Plant Biol* 54:183–206.

Curie C, Cassin G, Couch D et al. 2009. Metal movement within the plant: Contribution of nicotianamine and yellow stripe 1-like transporters. *Ann Bot* 103:1–11.

Curie C, Panaviene Z, Loulergue C et al. 2001. Maize yellow stripe1 encodes a membrane protein directly involved in Fe(III) uptake. *Nature* 409:346–349.

Cuypers A, Keunen E, Bohler S et al. 2012. Cadmium and copper stress induce a cellular oxidative challenge leading to damage versus signalling. In *Metal Toxicity in Plants: Perception, Signaling and Remediation*, eds. DK Gupta, LM Sandalio, pp. 65–89. Heidelberg, Germany: Springer.

De Michele R, Vurro E, Rigo C et al. 2009. Nitric oxide is involved in cadmium-induced programmed cell death in *Arabidopsis* suspension cultures. *Plant Physiol* 150:217–228.

Dechamps C, Noret N, Mozek R et al. 2008. Root allocation in metal-rich patch by *Thlaspi caerulescens* from normal and metalliferous soil—New insights into the rhizobox approach. *Plant Soil* 310:211–224.

Deinlein U, Weber M, Schmidt H et al. 2012. Elevated nicotianamine levels in *Arabidopsis halleri* roots play a key role in Zn hyperaccumulation. *Plant Cell* 24:708–723.

Delhaize E, Kataoka T, Hebb DM et al. 2003. Genes encoding proteins of the cation diffusion facilitator family that confer manganese tolerance. *Plant Cell* 15:1131–1142.

Dinneny JR, Long TA, Wang JY et al. 2008. Cell identity mediates the response of *Arabidopsis* roots to abiotic stress. *Science* 320:942–945.

Dobzhansky T. 1973. Nothing in biology makes sense except in the light of evolution. *Am Biol Teach* 35:125–129.

Duc C, Cellier F, Lobréaux S et al. 2009. Regulation of iron homeostasis in *Arabidopsis thaliana* by the clock regulator Time for Coffee. *J Biol Chem* 284:36271–36281.

Dučić T, Polle A. 2005. Transport and detoxification of manganese and copper in plants. *Brazil J Plant Physiol* 17:103–112.

Duffus JH. 2002. "Heavy metals"—A meaningless term? (IUPAC technical report). *Pure Appl Chem* 74:793–807.

Dutta S, Podile AR. 2010. Plant growth promoting rhizobacteria (PGPR): The bugs to debug the root zone. *Crit Rev Microbiol* 36:232–244.

Eide D, Broderius M, Fett J et al. 1996. A novel iron-regulated metal transporter from plants identified by functional expression in yeast. *Proc Natl Acad Sci U S A* 93:5624–5628.

Elobeid M, Göbel C, Feussner I et al. 2011. Cadmium interferes with auxin physiology and lignification in poplar. *J Exp Bot* 63:1413–1421.

El-Tayeb MA, El-Enany AE, Ahmed NL. 2006. Salicylic acid-induced adaptive response to copper stress in sunflower (*Helianthus annuus* L.). *Plant Growth Regul* 50:191–199.

Eskew DL, Welch RM, Cary EE. 1983. Nickel: An essential micronutrient for legumes and possibly all higher plants. *Science* 222:621–623.

Eun S-O, Shik Youn H, Lee Y. 2000. Lead disturbs microtubule organization in the root meristem of *Zea mays*. *Physiol Plant* 110:357–365.

Fargasova A. 1994. Effect of Pb, Cd, Hg, As, and Cr on germination and root growth of *Sinapis alba* seeds. *Bull Environ Contam Toxicol* 52:452–456.

Faucon MP, Colinet G, Mahy G et al. 2009. Soil influence on Cu and Co uptake and plant size in the cuprophytes *Crepidorhopalon perennis* and *C. tenuis* (Scrophulariaceae) in SC Africa. *Plant Soil* 317:201–212.

Filatov V, Dowdle J, Smirnoff N et al. 2006. Comparison of gene expression in segregating families identifies genes and genomic regions involved in a novel adaptation, zinc hyperaccumulation. *Mol Ecol* 15:3045–3059.

Finney LA, O'Halloran TV. 2003. Transition metal speciation in the cell: Insights from the chemistry of metal ion receptors. *Science* 300:931–936.

Foyer CH, Noctor G. 2009. Redox regulation in photosynthetic organisms: Signaling, acclimation, and practical implications. *Antiox Redox Signal* 11:861–905.

Frausto da Silva JJR, Williams RJP. 2001. *The Biological Chemistry of the Elements: The Inorganic Chemistry of Life*, 2nd edn. New York: Oxford University Press.

Freeman JL, Garcia D, Kim DG et al. 2005. Constitutively elevated salicylic acid signals glutathione-mediated nickel tolerance in *Thlaspi* nickel hyperaccumulators. *Plant Physiol* 137:1082–1091.

Freeman JL, Persans MW, Nieman K et al. 2004. Increased glutathione biosynthesis plays a role in nickel tolerance in *Thlaspi* nickel hyperaccumulators. *Plant Cell* 16:2176–2191.

Freeman JL, Salt DE. 2007. The metal tolerance profile of *Thlaspi goesingense* is mimicked in *Arabidopsis thaliana* heterologously expressing serine acetyl-transferase. *BMC Plant Biol* 7:63.

Frérot H, Faucon M-P, Willems G et al. 2010. Genetic architecture of zinc hyperaccumulation in *Arabidopsis halleri*: The essential role of QTL x environment interactions. *New Phytol* 187:355–367.

Gamalero E, Lingua G, Berta G et al. 2009. Beneficial role of plant growth promoting bacteria and arbuscular mycorrhizal fungi on plant responses to heavy metal stress. *Can J Microbiol* 55:501–514.

Ganeshan R, Manoharan. 1983. Effect of cadmium and mercury on germination, growth, and dry matter production of *Abelmoschus esculentus. Geobios* 10:9–12.

García MJ, Lucena C, Romera FJ et al. 2010. Ethylene and nitric oxide involvement in the up-regulation of key genes related to iron acquisition and homeostasis in *Arabidopsis. J Exp Bot* 61:3885–3899.

Garcia-Molina A, Andres-Colas N, Perea-Garcia A et al. 2011. The intracellular *Arabidopsis* COPT5 transport protein is required for photosynthetic electron transport under severe copper deficiency. *Plant J* 65:848–860.

Genc Y, Huang CY, Langridge P. 2007. A study of the role of root morphological traits in growth of barley in zinc-deficient soil. *J Exp Bot* 58:2775–2784.

Gendre D, Czernic P, Conéjéro G et al. 2007. TcYSL3, a member of the YSL gene family from the hyperaccumulator *Thlaspi caerulescens*, encodes a nicotianamine-Ni/Fe transporter. *Plant J* 49:1–15.

Gewin V. 2010. Food: An underground revolution. *Nature* 466:552–553.

Giehl RFH, Lima JE, von Wirén N. 2012. Localized iron supply triggers lateral root elongation in *Arabidopsis* by altering the AUX1-mediated auxin distribution. *Plant Cell* 24:33–49.

Gohre V, Paszkowski U. 2006. Contribution of the arbuscular mycorrhizal symbiosis to heavy metal phytoremediation. *Planta* 223:1115–1122.

Gong JM, Lee DA, Schroeder JI. 2003. Long-distance root-to-shoot transport of phytochelatins and cadmium in *Arabidopsis. Proc Natl Acad Sci U S A* 100:10118–10123.

Graziano M, Lamattina L. 2007. Nitric oxide accumulation is required for molecular and physiological responses to iron deficiency in tomato roots. *Plant J* 52:949–960.

Grotz N, Fox T, Connolly E et al. 1998. Identification of a family of zinc transporter genes from *Arabidopsis* that respond to zinc deficiency. *Proc Natl Acad Sci U S A* 95:7220–7224.

Guan TX, He HB, Zhang XD et al. 2011. Cu fractions, mobility and bioavailability in soil-wheat system after Cu-enriched livestock manure applications. *Chemosphere* 82:215–222.

Guerinot ML. 2010. Iron. In *Cell Biology of Metals and Nutrients*, eds. R Hell, R-R Mendel, pp. 75–94. Berlin, Germany: Springer.

Guerinot ML, Colangelo EP. 2004. The essential basic helix-loop-helix protein FIT1 is required for the iron deficiency response. *Plant Cell* 16:3400–3412.

Guimarães MdA, Gustin JL, Salt DE. 2009. Reciprocal grafting separates the roles of the root and shoot in zinc hyperaccumulation in *Thlaspi caerulescens. New Phytol* 184:323–329.

Gupta R. 1991. Toxic effects of mercury on seed germination of bean and mustard. *Comp Physiol Ecol* 16:43–45.

Gustin JL, Loureiro ME, Kim D et al. 2009. MTP1-dependent Zn sequestration into shoot vacuoles suggests dual roles in Zn tolerance and accumulation in Zn-hyperaccumulating plants. *Plant J* 57:1116–1127.

Hagemeyer J, Breckle S-W. 2002. Trace element stress in roots. In *Plant Roots: The Hidden Half*, eds. Y Waisel, A Eshel, U Kafkafi, 3rd edn., pp. 763–785. New York: Marcel Dekker, Inc.

Hagström J, James WM, Skene KR. 2001. A comparison of structure, development and function in cluster roots of *Lupinus albus* L. under phosphate and iron stress. *Plant Soil* 232:81–90.

Haines BJ. 2002. Zincophilic root foraging in *Thlaspi caerulescens. New Phytol* 155:363–372.

Hall JL. 2002. Cellular mechanisms for heavy metal detoxification and tolerance. *J Exp Bot* 53:1–11.

Hammond JP, Bowen HC, White PJ et al. 2006. A comparison of the *Thlaspi caerulescens* and *Thlaspi arvense* shoot transcriptomes. *New Phytol* 170:239–260.

Hanikenne M, Talke IN, Haydon MJ et al. 2008. Evolution of metal hyperaccumulation required cis-regulatory changes and triplication of HMA4. *Nature* 453:391–395.

Hansch R, Mendel RR. 2009. Physiological functions of mineral micronutrients (Cu, Zn, Mn, Fe, Ni, Mo, B, Cl). *Curr Opin Plant Biol* 12:259–266.

Hansel CM, Fendorf S. 2001. Characterization of Fe plaque and associated metals on the roots of mine-waste impacted aquatic plants. *Environ Sci Technol* 35:3863–3868.

Hao F, Wang X, Chen J. 2006. Involvement of plasma-membrane NADPH oxidase in nickel-induced oxidative stress in roots of wheat seedlings. *Plant Sci* 170:151–158.

Harmens H, Gusmão NGCPB, Den Hartog PR et al. 1993. Uptake and transport of zinc in zinc-sensitive and zinc-tolerant *Silene vulgaris. J Plant Physiol* 141:309–315.

Haydon MJ, Bell LJ, Webb AA. 2011. Interactions between plant circadian clocks and solute transport. *J Exp Bot* 62:2333–2348.

Haydon MJ, Cobbett CS. 2007. Transporters of ligands for essential metal ions in plants. *New Phytol* 174:499–506.

He J, Ren Y, Pan X et al. 2010. Salicylic acid alleviates the toxicity effect of cadmium on germination, seedling growth, and amylase activity of rice. *J Plant Nutr Soil Sci* 173:300–305.

Hermans C, Hammond JP, White PJ et al. 2006. How do plants respond to nutrient shortage by biomass allocation? *Trends Plant Sci* 11:610–617.

Heyno E, Klose C, Krieger-Liszkay A. 2008. Origin of cadmium-induced reactive oxygen species production: Mitochondrial electron transfer versus plasma membrane NADPH oxidase. *New Phytol* 179:687–699.

Hildebrandt U, Regvar M, Bothe H. 2006. Arbuscular mycorrhiza and heavy metal tolerance. *Phytochemistry* 68:139–146.

Horváth E, Szalai G, Janda T. 2007. Induction of abiotic stress tolerance by salicylic acid signaling. *J Plant Growth Regul* 26:290–300.

Hsu YT, Kao CH. 2003. Role of abscisic acid in cadmium tolerance of rice (*Oryza sativa* L.) seedlings. *Plan Cell Environ* 26:867–874.

Hsu YT, Kao CH. 2005. Abscisic acid accumulation and cadmium tolerance in rice seedlings. *Physiol Plant* 124:71–80.

Hussain D, Haydon MJ, Wang Y et al. 2004. P-type ATPase heavy metal transporters with roles in essential zinc homeostasis in *Arabidopsis*. *Plant Cell* 16:1327–1339.

Ide Y, Kusano M, Oikawa A et al. 2011. Effects of molybdenum deficiency and defects in molybdate transporter MOT1 on transcript accumulation and nitrogen/sulphur metabolism in *Arabidopsis thaliana*. *J Exp Bot* 62:1483–1497.

Ingle RA, Mugford ST, Rees JD et al. 2005. Constitutively high expression of the histidine biosynthetic pathway contributes to nickel tolerance in hyperaccumulator plants. *Plant Cell* 17:2089–2106.

Inoue Y, Konno M, Ooishi M. 2006. Temporal and positional relationships between Mn uptake and low-pH-induced root hair formation in *Lactuca sativa* cv. Grand Rapids seedlings. *J Plant Res* 119:439–447.

Ishimaru Y, Suzuki M, Tsukamoto T et al. 2006. Rice plants take up iron as an Fe^{3+}-phytosiderophore and as Fe^{2+}. *Plant J* 45:335–346.

Ivanov R, Brumbarova T, Bauer P. 2012. Fitting into the harsh reality: Regulation of iron-deficiency responses in dicotyledonous plants. *Mol Plant* 5:27–42.

Jain R, Srivastava S, Solomon S et al. 2010. Impact of excess zinc on growth parameters, cell division, nutrient accumulation, photosynthetic pigments and oxidative stress of sugarcane (*Saccharum* spp.). *Acta Physiol Plant* 32:979–986.

Jakoby M, Wang H-Y, Reidt W et al. 2004. FRU (BHLH029) is required for induction of iron mobilization genes in *Arabidopsis thaliana*. *FEBS Lett* 577:528–534.

Janik E, Maksymiec W, Mazur R et al. 2010. Structural and functional modifications of the major light-harvesting complex II in cadmium- of copper-treated Secale cereale. *Plant Cell Physiol* 51:1330–1340.

John R, Ahmad P, Gadgil K et al. 2009. Heavy metal toxicity: Effect on plant growth, biochemical parameters and metal accumulation by *Brassica juncea* L. *Int J Plant Prod* 3:1735–8043.

Jonak C, Nakagami H, Hirt H. 2004. Heavy metal stress. Activation of distinct mitogen-activated protein kinase pathways by copper and cadmium. *Plant Physiol* 136:3276–3283.

Kabata-Pendias A. 2010. *Trace Elements in Soils and Plants*, 4th edn. Boca Raton, FL: CRC.

Kaiser BN, Gridley KL, Brady JN et al. 2005. The role of molybdenum in agricultural plant production. *Ann Bot* 96:745–754.

Kampfenkel K, Kushnir S, Babiychuk E et al. 1995. Molecular characterization of a putative *Arabidopsis thaliana* copper transporter and its yeast homologue. *J Biol Chem* 270:28479–28486.

Kanady JS, Tsui EY, Day MW et al. 2011. A synthetic model of the MnCa subsite of the oxygen-evolving complex in photosystem II. *Science* 333:733–736.

Kawachi M, Kobae Y, Mori H et al. 2009. A mutant strain *Arabidopsis thaliana* that lacks vacuolar membrane zinc transporter MTP1 revealed the latent tolerance to excessive zinc. *Plant Cell Physiol* 50:1156–1170.

Khan GA, Declerck M, Sorin C et al. 2011. MicroRNAs as regulators of root development and architecture. *Plant Mol Biol* 77:47–58.

Kim SA, Guerinot ML. 2007. Mining iron: Iron uptake and transport in plants. *FEBS Lett* 581:2273–2280.

Kim Y-Y, Yang Y-Y, Lee Y. 2002. Pb and Cd uptake in rice roots. *Physiol Plant* 116:368–372.

Kobayashi Y, Kuroda K, Kimura K et al. 2008. Amino acid polymorphisms in strictly conserved domains of a P-Type ATPase HMA5 are involved in the mechanism of copper tolerance variation in *Arabidopsis*. *Plant Physiol* 148:969–980.

Kochian LV, Hoekenga OA, Pineros MA. 2004. How do crop plants tolerate acid soils?—Mechanisms of aluminum tolerance and phosphorous efficiency. *Annu Rev Plant Biol* 55:459–493.

Konno M, Ooishi M, Inoue Y. 2003. Role of manganese in low-pH-induced root hair formation in *Lactuca sativa* cv. Grand Rapids seedlings. *J Plant Res* 116:301–307.

Korshunova YO, Eide D, Clark WG et al. 1999. The IRT1 protein from *Arabidopsis thaliana* is a metal transporter with a broad substrate range. *Plant Mol Biol* 40:37–44.

Krämer U. 2010. Metal hyperaccumulation in plants. *Annu Rev Plant Biol* 61:517–534.

Krämer U, Cotter-Howells JD, Charnock JM et al. 1996. Free histidine as a metal chelator in plants that accumulate nickel. *Nature* 379:635–638.

Krämer D, Römheld V, Landsberg E et al. 1980. Induction of transfer-cell formation by iron deficiency in the root epidermis of *Helianthus annuus* L. *Planta* 147:335–339.

Krantev A, Yordanova R, Janda T et al. 2008. Treatment with salicylic acid decreases the effect of cadmium on photosynthesis in maize plants. *J Plant Physiol* 165:920–931.

Krznaric E, Verbruggen N, Wevers JHL et al. 2009. Cd-tolerant Suillus luteus: A fungal insurance for pines exposed to Cd. *Environ Pollut* 157:1581–1588.

Kwak JM, Mori IC, Pei ZM et al. 2003. NADPH oxidase AtrbohD and AtrbohF genes function in ROS-dependent ABA signaling in *Arabidopsis*. *EMBO J* 22:2623–2633.

Landsberg EC. 1984. Regulation of iron-stress response by whole-plant activity. *J Plant Nutr* 7:609–621.

Landsberg EC. 1986. Function of rhizodermal transfer cells in the Fe stress response mechanism of *Capsicum annuum* L. *Plant Physiol* 82:511–517.

Landberg T, Greger M. 2002. Differences in oxidative stress in heavy metal resistant and sensitive clones of *Salix viminalis. J Plant Physiol* 159:69–75.

Lanquar V, Ramos MS, Lelièvre F et al. 2010. Export of vacuolar manganese by AtNRAMP3 and AtNRAMP4 is required for optimal photosynthesis and growth under manganese deficiency. *Plant Physiol* 152:1986–1999.

Lasat MM, Baker AJM, Kochian LV. 1996. Physiological characterization of root Zn^{2+} absorption and translocation to shoots in Zn hyperaccumulator and nonaccumulator species of *Thlaspi. Plant Physiol* 112:1715–1722.

Lasat MM, Baker AJM, Kochian LV. 1998. Altered zinc compartmentation in the root symplasm and stimulated Zn^{2+} absorption into the leaf as mechanisms involved in zinc hyperaccumulation in *Thlaspi caerulescens. Plant Physiol* 118:875–883.

Lequeux H, Hermans C, Lutts S et al. 2010. Response to copper excess in *Arabidopsis thaliana*: Impact on the root system architecture, hormone distribution, lignin accumulation and mineral profile. *Plant Physiol Biochem* 48:673–682.

Li J-Y, Fu Y-L, Pike SM et al. 2010. The *Arabidopsis* nitrate transporter NRT1.8 functions in nitrate removal from the xylem sap and mediates cadmium tolerance. *Plant Cell* 22:1633–1646.

Lin YF, Liang HM, Yang SY et al. 2009. *Arabidopsis* IRT3 is a zinc-regulated and plasma membrane localized zinc/iron transporter. *New Phytol* 182:392–404.

Ling H-Q, Bauer P, Bereczky Z et al. 2002. The tomato fer gene encoding a bHLH protein controls iron-uptake responses in roots. *Proc Natl Acad Sci U S A* 99:13938–13943.

Lingam S, Mohrbacher J, Brumbarova T et al. 2011. Interaction between the bHLH transcription factor FIT and ETHYLENE INSENSITIVE3/ETHYLENE INSENSITIVE3-LIKE1 reveals molecular linkage between the regulation of iron acquisition and ethylene signaling in *Arabidopsis. Plant Cell* 23:1815–1829.

Liu DH, Jiang WS, Meng QM et al. 2009. Cytogenetical and ultrastructural effects of copper on root meristem cells of *Allium sativum* L. *Biocell* 33:25–32.

Liu X-M, Kim KE, Kim K-C et al. 2010a. Cadmium activates *Arabidopsis* MPK3 and MPK6 via accumulation of reactive oxygen species. *Phytochemistry* 71:614–618.

Liu J, Reid RJ, Smith FA. 2000. The mechanism of cobalt toxicity in mung beans. *Physiol Plant* 110:104–110.

Liu FJ, Tang YT, Du RJ et al. 2010b. Root foraging for zinc and cadmium requirement in the Zn/Cd hyperaccumulator plant *Sedum alfredii. Plant Soil* 327:365–375.

Long TA, Tsukagoshi H, Busch W et al. 2010. The bHLH transcription factor POPEYE regulates response to iron deficiency in *Arabidopsis* roots. *Plant Cell* 22:2219–2236.

Lucena C, Waters BM, Romera FJ et al. 2006. Ethylene could influence ferric reductase, iron transporter, and H^+-ATPase gene expression by affecting FER (or FER-like) gene activity. *J Exp Bot* 57:4145–4154.

Lugon-Moulin N, Ryan L, Donini P et al. 2006. Cadmium content of phosphate fertilizers used for tobacco production. *Agron Sustainable Dev* 26:151–155.

Lugtenberg B, Kamilova F. 2009. Plant-growth-promoting rhizobacteria. *Annu Rev Microbiol* 63:541–556.

Lux A, Martinka M, Vaculik M et al. 2011. Root responses to cadmium in the rhizosphere: A review. *J Exp Bot* 62:21–37.

Ma Z, Bielenberg DG, Brown KM et al. 2001. Regulation of root hair density by phosphorus availability in *Arabidopsis thaliana. Plant Cell Environ* 24:459–467.

Ma Y, Prasad MNV, Rajkumar M et al. 2011. Plant growth promoting rhizobacteria and endophytes accelerate phytoremediation of metalliferous soils. *Biotechnol Adv* 29:248–258.

Madejon P, Ramirez-Benitez JE, Corrales I et al. 2009. Copper-induced oxidative damage and enhanced antioxidant defenses in the root apex of maize cultivars differing in Cu tolerance. *Environ Exp Bot* 67:415–420.

Maksimovic I, Kastori R, Krsti L et al. 2007. Steady presence of cadmium and nickel affects root anatomy, accumulation and distribution of essential ions in maize seedlings. *Biol Plant* 51:589–592.

Maldonado-Magaña A, Favela-Torres E, Rivera-Cabrera F et al. 2011. Lead bioaccumulation in *Acacia farnesiana* and its effect on lipid peroxidation and glutathione production. *Plant Soil* 339:377–389.

Manohar M, Shigaki T, Hirschi KD. 2011. Plant cation/H^+ exchangers (CAXs): biological functions and genetic manipulations. *Plant Biol* 13:561–569.

Mari S, Gendre D, Katia Pianelli K et al. 2006. Root-to-shoot long-distance circulation of nicotianamine and nicotianamine-nickel chelates in the metal hyperaccumulator *Thlaspi caerulescens. J Exp Bot* 57:4111–4122.

Marschner H. 1995. *Mineral Nutrition of Higher Plants*, 2nd edn. London, U.K.: Academic Press.

Mastretta C, Taghavi S, van der Lelie D et al. 2009. Endophytic bacteria from seeds of *Nicotiana tabacum* can reduce cadmium phytotoxicity. *Int J Phytorem* 11:251–267.

McGrath SP, Micó C, Zhao FJ et al. 2010. Predicting molybdenum toxicity to higher plants: Estimation of toxicity threshold values. *Environ Pollut* 158:3085–3094.

Meda AR, Scheuermann EB, Prechsl UE et al. 2007. Iron acquisition by phytosiderophores contributes to cadmium tolerance. *Plant Physiol* 143:1761–1773.

Mei H, Cheng NH, Zhao J et al. 2009. Root development under metal stress in *Arabidopsis thaliana* requires the H^+/cation antiporter CAX4. *New Phytol* 183:95–105.

Meiser J, Lingam S, Bauer P. 2011. Posttranslational regulation of the iron deficiency basic helix-loop-helix transcription factor FIT is affected by iron and nitric oxide. *Plant Physiol* 157:2154–2166.

Mendoza-Cozatl DG, Jobe TO, Hauser F et al. 2011. Long-distance transport, vacuolar sequestration, tolerance, and transcriptional responses induced by cadmium and arsenic. *Curr Opin Plant Biol* 14:554–562.

Mesjasz-Przybylowicz J, Barnabas A, Przybyiowicz W. 2007. Comparison of cytology and distribution of nickel in roots of Ni-hyperaccumulating and non-hyperaccumulating genotypes of *Senecio coronatus*. *Plant Soil* 293:61–78.

Metwally A, Finkemeier I, Georgi M et al. 2003. Salicylic acid alleviates the cadmium toxicity in barley seedlings. *Plant Physiol* 132:272–281.

Meyer C-L, Peisker D, Courbot M et al. 2011. Isolation and characterization of *Arabidopsis halleri* and *Thlaspi caerulescens* phytochelatin synthases. *Planta* 234:83–95.

Micó C, Li HF, Zhao FJ et al. 2008. Use of Co speciation and soil properties to explain variation in Co toxicity to root growth of barley (*Hordeum vulgare* L.) in different soils. *Environ Pollut* 156:883–890.

Millaleo R, Reyes-Diaz M, Ivanov AG et al. 2010. Manganese as essential and toxic element for plants: Transport, accumulation and resistance mechanisms. *J Soil Sci Plant Nutr* 10:470–481.

Minglin L, Yuxiu Z, Tuanyao C. 2005. Identification of genes up-regulated in response to Cd exposure in *Brassica juncea* L. *Gene* 363:151–158.

Molassiotis A, Therios I, Dimassi K et al. 2005. Induction of Fe (III)-chelate reductase activity by ethylene and salicylic acid in iron-deficient peach rootstock explants. *J Plant Nutr* 28:669–682.

Monni S, Uhlig C, Hansen E et al. 2001. Ecophysiological responses of *Empetrum nigrum* to heavy metal pollution. *Environ Pollut* 112:121–129.

Moog PR, van der Kooij TA, Bruggemann W et al. 1995. Responses to iron deficiency in *Arabidopsis thaliana*: The Turbo iron reductase does not depend on the formation of root hairs and transfer cells. *Planta* 195:505–513.

Moore KL, Schröder M, Wu Z et al. 2011. High-resolution secondary ion mass spectrometry reveals the contrasting subcellular distribution of arsenic and silicon in rice roots. *Plant Physiol* 156:913–924.

Moradi AB, Conesa HM, Robinson BH et al. 2009. Root responses to soil Ni heterogeneity in a hyperaccumulator and a non-accumulator species. *Environ Pollut* 157:2189–2196.

Morel M, Crouzet J, Gravot A et al. 2009. AtHMA3, a P_{1B}-ATPase allowing Cd/Zn/Co/Pb vacuolar storage in *Arabidopsis. Plant Physiol* 149:894–904.

Morrissey J, Baxter IR, Lee J et al. 2009. The ferroportin metal efflux proteins function in Iron and Cobalt homeostasis in *Arabidopsis. Plant Cell* 21:3326–3338.

Mou D, Yao Y, Yang Y et al. 2011. Plant high tolerance to excess manganese related with root growth, manganese distribution and antioxidative enzyme activity in three grape cultivars. *Ecotoxicol Environ Saf* 74:776–786.

Mukherjee I, Campbell NH, Ash JS et al. 2006. Expression profiling of the *Arabidopsis* ferric chelate reductase (FRO) gene family reveals differential regulation by iron and copper. *Planta* 223:1178–1190.

Muller M, Schmidt W. 2004. Environmentally induced plasticity of root hair development in *Arabidopsis. Plant Physiol* 134:409–419.

Nakanishi H, Ogawa I, Ishimaru Y et al. 2006. Iron deficiency enhances cadmium uptake and translocation mediated by the Fe^{2+} transporters OslRT1and OslRT2 in rice. *Soil Sci Plant Nutr* 52:464–469.

Navari-Izzo F, Cestone B, Cavallini A et al. 2006. Copper excess triggers phospholipase D activity in wheat roots. *Phytochemistry* 67:1232–1242.

Nishida S, Tsuzuki C, Kato A et al. 2011. AtIRT1, the primary iron uptake transporter in the root, mediates excess nickel accumulation in *Arabidopsis thaliana. Plant Cell Physiol* 52:1433–1442.

Nouet C, Motte P, Hanikenne M. 2011. Chloroplastic and mitochondrial metal homeostasis. *Trends Plant Sci* 16:1360–1385.

Ortega-Villasante C, Hernández LE, Rellan-Alvarez R et al. 2007. Rapid alteration of cellular redox homeostasis upon exposure to cadmium and mercury in alfalfa seedlings. *New Phytol* 176:96–107.

Outten CE, O'Halloran TV. 2001. Femtomolar sensitivity of metalloregulatory proteins controlling zinc homeostasis. *Science* 292:2488–2492.

Pakrasi HB, Korshunova YO, Eide D et al. 1999. The IRT1 protein from *Arabidopsis thaliana* is a metal transporter with a broad substrate range. *Plant Mol Biol* 40:37–44.

Panou-Filotheou H, Bosabalidis AM. 2004. Root structural aspects associated with copper toxicity in oregano (*Origanum vulgare* subsp hirtum). *Plant Sci* 166:1497–1504.

Pasternak T, Rudas V, Potters G et al. 2005. Morphogenic effects of abiotic stress: Reorientation of growth in *Arabidopsis thaliana* seedlings. *Environ Exp Bot* 53:299–314.

Patra M, Bhowmik N, Bandopadhyay B et al. 2004. Comparison of mercury, lead and arsenic with respect to genotoxic effects on plant systems and the development of genetic tolerance. *Environ Exp Bot* 52:199–223.

Patra M, Sharma A. 2000. Mercury toxicity in plants. *Bot Rev* 66:379–422.

Perea-García A, Andrés-Colás N, Peñarrubia L. 2010. Copper homeostasis influences the circadian clock in *Arabidopsis. Plant Signal Behav* 5:1237–1240.

Persans MW, Nieman K, Salt DE. 2001. Functional activity and role of cation-efflux family members in Ni hyperaccumulation in *Thlaspi goesingense. Proc Natl Acad Sci U S A* 98:9995–10000.

Peto A, Lehotai N, Lozano-Juste J et al. 2011. Involvement of nitric oxide and auxin in signal transduction of copper-induced morphological responses in *Arabidopsis* seedlings. *Ann Bot* 108:449–457.

Phang IC, Leung DWM, Taylor H et al. 2010. Correlation of growth inhibition with accumulation of Pb in cell wall and changes in response to oxidative stress in *Arabidopsis thaliana* seedlings. *Plant Growth Regul* 64:17–25.

Pianelli K, Mari S, Marquès L et al. 2005. Nicotianamine over-accumulation confers resistance to nickel in *Arabidopsis thaliana. Transgenic Res* 14:739–748.

Pilet PE, Versel JM. 1981. Effect of mercury on growth and gravireaction of maize roots (*Zea mays*). *Z Pflanzenphysiol* 104:193–198.

Pilon-Smits EA, Quinn CF, Tapken W et al. 2009. Physiological functions of beneficial elements. *Curr Opin Plant Biol* 12:267–274.

Pittman JK. 2005. Managing the manganese: Molecular mechanisms of manganese transport and homeostasis. *New Phytol* 167:733–742.

Popova LP, Maslenkova LT, Yordanova RY et al. 2009. Exogenous treatment with salicylic acid attenuates cadmium toxicity in pea seedlings. *Plant Physiol Biochem* 47:224–231.

Potters G, Pasternak TP, Guisez Y et al. 2007. Stress-induced morphogenic responses: Growing out of trouble? *Trends Plant Sci* 12:98–105.

Potters G, Pasternak TP, Guisez Y et al. 2009. Different stresses, similar morphogenic responses: Integrating a plethora of pathways. *Plant Cell Environ* 32:158–169.

Pourrut B, Shahid M, Dumat C et al. 2011. Lead uptake, toxicity, and detoxification in plants. *Rev Environ Contam Toxicol* 213:113–136.

del Pozo T, Cambiazo V, Gonzalez M. 2010. Gene expression profiling analysis of copper homeostasis in *Arabidopsis thaliana. Biochem Biophys Res Commun* 393:248–252.

Puig S, Andres-Colas N, Garcia-Molina A et al. 2007. Copper and iron homeostasis in *Arabidopsis*: Responses to metal deficiencies, interactions and biotechnological applications. *Plant Cell Environ* 30:271–290.

Puig S, Peñarrubia L. 2009. Placing metal micronutrients in context: Transport and distribution in plants. *Curr Opin Plant Biol* 12:299–306.

Raskin I, Ensley BD. 2000. *Phytoremediation of Toxic Metals: Using Plants to Clean Up the Environment*. New York: John Wiley & Sons, Inc., pp. 53–70.

Ravet K, Reyt G, Arnaud N et al. 2012. Iron and ROS control of the DownSTream mRNA decay pathway is essential for plant fitness. *EMBO J* 31:175–186.

Remans T, Opdenakker K, Smeets K et al. 2010. Metal-specific and NADPH oxidase dependent changes in lipoxygenase and NADPH oxidase gene expression in *Arabidopsis thaliana* exposed to cadmium or excess copper. *Funct Plant Biol* 37:532–544.

Richard O, Pineau C, Loubet S et al. 2011. Diversity analysis of the response to Zn within the *Arabidopsis thaliana* species revealed a low contribution of Zn translocation to Zn tolerance and a new role for Zn in lateral root development. *Plant Cell Environ* 34:1065–1078.

Richau KH, Kozhevnikova AD, Seregin IV et al. 2009. Chelation by histidine inhibits the vacuolar sequestration of nickel in roots of the hyperaccumulator *Thlaspi caerulescens. New Phytol* 183:106–116.

Rivas-San Vicente M, Plasencia J. 2011. Salicylic acid beyond defence: Its role in plant growth and development. *J Exp Bot* 62:3321–3338.

Roberts LA, Pierson AJ, Panaviene Z et al. 2004. Yellow stripe1. Expanded roles for the maize iron-phytosiderophore transporter. *Plant Physiol* 135:112–120.

Rodriguez-Serrano M, Romero-Puertas MC, Zabalza A et al. 2006. Cadmium effect on oxidative metabolism of pea (*Pisum sativum* L.) roots. Imaging of reactive oxygen species and nitric oxide accumulation in vivo. *Plant Cell Environ* 29:1532–1544.

Rodriguez-Serrano M, Romero-Puertas MC, Pazmino DM et al. 2009. Cellular response of pea plants to cadmium toxicity: Cross talk between reactive oxygen species, nitric oxide, and calcium. *Plant Physiol* 150:229–243.

Rogers EE, Guerinot ML. 2002. FRD3, a member of the multidrug and toxin efflux family, controls iron deficiency responses in *Arabidopsis. Plant Cell* 14:1787–1799.

Romera FJ, Alcantara E. 2004. Ethylene involvement in the regulation of Fe-deficiency stress responses by Strategy I plants. *Funct Plant Biol* 31:315–328.

Romera FJ, Alcántara E, De La Guardia MD. 1992. Role of roots and shoots in the regulation of the Fe efficiency responses in sunflower and cucumber. *Physiol Plant* 85:141–146.

Römheld V, Marschner H. 1981. Iron deficiency stress induced morphological and physiological changes in root tips of sunflower. *Physiol Plant* 53:354–360.

Ruiz ON, Alvarez D, Torres C et al. 2011. Metallothionein expression in chloroplasts enhances mercury accumulation and phytoremediation capability. *Plant Biotechnol J* 9:609–617.

Sagardoy R, Morales F, Lopez-Millan AF et al. 2009. Effects of zinc toxicity on sugar beet (*Beta vulgaris* L.) plants grown in hydroponics. *Plant Biol* 11:339–350.

Sancenón V, Puig S, Mateu-Andres I et al. 2004. The *Arabidopsis* copper transporter COPT1 functions in root elongation and pollen development. *J Biol Chem* 279:15348–15355.

Sancenón V, Puig S, Mira H et al. 2003. Identification of a copper transporter family in *Arabidopsis thaliana. Plant Mol Biol* 51:577–587.

Schaaf G, Honsbein A, Meda AR et al. 2006. AtIREG2 encodes a tonoplast transport protein involved in iron-dependent nickel detoxification in *Arabidopsis thaliana* roots. *J Biol Chem* 281:25532–25540.

Schaaf G, Ludewig U, Erenoglu BE et al. 2004. ZmYS1 functions as a proton-coupled symporter for phytosiderophore- and nicotianamine-chelated metals. *J Biol Chem* 279:9091–9096.

Schat H, Llugany M, Vooijs R et al. 2002. The role of phytochelatins in constitutive and adaptive heavy metal tolerances in hyperaccumulator and non-hyperaccumulator metallophytes. *J Exp Bot* 53:2381–2392.

Schmidt W. 2008. Inner voices meet outer signals: The plasticity of rhizodermic cells. *Plant Sci* 174:239–245.

Schmidt W, Bartels M. 1996. Formation of root epidermal transfer cells in Plantago. *Plant Physiol* 110:217–225.

Schmidt W, Michalke W, Schikora A. 2003. Proton pumping by tomato roots. Effect of Fe deficiency and hormones on the activity and distribution of plasma membrane H^+-ATPase in rhizodermal cells. *Plant Cell Environ* 26:361–370.

Schmidt W, Schikora A. 2001. Different pathways are involved in phosphate and iron stress-induced alterations of root epidermal cell development. *Plant Physiol* 125:2078–2084.

Schmidt W, Tittel J, Schikora A. 2000. Role of hormones in the induction of iron deficiency responses in *Arabidopsis* roots. *Plant Physiol* 122:1109–1118.

Schmidt W, Yang TJW, Perry PJ et al. 2008. Manganese deficiency alters the patterning and development of root hairs in *Arabidopsis. J Exp Bot* 59:3453–3464.

Schützendübel A, Polle A. 2002. Plant responses to abiotic stresses: Heavy metal-induced oxidative stress and protection by mycorrhization. *J Exp Bot* 53:1351–1365.

Schwartz C, Morel JL, Saumier S et al. 1999. Root architecture of the zinc-hyperaccumulator plant *Thlaspi caerulescens* as affected by metal origin, content and localization in soil. *Plant Soil* 208:103–115.

Schwarz G, Mendel RR. 2006. Molybdenum cofactor biosynthesis and molybdenum enzymes. *Annu Rev Plant Biol* 57:623–647.

Seguela M, Briat JF, Vert G et al. 2008. Cytokinins negatively regulate the root iron uptake machinery in *Arabidopsis* through a growth-dependent pathway. *Plant J* 55:289–300.

Shahzad Z, Gosti F, Frérot H et al. 2010. The five AhMTP1 zinc transporters undergo different evolutionary fates towards adaptive evolution to zinc tolerance in *Arabidopsis halleri. PLoS Genet* 6:e1000911.

Shane MW, Lambers H. 2005. Cluster roots: A curiosity in context. *Plant Soil* 274:101–125.

Shanmugam V, Lo J-C, Wu C-L et al. 2011. Differential expression and regulation of iron-regulated metal transporters in *Arabidopsis halleri* and *Arabidopsis thaliana*—The role in zinc tolerance. *New Phytol* 190:125–137.

Shanmugam V, Tsednee M, Yeh K-C et al. 2012. ZINC TOLERANCE INDUCED BY IRON 1 reveals the importance of glutathione in the cross-homeostasis between zinc and iron in *Arabidopsis thaliana. Plant J* 69:1006–1017.

Sharma P, Dubey RS. 2005. Lead toxicity in plants. *Brazil J Plant Physiol* 17:35–52.

Sheldon AR, Menzies NW. 2005. The effect of copper toxicity on the growth and root morphology of Rhodes grass (*Chloris gayana* Knuth.) in resin buffered solution culture. *Plant Soil* 278:341–349.

Sheng X-F, Xia J-J, Jiang C-Y et al. 2008. Characterization of heavy metal-resistant endophytic bacteria from rape (*Brassica napus*) roots and their potential in promoting the growth and lead accumulation of rape. *Environ Pollut* 156:1164–1170.

Shin R, Berg RH, Schachtman DP. 2005. Reactive oxygen species and root hairs in *Arabidopsis* root response to nitrogen, phosphorus and potassium deficiency. *Plant Cell Physiol* 46:1350–1357.

Sinclair SA, Sherson SM, Jarvis R et al. 2007. The use of the zinc-fluorophore, Zinpyr-1, in the study of zinc homeostasis in *Arabidopsis* roots. *New Phytol* 174:39–45.

Smeets K, Ruytinx J, Semane B et al. 2008. Cadmium-induced transcriptional and enzymatic alterations related to oxidative stress. *Environ Exp Bot* 63:1–8.

Song WY, Choi KS, Kim DY et al. 2010a. *Arabidopsis* PCR2 is a zinc exporter involved in both zinc extrusion and long-distance zinc transport. *Plant Cell* 22:2237–2252.

Song WY, Park J, Mendoza-Cózatl DG et al. 2010b. Arsenic tolerance in *Arabidopsis* is mediated by two ABCC-type phytochelatin transporters. *Proc Natl Acad Sci U S A* 107:21187–21192.

Song WY, Sohn EJ, Martinoia E et al. 2003. Engineering tolerance and accumulation of lead and cadmium in transgenic plants. *Nat Biotechnol* 21:914–919.

Sunkar R, Chinnusamy V, Zhu J et al. 2007. Small RNAs as big players in plant abiotic stress responses and nutrient deprivation. *Trends Plant Sci* 12:301–309.

Sunkar R, Kapoor A, Zhu J-K. 2006. Posttranscriptional induction of two Cu/Zn superoxide dismutase genes in *Arabidopsis* is mediated by downregulation of miR398 and important for oxidative stress tolerance. *Plant Cell* 18:2051–2065.

Suzuki N. 2005. Alleviation by calcium of cadmium-induced root growth inhibition in *Arabidopsis* seedlings. *Plant Biotechnol* 22:19–25.

Taiz L, Zeiger E. 2010. *Plant Physiology*, 5th edn. Sunderland, MA: Sinauer Associates.

Takahashi R, Ishimaru Y, Senoura T et al. 2011. The OsNRAMP1 iron transporter is involved in Cd accumulation in rice. *J Exp Bot* 62:4843–4850.

Talke IN, Hanikenne M, Krämer U. 2006. Zinc-dependent global transcriptional control, transcriptional deregulation, and higher gene copy number for genes in metal homeostasis of the hyperaccumulator *Arabidopsis halleri. Plant Physiol* 142:148–167.

Tejada-Jimenez M, Galvan A, Fernandez E et al. 2009. Homeostasis of the micronutrients Ni, Mo and Cl with specific biochemical functions. *Curr Opin Plant Biol* 12:358–363.

Tewari RK, Kumar P, Sharma PN. 2008. Morphology and physiology of zinc-stressed mulberry plants. *Z Pflanzenernahr Bodenkd* 171:286–294.

Thomine S, Lelièvre F, Debarbieux E et al. 2003. AtNRAMP3, a multispecific vacuolar metal transporter involved in plant responses to iron deficiency. *Plant J* 34:685–695.

Thomine S, Wang R, Ward JM et al. 2000. Cadmium and iron transport by members of a plant metal transporter family in *Arabidopsis* with homology to Nramp genes. *Proc Natl Acad Sci U S A* 97:4991–4996.

Tomatsu H, Takano J, Takahashi H et al. 2007. An *Arabidopsis thaliana* high-affinity molybdate transporter required for efficient uptake of molybdate from soil. *Proc Natl Acad Sci U S A* 104:18807–18812.

Trampczynska A, Küpper H, Meyer-Klaucke W et al. 2010. Nicotianamine forms complexes with Zn(II) in vivo. *Metallomics* 2:57–66.

Ueno D, Milner MJ, Yamaji N et al. 2011. Elevated expression of TcHMA3 plays a key role in the extreme Cd tolerance in a Cd-hyperaccumulating ecotype of *Thlaspi caerulescens. Plant J* 66:852–862.

Ueno D, Yamaji N, Kono I et al. 2010. Gene limiting cadmium accumulation in rice. *Proc Natl Acad Sci U S A* 107:16500–16505.

Vacchina V, Mari S, Czernic P et al. 2003. Speciation of Nickel in a hyperaccumulating plant by high-performance liquid chromatography-inductively coupled plasma mass spectrometry and electrospray MS/MS assisted by cloning using yeast complementation. *Anal Chem* 75:2740–2745.

Válega M, Lima AIG, Figueira EMAP et al. 2009. Mercury intracellular partitioning and chelation in a salt marsh plant, *Halimione portulacoides* (L.) Aellen: Strategies underlying tolerance in environmental exposure. *Chemosphere* 74:530–536.

Van Hoof NA, Koevoets PLM, Hakvoort HWJ et al. 2001. Enhanced ATP-dependent copper efflux across the root cell plasma membrane in copper-tolerant *Silene vulgaris*. *Physiol Plant* 113:225–232.

Van de Mortel JE, Almar Villanueva L, Schat H et al. 2006. Large expression differences in genes for iron and zinc homeostasis, stress response, and lignin biosynthesis distinguish roots of *Arabidopsis thaliana* and the related metal hyperaccumulator *Thlaspi caerulescens*. *Plant Physiol* 142:1127–1147.

Van de Mortel JE, Schat H, Moerland PD et al. 2008. Expression differences for genes involved in lignin, glutathione and sulphate metabolism in response to cadmium in *Arabidopsis thaliana* and the related Zn/Cd-hyperaccumulator *Thlaspi caerulescens*. *Plant Cell Environ* 31:301–324.

Verbruggen N, Hermans C, Schat H. 2009a. Mechanisms to cope with arsenic or cadmium excess in plants. *Curr Opin Plant Biol* 12:364–372.

Verbruggen N, Hermans C, Schat H. 2009b. Molecular mechanisms of metal hyperaccumulation in plants. *New Phytol* 182:781–781.

Verbruggen N, Leduc D. 2009. Potential of plant genetic engineering for phytoremediation of toxic trace elements. In *Phytotechnologies Solutions for Sustainable Land Management*, ed. T Vanek, in Encyclopedia of Life Support Systems (EOLSS), Oxford, U.K.: EOLSS Publishers.

Verma S, Dubey RS. 2003. Lead toxicity induces lipid peroxidation and alters the activities of antioxidant enzymes in growing rice plants. *Plant Sci* 164:645–655.

Verret F, Gravot A, Auroy P et al. 2004. Overexpression of AtHMA4 enhances root- to-shoot translocation of zinc and cadmium and plant tolerance. *FEBS Lett* 576:306–312.

Vert G, Grotz N, Dedaldechamp F et al. 2002. IRT1, an *Arabidopsis* transporter essential for iron uptake from the soil and for plant growth. *Plant Cell* 14:1223–1233.

Veselov D, Kudoyarova G, Symonyan M et al. 2003. Effect of cadmium on ion uptake, transpiration and cytokinin content in wheat seedlings. *Bulg J Plant Physiol* 2003:353–359.

Voinnet O. 2009. Origin, biogenesis, and activity of plant microRNAs. *Cell* 136:669–687.

Von Wiren N, Marschner H, Romheld V. 1996. Roots of iron-efficient maize also absorb phytosiderophore-chelated zinc. *Plant Physiol* 111:1119–1125.

Wagner GJ, Sutton TG, Yeargan R. 1988. Root control of leaf cadmium accumulation in tobacco. *Tobacco Sci* 32:88–91.

Wang J, Ding H, Zhang A et al. 2010. A novel mitogen-activated protein kinase gene in maize (*Zea mays*), ZmMPK3, is involved in response to diverse environmental cues. *J Int Plant Biol* 52:442–452.

Ward JT, Lahner B, Yakubova E et al. 2008. The effect of iron on the primary root elongation of *Arabidopsis* during phosphate deficiency. *Plant Physiol* 147:1181–1191.

Waters BM, Blevins DG. 2000. Ethylene production, cluster root formation, and localization of iron(III) reducing capacity in Fe-deficient squash roots. *Plant Soil* 225:21–31.

Weber M, Harada E, Vess C et al. 2004. Comparative microarray analysis of *Arabidopsis thaliana* and *Arabidopsis halleri* roots identifies nicotianamine synthase, a ZIP transporter and other genes as potential metal hyperaccumulation factors. *Plant J* 37:269–281.

Werner T, Nehnevajova E, Kollmer I et al. 2010. Root-specific reduction of cytokinin causes enhanced root growth, drought tolerance, and leaf mineral enrichment in *Arabidopsis* and tobacco. *Plant Cell* 22:3905–3920.

Weyens N, van der Lelie D, Taghavi S et al. 2009. Exploiting plant-microbe partnerships to improve biomass production and remediation. *Trends Biotechnol* 27:591–598.

White PJ, Broadley MR. 2005. Biofortifying crops with essential mineral elements. *Trends Plant Sci* 10:1360–1385.

White PJ, Broadley MR. 2011. Physiological limits to zinc biofortification of edible crops. *Front Plant Sci* 2:80.

White JG, Zasoski RJ. 1999. Mapping soil micronutrients. *Field Crop Res* 60:11–26.

Whiting SN, Leake JR, McGrath SP et al. 2000. Positive responses to Zn and Cd by roots of the hyperaccumulator *Thlaspi caerulescens*. *New Phytol* 145:199–210.

Widodo JA, Broadley MR, Rose T et al. 2010. Response to zinc deficiency of two rice lines with contrasting tolerance is determined by root growth maintenance and organic acid exudation rates, and not by zinc-transporter activity. *New Phytol* 186:400–414.

Wightwick AM, Mollah MR, Partington DL et al. 2008. Copper fungicide residues in Australian vineyard soils. *J Agric Food Chem* 56:2457–2464.

Willems G, Drager DB, Courbot M et al. 2007. The genetic basis of zinc tolerance in the metallophyte *Arabidopsis halleri* ssp. halleri (Brassicaceae): An analysis of quantitative trait loci. *Genetics* 176:659–674.

Willems G, Frérot H, Gennen J et al. 2010. Quantitative trait loci analysis of mineral element concentrations in an *Arabidopsis halleri—Arabidopsis lyrata petraea* F2 progeny grown on cadmium contaminated soil. *New Phytol* 187:368–379.

Williams LE, Pittman JK. 2010. Dissecting pathways involved in manganese homeostasis and stress in higher plant cells. In *Cell Biology of Metal and Nutrients*, eds. R Hell, R-R Mendel, Plant Cell Monographs, Vol. 17, pp. 17–281. Berlin, Germany: Springer.

Wintz H, Fox T, Wu YY et al. 2003. Expression profiles of *Arabidopsis thaliana* in mineral deficiencies reveal novel transporters involved in metal homeostasis. *J Biol Chem* 278:47644–47653.

Wong MH, Bradshaw AD. 1982. A comparison of the toxicity of heavy metals, using root elongation of rye grass, *Lolium perenne. New Phytol* 91:255–261.

Wong CKE, Cobbett CS. 2009. HMA P-type ATPases are the major mechanism for root-to-shoot Cd translocation in *Arabidopsis thaliana. New Phytol* 181:71–78.

Woo E-J, Marshall J, Bauly J et al. 2002. Crystal structure of auxin-binding protein 1 in complex with auxin. *EMBO J* 21:2877–2885.

Wu J, Wang C, Zheng L et al. 2011. Ethylene is involved in the regulation of iron homeostasis by regulating the expression of iron-acquisition-related genes in *Oryza sativa. J Exp Bot* 62:667–674.

Wu T, Zhang H-T, Wang Y et al. 2012. Induction of root Fe(III) reductase activity and proton extrusion by iron deficiency is mediated by auxin-based systemic signalling in *Malus xiaojinensis. J Exp Bot* 63:859–870.

Xing J, Jiang R, Ueno D et al. 2008a. Variation on root-to-shoot translocation of cadmium and zinc among different accessions of the hyperaccumulators *Thlaspi caerulescens* and *Thlaspi praecox. New Phytol* 178:315–325.

Xing CH, Zhu MH, Cai MZ et al. 2008b. Developmental characteristics and response to iron toxicity of root border cells in rice seedlings. *J Zhejiang Univ Sci B* 9:261–264.

Xiong J, An L, Lu H et al. 2009a. Exogenous nitric oxide enhances cadmium tolerance of rice by increasing pectin and hemicellulose contents in root cell wall. *Planta* 230:755–765.

Xiong J, Lu H, Lu K et al. 2009b. Cadmium decreases crown root number by decreasing endogenous nitric oxide, which is indispensable for crown root primordia initiation in rice seedlings. *Planta* 230:599–610.

Xu J, Wang WY, Sun JH et al. 2011. Involvement of auxin and nitric oxide in plant Cd-stress responses. *Plant Soil* 346:107–119.

Xu J, Yin H, Li Y et al. 2010. Nitric oxide is associated with long-term zinc tolerance in *Solanum nigrum. Plant Physiol* 154:1319–1334.

Yamasaki H, Abdel-Ghany SE, Cohu CM et al. 2007. Regulation of copper homeostasis by micro-RNA in *Arabidopsis. J Biol Chem* 282:16369–16378.

Yamasaki H, Hayashi M, Fukazawa M et al. 2009. SQUAMOSA promoter binding protein-like7 is a central regulator for copper homeostasis in *Arabidopsis. Plant Cell* 21:347–361.

Yang M, Ding GD, Shi L et al. 2010. Quantitative trait loci for root morphology in response to low phosphorus stress in *Brassica napus. Theor Appl Gen* 121:181–193.

Yang X, Li T, Yang J et al. 2006. Zinc compartmentation in root, transport into xylem, and absorption into leaf cells in the hyperaccumulating species of *Sedum alfredii* Hance. *Planta* 224:185–195.

Yang TJ, Perry PJ, Ciani S et al. 2008. Manganese deficiency alters the patterning and development of root hairs in *Arabidopsis. J Exp Bot* 59:3453–3464.

Yang Y, Sun C, Yao Y, Zhang Y, Achal V. 2011. Growth and physiological responses of grape (*Vitis vinifera* "Combier") to excess zinc. *Acta Physiol Plant* 33:1483–1491.

Ye ZH, Baker AJM, Wong MH et al. 1997. Copper and nickel uptake, accumulation and tolerance in *Typha latifolia* with and without iron plaque on the root surface. *New Phytol* 136:481–488.

Yeh C-M, Chien P-S, Huang H-J. 2007. Distinct signalling pathways for induction of MAP kinase activities by cadmium and copper in rice roots. *J Exp Bot* 58:659–671.

Yeh C-M, Hung WC, Huang HJ. 2003. Copper treatment activates mitogen-activated protein kinase signalling in rice. *Physiol Plant* 119:392–399.

Yuan Y, Wu H, Wang N et al. 2008. FIT interacts with AtbHLH38 and AtbHLH39 in regulating iron uptake gene expression for iron homeostasis in *Arabidopsis. Cell Res* 18:385–397.

Yusuf M, Fariduddin Q, Hayat S et al. 2011. Nickel: An overview of uptake, essentiality and toxicity in plants. *Bull Environ Contam Toxicol* 86:1–17.

Zaïd EH, Arahou M, Diem HG et al. 2003. Is Fe deficiency rather than P deficiency the cause of cluster root formation in *Casuarina* species? *Plant Soil* 248:229–235.

Zelko I, Lux A, Czibula K. 2008. Difference in the root structure of hyperaccumulator *Thlaspi caerulescens* and non-hyperaccumulator *Thlaspi arvense. Int J Environ Pollut* 33:123–132.

Zengin FK. 2006. The effects of Co^{2+} and Zn^{2+} on the contents of protein, abscisic acid, proline and chlorophyll in bean (*Phaseolus vulgaris* cv. Strike) seedlings. *J Environ Biol* 27:441–448.

Zhang HM, Sun Y, Xie XT et al. 2009. A soil bacterium regulates plant acquisition of iron via deficiency-inducible mechanisms. *Plant J* 58:568–577.

Zhang H, Xia Y, Wang G et al. 2008. Excess copper induces accumulation of hydrogen peroxide and increases lipid peroxidation and total activity of copper-zinc superoxide dismutase in roots of *Elsholtzia haichowensis. Planta* 227:465–475.

Zhao FJ, McGrath SP. 2009. Biofortification and phytoremediation. *Curr Opin Plant Biol* 12:373–380.

Zhu C, Ding Y, Liu H. 2011. MiR398 and plant stress responses. *Physiol Plant* 143:1–9.

35

Maintaining Root Growth in Drying Soil: A Review of Progress and Gaps in Understanding

Eric S. Ober
Rothamsted Research

Robert E. Sharp
University of Missouri

I. Introduction

When there is little rainfall and upper soil layers are depleted of moisture, plants rely on the ability of root systems to proliferate throughout the soil profile to extract water. The patterns of root growth and soil drying are spatially heterogeneous, such that regions of the soil are dried unevenly, leaving patches or layers of moisture not yet explored by roots (Hodge et al. 2009; White and Kirkegaard 2010; Kano et al. 2011; Miyazawa et al. 2011). In some circumstances, to reach moisture, roots must pass through soil that is already dry because other roots—perhaps of neighboring plants—have previously extracted the soil water. In other circumstances, seed sown in dry soil germinates when surface layers are wetted, but seedling roots must penetrate through dry soil to find moisture in deeper layers. In such situations, roots must be able to continue growing through a soil matrix that is often at water potentials (ψ_w) that may be inhibitory to growth. For example, under drought conditions, the nodal root axes of maize (which are produced from the stem nodes) are challenged to grow into and through surface soil that may have become very dry. In these circumstances, the water for continued root growth can be supplied to the root tip via the phloem (Boyer et al. 2010). In addition, it was demonstrated that maize nodal roots are able to continue growing at tissue ψ_w lower than those that are inhibitory for the growth of leaves, stem, and reproductive structures (Figure 35.1; Sharp and Davies 1979; Westgate and Boyer 1985). Similarly, the primary root of several crop species is able to continue growing at low soil ψ_w that completely inhibits shoot growth (Sharp et al. 1988; Spollen et al. 1993; Yamaguchi et al. 2010). These findings indicate some form of internal regulation within the root growth zone that allows the maintenance of cell elongation under these conditions. The mechanisms underlying primary root growth maintenance at low ψ_w have been studied extensively, as reviewed previously (Sharp et al. 2004; Ober and Sharp 2007; Yamaguchi and Sharp 2010). However, large gaps remain in our understanding of how the growth of roots in response to low ψ_w is controlled and how the molecular physiology that regulates root growth and development at high ψ_w is modulated to enable continuation of growth when water is limiting.

In this chapter, we briefly review some of the important physiological features involved in root growth maintenance at low ψ_w, focusing on studies of the maize (*Zea mays* L.) primary root as a model system. We also speculate how putative key control factors of root growth and development under well-watered conditions may play a role in determining the response of root growth in drying soil. The growth and function of the entire root system also depend on the initiation and growth of lateral roots and root hairs, which are discussed elsewhere (Ingram and Malamy 2010; De Smet et al. 2012; Dubrovsky and Forde 2012; see also Chapters 4 and 6).

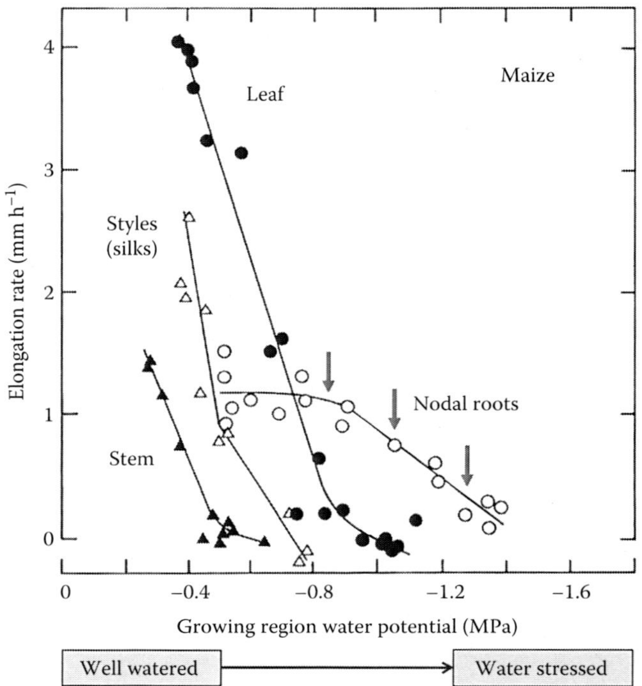

FIGURE 35.1 Elongation rate in different organs of maize as a function of the ψ_w of the tissues in which growth occurs, during the course of several days of soil drying. As soil water availability decreases, elongation rate decreases, but the sensitivity of elongation to the decline in tissue ψ_w differs depending on the organ. Note that nodal root elongation (arrows) continues at ψ_w that are completely inhibitory to growth of the leaves, stem, and silks. Therefore, the differential growth of root and shoot tissues is not due to variations in water availability to the different organs, but to the way the physiology of the cells is altered in response to low ψ_w. (Redrawn from Westgate, M.E. and Boyer, J.S., *Planta*, 164, 540, 1985.)

II. Root Elongation in the Soil Matrix

The bulk of water and nutrient uptake along root axes occurs in mature root tissue (mainly via the subtending lateral roots) some distance from the root tip (Melchior and Steudle 1993; Varney and Canny 1993); however, the placement of this mature tissue depends on the prior growth of the root tip through this region of the soil. Furthermore, uptake of water and nutrients diminishes with root age, so an effective root system depends on the continued generation of new root tissue. The elongation of roots depends on the balance between cell proliferation and differentiation (Tsukagoshi et al. 2010) and, in expanding cells, on the balance between cell wall loosening and wall stiffening (Cosgrove 2005; Boyer 2009). The key processes that determine the production of cells are the rate of cell division and the size of the meristem, which is determined by the number of cell divisions before a cell exits the meristem (Beemster and Baskin 2000; see also Chapter 3). Cell elongation entails extension of the cell wall matrix, deposition of new material and orientation, followed by the eventual stiffening of the cell wall that brings growth to an end (Fan et al. 2006; Boyer 2009).

A complex interplay of factors controls these processes, all occurring within a brief frame of time and space. For example, in the maize primary root growing in the absence of water deficit (high ψ_w), a newly formed cell is displaced from the base of the meristem to the base of the elongation zone, a distance of approximately 10 mm, in about 8 h (Sharp et al. 1988). During this time, the relative elongation rate of the cell greatly accelerates and then decelerates (Figure 35.2). The integrated elongation of the cells within the growth zone simultaneously pushes the root tip through the soil. When cell elongation ceases, the tissue is firmly fixed at a point in the soil matrix: the flux of water and nutrients and associations with rhizosphere microflora all occur at this local spot until the root dies. However, it is the molecular, biochemical, and biophysical changes that occur during the brief spatial and temporal passage of cells from the meristem through the growth zone that ultimately determines the growth of the entire root system. Accordingly, the physiology of this small volume of tissue deserves close scrutiny. Cell production was reported to be fairly robust in maize and soybean primary roots even under severe water deficit (Saab et al. 1992; Yamaguchi et al. 2010 [although a contrasting finding in maize was reported by Fraser et al. 1990]). In contrast, cell elongation rates are spatially modulated in water-stressed roots, as described in the following section.

A. Experimental Approaches Highlight Focal Points of Control

A great deal about root growth regulation has been learned by employing the analytical power of growth kinematics, a method that quantifies the spatial and temporal distribution of expansion rates within an organ (Silk 1984; Silk 2002; van der Weele et al. 2003; Wuyts et al. 2011). For example, in the maize primary root, kinematic analysis identified contiguous regions within the growth zone with distinct responses of cell elongation to low ψ_w (Figure 35.2; Sharp et al. 1988; Liang et al. 1997). Remarkably, local elongation rates are fully maintained in the early ontogenetic phases of growth (apical few mm) even under severe water deficits. However, deceleration and cessation of elongation occur closer to the apex than in well-watered roots, resulting in a shortened growth zone. Not surprisingly, different physiological mechanisms have been shown to underlie the spatially distinct growth responses to water stress (reviewed by Sharp et al. 2004; Ober and Sharp 2007). In particular, cell wall extensibility is enhanced in the apical region but decreased in the basal region of severely water-stressed roots (Wu et al. 1996). The enhancement of extensibility in the apical region is considered to be important to maintain cell elongation despite reduced turgor pressures resulting from incomplete osmotic adjustment (Sharp et al. 1990; Spollen and Sharp 1991). Recently, the kinematic approach was combined with transcriptomic and proteomic analyses, further revealing the complexity and coordination of processes involved in maize and soybean primary root growth regulation under water deficits (Poroyko et al. 2005, 2007; Zhu et al. 2007; Spollen et al. 2008; Yamaguchi et al. 2010;

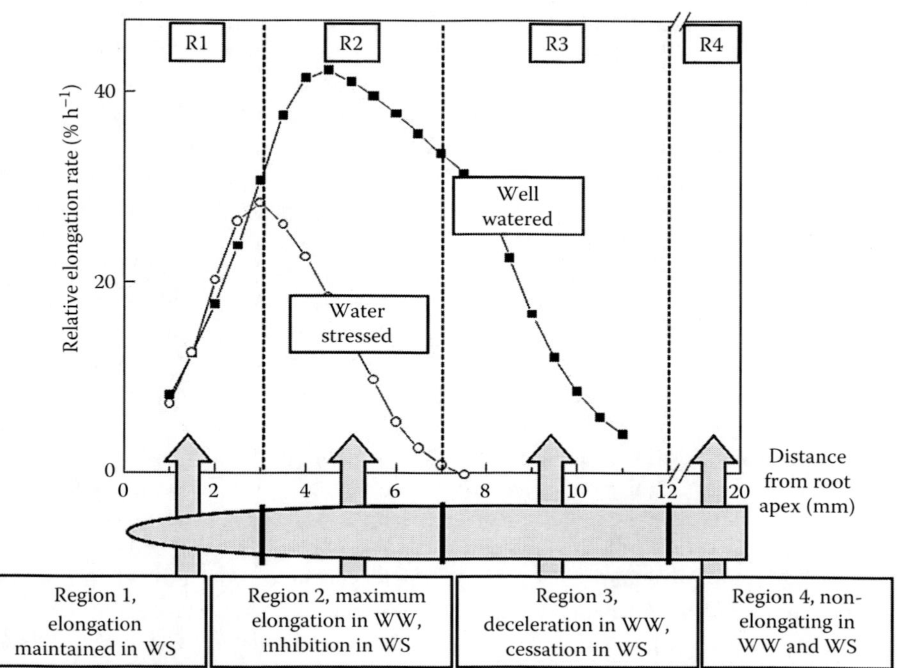

FIGURE 35.2 Relative elongation rate as a function of distance from the apex of the maize primary root growing under well-watered (WW, ψ_w of −0.03 MPa; filled symbols) or water-stressed (WS, ψ_w of −1.6 MPa; open symbols) conditions. Note that in Region 1 (R1), the local elongation rate in water-stressed roots is maintained at the well-watered rate even though turgor has decreased (Spollen and Sharp 1991). In Region 2 (R2), the high rate of elongation that is attained at high ψ_w is not achieved at low ψ_w because growth begins to decelerate prematurely. In Region 3 (R3), elongation rate at high ψ_w begins to decelerate as cell wall extensibility decreases and wall stiffening increases; at low ψ_w, cells in this region have already ceased growth. In Region 4 (R4), cells are no longer capable of elongation in either environment. Understanding the molecular physiology of the differential growth response to low ψ_w in these different regions provides insights into how cells sense and respond to the availability of water in the soil. (Modified from Liang, B.M. et al., *Plant Physiol.*, 115, 101, 1997.)

reviewed by Yamaguchi and Sharp 2010). The ability to analyze gene expression in specific root cell types (Birnbaum et al. 2003; Dinneny et al. 2008; see Chapter 2), and to image expression patterns using fluorescent tags and confocal microscopy, further enhances the spatial details that can be obtained in order to gain an understanding of the role of specific genes in growth regulation (Laplaze et al. 2005).

The pathways controlling root growth and development may indeed be highly complex, with redundancies, overlaps, and feedback loops, and an understanding of these networks is facilitated by using a systems approach (Moreno-Risueno et al. 2010; Lucas et al. 2011) as illustrated by the use of artificial neural network models (Gago et al. 2010). As growth control models become more sophisticated, they can incorporate physical connections (plasmodesmata) and mechanical stresses within tissues. Alternatively, the fundamental in vivo processes could be simpler than it appears from the mass of disparate reports in the literature, and the core players might become apparent once the myriad of experimental manipulations are distilled (Santner and Estelle 2009; Jaillais and Chory 2010; Ross et al. 2011).

A vital question that needs to be addressed is to what extent these control systems change when a root encounters dry soil. Is the fundamental structure of the regulatory network similar at both high and low ψ_w, only modulated in time and space within the root apex, or is there a qualitative change, bringing new factors into play that become increasingly pertinent as water deficit increases? The next sections include an examination of some recent findings in the regulation of root growth and development that may apply to roots growing at low ψ_w.

B. Hormonal Regulation of Root Growth

Although hormones are likely to play important roles in root growth regulation at low ψ_w, the roles of most of these compounds have not been elucidated. Abscisic acid (ABA) is often viewed as an inhibitor of growth, but this conclusion is mainly drawn from experiments in which high concentrations of exogenous ABA caused nonphysiological levels to accumulate in tissues (Trewavas and Jones 1991). In contrast, it was shown that endogenous accumulation of ABA, which occurs in most tissues under water-stressed conditions, is required for the maintenance of cell elongation in the apical region of the maize primary root growth zone at low ψ_w. Blocking the accumulation of ABA via either chemical agents or genetic lesions in the ABA biosynthetic pathway led to severe root growth inhibition (Saab et al. 1990, 1992). Exogenous ABA added back to the root to restore normal endogenous levels restored root growth rate (Sharp et al. 1994). It was further shown that in the absence of normal levels of endogenous ABA, ethylene production increased in maize primary root tissues during water deficit, leading to growth inhibition (Spollen et al. 2000). Thus, one way in which ABA helps to maintain root growth at low ψ_w is by keeping ethylene in check (reviewed by Sharp 2002). ABA also plays roles in regulating osmotic adjustment and redox balance in the root growth zone via proline accumulation (Ober and Sharp 1994; Sharma et al. 2011), in electrophysiological responses to low ψ_w including set points for ion homeostasis (Ober and Sharp 2003), and in cell wall extensibility (Wu et al. 1994). There is also some evidence that ABA accumulation can help maintain cell production rate in the maize primary root under water-stressed conditions (Saab et al. 1992).

In addition to ABA and ethylene, the concentrations of and sensitivity to other growth-regulating hormones are probably also modulated at low ψ_w. Moreover, hormonal interactions are the rule rather than the exception. For example, under normal conditions, interactions between auxin and cytokinins regulate the balance between cell proliferation and differentiation and affect the dimensions of the root meristem and elongation zone (Dello Ioio et al. 2008; Perilli et al. 2012). Cytokinins are negative regulators of root meristem activity and affect root growth by regulating the exit of cells from the meristem. Increased metabolism of cytokinins by overexpressing a cytokinin oxidase (CKX1), driven by a root-specific promoter to avoid deleterious effects on shoot growth, increased meristem size and promoted root growth rate, resulting in improved survival following severe water deficit (Werner et al. 2010). There is also evidence that the cytokinin status of root tissue can affect root responses to ABA (Nishiyama et al. 2011). This is by no means an exhaustive catalogue of the many reports describing how manipulation of hormone levels, ratios, and gradients affect root growth. There is still much to be learned about how the chemical signaling system that regulates growth in well-watered conditions is altered to maintain growth and development of roots when soil moisture availability becomes limiting.

C. Are DELLA Proteins a Central Integrator of Root Growth Regulation?

A common focal point of hormone action appears to be the DELLA protein family (Weiss and Ori 2007; Harberd et al. 2009; Santner and Estelle 2009). DELLA proteins are members of the GRAS family of transcription factors and act to restrain growth until they are degraded in a gibberellin (GA)-dependent fashion in proteasomes as part of a GA-receptor-ubiquitin complex (Harberd et al. 2009). Downstream, the DELLA protein RGA binds to the promoter of another GRAS transcription factor SCARECROW-LIKE 3 (SCL3), which is mainly localized to newly differentiated endodermal cells entering the elongation zone (Heo et al. 2011). SCL3 and DELLA proteins antagonize each other's action, resulting in a homeostatic control of growth and GA biosynthesis (Zhang et al. 2011). It has been shown that it is the endodermis that establishes the rate of root growth: expression of mutant DELLA protein resistant to GA degradation inhibits root growth only if expressed in cells of the endodermis, whereas expression in other cells results in uncontrolled cell expansion (Ubeda-Tomas et al. 2008). Normal root growth depends on the coordinated expansion of all cells within the tissue, but it appears that cells of the endodermis are the linchpin. This is important because any factors that restrict root growth at low ψ_w could be expressed solely in the endodermis and absent in other tissues.

Ethylene appears to stabilize DELLAs against GA-induced degradation (Achard et al. 2006), which may explain at least part of the growth-inhibitory role of ethylene in roots at low ψ_w. Ethylene also appears to inhibit root growth by stimulating auxin synthesis and its rootward transport (Růžička et al. 2007; Swarup et al. 2007). However, increased auxin also stimulates GA and the subsequent DELLA degradation (Perilli et al. 2012). These antagonistic effects are one example of feedback control that results in homeostasis of GA synthesis and action. It has been suggested that exogenous ABA stabilizes DELLA proteins against GA-mediated degradation, leading to growth inhibition (Achard et al. 2006). However, this could be explained by ethylene-induced DELLA stabilization, because excessive exogenous ABA can stimulate ethylene synthesis (Beaudoin et al. 2000). To our knowledge, the role of DELLA proteins in root growth responses to water deficit has not yet been explicitly examined.

D. Modulation of Cell Production, Cell Wall Extensibility, and Root Growth by Reactive Oxygen Species

Further downstream, modulation of ROS levels by DELLA proteins could be the way by which DELLAs eventually affect growth (Achard and Genschik 2009). There is evidence that ROS are involved in regulating the transition from cell proliferation to differentiation as well as cell wall extensibility (Marino et al. 2012). Increased expression of peroxidase via mutation of the UPBEAT1 transcription factor resulted in decreased levels of H_2O_2, leading to greater rates of cell production as a consequence of increased meristem size (Tsukagoshi et al. 2010). In addition, there was increased expression of NADPH oxidases in the *upb1* mutant, causing elevated levels of superoxide ($O_2\cdot^-$). It was proposed that $O_2\cdot^-$ stimulates cell proliferation, while H_2O_2 stimulates cell differentiation, and that the intersection of opposing gradients of these ROS within the root apex determines the point where cells stop dividing and begin to differentiate (Tsukagoshi et al. 2010). Any cellular or environmental factor that alters these ROS levels and shifts this developmental transition would have an effect on cell production rate and therefore root growth.

The effects of ROS on wall rheological properties have been studied extensively in tip-growing systems such as pollen tubes, root hairs, and fungal hyphae, and many of the processes that are understood to regulate growth in these unicellular systems probably also can be applied to the multicellular root growth zone. There is evidence that ROS may play a role in modulating cell wall extension properties in the maize primary root (Liszkay et al. 2004; Zhu et al. 2007). Analysis of the transcriptome (Spollen et al. 2008) and the cell wall proteome (Zhu et al. 2007) highlights the upregulation

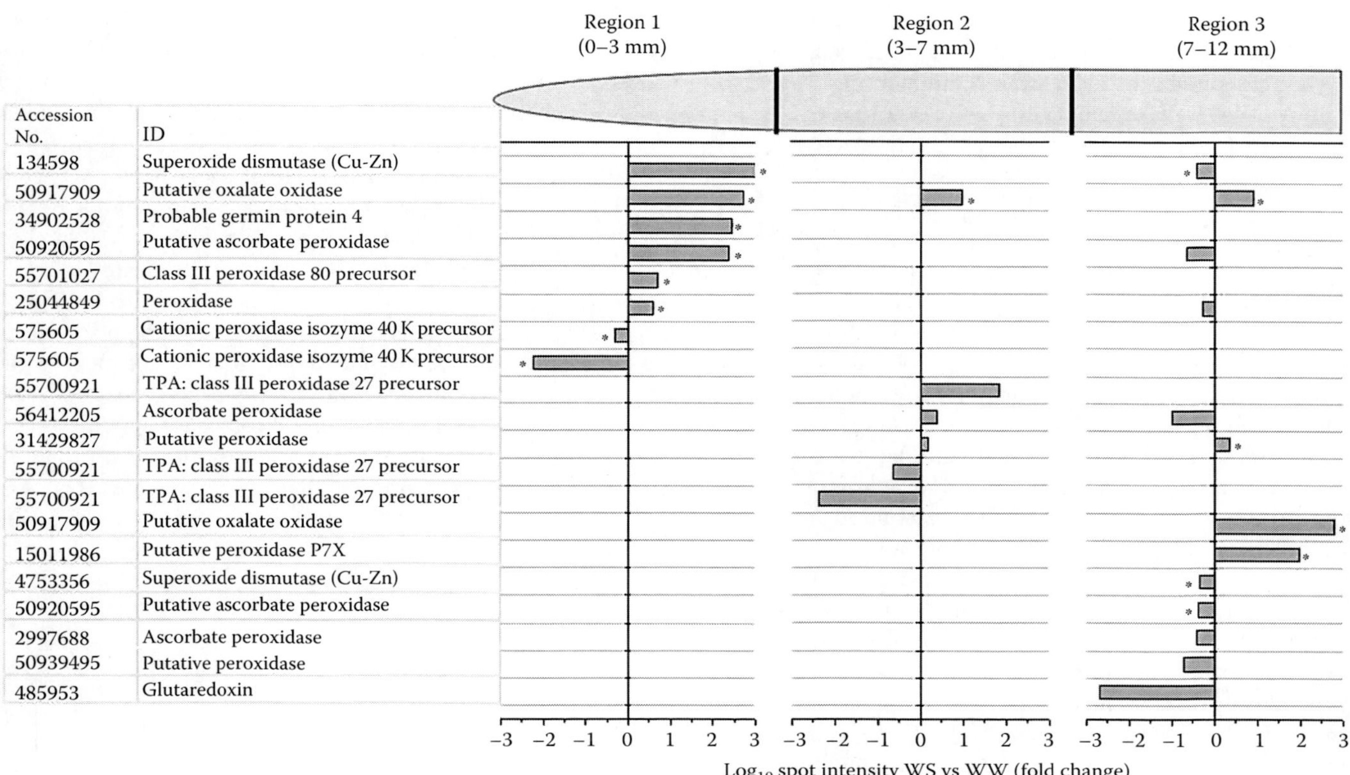

FIGURE 35.3 Cell wall proteins in the category of ROS metabolism that showed increased or decreased abundance in different regions of the maize primary root growth zone under water-stressed (WS, ψ_w of −1.6 MPa) compared with well-watered (WW, ψ_w of −0.03 MPa) conditions. This is one of five functional categories that in total comprised 152 cell wall proteins that showed primarily region-specific changes in abundance in response to water deficit. The changes that are marked with an asterisk are considered to be independent of developmental changes associated with the stress-induced shortening of the growth zone (Figure 35.2) and, therefore, likely to be true responses to water stress. (Modified from Zhu, J.M. et al., *Plant Physiol.*, 145, 1533, 2007.)

of a large number of ROS-related proteins and genes in response to water deficit. Prominent among these changes is the upregulation of oxalate oxidase (germin) in the apical region of the growth zone where cell elongation is maintained at low ψ_w (Figures 35.2 and 35.3). This enzyme family catalyzes the production of H_2O_2 from oxalate, and H_2O_2 can be metabolized to form hydroxyl radicals ($OH^•$) via the Fenton reaction. $OH^•$ may play an important role in altering the extensibility of the cell wall by cleavage of load-bearing wall polymer tethers (Fry 1998; Liszkay et al. 2004). ABA may be required to maintain an appropriate balance of ROS production and detoxification, in part via its control of ethylene, as uncontrolled ethylene production in roots of the ABA-deficient maize mutant *vp14* at low ψ_w leads to excess ROS and growth inhibition (Smith and Sharp, unpublished results). ABA has also been shown to increase expression of genes for antioxidant enzymes, for example, catalase in maize leaves (Guan et al. 2000).

E. Other Effects on Cell Wall Properties at Low ψ_w

In addition to ROS, there are examples of other factors that play important roles in modulating cell wall extensibility in roots at low ψ_w. Xyloglucan endotransglycosylase/hydrolases (XTH) and expansins are two key wall-modifying proteins that are upregulated at low ψ_w in the apical region of the maize primary root growth zone (Wu et al. 1994, 1996; Wu and Cosgrove 2000; Spollen et al. 2008). There is evidence that the response of XTH (Wu et al. 1994) but not of expansins (Wu et al. 2001) is regulated by ABA. In *Arabidopsis*, it has been shown that ABA can also affect wall chemistry during the transition from cell proliferation to cell differentiation, which is associated with a shift in relative abundance of LM5 galactan and LM6 arabinan epitopes (Talboys et al. 2011). Low concentrations (50 nM) of exogenous ABA, which increased root length and meristem size, increased the abundance of LM6 arabinan epitopes in accordance with delayed transition to differentiation. The response was blocked in the *abi4* ABA response mutant, but could not be replicated by treatment with the ethylene precursor ACC, suggesting the effect is independent of ethylene. Another ABA-regulated protein, HAP3b, enhanced cell elongation rates as well as rates of cell production when overexpressed in *Arabidopsis* root tissue (Ballif et al. 2011). In rice, transcriptomic analysis of the apical cm of roots highlighted that one upregulated gene under water deficit was a xyloglucan fucosyltransferase (most cell wall-related genes were downregulated) and expression was greater in drought-tolerant than in drought-susceptible lines

(Moumeni et al. 2011). Xyloglucan fucosyltransferase activity was associated with root growth in *Arabidopsis*, suggesting that fucose residues on the xyloglucan chain may affect the susceptibility of cell walls to remodeling by XTH (Perrin et al. 2003).

F. Perception of Low ψ_w and Intracellular Signaling

Another important element to consider in root growth in dry soil is how low ψ_w is sensed by the root and how this signal is transduced throughout the tissue. In this process, the plasma membrane and Ca^{2+} feature as key players. A decrease in ψ_w induces changes in root cell membrane potential that are maintained under steady-state conditions, long after the initial stimulus disappears (Ober and Sharp 2003). Interestingly, ABA, which is required for the maintenance of maize primary root growth at low ψ_w (see earlier text), is also required for regulation of the cell membrane potential and homeostatic set points for ions (Shabala and Lew 2002; Ober and Sharp 2003; MacRobbie 2006). Ion transport may be important for establishing growth patterns, as fluxes of Ca^{2+}, K^+, and H^+ occur in spatial association with rates of cell expansion within the root growth zone. A net flux of current enters the root near the tip and leaves the root near the base of the growth zone (Miller and Gow 1989; Fromm et al. 1997; Kiegle et al. 2000; Newman 2001). Current flux densities are greater in faster than in slower growing roots. The net influx of Ca^{2+}, which is greater in the growth zone than in the mature root, is mediated by channels that are sensitive to ROS and depend on NADPH oxidase activity (Demidchik et al. 2007; Mittler et al. 2011). Changes in ion transport could be a primary mechanism signaling changes in soil ψ_w. However, it is not certain if these flux patterns help determine the spatial dimensions of the growth zone or are a consequence of where growth is occurring. In the apoplast, Ca^{2+} binds to pectin, which affects the degree to which load-bearing wall polymers yield to turgor pressure (Boyer 2009). Ca^{2+} is also an important intracellular signal: Ca^{2+} regulates the activity of calcium-dependent kinases and other enzymes that affect growth (for details, see Chapter 20).

In addition to Ca^{2+}, it is clear that cell wall pH and its regulation play an important role in wall extensibility through direct effects on acidification of the wall matrix and on wall-loosening or wall-tightening proteins that have specific pH optima, such as expansins (Wu and Cosgrove 2000; Fan and Neumann 2004). Recent evidence suggests that Ca^{2+}-regulated proton flux into the apoplast helps establish the auxin gradient through the root tip (Monshausen et al. 2011), providing a clear example of the complexity and coordination of interacting factors involved in root growth regulation.

III. Root Growth in Drying Soil in the Field

In comparison with the amount of information on seedling root growth in laboratory experiments, there is considerably less information on how root growth responds to low soil ψ_w in the field. Due to the complexities of the field environment, experimentation is challenging, and extrapolation of laboratory results to the field must be done with caution. For instance, many soils also become hard as they dry; therefore, roots encounter both water stress and penetration resistance (Whitmore and Whalley 2009; Bengough et al. 2011; see also Chapter 37). Crop genetic improvement through manipulation of the root system will likely not come about through selections made in the field because of the effort required to do this on a large scale. However, there is renewed effort in finding ways to determine the phenotype of large populations at the seedling or young plant stages in order to uncover genotypic variation that can be observed in the field. There is some evidence to suggest that root behavior observed at the seedling stage under controlled conditions can indicate growth of mature root systems in the field. In Figure 35.4, an example is provided illustrating that primary root growth of sugar beet seedlings at high and low ψ_w can be correlated with the ability of the hybrids to yield relatively well under

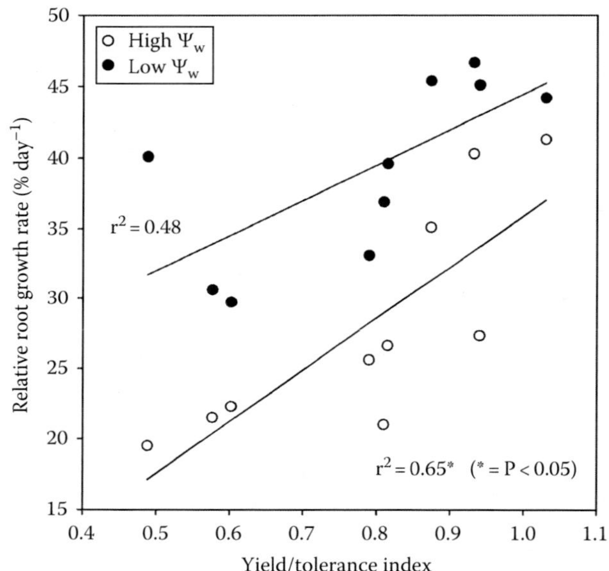

FIGURE 35.4 The correlation between sugar beet (*Beta vulgaris* L.) seedling relative root elongation rates at high (−0.03 MPa) and low (−1.0 MPa) ψ_w and the across-environment yield stability (Y/T index) measured under field conditions in mature plants. Symbols represent the mean values for different sugar beet hybrids. Seedlings were grown in vermiculite in the dark under nontranspiring conditions (to avoid confounding effects of shoot transpiration), and root lengths were measured at 4 days after transplanting germinated seeds. Under these conditions, root elongation rate was greater at low ψ_w than at high ψ_w because competition with shoot growth was greater at high ψ_w. The same genotypes were planted in the field and grown under fully irrigated or managed drought conditions (see Ober et al. 2004). The Y/T index (Fernandez 1992; Ober et al. 2004) combines relative yields under drought and relative yield potential, computed on the basis of sugar yield, which is highly correlated with total root biomass. The positive relationships suggest that genotypic rankings based on seedling screens can be used to identify vigorous, broadly adapted hybrids and that heritability is high (genotypic rankings are largely similar under low and high ψ_w).

both irrigated and drought conditions, reflecting broad adaptation and yield stability. The genotypes showed differential sensitivity of root and shoot growth to low ψ_w at the seedling stage, but this did not relate to field performance. The seedling test was most effective at high ψ_w as a screen for seedling vigor and yield potential in mature plants.

An alternative approach is to identify genotypes that have known differences in root growth responses to dry conditions in the field, then use these lines to devise a test that can be employed in controlled conditions on seedlings that will expose these genotypic differences. There is field evidence of altered rooting patterns in response to soil moisture supply. Genotypic differences in these patterns suggest that they are under genetic control (Hurd 1968; Sponchiado et al. 1989; Courtois et al. 2009). For instance, lines that exhibit greater proliferation of roots in deep soil layers can have a beneficial effect on crop performance (Figure 35.5; Kirkegaard et al. 2007; Courtois et al. 2009; Lopes and Reynolds 2010). However, in such plants, there is a risk that expenditure of carbon could be wasted if there is limited moisture at depth or if soil water supplies are exhausted too early in crop development (Sinclair et al. 2010). An examination of grape rootstocks that promoted high or low growth of vines showed that the high-vigor rootstocks had root systems with greater plasticity (i.e., could better exploit soil moisture when it became available), but did not differ in sensitivity to moisture stress (Bauerle et al. 2008). A comparison of winter wheat lines that differed in drought tolerance showed that a more drought-tolerant line (in terms of grain yield potential maintained under water-limited conditions) was better able to extract water from deep soil layers than a less drought-tolerant line (Figure 35.5). Similar results were observed in spring wheat lines (Lopes and Reynolds 2010). Clearly, at a macroscale, in order to manipulate roots so that they establish where they are needed, the cellular and molecular processes discussed in this chapter are vitally important to understand.

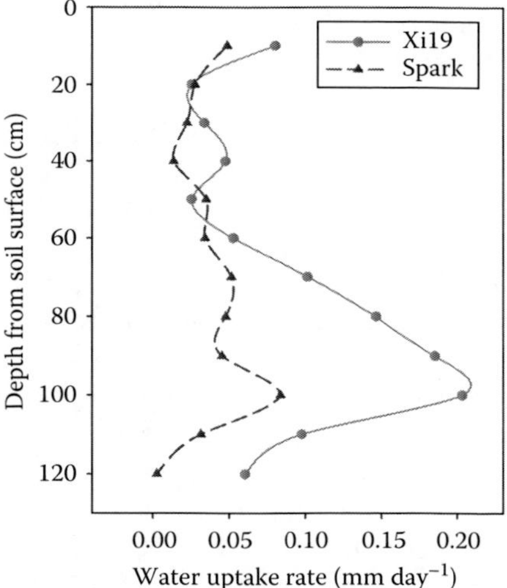

FIGURE 35.5 Comparison of water uptake rates at different depths in the soil profile for two contrasting winter wheat varieties. Field plots were grown under managed drought conditions using large polytunnel rainout shelters (Ober et al. 2010). Changes in soil water content during the period plots were covered (from 2 weeks before flowering until maturity) and were measured using a capacitance-type soil moisture meter inserted into PVC access tubes installed in each plot. Measurements were made at regular intervals at every 10 cm of soil depth to 120 cm from the soil surface. Smoothed lines were fit to mean values (n = 3) for the interval from May 20 to June 15, 2009. Genotypic differences were significant at 80–100 cm from the soil surface, indicating that the root system of Xi19 was better able to extract moisture from deep soil layers than Spark. Xi19 also exhibited greater drought tolerance (defined as the proportion of irrigated yield potential maintained under drought) than Spark.

IV. Conclusions

The rhizosphere environment is neither static nor homogeneous; root growth, therefore, must respond in dynamic ways to local challenges, such as lack of soil moisture. Root growth is determined by the balance between cell division and differentiation and between cell wall loosening and wall tightening. ABA accumulation is required for root growth maintenance at low ψ_w, but a comprehensive picture of the complex interactions between growth-regulating hormones in water-stressed roots remains elusive. One intersection of diverse pathways appears to be the DELLA protein family, which restricts root growth. ROS levels, cell wall pH, and apoplastic Ca^{2+} concentrations appear to play critical roles in directing wall extensibility, and ROS are also involved in regulating the transition from cell proliferation to differentiation. Regulation of ROS levels is determined by several key enzyme families (e.g., NADPH oxidases, oxalate oxidases, peroxidases, catalases) that affect redox poise and balance the production and detoxification of ROS.

However, there is still much to be learned about how putative key control factors of root growth and development under well-watered conditions play a role in determining the response of root growth in drying soil. Greater understanding is required to identify specific gene targets that can lead to improvements in root system function and crop yields in unfavorable environments. Experimental platforms ranging from petri plates to the field should be included in the endeavor. Specific challenges are to discover how roots sense low soil ψ_w and how these signals are transduced to effect genetic and physiological changes so that growth continues, thereby establishing new root tissue in areas with greater abundance of soil resources.

References

Achard P, Cheng H, De Grauwe L et al. 2006. Integration of plant responses to environmentally activated phytohormone signals. *Science* 311:91–94.

Achard P, Genschik P. 2009. Releasing the brakes of plant growth: How GAs shutdown DELLA proteins. *J Exp Bot* 60:1085–1092.

Ballif J, Endo S, Kotani M, MacAdam J, Wu Y. 2011. Overexpression of HAP3b enhances primary root elongation in *Arabidopsis. Plant Physiol Biochem* 49:579–583.

Bauerle TL, Smart DR, Bauerle WL, Stockert C, Eissenstat DM. 2008. Root foraging in response to heterogeneous soil moisture in two grapevines that differ in potential growth rate. *New Phytol* 179:857–866.

Beaudoin N, Serizet C, Gosti F, Giraudat J. 2000. Interactions between abscisic acid and ethylene signaling cascades. *Plant Cell* 12:1103–1115.

Beemster GTS, Baskin TI. 2000. STUNTED PLANT 1 mediates effects of cytokinin, but not of auxin, on cell division and expansion in the root of *Arabidopsis. Plant Physiol* 124:1718–1727.

Bengough AG, McKenzie BM, Hallett PD, Valentine TA. 2011. Root elongation, water stress, and mechanical impedance: A review of limiting stresses and beneficial root traits. *J Exp Bot* 62:59–68.

Birnbaum K, Shasha DE, Wang JY, Jung JW, Lambert GM, Galbraith DW, Benfey PN. 2003. A gene expression map of the *Arabidopsis* root. *Science* 302:1956–1960.

Boyer JS. 2009. Cell wall biosynthesis and the molecular mechanism of plant enlargement. *Funct Plant Biol* 36:383–394.

Boyer JS, Silk WK, Watt M. 2010. Path of water for root growth. *Funct Plant Biol* 37:1105–1116.

Cosgrove DJ. 2005. Growth of the plant cell wall. *Nat Rev Mol Cell Biol* 6:850–861.

Courtois B, Ahmadi N, Khowaja F et al. 2009. Rice root genetic architecture: Meta-analysis from a drought QTL database. *Rice* 2:115–128.

De Smet I, White PJ, Bengough AG et al. 2012. Analyzing lateral root development: How to move forward. *Plant Cell* 24:15–20.

Dello Ioio R, Linhares FS, Sabatini S. 2008. Emerging role of cytokinin as a regulator of cellular differentiation. *Curr Opin Plant Biol* 11:23–27.

Demidchik V, Shabala SN, Davies JM. 2007. Spatial variation in H_2O_2 response of *Arabidopsis thaliana* root epidermal Ca^{2+} flux and plasma membrane Ca^{2+} channels. *Plant J* 49:377–386.

Dinneny JR, Long TA, Wang JY et al. 2008. Cell identity mediates the response of *Arabidopsis* roots to abiotic stress. *Science* 320:942–945.

Dubrovsky JG, Forde BG. 2012. Quantitative analysis of lateral root development: Pitfalls and how to avoid them. *Plant Cell* 24:1–14.

Fan L, Linker R, Gepstein S et al. 2006. Progressive inhibition by water deficit of cell wall extensibility and growth along the elongation zone of maize roots is related to increased lignin metabolism and progressive stelar accumulation of wall phenolics. *Plant Physiol* 140:603–612.

Fan L, Neumann PM. 2004. The spatially variable inhibition by water deficit of maize root growth correlates with altered profiles of proton flux and cell wall pH. *Plant Physiol* 135:1–10.

Fernandez G. 1992. Effective selection criteria for assessing plant stress tolerance. In *Adaptation of Food Crops to Temperature and Water Stress*, Publ No., 93-410, ed. CG Kuo, pp. 257–270. Tainan, Taiwan: Asian Vegetable Research Development Center.

Fraser F, Silk WK, Rost TL. 1990. Effects of low water potential on cortical cell length in growing regions of maize roots. *Plant Physiol* 93:648–651.

Fromm J, Meyer AJ, Weisenseel MH. 1997. Growth, membrane potential and endogenous ion currents of willow (*Salix viminalis*) roots are all affected by abscisic acid and spermine. *Physiol Plant* 99:529–537.

Fry SC. 1998. Oxidative scission of plant cell wall polysaccharides by ascorbate-induced hydroxyl radicals. *Biochem J* 332:507–515.

Gago J, Landin M, Gallego PP. 2010. Artificial neural networks modeling the in vitro rhizogenesis and acclimatization of *Vitis vinifera* L. *J Plant Physiol* 167:1226–1231.

Guan LM, Zhao J, Scandalios JG. 2000. *Cis*-elements and transfactors that regulate expression of the maize *Cat1* antioxidant gene in response to ABA and osmotic stress: H_2O_2 is the likely intermediary signaling molecule for the response. *Plant J* 22:87–95.

Harberd NP, Belfield E, Yasumura Y. 2009. The angiosperm gibberellin-GID1-DELLA growth regulatory mechanism: How an "inhibitor of an inhibitor" enables flexible response to fluctuating environments. *Plant Cell* 21:1328–1339.

Heo J-O, Chang KS, Kim IA et al. 2011. Funneling of gibberellin signaling by the GRAS transcription regulator SCARECROW-LIKE 3 in the *Arabidopsis* root. *Proc Natl Acad Sci U S A* 108:2166–2171.

Hodge A, Berta G, Doussan C, Merchan, F Crespi M. 2009. Plant root growth, architecture and function. *Plant Soil* 321:153–187.

Hurd E. 1968. Growth of roots of seven varieties of spring wheat at high and low moisture levels. *Agron J* 60:201–205.

Ingram PA, Malamy JE. 2010. Root system architecture. In *Advances in Botanical Research*, eds. J-C Kader, M Delseny, pp. 75–117. New York: Academic Press.

Jaillais Y, Chory J. 2010. Unraveling the paradoxes of plant hormone signaling integration. *Nat Struct Mol Biol* 17:642–645.

Kano M, Inukai Y, Kitano H, Yamauchi A. 2011. Root plasticity as the key root trait for adaptation to various intensities of drought stress in rice. *Plant Soil* 342:117–128.

Kiegle E, Gilliham M, Haselhoff J, Tester M. 2000. Hyperpolarisation-activated calcium currents found only in cells from the elongation zone of *Arabidopsis thaliana* roots. *Plant J* 21:225–229.

Kirkegaard JA, Lilley JM, Howe GN, Graham JM. 2007. Impact of subsoil water use on wheat yield. *Aust J Agric Res* 58:303–315.

Laplaze L, Parizot B, Baker A et al. 2005. GAL4-GFP enhancer trap lines for genetic manipulation of lateral root development in *Arabidopsis thaliana. J Exp Bot* 56:2433–2442.

Liang BM, Sharp RE, Baskin TI. 1997. Regulation of growth anisotropy in well-watered and water-stressed maize roots. I. Spatial distribution of longitudinal, radial and tangential expansion rates. *Plant Physiol* 115:101–111.

Liszkay A, van der Zalm E, Schopfer P. 2004. Production of reactive oxygen intermediates ($O_2^{\cdot-}$, H_2O_2, and $^{\cdot}OH$) by maize roots and their role in wall loosening and elongation growth. *Plant Physiol* 136:3114–3123.

Lopes MS, Reynolds MP. 2010. Partitioning of assimilates to deeper roots is associated with cooler canopies and increased yield under drought in wheat. *Funct Plant Biol* 37:147–156.

Lucas M, Laplaze L, Bennett MJ. 2011. Plant systems biology: Network matters. *Plant Cell Environ* 34:535–553.

MacRobbie E. 2006. Osmotic effects on vacuolar ion release in guard cells. *Proc Natl Acad Sci U S A* 103:1135–1140.

Marino D, Dunand C, Puppo A, Pauly N. 2012. A burst of plant NADPH oxidases. *Trends Plant Sci* 17:9–15.

Melchior W, Steudle E. 1993. Water transport in onion (*Allium cepa* L.) roots. Changes in axial and radial hydraulic conductivities during root development. *Plant Physiol* 101:1305–1315.

Miller AL, Gow NAR. 1989. Correlation between root-generated ionic currents, pH, fusicoccin, indoleacetic acid, and growth of the primary root of *Zea mays*. *Plant Physiol* 89:1198–1206.

Mittler R, Vanderauwera S, Suzuki N et al. 2011. ROS signaling: The new wave? *Trends Plant Sci* 16:300–309.

Miyazawa Y, Yamazaki T, Moriwaki T, Takahashi H. 2011. Root tropism: Its mechanism and possible functions in drought avoidance. In *Advances in Botanical Research*, ed. T Ismail, pp. 349–375. New York: Academic Press.

Monshausen GB, Miller ND, Murphy AS, Gilroy S. 2011. Dynamics of auxin-dependent Ca^{2+} and pH signaling in root growth revealed by integrating high-resolution imaging with automated computer vision-based analysis. *Plant J* 65:309–318.

Moreno-Risueno MA, Busch W, Benfey PN. 2010. Omics meet networks—Using systems approaches to infer regulatory networks in plants. *Curr Opin Plant Biol* 13:126–131.

Moumeni A, Satoh K, Kondoh H et al. 2011. Comparative analysis of root transcriptome profiles of two pairs of drought-tolerant and susceptible rice near-isogenic lines under different drought stress. *BMC Plant Biol* 11:174.

Newman I. 2001. Ion transport in roots: Measurement of fluxes using ion-selective microelectrodes to characterize transporter function. *Plant Cell Environ* 24:1–14.

Nishiyama R, Watanabe Y, Fujita Y et al. 2011. Analysis of cytokinin mutants and regulation of cytokinin metabolic genes reveals important regulatory roles of cytokinins in drought, salt and abscisic acid responses, and abscisic acid biosynthesis. *Plant Cell* 23:2169–2183.

Ober ES, Clark CJAC, Le Bloa M, Royal A, Jaggard KW, Pidgeon JD. 2004. Assessing the genetic resources to improve drought tolerance in sugar beet: Agronomic traits of diverse genotypes under droughted and irrigated conditions. *Field Crops Res* 90:213–234.

Ober ES, Clark CJAC, Perry A. 2010. Traits related to genotypic differences in effective water use and drought tolerance in UK winter wheat. *Aspects Appl Biol* 105:13–22.

Ober ES, Sharp RE. 1994. Proline accumulation in maize (*Zea mays* L.) primary roots at low water potentials. I. Requirement for increased levels of abscisic acid. *Plant Physiol* 105:981–987.

Ober ES, Sharp RE. 2003. Electrophysiological responses of maize roots to low water potentials: Relationship to growth and ABA accumulation. *J Exp Bot* 54:813–824.

Ober ES, Sharp RE. 2007. Regulation of root growth responses to water deficit. In *Advances in Molecular-Breeding toward Drought and Salt Tolerant Crops*, eds. MA Jenks, PM Hasegawa, SM Jain, pp. 33–53. New York: Springer.

Perilli S, Di Mambro R, Sabatini S. 2012. Growth and development of the root apical meristem. *Curr Opin Plant Biol* 15:17–23.

Perrin RM, Jia ZH, Wagner TA et al. 2003. Analysis of xyloglucan fucosylation in *Arabidopsis*. *Plant Physiol* 132:768–778.

Poroyko V, Hejlek LG, Spollen WG et al. 2005. The maize root transcriptome by serial analysis of gene expression. *Plant Physiol* 138:1700–1710.

Poroyko V, Spollen WG, Hejlek LG et al. 2007. Comparing regional transcript profiles from maize primary roots under well-watered and low water potential conditions. *J Exp Bot* 58:279–289.

Ross JJ, Weston DE, Davidson SE, Reid JB. 2011. Plant hormone interactions: How complex are they? *Physiol Plant* 141:299–309.

Růžička K, Ljung K, Vanneste S et al. 2007. Ethylene regulates root growth through effects on auxin biosynthesis and transport-dependent auxin distribution. *Plant Cell* 19:2197–2212.

Saab IN, Sharp RE, Pritchard J. 1992. Effect of inhibition of abscisic acid accumulation on the spatial distribution of elongation in the primary root and mesocotyl of maize at low water potentials. *Plant Physiol* 99:26–33.

Saab IN, Sharp RE, Pritchard J, Voetberg GS. 1990. Increased endogenous abscisic acid maintains primary root growth and inhibits shoot growth of maize seedlings at low water potentials. *Plant Physiol* 93:1329–1336.

Santner A, Estelle M. 2009. Recent advances and emerging trends in plant hormone signalling. *Nature* 459:1071–1078.

Shabala SN, Lew RR. 2002. Turgor regulation in osmotically stressed *Arabidopsis* epidermal root cells. Direct support for the role of inorganic ion uptake as revealed by concurrent flux and cell turgor measurements. *Plant Physiol* 129:290–299.

Sharma S, Villamor JG, Verslues PE. 2011. Essential role of tissue-specific proline synthesis and catabolism in growth and redox balance at low water potential. *Plant Physiol* 157:292–304.

Sharp RE. 2002. Interaction with ethylene: Changing views on the role of abscisic in root and shoot growth responses to water stress. *Plant Cell Environ* 25:211–222.

Sharp RE, Davies WJ. 1979. Solute regulation and growth by roots and shoots of water-stressed maize plants. *Planta* 147:43–49.

Sharp RE, Hsiao TC, Silk WK. 1990. Growth of the maize primary root at low water potentials. II. The role of growth and deposition of hexose and potassium in osmotic adjustment. *Plant Physiol* 93:1337–1346.

Sharp RE, Poroyko V, Hejlek LG, Spollen WG, Springer GK, Bohnert HJ, Nguyen HT. 2004. Root growth maintenance during water deficits: Physiology to functional genomics. *J Exp Bot* 55:2343–2351.

Sharp RE, Silk W, Hsiao TC. 1988. Growth of the maize primary root at low water potentials. I. Spatial distribution of expansive growth. *Plant Physiol* 87:50–57.

Sharp RE, Wu Y, Voetberg GS, Saab IN, LeNoble ME. 1994. Confirmation that abscisic acid accumulation is required for maize primary root elongation at low water potentials. *J Exp Bot* 45:743–751.

Silk WK. 1984. Quantitative descriptions of development. *Ann Rev Plant Physiol* 35:479–518.

Silk WK. 2002. The kinematics of primary growth. In *Plant Roots: The Hidden Half*, ed. Y Waisel, A Eshel, U Kafkafi, 3rd edn., pp. 175–197, New York: Marcel Dekker, Inc.

Sinclair TR, Messina CD, Beatty A, Samples M. 2010. Assessment across the United States of the benefits of altered soybean drought traits. *Agron J* 102:475–482.

Spollen WG, LeNoble ME, Samuels TD, Bernstein N, Sharp RE. 2000. Abscisic acid accumulation maintains maize primary root elongation at low water potentials by restricting ethylene production. *Plant Physiol* 122:967–976.

Spollen WG, Sharp RE. 1991. Spatial distribution of turgor and root growth at low water potentials. *Plant Physiol* 96:438–443.

Spollen WG, Sharp RE, Saab IN, Wu Y. 1993. Regulation of cell expansion in roots and shoots at low water potentials. In *Water Deficits. Plant Responses from the Cell to the Community*, eds. JAC Smith, H Griffiths, pp. 37–52. Oxford: Bios Scientific.

Spollen W, Tao W, Valliyodan B et al. 2008. Spatial distribution of transcript changes in the maize primary root elongation zone at low water potential. *BMC Plant Biol* 8:32.

Sponchiado BN, White JW, Castillo JA, Jones PG. 1989. Root growth of four common bean cultivars in relation to drought tolerance in environments with contrasting soil types. *Exp Agric* 25:249–257.

Swarup R, Perry P, Hagenbeek D et al. 2007. Ethylene upregulates auxin biosynthesis in *Arabidopsis* seedlings to enhance inhibition of root cell elongation. *Plant Cell* 19:2186–2196.

Talboys PJ, Zhang HM, Knox JP. 2011. ABA signalling modulates the detection of the LM6 arabinan cell wall epitope at the surface of *Arabidopsis thaliana* seedling root apices. *New Phytol* 190:618–626.

Trewavas AJ, Jones HG. 1991. An assessment of the role of ABA in plant development. In *Abscisic Acid: Physiology and Biochemistry*, eds. WJ Davies, HG Jones, pp. 169–188. Oxford, U.K.: Bios Scientific.

Tsukagoshi H, Busch W, Benfey PN. 2010. Transcriptional regulation of ROS controls transition from proliferation to differentiation in the root. *Cell* 143:606–616.

Ubeda-Tomás S, Swarup R, Coates J et al. 2008. Root growth in *Arabidopsis* requires gibberellin/DELLA signalling in the endodermis. *Nat Cell Biol* 10:625–628.

Varney GT, Canny MJ. 1993. Rates of water uptake into the mature root system of maize. *New Phytol* 123:775–786.

van der Weele CM, Jiang HS, Palaniappan KK et al. 2003. A new algorithm for computational image analysis of deformable motion at high spatial and temporal resolution applied to root growth. Roughly uniform elongation in the meristem and also, after an abrupt acceleration, in the elongation zone. *Plant Physiol* 132:1138–1148.

Weiss D, Ori N. 2007. Mechanisms of cross talk between gibberellin and other hormones. *Plant Physiol* 144:1240–1246.

Werner T, Nehnevajova E, Kollmer I et al. 2010. Root-specific reduction of cytokinin causes enhanced root growth, drought tolerance, and leaf mineral enrichment in *Arabidopsis* and tobacco. *Plant Cell* 22:3905–3920.

Westgate ME, Boyer JS. 1985. Osmotic adjustment and the inhibition of leaf, root, stem and silk growth at low water potentials in maize. *Planta* 164:540–549.

White RG, Kirkegaard JA. 2010. The distribution and abundance of wheat roots in a dense, structured subsoil—Implications for water uptake. *Plant Cell Environ* 33:133–148.

Whitmore AP, Whalley WR. 2009. Physical effects of soil drying on roots and crop growth. *J Exp Bot* 60:2845–2857.

Wu Y, Cosgrove DJ. 2000. Adaptation of roots to low water potentials by changes in cell wall extensibility and cell wall proteins. *J Exp Bot* 51:1543–1553.

Wu Y, Sharp RE, Durachko DM, Cosgrove DJ. 1996. Growth maintenance of the maize primary root at low water potentials involves increases in cell wall extension properties, expansin activity and wall susceptibility to expansins. *Plant Physiol* 111:765–772.

Wu Y, Spollen WG, Sharp RE, Hetherington PR, Fry SC. 1994. Root growth maintenance at low water potentials. Increased activity of xyloglucan endotransglycosylase and its possible regulation by ABA. *Plant Physiol* 106:607–615.

Wu YJ, Thorne ET, Sharp RE, Cosgrove DJ. 2001. Modification of expansin transcript levels in the maize primary root at low water potentials. *Plant Physiol* 126:1471–1479.

Wuyts N, Bengough AG, Roberts TJ et al. 2011. Automated motion estimation of root responses to sucrose in two *Arabidopsis thaliana* genotypes using confocal microscopy. *Planta* 234:769–784.

Yamaguchi M, Sharp RE. 2010. Complexity and coordination of root growth at low water potentials: Recent advances from transcriptomic and proteomic analyses. *Plant Cell Environ* 33:590–603.

Yamaguchi M, Valliyodan B, Zhang J et al. 2010. Regulation of growth response to water stress in the soybean primary root. I. Proteomic analysis reveals region-specific regulation of phenylpropanoid metabolism and control of free iron in the elongation zone. *Plant Cell Environ* 33:223–243.

Zhang Z-L, Ogawa M, Fleet CM et al. 2011. SCARECROW-LIKE 3 promotes gibberellin signaling by antagonizing master growth repressor DELLA in *Arabidopsis. Proc Natl Acad Sci U S A* 108:2160–2165.

Zhu JM, Alvarez S, Marsh EL et al. 2007. Cell wall proteome in the maize primary root elongation zone. II. Region-specific changes in water soluble and lightly ionically bound proteins under water deficit. *Plant Physiol* 145:1533–1548.

36

Effects of Salinity on Root Growth

Nirit Bernstein
Agricultural Research Organization

I. Introduction

Salinity inhibits growth and development of most plants. Inhibition of shoot and root development is the primary response to the stress. Growth, morphology, anatomy, and physiology of roots are affected by salinity. Changes in water and ion uptake by the roots, production of hormonal signals that communicate information to the shoot, and changes in patterns of expression might induce changes in plant development. Since root growth is usually less sensitive to salt stress than shoot growth, an increased root/shoot ratio is often observed when plants are subjected to saline conditions (Cheeseman 1988; Cruz and Cuatreno 1990; Bernstein et al. 2004; Perica et al. 2008; Khayyat et al. 2009; Cécooli et al. 2011). Root growth of halophytes may be affected differently (Flowers and Colmer 2008). The restriction of root growth by salinity, which reduces the soil volume that can be explored by the root and hence the availability and uptake of water and essential minerals, diminishes the supply of nutrients to the shoot which may contribute to growth reduction (Pitman 1984; Bernstein et al. 1995; Lazof and Bernstein 1998). The increase in root/shoot ratio reduces the demand for element supply to the shoot and thereby has a potential to increase the ability of the root to supply those elements and present an adaptive advantage (Cheeseman 1988). A potentially negative effect of such a change is the decreased capacity of the shoot to supply assimilates to the root and the growing tissues, which is likely to affect plant development and survival particularly under long-term salinization.

The underlying mechanisms involved in the inhibition processes of root growth under salinity are not clearly established. Saline solutions impose on roots two types of stresses: osmotic stress resulting from lowered water potential in the root-growing medium and ionic stress induced by changes in concentrations of specific ions in the root medium and inside the growing tissue (Bernstein and Kafkafi 2002). A secondary induced oxidative stress affects the root cells when the increased production of reactive oxygen species (ROS) under salinity is not met by a suitable increase of the protective antioxidative system (Shoresh et al. 2011). The information available, concerning root responses to salinity and the mechanisms involved is limited compared to our current knowledge on shoot responses. A better understanding of the chemical, physiological, and molecular mechanisms involved in the different facets of root growth reduction would expand our understanding of root and whole-plant responses to saline conditions.

II. Root Growth under Salt Stress

Roots are in direct contact with the soil solution. As such, they are first to encounter the saline medium and are potentially the first site of damage or line of defense under salt stress. The shape and size of the root system are determined by extension growth of individual root tips as well as by the rate and location of lateral root development. Salinity affects these root developmental processes differently. While root elongation of many plants is severely inhibited by high concentrations of NaCl in the medium (Cramer et al. 1987; Zhong and Läuchli 1993b; Neumann et al. 1994; West et al. 2004; Fornes et al. 2007), lateral root formation is less affected (Waisel and Breckle 1987) or might even be stimulated by the stress (Kramer 1980).

Root fresh and dry biomass decrease as well with increasing salinity (Mauromicale and Licandro 2002; Bernstein et al. 2004, 2009; Khayyat et al. 2009), and the extent of decrease sometimes positively correlates with the extent of salt sensitivity of plant cultivars (Rewald et al. 2010).

In a single root system, the response to salinity of root elongation (Rahnama et al. 2011) and root branching (Zolla et al. 2010) can be root-type specific, inducing a change in root system architecture. For example, reduced length of the main root and increased branching in *Lycopersicum esculentum* Mill. resulted in a shorter and more branched root system, which might give rise to a larger root system (Karni et al. 2010). The degree of root system growth response to salinity can vary considerably within a family, genus, and species (Bernstein et al. 2001; Rahnama et al. 2011).

Growth of roots is restricted to a limited region of the root axis near its tip. Therefore, investigations into the physiological basis for root growth inhibition under stress involve examination of the root apical meristem and the adjoining elongation zone. Growth is not uniformly distributed throughout the apical growth region, but varies considerably along the root axis (see Chapters 3 and 6). Similar to leaves, response to salinity varies with location along the developing organ (Bernstein et al. 1993a,b; Zhong and Läuchli 1993b; Neves-Piestun and Bernstein 2001; West et al. 2004). Salinity imposes inhibition of root elongation rate by a combined effect on the size of the growth zone and the magnitude of localized tissue elongation. The length of the elongation zone of maize (Table 36.1; Pritchard 1994; Bustos et al. 2008), cotton (Zhong and Läuchli 1993b), sorghum (Koryo 1997), and *Arabidopsis* (West et al. 2004) roots was shortened under saline conditions. Reduction of growth intensity occurred throughout the elongation zone of cotton roots but only in the basal region, that is, more mature cells, of the growing region of the maize and *Arabidopsis* roots. Growth in the youngest cells (i.e., most apical zone) of maize- and *Arabidopsis*-growing regions was unaffected by salt stress. The spatial effect on root elongation was specific to NaCl and was not obtained with iso-osmotic concentration of mannitol (Pritchard 1994).

The stress-induced inhibition of root elongation may result from effects on the *rate of cell division, rate of cell expansion, duration of cell growth*, or *orientation of cell expansion*. In principle, any of these processes could underlie the root's growth response to salinity. Information about salt stress effect on root-*cell expansion* is available from three types of indirect studies and from growth kinematic analysis:

1. Salinity inhibits the elongation rate of root segments in the elongation zone (Zhong and Läuchli 1993b; West et al. 2004; Bustos et al. 2008). The reduction of tissue growth intensity (relative elemental growth rate, REGR) within the "elongation only" zone may be the result of the inhibition of cell elongation rate.
2. Swift changes in root growth rate following salinization or removal of salt. Such changes are due to cell expansion and not cell division because newly dividing cells do not contribute to organ elongation before several hours or even days have passed (Clarkson 1969; Powell et al. 1986). Plant organ growth rates were demonstrated to be reduced rapidly by salinity (Cramer and Bowman 1991) or increase rapidly after removal of the salt (Rawson and Munns 1984), suggesting that salinity decreases cell expansion rate.
3. Cell length measurements suggested that cell growth is inhibited. Mature root-cell size was reduced under salt stress (Kurth et al. 1986), and salinization resulted in shorter roots (Zidan et al. 1990; Azaizeh et al. 1992). However, differences in cell length do not necessarily reflect variations in cell elongation rates. Salinization may also affect cell length via an effect on the duration of cell elongation (Bernstein et al. 1993a; Lazof and Bernstein 1998).

Characterization of spatial and temporal aspects of cell displacement is required before conclusions can be drawn from elongating cell length data. Such information is available for the salt-stressed root of maize (Table 36.1) and *Arabidopsis* (calculated from tissue displacement: West et al. 2004). In maize, inferring cell growth rate from cell length data revealed that the elongation rate of cells located 2 mm from the tip of the root was lowered by 37% under salinity (Table 36.1; Bernstein and Ioffe, unpublished). In *Arabidopsis*, analysis of time-lapse photographs was used to determine the spatial velocity profile throughout the growth zone and to gain insight into cell expansion rates (West et al. 2004). The strain rate was reduced by salinity at the distal portion of the growing zone.

TABLE 36.1 Effect of Salt Stress on Elongation Rate of Maize Roots, Length of Root Elongation Zone, and Duration and Rate of Cortical Root-Cell Elongation[a]

Treatment	Root Elongation Rate (mm/h)	Length of Root Elongation Zone (mm)	Cell Elongation Rate (mm/h)	Duration of Cell Elongation (h)
1 mM NaCl	0.93 ± 0.0002	9.0	0.0078 ± 0.00014	17.15
80 mM NaCl	0.66 ± 0.0003	7.5	0.0049 ± 0.00015	25.5

Source: Ioffe and Bernstein, unpublished.

[a] Rates of cell elongation were calculated for cells located 2 mm from the root tip by combining results of cell elongation rates with their growth trajectory. Duration of cell elongation represents the length of time required for a cell initially located 05 mm from the root tip to be displaced past the end of the elongation zone. Data for first (3-day-old) adventitious roots of 12-day-old maize plants, 5 days following initiation of the salt treatment. Presented results are averages ±SE (n = 5 roots); results from 20–40 cells were averaged to give each individual root average.

Similar to cell expansion, only little evidence of salt effect on *cell division* rate has been reported. Salt stress inhibited cell number increase in roots (Samarajeewa et al. 1999) and apparent root-cell production rate (Kurth et al. 1986; Zidan et al. 1990). It induced nuclear deformation and degradation in meristematic root cells (Katsuhara and Kawasaki 1996), and the resulting cell death was suggested to be the cause of root growth inhibition. Evidence from some studies suggests, indirectly, a relationship between cell cycle regulation and salt growth tolerance. In *Arabidopsis*, the gain of AtHAL3a function, a protein involved in cell cycle regulation, induced altered growth rate and improved salt tolerance (Espinosa-Ruiz et al. 1999).

Trigonelline, a salt stress osmoregulator in legumes, was found to function as a cell cycle regulator (Tramontano and Jowe 1997). Kinematic analysis revealed as well that in the primary root of *Arabidopsis*, the overall growth reduction was due to a decrease in cell production and mature cell length (West et al. 2004). Since the average cell cycle duration was not affected, the reduced cell production was due to a smaller number of dividing cells, that is, meristem size reduction.

The cycling-dependent kinase activity, which catalyzed the cell cycle progression, and the promoter activity of the mitotic cyclin *CYCB1;2* were transiently reduced after exposure to salt stress. In *Arabidopsis* roots, the transcript level of the *CDC2aAt* gene, which encodes a cyclin-dependent kinase, decreased along with the inhibition of root growth (Burssens et al. 2000). In the seminal root of rye as well, salinity reduced the frequency of cell division and increased cell death in the meristem (Ogawa et al. 2006). Taken together, the available data suggest that root growth is inhibited under salinity by combined effects on cell production and elongation.

Salinity induced chromosomal abnormalities in mitotic root cells, which in some plants could be prevented by osmoprotectants. In *Allium cepa* L., chromosomal abnormalities in the mitotic cells at root tip were prevented by a prior treatment of the roots with myoinositol, a cyclic polyol osmoprotectant (Chatterjee and Majumder 2010), which has been correlated before with tolerance to drought and salinity (Sengupta et al. 2008). Also, overexpression of myoinositol-1-phosphate synthase coding genes in tobacco prevented DNA fragmentation under salinity (Chatterjee and Majumder 2010).

The *duration of root-cell elongation* in maize is lengthened under salt stress (Table 36.1). Cell growth kinematic analysis revealed that a nonstressed cell ceases elongation and reaches final length sooner than a salt-stressed cell, in spite of the reduction of growth zone length under stress. The lengthened duration of cell elongation does not compensate for the reduction in cell elongation rate. The elongation rate of the root was therefore slower under stress (Table 36.1). The number of elongating cells along the growth zone is another important factor in determination of organ elongation rate. The number of elongating cells in the root at any given time is a function of cell production rate in the meristem and the duration of individual cell elongation. It is still unknown how salt stress affects the number of elongating cells in the root.

In addition to the effect on cellular expansion rate, salinity might also change the *growth orientation* (growth anisotropy) of cells. Salinization was shown to cause thicker but shorter roots (barley: Huang and Redman 1995; cotton: Zhong and Läuchli 1993b; *Picea mariana* (Mill.) Britton, Sterns and Poggenb.: Croser et al. 2001; *Portulaca oleracea* L.: Franco et al. 2011) and to reduce cotton root elongation and change cell shape but not cell volume (Kurth et al. 1986). Tobacco tissue culture cells adapted over many generations to grow in a high-salt medium expanded less anisotropically than unadapted cells under nonstress conditions (Chang et al. 1996). Alteration of cell growth anisotropy implies an effect of NaCl on the cytoskeleton (discussed further in Section VII.C).

Root tissue development is also affected by salinity. Salt stress stimulated early development of cotton root endodermis and induced development of an exodermis (Reinhardt and Rost 1995). Stimulation of the vascular development under salinity occurs because the axial root elongation slows down to a greater extent than the xylem vessel differentiation, resulting in the mature vessels being found closer to the root tips of salt-stressed plants compared to nonstressed plants. In the halophyte *Suaeda maritima* (L.) Dumort., the Casparian strip was developed closer to the root apex (Hajibagheri et al. 1985), and in *Chloris gayana* Kunth roots, the cross-sectional area of the vascular cylinder decreased under salinity (Cécooli et al. 2011). These changes might be involved with altered regulation of water and solute transport under stress.

Salinity-induced changes in root morphology may be involved in the root response to salinity and extent of root and plant salt tolerance. For example, the *Thellungiella salsuginea* (Pall.) O.E. Schulz, the *Arabidopsis*-related halophyte, has a thickened primary root that acts as a sink for Na^+ sequestration, thereby preventing its accumulation in the lateral roots and young leaves (Vera-Estrella et al. 2005), and glycophytes can adjust root growth and root system architecture in response to salinity to avoid locally high salt concentration (Galvan-Ampudia and Testerink 2011).

Root system architecture, a result of lateral root positioning, initiation, and emergence, is determined by interplay between an endogenous developmental program controlled by accumulation of auxin and environmental signals. Abiotic stresses are known to modify the development of the root system by affecting hormone synthesis, transport, and sensitivity. Salinity seems to have a specific auxin-mediated effect on root system architecture which affects development of the lateral roots as well as the direction of root growth. The effect on lateral and adventitious root development is more complex than the inhibition of elongation usually seen for the primary root. In various experimental setups, stimulation or inhibition was documented (West et al. 2004; Wang et al. 2009; Zhao et al. 2010; Zolla et al. 2010), which might be related to different effects of the osmotic and anionic components of the salinity stress on these roots. While osmotic stress inhibits lateral root formation, ionic stress was found to stimulate both initiation and emergence (Zhao et al. 2010; Zolla et al. 2010; Galvan-Ampudia and Testerink 2011). This inhibition of lateral root initiation is mediated through inhibition of auxin transport from the shoot and reduced auxin accumulation in the meristematic cells (Zhao et al. 2010; Galvan-Ampudia and Testerink 2011). High salt concentration transiently repressed expression of PIN2, the auxin efflux carrier, and abundance of the PIN2 protein, suggesting that salt stress modulates this key regulator of auxin transport both transcriptionally and posttranscriptionally (Sun et al. 2008; Galvan-Ampudia and Testerink 2011).

III. Salinity Sensing and Signaling

In response to exposure of the root to salinity, the root needs to adjust to maintain its function under the altered conditions, as well as to signal to the shoot the occurrence of stress. Salinity-induced signals affect root development and ion homeostasis in the root. Plants responds rapidly and specifically to increase in Na^+ salinity (Tracy et al. 2008); however, the mechanism for sensing the change in Na^+ concentration is unknown. It is not yet known if the Na^+ is sensed extracellularly by a plasma membrane root-cell protein or intracellularly after it enters the root cells. An increase in cytoplasmic free calcium in the root cells is the earliest recorded event following root exposure to Na^+ salinity, which apparently results from Na^+-activated flux of Ca^{2+} across the plasma membrane and the tonoplast (Kiegle et al. 2000; Knight 2000; Moore et al. 2002; Tracy et al. 2008), and likely plays a role in the activation of signaling pathways (Tracy et al. 2008; Kudla et al. 2010). The extracellular salt composition modulates the changes in cytosolic Ca^{2+} which transmits the information intracellularly (Tracy et al. 2008), with a response that is cell specific—that is, was demonstrated to be lower in the root pericycle compared to other cell types (Kiegle et al. 2000).

The salt overly sensitive (SOS) pathway, which is the best characterized signaling pathway for salinity, seems to involve the Na^+-induced elevation of cytosolic free Ca^{2+}. The increase in the cytosolic free Ca^{2+} activates the calcium sensor SOS3, a calcineurin B-like protein (CBL4). Dimerization with SOS3 recruits SOS2, a CBL-interacting protein kinase to the plasma membrane, and thereby activation of the Na^+/H^+ antiporter SOS1 by phosphorylation (Tuteja 2007; Kudla et al. 2010). Not much is known about the involvement of this pathway in salinity tolerance; however, *sos* mutants of *Arabidopsis thaliana* (L.) Heynh. are less tolerant to salinity stress than wild-type plants, suggesting involvement of the pathway in some aspects of salinity tolerance (Zhu et al. 1998). Interestingly, the resulting increase in Na^+ efflux by the SOS pathway facilitates maintained root growth (Galvan-Ampudia and Testerink 2011). It was suggested that if this pathway is exhausted (under high Na^+ concentration), Na^+ levels in the root cells increase and can act as a signal for initiation of an avoidance agravitropic growth response, that is, root bending (Galvan-Ampudia and Testerink 2011). The SOS pathway seems also to be required for the activation of the mild-salinity-induced lateral root emergence. The analysis of auxin distribution in *sos3* mutants suggests that the SOS3 protein controls lateral root emergence by increasing auxin transport and production under salinity (Zhao et al. 2010).

Following salinization of the growth medium, a rapid reduction of leaf elongation rate occurs (Passioura and Munns 2000; Cramer 2002a). The mechanisms involved in this swift growth restriction are not clearly known, but are likely to be regulated by long-distance signaling from the root to the shoot by hormones or their precursors, since the reduction of leaf growth rate under salinity is not limited by carbohydrate availability or water status (Munns and Tester 2008).

The long-distance signal involved in regulating the salinity response is still not clear. Abscisic acid (ABA) is a reasonable candidate for this root signal since it is found in the xylem sap and increases under salinity stress (reviewed by Munns and Cramer 1996) and is known to be involved in root-shoot signaling under water stress (Zhu 2002; Davies et al. 2005). However, studies based on ABA analyses in the growing tissues of leaves and ABA-deficient mutants (Makela et al. 2003) do not support the ABA-control theory for saline conditions, and there is evidence that other hormones are also involved (Reviewed by Dodd 2005; Pérez-Alfocea et al. 2010).

IV. Possible Factors Affecting Root Growth under Salt Stress

It is generally accepted that salt stress may reduce root growth by osmotic effects, by specific ion effects (toxicity, deficiency, or ion imbalance), and/or by induction of oxidative stress. The degree at which these factors affect root growth depends on the inherent plant sensitivity, the duration of exposure to the stress, the concentration and type of salts involved, and environmental variables (such as physical and chemical properties of the growing medium, air humidity and composition, soil and air temperature, and light intensity).

A. Osmotic Effects

High salt concentrations in the root media result in low soil water potentials at the root zone and eventually may lead to a water deficit. Roots that are exposed to a sudden event of salinity or water stress lose their turgor and respond in an immediate cessation of elongation. Recovery of elongation can occur without full recovery of the turgor in the growing cells of the root, as long as the turgor exceeds the wall "yield threshold" (which depends on cell-wall properties that may change under stress) (Reviewed by Bernstein and Kafkafi 2002). Under long-term salinization, turgor of the growing root cells may or may not be reduced; therefore, changes in turgor alone cannot fully explain the effect of salinity on root elongation (Neuman et al. 1994; Pritchard 1994; and references in Bernstein and Kafkafi 2002).

Upon exposure to salinity, roots of many plants adjust osmotically. Biosynthesis of large quantities of organic osmolytes is metabolically expensive and potentially limiting the energy available for growth. The alternative, accumulation of inorganic ions from the root medium, is energetically less demanding (Yeo 1993) but may reduce root growth due to ion toxicity.

B. Specific Ion Effects

Generally, macronutrient concentration in roots is reduced under salinity, whereas some micronutrient concentration is increased (Izzo et al. 1991). Salinity-induced toxicity or deficiency of one or more ions may cause growth reduction under stress. The term *specific* means that the ion under consideration causes an additional depression of growth beyond what could be expected from its osmotic effect.

1. Toxicities

Although much research focused on ion-specific effects under salinization, it is yet unknown whether Na^+, Cl^-, or other ions are the predominant growth-limiting factors for root growth reduction. Since the causes of growth inhibition have to be found in the processes, regulatory events, and metabolic changes that occur in the growing cells, information concerning the relationships between cell elongation and concentration of elements in the growing tissue is needed for evaluation of their possible role in growth reduction (Bernstein et al. 1995). Salinization increases Na^+ content throughout the elongation region of cotton roots, but rates of sodium deposition are enhanced only in the region least affected by salinity (Zhong and Läuchli 1994). This suggests that in cotton roots, similar to leaves (Neves-Piestun and Bernstein 2005), sodium accumulation in the growing cells is not the main cause of root growth reduction under NaCl stress.

a. Na^+ and Cl^- Uptake

The root-cell plasma membrane is an important control point for the exclusion, or selective uptake, of essential ions. Na^+/K^+ discrimination and root internalization of Na^+, Cl^-, Ca^{2+}, and K^+ are often considered in relation to plant tolerance and sensitivity to salinity (Rengel 1992; Lazof and Bernstein 1998; Munns and Tester 2008).

Growth inhibition, the most common effect of salinity, is often correlated with high Na^+ concentration in the plant. For some plants, especially woody perennials (avocado, grapevine, and citrus), Na^+ is accumulated in the roots or stems, and Cl^- is transported to the shoot where it causes tissue damage. However, for most plants, Na^+ is considered to be the primary cause of ion-specific damage and processes which control uptake of Na^+ into the root, and its delivery and entry to the xylem are central for salt tolerance (Munns and Tester 2008). Root processes involved with net delivery of Na^+ to the xylem include influx of Na^+ from the soil solution into the root cells, efflux of Na^+ from these cells back to the soil solution, and efflux of Na^+ from inner root cells to the xylem. Sodium influx into root cells across the plasma membrane occurs down an electrical gradient and is at equilibrium at internal concentrations of about 10 mol m^{-3} Na^+ and ~0.2 mol m^{-3} Ca^{2+} (Bush 1995). It is therefore passive and rapid. Na^+ enters roots through voltage-independent (or weakly voltage-dependent) nonselective cation channels (Schachtman et al. 1991; Amtmann and Sanders 1999) and likely also by high-affinity K^+ transporter (HKT) family (Haro et al. 2005). Cyclic nucleotide-gated channels and ionotropic glutamate receptor-like channels are some of the possible candidates for selective cation channels in the root (Demidchik et al. 2002). Elevated root zone salinity increases the gradient that drives the passive movement of Na^+ across the plasma membrane. Ion homeostasis under saline environments relies on Na^+ transport from the cytoplasm to the cell-wall apoplast or to the vacuole. Part of the Na^+ which enters the root cells passively is pumped out by Na^+/H^+ antiporters (SOS1) on the plasma membrane, which are energized by P-type ATPases (Pardo et al. 2006). SOS1 is involved in salt resistance, and overexpression of SOS1 was shown to increase salinity resistance (*Arabidopsis*: Yang et al. 2009) and its downregulation to increase sensitivity to salinity (*Thellungiella halophila* (C.A. Mey.) O.E. Schulz: Oh et al. 2007). Na^+ efflux may also occur by transporters other than Na^+/H^+ antiporters, such as members of the CHX family (Pardo et al. 2006), or Na^+ translocating ATPases (Mennen et al. 1990). Compartmentation of Na^+ and Cl^- in the vacuoles prevents toxic effects in the cytosol and facilitates osmotic adjustment. The transport of Na^+ to the vacuole is mediated by a tonoplast Na^+/H^+ antiporter, which is energized by the V-type H^+ATPase and H^+-PPiase (Apse et al. 2003; Bao et al. 2008). The action of a Na^+/H^+ antiporter on the tonoplast and the plasma membrane reduces cytoplasmic Na^+, and although it increases the gradient which drives passive movement of Na^+ into the cell (Dupont 1992), it has an adaptive potential for salinity resistance. Overexpression of the vacuolar Na^+/H^+ antiporter indeed promotes sustained growth of *Arabidopsis thaliana* under elevated salinity (Apse et al. 1999). Salt tolerance of *Plantago* is associated with Na^+/H^+ antiporter activity (Prins 1995), and roots of salt-acclimated soybean have higher Na^+/H^+ activity than roots of nonstressed and nonacclimatized plants (Huang et al. 1998). The plasma membrane SOS1, Na^+/H^+ antiporter, which is expressed in cell of the stele as well (Munns and Tester 2008), may be involved in efflux of Na^+ from the stellar cells to the xylem.

The ability to compartmentalize Na^+ may result, in part, from stimulation of the H^+-ATPases of the plasmalemma (PM-ATPase) and the tonoplast (V-ATPase). Salinity increases ATPase activity in root cells, sometimes in correlation with salinity tolerance or ability for Na^+ compartmentation (Katsuhara et al. 1997; Lin et al. 1997; Zhang et al. 1997, 1999).

Cl^- influx into plant roots is thought to require energy in most situations and to be catalyzed by a $Cl^-/2H^+$ symporter (Felle 1994). The nature of the mechanisms of Cl^- exclusion by roots, which characterizes tolerance in some species (e.g., *Citrus*), is not understood. If chloride enters the cell against an electrical gradient by active processes, then sodium influx into the cell which depolarizes the plasma membrane should facilitate passive Cl^- entrance through anion channels (Logan et al. 1997). ATPase-driven anion channels in the tonoplast were suggested to facilitate subcellular compartmentation of Cl^- inside the vacuoles (Plant et al. 1994), but Cl^- loading into the xylem is thought to be passive and mediated by anion channels (Gilliham and Tester 2005). Na^+ and Cl^- can also cross membranes by diffusion

(Hille 1992), but this effect is probably small relative to the other transport mechanisms available for these ions.

2. Deficiencies and Ion Imbalance

Salinity-induced root growth restriction reduces the volume of soil that can be explored by the root system and therefore the quantities of nutrients available to the plant. Additionally, high concentrations of Na^+ and Cl^- near the root surface can affect membrane uptake processes and thereby reduce nutrient uptake. Antagonistic effects for influx into roots were reported for Na^+ and K^+ (Lynch and Läuchli 1984; Cramer et al. 1987), Na^+ and Ca^{2+} (Cramer et al. 1987), and Cl^- and NO_3^- (Xu et al. 2000), and salinity can also reduce phosphorous uptake (Martinez et al. 1996). Reduced Ca^{2+} and K^+ supply to the plant is often discussed in relation to salinity-induced growth inhibition.

The rate of water and nutrient transport from root to the shoots is reduced under salinity. Therefore, in addition to effects on mineral uptake, salinity may also affect nutrient transport and distribution in the plant (Lazof and Bernstein 1999). In pepper and tomato, the concentration of nutrient ions in the exudates increased by a factor of 2 or 3 following salinization, but the much reduced exudate flow (reduction by a factor of 17–20) resulted in lowered supply of nutrients to the shoots (Bernstein and Kafkafi 2002).

a. Reduced Ca^{2+} Supply

Calcium is an essential nutrient that is required for structural roles in the cell wall and membranes, as a countercation for inorganic and organic anions in the vacuole and as an intracellular messenger in the cytosol (White and Broadley 2003; see also Chapter 20). It plays a unique role in the response of plants to saline conditions (reviewed by Rengel 1992; Knight et al. 1997; Lazof and Bernstein 1998; Cramer 2002b). The importance of Ca^{2+} for the function of the plasma membrane of the root cells under salinity is well documented. High concentration of Na^+ can displace Ca^{2+} involved in pectin-associated cross-linking and Ca^{2+} present at the binding sites of the root cell membranes, thereby affecting cell-wall and membrane stability and cellular functions. NaCl salinity was demonstrated to displace membrane-associated Ca^{2+} from plasma membrane of root cells and to release Ca^{2+} from subcellular compartments that accumulate Ca^{2+} (reviewed by Bernstein and Kafkafi 2002).

The reduction in cell growth when Ca^{2+} of the cell membrane is replaced by Na^+ was attributed to potassium leakage from the cells (Leigh and Wyn Jones 1984). The salinity induced membrane depolarization (discussed in Section V.A), and the decrease in Ca^{2+} reduces K^+/Na^+ selectivity of the root cell membrane (further discussed in Section IV.2.b), resulting in increased Na^+ influx and K^+ leakage (Rubio et al. 2003; Horie et al. 2006; Tuna et al. 2007). The alleviating effects of supplemental Ca^{2+} may result from reduction of Ca^{2+} displacement and thereby preservation of plasma membrane and cell-wall integrity, K^+/Na^+ selectivity, and reduction of Na influx (Zhong and Läuchli 1993b; Cramer 2002b; Rubio et al. 2003). Elevated Ca^{2+} concentrations in the root zone have long been known to negate part of the salinity-induced root and shoot growth inhibition (LaHaye and Epstein 1971; Rengel 1992; Zhong and Läuchli 1993b; Bernstein et al. 1993a; Lazof and Bernstein 1998 and references therein; Reid and Smith 2000; Cramer 2002b; Shabala et al. 2003; Bolat et al. 2006; Melgar et al. 2006). Root growth of different species and different cultivars responds differently to supplement Ca^{2+} when salinized (Cramer 2002b). Salt-tolerant genotypes (barley: Brittisnich et al. 1989; melon: Yermiyahu et al. 1997) demonstrate lower displacement of membrane-associated Ca^{2+} by NaCl than salt-sensitive genotypes. This indicates that the partial alleviation of NaCl-induced root growth inhibition by Ca^{2+} may be related in part to the quantity of membrane-associated Ca^{2+} as well as to the intensity of Ca^{2+} membrane association.

Calcium plays an important role also in water transport in salt-stressed plants. Its concentration in the growing solution can determine the restoration of root hydraulic conductivity (Azaizeh and Steudle 1991; Martínez- Ballesta et al. 2003b; Cabañero et al. 2004). Results for melon and pepper plants suggest that NaCl decreases the passage of water through the membrane and roots by reducing the activity of Hg-sensitive aquaporins, and the ameliorative effect of Ca^{2+} on NaCl stress could be related to regulation of aquaporin function (Carvajal et al. 1999, 2000; Cabañero et al. 2004).

b. Reduced K^+ Supply

Appropriate concentration of K^+ in plant cells is essential for growth and metabolic processes (Leigh and Wyn Jones 1984). However, K^+ and Na^+ ions compete for entry into plant cells; thereby, under saline conditions, plants may suffer potassium deficiency (Maathuis and Amtmann 1999; Schachtman 2000). Increasing concentration of NaCl significantly inhibits K^+ influx into roots and increases the loss of intracellular K^+ (Lynch and Läuchli 1984; Cramer et al. 1987). Consequently, K^+ concentration in the plant is often reduced under salinity (Kronzucker et al. 2008). In barley, Na^+-activated net K^+ efflux was negatively correlated with salinity tolerance pointing at the relation to root K^+ status (Chen et al. 2005). Furthermore, most plants use K^+, rather than sodium, as a component for osmotic adjustment; therefore, reduced K^+ root intake under salinity may increase synthesis of organic osmolytes and have a negative effect on the energy balance in the plant.

The selectivity of K^+ over Na^+ for root uptake is of major importance for plant function under saline conditions and was the center of much research (reviewed by Zhang et al. 2010). The net root selectivity of K^+ over Na^+ differs between species and particularly between mono- and dicotyledonous halophytes (Flowers and Colmer 2008), with a mechanism which is not yet clear. Salt resistance in bread wheat is linked with a locus on the D genome that results in low Na^+ uptake and increased K^+/Na^+ discrimination (Munns et al. 2000), and the discrimination was also found in other species in the triticea which contain the D genome (e.g., Colmer et al. 2006; Gorham et al. 1990). The K^+/Na^+ discrimination trait seems to be controlled by a single gene locus, *Kna1* (Dvořák and Gorham 1992), and the gene has been cloned as HKT1;5 (Byrt et al. 2007). In some studies, no relationships were identified between K^+/Na^+ discrimination or K^+/Na^+ ratio and salt sensitivity, suggesting that high K^+/Na^+ selectivity may not be a reliable criterion for salinity resistance in some species (wheat: Schachtman et al. 1991; bread wheat: Genc et al. 2007; *Brassica*: He and Cramer 1993).

The effect of Ca^{+2} on K^+/Na^+ discrimination varies with root cell age. In young growing cells, salinity reduced the concentrations and deposition rates of K^+ and Ca^{2+} and increased Ca^{2+}supply enhanced K^+/Na^+ selectivity, thereby maintaining high K^+ concentration and root elongation. In more mature cells further away from the root tip, where the vacuole is more developed, the beneficial effects of Ca^{+2} diminish (Nakamura et al. 1990; Zhong and Läuchli 1994) probably because the selective properties of the tonoplast become more important.

3. Cellular Compartmentation

Salinity affects ion concentrations in subcellular compartments of root cells. Sequestration of Na^+ and Cl^- in the vacuoles is an adaptation which facilitates maintenance of lower concentrations in the cytosol. The concentration at which Na^+ and Cl^- become toxic in the cytoplasm of the root cell is not known. Direct measurements with ion-selective electrodes indicate the Na^+ concentration in the cytosol of salt-stressed root cells ranges from 10 to 30 mM (Carden et al. 2003). In some cases, the extent of stress-induced changes of ion concentration in subcellular compartments was found to differ among taxa differing in sensitivity to salinity. Specifically, reduced Na^+ and Cl^- accumulation in the cytoplasm were associated with salt tolerance (Hajibagheri et al. 1987, 1989); reduced P and K^+ concentrations in the cytoplasm were associated with salt sensitivity (Koyro and Stelzer 1988); increased Na^+, Cl^-, K^+, Mg^{2+}, Ca^{2+}, P, and S concentrations in vacuoles of growing root cells were associated with salt tolerance; and in mature cells of the halophyte *Suaeda maritima*, concentrations of Na^+ and Cl^- in the vacuole were higher than in the cytoplasm or the cell walls (Hajibagheri and Flowers 1989). Changes in the capacity for Na^+ compartmentation in the vacuole can be achieved by alteration of the expression levels of the tonoplast transporters (NHX, AVP1 H^+). Overexpression of AVP1 (Gaxiola et al. 2001) and NHX (Apse et al. 1999; Chen et al. 2007) resulted in increased tolerance to salinity suggesting a role for Na^+ compartmentation in the vacuole in salinity tolerance.

C. Oxidative Stress

Exposure to salinity increases the production of ROS such as $^1O_2$, $O^{\bullet}{}_2{}^-$, $OH^{\bullet}$, and H_2O_2 in the plant cells due to impaired electron transport processes in chloroplasts and mitochondria, induction of metabolic pathways such as photorespiration, and perturbation of the functioning of photosynthetic components (Dat et al. 2000; Sivakumar et al. 2000; Logan 2005). The chief toxicity of ROS has been attributed to an ability to initiate cascade reactions that result in the production of $OH^{\bullet}$ and other destructive species such as lipid peroxides.

The plant antioxidant defense system is comprised of enzymatic and nonenzymatic components which diminish the excess of ROS (Dat et al. 2000; Apel and Hirt 2004). Under salinity stress, ROS production can exceed the scavenging capacity and accumulate to levels that can damage cell components, and increased activity of enzymes that detoxify the ROS is often induced (Mitler et al. 2002). In addition to their damaging effects, ROS are also required for useful functions such as growth processes and signaling, and apoplastic ROS were suggested to play a role in cell expansion via effects on cell-wall loosening (Rodriguez et al. 2002; Foreman et al. 2003; Gomez et al. 2004; Rubio et al. 2009; Shoresh et al. 2011). Therefore, a tight regulation of ROS concentration is required for optimal function.

The effect of salt stress on the antioxidant responses in leaves was widely studied, and many studies found differences in levels of expression or activity of antioxidant enzymes, which are also associated with level of salinity tolerance of genotypes. Unlike leaves, only scarce information is available concerning the effects of salt stress on the oxidative and antioxidative response of roots (Panda and Upadhyay 2003; Bandeoglu et al. 2004; Mittova et al. 2004; Tsai et al. 2004; Kim et al. 2005; de Azevedo Neto et al. 2006; Cavalcanti et al. 2007; Seckin et al. 2009; Bernstein et al. 2010). Some of the results suggest relations between activity of the antioxidative system in the root and genotype tolerance to salinity (e.g., Demiral and Turkan 2005; de Azvedo Neto et al. 2006).

Since ROS level in the tissue is the net result of their production and detoxification, an increase in their concentration under salinity demonstrates inability of the protective antioxidant system to overcome the excess production by the stress and suggests contribution of oxidative damage to plant sensitivity. Concomitant with the variability that exists among plants in root sensitivity and tolerance mechanisms to salinity, accumulation of ROS in roots also varies with plant species and stress level. Hydrogen peroxide (H_2O_2) was shown to accumulate in the apoplast and the cytoplasm of young growing cells and mature root cells of broccoli already during the first 24 h of stress (Hernandez et al. 2010), as well as in roots of rice (Tsai et al. 2004) and *Lemna minor* L. (Panda and Upadhyay 2003). However, reduction or no change in its concentration in roots under salinity was documented in other studies (rice: Lee et al. 2001; barley: Kim et al. 2005). Like in the leaves of maize (Bernstein et al. 2010; Shoresh et al. 2011; Kravchik and Bernstein 2012), a higher accumulation of ROS under salinity was found in the growing root cells compared to the mature root cells (Hernandez et al. 2010), and the accumulation was localized to the mitochondria (Mittova et al. 2004; Leshem et al. 2007; Shi et al. 2007; Hernandez et al. 2010). This higher accumulation in the growing cells may represent higher sensitivity of the young growing cells or reflect requirements of ROS for cell expansion.

Accumulation of malondialdehyde (MDA), a secondary breakdown product of lipid peroxidation in the plant, is considered to be a marker of oxidative damage, reflecting lipid peroxidation. MDA increased under salinity in broccoli roots in correlation with accumulation of H_2O_2 (Hernandez et al. 2010), as well as in *Lotus japonicus* (Regel) K. Larsen despite the maintenance of antioxidant levels (Rubio et al. 2009), suggesting insufficient scavenging activity to overcome the salinity-induced overproduction. In maize, exposure to salinity reduced ROS content (Bustos et al. 2008), and the level of MDA was lower under salt stress compared to nonstressed roots throughout the cell developmental gradient in both the primary and adventitious roots (Bernstein et al. 2010), suggesting sufficient detoxification by the antioxidative machinery. Although these roots did not seem to experience oxidative damage under salinity, root activity was reduced by salinity indicating that other destructive salt effects distress roots (Bernstein et al. 2010). A similar reduction in MDA occurred also in a salt-tolerant maize line in parallel with the activity of antioxidant enzymes (de Azvedo Neto et al. 2006). An increase in MDA in a salt-sensitive, but not a salt-tolerant, variety of rice (Demiral and Turkan 2005) indicates as well a role for protection of roots from oxidative damage in salt sensitivity in maize.

The salt-stress-induced changes in the activity of antioxidant enzymes are spatially variable in the plant tissues and temporally regulated with duration of exposure to the stress or with the developmental stage of the root cells. For example, results for maize indicated variability in salinity-induced effects on activity of scavenging enzymes at different cell developmental stages of the root, between primary and adventitious roots and between leaves and roots (Bernstein et al. 2010). In barley, the increase in activity of the antioxidant enzymes was more significant and consistent in the root compared to the shoot (Kim et al. 2005), while in maize the increase under salinity was more pronounced in the shoot (Bernstein et al. 2010). In *Brassica oleracea* L., activity of antioxidant enzymes in the roots changed with duration of exposure to salinity and the level of salinity employed (Hernandez et al. 2010).

The dual role of ROS in growth stimulation, on one hand, and inhibition by oxidative damage on the other should be considered when evaluating the potential involvement of ROS and the antioxidant machinery in the physiology of the stressed plant. The observed differences in spatial and temporal induction of the antioxidant system between organs, cell development stages, duration of exposure, and the level of stress reflect the inherent variability in oxidative response to salinity. The capacity for antioxidant response seems to take part in determination of organ or plant sensitivity to the salinity-induced oxidative response. The physiological and molecular bases for the reduced sensitivity to salinity of roots compared to leaves are not known. In a recent study with maize, we have suggested that this reduced sensitivity reflects lower oxidative damage under salinity (Bernstein et al. 2010). Furthermore, contrary to leaves, Monshausen et al. (2007) demonstrated that ROS needs to be reduced in root tips to allow growth. Whether this difference in ROS requirement for growth by roots and leaves is the reason for their different sensitivity to salt-induced growth reduction needs to be further explored.

In the nonstressed plant, the balance between stimulatory and inhibitory activities depends on the type of ROS and its concentration (Knight 2007). Experimental evidence suggests that $O_2^{\bullet -}$ produced in the elongation zone of roots is required for cell elongation and consequently participates in the regulation of root elongation. However, salinity-induced growth inhibition was found not to be associated with changes in apoplastic $O_2^{\bullet -}$ levels (maize: Bustos et al. 2008).

V. Root Membrane Properties Involved in Sensitivity and Tolerance

The ability of plants to regulate shoot ion composition relies in part on uptake and transport processes in their roots. Cell membranes are the major sites for controlling active and passive solute and water flux. Membrane characteristics of root cells are therefore of special interest in the study of growth regulation under salt stress.

A. Initial Responses

The plasma membrane ATPase pumps generate and maintain an electrochemical gradient essential for ion and solute transport. Under salinity, specific salts and the total salt concentration in the root-growing solution can influence the H^+ gradients across the plasma membrane and thereby reduce the acquisition of nutrients that are transported into roots along these gradients. The membrane potential depolarizes rapidly under salinity but in many cases recovers within minutes (i.e., Cheeseman et al. 1985; Läuchli and Schubert 1989; Katsuhara and Tazawa 1990), suggesting that the change under salinity is usually transitory and, in the long term, the membrane potential remains unchanged. Salinization was demonstrated to induce membrane depolarization of about 70–80 mV (Shabala et al. 2005, 2006; Cuin and Shabala 2005), which results, for example, in the activation of depolarization-activated outward-rectifying K^+ channels (Véry and Sentenac 2003; Shabala et al. 2006), leading to an immediate loss of K^+ from the cell (Cuin and Shabala 2005; Shabala et al. 2006; Chen et al. 2007). Such depolarization-activated K^+ channels are primarily responsible for NaCl-induced K^+ efflux in barley (Chen et al. 2007) and *Arabidopsis* (Shabala et al. 2006). However, in wheat roots (Cuin et al. 2008), the extent of depolarization after the initial "shock" was lower (in the range of 40–65 mV) and did not correlate with the extent of the salt-induced K^+ efflux and demonstrated only a small recovery.

B. Membrane Composition

Mineral imbalances of the root medium, common in saline environments, often affect the chemical composition and structure of root cell membranes (Diepenbrock 1985; Wu et al. 1998, 2005; López-Péreza et al. 2009). Salinization caused a decrease in root membrane fluidity that was attributed to changes in the weight ratio of lipids to proteins but not to modifications in lipid composition (cf. Chung and Matsumoto 1989; Blits and Gallagher 1990; Borochov-Neori and Borochov 1991; Wu et al. 1998). The higher protein density may provide a change in membrane surface charge (Suhayda et al. 1990).

The selectivity of biological membranes to ions is determined by membrane parameters such as the size and charge of the polar head group, the length of the fatty acid chains, and the interaction of phospholipids with sterol. The fine structure and lipid composition of the membranes with their bound proteins are genetically encoded, and the expression of that code changes to a certain degree under the influence of salts in the external solution. Membrane fatty acid unsaturation increases under salinity (López-Péreza et al. 2009) and was suggested to affect cell salinity tolerance (Allakhverdiev et al. 1999). The increase of free sterol/phospholipid ratio and total sterol/phospholipid ratio under salinity might interfere with plasma membrane structural configuration (Mansour et al. 2002), and a shift to a more planar sterols might be advantageous in ion exclusion (Yahya et al. 1995). Since ion permeability and ATPase activity are influenced by membrane lipid composition, lipids have the capacity to regulate ion movement into the roots through their influence on both passive and active transport processes (Douglas and Walker 1984). Salt-induced changes in the root lipid composition have been correlated with the relative abilities of different plant taxa to tolerate salinity or adapt to it (Lin et al. 1996; Douglas and Walker 1984; López-Péreza et al. 2009).

VI. Root Hydraulic Conductivity

A common effect of several abiotic stresses including salinity is to cause tissue dehydration. Such dehydration is caused by an imbalance between root water uptake and leaf transpiration.

Water moves through roots in response to a water potential gradient mostly generated by transpiration. Knowledge of the driving forces and the resistance that control the movement of water through the soil–plant continuum is important for understanding the effect of salinity on root function and its integration with shoot responses.

Water uptake during growth maintains turgor pressure, which provides the driving force for cell expansion. The hydraulic conductivity of the water uptake pathway and the osmotic potential gradient between the growing cell and the source of water regulate the ability of the tissue to supply water to growing cells and might therefore be related to cell expansion growth and to overall root growth.

Some studies suggest that growth of plants is limited by the root ability to transport water to the shoot (Sánchez-Blanco et al. 1991; Alarcón et al. 1994) and that root biomass does not reflect the ability of the root system to absorb and conduct water (Krasowski and Caputa 2005). This issue is controversial. Changes in hydraulic conductivity of water pathways were suggested to play a role in inhibition of root growth following salinization. Root water uptake rate and hydraulic conductivity decrease under salinity (Silva et al. 2008), and salinity was demonstrated to reduce the hydraulic conductivity and growth of many plants including bean, lupin, maize, soybean, broccoli, and *Arabidopsis* (Munns and Passioura 1984; Joly 1989; Evlagon et al. 1990; Peyrano et al. 1997; Bastías et al. 2004; Muries et al. 2011; Sutka et al. 2011) but not of barley, sunflower, a different tomato cultivar, and olive (Shalhevet et al. 1976; Munns and Passioura 1984; Rewald et al. 2011). The conflicting reports may arise from inherent differences between roots of different cultivars or species or different types of roots in the plant. In olive, axial root hydraulics under salt stress reacted in a more plastic fashion than branch conductivities. Increased specific conductivity of roots, different plasticity of root hydraulics, and modifications of the mean conduit diameter could not account for the observed differences in salt resistance. Instead, a high within-population variability in root conductivity and increased variability in conduit size were suggested to represent favorable traits that enhance water uptake from soils with heterogeneous salinity (Rewald et al. 2011).

In addition to possible damaging effects on shoot–water relations, an increased root hydraulic resistance under salinity was suggested to be of adaptive nature, restricting the loss of water from the root to the surrounding medium (Rodrigez et al. 1997). It remains to be established whether reduced root conductivity is a significant factor for salt tolerance (Shannon et al. 1994) or if more salt-tolerant species have higher root hydraulic conductivities (An et al. 2003). Increased axial conductivity may balance an increase in radial resistance or compensate the decrease in root system size under salinity.

Parallel pathways—apoplastic, symplastic, and transmembrane—play a role during the passage of water through the root. Each of these pathways can contribute to overall root radial hydraulic resistance (Steudle and Peterson 1998). Reduction of root hydraulic conductance under salt stress seems to involve changes in the symplastic transcellular pathways. Movement through water channels (aquaporins) is an important component in the regulation of water movement in the transcellular pathway (Steudle and Peterson 1998; see also Chapter 24). They allow water to pass freely across cellular membranes, following osmotic or hydrostatic pressure gradients (Chrispeels and Maurel 1994). Reduced activity, or abundance, of water channels in the roots under salinity decreased sap flow and osmotic pressure-dependent hydraulic conductance in root systems (Carvajal et al. 1999; Martinez-Ballesta et al. 2003a). Moreover, increased expression of aquaporins of the PIP1 and PIP2 subfamilies and increased membrane lipid unsaturation in roots allowed to maintain membrane permeability to water (broccoli: López-Péreza et al. 2009). Aquaporins belonging to families other than the PIP aquaporins were observed as well to involve in the regulation of water uptake under salt stress (TIP aquaporin: Peng et al. 2007; Wang et al. 2011. NIP aquaporins: Gao et al. 2010).

The initial decrease in root hydraulic conductivity following salinization may be caused by an osmotic shock as a result of an aquaporin conformational change caused by negative pressures (Wan et al. 2004). Stepwise salinization reduced the root hydraulic conductivity value of maize root cortical cells to a lesser extent than one-step salinization that results in a higher osmotic shock (Wan 2010). At the same time, root hydraulic conductivity could decrease by a direct effect of Na^+

on aquaporin function (Carvajal et al. 1999). The initial reduction of root hydraulic conductivity correlated with a downregulation of PIP aquaporin genes (Martínez-Ballesta et al. 2003a; Boursiac et al. 2005). It resulted from internalization of plasma membrane vesicles containing PIP proteins (Boursiac et al. 2005) and was correlated with an increase in the percentage of water moving via the apoplastic path (Martínez-Ballesta et al. 2003a). Thereby, in the initial phase of salt stress (first few hours), the decrease in root hydraulic conductivity and water uptake is caused mainly by an osmotic shock (discussed by Aroca et al. 2012). At later stages of salt-stress exposure (after a few days), a partial or total recovery of root hydraulic conductivity was reported for some species (*Arabidopsis*: Martínez-Ballesta et al. 2003a; maize: Wan 2010) and is accompanied by an increase in suberin contents in root endodermis and/or exodermis cells (Schreiber et al. 2005). This suberization potentially diminishes apoplastic water flow and Na^+ and Cl^- entrance into the xylem (Zimmermann et al. 2000; Ranathunge and Schreiber 2011). Therefore, the partial recovery of the root hydraulic conductivity during long-term salinity exposure results from an enhancement of the cell-to-cell pathway since the apoplastic pathway is inhibited. This notion is supported by the accumulation of PIP proteins in roots under long-term salinization (from 3 to 15 days) (Marulanda et al. 2010; Muries et al. 2011), which should favor the cell-to-cell pathway (Aroca et al. 2012). The fact that salinity-reduced hydraulic conductivity of roots was not accompanied by changes in the content of lignin-like polymers or in the activity of syringaldazine oxide (an enzyme associated with lignification of cell walls) (Peyrano et al. 1997) also suggests that the NaCl-induced reduction in hydraulic conductance involves changes in the symplastic pathway.

Measurements with cell pressure probe revealed that salinization reduced the root cells' hydraulic conductivity by 30%–66%. From the hydrostatic and osmotic relaxation of turgor, the hydraulic conductivity of cortical cells was also found to decrease significantly by salinization (Azaizeh and Steudle 1991; Azaizeh et al. 1992).

VII. Root Cell-Wall Properties Involved in Salt Sensitivity and Tolerance

Plant growth may be defined as an irreversible increase in size resulting from cell division and cell expansion. Cell expansion is thought to be controlled by cell-wall mechanical properties and by cell turgor. Since root growth inhibition under salt stress is not necessarily caused by a decrease in turgor (see Section IV.A), information about effects of salinity on the cell wall is valuable for understanding growth inhibition and maintenance processes.

A. Mechanical Properties (Extensibility)

According to the combined growth equation of Lockhart (1965), cell expansion rate is a function of the wall extensibility coefficient, hydraulic conductivity, the osmotic potential gradient between the growing cell and the ambient water, and the wall yield threshold value.

Since long-term reduction of root elongation by salinity is not caused by salinity-induced loss of the capacity to maintain osmotic potential gradient, or turgor, in the growing cells of the root (Zimmermann et al. 1992; Neumann et al. 1994; Pritchard 1994, Section IV.A), it seems to involve hardening of cell walls of the expanding cells. The apparent yield threshold pressure is higher in salinized maize root tips than in nonsalinized ones (Neumann et al. 1994; Pritchard et al. 1994), and cell-wall extensibility is lower under long-term exposure to stress (Neumann et al. 1994; Nonami et al. 1995). Changes in physical properties of cell walls are specific to elongation cells and are influenced more by sodium than by calcium salts (Nonami et al. 1995).

Root elongation rate recovers rapidly after exposure to salinity. Recovery from a moderate osmotic shock is completed within an hour (Frensch and Hsiao 1994), and recovery from 150 mM NaCl takes occurs within a day (Munns 2002). Unlike leaves, these recoveries take place despite the turgor not being fully restored, indicating differences, yet unknown, in changes in cell-wall properties between the two organs.

B. Cell-Wall Composition and Ultrastructure

Changes in the mechanical properties of cell walls following salinization must be caused by modification in wall composition and structure. Salinity can increase or decrease biosynthesis of cell-wall material depending on the plant species and the stress level (reviewed by Bernstein and Kafkafi 2002). The extent of salinity-induced reduction in cell-wall polysaccharide content often correlates with the level of sensitivity to salinity (Suhayda et al. 1994). In maize roots, inhibition of biosynthesis of cellulosic material and its prevention by supplemental Ca^{+2} parallels the extent of growth restriction (Zhong and Läuchli 1988, 1993a); in halophytes, production of large amounts of cell-wall material might involve in growth protection by increasing the Ca-binding capacity of the wall (Binet 1985), and salinity-induced wall growth of xylem parenchyma cells in roots was suggested to allow the action of transfer cells (i.e., removal of Na from the xylem stream; Yeo et al. 1977). Proteomic analysis revealed upregulation under salinity of two cell-wall-related proteins putatively involved in wall reinforcement and biosynthesis in root cells of a salt-tolerant but not salt-sensitive genotype of tomato (caffeoyl-CoA *O*-methyltransferase 6 (AM37) which plays a role in lignin biosynthesis, and UDPglucose: protein transglucosylase-like protein: Manaa et al. 2011).

Salt-induced Ca^{2+} deficiency may account for the decrease in polysaccharide content in barley roots, and salt tolerance in wild barley was suggested to involve high pectic polysaccharide content in the cell wall, where polygalacturonan regions can cross-link by Ca^{2+}(Suhayda et al. 1994). Salinization increased the amount of polysaccharides of intermediate molecular size and decreased that of small-size molecules, indicating a possible inhibition of polysaccharide degradation (Zhong and Läuchli 1993a). This change was not observed with supplemental Ca^{2+}.

Salinity increased cell-wall protein and aromatic compounds in barley roots (Suhayda et al. 1994) and thereby the potential for oxidative coupling of proteins and phenolic entities. Since covalent bonding of wall residues is a mechanism by which cell-wall extensibility may be controlled, root growth inhibition under salinity may involve increased cross-linking of cell-wall polymers. Salinity-induced deposition of β-1,3-D-glucan, callose, in the walls of cortical root cells may hinder as well the expansion of root cells (Koryo 1997).

Salt stress affects the orientation of cell-wall cellulose microfibrils in roots (Koryo 1997). The possible involvement of the observed stress-induced changes in cell-wall ultrastructure in the processes of growth inhibition is unclear at this time.

C. Sorption Capacity of Cell Walls

The cell walls affect root and plant function also by their capacity for ion sorption. Cation binding to cell walls has been implicated in a range of physiological processes including wall extensibility and therefore cell elongation. Ion exchange between the medium and ionogenic groups in the polymeric cell-wall matrix modifies the composition of the microenvironment and therefore may influence the availability and entry of nutrients into plant roots. The sorption specificity of the wall for different cations, and its alterations under stress conditions, may therefore be of importance for growth sensitivity under saline conditions. Only few publications focused on plant cell walls as natural ion exchangers under salinity, and the accumulated data suggest a connection to salinity tolerance. Ca^{2+} and Na^+ are sorbed on identical sites of barley root-cell wall suggesting the possibility for Ca^{2+}/Na^+ competition for sorption (Stassart et al. 1981). The selectivity for Ca^{2+} of cell walls of bean, a glycophyte, is higher than in the halophyte *Cochlearia anglica* L. (Bigot and Binet 1986) in accordance with their growth sensitivity; the cation exchange capacity of the halophyte decreased under salt stress, but tended to rise in the glycophyte. Sorption of anions to cell walls, although low, may also change under saline conditions. Sorption of Cl^- to cell walls was found to increase with external NaCl concentrations in the 0–20 mM NaCl range (Richter and Dainty 1988). The apparent dissociation constant of ionogenic group in the root-cell walls strongly depends on a salt concentration in the solution (Meychik and Yermakov 2001), and high acidic properties of the cation exchangeable groups in cell walls are associated with optimal growth under salinity (Meychik et al. 2005).

VIII. Concluding Remarks

Salt stress imposes inhibition of root elongation in most plant species. The mechanism of root stress response and tolerance is complex and integrates numerous physiological aspects of regulation, metabolism, and biophysics. Furthermore, the variability in root sensitivity and tolerance to salinity between and within plant species is associated with a range of mechanisms of varied capacity to contribute to the integrated response. The complexity and variability of the response indicate that mechanistic understanding might need to be established independently for various plant systems.

Salinity is conventionally considered to reduce root growth by osmotic effects, specific ion effects, and/or oxidative stress. Therefore, investigations have centered on regulation of Na^+ uptake and compartmentation, interaction between Na^+ and acquisition of major mineral nutrients such as K^+ and Ca^{2+}, and turgor regulation. It is yet unknown whether Na^+, Cl^-, or other ions are the predominant growth-limiting factors for root growth, and the mechanism for sensing the change in Na^+ concentration is also unknown. Available data demonstrate that changes in turgor alone cannot fully explain the effect of salinity on root elongation. Unlike leaves, only scarce information is available concerning the effects of salt stress on the oxidative and antioxidative processes in roots. In accord with the variability that exists among plants in mechanisms of root response to salinity, accumulation of ROS in roots varies as well with plant species and stress level. The dual role of ROS in growth stimulation or induction of damage should be considered when evaluating the potential involvement of ROS and the antioxidant machinery in the physiology of the stressed root.

Regulatory effects of salinity on growth processes in the root have an effect on the dividing and expanding cells. Further spatial and temporal studies of growing root cells' response to salinity will aid to resolve the complexity of root responses to short- and long-term salt-stress exposures and its effect on whole-plant behavior.

References

Alarcón JJ, Sánchez-Blanco MJ, Bolarín MC, Torrecillas A. 1994. Growth and osmotic adjustment of two tomato cultivars during and after saline stress. *Plant Soil* 166:75–82.

Allakhverdiev SI, Nishiyama Y, Suzuki I, Tasakay Y, Murata N. 1999. Genetic engineering of the unsaturation of fatty acids in membrane lipids alters the tolerance of *Synechocystis* to salt stress. *Proc Natl Acad Sci U S A* 96:5862–5867.

Amtmann A, Sanders D. 1999. Mechanisms of Na^+ uptake by plant cells. *Adv Bot Res* 29:75–112.

An P, Inanaga S, Xiangjun L, Shimizu H, Tanimoto E. 2003. Root characteristics in salt tolerance. *Root Res* 12:125–132.

Apel K, Hirt H. 2004. Reactive oxygen species: Metabolism, oxidative stress and signal transduction. *Annu Rev Plant Biol* 55:373–399.

Apse MP, Aharon GS, Snedden WA, Blumwald E. 1999. Salt tolerance conferred by overexpression of a vacuolar Na^+/H^+ antiport in *Arabidopsis*. *Science* 285:1256–1258.

Apse MP, Sottosanto JB, Blumwald E. 2003. Vacuolar cation/H^+ exchange, ion homeostasis, and leaf development are altered in a T-DNA insertional mutant of AtNHX1, the *Arabidopsis* vacuolar Na^+/H^+ antiporter. *Plant J* 36:229–239.

Aroca R, Porcel R, Ruiz-Lozano JM. 2012. Regulation of root water uptake under abiotic stress Conditions. *J Exp Bot* 63:43–57.

Azaizeh H, Gunse B, Steudle E. 1992. Effects of NaCl and $CaCl_2$ on water transport across root cells of maize (*Zea mays* L.) seedlings. *Plant Physiol* 99:886–894.

Azaizeh H, Steudle H. 1991. Effects of salinity on water transport of excised maize (*Zea mays* L.) roots. *Plant Physiol* 97:1136–1145.

de Azevedo Neto ADD, Prisco JT, Eneas J, deAbreu CEB, Gomes E. 2006. Effect of salt stress on antioxidative enzymes and lipid peroxidation in leaves and roots of salt-tolerant and salt-sensitive maize genotypes. *Environ Exp Bot* 56:87–94.

Bandeoglu E, Eyidogan F, Yucel M, Oktem HA. 2004. Antioxidant responses of shoots and roots of lentil to NaCl-salinity stress. *Plant Growth Regul* 42:69–77.

Bao AK, Wang SM, Wu GQ, Xi JJ, Zhang JL, Wang CM. 2008. Overexpression of the *Arabidopsis* H^+-PPase enhanced the salt and drought tolerance in transgenic alfalfa (*Medicago sativa* L.). *Plant Sci* 176:232–240.

Bastías E, Fernandez-Garcia N, Carvajal M. 2004. Aquaporin functionality in roots of *Zea mays* in relation to the interactive effects of boron and salinity. *Plant Biol* 6:415–421.

Bernstein N, Ioffe M, Zilberstaine M. 2001. Salt-stress effects on avocado rootstock growth. I. Establishing criteria for determination of shoot growth sensitivity to the stress. *Plant Soil* 233:1–11.

Bernstein N, Kafkafi U. 2002. Root growth under salinity stress. In: *Plant Roots: The Hidden Half*, 3rd edn., eds. Y Waisel, A Eshel, U Kafkafi, pp. 787–819. New York: Marcel Dekker, Inc.

Bernstein N, Kravchik M, Dudai N. 2009. Salinity-induced changes in essential oil, pigments and salts accumulation in sweet basil (*Ocimum basilicum*), in relation to alterations of morphological development. *Ann Appl Biol* 156:167–177.

Bernstein N, Läuchli A, Silk WK. 1993a. Kinematics and dynamics of sorghum (*Sorghum bicolor* L.) leaf development at various Na/Ca salinities. I. Elongation growth. *Plant Physiol* 103:1107–1114.

Bernstein N, Meiri A, Zilberstaine M. 2004. Root growth of avocado [*Persea Americana* Mill] is more sensitive to salinity than shoot growth. *J Am Soc Hortic Sci* 129:188–192.

Bernstein N, Shoresh M, Xu Y, Huang B. 2010. Involvement of the plant antioxidative response in the differential growth sensitivity to salinity of leaves vs. roots during cell development. *Free Rad Biol Med* 49:1161–1171.

Bernstein N, Silk WK, Läuchli A. 1993b. Growth and development of sorghum leaves under conditions of NaCl stress: Spatial and temporal aspects of leaf growth inhibition. *Planta* 191:433–439.

Bernstein N, Silk WK, Läuchli A. 1995. Growth and development of sorghum leaves under conditions of NaCl stress: Possible role of some mineral elements growth inhibition. *Planta* 196:699–705.

Bigot J, Binet P. 1986. Exchange capacity and partial cation selectivity isolated from the roots of *Cochleria anglica* and *Phaseolus vulgaris* grown in media of different salinities. *Can J Bot* 64:955–958.

Binet P. 1985. Salt resistance and the environment of the cell wall of some halophytes. *Vegetatio* 61:241.

Bittisnich D, Robinson D, Whitecross M. 1989. Membrane-associated and intracellular free calcium levels in root cells under NaCl stress. In: *Plant Membrane Transport: The Current Position. Proceedings of 8th International Workshop on Plant Membrane*, ed. J Dainty, MI de Michelis, E Marré, F Rasi-Caldogno, pp. 681–682. Amsterdam, New York: Elsevier.

Blits KC, Gallagher JL. 1990. Salinity tolerance of *Kosteletzkya virginica*. II. Root growth, lipid content, ion and water relations. *Plant Cell Environ* 13:419–425.

Bolat I, Kaya C, Almaca A, Timucin S. 2006. Calcium sulfate improves salinity tolerance in rootstocks of plum. *J Plant Nutr* 29:553–564.

Borochov-Neori H, Borochov A. 1991. Response of melon plant to salt. 1 Growth, morphology and root membrane properties. *J Plant Physiol* 139:100–105.

Boursiac Y, Chen S, Luu DT, Sorieul M, van der Dries N, Maurel C. 2005. Early effects of salinity on water transport in *Arabidopsis* roots. Molecular and cellular features of aquaporin expression. *Plant Physiol* 139:790–805.

Burssens S, Himanen K, van de Cotte B et al. 2000. Expression of cell cycle regulatory genes and morphological alterations in response to salt stress in *Arabidopsis thaliana*. *Planta* 211:632–640.

Bush DS. 1995. Calcium regulation in plant cells and its role in signaling. *Annu Rev Plant Physiol Plant Mol Biol* 46:95–122.

Bustos D, Lascano R, Villasuso AL, Machado E, Senn ME, Cordoba A, Taleisnik E. 2008. Reductions in maize root-tip elongation by salt and osmotic stress do not correlate with apoplastic $O_2^{\bullet-}$ levels. *Ann Bot* 102:551–559.

Byrt CS, Platten JD, Spielmeyer W et al. 2007. HKT1;5-like cation transporters linked to Na^+ exclusion loci in wheat, Nax2 and Kna1. *Plant Physiol* 143:1918–1928.

Cabañero FJ, Martinez V, Carvajal M. 2004. Does calcium determine water uptake under saline conditions in pepper plants, or is it water flux which determines calcium uptake? *Plant Sci* 166: 443–450.

Carden DE, Walker DJ, Flowers TJ, Miller AJ. 2003. Single-cell measurements of the contributions of cytosolic Na^+ and K^+ to salt tolerance. *Plant Physiol* 131:676–683.

Carvajal M, Cerda A, Martínez V. 2000. Does calcium ameliorate the negative effect of NaCl on melon root water transport by regulating aquaporin activity? *New Phytol* 145:439–447.

Carvajal M, Martinez V, Alcaraz CF. 1999. Physiological function of water channels as affected by salinity in roots of paprika pepper. *Physiol Plant* 105:95–101.

Cavalcanti FR, Santos-Lima JPM, Ferreira-Silva SL, Viegas RA, Gomes-Silveira JA. 2007. Roots and leaves display contrasting oxidative response during salt stress and recovery in cowpea. *J Plant Physiol* 164:591–600.

Cécooli G, Ramos JC, Ortega LI, Acosta JM, Perreta MJ. 2011. Salinity induced anatomical and morphological changes in *Chloris gayana* Kunth roots. *Biocell* 35:9–17.

Chang PFL, Damsz B, Kononowicz AK, Reuveni M, Chen Zu Tang, Xu Yi, Hedges K, Tseng CC, Singh NK, Binzel ML, Narasimhan ML, Hasegawa PM, Bressan RA, Chen ZT, Xu-Y. 1996. Alterations in cell membrane structure and expression of a membrane-associated protein after adaptation to osmotic stress. *Physiol Plant* 98:505–516.

ChatterjeeJ, MajumderAL. 2010. Salt-induced abnormalities on root tip mitotic cells of *Allium cepa*: Prevention by inositol pretreatment. *Protoplasma* 245:165–172.

Cheeseman JM. 1988. Mechanisms of salinity tolerance in plants. *Annu Rev Plant Physiol* 87:547–550.

Cheeseman JM, Bloebaum PD, Wickens LK. 1985. Short term $^{22}Na^+$ and $^{42}K^+$ uptake in intact, mid-vegetative *Spergularia marina* plants. *Physiol Plant* 65:460–466.

Chen H, An R, Tang J-H et al. 2007. Over-expression of a vacuolar Na+/H+ antiporter gene improves salt tolerance in an upland rice. *Mol Breed* 19:215–225.

Chen Z, Newman I, Zhou M, Mendham N, Zhang G, Shabala S. 2005. Screening plants for salt tolerance by measuring K+ flux: A case study for barley. *Plant Cell Environ* 28:1230–1246.

Chen ZH, Pottosin II, Cuin TA et al. 2007. Root plasma membrane transporters controlling K^+/Na^+ homeostasis in salt stressed barley. *Plant Physiol* 145:1714–1725.

Chrispeels MJ, Maurel C. 1994. Aquaporins: The molecular basis of facilitated water movement through living plants? *Plant Physiol* 105:9–13.

Chung CG, Matsumoto H. 1989. Localization of the NaCl-sensitive membrane fraction in cucumber roots by centrifugation on sucrose density gradients. *Plant Cell Physiol* 30:1133–1138.

Clarkson DT. 1969. Metabolic aspects of aluminum toxicity and some possible mechanisms for resistance. *Br Ecol Soc Symp* 9:381–397 Oxford, U.K.: Blackwell Sci Pub.

Colmer TD, Flowers TJ, Munns R. 2006. Use of wild relatives to improve salt tolerance in wheat. *J Exp Bot* 57:1059–1078.

Cramer GR. 2002a. Response of abscisic acid mutants of *Arabidopsis* to salinity. *Func Plant Biol* 29:561–567.

Cramer GR. 2002b. Sodium-calcium interactions under salinity stress. In: *Salinity: Environment—Plants—Molecules*, eds. A Läuchli, U Lüttge, pp. 205–227. Dordrecht, the Netherlands: Kluwer.

Cramer GR, Bowman DC. 1991. Kinetics of maize leaf elongation. I. Increased yield threshold limits short-term, steady-state elongation rates after exposure to salinity. *J Exp Bot* 42:1417–1426.

Cramer GR, Lynch J, Läuchli A, Epstein E. 1987. Influx of Na^+, K^+, and Ca^{2+} into roots of salt-stressed cotton seedlings. Effects of supplemental Ca^{2+}. *Plant Physiol* 83:510–516.

Croser C, Renault S, Franklin J, Zwiazek JJ. 2001. The effect of salinity on the emergence and seedling growth of *Picea mariana*, *Picea glauca* and *Pinus banksiana*. *Environ Pollut* 115:9–16.

Cruz V, Cuartero J. 1990. Effects of salinity at several developmental stages of six genotypes of tomato (*Lycopersicon* spp.). In: *Eucarpia Tomato 90*, eds. J Cuartero, ML Gomes-Guilmon, R Fernández-Muñoz, pp. 81–86. *Proceedings of the XIth Eucarpia Meeting on tomato Genetics and Breeding*, Málaga, Spain.

Cuin TA, Shabala S. 2005. Exogenously supplied compatible solutes rapidly ameliorate NaCl-induced potassium efflux from barley roots. *Plant Cell Physiol* 46:1924–1933.

Cuin TC, Betts SA, Chalmandrier R, Shabala S. 2008. A root's ability to retain K^+ correlates with salt tolerance in wheat. *J Exp Bot* 59:2697–2706.

Dat J, Vandenabeele S, Vranova E, Van Montagu M, Inze D, Van Breusegen F. 2000. Dual action of the active oxygen species during plant stress responses. *Cell Mol Life Sci* 57:779–995.

Davies WJ, Kudoyarova G, Hartung W. 2005. Long-distance ABA signaling and its relation to other signaling pathways in the detection of soil drying and the mediation of the plant's response to drought. *J Plant Growth Regul* 24:285–295.

Demidchik V, Davenport RJ, Tester M. 2002. Nonselective cation channels. *Annu Rev Plant Biol* 53:67–107.

DemiralT, Türkan I. 2005. Comparative lipid peroxidation, antioxidant defense systems and proline content in roots of two rice cultivars differing in salt tolerance. *Environ Exp Bot* 53:247–257.

Diepenbrock W. 1985. The fatty acid composition of root membrane lipids from rape plants (*Brassica napus* L.) as affected by localized supply of K and Ca. *Agrochimica* 29:123–131.

Dodd I. 2005. Root-to-shoot signaling: Assessing the role of "up" in the up and down world of long-distance signaling in planta. *Plant Soil* 74:257–275.

Douglas TJ, Walker RR. 1984. Phospholipids, free sterols and adenosine triphosphate of plasma membrane-enriched preparations from roots of citrus genotypes differing in chloride exclusion ability. *Physiol Plant* 62:51–58.

Dupont FM. 1992. Salt-induced changes in ion transport: Regulation of primary pumps and secondary transporters. In: *Transport and Receptor Proteins of the Plant Membranes*, eds. DT Cooke, DT Clarkson, pp. 91–100. New York: Plenum Press.

Dvořák J, Gorham J. 1992. Methodology of gene transfer by homoeologous recombination into *Triticum turgidum*: Transfer of K^+/Na^+ discrimination from *Triticum aestivum*. *Genome* 35:639–646.

Espinosa-Ruiz A, Belles JM, Serrano R, Culianez-Macia FA. 1999. *Arabidopsis thaliana* AtHAL3: A flavoprotein related to salt and osmotic tolerance and plant growth. *Plant J* 20:529–539.

Evlagon D, Ravina I, Neumann PM. 1990. Interactive effects of salinity and calcium on hydraulic conductivity, osmotic adjustment and growth in primary roots of maize seedlings. *Isr J Bot* 39:239–247.

Felle H. 1994. The H^+/Cl^- symporter in root-hair cells of *Sinapis alba*. An electrophysiological study using ion-selective microelectrodes. *Plant Physiol* 106:1131–1136.

Flowers TJ, Colmer TD. 2008. Salinity tolerance in halophytes. *New Phytol* 179:945–963.

Foreman J, Demidchik V, Bothwell JHF et al. 2003. Reactive oxygen species produced by NADPH oxidase regulate plant cell growth. *Nature* 422:442–446.

Fornes F, María Belda R, Carrión C, Noguera V, García-Agustín P, Abad M. 2007. Pre-conditioning ornamental plants to drought by means of saline water irrigation as related to salinity tolerance. *J Hortic Sci* 113:52–59.

Franco JA, Cros V, Vicente M, Martínez-Sánchez JJ. 2011. Effects of salinity on the germination, growth, and nitrate contents of purslane (*Portulaca oleracea* L.) cultivated under different climatic conditions. *J Hortic Sci Biotechnol* 86:1–6.

Frensch J, Hsiao TC. 1994. Transient responses of cell turgor and growth of maize roots as affected by changes in water potential. *Plant Physiol* 104:247–254.

Galvan-Ampudia CS, TesterinkC. 2011. Salt stress signals shape the plant root. *Curr Opin Plant Biol* 14:296–302.

Gao Z, He X, Zhao B, Zhou C, Liang Y, Ge R, Shen Y, Huang Z. 2010. Overexpressing a putative aquaporin gene from wheat, TaNIP, enhances salt tolerance in transgenic *Arabidopsis*. *Plant Cell Physiol* 51:767–775.

Gaxiola RA, Li JS, Undurraga S et al. 2001. Drought- and salt tolerant plants result from overexpression of the AVP1 H^+-pump. *Proc Natl Acad Sci U S A* 98:11444–11449.

Genc Y, Mcdonald GK, Tester M. 2007. Reassessment of tissue Na^+ concentration as a criterion for salinity tolerance in bread wheat. *Plant Cell Environ* 30:1468–1498.

Gilliham M, Tester M. 2005. The regulation of anion loading to the maize root xylem. *Plant Physiol* 137:819–828.

Gorham J. 1990. Salt tolerance in the Triticeae: Ion discrimination in rye and Triticale. *J Exp Bot* 41:609–614.

Hajibagheri MA, Flowers TJ. 1989. X-ray microanalysis of ion distribution within root cortical cells of the halophyte *Suaeda maritima*. *Planta* 177:131–134.

Hajibagheri MA, Harvey DMR, Flowers TJ. 1987. Quantitative ion distribution within root cells of salt-sensitive and salt-tolerant maize varieties. *New Phytol* 105:367–379.

Hajibagheri MA, Yeo AR, Flowers TJ. 1985. Salt tolerance in *Suaeda maritime* (L.) Dum. Fine structure and ion concentrations in the apical region of roots. *New Phytol* 99:331–343.

Hajibagheri MA, Yeo AR, Flowers TJ, Collins JC. 1989. Salinity resistance in *Zea mays*: Fluxes of potassium, sodium and chloride, cytoplasmic concentrations and microsomal membrane lipids. *Plant Cell Environ* 12:753–757.

Haro R, Bañuelos MA, Senn MAE, Barrero-Gil J, Rodríguez-Navarro A. 2005. HKT1 mediates sodium uniport in roots. Pitfalls in the expression of HKT1 in yeast. *Plant Physiol* 139:1495–1506.

He T, Cramer GR. 1993. Salt tolerance of rapid-cycling brassica species in relation to potassium sodium ratio and selectivity at the whole plant and callus levels. *J Plant Nutr* 16:1263–1277.

Hernandez M, Fernandez-Garcia N, Diaz-Vivancos P, Olmos E. 2010. A different role for hydrogen peroxide and the antioxidative system under short and long salt stress in *Brassica oleracea* roots. *J Exp Bot* 61:521–535.

Hille B. 1992. *Ionic Channels of Excitable Membranes*, 2nd edn. Sunderland, MA: Sinauer Associates, Inc.

Horie T, Horie R, Chan WY, Leung HY, Schroeder JI. 2006. Calcium regulation of sodium hypersensitivities of sos3 and athkt1 mutants. *Plant Cell Physiol* 47:622–633.

Huang CY, Liau EC. 1998. The regulatory role of plasma membrane proton-pumping ATPase in salt tolerance of soybean plant growing under the salt-stress condition. *Taiwania* 43:225–234.

Huang J, Redman RE. 1995. Responses of growth, morphology and anatomy to salinity and calcium supply in cultivated and wild barley. *Can J Bot* 73:1859–1866.

Izzo R, Nayari-Izzo F, Quartacci MF. 1991. Growth and mineral absorption in maize seedlings as affected by increasing NaCl concentration. *J Plant Nutr* 14:678–699.

Joly RJ. 1989. Effects of sodium chloride on hydraulic conductivity of soybean root systems. *Plant Physiol* 91:1261–1265.

Karni L, Aktas H, Deveturero G, Aloni B. 2010. Involvement of root ethylene and oxidative stress-related activities in pre-conditioning of tomato transplants by increased salinity. *J Hortic Sci Biotechnol* 85:23–29.

Katsuhara M, Kawasaki T. 1996. Of stress induced nuclear and DNA degradation in meristematic cells of barley roots. *Plant Physiol* 37:169–173.

Katsuhara M, Mimura T, Tazawa M. 1990. ATP-regulated ion channels in the plasma membrane of Characeae alga, *Nitellopsis obtusa*. *Plant Physiol* 93:343–346.

Katsuhara M, Tazawa M. 1990. Mechanism of calcium-dependent salt tolerance in cells of *Nitellopsis obtusa*: Role of intracellular adenine nucleotides. *Plant Cell Environ* 13:179–184.

Katsuhara M, Yazaki Y, Sakano K, Kawasaki T. 1997. Intracellular pH and proton-transport in barley root cells under salt stress: In vivo ^{31}P-NMR study. *Plant Cell Physiol* 38:2155–2160.

Khayyat M, Rajaee S, Sajjadinia S, Saied E, Enayatollaha T. 2009. Calcium effects on changes in chlorophyll contents, dry weight and micronutrients of strawberry (*Fragaria × ananassa* Duch.) plants under salt-stress conditions. *Fruits* 64:53–59.

Khayyat M, Kiegle E, Moore C, Haseloff J, Tester M, Knight M. 2000. Cell-type specific calcium responses to drought, NaCl, and cold in *Arabidopsis* root: A role for endodermis and pericycle in stress signal transduction. *Plant J* 23:267–278.

Kim SY, Lim JH, Park MR et al. 2005. Enhanced antioxidant enzymes are associated with reduced hydrogen peroxide in barley roots under saline stress. *J Biochem Mol Biol* 38:218–224.

Knight H. 2000. Calcium signaling during abiotic stress in plants. *Int Rev Cytol* 192:269–324.

Knight MR. 2007. New ideas on root hair growth appear from the flanks. *Proc Natl Acad Sci U S A* 104:20649–20650.

Knight H, Trewavas AJ, Knight MR. 1997. Calcium signaling in *Arabidopsis thaliana* responding to drought and salinity. *Plant J* 12:1067–1078.

Koryo HW. 1997. Ultrastructural and physiological changes in root cells of sorghum plants (*Sorghum bicolor* S. Sudanensis cv. Sweet Sioux) induced by NaCl. *J Exp Bot* 308:693–706.

Koyro WW, Stelzer R. 1988. Ion concentrations in the cytoplasm and vacuoles of rhizodermis cells from NaCl treated sorghum, *Spartina* and *Puccinellia* plants. *J Plant Physiol* 133:441–446.

Kramer D. 1980. Transfer cells in the epidermis of roots. In: *Plant Membrane Transport: Current Conceptual Issues*, ed. RM Spanswick, WJ Lucas, J Dainty, pp. 393–394. Amsterdam, the Netherlands: Elsevier.

Krasowski M, Caputa A. 2005. Relationships between the root system size and its hydraulic properties in white spruce seedlings. *New Forests* 30:127–146.

Kravchik M, Bernstein N. 2012. Effects of salinity on the transcriptome of growing maize leaf cells points at differential involvement of the antioxidative response in cell growth restriction. *BMC Genomics*.

Kronzucker HJ, Szczerba MW, Schulze LM, Britto DT. 2008. Non-reciprocal interactions between K^+ and Na^+ ions in barley (*Hordeum vulgare* L.). *J Exp Bot* 59:2793–2801.

Kudla J, Batistic O, Hashimoto K. 2010. Calcium signals: The lead currency of plant information processing. *Plant Cell* 22:541–563.

Kurth E, Cramer GR, Läuchli A, Epstein E. 1986. Effects of NaCl and $CaCl_2$ on cell enlargement and cell production in cotton roots. *Plant Physiol* 82:1102–1106.

LaHaye PA, Epstein E. 1971. Calcium and salt toleration by bean plants. *Physiol Plant* 25:213–218.

Läuchli A, Schubert S. 1989. The role of calcium in the regulation of membrane and cellular growth processes under salt stress. In: *Environmental Stress in Plants*, ed. JH Cherry, NATO ASI Ser 19, pp. 131–138. Berlin, Germany: Springer.

Lazof DB, Bernstein N. 1998. The NaCl-induced inhibition of shoot growth: The case for disturbed nutrition with special consideration of calcium nutrition. *Adv Bot Res* 29:113–189.

Lazof DB, Bernstein N. 1999. Effect of salinization on nutrient transport to lettuce leaves: Consideration of leaf developmental stage. *New Phytol* 144:85–94.

Lee DH, Kim YS, Lee CB. 2001. The inductive responses of the antioxidant enzymes by salt stress in the rice (*Oryza sativa* L.). *J Plant Physiol* 158:737–745.

Leigh RA, Wyn Jones RG. 1984. A hypothesis relating critical potassium concentrations for growth to the distribution and functions of this ion in plant cell. *New Phytol* 97:1–13.

Leshem Y, Seri L, Levine A. 2007. Induction of phosphatidylinositol 3-kinase-mediated endocytosis by salt stress leads to intracellular production of reactive oxygen species and salt tolerance. *Plant J* 51:185–197.

Lin H, Salus SS, Schumaker KS. 1997. Salt sensitivity and the activities of the H^+-ATPases in cotton seedlings. *Crop Sci* 37:190–197.

Lin H, Wu L. 1996. Effects of salt stress on root plasma membrane characteristics of salt-tolerant and salt-sensitive buffalo grass clones. *Environ Exp Bot* 36:239–254.

Lockhart JA. 1965. An analysis of irreversible plant cell elongation. *J Theor Biol* 8:264–275.

Logan BA. 2005. Reactive oxygen species and photosynthesis. In: *Antioxidants and Reactive Oxygen Species in Plants*, ed. N Smirnoff, pp. 250–267. Oxford, U.K.: Blackwell.

Logan H, Basset M, Véry AA, Sentenac H. 1997. Plasma membrane transport systems in higher plants: From black boxes to molecular physiology. *Physiol Plant* 100:1–5.

López-Péreza L, Martínez-Ballestaa MdC, Maurelb C, Carvajal M. 2009. Changes in plasma membrane lipids, aquaporins and proton pump of broccoli roots, as an adaptation mechanism to salinity. *Phytochemistry* 70:492–500.

Lynch J, Läuchli A. 1984. Potassium transport in salt-stresses barley roots. *Planta* 161:295–301.

Maathuis FJM, Amtmann A. 1999. K^+ nutrition and Na^+ toxicity: The basis of cellular K^+/Na^+ ratios. *Ann Bot* 84:123–133.

Makela P, Munns R, Colmer TD, Peltonen-Sainio P. 2003. Growth of tomato and its ABA-deficient mutant (*sitiens*) under saline conditions. *Plant Physiol* 117:58–63.

Manaa A, Ben Ahmed H, Valot B, Bouchet JP, Aschi-Smiti S, Causse M, Faurobert M. 2011. Salt and genotype impact on plant physiology and root proteome variations in tomato. *J Exp Bot* 62:2797–2813.

Mansour MMF, Salama KHA, Al-Mutawa MM, Abou Hadid AF. 2002. Effect of NaCl and polyamines on plasma membrane lipids of wheat roots. *Biol Plant* 45:235–239.

Martínez-Ballesta MC, Aparicio F, Pallas V, Martínez V, Carvajal M. 2003a. Influence of saline stress on root hydraulic conductance and PIP expression in *Arabidopsis*. *J Plant Physiol* 160:689–697.

Martínez-Ballesta MC, Martinez V, Carvajal M. 2003b. Aquaporin functionality in relation to H^+-ATPase activity in root cells of *Capsicum annuum* grown under salinity. *Physiol Plant* 117:413–420.

Martinez V, Bernstein N, Läuchli A. 1996. Salt induced inhibition of phosphorus transport in lettuce leaves. *Physiol Plant* 97:118–122.

Mauromicale G, Licandro P. 2002. Salinity and temperature effects on germination, emergence and seedling growth of globe artichoke. *Agronomie* 22:443–450.

Melgar JC, Benlloch M, Fernandez-Escobar R. 2006. Calcium increases sodium exclusion in olive plants. *Sci Hortic* 109:303–305.

Mennen H, Jacoby B, Marschner H. 1990. Is sodium proton antiport ubiquitous in plant cells? *J Plant Physiol* 137:180–183.

Meychik NR, Nikolaeva JI, Yermakov IP. 2005. Ion exchange properties of the root cell walls isolated from the Halophyte Plants (*Suaeda altissima* L.) grown under conditions of different salinity. *Plant Soil* 277:163–174.

Meychik NR, Yermakov IP. 2001. Ion exchange properties of plant root cell walls. *Plant Soil* 234:181–193.

Mittler R. 2002. Oxidative stress, antioxidants and stress tolerance. *Trends Plant Sci* 7:406–410.

Mittova V, Guy M, Tal M, Volokita M. 2004. Salinity up-regulates the antioxidative system in root mitochondria and peroxisomes of the wild salt-tolerant tomato species *Lycopersicon pennellii*. *J Exp Bot* 55:1105–1113.

Monshausen GB, Bibikova TN, Messerli MA, Shi C, Gilroy S. 2007. Oscillations in extracellular pH and reactive oxygen species modulate tip growth of *Arabidopsis* root hairs. *Proc Natl Acad Sci U S A* 104:20996–21001.

Moore CA, Bowden HC, Scrase-Field S, Knight MR, White PJ. 2002. The deposition of suberin lamellae determines the magnitude of cytosolic Ca^{2+} elevations in root endodermal cells subjected to cooling. *Plant J* 30:457–465.

Munns R. 2002. Comparative physiology of salt and water stress. *Plant Cell Environ* 25:239–250.

Munns R, Cramer GR. 1996. Is coordination of leaf and root growth mediated by abscisic acid? *Plant Soil* 185:33–49.

Munns R, Hare RA, James RA, Rebetzke GJ. 2000. Genetic variation for improving the salt tolerance of durum wheat. *Aust J Agric Res* 51:69–74.

Munns R, Passioura JB. 1984. Hydraulic resistance of plants. III. Effects of NaCl in barley and lupin. *Aust J Plant Physiol* 11:351–359.

Muns R, Tester M. 2008. Mechanisms of salinity tolerance. *Annu Rev Plant Biol* 59:651–681.

Muries B, Faize M, Carvajal M, Martínez-Ballesta MC. 2011. Identification and differential induction of the expression of aquaporins by salinity in broccoli plants. *Mol Biosyst* 7:1322–1335.

Nakamura Y, Tanaka K, Ohta E, Sakata M. 1990. Protective effect of external Ca^{2+} on elongation and the intracellular concentration of K^+ in intact mung bean root under high NaCl stress. *Plant Cell Physiol* 31:815–821.

Neumann PM, Azaizeh H, Leon D. 1994. Hardening of root cell walls: A growth inhibitory response to salinity stress. *Plant Cell Environ* 17:303–309.

Neves-Piestun BG, Bernstein N. 2001. Salinity-induced inhibition of leaf elongation is not mediated by changes in cell-wall acidification capacity. *Plant Physiol* 125:1419–1428.

Neves-Piestun BG, Bernstein N. 2005. Salinity induced changes in the nutritional status of expanding cells may impact leaf growth inhibition in Maize. *Funct Plant Biol* 32:141–152.

Nonami H, Tanimoto K, Tabuchi A, Fukuyama T, Hashimoto Y. 1995. Salt stress under hydroponic conditions causes changes in cell wall extension during growth. *Acta Hortic* 396:91–98.

Ogawa A, Kitamichi K, Toyofuku K, Kawashima C. 2006. Quantitative analysis of cell division and cell death in seminal root of rye under salt stress. *Plant Prod Sci* 9:56–64.

Oh DH, Gong QQ, Ulanov A et al. 2007. Sodium stress in the halophyte *Thellungiella halophila* and transcriptional changes in a thsos1-RNA interference line. *J Integr Plant Biol* 49:1484–1496.

Panda SK, Upadhyay RK. 2003. Salt stress injury induces oxidative alterations and antioxidative defence in the roots of. *Lemna minor. Biol Plant* 48:249–253.

Pardo JM, Cubero B, Leidi EO, Quintero FJ. 2006. Alkali cation exchangers: Roles in cellular homeostasis and stress tolerance. *J Exp Bot* 57:1181–1199.

Passioura JB, Munns R. 2000. Rapid environmental changes that affect leaf water status induce transient sugars or pauses in leaf expansion rate. *Aust J Plant Physiol* 27:941–948.

Pérez-Alfocea F, Albacete A, Ghanem ME, Dodd IC. 2010. Hormonal regulation of source-sink relations to maintain crop productivity under salinity: A case study of root-to-shoot signalling in tomato. *Funct Plant Biol* 37:592–603.

Perica S, Goreta S, Selak GV. 2008. Growth biomass allocation and leaf ion concentration of seven olive (*Olea europaea* L.) cultivars under increased salinity. *Sci Hortic* 117:123–129.

Peng YH, Lin WL, Cai WM, Arora R. 2007. Overexpression of a *Panax ginseng* tonoplast aquaporin alters sat tolerance, drought tolerance and cold acclimation ability in transgenic *Arabidopsis* plants. *Planta* 226:729–740.

Peyrano G, Taleisnik E, Quiroga M, De Forchetti SM, Tigier H. 1997. Salinity effects of hydraulic conductance, lignin content and peroxidase activity in tomato roots. *Plant Physiol Biochem* 35:387–393.

Pitman MG. 1984. Transport across the root and shoot/root interactions. In: *Salinity Tolerance in Plants—Strategies for Crop Improvement*, eds. RC Staples, GH Toenniessen, pp. 93–123. New York: Wiley and Sons.

Plant P, Gelli A, Blumwald E. 1994. Vacuolar chloride regulation of an anion-selective tonoplast channel. *J Membr Biol* 140:1–12.

Powell MJ, Davies MS et al. 1986. The influence of zinc on the cell cycle in the root meristem of a zinc-tolerant and a non-tolerant cultivar of *Festuca rubra* L. *New Phytol* 102:419–428.

Prins HBA. 1995. Salt stress in *Plantago*. The role of membranes, channels and pumps. *Acta Phytopathol Entomol Hung* 30:1–2, 5–13.

Pritchard J. 1994. The control of cell expansion in roots. *Tansley Rev* 68; *New Phytol* 127:3–26.

Quan LJ, Zhang B, Shi WW, Li HY. 2008. Hydrogen peroxide in plants: A versatile molecule of the reactive oxygen species network. *J Integr Plant Biol* 50:2–18.

Rahnama A, Munns R, Poustini K, Michelle W. 2011. A screening method to identify genetic variations in root growth response to salinity gradient. *J Exp Bot* 62:69–77.

Ranathunge K, Schreiber L. 2011. Water and solute permeabilities of *Arabidopsis* roots in relation to the amount and composition of aliphatic suberin. *J Exp Bot* 62:1961–1974.

Rawson HM, Munns R. 1984. Leaf expansion in sunflower as influenced by salinity and short-term changes in carbon fixation. *Plant Cell Environ* 7:207–213.

Reid RJ. Smith FA. 2000. The limits of sodium/calcium interactions in plant growth. *Aust J Plant Physiol* 27:709–715.

Reinhardt DH, Rost TL. 1995. Salinity accelerated endodermal development and induces an exodermis in cotton seedling roots. *Environ Exp Bot* 35:563–574.

Rengel Z. 1992. The role of calcium in salt toxicity. *Plant Cell Environ* 15:625–632.

Rewald B, Leuschner C, Wiesman Z, Ephrath JE. 2011. Influence of salinity on root hydraulic properties of three olive varieties. *Plant Biosyst* 145:12–22.

Rewald B, Rachmilevitch S, Ephrath, JE. 2010. Salt stress effects on root systems of two mature olive cultivars. *Acta Hortic* 888:109–118.

Richter C, Dainty J. 1988. Ion behavior in plant cell walls. II. Measurement of the Donnan free space, anion-exclusion space, anion-exchange capacity and cation-exchange capacity in delignified *Sphagnum russowi* cell walls. *Can J Bot* 67:460–464.

Rodriguez P, Dell Amico J, Morales D, Sanchez Blanco MJS, Alarcon JJ. 1997. Effects of salinity on growth, shoot water relations and root hydraulic conductivity in tomato plants. *J Agric Sci* 4:439–444.

Rodriguez AA, Grunberg KA, Taleisnik EL. 2002. Reactive oxygen species in the elongation zone of maize leaves are necessary for leaf extension. *Plant Physiol* 129:1627–1632.

Rubio MC, Bustos-Sammamed P, Clemente MR, Becana M. 2009. Effects of salt stress on expression of antioxidant genes and proteins in the model legume *Lotus japonicus. New Phytol* 181:851–859.

Rubio F, Flores P, Navarro JM, Martínez V. 2003. Effects of Ca^{2+}, K^+ and cGMP on Na^+ uptake in pepper plants. *Plant Sci* 165:1043–1049.

Samarajeewa PK, Barrero RA, Umeda-Hara C, Kawai M, Uchimiya H. 1999. Cortical cell death, cell proliferation, macromolecular movements and *r*Tipl expression pattern in roots of rice (*Oryza sativa* L.) under salt-stress. *Planta* 207:354–361.

Sánchez-Blanco MJ, Bolarín MC, Alarcón JJ, Torrecillas A. 1991. Salinity effects on water relations in *Lycopersicon esculentum* and its wild salt-tolerant relative species *L. pennellii. Physiol Plant* 83:269–274.

Schachtman DP. 2000. Molecular insights into the structure and function of plant K^+ transport mechanisms. *Biochim Biophys Acta* 1465.:127–139.

Schachtman DP, Tyerman SD, Terry BR. 1991. The K^+/Na^+ selectivity of a cation channel in the plasma membrane of root cells does not differ in salt-tolerant and salt-sensitive wheat species. *Plant Physiol* 97:598–605.

Seckin B, Sekmen AH, Turkan I. 2009. An enhancing effect of exogenous mannitol on the antioxidant enzyme activities in roots of wheat under salt stress. *J Plant Growth Regul* 28:12–20.

Sengupta S, Patra B, Ray S, Majumder AL. 2008. Inositol methyl transferase from a halophytic wild rice *Porteresia coarctata* Roxb. (Tateoka): Its expression pattern under abiotic stress. *Plant Cell Environ* 31:1442–1459.

Shabala L, Cuin TA, Newman IA, Shabala S. 2005. Salinity induced ion flux patterns from the excised roots of *Arabidopsis* SOS mutants. *Planta* 222:1041–1050.

Shabala S, Demidchik V, Shabala L, et al. 2006. Extracellular Ca^{2+} ameliorates NaCl-induced K^+ loss from *Arabidopsis* root and leaf cells by controlling plasma membrane K^+-permeable channels. *Plant Physiol* 141:1653–1665.

Shabala S, Shabala L, Van Volkenburgh E. 2003. Effect of calcium on root development and root ion fluxes in salinised barley seedlings. *Func Plant Biol* 30:507–514.

Shalhevet J, Maas EV, Hoffman GJ, Ogata G. 1976. Salinity and hydraulic conductance of roots. *Physiol Plant* 38:224–232.

Shannon MC, Grieve CM, Francois LE. 1994. Whole-plant response to salinity. In: *Plant-Environment Interactions*, ed. RE Wilkinson, pp. 199–244. New York: Marcel Dekker, Inc.

Shi Q, Ding F, Wang X, Wei M. 2007. Exogenous nitric oxide protects cucumber roots against oxidative stress induced by salt stress. *Plant Physiol Biochem* 45:542–550.

Shoresh M, Spivak M, Bernstein N. 2011. Involvement of calcium-mediated effects on ROS metabolism in regulation of growth improvement under salinity. *Free Rad Biol Med* 51:1221–1234.

Silva C, Martínez V, Carvajal M. 2008. Osmotic versus toxic effects of NaCl on pepper plants. *Biol Plant* 52:72–79.

Sivakumar P, Sharmila P, Saradhi PP. 2000. Proline alleviates salt-stress-induced enhancement in ribulose-1,5-bisphosphate oxygenase activity. *Biochem Biophys Res Commun* 279:512–515.

Stassart JM, Neirinckx L, Dejaegere R. 1981. The interactions between monovalent cations and calcium during their adsorption on isolated cell walls and absorption by intact barley roots. *Ann Bot* 47:647–652.

Steudle E, Peterson CA. 1998. How does water get through roots. *J Exp Bot* 49:775–788.

Suhayda CG, Giannini JL, Briskin DP, Shannon MC. 1990. Electrostatic changes in *Lycopersicon esculentum* root plasma membrane resulting from salt stress. *Plant Physiol* 93:471–478.

Suhayda CG, Redmann RE, Wang X. 1994. Salinity alters root cell wall properties and trace metal uptake in barley. In: *Biochemistry of Metal Micronutrients in the Rhizosphere*, eds. JA Manthey, DE Crowley, DG Luster, pp. 325–342. Boca Raton, FL: Lewis Publishers.

Sun F, Zhang W, Hu H et al. 2008. Salt modulates gravity signaling pathway to regulate growth direction of primary roots in *Arabidopsis. Plant Physiol* 146:178–188.

Sutka M, Li G, Boudet J, Boursiac Y, Doumas P, Maurel C. 2011. Natural variation of root hydraulics in *Arabidopsis* grown in normal and salt-stressed conditions. *Plant Physiol* 155:1264–1276.

Tracy FE, Gilliham M, Dodd A.N, Webb AA, Tester M. 2008. NaCl-induced changes in cytosolic free Ca^{2+} in *Arabidopsis* thaliana are heterogeneous and modified by external ionic composition. *Plant Cell Environ* 31:1063–1073.

Tsai YC, Hong CY, Liu LF, Kao CH. 2004. Relative importance of Na^+ and Cl^- in NaCl-induced antioxidant systems in roots of rice seedlings. *Physiol Plant* 122:86–94.

Tramontano WA, Jowe D. 1997. Trigonelline accumulation in salt-stressed legumes and the role of other osmoregulators as cell cycle control agents. *Phytochemistry* 446:1037–1040.

Tuna AL, Kaya C, Ashraf M, Altunlu H, Yokas I, Yagmur B. 2007. The effects of calcium sulphate on growth, membrane stability and nutrient uptake of tomato plants grown under salt stress. *Environ Exp Bot* 59:173–178.

Tuteja N. 2007. Mechanisms of high salinity tolerance in plants. *Methods Enzymol* 428:419–438.

Vera-Estrella R, Barkla BJ, Garcia-Ramirez L, Pantoja O. 2005. Salt stress in *Thellungiella halophila* activates Na^+ transport mechanisms required for salinity tolerance. *Plant Physiol* 139:1507–1517.

Véry AA, Sentenac H. 2003. Molecular mechanisms and regulation of K^+ transport in higher plants. *Annu Rev Plant Biol* 54:575–603.

Waisel Y, Breckle SW. 1987. Differences in responses of various radish roots to salinity. *Plant Soil* 104:191–194.

Wan XC. 2010. Osmotic effects of NaCl on cell hydraulic conductivity of corn roots. *Acta Biochim Biophys Sin* 42:351–357.

Wan XC, Steudle E, Hartung W. 2004. Gating of water channels (aquaporins) in cortical cells of young corn roots by mechanical stimuli (pressure pulses): Effects of ABA and $HgCl_2$. *J Exp Bot* 55:411–422.

Wang X, Li Y, Ji W et al. 2011. A novel *Glycine soja* tonoplast intrinsic protein gene responds to abiotic stress and depresses salt and dehydration tolerance in transgenic *Arabidopsis thaliana*. *J Plant Physiol* 168:1241–1248.

Wang Y, Li K, Li X. 2009. Auxin redistribution modulates plastic development of root system architecture under salt stress in *Arabidopsis* thaliana. *J Plant Physiol* 166:1637–1645.

West G, Inze D, Beemster GT. 2004. Cell cycle modulation in the response of the primary root of *Arabidopsis* to salt stress. *Plant Physiol* 135:1050–1058.

White PJ, Broadley MR. 2003. Calcium in plants. *Ann Bot* 92:487–511.

Wu J, Seliskar DM, Gallagher JL. 1998. Stress tolerance in the salt marsh plant *Spartina patens*: Impact of NaCl on growth and root plasma membrane lipid composition. *Physiol Plant* 102:307–317.

Wu J, Seliskar DM, Gallagher JL. 2005. The response of plasma membrane lipid composition in callus of the halophyte *Spartina patens* (Poaceae) to salinity stress. *Am J Bot* 92:852–858.

Xu G, Magen H, Tarchitzky J, Kafkafi U. 2000. Advances in chloride nutrition of plants. *Adv Agron* 68:97–159.

Yahya A, Liljenberg C, Nilsson R, Lindberg S, Banas A. 1995. Effects of pH and mineral supply on lipid composition and protein pattern of plasma membranes from sugar beet roots. *J Plant Physiol* 146:81–87.

Yang Q, Chen ZZ, Zhou XF et al. 2009. Overexpression of SOS (Salt Overly Sensitive) genes increases salt tolerance in transgenic *Arabidopsis*. *Mol Plant* 2:22–31.

Yeo AR. 1993. Salinity resistance: Physiologies and prices. *Physiol Plant* 58:214–222.

Yeo AR, Kramer D, Läuchli A, Gullasch J. 1977. Ion distribution in salt-stressed mature *Zea mays* roots in relation to ultrastructure and retention of sodium. *J Exp Bot* 28:17–29.

Yermiyahu U, Nir S, Ben Hayyim G, Kafkafi U, Kinraide TB. 1997. Root elongation in saline solution related to calcium binding to root cell plasma membranes. *Plant Soil* 191:67–76.

Zhang JL, Flowers TJ, Wang SM. 2010. Mechanisms of sodium uptake by roots of higher plants. *Plant Soil* 326:45–60.

Zhang JS, Xie C, Li ZY, Chen Y. 1999. Expression of the plasma membrane H^+-ATPase gene in response to salt stress in a rice salt-tolerant mutant and its original variety. *Theor App Genet* 99:1006–1011.

Zhang H, Zhou JM, Gguo Y, Chen SY. 1997. A physiological study on the salt-tolerant mutant of rice. *Acta Physiol Sin* 23:181–186.

Zhao Y, Wang T, Zhang W, Li X. 2010. SOS3 mediates lateral root development under low salt stress through regulation of auxin redistribution and maxima in *Arabidopsis*. *New Phytol* 189:1122–1134.

Zhong H, Läuchli A. 1988. Incorporation of [^{14}C]glucose into cell wall polysaccharides of cotton roots: Effects of NaCl and $CaCl_2$. *Plant Physiol* 88:511–514.

Zhong H, Läuchli A. 1993a. Changes in cell wall composition and polymer size in primary roots of cotton seedlings under high salinity. *J Exp Bot* 44:773–778.

Zhong H, Läuchli A. 1993b. Spatial and temporal aspects of growth in the primary root of cotton seedlings: Effect of NaCl and $CaCl_2$. *J Exp Bot* 44:763–771.

Zhong H, Läuchli A. 1994. Spatial distribution of solutes, K, Na, Ca and their deposition rates in the growth zone of primary cotton roots: Effects of NaCl and $CaCl_2$. *Planta* 194:34–41.

Zhu JK. 2002. Salt and drought signal transduction in plants. *Annu Rev Plant Biol* 53:247–273.

Zhu JK, Liu JP, Xiong LM. 1998. Genetic analysis of salt tolerance in *Arabidopsis*: Evidence for a critical role of potassium nutrition. *Plant Cell* 10:1181–1191.

Zidan I, Azaizeh H, Neumann PM. 1990. Does salinity reduce growth of maize root epidermal cells by inhibiting their capacity for cell wall acidification? *Plant Physiol* 93:7–11.

Zimmermann HM, Hartmann K, Schreiber L, Steudle E. 2000. Chemical composition of apoplastic transport barriers in relation to radial hydraulic conductivity of corn roots (*Zea mays* L.). *Planta* 210:302–311.

Zimmermann U, Rygol J, Balling A, Klock G, Metzler A, Haase A. 1992. Radial turgor and osmotic pressure profiles in intact and excised roots of *Aster tripolium*. Pressure probe measurements and nuclear magnetic resonance imaging analysis. *Plant Physiol* 99:186–196.

Zolla G, Heimer YM, Barak S. 2010. Mild salinity stimulates a stress-induced morphogenic response in *Arabidopsis* thaliana roots. *J Exp Bot* 61:211–224.

37

Soil Mechanical Resistance and Root Growth and Function

W. Richard Whalley
Rothamsted Research

A. Glyn Bengough
The James Hutton Institute
University of Dundee

I. Introduction

Roots experience mechanical impedance due to the force required to displace soil particles as they elongate. Mechanical impedance (or soil strength) is the most ubiquitous constraint to root growth. Strong soil can be a serious agricultural problem, as the ability of the root system to access water and nutrients from the deeper soil layers is restricted (Barraclough and Weir 1988). Soil is a complicated material and its properties are affected by land management and also by the roots themselves. This chapter has two objectives: first, to explain how soil type, management, and water status affect soil strength and, second, to describe how roots are able to penetrate strong soil and how root traits enable good penetration of strong soils. The idea of soil strength is communicated in a number of ways and often "compact" or "hard" is used to describe soil that is difficult for roots to deform and penetrate. In this chapter we refer to strong and weak soil. To clarify our definition of strong and weak soil, we first describe its measurement in a way that is appropriate to root growth studies: "When you can measure what you are speaking about and express it in numbers, you know something about it," Lord Kelvin, speaking to the Institution of Civil Engineers, May 3, 1883.

II. Measuring Soil Strength

Penetrometers are commonly used to estimate the mechanical impedance to root elongation. Penetrometers are usually based on steel cones that are pushed down through the soil, and the force needed to do this is recorded as function of depth and converted to a pressure by dividing it with the area of the base of the cone. A penetrometer is shown in use in Figure 37.1.

The early use of penetrometers was frequently associated with the problem of trafficability, of heavy machinery, but recently penetrometer resistance has become an important measurement for assessing whether roots can penetrate soil (Bengough and Mullins 1990; To and Kay 2005; Whalley et al. 2005). However, the resistance of a soil to a penetrometer can be a factor of up to eight times greater than the force that a root needs to exert to penetrate the same soil (Bengough and Mullins 1990). The greater resistance of soil to a penetrometer in comparison with a root can be explained largely by a lower frictional resistance between roots and soil in comparison with metal and soil (Bengough and McKenzie 1997). Although lubricated penetrometers have been shown to have lower penetrometer resistance (Tollner and Verma 1987), a more practicable solution to take account of the soil to metal friction when measuring penetrometer resistance is to rotate

FIGURE 37.1 The photograph shows a penetrometer being used to measure the strength of the soil in the field on the long-term experiments at Rothamsted Research. The insert shows a cone of the penetrometer being pushed into soil. The shaft of the penetrometer has a smaller diameter than the cone so the force measured depends only on the deformation of soil by the cone. The penetrometer resistance can be calculated by dividing the force required to push the penetrometer through the soil by the cross-sectional area of the base of the cone. (From Whalley, W.R. et al., *Soil Tillage Res.*, 84, 18, 2005. With permission.)

the penetrometer (Bengough et al. 1997). Typically the fixed penetrometer has a resistance to penetration two to eight times higher than that of a rotating penetrometer (Whalley et al. 2005). This is close to the ratio between the resistance to a fixed penetrometer and the resistance a root exerts to penetrate the same soil (Bengough and Mullins 1991). It is now generally accepted that a fixed penetrometer resistance in excess of 2 MPa will seriously restrict root elongation (e.g., Groenevelt et al. 2001). However, it is misleading to think of this as a threshold value, since the elongation rate of roots decreases in an approximately linear fashion until high penetrometer resistances are encountered.

III. Physical Basis for Mechanical Resistance in Soil

A. Soil Strength due to Soil Drying

As a first approximation, in loose soil there is a common relationship between matric potential (the energy status of water held, under tension, by capillary forces in soil pores) and the resistance to a penetrometer for a range of different soils. In loose agricultural soils, such as seedbeds, it is the capillary forces between particles that give soil its strength. Soil strength is generally large in soils that have many fine water-filled capillaries held under large tension. In relatively loose soil, if the matric potential, ψ, is then multiplied by the degree of saturation S (defined as θ/θ_{max} where θ is the water content and θ_{max} is the maximum water content or soil porosity), the function $|\psi S|$ is proportional to penetrometer resistance (see Figure 37.2a). In Figure 37.2b the resistance to affixed and to a rotating penetrometer for loamy sand is compared. Rotation of the penetrometer alters the direction of the soil-metal friction force vector so that it is not recorded by the force meter, which is only sensitive to the vertical force on the penetrometer shaft. The data in Figure 37.2b show that even in loose soil, only a small amount of drying (to a matric potential of −100 kPa in this case) can result in substantial mechanical constraints to root elongation.

A seemingly anomalous outcome of a regression of penetrometer resistance against $|\psi S|$ is that as soil increasingly dries and S approaches zero, the soil might be expected to become weaker. Indeed for sand-koalinite mixtures, this has been shown to be the case (Mullins and Panayiotopoulos 1984) and beach sand behaves in a similar manner. However, the strength of very dry agricultural soils is not usually low, probably because of cementation by biological exudates and mineral precipitation as soil dries. Provided it is applied within the range of the agricultural soils from which it is derived, a more general relationship between water stress and soil strength for soils of various densities was proposed by Whalley et al. (2007):

$$\mathrm{Log}_{10}Q = 0.35\,\mathrm{Log}_{10}|\psi S| + 0.93\rho + 1.26 \quad (37.1)$$

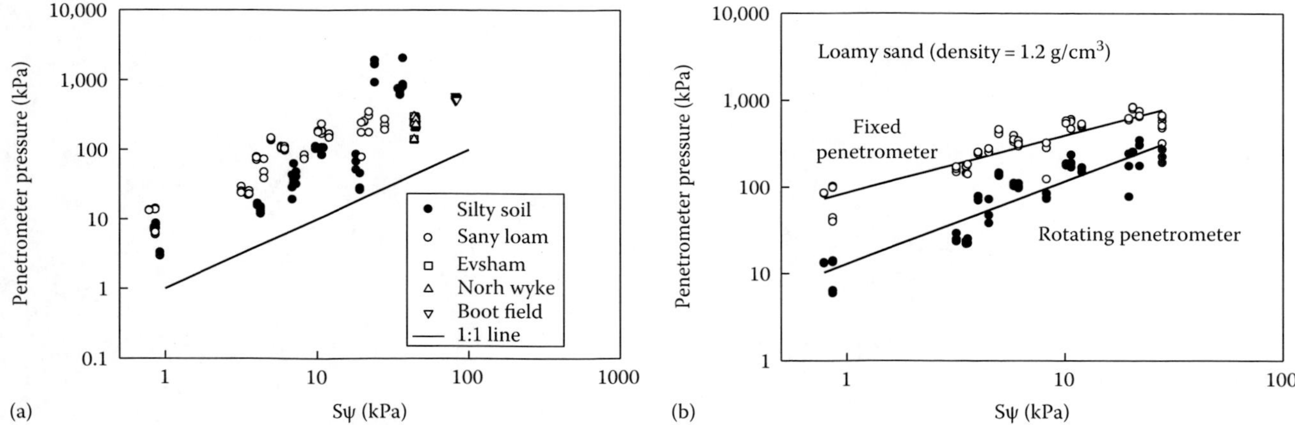

FIGURE 37.2 (a) The resistance to a rotating penetrometer is plotted against the products of degree of saturation and matric potential. This resistance is thought to be an approximation of the resistance to penetration that roots experience. (b) Plotted against time. Three different irrigation treatments are shown: rainfed (open circles and heavy line), drying at a depth of 20 cm limited to a matric potential of −80 kPa (closed circles and dotted line) and well-watered (open squares and light solid line). The resistance to a fixed and rotating penetrometer is compared. Nonrotating (or fixed) penetrometers have a much higher penetrometer resistance. These were plotted using data from Whalley et al. (2005).

This pedotransfer function was developed for a large number of Canadian soils and it has been tested on data from U.K. soils. Two important messages emerge from Equation 37.1. First, even wet soils can be strong as illustrated in Figure 37.2 and supported by field data (Figure 37.3) indicating that at matric potentials of −80 kPa or wetter, the penetrometer resistance could be more than 2 MPa. Second, dense soils become too strong for roots to penetrate at relatively high matric potentials (i.e., in well-watered soils). An important feature of Equation 37.1 is that it is not specific to soil type, which has been achieved in part by expressing soil water status as the product $|\psi S|$. Although it is possible to calculate penetrometer pressure precisely (e.g., Farrell and Greacen 1966), too much information is needed from geotechnical soil tests, which are too time consuming to make it practical. Moreover, few laboratories are sufficiently well equipped to make these tests. Thus Equation 37.1 is a compromise that can be improved as more data become available but provides a relatively simple and robust method to explore the interaction between water stress, soil density, and soil strength.

B. Soil Strength due to Compaction

Equation 37.1 includes the effects of soil density on soil strength. In practice the effects of compaction and soil water are not independent. As shown by Gregory et al. (2010), compaction and

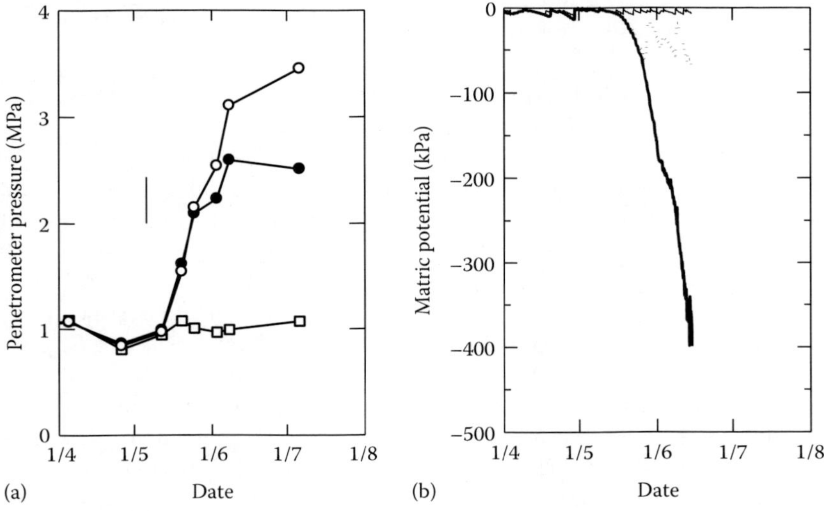

FIGURE 37.3 The mean penetrometer resistance at a depth of 20 cm (a), the matric potential at a depth of 20 cm (b) plotted against time. Three different irrigation treatments are shown: rainfed (open circles and heavy line), drying at a depth of 20 cm limited to a matric potential of −80 kPa (closed circles and dotted line) and well-watered (open squares and light solid line). Note that when soil dries to −80 kPa, the penetrometer resistance exceeds 2 MPa. (Data redrawn from Whalley, W.R., et al., *Plant Soil*, 280, 279, 2006. With permission.)

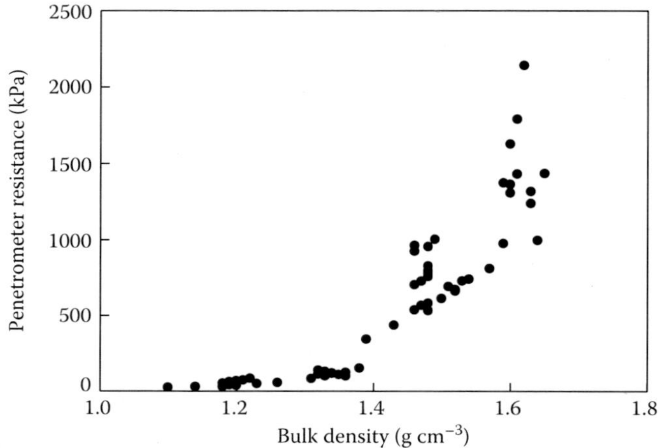

FIGURE 37.4 Penetrometer resistance plotted against soil bulk density. Penetrometer resistance becomes increasingly sensitive to bulk density at higher values of bulk density (>1.4 g cm^{-3} in this example). (Data from Whalley, W.R. et al., *Soil Tillage Res.*, 84, 18, 2005. With permission.)

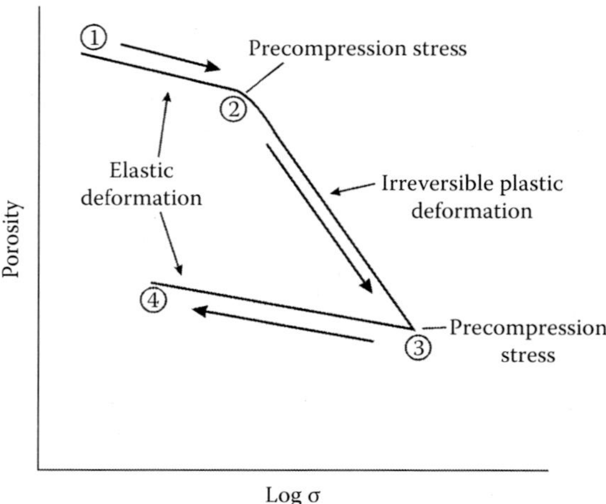

FIGURE 37.5 The relationship between soil porosity and the logarithm of the mean compaction pressure σ. When loose soil (1) is compressed initially, it will deform elastically along an elastic rebound line (1 to 2) until the precompression stress (2) is reached. If the compaction pressure is increased, the soil will fail plastically (2 to 3). When the compaction pressure is reduced (3), the soil will recover along a new elastic rebound curve (3 to 4) and there will be a permanent decrease in porosity (4).

deformation of soil can change the relationship between water content and matric potential. Thus the use of Equation 37.1 to estimate the effects of a change in density on soil strength can only be an approximation because the effect of densification on matric potential is not taken into account. Experimental data confirm the prediction of Equation 37.1 that penetrometer resistance is very sensitive to soil density (Figure 37.4).

Compaction is widely cited as one of the more important factors that contribute to poor soil structure and strong soil (e.g., Fritton 2008). However, compaction is not a stress but a condition that alters the balance between different physical stresses. In pot experiments, poor crop growth is often associated with increased soil strength (Bingham and Bengough 2003). However, in the field, soil compaction does not always reduce yields (Whalley et al. 2008). In some cases the effects of compaction have been found to be cumulative in the sense that the observation of lower yields becomes apparent following repeated compaction of the same field (Sweeney et al. 2006). Håkansson (1990) proposed that the degree of soil compaction significantly influenced the yield of crop. Water and oxygen availability are also likely to be determined by both compaction-induced changes in water release characteristic of soil (Gregory et al. 2010) as well as lower hydraulic saturated conductivity in dense soils (Matthews et al. 2010).

Soil compactibility is very sensitive to soil water status and there is an optimum water content at which soils are most compactable (Gregory et al. 2006). It is difficult to compact wet soils because the compressive forces are partly supported by the positive water pressure within the soil pores, while very dry soils are also strong due to large capillary tension in water films increasing the effective stress between soil particles. At any given water content, the compression characteristic is a useful description of the soil deformation behavior as a function of the applied compaction pressure (Figure 37.5).

When the pressure applied to the soil increases, initially there is limited elastic deformation along the elastic rebound line. When the precompaction stress is exceeded, the soil will deform with increasing pressure more rapidly and irreversibly (i.e., plastic deformation). When the compaction pressure is removed, the soil will recover along a new elastic rebound line. The compression characteristic is useful because it provides a simplification of a highly complex process (see Mitchell and Soga 2005).

C. Effects of Soil Type on Soil Strength

While there is much discussion on the effects of different soil types on soil strength that may be useful for descriptive purposes, it can be confusing. Clay soils that are widely referred to as "heavy" soils maintain relatively low penetrometer pressure as they dry (Gregory et al. 2007). As roots take up water from clay soils, the soil shrinks and the matric potential stays high, although clay soils do increase in dry bulk density as they shrink, typically by up to 20%. In Figure 37.6, we have calculated the penetrometer resistance, with Equation 37.1, for soils at two matric potentials as a function of density. The range of densities that typically occur in drying clays and those commonly found in sandy soils is indicated in Figure 37.6. The benefit of maintaining a high (wet) water potential outweighs the negative effect of density increase in drying clays. Typical densities for sandy soils, often referred to as "light," also indicated in Figure 37.6, are much higher than those found in clays. It is preferable to relate root growth to stresses in the soil and their effects on the root, rather than soil type.

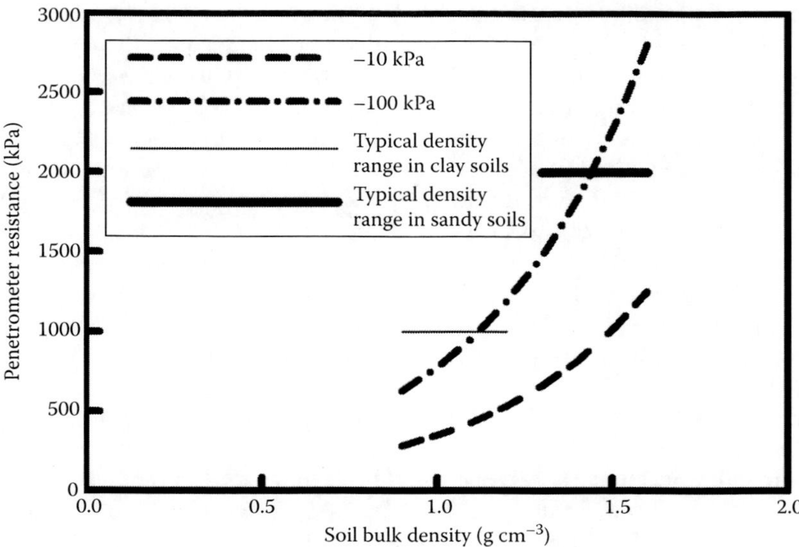

FIGURE 37.6 The penetrometer resistance of soil at two matric potentials (−10 and −100 kPa) calculated with Equation 37.1 plotted against soil bulk density.

IV. Growth Pressure Generated by Roots

Measurements of maximum axial growth pressure (σ_{max}), which is the maximum force per unit area the root can exert, (σ_{max}), are reported widely (Pfeffer 1893; Gill and Bolt 1955; Eavis et al. 1969; Misra et al. 1986; Souty and Stępniewski 1988; Whalley and Dexter 1993; Whalley et al. 1994; Clark et al. 1996). Several different designs of apparatus for measuring σ_{max} have been reported, all of which are "homemade," and these are compared by Clark et al. (1999). The principle of the measurement is to allow a root to elongate into a small blind hole in a piece of ceramic that will completely impede the root. If the root is anchored securely, the maximum growth force can be measured using a force transducer attached to the ceramic (Figure 37.7). The rate of growth force development is dependent on temperature (Bengough et al. 1994; Whalley et al. 1994). At 25°C, the roots of pea approach a maximum growth force in about 24 h. Although growth force may continue to increase gradually beyond that point, the value of σ_{max} is stable with time. The value of σ_{max} in pea was not found to be sensitive to temperature over the range 5°C–30°C (Bengough et al. 1994; Whalley et al. 1994). A limitation of this approach to measuring root growth pressure is that it can only be easily used with seedlings, although Misra (1997) did measure σ_{max} in both lateral and primary roots of pea and eucalypt (*Eucalyptus nitens* Maiden) seedlings. As far as we are aware, measurements of σ_{max} have not been made on the mature roots and they are restricted to seedlings no more than a few days old.

One of the main outcomes of maximum growth pressure measurements has been an understanding of how individual roots respond to environmental stresses. Some of the most useful data have been provided by experiments, in which σ_{max} was measured as a function of oxygen, water, and temperature stress. Eavis et al. (1969) have shown that, in the case of pea, σ_{max} was not significantly affected when the oxygen partial pressure was reduced from 21 to 3 kPa and was in the range 1.1–1.4 MPa. In the case of cotton (*Gossypium hirsutum* L.), σ_{max} was reduced from 1.1 to 0.5 MPa when the oxygen

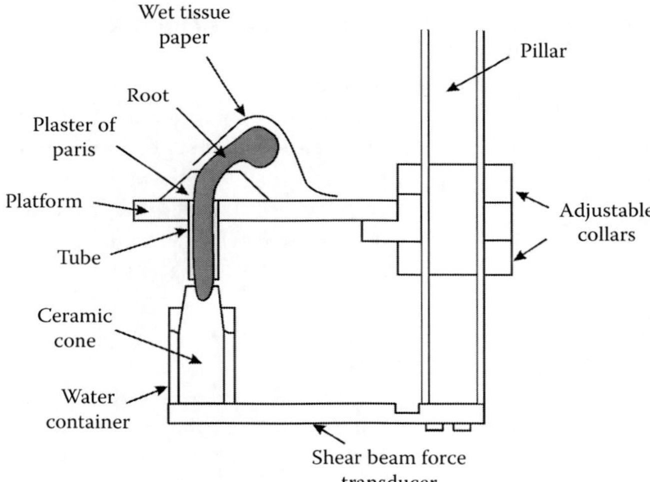

FIGURE 37.7 An example of an apparatus that can be used to measure the maximum force that completely impeded roots can exert. The root is cemented into a hole with plaster of Paris so that it elongates down a narrow tube into a small hole in a moist ceramic. The growth force can be measured using a shear beam force transducer. The maximum root growth pressure can be estimated by dividing the maximum growth force by the area of the root between the end of the tube and the top of the ceramic. (Redrawn from Clark, L.J. et al., *Plant Cell Environ.*, 19, 1099, 1996.)

partial pressure was reduced from 21 to 3 kPa. In experiments on young maize, Souty and Stępniewski (1988) varied oxygen partial pressures from 0 to 20 kPa either around the whole seedling or just the radicle. The values of σ_{max} decreased with oxygen partial pressure in both cases, although the decrease was less severe when only the radicle was subjected to oxygen shortage, probably because oxygen could diffuse inside the root. Eavis et al. (1969) measured a significant decrease in σ_{max} for pea and cotton roots when they were subjected to matric potentials of −0.03 or −0.01 MPa. Whalley et al. (1998) extended the range of water stress for pea down to −0.45 MPa. Over this larger range, the value of σ_{max} almost halved (i.e., from 0.66 to 0.35 MPa).

Clark et al. (1996) combined measurements of maximum growth pressure with micropressure probe measurements of cell turgor. The micropressure probe technique is described by Pritchard et al. (1990) and is based on the use of a micropipette, which is connected to a pressure transducer. By making turgor pressure measurements in impeded pea roots, mounted in a specially adapted device for measuring σ_{max}, Clark et al. (1996) were able to demonstrate that when pea roots become impeded, turgor increased from 0.55 to 0.78 MPa and confining pressure due to the cell wall tension decreased from 0.55 to 0.26 MPa.

There is no evidence of any genotypic variability in σ_{max} generated by the roots of different plant species. Clark and Barraclough (1999) found that although roots of dicotyledons have often been observed to grow better in strong soil than monocotyledons (e.g., Materechera et al. 1992), the mean σ_{max} for dicots was 0.41 MPa compared to 0.44 MPa for monocots.

V. Root Penetration and Bending Stiffness

To elongate through soil where existing channels are smaller than the root diameter, roots must exert a growth pressure (which results from turgor pressure and cell wall relaxation) to deform the soil around the root (Greacen and Oh 1972; Clark et al. 1996). When roots approach a layer of strong soil, they may either penetrate the layer or may get deflected from their original direction (Dexter and Hewitt 1978). The mechanisms that determine the outcome of such an encounter are not fully understood. If the root does not have sufficient lateral support, bending of the root may occur above the strong layer.

It has long been known that roots of some species are better at penetrating strong soil than others (Materechera et al. 1992). In particular, dicots have better root penetration than monocots. But as we have discussed, dicots do not have greater maximum axial growth pressures than monocots (Clark and Barraclough 1999), which implies that differences in growth pressure do not account for differences in root penetration ability. It has been suggested that species with thicker roots gave better penetration because they were more resistant to buckling (Whiteley et al. 1982) or because of axial stress relief (Kirby and Bengough 2002). There is evidence for differences in root penetration ability between cultivars as well as between species. Yu et al. (1995) used wax layers to assess root penetration in rice, following the approach described by Taylor and Gardner (1960). Wax layers were made by mixing hard paraffin wax and white soft paraffin (petrolatum) to give a thin layer with a predetermined mechanical strength. By installing these layers beneath the surface of a low-impedance growing medium and counting the number of roots that penetrated the strong wax layer, an index of the relative ability of roots to penetrate strong layers can be obtained.

Cultivar differences in the ability of rice roots to penetrate strong media are much smaller without a high spatial gradient in strength (Clark et al. 2008a,b). This was illustrated most clearly by Clark et al. (2008a,b) who found that root penetration was greatly increased when soil strength increased with depth at a scale of only a few millimeters (Figure 37.8) to the extent that the effects of the gradient in soil strength were comparable with the genotypic variability in root penetration.

Interestingly, cultivars with good root penetration of hard wax layers did not have longer roots in uniformly strong sand, but they did have greater root diameters (Clark et al. 2002). Differences in the ability of rice roots to penetrate strong layers appear therefore not to be related to their ability to elongate through strong soil but rather to their ability to penetrate a

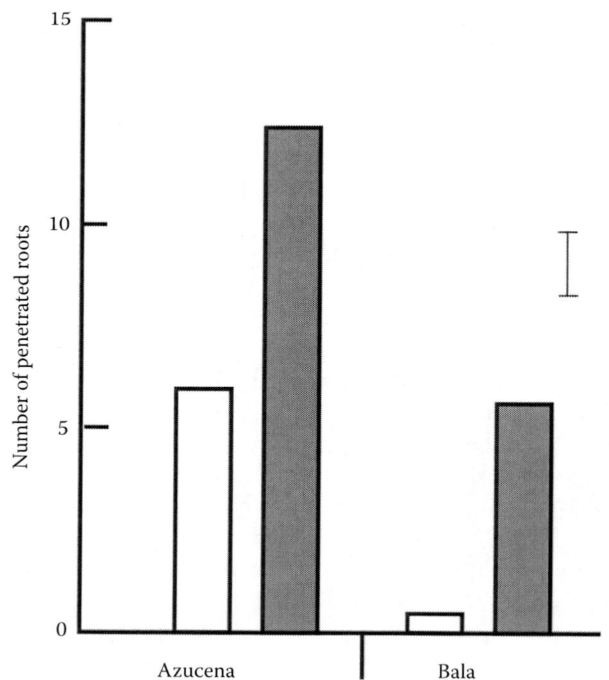

FIGURE 37.8 The number of root axes that penetrated a multilayered wax disk with strength increasing with depth (shaded bars) or a control with strength decreasing with depth (open bars). (Error bar shows the SED, 21 df.) (Data replotted from Clark, L.J. et al., *Plant Soil*, 307, 235, 2008. With permission.)

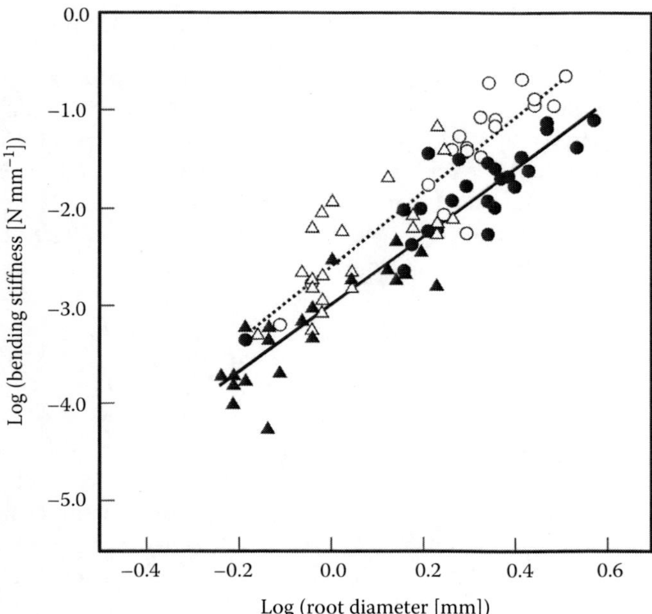

FIGURE 37.9 Log-log plot of bending stiffness against root diameter for roots of Azucena (●, ○) and Bala (▲, △) rice cultivars that either had not reached (○, Δ, dotted regression line) or had penetrated (●, ▲, solid regression line) a strong layer. Each point represents a single root. The roots that had penetrated the strong layer were less stiff in bending than those that had not. This was thought to be cell wall loosening needed to deform strong layers. (Data are replotted from Clark, L.J. et al., *Funct Plant Biol*, 35, 1163, 2008. With permission.)

boundary when there is a rapid increase in mechanical impedance. This suggests that good root penetration results not from high growth pressures (see previous section) but from the roots resisting bending as they encounter a hard layer.

There is an approximately fourth-power relationship between root diameter and bending stiffness (Figure 37.9). For a given root diameter, penetrated roots were less stiff than roots that had not reached the wax disk, suggesting that an encounter with a strong layer decreases bending stiffness (Clark et al. 2008a,b). When Azucena and Bala rice cultivars were grown in both weak (*control*) and strong (*impeded*) sand, in a system that is designed to allow mechanical impedance to be varied independently of soil aeration and water status (see Section VI), the bending stiffness of roots removed from strong sand was significantly lower ($P < 0.001$) than those removed from weak sand. It was assumed greater cell wall relaxation in roots grown in strong sand, which is required to increase growth pressure (Clark et al. 1996), resulted reduced stiffness of roots encountering strong soil or strong layers.

VI. Experimental Approaches to Investigate the Effects of Strong Soil

It is tempting to think that if soil is used in experimental systems to study the effects of soil strength, the results will have direct field relevance. However, the difficulty with this approach is that interpretation of plant responses remain specific to that soil type because they are pertinent to a certain and often unique combination of stresses that arise because of the soil type, water status, and density as described earlier. As shown in Figure 37.3, even relatively moist soils can have high soil strength. The sand culture system allows the mechanical strength of the root growth environment to be increased without affecting aeration or the hydraulic environment. When a weight is placed on the sand surface, the mechanical impedance of the medium is increased as the resistance of sand grains to displacement is increased, but there is negligible compaction of the sand core (Figure 37.10).

This allows mechanical impedance to be varied independently of aeration and water status of the growing medium and uses non-flooded conditions.

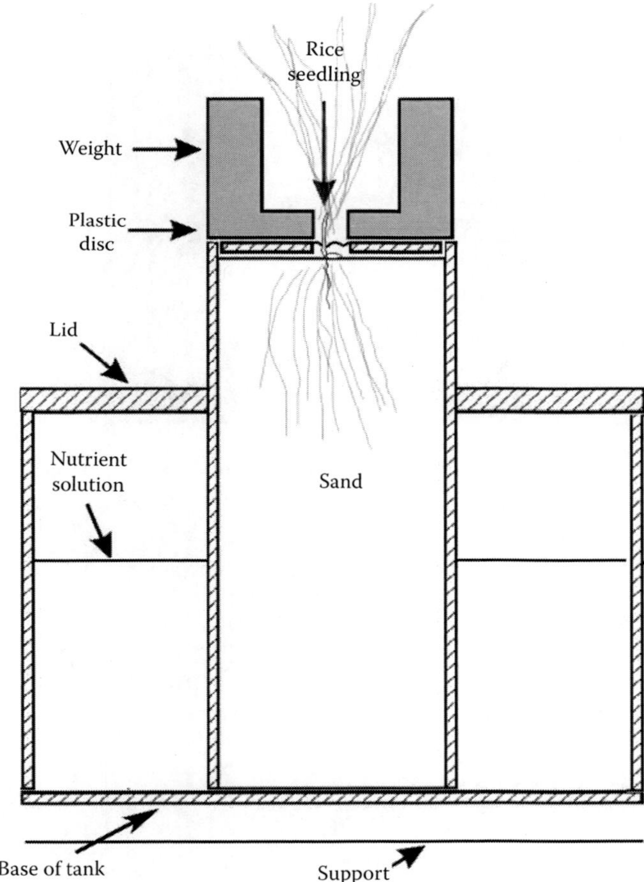

FIGURE 37.10 Schematic diagram and photo of an experimental system designed to explore the effects of high soil strength on the growth of both roots and shoots. The system is based on sand culture and the sand is either confined by the load from a large steel weight or unconfined in the control (white foam of negligible weight). Water availability and nutrient availability are similar in both impeded and control sand columns, with only the mechanical impedance to root growth differing. (From Clark, L.J. et al., *Field Crops Res.*, 76, 189, 2002. With permission.)

VII. Effect of Soil Strength on Root Morphology: Elongation and Cell Expansion

A recent survey of 19 U.K. arable soils showed that 50% of them had a penetrometer resistance of >2 MPa at a matric potential of −200 kPa (Bengough et al. 2011). Root elongation slows when roots encounter hard soil that exerts a large reaction force on the expanding root tissue (Figure 37.11). A penetrometer resistance of between 0.8 and 2 MPa is sufficient to halve the elongation rate of seedlings in repacked soil, in the absence of water stress (Bengough et al. 2011). This penetrometer resistance is greater than the turgor pressure within the expanding root cells (typically 0.4–0.7 MPa; e.g., Pritchard 1994) because penetrometers encounter between two and eight times greater resistance than plant roots (Bengough and Mullins 1991; Iijima et al. 2003). Indeed, peanut roots have been observed to penetrate soil with penetrometer resistances of more than 5 MPa in a loamy sand soil (Taylor and Ratliff 1969).

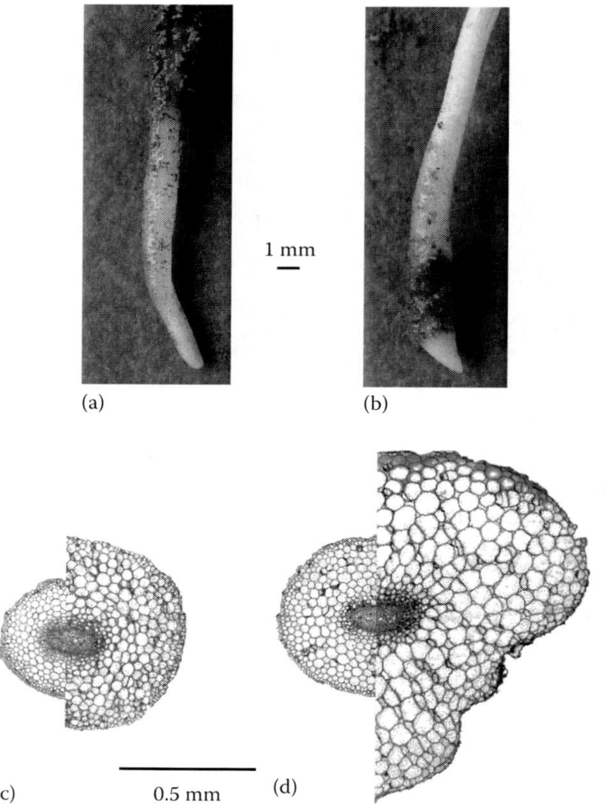

FIGURE 37.11 Tips of the primary roots of maize seedlings grown in the field either along a continuous preexisting channel (a; unimpeded) or into bulk soil (b; mechanically impeded), causing root elongation to slow and the root to thicken and they show cross sections through lupin roots at either (c) 4–5 mm behind the tip or (d) 10–11 mm behind the root tip. Left halves of sections are for unimpeded roots; right halves are for mechanically impeded roots. (c and b: Reproduced from Hanbury, C.D. and Atwell, B.J., Growth dynamics of mechanically impeded lupin roots: Does altered morphology induce hypoxia? *Ann. Bot.*, 96, 913–924, 2005, Figure 4, by permission of Oxford University Press.)

This reaction force of the soil on the root (σ), coupled with the yield threshold of the cell walls (Y), opposes the expanding turgor pressure (P) within the root cells. The rate of expansion of the length (l) of tissue in the elongation zone is also regulated by the extensibility of the cell walls in the elongation zone (m) and can be described by the modified Lockhart equation (Greacen 1986; Bengough et al. 2006):

$$l^{-1}\left(\frac{dl}{dt}\right) = m(\sigma)\big(P - Y(\sigma) - \sigma\big) \tag{37.2}$$

The parameters m and Y are written as functions of σ to emphasize that they vary with the mechanical impedance experienced by the root, making this equation harder to apply than it may first appear (Passioura and Fry 1992). Cell wall extensibility and yield threshold depend on the relative rates of making and breaking cross-linking tethers between microfibrils in the cell wall, which is likely to be under enzymatic control. The mechanical stress causes physiological changes in the root and reorientation of microfibril deposition in the cell walls from radial to longitudinal, similar to the reorientation that occurs in response to ethylene (Veen 1982). These changes were observed using polarized light microscopy and remain the only observations of this important phenomenon to date. The anisotropic stiffening of the cell walls decreases the pressure required for a cell to expand radially while increasing that required for axial extension. This results in shorter radially expanded cells (Figure 37.11), shortening of the elongation zone and slowing of the root elongation rate of impeded roots (Croser et al. 1999; Hanbury and Atwell 2005). Cells cease expansion closer to the root tip in impeded roots and root hairs emerge closer to the root tip from the maturing epidermal cells. The rate of cell flux (the rate at which new cells are added onto an individual cell file) is also slowed and is also associated with shortening of the elongation zone. In pea roots where root elongation rate was slowed to only 20% of the unimpeded rate, cortical cell length and cell flux were both halved, contributing to an elongation zone of 8 mm as compared with 13 mm in the unimpeded roots (Croser et al. 1999, 2000).

The elongation rate of maize roots subject to mechanical impedance has been restored to 90% of the unimpeded rate by addition of 10 μM aminoethoxyvinyl glycine plus 1 μM silver thiosulfate, for maize roots grown in glass beads with triaxial cells subject to an external confining pressure (Sarquis et al. 1991). This provides evidence that ethylene plays a role in the root response to soil strength, in support of observations by Kays et al. (1974): ethylene was released from the roots of *Vicia faba* L. when they encountered a mechanical barrier, and evolution of ethylene diminished to control rates when the barrier was removed.

There are only a few studies where both the root penetration resistance (root force divided by root cross-sectional area) have been measured alongside root elongation rate (Figure 37.12).

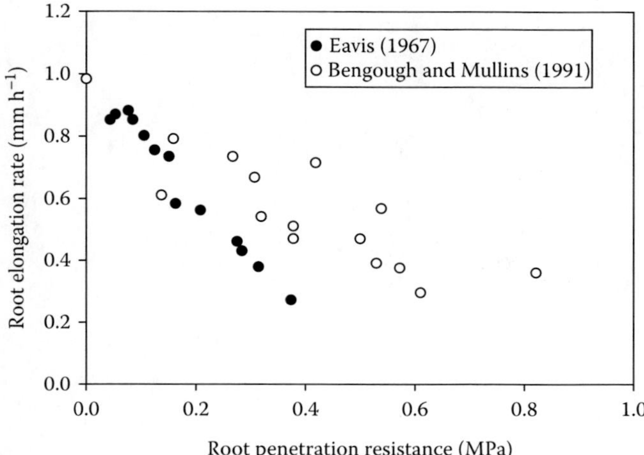

FIGURE 37.12 Elongation rate as a function of root penetration resistance. It should be noted that penetrometer resistance is between 2 and 8 times greater than the root penetration resistance. (Data replotted from Eavis 1967; Bengough, A.G. and Mullins, C.E. *Plant Soil*, 131, 59, 1991; Bengough, A.G. and Mullins, C.E., *Soil Sci.*, 4, 341, 1990. With permission.)

Elongation rate decreased approximately linearly with increasing root penetration resistance, presumably ceasing when the reaction force of the soil on the root exceeds the turgor pressure in the cells—a maximum of approximately 1 MPa. It is important to consider how root elongation rates respond to variations in soil strength, as occur naturally in structured soils. When roots grow from hard soil to loose soil, their elongation rate takes a period of 2–5 days to recover to unimpeded rates (Bengough and Young 1993; Hanbury and Atwell 2005). This is due to physiological changes occurring in the root in response to the mechanical impedance that cannot be reversed rapidly. The slowing of cell flux and stiffening of cell walls at the rear of the elongation zone associated with the slower elongation rate (Equation 37.2) may not be reversed until sufficient new cells enter the elongation zone (Croser et al. 1999, 2000). There appear to be two components to the elongation response to mechanical impedance—a substantial rapid change in elongation rate associated with the immediate change in the balance between turgor pressure and opposing soil reaction pressure and a slower adaptation to the new environment associated with changes in the cell wall properties and cell flux in the elongation zone (Bengough and Mackenzie 1994; Hanbury and Atwell 2005). Thus the elongation rate of a root depends substantially on its stress history during the previous few days, and not solely on the stress experienced by the root at any given time. There is interesting evidence that the recent elongation rate of excavated roots can be estimated from the length of the root elongation zone (approximated by the distance from root tip to root hairs; Watt et al. 2003; Pages et al. 2010), although this has only recently been verified for roots grown in a range of soil densities and matric potentials (Schmidt et al. submitted).

New imaging methods are enabling in situ studies of root elongation with high temporal and spatial resolution. Digital photography with image analysis using optical flow methods, such as particle image velocimetry, provides methods of tracking the displacement of patches of image texture in a series of time-lapse images. Deformation patterns around growing maize root tips have quantified sand displacements as small as 0.5 μm and shown patterns of soil particle movement that are influenced by the nature of the root cap-soil interface (Vollsnes et al. 2010). When a functional root cap was present, there was little axial displacement of sand particles ahead of the root tip, with most sand deformation occurring radially (Figure 37.13).

In contrast, in decapped roots, there was substantial deformation of the sand ahead of the root tip, and this increased resistance to growth, slowed root elongation, and increased radial thickening of the root. In loose soil, decapped roots elongated at the same rate as roots with intact root caps, but this was not the case in compacted soil. Direct measures of the root penetration resistance in compacted soil suggested that decapped roots experienced approximately double the root penetration resistance of intact roots and elongated at only half the rate of roots with an intact cap (Iijima et al. 2003). The reason for this is probably the presence of a slippery coating of detached border cells and mucilage that provides a low-friction interface between the intact root cap and the soil (Figure 37.14).

Whereas the cell flux to the root proper decreased in mechanically impeded roots, the number of detached border cells increased in maize roots grown in compacted sand (Iijima et al. 2000). The number of border cells released increased from 1930 to 3220 per day; this represented a 12-fold increase in border cell release per unit root extension. The production rate of border cells was sufficient to completely cover the surface of mechanically impeded maize root tips. These border cells can be thought of as a disposable drilling tissue that surrounds the root cap, with the border cells providing a turgid cushion between the rough soil particles and the root cap and epidermis surface. The release of border cells was greatest for the seedling roots of brome grass (*Bromus carinatus* Hook. & Arn.), maize, and cucumber seedlings and decreases from thousands to hundreds as plants age (Odell et al. 2008). Seedling establishment is a crucial time for rapid root extension to enable roots to access sufficient soil volume for water and nutrient uptake, especially if rainfall is intermittent and in a drying soil profile. It must also be noted that border cells serve biological as well as physical functions in soil, making the root tip harder to colonize—by both pathogenic and benign soil organisms (Hawes et al. 2000; Humphris et al. 2005).

Border cell release from root tips is particularly apparent when root tips are immersed in water. Rapid expansion of the mucilage layer around roots occurs as it absorbs water to many times its own dry mass (Sealey et al. 1995; McCully and Boyer 1997; Read et al. 1999). Despite the relatively small quantity of water held by mucilage at dry potentials, it is likely that both mucilage and other exudates significantly alter the wetting and drying properties of the rhizosphere and hence the hydraulic conductivity of this crucial zone of soil through which virtually

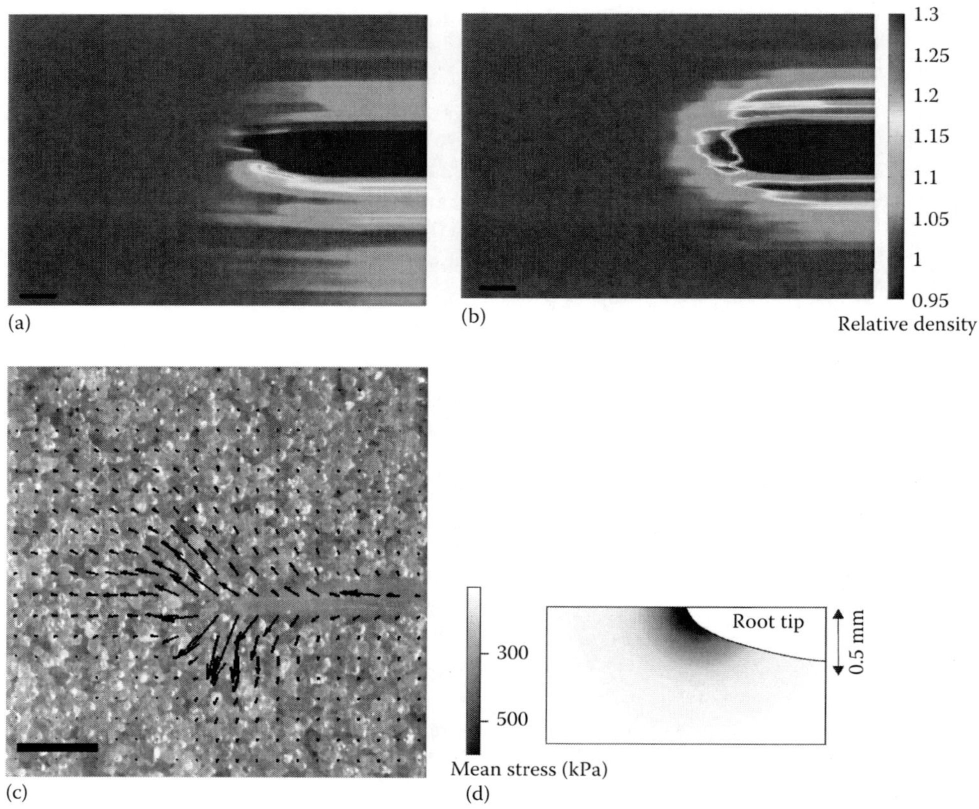

FIGURE 37.13 **(See color insert.)** Soil displacement patterns around growing root tips. (a and b) Soil compression around wild type (KYS; a) or decapped mutant (*agt1*$_{dec}$; b) maize roots (Redrawn from Vollsnes, A.V. et al., *Eur. J. Soil Sci.*, 61, 926, 2010, Figures 9a and b.) estimated from (c) displacement fields in sand measured with particle image velocimetry. (Redrawn from Vollsnes, A.V. et al., *Eur. J. Soil Sci.*, 61, 926, 2010, Figures 6a.) (d) Predicted stress distribution around root tips of pea growing in sandy loam. Black scale bars are 2 mm long. (Redrawn Kirby, J.M. and Bengough, A.G., *Eur. J. Soil Sci.*, 53, 119, 2002, Figure 4. With permission.)

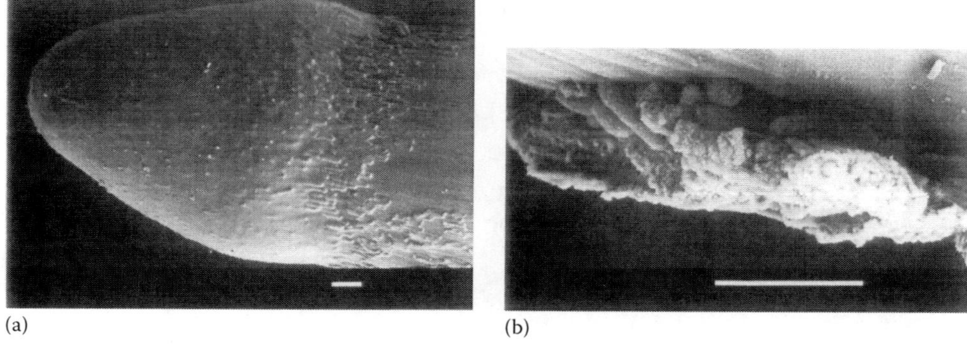

FIGURE 37.14 Scanning electron micrographs showing root border cells (sloughed root cap cells) of maize grown on (a) filter paper showing thick mucilage layer coating root cap with border cells exposed behind, (b) border cells separating root epidermis from soil particles. Scale bar is 100 μm. (Reproduced from Bengough, A.G. and McKenzie, B.M., Sloughing of root cap cells decreases the frictional resistance to maize (*Zea mays* L.) root growth, *J. Exp. Bot.*, 48, 885–893, 1997, Plate 1A and B, by permission of Oxford University Press.)

all water and nutrients flow (Read et al. 2003). The shear properties of polysaccharide-soil mixtures may be substantially different from unamended soil, with extracellular polysaccharides tending to increase viscosity (Barre and Hallett 2009). Recent neutron imaging of water in the rhizosphere suggests that it displays hysteretic-like behavior, remaining wetter than the bulk soil in a drying environment but being harder to rewet following irrigation (Carminati et al. 2010). An interesting and as yet unexplored possibility concerns whether the release of mucilage from the root cap significantly alters the deformation behavior of the soil and hence the mechanical impedance to root growth. Indeed, it has long been known that mechanical impedance substantially increases the release of mucilage from root tips (Barber and Gunn 1974; Boeuf-Tremblay et al. 1995).

In maize roots grown in ballotini subject to an external confining pressure, the dry mass of root exudate doubled as compared with unimpeded controls (Barber and Gunn 1974). Much of this exudate is closely adhered to the surface of the glass beads in the medium (Boeuf-Tremblay et al. 1995). In its hydrated condition, the combined layer of mucilage and border cells associated with the root cap has only a tiny coefficient of friction, 0.04 (Bengough and Kirby 1999). The frictional properties of the mucilage under drier conditions are less certain, and it is likely that dry mucilage will represent a less-slippery interface with the soil (McCully 1999). The degree of lubrication that mucilage and border cells proffer to roots in dry soil is currently uncertain, though of considerable importance, given the large increase in strength as soil dries.

VIII. Root System Growth and Function in Heterogeneous Soil

In the previous section we have been mainly concerned with the growth of individual root tips in hard soil, frequently the primary roots of seedlings. Lateral root length increases rapidly as plants age, such that lateral roots rapidly represent the bulk of the total root length. The total length of first- and second-order lateral roots of a 3-week-old barley plant was predicted to exceed the length of main axes by more than 10-fold (Hackett and Rose 1972). Thus we must consider how the root system as a whole responds to soil strength and the likely effects of variations in soil strength on root system growth and function.

There have been relatively few measurements of lateral root growth in hard soils. Both lateral roots and the primary root of pea were found to exert a maximum root growth pressure of approximately 0.3 MPa, suggesting there is little difference in the pressures that these roots can generate (Misra 1997). In hard soils, root systems appear stunted and may also have more branches per unit length of main axis (Table 37.1).

This visual appearance is largely due to the slower elongation rate of the main axes, sometimes resulting in a shorter distance between lateral roots, as generally the total number of lateral roots either remains unchanged or decreases with increased mechanical impedance. The initiation of individual lateral roots may depend on the small-scale variation in mechanical stress experienced by the root tip—as observed when pea root axes remained unbranched when traversing large soil pores (Tsegaye and Mullins 1994). Lateral roots tend to emerge on the convex bend in a main root axis when a root has to grow around an obstruction in the soil (see Chapters 6 and 20, for description of the molecular basis of this phenomenon), indicating again that the mechanical force encountered by the root tip influences lateral root production as well as elongation (Goss and Russell 1980).

Mechanical impedance may constrain root growth to biopores and cracks between soil peds in hard soils, especially compacted subsoils (Passioura 1991; White and Kirkegaard 2010). Many roots may cluster within a single biopore, decreasing the relative effectiveness of these roots as sinks for water and nutrients, as the time constant for water extraction depends

TABLE 37.1 Effects of Mechanical Impedance on Elongation of First-Order Lateral Roots in Various Growth Media

Species	Growth Medium	Effects of Mechanical Impedance on Lateral Root Growth	References
I. Unimpeded versus main axes and laterals impeded			
Maize	Loamy sand	Decreased lateral root number with increasing soil bulk density (especially fewer laterals at depth). Length per lateral root unaffected	Kuchenbuch and Ingram (2004)
Eucalyptus	Clay aggregates	Fewer lateral roots in total, more closely spaced per length main axis. Length of individual lateral roots increased	Misra and Gibbons (1996)
Pea	Sandy clay loam	Depended on pea species and the presence of soil structural features (large pores)	Tsegaye and Mullins (1994)
Barley and wheat	Sandy loam	Lateral length restricted less than main axes in barley, but not in wheat. Number of first-order laterals per length main axes unchanged	Bingham and Bengough (2003)
II. Unimpeded versus main axes impeded (laterals unimpeded)			
Barley	Glass beads under external confining pressure	Same number of laterals but more closely spaced and longer individual lateral roots (same total lateral length)	Goss (1977)

strongly on the degree of root clustering within any given geometry of biopores, cracks, or soil peds (Passioura 1991). In a hard Australian subsoil, >85% of wheat roots were located within biopores or cracks, with 44% of these roots clumped together in groups of three or more (White and Kirkegaard 2010). Such detailed studies of root distribution in structured soil are crucial to our understanding of root growth and function in field soils but remain rare. In situ imaging of soil pore networks is rapidly becoming a powerful technique that yields information on pore size and distribution in 3 dimensions, including measures of tortuosity and continuity that cannot be readily obtained by other methods (Jassogne et al. 2007). Two classes of pores could be identified from these images—firstly, tubular biopores likely to have been formed by roots and, secondly, pores of much more variable shape and size. The ability of roots to find and exploit such pores is likely to be of substantial advantage where deep water extraction is important and has recently resulted in the development of a field screen to investigate this rooting trait (McKenzie et al. 2009).

The response of roots to soil of variable strength has been studied for root systems growing in soil that has either a vertical or horizontal contrast in soil strength (Bingham and Bengough 2003). The plasticity of plant root systems to exhibit compensatory growth of roots in the unimpeded soil zone depended on species. Broccoli, but not lettuce, showed compensatory root growth in loose topsoil (Montagu et al. 1998), while barley, but not wheat, showed compensatory root growth in the loose soil part of a vertically divided root system (Bingham and Bengough 2003). Such genotypic variation makes it harder to generalize root response to mechanical stress but also indicates genetic potential for selecting root system phenotypes with traits that display characteristics that may benefit crop performance.

IX. Evidence for the Effects of Strong Soil on Yield

The effects of soil strength on plant growth should be considered within the context of a multiple stress environment. As we have described, when soil water status or density is altered, then strength and water availability will change in some coordinated way depending on soil type. Recently Dodd et al. (2010) found that the multiple stress environment in different root growth media appeared to be integrated by root water potential. Root-sourced signaling has become an accepted mechanism that determines how root stresses in the rhizosphere control shoot elongation and stomatal conductance (reviewed by Davies et al. 2000; Clark et al. 2005; Dodd 2005). A key question to resolve in a multistress environment is what physical stress or combinations of stresses mediate the signaling processes. Although most work has been directed at the effects of water stress, the impacts of high soil strength on shoot growth are well documented (e.g., Masle and Passioura 1987) and they can be observed at matric potentials much higher (i.e., less negative) than those frequently used by plant scientists to investigate the effects of water stress. Blum et al. (1991) concluded that shoot growth was limited not by water availability *per se* but by drying of the surface soil. In field experiments, the increased strength of the soil surface (Whalley et al. 2008) appeared to give the best explanation for reduced yield of winter wheat and by comparison with previous work (Whalley et al. 2006), yield loss appeared to be related to soil strength rather than low matric potential. Whalley et al. (2006) grew wheat in sand culture (Figure 37.10) with either mechanically weak or strong root growth environments. They found that simply by increasing the mechanical strength of the root zone, shoot growth was approximately halved. Since there was no change in the stomatal conductance, this suggests that the reduction in shoot growth was mediated by chemical signaling.

Deeper roots are advantageous in water-limited environments allowing greater access to water at depth (Lopes and Reynolds 2010). It seems likely that the ability to penetrate strong soil will be a component of deep rooting, although the ability to penetrate strong layers and deep rooting have never been clearly associated with each other in the field environment. It is likely that deep water extraction by roots is affected by interactions with the plant, environment, and soil. This has resulted in a lack of consensus on whether a large root system contributes to wheat adaptation in water-limited environments (e.g., Palta et al. 2011), although studies such as that of Lopes and Reynolds (2010) support such an adaptation. Irrespective of the role of roots in deep water or nutrient uptake, there is clear evidence that strong soil initiates chemical signaling, in the impeded roots, which can reduce leaf elongation.

X. Conclusions

In this chapter we have discussed the origin of mechanical impedance to root elongation in agricultural soils and the consequences for roots. Even in relatively in well-watered soils, mechanical impedance to root elongation often reduces root elongation rates (Bengough et al. 2011). In contrast to the common assumption that only compacted soils are strong, we show that even relatively loose soils often become strong as they dry. Soil type (e.g., sand or clay) and condition (e.g., loose or compact) determine the relative balance of water stress and soil strength and the extent that these limit root and crop growth for any given season. Although roots exert surprisingly high pressures, there is little evidence for genotypic variability in these pressures. Instead, variability in root penetration of strong soils is probably linked to variation in the biomechanical properties of individual root tips (e.g., root bending stiffness, root thickening, frictional properties of the root cap). The effects of high soil strength on root growth can be observed in relatively wet soils with little limitation on water availability. Thus the effects of soil strength on root elongation and plant growth are frequently overlooked, despite soil strength often being the major limitation to root growth in agricultural soils.

References

Barber DA, Gunn KB. 1974. The effect of mechanical forces on the exudation of organic substances by the roots of cereal plants. *New Phytol* 73:39–45.

Barraclough PB, Weir AH. 1988. Effects of a compacted subsoil layer on root and shoot growth, water use and nutrient uptake of winter wheat. *J Agric Sci* 110:207–216.

Barre P, Hallett PD. 2009. Rheological stabilization of wet soils by model root and fungal exudates depends on clay mineralogy. *Eur J Soil Sci* 60:525–538.

Bengough AG, Bransby MF, Hans J, McKenna SJ, Roberts TJ, Valentine TA. 2006. Root responses to soil physical conditions; growth dynamics from field to cell. *J Exp Bot* 57:437–447.

Bengough AG, Kirby JM. 1999. Tribology of the root cap in maize (*Zea mays*) and peas (*Pisum sativum*). *New Phytol* 142:421–425.

Bengough AG, Mackenzie CJ. 1994. Simultaneous measurement of root force and elongation for seedling pea roots. *J Exp Bot* 45:95–102.

Bengough AG, Mackenzie CJ, Elangwe HE. 1994. Biophysics of the growth responses of pea roots to changes in penetration resistance. *Plant Soil* 167:135–141.

Bengough AG, McKenzie BM. 1997. Sloughing of root cap cells decreases the frictional resistance to maize (*Zea mays* L.) root growth. *J Exp Bot* 48:885–893.

Bengough AG, McKenzie BM, Hallett PD, Valentine TA. 2011. Root elongation, water stress, and mechanical impedance: A review of limiting stresses and beneficial root tip traits. *J Exp Bot* 62:59–68.

Bengough AG, Mullins CE. 1990. Mechanical impedance to root growth: A review of experimental techniques and root growth responses. *Soil Sci* 4:341–358.

Bengough AG, Mullins CE. 1991. Penetrometer resistance, root penetration resistance and root elongation rate in 2 sandy loam soils. *Plant Soil* 131:59–66.

Bengough AG, Mullins CE, Wilson G. 1997. Estimating soil frictional resistance to metal probes and its relevance to the penetration of soil by roots. *Eur J Soil Sci* 48:603–612.

Bengough AG, Young IM. 1993. Root elongation of seedling peas through layered soil of different penetration resistances. *Plant Soil* 149:129–139.

Bingham IJ, Bengough AG. 2003. Morphological plasticity of wheat and barley roots in response to spatial variation in soil strength. *Plant Soil* 250:273–282.

Blum A, Johnson JW, Ramseur EL, Tollner EW. 1991. The effect of a drying topsoil and possible non-hydraulic signal on wheat growth and yield. *J Exp Bot* 42:1225–1231.

Boeuf-Tremblay V, Plantureux S, Guckert A. 1995. Influence of mechanical impedance on root exudation of maize seedlings at two development stages. *Plant Soil* 172:279–287.

Carminati A, Moradi AB, Vetterlein D et al. 2010. Dynamics of soil water content in the rhizosphere. *Plant Soil* 332:163–176.

Clark LJ, Barraclough PB. 1999. Do dicotyledons generate greater maximum axial root growth pressures than monocotyledons? *J Exp Bot* 50:1263–1266.

Clark LJ, Bengough AG, Whalley WR, Dexter AR, Barraclough PB. 1999. Maximum growth pressure in pea seedlings: Effects of measurement techniques and cultivars. *Plant Soil* 209:101–109.

Clark LJ, Cope RE, Whalley WR, Barraclough PB, Wade LJ. 2002. Root penetration of strong soil in rainfed lowland rice: Comparison of laboratory screens with field performance. *Field Crops Res* 76:189–198.

Clark LJ, Ferraris S, Price AH, Whalley WR. 2008a. A gradual rather than abrupt increase in strength gives better root penetration of strong layers. *Plant Soil* 307:235–242.

Clark LJ, Gowing DJ, Lark RM et al. 2005. Sensing the physical and nutritional status of the root environment in the field: A review of progress and opportunities. *J Agric Sci* 143:347–358.

Clark LJ, Price AH, Steele KA, Whalley WR. 2008b. Evidence from near-isogenic lines that root penetration increases with root diameter and bending stiffness in rice. *Funct Plant Biol* 35:1163–1171.

Clark LJ, Whalley WR, Dexter AR, Barraclough PB, Leigh RA. 1996. Turgor pressure is increased in the apex of pea roots by complete mechanical impedance. *Plant Cell Environ* 19:1099–1102.

Croser C, Bengough AG, Pritchard J. 1999. The effect of mechanical impedance on root growth in pea (*Pisum sativum*). I. Rates of cell flux, mitosis, and strain during recovery. *Physiol Plant* 107:277–286.

Croser C, Bengough AG, Pritchard J. 2000. The effect of mechanical impedance on root growth in pea (*Pisum sativum*). II. Cell expansion and wall rheology during recovery. *Physiol Plant* 109:150–159.

Davies WJ, Bacon MA, Thompson DS, Sobeih W, Rodríguez, LG. 2000. Regulation of leaf and fruit growth in plants growing in drying soil: Exploitation of the plants' chemical signalling system and hydraulic architecture to increase the efficiency of water use in agriculture. *J Exp Bot* 51:1617–1626.

Dexter AR, Hewitt JS. 1987. The deflection of plant roots. *J Agric Eng Res* 23:17–22.

Dodd IC. 2005. Root-to-shoot signalling: Assessing the roles of 'up' in the up and down world of long-distance signalling *in planta*. *Plant Soil* 274:251–270.

Dodd IC, Egea G, Watts CW, Whalley WR. 2010. Root water potential integrates discrete soil physical properties to influence ABA signalling during partial rootzone drying. *J Exp Bot* 16:3543–3551.

Eavis BE, Ratliff LF, Taylor HM. 1969. Use of a dead-load technique to determine axial root growth pressure. *Agron J* 61:640–643.

Farrell DA, Greacen EL. 1966. Resistance to penetration of fine probes in compressible soil. *Aust J Soil Res* 4:1–17.

Fritton DD. 2008. Evaluation of pedotransfer and measurement approaches to avoid soil compaction. *Soil Tillage Res* 99:268–287.

Gill WR, Bolt GH. 1955. Pfeffer's studies on the root growth pressures exerted by plants. *Agron J* 47:166–168.

Goss MJ. 1977. Effects of mechanical impedance on root-growth in barley (*Hordeum vulgare*) 1. Effects on elongation and branching of seminal root axes. *J Exp Bot* 28:96–111.

Goss MJ, Russell RS. 1980. Effects of mechanical impedance on root growth in barley (*Hordeum vulgare* L.). III. Observation on the mechanism of response. *J Exp Bot* 31:577–588.

Greacen EL. 1986. Root response to soil mechanical properties. *Transactions of the 13th Congress of International Society of Soil Science*, Hamburg, Germany, Vol. 5, pp. 20–47.

Greacen EL, Oh JS. 1972. Physics of root growth. *Nat New Biol* 235:24–25.

Gregory AS, Bird NRA, Whalley WR, Matthews GP, Young IM. 2010. Deformation and shrinkage effects on the soil water release characteristic. *Soil Sci Soc Am J* 74:1104–1112.

Gregory AS, Watts CW Whalley WR et al. 2007. Physical resilience of soil to field compaction and the interactions with plant growth and microbial community structure. *Eur J Soil Sci* 58:1221–1232.

Gregory AS, Whalley WR, Watts CW, Bird NRA, Hallett PD, Whitmore AP. 2006. Calculation of the compression index and the precompression stress from soil compression test data. *Soil Tillage Res* 89:45–57.

Groenevelt PH, Grant CD, Semesta S. 2001. A new procedure to determine water availability. *Aust J Soil Res* 39:577–598.

Hackett C, Rose DA. 1972. A model of the extension and branching of a seminal root of barley, and its use in studying relations between root dimensions I. The model. *Aust J Biol Sci* 25:669–679.

Håkansson I. 1990. A method for characterizing the state of compactness of the plough layer. *Soil Tillage Res* 16:105–120.

Hanbury CD, Atwell BJ. 2005. Growth dynamics of mechanically impeded lupin roots: Does altered morphology induce hypoxia? *Ann Bot* 96:913–924.

Hawes MC, Gunawardena U, Miyasaka S, Zhao XW. 2000. The role of root border cells in plant defense. *Trends Plant Sci* 5:128–133.

Humphris SN, Bengough AG, Griffiths BS et al. 2005. Root cap influences root colonisation by *Pseudomonas fluorescens* SBW25 on maize. *FEMS Microbiol Ecol* 54:123–130.

Iijima M, Higuchi T, Barlow PW, Bengough AG. 2003. Root cap removal increases root penetration resistance in maize (*Zea mays* L.). *J Exp Bot* 54:2105–2109.

Iijima M, Griffiths B, Bengough AG. 2000. Sloughing of cap cells and carbon exudation from maize seedling roots in compacted sand. *New Phytol* 145:477–482.

Jassogne L, McNeill A, Chittleborough D. 2007. 3D-visualization and analysis of macro- and meso-porosity of the upper horizons of a sodic, texture-contrast soil. *Eur J Soil Sci* 58:589–598.

Kays SJ, Nicklow CW, Simons DH. 1974. Ethylene in relation to response of roots to physical impedance. *Plant Soil* 40:565–571.

Kirby JM, Bengough AG. 2002. Influence of soil strength on root growth: Experiments and analysis using a critical-state model. *Eur J Soil Sci* 53:119–127.

Kuchenbuch RO, Ingram KT. 2004. Effects of soil bulk density on seminal and lateral roots of young maize plants (*Zea mays* L.). *J Plant Nutr Soil Sci* 167:229–235.

Lopes MS, Reynolds MP. 2010. Partitioning of assimilates to deeper roots is associated with cooler canopies and increased yield under drought in wheat. *Funct Plant Biol* 37:147–156.

Masle J, Passioura JB. 1987. The effect of soil strength on the growth of young wheat plants. *Aust J Plant Physiol* 14:643–656.

Materechera SA, Dexter AR, Alston AM, Kirby JM. 1992. Growth of seedling roots in response to external osmotic stress by polyethylene glycol 20,000. *Plant Soil* 143:85–91.

Matthews GP, Laudone GM, Gregory AS, Bird NRA, Matthews AGDG, Whalley WR. 2010. Measurement and simulation of the effect of compaction on the pore structure and saturated hydraulic conductivity of grassland and arable soil. *Water Resour Res* 46:W05501.

McCully ME. 1999. Roots in soil: Unearthing the complexities of roots and their rhizospheres. *Annu Rev Plant Physiol Plant Mol Biol* 50:695–718.

McCully ME, Boyer JS. 1997. The expansion of maize root-cap mucilage during hydration. 3. Changes in water potential and water content. *Physiol Plant* 99:169–177.

McKenzie BM, Bengough AG, Hallett PD, Thomas WTB, Forster B, McNicol JW. 2009. Deep rooting and drought screening of cereal crops: A novel field-based method and its application. *Field Crop Res* 112:165–171.

Misra RK. 1997. Maximum axial growth pressures of the lateral roots of pea and eucalypt. *Plant Soil* 188:161–170.

Misra RK, Dexter AR, Alston A. 1986. Maximum axial and radial growth pressures of plant roots. *Plant Soil* 95:315–326.

Misra RK, Gibbons AK. 1996. Growth and morphology of eucalypt seedling-roots, in relation to soil strength arising from compaction. *Plant Soil* 182:1–11.

Mitchell JK, Soga K. 2005. *Fundamentals of Soil Behaviour*. Hoboken, NJ: Wiley & Sons.

Montagu KD, Conroy JP, Francis GS. 1998. Root and shoot response of field-grown lettuce and broccoli to a compact subsoil. *Aust J Agric Res* 49:89–97.

Mullins CE, Panayiotopoulos KP. 1984. The strength of unsaturated mixtures of sand and kaolin and the concept of effective stress. *J Soil Sci* 35:459–468.

Odell RE, Dumlao MR, Samar D, Silk WK. 2008. Stage-dependent border cell and carbon flow from roots to rhizosphere. *Am J Bot* 95:441–446.

Pages L, Serra V, Draye X, Doussan C, Pierret A. 2010. Estimating root elongation rates from morphological measurements of the root tip. *Plant Soil* 328:35–44.

Palta JA, Chen X, Milroy SP, Rebetzke GJ, Dreccer MF, Watt M. 2011. Large root systems: Are they useful is adapting wheat to dry environments. *Funct Plant Biol* 38:347–354.

Passioura JB. 1991. Soil structure and plant-growth. *Aust J Soil Res* 29:717–728.

Passioura JB, Fry SC. 1992. Turgor and cell expansion: Beyond the Lockhart equation. *Aust J Plant Physiol* 19:565–576.

Pfeffer W. 1893. Druck und Arbeitsleistung durch Wachsende Pflanzen. Abhandlungen der Königlich Sächsiscten Gesellschaft der Wissenschaften 33:235–474.

Pritchard J. 1994. The control of cell expansion in roots. *New Phytol* 127:3–26.

Pritchard J, Jones RGW, Tomos AD. 1990. Measurement of yield threshold and cell-wall extensibility of intact wheat roots under different ionic, osmotic and temperature treatments. *J Exp Bot* 41:669–675.

Read DB, Bengough AG, Gregory PJ et al. 2003. Plant roots release phospholipid surfactants that modify the physical and chemical properties of soil. *New Phytol* 157:315–326.

Read DB, Gregory PJ, Bell AE. 1999. Physical properties of axenic maize root mucilage. *Plant Soil* 211:87–91.

Sarquis JI, Jordan WR, Morgan PW. 1991. Ethylene evolution from maize (*Zea-mays* L.) seedling roots and shoots in response to mechanical impedance. *Plant Physiol* 96:1171–1177.

Sealey LJ, McCully ME, Canny MJ. 1995. The expansion of maize root-cap mucilage during hydration 1. Kinetics. *Physiol Plant* 93:38–46.

Souty N, Stępniewski W. 1988. The influence of external oxygen concentration on axial root growth force of maize radicles. *Agronomie* 8:295–300.

Sweeney DW, Kirkham MB, Sisson JB. 2006. Crop and soil response to wheel-track compaction of a claypan soil. *Agron J* 98:637–643.

Taylor HM, Gardner HR. 1960. Use of wax substrates in root penetration studies. *Soil Sci Soc Am Proc* 24:287–496.

Taylor HM, Ratliff LH. 1969. Root elongation rates of cotton and peanuts as a function of soil strength and water content. *Soil Sci* 108:113–119.

To J, Kay BD. 2005. Variation in penetrometer resistance with soil properties: The contribution of effective stress and implications for pedotransfer functions. *Geoderma* 126:161–276.

Tollner EW, Verma BV. 1987. Lubricated and non-lubricated cone penetrometer performance comparison in six soils. *Trans Am Soc Agric Eng* 30:1611–1618.

Tsegaye T, Mullins CE. 1994. Effect of mechanical impedance on root-growth and morphology of 2 varieties of pea (*Pisum-sativum* L.). *New Phytol* 126:707–713.

Veen BW. 1982. The influence of mechanical impedance on the growth of maize roots. *Plant Soil* 66:101–109.

Vollsnes AV, Futsaether CM, Bengough AG. 2010. Quantifying rhizosphere particle movement around mutant maize roots using time-lapse imaging and particle image velocimetry. *Eur J Soil Sci* 61:926–939.

Whalley WR, Bengough AG, Dexter AR. 1998. Water stress induced by PEG decreases the maximum growth pressure of the roots of pea seedlings. *J Exp Bot* 49:1689–1694.

Whalley WR, Clark LJ, Dexter AR. 1994. The temperature dependence of the maximum axial growth pressure of the roots of pea (*Pisum sativum* L.). *Plant Soil* 163:211–215.

Whalley WR, Clark LJ, Gowing DJG, Cope RE, Lodge RJ, Leeds-Harrison PB. 2006. Does soil strength play a role in wheat yield losses caused by soil drying? *Plant Soil* 280:279–290.

Whalley WR, Dexter AR. 1993. The maximum axial growth pressure of roots of spring and autumn cultivars of lupin. *Plant Soil* 157:313–318.

Whalley WR, Leeds-Harrison PB, Clark LJ, Gowing DJG. 2005. The use of effective stress to predict the penetrometer resistance of unsaturated soils. *Soil Tillage Res* 84:18–27.

Whalley WR, To J, Kay BD, Whitmore AP. 2007. Prediction of the penetrometer resistance of soils with models with few parameters. *Geoderma* 137:370–377.

Whalley WR, Watts CW, Gregory AS, Mooney SJ, Clark LJ, Whitmore AP. 2008. The effect of soil strength on the yield of wheat. *Plant Soil* 306:237–247.

Watt M, McCully ME, Kirkegaard JA. 2003. Soil strength and rate of root elongation alter the accumulation of *Pseudomonas* spp. and other bacteria in the rhizosphere of wheat. *Funct Plant Biol* 30:483–491.

White RG, Kirkegaard JA. 2010. The distribution and abundance of wheat roots in a dense, structured subsoil—Implications for water uptake. *Plant Cell Environ* 33:133–148.

Whiteley GM, Hewitt JS, Dexter AR. 1982. The buckling of plant roots. *Physiol Plant* 54:333–342.

Yu L-X, Ray JD, O'Toole JC, Nguyen HT. 1995. Use of wax-petrolatum layers for screening rice root penetration. *Crop Sci* 35:684–687.

VI

Root–Rhizosphere Interactions

38

Fungal Root Endophytes

Thomas N. Sieber
ETH (Swiss Federal Institute of Technology)

Christoph R. Grünig
Microsynth AG

I. Introduction

The peripheral root tissues form a morphologically, physically, and chemically complex microcosm that provides a broad selection of different habitats for a myriad of microorganisms: bacteria, actinomycetes, protozoa, nematodes, microalgae, and fungi. The boundary between roots and soil changes all the time because roots constantly modify the nearby soil structure by their mechanical and metabolic activity (Foster et al. 1983). The rhizoplane, the epidermis, and the outer cortex are colonized by microorganisms in a nonrandom manner. Heavy microbial growth can occur on some individual cells, while neighboring cells are almost devoid of microorganisms (Bowen and Rovira 1976). Patchiness of microbial root colonization probably reflects the uneven distribution of organic debris in the soil, which serves as food base for the microorganisms. Many soil bacteria and fungi are able to colonize inter- and or intracellularly the epidermal and the outer cortical cells (OCO) of healthy roots. Only a comparatively small number of organisms, for example, mycorrhizal fungi, endophytes, and pathogens, possess, however, the ability to cross the inner boundary of the rhizosphere and to colonize the inner of root tissues (Bazin et al. 1990).

An endophyte is literally defined as one plant living within another organism. Applying the generally accepted five-kingdom system of Whittaker (1969), the term "endomycete" would, thus, seem more appropriate for a fungus living internally in a plant. This term is, however, ambiguous because it may lead to confusion with members of the Endomycetes, a class introduced by von Arx (1967) to accommodate ascomycetous yeasts and fungi with an yeast like growth phase. De Bary (1866) coined the term "endophyte" to distinguish organisms that invade and reside within host tissues or cells from "epiphytes," those fungi living on the outer surfaces of plants. Various plant/fungus symbioses constitute a continuum from antagonism to mutualism, and the type of symbiosis may change over time and space. Diseases caused by highly virulent pathogens or well-developed ectomycorrhizae are immediately obvious also to the nonspecialist, whereas the result of a plant/fungus association in between these two extremes may escape observation even by a specialist. The impossibility to unequivocally define the behavior of an endophyte as antagonistic or mutualistic becomes obvious by the findings of Freeman and Rodriguez (1993) who observed a fungal plant pathogen to convert to a nonpathogenic, endophytic mutualist by mutation at a single locus. Thus, a pragmatic definition of endophytism should be applied, which includes all organisms located within apparently healthy, functional root tissues at the moment of sample collection. The currently available methods to detect endophytes are destructive and, thus, the addendum "at the moment of sample collection" was necessary to account for the dynamic nature of plant–fungus interactions. Only fungi will be considered in this chapter although bacteria can live endophytically too (Chanway 1996; Schulz et al. 2006). Endophytic bacteria have been shown to be of vital importance in some symbioses, for example, actinorhizal symbiosis between tree roots (*Alnus* spp.) and *Frankia* spp., symbiosis between sugarcane (*Saccharum officinarum* L.) and *Acetobacter diazotrophicus*, a nitrogen-fixing bacterial endophyte (Dong et al. 1994), or plant growth–promoting rhizobacteria (Wei et al. 1994). The fungal partners in mycorrhizal associations are clearly also endophytes because part or all of their thallus is localized within the roots. Comprehensive articles, reviews, and books have been provided for many of the groups of classical mycorrhizal associations, namely, ecto-(ECM), ectendo-(EEM), or arbuscular mycorrhizae (AM) and the ericoid, arbutoid, orchid, or monotropoid mycorrhizae (Harvais and Hadley 1967; Read 1983; Mikola 1988; Allen et al. 1991; Egger et al. 1991; Allen and Allen 1992; Currah and Zelmer 1992; Read 1992, 1996; Brundrett 2004; Smith and Read 2008; Brundrett 2009). Emphasis in this chapter will be laid on root–fungus associations that are not regarded as mycorrhizae in the classical sense although literally they also form "fungus–root" entities of tightly interwoven fungus-plant tissues. Fungal root endophytes are ubiquitous and probably more abundant than classical mycorrhizae because they are not confined to the root tips and can occur everywhere in the root system. They are regularly isolated during studies about classical mycorrhizae but are then rather regarded as nuisance than interesting research objects. Since fungal root endophytes have been largely neglected, information about their functions is rare. After all, the number of publications about fungal root endophytes started to increase exponentially since about a decade ago, and there is hope that this group of organisms will eventually obtain the attention it deserves. In addition to mycorrhizae, root-colonizing obligate biotrophs with a prolonged latent phase such as certain smuts (Garcia-Guzman et al. 1996) or rust fungi, that is, *Tranzschelia fusca* (G. Winter) Dietel in the rhizomes of *Anemone nemorosa* L., will not be discussed in this chapter.

This chapter is based in part upon information in several previous reviews (which it does not supersede): (Melin 1923; Pearson and Read 1973; Garrett 1981; Foster et al. 1983; Read 1983; Wilcox 1983; Duddrige 1985; Foster 1986; Deacon 1987; Bazin et al. 1990; Petrini 1991; Newell 1992; Cannon and Hawksworth 1995; Bills 1996; Wilcox 1996; Jumpponen and Trappe 1998a; Sieber 2002; Addy et al. 2005; Mandyam and Jumpponen 2005; Sieber and Grünig 2006; Grünig et al. 2008b; Peterson et al. 2008; Newsham et al. 2009). Our referencing is not comprehensive, but we tried to include key references that can provide access to more literature on fungal root endophytes. In this chapter, we will concentrate on the taxonomy, diversity, physiology, and ecology of root endophytes and conclude with some ideas about interesting avenues for future research. Nomenclature of fungal names follows the system applied in the Index Fungorum (http://www.indexfungorum.org/Names/Names.asp). If both the teleomorph (sexual reproductive stage) and the anamorph (asexual reproductive stage) are produced, the name of the teleomorph will be used. Names of the anamorph(s) can be retrieved from the list of synonyms provided for teleomorphs with known anamorph(s) in the Index Fungorum.

Morphology of the endophyte–root symbioses is very variable and changes over time. It depends on both host and endophyte species; developmental stage of the tree, type, and age of the roots; edaphic and climatic conditions; and the microbial community in the rhizosphere. Sections IV.A1, IV.B1, IV.C1, V.A1, V.B1, and V.C1 about the histology of the endophyte–root interface will be included into this chapter of each host taxon.

II. Species and Hosts of Root Endophytes

Endophytic fungi were detected in the roots of all examined plant species. Grasses, orchids, and species of the Ericaceae and Pinaceae are the best-studied plants (Figure 38.1). Diversity, frequency, and population density of endophyte species depend on the edaphic and climatic conditions, on the heterogeneity of habitats and niches present within the host tissues, and on the competing organisms. Dark septate endophytes (DSE) probably occur in the roots of any plant species. They are the most frequent and most widespread root endophytes (Stoyke et al. 1992; Ahlich and Sieber 1996; Jumpponen and Trappe 1998a; Grünig et al. 2008b; Figure 38.1). Species of *Cylindrocarpon, Fusarium, Gibberella, Ilyonectria*, and *Neonectria* and the Sebacinales are frequent non-DSE. Fungi of the Sebacinales form mycorrhizae with ECM plants and orchids (OM) and associations of uncertain status with many other plant species (Warcup 1988; Weiss and Oberwinkler 2001; Weiss et al. 2011). The genera *Microdochium* and *Cryptosporiopsis* comprise both DSE and non-DSE species. *Microdochium nivale* (Fr.) Samuels & I.C. Hallett (anamorph of *Monographella nivalis* (Schaffnit) E. Müll.) and *Microdochium bolleyi* (R. Sprague) de Hoog & Herm.-Nijh. are both endophytes in grasses, but colonies of *M. nivale* are white changing to orange where sporodochia form, whereas those of *M. bolleyi* are black.

A. Dark Septate Endophytes

DSE is a form taxon and serves primarily to differentiate these fungi from endophytes with septate, hyaline hyphae and from fungi with sparsely septate, hyaline hyphae, which are characteristic for AM fungi. However, the distinction based on the degree of pigmentation is often arbitrary because melanization varies greatly in some species (Addy et al. 2005; Hambleton and Sigler 2005) especially that of mycelia in colonized roots (Yu et al. 2001; Barrow 2003). We propose to define DSE as those endophytic fungi, which form either at least partly dark brown, dark gray, or black colonies on 2% (w/v) malt extract agar (MEA) when incubated at 20°C and/or distinctly melanized structures in roots. Most DSE are ascomycetes, and consequently dematiaceous ascomycetes that are able to colonize the interior of plant tissues without inducing disease are potential DSE. However, the most frequent root-colonizing DSE belong to the genera *Microdochium* (anamorphic *Monographella* (Xylariales)), *Periconia* (anamorphic Pleosporales), *Harpophora* (anamorphic *Gaeumannomyces*), *Cadophora* (anamorphic Helotiales), *Cryptosporiopsis* (anamorphic *Pezicula* (Helotiales)), *Phialocephala*, and *Acephala* (both anamorphic Helotiales) (Figure 38.1). Some of the best-studied DSE are *Microdochium bolleyi*, *Periconia macrospinosa* Lefebvre & Aar.G. Johnson, *Gaeumannomyces* and *Harpophora* species, *Acephala applanata* Grünig & T.N. Sieber, and *Phialocephala fortinii* s.l. C.J.K. Wang & H.E. Wilcox. *P. fortinii* s.l. consists of several morphologically indistinguishable but reproductively isolated cryptic species (CSP) (Grünig 2004; Grünig et al. 2007) that together with *A. applanata* form the *P. fortinii* s.l.–*A. applanata*

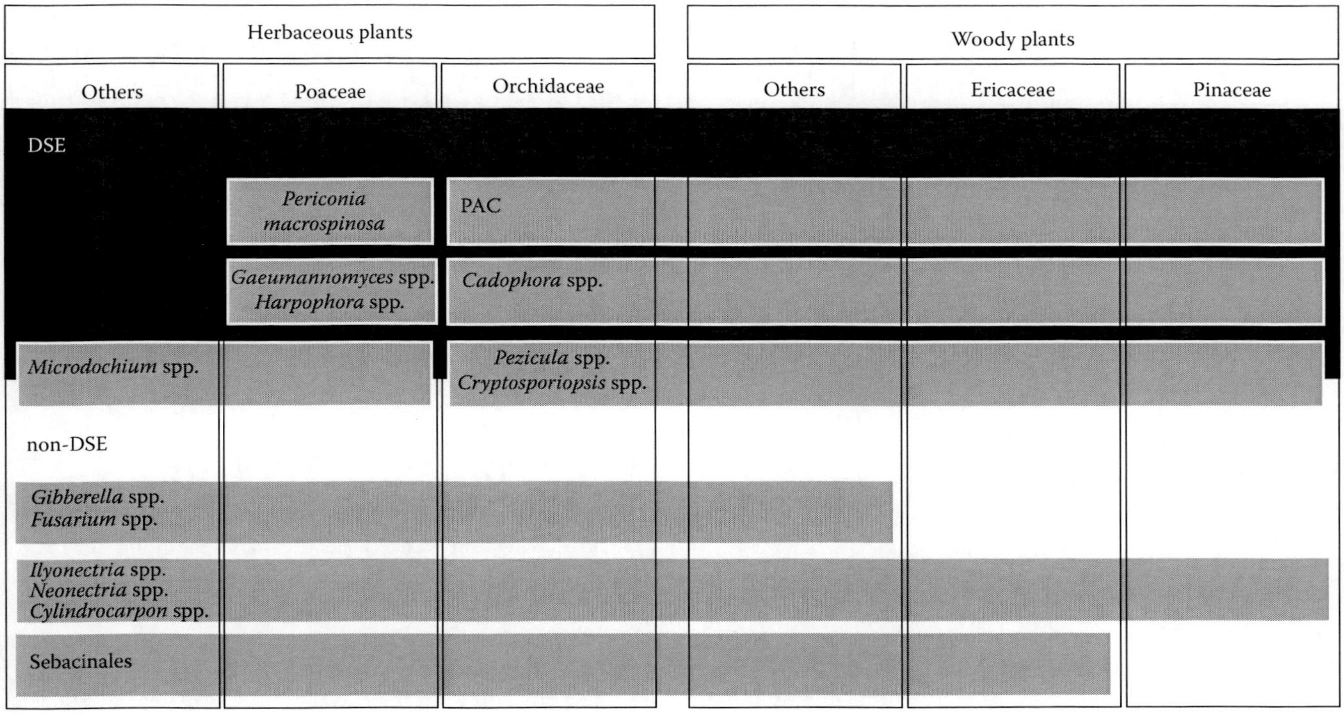

Herbaceous plants

Woody plants

Others | Poaceae | Orchidaceae | Others | Ericaceae | Pinaceae

DSE

Periconia macrospinosa

PAC

Gaeumannomyces spp.
Harpophora spp.

Cadophora spp.

Microdochium spp.

Pezicula spp.
Cryptosporiopsis spp.

non-DSE

Gibberella spp.
Fusarium spp.

Ilyonectria spp.
Neonectria spp.
Cylindrocarpon spp.

Sebacinales

FIGURE 38.1 Simplified schematic display of the main taxa of root endophytes and the major groups of their plant hosts. The black upper box indicates occurrence of DSE, the blank lower box occurrence of non-DSE; species of taxa positioned in the black box are DSE, those in the blank box non-DSE; the genera *Microdochium* and *Pezicula* (*Cryptosporiopsis*) comprise both DSE and non-DSE species.

species complex (PAC) (Grünig et al. 2008b). PAC are among the best-characterized DSE. They are very widespread and abundant in roots of woody plants, especially conifers and ericaceous plants, all over the Northern Hemisphere (Tables 38.4 through 38.6). Section II.A1 is therefore dedicated to them (see below).

Host specificity of DSE is the exception, but some species prefer certain host taxa. *Microdochium*, *Periconia*, and *Harpophora* species prefer grasses, whereas *Cadophora*, *Cryptosporiopsis*, and PAC species preferentially occur on orchids and woody plant species (Grünig et al. 2008b; Figure 38.1). PAC seem to be confined to forest ecosystems because reports about their presence in arable soils or in roots of plants growing in these soils are very rare (Ahlich-Schlegel 1997; Brenn et al. 2008).

1. *Phialocephala fortinii s.l.–Acephala applanata Species* Complex

A. applanata is the only morphologically distinct PAC species, because its colonies on MEA differ from those of *Phialocephala fortinii* s.l. by the absence of aerial mycelium (Grünig and Sieber 2005). The structures of conidiophores are similar among species and highly variable within species and, consequently, do not allow to discriminate species (Grünig et al. 2008a). Differentiation of PAC species is only possible using molecular methods. Seven of the *Phialocephala fortinii* s.l. species were formally described based on population differentiation and differences at five sequence loci (Grünig et al. 2007, 2008a): *Phialocephala fortinii* s.s., *Phialocephala subalpina*, *Phialocephala letzii* Grünig & T. N. Sieber, *Phialocephala uotilensis* Grünig & T. N. Sieber, *Phialocephala turicensis* Grünig & T. N. Sieber, *Phialocephala europaea* Grünig & T.N. Sieber, and *Phialocephala helvetica* Grünig & T.N. Sieber. Identification of members of PAC is possible using ITS sequencing (Grünig et al. 2008b). However, species identification within PAC to the species level requires sequencing of at least three of the sequence loci (Grünig et al. 2008a). Usually, sequencing of the pPF-076, pPF-018, and beta-tubulin locus using the primers listed in Grünig et al. (2007) is enough to identify the species. Alternatively, identification can occur by multiplex-polymerase-chain-reaction (PCR)-amplified microsatellite (MS) loci using the primers developed by Queloz et al. (2008, 2010).

The geographical distribution of PAC species ranges from arctic to subtropical regions throughout the Northern Hemisphere (Piercey et al. 2004; Zhang et al. 2009; Queloz et al. 2011). The wide geographic distribution of PAC contrasts with the results obtained from baiting experiments, which provided no evidence for aerial dispersal of PAC (Bachmann 2010). Nevertheless, strains with identical inter-simple sequence repeats (ISSR) fingerprint and single-copy restriction fragment length polymorphism (RFLP) haplotype were repeatedly found in forest stands situated a few km apart, and migration rates measured between continents were low but still sufficient to prevent speciation (Grünig et al. 2004, 2011; Grünig and Sieber 2005; Queloz 2010). This is at least surprising, since PAC do not or only rarely sporulate after prolonged incubation at low temperature, and germination of conidia was never observed. Probably, anthropogenic gene and genotype flow occurs by the movement of nursery plants colonized by PAC locally and around the globe (Brenn et al. 2008).

Abundance distribution curves of PAC species in local communities are hyperbolic with a few abundant species and many "rare" species (Grünig et al. 2006), consistent with the community structures observed in many other biological systems (McGill et al. 2007). In a recent study, 44 communities from the Northern Hemisphere comprising more than 5000 PAC strains were analyzed (Queloz et al. 2011). Species diversity and community structure were neither correlated with the tree community, geographical location, soil properties, management practices, precipitation, nor temperature, supporting the hypothesis of "everything is everywhere" (Baas-Becking 1934). Indeed, it is known that host specificity of PAC species is low or lacking because most species can be isolated from a broad range of woody plant species (Grünig et al. 2008b). For example, the most abundant PAC species, *Phialocephala subalpina* Grünig & T.N. Sieber, was isolated from 16 plant species compared to many fungal pathogens that often have narrow host ranges (Tables 38.4 through 38.6).

PAC species were shown to be a genetically highly variable on a regional level (Harney et al. 1997; Grünig et al. 2001). Within forest plots of 200 m^2, PAC can form communities of up to 13 sympatrically occurring species (Grünig et al. 2004; Queloz et al. 2010). More than 25 genets were present in an area of only 9 m^2 of forest soil, and this community remained stable for several years (Queloz et al. 2005). More homogenous population structures would be expected for these supposedly mitotic PAC fungi. There are at least two possibilities to explain this phenomenon. Either the assumption of asexuality is wrong, and PAC enjoy some hidden sex such as parasexuality, or it is a polyphyletic species complex, and what we see today is the result of a convergent evolutionary process. From a molecular genetics point of view, PAC meet the requirements for sex, because they possess mating-type genes (Zaffarano et al. 2010), and probably, they are also functional because no gametic disequilibrium can be found in PAC species based on single-copy RFLP data (Grünig et al. 2007). Whereas mating-type genes of *Acephala applanata* are organized in a homothallic fashion (MAT1-1 and MAT1-2 in the same individual interrupted by a transposable element), other PAC species are heterothallic possessing only either the MAT1-1 or MAT1-2 idiomorph in their genomes (Zaffarano et al. 2010, 2011). Zaffarano et al. (2010) showed that in more than 80% of the populations of PAC species a 1:1 mating-type ratio and gametic equilibrium can be observed. In addition, MAT genes were shown to evolve under strong purifying selection. All three observations support the assumption that cryptic sexual reproduction occurs in PAC species. Moreover, the presence of sexual reproduction is also supported by the finding of the teleomorphic state in *Phaeomollisia piceae* T.N. Sieber & Grünig, a phylogenetically closely related species (Grünig et al. 2009).

Another secret is whether PAC fungi are able to grow through soil. PAC can readily be baited from forest soil using Norway-spruce seedlings as bait (Ahlich et al. 1998; Trüssel 2011). However, it is not known if PAC actively grow through soil toward host roots or are passively waiting as dormant propagules, for example, microsclerotia in or on root debris, until a susceptible root "grows past." Trüssel (2011) tested mycelial growth through soil substrate in nonsterilized and gamma-sterilized forest soil and peat–vermiculite. Norway-spruce seedlings were used as bait separated from the PAC inoculated substrate by a mesh screen that did not allow passage of roots. Growth of mycelium occurred only in the sterilized substrates and was slow with maximal distances of 2 cm covered in 3 months at 20°C, although the pH of 3.8 was optimal (Trüssel 2011). No growth occurred in nonsterilized forest soil probably due to inhibition by other soil microorganisms. Thus, it remains dubious if PAC can actively grow through soil or are transmitted only by root contacts.

III. Methods of Detection

Isolation from plant tissues (1), histological examination (2), assays of fungal-specific molecules, for example, ergosterol, a sterol characteristic of fungal membranes (3), and culture-independent DNA extraction and sequencing of specific loci (4) are the four major groups of methods used to detect and describe communities of endophytic fungi (Parsons 1981; Savage and Sall 1981; El-Nashaar et al. 1986; Newell et al. 1988; Hampton et al. 1990; Newell 1992; Reissinger et al. 2001; Sieber 2002; Schulz and Boyle 2005; Jumpponen and Jones 2009). Each of these methods has its advantages and disadvantages. Methods (1) and (4) but not (2) and (3) allow identification of the fungi. Only culturable fungi can be detected with method (1), and the number and kind of species retrieved depend on the strength of surface sterilization, the medium used for incubation, and the sample size. With method (4), it remains obscure whether the DNAs originated from living or dead fungi, and consequently, the surface-sterilization method used in combination with method (4) must not only kill organisms on the plant surface but also remove their DNA. Furthermore, method (4) strongly depends on the quality of the databases used for sequence identification. Consequently, only the application of several of these methods combined allows obtaining a complete picture of the endophytic mycobiota in a plant tissue.

Methods were developed to specifically detect certain species. Quantitative PCR was developed to detect and quantify PAC in root tissues (Tellenbach et al. 2010). Quantification by qPCR is, however, problematic because the amount of DNA is measured and used to estimate biomass. Even if it is assumed that each cell contains one nucleus, type and size of fungal cells can vary considerably within and among isolates making estimates inaccurate. For example, the method worked fine for PAC strains that formed uniform, regular mycelia in the host but not for those that produced lots of microsclerotia (Tellenbach et al. 2010). Although based on DNA, microsatellites provide an even more sensitive detection and quantification method because genets of the same species can be differentiated and quantified provided that genets possess different alleles at at least one locus (Reininger et al. 2011b).

IV. Root Endophytes of Herbaceous Plants

Agricultural plants are among the best-studied herbaceous plants in regard to root endophytes. Nonagricultural herbaceous plants have only rarely been examined, and the detection of non-mycorrhizal root endophytes was mostly a by-product during studies of OM or AM fungi.

A. Grass Endophytes

Members of the Clavicipitaceae are the best-known grass endophytes. They have received a lot of attention in the past due to their beneficial effects upon their hosts (Clay 1988; Schardl 2010). Some function as biocontrol agents against insects and other herbivores, others produce growth-promoting metabolites. Most effects are based on various kinds of alkaloids produced by these endophytes. Clavicipitaceous grass endophytes are considered to colonize their hosts systemically. However, roots are usually not colonized. Azevedo and Welty (1995) inoculated axenically grown tall fescue (*Festuca arundinacea* Schreb.) seedlings with *Neotyphodium coenophialum* (Morgan-Jones & W. Gams) Glenn, C.W. Bacon & Hanlin but could not observe direct penetration of the fungus into intact root cortex cells. Consequently, improved growth and biomass accumulation of roots as well as altered root morphology of infected plants grown at low nutrient availability (e.g., phosphorus) are mediated by the endophyte activity in the aerial plant parts (Malinowski et al. 1998; Malinowski and Belesky 1999). However, roots of grasses are habitats for many non-clavicipitalean endophytes (Table 38.1).

The most frequent genera of grass-root endophytes are *Fusarium, Gaeumannomyces, Gibberella, Harpophora, Microdochium, Monographella,* and *Periconia* (Figure 38.1). For example, Riesen and Sieber (1985) and Sieber et al. (1988) isolated more than 100 species from roots of winter wheat. *Fusarium culmorum* (W.G. Sm.) Sacc., *Fusarium oxysporum* Schltdl., *Gibberella zeae* (Schwein.) Petch, *Microdochium bolleyi, Monographella nivalis, Periconia macrospinosa* Lefebvre & Aar.G. Johnson, and *Phaeosphaeria nodorum* (E. Müll.) Hedjar., the causal agent of glume blotch, were the dominant fungal species. *M. bolleyi* was also regularly recovered from the roots of healthy field-grown barley (*Hordeum vulgare* L.), oats (*Avena sativa* L.), and pasture grass (Murray and Gadd 1981).

Stoyke and Currah (1991) isolated DSE from an unidentified grass species in an alpine habitat of the Rocky Mountains in Alberta, Canada. Similarly, dark septate endophytic hyphae were observed in grasses collected on various islands of the Southern Atlantic Ocean and the Antarctic Peninsula (Christie and Nicolson 1983), and Upson et al. (2009) isolated 243 DSE strains from the roots of *Deschampsia antarctica* E. Desv. and *Colobanthus quitensis* (Kunth) Bartl. (Caryophyllaceae) collected from 17 sites across a 1470 km transect through maritime and sub-Antarctica. Most DSE belonged to the Helotiales. *Leptodontidium orchidicola* Sigler & Currah, *Pezoloma ericae* (D.J. Read) Baral, and species of *Tapesia* and *Mollisia* could be identified by ITS sequence comparison. Li et al. (2005) detected considerable colonization of roots of several grass species collected in Kunming, China, by DSE.

Grass and cereal roots are often colonized by dark septate *Gaeumannomyces* species and relatives. Take-all, caused by *Gaeumannomyces graminis* var. *tritici J.* Walker and *Gaeumannomyces graminis* var. *avenae* (E.M. Turner) Dennis, is the most important root disease of wheat and oat worldwide (Freeman and Ward 2004), but many *Gaeumannomyces* species are nonpathogenic (Deacon 1981; Sieber 1985; Deacon 1987; Skipp and Christensen 1989; Blaschke 1991; Crous et al. 1995; Gutteridge et al. 2007). *Gaeumannomyces* species possess *Phialophora*-like anamorphs, which have been accommodated in the genus *Harpophora* by Gams (2000). Whereas it is accepted that *Harpophora graminicola* (Deacon) W. Gams is the anamorph of *Gaeumannomyces cylindrosporus* Hornby, Slope, Gutter. & Sivan. (Ward and Bateman 1999; Freeman and Ward 2004), there is some debate about other *Harpophora–Gaeumannomyces* connections, especially when it comes to the anamorphs of the various varieties of *G. graminis* (Sacc.) Arx & D.L. Olivier. A *Harpophora* species isolated by McKeen (1952) from corn roots was described as *Harpophora radicicola* (Cain) W. Gams by Cain (1952). Some workers interpreted *H. radicicola* as the anamorphic state of *G. graminis* var. *tritici* (Lemaire and Ponchet 1963; Simonsen 1971). However, this was considered not tenable by Walker (1981). In fact, identity of *H. radicicola* with *Gaeumannomyces graminis* var. *maydis* J.M. Yao, Yong C. Wang & Y.G. Zhu, the maize take-all fungus from China (Yao et al. 1992), was demonstrated using molecular genetics methods (Ward and Bateman 1999). In addition, the same authors found *H. radicicola* to be almost identical to *Harpophora zeicola* (Deacon & D.B. Scott) W. Gams, which was described as a weak parasite of drought or otherwise stressed maize plants in South Africa and France (Deacon and Scott 1983). *G. graminis* var. *maydis* was also observed in the roots of three alpine grasses (*Deschampsia caespitosa* (L.) P. Beauv., *Festuca pumila* Chaix, and *Poa alpina* L.) growing at timberline (1900 m asl.) in Bavaria (Blaschke 1991). Colonization of roots of *Deschampsia flxuosa* (L.) Trin. by an unidentified DSE increased with increasing temperature under a global warming scenario but was not affected by elevated CO_2 levels (Olsrud et al. 2010; Table 38.1).

1. Anatomy

Epidermal and cortical root cells of cereal and grass roots filled with heavily melanized microsclerotia are characteristic for colonization by *M. bolleyi* (Murray and Gadd 1981). In contrast, inter- and intracellular hyphae of this fungus are hyaline. Colonization of the stelar tissues of healthy roots does usually not occur. These patterns of colonization correspond well with those observed by Rasmann et al. (2009) for *M. bolleyi* in tomato roots

TABLE 38.1 Endophytes in Grasses (Family: Poaceae), Sedges (Family: Cyperaceae) and Bromeliads (Family: Bromeliaceae)

Endophyte				Host					
Genus[a]	Species[a]	Fungus Order[a]	DSE/Not DSE	Species	Plant Order	Woody/ Herbaceous	Type of Experiment[b]	Special Effect[c]	References
Acremonium	sp.	Hypocreales	Not DSE	*Holcus lanatus*	Poales	Grass	Isol		Sánchez Márquez et al. (2010)
Acremonium	*strictum*	Hypocreales	Not DSE	*Ammophila arenaria*	Poales	Grass	In vitro exp	a	Hol et al. (2007)
Alternaria	sp.	Pleosporales	DSE	*Bouteloua gracilis*	Poales	Grass	Isol		Herrera et al. (2010)
Alternaria	sp.	Pleosporales	DSE	*Holcus lanatus*	Poales	Grass	Isol		Sánchez Márquez et al. (2010)
Aspergillus	sp.	Eurotiales	Not DSE	*Stipa grandis*	Poales	Grass	Isol		Su et al. (2010)
Bipolaris	sp.	Pleosporales	DSE	*Bouteloua gracilis*	Poales	Grass	Isol		Herrera et al. (2010)
Campanella	sp.	Agaricales	DSE	*Bouteloua gracilis*	Poales	Grass	Isol		Herrera et al. (2010)
Chaetomium	*funicola*	Sordariales	Not DSE	*Hordeum vulgare*	Poales	Grass	In vitro exp	b	Vilich et al. (1998)
Chaetomium	*globosum*	Sordariales	Not DSE	*Hordeum vulgare*	Poales	Grass	In vitro exp	b	Vilich et al. (1998)
Cladorrhinum	*foecundissimum*	Sordariales	N.A.	*Elytrigia repens*	Poales	Grass	Isol & in vitro exp	c	Gasoni and Stegamn de Gurfinkel (2009)
Cladosporium	*cladosporiodes*	Capnodiales	DSE	*Bothriochloa macra*	Poales	Grass	Isol		White and Backhouse (2007)
Cladosporium	*cladosporiodes*	Capnodiales	DSE	*Hyparrhenia hirta*	Poales	Grass	Isol		White and Backhouse (2007)
Coniothyrium	sp.	Pleosporales	Not DSE	*Triticum aestivum*	Poales	Grass	Isol		Crous et al. (1995)
Cryptosporiopsis	*rhizophila*	Helotiales	DSE	*Deschampsia flexuosa*	Poales	Grass	In vitro exp	d	Zijlstra et al. (2005)
Curvularia	*inaequalis*	Pleosporales	DSE	*Holcus lanatus*	Poales	Grass	Isol		Sánchez Márquez et al. (2010)
Cylindrocarpon	*didymum*	Hypocreales	Not DSE	*Triticum aestivum*	Poales	Grass	Isol		Sieber (1985)
Dictyochaeta	*fertilis*	Chaetosphaeriales	Not DSE	*Lolium perenne*	Poales	Grass	Isol		Skipp and Christensen (1989)
Drechslera	sp.	Pleosporales	DSE	*Holcus lanatus*	Poales	Grass	Isol		Sánchez Márquez et al. (2010)
Embellisia	*chlamydospora*	Pleosporales	DSE	*Carex stenophylla*	Poales	Sedge	Isol		Graf, pers. comm. (2008)
Epicoccum	sp.	Pleosporales	Not DSE	*Holcus lanatus*	Poales	Grass	Isol		Sánchez Márquez et al. (2010)
Fusarium	*culmorum*	Hypocreales	Not DSE	*Triticum aestivum*	Poales	Grass	Isol		Sieber et al. (1988)
Fusarium	*oxysporum*	Hypocreales	Not DSE	*Bothriochloa macra*	Poales	Grass	Isol		White and Backhouse (2007)
Fusarium	*oxysporum*	Hypocreales	Not DSE	*Holcus lanatus*	Poales	Grass	Isol		Sánchez Márquez et al. (2010)
Fusarium	*oxysporum*	Hypocreales	Not DSE	*Hyparrhenia hirta*	Poales	Grass	Isol		White and Backhouse (2007)
Fusarium	*oxysporum*	Hypocreales	Not DSE	*Lolium perenne*	Poales	Grass	Isol		Skipp and Christensen (1989)
Fusarium	*oxysporum*	Hypocreales	Not DSE	*Oryza sativa*	Poales	Grass	Isol		Fisher and Petrini (1992)
Fusarium	*oxysporum*	Hypocreales	Not DSE	*Stipa grandis*	Poales	Grass	Isol		Su et al. (2010)
Fusarium	*redolens*	Hypocreales	Not DSE	*Stipa grandis*	Poales	Grass	Isol		Su et al. (2010)
Fusarium	sp.	Hypocreales	DSE	*Bouteloua gracilis*	Poales	Grass	Isol		Herrera et al. (2010), Khidir et al. (2010)
Fusarium	sp.	Hypocreales	DSE	*Sporobolus cryptandrus*	Poales	Grass	Isol		Khidir et al. (2010)
Fusarium	spp.	Hypocreales	Not DSE	*Stipa grandis*	Poales	Grass	Isol		Su et al. (2010)
Gaeumannomyces	*cylindrosporus*[d]	Magnaporthaceae	DSE	*Poa pratensis*	Poales	Grass	Isol		Saleh and Leslie (2004)
Gaeumannomyces	*cylindrosporus*[d]	Magnaporthaceae	DSE	Various grass species	Poales	Grass	Isol	e	Deacon (1981)
Gaeumannomyces	*cylindrosporus*[d]	Magnaporthaceae	DSE	*Vulpia ciliata*	Poales	Grass	In vitro exp	f	Newsham (1999)
Gaeumannomyces	*graminis* var. *graminis*	Magnaporthaceae	DSE	*Oryza sativa*	Poales	Grass	Isol		Saleh and Leslie (2004)

Gaeumannomyces	*graminis* var. *maydis*[e]	Magnapothaceae	DSE	*Lolium perenne*	Poales	Grass	Isol		Skipp and Christensen (1989)
Gaeumannomyces	*graminis* var. *maydis*[e]	Magnaporthaceae	DSE	*Zea mays*	Poales	Grass	Isol		Cain (1952)
Gaeumannomyces	*graminis* var. *maydis*[e]	Magnaporthaceae	DSE	*Zea mays*	Poales	Grass	Isol		Ward and Bateman (1999)
Gaeumannomyces	sp.	Magnaportaceae	DSE	*Bouteloua gracilis*	Poales	Grass	Isol		Herrera et al. (2010)
Gibberella	*avenacea*	Hypocreales	Not DSE	*Triticum aestivum*	Poales	Grass	Isol		Crous et al. (1995)
Gibberella	*zeae*	Hypocreales	Not DSE	*Triticum aestivum*	Poales	Grass	Isol		Sieber et al. (1988)
Harpophora	*oryzae*	Magnaporthaceae	DSE	*Oryza sativa*	Poales	Grass	Isol		Yuan et al. (2010)
Idriella	*lunata*	Helotiales	Not DSE	*Stipa grandis*	Poales	Grass	Isol		Su et al. (2010)
Ilyonectria	*radicicola*	Hypocreales	Not DSE	*Triticum aestivum*	Poales	Grass	Isol		Sieber (1985)
Leptodontidium	sp.	Helotiales	Not DSE	*Holcus lanatus*	Poales	Grass	Isol		Sánchez Márquez et al. (2010)
Meliniomyces	*variabilis* LtVB3	Leotiomycetes	DSE	*Hordeum vulgare*	Poales	Grass	Isol		Narisawa et al. (2004)
Microdochium	*bolleyi*	Xylariales	DSE	*Elymus farctus*	Poales	Grass	Isol		Sánchez Márquez et al. (2008)
Microdochium	*bolleyi*	Xylariales	DSE	*Poa alpigena*	Poales	Grass	Isol		Väre et al. (1992)
Microdochium	*bolleyi*	Xylariales	DSE	*Triticum aestivum*	Poales	Grass	Isol		Sieber et al. (1988)
Microdochium	*bolleyi*	Xylariales	DSE	*Triticum aestivum*	Poales	Grass	In vitro & field exp	g	Reinecke (1978)
Microdochium	*bolleyi*	Xylariales	DSE	*Triticum aestivum*	Poales	Grass	Isol		Riesen and Sieber (1985)
Microdochium	*bolleyi*	Xylariales	DSE	*Triticum aestivum*	Poales	Grass	In vitro exp	h	Kirk and Deacon (1987)
Microdochium	sp.	Xylariales	DSE	*Andropogon gerardii*	Poales	Grass	In vitro exp	i	Mandyam et al. (2010)
Microdochium	sp.	Xylariales	DSE	*Bouteloua gracilis*	Poales	Grass	Isol		Herrera et al. (2010)
Microdochium	sp.	Xylariales	DSE	Grasses	Poales	Grass	Isol		Mandyam et al. (2010)
Microdochium	spp.	Xylariales	DSE	*Stipa grandis*	Poales	Grass	Isol		Su et al. (2010)
Mollisia	sp.	Helotiales	DSE	*Deschampsia antarctica*	Poales	Grass	Isol		Upson et al. (2009)
Moniliophthora	sp.	Agaricales	DSE	*Bouteloua gracilis*	Poales	Grass	Isol		Herrera et al. (2010), Khidir et al. (2010)
Moniliophthora	sp.	Agaricales	DSE	*Sporobolus cryptandrus*	Poales	Grass	Isol		Khidir et al. (2010)
Monodictys	sp.	Dothideomycetes	DSE	*Carex stenophylla*	Poales	Sedge	Isol		Graf, pers. comm. (2008)
Monographella	*nivalis*	Xylariales	Not DSE	*Triticum aestivum*	Poales	Grass	Isol		Sieber et al. (1988)
Monographella	*nivalis*	Xylariales	Not DSE	*Triticum aestivum*	Poales	Grass	Isol		Sieber et al. (1988)
Monosporascus	sp.	Xylariales	Not DSE	*Stipa grandis*	Poales	Grass	Isol		Su et al. (2010)
Neurospora	sp.	Sordariales	Not DSE	*Stipa grandis*	Poales	Grass	Isol		Su et al. (2010)
Paraphaeosphaeria	sp.	Pleosporales	DSE	*Bouteloua gracilis*	Poales	Grass	Isol		Herrera et al. (2010), Khidir et al. (2010)
Paraphaeosphaeria	sp.	Pleosporales	DSE	*Sporobolus cryptandrus*	Poales	Grass	Isol		Khidir et al. (2010)
Paraphoma	*fimeti*	Pleosporales	DSE	*Vulpia ciliata*	Poales	Grass	In vitro exp	f	Newsham (1994)
Penicillium	sp.	Eurotiales	Not DSE	*Holcus lanatus*	Poales	Grass	Isol		Sánchez Márquez et al. (2010)
Penicillium	sp.	Eurotiales		*Stipa grandis*	Poales	Grass	Isol		Su et al. (2010)
Periconia	*macrospinosa*	Pleosporales	DSE	*Andropogon gerardii*	Poales	Grass	In vitro exp	j	Mandyam et al. (2010)

(continued)

TABLE 38.1 (continued) Endophytes in Grasses (Family: Poaceae), Sedges (Family: Cyperaceae) and Bromeliads (Family: Bromeliaceae)

Endophyte				Host					
Genus[a]	Species[a]	Fungus Order[a]	DSE/Not DSE	Species	Plant Order	Woody/ Herbaceous	Type of Experiment[b]	Special Effect[c]	References
Periconia	*macrospinosa*	Pleosporales	DSE	*Bothriochloa macra*	Poales	Grass	Isol		White and Backhouse (2007)
Periconia	*macrospinosa*	Pleosporales	DSE	*Holcus lanatus*	Poales	Grass	Isol		Sánchez Márquez et al. (2010)
Periconia	*macrospinosa*	Pleosporales	DSE	*Hyparrhenia hirta*	Poales	Grass	Isol		White and Backhouse (2007)
Periconia	*macrospinosa*	Pleosporales	DSE	*Stipa grandis*	Poales	Grass	Isol		Su et al. (2010)
Periconia	*macrospinosa*	Pleosporales	DSE	*Andropogon gerardii*	Poales	Grass	In vitro exp	k	Mandyam et al. (2010)
Periconia	*macrospinosa*	Pleosporales	DSE	Grasses	Poales	Grass	Isol		Mandyam et al. (2010)
Periconia	*macrospinosa*	Pleosporales	DSE	*Holcus lanatus*	Poales	Grass	Isol		Sánchez Márquez et al. (2010)
Periconia	sp.	Pleosporales	DSE	*Bouteloua gracilis*	Poales	Grass	Isol		Herrera et al. (2010)
Phaeocytostroma	*plurivorum*	Pezizomyc incert	N.A.	*Stipa grandis*	Poales	Grass	Isol		Su et al. (2010)
Phaeosphaeria	*nodorum*	Pleosporales	Not DSE	*Triticum aestivum*	Poales	Grass	Isol		Sieber et al. (1988)
Phaeosphaeria	*nodorum*	Pleosporales	Not DSE	*Triticum aestivum*	Poales	Grass	Isol		Sieber et al. (1988)
Phialocephala	*fortinii s.l.*	Helotiales	DSE	*Deschampsia flexuosa*	Poales	Grass	Isol		Tejesvi and Ruotsalainen (2010)
Phialocephala	*fortinii s.l.*	Helotiales	DSE	*Deschampsia flexuosa*	Poales	Grass	In vitro exp	l	Zijlstra et al. (2005)
Phialocephala	*fortinii s.l.*	Helotiales	DSE	*Poa alpigena*	Poales	Grass	Isol		Väre et al. (1992)
Phialocephala	*fortinii s.l.*	Helotiales	DSE	*Poa alpigena*	Poales	Grass	Isol		Väre et al. (1992)
Phialocephala	sp. 8	Helotiales	DSE	*Carex aquatilis*	Poales	Sedge	Isol		Grünig et al. (2009)
Phialophora	sp.	N.A.	DSE	*Carex curvula*	Poales	Sedge	Isol		Haselwandter and Read (1982)
Phialophora	sp.	N.A.	DSE	*Carex firma*	Poales	Sedge	Isol	f	Haselwandter and Read (1982)
Phialophora	sp.	Helotiales	N.A.	*Deschampsia flexuosa*	Poales	Grass	Isol		Tejesvi and Ruotsalainen (2010)
Phialophora	sp.	Helotiales	N.A.	*Stipa grandis*	Poales	Grass	Isol		Su et al. (2010)
Phialophora	sp. 1	Helotiales	DSE	*Triticum aestivum*	Poales	Grass	Isol		Rooden, unpublished
Phialophora	sp. 1	Helotiales	DSE	*Triticum aestivum*	Poales	Grass	Isol		Rooden, unpublished
Phoma	sp.	Pleosporales	DSE	*Bouteloua gracilis*	Poales	Grass	Isol		Herrera et al. (2010)
Phoma	*glomerata*	Pleosporales	DSE	*Triticum aestivum*	Poales	Grass	Isol		Crous et al. (1995)
Podospora	sp.	Sordariales	Not DSE	*Holcus lanatus*	Poales	Grass	Isol		Sánchez Márquez et al. (2010)
Polymyxa	sp.	Plasmodiophoraceae	Not DSE	*Sorghum bicolor*	Poales	Grass	In vitro exp		Galamay et al. (1992)
Pyrenochaeta	sp. 1	Pleosporales	DSE	*Holcus lanatus*	Poales	Grass	Isol		Sánchez Márquez et al. (2010)
Rhizoctonia	sp.	Cantharellales	DSE	*Carex curvula*	Poales	Sedge	Isol		Haselwandter and Read (1982)
Rhizoctonia	sp.	Cantharellales	DSE	*Carex firma*	Poales	Sedge	Isol	f	Haselwandter and Read (1982)

Rhizoctonia	sp.	Cantharellales	DSE	*Carex lasiocarpa, C. aquatilis*	Poales	Sedge	Isol		Thormann et al. (1999)
Rhizoctonia	spp.	Cantharellales	Not DSE	*Triticum aestivum*	Poales	Grass	Isol		Sieber et al. (1988)
Tapesia	sp.	Helotiales	DSE	*Deschampsia antarctica*	Poales	Grass	Isol		Upson et al. (2009)
Thielavia	*appendiculata*	Sordariales	Not DSE	*Stipa grandis*	Poales	Grass	Isol		Su et al. (2010)
Trematosphaeria	*clarkii*	Pleosporales	DSE	*Oryza sativa*	Poales	Grass	Isol		Fisher and Petrini (1992)
Trematosphaeria	*clarkii*	Pleosporales	DSE	*Oryza sativa*	Poales	Grass	Isol		Fisher and Petrini (1992)
Unidentified DSE			DSE	*Arrhenatherum elatius*	Poales	Grass	Micros		Deram et al. (2008)
Unidentified DSE			DSE	*Bouteloua eriopoda*	Poales	Grass	In vitro exp		Barrow et al. (2004)
Unidentified DSE			DSE	*Bouteloua gracilis*	Poales	Grass	ITS		Green et al. (2008)
Unidentified DSE			DSE	*Bouteloua gracilis*	Poales	Grass	Micros		Medina-Roldan et al. (2008)
Unidentified DSE			DSE	*Cynodon dactylon*	Poales	Grass	Micros		Li et al. (2005)
Unidentified DSE			DSE	*Cyperus rotundus*	Poales	Grass	Micros		Li et al. (2005)
Unidentified DSE			DSE	*Deschampsia flexuosa*	Poales	Grass	Open-top chambers in forest	m	Olsrud et al. (2010)
Unidentified DSE			DSE	*Deuterocohnia longipetala*	Poales	Bromelia	Micros		Lugo et al. (2009)
Unidentified DSE			DSE	*Digitaria cruciata*	Poales	Grass	Micros		Li et al. (2005)
Unidentified DSE			DSE	*Dyckia* spp.	Poales	Bromelia	Micros		Lugo et al. (2009)
Unidentified DSE			DSE	*Paspalum distichum*	Poales	Grass	Micros		Li et al. (2005)
Unidentified DSE			DSE	*Phragmites australis*	Poales	Grass	Micros		Dolinar and Gaberscik (2010)
Unidentified DSE			DSE	*Poa annua*	Poales	Grass	Micros		Li et al. (2005)
Unidentified DSE			DSE	*Tillandsia* spp.	Poales	Bromelia	Micros		Lugo et al. (2009)

[a] Genus, species and order (family) names according to index fungorum (http://www.indexfungorum.org/Names/Names.asp, December 20, 2011). If both the teleomorph (sexual reproductive stage) and the anamorph (asexual reproductive stage) are produced, the name of the teleomorph is given. Names of anamorph(s) can be retrieved from the list of the teleomorph's synonyms provided in the index. The next lower taxon is given if the fungus Order is not known with certainty ("incertae sedis").

[b] Isol, isolation from surface-sterilized roots; in vitro exp, fungus used in in-vitro experiment(s); ITS, detection by DNA extraction, sequencing of the ITS regions and sequence comparison with sequence databases; Micros, detection by microscopy.

[c] Special effect details: (a) Plant growth stimulation; improved sand stabilizing role of host in coastal dunes; control of nematodes. (b) Growth stimulation of roots; control of *Blumeria graminis*. (c) Biocontrol of damping-off, root and stem rot, caused by *Thanatephorus cucumeris*, in cotton. (d) Enhanced nitrogen uptake. (e) Control of *Gaeumannomyces graminis*. (f) Plant growth stimulation. (g) Control of *Phaeosphaeria nodorum*, *Fusarium* and *Gibberella* species. (h) Control of *Gaeumannomyces graminis* var. *tritici* and *Gaeumannomyces cylindrosporus*. (i) Plant growth stimulation by one isolate. (j) Plant growth stimulation by two of three isolates. (k) Plant growth stimulation by one of two isolates. (l) Increase of nitrogen uptake. (m) Increased colonization at higher temperatures, but not at elevated CO_2.

[d] Anamorph: *Harpophora graminicola*.

[e] Anamorph: *Harpophora radicicola*.

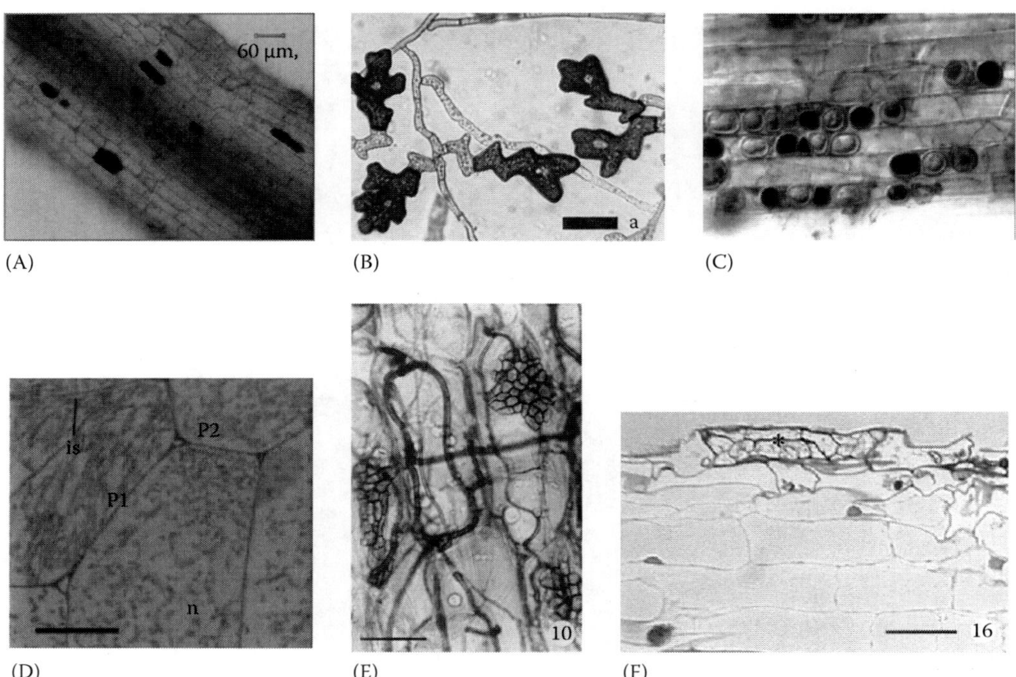

FIGURE 38.2 (A) Multicellular microsclerotia of *Microdochium bolleyi* filling single cortex cells in a tomato root. (From Rasmann, C. et al., *Appl. Soil Ecol.*, 43, 22, 2009.) (B) Hyphopodia of *Gaeumannomyces graminis* var. *graminis* wild-type strain JH2033 formed on Mylar films. (From Money, N.P. et al., *Fungal Genet. Biol.*, 24, 240, 1998.) Scale bar = 20 μm. (C) Intracellular spherical to pear-shaped chlamydospores of *Piriformospora indica* in cortical cells of a barley root. (From Waller, F. et al., *Proc. Natl. Acad. Sci. U S A*, 102, 13386, 2005.) (D) Morphology and fungal colonization (most probably *Inocybe* sp.) of coralloid rhizomes of the orchid *Epipogium aphyllum*. Magnification detail of infected cells with view of elongated fungal pelotons cut transversally in P1 and longitudinally in P2 (n, orchid cell nucleus, is, inflorescence shoots). (From Roy, M. et al., *Ann. Bot.*, 104, 595, 2009.) Scale bar = 50 μm. (E) Squash mount of mycorrhizal root of *Calypso bulbosa* stained with chlorazol black showing sclerotium-like structures of a DSE in cortical cells. (From Currah, R.S. et al., *Am. J. Bot.*, 75, 739, 1988.) Scale bar = 13 μm. (F) Epidermal microsclerotium (*) of *Phialocephala fortinii* s.l. (CSP 12, strain UAMH 9525) in a root of *Asparagus officinalis*. (From Yu, T. et al., *Can. J. Microbiol.*, 47, 741, 2001.) Scale bar = 15 μm.

(Figure 38.2A). Microsclerotia like those formed by *M. bolleyi* were also observed in roots of *Poa alpina* (cited as *Poa alpigena*) in Spitsbergen (Väre et al. 1992).

Colonization by *Harpophora* spp. (anamorphic *Gaeumannomyces* spp.) is usually recognized by dark septate runner hyphae growing on the root surface. The surface mycelium of some species also possesses hyphopodia that can be lobed or unlobed (Deacon 1981), a feature that is considered suitable to discriminate the varieties *tritici*, *avenae*, and *maydis* of *G. graminis*, which possess simple hyphopodia from the variety *graminis* that has lobed hyphopodia (Freeman and Ward 2004; Figure 38.2B). Depth of penetration into the root cortex and the stele and, thus, virulence depend on the fungus and host species, the environmental conditions, and the age and type of roots.

G. graminis var. *maydis* forms a net of stout brown runner hyphae with lateral hyphopodia on the root surface (McKeen 1952). Infection of the cortical cells of primary, seminal, and adventitious roots occurred by slender colorless microhyphae, but the roots appeared usually healthy. Sometimes individual cells were completely filled with enlarged fungus cells that later assumed thick, brown walls typical for microsclerotia (McKeen 1952). Skipp and Christensen (1989) observed microsclerotia similar to those observed by McKeen (1952) in cortical cells of roots of *Lolium perenne* L. from various sites in New Zealand. Isolations from surface-sterilized root segments confirmed presence of *G. graminis* var. *maydis*. *G. cylindrosporus* is a nonpathogenic endophyte (Deacon 1987). Mycelia of *G. cylindrosporus* and *G. graminis* develop similarly on roots, but *G. cylindrosporus* does not penetrate the vascular system. Whereas in young regions of the roots, the dark, lysis-resistant runner hyphae of *G. cylindrosporus* are confined to the root surface and the fungus penetrates only little into the cortex, on older regions, the runner hyphae grow intercellularly deep within the cortex (Deacon 1987).

Intracellular pycnidia of *Phoma fimeti* Brunaud were observed in the root cortex of the annual grass *Vulpia ciliata* Dumort (Newsham 1994). Intracellular sporulation may be ecologically advantageous because conidia can spread rapidly right after rupture of the cell wall in sloughed-off cortical cells.

Galamay et al. (1992) observed blackberry-like sclerotia of an endophyte in the epidermal and hypodermal cells of nodal and first-order lateral roots of *Sorghum bicolor* (L.) Moench. The fungus was tentatively identified as *Polymyxa* sp. (Plasmodiophoromycetes). Since tissues internal to the hypodermis were never colonized, it was assumed that the hypodermis functions as a barrier to protect the inner tissues from colonization by the fungus.

In substrate inoculated with *Piriformospora indica* Sav. Verma, Aj. Varma, Rexer, G. Kost & P. Franken, the fungus entered barley (*H. vulgare* L.) roots primarily via root hairs and proceeded intracellularly into rhizodermal cells and later on into the root cortex (Waller et al. 2005). It forms spherical to pear-shaped chlamydospores in both root hairs and cortical cells (Figure 38.2C). Hyphae were detected neither in the central part of the roots beyond the endodermis nor in stems or leaves.

2. Endophyte–Pathogen Interactions, Biological Control

Antagonism between *Phaeosphaeria nodorum* and *M. bolleyi* was observed by Reinecke (1978) and Reinecke and Fokkema (1981). *M. bolleyi* behaved also as an antagonist against *Fusarium* species (Reinecke 1978; Sieber 1985) and was reported to reduce damage to or colonization of cereal roots by *G. graminis* var. *tritici* (Kirk and Deacon 1987). Similarly, *G. cylindrosporus* controlled pathogens in *Vulpia ciliata* (Newsham 1999) and gave significant control of take-all by competition for senescing root tissues (Deacon 1981). *G. cylindrosporus* is found in relatively low amounts on cereal roots unless the cereal crop follows 1 or 2 years of grass (or cereal crop) (Deacon 1987). Crop rotation, a very effective form of disease management with a long tradition, seems to be disadvantageous for *G. cylindrosporus* and, thus, for control of take-all. Several other endophytic fungi possess antagonistic activity against *G. graminis* var. *tritici* (Macia-Vicente et al. 2008): *Acremonium blochii* (Matr.) W. Gams, *Acremonium furcatum* (Moreau & V. Moreau) ex W. Gams, *Aspergillus fumigatus* Fresen., *Cyclindrocarpon* sp., *Dactylaria* sp., *Gibberella intricans* Wollenw. (teleomorph of *Fusarium equiseti* (Corda) Sacc.), *Ilyonectria radicicola* (Gerlach & L. Nilsson) Chaverri & C. Salgado (teleomorph of *Cylindrocarpon destructans* (Zinsm.) Scholten), *Phoma herbarum* Westend., *P. leveillei* Boerema & G.J. Bollen. *A. blochii*, *Aspergillus fumigatus*, *Dactylaria* sp., *G. intricans*, and *Phoma herbarum* reduced colonization of barley roots by this fungus. *Cladorrhinum foecundissimum* Sacc. & Marchal, isolated from the grass *Elytrigia repens* (L.) Desv. ex Nevski, controlled root and stem rot caused by *Rhizoctonia solani* J.G. Kühn in cotton (*Gossypium hirsutum* L.) (Gasoni and Stegman de Gurfinkel 2009). Disease severity of powdery mildew (*Blumeria graminis* (DC.) Speer f. sp. hordei Jacz.) on primary leaves of *H. vulgare* was reduced when seeds had been treated with spore suspensions of *Chaetomium globosum* Kunze ex Fr. and *Chaetomium funicola* Cooke (Vilich et al. 1998). Endophytic *Acremonium strictum* W. Gams in roots of the grass *Ammophila arenaria* (L.) Link was nematicidal and reduced the adverse effects of nematodes (Hol et al. 2007).

3. Endophyte–Plant Interactions

Root inoculations with a *Microdochium* species (DSE) and *Periconia macrospinosa* Lefebvre & Aar.G. Johnson (DSE) from grass roots collected in a mesic tallgrass prairie increased the biomass of the grass *Andropogon gerardii* Vitman (Mandyam et al. 2010). Nitrogen uptake of *D. flexuosa* was enhanced by the DSE *Cryptosporiopsis rhizophila* Verkley & Zijlstra and *Phialocephala fortinii* s.l. (Zijlstra et al. 2005). Presence of the DSE *Phoma fimeti* in the roots of the annual grass *V. ciliata* increased root and shoot biomass, root lengths, and tiller numbers (Newsham 1994). Similarly, *G. cylindrosporus* increased tiller number and biomass production of *V. ciliata* (Newsham 1999). *Metacordyceps chlamydosporia* (H.C. Evans) G.H. Sung, J.M. Sung, Hywel-Jones & Spatafora, a clavicipitalean nematode parasite colonized barely roots endophytically and promoted plant growth (Macia-Vicente et al. 2008). *Acremonium strictum* was shown to increase root biomass and the number of tillers in *Ammophila arenaria* (Hol et al. 2007). Increased root fresh weight was observed after seeds of barley (*H. vulgare*) were treated with spore suspensions of *C. globosum* or *C. funicola* (Vilich et al. 1998). Similarly, inoculations with *Piriformospora indica* enhanced yield in barley (Waller et al. 2005).

B. Orchid Endophytes

Classical OM fungi of the *Rhizoctonia*-type are polyphyletic and belong to several different orders of the Basidiomycotina (Sebacinales, Ceratobasidiales, Tulasnellales) (Currah and Zelmer 1992; Smith and Read 2008; Weiss et al. 2011). Orchids, especially achlorophyllous, mycoheterotrophic ones, often associate also with basidiomycetes that form ECM with woody plants, for example, *Inocybe, Hebeloma, Xerocomus, Lactarius*, or *Thelephora* species and, thus, probably indirectly exploit the neighboring trees as carbon sources (Roy et al. 2009; Table 38.2). In addition, many orchids host members of the ascomycotina. *Leptodontidium orchidicola* Sigler & Currah, *Phialocephala fortinii* s.l., *Trichocladium opacum* (Corda) S. Hughes, and *Trichosporiella multisporum* Sigler & Currah were frequently isolated from terrestrial orchids (Currah et al. 1987, 1988, 1990; Currah and Sherburne 1992). Tao et al. (2008) examined the endophytes of the terrestrial orchid *Bletilla ochracea* Schltr. using denaturing gradient gel electrophoresis (DGGE) and random cloning analysis and detected 10 operational taxonomic units (OTUs) of endophytic fungi. Two *Mycosphaerella* species, an unknown ascomycete species, and an *Alternaria* species dominated. *Oidiodendron* spp. may be significant symbionts in orchids because they are also known to form endomycorrhizae with ericaceous plants (Currah and Zelmer 1992). Vujanovic et al. (2000) described *Phialocephala victorinii* Vujan. & St-Arn., an orchid endophyte with dark septate hyphae. Epiphytic and lithophytic orchids in the tropics often yield large numbers of isolates of fungi belonging to the Xylariales (Bayman et al. 1997; Table 38.2). Members of the Xylariaceae are also the most frequently isolated endophytes of leaves of tropical palms (Rodrigues 1996). Epiphytic orchids may, thus, harbor the same endophytes as the trees they grow on.

TABLE 38.2 Endophytes in Orchids (Family Orchidaceae)

Endophyte				Host					
Genus[a]	Species[a]	Fungus Order[b]	DSE/Not DSE	Species	Plant Order	Woody/Herbaceous	Type of Experiment[b]	Special Effect[c]	References
Acremonium	sp.	Hypocreales	Not DSE	*Dendrobium loddigesii*	Asparagales	Orchid	Isol		Chen et al (2010b)
Alternaria	sp.	Pleosporales	DSE	*Oncidium warmingii*	Asparagales	Orchid	Isol and in vitro exp with crude extracts	a	Vaz et al. (2009)
Alternaria	sp.	Pleosporales	DSE	*Bletilla ochracea*	Asparagales	Orchid	Mol		Tao et al. (2008)
Armillaria	*mellea*	Agaricales	Not DSE	*Gastrodia elata*	Asparagales	Orchid	Isol		Xu and Guo (2000)
Aspergillus	spp.	Eurotiales	Not DSE	*Lepanthes* spp.	Asparagales	Orchid	Isol		Bayman et al. (1997)
Cadophora	*malorum*	Helotiales	DSE	*Bletilla striata*	Asparagales	Orchid	Isol		Chen et al. (2010a)
Cadophora	sp.	Helotiales	Not DSE	*Gymnadenia conopsea*	Asparagales	Orchid	Isol		Stark et al. (2009)
Cenococcum	*geophilum*	Dothideomycetes	Not DSE	*Gymnadenia conopsea*	Asparagales	Orchid	Isol		Stark et al. (2009)
Colletotrichum	sp.	Sordariomycetes	Not DSE	*Dendrobium nobile*	Asparagales	Orchid	Isol		Yuan et al. (2009)
Colletotrichum	sp.	Sordariomycetes	Not DSE	*Lepanthes* spp.	Asparagales	Orchid	Isol		Bayman et al. (1997)
Cryptococcus	*carnescens*	Tremellales	Not DSE	*Gymnadenia conopsea*	Asparagales	Orchid	Isol		Stark et al. (2009)
Cryptosporiopsis	*ericae*	Helotiales	DSE	*Spiranthes sinensis*	Asparagales	Orchid	Isol		Chen et al. (2010a)
Cylindrocarpon	sp.	Hypocreales	Not DSE	*Cremastra appendiculata*	Asparagales	Orchid	Isol		Zhang et al. (2006)
Cylindrocarpon	sp.	Hypocreales	Not DSE	*Sophronitis fournieri*	Asparagales	Orchid	Isol		Vaz et al. (2009)
Epicoccum	*nigrum*	Pleosporales	DSE	*Maxillaria rigida*	Asparagales	Orchid	Isol and in vitro exp with crude extracts	b	Vaz et al. (2009)
Exophiala	sp.	Chaetothyriales	DSE	*Gymnadenia conopsea*	Asparagales	Orchid	Isol		Stark et al. (2009)
Fusarium	*chlamydosporum*	Hypocreales	Not DSE	*Dendrobium crumenatum*	Asparagales	Orchid	Isol		Siddiquee et al. (2010)
Fusarium	*oxysporum*	Hypocreales	Not DSE	*Gymnadenia conopsea*	Asparagales	Orchid	Isol		Stark et al. (2009)
Fusarium	sp.	Hypocreales	Not DSE	*Anacamptis pyramidalis*	Asparagales	Orchid	Isol		Gezgin and Eltem (2009)
Fusarium	sp.	Hypocreales	Not DSE	*Bulbophyllum involutum*	Asparagales	Orchid	Isol and in vitro exp with crude extracts	b	Vaz et al. (2009)
Fusarium	sp.	Hypocreales	Not DSE	*Dendrobium loddigesii*	Asparagales	Orchid	Isol		Chen et al. (2010b)
Fusarium	sp.	Hypocreales	Not DSE	*Isochilus linearis*	Asparagales	Orchid	Isol		Vaz et al. (2009)
Fusarium	sp.	Hypocreales	Not DSE	*Ophrys fusca*	Asparagales	Orchid	Isol		Gezgin and Eltem (2009)
Fusarium	sp.	Hypocreales	Not DSE	*Orchis sancta*	Asparagales	Orchid	Isol		Gezgin and Eltem (2009)
Fusarium	sp.	Hypocreales	Not DSE	*Serapias vomeracea* subsp. *orientalis*	Asparagales	Orchid	Isol		Gezgin and Eltem (2009)
Geopyxis	sp.	Pezizales	Not DSE	*Gymnadenia conopsea*	Asparagales	Orchid	Isol		Stark et al. (2009)
Gibberella	*fujikuroi* var. *fujikuroi*	Hypocreales	Not DSE	*Epidendrum secundum*	Asparagales	Orchid	Isol		Vaz et al. (2009)
Guignardia	*mangifera*	Botryosphaeriales	Not DSE	*Dendrobium nobile*	Asparagales	Orchid	Isol		Yuan et al. (2009)
Gymnopus	sp.	Agaricales	Not DSE	*Wullschlaegelia aphylla*	Asparagales	Orchid	Isol	c	Martos et al. (2009)
Hebeloma	sp.	Agaricales	Not DSE	*Epipogium aphyllum*	Asparagales	Orchid	Isol		Roy et al. (2009)
Hypocrea	sp.	Hypocreales	not DSE	*Sophronitis longipes*	Asparagales	Orchid	Isol		Vaz et al. (2009)

Hypoxylon	sp	Xylariales	not DSE	*Dendrobium nobile*	Asparagales	Orchid	Isol		Yuan et al. (2009)
Ilyonectria	*radicicola*	Hypocreales	not DSE	*Gymnadenia conopsea*	Asparagales	Orchid	Isol		Stark et al. (2009)
Inocybe	sp.	Agaricales	not DSE	*Epipogium aphyllum*	Asparagales	Orchid	Isol		Roy et al. (2009)
Lactarius	*pubescens*	Agaricales	not DSE	*Gymnadenia conopsea*	Asparagales	Orchid	Isol		Stark et al. (2009)
Lactarius	sp.	Agaricales	not DSE	*Epipogium aphyllum*	Asparagales	Orchid	Isol		Roy et al. (2009)
Lecanora	sp.	Lecanorales	not DSE	*Gymnadenia conopsea*	Asparagales	Orchid	Isol		Stark et al. (2009)
Leptodontidium	*orchidicola*	Helotiales	DSE	Orchids	Asparagales	Orchid	Isol		Currah et al. (1987, 1988, 1990)
Leptodontidium	*orchidicola*	Helotiales	not DSE	*Gymnadenia conopsea*	Asparagales	Orchid	Isol		Stark et al. (2009)
Leptodontidium	sp.	Helotiales	DSE	*Cephalanthera longifolia*	Asparagales	Orchid	Isol		Abadie et al. (2006)
Leptodontidium	sp.	Helotiales	DSE	*Dendrobium nobile*	Asparagales	Orchid	Isol and in vitro exp	d	Hou and Guo (2009)
Morchella	sp.	Pezizales	Not DSE	*Gymnadenia conopsea*	Asparagales	Orchid	Isol		Stark et al. (2009)
Mycena	*alphitophora*	Agaricales	Not DSE	*Gastrodia elata*	Asparagales	Orchid	Isol		Xu and Guo (2000)
Mycena	sp.	Agaricales	Not DSE	*Wullschlaegelia aphylla*	Asparagales	Orchid	Isol	c	Martos et al. (2009)
Mycosphaerella	sp.		N.A.	*Bletilla ochracea*	Asparagales	Orchid	Mol		Tao et al. (2008)
Penicillium	*griseofulvum*	Ascomycota incert	Not DSE	*Dendrobium nobile*	Asparagales	Orchid	Isol		Yuan et al. (2009)
Penicillium	spp.	Eurotiales	Not DSE	*Lepanthes* spp.	Asparagales	Orchid	Isol		Bayman et al. (1997)
Pestalotia	spp.	Xylariales	Not DSE	*Lepanthes* spp.	Asparagales	Orchid	Isol		Bayman et al. (1997)
Peziza	sp.	Pezizales	Not DSE	*Gymnadenia conopsea*	Asparagales	Orchid	Isol		Stark et al. (2009)
Phialocephala	*fortinii s.l.*	Helotiales	DSE	Orchids	Asparagales	Orchid	Isol		Currah et al. (1987, 1988, 1990)
Phialocephala	*victorinii*	Helotiales	DSE	Orchids	Asparagales	Orchid	Isol		Vujanovic et al. (2000)
Phialocephala	*europaea*	Helotiales	DSE	*Calypso bulbosa*	Asparagales	Orchid	Isol		Currah et al. (1988)
Phialophora	sp.	Chaetothyriales	Not DSE	*Gymnadenia conopsea*	Asparagales	Orchid	Isol		Stark et al. (2009)
Phialophora	sp.	Helotiales	Not DSE	*Cypripedium* spp.	Asparagales	Orchid	Isol		Shefferson et al. (2005)
Resinicium	sp.	Aphyllophorales	Not DSE	*Gastrodia similis*	Asparagales	Orchid	Isol	c	Martos et al. (2009)
Rhizoctonia	spp.	Cantharellales	Not DSE	*Lepanthes* spp.	Asparagales	Orchid	Isol		Bayman et al. (1997)
Rhodotorula	*mucilaginosa*	Sporidiobolales	Not DSE	*Acianthera hamosa*	Asparagales	Orchid	Isol		Vaz et al. (2009)
Russula	*brevipes*	Agaricales	Not DSE	*Limodorum abortivum*	Asparagales	Orchid	Isol		Girlanda et al. (2006)
Russula	*chloroides*	Agaricales	Not DSE	*Limodorum abortivum*	Asparagales	Orchid	Isol		Girlanda et al. (2006)
Russula	*delica*	Agaricales	Not DSE	*Limodorum abortivum*	Asparagales	Orchid	Isol		Girlanda et al. (2006)
Russula	*exalbicans*	Agaricales	Not DSE	*Gymnadenia conopsea*	Asparagales	Orchid	Isol		Stark et al. (2009)
Russula	*lepidicolor*	Agaricales	Not DSE	*Dipodium hamiltonianum*	Asparagales	Orchid	Isol		Dearnaley and Le Brocque (2006)
Russula	*lilacea*	Agaricales	Not DSE	*Dipodium hamiltonianum*	Asparagales	Orchid	Isol		Dearnaley and Le Brocque (2006)
Russula	sp.	Agaricales	Not DSE	*Cypripedium parviflorum*	Asparagales	Orchid	Isol		Shefferson et al. (2005)
Serendipita	*vermifera*	Auriculariales	Not DSE	*Orchids*	Asparagales	Orchid	Isol & in vitro exp	d	Weiss et al. (2011)
Terfezia	sp.	Pezizales	Not DSE	*Gymnadenia conopsea*	Asparagales	Orchid	Isol		Stark et al. (2009)
Tetracladium	sp.	Helotiales	DSE	*Cephalanthera longifolia*	Asparagales	Orchid	Isol		Abadie et al. (2006)
Tetracladium	sp.	Helotiales	DSE	*Orchis militaris*	Asparagales	Orchid	Isol		Vendramin et al. (2010)
Thelephora	sp.	Thelephorales	Not DSE	*Epipogium aphyllum*	Asparagales	Orchid	Isol		Roy et al. (2009)
Titaea	*maxilliforme*	Helotiales	Not DSE	*Gymnadenia conopsea*	Asparagales	Orchid	Isol		Stark et al. (2009)

(*continued*)

TABLE 38.2 (continued) Endophytes in Orchids (Family Orchidaceae)

Endophyte				Host					
Genus[a]	Species[a]	Fungus Order[b]	DSE/Not DSE	Species	Plant Order	Woody/Herbaceous	Type of Experiment[b]	Special Effect[c]	References
Trichocladium	*opacum*	Sordariales	DSE	Orchids	Asparagales	Orchid	Isol		Currah et al. (1987, 1988, 1990)
Trichoderma	*asperellum*	Hypocreales	Not DSE	*Epidendrum secundum*	Asparagales	Orchid	Isol		Vaz et al. (2009)
Trichosporiella	*multisporum*	Helotiales	Not DSE	*Orchids*	Asparagales	Orchid	Isol		Currah et al. (1987, 1988, 1990)
Verpa	*conica*	Pezizales	Not DSE	*Gymnadenia conopsea*	Asparagales	Orchid	Isol		Stark et al. (2009)
Wilcoxina	*rehmii*	Pezizales	Not DSE	*Gymnadenia conopsea*	Asparagales	Orchid	Isol		Stark et al. (2009)
Wilcoxina	sp.	Pezizales	DSE	*Cephalanthera longifolia*	Asparagales	Orchid	Isol		Abadie et al. (2006)
Xerocomus	sp.	Agaricales	Not DSE	*Epipogium aphyllum*	Asparagales	Orchid	Isol		Roy et al. (2009)
Xylaria	spp.	Xylariales	Not DSE	*Dendrobium nobile*	Asparagales	Orchid	Isol		Yuan et al. (2009)
Xylaria	spp.	Xylariales	Not DSE	*Lepanthes* spp.	Asparagales	Orchid	Isol		Bayman et al. (1997)
Unidentified DSE			DSE	*Maianthemum bifolium*	Asparagales	Herb	Isol		Postma et al. (2007)

[a] Genus, species and order (family) names according to index fungorum (http://www.indexfungorum.org/Names/Names.asp, December 20, 2011). If both the teleomorph (sexual reproductive stage) and the anamorph (asexual reproductive stage) are produced, the name of the teleomorph is given. Names of anamorph(s) can be retrieved from the list of the teleomorph's synonyms provided in the index. The next lower taxon is given if the Fungus order is not known with certainty ("incertae sedis").

[b] Isol, isolation from surface-sterilized roots; in vitro exp, fungus used in in-vitro experiment(s); Mol, culture-free, molecular detection method.

[c] Special effect details: (a) Antibacterial against *Escherichia coli* and *Staphylococcus aureus*. (b) Antimycotic against pathogenic yeasts *Candida krusei* and *C. albicans*. (c) Food chain link between dead leaves and orchid. (d) Plant growth stimulation.

Confirmation of this assumption would evoke the question of whether the infection occurs by hyphae growing directly from one host to the other or by inoculi (spores, conidia) on each host independently.

1. Anatomy

In contrast to the complex globular masses (hyphal coils) or pelotons of branched and anastomosed hyphae formed by classical orchid mycorrhizae within cortical cells (Roy et al. 2009; Figure 38.2D), *L. orchidicola* and *Phialocephala fortinii* were observed to form small sclerotia (Currah et al. 1988; Figure 38.2E).

2. Endophyte–Pathogen and Endophyte–Plant Interactions

Crude extracts from *Alternaria* sp. (DSE) from the roots of *Oncidium warmingii* Rchb. f. showed significant antibacterial activity against *Escherichia coli* and *Staphylococcus aureus* (Vaz et al. 2009). Similarly, *Epicoccum nigrum* Link (DSE) from *Maxillaria rigida* Barb. Rodr. and an unidentified *Fusarium* species from *Bulbophyllum involutum* Borba, Semir & F. Barros controlled the human pathogenic yeasts *Candida krusei* (Castell.) Berkhout and *Candida albicans* (C.P. Robin) Berkhout (Vaz et al. 2009). A *Leptodontidium* species associated with a subtropical *Dendrobium* species stimulated growth of *Dendrobium nobile* Lindl. seedlings in vitro (Hou and Guo 2009).

C. Other Herbaceous Plant Hosts

A plethora of fungi were detected in asymptomatic, healthy roots of many non-graminiculous, non-orchid herbaceous plant species growing in arctic, alpine (Christie and Nicolson 1983; Stoyke and Currah 1991; Väre et al. 1992; Ruotsalainen et al. 2004; Schmidt et al. 2008; Lv et al. 2010), aquatic (Kai and Zhao 2006), arid (Lugo et al. 2009), or neotropical habitats (Lehnert et al. 2009; Table 38.3). Terrestrial Bromeliaceae were colonized by both AM fungi and DSE except *Bromelia urbaniana* (Mez) L.B. Sm., which was not colonized by either one of them. In contrast, epiphytic Bromeliaceae were colonized only by DSE (Lugo et al. 2009). AM structures were observed in seven and DSE in four of 32 hydrophyte species from lakes and streams in Southwestern China (Kai and Zhao 2006). The frequency of hydrophytes hosting AM fungi and/or DSE was higher in water bodies of the Upper Parana in South America where AM was detected in nine species and DSE in 16 of 24 species (De Marins et al. 2009). Fourteen species of endophytes were isolated from the roots of *Saussurea involucrata* Kar. & Kir. collected at >2600 m asl in the Tianshan Mountains in China. Species of *Cylindrocarpon*, *Phoma*, and *Fusarium* dominated the endophytic fungal community (Lv et al. 2010). Fernandez et al. (2008) detected DSE in the roots of *Lycopodium paniculatum* Desv. ex Poir. and *Equisetum bogotense* Kunth in a Valdivian temperate forest of Patagonia, Argentina. DSE were also detected in Gentianaceae *Gentianella magellanica* Gaudich., *Gentianella parviflora* (Griseb.) T.N. Ho, *Gentianella multicaulis* (Gillies ex Griseb.) Fabris, and *Gentiana prostrata* Haenke indigenous in Argentina (Salvarredi et al. 2010). *Trichoderma asperellum* Samuels, Lieckf. & Nirenberg, *Gliocladium virens* J.H. Mill., Giddens & A.A. Foster, and *Hypocrea lixii* Pat. were endophytic in and epiphytic on banana roots (Xia et al. 2011), whereas *Trichoderma brevicompactum* G.F. Kraus, C.P. Kubicek & W. Gams was isolated only from inside of the roots. Genetic diversity of endophytic *T. asperellum* and *G. virens* was lower than that of epiphytic ones, suggesting that only selected genotypes are able to infect the roots. Väre et al. (1992) studied the fungi associated with roots of 76 (72 herbaceous) plant species in Spitsbergen. Ectomycorrhizae and AM fungi were absent from all herbaceous plant species, except *Pedicularis dasyantha* Hadač that showed slight ectomycorrhizal colonization. In contrast, root endophytes were commonly isolated. As already mentioned earlier, Upson et al. (2009) isolated 243 mostly helotialean DSE strains from the roots of *Deschampsia antarctica* (grass) and *Colobanthus quitensis* (Kunth) Bartl. (Caryophyllaceae) along a transect across sub-Antarctica.

1. Anatomy

Rodríguez-Gálvez and Mendgen (1995) studied the ultrastructure of the infection process by *Fusarium oxysporum* f. sp. *vasinfectum*, causal agent of tracheomycosis of species of the Malvaceae, for example, cotton. High-pressure freezing of infected cortical cells revealed that *F. oxysporum* penetrates and grows within the host cells without inducing damages such as plasmolysis, cell degeneration, or host necrosis. It was suggested, therefore, that *F. oxysporum* f. sp. *vasinfectum* has an endophytic, biotrophic phase during colonization of the root tips.

The most frequently observed fungal structures in the roots of 30 of 72 examined herbaceous plant species in Spitsbergen were inter- and intracellularly growing melanized, septate mycelia (DSE) (Väre et al. 1992). In addition, cortical cells of *Polemonium boreale* Adams, a herbaceous Ericaceae, were filled with dark microsclerotia that resembled those formed by *M. bolleyi* and *Phialocephala fortinii* s.l. Similarly, Treu et al. (1996) detected dark microsclerotia in root cells of various alpine plants collected in Denali National Park, Alaska. A strain of *P. fortinii* s.l. CSP 12 (UAMH 9525) isolated from *Vaccinium vitis-idaea* L. formed microsclerotia having a "puzzle-like" appearance in the roots of *Asparagus officinalis* L. similar to those observed in conifer roots (Yu et al. 2001; Figure 38.2F). In the cortex of young roots of Chinese cabbage, dark septate hyphae of *Cladophialophora (Heteroconium) chaetospira* (Grove) Crous & Arzanlou were abundant (Figure 38.3A), whereas the cortex of older roots was filled with microsclerotia (Yonezawa et al. 2004). Similarly, hyphae and chlamydospores of DSE were observed in primary roots of *Xenophyllum rosenii* (R.E. Fr.) V.A. Funk (Asteraceae) collected at 5389 m asl (Schmidt et al. 2008; Figure 38.3B). Sclerotia and hyphae growing very close to vascular bundles were also observed in *Equisetum pratense* Ehrh. roots (Hodson et al. 2009).

TABLE 38.3 Endophytes in Other Herbaceous Plants

Endophyte				Host					
Genus[a]	Species[a]	Fungus Order[a]	DSE/Not DSE	Species	Plant Order	Woody/ Herbaceous	Type of Experiment[b]	Special Effect[c]	References
Acremonium	*alternatum*	Hypocreales	Not DSE	*Brassica oleracea* var. *gemmifera*	Brassicales	Herb	In vitro exp	a	Raps and Vidal (1998), Dugassa-Gobena et al. (1998)
Acremonium	*strictum*	Hypocreales	Not DSE	*Solanum lycopersicum*	Solanales	Herb	In vitro exp	b	Raps and Vidal (1996), Vidal (1996)
Aureobasidium	*pullulans*	Dothideales	Not DSE	*Pteridium aquilinum*	Pteridophyta	Perennial fern	Isol		Petrini et al. (1992)
Cladophialophora	*chaetospira*	Chaetothyriales	DSE	*Brassica campestris*	Brassicales	Herb	In vitro exp	c	Narisawa et al. (1998)
Cladosporium	sp.	Capnodiales	DSE	*Saussurea involucrata*	Asterales	Herb	Isol		Lv et al. (2010)
Cylindrocarpon	sp.	Hypocreales	Not DSE	*Saussurea involucrata*	Asterales	Herb	Isol		Lv et al. (2010)
Discocistella	*grevillei*	Helotiales		*Saussurea involucrata*	Asterales	Herb	Isol		Lv et al. (2010)
Fusarium	*oxysporum*	Hypocreales	Not DSE	*Brassica oleracea* var. *capitata*	Brassicales	Herb	In vitro exp		Davis (1967), Matta (1989)
Fusarium	*oxysporum*	Hypocreales	Not DSE	*Citrullus vulgaris*	Cucurbitales	Herb	In vitro exp	d	Davis (1967), Matta (1989)
Fusarium	*oxysporum*	Hypocreales	Not DSE	*Dianthus caryophyllus*	Caryophyllales	Herb	In vitro exp	d	Davis (1967), Matta (1989)
Fusarium	*oxysporum*	Hypocreales	Not DSE	*Linum usitatissimum*	Malpighiales	Herb	In vitro exp	d	Davis (1967), Matta (1989)
Fusarium	*oxysporum*	Hypocreales	Not DSE	*Lycopersicon esculentum*	Solanales	Herb	In vitro exp	d	Davis (1967), Matta (1989)
Hypocrea	*lixii*	Hypocreales	Not DSE	*Musa acuminata*	Zingiberales	Herb	Isol		Xia et al. (2011)
Ilyonectria	*radicicola*	Hypocreales	Not DSE	*Pteridium aquilinum*	Pteridophyta	Perennial fern	Isol		Petrini et al. (1992)
Leptodontidium	*orchidicola*	Helotiales	DSE	*Colobanthus quitensis*	Caryophyllales	Herb	Isol		Upson et al. (2009)
Leptodontidium	*orchidicola*	Helotiales	DSE	*Pedicularis bracteosa*	Lamiales	Herb	Isol		Currah et al. (1987)
Leptodontidium	*orchidicola*	Helotiales	DSE	*Saussurea involucrata*	Asterales	Herb	Isol		Lv et al. (2010)
Leptosphaeria	sp.	Pleosporales	DSE	*Saussurea involucrata*	Asterales	Herb	Isol		Lv et al. (2010)
Meliniomyces	*variabilis* LtVB3	Leotiomycetes	DSE	*Brassica rapa*	Brassicales	Herb	In vitro exp	e	Ohtaka and Narisawa (2008)
Meliniomyces	*variabilis* LtVB3	Leotiomycetes	DSE	*Solanum lycopersicum*	Solanales	Herb	In vitro exp	f	Ohtaka and Narisawa (2008)
Microdochium	*bolleyi*	Xylariales	DSE	*Lycopersicon esculentum*	Solanales	Herb	Isol and Micros	g	Rasmann et al. (2009)
Mortierella	sp.	Zygomycota	Not DSE	*Pteridium aquilinum*	Pteridophyta	Perennial fern	Isol		Petrini et al. (1992)
Mycocentrospora	*acerina*	Pleosporales	DSE	*Saussurea involucrata*	Asterales	Herb	In vitro exp	h	Wu et al. (2010)
Mycocentrospora	*acerina*	Pleosporales	DSE	*Saussurea involucrata*	Asterales	Herb	Isol		Lv et al. (2010)
Phaeosphaeria	*avenaria*	Pleosporales	DSE	*Saussurea involucrata*	Asterales	Herb	Isol		Lv et al. (2010)
Phialophora	*cyclaminis*	Chaetothyriales	DSE	*Cyclamen persicum*	Ericales	Herb	Isol		Schol-Schwarz (1970)
Phoma	*chrysanthemicola*	Pleosporales	DSE	*Chrysanthemum morifolium*	Asterales	Herb	Isol		Aveskamp et al. (2009)
Phoma	*chrysanthemicola*	Pleosporales	DSE	*Heteropappus semiprostratus*	Asterales	Herb	Isol		de Graaf (unpublished)
Phoma	*chrysanthemicola*	Pleosporales	DSE	*Saussurea involucrata*	Asterales	Herb	Isol		Lv et al. (2010)
Phoma	*glomerata*	Pleosporales	DSE	*Saussurea involucrata*	Asterales	Herb	Isol		Lv et al. (2010)
Phoma	*sclerotioides*	Pleosporales	DSE	*Saussurea involucrata*	Asterales	Herb	Isol		Lv et al. (2010)

Piriformospora	*indica*	Sebacinales	Not DSE	Various spp.	N.A.	Various	In vitro exp	i	Varma et al. (2001)
Tapesia	sp.	Helotiales	DSE	*Colobanthus quitensis*	Caryophyllales	Herb	Isol		Upson et al. (2009)
Trichoderma	*asperellum*	Hypocreales	Not DSE	*Musa acuminata*	Zingiberales	Herb	Isol		Xia et al. (2011)
Trichoderma	*virens*	Hypocreales	Not DSE	*Musa acuminata*	Zingiberales	Herb	Isol		Xia et al. (2011)
Trichoderma	*brevicompactum*	Hypocreales	Not DSE	*Musa acuminata*	Zingiberales	Herb	Isol		Xia et al. (2011)
Triscelophorus	*monosporus*	Pezizomycotina	Not DSE	*Angiopteris evecta*	Pteridophyta	Perennial fern	Isol		Raviraja et al. (1996)
Triscelophorus	*monosporus*	Pezizomycotina	Not DSE	*Christela dentata*	Pteridophyta	Small fern	Isol		Raviraja et al. (1996)
Unidentified DSE			DSE	*Astragalus cf. arequipensis*	Fabales	Herb	Isol		Schmidt et al. (2008)
Unidentified DSE			DSE	*Bartsia pumila*	Lamiales	Herb	Isol		Schmidt et al. (2008)
Unidentified DSE			DSE	*Ceradenia* spp.	Polypodiales	Herb	Micros		Lehnert et al. (2009)
Unidentified DSE			DSE	*Cochlidium serrulatum*	Polypodiales	Herb	Micros		Lehnert et al. (2009)
Unidentified DSE			DSE	*Elaphoglossum* spp.	Polypodiales	Herb	Micros		Lehnert et al. (2009)
Unidentified DSE			DSE	*Equisetum bogotense*	Equisetales	Herb	Micros		Fernandez et al. (2008)
Unidentified DSE			DSE	*Galium odoratum*	Gentianales	Herb	Isol		Postma et al. (2007)
Unidentified DSE			DSE	*Gramitis paramicola*	Polypodiales	Herb	Micros		Lehnert et al. (2009)
Unidentified DSE			DSE	*Hydrilla verticillata*	Alismatales	Herb	Micros		Kai and Zhao (2006)
Unidentified DSE			DSE	*Hygrophila cf. costata*	Lamiales	Herb	Micros		De Marins et al. (2009)
Unidentified DSE			DSE	*Hymenophyllum* spp.	Polypodiales	Herb	Micros		Lehnert et al. (2009)
Unidentified DSE			DSE	*Lellingeria* spp.	Polypodiales	Herb	Micros		Lehnert et al. (2009)
Unidentified DSE			DSE	*Limnobium laevigatum*	Apiales	Herb	Micros		De Marins et al. (2009)
Unidentified DSE			DSE	*Lycopodium* sp.	Lycopodiales	Herb	Isol		Schmidt et al. (2008)
Unidentified DSE			DSE	*Lycopodium paniculatum*	Lycopodiales	Herb	Micros		Fernandez et al. (2008)
Unidentified DSE			DSE	*Melpomene* spp.	Polypodiales	Herb	Micros		Lehnert et al. (2009)
Unidentified DSE			DSE	*Mercurialis perennis*	Malpighiales	Herb	Isol		Postma et al. (2006)
Unidentified DSE			DSE	*Micropolypodium* sp.	Polypodiales	Herb	Micros		Lehnert et al. (2009)
Unidentified DSE			DSE	*Mnioides* sp.	Asterales	Herb	Isol		Schmidt et al. (2008)
Unidentified DSE			DSE	*Myriophyllum brasiliense*	Saxifragales	Herb	Micros		De Marins et al. (2009)
Unidentified DSE			DSE	*Oenanthe decumbens*	Apiales	Herb	Micros		Kai and Zhao (2006)
Unidentified DSE			DSE	*Pedicularis* spp.	Lamiales	Herb	Isol		Li and Guan (2007)
Unidentified DSE			DSE	*Perezia coerulescens*	Asterales	Herb	Isol		Schmidt et al. (2008)
Unidentified DSE			DSE	*Plantago asiatica*	Lamiales	Herb	Micros		Li et al. (2005)
Unidentified DSE			DSE	*Polygonum* spp.	Caryophyllales	Herb	Micros		De Marins et al. (2009)
Unidentified DSE			DSE	*Potamogeton tepperi*	Alismatales	Herb	Micros		Kai and Zhao (2006)
Unidentified DSE			DSE	*Rotala rotundifolia*	Myrtales	Herb	Micros		Kai and Zhao (2006)
Unidentified DSE			DSE	*Saxifraga aizoides*	Saxifragales	Herb	Isol	j	Ruotsalainen et al. (2004)
Unidentified DSE			DSE	*Senecio* sp.	Asterales	Herb	Isol		Schmidt et al. (2008)
Unidentified DSE			DSE	*Sibbaldia procumbens*	Rosales	Herb	Isol	j	Ruotsalainen et al. (2004)
Unidentified DSE			DSE	*Solidago virgaurea*	Asterales	Herb	Isol	j	Ruotsalainen et al. (2004)
Unidentified DSE			DSE	*Stellaria nemorum*	Caryophyllales	Herb	Isol		Postma et al. (2006)
Unidentified DSE			DSE	*Stylidium productum*	Asterales	Herb	Isol		Chambers et al. (2008)
Unidentified DSE			DSE	*Terpsichore* spp.	Polypodiales	Herb	Micros		Lehnert et al. (2009)

(continued)

TABLE 38.3 (continued) Endophytes in Other Herbaceous Plants

Endophyte				Host					
Genus[a]	Species[a]	Fungus Order[a]	DSE/Not DSE	Species	Plant Order	Woody/ Herbaceous	Type of Experiment[b]	Special Effect[c]	References
Unidentified DSE			DSE	*Trichomanes* spp.	Polypodiales	Herb	Micros		Lehnert et al. (2009)
Unidentified DSE			DSE	*Trientalis europaea*	Ericales	Herb	Isol	j	Ruotsalainen et al. (2004)
Unidentified DSE			DSE	*Trifolium repens*	Fabales	Herb	Micros		Li et al. (2005)
Unidentified DSE			DSE	*Viola biflora*	Malpighiales	Herb	Isol	j	Ruotsalainen et al. (2004)
Unidentified DSE			DSE	*Werneria orbignyana*	Asterales	Herb	Isol		Schmidt et al. (2008)
Unidentified DSE			DSE	*Werneria* sp.	Asterales	Herb	Isol		Schmidt et al. (2008)
Unidentified DSE			DSE	*Xenophyllum rosenii*	Asterales	Herb	Isol		Schmidt et al. (2008)

[a] Genus, species and order (family) names according to index fungorum (http://www.indexfungorum.org/Names/Names.asp, December 20, 2011). If both the teleomorph (sexual reproductive stage) and the anamorph (asexual reproductive stage) are produced, the name of the teleomorph is given. Names of anamorph(s) can be retrieved from the list of the teleomorph's synonyms provided in the index. The next lower taxon is given if the Fungus order is not known with certainty ("incertae sedis").

[b] Isol, isolation from surface-sterilized roots; in vitro exp, fungus used in in-vitro experiment(s); Micros, detection by microscopy.

[c] Special effect details: (a) Control of diamondback moth and cabbage aphid *Brevicoryne brassicae*. (b) Control of nematodes and greenhouse whitefly larvae. (c) Control of clubroot (*Plasmodiophora brassicae*). (d) Induction of resistance. (e) Control of *Verticillium longisporum*. (f) Control of *Fusarium* wilt. (g) Increased colonization by *Microdochium bolleyi* like fungi in organically grown tomatoes. (h) Stimulation of plant root development. (i) Plant growth stimulation. (j) No correlation with altitude.

2. Endophyte–Pathogen, Endophyte–Nematode, and Endophyte–Insect Interactions

More than 70 formae speciales of *Fusarium oxysporum* are pathogenic on different hosts (Armstrong and Armstrong 1981). In general, each form causes symptoms only on one or a few related plant species, whereas they occur as nonpathogenic endophytes in other plant species. Absence of adverse effects of *F. oxysporum* isolates in infection experiments is in line with this observation (Davis 1967; Gessler and Kuc 1982; Rodríguez-Gálvez and Mendgen 1995). Nonpathogenic isolates of *F. oxysporum* were shown to be able to penetrate and colonize the tissues of carnation and tomato plants without symptom expression (Postma and Rattink 1991; Hallmann and Sikora 1994). *Fusarium* wilt disease is inhibited in plants inoculated with formae speciales of *F. oxysporum* to which they are not susceptible prior to inoculation with forms to which they are susceptible. This phenomenon is called "cross-protection" or "induced resistance" (Davis 1967; Matta 1989). Inoculation of individual tomato, flax, carnation, cabbage, and watermelon seedlings with any one of nine formae speciales of *F. oxysporum* markedly reduced susceptibility to forms that they are susceptible to (Davis 1967). Similarly, resistance of cucumber (*Cucumis sativus* L.) against *F. oxysporum* f. sp. *cucumerinum* J.H. Owen was induced by inoculation of formae speciales that are not pathogenic on cucumber (Gessler and Kuc 1982). Cross-protection can be based on an indirect mechanism mediated by the plant or on direct interactions between inducer and challenger (Matta 1989). Chitin synthase–deficient mutants of the tomato root pathogen *F. oxysporum* f. sp. *lycopersici* (Sacc.) W.C. Snyder & H.N. Hansen elicited plant defense response and protected the plants against wild-type infections (Pareja-Jaime et al. 2010). *Fusarium solani* colonized tomato roots endophytically and protected them against

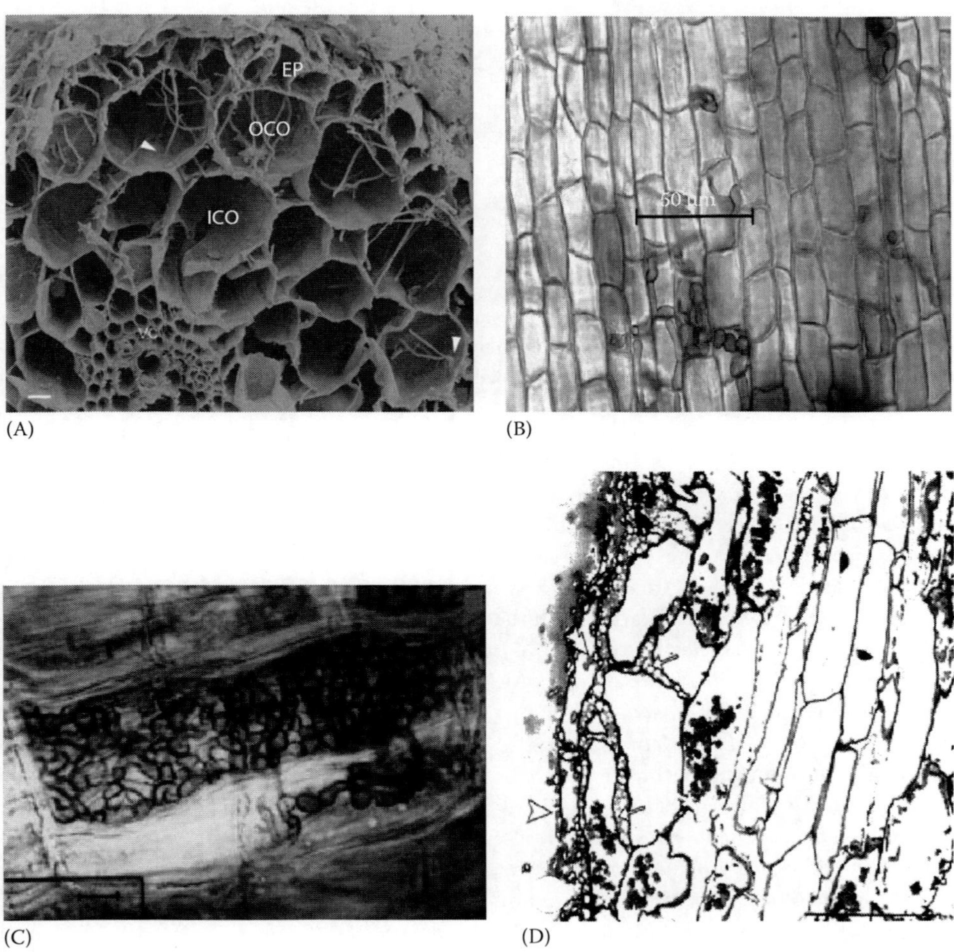

FIGURE 38.3 (A) Scanning electron micrographs of cross section of a Chinese cabbage root infected by *Cladophialophora chaetospira*. Abundant fungal hyphae developing in EPs and within OCO (arrowheads, appressorium-like swollen structures; ICO, inner cortical cells; VC, vascular cylinder). (From Yonezawa, M. et al., *Mycoscience*, 45, 367, 2004.) Scale bar = 5 µm. (B) Dark septate endophytic fungi in roots of *Xenophyllum rosenii* at 5389 m. (From Schmidt, S.K. et al., *Arct. Antarct. Alp. Res.*, 40, 576, 2008.) (C) *Gaultheria poeppiggi* EP of hair root filled with sclerotia of the *Phialocephala* type. (From Urcelay, C., *Mycorrhiza*, 12, 89, 2002.) Scale bar = 10 µm. (D) Roots of *Pinus strobus* colonized by *Phialocephala fortinii* s.l. (strain UAMH 10266). Longitudinal section of colonized root showing surface hyphae (arrowhead) and Hartig net (arrows) and intercellular hyphae. (From Peterson, R.L. et al., *Botany*, 86, 445, 2008.) Scale bar = 50 µm.

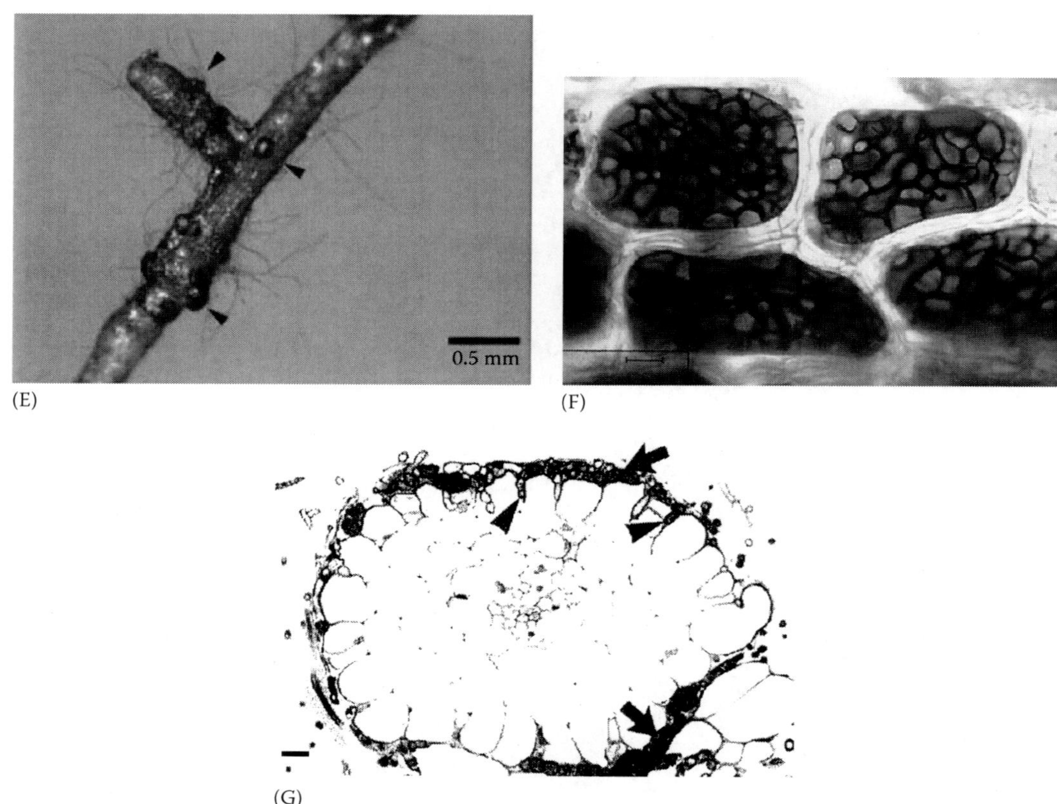

FIGURE 38.3 (continued) (E) Ectomycorrhiza formed by *Acephala macrosclerotiorum* on *Pinus sylvestris*. The mantle appears verrucous due to characteristic, melanized wartlike sclerotia (arrowheads). (From Münzenberger, B. et al., *Mycorrhiza*, 19, 481, 2009.) (F) Cells of the phlobaphene cork of a Norway-spruce (*Picea abies*) fine root filled with multicellular sclerotia composed of bubble-shaped, thick-walled, melanized cells of *Phialocephala fortinii* s.l. Scale bar = 10 μm. (Courtesy of O. Holdenrieder). (G) Section of root of axenically grown *Salix glauca* inoculated with *P. fortinii* s.l. (CSP13, strain UAMH 8148). Transverse section of a lateral root with a patchy fungal mantle (arrows) associated with the epidermal layer and Hartig net initials (arrowheads). (From Fernando, A.A. and Currah, R.S., *Can. J. Bot.*, 74, 1071, 1996.) Scale bar = 10 μm.

the root pathogen *F. oxysporum* f. sp. *radicis-lycopersici* Jarvis & Shoemaker (Kavroulakis et al. 2007). Vascular discolorations were reduced compared to the controls when wounded roots of tomato seedlings were treated with an unidentified *Acremonium* species prior to inoculation with the pathogen, *F. oxysporum* f. sp. *lycopersici* (Phillips et al. 1967). Inoculations with nonpathogenic *Fusarium* forms can provide protection not only against other forms of *Fusarium* but also against other diseases. Metabolites of *F. oxysporum* can also significantly reduce growth of soilborne pathogens such as *Phytophthora cactorum* (Lebert & Cohn) J. Schröt., *Pythium ultimum* Trow, and *Thanatephorus cucumeris* (A.B. Frank) Donk in vitro (Hallmann and Sikora 1996), and *F. oxysporum* strain EF 119 was efficient against tomato late blight caused by *Phytophthora infestans* (Mont.) de Bary and inhibited growth of *Pythium ultimum* and *Phytophthora capsici* Leonian (Kim et al. 2007). Inoculations of tomato plants with nonpathogenic *F. oxysporum* strains significantly reduced colonization by the plant parasitic root-knot nematode *Meloidogyne incognita* Kofoid & White without adversely affecting plant health (Hallmann and Sikora 1994; Dababat and Sikora 2007; Dababat et al. 2008). Similarly, endophytic *F. oxysporum* can control the burrowing nematode *Radopholus similis* Cobb, which is one of the key pests of banana (Athman et al. 2006; Mendoza and Sikora 2009; Paparu et al. 2009).

Biological control of diseases, nematode, or insect pests was also reported for root endophytes other than *Fusarium* spp. Root inoculations of tomato plants with the soilborne endophyte *Acremonium strictum* significantly reduced the frequency of root-knots induced by the nematode *Meloidogyne hapla* Chitwood (Raps and Vidal 1996). Altered plant growth of endophyte-colonized roots and direct parasitism of *A. strictum* on the nematode eggs were assumed to be responsible for nematode control. Culture filtrates of endophytic fungi isolated from roots of tomato and banana plants in Kenya and Uganda caused a significant reduction of the activity of several nematode species (Schuster et al. 1995). Mortality of greenhouse whiteflies (*Trialeurodes vaporariorum* Westwood [Homoptera]) on tomato plants inoculated with the endophyte *A. strictum* was only increased if the plants suffered from drought stress. Interestingly, the insects were preferentially feeding on plants with inoculated roots (Vidal 1996). Similarly, the insect also preferred French beans (*Phaseolus vulgaris* L.) inoculated with *A. strictum* to uninoculated ones (Moll and Vidal 1995). In addition, females laid more eggs on inoculated than on uninoculated plants, and performance of the larvae on endophyte-infected plants was slightly better. The observed changes of the sugar and amino acid composition of the phloem sap were assumed responsible for the behavior of the insects.

Root inoculations of Brussels sprouts (*Brassica oleracea* var. *gemmifera* DC.) with the soilborne endophyte *Acremonium alternatum* Link had a significant effect on the performance of diamondback moth (*Plutella xylostella* L.) (Dugassa-Gobena et al. 1998; Raps and Vidal 1998). Larvae fed with leaves of endophyte-inoculated plants experienced a higher mortality or showed delayed growth, development, and pupation compared to controls. The observed changes of the phytosterol composition of the host plants infected by endophytes were suspected to confer control of the insect, because these changes altered the suitability of certain phytosterols to be enzymatically dealkylated to cholesterol in insects. This subsequently interfered with the molting processes of the larvae (Dugassa-Gobena et al. 1998; Raps and Vidal 1998).

Sixteen endophytic fungal isolates from roots of Chinese cabbage (*Brassica campestris* L.) almost completely suppressed clubroot, caused by the soilborne fungus *Plasmodiophora brassicae* Woronin, in sterile soil (Narisawa et al. 1998). Two of these isolates were also effective in nonsterile soil and could be identified as *Cladophialophora chaetospira* (DSE), a species commonly found on rotting wood in Europe (Ellis 1976) and isolated from roots of *Picea abies* (L.) Karst. (Crous et al. 2007). Similarly, DSE isolate LtVB3 from barley roots was shown to colonize roots of Chinese cabbage and tomato endophytically and suppress *Verticillium* yellows of Chinese cabbage and *Fusarium* wilt of tomato (Narisawa et al. 2004). LtVB3 could meanwhile be identified as *Meliniomyces variabilis* Hambl. & Sigler (Ohtaka and Narisawa 2008). Several endophytes (*Phaeosphaeria avenaria* (G.F. Weber) O.E. Erikss., *Leptosphaeria* sp., *Phoma chrysanthemicola* Hollós, *Cladosporium* sp., and *Cylindrocarpon* sp.) isolated from the roots of the alpine plant *Saussurea involucrata* (Kar. & Kir.) Sch. Bip. in the Tianshan Mountains in China possessed antimicrobial activity against human pathogenic fungi and bacteria (Lv et al. 2010).

Piriformospora indica Sav. Verma, Aj. Varma, Rexer, G. Kost & P. Franken (Sebacinales, Basidiomycota), a fungus isolated from an AM spore collected from desert soil in India, is able to colonize the roots of various plant species as an endophyte (Verma et al. 1998; Figure 38.2C). Inoculation with the fungus and application of fungal culture induced resistance of their hosts against biotic and abiotic stress (Varma et al. 2001; Deshmukh and Kogel 2007; Serfling et al. 2007; Achatz et al. 2010). Due to its ease of culture, this fungus serves as a model organism for the study of beneficial plant–microbe interactions and a new tool for improving plant production systems. Colonization was inter- and intracellular with coils and branches or round chlamydospore-like structures and no arbuscules. Neither shoot nor stelar tissues were colonized. Recently, sebacinalean endophytes were found to be universally present as symptomless endophytes in many liverworts, bryophytes, pteridophytes, and herbaceous angiosperms, for example, wheat, maize, and the model plant *Arabidopsis thaliana* (Weiss et al. 2011).

3. Endophyte–Plant Interactions

Chinese cabbage seedlings from seed treated with the two *C. chaetospira* strains isolated by Narisawa et al. (1998) appeared healthy, and inoculation with one of the isolates promoted plant growth. Similarly, *Mycocentrospora acerina* (R. Hartig) Deighton, another DSE, promoted root development of *S. involucrata* (Wu et al. 2010). Inoculation with *Piriformospora indica* and application of fungal culture filtrate promoted plant growth and seed yield (Varma et al. 2001; Deshmukh and Kogel 2007; Serfling et al. 2007; Achatz et al. 2010).

Growth promotion and improved uptake of phosphorus were observed also in cotton when the roots were colonized endophytically by *C. foecundissimum* Sacc. & Marchal (Gasoni and Stegman De Gurfinkel 1997). Intercellular hyphae formed dense layers on the distal (outer) side of the endodermis. The fungus grew also intracellularly in root hairs.

4. Endophyte–Environment Interactions

Inoculation of DSE (*Cadophora* and *Rhizoctonia* species) onto aseptically grown seedlings of the alpine sedges *Carex curvula* All. and *Carex firma* Host. resulted in a significant increase of dry matter production in *C. firma* but not in *C. curvula* compared to uninoculated controls (Haselwandter and Read 1982). The authors concluded that the function of the relationship between *Carex* roots and their DSE may be comparable to that found between plants and AM fungi. Like AM, DSE provided improved phosphorus supply to the plant. DSE may replace AM in stressed environments. This hypothesis is supported by the findings of Currah and Van Dyk (1986) who observed roots growing in alpine soils with little organic matter to have DSE and those growing in habitats with visibly more organic matter to have AM. Three alpine species of the Fabaceae (*Astragalus alpinus* L., *A. vexilliflexus* E. Sheld., *Oxytropis jordalii* A.E. Porsild), a family that is usually endomycorrhizal in other habitats, lacked AM but were colonized by DSE (Currah and Van Dyk 1986), an observation similar to that made by Christie and Nicolson (1983) for two grass species in the Antarctic region. Conversely, O'Dell and Trappe (1992) detected AM in *Astragalus cottonii* M.E. Jones, *Lupinus latifolius* Lindl. ex J. Agardh, *L. lepidus* Douglas ex Lindl., and *Oxytropis campestris* (L.) DC. from alpine habitats. The two *Lupinus* species and *O. campestris* were also colonized by septate endophytes, whereas *A. cottonii* was not. One of the DSE observed in the roots of *L. latifolius* was later on identified as *Phialocephala fortinii* (O'Dell et al. 1993).

AM fungi were absent in *Xenophyllum rosenii* (R.E. Fr.) V.A. Funk and *Perezia coerulescens* Wedd. (both Asteraceae), but roots of both plant species were heavily colonized by DSE fungi at 5391 m asl in the Peruvian Andes (Schmidt et al. 2008; Figure 38.3B). At slightly lower elevations (5240–5250 m), AM fungi were present while DSE fungi were rare except in

Bartsia pumila Benth. (Scrophulariaceae) and plants outside of the Asteraceae. In Colorado at 4300 m asl, AM fungi were rare, but all examined plants were colonized by DSE fungi (Schmidt et al. 2008). Absence of AM fungi and presence of dark septate hyphae were also observed for roots of various herbaceous plant species collected in Arctic Canada (Bledsoe et al. 1990). DSE occurred also in several alpine plants in the Canadian Rocky Mountains (Stoyke and Currah 1991). In contrast, Kohn and Stasovski (1990) found no DSE in roots collected in arctic Canada, but AM was observed in one plant species. Ruotsalainen et al. (2004) could not find any consistent correlation between the colonization of various Arctic–alpine plants by AM or DSE and altitude in Northern Norway. DSE might not only replace AM at high altitudes and latitudes but also in acidic soils as suggested by Postma et al. (2007) who found that colonization of herbaceous plants (*Galium odoratum* (L.) Scop., *Mercurialis perennis* L., *Stellaria nemorum* L.) by DSE increased with decreasing pH whereas colonization by AM decreased. The same authors also detected a positive correlation between leaf magnesium concentrations and the presence of DSE in *G. odoratum*. DSE are probably the most widespread root–fungus association in polar regions, but their roles in plant nutrition and survival are still poorly understood (Newsham et al. 2009). Li et al. (2005) detected considerable colonization of roots of *Trifolium repens* L. and *Plantago asiatica* L. by DSE. Colonization by DSE correlated with relative humidity and sunlight hours but not with AM colonization, temperature, rainfall, nitrogen, phosphorus potassium, or organic matter content in the soil.

V. Root Endophytes of Woody Plant Species

Many shrub and trees species belonging to various plant families were examined for the presence of root endophytes (Tables 38.4 through 38.6). Members of the Cupressaceae (Lihnell 1939; Hennon et al. 1990), Ericaceae (including Epacridaceae) (Oberholzer-Tschütscher 1982; Widler and Müller 1984; Hutton et al. 1994; Ahlich and Sieber 1996; Steinke et al. 1996; Hambleton and Currah 1997; Vodnik et al. 1997; Verkley et al. 2003; Ruotsalainen et al. 2004; Hambleton and Sigler 2005; Sigler et al. 2005; Grünig et al. 2009; Newsham et al. 2009; Bagyalakshmi et al. 2010; Kjoller et al. 2010; Tian et al. 2011), and Pinaceae (Bloomberg 1966; Parkinson and Crouch 1969; Galaaen and Venn 1979; Kowalski 1982a,b; Courtois and Ruschen 1987; Summerbell 1989; Courtois 1990b; Danielson and Visser 1990; Kattner and Schönhar 1990; Fisher et al. 1991a; Manka and Mroczkiewicz 1991; Heslin et al. 1992; Holdenrieder and Sieber 1992; Kattner 1992a; Ahlich and Sieber 1996; Horton et al. 1998; Grönberg et al. 2006; Kaparakis and Sen 2006; Menkis et al. 2006; Brenn et al. 2008; Grünig et al. 2008b, 2009; Bachmann 2010; Kandalepas et al. 2010; Rivera-Orduna et al. 2011; Wagg et al. 2011) are among the most intensively studied plant species. Some broadleaf tree and shrub species have been examined as well (Currah and Van Dyk 1986; Domanski and Kowalski 1987; Summerbell 1989; Sridhar and Bärlocher 1992; Cother and Gilbert 1994; Fisher et al. 1995a, 1995b; Iqbal et al. 1995; Ahlich and Sieber 1996; Raviraja et al. 1996; Werner et al. 1997; Barrow et al. 2004; Beauchamp et al. 2005; Newsham et al. 2009; Bagyalakshmi et al. 2010; Kandalepas et al. 2010).

Since presence and function of fungal root endophytes have most intensively been studied for members of the Ericales and Pinales, Sections V.A and V.B will be dedicated to these two groups of hosts.

A. Root Endophytes of Woody Ericales

Non-mycorrhizal endophytes are often isolated from ericaceous hosts in addition to the classical mycorrhizal fungi *Pezoloma (Hymenoscyphus) ericae* (D.J. Read) Baral (anamorph: *Scytalidium vaccinii* Dalpé, Litten & Sigler) and *Oidiodendron* spp. Species of the genera *Acephala*, *Cryptosporiopsis, Cylindrocarpon, Meliniomyces, Phialocephala*, and *Trichocladium* are most often observed (Oberholzer-Tschütscher 1982; Widler and Müller 1984; Väre et al. 1992; Hambleton and Currah 1997; Vodnik et al. 1997; Verkley et al. 2003; Hambleton and Sigler 2005; Sigler et al. 2005; Grünig et al. 2006; Grünig et al. 2008b, 2009, 2011; Queloz et al. 2010; Table 38.4). *Phialocephala fortinii* s.l. were present in 17 of 19 ericaceous hosts from boreal and alpine sites in Alberta, Canada (Hambleton and Currah 1997), in *Gaultheria shallon* Pursh from coastal British Columbia, Canada, and in *Calluna vulgaris* (L.) Hull as well as *Vaccinium myrtillus* L. from a subalpine site in Switzerland (Ahlich and Sieber 1996; Grünig et al. 2006). A fungus isolated by Vodnik et al. (1997) from the roots of *Erica carnea* L. from Slovenia and deposited as *Phialocephala fortinii* at UAMH as strain number 8433 could meanwhile be identified as *Phialocephala subalpina* Grünig & T.N. Sieber (Grünig et al. 2008b). A hitherto unknown species of *Phialocephala* was isolated from *Vaccinium vitis-idaea*, *V. myrtillus* and *V. uliginosum* growing on permafrost soil in the Jura mountains of Switzerland and described as *Phialocephala glacialis* Grünig & T.N. Sieber (Grünig et al. 2009). Interestingly, the same species was also detected in healthy needles of Norway-spruce trees nearby. All seven described species of *P. fortinii* s.l. and *A. applanata* could be detected in roots of *V. myrtillus* and other *Vaccinium* species in many countries in Europe and some of them also in North America (Queloz 2010; Table 38.4). Unidentified DSE were also isolated from *Symplocos cochinchinensis* (Lour.) S. Moore and several other plant species from shola vegetation in the Western Ghats region, southern India (Bagyalakshmi et al. 2010).

Cryptosporiopsis species (anamorphic *Pezicula* species) are regularly isolated from rhizomes and fine roots of ericaceous plants. *Cryptosporiopsis ericae* Sigler from roots of *Vaccinium membranaceum* Douglas ex Torr., *V. ovalifolium* Sm. and *G. shallon* Pursh, and *C. brunnea* Sigler from *G. shallon* from northwestern North America were described as new species (Sigler et al. 2005), and a *Cryptosporiopsis* species was isolated

TABLE 38.4 Endophytes in Ericoid Plants (Family Ericaceae)

Endophyte				Host					
Genus[a]	Species[a]	Fungus Order[a]	DSE/Not DSE	Species	Plant Order	Woody/ Herbaceous	Type of Experiment[b]	Special Effect[c]	References
Acephala	*applanata*	Helotiales	DSE	*Vaccinium myrtillus*	Ericales	Woody shrub	Isol		Queloz (2010)
Acephala	sp. 1	Helotiales	DSE	*Cassiope mertensiana*	Ericales	Woody shrub	Isol		Grünig et al. (2009)
Acephala	sp. 3	Helotiales	DSE	*Vaccinium myrtillus*	Ericales	Woody shrub	Isol		Grünig et al. (2009)
Acephala	sp. 7	Helotiales	DSE	*Calluna vulgaris*	Ericales	Woody shrub	Isol		Pietrowski (unpublished)
Cryptocline	*dubia*	Helotiales	Not DSE	*Arctostaphylos uva-ursi*	Ericales	Woody shrub	Isol		Widler and Müller (1984)
Cryptosporiopsis	sp.	Helotiales	DSE	*Arctostaphylos uva-ursi*	Ericales	Woody shrub	Isol		Widler and Müller (1984)
Cryptosporiopsis	sp.	Helotiales	DSE	*Erica carnea*	Ericales	Woody shrub	Isol		Oberholzer-Tschütscher (1982)
Cryptosporiopsis	*brunnea*	Helotiales	DSE	*Gaultheria shallon*	Ericales	Woody shrub	Isol		Sigler et al. (2005)
Cryptosporiopsis	*ericae*	Helotiales	Variable	*Vaccinium membranaceum*	Ericales	Woody shrub	Isol		Sigler et al. (2005)
Cryptosporiopsis	*rhizophila*	Helotiales	DSE	*Erica tetralix*	Ericales	Woody shrub	Isol		Verkley et al. (2003)
Cylindrocarpon	*didymum*	Hypocreales	Not DSE	*Arctostaphylos uva-ursi*	Ericales	Woody shrub	Isol		Widler and Müller (1984)
Cylindrocarpon	*didymum*	Hypocreales	Not DSE	*Erica carnea*	Ericales	Woody shrub	Isol		Oberholzer-Tschütscher (1982)
Cystodendron	*dryophilum*	Helotiales	Not DSE	*Arctostaphylos uva-ursi*	Ericales	Woody shrub	Isol		Widler and Müller (1984)
Gliocladium	*solani f. nigrovirens*	Hypocreales	Variable	*Arctostaphylos uva-ursi*	Ericales	Woody shrub	Isol		Widler and Müller (1984)
Herpotrichia	sp.	Pleosporales	DSE	*Vaccinium* sp.	Ericales	Woody shrub	Isol		Müller (unpublished)
Ilyonectria	*radicicola*	Hypocreales	Not DSE	*Erica carnea*	Ericales	Woody shrub	Isol		Vodnik et al. (1997)
Marasmius	*scorodonius*	Agaricales	Not DSE	*Arctostaphylos uva-ursi*	Ericales	Woody shrub	Isol		Widler and Müller (1984)
Meliniomyces	sp. 3	Leotiomycetes	Variable	*Gaultheria shallon*	Ericales	Woody shrub	Isol		Hambleton and Sigler (2005)
Meliniomyces	sp. 3	Leotiomycetes	Variable	*Vaccinium myrtillus*	Ericales	Woody shrub	Isol		Hambleton and Sigler (2005)
Meliniomyces	*variabilis*	Leotiomycetes	DSE	*Andromeda polifolia*	Ericales	Woody shrub	Isol		Kjoller et al. (2010)
Meliniomyces	*variabilis*	Leotiomycetes	DSE	*Empetrum hermaphroditum*	Ericales	Woody shrub	Isol		Kjoller et al. (2010)
Meliniomyces	*variabilis*	Leotiomycetes	DSE	*Rhododendron albiflorum*	Ericales	Woody shrub	Isol		Hambleton and Sigler (2005)
Meliniomyces	*variabilis*	Leotiomycetes	DSE	*Vaccinium uliginosum*	Ericales	Woody shrub	Isol		Kjoller et al. (2010)
Meliniomyces	*variabilis*	Leotiomycetes	DSE	*Vaccinium vitis-idaea*	Ericales	Woody shrub	Isol		Kjoller et al. (2010)

(continued)

TABLE 38.4 (continued) Endophytes in Ericoid Plants (Family Ericaceae)

Endophyte				Host					
Genus[a]	Species[a]	Fungus Order[a]	DSE/Not DSE	Species	Plant Order	Woody/Herbaceous	Type of Experiment[b]	Special Effect[c]	References
Microdochium	*bolleyi*	Xylariales	DSE	*Polemonium boreale*	Ericales	Herb	Isol		Väre et al. (1992)
Monodictys	*putredinis*	Dothideomycetes	Not DSE	*Erica carnea*	Ericales	Woody shrub	Isol		Oberholzer-Tschütscher (1982)
Oidiodendron	*griseum*	Dothideomycetes	DSE	*Loiseleuria procumbens*	Ericales	Woody shrub	Isol		Stoyke and Currah (1991), Hambleton and Currah (1997)
Oidiodendron	*griseum*	Dothideomycetes	DSE	*Phyllodoce glanduliflora*	Ericales	Woody shrub	Isol		Stoyke and Currah (1991), Hambleton and Currah (1997)
Oidiodendron	*griseum*	Dothideomycetes	DSE	*Vaccinium myrtilloides*	Ericales	Woody shrub	Isol		Stoyke and Currah (1991), Hambleton and Currah (1997)
Oidiodendron	*griseum*	Dothideomycetes	DSE	*Vaccinium vitis-idaea*	Ericales	Woody shrub	Isol		Stoyke and Currah (1991), Hambleton and Currah (1997)
Oidiodendron	*maius*	Dothideomycetes	DSE	Various spp.	Ericales	Woody shrub	Isol		Hambleton and Currah (1997)
Pezoloma	*ericae*	Leotiomycetes	DSE	*Calluna vulgaris*	Ericales	Woody shrub	Isol		Read (1974)
Pezoloma	*ericae*	Leotiomycetes	DSE	*Ledum groenladicum*	Ericales	Woody shrub	Isol		Hambleton et al. (1999)
Phialocephala	*fortinii s.l.*	Helotiales	DSE	*Polemonium boreale*	Ericales	Herb	Isol		Väre et al. (1992)
Phialocephala	*fortinii s.l.*	Helotiales	DSE	Various spp.	Ericales	N.A.	Isol		Hambleton and Currah (1997)
Phialocephala	*europaea*	Helotiales	DSE	*Vaccinium myrtillus*	Ericales	Woody shrub	Isol		Queloz (2010)
Phialocephala	*europaea*	Helotiales	DSE	*Vaccinium uliginosum*	Ericales	Woody shrub	Isol		Queloz (2010)
Phialocephala	*europaea*	Helotiales	DSE	*Vaccinium vitis-idaea*	Ericales	Woody shrub	Isol		Queloz (2010)
Phialocephala	*fortinii s.s.*	Helotiales	DSE	*Cassiope mertensiana*	Ericales	Woody shrub	Isol		Queloz (2010)
Phialocephala	*fortinii s.s.*	Helotiales	DSE	*Kalmia microphylla*	Ericales	Woody shrub	Isol		Queloz (2010)
Phialocephala	*fortinii s.s.*	Helotiales	DSE	*Luteka pectinata*	Ericales	Woody shrub	Isol		Queloz (2010)
Phialocephala	*fortinii s.s.*	Helotiales	DSE	*Vaccinium myrtillus*	Ericales	Woody shrub	Isol		Grünig et al. (2008a)
Phialocephala	*fortinii s.s.*	Helotiales	DSE	*Vaccinium uliginosum*	Ericales	Woody shrub	Isol		Queloz (2010)
Phialocephala	*fortinii s.s.*	Helotiales	DSE	*Vaccinium vitis-idaea*	Ericales	Woody shrub	Isol		Queloz (2010)
Phialocephala	*glacialis*	Helotiales	DSE	*Vaccinium myrtillus*	Ericales	Woody shrub	Isol		Grünig et al. (2009)

Phialocephala	*helvetica*	Helotiales	DSE	*Erica carnea*	Ericales	Woody shrub	Isol		Queloz (2010)
Phialocephala	*helvetica*	Helotiales	DSE	*Vaccinium myrtillus*	Ericales	Woody shrub	Isol		Queloz (2010)
Phialocephala	*letzii*	Helotiales	DSE	*Vaccinium myrtillus*	Ericales	Woody shrub	Isol		Queloz (2010)
Phialocephala	*letzii*	Helotiales	DSE	*Vaccinium vitis-idaea*	Ericales	Woody shrub	Isol		Queloz (2010)
Phialocephala	*subalpina*	Helotiales	DSE	*Arctostaphylos uva-ursi*	Ericales	Woody shrub	Isol		Queloz (2010)
Phialocephala	*subalpina*	Helotiales	DSE	*Empetrum hermaphroditum*	Ericales	Woody shrub	Isol		Queloz (2010)
Phialocephala	*subalpina*	Helotiales	DSE	*Erica carnea*	Ericales	Woody shrub	Isol		Vodnik et al. (1997)
Phialocephala	*subalpina*	Helotiales	DSE	*Vaccinium myrtillus*	Ericales	Woody shrub	Isol		Grünig et al. (2008a)
Phialocephala	*subalpina*	Helotiales	DSE	*Vaccinium uliginosum*	Ericales	Woody shrub	Isol		Queloz (2010)
Phialocephala	*subalpina*	Helotiales	DSE	*Vaccinium vitis-idaea*	Ericales	Woody shrub	Isol		Queloz (2010)
Phialocephala	*turicensis*	Helotiales	DSE	*Vaccinium myrtillus*	Ericales	Woody shrub	Isol		Queloz (2010)
Phialocephala	*uotilensis*	Helotiales	DSE	*Vaccinium myrtillus*	Ericales	Woody shrub	Isol		Queloz (2010)
Phialophora	*bubakii*	Chaetothyriales	Not DSE	*Erica carnea*	Ericales	Woody shrub	Isol		Oberholzer-Tschütscher (1982)
Rhizoctonia	sp.	Cantharellales	DSE	*Erica carnea*	Ericales	Woody shrub	Isol		Vodnik et al. (1997)
Trichocladium	*opacum*	Sordariales	DSE	*Arctostaphylos uva-ursi*	Ericales	Woody shrub	Isol		Widler and Müller (1984)
Trichocladium	*opacum*	Sordariales	Not DSE	*Erica carnea*	Ericales	Woody shrub	Isol		Oberholzer-Tschütscher (1982)
Varicosporium	sp.	Helotiales	DSE	*Arctostaphylos uva-ursi*	Ericales	Woody shrub	Isol		Widler and Müller (1984)
Unidentified DSE			DSE	*Empetrum hermaphroditum*	Ericales	Woody shrub	Isol	a	Ruotsalainen et al. (2010)
Unidentified DSE			DSE	*Symplocos cochinchinensis*	Ericales	Tree	Isol		Bagyalakshmi et al. (2010)

[a] Genus, species and order (family) names according to index fungorum (http://www.indexfungorum.org/Names/Names.asp, December 20, 2011). If both the teleomorph (sexual reproductive stage) and the anamorph (asexual reproductive stage) are produced, the name of the teleomorph is given. Names of anamorph(s) can be retrieved from the list of the teleomorph's synonyms provided in the index. The next lower taxon is given if the Fungus order is not known with certainty ("incertae sedis").

[b] Isol, isolation from surface-sterilized roots.

[c] Special effect details: (a) Colonization frequency positively correlated with light intensity.

from *C. vulgaris*, *Erica tetralix* L., *V. vitis-idaea*, and *V. myrtillus* in the Netherlands and described as *C. rhizophila* Verkley & Zijlstra (Verkley et al. 2003).

Meliniomyces species seem to be widespread endophytes in roots of *Andromeda polifolia* L., *Empetrum hermaphroditum* Hagerup, *Phyllodoce empetriformis* (Sm.) D. Don, *Rhododendron albiflorum* Hook., *V. membranaceum*, *V. uliginosum* L., and *V. vitis-idaea* in alpine heathlands and subarctic mires (Hambleton and Sigler 2005; Kjoller et al. 2010).

Dark-colored, sterile, and slow-growing mycelia were also isolated in Australia from Epacridaceae, a family close to the Ericaceae but distributed in the Southern Hemisphere (Hutton et al. 1994). Pectic zymogram analysis revealed that none of the Australian isolates matched fungi known to be infective with Ericaceae in the Northern Boreal and Alpine Zone, for example, *Pezoloma ericae* and *Oidiodendron* spp., even though all isolates appeared morphologically similar and had similar growth rates. Two-thirds of the endophytes isolated from *Leucopogon parviflorus* (H. Andrews) Lindl., another representative of Epacridaceae, were sterile and had slow growth rates. Some of them were dark and similar to *P. ericae*. Inspection of roots under the light microscope revealed that there were at least two different fungi that consistently formed ericoid mycorrhizal structures at these sites, possibly *P. ericae* and *Oidiodendron* sp. (Steinke et al. 1996).

1. Anatomy

Dark septate inter- and intracellular hyphae and microsclerotia were observed in epidermal cells (EPs) of ericaceous "hair roots" of *Gaultheria poeppiggi* DC. collected in the Cordoba mountains of central Argentina (Urcelay 2002; Figure 38.3C). Similarly, DSE were present in hair roots of *Woollsia pungens* (Cav.) F. Muell. in New South Wales (Chambers et al. 2008). *Phialocephala fortinii* s.s. occurred as extensive wefts of dark, septate hyphae on the root surface and as intracortical sclerotia of compact, darkly pigmented and irregularly lobed, thick-walled hyphae in axenic seedlings of *Menziesia ferruginea* Sm. (Ericaceae) (Stoyke and Currah 1991, 1993). Intracellular coils and colonization of the vascular tissues were not observed. This association differs from the ericoid mycorrhizal type, in which elaborate intracellular hyphal branches are ensheathed by invaginations of the host plasma membrane but represents a fungus–root association that seems to be common in alpine plants. Similarly, the pattern of root colonization of *Rhododendron brachycarpum* D.Don by *Phialocephala fortinii* s.l. differed from that observed in ericoid mycorrhizae (Currah et al. 1993). *Phialocephala fortinii* s.l. colonized the EPs, in which it sometimes formed black sclerotia, but was not able to colonize the thick-walled, phenol-rich exodermal layer of hair roots. Intercellular colonization between epidermis and exodermis layers was, however, common.

2. Endophyte–Plant Interactions

Presence of *Phialocephala fortinii* s.s. in axenic cultures of *M. ferruginea* (Ericaceae) caused a 10-fold increase of seedling mortality compared to mortality of control plants (Stoyke and Currah 1993). Established plants and growth rates were, however, not affected. *Phialocephala fortinii*-like DSE strains were shown to differ in virulence against *R. brachycarpum*. In resynthesis experiment, one strain had a significant negative effect on dry weight accumulation of seedlings, but plants looked healthy, whereas a second one had no effect (Currah et al. 1993).

3. Endophyte–Environment Interactions

Phialocephala fortinii s.l. was only rarely detected in ericaceous roots from an acidic wetland but occurred quite frequently in roots from an alpine site in Alberta (Hambleton and Currah 1997). *Phialocephala subalpina* and a *Rhizoctonia* sp. were shown to be sensitive to heavy metals and mainly occurred in roots of *E. carnea* growing in non-polluted soil, whereas *Cladosporium herbarum* (Pers.) Link and *Ilyonectria radicicola* were the fungi most frequently isolated from plant roots originating from lead contaminated soil (Vodnik et al. 1997). Stress levels along three abiotic gradients (pollution, elevation, distance from seashore) on Kola Peninsula in Russia had no effect on root colonization of crowberry (*E. hermaphroditum*) by DSE, but distance to mountain birch (*Betula pubescens* ssp. *czerepanovii* (Orlova) Hämet-Ahti) had an effect; colonization was higher outside the canopy area (Ruotsalainen et al. 2010). Substitution of AM fungi at higher altitude could not be observed for *Trientalis europaea* L. and several other alpine plants collected along a gradient from 0 to 1400 m at Mt. Paras, North Norway (Ruotsalainen et al. 2004).

B. Root Endophytes of Pinales

Endophyte species diversity strongly depends on the degree of surface sterilization and the sample size; the weaker the sterilization and the more samples, the more species are detected. For example, Holdenrieder and Sieber (1992) could detect 120 taxa in >500 serially washed Norway-spruce root segments. However, if surface sterilization is applied, diversity of fungal communities steeply decreases. Moreover, the diversity of endophyte communities in roots is also much lower than that of endophyte communities in aboveground plant parts (Sieber 1988, 1989, 2007; Sridhar and Bärlocher 1992; Kowalski 1993; Sieber-Canavesi and Sieber 1993; Ahlich and Sieber 1996; Menkis 2004; Rivera-Orduna et al. 2011). Only five species of endophytes (*Alternaria* sp., *Cochliobolus* sp., *Penicillium* sp., *Phoma medicaginis*, and a species of the Xylariaceae) could be isolated from roots of *Taxus globosa* Schltdl. in Mexico, whereas bark contained at least 17 different species (Rivera-Orduna et al. 2011). More than 90% of the isolates collected from root tips of *Picea abies* and *Pinus sylvestris* L. belonged to the genus *Phialocephala* (Menkis 2004; Table 38.5). Non-mycorrhizal fine roots of healthy *Abies alba* Mill., *Picea abies*, and *Pinus sylvestris* in Europe are heavily colonized by endophytic fungi. Depending on the site, between 88% and 100%

TABLE 38.5 Endophytes in Conifers

Endophyte				Host					
Genus[a]	Species[a]	Fungus Order[a]	DSE/Not DSE	Species	Plant Order	Woody/ Herbaceous	Type of Experiment[b]	Special Effect[c]	References
Acephala	*applanata*	Helotiales	DSE	*Abies alba*	Pinales	Tree	Isol		Queloz (2010)
Acephala	*applanata*	Helotiales	DSE	*Picea abies*	Pinales	Tree	Isol		Grünig and Sieber (2005)
Acephala	*applanata*	Helotiales	DSE	*Pinus mugo*	Pinales	Tree	Isol		Queloz (2010)
Acephala	*applanata*	Helotiales	DSE	*Pinus sylvestris*	Pinales	Tree	Isol		Queloz (2010)
Acephala	*macrosclerotiorum*	Helotiales	DSE	*Picea abies*	Pinales	Tree	Isol		Menkis et al. (2004)
Acephala	*macrosclerotiorum*	Helotiales	DSE	*Pinus sylvestris*	Pinales	Tree	Isol		Münzenberger et al. (2009)
Acephala	sp. 2	Helotiales	DSE	*Pinus sylvestris*	Pinales	Tree	Isol		Grünig et al. (2009)
Acephala	sp. 4	Helotiales	DSE	*Pinus banksiana*	Pinales	Tree	Isol		Grünig et al. (2009)
Alternaria	sp.	Pleosporales	DSE	*Taxus globosa*	Pinales	Tree	Isol		Rivera-Orduna et al. (2011)
Anguillospora	*filiformis*	Pleosporales	Not DSE	*Picea glauca*	Pinales	Tree	Isol		Sridhar and Bärlocher (1992)
Aspergillus	*versicolor*	Eurotiales	Not DSE	*Picea abies*	Pinales	Tree	Isol		Courtois (1990a,b)
Cadophora	*finlandica*	Helotiales	DSE	*Picea abies*	Pinales	Tree	Isol		Menkis (unpublished)
Cadophora	*finlandica*	Helotiales	DSE	*Pinus resinosa*	Pinales	Tree	in vitro exp	a	Alberton et al. (2010)
Cadophora	*finlandica*	Helotiales	DSE	*Pinus sylvestris*	Pinales	Tree	Isol		Wang and Wilcox (1985)
Cadophora	*malorum*	Helotiales	DSE	*Picea abies*	Pinales	Tree	Isol		Sieber (unpublished)
Cadophora	sp. 1	Helotiales	DSE	*Picea abies*	Pinales	Tree	Isol		Queloz (2010)
Cadophora	sp. 2	Helotiales	DSE	*Abies alba*	Pinales	Tree	Isol		Ahlich and Sieber (1996)
Cadophora	sp. 2	Helotiales	DSE	*Pinus sylvestris*	Pinales	Tree	Isol		Bachmann (2010)
Chloridium	*paucisporum*	Chaetosphaeriales	DSE	*Pinus resinosa*	Pinales	Tree	In vitro exp	a	Alberton et al. (2010)
Cladophialophora	*chaetospira*	Chaetothyriales	DSE	*Picea abies*	Pinales	Tree	Isol		Crous et al. (2007)
Cochliobolus	sp.	Pleosporales	DSE	*Taxus globosa*	Pinales	Tree	Isol		Rivera-Orduna et al. (2011)
Coniothyrium	sp.	Pleosporales	Not DSE	*Pinus sylvestris*	Pinales	Tree	Isol		Görke (1998)
Cryptosporiopsis	*abietina*	Helotiales	Not DSE	*Picea abies*	Pinales	Tree	Isol		Kattner and Schönhar (1990)
Cryptosporiopsis	*radicicola*	Helotiales	DSE	*Abies alba*	Pinales	Tree	Isol		Ahlich and Sieber (1996)
Cryptosporiopsis	*radicicola*	Helotiales	DSE	*Pinus sylvestris*	Pinales	Tree	Isol		Ahlich and Sieber (1996)
Cryptosporiopsis	sp.	Helotiales	DSE	*Chamaecyparis nootkatensis*	Pinales	Tree	Isol		Hennon et al. (1990)
Cryptosporiopsis	*cf melanigena*	Helotiales	DSE	*Picea abies*	Pinales	Tree	Isol		Queloz (2010)
Cryptosporiopsis	*ericae*	Helotiales	Variable	*Picea abies*	Pinales	Tree	Isol		Sigler et al. (2005)
Cylindrocarpon	*aquaticum*	Hypocreales	Not DSE	*Picea glauca*	Pinales	Tree	Isol		Sridhar and Bärlocher (1992)
Cylindrocarpon	*didymum*	Hypocreales	Not DSE	*Chamaecyparis nootkatensis*	Pinales	Tree	Isol		Hennon et al. (1990)
Cylindrocarpon	*didymum*	Hypocreales	Not DSE	*Pinus sylvestris*	Pinales	Tree	Isol		Ahlich and Sieber (1996)
Cylindrocarpon	*obtusisporum*	Hypocreales	Variable	*Picea abies*	Pinales	Tree	Isol		Queloz (2010)
Didymella	*bryoniae*	Pleosporales	DSE	*Picea abies*	Pinales	Tree	Isol		Queloz (2010)
Didymosphaeria	sp.	Pleosporales	DSE	*Picea abies*	Pinales	Tree	Isol		Brenn et al. (2008)
Gelatinosporium	sp.	Helotiales	Not DSE	*Chamaecyparis nootkatensis*	Pinales	Tree	Isol		Hennon et al. (1990)

(continued)

TABLE 38.5 (continued) Endophytes in Conifers

Endophyte				Host					
Genus[a]	Species[a]	Fungus Order[a]	DSE/Not DSE	Species	Plant Order	Woody/Herbaceous	Type of Experiment[b]	Special Effect[c]	References
Herpotrichia	sp.	Pleosporales	DSE	*Picea abies*	Pinales	Tree	Isol		Grünig (unpublished)
Humicolopsis	*cephalosporioides*	Pezizomycotina	DSE	*Picea abies*	Pinales	Tree	Isol		Courtois (1990a,b)
Ilyonectria	*radicicola*	Hypocreales	Not DSE	*Picea abies*	Pinales	Tree	Isol		Görke (1998)
Ilyonectria	*radicicola*	Hypocreales	Variable	*Pinus sylvestris*	Pinales	Tree	Isol		Bachmann (2010)
Leptosphaeria	sp. 1	Pleosporales	DSE	*Pinus sylvestris*	Pinales	Tree	Isol		Bachmann (2010)
Macrophomina	*phaseolina*	Botryosphaeriales	DSE	*Pinus sylvestris*	Pinales	Tree	Isol		Bachmann (2010)
Meliniomyces	sp. 2	Leotiomycetes	Variable	*Pinus sylvestris*	Pinales	Tree	Isol		Hambleton and Sigler (2005)
Meliniomyces	sp. 4	Leotiomycetes	Variable	*Pinus sylvestris*	Pinales	Tree	Isol		Hambleton and Sigler (2005)
Meliniomyces	*variabilis*	Leotiomycetes	DSE	*Pinus resinosa*	Pinales	Tree	In vitro exp	a	Alberton et al. (2010)
Meliniomyces	*variabilis*	Leotiomycetes	DSE	*Tsuga heterophylla*	Pinales	Tree	Isol		Hambleton and Sigler (2005)
Meliniomyces	*vraolstadiae*	Leotiomycetes	DSE	*Pinus resinosa*	Pinales	Tree	In vitro exp	a	Alberton et al. (2010)
Paecilomyces	sp.	Eurotiales	Not DSE	*Pinus radiata*	Pinales	Tree	Isol		Sieber and Langenegger (unpublished)
Penicillium	*nigricans*	Eurotiales	DSE	*Picea abies*	Pinales	Tree	Isol		Courtois (1990a,b)
Phialocephala	*europaea*	Helotiales	DSE	*Abies alba*	Pinales	Tree	Isol		Queloz (2010)
Phialocephala	*europaea*	Helotiales	DSE	*Larix kaempferi*	Pinales	Tree	Isol		Queloz (2010)
Phialocephala	*europaea*	Helotiales	DSE	*Picea abies*	Pinales	Tree	Isol		Grünig et al. (2008a)
Phialocephala	*europaea*	Helotiales	DSE	*Picea glauca*	Pinales	Tree	Isol		Queloz (2010)
Phialocephala	*europaea*	Helotiales	DSE	*Pinus sylvestris*	Pinales	Tree	Isol		Queloz (2010)
Phialocephala	*fortinii s.l.*	Helotiales	DSE	*Abies alba*	Pinales	Tree	Isol		Ahlich and Sieber (1996)
Phialocephala	*fortinii s.l.*	Helotiales	DSE	*Picea abies*	Pinales	Tree	Isol		Ahlich and Sieber (1996)
Phialocephala	*fortinii s.l.*	Helotiales	DSE	*Pinus resinosa*	Pinales	Tree	In vitro exp	a	Alberton et al. (2010)
Phialocephala	*fortinii s.l.*	Helotiales	DSE	*Larix decidua*	Pinales	Tree	Isol		Grünig et al. (2008a)
Phialocephala	*fortinii s.s.*	Helotiales	DSE	*Pinus sylvestris*	Pinales	Tree	Isol		Ahlich and Sieber (1996)
Phialocephala	*fortinii s.s.*	Helotiales	DSE	*Abies alba*	Pinales	Tree	Isol		Queloz (2010)
Phialocephala	*fortinii s.s.*	Helotiales	DSE	*Larix decidua*	Pinales	Tree	Isol		Queloz (2010)
Phialocephala	*fortinii s.s.*	Helotiales	DSE	*Picea abies*	Pinales	Tree	Isol		Grünig et al. (2008a)
Phialocephala	*fortinii s.s.*	Helotiales	DSE	*Picea glauca*	Pinales	Tree	Isol		Queloz (2010)
Phialocephala	*fortinii s.s.*	Helotiales	DSE	*Picea mariana*	Pinales	Tree	Isol		Queloz (2010)
Phialocephala	*fortinii s.s.*	Helotiales	DSE	*Pinus cembra*	Pinales	Tree	Isol		Queloz (2010)
Phialocephala	*fortinii s.s.*	Helotiales	DSE	*Pinus mugo*	Pinales	Tree	Isol		Queloz (2010)
Phialocephala	*fortinii s.s.*	Helotiales	DSE	*Pinus sylvestris*	Pinales	Tree	Isol		Grünig et al. (2008a)
Phialocephala	*helvetica*	Helotiales	DSE	*Juniperus* sp.	Pinales	Tree	Isol		Queloz (2010)
Phialocephala	*helvetica*	Helotiales	DSE	*Picea abies*	Pinales	Tree	Isol		Grünig et al. (2008a)
Phialocephala	*helvetica*	Helotiales	DSE	*Pinus leucodermis*	Pinales	Tree	Isol		Queloz (2010)
Phialocephala	*helvetica*	Helotiales	DSE	*Pinus sylvestris*	Pinales	Tree	Isol		Queloz (2010)
Phialocephala	*letzii*	Helotiales	DSE	*Abies alba*	Pinales	Tree	Isol		Queloz (2010)
Phialocephala	*letzii*	Helotiales	DSE	*Picea abies*	Pinales	Tree	Isol		Grünig et al. (2008a)

Phialocephala	*sphaeroides*	Helotiales	DSE	*Picea abies*	Pinales	Tree	Isol		Grünig et al. (2009)
Phialocephala	*subalpina*	Helotiales	DSE	*Abies alba*	Pinales	Tree	Isol		Queloz (2010)
Phialocephala	*subalpina*	Helotiales	DSE	*Abies spectabilis*	Pinales	Tree	Isol		Queloz (2010)
Phialocephala	*subalpina*	Helotiales	DSE	*Picea abies*	Pinales	Tree	Isol		Grünig et al. (2008a)
Phialocephala	*subalpina*	Helotiales	DSE	*Picea glauca*	Pinales	Tree	Isol		Queloz (2010)
Phialocephala	*subalpina*	Helotiales	DSE	*Pinus mugo*	Pinales	Tree	Isol		Queloz (2010)
Phialocephala	*subalpina*	Helotiales	DSE	*Pinus strobus*	Pinales	Tree	Isol		Queloz (2010)
Phialocephala	*subalpina*	Helotiales	DSE	*Pinus sylvestris*	Pinales	Tree	Isol		Grünig et al. (2008a)
Phialocephala	*subalpina*	Helotiales	DSE	*Tsuga dumosa*	Pinales	Tree	Isol		Queloz (2010)
Phialocephala	*turicensis*	Helotiales	DSE	*Abies alba*	Pinales	Tree	Isol		Queloz (2010)
Phialocephala	*turicensis*	Helotiales	DSE	*Picea abies*	Pinales	Tree	Isol		Queloz (2010)
Phialocephala	*turicensis*	Helotiales	DSE	*Taxus baccata*	Pinales	Tree	Isol		Queloz (2010)
Phialocephala	*uotilensis*	Helotiales	DSE	*Abies alba*	Pinales	Tree	Isol		Queloz (2010)
Phialocephala	*uotilensis*	Helotiales	DSE	*Picea abies*	Pinales	Tree	Isol		Grünig et al. (2008a)
Phialophora	*melinii*	Helotiales	Not DSE	*Chamaecyparis nootkatensis*	Pinales	Tree	Isol		Hennon et al. (1990)
Phoma	sp. 1	Pleosporales	DSE	*Pinus sylvestris*	Pinales	Tree	Isol		Bachmann (2010)
Phoma	*exigua*	Pleosporales	DSE	*Picea abies*	Pinales	Tree	Isol		Queloz (2010)
Phoma	*medicaginis*	Pleosporales	DSE	*Taxus globosa*	Pinales	Tree	Isol		Rivera-Orduna et al. (2011)
Phoma	*radicina*	Pleosporales	DSE	*Pinus sylvestris*	Pinales	Tree	Isol		Bachmann (2010)
Phomopsis	sp.	Diaporthales	Not DSE	*Pinus radiata*	Pinales	Tree	Isol		Sieber and Langenegger (unpublished)
Rhizoctonia	sp.	Cantharellales	Not DSE	*Pinus sylvestris*	Pinales	Tree	in vitro exp	b	Grönberg et al. (2006)
Sphaeropsis	*sapinea*	Pezizomycotina	DSE	*Pinus radiata*	Pinales	Tree	Isol		Sieber and Langenegger (unpublished)
Sporidesmium	sp.	Pleosporales	Not DSE	*Chamaecyparis nootkatensis*	Pinales	Tree	Isol		Hennon et al. (1990)
Sydowia	*polyspora*	Dothideales	DSE	*Pinus sylvestris*	Pinales	Tree	Isol		Görke (1998)
Trichoderma	*hamatum*	Hypocreales	Not DSE	*Picea abies*	Pinales	Tree	Isol		Kattner and Schönhar (1990)
Trichoderma	*polysporum*	Hypocreales	Not DSE	*Picea abies*	Pinales	Tree	Isol		Kattner and Schönhar (1990)
Trichoderma	*viride*	Hypocreales	Not DSE	*Picea abies*	Pinales	Tree	Isol		Kattner and Schönhar (1990)

[a] Genus, species and order (family) names according to index fungorum (http://www.indexfungorum.org/Names/Names.asp, December 20, 2011). If both the teleomorph (sexual reproductive stage) and the anamorph (asexual reproductive stage) are produced, the name of the teleomorph is given. Names of anamorph(s) can be retrieved from the list of the teleomorph's synonyms provided in the index. The next lower taxon is given if the Fungus order is not known with certainty ("incertae sedis").

[b] Isol, isolation from surface-sterilized roots; in vitro exp, fungus used in in-vitro experiment(s).

[c] Special effect details: (a) Plant growth stimulation at elevated CO_2. (b) Stimulation of early root development.

of the trees examined in Finland, Germany, and Switzerland had roots colonized by endophytes, and DSE were most abundant (Ahlich and Sieber 1996; Görke 1998). They were present in between 53% and 90% of the trees, and the majority of the isolates belonged to PAC. Up to 70% of the fine roots of a single root system of Norway spruce (*Picea abies*) can be colonized by DSE (Holdenrieder and Sieber 1992). *Taxodium distichum* (L.) Rich. seems to be an exception regarding dominance of DSE, because DSE were very rare in roots of this tree species and occurred in only 0.33% of the root samples (Kandalepas et al. 2010). Whereas DSE, especially PAC, dominate the endophyte communities, species of *Cryptosporiopsis* and *Cylindrocarpon* occur sometimes also quite frequently in conifer roots (Courtois 1990a; Kattner and Schönhar 1990; Holdenrieder and Sieber 1992; Ahlich and Sieber 1996; Table 38.5). A special community was observed in submersed roots of *Picea glauca* (Moench) Voss (Sridhar and Bärlocher 1992). The community was dominated by aquatic hyphomycetes, for example, *Anguillospora filiformis* Greath. and *Heliscus lugdunensis* Sacc. & Therry (Table 38.5).

Non-mycorrhizal endophytes are often associated with ectomycorrhizae of conifers. DSE, *Umbelopsis isabellina* (Oudem.) W. Gams, *Penicillium spinulosum* Thom, and *Penicillium montanense* M. Chr. & Backus were most frequently associated with ectomycorrhizae of *Picea mariana* (Mill.) Britton, Sterns & Poggenb. (Summerbell 1989). *Oidiodendron* sp. considered non-mycorrhizal on conifers, and DSE were most frequently isolated from ectomycorrhizae of Sitka spruce (*Picea sitchensis* (Bong.) Carrière) in Irish forest mixed stands (Schild et al. 1988; Heslin et al. 1992). *Oidiodendron* sp. and DSE appeared to be antagonistic. DSE were only abundant when *Oidiodendron* sp. was rare and vice versa. Douglas-fir seedlings preinoculated with the ectomycorrhizal fungus *Rhizopogon vinicolor* A.H. Sm. and outplanted on eastern Vancouver Island were to some extend colonized by DSE after one growing season (Berch and Roth 1993). DSE were shown to survive forest fires as resident inoculum and to colonize *Pinus muricata* D. Don seedlings right after germination (Horton et al. 1998). Eighteen percent of the seedlings sampled until 5 months after germination were colonized by AM fungi, ectomycorrhizal fungi, and DSE simultaneously.

Except for members of the Epacridaceae, root endophytes of woody plant species have not been examined intensively in the Southern Hemisphere. This is especially true for conifers. Root samples of *Pinus radiata* D. Don collected at four sites in Western Australia revealed *Sphaeropsis sapinea* (Fr.) Dyko & B. Sutton, a fungus usually confined to pine needles in the Northern Hemisphere, to be the main colonizer at two of the four sites (Sieber and Langenegger, unpublished). A *Paecilomyces* sp. and a *Phomopsis* sp. dominated at each one of the other two sites.

DSE are prevalent in roots of conifer seedlings in nurseries and naturally regenerating seedlings in the forest. Primary roots of 1-year-old nursery-grown Douglas-fir (*Pseudotsuga menziesii* (Mirb.) Franco) seedlings were most frequently colonized by DSE, *Fusarium* spp., and *Cylindrocarpon* spp. (Bloomberg 1966). Similarly, DSE were among the earliest and most abundant root colonizers of nursery-grown seedlings of *Pinus banksiana* Lamb., *Pinus contorta* Douglas ex Loudon, and *Picea glauca* (Moench) Voss (Danielson and Visser 1990). The frequency of root segments of nursery seedlings of *Picea abies* and *Pinus sylvestris* L. colonized by endophytic fungi in general varied between 0.5% and 10% and that by PAC between 0% and 4%, that is, colonization density was significantly lower than expected from densities observed in spruce forests (Brenn et al. 2008). Interestingly, PAC could only be detected in plants from nurseries situated directly in forests or very close to forests but not in plants from nurseries on agricultural plots, because PAC are very rare in arable soils (Ahlich-Schlegel 1997). Endophytic fungi were isolated from 1- to 5-year-old seedlings of *Pinus nigra* var. *laricio* Maire sampled in an area of natural tree regeneration (Parkinson and Crouch 1969). DSE were most frequently isolated from roots of 5-year-old seedlings and *Penicillium* spp. dominated in roots of younger plants. The frequency of colonization by DSE varied between 30% and 100% among plant individuals. DSE did not show any preference for specific parts in the root system. *Penicillium* spp. were, however, most frequently isolated from the upper tap and upper lateral roots, whereas *Ilyonectria radicicola* showed a preference for the lower vertical roots.

1. Anatomy

Structural features of the endophyte–root symbioses of conifers are best studied for symbioses with PAC fungi. In contrast to mycorrhizal fungi, PAC are not confined to the root tips but can occur everywhere in the root system from root tips to the bark of coarse roots at the stem base. First- and second-order laterals of primary roots of *Pinus contorta* colonized by *Phialocephala fortinii* s.l. were characterized by surface patches of sclerotia and loose wefts of hyphae growing along the root surface and hyphae as well as sclerotia growing inter- and intracellularly in the outer cortex (O'Dell et al. 1993). Occasionally, patches of intercellular labyrinthine fungal tissue, similar to Hartig net tissue, were formed on the surface of primary pine roots, and proximal portions of lateral roots frequently had a sporadic mantle. Similarly, roots of *Pinus strobus* L. showed varying amounts of surface hyphae, Hartig net structures, and intracellular hyphae (Peterson et al. 2008; Figure 38.3D). *Acephala macrosclerotiorum* Münzenberger & Bubner, a recently described close relative of PAC, formed thin mantles and a distinct Hartig net on *Pinus sylvestris* and *Picea abies* (Münzenberger et al. 2009). *A. macrosclerotiorum* forms characteristic, melanized wartlike sclerotia giving the mantle a verrucous appearance (Figure 38.3E). DSE formed good mantles and Hartig nets on some of the best nursery stocks of pines and white spruce (Danielson and Visser 1990). It persisted on the stock for as long as the seedlings were in the nursery, but once outplanted, it was replaced by mycorrhizal fungi, especially *Thelephora terrestris* Ehrh. and E-strain fungi. Hartig net structures could not be observed in resynthesis experiments between *Picea abies* and various PAC strains, and microsclerotia production was strongly strain dependent (Tellenbach et al. 2010).

Colonization of fine roots (diameter <5 mm) of Norway spruce and Scots pine by DSE was examined in field samples from Finland and Switzerland (Sieber, unpublished). Density of surface mycelia and intracortical fungal structures are much lower in field samples than in roots from in vitro synthesis experiments, and thus, the endophytes are more difficult to locate. DSE were observed to colonize primary roots down to the innermost cortical cells. Colonization of the endodermis or the expanded pericycle did not occur. DSE most frequently occurred as intra- and intercellular hyphae and small intracellular microsclerotia composed of cells with thin, melanized walls. The rhytidome of older conifer roots is frequently colonized by dark, septate hyphae down to the youngest layer of phellem cells. Intensely colonized roots can be recognized already in the field by the black appearance of the rhytidome; sometimes, it is necessary to remove the loose bark scales to see the more proximal phellem tissues (Grünig et al. 2008b). The black appearance of the rhytidome is due to cells of the phlobaphene cork filled with multicellular sclerotia composed of bubble-shaped, thick-walled, melanized cells (Holdenrieder, personal communication, 1997; Figure 38.3F). These sclerotia might serve as inoculi and food bases from where mycelia can grow to colonize new substrates. Holdenrieder (1989) presented excellent SEM pictures of hyphal aggregations of DSE filling entire phellem cells of Norway-spruce roots. Endophytic colonization of cells proximal to the phellem has, so far, not conclusively been demonstrated but is certainly possible considering the frequent colonization of the xylem in forest trees by DSE (including PAC) (Görke 1998).

Intracellular colonization of Norway-spruce seedlings inoculated with DSE and *Cryptosporiopsis* cf. *abietina* Petr., possibly conspecific with *C. radiciola* Kowalski & Kehr, was observed in root cortex cells (Haug et al. 1988). Whereas the infection by DSE was confined to the cortex, *C.* cf. *abietina* colonized also the vascular tissue and caused decline of the seedlings. The older seedlings (5-month-old compared to 3-week-old) resisted for a longer period of time. Interestingly, addition of malt caused DSE to become pathogenic and to colonize the vascular tissues.

2. Endophyte–Pathogen Interactions, Biological Control

Some endophytes were demonstrated to confer biological control against other microorganisms. Duda and Sierota (1987) successfully used *Trichoderma viride* Pers., which was found highly pathogenic on Norway-spruce seedlings (Forbrig 1989) to control damping-off of *Pinus sylvestris* seedlings caused by *Fusarium oxysporum* or *Rhizoctonia solani*. DSE were shown to protect roots of *Picea abies*, *Pinus sylvestris*, and *Quercus robur* L. against other organisms, for example, *F. oxysporum* and *R. solani* (Manka 1960; Manka and Przezbòrski 1978, 1987). DSE had only an adverse effect onto the trees when these had been planted on an unsuitable site or after an *Armillaria* attack. Some isolates of *Phialocephala subalpina* increased survival of Norway-spruce seedlings inoculated with either one of the two root pathogens *Phytophthora plurivora* T. Jung & T.I. Burgess (syn. *P. citricola* Sawada) or *Elongisporangium undulatum* (H.E. Petersen) Uzuhasi, Tojo & Kakish. (syn. *Pythium undulatum* H.E. Petersen) (Tellenbach and Sieber 2012). *Phialocephala subalpina* seems to confer an indirect benefit to its host and might therefore be tolerated in natural spruce populations, despite negative effects on plant performance.

Biological control seems to be effective not only among fungal species but also among genotypes (strains) of the same species. Pathogenicity is often strain dependent (Tellenbach et al. 2011) (see text in the following paragraph). Nonpathogenic PAC strains significantly reduced density of colonization of Norway-spruce (*Picea abies*) roots by pathogenic PAC strains, attenuating adverse effects on plant growth (Reininger et al. 2011a, 2012). This highlights the importance of high genotypic diversity of PAC fungi colonizing the same root system.

3. Endophyte–Plant Interactions

DSE had very marked pathogenic properties toward *Pinus sylvestris* under pure culture conditions. However, DSE could achieve dominance on root surfaces of healthy elongating roots of pines growing under natural conditions, when its pathogenic effects appear to be reduced (Robertson 1954). De la Bastide and Kendrick (1990) treated white-pine seedlings (*Pinus strobus*) with benomyl to reduce the pathogenic effect of DSE. Adverse effects caused by DSE occurred also in experiments of Richard et al. (1971) with *Picea mariana*, Wilhelm et al. (1969) with *Pinus pinea* L. seedlings, Wilcox and Wang (1987) with *Pinus resinosa* Aiton, *Picea rubens* Sarg., and *Betula alleghaniensis* Britton, and Melin (1923) with *Pinus sylvestris* and *Picea abies*.

General health status as expressed by needle color, presence or absence of needle-tip chlorosis, mortality rate, and dry weight of Norway-spruce seedlings used to bait DSE from forest soils was not correlated with the colonization of the roots by DSE (mainly PAC) (Ahlich et al. 1998). Hennon et al. (1990) frequently isolated DSE, *Cryptosporiopsis* sp., *Gelatinosporium* sp., *Sporidesmium* sp., *Cylindrocarpon didymum* (Harting) Wollenw., and *Cadophora melinii* Nannf. from healthy and declining cedar roots during their studies on declining *Chamaecyparis nootkatensis* (D. Don) Spach in Southeast Alaska. DSE were most frequently isolated. Healthy and declining trees were equally frequently colonized. Except for *C. didymum*, none of the fungi proved to be pathogenic in infection experiments with cedar seedlings. None of the seedlings died. Similarly, DSE were isolated from diseased and healthy seedlings of Norway spruce equally frequently and were not pathogenic in infection experiments (Galaaen and Venn 1979). This contrasts with *Pythium sylvaticum* W.A. Campb. & F.F. Hendrix that was isolated almost only from diseased seedlings and was highly virulent when inoculated onto axenically grown seedlings. The endophytic mycobiota in roots of European silver fir (*A. alba*) seedlings was examined at a site with natural regeneration and another site where fir did not regenerate (Kowalski 1982a). *Ilyonectria radicicola* (teleomorph of *C. destructans*) was the most frequently isolated pathogenic fungus, especially in young, 1-year-old seedlings. In natural regenerations, DSE were very common and *I. radicicola* occurred only sporadically.

Likewise, *I. radicicola* was demonstrated to be pathogenic on fir and pine seedlings in vitro (Dahm et al. 1987). Virulence was affected by pH, temperature, light intensity, and also by associated bacteria and actinomycetes. In contrast, Bloomberg (1966) isolated DSE and *Cylindrocarpon* spp. more frequently from healthy Douglas-fir seedlings. The frequency of *Fusarium* spp. did, however, not depend on the health status of the plants. DSE, *Trichoderma viride*, and *I. radicicola* were pathogenic when inoculated onto 2 to 3-week-old axenically grown seedlings of *Picea abies* (Forbrig 1987, 1989). *T. viride* and *I. radicicola* killed the seedlings rapidly. DSE were suspected to infect only weakened trees. The interactions with Norway spruce of more than 30 isolates of four different PAC species and from three geographical regions were strongly isolate dependent and ranged from neutral to highly virulent, but no strain had a stimulating effect on plant growth (Tellenbach et al. 2011). Variation in virulence was much higher within than among PAC species, but only isolates of *Phialocephala subalpina*, one of the most frequent PAC species, were highly virulent. Disease caused by *Phialocephala subalpina* genotypes from the native range of Norway spruce was more severe than that induced by genotypes from outside the range. Virulence was not correlated with the phylogenetic relatedness of the isolates but was positively correlated with the extent of fungal colonization as measured by quantitative real-time PCR. The results obtained by Tellenbach et al. (2011) contrast with those reported by Peterson et al. (2008) who observed the production of lateral roots of *Pinus strobus* seedlings inoculated with *Phialocephala fortinii* s.l. being greater than that of uninoculated controls.

Chloridium paucisporum C.J.K. Wang & H.E. Wilcox from 3-year-old nursery seedlings of *Pinus resinosa* was found to stimulate *Pinus resinosa* seedlings growth (Wilcox and Ganmore-Neumann 1974). The plant–fungus association was found to be EEM but different from that of E-strain fungi. Jumpponen and Trappe (1998b) showed that the culture system under which an association is studied may also affect the host–fungus interaction. The effects of various glucose concentrations on *Pinus contorta* seedlings inoculated with *Phialocephala fortinii* s.l. were studied in an axenic and an open pot system. Inoculation resulted in substantial increase of biomass in the axenic system, and host biomass increased with increasing glucose concentration. Glucose alone did not significantly affect host biomass. In the open pot cultures, inoculation did not affect biomass. The observed growth stimulation in the closed culture system was probably due to CO_2 fertilization as a consequence of fungal respiration (Jumpponen and Trappe 1998b). Growth stimulation and inhibition were observed also in cell-free extracts of endophytes. Culture filtrates of *T. viride* reduced and filtrates of DSE stimulated elongation of excised roots of *Pinus sylvestris*, whereas *I. radicicola* reduced formation of lateral roots (Turner 1962). Binucleate Rhizoctonia endophytes promoted formation of adventitious roots of Scots pine (*Pinus sylvestris* L.) hypocotyl cuttings (Kaparakis and Sen 2006). Similarly, Rhizoctonia stimulated early seedling growth in a nitrogen-limited soil (Grönberg et al. 2006).

4. Endophyte–Environment Interactions

Knowledge about the influence of environmental factors on frequency and colonization of root endophytes is sparse. Although some factors were identified in the past, more efforts are needed to get a more complete picture of the key factors and how they interact with each other and with the endophyte–plant system. The proton [H^+] concentration (pH) in the soil and the type of biogeoclimatic zone were identified as factors affecting root endophyte communities. Statistically significant correlations existed between soil pH and the frequency of colonization by *Cryptosporiopsis radicicola* Kowalski & C. Bartnik, *Cylindrocarpon didymum*, and DSE (Ahlich and Sieber 1996). The correlations with *C. radicicola* and *C. didymum* were positive, and the correlation with DSE was negative. Some *Cylindrocarpon* species are well known to prefer alkaline conditions (Matturi and Stenton 1964; Domsch et al. 1980; Schönhar 1987). DSE were, however, considered not to be influenced by soil pH according to Melin (1924). Although DSE were absent in the soil samples of only 2 of 72 sites, DSE occurred less frequently in soils with a high pH value (Ahlich et al. 1998). Maximum isolation was from soils with pH values ranging from 3.5 to 4.5. Similarly, Manka (1960) gives pH 4 as the optimum for growth of DSE. Danielson and Visser (1989) considered a pH value of 3.1 the minimum for growth of DSE contrasting with the results of Ahlich et al. (1998) who detected DSE in almost 90% of the root segments of the Norway-spruce seedlings used to bait endophytes from a forest soil with pH 3.0. A relationship between altitude and pH value was reported by Holdenrieder and Sieber (1992) who found the most prominent difference in fungal associations in Norway-spruce roots to exist between roots colonized mainly by *I. radicicola* from alkaline soils at low altitude and roots dominated by DSE from acidic soil (peat bog), as well as from alkaline soil at high altitude.

In Switzerland, *Phialocephala fortinii* s.l. proofed to occur preferentially on high altitudes (Ahlich and Sieber 1996). *Trichoderma viride* occurred preferentially in roots from extremely acidic soils, *T. hamatum* (Bonord.) Bainier from moderately acidic, and *T. polysporum* (Link) Rifai from weakly acidic soils (Kattner and Schönhar 1990). *Cryptosporiopsis abietina* (perhaps conspecific with *C. radicicola*) and *I. radicicola* preferred a neutral environment. Liming reduced *T. viride* and a *Penicillium* sp., whereas the frequency of DSE, *Penicillium spinulosum* Thom, *T. polysporum*, and some *Mortierella* spp. increased (Kattner 1992b). Inoculations with root endophytes may enhance survival rates of seedlings planted at polluted sites. LoBuglio and Wilcox (1988) inoculated *Pinus resinosa* seedlings with *Cadophora finlandica* (C.J.K. Wang & H.E. Wilcox) T.C. Harr. & McNew and planted them onto iron tailings in an old iron mine. *C. finlandica* inoculated seedlings had a higher survival rate than uninoculated controls. In addition, mortality of seedlings treated with *C. finlandica* was lower than that of seedlings treated with ectomycorrhizal fungi.

Data about the influence of phosphorus (P) and nitrogen (N) availability on colonization by endophytes and plant performance are sparse. *Pinus contorta* seedlings inoculated with *Phialocephala fortinii* s.l. showed enhanced P uptake and increased growth (Jumpponen et al. 1998). In addition, N uptake was increased when *Phialocephala fortinii* s.l. inoculation was combined with addition of N. N fertilization of a 20-year-old stand of western hemlock (*Tsuga heterophylla* (Raf.) Sarg.) with urea did not change the total number of fine roots and the number of basidiomycetous mycorrhizae (Kernaghan et al. 1995). However, *Cenococcum geophilum* mycorrhizae decreased slightly, whereas those lacking mantels such as EEM and DSE increased.

Pinus sylvestris seedlings inoculated with DSE (*Phialocephala fortinii* s.l., *Cadophora finlandica*, *Chloridium paucisporum*, *Pezoloma ericae*, *Meliniomyces variabilis*, and *M. vraolstadiae* Hambl. & Sigler) accumulated on average 17% more biomass than control seedlings under elevated CO_2 combined with N limitation, indicating that DSE fungi increase plant nutrient use efficiency and are, thus, more beneficial to the plant under elevated CO_2 (Alberton et al. 2010).

5. Extracellular Enzymes and Fungicide Resistance

The production of biocidal metabolites, plant growth hormones, exoenzymes, and siderophores or the resistance against noxious substances can be advantageous when competing with other soil microorganisms for space, nutrients, and/or infection sites (Currah and Tsuneda 1993; Ahlich-Schlegel 1997; Caldwell et al. 2000; Bartholdy et al. 2001; Schulz et al. 2002; Grünig et al. 2008b). *C. finlandica* and several PAC isolates utilized cellulose, laminarin, starch, and xylan as sole carbon source and proteins and nucleic acids as sole nitrogen and phosphorus sources (Caldwell et al. 2000). However, lignolytic activity was not observed. *T. viride*, *Mortierella nana* Linnem., and DSE exhibited higher pectolytic than cellulolytic activity, and the activity was strain dependent (Dahm 1987). *Trichoderma* isolates were the most active, DSE the least. Presence of cellulolytic activity of DSE was demonstrated by Gams (1963), Levisohn (1954), and Melin (1923). The opposite observation by Schelling (1952) may be due to the fact that she used neither trace elements nor growth factors in her experiments. Ahlich-Schlegel (1997) tested more than 200 DSE isolates for fungicide resistance and the presence of extracellular enzymes and was able to demonstrate that enzyme activity is strain dependent. One-third of all strains produced proteases. Some *Acephala applanata* strains were able to produce amylases, but other PAC species were not. Laccase activity, an oxidase used to degrade lignin, was observed in all *A. applanata* strains but was present only in half of the other PAC strains. Most strains produced phenoloxidases, but activity was much higher in *A. applanata* isolates. Similar observations were made by Grünig et al. (2008b) who found *A. applanata* to be a good producer of amylases, laccases, and proteases in contrast to other PAC species. Benomyl at a concentration of ≥10 mg L^{-1} inhibited all strains, whereas thiabendazole did so only at a concentration of ≥100 mg L^{-1} (Ahlich-Schlegel 1997). At 10 mg L^{-1} thiabendazole, all *A. applanata* isolates were inhibited; the reaction of the other DSE was variable. Cycloheximide at ≥100 mg L^{-1} inhibited *A. applanata*, *Phialocephala letzii*, *Phialocephala helvetica*, and *Phialocephala subalpina* the most and *Phialocephala uotilensis* and *Phialocephala turicensis* the least (Ahlich-Schlegel 1997; Grünig et al. 2008a,b). Toxins were responsible for pathogenicity of *I. radicicola* but the high pectinase and cellulase production of this fungus were assumed to enhance virulence (Lyr and Kluge 1968).

Bartholdy et al. (2001) studied the siderophore profiles of five PAC strains. Siderophores controlling the uptake of iron ions by fungi play a crucial role in the infection process in some plant–fungus systems (Johnson 2008). The siderophore profiles of two *Phialocephala fortinii* s.s. strains were highly similar. They produced almost exclusively ferricrocin. In contrast, a strain each of *Phialocephala subalpina* and *Phialocephala europaea* possessed unique siderophore profiles and produced ferrirubin in addition to ferricrocin, but the amount of ferricrocin produced was significantly lower in *Phialocephala subalpina* and significantly higher in *Phialocephala europaea* than in *Phialocephala fortinii* s.s. A strain of CSP9 isolated from *Larix decidua* Mill. in Germany produced the plant growth hormone IAA (Schulz et al. 2002). The vacuolar system in hyphae of PAC strain UAMH_9608 (*Phialocephala fortinii* s.s.) was investigated, and the presence of polyphosphate in the vacuoles was confirmed (Saito et al. 2006). An endophytic strain of *Phialocephala scopiformis* T. Kowalski & Kehr, a species with high affinities to the PAC, was recently reported to produce rugulosin—a toxic metabolite against some herbivores of *Picea glauca* needles (Miller et al. 2008; Sumarah et al. 2008). Whether PAC members also produce secondary metabolites toxic to some herbivores feeding on roots remains to be tested.

6. Stress Tolerance and Oligotrophic Growth of DSE

DSE are resistant against repeated thawing–freezing and against permanent drought and are able to grow under oligotrophic conditions. MEA discs and cellophane pieces colonized by PAC isolates were frozen at −20°C and thawn and refrozen daily (Ahlich-Schlegel 1997). Most isolates survived for at least 10 days. *Phialocephala fortinii* s.l. survived on average longer than *A. applanata*. One *Phialocephala fortinii* s.l. isolate survived for more than 50 days. Drought resistance was tested by putting cellophane sheets colonized by PAC isolates in a desiccator. All isolates survived for at least 8 months. PAC isolates were grown on nitrogen-free water agar and subcultured four times every 21 days. Mycelium was very sparse in all isolates, but growth was not reduced even after the fourth subculture. Colony diameters ranged from 29 to 43 mm for *Acephala applanata* and 43–63 for *Phialocephala fortinii* s.l. Growth under oligotrophic conditions was studied in plastic Petri dishes that may have released some nitrogen into the medium. Thus, the experiment should be repeated using glass Petri dishes.

C. Root Endophytes in Other Woody Plant Species

Members of the Betulaceae, Fagaceae, and Salicaceae are the best studied for the presence of root endophytes (Table 38.6). Species of *Cryptosporiopsis*, *Cylindrocarpon*, *Meliniomyces*, and

TABLE 38.6 Endophytes in Other Woody Plants

Endophyte				Host					
Genus[a]	Species[a]	Fungus Order[a]	DSE/Not DSE	Species	Plant Order	Woody/Herbaceous	Type of Experiment[b]	Special Effect[c]	References
Acephala	*applanata*	Helotiales	DSE	*Betula pubescens*	Fagales	Tree	Isol		Queloz (2010)
Acephala	sp. 2	Helotiales	DSE	*Sorbus aucuparia*	Rosales	Tree	Isol		Grünig et al. (2009)
Alternaria	*alternata*	Pleosporales	DSE	*Atriplex vesicaria*	Caryophyllales	Woody shrub	Isol		Cother and Gilbert (1994)
Alternaria	*chlamydospora*	Pleosporales	DSE	*Atriplex vesicaria*	Caryophyllales	Woody shrub	Isol		Cother and Gilbert (1994)
Anguillospora	*filiformis*	Pleosporales	Not DSE	*Acer spicatum*	Sapindales	Tree	Isol		Sridhar and Bärlocher (1992)
Anguillospora	*filiformis*	Pleosporales	Not DSE	*Betula papyrifera*	Fagales	Tree	Isol		Sridhar and Bärlocher (1992)
Anguillospora	*longissima*	Pleosporales	Not DSE	*Salix babylonica*	Malpighiales	Tree	Isol		Iqbal et al. (1995)
Articulospora	*proliferata*	Helotiales	Not DSE	*Mangifera indica*	Sapindales	Tree	Isol		Iqbal et al. (1995)
Articulospora	*proliferata*	Helotiales	Not DSE	*Salix babylonica*	Malpighiales	Tree	Isol		Iqbal et al. (1995)
Ascochyta	sp.	Pleosporales	DSE	*Atriplex vesicaria*	Caryophyllales	Woody shrub	Isol		Cother and Gilbert (1994)
Cadophora	*fastigiata*	Helotiales	DSE	*Quercus robur*	Fagales	Tree	Isol		Halmschlager and Kowalski (2004)
Cadophora	sp.	Helotiales	DSE	*Quercus petraea*	Fagales	Tree	Isol		Halmschlager and Kowalski (2004)
Cadophora	sp. 3	Helotiales	DSE	*Myricaria prostrata*	Caryophyllales	Woody shrub	Isol		Graf, pers. comm. (2008)
Chaetomium	*cochliodes*	Sordariales	Not DSE	*Aphelandra tetragona*	Lamiales	Woody shrub	Isol		Werner et al. (1997)
Cladosporium	*tenuissimum*	Capnodiales	DSE	*Alnus glutinosa*	Fagales	Tree	Isol		Fisher et al. (1991b)
Clavariopsis	*aquatica*	Pleosporales	Not DSE	*Salix babylonica*	Malpighiales	Tree	Isol		Iqbal et al. (1995)
Coniothyrium	sp.	Pleosporales	DSE	*Atriplex vesicaria*	Caryophyllales	Woody shrub	Isol		Cother and Gilbert (1994)
Cryptosporiopsis	*radicicola*	Helotiales	DSE	*Fagus sylvatica*	Fagales	Tree	Isol		Ahlich and Sieber (1996)
Cryptosporiopsis	*radicicola*	Helotiales	DSE	*Quercus petraea*	Fagales	Tree	Isol		Halmschlager and Kowalski (2004)
Cryptosporiopsis	*melanigena*	Helotiales	DSE	*Quercus petraea*	Fagales	Tree	Isol		Kowalski et al. (1998)
Cryptosporiopsis	*radicicola*	Helotiales	DSE	*Quercus robur*	Fagales	Tree	Isol		Kowalski and Bartnik (1995)
Cylindrocarpon	*didymum*	Hypocreales	Not DSE	*Fagus sylvatica*	Fagales	Tree	Isol		Ahlich and Sieber (1996)
Cylindrocarpon	*magnusianum*	Hypocreales	Not DSE	*Fagus sylvatica*	Fagales	Tree	Isol		Görke (1998)
Cystodendron	sp.	Helotiales	Variable	*Quercus petraea*	Fagales	Tree	Isol		Halmschlager and Kowalski (2004)
Embellisia	*chlamydospora*	Pleosporales	DSE	*Myricaria prostrata*	Caryophyllales	Woody shrub	Isol		Graf, pers. comm. (2008)
Flagellospora	*curvula*	Hypocreales	Not DSE	*Mangifera indica*	Sapindales	Tree	Isol		Iqbal et al. (1995)
Flagellospora	*curvula*	Hypocreales	Not DSE	*Populus hybrida*	Malpighiales	Tree	Isol		Iqbal et al. (1995)
Flagellospora	*curvula*	Hypocreales	Not DSE	*Salix babylonica*	Malpighiales	Tree	Isol		Iqbal et al. (1995)
Flagellospora	*fusaroides*	Hypocreales	Not DSE	*Mangifera indica*	Sapindales	Tree	Isol		Iqbal et al. (1995)
Flagellospora	*fusaroides*	Hypocreales	Not DSE	*Salix babylonica*	Malpighiales	Tree	Isol		Iqbal et al. (1995)
Gibberella	*intricans*	Hypocreales	Not DSE	*Atriplex vesicaria*	Caryophyllales	Woody shrub	Isol		Werner et al. (1997)
Fusarium	*oxysporum*	Hypocreales	Not DSE	*Atriplex vesicaria*	Caryophyllales	Woody shrub	Isol		Cother and Gilbert (1994)
Fusarium	sp.	Hypocreales	Not DSE	*Mangifera indica*	Sapindales	Tree	Isol		Iqbal et al. (1995)
Fusarium	sp.	Hypocreales	Not DSE	*Populus hybrida*	Malpighiales	Tree	Isol		Iqbal et al. (1995)
Fusarium	sp.	Hypocreales	Not DSE	*Salix babylonica*	Malpighiales	Tree	Isol		Iqbal et al. (1995)
Gibberella	*baccata*	Hypocreales	Not DSE	*Atriplex vesicaria*	Caryophyllales	Woody shrub	Isol		Cother and Gilbert (1994)
Gibberella	*nygamai*	Hypocreales	Not DSE	*Atriplex vesicaria*	Caryophyllales	Woody shrub	Isol		Cother and Gilbert (1994)
Haematonectria	*haematococca*	Hypocreales	Not DSE	*Aphelandra tetragona*	Lamiales	Woody shrub	Isol		Werner et al. (1997)

Ilyonectria	*radicicola*	Hypocreales	Not DSE	*Alnus glutinosa*	Fagales	Tree	Isol	Fisher et al. (1991b)
Ilyonectria	*radicicola*	Hypocreales	Not DSE	*Aphelandra tetragona*	Lamiales	Woody shrub	Isol	Werner et al. (1997)
Ilyonectria	*radicicola*	Hypocreales	Not DSE	*Betula pendula*	Fagales	Tree	Isol	Görke (1998)
Ilyonectria	*radicicola*	Hypocreales	Not DSE	*Fagus sylvatica*	Fagales	Tree	Isol	Görke (1998)
Ilyonectria	*radicicola*	Hypocreales	Not DSE	*Gynoxys oleifolia*	Asterales	Tree	Isol	Fisher et al. (1995b)
Ilyonectria	*radicicola*	Hypocreales	Variable	*Quercus robur*	Fagales	Tree	Isol	Halmschlager and Kowalski (2004)
Ilyonectria	*radicicola*	Hypocreales	Variable	*Tilia petiolaris*	Malvales	Tree	Isol	Schroers et al. (2008)
Nectria	*lugdunensis*	Hypocreales	Not DSE	*Acer spicatum*	Sapindales	Tree	Isol	Sridhar and Bärlocher (1992)
Nectria	*lugdunensis*	Hypocreales	DSE	*Alnus glutinosa*	Fagales	Tree	Isol	Fisher et al. (1991b)
Leptodontidium	sp.	Helotiales	DSE	*Eucalyptus regnans*	Myrtales	Tree	Isol	Tedersoo et al. (2009)
Leptodontidium	sp.	Helotiales	DSE	*Nothofagus cunninghamii*	Fagales	Tree	Isol	Tedersoo et al. (2009)
Leptodontidium	sp.	Helotiales	DSE	*Pomaderris apetala*	Rosales	Tree	Isol	Tedersoo et al. (2009)
Libertella	spp.	Xylariales	Not DSE	*Atriplex vesicaria*	Caryophyllales	Woody shrub	Isol	Cother and Gilbert (1994)
Lunulospora	*curvula*	Pezizomycotina	Not DSE	*Populus hybrida*	Malpighiales	Tree	Isol	Iqbal et al. (1995)
Lunulospora	*curvula*	Pezizomycotina	Not DSE	*Salix babylonica*	Malpighiales	Tree	Isol	Iqbal et al. (1995)
Meliniomyces	*bicolor*	Leotiomycetes	DSE	*Nothofagus procera*	Fagales	Tree	Isol	Hambleton and Sigler (2005)
Meliniomyces	*bicolor*	Leotiomycetes	DSE	*Quercus robur*	Fagales	Tree	Isol	Hambleton and Sigler (2005)
Meliniomyces	sp. 1	Leotiomycetes	Variable	*Betula pubescens*	Fagales	Tree	Isol	Hambleton and Sigler (2005)
Meliniomyces	*vraolstadiae*	Leotiomycetes	Variable	*Betula pubescens*	Fagales	Tree	Isol	Hambleton and Sigler (2005)
Meliniomyces	*vraolstadiae*	Leotiomycetes	Variable	*Betula pubescens*	Fagales	Tree	Isol	Hambleton and Sigler (2005)
Monodictys	*arctica*	Dothideomycetes	DSE	*Saxifraga oppositifolia*	Saxifragales	Woody shrub	Isol	Day et al. (2006)
Oidiodendron	sp.	Dothideomycetes	DSE	*Eucalyptus regnans*	Myrtales	Tree	Isol	Tedersoo et al. (2009)
Oidiodendron	sp.	Dothideomycetes	DSE	*Nothofagus cunninghamii*	Fagales	Tree	Isol	Tedersoo et al. (2009)
Oidiodendron	sp.	Dothideomycetes	DSE	*Pomaderris apetala*	Rosales	Tree	Isol	Tedersoo et al. (2009)
Penicillium	*pinetorum*	Eurotiales	Not DSE	*Aphelandra tetragona*	Lamiales	Woody shrub	Isol	Werner et al. (1997)
Penicillium	*restrictum*	Eurotiales	Not DSE	*Betula pendula*	Fagales	Tree	Isol	Görke (1998)
Penicillium	*restrictum*	Eurotiales	Not DSE	*Fagus sylvatica*	Fagales	Tree	Isol	Görke (1998)
Phialocephala	*europaea*	Helotiales	DSE	*Fagus sylvatica*	Fagales	Tree	Isol	Queloz (2010)
Phialocephala	*europaea*	Helotiales	DSE	*Salix* sp.	Malpighiales	Tree	Isol	Queloz (2010)
Phialocephala	*europaea*	Helotiales	DSE	*Sorbus aucuparia*	Rosales	Woody shrub	Isol	Queloz (2010)
Phialocephala	*fortinii s.l.*	Helotiales	DSE	*Betula pendula*	Fagales	Tree	Isol	Görke (1998)
Phialocephala	*fortinii s.l.*	Helotiales	DSE	*Fagus sylvatica*	Fagales	Tree	Isol	Ahlich and Sieber (1996)
Phialocephala	*fortinii s.s.*	Helotiales	DSE	*Betula pendula*	Fagales	Tree	Isol	Queloz (2010)
Phialocephala	*fortinii s.s.*	Helotiales	DSE	*Betula pubescens*	Fagales	Tree	Isol	Queloz (2010)
Phialocephala	*fortinii s.s.*	Helotiales	DSE	*Sorbus aucuparia*	Rosales	Woody shrub	Isol	Queloz (2010)
Phialocephala	*helvetica*	Helotiales	DSE	*Fagus sylvatica*	Fagales	Tree	Isol	Queloz (2010)
Phialocephala	*letzii*	Helotiales	DSE	*Fagus sylvatica*	Fagales	Tree	Isol	Queloz (2010)
Phialocephala	*letzii*	Helotiales	DSE	*Helianthemum* sp.	Malvales	Woody shrub	Isol	Queloz (2010)
Phialocephala	*letzii*	Helotiales	DSE	*Sorbus aucuparia*	Rosales	Woody shrub	Isol	Queloz (2010)

(*continued*)

TABLE 38.6 (continued) Endophytes in Other Woody Plants

Endophyte				Host					
Genus[a]	Species[a]	Fungus Order[a]	DSE/Not DSE	Species	Plant Order	Woody/ Herbaceous	Type of Experiment[b]	Special Effect[c]	References
Phialocephala	sp. 9	Helotiales	DSE	*Myricaria prostrata*	Caryophyllales	Woody shrub	Isol		Graf, pers. comm. (2008)
Phialocephala	*sphaeroides*	Helotiales	DSE	*Aralia nudicaulis*	Apiales	Woody shrub	Isol		Wilson et al. (2004)
Phialocephala	*subalpina*	Helotiales	DSE	*Betula pendula*	Fagales	Tree	Isol		Queloz (2010)
Phialocephala	*subalpina*	Helotiales	DSE	*Betula tortuosa*	Fagales	Tree	Isol		Queloz (2010)
Phialocephala	*subalpina*	Helotiales	DSE	*Sorbus aucuparia*	Rosales	Woody shrub	Isol		Queloz (2010)
Phialocephala	*turicensis*	Helotiales	DSE	*Fagus sylvatica*	Fagales	Tree	Isol		Queloz (2010)
Phoma	*prunicola*	Pleosporales	DSE	*Atriplex vesicaria*	Caryophyllales	Woody shrub	Isol		Cother and Gilbert (1994)
Phoma	*variospora*	Pleosporales	DSE	*Atriplex vesicaria*	Caryophyllales	Woody shrub	Isol		Cother and Gilbert (1994)
Phomopsis	*alnea*	Diaporthales	Not DSE	*Alnus glutinosa*	Fagales	Tree	Isol		Fisher et al. (1991b)
Phomopsis	spp.	Diaporthales	DSE	*Atriplex vesicaria*	Caryophyllales	Woody shrub	Isol		Cother and Gilbert (1994)
Pleospora	*herbarum*	Pleosporales	DSE	*Atriplex vesicaria*	Caryophyllales	Woody shrub	Isol		Cother and Gilbert (1994)
Pleospora	*obrusa*	Pleosporales	DSE	*Atriplex vesicaria*	Caryophyllales	Woody shrub	Isol		Cother and Gilbert (1994)
Pleospora	*phaeocomoides*	Pleosporales	DSE	*Atriplex vesicaria*	Caryophyllales	Woody shrub	Isol		Cother and Gilbert (1994)
Pyrenochaeta	sp. 1	Pleosporales	DSE	*Caragana versicolor*	Fabales	Woody shrub	Isol		Graf, pers. comm. (2008)
Serendipita	*vermifera*	Auriculariales	Not DSE	*Tilia* sp.	Malvales	Tree	In vitro exp	a	Oberwinkler (1964)
Sesquicillium	*candelabrum*	Hypocreales	Not DSE	*Betula pendula*	Fagales	Tree	Isol		Görke (1998)
Sesquicillium	*candelabrum*	Hypocreales	Not DSE	*Fagus sylvatica*	Fagales	Tree	Isol		Görke (1998)
Sporormiella	*intermedia*	Pleosporales	DSE	*Atriplex vesicaria*	Caryophyllales	Woody shrub	Isol		Cother and Gilbert (1994)
Tetracladium	*marchalianum*	Helotiales	Not DSE	*Mangifera indica*	Sapindales	Tree	Isol		Iqbal et al. (1995)
Tetracladium	marchalianum	Helotiales	Not DSE	*Populus hybrida*	Malpighiales	Tree	Isol		Iqbal et al. (1995)
Tetracladium	*marchalianum*	Helotiales	Not DSE	*Salix babylonica*	Malpighiales	Tree	Isol		Iqbal et al. (1995)
Trichoderma	*viride*	Hypocreales	Not DSE	*Aphelandra tetragona*	Lamiales	Woody shrub	Isol		Werner et al. (1997)
Triscelophorus	*konajensis*	Pezizomycotina	Not DSE	*Coffea arabica*	Gentianales	Woody shrub	Isol		Raviraja et al. (1996)
Triscelophorus	*monosporus*	Pezizomycotina	Not DSE	*Coffea arabica*	Gentianales	woody shrub	Isol		Raviraja et al. (1996)
Unidentified DSE			DSE	*Atriplex canescens*	Caryophyllales	Woody shrub	In vitro exp		Barrow et al. (2004)
Unidentified DSE			DSE	*Daphniphyllum neilgherrense*	Saxifragales	Tree	Isol		Bagyalakshmi et al. (2010)
Unidentified DSE			DSE	*Elaeocarpus munronii*	Oxalidales	Tree	Isol		Bagyalakshmi et al. (2010)
Unidentified DSE			DSE	*Euodia roxburghiana*	Sapindales	Tree	Isol		Bagyalakshmi et al. (2010)
Unidentified DSE			DSE	*Syzygium arnottianum*	Myrtales	Tree	Isol		Bagyalakshmi et al. (2010)
Unidentified DSE			DSE	*Syzygium montanum*	Myrtales	Tree	Isol		Bagyalakshmi et al. (2010)
Unidentified DSE			DSE	*Tamarix ramosissima*	Caryophyllales	Woody shrub	Isol		Beauchamp et al. (2005)

[a] Genus, species and order (family) names according to index fungorum (http://www.indexfungorum.org/Names/Names.asp, December 20, 2011). If both the teleomorph (sexual reproductive stage) and the anamorph (asexual reproductive stage) are produced, the name of the teleomorph is given. Names of anamorph(s) can be retrieved from the list of the teleomorph's synonyms provided in the index. The next lower taxon is given if the Fungus order is not known with certainty ("incertae sedis").

[b] Isol, isolation from surface-sterilized roots; in vitro exp, fungus used in in-vitro experiment(s).

[c] Special effect details: (a) Plant growth stimulation.

Phialocephala were most frequently isolated. *Cryptosporiopsis radicicola*, a species often isolated from roots of *Quercus robur* (Kowalski and Bartnik 1995; Halmschlager and Kowalski 2004), was the dominant endophyte in roots of European beech (*Fagus sylvatica*) in Switzerland (Ahlich and Sieber 1996), and *Cryptosporiopsis melanigena* T. Kowalski & Halmschl. was often isolated from roots of *Q. robur* and *Q. petraea* (Kowalski et al. 1998). According to Kowalski (1983), the wood and bark endophyte *Pezicula cinnamomea* (DC.) Sacc., with its anamorph *Cryptosporiopsis grisea* (Pers.) Petr., can also spread into the roots of dying trees. The main colonizer of *Dryas octopetala* L. roots sampled at Spitsbergen was a *Cryptosporiopsis* species (Fisher et al. 1995a). DSE (including *Phialocephala fortinii* s.l.) were demonstrated to be the most frequent endophytes in apparently healthy xylem of roots of *Betula pendula* and *F. sylvatica* in Germany (Görke 1998; Table 38.6). Similarly, DSE were observed in 42% and 40% of the root samples from *Iva frutescens* L. and *Triadica sebifera* (L.) Small, respectively, collected in the southeastern marshes of Louisiana (Kandalepas et al. 2010). DSE also dominated the endophyte community in woody shrubs collected in Kailash-Manasarovar Region on the Tibetan plateau at 4700 m (F. Graf, personal communication, July 2008; Table 38.6). Desertification by wind and water erosion is a serious threat in this region, and the mycelium of mycorrhizal and other endophytic fungi could have a stabilizing effect (Burri et al. 2009).

However, DSE do not always dominate the non-mycorrhizal endophyte community in roots of woody plants. DSE were absent or rare in roots of *Gynoxis oleifolia* Muchler (Compositae) collected in Ecuador at 3300 m asl. *I. radicicola, Epicoccum nigrum* Link, *Trichoderma harzianum* Rifai, and *Truncatella angustata* (Pers.) S. Hughes were the endophytic fungi most frequently observed (Fisher et al. 1995b).

Very special endophyte communities are found on roots constantly submersed in water (Sengupta et al. 1988; Fisher et al. 1991b; Sridhar and Bärlocher 1992; Iqbal et al. 1995; Raviraja et al. 1996). Submersed roots are often colonized by aquatic hyphomycetes in addition to "terrestrial" fungi (Table 38.6). The effects of aquatic hyphomycetes on their host are probably small. Submersed roots may, however, constitute refuge and permanent sources of inoculum for the aquatic hyphomycetes to persist in streams (Raviraja et al. 1996). Two septate, dematiaceous endophytes, a coelomycete and a *Rhizoctonia*-like fungus, were frequently isolated from roots of some mangroves and saline-resistant plants growing in the seabound delta region of West Bengal (Sengupta et al. 1988). They grew inter- and intracellularly either alone or simultaneously with AM and formed sclerotia in culture. Absence of AM in some halophytes indicates that endophytes may replace AM in extreme habitats, similar to what has been observed for arctic and alpine herbaceous plants.

Some woody plant species were examined also on the Southern Hemisphere. Roots of *Nothofagus cunninghamii* (Hook.) Oerst., *Eucalyptus regnans* F. Muell., and *Pomaderris apetala* Labill. from Tasmania were colonized by species of *Leptodontidium* and *Oidiodendron* (Tedersoo et al. 2009). These helotialean fungi are phylogenetically closely related to root endophytes and ericoid mycorrhizal fungi of the Northern Hemisphere, suggesting strong ecological and evolutionary links. Seventy-one fungal endophytes could be isolated from bladder saltbush (*Atriplex vesicaria* Heward ex Benth.) in Southern Australia (Cother and Gilbert 1994). *Fusarium* spp., *Alternaria chlamydospora* Mouch., *Libertella* spp., *Phoma variospora* Shreem., and *Sporormiella intermedia* (Auersw.) S.I. Ahmed & Cain ex Kobayasi were isolated as root endophytes for the first time in Australia. Species of known pathogenic genera (*Ascochyta, Coniothyrium, Phomopsis, Pleospora*) occurred also quite frequently.

1. Anatomy

Phialocephala fortinii CSP 13 (CSP 13 according to Grünig et al. (2008b)) formed a Hartig net and a thin, patchy mantle in axenic culture with *Salix glauca* L. seedlings (Fernando and Currah 1996; Figure 38.3G). An interesting fungus–root association was described by Sequerra et al. (1995). *Penicillium nodositatum* Valla was observed to penetrate and colonize cortical cells of roots of *Alnus incana* (L.) Moench and to induce myconodules similar to those formed by actinorhizal bacteria (*Frankia* spp.). Host defense was minimal, and the host plasma membrane, invaginated around the endophyte, kept its integrity as it does in symbiotic associations. *Penicillium nodositatum* was, thus, considered as a neutral microsymbiont similar to a compatible but ineffective Frankia strain.

2. Endophyte–Pathogen Interactions

Trichoderma spp. and *Gliocladium* spp. were shown to control *Phytophthora cactorum* root and crown rot of apple (Smith et al. 1990). Some isolates significantly reduced root rot and increased plant weight. DSE were shown to protect roots of *Picea abies, Pinus sylvestris*, and *Q. robur* against other organisms, for example, *F. oxysporum* and *R. solani* (Manka 1960; Manka and Przezbòrski 1978, 1987).

3. Endophyte–Plant Interactions

Leptodontidium orchidicola caused a marked increase in host root length but also invaded the stele, causing extensive cellular lysis in axenic culture with *Salix glauca* seedlings (Fernando and Currah 1996). In pot monocultures with various subalpine plant species, the effects of *L. orchidicola* strains on host dry weight were strain and host specific (Fernando and Currah 1996). The effects of *Phialocephala fortinii* CSP 13 were also host specific. In pot combination cultures (*Dryas octopetala, Picea abies, Potentilla fruticosa* L., and *S. glauca* in the same pot), the *Phialocephala fortinii* s.l.–*Potentilla fruticosa* symbiosis resulted in a significant increase in shoot weight in contrast to the results of the same symbiosis in monoculture resynthesis. DSE isolated from mangroves stimulated growth of *Cajanus cajan* (L.) Millsp. in the absence of easily available phosphorus (Sengupta et al. 1988).

4. Endophyte–Environment Interactions

Caldwell et al. (2000) studied the utilization of major forms of carbon, nitrogen, and phosphorus commonly present in plant litter and detritus for two dark septate, root endophyte isolates from alpine *Salix* species. Both isolates utilized cellulose, laminarin, starch, and xylan as sole carbon source and proteins and ribonucleic acids as sole nitrogen and phosphorus sources. However, lignolytic activity could not be detected.

VI. Conclusions and Outlook

The presence of non-mycorrhizal fungal endophytes in plant roots is a worldwide phenomenon. There is probably no plant species without root endophytes. Species diversity of endophyte communities in roots is low compared to the one in aerial tissues. However, the within-species genotypic diversity of root endophytes can be high. Fungi with melanized, septate hyphae are prevalent in many of the communities. These fungi usually do not sporulate readily in culture and are, therefore, assigned to the form taxon DSE. DSE are a very diverse group of fungi. They can be separated into two subgroups. The first includes DSE, which occur mainly on herbaceous plants such as species of the *Harpophora–Gaeumannomyces* complex, *Microdochium* or *Periconia*. The second subgroup is formed by PAC, which occur preferentially in woody plant roots. PAC are rare or absent in plant roots from arable soils or pastures. Antagonism with species of the *Harpophora–Gaeumannomyces* complex and/or other microorganisms could be the reason. *Harpophora* species were shown to confer induced resistance (cross-protection) against closely related species or varieties in agricultural plants, especially cereals. Similarly, PAC were shown to protect forest trees in a similar way, for example, against oomycetous root pathogens. The system of induced resistance is expected to be even more balanced in undisturbed forest ecosystems than in agricultural systems because the microbial community in the rhizosphere and in the roots was allowed to equilibrate for decades if not centuries. The study of these systems is hampered by the high genotypic variability of the organisms involved.

With the availability of many different molecular genetics markers, it is now possible to identify and characterize species and genets of PAC allowing experimentation with clearly defined strains. Several strains are available from international culture collections. The high variability of DSE and PAC also explains, at least in part, the contradictory results obtained in endophyte–plant interaction experiments. Whereas there are several examples of non-PAC DSE with beneficial effects on plant growth, the majority of PAC revealed no effect or only slightly positive effects. More recent in vitro multi-strain experiments with defined PAC strains indicate that PAC function along a continuum from neutral to pathogenic. Plant growth promotion has not been observed. However, though PAC are costly, they seem to provide control of more serious pathogens and perhaps also herbivores. Future experiments should be shifted more toward natural conditions.

The genome of strain UAMH11012 of *Phialocephala subalpina* was sequenced, and gene models are currently manually validated and annotated. First estimates based on 10 Mbp manually validated gene models show that the *Phialocephala subalpina* genome will include roughly 20,000 gene models. Special emphasis is given to annotation of gene clusters related to secondary metabolites. The availability of the annotated genome of *Phialocephala subalpina* will allow performing transcriptomic analysis of host–PAC interactions and comparative genomic studies with ectomycorrhizal and pathogenic fungi in the near future. In addition, the completely sequenced and annotated mitochondrial genome of *Phialocephala subalpina* is already available (Duò et al. 2012). Several mitochondrial loci proved to be suitable for species diagnosis in PAC and can be used for community studies.

We do not know the functions root endophytes had in the past nor which functions they will have in the future. We only know that nature is a dynamic system. What we see today is one picture in the evening-filling film entitled "Evolution," but it is fascinating to speculate about the next step.

References

Abadie J-C, Puttsepp U, Gebauer G, Faccio A, Bonfante P, Selosse M-A. 2006. *Cephalanthera longifolia* (Neottieae, Orchidaceae) is mixotrophic: A comparative study between green and nonphotosynthetic individuals. *Can J Bot* 84:1462–1477.

Achatz B, von Rueden S, Andrade D et al. 2010. Root colonization by *Piriformospora indica* enhances grain yield in barley under diverse nutrient regimes by accelerating plant development. *Plant Soil* 333:59–70.

Addy HD, Piercey MM, Currah RS. 2005. Microfungal endophytes in roots. *Can J Bot* 83:1–13.

Ahlich K, Rigling D, Holdenrieder O, Sieber TN. 1998. Dark septate hyphomycetes in Swiss conifer forest soils surveyed using Norway-spruce seedlings as bait. *Soil Biol Biochem* 30:1069–1075.

Ahlich K, Sieber TN. 1996. The profusion of dark septate endophytic fungi in non-ectomycorrhizal fine roots of forest trees and shrubs. *New Phytol* 132:259–270.

Ahlich-Schlegel K. 1997. Vorkommen und Charakterisierung von dunklen, septierten Hyphomyceten (DSH) in Gehölzwurzeln. PhD thesis, Zürich, Switzerland: ETH Zürich.

Alberton O, Kuyper TW, Summerbell RC. 2010. Dark septate root endophytic fungi increase growth of Scots pine seedlings under elevated CO_2 through enhanced nitrogen use efficiency. *Plant Soil* 328:459–470.

Allen MF, Allen EB. 1992. Mycorrhizae and plant community development: Mechanisms and patterns. In *The Fungal Community*, eds. GC Carroll, DT Wicklow, pp. 455–479. New York: Marcel Dekker, Inc.

Allen EA, Hoch HC, Steadman JR, Stavely RJ. 1991. Influence of leaf surface features on spore deposition and the epiphytic growth of phytopathogenic fungi. In *Microbial Ecology of Leaves*, eds. JH Andrews, SS Hirano, pp. 87–110. New York: Springer.

Armstrong GM, Armstrong JK. 1981. Formae speciales and races of *Fusarium oxysporum* causing wilt diseases. In *Fusarium: Diseases, Biology, and Taxonomy*, eds. PE Nelson, TA Tousson, RJ Cook, pp. 391–399. University Park, PA: The Pennsylvania State University Press.

von Arx JA. 1967. Pilzkunde—Ein kurzer Abriss der Mykologie unter besonderer Berücksichtigung der Pilze in Reinkultur. Vaduz, Fürstentum Liechtenstein: J. Cramer.

Athman SY, Dubois T, Viljoen A et al. 2006. In vitro antagonism of endophytic *Fusarium oxysporum* isolates against the burrowing nematode *Radopholus similis. Nematology* 8:627–636.

Aveskamp MM, Verkley GJM, de Gruyter J et al. 2009. DNA phylogeny reveals polyphyly of *Phoma* section Peyronellaea and multiple taxonomic novelties. *Mycologia* 101:363–382.

Azevedo MD, Welty RE. 1995. A study of the fungal endophyte *Acremonium coenophialum* in the roots of tall fescue seedlings. *Mycologia* 87:289–297.

Baas-Becking LGM. 1934. *Geobiologie of inleiding tot de milieukunde*. Den Haag, the Netherlands: W.P. van Stockum and Zoon.

Bachmann S. 2010. Zur Ausbreitungsbiologie von endophytischen Wurzelpilzen des *Phialocephala fortinii* s.l.—*Acephala applanata* Artenkomplexes (PAC). BSc Thesis, Zürich, Switzerland: ETH Zürich.

Bagyalakshmi G, Muthukumar T, Sathiyadash K, Muniappan V. 2010. Mycorrhizal and dark septate fungal associations in shola species of Western Ghats, southern India. *Mycoscience* 51:44–52.

Barrow JR. 2003. Atypical morphology of dark septate fungal root endophytes of *Bouteloua* in arid southwestern USA rangelands. *Mycorrhiza* 13:239–247.

Barrow JR, Osuna-Avila P, Reyes-Vera I. 2004. Fungal endophytes intrinsically associated with micropropagated plants regenerated from native *Bouteloua eriopoda* torr. and *Atriplex canescens* (Pursh) Nutt. *In Vitro Cell Dev Biol Plant* 40:608–612.

Bartholdy BA, Berreck M, Haselwandter K. 2001. Hydroxamate siderophore synthesis by *Phialocephala fortinii*, a typical dark septate fungal root endophyte. *BioMetals* 14:33–42.

Bayman P, Lebron LL, Tremblay RL, Lodge DJ. 1997. Variation in endophytic fungi from roots and leaves of *Lepanthes* (Orchidaceae). *New Phytol* 135:143–149.

Bazin MJ, Markham P, Scott EM, Lynch JM. 1990. Population dynamics and rhizosphere interactions. In *The Rhizosphere*, ed. JM Lynch, pp. 99–127. Chichester, UK: John Wiley & Sons.

Beauchamp VB, Stromberg JC, Stutz JC. 2005. Interactions between *Tamarix ramosissima* (saltcedar), *Populus fremontii* (cottonwood), and mycorrhizal fungi: Effects on seedling growth and plant species coexistence. *Plant Soil* 275:221–231.

Berch SM, Roth AL. 1993. Ectomycorrhizae and growth of Douglas fir seedlings preinoculated with *Rhizopogon vinicolor* and outplanted on eastern Vancouver Island. *Can J For Res* 23:1711–1715.

Bills GF. 1996. Isolation and analysis of endophytic fungal communities from woody plants. In *Endophytic Fungi in Grasses and Woody Plants*, eds. SC Redlin, LM Carris, pp. 31–65. St. Paul, MN: APS Press.

Blaschke H. 1991. Distribution, mycorrhizal infection, and structure of roots of calcicole floral elements at treeline, Bavarian Alps, Germany. *Arct Alp Res* 23:444–450.

Bledsoe C, Klein P, Bliss LC. 1990. A survey of mycorrhizal plants on Truelove Lowland, Devon Island, N. W. T., Canada. *Can J Bot* 68:1848–1856.

Bloomberg WJ. 1966. The occurrence of endophytic fungi in Douglas-fir seedlings and seed. *Can J Bot* 44:413–420.

Bowen GD, Rovira AD. 1976. Microbial colonization of plant roots. *Annu Rev Phytopathol* 14:121–144.

Brenn N, Menkis A, Grünig CR, Sieber TN, Holdenrieder O. 2008. Community structure of *Phialocephala fortinii* s. lat. in European tree nurseries, and assessment of the potential of the seedlings as dissemination vehicles. *Mycol Res* 112:650–662.

Brundrett M. 2004. Diversity and classification of mycorrhizal associations. *Biol Rev* 79:473–495.

Brundrett MC. 2009. Mycorrhizal associations and other means of nutrition of vascular plants: Understanding the global diversity of host plants by resolving conflicting information and developing reliable means of diagnosis. *Plant Soil* 320:37–77.

Burri K, Graf F, Böll A. 2009. Revegetation measures improve soil aggregate stability: A case study of a landslide area in Central Switzerland. *For Snow Landsc Res* 82:45–60.

Cain RF. 1952. Studies of fungi imperfecti. I. *Phialophora. Can J Bot* 30:338–343.

Caldwell BA, Jumpponen A, Trappe JM. 2000. Utilization of major detrital substrates by dark-septate, root endophytes. *Mycologia* 92:230–232.

Cannon PF, Hawksworth DL. 1995. The diversity of fungi associated with vascular plants: The known, the unknown and the need to bridge the knowledge gap. *Adv Plant Pathol* 11:277–302.

Chambers SM, Curlevski NJA, Cairney JWG. 2008. Ericoid mycorrhizal fungi are common root inhabitants of non-Ericaceae plants in a south-eastern Australian sclerophyll forest. *FEMS Microbiol Ecol* 65:263–270.

Chanway CP. 1996. Endophytes: They're not just fungi! *Can J Bot* 74:321–322.

Chen XM, Dong HL, Hu KX, Sun ZR, Chen JA, Guo SX. 2010a. Diversity and antimicrobial and plant-growth-promoting activities of endophytic fungi in *Dendrobium loddigesii* Rolfe. *J Plant Growth Regul* 29:328–337.

Chen JA, Dong HL, Meng ZX, Guo SX. 2010b. *Cadophora malorum* and *Cryptosporiopsis ericae* isolated from medicinal plants of the Orchidaceae in China. *Mycotaxon* 112:457–461.

Christie P, Nicolson TH. 1983. Are mycorrhizas absent from the antarctic? *Trans Br Mycol Soc* 80:557–560.

Clay K. 1988. Fungal endophytes of grasses: A defensive mutualism between plants and fungi. *Ecology* 69:10–16.

Cother EJ, Gilbert RL. 1994. The endophytic mycoflora of bladder saltbush (*Atriplex vesicaria* Hew. ex Benth.) and its possible role in the plant's periodic decline. *Proc Linn Soc NSW* 114:149–169.

Courtois H. 1990a. Die Pilzflora im Kronenbereich und in der Rhizosphäre von Fichten (*P. abies* Karst.) und ihre Bedeutung zur Interpretation von "Waldsterbe"-Symptomen. *Angew Bot* 64:381–392.

Courtois H. 1990b. Endophytische Mikropilze in Fichtenfeinwurzeln. Zur Grundbelastung des Baumsterbens. *Allg Forst- J-Ztg* 161:189–198.

Courtois H, Ruschen G. 1987. Haben wurzelinfizierende Mikropilze Einfluss auf das "Waldsterben"? *Allg Forst- J-Ztg* 158:189–194.

Crous PW, Petrini O, Marais GF, Pretorius ZA, Rehder F. 1995. Occurrence of fungal endophytes in cultivars of *Triticum aestivum* in South Africa. *Mycoscience* 36:105–111.

Crous PW, Schubert K, Braun U et al. 2007. Opportunistic, human-pathogenic species in the Herpotrichiellaceae are phenotypically similar to saprobic or phytopathogenic species in the Venturiaceae. *Studies Mycol* 53:185–234.

Currah RS, Hambleton S, Smreciu A. 1988. Mycorrhizae and mycorrhizal fungi of *Calypso bulbosa. Am J Bot* 75:739–752.

Currah RS, Sherburne R. 1992. Septal ultrastructure of some fungal endophytes from boreal orchid mycorrhizas. *Mycol Res* 96:583–587.

Currah RS, Sigler L, Hambleton S. 1987. New records and new taxa of fungi from the mycorrhizae of terrestrial orchids of Alberta. *Can J Bot* 65:2473–2482.

Currah RS, Smreciu AE, Hambleton S. 1990. Mycorrhizae and mycorrhizal fungi of boreal species of *Platanthera* and *Coeloglossum* (Orchidaceae). *Can J Bot* 68:1171–1181.

Currah RS, Tsuneda A. 1993. Vegetative and reproductive morphology of *Phialocephala fortinii* (Hyphomycetes, *Mycelium radicis atrovirens*) in culture. *Trans Mycol Soc Jpn* 34:345–356.

Currah RS, Tsuneda A, Murakami S. 1993. Morphology and ecology of *Phialocephala fortinii* in roots of *Rhododendron brachycarpum*. *Can J Bot* 71:1639–1644.

Currah RS, Van Dyk M. 1986. A survey of some perennial vascular plant species native to Alberta for occurrence of mycorrhizal fungi. *Can Field-Nat* 100:330–342.

Currah RS, Zelmer C. 1992. A key and notes for the genera of fungi mycorrhizal with orchids and a new species in the genus *Epulorhiza. Rep Tottori Mycol Inst* 30:43–59.

Dababat AE-FA, Selim ME, Saleh AA, Sikora RA. 2008. Influence of *Fusarium* wilt resistant tomato cultivars on root colonization of the mutualistic endophyte *Fusarium oxysporum* strain 162 and its biological control efficacy toward the root-knot nematode *Meloidogyne incognita. J Plant Dis Prot* 115:273–278.

Dababat AE-FA, Sikora RA. 2007. Induced resistance by the mutualistic endophyte, *Fusarium oxysporum* strain 162, toward *Meloidogyne incognita* on tomato. *Biocontrol Sci Technol* 17:969–975.

Dahm H. 1987. Cellulolytic and pectolytic activity of nonmycorrhizal fungi associated with roots of forest trees. *Acta Microbiol Pol* 36:317–324.

Dahm H, Strzelczyk E, Michniewicz M. 1987. Effect of pH, temperature and light on the pathogenicity of *Cylindrocarpon destructans* to pine seedlings in associative cultures with bacteria and actinomycetes. *Eur J For Pathol* 17:141–148.

Danielson RM, Visser S. 1989. Effects of forest soil acidification on ectomycorrhizal and vesicular-arbuscular mycorrhizal development. *New Phytol* 112:41–47.

Danielson RM, Visser S. 1990. The mycorrhizal and nodulation status of container-grown trees and shrubs reared in commercial nurseries. *Can J For Res* 20:609–614.

Davis D. 1967. Cross-protection in *Fusarium* wilt diseases. *Phytopathology* 57:311–314.

Day MJ, Gibas CFC, Fujimura KE, Egger KN, Currah RS. 2006. *Monodictys arctica*, a new hyphomycete from the roots of *Saxifraga oppositifolia* collected in the Canadian High Arctic. *Mycotaxon* 98:261–272.

De Bary A. 1866. *Morphologie und Physiologie der Pilze, Flechten und Myxomyceten*. Leipzig, Germany: Engelmann.

De la Bastide PY, Kendrick B. 1990. The in vitro effects of benomyl on disease tolerance, ectomycorrhiza formation, and growth of white pine (*Pinus strobus*) seedlings. *Can J Bot* 68:444–448.

De Marins JF, Carrenho R, Thomaz SM. 2009. Occurrence and coexistence of arbuscular mycorrhizal fungi and dark septate fungi in aquatic macrophytes in a tropical river-floodplain system. *Aquat Bot* 91:13–19.

Deacon JW. 1981. Ecological relationships with other fungi: Competitors and hyperparasites. In *Biology and Control of Take-All*, eds. MJC Asher, PJ Shipton, pp. 75–101. London, U.K.: Academic Press.

Deacon JW. 1987. Programmed cortical senescence: A basis for understanding root infection. In *Fungal Infection of Plants*, eds. GF Pegg, PG Ayres, pp. 285–297. Cambridge, U.K.: Cambridge University Press.

Deacon JW, Scott DB. 1983. *Phialophora zeicola* sp. nov., and its role in the root-rot stalk-rot complex of maize. *Trans Br Mycol Soc* 81:247–262.

Dearnaley JDW, Le Brocque AF. 2006. Molecular identification of the primary root fungal endophytes of *Dipodium hamiltonianum* (Orchidaceae). *Aust J Bot* 54:487–491.

Deram A, Languereau-Leman F, Howsam M, Petit D, Van Haluwyn C. 2008. Seasonal patterns of cadmium accumulation in *Arrhenatherum elatius* (Poaceae): Influence of mycorrhizal and endophytic fungal colonisation. *Soil Biol Biochem* 40:845–848.

Deshmukh SD, Kogel KH. 2007. *Piriformospora indica* protects barley from root rot caused by *Fusarium graminearum. J Plant Dis Prot* 114:263–268.

Dolinar N, Gaberscik A. 2010. Mycorrhizal colonization and growth of *Phragmites australis* in an intermittent wetland. *Aquat Bot* 93:93–98.

Domanski S, Kowalski T. 1987. Fungi occurring on forests injured by air pollutants in the Upper Silesia and Cracow industrial regions. X. Mycoflora of dying young trees of *Alnus incana. Eur J For Pathol* 17:337–348.

Domsch KH, Gams W, Anderson TH. 1980. *Compendium of Soil Fungi*. London, U.K.: Academic Press.

Dong Z, Canny MJ, McCully ME et al. 1994. A nitrogen-fixing endophyte of sugarcane stems. *Plant Physiol* 105:1139–1147.

Duda B, Sierota ZH. 1987. Survival of Scots pine seedlings after biological and chemical control of damping-off in plastic greenhouses. *Eur J For Pathol* 17:110–117.

Duddrige JA. 1985. A comparative ultrastructural analysis of the host-fungus interface in mycorrhizal and parasitic associations. In *Developmental Biology of Higher Fungi*, eds. D Moore, LA Casselton, DA Wood, JC Frankland, pp. 141–173. Cambridge, U.K.: Cambridge University Press.

Dugassa-Gobena D, Raps A, Vidal S. 1998. Influence of fungal endophytes on allelochemicals of their host plants and the behaviour of insects. *Med Fac Landbouww Univ Gent* 63:333–337.

Duò A, Bruggmann R, Zoller S, Grünig CR. 2012. Mitochondrion genome evolution in species belonging to the *Phialocephala fortinii* s.l.—*Acephala applanata* species complex. *BMC Genomics* 13:166.

Egger KN, Danielson RM, Fortin JA. 1991. Taxonomy and population structure of E-strain mycorrhizal fungi inferred from ribosomal and mitochondrial DNA polymorphisms. *Mycol Res* 95:866–872.

El-Nashaar HM, Moore LW, George RA. 1986. Enzyme-linked immunosorbent assay quantification of initial infection of wheat by *Gaeumannomyces graminis* var. *tritici* as moderated by biocontrol agents. *Phytopathology* 76:1319–1322.

Ellis MB. 1976. *More Dematiaceous Hyphomycetes*. Kew, Surrey, U.K.: CAB International Mycological Institute (IMI).

Fernandez N, Messuti MI, Fontenla S. 2008. Arbuscular mycorrhizas and dark septate fungi in *Lycopodium paniculatum* (Lycopodiaceae) and *Equisetum bogotense* (Equisetaceae) in a valdivian temperate forest of Patagonia, Argentina. *Am Fern J* 98:117–127.

Fernando AA, Currah RS. 1996. A comparative study of the effects of the root endophytes *Leptodontidium orchidicola* and *Phialocephala fortinii* (fungi imperfecti) on the growth of some subalpine plants in culture. *Can J Bot* 74:1071–1078.

Fisher PJ, Graf F, Petrini LE, Sutton BC, Wookey PA. 1995a. Fungal endophytes of *Dryas octopetala* from a high arctic polar semidesert and from the Swiss Alps. *Mycologia* 87:319–323.

Fisher PJ, Petrini O. 1992. Fungal saprobes and pathogens as endophytes of rice (*Oryza sativa* L.). *New Phytol* 120:137–143.

Fisher PJ, Petrini O, Petrini LE. 1991a. Endophytic Ascomycetes and Deuteromycetes in roots of *Pinus sylvestris. Nova Hedwigia* 52:11–15.

Fisher PJ, Petrini LE, Sutton BC, Petrini O. 1995b. A study of fungal endophytes in leaves, stems and roots of *Gynoxis oleifolia* Muchler (Compositae) from Ecuador. *Nova Hedwigia* 60:589–594.

Fisher PJ, Petrini O, Webster J. 1991b. Aquatic hyphomycetes and other fungi in living aquatic and terrestrial roots of *Alnus glutinosa. Mycol Res* 95:543–547.

Forbrig R. 1987. Anatomische und histologische Untersuchungen an pilzinfizierten Fichtenkeimlingen (*Picea abies* Karst.). *Allg Forst- J-ztg* 158:222–229.

Forbrig R. 1989. Anatomische und biologische Untersuchungen an pilzinfizierten Fichtenkeimlingen (*Picea abies* Karst.). II. *Allg Forst- J-ztg* 160:137–144.

Foster RC. 1986. The ultrastructure of the rhizoplane and rhizosphere. *Annu Rev Phytopathol* 24:211–234.

Foster RC, Rovira AD, Cock TW. 1983. *Ultrastructure of the Root-Soil Interface*. St. Paul, MN: APS Press.

Freeman S, Rodriguez RJ. 1993. Genetic conversion of a fungal plant pathogen to a nonpathogenic, endophytic mutualist. *Science* 260:75–78.

Freeman J, Ward E. 2004. *Gaeumannomyces graminis*, the take-all fungus and its relatives. *Mol Plant Pathol* 5:235–252.

Galaaen R, Venn K. 1979. *Pythium sylvaticum* Campbell & Hendrix and other fungi associated with root dieback of 2/0 seedlings of *Picea abies* (L.) Karst. in Norway. *Rep Norw For Res Inst* 34:265–280.

Galamay TO, Yamauchi A, Kono Y, Hioki M. 1992. Specific colonization of the hypodermis of sorghum roots by an endophyte, *Polymyxa* sp. *Soil Sci Plant Nutr* 38:573–578.

Gams W. 1963. *Mycelium radicis atrovirens* in forest soils, isolation from soil microhabitats and identification. In *Soil Organisms*, eds. J Dockson, J van der Drift, pp. 176–182. Amsterdam, the Netherlands: North Holland Publications.

Gams W. 2000. *Phialophora* and some similar morphologically little-differentiated anamorphs of divergent ascomycetes. *Stud Mycol* 45:187–199.

Garcia-Guzman G, Burdon JJ, Nicholls AO. 1996. Effects of the systemic flower infecting-smut *Ustilago bullata* on the growth and competitive ability of the grass *Bromus catharticus. J Ecol* 84:657–665.

Garrett SD. 1981. *Soil Fungi and Soil Fertility*, 2nd edn. Oxford, U.K.: Pergamon Press.

Gasoni L, Stegman De Gurfinkel B. 1997. The endophyte *Cladorrhinum foecundissimum* in cotton roots: Phosphorus uptake and host growth. *Mycol Res* 101:867–870.

Gasoni L, Stegman de Gurfinkel B. 2009. Biocontrol of *Rhizoctonia solani* by the endophytic fungus *Cladorrhinum foecundissimum* in cotton plants. *Aust Plant Pathol* 38:389–391.

Gessler C, Kuc J. 1982. Induction of resistance to *Fusarium* wilt in cucumber by root and foliar pathogens. *Phytopathology* 72:1439–1441.

Gezgin Y, Eltem R. 2009. Diversity of endophytic fungi from various Aegean and Mediterranean orchids (saleps). *Turk J Bot* 33:439–445.

Girlanda M, Selosse MA, Cafasso D et al. 2006. Inefficient photosynthesis in the Mediterranean orchid *Limodorum abortivum* is mirrored by specific association to ectomycorrhizal Russulaceae. *Mol Ecol* 15:491–504.

Görke C. 1998. Mykozönosen von Wurzeln und Stamm von Jungbäumen unterschiedlicher Bestandsbegründungen. *Biblio Mycol* 173:1–462.

Green LE, Porras-Alfaro A, Sinsabaugh RL. 2008. Translocation of nitrogen and carbon integrates biotic crust and grass production in desert grassland. *J Ecol* 96:1076–1085.

Grönberg H, Kaparakis G, Sen R. 2006. Binucleate *Rhizoctonia* (*Ceratorhiza* spp.) as non-mycorrhizal endophytes alter *Pinus sylvestris* L. seedling root architecture and affect growth of rooted cuttings. *Scan J For Res* 21:450–457.

Grünig CR. 2004. Population biology of the tree-root endophyte *Phialocephala fortinii*. PhD thesis no. 15313, Zürich, Switzerland: ETH Zürich, http://e-collection.ethbib.ethz.ch/cgi-bin/show.pl?type = diss&nr = 15313

Grünig CR, Brunner PC, Duò A, Sieber TN. 2007. Suitability of methods for species recognition in the *Phialocephala fortinii–Acephala applanata* species complex using DNA analysis. *Fung Genet Biol* 44:773–788.

Grünig CR, Duò A, Sieber TN. 2006. Population genetic analysis of *Phialocephala fortinii* s.l. and *Acephala applanata* in two undisturbed forests in Switzerland and evidence for new cryptic species. *Fung Genet Biol* 43:410–421.

Grünig CR, Duò A, Sieber TN, Holdenrieder O. 2008a. Assignment of species rank to six reproductively isolated cryptic species of the *Phialocephala fortinii* s. l.-*Acephala applanata* species complex. *Mycologia* 100:47–67.

Grünig CR, McDonald BA, Sieber TN, Rogers SO, Holdenrieder O. 2004. Evidence for subdivision of the root-endophyte *Phialocephala fortinii* into cryptic species and recombination within species. *Fung Genet Biol* 41:676–687.

Grünig CR, Sieber TN. 2005. Molecular and phenotypic description of the widespread root symbiont *Acephala applanata* gen. et sp. nov., formerly known as "Dark Septate Endophyte Type 1." *Mycologia* 97:628–640.

Grünig CR, Sieber TN, Holdenrieder O. 2001. Characterisation of dark septate endophytic fungi (DSE) using inter-simple-sequence-repeat-anchored polymerase chain reaction (ISSR-PCR) amplification. *Mycol Res* 105:24–32.

Grünig CR, Queloz V, Duo A, Sieber TN. 2009. Phylogeny of *Phaeomollisia piceae* gen. sp. nov.: A dark, septate, conifer-needle endophyte and its relationships to *Phialocephala* and *Acephala*. *Mycol Res* 113:207–221.

Grünig CR, Queloz V, Sieber TN. 2011. Structure of diversity in dark septate endophytes: From species to genes. In *Endophytes of Forest Trees: Biology and Applications*, eds. AM Pirttilä, AC Frank, pp. 3–30. Forestry Sciences 80, Berlin, Germany: Springer Science & Business Media.

Grünig CR, Queloz V, Sieber TN, Holdenrieder O. 2008b. Dark septate endophytes (DSE) of the *Phialocephala fortinii* s.l.—*Acephala applanata* species complex in tree roots: Classification, population biology, and ecology. *Botany* 86:1355–1369.

Gutteridge RJ, Jenkyn JF, Bateman GL. 2007. The potential of non-pathogenic *Gaeumannomyces* spp., occurring naturally or introduced into wheat crops or preceding crops, for controlling take-all in wheat. *Ann Appl Biol* 150:53–64.

Hallmann J, Sikora RA. 1994. Occurrence of plant parasitic nematodes and non-pathogenic species of *Fusarium* in tomato plants in Kenya and their role as mutualistic synergists for biological control of root-knot nematodes. *Int J Pest Manage* 40:321–325.

Hallmann J, Sikora RA. 1996. Toxicity of fungal endophyte secondary metabolites to plant parasitic nematodes and soil-borne plant pathogenic fungi. *Eur J Plant Pathol* 102:155–162.

Halmschlager E, Kowalski T. 2004. The mycobiota in nonmycorrhizal roots of healthy and declining oaks. *Can J Bot* 82:1446–1458.

Hambleton S, Currah RS. 1997. Fungal endophytes from the roots of alpine and boreal Ericaceae. *Can J Bot* 75:1570–1581.

Hambleton S, Sigler L. 2005. *Meliniomyces*, a new anamorph genus for root-associated fungi with phylogenetic affinities to *Rhizoscyphus ericae* (*Hymenoscyphus ericae*), Leotiomycetes. *Stud Mycol* 53:1–27.

Hambleton S, Huhtinen S, Currah RS. 1999. *Hymenoscyphus ericae*: A new record from western Canada. *Mycological Research* 103:1391–1397.

Hampton R, Ball E, De Boer S. 1990. *Serological Methods*. St. Paul, MN: APS Press.

Harney SK, Rogers SO, Wang CJK. 1997. Molecular characterization of dematiaceous root endophytes. *Mycol Res* 101:1397–1404.

Harvais G, Hadley G. 1967. The relation between host and endophyte in orchid mycorrhiza. *New Phytol* 66:205–215.

Haselwandter K, Read DJ. 1982. The significance of a root-fungus association in two *Carex* species of high-alpine plant communities. *Oecologia* 53:352–354.

Haug I, Weber G, Oberwinkler F. 1988. Intracellular infection by fungi in mycorrhizae of damaged spruce trees. *Eur J For Pathol* 18:112–120.

Hennon PE, Shaw III CG, Hansen EM. 1990. Symptoms and fungal associations of declining *Chamaecyparis nootkatensis* in Southeast Alaska. *Plant Dis* 74:267–273.

Herrera J, Khidir HH, Eudy DM, Porras-Alfaro A, Natvig DO, Sinsabaugh RL. 2010. Shifting fungal endophyte communities colonize *Bouteloua gracilis*: Effect of host tissue and geographical distribution. *Mycologia* 102:1012–1026.

Heslin MC, Blasius D, McElhinney C, Mitchell DT. 1992. Mycorrhizal and associated fungi of Sitka spruce in Irish forest mixed stands. *Eur J For Pathol* 22:46–57.

Hodson E, Shahid F, Basinger J, Kaminskyj S. 2009. Fungal endorhizal associates of *Equisetum* species from Western and arctic Canada. *Mycol Prog* 8:19–27.

Hol WHG, de la Pena E, Moens M, Cook R. 2007. Interaction between a fungal endophyte and root herbivores of *Ammophila arenaria*. *Basic Appl Ecol* 8:500–509.

Holdenrieder O. 1989. Endophytes and rhizoplane fungi of Norway spruce. In *Proceedings of the Seventh International Conference on Root and Butt Rots*, ed. DJ Morrison, pp. 531–545. August 9–16, 1988. Victoria, British Columbia, Canada: Forestry Canada, Pacific Forestry Centre.

Holdenrieder O, Sieber TN. 1992. Fungal associations of serially washed healthy non-mycorrhizal roots of *Picea abies*. *Mycol Res* 96:151–156.

Horton TR, Cazares E, Bruns TD. 1998. Ectomycorrhizal, vesicular-arbuscular and dark septate fungal colonization of bishop pine (*Pinus muricata*) seedlings in the first 5 months of growth after wildfire. *Mycorrhiza* 8:11–18.

Hou X-Q, Guo S-X. 2009. Interaction between a dark septate endophytic isolate from *Dendrobium* sp. and roots of *D. nobile* seedlings. *J Integr Plant Biol* 51:374–381.

Hutton BJ, Dixon KW, Sivasithamparam K. 1994. Ericoid endophytes of Western Australian heaths (Epacridaceae). *New Phytol* 127:557–566.

Iqbal SH, Firdaus EB, Yousaf N. 1995. Freshwater hyphomycete communities in a canal: 1. Endophytic hyphomycetes of submerged roots of trees sheltering a canal bank. *Can J Bot* 73:538–543.

Johnson L. 2008. Iron and siderophores in fungal-host interactions. *Mycol Res* 112:170–183.

Jumpponen A, Jones KL. 2009. Massively parallel 454 sequencing indicates hyperdiverse fungal communities in temperate *Quercus macrocarpa* phyllosphere. *New Phytol* 184:438–448.

Jumpponen A, Mattson KG, Trappe JM. 1998. Mycorrhizal functioning of *Phialocephala fortinii* with *Pinus contorta* on glacier forefront soil: Interactions with soil nitrogen and organic matter. *Mycorrhiza* 7:261–265.

Jumpponen A, Trappe JM. 1998a. Dark septate endophytes: A review of facultative biotrophic root-colonizing fungi. *New Phytol* 140:295–310.

Jumpponen A, Trappe JM. 1998b. Performance of *Pinus contorta* inoculated with two strains of root endophytic fungus, *Phialocephala fortinii*: Effects of synthesis system and glucose concentration. *Can J Bot* 76:1205–1213.

Kai W, Zhao ZW. 2006. Occurrence of arbuscular mycorrhizas and dark septate endophytes in hydrophytes from lakes and streams in southwest China. *Int Rev Hydrobiol* 91:29–37.

Kandalepas D, Stevens KJ, Shaffer GP, Platt WJ. 2010. How abundant are root-colonizing fungi in Southeastern Louisiana's degraded marshes? *Wetlands* 30:189–199.

Kaparakis G, Sen R. 2006. Binucleate *Rhizoctonia* (*Ceratorhiza* spp.) induce adventitious root formation in hypocotyl cuttings of *Pinus sylvestris* L. *Scan J For Res* 21:444–449.

Kattner D. 1992a. Der Einfluss von Trockenstress auf die Besiedlung von Fichtenfeinwurzeln (*Picea abies* Karst.) durch *Trichoderma viride* und andere endophytische Mikropilze. *Forstwissenschaftliches Centralblatt* 111:383–389.

Kattner D. 1992b. Langzeitwirkung einer Calciumdüngung auf die Besiedlung von Fichtenwurzeln durch Mikropilze. *Allg Forst- J-ztg* 163:138–142.

Kattner D, Schönhar S. 1990. Untersuchungen über das Vorkommen mikroskopischer Pilze in Feinwurzeln optisch gesunder Fichten (*Picea abies* Karst.) auf verschiedenen Standorten. *Mitteilungen des Vereins Forstlicher Standortskunde und Forstpflanzenzüchtung* 35:39–43.

Kavroulakis N, Ntougias S, Zervakis GI, Ehaliotis C, Haralampidis K, Papadopoulou KK. 2007. Role of ethylene in the protection of tomato plants against soil-borne fungal pathogens conferred by an endophytic *Fusarium solani* strain. *J Exp Bot* 58:3853–3864.

Kernaghan G, Berch S, Carter R. 1995. Effect of urea fertilization on ectomycorrhizae of 20-year-old *Tsuga heterophylla*. *Can J For Res* 25:891–901.

Khidir HH, Eudy DM, Porras-Alfaro A, Herrera J, Natvig DO, Sinsabaugh RL. 2010. A general suite of fungal endophytes dominate the roots of two dominant grasses in a semiarid grassland. *J Arid Environ* 74:35–42.

Kim HY, Choi GJ, Lee HB et al. 2007. Some fungal endophytes from vegetable crops and their anti-oomycete activities against tomato late blight. *Lett Appl Microbiol* 44:332–337.

Kirk JJ, Deacon JW. 1987. Control of the take-all fungus by *Microdochium bolleyi*, and interactions involving *M. bolleyi*, *Phialophora graminicola* and *Periconia macrospinosa* on cereal roots. *Plant Soil* 98:231–237.

Kjoller R, Olsrud M, Michelsen A. 2010. Co-existing ericaceous plant species in a subarctic mire community share fungal root endophytes. *Fungal Ecol* 3:205–214.

Kohn LM, Stasovski E. 1990. The mycorrhizal status of plants at Alexandra Fiord, Ellesmere Island, Canada, a high arctic site. *Mycologia* 82:23–35.

Kowalski S. 1982a. Role of mycorrhiza and soil fungi in natural regeneration of fir (*Abies alba* Mill.) in Polish Carpathians and Sudetes. *Eur J For Pathol* 12:107–112.

Kowalski T. 1982b. Vorkommen von Pilzen in durch Luftverunreinigung geschädigten Wäldern im Oberschlesischen und Krakauer Industriegebiet. VIII. Mykoflora von *Larix decidua* an einem Standort mit mittlerer Immissionsbelastung. *Eur J For Pathol* 12:262–272.

Kowalski T. 1983. Fungi occurring in forest injured by air-pollutants in the upper Silesi and Cracow industrial regions. 9. Mycoflora on *Quercus robur* L and *Quercus rubra* L. located in a zone with medium-high air-pollution damage. *Eur J For Pathol* 13:46–59.

Kowalski T. 1993. Fungi in living symptomless needles of *Pinus sylvestris* with respect to some observed disease processes. *J Phytopathol* 139:129–145.

Kowalski T, Bartnik C. 1995. *Cryptosporiopsis radicicola* sp. nov. from roots of *Quercus robur*. *Mycol Res* 99:663–666.

Kowalski T, Halmschlager E, Schrader K. 1998. *Cryptosporiopsis melanigena* sp. nov., a root-inhabiting fungus of *Quercus robur* and *Q. petraea*. *Mycol Res* 102:347–354.

Lehnert M, Kottke I, Setaro S, Pazmino LF, Pablo Suarez J, Kessler M. 2009. Mycorrhizal associations in ferns from Southern Ecuador. *Am Fern J* 99:292–306.

Lemaire JM, Ponchet J. 1963. *Phialophora radicicola* Cain, forme conidienne du *Linocarpon cariceti* B. et Br. *Comptes Rendus Hebdomadaires des Séances de l'Académie d'Agriculture de France* 49:1067–1069.

Levisohn I. 1954. Test for the pseudomycorrhizal group of soil fungi. *Nature* 174:408–409.

Li A-R, Guan K-Y. 2007. Mycorrhizal and dark septate endophytic fungi of *Pedicularis* species from northwest of Yunnan Province, China. *Mycorrhiza* 17:103–109.

Li LF, Yang A, Zhao ZW. 2005. Seasonality of arbuscular mycorrhizal symbiosis and dark septate endophytes in a grassland site in southwest China. *FEMS Microbiol Ecol* 54:367–373.

Lihnell D. 1939. Untersuchungen über die Mykorrhizen und die Wurzelpilze von *Juniperus communis*. *Symb Bot Upsal* 3:1–141.

LoBuglio KF, Wilcox HE. 1988. Growth and survival of ectomycorrhizal and ectendomycorrhizal seedlings of *Pinus resinosa* on iron tailings. *Can J Bot* 66:55–60.

Lugo MA, Molina MG, Crespo EM. 2009. Arbuscular mycorrhizas and dark septate endophytes in bromeliads from South American arid environment. *Symbiosis* 47:17–21.

Lv Y-L, Zhang F-S, Chen J et al. 2010. Diversity and antimicrobial activity of endophytic fungi associated with the alpine plant *Saussurea involucrata*. *Biol Pharm Bull* 33:1300–1306.

Lyr H, Kluge E. 1968. Zusammenhänge zwischen Pathogenität, Enzym- und Toxinproduktion bei *Cylindrocarpon radicicola*. *Phytopathol Z* 62:220–231.

Macia-Vicente JG, Jansson H-B, Mendgen K, Lopez-Llorca LV. 2008. Colonization of barley roots by endophytic fungi and their reduction of take-all caused by *Gaeumannomyces graminis* var. *tritici*. *Can J Microbiol* 54:600–609.

Malinowski DP, Alloush GA, Belesky DP. 1998. Evidence for chemical changes on the root surface of tall fescue in response to infection with the fungal endophyte *Neotyphodium coenophialum*. *Plant Soil* 205:1–12.

Malinowski DP, Belesky DP. 1999. *Neotyphodium coenophialum*-endophyte infection affects the ability of tall fescue to use sparingly available phosphorus. *J Plant Nutr* 22:835–853.

Mandyam K, Jumpponen A. 2005. Seeking the elusive function of the root-colonising dark septate endophytic fungi. *Stud Mycol* 53:173–189.

Mandyam K, Loughin T, Jumpponen A. 2010. Isolation and morphological and metabolic characterization of common endophytes in annually burned tallgrass prairie. *Mycologia* 102:813–821.

Manka K. 1960. O grzybie korzenowym *Mycelium radicis atrovirens* Melin. *Monogr Bot* 10:147–158.

Manka K, Mroczkiewicz M. 1991. A contribution to *Mycelium radicis atrovirens* Melin occurrence in scots pine (*Pinus silvestris* L.) roots. *Phytopathol Polonica* 2:102–105.

Manka K, Przezbòrski A. 1978. Biologiczne zwalczanie zgorzeli siewek sosny zwyczajnej za pomoca grzyba *Mycelium radicis atrovirens* Melin. *Zeszyty Problemowe Postepow Nauk Rolniczych* 213:173–180.

Manka K, Przezbòrski A. 1987. W kierunku biologicznej ochrony siewek drzew lesnych przed pasozytnicza zgorzela. *Sylwan* 131:53–59.

Martos F, Dulormne M, Pailler T et al. 2009. Independent recruitment of saprotrophic fungi as mycorrhizal partners by tropical achlorophyllous orchids. *New Phytol* 184:668–681.

Matta A. 1989. Induced resistance to *Fusarium* wilt diseases. In *Vascular Wilt Diseases of Plants—Basic Studies and Control*, eds. EC Tjamos, CH Beckman, pp. 175–96. Berlin, Geramny: Springer.

Matturi ST, Stenton H. 1964. Distribution and status in the soil of *Cylindrocarpon* species. *Trans Br Mycol Soc* 47:577–587.

McGill BJ, Etienne RS, Gray JS et al. 2007. Species abundance distributions: Moving beyond single prediction theories to integration within an ecological framework. *Ecol Lett* 10:995–1015.

McKeen WE. 1952. *Phialophora radicicola* Cain, a corn rootrot pathogen. *Can J Bot* 30:344–347.

Medina-Roldan E, Arredondo JT, Huber-Sannwald E, Chapa-Vargas L, Olalde-Portugal V. 2008. Grazing effects on fungal root symbionts and carbon and nitrogen storage in a shortgrass steppe in Central Mexico. *J Arid Environ* 72:546–556.

Melin E. 1923. Experimentelle Untersuchungen über die Konstitution und Ökologie der Mykorrhizen von *Pinus sylvestris* L. und *Picea abies* (L.) Karst. In *Mykologische Untersuchungen und Berichte 2*, ed. R Flack, pp. 334–373. Kassel, Germany: Druck und Verlag G. Gottheilt.

Melin E. 1924. Über den Einfluss der Wasserstoffionenkonzentration auf die Virulenz der Wurzelpilze von Kiefer und Fichte. *Botaniska Notiser*1924:38–48.

Mendoza AR, Sikora RA. 2009. Biological control of *Radopholus similis* in banana by combined application of the mutualistic endophyte *Fusarium oxysporum* strain 162, the egg pathogen *Paecilomyces lilacinus* strain 251 and the antagonistic bacteria *Bacillus firmus*. *Biocontrol* 54:263–272.

Menkis A. 2004. Ecology and molecular characterization of dark septate fungi from roots, living stems, coarse and fine woody debris. *Mycol Res* 108:965–973.

Menkis A, Vasiliauskas R, Taylor AFS, Stenstrom E, Stenlid J, Finlay R. 2006. Fungi in decayed roots of conifer seedlings in forest nurseries, afforested clear-cuts and abandoned farmland. *Plant Pathol* 55:117–129.

Mikola P. 1988. Ectendomycorrhiza of conifers. *Silva Fennica* 22:19–27.

Miller JD, Sumarah MW, Adams GW. 2008. Effect of a rugulosin-producing endophyte in *Picea glauca* on *Choristoneura fumiferana*. *J Chem Ecol* 34:362–368.

Moll M, Vidal S. 1995. Einfluss des wurzelbürtigen Endophyten *Acremonium strictum* Gams auf das Wirtswahlverhalten und die Entwicklung von *Trialeurodes vaporariorum* (Westwood) (Homoptera, Aleyrodidae) an Buschbohnen. *Mitteilungen der Deutschen Gesellschaft für allgemeine und angewandte Entomologie* 10:445–448.

Money NP, Caesar-TonThat T-C, Frederick B, Henson JM. 1998. Melanin synthesis is associated with changes in hyphopodial turgor, permeability, and wall rigidity in *Gaeumannomyces graminis* var. *graminis*. *Fungal Genet Biol* 24:240–251.

Münzenberger B, Bubner B, Wöllecke J et al. 2009. The ectomycorrhizal morphotype *Pinirhiza sclerotia* is formed by *Acephala macrosclerotiorum* sp nov., a close relative of *Phialocephala fortinii*. *Mycorrhiza* 19:481–492.

Murray DIL, Gadd GM. 1981. Preliminary studies on *Microdochium bolleyi* with special reference to colonization of barley. *Trans Br Mycol Soc* 76:397–403.

Narisawa K, Tokumasu S, Hashiba T. 1998. Suppression of clubroot formation in Chinese cabbage by the root endophytic fungus, *Heteroconium chaetospira. Plant Pathol* 47:206–210.

Narisawa K, Usuki F, Hashiba T. 2004. Control of verticillium yellows in Chinese cabbage by the dark septate endophytic fungus LtVB3. *Phytopathology* 94:412–418.

Newell SY. 1992. Estimating fungal biomass and productivity in decomposing litter. In *The Fungal Community*, eds. GC Carroll, DT Wicklow, pp. 521–561. New York: Marcel Dekker, Inc.

Newell SY, Arsuffi TL, Fallon RD. 1988. Fundamental procedures for determining ergosterol content of decaying plant material by liquid chromatography. *Appl Environ Microbiol* 54:1876–1879.

Newsham KK. 1994. First record of intracellular sporulation by a coelomycete fungus. *Mycol Res* 98:1390–1392.

Newsham KK. 1999. *Phialophora graminicola*, a dark septate fungus, is a beneficial associate of the grass *Vulpia ciliata* ssp. *ambigua. New Phytol* 144:517–524.

Newsham KK, Upson R, Read DJ. 2009. Mycorrhizas and dark septate root endophytes in polar regions. *Fungal Ecol* 2:10–20.

O'Dell TE, Massicotte HB, Trappe JM. 1993. Root colonization of *Lupinus latifolius* Agardh. and *Pinus contorta* Dougl. by *Phialocephala fortinii* Wang & Wilcox. *New Phytol* 124:93–100.

O'Dell TE, Trappe JM. 1992. Root endophytes of lupin and some other legumes in northwestern USA. *New Phytol* 122:479–485.

Oberholzer-Tschütscher B. 1982. Untersuchungen über endophytische Pilze von *Erica carnea* L., Zürich, Switzerland: Institute of Microbiology, Swiss Federal Institute of Technology.

Oberwinkler F. 1964. Intrahymenial Heterobasidiomycetes. *Sebacina* strains without fruiting bodies and their systematic placement. *Nova Hedwigia* 7:489–499.

Ohtaka N, Narisawa K. 2008. Molecular characterization and endophytic nature of the root-associated fungus *Meliniomyces variabilis* (LtVB3). *J Gen Plant Pathol* 74:24–31.

Olsrud M, Carlsson BA, Svensson BM, Michelsen A, Melillo JM. 2010. Responses of fungal root colonization, plant cover and leaf nutrients to long-term exposure to elevated atmospheric CO_2 and warming in a subarctic birch forest understory. *Global Change Biol* 16:1820–1829.

Paparu P, Dubois T, Coyne D, Viljoen A. 2009. Dual inoculation of *Fusarium oxysporum* endophytes in banana: Effect on plant colonization, growth and control of the root burrowing nematode and the banana weevil. *Biocontrol Sci Technol* 19:639–655.

Pareja-Jaime Y, Martin-Urdiroz M, Gonzalez Roncero MI, Antonio Gonzalez-Reyes J, Ruiz Roldan MDC. 2010. Chitin synthase-deficient mutant of *Fusarium oxysporum* elicits tomato plant defence response and protects against wild-type infection. *Mol Plant Pathol* 11:479–493.

Parkinson D, Crouch R. 1969. Studies on fungi in a pinewood soil. V. Root mycofloras of seedlings of *Pinus nigra* var. *laricio. Revue d' Écologie et de Biologie du Sol* 6:263–275.

Parsons JW. 1981. Chemistry and distribution of amino sugars in soils and soil organisms. In *Soil Biochemistry*. Vol. 5, eds. EA Paul, JN Ladd, pp. 197–227. New York: Marcel Dekker, Inc.

Pearson V, Read DJ. 1973. The biology of mycorrhiza in the Ericaceae. *New Phytol* 72:371–379.

Peterson RL, Wagg C, Pautler M. 2008. Associations between microfungal endophytes and roots: Do structural features indicate function? *Botany* 86:445–456.

Petrini O. 1991. Fungal endophytes of tree leaves. In *Microbial Ecology of Leaves*, eds. JH Andrews, SS Hirano, pp. 179–197. New York: Springer.

Petrini O, Fisher PJ, Petrini LE. 1992. Fungal endophytes of bracken (*Pteridium aquilinum*), with some reflections on their use in biological control. *Sydowia* 44:282–293.

Phillips DV, Leben C, Allison CC. 1967. A mechanism for the reduction of *Fusarium* wilt by a *Cephalosporium* species. *Phytopathology* 57:916–919.

Piercey MM, Graham SW, Currah RS. 2004. Patterns of genetic variation in *Phialocephala fortinii* across a broad latitudinal transect in Canada. *Mycol Res* 108:955–964.

Postma J, Rattink H. 1991. Biological control of *Fusarium* wilt of carnation with a nonpathogenic isolate of *Fusarium oxysporum. Can J Bot* 70:1199–1205.

Postma JWM, Olsson PA, Falkengren-Grerup U. 2007. Root colonisation by arbuscular mycorrhizal, fine endophytic and dark septate fungi across a pH gradient in acid beech forests. *Soil Biol Biochem* 39:400–408.

Queloz V. 2010. Biogeography and evolution of the *Phialocephala fortinii* s.l.—*Acephala applanata* species complex (PAC). PhD thesis, Zürich, Switzerland: ETH Zürich.

Queloz V, Duo A, Grünig CR. 2008. Isolation and characterization of microsatellite markers for the tree-root endophytes *Phialocephala subalpina* and *Phialocephala fortinii* s.s. *Mol Ecol Resour* 8:1322–1325.

Queloz V, Duo A, Sieber TN, Grünig CR. 2010. Microsatellite size homoplasies and null alleles do not affect species diagnosis and population genetic analysis in a fungal species complex. *Mol Ecol Resour* 10:348–367.

Queloz V, Grünig CR, Sieber TN, Holdenrieder O. 2005. Monitoring the spatial and temporal dynamics of a community of the tree-root endophyte *Phialocephala fortinii* sl. *New Phytol* 168:651–660.

Queloz V, Sieber TN, Holdenrieder O, McDonald BA, Grünig CR. 2011. No biogeographical pattern for a root-associated fungal species complex. *Global Ecol Biogeogr* 20:160–169.

Raps A, Vidal S. 1996. The influence of a root-colonizing endophyte on the development of root-knot nematodes on tomato. *Mitteilungen aus der Biologischen Bundesanstalt für Land- und Forstwirtschaft (Berlin-Dahlem)* 321:453.

Raps A, Vidal S. 1998. Indirect effects of an unspecialized endophytic fungus on specialized plant-herbivorous insect interactions. *Oecologia* 114:541–547.

Rasmann C, Graham JH, Chellemi DO, Datnoff LE, Larsen J. 2009. Resilient populations of root fungi occur within five tomato production systems in southeast Florida. *Appl Soil Ecol* 43:22–31.

Raviraja NS, Sridhar KR, Bärlocher F. 1996. Endophytic aquatic hyphomycetes of roots of plantation crops and ferns from India. *Sydowia* 48:152–160.

Read DJ. 1974. *Pezizella ericae* sp nov, perfect state of a typical mycorrhizal endophytes of ericaceae. *Trans Br Mycol Soc* 63:381.

Read DJ. 1983. The biology of micorrhiza in the Ericales. *Can J Bot* 61:985–1004.

Read DJ. 1992. The mycorrhizal fungal community with special reference to nutrient mobilization. In *The Fungal Community*, eds. GC Carroll, DT Wicklow, pp. 631–652. New York: Marcel Dekker, Inc.

Read DJ. 1996. The structure and function of the ericoid mycorrhizal root. *Ann Bot* 77:365–374.

Reinecke P. 1978. *Microdochium bolleyi* at the stem base of cereals. *Z Pflanzenkrank Pflanzensch* 85:679–685.

Reinecke P, Fokkema NJ. 1981. An evaluation of methods of screening fungi from the haulm base of cereals for antagonism to *Pseudocercosporella herpotrichoides* in wheat. *Trans Br Mycol Soc* 77:343–350.

Reininger V, Grünig CR, Sieber TN. 2011a. Are plant communities shaped by fungal root endophytes? *Phytopathology* 101:S151–S152.

Reininger V, Grünig CR, Sieber TN. 2011b. Microsatellite-based quantification method to estimate biomass of endophytic *Phialocephala* species in strain mixtures. *Microb Ecol* 61:676–683.

Reininger V, Grünig CR, Sieber TN. 2012. Host species and strain combination determine growth reduction of spruce and birch seedlings colonized by root-associated dark septate endophytes. *Environ Microbiol* 14:1064–1076.

Reissinger A, Vilich V, Sikora RA. 2001. Detection of fungi in planta: Effectiveness of surface sterilization methods. *Mycol Res* 105:563–566.

Richard C, Fortin J-A, Fortin A. 1971. Protective effect of an ectomycorrhizal fungus against the root pathogen *Mycelium radicis atrovirens*. *Can J For Res* 1:246–251.

Riesen T, Sieber TN. 1985. *Endophytic Fungi in Winter Wheat (Triticum aestivum L.)*. Zürich, Switzerland: Institute of Microbiology, Swiss Federal Institute of Technology.

Rivera-Orduna FN, Suarez-Sanchez RA, Flores-Bustamante ZR, Gracida-Rodriguez JN, Flores-Cotera LB. 2011. Diversity of endophytic fungi of *Taxus globosa* (Mexican yew). *Fungal Divers* 47:65–74.

Robertson NF. 1954. Studies on the mycorrhiza of *Pinus silvestris*. I. Pattern of development of mycorrhizal root and its significance for experimental studies. *New Phytol* 53:253–283.

Rodrigues KF. 1996. Fungal endophytes of palms. In *Endophytic Fungi in Grasses and Woody Plants*, eds. SC Redlin, LM Carris, pp. 121–132. St. Paul, MN: APS Press.

Rodríguez-Gálvez E, Mendgen K. 1995. The infection process of *Fusarium oxysporum* in cotton root tips. *Protoplasma* 189:61–72.

Roy M, Yagame T, Yamato M et al. 2009. Ectomycorrhizal *Inocybe* species associate with the mycoheterotrophic orchid *Epipogium aphyllum* but not its asexual propagules. *Ann Bot* 104:595–610.

Ruotsalainen AL, Markkola AM, Kozlov MV. 2010. Birch effects on root fungal colonisation of crowberry are uniform along different environmental gradients. *Basic Appl Ecol* 11:459–467.

Ruotsalainen AL, Vare H, Oksanen J, Tuomi J. 2004. Root fungus colonization along an altitudinal gradient in North Norway. *Arct Antarct Alp Res* 36:239–243.

Saito K, Kuga-Uetake Y, Saito M, Peterson RL. 2006. Vacuolar localization of phosphorus in hyphae of *Phialocephala fortinii*, a dark septate fungal root endophyte. *Can J Microbiol* 52:643–650.

Saleh AA, Leslie JE. 2004. *Cephalosporium maydis* is a distinct species in the *Gaeumannomyces-Harpophora* species complex. *Mycologia* 96:1294–1305.

Salvarredi LA, Crespo EM, Menoyo E, Filippa EM, Barboza GE, Lugo MA. 2010. Arbuscular mycorrhizas and dark septates endophytes in native Gentianaceae from Argentina. *Boletin De La Sociedad Argentina De Botanica* 45:223–229.

Sánchez Márquez S, Bills GF, Domínguez Acuña D, Zabalgogeazcoa I. 2010. Endophytic mycobiota of leaves and roots of the grass *Holcus lanatus*. *Fungal Divers* 41:115–123.

Sánchez Márquez S, Bills GF, Zabalgogeazcoa I. 2008. Diversity and structure of the fungal endophytic assemblages from two sympatric coastal grasses. *Fungal Divers* 33:87–100.

Savage SD, Sall MA. 1981. Radioimmunosorbent assay for *Botrytis cinerea*. *Phytopathology* 71:411–415.

Schardl CL. 2010. The *Epichloë*, symbionts of the grass subfamily Pooideae. *Ann Missouri Bot Gard* 97:646–665.

Schelling CL. 1952. Zur Kenntnis von *Mycelium radicis atrovirens* Melin mit besonderer Berücksichtigung der Verwertbarkeit verschiedener Kohlenstoffquellen. ed. Zürich, Switzerland: Swiss Federal Institute of Technology.

Schild DE, Kennedy A, Stuart MR. 1988. Isolation of symbiont and associated fungi from ectomycorrhizas of Sitka spruce. *Eur J For Pathol* 18:51–61.

Schmidt SK, Sobieniak-Wiseman LC, Kageyama SA, Halloy SRP, Schadt CW. 2008. Mycorrhizal and dark-septate fungi in plant roots above 4270 meters elevation in the Andes and Rocky Mountains. *Arct Antarct Alp Res* 40:576–583.

Schol-Schwarz MB. 1970. Revision of the genus *Phialophora* (Moniliales). *Persoonia* 6:59–94.

Schönhar S. 1987. Untersuchungen über das Vorkommen pilzlicher Parasiten an Feinwurzeln 70–90jähriger Fichten (*Picea abies* Karst.). *Mitteilungen des Vereins Forstlicher Standortskunde und Forstpflanzenzüchtung* 33:77–80.

Schroers HJ, Zerjav M, Munda A, Halleen F, Crous PW. 2008. *Cylindrocarpon pauciseptatum* sp nov., with notes on *Cylindrocarpon* species with wide, predominantly 3-septate macroconidia. *Mycol Res* 112:82–92.

Schulz B, Boyle C. 2005. The endophytic continuum. *Mycol Res* 109:661–686.

Schulz B, Boyle C, Draeger S, Römmert AK, Krohn K. 2002. Endophytic fungi: A source of novel biologically active secondary metabolites. *Mycol Res* 106:996–1004.

Schulz B, Boyle C, Sieber TN. 2006. *Microbial Root Endophytes*. Berlin, Germany: Springer.

Schuster RP, Sikora RA, Amin N. 1995. Potential of endophytic fungi for the biological control of plant parasitic nematodes. *Mededelingen Faculteit Landbouwkundige en Toegepaste Biologische Wetenschappen Universiteit Gent* 60:1047–1052.

Sengupta A, Chakraborty DC, Chaudhuri S. 1988. Do septate endophytes also have a mycorrhizal function for plants under stress? In *Mycorrhizae for Green Asia: First Asian Conference on Mycorrhizae*, eds. A Mahadevan, N Raman, K Natarajan, pp. 169–174. Madras, India: Centre of Advanced Studies in Botany, University of Madras, Guindy Campus.

Sequerra J, Capellano A, Gianinazzi-Pearson V, Moiroud A. 1995. Ultrastructure of cortical root cells of *Alnus incana* infected by *Penicillium nodositatum*. *New Phytol* 130:545–555.

Serfling A, Wirsel SGR, Lind V, Deising HB. 2007. Performance of the biocontrol fungus *Piriformospora indica* on wheat under greenhouse and field conditions. *Phytopathology* 97:523–531.

Shefferson RP, Weiss M, Kull T, Taylors DL. 2005. High specificity generally characterizes mycorrhizal association in rare lady's slipper orchids, genus *Cypripedium*. *Mol Ecol* 14:613–626.

Siddiquee S, Yusuf UK, Zainudin NAIM. 2010. Morphological and molecular detection of *Fusarium chlamydosporum* from root endophytes of *Dendrobium crumenatum*. *Afr J Biotechnol* 9:4081–4090.

Sieber TN. 1985. Endophytische Pilze von Winterweizen (*Triticum aestivum* L.). Zürich, Switzerland: Institute of Microbiology, Swiss Federal Institute of Technology.

Sieber T. 1988. Endophytic fungi in needles of healthy-looking and diseased Norway spruce (*Picea abies* L. Karsten). *Eur J For Pathol* 18:321–342.

Sieber TN. 1989. Endophytic fungi in twigs of healthy and diseased Norway spruce and white fir. *Mycol Res* 92:322–326.

Sieber TN. 2002. Fungal root endophytes. In *Plant Roots: The Hidden Half*, eds. Y Waisel, A Eshel, U Kafkafi, 3rd edn., pp. 887–917. New York: Marcel Dekker, Inc.

Sieber TN. 2007. Endophytic fungi in forest trees: Are they mutualists? *Fungal Biol Rev* 21:75–89.

Sieber TN, Grünig CR. 2006. Biodiversity of fungal root-endophyte communities and populations in particular of the dark septate endophyte *Phialocephala fortinii*. In *Microbial Root Endophytes*, eds. B Schulz, C Boyle, TN Sieber, pp. 107–132. Berlin, Germany: Springer.

Sieber TN, Riesen TK, Müller E, Fried PM. 1988. Endophytic fungi in four winter wheat cultivars (*Triticum aestivum* L.) differing in resistance against *Stagonospora nodorum* (Berk.) Cast. & Germ. = *Septoria nodorum* (Berk.) Berk. *J Phytopathol* 122:289–306.

Sieber-Canavesi F, Sieber TN. 1993. Successional patterns of fungal communities in needles of European silver fir (*Abies alba* Mill.). *New Phytol* 125:149–161.

Sigler L, Allan T, Lim SR, Berch S, Berbee M. 2005. Two new *Cryptosporiopsis* species from roots of ericaceous hosts in western North America. *Stud Mycol* 53:53–62.

Simonsen J. 1971. *Phialophora radicicola* Cain, the conidial state of *Gaeumannomyces graminis* in Denmark. *Friesia* 9:361–368.

Skipp RA, Christensen MJ. 1989. Fungi invading roots of perennial ryegrass (*Lolium perenne* L.) in pasture. *NZ J Agric Res* 32:423–431.

Smith SE, Read DJ. 2008. *Mycorrhizal Symbiosis*, 3rd edn. San Diego, CA: Academic Press.

Smith VL, Wilcox WF, Harman GE. 1990. Potential biological control of *Phytophthora* root and crown rots of apple by *Trichoderma* and *Gliocladium* spp. *Phytopathology* 80:880–885.

Sridhar KR, Bärlocher F. 1992. Endophytic aquatic hyphomycetes of roots of spruce, birch and maple. *Mycol Res* 96:305–308.

Stark C, Babik W, Durka W. 2009. Fungi from the roots of the common terrestrial orchid *Gymnadenia conopsea*. *Mycol Res* 113:952–959.

Steinke E, Williams PG, Ashford AE. 1996. The structure and fungal associates of mycorrhizas in *Leucopogon parviflorus* (Andr.) Lindl. *Ann Bot* 77:413–419.

Stoyke G, Currah RS. 1991. Endophytic fungi from the mycorrhizae of alpine ericoid plants. *Can J Bot* 69:347–352.

Stoyke G, Currah RS. 1993. Resynthesis in pure culture of a common subalpine fungus-root association using *Phialocephala fortinii* and *Menziesia ferruginea* (Ericaceae). *Arct Alp Res* 25:189–193.

Stoyke G, Egger KN, Currah RS. 1992. Characterization of sterile endophytic fungi from the mycorrhizae of subalpine plants. *Can J Bot* 70:2009–2016.

Su Y-Y, Guo L-D, Hyde KD. 2010. Response of endophytic fungi of *Stipa grandis* to experimental plant function group removal in Inner Mongolia steppe, China. *Fungal Divers* 43:93–101.

Sumarah MW, Puniani E, Blackwell BA, Miller JD. 2008. Characterization of polyketide metabolites from foliar endophytes of *Picea glauca*. *J Nat Prod* 71:1393–1398.

Summerbell RC. 1989. Microfungi associated with the mycorrhizal mantle and adjacent microhabitats within the rhizosphere of black spruce. *Can J Bot* 67:1085–1095.

Tao G, Liu ZY, Hyde KD, Liu XZ, Yu ZN. 2008. Whole rDNA analysis reveals novel and endophytic fungi in *Bletilla ochracea* (Orchidaceae). *Fungal Divers* 33:101–122.

Tedersoo L, Paertel K, Jairus T, Gates G, Poldmaa K, Tamm H. 2009. Ascomycetes associated with ectomycorrhizas: Molecular diversity and ecology with particular reference to the Helotiales. *Environ Microbiol* 11:3166–3178.

Tejesvi MV, Ruotsalainen AL, Markkola AM, Pirttila AM. 2010. Root endophytes along a primary succession gradient in northern Finland. *Fungal Divers* 41:125–134.

Tellenbach C, Sieber TN. 2012. Do colonization by dark septate endophytes and elevated temperature affect pathogenicity of oomycetes? *FEMS Microbiology Ecology* 82:157–168.

Tellenbach C, Grünig CR, Sieber TN. 2010. Suitability of quantitative real-time PCR to estimate the biomass of fungal root endophytes. *Appl Environ Microbiol* 76:5764–5772.

Tellenbach C, Grünig CR, Sieber TN. 2011. Negative effects on survival and performance of Norway spruce seedlings colonized by dark septate root endophytes are primarily isolate-dependent. *Environ Microbiol* 13:2508–2517.

Thormann MN, Currah RS, Bayley SE. 1999. The mycorrhizal status of the dominant vegetation along a peatland gradient in southern boreal Alberta, Canada. *Wetlands* 19:438–450.

Tian W, Zhang CQ, Qiao P, Milne R. 2011. Diversity of culturable ericoid mycorrhizal fungi of *Rhododendron decorum* in Yunnan, China. *Mycologia* 103:703–709.

Treu R, Laursen GA, Stephenson SL, Landolt JC, Densmore R. 1996. Mycorrhizae from Denali National Park and Preserve, Alaska. *Mycorrhiza* 6:21–29.

Trüssel D. 2011. Zur Verbreitungsbiologie von *Phialocephala subalpina, P. fortinii* s.s. und *Acephala applanata*—Vegetative Ausbreitung im Boden. MSc thesis, Zürich, Switzerland: ETH Zürich.

Turner PD. 1962. Morphological influence of exudates of mycorrhizal and non-mycorrhizal fungi on excised root cultures of *Pinus sylvestris* L. *Nature* 194:551–552.

Upson R, Newsham KK, Bridge PD, Pearce DA, Read DJ. 2009. Taxonomic affinities of dark septate root endophytes of *Colobanthus quitensis* and *Deschampsia antarctica,* the two native antarctic vascular plant species. *Fungal Ecol* 2:184–196.

Urcelay C. 2002. Co-occurrence of three fungal root symbionts in *Gaultheria poeppiggi* DC in Central Argentina. *Mycorrhiza* 12:89–92.

Väre H, Vestberg M, Eurola S. 1992. Mycorrhiza and root associated fungi in Spitsbergen. *Mycorrhiza* 1:93–104.

Varma A, Singh A, Sudha et al. 2001. *Piriformospora indica:* An axenically culturable mycorrhiza-like endosymbiotic fungus. In *The Mycota IX. Fungal Associations*, ed. B Hock, pp. 125–150. Berlin, Germany: Springer.

Vaz ABM, Mota RC, Bomfim MRQ et al. 2009. Antimicrobial activity of endophytic fungi associated with Orchidaceae in Brazil. *Can J Microbiol* 55:1381–1391.

Vendramin E, Gastaldo A, Tondello A, Baldan B, Villani M, Squartini A. 2010. Identification of two fungal endophytes associated with the endangered orchid *Orchis militaris* L. *J Microbiol Biotechnol* 20:630–636.

Verkley GJM, Zijlstra JD, Summerbell RC, Berendse F. 2003. Phylogeny and taxonomy of root-inhabiting *Cryptosporiopsis* species, and *C. rhizophila* sp nov., a fungus inhabiting roots of several Ericaceae. *Mycol Res* 107:689–698.

Verma S, Varma A, Rexer KH et al. 1998. *Piriformospora indica*, gen. et sp. nov., a new root-colonizing fungus. *Mycologia* 90:896–903.

Vidal S. 1996. Changes in suitability of tomato for whiteflies mediated by a non-pathogenic endophytic fungus. *Entomol Exp Appl* 80:272–274.

Vilich V, Dolfen M, Sikora RA. 1998. *Chaetomium* spp. colonization of barley following seed treatment and its effect on plant growth and *Erysiphe graminis* f. sp. *hordei* disease severity. *Z Pflanzenkrank Pflanzensch* 105:130–139.

Vodnik D, Mihelcic M, Gogala N. 1997. Isolation of root associated fungi from *Erica herbacea* and their tolerance to lead. *Acta Biol Slovenica* 41:35–42.

Vujanovic V, St-Arnaud M, Barabé D, Thibeault G. 2000. *Phialocephala victorinii* sp. nov., endophyte of *Cypripedium parviflorum. Mycologia* 92:571–576.

Wagg C, Husband BC, Green DS, Massicotte HB, Peterson RL. 2011. Soil microbial communities from an elevational cline differ in their effect on conifer seedling growth. *Plant Soil* 340:491–504.

Walker J. 1981. Taxonomy of take-all fungi and related genera and species. In *Biology and Control of Take-All*, eds. MJC Asher, PJ Shipton, pp. 15–74. London, U.K.: Academic Press.

Waller F, Achatz B, Baltruschat H et al. 2005. The endophytic fungus *Piriformospora indica* reprograms barley to salt-stress tolerance, disease resistance, and higher yield. *Proc Natl Acad Sci U S A* 102:13386–13391.

Wang CJK, Wilcox HE. 1985. New species of ectendomycorrhizal and pseudomycorrhizal fungi: *Phialophora finlandia, Chloridium paucisporum,* and *Phialocephala fortinii. Mycologia* 77:951–958.

Warcup JH. 1988. Mycorrhizal associations of isolates of *Sebacina vermifera. New Phytol* 110:2272–2231.

Ward E, Bateman GL. 1999. Comparison of *Gaeumannomyces*- and *Phialophora*-like fungal pathogens from maize and other plants using DNA methods. *New Phytol* 141:323–331.

Wei G, Kloepper JW, Tuzun S. 1994. Induced systemic resistance to cucumber diseases and increased plant growth by plant growth-promoting rhizobacteria under field conditions. In *Improving Plant Productivity with Rhizosphere Bacteria*, eds. MH Ryder, PM Stephens, GD Bowen, pp. 70–71. Glen Osmond, Australia: CSIRO Division of Soils.

Weiss M, Oberwinkler F. 2001. Phylogenetic relationships in Auriculariales and related groups—Hypotheses derived from nuclear ribosomal DNA sequences. *Mycol Res* 105:403–415.

Weiss M, Sykorova Z, Garnica S et al. 2011. Sebacinales everywhere: Previously overlooked ubiquitous fungal endophytes. *PLOS One* 6:e16793.

Werner C, Petrini O, Hesse M. 1997. Degradation of the polyamine alkaloid aphelandrine by endophytic fungi isolated from *Aphelandra tetragona. FEMS Microbiol Lett* 155:147–153.

White IR, Backhouse D. 2007. Comparison of fungal endophyte communities in the invasive panicoid grass *Hyparrhenia hirta* and the native grass *Bothriochloa macra. Aust J Bot* 55:178–185.

Whittaker RH. 1969. New concepts of kingdoms of organisms. *Science* 163:150–160.

Widler B, Müller E. 1984. Untersuchungen über endophytische Pilze von *Arctostaphylos uva-ursi* (L.) Sprengel (Ericaceae). *Botanica Helvetica* 94:307–337.

Wilcox HE. 1983. Fungal parasitism of woody plant roots from mycorrhizal relationships to Plant Dis. *Annu Rev Phytopathol* 21:221–242.

Wilcox HE. 1996. Mycorrhizae. In *The Hidden Half*, eds. Y Waisel, A Eshel, U Kafkafi, pp. 689–721. New York: Marcel Dekker, Inc.

Wilcox HE, Ganmore-Neumann R. 1974. Ectendomycorrhizae in *Pinus resinosa* seedlings. I. Characteristics of mycorrhizae produced by a black imperfect fungus. *Can J Bot* 52:2145–2155.

Wilcox HE, Wang CJK. 1987. Mycorrhizal and pathological associations of dematiaceous fungi in roots of 7-month-old tree seedlings. *Can J For Res* 17:884–899.

Wilhelm S, Nelson PE, Ford DH. 1969. A gray sterile fungus pathogenic on strawberry roots. *Phytopathology* 59:1525–1529.

Wilson BJ, Addy HD, Tsuneda A, Hambleton S, Currah RS. 2004. *Phialocephala sphaeroides* sp nov., a new species among the dark septate endophytes from a boreal wetland in Canada. *Can J Bot* 82:607–617.

Wu LQ, Lv YL, Meng ZX, Chen J, Guo SX. 2010. The promoting role of an isolate of dark-septate fungus on its host plant *Saussurea involucrata* Kar. et Kir. *Mycorrhiza* 20:127–135.

Xia X, Lie TK, Qian X, Zheng Z, Huang Y, Shen Y. 2011. Species diversity, distribution, and genetic structure of endophytic and epiphytic *Trichoderma* associated with banana roots. *Microb Ecol* 61:619–625.

Xu JT, Guo SX. 2000. Retrospect on the research of the cultivation of *Gastrodia elate* Bl, a rare traditional Chinese medicine. *Chin Med J* 113:686–692.

Yao JM, Wang YC, Zhu YG. 1992. A new variety of the pathogen of maize take-all. *Acta Mycol Sin* 11:99–104.

Yonezawa M, Usuki F, Narisawa K, Takahashi J, Hashiba T. 2004. Anatomical study on the interaction between the root endophytic fungus *Heteroconium chaetospira* and Chinese cabbage. *Mycoscience* 45:367–371.

Yu T, Nassuth A, Peterson RL. 2001. Characterization of the interaction between the dark septate fungus *Phialocephala fortinii* and *Asparagus officinalis* roots. *Can J Microbiol* 47:741–753.

Yuan Z-L, Chen Y-C, Yang Y. 2009. Diverse non-mycorrhizal fungal endophytes inhabiting an epiphytic, medicinal orchid (*Dendrobium nobile)*: Estimation and characterization. *World J Microbiol Biotechnol* 25:295–303.

Yuan ZL, Lin FC, Zhang CL, Kubicek CP. 2010. A new species of *Harpophora* (Magnaporthaceae) recovered from healthy wild rice (*Oryza granulata)* roots, representing a novel member of a beneficial dark septate endophyte. *FEMS Microbiol Lett* 307:94–101.

Zaffarano PL, Duò A, Grünig CR. 2010. Characterization of the mating type (MAT) locus in the *Phialocephala fortinii* s.l.—*Acephala applanata* species complex. *Fungal Genet Biol* 47:761–772.

Zaffarano PL, Queloz V, Duò A, Grünig CR. 2011. Sex in the PAC: A hidden affair in dark septate endophytes? *BMC Evol Biol* 11:282.

Zhang C, Yin L, Dai S. 2009. Diversity of root-associated fungal endophytes in *Rhododendron fortunei* in subtropical forests of China. *Mycorrhiza* 19:417–423.

Zhang MS, Wu SJ, Jie XJ et al. 2006. Effect of endophyte extract on micropropagation of *Cremastra appendiculata* (D. Don.) Makino (Orchidaceae). *Propag Ornament Plants* 6:83–89.

Zijlstra JD, Van't Hof P, Baar J et al. 2005. Diversity of symbiotic root endophytes of the Helotiales in ericaceous plants and the grass, *Deschampsia flexuosa. Stud Mycol* 53:147–162.

39

Molecular Physiology of Tree Ectomycorrhizal Interactions

Mohammad Tanbir Habib
Georg-August University of Göttingen

Till Heller
Georg-August University of Göttingen

Andrea Polle
Georg-August University of Göttingen

I. Introduction

Roots of most land plant species are colonized by mycorrhizal fungi. This association serves mutual benefits: plant-derived carbohydrates are allocated to the fungus and in return the fungus delivers nutrients, in particular phosphorus and nitrogen to its host plant (Smith and Read 2008). It has been estimated that about 80% of the plant P and N contents have been acquired via mycorrhizas underlining their importance for plant nutrition (van der Heijden et al. 2008). However, it should not be forgotten that mycorrhizal fungi provide additional ecological benefits to their hosts such as increased drought tolerance, decreased susceptibility to biotic diseases, improved tolerance to heavy metals, excess salinity, and other environmental cues (Schützendübel and Polle 2002; Polle and Schützendübel 2003; Luo et al. 2009; Beniwal et al. 2010; see also Chapter 40).

Among different mycorrhizal lifestyles, arbuscular and ectomycorrhizal fungi (AM and EM, respectively) have been most studied. While AMs form treelike structures (arbuscules), coils, and vesicles within the cells, EMs colonize only the interface between root cortex cells forming a netlike structure (Hartig net) for mutual nutrient exchange. EM hyphae ensheath root tips resulting in typical mantle structures (called "morphotype" or "phylotype") (Figure 39.1A and B) characterized by differences in thickness, color, presence, and lengths of emanating hyphae. These structures are fundamental to the uptake and allocation of nutrients as discussed in Section III.

While AMs are formed with roots of the majority of higher plant species including grasses, herbs, and trees, EMs have hitherto only been detected in woody species (Brundrett 2009). Most tree species of the boreal and temperate climatic zone are associated with EM fungi, whereas in tropical ecosystem, AM associations are prevalent. However, there are exceptions, e.g., members of the family of Dipterocarpaceae and Fagaceae form EM (Brundrett 2009). Notably, tropical mono-dominance, i.e., mixed forests dominated by one particular species, is often caused by EM-forming tree species (McGuire et al. 2008). On a global scale, more than 7500 different EM species have been identified and estimates suggest that the total species numbers will be in the range from 25,000 to 30,000 (Rinaldi et al. 2008). New sequencing techniques offer the possibility for undirected, encompassing analyses of fungal species richness (Bueé et al. 2009). With the advent of these methods, knowledge on the diversity and composition of mycorrhizal species in the different ecosystems is currently rapidly increasing suggesting that EM diversity is similar in tropical, temperate, and boreal forests (Peay et al. 2010; Tedersoo and Nara 2010; Tedersoo et al. 2010a,b; Lang et al. 2011). Apart from significant diversity within ecosystems, there is also significant temporal variation in the colonization pattern of different fungal species (Koide et al. 2007). Furthermore, a single tree species can be colonized by more than 100 different EM fungi (Lang et al. 2011). To unravel the significance of EM diversity, central questions are whether root colonization of different EM fungi involves the same molecular events

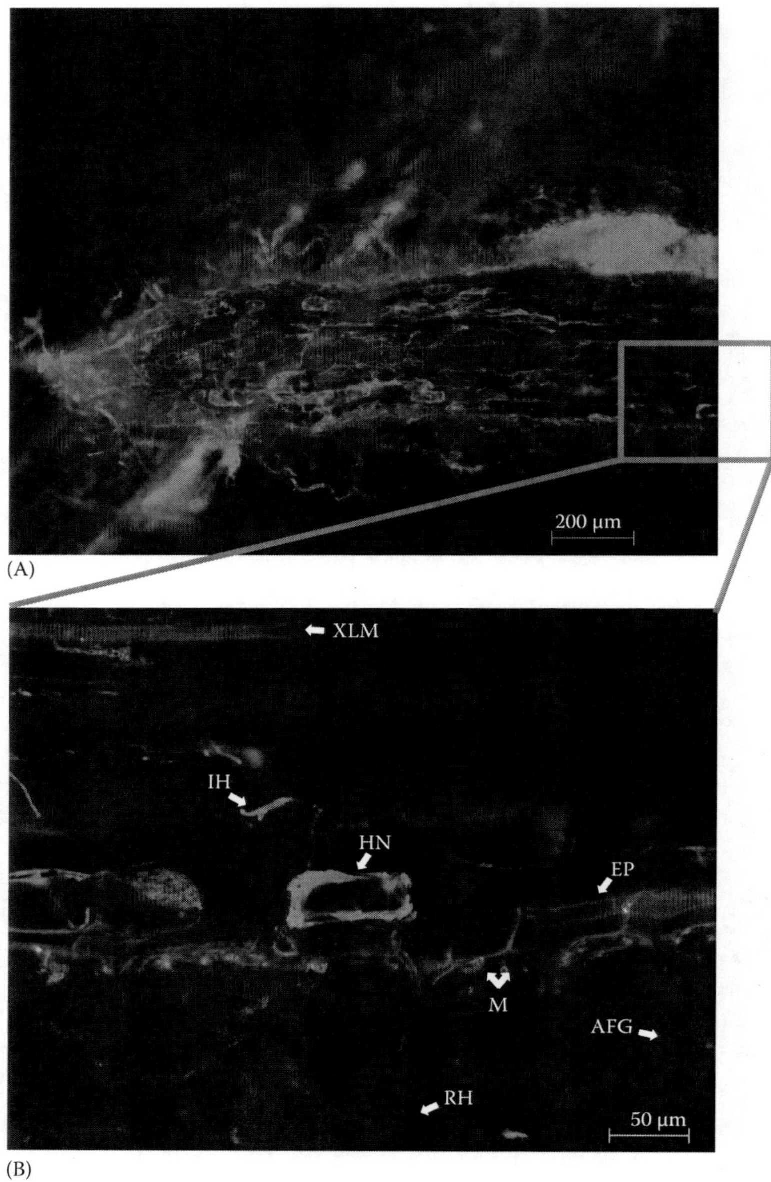

FIGURE 39.1 **(See color insert.)** Morphological stages of *L. bicolor* on and in *Populus* x *canescens* roots. (A) Root tip ensheathed by a fungal mantle with emanating hyphae, (B) details of root cells in contact with hyphae forming a Hartig net. The organisms were cocultured for 5 weeks in a Petri dish on agar medium. Infected roots were fixed and 65–70 μm thick longitudinal sections were prepared with a sliding microtome. The fungal hyphae were stained with the chitin-specific florescent dye WGA-Alexa Fluor-488 (green), and the plant cell walls were stained with propidium iodide (red). Sample was visualized under a Zeiss fluorescent microscope. M = mantle, HN = Hartig net, RH = root hair, EP = epidermis, XLM = xylem, IH = invading hyphae, AFG = autofluorescence of glue.

and whether the functional traits of different EM species differ, thus providing complementary benefits to their host.

In this chapter, we place a focus on the molecular physiology of EM fungal tree interaction. With respect to functions and ecology of AM fungi, we refer the reader to several excellent recent reviews (Bucher 2007; Smith and Smith 2011). Here, we summarize knowledge on the steps required for formation of EM, consider the consequences for root physiology and morphology, and discuss aspects of the impact of EM fungal diversity on these processes. The structures formed by this interaction eventually serve mutual benefits, for plants particularly facilitated access to P and N (Read and Perez-Moreno 2003; Plassard and Dell 2010; Wu 2010; Cairney 2011). The detailed consideration of the molecular mechanisms involved in the regulation of N and P uptake, transformation, and transport is beyond the scope of this chapter (for N uptake mechanisms, see Chapter 25). Instead, we provide an overview on extracellular EM fungal enzyme activities, which are the prerequisite to increase plant supply with P and N and which may be key factors to understanding functional diversity of EM.

II. Molecular and Physiological Changes in Roots In Response To Ectomycorrhizal Colonization

A. Transcriptional, Physiological, and Morphological Changes in Roots during Mycorrhizal Formation

The formation of a mature, functional EM takes place in four stages—preinfection, colonization, differentiation, and functioning (Martin et al. 1997)—which are governed by a series of transcriptional and physiological rearrangements in both partners. At the preinfection stage, there is no actual physical contact between the tree root and its potential ectosymbiont, but they can recognize each other via yet unknown long-distance communicating molecules (Martin et al. 2007). One potential candidate is ethylene (Splivallo et al. 2009; Labbe et al. 2011), but other water-soluble compounds as in case of AM- and nodule-forming bacteria may also be involved (Dénarié et al. 1996; Maillet et al. 2011). Although at this earliest stage of EM development the partners do not yet support each other with nutrients, the preinfection stage is crucial for compatibility determination and preparation of metamorphic changes.

Several in vitro studies have been carried out to study transcriptional patterns without direct contact by separating the EM partners by a semipermeable membrane (Menotta et al. 2004; Felten et al. 2009; da Silva Coelho et al. 2010). These studies employed different plant–fungal combinations using the ascomycetes *Tuber borchii* Vittad. with *Tilia americana* L. (Menotta et al. 2004) and *Hydnangium* spp. Wallr. with *Eucalyptus grandis* W. Hill ex Maid. (da Silva Coelho et al. 2010) and the basidiomycete *Laccaria bicolor* (Maire) P.D. Orton with *Populus tremula* L. x *P. alba* L. (Felten et al. 2009). In other studies, plant or fungal tissues were investigated at an early stage after physical contact between hyphae and roots (*Eucalyptus gobulus* Labill. with *Pisolithus microcarpus* (Cooke and Massee) G. Cunn., Duplessis et al. 2005; *Betula pendula* Roth with *Paxillus involutus* (Fries) Fries, Le Quéré et al. 2005). In roots, early symbiotic responses were activation of genes for pathogen-related proteins, heat shock factors, glutathione S-transferases, and metallothioneins (Johansson et al. 2004; Duplessis et al. 2005; Felten et al. 2009), suggesting that fungal recognition by the plants involves redox regulation and defense activation.

Early developmental stages of mycorrhizal formation, when plant and fungus are in contact but have not yet formed the Hartig net, are also characterized by increased expression of defense- and pathogen-related plant genes (Voiblet et al. 2001; Podila et al. 2002; Duplessis et al. 2005; Frettinger et al. 2007; Heller et al. 2008). For example, activation of metallothioneins was found in early interaction of *E. gobulus–P. microcarpus* (Duplessis et al. 2004), *B. pendula–P. involutus* (Johansson et al. 2004), and *Quercus robur–Piloderma croceum* J. Erikss and Hjortstam (Krüger et al. 2004; Frettinger et al. 2007). Glutathione S-transferase expression was increased in *P. involutus–B. pendula* (Johansson et al. 2004) as well as in *Pinus sylvestris* L.–*L. bicolor* (Heller et al. 2011). However, with the establishment of a functional EM structure, downregulation of defense responses has been reported (Voiblet et al. 2001; Felten et al. 2009; Heller et al. 2011). Indeed, reanalyzing the transcriptome data published by Luo et al. (2009) for fully functional *P. tremula* x *P. alba* –*P. involutus* mycorrhizas revealed that the category "response to stimulus" was significant and contained only genes with decreased transcript levels (using data in supplementary Table S1 in Luo et al. 2009 for AgriGO analysis/bioinfo.cau.edu.cn/agriGO/). Altogether this indicates that EMs overcome the plants' shield against invading microorganisms.

There is now evidence that small secreted fungal peptides are involved in EM formation (Martin et al. 2008, 2010). In *L. bicolor*–poplar interaction mycorrhizal induced small secreted protein 7 (MiSSP7) is the most highly regulated fungal transcript in symbiotic tissue that enters the root cell via endocytosis and accumulates in the plant nucleus. MiSSP7 RNAi mutant lines were unable to penetrate between the root cells, indicating that the fungus is also actively participating in the establishment of the symbiosis (Plett et al. 2011). Moreover, MiSSP7 of *L. bicolor* reprograms the host transcriptome, similar to what has been described for effectors from plant pathogenic fungi (Doehlemann et al. 2009; Plett et al. 2011). This suggests that mycorrhizal and pathogenic fungi as well as pathogenic oomycetes use similar mechanism to suppress the host defense (Dou et al. 2008; Rafiqi et al. 2010; van West et al. 2010; Kloppholz et al. 2011; Plett et al. 2011).

To enable formation of EM structures, root development is also reprogrammed. Early responses of plants to the presence of EM fungi were upregulation of many ethylene-responsive and some ethylene-producing genes and also transcripts related to auxin signaling (Duplessis et al. 2005; Felten et al. 2009). These hormones may also be of fungal origin, e.g., *T. borchii*, *Tuber melanosporum* Vittad., and *Hebeloma cylindrosporum* Romagn. (Gay et al. 1994; Splivallo et al. 2009). Applying exogenous auxin to the EM system of *P. croceum* J. Erikss. and Hjortstam–*Q. robur* L. resulted in more intense EM colonization of the host compared to untreated controls (Herrmann et al. 2004). A stimulative role of auxin during host infection has also been proposed for AMs and root pathogenic fungi (Hanlon and Coenen 2011; Kidd et al. 2011). Moreover, when poplar roots were incubated with MiSSP7 protein, auxin-responsive genes were induced as early as 1 h after exposure (Plett et al. 2011).

Auxin and ethylene have been implicated in the early phase of EM formation and probably lead eventually to changes in root morphology (Felten et al. 2009; Splivallo et al. 2009). Whether other volatile compounds are involved in the induction of changes in root morphology remains to be studied. Diffusible signal(s) from *L. bicolor* is able to alter the endogenous plant auxin distribution, followed by the simulation of lateral root development, even when plant and fungus are separated by a semipermeable membrane (Felten et al. 2009). Transcriptomic study of root tissues to identify gene networks that regulate lateral root development in poplar revealed that many auxin biosynthesis or auxin transport-related transcripts were regulated during indirect interaction between poplar root and the EM fungus *L. bicolor* (Felten et al. 2009). Among these are members of the auxin-responsive transcription factor gene family PtaIAA, IAA-amido synthetases gene family PtaGH3, auxin influx carriers PtaAUX, and auxin efflux carriers PtaPIN (Felten et al. 2009). It has been proposed that the effect on lateral root induction in poplar is due to unknown fungal signals that cause an auxin accumulation at the root apex and increase the polar auxin transport through PtaPIN9 (Felten et al. 2009). Furthermore, *Clv1-like* gene and *Nodulin21-like* genes, which may function in lateral root initiation, are regulated in different developmental stages during the interaction of *P. sylvestris* with *L. bicolor* (Heller et al. 2008). The formation of lateral roots is ecologically important because the newly emerged roots increase the nutrient absorption area for the plant and provide additional habitats for the ectosymbiont.

While the presence of EM is stimulating formation of new lateral roots, the fungus needs to block further growth of the root tip after ensheathing to prevent the breakthrough of the growing root tip through the fungal mantle. Indeed, in mature mycorrhizas, auxin is diminished and no active auxin could be localized in root tips of mycorrhizal compared to non-mycorrhizal plants (Luo et al. 2009). Transcript levels of auxin-sensitive *GH3* were also suppressed in fully developed mycorrhizal pine roots (Reddy et al. 2006). This underlines that the auxin homeostasis gene GH3 is a molecular marker of EM-induced early reprogramming of plant transcriptome as suggested by Heller et al. (2011), but not of a fully developed EM. In fully developed EM, a large number of genes showed differential regulation compared to non-EM roots (see supplement Table S1 in Luo et al. 2009). Notably, analysis of gene ontologies with AgriGO (bioinfo.cau.edu.cn/agriGO) revealed significant enrichment of genes with decreased transcript levels in the categories "system development" and "organ development." These categories contain transcription factors important for cell fate determination (NF-Y = nuclear transcription factor Y; AS1 = asymmetric leaves [AS1] that a gene involved in specifying cell boundaries; AML4 = ARABIDOPSIS MEI2-LIKE 4, a transcription factor that plays a role in meiosis; TOPLESS-RELATED 3 that mediates auxin-dependent transcriptional repression), SNF2 (denominated after a sucrose on-fermentation mutant of yeast) that encodes a putative chromatin remodeling ATPase, and phospholipase D, which may function in vesicle trafficking required for auxin transport (see Luo et al., supplementary Table 1). The decreases in these plant developmental regulators suggest that the fungus induces growth arrest of the colonized root tip by yet unknown signaling pathways.

B. Transcriptional, Physiological, and Morphological Changes in EM-Forming Fungi during Mycorrhizal Establishment and Functioning

EM fungi occur as free-living saprophytes in the soil. This is a major difference from AM fungi, which do not proliferate outside their host plants. In vitro growth experiments reveal morphological changes in free mycelium of EM-forming fungi when a potential host is present. For example, *T. borchii* grown in presence of its natural tree host *T. americana*, but without physical contact, shows rapid apical growth of fungal hyphae (Menotta et al. 2004). The apical growth of mycelium stops after getting close to the host root and the subapical hyphae start to differentiate (Menotta et al. 2004). The preinfection stage is also accompanied by reprogramming of the fungal transcriptome, in particular of the expression of genes for cell structure-related proteins such as a GAS-2 homologue, actin-associated proteins, and centractin-like cell nuclear migration protein (Menotta et al. 2004). Similar transcriptional changes occur in the EM basidiomycete *Hydnangium* spp. during indirect interaction with *E. grandis* (da Silva Coelho et al. 2010). Furthermore, in the ectosymbionts, transcripts related to energy metabolism, cell structure, and cellular detoxification were increased including, e.g., regulation of cytochrome P450 and stimulation of glutathione S-transferase (Menotta et al. 2004; da Silva Coelho et al. 2010).

While plants show regulation of defense- and auxin-responsive genes during early interactions in different associations such as *Pinus resinosa* Ait.–*L. bicolor, B. pendula–P. involutus, E. gobulus–P. microcarpus*, and *E. gobulus–P. tinctorius*, the fungal partners regulate mainly genes for structural components like actin and hydrophobin (Tagu et al. 1996; Voiblet et al. 2001; Podila et al. 2002; Johansson et al. 2004; Table 1 of Duplessis et al. 2005; Le Quéré et al. 2005). Among early downregulated fungal transcripts are the translation elongation factor EF-1 gamma and several signaling genes (Ras-related GTPase rho-type, GTP binding proteins ypt1 and SEC4; resp., Johansson et al. 2004; Le Quéré et al. 2005), while *HydPt-2* of *Pisolithus* spp. encoding a structural component of the cell wall is increased during early symbiosis development (Tagu et al. 1996; Voiblet et al. 2001; Acioli-Santos et al. 2008; da Silva Coelho et al. 2010). Although our understanding of the mechanistic events is still very limited, emerging data indicate that restructuring the fungal shape involves similar molecular events during the establishment of different fungal–plant interactions.

The largest group of transcripts regulated during early interaction represents structural subunits of the ribosome and other components of protein synthesis machinery. These transcripts were mainly downregulated at an early time point during EM interaction and gradually reached their normal expression levels at later time points. For example, in the EM fungi *P. involutus* and *P. microcarpus*, several early repressed and late balanced/induced transcripts in contact with host show sequence similarity to a *L. bicolor* translation elongation factor EF-1 gamma (protein ID 677254) (Duplessis et al. 2005; Le Quéré et al. 2005). An exception is the upregulation of *L. bicolor* EF-1 alpha/Tu (protein ID 568550) and the corresponding transcript in *P. involutus* and *P. tinctorius* during interaction with their natural hosts (Voiblet et al. 2001; Podila et al. 2002; Johansson et al. 2004). Generally, downregulation of transcripts representing protein synthesis appears to indicate fungal alteration of saprotrophic to biotrophic status.

EM formation between different fungi and plant species involves stimulation of fungal GSTs (Podila et al. 2002; Tagu et al. 2003; Le Quéré et al. 2005). GSTs are also induced in *L. bicolor* interacting with its helper bacteria *Pseudomonas fluorescens* (Deveau et al. 2007). Since differences in GST activation were observed between *P. involutus* strains that formed a functional EM with birch and those that were not able to colonize the host, it was assumed that these enzymes may be involved in host recognition (Le Quéré et al. 2005). However, in different organisms, GSTs form large families with many members whose functions are not yet well understood (Edwards et al. 2000). Therefore, it will be difficult to assign functional roles to these enzymes in EM formation.

The primary metabolism of the fungi also undergoes changes during EM formation. In addition to carbohydrate (hexokinase, pyruvate kinase) and amino acid metabolism, transcripts for various enzymes of the fatty acid metabolism were affected (Table 39.1). For example, in *L. bicolor*, transcripts representing the members of thiolase family (3-ketoacyl-CoA thiolase) are differentially regulated (Podila et al. 2002; Reich et al. 2009). While *THIK1* was induced in early EM stages compared to free-living mycelium in *P. resinosa—L. bicolor* interaction (Podilaetal.2002), *THIK5* was downregulated in 9-month-old greenhouse *Pseudotsuga menziesii* (Mirb.) Franco–*L. bicolor* mycorrhizas compared to free-living mycelium (Reich et al. 2009). In *Eucalyptus globules–P. microcarpus* interaction, fungal transcripts related to mitochondrial energy metabolism were increased (Duplessis et al. 2005), indicating that the transition from the free to the mutualistic lifestyle involves the investment of fungal carbon resources.

The fully developed EM consists of root cells in close contact with the fungal cells forming the Hartig net, which are connected with the fungal hyphae ensheathing the roots' tip as a mantle and also the hyphae emanating into the soil (Figure 39.1A). These specific fungal tissues fulfill different functions with respect to nutrient acquisition from the soil and translocation of compounds to plant cells. Therefore, it would be interesting to identify fungal morphogenes that lead to the development of specific structures. To date, very little is known about these processes. Table 39.2 compiles a list of fungal genes that were differentially regulated in the mantle compared to rhizomorphs (= long-reaching hyphae that are connected with each other forming rootlike structures) as well as of genes showing differences between free-living mycelium and rhizomorphs. Most genes identified are also found during EM formation. However, some candidates like *lecA* of *P. involutus* and *L. bicolor*, representing a gene encoding a fungal fruiting body lectin, may be important for the establishment of structural differences. Lectins bind to sugar moieties and, therefore, may be involved in recognition and adhesion (Bovi et al. 2011). For example, in poplar-*L. bicolor* interaction, the expression of *lecA* was 12-fold upregulated in the fully developed mycorrhizas compared to the free-living mycelium (Martin et al. 2008). In birch–*P. involutus*, *lecA* was downregulated at the early stage but was gradually restored at later time points (Le Quéré et al. 2005). Furthermore, *lecA* induction was missing in a *P. involutus* strain (Le Quéré et al. 2004) unable to form a functional mycorrhiza (Gafur et al. 2004).

Among other early repressed and late induced fungal transcripts in *P. involutus*–birch interaction are *cipC1* and *cipC2* (Le Quéré et al. 2005). CipC-like transcripts are among the most abundant transcripts in the free-living mycelium of both EM fungi, *L. bicolor* and *P. microcarpus* (Peter et al. 2003). CipC has initially been found as a concanavalin-induced protein, whose functions have not yet been clarified (Melin et al. 2002). CipC transcripts are more abundant in *P. involutus* cells within the EM root tip than in the extraradical mycelium (Morel et al. 2005). Thus, it was hypothesized that the cipC genes of *P. involutus* might be linked to the changes in morphology during mycorrhization (Morel et al. 2005). Their orthologue is also regulated during *L. bicolor* interaction with its helper bacterial strain *P. fluorescens* (Deveau et al. 2007). All these findings suggest that they might play a role during the shift from saprotrophic to biotrophic lifestyle. Comparison between the genomic sequence of compatible and incompatible *P. involutus* isolates and the respective transcript revealed that cipC genes are among the rapidly evolving genes within the species and therefore have a potential role in host specificity (Le Quéré et al. 2004).

With regard to the beneficial effects of EM for plants, the regulation of genes involved in nutrient acquisition is of particular interest. EM fungi obtain plant carbohydrates and constitute a significant sink for photosynthetic products (Lamhamedi et al. 1994; Nehls et al. 2010; Teramoto et al. 2011). Indeed, expression of sugar transporters and hexokinases was increased in *P. involutus*, *P. microcarpus*, *L. bicolor*, and *T. melanosporum* Vittad. during interaction with their corresponding host in comparison with free-living mycelium (Johansson et al. 2004; Duplessis et al. 2005; Martin et al. 2010). When plant carbohydrate supply to the fungi was manipulated by shading or girdling, both colonization and EM diversity decreased (Druebert et al. 2009; Pena et al. 2010). Furthermore, plant varieties with lower inherent growth rates such as inland of *P. menziesii* displayed lower EM fungal diversity than fast-growing coastal varieties (Ducic et al. 2009). This indicates that plant carbon productivity drives EM formation. Apparently, the plant somehow recognizes the potential or realized benefits it obtains by specific fungi. Recent studies employing two different AM fungal species that provided high or low benefits to a given host plant showed that more carbohydrates were allocated to the more efficient than to the less efficient fungus (Kiers et al. 2011). When the nitrate reductase gene (protein ID 291348) of *L. bicolor* was knocked down, the capacity of the fungus for root colonization was compromised (Kemppainen et al. 2009). This underlined that the host was able to monitor the nutritional status of candidate ectosymbionts and avoided the establishment of an inefficient interaction.

The colonization with EM generally increases the phosphorus content of the host (Plassard and Dell 2010). The genome of *L. bicolor* contains five Pi transporters, which have not yet been characterized, but they are not constitutively upregulated in functional EMs (Martin et al. 2008). Studies with *HcPT1* and *HcPT2*, two orthologues in *H. cylindrosporum* Romagn., indicate that

TABLE 39.1 Fungal Transcripts That Are Differentially Expressed during Ectomycorrhiza Formation in Axenic Cultures with Different Host Plants

Fungus	Host Plant	NCBI ID	Lb ID	Annotation	Regulation	References
				Metabolism		
Amino acid transport and metabolism						
P. involutus	*B. pendula*	T143A00784F	671723	Cystathionine beta-lyases/ cystathionine gamma-synthases	25 days ↑	Johansson et al. (2004)
P. microcarpus	*E. gobulus*	St187	188701	Delta-1-pyrroline-5-carboxylate dehydrogenase	12 days ↑	Duplessis et al. (2005)
P. involutus	*B. pendula*	T143A02145F	311861	Isocitrate dehydrogenase	2–14 days ↑	Le Quere et al. (2005)
Carbohydrate transport and metabolism						
P. involutus	*B. pendula*	T143A02411F	312018	Hexokinase	25 days ↑	Johansson et al. (2004)
P. microcarpus	*E. gobulus*	P061D08	312018	Hexokinase	7 days ↑	Duplessis et al. (2005)
P. microcarpus	*E. gobulus*	Lb02E13	301694	Pyruvate kinase	7 days ↑	Duplessis et al. (2005)
Catalytic activity						
P. microcarpus	*E. gobulus*	6B4	709187	Histidine triad-like motif	12 days ↑	Duplessis et al. (2005)
Eicosanoid and glutathione metabolism						
P. involutus	*B. pendula*	T143B05234F	379253	MAPEG	25 days ↓	Johansson et al. (2004)
Energy production and conversion						
P. involutus	*B. pendula*	T143B00685F	673699	Cytochrome c oxidase	4–8 days ↑	Le Quere et al. (2005)
P. microcarpus	*E. gobulus*	Lb03F15	710288	Cytochrome c oxidase, subunit I	12 days ↑	Duplessis et al. (2005)
P. microcarpus	*E. gobulus*	Lb01K02	678672	Dihydrolipoamide acetyltransferase	12 days ↑	Duplessis et al. (2005)
P. microcarpus	*E. gobulus*	EP2102M07	693318	Glycolate oxidase	7–12 days ↓	Duplessis et al. (2005)
P. microcarpus	*E. gobulus*	8C6	683738	Kynurenine 3-monooxygenase and related flavoprotein monooxygenases	12 days ↑	Duplessis et al. (2005)
P. involutus	*B. pendula*	T143B04084F	301012	Mitochondrial carnitine-acylcarnitine carrier protein	25 days ↓	Johansson et al. (2004)
P. microcarpus	*E. gobulus*	Lb01D10	477251	Mitochondrial F1F0-ATP synthase, subunit c/ATP9/proteolipid	7–12 days ↑	Duplessis et al. (2005)
P. involutus	*B. pendula*	T143B00480	695354	Mitochondrial solute carrier protein	2–8d ↓	Le Quere et al. (2005)
P. microcarpus	*E. gobulus*	Lb01D07	627608	Monooxygenase	7–12 days ↑	Duplessis et al. (2005)
P. involutus	*B. pendula*	T143A01066F	326114	NAD-malate dehydrogenase	4–8 days ↓	Le Quere et al. (2005)
P. tinctorius	*E. gobulus*	BE704435	185536	NAD-malate dehydrogenase	4 days ↑	Voiblet et al. (2001)
P. involutus	*B. pendula*	T143B00262	191230	Predicted mitochondrial carrier protein	25 days ↓	Johansson et al. (2004)
P. microcarpus	*E. gobulus*	7A5	676904	Predicted quinone oxidoreductase	12 days ↑	Duplessis et al. (2005)
P. involutus	*B. pendula*	T143B02973F	467809	Ubiquinol cytochrome c reductase	2–8 days ↑, 14 days ↓	Le Quere et al. (2005)
P. microcarpus	*E. gobulus*	EP402D12	313187	Ubiquinol-cytochrome c reductase	4 days ↑	Duplessis et al. (2005)
P. microcarpus	*E. gobulus*	Lb03N24	491148	Vacuolar H+-ATPase V1 sector, subunit G	12 days ↑	Duplessis et al. (2005)
Inorganic ion transport and metabolism						
P. microcarpus	*E. gobulus*	St94	313939	PAFC16201, RNA polymerase II-associated protein	12 days ↑	Duplessis et al. (2005)
Lipid transport and metabolism						
P. involutus	*B. pendula*	T143A02319F	185962	Cytochrome P450 CYP4	4 days ↑, 8 days ↑	Le Quere et al. (2005)
L. Bicolor	*P. resinosa*	LbSSH00051	392649	Cytochrome P450, E-class P450, group I	6–72 h ↑	Podila et al. (2002)
L. Bicolor	*P. resinosa*	LbSSH00109	185981	THIK-1, 3-ketoacyl-CoA thiolase	6–72 h ↑	Podila et al. (2002)
Metallothionein, Est-supported						
P. microcarpus	*E. gobulus*	P062D01	399677	Metallothionein	4 days ↓	Duplessis et al. (2006)
Phosphorylcholine transferase						
P. microcarpus	*E. gobulus*	5D10	305504	Lipid transport and metabolism	12 days ↑	Duplessis et al. (2005)

TABLE 39.1 (continued) Fungal Transcripts That Are Differentially Expressed during Ectomycorrhiza Formation in Axenic Cultures with Different Host Plants

Fungus	Host Plant	NCBI ID	Lb ID	Annotation	Regulation	References
				Metabolism		
Posttranslational modification, protein turnover, chaperones						
P. microcarpus	*E. gobulus*	P061F09	660461	FKBP-type peptidyl-prolyl *cis–trans* isomerase	4–21 days ↓	Duplessis et al. (2005)
P. microcarpus	*E. gobulus*	Lb01E18	674829	Glutathione S-transferase	12 days ↑	Duplessis et al. (2005)
P. microcarpus	*E. gobulus*	Lb04A03	658386	Glutathione S-transferase	12 days ↑	Duplessis et al. (2005)
P. microcarpus	*E. gobulus*	St82	631723	Gpi-anchor transamidase	12 days ↑	Duplessis et al. (2005)
L. Bicolor	*P. resinosa*	LbSSH00007	189966	Hsp70 chaperone BiP	6–72 h ↑	Podila et al. (2002)
P. tinctorius	*E. gobulus*	AW600823	245357	Thioredoxin reductase	4 days ↑	Voiblet et al. (2001)
P. microcarpus	*E. gobulus*	8C7	245357	Thioredoxin reductase LBTR	12 days ↑	Duplessis et al. (2005)
P. microcarpus	*E. gobulus*	Lb02E11	692536	Thioredoxin-like protein	7–12 days ↑	Duplessis et al. (2005)
P. microcarpus	*E. gobulus*	Lb02E18	181297	UBCc, Ubiquitin-conjugating enzyme E2	7–12 days ↑	Duplessis et al. (2005)
P. involutus	*B. pendula*	T143B01152F	192623	Ubiquitin	2–8 days ↓	Le Quere et al. (2005)
P. involutus	*B. pendula*	T143B01795F	192623	Ubiquitin	2–8 days ↓	Le Quere et al. (2005)
P. microcarpus	*E. gobulus*	5D7	605856	Ubiquitin C-terminal hydrolase	12 days ↑	Duplessis et al. (2005)
P. involutus	*B. pendula*	T143B00558	473733	Ubiquitin-protein ligase	2–8 days ↓	Le Quere et al. (2005)
P. involutus	*B. pendula*	T143B02582F	657951	Ubiquitin-protein ligase	2–8 days ↓	Le Quere et al. (2005)
P. involutus	*B. pendula*	T143A02388F	691182	Ubiquitin-protein ligase	2–8 days ↓	Le Quere et al. (2005)
P. microcarpus	*E. gobulus*	St46	652632	Ubiquitin-protein ligase	4 days ↑	Duplessis et al. (2005)
P. microcarpus	*E. gobulus*	Lb06D12	483673	Ubiquitin-specific protease UBP14	12 days ↑	Duplessis et al. (2005)
L. Bicolor	*P. resinosa*	LbSSH00058	674829	Glutathione S-transferase	6–72 h ↑	Podila et al. (2002)
L. Bicolor	*P. resinosa*	LbSSH00019	674829	Glutathione S-transferase	6–72 h ↑	Podila et al. (2002)
P. involutus	*B. pendula*	T143B01510F	571619	Glutathione S-transferase	25 days ↓	Johansson et al. (2004)
P. involutus	*B. pendula*	T143B01194F	438154	Glutathione S-transferase	2–8 days ↓	Le Quere et al. (2005)
P. involutus	*B. pendula*	T143A00638F	172363	LbPrx1-Cys, Alkyl hydroperoxide reductase	2–8 days ↓, 14 days ↑	Le Quere et al. (2005)
P. involutus	*B. pendula*	T143B00811F	379253	MAPEG	4–8d ↑	Le Quere et al. (2005)
P. involutus	*B. pendula*	T143A04315F	295142	Thioredoxin, LbTrx3	2–8 days ↓	Le Quere et al. (2005)
P. involutus	*B. pendula*	T143B04641F	185202	PHO88-related membrane protein	25 days ↑	Johansson et al. (2004)
Proteolysis and peptidolysis						
P. microcarpus	*E. gobulus*	11D7	703702	Metallopeptidase	12 days ↑	Duplessis et al. (2005)
P. tinctorius	*E. gobulus*	AW600817	637700	Neutral zinc metallopeptidases	4 days ↑	Voiblet et al. (2001)
P. microcarpus	*E. gobulus*	11A7	184269	Prolyl aminopeptidase	12 days ↑	Duplessis et al. (2005)
Secondary metabolites biosynthesis, transport and catabolism						
P. involutus	*B. pendula*	T143B03447F	638005	Cytochrome P450 CYP2 subfamily	25 days ↓	Johansson et al. (2004)
P. microcarpus	*E. gobulus*	EP402C18	295462	Cytochrome P450 CYP2 subfamily	7–14 days ↑	Duplessis et al. (2005)
Stress						
L. Bicolor	*P. resinosa*	LbSSH00059	488195	Peroxidase activity, electron transport, response to oxidative stress	6–72 h ↑	Podila et al. (2002)
Unknown function						
P. involutus	*B. pendula*	T143B00870F	705012	Predicted alpha/beta hydrolase BEM46	25 days ↑	Johansson et al. (2004)
				Regulation		
Endonuclease activity						
P. tinctorius	*E. gobulus*	BE704426	621647	LAGLIDADG DNA endonuclease	4 days ↓	Voiblet et al. (2001)
Information storage and processing						
P. tinctorius	*C. sativa*	ES343845	656443	Heat shock protein Hsp70	12 h ↓	Acioli-Santos et al. (2008)
P. tinctorius	*C. sativa*	ES343842	659763	Heat shock protein Hsp90	12 h ↓	Acioli-Santos et al. (2008)
L. Bicolor	*P. resinosa*	LbSSH00025	701022	Hsp27-ERE-TATA-binding protein	6–72 h ↑	Podila et al. (2002)

(*continued*)

TABLE 39.1 (continued) Fungal Transcripts That Are Differentially Expressed during Ectomycorrhiza Formation in Axenic Cultures with Different Host Plants

Fungus	Host Plant	NCBI ID	Lb ID	Annotation	Regulation	References
				Regulation		
Meiotic recombination-related protein						
P. tinctorius	*E. gobulus*	BE704429	399758	LIM15/DMC1 homolog	4 days ↓	Voiblet et al. (2001)
Nuclear structure						
P. microcarpus	*E. gobulus*	8D10	637483	Protein tyrosine phosphatase SHP1	12 days ↑	Duplessis et al. (2005)
P. tinctorius	*E. gobulus*	AW600836	637483	Protein tyrosine phosphatase SHP1	4 days ↑	Voiblet et al. (2001)
RNA processing and modification						
P. microcarpus	*E. gobulus*	5B8	471190	Small nuclear ribonucleoprotein Sm D3	4–21 days ↓	Duplessis et al. (2005)
P. tinctorius	*E. gobulus*	AW600907	599334	WD40-repeat-containing subunit of the 18S rRNA processing complex	4 days ↑	Voiblet et al. (2001)
Signal transduction						
P. involutus	*B. pendula*	T143B05214F	183731	GTP-binding protein ypt1	25 days ↓	Johansson et al. (2004)
P. involutus	*B. pendula*	T143B01572F	179004	Ras-related small GTPase, Rho type	25 days ↓	Johansson et al. (2004)
P. involutus	*B. pendula*	T143A01259F	387559	GTP-binding protein, SEC4	2–8 days ↓	Le Quere et al. (2005)
P. involutus	*B. pendula*	T143A03762F	661463	Calmodulin	2–8 days ↓	Le Quere et al. (2005)
Threonyl-tRNA synthetase						
P. microcarpus	*E. gobulus*	EP402D17	641282	Translation, ribosomal structure and biogenesis	7 days ↑	Duplessis et al. (2005)
Transcription						
P. involutus	*B. pendula*	T143A03157F	318844	Elongation factor 1 beta/delta chain	2–8 days ↓	Le Quere et al. (2005)
P. microcarpus	*E. gobulus*	P061E06	318844	Elongation factor 1 beta/delta chain	4 days ↑	Duplessis et al. (2005)
L. bicolor	*P. resinosa*	LbSSH00013	469385	PF6.2.1 (Symbiosis related TF)	6–72 h ↑	Podila et al. (2002)
Translation, ribosomal structure and biogenesis						
L. bicolor	*P. resinosa*	LbSSH00047	568550	Translation elongation factor EF-1 alpha/Tu	6–72 h ↑	Podila et al. (2002)
P. involutus	*B. pendula*	T143C02273F	568550	Translation elongation factor EF-1 alpha/Tu	25 days ↑	Johansson et al. (2004)
P. tinctorius	*E. gobulus*	AW600874	568550	Translation elongation factor EF-1 alpha/Tu	4 days ↑	Voiblet et al. (2001)
P. involutus	*B. pendula*	T143B01464F	677254	Translation elongation factor EF-1 gamma	2–8 days ↓	Le Quere et al. (2005)
P. involutus	*B. pendula*	T143B03045F	677254	Translation elongation factor EF-1 gamma	2–8 days ↓, 14 days ↑	Le Quere et al. (2005)
P. involutus	*B. pendula*	T143B00505	677254	Translation elongation factor EF-1 gamma	4d ↓	Le Quere et al. (2005)
P. microcarpus	*E. gobulus*	Lb03E08	677254	Translation elongation factor EF-1 gamma	4–12 days ↓, 21 days ↑	Duplessis et al. (2005)
P. microcarpus	*E. gobulus*	7A7	677254	Translation elongation factor EF-1 gamma	4 days ↓	Duplessis et al. (2005)
P. microcarpus	*E. gobulus*	9B2	677254	Translation elongation factor EF-1 gamma	7 days ↓	Duplessis et al. (2005)
P. involutus	*B. pendula*	T143B03743F	475910	Translation initiation factor 1 (eIF-1/SUI1)	25 days ↓	Johansson et al. (2004)
P. involutus	*B. pendula*	T143B03743F	475910	Translation initiation factor 1 (eIF-1/SUI1)	2–8 days ↓	Le Quere et al. (2005)
P. involutus	*B. pendula*	CN072186	708926	Translation initiation factor 4F	14 days ↑	Morel et al. (2005)
Zinc ion binding						
P. tinctorius	*E. gobulus*	BE704427	672265	SCF ubiquitin ligase	4 days ↓	Voiblet et al. (2001)
P. tinctorius	*E. gobulus*	AW600812	479280	Zinc finger, DksA/TraR C4-type	4 days ↑	Voiblet et al. (2001)

TABLE 39.1 (continued) Fungal Transcripts That Are Differentially Expressed during Ectomycorrhiza Formation in Axenic Cultures with Different Host Plants

Fungus	Host Plant	NCBI ID	Lb ID	Annotation	Regulation	References
				Structure		
Cytoskeleton						
P. involutus	*B. pendula*	T143B01146F	192701	Actin 1 (β-actin)	2–8 days ↓	Le Quere et al. (2005)
P. involutus	*B. pendula*	T143B00767F	292399	Actin depolymerizing factor	2–8 days ↓	Le Quere et al. (2005)
P. microcarpus	*E. gobulus*	3D10	192701	Actin-1	4 days ↓, 7 days ↓	Duplessis et al. (2005)
P. involutus	*B. pendula*	T143A02917F	399439	Actin-related protein ARPC3	2 days ↓	Le Quere et al. (2005)
P. microcarpus	*E. gobulus*	P052E04	294176	Actin-related protein ARPC3	4 days ↓, 7 days ↓	Duplessis et al. (2005)
P. tinctorius	*E. gobulus*	AW600818	399446	actin-related protein ARPC3	4 days ↑	Voiblet et al. (2001)
L. bicolor	*P. resinosa*	LbSSH00014	192523	Alpha tubulin	6–72 h ↑	Podila et al. (2002)
P. involutus	*B. pendula*	T143A00886F	192523	Alpha tubulin	25 days ↑	Johansson et al. (2004)
P. involutus	*B. pendula*	T143A00655F	192524	Alpha tubulin	4 days ↓	Le Quere et al. (2005)
P. involutus	*B. pendula*	T143B00990F	659723	Microtubule-associated anchor protein involved in autophagy and membrane trafficking	2–8 days ↓	Le Quere et al. (2005)
Intracellular trafficking, secretion, and vesicular transport						
P. involutus	*B. pendula*	T143C01496F	453088	GTP-binding ADP-ribosylation factor-like protein ARL1	25 days ↓	Johansson et al. (2004)
L. bicolor	*P. resinosa*	LbSSH00064	317058	Nuclear pore complex, Nup98 component	6–72 h ↑	Podila et al. (2002)
Membrane						
P. tinctorius	*E. gobulus*	AW600908	295862	FUN34 transmembrane protein	4 days ↑	Voiblet et al. (2001)
P. tinctorius	*E. gobulus*	AW731609	295862	FUN34 transmembrane protein	4 days ↑	Voiblet et al. (2001)
N-acetyl galactosamine-binding						
P. involutus	*B. pendula*	T143B04748F	185716	Lectin	2–8 days ↓	Le Quere et al. (2005)
Structural constituent of cell wall						
P. involutus	*B. pendula*	T143A05011F	241509	LbH12_Class I hydrophobin	25 days ↑	Johannson et al. (2004)
P. tinctorius	*C. sativa*	ES343840	241509	LbH12_Class I hydrophobin	12 h ↓	Acioli-Santos et al. (2008)
Hydnangium sp.	*E. grandis*	ABA46363	335058	LbH13_Class I hydrophobin	15 days ↑	da silva Coelho et al. (2010)
P. involutus	*B. pendula*	T143B03881F	335058	LbH13_Class I hydrophobin	25 days ↑	Johannson et al. (2004)
P. microcarpus	*E. gobulus*	EP401L13	335058	LbH13_Class I hydrophobin	12–21 days ↓	Duplessis et al. (2005)
P. microcarpus	*E. gobulus*	EP2102D13	335058	LbH13_Class I hydrophobin	4–12 days ↑	Duplessis et al. (2005)
P. microcarpus	*E. gobulus*	5B9	335058	LbH13_Class I hydrophobin	4–21 days ↓	Duplessis et al. (2005)
P. microcarpus	*E. gobulus*	EP402A12	335058	LbH13_Class I hydrophobin	4 days ↑	Duplessis et al. (2005)
P. microcarpus	*E. gobulus*	P062A06	335058	LbH13_Class I hydrophobin	7–21 days ↓	Duplessis et al. (2005)
P. microcarpus	*E. gobulus*	EP402D21	335058	LbH13_Class I hydrophobin	7 days ↑	Duplessis et al. (2005)
P. tinctorius	*E. gobulus*	AAC49308.1	335058	LbH13_Class I hydrophobin	12 h–7 days ↑	Tagu et al. (1996)
P. tinctorius	*E. gobulus*	AW600900	335058	LbH13_Class I hydrophobin	4 days ↑	Voiblet et al. (2001)
P. tinctorius	*E. gobulus*	AAC49307.1	624234	LbH2_Class I hydrophobin	12 h–7 days ↑	Tagu et al. (1996)
P. tinctorius	*C. sativa*	ES343843	624234	LbH2_Class I hydrophobin	12 h ↓	Acioli-Santos et al. (2008)
				Unknown		
P. involutus	*B. pendula*	T143B00658F	399510	CipC1, Concanamycin Induced Protein C	2–8 days ↓	Le Quere et al. (2005)
P. involutus	*B. pendula*	T143B03675F	470792	Predicted protein	2–8 days ↓	Le Quere et al. (2005)
P. involutus	*B. pendula*	T143B03793F	470792	Predicted protein	2–8 days ↓	Le Quere et al. (2005)
P. tinctorius	*E. gobulus*	BE704428	655810	NIPSNAP1 protein	4 days ↓	Voiblet et al. (2001)
P. microcarpus	*E. gobulus*	P061F11	620912	Predicted protein	4–21 days ↓	Duplessis et al. (2005)

The table is arranged by Gene Annotation Ontology.
NCBI, National center for biotechnology information; GOA, Gene ontology annotation; LB, *L. bicolor*.

TABLE 39.2 Transcripts That Are Differentially Expressed in Different Tissues of *P. involutus* Colonizing *B. pendula*

NCBI ID	Lb ID	Putative Function	Time	Tissue	References
Amino acid transport and metabolism					
T143B02460F	176573	Arylformamidase	28 days	TIP vs RZM or PCH	Wright et al. (2005)
CN072176	659644	Translation elongation factor EF-3b	14 days ↑	M vs PCH	Morel et al. (2005)
Carbohydrate binding					
T143A03455F	295130	CBM12 containing protein	28 days ↓	TIP vs RZM or PCH	Wright et al. (2005)
Carbohydrate transport					
T143A02148F	690162	Predicted suger transporter	28 days ↑	TIP vs RZM or PCH	Wright et al. (2005)
Cytoskeleton					
CN072163	659723	Microtubule-associated anchor protein, autophagy, membrane trafficking	14 days ↑	M vs PCH	Morel et al. (2005)
Cytoskeleton					
T143A00655F	192524	Alpha tubulin	28 days ↓	TIP vs RZM or PCH	Wright et al. (2005)
T143B01146F	192701	Actin 1 (β-actin)	28 days ↓	TIP vs RZM or PCH	Wright et al. (2005)
T143B00767F	292399	Actin depolymerizing factor	28 days ↓	TIP vs RZM or PCH	Wright et al. (2005)
Energy production and conversion					
T143B00480	695354	Mitochondrial solute carrier protein	28 days ↓	TIP vs RZM or PCH	Wright et al. (2005)
Inorganic ion transport and metabolism					
N-acetyl galactosamine-binding lectin					
T143A04543F	464553	Ca^{2+}-modulated nonselective cation channel polycystin	28 days ↓	TIP vs RZM or PCH	Wright et al. (2005)
T143B04748F	185716	Lectin	28 days	TIP vs RZM or PCH	Wright et al. (2005)
Posttranslational modification, protein turnover, chaperones					
T143B01194F	438154	Glutathione S-transferase	28 days	TIP vs RZM or PCH	Wright et al. (2005)
CN072158	192623	Ubiquitin	14 days ↑	M vs PCH	Morel et al. (2005)
CN072180	659763	Molecular chaperone (HSP90 family)	14 days ↑	M vs PCH	Morel et al. (2005)
T143B04641F	185202	PHO88-related membrane protein	28 days ↑	TIP vs RZM or PCH	Wright et al. (2005)
Signal transduction					
T143B03506F	659682	Ras-related small GTPase, Rho type	28 days ↓	TIP vs RZM or PCH	Wright et al. (2005)
Transcription					
CN072185	493900	Transcription factor containing NAC and TS-N domains	14 days ↑	M vs PCH	Morel et al. (2005)
Translation, ribosomal structure and biogenesis					
T143B01464F	677254	Translation elongation factor EF-1 gamma	28 days ↓	TIP vs RZM or PCH	Wright et al. (2005)
CN072166	475910	Translation initiation factor 1 (eIF-1/SUI1)	14 days ↑	M vs PCH	Morel et al. (2005)
CN072181	700222	Mitochondrial translation elongation factor Tu	14 days ↑	M vs PCH	Morel et al. (2005)
Unknown					
T143B00658F	399510	CipC1, Concanamycin Induced Protein C	28 days ↓	TIP vs RZM or PCH	Wright et al. (2005)

The table is arranged by gene annotation ontology.
M, mycorrhiza; FLM, free-living mycelium; TIP, mycorrhizal root tip; RZM, rhizomorphs; PCH, extramatrical mycelium; vs, versus.

expression of these transporters depends strongly on P availability and, thus, increases only during Pi starvation (Tatry et al. 2009). Ecological studies with two varieties of *P. menziesii* with increasing degree of colonization by EM showed that a tight positive correlation existed between plant net primary production and, thus, plant P content, but not with phosphorus use efficiency (Ducic et al. 2009). In contrast, plant nitrogen use efficiency, i.e., the amount of N used per unit of plant biomass produced, increased with the extent of root colonization (Ducic et al. 2009). This indicates that EM fungi stimulate biomass production leading to more efficient N utilization, whereas higher P uptake does not stimulate growth correspondingly. One reason may be that fungi store large quantities of P as polyphosphates that are not easily available to the plant (Ashforda et al. 1999). Furthermore, a surplus of P translocated to the plant can be sequestered in vacuoles together with cations (Ducic and Polle 2007). A recent systems biology study, in which the transcriptome of *P. tremuloides–L. bicolor* mycorrhizas was sequenced, showed that transcripts for inorganic P transporters were enriched neither in the fungal nor in the plant transcriptome (Larsen et al. 2011). Also, fungal ammonium transporters were not enriched (Larsen et al. 2011). In contrast to the fungus, in the host plant, ammonium transporters increased in EM roots (Couturier et al. 2007; Luo et al. 2009). Notably, both amino acid transporters of fungal and plant origin were strongly enriched in EM (Larsen et al. 2011), suggesting increased uptake and translocation of amino acid in both partners. Moreover, the concentration of *P. involutus* mRNA coding for hexose transporter and inorganic phosphate transporter was higher in *B. pendula* mycorrhizal root tips compared to rhizomorphic tissue and mycelia patch (Wright et al. 2005). However, as discussed earlier, the transcriptional regulation of these processes is complex and most likely context-dependent since studies investigating other EM types also reported decreases in these transport systems (see Tables 39.1 and 39.2).

In conclusion, this section shows that EM formation involves diffusible signals that appear to act via auxin signaling and affect root morphology in an interaction that does not require physical contact. Small secreted fungal molecules can travel into the plant nucleus and induce reprogramming that enables EM formation. The establishment of EM is characterized by suppression of massive defense reactions and, thus, resembles that of the colonization of maize with the biotrophic fungus *Ustilago maydis* (Doehlemann et al. 2008). Ecological and genetic analysis revealed that the plant is able to monitor putative benefits and does not foster colonization of inefficient fungi. Transcriptional analyses of various fungal–plant EM systems showed reorganization and differentiation of the fungal shape leading to functional differences. The situation regarding plant P and N nutrition is complex. Regulation at the transcript level appears to be strongly influenced by plant demand and carbon supply to the fungus. Thus, it is difficult to recognize consistent regulatory patterns. Both plants and fungi exude enzymes to facilitate nutrient acquisition, thereby actively influencing their environment. Therefore, understanding of the regulation of nutrient exchange in the EM mutualistic interactions most likely requires consideration of the abilities of EM to modify nutrient relations in their immediate environment.

III. Consequences of Ectomycorrhizas for Extracellular Enzyme Activities with Emphasis on Plant Nutrition

Nutrient acquisition from mineral substances is achieved by exudation of organic acids and siderophores resulting in the solubilization of recalcitrant compounds (Rosling 2009). Nutrients from organic soil particles are mainly obtained through the secretion of specific EM enzymes (Smith and Read 2008). These enzymes are involved in nutrient exploitation from litter and other organic matter in natural forest soils. In these soils, the major limiting nutrients are phosphorus and nitrogen. Macromolecules like proteins, chitin, and phytate are the major sources of these nutrients and occur in dead plant and fungal tissues (Leake et al. 2002; Read and Perez-Moreno 2003). Enzymes able to break down cell wall components of dead plant tissues, like cellulases, hemicellulases, pectinases, and lignin-degrading enzymes, are, therefore, important to get access to N- and P-containing macromolecules (Perez-Moreno and Read 2000). The latter need to be degraded to compounds that can be taken up by EM fungi.

A. Secreted Ectomycorrhizal Enzymes Involved in Exploitation of Phosphorus

Organic phosphorus complexes make up 30%–70% of the total phosphorus content in soils (Stevenson and Cole 1999). Extracellular enzymes such as acid phosphatases, phosphomonoesterases, and phosphodiesterases, which can release P from complex substrates, are very widespread across different taxa of EM fungi (Bae and Barton 1989; Antibus et al. 1992; McElhinney and Mitchell 1993; Conn and Dighton 2000; Alvarez et al. 2006; Courty et al. 2006; Mosca et al. 2007; Hrynkiewicz et al. 2009; see also Table 39.3). Acid phosphatases are also secreted by plants, but their transcription is strongly reduced in EM roots (Luo et al. 2009). After uptake of the mobilized phosphorus, P is stored in EM fungal hyphae in form of polyphosphates. Orthophosphate and short-chain polyphosphates are the predominant transport forms of P toward the host plant (Smith and Read 2008). Many tree species, especially those producing cluster roots, exude carboxylates to release phosphate (Radersma and Grierson 2004). Still, acid phosphatases from EM mutualists are important for the utilization of organic phosphorus compounds because their hyphae explore a much wider range of the surrounding media than the trees' rhizosphere and, thus, reach a much larger soil volume from which they can mobilize phosphorus in nutrient-deficient soils (Häussling and Marschner 1989). In addition to increasing the uptake of Pi, mycorrhizal fungi increase the spectrum of P sources utilized by tree roots by mediating the dissolution of insoluble metallophosphate salts and the hydrolysis of organic phosphorus compounds (Cumming and Weinstein 1990a,b; Cumming 1992; Cumming and Silverman 1996).

Different EM fungi grown in pure culture also showed extracellular acid phosphomonoesterase activity, indicating that this function is also important during saprotrophic life (Tibbett et al. 1998; Nygren and Rosling 2009; Table 39.3). In a comprehensive study, basidiomycetes belonging to the species *H. cylindrosporum*, *Laccaria laccata* (Scop.) Cooke, *P. involutus*, *Rhizopogon rubescens* Tul. and C. Tul., *Suillus collinitus* (Fr.) Kuntze, *Suillus granulatus* (L.) Roussel, and *Suillus luteus* A.H. Smith and Thiers were cultured in a low-phosphate medium to test activities of enzymes involved in P mobilization

TABLE 39.3 Selection of Secreted Fungal Enzymes with Functions in Nutrient Acquisition: Fungal Enzymes with Functions in Phosphorus Acquisition

Host Plant	EM-Fungus	Enzyme-Name	EC-Nr.	Regulation	References
Quercus petraea *Q. robur*	*L. quietus, Cortinarius anomalus, Xerocomus chrysenteron*	Acid phosphatase	3.1.3.2	↑	Courty et al. (2005)
Picea abies *Fagus sylvatica*	*C. geophilum, Russula fellea, R. nigricans, Xerocomus* sp.	Acid phosphatase	3.1.3.2	Present	Courty et al. (2006)
Quercus petraea *Q. robur* *Carpinus betulus*	*Byssocorticium atroviren, Cortinarius olivaceofuscus,L. quietus, L. subdulcis, Tomentella* sp.	Acid phosphatase	3.1.3.2	Present	Courty et al. (2006)
Quercus petraea *Q. robur* *Carpinus betulus*	*Amanita rubescens, Byssocorticium atrovirens, C. geophilum, Clavulina cristata, Cortinarius anomalus, Lactarius chrysoreus, L. helvus, L. quietus, Peziza depressa, P. croceum, Russula atropurpurea, R. cyanoxantha, R. nigricans, R. ochroleuca, Sebacina helvelloides, Tomentella badia, T. botryoides, T. ellisii, T. punicea, T. sublilacina, Tuber puberulum, Xerocomus chrysenteron*	Acid phosphatase	3.1.3.2	Present	Buée et al. (2007)
Pinus contorta	*S. granulatus*	Phosphatase	3.1.3.2	Present	Cullings et al. (2008)
Nothofagus obliqua	*C. geophilum, P. involutus Pisolithus tinctorius*	Surface-bound Phosphomonoesterase	3.1.3.2	↑	Alvarez et al. (2006)
Salix Polaris	*Cenococcum* sp.	Alkaline phosphatase Acid phosphatase	3.1.3.1 3.1.3.2	Present	Hrynkiewicz et al. (2009)
P. sylvestris	*A. muscaria, C. geophilum, L. bicolor, Rhizopogon roseolus, Suillus bovinus, S. luteus, S. variegates*	Phosphomonoesterase	3.1.3.2	Present	Nygren and Rosling (2009)
Picea abies	*C. geophilum, Cortinarius glaucopus, L. bicolor, Meliniomyces bicolor*	Phosphomonoesterase	3.1.3.2	Present	Nygren and Rosling (2009)
Quercus sp.	*Lactarius chrysorrheus, Xerocomus communis*	Phosphomonoesterase	3.1.3.2	Present	Nygren and Rosling (2009)
B. pendula	*Tricholoma scalpturatum*	Phosphomonoesterase	3.1.3.2	Present	Nygren and Rosling (2009)
n.a. (culture filtrates)	*Amanita spissa, Piloderma aff. fallax, Russula sanguinea, Tricholoma fulvum*	Phosphomonoesterase	3.1.3.2	Present	Nygren and Rosling (2009)
n.a. (culture filtrates)	*H. cylindrosporum, L. laccata, P. involutus, R. rubescens, S. collinitus, S. granulates, S. luteus*	p-Nitrophenyl Phosphatase Phytase	3.1.3.41 3.1.3.8	Present	Quiquampoix and Mousain (2005)
n.a. (genome analysis)	*L. bicolor*	Acid phosphatase	3.1.3.2	Present	Martin et al. (2008)
P. pinaster	*Rhizopogon luteolus Sphaerosporella brunea*	Acid phosphatase	3.1.3.2	Present	Ali et al. (2009)
P. pinaster	*H. cylindrosporum*	Alkaline phosphatase Acid phosphatase	3.1.3.1 3.1.3.2	↑	van Aarle and Plassard (2010)
Quercus petraea	ECM community	Phosphatase	3.1.3.2	↑	Diedhiou et al. (2010)

(Quiquampoix and Mousain 2005). *R. rubescens*, *S. collinitus*, *S. granulatus*, and *S. luteus* produced free extracellular phosphomonoesterases, whereas *H. cylindrosporum*, *S. collinitus*, and *S. granulatus* exhibited free extracellular phytase activity (Quiquampoix and Mousain 2005), required to mobilize P from phytate, a dominant form of organic phosphorus in soil (Turner et al. 2002). *H. cylindrosporum* secretion also brought about high levels of acid phosphatase activity in the external medium. Addition of these enzymes to podzol resulted in increased release of organic P in the absence of the fungus (Louche et al. 2010). This underlines the importance of secreted acid phosphatases for P cycling between plant and soil.

Sequencing of fungal genomes revealed that *L. bicolor* but not the ascomycete *T. melanosporum* contained genes for acid phosphatase (Martin et al. 2008, 2010). However, this lack is not typical of EM-forming ascomycetes since *Cenococcum geophilum* Fr. displayed acid phosphatase activity (Hrynkiewicz et al. 2009) and an orthologue to the *L. bicolor* gene is present in the genome of a human pathogen, the ascomycete *Aspergillus fumigatus* Fresen. (Nierman et al. 2005).

Emanating hyphae as well as EM roots of *Pinus pinaster* Aiton trees grown in soil with low phosphorus availability showed high activities of surface-bound acid phosphomonoesterase activity (Ali et al. 2009; van Aarle and Plassard 2010). However, in other studies, activation of these enzymes was not detected after growth of different EM fungi on either inorganic or organic P sources (Antibus et al. 1992; Nygren and Rosling 2009). Obviously, the presence of organic phosphorus has no activating effect on the production of phosphatases in the examined EM fungi. In summary, the studies on secreted fungal phosphatases suggest that the biosynthesis of the enzymes is stimulated when the surrounding media have low phosphorus concentrations and inorganic phosphorus is unavailable in the external solution.

B. Secreted Enzymes Involved in Nitrogen Acquisition

In mycorrhizal roots, uptake of NO_3^- and NH_4^+ can be regulated at the transcriptional level by adjusting the amount of membrane transporters (see Section II.B; Table 39.4). However, temperate and boreal forest ecosystems are generally N-limited (LeBauer and Treseder 2008), and therefore, additional sources of N need to be tapped. Forest trees can utilize organic N sources in the form of simple amino acids (Näsholm and Persson 2001), whereas more complex compounds present in typical forest soil such as proteins and chitin have to be degraded before they become available for plant nutrition. EMs play essential roles in the degradation of these compounds and are, therefore, essential for N cycling in N-limited forest ecosystems (Smith and Read 2008). A vast range of EM fungi species with different host tree species has the ability to utilize proteins as their sole source of nitrogen (Abuzinadah and Read 1986; Abuzinadah et al. 1986; Finlay et al. 1992; Keller 1996; Lilleskov et al. 2002; Rangel-Castro et al. 2002; Guidot et al. 2005; Nygren et al. 2007). Furthermore, considerable quantities of the assimilated N are transferred to the host plant (Finlay et al. 1992; Taylor et al. 2004).

Most studies on the saprotrophic abilities of EM fungi to mobilize N from organic sources have focused on the production of extracellular proteases (Leake and Read 1997; Lindahl et al. 2005). For example, Taylor et al. (2000) used milk powder plates to detect protease activity in several slow-growing EM fungi. Maijala et al. (1991) provided evidence for the production of two proteases by *Amanita regalis* (Fr.) Michael. Nehls et al. (2001a) found that the EM fungus *Amanita muscaria* (L.) Lam. produced two aspartic proteases, AmProt1 and AmProt2, which were important for both N and C nutrition. *AmProt1* was strongly upregulated under carbohydrate or nitrogen starvation or when fungal hyphae were grown under low pH conditions but suppressed in the presence of amino acids (Nehls et al. 2001a).

Fungal proteases also play roles in climatic adaptation of N nutrition. Strains of *Hebeloma* spp. from different climatic zones were grown in axenic culture at different temperatures (Tibbett et al. 1999). Growth at low temperature induced greater proteolytic activity than that at higher temperatures. Many of the strains produced proteases, which retained significant activity at temperatures as low as 0°C with a thermal optimum between 0°C and 6°C. This suggested that the contribution of cold-active proteases to the N acquisition potential of EM fungi was more significant than previously known.

Proteins often form protein–phenol complexes with tannins resulting in extremely recalcitrant molecules. In order to get access to these proteins, tannin–protein complexes need to be digested by phenoloxidases and peroxidases (Bending and Read 1996; Wu et al. 2003). These enzyme classes are also involved in the breakdown of lignin and have been detected not only in saprotrophs (Kellner et al. 2009) but also in several EM taxa (Chen et al. 2003; Luis et al. 2005; Martin et al. 2008, 2010). Secreted polyphenol oxidase activities and laccase activities have frequently been documented (see Section II.C; Table 39.4), but compared to the numerous publications concerning these enzymes, there is rather limited evidence for the occurrence of other secreted lignolytic enzymes, like lignin peroxidases or manganese peroxidases (Cairney et al. 2003; Baldrian 2006). Overall, the release of nitrogen from plant litter by EM-derived enzymes with saprotrophic functions is an important link in the N cycle between plants and soil (Wurzburger and Hendrick 2009).

Another source of organic nitrogen in soils is chitin, derived from dead fungal or arthropod material. Chitin is a long-chain polymer of an N-acetylglucosamine, a derivative of glucose. The potential of EM fungi to use chitin as a nitrogen source is widely distributed among different species (Leake and Read 1990; Lindahl and Taylor 2004). Chitin is degraded by different chitinases in several steps: chitin hydrolases (chitinase A, EC 3.2.1.14) degrade the polymer to chitin oligomers. Chitinase B (EC 3.2.2.14) hydrolyses the oligomers to chitinbiose, a disaccharide of N-acetylglucosamine. In the last step, chitinbiose is degraded by chitobiase (EC 3.2.1.52) to N-acetylglucosamine (Antranikian 2005). EM fungi secrete chitinases into their environment (Table 39.4), but the released activities strongly depend on the prevailing conditions, i.e., organic soil, mineral soil, or dead woody debris (Bueé et al. 2007). Dead woody debris was predominantly colonized by the EM genera *Lactarius* and *Tomentella* displaying high chitinase activities (Bueé et al. 2007). This suggests that these EM can obtain N via the chitin degradation of dead or living saprotrophs co-colonizing dead wood. Different EM species exhibited different chitinase activities, when co-occurring in the same environment. Chitinase activities also changed within species between niches (Bueé et al. 2007). Another study showed that the EM community of oak trees displayed higher chitinase and protease activities, when their soil habitat was enriched in N and P, an effect that that was due to few specialized EM fungal species (Diedhiou et al. 2010). In conclusion, these reports underline that EM fungi have different abilities for degradation of complex compounds, which can—at least within given limits—be optimized to access recalcitrant nitrogen sources in their environment.

C. Ectomycorrhizal Enzymes Involved in Degradation of Complex C-Bearing Compounds

A hallmark of EM lifestyle is that root-associated EM fungi obtain carbon directly from the plant. However, when living free in the soil or when access to valuable N source requires degradation of complex polymers, carbon can also be gained from organic litter (Treseder et al. 2005). This ability originates from a large array of genes encoding enzymes for the decomposition of cellulose, lignin, tannins, etc. (Martin et al. 2008, 2010; Table 39.5). For example, Chambers et al. (1999) detected manganese-dependent peroxidase activity in *Tylospora fibrillosa* (Burt)

TABLE 39.4 Selection of Secreted Fungal Enzymes with Functions in Nutrient Acquisition: Fungal Enzymes with Functions in Nitrogen Acquisition

Host Plant	EM-Fungus	Enzyme-Name	EC-Nr.	Regulation	References
n.a. (DNA templates extracted from sporocarps)	*Amanita crocea, A. Virosa,Coltrichia perennis, Cortinarius balteatus, C. Anthracinus, Hygrophorus erubescens, P. involutus, Piloderma byssinum, P. fallax, Russula badia, R. Roseipes, Sarcodon imbricatum, Suillus flavidus, S. luteus, Tricholoma focale, T. fulvum*	N-acetyl-hexosaminidase	3.2.1.52	Present	Lindahl and Taylor (2004)
Picea abies *Fagus sylvatica*	*Russula fellea, R. nigricans, C. geophilum, Xerocomus* sp.	Chitinase Leucine amino-peptidase	3.2.1.14 3.4.11.1	Present	Courty et al. (2006)
Quercus petraea *Q. robur* *Carpinus betulus*	*Byssocorticium atrovirens, Cortinarius olivaceofuscus, L. quietus, L. subdulcis, Tomentella* sp.	Chitinase Leucine amino-peptidase	3.2.1.14 3.4.11.1	Present	Courty et al. (2006)
Quercus petraea *Q. robur* *Carpinus betulus*	*Amanita rubescens, Byssocorticium atrovirens, C. geophilum, Clavulina cristata, Cortinarius anomalus, Lactarius chrysoreus, L. helvus, L. quietus, Peziza depressa, P. croceum, Russula atropurpurea, R. cyanoxantha, R. nigricans, R. ochroleuca, Sebacina helvelloides, Tomentella badia, T. botryoides, T. ellisii, T. punicea, T. sublilacina, Tuber puberulum, Xerocomus chrysenteron*	Chitinase Leucine amino-peptidase	3.2.1.14 3.4.11.1	Present	Buée et al. (2007)
Pinus contorta	*S. granulatus*	Protease	3.4.11.6	Present	Cullings et al. (2008)
n.a. (genome analysis)	*L. bicolor*	Protease	3.4.11.6	Present	Martin et al. (2008)
n.a. (culture filtrates)	*Lactarius controversus, P. involutus, P. croceum, Suilius bovinus*	Protease	3.4.11.6	Present	Bending and Read (1996)
n.a. (culture filtrates)	*Lacctarius corttroversus*	Polyphenol oxidase	1.10.3.1	Present	Bending and Read (1996)
n.a. (liquid cultures)	*P. involutus, Rhizopogon roseolus* Fr	Chitinase	3.2.1.14	Present	Leake and Read (1990)
P. sylvestris	*A. muscaria, Hydnum rufescens* schaeff, *L. bicolor, Lactarius deliciosus, L. deterrimus, L. quieticolor, L. semisanguifluus*	Protease	3.4.11.6	Present	Nygren et al. (2007)
n.a.	*Amanita spissa, Lactarius rufus*	Protease	3.4.11.6	Present	Nygren et al. (2007)
Picea abies	*C. geophilum, Cortinarius glaucopus, Hydnum rufescens, L. bicolor, Lactarius auriolla, L. deterrimus, L. quieticolor*	Protease	3.4.11.6	Present	Nygren et al. (2007)
Tilia cordata	*Boletus luridus*	Protease	3.4.11.6	Present	Nygren et al. (2007)
Q. robur	*Lactarius chrysorrheus, L. evosmus*	Protease	3.4.11.6	Present	Nygren et al. (2007)
Quercus sp.	*Lactarius acerrimus, L. quietus, L. zonarius.*	Protease	3.4.11.6	Present	Nygren et al. (2007)
Populus sp.	*Lactarius controversus*	Protease	3.4.11.6	Present	Nygren et al. (2007)
Salix repens	*Lactarius controversus*	Protease	3.4.11.6	Present	Nygren et al. (2007)
Betula sp.	*Lactarius pubescens* Fr.	Protease	3.4.11.6	Present	Nygren et al. (2007)
n.a. (Mycelia grown on Petri dishes or in liquid culture)	*A. muscaria*	Aspartic protease (AmProt1)	3.4.23.24	↑	Nehls et al. (2001)
n.a. (plate assays)	*A. regalis, Paxillus invohitus, Suilliis bovinus*	Protease	3.4.11.6	Present	Maijala et al. (1991)
n.a (axenic cultures)	*Hebeloma* sp.	Protease	3.4.11.6	↑	Tibbett et al. (1999)
Quercus petraea	ECM community	Chitinase Protease	3.2.1.14 3.4.11.6	↑	Diedhiou et al. (2010)
n.a. genome analysis	*L. bicolor*	Chitinase Protease	3.2.1.14 3.4.11.6	Present	Martin et al. (2008)

Bourdot and Galzin. Lignin peroxidase-encoding genes were found in a phylogenetically wide range of EM fungi (Bodeker et al. 2009). In addition, laccase-like genes and activities are widespread in EM basidiomycetes, e.g., encompassing *Amanita*, *Cortinarius*, *Hebeloma*, *Lactarius*, *Paxillus*, *Piloderma*, *Russula*, *Tylospora*, and *Xerocomus* (Chen et al. 2003; Luis et al. 2005). Laccases decompose phenolic compounds by oxidation, making polyphenol-bound soil proteins more accessible to fungal proteases (Ramstedt and Söderhäll 1983). A peak of laccase activity is found in spring, supporting active saprotrophic growth of the

TABLE 39.5 Selection of Secreted Fungal Enzymes with Functions in Nutrient Acquisition: Fungal Enzymes with Functions in Carbon Acquisition

Host Plant	EM-Fungus	Enzyme-Name	EC-Nr.	Regulation	References
n. a. (culture filtrates)	*T. fibrillosa*	H3-like manganese-dependent peroxidase	1.11.1.13	Present	Chambers et al. (1999)
n.a. (extracted DNA from dried fruit bodies, cultured material, and fresh, whole sporocarps)	*Cortinarius armillatus, C. traganus, C. hinnuleus, C. infractus, C. malachius, Hygrophorus agathosmus, Russula sardonia, R. xerampelina, Lactarius fulvissimus, L. rufus, Gomphus clavatus*	Different currently sequenced ClassII peroxidases from basidiomycetes	1.11.1	Present	Bodeker et al. (2009)
n.a. (extracted DNA from dried basidiome material and cultured mycelia)	*Lactarius clarkeae, L. rufu, L. scrobiculatus, Piloderma fallax, Rhizopogon luteolus, Russula adusta, R. puellaris, T. fibrillosa*	White rot fungal like laccases	1.10.3.2	Present	Chen et al. (2003)
n.a. (extracted DNA from dried basidiome material and cultured mycelia)	*Piloderma byssinum J.* Erikss. & Ryvarden	White rot fungal like laccases	1.10.3.2	🡅	Chen et al. (2003)
Fagus sylvatica *Q. robur*	*Hebeloma radicosum, Lactarius subdulcis, Russula nigricans, Russula mairei, Boletaceae* sp.	Laccase	1.10.3.2	Present	Luis et al. (2005)
Quercus petraea *Q. robur*	*L. quietus, Cortinarius anomalus, Xerocomus chrysenteron*	Laccase	1.10.3.2	↑	Courty et al. (2009)
Picea abies *Fagus sylvatica*	*C. geophilum, Russula fellea, R. nigricans, Xerocomus* sp.	ß-Glucosidase Laccase 1.10.3.2 Cellobiohydrolase Glucuronidase Xylosidase	3.1.3.2 3.2.1.91 3.2.1.31 3.2.1.37	Present	Courty et al. (2006)
Quercus petraea *Q. robur* *Carpinus betulus*	*Cortinarius olivaceofuscus* *Byssocorticium atrovirens, L. quietus, L. subdulcis, Tomentella* sp.	ß-Glucosidase Laccase Cellobiohydrolase Glucuronidase Xylosidase	3.1.3.2 1.10.3.2 3.2.1.91 3.2.1.31 3.2.1.37	Present	Courty et al. (2006)
Quercus petraea *Q. robur* *Carpinus betulus*	*Amanita rubescens, Byssocorticium atrovirens, C. geophilum, Clavulina cristata, Cortinarius anomalus, Lactarius chrysoreus, L. helvus, L. quietus, Peziza depressa, P. croceum, Russula atropurpurea, R. cyanoxantha, R. nigricans, R. ochroleuca, Sebacina helvelloides, Tomentella badia, T. botryoides, T. ellisii, T. punicea, T. sublilacina, Tuber puberulum, Xerocomus chrysenteron*	ß-Glucosidase Laccase 1.10.3.2 Cellobiohydrolase Glucuronidase Xylosidase	3.1.3.2 3.2.1.91 3.2.1.31 3.2.1.37	Present	Buée et al. (2007)
n.a. (culture filtrates)	*T. matsutake*	ß-Glucosidase	3.1.3.2	present	Kusuda et al. (2006)
n.a. genome analysis	*L. bicolor*	GH5 cellulase family	3.2.1.4	Present	Martin et al. (2008)
Pinus contorta	*S. granulatus*	ß-Glucosidase Laccase Endocellulase Manganese peroxidase Lignin-peroxidase	3.1.3.2 1.10.3.2 3.2.1.4 1.11.1.13	Present	Cullings et al. (2008)

n.a., no information available; 🡅, increased transcript levels; 🡇, decreased transcript levels under the conditions indicated under comments; ↑, elevated enzyme activity under the conditions indicated under comments.

fungus when most photosynthates are being used by the tree to build new plant tissues (Courty et al. 2007). This supports the hypothesis that extracellular fungal enzymes play important roles in adapting nutrition to changing environments, which may be particularly useful for life in temperate and boreal ecosystems with their large fluctuations in temperature.

Genes encoding for N-acetyl hexosaminidases, acting on glucosides, galactosides, and several oligosaccharides, were also detected in representatives of the class of EM basidiomycetes (Lindahl and Taylor 2004). Assays of enzyme activities of isolated EM root tips showed that laccase, β-glucosidase, and cellobiohydrolase activities differed depending on the examined fungal species (Courty et al. 2005, 2006; Buée et al. 2007). As described earlier for nitrogen acquisition, the enzyme activity profiles were influenced by EM species, location, and soil horizon although occurring in the same habitat (Courty et al. 2005, 2006;

Buée et al. 2007). Many EM fungi such as *Peziza* spp. (Egger 2006) or *Tricholoma matsutake* (S. Ito and S. Imai) Singer (Kusuda et al. 2006) possess oxidative and cellulolytic enzyme activities (laccase, β-glucosidase, and cellobiohydrolase) in extracellular compartments (Bending and Read 1995).

Although the capacity to attack plant cell walls is well developed in some EM fungi, the sequenced genome of *L. bicolor* shows that the diversity of these genes is lower than in typical saprotrophs. For example, the cellulase gene family was strongly downsized compared to typical saprotrophic fungal species, because only the GH5 cellulase subfamily was found, but not GH6 and GH7 (Martin et al. 2008; Martin and Nehls 2009). Furthermore, a reduced number of genes for hemicellulose- and pectin-degrading enzymes were present pointing to lower saprotrophic abilities of *L. bicolor* compared with saprotrophs. These recent findings led to the assumption that genes encoding degrading enzymes with saprotrophic functions could probably be interpreted as evolutionary orthologues, which become active when carbon acquisition from their host plant is limited (Baldrian 2009). This assumption is supported by the observation that activities of the secreted enzymes laccase, glucuronidase, cellobiohydrolase, and β-glucosidase were strongly enhanced in *Lactarius quietus* Fr. during the period before budbreak in spring showing that the EM fungus responded to C shortage in the tree by temporarily switching to saprotrophic behavior (Courty et al. 2007). When carbon supply to the roots was decreased by partial defoliation of pine, saprotrophic enzyme activities of endocellulase, D-glucosidase, laccase, manganese peroxidase, lignin peroxidase, phosphatase, and protease in *Suillus granulates* also increased (Cullings et al. 2008). Altogether these results document that many EM fungi are able to adapt their nutrition between the saprotrophic and biotrophic lifestyle.

IV. Conclusion and Outlook

EM fungi show a worldwide distribution and are mainly associated with roots of tree species. Their diversity is high and many of them have distinctive host preferences (Lang et al. 2011). Therefore, the question arose if colonization by different EM fungal species involves similar molecular events. Although it is currently too early to answer this question conclusively, the data compiled for different host plant–fungal associations indicate that reprogramming of auxin signaling in plant roots is an important step toward successful colonization for several EM species. Thereby, EM fungi not only manipulate the root to produce higher surface area but also create additional "habitats" for their own further colonization. Furthermore, the data collected for different plant–fungal associations indicate that EMs influence the root cellular redox state and succeed in suppression of massive defense reactions. In contrast to plants, in most EM fungi analyzed to date, mainly the expression of genes encoding structural components was affected.

Very little is known about factors mediating host specificity and recognition. However, first genetic and a range of ecological studies indicate that the mutual potential benefit can be sensed by both interacting organisms. A first example has shown that a small secreted fungal peptide is important for the establishment of the interaction by directly entering the plant cell nucleus (Plett et al. 2011). Since the peptide is specific to *L. bicolor*, the question whether general mechanisms underlie EM–plant interactions remains unanswered. It is also unclear how the fungus senses its host plant.

Although a coarse pattern emerges with respect to the process of root colonization that eventually leads to improved plant performance, the mechanisms responsible for EM achieving "upgrading" of plants appear surprisingly complex. Only few general molecular responses with regard to improved nutrient supply have been identified. For example, in most cases, increase in plant ammonium transporters was found in EM roots. However, it was unexpected that the widespread positive influence of EM on P nutrition was not immediately apparent from the analysis of the genome or transcriptome of *L. bicolor* (Martin et al. 2008). The data compiled here show that the sources of P as well as of N are complex, especially in boreal and temperate forest ecosystems. The abilities of different EM fungi to secrete enzymes that mobilize these resources will inevitably change the nutrient concentrations in the immediate root environment. This in turn may result in feedback responses in the fungus as well as in the plant. As a consequence, the differences observed and contrasting observations regarding plant or fungal uptake systems in mutualistic interactions are currently difficult to interpret. The key to improved understanding how fungi affect plant nutrient supply is most likely the different ability of EM fungal species to secrete enzymes that can decompose complex organic compounds. Field studies of enzyme activities of different EM species highlighted an enormous flexibility to adapt enzyme production to demand. There is now emerging evidence that functional traits of different EM species differ and that they provide complementary benefits to their host. In the near future, genome projects, which have currently been launched, will help us to understand better to which extent different fungal toolboxes of secretory enzymes contribute to niche differentiation and complementary ecological functions and if contrasting fungal functions result in different feedback responses of the molecular biology of their host plants.

Acknowledgments

We are grateful to the Federal State of Lower Saxony (Ministerium für Wissenschaft und Kultur and Niedersächsisches Vorab) for funding our department by the Excellence Cluster Functional Biodiversity Research (FBR) in the project areas POPDIV and MICRORHIZO and to the German Science Foundation (DFG) for supporting the project Poplar Communication (Po362/20).

References

van Aarle IM, Plassard C. 2010. Spatial distribution of phosphatase activity associated with ectomycorrhizal plants is related with soil type. *Soil Biol Biochem* 42: 324–330.

Abuzinadah RA, Finlay RD, Read DJ. 1986. The role of proteins in the nitrogen nutrition of ectomycorrhizal plants II. Utilization of protein by mycorrhizal plants of *Pinus contorta*. *New Phytol* 103: 495–506.

Abuzinadah RA, Read DJ. 1986. The role of proteins in the nitrogen nutrition of ectomycorrhizal plants. I. Utilization of peptides and proteins by ectomycorrhizal fungi. *New Phytol* 103: 481–493.

Acioli-Santos B, Sebastiana M, Pessoa F et al. 2008. Fungal transcript pattern during the preinfection stage (12 h) of ectomycorrhiza formed between *Pisolithus tinctorius* and *Castanea sativa* roots, identified using cDNA microarrays. *Curr Microbiol* 57: 620–625.

Ali MA, Louche J, Legname E, Duchemin M, Plassard C. 2009. *Pinus pinaster* seedlings and their fungal symbionts show high plasticity in phosphorus acquisition in acidic soils. *Tree Physiol* 29: 1587–1597.

Alvarez M, Gieseke A, Godoy R, Härtel S. 2006. Surface-bound phosphatase activity in ectomycorrhizal fungi: A comparative study between a colorimetric and a microscope-based method. *Biol Fertil Soils* 42: 561–568.

Antibus RK, Sinsabaugh RL, Linkins AE. 1992. Phosphatase activities and phosphorus uptake from inositol phosphate by ectomycorrhizal fungi. *Can J Bot* 70: 794–801.

Antranikian G. 2005. *Angewandte Mikrobiologie*. Berlin, Germany: Springer.

Ashforda AE, Veskb PA, Orlovicha DA, Markovinab AL, Allaway WG. 1999. Dispersed polyphosphate in fungal vacuoles in *Eucalyptus pilularis/Pisolithus tinctorius* ectomycorrhizas. *Fungal Genet Biol* 28: 21–33.

Bae KS, Barton LL. 1989. Alkaline phosphatase and other hydrolyases produced by *Cenococcum graniforme*, an ectomycorrhizal fungus. *Appl Environ Microbiol* 55: 2511–2516.

Baldrian P. 2006. Fungal laccases: Occurrence and properties. *FEMS Microbiol Rev* 30: 215–242.

Baldrian P. 2009. Ectomycorrhizal fungi and their enzymes in soils: Is there enough evidence for their role as facultative soil saprotrophs? *Oecologia* 161: 657–660.

Bending GD, Read DJ. 1995. The structure and function of the vegetative mycelium of ectomycorrhizal plants V. Foraging behaviour and translocation of nutrients from exploited litter. *New Phytol* 130: 401–409.

Bending GD, Read DJ. 1996. Nitrogen mobilization from protein–polyphenol complex by ericoid and ectomycorrhizal fungi. *Soil Biol Biochem* 28: 1603–1612.

Beniwal RS, Langenfeld-Heyser R, Polle A. 2010. Ectomycorrhiza and hydrogel protect hybrid poplar from water deficit and unravel plastic responses of xylem anatomy. *Environ Exp Bot* 69: 189–197.

Bodeker ITM, Nygren CMR, Taylor AFS, Olson A, Lindahl BD. 2009. Class II peroxidase-encoding genes are present in a phylogenetically wide range of ectomycorrhizal fungi. *ISME J* 3: 1105–1115.

Bovi M, Carrizo ME, Capaldin S et al. 2011. Structure of a lectin with antitumoral properties in king bolete (*Boletus edulis*) mushrooms. *Glycobiology* 21: 1000–1009.

Brundrett MC. 2009. Mycorrhizal associations and other means of nutrition of vascular plants: Understanding the global diversity of host plants by resolving conflicting information and developing reliable means of diagnosis. *Plant Soil* 320: 37–77.

Bucher M. 2007. Functional biology of plant phosphate uptake at root and mycorrhiza interfaces. *New Phytol* 173: 11–26.

Buée M, Courty PE, Mignot D, Garbaye J. 2007. Soil niche effect on species diversity and catabolic activities in an ectomycorrhizal fungal community. *Soil Biol Biochem* 39: 1947–1955.

Buée M, Reich M, Murat C et al. 2009. 454 Pyrosequencing analyses of forest soils reveal an unexpectedly high fungal diversity. *New Phytol* 184: 449–456.

Cairney JWG. 2011. Ectomycorrhizal fungi: The symbiotic route to the root for phosphorus in forest soils. *Plant Soil* 344: 51–71.

Cairney JWG, Taylor AFS, Burke RM. 2003. No evidence for lignin peroxidase genes in ectomycorrhizal fungi. *New Phytol* 160: 461–462.

Chambers SM, Burke RM, Brooke PR, Cairney JWG. 1999. Molecular and biochemical evidence for manganese-dependent peroxidase activity in *Tylospora fibrillosa*. *Mycol Res* 103: 1098–1102.

Chen DM, Bastias BA, Taylor AFS, Cairney JWG. 2003. Identification of laccase-like genes in ectomycorrhizal basidiomycetes and transcriptional regulation by nitrogen in *Piloderma byssinum*. *New Phytol* 157: 547–554.

Conn C, Dighton J. 2000. Litter quality influences on decomposition, ectomycorrhizal community structure and mycorrhizal root surface acid phosphatase activity. *Soil Biol Biochem* 32: 489–496.

Courty PE, Breda N, Garbaye J. 2007. Relation between oak tree phenology and the secretion of organic matter degrading enzymes by *Lactarius quietus* ectomycorrhizas before and during bud break. *Soil Biol Biochem* 39: 1655–1663.

Courty PE, Pouysegur R, Bueé M, Garbaye J. 2006. Laccase and phosphatase activities of the dominant ectomycorrhizal types in a lowland oak forest. *Soil Biol Biochem* 38: 1219–1222.

Courty PE, Pritsch K, Schloter M, Hartmann A, Garbaye J. 2005. Activity profiling of ectomycorrhiza communities in two forest soils using multiple enzymatic tests. *New Phytol* 167: 309–319.

Couturier J, Montanini B, Martin F, Brun A, Blaudez D, Chalot M. 2007. The expanded family of ammonium transporters in the perennial poplar plant. *New Phytol* 174: 137–150.

Cullings KW, Ishkhanova G, Henson J. 2008. Defoliation effects on enzyme activities of the ectomycorrhizal fungus *Suillus granulatus* in a *Pinus contorta* (lodgepole pine) stand in Yellowstone National Park. *Oecologia* 158: 77–83.

Cumming JR. 1992. Developing low pH media to study the response of mycorrhizal fungi to phosphorus source and metal ion exposure. In *Proceedings of the 12th North American Forest Biology Workshop*, eds., SJ Colombo, G Hogan, V Wearn, p. 155. Sault Ste Marie, Ontario, Canada: Ministry of Natural Resources.

Cumming JR, Silverman TS. 1996. Citrate modulates acid phosphatase activity in *Laccaria bicolor. Plant Physiol* 111: 272–272.

Cumming JR, Weinstein LH. 1990a. Aluminium-mycorrhizal interactions in the physiology of pitch pine seedlings. *Plant Soil* 125: 7–18.

Cumming JR, Weinstein LH. 1990b. Utilization of $AlPO_4$ as a phosphorus source by ectomycorrhizal *Pinus rigida* Mill. seedlings. *New Phytol* 116: 99–106.

Dénarié J, Debellé F, Promé JC. 1996. Rhizobium lipo-chitooligosaccharide nodulation factors: Signaling molecules mediating recognition and morphogenesis. *Annu Rev Biochem* 65: 503–535.

Deveau A, Palin B, Delaruelle C et al. 2007. The mycorrhiza helper *Pseudomonas fluorescens* BBc6R8 has a specific priming effect on the growth, morphology and gene expression of the ectomycorrhizal fungus *Laccaria bicolor* S238N. *New Phytol* 175: 743–755.

Diedhiou AG, Dupouey JL, Buee M. 2010. The functional structure of ectomycorrhizal communities in an oak forest in central France witnesses ancient Gallo-Roman farming practices. *Soil Biol Biochem* 42: 860–862.

Doehlemann G, van der Linde K, Assmann D et al. 2009. *Pep1*; a secreted effector protein of *Ustilago maydis*, is required for successful invasion of plant cells. *PLoS Pathog* 5: e1000290.

Dou D, Kale SD, Wang X et al. 2008. RXLR-mediated entry of *Phytophthora sojae* effector *Avr1b* into soybean cells does not require pathogen-encoded machinery. *Plant Cell* 20: 1930–1947.

Druebert C, Lang C, Valtanen K, Polle A. 2009. Beech carbon productivity as driver of ectomycorrhizal abundance and diversity. *Plant Cell Environ* 32: 992–1003.

Ducic T, Berthold D, Langenfeld-Heyser R, Beese F, Polle A. 2009. Mycorrhizal communities in relation to biomass production and nutrient use efficiency in two varieties of Douglas fir (*Pseudotsuga menziesii* var. menziesii and var. glauca) in different forest soils. *Soil Biol Biochem* 41: 742–753.

Ducic T, Polle A. 2007. Manganese toxicity in two varieties of Douglas fir (*Pseudotsuga menziesii* var. viridis and glauca) seedlings as affected by phosphorus supply. *Funct Plant Biol* 34: 31–40.

Duplessis S, Courty PE, Tagu D, Martin F. 2005. Transcript patterns associated with ectomycorrhiza development in *Eucalyptus globules* and *Pisolithus microcarpus*. *New Phytol* 165: 599–611.

Edwards E, Dixon DP, Walbot V. 2000. Plant glutathione S-transferases: Enzymes with multiple functions in sickness and in health. *Trends Plant Sci* 5: 193–198.

Egger KN. 2006. The surprising diversity of ascomycetous mycorrhizas. *New Phytol* 170: 421–423.

Felten J, Kohler A, Morin E et al. 2009. The ectomycorrhizal fungus *Laccaria bicolor* stimulates lateral root formation in poplar and *Arabidopsis* through auxin transport and signaling. *Plant Physiol* 151: 1991–2005.

Finlay RD, Frostegard A, Sonnerfeldt A-N. 1992. Utilization of organic and inorganic nitrogen sources by ectomycorrhizal fungi in pure culture und in symbiosis with *Pinus contorta* Dougl. Ex Loud. *New Phytol* 120: 105–116.

Frettinger P, Derory J, Herrmann S et al. 2007. Transcriptional changes in two types of pre-mycorrhizal roots and in ectomycorrhizas of oak microcuttings inoculated with *Piloderma croceum*. *Planta* 225: 331–340.

Gafur A, Schützendübel A, Langenfeld-Heyser R, Fritz E, Polle A. 2004. Compatible and incompetent *Paxillus involutus* isolates for ectomycorrhiza formation in vitro with poplar (Populus x canescens) differ in H_2O_2 production. *Plant Biol* 6: 91–99.

Gay G, Normand L, Marmeisse R, Sotta B, Debaud JC. 1994. Auxin overproducer mutants of *Hebeloma cylindrosporum* Romagnesi have increased mycorrhizal activity. *New Phytol* 128: 645–657.

Guidot A, Verner M-C, Debaud J-C, Marmeisse R. 2005. Intraspecific variation in use of different organic nitrogen sources by the ectomycorrhizal fungus *Hebeloma cylindrosporum*. *Mycorrhiza* 15: 167–177.

Hanlon MT, Coenen C. 2011. Genetic evidence for auxin involvement in arbuscular mycorrhiza initiation. *New Phytol* 189: 701–709.

Häussling M, Marschner H. 1989. Organic and inorganic soil phosphates and acid phosphatase activity in the rhizosphere of 80-year-old Norway spruce [*Picea abies* (L.) Karst.] trees. *Biol Fertil Soil* 8: 128–133.

Heller G, Adomas A, Li G et al. 2008. Transcriptional analysis of *Pinus sylvestris* roots challenged with the ectomycorrhizal fungus *Laccaria bicolor*. *BMC Plant Biol* 8: 19.

Heller G, Lundén K, Finlay RD, Asiegbu FO, Elfstrand M. 2011. Expression analysis of *Clavata1*-like and *Nodulin21*-like genes from *Pinus sylvestris* during ectomycorrhiza formation. *Mycorrhiza* 22: 271–277.

Herrmann S, Oelmüller R, Buscot F. 2004. Manipulation of the onset of ectomycorrhiza formation by indole-3-acetic acid, activated charcoal or relative humidity in the association between oak microcuttings and *Piloderma croceum*: Influence on plant development and photosynthesis. *J Plant Physiol* 161: 509–517.

Hrynkiewicz K, Baum C, Leinweber P. 2009. Mycorrhizal community structure, microbial biomass P and phosphatase activities under *Salix polaris* as influenced by nutrient availability. *Eur J Soil Biol* 45: 168–175.

Johansson T, Le Quere A, Ahren D et al. 2004. Transcriptional responses of *Paxillus involutus* and *Betula pendula* during formation of ectomycorrhizal root tissue. *Mol Plant Microbe Interact* 17: 202–215.

Keller G. 1996. Utilization of inorganic and organic nitrogen sources by high-subalpine ectomycorrhizal fungi of *Pinus cembra* in pure culture. *Mycol Res* 100: 989–998.

Kellner H, Luis P, Schlitt B, Buscot F. 2009. Temporal changes in diversity and expression patterns of fungal laccase genes within the organic horizon of a brown forest soil. *Soil Biol Biochem* 41: 1380–1389.

Kemppainen M, Duplessis S, Martin F, Pardo AG. 2009. RNA silencing in the model mycorrhizal fungus *Laccaria bicolor*: Gene knock-down of nitrate reductase results in inhibition of symbiosis with Populus. *Environ Microbiol* 11: 1878–1896.

Kidd BN, Kadoo NY, Dombrecht B et al. 2011. Auxin signaling and transport promote susceptibility to the root-infecting fungal pathogen *Fusarium oxysporum* in *Arabidopsis*. *Mol Plant Microbe Interact* 24: 733–748.

Kiers ET, Duhamel M, Beesetty Y et al. 2011. Reciprocal rewards stabilize cooperation in the mycorrhizal symbiosis. *Science* 333: 880–882.

Kloppholz S, Kuhn H, Requena N. 2011. A secreted fungal effector of *Glomus intraradices* promotes symbiotic biotrophy. *Curr Biol* 21: 1204–1209.

Koide RT, Shumway DL, Xu B, Sharda JN. 2007. On temporal partitioning of a community of ectomycorrhizal fungi. *New Phytol* 174: 420–429.

Krüger A, Peskan-Berghöfer T, Frettinger P, Herrmann S, Buscot F, Oelmüller R. 2004. Identification of premycorrhiza-related plant genes in the association between *Quercus robur* and *Piloderma croceum. New Phytol* 163: 149–157.

Kusuda M, Ueda M, Konishi Y, Araki Y, Yamanaka K, Nakazawa M, Miyatake K, Terashita T. 2006. Detection of β-glucosidase as saprotrophic ability from an ectomycorrhizal mushroom, *Tricholoma matsutake. Mycoscience* 47: 184–189.

Labbe J, Jorge V, Kohler A et al. 2011. Identification of quantitative trait loci affecting ectomycorrhizal symbiosis in an interspecific F_1 poplar cross and differential expression of genes in ectomycorrhizas of the two parents: *Populus deltoides* and *Populus trichocarpa. Tree Genet Gen* 7: 617–627.

Lamhamedi MS, Godbout C, Fortin JA. 1994. Dependence of *Laccaria bicolor* basidiome development on current photosynthesis of *Pinus strobus* seedlings. *Can J For Res* 24: 1797–1804.

Lang C, Seven J, Polle A. 2011. Host preferences and differential contributions of deciduous tree species shape mycorrhizal species richness in a mixed Central European forest. *Mycorrhiza* 21: 297–308.

Larsen PE, Sreedasyam A, Trivedi G, Podila GK, Cseke LJ, Collart FR. 2011. Using next generation transcriptome sequencing to predict an ectomycorrhizal metabolome. *BMC Syst Biol* 5: 70.

Le Quéré A, Schützendübel A, Rajashekar B et al. 2004. Divergence in gene expression related to variation in host specificity of an ectomycorrhizal fungus. *Mol Ecol* 13: 3809–3819.

Le Quéré A, Wright DP, Söderström B, Tunlid A, Johansson T. 2005. Global patterns of gene regulation associated with the development of ectomycorrhiza between birch (*Betula pendula* Roth.) and *Paxillus involutus* (Batsch) Fr. *Mol Plant Microbe Interact* 18: 659–673.

Leake JR, Donnelly DP, Boddy L. 2002. Interactions between ectomycorrhizal and saprotrophic fungi. In *Mycorrhizal Ecology*, eds., MGA van der Heijden, IR Sanders, pp. 346–372. Berlin, Germany: Springer.

Leake JR, Read DJ. 1990. Chitin as a nitrogen source for mycorrhizal fungi. *Mycol Res* 94: 993–995.

Leake J, Read DJ. 1997. Mycorrhizal fungi in terrestrial habitats. *Mycota* 4: 281–301.

LeBauer DS, Treseder KK. 2008. Nitrogen limitation of net primary productivity in terrestrial ecosystems is globally distributed. *Ecology* 89: 371–379.

Lilleskov EA, Hobbie EA, Fahey TJ. 2002. Ectomycorrhizal fungal taxa differing in response to nitrogen deposition also differ in pure culture organic nitrogen use and natural abundance of nitrogen isotopes. *New Phytol* 154: 219–231.

Lindahl BD, Finaly RD, Cairney JWG. 2005. Enzymatic activities of mycelia in mycorrhizal fungal communities. In *The Fungal Community-its Organization and Role in the Ecosystem*, eds., J Dighton, JF White, P Oudemans, pp. 331–348. Boca Raton, FL: Taylor & Francis Group.

Lindahl BD, Taylor AFS. 2004. Occurrence of N-acetylhexosaminidase encoding genes in ectomycorrhizal basidiomycetes. *New Phytol* 164: 193–199.

Louche J, Ali MA, Cloutier-Hurteau B, Sauvage F-X, Quiquampoix H, Plassard C. 2010. Efficiency of acid phosphatases secreted from the ectomycorrhizal fungus *Hebeloma cylindrosporum* to hydrolyse organic phosphorus in podzols. *FEMS Microbiol Ecol* 73: 323–335.

Luis P, Kellner H, Zimdars B, Langer U, Martin F, Buscot F. 2005. Patchiness and spatial distribution of laccase genes of ectomycorrhizal, saprotrophic, and unknown basidiomycetes in the upper horizons of a mixed forest cambisol. *Microb Ecol* 50: 570–579.

Luo ZB, Janz D, Jiang XN et al. 2009. Upgrading root physiology for stress tolerance by ectomycorrhizas: Insights from metabolite and transcriptional profiling into reprogramming for stress anticipation. *Plant Physiol* 151: 1902–1917.

Maijala P, Fagerstedt KV, Raudaskoski M. 1991. Detection of extracellular cellulolytic and proteolytic activity in ectomycorrhizal fungi and *Heterobasidion annosum* (Fr.). *Bref. New Phytol* 117: 643–648.

Maillet F, Poinsot V, André O et al. 2011. Fungal lipochitooligosaccharide symbiotic signals in arbuscular mycorrhiza. *Nature* 469: 58–63.

Martin F, Aerts A, Ahren D et al. 2008. The genome of *Laccaria bicolor* provides insights into mycorrhizal symbiosis. *Nature* 452: 88–92.

Martin F, Kohler A, Duplessis S. 2007. Living in harmony in the wood underground: Ectomycorrhizal genomics. *Curr Opin Plant Biol* 10: 204–210.

Martin F, Kohler A, Murat C et al. 2010. Périgord black truffle genome uncovers evolutionary origins and mechanisms of symbiosis. *Nature* 464: 1033–1038.

Martin F, Lapeyrie F, Tagu D. 1997. Altered gene expression during ectomycorrhiza development. *Mycota* 6: 223–242.

Martin F, Nehls U. 2009. Harnessing ectomycorrhizal genomics for ecological insights. *Curr Opin Plant Biol* 12: 508–515.

McElhinney C, Mitchell DT. 1993. Phosphatase activity of four ectomycorrhizal fungi in a Sitka spruce-Japanese larch plantation in Ireland. *Mycol Res* 97: 725–732.

McGuire KL, Henkel TW, de la Cerda GI, Villa G, Edmund F, Andrew C. 2008. Dual mycorrhizal colonization of forest-dominating tropical trees and the mycorrhizal status of non-dominant tree and liana species. *Mycorrhiza* 18: 217–222.

Melin P, Schnürer J, Wagner EG. 2002. Proteome analysis of *Aspergillus nidulans* reveals proteins associated with the response to the antibiotic concanamycin A, produced by *Streptomyces* species. *Mol Genet Gen* 267: 695–702.

Menotta M, Amicucci A, Sisti D, Gioacchini AM, Stocchi V. 2004. Differential gene expression during pre-symbiotic interaction between *Tuber borchii* Vitad. and *Tilia americana* L. *Curr Genet* 46: 158–165.

Morel M, Jacob C, Kohler A et al. 2005. Identification of genes differentially expressed in extraradical mycelium and ectomycorrhizal roots during *Paxillus involutus-Betula pendula* ectomycorrhizal symbiosis. *Appl Environ Microbiol* 71: 382–391.

Mosca E, Montecchio L, Scattolin L, Garbaye J. 2007. Enzymatic activities of three ectomycorrhizal types of *Quercus robur* L. in relation to tree decline and thinning. *Soil Biol Biochem* 39: 2897–2904.

Näsholm T, Persson J. 2001. Plant acquisition of organic nitrogen in boreal forests. *Plant Physiol* 111: 419–426.

Nehls U, Bock A, Einig W, Hampp R. 2001a. Excretion of two proteases by ectomycorrhizal fungus *Amanita muscaria*. *Plant Cell Environ* 24: 741–747.

Nehls U, Göhringer F, Wittulsky S, Dietz S. 2010. Fungal carbohydrate support in the ectomycorrhizal symbiosis: A review. *Plant Biol* 12: 292–301.

Nierman WC, Pain A, Anderson MJ et al. 2005. Genomic sequence of the pathogenic and allergenic filamentous fungus *Aspergillus fumigatus*. *Nature* 438: 1151–1156.

Nygren CM, Edqvist J, Elfstrand M, Heller G. 2007. Detection of extracellular protease activity in different species and genera of ectomycorrhizal fungi. *Mycorrhiza* 17: 241–248.

Nygren C, Rosling A. 2009. Localisation of phosphomonoesterase activity in ectomycorrhizal fungi grown on different phosphorus sources. *Mycorrhiza* 19: 197–204.

Peay KG, Kennedy PG, Davies JS, Tan S, Bruns TD. 2010. Potential link between plant and fungal distributions in a dipterocarp rainforest: Community and phylogenetic structure of tropical ectomycorrhizal fungi across a plant soil ecotone. *New Phytol* 185: 529–542.

Pena R, Offermann C, Simon J et al. 2010. Carbon limitations after girdling affect ectomycorrhizal diversity and reveal functional differences of EM community composition in a mature beech forest (*Fagus sylvatica*). *Appl Environ Microbiol* 76: 1831–1841.

Perez-Moreno J, Read DJ. 2000. Mobilization and transfer of nutrients from litter to tree seedlings via the vegetative mycelium of ectomycorrhizal plants. *New Phytol* 145: 301–309.

Peter M, Courty PE, Kohler A et al. 2003. Analysis of expressed sequence tags from the ectomycorrhizal basidiomycetes *Laccaria bicolor* and *Pisolithus microcarpus*. *New Phytol* 159: 117–129.

Plassard C, Dell B. 2010. Phosphorus nutrition of mycorrhizal trees. *Tree Physiol* 30: 1129–1139.

Plett JM, Kemppainen M, Kale SD et al. 2011. A secreted effector protein of *Laccaria bicolor* is required for symbiosis development. *Curr Biol* 21: 1197–1203.

Podila GK, Zheng J, Balasubramanian S et al. 2002. Fungal gene expression in early symbiotic interactions between *Laccaria bicolor* and red pine. *Plant Soil* 244: 117–128.

Polle A, Schützendübel A. 2003. Heavy metal signalling in plants: Linking cellular and organismic responses. In *Plant Responses to Abiotic Stresses, Topics in Current Genetics*, eds., H Hirt, K Shinozaki, Vol. 4, pp. 167–215. Berlin, Germany: Springer.

Quiquampoix H, Mousain D. 2005. Enzymatic hydrolysis of organic phosphorus. In *Organic Phosphorus in the Environment*, eds., BL Turner, E Frossard, DS Baldwin, pp. 89–112. Wallingford, U.K.: CAB International.

Radersma S, Grierson PF. 2004. Phosphorus mobilization in agroforestry: Organic anions, phosphatase activity and phosphorus fractions in the rhizosphere. *Plant Soil* 259: 209–219.

Rafiqi M, Gan PH, Ravensdale M et al. 2010. Internalization of flax rust avirulence proteins into flax and tobacco cells can occur in the absence of the pathogen. *Plant Cell* 22: 2017–2032.

Ramstedt M, Söderhäll K. 1983. Proteases, phenoloxidase and pectinase activities in mycorrhizal fungi. *Trans Br Mycol Soc* 81: 157–160.

Rangel-Castro JI, Danell E, Taylor AFS. 2002. Use of different nitrogen sources by the edible ectomycorrhizal mushroom *Cantharellus cibarius*. *Mycorrhiza* 12: 131–137.

Read DJ, Perez-Moreno J. 2003. Mycorrhizas and nutrient cycling: A journey towards relevance. *New Phytol* 157: 475–492.

Reddy SM, Hitchin S, Melayah D et al. 2006. The auxin-inducible GH3 homologue Pp-GH3.16 is down regulated in *Pinus pinaster* root systems on ectomycorrhizal symbiosis establishment. *New Phytol* 170: 391–400.

Reich M, Göbel C, Kohler A et al. 2009. Fatty acid metabolism in the ectomycorrhizal fungus *Laccaria bicolor*. *New Phytol* 182: 950–964.

Rinaldi AC, Comandini O, Kuyper TW. 2008. Ectomycorrhizal fungal diversity: Separating the wheat from the chaff. *Fungal Divers* 33: 1–45.

Rosling A. 2009. Trees, mycorrhiza and minerals-field relevance of in vitro experiments. *Geomicrobiol J* 26: 389–401.

Schützendübel A, Polle A. 2002. Plant responses to abiotic stresses: Heavy metal-induced oxidative stress and protection by mycorrhization. *J Exp Bot* 53: 1351–1365.

da Siva Coelho I, de Queiroz MV, Costa MD, Kasuya MCM, de Araújo EF. 2010. Identification of differentially expressed genes of the fungus *Hydnangium* sp. during the pre-symbiotic phase of the ectomycorrhizal association with *Eucalyptus grandis*. *Mycorrhiza* 20: 531–540.

Smith SE, Read DJ. 2008. *Mycorrhizal Symbiosis*, 3rd edn. London, U.K.: Academic Press.

Smith SE, Smith FA. 2011. Roles of arbuscular mycorrhizas in plant nutrition and growth: New paradigms from cellular to ecosystem scales. *Annu Rev Plant Biol* 62: 227–250.

Splivallo R, Fischer U, Gobel C, Feussner I, Karlovsky P. 2009. Truffles regulate plant root morphogenesis via the production of auxin and ethylene. *Plant Physiol* 150: 2018–2029.

Stevenson FJ, Cole MA. 1999. Phosphorus. In *Cycles of Soil: Carbon, Nitrogen, Phosphorus, Sulfur, Micronutrients*, eds., FJ Stevenson, MA Cole, pp. 279–329, New York: John Wiley & Sons.

Tagu D, Nasse B, Martin F. 1996. Cloning and characterization of hydrophobins-encoding cDNAs from the ectomycorrhizal basidiomycete *Pisolithus tinctorius. Gene* 168: 93–97.

Tagu D, Palin B, Balestrini R et al. 2003. Characterization of a symbiosis- and auxin-regulated glutathione-S-transferase from *Eucalyptus globulus* roots. *Plant Physiol Biochem* 41: 611–618.

Tatry MV, Kassis EE, Lambilliotte R et al. 2009. Two differentially regulated phosphate transporters from the sym biotic fungus *Hebeloma cylindrosporum* and phosphorus acquisition by ectomycorrhizal *Pinus pinaster. Plant J* 57: 1092–1102.

Taylor AFS, Gebauer G, Read DJ. 2004. Uptake of nitrogen and carbon from double-labelled 15N and 13C glycine by mycorrhizal pine seedlings. *New Phytol* 164: 383–388.

Taylor AFS, Martin F, Read DJ. 2000. Fungal diversity in ectomycorrhizal communities of Norway spruce (*Picea abies* (L.) Karst.) and Beech (*Fagus sylvatica* L.) along north-south transects in Europe. *Ecol Stud* 142: 343–365.

Tedersoo L, May TW, Smith ME. 2010a. Ectomycorrhizal lifestyle in fungi: Global diversity, distribution, and evolution of phylogenetic lineages. *Mycorrhiza* 20: 217–263.

Tedersoo L, Nara K. 2010. General latitudinal gradient of biodiversity is reversed in ectomycorrhizal fungi. *New Phytol* 185: 351–354.

Tedersoo L, Nilsson RH, Abarenkov K et al. 2010b. 454 Pyrosequencing and Sanger sequencing of tropical mycorrhizal fungi provide similar results but reveal substantial methodological biases. *New Phytol* 188: 291–301.

Teramoto M, Wu B, Hogetsu T. 2011. Transfer of ^{14}C-photosynthate to the sporocarp of an ectomycorrhizal fungus *Laccaria amethystina. Mycorrhiza* 22: 219–225.

Tibbett M, Sanders FE, Cairney JWG. 1998. The effect of temperature and inorganic phosphorus supply on growth and acid phosphatase production in arctic and temperate strains of ectomycorrhizal *Hebeloma* spp. in axenic culture. *Mycol Res* 102: 129–135.

Tibbett M, Sanders FE, Cairney JWG, Leake JR. 1999. Temperature regulation of extracellular proteases in ectomycorrhizal fungi (*Hebeloma* spp.) grown in axenic culture. *Mycol Res* 103: 707–714.

Treseder KK, Allen MF, Ruess RW, Pregitzer KS, Hendrick RL. 2005. Lifespans of fungal rhizomorphs under nitrogen fertilization in a pinyon-juniper woodland. *Plant Soil* 270: 249–255.

Turner BL, Paphazy MJ, Haygarth PM, McKelvie ID. 2002. Inositol phosphates in the environment. *Philos Trans R Soc B* 57: 449–469.

Van der Heijden MGA, Bardgett RD, vanStraalen NM. 2008. The unseen majority: Soil microbes as drivers of plant diversity and productivity in terrestrial ecosystems. *Ecol Lett* 11: 296–310.

Voiblet C, Duplessis S, Encelot N, Martin F. 2001. Identification of symbiosis-regulated genes in *Eucalyptus globules–Pisolithus tinctorius* ectomycorrhiza by differential hybridization of arrayed cDNAs. *Plant J* 25: 181–191.

van West P, de Bruijn I, Minor KL et al. 2010. The putative RxLR effector protein SpHtp1 from the fish pathogenic oomycete *Saprolegnia parasitica* is translocated into fish cells. *FEMS Microbiol Lett* 310: 127–137.

Wright DP, Johansson T, Le Quéré A, Söderström B, Tunlid A. 2005. Spatial patterns of gene expression in the extrametrical mycelium and mycorrhizal root tips formed by the ectomycorrhizal fungus *Paxillus involutus* in association with birch (*Betula pendula*) seedlings in soil microcosms. *New Phytol* 167: 579–596.

Wu T. 2010. Can ectomycorrhizal fungi circumvent the nitrogen mineralization for plant nutrition in temperate forest ecosystems? *Soil Biol Biochem* 43: 1109–1117.

Wu T, Sharda JN, Koide RT. 2003. Exploring interactions between saprotrophic microbes and ectomycorrhizal fungi using a protein-tannin complex as an N source by red pine (*Pinus resinosa*). *New Phytol* 159: 131–139.

Wurzburger N, Hendrick RL. 2009. Plant litter chemistry and mycorrhizal roots promote a nitrogen feedback in a temperate forest. *J Ecol* 97: 528–536.

40

Mycorrhizae—Rhizosphere Determinants of Plant Communities: What Can We Learn from the Tropics?

Ingrid Kottke
Eberhard-Karls University Tübingen

Gábor M. Kovács
Eötvös Loránd University

I. Introduction

Plant communities are characterized by species composition and abundances. Local occurrence of communities was explained by assuming the abiotic factors and plant–plant interactions as the most important drivers (Ellenberg 1996). Recently, phylogeny and paleobiogeography of plants became more relevant to understand, for example, disparity of species richness among tropical, temperate, and boreal biomes (Fine et al. 2008). Biotic interactions came into interest as shaping tropical plant communities (Burslem et al. 2005). Here, we focus on soil fungi that mutually associate with fine roots of more than 90% of land plants by forming mycorrhizae (Smith and Read 2008). By the nonrandom, phylogenetic conservative associations of mycobionts and plants established in coevolution since appearance of first land plants, local presence or absence of the distinct mycobionts appeared as an important determinant upon establishment of land plants and their communities (Smith and Read 2008).

In the previous edition (Kottke 2002), we stressed on the hypothesis that species-rich plant communities are supported by broad sharing of mycobionts, while specific plant–fungus associations may restrict plant diversity (Molina et al. 1992). We exemplified these ideas by then published data and own results mostly obtained from northern hemisphere pasture and forest plant communities. Meanwhile, a body of important information came from investigations in Neotropical and East Asian rain forests and from mycorrhizal associations of orchids not considered in the previous edition. The progress was mainly achieved by molecular study of the mycobionts. Thus, bias from sampling of spores (Glomeromycota in arbuscular mycorrhizas/AMF), fruit bodies, or mycorrhiza morphotypes (Basidiomycota and Ascomycota in ectomycorrhizas/ECMF) was circumvented, as well as dependence on mostly unsuccessful mycelium isolation (orchid mycorrhizal fungi/OMF).

The amount of molecular diversity studies of mycorrhizal fungi has increased by magnitudes already during the 1990s (Horton and Bruns 2001), but in the last decade, the decreasing costs of Sanger sequencing and the advent of pyrosequencing-based next-generation sequencing (NGS) techniques have revolutionized the fungal diversity studies (e.g., see references in Hibbett et al. 2011). With these methods, the fungal community of a certain area could more precisely be studied than before. Moreover, areas hard to visit and to sample regularly, like the tropics, can be easier investigated. The new techniques, however, have raised other problems regarding the sampling, the wet lab (e.g., studied loci, biased primers), and the analyses (e.g., reference datasets, species delimitation, clustering to molecular operational taxonomic units—MOTUs). The obvious need for consensus-based standardized methods was emphasized several times (e.g., Horton and Bruns 2001; Lilleskov and Parrent 2007) but has only been partly achieved so far. Keeping these problems in mind, we collected and discussed data from literature and our own investigations focusing on the impact of narrow or broad sharing of mycobionts on occurrence and diversity of trees in tropical rain forests and on orchids in open landscapes and terrestrial and epiphytic habitats.

II. Tree Species Richness of Neotropical Rain Forests as Supported by Common Mycorrhizal Fungi

Neotropical rain forests are extraordinary rich in tree species forming mixed stands (Gentry 1995; Valencia et al. 2004; Homeier et al. 2009). The vast majority of these tree species are symbiotically associated with mycorrhizal fungi of phylum Glomeromycota forming arbuscular mycorrhizae, termed AMF further on (Kottke et al. 2004). Anthropogenic threats especially by slash and burn turn vast Neotropical forests into pastures or fields, which, after getting abandoned, recover slowly and turn mostly into secondary forests (Peterson and Carson 2008). Besides other factors, potential loss in diversity or turnover in AMF composition could presumably impede long-term recovery of primary rain forests.

Studies of AMF communities in relation to plant diversity in Neotropical rain forests were carried out, based on AMF spore counts, in lowland evergreen forests in Nicaragua and Costa Rica (Picone 2000; Lovelock et al. 2003) and in Panama (Mangan et al. 2004). In the latter forest, AMF of seedlings of three tree species were additionally studied by molecular techniques (Husband et al. 2002a,b). Molecular identification was also applied to investigate AMF from roots of adult trees of 30 species and from planted seedlings of four native tree species in the tropical mountain rain forest area of Southern Ecuador (Haug et al. 2010). Lovelock et al. (2003) and Husband et al. (2002a) compared AMF from two sites in primary forest, while Mangan et al. (2004) studied impact of forest fragmentation on AMF communities in mainland and island forest. Picone (2000) and Haug et al. (2010) compared AMF communities in primary forests with those of reforested, abandoned pastures.

Investigators congruently found broad sharing of AMF by diverse trees in the respective sites, although comparable molecular delimitation of AMF species differed among studies and correct microscopic identification and estimation of spore numbers is problematic. Differences of AMF community composition were only found from spore abundance data between both co-occurring tree species and similar sampling sites (Lovelock et al. 2003; Mangan et al. 2004). However, based on molecular data, high turnover of members of the AMF Glomus group A was found when abandoned pastures and primary mountain forests were compared (Haug et al. 2010) (Figure 40.1). No decline in number of AMF taxa by forest fragmentation or conversion to pastures was observed. Number of defined AMF taxa ranged

FIGURE 40.1 Scheme indicating tree species–rich Neotropical forest (a) and regenerating or planted seedlings on abandoned pastures (b) as promoted by distinct networks of Glomeromycota, forest specialists, and open area-adapted species, respectively (in light grey: mycelia and vegetative spores of Glomeromycota).

from 13 to 28 per site, but number of AMF-associated trees were about ten times larger. Thus, diversity of trees and AMF within sites was not correlated.

Differences in the AMF community composition may probably be best explained by what was called "site-dependent host preferences" (Husband et al. 2002a) or "host ecological group-related AMF" (Öpik et al. 2009). The cited authors interpreted these preferences as having potentially important impact on tree diversity in forest communities, assuming that preferences in mycorrhizal associations would indicate specific support of coexisting, competing trees in microhabitats. Although there is evidence from pot experiments that growth of tropical tree seedlings may differ among tree species and AMF combinations (Kiers et al. 2000), an assortment effect on tree community in tropical rain forest is most likely negligible. A multitude of AMF associate with each individual tree, and negative effects of one AMF species will be balanced by positive feedback of another fungus (Bever et al. 2010). In our opinion, the obvious disposition of tropical rain forest AMF to associate with a broad range of tree species appears to be crucial for recruitment and maintenance of high tree diversity. Kiers et al. (2000) found that all inocula were able to colonize the roots of all seedlings including early and late successional species. Husband et al. (2002b) observed that seedlings were first colonized by dominant fungal types. Planted seedlings of primary forest species on abandoned pastures were associated with a number of widespread fungi (Haug et al. 2010). Thus, seedling regeneration is seemingly not limited by lack of AMF on abandoned pastures in Neotropical rain forest areas. Species-rich herbaceous and shrubby vegetation established as a consequence of forest disturbance might be important for high diversity and turnover of AMF on abandoned pastures (Mangan et al. 2004; Günter et al. 2009).

However, Husband et al. (2002b) found replacement of early colonizing AMF by more rare fungal taxa after 2 years of seedling growth in lowland forest sites, and Haug et al. (2010) stated that the majority of AMF taxa encountered in the primary forest were restricted to the forest and not found on tree seedlings in abandoned pastures. Interestingly, Öpik et al. (2009) analyzing AMF of boreo-nemoral herbs found forest-specific lineages colonizing shade-tolerant herbs. The question of "forest specialist AMF" as substantial prerequisite for restoration of primary tropical rain forest and their dependence on late forest succession soil conditions may be of crucial importance for restoration and conservation programs.

III. From Individual ECM Trees to Monodominant ECM Stands in the Wet Tropics

Ectomycorrhiza (ECM) forming Nyctaginaceae of genera *Neea* and *Guapira* occur as scattered individuals among a multitude of AM-forming trees in the mountain rain forest at Reserva Biológica San Francisco, Southern Ecuador (Kottke et al. 2004). The ECM-forming fungi (ECMF) were identified by morphotyping and molecular sequencing of internal transcribed spacer rDNA (*ITS rDNA*, Haug et al. 2005). *Guapira* sp. and *Neea* sp. from inside the primary forest came out as associated with only one mycorrhizal fungus, both closely related and belonging to the Thelephoraceae (Table 40.1). Five fungal phylotypes belonging to three taxonomic groups were found with another *Neea* sp. on the border of the primary forest (Table 40.1; Figure 40.2a). In lowland forest of the Yasuni National Park, Western Amazonia, Ecuador, a study site of 25 ha comprised 15 species of *Neea*, two species of *Guapira*, and nine species of *Coccoloba* (Polygonaceae), which were associated, respectively, with 11, 9, and 25 ECMF phylotypes from up to five taxonomic groups per plant genus (Table 40.1). Only, respectively, 3, 6, and 5 phylotypes were shared among two tree genera (Fig. 2 in Tedersoo et al. 2010b). *Neea* "commun" tended to aggregate in sparse patches of up to 10 individuals, while the other trees occurred well segregated.

The Guayana Region in northeastern South America is characterized by a flora of high species richness, among them only few ECM-forming tree species of Caesalpiniaceae and Dipterocarpaceae such as *Pakaraimaea dipterocarpacea* Mag. and Ashton showing a dispersed occurrence with individuals mostly separated between 400 and 1100 m from each other (Henkel et al. 2002; Moyersoen 2006). Ectomycorrhizal state was shown by molecular evidence from mycorrhizas of *P. dipterocarpacea* (Moyersoen 2006) and assessed from fruit body communities under *Dicymbe* spp. (Caesalpiniaceae; Henkel et al. 2002; Table 40.1). *Dicymbe corymbosa* Benth. is a striking example of monodominant ECM stands ranging from one to many hectares in extent, immersed in a matrix of tree species–rich AM forests (Henkel et al. 2005). The ECM state and mast fruiting years were considered as the main drivers of monodominance. Field experiments showed that seedlings survived best when germinating within the "hyphal network" of the adult trees (McGuire 2007).

The lowland rain forests in West Africa are mixed stands of more than 400 tree species mostly forming AM. However, these forests are definitely dominated by a number of caesalpinioid species within nine genera of tribe Amherstieae and one genus of tribe Detariae (*Afzelia*) all forming ECM (Newberry et al. 1988). Large, emergent trees may occur in groves or dominate in large areas. A recent study using molecular tools for identification of ECMF in one of the established Caesalpiniaceae plots of Korup National Park, Cameroon, found 111 or 124 ECMF species belonging to 16 fungal lineages, using Sanger sequencing and NGS (next-generation sequencing) technique, respectively (Table 40.1; Tedersoo et al. 2010a). These numbers are distinctly higher than the fungal taxa in the earlier-described Neotropical forests. Identification of ECMF and mycorrhizas from tree species by use of molecular tools in Guinea tropical rain forest revealed 39 taxa from 7 fungal groups (Table 40.1; Diédhiou et al. 2010). Twenty of the taxa were only associated with a single host of the 4 Caesalpiniaceae species or with *Uapaca esculenta* A. Chev. ex Aubrév. and Leandri (Phyllanthaceae, Euphorbiales), 19 were shared by 2, 6 by all the 5 tree species. Seedlings also shared ECMF among each other and with multiple trees, but

TABLE 40.1 Fungal Groups Associated to Ectomycorrhiza Forming Trees in the Wet Tropics Displaying Correlated Increase of Diversity

			Groups of Ectomycorrhizal Fungi																
Plant Family	Region	Investigated Species	Thelephoraceae	Russulaceae	Sebacinaceae	Cantharellaceae	Clavulinaceae	Amanitaceae	Inocybe	Cortinariaceae	Boletaceae	Sclerodermataceae	Tricholomataceae	Atheliales	Hymenochaetales	Hysterangiales	Hygrophoraceae	Ascomycota	References
Nyctaginaceae	Andes of Southern Ecuador	*Guapira* sp.	+																Haug et al. (2005)
		Neea sp. 1	+	+														+	
		Neea sp. 2	+																
	Amazonia, Ecuador	*Guapira* 2 spp.	+	+		+	+		+										Tedersoo et al. (2010b)
		Neea 15 spp. (one in clumps)	+	+			+	+											
Polygonaceae	Amazonia, Ecuador	*Coccoloba* 9 spp.	+	+	+		+		+										Tedersoo et al. (2010b)
Caesalpiniaceae	Guyana	*Dicymbe* 3 spp, *Aldinia insignis*		+	+	+	+	+		+	+								Henkel et al. (2002, 2004)
Dipterocarpaceae	Guyana	*Pakaraimaea dipterocarpacea*			+		+	+	+	+									Moyersoen (2006)
Phyllanthaceae	Senegal	*Uapaca guineensis*		+		+		+			+	+							Thoen and Ba (1989)
Caesalpiniaceae	Senegal	*Afzelia africana*		+		+		+	+		+	+							Thoen and Ba (1989)
	Congo Basin	*Gilbertiodendron dewevrei*		+		+		+		+	+								Fassi and Moser (1991)
	Guinea	*Cryptosepalum tetraphyllum* and three other species	+	+			+				+	+	+						Riviere et al. (2007); Diédhiou et al. (2010)
	Cameroon	*Microberlinia bisulcata* and *Tetraberlinia* 3 spp., 10 other spp.	+	+	+	+	+	+	+	+	+	+		+				+	Tedersoo et al. (2010b)
Dipterocarpaceae	West Ghats, India	*Vateria indica, Dipterocarpus indicus, Kingiodendron pinnatum*		+				+			+								Riviere et al. (2007)
	Malaysia	Eight species	+	+		+		+		+	+	+					+		Sirikantaramas et al. (2003)
	Seychelles	*Vateriopsis seychellarum and Intsia bijuga*	+			+			+	+	+	+			+				Tedersoo et al. (2007)
	Borneo	87 species	+	+	+	+	+	+		+	+	+	+	+	+	+		+	Peay et al. (2010a)

(a)

(b)

FIGURE 40.2 Scheme displaying (a) isolated ECM-forming trees with few ECMF in Neotropical mountain rain forest and (b) clumped stands with diverse ECMF as promoting monodominance of ECM trees in highly diverse AM forests of tropical Africa and Asia. Fungal groups, see Table 40.1.

there was a significant turnover among seedlings and adults. Adults served as "nurse trees" for conspecific and non-conspecific seedlings. Seedlings thus tended to be more generalists as compared to adults. Thoen and Ba (1989), by investigating basidiomes in Senegal under *Afzelia africana* Sm. Ex Pers. (Caesalpiniaceae) and *Uapaca guineensis* Müll.-Arg., found quite distinct ECMF communities; only 6 fungal species were shared, while 24 or 13 ECMF species, respectively, were only associated with one of the tree species, indicating taxonomic selectivity in the adult phase.

Borneo lowland rain forest is characterized by high numbers of abundant species (87) in the family Dipterocarpaceae accounting for 42% of basal area. All members of this family form ECM (Peay et al. 2010b). Investigating the mycobiont communities on two sites in Borneo lowland rainforest using molecular techniques for root extracted DNA sequencing, 105 ECMF taxa were found covering similar fungal lineages as in the before mentioned biomes (Table 40.1). Tree and fungal species covaried among the plots and sites on sand or clay soil, 14 out of 21 plots sharing no fungal species in common, only one ECMF taxon (Atheliales) occurring in both soil types. Tedersoo et al. (2007) investigated ECMF by molecular tools on the Seychelles from two indigenous tree species, *Vateriopsis seychellarum* (Dyer) Heim (Dipterocarpaceae) and *Intsia bijuga* (Coleb.) Kuntze (Caesalpiniaceae) and found 30 ECMF phylotypes, only three overlapping on these two tree species. Thus, specific associations of ECMF were prominent, supporting the earlier statement by Smits (1983) of high specificity on species level in Dipterocarpaceae. Natural regeneration of Dipterocarpaceae after logging is considered very poor. Absence of Dipterocarpaceae in secondary forests of the nonseasonal zone was observed, probably due to short-distance dispersal strategy of seeds and confinement to the ECMF network of mother trees (Smits 1983). Restriction for establishment of ECM seedlings because of missing fungal partner is generally strong because specificity of ECMF on species level is a global phenomenon (Dickie and Reich 2005 and literature therein). Seeds of ECM trees are mostly large, limiting dispersal, and Basidiomycota ECMF colonize via dikaryotic hyphae not via single, monokaryotic spores. Thus, colonization is best guaranteed within the already existing fungal networks of the mother trees.

A very special case of support by ECMF was described for *Graffenrieda emarginata* (Ruiz and Pav.) Triana (Melastomataceae; Haug et al. 2004). In the comprehensively studied 12 ha primary forest at Reserva Biológica San Francisco, Southern Ecuador, more than 190 tree species over 5 cm diameter at breast height were identified. Only *G. emarginata* occurred frequently but restricted to the mountain ridges that are covered by a thick humus layer (Homeier et al. 2009). Studies using transmission electron microscopy and molecular techniques revealed that *G. emarginata* forms AM but is additionally associated with a *Rhizoscyphus* sp. (Ascomycota) and a *Tulasnella* sp. (Basidiomycota), both fungi forming superficial ectomycorrhizae (Haug et al. 2004). The authors hypothesized that the double mycorrhization may be of competitive advantage for the rather slow-growing *G. emarginata*. *Rhizoscyphus ericae* (D. J. Read) W. Y. Zhuang and Korf, a closely related mycobiont of Ericaceae, was shown to be efficient in nitrogen acquisition from organic soil compounds (Smith and Read 2008 and literature therein), and similar facilities might be expected from the *G. emarginata* associate.

Summing up, wet tropical ECM trees associate with members of similar fungal lineages (Table 40.1) that show a very restricted overlap on the fungal species level. A corresponding situation was found for temperate areas (Smith and Read 2008). Evaluation of specificity in associations on the species–species level is, however, hampered by frequent lack of precise information in the published datasets, and our interpretation is, therefore, preliminary. In the temperate regions, as a general rule, high number of ECMF is correlated with a low number of tree species, while in the wet tropics, situation is quite different. The ECM trees—either simple individuals or patches—could be considered as islands for ECMF and island biogeography models could be used to interpret the observed trend of parallel increase of tree and fungal species numbers (Peay et al. 2007, 2010a). The size of the ECM islands, correlating to isolated or clumped to dominant occurrence of ECM trees in AM forests, could be seen as a reason for the observed correlation among number of ECMF and number and abundance of ECM-tree species (Table 40.1; Figure 40.2).

IV. Impact of Mycobionts on Establishment of Orchids in Open Habitats, Tropical AM Forests, and ECM Forests

Orchids are probably the most species-rich plant group occurring worldwide as terrestrials and epiphytes, habitats ranging from deserts and open grasslands to temperate and tropical forests. In most habitats, orchids occur as scattered individuals or in small bunch. In the tropical mountain rainforest, they may become extraordinary abundant as terrestrials and epiphytes. Also regenerating man-made landslides are crowded by orchids (Bussmann et al. 2008). Orchids have tiny seeds that can easily be spread by the wind, and it might thus be expected that orchids are to be found everywhere. However, studies showed that seed dispersal is most intensive only up to 1 m in the surroundings of a fruiting individual (Jacquemyn et al. 2007; Jersáková and Malinová 2007). Furthermore, orchid seed germination and early, achlorophyllous development depend on fungi for carbon supply (Smith 1967; Hadley 1969). Orchid establishment is, thus, correlated with presence of nearby adult plants that preserve the appropriate fungi in their mycorrhizas for further generations (Batty et al. 2001; McKendrick et al. 2002; Diez 2007).

Molecular identification of orchid mycobionts and investigation of orchid carbon gain revealed different, unexpected fungal strategies for carbon supply in the respective communities (Selosse and Roy 2009 and references therein). Members of Tulasnellales, Sebacinales, and Ceratobasidiales (Agaricomycotina) associate with green orchids in open habitats worldwide (Kottke and Suárez 2009; Yukawa et al. 2009) and with green, terrestrial, and epiphytic orchids in tropical mountain forests (Otero et al. 2002; Suárez et al. 2006, 2008; Kottke et al. 2010). Recently, Atractiellomycetes (rust-related Basidiomycota) were found to form orchid mycorrhiza structures in Andean orchids (Kottke et al. 2010). All these fungi fructificate inconspicuously on bark, rotten wood, or debris and, thus, display a saprophytic growth phase. Carbon acquisition from the organic sources to nourish the orchid protocorm should be expected (Midgley et al. 2006), but saprophytic capacities are probably low (Batty et al. 2001; Liebel et al. 2010). These fungi do not promote achlorophyllous orchids in the adult stage, but most likely obtain carbon from their green hosts. This was experimentally shown for *Ceratobasidium cornigerum* (Bourdot) D. P. Rogers–*Goodyera repens* (L.) R. Br. association (Cameron et al. 2006, 2008), Tulasnella (*Epulorhiza*) isolate–*Serapias strictiflora* Welwitsch ex. Veiga (Látalová and Baláz 2010), and was indicated by field measurements (Liebel et al. 2010).

Tulasnellales are the most frequent orchid mycobionts spanning from associations with basal Apostasiodeae (Yukawa et al. 2009) and Cypripedioideae (Shefferson et al. 2005; Shimura et al. 2009) onto Orchidoideae (Jacquemyn et al. 2010) and higher Epidendroideae (Suárez et al. 2006). *Tulasnella* species diversity, defined as molecular operational taxonomic units (MOTUs), was found high, but turnover among host types (epiphytes and terrestrials) and habitats (pristine forest and recovering landslides) was low in the tropical mountain rain forest of the Northern Andes (Figure 40.3; Suárez et al. 2006; Herrera et al. unpublished). Thus, these mycobionts most likely support the extraordinary species richness and abundance of orchids as terrestrials and epiphytes in the Andean tropical rain forest. *Tulasnella* mycobionts were further found to be shared by *Chiloglottis* species, sampled through New South Wales Australia (Roche et al. 2010), by different orchid genera (*Liparis*, *Tipularia*, *Goodyera pubescens* [Willd.] R. Brown) sampled in the United States (McCormick et al. 2004), among different orchid genera (*Epipactis*, *Orchis*) from pristine forest and oil shale ash-polluted habitats in Estonia (Shefferson et al. 2008) and among five *Orchis* spp. throughout Europe (Jacquemyn et al. 2010). It is also interesting that some sequences of *Tulasnella* isolates from Australian orchids are close to *Tulasnella* from Europe or South America (Suárez et al. 2006), indicating sharing of historically related, but not identical, supposedly locally radiated taxa.

Sebacina spp. and *Ceratobasidium* spp. are also worldwide spread orchid mycobionts. However, *Ceratobasidium* is seemingly preferentially associated and shared within distinct orchid genera in a wide geographical range, like *Goodyera* spp.

FIGURE 40.3 (**See color insert.**) Scheme indicating the extraordinary species-rich orchid communities as epiphytes (e.g., *Stelis* spp.) and terrestrials (e.g., *Epidendrum* spp., *Prostechea vespa*, *Sobralia rosea*) in the Andean tropical mountain rain forest supported by multiple, shared Tulasnellales (basidiomes on bark and rotten branches).

(Shefferson et al. 2010) or *Apostasia* spp. (Yukawa et al. 2009), and rarely encountered in other genera and biomes (Otero et al. 2002; Kottke et al. 2010). *Sebacina* spp. (subgroup B) and Atractiellomycetes were shared among the terrestrial and epiphytic orchids in the Neotropical pristine forest and on a landslide, similar to *Tulasnella* spp. but were less frequent (Kottke et al. 2010).

Achlorophyllous and green orchids that occur in shady forests dominated by ectomycorrhiza-forming trees associate with the ECMF linking them via mycelia to ectomycorrhizas of the surrounding trees (Table 40.2; Figure 40.4), from where they derive all or part of their carbon as revealed by stable isotopes analyses (Gebauer and Meyer 2003; Girlanda et al. 2006). There is a consistent agreement that the switch to the efficient carbon supply by ECMF was important to the evolutionary loss of chlorophyll and full mycotrophy of orchids throughout their lifetime. The switch is, however, gradual, and amounts of carbon derived by green orchid individuals via fungi or by own photosynthesis may depend on light availability (Preiss et al. 2010).

Individual achlorophyllous orchid species were found to depend on certain ectomycorrhizal genera (Table 40.2) and may even prefer closely related taxa (Taylor and Bruns 1999; Selosse et al. 2002). Abundance of preferred fungi is additionally crucial for establishment of achlorophyllous orchid species (McCormick et al. 2009). *Cypripedium calceolus* L. was found to be autotrophic (Preiss et al. 2010) and was associated with *Tulasnella* spp., but mixotrophic *Epipactis* spp., *Cephalanthera* spp. and *Cypripedium fasciculatum* Kellog ex S. Watson, while still associated with *Tulasnella* spp., switched additionally or predominantly to ECMF (Table 40.2). Most *Tulasnella* spp. associating with *Epipactis* and *Cypripedium* fall into a distinct clade irrespective of geographical separation in Europe, North America, and Western China (Suárez et al. 2006; Shefferson et al. 2007, 2008; Shimura et al. 2009; Yuan et al. 2010). *Tulasnella* spp. from another clade were already shown to form ectomycorrhizas in European forests (Bidartondo et al. 2003). An unidentified *Tulasnella* was found to form superficial ECM with *G. emarginata* (Melastomataceae) in the Neotropical mountain rain forest (Haug et al. 2004). Thus, we may assume that some *Tulasnella* spp. take part in ECMF hyphal networks. The occurrence of Neottieae and other orchid groups in shady ECM forests would then be consistently explained by their ECMF mycobionts (Figure 40.4).

Mycoheterotrophic orchids occur also in forests dominated by AM-forming trees and were studied in tropical Asia and Southern Africa. These orchids associate with saprophytic fungi belonging to Agaricales and Hymenochaetales in Basidiomycota (Selosse et al. 2010). *Gastrodia similis* Bosser (La Réunion, Macarenen, Indian Ocean) associates with

TABLE 40.2 Fungal Groups Associated with Orchids in ECM Forests

	Fungal Group										
	Sebacinales group A (*S. incrustans* group)	Thelephorales	Russulales	*Inocybe*	*Clavulina*	*Tuber*	Other ECMF	Tulasnellales	Sebacinales group B (*S. vermifera* group)	Ceratobasidiales	References
Mycoheterotrophic orchids											
Neottia nidus-avis	+++										Selosse et al. (2002); McKendrick et al. (2002); Bidartondo et al. (2004)
Hexalectris spicata, H. revoluta	+++										Taylor et al. (2003)
Cephalanthera austine		+++									Taylor and Bruns (1997); McCormick et al. (2004)
Corallorhiza trifida		+++									McKendrick et al. (2000)
Corallorhiza maculata, C. mertensiana			+++								Taylor and Bruns (1999); Taylor et al. (2004)
Dipodium hamiltonianum, D. variegatum			+++								Bougoure and Dearnaley (2005); Dearnaley and Le Brocque (2006)
Epipogium aphyllum		+		+++			+				Roy et al. (2009a)
Aphyllorchis montana, A. caudata	+	+++	+++		+++		+				Roy et al. (2009b)
Cephalanthera exigua		+++					+				Roy et al. (2009b)
Mixotrophic orchids											
Cephalanthera damasonium,[a] *C. rubra, C. longifolia*[a]	+	+++		+		+	+			+	Julou et al. (2005); Bidartondo et al. (2004); Abadie et al. (2006)
Cypripedium fasciculatum	+		+				+	+++			Shefferson et al. (2005, 2007); Whitridge and Southworth (2005); Suárez et al. (2008)
Epipactis helleborine, E. microphylla, E. atrorubens	+	+	+			+++	+	+			Selosse et al. (2004); Bidartondo et al. (2004); Bidartondo and Read (2008); Suárez et al. (2006); Shefferson et al. (2008); Ogura-Tsujita and Yakawa (2008a)
Limodorum abortivum, L. brulloi, L. trabutianum			+			+					Girlanda et al. (2006)
Green orchids in ECM forest											
Cypripedium calceolus[b]								+++			Shefferson et al. (2005, 2007)
Cypripedium candidum, C. montanum		+						+++			Shefferson et al. (2005, 2007)
Cypripedium californicum								+	+	+	Shefferson et al. (2005, 2007)
Cypripedium parviflorum			+				+	+++	+		Shefferson et al. (2005)
Platanthera chlorantha[b]								+			Bidartondo et al. (2004)

+, fungi present; +++, fungi dominating.
[a] Including heterotrophic individuals.
[b] Autotrophic or of unknown state.

wood-decaying *Resinicium* spp., *Gastrodia confusa* Honda and Tuyama in bamboo forests of Japan associates with litter-decaying *Mycena* spp. (Ogura-Tsujita et al. 2009), and *Wullschlaegelia aphylla* (Sw.) Rchb. f. (Guadeloupe, Caribbeans) associates with *Gymnopus* and *Mycena* spp. (Martos et al. 2009). *Eulophia zollingeri* (Rchb. F.) J. J. Sm. is associated with *Psathyrella candolleana* (Fries) Maire (Japan; Ogura-Tsujita and Yakawa 2008b). *Coprinellus* (*Coprinus* by the authors) and *Psathyrella* promote *Epipogium roseum* (D. Don) Lindl. (Yamato et al. 2005; Yagame et al. 2007). These saprophytic fungi might efficiently derive carbon from organic resources and replace the ECMF in AM tropical forests.

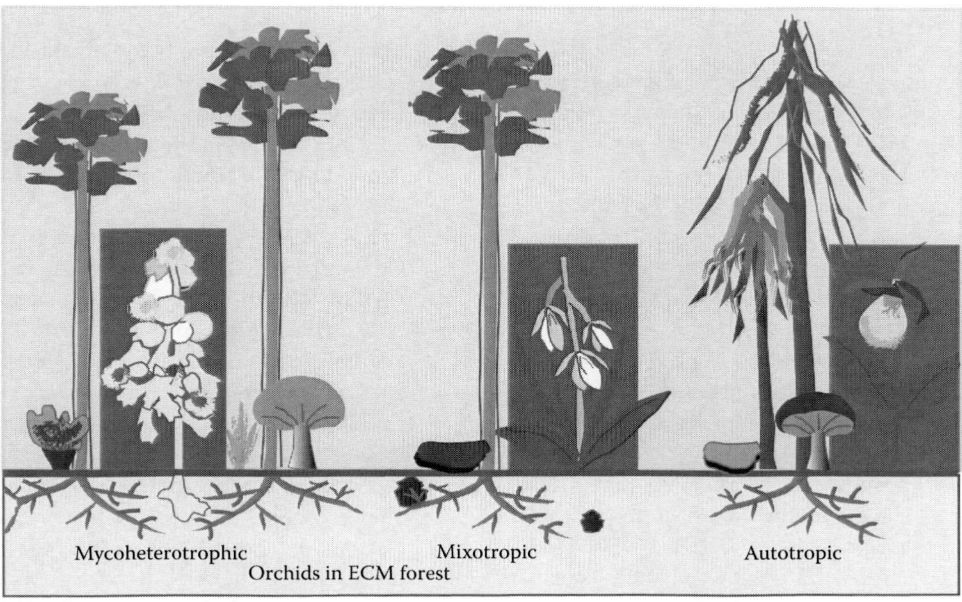

FIGURE 40.4 (**See color insert.**) Scheme displaying orchids in shady ECM forests (Fagales, Caesalpiniaceae, Dipterocarpaceae, Pinaceae) as supported by ECMF mycobionts profiting from carbon supply by trees. Fungi indicate main groups as given in Table 40.2 (Thelephorales, Sebacinales group A, Russulales, Tuberaceae, Tulasnellales).

V. Conclusions

In the previous edition (Kottke 2002), we compiled data obtained by identification of mycorrhizal fungi mostly from the northern hemisphere and came to the conclusion that AM- and ECM-associated plant communities rely on different life strategies. Focusing on tropical wet forests and using the results of molecular delimitation of fungal taxa, the earlier statement is principally corroborated and requires an extension. High tree species richness in tropical wet forests is linked to broad sharing of AMF similar to the situation found in temperate grassland communities. The number of AMF taxa detected by sequencing the root-associated mycobionts of tropical trees is, however, much higher than previously thought, and a turnover among habitats was found, indicating forest specialist fungi.

ECM-associated trees tend to form monodominant stands in temperate forests but only in few areas of the wet tropics. The tendency to monodominance is intrinsically linked with narrow host range of mycobionts. Tropical wet forests are always rich in AM tree species, and ECM trees may occur as either single individuals or small groups. Both could be interpreted as islands for ECMF. Population and phylogenetic studies of EMF and trees have revealed dispersal limitation at both landscape and continental scales. While in temperate areas number of tree species are low in comparison to the number of mycobionts, a correlated increase of fungal and tree species richness is observed in the wet tropical forests.

Orchids were not considered in the previous edition, but recent molecular identification of mycobionts from orchid roots sampled worldwide and especially in tropical habitats revealed somewhat similar situation as mentioned earlier. Orchids growing in open areas and as epiphytes are seemingly associated with widespread fungi and locally share mycobionts on the species level. Broad sharing of mycobionts is presumably supporting the extraordinary richness of epiphytic and terrestrial orchids in tropical mountain forests and probably essential for establishment of epiphytes. On contrast, ECMF-associated orchids, including most of the achlorophyllous taxa, are rare and mostly specifically linked to their mycobionts, restricted to the respective ECM forest. An even lower number of achlorophyllous orchid species is associated to saprophytic fungi in AM forests.

The examples of the tropics presented here and the northern hemisphere presented in the previous edition show us how characteristic the composition of root colonizing fungal communities can be and how this distinctiveness correlates with and influences the plant communities. The tropical examples revealed similar general patterns as studies of other habitats. Determinism of plant communities by mycorrhizal associations can be interpreted as the question of inherited specificity—either specificity of phylogenetic lineages (e.g., species or genera of ECM trees and their fungal partners), specificity of habitats (e.g., forest versus disturbed areas), or specificity of life strategies (e.g., green versus mycotrophic orchids). Thus, the new insights from tropical communities revealed that determinism of mycorrhizal interaction in plant communities is the outcome of long-lasting functional evolution, biogeography, and adaptation of plants and fungi to certain habitats.

Acknowledgments

Support by research grants of Deutsche Forschungsgemeinschaft (DFG FOR 420; RU 816) to I. Kottke and Hungarian Research Fund (OTKA K72776; N181157) and an Alexander von Humboldt Fellowship to G. M. Kovács are greatly appreciated. We thank Verena Uhle-Schneider for skillfully preparing the figures.

References

Abadie LC, Puttsepp Ü, Gebauer G, Faccio A, Bonfante P, Selosse MA. 2006. *Cephalanthera longifolia* (Neottieae, Orchidaceae) is mixotrophic: A comparative study between green and nonphotosynthetic individuals. *Can J Bot* 84:1462–1477.

Batty AL, Dixon KW, Brundrett M, Sivasithamparam K. 2001. Constraints to symbiotic germination of terrestrial orchid seed in a Mediterranean bushland. *New Phytol* 152:511–520.

Bever JD, Dickie IA, Facelli E et al. 2010. Rooting theories of plant community ecology in microbial interactions. Review. *Trends Ecol Evol* 25:468–478.

Bidartondo MI, Bruns TD, Weiß M, Sérgio C, Read DJ. 2003. Specialized cheating of the ectomycorrhizal symbiosis by an epiparasitic liverwort. *Proc R Soc Lond B* 270:835–842.

Bidartondo MI, Burghardt B, Gebauer G, Bruns TD, Read DJ. 2004. Changing partners in the dark: Isotopic and molecular evidence of ectomycorrhizal liaison between forest orchids and trees. *Proc R Soc Lond B* 271:1799–1806.

Bidartondo MI, Read DJ. 2008. Fungal specificity bottlenecks during orchid germination and development. *Mol Ecol* 17:3707–3716.

Bougoure JJ, Dearnaley JD. 2005. The fungal endophytes of *Dipodium variegatum* (Orchidaceae). *Australas Mycol* 24:15–19.

Burslem D, Pinard M, Hartley S. 2005. *Biotic Interactions in the Tropics. Their Role in the Maintenance of Species Diversity.* Cambridge, U.K.: Cambridge University Press.

Bussmann RW, Wilcke W, Richter M. 2008. Landslides as important disturbance regimes—Causes and regeneration. In *Gradients in a Tropical Mountain Ecosystem of Ecuador*, eds. E Beck, J Bendix, I Kottke, F Makeschin, R Mosandl, Series Ecol. Stud., Vol. 198, pp. 319–330. Berlin, Germany: Springer.

Cameron DD, Johnson I, Read DJ, Leake JR. 2008. Giving and receiving: Measuring the carbon cost of mycorrhizas in the green orchid, *Goodyera repens. New Phytol* 180:176–184.

Cameron DD, Leake JR, Read DJ. 2006. Mutualistic mycorrhiza in orchids: Evidence from plant-fungus carbon and nitrogen transfer in the green-leaved terrestrial orchid *Goodyera repens. New Phytol* 171:405–416.

Dearnaley JD, Le Brocque AF. 2006. Molecular identification of the primary root fungal endophytes of *Dipodium hamiltonianum* (Yellow hyacinth orchid). *Aust J Bot* 54:487–491.

Dickie IA, Reich P. 2005. Ectomycorrhizal fungal communities at forest edges. *J Ecol* 93:244–255.

Diédhiou A, Selosse M-A, Galiana A et al. 2010. Multi-host ectomycorrhizal fungi are predominant in a Guinean tropical rainforest and shared between canopy trees and seedlings. *Environ Microbiol* 12:2219–2232.

Diez JM. 2007. Hierarchical patterns of symbiotic orchid germination linked to adult proximity and environmental gradients. *J Ecol* 95:159–170.

Ellenberg H. 1996. Vegetation Mitteleuropas mit den Alpen in ökologischer, dynamischer und historischer Sicht. 5. Auflg., UTB für Wissenschaft, Stuttgart, Eugen Ulmer.

Fassi B, Moser M. 1991. Mycorrhizae in the natural forests of tropical Africa and the neotropics. In *Fungi, Plants and Soil*, ed. A Fontana, pp. 157–202. Torino, Italy: Centro di Studio sulla Micologia.

Fine PV, Ree RH, Burnham RJ. 2008. The disparity in tree species richness among tropical, temperate, and boreal biomes: The geographic area and age hypothesis. In *Tropical Forest Community Ecology*, eds. WP Carson, SA Schnitzer, pp. 31–45. Chichester, U.K.: Wiley-Blackwel.

Gebauer G, Meyer M. 2003. ^{15}N and ^{13}C natural abundance studies of autotrophic and myco-heterotrophic orchids provides insight into nitrogen and carbon gain from fungal associations. *New Phytol* 160:209–223.

Gentry AH. 1995. Patterns of diversity and floristic composition in neotropical montane forests. In *Biodiversity and Conservation of Neotropical Montane Forests*, eds. SP Churchill, H Balslev, E Forero, JL Luteyn. New York: New York Botanical Garden, pp. 103–126.

Girlanda M, Selosse MA, Cafasso D et al. 2006. Inefficient photosynthesis in the Mediterranean orchid *Limodorum abortivum* is mirrored by specific associations to ectomycorrhizal Russulaceae. *Mol Ecol* 15:491–504.

Günter S, Gonzalez P, Ivarez GA et al. 2009. Determinants for successful reforestation of abandoned pastures in the Andes: Soil conditions and vegetation cover. *For Ecol Manage* 258:81–91.

Hadley G. 1969. Cellulose as a carbon source for orchid mycorrhiza. *New Phytol* 68:933–939.

Haug I, Lempe J, Homeier J et al. 2004. *Graffenrieda emarginata* (Melastomataceae) forms mycorrhizas with Glomeromycota and with a member of *Hymenoscyphus ericae* aggr. in the organic soil of a neotropical mountain rain forest. *Can J Bot* 82:340–356.

Haug I, Weiß M, Homeier J, Oberwinkler F, Kottke I. 2005. Russulaceae and Thelephoraceae form ectomycorrhizas with members of the Nyctaginaceae (Caryophyllales) in the tropical mountain rain forest of southern Ecuador. *New Phytol* 165:923–936.

Haug I, Wubet T, Weiß M et al. 2010. Species-rich but distinct arbuscular mycorrhizal communities in reforestation plots on degraded pastures and in neighboring primary tropical mountain rain forest. *J Trop Ecol* 51:125–148.

Henkel TW, Mayor JR, Woolley LP. 2005. Mast fruiting and seedling survival of the ectomycorrhizal, monodominant *Dicymbe corymbosa* (Caesalpiniaceae) in Guyana. *New Phytol* 167:543–556.

Henkel TW, Roberts P, Aime MC. 2004. Sebacinoid species from the Pakaraima Mountains of Guyana. *Mycotaxon* 89:433–439.

Henkel TW, Terborgh J, Vilgalys RJ. 2002. Ectomycorrhizal fungi and their hosts in the Pakaraima Mountains of Guyana. *Mycol Res* 106:515–531.

Hibbett DS, Ohman A, Glotzer D, Nuhn M, Kirk P, Nilsson RH. 2011. Progress in molecular and morphological taxon discovery in Fungi and options for formal classification of environmental sequences. *Fung Biol Rev* 25:38–47.

Homeier J, Breckle S-W, Günter S, Rollenbeck RT, Leuschner C. 2009. Tree diversity, forest structure and productivity along altitudinal and topographical gradients in a species-rich ecuadorian montane rain forest. *Biotropica* 42:140–148.

Horton TR, Bruns TD. 2001. The molecular revolution in ectomycorrhizal ecology: Peeking into the black-box. *Mol Ecol* 10:1855–1871.

Husband R, Herre EA, Turner SL, Gallery R, Young JP. 2002a. Molecular diversity of arbuscular mycorrhizal fungi and patterns of host association over time and space in a tropical forest. *Mol Ecol* 11:2669–2678.

Husband R, Herre EA, Young JP. 2002b. Temporal variation in the arbuscular mycorrhizal communities colonising seedlings in a tropical forest. *FEMS Microbiol Ecol* 42:131–136.

Jacquemyn H, Brys R, Vandepitte K, Honnay O, Roldán I, Wigand T. 2007. A spatially explicit analysis of seedling recruitment in the terrestrial orchid *Orchis purpurea. New Phytol* 176:448–459.

Jacquemyn H, Honnay O, Cammue BP, Brys R, Lievens B. 2010. Low specificity and nested subset structure characterize mycorrhizal associations in five closely related species of the genus *Orchis. Mol Ecol* 19:4086–4095.

Jersáková J, Malinová T. 2007. Spatial aspects of seed dispersal and seedling recruitment in orchids. *New Phytol* 176:237–241.

Julou T, Burghardt B, Gebauer G, Berveilleir D, Damesin C, Selosse MA. 2005. Mixotrophy in orchids: Insight from a comparative study of green individuals and nonphotosynthetic individuals of *Cephalanthera damasonium. New Phytol* 166:639–653.

Kiers ET, Lovelock CE, Krieger EL, Herre EA. 2000. Differential effects of tropical arbuscular mycorrhizal fungal inocula on root colonization and tree seedlings growth: Implications for tropical forest diversity. *Ecol Lett* 3:106–113.

Kottke I. 2002. Mycorrhizae-Rhizosphere determinants of plant communities. In *Plant Roots: The Hidden Half*, 3rd edn., eds. Y Waisel, A Eshel, U Kafkafi, pp. 919–932. New York: Marcel Dekker, Inc.

Kottke I, Beck A, Oberwinkler F, Homeier J, Neill D. 2004. Arbuscular endomycorrhizas are dominant in the organic soil of a neotropical montane cloud forest. *J Trop Ecol* 20:125–129.

Kottke I, Suárez JP. 2009. Mutualistic, root-inhabiting fungi of orchids—Identification and functional types. In *Proceedings of the Second Scientific Conference on Andean Orchids*, eds. AM Pridgeon, JP Suárez, pp. 84–99. Loja, Ecuador: Universidad Técnica Particular de Loja.

Kottke I, Suárez JP, Herrera P et al. 2010. Atractiellomycetes belonging to the 'rust' lineage (Pucciniomycotina) form mycorrhizae with terrestrial and epiphytic Neotropical orchids. *Proc R Soc B* 277:1289–1296.

Látalová K, Baláz M. 2010. Carbon nutrition of mature green orchid *Serapias strictiflora* and its mycorrhizal fungus *Epulorhiza* sp. *Biol Plant* 54:97–104.

Liebel HA, Bidratondo MI, Preiss K et al. 2010. C and N isotope signatures reveal constraints to nutritional modes in orchids from the Mediterranean and Macronesia. *Am J Bot* 97:903–912.

Lilleskov EA, Parrent JL. 2007. Can we develop general predictive models of mycorrhizal fungal community–environment relationships? *New Phytol* 174:250–256.

Lovelock CA, Andersen K, Morton JB. 2003. Arbuscular mycorrhizal communities in tropical forests are affected by host tree species and environment. *Oecologia* 135:268–279.

Mangan SA, Eom AH, Adler GH, Yavitt JB, Herre EA. 2004. Diversity of arbuscular mycorrhizal fungi across a fragmented forest in Panama: Insular spore communities differ from mainland communities. *Oecologia* 141:687–700.

Martos F, Dulormne M, Pailler T et al. 2009. Independent recruitment of saprotrophic fungi as mycorrhizal partners by tropical achlorophyllous orchids. *New Phytol* 184:668–681.

McCormick MK, Whigham DF, O'Neill J. 2004. Mycorrhizal diversity in photosynthetic terrestrial orchids. *New Phytol* 163:425–438.

McCormick MK, Whigham DF, O'Neill JP et al. 2009. Abundance and distribution of *Corallorhiza odontorhiza* reflect variations in climate and ectomycorrhizae. *Ecol Monogr* 79:619–635.

McGuire KL. 2007. Common ectomycorrhizal networks may maintain monodominance in a tropical rain forest. *Ecology* 88:566–574.

McKendrick SL, Leake JR, Taylor DL, Read DJ. 2000. Symbiotic germination and development of myco-heterotrophic plants in nature: Ontogeny of *Corallorhiza trifida* and characterization of its mycorrhizal fungi. *New Phytol* 145:523–537.

McKendrick SL, Leake DJ, Taylor DL, Read DJ. 2002. Symbiotic germination and development of myco-heterotrophic orchid *Neottia nidus-avis* in nature and its requirement for locally distributed *Sebacina* spp. *New Phytol* 154:233–247.

Midgley DJ, Jordan LA, Saleeba JA, McGee PA. 2006. Utilisation of carbon substrates by orchid and ericoid mycorrhizal fungi from Australian dry sclerophyll forests. *Mycorrhiza* 16:175–182.

Molina R, Massicotte H, Trappe J. 1992. Specificity phenomena in mycorrhizal symbiosis: Community-ecological consequences and practical implication. In *Mycorrhizal Functioning. An Integrative Plant-Fungal Process*, ed. MJ Allen, pp. 357–423. New York: Chapman and Hall.

Moyersoen B. 2006. *Pakaraimaea dipterocarpacea* is ectomycorrhizal, indicating an ancient Gondwanaland origin for the ectomycorrhizal habit in Dipterocarpaceae. *New Phytol* 172:753–762.

Newberry DM, Alexander IJ, Thomas DW, Gartlan JS. 1988. Ectomycorrhizal rain-forest legumes and soil phosphorous in Korup National Park, Cameroon. *New Phytol* 109:433–450.

Ogura-Tsujita Y, Gebauer G, Hashimoto T, Umata H, Yukawa T. 2009. Evidence for novel and specialized parasitism: The orchid *Gastrodia confusa* gains carbon from saprotrophic *Mycena*. *Proc R Soc Lond B* 276:761–767.

Ogura-Tsujita Y, Yakawa T. 2008a. *Epipactis helleborine* shows strong mycorrhizal preference towards ectomycorrhizal fungi with contrasting geographic distributions in Japan. *Mycorrhiza* 18:331–338.

Ogura-Tsujita Y, Yakawa T. 2008b. High mycorrhizal specificity in a widespread mycoheterotrophic plant, *Eulophia zollingeri* (Orchidaceae). *Am J Bot* 95:93–97.

Öpik M, Metsis M, Daniell TJ, Zobel M, Moora M. 2009. Large-scale parallel 454 sequencing reveals host ecological group specificity of arbuscular mycorrhizal fungi in a boreonemoral forest. *New Phytol* 184:424–437.

Otero JT, Ackermann JD, Bayman P. 2002. Diversity and host specificity of endophytic Rhizoctonia-like fungi from tropical orchids. *Am J Bot* 89:1852–1858.

Peay KG, Bruns TD, Kennedy PG, Bergemann SE, Garbelotto M. 2007. A strong species–area relationship for eukaryotic soil microbes: Island size matters for ectomycorrhizal fungi. *Ecol Lett* 10:470–480.

Peay KG, Garbelotto M, Bruns TD. 2010a. Evidence of dispersal limitation in soil microorganisms: Isolation reduces species richness on mycorrhizal tree islands. *Ecology* 91:3631–3640.

Peay KG, Kennedy PG, Davies SJ, Tan S, Bruns TD. 2010b. Potential link between plant and fungal distribution in a dipterocarp rainforest: Community and phylogenetic structure of tropical ectomycorrhizal fungi across a plant and soil ecotone. *New Phytol* 185:529–542.

Peterson CJ, Carson WP. 2008. Processes constraining woody species succession on abandoned pastures in the tropics: On the relevance of temperate models of succession. In *Tropical Forest Community Ecology*, eds. WP Carson, SA Schnitzer, pp. 365–383. Chichester, U.K.: Wiley-Blackwell.

Picone C. 2000. Diversity and abundance of arbuscular-mycorrhizal fungus spores in tropical forest and pastures. *Biotropica* 32:734–750.

Preiss K, Adam IK, Gebauer G. 2010. Irradiance governs exploitation of fungi: Fine-tuning of carbon gain by two partially myco-heterotrophic orchids. *Proc R Soc Lond B* 277:1333–1336.

Riviere T, Diédhiou AG, Diabate M et al. 2007. Genetic diversity of ectomycorrhizal Basidiomycetes from African and Indian tropical rain forests. *Mycorrhiza* 17:415–428.

Roche SA, Carter RJ, Peakall R, Smith LM, Whitehead MR, Linde CC. 2010. A narrow group of monophyletic *Tulasnella* (Tulasnellaceae) symbiont lineages are associated with multiple species of *Chiloglottis* (Orchidaceae): Implications for orchid diversity. *Am J Bot* 97:1313–1327.

Roy M, Takahiro Y, Msahide Y et al. 2009a. Ectomycorrhizal *Inocybe* species associate with the mycoheterotrophic orchid *Epipogium aphyllum* Sw., but not with its asexual propagules. *Ann Bot* 104:595–610.

Roy M, Watthana S, Stier A, Richard F, Vessabutr S, Selosse M-A. 2009b. Mycoheterotrophic orchids from Thailand tropical dipterocarpacean forests associate with a broad diversity of ectomycorrhizal fungi. *BMC Biol* 7:51.

Selosse M-A, Faccio A, Scappaticci G, Bonfante P. 2004. Chlorophyllous and achlorophyllous specimens of *Epipactis microphylla* (Neottieae, Orchidaceae) are associated with ectomycorrhizal septomycetes, including truffles. *Microb Ecol* 47:416–426.

Selosse M-A, Martos F, Perry BA, Padamsee M, Roy M, Pailler T. 2010. Saprotrophic fungal mycorrhizal symbionts in achlorophyllous orchids. *Plant Signal Behav* 5:1–5.

Selosse M-A, Roy M. 2009. Green plants that feed on fungi: Facts and questions about mixotrophy. *Trends Plant Sci* 14:64–70.

Selosse M-A, Weiß M, Jany J-L, Tiller A. 2002. Communities and populations of sebacinoid Basidiomycetes associated with the achlorophyllous orchid *Neottia nidus-avis* (L.) L.C.M. Rich. and neighbouring tree ectomycorrhizae. *Mol Ecol* 11:1831–1844.

Shefferson RP, Cowden CC, McCormick MK, Yukawa T, Ogura-Tsujita Y, Hashimoto T. 2010. Evolution of host breadth in broad interactions: Mycorrhizal specificity in East Asian and North American rattlesnake plantains (*Goodyera* spp.) and their fungal hosts. *Mol Ecol* 19:3008–3017.

Shefferson RP, Kull T, Tali K. 2008. Mycorrhizal interactions of orchids colonizing Estonian mine tailing hills. *Am J Bot* 95:156–164.

Shefferson RP, Taylor DL, Weiß M et al. 2007. The evolutionary history of mycorrhizal specificity among lady's slipper orchids. *Evolution* 61:1380–1390.

Shefferson RP, Weiß M, Kull T, Taylor DL. 2005. High specificity generally characterizes mycorrhizal association in rare lady's slipper orchids, genus *Cypripedium*. *Mol Ecol* 14:613–626.

Shimura H, Sadamoto M, Matsuura M, Kawahara T, Naito S, Koda Y. 2009. Characterization of mycorrhizal fungi from the threatened *Cypripedium macranthos* in a northern island of Japan: Two phylogenetically distinct fungi associated with the orchid. *Mycorrhiza* 19:525–534.

Sirikantaramas S, Sugiokai N, Lee S et al. 2003. Molecular identification of ectomycorrhizal fungi associated with Dipterocarpaceae. *Tropics* 13:69–77.

Smith SE. 1967. Carbohydrate translocation into orchid mycorrhizas. *New Phytol* 66:371–378.

Smith SE, Read DJ. 2008. *Mycorrhizal Symbiosis*, 3rd edn. San Diego, CA: Academic Press.

Smits WT. 1983. Dipterocarpaceae and mycorrhiza: An ecological adaptation and a factor in forest regeneration. *Flora Malaysia Bull* 36:3926–3937.

Suárez JP, Weiß M, Abele A, Garnica S, Oberwinkler F, Kottke I. 2006. Diverse tulasnelloid fungi form mycorrhizas with epiphytic orchids in an Andean cloud forest. *Mycol Res* 110:1257–1270.

Suárez JP, Weiß M, Abele A, Oberwinkler F, Kottke I. 2008. Members of Sebacinales subgroup B form mycorrhizae with epiphytic orchids in a neotropical mountain rain forest. *Mycol Progress* 7:75–85.

Taylor DL, Bruns TD. 1997. Independent, specialized invasion of ectomycorrhizal mutualism by two non photosynthetic orchids. *Proc Natl Acad Sci USA* 94:4510–4515.

Taylor DL, Bruns TD. 1999. Population, habitat and genetic correlates of mycorrhizal specialization in the "cheating" orchids *Corallorhiza maculata* and *C. mertensiana*. *Mol Ecol* 8:1719–1732.

Taylor DL, Bruns TD, Hodges SA. 2004. Evidence for mycorrhizal races in a cheating orchid. *Proc R Soc Lond B* 271:35–43.

Taylor DL, Bruns TD, Szaro TM, Hodges SA. 2003. Divergence in mycorrhizal specialization within *Hexalectris spicata* (Orchidaceae), a non-photosynthetic desert orchid. *Am J Bot* 90:1168–1179.

Tedersoo L, Nilsson RH, Abarenkov K et al. 2010a. 454 Pyrosequencing and Sanger sequencing of tropical mycorrhizal fungi provide similar results but reveal substantial methological biases. *New Phytol* 188:291–301.

Tedersoo L, Sadam A, Zambrano M, Valencia R, Bahram M. 2010b. Low diversity and high host preference of ectomycorrhizal fungi in Western Amazonia, a neotropical diversity hotspot. *ISME J* 4:465–471.

Tedersoo L, Suvi T, Beaver K, Kõljalg U. 2007. Ectomycorrhizal fungi of the Seychelles: Diversity patterns and host shifts from the native *Vateriopsis seychellarum* (Dipterocarpaceae) and *Intsia bijuga* (Caesalpiniaceae) to the introduced *Eucalyptus robusta* (Myrtaceae), but not *Pinus caribea* (Pinaceae). *New Phytol* 175:321–333.

Thoen D, Ba AM. 1989. Ectomycorrhizas and putative ectomycorrhizal fungi of *Afzelia africana* Sm. and *Uapaca guinensis* Müll. Arg. in southern Senegal. *New Phytol* 113:549–559.

Valencia R, Foster RB, Villa G et al. 2004. Tree species distributions and local habitat variation in the Amazon: Large forest plot in eastern Ecuador. *J Ecol* 92:214–229.

Whitridge H, Southworth D. 2005. Mycorrhizal symbionts of the terrestrial orchid *Cypripedium fasciculatum*. *Selbyana* 26:254–260.

Yagame T, Yamato M, Mii M, Suzuki A, Iwase K. 2007. Developmental process of achlorophyllous orchid, *Epipogium roseum*: From seed germination to flowering under symbiotic cultivation with mycorrhizal fungi. *J Plant Res* 120:229–236.

Yamato M, Yagame T, Suzuki A, Iwase K. 2005. Isolation and identification of mycorrhizal fungi associated with an achlorophyllous plant, *Epipogium roseum* (Orchidaceae). *Mycoscience* 46:73–77.

Yuan L, Yang ZL, Li SY, Hu H, Huang JL. 2010. Mycorrhizal specificity, preference, and plasticity of six slipper orchids from South Western China. *Mycorrhiza* 20:559–568.

Yukawa T, Ogura-Tsujita Y, Shefferson RP, Yokoyama J. 2009. Mycorrhizal diversity in *Apostasia* (Orchidaceae) indicates the origin and evolution of orchid mycorrhiza. *Am J Bot* 96:1997–2009.

41

Response of Soybean Roots to Soybean Cyst Nematode at the Molecular Level

Benjamin F. Matthews
United States Department of Agriculture

Heba M.M. Ibrahim
United States Department of Agriculture
Cairo University

Parsa Hosseini
United States Department of Agriculture
Towson University

Nadim Alkharouf
Towson University

Savithiry S. Natarajan
United States Department of Agriculture

I. Introduction

Plant parasitic nematodes are major pests of agronomically important crop plants and cause an estimated $100 billion in damage annually worldwide (Koenning et al. 1999). These microscopic invertebrate pests attack all types of plants. These include the major cereals, rice, wheat and corn, potato, tomato, vegetable crops, trees, ornamentals, turf grass, and other plants. In fact, more than 200 species of plant parasitic nematode are reported to attack rice alone (Prot et al. 1994)! Crop rotation is often used to help alleviate nematode damage to crops. Poor and non-hosts must be alternated with the susceptible crop in the cropping sequence to minimize nematode damage. Chemical control can be costly and can damage the environment. Nematodes can have a large and broad host range, but many have a restricted host range. For example, in the genus *Meloidogyne*, some members can reproduce on a wide range of vegetable crops, such as potato and tomato, while only some populations can reproduce on a few hosts (Moens et al. 2009).

The soybean cyst nematode (SCN; *Heterodera glycines*) is a sedentary, obligate plant parasitic nematode with a narrower host range than the root-knot nematode, *Meloidogyne incognita*. SCN is of particular importance to soybean as it is the most devastating pest of soybean in the United States, causing an estimated $1 billion in losses annually (Wrather and Koenning 2006). Eggs of SCN can remain viable in the soil for up to 9 years. Thus, although crop rotation can reduce the number of viable SCN in the soil, it does not eliminate the problem. There are some soybean genotypes that are resistant to specific, genetically diverse populations of SCN (Niblack et al. 2002). In early literature, these SCN populations were referred to as "races." Because there are numerous genetically diverse populations of SCN, there is no commercial cultivar of soybean with resistance to all SCN. Therefore, soybean breeding and genetic engineering continue in an effort to develop elite soybean cultivars with broad resistance to SCN.

The life cycle of the SCN and its background have been recounted numerous times in papers and reviews (e.g., Bird and Koltai 2000; Goverse et al. 2000; Williamson and Gleason 2003; Lilley et al. 2005; Niblack et al. 2006; Klink et al. 2007a, 2009a; Mitchum and Baum 2008; Abad and Williamson 2010). Therefore, only a brief accounting of the life cycle of SCN is given here. The life cycle of the SCN includes the egg, four juvenile stages, and the adult stage. It is the second-stage juvenile (J-2) that hatches from the egg and migrates through the soil to enter the root. The SCN J-2 prefers entering just above the root tip but also may enter the root at other points along its entire length. The J-2 migrates intracellularly to the vascular cylinder, secreting cell wall–degrading enzymes along the way. It selects a feeding site and pierces a cell with its stylet and injects glandular secretions into the cell. The cell becomes metabolically hyperactive, and its cell wall degrades along with that of adjacent cells. The nucleus enlarges and mitochondria and plastids proliferate, while the central vacuole disperses into small vesicles during cytoplasmic condensation. Neighboring cells merge with the initial feeding cell to form a syncytium, which can be composed of 200 cells or more in a compatible host by the time the nematode is mature (Jones 1981; Jung and Wyss 1999). As the morphology of the soybean cells at the feeding site change, so does the physiology. There are also dramatic changes in the physiology of the nematode when it shifts from a motile to sedentary life style. In roots exhibiting an incompatible interaction with the nematode, the syncytia degrade. In the interaction of SCN with *Arabidopsis*, H_2O_2 is produced, and there are symptoms of a hypersensitive response (Waetzig et al. 1999). Similarly, H_2O_2 accumulates, and there is a hypersensitive response in incompatible reaction of tomato roots with the root-knot nematode, *M. incognita* (Melillo et al. 2006).

Numerous scientists have examined the interaction of soybean with SCN at the molecular level with special emphasis on gene expression. Earlier studies examined one or a few soybean genes (Hermsmeier 1998; Vaghchhipawala et al. 2001; Bird and Kaloshian 2003; Alkharouf et al. 2004; Williamson and Gleason 2004). A first attempt to broaden the spectrum of genes analyzed during interactions of soybean with SCN was made by Khan et al. (2004) using noncommercial microarrays made from approximately 1300 cDNAs printed on glass microscope slides. The expression of soybean genes expressed in compatible roots of cv. Kent was studied 2 days after infection (dai). Although these data were limited by the number of cDNAs printed on the array, many of the genes identified in this chapter as having increased expression were confirmed in later studies when larger-scale commercial microarrays became available. Genes encoding extensin, peroxidase, aspartic protease, fructose-1,6-bisphosphate aldolase, basic chitinase, β-1,3-endoglucanase, and sucrose synthase were found to be upregulated. A second, larger noncommercial microarray of 6000 printed cDNA inserts was used to measure transcript abundance in roots of soybean infected with SCN over a time course of 6 and 12 h after infection (hai) and 1,2,4,6, and 8 dai (Alkharouf et al. 2006). A broader range of up- and downregulated genes was identified, including the genes identified by Khan et al. (2004) and genes encoding enzymes in the glycolysis pathway, the phenylpropanoid metabolism, PR genes, cell wall enzymes, and others.

The next step in studying gene expression during interaction of soybean with SCN was the application of laser capture microdissection to collect syncytia formed by the nematode, so genes expressed specifically at the feeding site could be identified. Precise capture of syncytial cells using laser capture microdissection dramatically reduced the number of unwanted cells, so gene expression in specific cell types could be conducted with low background arising from unwanted cell types (Figure 41.1A and B; Kerk et al. 2003; Ramsay et al. 2004; Klink et al. 2005, 2007b, 2008, 2009a–c, 2010). A cDNA library was generated from the mRNA isolated using LCM of syncytia from compatible roots (cv. Kent) 8 dai followed by one-pass sequencing of cDNA clones. It allowed for identification of 800 genes expressed in the syncytium (Alkharouf et al. 2004). The results were checked using RT-PCR. ESTs representing genes encoding extensin, xylose synthase, protein kinases, α- and β-tubulin, peroxidase, asparaginyl endopeptidase, and ribosomal proteins were among the most highly represented.

Once commercial microarrays for soybean became available, the technology was more accessible to small laboratories. Thus, it was easier to obtain data for the abundance of transcripts of a large number of soybean genes in one experiment. For example, the Affymetrix soybean microarray contained more than 37,500 probe sets to measure transcript abundance. This array also conveniently had approximately 7500 probe sets to measure transcripts of SCN. Thus, transcript abundance of a large set of genes from soybean and SCN could be measured in parallel using the same experimental material (Ithal et al. 2007a,b; Klink et al. 2007a,b, 2009a, 2010; Mazarei et al. 2011). When transcript abundance was compared in compatible and incompatible roots through a time series of 12 hai, 3 dai, and 8 dai, 2000–8000 gene probe sets indicated a changed in abundance more than

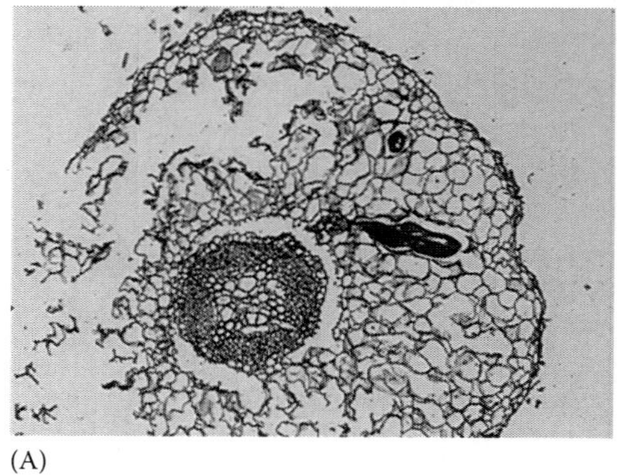
(A)

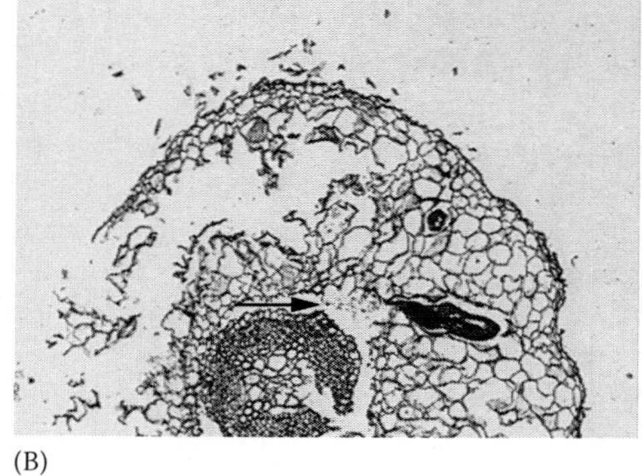
(B)

FIGURE 41.1 Laser capture microdissection of the syncytium formed by SCN at 6 dai. A cross section of a soybean root infected with SCN (N) is shown. (A) The nematode feeding site is intact. (B) The syncytium (arrow) has been removed.

1.5-fold at each time point (Klink et al. 2007a). This suggested that approximately 5%–21% of the genes in the roots that were represented on the microarray were altered in expression 1.5-fold or more. Numerous resistance-related genes were identified with increased transcript abundance at 3 and 8 dai, representing stages wherein nematodes were beginning to feed and later, when the nematode feeding site was better established. Among the genes having elevated transcripts in the incompatible interaction at 12 hai, 3 dai, and 8 dai were genes encoding pathogenesis-related gene 5 (PR5), PR6, cysteine protease inhibitor, several transcription factors, and β-fructosidase (Klink et al. 2007). At 12 hai and 3 dai, transcripts encoding peroxidase, pectinesterase, lipoxygenase, expansin, and others were elevated but declined by 8 dai (Klink et al. 2007a). In compatible roots, transcripts for genes encoding expansin, pectate lyase, peroxidase, serine carboxypeptidase, a hydroxyproline-rich protein, and others were elevated at later time points, 8, 12, and 16 dai (Putoff et al. 2007). Mazarei et al. (2011) examined two sister lines, one resistant and the other compatible to SCN race 2. Their microarray and qRT-PCR data correlated well and demonstrated that polygalacturonase-inhibiting protein (PGIP), stress-induced protein (SAM22), salicylic acid methyl transferase protein (SABATH2), and other genes were induced at 3, 6, and 9 dai in the resistant interaction but remained constant in the susceptible interaction.

More recently, microarrays and laser capture microdissection have been used concurrently to provide insights into gene expression specifically in syncytial cells during compatible and incompatible reactions. This powerful combination provided syncytial cells removed from what would be an overwhelming background of non-syncytial cells and allowed the measurement of transcripts of thousands of genes in parallel from theses small samples (Ithal et al. 2007a,b; Klink et al. 2007b, 2009a, 2010).

In the future, data from proteomics will be integrated with data from gene expression studies, so changes in transcript levels and protein abundance can be used to provide a better picture of events occurring within the plant and within particular cells during nematode infection. Proteomic technologies are powerful tools for examining alterations in protein profiles caused by mutations, introduction or silencing of genes, or responses to various stress stimuli including nematode infection (Görg et al. 2000; Dubey and Grover 2001; Afzal, et al. 2009). Gel-based proteomics with mass spectrometric characterization is widely used for determining protein profiles in plants (Porubleva et al. 2001; Herman et al. 2003; Natarajan et al. 2006, 2007; Xu et al. 2006, 2007) and pathogen reference maps (Kaji et al. 2000; Schrimpf et al. 2001; Mawuenyega et al. 2003; Tabuse et al. 2005), as well as for investigating temperature and oxidative stress responses (Sule et al. 2004; Wang et al. 2004; Yan et al. 2006), pathogen–plant relationships (Riggs et al. 1982; Wan et al. 2005; Roggero and Pennazio 2008; Afzal et al. 2009), root symbiotic interactions (Mathesius et al. 2003; Hoa et al. 2004; Amiour et al. 2006; Valot et al. 2006; Segarra et al. 2007; Larrainzar et al. 2007), and parasitic relationships (De Meutter et al. 2001). Recent advances in the availability of commercial immobilized pH gradient (IPG) strips with different pH ranges (wide and narrow), image analysis software, and ionization technology using different modern mass spectrometers, together with the establishment of protein databases, have substantially increased the accuracy of protein profile characterization from complex protein mixtures and offer high-throughput analysis.

There are many reports in the literature describing the physiological and histological basis of nematode–host interactions. Relatively, fewer reports are available on the proteomics of the nematode and its pathogenesis. Basic information about the nematode's proteome would be useful for devising strategies to control SCN and developing soybeans with broad resistance against SCN. Such a proteome profile could considerably improve our understanding of SCN parasitism and its virulence.

Studies on the proteomics of SCN have been limited by the difficulty of collecting enough nematode and host material for protein extraction of sufficient quantity and adequate quality to analyze by 2-dimensional polyacrylamide gel electrophoresis (2D-PAGE). In addition, a paucity of sequence information from pathogen and host has also limited the number of proteins that can be identified. Several techniques have been used to concentrate the nematode and host proteins. Navas et al. (2002) precipitated proteins from 18 isolates of *Meloidogyne* spp. with acetone and examined protein variation using 2D-PAGE. They analyzed 86–203 protein positions from these gels and used the results for phylogenetic analysis. Mbeunkui et al. (2010) concentrated *Meloidogyne hapla* lysate using a 10 kDa molecular weight cutoff concentrator, separated the resulting protein with SDS-PAGE, and used liquid chromatography/mass spectrometry (LC/MS) to identify 516 proteins.

To understand the biology of the parasitic relationship between a nematode and its host plant, it is important to understand the proteomic role of the nematode's pharyngeal secretions because they contain cell wall–degrading enzymes such as cellulases, which facilitate the migration of the nematode through the root and aid in maintaining the feeding site (Smant et al. 1998). Identification of a number of secretory proteins produced in the pharyngeal glands, amphids, and cuticle using monoclonal antibody and analysis of mRNAs has been reported, and it is generally accepted that secretions from pharyngeal glands contain the pathogenicity factors responsible for feeding site establishment (Hussey 1989; Williamson and Hussey 1996). Secreted proteins from *Globodera rostochiensis*, such as β-1, 4-endoglucanases, and peroxiredoxin, were identified by Smant et al. (1998) and Robertson et al. (2000) using monoclonal antibodies. Meutter et al. (2001) analyzed isolated pharyngeal glands proteins of cyst nematode *Heterodera schachtii* using both 1D and 2D-PAGE. They reported about 30 protein bands or spots visible on Coomassie blue-stained gels and several additional spots visible using silver staining. Most of the proteins from 2D-PAGE were below 55 kDa with basic isoelectric point. They identified spot clusters that included cell wall–degrading enzymes. In addition, the authors identified two endoglucanases and a novel protein in the gland secretions. In secretions of the potato cyst nematode, Robertson et al. (1999) visualized only 10 proteins using SDS-PAGE ranging in size from 15–70 kDa that included proteases and the antioxidant enzyme, superoxide dismutase, (SOD). SOD may have a role in protecting the nematode from the plant resistance response, neutralizing oxygen free radicals produced in the oxidative burst associated with nematode infection (Jones et al.1997; Robertson et al. 2000). These protein studies were complemented by molecular studies of the mRNAs present in the esophageal glands of SCN. These studies reported that more than sixty genes were expressed in these glands, some encoding enzymes that digest components of the plant cell wall, including β-1,4-endoglucanse and pectinase (Smant et al. 1998). Also, a product of a gene encoding chorismate mutase (Lambert et al. 1999; Bekal et al. 2003) and a CLAVATA-like protein (Wang et al. 2005; Replogle et al. 2011) were produced. Some of these gene products were targeted to the plant nucleus, while others remained in the cytoplasm (Elling et al. 2007). These and other nematode proteins were secreted by the nematode in an attempt to reprogram the plant cell to facilitate nematode feeding. The plant responds in kind in an attempt to combat the nematode.

II. Carbohydrate Synthesis and Metabolism at the Nematode Feeding Site

Several studies on syncytia formed by *H. schachtii* in *Arabidopsis* roots, including analysis of metabolic profiles of sugars and amino acids, indicate that syncytia accumulate starch, which would be a source of carbohydrate for the nematode (Hofmann and Grundler 2008; Hofmann et al. 2010). Furthermore, specific sugar transporters are induced during nematode feeding (Hofmann et al. 2009). In addition, transcripts encoding starch synthase (EC 2.4.2.21), starch phosphorylase (EC 2.4.1.1), and ADP-glc pyrophosphorylase are elevated in syncytia formed in *Arabidopsis* by *H. schachtii* at 5 and 15 dai (Szakasits et al. 2009).

This is further supported by gene expression analysis of syncytia formed by SCN in soybean roots in the compatible interaction at 2 and 10 dai and at 6 dai in the incompatible reaction, because transcripts for genes encoding enzymes important to accumulation of starch and sugars in the syncytia were elevated (Ithal et al. 2007a; Klink et al. 2009a). For example, in compatible roots at 10 dai by SCN, the expression of genes encoding enzymes in gluconeogenesis/glycolysis increased from α-D-glucose and β-D-glucose (EC 2.7.1.1) through the pathway including genes encoding pyruvate dehydrogenase (EC 1.2.4.1) in roots of soybean cv. Williams 82 during the compatible reactions (Ithal et al. 2007a; Figure 41.2). Thus, carbohydrate metabolism is increased throughout gluconeogenesis/glycolysis. It is noteworthy that the gene encoding fructose bisphosphatase (EC 3.1.3.11) is one of the most highly expressed genes in soybean syncytia at 9 dai in a compatible interaction. Yet in an incompatible interaction at 9 dai, when syncytia in incompatible roots collapse, expression of genes encoding fructose bisphosphatase is at the control (uninfected cells) levels (Klink et al. 2009a). Fructose bisphosphatase catalyzes a unidirectional reaction toward starch and sucrose synthesis. Fructose bisphosphatase converts D-fructose 1,6 bisphosphate and H_2O into D-fructose 6-phosphate and phosphate. Thus, elevation of the level of expression of genes encoding this enzyme supports the data of Hoffmann and Grundler (2008) indicating that starch serves as a major sugar reservoir in syncytia. Furthermore, in the compatible and incompatible interactions, transcripts encoding enzymes in the steps between pyruvate to D-fructose 6-phosphate are all elevated, specifically those encoding fructose-bisphosphate aldolase (EC 4.2.1.13), glyceraldehydes-3-phosphate dehydrogenase (EC 1.2.1.12), phosphoglycerate kinase (EC 2.7.2.3), phosphoglycerate mutase (EC 5.4.2.1), and pyruvate kinase (EC 2.7.1.40; Ithal et al. 2007a; Klink et al. 2007b, 2009a). Furthermore, during the compatible interaction at 10 dai (Ithal et al. 2007a), in addition to elevated transcripts encoding fructose bisphosphatase, other transcripts are elevated for genes encoding fructose-bisphosphate aldolase (EC 4.1.2.13), 6-phosphofructokinase (EC 2.7.1.11), and glucose-6-phosphate isomerase (EC 5.3.1.9), but transcripts of these genes are not elevated in the incompatible reaction at 8 dai (Klink et al. 2009a). A gene encoding phosphoglycerate mutase is induced in soybean (Hermsmeier et al. 1998) by SCN and in *Arabidopsis* by *H. schachtii* and *M. incognita*.

Interestingly, a gene encoding aldose-1-epimerase (galactomutarotase, EC 5.1.3.3) is strongly downregulated in syncytia in the compatible interaction (Ithal et al. 2007a), whereas it is moderately upregulated in the incompatible interaction (Klink et al. 2009a). Increased activity of this enzyme has been correlated with decreased cellulose production (Fekete et al. 2008). Thus, decreasing the amount of transcript of this enzyme may decrease the amount of enzyme produced leading to a possible increase in cellulose production, which is needed to fortify the cell walls adjacent to the syncytium.

In syncytia from the incompatible reaction at 9 dai, the transcript abundance of genes encoding enzymes on the path from acetate to ethanol are moderately increased, i.e., the genes encoding aldehyde dehydrogenase (EC 1.2.1.3) and alcohol dehydrogenase (1.1.1.1) (Klink et al. 2009a). In contrast, in compatible syncytia at 10 dai, transcripts of one gene encoding aldehyde dehydrogenase are elevated, while transcripts of another are decreased. Furthermore, a gene encoding alcohol dehydrogenase is strongly downregulated. As a result of these changes, two possible modification of metabolism might occur: (1) There may be an increased flow of carbon in the direction of starch production, and (2) there may be increased acetyl-coA synthesis leading to an increase in the synthesis of fatty acids, leading to the production of lipids. The expression of several genes encoding enzymes of the TCA cycle (Figure 41.3) is also elevated in. This may allow more energy to be produced in the syncytium and provide carbon to the nematode. For example, the genes encoding citrate (Si)-synthase (EC 2.3.3.1), aconitate hydratase (EC 4.2.1.3), succinyltransferase (EC 2.3.1.61), fumarate hydratase (EC 4.2.1.2), and malate dehydrogenase (EC 1.1.1.37) are all moderately upregulated.

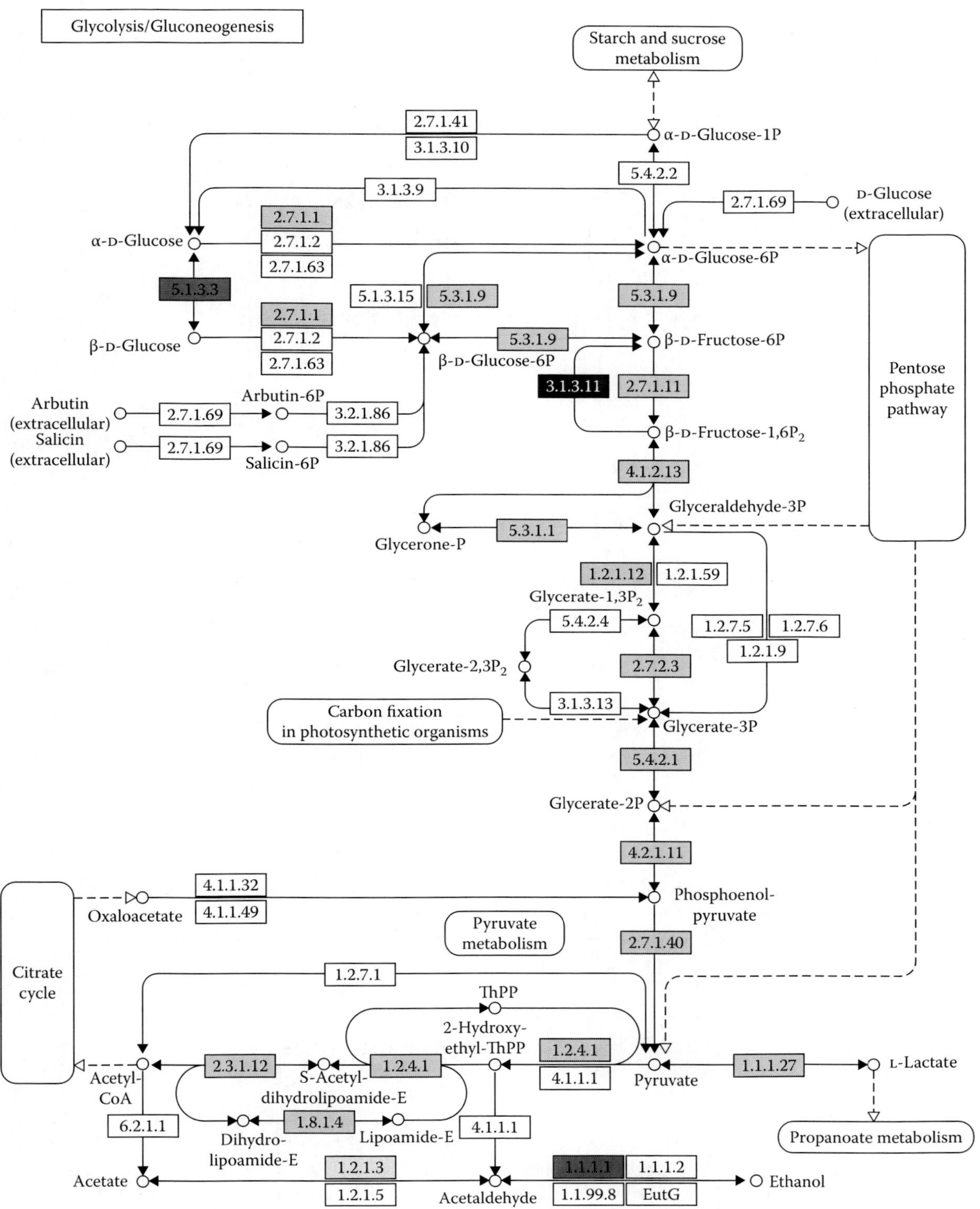

FIGURE 41.2 **(See color insert.)** Expression profiles of the RNAs encoding enzymes in glycolysis/gluconeogenesis pathway in compatible syncytia 10 dai derived from data from Ithal et al. (2007a). Enzymes colored in red are encoded by downregulated genes. Enzymes colored in yellow are encoded by more than one gene and those different gene copies are up- and downregulated, respectively. Genes encoding enzymes colored in green are upregulated. Dark green indicates genes with transcripts accumulating more than 75% of the upregulated genes.

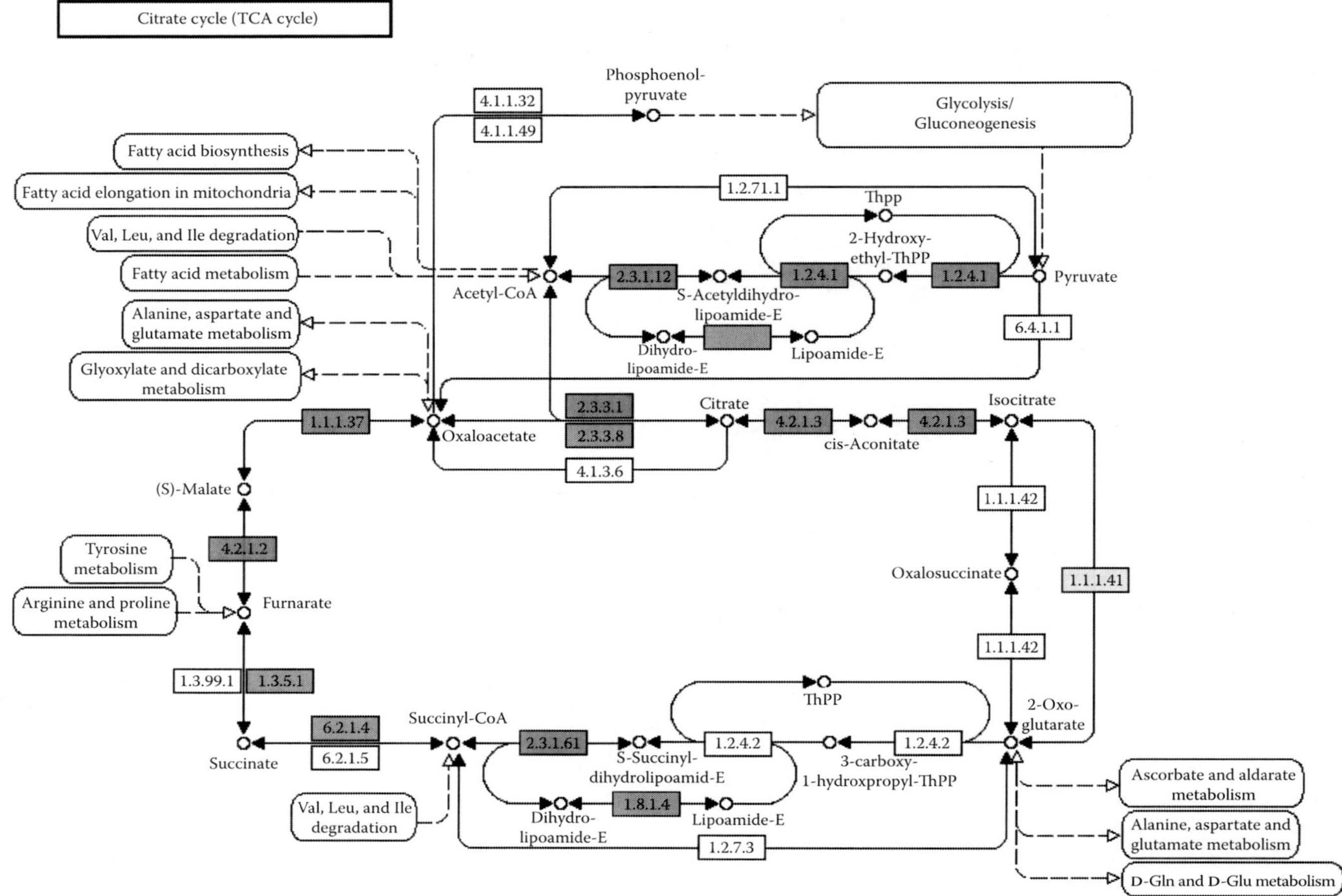

FIGURE 41.3 **(See color insert.)** Expression profiles of the RNAs encoding enzymes in the TCA cycle as found in syncytia formed during a compatible reaction 10 dai. Enzymes colored in yellow are encoded by more than one gene and those different gene copies are up- and downregulated, respectively. Genes encoding enzymes colored in green are upregulated. (Data from Ithal N et al., *Mol. Plant-Microbe. Interact.*, 20, 293, 2007a.)

III. Fatty Acid Biosynthesis

As syncytia enlarge during the compatible interaction, lipid bodies accumulate, and there is a proliferation of smooth endoplasmic reticulum (Golinowski et al. 1997), while in degenerating syncytia of incompatible reactions, lipid-like globules are also present (Riggs et al. 1973). In compatible syncytia at 5 and 10 dai, transcripts of several genes involved in fatty acid biosynthesis are upregulated, including genes encoding FabH (EC 2.3.1.180; β–ketoacyl-acyl-carrier-protein synthase III), FabG (EC 1.1.1.100; 3-oxoacyl-[acyl-carrier-protein] reductase), FabI (EC 1.3.1.9; enoyl-[acyl-carrier-protein] reductase), FabK (EC 1.3.1.-; enoyl-[acyl-carrier-protein] reductase II), FabL (1/3/1/-; enoyl-[acyl-carrier-protein] reductase III), and FabB (EC 2.3.1.41; β-ketoacyl-[acyl-carrier-protein] synthase I; Figure 41.4). However, also in the incompatible reaction, transcripts of genes encoding several enzymes involved in fatty acid biosynthesis are upregulated in the syncytia at 8 dai, including genes encoding acetyl-coA carboxylase (EC: 6.4.1.2), FabG, FabZ, and FabI. Acetyl-CoA carboxylase irreversibly catalyzes the carboxylation of acetyl-CoA to form malonyl-CoA, a precursor to fatty acid biosynthesis. This is also supported by reports of increased formation of endoplasmic reticulum and lipid bodies (Riggs et al. 1973; Golinowski et al. 1977).

IV. Cell Walls

Several genes involved in cell wall synthesis, restructuring, and dissolution were differentially expressed in roots of soybean cv. Williams 82 during the incompatible and compatible reactions with SCN. Expansins are nonenzymatic proteins involved in plant cell expansion. Several genes encoding expansins were greatly increased in expression in syncytia during the incompatible reaction at 3, 6, and 9 dai, particularly the genes coinciding with GenBank numbers CF805822, CA785167, and CD394837 (Figure 41.5A). At 8 dai in syncytia from the incompatible reaction, transcripts of CA785167 and CF805822 were over 250-fold increased in expression compared to noninfected cells (Klink

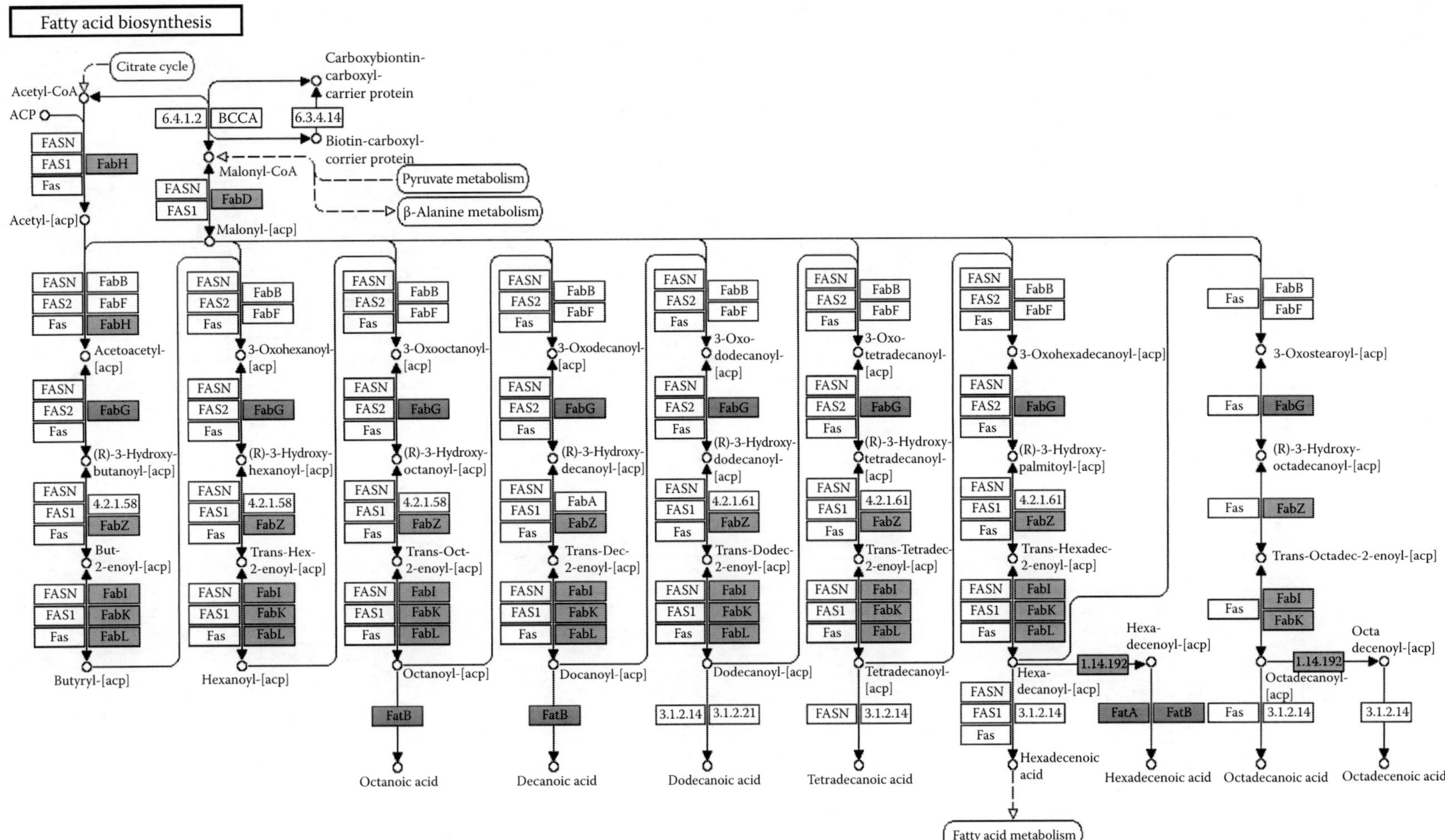

FIGURE 41.4 Induced genes encoding enzymes involve in fatty acid biosynthesis in syncytia of the compatible interaction at 5 dai. Genes encoding enzymes colored in green are upregulated. (Data from Ithal N. et al., *Mol Plant-Microbe. Interact.*, 20, 293, 2007a.)

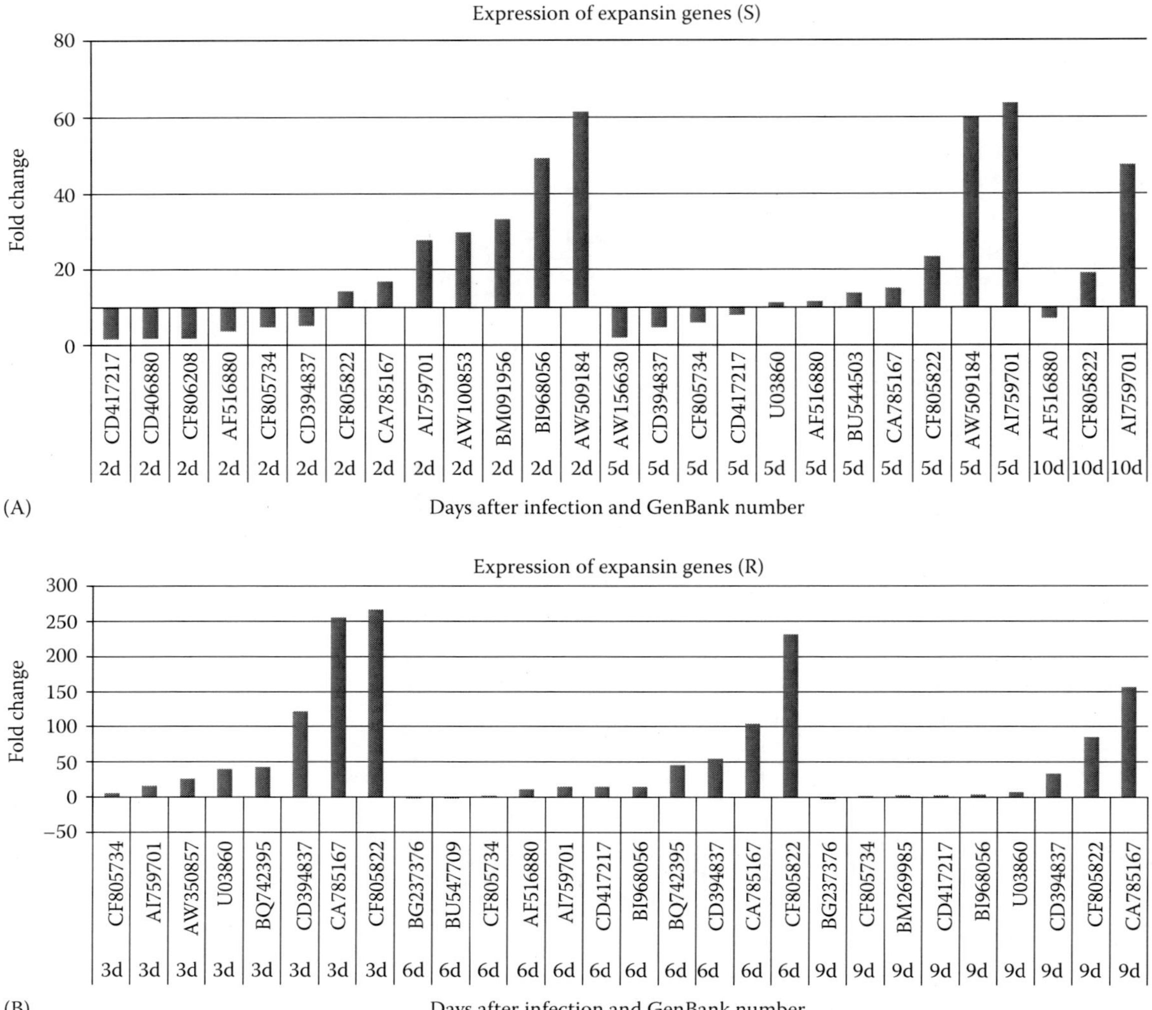

FIGURE 41.5 Histogram showing the expression level of expansins enzymes in SCN-infected soybean cv. Peking plants displaying (A) a susceptible (S) and (B) a resistant (R) interaction at different days after infection. The genes with the highest transcript abundance in syncytia from the incompatible reaction coincided with GenBank numbers CF805822, CA785167, and CD394837. (Data from Klink, V.P. et al., *Plant Mol. Biol.*, 71, 525, 2009a.)

et al. 2007b, 2009a). In syncytia from the compatible reaction at 10 dai, transcripts of genes encoding expansins accumulated up to 60-fold more than in control cells (Figure 41.5B). Thus, some genes encoding expansin were expressed more than five-fold higher in syncytia during the incompatible reaction as compared with during the compatible reaction at 3 dai and remained expressed at higher levels at 6 and 9 dai.

Plant parasitic nematodes such as SCN and RKN (root-knot nematodes) produce cellulases to soften the plant cell wall, so they can penetrate the root of the host and migrate toward the vascular tissue (Smant et al. 1998; Wang et al. 1999; Ibrahim et al. 2011). Genes encoding cellulases are also upregulated in syncytia formed by SCN, especially the cellulose genes represented by BI785739 and BI969418 at 3 dai and CF806812 at 6 and 9 dai (data from Klink et al. 2007a). Cellulases in the syncytia during the susceptible interaction are both up- and downregulated (data from Ithal et al. 2007a).

V. Defense Response

A. Hypersensitive Response

The hypersensitive response has been associated with the incompatible interaction of soybean with nematodes for many years. Histological studies of the interaction of the RKN with soybean indicated that a hypersensitive response took place in the incompatible interaction (Kaplan et al. 1979). The hypersensitive response has been described in other interactions of plants with plant parasitic nematodes. Plant cell necrosis around the head of the nematode was observed in interactions of tomato with RKN (Dropkin et al. 1969). The hypersensitive reaction repressed the production of giant cells in the incompatible response of the RKN with tomato (Paulson and Webster 1972). The accumulation of hydrogen peroxide was detected in incompatible interactions of tomato and RKN (Melillo et al. 2006). Gene expression analysis indicates that the gene represented by BG237063, a component of the necrotic lesion membrane attack complex, that is involved in the respiratory burst is downregulated along with BG406413, respiratory burst oxidase homolog, and BI971993, respiratory burst oxidase protein F, in syncytia during the susceptible interaction (Figure 41.6A). In syncytia formed during the resistant interaction transcript, abundance of BG237063 was elevated fourfold at 6 and 9 dai, transcript abundance of BI967244 was elevated over twofold at 3 and 6 dai and over threefold at 9 dai, and transcript levels of BE822942 were elevated fivefold at 6 dai. These gene expression data support earlier work indicating that a stronger hypersensitive response occurs during the resistant interaction than is invoked during the susceptible interaction.

Along with the increased abundance of some of these transcripts in syncytia displaying the resistant interaction and a decrease in abundance in those displaying a susceptible reaction, there was a concomitant increase in the abundance of transcripts for many different peroxidases in the resistant interaction with genes represented by AW309606, BU548599, and CF809087 increased over 100-fold over the time course (data from Klink et al. 2007a).

B. Jasmonic Acid and Salicylic Acid

Reports by Klink et al. (2007a,b, 2009a) and Ithal et al. (2007a) indicate that there are changes in the expression of genes that encode enzymes involved in jasmonic acid synthesis during the compatible interaction of soybean roots with SCN. In both cases, the authors used laser capture microdissection to collect syncytia formed by SCN for gene expression analysis using Affymetrix microarrays. An overview of the expression pattern of some of the genes involved in jasmonic acid synthesis in incompatible syncytia at 9 dai is given in Figure 41.7. Although two genes encoding allene oxide synthase (CA802684, CD397298) are downregulated in syncytia during the incompatible reaction, one (CA819306) is upregulated at 6 and 9 dai. There are two probes representing an EST of a gene encoding allene oxide cyclase (BG789780) on the microarray. In all instances, the trends were the same. The transcripts of this gene were greatly elevated at 3 dai and decreased but were still strongly elevated at 6 and 9 dai. Transcripts of genes encoding OPD reductase, represented by BU765938 and CF808146, were slightly upregulated at 3 and 6 dai. However, in the compatible reaction, only BI968944 encoding 12-oxophytodienoate reductase (OPR) was upregulated.

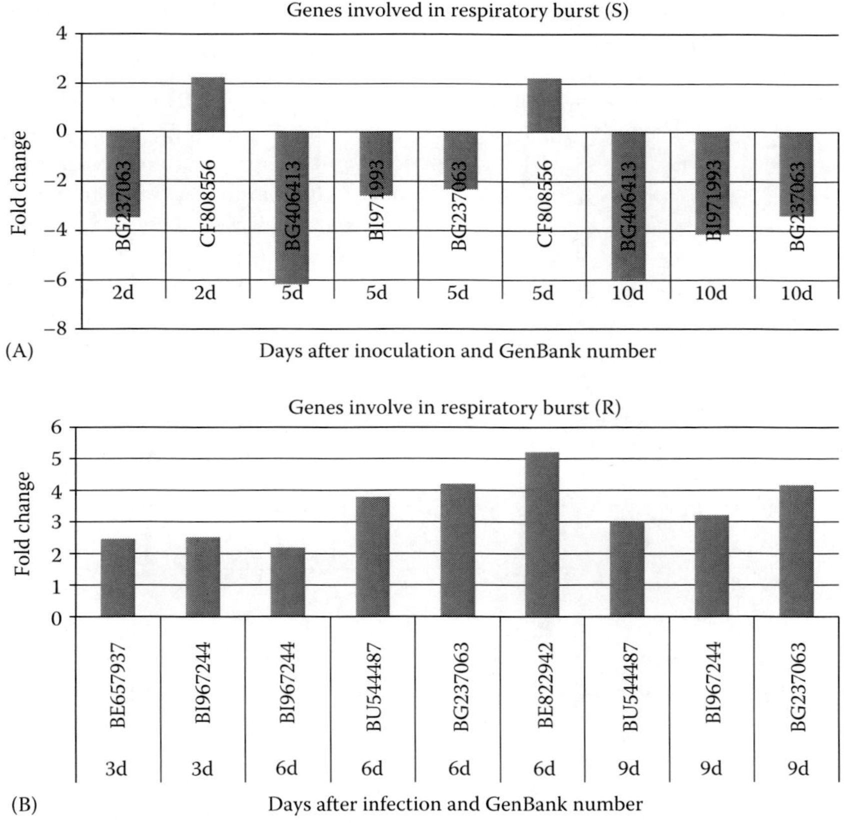

FIGURE 41.6 Expression of genes involved in production of the respiratory burst in syncytia. (A) during the susceptible (S) interaction (Data from Ithal, N. et al., *Mol. Plant-Microbe. Interact.*, 20, 293, 2007a.) and (B) during the resistant interaction. (Data from Klink, V.P. et al., *Planta*, 226, 1423, 2007a.)

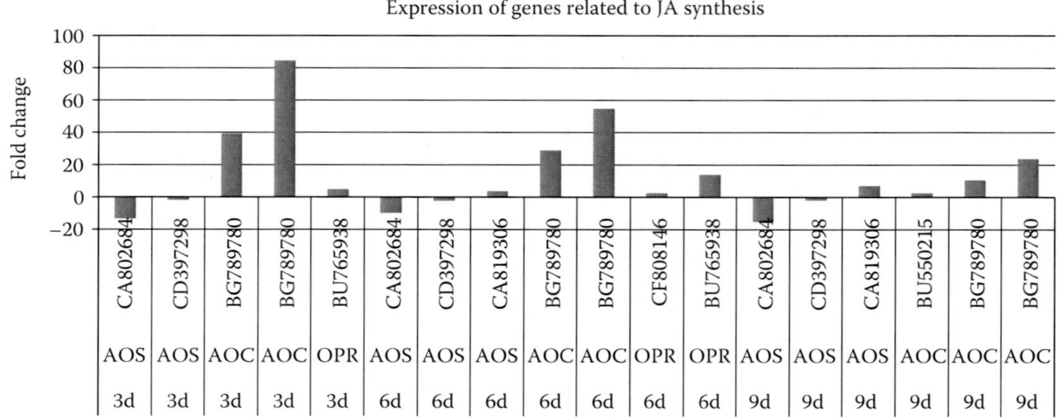

FIGURE 41.7 Graph showing the expression level of intermediate genes to JA in cyst nematode-infected incompatible soybean plants at 3, 6, and 9 dai. Transcript levels of allene oxide synthase (AOS), allene oxide cyclase (AOC), and 12-oxophytodienoate reductase (OPR) are given. Allene cyclase (AOC, BG789780) was represented by two probes on the microarrays. (Data from Klink, V.P. et al., *Plant Mol. Biol.*, 71, 525, 2009a.)

C. Other Defense-Related Genes

The genes encoding respiratory burst oxidase homologues B and F were upregulated in the incompatible reaction between Soybean cv. Peking and SCN at 3, 6, and 9 dai, while in the compatible interaction, there was no change in the expression of the gene encoding respiratory burst oxidase homologue B and there is a fivefold downregulation of the gene encoding respiratory burst oxidase homologue F (Klink et al. 2009a). Genes encoding PR-2, PR-3, PR-4, PR-5, and PR-10 are upregulated in the incompatible interaction (Klink et al. 2009a), while a mixed reaction of gene expression was apparent at 5 and 10 dai in the compatible reaction (Ithal et al. 2007a). Although several genes encoding PR10 are overexpressed, transcript levels of genes represented by CF805736, CF921432, and X60043 are particularly elevated (Figure 41.8) However, genes encoding PR-1 and PR-15 are upregulated in both compatible and incompatible reactions (Ithal et al. 2007a; Klink et al. 2009a). Transcripts of CF806709 representing PR15 are greatly abundant in the incompatible interaction at 2, 5, and 10 dai and are much higher than those of any PR15 gene at any timepoint in the incompatible interaction.

VI. Changes of Protein Expression in Plants Upon Infection by Nematodes

The proteomics of SCN–host interactions were reported by Afzal et al. (2009). The authors used 2D-PAGE and mass spectrometry and compared proteomic data from 10 days SCN-infected and SCN-noninfected soybean roots. They identified 28 protein spots that were different and grouped these proteins into 6 functional groups. Metabolite analysis identified 131 metabolites among which 58 were altered by one or more treatment. Based on the proteomic and metabolomic analyses, the authors concluded that 17 pathways were altered by the SCN resistance

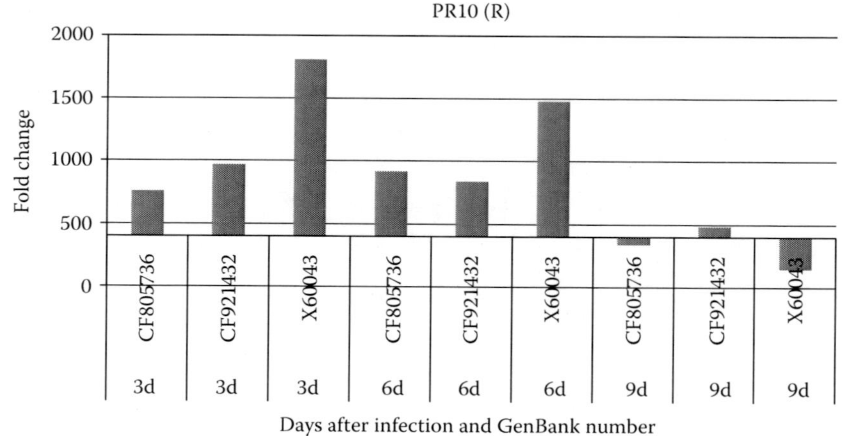

FIGURE 41.8 Graph showing the expression level of three highly expressed genes encoding PR10 in resistant syncytia at 3, 6, and 9 dai. (Data from Klink, V.P. et al., *Plant Mol. Biol.*, 71, 525, 2009a.)

alleles of the resistance locus, *rhg1*. These changes included systemic acquired resistance-like responses, such as xenobiotic, phytoalexin, ascorbate, and inositol metabolism, as well as primary metabolism pathways like amino acid synthesis and glycolysis. A gene (10A06) encoding a protein that is most likely secreted as an effector into the developing syncytia during early plant parasitism was cloned from the sugar beet cyst nematode, *H. schachtii* (Hewezi et al. 2010). When 10A06 was constitutively expressed in *Arabidopsis*, plant morphology was affected and its susceptibility to *H. schachtii* was increased. The authors used yeast two-hybrid assays to identify spermidine synthase 2 (SPDS2), involved in polyamine biosynthesis as a key protein that interacts with 10A06. Transgenic plants expressing 10A06 exhibited elevated SPDS2 mRNA abundance, significantly higher spermidine (Spd) content, increased polyamine oxidase (PAO) activity, and enhanced plant susceptibility to *H. schachtii*. Increased PAO activity stimulates the induction of the cellular antioxidant machinery in nematode syncytia.

Jaubert et al. (2002) employed 2D-PAGE and microsequencing to identify stylet-secreted proteins from *M. incognita*. They separated 40 proteins whose molecular weight and isoelectric point ranged from 6.5 to 83 kDa and 5.0 to 7.5, respectively. The seven most abundant proteins were internally micro-sequenced and identified as the β-chain of ATP synthase, two tropomyosins, a troponin-like protein, a myosin regulatory light chain (MRLC), a 14-3-3 protein, and a calreticulin (CRT). The gene for CRT, which is a calcium-binding protein involved in multiple functions including intracellular calcium homeostasis and protein maturation was cloned and shown to be actively expressed in the esophageal glands of J2s of *M. incognita* (Jaubert et al. 2002). Bellafiore et al. (2008) employed multidimensional protein identification technology (MudPIT) to identify 486 secreted proteins from *M. incognita*. Earlier studies had visualized nine bands on SDS gels in the protein from the stylet exudates from adult females of *M. incognita* (Veech et al. 1987).

By comparison of 1D and 2D protein patterns, Callahan et al. (1997) reported that many polypeptides were differentially expressed in roots from root-knot nematode resistant and susceptible cotton (*Gossypium hirsutum*) lines at 8 days after inoculation. In addition, a relatively abundant 14 kDa plant protein was expressed only in the resistant isoline 81–249 at 8 dai and was localized to the nematode-induced galls. Digestion with cyanogens bromide (CNBr) yielded two major fragments of 9 and 4 kDa, from which partial amino acid sequences were obtained. Comparison of these partial sequences with available gene databases did not reveal strong homologies with other sequences.

Differentially expressed proteins in resistant cultivars of cotton and coffee were analyzed at 6 and 10 days after phytonematode infection with *M. incognita and M. paranaensis* (Franco et al. 2010). The authors found 100 protein spots in the Coomassie Brilliant Blue (CBB)–stained coffee root 2D gels, which revealed many differentially expressed proteins at 6 days after inoculation of nematodes, including 7 upregulated and 3 downregulated. The same proteins, when analyzed at 10 days after inoculation, showed different localization patterns on the gels, suggesting that the period of nematode infection could be related to protein pattern modification. The 2D-PAGE protein maps of cotton root had 150 protein spots from CBB stained gels, and 50 additional spots were visualized using silver staining. At 6 days after *M. incognita* inoculation, they found four up- and five downregulated proteins. At 10 dai, the authors saw additional differential expression of proteins, including 13 up- and 2 downregulated. Two proteins were identified from coffee: a chitinase that is involved in defense responses to pathogens and a pathogenesis-related protein. One protein, a quinone reductase 2 that is known to protect plant cells from oxidative damages, was identified from cotton root.

De Boer et al. (1992) used 2D-PAGE to examine the differences in total proteins between two motile stages and two sedentary stages of the potato cyst nematode *G. rostochiensis* and reported 542 reproducible protein spots using a sensitive silver stain. Of these, 401 protein spots change their intensity or presence in one or more of the four developmental stages. From this, they concluded that the potato cyst nematode has a very dynamic protein metabolism.

VII. Conclusions

Our understanding of the interaction between the plant and the nematode has been greatly enhanced over the past two decades through molecular, gene expression and proteomic studies. Yet much more needs to be resolved. A full understanding of the compatible and incompatible interactions between soybean and SCN, including signaling and control mechanisms, remains elusive. Application of this new knowledge remains to be applied to successfully develop soybean with broad resistance to nematodes.

Acknowledgments

The authors gratefully acknowledge support from United Soybean Board and from the BioGreen 21 Program (no. PJ007031), Rural Development Administration, Republic of Korea.

References

Abad P, Williamson VM. 2010. Plant nematode interaction: A sophisticated dialogue. *Adv Bot Res* 53:147–192.

Afzal AJ, Natarajan A, Saini N et al. 2009. The nematode resistance allele at the rhg1 locus alters the proteome and primary metabolism of soybean roots. *Plant Physiol* 151:1264–1280.

Alkharouf N, Khan R, Matthews BF. 2004. Analysis of expressed sequence tags from roots of resistant soybean infected by the soybean cyst nematode. *Genome* 47:380–388.

Alkharouf NW, Klink V, Chouikha IB et al. 2006. Microarray analyses reveal global changes in gene expression of susceptible *Glycine max* (soybean) roots during infection by *Heterodera glycines* (soybean cyst nematode). *Planta* 224:838–852.

Amiour N, Recorbet G, Robert F, Gianinazzi S, Dumas-Gaudot E. 2006. Mutations in DMI3 and SUNN modify the appressorium-responsive root proteome in arbuscular mycorrhiza. *Mol Plant-Microbe Interact* 19:988–997.

Bekal S, Niblack TL, Lambert KN. 2003. A chorismate mutase from the soybean cyst nematode *Heterodera glycines* shows polymorphisms that correlate with virulence. *Mol Plant-Microbe Interact* 16:439–446.

Bellafiore S, Shen ZX, Rosso MN, Abad P, Shih P, Briggs SP. 2008. Direct identification of the *Meloidogyne incognita* secretome reveals proteins with host cell reprogramming potential. *PLoS Pathog*. 4:e1000192.

Bird DMcK, Kaloshian I. 2003. Are roots special? Nematodes have their say. *Physiol Mol Plant Pathol* 62:115–123.

Bird DMcK, Koltai H. 2000. Plant parasitic nematodes: Habitats, hormones, and horizontally-acquired genes. *J Plant Growth Regul* 19:183–194.

Davis EL, Caplan DT, Dickson, DW, Mitchell DJ. 1989. Root tissue response of two related soybean cultivars to infection by lectin-treated *Meloidogyne* spp. *J Nematol* 21:219–228.

De Boer JM, Overmars HA, Bakker J, Gommers FJ. 1992. Analysis of two-dimensional protein patterns from developmental stages of the potato cyst-nematode, *Globodera rostochiensis. Parasitology* 105:461–474.

De Meutter J, Vanholme B, Bauw G, Tytgat T, Gheysen G. 2001. Preparation and sequencing of secreted proteins from the pharyngeal glands of the plant parasitic nematode *Heterodera schachtii. Mol Plant Pathol* 2:297–301.

Dropkin VH. 1969. Cellular responses of plants to nematode infections. *Annu Rev Phytopathol* 7:101–122.

Dubey H, Grover A. 2001. Current initiatives in proteomics research: The plant perspective. *Curr Sci* 80:262–269.

Elling AA, Davis EL, Hussey RS, Baum RS. 2007. Active uptake of cyst nematode parasitism proteins into the plant cell nucleus. *Int J Parasitol* 37:1269–1279.

Fekete E, Seiboth B, Kubicek C, Szentirmai A, Karaffa L. 2008. Lack of aldose 1-epimerase in *Hypocrea jecorina* (anamorph *Trichoderma reesei*): A key to cellulose gene expression on lactose. *Proc Natl Acad Sci USA* 105:7141–7146.

Franco OL, Pereira JL, Costa PHA et al. 2010. Methodological evaluation of 2-De to study root proteomics during nematode infection in cotton and coffee plants. *Prep Biochem Biotechnol* 40:152–163.

Golinowski W, Sobczak M, Kurek W, Grymaszewska G. 1997. The structure of syncytia. In: *Cellular and Molecular Aspects of Plant-Nematode Interactions in Plant Pathology*, eds. C Fenoll, FMW Grundler, SA Ohl, Vol. 10, pp. 80–97. Dordrecht, the Netherlands: Kluwer.

Gorg A, Obermaier C, Boguth G, Harder A, Scheibe B, Wildgruber R, Weiss W. 2000. The current state of two-dimensional electrophoresis with immobilized pH gradients. *Electrophoresis* 21:1037–1053.

Goverse A, Engler JDA, Verhees J, Krol SVD, Helder J, Gheysen G. 2000. Cell cycle activation by plant parasitic nematodes. *Plant Mol Biol* 43:747–761.

Herman EM, Helm RM, Jung R, Kinney AJ. 2003. Genetic modification removes an immunodominant allergen from soybean. *Plant Physiol* 132:36–43.

Hermsmeier D, Mazarei M, Baum TJ. 1998. Differential display analysis of the early compatible interaction between soybean and soybean cyst nematode. *Mol Plant Microbe Interact* 11:1258–1263.

Hewezi T, Howe PJ, Maier TR, Hussey RS, Mitchum MG, Davis, EL, Baum TJ. 2010. *Arabidopsis spermidine* synthase is targeted by an effector protein of the cyst nematode *Heterodera schachtii. Plant Physiol* 152:968–984.

Hoa LTP, Nomura M, Kajiwara H, Day DA, Tajima S. 2004. Proteomic analysis on symbiotic differentiation of mitochondria in soybean nodules. *Plant Cell Physiol* 45:300–308.

Hofmann J, El Ashry AEN, Anwar S, Erban A, Kopka J, Grudler F. 2010. Metabolic profiling reveals local and systemic responses of host plants to nematode parasitism. *Plant J* 62:1058–1071.

Hofmann J, Grundler FMW. 2008. Starch as a sugar reservoir for nematode-induced syncytia. *Plant Signal Behav* 3:961–962.

Hofmann J, Szakasits D, Blochl A et al. 2008. Starch serves as carbohydrate storage in nematode-induced syncytia. *Plant Physiol* 146:228–235.

Hussey RS, Boerma HR. 1989. Tolerance in maturity groups V-VIII soybean cultivars to *Heterodera glycines. J Nematol* 21:686–692.

Ibrahim HMM, Hosseini P, Alkharouf NW et al. 2011. Analysis of gene expression in soybean roots in response to root knot nematode using microarray and KEGG pathways. *BMC Genom*. DOI: 10.1186/1471-2164-12-220.

Ithal N, Recknor J, Nettleston D, Hearne L, Maier T, Baum TJ, Mitchum MG. 2007a. Developmental transcript profiling of cyst nematode feeding cells in soybean roots. *Mol Plant-Microbe Interact* 20:293–305.

Ithal N, Recknor J, Nettleston D, Hearne L, Maier T, Baum TJ, Mitchum MG. 2007b. Parallel genome-wide expression profiling of host and pathogen during soybean cyst nematode infection of soybean. *Mol Plant-Microbe Interact* 20:293–305.

Jaubert S, Ledger TN, Laffaire JB, Piotte C, Abad P, Rosso MN. 2002. Direct identification of stylet secreted proteins from root-knot nematodes by a proteomic approach. *Mol Biochem Parasitol* 121:205–211.

Jones MGK. 1981. The development and function of plant cells modified by endoparasitic nematodes. In: *Plant Parasitic Nematodes*, eds. IBM Zuckerman, RA Rohde, Vol. II, pp. 255–279. New York: Academic Press.

Jones JT, Roberston WM. 1997. Nematode secretions. In: *Cellular and Molecular Aspects of Plant-Nematode Interactions*, eds. C Fenoll, FMW Grundler, SA Ohl, pp. 98–106. Dordrecht, the Netherlands: Kluwer.

Jung C, Wyss U. 1999. New approaches to control plant parasitic nematodes. *Appl Microbiol Biotechnol* 51:439–446.

Kaji H, Tsuji T, Mawuenyega KG, Wakamiya A, Taoka M, Isobe T. 2000. Profiling of *Caenorhabditis elegans* proteins using two-dimensional gel electrophoresis and matrix assisted laser desorption/ionization-time of flight-mass spectrometry. *Electrophoresis* 21:1755–1765.

Kaplan DT, Thomason IJ, Van Gundy SD. 1979. Histological study of the compatible and incompatible interaction of soybeans and *Meloidogyne incognita. J Nematol* 11:338–343.

Kerk N, Ceserani T, Tausta SL, Sussex IM, Nelson TM. 2003. Laser capture microdissection of cells from plant tissues. *Plant Physiol* 132:27–35.

Khan R, Alkharouf N, Beard H et al. 2004. Microarray analysis of gene expression in soybean roots compatible to the soybean cyst nematode two days post invasion. *J Nematol* 36:241–248.

Klink VP, Alkharouf N, MacDonald M, Matthews BF. 2005. Laser capture microdissection (LCM) and expression analysis of *Glycine max* (soybean) syncytium containing root regions formed by the plant pathogen *Heterodera glycines* (soybean cyst nematode). *Plant Mol Biol* 59:965–979.

Klink VP, Hosseine P, Matsye P, Alkharouf NW, Matthews BF. 2009a. A gene expression analysis of syncytia laser microdissected from the roots of the *Glycine max* (soybean) genotype PI 548402 (Peking) undergoing a resistant reaction after infection by *Heterodera glycines* (soybean cyst nematode). *Plant Mol Biol* 71:525–567.

Klink VP, Hosseine P, Matsye P, Alkharouf NW, Matthews BF. 2009b. A gene expression analysis of syncytia laser microdissected from the roots of the *Glycine mzx* (soybean) genotype PI 548402 undergoing a resistant reaction after infection by *Heterodera glycines* (soybean cyst nematode). *Plant Mol Biol* 71:525–567.

Klink VP, Hosseine P, Matsye P, Alkharouf NW, Matthews BF. 2010. Syncytium gene expression in *Glycine max* [PI88788] roots undergoing a resistant reaction to the parasitic nematode *Heterodera glycines. Plant Physiol Biochem* 48:176–193.

Klink VP, Matthews BF. 2008. The use of laser capture microdissection to study the infection of *Glycine max* (soybean) by *Heterodera glycines* (soybean cyst nematode). *Plant Signal Behav* 3:1–3.

Klink VP, Overall CC, Alkharouf NW, MacDonald MH, Matthews BF. 2007a. A time-course comparative microarray analysis of an incompatible and compatible response by *Glycine max* (soybean) to *Heterodera glycines* (soybean cyst nematode) infection. *Planta* 226:1423–1447.

Klink VP, Overall CC, Alkharouf NW, MacDonald MH, Matthews BF. 2007b. Laser capture microdissection (LCM) and comparative microarray expression analysis of syncytial cells isolated from incompatible and compatible soybean (*Glycine max*) roots infected by the soybean cyst nematode (*Heterodera glycines*). *Planta* 226:1389–1409.

Klink VP, Overall C, Matthews BF. 2007c. Developing a systems biology approach to study disease progression caused by *Heterodera glycines* in *Glycines max. Gene Regul Syst Biol* 2:17–33.

Koenning SR, Overstreet C, Noling JW, Donald PA, Becker JO, Fortnum BA. 1999. Survey of crop losses in response to phytoparasitic nematodes in the United States for 1994. *J Nematol* 31:587–618.

Lambert KN, Allen KD, Sussex IM. 1999. Cloning and characterization of an esophageal-gland specific chorismate mutase from the phytopathogenic nematode *Meloidogyne javanica. Mol Plant Microbe Interact* 12:328–336.

Larrainzar E, Wienkoop S, Weckwerth W, Ladrera R, Arrese-Igor C, Gonzalez EM. 2007. Medicago truncatula root nodule proteome analysis reveals differential plant and bacteroid responses to drought stress. *Plant Physiol* 144:1495–1507.

Lilley CH, Atkinson HJ, Urwin PE. 2005. Molecular aspects of cyst nematodes. *Mol Plant Pathol* 6:577–588.

Mathesius U. 2003. Conservation and divergence of signalling pathways between roots and soil microbes - The rhizobium-legume symbiosis compared to the development of lateral roots, mycorrhizal interactions and nematode-induced galls. *Plant Soil* 255:105–119.

Mawuenyega KG, Kaji H, Yamauchi Y et al. 2003. Large-scale identification of *Caenorhabditis elegans* proteins by multidimensional liquid chromatography - Tandem mass spectrometry. *J Proteome Res* 2:23–35.

Mazarei M, Puthoff DP, Hart JK, Rodermel SR, Baum TJ. 2002. Identification and characterization of a soybean ethylene-responsive element-binding protein gene whose mRNA expression changes during soybean cyst nematode infection. *Mol Plant Microbe Interact* 6:577–586.

Mbeunkui F, Scholl EH, Opperman CH, Goshe MB, Bird DM. 2010. Proteomic and bioinformatic analysis of the root-knot nematode *Meloidogyne hapla*: The basis for plant parasitism. *J Proteome Res* 9:5370–5381.

Melillo MT, Leonetti P, Bongiovanni M, Castagnone-Sereno P, Bleve-Zacheo T. 2006. Modulation of reactive oxygen species activities and H_2O_2 accumulation during compatible and incompatible tomato—Root-knot nematode interactions. *New Phytol* 170:501–512.

Mitchum MG, Baum TJ. 2008. Genomics of the soybean cyst nematode-soybean interaction. In: *Genetics and Genomics of Soybean*, ed. G Stacey, pp. 321–341. New York: Springer.

Moens M, Perry RN, Starr JL. 2009. *Meloidogyne* species - A diverse group of novel and important plant parasites. In: *Root-Knot Nematodes*, eds. RN Perry, M Moens, JL Starr, pp. 1–22. Wellingford, U.K.: CAB International.

Natarajan SS, Xu CP, Bae HH, Bailey BA. 2007. Proteomic and genomic characterization of Kunitz trypsin inhibitors in wild and cultivated soybean genotypes. *J Plant Physiol* 164:756–763.

Natarajan SS, Xu CP, Bae HH, Caperna TJ, Garrett WA. 2006. Characterization of storage proteins in wild (*Glycine soja*) and cultivated (*Glycine max*) soybean seeds using proteomic analysis. *J Agric Food Chem* 54:3114–3120.

Navas A, Lopez JA, Esparrago G, Camafeita E, Albar JP. 2002. Protein variability in *Meloidogyne* spp. (Nematoda: Meloidogynidae) revealed by two-dimensional gel electrophoresis and mass spectrometry. *J Proteome Res* 1:421–427.

Niblack TL, Arelli PR, Noel GR et al. 2002. A revised classification scheme for genetically diverse populations of *Heteroderaglycines. J Nematol* 34:279–288.

Niblack TL, Lambert KN, Tylka GL. 2006. A model plant pathogen from the kingdom Animalia: *Heterodera glycines*, the soybean cyst nematode. *Annu Rev Phytopathol* 44:283–303.

Paulson RE, Webster JM. 1972. Ultrastructure of the hypersensitive reaction in roots of tomato, *Lycopersicon esculentum* L. to infection by the root-knot nematode, *Meloidogyne incognita. Physiol Plant Pathol* 2:227–234.

Porubleva L, Velden KV, Kothari S, Oliver DJR, Chitnis PR. 2001. The proteome of maize leaves: Use of gene sequences and expressed sequence tag data for identification of proteins with peptide mass fingerprints. *Electrophoresis* 22:1724–1738.

Prot JC, Sorianoo IRS, Matias D. 1994. Major root-parasitic nematodes associated with irrigated rice in the Philippines. *Fundam Appl Nematol* 17:75–78.

Putoff DP, Ehrenfried ML, Vinyard BT, Tucker ML. 2007. GeneChip profiling of transcriptional responses to soybean cyst nematode, *Heterodera glycines*, colonization of soybean roots. *J Exp Bot* 58:3407–3418.

Ramsay K, Wang Z, Jones MGK. 2004. Using laser capture microdissection to study gene expression in early stages of giant cells induced by root-knot nematodes. *Mol Plant Pathol* 5:587–592.

Replogle A, Wang J, Bleckmann A et al. 2011. Nematode CLE signaling in *Arabidopsis* requires CLAVATA2 and CORYNE. *Plant J* 65:430–440.

Riggs RD, Kim KS, Gipson I. 1973. Ultrastructural changes in Peking soybeans infected with *Heterodera glycines. Phytopathology* 63:76–84.

Riggs RD, Rakes L, Hamblen ML. 1982. Morphometric and serologic comparisons of a number of populations of cyst nematodes. *J Nematol* 14:188–199.

Robertson L, Robertson WM, Jones JT. 1999. Direct analysis of the secretions of the potato cyst nematode *Globodera rostochiensis. Parasitology* 119:167–176.

Robertson L, Robertson WM, Sobczak JH et al. 2000. Cloning, expression and functional characterisation of a peroxiredoxin from the potato cyst nematode *Globodera rostochiensis. Mol Biochem Parasitol* 111:41–49.

Roggero P, Pennazio S. 2008. The extracellular acidic and basic pathogenesis-related proteins of soybean induced by viral infection. *J Phytopathol* 127:274–280.

Schrimpf SP, Langen H, Gomes AV, Wahlestedt C. 2001. A two-dimensional protein map of *Caenorhabditis elegans. Electrophoresis* 22:1224–1232.

Segarra G, Casanova E, Bellido D, Odena MA, Oliveira E, Trillas I. 2007. Proteome, salicylic acid, and jasmonic acid changes in cucumber plants inoculated with *Trichoderma asperellum* strain T34. *Proteomics* 7:3943–3952.

Smant G, Stokkermans JPWG, Yan Y et al. 1998. Endogenous cellulases in animals: Isolation of beta-1,4-endoglucanase genes from two species of plant-parasitic cyst nematodes. *Proc Natl Acad Sci USA* 95:4906–4911.

Sule A, Vanrobaeys F, Hajos G, Van Beeumen J, Devreese B. 2004. Proteomic analysis of small heat shock protein isoforms in barley shoots. *Phytochemistry* 65:1853–1863.

Szakasits D, Heinen P, Wieczorek K et al. 2009. The transcriptome of syncytia induced by the cyst nematode *Heterodera schachtii* in *Arabidopsis* roots. *Plant J* 57:771–784.

Tabuse Y, Nabetani T, Tsugita A. 2005. Proteomic analysis of protein expression profiles during *Caenorhabditis elegans* development using two-dimensional difference gel electrophoresis. *Proteomics* 5:2876–2891.

Vaghchhipawala Z, Bassuner R, Clayton K, Lewers K, Shoemaker R, Mackenzie S. 2001. Modulations in gene expression and mapping of genes associated with cyst nematode infection of soybean. *Mol Plant Microbe Interact* 14:42–54.

Valot B, Negroni L, Zivy M, Gianinazzi S, Dumas-Gaudot E. 2006. A mass spectrometric approach to identify arbuscular mycorrhiza-related proteins in root plasma membrane fractions. *Proteomics* 6:S145–S155.

Veech JA, Starr JL, Nordgren RM. 1987. Production and partial characterization of stylet exudates from adult females of *Meloidogyne incognita. J Nematol* 19:463–468.

Waetzig GH, Sobczak M, Grundler FMW. 1999. Localization of hydrogen peroxide during the defence response of *Arabidopsis thaliana* against the plant-parasitic nematode *Heterodera glycines. Nematology* 1:681–686.

Wan JR, Torres M, Ganapathy A et al. 2005. Proteomic analysis of soybean root hairs after infection by *Bradyrhizobium japonicum. Mol Plant Microbe Interact* 18:458–467.

Wang S, Meyers D, Yan Y, Baum T, Smant G, Hussey R, Davis E. 1999. In planta localization of a beta-1,4-endoglucanase secreted by *Heterodera glycines. Mol Plant Microbe Interact* 12:64–67.

Wang X, Mitchum MG, Gao B et al. 2005 A parasitism gene from a plant-parasitic nematode with function similar to *CLAVATA3/ESR (CLE)* of *Arabidopsis thaliana. Mol Plant Pathol* 6:187–191.

Wang KJ, Yamashita T, Watanabe MR, Takahata Y. 2004. Genetic characterization of a novel Tib-derived variant of soybean Kunitz trypsin inhibitor detected in wild soybean (*Glycine soja*). *Genome* 47:9–14.

Williamson VM, Gleason GA. 2004. Plant-nematode interactions. *Curr Opin Plant Biol* 6:327–333.

Williamson VM, Hussey RS. 1996. Nematode pathogenesis and resistance in plants. *Plant Cell* 8:1735–1745.

Wrather JA, Koenning SR. 2006. Estimates of disease effects on soybean yields in the United States 2003 to 2005. *J Nematol* 38:173–180.

Xu CP, Caperna TJ, Garrett WM, Cregan P, Bae HH, Luthria DL, Natarajan SS. 2007. Proteomic analysis of the distribution of the major seed allergens in wild, landrace, ancestral, and modern soybean genotypes. *J Sci Food Agric* 87:2511–2518.

Xu CP, Garrett WM, Sullivan J, Caperna TJ, Natarajan SS. 2006. Separation and identification of soybean leaf proteins by two-dimensional gel electrophoresis and mass spectrometry. *Phytochemistry* 67:2431–2440.

Yan SP, Zhang QY, Tang ZC, Su WA, Sun WN. 2006. Comparative proteomic analysis provides new insights into chilling stress responses in rice. *Mol Cell Proteome* 5:484–496.

VII

Modern Research Techniques

42

Minirhizotron Techniques

Boris Rewald
University of Natural Resources and Life Sciences

Jhonathan E. Ephrath
Ben-Gurion University of the Negev

I. Introduction

Special techniques are required to investigate root systems since they are hidden in the soil. Traditionally, destructive techniques like coring, trenching, and excavating have been used to access roots in situ. More recently, nondestructive techniques including rhizotrons and minirhizotrons (MRs) were developed in order to allow direct and repeated observations of the roots within the rhizosphere.

Installations using transparent "walls" to study roots in soil are termed rhizotrons (Böhm 1979). Walk-in rhizotron facilities or smaller-sized rhizotron chambers can also be used as lysimeters, and they may also include sensors that monitor soil conditions (Karnok and Kucharski 1982; Pan et al. 2001; Meier and Leuschner 2008). However, large rhizotrons have several disadvantages, with setup and operational cost being the primary ones; therefore, a very limited number of these facilities were built worldwide. In need for continuous nondestructive measurements of root traits in agricultural, silvicultural, and pristine ecosystems, the MR system was developed and has ever since gained wide acceptance. While glass plates ("root windows") have been used since the early 1900s (e.g., McDougall 1916), the MR concept was originally proposed by Bates (1937). In a work on fruit trees, he designed observation trenches in form of a walled chamber fitted with "root windows." However, what can probably be considered the first study with MRs as we know them today, using transparent tubes and an imaging device, was conducted by Waddington (1971).

MRs have helped improve our understanding of root systems, for example, in respect of standing stock, root production and longevity, root–parasite and root–hyphae interactions, and root phenology and distribution (e.g., Upchurch and Ritchie 1983; McMichael and Taylor 1987; Aerts et al. 1992; Hendrick and Pregitzer 1992; Hooker et al. 1995; Kosola et al. 1995; Eissenstat et al. 2000; Treseder et al. 2005; Vargas and Allen 2008; Ephrath and Eizenberg 2010).

Although reviews have previously discussed how to install the MR tubes and how to collect and use the obtained images (Taylor 1987; Box 1996; Hendrick and Pregitzer 1996a,b; Majdi 1996; Johnson et al. 2001; Mainiero 2006; McMichael and Zak 2006), there is an ongoing need to point out the proper use and possible pitfalls of MR systems to new users and to promote "good practice" standards. This chapter addresses five specific topics: (1) installation of MR observation tubes (MR-OTs), (2) MR image capturing systems, (3) image acquisition and analysis, (4) application of the MR technique, and (5) an outlook on recent and future developments, which could extend the range of applications of this technique.

II. Installation Protocols and Materials Used for Minirhizotron Observation Tubes

A. Installation of Minirhizotron Observation Tubes

Whichever MR system is used, it requires that transparent MR-OTs be installed in the soil. Because MR studies are conducted in a wide range of natural and artificial soil environments, soil type and species-specific factors have to be taken into account during MR-OT installation. In order to minimize soil compaction and plant damage in situ, trampling must be avoided during installation. Installing MRs before planting or when root and shoot biomass is at its annual low is recommended.

Most MR studies are conducted in rather homogeneous soils, while very stony soils are only rarely addressed because of the difficult installation process (but see Phillips et al. 2000). Holes to insert the MR-OT are usually made using an auger (Kage et al. 2000), a soil corer (Hummel et al. 1989), or a combination of both (see Johnson et al. 2001). Depending on the bulk soil density and the required depth, researchers may install tubes manually or by mechanical drilling devices (Brown and Upchurch 1987). Manual MR-OT installation is mostly conducted down to less than 100 cm depth, while greater depths are accessed by tractor-mounted or portable auger systems (Kloeppel and Gower 1995). For using a manual auger, a supporting stand should be fixed on the topsoil to guide the drill at the desired position and angle and may cause additional soil disturbance. A soil corer has the advantage of a more smooth soil interface and no additional disturbance by a supporting stand when MR-OTs are installed vertically; however, using a hammer or a ram is tedious, and the soil might become asymmetrically compacted (Figure 42.1A).

Ideally, MR tubes should be installed in such a way that they are in close contact with the soil matrix, affecting root growth only in the way other large objects such as stones do. However, it is extremely difficult to ensure complete and uniform contact of the OT surface with the soil with no gap. Contrasting considerations must be taken into account while deciding of the drill size. A tight fit will prevent the formation of a gap and will reduce the risk of tube rotation (a problem common for short MR-OT in shrinking soils; see Johnson et al. [2001] for "anchoring devices"). But it can cause soil compression that might lead to reduced root growth near the tube (McMichael and Taylor 1987). If the hole is oversized, MR-OT installation is easier and scratches at the tube surface can be avoided; however, even narrow voids between the wall of the OT and the soil will constitute a low-resistance path that can artificially increase root growth, branching, and survival (van Noordwijk et al. 1985; Volkmar 1993 and references therein; Figures 42.1B and 42.2). They are also prone to moisture condensation that may interfere with root observation (Figure 42.2). While backfilling of oversized holes with sieved soil material after tube placement makes MR-OT installation easier (Kloeppel and Gower 1995), the unnatural density and structure of the soil will most likely influence root traits. For a discussion of the appropriate installation and use of MR-OTs in wetland ecosystems see Iversen et al. (2012).

B. Angle of Installation

Commonly MR-OTs are installed either vertically (90°) or angled. Many of the angled MR-OTs are installed at 30° or 45°, but different angles are common (see Johnson et al. 2001). It was proposed that angled MRs can estimate root depth distribution of herbaceous/crop plants better than vertical tubes and to reduce the artificial funneling of roots down the root/MR tube interface (Bragg et al. 1983; Merrill et al. 1994; Pagès and Bengough 1997). It can be speculated that funneling is related to the persistence of a gap around the tube that may increase water infiltration and consist a low-resistance path that roots tend to follow due to gravitropism. However, Ephrath et al. (1999) found no preferential growth of wheat roots as a result of steep insertion angle in a sandy soil with homogeneous bulk density. No studies addressing the influence of different OT installation angles on MR results for woody species are known to these authors. Further studies with different plant soil types are needed for a conclusive answer. These authors expect differences in root-growth pattern between angled and vertically installed tubes to be soil- and species-specific but highly related to installation protocols.

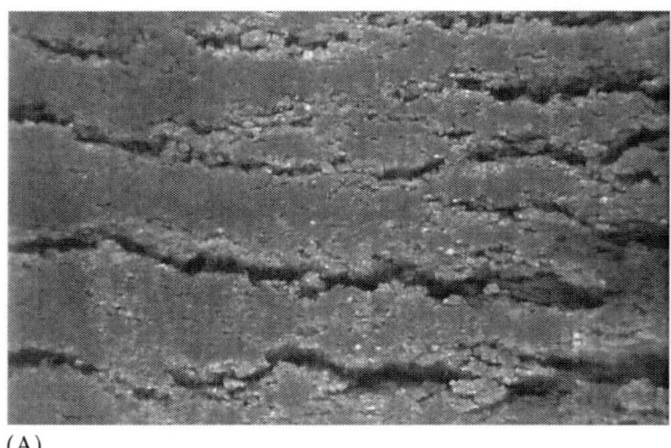
(A)

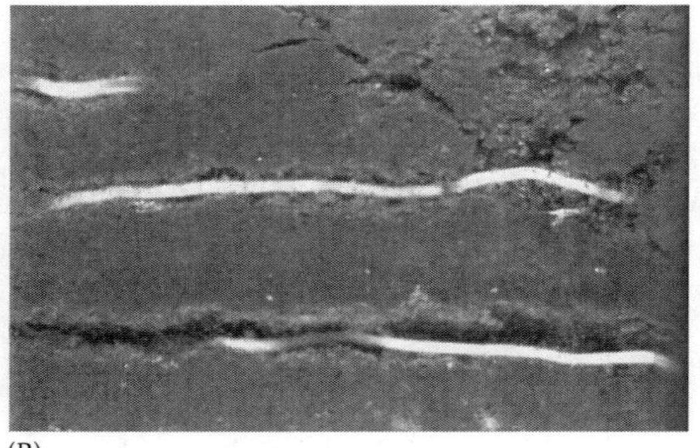
(B)

FIGURE 42.1 (A) Image of soil immediately after coring and tube installation into a sandy loam. Note the cracks in the soil profile resulting from hammering the coring tool into place. (B) Cotton roots preferentially exploiting resulting cracks in soil profile. (Images courtesy of Dennis Gitz, USDA-ARS, Lubbock, TX; pictures were taken with a camera MR, Bartz, Carpinteria, CA.)

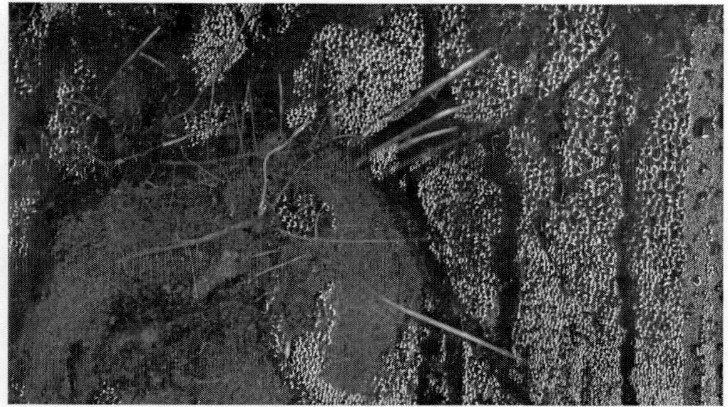

FIGURE 42.2 Picture section within an inappropriately installed MR-OT (90°) 3 months after installation (soil corer). The picture was captured with a scanner MR (CID, Cedar Rapids, IA). The roots of *Fagus sylvatica* can be seen to grow toward soil voids; many regions of the picture are obscured by condensed water in alternation to voids.

However, while vertical MR-OTs are more easily installed and depth at each recording point is more easily determined, angled tubes can reach underneath individual plants, which might be important in low plant density and/or in agricultural systems (i.e., in rows; Figure 42.3).

Installation of MR-OT in pots, lysimeters, and phytotrons with rather artificial, homogeneous soil environments seems to be less problematic because soil can be equally distributed

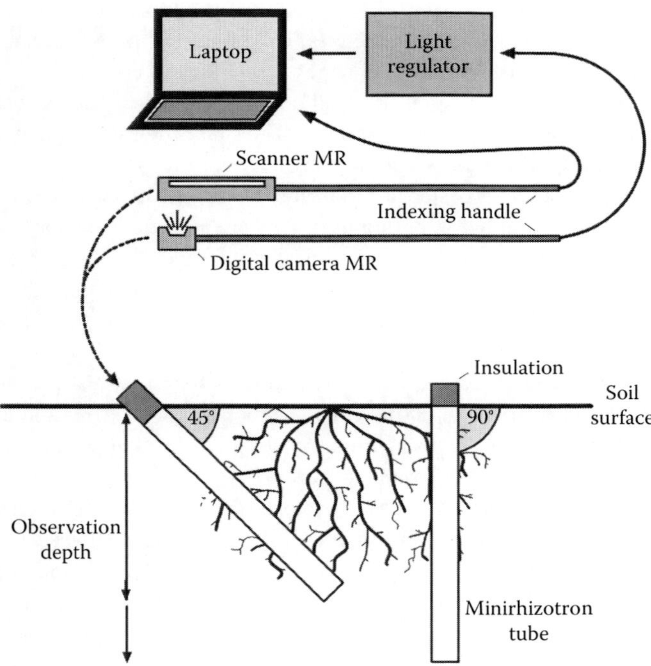

FIGURE 42.3 Setup of MR-OTs in both angled (e.g., 45°) and vertical positions (90°); aboveground light and temperature insulation and the observation depths are indicated. Images are captured by either digital camera–based or scanner-based MR systems connected to a laptop. Indexing handles allow for exact positioning of the devices in the OT; the light intensity of camera MR systems can be regulated.

around tubes during setup, reducing problems of air gaps and disturbance of root systems. MRs are often installed horizontally in such artificial sites (Liedgens 1998; Meier and Leuschner 2008). Horizontal tubes have the advantage of maximizing the observation area per soil depth (Smucker 1993), but a study by Dubach and Russelle (1995) revealed that there was a large variation between the root numbers on upper and lower sides of horizontally installed MR-OTs and that neither side was well correlated with the root counts on both horizontal sides.

C. Tube Protection from Light and Weather

Polycarbonate and PVC plugs (Box 1996; Phillips et al. 2000), rubber bungs (Majdi and Kangas 1997), or plastic end caps (Meier and Leuschner 2008) were used to seal the lower end of MR-OT that is in the ground. This is most important in moist soils in order to prevent water accumulation on the tube's inner surface, while it is of less concern in dry environments. However, the aboveground portion of MR-OT should always be insulated and covered with a lighttight cap and painted or covered with opaque tape to reduce thermal fluctuations and exclude light that can affect roots (Levan et al. 1987) and root-associated microbes (Klironomos and Allen 1995); special care has to be taken in soils that develop cracks while drying (Dubach and Russelle 1995). Especially in high-solar-radiation environments and without canopy cover, it is recommended to choose reflective colors for the tube cover and to reduce the protruding length of MR-OT to the minimum in order to avoid excessive heating (Figures 42.3 and 42.4A and B); insulation material, placed inside the protruding end of the tube, might further reduce temperature fluctuation of the soil around it. In areas with high snowfall, a support stand may be needed for angled MR-OT to prevent cracks on tubes' aboveground caused by the snow weight (Johnson et al. 2001).

D. Time Lag before First Measurement

Insertion of MR-OTs causes disturbances of the soil and root systems as discussed in Section II. A. It is unclear to date how fast different soil types and root systems return to equilibrium conditions; most researchers allow for a time period of 6–12 months, while some start their measurements immediately or within a few months (see Johnson et al. 2001). A timely start seems more unproblematic if MR-OTs are installed before planting, for example, in agricultural ecosystems or phytotrons; however, there may be a release of nutrients near recently installed tubes (Joslin and Wolfe 1999). In ecosystems with established root systems, first year's data were often found to be atypical as compared to subsequent years (Aerts et al. 1989; Burke and Raynal 1994). According to a meta-analysis by Strand et al. (2008), estimated longevity of tree fine root increased up to 40% with increasing time since OT installation (Figure 42.5). They concluded that tree root systems needed up to 3 years to return to equilibrium and that the longevity of fine roots established during the "pre-equilibrium" period was 50% shorter as compared to roots that developed in the "post-equilibration" period. The results indicate that short-term MR studies have contributed to the overestimation of fine-root turnover rates (Strand et al. 2008).

(A)

(B)

FIGURE 42.4 Aboveground view on two differently installed MR-OTs. (A) A too long protruding tube installed at 45° with severed insulation; (B) a vertically installed tube. A short aboveground tube length and sufficient insulation to prevent light penetration and reduce tube heating.

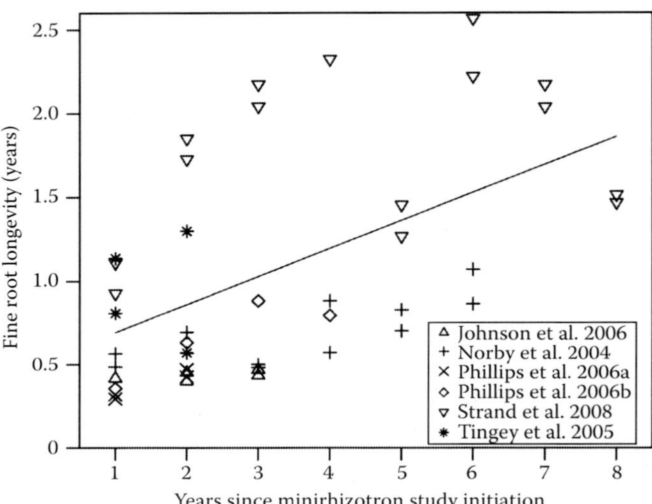

FIGURE 42.5 A meta-analysis by Strand et al. (2008) showing fine-root longevity as a function of the time since MR study initiation; six studies conducted in North American forest ecosystems are displayed. The increase in determined root longevity by time since MR-OT installation is indicated by a solid line. See Strand et al. (2008) for details.

Thus, in order to avoid having the disturbances during MR-OT installation affect the measurements of root traits, a sufficiently long soil- and species-specific equilibration time has to be taken into account. However, data collected during the "pre-equilibration" period could be used to determine the phenological pattern of root growth in general (Burke and Raynal 1994) or to determine the regrowth potential of root systems after disturbance (similar to ingrowth cores).

E. Material and Types of Minirhizotron Observation Tubes

A wide variety of materials have been used for MR-OT. Transparent and rigid MR tubes made of materials such as glass (Richards 1984; Eissenstat and Caldwell 1988; Fitter et al. 1999), polycarbonate (PC, known as Lexan; van Noordwijk et al. 1985; Box and Johnson 1987), polymethyl 2-methylpropenoate (or polymethyl methacrylate [PMMA], known as acrylic, Perspex, Plexiglas, or Acrylite; Itoh 1985; Vos and Groenwold 1987; Kloeppel and Gower 1995), and cellulose acetate butyrate (CAB or butyrate; Box et al. 1989; Hendrick and Pregitzer 1992; Wells and Eissenstat 2001; Yang et al. 2003) have been used. Inner diameter of the tubes used ranged from 13 mm (Boroscope; Upchurch and Ritchie 1983) to 64 mm (Scanner MR; Gaul et al. 2009); square tubes were used rarely (van Noordwijk et al. 1985). A future standardization of MR-OT diameters is desirable to allow for a greater flexibility in using different image capturing devices.

Rigid tubes made out of plastic have established themselves as the most common type of MR-OT because they are of greater durability than glass tubes, especially in rocky, swelling, or freezing soils, of easier use than flexible/inflated tubes (see the text in the following paragraph), and often the least expensive option. However, there are significant differences in the scratch resistance and transmissibility (e.g., for UV light) of rigid tubes; thus, on the one hand, the choice of material has to be made according to image capturing system, soil type, expected time, and intensity of use, as well as cost and availability (Wang et al. 1995; Johnson et al. 2001). On the other hand, the tube material might influence root traits. Although addressed as early as 1976 (Taylor and Böhm 1976), only a few studies have evaluated the influence of the clear materials on root growth. While Brown and Upchurch (1987) found no differences between the tube materials they tested, Withington et al. (2003) found the OT material to cause changes in root production and phenology. The production of apple roots was greatest around glass tubes, and these roots became pigmented later and lived longer than roots that grew near acrylic and CAB MR-OT. Furthermore, roots became pigmented faster next to CAB tubes than next to acrylic tubes; root survival was shorter near CAB tubes in three of four deciduous hardwood species but shorter near acrylic tubes for three conifer species. The comparison of root length density (RLD) with root standing crop from cores showed that the data gained from acrylic tubes matched more closely than the data from butyrate tubes (Withington et al. 2003). Future studies should seek to further clarify the impact and underlying mechanisms of different MR-OT materials on root traits of various species to prevent artifacts.

MRs with inflatable/flexible walls have been developed to address special needs (as sampling of soil and roots) and to optimize soil–tube contact. Especially in soils that tend to shrink during drying (e.g., clayey soils), inflatable tubes were meant to solve the problem of voids that form in such cases (Merrill 1992). Materials used to make MR-OT with flexible walls include cellulose acetate (Merrill et al. 1987), polyvinyl (Merrill 1992; Merrill et al. 2005), fluoroethylene propylene (known as Teflon FEP, Kosola 1999), and rubber (Gijsman et al. 1991; López et al. 1996). As very low-cost alternative standard (opaque), drain pipes with an opening ("cut out") have also been used; soil was prevented from collapse, and roots were prevented from growing into the opening by inserting an inflatable inner tube between readings (Harun and Roslan 2003).

F. Number of Replicate Tubes

To the best of our knowledge, there is to date no study that dealt sufficiently with the required number of MR-OT. There is no doubt that this number has to take into account several aspects besides resource availability. Taylor et al. (1990) suggested that eight tubes were required in order to estimate the RLD of a plot, a number similar to that recommended for soil coring approaches. Horizontally installed tubes were suggested to reduce the number of required MR-OT (Smucker 1993); however, this seems to be due to the effect of the higher tube length per soil depth rather than of the installation angle *per se*. According to our experience, the number of tubes should reflect the variability of the soil and the root systems. If rooting patterns are known, for example, in studies of agricultural fields or orchards, where most roots can be found close to an irrigation system (Rewald et al. 2011b), fewer tubes (e.g., five to six tubes) seem to be sufficient. In ecosystems where unknown or heterogeneous root distribution is studied or experimental manipulations are conducted (e.g., drought treatments or FACE experiments), a larger quantity of MR-OT should be installed per plot (e.g., ≥12 tubes; Smucker 1993; Vogt at al. 1998; Johnson et al. 2001).

III. Image Capturing by Minirhizotron Systems

A. Image Capturing Devices

In the early MR systems, the use of simple mirrors and a light source gave reasonable correlations between MR measurements and washed root lengths (Gregory 1979), but the low image quality and its limited size generally restricted the accuracy (Keng and Kusaka 1988). Since the early 1970s, several types of image capturing devices, that is, fiber optics, endoscopes, boroscopes, root periscopes, and telescopes, have been developed to increase the quality of images and to facilitate image capturing (e.g., Sanders and Brown 1978; Richards 1984; Rush et al. 1984; Itoh 1985; van Noordwijk et al. 1985; Poelman et al. 1996). Notably, the use of miniature (color) video cameras improved the operation of MRs by using the microphone to record the camera location and other information on the audio track (Upchurch and Ritchie 1983, 1984; Johnson et al. 2001).

In the last decade, the MR technology has advanced considerably; in particular, the development of digital image capturing technologies made it possible to conduct faster and more comprehensive measurements. Although boroscope-based MRs are still being used to study very small root systems, and video camera MRs are widely used because of their availability, the two commonly used MR systems that exist today are (1) digital (video) camera–based MR and (2) scanner-based MR systems. Both systems store the images on a (mobile) computer equipped with software to capture and label pictures. They also usually have an indexed handle (e.g., Ferguson and Smucker 1989) that enables the user to take repeated pictures at the same soil location (Figure 42.3).

1. *Digital (video) camera MR*: Different sizes of cameras exist, and the diameter of the MR-OTs has to be selected according to the diameter of the camera housing. The most commonly used MR systems are produced by Bartz Technology Corporation (subsequently named "Bartz"; Carpinteria, CA), but basic camera MR systems can be easily custom made using webcams and LED lighting (e.g., Faget et al. 2010). The imaging qualities of MR cameras differ, often restraining digital zooming; but, some systems allow for optical zooming up to 100× to study root details, soil fauna, or single hyphae (Allen et al. 2007). Most camera MR systems further allow manual setting of the focus either via software- or hardware-based lens focusing and the capture of video sequences; both features enable studies of root–soil fauna interactions in soil voids (Lussenhop and Fogel 1993).

   The image recorded by the camera covers a narrow section of the tube perimeter (<2 cm wide); cameras with a wider view cannot be used because the soil–tube boundary gets blurry at the edges because of the curving tube. In order to capture a wider soil profile (on several pictures), the camera has to be rotated in each measuring depth.

   All camera MRs employ a special light source; the lighting intensity often can be changed by a control box (Figure 42.3), a feature that is rarely needed because of the exposure correction capabilities of the image capturing software. However, it is important that the light intensity will be the same over the whole viewed area. In general, the external light source and the exposed lens of MR cameras make it easier to customize the emitted/recorded wavelength by filters, for example, to take pictures of white light or green fluorescence emitted by GFP (see Section VI.A).
2. *Scanner MR*: By now, the only commercial scanner-based MR system is produced by CID BioScience Inc. (subsequently named "CID"; Camas, WA). In this device,

a modified CCD (charge-coupled device) flatbed scanner is used to capture the images. CCD-type scanners are preferred due to their larger depth of field (Dannoura et al. 2008). The scanner can take a 360° picture of the soil–tube boundary, thus recording the whole soil profile (picture size approx. 20 cm wide × 22 cm high). These bigger pictures reduce the number of pictures to be taken and annul the need for measurements at different directions. They may facilitate correct data interpretation because larger parts of the branching root system can be seen. However, in densely rooted soils, the larger picture size increases the time for data analysis significantly, and the user might consider analyzing only narrow sections of the soil profile.

Before the scanner can be used, white balancing has to be done within a calibration tube. The maximum scanner resolution is 1200 dpi enabling moderate and post–image capturing digital zooming on picture details; however, in routine use, lower resolutions (approx. 300–600 dpi) will likely be chosen to reduce measuring time. Homogeneous lighting and automatic focusing are intrinsic features of scanner MR; the autofocus is convenient during standard use but limits observations in soil voids. The inner diameter of MR-OT for the currently available scanner MR must be 64 mm, thus "regular" tubes (as used by camera MR systems) cannot be used due to smaller diameters.

Each MR system has its own advantages and disadvantages; most pronounced differences exist in picture size and resolution. Future users should choose a device that suits their research needs, but the availability/costs of OTs should also be considered.

B. Temperature Increase by the Minirhizotron Lighting System

The lighting systems of MR image capturing devices can cause an increase in soil temperature up to 3.5°C (Van Rees 1998). It has been argued that short temperature increases from MR systems are insignificant when considering the surface soil temperature fluctuations in the field (McMichael and Burke 1996). However, diurnal fluctuations in soil temperatures at greater soil depth are much smaller; therefore, the effect of short-term temperature increases caused by stationary light systems on root growth and development may be more pronounced at depth than at the surface, especially in the winter when soil temperatures are low (Gaul et al. 2008). Because of the possible effects of increased soil temperatures from MR light on root growth and development as well as on soil fauna activity, lighting intensity and exposure time should be reduced as much as possible.

IV. Image Acquisition and Analysis

A. Frequency of Image Capturing

One of the main advantages of the MR technique is the possibility to conduct continuous measurements. Tingey et al. (2003) demonstrated that MR sampling frequencies had major effects on estimated fine-root production and mortality in both evergreen and deciduous tree species. Because fine roots are short-lived and particularly prone to herbivory, they may appear and disappear between image capturing events (Hendrick and Pregitzer 1996a,b; Eissenstat and Yanai 1997; Vogt et al. 1998), leading to underestimation of fine-root production and mortality (Johnson et al. 2001). Most MR studies choose capturing frequencies between 2 and 4 weeks with lower sampling rates during expected root dormancy (e.g., winter). However, the main factor that will determine the frequency of image collection in MR studies is the studied root trait (Taylor 1987). For example, in studies of root turnover rates, the image capturing rate will depend on an approximated mortality rate, while for documentation of standing stocks or rooting depth, lower sampling frequencies can be chosen (see Mainiero [2006] for temperature dependence).

B. Effect of Imaging Direction

Potential errors in MR studies might be caused by the spatial orientation of image capturing. A study by Dubach and Russelle (1995) found a large variation in root numbers on different sides of horizontally installed MR-OT. Furthermore, root distribution is highly influenced by the irrigation regime; for example, in agricultural systems, roots are especially concentrated close to water emitters (e.g., Shani et al. 1995; Rewald et al. 2011b). While methodological studies are virtually absent, the need for careful selection of the imaging direction is obvious, unless 360° are scanned.

C. Image Analysis

The first comprehensive analyses of MR pictures were based on manual tracing of roots on transparent sheets (Cheng et al. 1991). For more detailed analyses, special computer programs are used. Taking into account that the number of images taken on a single MR experiment will be in the magnitude of thousands, image file handling is of great importance. Most MR image analysis programs require files to be named in a way that allows the program to distinguish between different experiments, tubes, depths, and dates. The most commonly used naming convention is ICAP (Bartz, Carpinteria, CA); the names are either given automatically (e.g., BTC I-CAP Image Capturing System; Bartz, Carpinteria, CA) or manually.

Various commercial and freeware computer programs are used to analyze MR images, examples are Rootfly (Birchfield and Wells 2006), RooTracker (Duke University, Durham, NC), Root Measurement System (Ingram and Leers 2001), and WinRHIZO Tron (Régent Instruments, Quebec, Canada). In all programs, the user must trace the roots manually; the process involves marking roots on a computer screen by moving the mouse along the roots and setting nodes and diameters. The overlaid marks can be copied to the consecutive MR picture taken later at the same position so differences between the two pictures can be determined, for example, increases in length, in diameter, or the death of root segments. Commonly, a function that allows recording of root segment–specific information like color and mycorrhizal status is also available. A common shortcoming of many analysis software types is that each root segment in the branching root system is labeled as single roots; a feature that allows grouping of roots according to orders is rare (WinRhizo Tron 2011a; Regent, Canada). Only recently, CID Bio-Science Inc., Camas, WA, developed a program (CI-690) that allows the user to trace roots using a touch screen input; although this is an interesting approach believed to make root tracing easier, no published report of end-user experience is available at the time of publication.

As mentioned previously, a major difference between the camera MR and the scanner MR system is the image size. Larger images (i.e., especially higher images, paralleling several images taken with camera MR systems) might allow for more accurate measurements since certain errors that result from difficulties in tracing roots in small images, for example, measuring the same root twice in different frames when analyzing overlapping roots in multiframe pictures, can be avoided.

D. Automated Image Analysis

Since manual analysis of MR images is very time consuming, its automation is sought for a long time (e.g., Richner et al. 2000). Although still in early stages, some computer programs allow for automatic detection of roots in MR images. The methods are generally based on image thresholding and region- or contour-based techniques to distinguish roots from the soil background, which often includes extraneous objects (Erz et al. 2005; Zeng et al. 2010). However, until now automatic root detection is limited since a low contrast between roots and background often results in systematically lower RLD compared to manual analyses (Vamerali et al. 1999). For root–soil systems with a high contrast, Zeng et al. developed a system that can detect, label, and measure individual roots, thereby setting the stage for automated tracking of roots through time (Zeng et al. 2008). Future approaches will likely involve the use of advanced imaging techniques like combinations of visual light and near-infrared reflectance to distinguish automatically between living and dead roots, organic matter, and mineral soil (Nakaji et al. 2008; Lei and Bauhus 2010 and references within).

V. Applications of the Minirhizotron Technique

If installed and analyzed carefully, the most serious limitations to the MR technique seem to be the initial costs of hard- and software and the time lag until soil and root dynamics come back to steady-state conditions after tube installation. Furthermore, while labor costs for tube installation and picture capturing are moderate, image analysis can become very time consuming and sufficient resources have to be allocated for this purpose (e.g., Coupe et al. 2009).

A. Minirhizotrons and Measurements of Standing Stock and Root Depth Distribution

MRs have been used extensively in assessing RLD (Bland and Dugas 1988; Franco and Abrisqueta 1997) and rooting depth (Hendrick and Pregitzer 1992; Majdi et al. 1992; Baumann et al. 2005). Fewer studies have related the measured RLD to root biomass (Johnson et al. 2001; Brown et al. 2009). The most commonly used method in root research is destructive soil coring (e.g., Rose et al. 2011); unlike MRs, soil cores reveal root masses directly without the need for calibration (Ephrath et al. 1999). Furthermore, roots from soil cores can often be reliably sorted into live and dead masses according to visual/mechanical or chemical criteria (Rewald et al. 2011c). Many studies found correlations between MR data and root biomass determined by soil coring, although the level of correlation varied between studies and species (e.g., Box and Ramseur 1993; Murphy and Smucker 1995; Jose et al. 2001; Gaul et al. 2009). At high soil bulk density, for example, rigid tube measurements consistently overestimated actual rooting density (as determined by soil coring) of both wheat and beans, while in the case of flexible MR-OT, the two measurements did not differ significantly (De Ruijter et al. 1996). Thus, soil coring might be preferred to study effects of different bulk densities instead of rigid MR-OT.

Contradictory reports were published concerning the underestimation of rooting frequency in the surface soil layers. While many studies on crops concluded that MRs underestimate RLD especially in the surface soil layer (Gregory 1979; Bragg et al. 1983; Upchurch and Ritchie 1983; Parker et al. 1991; Samson and Sinclair 1994), others did not. For example, Jose et al. (2001) found that the root biomass of *Zea mays* was slightly underestimated by the MR technique in the top 30 cm of the soil but for two tree species no significant difference occurred in surface or deeper soil layers. In general, there are more studies reporting underestimation of crop rooting density than that of trees by MRs (e.g., Rytter and Hanson 1996; Franco and Abrisqueta 1997; Ephrath et al. 1999; Jose et al. 2001; Gaul et al. 2009). These conflicting reports may be the result of a number of factors that can influence MR data such as species, soil type and density, tube installation technique, replicate numbers, and sampling errors. While in short-term experiments with slow-growing species or in dense soils, the soil cores could be more suitable to quantify fine-root abundance and distribution, the MR technique is a reliable method with a minimal site disturbance.

Two methods are used to relate the MR RLD data to the soil volume and to convert the RLD to biomass (see Johnson et al. [2001] for details). Merrill and Upchurch (1994) and Merrill et al. (1994) used the number of roots that intersect the MR picture to calculate the expected root length within the soil volume taken by the MR-OT. In the other approach, assumptions regarding depth of view must be made when root length or root surface area is converted to biomass per unit of soil volume. Typically, the values used for depth of field range from 2 to 3 mm (Sanders and Brown 1978; Itoh 1985; Steele et al. 1997; Brown et al. 2009). Specific root length (SRL) data are used in order to convert RLD to biomass values. Since SRL varies with diameter or root branching order, it should be determined for the different root classes by destructive sampling (Rewald et al. 2011a). These volumetric data sets can also be expressed on a ground surface area basis by relating the volume to the length of the MR-OT. In homogeneous rooting systems, a direct "biomass calibration" of MR data may be done with root biomass density data obtained from nearby soil cores (Ephrath et al. 1999; Johnson et al. 2001).

B. Minirhizotrons for Estimation of Root Production and Demography

A number of independent methods are available for estimating fine-root dynamics and turnover (Tierney and Fahey 2001; Majdi et al. 2005). Besides MRs, methods to determine fine-root production and turnover include (1) indirect mass-related techniques (by coring), (2) experimental setups including ingrowth cores and miniature root-growth chambers, and (3) changes in carbon isotopic ratios (Powell and Day 1991; Hendrick and Pregitzer 1992, 1993; Majdi 1996; Gaudinski et al. 2000; Rewald and Leuschner 2009). However, sequential coring will only reflect root growth and death during the period prior to sampling, while MRs, provided that root system was allowed to return to equilibrium growth after soil disturbance (see, e.g., Gaul et al. 2009), can determine short-term changes in root dynamics. Thus, the MR technique is suggested to provide more realistic results of fine-root dynamics than sequential coring (Publicover and Vogt 1993; Hendricks et al. 2006; Majdi et al. 2007). Two major limitations of the ingrowth core method compared to MR are (1) that no information on time of root ingrowth or mortality is obtained when the amount of living and dead roots is measured and (2) that the homogenized/replaced soil of the reconstructed ingrowth core will present a physical and chemical artificial and less competitive soil environment, which will be colonized at different rates than other parts of the rooting volume (Majdi 1996). Furthermore, destructive methods cause repeated and prolonged soil disturbance, so they are not suitable for long-term research on plots of limited size.

The reliability of the MR method depends inter alia on the accuracy of assessing the physiological status of roots, that is, whether roots are dead or alive. For using Kaplan–Meier statistics for root longevity calculations (Majdi et al. 2001; Tierney and Fahey 2001; Pritchard et al. 2008a; Strand et al. 2008), the observed roots have to be pooled into two groups: roots that are alive and roots that died during the observation period. However, to date three different criteria have been used for determining the death of roots in MR studies: color changes (Cheng et al. 1991; Hendrick and Pregitzer 1992), signs of disintegration (Repo et al. 2008), and disappearance (Comas et al. 2000).

When roots become dark brown or black, they are considered dead. However, root color may darken as a result of secondary growth or suberization, and this can complicate visual estimates of the physiological status (Figure 42.6). Furthermore, under visible light, it is sometimes difficult to distinguish between roots and organic debris, which might influence the determination of root disintegration. To aid in separation of live and dead roots, some MR systems included both visible and ultraviolet (UV) lights (Bartz, Carpinteria, CA). Under UV light, live roots are supposed to fluoresce more strongly than dead roots (Dyer and Brown 1983). However, Wang et al. (1995) found that some dead root still fluoresced under UV light; consequently, no significant differences between estimates of live root proportions by visible or UV light were found.

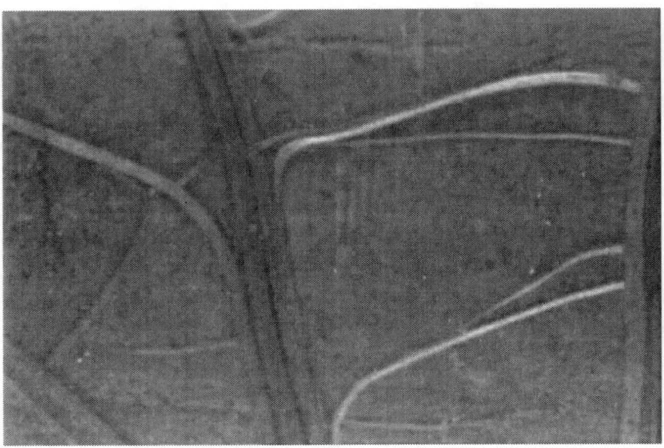

FIGURE 42.6 Root branches of *Tamarix aphylla*. Older roots (dark) can be clearly distinguished from newer (white) roots. The picture was taken with a camera MR, Bartz, Carpinteria, CA.

Moreover, the disappearance of roots from the MT picture might not always be a result of root death and decay, but disappeared roots may remain alive and only vanished from the field of view because of a change in growth angle or superposition by other roots, biofilms, hyphae, or moving soil particles. The criteria used for classifying roots to "dead" or "alive" fractions as well as the methods used for calculations may lead to significant differences in estimated root turnover rates (see Satomura et al. 2007 and references therein).

Furthermore, the discrepancies in estimated fine-root turnover rates between different approaches are illustrated by several studies based on changes in carbon isotopic ratios, which have reported higher turnover rates than usually measured by MRs (Gaudinski et al. 2000; Trumbore et al. 2006). The model of Guo et al. (2008) indicated that median-based longevity estimates made by MR studies underestimated actual longevity, whereas simulated mean residence time of carbon from isotopic studies overestimated longevity. Longevity distributions of fine roots are often positively skewed, indicating different fine-root longevities within one root system (Tierney and Fahey 2002; Joslin et al. 2006; Trumbore et al. 2006). The model of Guo et al. (2008) considers the heterogeneity of root systems by assuming that the most dynamic pool is dominated by first-order (i.e., the root tip) and second-order roots, while longer-lived roots are mostly roots of higher orders (Eissenstat et al. 2000; Wells et al. 2002; Guo et al. 2008). Because isotopic studies are based on residence time of root mass, they are biased by larger and older roots that are less numerous but contain more carbon and live longer. On the other hand, since MR estimates are rather number based, they are biased by the frequent first- and second-order roots, which have the fastest growth and turnover but contain less carbon. Pritchard and Strand (2008) suggested conducting root survival analyses based on volume, instead of the individual roots themselves, to improve the quantification of turnover of fine-root mass in MR studies. Furthermore, a longer duration of MR studies and the classification of root orders during picture analysis might increase data accuracy. In the absence of one standard method to determine root longevity, combined approaches including MRs and coring are widely recommended (Nadelhoffer 2000; Hertel and Leuschner 2002; Hendricks et al. 2006).

C. Minirhizotrons for Studying Root Morphology

MRs have been used for assessing root morphology (Upchurch 1985; Withington et al. 2003; Basile et al. 2007; Figure 42.7). The most common morphological parameter assessed is root diameter; fewer studies have addressed root pigmentation and branching. However, care has to be taken when choosing the material of MR-OT; Withington et al. (2003) showed that the root morphology of different tree species was significantly different between plastic and glass MR-OT. Compared to glass tubes, the mean root diameter of roots observed through plastic tubes was higher in two of six species, and the time from birth to pigmentation was significantly decreased against CAB tubes in four out of six tree species.

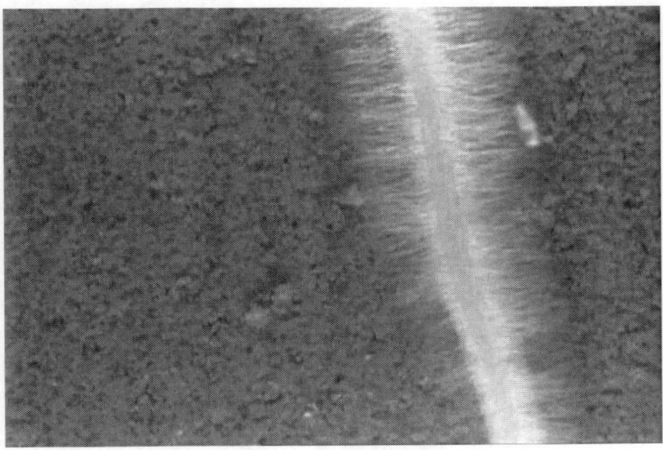

FIGURE 42.7 Root of *Cakile maritima* featuring root hairs. Although no soil voids can be detected, root hairs seem to concentrate at the tube–soil interface. The picture was taken with a camera MR, Bartz, Carpinteria, CA.

D. Minirhizotrons for Studying Belowground Interactions

MRs have so far been used for studying belowground interactions between roots and their mycorrhizal partners, roots and soil fauna/plant parasites, and competitive interactions.

Until now, most of the studies aimed at understanding the dynamics of mycorrhizal colonization have used soil cores (Mukerji et al. 2006). Fungal structures down to single hyphae can be studied with magnifying MR image capturing devices (Figure 42.8), allowing density estimates of ectomycorrhizae, rhizomorphs, and colonies of saprophytic fungi (Treseder et al. 2005; Pritchard et al. 2008b; Hasselquist et al. 2010). Direct observation of invertebrates in MRs is appealing because the methods used to extract invertebrates from soil select mobile, desiccation-resistant species. However, unresolved issues are high light or UV intensities, which might drive away invertebrates from the viewing area, and increased soil temperatures, which could affect soil fauna activity (Snider et al. 1990; Lussenhop and Fogel 1993). Observations of mycorrhizae or soil fauna by means of MRs are still rare, and future studies should seek to take full advantage of this direct, in situ observation technique.

Belowground resource competition, mediated through root–root interaction, is of wide importance in plant ecosystems (Rewald and Leuschner 2009). Since MRs can estimate root biomass and distribution, they were suggested to be able to assess the degree of competition (Jose et al. 2001; Båth et al. 2008). However, until now studies are restricted to tree–crop interactions (e.g., Campbell et al. 1994; Gillespie et al. 2000) due to difficulties in distinguishing roots of different species in situ (but see Section VI.A). In addition to root

FIGURE 42.8 Roots of *Fagus sylvatica* and hyphae on a sufficiently installed MR-OT surface (i.e., only marginal soil voids); three root orders can be distinguished. The image was captured with a scanner MR system, CID, Cedar Rapids, IA. (Image courtesy of M. Lukac and D.L. Godbold, Bangor FACE, Bangor, U.K.)

FIGURE 42.9 **(See color insert.)** *Solanum lycopersicum* roots infested with *Orobanche aegyptiaca* tubercles. The picture was taken with a magnifying camera MR, Bartz, Carpinteria, CA.

FIGURE 42.10 **(See color insert.)** The fluorescent green roots belong to a genetically transformed *Zea mays* genotype expressing the GFP. The dark nonfluorescent roots belong to a non-GFP maize variety and can be clearly distinguished from the GFP roots. The picture was taken with a custom-made UV-MR camera system. (see Faget et al. [2009] for details; Image courtesy of M. Faget, M. Liedgens, P. Stamp, P. Flutsch and J.M. Herrera, Zurich, Switzerland.)

competition, belowground parasitic interactions can also be studies by means of MRs. Eizenberg et al. (2005) successfully used MRs for in situ monitoring of the early stages of the root parasite *Orobanche* development (Figure 42.9) and for detecting herbicide effect on the underground stages of the parasitic interaction.

VI. Recent and Future Developments in Minirhizotron Systems

A. Distinguishing between Species

Understanding of plant interactions is greatly limited by the inability to identify and quantify roots associated with different species; thus, developing such capabilities would greatly improve the MR technique. Discrimination of root fragments by visual morphology inspection is difficult and time consuming even for experienced taxonomists (Hertel and Leuschner 2006; Li et al. 2006), especially if picture resolution and visual view are limited. To overcome this limitation, Faget et al. recently used transgenic plants, expressing green fluorescent protein (GFP), combined with a modified MR imaging system (Faget et al. 2009, 2010). By illuminating the roots with white light, they could see all roots, but when switching to the appropriate UV excitation wavelength and detecting the appropriate fluorescence, only the roots that contain the GFP showed in the picture. This allowed the users to distinguish between transgenic and nontransgenic roots (Figure 42.10). Other promising technologies, aiming to distinguish between root species identities, are the near-infrared reflectance spectroscopy (NIRS) and the Fourier transform infrared spectroscopy (FTIR), which provide information about the presence, character, and number of functional chemical groups (Nakaji et al. 2008; Naumann et al. 2010); similar technologies are envisaged to be operative in MR-OTs.

B. Automatic Imaging Systems

Automatic image capturing systems would allow increased sampling frequency. Smucker et al. (1987) have designed an automatic device (to move a camera along the MR tube one screen distance at a time) many years ago, but it was not widely used. Recently, Allen et al. (2007) equipped MR tubes with a custom-made robotic automated MR system; however, only few automatic scanning systems are commercially available to date (e.g., CID, Cedar Rapids, IA and RhizoSystems, Idlyllwild, CA), allowing for continuous observation of root growth in surface soil layers. Having multiple automatic MR that respond to remote commands or environmental triggers (e.g., rainfall) will allow simultaneous data collection at multiple points in space and time and reduce the risk of missing short-lived roots.

C. Measurement of Environmental Parameters

The combination of the current MR techniques with other optical measurements could allow for measurements of soil traits (e.g., soil water content and temperature, pH values, and nutrients) that influence root growth. Standardization in MR-OT diameter and a careful selection of tube materials (e.g., low ion content) could allow for their multiple use for other techniques such as frequency domain reflectometry (FDR) and capacitance or neutron probes for determining soil water contents (e.g., Andrén et al. 1991; Kirkham et al. 1998). While a dual use of tubes would directly allow determining the influence of soil moisture on the observed rooting pattern, probe diameters and tube requirements are often not matched with MR imaging devices, preventing wide use.

Another promising technology that could be combined with MR systems are optical sensors ("optodes"). Optodes are growing in popularity due to the low-cost and long-term stability but are currently used in "rhizoboxes" only (Gansert and Blossfeld 2008). The fundamental principle is based on the ability of selected substances (embedded in foils) to act as dynamic luminescence quenchers. In the case of oxygen, if a ruthenium complex is illuminated with blue light, it will be excited and emit a red luminescent light with an intensity, or lifetime, that depends on the ambient oxygen concentration. The emitted light can be recorded with a photo sensor. Currently, optodes exist for O_2, CO_2, pH, Ca, P, and N determinations (e.g., Grunth et al. 2008; Strömberg 2008). Transparent, planar optode foil on the surface of MR-OT would provide valuable information about root/soil interactions including root exudation and presents an alternative to electrode-based sensors or other more soil disturbing analytical instrumentation.

VII. Concluding Remarks

MR systems have proven to be very useful for studying rhizosphere processes like fine-root growth and turnover, root morphology, and belowground interactions. MRs are composed of two constituents: clear OTs, which are installed in matching soil cores, and an image capturing device. Installed properly and after an equilibration period, they allow for studying the rhizosphere in a continuous, nondestructive manner. Although MR systems have provided many insights into rhizosphere processes, our review outlined wide variations in tube installation procedures, image acquisition, and data processing techniques. Since the MR image capturing devices improved constantly during the last decades, the main issues that have to be taken under consideration by future methodological studies and when using the MR system are the MR-OT installation (e.g., installation procedure *per se*, angle, tube material), the time lag between installation and the start of the study, the frequency of measurements, and the image analysis. To facilitate the utilization of the MR technique, extensive future research is needed on automatic image analysis and on imaging techniques beyond visible light.

We have no doubt that as root research continues to gain importance as a research field, this will result in future large-scale development of the MR technique.

References

Aerts R, Bakker C, de Caluwe H 1992 Root turnover as determinant of the cycling of C, N, and P in a dry heathland ecosystem. *Biogeochemistry* 15:175–190.

Aerts R, Berendse F, Klerk NM, Bakker C 1989 Root production and root turnover in two dominant species of wet heathlands. *Oecologia* 81:374–378.

Allen MF, Vargas R, Graham EA, Swenson W, Hamilton M, Taggart M, Harmon TC, Rat'ko A, Rundel P, Fulkerson B, Estrin D 2007 Soil sensor technology: Life within a pixel. *Bioscience* 57:859–867.

Andrén O, Rajkai K, Kätterer T 1991 A nondestructive technique for studies of root distribution in relation to soil-moisture. *Agric Ecosyst Environ* 34:269–278.

Basile B, Bryla DR, Salsman ML, Marsal J, Cirillo C, Johnson RS, Dejong TM 2007 Growth patterns and morphology of fine roots of size-controlling and invigorating peach rootstocks. *Tree Physiol* 27:231–241.

Bates GH 1937 A device for the observation of root growth in the soil. *Nature* 139:966–967.

Båth B, Kristensen HL, Thorup-Kristensen K 2008 Root pruning reduces root competition and increases crop growth in a living mulch cropping system. *J Plant Int* 3:211–221.

Baumann DL, Workmaster BA, Kosola KR 2005 'Ben Lear' and 'Stevens' cranberry root and shoot growth response to soil water potential. *Hortscience* 40:795–798.

Birchfield S, Wells CE 2006 Rootfly: Software for minirhizotron image analysis. http://www.ces.clemson.edu/stb/rootfly/

Bland WL, Dugas WA 1988 Root length density from minirhizotron observations. *Agron J* 80:271–275.

Böhm W 1979 *Methods of Studying Root Systems*. Berlin, Germany: Springer.

Box JE 1996 Modern methods for root investigations. In *Plant Roots: The hidden Half*, eds. Y Waisel, A Eshel, U Kafkafi, pp. 193–237. New York: Marcel Dekker, Inc.

Box JE, Johnson JW 1987 Minirhizotron rooting comparisons of three wheat cultivars. In *Minirhizotron Observation Tubes: Methods and Applications for Measuring Rhizosphere Dynamics*, ed. HM Taylor, pp. 123–130. Madison, WI: American Society of Agronomy.

Box JE, Ramseur EL 1993 Minirhizotron wheat root data—Comparisons to soil core root data. *Agron J* 85:1058–1060.

Box JE, Smucker AJM, Ritchie JT 1989 Minirhizotron installation techniques for investigating root responses to drought and oxygen stresses. *Soil Sci Soc Amer J* 53:115–118.

Bragg PL, Govi G, Cannell RQ 1983 A comparison of methods, including angled and vertical minirhizotrons, for studying root-growth and distribution in a spring oat crop. *Plant Soil* 73:435–440.

Brown ALP, Day FP, Stover DB 2009 Fine root biomass estimates from minirhizotron imagery in a shrub ecosystem exposed to elevated CO_2. *Plant Soil* 317:145–153.

Brown DA, Upchurch DR 1987 Minirhizotrons: A summary of methods and instruments in current use. In *Minirhizotron Observation Tubes: Methods and Applications for Measuring Rhizosphere Dynamics*, ed. HM Taylor, pp. 15–30. Madison, WI: American Society of Agronomy.

Burke MK, Raynal DJ 1994 Fine root growth phenology, production, and turnover in a northern hardwood forest ecosystem. *Plant Soil* 162:135–146.

Campbell CD, Mackie-Dawson LA, Reid EJ, Pratt SM, Duff EI, Buckland ST 1994 Manual recording of minirhizotron data and its application to study the effect of herbicide and nitrogen fertiliser on tree and pasture root growth in a silvopastoral system. *Agrofor Syst* 26:75–87.

Cheng WX, Coleman DC, Box JE 1991 Measuring root turnover using the minirhizotron technique. *Agr Ecosyst Environ* 34:261–267.

Comas LH, Eissenstat DM, Lakso AN 2000 Assessing root death and root system dynamics in a study of grape canopy pruning. *New Phytol* 147:171–178.

Coupe MD, Stacey JN, Cahill JF 2009 Limited effects of above- and belowground insects on community structure and function in a species-rich grassland. *J Veg Sci* 20:121–129.

Dannoura M, Kominami Y, Oguma H, Kanazawa Y 2008 The development of an optical scanner method for observation of plant root dynamics. *Plant Root* 2:14–18.

De Ruijter FJ, Venn BW, van Oijen M 1996 A comparison of soil core sampling and minirhizotrons to quantify root development of field-grown potatoes. *Plant Soil* 182:301–312.

Dubach M, Russelle MP 1995 Reducing the cost of estimating root turnover with horizontally installed minirhizotrons. *Agron J* 87:258–263.

Dyer D, Brown DA 1983 Relationship of fluorescent intensity to ion uptake and elongation rates of soybean roots. *Plant Soil* 72:127–134.

Eissenstat DM, Caldwell MM 1988 Seasonal timing of root-growth in favorable microsites. *Ecology* 69:870–873.

Eissenstat DM, Wells CE, Yanai RD, Whitbeck JL 2000 Building roots in a changing environment: Implications for root longevity. *New Phytol* 147:33–42.

Eissenstat DM, Yanai RD 1997 The ecology of root lifespan. *Adv Ecol Res* 27:1–60.

Eizenberg H, Shtienberg D, Silberbush M, Ephrath JE 2005 A new method for *in-situ* monitoring of the underground development of *Orobanche cumana* in sunflower (*Helianthus annuus*) with a mini-rhizotron. *Ann Bot* 96:1137–1140.

Ephrath JE, Eizenberg H 2010 Quantification of the dynamics of *Orobanche cumana* and *Phelipanche aegyptiaca* parasitism in confectionery sunflower. *Weed Res* 50:140–152.

Ephrath JE, Silberbush M, Berliner PR 1999 Calibration of minirhizotron readings against root length density data obtained from soil cores. *Plant Soil* 209:201–208.

Erz G, Veste M, Anlauf H, Breckle SW, Posch S 2005 A region and contour based technique for automatic detection of tomato roots in minirhizotron images. *J Appl Bot Food Qual* 79:83–88.

Faget M, Herrera JM, Stamp P, Aulinger-Leipner I, Frossard E, Liedgens M 2009 The use of green fluorescent protein as a tool to identify roots in mixed plant stands. *Funct Plant Biol* 36:930–937.

Faget M, Liedgens M, Stamp P, Flütsch P, Herrera JM 2010 A minirhizotron imaging system to identify roots expressing the green fluorescent protein. *Comput Electron Agric* 74:163–167.

Ferguson JC, Smucker AJM 1989 Modifications of the minirhizotron video camera system for measuring spatial and temporal root dynamics. *Soil Sci Soc Am J* 53:1601–1605.

Fitter AH, Self GK, Brown TK, Bogie DS, Graves JD, Benham D, Ineson P 1999 Root production and turnover in an upland grassland subjected to artificial soil warming respond to radiation flux and nutrients, not temperature. *Oecologia* 120:575–581.

Franco JA, Abrisqueta JM 1997 A comparison between minirhizotron and soil coring methods of estimating root distribution in young almond trees under trickle irrigation. *J Hortic Sci* 72:797–805.

Gansert D, Blossfeld S 2008 The application of novel optical sensors (optodes) in experimental plant ecology. In *Progress and Perspectives in Non-Invasive Bioprocess Analysis and Biogeochemical Interaction*, eds. U Lüttge, W Beyschlag, J Murata, pp. 333–358. Berlin, Germany: Springer.

Gaudinski JB, Trumbore SE, Davidson EA, Zheng SH 2000 Soil carbon cycling in a temperate forest: Radiocarbon-based estimates of residence times, sequestration rates and partitioning of fluxes. *Biogeochemistry* 51:33–69.

Gaul D, Hertel D, Leuschner C 2008 Effects of experimental soil frost on the fine-root system of mature Norway spruce. *J Plant Nutr Soil Sci* 171:690–698.

Gaul D, Hertel D, Leuschner C 2009 Estimating fine root longevity in a temperate Norway spruce forest using three independent methods. *Funct Plant Biol* 36:11–19.

Gijsman AJ, Floris J, van Noordwijk M, Brouwer G 1991 An inflatable minirhizotron system for root observations with improved soil/tube contact. *Plant Soil* 134:261–269.

Gillespie AR, Jose S, Mengel DB et al. 2000 Defining competition vectors in a temperate alley cropping system in the Midwestern USA—1. Production physiology. *Agrofor Syst* 48:25–40.

Gregory PJ 1979 Periscope method for observing root-growth and distribution in field soil. *J Exp Bot* 30:205–214.

Grunth NL, Askaer L, Elberling B 2008 Oxygen depletion and phosphorus release following flooding of a cultivated wetland area in Denmark. *Danish J Geogr* 108:17–25.

Guo D, Li H, Mitchell RJ, Han W, Hendricks JJ, Fahey TJ, Hendrick RL 2008 Fine root heterogeneity by branch order: Exploring the discrepancy in root turnover estimates between minirhizotron and carbon isotopic methods. *New Phytol* 177:443–456.

Harun MH, Roslan M 2003 An improved minirhizotron for observing oil palm root development and turnover. Malaysian Palm Oil Board, Ministry of Primary Industries, Kuala Lumpur, Malaysia, Report 179, pp. 1–2.

Hasselquist N, Vargas R, Allen M 2010 Using soil sensing technology to examine interactions and controls between ectomycorrhizal growth and environmental factors on soil CO_2 dynamics. *Plant Soil* 331:17–29.

Hendricks JJ, Hendrick RL, Wilson CA, Mitchell RJ, Pecot SD, Guo DL 2006 Assessing the patterns and controls of fine root dynamics: An empirical test and methodological review. *J Ecol* 94:40–57.

Hendrick RL, Pregitzer KS 1992 Spatial variation in tree root distribution and growth associated with minirhizotrons. *Plant Soil* 143:283–288.

Hendrick RL, Pregitzer KS 1993 The dynamics of fine root length, biomass, and nitrogen content in two northern hardwood ecosystems. *Can J For Res* 23:2507–2520.

Hendrick RL, Pregitzer KS 1996a Applications of minirhizotrons to understand root function in forests and other natural ecosystems. *Plant Soil* 185:293–304.

Hendrick RL, Pregitzer KS 1996b Temporal and depth-related patterns of fine root dynamics in northern hardwood forest. *J Ecol* 84:167–176.

Hertel D, Leuschner C 2002 A comparison of four different fine root production estimates with ecosystem carbon balance data in a *Fagus-Quercus* mixed forest. *Plant Soil* 239:237–251.

Hertel D, Leuschner C 2006 The in situ root chamber: A novel tool for the experimental analysis of root competition in forest soils. *Pedobiologia* 50:217–224.

Hooker JE, Black KE, Perry RL, Atkinson D 1995 Arbuscular mycorrhizal fungi induced alteration to root longevity of poplar. *Plant Soil* 172:327–329.

Hummel JW, Levan MA, Sudduth KA 1989 Minirhizotron installation in heavy soils. *Trans Am Soc Agric Biol Eng* 32:770–776.

Ingram KT, Leers GA 2001 Software for measuring root characters from digital images. *Agron J* 93:918–922.

Itoh S 1985 In situ measurement of rooting density by micro-rhizotron. *Soil Sci Plant Nutr* 31:653–656.

Iversen C, Murphy M, Allen M, Childs J, Eissenstat D, Lilleskov E, Sarjala T, Sloan V, Sullivan P 2012 Advancing the use of minirhizotrons in wetlands. *Plant Soil* 352:23–39.

Johnson MG, Tingey DT, Phillips DL, Storm MJ 2001 Advancing fine root research with minirhizotrons. *Environ Exp Bot* 45:263–289.

Jose S, Gillespie AR, Seifert JR, Pope PE 2001 Comparison of minirhizotron and soil core methods for quantifying root biomass in a temperate alley cropping system. *Agrofor Syst* 52:161–168.

Joslin JD, Gaudinski JB, Torn MS, Riley WJ, Hanson PJ 2006 Fine-root turnover patterns and their relationship to root diameter and soil depth in a ^{14}C-labeled hardwood forest. *New Phytol* 172:523–535.

Joslin JD, Wolfe MH 1999 Disturbances during minirhizotron installation can affect root observation data. *Soil Sci Soc Am J* 63:218–221.

Kage H, Kochler M, Stutzel H 2000 Root growth of cauliflower (*Brassica oleracea* L. *botrytis*) under unstressed conditions: Measurement and modelling. *Plant Soil* 223:131–145.

Karnok KJ, Kucharski RT 1982 Design and construction of a rhizotron-lysimeter facility at the Ohio State University. *Agron J* 74:152–156.

Keng JCW, Kusaka T 1988 Application of fiber optic techniques in underground observations. *Soil Technol* 1:157–167.

Kirkham MB, Grecu SJ, Kanemasu ET 1998 Comparison of minirhizotrons and the soil-water-depletion method to determine maize and soybean root length and depth. *Eur J Agron* 8:117–125.

Klironomos JN, Allen MF 1995 UV-B-mediated changes on below-ground communities associated with the roots of *Acer saccharum*. *Funct Ecol* 9:923–930.

Kloeppel BD, Gower ST 1995 Construction and installation of acrylic minirhizotron tubes in forest ecosystems. *Soil Sci Soc Am J* 59:241–243.

Kosola KR 1999 Laparoscopic sampling of roots of known age from an expandable-wall minirhizotron system. *Agron J* 91:876–879.

Kosola KR, Eissenstat DM, Graham JH 1995 Root demography of mature citrus trees—The influence of *Phytophthora nicotianae*. *Plant Soil* 171:283–288.

Lei P, Bauhus J 2010 Use of near-infrared reflectance spectroscopy to predict species composition in tree fine-root mixtures. *Plant Soil* 333:93–103.

Levan MA, Ycas JW, Hummel JW 1987 Light leak effects on near-surface soybean rooting observed with minirhizotrons. In *Minirhizotron Observation Tubes: Methods and Applications for Measuring Rhizosphere Dynamics*, ed. HM Taylor, pp. 89–98. Madison, WI: American Society of Agronomy.

Li L, Sun JH, Zhang FS, Guo TW, Bao XG, Smith FA, Smith SE 2006 Root distribution and interactions between intercropped species. *Oecologia* 147:280–290.

Liedgens M 1998 Seasonal development of the maize root system in minirhizotron-equipped lysimeters. PhD thesis. Zürich, Switzerland: Federal Institute of Technology.

López B, Sabaté S, Gracia C 1996 An inflatable minirhizotron system for stony soils. *Plant Soil* 179:255–260.

Lussenhop J, Fogel R 1993 Observing soil biota *in situ*. *Geoderma* 56:25–36.

Mainiero R 2006 Fine root dynamics by minirhizotron. In *Handbook of Methods Used in Rhizosphere Research–Online edition*, eds. J Luster, R Finlay. Birmensdorf, Switzerland: Swiss Federal Research Institute WSL.

Majdi H 1996 Root sampling methods—Applications and limitations of the minirhizotron technique. *Plant Soil* 185:255–258.

Majdi H, Damm E, Nylund JE 2001 Longevity of mycorrhizal roots depends on branching order and nutrient availability. *New Phytol* 150:195–202.

Majdi H, Kangas P 1997 Demography of fine roots in response to nutrient applications in a Norway spruce stand in southwestern Sweden. *Ecoscience* 4:199–205.

Majdi H, Nylund JE, Agren GI 2007 Root respiration data and minirhizotron observations conflict with root turnover estimates from sequential soil coring. *Scand J For Res* 22:299–303.

Majdi H, Pregitzer KS, Moren AS, Nylund JE, Agren GI 2005 Measuring fine root turnover in forest ecosystems. *Plant Soil* 276:1–8.

Majdi H, Smucker AJM, Persson H 1992 A comparison between minirhizotron and monolith sampling methods for measuring root-growth of maize (*Zea mays* L.). *Plant Soil* 147:127–134.

McDougall WB 1916 The growth of forest tree roots. *Am J Bot* 3:384–392.

McMichael BL, Burke JJ 1996 Temperature effects on root growth. In *Plant Roots: The Hidden Half*, eds. Y Waisel, A Eshel, U Kafkafi, pp. 383–396. New York: Marcel Dekker, Inc.

McMichael BL, Taylor HM 1987 Applications and limitations of rhizontrons and minirhizotrons. In *Minirhizotron Observation Tubes: Methods and Applications for Measuring Rhizosphere Dynamics*, ed. HM Taylor, pp. 1–14. Madison, WI: American Society of Agronomy.

McMichael B, Zak J 2006 The role of rhizotrons and minirhizotrons in evaluating the dynamics of rhizoplane-rhizosphere microflora. In *Microbial Activity in the Rhizoshere*, eds. KG Mukerji, C Manocharachary, J Singh, Vol. 7, pp. 71–87. Berlin, Germany: Springer.

Meier IC, Leuschner C 2008 Genotypic variation and phenotypic plasticity in the drought response of fine roots of European beech. *Tree Physiol* 28:297–309.

Merrill SD 1992 Pressurized-wall minirhizotron for field observation of root-growth dynamics. *Agron J* 84:755–758.

Merrill SD, Doering EJ, Reichman GA 1987 Application of a minirhizotron with flexible, pressurized walls to a study of corn root growth. In *Minirhizotron Observation Tubes: Methods and Applications for Measuring Rhizosphere Dynamics*, ed. HM Taylor, pp. 131–143. Madison, WI: American Society of Agronomy.

Merrill SD, Tanaka DL, Hanson JD 2005 Comparison of fixed-wall and pressurized-wall minirhizotrons for fine root growth measurements in eight crop species. *Agron J* 97:1367–1373.

Merrill SD, Upchurch DR 1994 Converting root numbers observed at minirhizotrons to equivalent root length density. *Soil Sci Soc Am J* 58:1061–1067.

Merrill SD, Upchurch DR, Black AL, Bauer A 1994 Theory of minirhizotron root directionality observation and application to wheat and corn. *Soil Sci Soc Am J* 58:664–671.

Mukerji KG, Manoharachary C, Singh J 2006 *Microbial Activity in the Rhizosphere*. New York: Springer.

Murphy SL, Smucker AJM 1995 Evaluation of video image-analysis and line-intercept methods for measuring root systems of alfalfa and ryegrass. *Agron J* 87:865–868.

Nadelhoffer KJ 2000 The potential effects of nitrogen deposition on fine-root production in forest ecosystems. *New Phytol* 147:131–139.

Nakaji T, Noguchi K, Oguma H 2008 Classification of rhizosphere components using visible-near infrared spectral images. *Plant Soil* 310:245–261.

Naumann A, Heine G, Rauber R 2010 Efficient discrimination of oat and pea roots by cluster analysis of Fourier transformed infrared (FTIR) spectra. *Field Crop Res* 119:78–84.

Norby RJ, Ledford J, Reilly CD, Miller NE, O'Neill EG 2004 Fine-root production dominates response of a deciduous forest to atmospheric CO2 enrichment. *Proc Natnl Acad Sci* 101:9689–9693.

van Noordwijk M, de Jager A, Floris J 1985 A new dimension to observations in minirhizotrons: A stereoscopic view on root photographs. *Plant Soil* 86:447–453.

Vos J, Groenwold J 1987 The relation between root growth along observation tubes and in bulk soil. In *Minirhizotron observation tubes: Methods and applications for measuring rhizosphere dynamics*, ed. HM Taylor, pp. 39–49. Madison, WI: American Society of Agronomy.

Pagès L, Bengough AG 1997 Modelling minirhizotron observations to test experimental procedures. *Plant Soil* 189:81–89.

Pan WL, Young FL, Bolton RP 2001 Monitoring Russian thistle (*Salsola iberica*) root growth using a scanner-based, portable mesorhizotron. *Weed Technol* 15:762–766.

Parker CJ, Carr MKV, Jarvis NJ, Puplampu BO, Lee VH 1991 An evaluation of the minirhizotron technique for estimating root distribution in potatoes. *J Agric Sci* 116:341–350.

Phillips DL, Johnson MG, Tingey DT, Biggart C, Nowak RS, Newsom JC 2000 Minirhizotron installation in sandy, rocky soils with minimal soil disturbance. *Soil Sci Soc Am J* 64:761–764.

Phillips DL, Johnson MG, Tingey DT, Catricala CE, Hoyman TL, Nowak RS 2006a Effects of elevated CO2 on fine root dynamics in a Mojave Desert community: a FACE study. *Global Ch Biol* 12:61–73.

Phillips DL, Johnson MG, Tingey DT, Storm MJ, Ball JT, Johnson DW 2006b CO2 and N-fertilization effects on fine-root length, production, and mortality: a 4-year Ponderosa pine study. *Oecologia* 148:517–525.

Poelman G, van de Koppel J, Brouwer G 1996 A telescopic method for photographing within 8×8 cm minirhizotrons. *Plant Soil* 185:163–167.

Powell SW, Day Jr. FP 1991 Root production in four communities in the Great Dismal swamp. *Am J Bot* 78:288–297.

Pritchard SG, Strand AE 2008 Can you believe what you see? Reconciling minirhizotron and isotopically derived estimates of fine root longevity. *New Phytol* 177:287–291.

Pritchard SG, Strand AE, McCormack ML et al. 2008a Fine root dynamics in a loblolly pine forest are influenced by free-air-CO_2-enrichment: A six-year-minirhizotron study. *Global Change Biol* 14:588–602.

Pritchard SG, Strand AE, McCormack ML, Davis MA, Oren R 2008b Mycorrhizal and rhizomorph dynamics in a loblolly pine forest during five years of free-air-CO_2-enrichment. *Global Change Biol* 14:1252–1264.

Publicover DA, Vogt KA 1993 A comparison of methods for estimating forest fine-root production with respect to sources of error. *Can J For Res* 23:1179–1186.

Rachmilevitch S, Ephrath JE 2011c Salt stress effects on root systems of two mature olive cultivars. *Acta Hortic*, 888:109–118.

Repo T, Lehto T, Finer L 2008 Delayed soil thawing affects root and shoot functioning and growth in Scots pine. *Tree Physiol* 28:1583–1591.

Rewald B, Ephrath JE, Rachmilevitch S 2011a A root is a root is a root?—Water uptake rates of *Citrus* root orders. *Plant Cell Environ* 34:33–42.

Rewald B, Leuschner C 2009 Belowground competition in a broad-leaved temperate mixed forest: Pattern analysis and experiments in a four-species stand. *Eur J For Res* 128:387–398.

Rewald B, Rachmilevitch S, McCue MD, Ephrath JE 2011b Influence of saline drip-irrigation on fine root and sap-flow densities of two mature olive varieties. *Environ Exp Bot* 72:107–114.

Richards JH 1984 Root growth response to defoliation in two *Agropyron* bunchgrasses: Field observations with an improved root periscope. *Oecologia* 64:21–25.

Richner W, Liedgens M, Bürgi H, Soldati A, Stamp P 2000 Root image analysis and interpretation. In *Root Methods: A Handbook*, eds. AL Smit, AG Bengough, C Engels, M van Noordwijk, S Pellerin, SC van de Geijn, pp. 305–342. Berlin, Germany: Springer.

Rose L, Coners H, Leuschner C 2011 Effects of fertilization and cutting frequency on the water balance of a temperate grassland. *Ecohydrology* 5:64–72.

Rush CM, Upchurch DR, Gerik TJ 1984 In situ observations of *Phymatotrichum omnivorum* with a borescope minirhizotron system. *Phytopathology* 74:104–105.

Rytter RM, Hansson AC 1996 Seasonal amount, growth and depth distribution of fine roots in an irrigated and fertilized *Salix viminalis* L. plantation. *Biomass Bioenergy* 11:129–137.

Samson BK, Sinclair TR 1994 Soil core and minirhizotron comparison for the determination of root length density. *Plant Soil* 161:225–232.

Sanders JL, Brown DA 1978 A new fiber optic technique for measuring root growth of soybeans under field conditions. *Agron J* 70:1073–1076.

Satomura T, Fukuzawa K, Horikoshi T 2007 Considerations in the study of tree fine-root turnover with minirhizotrons. *Plant Root* 1:34–45.

Shani U, Waisel Y, Eshel A 1995 The development of melon roots under trickle irrigation: Effects of the location of emitters. In *Structure and Function of Roots*, ed. F Baluska, pp. 223–225. Amsterdam, the Netherlands: Kluwer.

Smucker AJM 1993 Soil environmental modifications of root dynamics and measurement. *Annu Rev Phytopathol* 31:191–216.

Smucker AJM, Ferguson JC, Debruyn WP, Belford LR, Ritchie JT 1987 Image analysis of video recorded plant root systems. In *Minirhizotron Observation Tubes: Methods and Applications for Measuring Rhizosphere Dynamics*, ed. HM Taylor, pp. 67–80. Madison, WI: American Society of Agronomy.

Snider RJ, Snider R, Smucker AJM 1990 Collembolan populations and root dynamics in Michigan agroecosystems. In *Rhizoplane Dynamics*, eds. JE Box, LC Hammond, pp. 168–191. Boulder, CO: Westview.

Steele SJ, Gower ST, Vogel JG, Norman JM 1997 Root production, net primary production and turnover in aspen, Jack pine and Black spruce forests in Saskatchewan and Manitoba. *Tree Physiol* 17:577–587.

Strand AE, Pritchard SG, McCormack ML, Davis MA, Oren R 2008 Irreconcilable differences: Fine-root life spans and soil carbon persistence. *Science* 319:456–458.

Strömberg N 2008 Determination of ammonium turnover and flow patterns close to roots using imaging optodes. *Environ Sci Technol* 42:1630–1637.

Taylor HM 1987 *Minirhizotron Observation Tubes: Methods and Applications for Measuring Rhizosphere Dynamics*. Madison, WI: American Society of Agronomy.

Taylor HM, Böhm W 1976 Use of acrylic plastic as rhizotron windows. *Agron J* 68:693–694.

Taylor HM, Upchurch DR, McMichael BL 1990 Applications and limitations of rhizotrons and minirhizotrons for root studies. *Plant Soil* 129:29–35.

Tierney GL, Fahey TJ 2001 Evaluating minirhizotron estimates of fine root longevity and production in the forest floor of a temperate broadleaf forest. *Plant Soil* 229:167–176.

Tierney GL, Fahey TJ 2002 Fine root turnover in a northern hardwood forest: A direct comparison of the radiocarbon and minirhizotron methods. *Can J For Res* 32:1692–1697.

Tingey DT, Phillips DL, Johnson MG 2003 Optimizing minirhizotron sample frequency for an evergreen and deciduous tree species. *New Phytol* 157:155–161.

Tingey DT, Phillips DL, Johnson MG, Rygiewicz PT, Beedlow PA, Hogsett WE 2005 Estimates of Douglas-fir fine root production and mortality from minirhizotrons. For *Ecol Mngmnt* 204:359–370.

Treseder KK, Allen MF, Ruess RW, Pregitzer KS, Hendrick RL 2005 Lifespans of fungal rhizomorphs under nitrogen fertilization in a pinyon-juniper woodland. *Plant Soil* 270:249–255.

Trumbore S, Da Costa ES, Nepstad DC et al. 2006 Dynamics of fine root carbon in Amazonian tropical ecosystems and the contribution of roots to soil respiration. *Global Change Biol* 12:217–229.

Upchurch DR 1985 Relationship between observations in minirhizotrons and true root length density. PhD thesis. Lubbock, TX: Texas Tech University.

Upchurch DR, Ritchie JT 1983 Root observations using a video recording system in mini-rhizotrons. *Agron J* 75:1009–1015.

Upchurch DR, Ritchie JT 1984 Battery-operated video camera for root observations in minirhizotrons. *Agron J* 76:1015–1017.

Vamerali T, Ganis A, Bona S, Mosca G 1999 An approach to minirhizotron root image analysis. *Plant Soil* 217:183–193.

Van Rees KCJ 1998 Soil temperature effects from minirhizotron lighting systems. *Plant Soil* 200:113–118.

Vargas R, Allen MF 2008 Dynamics of fine root, fungal rhizomorphs, and soil respiration in a mixed temperate forest: Integrating sensors and observations. *Vadose Zone J* 7:1055–1064.

Vogt KA, Vogt DJ, Bloomfield J 1998 Analysis of some direct and indirect methods for estimating root biomass and production of forests at an ecosystem level. *Plant Soil* 200:71–89.

Volkmar KM 1993 A comparison of minirhizotron techniques for estimating root length density in soils of different bulk densities. *Plant Soil* 157:239–245.

Waddington J 1971 Observation of plant roots *in situ*. *Can J Bot* 49:1850–1852.

Wang ZQ, Burch WH, Mou P, Jones RH, Mitchell RJ 1995 Accuracy of visible and ultra-violet light for estimating live root proportions with minirhizotrons. *Ecology* 76:2330–2334.

Wells CE, Eissenstat DM 2001 Marked differences in survivorship among apple roots of different diameters. *Ecology* 82:882–892.

Wells CE, Glenn DM, Eissenstat DM 2002 Changes in the risk of fine-root mortality with age: A case study in peach, *Prunus persica* (Rosaceae). *Am J Bot* 89:79–87.

Withington JM, Elkin AD, Bulaj B, Olesinski J, Tracy KN, Bouma TJ, Oleksyn J, Anderson LJ, Modrzynski J, Reich PB, Eissenstat DM 2003 The impact of material used for minirhizotron tubes for root research. *New Phytol* 160:533–544.

Yang J, Hammer RD, Blanchar RW 2003 Minirhizotron quantification of soybean root growth as affected by reduced—A horizon in soil. *J Plant Nutr Soil Sci* 166:708–711.

Zeng G, Birchfield ST, Wells CE 2008 Automatic discrimination of fine roots in minirhizotron images. *New Phytol* 177:549–557.

Zeng G, Birchfield ST, Wells CE 2010 Rapid automated detection of roots in minirhizotron images. *Mach Vision Appl* 21:309–317.

43

Noninvasive Tools for Measuring Metabolism and Biophysical Analyte Transport at the Root–Rhizosphere Interface: Self-Referencing Electrochemical and Optical Sensors

Eric S. McLamore
University of Florida

D. Marshall Porterfield
Purdue University

I. Introduction

The interface between roots and the surrounding rhizosphere is the boundary layer formed at the root surface, and the understanding of spatial and temporal variations of analyte activity within this unstirred layer provides critical information regarding root physiology. Thus, root biologists have been exploring this mass boundary layer using a wide variety of different techniques. Among these techniques, microsensors have contributed a large dataset regarding spatial and temporal variations in soil chemistry.

In the third edition of *Plant Roots: The Hidden Half* (Porterfield 2002), we introduced the self-referencing microsensor, including a brief history of the governing theory, general methods for constructing ion-selective and polarographic electrochemical microsensors, the governing theory behind the self-referencing technique, and noted the potential for utilizing self-referencing biosensors (see Section III for a synopsis). In the last decade, the field has grown considerably, in large part driven by the development of optical microsensors for measuring oxygen flux (Porterfield et al. 2006; Sanchez et al. 2008; Chatni et al. 2009; Chatni and Porterfield 2009; McLamore et al. 2010a,b), enzyme-based microbiosensors (McLamore et al. 2010c, 2011; Shi et al. 2011a,b,c), and microelectrodes for biomolecules such as nitric oxide (Prado et al. 2004; Salmi et al. 2007) and auxin (McLamore et al. 2010d). Two recent reviews discuss developments in the technique over the last decade (Porterfield 2007; McLamore and Porterfield 2011) and provide a number of examples of how the noninvasive technique can provide real-time data describing

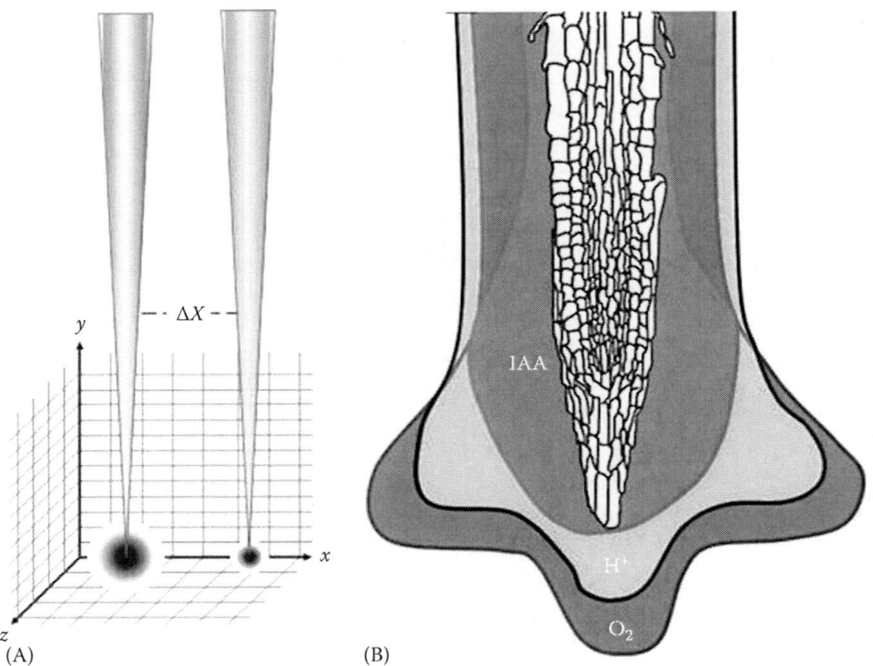

FIGURE 43.1 (A) Conceptual model of self-referencing microsensor technique for real-time monitoring of concentration gradients. (B) Oxygen influx, proton efflux, and induced auxin uptake in *Z. mays* roots based on measurements using self-referencing microsensors. (Data from McLamore, E.S. et al., *Planta*, 232, 1087, 2010b; McLamore, E.S. et al., *Plant J.*, 63, 1004, 2010d.)

physiological transport in cells, tissues, organs, and whole organisms. In this chapter, we discuss how these tools can be used to probe the physiological activity of plant roots, laying the foundation for increasing our knowledge of real-time, spatially resolved analyte transport (Figure 43.1).

II. Tools for Studying Root Physiology

A. A Note on Nomenclature: Sensors, Detectors, and Assays

Since the development of the polarographic electrode by Clark et al. (1953), there has been much progress in the area of sensors, microsensors, and recently nanosensors. Nanomaterials research has energized sensor development efforts by enhancing signal acquisition for both electrochemical (Wang 2005) and optical devices (Bonaccorso et al. 2010). This explosion of technologies includes a large collection of sensors, detectors, and assays for physiological recordings. The nomenclature used for describing these devices is of vital importance, as the attributes and features of sensors, detectors, and assays are quite different. The most important characteristics describing performance of these tools are sensitivity, limit of detection, and hysteresis (Table 43.1). Use of the correct terminology is critical, as tools for measuring biophysical transport of molecules at the root–rhizosphere interface must be highly sensitive and selective, with a physiologically relevant limit of detection and little to no hysteresis. The quantitative nature of sensors (which is a major advantage over detectors and detector arrays) is due to reversible molecular interaction(s) between the recognition element within the sensor and the target analyte.

TABLE 43.1 Operating Characteristics of Detectors, Detector Arrays and Sensors

Device	Sensitivity	Limit of Detection	Hysteresis
Detector	Qualitative	Must be defined by user	High
Detector array	Semiquantitative	Must be defined by user	High
Sensor	Quantitative	Mathematically/ empirically defined	Low

There are many detectors, detector arrays, and sensors that have been valuable tools for studying roots and root cells. These technologies have provided a wealth of fundamental information and new knowledge. As with any technology, the advantages and caveats of each device should be well documented to ensure that the tool can provide data that can be used in order to investigate hypothesis-driven research questions. The following summaries of detectors, detector arrays, and sensors are not intended to be exhaustive, but rather highlight the unique signals obtained using self-referencing microsensors, which is covered in detail in the subsequent section.

B. Use of Microassays in Root Studies

Microassays are detector arrays that have been used to study roots and root cells by means of fluorescence intensity or electrochemical activity, which is usually measured in microtiter plates (Ludwig-Müller et al. 1997; O'Riordan et al. 2000; Alderman et al. 2004; Kupper et al. 2004; Serrano et al. 2007). Although many attempts have been made at quantifying analyte uptake/production rates, microassays are semiquantitative due to significant signal hysteresis. Additionally, most microassays require invasive sampling of roots and/or addition of extracting agents, limiting the broad applicability of the results.

C. Ion-Selective Electrodes in Root Studies

As reviewed by Porterfield (2002), ion-selective electrodes (ISE) are sensors that have been used in a wide variety of plant root studies. ISEs have been used to measure the activity of ions along the surface of roots, including H^+ (Peters and Felle 1999), Ca^{2+} (Felle 1994), Cl^- (Herrmann and Felle 1995), and K^+ (Maathuis and Sanders 1993) (among many others). While these static ISEs provided a wealth of data concerning the activity near plant roots, they were not capable of providing quantitative data concerning the polarity of ion movement within the boundary layer. Invasive studies in plant roots have also been conducted using ISEs (Felle 1998; Felle et al. 1999a,b). While these invasive measurements are experimentally robust and can be used in situ, they induce an experimental error by causing local tissue damage and inducing oxidative bursts (see Porterfield 2002).

D. Techniques for Measuring Oxygen in Root Studies

Respirometry (measurement of gas phase oxygen) has been used extensively for measuring physiology of plants, developing roots, and root cells (Lamboursain et al. 2002; De Dobbeleer et al. 2006; Cloutier et al. 2009; Gupta et al. 2009). However, respirometry suffers from many disadvantages, including (1) underestimation of metabolic rates due to oxygen limitation in the measurement chamber, (2) requirement of oxygen/ADP microinjection, (3) significant background artifacts, (4) poor detection limit due to oxygen consumption by the electrode, (5) low signal-to-noise ratio, and (6) autoxidation of commonly used reagents (e.g., ascorbate and phenylenediamine dihydrochloride). Analysis of respirometry often requires "data smoothing," "time correction," and best-of-fit correlation for noise reduction (Gnaiger et al. 2000). Respirometry data are often reported as flux, although the magnitude and the direction of analyte transport are not directly measured (i.e., values are *ad hoc* calculation of the negative time derivative of oxygen concentration) (Armstrong et al. 2009; Gupta et al. 2009). Respirometry does not provide spatial resolution, which limits the types of hypothesis-driven research questions that can be posed regarding the form and function of roots.

Polarographic oxygen microelectrodes (reviewed by Porterfield 2002) overcome many of the disadvantages faced with respirometry and microassays and provide direct measurement of oxygen concentrations at the micro- or macroscale (Verslues et al. 1998; Porterfield et al. 1999; Porterfield 2002; Buerk 2004; Land et al. 1999). Polarographic electrodes do however have many disadvantages, including a low signal-to-noise ratio, excessive drift/noise, requirement of a reference electrode, consumption of extracellular oxygen (significantly reducing sensitivity), and (depending on the design) poor selectivity.

Development of fiber-optic microsensors (optrodes) has resolved many of the problems experienced by polarographic oxygen electrodes, including no oxygen consumption at the tip of the optrode (Kühl and Jørgensen 1992), rapid response time (Lee and Okura 1997), no requirement of a reference electrode (Chatni and Porterfield 2009), high reproducibility, and excellent selectivity (Miller and Dunton 2007). Signal-to-noise ratio for optrodes (Chatni and Porterfield 2009; McLamore et al. 2010b) is orders of magnitude higher than polarographic microelectrodes (Mancuso et al. 2000). The low signal-to-noise ratio for polarographic electrodes is in part due to the use of small tip diameters, which induces antenna effects on the measured signal. In optrodic sensing, on the other hand, small tip diameters do not cause an increase in measured noise. Among various designs in the current literature, optrodes utilizing luminescent dyes and frequency-modulated lifetime are preferred (Kuhl and Jørgensen 1992; McEvoy et al. 2003; Chen et al. 2009; Roche et al. 2010). By modulating the excitation intensity, phase shifts can be monitored using a lock-in amplifier (see Figure 43.5), reducing noise, drift, and eliminating calibration shifts associated with photobleaching commonly encountered when using fluorescence intensity-based optrodes (Wolfbeis 2004; Borisov and Wolfbeis 2008; Chatni and Porterfield 2009; Chatni et al. 2009).

While the devices described earlier have benefits for monitoring biochemical markers in/near roots, many are invasive/destructive, have low spatial resolution and low signal-to-noise ratio, and are not capable of monitoring the dynamic flux of molecules in the boundary layer formed at the root–rhizosphere interface.

III. Self-Referencing Physiological Sensors

Comprehensive reviews of the self-referencing (SR) technique may be found in Porterfield (2002), Porterfield (2007), McLamore and Porterfield (2011), Newman (2001) and Newman et al. (2012). These reviews include details on the governing theory, hardware requirements, new developments in types of sensors, and recent enhancements of transduction/acquisition schemes. Here, we will briefly review the basics of the technique, governing equations, and theory for SR ion-selective electrodes, SR amperometric microsensors, and SR optical microsensors.

SR is a microsensor technique used to filter noise in the extracellular environment. This noise is due to a combination of "antenna effects," relatively small active surface area of microsensors, and dynamic spatial and temporal transport in the boundary layer near living cells/tissues. Biophysical transport of molecules in this diffusive boundary layer is governed by Fick's law of diffusion and can be predicted using diffusion–reaction

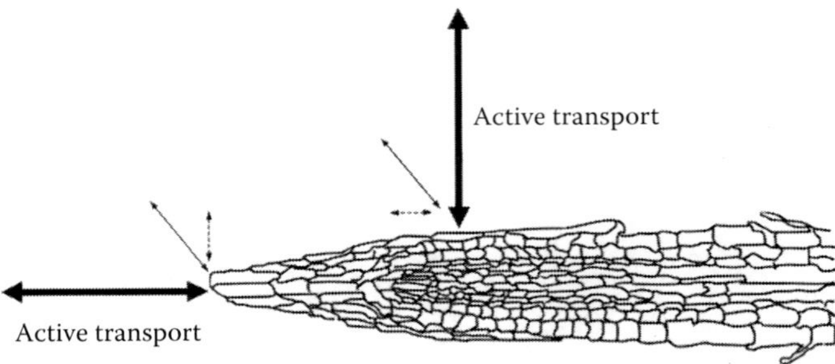

FIGURE 43.2 Conceptual 3D transport at the rhizosphere–root interface. All three dimensions of transport are shown (thickness and length of arrows indicate relative magnitude). Biological transport occurs in the direction perpendicular to the tangent of the root surface and is due to cellular activity within the root.

kinetic models. For roots, this diffusion layer is approximately 100–1200 μm in length (McLamore and Porterfield 2011). Transport at the root–rhizosphere interface is driven by concentration gradients that are regulated by living cells. The direction of transport at the root surface is in the vector perpendicular to the tangent of the root surface (Figure 43.2). Transport in the remaining two dimensions (parallel to the root surface) occurs at relatively short length and time scales (McLamore and Porterfield 2011). The transport of molecules in the unstirred boundary layer near developing roots holds many of the secrets to the underlying biochemistry in roots.

The SR microsensor technique is based on the biophysical principle of cell-directed gradient transport described earlier and involves computer-controlled oscillation of a single microelectrode in the primary direction of biological transport (see Figure 43.2). A microsensor is positioned within 1 μm of the root surface using a camera/zoomscope and stepper motors. The microsensor is then oscillated between two positions (known as "near pole" and "far pole") at a constant excursion distance (ΔX); oscillation within the 0.2–0.5 Hz range does not induce mixing of the unstirred layer. Concentration gradients (ΔC) are recorded in real time during microsensor oscillation. If the value of the molecular diffusion coefficient is known (or can be estimated), biophysical flux can be recorded in real time based on Fick's first law of diffusion (Equation 43.1). SR uses a move-wait-measure technique for detection of real-time changes in differential analyte concentration over a fixed excursion distance (Porterfield 2002).

$$J = -D\frac{\Delta C}{\Delta X} = -D\frac{C_{(x)} - C_{(x+\Delta x)}}{\Delta X} \tag{43.1}$$

where

J is the analyte flux
D is the molecular diffusion constant
$C_{(x)}$ is the analyte concentration at location (x), termed "near pole"
$C_{(x+\Delta X)}$ is the analyte concentration at location ($x + \Delta x$), termed "far pole"

Calculation of flux via Equation 43.1 depends on a known value of the molecular diffusion coefficient (and/or effective diffusion coefficient).

The SR microsensor technique converts static micro-/nanosensors with otherwise low signal-to-noise ratios into dynamic flux sensors capable of filtering out signals not associated with biologically active transport. SR significantly improves sensitivity, selectivity, and signal-to-noise ratio (Jaffe and Nuccitelli 1974). Porterfield (2002) discussed an important hardware development, which involves the use of a DC-coupled amplification scheme (Figure 43.3), significantly improving system performance. The use of a single microelectrode is an extremely important concept for noise filtration, as use of sensor arrays for measuring concentration gradients does not provide phase-sensitive detection. In other words, noise from individual sensors "washes out" differential signals, and measuring biologically relevant differential signals (1–1000 pmol cm^{-2} s^{-1}) is not possible.

If the excursion distance (ΔX) is within the linear portion of the concentration gradient (typically ≈10% of the total boundary layer thickness), Fick's first law of diffusion may be used to directly calculate flux based on measured values of $\Delta C/\Delta X$ using Equation 43.1. Multidimensional measurements (i.e., square wave oscillation) are possible (Kunkel et al. 2006), although this approach should be reserved for highly specialized applications due to temporal limitations. McLamore et al. (2009) developed a technique for accounting for drift artifacts during measurement using SR microsensors.

A. SR Ion-Selective Microelectrodes

SR micro-ISE (μISE) operate according to Nernst Law, where voltage is measured versus a reference electrode (commonly 3 M KCl in contact with a Ag/AgCl pellet). According to the Nernst equation, a decade difference in ion concentration produces a measured potential of 59.16 mV for monovalent ions and 29.58 mV for divalent ions (Figure 43.4). For SR ion-selective electrodes, the Fick equation is combined with the Nernst or Nikolsky–Eisenmann equation, and simplified to Equation 43.2.

$$J = -D\left(\frac{\Delta C}{\Delta X}\right) = -D\left(\frac{\left(10^{E_{(x)}-b/s} - 10^{E_{(x+\Delta x)}-b/s}\right)}{\Delta X}\right) \tag{43.2}$$

where

J is the primary ion flux (μmol cm^{-2} s^{-1})
$E_{(x)}$ is the electrode potential at near pole (mV)
$E_{(x+\Delta x)}$ is the electrode potential at far pole (mV)
s is the Nernst slope (mv/log$_{10}$ C)
b is the y-intercept of Nernst calibration plot (mV)

For measuring H^+ flux, the SR μISE technique has been extended to account for buffering effects using a technique in Porterfield et al. (2009). The user friendly technique requires only the input of the total buffer concentration and pK value by the user to correct for this artifact.

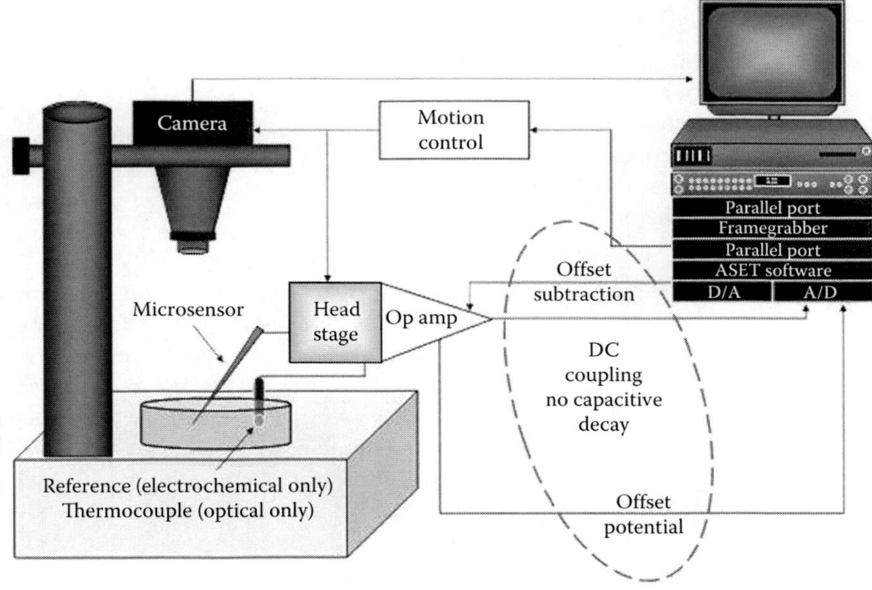

FIGURE 43.3 Basic hardware required for self-referencing technique, highlighting the DC coupling scheme, which resolves the capacitive decay problem experienced when using AC-coupled approaches. All hardware may be obtained from Applicable Electronics, Inc. (Sandwich, MA) and Science Wares (Fallmouth, MA).

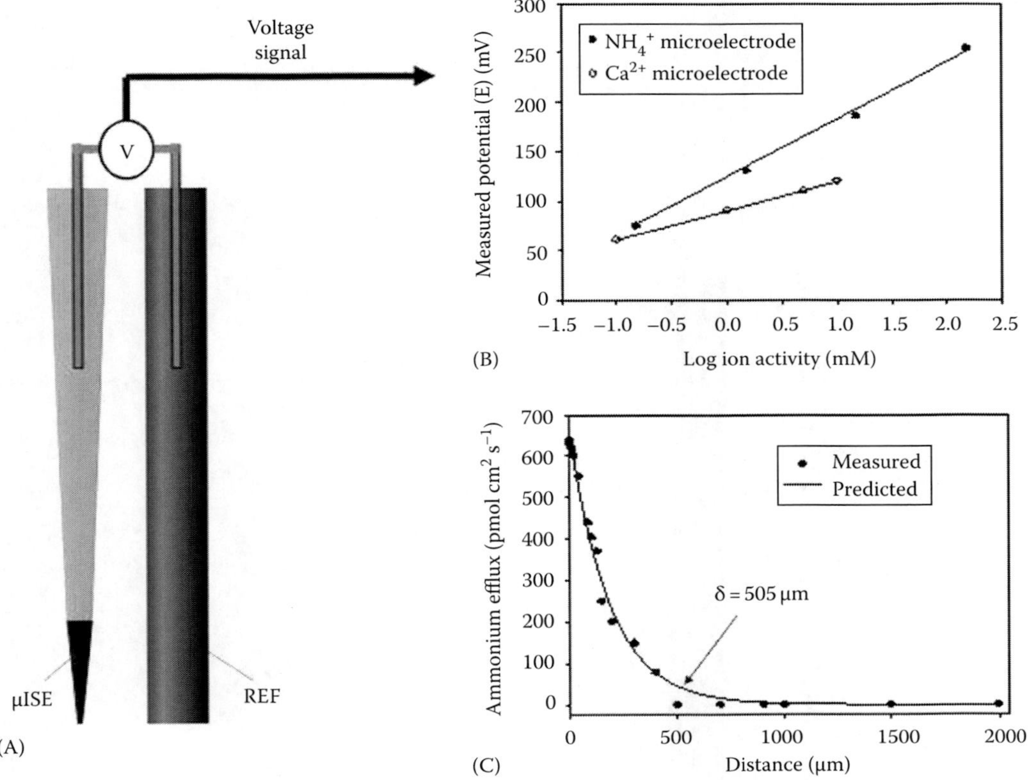

FIGURE 43.4 (A) Schematic of basic transduction scheme for μISE (where REF = reference electrode). (B) Calibration of NH_4^+ and Ca^{2+} electrodes in ¼ MS media at 20°C and (C) quantification of NH_4^+ boundary layer formed near an abiotic point source (1 mM NH_4Cl alginate bead) in ¼ MS media using self-referencing technique. The thickness of the mass boundary layer (δ) was estimated to be 505 μm. The excursion distance (ΔX) was 30 μm, and the microsensor oscillation frequency (F) was 0.33 Hz for all measurements.

B. SR Amperometric Microsensors

Amperometric sensors operate using a three electrode scheme (or a two electrode scheme with a polarizable electrode). Voltage is applied that is equivalent to the oxidation or reduction potential of the target analyte, and electron flow (current) is measured between the target electrode (noted as μSensor in Figure 43.5) and a reference electrode. Combining the first-order Fick equation with the equation for calibration of an amperometric sensor, the equation describing SR amperometric sensors is

$$J = -D\frac{\Delta C}{\Delta X} = -D\frac{\left[\left(i_{(x)} - i_{(x+\Delta x)}\right)/m\right]}{\Delta X} \quad (43.3)$$

where

$i_{(x)}$ is the electrode current near pole (pA)
$i_{(x+\Delta x)}$ is the electrode current at far pole (pA)
m is the linear slope of calibration plot (pA/μM)

C. SR Fiber-Optic Microsensors

Optical microelectrodes (or optrodes) are fiber-optic cables, where the tip has been modified by immobilization of an analyte-sensitive dye (the dye used in the examples mentioned in this chapter is the O_2-quenched luminophore platinum tetrakis (pentafluorophenyl) porphyrin). Most approaches utilize a thermocouple for temperature corrections (Figure 43.6), although recent advancements have allowed the use of optrodes that do not require a thermocouple (Wolfbeis 2004). The luminophore is excited by a short pulse of light through the fiber optic, and luminescence lifetime is measured. Measurement of luminescence lifetime has many advantages over intensity-based methods, including reduced photobleaching due to short excitation periods, little or no source/detector drift, and reduced luminophore degradation (Wolfbeis 2004). These advantages are further enhanced by using frequency-modulated excitation, where oxygen quenching induces phase shifts that can easily be measured using a photomultiplier tube and lock-in amplifier without the use of a reference electrode (Chatni and Porterfield 2009; Chatni et al. 2009). For SR oxygen optrode analysis using phosphorescent dyes, the equation describing ΔC based on phase angle measurements at the near pole and far pole is based on the Stern–Volmer relationship:

$$J = -D\frac{\left[\left(\phi_1 - \phi_2\right)/b\right]}{\Delta X} \quad (43.4)$$

where

ϕ_1 represents phase angle at near pole
ϕ_2 represents phase angle at far pole
b represents slope of calibration plot (0%–21% oxygen)

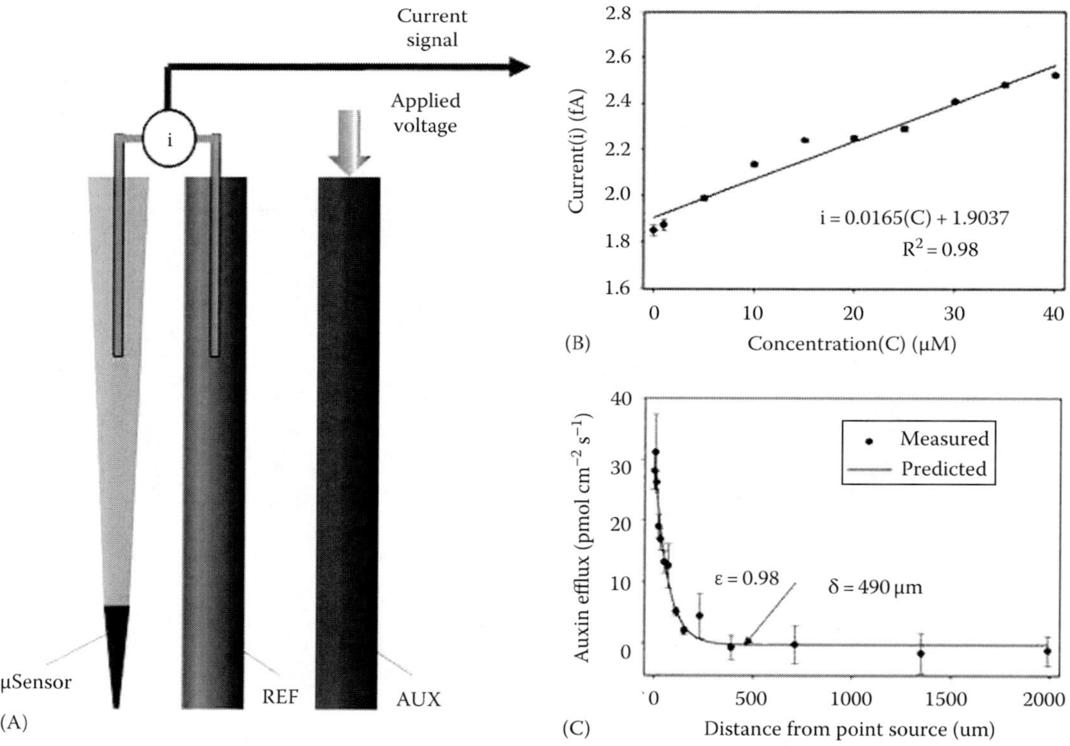

FIGURE 43.5 (A) Basic transduction scheme for amperometric sensors (where REF = reference electrode and AUX = auxiliary electrode). (B) Calibration of amperometric sensor for detecting IAA in ¼ MS media at 20°C and (C) quantification of IAA boundary layer formed near an abiotic point source in ¼ MS media using self-referencing technique with a δ value of 490 μm (ΔX = 20 μm, F = 0.30 Hz).

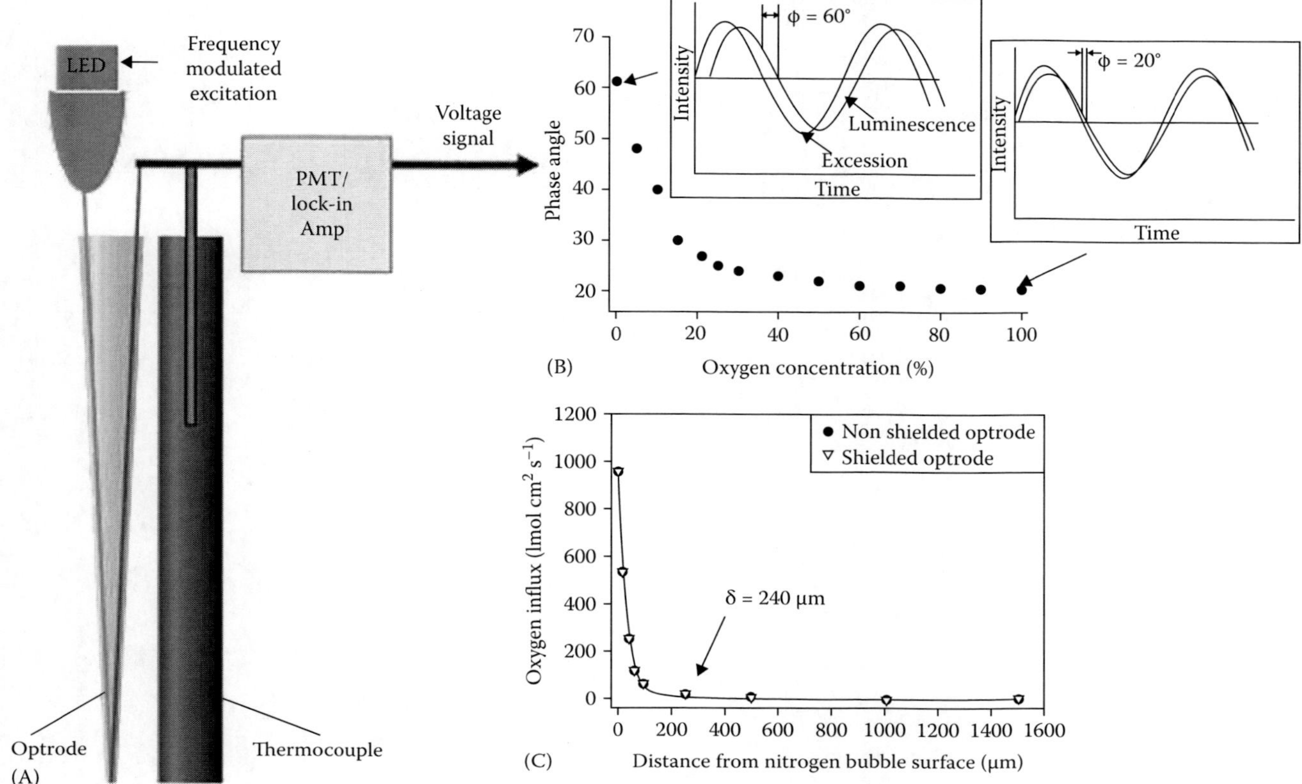

FIGURE 43.6 (A) Transduction scheme for optrodes using an oxygen-quenched luminophore (where PMT = photomultiplier tube). (B) Frequency-modulated calibration of a PtTFPP optrode in ¼ MS media (pH = 7.2) at 20°C and (C) quantification of boundary layer formed near an abiotic O_2 sink ($N_2(g)$ bubble) in ¼ MS media using self-referencing technique. The measured δ value was 240 μm (ΔX = 35 μm, F = 0.33 Hz).

The transport of metabolites and biomolecules at the root–rhizosphere interface is complex. Use of SR microsensors allows us to begin to fill this gap in the knowledge by noninvasively probing real-time flux. The following sections review some recent examples of noninvasive, real-time studies on various plant roots using SR physiological sensors, highlighting ultradian oscillatory transport, biochemical pathway and pharmacology studies, and spatially resolved measurements of form–function relationships for intact roots.

IV. Spatially Resolved Measurements at the Rhizosphere–Root Interface

Quantification of structure–function relationships of roots is vitally important to our understanding of root behavior under physiological and pathophysiological conditions. Most of these studies focus on cellular function within well-known spatial zones of metabolically distinct cells (e.g., meristematic and elongation zones).

A. Oxygen Flux in Developing Roots

Using a SR optrode, oxygen flux along the surface of soybean (*Glycine max* L.), maize (*Zea mays*), and kidney bean (*Phaseolus vulgaris* L.) roots was measured (Figure 43.7). Optrode excursion distance (50 μm) remained perpendicular to the tangent of the root surface, and distance from the root tip was linearized to account for root curvature. Flux values were based on excursion distance (ΔX) from the root surface, as opposed to cylindrical coordinates extrapolating to stele tissues. Kochian et al. (1992) compared root–rhizosphere ion flux using radial and planar coordinate systems and found that analyses using the planar coordinate system were best suited for relatively large roots (such as young maize roots), while cylindrical coordinate systems were best for small roots such as *Arabidopsis*. For all roots (n = 5), the maximum O_2 flux occurred near the elongation zone (McLamore et al. 2010b), where O_2 influx for *G. max* (75 ± 1 pmol cm^{-2} s^{-1}), *P. vulgaris* (106 ± 1 pmol cm^{-2} s^{-1}), and *Z. mays* (117 ± 4 pmol cm^{-2} s^{-1}) was higher than average influx at root tips (38 ± 2 pmol cm^{-2} s^{-1}). These results are consistent with previously reported trends and indicate that the peak in O_2 influx near the elongation zone/meristematic tissue corresponds to high cell growth rate and nutrient/energy demand. All of the measured O_2 flux was due to respiratory activity, as the young seedlings did not have mature aerenchyma or shoots. A characteristic "dip" in O_2 flux between the tip and zone of elongation for *Z. mays* has been observed in numerous studies (Mancoso and Boselli 2002; Porterfield 2002; McLamore et al. 2010b). However, this

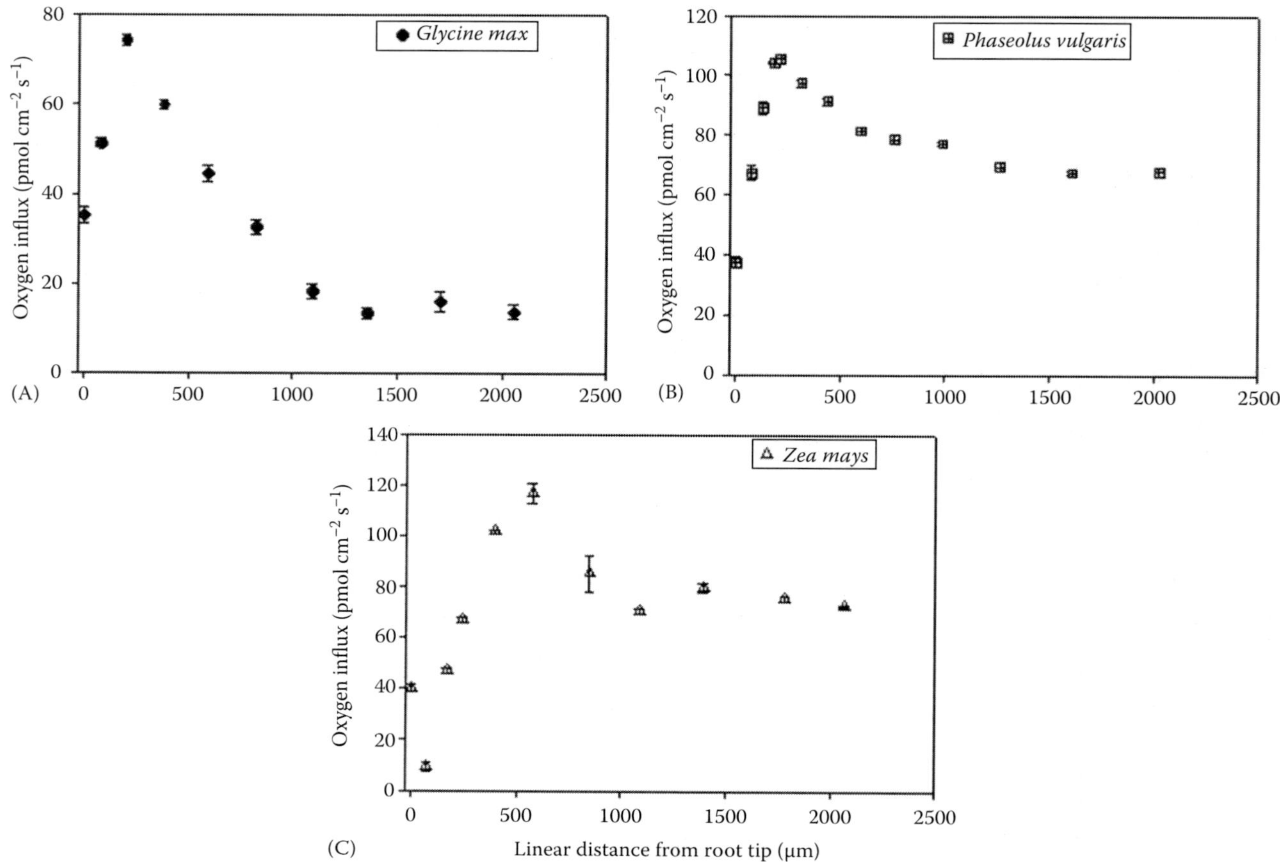

FIGURE 43.7 Rhizosphere oxygen flux profiles at the surface of 7-day-old (A) *G. max*, (B) *P. vulgaris*, and (C) *Z. mays* roots ($n = 5$ roots). For all roots, the maximum oxygen flux was near the meristematic/elongation zone (average root diameter was 820 ± 60 μm ($\Delta X = 50$ μm, $F = 0.25$ Hz). (Reproduced from McLamore, E.S. et al., *Planta*, 232, 1087, 2010b.)

structure–function relationship has not been observed in the two dicots in Figure 43.7, and there are a number of research groups currently investigating the physiological relevance of this respiratory pattern.

B. Induced Auxin Uptake in Developing Maize Roots

Uptake of exogenously applied IAA across the root tissue surface has been attributed to a combination of diffusion and carrier-mediated transport (Martin and Pilet 1986; Bennett et al. 1996; Mancuso et al. 2005). McLamore et al. (2010d) measured induced IAA flux near the surface of 3- to 5-day B73 inbred maize roots using a SR IAA microsensor (Figure 43.8). A tapered parylene-insulated platinum/iridium microelectrode (2–4 μm tip diameter) was modified be electrodepositing amorphous platinum nanoclusters (i.e., Pt black) and subsequently immobilizing a layer of multiwalled carbon nanotubes on top. By polarizing the amperometric electrode at 750 mV, IAA in the rhizosphere can be detected using oxidative amperometry catalyzed by the immobilized nanomaterials. The IAA microsensor was calibrated between 0.1 and 40 μM IAA in ¼ Murashige and Skoog (MS) media (pH = 7.4); the lower limit of detection of the microelectrode was calculated to be 400 nM IAA (McLamore et al. 2010d). Performance of microsensors was unaffected (<3% change in output) by compounds commonly found in plant growth media such as $Ca(NO_3)_2$, NaH_2PO_4, $MgSO_4$, KCl, $CuSO_4$, KH_2PO_4, KNO_3, $MnCl_2$, NaN_3, sucrose, glucose, citrate, oxalate, malate, ascorbate, nitric oxide, glucose, and 2,4-dichlorophenoxy acetic acid, as well as NPA and NOA (Mancuso et al. 2005; McLamore et al. 2010d). In B73, the maximum influx occurred between the root apical meristem and elongation zone (i.e., distal elongation zone). This structure–function relationship was consistent with radiotracer assays of basipetal IAA transport in roots (Jones 1990; Geisler et al. 2005; Peer and Murphy 2007), analyses of root gravitropic bending in mutants and wild type plants (Mullen et al. 1998; Swarup et al. 2005), and subcellular localization studies of auxin transporters (reviewed by Zazimalova et al. 2010).

C. Proton Flux in Developing *Typha latifolia* Roots

Oxygen transport in wetland plants has been studied using a wide variety of techniques. In addition to the oxygen released from the roots, rhizosphere acidification associated with root

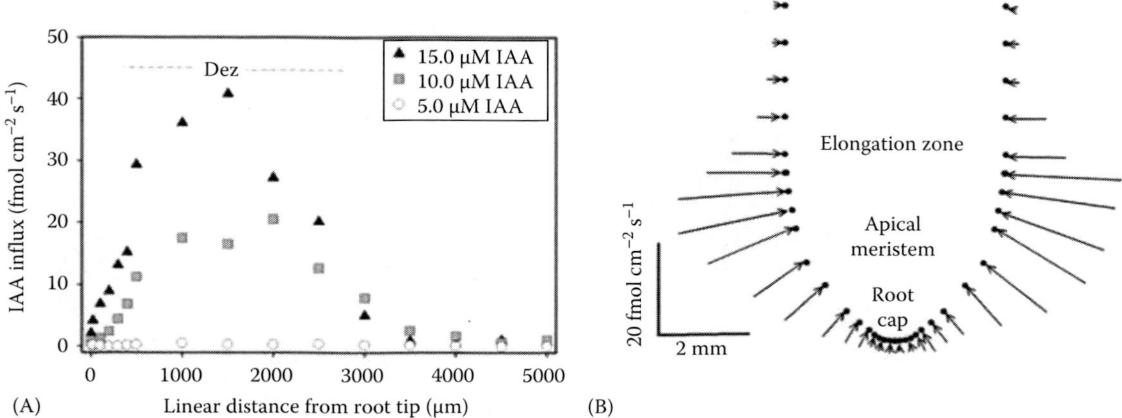

FIGURE 43.8 Measurement of induced IAA influx in maize roots using a SR IAA microsensor. (A) Induced IAA influx for 4-day B73-inbred maize roots in ¼ MS media following external addition of IAA. (B) Representative model of root IAA influx profile following external addition of 15 μM IAA. Arrows indicate the magnitude of average IAA uptake at various locations (graph to the left of figure B shows the scale bar for the model) ($\Delta X = 30$ μm, $F = 0.30$ Hz). (Reproduced from McLamore, E.S. et al., *Plant J.*, 63, 1004, 2010d.)

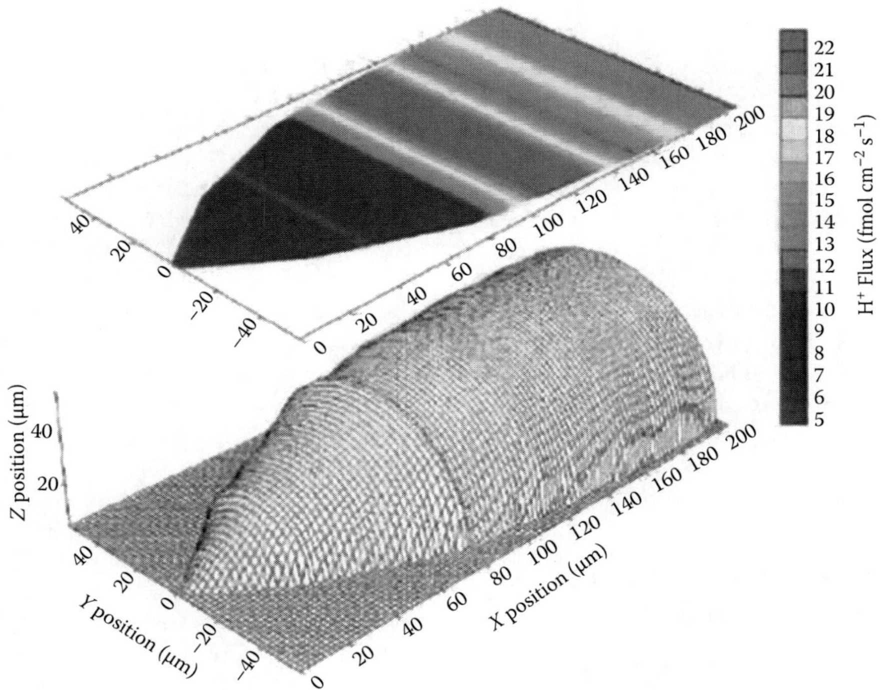

FIGURE 43.9 **(See color insert.)** 3D profile of proton efflux from a *T. latifolia* root. Increased proton flux occurs near the zone of elongation, where cells are rapidly metabolizing and driving polar root elongation ($\Delta X = 10$ μm, $F = 0.30$ Hz). (Reproduced from Porterfield, D.M., Use of microsensors for studying the physiological activity of plant roots, in *Plant Roots: The Hidden Half*, eds. Y. Waisel, A. Eshel, U. Kafkafi, 3rd edn., Marcel Dekker, Inc., New York, pp. 333–347, 2002.)

cell H^+ pumping is an important consideration for understanding the local biogeochemistry. Seasonal fluctuations of redox potential, phytotransport, and acidification in the rhizosphere of wetland plants and crops (e.g., rice) lead to dynamic soil conditions and affect both the plant root viability and the symbiotic microbes on or near roots. As we discussed previously, spatial profiles of proton efflux from *T. latifolia* L. roots in ¼ MS solution were probed (Porterfield 2002). Proton efflux was lowest near the root tip and maximum near the apical meristem/elongation zone approximately 140 μm from the root tip (linearized distance) (Figure 43.9).

D. Nitrogen Flux in Maize Roots

Assimilation of NO_3^- in plant roots requires significantly higher metabolic rates than NH_4^+, as NO_3^- must be reduced to nitrite

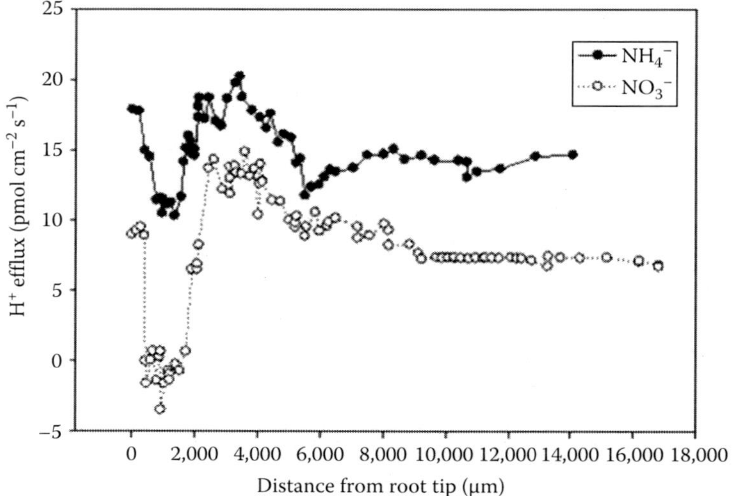

FIGURE 43.10 Proton flux patterns along the roots of *Z. mays* exposed to different forms of nitrogen (NH_4^+ or NO_3^-). Roots supplied with ammonium in the hydroponic solution had significantly more rhizosphere H^+ efflux activity compared to plants provided with nitrate (ΔX = 25 μm, F = 0.30 Hz). (Reproduced from McLamore, E.S. and Porterfield, D.M., *Chem. Soc. Rev.*, 40, 5308, 2011.)

and then to ammonium via nitrate reductase and nitrite reductase, respectively. This metabolic cost requires increased proton pumping, and electrophysiology studies link nitrate and ammonia assimilation with trans-plasma membrane proton pumping. Proton efflux was measured using a SR H^+-selective electrode for *Z. mays* roots using ammonium or nitrate as the sole nitrogen source in hydroponic solutions (Figure 43.10). The ammonium solution contained 200 μM $(NH_4)_2SO_4$ and 100 μM Ca_2SO_4, while the NO_3^- solution contained 200 μM $Ca(NO_3)_2$ and 100 μM Ca_2SO_4. In profiles of root H^+ transport, flux was largest near the elongation zone/apical meristem, although a net influx of protons occurred near the apical meristem of roots grown in NO_3^-. Among all samples tested, roots supplied with ammonium were measured to have between 30% and 100% more rhizosphere H^+ efflux activity as compared to nitrate provided plants.

V. Real-Time Measurements at the Rhizosphere–Root Interface

A. Ultradian Oscillatory Transport in Roots

Use of noninvasive SR microsensors has allowed many researchers to study the oscillatory transport of ions, metabolites, and biomolecules (Shabala et al. 2006). These ultradian oscillations (i.e., regular cycles with a period less than 12–24 h) in root–rhizosphere transport are believed to be due to the dynamic regulation of homeostatic/metabolic transport via membrane pumps and channels. Oscillation of ionic transport (Ca^{2+}, H^+, K^+) has been measured using the SR technique in *Z. mays* roots (Shabala et al. 1997), *Z. mays* protoplast (Shabala and Newman 1998), and pollen tubes (Feijó et al. 2001; Holdaway-Clarke and Hepler 2003). Oxygen oscillations have been measured for *Olea europaea* L. roots (Mancuso et al. 2000) and *Z. mays* roots (Porterfield 2002; Shabala et al. 2006; McLamore et al. 2010b).

Using a SR optrode, oscillatory transport of oxygen at the root tip of 7-day-old *G. max*, *Z. mays*, and *P. vulgaris* had an average period of 3.9 ± 0.3, 4.0 ± 0.3, and 3.3 ± 0.3 min, respectively (n = 5 roots) (Figure 43.11). No significant difference was measured between the legume roots, although the period of oscillation in *Z. mays* roots was significantly shorter than that of legumes (McLamore et al. 2010b). This technology has numerous advantages over SR polarographic microsensors (Mancuso et al. 2000), including no oxygen consumption by the sensor, no measureable sensor hysteresis/drift, facile sensor fabrication, high reproducibility, and excellent selectivity (McLamore et al. 2010). For demonstration purposes, a background measurement was taken 3 mm from the root surface, where no significant O_2 flux was recorded.

Shabala et al. (2006) recently reviewed oscillatory transport in roots, including a proposed model of the biochemical mechanisms regulating ultradian transport. Root–rhizosphere H^+ flux in 3 days barley roots (*Hordeum vulgare* L.) had a regular oscillation period of approximately 1.8 min (Shabala et al. 2006). Increases in rhizosphere salinity (5–100 mM NaCl) correlated with increased oscillation period at an average rate of approximately 82 min μM^{-1} NaCl (Figure 43.12). In addition to the effect of salt stress on ultradian transport, Shabala et al. (2006) also noted that the oscillation period is affected by age, growth rate, temperature, light, and nutrient availability.

In a recent study, ultradian oscillations in IAA root–rhizosphere flux near the distal elongation zone were measured in 4d B73-inbred *Z. mays* roots using a SR IAA microsensor (Figure 43.13) (McLamore et al. 2010d). These oscillations (period of 3.6–5.7 min) are thought to be a direct result of changes in rhizosphere IAA levels that are derived from apoplastic pools, as IAA was recently shown to diffuse into external media in a study by Blakeslee et al. (2007). These studies suggest that changes in apoplastic IAA pools are in turn regulated by uptake transport

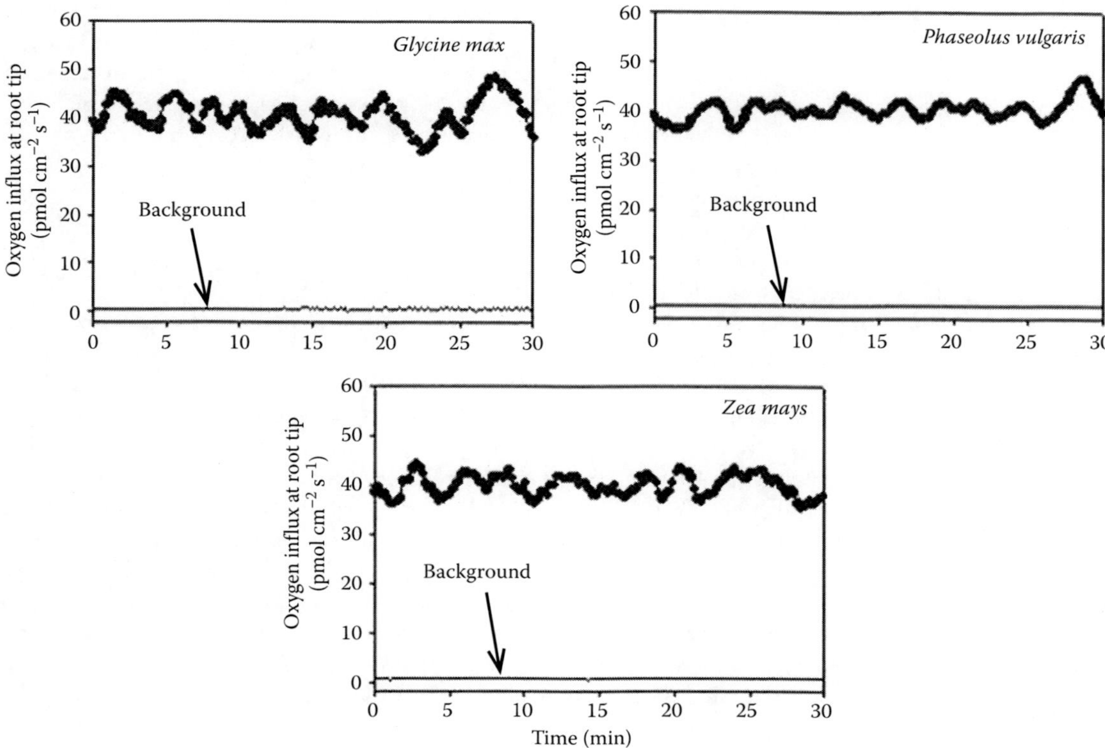

FIGURE 43.11 Oscillation of oxygen uptake measured at the tip of 7-day-old roots of *G. max*, *P. vulgaris*, and *Z. mays* measured using a SR optrode (ΔX = 50 μm, F = 0.25 Hz). (Reproduced from McLamore, E.S. et al., *Planta*, 232, 1087, 2010b.)

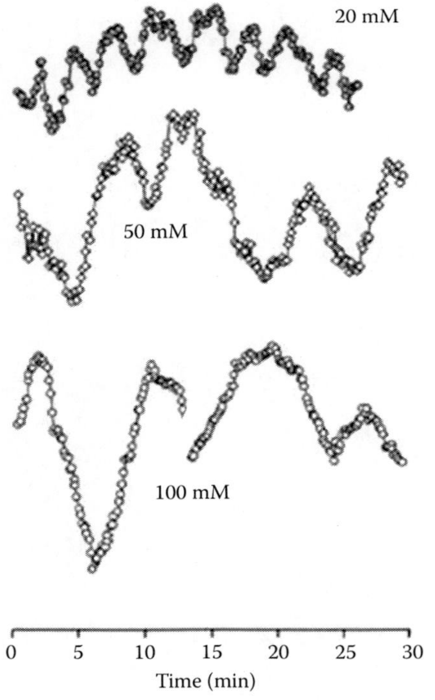

FIGURE 43.12 Root–rhizosphere H^{+} flux in 3 days barley roots (*H. vulgare*) indicated an oscillation period of approximately 1.8 min (Shabala et al. 2006). Values above each trace indicate the supplemental concentration of NaCl in the growth media. (Reproduced from Shabala et al. 2006; McLamore, E.S. et al., *Planta*, 232, 1087, 2010b.)

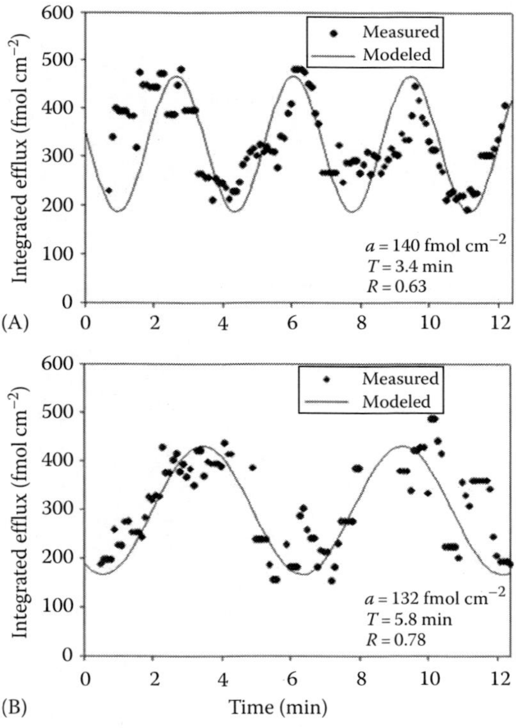

FIGURE 43.13 Oscillation of IAA transport in the DEZ measured using a SR IAA microsensor (ΔX = 20 μm, F = 0.30 Hz). (A) B73 and (B) BR2 *Z. mays* roots. (Reproduced from McLamore, E.S. et al., *Plant J.*, 63, 1004, 2010d.)

proteins (e.g., AUX1) and efflux transporters (e.g., PIN/ABCB). IAA transport was also measured in maize IAA transport mutants (BRACHYTIC2 or BR2), and flux in the DEZ was significantly lower than B73 roots (Blakeslee et al. 2007). Oscillation of IAA flux in BR2 roots (6.3–7.9 min) was significantly longer than B73 roots, and current ongoing studies are investigating the kinetics of IAA transport in the distal elongation zone of inbred B73 and BR2 *Z. mays* roots (McLamore et al. 2010d).

Oscillations in metabolism/ionic flux at root surfaces have been reported numerous times (Porterfield and Smith 2000; Shabala et al. 2006), and models/predictions based on steady-state concentration must be updated to consider these spatially and temporally dynamic variations in cell transport. These oscillations are thought to be driven by diffusion-limited uptake of metabolites, which is a function of the local oxygen concentration, ATP/ADP turnover, and transmembrane H^+ transport (Shabala and Newman 1998; Shabala et al. 1997; Shabala and Knowles 2002; McLamore et al. 2010b,d). Oscillation of respiratory oxygen flux in roots may suggest that observed ionic oscillations are an artifact of metabolic oscillations, although further studies are needed to tease out the relationship between metabolism, molecular diffusion, and ion transport. Shabala et al. (2006) developed a feedback-controlled oscillatory model to describe these oscillations in membrane transport activity and suggested a possible role for this behavior in adaptive response(s) to salinity, temperature, osmotic changes, hypoxia, and pH stresses. Details on the mechanisms driving these oscillations (and those noted in other organisms) are not yet completely understood, though use of noninvasive real-time tools for analyzing these oscillations is improving our understanding of the underlying cell biology.

B. Pharmacological Studies in Roots

During real-time measurement of analyte flux, exposure of roots to pharmacological inhibitors activating or deactivating specific biochemical pathways allows one to understand the temporal dynamics of physiological transport. These pharmacological studies have been used for understanding root–rhizosphere transport of H^+, K^+, Ca^{2+}, O_2, and IAA (Cárdenas et al. 1999; Shabala et al. 2000; McLamore et al. 2010b,d). The ability to noninvasively monitor flux to/from metabolic pathways using real-time techniques adds to our understanding of the complex and dynamic nature of root physiology. During real-time measurement of oxygen transport to roots using a SR optrode for intact *G. max*, *P. vulgaris*, and *Z. mays* roots, addition of 1.5 mM potassium cyanide KCN caused a reduction of oxygen influx at the root tip by 57 ± 1, 56 ± 2, and 53 ± 2%, respectively (Figure 43.14).

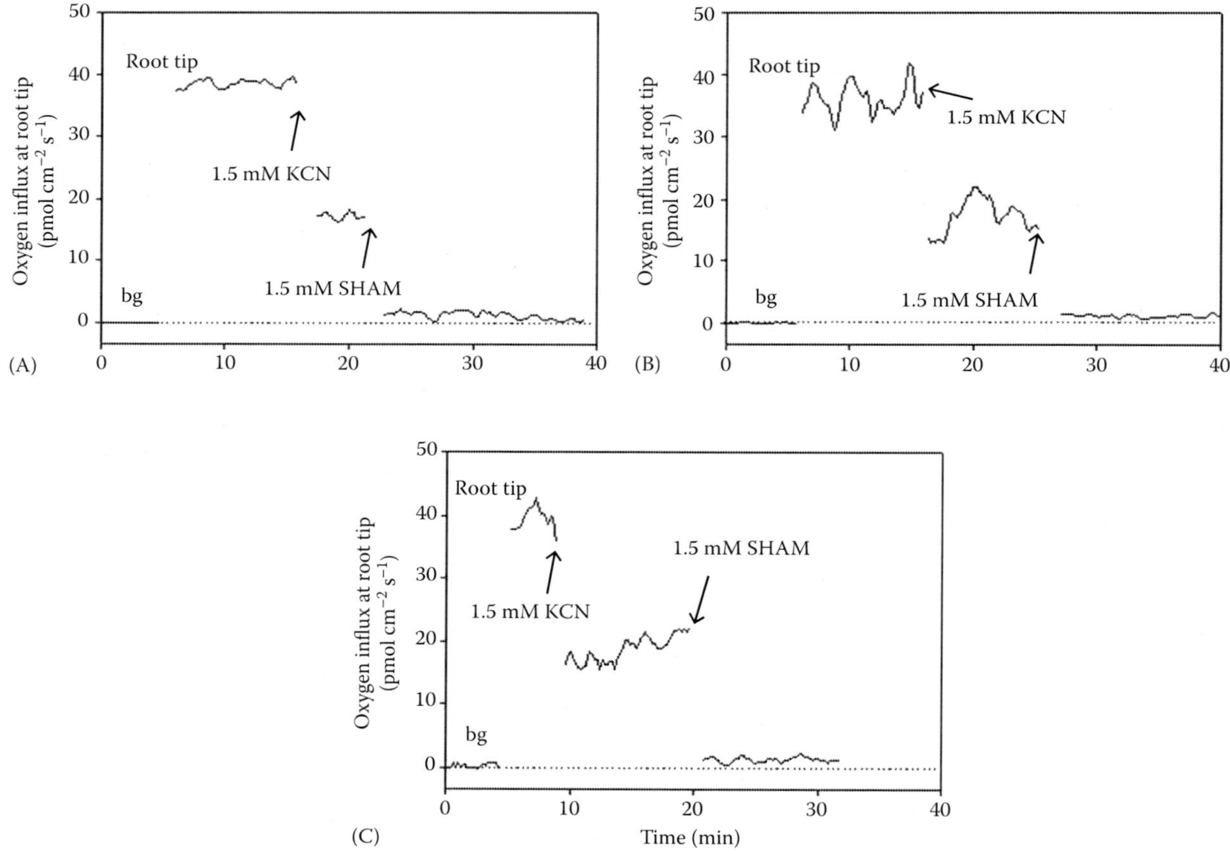

FIGURE 43.14 Real-time oxygen influx during addition of KCN and SHAM for (A) *G. max*, (B) *P. vulgaris*, and (C) *Z. mays*. Background measurements (bg) were taken 3 mm from the root surface (ΔX = 50 µm, F = 0.25 Hz). (Reproduced from McLamore, E.S. et al., *Planta*, 232, 1087, 2010b.)

This reduction in oxygen uptake was due to inhibition of the cytochrome (cyt) pathway and was concentration dependent with maximum inhibition occurring at approximately 1.5 mM KCN (McLamore et al. 2010b). After inhibition of the cyt pathway, subsequent addition of 1.5 mM salicyl hydroxyamic (SHAM, an inhibitor of the alternative oxidase pathway—Aox) reduced oxygen flux by approximately 97% ± 1% from basal levels for all roots measured. This reduction in O_2 flux is expected, as inhibition of both the cyt and alternative oxidase pathways reduces all metabolic electron transport. This pharmacological experiment was designed to observe "spillover" O_2 flux associated with the Aox pathway (by first inhibiting the cyt pathway), although the experiment could also be conducted by adding SHAM first, followed by KCN for observing specific electron and oxygen flux due to the Aox pathway. Following inhibition of the cyt pathway, the period and amplitude of oxygen oscillations was significantly reduced for all roots (McLamore et al. 2010b). Inhibition of the Aox pathway caused a subsequent decrease in period and amplitude of respiratory oscillations at root tips. Following addition of inhibitors, oscillations were stable for a minimum of 30 min (minimum of 12 oscillations recorded for each root). This alteration of oscillatory behavior is consistent with predictions of transport in oscillating feedback systems (Gradmann and Slayman 1975; Shabala et al. 2006) and is indicative of a reestablishment of membrane gradient transport.

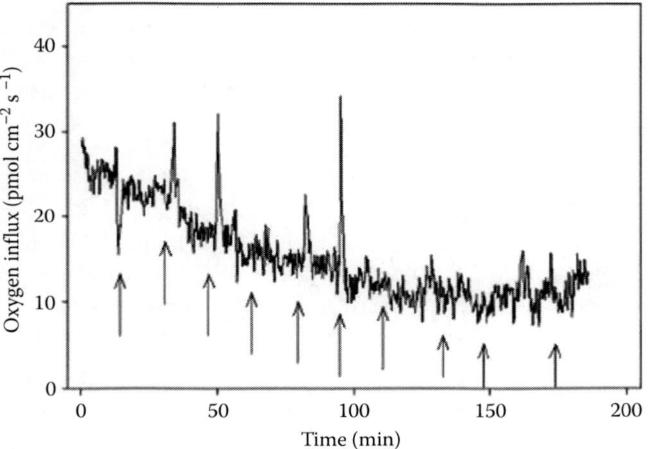

FIGURE 43.15 Real-time oxygen influx for a 4 days B73 *Z. mays* root in ¼ MS medium following increasing external addition of IAA; arrows indicate external addition of IAA in increments of 1.5 μM (ΔX = 50 μm, F = 0.25 Hz). (Reproduced from McLamore, E.S. et al., *Plant J.*, 63, 1004, 2010d.)

To study saturation of transport proteins in B73-inbred maize roots, the growth phytohormone IAA is commonly added to growth media (Mancuso et al. 2005), affecting uptake across the root tissue surface due to a combination of diffusion and carrier-mediated transport. While this is a useful technique for studying root–rhizosphere transport, a drawback of the addition of external IAA for experimental purposes in the 5–15 μM range is inhibition of root elongation (McLamore et al. 2010d). Using a SR optrode, average O_2 influx at the distal elongation zone of B73 roots was 26.1 ± 2.3 pmol cm^{-2} s^{-1}. Addition of external IAA significantly decreased oxygen flux, and concentrations as low as 1.5 μM-IAA caused a significant reduction in root oxygen flux to root meristematic tissue within 10 min (Figure 43.15), which was confirmed with trypan blue and carboxy-H2DCFDA staining (Geisler et al. 2005).

In a related study, IAA flux was measured near the distal elongation zone using a SR IAA microsensor (McLamore et al. 2010d), and auxin efflux and influx transport proteins were inhibited using 1-*N*-naphthylphthalamic acid (NPA) and 2-naphthoxyacetic acid (NOA), respectively. Addition of 2.5 μM NPA reduced IAA efflux to undetectable levels within approximately 30 min, and in addition, influx of IAA was reduced by 25% ± 3%. This indicates that apoplast auxin pools were depleted due to inhibition by NPA (Figure 43.16) (McLamore et al. 2010d). In a parallel experiment, auxin uptake was inhibited in B73 roots using 2.5 μM NOA after approximately 20 min. Over the time scale of this experiment, efflux increased by an average of 170% ± 16% after a relatively short period (approximately

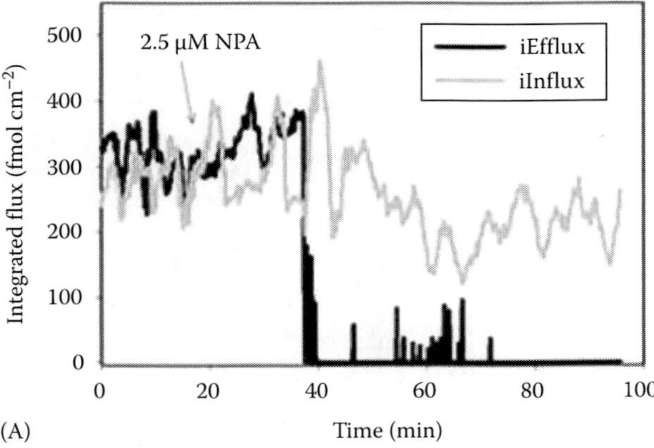

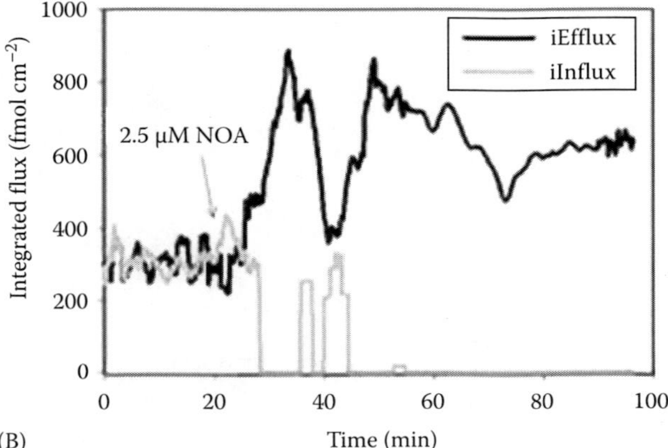

FIGURE 43.16 Endogenous root–rhizosphere IAA flux in root meristematic tissue of 4 days B73-inbred *Z. mays* roots. (A) Addition of 2.5 μM NPA caused a decrease in IAA efflux. (B) Addition of 2.5 μM NOA caused influx to diminish to approximately zero, while efflux from the apoplastic pools increased (ΔX = 30 μm, F = 0.30 Hz). (Reproduced from McLamore, E.S. et al., *Plant J.*, 63, 1004, 2010d.)

30 min), validating previous data implicating NOA as a carrier influx inhibitor (Parry et al. 2001; Rahman et al. 2002).

SR microsensors are a very useful tool for dissecting relative ratios of respiratory and/or photosynthetic oxygen flux in roots. Uptake and/or release can be monitored during oxidative phosphorylation, the alternative oxidase pathway, and many other pathways using real-time pharmacological inhibition studies. A major advantage of this technique is the noninvasive nature of the measurements. Using this approach, we are able to perform the real-time measurements while leaving the root(s) intact for subsequent analysis (e.g., real-time qPCR and fluorescent staining) or continued growth monitoring.

VI. Conclusions

A summary of some manuscripts utilizing SR to describe physiological processes in roots is presented in Table 43.2. These efforts are categorized into studies utilizing probes for measuring net ion flux (i.e., biocurrent), transport of specific ions, flux of biological molecules such as IAA, and root oxygen flux. While most efforts have focused on ions and oxygen, recent development in sensors for measuring other important analytes will no doubt expand this list considerably, and we are already seeing the application of these tools (e.g., enzyme-based microbiosensors) for measuring compounds such as nitric oxide, NADH, glucose, ethanol, and ATP in other research applications (Porterfield 2007; McLamore et al. 2010a,c, reviewed by McLamore and Porterfield 2011). Inclusion of catalytic materials such as nanomaterials as active components within microsensors has greatly enhanced signal acquisition (Wang 2005; Bonaccorso et al. 2010), and these developments have had a significant impact on the field of self-referencing (McLamore and Porterfield 2011). We look forward to the next body of experimental data using these novel tools in the study of root physiology.

There are numerous excellent reviews highlighting the concerted efforts of various groups using SR sensors for studying pollen tube physiology (Heper et al. 2006), ion transporters in roots (Newman 2001), and oscillatory transport at the root–rhizosphere interface (Shabala et al. 2006). We anticipate that current and ongoing research in root physiology will benefit greatly from the self-referencing technique, allowing us to probe the hidden half, utilizing novel microsensors for detecting molecules such as citrate, malate, ATP, and ethanol. In particular, the noninvasive nature of the measurements and lack of externally applied reagents allow experimentation for a variety of hypothesis-driven questions ranging from basic metabolism to growth, development, and stress adaptation. These new tools will allow root physiologists to tackle current issues such as climate change, bioenergy production, and maintaining a safe and productive supply of food and water.

VII. Future Directions

The last few decades has seen an explosion in the development of sensor technologies for life sciences research, and we expect these new and exciting micro-/nanoscale technologies will allow root physiologists to probe and interrogate cells and tissues in

TABLE 43.2 Summary of Some Manuscripts Utilizing SR to Study Root Physiology

	Analyte(s)	Specimen(s)	References
Current	Net ionic current	Maize (*Zea mays*) roots	Bjorkman et al. (1985); Bjorkman and Leopold (1987a); Bjorkman and Leopold (1987b)
Ions	Ca^{2+}	Wheat (*Triticum aestivum*) roots	Huang et al. (1992)
	Ca^{2+}	*Arabidopsis thaliana* root hair cells	Schiefelbein et al. (1992)
	Ca^{2+}, H^+	Alfalfa (*Medicago sativa*) root hair	Bennett et al. (1994)
	Ca^{2+}	Corn poppy (*Papaver rhoeas*) roots	Franklin-Tong et al. (2002)
	Ca^{2+}	Bean (*Phaseolus vulgaris*) root hairs	Cardenas et al. (1999)
	K^+, Ca^{2+}	Wheat (*Triticum aestivum*) root protoplasts	Gilliham et al. (2006)
	NH_4^+, NO_3^-	Barley (*Hordeum vulgare*) roots	Henriksen et al. (1992)
	K^+, H^+, Ca^{2+}	Maize (*Zea mays*) roots	Kochian et al. (1992)
	H^+, Ca+	Maize (*Zea mays*) roots	Shabala (1997)
	K^+, H^+, Ca^{2+}	Maize (*Zea mays*) root protopolasts	Shabala (1998)
	H^+, Ca^{2+}	Maize (*Zea mays*) roots	Shabala and Newman (1997)
	H^+, K^+	Maize (*Zea mays*) roots	Newman et al. (1987)
	Cd^{2+}	Wheat (*Triticum aestivum*) roots	Pineros et al. (1998)
Biomolecules	IAA	Maize (*Zea mays*) roots	Mancuso et al. (2005)
	IAA	Maize (*Zea mays*) roots	McLamore et al. (2010b)
Oxygen	O_2	Olive tree (*Olea europaea*) roots	Mancuso et al. (2000)
	O_2	Grape vine (*Vitis*) roots	Mancuso and Bosellil (2002)
	O_2	Soybean (Glycine max), maize (Zea mays), kidney bean (*Phaseolus vulgaris*) roots	McLamore et al. (2010a)
	O_2	Maize (*Zea mays*) roots	Porterfield (2002); Shabala (2006)

ways which were not previously possible. As discussed in this chapter, many researchers are aiming both to develop new sensors and use these tools in novel modalities for measuring root physiology. Although field applications of the SR microsensor technique have not yet been demonstrated, the lessons learned from noninvasively probing the functional realm of roots have been invaluable. As research in the life sciences becomes more interdisciplinary, the separation between science, engineering, and technology becomes nonexistent, allowing researchers to design tools based on specific hypothesis-driven research questions, rather than formulating hypotheses within the confines of available technology. In the next decade, we expect continued use of existing SR microsensors described here, and development of new SR microsesors (e.g., malate, citrate) will allow us to couple real-time measurements of physiological analyte flux with genomic analysis using techniques such as real-time quantitative polymerase chain reaction. This experimentation capability will allow us to begin to bridge the gap between the genome and the physiome, coupling real-time root–rhizosphere flux to the central dogma.

Acknowledgments

The authors wish to thank the many funding sources for this work, including the National Science Foundation, Office of Naval Research, Environmental Protection Agency, Purdue University, and the University of Florida. We would also like to thank Al Shipley (Applicable Electronics Inc.) and Eric Karplus (Science Wares Inc.) for their ongoing and continued technical support and innovation.

References

Alderman J, Hynes J, Floyd SM, Kruger J, O'Connor R, Papkovsky DB 2004 A low-volume platform for cell-respirometric screening based on quenched-luminescence oxygen sensing. *Biosens Bioelectron* 19:1529–1535.

Armstrong W, Webb T, Darwent M, Beckett PM 2009 Measuring and interpreting respiratory critical oxygen pressures in roots. *Ann Bot* 103:281–293.

Bennett ANS, Cox DN, Shipley A, Ehrhardt DW, Long SR 1994 Effects of NOD factors on alfalfa root hair Ca^{++} and H^+ currents and on cytoskeletal behavior. In *Advances in Molecular Genetics of Plant–Microbe Interactions*, eds. MJ Daniels, JA Donnie, AE Osbourn, Vol. 3, pp. 107–113. Dordrecht, the Netherlands: Kluwer.

Bennett MJ, Marchant A, Green HG et al. 1996 *Arabidopsis* AUX1 gene: A permease-like regulator of root gravitropism. *Science* 273:948–950.

Bjorkman T, Leopold, AC 1987a An electric current associated with gravity sensing in maize roots. *Plant Physiol* 84:841–846.

Bjorkman, T, Leopold, AC 1987b Effects of inhibitors on auxin transport and of calmodulin on a gravisensing-dependent current in maize roots. *Plant Physiol* 84:847–850.

Bjorkman T, Leopold AC, Scheffey C, Jaffe LF 1985 Gravistimulation-induced changes in current patterns around corn root caps. *Physiologist* 28:297.

Blakeslee JJ, Bandyopadhyay A, Lee OR et al. 2007 Interactions among PINFORMED (PIN) and P-glycoprotein (PGP) auxin transporters in *Arabidopsis thaliana*. *Plant Cell* 19:131–147.

Bonaccorso F, Sun Z, Hasan T, Ferrari AC 2010 Graphene photonics and optoelectronics. *Nat Photon* 4:611–622.

Borisov SM, Wolfbeis OS 2008 Optical biosensors. *Chem Rev* 108:423–461.

Buerk DG 2004 Oxygen sensing. *Methods Enzymol* 381:665–690.

Cárdenas L, Feijó JA, Kunkel JG et al. 1999 *Rhizobium* nod factors induce increases in intracellular free calcium and extracellular calcium influxes in bean root hairs. *Plant J* 19:347–352.

Chatni MR, Maier DE, Porterfield DM 2009 Evaluation of microparticle materials for enhancing the performance of fluorescent lifetime based optrodes. *Sens Actuat. B* 141:471–477.

Chatni MR, Porterfield DM 2009 Self-referencing optrode technology for non-invasive real-time measurement of biophysical flux and physiological sensing. *Analyst* 134:2224–2232.

Chen R, Farmery AD, Obeid AN, Hahn CEW 2009 Plastic fibre optic oxygen sensors based on a polymer matrix doped with Pt (II) complexes. *Proc SPIE*. 7503:1–4.

Clark LC, Wolf R, Granger D, Taylor Z 1953 Continuous recording of blood oxygen tensions by polarography. *J Appl Physiol* 6:189–193.

Cloutier M, Chen J, Tatge F, McMurray-Beaulieu V, Perriera, Jolicoeur M 2009 Kinetic metabolic modelling for the control of plant cells cytoplasmic phosphate. *J Theor Biol* 259:118–131.

De Dobbeleer C, Cloutier M, Fouilland M, Legros R, Jolicoeur MA 2006 High-rate perfusion bioreactor for plant cells. *Biotechnol Bioeng* 95:1126–1137.

Feijó JA, Sainhas J, Holdaway-Clarke T, Cordeiro MS, Kunkel JG, Hepler PK 2001 Cellular oscillations and the regulation of growth: The pollen tube paradigm. *Bioassays* 23:86–94.

Felle HH 1994 The H+/Cl– symporter in root-hair cells of *Sinapis alba*. An electrophysiological study using ion-selective microelectrodes. *Plant Physiol* 106:1131–1136.

Felle HH 1998 The apoplastic pH of the *Zea mays* root cortex as measured with pH-sensitive microelectrodes: aspects of regulation. *J Exp Bot* 49:987–995.

Felle HH, Kondorosi E, Kondorosi A, Schultze M 1999a Elevation of the cytosolic free (Ca^{2+}) is indispensable for the transduction of the Nod factor signal in alfalfa. *Plant Physiol* 121:273–279.

Felle HH, Kondorosi E, Kondorosi A, Schultze M 1999b Nod factors modulate the concentration of cytosolic free calcium differently in growing and non-growing root hairs of *Medicago sativa* L. *Planta* 209:207–212.

Franklin-Tong VE, Holdaway-Clarke TL, Straatman KR, Kunkel JG, Hepler PK 2002 Involvement of extracellular calcium influx in the self-incompatibility response of *Papaver rhoeas*. *Plant J* 29:333–345.

Geisler M, Blakeslee JJ, Bouchard R 2005 Cellular efflux of auxin catalyzed by the *Arabidopsis* MDR/PGP transporter AtPGP1. *Plant J* 44:179–194.

Gilliham M, Sullivan W, Tester M, Tyerman SD 2006 Simultaneous flux and current measurement from single plant protoplasts reveals a strong link between K^+ fluxes and current, but no link between Ca^{2+} fluxes and current. *Plant J* 46:134–144.

Gnaiger E, Mendez G, Hand SC 2000 High phosphorylation efficiency and depression of uncoupled respiration in mitochondria under hypoxia. *Proc Natl Acad Sci USA* 97:11080–11085.

Gradmann D, Slayman CL 1975 Oscillations of an electrogenic pump in the plasma membrane of *Neurospora. J Membrane Biol* 23:181–212.

Gupta KJ, Zabalza A, van Dongen JT 2009 Regulation of respiration when the oxygen availability changes. *Physiol Plant* 137:383–391.

Henriksen GH, Raman DR, Walker LP, Spanswick RM 1992 Measurement of net fluxes of ammonium and nitrate at the surface of barley roots using ion-selective microelectrodes: Patterns of uptake along the root axis and evaluation of the microelectrode flux estimation technique. *Plant Physiol* 99:734–747.

Herrmann A, Felle HH 1995 Tip growth in root hair cells of *Sinapis alba* L.: Significance of internal and external Ca^{2+} and pH. *New Phytol* 129:523–533.

Holdaway-Clarke TL, Hepler PK 2003 Control of pollen tube growth: Role of ion gradients and fluxes. *New Phytol* 159:539–563.

Huang JW, Grunes DL, Kochian LV 1992 Aluminum effects on the kinetics of calcium uptake into cells of the wheat root apex. *Planta* 188:414–421.

Jaffe LF, Nuccitelli R 1974 An ultrasensitive vibrating probe for measuring steady extracellular currents. *J Cell Biol* 63:614–628.

Jones AM 1990 Location of transported auxin in etiolated maize shoots using 5-azidoindoleacetic acid. *Plant Physiol* 93:1154–1161.

Kochian LV, Shaff JE, Kühtreiber WM, Jaffe LF, Lucas WJ 1992 Use of an extracellular, ion-selective, vibrating microelectrode system for the quantification of K^+, H^+, and Ca^{2+} fluxes in maize roots and maize suspension cells. *Planta* 188:601–610.

Kühl M, Jørgensen BB 1992 Spectral light measurements in microbenthic phototrophic communities with a fiber-optic microprobe coupled to a sensitive diode array detector. *Limnol Ocean* 37:1813–1823.

Kunkel JG, Cordeiro S, Xu Y, Shipley AM, Feijó JA 2006 Electrochemical sensor applications to the study of molecular physiology and analyte flux in plants. In *Plant Electrophysiology: Theory and Methods*, eds. AG Volkov, pp. 109–137. Berlin, Germany: Springer.

Kupper H, Setlik I, Hlasek M 2004 A versatile chamber for simultaneous measurements of oxygen exchange and fluorescence in filamentous and thallous algae as well as higher plants. *Photosynthetica* 42:579–583.

Lamboursain L, St-Onge F, Jolicoeur M 2002 A lab-built respirometer for plant and animal cell culture. *Biotechnol Progr* 18:1377–1386.

Land SC, Portefield DM, Sanger RH, Smith PJS 1999 The self-referencing oxygen-selective microelectrode: Detection of transmembrane oxygen flux from single cells. *J Exp Biol* 202:211–218.

Lee SK, Okura I 1997 Photostable optical oxygen sensing material: Platinum tetrakis (pentafluorophenyl) porphyrin immobilized in polystyrene. *Anal Commun* 34:185–188.

Ludwig-Müller J, Kaldorf M, Sutter EG, Epsteind E 1997 Indole-3-butyric acid (IBA) is enhanced in young maize (*Zea mays* L.) roots colonized with the arbuscular mycorrhizal fungus *Glomus intraradices. Plant Sci* 125:153–162.

Maathuis FJM, Sanders D 1993 Energization of potassium uptake in *Arabidopsis thaliana. Planta* 191:302–307.

Mancuso S, Boselli M 2002 Characterisation of the oxygen fluxes in the division, elongation and mature zone of *Vitis* roots: Influence of oxygen availability. *Planta* 214:767–774.

Mancuso S, Marras AM, Magnus V, Baluska F 2005 Noninvasive and continuous recordings of auxin fluxes in intact root apex with a carbon nanotube-modified and self-referencing microelectrode. *Anal Biochem* 341:344–351.

Mancuso S, Papeschi G, Marras AM 2000 A polarographic, oxygen-selective, vibrating-microelectrode system for the spatial and temporal characterization of transmembrane oxygen fluxes in plants. *Planta* 211:384–389.

Martin HV, Pilet PE 1986 Saturable uptake of indol-3yl-acetic acid by maize roots. *Plant Physiol* 81:889–895.

McEvoy AK, Von Bültzingslöwen C, McDonagh C, MacCraith BD, Klimant I, Wolfbeis OS 2003 Optical sensors for application in intelligent food packaging technology. *Proc SPIE* 4876:806–815.

McLamore ES, Diggs A, Calvo Marzal P, Shi J, Claussen J, Murphy A, Porterfield DM 2010d Non-invasive quantification of endogenous root auxin transport using an integrated flux microsensor technique. *Plant J* 63:1004–1016.

McLamore ES, Jaroch D, Chatni MR, Porterfield DM 2010b Self-referencing optrodes for measuring real time oxygen flux in plant roots and photosynthetic microbial mats. *Planta* 232:1087–1099.

McLamore ES, Mohanty S, Shi J, Rickus JL, Porterfield DM 2010c Real time neuronal glutamate flux during potassium stimulation. *J Neurosci Meth* 189:14–22.

McLamore ES, Porterfield DM 2011 Non-invasive tools for measuring metabolism and biophysical analyte transport: Self-referencing physiological sensing. *Chem Soc Rev* 40:5308–5320.

McLamore ES, Porterfield DM, Banks MK 2009 Non-invasive self-referencing electrochemical sensors for quantifying real time biofilm analyte flux. *Biotechnol Bioeng* 102:791–799.

McLamore ES, Shi J, Jaroch D et al. 2011 A self referencing enzyme-based microbiosensor for real time measurement of physiological glucose transport. *Biosens Bioelectron* 26:2237–2245.

McLamore ES, Zhang W, Porterfield DM, Banks MK 2010a Real-time, non-invasive biofilm physiology during chemical toxin exposure. *Environ Sci Technol* 44:7050–7057.

Miller HL, Dunton KH 2007 Stable isotope (C-13) and O_2 micro-optode alternatives for measuring photosynthesis in seaweeds. *Mar Ecol Prog Ser* 329:85–97.

Mullen JL, Ishikawa H, Evans ML 1998 Analysis of changes in relative elemental growth rate patterns in the elongation zone of *Arabidopsis* roots upon gravistimulation. *Planta* 206:598–603.

Newman IA 2001 Ion transport in roots: Measurement of fluxes using ion-selective microelectrodes to characterize transporter function. *Plant Cell Environ* 24:1–14.

Newman IA, Kochian LV, Grusak MA, Lucas WJ 1987 Fluxes of H^+ and K^+ in corn roots: Characterization and stoichiometries using ion selective electrodes. *Plant Physiol* 84:1177–1184.

Newman I, Chen SL, Porterfield DM, Sun J 2012 Non-invasive flux measurements using microsensors: Theory, limitations, and systems. *Methods Mol Biol* 913:101–117.

O'Riordan TC, Buckley D, Ogurtsov VI, O'Connor R, Papkovsky DB 2000 A cell viability assay based on monitoring respiration by optical oxygen sensing. *Anal Biochem* 278:221.

Parry G, Delbarre A, Marchant A, Swarup R, Napier R, Perrot-Rechenmann C, Bennett MJ 2001 Novel auxin transport inhibitors phenocopy the auxin influx carrier mutation aux1. *Plant J* 25:399–406.

Peer WA, Murphy AS 2007 Flavonoids and auxin transport: Modulators or regulators? *Trends Plant Sci* 12:556–563.

Peters W, Felle HH 1999 The correlation of profiles of surface pH and elongation growth in maize roots. *Plant Physiol* 121:905–912.

Pineros MA, Shaff JE, Kochian LV 1998 Development, characterization, and application of a cadmium-selective microelectrode for the measurement of cadmium fluxes in roots of *Thlaspi* species and wheat. *Plant Physiol* 116:1393–1401.

Porterfield DM 2002 Use of microsensors for studying the physiological activity of plant roots. In *Plant Roots: The Hidden Half*, ed. Y Waisel, A Eshel, U Kafkafi, 3rd edn., pp. 333–347. New York: Marcel Dekker, Inc.

Porterfield DM 2007 Measuring metabolism, biophysical flux in the tissue, cellular, sub-cellular domains: Recent developments in self-referencing amperometry for physiological sensing. *Biosens Bioelectron* 22:1186–1196.

Porterfield DM, Kuang A, Smith PJS, Crispi ML, Musgrave ME 1999 Oxygen-depleted zones inside reproductive structures of Brassicaceae: Implications for oxygen control of seed development. *Can J Bot* 77:1439–1446.

Porterfield DM, McLamore ES, Banks MK 2009. Microsensor technology for measuring H^+ flux in buffered media. *Sens Actuat B* 136:383–387.

Porterfield DM, Rickus JL Kopelman R 2006 Non-invasive approaches to measuring respiratory patterns using a PtTFPP based, phase-lifetime, self-referencing oxygen optrode. *Proc SPIE: Smart Med Biomed Sensor Technol* 6380:1–8.

Porterfield DM, Smith PJS 2000 Single-cell, real-time measurements of extracellular oxygen, proton fluxes from *Spirogyra grevilleana. Protoplasma* 212:80–88.

Prado AM, Porterfield DM, Feijó JA 2004 Nitric oxide is involved in growth regulation, re-orientation of pollen tubes. *Development* 131:2707–2714.

Rahman A, Hosokawa S, Oono Y, Amakawa T, Goto N, Tsurumi S 2002 Auxin and ethylene response interactions during Arabidopsis root hair development dissected by auxin influx modulators. *Plant Physiol* 130:1–10.

Roche PJR, Cheung MCK, Yao L, Kirk AG, Chodavarapu VP 2010 Enhancement of luminescent quenching based oxygen sensing by gold nanoparticles: Comparison between luminophore: matrix: nanoparticle thin films on glass, gold coated substrates. *J Nanophoton* 4:043521.

Salmi ML, Morris KE, Roux SJ, Porterfield DM 2007 Nitric oxide, cGMP signaling in calcium-dependent development of cell polarity in *Ceratopteris richardii. Plant Physiol* 144:94–104.

Sanchez B, Ochoa-Acuna H, Porterfield DM, Sepulveda MS 2008 Oxygen flux as an indicator of physiological stress in fathead minnow (*Pimephales promelas*) embryos: A real-time biomonitoring system of water quality. *Environ Sci Technol* 42:7010–7017.

Schiefelbein JW, Shipley A, Rowse P 1992 Calcium influx at the tip of growing root-hair cells of *Arabidopsis thaliana. Planta* 187:455–459.

Serrano M, Robatzek S, Torres M et al. 2007 Chemical interference of pathogen-associated molecular pattern-triggered immune responses in *Arabidopsis* reveals a potential role for fatty-acid synthase type II complex-derived lipid signals. *J Biol Chem* 282:6803–6811.

Shabala S, Babourina O, Newman I 2000 Ion-specific mechanisms of osmoregulation in bean mesophyll cells. *J Exp Bot* 151:1243–1253.

Shabala SN, Newman IA 1998 Osmotic sensitivity of Ca^{2+}, H^+ transporters in corn roots: Effect on fluxes, their oscillations in the elongation region. *J Membr Biol* 161:45–54.

Shabala SN, Newman IA, Morris J 1997 Oscillations in H^+, Ca^{2+} ion fluxes around the elongation region of corn roots, effects of external pH. *Plant Physiol* 113:111–118.

Shabala L, Ross T, McMeekin T, Shabala S 2006 Non-invasive microelectrode ion flux measurements to study adaptive responses of microorganisms to the environment. *FEMS Microbiol Rev* 30:472–486.

Shabala S, Knowles A 2002 Rhythmic patterns of nutrient acquisition by wheat roots. *Funct Plant Biol* 29:595–605.

Shi J, Cha T-G, Claussen JC, Diggs AR, Choi JH, Porterfield DM 2011c Microbiosensors based on DNA modified single-walled carbon nanotube, Pt black nanocomposites. *Analyst* 136:4916–4924.

Shi J, Claussen JC, McLamore ES et al. 2011b A comparative study of enzyme immobilization strategies for multi-walled carbon nanotube glucose biosensors. *Nanotechnology* 22:355502.

Shi J, McLamore ES, Jaroch D, Claussen JC, Rickus JL, Porterfield DM 2011a Oscillatory glucose flux in INS 1 pancreatic β cells: A self-referencing microbiosensor study. *J Anal Biochem* 411:185–193.

Swarup R, Kramer EM, Perry P et al. 2005 Root gravitropism requires lateral root cap, epidermal cells for transport, response to a mobile auxin signal. *Nat Cell Biol* 7:1057–1065.

Verslues PE, Ober ES, Sharp RE 1998 Root growth, oxygen relations at low water potentials. Impact of oxygen availability in polyethylene glycol solutions. *Plant Physiol* 116:1403–1412.

Wang J 2005 Carbon-nanotube based electrochemical biosensors: A review. *Electroanalysis* 17:7–14.

Wolfbeis OS 2004 Fiber optic chemical sensors, biosensors. *Anal Chem* 76:3269–3284.

Zazimalova E, Murphy AS, Yang H, Hoyerova K, Hosek P 2010 Auxin transporters—Why so many? *Cold Spring Harb Perspect Biol* 2:2–14.

Genes and Mutants Index

B

C

D

E

F

G

H

I

J

K

L

M

N

P

Q

R

S

T

U

V

W

X

Y

Z

Organism Index

D

E

F

M

N

O

Q

R

S

T

U

Terms Index

D

E

F

T

U

V

W

X

Z